REVIEWS in MINERALOGY and GEOCHEMISTRY

Volume 87   2022

Geological Melts

EDITORS

Daniel R. Neuville
Université de Paris, France

Grant S. Henderson
University of Toronto, Canada

Donald B. Dingwell
Ludwig-Maximilians-Universität München, Germany

Series Editor: **Ian Swainson**

MINERALOGICAL SOCIETY of AMERICA
GEOCHEMICAL SOCIETY

COVER ILLUSTRATIONS

Background: Simultaneous effusive/explosive activity of the 2021 Cumbre Vieja euption, La Palma (Photo credit: Ulrich Kueppers: 7 October 2021).

Lower right inset: Glass structure visualized from molecular dynamics simulations (Le Losq et al. 2017).

Upper left inset: SEM photo of volcanic ash from the submarine eruption of Serreta off Terceira, Azores (Kueppers and Cimarelli 2018).

Lower left inset: Scanning calorimetric glass transition determination of a Pantellerite glass. (redrawn schematically from Gottsmann and Dingwell 2002).

Upper right inset: Raman spectra of a basaltic glass (redrawn from Amalberti et al. 2021).

Amalberti J et al. (2021) Raman spectroscopy to determine CO_2 solubility in mafic silicate melts at high pressure: Haplobasaltic, haploandesitic and approach of basaltic compositions (2021) Chem Geol 582:120413. https://doi.org/10.1016/j.chemgeo.2021.120413

Gottsmann J, Dingwell DB (2002) The thermal history of a rheomorphic air-fall deposit: The 8 ka pantellerite flow of Mayor Island, New Zealand. Bull Volcanol 64:410–422

Le Losq C et al. (2017) Percolation channels: a universal idea to describe the atomic structure and dynamics of glasses and melts. Sci Rep 7:16490. https://doi.org/10.1038/s41598-017-16741-3 (CC-BY-4.0, http://creativecommons.org/

Reviews in Mineralogy and Geochemistry, Volume 87
Geological Melts

ISSN 1529-6466 (print)
ISSN 1943-2666 (online)
ISBN 978-1-946850-08-9
COPYRIGHT 2022

THE MINERALOGICAL SOCIETY OF AMERICA
3635 CONCORDE PARKWAY, SUITE 500
CHANTILLY, VIRGINIA, 20151-1125, U.S.A.
WWW.MINSOCAM.ORG

Geological Melts

87   *Reviews in Mineralogy and Geochemistry*   **87**

PREFACE

The initiative for the development of this volume arose from preliminary discussions between the editors that it had been 25 years, and a whole generation, since RIMG had dedicated a special issue to the topic of silicate melts and their many roles in materials and geological sciences. Back in 1995 silicate melts were being rediscovered as a vital component of the earth sciences in the wake of an increasing awareness of their structural complexity, the development of new experimental techniques and their application to the determination of structure and properties of these melts, as well as, the partially unexpected results of such studies. A further overriding theme was the emerging consensus of the concept of the glass transition as a key to the interpretation and extrapolation of structure and property determinations and as a fundamental physical transition governing many phenomena in petrology and volcanology.

The result of that confluence of interests and ambitions has been a remarkable quarter century of opportunities for the study of the molten state in both materials and geological sciences and involving national programs and international collaborations on an unprecedented scale.

Collected in this volume are a compact set of chapters covering fundamental aspects of the nature of silicate melts and the implications for the systems in which they participate, both technological and natural. The contents of this volume may perhaps best be summarized as structure – properties – dynamics. The volume contains syntheses of short and medium range order, structure-property relationships, and computation-based simulations of melt structure. It continues with analyses of the properties (mechanical, diffusive, thermochemical, redox, nucleation, rheological) of melts. The dynamic behavior of melts in magmatic and volcanic systems, is then treated in the context of their behavior in magma mixing, strain localization, frictional melting, magmatic fragmentation, and hot sintering. Finally, the non-magmatic, extraterrestrial and prehistoric roles of melt and glass are presented in their respective contexts.

This volume is the cumulative effort of many people whom we gratefully thank, especially the authors of the chapters for their contributions and patience, and the reviewers for their comments and suggestions. We thank Ian Swainson and Rachel Russell for all their work. Finally, we hope graduate students and researchers new to the field will find the volume helpful as a starting point to these fascinating areas of such great importance for understanding the workings of the Earth System and realms beyond.

D.B. Dingwell (Munich)

Grant Henderson (Toronto)

Daniel -R. Neuville (Paris)

1529-6466/22/0087-0000$00.00 (print)
1943-2666/22/0087-0000$00.00 (online)

http://dx.doi.org/10.2138/rmg.2022.87.00

Geological Melts

87   *Reviews in Mineralogy and Geochemistry*   **87**

TABLE OF CONTENTS

1   The Short-Range Order (SRO) and Structure

GS Henderson, JF Stebbins

INTRODUCTION 1
ACRONYMS AND ABBREVIATIONS 2
WHAT IS SHORT-RANGE ORDER (SRO)? 3
    The Si–O bond, O–Si–O and Si–O–Si angles 4
    Techniques used to probe the SRO in glasses and melts 6
NETWORK-FORMING OXIDES AND CATIONS 15
    Silicon coordination in oxide glasses and melts 16
    Aluminum coordination 19
OXYGEN ANION SPECIATION 22
    Order/disorder among bridging oxygens 22
    Effects of modifier oxides: non-bridging oxygens 25
    Measurements of distributions of non-bridging oxygens 26
    The connection between network cation coordination and non-bridging oxygens 28
    Higher oxygen anion coordination numbers 30
    "Free" oxide anions 31
DISTRIBUTIONS OF NBO AND BO AROUND NETWORK CATIONS:
   Q SPECIES 34
    Variations on a theme: issues of fitting and data analysis 38
    Modifier and charge compensating cations 40
ACKNOWLEDGEMENTS 43
REFERENCES 43

2   From Short to Medium Range Order in Glasses and Melts by Diffraction and Raman Spectroscopy

JWE Drewitt, L Hennet, DR Neuville

INTRODUCTION 55
X-RAY, NEUTRON, AND RAMAN SCATTERING OF GLASSES AND MELTS 57
    X-ray and neutron diffraction 57
    Raman spectroscopy 59
    The boson peak: A signature of SRO or MRO? 61
MULTICOMPONENT SILICATE GLASSES 62
    Binary silicate glasses 62

Diffraction measurements 63
Raman spectroscopy measurements 65
Lead silicate glasses 68
Aluminosilicate glasses 68
Diffraction and Raman measurements along the tectosilicate join 69
Charge compensator versus network modifier 71
GLASS AND MELT STRUCTURE UNDER EXTREME CONDITIONS........................72
High temperature containerless processing..73
Aluminate melts 73
Aluminosilicate melts...79
 Iron silicate melts and glasses 80
Glasses and melts at high pressure 83
Pressure induced modifications in MRO and SRO in SiO_2 and GeO_2 glass...............84
Silicate melt structure at high pressure..86
Al coordination change at high pressure 87
SUMMARY AND FUTURE PERSPECTIVES 88
ACKNOWLEDGEMENTS 89
REFERENCES 89

3   Link between Medium and Long-range Order and Macroscopic Properties of Silicate Glasses and Melts

DR Neuville, C Le Losq

INTRODUCTION 105
Glass structure versus macroscopic properties.......................................105
Thermodynamic approach to glass transition..109
Viscosity and glass transition 109
Configurational properties and glass structure 110
Pressure–temperature space..113
SILICATE GLASSES AND MELTS..113
Alkali or earth alkaline silicate glasses and melts 113
Viscosity of silicate melts...115
Ideal mixing: mixing alkali or alkaline-earth in silicate glasses and melts.............116
Mixing alkali and alkaline-earth elements in silicate glasses................120
Silicate glasses and others network formers..125
Silicate melts: how can we use existing structural knowledge to model
   melt properties...127
ALUMINOSILICATE GLASSES AND MELTS..131
Molar volume, coordination number and structure 132
Proportion of Al in five-fold coordination..133
Molar volume, AlO_4^{4-} and implications for the coordination number of
   metal cations..135
Glass transition temperature 136
Link between structure and properties of aluminosilicate melts: example of the
   $CaO–Al_2O_3–SiO_2$ system...136
From the CAS system to other chemical systems 140
Alkaline-earth mixing in aluminosilicate glasses and melts 142
Models for alkali aluminosilicate melts 144
ALUMINATE GLASSES AND MELTS 147

Al_2O_3...148
Al_2O_3–$CaAl_2O_4$ compositions..148
CA–C3A compositions..149
C3A–CaO compositions...149
Link with observations in other binary systems 149
CONCLUSION AND PERSPECTIVES 150
REFERENCES 152

4 Topology and Rigidity of Silicate Melts and Glasses

M Micoulaut, M Bauchy

INTRODUCTION 163
ROLE OF NETWORK RIGIDITY 164
Rigidity theory of network glasses...164
The situation in silicate glasses...165
Rigidity Hamiltonians and floppy modes 169
Temperature dependent constraints 170
MOLECULAR DYNAMICS AND RIGIDITY 172
Bond-bending and bond-stretching..173
Behavior in the liquid phase 174
ISOSTATIC RELAXATION..175
Reversibility windows 176
OTHER APPLICATIONS 180
Diffusivity anomalies 180
Prediction of glass hardness 181
Prediction of glass stiffness 182
Origin of fracture toughness anomalies....................................183
Prediction of dissolution kinetics 184
Other applications and conclusion...185
REFERENCES 185

5 Molecular Simulations of Oxide and Silicate Melts and Glasses

S Jahn

INTRODUCTION 193
SIMULATION METHODS..193
Classical potentials..194
Electronic structure methods 196
Molecular dynamics simulations..197
Monte-Carlo simulations 198
Data analysis...198
SIMULATIONS OF OXIDE AND SILICATE MELTS AND GLASSES ...203
Oxide melts and glasses at ambient pressure 203
Silicate melts and glasses at ambient pressure 206
Melts and glasses at high pressure...212
CONCLUSIONS...218
REFERENCES 219

6   Mechanical Properties of Oxide Glasses

BP Rodrigues, T To, MM Smedskjaer, L Wondraczek

INTRODUCTION 229
ELASTIC CONSTANTS AND POISSON'S RATIO 230
FRACTURE AND BRITTLENESS 232
EXPERIMENTAL METHODS FOR STRENGTH AND TOUGHNESS TESTING.........236
    Biaxial test..237
    Uniaxial test..238
    Fracture toughness test 239
WEIBULL DISTRIBUTION AND PROBABILITY PLOTS 240
STRESS CORROSION AND FATIGUE...243
    Region 0: environmental limit...244
    Region I: stress corrosion 244
    Region II: transport kinetics 245
    Region III: inert crack growth 246
    Stress corrosion mechanism 246
INDENTATION HARDNESS, SCRATCH RESISTANCE AND CRACKING 246
    Stress fields...247
    Deformation mechanism 248
    Hardness 249
    Strain-rate sensitivity..249
    Indentation cracking 250
    Scratch resistance 253
FRACTOGRAPHY..254
    Crack branching pattern and angle 254
    Fracture surfaces (mirrors, mist and hackle) 255
RESIDUAL STRESS AND METHODS FOR ENHANCING
  THE PRACTICAL STRENGTH OF GLASSES 256
    Thermal strengthening...257
    Chemical strengthening..258
MECHANICAL PROPERTY EXAMPLES...260
    Fracture toughness versus Young's modulus 260
    Fracture energy versus Poisson's ratio 260
    Properties related to Poisson's ratio 261
    Stress optical coefficient and persistent anisotropy 262
PERSPECTIVE: TOPOLOGICAL CONSTRAINT THEORY 262
PERSPECTIVE: STRUCTURAL HETEROGENEITY
  AND NON-AFFINE DEFORMATION 264
PERSPECTIVE: MODELING AND SIMULATION 266
    Finite element methods 266
    Peridynamics 268
    Molecular dynamics 268
    Machine learning...270
OUTLOOK...270
ACKNOWLEDGEMENT 271
REFERENCES 272

7   Diffusion in Melts and Magmas

Y Zhang, T Gan

INTRODUCTION 283
FUNDAMENTALS OF DIFFUSION 284
    Fick's laws 284
    Various kinds of diffusion and diffusivities...286
    Dependence of D on temperature, pressure, and melt composition 287
SOME USEFUL SOLUTIONS TO THE DIFFUSION EQUATION AND
  EXPERIMENTAL DESIGNS FOR OBTAINING DIFFUSIVITY 289
    Diffusion couples 289
    Sorption or desorption 290
    Diffusion in melts during diffusive mineral dissolution 291
    Thin-source diffusion 292
    Isotropic diffusion in spheres 293
    Variable diffusivity along a profile 293
    Diffusion distance and square root of time relation 295
MULTICOMPONENT DIFFUSION 295
    Effective binary diffusion 296
    Multicomponent diffusion theory...298
    Recent studies of multicomponent diffusion...300
TRACER AND EFFECTIVE BINARY DIFFUSION DATA 307
    H_2O diffusion 308
    Diffusion of alkalis...312
    Cu diffusion 313
    Diffusion of Sc, Y, and REE 315
    Diffusivities of Li, Rb, Sr, Ba, Sn, V, Zr, Hf, Th, U, Nb and Ta...................317
    SiO_2 diffusion 318
    Self diffusion of O, Si, Mg and Ca, and interdiffusivity of Ni and Co in a
      peridotite melt..319
    Diffusion of Mo and W...320
    Diffusion of F, Cl, and S..320
    Major and trace element diffusion (OEBD) in shoshonite–rhyolite
      diffusion couple 321
DIFFUSIVE ELEMENTAL AND ISOTOPE FRACTIONATION DURING
  MAGMATIC PROCESSES 322
DIFFUSIVITY IN CRYSTAL-BEARING AND BUBBLE-BEARING MAGMAS..........325
    Crystal-bearing magmas...328
    Bubble-bearing magmas 329
CONCLUSIONS..331
ACKNOWLEDGEMENT 331
REFERENCES 331

8   Silicate Melt Thermochemistry and the Redox State of Magmas

R Moretti, G Ottonello

INTRODUCTION AND RATIONALE...339
THE EMERGENCE AND RISE OF THE CONCEPT OF OXYGEN FUGACITY 345
    Oxygen exchange between metals and their oxides...............................345
    Oxygen fugacity and iron-bearing solid phases 348
MODELLING THE REDOX STATE OF SILICATE MELTS...................................350
    Non-reactive species-based approaches to iron redox...............................350
    The ionic approach and the role of the ligand 357
    Oxygen fugacity, sulfur and joint Fe–S redox exchanges 368
THE (FULL) *AB-INITIO* PERSPECTIVE:
    THE CASE OF IRON REDOX 374
    Electron transfer and solute–solvent interactions in melts....................374
    The normal oxygen electrode 376
    Ab-initio iron redox 382
REMARKS AND PERSPECTIVES..387
CODE AVAILABILITY...390
APPENDIX 391
    Isolated molecules (gaseous state)..391
    Neutral and charged species in solution 392
ACKNOWLEDGEMENTS 397
REFERENCES 397

9   Nucleation, Growth, and Crystallization in Oxide Glass-formers. A Current Perspective

M Montazerian, ED Zanotto

INTRODUCTION 405
TABLE OF SYMBOLS 407
CRYSTAL NUCLEATION AND CLASSICAL NUCLEATION THEORY 408
    Recent findings that warrant further research: Examples of experimental tests........413
BASIC MODELS OF CRYSTAL GROWTH IN SUPERCOOLED LIQUIDS..................416
    Experimental tests 420
OVERALL CRYSTALLIZATION AND GLASS-FORMING ABILITY:
    THE JOHNSON–MEHL–AVRAMI–KOLMOGOROV APPROACH............................422
    Glass stability against crystallization 426
PERSPECTIVES 426
ACKNOWLEDGMENTS...427
REFERENCES 428

10   Thermodynamics of Multi-component Gas–Melt Equilibrium in Magmas: Theory, Models, and Applications

P Papale, R Moretti, A Paonita

INTRODUCTION 431
FUNDAMENTAL EQUATIONS OF EQUILIBRIUM THERMODYNAMICS 435
    Fundaments of thermodynamic equilibrium 435
    Terms in the equilibrium equations 440
REGULAR SOLUTION MODELING APPROACH TO
VOLATILE–MELT THERMODYNAMICS 444
    Excess Gibbs energy..444
    Early models...446
    The SOLWCAD model (Papale et al. 2006) 447
    The MagmaSat model (Ghiorso and Gualda 2015)..451
    Performance of the SOLWCAD and MagmaSat models 452
REACTIVE SPECIES-BASED APPROACHES TO VOLATILE–MELT EQUILIBRIA ...455
    Making solubility models (nearly) ideal: the Burnham model for water solubility ..455
    General aspects of the reactive species-based approaches to
        volatile–melt equilibria..458
    H_2O models revisited: the role of speciation in ionic polymeric-approaches 459
    Speciation-based CO_2 models 469
    The VolatileCalc model 473
    Acid–base compositional control on volatile speciation 476
    Sulfur solubility..482
    Modeling sulfur solubility in silicate melts: the CTSFG model 485
    Additional models for sulfur solubility in silicate melts 494
    Combined C–H–O–S–(±Cl) saturation models 497
    Halogen solubility 503
    First principles and molecular dynamics approaches...505
NOBLE GASES..507
    General aspects of noble gas solubility in magmas..507
    Basic thermodynamics of noble gas solubility in silicate melts............................508
    Pure noble gases and noble gas mixtures 509
    Reference state and activity–composition relationships..510
    Models based on statistical mechanics...514
    Models accounting for melt composition 515
    Modeling mixed H_2O–CO_2–noble gases 518
    Ar solubility as a proxy for N_2 solubility 521
    Applications..521
    Determination of magma storage conditions..522
    Interpretation of volcano degassing data 529
    Using noble gases with major volatiles 533
    Constraining and modeling magma and eruption dynamics 537
CONCLUDING REMARKS..543
REFERENCES 543

11   High Pressure Melts

T Sakamaki, E Ohtani

INTRODUCTION 557
EXISTENCE OF MAGMA IN THE INTERIOR OF THE EARTH.................................557
    Evidence for mantle melting 557
    H_2O in Earth's mantle...558
    Mantle transition zone 558
    Ultralow velocity zone at the base of lower mantle 558
PHYSICAL PROPERTY OF HIGH PRESSURE MELTS 559
    Density of melt at high pressure...559
    Viscosity of melt at high pressure 560
H_2O-BEARING MELTS...562
    Generation of hydrous magma in the interior of the Earth......................................562
    H_2O effect on the melt density...563
    The behavior of hydrous magmas above and below the mantle transition zone 564
    H_2O effect on the melt viscosity...565
CO_2-BEARING MELTS...566
    CO_2-rich magma in the interior of the Earth 566
    CO_2 effect on the melt density...566
    CO_2 effect on the melt viscosity 567
    Viscosity of carbonate melt 567
MAGMA OCEAN IN THE EARLY EARTH 568
    Melting in the early Earth's mantle 568
    Dynamics of the magma ocean crystallization...568
REFERENCES 569

12   Volatile-bearing Partial Melts in the Lithospheric and Sub-Lithospheric Mantle on Earth and Other Rocky Planets

R Dasgupta, P Chowdhury, J Eguchi, C Sun, S Saha

INTRODUCTION 575
THE EFFECTS OF MAJOR VOLATILES ON MELTING
  OF THE EARTH'S MODERN MANTLE 576
    Solidus of CO_2, H_2O, and CO_2+H_2O-bearing peridotite...............................577
    The effects of CO_2 and H_2O contents and major element bulk composition
      on the solidus..578
    Carbonated peridotite solidus during open system processes 581
    The composition of H_2O–CO_2-bearing partial melts from nominally
      volatile-free mantle—beneath oceans and continents 582
THE EFFECTS OF VOLATILE STORAGE IN ACCESSORY PHASES
  VERSUS MAJOR MANTLE MINERALS...586
    Mobilization of carbon versus sulfur-bearing accessory phase during melting
      of the Earth's mantle..587
    Sulfur mobilization by incipient melt in the Earth's mantle 588
    Extraction of C–S–H volatiles from other rocky mantles in the Solar System 591

Carbon contents of graphite-saturated mantle melts—application to
planetary mantles 593
Sulfur concentrations at sulfide saturation—application to planetary mantles 596
Fractionation of carbon and sulfur during mantle melting at
varying oxygen fugacity 596
Carbon, sulfur and water as a function of melting degree and mantle redox 597
CONCLUDING REMARKS...600
ACKNOWLEDGMENTS...601
REFERENCES 601

13   Decrypting Magma Mixing in Igneous Systems

D Morgavi, M Laumonier, M Petrelli, DB Dingwell

INTRODUCTION 607
HISTORICAL PERSPECTIVE (1851 TO MODERN TIMES)..607
DEFINITIONS AND FIELD EVIDENCE OF MAGMA MIXING 610
Definitions of mixing and mingling 610
Mixing and mingling structures 610
SCALING RULES AND NUMERICAL MODELLING..612
Stretching and folding plus diffusion: kinematic description of magma mixing 612
Complete fluid dynamic description of magma mixing 615
EXPERIMENTAL STUDIES FOR DECIPHERING THE COMPLEXITY OF
MAGMA MIXING 623
Reproducing the textural evidences of magma mixing 623
Rheological constraints 625
Flow regimes in magma chambers and mixing 627
Experiments involving crystal disequilibrium during magma mixing 629
CHEMICAL INVESTIGATIONS: FROM A SIMPLE LINEAR RELATION
TO A COMPLEX INTERPLAY WITH SYSTEM DYNAMICS 629
IMPLICATION OF MAGMA MIXING AND FUTURE DEVELOPMENTS 630
ACKNOWLEDGEMENTS 631
REFERENCES 632

14   Magma / Suspension Rheology

S Kolzenburg, MO Chevrel, DB Dingwell

THEORETICAL CONSIDERATIONS..639
CONVENTIONAL DESCRIPTIONS OF RHEOLOGICAL DATA 640
EXPERIMENTS ON ANALOGUE MATERIALS...646
Particle suspension analogues 647
Bubble suspension analogues..648
Experimental materials and measurement strategies..649
EXPERIMENTS ON HIGH TEMPERATURE SILICATE MELT SUSPENSIONS 650
Concentric cylinder experiments 652
Bubble bearing suspensions 662
Parallel plate experiments..663
Torsion experiments 668
FIELD RHEOLOGY 672

Falling sphere 674
Penetrometers 674
Rotational viscometers 675
Toward parameterization: requirements for future field viscometry 677
PARAMETERIZATION STRATEGIES 679
Particle suspensions..679
Parameterization of the relative suspension viscosity η_r 679
Parameterization of the yield stress τ_y..689
Bubble suspensions 692
The yield stress dilemma..702
TECHNOLOGICAL ADVANCES..704
OUTSTANDING CHALLENGES 705
Models for multiphase rheology..705
Reactive flow and phase dynamics 706
Filling data gaps 707
Connecting magma rheology and rock mechanics..707
Exploiting multidisciplinary datasets 708
Characterizing nanoscale processes 708
ACKNOWLEDGMENTS..709
REFERENCES 709

15   Strain Localization in Magmas

Y Lavallée, JE Kendrick

INTRODUCTION 721
MATERIAL DEFORMATION AND STRAIN LOCALIZATION....................................722
Evidence for strain localization in magmas: the geologic record............................722
Strain regimes..724
Stress and strain regimes in magmatic environments..724
Evolution of properties and conditions during magma transport 725
Deformation modes: ductile vs brittle 726
STRAIN LOCALIZATION: AN INTERPLAY BETWEEN
DEFORMATION MECHANISMS 728
Viscous flow and energy dissipation 728
Bubble deformation and alignment 729
Crystal alignment and deformation 734
Multiphase magma rupture..739
Fault processes..748
CONSEQUENCES OF STRAIN LOCALIZATION IN MAGMAS 750
Brittle versus ductile deformation modes in rocks and magmas..............................750
Construction versus destruction of permeability 751
Implications for magma seismicity and tilt at volcanoes 752
Transport in volcanic conduits and during volcanic eruptions: a model..................754
ACKNOWLEDGEMENTS 757
REFERENCES 757

16   Magma Fragmentation

B Scheu, DB Dingwell

INTRODUCTION 767
SILICATE LIQUIDS—THE BASIS OF MAGMA...768
FRAGMENTATION—A MATERIALS TRIGGER 769
EXPERIMENTAL VOLCANOLOGY OF MAGMATIC FRAGMENTATION.................773
    Early experimental approaches...774
    Fragmentation experiments using magma analogues..776
    Fragmentation experiments using silicate melts and volcanic rocks.......................777
THEORETICAL MODELS AND CRITERIA FOR MAGMA FRAGMENTATION.........778
  MAGMA FRAGMENTATION BEHAVIOR 780
    Fragmentation threshold..780
    Influence of permeability on magma fragmentation 782
    Speed of magma fragmentation...783
    Timescales of magma ascent and fragmentation—Implications for eruption styles. 785
PRODUCTS OF MAGMA FRAGMENTATION...787
    Energetic considerations of magma fragmentation 787
    Grain size distribution of fragmentation products...788
SECONDARY FRAGMENTATION 790
EXPLOSIVE AND NON-EXPLOSIVE MAGMA WATER INTERACTION 791
NON-MAGMATIC FRAGMENTATION: STEAM-DRIVEN ERUPTIONS.................792
CONCLUDING REMARKS AND FUTURE PERSPECTIVES.............................793
ACKNOWLEDGEMENTS 794
REFERENCES 794

17   Hot Sintering of Melts, Glasses and Magmas

FB Wadsworth, J Vasseur, EW Llewellin, DB Dingwell

INTRODUCTION TO SINTERING IN VOLCANIC ENVIRONMENTS 801
    Families of sintering phenomena...803
    Phenomenology and internal texture 804
    Conceptual approaches...807
THEORETICAL MODELS FOR SINTERING UNDER NO EXTERNAL LOAD 808
    'Free sintering' of many particles or sintering under zero applied load...................810
    Extending sintering models to polydisperse systems 813
'PRESSURE SINTERING' OR SINTERING UNDER EXTERNAL LOAD....................813
    The extended Mackenzie–Shuttleworth model (Wadsworth et al. 2019).................814
    The extended Scherer model (Scherer 1986) 814
    The Quane and Russell (2005) model. 814
PORE SIZES, SURFACE AREA AND INTER-PARTICLE DISTANCES
  IN SINTERING SYSTEMS 815
    Pore sizes...815
    Specific surface area..818
    Inter-particle distance...818
EXPERIMENTAL APPROACHES...818
    Sintering under equilibrium pressures..820
    Sintering under differential pressures...822

EMPIRICAL DATA AND ANALYSIS 822
    Fundamental quantities specific to the data used here............................822
    Sintering of two droplets or particles 823
    Sintering of large systems of many droplets 824
SINTERING OF MORE COMPLEX SYSTEMS...826
    The effect of rigid crystals...827
    Sintering with diffusive hydration or dehydration...................................828
PHYSICAL PROPERTIES OF SINTERED SYSTEMS 830
    The sintered filter: permeability during sintering (Wadsworth et al. 2021) 830
    Elastic properties 831
    Compressive strength of sintered materials..832
SINTERING DYNAMICS MAPS..833
RECIPES FOR SINTERING AND APPLICATIONS 834
OUTLOOK 836
ACKNOWLEDGMENTS..836
REFERENCES 837

18   Models for Viscosity of Geological Melts

JK Russell, K-U Hess, DB Dingwell

INTRODUCTION 841
EXPERIMENTAL DETERMINATION OF MELT VISCOSITY 842
    Measurement techniques 843
    Viscosity data for natural silicate melts...846
MODEL CONSIDERATIONS 846
TEMPERATURE DEPENDENCE OF MELT VISCOSITY 848
    Non-Arrhenian functions for melt viscosity..848
THE HIGH-TEMPERATURE LIMIT TO MELT VISCOSITY (A) 851
    Practical application of a high-T limit 852
    Implications of a common A...853
EARLY MODELS AND MODELLING OF GEOLOGICAL MELTS 856
MODELS FOR HYDROUS SILICIC MELTS 861
    Hydrous low-P models for silicic melts 861
    Hydrous high-P models for silicic melts...863
OTHER MODELS FOR RESTRICTED COMPOSITIONAL RANGES 866
    Models for melts from the Phlegrean Field..866
    A model for natural Fe-bearing silicate melts 867
    A model for extraterrestrial melts...868
MULTICOMPONENT MELT MODELS FOR GEOLOGICAL SYSTEMS.....871
    Non-Arrhenian multicomponent melt models (anhydrous) 871
    Non-Arrhenian multicomponent melt models (hydrous) 873
GRD MODEL (2008) 874
    Attributes 875
    Weaknesses...878
QUO VADIMUS 880
CLOSING THOUGHTS...881
ACKNOWLEDGEMENTS 882
REFERENCES 882

19   Non-terrestrial Melts, Magmas and Glasses

G Libourel, P Beck, J-A Barrat

INTRODUCTION 887
CHONDRULES: THE EARLIEST MOLTEN DROPLETS
  OF THE SOLAR SYSTEM..888
    Extraterrestrial igneous droplets...889
    Chondrule thermal history..891
    Evidence for gas–melt interaction during chondrule formation...............................894
    Astrophysical implications 896
MAGMATIC MELTS FROM PROTOPLANETS...896
    Alkali-depleted protoplanetary basalts. 897
    Andesitic or trachyandesitic achondrites..901
IMPACT-RELATED MELTS AND GLASSES..902
    Impact induced melting and vaporisation, lessons from terrestrial impactites.........902
    Impact-related amorphization of minerals: dense glass or not.................................903
    High-pressure melting, and high-pressure phase bearing melt. 904
    Transported glass fragments and glass beads in extra-terrestrial regolith and
      regolith breccias...906
    Prospects..911
ACKNOWLEDGEMENTS 912
REFERENCES 912

20   Frictional Melting in Magma and Lava

JE Kendrick, Y Lavallée

INTRODUCTION 919
FAULT FRICTION 921
    Identifying pseudotachylytes..921
    Fault slip in tectonic and volcanic environments 923
    Dynamics of frictional sliding..926
FRICTIONAL MELTING 928
    The history of experimental approaches..928
    Mechanical response to melting 929
    Selective melting 932
    Frictional melt rheology 937
    Viscous remobilization 944
    Thermal vesiculation 945
    Elevated ambient temperature 947
    Fault healing and cyclic rupture 951
SUMMARY AND CONCLUDING REMARKS 952
ACKNOWLEDGEMENTS 953
REFERENCES 953

21   Non-Magmatic Glasses

MR Cicconi, JS McCloy, DR Neuville

OVERVIEW 965
ACRONYMS AND GLOSSARY 965
INTRODUCTION 966
    The composition and origin of natural glasses 966
METAMORPHIC GLASSES 968
    Pyrometamorphic glasses 968
    Pseudotachylite or frictionites 972
GLASSES FROM HIGHLY ENERGETIC EVENTS 974
    Impactites 974
    Tektites and microtektites 977
    Enigmatic impact glasses 982
    Fulgurite 987
    Trinitite or nuclear glass 993
GLASS PROPERTIES 996
    Liquid immiscibility 996
    Reduced iron species 997
    Glass structure 1000
CONCLUDING REMARKS 1005
ACKNOWLEDGMENTS 1005
REFERENCES 1005

22   Silicate Glasses and Their Impact on Humanity

RE Youngman

INTRODUCTION 1015
HISTORY AND IMPACT OF SILICATE GLASSES 1016
    Glass before and during the Age of Antiquity 1016
    Silicate glasses during the Middle Ages 1019
    Silicate glasses in modern and contemporary history 1022
    Summary and perspectives on the Glass Age 1035
ACKNOWLEDGEMENTS 1037
REFERENCES 1038

23   Glass as a State of Matter—The "newer" Glass Families from Organic, Metallic, Ionic to Non-silicate Oxide and Non-oxide Glasses

D Möncke, B Topper, AG Clare

OVERVIEW 1039
INTRODUCTION 1039
    The glassy state of matter 1042
    The glass family tree 1042
    Glass models 1043

ORGANIC GLASSES 1044
    Man-made polymeric glasses 1046
    Amber 1047
    Organic metal framework glasses 1047
INORGANIC METALLIC GLASSES 1048
INORGANIC NON-METALLIC NON-OXIDE GLASSES 1049
    Chalcogen(ide) glasses 1049
    Fluoride glasses 1052
    Other ionic and molecular glasses 1055
    Nitride glasses 1055
SIMPLE INORGANIC OXIDE GLASSES 1058
    Phosphate glasses 1058
    Borate glasses 1066
    Tellurite glasses 1071
    Germanate glasses 1073
    Other glasses 1075
GLASS FAMILIES IN COMPARISON 1076
CONCLUDING REMARKS 1078
ACKNOWLEDGEMENTS 1078
REFERENCES 1079

Reviews in Mineralogy & Geochemistry
Vol. 87 pp.1-53, 2022
Copyright © Mineralogical Society of America

1

The Short-Range Order (SRO) and Structure

Grant S. Henderson

Department of Earth Sciences
University of Toronto
22 Russell St
Toronto
Ontario, M5S 3B1
Canada

henders@es.utoronto.ca

Jonathan F. Stebbins

Department of Geological Sciences
Stanford University
Stanford
California 94305–2115
USA

stebbins@stanford.edu

INTRODUCTION

Magmas are mixtures of melt, crystals and dissolved gasses with or without crystalline material (Mysen and Richet 2019). They play an important role in igneous processes particularly those influenced by their viscosity, density, and other physical properties. Much of the research to date has used quenched melts (or glasses) as proxies for studying the melt portion of magmas. Studies of glasses and melts at elevated temperatures or in-situ have been carried out since the late 1970s (cf., Waseda and Egami 1979; Exarhos et al. 1988) but remain difficult experiments to perform due to technical difficulties dealing with high temperatures and molten liquids. Problems can arise from, for example, thermal broadening of spectroscopic peaks (often reflecting important dynamics in the liquid state), black body radiation, and motional averaging, as well as challenges of thermal gradients, volatilization, accurate P/T measurement, etc. Another complication is the complex chemical compositions of natural melts and glasses which can make it difficult to analyze and interpret data. Therefore, the glasses studied are usually simple synthetic, rather than natural compositions. The assumption is that the structure of the glass resembles that of the melt at the glass transition temperature, the temperature at which there is a transition from liquid-like to solid-like properties and behavior. One of the key issues is the quality and information content of data obtainable on glasses vs. liquids.

A more complete definition of a glass is given by Zanotto and Mauro (2017). Several recent books provide greater detail on glass and melt structure and properties than can be included here (e.g., Greaves and Sen 2007; Le Losq et al. 2019b; Musgraves et al. 2019; Mysen and Richet 2019; Varshneya 2019; Greaves 2020; Richet 2020) including high pressure studies (Kono and Sanloup 2018). Other RiMG volumes on silicate melts (Stebbins et al. 1995) and on spectroscopic methods (Henderson et al. 2014b) as well as other chapters in this volume provide extended and complementary information as well.

1529-6466/22/0087-0001$10.00 (print)
1943-2666/22/0087-0001$10.00 (online)   http://dx.doi.org/10.2138/rmg.2022.87.01

ACRONYMS AND ABBREVIATIONS

AWAXS	Anomalous wide-angle X-ray scattering	MAS	Magic angle spinning (NMR)
Ab Initio	First principles	MBO	Bridging oxygen (Si–O–Si) with one or more alkali or alkaline-earth cations attached to it.
BE	Binding energy	MD	Molecular dynamics
BO	Bridging oxygen	MQ	Multiple quantum (NMR)
CA	$CaO–Al_2O_3$ or $CaAl_2O_4$	NBO	Non-bridging oxygen
CA2	$CaO–Al_2O_3$ or $CaAl_4O_7$	NMR	Nuclear magnetic resonance
C3A	$3CaO–Al_2O_3$ or $Ca_3Al_2O_6$	NRIXS	Non resonant inelastic X-ray scattering (same as XRS)
C2A	$2CaO–Al_2O_3$ or $Ca_2Al_2O_5$	O^{2-}	Free oxide (same as FO)
CMAS	$CaO–MgO–Al_2O_3–SiO_2$ or $CaMgAl_2SiO_7$	PCA	Principle component analysis
CN	Coordination number	PCF	Pair correlation function
CQ	quadrupolar coupling constant (NMR)	PSF	Partial structure factor
CSA	Chemical Shift Anisotropy (NMR)	RDF	Radial distribution function
DAS	Dynamic Angle Spinning (NMR)	RMC	Reverse Monte Carlo
DAXS	Diffraction anomalous X-ray scattering	SRO	Short-range order
DFT	Density functional theory	tricluster	A group of three network cations, each of which has three or four oxygen first neighbors which share a single oxygen i.e., OAl_3
DOR	Double Rotation (NMR)	UV	Ultraviolet
ELNES	Energy loss near-edge spectroscopy	XANES	X-ray absorption near-edge structure
EPSR	Empirical potential structure refinement	XAFS	X-ray absorption fine structure spectroscopy (same as EXAFS)
eV	Electron volt	XAS	X-ray absorption spectroscopy
EXAFS	Extended X-ray absorption fine structure	XPS	X-ray photoelectron spectroscopy
FO	Free Oxide (same as O^{2-})	XRS	X-ray Raman spectroscopy (same as NRIXS)
FWHM	Full width at half maximum	2D	Two-dimensional (NMR)
IR	Infrared also NIR (Near IR) and MIR (Mid IR)	3Q	Triple quantum (NMR)
keV	Kilo electron volt		

WHAT IS SHORT-RANGE ORDER (SRO)?

The interatomic forces that link atoms in glasses and melts are the same as those in crystalline materials (Zachariasen 1932). Crystal structures can be defined by a unit cell, lattice type, atom positions, point and space group symmetry. Glasses form extended three-dimensional structures like crystalline materials, but in a glass, there is no lattice and no translational symmetry. No formal unit cell can be defined. The lack of well-ordered periodic structure like that of a crystal means that the structure of a glass must be defined using an alternative approach. The most common way of defining the structure of glasses is to use concepts of order even though the presence of order in glasses cannot be precisely defined. The degree of order is usually interpreted in terms of *short, intermediate* or *medium range order*. It is the lack of long-range order that differentiates a glass from a crystal. Drewitt et al. (2022, this volume) discuss the intermediate-range order in glasses and melts.

Wright (1994) defined four ranges: (I) the structural unit, (II) the interconnection of adjacent structural units, (III) the network topology (medium-range order), and (IV) long-range fluctuations in density (long-range order). The first two ranges of order defined by Wright (1994) are generally grouped together to define short-range order.

In simple silicate glasses the basic building block is the SiO_4 tetrahedron. The arrangement of atoms around the tetrahedron and how it is linked to its immediate tetrahedral neighbours constitutes one key aspect of the short range order (SRO). Other factors such as coordination of non-network cations also play a role. SiO_4 tetrahedra are made up of 4 Si–O bonds and 4 O–Si–O angles. The latter are generally close to $109.4°$ and tend to be relatively limited (ranging from $98-122°$) except at high pressures (*P*) and temperatures (*T*). If the oxygens are shared with the Si atoms of the adjacent tetrahedron they are referred to as **bridging oxygens (BO)** and link the tetrahedra together. The number of bridging oxygens associated with a tetrahedron defines what is termed a Q^n **species** (based on chemical nomenclature for a 'quaternary' group with four bonds) where *n* is the number of bridging oxygens and can vary between 0–4 (Fig. 1).

If an oxygen is bonded to only one Si atom (or other element such as Al, B, or P, with high charge and small coordination number) it is termed a **non-bridging oxygen (NBO)**. This type of bond has excess negative charge on the oxygen resulting in the need for the charge to be balanced by the close approach of one or usually several "charge balancing cations" such as an alkalis or alkaline-earths. There are a number of parameters that can be defined which fully describe the nature of the tetrahedron and its linkage to adjacent tetrahedra.

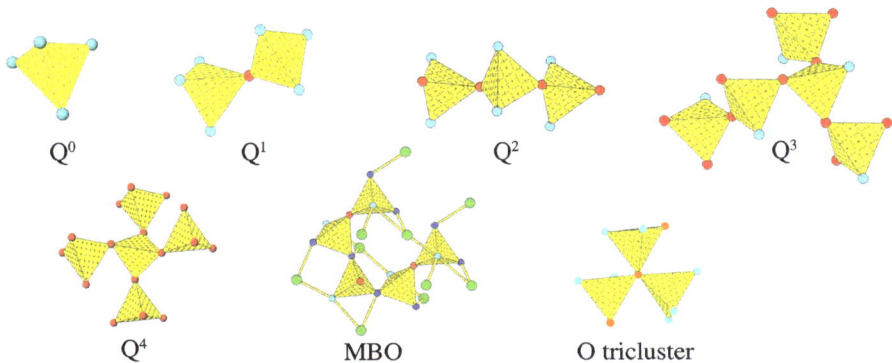

Figure 1. Q^n species: NBO (sky blue), BO (red), MBO (blue). Q^0 tetrahedron with no BO; Q^1 with 1 BO; Q^2 with 2 BO; Q^3 with 3 BO; partial structure of α-$Na_2Si_2O_5$ showing Q^3 tetrahedra with regular BO (red) and MBO (blue) with Na (green) attached; and an oxygen tricluster (orange) being shared with 3 AlO_4 tetrahedra.

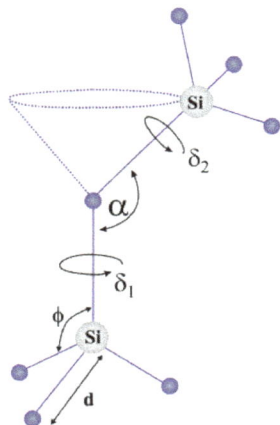

Figure 2. Parameters that define the SRO: α intertetrahedral angle, δ bond torsion angles, d Si–O distance, Φ O–Si–O angle. Reprinted from The structure of silicate melts: a glass perspective. Henderson GS Canadian Mineralogist 43: 1921–1958. Copyright Mineralogical Association of Canada (2005).

These are: the Si–O bond length (d); the tetrahedral (O–Si–O) bond-angle (ϕ); the inter-tetrahedral (Si–O–Si) or dihedral bond-angle (α); the bond-torsion angles (δ_1 and δ_2), and the Si coordination number (N). These parameters are shown in Figure 2. There is little difference between a tetrahedron in a glass and a crystal. This is because the parameters associated with individual tetrahedra are relatively fixed. However, the wide variation in α (116–180° in silica glass), and δ_1 and δ_2 distinguishes a glass from its corresponding crystalline analogue: disorder begins with the interconnection of adjacent tetrahedra (Mozzi and Warren 1969).

The Si–O bond, O–Si–O and Si–O–Si angles

Excellent reviews of the nature and behaviour of the Si–O bonds and angles found in crystalline silicates and in glasses are given in Liebau (1985) and Mysen and Richet (2019). Only a brief overview will be given here. It is important to remember that the bond distances and angles determined for glasses are generally average distances and not explicit distances as determined for crystalline materials. This is because of the nature of the disorder in glasses. In silica glass the average Si–O bond length is ~1.59–1.62 Å, with a Si–O–Si angle varying from ~120–180° (Mozzi and Warren 1969; Mei et al. 2007). An often stated Si–O–Si average angle is 144° as determined by Mozzi and Warren. However, this value is significantly different from nearly every other study (using multiple techniques) of the Si–O–Si angle in SiO_2 glass where the average angle is around 147–151°. The difference can be attributed to the manner in which Mozzi and Warren (1969) calculated their Si–Si bond distance distribution. A discussion of this and further references are given in Wright (1994); Henderson (2005); Mei et al. (2007) and Trease et al. (2017). The average O–O distance is ~2.62–2.65 Å and Si–Si distance ~3.06–3.12 Å (Mozzi and Warren 1969; Konnert and Karle 1973; Wright 1994). The bridging oxygen bond lengths in silicate glasses and crystals lengthens with decrease in the Si–O–Si angle and with increase in coordination of Si (Liebau 1985). Furthermore, Si–NBO distances are shorter than Si–BO distances because of the weaker attraction to lower-charged cations. In crystalline silicates the Si–NBO distance is typically 1.58 Å and Si–BO about 1.63 Å. The coordination of the oxygen atoms also influences the Si–O distance and Si–O–Si angle: Si–O bond lengths increase as the coordination of O increases while the Si–O–Si angle tends to decrease (cf., Liebau 1985; Gibbs et al. 1997).

Recently, Nesbitt et al. (2015a) were able to detect at least two types of Si–O–Si. One type of BO was connected to 2 Si atoms with a coordination of 2 while another termed **MBO** (where M = alkali or alkaline-earth) had alkali or alkaline-earth cations bonded to the BO resulting in the oxygen being 3-fold or higher coordinated. Such bonds must be present because while one Na atom produces 1 NBO its coordination number is around 5 and consequently some of the 5 bonds must be to bridging oxygens. Discriminating between the close approach of an alkali to the BO as opposed to actual formation of a bond with the BO, however, remains a challenging issue. Extensive molecular orbital calculations do show that such bonds are likely and energetically favoured (Gibbs 1982; Gibbs et al. 1997, 2009) Furthermore, they are very common in alkali silicate crystal structures such as α-$Na_2Si_2O_5$ where chains of SiO_4 tetrahedra consist exclusively of Q^3 tetrahedra but with 2 MBOs and one regular BO (Pant and Cruickshank 1968). The presence of these MBOs can influence the peaks or bands observed in the Raman spectra of alkali and alkaline-earth silicate glasses and alter binding energies (BE) in O 1s XPS spectra (see below).

While there are many models of the possible structure of SiO_2 and silicate glasses (see Henderson 2005 for a more extensive discussion and Wright 2019) there are two principal models that are widely referred to when describing the structure of such glasses. These are the crystallite theory (cf., Lebedev 1921, Randall et al. 1930 and Porai-Koshits 1958) and the random network model (cf., Zachariasen 1932, Warren 1933) or modified random network (MRN) model of Greaves et al. (1981), Greaves (1985) and Greaves and Sen (2007). Briefly the crystallite hypothesis treats the extended structure of the glass as regions of crystal-like structure (crystallites, 8–15 Å in size) which have more disordered areas at the interface between the crystallites. The random network model treats the glass as a continuous random network of SiO_4 tetrahedra which are fully bonded through BOs to adjacent tetrahedra making up the network. Zachariasen (1932) recognised that certain elements made up such networks, and termed them **network formers** (Si, Al, P, As, B). On the other hand, addition of oxides of other elements tended to break the network up. The latter he termed **network modifiers** and includes the alkalis and alkaline-earths. The added oxide ions convert the BOs to NBOs when added to the network in excess of T^{3+} cations. Some elements (e.g., Fe, Ti) were termed "intermediate" since they may act as either a modifier or former depending upon the composition of the glass and/or T and P conditions (Varshneya 2019). Greaves et al. (1981) suggested that modifier cations were not homogeneously distributed in silica glass and did not randomly break up the fully connected network. They suggested that the alkalis (Na in the study) were heterogeneously distributed in the glass with the Na–NBO linkages tending to form "percolation" channels within a silica rich network (Fig. 3). This model is the most widely accepted model in geological glass and melt research. However, it should be mentioned that direct structural evidence of channels remains elusive although classical (Cormack et al. 2003; Du and Cormack 2003) and first principles or *ab initio* molecular dynamics simulations do suggest their presence (Fig. 3). However, the "channels" in melt simulations are not channels in the sense suggested by Greaves (1985) but "dynamic trajectories" of the diffusion of Na through the simulated structure.

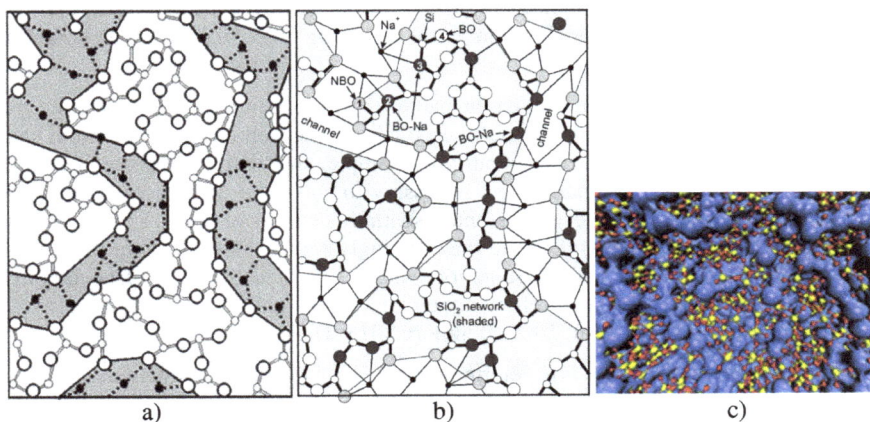

a)   b)   c)

Figure 3. Examples of a modified random network containing **a)** channels (reprinted from, The structure of silicate melts: a glass perspective. Henderson GS Canadian Mineralogist 43:1921–1958. Copyright Mineralogical Association of Canada (2005), **b)** with MBOs present near the channels (Reprinted from Journal of Non-crystalline Solids, Nesbitt HW, Henderson GS, Bancroft GM, Ho R 409:139–148), Copyright (2015), with permission from Elsevier and **c)** from a computer simulation showing dynamic trajectories of the Na atoms (Reprinted with permission from Meyer A, Horbach J ,Kolb W, Kargl F, Schober H Physical Review Letters 93:027801(2004) Copyright (2004) by the American Physical Society. http://dx.doi.org/10.1103/PhysRevLett.93.027801).

Techniques used to probe the SRO in glasses and melts

There are a wide range of techniques that are employed in studying the structure of glasses and melts (cf., Stebbins et al. 1995; Henderson et al. 2014b; Affatigato 2015; Henderson 2020). Some are direct structural probes, such as neutron and X-ray diffraction (scattering) while others investigate specific types of atoms such as Nuclear Magnetic Resonance (NMR), X-ray absorption spectroscopy (XAS), Mössbauer and X-ray photoelectron spectroscopy (XPS) or structural groups and vibrational properties (Raman spectroscopy, infrared (IR) and Brillouin spectroscopy). Regardless of the technique employed, the data represent the contributions from all of the local structural configurations. Furthermore, these contributions may be highly overlapping in the observed spectrum and thus difficult to uniquely evaluate. Unlike crystalline materials, in which individual site occupancies and geometries can be determined, information on specific atom sites is not possible in glasses and melts. In addition, for melts, temperature-induced structural changes with changing T are important to obtain. However, one needs to be aware of the time scale of the experimental technique used to probe the structure and the rate of changes in the melt. A large discrepancy between the two provides information on a time-averaged structure and not individual structural features.

Space is too limited to give overviews of every technique and interested readers are referred to Henderson (2020) or the specialized chapters given in Henderson et al. (2014b), Affatigato (2015), Musgraves et al. (2019), and Richet (2020). Nevertheless, we will briefly describe the most commonly used techniques.

X-ray and neutron diffraction techniques. Neutron (Hannon 2015) and X-ray diffraction (Benmore 2015) (also called *scattering* when applied to amorphous materials like glasses and melts) are extensively used to determine average bond lengths, coordination numbers and angles over both short and intermediate range length scales. Neutron diffraction and its variant "isotope substituted" neutron diffraction (see below) is frequently used to study lighter elements and elements that are close to each other in the periodic table for which X-rays have difficulty discriminating, because X-rays interact with the electron cloud of an element while neutrons interact with nuclei and their neutrons. Consequently, neutrons provide better spatial resolution than X-rays. However, neutron scattering experiments generally require much larger samples than X-ray scattering (e.g., 50 g vs. 1 mg), sometimes limiting applications.

Figure 4 shows the powder diffraction pattern for crystalline quartz versus the glass slide on which the sample sits. The crystalline sample shows well defined diffraction peaks, or "reflections", which can be used to identify the sample, and, with Rietveld techniques, to determine the crystallographic parameters necessary to solve the crystal structure. On the other hand, the glass pattern is very weak and essentially consists of one or more very broad humps in the trace (Fig. 4 inset). Despite the seemingly low resolution, useful information about bond distances, angles and coordination can be extracted from the glass data using appropriate methods. One of the following references should be consulted for a more thorough discussion (Fischer et al. 2006; Benmore 2015; Hannon 2015).

The most common approach is to determine the Radial Distribution Functions (RDF), and the total and pair correlation functions (PCF). The RDF is a one-dimensional representation of the three-dimensional structure which is averaged over the entire system. It represents the probability of finding a given atom at any distance r from an atom at the centre of the system. As in diffraction from crystals the angle between incident and scattered beams is 2θ and λ the X-ray or neutron wavelength, the wave or scattering vector is then given by $4\pi \sin \theta/\lambda$. This vector is denoted by s or k in X-ray diffraction and Q in neutron scattering.

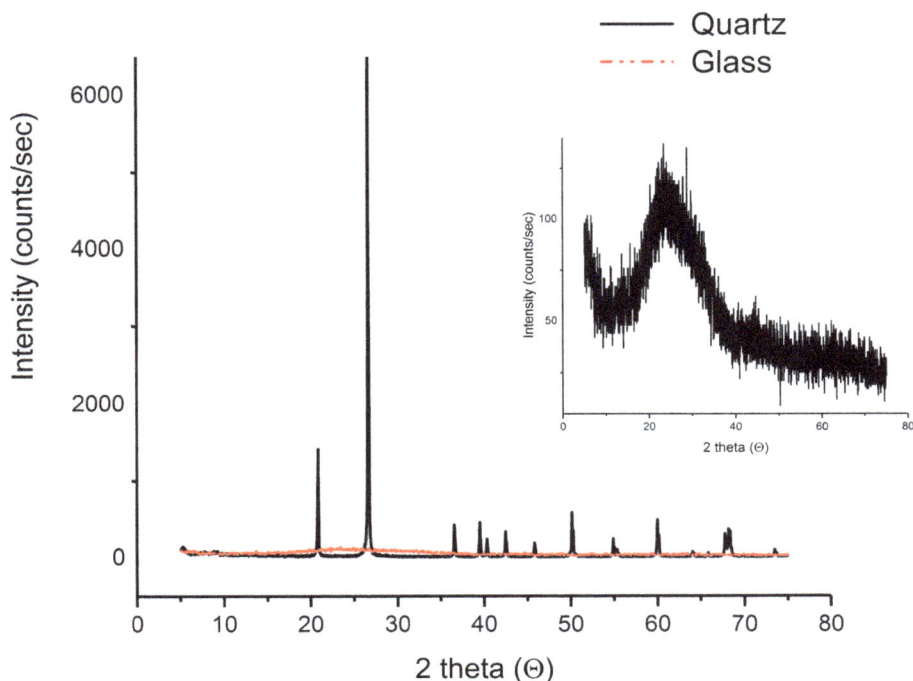

Figure 4. Powder diffraction traces for crystalline SiO_2 (quartz) and the glass slide on which the sample was mounted (red, dash/dot). The inset shows the glass diffraction trace enlarged to give an indication of the difference in intensity between a glass and crystalline sample.

Since diffraction data provides information about the average structure of the glass, the RDF peaks indicate average interatomic distances and average bond distance variations (peak widths). The areas under the peaks are related to average coordination number. RDF peaks are relatively high and narrow at low radial distances, but broader and of lower intensity at longer distances. This indicates the structure is less variable at shorter distances than longer ones. This enables the identification of the interatomic distances at low r but longer distance peaks are more ambiguous as there are more than one contribution to the peaks. By collecting data to high k space one can minimize the overlapping contribution: Neutron data is often collected out to Q (k) ranges of ~40–50 Å^{-1} while X-ray data is obtained using X-ray photons (>40 keV) with wavelengths lower than 0.03 nm which improves k to ≥ 30 Å^{-1}. An alternative approach is to use techniques that can separate the individual PSFs (e.g., anomalous X-ray scattering or isotope substitution, see below). Analysis and interpretation of X-ray and neutron diffraction data is often performed in conjunction with molecular dynamics (MD) simulations. The simulations in turn may use some sort of model such as empirical potential structure refinement (EPSR) to interpret the data. Care must be taken as to what is reported, the actual pair correlation functions and experimental distances, versus MD or EPSR distances.

Isotope substituted neutron diffraction (c.f., Cormier 2019) uses the dependence on number of neutrons in the sample. This enables one to more clearly discriminate the different atom pairs contributing to the RDF and enables a more unambiguous interpretation. There are, however, some caveats that must be considered (cf., Hannon 2015; Cormier 2019).

Vibrational techniques (Raman, IR, Brillouin). The reader is referred to relatively recent reviews of Raman (Dubessy et al. 2012; Rossano and Mysen 2012; Neuville et al. 2014; Almeida and Santos 2015), Infrared (McMillan and Hofmeister 1988; Kamitsos 2015) and Brillouin (Speziale et al. 2014; Kieffer 2015) spectroscopy for more comprehensive discussions of these techniques, or Henderson (2020) for an overview. Infrared (IR), Raman and Brillouin spectroscopy each probe various vibrational properties of a glass. Incident electromagnetic radiation of appropriate frequencies (IR–Visible regions) interact with the vibrational and rotational modes of structural units making up the structure of the glass. IR spectroscopy involves the absorption of photons in the infrared region of the EM spectrum while Raman is based on photon inelastic scattering. The latter is by far the most common vibrational spectroscopic technique used to probe the Q^n speciation, as well as the intermediate-range structure of glasses. Brillouin spectroscopy is primarily used to investigate the elastic properties of glasses and involves the interaction of photons with acoustic waves or phonons in the glass.

Infrared spectroscopy (IR). In IR spectroscopy incident photons from an IR source are absorbed by the vibrating atoms of the sample. Two energy ranges are usually recognised: far IR (~ 10–400 cm^{-1}) and mid-IR regions (~ 400–5000 cm^{-1}). Absorption occurs when there is a change in the induced dipole moment of the bonds that are vibrating. It is the change in the incident versus transmitted or reflected IR radiation that is measured, and these changes are strongest in polar molecules and asymmetric vibrations. Currently, reflectance IR is probably the most common IR spectroscopy method used to investigate glass structure while absorption studies are generally used to investigate volatiles such as water (see Fig. 5 and Behrens et al. 2009) and CO_2 in glasses (cf., Kamitsos 2015). The peaks observed in IR spectra are characteristic of molecular groups or atomic vibrational motions that make up the SRO of the glass.

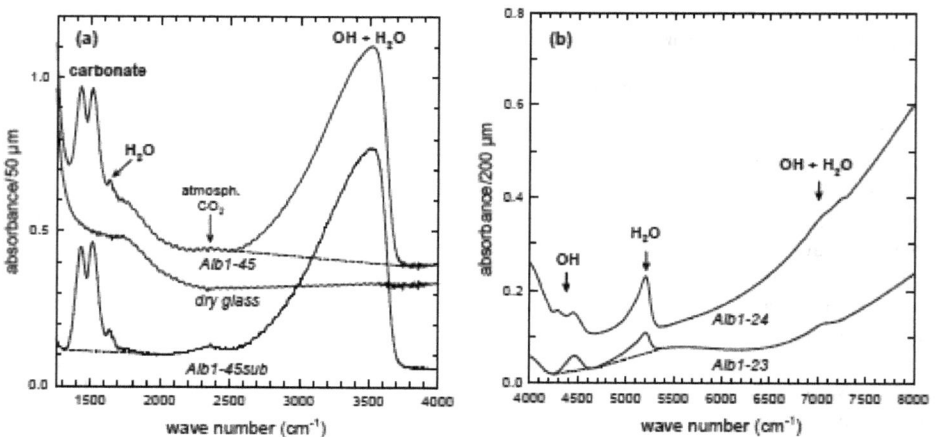

Figure 5. IR absorption spectra of albite glass (Alb1) from Behrens et al. (2009). Spectra are plotted with an offset for clarity. **(a)** MIR. Bold lines represent measured spectra. To evaluate the carbonate band intensity a spectrum of the starting glass scaled to same thickness was subtracted from the sample spectrum. A linear baseline (dashed) was fitted to the raw spectrum to quantify the 3550 cm^{-1} band intensity and subtraction spectrum to measure the peak heights of the carbonate peaks. **(b)** NIR. Linear baselines (dashed) were employed to evaluate the peak heights of the NIR combination bands. Note that the Fe-related band at 5600 cm^{-1} disappeared and the OH combination bands splits into two peaks in the spectrum of the water-rich sample Alb1_24. This observation indicates alteration during quench.

Raman spectroscopy. As noted above, Raman spectroscopy is widely used to investigate both the short and intermediate range structure of glasses and melts. It is particularly sensitive to elucidating the different types of Q^n species in glasses although quantification of the different species remains problematic. The reader should refer to one of the references noted above, as well as, Dubessy et al. (2012) for descriptions of the different types of spectrometers. Essentially, Raman spectroscopy relies on measuring the change in photon energy when photons from a laser are scattered from the vibrating atoms or molecular groups in the sample. During the collisional process, a few of the incident photons can gain or lose energy (most are scattered with no change in energy). This gain or loss in energy is detected and is characteristic of the atoms and molecular groups involved in the vibrational motion. The difference in energy is termed the Raman shift (Δv) and is reported in terms of cm^{-1}. The Raman shift results from inelastic interactions between the incident photons and the electron cloud around the vibrating molecules. When the electron cloud is easily deformed it is said to be Raman active and the deformation is termed the polarizability. Vibrations and molecular motions that are polarizable are Raman active.

In mineralogical applications, Raman spectroscopy is primarily used to characterize mineralogical phases, as well as, inclusions contained within gems and crystals (See Beyssac and Pasteris 2020). This approach uses Raman spectra as a fingerprinting technique comparing an unknown spectrum with some sort of crystalline standard. With glasses and melts the Raman spectrum is more commonly used to discriminate different vibrations and vibrational groups contained within the glass. In particular for SRO, it is the NBO and BO vibrations in the 800–1300 cm^{-1} region of the spectrum that are important. Figure 6a shows the Raman spectrum for SiO_2 glass and for SiO_2 glass with 5 mol% Na_2O added. Clearly there are significant changes in the high frequency region as network modifier (Na_2O) is added to the fully connected SiO_2 network composed entirely of Q^4 tetrahedra where all the oxygens attached to Si are BOs. The addition of a modifier creates NBOs and other types of Q^n ($n = 1$–3) species. Figure 6b (O'Shaughnessy et al. 2020) is a plot of a series of sodium silicate glasses with increasing modifier content. It shows the progressive change in shape and intensity of the bands in the high frequency region of the spectrum. Both Q^3 (at 1100 cm^{-1}) and Q^2 bands (at 980 cm^{-1}) are clearly observed at higher alkali contents.

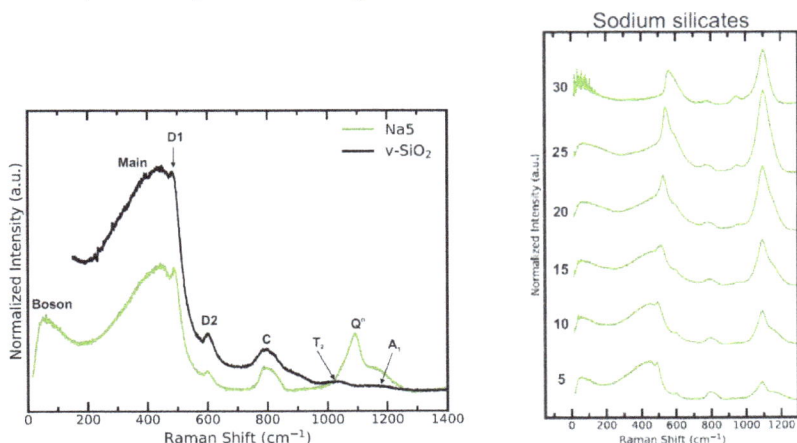

Figure 6. Unpolarized Raman spectra for **a)** SiO_2 glass and of Na-silicate glass with 5 mol% Na_2O added. Note the increase in intensity in the 1100 cm^{-1} region due to the presence of NBOs, as well as, increased intensity of the A_1 band of pure SiO_2; **b)** Full range of spectra for 5–30 mol% Na_2O added showing the rapid increase in Q^n bands as Na_2O increases. Republished with permission of John Wiley & Sons - Books from The influence of modifier cations on the Raman stretching modes of Q^n species in alkali silicate glasses, C O'Shaughnessy, GS Henderson, HW Nesbitt, GM Bancroft, DR Neuville, Journal of the American Ceramic Society 00:1–11, https://doi.org/10.1111/jace.17081, copyright (2020) permission conveyed through Copyright Clearance Center, Inc.

Raman peak or band positions depend primarily on the types of atoms undergoing vibration and the nature of the bonding between the atoms involved. The intensity is primarily related to the polarizability of the interatomic bonding as well as the types of atoms involved. The SiO_2 spectrum shown in Figure 6a can be used to aid interpretation of more complex compositions. The Boson peak at 10–200 cm^{-1} is characteristic of glasses and while still somewhat controversial is related to longer-range interactions. The broad feature labelled "main" in Figure 6a is at around 440 cm^{-1} and primarily related to asymmetric vibrations associated with BOs in rings containing > 5 SiO_4 tetrahedra with a mean of around 6 tetrahedra per ring. The sharp features labelled D1 and D2 (for defect band 1 and defect band 2) at 490 and 606 cm^{-1}, respectively are due to the breathing motion of oxygens in small 4- and 3-membered rings. The feature labelled C at ~800 cm^{-1} is a combination of two peaks whose origin is rather vague. They have been assigned to various "cage rattling" motions of the Si and O, as well as BO bending and symmetric stretch. The features labelled T_2 and A_1 are both due to motions of the BO in Q^4 tetrahedra (Fig. 7). T_2 involves the motion of 2 BO toward the Si atoms while the other 2 BOs move away from the Si, while the A_1 mode is the motion of all the BO toward and away from the Si. The A_1 band is actually 2 bands whose origin is unclear. Furthermore, it remains present over an extensive compositional range when modifiers are added to the glass. On the other hand, the T_2 band is somewhat controversial in that some authors propose that it is present in all glasses while others suggest that it becomes inactive upon modifier addition (see discussion of curve fitting and text below).

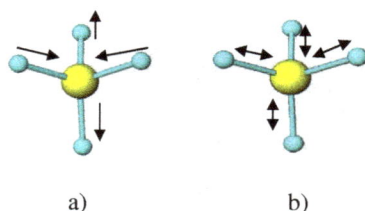

a)   b)

Figure 7. Q^4 vibrational modes. **a)** T_2 mode where pairs of oxygens vibrate toward and away from the Si atom and **b)** A_1 mode where all the oxygen move toward and away from the Si atom.

With the addition of a network modifier, additional bands are observed in the 850–1300 cm^{-1} region. These additional bands are due to the presence of NBO symmetric stretch vibrations and they occur at relatively fixed positions for the different Q^n (where $n = 0$–3) species. However, there are a number of issues that are worth pointing out. In order to quantify the relative proportions of the different Q species the high frequency envelope is often "curve fit" or "deconvolved" by fitting a series of Gaussian curves to the envelope. However, in the vast majority of studies the number of bands observable in the high frequency envelope exceeds the number of Q^n species present, and, consequently, additional bands must be added to obtain reasonable fits. These additional bands have been assigned to a range of possible structural species from two Q^3 and Q^2 bands (Matson et al. 1983) to the presence of the T_2 band (cf., Le Losq et al. 2014), to assignments such as Si–O^0 (cf., Franz and Mysen 1995).

Use of pure Gaussian line shapes may be responsible for some of these ambiguities. For example, Kamitsos and Risen (1984) noted that Raman bands of silicate glasses were predominantly Lorentzian rather than Gaussian. And more recently, Bancroft et al. (2018) and Nesbitt et al. (2018) have explored the lineshapes and linewidths of bands in the high frequency envelope of alkali and alkaline-earth silicate glasses. Bancroft et al. (2018) and O'Shaughnessy et al. (2020) point out that the additional bands are required because a) Gaussian line shapes do not fit the side wings of the high frequency Raman bands well;

b) there are additional bands in the spectral envelope due to MBO vibrations; c) the additional MBO bands make the Q^3 vibrational peak asymmetric; and d) the A_1 mode of the Q^4 tetrahedra is present even to high alkali contents and occurs as two peaks which merge together at 20 mol% added alkali (O'Shaughnessy et al. 2020). An example of their fits for two sodium silicate glass compositions with 15 and 30 mol% added Na_2O are shown in Figure 8.

However, it should be noted that these findings are only for alkali and alkaline-earth containing silicate glasses. When Al or other elements (e.g., Ti, P) are present, the high frequency envelope is more complex. Line shapes and assignments may well be different (cf., Le Losq et al. 2014). Care should be used when interpreting data for aluminosilicate glasses and other compositions: interpretations based on studies of alkali or alkaline-earth silicate glasses may not be applicable.

It is important to note that when curve fitting the high frequency envelope, care should be taken as to the nature of the vibrations. Authors may refer to the same vibrational motion using different terminology and it is important to have a good understanding of the terminology and nature of the motions i.e., rocking, bending, asymmetric, symmetric, rattling, twisting, torsional.

Furthermore, in both infrared and Raman spectra, the area of a fitted component is not necessarily directly proportional to the actual fraction of a given structural species, as absorption or scattering cross sections are expected to be affected to some degree by the structure itself. Independent constraints of mass-balance, or 'calibration' from independent measurements by other methods (e.g., NMR spectroscopy), have been applied to this problem (Mysen and Richet 2019). Full, quantitative calculation of vibrational spectra for glasses is potentially very important for resolving such empirical issues but remains a challenging problem. Recent Raman spectral simulations on alkali silicate glass structures (determined from classical MD simulations) suggest that component lineshapes for some Q^n species may be highly non-Gaussian and more complex, than conventional deconvolution functions (cf., Kilymis et al. 2019).

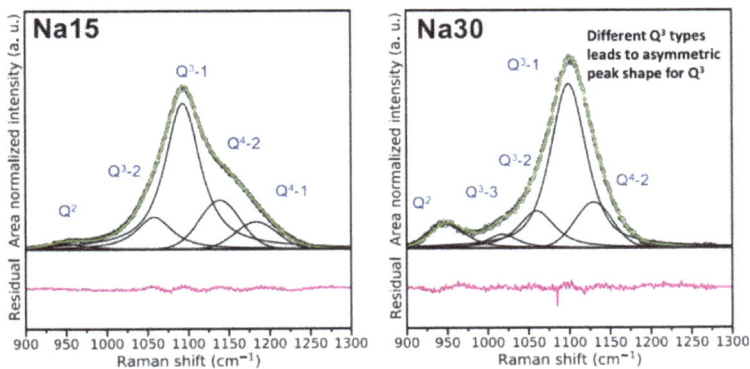

Figure 8. Fits for sodium silicate glasses with 15 mol% (Na15) and 30 mol% Na_2O (Na30). Q^3–1 is Q^3 tetrahedra without an alkali attached to the BO while Q^3–2 and Q^3–3 have 1 or 2 alkalis attached to the BO. Similarly, for the Q^4 and Q^2 tetrahedra. Note the asymmetry in the most intense band in Na30 due to the presence of multiple types of Q^3 species (BO, MBO). An alternative explanation for the additional peaks is second neighbour effects on the Q^n species (Olivier et al. 2001). In this model the Q^n units may be clustered with $Q^{n,ijkl}$ where *ijkl* are adjacent Q species of varying character i.e., Q^0, Q^1, Q^2, Q^3 etc.

Brillouin Spectroscopy. Brillouin spectroscopy is a technique (cf., Speziale et al. 2014) that uses the interaction between optical photons from an incident laser and phonon vibrations in a solid, to probe the elastic properties and acoustic velocities of the sample. It measures the scattered light within 10^{-2} to <10 cm^{-1} (usually $\pm$ 1–2 cm^{-1}) of the incident laser light.

Thus, one needs very high discrimination of photon energies which is usually obtained using a Fabry-Perrot spectrometer. There are two common types of measurements (cf., Speziale et al. 2014) which are used to obtain the sound velocity via

$$v_{s,p} = \lambda c \, \Delta\sigma / 2 \sin(\theta/2) \tag{1}$$

λ and c are the wavelength and velocity of light, and $v_{s,p}$ is the sound velocity. The Brillouin shift is $\Delta\sigma$ and the angle between the incident and scattered light is θ. Brillouin spectroscopy is used to observe changes in viscoelastic behaviour, acoustic velocities, coupling coefficients (between fluctuations and the electromagnetic field), the lifetime of the interactions, the anisotropy of the interactions, as well as, detecting polyamorphism, at high T and P.

Nuclear magnetic resonance (NMR) techniques. Nuclear magnetic resonance (NMR) spectroscopy records signals from nuclear spin transitions specific to a given nuclide, generally in the radiofrequency range (10's to 100's of MHz), which are observable when a sample is placed in a high magnetic field. The energies of these transitions are slightly perturbed by the electronic environment surrounding the nucleus, and thus by the chemical structure. Isotopes of many elements have readily measurable NMR spectra (MacKenzie and Smith 2002; Stebbins and Xue 2014), which can yield unique information about short to intermediate range structure in crystalline and glassy solids as well as in liquids. NMR of nuclides including ^{29}Si, ^{27}Al, ^{23}Na, and, with isotopic enrichment, ^{17}O, has been widely applied to silicate and aluminosilicate glasses of most central interest to the geosciences. ^{1}H, ^{13}C, ^{19}F, and even ^{33}S and ^{37}Cl NMR have provided further information on structural environments of geochemically important volatile species; a number of nuclides that are more difficult to observe (e.g., ^{25}Mg, ^{39}K, ^{43}Ca) are beginning to be informative. NMR studies of all of these and of other key components of technological glasses, especially ^{11}B and ^{31}P but including many others, have revolutionized our understanding of oxide glass structure.

The physics of the relationships of NMR observables to structure and dynamics is complex, even more so because modern NMR experiments often take advantage of interactions between like and unlike nuclear spins to gain unique information about proximity and interconnections between cations and ions. We can only begin to introduce the basics in this section. A more complete view was presented in several RiMG volumes on spectroscopic methods (Kirkpatrick 1988; Stebbins 1988, 1995; Stebbins and Xue 2014); several excellent books focus on the broader scope of solid state NMR of inorganic materials (Engelhardt and Michel 1987; MacKenzie and Smith 2002; Wasylishen et al. 2012) and on modern NMR experiments in general (Levitt 2001; Duer 2002, 2004). A number of extensive reviews are focused on NMR of glasses (Eckert 1992, 1994, 2017; Kirkpatrick et al. 1986; Stebbins 2002; Kohn 2004; Massiot et al. 2008; Zwanziger et al. 2019); in turn, many detailed reviews of glass structure refer extensively to NMR studies (Greaves and Sen 2007; Mysen and Richet 2019).

As for other spectroscopic methods, NMR spectra are generally presented as intensity vs. frequency (energy). Scales are most commonly expressed as the shift in frequency relative to a standard, in parts per million (ppm); for historical reasons plotted frequencies usually *decrease* from left to right. Intensities are most commonly shown in arbitrary units and analysis of data done in terms of relative contributions of different spectral components ("peaks" or "resonances"). However, in an ideal NMR experiment, intensities are actually quantitative with no corrections for 'cross sections' of different molecular groups or bonding environments, with the signal strictly proportional to the number of spins of a given nuclide in the sample. This can allow NMR to be used to quantitatively test or to calibrate other methods. Of course, as in any experiment, there are many ways that NMR results can be 'non-ideal' (Stebbins and Xue 2014). An important practical limitation for most NMR experiments on solid materials is that concentrations of ions with unpaired electron spins (e.g., many transition metal and rare earth cations) greater than a few tenths of 1% can cause serious spectral broadening and loss of information. Although there has been much recent progress in NMR of such 'paramagnetic' systems in crystalline oxides and silicates (Grey and Dupré, 2004; Strobridge et al. 2014; Kim et al. 2015a,b; Pecher et al. 2017; Stebbins et al. 2017, 2018), this has not yet extended very far into studies of glass structure.

A number of important NMR-active nuclides (e.g., ^1H, ^{13}C, ^{29}Si, ^{31}P) have nuclear spins of 1/2, and the primary effect on spectra is through the 'chemical shift'. This is the small change in frequency dominated by the short-range structure, most notably first-shell bond distances, and, especially for cations, the coordination number by oxygen. For example, increasing the oxygen coordination (with accompanying increase in mean bond distance) systematically decreases the chemical shift (to lower frequency) for most cations observable by NMR. Bond angles (e.g., mean Si–O–Si angle around Si), and second neighbors (e.g., Na^+ vs. Ca^{2+} vs. Al^{3+} vs. Si^{4+} first cation neighbors to Si) also have well-studied effects on chemical shift. All of these have been systematized by decades of study on crystalline model compounds, and, more recently, increasingly accurate first-principles calculations (e.g., Charpentier et al. 2004; Ashbrook et al. 2007; Pedone et al. 2010; Angeli et al. 2011).

Many other NMR nuclides (e.g., ^{11}B, ^{17}O, ^{23}Na, ^{27}Al) have nuclear spins > 1/2 and are thus also subject to nuclear 'quadrupolar' interactions (MacKenzie and Smith 2002; Stebbins and Xue 2014). The most important parameter, the 'quadrupolar coupling constant' (C_Q) depends on the distortion of the local chemical environment from purely cubic local symmetry. In an ordered crystal, this produces a characteristic peak shape and shift down in frequency that can be analyzed to give structural information; for comparable sites in glasses, structural disorder generally yields a shifted, broadened, asymmetric peak that can only be approximated by fitting with several adjustable variables and distributions of parameters. Fortunately, quadrupolar shifts and broadening decrease greatly with ever higher magnetic field strengths, giving increasing resolution, quantitation, and information content. At higher fields, the chemical shift contribution to the spectrum begins to dominate, often simplifying extraction of structural information.

Nuclear magnetic interactions (like those observed in some other spectroscopies) generally depend on the orientation of a molecule, crystal, or structural group in the external magnetic field. In a polycrystalline or an amorphous solid, this can lead to a large spreading out of the signal and loss in resolution and reduction in signal to noise ratio. Some such 'anisotropic' effects (chemical shift anisotropy and nuclear dipole couplings) can be averaged out by rapid rotation of the sample on an axis at the 'magic angle' (about 54.7°, 'magic angle spinning' or MAS NMR) to the external field, usually resulting in much narrower peaks (at the 'isotropic chemical shift') and much enhanced resolution. In some cases, the anisotropy itself can carry useful structural information, as sometimes seen in 'static' NMR experiments (non-spinning) or in advanced experiments such as 'Magic Angle Flipping' (Davis et al. 2010, 2011). Quadrupolar broadening is only partially averaged by MAS but can be eliminated by 2-dimensional 'multiple quantum' (MQMAS) NMR. Accurate quantitation of intensities in the latter can be complicated, however. Dynamic Angle Spinning (DAS) and Double Rotation (DOR) NMR can also eliminate quadrupolar effects but may be technically more difficult to implement (Florian et al. 1996).

In liquids (and even in some mobile solids), reorientation may be rapid about multiple, random axes, completely averaging all anisotropic effects, and yielding the very narrow NMR lines typical of molecular liquids. This effect is routinely exploited in ^1H NMR in organic chemistry. If, at the same time, bond-breaking and exchange among different structural environments is rapid compared to the frequency width of NMR spectrum, only a single, averaged NMR signal will be observed. This situation is common in very high temperature, *in situ* NMR experiments on oxide liquids (Stebbins and Farnan 1992; Florian et al. 1995; Massiot et al. 1995, 2008; Capron et al. 2001; Kanehashi and Stebbins 2007), for example when Al^{3+} cations may have 4 vs. 5 vs. 6 oxygen neighbors. Spectra still will contain useful, quantitative information about the average local structure, and effects on temperature and composition.

In the intermediate dynamical regime, typically up to a few hundred °C above the glass transition temperature, viscosity is still high enough, and structural exchange slow enough, so that in some cases the dynamics of bond breaking in the liquid can actually be measured by NMR (Farnan and Stebbins 1990a,b, 1994).

Many advanced NMR experiments take advantage of the interactions between like or between unlike nuclear spins to gain unique information about ordering and structure. Indeed, it is this potential to directly probe element-specific, interatomic interactions that may be the most unique capability of NMR spectroscopy. Such interactions can be through-space and highly sensitive to distance (dipole-dipole interactions) or through-bond and highly sensitive to bonded connections ('J' couplings). In such experiments, which are described by a plethora of acronyms (Duer 2004), two or more nuclear spin systems may be perturbed simultaneously or sequentially. Data are often recorded as two-dimensional spectra, with two frequency axes derived from variation of some time interval during the experiment. In oxide glasses, most such 'double (or even triple) resonance' experiments rely on pairs of nuclides with strong nuclear dipolar couplings, such as ^{1}H with ^{29}Si or ^{27}Al, or other pairs such as ^{11}B–^{27}Al and ^{27}Al–^{31}P (van Wüllen et al. 1996a,b; Chan et al. 1999; Zhang and Eckert 2006). However, recent advances have yielded results with more weakly coupled pairs such as ^{17}O–^{27}Al and ^{17}O–^{29}Si (Jaworski et al. 2015; LaComb et al. 2016; Sukenaga et al. 2017). Other advanced NMR experiments, relying on interactions between like pairs of nuclides (e.g., ^{29}Si–^{29}Si) can provide unique information on longer range structure, including the extent of phase separation (Martel et al. 2011, 2014). As in any complex experiment, careful testing and even calibration of results on crystalline materials of known structure is very helpful in validation.

X-ray absorption and X-ray Raman. X-ray absorption spectroscopy (XAS) consists of two principle techniques: X-ray absorption near-edge structure (XANES) spectroscopy (cf., Henderson et al. 2014a) and Extended X-ray absorption spectroscopy (EXAFS) (cf., Newville 2014). The XAS spectrum is composed of 3 regions: XANES (0–50 eV above the X-ray absorption edge); EXAFS (~ 50–1500 eV above the X-ray edge), and the pre-edge at ~20–30 eV below the edge (Fig. 9). Since both EXAFS and XANES are element specific, they can potentially be obtained from any element at concentrations as low as ~ 100 ppm.

Both techniques probe the structural environment of specific types of atoms but of course the information extracted is for the average structure of the element over all possible sites in

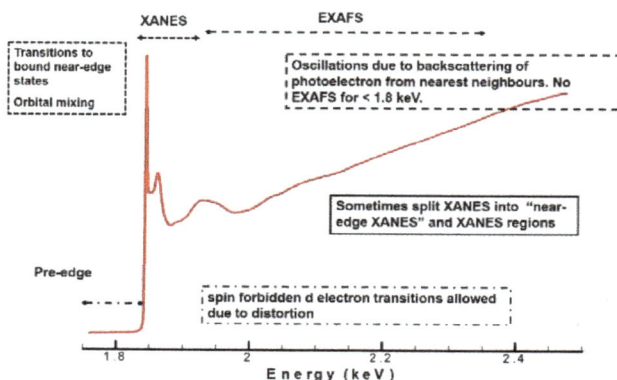

Figure 9. XAS spectrum for Si *K*-edge showing the XANES and EXAFS regions, as well as the pre-edge which is used to discriminate different transition metal coordination states. It is not possible to obtain EXAFS for elements with excitation energies below 1.8 keV because of interference effects from overlapping elemental edges.

the glass. EXAFS is primarily used to determine the first shell or nearest neighbour interatomic distances and angles around the element of interest. It can also determine features out to 2nd and 3rd neighbours (it is limited to maximum distances of ~10 Å), as well as the coordination of the element. XANES provides information about the electronic structure of the element, the oxidation state, and for transition metals, the average coordination. XANES data are qualitative while interatomic distances and static disorder or Debye–Waller parameters determined by EXAFS are quantitative. X-ray Raman spectroscopy (XRS) or Near-Edge Inelastic X-ray scattering (NRIXS) is similar to XANES producing spectra that are qualitatively similar. However, the excitation mechanism is different and uses momentum transfer (similar to optical Raman) of hard X-rays (>10 keV) to excite the electronic transitions of interest. The great advantage of X-ray Raman is that it excites all the elements in the sample and can be used for high T and P *in-situ* studies. It is currently the only technique that can probe light elements (<2 keV) at high T and P as it does not require the sample to be under high vacuum. The disadvantage is that it can take a long time to collect data, although this is improving as the number of detectors on synchrotron X-ray Raman beamlines used to measure the emitted photons are increasing. The reader should see Lee et al. (2014) for a more comprehensive discussion of X-ray Raman.

Another technique that has been used to investigate the SRO of glasses is Electron Loss Near-Edge spectroscopy or ELNES. This technique provides information similar to XANES (cf., Brydson et al. 2014).

The role of transition metals in the structure of glasses is particularly important. The pre-edge of the XAS spectrum can be used to identify the oxidation state and coordination of many transition metals. The peaks in this region are from spin forbidden transitions of the d-electrons which are allowed due to site distortion. Application of curve fitting techniques and comparison to crystalline standards can be used to determine the likely coordination and oxidation states in the glass. Another useful method for determining transition metal coordination and oxidation state is to use *L*- or *M*-edge XANES (cf., Henderson et al. 2002; Höche et al. 2013, Smythe et al. 2013) but it is more difficult to interpret. This is because the electronic transitions are due to 2p to 3d states. Interpretation is greatly facilitated by first principles calculations in which the partial density of states are calculated for the different orbitals.

The XAS edge position is also dependent upon both oxidation state and coordination while the XANES depends on the next nearest neighbours and the electronic interactions contributing to the spectral envelope. Consequently, a finger printing technique can be used to compare glass XANES with crystalline analogues, as well as some sort of linear combination modelling to obtain relative proportions of different CN and oxidation states.

Other techniques. There are a number of other techniques that are employed to investigate the SRO of glasses. These include Mössbauer Spectroscopy (primarily for Fe, Sn, Au oxidation state and coordination) (Johnson and Johnson 2005), X-ray photoelectron spectroscopy (XPS) (Nesbitt and Bancroft 2014), and UV/visible spectroscopy (Rossman 2014). XPS is used to probe the nature of the oxygen environments in glasses (Nesbitt et al. 2015c) while the latter is mainly used to probe the role of transition elements in color in glasses (cf., Calas et al. 2020).

NETWORK-FORMING OXIDES AND CATIONS

Pure SiO_2, GeO_2, and B_2O_3 readily form glasses on quenching from the high-temperature liquid (melt) phase. The cations in these glasses have high valences and relatively small radii, leading to short-range structures with small numbers of oxygens in the first shell—four for silica and germania, three for boron oxide, as in corresponding low-pressure crystalline phases (or at least one of those phases in the case of GeO_2). The short cation–oxygen bonds are especially strong; the liquid phases have relatively high viscosities, low diffusivities, and are slow to crystallize.

In the crystals as well as the glasses, and probably to a large degree in the high temperature liquids, the SiO_4, GeO_4, or BO_3 polyhedra are connected to each other through bridging oxygens, forming strong three-dimensional networks. Hence the name 'network forming oxides', and 'network forming cations'. Other cations can play this role in some more complex oxide melt compositions, including Al^{3+} (especially important in magmatic liquids and glasses, see below), P^{5+}, and even geologically rare Sb^{5+}, As^{5+}, Mo^{6+}, and others. For this chapter, our discussion of short-range structure around such network cations will focus primarily on geochemically abundant Si^{4+} and Al^{3+}, with a few comparisons to Ge^{4+} and B^{3+} for context.

Silicon coordination in oxide glasses and melts

Pure SiO_2 glass has been extensively studied as an endmember for silicates, and because silica is a key material for advanced technologies such as data transmission fibers and UV-transmitting optical components in semiconductor fabrication. All evidence points to four-fold oxygen coordination (SiO_4 tetrahedra, $^{[4]}Si$) as by far the predominant structural group in pure silica: for example, ^{29}Si NMR studies have detected only this species with a detection limit for $^{[5]}Si$ and $^{[6]}Si$ of a fraction of one percent. This remains the case even for glasses quenched from pressures up to about 15 GPa, even when samples are recovered with up to about 15% increased density (Devine et al. 1987; Xue et al. 1991; Trease et al. 2017).

Given the similarity of Si coordination in the pure glass phase and in low-pressure crystalline forms of silica, structural studies have concentrated on measuring mean values and statistical distributions of slightly longer-range variables, in particular Si–O–Si bond angles, Si–Si distances, and proportions of rings of various sizes (see above). Most available methods have been brought to bear on this problem, for example X-ray scattering (Neuefeind and Liss 1996; Benmore 2015), Raman (Martinet et al. 2015), simple ^{29}Si MAS NMR (Devine et al. 1987) to advanced, 2-dimensional ^{17}O NMR (Trease et al. 2017). There is now a reasonably good consensus on at least an overall, statistical description in ambient pressure and a few high-pressure silica glasses, that correlates well with more 'direct' methods such as X-ray scattering.

With increasing pressure, the main structural response of the glass network involves deformation of the SiO_4 tetrahedron (Inamura et al. 2001) accompanied by changes in the Si–O–Si angle. However, at around 7 GPa there is evidence for 'polyamorphism' (Inamura et al. 2004). Polyamorphism is similar to polymorphism in crystalline materials where the structure of the mineral undergoes a displacive (α-quartz to β-quartz) or reconstructive (coesite to stishovite) phase transformation resulting in a different mineral structure or polymorph. The polymorph has a well-defined P, T and composition range over which it is found. Glasses may also undergo similar behaviour with different structures being observed over relatively well-defined P, T and composition ranges (cf., Majérus et al. 2004a). Polyamorphism has now been observed in a number of glasses including SiO_2, GeO_2, as well as, alkali and alkaline-earth silicates and aluminosilicate glasses (Majérus et al. 2004b).

At much higher pressures, *in situ* observations by X-ray and neutron scattering, and other methods, clearly show that Si coordination in pure silica glass does eventually increase to 5 and 6 up to 100 GPa (Sato and Funamori 2008, 2010). The onset of the 4- to 6-fold coordination change is around 15 GPa (Benmore et al. 2010) and increases linearly above this pressure with 5-fold Si acting as an intermediate during the transformation (cf., Zeidler et al. 2014). Below 15 GPa the SiO_4 tetrahedra shrink slightly. A transformation from 6 to higher coordinated Si may occur above 140 GPa (cf., Murakami and Bass 2010). With increasing T there is little change in SiO_2 SRO. The Si–O bond lengthens but the overall structural unit remains the SiO_4 tetrahedron which expands slightly. The Si coordination remains 4 even into the melt phase and the Si–O–Si angle undergoes a slight decrease (Mei et al. 2007). Such studies have demonstrated large structural changes at very high pressures in a wider range of glass compositions, for example the geophysically more interesting $MgSiO_3$ (Kono et al. 2018).

Variations in Si coordination become more readily measurable in two- or multicomponent-silicate compositions, most obviously in binary potassium and sodium silicates such as $K_2Si_4O_9$. [5]Si and [6]Si were first detected, and quantified, in such compositions by ^{29}Si MAS NMR in glasses quenched from melts at pressures up to 12 GPa and decompressed for study at ambient pressure (Stebbins and McMillan 1989; Xue et al. 1991) (Fig. 10). About 5 to 10% of such species were seen but their abundances decrease to low but still measurable values (<0.1%) in glasses quenched at ambient P (Stebbins and McMillan 1993). Raman spectra of these glasses showed obvious changes, but these did not readily correlate with cation coordination increases. Later NMR studies also detected high-coordinated Si in other compositions including Ca and Mg silicates and Na aluminosilicates (Stebbins and Poe 1999; Gaudio et al. 2008; Kelsey et al. 2009a). Data for the metastable, high pressure crystalline phase $CaSi_2O_5$, which contains [4]Si, [5]Si, and [6]Si sites, confirmed that ^{29}Si chemical shifts near to −150 ppm are indeed due to [5]Si (Stebbins and Poe 1999).

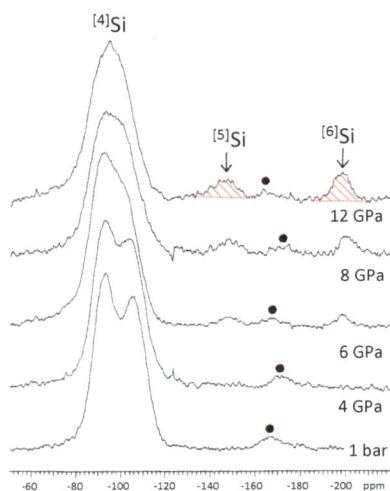

Figure 10. ^{29}Si MAS NMR spectra of $Na_2Si_4O_9$ glass, quenched from melts at pressures shown. The predominant intensity is from [4]Si, but peaks for both [5]Si and [6]Si are clearly visible, and are more intense in higher pressure samples. Black dots mark spinning sidebands (Xue et al. 1991).

More recent work on pressure effects on recoverable glass structure found that, in typical experiments in solid-medium high pressure apparatus, significant transient pressure drops during cooling may occur, leading to underestimation of both density increases and structural changes in recovered glasses (Gaudio et al. 2009; Bista et al. 2015; Stebbins and Bista 2018). Bringing a glass to a metastable equilibrium supercooled liquid state, just above its glass transition temperature, yields more accurate results with higher coordination levels for Si as well as for Al (see below) and B in borosilicates (Bista et al. 2017) when compared to samples quenched from above the liquidus. Possible change in network cation coordination during decompression of such quenched glasses is still an incompletely understood problem, but for multicomponent glasses is thought to be relatively unimportant (Malfait et al. 2012) at least to pressures of about 10 GPa. In any case, measured structural changes in recovered high pressure glasses can be correlated with measured property changes, most notably densification, to constrain mechanisms. For example, it is clear that in the aluminosilicates and borosilicates studied, recoverable densification is considerably greater than can be readily explained by observed increases in network cation coordination, indicating that much of the bulk density change is occurring by compression of 'soft' network modifier sites.

^{29}Si MAS NMR studies have also measured low concentrations (up to a few tenths of 1%) of [5]Si in alkali silicate glasses with high (>70%) silica contents formed at 1 bar pressure (Stebbins 1991, 2019; Stebbins and McMillan 1993). Glasses quenched more rapidly, and thus recording the melt structure at higher fictive temperatures (fictive temperature is the point at which the liquid structure 'freezes in' for a given cooling rate; it can be varied by about 100 to 200 K by common lab methods), have more [5]Si, consistent with the hypothesized role of this species as a high-energy 'transition complex' in mechanisms of bond swapping and network diffusion that may control transport properties in the liquid state (Fig. 11). However, this hypothesis has mostly been based on molecular dynamics simulations and other theoretical approaches (Angell et al. 1982; Brawer 1985; Yuan and Cormack 2001; Machacek et al. 2007; Ohkubo et al. 2017) and has not yet been directly validated by experiment.

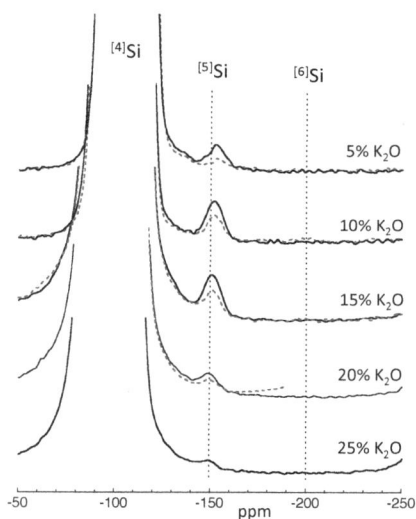

Figure 11. ^{29}Si MAS NMR spectra of K-silicate glasses formed at 1 bar pressure, with mole% K_2O as shown. The vertical scales are greatly enlarged to show [5]Si peaks. [6]Si is too low in concentration to detect in these samples. Solid lines show data for rapidly quenched glasses (higher fictive temperatures); dashed lines for annealed, slowly cooled glasses (lower fictive temperatures). Systematic effects of composition can be seen in changing [5]Si peak intensities. Reprinted from Pentacoordinate silicon in ambient pressure potassium and lithium silicate glasses: Temperature and compositional effects and analogies to alkali borate and germanate systems. Stebbins JF, Journal of Non-Crystalline Solids: X 1: 100012 Published by Elsevier (CC BY-NC-ND 4.0).

The first systematic study of compositional control on [5]Si concentrations, again in the ambient-pressure K_2O–SiO_2 system, showed that the high coordinate species starts off very low at the high-silica end of the binary, rises to a maximum at about 10 mole% K_2O, then drops below detection above about 25% K_2O (Stebbins 2019). This complex pattern is qualitatively similar to what has long been known from ^{11}B NMR of alkali borate glasses ([3]B + [4]B species), and which has long been thought to be the case (somewhat less definitively) for alkali germanate glasses ([4]Ge + [5]Ge + [6]Ge), although in these systems the concentrations of higher-coordinated cations are orders of magnitude higher than SiO_5 in silicates. This suggests that there may be underlying energetic similarities among all three types of systems, perhaps involving a fundamental mechanism of interaction between cation coordination and oxygen speciation (see below).

Aluminum coordination

Pure, molten Al_2O_3 cannot be quenched to a glass. Remarkably, given its high melting point (2072 °C), the stable liquid has been studied extensively by *in-situ* methods, including X-ray and neutron scattering (Ansell et al. 1997; Landron et al. 2001; Weber et al. 2008; Shi et al. 2019) and ^{27}Al NMR (Florian et al. 1995; Skinner et al. 2014). Ansell et al. (1997) found Al was predominantly tetrahedral [4]Al in Al_2O_3 liquid as compared to being octahedral ([6]Al) in corundum. On the other hand, Landron et al. (2001) using neutron diffraction suggested that a significant proportion (25%) of the Al sites were [5]Al. They also suggested that oxygen triclusters also formed (cf., Lacey 1963).

Results for pure liquid alumina do not uniquely quantify the proportions of [4]Al, [5]Al and [6]Al (for example, the NMR sees only the motional average of all three), but it is clear that the mean coordination number is between 4 and 5, far lower than in the six-fold Al coordination in crystalline corundum. Amorphous solid aluminum oxide can be made by sputtering methods employed in the fabrication of optical and other thin-film coatings. In such solids, ^{27}Al MAS and MQMAS NMR can measure speciation, which appears to give a somewhat higher mean value of Al coordination than in the equilibrium stable liquid (Kim et al. 2014). It has been suggested that such thin-film oxides might represent the structure of metastable, deeply supercooled liquids. If that is the case, then here the effect of increasing temperature is to decrease the mean network cation coordination.

Al^{3+} in silicate and oxide minerals is commonly 4- or 6- coordinated, with a few structures (e.g., andalusite) containing [5]Al. In most low-pressure aluminosilicate glasses and melts, [4]Al groups predominate, although [5]Al and [6]Al become more abundant in peraluminous compositions and in non-geological compositions with high borate or phosphate contents. Up to about 10% of [5]Al, with a few % [6]Al, are present in most metaluminous and peralkaline aluminosilicate compositions. These are readily detected by ^{27}Al MAS NMR at high magnetic fields (required to reduce the effects of quadrupolar coupling), or by 2-dimensional, MQMAS NMR (McMillan and Kirkpatrick 1992; Toplis et al. 2000; Neuville et al. 2004, 2006, 2008) (Fig. 12). Accurate quantification of these species may be somewhat model-dependent because of overlap of peaks for the three species, but different studies are more-or-less consistent.

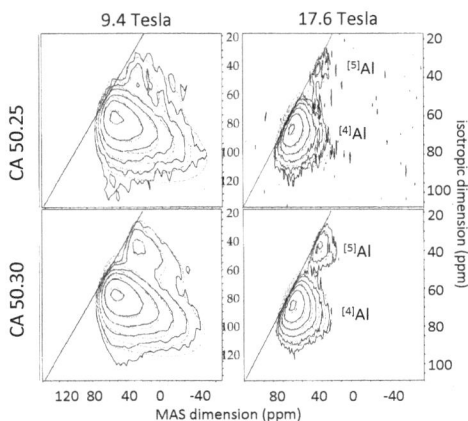

Figure 12. ^{27}Al MQMAS NMR spectra for two Ca-aluminosilicate glasses, collected at two different magnetic fields as labeled. CA50.25 is 50 mol% SiO_2, 25 mol% Al_2O_3, 25 mol% CaO (=$CaAl_2Si_2O_8$); CA50.30 is 50 mol% SiO_2, 30 mol% Al_2O_3, 20 mol% CaO (peraluminous). [5]Al signals are clearly visible in both but are more intense for the peraluminous composition. Reprinted from Neuville DR, Cormier L, Massiot D, ^{27}Al MQMAS NMR spectra for two Ca-aluminosilicate glasses, Al environment in tectosilicate and peraluminous glasses: a Al-27 MQ-MAS NMR, Raman, and XANES investigation. Geochimica et Cosmochimica Acta 68: 5071–5079, copyright (2004), with permission from Elsevier.

Within a given data set, systematic effects of composition or pressure may be better constrained than absolute speciation, especially for more disordered compositions with broader peaks (e.g., Mg aluminosilicates). Compositional effects on Al coordination are complex, but it is clear that network modifier/charge compensator cations with higher field strength (defined here as the ratio of the cation valence to the square of the mean cation–oxygen bond distance), systematically increase the Al coordination ($K^+ < Na^+ < Ca^{+2} < Mg^{+2}$) (Allwardt et al. 2007; Kelsey et al. 2009b).

As in crystals, increases in Al coordination in melts and glasses with higher pressure are more rapid than for Si, although only a few studies have been made of both cations in the same high-pressure glasses (Yarger et al. 1995; Kelsey et al. 2009a). Indeed, in some Ca aluminosilicates quenched from 10 GPa, [4]Al becomes the least abundant Al species (Allwardt et al. 2005b), whereas it still predominates in analogous Na and K compositions. The effect of non-network cation field strength noted above is thus even more apparent in high pressure glasses (Fig. 13). Also important is a strong effect of non-bridging oxygen (NBO) content, as NBO provides one relatively 'easy' pathway to the formation of [5]Al and [6]Al (see below). Low-NBO alkali aluminosilicate glasses, for example oft-studied 'charge balanced' compositions such as $NaAlSi_3O_8$ (albite) and $NaAlSi_2O_6$ (jadeite), thus show the least changes in Al coordination with pressure (Bista et al. 2015; Bista and Stebbins 2017). These compositions may be good representations of what happens in granitic compositions but are poor models for glasses rich in NBO and high field strength cations such as Mg^{2+}, Fe^{2+}, and Ca^{2+}. The effect of NBO content decreases, however, for higher field strength modifier cations, e.g., it is less apparent in Mg aluminosilicates (Bista and Stebbins 2017). This is to be expected as the length and strength of the modifier-oxygen bond approaches that of Al–O, decreasing the energetic distinction between BO and NBO.

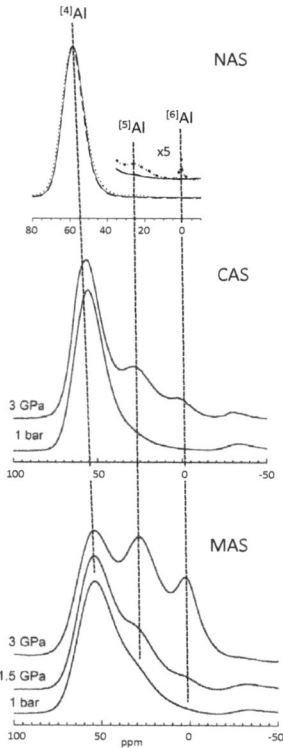

Figure 13. ^{27}Al MAS NMR spectra (18.8 Tesla field) for glasses quenched at 1 bar, and from just above the glass transition temperatures at high pressures. All three compositions are on 'charge balanced' joins (1:1 ratio of modifier oxide to alumina): NAS is 67% SiO_2 ($NaAlSi_2O_6$), CAS and MAS are 60% SiO_2. For NAS, spectra for lower pressure (solid and dashed) and high pressure (3 GPa, dotted) glasses are superimposed; a small amount of [5]Al at 3 GPa only is visible with a 5× increase in vertical scale. The tiny [6]Al peak in that spectrum is from a trace of crystalline jadeite. Note the large effect of increasing modifier cation field strength ($Na^+ < Ca^{2+} < Mg^{2+}$) on [5]Al and [6]Al concentrations, both at low and high pressure. The development of [5]Al and [6]Al in the NAS and CAS glasses is limited by the low NBO contents of these compositions; this mechanism is much less important in the MAS system (Bista et al. 2015; Bista and Stebbins 2017).

As for Si coordination in alkali silicates, recent work has shown that earlier studies on glasses quenched from high temperature melts at high pressure may under-represent both recoverable Al coordination increases and density changes because of pressure drop effects, sometimes avoidable by working near to the glass transition temperature (Gaudio et al. 2009; Bista et al. 2015). It is now clear that in simple three-component analogs of basaltic liquids, [5]Al and [6]Al become important components (totalling >30% of Al) even at typical source-region pressures of 3 GPa, suggesting that models of properties vs. composition may need to be more complex to account for structural change even in this regime (Ghiorso 2004).

A few ^{27}Al MAS NMR studies of aluminosilicate glasses with varying fictive temperature have shown that, as for [5]Si, [5]Al concentrations increase with temperature, correlating with what is measured from averaged ^{27}Al chemical shifts in high temperature, in-situ spectroscopy (Stebbins et al. 2008; Le Losq et al. 2014). However, other work on compositions with multiple different network cations (e.g., aluminoborosilicates) has shown that complex interactions among different cationic and anionic species can lead to either increases or decreases with temperature in coordination, depending on composition (Morin and Stebbins 2016). *In-situ*, high T NMR studies of a number of aluminate liquids have provided further insights into these effects as well as the dynamics controlling viscosity and diffusion (Florian et al. 1995; Massiot et al. 1995; Capron et al. 2001; Massiot et al. 2008).

Other experimental approaches, for example X-ray and neutron scattering, Raman and X-ray spectroscopies, tend to be less quantitative in measuring relatively small concentrations of high-coordinate Si and Al, but are often more feasible to implement for in-situ high temperature and/or high pressure studies, which are potentially very important in avoiding uncertainties introduced by quenching and/or decompression. Some of the earliest studies on aluminosilicate glasses at high P were those of Kushiro (1976, 1978) and colleagues (Kushiro et al. 1976) who investigated the pressure effects on the viscosity of albite, jadeite and andesite and basaltic melts. They concluded that the viscosity behaviour of the melts was due in large part to changes in the coordination of Al from [4]Al to [6]Al although this was questioned by a number of other studies. Since then there have been many studies on the structure of aluminosilicate glasses and melts at both ambient (Weigel et al. 2008a; Neuville et al. 2010) and high P (cf., Drewitt et al. 2011, 2015; Bernasconi et al. 2016), high T (Hennet et al. 2016; Allu et al. 2018), and high P and T (Drewitt et al. 2011; Sakamaki et al. 2012), using a number of techniques (X-ray and neutron diffraction, Al K- and $L_{2,3}$-edge XANES, Raman, NMR and MD simulations).

For example, Neuville et al. (2010) looked at glasses, liquids and crystals along the CaO–Al_2O_3 join and found that Al was predominantly [6]Al in high alumina (>80% mol% Al_2O_3) melts, while at lower Al_2O_3 content, (~25 mol% Al_2O_3) Al was predominantly [4]Al and in Q^2 species. Glasses along the $(CaO)_x–(Al_2O_3)_{1-x}$ binary have been investigated by Drewitt et al. (2011, 2012) at high T. The results show Al is predominantly [4]Al in the melt with approximately 20% [5]Al at $x = 0.33$ but this reduces with increasing CaO concentration. Ca is predominantly in 6-fold coordination while oxygen triclusters occur in $CaAl_4O_7$ (CA2) (13%) $CaAl_2O_4$ (CA) (7%) and $Ca_3Al_2O_6$ (C3A) (0.6%) composition liquids with numbers decreasing with increasing CaO. High pressure studies of CA and anorthite ($CaAl_2Si_2O_8$) composition glasses up to 32.4 GPa have been performed by Drewitt et al. (2015) and suggest a continuous change in Al coordination with P with [5]Al dominating at 5 GPa and [6]Al at higher pressures.

Weigel et al. (2008a) investigated a number of sodium aluminosilicate glasses using Al $L_{2,3}$ XANES and showed that this edge was sensitive to both Al coordination and electronic properties. Albite ($NaAlSi_3O_8$), jadeite ($NaAlSi_2O_6$) and nepheline ($NaAlSiO_4$) composition glasses all showed Al predominantly in 4-fold coordination and unchanged with the replacement of Al by Fe. Similar results were obtained by Bernasconi et al. (2016) on a series of Ca and Na aluminosilicate glasses. Average coordinations from their EPSR analysis showed Si and Al both 4-fold coordinated, and Na 6–7-fold and Ca 6-fold coordinated. Sakamaki et al. (2012)

investigated jadeite composition melts up to 4.9 GPa and also found Al predominantly in 4-fold coordination. Pressure-induced structural changes mainly involved a systematic reduction in the T–O–T (T = Si, Al) angle, shortening of the T–O and T–T bond distances. Hennet et al. (2016) looked at the $CaO–Al_2O_3–SiO_2$ system at high temperature using neutron diffraction and aerodynamic levitation heating. Their conclusions were that Al is predominantly [4]Al with some [5]Al at high T with little change in the Ca coordination (predominantly 6-fold). Recently, Moulton et al. (2019) have investigated anorthite composition glasses to 20 GPa using Raman and Brillouin spectroscopy. They suggest that the Brillouin spectroscopy exhibits distinct changes at 2 and 5 GPa with the latter indicative of a change of coordination of Si to [5]Si (cf., Moulton et al. 2019 for a complete discussion). Finally, the CMAS system has been studied recently by Allu et al. (2018). They used a variety of techniques to explore the structure of Mg and Ca containing aluminosilicate glasses and glass ceramics and its relationship to aluminum avoidance and crystalline phases.

OXYGEN ANION SPECIATION

Most geoscientists first learn about oxide and silicate mineral structures from models of interconnected cation polyhedra, which as noted above form the starting point for many discussions of glass and melt structure. This is, of course, a useful approach, but can be somewhat limiting, as various forms of oxide anions are generally the numerically and volumetrically predominant ions in any oxide material. Changes in coordination (speciation) of oxide ions are intimately connected to changes in the coordination of the cations: changes in one must lead to changes in the other. When both can actually be measured then quantitative assessments of mechanisms linking structure and properties may be derived (Stebbins 2020).

Order/disorder among bridging oxygens

A mineralogically familiar example of the importance of oxygen anion speciation is that of order/disorder among bridging oxygens in framework aluminosilicates, e.g., feldspars and feldspathoids and their glassy and molten equivalents along joins such as $Na_2Al_2O_4–SiO_2$ and $CaAl_2O_4–SiO_2$. Given entirely tetrahedral coordination of Al and Si in such compositions (exact in crystals, a starting approximation in many glasses), all of the network forming cations are linked to others, and all of the oxygens form bridges between them, either Si–O–Si, Si–O–Al, or Al–O–Al. Simple bond valence and field strength considerations indicate a systematic increase in residual negative charge on bridging oxygens in the order given, requiring more interaction with 'charge compensating' cations, typically K^+, Na^+, or Ca^{2+} in feldspars and corresponding glasses.

An equilibrium among the three BO species can be written:

$$Si–O–Si + Al–O–Al = 2(Si–O–Al) \qquad (2)$$

In typical aluminosilicate minerals with Al/Si less than or equal to 1, Al–O–Al species are generally considered to be energetically unfavorable because of the difficulty of bringing in a sufficient number of relatively large charge compensating cations (the 'aluminum avoidance' that is prevalent in K-, Na- and Ca-aluminosilicate minerals), leading to Al/Si ordering and shifting Equilibrium (2) to the right. Higher temperatures will favor the disorder that is generated by shifting this equilibrium to the left in most geologically interesting compositions (Si/Al > 1), as is suspected in disordered crystalline anorthite formed near the melting point (Phillips et al. 1992). Higher field strength compensator cations clearly also can lead to substantial Al/Si disorder and violation of 'aluminum avoidance', as clearly shown by ^{29}Si MAS NMR studies of high temperature crystalline $LiAlSiO_4$ (β-eucryptite) and $Mg_2Al_4Si_5O_{18}$ (synthetic cordierite), both of major importance in advanced ceramic materials (Putnis et al. 1985; Phillips et al. 2000). These same energetic considerations are a starting point for thinking about glass and melt structure.

^{29}Si MAS NMR has long been an important tool for assessing Al/Si order/disorder in aluminosilicate minerals because varying the number of Al first neighbors to an SiO_4 group systematically changes the chemical shift (Engelhardt and Michel 1987; Kirkpatrick 1988), often leading to resolvable peaks in spectra. In glasses, unfortunately, ranges in chemical shift, due to disorder, broaden peaks and eliminate such resolution. However, with some assumptions required for fitting of overlapping components, and simultaneous analysis of spectra from multiple compositions, it may be feasible to estimate the extent of disorder in glasses such as the $NaAlSiO_4$–SiO_2 and $CaAl_2Si_2O_8$–SiO_2 systems (Murdoch et al. 1985; Lee and Stebbins 1999). Partial ordering (i.e., more Al–O–Al than predicted by 'aluminum avoidance', but less than a fully random distribution) was determined, as well as a greater extent of disorder in the Ca- vs. Na-aluminosilicate system. The results for the alkali aluminosilicates were independently confirmed by 2-dimensional ^{17}O MQMAS NMR, which can greatly enhance resolution among the bridging oxygen species in some compositions (Dirken et al. 1997), and demonstrated the clear presence of Al–O–Al (Lee and Stebbins 2000a,b; Fig. 14).

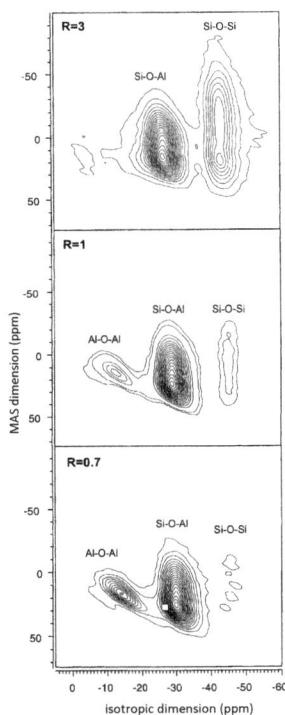

Figure 14. ^{17}O MQMAS NMR spectra for glasses on the $Na_2Al_2O_4$–SiO_2 ('charge balanced') join, with R = Si/Al as labeled. Peaks for Al–O–Al, Al–O–Si, and Si–O–Si bridging oxygens are well-resolved in the 'isotropic' dimension (largely because of differences in quadrupolar coupling), and not resolved at all in the conventional MAS dimension of the 2-D plots. The clear presence of all three peaks for R = 1 composition ($NaAlSiO_4$, 'nepheline') demonstrates significant, but not fully random, Al/Si disorder in the network. Reprinted from Lee SK, Stebbins JF, Al–O–Al and Si–O–Si sites in framework aluminosilicate glasses with Si/Al = 1: quantification of framework disorder, Journal of Non-Crystalline Solids 270:260-264, copyright (2000), with permission from Elsevier.

The higher field strength of Ca^{2+} vs. Na^+ pushes Reaction (2) to the left in glasses and melts, presumably because the concentration of negative charge on the Al–O–Al oxygen more effectively balances the higher cation charge (Lee and Stebbins 1999). It is also more difficult to fit enough larger, low charged cations around a relatively highly charged anion such as Al–O–Al ('steric hindrance', Mysen and Richet 2019), leading to the same compositional shift. This leads to greater disorder, and in common compositional ranges, a correspondingly greater concentration of Si–O–Si groups. This is in turn correlated with a higher thermodynamic activity of silica in aluminosilicates with higher field strength cations as measured by phase equilibria (Ryerson 1985), suggesting a connection between this key petrological variable and the fraction of 'silica-like' structural units. Subsequent work on glasses with varying fictive temperatures confirmed the expected increase in Al/Si disorder at higher temperature and allowed an estimate of the enthalpy change associated with Reaction (2) that were consistent with calorimetric estimates (Dubinsky and Stebbins 2006). The mixing of three bridging oxygen species contributes significantly to overall configurational entropy in the liquids but cannot account for all of it (Stebbins 2008). These studies of ambient pressure glasses also opened the doorway for a number of subsequent measurements of changes in oxygen speciation in a wide range of high-pressure aluminosilicate glasses (Lee et al. 2004, 2005, 2006; Lee 2005). As discussed below, as Si and Al coordinations increase at higher pressure, oxygen speciation becomes considerably more complex as a variety of structures develop with coordinations of more than two network cations.

Glasses rich in other network cations, notably B and P, are outside the scope of this chapter but are of considerable interest in technological materials (Greaves and Sen 2007; Musgraves et al. 2019; Richet 2020). Like ^{27}Al, ^{11}B and ^{31}P are abundant, easy-to–observe NMR nuclides with strong nuclear dipolar interactions. Such compositions have thus been the subject of many advanced double-resonance NMR studies, which can define in considerable detail the interconnections, and thus order/disorder, among borate, aluminate, phosphate (and in some cases silicate) polyhedra (Van Wüllen et al. 1996a,b; Züchner et al. 1998; Chan et al. 1999; Zhang and Eckert 2006). Further intriguing progress has been made using NMR methods based on both through-space and through-bond interactions ("J-couplings") between ^{17}O and cations such as ^{23}Na, ^{27}Al, and ^{11}B (Lee et al. 2009; Jaworski et al. 2015). In Na–Ca aluminosilicate glasses, for example, the latter approach has clearly resolved ^{17}O NMR signals for bridging oxygens connected to Al, and for both bridging and non-bridging oxygens bonded to Na^+ cations (Fig. 15) (Sukenaga et al. 2017). In contrast, NBO on Al cations are low enough in concentration to be undetectable in these compositions.

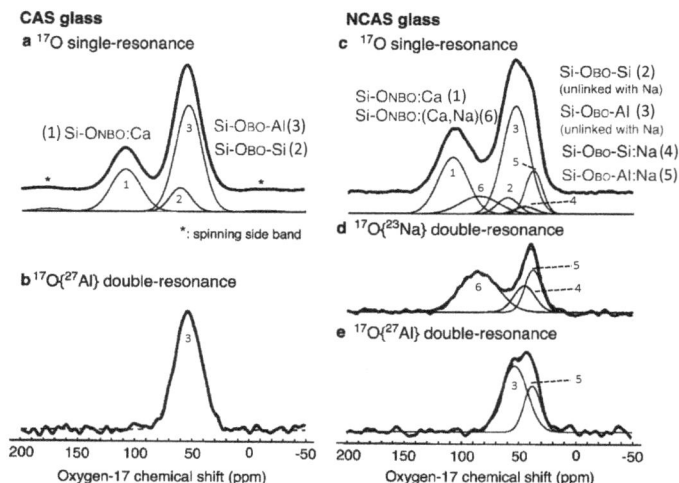

Figure 15. ^{17}O MAS NMR spectra for Ca aluminosilicate (CAS) and Ca–Na aluminosilicate (NCAS) glasses collected at the very high magnetic field of 20 Tesla, which nearly eliminates effects of quadrupolar coupling for this nuclide. The single resonance spectra (**a** and **c**) show all of the oxygen sites; the double resonance spectra (**b, d, e**), in contrast, show only oxygen sites that are directly connected to either ^{23}Na or to ^{27}Al, including both NBO and BO. Overlapping resonances can thus be unambiguously quantified. Peaks for each species are numbered here; these are marked in color in the original publication. (Adapted with permission from Sukenaga S, Florian P, Kanehashi K, Shibata H, Saito N, Nakashima K, Massiot D. Oxygen speciation in multicomponent silicate glasses using through bond double resonance NMR spectroscopy. Journal of Physical Chemistry Letters 8:2274–2279. Copyright (2017) American Chemical Society.

Effects of modifier oxides: non-bridging oxygens

In silicates, the most obvious, and fundamental, connection, between composition and oxide ion speciation can be described by what happens when an oxide of a relatively large, low-charge cation (network modifier oxides, e.g., K_2O, Na_2O, CaO, MgO etc.) is added to silica:

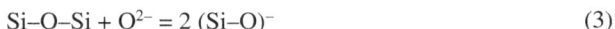

$$Si–O–Si + O^{2-} = 2 \text{ (Si–O)}^- \tag{3}$$

or,

$$BO + FO = 2 \text{ NBO} \tag{4}$$

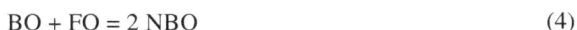

Here, the O^{2-} species is defined as an oxygen anion bonded to no small, high-charged network cation such as Si^{4+} or Al^{3+}. Especially in the context of glass and melt structure, this is often denoted as a 'free' oxide ion, 'FO'. The distinctions among BO, NBO and FO become blurred as network cation coordinations increase, especially at high pressure. The overall transition in oxygen ion species summarized by Reaction (4) is that familiar in silicate crystal structures, for example the reaction of 1 mole of quartz (all BO) with 1 mole of MgO (all O^{2-}) to form 1 mole of enstatite, $MgSiO_3$ (each Si having 2 NBO and 2 BO). The bridging oxygens can of course be other species such as Si–O–Al in more complex compositions.

In normal glass-forming compositions, in which overall O/Si < 4, Reaction (4) is often assumed (at least as a first approximation) to go to completion (but see discussion below), minimizing the concentration of FO. At higher ratios of O/Si, e.g., in the Mg_2SiO_4–MgO binary, there are too many oxide ions to all have a Si neighbor and FO must thus become abundant, but such liquids can rarely be quenched to glasses. Assuming for now a composition that allows all oxygen anions to be either BO or NBO, and that all Si (and Al) have exactly 4 neighbors, the overall proportions of BO and NBO can be readily calculated from composition alone, yielding the useful descriptive ratio of mean NBO per tetrahedral cation (NBO/T) (Mysen and Richet 2019), which has often been used to parameterize melt property variations with composition. In systems containing Al_2O_3, some of the O^{2-} will go into forming AlO_4 groups, which can be accounted for with reactions such as:

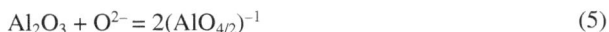

$$Al_2O_3 + O^{2-} = 2(AlO_{4/2})^{-1} \tag{5}$$

(The subscript in the latter denotes four coordinated Al, with 4 bridging oxygens each shared by one other network cation. This notation accounts for both charge and mass balance.) Considering mole fractions (or simply numbers of moles) of the standard major oxide components, we can define the sum of modifier oxide anions as

$$M = Na_2O + K_2O + CaO + MgO + FeO \tag{6}$$

and the sum of tetrahedral cations as

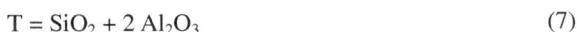

$$T = SiO_2 + 2 Al_2O_3 \tag{7}$$

With the assumptions given above,

$$NBO/T = 2 \times (M - Al_2O_3)/T \tag{8}$$

This is surely a useful first approximation, but, as noted in the previous section, the assumption about all-tetrahedral network cation coordination may be inaccurate even in 1 bar glasses and certainly in high pressure glasses. Also, the added assumption about minimal FO has been called into question by a number of recent experimental studies as well (see below).

Reaction (4) is generally discussed in the context of the 'depolymerization' of silica-rich network liquids by the addition of network modifying oxides, and clearly plays a central role in accompanying drastic reductions in viscosity, glass transition temperature, liquidus temperatures and effects on all other melt and glass properties, as discovered millennia ago by ancient glass makers. However, it is essentially the same as the 'polymerization' reaction considered in early models of silicate melt thermodynamics, developed for metallurgical slags and based on considering what happens when silica is added to a liquid such as FeO or CaO rather than the opposite (Masson 1977; Hess 1980), although this is often written in the reverse direction resulting in polymerization and production of FO.

Written in the form as in Reaction (4), the role of the oxide ion in 'modifying' the network is emphasized, not that of the 'modifier cation'. The size, charge, and electronic structure of the latter certainly do play major roles in the distribution of the resulting anionic species, bond strengths, order/disorder, and subsequent effects on glass and melt properties. 'Modifier' or 'charge balancing' cations generally have charges of +1 or +2 and coordination numbers of 5 or above (although sometimes 4). Each such cation can thus contribute only a fraction $\ll 1$ (e.g., 2/6 for Ca^{2+} with six coordination) to balance the -1 formal charge on an NBO anion. The latter must therefore be coordinated by multiple cations (in glasses and melts as is well-known in silicate crystal structures). In silica-rich melts, the system may have difficulty reaching this distribution, leading to clustering of NBO and modifier cations (the 'modified random network model', Greaves and Ngai 1994; Greaves and Sen 2007), large effects on ionic diffusivity, and often to liquid-liquid phase separation in binary silicate compositions, which is systematically more important as the field strength of the modifier increases (e.g., $K^+ < Na^+ < Ca^{+2} < Mg^{+2}$).

Measurements of distributions of non-bridging oxygens

Spectroscopic and other structural methods have long been applied to silicate glasses of simple compositions to test the validity and extent of reactions such as Reaction (4). In some binary compositions (e.g., $K_2O–SiO_2$ and $CaO–SiO_2$), for example, peaks for BO and NBO are relatively well resolved in 1-dimensional ^{17}O NMR spectra (Fig. 16) as well as O1s XPS spectra. A properly designed NMR experiment can yield quantitative measures of NBO and BO concentrations that generally are consistent with reactions such as Reaction (4) going to completion or nearly so (Xue et al. 1994; Stebbins et al. 2001b; Thompson et al. 2012; Stebbins and Sen 2013). However, analysis of such data may require fitting of partially overlapping peaks in spectra. This can be complicated by asymmetrical line shapes especially

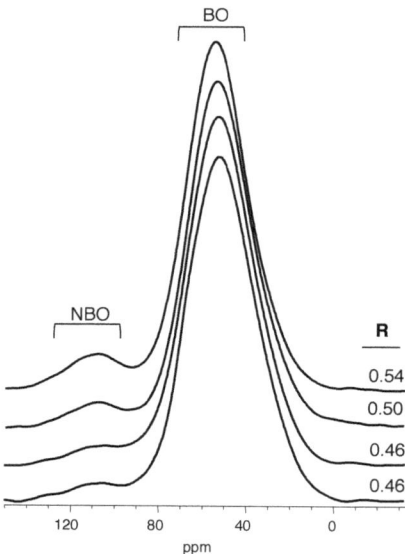

Figure 16. ^{17}O MAS NMR spectra (14.1 Tesla field) for Ca aluminosilicate glasses, all with 30% SiO_2, and having molar $CaO/(CaO+Al_2O_3)$ ratios 'R' as labeled. $R=0.5$ thus marks the 'charge balanced' composition, which, in conventional models, should have no NBO. However, NBO are clearly detected, well into the peraluminous range ($R<0.5$); (Thompson and Stebbins 2011).

for the BO, which have relatively large quadrupolar coupling constants, as well as by the broadening caused by structural disorder. As in any such data analysis, the resulting precisions may not be sufficient to rule out the presence of relatively low concentrations of other species when estimated by difference (see discussion below of FO concentrations, for example): direct detection of such species can thus lead to more robust conclusions. In other compositions of major interest to geochemistry, notably Na and Mg silicates, resolution in 1-D ^{17}O NMR spectra is poor because of overlapping ranges in chemical shifts, but 2-dimension methods such as MQMAS (Allwardt and Stebbins 2004) and DAS NMR (Florian et al. 1996) can separate peaks, eliminate effects of quadrupolar coupling and often improve quantitation.

Because of strong effects of the modifier cation charge and radius on ^{17}O chemical shifts, in glasses with more than one type of modifier, NMR spectra can provide information on cation distributions around NBO and hence on a key aspect of order/disorder. For example, NBO in $CaSiO_3$–$MgSiO_3$ glasses probably have on average 3 modifier cation neighbors as is typical of corresponding crystals, distributed among 3Ca, 2Ca1Mg, 1Ca2Mg, and 3Mg permutations. Variations in NBO peak position and width with Ca/Mg ratio in ^{17}O NMR spectra of such glasses are consistent with a random Ca-Mg mixing (Allwardt and Stebbins 2004), and thus with early studies of viscosity and configurational entropy (Neuville and Richet 1991). The same has been seen in $K_2Si_2O_5$–$Na_2Si_2O_5$ glasses (Florian et al. 1996). When modifier cations differ more greatly, for example K^+ vs. Mg^{2+}, ^{17}O NMR has demonstrated strong cation ordering, with NBO coordinated primarily by Mg^{2+} and thus requiring K^+ to have mostly BO neighbors (Allwardt and Stebbins 2004), as in the mineral phlogopite.

Similarly, ^{17}O NMR can sometimes measure concentrations of NBO attached to different network cations, as these may yield well-resolved peaks in spectra. For example, it is clear in Ca aluminosilicates, Al–NBO are abundant only in glasses with very high Al/Si, with strong favoring of Si–NBO in most compositions (Fig. 17) (Allwardt et al. 2003). This confirmed earlier inferences from Raman spectroscopy (Mysen 1988; Mysen and Richet 2019). In glasses with even higher field strength modifiers (e.g., La^{3+} and Y^{3+}), Al–NBO are more easily stabilized (Jaworski et al. 2015).

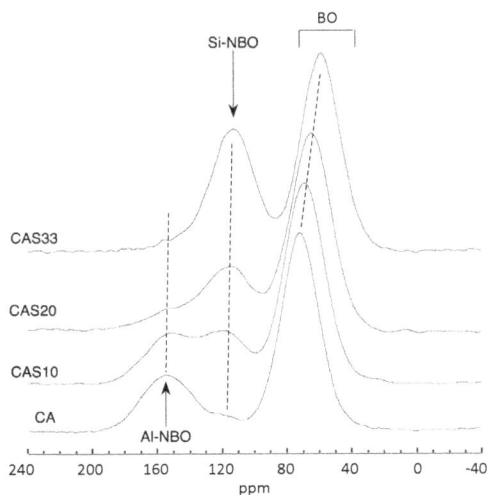

Figure 17. ^{17}O MAS NMR spectra (14.1 Tesla field) for low-silica Ca aluminosilicate glasses with molar $CaO > Al_2O_3$, showing clear distinction between NBO on Si and NBO on Al. The latter are significant only at very low Si/Al ratios, and are strongly ordered onto Si in most aluminosilicates. From lowest to uppermost spectra, Si/Al ratios are 0.01, 0.2, 0.36, and 0.97 (Allwardt et al. 2003). The BO peak position shifts with changing proportions of Al–O–Al, Al–O–Si, and Si–O–Si.

As described in the previous section, in aluminosilicate glasses the larger, lower charged cations also play the role of 'charge compensator' for partially charged BO such as Si–O–Al. Compositions in which molar $M^{2+}O$ and/or M^{1+}_2O are equal to that of Al_2O_3 are often described as 'charge compensated' and fall on joins that include the feldspar and feldspathoid stoichiometries, e.g., $KAlSiO_4$–SiO_2 and $CaAl_2Si_2O_8$–SiO_2. As in these minerals, the glasses are often described as 'fully polymerized', with no NBO and all BO. This may be a good initial approximation, and such liquids do have viscosities near to maxima, implying high proportions of strong BO bonds. However, ^{17}O NMR measurements of NBO concentrations in glasses of fixed silica contents and varying modifier/alumina ratios have clearly shown the presence of a few% NBO in 'charge compensated' compositions (Stebbins and Xu 1997; Thompson and Stebbins 2011) (Figs. 16 and 18). This deviation again increases with cation charge, at least with $Ca^{2+} > K^+$, and was previously hypothesized from detailed variations with composition in melt viscosities (Toplis et al. 1997; Toplis and Dingwell 2004). It must be concluded that one or more of the assumptions commonly applied to Reaction (4) to estimate NBO is inaccurate. The measurement of significant contents of $^{[5]}Al$ by NMR can explain much of this variation; another 'non-standard' oxygen species has also long been suggested, namely oxygen ions with 3 instead of 2 network cation neighbors, commonly denoted 'triclusters' (see section below).

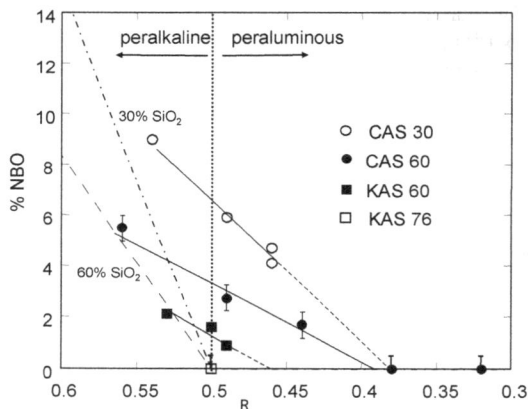

Figure 18. Non-bridging oxygen concentrations (out of total O) in Ca-aluminosilicate (CAS) and K-aluminosilicate (KAS) glasses, derived from ^{17}O MAS NMR spectra as in Figure 16, vs. $R = CaO/(CaO+Al_2O_3)$ or $K_2O/(K_2O+Al_2O_3)$. Dotted vertical line divides peraluminous from peralkaline compositions; dashed lines are predictions from standard models for 30% and 60% SiO_2 compositional joins, in which NBO goes to zero at $R=0.5$ (Thompson and Stebbins 2011).

The connection between network cation coordination and non-bridging oxygens

Reactions (3) or (4) are the common textbook description of modification of an oxide glass or liquid network by addition of a modifier oxide, and clearly is a good starting point for low-pressure silicates. However, in the other well-studied simple oxide systems, borates and germanates, standard models appear to be very different. Although details of these systems are outside the scope of this chapter, a brief description is useful especially in thinking about geological melts at high pressure.

Decades of research using most known structural tools have shown that pure B_2O_3 glass and liquid are comprised of corner-shared networks of BO_3 triangles. As a modifier oxide is added, viscosities and glass transition temperatures initially increase to maxima, then decrease. Molar volumes are also varying highly non-linearly with composition. ^{11}B NMR accurately measured proportions of $^{[3]}B$ and $^{[4]}B$: the latter first increases (leading to a more connected network) then decreases (Silver and Bray 1958; Dell et al. 1983; Aguiar and Kroeker 2007).

In the initial stages of this change, the reaction below, at least at temperatures near the glass transition where the structure is quenched in, goes nearly to completion:

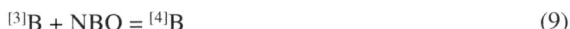

$$[3]B + NBO = [4]B \qquad (9)$$

Only at high modifier content (with Na/B > 0.5), when the $[4]B/$ $[3]B$ declines, do NBO become important. Here, a reaction analogous to Reaction (4) becomes important. Increasing temperature shifts Reaction (9) to the left, higher pressure shifts it to the right (Gupta et al. 1985; Du et al. 2004; Wu and Stebbins 2013).

In alkali germanates, known crystal structures and variations in glass and melt properties with composition have long suggested that an analogous process may occur. Pure GeO_2 glass, like SiO_2, appears to be a network of corner-shared tetrahedra (cf., Micoulaut et al. 2006), but as a modifier oxide is added the coordination number of some of the Ge initially increases to 5 (or 5 + 6; see Henderson 2007 for a summary). Again, an equilibrium reaction can be written (two NBO would be used for $[6]Ge$):

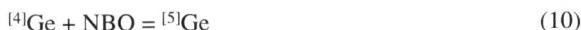

$$[4]Ge + NBO = [5]Ge \qquad (10)$$

As in the borates, coordination goes up, then decreases back to 4 at high modifier contents in both glasses and crystals-again with NBO becoming abundant; glass and liquid molar volumes again are varying non-linearly with composition. Again, many methods have been used to constrain the extent of such reactions, including X-ray and neutron scattering (Hoppe et al. 2000), ^{17}O NMR (Du and Stebbins 2006; Lee and Lee 2006; Peng and Stebbins 2007), and vibrational spectroscopy (Henderson and Wang 2002; Henderson et al. 2010). Given that there is not an experimental tool that can directly and accurately measure Ge coordination (unlike the cases of ^{29}Si and ^{11}B NMR), considerable uncertainty remains in the details, but reactions like Reaction (10) are generally thought to capture much of what happens at high GeO_2 contents.

It has long been suggested that an analogous reaction could account for at least part of the Si coordination increases with pressure known to occur in NBO-rich glasses:

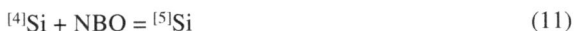

$$[4]Si + NBO = [5]Si \qquad (11)$$

^{29}Si and ^{17}O NMR studies of the same high pressure alkali silicate glasses have supported its validity in these systems (Xue et al. 1994; Allwardt et al. 2004) (Fig. 19) and, with ^{27}Al NMR, have extended it to aluminosilicates (Lee et al. 2004, 2006; Allwardt et al. 2005a). In silicates, simple considerations of equilibrium constants thus suggest that Si cation coordination might increase monotonically with NBO concentration. As noted above, however, this is far from the case in the better-studied borates and germanates; new data have now shown a similar pattern of coordination increase, then decrease in ambient pressure alkali silicates as well (Stebbins 2019a). A very simple thermodynamic model, considering only non-ideal mixing of oxygen anionic species, can at least qualitatively reproduce these patterns in all three types of system, suggesting a greater commonality in underlying structural mechanisms than previously expected. These findings suggest that just as in the well-studied ambient pressure borate and germanate systems (Stebbins et al. 2013), physical properties of silicate glasses and melts at high pressures (where $[4]Si$, $[5]Si$ and $[6]Si$ may all be abundant) may vary in a strongly non-linear fashion with composition. As yet, however, few property data are available on compositionally simple melts to constrain such models.

Apart from obvious interactions of increased pressure with increased network coordination and effects on melt properties, there is a long-standing interest in $[5]Si$ as a possible transient species in the dynamics of bond-breaking, diffusion, viscous flow, and even melting and crystallization. Most insights into this role come from theory and computer simulations, but new experimental data have renewed such discussions and provided important new constraints (Nesbitt et al. 2019; Stebbins 2019, 2020).

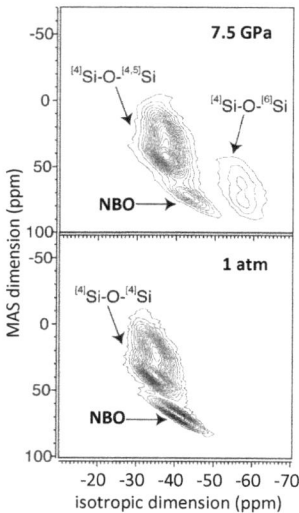

Figure 19. ^{17}O MQMAS spectra for $K_2Si_4O_9$ glasses quenched from melts at 1 bar and at 7.5 GPa. The loss of intensity in the NBO peak, and development of signal for $^{[4]}Si–O–^{[6]}Si$ groups, is clear. Signal for oxygens bonded to $^{[5]}Si$ overlaps with that for conventional bridging oxygens, however. Reprinted from Allwardt JR, Schmidt BC, Stebbins JF, Structural mechanisms of compression and decompression in high pressure $K_2Si_4O_9$ glasses: An investigation utilizing Raman and NMR spectroscopy of high-pressure glasses and crystals, Chemical Geology 213:137–151, copyright (2004), with permission from Elsevier.

Higher oxygen anion coordination numbers

Bridging oxygens in framework silicate crystals and glasses are often described as 'two coordinated', meaning two network cation neighbors, and ignoring the charge compensating cations. Oxygens bonded to three tetrahedral network cations are often called 'triclusters', and have sometimes been hypothesized to account for charge balance and property variations in peraluminous glasses and liquids (Lacey 1963; Toplis et al. 1997; Toplis and Dingwell 2004). OAl_3 (or OB_3) triclusters do occur in a small number of known crystal structures e.g., $CaAl_4O_7$ (grossite) and $SrAl_4O_7$ and their borate analogs. The O anion is somewhat overbonded, leading to somewhat lengthened Al–O distances. Tetrahedral triclusters involving one or more Si^{4+} cations would be even more overbonded and less likely to be stable in melts and glasses. ^{17}O and ^{27}Al NMR studies have characterized such sites in aluminate and borate crystals (Stebbins et al. 2001a; LaComb and Stebbins 2015), but unfortunately in most aluminosilicate glasses of interest to geosciences, these resonances have considerable overlap with other, majority species, making quantitation non-unique (Stebbins et al. 2001a). In at least one aluminate glass ($CaAl_2O_4$), double resonance NMR has more definitively shown the presence of a few% of OAl_3 groups, as ^{17}O chemical shift separations are greater than in aluminosilicates (Iuga et al. 2005). Diffraction data on aluminate liquids, when modeled by reverse Monte Carlo (RMC) simulations, has also suggested significant fractions of oxide anions coordinated by three network cations having various coordination numbers (Skinner et al. 2012, 2014), as has ^{17}O NMR on amorphous alumina (Lee and Ryu 2017).

As noted above, bridging oxygens in glasses that carry a nominal negative charge (Si–O–Al and Al–O–Al) are expected to also have one or more 'charge compensating' cations in their first coordination shell, as is well-known in aluminosilicate crystals. Nominally 'uncharged' Si–O–Si groups may also interact with such cations, as is also well-known in many silicate crystal structures. Thus, a second type of bridging oxygen has been shown to occur in Na silicate glasses. Using O 1s XPS Nesbitt et al. (2015a) demonstrated the presence of bridging oxygens with an alkali "attached" and termed these types of BO, MBO where M = alkali oxide. They suggested that these types of bridging oxygens must be present in areas where the glass structure is heterogeneous with a non uniform distribution of Na atoms, near for example, percolation channels. Classical MD simulations (e.g., Mead and Mountjoy 2005, 2006; Tilocca and de Leeuw 2006; Mountjoy 2007; Machacek et al. 2010) have also shown their presence in Na and Na–Ca glasses. While alkali cations are needed to "charge balance" the NBOs the alkali and alkaline-earth cation coordination numbers are generally around 5–7. This implies that some of the alkalis must also interact with BOs and in many compositions the numbers of MBO should actually exceed the numbers of BO. This is certainly true for crystalline alkali silicate structures. For example, in α-$Na_2Si_2O_5$ 2 of every 3 BOs on a Q^3 tetrahedron is also bonded to a Na (Pant and Cruickshank 1968), while in $K_2Si_2O_5$ (de Jong et al. 1998) the BOs may have 1, 2 or 3 K atoms attached (dependent upon the K–O cutoff length being 3.25 Å) like similar composition glasses (cf., Nesbitt et al. 2017a). The crystalline structures show the coordination of the BO increasing from 2–3–4–5 as the structure become increasingly depolymerized ($Q^3 \rightarrow Q^2$). The presence of such MBOs must have a significant impact on the behaviour and physical properties of glasses and melts. Unfortunately, this aspect of the structure and coordination around the BOs has been largely unexplored, mainly due to the difficulty in "seeing" the alkalis and alkaline-earths and determining the nature of their interaction with the BOs in disordered materials.

Methods such as ^{17}O NMR, for example, have noted effects on chemical shifts for Si–O–Si groups in glasses that vary with the nature of the modifier cation, but have been unable to resolve distinct BO species with varying coordination numbers (Lee and Stebbins 2009; Allwardt and Stebbins 2004). Recently, high-field, double resonance $^{17}O\{^{23}Na\}$ NMR has unambiguously resolved both Si–O–Si and Si–O–Al BO that have one or more Na^+ neighbors through the exploitation of Na–O through-bond coupling (Fig. 15) (Sukenaga et al. 2017).

In ambient-pressure glasses (Le Losq et al. 2014), and especially in high pressure aluminosilicate liquids and glasses in which substantial fractions of the network cations become 5- and 6-fold coordinated, oxygen coordination must increase as well simply from consideration of local charge balance and bond counting. For example, oxygens with two [5]Al and one [4]Al or [4]Si neighbor are less overbonded and more likely to be stable than tetrahedral "triclusters' discussed above. Some of these may have coordinations analogous to oxygen anions in the mineral sillimanite (Al_2SiO_5), with one [4]Si, one [4]Al, and one [6]Al first neighbor and probably continue to resemble bridging oxygens. Others may have bond lengths and energetic roles that are quite different, and a great variety of different species can be expected or imagined. At very high pressures, the oxygen coordinations seen in minerals such as stishovite (3 [6]Si neighbors) and corundum (4 [6]Al neighbors) may be better analogs for melt and glass structure. A number of ^{17}O MQMAS NMR studies of high-pressure glasses have begun to reveal such complexities (Lee 2005; Lee et al. 2006). In any case, the simple distinctions among bridging, non-bridging, and 'free' oxygen anions that form the basis for much of our thinking about glass structure will break down in such high-pressure systems, requiring more complex models and ways describing structures.

"Free" oxide anions

As mentioned above, various forms of Reaction (4) have long been used not only to describe the 'depolymerization' of silica-rich liquids with the addition of modifier oxides, but to model the thermodynamics (solid–liquid–vapor equilibria) of the very low-silica liquids of primary importance in metallurgical slags (Toop and Samis 1962; Masson 1977; Hess 1980). In compositions with O/Si ratios >4, the 'free' oxide species (FO) must be present by stoichiometry, as there are simply not enough SiO_4 groups to bond to all of the oxygen anions. In such liquids, FO species can be quite abundant (e.g., 100 % in liquids such as pure MgO) and their concentration in the MO–SiO_2 binaries can shift considerably depending on the field strength of the modifier cation, reaching greatest concentrations in liquids dominated by the divalent transition metal cations (e.g., Fe^{2+}, Ni^{2+}, Co^{2+}), lower for Mg^{2+}, and lower still in Ca- and alkali-silicates. In the latter, additional constraints on oxide anion activities can be derived from vapor pressure and electrochemical measurements (Semkow and Haskin 1985; Abdelouhab et al. 2008). Because of their weak average bond strengths and resulting high fragility

Henderson & Stebbins

(low viscosity at high temperature), such FO-rich liquids generally do not quench to form glasses, limiting the extent of direct measurements of their structures and comparisons to more silica-rich range of interest to geochemistry. When extrapolated to compositions more likely to form glasses (e.g., typically >50% SiO_2), however, FO concentrations predicted by such models are generally very low ($\ll 1\%$). It was thus especially interesting and surprising that percent-levels of FO anions were reported in alkali silicate glasses containing as much as 50 to 60 mol% SiO_2, based on fits of partially resolved components for NBO and BO in oxygen 1s XPS spectra: the signal for FO itself is apparently unresolvable in such data (Nesbitt and Dalby 2007; Nesbitt et al. 2011; Sawyer et al. 2015). The %FO was calculated from the proportion of BO and NBO determined from the XPS spectra. The FO signal itself lies within the NBO XPS peak. This has been clearly shown by Nesbitt et al. (2015a) using PbO–SiO_2 glasses where the FO peak can be detected at very high PbO contents (see Fig. 20). The amounts of FO determined from the relative peak areas are fully consistent with the amounts of FO detected by ^{17}O NMR (Lee and Kim 2015) and determined from equilibrium thermodynamic calculations. Although these conclusions about FO in alkali-silicate glasses are controversial and depend (like many spectroscopic studies) on assumptions used in fitting, they have led to considerable renewed interest in this oxygen anionic species (Malfait 2015; Nesbitt et al. 2015b).

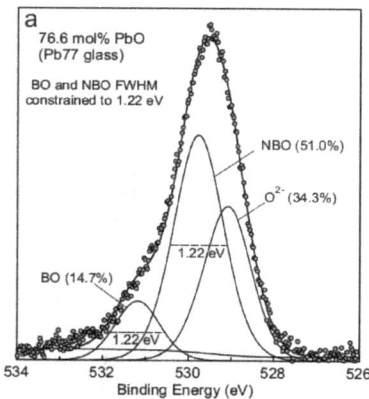

Figure 20.: Fits to the O 1s XPS spectrum for a PbO–SiO_2 glass with 76.6 mol% PbO from Nesbitt et al. (2015c). The NBO peak clearly has 2 contributions: one from NBO and one from FO or O^{2-}.

Direct NMR spectroscopic measurements of FO concentrations in glasses are sometimes feasible, as for BO and NBO species as described above. The gap between silica-poor, FO-rich liquids and glass-forming compositions can be bridged in a few cases when low silica glasses can be made by containerless (levitation) melting and rapid quenching of compositions with low liquidus temperatures, most notably Ca–Mg silicates. These can form glasses even at 'sub-orthosilicate' compositions, down to about 28 mol% SiO_2, with O/Si=4.4 (Nasikas et al. 2012). Here, a double-resonance, 2-D NMR method ('heteronuclear correlation', HETCOR) has been applied, which distinguished oxygen sites with and without Si first neighbors by taking advantage of the ^{17}O–^{29}Si nuclear dipolar couplings. Although results were of relatively low precision, about 10 ± 3 %FO was observed in the 29 mol% silica glass, and about 8 ± 3 %FO in a glass close to the olivine stoichiometry (33 mol% SiO_2) (Hung et al. 2016) (Fig. 21). The latter result confirms that, as in models of very low silica slag compositions, reactions such as Reaction (4) do not always 'go to completion'. By mass balance, excess FO at, say, the olivine composition, requires as well the presence of excess BO, i.e., more linkages between SiO_4 groups than expected if FO were absent. Considerations of the enthalpy changes associated with such reactions suggest that FO oxide species may become considerably more abundant at high temperatures in low silica liquids, especially those rich in Mg^{2+} and Fe^{2+} that are most relevant to ultramafic magmas in the Earth (Stebbins 2017).

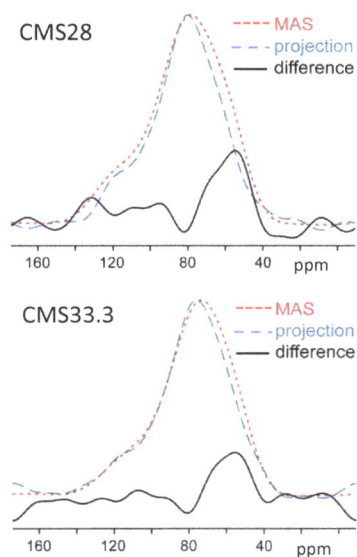

Figure 21. $^{17}O \rightarrow ^{29}Si$ cross-polarization heteronuclear correlation (CP-HETCOR) NMR spectra (18.8 Tesla field) for Ca-Mg silicate (CMS) glasses with very low silica contents of 28 and 33.3 mol%. The MAS spectra (short dashed lines) show signals for all of the oxygens, while the 'projections' of the CP-HETCOR 2-dimensional spectra (long dashed lines) show signal only for oxygen sites adjacent to Si. The difference spectra (solid lines) thus highlight the presence of oxygen anions ('free' oxide) that are not bonded to any Si. Reprinted from Hung I, Gan Z, Gor'kov PL, Kaseman DC, Sen S, LaComb M, Stebbins JF, Detection of "free" oxide ions in low-silica Ca/Mg silicate glasses: Results from $^{17}O \rightarrow ^{29}Si$ HETCOR NMR. Journal of Non-Crystalline Solids 445–446:1–6, copyright (2016), with permission from Elsevier.

In several glasses with compositions far from those of geological materials, conventional ^{17}O MAS NMR has readily and directly quantified FO concentrations, as peaks for this species are relatively narrow, with large chemical shift differences, and thus are well resolved. For example, the PbO–SiO$_2$ binary system has an especially wide range of glass-forming compositions. Here, even at silica contents considerably higher than the olivine stoichiometry (e.g., O/Si of about 3.5, about 40 mol% SiO$_2$), a few%FO were directly observed (Lee and Kim 2015). In an even more exotic system, SiO$_2$–HfO$_2$ amorphous thin films made by sputtering, large concentrations of FO were directly observed by ^{17}O NMR in compositions up to 41 mole% silica (Kim et al. 2015b). A thermodynamic analysis of these data suggested that these films could represent the structure of quenched, deeply supercooled liquids metastable with respect to both crystallization and to liquid–liquid phase separation.

Conventional 1-D ^{17}O MAS NMR can clearly detect FO when it is present in crystals such as CaO and Ca$_3$SiO$_5$, again because of relatively narrow peaks (small quadrupolar coupling constants) and good chemical shift separation from BO and NBO (Thompson et al. 2012); ^{17}O MQMAS NMR also readily detected this type of site in wadsleyite (high-pressure Mg$_2$SiO$_4$) (Ashbrook et al. 2007). However, no signal for FO was detected (<1%) in a CaO–SiO$_2$ glass with 44 mol% silica (Thompson et al. 2012) (Fig. 22). Similarly, no FO signals were seen for K$_2$O–SiO$_2$ glasses down to 60 mol% silica, based on a chemical shift range for this species estimated from a theoretical calculation (Stebbins and Sen 2013). It's important to note that, as mentioned above for small but sometimes measurable concentrations of $^{[5]}Si$ groups, even very low concentrations of FO groups may potentially be important for dynamical processes in melts (Nesbitt et al. 2017c, 2019). Such inferences still rely largely on theory and simulations but can point the way to future experimental priorities.

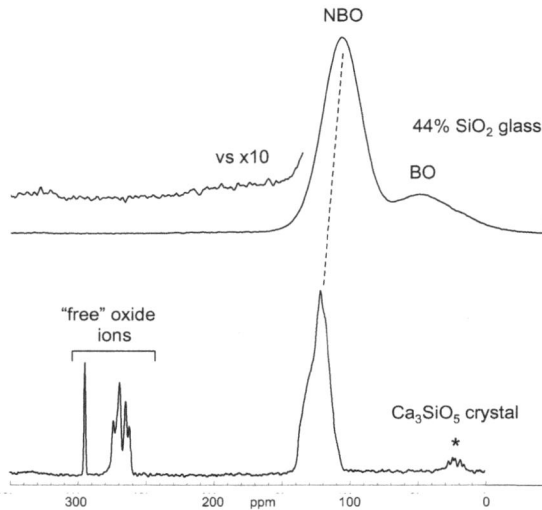

Figure 22. ^{17}O MAS NMR spectra for crystalline Ca_3SiO_5 and a Ca-silicate glass with 44% SiO_2. The crystal contains several oxygen anion sites with no Si neighbors ('free' oxide ions); the sample also contains a small amount of residual CaO with similar O^{2-} sites (narrow peak at 295 ppm). The complex NBO line shape is due to the presence of several distinct NBO sites as well. The glass (near the low end of the range of silica contents obtainable by conventional quench methods) contains BO and NBO sites as expected, but no detectable 'free' oxide signals. Reprinted from: Thompson LM, McCarty RJ, Stebbins JF, Estimating accuracy of ^{17}O NMR measurements in oxide glasses: Constraints and evidence from crystalline and glassy calcium and barium silicates. Journal of Non-Crystalline Solids 358:2999–3006, copyright (2012), with permission from Elsevier.

DISTRIBUTIONS OF NBO AND BO AROUND NETWORK CATIONS: Q SPECIES

As noted above, in aluminosilicate glass and melt compositions with low concentrations of FO, and of high-coordinated Si and Al, the overall fractions of BO and NBO can be calculated from composition alone. For example, a glass of $Na_2Si_2O_5$ composition would have an average of 1 NBO per Si, as the corresponding sheet silicate crystal has exactly 1 NBO on each SiO_4 tetrahedron; an $MgSiO_3$ glass should have an average of 2 NBO per Si, as crystalline enstatite has exactly 2 NBO on each SiO_4 group. Crystalline $Na_2Si_2O_5$ is thus comprised entirely of Q^3 groups, pyroxenes such as enstatite are all Q^2, silica polymorphs all Q^4, etc. In the corresponding glasses, composition alone gives at best only the mean value of NBO/T; if FO concentrations (through a shift in Reaction (4) to its left, as suggested for some high-alkali silicates, Nesbitt et al. 2015a) or higher-coordinate Si or Al become significant, such calculations become less accurate approximations, as discussed below. Decades of spectroscopic studies have provided extensive information on actual tetrahedral species distributions and how they vary with composition, temperature, and pressure, but some details remain incompletely resolved.

The alkali disilicates (e.g., $Na_2Si_2O_5$), for example, are good glass formers and low-melting liquids, and have long served as revealing experimental subjects. As in the crystals, all of the tetrahedra could be Q^3 groups, but early Raman spectra showed that Q^2 groups must also be present, requiring by mass balance the presence of Q^4 species as well (cf., Mysen and Richet 2019); work on wide ranges of binary and ternary glasses provided the first models of compositional effects on Q^n species distributions in these systems. ^{29}Si NMR confirmed these findings and provided more directly quantitative results, although for both types of spectroscopy partially overlapping peaks may require assumptions about component peak shapes, most often taken as Gaussian in form (Murdoch et al. 1985; Emerson et al. 1989; Maekawa et al. 1991; Malfait et al. 2007a) (Fig. 23). NMR results have been used to 'calibrate' Raman cross sections in glass spectra as well (Mysen 1990; Mysen and Frantz 1993; Mysen and Richet 2019).

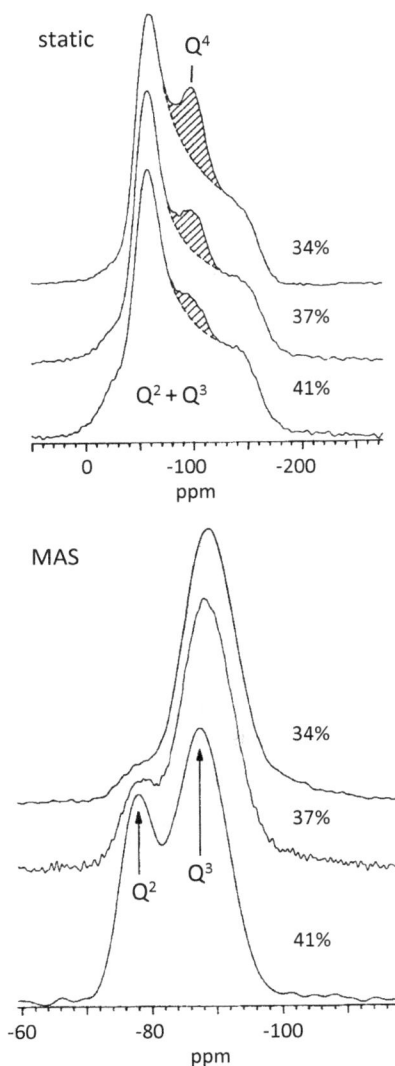

Figure 23. ^{29}Si NMR spectra for Na silicate glasses, with% Na_2O as labeled. Upper group are for 'static' (non-spinning) samples and show overlapping spectra for Q^2 and Q^3 components broadening by chemical shift anisotropy (CSA). In contrast, highly symmetrical Q^4 sites (cross-hatched) have low CSA and narrow peaks and are thus readily quantified without the need for fitting. In the MAS spectra (note much smaller frequency range plotted), Q^2 and Q^3 components are relatively well resolved and the Q^4 signals are more difficult to detect. The presence of significant Q^4 intensity for glasses with $Na_2O > 33$ mol?% ($Na_2Si_2O_5$), and of Q^2 intensity near to this composition, show definitively that there are more Q^n species present than required by stoichiometry, contributing to disorder in the glass. Reproduced from Stebbins JF, Identification of multiple structural species in silicate glasses by Si-29 NMR. Nature 330:465–467. Copyright Stebbins JF (1987).

Results on Q^n speciation in the binary silicate systems have often been analyzed by thermodynamic approaches, where equilibria are defined for a series of 'disproportionation' reactions such as:

$$2Q^3 = Q^2 + Q^4 \tag{12}$$

or, more generally for $n = 1, 2, 3$:

$$2Q^n = Q^{n+1} + Q^{n-1}$$

"Apparent" equilibrium constants (i.e., ignoring activity coefficients) can be defined in terms of the mole fractions of the species, for example as

$$k_3 = [Q^2] \times [Q^4] / [Q^3]^2 \tag{13}$$

As noted above, if FO are abundant, other equilibria involving this species will also be important, as has long been modeled in low-silica slag compositions and which has recently been proposed in some high-alkali silicate glasses (Nesbitt et al. 2011, 2015a); in this case, calculation of Q^n species abundance from fits of spectra will also become more complicated. Reactions involving FO may also be especially interesting in melting of silicates such as Na_2SiO_3 where both polymerization and depolymerization reactions may occur (Nesbitt et al. 2017b, 2019). However, in lithium silicate glasses with extremely high modifier contents (Li_2O fractions as high as 64%, prepared by special rapid quench methods), where FO contents might be expected to be especially high, fits for Q^n components of ^{29}Si NMR spectra yielded good agreement with expected compositions without the need for additional oxide ion species (Larson et al. 2006).

Given the vagaries (and varying assumptions) in fitting of spectroscopic data, there is some consensus on approximate values of k_n for alkali silicates, with increasing values for smaller, higher field strength modifiers ($Li^+ > Na^+ > K^+$) promoted by the increasing degree of concentration of negative charge around the modifier (Brandriss and Stebbins 1988; Maekawa et al. 1991; Stebbins et al. 1992), analogous to that noted above for bridging oxygen species. For example, well-studied $Na_2Si_2O_5$ glass has about 10% Q^2 groups, a corresponding fraction of Q^4 groups, and about 80% Q^3 groups. This is more disordered than in the crystal, but less so than predicted by a random model of NBO distribution over the tetrahedra (25% Q^2). For a given silica content, the resulting higher Q^4 concentrations in the liquids with higher field strengths correlates with the higher thermodynamic activity of silica (the Q^4-like component), the latter long known by comparisons of the binary phase diagrams (Ryerson 1985; Stebbins 2016). As mentioned above for NBO distributions, this effect of modifier field strength on speciation and activities is correlated with the energetics that drive liquid–liquid phase separation.

NMR studies of glasses with varying cooling rates, and thus fictive temperature, show that k_3 increases with higher temperature (Reaction (12) is pushed to the right by a positive reaction enthalpy, Brandriss and Stebbins 1988). This finding has been confirmed and greatly extended by high temperature, *in situ* Raman spectroscopy (Mysen and Frantz 1993; Malfait et al. 2007b, 2008). The mixing of Q^n species contributes significantly to overall configurational entropy and heat capacity of melts (e.g., Le Losq and Neuville 2017) but is far from explaining totals.

In binary alkaline earth silicate systems, the range of compositions that can be quenched into single-phase glasses is narrower than in the alkali silicates, and, in both NMR and Raman spectra, component peaks assignable to different Q^n groups are even less well resolved because of greater overall disorder. Fitting of conventional 1-D ^{29}Si MAS NMR spectra become highly model-dependent, for example. ^{29}Si NMR, using the advanced 'magic angle flipping' method and 2-D data collection, has allowed greater distinctions to be made for Q^n species, based on their local asymmetry (chemical shift anisotropy) instead of simply their isotropic chemical shifts (Davis et al. 2010, 2011). In compositions such as $MgSiO_3$ (Davis et al. 2011) and $CaSiO_3$ (Zhang et al. 1997) glass, the available data indicate that k_3 (as well as the analogous k_2) values are highest values are higher than in the alkali systems, as the divalent modifier cations are even more effective at concentrating the negative charges on the NBO. In the latter composition (Fig. 24), total NBO/Si, as estimated from a fitted model, was slightly lower than expected from composition and the conventional assumptions discussed above, possibly allowing for about 1% 'free' oxide ion; whether this is significantly different from zero is not certain given uncertainties in the fitting process. In $MgSiO_3$ glass and in a composition close to Mg_2SiO_4, the estimated NBO/T values were also below those expected from composition, again suggesting the presence of a few % 'free' oxide ion in these low silica, high field-strength compositions.

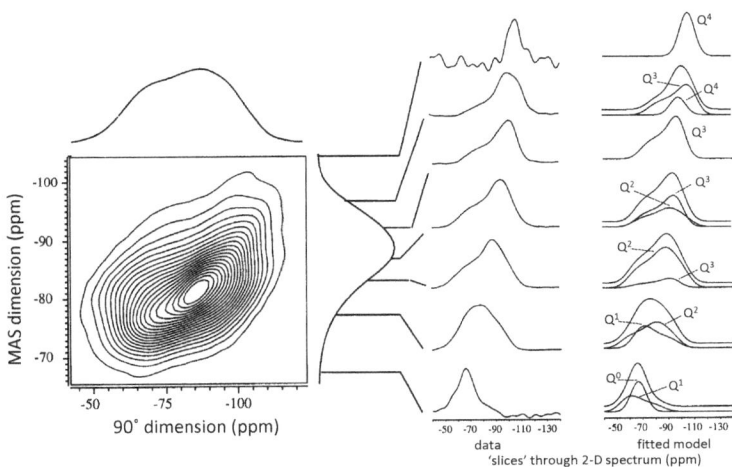

Figure 24. Two-dimensional 'magic angle flipping' (MAF) ^{29}Si NMR spectra for $CaSiO_3$ glass. In the 90° dimension, the line shapes are influenced by the chemical shift anisotropy (CSA), which helps in distinguishing among the Q^n species. A fitted model for successive slices in the MAS dimension is shown at the right. Each fit includes components for no more than two Q^n species with distinct CSA's, although low intensities of other components must make minor contributions to the line shapes. Reprinted with permission from Anionic species determination in $CaSiO_3$ glass using two-dimensional ^{29}Si NMR. Zhang P, Grandinetti PJ, Stebbins JF. Journal of Physical Chemistry B 101:4004–4008. Copyright (1997) American Chemical Society.

As well as statements of equilibria among Q^n species, reactions such as Reaction (12) are often considered in terms of accounting for mass and charge balance as melt and glass structure change with composition, temperature and pressure. For example, with the simple approximations mentioned above (negligible FO and constant network coordination numbers), an increase in Q^4 species is expected to be accompanied by an increase in Q^2 (or other species with $n < 3$). When 'observations' suggest otherwise, then these assumptions may be breaking down, such as in reported Q^n speciation in MD simulations of an Na-silicate liquid at high pressure (Noritake et al. 2012; Mysen and Richet 2019). However, it is important, especially in the case of simulations, that counting of all species is done in a consistent fashion and includes changes in both anion and cation coordination.

In aluminosilicates and other multicomponent glasses, analysis of both NMR and Raman spectra to derive Q^n speciation becomes increasingly complex and model-dependent, in large part because of the much wider variety of short-range structural species with overlapping spectral contributions. Such data remain useful at testing structural models, and in detecting relative effects of composition, temperature, and pressure on structure.

Variations on a theme: issues of fitting and data analysis

In more detail, various approaches have been taken to analyzing NMR and Raman (and other) spectra of glasses, that may yield somewhat disparate results. Such issues can be related to assumptions of nominal vs. measured compositions, which are often not reported or are potentially affected by unanalyzed constituents such as CO_2 and H_2O. Data quality (signal to noise ratios, accurate baselines, etc.) have generally improved from early to more recent studies, but no data set is perfect. Even in the simplest, most well-analyzed binary compositions, in the best-resolved and least disordered systems (alkali silicates), derivation of Q^n species populations usually requires fitting of spectra. This in turn requires assumptions about component peak shapes, numbers of component peaks and possibly of other constraints. Vibrational spectra are further complicated by potential variation in absorption cross sections between different structural groups. In ^{29}Si MAS NMR and Raman spectra, most commonly symmetrical Gaussian ('bell-shaped') are chosen, based on the assumption that a random (or at least disordered) distribution of some structural variable such as bond length will map linearly into an observable parameter such as chemical shift or vibrational frequency. There may or may not be good physical basis for this assumption: it has rarely been explored in detail (cf., Bancroft et al. 2018) but recent theoretical work holds considerable promise in this question (Angeli et al. 1999, 2011; Charpentier et al. 2004; Kilymis et al. 2019). In some cases, observed peak shapes seem more consistent with a "Lorentzian" (exponential) line shape (or a mix of Gaussian and Lorentzian), which can have a physical justification when the line width is controlled by lifetime effects: common in liquids but unusual in glasses that are low in magnetic impurities.

A given set of results for Q^n species concentrations, resulting from an unconstrained fit of, say, a ^{29}Si MAS NMR spectrum, can be independently tested by summing up the total NBO for each species and comparing with that predicted from composition, with assumptions as noted above in considering Reaction (4). In many cases, results are consistent given significant uncertainties in composition and in the fitting process. A calculation of too many total NBO per Si from such analysis can't be explained by a simple structural variation and suggests a problem with assumptions about composition, with the data themselves, or with the fitting procedure. A calculation of too few NBO per Si may suggest similar problems if it assumed that FO are negligible, or may suggest an unexpectedly high concentration of FO, i.e., Reaction (4) is 'incomplete'. Authors have also refined results by constraining the relative areas of the Q^n components such that experimental (cf., Sawyer et al. 2015) and predicted total NBO/Si ratios are identical, and testing results against unconstrained fits (Malfait et al. 2007a). In some cases, it is obvious that simple models with a single Gaussian component for each Q^n species are inadequate, requiring two components for at least one species to reproduce experimental spectra (Sen and Youngman 2003; Malfait et al. 2007a). Spectral resolution may support such models, and two components can be physically justified, for example by second cation neighbor effects; in other cases, the addition of an added fitting component may be a proxy for a more complex, asymmetrical distribution of whatever structural variable is controlling the observed frequency. Fits with more components are more open to re-interpretation by subsequent authors, as has been done recently for some NMR spectra of alkali silicate glasses, resulting in different conclusions about issues such as FO concentrations (Malfait 2015; Nesbitt et al. 2015b; Sawyer et. al 2015).

Raman spectra, like NMR, have routinely been fitted by Gaussian line shapes (see Hawthorne 1988 for a discussion of basic fitting methods); most commonly the fits have few or no constraints on linewidths, leading to widely varying results among publications. However, as noted above, recent studies indicate that the peaks for some Q^n species in alkali silicate glasses may be better matched by Lorentzian instead of purely Gaussian models, or lineshapes that are mixtures of both, as noted by Kamitsos and Risen (1984). The best-fit Lorentzian/Gaussian proportion (typically 90–70% of the former) may furthermore change systematically with composition, introducing additional adjustable parameters to modelling of spectra (Bancroft et al. 2018).

This approach to fitting Raman spectra in the high frequency envelope (where the Q^n vibrations lie) may lead to more accurate estimates of Q^n proportions, with fewer extraneous fitted components. Component peak shapes for some Q^n bands may also actually be asymmetrical, requiring further care in fitting to produce the physically most meaningful results (McMillan 1984; Bancroft et al. 2018; O'Shaughnessy et al. 2020); constraints on peak widths may also be helpful.

Another approach that is gaining favour is to use Principle Component Analysis (PCA) and statistical algorithms such as Non-negative Matrix factorization to fit spectra (cf., Herzog and Zakaznova-Herzog 2011; Woelffel et al. 2015). These use a purely statistical method applied simultaneously to a set of spectra for glasses of varying composition to deconvolute (fit) the various components contributing to the spectral envelope as well as which components are correlated to each other. While interesting, like all fitting techniques the results depend on the assumptions made at the beginning of the fits and in their subsequent interpretations. In any case, a calculated experimental envelope, even with a minimized statistical measure such as χ^2, that closely matches the experimental one is not necessarily the most "correct" fit if the assumptions and analysis are physically unrealistic or incorrect.

In summary it can be said that almost any empirical analysis of spectra of glasses requires making assumptions that are dependent upon some model of the structure, even if this is simply a conception of the relationship of disorder to resulting line shape. In the most convincing results, assumptions are explicitly stated and tested when possible on model systems, whether amorphous, liquid, or crystalline. More adjustable variables, e.g., number of fitted peaks or varying peak shape, can improve statistical quality of fits but may or may not be physically justifiable. More constraints on fits (relative peak areas, positions, etc.) may bring results closer to an expected model of the structure but can also build in potentially inaccurate assumptions. Simultaneous fitting of spectra for multiple glasses over a wide composition range can be helpful but may require further assumptions about the constancy of parameters such as peak position, shape and width (Lee and Stebbins 1999; Zakaznova-Herzog et al. 2007). In any case, a realistic assessment of uncertainty (not simply the statistical reproducibility of a given model), including possible variations due to assumptions used in fitting, is most helpful in testing the robustness of conclusions about structure.

One pathway towards significant improvement in the analysis of spectroscopic (and other) data on glass structure is the improvement in theoretical tools to more quantitatively connect structure with observables. This is a large subject largely out of the scope of this chapter, but a few example can be given. Recent work on 2-dimensional, ^{17}O MQMAS of multicomponent glasses combined high quality experimental data with molecular dynamics simulations and DFT calculations to produce the first accurate quantitative models of distributions of species and the extent of disorder in such systems, greatly expanding our knowledge (Charpentier et al. 2004; Angeli et al. 2011). Theoretical calculations of full vibrational spectra (including cross sections, intensities, effects of disorder, and not simply mean frequencies) could potentially allow more complete extraction of information from the available experimental data, which may be especially critical for high temperature and high-pressure measurements. Raman vibrational spectra are particularly difficult to simulate due to the complex nature of the polarizability which influences peak intensities. However, progress is being made using DFT calculations (cf., Kilymis et al. 2019).

Analysis of X-ray and neutron diffraction data on glasses and especially on high temperature melts has progressed dramatically with computer simulations and 'reverse Monte Carlo' methods to generate structures that match experiment (Skinner et al. 2012, 2014). However, one should note that X-ray and neutron correlation functions can often be fit with multiple structural models. This is because nearly any model structure involving SiO_4 tetrahedra will fit the first 3 interatomic distances in the RDFs reasonably well. The peaks are the Si–O, O–O and Si–Si distance. In compositions that contain Al, Fe or Ti it becomes difficult to distinguish differences in the T–O (where T = Si, Al, Fe or Ti) bond lengths and partial correlation modelling using

Gaussian curve fits are often employed (see Hannon 2014). Beyond the Si–Si distance there are multiple interatomic distances contributing to the peaks and it in this distance range (3.5–10 Å or higher) where comparison of simulations with experimental data breaks down. As DFT and other first principles algorithms and computational techniques improve, it should be possible to develop more realistic simulations of the structure beyond the SRO.

Modifier and charge compensating cations

In the above discussions we have reserved the term 'network forming' cation for those with +3 or +4 valence with only 3 or 4 strong, partially covalent bonds to oxygen. Already when these coordinations increase with pressure (e.g., for Si and Al), the structural role of such cations may become less well-defined. 'Network modifying' cations usually have +1 or +2 valence, coordination numbers of 5 to 8 or even more, and longer, weaker, and more ionic bonds. In magmas these are most commonly represented by K^+, Na^+, Ca^{2+}, and Fe^{2+}. The smaller Mg^{2+} and Li^+ may have coordination numbers as low as 4 in some systems (Ildefonse et al. 1995; Li et al. 1999), but its role in melt and glass properties remains that of a 'modifier' (Fiske and Stebbins 1994; George and Stebbins 1998). Fe^{3+} may also often have tetrahedral coordination in glasses, as determined by Mössbauer and other spectroscopic studies (Mysen and Richet 2019), and its higher valence may result in a structural role more like that of Al^{3+} (indeed, crystals such as $KFeSiO_4$ and $KFeSi_2O_6$ have tetrahedral framework structures similar to their aluminosilicate analogs, kalsilite and leucite). Other larger, high-valence cations that are much less abundant in geological glasses and melts (e.g., REE^{3+}, Ti^{4+}, Zr^{4+}, Mo^{6+}, etc.) have often been described as having behavior 'intermediate' between network formers and modifiers, reflecting their complex effects on properties (Varshneya 1994). Short range structure around such cations may be very interesting for controlling thermodynamic activities and thus phase equilibria of minor minerals that control concentrations of trace elements and radiogenic isotopes, such as rutile, zircon, and monazite.

Network modifying cations in crystalline silicates (apart from orthosilicates, containing only Q^0 groups and only NBO) are often coordinated by mixtures of both bridging and non-bridging oxygens, simply because there are not enough of the latter to fill the entire coordination shell. In mixed cation systems, the smaller, higher field strength modifiers may concentrate NBO, for example Ca^{2+} with 4 BO and 4 NBO vs. Mg^{2+} with 6 NBO in crystalline diopside ($CaMgSi_2O_6$). The same is expected in molten and glassy silicates, complicating structural models. Because Reaction (4) produces only 2 moles of NBO for each mole of M^+O or M^{2+}_2O added to silica, and cation coordination numbers are at least 4 (often 6 to 8), higher field strength modifiers may contribute to clustering of NBO and hence of modifier cations in glasses and liquids.

In glasses and melts, it is often expected that modifier coordination numbers will be lower than in corresponding crystals because of the higher bulk density of the latter, and there are some data to support this presumption (Brown et al. 1995). However, caution is often needed because of the challenges of measuring coordination in a highly disordered structure with a potentially wide range of first-shell cation–oxygen distances. EXAFS techniques are often used to determine coordination numbers in glasses. But, the CNs determined are averages and have fairly high uncertainties e.g., ±20% (cf., Newville 2014; Dalba and Rocca 2015). The pre-edge XANES peaks can also be used extract coordination numbers (cf., Henderson et al. 2014a and references therein) and have been used extensively to determine Fe^{3+}/Fe^{2+} and Ti^{4+} coordination in a variety of glass compositions.

The structural environment of Fe is complex since it may exist in two oxidation states (Fe^{3+} and Fe^{2+}) as well as, in 4-, 5- and 6-fold coordination. Titanium is similar in that while Ti^{3+} is not likely in terrestrial magmas, Ti^{4+} can exist in 4-, 5-, and 6-fold coordination. For both transition metals, 5-fold coordination seems to more common in glasses and melts, but the proportions of the different coordination states depend upon composition, *T* and *P*. In addition, it remains difficult to discriminate between the presence of 5-fold coordination versus a mixture of 4- and 6-fold coordination. Mysen and Richet (2019) give a summary of the characteristic coordination and oxidation states of Fe (Montazerian and Zanatto 2022; Sakamaki and Ohtani 2022, both this volume) and Ti (Dasgupta et al. 2022, this volume) in common magmatic compositions as does Henderson et al. (2014a).

While Raman spectroscopy is predominantly used to probe the Q^n species in silicate glasses and melts, it has also been used to investigate the likely CN of both Fe (Welsch et al. 2017) and Ti (Henderson and Fleet 1995; Scannell and Huang 2016; Scannell et al. 2016a,b). It should be noted however that trying to determine the coordination of these cations by Raman spectroscopy is not simple. It requires careful analysis of the spectra and it is often subtle changes in the spectral envelope that are important. Addition of Fe and Ti oxides to alkali silicate composition glasses appear to generate Q^2 and Q^3 species (cf., Welsch et al. 2017; Henderson and Fleet 1995; Mysen and Neuville 1995; Nayak et al. 2019a) with the Q^2 and Q^3 band intensities exhibiting a strong dependence upon the Fe and Ti concentrations

Nevertheless, for Fe containing glasses, because of the strong correlation between Q species' intensity and the addition of Fe oxide, they can be used to determine the redox state of iron in glasses (Cochain et al. 2012; Le Losq et al. 2019a). The reader is referred to Le Losq et al. (2019a) for a more comprehensive discussion of the different procedures for redox determination in Fe containing glasses.

X-ray and neutron diffraction (cf., Weigel et al. 2006, 2008b) and Fe or Ti K- and L-edge XANES (cf., Henderson et al. 2014a for a comprehensive discussion of XANES) have been extensively used to determine the coordination of Fe and Ti cations in glasses. The majority of diffraction studies have found that in alkali silicate glasses the majority of Fe^{3+} is in 4-fold coordination with some Fe^{3+} in 5-fold coordination. Ferrous iron (Fe^{2+}) is also in 4-fold coordination. The $^{[4]}Fe^{3+}$–O distance is 1.87 Å while $^{[5]}Fe^{3+}$–O is 2.01 Å (Weigel et al. 2008b).

The pre-edge features of Fe and Ti K-edge XANES have been extensively used to investigate Fe and Ti coordination in glasses and is preferable to Mössbauer spectroscopy for iron contents $<0.1\%$ (see Henderson et al. (2014a) and references therein for a more comprehensive discussion). Recently Nayak et al. (2019a,b) have studied Li_2O, Na_2O and K_2O–silicate glasses at fixed iron contents using EXAFS, Raman and Mössbauer. They show that in Na_2O–silicate glasses Q^2 increases as Fe oxide is added as noted above, while in K_2O-containing glasses Q^2 is abundant in lower K_2O composition glasses but Q^3 and Q^4 become more abundant at high K_2O contents.

In many cases, more robust conclusions can be drawn from *relative*, rather than *absolute* changes in modifier cation coordination, or mean cation-oxygen bond distances, as a function of composition, temperature, or pressure. For example, easily-observable ^{23}Na MAS NMR spectra of Na-rich glasses always have relatively broad peaks with no resolution among signals for cations with varying coordination numbers, as well as asymmetry related to quadrupolar coupling. Nonetheless, collection of spectra at multiple magnetic fields, and application of MQMAS NMR, can yield useful results on compositional effects on Na coordination environment (Gee et al. 1997; Kohn et al. 1998; Angeli et al. 2011). Decreasing the bulk NBO/T, for example in a Na silicate–Na aluminosilicate binary, does shift the NMR peak in a direction suggesting longer mean Na–O distances and likely higher coordination, as calibrated by observations of crystalline silicates with known structures (Xue and Stebbins 1993; George and Stebbins 1995; Stebbins 1998). This is expected as the ratio of highly charged NBO to lesser charged BO (Si–O–Al in this example) decreases, requiring more total O first neighbors. A shift in the opposite direction (towards shorter Na–O distances) has been reported in high pressure glasses, consistent with the compression of the 'soft' alkali ion sites (Allwardt et al. 2005b; Lee et al. 2006; Kelsey et al. 2009a) (Fig. 25).

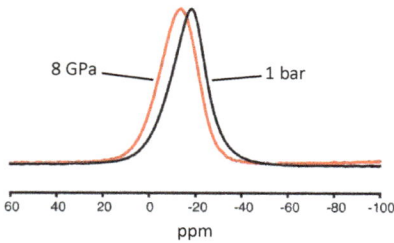

Figure 25. ^{23}Na MAS NMR spectra for NaAlSi$_3$O$_8$ glasses, showing a shift to higher frequency (left side) with higher quench pressure, consistent with reduction in mean Na–O bond length. Reprinted from The effect of Na/Si ratio on the structure of sodium silicate and aluminosilicate glasses quenched from melts at high pressure: A multi-nuclear (Al-27, Na-23, O-17) 1D and 2D solid-state NMR study, Lee SK, Cody GD, Fei Y, Mysen BO, Chemistry of Geology 229:162–172, copyright (2006), with permission from Elsevier.

A unique aspect of NMR can be its ability to measure interactions between pairs of like or of unlike nuclides, through experiments relying on through-space nuclear dipolar couplings or through bond 'J' couplings. The former are very sensitive to interatomic distance, the latter to actual bond connections and orbital perturbations. Among magmatically abundant modifier cations, Na–Na dipolar couplings are especially strong, and have been used effectively to estimate mean Na–Na distances in glasses and thus test models of homogeneous vs. clustered cation distributions (Gee and Eckert 1995). Na interactions with BO and NBO in aluminosilicates have been explored with through-bond double resonance NMR methods, again testing models of cation distributions (Sukenaga et al. 2017).

Other geologically common modifier cations (^{39}K, ^{43}Ca, ^{25}Mg) are less amenable to NMR spectroscopy for reasons of low resonant frequency, low natural abundance, and/or large quadrupolar coupling, but some progress has been made using very high magnetic fields and advanced 2D methods such as MQMAS (Kroeker and Stebbins 2000; Mackenzie and Smith 2002; Stebbins et al. 2002; Angeli et al. 2007; Pedone et al. 2010). The highly magnetic, unpaired electronic spins on Fe^{2+} and Fe^{3+} cause major technical complications for the observation of ^{57}Fe NMR in solids, as well as for obtaining and interpreting NMR spectra of other more commonly studied nuclides (Stebbins et al. 2018). In some cases, *in situ* high temperature NMR in low-viscosity silicate liquids can observe very narrow resonances for 'difficult' nuclides such as ^{25}Mg (Fiske and Stebbins 1994; George and Stebbins 1998). Such spectra show only the motionally averaged cation environment, but still contain information on temperature and compositional effects on mean modifier cation coordination.

Network modifiers such as the alkali and alkaline-earth elements are also difficult to probe by X-ray, neutron and XAS methods. This is primarily due to a number of factors including low atomic numbers, volatilization of the alkalis and alkaline-earths, core hole lifetime broadening and interference effects from other elemental edges in XAS. The latter prevents, for example, obtaining the EXAFS signal of elements below the K-edge of Si. In general, one needs data out to a Å^{-1} range of at least 10 Å^{-1} or greater to discriminate bond distance variations. Greaves et al. (1981) were able to obtain a Na EXAFS signal out to ~400 eV (10 Å^{-1}) due to modification of the soft X-ray beamline used in their study. The majority of X-ray structural studies of light elements in glasses are XANES. Henderson et al. (2014a) reviews the XANES data for both the alkalis and alkaline-earths, as well as, many other elements (down to Br) in the periodic table and should be consulted for more detailed information.

Of increasing interest is the structural role of network modifiers in glasses (e.g., Li O'Shaughnessy et al. 2018a,b; Cs, Li et al. 1999; O'Shaughnessy et al. 2017) and melts at high T and/or P. As noted previously, it is difficult to obtain such information but the technique of X-ray Raman spectroscopy (XRS; Lee et al. 2014; de Clermont Gallerande et al. 2018) is becoming increasingly popular. Recent XRS studies have looked at the O K-edge in synthetic basalt compositions (haplobasalt) at high P (Moulton et al. 2016a,b), O and B K-edges in borosilicate glasses (Lelong et al. 2017) and the Li K-edge in melts (Cicconi et al. 2019, in prep) to name a few.

ACKNOWLEDGEMENTS

We thank Charles Le Losq and an anonymous reviewer for helpful comments on the original manuscript, and Ian Swainson for enormous help in editing and layout throughout this chapter and volume. JFS thanks the NSF for support through grant number EAR-1521055. GSH thanks NSERC for financial support through an NSERC discovery grant.

REFERENCES

Abdelouhab S, Podor R, Rapin C, Toplis MJ, Berthoud P, Vilasi M (2008) Determination of Na_2O activities in silicate melts by EMF measurements. J Non-Cryst Solids 354:3001–3011

Affatigato, M (ed) (2015) Modern Glass Characterisation. John Wiley and Sons, USA

Aguiar PM, Kroeker S (2007) Boron speciation and non-bridging oxygens in high-alkali borate glasses. J Non-Cryst Solids 353:1834–1839

Almeida RM, Santos LF (2015) Raman spectroscopy of glasses. *In:* Modern Glass Characterisation, M Affatigato (Ed), John Wiley and Sons, USA, 74–106

Allu AR, Gaddam A, Ganisetti S, Balaji S, Siegel R, Mather GC, Fabian M, Pascual MJ, Ditaranto N, Milius W, Senker J, Agarkov DA, Kharton VV, Ferreira JMF (2018) Structure and crystallization of alkali-earth aluminosilicate glasses: Prevention of the alumina-avoidance principle. J Phys Chem B 122:4737–4747

Allwardt JR, Stebbins JF (2004) Ca–Mg and K–Mg mixing around non-bridging oxygens in silicate glasses: An investigation using oxygen-17 MAS and 3QMAS NMR. Am Mineral 89:777–784

Allwardt JR, Lee SK, Stebbins JF (2003) Bonding preferences of non-bridging oxygens in calcium aluminosilicate glass: evidence from ^{17}O MAS and 3QMAS NMR on calcium aluminate and low-silica Ca-aluminosilicate glasses. Am Mineral 88:949–954

Allwardt JR, Schmidt BC, Stebbins JF (2004) Structural mechanisms of compression and decompression in high pressure $K_2Si_4O_9$ glasses: An investigation utilizing Raman and NMR spectroscopy of high-pressure glasses and crystals. Chem Geol 213:137–151

Allwardt JR, Stebbins JF, Schmidt BC, Frost DJ (2005a) The effect of composition, compression, and decompression on the structure of high pressure aluminosilicate glasses: an investigation utilizing ^{17}O and ^{27}Al NMR. *In:* Frontiers of High Pressure Research. Chen J, Wang Y, Duffy T S, Shen G, Dobrzhinetskaya LF (eds) Elsevier, Amsterdam, p 211–240

Allwardt JR, Stebbins JF, Schmidt BC, Frost DJ, Withers AC, Hirschmann MM (2005b) Aluminum coordination and the densification of high-pressure aluminosilicate glasses. Am Mineral 90:1218–1222

Allwardt JA, Stebbins JF, Terasaki H, Du L-S, Frost DJ, Withers AC, Hirschmann MM, Susuki A, Ohtani E (2007) Effect of structural transitions on macroscopic properties of high-pressure silicate melts: ^{27}Al MAS NMR and density measurements of glasses and falling sphere viscometry measurements. Am Mineral 92:1093–1104

Angeli F, Charpentier T, Faucon P, Petit JC (1999) Structural characterization of glasses from the inversion of ^{23}Na and ^{27}Al 3Q-MAS NMR spectra. J Phys Chem B 103:10356–10364

Angeli F, Gaillard M, Jollivet P, Carpentier T (2007) Contribution of [43]Ca MAS NMR for probing the structural configuration of calcium in glass. Chem Phys Let 440:324–328

Angeli F, Villain O, Schuller S, Ispas S, Charpentier T (2011) Insight into sodium silicate glass structural organization by multinuclear NMR combined with first-principles calculations. Geochim Cosmochim Acta 75:2453–2469

Angell CA, Cheeseman PA, Tamaddon S (1982) Pressure enhancement of ion mobilities in liquid silicates from computer simulation studies to 800 kilobars. Science 218:885–887

Ansell S, Krishnan S, Weber JKR, Felten JJ, Nordine PC, Beno MA, Price DL, Saboungi M-L (1997) Structure of liquid aluminum oxide. Phys Rev Lett 78:464–466

Ashbrook SE, Le Polles L, Pickard CJ, Berry AJ, Wimperis S, Farnan I (2007) First-principles calculations of solid-state ^{17}O and ^{29}Si NMR spectra of Mg_2SiO_4 polymorphs. Phys Chem Chem Phys 9:1587–1598

Bancroft GM, Nesbitt HW, Henderson GS, O'Shaughnessy C, Withers AC, Neuville DR (2018) Lorentzian dominated lineshapes and linewidths for Raman symmetric stretch peaks (800–1200 cm^{-1}) in Q^n ($n = 1$–3) species of alkali silicate glasses/melts. J Non-Crys Solids 484:72–83

Behrens H, Misiti V, Freda C, Vetere F, Botcharnikov RE, Scarlato P (2009) Solubility of H_2O and CO_2 in ultrapotassic melts at 1200 and 1250 degrees C and pressure from 50 to 500 MPa. Am Mineral 94:105–120

Benmore CJ (2015) X-ray diffraction from glass, *In:* Modern Glass Characterisation, M Affatigato (Ed), John Wiley and Sons, USA, p 241–270

Benmore CJ, Soinard E, Amin SA, Guthrie M, Shastri S, Lee PL, Yarger JL (2010) Structural and topological changes in silica glass at pressure. Phys Rev B 81:054105

Bernasconi A, Dapiaggi M, Bowron D, Ceola S, Maurina S (2016) Aluminosilicate-based glasses structural investigation by high-energy X-ray diffraction. J Mater Sci 51:8845–8860

Beyssac O, Pasteris JD (eds) (2020) Raman spectroscopy in earth and planetary sciences. Elements 16:2

Bista S, Stebbins JF (2017) The role of modifier cations in network modifier cation coordination increases with pressure in aluminosilicate glasses and melts from 1 to 3 GPa. Am Mineral 102:1657–1666

Bista S, Stebbins JF, Hankins B, Sisson TW (2015) Aluminosilicate melts and glasses at 1 to 3 GPa: Temperature and pressure effects on recovered structural and density changes. Am Mineral 100:2298–2307

Bista S, Stebbins JF, Wu J, Gross TM (2017) Structural changes in calcium aluminoborosilicate glasses recovered from pressures of 1.5 to 3.0 GPa: Interactions of two network species with coordination number increases. J Non-Cryst Solids 478:50–57

Brandriss ME, Stebbins JF (1988) Effects of temperature on the structures of silicate liquids: ^{29}Si NMR results. Geochim Cosmochim Acta 52:2659–2670

Brawer S (1985) Relaxation in Viscous Liquids and Glasses. American Ceramic Society, Inc., Columbus

Brown GE, Jr, Farges F, Calas G (1995) X-ray scattering and X-ray spectroscopy studies of silicate melts. Rev Mineral 32:317–410

Brydson RK, Brown A, Benning LG, Livi K (2014) Analytical transmission electron microscopy. Rev Mineral Geochem 78:219–269

Calas G, Galoisy L, Cormier L (2020) The color of glass. *In:* Richet P (Ed.), Encyclopedia of Glass Science, Technology, History, and Culture, J Wiley and Sons, Hoboken, NJ

Capron M, Florian P, Fayon F, Trumeau D, Hennet L, Gaihlanou M, Thiaudiere D, Landron C, Douy A, Massiot D (2001) Local structure and dynamics of high temperature SrO–Al_2O_3 liquids studied by ^{27}Al NMR and Sr K-edge XAS spectroscopy. J Non-Cryst Solids 293–295:496–501

Chan JCC, Bertmer M, Eckert H (1999) Site connectivities in amorphous materials studied by double-resonance NMR of quadrupolar nuclei: High-resolution ^{11}B–^{27}Al spectroscopy of aluminoborate glasses. J Am Chem Soc 121:5238–5248

Charpentier T, Ispas S, Profeta M, Mauri F, Pickard CJ (2004) First-principles calculation of O-17, Si-29, and Na-23 NMR spectra of sodium silicate crystals and glasses. J Phys Chem B 108:4147–4161

Cicconi MR, O'Shaughnessy C, Hennet L, Henderson G, Neuville DR (2019) X-ray Raman scattering study of Li-based compounds at high temperature. presented at X-ray Raman scattering spectroscopy Workshop, ESRF, April 2019

Cochain B, Neuville DR, Henderson GS, McCammon CA, Pinet O, Richet P (2012) Effects of iron content and redox sate on the structure of sodium borosilicate glasses: A Raman, Mössbauer and boron K-edge XANES spectroscopy study. J Am Ceram Soc 95:962–971

Cormack AN, Du J, Zeitler TR (2003) Sodium ion migration mechanisms in silicate glasses probed by molecular dynamics simulations. J Non-Cryst Solids 323:147–154

Cormier L (2019) Neutron and X-ray diffraction of glass. *In:* Springer Handbook of Glass. Springer Nature, Cham, Switzerland, 1045–1083

Dalba G, Rocca F (2015) XAFS spectroscopy and glass structure. *In:* Modern Glass Characterisation, M Affatigato (Ed), John Wiley and Sons, USA, p 271–314

Dasgupta R, Chowdhury P, Eguchi J, Sun C, Saha S (2022) Volatile-bearing partial melts in the lithospheric and sub-lithospheric mantle on Earth and other rocky planets. Rev Mineral Geochem 87:575–606

Davis MC, Kaseman DC, Parvani SM, Sanders KJ, Grandinetti PJ, Massiot D, Florian P (2010) $Q^{(n)}$ species determination in K_2O–$2SiO_2$ glass by ^{29}Si magic angle flipping NMR. J Phys Chem A 114:5503–5508

Davis MC, Sanders KJ, Grandinetti PJ, Gaudio SJ, Sen S (2011) Structural investigations of magnesium silicate glasses by ^{29}Si 2D magic-angle flipping NMR. J Non-Cryst Solids 357:2787–2795

de Clermont Gallerande E, Cabaret D, Lelong G, Brouder C, Attaia, M-B, Paulatto L, Gillmore K, Shale Ch J, Radtke G (2018) First-principles modeling of X-ray Raman scattering spectra. Phys Rev B 98:21404

de Jong BHWS, Supèr HTJ, Spek AL, Nachtegaal G, Fischer JC (1998) Mixed alkali systems: Structure and ^{29}Si MASNMR of $Li_2Si_2O_5$ and $K_2Si_2O_5$. Acta Cryst B 54:568–577

Dell WJ, Bray PJ, Xiao SZ (1983) ^{11}B NMR studies and structural modeling of Na_2O–B_2O_3–SiO_2 glasses of high soda content. J Non-Cryst Solids 58:1–16

Devine RAB, Dupree R, Farnan I, Capponi JJ (1987) Pressure-induced bond-angle variation in amorphous SiO_2. Phys Rev B 35:2560–2562

Dirken PJ, Kohn SC, Smith ME, van Eck ERH (1997) Complete resolution of Si–O–Si and Si–O–Al fragments in an aluminosilicate glass by ^{17}O multiple quantum magic angle spinning NMR spectroscopy. Chem Phys Let 266:568–574

Drewitt JWE, Jahn S, Sanloup C, Cristiglio V, Bytchkov A, Leydier M, Brassamin S, Fischer HE, Hennet L (2011) The structure of liquid calcium aluminates as investigated using neutron and high energy X-ray diffraction in combination with molecular dynamics simulation methods. J Phys Condens Matter 23:155101

Drewitt JWE, Hennet L, Zeidler A, Jahn S, Salmon PS, Neuville DR, Fischer HE (2012) Structural transformations on vitrification in the fragile glass forming system $CaAl_2O_4$. Phys Rev Lett 109:235501

Drewitt JWE, Jahn S, Sanloup C, de Grouchy C, Garbarino G, Hennet L (2015) Development of chemical and topological structure in aluminosilicate liquids and glasses at high pressure. J Phys Condens Matter 27:105103

Drewitt JWE, Hennet L, Neuville DR (2022) From short to medium range order in glasses and melts by diffraction and Raman spectroscopy. Rev Mineral Geochem 87:55–103

Du J, Cormack AN (2003) The medium range structure of sodium silicate glasses: a molecular dynamics simulation. J Non-Cryst Solids 323:66–79

Du LS, Stebbins JF (2006) Oxygen sites and network coordination in sodium germanate glasses and crystals: high-resolution oxygen-17 and sodium-23 NMR. J Phys Chem B 110:12427–12437

Du LS, Allwardt JR, Schmidt BC, Stebbins JF (2004) Pressure-induced structural changes in a borosilicate glass-forming liquid: boron coordination, non-bridging oxygens, and network ordering. J Non-Cryst Solids 337:196–200

Dubessy J, Caumon MC, Rull F (eds) (2012) Raman spectroscopy applied to Earth Sciences and Cultural Heritage. EMU Volume 12, Aberystwyth, UK

Dubinsky EV, Stebbins JF (2006) Quench rate and temperature effects on framework ordering in aluminosilicate melts. Am Mineral 91:753–761

Duer MJ (2002) Solid-state NMR spectroscopy: principles and applications. Blackwell, Oxford, UK

Duer MJ (2004) Introduction to solid-state NMR spectroscopy. Blackwell, Oxford, UK

Eckert H (1992) Structural characterization of noncrystalline solids and glasses using solid state NMR. Prog Nucl Mag Res 24:159–293

Eckert H (1994) Structural studies of non-crystalline solids using solid state NMR. New experimental approaches and results. *In:* Blümich, B (Ed.), Solid-State NMR IV. Methods and Applications of Solid-State NMR. Springer-Verlag, Berlin, p 127–198

Eckert H (2017) Spying spins on messy materials: 60 years of glass structure elucidation by NMR spectroscopy. Int J Appl Glass Sci 2017:1–21

Emerson JF, Stallworth PE, Bray PJ (1989) High-field ^{29}Si NMR studies of alkali silicate glasses. J Non-Cryst Solids 113:253–259

Engelhardt G, Michel D (1987) High-Resolution Solid-State NMR of Silicates and Zeolites. Wiley, New York

Exarhos GJ, Frydrych WS, Walrafen GE, Fisher M, Pugh E, Garofilini SH (1988) Vibrational spectra of silica near 2400°K: Measurement and molecular dynamics simulations, *In:* Clark RJH, Long DA (Eds), Proceeding of 11th International Conference Raman Spectroscopy, Wiley p. 503–504

Farnan I, Stebbins JF (1990a) A high temperature ^{29}Si NMR investigation of solid and molten silicates. J Am Chem Soc 112:32–39

Farnan I, Stebbins JF (1990b) Observation of slow atomic motions close to the glass transition using 2-D ^{29}Si NMR. J Non-Cryst Solids 124:207–215

Farnan I, Stebbins JF (1994) The nature of the glass transition in a silica-rich oxide melt. Science 265:1206–1209

Fischer HE, Barnes AC, Salmon PS (2006) Neutron and x-ray diffraction studies of liquids and glasses. Rep Prog Phys 69:233–299

Fiske P, Stebbins JF (1994) The structural role of Mg in silicate liquids: a high-temperature ^{25}Mg, ^{23}Na, and ^{29}Si NMR study. Am Mineral 79:848–861

Florian P, Massiot D, Poe B, Farnan I, Coutures JP (1995) A time-resolved ^{27}Al NMR study of the cooling process of liquid alumina from 2450 °C to crystallisation. Solid Stat Nucl Magnet Reson 5:233–238

Florian P, Vermillion KE, Grandinetti PJ, Farnan I, Stebbins JF (1996) Cation distribution in mixed alkali disilicate glasses. J Am Chem Soc 118:3493–3497

Franz JD, Mysen BO (1995) The Raman spectra and structure of BaO–SiO$_2$, SrO–SiO$_2$ and CaO–SiO$_2$ melts at 1600°C. Chem Geol 121:155–175

Gaudio SJ, Sen S, Lesher CE (2008) Pressure-induced structural changes and densification of vitreous MgSiO$_3$. Geochim Cosmochim Acta 72:1222–1230

Gaudio SJ, Lesher CE, Sen S (2009) Structural relaxation phenomena in amorphous silicates at high pressure and temperature. COMPRES ann mtg (abst)

Gee B, Eckert H (1995) ^{23}Na nuclear magnetic resonance spin echo decay spectroscopy of sodium silicate glasses and crystalline model compounds. Solid State NMR 5:113–122

Gee B, Janssen M, Eckert H (1997) Local cation environments in mixed alkali silicate glasses studied by multinuclear single and double resonance magic-angle spinning NMR. J Non-Cryst Solids 215:41–50

George AM, Stebbins JF (1995) High temperature ^{23}Na NMR data for albite: comparison to chemical shift models. Am Mineral 80:878–884

George AM, Stebbins JF (1998) Structure and dynamics of magnesium in silicate melts: a high temperature ^{25}Mg NMR study. Am Mineral 83:1022–1029

Ghiorso MS (2004) An EOS for silicate melts. III. Analysis of stoichiometric liquids at elevated pressure: shock compression data, molecular dynamics simulations and mineral fusion curves. Am J Sci 304:752–810

Gibbs GV (1982) Molecules a models for bonding in silicates. Am Mineral 67:421–450

Gibbs GV, Hill FC, Boisen Jr. MB (1997) The Si–O bond and electron density distributions. Phys Chem Mineral 24:167–178

Gibbs GV Wallace AF, Cox DF, Downs RT, Ross NL, Rosso KM (2009) Bonded interactions in silica polymorphs, silicates and siloxane molecules. Am Mineral 94:1085–1102

Greaves GN (1985) EXAFS and the structure of glass. J Non-Cryst Solids 71:203–217

Greaves GN (2020) The extended structure of Glass. *In:* Richet P (Ed.), Encyclopedia of Glass Science, Technology, History, and Culture, J Wiley and Sons, Hoboken, NJ

Greaves GN, Ngai KL (1994) Ionic transport properties in oxide glasses derived from atomic structure. J Non-Cryst Solids 172:1378–1388

Greaves GN, Sen S (2007) Inorganic glasses, glass-forming liquids and amorphizing solids. Adv Phys 56:1–166

Greaves GN, Fontaine A, Lagarde P, Raoux D, Gurman SJ (1981) Local structure of silicate glasses. Nature 293:611–616

Grey CP, Dupré N (2004) NMR studies of cathode materials for lithium-ion rechargeable batteries. Chem Rev 104:4493–4512

Gupta PK, Lui ML, Bray PJ (1985) Boron coordination in rapidly cooled and in annealed aluminum borosilicate glass fibers. J Am Ceram Soc 68:C82

Hannon AC (2015) Neutron diffraction techniques for structural studies of glasses. *In:* Modern Glass Characterisation, M Affatigato (Ed), John Wiley and Sons, USA, p 158–240

Hawthorne FC (1988) Spectrum fitting methods. Rev Mineral 18:63–98

Henderson GS (2005) The structure of silicate melts: A glass perspective. Can Mineral 43:1921–1958

Henderson GS (2007) The germanate anomaly: What do we know? J Non-Cryst Solids 353:1695–1704

Henderson GS (2020) Structural Probes. *In:* Richet P (Ed.), Encyclopedia of Glass Science, Technology, History, and Culture. J Wiley and Sons, Hoboken, NJ

Henderson GS, Fleet ME (1995) The structure of titanium silicate glasses. Can Mineral 33:399–408

Henderson GS, Wang HM (2002) Germanium coordination and the germanium anomaly. Eur J Mineral 14:733–744

Henderson GS, Xiaoyang Liu, Fleet ME (2002) A Ti L-edge X-ray Absorption study of Ti-silicate glasses. Phys Chem Mineral 29:32–42

Henderson GS, Soltay LG, Wang HM (2010) Q speciation in germanate glasse. J Non-Cryst Solids 356:2480–2485

Henderson GS, de Groot FMF, Moulton BJA (2014a) X-ray Absorption Near-Edge Structure (XANES) spectroscopy. Rev Mineral Geochem 78:75–138

Henderson GS, Neuville DR, Downs RT (eds) (2014b) Spectroscopic methods in Mineralogy and Materials Sciences. Reviews in Mineralogy and Geochemistry Vol 78, Min Soc Am, Washington DC

Hennet L, Drewitt, JWE, Neuville DR, Cristiglio V, Kozaily J, Brassamin S, Zanghi D, Fischer HE (2016) Neutron diffraction of calcium aluminosilicate glasses and melts. J Non-Cryst Solids 451:89–93

Hess PC (1980) Polymerization model for silicate melts. *In:* Physics of Magmatic Processes. Hargraves R B (ed) Princeton University Press, Princeton, NJ, p 3–48

Höche T, Ikeno H, Mäder M, Henderson GS, Blyth RIR, Sales BC, Tanaka I (2013) Vanadium $L_{2,3}$ XANES experiments and first principles multielectron calculations: Impact of second-nearest neighbouring cations on Vanadium-bearing Fresnoites. Am Mineral 97:665–670

Hoppe U, Kranold R, Weber H-J, Neuefeind J, Hannon AC (2000) The structure of potassium germanate glasses-a combined x-ray and neutron scattering study. J Non-Cryst Solids 278:99–114

Hung I, Gan Z, Gor'kov PL, Kaseman DC, Sen S, LaComb M, Stebbins JF (2016) Detection of "free" oxide ions in low-silica Ca/Mg silicate glasses: Results from ^{17}O–^{29}Si HETCOR NMR. J Non-Cryst Solids 445–446:1–6

Inamura Y, Arai M, Nakamura M, Otomo T, Kitamura N, Bennington SM, Hannon AC, Buchenau U (2001) Intermediate range structure and low energy dynamics of densified vitreous silica. J Non-Cryst Solids 293–295:389–393

Inamura Y, Katayama Y, Utsumi W, Funakoshi K-I (2004) Transformations in the intermediate-range structure of SiO_2 glass under high pressure and temperature. Phys Rev Lett 93:015501

Ildefonse P, Calas G, Flank AM, Lagarde P (1995) Low Z elements (Mg, Al and Si) K-edge X-ray absorption spectroscopy in minerals and disordered systems. Nucl Instrum Methods Phys Res B 97:172–175

Iuga D, Morais C, Gan ZH, Neuville DR, Cormier L, Massiot D (2005) NMR heteronuclear correlation between quadrupolar nuclei in solids. J Am Chem Soc 127:11540–11541

Jaworski A, Stevensson B, Edén M (2015) Direct ^{17}O NMR experimental evidence for Al–NBO bonds in Si-rich and highly polymerized aluminosilicate glasses. Phys Chem Chem Phys 17:18269–18272

Johnson JA, Johnson CE (2005) Mössbauer spectroscopy as a probe of silicate glasses. J Phys Condens Matter 17:R318–R412

Kamitsos EI, (2015) Infrared spectroscopy of glasses. *In:* Modern Glass Characterisation, M Affatigato (ed), John Wiley and Sons, USA, p32–73

Kamitsos EI, Risen Jr WM (1984) Vibrational spectra of alkali and mixed alkali pentasilicate glasses. J Non-Cryst Solids 65:333–354

Kanehashi K, Stebbins JF (2007) *In-situ* high temperature ^{27}Al NMR study on dynamics of calcium aluminosilicate glass and melt. J Non-Cryst Solids 353:4001–4010

Kelsey KE, Stebbins JF, Mosenfelder JL, Asimow PD (2009a) Simultaneous aluminum, silicon, and sodium coordination changes in 6 GPa sodium aluminosilicate glasses. Am Mineral 94:1205–1215

Kelsey KE, Stebbins JF, Singer DM, Brown GE, Jr., Mosenfelder JL, Asimow PD (2009b) Cation field strength effect on high pressure aluminosilicate glass structure: Multinuclear NMR and La XAFS results. Geochim Cosmochim Acta 73:3914–3933

Kieffer J (2015) Brillouin Light Scattering. *In:* Modern Glass Characterisation, M Affatigato (Ed), John Wiley and Sons, USA, p 107–157

Kilymis D, Ispas, S, Helhen B, Peuget S, Delaye J-M (2019) Vibrational properties of sodosilicate glasses from first principles calculations. Phys Rev B 99:054209

Kim N, Bassiri R, Fejer MM, Stebbins JF (2014) The structure of ion beam sputtered amorphous alumina films and effects of Zn doping: high-resolution ^{27}Al NMR. J Non-Cryst Solids 405:1–6

Kim J, Ilott AJ, Middlemiss DS, Chenova NA, Pinney N, Morgan D, Grey CP (2015a) ^{2}H and ^{27}Al solid-state NMR study of the local environments in Al-doped 2-line ferrihydrite, goethite, and lepidocrocite. Chem Mat 27:3966–3978

Kim N, Bassiri R, Fejer MM, Stebbins JF (2015b) Structure of amorphous silica–hafnia and silica–zirconia thin-film materials: The role of a metastable equilibrium state in non-glass-forming oxide systems. J Non-Cryst Solids 429:5–12

Kirkpatrick RJ (1988) MAS NMR spectroscopy of minerals and glasses. Rev Mineral 18:341–403

Kirkpatrick RJ, Dunn T, Schramm S, Smith KA, Oestrike R, Turner G (1986) Magic-angle sample-spinning nuclear magnetic resonance spectroscopy of silicate glasses: a review. *In:* Structure and Bonding in Noncrystalline Solids, Walrafen GE, Revesz AG (eds.) Plenum Press, New York, p 302–327

Kohn SC (2004) NMR studies of silicate glasses. EMU Notes in Mineralogy 6:399–419

Kohn SC, Smith ME, Dirken PJ, van Eck ERH, Kentgens APM, Dupree R (1998) Sodium environments in dry and hydrous albite glasses; improved ^{23}Na solid state NMR data and their implications for water dissolution mechanisms. Geochim Cosmochim Acta 62:79–87

Konnert KH, Karle J (1973) Crystalline ordering in silica and germania glasses. Science 179:177–179

Kono Y, Sanloup C (2018) Magmas Under Pressure: Advances in High-Pressure Experiments on Structure and Properties of Melts. Amsterdam, Elsevier

Kono Y, Shibazaki Y, Kenney-Benson C, Wang Y, Shen G (2018) Pressure-induced structural change in MgSiO$_3$ glass at pressures near the Earth's core-mantle boundary. PNAS 115:1742–1747

Kroeker S, Stebbins JF (2000) Magnesium coordination environments in glasses and minerals: New insight from high field magnesium-25 MAS NMR. Am Mineral 85:1459–1464

Kushiro I (1976) Changes in viscosity and structure of melt of NaAlSi$_2$O$_6$ composition at high pressure. J Geophys Res 81:6347–6350

Kushiro I (1978) Viscosity and structural changes of albite (NaAlSi$_3$O$_8$) melt at high pressures. Earth Planet Sci Lett 41:87–90

Kushiro I, Yoder Jr HS, Mysen BO (1976) Viscosities of basalt and andesite melts at high pressures. J Geophys Res 81:6351–6356

Lacey ED (1963) Aluminum in glasses and melts. Phys Chem Glasses 4:234–238

LaComb M, Stebbins JF (2015) Oxygen triclusters in crystalline and glassy SrB$_4$O$_7$: high resolution ^{11}B and ^{17}O NMR analysis. J Non-Cryst Solids 428:105–111

LaComb, M, Rice D, Stebbins JF (2016) Network oxygen sites in calcium aluminoborate glasses: Results from ^{17}O{^{27}Al} and ^{17}O{^{11}B} double resonance NMR. J Non-Cryst Solids 447:248–254

Landron C, Hennet L, Jenkins TE, Greaves GN, Coutures JP, Soper AK (2001) Liquid alumina: Detailed atomic coordination determined from neutron diffraction data using empirical structure refinement. Phys Rev Lett 21:4839–4842

Larson C, Doerr J, Affatigato M, Feller S, Holland D, Smith ME (2006) A ^{29}Si MAS NMR study of silicate glasses with a high lithium content. J Phys Condens Matter 18:11323–11331

Lebedev AA (1921) On the polymorphism and annealing of glass. Proc State Optical Institute 2:1–20

Lee SK (2005) Microscopic origins of macroscopic properties of silicate melts and glasses at ambient and high pressure: implications for melt generation and dynamics. Geochim Cosmochim Acta 69:3695–3710

Lee SK, Lee BH (2006) Atomistic origin of germanate anomaly in GeO$_2$ and Na-germanate glasses: insights from two-dimensional ^{17}O NMR and quantum chemical calculations. J Phys Chem B 110:16408–16412

Lee SK, Kim EJ (2015) Probing metal-bridging oxygen and configurational disorder in amorphous lead silicates: Insights from ^{17}O solid-state nuclear magnetic resonance. J Phys Chem C 119:748–756

Lee SK, Ryu S (2017) Probing triply coordinated oxygen in amorphous Al$_2$O$_3$. J Phys Chem Lett 9:150–156

Lee SK, Stebbins JF (1999) The degree of aluminum avoidance in aluminosilicate glasses. Am Mineral 84:937–945

Lee SK, Stebbins JF (2000a) Al–O–Al and Si–O–Si sites in framework aluminosilicate glasses with Si/Al=1: quantification of framework disorder. J Non-Cryst Solids 270:260–264

Lee SK, Stebbins JF (2000b) The structure of aluminosilicate glasses: high-resolution ^{17}O and ^{27}Al MAS and 3QMAS NMR study. J Phys Chem B 104:4091–4100

Lee SK, Stebbins JF (2009) Effects of the degree of polymerization on the structure of sodium silicate and aluminosilicate glasses: an ^{17}O NMR study. Geochim Cosmochim Acta 73:1109–1119

Lee SK, Cody GD, Fei Y, Mysen BO (2004) Nature of polymerization and properties of silicate melts and glasses at high pressure. Geochim Cosmochim Acta 68:4189–4200

Lee SK, Cody GD, Mysen BO (2005) Structure and extent of disorder in quaternary (Ca–Mg and Ca–Na) aluminosilicate glasses and melts. Am Mineral 90:1393–1401

Lee SK, Cody GD, Fei Y, Mysen BO (2006) The effect of Na/Si ratio on the structure of sodium silicate and aluminosilicate glasses quenched from melts at high pressure: A multi-nuclear (Al-27, Na-23, O-17) 1D and 2D solid-state NMR study. Chem Geol 229:162–172

Lee SK, Deschamps M, Hiet J, Massiot D, Park SY (2009) Connectivity and proximity between quadrupolar nuclides in oxide glasses: insights from through-bond and through-space correlations in solid-state NMR. J Phys Chem B 113:5162–5167

Lee SK, Eng PJ, Mao H-K, (2014) Probing of pressure-induced bonding transitions in crystalline and amorphous earth materials: insights from X-ray Raman scattering at high pressure Rev Mineral Geochem 78:139–174

Le Losq C, Neuville DR (2017) Molecular structure, configurational entropy and viscosity of silicate melts: Link through the Adam and Gibbs theory of viscous flow. J Non-Cryst Solids 463:175–188

Le Losq C, Neuville DR, Florian P, Henderson GS, Massiot D (2014) The role of Al^{3+} on rheology and structural changes in sodium silicate and aluminosilicate glasses and melts. Geochim Cosmochim Acta 126:495–517

Le Losq C, Berry AJ, Kendrick MA, Neuville DR, O'Neill HStC (2019a) Determination of the oxidation state of iron in Mid-Ocean Ridge basalt glasses by Raman spectroscopy, Am Mineral 104:1032–1042

Le Losq C, Cicconi MR, Greaves GN, Neuville DR (2019b) Silicate glasses. *In:* Springer Handbook of Glass. Springer Nature, Cham, Switzerland, p 441–503

Lelong G, Cormier L, Hennet L, Michel F, Rueff J-P, Ablett JM, Monaco G (2017) Lithium borate crystals and glasses: How similar are they? A non-resonant inelastic X-ray scattering study around the B and O *K*-edge. J Non-Cryst Solids 472:1–8

Levitt, MH (2001) Spin Dynamics, Basics of Nuclear Magnetic Resonance. Wiley, New York

Liebau F (1985) Structural Chemistry of Silicates. Springer-Verlag, New York

Li D, Peng M, Murata T (1999) Coordination and local structure of magnesium in silicate minerals and glasses: Mg K-edge XANES study. Can. Mineral. 37:199-206

Machacek J, Gedeon O, Liska M (2007) Molecular approach to the 5-coordinated silicon atoms in silicate glasses. Phys Chem Glasses 48:345–353

Machacek J, Gedeon O, Liska M (2010) The MD study of mixed alkali effect in alkali silicate glasses. Phys Chem Glasses: Eu J Glass Sci Tech 1:65–68

MacKenzie KJD, Smith ME (2002) Multinuclear Solid-State NMR of Inorganic Materials. Pergamon, New York

Maekawa H, Maekawa T, Kawamura K, Yokokawa T (1991) The structural groups of alkali silicate glasses determined from ^{29}Si MAS-NMR. J Non-Cryst Solids 127:53–64

Majérus O, Cormier L, Calas G, Beuneu B (2004a) A neutron diffraction study of temperature-induced structural changes in potassium disilicate glass and melt. Chem Geol 213:89–102

Majérus O, Cormier L, Itié JP, Galoisy L, Neuville DR, Calas G (2004b) Pressure-induced Ge coordination change and polyamorphism in SiO_2–GeO_2 glasses. J Non-Cryst Solids 345/346:34–38

Malfait WJ (2015) Comment on "Spectroscopic studies of oxygen speciation in potassium silicate glasses and melts." Can J Chem 93:578–580

Malfait WJ, Halter WE, Morizet Y, Meier BH, Verel R (2007a) Structural control on bulk melt properties: Single and double quantum ^{29}Si NMR spectroscopy on alkali-silicate glasses. Geochim Cosmochim Acta 71:6002–6018

Malfait WJ, Zakaznova-Herzog VP, Halter WE (2007b) Quantitative Raman spectroscopy: High-temperature speciation of potassium silicate melts. J Non-Cryst Solids 353:4029–4042

Malfait WJ, Zakaznova-Herzog VP, Halter, WE (2008) Quantitative Raman spectroscopy: speciation of Na-silicate glasses and melts. Am Mineral 93:1505–1518

Malfait WJ, Verel R, Ardia P, Sanchez-Valle C (2012) Aluminum coordination in rhyolite and andesite glasses and melts: Effect of temperature, pressure, composition and water content. Geochim Cosmochim Acta 77:11–26

Martel L, Allix M, Millot F, Sarous-Kanian V, Véron E, Ory S, Massiot D, Deschamps M (2011) Controlling the size of nanodomains in calcium aluminosilicate glasses. J Phys Chem C 115:18935–18945

Martel L, Massiot D, Deschamps M (2014) Phase separation in sodium silicates observed by solid-state MAS-NMR. J Non-Cryst Solids 390:37–44

Martinet C, Kassir-Bodon A, Deschamps T, Cornet A, Le Floch S, Martinez V, Champagnon B (2015) Permanently densified SiO_2 glasses: A structural approach. J Phys Condens Matter 27:325401

Matson DW, Sharma SK, Philpotts JA (1983) The structure of high silica alkali-silicate glasses. A Raman spectroscopic investigation. J Non-Cryst Solids 58:323–352

Massiot D, Trumeau D, Touzo B, Farnan I, Rifflet JC, Douy A, Coutures JP (1995) Structure and dynamics of $CaAl_2O_4$ from liquid to glass: A high-temperature ^{27}Al NMR time-resolved study. J Phys Chem 99:16455–16459

Massiot D, Fayon F, Montouillout V, Pellerin N, Hiet J, Roiland C, Florian P, Coutures J-P, Cormier L, Neuville DR (2008) Structure and dynamics of oxide melts and glasses: A view from multinuclear and high temperature NMR. J Non-Cryst Solids 354:249–254

Masson CR (1977) Anionic constitution of glass-forming melts. J Non-Cryst Solids 25:3–41

McMillan PF (1984) Structural studies of silicate glasses and melts—Applications and limitations of Raman spectroscopy. Am Mineral 69:627–644

McMillan P, Hofmeister AM (1988) Infrared and Raman spectroscopy. Rev Mineral 18:99–159

McMillan PF, Kirkpatrick RJ (1992) Al coordination in magnesium aluminosilicate glasses. Am Mineral 77:898–900

Mead RN, Mountjoy G (2005) The structure of $CaSiO_3$ glass and the modified random network model. Phys Chem Glasses 46:311–314

Mead RN, Mountjoy G (2006) A molecular dynamics study of the atomic structure of $(CaO)_x(SiO_2)_{1-x}$ glasses. J Phys Chem B 110:14273–14278

Mei Q, Benmore CJ, Weber JKR (2007) Structure of liquid SiO_2: A measurement buy high-energy X-ray diffraction. Phys Rev Lett 98:057802

Meyer A, Horbach J ,Kolb W, Kargl F, Schober H (2004) Channel formation and intermediate range order in sodium silicate melts and glasses. Phys Rev Lett 93:027801

Micoulaut M, Cormier L, Henderson GS (2006) The structure of crystalline, amorphous and liquid GeO_2. J Phys Condens Matter 18:R753–R784

Montazerian M, Zanotto ED (2022) Nucleation, growth, and crystallization in oxide glass-formers. A current perspective. Rev Mineral Geochem 87:405–429

Morin EI, Stebbins JF (2016) Separating the effects of composition and fictive temperature on Al and B coordination in Ca, La, Y aluminosilicate, aluminoborosilicate, and aluminoborate glasses. J Non-Cryst Solids 432:384–392

Moulton BJA, Henderson GS, Sonneville C, O'Shaughnessy CA, Zuin L, Regier T, de Ligny D (2016a) The structure of haplobasaltic glasses investigated using X-ray absorption near edge structure (XANES) spectroscopy at the Si, Al, Mg, and O K-edges and Ca, Si and Al $L_{2,3}$-edges. Chem Geol 420:213–230

Moulton BJA, Henderson GS, Fukui H, Hiraoka N, de Ligny D, Sonneville C, Kanzaki M (2016b) *In-situ* structural changes of amorphous diopside $(CaMgSi_2O_6)$ up to 20 GPa: A Raman and O K-edge X-ray Raman spectroscopic study. Geochim Cosmochim Acta 178:41–61

Moulton BJA, Henderson GS, de Ligny D, Martinet, C, Martinez V, Sonneville C (2019) Structure-longitudinal sound velocity relationships in glassy anorthite $(CaAl_2Si_2O_8)$ up to 20GPa: An *in-situ* Raman and Brillouin spectroscopy study. Geochim Cosmochim Acta 261:132–144

Mountjoy G (2007) The local environment of oxygen in silicate glasses from molecular dynamics. J Non-Cryst Solids 353:1849–1853

Mozzi RL,Warren BE (1969) The structure of vitreous silica. J Appl Crystallogr 2:164–172

Murakami M, Bass JD (2010) Spectroscopic evidence for ultrahigh-pressure polymorphism in SiO_2 glass. Phys Rev Lett 104:025504

Murdoch JB, Stebbins JF, Carmichael ISE (1985) High-resolution ^{29}Si NMR study of silicate and aluminosilicate glasses: the effect of network-modifying cations. Am Mineral 70:332–343

Musgraves JD, Hu, J, Calvez L (Eds.) (2019) Springer Handbook of Glass. Springer Nature, Cham, Switzerland

Mysen BO (1988) Structure and Properties of Silicate Melts. Elsevier, Amsterdam

Mysen BO (1990) Effect of pressure, temperature and bulk composition on the structure and species distribution in depolymerized alkali aluminosilicate melts and quenched glasses. J Geophys Res 95:15733–15744

Mysen BO, Frantz JD (1993) Structure and properties of alkali silicate melts at magmatic temperatures. Eur J Mineral 5:393–407

Mysen BO, Neuville DR (1995) Effect of temperature and TiO_2 content on the structure of $Na_2Si_2O_5$–$Na_2Ti_2O_5$ melts and glasses. Geochem Cosmochem Acta 59:325–342

Mysen BO, Richet P (2019) Silicate Glasses and Melts, Properties and Structure, 2nd edition. Elsevier, Amsterdam

Nasikas NK, Edwards TG, Sen S, Papatheodorou GN (2012) Structural characteristics of novel Ca–Mg orthosilicate and suborthosilicate glasses: results from ^{29}Si and ^{17}O NMR spectroscopy. J Phys Chem B 116:2696–2702

Nayak MT, Desa JAE, Reddy VR, Nayak C, Bhattacharyya D, Jha SN (2019a) Structure of silicate glasses with varying sodium and fixed iron contents. J Non-Cryst Solids 509:42–47

Nayak MT, Desa JAE, Reddy VR, Nayak C, Bhattacharyya D, Jha SN (2019b) Structural studies of potassium silicate glass with fixed iron content and their relation to similar alkali silicates. J Non-Cryst Solids 518:85–91

Nesbitt HW, Bancroft GM (2014) High resolution core- and valence-level XPS studies of the properties (structural, chemical and bonding) of silicate minerals and glasses. Rev Mineral Geochem 78:271–329

Nesbitt HW, Dalby KN (2007) High resolution O 1s XPS spectral, NMR, and thermodynamic evidence bearing on anionic silicate moieties (units) in $PbO–SiO_2$ and $Na_2O–SiO_2$ glasses. Can J Chem 85:782–792

Nesbitt HW, Bancroft GM, Henderson GS, Ho R, Dalby KN, Huang Y, Yan Z (2011) Bridging, non-bridging and free (O^{2-}) oxygen in $Na_2O–SiO_2$ glasses: An X-ray photoelectron spectroscopic (XPS) and nuclear magnetic resonance (NMR) study. J Non-Cryst Solids 357:170–180

Nesbitt HW, Henderson GS, Bancroft GM, Ho R (2015a) Experimental evidence for Na coordination to bridging oxygen in Na-silicate glasses: Implications for spectroscopic studies and for the modified random network model. J Non-Cryst Solids 409:139–148

Nesbitt HW, Bancroft, GM, Thibault, Y, Sawyer, R, Secco, RA (2015b) Reply to the comment by Malfait on "Spectroscopic studies of oxygen speciation in potassium silicate glasses and melts." Can J Chem 93:581–587

Nesbitt HW, Bancroft GM, Henderson GS, Sawyer R, Secco R (2015c) Direct and indirect evidence for free oxygen (O^{2-}) in MO–silicate glasses. Am Mineral 100:2566–2578

Nesbitt HW, Henderson GS, Bancroft GM, O'Shaughnessy C (2017a) Electron densities over Si and O atoms of tetrahedral and their impact on Raman stretching frequencies and Si–NBO force constants. Chem Geol 461:65–74

Nesbitt HW, Bancroft GM, Henderson GS, Richet P, O'Shaughnessy C (2017b) Silicate mineral melting, crystallization and the glass transition: Toward a unified description for silicate phase transitions. Am Mineral 102:412–420

Nesbitt HW, Henderson GS, Bancroft GM, Sawyer R, Secco RA (2017c) Bridging oxygen speciation in K-silicate glasses and melts, with implications for spectroscopic studies and glass structure. Chem Geol 461:13–22

Nesbitt HW, Bancroft GM, Henderson GS (2018) Temperature dependence of Raman shifts and linewidths for Q^0 and Q^2 crystals of silicates, phosphates and sulfates. Am Mineral 424:72–83

Nesbitt HW, O'Shaughnessy C, Henderson GS, Bancroft GM, (2019) Factors affecting line shapes and intensities of Q^3 and Q^4 Raman bands of Cs glasses. Chem Geol 505:1–11

Neufeind J, Liss KD (1996) Bond angle distribution in amorphous germania and silica. Ber Bunsen-Gesell Phys Chem 100:1341–1349

Neuville DR, Richet P (1991) Viscosity and mixing in molten (Ca, Mg) pyroxenes and garnet. Geochim Cosmochim Acta 55:1011–1019

Neuville DR, Cormier L, Massiot D (2004) Al environment in tectosilicate and peraluminous glasses: A Al-27 MQ-MAS NMR, Raman, and XANES investigation. Geochim Cosmochim Acta 68:5071–5079

Neuville DR, Cormier L, Massiot D (2006) Al coordination and speciation in calcium aluminosilicate glasses: effects of composition determined by Al-27 MQ-MAS NMR and Raman spectroscopy. Chem Geol 229:173–185

Neuville DR, Cormier L, Montouillout V, Florian P, Millot F, Rifflet JC, Massiot D (2008) Structure of Mg- and Mg/Ca aluminosilicate glasses: ^{27}Al NMR and Raman spectroscopy investigations. Am Mineral 93:1721–1731

Neuville DR, Henderson GS, Cormier L, Massiot D (2010) The structure of crystals, glasses, and melts along the CaO–Al$_2$O$_3$ join: Results from Raman, Al *L*- and *K*-edge X-ray absorption, and ^{27}Al NMR spectroscopy. Am Mineral 95:1580–1589

Neuville DR, de Ligny D, Henderson GS (2014) Advances in Raman spectroscopy applied to Earth and materials sciences. Rev Mineral Geochem 78:509–549

Newville M (2014) Fundamentals of XAFS. Rev Mineral Geochem 78:33–74

Noritake F, Kawamura K, Yoshino T (2012) Molecular dynamics simulation and electrical conductivity measurement of Na$_2$O.3SiO$_2$ melt under high pressure; relationship between its structure and properties. J. Non-Cryst Solids, 358: 3109–3118

Ohkubo T, Tsuchida E, Deguchi K, Ohki S, Shimuzu T, Otomo T, Iwadate Y (2017) Insights from ab initio molecular dynamics simulations for a multicomponent oxide glass. J Am Ceram Soc 101:1122–1134

O'Shaughnessy C, Henderson GS, Nesbitt HW, Bancroft GM, Neuville DR (2017) The structure of cesium silicate glasses and liquids. Chem Geol 461:82–95

O'Shaughnessy C, Henderson GS, Moulton BJ, Zuin L, Neuville DR (2018a) A Li *K*-edge XANES study of salts and minerals. J Synchrotron Radiat 25:1–9

O'Shaughnessy C, Henderson GS, Moulton BJ, Zuin L, Neuville DR (2018b) The effect of alkaline-earth substitution on the Li *K*-edge of lithium silicate glasses. J Non-Cryst Solids 500:417–421

O'Shaughnessy C, Henderson GS, Nesbitt HW, Bancroft GM, Neuville DR (2020) The influence of modifier cations on the Raman stretching modes of Q^n species in alkali silicate glasses. J Am Ceram Soc 103:3991–4001

Olivier L, Yuan X, Cormack AN, Jäger C (2001) Combined ^{29}Si double quantum NMR and MD simulation studies of network connectivities of binary Na$_2$O.SiO$_2$ glasses: new prospects and problems. J Non-Cryst Solids 293–295:53–66

Pant AK, Cruickshank DWJ (1968) The crystal structure of α-Na$_2$Si$_2$O$_5$. Acta Cryst B 24:13–19

Pedone A, Charpentier T, Menziani MC (2010) Multinuclear NMR of CaSiO$_3$ glass: simulation from first-principles. Phys Chem Chem Phys 12:6054–6066

Pecher O, Carretero-Gonzalez J, Griffit KJ, Grey CP (2017) Materials' methods: NMR in battery research. Chem Mater 29:213–242

Peng L, Stebbins JF (2007) Sodium germanate glasses and crystals: NMR constraints on variation in structure with composition. J Non-Cryst Solids 353:4732–4742

Phillips BL, Kirkpatrick RJ, Carpenter MA (1992) Investigation of short-range Al,Si order in synthetic anorthite by ^{29}Si MAS NMR spectroscopy. Am Mineral 77:484–495

Phillips BL, Xu HW, Heaney PJ, Navrotsky A (2000) Si-29 and Al-27 MAS-NMR spectroscopy of beta-eucryptite (LiAlSiO$_4$): The enthalpy of Si,Al ordering. Am Mineral 85:181–188

Porai-Koshits EA (1958) The Possibilities and Results of X-ray Methods for Investigation of Glassy Substances. *In:* The Structure of Glass, Lebedev AA, (Ed) New York: Consultants Bureau, p 25–35

Putnis A, Fyfe CA, Gobbi GC (1985) Al,Si ordering in cordierite using "magic angle spinning" NMR. Phys Chem Mineral 12:211–216

Randall JT, Rooksby HP, Cooper BS (1930) X-ray diffraction and the structure of vitreous solids. Z Kristall 75:196–214

Richet P (ed) (2020) Encyclopedia of Glass Science, Technology, Culture and History. J Wiley and Sons, Hoboken, NJ

Rossano S, Mysen BO (2012) Raman spectroscopy of silicate glasses and melts in geological systems. *In:* Raman spectroscopy applied to Earth sciences and cultural heritage, Dubessy J, Caumon M-C, Rull F (eds) EMU Volume 12, Aberystwyth, UK, p 321–361

Rossman GR (2014) Optical spectroscopy. Rev Mineral Geochem 78:371–398

Ryerson FJ (1985) Oxide solution mechanisms in silicate melts: systematic variations in the activity coefficient of SiO_2. Geochim Cosmochim Acta 49:637–650

Sakamaki T, Ohtani E (2022) High pressure melts. Rev Mineral Geochem 87:557–574

Sakamaki T, Wang Y, Park C, Yu T, Shen G (2012) Structure of jadeite melt at high pressure up to 4.9 GPa. J Appl Phys 111:112623

Sato T, Funamori N (2008) Sixfold-coordinated amorphous polymorph of SiO_2 under high pressure. Phys Rev Lett 101:255502

Sato T, Funamori N (2010) High pressure structural transformation of SiO_2 glass up to 100 GPa. Phys Rev B 82:184102

Sawyer R, Nesbitt HW, Bancroft GM, Thibault Y, Secco RA (2015) Spectroscopic studies of oxygen speciation in potassium silicate glasses and melts. Can J Chem 93:60–73

Scannell G, Huang L (2016) Structure and thermo-mechanical response of $Na_2O–TiO_2–SiO_2$ glasses to temperature. J Non-Cryst Solids 453:46–58

Scannell G, Barra S, Huang L (2016a) Structure and properties of $Na_2O–TiO_2–SiO_2$ glasses: Role of Na and Ti on modifying the silica network. J Non-Cryst Solids 448:52–61

Scannell G, Koike A, Huang L (2016b) Structure and thermo-mechanical response of $TiO_2–SiO_2$ glasses to temperature. J Non-Cryst Solids 447:238–247

Semkow KW, Haskin LA (1985) Concentrations and behavior of oxygen and oxide ions melts of composition CaO–MgO–$x$$SiO_2$. Geochim Cosmochim Acta 49, 1897–1908

Sen S, Youngman RE (2003) NMR study of Q-speciation and connectivity in $K_2O–SiO_2$ glasses with high silica content. J Non-Cryst Solids 331:100–107

Shi C, Alderman OLG, Berman D, Du J, Neuefeind J, Tamalonis A, Weber RJK, You J, Benmore CJ (2019) The structure of amorphous and deeply supercooled liquid alumina. Front Mat 6:1–15

Silver AH, Bray PJ (1958) Nuclear magnetic resonance absorption in glass .1. Nuclear quadrupolar effects in boron oxide, soda-boric oxide, and borosilicate glass. J Chem Phys 29:984–990

Skinner LB, Barnes AC, Salmon PS, Fischer HE, Drewitt JWE, Honkimäki V (2012) Structure and triclustering in Ba–Al–O glass. Phys Rev B 85:06420

Skinner LB, Benmore CJ, Weber JKR, Du J, Neuefeind J, Tumber S, Parise JB (2014) Low cation coordination in oxide melts. Phys Rev Lett 112:157801–157801–157805

Smythe DJ, Brenan JM, Bennett NR, Regier T, Henderson GS (2013) Quantitative determination of cerium oxide states in alkali-aluminosilicate glasses using $M_{4,5}$-edge XANES. J Non-Cryst Solids 378:258–264

Speziale S, Marquardt H, Duffy T (2014) Brillouin spectroscopy and its applications in geosciences. Rev Mineral Geochem 78:543–603

Stebbins JF (1987) Identification of multiple structural species in silicate glasses by Si-29 NMR. Nature 330:465–467

Stebbins JF (1988) NMR spectroscopy and dynamic processes in mineralogy and geochemistry. Rev Mineral 18: 405–430

Stebbins JF (1991) Experimental confirmation of five-coordinated silicon in a silicate glass at 1 atmosphere pressure. Nature 351:638–639

Stebbins JF (1995) Dynamics and structure of silicate and oxide melts: nuclear magnetic resonance studies. Rev Mineral 32:191–246

Stebbins JF (1998) Cation sites in mixed-alkali oxide glasses: correlations of NMR chemical shift data with site size and bond distance. Solid State Ionics 112:137–141

Stebbins JF (2002) NMR studies of oxide glass structure, in: Duer, M (Ed), Solid State NMR: Theory and Applications. Blackwell Scientific, Oxford, p 391–436

Stebbins JF (2008) Temperature effects on the network structure of oxide melts and their consequences for configurational heat capacity. Chem Geol 256:80–91

Stebbins JF (2016) Glass structure, melt structure and dynamics: some concepts for petrology. Am Mineral 101:753–768

Stebbins JF (2017) "Free" oxide ions in silicate melts: thermodynamic considerations and probable effects of temperature. Chem Geol 461:2–12

Stebbins JF (2019) Pentacoordinate silicon in ambient pressure potassium and lithium silicate glasses: Temperature and compositional effects and analogies to alkali borate and germanate systems. J Non-Cryst Solids: X 1:100012–1–12

Stebbins JF (2020) Short-range structure and order in oxide glasses. *In:* Richet P (Ed.), Encyclopedia of Glass Science, Technology, History, and Culture. J Wiley and Sons, Hoboken, NJ

Stebbins JF, Bista S (2018) Pentacoordinated and hexacoordinated silicon cations in a potassium silicate glass: effects of pressure and temperature. J Non-Cryst Solids 505:234–240

Stebbins JF, Farnan I (1992) The effects of temperature on silicate liquid structure: a multi-nuclear, high temperature NMR study. Science 255:586–589

Stebbins JF, McMillan P (1989) Five- and six- coordinated Si in $K_2Si_4O_9$ glass quenched from 1.9 GPa and 1200°C. Am Mineral 74:965–968

Stebbins JF, McMillan P (1993) Compositional and temperature effects on five coordinated silicon in ambient pressure silicate glasses. J Non-Cryst Solids 160:116–125

Stebbins JF, Poe BT (1999) Pentacoordinate silicon in high-pressure crystalline and glassy phases of calcium disilicate (CaSi$_2$O$_5$). Geophys Res Lett 26:2521–2523

Stebbins JF, Sen S (2013) Oxide ion speciation in potassium silicate glasses: new limits from ^{17}O NMR. J Non-Cryst Solids 368:17–22

Stebbins JF, Xu Z (1997) NMR evidence for excess non-bridging oxygens in an aluminosilicate glass. Nature 390:60–62

Stebbins JF, Xue X (2014) NMR spectroscopy of inorganic Earth materials. Rev Mineral Geochem 78:605–653

Stebbins JF, Farnan I, Xue X (1992) The structure and dynamics of alkali silicate liquids: a view from NMR spectroscopy. Chem Geol 96:371–386

Stebbins JF, McMillan PF, Dingwell DB (Eds) (1995) Structure, Dynamics, and Properties of Silicate Melts. Rev Mineral Vol. 32, Mineral Soc Am, Washington, DC

Stebbins JF, Oglesby JV, Kroeker S (2001a) Oxygen triclusters in crystalline CaAl$_4$O$_7$ (grossite) and calcium aluminosilicate glasses: oxygen-17 NMR. Am Mineral 86:1307–1311

Stebbins JF, Oglesby JV, Lee SK (2001b) Oxygen sites in silicate glasses: a new view from oxygen-17 NMR. Chem Geol 174:63–75

Stebbins JF, Du L-S, Kroeker S, Neuhoff P, Rice D, Frye J, Jakobsen HJ (2002) New opportunities for high-resolution solid-state NMR spectroscopy of oxide materials at 21.1 and 18.8 Tesla fields. Solid State NMR 21:105–115

Stebbins JF, Dubinsky EV, Kanehashi K, Kelsey KE (2008) Temperature effects on non-bridging oxygen and aluminum coordination number in calcium aluminosilicate glasses and melts. Geochim Cosmochim Acta 72:910–925

Stebbins JF, Wu J, Thompson LM (2013) Interactions between network cation coordination and non-bridging oxygen abundance in oxide melts and glasses: insights from NMR spectroscopy. Chem Geol 346:34–46

Stebbins JF, McCarty RJ, Palke AC (2017) Solid-state NMR and short-range order in crystalline oxides and silicates: a new tool in paramagnetic resonances. Acta Cryst C 73:128–136

Stebbins JF, McCarty RJ, Palke AC (2018) Toward the wider application of ^{29}Si NMR spectroscopy to paramagnetic transition metal silicate minerals and glasses: Fe(II), Co(II), and Ni(II) silicates. Am Mineral 103:776–291

Strobridge FC, Middlemiss DS, Pell AJ, Leskes M, Clément RJ, Pourpoint F, Lu Z, Hanna JV, Pintacuda G, Emsley L, Samosen A, Grey CP (2014) Characterizing local environments in high energy density Li-ion battery cathodes: a combined NMR and first principles study of LiFe$_x$Co$_{1-x}$PO$_4$. J Mat Chem A 2:11948–11957

Sukenaga S, Florian P, Kanehashi K, Shibata H, Saito N, Nakashima K, Massiot D (2017) Oxygen speciation in multicomponent silicate glasses using through bond double resonance NMR spectroscopy. J Phys Chem Let 8:2274–2279

Thompson LM, Stebbins JF (2011) Non-bridging oxygen and high-coordinated aluminum in metaluminous and peraluminous calcium and potassium aluminosilicate glasses: High-resolution ^{17}O and ^{27}Al MAS NMR results. Am Mineral 96:841–853

Thompson LM, McCarty RJ, Stebbins JF (2012) Estimating accuracy of ^{17}O NMR measurements in oxide glasses: Constraints and evidence from crystalline and glassy calcium and barium silicates. J Non-Cryst Solids 358:2999–3006

Tilocca A, de Leeuw NH (2006) Structural and electronic properties of modified sodium and soda-lime silicate glasses by Car-Parrinello molecular dynamics. J Mat Chem 16:1950–1955

Toop GW, Samis CS (1962) Some new ionic concepts of silicate slags. Can Met Quart 1:129–152

Toplis MJ, Dingwell DB (2004) Shear viscosities of CaO–Al$_2$O$_3$–SiO$_2$ and MgO–Al$_2$O$_3$–SiO$_2$ liquids: Implications for the structural role of aluminum and the degree of polymerization of synthetic and natural aluminosilicate melts. Geochim Cosmochim Acta 68:5169–5188

Toplis MJ, Dingwell DB, Lenci T (1997) Peraluminous viscosity maxima in Na$_2$O–Al$_2$O$_3$–SiO$_2$ liquids: the role of triclusters in tectosilicate melts. Geochim Cosmochim Acta 61:2605–2612

Toplis MJ, Kohn SC, Smith ME, Poplett IJF (2000) Four coordinate aluminum in tectosilicate glasses observed by triple quantum MAS NMR. Am Mineral 85:1556–1560

Trease NM, Clark TM, Grandinetti PJ, Stebbins JF, Sen S (2017) Bond length-bond angle correlation in densified silica-results from ^{17}O NMR spectroscopy. J Chem Phys 146:184505–184501–184509

van Wüllen L, Gee B, Zuchner L, Bertmer M, Eckert H (1996a) Connectivities and cation distributions in oxide glasses: new results from solid-state NMR. Ber Bunsen-Gesell Phys Chem 100:1539–1549

van Wüllen L, Züchner L, Müller-Warmuth W, Eckert H (1996b) ^{11}B{^{27}Al} and ^{27}Al{^{11}B} double resonance experiments on a glassy sodium aluminoborate. Solid State NMR 6:203–212

Varshneya AK (2019) Fundamentals of Inorganic Glasses, 3rd edition. Elsevier, Amsterdam

Waseda Y, Egami T (1979) Effect of low-temperature annealing and deformation on the structure of metallic glasses by X-ray diffraction. J Mater Sci 14:1249–1253

Wasylishen RE, Ashbrook SE, Wimperis S (2012) NMR of Quadrupolar Nuclei in Solid Materials. John Wiley and Sons Ltd., Chichester, UK

Weber R, Sen S, Youngman RE, Hart RT, Benmore CJ (2008) Structure of high alumina content Al$_2$O$_3$–SiO$_2$ composition glasses. J Phys Chem B 112:16726–16733

Weigel C, Cormier L, Galoisy L, Calas, G, Bowron D, Beuneu B (2006) Determination of Fe^{3+} sites in a NaFeSi$_2$O$_6$ glass by neutron diffraction with isotopic substitution coupled with numerical simulation. App Phys Lett 89:141911

Weigel C, Calas G, Cormier L, Galoisy L, Henderson GS (2008a) High-resolution Al $L_{2,3}$-edge X-ray absorption near edge structure spectra of Al-containing crystals and glasses: Coordination number and bonding information from edge components. J Phys Condens Matter 20:135219

Weigel C, Cormier L, Calas G, Galoisy L, Bowron DT (2008b) Nature and distribution of iron sites in a sodium silicate glass investigated by neutron diffraction and EPSR simulation. J Non-Cryst Solids 354:5378–5385

Welsch A-M, Knipping JL, Beherens H, (2017) Fe-oxidation state in alkali-trisilicate glasses–A Raman spectroscopic study. J Non-Cryst Solids 471:28–38

Woelffel W, Claireaux C, Toplis M, Burov E, Barthel E, Shukla A, Biscaras J, Chopinet MH, Gouillart E (2015) Analysis of soda-lime glasses using non-negative matrix factor deconvolution of Raman spectra. J Non-Cryst. Solids 428:121–132

Wright AC (1994) Neutron scattering from vitreous silica.V. The structure of vitreous silica: what have we learned from 60 years of diffraction studies? J Non-Cryst Solids 179:84–115

Wright AC (2019) Density fluctuations in vitreous SiO_2 and GeO_2. Phys Chem Glasses: E J Glass Sci Tech B 60:33–48

Wu J, Stebbins JF (2013) Temperature and modifier cation field strength effects on aluminoborosilicate glass network structure. J Non-Cryst Solids 362:73–81

Xue X, Stebbins JF (1993) ^{23}Na NMR chemical shifts and the local Na coordination environments in silicate crystals, melts, and glasses. Phys Chem Mineral 20:297–307

Xue X, Stebbins JF, Kanzaki M, McMillan PF, Poe B (1991) Pressure-induced silicon coordination and tetrahedral structural changes in alkali silicate melts up to 12 GPa: NMR, Raman, and infrared spectroscopy. Am Mineral 76:8–26

Xue X, Stebbins JF, Kanzaki M (1994) Correlations between O-17 NMR parameters and local structure around oxygen in high-pressure silicates and the structure of silicate melts at high pressure. Am Mineral 79:31–42

Yarger JL, Smith KH, Nieman RA, Diefenbacher J, Wolf GH, Poe BT, McMillan PF (1995) Al coordination changes in high-pressure aluminosilicate liquids. Science 270:1964–1967

Yuan X, Cormack AN (2001) Local structures of MD-modeled vitreous silica and sodium silicate glasses. J Non-Cryst Solids 283:69–87

Zachariasen WH (1932) The atomic arrangement of glass. J Am Chem Soc 54:3841–3850

Zakaznova-Herzog VP, Malfait WJ, Herzog F, Halter WE (2007) Quantitative Raman spectroscopy: Principles and application to potassium silicate glasses. J Non-Cryst Solids 353:4015–4028

Zanotto ED Mauro JC (2017) The glassy state of matter: Its definition and ultimate fate. J Non-Cryst Solids 471:490–495

Zeidler A, Wezka K, Rowlands RF, Whittaker DAJ, Salmon P, Polidori A, Drewitt JWE, Klotz S, Fischer HE, Wilding MC, Bull CL, Tucker MG, Wilson M (2014) High-pressure transformation of SiO_2 glass from a tetrahedral to an octahedral network: A joint approach using neutron diffraction and molecular dynamics. Phys Rev Lett 113:135501

Zhang L, Eckert H (2006) Short- and medium-range order in sodium metaphosphate glasses: new insights from high-resolution dipolar solid-state NMR. J Phys Chem B 110:8946–8958

Zhang P, Grandinetti PJ, Stebbins JF (1997) Anionic species determination in $CaSiO_3$ glass using two-dimensional ^{29}Si NMR. J Phys Chem B101: 4004–4008

Züchner L, Chan JCC, Müller-Warmuth W, Eckert H (1998) Short-range order and site connectivities in sodium aluminoborate glasses: I. Quantification of local environments by high-resolution ^{11}B, ^{23}Na, and ^{27}Al solid-state NMR. J Phys Chem B 102:4495–4506

Zwanziger JW, Werner-Zwanziger U, Calahoo C, Paterson AL (2019) Nuclear magnetic resonance and electron paramagnetic resonance studies of glass. *In:* Springer Handbook of Glass. Springer Nature, Cham, Switzerland, 953–993

Reviews in Mineralogy & Geochemistry
Vol. 87 pp. 55-103, 2022
Copyright © Mineralogical Society of America

2

From Short to Medium Range Order in Glasses and Melts by Diffraction and Raman Spectroscopy

James W. E. Drewitt

School of Physics
University of Bristol
H H Wills Physics Laboratory
Tyndall Avenue
Bristol, BS8 1TL
United Kingdom

james.drewitt@bristol.ac.uk

Louis Hennet

Conditions Extrêmes et Matériaux: Haute Température et Irradiation
1d avenue de la Recherche Scientifique
Interfaces, Confinement, Matériaux et Nanostructures
1b rue de la Férollerie
CNRS, Université d'Orléans
45071 Orléans cedex 2
France

louis.hennet@cnrs-orleans.fr

Daniel R. Neuville

Géomatériaux
CNRS-Institut de physique du globe de Paris
Université de Paris
1 rue Jussieu
75005 Paris
France

neuville@ipgp.fr

INTRODUCTION

The structure of glasses and the melts from which they are formed is intrinsically disordered, making their structural characterization difficult. Whilst the structure of solid minerals can be readily determined from the periodicity of the crystallographic unit cell, the absence of long-range symmetry in glasses and melts means their "unit cell" is essentially of infinite extent. However, although structurally disordered, the atomic-scale arrangements in glasses and melts are not strictly random. As detailed in Henderson and Stebbins (2022, this volume), glasses typically exhibit a high degree of local chemical short-range order (SRO) in the form of well-defined coordination polyhedra and chemical bond lengths, often similar to the corresponding crystal. This is illustrated in Figure 1a for pure silica (SiO_2) glass where, as for crystalline polymorphs of SiO_2, every silicon atom is bonded to four oxygen atoms in tetrahedral units bonded to the next tetrahedron by a bridging oxygen atom.

1529-6466/22/0087-0002$05.00 (print)
1943-2666/22/0087-0002$05.00 (online)

http://dx.doi.org/10.2138/rmg.2022.87.02

In a glass, periodicity breaks down beyond the scale of a few atoms. Nevertheless, different patterns of medium range order (MRO), also called intermediate range order, may occur beyond the next nearest neighbour length-scale at ~5–20 Å (1 Ångstrom = 10^{-10} m) related to the topology of the glass network (Elliot 1991; Price 1996). In the silica glass example in Figure 1a, substantial MRO arises from ring structures that enclose open regions in the network of corner shared SiO_4 tetrahedra. Topological and chemical ordering may also persist in some network glasses on extended length scales up to nanometer distances (Salmon and Zeidler 2013).

The addition of alkali, alkaline earth, or metal "modifier" cations to SiO_2 glass disrupts the silicate network by breaking Si–O bonds leading to the formation of non-bridging oxygen (NBO) atoms (see Henderson and Stebbins 2022, this volume). Two competing theories have been proposed to describe the MRO structure and distribution of modifier cations in silicate glasses: the perturbed cation distribution (PCD) model (Lee and Stebbins 2003), and the modified random network (MRN) model (Greaves 1985). The PCD model assumes a relatively homogeneous local distribution of modifier cations in the glass network. In contrast, the MRN model predicts a heterogenous distribution of modifier cations forming diffusion channels which percolate throughout the disrupted tetrahedral silicate network (see Fig. 1b). Although evidence from experiments and simulations appears to substantiate the Greaves MRN model in alkali silicate glasses (Meyer et al. 2004; Du and Cormack 2004; Cormier and Cuello 2011; Le Losq et al. 2017), the question of whether cation diffusion channels are a general feature of MRO in modified silicate glasses remains unanswered.

Diffraction (x-ray and neutron) and vibrational spectroscopies (Raman or infrared) provide detailed information on SRO in glasses. Raman spectroscopy is also sensitive to MRO and can be exploited to identify ring structures in glasses. Pre-peaks observed in diffraction measurements of glasses can also be highly indicative of MRO. Combining the strengths of diffraction measurements, Raman spectroscopy, and computer simulation (Jahn 2022, this volume) enables detailed structural insights to be obtained on SRO and MRO in non-crystalline materials including glasses and their melts, offering a complementary perspective to the cationic environment probed by Nuclear Magnetic Resonance (NMR) or x-ray absorption spectroscopy.

In this chapter, we provide an overview of x-ray or neutron diffraction and Raman spectroscopy measurements to characterize the structure of geologically relevant silicate and aluminate glasses and their melts, including the considerable reorganization in SRO and MRO that can take place as a function of composition and high-temperature (T) and -pressure (p) conditions experienced in deep planetary interiors.

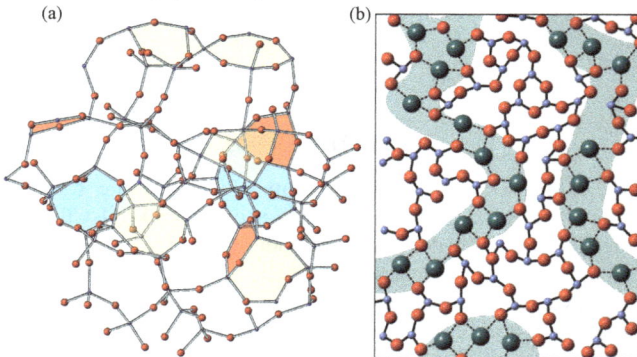

Figure 1. (a) Structural model of pure silica (SiO_2) glass. 3-, 4-, and 5-membered rings of SiO_4 tetrahedra are shown by the **red, yellow,** and **blue shaded regions,** respectively. Si = **small blue spheres;** O = **large red spheres**. Created using the SiO_2 glass configuration from Le Roux and Petkov (2010a). **(b)** The Greaves modified random network (MRN) model (after Greaves 1985). Modifier cations (**large green spheres with dashed lines** denoting ionic bonds) form percolation channels in the covalently bonded glass network of Si (**blue spheres**) and O (**red spheres**).

X-RAY, NEUTRON, AND RAMAN SCATTERING OF GLASSES AND MELTS

A brief overview of the theory of diffraction and Raman spectroscopy to obtain information on SRO and MRO in the atomic-scale structure in glasses and melts is provided in the following by reference to pure SiO_2 (silica), one of the principal components of all-natural and most commercial silicate glasses. For more exhaustive theoretical treatment of x-ray and neutron diffraction studies of liquids and glasses we refer the reader to the reviews by Fischer et al. (2006), Hannon (2015), Benmore (2015), and Cormier (2019). For the theory of Raman scattering and required data treatment and normalization procedures for oxide glasses we recommend Rull (2012), Rossano and Mysen (2012), Neuville et al. (2014a), and Almeida and Santos (2015).

X-ray and neutron diffraction

X-ray diffraction (XRD), pioneered by M. von Laue, W. H. Bragg, and L. Bragg in the early 20th century, is a powerful tool for characterizing the structure of materials at the atomic level. Although less widely available and requiring large-scale reactor or accelerator sources, diffraction by neutrons provides highly complementary information: while x-ray sensitivity increases with atomic number, the scattering power of neutrons varies between elements and isotopes such that neutron diffraction has greater sensitivity to some light elements (e.g. oxygen) compared to x-rays.

Consider first a crystalline solid composed of atoms organized in a highly ordered repeating periodic arrangement in three dimensions (e.g. crystalline SiO_2, Fig. 2a). The interatomic distances are of the order of Ångstroms, the same order of magnitude as the wavelength of x-rays or high energy thermal neutrons: a crystal thus constitutes a three-dimensional network which can diffract x-rays or neutrons in accordance to Bragg's law, producing sharp constructive interference peaks associated with the crystalline lattice spacings. Similarly, inelastic light scattering from crystalline materials typically produces intense narrow bands corresponding to specific vibrational modes or chemical bonds (see the next section on Raman spectroscopy). In the case of a non-crystalline solid (e.g. SiO_2 glass, Fig. 2b),

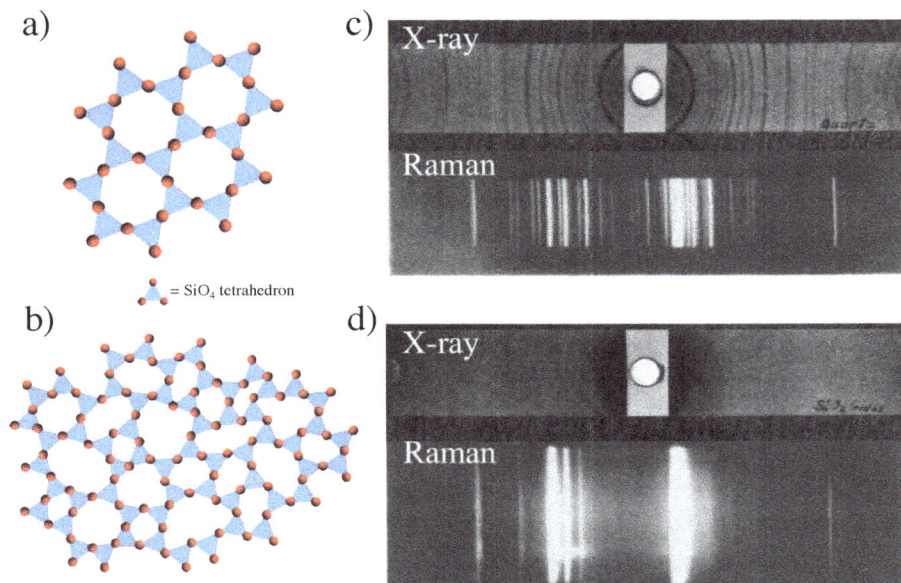

Figure 2. Two dimensional representations of the **a)** crystalline and **b)** amorphous structures of SiO_2 (after Zachariasen 1932) shown together with their corresponding XRD (Warren 1934a) and Raman (Gross and Ramanova 1929) photographs in **c)** and **d)**.

the structure is disordered with no long-range translational periodicity such that diffraction patterns exhibit broad diffuse peaks. This is demonstrated by the first XRD patterns (Warren 1934a,b,c) measured for SiO_2 crystal (quartz) (Fig. 2c) and glass (Fig. 2d), with sharp intense peaks observed for the crystal and diffuse features for the glass measurements. However, despite the structural disorder in glasses and melts, chemical bonding constraints give rise to a high degree of SRO which, together with MRO, is encoded in the diffuse scattering signal.

Diffraction provides information on liquid and glass structure by measurement of the total structure factor, $S(Q)$, obtained after processing the diffracted intensity, where the magnitude of the reciprocal space scattering vector $Q = 4\pi \sin\theta/\lambda$ where θ is one half the scattering angle and λ denotes the wavelength of the incident x-ray or neutron beams (Fischer et al. 2006). The real space total pair distribution function, $G(r)$, provides a measure of the probability of finding two atoms a distance r apart and is given by the Fourier transform relation

$$G(r) - 1 = \frac{1}{2\pi^2 r\rho_0} \int_0^\infty Q[S(Q) - 1]\sin(Qr)\,dQ, \tag{1}$$

where ρ_0 denotes the atomic number density. For a glass containing n different chemical species, the $S(Q)$ and $G(r)$ are comprised of a weighted sum of the Faber–Ziman partial structure factors $S_{\alpha\beta}(Q)$ (Faber and Ziman 1965) or partial pair distribution functions $g_{\alpha\beta}(r)$ for $n(n+1)/2$ atomic pairs for chemical species α or β:

$$S(Q) = \sum_{\alpha=1}^{n}\sum_{\beta=1}^{n} W_{\alpha\beta}^S S_{\alpha\beta}(Q) \tag{2}$$

$$G(r) = \sum_{\alpha=1}^{n}\sum_{\beta=1}^{n} W_{\alpha\beta}^G g_{\alpha\beta}(r) \tag{3}$$

For neutrons $W_{\alpha\beta}^G = W_{\alpha\beta}^S = c_\alpha c_\beta b_\alpha b_\beta / \left(\sum_\alpha c_\alpha b_\alpha\right)^2$, where c denotes concentration and b is the neutron scattering length (Sears 1992). X-ray form-factors are Q-dependent and the $S_{\alpha\beta}(Q)$ weighting factors in XRD are $W_{\alpha\beta}^S(Q) = c_\alpha c_\beta f_\alpha(Q) f_\beta^*(Q) / \left(\sum_\alpha c_\alpha f_\alpha(Q)\right)^2$, where $f(Q)$ and $f^*(Q)$ denote the complex atomic scattering factor and its conjugate (e.g. Waasmaier and Kirfel 1995). As such, the x-ray $G(r)$ cannot be described in terms of simple weighted linear combination of $g_{\alpha\beta}(r)$. The Warren–Krutter–Morningstar (WKM) approximation (Warren 1936a) in which $W_{\alpha\beta}^S(Q)$ is assumed to be independent of Q such that the weighting factor $W_{\alpha\beta}^G = W_{\alpha\beta}^S(Q=0)$, can be used to eliminate this complication. In most cases, the WKM is a good approximation, however an expansion to higher order can be made for improved accuracy (Masson and Thomas 2013). Alternatively, if a real-space peak of interest arises solely from one distinct partial pair distribution function then its real-space weighting factor can be eliminated by division prior to Fourier transformation (Zeidler et al. 2009).

The average number of neighboring atoms in a coordination shell between r_1 and r_2 can be calculated by integrating the radial distribution function, $RDF(r)$ (Fischer et al. 2006):

$$\bar{n}_\alpha^\beta = \int_{r_1}^{r_2} RDF(r) = 4\pi\rho_0 \int_{r_1}^{r_2} r^2 G(r). \tag{4}$$

The $S(Q)$ and $G(r)$ functions measured for SiO_2 (silica) glass by XRD are shown in Figure 3 (Kohara and Suzuya 2005). The pronounced first sharp diffraction peak (FSDP) in $S(Q)$ at $Q_1 = 1.55 \text{ Å}^{-1}$ is observed in many different glasses and is a signature of MRO in the glass with a periodicity $2\pi/Q_1$ (Price et al. 1989; Elliot 1991; Salmon 1994). The coherence length, which estimates the distance in real space over which the intermediate range ordering persists, is given by $2\pi/\Delta Q_1$, where ΔQ_1 full width at half maximum of the FSDP, estimates the distance in real space over which the MRO persists (Salmon 1994). Hence, a lower position Q_1 reflects increasingly longer-range order, and a sharper peak reflects MRO persisting for a longer distance in real space. The FSDP is highly sensitive to variations in MRO induced by p, T, and compositional changes (discussed later).

Under ambient conditions, the FSDP observed in the $S(Q)$ for SiO_2 glass is attributed to tetrahedral units in ring arrangements (Fig. 1a). While information on the ring structure can potentially be extracted directly from its real-space representation (Shi et al. 2019), in practice analysis of molecular dynamics simulation models (Jahn 2022, this volume) using e.g. ring statistics (LeRoux and Jund 2010b) or more recently persistent homology (Hiraoka et al. 2016; Hosokawa et al. 2019, Onodera et al. 2020) techniques are required to quantify the full nature of MRO arising from glass and melt network connectivity.

The oscillatory features in $S(Q)$ at high-Q relate to local structural ordering, as manifested in real-space by peaks in $G(r)$ at the characteristic interatomic distances of short-range chemical bonds. The Fourier transform integral in Equation (1) requires $S(Q)$ to be known to infinitely high Q but in practice diffraction measurements have a finite maximum scattering vector, Q_{max}. Termination ripples can be minimized by multiplying $S(Q)$ by a smooth modification function that decays to zero at Q_{max} prior to Fourier transformation (Fischer et al. 2006). In general, high-energy x-ray or neutron beams and detectors that cover a wide range of scattering angles are required to probe the bulk structure of glasses and melts with a sufficiently high Q_{max} for good resolution in $G(r)$. For this reason, diffraction measurements of glasses and melts are typically performed at large-scale synchrotron x-ray or spallation/reactor source neutron user facilities.

The $G(r)$ for pure silica glass (Fig. 3b) is dominated by the first peak at 1.60 Å, followed by a second peak at 2.61 Å, corresponding to the nearest neighbor Si–O and O–O bond distances in a SiO_4 tetrahedron, respectively. The angle O–Si–O calculated from these two distances using the law of cosines is very close to the perfect intra tetrahedral angle of 109.5°. The $G(r)$ becomes increasingly featureless at distances greater than ~6 Å due to the lack of long-range ordering.

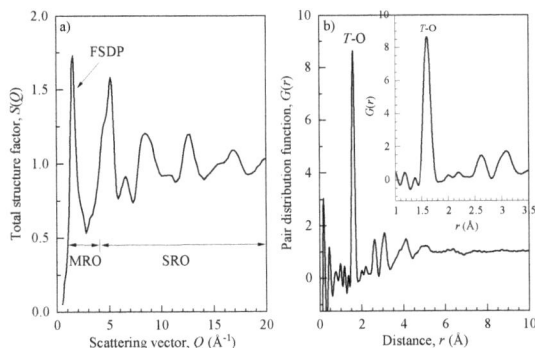

Figure 3. a) Total structure factor $S(Q)$ and **b)** pair distribution function, $G(r)$, for SiO_2 glass, from synchrotron XRD measurements (data from Kohara and Suzuya 2005). Regions of $S(Q)$ associated with SRO and MRO are indicated including the first sharp diffraction peak (FSDP).

Raman spectroscopy

Raman spectroscopy, discovered by C. V. Raman and K. S. Krishnan in 1928 using filtered sunlight, relies on the inelastic scattering of light and provides information on the vibrational modes in the system. In principle, vibrational spectroscopy is possible using any source of monochromatic light, including x-rays. Raman spectroscopy in most modern laboratory settings typically employs laser light from near ultraviolet to visible and near infrared wavelengths in the order of 10^{-6}–10^{-7} m (energy ~1–12 eV, frequency ~3×10^{14}–3×10^{15} Hz).

In a Raman experiment, a sample is irradiated by a focused beam of monochromatic laser light with wavelength λ_0 and energy $E_0 = hc / \lambda_0$, where h is Planck's constant and c the speed of light. Most of the incident photons undergo an elastic scattering process (Rayleigh scattering) with the energy of the scattered light $E = E_0$. A small fraction of the incident photons exchange energy with molecular bond vibrations producing very weak inelastic Raman scattering intensity with energy $E = E_0 \pm E_m$. Here $E_m = hc / \lambda_m$ corresponds to the energy of an elementary excitation of a molecule measured in Raman spectroscopy relative to the incident energy E_0. If a photon is absorbed, exciting the molecule to a higher vibrational mode, then $E < E_0$ and the measured intensities are referred to as 'Stokes' lines. Alternatively, if the molecule is de-excited into a lower vibrational mode then a photon with $E > E_0$ is emitted giving rise to 'Anti-Stokes' lines. The 'Anti-Stokes' intensity is typically weaker and normally only the 'Stokes' side of the Raman spectrum is measured.

In practice, Raman spectra are expressed as a function of wavenumber (in cm^{-1}) instead of the energy with the Raman shift of the vibrational mode given by:

$$\Delta\omega = \frac{1}{\lambda_0} - \frac{1}{\lambda_m}$$

Raman active vibration modes are determined by selection rules, where only molecular excitations that result in a change in polarizability will produce a Raman signal (Rull 2012).

Raman spectroscopy provides a spectral signature of a given material. The position of the Raman lines or bands are characteristic of the composition and the intensities correlate with the concentration of the chemical species. The line width can provide information including the size of crystallites in polycrystalline materials or the degree of structural disorder. Like diffraction measurements, glasses and disordered materials yield broad diffuse Raman peaks (Fig. 2d). The Raman spectrum of SiO_2 glass and two crystalline polymorphs, quartz and cristobalite, are shown for comparison on Figure 4. The Raman spectrum of silica glass has been well characterized and, as described in chapter 2 (Henderson and Stebbins 2022, this volume) the specific Raman bands annotated in Figure 4 can be attributed to distinct vibrational features.

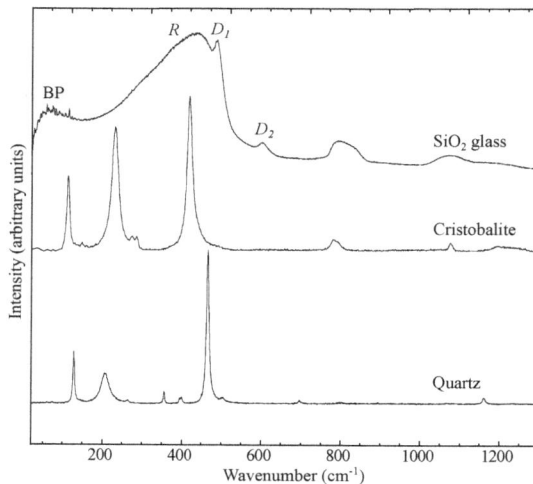

Figure 4. Raman spectra of SiO_2 polymorphs: quartz, cristobalite and glass (redrafted from Neuville et al. 2014). See the text for the assignment of the BP, R, D1, D2, peaks in the glass spectrum.

The Boson peak (BP) at ~80 cm^{-1}, so-called because the temperature dependence of the peak intensity obeys Bose-Einstein statistics, is a ubiquitous yet controversial feature of glass Raman spectra (Nakayama 2002) which may be a signature of longer-range vibrational structure (discussed in more detail below). The Raman spectrum below ~1000 cm^{-1} has a strong sensitivity to MRO, typically related to Si–O–Si breathing modes from SiO$_4$ tetrahedra in ring arrangements (Sharma et al. 1981; Galeener 1982a,b, 1983; Geissberger and Galeener 1983; Barrio et al. 1993; McMillan et al. 1994; Pasquerello and Car 1998; Umari and Pasquerello 2002; Umari et al. 2003; Rahmani et al. 2003; Kalampounias et al. 2006). The broad peak at 440 cm^{-1} (R) corresponds to bending vibrations associated with the broad distribution of inter tetrahedral Si–O–Si angles in n-membered rings with $n \geq 5$. The narrow D_1 (~495 cm^{-1}) and D_2 (~606 cm^1) bands are associated with the well–defined Si–O–Si breathing modes in 4- and 3-membered rings, respectively. The asymmetric peaks centered at ~800 cm^1 and ~1070 cm^{-1} are attributed Si–O stretching vibrations in SiO$_4$ tetrahedra (McMillan 1984; Sarnthein et al. 1997; Kalampounias et al. 2006).

The boson peak: A signature of SRO or MRO?

The BP observed in Raman, inelastic neutron, and infrared spectroscopy arises from excess low frequency, ω, contributions in THz range of the vibrational density of states (VDOS), g(ω) compared to the Debye elastic continuum model [g(ω) $\propto \omega^2$] obeyed by crystalline materials. The BP appears to be a universal feature of inorganic glasses; however, its origin remains controversial and several competing theories have been developed (Nakayama 2002; Greaves and Sen 2007; Brink et al. 2016; Wang et al. 2018; Baggioli and Zaccone 2019).

The soft-potential model has been used to explain the BP (Buchenau et al. 1992; Gurevich et al. 1993; Parshin 1993), in which quasi-localized vibrational modes are associated with "defects" in the disordered structure (Laird and Schober 1991). Quasi-local vibrations caused by linked "rigid" SiO$_4$ tetrahedra which rotate against each other without distortion are offered as an explanation for the BP in the Raman spectrum of pure SiO$_2$ glass (Buchenau et al. 1986; Taraskin and Elliot 1999; Hehlen et al. 2000, 2002). However, the ubiquity of the BP in a wide range of glasses indicates the requirement for a more generalized interpretation.

Experimental evidence for a correlation between the positions of the FSDP and BP in some glasses is strongly suggestive of a connection to MRO (Novikov and Sokolov 1991; Sokolov et al. 1992; Börjesson et al. 1993; Price 1996). Several interpretations of the BP invoke a degree of MRO or extended range ordering, including nano-structured clusters (Duval 1986; Mermet et al. 1996), density fluctuations intrinsic to the disordered structure (Elliot 1992), or heterogeneity in the elastic constant producing "soft" regions in the material (Sokolov 1999; Schirmacher et al. 2007, 2015; Schirmacher 2013; Marruzzo et al. 2013). Recent Raman measurements showing an increase in BP intensity on substitution of Na by K in aluminosilicate glasses indicates that the BP may be related to the development of nano-structured percolation channels (Le Losq et al. 2017).

Another explanation indicates there is a relationship between the BP and the glass transition temperature (T_g) associated with marked changes in MRO at T_g (Levelut et al. 1995; Takahashi et al. 2009; Stavrou et al. 2010; Tomoshige et al. 2019), where theoretical interpretations predict a phonon-saddle transition (Grigera et al. 2003). Alternatively, it has been suggested that the BP simply originates from a broadening or shift of the lowest Van Hove singularity of the corresponding crystalline state (Taraskin et al. 2001; Chumakov et al. 2011) and as such relates to SRO in the glass.

MULTICOMPONENT SILICATE GLASSES

Most natural magmatic glasses and melts contain high (rhyolites) or moderate (basalts) concentrations of SiO_2 with a variety of additional oxide components such as Al_2O_3, MgO, CaO, FeO, and other metal oxides. Metamorphic glasses may also be formed by vitrification of crustal silicates during extreme natural events including hypervelocity meteorite or comet impacts (impactites, tektites), lightning strikes (fulgurites), and natural combustion of coal beds (Buchite) (McCloy 2019; Cicconi et al. 2022, this volume). A structural model of a typical multi-component silicate glass is illustrated in Figure 5a. Here, the small Si^{4+} or Al^{3+} cations are the network formers, denoted T, bonded to 4 oxygen atoms in tetrahedral units interlinked by bridging oxygen (BO) atoms to form the backbone of a network structure with three, four, five, and more membered rings. Larger lower-charged cations, denoted M (e.g. alkaline, alkaline-earth, other metal cations), adopt varying roles as either charge compensators or network modifiers forming larger structural motifs and perturbing the network structure via the formation of non-bridging oxygen (NBO) atoms.

An example of a pair distribution function $G(r)$ for this silicate glass model is shown in Figure 5b featuring three distinct peaks which can be attributed to local structural ordering; the first from nearest neighbour intra tetrahedral bonds (T–O), the second peak arising from cation-oxygen (M–O) or oxygen-oxygen (O–O) distances, and the third peak and features at higher distances due to tetrahedra-cation (T–M), tetrahedra-tetrahedra (T–T), and cation-cation (M–M) correlations. As discussed earlier, for a glass containing n different chemical species the $G(r)$ comprises a weighted sum of $n(n+1)/2$ individual partial pair distribution functions. Due to the complexity of these overlapping atom-atom correlations it becomes increasingly difficult to assign peaks to specific structural features in higher-order coordination shells at length scales greater than ~3 Å. However, information on these correlations may be obtained by making a systematic survey of simple systems, or individual partial pair correlations may be resolved directly by using specialist element selective techniques such as neutron diffraction with isotope subsition (NDIS). This method exploits the contrast in neutron scattering length for isotopes of the same element by measuring the diffraction patterns of two or more samples that are identical in every respect, except for the isotopic enrichment of one or more element. By taking linear combinations of the measured diffraction patterns, correlations involving the substituted species can be extracted or eliminated. The resolution of structural information on a partial pair distribution function level using NDIS, as well as conventional total scattering measurements, offer a rigourous test of the efficacy of molecular dynamics (MD) simulations (Jahn 2022, this volume) which can provide a full-picture of glass and melt structure and properties.

An example Raman spectrum for the model multicomponent silicate glass structure is shown in Figure 5c. The biggest difference compared to the spectrum of pure SiO_2 glass (Fig. 4) is in the appearance of a strong peak in the high wavenumber region centered at around 1100 cm^{-1}. This intensity arises from T–O stretching vibrations in different tetrahedral species with n BO and $4n$ NBO: i.e. the Q^n units, from isolated (Q^0) to fully polymerized tetrahedra (Q^4) (see chapter 2 by Henderson and Stebbins (2022, this volume) and discussion therein).

Binary silicate glasses

In the following, we present some XRD and Raman measurements of binary silicate glass compositions in the system $(MO)_x(SiO_2)_{1-x}$ for M= Li$_2$, Na$_2$, K$_2$, Mg, Ca, Ba, Pb, with mole fraction x from 0 to 0.7 and discuss the effect of the network modifier on the silicate glass network. The results presented illustrate all the salient features observed in the $G(r)$ functions and Raman spectra of binary silicate glasses and melts. Specific compositions are denoted by concatenating the element M with the mole fraction in per cent X (MX, e.g. Ca50 = $(CaO)_{0.5}(SiO_2)_{0.5}$).

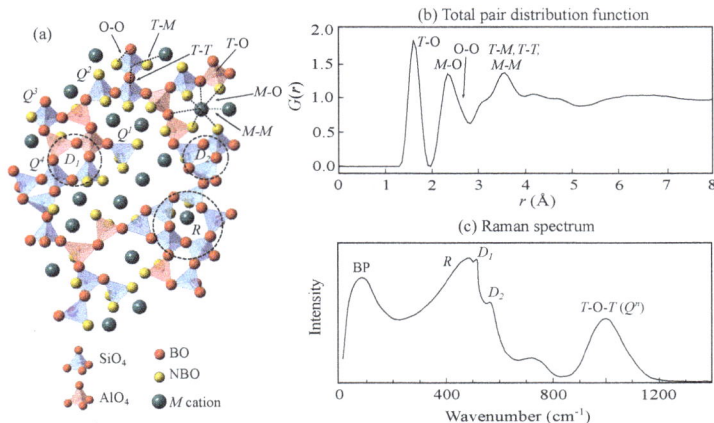

Figure 5. (a) Schematic of a model aluminosilicate glass structure, where network formers T=Si, Al are represented by the **small spheres** in the center of tetrahedral units, bridging and non-bridging oxygen atoms by the **red and gold spheres**, and modifier cations M in **green**. Representative **(b)** pair distribution function $G(r)$ and **(c)** Raman spectrum for this model glass are shown with peaks corresponding to specific SRO and MRO structural features highlighted.

Diffraction measurements

First, we consider x-ray pair distribution functions $G(r)$ measured for Ca- and Pb-silicate glasses shown in Figure 6 which have features typical for all alkali, alkaline earth, and metal–oxide silicate glasses. The first peak between 1.6 and 1.7 Å arises from the nearest neighbour Si–O bond, with its intensity and variation dependent upon the nature of the M cation and the proportion of silica. On addition of CaO we observe a peak at 2.3 Å arising from Ca–O bonds in Ca-centered polyhedra with an average 6-fold coordination by oxygen (Cormack and Du 2001). New Ca–Ca correlations also give rise to a peak at ~3.5 Å. With increasing CaO fraction, both the Ca–O and Ca–Ca peaks become progressively more significant with a corresponding reduction in height of the Si–O peak. This is even more evident in the $G(r)$ functions measured for Pb-silicate glasses, due to the high atomic number of Pb (Z=82) compared to Si, O, or Ca, leading to a higher x-ray scattering cross-section. At high Pb concentrations ($x > 0.5$) two clear Pb–O peaks at 2.26 Å and 2.66 Å are resolved, corresponding to Pb in 4- and 6-fold coordination, respectively. Takaishi et al. (2005) report a higher Pb–O bond length of 2.78 Å in Pb-silicate glasses from neutron diffraction measurements. However, these authors failed to consider correlations at longer distances, especially Pb–Si and Pb–Pb (Fig. 7), which have a strong influence on the position of the Gaussian functions fitted to their real-space pair distribution functions. Takaishi et al. (2005) also assign a peak at 2.65 Å to O–O bonds in their Gaussian fit. In our XRD measurements of Pb-silicate glasses, the O-O correlations are poorly weighted ($W_{OO} = 0.090$ ($x = 0.3$), 0.046 ($x = 0.5$) and 0.025 ($x = 0.7$)) and contribute little to the measured signal. Since the intensity of the peak at 2.66 Å is strongly influenced by the fraction of Pb present in the glass, we assign this peak to mainly Pb–O bonds in 6-fold coordination. We note that this relationship is also visible in the data published by Takaishi et al. (2005), which clearly show an increase in intensity of the peak at 2.65 Å at the highest Pb content when the weighting of Pb–O correlations becomes higher than for O–O correlations.

The Gaussian fits presented in Figure 7 reproduce very well the experimental total correlation functions $T(r)$ considering the two well defined PbO_4 and PbO_6 contributions. While the presence of PbO_5 units cannot be ruled out, their contribution will lie between the PbO_4 and PbO_6 and are difficult to determine unambiguously.

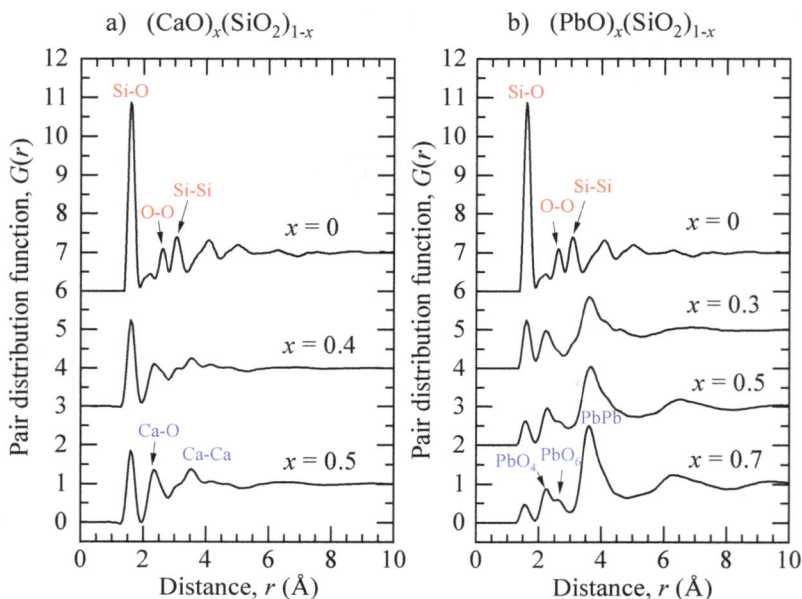

Figure 6. Total pair distribution functions $G(r)$ SiO_2 and silicate glasses in the system $(MO)_x(SiO_2)_{1-x}$ for **a)** M = Ca and **b)** M = Pb (previously unpublished XRD measurements made at beamline ID11 at the European Synchrotron Radiation Facility (ESRF), Grenoble, France).

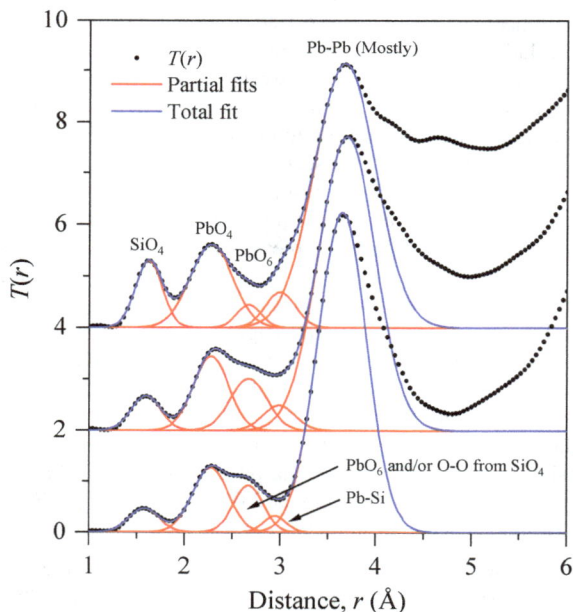

Figure 7. Total correlation functions $T(r)$ obtained from the XRD data for $(PbO)_x(SiO_2)_{1-x}$ for x = 0.3 (**top**), 0.5 (**middle**), 0.7 (**bottom**) from Figure 6, presented together with partial Gaussian fits (**red curves**) and their combined totals (**blue curves**).

Although Pb and other heavy elements are strongly probed by XRD, lighter elements such as Ca are probed less strongly. Neutron diffraction with isotope subsitution (NDIS) is a powerful tool for isolating these weaker or heavily overlapped pair correlations. Figure 8 shows the first order (Eckersley et al. 1988) and second order "double" (Gaskell et al. 1991) difference functions from $^{44}Ca/^{nat}Ca$ NDIS measurements of $CaSiO_3$ (Ca50) glass (with 3 mol.% Al_2O_3 added to prevent nucleation, Eckersley et al. 1988). The results reveal well defined Ca–O and Ca–Ca distances indicating ordering of the modifier Ca^{2+} cations up to 10 Å. Originally attributed by Gaskell et al. (1991) to sheets of edge shared CaO_6 octahedra, subsequent MD simulations have revised this model of medium range cationic ordering to more chainlike clusters or channels of Ca-centered polyhedra (Mead and Mountjoy 2006; Benmore et al. 2010a; Skinner et al. 2012a; Cormier and Cuello 2013). A similar distribution of Mg-centered polyhedra contributing to MRO is observed from a $^{25}Mg/^{nat}Mg$ NDIS study of $MgSiO_3$ glass (Cormier and Cuello 2011). These inhomogeneous distributions of cation centered polyhedra pervading the silicate network appears to substantiate the Greaves MRN model (Fig. 1b). Further NDIS measurements revealing the development of cationic MRO on vitrification of oxide melts are discussed later in this chapter.

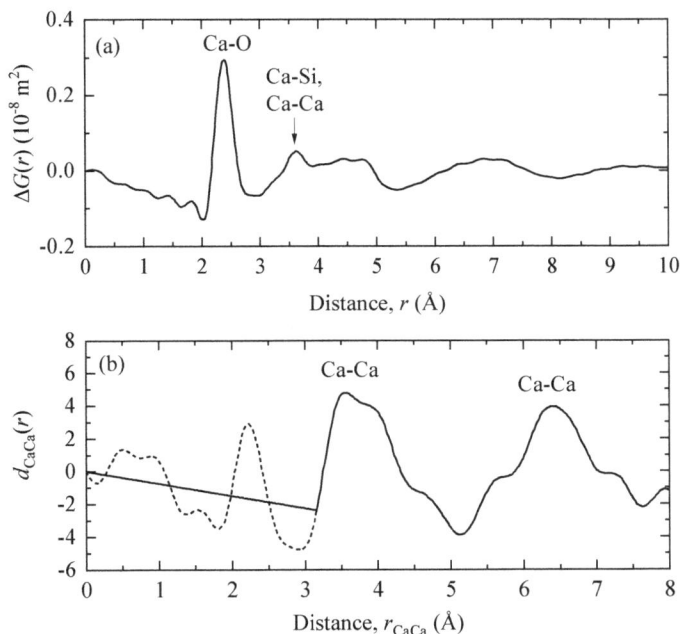

Figure 8. (a) First order difference function $\Delta G(r)$ and **(b)** reduced partial pair distribution function $d_{CaCa}(r) = 4\pi r\rho_0 [g_{CaCa}(r) - 1]$ by double difference from the $^{44}Ca/^{nat}Ca$ NDIS results reported in Eckersley et al. (1988) and Gaskell et al. (1991) for $(CaO)_{0.48}(SiO_2)_{0.49}(Al_2O_3)_{0.03}$ glass. The peak at ~2.2 Å in the unphysical low-r features below the first Ca–Ca interatomic distance (**dashed curve**) in (b) arises from an incomplete subtraction of partials $g_{\alpha\beta\neq CaCa}(r)$.

Raman spectroscopy measurements

We now consider unpolarized Raman spectra for the full range of M-silicate compositions shown in Figure 9, for which distinct and progressive changes are observed with the type and proportion of M cation content. In the following, we discuss the structural information provided by these spectra with respect to the four key spectral regimes; the boson (20–200 cm^{-1}), low (200–600 cm^{-1}), intermediate (600–800 cm^{-1}), and high (800–1200 cm^{-1}) wavenumber regions.

The boson region: At low wavenumbers, the BP generally experiences an increase in frequency with reducing SiO_2 concentrations, indicating a correlation between the BP frequency and glass depolymerization (Neuville 2006, 2014a; Richet 2012). The BP is also influenced by the introduction of different network modifiers, where at constant SiO_2 concentration the BP shifts to higher frequency with the addition of small or medium sized ions and lower frequency with the introduction of larger heavier ions (Neuville 2005, 2006). These changes are accompanied by a narrowing of the BP width for glasses containing heavier elements (Ba, Pb) compared to lighter elements (Na, Li, Mg, Ca), consistent with a very narrow boson peak observed in the Raman spectra of cesium silicate glasses (O'Shaughnessy et al. 2017). This appears to point towards a modification of MRO in the silicate glasses, and perhaps the development of ordering of the modifier cations. Changes in intensity and position of the BP may also be explained by distortion of the SiO_4 tetrahedra (Hehlen et al. 2002).

The low frequency region: With the addition of modifier oxides, the R, D_1, and D_2 bands in the spectral domain between ~250 and 600 cm^{-1} originating from $\geq$ 5-, 4-, or 3-membered rings, respectively, evolve into a broad peak in the region of the D_2 vibration at ~580 to 680 cm^{-1}, reflecting the breaking of Si–O–Si bonds and introduction of NBO leading to a reduction in the size of the ring structures formed from interconnected SiO_4 tetrahedra. However, for lighter modifier elements (e.g. Li-silicate glass) the spectrum retains significant resemblance to the pure SiO_2 glass spectrum indicating the presence of silica-rich regions in the glass (Matson et al. 1983). This has been interpreted as consistent with the MRN model, where Li atoms form MRO clusters within an underlying silicate network (Le Losq et al. 2019a).

Although Figure 9 shows only unpolarized spectra, the R, D_1, and D_2 bands are highly polarized. Polarized Raman measurements of Na-silicate glasses, where spectra are recorded with parallel (VV) or perpendicular (VH) polarizations of incident and scattered light, reveal a strong reduction in intensity of the R-band accompanied by the development of a strong narrow band between ~540 to 600 cm^{-1} in VV polarized spectra with increasing Na_2O fraction (Hehlen and Neuville 2015). This is consistent with a reduction in the Si–O–Si angle due to a reduction in ring size and increasing preponderance of D_2 modes. The R, D_1, and D_2 bands are inactive in the VH polarized spectra which instead reveal the development of a peak at ~350 cm^{-1} and a lower frequency peak at ~175 cm^{-1} with increasing cation concentration attributed to cation motions (Hehlen and Neuville 2015). This will be discussed in more detail below, in relation to the dual role of cations as network modifiers or compensators in aluminosilicate compositions.

The intermediate frequency region: The asymmetric band observed in silica and silica-rich glasses at around 800 cm^{-1} is attributed to bending vibrations in SiO_4 tetrahedra (Sarnthein et al. 1997; Taraskin and Elliott 1997; Spiekermann et al. 2013). It is, however, interesting to note the presence of a high frequency shoulder in this peak for pure SiO_2. This asymmetry has been attributed to two different structures of SiO_4 units with distinct inter-tetrahedral Si–O–Si angles coexisting in the glass (Seifert et al. 1982; Neuville and Mysen 1996; Kalampounias et al. 2006). However, direct structural studies have not observed a bimodal distribution of the Si–O–Si angle in SiO_4 units. We note that different n-membered rings exhibit different Si–O–Si angles such that this observation could be simply explained by the variation in Si–O stretching frequency in tetrahedral units as a function of ring size (Le Losq et al. 2019a). This is supported by the loss of this high-frequency shoulder with the addition of modifier cations and hence reduction in the ring size and Si–O–Si distributions. However, since the Si–NBO frequency is related to the bond force constant other explanations are also possible, including e.g. a change in Si–O–Si angle as alkali atoms bond to the BO or by the presence of an alkali as a charge compensator.

The high frequency region: The high frequency Raman bands for silicate glasses centered at ~1100 cm^{-1} are associated with Si–O stretching vibrations in different Q^n tetrahedral units and are typically interpreted by spectral deconvolution (Mysen et al. 1982a,b; McMillan 1984), as discussed in detail in Henderson and Stebbins (2022, this volume). Here we provide additional details for selected glass compositions.

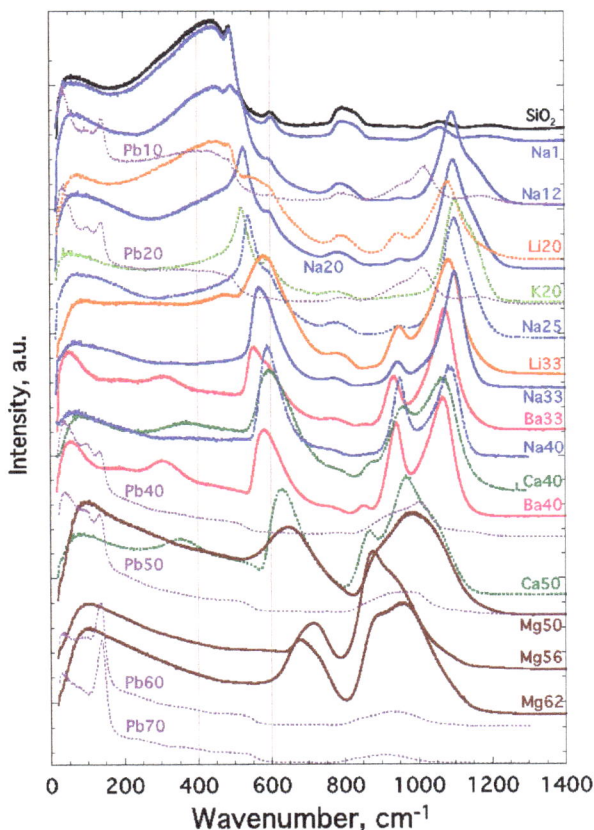

Figure 9. Unpolarized Raman spectra of SiO_2 and silicate glasses in the system $(MO)_x(SiO_2)_{1-x}$ for $M = Li_2$, Na_2, K_2, Mg, Ca, Ba, Pb. The spectra are labelled *MX* for cation fraction *X* in mole%. Original data are in Neuville et al. (2014) and Ben Kacem et al. (2017).

Spectral deconvolution of these high-frequency bands into their constituent components by Gaussian curve fitting is illustrated in Figure 10 for Ca40 and Na40 glass spectra (Neuville 2006). Here the Gaussian bands at ~900 cm^{-1}, 960 cm^{-1}, and 1080 cm^{-1} can be attributed to the vibrations of Q^1, Q^2, and Q^3 species, respectively. To fully model the high-frequency region, it is necessary to include an additional band at 1020 cm^{-1} for Ca40 and 1040 cm^{-1} for Na$_{40}$. This contribution can be compared to the band observed at ~1070 cm^{-1} in the pure SiO_2 glass spectrum (Fig. 4). This band has been attributed to the T_2 stretching mode of Q^4 units (Neuville et al. 2014a; Henderson and Stebbins 2022, this volume) although recent studies (Bancroft et al. 2018, O'Shaughnessy et al. 2020) assigned this band to Q^3 tetrahedra close to alkali cations. The Raman spectrum of the Na1 glass (Fig. 9) shows that the addition of only 1 mol% of Na_2O in silica leads to a slight increase of the intensity of the peak at 1070 cm^{-1}. Further addition of modifier ions, however, leads to a large increase in the intensity corresponding to the higher frequency $Q^4 A_1$ mode for higher alkali composition silicate glasses (Bancroft et al. 2018; O'Shaughnessy et al. 2020; Henderson and Stebbins 2022, this volume). At low alkali oxide contents, the A_1 mode is composed of two bands. These bands merge into a single band after the addition of 20 mol% of Cs oxide (O'Shaughnessy et al. 2020).

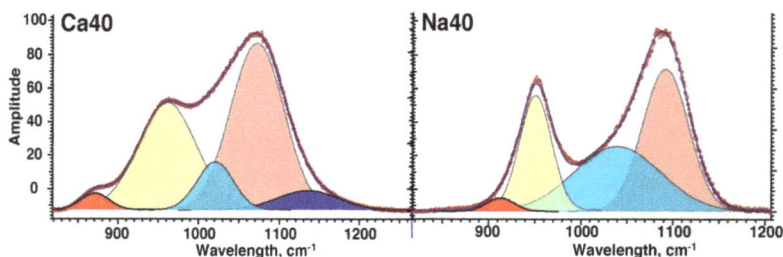

Figure 10. Gaussian deconvolution of Raman spectra for Ca_{40} and Na_{40}, redrafted from (Neuville 2006). Band assignments (from low to high frequency) are: Q^1, Q^2, T_{2s}, Q^3 and Q^4. No Q^4 species are present in the Na_{40} glass.

The band at 1135 cm^{-1} in the Ca40 spectrum is attributed to the A_1 stretching mode of Q^4 units. Substitution of Ca by Na reduces the network polymerization as evident by the complete absence of this Q^4 band in the Na40 glass spectrum. With increasing modifier cation concentration, increasingly depolymerized species are present, including Q^0 and Q^1 species, leading to a significant shift in the intensity maximum of the Si–O stretching bands to lower frequencies (see e.g. the spectra for Mg_{62} and Mg_{56} and Fig. 12 of Neuville et al. 2014a).

Lead silicate glasses

The high-frequency domain, 850–1300 cm^{-1}, in the Raman spectra of lead silicate glasses appears less intense compared to other glass compositions. This is a result of the normalization process as Pb-silicate glasses exhibit the highest intensity peaks at low frequency. Two very intense low-frequency peaks at 100 cm^{-1} and 141 cm^{-1} are dominant features in Raman spectra of Pb-silicate glasses (Worrell and Henshall 1978; Furukawa et al. 1978; Ohno et al. 1991; Zahra et al. 1993; Feller et al. 2010, Ben Kacem et al. 2017). The peak at 141 cm^{-1} is attributed to covalent Pb–O–Pb bonding in interconnected tetrahedral PbO_4 units (Furukawa et al.1978; Worrell and Henshall (1978); Zahra et al.1993) and correlates with NMR spectra which show the proportion of Pb–O–Pb linkages increases with PbO content (Lee and Kim 2015). The Raman band at 100 cm^{-1} has been attributed to ionic Pb–O bonds associated with NBO atoms in PbO_6 units (Worrell and Henshall 1978; Ohno et al. 1991)

A variety of charge balancing models have been reported which predict MRO in the form of different extended networks of interconnected Pb-centered units in the glass, including dimeric zigzag chains of covalently bonded PbO_4 tetrahedra (Morikawa et al. 1982), screw chains of PbO_n polyhedra ($n = 3$ or 4) (Imaoka et al. 1986), Pb cations in predominantly edge-shared PbO_3 trigonal pyramid arrangements (Takaishi et al. 2005), and pyramidal PbO_n units with a mix of corner and edge sharing with electron lone-pairs organizing to form voids in the glass (Alderman et al. 2013).

XRD reveals Pb can adopt both 4- and 6-fold coordinated sites in Pb-silicate glasses (Figs. 6 and 7), where Pb in 4-fold coordination forms an interconnected sub-network mixing mechanically with the SiO_4 tetrahedral network without chemical interaction (Neuville and Le Losq 2022, this volume). This is supported by the glass transition temperature for $PbO–SiO_2$ glasses which varies almost linearly between 30 and 90 mol.% of SiO_2 fraction (Ben Kacem et al. 2017) implying heterogenous mixing. Pb in 6-fold coordination plays a similar role to alkali or alkaline-earth elements breaking up the silicate network polymerization by forming NBOs.

Aluminosilicate glasses

Aluminosilicate glasses and their melts are of interest due to their relevance to natural magmas. Aluminum is classed as an intermediate glass former due to its ability to both compete with Si to form a 4-fold coordinated network structure or to behave as a network modifier assuming 5- or 6-fold coordination and reducing the glass network connectivity via the formation of non-bridging oxygens (Sun 1947). This behavior is particularly sensitive to the relative proportions of Al and other modifier cations (Bista and Stebbins 2017). To form a perfectly connected network of AlO_4 tetrahedra the Al:O ratio needs to be precisely 1:2 such that any two Al atoms are connected by a single bridging oxygen. For the $(MO)_x(Al_2O_3)_{1-x}$ compositions this occurs at the ratio $R = MO/Al_2O_3 = 1$ ($x = 0.5$), where modifier cations, assuming a uniform distribution, will perfectly charge compensate AlO_4 units which have an overall charge of -1. Aluminosilicate compositions along this join are denoted tectosilicate or meta-aluminosilicate glasses. Peraluminous aluminosilicates, classified as glasses with the ratio $R = MO/Al_2O_3 < 1$, have an excess of Al and are unable to form an ideal charge compensated corner-sharing network of AlO_4 tetrahedra (Mysen and Richet 2019). This oxygen deficiency may be compensated for by the formation of highly coordinated AlO_5 and AlO_6 units and/or a change in network connectivity. In peralkaline glasses ($R = MO/Al_2O_3 > 1$) there are excess modifier cations which will act to charge compensate AlO_4 tetrahedra and potentially depolymerize the network. There is, however, increasing evidence for deviations from this simple model, with significant proportions AlO_5 and even 5-fold coordinated SiO_5 units observed in glasses that are sufficiently charge compensated by metal cations (Stebbins 1991, et al. 1997; Toplis et al. 2000). Significant fractions of AlO_4 have also been observed in the alumina-rich glasses (Mysen and Richet 2005) in which charge neutrality may be accomplished by the formation of $O:(Al/Si)_3$ triclusters in which one O atom is shared by three Al/Si tetrahedral units (Lacy 1963; Toplis et al. 1997, 2000; Stebbins and Xu 1997; Stebbins et al. 2001; Mysen and Toplis 2007).

Diffraction and Raman measurements along the tectosilicate join

The $G(r)$ functions from neutron diffraction experiments and Raman spectra for MO–Al_2O_3–SiO_2 (M = Ca, Sr) glasses along the tectosilicate join ($R = MO/Al_2O_3 = 1$) are shown in Figures 11 and 12. Compositions are denoted by concatenating the modifier element M with the mole fraction X of SiO_2 and Y of Al_2O_3 in per cent ($MX.Y$). With increasing substitution of Si by Al the nearest neighbor peak T–O in $G(r)$ is broadened and shifts to higher distances from 1.60 Å in pure SiO_2 to 1.76 Å in silica-free Ca0.50 glass. This results from Al–O correlations with longer nearest neighbour lengths overlapping the shorter Si–O correlations (Cormier et al. 2000, 2003; Hennet et al. 2016). For the Ca-aluminosilicate glasses, the peak at ~2.30 to 2.35 Å

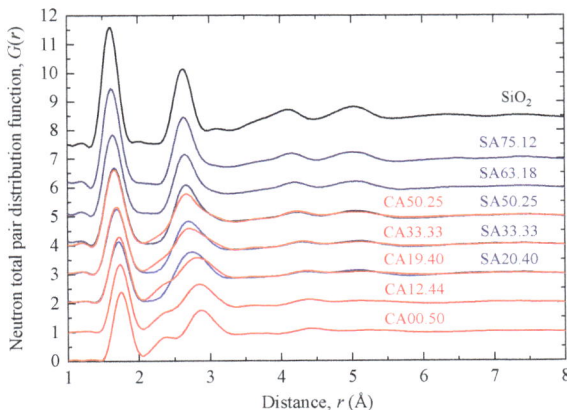

Figure 11. Neutron total pair distribution functions for M-aluminosilicate (M = Ca,Sr) glasses along the tectosilicate ($R = 1$) join (data from Bowron 2008; Kozaily 2012; Hennet et al. 2016; Charpentier et al. 2018). Compositions are denoted by $MX.Y$ for mole per cent fraction X of SiO_2 and Y of Al_2O_3, where the fraction $MO = 100 - (X + Y)$.

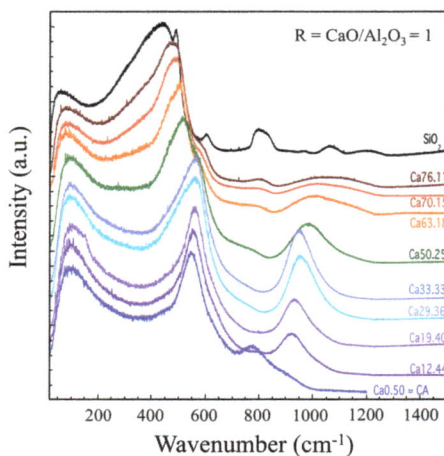

Figure 12. Raman spectra for calcium aluminosilicate glasses along the tectosilicate (R = 1) join (data from Neuville et al. 2004, 2010, 2014a). Compositions are denoted by the notation Ca$X.Y$ for mole per cent fraction X of SiO_2 and Y of Al_2O_3, where the fraction CaO = 100 − ($X + Y$).

is attributed to the first Ca–O bond and becomes increasingly conspicuous with reducing SiO_2 concentration due to higher weighting on the $g_{CaO}(r)$ partial pair distribution functions. It is difficult to precisely determine the Ca–O coordination environment using conventional x-ray or neutron diffraction due to considerable overlap by the $g_{OO}(r)$ partial pair distribution functions (Hannon and Parker 2000; Benmore et al. 2003; Mei et al. 2008a,b; Drewitt et al. 2011; Hennet et al. 2016). Direct measurements of the first Ca–O coordination shell in both calcium-silicate (as previously discussed) and -aluminate glasses by NDIS provide average Ca–O coordination numbers of ~6.2 to 6.5 (Eckersley et al. 1988; Drewitt et al. 2012), with MD simulations revealing a broad distribution of Ca–O polyhedra centered on ~6 (Drewitt et al. 2012; Jakse et al. 2012).

Despite the larger neutron scattering length b_{Sr} = 0.702(2) fm, compared to b_{Ca} = 0.470(2) fm (Sears 1992), the nearest neighbour Sr–O correlations are not discernible in the $G(r)$ functions for Sr-aluminosilicate glasses. This is due to the larger ionic radius of Sr^{2+} cf. Ca^{2+} (Shannon 1976) due to both higher atomic number and larger Sr–O coordination number of ~8 to 9 (Novikov et al. 2017; Charpentier et al. 2018). Furthermore, the O–O correlations associated with these highly coordinated Sr-centered polyhedra are shifted to lower distances compared to Ca-aluminosilicate measurements. As a result, the nearest neighbour peak in $g_{SrO}(r)$ is completely overlapped by the $g_{OO}(r)$ correlations (Florian et al. 2018). This highlights a need for future diffraction studies of Sr-aluminosilicate glasses using element specific techniques.

In Raman spectra of aluminosilicate glasses, similar spectral features and vibrations are observed for Al as for Si, where Al can adopt Q^4, Q^3 and Q^2 speciation depending on silica fraction (McMillan and Piriou 1983; Neuville et al. 2008a,b, 2010; Licheron et al. 2011). The substitution of Al for Si in tetrahedral positions leads to detectable shifts in frequency, broadening, and reduction in spectral resolution of the Raman bands relative to the SiO_2 glass spectrum (Mysen et al. 1981; Seifert et al. 1982; McMillan et al. 1982; McMillan and Piriou 1982, 1983; McMillan 1984; Neuville and Mysen 1996; Neuville et al. 2004, 2006, 2008a,b; Le Losq and Neuville 2013; Le Losq et al. 2017). In particular, the T–O stretching vibrations shift to lower wavenumber, with a corresponding increase in position of T–O–T bending modes, as a result of a reduction in the (Si,Al)–O force constant and/or Si,Al coupling (Rossano and Mysen 2012). These changes are illustrated for calcium aluminosilicate glasses in Figure 12 showing the Raman spectra along the tectosilicate join. Along this join (tectosilicate), glasses are fully polymerized and Al^{3+} substitutes for Si^{4+} in Q^4 tetrahedral sites regardless of the modifier element (Neuville and Mysen 1996; Neuville et al. 2004, 2006, 2008a,b; Le Losq and Neuville 2013; Novikov et al. 2017; Le Losq et al. 2017; Ben Kacem 2017). This is illustrated in Figure 13a, where the Q^{4I}, Q^{4II} vibrational modes (assigned to Q^4 tetrahedra with 2 different Si–O–Si angles, see Henderson and Stebbins (2022, this volume) and T_{2s} mode (the T_2 mode associated with Q^4 tetrahedra observed in SiO_2 glass, see Henderson and Stebbins 2022, this volume) determined by Gaussian deconvolution, experience a linear shift to lower wavenumbers with increasing substitution of Si for Al associated with a continuous shift from the Si–O–Si vibration to pure Al–O–Al vibrations in the CA50.00 glass (McMillan and Piriou 1983; Neuville and Mysen 1996; Neuville et al. 2004, 2006, 2008a,b, 2010; Licheron et al. 2011; Novikov et al. 2017). This variation correlates well with the ^{27}Al NMR chemical shift for Al in four-fold coordination and the T–O bond length measured by diffraction (Fig. 13b), which both increase linearly with increasing Al fraction (Neuville et al. 2004, 2006, 2008a,b). Modifier cations nano-segregate into percolation channels at high Al/Si and high modifier content, with their presence indicated by an increase of the D_1 and D_2 peak intensity (Le Losq et al. 2017).

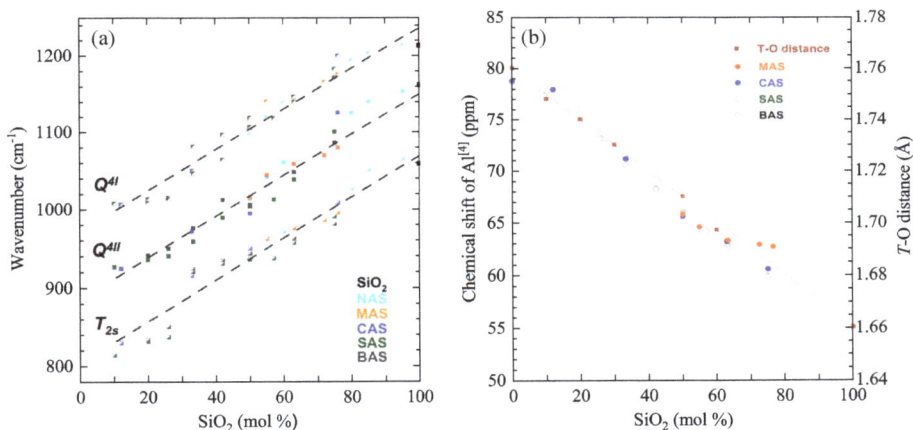

Figure 13. a) Wavenumber of the 3 Gaussian bands as a function of SiO_2 for NAS, MAS, CAS, SAS and BAS tectosilicate glasses (Neuville and Mysen 1996; Neuville et al. 2004, 2006, 2008; Novikov et al. 2017) and **b)** ^{27}Al NMR chemical shift, δ_{iso} and T–O distance obtained from the $G(r)$ as a function of SiO_2 for CAS glass system (Fig. 11).

Charge compensator versus network modifier

As noted above the substitution of an alkali or alkaline-earth element by Al has a significant influence on the Raman spectra of aluminosilicate glasses. Some of the changes are related to whether the added alkali or alkaline-earth cations behave as charge compensators or network modifiers. The dual role of the cations can be explored through analysis of the VH component of polarized Raman spectra. (Hehlen and Neuville 2015, 2020).

The VH Raman spectra of calcium aluminosilicate glasses along the 50% SiO_2 join are shown in Figure 14. A strong band is apparent at ~350 cm^{-1} for the alumina-free calcium silicate glass (Ca50.00) which is not visible in unpolarized Raman spectra of the Ca50.X series (Fig. 9). The intensity of this peak reduces with decreasing CaO/Al_2O_3 ratio and its position is sensitive to the mass of the cation (Na: 350 cm^{-1}, Mg: 360 cm^{-1}, Ca: 352 cm^{-1}, Sr: 334 cm^{-1}, Ba: 310 cm^{-1}, Hehlen and Neuville 2015, 2020). This low frequency vibration is attributed to cations acting as network modifiers. When cations adopt a charge compensation role for the AlO_4^- tetrahedra this vibration disappears.

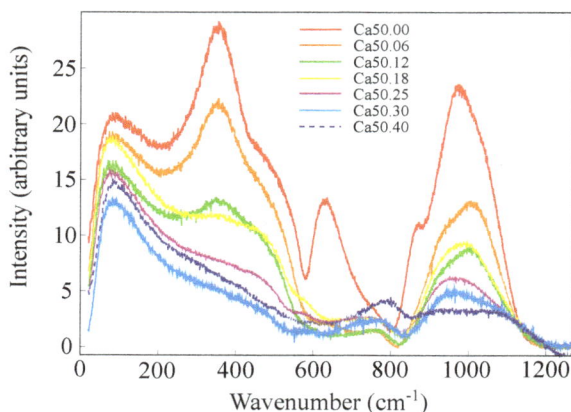

Figure 14. VH Raman spectra of calcium aluminosilicate glasses along the 50% SiO_2 join (redrafted from Hehlen and Neuville 2015).

This change in role of modifier cations with varying CaO/Al_2O_3 ratio is supported by a pre-edge shift observed in Ca K-edge x-ray absorption near edge structure (XANES) measurements of Ca-aluminosilicate glasses from Ca50 to Ca50.25 compositions (Cicconi et al. 2016). A similar change in the structural role of Na^+ ions from network modifier to charge compensator are indicated by an evolution of the ^{23}Na NMR chemical shift of Na-aluminosilicate glasses (Le Losq et al. 2014). This also appears to correlate with the disappearance of the peak at $350 \, cm^{-1}$ observed in the VH Raman spectra between sodium silicate and sodium tectosilicate glasses (Hehlen and Neuville 2015). Being able to determine the behavior of alkali and alkaline-earth cations as charge compensators or network modifiers is very useful. Clearly, there is potential in examining the VH spectra and more work needs to be done.

GLASS AND MELT STRUCTURE UNDER EXTREME CONDITIONS

Despite decades of research focused on aluminosilicate glasses, much less direct information is available on the structure of magmas in their high temperature liquid state (Henderson 2005) or under the high p conditions of deep planetary interiors (Kono and Sanloup 2018). This is due to the challenges associated with making measurements above the high melting temperatures of aluminosilicates (> 1500 K), where conventional furnaces present a high risk of chemical reaction with a sample, spectroscopy measurements are inhibited by thermal radiation (Papatheodorou et al. 2010), and the thermal motions and structural disorder lead to weaker spectroscopy or diffraction signals compared to the solid glass. Measurements at high-p and/or -T are doubly challenging, requiring specialized instrumentation to generate these extreme conditions while simultaneously allowing good accessibility to the sample and minimizing unwanted contributions from the sample environment. As a result, quenched glasses have long been used as analogues for the study of melts (Henderson 2005). However, while important, information obtained on glass structure is not always representative of melts in their natural liquid state (Wilding et al. 2010; Drewitt et al. 2012). This is particularly apparent for so-called *fragile* liquids such as depolymerized silicate melts (Giordano and Dingwell 2003) in which dynamical properties and viscosity exhibit a marked deviation from Arrhenius behavior compared to more traditional *strong* glass forming liquids such as pure silica (Angell 1991, 1995). Glasses and melts also experience significant reorganization in SRO and MRO under compression such that measurements made at high-p conditions are necessary to understand the full nature of magmas at depth.

High temperature containerless processing

In recent years, containerless techniques such as aerodynamic-levitation with laser-heating have enabled *in situ* diffraction and spectroscopy measurements of the structure of liquid oxides (Price 2010; Papatheodorou et al. 2010; Hennet et al. 2011a; Benmore and Weber 2017). Levitating liquids on a stream of inert gas eliminates the possibility of chemical reaction with containment material and experiments provide clean data sets of the liquid sample enabling advanced techniques such as NDIS to be applied (Drewitt et al. 2012, 2017, 2019; Skinner et al. 2012a). The high flux of synchrotron sources means that XRD measurements using ultra-fast millisecond acquisition times are possible with large area detectors. This enables diffraction patterns to be measured as an oxide liquid is supercooled into the solid glass state (Hennet et al. 2007, 2011b; Benmore et al. 2010a; Skinner et al. 2013a; Drewitt et al. 2019). Another key advantage of containerless processing is the suppression of heterogenous nucleation thereby promoting deep supercooling which, combined with rapid quenching, facilitates the formation of glass compositions such as peridotite (Auzende et al. 2011) or pure aluminates (Drewitt et al. 2019) that cannot ordinarily be formed using conventional methods.

Aluminate melts

Pure liquid alumina has a poor glass forming ability and cannot be quenched to a glass (Skinner et al. 2013b; Shi et al. 2019). However, the glass forming ability of aluminate compositions is enhanced significantly by the addition of alkali, alkaline earth, or metal oxides in combination with containerless processing or other fast quenching techniques. Extensive experimental and simulation studies have been made to understand the structure of silica-free calcium aluminate glasses and their melts (McMillan and Piriou 1983; Morikawa et al. 1983; Poe et al. 1993, 1994; Massiot et al. 1995; McMillan et al. 1996; Daniel et al. 1996; Hannon and Parker 2000; Weber et al. 2003; Benmore et al. 2003; Iuga et al. 2005; Kang et al. 2006; Thomas et al. 2006; Hennet et al. 2007; Mei et al. 2008a,b; Cristiglio et al. 2010; Neuville et al. 2010; Mountjoy et al. 2011; Licheron et al. 2011; Drewitt et al. 2011, 2012, 2017, 2019; Liu et al. 2020). Other compositions have also been investigated, including barium (Licheron et al. 2011; Skinner et al. 2012b), strontium (Weber et al. 2003; Licheron et al. 2011; Novikov et al. 2017), lead (Barney et al. 2007), and rare-earth aluminates (Wilding et al. 2002; Weber et al. 2004a; Du and Corrales 2007; Barnes 2015). A liquid–liquid phase transition (LLPT) has even been proposed in the yttria–aluminate system, inferred from the observation of coexisting low-density and high-density phases in the supercooled melts (Aasland and McMillan 1994; Wilding et al. 2002, 2005, 2015; Weber et al. 2004a,b; Wilson and McMillan 2004; Greaves et al. 2008), although this result is disputed by subsequent measurements which find no structural or thermal signatures consistent with a LLPT (Barnes et al. 2009) and the observations may instead be attributed to nanocrystalline inclusions forming within the glassy material (Nagashio and Kuribayashi 2002; Tangeman et al. 2004; Skinner et al. 2008; Barnes et al. 2009).

Molten calcium aluminates are very *fragile* refractory liquids. Fragile liquids tend to exhibit greater variation in SRO and MRO compared to *strong* glass forming liquids and encounter high potential energy barriers during supercooling such that they become trapped in deep local energy minima on approach to the glass transition (Drewitt et al. 2019). The comprehensive range of techniques applied to the calcium aluminate system reveal significant transformations in SRO and MRO structure taking place in these liquids during glass formation.

Neutron and XRD $G(r)$ functions for $(CaO)_x(Al_2O_3)_{1-x}$ glasses and melts with $x = 0.5$ ($CaAl_2O_4$, CA) and $x = 0.75$ ($Ca_3Al_2O_6$, C3A) are shown in Figure 15 (Drewitt et al. 2011, 2012, 2017, 2019). The results of MD simulations made using aspherical ion model (AIM) potentials, which consider polarization effects up to the quadrupolar level (see Jahn 2022, this volume), are also shown (Drewitt et al. 2011, 2012, 2019). The first peak in $G(r)$ for the melts at ~1.78 Å is attributed to the nearest neighbour Al–O bond, while the second peak at ~2.3 Å

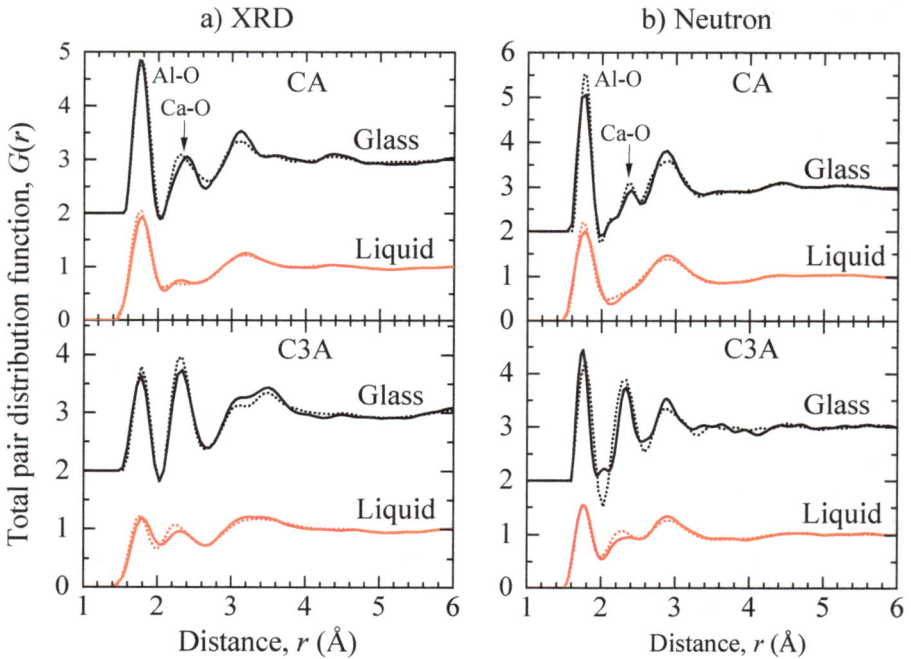

Figure 15. Total pair distribution functions $G(r)$ from **a)** x-ray and **b)** neutron diffraction of CA and C3A glasses (**solid black curves**) and liquids (**solid red curves**) together with the corresponding functions computed directly from MD simulations (**dotted curves**) (data from Drewitt et al. 2011, 2012, 2017, 2019).

arises from Ca–O correlations (Drewitt et al. 2011). On vitrification, the intensity and resolution of the Al–O and Ca–O peaks is significantly enhanced, reflecting the thermal motions and higher overall structural disorder in the liquid state. The Al–O and Ca–O peak positions experience shifts to ~1.75 Å and ~2.35 Å, consistent with a reduction and increase in Al–O and Ca–O coordination numbers, respectively. However, accurate quantification of these coordination numbers is hindered by the penetration of Ca–O correlations into the first Al–O coordination shell and considerable overlap with other atom-atom interactions at higher bond lengths. This is particularly acute for the liquids, but also affects the glass measurements (Hannon and Parker 2000; Mei et al. 2008a,b; Drewitt et al. 2011). Element selective techniques are, therefore, required to isolate the Al and Ca coordination environments. This represents a significant challenge at high temperatures: ^{27}Al NMR spectroscopy measurements observe the fast exchange limit such that individual coordination populations cannot be resolved in the high temperature melts (Coté et al. 1992; Poe et al. 1993, 1994; Massiot et al. 1995; Florian et al. 2018), ^{43}Ca NMR is limited by low sensitivity and natural abundance (Dupree et al. 1997), x-ray absorption measurements of high temperature levitated melts are difficult to perform, and NDIS is limited by the sample size and neutron flux.

Despite the small size (~2–3 mm diameter) of levitated melt droplets, neutron diffraction with isotope subsitution (NDIS) has nevertheless been successfully applied to precisely measure the Al–O and Ca–O coordination environments in levitated liquid CA and C3A (Drewitt et al. 2012, 2017). The pseudo-binary pair distribution functions $g_{\mu\mu}(r)$ and $g_{Ca\mu}(r)$ for the CA and C3A melts and the quenched CA glass are shown in Figure 16. In $g_{\mu\mu}(r)$ all interactions involving Ca are eliminated and the function contains contributions from μ–μ (μ = Al, O) pair

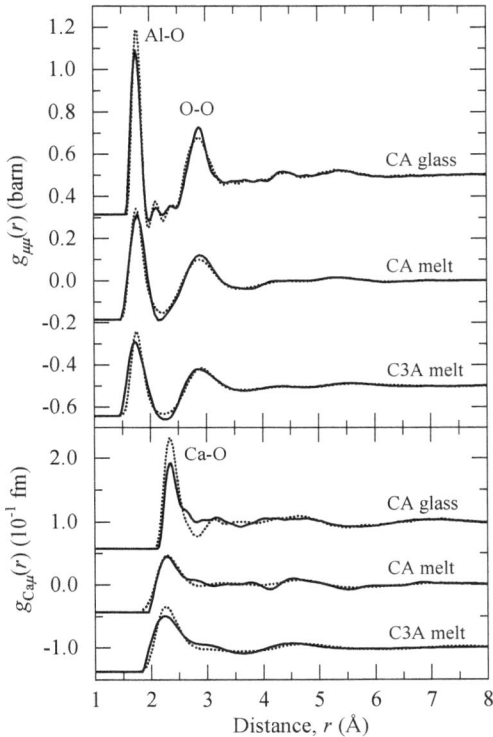

Figure 16. The real-space pseudo-binary pair distribution functions $g_{\mu\mu}(r)$ and $g_{Ca\mu}(r)$ for the CA (Drewitt et al. 2012) and C3A (Drewitt et al. 2017) melts from NDIS measurements (**solid curves**) and AIM-MD (**dotted curves**) simulations (Drewitt et al. 2011, 2012). (Reprinted from Drewitt et al. 2019, doi:10.1088/1742-5468/ab47fc © SISSA Medialab Srl. Reproduced by permission of IOP Publishing. All rights reserved).

correlations only, showing well resolved peaks corresponding to Al–O nearest neighbors and O–O bonds. In contrast, $g_{Ca\mu}(r)$ is comprised solely from Ca–μ pair correlations and the main peak corresponds to the Ca–O nearest-neighbors. The NDIS experiments for liquid C3A and CA glass go one step further to provide direct measurement of the $g_{CaCa}(r)$ partial pair distribution function, as shown in Figure 17.

The AIM-MD model is in good general agreement with the neutron and XRD total structure measurements and the NDIS difference functions. The CA composition has an Al:O ratio of 1:2 such that it is feasible to form a fully polymerized network of Q^4-species in which any two Al atoms are connected by a single bridging oxygen. The C3A composition on the other hand is significantly depolymerized with a theoretical mean number of 2 bridging oxygens per Al atom. The combined NDIS, XRD, and AIM-MD results reveal that the structure of liquid CA deviates from this simple network model containing 83% AlO_4 tetrahedra with 12% non-bridging oxygens and 15% AlO_5 units, many of which share edges (Drewitt et al. 2012). Considering all Al–O pairs, 18% oxygen atoms have a coordination > 2, with 7% "formal" triclusters involving AlO_4 tetrahedra only. Liquid C3A is composed of a higher fraction of AlO_4 tetrahedra (93%) with 60% non-bridging oxygen and 36% bridging oxygen atoms (Drewitt et al. 2017), where the latter is slightly higher than expected from the O:Al ratio (Skinner et al. 2012b). This is accounted for by the presence of 3 to 4% "free oxygen" ions, which do not participate in Al–O bonding, and ~1% oxygen triclusters. While the majority of AlO_4 tetrahedra in C3A melt belong to a single infinitely connected network, 15 to 20% belong to smaller clusters with ~10% forming Al_2O_7 dimers or isolated AlO_4 tetrahedra.

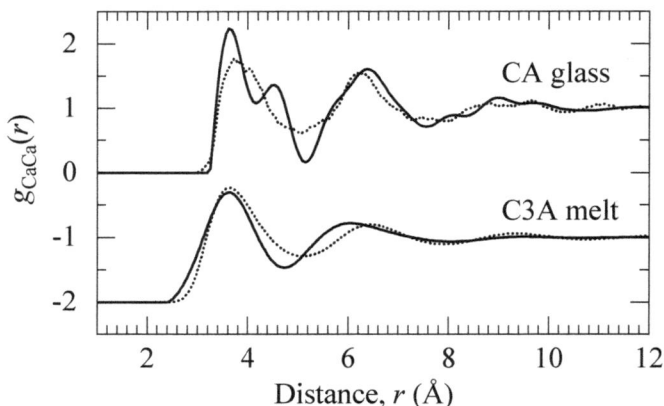

Figure 17. The partial pair distribution function $g_{CaCa}(r)$ for CA glass (Drewitt et al. 2012) and C3A melt (Drewitt et al. 2017) determined directly from NDIS experiments (**solid curves**) and generated from the AIM-MD simulations (Drewitt et al. 2011, 2012). (Reprinted from Drewitt et al. 2019, doi:10.1088/1742-5468/ab47fc © SISSA Medialab Srl. Reproduced by permission of IOP Publishing. All rights reserved).

On vitrification, the over coordinated CA glass structure reorganizes to form a predominantly corner shared network of AlO_4 tetrahedra. The C3A glass is also characterised by a preponderance (99%) of aluminium in AlO_4 tetrahedra with 85% belonging to a single infinite network. This is consistent with ^{27}Al NMR (Neuville et al. 2006, 2007) and XANES spectroscopy at the L and K-edges (Neuville et al. 2008b, 2010) which show Al is predominantly four-fold coordinated by oxygen in glasses with compositions $x = 0.5$ to 0.75. There is, however, a residual 3.5% Al in AlO_5 units remaining in CA glass (Neuville et al. 2006) and the AIM-MD results indicate the presence of ~4% residual AlO_5 units in CA glass, accompanied by 7% non-bridging oxygens and 5% "formal" oxygen triclusters (Drewitt et al. 2012). This latter result is consistent with 5% oxygen triclusters detected in the glass by heteronuclear correlation NMR spectroscopy (Iuga et al. 2005).

Differences in network connectivity are apparent from the Raman spectra of CA, C12A7 ($x = 0.632$, $Ca_{12}Al_{14}O_{33}$) and C3A glasses shown in Figure 18. All spectra exhibit a Boson peak at ~90 cm^{-1}, a strong band at ~560 cm^{-1} with significant asymmetry at higher wavenumbers, and another strong band at ~780 cm^{-1}, which increases in intensity relative to the 560 cm^{-1} band with increasing CaO content. The 560 cm^{-1} band is attributed to transverse motions of bridging oxygens in Al–O–Al linkages, with the high frequency asymmetry caused by Al–O stretching vibrations of the fully polymerized tetrahedral aluminate groups (McMillan and Piriou 1983). This band experiences a shift to higher frequencies and reduction in intensity with increasing CaO content associated with changes in polarizability of the Al–O–Al vibrations resulting from an increase in the Al–O force constant as Q^2 species are introduced into the glass network. This is consistent with a shift to lower energies of the $L_{2,3}$-edge in Al XANES spectroscopy measurements attributed to the presence of Q^2 species and associated stronger hybridization of non-bridging oxygens (Neuville et al. 2010). The weak band in the CA glass spectrum at 780 cm^{-1}, and shoulder at 910 cm^{-1}, are attributed to Al–O stretching vibrations of Q^4 Al species. The increase in intensity and shift to lower frequencies of the 780 cm^{-1} band with increasing CaO content is attributed to increasing fractions of non-bridging oxygens giving rise to more intense vibrations compared to bridging oxygens (McMillan and Piriou 1983). This is consistent with ^{27}Al NMR (Neuville et al. 2006) and Al K-edge XANES results (Neuville et al. 2008a,b), which suggest Al units predominantly in Q^2 and Q^3 speciation in C3A glass. It also implies that the C12A7 glass contains a significant fraction of NBO.

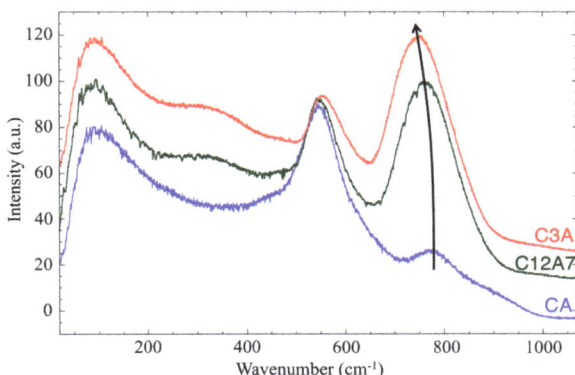

Figure 18. Raman spectra of CA, C12A7, C3A glasses (data from Neuville et al. 2006, 2010; Licheron et al. 2011). The arrow indicates the increase in intensity of the 780 cm^{-1} band and corresponding shift to lower frequencies with increasing CaO concentration.

Calcium has a broad distribution of 4- to 9-fold Ca–O coordination sites, with average Ca–O coordination numbers of 6.2 in CA (Drewitt et al. 2012) and 5.6 in C3A (Drewitt et al. 2017) melts, compared to ~5 in liquid CaSiO$_3$ (Skinner et al. 2012a). In CA melt, edge- and face-sharing Ca-centered polyhedra form small clusters contributing to MRO. In C3A melt all Ca-centered units are connected by corners to a single network with 90% edge- and face-sharing. The direct measurement of $g_{CaCa}(r)$ by NDIS (Fig. 17) gives an average Ca–Ca coordination number of ~8 (~8.5 from AIM-MD) compared to a value ~5 in CA melt, reflecting the greater fraction of Ca in the C3A composition and hence greater tendency towards clustering. The average Ca–O coordination in C3A glass reduces slightly to 5.5 (Drewitt et al. 2019). However, on vitrification, the CA glass experiences a remarkable development of cationic MRO with edge- and face-sharing Ca-centered polyhedra with an average Ca–O coordination of 6.4 which form large branched chainlike clusters that weave through the glass network (Fig. 19) (Drewitt et al. 2012). Similar clusters of Ca-centered octahedra are observed in CaSiO$_3$ glass (Benmore et al. 2010a; Skinner et al. 2012a) and it has been argued that Ca-clustering is in fact an essential requirement for glass formation and is responsible for the fragile behavior of the melt (Skinner et al. 2012a).

(a) Melt   (b) Glass

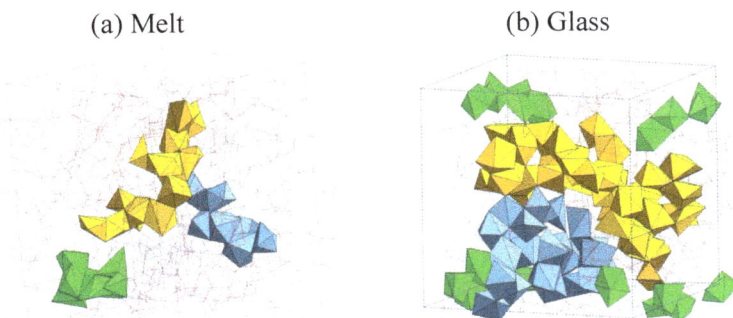

Figure 19. Snapshots illustrating the largest clusters of edge-sharing Ca-centred polyhedra in the AIM-MD simulations of CaAl$_2$O$_4$ from Drewitt et al. (2012) for **(a)** the melt 2500 K and **(b)** the glass at 350 K. The clusters are represented by the light (**yellow**), dark (**blue**), and medium (**green**) shaded units which involve 16, 9, and 8 Ca atoms in the liquid or 44, 24, and 19 Ca atoms in the glass, respectively. Reprinted figure with permission from Drewitt et al. Physical Review Letters 109:235501 (2012) doi:10.1088/0953-8984/24/9/099501. Copyright (2012) by the American Physical society.

In summary, at the Al_2O_3-rich limit of the glass forming region of calcium aluminate melts up to 20% of the Al reside in AlO_5 coordination sites (Drewitt et al. 2011), increasing to over 30% in pure alumina (Skinner et al. 2013b). These highly coordinated Al units are incompatible with glass formation. However, although the equimolar (CA) melt contains 15% AlO_5 polyhedra (Drewitt et al. 2011), these highly coordinated units and oxygen triclusters breakdown on vitrification and the supercooled melt structure reorganizes to form a predominantly corner-shared AlO_4 tetrahedral glass network (Drewitt et al. 2012). This is accompanied by the development of medium range cationic ordering associated with the formation of long clusters of edge- and face-sharing Ca-centered polyhedra (Fig. 19). While significantly depolymerized, the structure of C3A melt remains largely composed of AlO_4 tetrahedra (Drewitt et al. 2011). Although many of these tetrahedra belong to a single infinite network, ~20% aluminum in C3A melt belong to smaller clusters with ~10% unconnected Al_2O_7 dimers and AlO_4 monomers (Drewitt et al. 2017). On vitrification, C3A glass is characterized by a tetrahedral aluminate network structure, where 85% of AlO_4 units belong to a single infinite structure with only 5% isolated tetrahedra or dimers (Drewitt et al. 2019). Beyond the CaO-rich limit of the glass forming region, the number of isolated AlO_4 units is expected to increase such that the glass can no longer support the formation of an infinitely connected network of AlO_4 units (Drewitt et al. 2017).

The significant transformations that take place in SRO and MRO during glass formation in the calcium aluminate glasses are captured by time-resolved measurements of aerodynamically levitated liquid calcium aluminates recorded during solidification of the high temperature melts through their glass transition (Hennet et al. 2007; Drewitt et al. 2019). In particular, the changes in MRO associated with the development of cationic ordering is indicated by the evolution of the FSDP in $S(Q)$ attributed to cation–cation correlations (Fig. 20). The work reviewed here for a representative fragile glass-forming system demonstrates that caution is required when considering glasses as analogues for geological melts. Similar differences in the local structural environment have also been observed in fragile MgO–SiO_2 liquids (Wilding et al. 2010) and natural silicate melts in the CaO–MgO–Al_2O_3–SiO_2 (CMAS) system encompass a range of kinetic fragilities (Giordano and Dingwell 2003; Giordano and Russell 2007; Giordano et al. 2008). As such, measurements made on glasses cannot be assumed to be representative of melts in their natural liquid state and models should be based, where possible, on liquid-state measurements.

(a) $x = 0.5$ (CA)   (b) $x = 0.75$ (C3A)

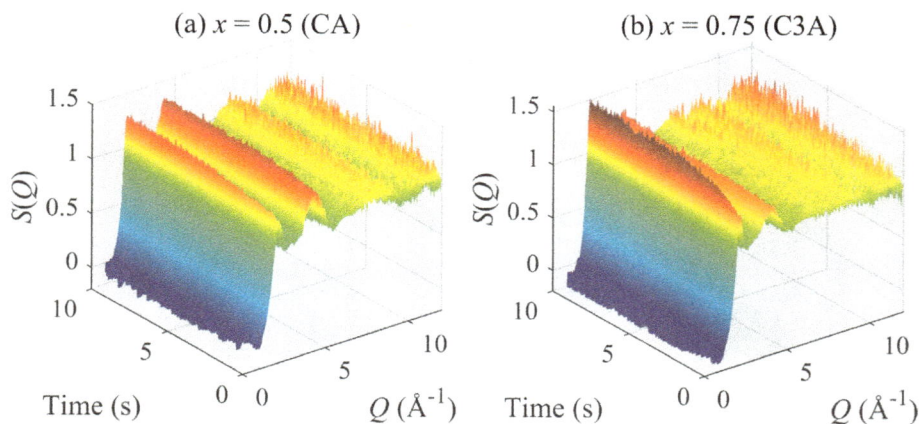

Figure 20. Time resolved synchrotron XRD measurements of the structure factors $S(q)$ of aerodynamically levitated liquid **(a)** CA and **(b)** C3A during glass formation with 30 ms acquisition times. (Reprinted from Drewitt et al. 2019 doi:10.1088/1742-5468/ab47fc © SISSA Medialab Srl. Reproduced by permission of IOP Publishing. All rights reserved).

Aluminosilicate melts

As for aluminates, aluminosilicate melts have high melting points and only a few diffraction studies have been reported, mainly using laser heating with aerodynamic levitation at neutron sources (Jakse et al. 2012, Hennet et al. 2016, Florian et al. 2018) or levitation inside the bore of a NMR spectrometer (Poe et al. 1992; Florian et al. 2018). Alternatively, wire furnace methods, in which a small quantity of sample is placed in a small hole (~500 μm diameter) in a Pt–Ir alloy wire and melted by resistive heating (Mysen and Frantz 1992; Richet et al. 1993; Neuville et al. 2014b), have been adopted for x-ray absorption spectroscopy (XAS) (Neuville et al. 2008b), Raman spectroscopy (Daniel et al. 1995; Neuville and Mysen 1996), XRD (Neuville et al. 2014b), and small angle x-ray scattering (SAXS) (di Genova et al. 2020) of aluminosilicate melts.

In contrast to aluminate melts, which can be stably levitated for days without significant mass loss (Drewitt et al. 2012, 2017), aerodynamically levitated laser heated silicate melts experience significant sample vaporization at high SiO_2 fractions. This leads to significant changes in composition, as experienced by *in situ* XRD measurements of liquid SiO_2 (Mei et al. 2007; Skinner et al. 2013a) and NDIS measurements of $CaSiO_3$ melt (Skinner et al. 2012a) for which sample mass loss is particularly detrimental due to the requirement for the different isotopically enriched samples to have identical composition. Thus, neutron diffraction experiments of aluminosilicate melts, which require counting times of several hours, have been limited to compositions up to about 33% of silica in melts containing CaO (Hennet et al. 2016) and 42% in melts containing SrO (Florian et al. 2018).

Figure 21a shows a comparison of the neutron structure factors measured for a range of calcium aluminosilicates melts along the charge compensator line together with the corresponding glass. On melting, the FSDP becomes less distinct with lower intensity and increased broadening, which is indicative of increased disorder in the organization of polyhedra on medium range length scales. This augmentation of disorder is also evident in the broadening and reduction in structure of the peaks over the full Q-range. Similar changes are observed for other CaO/Al_2O_3 ratios (Hennet et al. 2016). Along the join $R = 1$, this disorder is mainly related to an increase of the proportion of AlO_5, which is apparent from the asymmetric high-r broadening of the first peak in $G(r)$ (Fig. 21b), related to a slight increase of the Al–O distance (Drewitt et al. 2012).

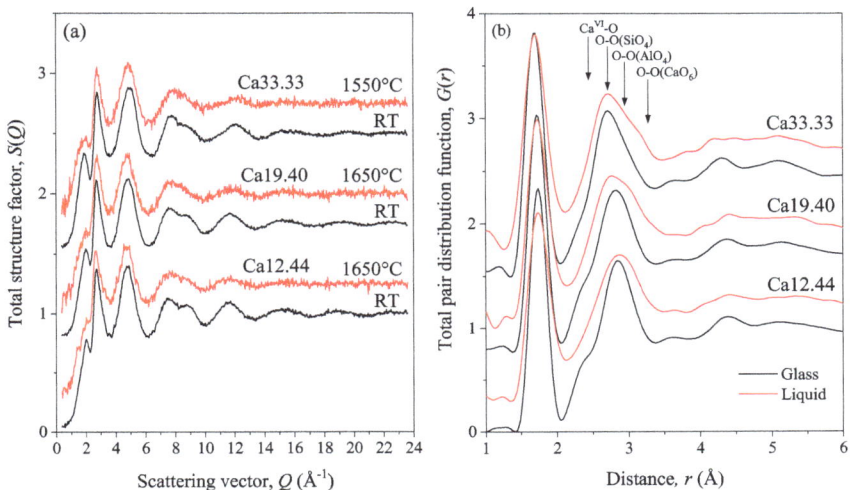

Figure 21. Total structure factors $S(Q)$ of some CAS glasses measured using neutron scattering **(a)** and corresponding total pair distribution functions **(b)**. The liquid temperatures are reported in the figure. (Data from Hennet et al. 2016).

As previously observed for glasses, the CaO correlation, which was largely overlapped with the O–O correlations belonging to SiO_4 tetrahedra, is now fully overlapped at high temperature complicating the interpretation of the measurements. Thus, molecular dynamics simulations have been employed to provide more insight on the liquid structure, revealing the AlO_5 units are accompanied by increased proportions of NBO and oxygen triclusters (Jakse et al. 2012).

The MD simulations reveal that the addition of silica to aluminate compositions reduces the quantity of AlO_5 present from ~18% in Ca0.50 to ~10% in Ca12.44 and Ca19.40 melts. These melts contain ~14% OAl_3 triclusters and ~12% NBO. The simulations reveal a continuous development in the AlO_4 network and corresponding reduction in AlO_5 units, oxygen triclusters, and NBOs during supercooling to a glass. Calcium occupies a distribution of coordination sites ranging from 4 to 8 with an average Ca–O coordination number of ~6. The potentials used to describe this structural environment also provide a relatively accurate model of the dynamical properties of these melts (Bouhadja et al. 2013).

Neutron diffraction measurements of $SrO–Al_2O_3–SiO_2$ melts reveal similar SRO associated with the Al and Si atoms as for calcium aluminates with an increase in the fraction of AlO_5 units in the melts estimated at 5–6% (Florian et al. 2018). As for the glasses, it is difficult to extract information from the SrO contribution, which is overlapped by O–O correlations arising from Si–O and Al–O tetrahedra and highly coordinated Sr–O polyhedra (cf. Fig. 11). By looking at the possible O–O correlations attributed to these Sr–O polyhedra, the Sr–O coordination number was estimated to be ~8 for the join $R = 3$ and somewhat less for $R = 1$ (Florian et al. 2018).

The *in situ* high temperature NMR experiments (Florian et al. 2018) provide more information on the network structure. With increasing ratio $X = Si/(Al+Si)$, along the $R = 1$ join, the network composed mainly from AlO_4 units at low Si content is gradually replaced by a network of SiO_4 tetrahedra. For $X > 0.7$, the same behavior is observed along the $R = 3$ join. However, below this ratio, AlO_4 units contain NBOs which break up the aluminosilicate network. Consequently, for a given silica content, the addition of SrO reduces the medium range order (smaller ring sizes for $R = 3$ than for $R = 1$). For per-aluminous compositions, large amounts of 5- and 6-fold coordinated aluminum are found in the melts. Different mechanisms than those observed along the joins $R = 1$ and 3, especially at low silica content could also suggest the presence of oxygen triclusters.

Iron silicate melts and glasses

As the fourth most abundant element in the Earth's crust, iron plays an important role due to its high mass having a strong influence on the density and viscosity of natural volcanic magmas. Iron adopts a dual character as both a network former and modifier in silicate glasses. This behavior is modulated by the coexistence of ferrous (Fe^{2+}) and ferric (Fe^{3+}) iron cations which adopt different structural roles. The redox ratio $Fe^{3+}/\Sigma Fe$ is strongly influenced by chemical composition, temperature, and oxygen fugacity (Wilke 2005). This mixed valency complicates the interpretation of the local coordination environment in structural measurements. In general, Fe^{3+} is predominantly a network former with 4-fold coordination in silicate glasses (Burkhard 2000), although this is highly dependent on composition and oxidation state, and higher coordinated Fe^{3+} sites have been observed (Virgo and Mysen 1985; Hannoyer et al. 1992; Keppler 1992; Kim and Lee 2020). In early models of iron silicate glasses, Fe^{2+} was assumed to occupy 6-fold coordinated sites as in corresponding silicate minerals (Boon and Fyfe 1972; Binsted et al. 1986). However, a wide distribution of 4-, 5-, and 6-fold coordination with oxygen has been reported in synthetic and natural iron silicate glasses by using x-ray, Mössbauer and optical absorption spectroscopies (Calas and Petiau 1983; Virgo and Mysen 1985; Binsted et al. 1986; Iwamoto et al. 1987; Dingwell and Virgo 1988; Hannoyer et al. 1992; Keppler 1992; Wang et al. 1993; Rossano et al. 1999, 2000a,b; Wu et al. 1999; Galoisy et al. 2001; Giuli et al. 2002; Farges et al. 2004; Wilke et al. 2004, 2007;

Jackson et al. 2005; Mysen 2006; Bingham et al. 2007; Giuli et al. 2011, 2012; Nyrow et al. 2014; Zhang et al. 2016, 2018; Alderman et al. 2017a; Cottrell et al. 2018 Nayak et al. 2019) and x-ray and neutron diffraction (Johnson et al. 1999; Holland et al. 1999; Weigel et al. 2006, 2008a,b; Drewitt et al. 2013; Wright et al. 2014).

More limited direct information is available on the local coordination environment of iron in silicate melts by comparison with the quenched glass. Using a conventional XRD source, Waseda et al. were the first to report the structure of iron silicate melts suggesting a reduction in Fe–O coordination from 6- to 4-fold with increasing SiO_2 fraction (Waseda and Toguri 1978; Waseda et al. 1980). However, serious inconsistencies have been found in this data including the assignment of unrealistically small Fe–Si bond distances and Fe–O–Si bond angles (Jackson et al. 1993; Drewitt et al. 2013). High temperature XAS measurements have indicated Fe^{2+} residing solely in 4-fold sites in alkali silicate (Waychunas et al. 1988) melts, and a complete conversion from 6-fold Fe^{2+} sites in crystalline fayalite (Fe_2SiO_4) to 4-fold in the melt (Jackson et al. 1993). A subsequent XAS study found only slightly higher quantities of low coordinated Fe^{2+} in reduced silicate melts (Wilke et al. 2007). This is confirmed by *in situ* synchrotron XRD measurements of laser heated silicate liquids, including Fe_2SiO_4 and $FeSiO_3$, levitated on an Ar gas stream in air (Drewitt et al. 2013). Here, two distinct Fe–O peaks are resolved (Fig. 22) with bond lengths 1.93 and 2.20 Å corresponding to FeO_4 and FeO_6 units, respectively, with average Fe–O coordination numbers of 4.8 (Fe_2SiO_4) and 5.1 ($FeSiO_3$). Both Fe–O peaks are also apparent in the measurements of a basalt glass and melt. Although only two Fe–O peaks are

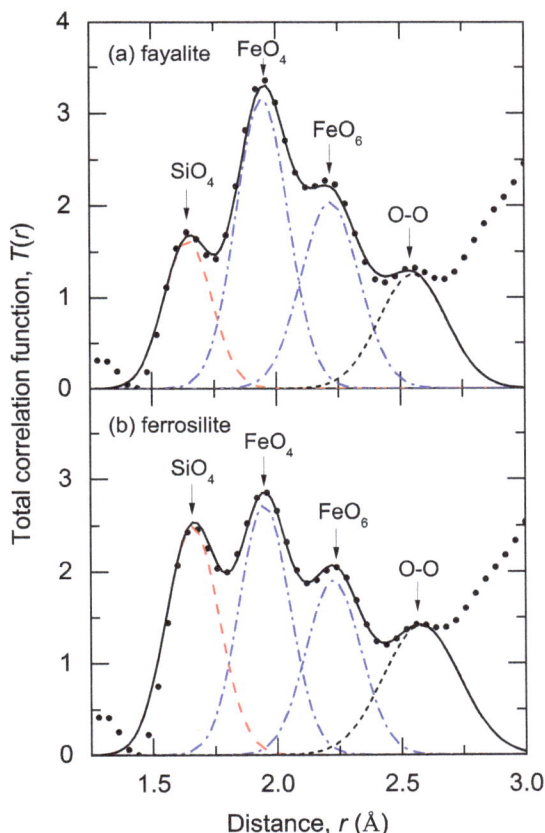

Figure 22. Gaussian fits to the main peaks in $T(r) = 4\pi r \rho_0 G(r)$ from synchrotron XRD measurements by Drewitt et al. (2013) for **(a)** liquid fayalite (Fe_2SiO_4) and **(b)** liquid ferrosilite ($FeSiO_3$). The **black circles** are the measured data and the **solid black curves** are the superposition of the fitted Gaussian functions. Reprinted figure with permission from Drewitt et al. Physical Review B 87:224201 (2013) http://dx.doi.org/10.1103/PhysRevB.87.224201. Copyright (2013) by the American Physical Society.

apparent, the results do not exclude the presence of up to ~10–25% FeO_5 units in the iron silicate melts and glasses studied (Drewitt et al. 2013). The coexistence of 4-, 5-, and 6-fold coordinated iron in silicate melts has important implications for the partitioning behaviour of iron in natural melts or transport properties of magmas. The ratio of these different units is affected by oxygen fugacity, with *in situ* Fe *K*-edge XANES and synchrotron XRD measurements of levitated fayalitic melts revealing average Fe–O coordination numbers increasing from 4.4 in the reduced melt to 4.7 at higher $Fe^{3+}/\Sigma Fe$ (Alderman et al. 2017b).

Raman spectroscopy has also been used to interpret the structural role of iron in silicate glasses and melts and link the redox equilibrium to the silicate structure (Mysen et al. 1980, 1984, 1985; Fox et al. 1982; Wang et al. 1993, 1995), and more recently to quantify the Fe redox ratio (Magnien et al. 2006, 2008; Roskosz et al. 2008; Di Muro et al. 2009; Cochain et al. 2012; Di Genova et al. 2015, 2016, 2017; Le Losq et al. 2019b). Raman spectra of iron silicate glasses and melts vary systematically with varying $Fe^{3+}/\Sigma Fe$ ratio due to the different roles of Fe^{2+} and Fe^{3+} influencing the network structure and Q^n speciation in different ways (Le Losq et al. 2019b). Samples with high $Fe^{3+}/\Sigma Fe$ ratio have characteristically strong intensity at ~900 cm^{-1} assigned to Fe^{3+}-O stretching vibrations (Wang et al. 1995; Magnien et al. 2004, 2006, 2008; Di Muro et al. 2009; Cochain et al. 2012; Le Losq et al. 2019b). This is illustrated in Figure 23 for an iron-pyroxene glass composition with variation of the redox state $Fe^{3+}/\Sigma Fe$ from 0.09 to 0.97. Here the intensity of the Fe^{3+}–O stretching band at 910 cm^{-1} increases progressively with increasing Fe^{3+} fraction consistent with the formation of tetrahedral ferric iron, as indicated from the Fe pre-edge XANES (Wilke et al. 2001).

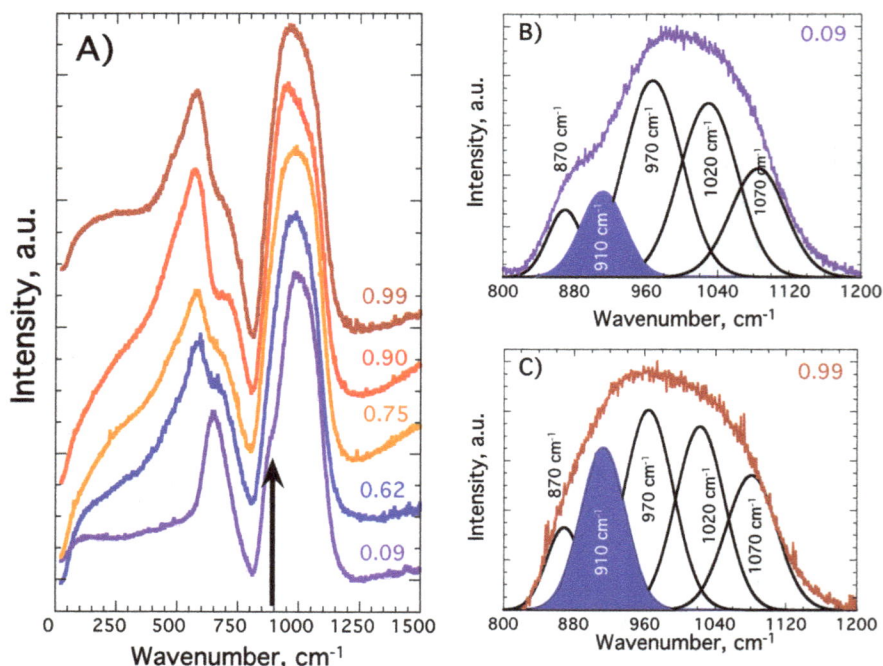

Figure 23. a) Raman spectra at room temperature for a series of iron-pyroxene glasses (redrafted from Magnien et al. 2006). The $Fe^{3+}/\Sigma Fe$ redox ratio is indicated and the spectra are displaced vertically for clarity. The spectra were normalized to the maximum intensity of the spectra corrected from the excitation lines (see Neuville et al. 2014). **b)** Deconvolution of Raman spectrum of pyroxene glass with redox ratio $Fe^{3+}/\Sigma Fe = 0.09$, and **c)** $Fe^{3+}/\Sigma Fe = 0.99$.

Glasses and melts at high pressure

Pressure can lead to considerable reorganization of SRO and MRO in glasses and melts. The structural response of melts to high p determines the density and dynamical properties of natural magmas in deep planetary interiors. Understanding the link between structure and properties of melts at high-p is, therefore, integral to understanding a range of problems including the distribution and causes of chemical alteration (metasomatism) of the mantle due to upward or lateral migration of melts, large scale differentiation from an early magma ocean, and mantle melting at mid-ocean ridges and subduction zones. Gravity driven melt transportation is proportional to melt mobility, which is dependent upon the viscosity of the melt and the density contrast between the melt and solid mantle. Early equation of state data for melts indicated that basaltic magmas are more compressible than mantle minerals leading to the possible existence of a melt–crystal density crossover at depth (Stolper et al. 1981; Rigden et al. 1984; Agee and Walker 1988).

Many network glasses experience permanent densification upon recovery from cold (i.e. room temperature) compression associated with irreversible changes due to a collapse in MRO and void space, with the local coordination environments typically returning to their ambient structures (Bridgman and Šimon 1953; Grimsditch 1984; Polian and Grimsditch 1990; Susman et al. 1991; Ishihara et al. 1999; Inamura et al. 2001; Trachenko and Dove 2002; Sampath et al. 2003; Huang and Kieffer 2004a,b; Champagnon et al. 2008; Rouxel et al. 2008). Heat treatment during compression promotes rebonding and further permanent densification of recovered samples (Polian and Grimsditch 1990; Ishihara et al. 1999; Trachenko and Dove 2002; Inamura et al. 2004; Martinet et al. 2015), where highly coordinated local environments reminiscent of the high-p state are recoverable to ambient conditions in glasses equilibrated at high-p–T above the liquidus (Stebbins and McMillan 1989; Li et al. 1995; Yarger et al. 1995; Ohtani et al. 1985; Xue et al. 1989, 1991; Stebbins and Poe 1999; Allwardt et al. 2004; Kelsey et al. 2009) or at high-p close to the glass transition (Allwardt et al. 2005a,b; Gaudio et al. 2008, 2015; Guerette et al. 2015; Stebbins and Bista 2019). Nevertheless, in order to unequivocally capture the evolution in SRO and MRO in glasses and melts at high p there is no substitute for *in situ* experiments, and such studies are clearly essential for melts which do not quench to a glass.

Recent advances in experimental and analytical techniques now allow for the measurement of the structure and properties of glasses and melts to be measured *in situ* at high-p. Pressure cells used for high-p research can be classified into two main types: large volume devices, such as a piston cylinder devices and multi anvil (MA) apparatus (Ito 2007), or the much smaller volume diamond anvil cell (DAC) (Mao and Mao 2007).

Large volume devices are typically used for *ex situ* phase equilibrium experiments under simultaneous high-p–T conditions using resistive heating. Piston cylinder devices can achieve moderate p up to 5–6 GPa and 1800 °C (Holloway and Wood 1988), corresponding to upper mantle or lunar core conditions. As the name suggests, the MA apparatus utilizes multiple anvils to compress a sample from four or more directions to generate conditions of up to 30 GPa and 3000 °C (Holloway and Wood 1988), corresponding to mid to lower mantle conditions. Use of sintered diamond MA cubes, in place of standard tungsten carbide cubes, allows p as high as 90 GPa to be generated (Zhai and Ito 2011), providing access to deep lower mantle conditions. MA devices with x-ray transparent apertures are now commonplace at most third-generation synchrotron radiation sources (Liebermann 2011) and have been exploited for *in situ* high-p measurements of the structure and properties of silicate glasses and melts (Funamori et al. 2004; Ohtani et al. 2005). However, MA devices are expensive to build and operate and sample accessibility is limited by the large-scale bulky nature of the apparatus. The Paris–Edinburgh (PE) cell is a large volume press specifically designed to optimize sample accessibility for *in situ* powder neutron diffraction experiments at p up to ~25 GPa (Besson et al. 1992; Klotz et al. 1995) where developments made within the last decade allow for high-quality *in situ* neutron diffraction measurements to be made of cold-compressed glass in the gigapascal regime

(Drewitt et al. 2010; Salmon et al. 2012; Salmon and Zeidler 2015). The PE cell has been adapted for high-p–T synchrotron XRD experiments (Besson et al. 1996; Mezouar et al. 1999; Crichton and Mezouar 2005) and used for *in situ* x-ray scattering measurements of silicate melts (Yamada et al. 2011; Sakamaki et al. 2012; Kono et al. 2014). Recently, a double-stage large volume cell was developed by incorporating a pair of secondary diamond anvils into a primary-stage PE cell to generate pressures in the megabar (1 Mbar = 100 GPa) regime (Kono et al. 2019) and has been used for measuring structure of different glasses at ultrahigh pressure (Kono et al. 2016, 2018; Ohira et al. 2019; Shibazaki et al. 2019; Wilding et al. 2019).

The DAC generates extremely high static pressures by compressing a sample between two opposing single crystal diamond anvils (Bassett 2009). High-p experiments up to ~1 megabar (1 Mbar = 100 GPa) are routine in the DAC (Li et al. 2018), with multimegabar pressures feasible by employing beveled (Bell et al. 1984) or toroidal (Dewaele et al. 2018) anvils to provide access to static p corresponding to Earth's core and beyond, with p in the Terapascal (1000 GPa) regime achievable using secondary micro-ball nanocrystalline diamond anvils (Dubrovinsky et al. 2012; Dubrovinskaia et al. 2016). The DAC is an extremely versatile and portable device with relatively low manufacturing costs. The transparency of diamonds across the electromagnetic spectrum offers a key advantage over other high-p methods, including the ability to observe a sample *in situ* under compression using optical microscopy, spectroscopy, and x-ray scattering techniques. Laser heating techniques can also be employed to heat samples at high-p to temperatures up to ~5000 K. Dedicated laser heated diamond anvil cell (LHDAC) setups for *in situ* high-p–T structure measurements are now widely available at synchrotron sources around the world (Watanuki et al. 2001; Shen et al. 2001; Meng et al. 2006; Caldwell et al. 2007; Liermann et al. 2010; Petitgirard et al. 2014). The LHDAC method combined with synchrotron XRD has been exploited in the Earth Sciences to study metallic (Shen et al. 2004; Morard et al. 2013, 2017) and silicate (Sanloup et al. 2013b; Drewitt et al. 2015) melt structure under deep mantle and core conditions. Simultaneous high-p–T conditions may also be achieved in a resistively heated (RH)-DAC. Although the maximum attainable temperature is lower than for the LHDAC, the RHDAC has the distinct advantage of providing homogeneous heating over the whole sample allowing the measurement of larger wholly molten samples without contamination from thermal insulation media or laser coupling material (de Grouchy et al. 2017; Louvel et al. 2020; Drewitt et al. 2020).

Regardless of whether large volume or DAC apparatus are used, measuring glass or melt structure *in situ* at high-p requires accurate characterization of the p-dependent background scattering from the cell to extract a precise measurement of the diffuse glass or melt signal (Shen et al. 2003; Drewitt et al. 2010). This background scattering can be reduced using spatial collimation (Mezouar et al. 2002) or by reducing the fraction of non-sample components within the path of the incident beam: e.g. much of the x-ray Compton scattering from DAC anvils can be eliminated by using perforated diamonds (Soignard et al. 2010).

Pressure induced modifications in MRO and SRO in SiO_2 and GeO_2 glass

In situ high-p spectroscopy (Grimsditch 1984; Hemley et al. 1986; Williams and Jeanloz 1988; Zha et al. 1994; Polsky et al. 1999; Lin et al. 2007; Champagnon et al. 2008; Murakami and Bass 2010), and x-ray (Meade et al. 1992; Inamura et al. 2004; Sato and Funamori 2008, 2010; Benmore et al. 2010b; Shen et al. 2011; Prescher et al. 2017; Kono et al. 2020) and neutron (Zeidler et al. 2014a) diffraction have been used extensively to determine the p-induced structural modifications of pure SiO_2 glass. Extensive spectroscopy (Itié et al. 1989; Durben and Wolf 1991; Polsky et al. 1999; Hong et al. 2007; Vaccari et al. 2009; Baldini et al. 2010; Dong et al. 2017; Spiekermann et al. 2019), and x-ray (Guthrie et al. 2004; Hong et al. 2007, 2014; Mei et al. 2010; Kono et al. 2016) and neutron (Guthrie et al. 2004; Drewitt et al. 2010; Salmon et al. 2012; Wezka et al. 2012) diffraction measurements reveal GeO_2 glass undergoes analogous p-induced transformations to SiO_2 but over a much lower-p regime.

There is firm consensus from these studies for the operation of two key densification mechanisms in pure SiO_2 and GeO_2 glass starting with a steady collapse in MRO associated with the closure of open ring structures which is accompanied at higher pressures by a continuous transformation from a tetrahedral to octahedral glass network. The changes in MRO are evident in the reciprocal space diffraction patterns from a shift in position of the FSDP at 1.55 Å^{-1} at ambient-p to higher Q values with increasing p (Fig. 24). This change is accompanied by the development of a second peak at ~3 Å^{-1}. The emergence of a new peak observed at ~3 Å^{-1} in $S(Q)$ for SiO_2 and GeO_2 glass at high-p is observed for many other types of oxide glasses and melt compositions including calcium aluminate ($CaAl_2O_4$) and anorthite ($CaAl_2Si_2O_8$) (Drewitt et al. 2015), forsterite (Mg_2SiO_4) (Benmore et al. 2011; Adjaoud et al. 2008), enstatite and wollastonite (Salmon et al. 2019), jadeite ($NaAlSi_2O_6$), albite ($NaAlSi_2O_6$) and diopside ($CaMgSi_2O_6$) (Sakamaki et al. 2014a,b). In SiO_2 glass, this peak has been attributed to the breakdown of MRO and emergence of SiO_6 octahedra (Meade et al. 1992; Benmore et al. 2010b). More generally, this peak is consistent with the development of short-range topological ordering arising from a more densely packed structure (Elliot 1995; Salmon et al. 2005, 2006; Sakamaki et al. 2014b; Drewitt et al. 2015). Advances made in neutron diffraction methodologies for high precision *in situ* measurements of glasses at high p (Drewitt et al. 2010) have enabled NDIS measurements of GeO_2 glass to directly measure the individual Ge and O environments (Wezka et al. 2012). From these measurements, the p-dependence of the mean Ge–O–Ge bond angle was found, which reduces with increasing p consistent with compaction of the open corner shared tetrahedral network structure.

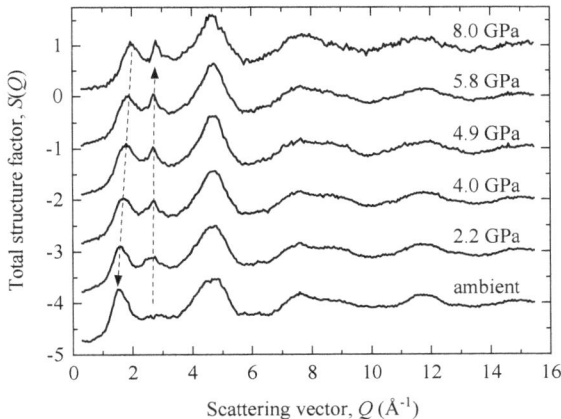

Figure 24. Total structure factors $S(Q)$ for GeO_2 glass as measured by in situ neutron diffraction at high pressure in the PE cell from ambient to 8.0 GPa. The changes in intensity and position of the FSDP and principal peak are indicated by the arrows. (Data from Drewitt et al. 2010).

The transformation in SRO from a tetrahedral to octahedral glass network occurs over the p range of ~10 to ~40 GPa (SiO_2) and ~5 to 30 GPa (GeO_2). The onset of coordination change in network forming structural motifs for a wide range of oxide glasses can be rationalised in terms of the oxygen packing fraction (Wang et al. 2014; Zeidler et al. 2014b; Zeidler and Salmon 2016; Salmon 2018). The upper limit of stability for tetrahedral motifs in both SiO_2 and GeO_2 glass occurs at and oxygen packing fraction of ~0.58, consistent with random loose packing of hard spheres. The conversion to an octahedral network is largely completed at an oxygen packing fraction of ~0.64 which corresponds to random close packing of hard spheres (Zeidler et al. 2014b). Molecular dynamics simulations reveal that GeO_5 and SiO_5 units act as important intermediaries on transformation towards an octahedral glass, where the coordination changes primarily from 4 ⊠ 5 and 5 ⊠ 6 (Wezka et al. 2012; Zeidler et al. 2014a).

This is supported by the measured mean O–Ge–O bond angle from NDIS (Wezka et al. 2012) and can be understood in terms of a 'zipper' mechanism for ring closure developed for SiO_2 glass in which a tetrahedral SiO_4 unit belonging to a ring structure forms a new bond with a bridging oxygen within the ring, leading to the formation of a distorted square pyramidal SiO_5 unit and a threefold coordinated oxygen atom (Wezka et al. 2012; Zeidler et al. 2014a). This demonstrates a striking interrelation between the changes experienced in SRO and MRO which is underscored by a correlation between the position of the FSDP in $S(Q)$, relative to its position at ambient-p, and the onset of coordination change (Zeidler and Salmon 2016; Salmon 2018).

Some studies using *in situ* XRD (Sato and Funamori 2010), Brillouin spectroscopy (Murakami and Bass 2010) and $K\beta''$ x-ray emission spectroscopy (Spiekermann et al. 2019) suggest a plateau in octahedral coordination persisting in SiO_2 and GeO_2 glass beyond 100 GPa. However, molecular dynamics simulations of both SiO_2 and GeO_2 glass (Brazhkin et al. 2011) and SiO_2 melt (Karki et al. 2007) indicate a continuous increase in coordination number > 6 at 100 GPa. Recent *in situ* XRD measurements for SiO_2 in a DAC up to 172 GPa (Prescher et al. 2017) and in a double-stage large-volume cell to 120 GPa (Kono et al. 2020), as well as GeO_2 glass up to 91.7 GPa (Kono et al. 2016), appear to confirm this.

Silicate melt structure at high pressure

In situ diffraction measurements of the structure of silicate melts at high p have advanced significantly in the last decade. Exploratory *in situ* synchrotron XRD measurements of molten silicates were first reported for $CaSiO_3$ and $MgSiO_3$ melts up to 6 GPa using a cubic multi-anvil device (Funamori et al. 2004). In the last decade, the structure of a range of silicate melts have been measured by diffraction using large volume pressure cells at p up to ~10 GPa including forsterite–enstatite (Mg_2SiO_4–$MgSiO_3$) (Yamada et al. 2007, 2011), albite ($NaAlSi_3O_8$) (Yamada et al. 2011), jadeite ($NaAlSi_2O_6$) (Sakamaki et al. 2012, 2014a; Wang et al. 2014), fayalite (Fe_2SiO_4) (Sanloup 2013a), diopside ($CaMgSi_2O_6$) (Wang et al. 2014), and basaltic melts (Sakamaki et al. 2013; Crépisson et al. 2014).

A compilation of structure, density, and viscosity data of silicate melts at high pressure reveals a reduction in isothermal viscosity of polymerized melts up to ~3–5 GPa where it encounters a turnover to a normal (positive) p-dependence (Schmelzer et al. 2005; Wang et al. 2014). Wang et al. (2014) attribute this viscosity turnover to the tetrahedral packing limit, below which the structure is highly compressible with transformations dominated by changes in network connectivity and the collapse of MRO resulting in a reduction in inter-tetrahedral bond angle, and above which the structural response is dominated by the increase in nearest neighbor atom-atom coordination numbers. Structural measurements of molten fayalite reveal an increase in average Fe–O coordination from 4.8 GPa at ambient-p (Drewitt et al. 2013) to 7.2 at 7.5 GPa (Sanloup et al. 2013a). It is suggested that this rapid increase in Fe coordination is responsible for the higher compressibility of fayalite melt compared to molten Mg-rich San Carlos olivine (Guillot and Sator 2007, Sanloup et al. 2013a).

To date, only two studies have been reported for the *in situ* structure of silicate melts beyond 10 GPa. Both studies utilise the LHDAC combined with synchrotron XRD, with molten basalt reported to lower mantle conditions up to 60 GPa using Yb-fibre laser heating (Sanloup et al. 2013b) and liquid anorthite reported to 32.5 GPa using CO_2 laser heating (Drewitt et al. 2015). The measurements of molten basalt reveal an increase in Si–O coordination change from 4 at ambient-p (Drewitt et al. 2013) to 6 at 35 GPa, consistent with the results for pure SiO_2 glass discussed above. By normalising the x-ray intensities using an iterative procedure that minimises the unphysical low-r oscillations below the first interatomic bond distance (Eggert et al. 2002) a converged solution for melt density was determined directly from the liquid diffraction measurements. The results reveal that molten basalt experiences a density crossover with the solid mantle and may be gravitationally stable in the deep lower mantle

(Sanloup et al. 2013b). *In situ* density measurements of silicate glasses using x-ray absorption contrast techniques similarly predict the accumulation of Fe-rich silicate melts above the core mantle boundary (Petitgirard et al. 2015, 2017), consistent the formation of a dense basal magma ocean in the Earth's early mantle (Labrosse et al. 2007).

Al coordination change at high pressure

NMR and x-ray spectroscopy measurements of permanently densified aluminosilicate glasses, synthesised from high-p melts (up to 12 GPa) in multi anvil apparatus and recovered to ambient conditions, revealed the development of highly coordinated AlO_5 and AlO_6 populations occur at far lower pressures than changes observed for Si–O coordination and as such has a greater influence on controlling melt mobility and magmatic processes in the shallow upper mantle (Li et al. 1995; Yarger et al. 1995; Allwardt et al. 2005a,b; Kelsey et al. 2009). It is likely, however, that these early *ex situ* studies significantly underestimate the population of highly coordinated units recoverable from high-p due to large transient drops in pressure during temperature quench (Stebbins and Bista 2019).

Direct *in situ* synchrotron XRD measurements of Al coordination change in $CaAl_2O_4$ and $CaAl_2Si_2O_8$ (anorthite) glasses to 30 GPa reveal a continuous increase in Al–O coordination reaching an average value of ~6 by 25 GPa (Drewitt et al. 2015). The only reported *in situ* measurements of Al–O coordination change in a melt at high-p, made for liquid $CaAl_2Si_2O_8$ to 32.4 GPa using synchrotron XRD in a DAC with CO_2 laser heating, reveals the change in local Al structure is comparable to the corresponding glass with an average Al–O coordination of 5 by ~10 GPa and 6 by ~30 GPa (Drewitt et al. 2015).

The increase in Al–O coordination from 4-, to 5-, to 6-fold up to 30 GPa measured for molten anorthite is consistent with the results of MD simulations (de Koker 2010; Drewitt et al. 2015). At ambient pressure, chemical ordering dominates in the form of AlO_4 and SiO_4 tetrahedra. On initial pressurization, the relatively open network structure is compressed and destroyed (Drewitt et al. 2015). This is supported by *in situ* Raman measurements of the glass which reveal a rapid reduction in the inter-tetrahedral bond angle below 2 GPa (Moulton et al. 2019). The increase in Al–O coordination number increases almost immediately on compression with the formation of AlO_5 motifs and increasing fractions of AlO_6 units after ~5 GPa (Drewitt et al. 2015). At 10–15 GPa, the fraction of AlO_5 units reaches a maximum fraction of ~50% and are increasingly replaced by AlO_6 units to become the dominant feature at $\geq$ 30 GPa (Fig. 25a). These changes in coordination are accompanied by a reduction in NBO fraction to only a few per cent by 10–15 GPa (Drewitt et al. 2015), consistent with Raman spectroscopy measurements which indicate that depolymerized Q^2 and Q^3 AlO_4 units are the first to transform to higher coordination (Muniz et al. 2016). The Si–O coordination number follows a similar behaviour as for pure SiO_2 glass. The Ca–O coordination begins to increase immediately on compression, rising continuously to an average value of ~10 at 30 GPa. The increase in highly coordinated Al and Ca units is accompanied by a development in polyhedral connectivity. At 10 GPa, ~95% Ca and 50% Al atoms belong to single large clusters, increasing to 100% Ca and 90% Al atoms by 30 GPa (Fig. 25b).

At higher pressures, recent *in situ* synchrotron XRD measurements of an aluminosilicate glass using a double-stage PE cell suggest the Al–O coordination number remains constant at ~6 up to 110 GPa (Ohira et al. 2019). However, MD simulations indicate a continuous increase in Al–O coordination number in aluminosilicate glasses and melts, approaching values of ~7 at 100 GPa (de Koker 2010; Bajgain et al. 2015; Ghosh and Karki 2018). We note that it is not possible to separate the Si–O and Al–O nearest neighbour contributions in the total pair distribution functions of aluminosilicate glass measured by Ohira et al. (2019). This contrasts with the measurements reported by Drewitt et al. (2015) where the Al–O coordination environment was quantified directly from the measurement of silica-free $CaAl_2O_4$ glass.

(a) AlO$_x$ coordination

(b) Clustering of edge- and face-sharing AlO$_x$ polyhedra

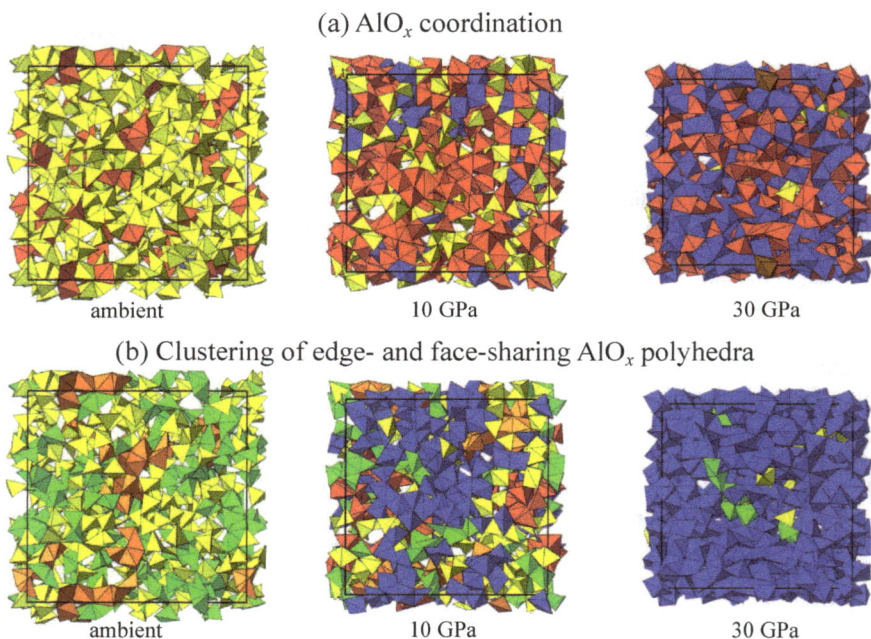

Figure 25. Snapshots from MD simulations of liquid CaAl$_2$O$_4$ at 2500 K at ambient, 10 GPa, and 30 GPa. **(a)** AlO$_x$ coordination with $x = 4$ (**yellow**), 5 (**red**), 6 (**blue**), and 7 (**brown**). **(b)** clusters of > 100 (**blue**), 10–100 (**red**), 5–9 (**orange**), 2–4 (**green**), and 1 (**yellow**) of edge- and face-sharing AlO$_x$ polyhedra. Reprinted figure with permission from Drewitt et al. Journal of Physics: Condensed Matter 27:105103 (2015) http://dx.doi.org/10.1088/0953-8984/27/10/105103 Copyright (2015) by the Institute of Physics.

SUMMARY AND FUTURE PERSPECTIVES

In this chapter we demonstrate the high complementarity between x-ray or neutron diffraction and Raman spectroscopy techniques for measuring SRO and MRO in geologically relevant silicate and aluminate glasses and their melts. Nevertheless, despite decades of research important questions remain unresolved. For example, the origin of the Boson peak, a ubiquitous feature in Raman spectra of glass, remains a mystery with competing theories invoking both MRO and SRO structure (Greaves and Sen 2007). Secondly, understanding the incorporation of modifier cations into aluminosilicate glasses and melts is important for understanding the viscous behavior of natural melts. While cationic percolation channels appear to be a general feature of MRO in alkali silicates (Greaves 1985; Meyer et al. 2004; Du and Cormack 2004; Cormier and Cuello 2011; Le Losq et al. 2017), more research is needed to elucidate if these cation diffusion channels are a general feature of alkaline-earth and transition-metal aluminosilicate melts and to understand their influence on the dynamic behavior of natural melts. This work will benefit from the further application of element selective diffraction techniques such as NDIS of glasses and their high-T melts (Drewitt et al. 2012, 2017) combined with advanced molecular dynamics simulation techniques (Jahn 2022, this volume).

Recent technological developments have facilitated diffraction and spectroscopy measurements of aluminosilicate glasses and their melts to be made *in situ* under extreme high-T and/or -p conditions experienced in planetary interiors. *In situ* high pressure measurements of silicate glasses and their germanate analogues have revealed a collapse in MRO associated with ring structures at ambient conditions followed by a development in SRO associated with increasing local coordination number. These changes are intimately linked to magmatic transport properties at depth, with an interplay between their viscosity and compressibility arising from collapsing MRO and subsequent increase in SRO at depth (Wang et al. 2013). There exists a striking interrelation between the destruction of MRO via ring closure events and the development of highly coordinated network forming units at high-p, underscored by the correlation between the relative position of the FSDP in $S(Q)$ measured at high-p and the onset of coordination change (Zeidler and Salmon 2016; Salmon 2018). However, while these measurements have improved our understanding of the structural modifications in simple silicate glasses at high p, much less information is available on other cations. Also, although important, information obtained on glass structure is not always representative of melts in their natural liquid state (Wilding et al. 2010; Drewitt et al. 2012) and as such *in situ* high-p–T measurements of the melts are highly desirable.

The last decade has seen a significant improvement in high-p–T technology and a wide range of *in situ* measurements of silicate melts have been reported (Kono et al. 2014). Nevertheless, despite these developments only two measurements of the *in situ* structure of silicate melts have been reported beyond 10 GPa, both utilising the LHDAC and synchrotron XRD (Sanloup et al. 2013b; Drewitt et al. 2015). The next decade is, therefore, likely to see an increase in LHDAC or RHDAC (Louvel et al. 2020) experiments applied to melts assisted by the increased flux and micro/nano focus capabilities of next generation high-energy synchrotron radiation facilities. The LHDAC methods used in XRD studies could also become more routinely adopted for Raman spectroscopy measurements to monitor changes in MRO in aluminosilicate melts at $T > 2000$ K by applying methods such as Coherent Anti-Stokes Raman Scattering Microscopy (CARS) (Cheng and Xie 2004; Baer and Yoo 2005) to overcome the deleterious effect of blackbody radiation on the Raman scattering signal. Finally, future developments in PE cell technology could include neutron transparent heating assemblies so that NDIS can be employed at high-p–T to reveal unprecedented resolution of the short and medium range cationic ordering in geological melts at high-p.

ACKNOWLEDGEMENTS

We thank Grant Henderson and an anonymous reviewer for their helpful comments on the original manuscript. JD acknowledges funding from NERC (NE/P002951/1) and EPSRC (EP/V001736/1).

REFERENCES

Aasland S, McMillan PF (1994) Density-driven liquid–liquid phase separation in the system Al_2O_3–Y_2O_3. Nature 369:633–636

Adjaoud O, Steinle-Neumann G, Jahn S (2008) Mg_2SiO_4 liquid under high pressure from molecular dynamics. Chem Geol 256:185–192

Alderman OLG, Hannon AC, Holland D, Feller S, Lehr G, Vitale AJ, Hoppe U, Zimmerman M, Watenphul A (2013) Lone-pair distribution and plumbite network formation in high lead silicate glass, $80PbO$–$20SiO_2$ Phys Chem Chem Phys 15:8506–8519

Alderman OLG, Wilding MC, Tamalonis A, Sendelbach S, Heald SM, Benmore CJ, Johnson CE, Johnson JA, Hah H-Y, Weber JKR (2017a) Iron K-edge x-ray absorption near-edge structure spectroscopy of aerodynamically levitated silicate melts and glasses. Chem Geol 453:169–185

Alderman OLG, Lazareva L, Wilding MC, Benmore CJ, Heald SM, Johnson CE, Johnson JA, Hah H-Y, Sendelbach S, Tamalonis A, Skinner LB, Parise JB, Weber JKR (2017b) Local structural variation with oxygen fugacity in Fe_2SiO_{4+x} fayalitic iron silicate melts. Geochim Cosmochim Acta 203:15–36

Allwardt JR, Schmidt BC, Stebbins JF (2004) Structural mechanisms of compression and decompression in high-pressure $K_2Si_4O_9$ glasses: an investigation utilizing Raman and NMR spectroscopy of glasses and crystalline materials. Chem Geol 213:137–151

Allwardt JR, Stebbins JF, Schmidt BC, Frost DJ, Withers AC, Hirschmann MM (2005) Aluminum coordination and the densification of high-pressure aluminosilicate glasses. Am Mineral 90:1218–1222

Allwardt JR, Poe BT, Stebbins JF (2005b) The effect of fictive temperature on Al coordination in high-pressure (10 GPa) sodium aluminosilicate glasses. Am Mineral 90:1453–1457

Almeida RM, Santos LF (2015) Raman spectroscopy of glasses. *In:* Modern Glass Characterisation, Affatigato M (Ed), John Wiley and Sons, USA, p 158–240

Agee CB, Walker D (1988) Static compression and olivine flotation in ultrabasic silicate liquid. J Geophys Res 93:3437–3449

Angell CA (1991) Relaxation in liquids, polymers and plastic crystals—strong/fragile patterns and problems. J Non-Cryst Solids 131–133:13–31

Angell CA (1995) Formation of glasses from liquids and biopolymers. Science 267:1924–1935

Auzende A-L, Gillot J, Coquet A, Hennet L, Ona-Nguema G, Bonnin D, Esteve I, Roskosz M, Fiquet G (2011) Synthesis of amorphous MgO-rich peridotitic starting material for laser-heated diamond anvil cell experiments—application to iron partitioning in the mantle. High Press Res 31:199–213

Baer BJ, Yoo C-S (2005) Laser heated high density fluids probed by coherent anti-Stokes Raman spectroscopy. Rev Sci Instr 76:013907

Bajgain S, Ghosh DB, Karki BB (2015) Structure and density of basaltic melts at mantle conditions from first-principles simulations. Nature Commun 6:8578

Baldini M, Aquilanti G, Mao H-k, Yang W, Shen G, Pascarelli S, Mao WL (2010) High-pressure EXAFS study of vitreous GeO_2 up to 44 GPa. 81:024201

Baggioli M, Zaccone A (2019) Universal origin of boson peak vibrational anomalies in ordered crystals and in amorphous materials. Phys Rev Lett 122:145501

Bancroft GM, Nesbitt HW, Henderson GS, O'Shaughnessy C, Withers AC, Neuville DR (2018) Lorentzian dominated lineshapes and linewidths for Raman symmetric stretch peaks (800–1200 cm^{-1}) in Q^n ($n = 1$–3) species of alkali silicate glasses/melts. J Non-Cryst Solids 484:72–83

Barnes AC (2015) A comparison of structural models of $Tb_3Al_5O_{12}$ and $Nd_3Al_5O_{12}$ glasses obtained when using x-ray data alone and when x-ray and neutron data are combined. Z Phys Chem 230:387–415

Barnes AC, Skinner LB, Salmon PS, Bytchkov A, Pozdnyakova I, Farmer TO, Fischer HE (2009) Liquid–liquid phase transition in supercooled yttria–alumina. Phys Rev Lett 103:225702

Barney ER, Hannon AC, Holland D, Winslow D, Rijal B, Affatigato M, Feller SA (2007) Structural studies of lead aluminate glasses. J Non-Cryst Solids 353:1741–1747

Barrio RA, Galeener FL, Martínez E, Elliot RJ (1993) Regular ring dynamics in AX_2 tetrahedral glasses. Phys Rev B 48:15672–15689

Bassett WA (2009) Diamond anvil cell, 50th birthday. High Press Res 29:163–186

Bell PM, Mao H-k, Goettel K (1984) Ultrahigh pressure: beyond 2 megabars and the ruby fluorescence scale. Science 226:542–544

Ben Kacem I, Gautron L, Coillot D, Neuville DR (2017) Structure and properties of lead silicate glasses and melts. Chem Geol 461:104–114

Benmore CJ (2015) X-ray diffraction from glass. *In:* Modern Glass Characterisation, Affatigato M (Ed), John Wiley and Sons, USA, p 241–270

Benmore CJ, Weber JKR (2017) Aerodynamic levitation, supercooled liquids and glass formation. Adv Phys X 2:717–736

Benmore CJ, Weber JKR, Sampath S, Siewenie J, Urquidi J, Tangeman JA (2003) A neutron and x-ray diffraction study of calcium aluminate glasses. J Phys: Condens Matter 15:S2413–S2423

Benmore CJ, Weber JKR, Wilding MC, Du J, Parise JB (2010a) Temperature-dependent structural heterogeneity in calcium silicate liquids. Phys Rev B 82:224202

Benmore CJ, Soignard E, Amin SA, Guthrie M, Shastri SD, Lee PL, Yarger JL (2010b) Structural and topological changes in silica glass at pressure. Phys Rev B 81:054105

Benmore CJ, Soignard E, Guthrie M, Amin SA, Weber JKR, McKiernan K, Wilding MC, Yarger JL (2011) High pressure x-ray diffraction measurements on Mg_2SiO_4 glass. J Non-Cryst Solids 357:2632–2636

Besson JM, Nelmes RJ, Hamel G, Loveday JS, Weill G, Hull S (1992) Neutron powder diffraction above 10 GPa. Physica B 180–181:907–910

Besson JM, Grima P, Gauthier M, Itié JP, Mézouar M, Häusermann D, Hanfland M (1996) Pretransitional behavior in zincblende HgTe under high pressure and temperature. Phys Stat Sol 198:419–425

Binsted N, Greaves GN, Henderson CMB (1986) Fe K-edge x-ray absorption spectroscopy of silicate minerals and glasses. J Phys Colloques 47:C8-837–840

Bingham PA, Parker JM, Searle TM, Smith I (2007) Local structure and medium range ordering of tetrahedrally coordinated Fe^{3+} ions in alkali–alkaline earth–silica glasses. J Non-Cryst Solids 353:2479–2494

Bista S, Stebbins JF (2017) The role of modifier cations in network cation coordination increases with pressure in aluminosilicate glasses and melts from 1 to 3 GPa. Am Mineral 102:1657–1666

Brink T, Koch L, Albe K (2016) Structural origins of the boson peak in metals: From high-entropy alloys to metallic glasses. Phys Rev B 94:224203

Boon JA, Fyfe WS (1972) The coordination number of ferrous ions in silicate glasses. Chem Geol 10:287–298

Börjesson L, Hassan AK, Swenson J, Torell LM, Fontana A (1993) Is there a correlation between the first sharp diffraction peak and the low frequency vibrational behavior of glasses? Phys Rev Lett 70:1275–1278

Bouhadja M, Jakse N, Pasturel A (2013) Structural and dynamic properties of calcium aluminosilicate melts: a molecular dynamics study. J Chem Phys 138:224510

Bowron DT (2008) An experimentally consistent atomistic structural model of silica glass. Mater Sci Eng B 149:166–170

Brazhkin VV, Lyapin AG, Trachenko K (2011) Atomistic modeling of multiple amorphous-amorphous transitions in SiO_2 and GeO_2 glasses at megabar pressures. Phys Rev B 83:132103

Bridgman PW, Šimon I (1953) Effects of very high pressures on glass. J Appl Phys 24:405–413

Buchenau U, Prager M, Nücker N, Dianoux AJ, Ahmad N, Phillips WA (1986) Low-frequency modes in vitreous silica. Phys Rev B 34:5665–5673

Buchenau U, Galperin YM, Gurevich VL, Schober HR (1991) Anharmonic potentials and vibrational localization in glasses. Phys Rev B 43:5039

Buchenau U, Galperin YM, Gurevich VL, Parshin DA, Ramos MA, Schober HR (1992) Interaction of soft modes and sound waves in glasses Phys Rev B 46:2798–2808

Burkhard DJM (2000) Iron-bearing silicate glasses at ambient conditions. J Non-Cryst Solids 275:175–188

Calas G, Petiau J (1983) Coordination of iron in oxide glasses through high-resolution K-edge spectra: Information from the pre-edge. Solid State Commun 48:625–629

Caldwell WA, Kunz M, Celestre RS, Domning EE, Walter MJ, Walker D, Glossinger J, MacDowell AA, Padmore HA, Jeanloz R, Clark SM (2007) Laser-heated diamond anvil cell at the advanced light source beamline 12.2.2. Nucl Instr Meth Phys Res A 582:221–225

Champagnon B, Martinet C, Boudeulle M, Vouagner D, Coussa C, Deschamps T, Grosvalet L (2008) High pressure elastic and plastic deformations of silica: In situ diamond anvil cell Raman experiments. J Non-Cryst Solids 354:569–573

Charpentier T, Okhotnikov K, Novikov AN, Hennet L, Fischer HE, Neuville DR, Florian P (2018) Structure of strontium aluminosilicate glasses from molecular dynamics simulation, neutron diffraction, and nuclear magnetic resonance studies. J Phys Chem B 122:9567–9583

Cheng J-X, Xie XS (2004) Coherent anti-Stokes Raman scattering microscopy: instrumentation, theory, and applications. J Phys Chem B 108:827–840

Crichton WA, Mezouar M (2005) Methods and application of the Paris–Edinburgh Press to X-ray diffraction structure solution with large-volume samples at high pressures and temperatures. Chapter 17 *In*: Advances in High-Pressure Technology for Geophysical Applications, Chen J, Wang Y, Duffy TS, Shen G, Dobrzhinetskaya LF (eds) Elsevier, Holland, p 353–369

Chumakov AI, Monaco G, Monaco A, Crichton WA, Bosak A, Rüffer R, Meyer A, Kargl F, Comez L, Fioretto D, Giefers H, Roitsch S, Wortmann G, Manghnani MH, Hushur A, Williams Q, Balogh J, Parliński K, Jochym P, Piekarz P (2011) Equivalence of the Boson Peak in glasses to the transverse acoustic van Hove singularity in crystals. Phys Rev Lett 106:225501

Cicconi MR, de Ligny D, Gallo TM, Neuville DR (2016) Ca neighbors from XANES spectroscopy: a tool to investigate structure, redox and nucleation processes in silicate glasses, melts and crystals. Am Mineral 101:1232–1235

Cicconi MR, McCloy JS, Neuville DR (2022) Non-magmatic glasses. Rev Mineral Geochem 87:965–1014

Cochain B, Neuville DR, Henderson GS, McCammon CA, Pinet O, Richet P (2012) Effects of iron content, redox state and structure of sodium borosilicate glasses: A Raman, Mössbauer and boron K-edge XANES spectroscopy study. J Am Ceram Soc 95:962–971

Cormack AN, Du J (2001) Molecular dynamics simulations of soda–lime silicate glasses. J Non-Cryst Solids 293–295:283–289

Cormier L (2019) Neutron and X-ray diffraction of glass. *In*: Handbook of Glass, Musgraves JD, Hu J, Calvez L (eds) Springer Nature, Switzerland pp 1045–1092

Cormier L, Cuello GJ (2011) Mg coordination in a $MgSiO_3$ glass using neutron diffraction coupled with isotopic substitution. Phys Rev B 83:224204

Cormier L, Cuello GJ (2013) Structural investigation of glasses along the $MgSiO_3$–$CaSiO_3$ join: Diffraction studies. Geochim Cosmochim Acta 122:498–510

Cormier L, Neuville DR, Calas G (2000) Structure and properties of low-silica calcium aluminosilicate glasses. J Non-Cryst Solids 274:110–114

Cormier L, Ghaleb D, Neuville DR, Delaye J-M, Calas G (2003) Chemical dependence of network topology of calcium aluminosilicate glasses: a computer simulation study. J Non-Cryst Solids 332:255–270

Coté B, Massiot D, Taulelle F, Coutures J-P (1992) ^{27}Al NMR spectroscopy of aluminosilicate melts and glasses. Chem Geol 96:367–370

Cottrell E, Lanzirotti A, Mysen B, Birner S, Kelley KA, Botcharnikov R, Davis FA, Newville M (2018) A Mössbauer-based XANES calibration for hydrous basalt glasses reveals radiation-induced oxidation of Fe. Am Mineral 103:489–501

Crépisson C, Morard G, Bureau H, Prouteau G, Morizet Y, Petitgirard S, Sanloup C (2014) Magmas trapped at the continental lithosphere-asthenosphere boundary. Earth Planet Sci Lett 393:105–112

Cristiglio V, Cuello GJ, Hennet L, Pozdnyakova I, Leydier M, Kozaily J, Fischer HE, Johnson MR, Price DL (2010) Neutron diffraction study of molten calcium aluminates. J Non-Cryst Solids 356:2492–2496

Daniel I, Gillet P, McMillan PF, Richet P (1995) An in situ high-temperature structural study of stable and metastable CaAl$_2$Si$_2$O$_8$ polymorphs. Mineral Mag 59:25–33

Daniel I, McMillan PF, Gillet P, Poe BT (1996) Raman spectroscopic study of structural changes in calcium aluminate (CaAl$_2$O$_4$) glass at high pressure and high temperature. Chem Geol 128:5–15

de Grouchy CJL, Sanloup C, Cochain B, Drewitt JWE, Kono Y, Crépisson C (2017) Lutetium incorporation in magmas at depth: Changes in melt local environment and the influence on partitioning behaviour. Earth Planet Sci Lett 464:155–165

de Koker N (2010) Structure, thermodynamics, and diffusion in CaAl$_2$Si$_2$O$_8$ liquid from first-principles molecular dynamics. Geochim Cosmochim Acta 74:5657–5671

Dewaele A, Loubeyre P, Occelli F, Marie O, Mezouar M (2018) Toroidal diamond anvil cell for detailed measurements under extreme static pressures. Nat Commun 9:2913

Di Genova D, Morgavi D, Hess K-U, Neuville DR, Borovkov N, Perugini D, Dingwell DB (2015) Approximate chemical analysis of volcanic glasses using Raman spectroscopy. J Raman Spectr 46:1235–1244

Di Genova D, Kolzenburg S, Vona A, Chevrel MO, Hess K-U, Neuville DR, Ertel-Ingrisch W, Romano C, Dingwell DB (2016) Raman spectra of Martian glass analogues: A tool to approximate their chemical composition. J Geophys Res Planets 121:740–752

Di Genova D, Vasseur J, Hess K-U, Neuville DR, Dingwell DB (2017) Effect of oxygen fugacity on the glass transition, viscosity and structure of silica- and iron-rich magmatic melts. J Non-Cryst Solids 470:78–85

Di Genova D, Brooker RA, Mader HM, Drewitt JWE, Longo A, Deubener J, Neuville DR, Fanara S, Shebanova O, Anzellini S, Arzilli F, Bamber EC, Hennet L, La Spina G, Miyajima N (2020) In situ observation of nanolite growth in volcanic melt : A driving force for explosive eruptions. Sci Adv 6:eabb0413

Di Muro A, Métrich M, Mercier M, Giordano D, Massare D, Montagnac G (2009) Micro-Raman determination of iron redox state in dry natural glasses: Application to peralkaline rhyolites and basalts. Chem Geol 259:78–88

Dingwell DB, Virgo D (1988) Viscosities of melts in the Na$_2$O–FeO–Fe$_2$O$_3$–SiO$_2$ system and factors controlling relative viscosities of fully polymerized silicate melts. Geochim Cosmochim Acta 52:395–403

Dong J, Yao H, Guo Z, Jia Q, Wang Y, An P, Gong Y, Liang Y, Chen D (2017) Revisiting local structural changes in GeO$_2$ glass at high pressure. J Phys Condens Matter 29:465401

Drewitt JWE, Salmon PS, Barnes AC, Klotz S, Fischer HE, Crichton WA (2010) Structure of GeO$_2$ glass at pressures up to 8.6 GPa. Phys Rev B 81:014202

Drewitt JWE, Jahn S, Cristiglio V, Bytchkov A, Leydier M, Brassamin S, Fischer HE, Hennet L (2011) The structure of liquid calcium aluminates as investigated using neutron and high energy x-ray diffraction in combination with molecular dynamics simulation methods. J Phys Condens Matter 23:155101

Drewitt JWE, Hennet L, Zeidler A, Jahn S, Salmon PS, Fischer HE (2012) Structural transformations on vitrification in the fragile glass-forming system CaAl$_2$O$_4$. Phys Rev Lett 109:235501

Drewitt JWE, Sanloup C, Bytchkov A, Brassamin S, Hennet L (2013) Structure of (Fe$_x$Ca$_{1-x}$O)$_y$(SiO$_2$)$_{1-y}$ liquids and glasses from high-energy x-ray diffraction: Implications for the structure of natural basaltic magmas. Phys Rev B 87:224201

Drewitt JWE, Jahn S, Sanloup C, de Grouchy C, Garbarino G, Hennet L (2015) Development of chemical and topological structure in aluminosilicate liquids and glasses at high pressure. J Phys: Condens Matter 27:105103

Drewitt JWE, Barnes AC, Jahn S, Kohn SC, Walter MJ, Novikov AN, Neuville DR, Fischer HE, Hennet L (2017) Structure of liquid tricalcium aluminate. Phys Rev B 95:064203

Drewitt JWE, Jahn S, Hennet L (2019) Configurational constraints on glass formation in the liquid calcium aluminate system. J Stat Mech 104012

Drewitt JWE, Turci F, Heinen BJ, Macleod SG, Qin F, Kleppe AK, Lord OT (2020) Structural ordering in liquid gallium under extreme conditions. Phys Rev Lett 124:145501

Du J, Cormack AN (2004) The medium range structure of sodium silicate glasses: a molecular dynamics simulation. J Non-Cryst Solids 349:66–79

Du J, Corrales R (2007) Understanding lanthanum aluminate glass structure by correlating molecular dynamics simulation results with neutron and x-ray scattering data. J Non-Cryst Solids 353:210–214

Dubrovinskaia N, Dubrovinsky L, Solopova NA, Abakumov A, Turner S, Hanfland M, Bykova E, Bykov M, Prescher C, Prakapenka VB, Petitgirard S, Chuvashova I, Gasharova B, Mathis Y-L, Ershov P, Snigireva I, Snigirev A (2016) Terapascal static pressure generation with ultrahigh yield strength nanodiamond. Science Adv 2:e1600341

Dubrovinsky L, Dubrovinskaia N, Prakapenka VB, Abakumov AM (2012) Implementation of micro-ball nanodiamond anvils for high-pressure studies above 6 Mbar. Nat Commun 3:1163

Dupree R, Howes AP, Kohn SC (1997) Natural abundance solid state ^{43}Ca NMR. Chem Phys Lett 276:399–404

Durben DJ, Wolf GH (1991) Raman spectroscopic study of the pressure-induced coordination change in GeO$_2$ glass. Phys Rev B 43:2355

Duval E, Boukenter A, Champagnon B (1986) Vibration eigenmodes and size of microcrystallites in glass: observation by very-low-frequency Raman scattering. Phys Rev Lett 56:2052–2055

Eckersley MC, Gaskell PH, Barnes AC, Chieux P (1988) Structural ordering in a calcium silicate glass. Nature 335:525–527

Eggert JH, Weck G, Loubeyre P, Mezouar M (2002) Quantitative structure factor and density measurements of high-pressure fluids in diamond anvil cells by x-ray diffraction: Argon and water. Phys Rev B 65:174105

Elliot S (1991) Medium-range structural order in covalent amorphous solids. Nature 354:445–452

Elliot S (1992) A unified model for the low-energy vibrational behaviour of amorphous solids. Europhys Lett 19:201–206

Elliot SR (1995) Second sharp diffraction peak in the structure factor of binary covalent network glasses. Phys Rev B 51:8599–8601

Faber TE, Ziman JM (1965) A theory of the electrical properties of liquid metals: III. the resistivity of binary alloys. Phil Mag 11:153–173

Farges F, Lefrère Y, Rossano S, Berthereau A, Calas G, Brown Jr GE (2004) The effect of redox state on the local structural environment of iron in silicate glasses: a combined XAFS spectroscopy, molecular dynamics, and bond valence study. J Non-Cryst Solids 344:176–188

Feller S, Lodden G, Riley A, Edwards T, Croskrey J, Schue A, Liss D, Stentz D, Blair S, Kelley M, Smith G, Singleton S, Affatigato M, Holland D, Smith ME, Kamitsos EI, Varsamis CPE, Loannou E (2010) A multispectroscopic structural study of lead silicate glasses over an extended range of compositions. J Non-Cryst Solids 356:304–313

Fischer HE, Barnes AC, Salmon PS (2006) Neutron and x-ray diffraction studies of liquids and glasses. Rep Prog Phys 69:233–299

Florian P, Novikov A, Drewitt JWE, Hennet L, Sarou-Kanian V, Massiot D, Fischer HE, Neuville DR (2018) Structure and dynamics of high-temperature strontium aluminosilicate melts. Phys Chem Chem Phys 20:27865–27877

Fox KE, Furukawa Y, White WB (1982) Transition metal ions in silicate melts. Part 2. Iron in sodium silicate glasses. Phys Chem Glasses 23:169–178

Funamori N, Yamamoto S, Yagi T, Kikegawa T (2004) Exploratory studies of silicate melt structure at high pressures and temperatures by in situ X-ray diffraction. J Geophys Res 109:B03203

Furukawa T, Brawer SA, White WB, (1978) The structure of lead silicate glasses determined by vibrational spectroscopy. J Mater Sci 13:268–282

Galeener FL (1982a) Planar rings in vitreous silica. J Non-Cryst Solids 49:53–62

Galeener FL (1982b) Planar rings in glasses. Solid State Commun 44:1037–1040

Galeener FL, Leadbetter AJ, Stringfellow MW (1983) Comparison of the neutron, Raman, and infrared vibrational spectra of vitreous SiO_2, GeO_2, and BeF_2. Phys Rev B 27:1052–1078

Galoisy L, Calas G, Arrio MA (2001) High-resolution XANES spectra of iron in minerals and glasses: structural information from the pre-edge region. Chem Geol 174:307–319

Gaskell PH, Eckersley MC, Barnes AC, Chieux P (1991) Medium-range order in the cation distribution of a calcium silicate glass. Nature 350:675–677

Gaudio SJ, Sen S, Lesher CE (2008) Pressure-induced structural changes and densification of vitreous $MgSiO_3$. Geochim Cosmochim Acta 72:1222–1230

Gaudio SJ, Lesher CE, Maekawa H, Sen S (2015) Linking high-pressure structure and density of albite liquid near the glass transition. Geochim Cosmochim Acta 157:28–38

Geissberger AE, Galeener FL (1983) Raman studies of vitreous SiO_2 versus fictive temperature. Phys Rev B 28:3266–3271

Ghosh DB, Karki BB (2018) First-principles molecular dynamics simulations of anorthite ($CaAl_2Si_2O_8$) glass at high pressure. Phys Chem Mater 45:575–587

Giordano D, Dingwell DB (2003) The kinetic fragility of natural silicate melts. J Phys: Condens Matter 15:S945

Giordano D, Russell JK (2007) A rheological model for glassforming silicate melts in the systems CAS, MAS, MCAS. J Phys: Condens Matter 19:205148

Giordano D, Russell JK, Dingwell DB (2008) Viscosity of magmatic liquids: A model. Earth Planet Sci Lett.271:123–134

Giuli G, Pratesi G, Cipriani C, Paris E (2002) Iron local structure in tektites and impact glasses by extended x-ray absorption fine structure and high-resolution x-ray absorption near-edge structure spectroscopy. Geochim Cosmochim Acta 66:4347–4353

Giuli G, Paris E, Hess K-U, Dingwell DB, Cicconi MR, Eeckhout SG, Fehr KT, Valenti P (2011) XAS determination of the Fe local environment and oxidation state in phonolite glasses. Am Mineral 96:631–636

Giuli G, Alonso-Mori R, Cicconi MR, Paris E, Glatzel P, Eeckhout SG, Scaillet B (2012) Effect of alkalis on the Fe oxidation state and local environment in peralkaline rhyolitic glasses. Am Mineral 97:468–475

Goldstein M (1969) Viscous liquids and the glass transition: a potential energy barrier picture. J Chem Phys 51:3728–3739

Greaves GN (1985) EXAFS and the structure of glass. J Non-Cryst Solids 71:203–217

Greaves GN, Sen S (2007) Inorganic glasses, glass-forming liquids and amorphizing solids. Adv Phys 56:1–166

Greaves GN, Wilding MC, Fearn S, Langstaff D, Kargl F, Cox S, Van QV, Majérus O, Benmore CJ, Weber R, Martin CM, Hennet L (2008) Detection of first-order liquid/liquid phase transitions in yttrium oxide–aluminum oxide melts. Science 322:566–570

Grigera TS, Martin-Mayer V, Parisi G, Verrocchio P (2003) Phonon interpretation of the 'boson peak' in supercooled liquids. Nature 422:289–292

Grimsditch M (1984) Polymorphism in amorphous SiO_2. Phys Rev Lett 52:2379–2381

Gross E, Romanova M (1929) Über die Lichtzerstreuung in Quarz und festen amorphen Stoffen, welche die Gruppe SiO_2 enthalten. Z Physik 55:744–752

Guillot B, Sator N (2007) A computer simulation study of natural silicate melts. Part II: High pressure properties. Geochim Cosmochim Acta 71:4538–4556

Guerette M, Ackerson MR, Thomas J, Yuan F, Watson EB, Walker D, Huang L (2015) Structure and properties of silica glass densified in cold compression and hot compression. Sci Rep 5:15343

Gurevich VL, Parshin DA, Pelous J, Schober HR (1993) Theory of low-energy Raman scattering in glasses. Phys Rev B 48:16318–16331

Guthrie M, Tulk CA, Benmore CJ, Xu J, Yarger JL, Klug DD, Tse JS, Mao H-k, Hemley RJ (2004) Formation and structure of a dense octahedral glass. Phys Rev Lett 93:115502

Hannon AC, Parker JM (2000) The structure of aluminate glasses by neutron diffraction. J Non-Cryst Solids 274:102–109

Hannon AC (2015) Neutron diffraction techniques for structural studies of glasses. *In:* Modern Glass Characterisation, Affatigato M (Ed), John Wiley and Sons, USA, p 158–240

Hannoyer B, Lenglet M, Dürr J, Cortes R (1992) Spectroscopic evidence of octahedral iron (III) in soda-lime silicate glasses. J Non-Cryst Solids 151:209–216

Hehlen B, Neuville DR (2015) Raman response of network modifier cations in alumino–silicate glasses. J Phys Chem B 119:4093–4098

Hehlen B, Neuville DR (2020) Non network-former cations in oxide glasses spotted by Raman scattering. Phys Chem Chem Phys 22:12724–12731

Hehlen B, Courtens E, Vacher R, Yamanaka A, Kataoka M, Inoue K (2000) Hyper-Raman scattering observation of the Boson peak in vitreous silica. Phys Rev Lett 84:5355–5358

Hehlen B, Courtens E, Yamanaka A, Inoue K (2002) Nature of the Boson peak of silica glasses from hyper-Raman scattering. J Non-Cryst Solids 307–310:87–91

Hemley RJ, Mao HK, Bell PM, Mysen BO (1986) Raman spectroscopy of SiO_2 glass at high pressure. 57:747–750

Henderson GS (2005) The structure of silicate melts: a glass perspective. Can Mineral 43:1921–1958

Henderson GS, Stebbins JF (2022) The short-range order (SRO) and structure. Rev Mineral Geochem 87:1–53

Hennet L, Pozdnyakova I, Bytchkov A, Price DL, Greaves GN, Wilding M, Fearn S, Martin CM, Thiaudière D, Bérar J-F, Boudet N, Saboungi M-L (2007) Development of structural order during supercooling of a fragile oxide melt. J Chem Phys 126:074906

Hennet L, Cristiglio V, Kozaily J, Pozdnyakova I, Fischer HE, Bytchkov A, Drewitt JWE, Leydier M, Thiaudière D, Gruner S, Brassamin S, Zanghi D, Cuello GJ, Koza M, Magazù S, Greaves GN, Price DL (2011a) Aerodynamic levitation and laser heating: Applications at synchrotron and neutron sources. Eur Phys J 196:151–165

Hennet L, Pozdnyakova I, Bytchkov A, Drewitt JWE, Kozaily J, Leydier M, Brassamin S, Zanghi D, Fischer HE, Greaves GN, Price DL (2011b) Application of time resolved x-ray diffraction to study the solidification of glass-forming melts. High Temp High Press 40:263–270

Hennet L, Drewitt JWE, Neuville DR, Cristiglio V, Kozaily J, Brassamin S, Zanghi D, Fischer HE (2016) Neutron diffraction of calcium aluminosilicate glasses and melts. J Non-Cryst Solids 451:89–93

Hiraoka Y, Nakamura T, Hirata A, Escolar EG, Matsue K, Nishiura Y (2016) Hierarchical structures of amorphous solids characterized by persistent homology. PNAS113:7035–7040

Holland D, Mekki A, Gee IA, McConville CF, Johnson JA, Johnson CE, Appleyard P, Thomas M (1999) The structure of sodium iron silicate glass—a multi-technique approach. J Non-Cryst Solids 253:192–202

Holloway JR, Wood BJ (1988) Simulating the Earth: Experimental Geochemistry. Boston: Unwin Hyman.

Hong X, Shen G, Prakapenka VB, Newville M, Rivers ML, Sutton SR (2007) Intermediate states of GeO_2 glass under pressures up to 35GPa. Phys Rev B 75:104201

Hong X, Ehm L, Duffy TS (2014) Polyhedral units and network connectivity in GeO_2 glass at high pressure: An X-ray total scattering investigation. Appl Phys Lett 105:081904

Hosokawa S, Bérar J-F, Boudet N, Pilgrim W-C, Pusztai L, Hiroi S, Maruyama K, Kohara S, Kato H, Fischer HE, Zeidler A (2019) Partial structure investigation of the traditional bulk metallic glass $Pd_{40}Ni_{40}P_{20}$. Phys Rev B 100:054204

Huang L, Kieffer J (2004a) Amorphous-amorphous transitions in silica glass. I. Reversible transitions and thermomechanical anomalies. Phys Rev B 69:224203

Huang L, Kieffer J (2004b) Amorphous-amorphous transitions in silica glass. II. Irreversible transitions and densification limit. Phys Rev B 69:224204

Imaoka M, Hasegawa H, Yasui I (1986) X-ray diffraction analysis on the structure of the glasses in the system PbO–SiO_2. J Non-Cryst Solids 85:393–412

Itié JP, Calas G, Petiau J, Fontaine A, Tolentino H (1989) Pressure-induced coordination changes in crystalline and vitreous GeO_2. Phys Rev Lett 63:398

Ishihara T, Shirakawa Y, Iida T, Kitamura N, Matsukawa M, Ohtori N, Umesaki N (1999) Brillouin scattering in densified GeO_2 glasses. Jpn J Appl Phys 38:3062–3065

Inamura Y, Arai M, Nakamura M, Otomo T, Kitamura N, Bennington SM, Hannon AC, Buchenau U (2001) Intermediate range structure and low-energy dynamics of densified vitreous silica. J Non-Cryst Solids 293–295:389–393

Inamura Y, Katayama Y, Utsumi W, Funakoshi K (2004) Transformations in the intermediate-range structure of SiO_2 glass under high pressure and temperature. Phys Rev Lett 93:015501

Ito E (2007) Multianvil cells and high-pressure experimental methods. *In:* Treatise on Geophysics: Theory and Practice (second edition), Elsevier, Holland 2:197–230

Iuga D, Morais C, Gan Z, Neuville DR, Cormier L, Massiot D (2005) NMR heteronuclear correlation between quadrupolar nuclei in solids. J Am Chem Soc 127:11540–11541

Iwamoto N, Umesaki N, Atsumi T (1987) EXAFS and x-ray diffraction studies of iron ions in a $0.2(Fe_2O_3) \cdot 0.8(Na_2O \cdot 2SiO_2)$ glass. J Mater Sci Lett 6:271–273

Jackson WE, Mustre de Leon J, Brown Jr GE, Waychunas GA, Conradson SD, Combes J-M (1993) High-temperature XAS study of Fe_2SiO_4 liquid: reduced coordination of ferrous iron. Science 262:229–233

Jackson WE, Farges F, Yeager M, Mabrouk PA, Rossano S, Waychunas GA, Solomon EI, Brown Jr GE (2005) Multi-spectroscopic study of Fe(II) in silicate glasses: Implications for the coordination environment of Fe(II) in silicate melts. Geochim Cosmochim Acta 69:4315–4332

Jakse N, Bouhadja M, Kozaily J, Drewitt JWE, Hennet L, Neuville DR, Fischer HE, Cristiglio V, Pasturel A (2012) Interplay between non-bridging oxygen, triclusters, and fivefold Al coordination in low silica content calcium aluminosilicate melts. Appl Phys Lett 101:201903

Jahn S (2022) Molecular simulations of oxide and silicate melts and glasses. Rev Mineral Geochem 87:193-227

Johnson JA, Johnson CE, Holland D, Mekki A, Appleyard P, Thomas MF (1999) Transition metal ions in ternary sodium silicate glasses: a Mössbauer and neutron study. J Non-Cryst Solids 246:104–114

Kalampounias AG, Yannopoulos SN, Papatheodorou GN (2006) Temperature-induced structural changes in glassy, supercooled, and molten silica from 77 to 2150 K. J Chem Phys 124:014504

Karki BB, Bhattarai D, Stixrude L (2007) First-principles simulations of liquid silica: Structural and dynamical behavior at high pressure. Phys Rev B 76:104205

Kang E-T, Lee S-J, Hannon AC (2006) Molecular dynamics simulations of calcium aluminate glasses. J Non-Cryst Solids 352:725–736

Kelsey KE, Stebbins JF, Mosenfelder JL, Asimow PD (2009) Simultaneous aluminum, silicon, and sodium coordination changes in 6 GPa sodium aluminosilicate glasses. Am Mineral 94:1205–1215

Keppler H (1992) Crystal field spectra and geochemistry of transition metal ions in silicate melts and glasses. Am Mineral 77:62–75

Kim H, Lee SK (2020) Extent of disorder in iron-bearing albite and anorthite melts: Insights from multi-nuclear (^{29}Si, ^{27}Al, and ^{17}O) solid-state NMR study of iron-bearing $NaAlSi_3O_8$ and $CaAl_2Si_2O_8$ glasses. Chem Geol 538:119498

Kohara S, Suzuya K (2005) Intermediate-range order in vitreous SiO_2 and GeO_2. J Phys: Condens Matter 17:S77-S86

Kono Y, Sanloup C (Eds.) (2018) Magmas Under Pressure, Elsevier, Holland

Kono Y, Park C, Kenney-Benson C, Shen G, Wang Y (2014) Toward comprehensive studies of liquids at high pressures and high temperatures: Combined structure, elastic wave velocity, and viscosity measurements in the Paris–Edinburgh cell. Phys Earth Planet Inter 228:269–280

Kono Y, Kenney-Benson C, Ikuta D, Shibazaki Y, Wang Y, Shen G (2016) Ultrahigh-pressure polyamorphism in GeO_2 glass with coordination number > 6. PNAS 113:3436–3441

Kono Y, Shibazaki Y, Kenney-Benson C, Wang Y, Shen G (2018) Pressure-induced structural change in $MgSiO_3$ glass at pressures near the Earth's core–mantle boundary. PNAS 115:1742–1747

Kono Y, Kenney-Benson C, Shen G (2019) Opposed type double stage cell for Mbar pressure experiment with large sample volume. High Press Res 40:175–183

Kono Y, Shu Y, Kenney-Benson C, Wang Y, Shen G (2020) Structural evolution of SiO_2 glass with Si coordination number greater than 6. Phys Rev Lett 125:205701

Kozaily J (2012) Structure and dynamics of calcium aluminosilicate melts. Ph.D. thesis, University of Orléans http://www.theses.fr/2012ORLE2003/document

Klinger MI (2010) Soft atomic motion modes in glasses: Their role in anomalous properties. Phys Rep 492:111–180

Klotz S, Besson JM, Hamel G, Nelmes RJ, Loveday JS, Marshall WG, Wilson RM (1995) Neutron powder diffraction at pressures beyond 25 GPa. Appl Phys Lett 66:1735–1737

Labrosse S, Hernlund JW, Coltice N (2007) A crystallizing dense magma ocean at the base of the Earth's mantle. Nature 450:866–869

Lacy ED (1963) Aluminium in glasses and melts. Phys Chem Glasses 4:234–238

Laird BB, Schober HR (1991) Localized low-frequency modes in a simple model glass. Phys Rev Lett 66:636–369

Le Losq C, Neuville DR (2013) Effect of Na/K mixing on the structure and rheology of tectosilicate silica-rich melts. Chem Geol 346:57–71

Le Losq C, Neuville DR, Florian P, Henderson GS, Massiot D (2014) The role of Al^{3+} on rheology and structural changes in sodium silicate and aluminosilicate glasses and melts. Geochim Cosmochim Acta 126:495–517

Le Losq C, Neuville DR, Chen W, Florian P, Massiot D, Zhou Z, Greaves GN (2017) Percolation channels: a universal idea to describe the atomic structure of glasses and melts. Sci Rep 7:16490

Le Losq C, Cicconi MR, Greaves GN, Neuville DR (2019a) Silicate glasses. *In:* Handbook of Glass, Musgraves JD, Hu J, Calvez L (eds) Springer Nature, Switzerland pp 441–503

Le Losq C, Berry AJ, Kendrick MA, Neuville DR, O'Neill HStC (2019b) Determination of the oxidation state of iron in Mid-Ocean Ridge basalt glasses by Raman spectroscopy. Am Mineral 104:1032–1042

Lee SK, Kim EJ (2015) Probing metal-bridging oxygen and configurational disorder in amorphous lead silicates: insights from [17]O solid-state nuclear magnetic resonance. J Phys Chem C 119:748–756

Lee SK, Stebbins JF (2003) The distribution of sodium ions in aluminosilicate glasses: a high-field Na-23 MAS and 3Q MAS NMR study. Geochim Cosmochim Acta 67:1699–1709

Le Roux S, Jund P (2010) Ring statistics analysis of topological networks: New approach and application to amorphous GeS_2 and SiO_2 systems. Comp Mater Sci 49:70–83

Le Roux S, Petkov V (2010) ISAACS—interactive structure analysis of amorphous and crystalline systems. J Appl Cryst 43:181–185

Levelut C, Gaimes N, Terki F, Cohen-Sohal G, Pelous J, Vacher R (1995) Glass-transition temperature: Relation between low-frequency dynamics and medium-range order. Phys Rev B 51:8606–8609

Li D, Secco RA, Bancroft GM, Fleet ME (1995) Pressure induced coordination change of Al in silicate melts from Al K edge XANES of high pressure $NaAlSi_2O_6$–$NaAlSi_3O_8$ glasses. Geophys Res Lett 22:3111–3114

Li B, Ji C, Yang W, Wang J, Yang K, Xu R, Liu W, Cai Z, Chen J, Mao H-k (2018) Diamond anvil cell behavior up to 4 Mbar. PNAS 115:1713–1717

Licheron M, Montouillout V, Millot F, Neuville DR (2011) Raman and ^{27}Al NMR structure investigations of aluminate glasses: $(1-x)Al_2O_3$-$_x$ MO, with M=Ca, Sr, Ba and $0.5 < x < 0.75$). J Non Cryst Solids 357:2796–2801

Liebermann RC (2011) Multi-anvil, high pressure apparatus: a half-century of development and progress. High Press Res 31:493–532

Liermann H-P, Morgenroth W, Ehnes A, Berghäuser A, Winkler B, Franz H, Weckert E (2010) The Extreme Conditions Beamline at PETRA III, DESY: Possibilities to conduct time resolved monochromatic diffraction experiments in dynamic and laser heated DAC. J Phys: Conf Ser 215:012029

Lin J-F, Fukui H, Prendergast D, Okuchi T, Cai YQ, Hiraoka N, Yoo C-S, Trave A, Eng P, Hu MY, Chow P (2007) Electronic bonding transition in compressed SiO_2 glass. Phys Rev B 75:012201

Liu H, Chen W, Pan R, Shan Z, Qiao A, Drewitt JWE, Hennet L, Jahn S, Langstaff DP, Chass GA, Tao H, Yue Y, Greaves GN (2020) From molten calcium aluminates through phase transitions to cement phases. Adv Sci 7:1902209

Louvel M, Drewitt JWE, Ross A, Thwaites R, Heinen BJ, Keeble DS, Beavers CM, Walter MJ, Anzellini S (2020) The HXD95: a modified Bassett-type hydrothermal diamond-anvil cell for in situ XRD experiments up to 5 GPa and 1300 K. J Synchrotron Rad 27:529–537

Mao H-K, Mao WL (2007) Diamond-anvil cells and probes for high *P–T* mineral physics studies. *In:* Treatise on Geophysics: Theory and Practice (second edition), Elsevier 2:263–291

Magnien V, Neuville DR, Cormier L, Mysen BO, Briois V, Belin S, Pinet O, Richet P (2004) Kinetics of iron oxidation in silicate melts: A preliminary XANES study. Chem Geol 213:253–263

Magnien V, Neuville DR, Cormier L, Roux J, Hazemann J-L, Pinet O, Richet P (2006) Kinetics of iron redox reactions: A high-temperature X-ray absorption and Raman spectroscopy study. J Nucl Mater 352:190–195

Magnien V, Neuville DR, Cormier L, Roux J, Hazemann J-L, de Ligny D, Pascarelli S, Vickridge I, Pinet O, Richet P (2008) Kinetics and mechanisms of iron redox reactions in silicate melts: The effects of temperature and alkali cations. Geochim Cosmochim Acta 72:2157–2168

Marruzzo A, Schirmacher W, Fratalocchi A, Ruocco G (2013) Heterogeneous shear elasticity of glasses: the origin of the boson peak. Sci Rep 3:1407

Martinet C, Kassir-Bodon A, Deschamps T, Cornet A, Le Floch S, Martinez V, Champagnon B (2015) Permanently densified SiO_2 glasses: a structural approach. J Phys: Condens Matter 27:325401

Massiot D, Trumeau D, Touzo B, Farnan I, Rifflet J-C, Douy A, Coutures J-P (1995) Structure and dynamics of $CaAl_2O_4$ from liquid to glass: A high-temperature ^{27}Al NMR time-resolved study J Phys Chem 99:16455–16459

Masson O, Thomas P (2013) Exact and explicit expression of the atomic pair distribution function as obtained from X-ray total scattering experiments. J Appl Crystallogr 46:461–465

Matson DW, Sharma SK, Philpotts JA (1983) The structure of high-silica alkali-silicate glasses. A Raman spectroscopic investigation. J Non-Cryst Solids 58:323–352

McCloy JS (2019) Frontiers in natural and un-natural glasses: An interdisciplinary dialogue and review. J Non-Cryst Solids: X 4:100035

McMillan PF (1984) Structural studies of silicate glasses and melts—Applications and limitations of Raman spectroscopy. Am Mineral 69:622–644

McMillan P, Piriou B (1982) The structures and vibrational spectra of crystals and glasses in the silica–alumina system. J Non-Cryst Solids 53:279–298

McMillan P, Piriou B (1983) Raman spectroscopy of calcium aluminate glasses and crystals. J Non-Cryst Solids 55:221–242

McMillan P, Piriou B, Navrotsky A (1982) A Raman spectroscopic study of glasses along the joins silica–calcium aluminate, silica–sodium aluminate, and silica–potassium aluminate. Geochim Cosmochim Acta 46:2021–2037

McMillan PF, Poe BT, Gillet P, Reynard B (1994) A study of SiO_2 glass and supercooled liquid to 1950 K via high-temperature Raman spectroscopy. Geochim Cosmochim Acta 58:3653–3664

McMillan PF, Petuskey WT, Cote B, Massiot D, Landron C, Coutures JP (1996) A structural investigation of CaO–Al_2O_3 glasses via ^{27}Al MAS-NMR. J Non-Cryst Solids 195:261–71

Mead RN, Mountjoy G (2006) A molecular dynamics study of the atomic structure of $(CaO)_x(SiO_2)_{1-x}$ Glasses. J Phys Chem B 110:14273–14278

Meade C, Hemley RJ, Mao HK (1992) High-pressure x-ray diffraction of SiO_2 glass. Phys Rev Lett 69:1387

Mei Q, Benmore CJ, Weber JKR (2007) Structure of liquid SiO_2: A measurement by high-energy x-ray diffraction. Phys Rev Lett 98:057802

Mei Q, Benmore CJ, Siewenie J, Weber JKR, Wilding M (2008a) Diffraction study of calcium aluminate glasses and melts: I. High energy x-ray and neutron diffraction on glasses around the eutectic composition. J Phys Condens Matter 20:245106

Mei Q, Benmore CJ, Weber JKR, Wilding M, Kim J, Rix J (2008b) Diffraction study of calcium aluminate glasses and melts: II. High energy x-ray diffraction on melts. J Phys Condens Matter 20:245107

Mei Q, Sinogeikin S, Shen G, Amin S, Benmore CJ, Ding K (2010) High-pressure x-ray diffraction measurements on vitreous GeO2 under hydrostatic conditions. Phys Rev B 81:174113

Meng Y, Shen G, Mao H-k (2006) Double-sided laser heating system at HPCAT for in situ x-ray diffraction at high pressures and high temperatures. J Phys: Condens Matter 18:S1097-S1103

Mermet A, Duval E, Etienne S, G'Sell C (1996) Low frequency Raman scattering study of the nanostructure of plastically deformed polymer glasses. J Non-Cryst Solids 196:227–232

Meyer A, Horbach J, Kob W, Kargl F, Schober H (2004) Channel formation and intermediate range order in sodium silicate melts and glasses. Phys Rev Lett 93:027801

Mezouar M, Le Bihan T, Libotte H, Le Godec Y, Häusermann D (1999) Paris Edinburgh large-volume cell coupled with a fast imaging-plate system for structural investigation at high pressure and high temperature. J Synchrotron Rad 6:1115–1119

Mezouar M, Faure P, Crichton W, Rambert N, Sitaud B, Bauchau S, Blattmann G (2002) Multichannel collimator for structural investigation of liquids and amorphous materials at high pressures and temperatures. Rev Sci Instr 73:3570–3574

Morard G, Siebert J, Andrault D, Guignot N, Garbarino G, Guyot F, Antonangeli D (2013) The Earth's core composition from high pressure density measurements of liquid iron alloys. Earth Planet Sci Lett 373:169–178

Morard G, Nakajima Y, Andrault D, Antonangeli D, Auzende AL, Boulard E, Cervera S, Clark AN, Lord OT, Siebert J, Svitlyk V, Garbarino G, Mezouar M (2017) Structure and density of Fe–C liquid alloys under high pressure. J Geophys Res Solid Earth 122:7813–7823

Morikawa H, Takagi Y, Ohno H (1982) Structural analysis of 2PbO–SiO_2 glass. J Non-Cryst Solids 53:173–182

Morikawa H, Marumo F, Koyama T, Yamane M, Oyobe A (1983) Structural analysis of $12CaO.7Al_2O_3$ glass. J Non-Cryst Solids 56:355–360

Moulton BJA, Henderson GS, Martinet C, Martinez V, Sonneville C, de Ligny D (2019) Structure—longitudinal sound velocity relationships in glassy anorthite (CaAl2Si2O8) up to 20 GPa: An in situ Raman and Brillouin spectroscopy study. Geochim Cosmochim Acta 261:132–144

Mountjoy G, Al-Hasni BM, Storey C (2011) Structural organisation in oxide glasses from molecular dynamics modelling. J Non-Cryst Solids 357:2522–2529

Muniz RF, de Ligny D, Martinet C, Sandrini M, Medina AN, Rohling JH, Baesso ML, Lima SM, Andrade LHC, Guyot Y (2016) In situ structural analysis of calcium aluminosilicate glasses under high pressure. J Phys Condens Matter 28:315402

Murakami M, Bass JD (2010) Spectroscopic evidence for ultrahigh-pressure polymorphism in SiO_2 Glass. Phys Rev Lett 104:025504

Mysen BO (2006) The structural behavior of ferric and ferrous iron in aluminosilicate glass near meta-aluminosilicate joins. Geochim Cosmochim Acta 70:2337–2353

Mysen BO, Frantz JD (1992) Raman spectroscopy of silicate melts at magmatic temperatures: Na_2O–SiO_2, K_2O–SiO_2 and Li_2O–SiO_2 binary compositions in the temperature range 25–1475°C, Chem Geol 96:321–332

Mysen BO, Richet P (2005) Silicate Glasses and Melts. Developments in Geochemistry Volume 10, Elsevier, Holland, 1st edition

Mysen BO, Richet P (2019) Silicate Glasses and Melts, Elsevier, Holland, 2nd edition

Mysen BO, Toplis MJ (2007) Structural behavior of Al^{3+} in peralkaline, metaluminous, and peraluminous silicate melts and glasses at ambient pressure. Am Mineral 92:933–946

Mysen BO, Seifert FA, Virgo D (1980) Structure and redox equilibria of iron-bearing silicate melts. Am Mineral 65:867–884

Mysen BO, Virgo D, Kushiro I (1981) The structural role of aluminum in silicate melts—a Raman spectroscopic study at 1 atmosphere. Am Mineral 66:678–70

Mysen BO, Finger LW, Virgo D, Seifert FA (1982a) Curve-fitting of Raman spectra of silicate glasses. Am Min 67:686–695

Mysen BO, Virgo D, Seifert FA (1982b) The structure of silicate melts: implications for chemical and physical properties of natural magma. Rev Geophys Space Phys 20:353–383

Mysen BO, Virgo D, Seifert FA (1984) Redox equilibria of iron in alkaline earth silicate melts: relationships between melt structure, oxygen fugacity, temperature and properties of iron-bearing silicate liquids. Am Mineral 69:834–847

Mysen BO, Virgo D, Neumann E-R, Seifert FA (1985) Redox equilibria and the structural states of ferric and ferrous iron in melts in the system CaO–MgO–SiO_2–Fe–O: relationships between redox equilibria, melt structure and liquidus phase equilibria. Am Mineral 70:317–331

Nagashio K, Kuribayashi K (2002) Spherical yttrium aluminum garnet embedded in a glass matrix. J Am Ceram Soc 85:2353–2358

Nakayama T (2002) Boson peak and terahertz frequency dynamics of vitreous silica. Rep Prog Phys 65:1195–1242

Nayak MT, Desa JAE, Reddy VR, Nayak C, Bhattacharyya D, Jha SN (2019) Structures of silicate glasses with varying sodium and fixed iron contents. J Non-Cryst Solids 509:42–47

Neuville DR (2005) Structure and properties in (Sr,Na) silicate glasses and melts. Phys Chem Glasses 46:112–118

Neuville DR (2006) Viscosity, structure and mixing in (Ca, Na) silicate melts. Chem Geol 229:28–42

Neuville DR, Le Losq C (2022) Link between medium and long-range order and macroscopic properties of silicate glasses and melts. Rev Mineral Geochem 87:105–162

Neuville DR, Mysen BO (1996) Role of aluminum in the silicate network: In situ, high-temperature study of glasses and melts on the join SiO_2–$NaAlO_2$. Geochim Cosmochim Acta 60:1727–1737

Neuville DR, Cormier L, Massiot D (2004) Al environment in tectosilicate and peraluminous glasses: A ^{27}Al MQ-MAS NMR, Raman, and XANES investigation. Geochim Cosmochim Acta 68:5071–5079

Neuville DR, Cormier L, Massiot D (2006) Al coordination and speciation in calcium aluminosilicate glasses: Effects of composition determined by ^{27}Al MQ-MAS NMR and Raman spectroscopy. Chem Geol 229:173–185

Neuville DR, Cormier L, Montouillout V, Massiot D (2007) Local Al site distribution in aluminosilicate glasses by ^{27}Al MQMAS NMR. J Non-Cryst Solids 353:180–184

Neuville DR, Cormier L, Montouillout V, Florian P, Millot F, Rifflet JC, Massiot D (2008a) Structure of Mg- and Mg/Ca aluminosilicate glasses: ^{27}Al NMR and Raman spectroscopy investigations. Am Mineral 93:1721–1731

Neuville DR, Cormier L, Flank AM, de Ligny D, Roux J, Flank A-M, Lagarde P (2008b) Environments around Al, Si and Ca in aluminate and aluminosilicate melts by x-ray absorption spectroscopy at high temperature. Am Mineral 93:228–234

Neuville DR, Henderson GS, Cormier L, Massiot D (2010) The structure of crystals, glasses, and melts along the CaO–Al_2O_3 join: Results from Raman, Al L- and K-edge X-ray absorption, and ^{27}Al NMR spectroscopy. Am Mineral 95:1580–1589

Neuville DR, de Ligny D, Henderson GS (2014a) Advances in Raman spectroscopy applied to earth and material sciences. Rev Mineral Geochem 78:509–541

Neuville DR, Hennet L, Florian P, de Ligny D (2014b) In situ high-temperature experiments. Rev Mineral Geochem 78:779–800

Nyrow A, Sternemann C, Wilke M, Gordon RA, Mende K, Yavaꭓ H, Simonelli L, Hiraoka N, Sahle ChJ, Huotari S, Andreozzi GB, Woodland AB, Tolan M, Tse JS (2014) Iron speciation in minerals and glasses probed by $M^{2/3}$-edge X-ray Raman scattering spectroscopy. Contrib Mineral Petrol 167:1012

Novikov VN, Sokolov AP (1991) A correlation between low-energy vibrational spectra and first sharp diffraction peak in chalcogenide glasses. Solid State Commun 77:243–247

Novikov AN, Neuville DR, Hennet L, Gueguen Y, Thiaudière D, Charpentier T, Florian P (2017) Al and Sr environment in tectosilicate glasses and melts: viscosity, Raman and NMR investigation. Chem Geol 461:115–127

Ohira I, Kono Y, Shibazaki Y, Kenney-Benson C, Masuno A, Shen G (2019) Ultrahigh pressure structural changes in a 60 mol.% Al_2O_3–40 mol.% SiO_2 glass. Geochem Perspect Lett 10:41–45

Ohno H, Nagasaki T, Igawa N, Kawamura H (1991) Neutron irradiation effects of PbO–SiO_2 glasses. J Nucl Mater 179–181:473–476

Ohtani E, Taulelle F, Angell CA (1985) Al^{3+} coordination changes in liquid aluminosilicates under pressure. Nature 314:78–81

Ohtani E, Suzuki A, Ando R, Urakawa S, Funakoshi K, Katayama Y (2005) Viscosity and density measurements of melts and glasses at high pressure and temperature by using the multi-anvil apparatus and synchrotron X-ray radiation. Chapter 10 *In:* Advances in High-Pressure Technology for Geophysical Applications, Chen J, Wang Y, Duffy TS, Shen G, Dobrzhinetskaya LF (eds) Elsevier, Holland, p 195–209

Onodera Y, Kohara S, Salmon PS, Hirata A, Nishiyama N, Kitani S, Zeidler A, Shiga M, Masuno A, Inoue H, Tahara S et al. (2020) Structure and properties of densified silica glass: characterizing the order within disorder. NPG Asia Mater 12:85

O'Shaughnessy C, Henderson GS, Nesbitt HW, Bancroft GM, Neuville DR (2017) Structure-property relations of caesium silicate glasses from room temperature to 1400 K: Implications from density and Raman spectroscopy. Chem Geol 461:82–95

O'Shaughnessy C, Henderson GS, Nesbitt HW, Bancroft GM, Neuville DR (2020) The influence of modifier cations on the Raman stretching modes of Q^n species in alkali silicate glasses. J Am Ceram Soc 103:3991–4001

Papatheodorou GN, Kalampounias AG, Yannopoulos (2010) Raman spectroscopy of high temperature melts. Chapter 20 *In:* Molten Salts and Ionic Liquids: Never the Twain? Gaune-Escard M, Seddon KR (Eds) John Wiley and Sons, Inc. NJ, USA, p 301–340

Parshin DA (1993) Soft potential model and universal properties of glasses. Physica Scripta T49:180–185

Pasquarello A, Car R (1998) Identification of Raman defect lines as signatures of ring structures in vitreous silica. Phys Rev Lett 80:5145–5147

Petitgirard S, Salamat A, Beck P, Weck G, Bouvier P (2014) Strategies for in situ laser heating in the diamond anvil cell at an X-ray diffraction beamline. J Synchrotron Rad 21:89–96

Petitgirard S, Malfait WJ, Sinmyo R, Kupenko I, Hennet L, Harries D, Dane T, Burghammer M, Rubie DC (2015) Fate of $MgSiO_3$ melts at core–mantle boundary conditions. PNAS 112:14186–14190

Petitgirard S, Malfait WJ, Journaux B, Collings IE, Jennings ES, Blanchard I, Kantor I, Kurnosov A, Cotte M, Dane T, Burghammer M, Rubie DC (2017) SiO_2 glass density to lower-mantle pressures. Phys Rev Lett 119:215701

Phillips JC (1984) Microscopic origin of anomalously narrow Raman lines in network glasses. J Non-Cryst Solids 63:347–355

Poe BT, McMillan PF, Coté B, Massiot D, Coutures J-P (1992) Silica–alumina liquids: in-situ study by high-temperature ^{27}Al NMR spectroscopy and molecular dynamics simulation. J Phys Chem 96:8220–8224

Poe BT, McMillan PF, Coté B, Massiot D, Coutures JP (1993) Magnesium and calcium aluminate liquids: in situ high-temperature ^{27}Al NMR spectroscopy. Science 259:786–788

Poe BT, McMillan PF, Coté B, Massiot D, Coutures J-P (1994) Structure and dynamics in calcium aluminate liquids: high-temperature ^{27}Al NMR and Raman spectroscopy. J Am Ceram Soc 77:1832–1838

Polian A, Grimsditch M (1990) Room-temperature densification of a-SiO_2 versus pressure. Phys Rev B 41:6086–6087

Polsky CH, Smith KH, Wolf GH (1999) Effect of pressure on the absolute Raman scattering cross section of SiO_2 and GeO_2 glasses. J Non-Cryst Solids 248:159–168

Prescher C, Prakapenka VB, Stefanski J, Jahn S, Skinner LB, Wang Y (2017) Beyond sixfold coordinated Si in SiO_2 glass at ultrahigh pressures. PNAS 114:10041–10046

Price DL (1996) Intermediate-range order in glasses. Curr Opin Solid State Mater Sci 1:572–577

Price DL (2010) High-Temperature Levitated Materials. Cambridge University Press

Price DL, Moss SC, Reijers R, Saboungi M-L, Susman S (1989) Intermediate-range order in glasses and liquids. J Phys: Condens Matter 1:1005–1008

Rahmani A, Benoit M, Benoit C (2003) Signature of small rings in the Raman spectra of normal and compressed amorphous silica: A combined classical and ab initio study. Phys Rev B 68:184202

Richet P (2012) Boson peak of alkali and alkaline earth silicate glasses: Influence of the nature and size of the network-modifying cation. J Chem Phys 136:034703

Richet P, Gillet P, Pierre A, Bouhifd MA, Daniel I, Fiquet G (1993) Raman spectroscopy, x-ray diffraction, and phase relationship determinations with a versatile heating cell for measurements up to 3600 K (or 2700 K in air). J Appl Phys 74:5451–5456

Rigden SM, Ahrens TJ, Stolper EM (1984) Densities of liquid silicates at high pressures. Science 226:1071–1074

Roskosz M, Toplis MJ, Neuville DR, Mysen BO (2008) Quantification of the kinetics of iron oxidation in silicate melts using Raman spectroscopy and assessment of the role of oxygen diffusion. Am Mineral 93:1749–1759

Rossano S, Mysen B (2012) Raman spectroscopy of silicate glasses and melts in geological systems. *In:* Raman spectroscopy applied to Earth sciences and cultural heritage, Dubessy J, Caumon M-C, Rull F (eds) EMU Notes Mineral 12:321–366

Rossano S, Balan E, Morin G, Bauer J-P, Calas G, Brouder C (1999) ^{57}Fe Mössbauer spectroscopy of tektites. Phys Chem Mineral 26:530–538

Rossano S, Ramos A, Delaye J-M, Creux S, Filippono A, Brouder C, Calas G (2000a) EXAFS and molecular dynamics combined study of CaO−FeO−2SiO_2 glass. New insight into site significance in silicate glasses. Europhys Lett 49:597–602

Rossano S, Ramos AY, Delaye J-M (2000b) Environment of ferrous iron in $CaFeSi_2O_6$ glass; contributions of EXAFS and molecular dynamics. J Non-Cryst Solids 273:48–52

Rouxel T, Ji H, Hammouda T, Moréac A (2008) Poisson's Ratio and the densification of glass under high pressure. Phys Rev Lett 100:225501

Rull F (2012) The Raman effect and the vibrational dynamics of molecules and crystalline solids. *In:* Raman spectroscopy applied to Earth sciences and cultural heritage, Dubessy J, Caumon M-C, Rull F (eds) EMU Notes Mineral 12:1–60

Sakamaki T, Wang Y, Park C, Yu T, Shen G (2012) Structure of Jadeite Melt at High Pressures up to 4.9 GPa. J Appl Phys 111:112623

Sakamaki T, Suzuki A, Ohtani E, Terasaki H, Urakawa S, Katayama Y, Funakoshi K-I, Wang Y, Hernlund JW, Ballmer MD (2013) Ponded melt at the boundary between the lithosphere and asthenosphere. Nat Geosci 6:1041

Sakamaki T, Wang Y, Park C, Yu T, Shen G (2014a) Contrasting behavior of intermediate-range order structures in jadeite glass and melt. Phys Earth Planet Inter 228:281–286

Sakamaki T, Kono Y, Wang Y, Park C, Yu T, Jing Z, Shen G (2014b) Contrasting sound velocity and intermediate-range structural order between polymerized and depolymerized silicate glasses under pressure. Earth Planet Sci Lett 391:288–295

Salmon PS (1994) Real space manifestation of the first sharp diffraction peak in the structure factor of liquid and glassy materials. Proc R Soc Lond A 445:351–365

Salmon PS (2018) Densification mechanisms of oxide glasses and melts. Chapter 13 *In:* Magmas Under Pressure, Kono Y, Sanloup C (Eds.), Elsevier, Holland, p 343–369

Salmon PS, Zeidler A (2013) Identifying and characterizing the different structural length scales in liquids and glasses: an experimental approach. Phys Chem Chem Phys 15:15286

Salmon PS, Zeidler A (2015) Networks under pressure: the development of in situ high-pressure neutron diffraction for glassy and liquid materials. J Phys: Condens Matter 27:133201

Salmon PS, Martin RA, Mason PE, Cuello GJ (2005) Topological versus chemical ordering in network glasses at intermediate and extended length scales. Nature 435:75–78

Salmon PS, Barnes AC, Martin RA, Cuello GJ (2006) Glass fragility and atomic ordering on the intermediate and extended range. Phys Rev Lett 96:235502

Salmon PS, Drewitt JWE, Whittaker DAJ, Zeidler A, Wezka K, Bull CL, Tucker MG, Wilding MC, Guthrie M, Marrocchelli D (2012) Density-driven structural transformations in network forming glasses: a high-pressure neutron diffraction study of GeO_2 glass up to 17.5 GPa. J Phys: Condens Matter 24:439601

Salmon PS, Moody GS, Ishii Y, Pizzey KJ, Polidori A, Salanne M, Zeidler A, Buscemi M, Fischer HE, Bull CL, Klotz S, Weber R, Benmore CJ, Macleod SG (2019) Pressure induced structural transformations in amorphous $MgSiO_3$ and $CaSiO_3$. J Non-Cryst Solids: X 3:100024

Sampath S, Benmore CJ, Lantzky KM, Neuefeind J, Leinenweber K, Price DL, Yarger JL (2003) Intermediate-range order in permanently densified GeO_2 Glass. Phys Rev Lett 90:115502

Sanloup C, Drewitt JWE, Crépisson C, Kono Y, Park C, McCammon C, Hennet L, Brassamin S, Bytchkov A (2013a) Structure and density of molten fayalite at high pressure. Geochim Cosmochim Acta 118:118–128

Sanloup C, Drewitt JWE, Konopkova Z, Dalladay-Simpson P, Morton DM, Rai N, van Westrenen W, Morgenroth W (2013b) Structural change in molten basalt at deep mantle conditions. Nature 503:104–107

Sarnthein J, Pasquarello A, Car R (1997) Origin of the High-Frequency Doublet in the Vibrational Spectrum of Vitreous SiO_2. Science 275:1925–1927

Sato T, Funamori N (2008) Sixfold-coordinated amorphous polymorph of SiO_2 under high pressure. Phys Rev Lett 101:255502

Sato T, Funamori N (2010) High-pressure structural transformation of SiO_2 glass up to 100 GPa. Phys Rev B 82:184102

Schirmacher W (2013) The boson peak. Physica Status Solidi (b) 250:937–253

Schirmacher W, Ruocco G, Scopigno T (2007) Acoustic attenuation in glasses and its relation with the boson peak. Phys Rev Lett 98:025501

Schirmacher W, Scopigno T, Ruocco G (2015) Theory of vibrational anomalies in glasses. J Non-Cryst Solids 407:133–140

Schmelzer JWP, Zanotto ED, Fokin VM (2005) Pressure dependence of viscosity. J Chem Phys 112:074511

Sears VF (1992) Neutron scattering lengths and cross sections. Neutron News 3:26–37

Seifert F, Mysen BO, Virgo D (1982) 3-dimensional network structure of quenched melts (glass) in the systems SiO_2–$NaAlO_2$, SiO_2–$CaAl_2O_4$ and SiO_2–$MgAl_2O_4$. Am Mineral 67:696–717

Shannon RD (1976) Revised effective ionic radii and systematic studies of interatomic distances in halides and chalcogenides. Acta Cryst A32:751–767

Sharma SK, Mammone JF, Nicol MF (1981) Raman investigation of ring configurations in vitreous silica. Nature 292:140–141

Shen G, Rivers ML, Wang Y, Sutton SR (2001) Laser heated diamond cell system at the Advanced Photon Source for in situ x-ray measurements at high pressure and temperature. Rev Sci Instr 72:1273–1282

Shen G, Prakapenka VB, Rivers ML, Sutton SR (2003) Structural investigation of amorphous materials at high pressures using the diamond anvil cell. Rev Sci Instr 74:3021–3026

Shen GY, Prakapenka VB, Rivers ML, Sutton SR (2004) Structure of liquid iron at pressures up to 58 GPa. Phys Rev Lett 92:185701

Shen G, Mei Q, Prakapenka VB, Lazor P, Sinogeikin S, Meng Y, Park C (2011) Effect of helium on structure and compression behavior of SiO_2 glass. PNAS 108:6004–6007

Shi Y, Neuefeind J, Ma D, Page K, Lambersona LA, Smith NJ, Tandia A, Song AP (2019) Ring size distribution in silicate glasses revealed by neutron scattering first sharp diffraction peak analysis. J Non-Cryst Solids 516:71–81

Shi C, Alderman OLG, Berman D, Du J, Neuefeind J, Tamalonis A, Weber JKR, You J, Benmore CJ (2019) The structure of amorphous and deeply supercooled liquid alumina. Front Mater 6:38

Shibazaki Y, Kono Y, Shen G (2019) Compressed glassy carbon maintaining graphite-like structure with linkage formation between graphene layers. Sci Rep 9:7531

Skinner LB, Barnes AC, Salmon PS, Crichton WA (2008) Phase separation, crystallization and polyamorphism in the Y_2O_3–Al_2O_3 system. J Phys: Condens Matter 20:205103

Skinner LB, Benmore CJ, Weber JKR, Tumber S, Lazareva L, Neuefeind J, Santodonato L, Du J, Parise JB (2012a) Structure of molten $CaSiO_3$: Neutron diffraction isotope substitution with aerodynamic levitation and molecular dynamics study. J Phys Chem B 116:13439–13447

Skinner LB, Barnes AC, Salmon PS, Fischer HE, Drewitt JWE, Honkimäki V (2012b) Structure and triclustering in Ba-Al–O glass. Physical Review B 85:064201

Skinner LB, Benmore CJ, Weber JKR, Wilding MC, Tumber SK, Parise JB (2013a) A time resolved high energy x-ray diffraction study of cooling liquid SiO_2. Phys Chem Chem Phys 15:8566–8572

Skinner LB, Barnes AC, Salmon PS, Hennet L, Fischer HE, Benmore CJ, Kohara S, Weber JKR, Bytchkov A, Wilding MC, Parise JB, Farmer TO, Pozdnyakova I, Tumber SK, Ohara K (2013b) Joint diffraction and modelling approach to the structure of liquid alumina. Phys Rev B 87:024201

Soignard E, Benmore CJ, Yarger JL (2010) A perforated diamond anvil cell for high-energy x-ray diffraction of liquids and amorphous solids at high pressure. Rev Sci Instr 81:035110

Sokolov AP (1999) Vibrations at the boson peak: random- and coherent-phase contributions. J Phys:Condens Matter 11:A213–A218

Sokolov AP, Kisliuk A, Soltwisch M, Quitmann D (1992) Medium-range order in glasses: comparison of Raman and diffraction measurements. Phys Rev Lett 69:1540–1543

Spiekermann, G, Steele-MacInnis, M, Kowalski PM, Schmidt C, Jahn S (2013) Vibrational properties of silica species in MgO–SiO$_2$ glasses obtained from ab initio molecular dynamics. Chem Geol 346:22–33

Spiekermann G, Harder M, Gilmore K, Zalden P, Sahle CJ, Petitgirard S, Wilke M, Biedermann N, Weis C, Morgenroth W, Tse JS, Kulik E, Nishiyama N, Yava⊠ H, Sternemann C (2019) Persistent octahedral coordination in amorphous GeO$_2$ Up to 100 GPa by K⊠'' X-Ray Emission Spectroscopy. Phys Rev X 9:011025

Stavrou E, Tsiantos C, Tsopouridou RD, Kripotou S, Kontos AG, Raptis C, Capoen B, Bouazaoui M, Turrell S, Khatir S (2010) Raman scattering boson peak and differential scanning calorimetry studies of the glass transition in tellurium–zinc oxide glasses. J Phys: Condens Matter 22:195103

Stebbins JF (1991) NMR evidence for five-coordinated silicon in a silicate glass at atmospheric pressure. Nature 351:638–639

Stebbins JF, Bista S (2019) Pentacoordinated and hexacoordinated silicon cations in a potassiumsilicate glass: Effects of pressure and temperature. J Non-Cryst Solids 505:234–240

Stebbins JF, McMillan P (1989) Five- and six-coordinated Si in K$_2$Si$_4$O$_9$ glass quenched from 1.9 GPa and 1200 °C. Am Mineral 74:965–968

Stebbins JF, Poe BT (1999) Pentacoordinate silicon in high-pressure crystalline and glassy phases of calcium disilicate (CaSi$_2$O$_5$). Geophys Res Lett 26:2521–2523

Stebbins JF, Xu Z (1997) NMR evidence for excess non-bridging oxygen in an aluminosilicate glass. Nature 390:60–62

Stebbins JF, Oglesby JV, Xu Z (1997) Disorder among network-modifier cations in silicate glasses: New constraints from triple-quantum 17O NMR. Am Mineral 82:1116–1124

Stebbins JF, Oglesby JV, Kroeker S (2001) Oxygen triclusters in crystalline CaAl$_4$O$_7$ (grossite) and in calcium aluminosilicate glasses: ^{17}O NMR. Am Mineral 86:1307–1311

Stolper E, Walker D, Hager BH, Hays JF (1981) Melt segregation from partially molten source regions: the importance of melt density and source region size. J Geophys Res 86:6261–6271

Sun K-H (1947) Fundamental condition of glass formation. J Am Ceram Soc 30:277–281

Susman S, Volin KJ, Price DL, Grimsditch M, Rino JP, Kalia RK, Vashishta P, Gwanmesia G, Wang Y, Liebermann RC (1991) Intermediate-range order in permanently densified vitreous SiO$_2$. A neutron-diffraction and molecular-dynamics study. Phys Rev B 43:1194–1197

Takaishi T, Takahashi M, Jin J, Uchino T, Yoko T, Takahashi M (2005) Structural study on PbO–SiO$_2$ glasses by x-ray and neutron diffraction and ^{29}Si MAS NMR measurements. J Am Ceram Soc 88:1591–1596

Takahashi Y, Osada M, Masai H, Fujiwara T (2009) Crystallization and nanometric heterogeneity in glass: In situ observation of the boson peak during crystallization. Phys Rev B 79:214204

Tangeman JA, Philips BL, Nordine PC, Weber JKR (2004) Thermodynamics and structure of single- and two-phase yttria–alumina glasses. J Phys Chem B 108:10663–10671

Taraskin SN, Elliott SR (1997) Nature of vibrational excitations in vitreous silica. Phys Rev B 56:8605–8622

Taraskin SN, Elliot SR (1999) Low-frequency vibrational excitations in vitreous silica: the Ioffe–Regel limit. J Phys Condens Matter 11:A219–A227

Taraskin SN, Loh YH, Nataranjan G, Elliott S (2001) Origin of the boson peak in systems with lattice disorder. Phys Rev Lett 86:1255–1261

Thomas BWM, Mead RN, Mountjoy G (2006) A molecular dynamics study of the atomic structure of (CaO)$_x$(Al$_2$O$_3$)$_{1-x}$ glass with $x = 0.625$ close to the eutectic. J Phys: Condens Matter 18:4697–4708

Tomoshige N, Mizuno H, Mori T, Kim K, Matubayasi N (2019) Boson peak, elasticity, and glass transition temperature in polymer glasses: Effects of the rigidity of chain bending. Sci Rep 9:19514

Toplis MJ, Dingwell DB, Lenci T (1997) Peraluminous viscosity maxima in Na$_2$O–Al$_2$O$_3$–SiO$_2$ liquids: The role of triclusters in tectosilicate melts. Geochim Cosmochim Acta 61:2605–2612

Toplis MJ, Kohn SC, Smith ME, Poplett IJF (2000) Fivefold-coordinated aluminum in tectosilicate glasses observed by triple quantum MAS NMR. Am Mineral 85:1556–1560

Trachenko K, Dove MT (2002) Densification of silica glass under pressure. J Phys: Condens Matter 14:7449

Umari P, Pasquarello A (2002) Modeling of the Raman spectrum of vitreous silica: concentration of small ring structures. Physica B 316–317:572–574

Umari P, Gonze X, Pasquarello A (2003) Concentration of small ring structures in vitreous silica from a first-principles analysis of the Raman spectrum. Phys Rev Lett 90:02741

Vaccari M, Aquilanti G, Pascarelli S, Mathon O (2009) A new EXAFS investigation of local structural changes in amorphous and crystalline GeO2 at high pressure. J Phys Condens Matter 21:145403

Virgo D, Mysen BO (1985) The structural state of iron in oxidized vs. reduced glasses at 1 atm: A ^{57}Fe Mössbauer study. Phys Chem Min 12:65–76

Waasmaier D, Kirfel A (1995) New analytical scattering-factor functions for free atoms and ions. Acta Cryst. A51: 416–431

Wang Z, Cooney TF, Sharma SK (1993) High temperature structural investigation of Na$_2$O·0.5Fe$_2$O$_3$·3SiO$_2$ and Na$_2$O·FeO·3SiO$_2$ melts and glasses. Contrib Mineral Pet 115:112–122

Wang Z, Cooney TF, Sharma SK (1995) In situ structural investigation of iron-containing silicate liquids and glasses. Geochim Cosmochim Acta 59:1571–1577

Wang Y, Sakamaki T, Skinner LB, Jing Z, Yu T, Kono Y, Park C, Shen G, Rivers ML, Sutton SR (2014) Atomistic insight into viscosity and density of silicate melts under pressure. Nature Commun 5:3241

Wang Y, Hong L, Wang Y, Schirmacher W, Zhang J (2018) Disentangling boson peaks and Van Hove singularities in a model glass. Phys Rev B 98:174207

Warren BE (1934a) Identification of crystalline substances by means of x-rays. J Am Ceram Soc 17:73–77

Warren BE (1934b) The diffraction of x-rays in glass. Phys Rev B 45:657–661

Warren BE (1934c) X-ray determination of the structure of glass. J Am Ceram Soc 17:249–254

Warren BE, Krutter H, Morningstar O (1936) Fourier analysis of x-ray patterns of vitreous SiO$_2$ and B$_2$O$_3$. J Am Ceram Soc 19:202–206

Waseda Y, Toguri JM (1978) The structure of the molten FeO–SiO$_2$ system. Metall Trans B 9:595–601

Waseda Y, Shiraishi Y, Toguri JM (1980) Determination of the molten FeO–Fe$_2$O$_3$–SiO$_2$ system by X-ray diffraction. Trans Jpn Inst Met 21:51–62

Watanuki T, Shimomura O, Yagi T, Kondo T, Isshiki M (2001) Construction of laser-heated diamond anvil cell system for in situ x-ray diffraction study at SPring-8. Rev Sci Instr 72:1289–1292

Waychunas GA, Brown Jr GE, Ponader CW, Jackson WE (1988) Evidence from X-ray absorption for network-forming Fe^{2+} in molten alkali silicates. Nature 332:251–253

Weber JKR, Benmore CJ, Tangeman JA, Siewenie J, Hiera KJ (2003) Structure of binary CaO–Al$_2$O$_3$ and SrO–Al$_2$O$_3$ liquids by combined levitation-neutron diffraction. J Neutron Res 11:113–121

Weber JKR, Abadie JG, Hixson AD, Nordine PC, Jerman GA (2004a) Glass formation and polymorphism in rare-earth oxide–aluminum oxide compositions. J Am Ceram Soc 83:1868–1872

Weber JKR, Benmore CJ, Siewenie J, Urquidi J, Key TS (2004b) Structure and bonding in single- and two-phase alumina-based glasses. Phys Chem Chem Phys 6:2480–2483

Weigel C, Cormier L, Galoisy L, Calas G (2006) Determination of Fe^{3+} sites in a NaFeSi$_2$O$_6$ glass by neutron diffraction with isotopic substitution coupled with numerical simulation. Appl Phys Lett 89:141911

Weigel C, Cormier L, Calas G, Galoisy L, Bowron DT (2008a) Intermediate-range order in the silicate network glasses NaFe$_x$Al$_{1-x}$Si$_2$O$_6$ ($x = 0$:0.5:0.8:1): A neutron diffraction and empirical potential structure refinement modeling investigation. Phys Rev B 78:064202

Weigel C, Cormier L, Calas G, Galoisy L, Bowron DT (2008b) Nature and distribution of iron sites in a sodium silicate glass investigated by neutron diffraction and EPSR simulation. J Non-Cryst Solids 354:5378–5385

Wilding M, Bingham PA, Wilson M, Kono Y, Drewitt JWE, Brooker RA, Parise JB (2019) CO$_{3+1}$ network formation in ultra-high pressure carbonate liquids. Sci Rep 9:15416

Wezka K, Salmon PS, Zeidler A, Whittaker DAJ, Drewitt JWE, Klotz S, Fischer HE, Marrocchelli D (2012) Mechanisms of network collapse in GeO$_2$ glass: high-pressure neutron diffraction with isotope substitution as arbitrator of competing models. J Phys: Condens Matter 24:502101

Wilding MC, McMillan PF, Navrotsky A (2002) Thermodynamic and structural aspects of the polyamorphic transition in yttrium and other rare-earth aluminate liquids. Physica A 314:379–390

Wilding MC, Wilson M, McMillan PF (2005) X-ray and neutron diffraction studies and MD simulation of atomic configurations in polyamorphic Y$_2$O$_3$–Al$_2$O$_3$ systems. Phil Trans R Soc A 363:589–607

Wilding MC, Benmore CJ, Weber JKR (2010) Changes in the local environment surrounding magnesium ions in fragile MgO–SiO$_2$ liquids. Europhys Lett 89:26005

Wilding MC, Wilson M, McMillan PF, Benmore CJ, Weber JKR, Deschamps T, Champagnon B (2015) Structural properties of Y$_2$O$_3$–Al$_2$O$_3$ liquids and glasses: An overview. J Non-Cryst Solids 407:228–234

Wilke M (2005) Fe in magma—An overview. Ann Geophys 48:609–617

Wilke M, Farges F, Petit P-E, Brown GE, Martin F (2001) Oxidation state and coordination of Fe in minerals: An Fe K-XANES spectroscopic study. Am Mineral 86:714–730

Wilke M, Partzsch GM, Bernhardt R, Lattard D (2004) Determination of the iron oxidation state in basaltic glasses using XANES at the K-edge. Chem Geol 213:71–87

Wilke M, Farges F, Partzsch GM, Schmidt C, Behrens H (2007) Speciation of Fe in silicate glasses and melts by in-situ XANES spectroscopy. Am Mineral 92:44–56

Wilson M, McMillan PF (2004) Interpretation of x-ray and neutron diffraction patterns for liquid and amorphous yttrium and lanthanum aluminum oxides from computer simulation. Phys Rev B 69:054206

Williams Q, Jeanloz R (1988) Spectroscopic evidence for pressure-induced coordination changes in silicate glasses and melts. Science 239:902–905

Worrell CA, Henshall T (1978) Vibrational spectroscopic studies of some lead silicate glasses. J Non-Cryst Solids 29:283–299

Wright AC, Clarke SJ, Howard CK, Bingham PA, Forder SD, Holland D, Martlew D, Fischer HE (2014) The environment of Fe^{2+}/Fe^{3+} cations in a soda–lime–silica glass. Phys Chem Glasses: Eur J Glass Sci Technol B 55:243–252

Wu Z, Bonnin-Mosbah M, Duraud JP, Métrich N, Delaney JS (1999) XANES studies of Fe-bearing glasses. J Synchrotron Rad 6:344–346

Xue X, Stebbins RG, Stebbins JF (1989) Silicon coordination and speciation changes in a silicate liquid at high pressures. Science 245:962–964

Xue X, Stebbins JF, Kanzaki M, McMillan PF, Poe B (1991) Pressure-induced silicon coordination and tetrahedral structural changes in alkali oxide–silica melts up to 12 GPa: NMR, Raman, and infrared spectroscopy. Am Mineral 76:8–26

Yamada A, Inoue T, Urakawa S, Funakoshi K-I, Funamori N, Kikegawa T, Ohfuji H, Irifune T (2007) In situ X-ray experiment on the structure of hydrous Mg-silicate melt under high pressure and high temperature. Geophys Res Lett 34:L10303

Yamada A, Wang Y, Inoue T, Yang W, Park C, Yu T, Shen G (2011) High-pressure x-ray diffraction studies on the structure of liquid silicate using a Paris–Edinburgh type large volume press. Rev Sci Instr 82:015103

Yarger JL, Smith KH, Nieman RA, Diefenbacher J, Wolf GH, Poe BT, McMillan PF (1995) Al coordination changes in high-pressure aluminosilicate liquids. Science 270:1964–1967

Zachariasen WH (1932) The atomic arrangement in glass. J Am Chem Soc 54:3841–3851

Zahra AM, Zahra CY, Piriou B (1993) DSC and Raman studies of lead borate and lead silicate glasses. J Non-Cryst Solids 155:45–55

Zeidler A, Salmon PS (2016) Pressure-driven transformation of the ordering in amorphous network-forming materials. Phys Rev B 93:214204

Zeidler A, Drewitt JWE, Salmon PS, Barnes AC, Crichton WA, Klotz S, Fischer HE, Benmore CJ, Ramos S, Hannon AC (2009) Establishing the structure of GeS_2 at high pressures and temperatures: a combined approach using x-ray and neutron diffraction. J Phys: Condens Matter 21:474217

Zeidler A, Wezka K, Rowlands RF, Whittaker DAJ, Salmon PS, Polidori A, Drewitt JWE, Klotz S, Fischer HE, Wilding MC, Bull CL, Tucker MG, Wilson M (2014a) High-pressure transformation of SiO_2 glass from a tetrahedral to an octahedral network: a joint approach using neutron diffraction and molecular dynamics. Phys Rev Lett 113:135501

Zeidler A, Salmon PS, Skinner LB (2014b) Packing and the structural transformations in liquid and amorphous oxides from ambient to extreme conditions. PNAS 111:10045–10048

Zha C-s, Hemley RJ, Mao H-k, Duffy TS, Meade C (1994) Acoustic velocities and refractive index of SiO_2 glass to 57.5 GPa by Brillouin scattering. Phys Rev B 50:13105–13112

Zhai S, Ito E (2011) Recent advances of high-pressure generation in a multianvil apparatus using sintered diamond anvils. Geosci Front 2:101–106

Zhang HL, Hirschmann MM, Cottrell E, Newville M, Lanzirotti A (2016) Structural environment of iron and accurate determination of Fe^{3+}/⊠Fe ratios in andesitic glasses by XANES and Mössbauer spectroscopy. Chem Geol 428:48–58

Zhang HL, Cottrell E, Solheid PA, Kelley KA, Hirschmann MM (2018) Determination of Fe^{3+}/⊠Fe of XANES basaltic glass standards by Mössbauer spectroscopy and its application to the oxidation state of iron in MORB. Chem Geol 479:166–175

Reviews in Mineralogy & Geochemistry
Vol. 87 pp. 105-162, 2022
Copyright © Mineralogical Society of America

3

Link between Medium and Long-range Order and Macroscopic Properties of Silicate Glasses and Melts

Daniel R. Neuville and Charles Le Losq

Géomatériaux, Institut de physique du globe de Paris
CNRS, Université de Paris Cité
Paris, France

neuville@ipgp.fr

lelosq@ipgp.fr

INTRODUCTION

The two first chapters of this volume focus on the structure of magmas at different scales: short and medium range order. A quick review of the glass literature, from the 90's or older, reveals that most published papers focused on the structure of glasses, or on properties of glasses or melts. Depending on their original discipline, the work of the scientific community in those areas are different. Physicists published papers on glass structure, including the structure of SiO_2, GeO_2, B_2O_3 as seen by X-ray or neutron diffraction and Raman spectroscopy for example (Galeener et al. 1983; Wright 1990; Gaskell et al. 1991). Chemists worked on more complex glass compositions, like chalcogenide (Poulain and Lucas 1970; Poulain 1983), or focused their work on specific properties of glasses like refractive index, density or the oxidation state of multivalent elements (Schreiber 1986). Earth scientists worked on properties of melts involved in geologic phenomena, like diffusivity, heat capacity, density or viscosity (Bottinga and Weill 1970, 1972; Bacon 1977; Carmichael et al. 1977; Robie et al. 1979; Ryan and Blevins 1987). This has changed over time. Since the 2000's, it is easier to simultaneously investigate the structure and properties of glasses. Industrial and technological needs evolved, and the interest of the Earth sciences community became more and more focused on the acquisition and interpretation of in situ data to address problems requiring insights about silicate melts at high temperature and high pressure. As a result, the scientific community became more unified, performing more and more studies that simultaneously investigated the structure and properties of glasses and melts. Presently, we have reached a point where it is possible to link together experimental and structural/thermodynamic data to build models for solving industrial or Earth sciences problems. The aim of this chapter is to show how this is possible.

Glass structure versus macroscopic properties

Among the most important questions about glasses and melts, one is critical for many applications and studies: how the melt/glass structure affects macroscopic properties? For Doremus (1973), a glass is an amorphous solid. The term solid implies a high viscosity, usually greater than 10^{10} Pa·s. This viscosity therefore limits the flow of the body. The amorphous term implies the absence of long-range order (see Fig. 5A in Drewitt et al. 2022, this volume), which reveals an analogy with the liquid state. So, a glass is a solid whose properties are similar to those of liquids. Parks and Huffman (1926) even talk about "a fourth state of matter". However, no consensus exists regarding glass, and the nature of glass and the glass transition are two fundamental questions that remain open in condensed matter physics (e.g., see the different definitions in and debate between Zanotto and Mauro 2017; Popov 2018; Schmelzer and Tropin 2018).

1529-6466/22/0087-0003$10.00 (print)
1943-2666/22/0087-0003$10.00 (online)

http://dx.doi.org/10.2138/rmg.2022.87.03

An important point is that glasses can exhibit very different chemical compositions and can be prepared by different means. Mineral (i.e., silicate), oxide, non-oxide, metal and organic glasses exist (see for an exhaustive description in Musgraves et al. 2019). In fact, glasses are found regardless of the type of chemical bonds that link the atoms they contain: covalent, ionic, metallic, Van der Waals or hydrogen bonding. The glassy state is a characteristic of condensed matter. Glasses can be obtained by the rapid quench of liquids, or of a gas by solid condensation, by amorphization of a crystalline phase or by sol-gel methods (Descamps 2017). The most common method for obtaining a glass is by the rapid quench of a liquid[1]. During the quench, viscosity continuously increases up to a value so high that the final state can be considered as solid. One can therefore imagine the structure of the glass as being similar to that of a liquid whose movements are hindered. Glasses, like undercooled liquids, have a disordered structure that was frozen in at the glass transition temperature.

A fundamental question is *what happens when a liquid crosses the glass transition temperature?* The latter is the temperature (often abbreviated T_g, and more realistically it is a small temperature interval) at which second order thermodynamic properties like heat capacity show a transition, revealing the locking-in of the atoms in fixed (but undetermined) positions. In essence, T_g is the temperature at which a number of thermodynamic properties go from liquid-like to solid-like values.

Unlike a crystallized solid, it is not possible to introduce the term of melting temperature for the glassy state. The solidification of a liquid into glass is accompanied, in fact, by a continuous and gradual increase in viscosity upon cooling, without the appearance of a crystalline structure (Fig. 1). This behavior remains true regardless of the glass/melt chemical composition (silicate, aluminosilicate, borate, borosilicate, germanate, chalcogenate, tellurate, metallic, organic... see for details regarding each glass family the Musgraves et al. 2019). This shows the continuous transition from glass to liquid for a property such as viscosity (Tamman 1925; see Fig. 1). The glass transition is a dynamic phenomenon that characterizes the loss of internal thermodynamic equilibrium. Indeed, the properties of a glass no longer depend solely on pressure and temperature, but also on the temperature at which the glass transition occurs. Finally, Tamman (1925) sums the concept as "The viscosity of a liquid increases with increasing undercooling, and in a rather *narrow temperature interval it increases very rapidly to values characteristic of solid crystals. A brittle glass is thus formed from an easily mobile liquid. This change in viscosity does not correspond to the behavior of the other properties, which in this temperature interval change relatively only slightly. The change in viscosity is a continuous one and no temperature can be chosen as the freezing-point, the point at which the liquid becomes solid. Glasses are undercooled liquids."*

The simplest and earliest characterization of the glass transition is due to Parks and Huffmann (1927), who investigated organic liquids. They note Nernst (1911) has stated that heating a glass *"externally it has the properties of a solid, owing to great viscosity and considerable rigidity, produced by strong mutual action of the molecules. An amorphous body differs from a crystal, however, in its complete isotropy and absence of a melting point; on heating, it passes continuously from the amorphous to the usual liquid state, as its properties show steady change with rise of temperature, and no breaks anywhere."*

In fact, Parks and Huffmann (1927) stated *"While there is no definite temperature, comparable to the melting point of a crystal, at which all properties undergo a sharp change, there is nevertheless a temperature interval, definite and reproducible, in which a number of properties change with a rapidity approaching that observed in the case of the melting process of a crystal. In brief, there is a softening region instead of a melting point. The glass as it exists below this softening region differs so markedly from the liquid existing above that it might well be considered as a different state of the substance."*

---

[1] "Rapid cooling rate" means 15°/min, the term "classic cooling rate" is also used.

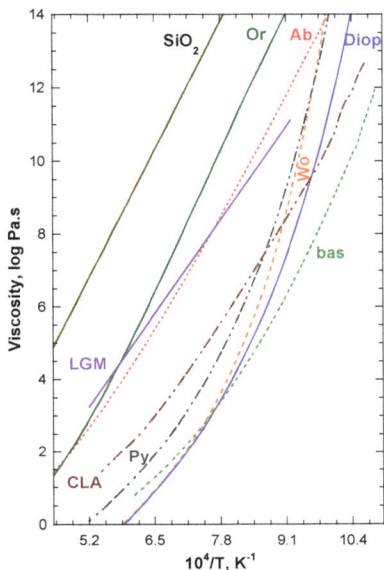

Figure 1. Viscosity versus 1/T curves for different melt compositions. LGM and CLA are rhyolite and andesite melts from Neuville et al. (1993), bas is a basalt melt from Villeneuve et al. (2008), SiO_2 is silica melt from Heterington et al. (1964) and Urbain et al. (1982), Ab and Or are albite, $NaAlSi_3O_8$, and orthoclase, $KAlSi_3O_8$ melts from (Le Losq and Neuville 2013; Le Losq et al. 2017); Py, Wo, Diop, are pyrope, $Mg_3Al_2Si_3O_{12}$, wollastonite, $CaSiO_3$, and diopside, $(CaMg)SiO_3$, melt compositions (Neuville and Richet 1991).

From an energetic point of view, Moyniham et al. (1974) showed that the variations in relative enthalpy and heat capacity with temperature for two different cooling or heating speeds exhibit a sudden change in temperature and corresponds to a dampened variation of these two properties. They showed that for the same liquid, the glass transition temperature varies with the cooling or heating rate. The higher the cooling rate, the higher relative enthalpy or heat capacity (Fig. 2). The glass transition thus corresponds to a small temperature and pressure interval upon which properties such as heat capacity undergo a second order transition (Moyniham et al. 1974). This interval is at higher temperatures at high cooling rates (situation A in Fig. 2), and moves to lower temperatures at slower cooling rates (situation B in Fig. 2).

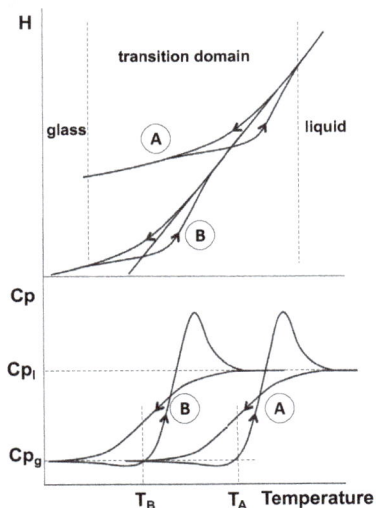

Figure 2. Relative enthalpy and heat capacity versus temperature for two different cooling rates, A and B correspond respectively to fast and low temperature change (redrafted from Moyniham et al. 1974). C_{p_l} and C_{p_g} correspond to the heat capacities of the liquid and the glass, respectively. T_B and T_A correspond to the glass transition at the two different rates A and B, respectively.

It is theoretically possible to obtain identical variations for all properties as a function of time at the same temperature (thermal expansion, viscosity for example) but it is not necessary (Moyniham, et al. 1974). With sudden changes in temperature, variations of the liquid properties linked to atomic mobility require some time to reach a new equilibrium. This time is called the relaxation time, τ. A liquid has a large number of different configurational states (Goldstein 1969, 1976). Each state corresponds to a minimum in potential energy and when temperature decreases, the number of possible configurations decreases as the domains for structural rearrangement become larger and larger. As a result, the time for structural relaxation, τ, of the liquid increases. Crossing the glass transition leads to the atoms trapped in given but disorganised positions. This results in quenching the liquid into a glass. The configuration state of the glass does not change from T_g to 0 K, as shown by the fact that the residual entropy of glass remains constant below T_g (e.g., see Richet et al. 1986, 1991; Richet and Neuville 1992; Tequi et al. 1993; Goldstein 2011; Schmelzer et al. 2018 and references cited therein). It should be noted that Raman and infrared vibrational spectroscopy data suggest that the structure of a glass is an image of the instantaneous configuration of the liquid at T_g (Sharma et al. 1978; Shevyakov et al. 1978; Kashio et al 1980; Kusabiraki and Shiraishi 1981; Kusabiraki 1986; Neuville and Mysen 1996).

The uniform cooling of a liquid at a rate $q = dT/dt$ can be likened to a series of instant temperature jumps $\Delta T = T_{fi} - T_i$, where T_{fi} and T_i are the final and initial temperatures, each jump being followed by a Δt period during which the temperature remains constant and equal to T_{fi}. Just after each jump, the viscosity of the liquid, η_l, is still worth η_{T_i} and differs from $\eta_{T_{fi}} - \eta_{T_i}$ from the new balance value, $\eta_{T_{fi}}$. This is called relaxation, the process by which the liquid tends to reach the state of equilibrium associated with the final temperature T_{fi}. To characterize the kinetics of this evolution, we can define a viscous relaxation time τ_η as equal to:

$$\tau_\eta = (\eta_l - \eta_{T_{fi}}) / (\delta\eta / \tau_\eta) \tag{1}$$

Experimentally, τ_η increases when T decreases. Three cases can be distinguished:

1. $\Delta t \gg \tau_\eta$, the substance has a relatively long time to equilibrate its structure at the new temperature. We are in the liquid state or in a glass state at the thermodynamic equilibrium for which the equilibrium viscosity is reached almost instantaneously. During cooling, the viscosity increases to its equilibrium value.
2. $\Delta t \sim \tau_\eta$, this corresponds to the glass transition domain. Viscosity depends on both the time and temperature to which the glass was previously subjected.
3. $\Delta t \ll \tau_\eta$, the relaxation time is much higher than the measurement time. No configurational rearrangement is then possible. The liquid freezes. Only the vibrational part of the heat capacity remains, which is close to the heat capacity of the crystal.

Relaxation times depend heavily on temperature (Simmons et al. 1970, 1974; Rekhson 1975, 1989; Dingwell and Webb 1990), as well as on the structure and therefore glass composition; e.g., borosilicate glasses have higher relaxation times than silicate or aluminosilicate glass compositions (Sipp et al. 1997). Figure 3 shows, for a soda-lime silicate glass, that relaxation time increases with decreasing temperature: e.g., equilibrium is reached after 500, 1200, 1600 min respectively at 795, 788 and 777 K.

However, it is important to note that there is no reason why the relaxation times of the various properties should be identical at equal cooling speeds. Moynihan et al. (1976a) indicate that *"for example, the enthalpy changes occurring during the approach to equilibrium of a network glass following a change in temperature could be considered to arise from the breaking of some of the network bonds, while the volume changes are due to rearrangement of the structure into less densely packed configurations. One cannot say a priori, however, which of these two processes should on the average occur earlier in time. That is, bond breaking might be a necessary precursor to volume-changing rearrangements of the structure, so that one would expect H to relax faster than V. On the other hand, configurational rearrangements leading to large volume changes might occur early in time, stressing the network and leading to subsequent bond breaking, so that one would expect V to relax faster than H."* In practice, it appears that the differences in relaxation time between volumetric and calorimetric relaxation times are relatively small, if not almost non-existent, and are essentially due to differences in cooling or heating rates (Sasabe et al. 1977; Moynihan and Gupta 1978; Dingwell and Webb 1990). Differences between relaxation times for the same material but determined with methods sensitive to different properties are clearly visible when comparing data from dilatometry, calorimetry and viscosimetry. For example, in the case of alkaline-earth aluminosilicate glass compositions, drop calorimetry T_g are generally found at temperatures at which viscosities are of 10.7 ± 0.5 log Pa·s, a value lower than that of ~12 log Pa·s that is known to be the reference for the determination of the viscous T_g (Neuville and Richet 1991; Tequi et al. 1991).

Thermodynamic approach to glass transition

We have seen that temperature and pressure are not sufficient parameters to characterize the state of glass. Tool and Eichlin (1931) introduced the concept of fictive temperature, T_{fic}, to characterize a glass at constant pressure. It can be defined as the temperature at which the glass would be in an equilibrium state (as a melt) if it could be heated and measured instantly. The fictive temperature can be defined during cooling in dilatometry or calorimetry, as the glassy transition temperature.

It is possible to give a formal thermodynamic definition of the fictive temperature, considering it as an order parameter. Moynihan et al. (1974, 1976a,b) discussed fictive temperatures, their influence on enthalpy properties and order parameters. It can be recalled, that the glass transition looks like a second-order transition for which volume, enthalpy and viscosity are continuous functions of T but not heat capacity, dilatation and compressibility coefficients. For the latter parameters, a steep transition is observed when temperature crosses the glass transition.

In reality, fictive temperature is a critical parameter as it reflects how viscosity, hence relaxation time, can show non-equilibrium behavior and can depend on time in the supercooled domain. In Figure 3, the effect of fictive temperature is clearly visible. The downward curve is obtained from a sample that has a fictive temperature below the measurement temperature. Similar behaviors are visible for density, or refractive index (Winter 1943; Ritland 1954). When the fictive temperature of the glass is higher than the measurement temperature, the glass has a more disordered configurational state than the state it should have if the glass were in thermodynamic equilibrium, i.e., its "fictive" configuration entropy is too high. It has a lower viscosity than its equilibrium viscosity, so its viscosity increases over time until the fictive temperature equals the measurement temperature. For a fictive temperature lower than that of measurement, the viscosity decreases until the establishment of the thermal equilibrium characterized by the equality of the two temperatures. In Figure 2, the relaxation curves over time for the cases of cooling or heating are asymmetrical. This is because a liquid increases its entropy faster with an increase in temperature (Rekhson 1980; Dingwell and Webb 1990; Sipp et al. 1997)[2].

Viscosity and glass transition

The glass transition is closely related to transport phenomena and more specifically to viscosity (Gibbs and Dimarzio 1958; Adam and Gibbs 1965; Goldstein 1969; Grest and Cohen 1980; Richet 1984; Scherer 1984), one of the fundamental properties of liquids that intervenes in transport processes. For the experimentalist, liquid viscosity also can be considered as a structural probe. Many models attempt to describe the variation of viscosity with temperature.

---

[2] A fictive pressure, P_{fic}, can be defined as the pressure at which the glass would be in an equilibrium state if it were subjected instantly to pressure equal to P_{fic}, at constant temperature.

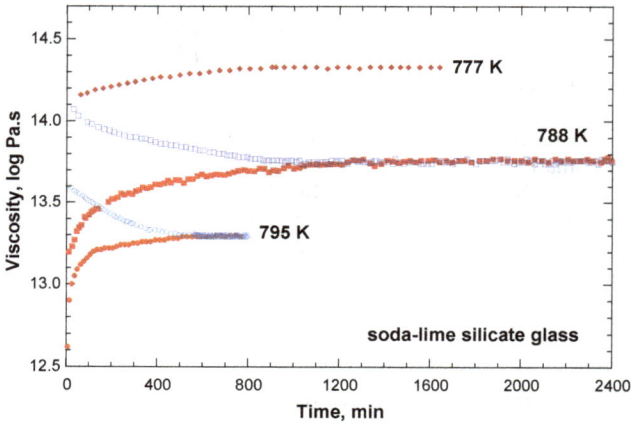

Figure 3. Viscosity versus time for a soda-lime silicate glass at different temperatures (redrafted from Sipp et al. 1997). We can note that increasing viscosities represent relaxation after previous measurements performed at higher temperatures, whereas decreasing viscosities indicate relaxation after annealing at lower temperatures.

Among them, we can cite the empirical Arrhenius and Tamman–Vogel–Flucher (TVF) equations, or the thermodynamic Adam–Gibbs (AG) model (see also the section *Silicate glasses and melts*) allows linking melt mobility to its thermodynamic properties such as heat capacity and configurational entropy, and can be used for viscosity predictions and extrapolations (Neuville and Richet 1990). Le Losq and Neuville (2017) further showed that the AG model allows linking melt structural knowledge to thermodynamic properties and viscosity, in order to build complete, extensive models for property predictions.

Configurational properties and glass structure

Configuration properties are the key to understand the difference between liquids and glasses. To illustrate the importance of such configurational aspects, consider the second order thermodynamic properties of crystal, glass and liquid pyrope ($Mg_3Al_2Si_3O_{12}$).

Table 1. Heat capacity of pyrope and $Mg_3Al_2Si_3O_{12}$, glass and liquid in $J \cdot mol^{-1} \cdot K^{-1}$. Data are from Tequi et al. (1991).

$T(K)$	Crystal	Glass	Liquid
298	325.6	327.8	
1020	492.2	504.4	665.7
1700	522.6		689.5

Above room temperature, the glass and crystalline phase have similar heat capacities, C_p. On the contrary, the C_p of the liquid is 25% higher than that of solids (crystal or glass). Moreover, it should also be noted that the C_p of the liquid usually increases with temperature of the liquid and cannot be considered as a constant. These differences can have a significant impact when you integrate with temperature or pressure, as shown by the example of pyrope and $Mg_3Al_2Si_3O_{12}$ liquid. For instance, a simple extrapolation of the heat capacity of the $Mg_3Al_2Si_3O_{12}$ liquid to low temperatures would produce a liquid with a lower entropy than that of its crystalline form below 694 K (Fig. 4A). This temperature actually corresponds to the Kauzmann temperature, a temperature below which the liquid entropy would become lower than that of a crystal with the same composition. Such a system does not exist because

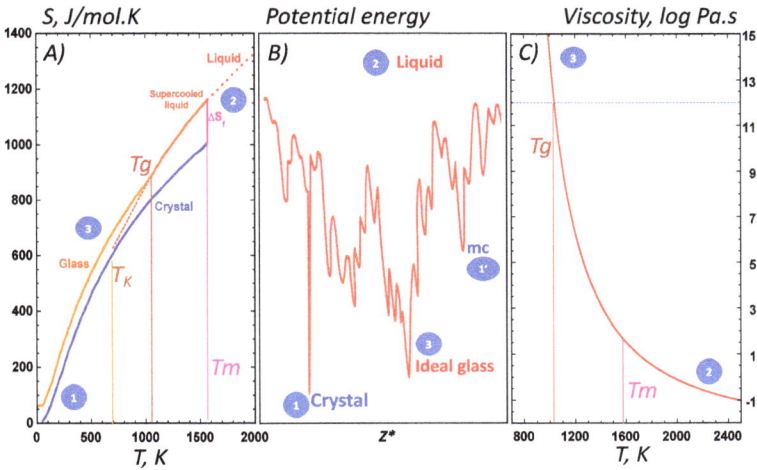

Figure 4. A) entropy versus temperature (Tequi et al. 1991), **B)** schematic representation of the complex higher dimensional potential energy hypersurface of an N-particle system in terms of a two-dimensional diagram of chemical potential versus some collective configuration coordinate Z^*, which defies precise definition, but the width of the funnel represents entropy (Angell 1991; Mossa et al . 2002), and **C)** viscosity versus temperature for the $Mg_3Al_2Si_3O_{12}$ composition (Neuville and Richet 1991). The same numbers are all connected together in the three figures. 1 is the Pyrope crystal that presents the minimum potential energy. 1′ is a metastable crystal, mc, that can have a potential energy higher than that observed for the glass. 2 is the liquid with a low viscosity, high entropy and high potential energy, in which atoms can move rapidly from one position to one other. 3 is the glass, in which atoms can move very slowly, there is a relaxation time, viscosity increase and entropy decrease, and the atom can fall in different potential energy minima and stay inside or jump to another with a lower potential energy. T_m is melting temperature, T_g, glass transition temperature, T_K Kauzmann temperature, ΔS_f entropy of melting at the melting temperature. Supercooled liquids are liquids between melting temperature and glass transition temperature.

a property of a liquid like the entropy should always remain higher than that of its crystalline form at a given temperature. The violation of this idea is known as the Kauzmann paradox (Kauzmann 1948)[3]. The pyrope example shows that configuration terms play an important role on the properties of liquids. It is therefore necessary to understand how configurational properties vary, and their link to melt and glass structure. Figure 4A shows the variation of the entropy of a crystal, glass and liquid for the $Mg_3Al_2Si_3O_{12}$ composition as a function of temperature. It is clear that the residual entropy corresponds to the difference between the entropy of the glass and that of the crystal at 0 K. This residual entropy is very small compared to the entropy of glass, crystal and liquid. This residual entropy is called configurational entropy, S^{conf}, and it is the key to understanding glasses and liquids. To determine S^{conf}, it is necessary to solve the entropic cycle represented in Figure 4A. However, this is only possible for minerals that melt congruently, and with a composition that can exist in a stable form as crystal, liquid and glass. Measuring the heat capacity of the crystal, C_{p_c}, from 0 K to the melting temperature, T_m, allows determining its absolute entropy S_c at T_m:

$$S_c = \int_0^{T_m} \frac{C_{p_c}}{T}\,dt \qquad (2)$$

By adding the entropy of fusion of the crystal, ΔS_f, to S_c, the entropy of the liquid S_l is then obtained at T_m. Measuring the heat capacity between the melting temperature, T_m, and the glass transition temperature, T_g, allows determining the heat capacity of the liquid and the

---

[3] This well-known paradox in temperature is recently proposed via pressure, but without any experimental basis, see the papers of Schmelzer et al. (2016) and Mauro (2011).

supercooled liquid, C_{p_l}. The measurement of the heat capacity of the glass between T_g and 0 K provides C_{p_g}. Finally, the residual entropy, also known as the so-called configurational entropy S^{conf}, is obtained at 0 K via:

$$S^{conf} (0 \text{ K}) = \int_0^{T_m} \frac{C_{p_c}}{T} dt + \Delta S_f + \int_{T_m}^{T_g} \frac{C_{p_l}}{T} dt + \int_{T_g}^0 \frac{C_{p_g}}{T} dt \qquad (3)$$

This calculation is illustrated in Figure 4A for a composition of $Mg_3Al_2Si_3O_{12}$ (Tequi et al. 1991). As this Figure shows, S^{conf} represents a small difference between large numbers, and its value will be affected by a very large error if calorimetric measurements are not made with the highest possible accuracy. For this purpose, the studies focused on determining S^{conf} from calorimetry measurements require high-precision measurements of thermal capacity: heat capacities need to be determined better than 0.2% by adiabatic calorimetry between 0 K and ambient, and better than 0.5% by high-temperature drop calorimetry (e.g., Tequi et al. 1991). Aside from the time and the difficulty of performing such high precision heat capacity measurements, one of the main drawbacks of the calculation of the configurational entropy through the thermodynamic cycle is the necessity to study minerals that melt congruently, with compositions existing as glass, liquid and crystal as stated previously. Only a few chemical compositions can be studied in this way. Fortunately, as we shall see later, S^{conf} can be determined through the use of the Adam and Gibbs model combined with viscosity measurements.

Figure 4B schematically illustrates the positions of atoms in a crystal as a function of the collective configuration coordinate, as determined by local minima of interatomic potentials that are at specific positions, allowing the building of a repetitive pattern characterized by long range order. In a glass, the bond angles and interatomic distances are not constant but extend over a range of relatively close values. Long-range order does not exist, and one can represent the interatomic potentials as a plane showing minima separated by barriers with varying shapes and heights. As a result, glass entropy is higher than that of crystal at given temperature (point 3, Fig. 4). The ideal glass with the minimum potential exists, but also other minima can exist with higher potential for the glass state or even the metastable crystal (point 1'). Now suppose that a certain amount of heat is brought instantly to the glass. Below the glass transition, the thermal energy brought upon increasing temperature is accommodated as vibrations, via an increase in average vibration amplitudes. The specific heat is vibrational in nature and the material behaves like a solid. At and above T_g, the thermal energy becomes important enough to allow the atoms to cross the energy barriers that separate the different configurational states (Richet and Neuville 1992). This configurational contribution is necessarily positive, and the liquid or glass states can be defined by the existence or absence of this contribution (Davis and Jones 1953). In a way, the glass transition can be seen as the beginning of the exploration by atoms of positions corresponding to the highest values of interatomic potential (Goldstein 1969). This distribution of configurations over increasingly high potential energy states is the main characteristic of atomic mobility and low viscosity or relaxation time (point 2, Fig. 4).

If we now look at the changes in volume in an amorphous material, we notice that a general characteristic of interatomic potentials is their anharmonic nature, i.e., that the forces applied to the vibrational atoms are not exactly proportional to the movements in relation to the equilibrium positions of these atoms (Richet and Neuville 1992). An increase in vibration amplitudes therefore leads to an increase in interatomic distances. Like solids, liquids also have such anharmonic vibrational expansion, but higher-energy patterns that begin to be explored over the glass transition are generally associated with increased interatomic distances. This is why the thermal expansion coefficient generally increases significantly at the glass transition (Richet and Neuville 1992).

Pressure–temperature space

As we have mentioned before, there is a glassy transition pressure (Rosenhauer et al. 1979), which allows one to place the glass transition in a pressure-temperature plane for a given cooling or heating rate. However, the effects of pressure and temperature on the properties of glasses are actually different. High pressure can produce irreversible configurational changes. So, for given experimental quench rates or measurement time scales, the kinetics of configurational changes are markedly different depending on whether they are caused by pressure or temperature. The main reason for this difference is that the shape of potential energy wells varies little with temperature but significantly with pressure. If high kinetic energy is required to cross the potential barriers at constant pressure, changes in these barriers with pressure can lead to new configurational states at lower temperatures, if the pressure is high enough. The processes occurring at high pressure in glasses are likely similar to those that occur in liquids. Since the 2000's, many papers have shown significant changes in the glass structure with increasing pressure, more or less linked to changes in their properties (e.g., Poe et al. 2001; Suzuki et al. 2002; Wang et al. 2014). In this volume, an exhaustive chapter can be found on the effect of pressure on the structure and properties of melts and glasses by Sakamaki and Ohtani (2022, this volume).

SILICATE GLASSES AND MELTS

Alkali or earth alkaline silicate glasses and melts

As depicted in the previous two chapters, silica glass is a material well-connected at the molecular scale, built from a three-dimensional network of tetrahedral units composed of covalently bonded central Si and apical O atoms (a.k.a. Bridging Oxygens or BO). However, contrary to α-quartz that forms perfect trigonal crystals, SiO_2 glass is built by SiO_4 tetrahedra distributed disorderly yet not totally randomly (e.g., see Fig. 5A in Drewitt et al. 2022, this volume). The relative density d of silica glass equals 2.20 (molar volume $V_m = 27.311$ cm^3), a value lower than that observed for α-quartz ($d = 2.65$ gcm^{-3}, $V_m = 22.673$ cm^3) that presents a compact and well-ordered structure. This implies that silica glass presents a rather porous structure, with possibly a minor amount of Non-Bridging Oxygen (NBO, see Henderson and Stebbins 2022, this volume) atoms not connected to 2 Si atoms (Brückner 1970; Fanderlik 1990). The fraction of NBO in silica is usually considered as negligible, making the silica structure a very strong one.

As an alkali oxide is added, Na_2O for example, more and more NBO are created until reaching the possibility to create isolated SiO_4 tetrahedra linked together by only ionic Na–O bonds. For alkali silicates, this corresponds to a theoretical view because Na_4SiO_4 glasses do not exist. However, in the Ba-silicate system, it is possible to obtain glasses with 33% and 37% silica with classic cooling rate (Bender et al. 2002), while in the Mg-silicate system it is possible to obtain glasses up to 62 MgO in mole percent (38% SiO_2) but below 50 percent of silica, it is necessary to use a fast cooling rate as shown in Raman spectra (Neuville et al. 2014a). Glass formation in the SiO_2–Na_2O system is actually not possible for amounts of Na_2O lower than that of the metasilicate composition Na_2SiO_3 (Schairer and Bowen 1956). At higher Na_2O contents, up to 58 mole percent, it is possible to obtain Na silicate glasses only via high-speed quenching of a small quantity of melt (a few grams; Imaoka and Yamazaki 1963). Stable, large pieces of glass cannot be formed below 60 mol% of SiO_2 in this system (Neuville 2006). For other alkali-silicate glasses, it is possible to make glasses up to 35 and 55 mole% of Li_2O and K_2O, respectively (Imaoka and Yamazaki 1963; Levin et al. 1964).

To summarize, alkali silicate glasses can be easily obtained at normal cooling rates in the 60 to 99 mole% SiO_2 range (see Raman spectra of silicate glasses in Neuville et al. 2014), and alkaline-earth silicate glasses between 33 and 75 mol% SiO_2, particularly with heavy alkaline-earth elements like Ba. For Mg, the glass forming domain is smaller, and ranges between

37 and 55 mole% SiO_2 (Imaoka and Yamazaki 1963; Richet et al. 2009). In the $CaO–SiO_2$ system between 99 and 62 mole% SiO_2, the melts at temperatures just above the liquidus consists of two immiscible phases (Imaoka and Yamazaki 1963; Levin et al. 1964; Hudon and Baker 2002a,b; Neuville 2006).

Alkali or alkaline-earth silicate glasses show almost linear variations in molar volume with addition of $M^{2+}O$ in silica glasses (Fig. 5). Molar volumes obtained from density measurements for $CaO–Al_2O_3–SiO_2$ and $Na_2O–Al_2O_3–SiO_2$ are also available in the literature (Seifert et al. 1982; Doweidar 1998, 1999) but we have chosen not to report them on the figures for two reasons: i) they show similar trends to ours, ii) they are made with different cooling rate than ours (not mentioned in the articles). The molar volume (V_m) of silicate glasses increases or decreases as a function of the cation size[4]. In the case of cations presenting an ionic radius larger than O^{2-} (132 pm), like Cs^+, Rb^+ or K^+, the glass molar volume increases with cation addition. For cations presenting an ionic radius lower than that of O^{2-}, like Na^+ and Li^+, V_m of silicate glasses decreases both with the cation size and its amount. A similar behavior is observed in alkaline-earth silicate glasses. In Ca and Mg silicate glasses, V_m varies almost linearly with the glass SiO_2 (Fig. 5). In Figure 5, a dotted straight line is drawn between SiO_2 and MgO with Mg in 6-fold coordination, and it is clearly observed that the V_m of Mg-silicate glasses are below this line, which suggests that the CN of Mg is significantly less than 6 in amorphous magnesium silicates. Such an interpretation agrees with ^{25}Mg Nuclear Magnetic Resonance (NMR) spectroscopy that shows that Mg is essentially in four-fold coordination in silicate glasses (Fiske and Stebbins 1994; Georges and Stebbins 1998; Kroeker and Stebbins 2000; Shimoda et al. 2007a,b), as well as with XANES at the Mg K-edge (Trcera et al. 2009) while Ca is essentially in six fold coordination in silicate glasses as shown by XANES at the Ca K-edge (Cormier and Neuville 2004; Neuville et al. 2004b; Ispa et al. 2010; Cicconi et al. 2016). These gentle variations may indicate that Ca and Mg remain almost in the same coordination number, CN, in silicate glasses, this CN being slowly affected by the SiO_2 content or by the variation in the number of bridging oxygens around Si.

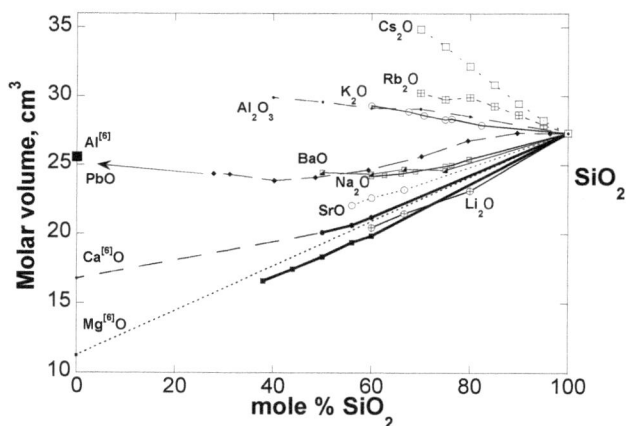

Figure 5. Molar volume of glasses at room temperature: data for Ca, Mg, Na, K, Li silicate glasses are original data from Neuville, Sr and Ba Novikov (2017) Rb, Cs from O'Shaughnessy et al. (2017, 2020); Pb-silicate from Ben Kacem et al. (2017) and Al-silicate glasses from Wang et al. (2020). Molar volume of $^{[6]}Al$ corresponds to Al_2O_3, and MgO and CaO are from Robie et al. (1979). Density of MA44.00 and MA38.00 glasses are respectively 2,807 and 2,881, Raman spectra of these two glasses are given by Neuville et al. (2014).

---

[4] Cation sizes are given in Whittaker and Muntus (1970) and molar volumes are from Robie et al. (1979).

Viscosity of silicate melts

In Figure 6A, the viscosity of SiO_2 and soda silicate melts are plotted as a function of reciprocal temperature. Silica shows the highest viscosity of all glass compositions, as well as a strongly linear behavior indicating an Arrhenian behavior (slope is equal to an activation energy).

The fact that silica easily forms a glass, albeit a strong one with a very high viscosity, directly indicates that Si is a network former cation. Other elements, presenting lower valence and larger ionic radius, can play different roles in the glass structure. For instance, the introduction of Na_2O in silica glass yields to breaking the covalent Si–O–Si bridges, and, hence, to the formation of Non-Bridging Oxygens, NBO. Na is considered, in this case, as a network modifier cation. Na fits into the glass structure as cations linked to the surrounding oxygens by bonds that are much more ionic and thus weaker than Si–O covalent bonds. Thus, the structure of sodium silicate glasses is weaker than that of vitreous silica. This translates into a lowering of the viscosity at given temperature when adding Na_2O in SiO_2 (Fig. 6A). Furthermore, the introduction of Na_2O in silica glass produces a non-Arrhenian behavior that increases with x for the NSx glass; x corresponds to the ratio of SiO_2/Na_2O in mole percent (Fig. 5A). When viscosity varies with $1/T$ following an Arrhenian behavior, it follows this equation:

$$\log \eta = A + B/T \qquad (4)$$

with A the viscosity at infinite temperature and B an activation energy. This equation can only be used for a few chemical compositions like SiO_2, GeO_2, $NaAlSi_3O_8$ or $KAlSi_3O_8$. An Arrhenian behavior implies that the activation energy, B, is independent of temperature,

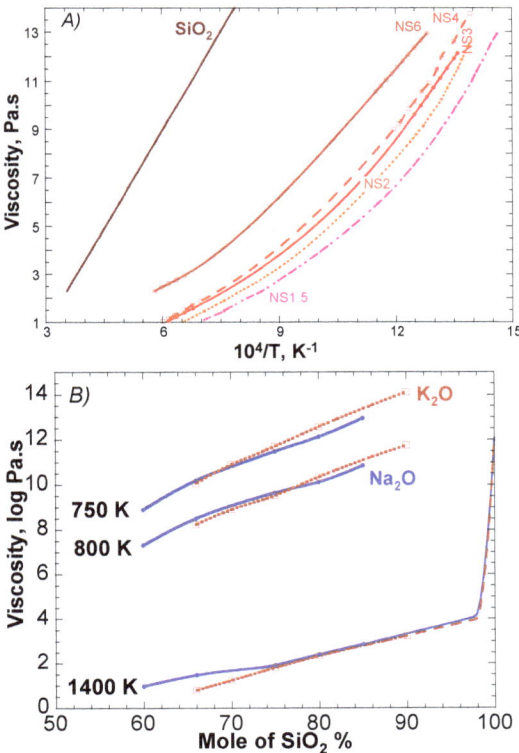

Figure 6. A) viscosity versus reciprocal temperature (SiO_2 NS6, NS4, NS3, NS2, NS1.5, data from Neuville 2006 and Le Losq et al. 2014). NSx: x corresponds to the ratio of SiO_2/Na_2O in mole percent (e.g., NS4: $x = 4 = 80/20$: 80% SiO_2 and 20% Na_2O); B) viscosity versus SiO_2 content for Na- and K-silicate glasses (data from Poole 1948; Bockris et al. 1955; Neuville 1992, 2006).

and, hence, that the process underlying melt viscous flow remains the same regardless of temperature. For melts showing a strong non-Arrhenian behavior like NS3, the activation energy term, B, decreases from 2000 kJ mol^{-1} at 1000 K down to 300 kJ mol^{-1} at 1800 K. In such conditions, it is no longer possible to use the Arrhenian equation. For this reason, Tamman, Vogel and Fulcher proposed the so-called TVF equation (Vogel 1921; Fulcher 1925; Tamman and Hesse 1926) to link viscosity and temperature:

$$\log \eta = A + B/(T - T_1) \qquad (5)$$

A, B, and T_1 are just fitting parameters without physical meaning. The TVF equation is very useful to fit and interpolate viscosity data at low and/or at high temperature (Neuville and Richet 1990). This equation has been used in many different papers and is at the basis of several viscosity models (Bottinga and Weill 1971; Shaw 1972; Persikov 1991; Giordano et al. 2008; Giordano and Russell 2018). The empirical parameters A, B, T_1 can be linked to glass structure (Giordano and Russell 2018). However, while finding very practical applications, this equation remains purely empirical.

Aside from the empirical Arrhenius or TVF equations, the Adam–Gibbs equation, derived from the theory of Adam and Gibbs (1965), is one of the best candidates to fit and extrapolate viscosity measurements for silicate melts (Urbain 1972; Wong and Angel 1976; Richet 1984; Scherer 1984; Neuville and Richet 1990, 1991; Mauro et al. 2009). The strength of this equation is that it allows relating viscosity measurements to heat capacity and configurational entropy data (Urbain 1972; Wong and Angell 1976; Richet 1984; Scherer 1984). The Adam–Gibbs theory and equation will be discussed in more details in the next section. No other equation allows relating thermodynamic to dynamic variables in a simple way.

The variations observed and described previously for sodium silicate glasses are similar for all alkali oxides, as for instance visible for K_2O in Figure 6B. Generally, the addition of a few percent of alkali in silica glass produces an important viscosity decrease of a few orders of magnitude (Lecko et al. 1977). Ten mole percent of alkali oxides decreases the viscosity by ~ 10 orders of magnitude at 1400 K, this decrease being larger at low temperature near the glass transition temperature. After 10 mol% of added alkali oxides, the decrease in viscosity with increasing alkali content becomes less pronounced and the viscosity varies almost linearly at high temperature. These viscosity variations for more than 10% of added alkali are well correlated with increases in thermal expansion and heat capacity of the liquid which vary almost linearly as a function of chemical change at high temperature (Bockris et al. 1955; Lange and Carmichael 1987; Lange and Navrotsky 1992).

Ideal mixing: mixing alkali or alkaline-earth in silicate glasses and melts

In the section *Viscosity of silicate melts*, we observed that molar volumes, as well as viscosity and glass transition temperature of alkali and alkaline-earth silicate glasses all changes with the addition of network modifiers. Mixing different metal cations (like Ca and Mg, or Na and K) in silicate glasses results in large variations in their properties, such as large decreases of their glass transition temperatures or large increases in their electrical conductivity (Day 1976). This is the so-called mixed alkali effect, MAE, reviewed many times (Richet 1984; Neuville and Richet 1991; Allward and Stebbins 2004; Cormier and Cuello 2013; Le Losq and Neuville 2013; Le Losq et al. 2017; Bødker et al. 2020) in the literature for alkali silicate glasses since the work of Day (1976).

In this section, we show and introduce some terminology about mixing elements and their effect on the structure and macroscopic properties of melts and glasses. Ca and Mg are very important elements in Earth and material sciences, and it thus it is particularly important to understand how they mix in silicate melts. To understand this, we can look at how viscosity varies along the $CaSiO_3$–$MgSiO_3$ binary. Neuville and Richet (1991) have shown that at constant temperature, the viscosity of $(Ca,Mg)SiO_3$ composition is always lower than the viscosity of the end-member, $CaSiO_3$ or $MgSiO_3$ (Fig. 7A).

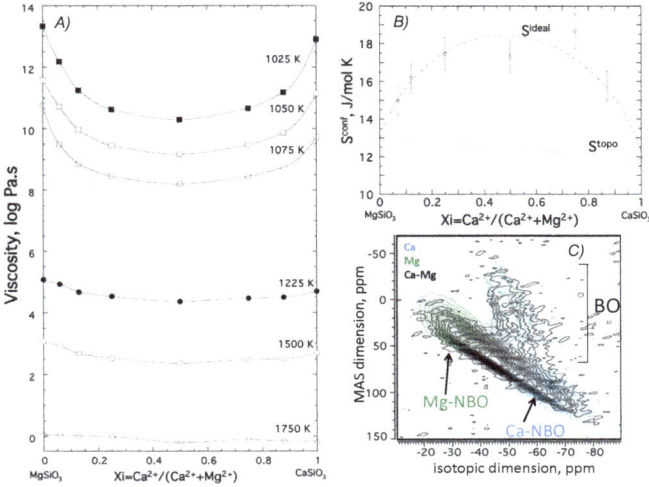

Figure 7. A) Viscosity versus $x_i = Ca^{2+}/(Ca^{2+}+Mg^{2+})$ at constant temperature and **B)** configurational entropy for Ca/Mg mixing data from dashed line correspond to the total configurational entropy and full line to the topological variation of the entropy see text (Neuville and Richet 1991) **C)** ^{17}O triple-quantum NMR spectrum redrafted from Allwardt and Stebbins (2004).

By using the Adam–Gibbs equation, it is possible to determine the configurational entropy, S^{conf}. Adam and Gibbs (1965) theory of relaxation processes is based on the idea that transporting matter in a viscous liquid requires a cooperative change in the fluid configuration. A liquid with zero configuration entropy would be analogous to a perfect crystal, and no material displacement could occur since a single configuration would be available. The viscosity would then be infinite. If only two configurations were possible, a movement of matter could only occur if all the atoms of the liquid changed position simultaneously. The probability of such a cooperative movement would be very low and viscosity extremely high, but no longer infinite. More generally, as S^{conf} grows, cooperative configurational changes can occur independently of each other in increasingly smaller volumes of the liquid. At the same time, viscosity decreases and is predicted by:

$$\log \eta = A_e - B_e/TS^{conf}(T) \tag{6}$$

where A_e is a pre-exponential term and B_e a measure of the Gibbs free energy barriers hindering configurational rearrangements in the liquid. $S^{conf}(T)$ can be written as:

$$S^{conf}(T) = S^{conf}(T_g) + \int_{T_g}^{T} \frac{C_p^{conf}(T)}{T} dt \tag{7}$$

and

$$C_p^{conf}(T) = C_{p_l}(T) - C_{p_g}(T_g) \tag{8}$$

where C_{p_l} and C_{p_g} are the heat capacity of the liquid and the glass.

C_p^{conf} is the melt configurational heat capacity, equal to the difference between the heat capacity of the liquid C_{p_l} and the heat capacity of the glass at T_g, $C_{p_g}(T_g)$ (Richet et al. 1986). Those values can be measured or modelled with the existing parametric equations (Richet and Bottinga 1984; Stebbins et al. 1984; Richet 1987; Tangeman and Lange 1998; Russell and Giordano 2017), see for examples and details Neuville (2005, 2006), Le Losq et al. (2014) and Le Losq and Neuville (2017). One may also note that $C_{p_g}(T_g) \sim 3R$, the Dulong and Petit limit (Petit and Dulong 1819), such that it can be easily estimated from this simple calculation (Richet 1987). The configurational heat capacity is a very important term, in particular, the higher C_p^{conf} is, the faster S^{conf} increases with T, and the lower the B_e/S^{conf} ratio.

The importance of these differences between configurational C_p is illustrated in Figure 1, where in the case of $NaAlSi_3O_8$, the liquid C_p^{conf} represents less than 10% of that of the glass, and deviations from Arrhenius behavior are small. On the contrary, $CaSiO_3$ or $Mg_3Al_2Si_3O_{12}$ have a C_p^{conf} representing 30% of the C_p of the glass, and the viscosity versus $1/T$ curves depart significantly from a straight line. Because of the importance of C_p^{conf}, it is therefore imperative to know the heat capacity of liquids and glasses for a given temperature and chemical composition. We can note that the configurational heat capacity $C_p^{conf}(T)$ depends essentially on the heat capacity of the liquid, C_{p_l}, and in the case of alkaline-earth silicate melts, C_{p_l} is independent of T and varies mostly linearly with melt composition (Stebbins et al. 1984; Richet and Bottinga 1985; Lange and Navrotsky 1992). This explains the pseudo-linear variations of viscosity observed at high temperature (1750 K, Fig. 7A).

However, near T_g, the melt viscosity depends strongly on $S^{conf}(T_g)$, which shows large non-linear variations with melt composition. This explains the large, non-linear viscosity changes at supercooled temperature upon mixing alkaline-earth elements in silicate melts (Fig. 7A). Using Equation (6) in conjunction with viscosity data and heat capacity values (from data or models), S^{conf} can be calculated. S^{conf} along the $CaSiO_3$–$MgSiO_3$ binary is plotted in Figure 7B. To reproduce S^{conf} variations with composition, it is necessary to look at the contributions to $S^{conf}(T_g)$ in Equation (7). S^{conf} records topological contributions from the melt structure (bond angle and interatomic distance distributions, etc.) as well as excess entropy arising from chemical mixing effects. It thus is common to express S^{conf} as the addition of those two sources (Neuville and Richet 1991):

$$S^{conf}(T_g) = S^{topo} + S^{mix} \tag{9}$$

with S^{mix} the term embedding all mixing contributions and S^{topo} the entropy arising from the topology of the glass network. The topology of the network for the two end-members is distinctly different and is also different compared to the crystalline analogues.

S^{topo}, the topological configuration entropy can be approximated as a linear variation of the configurational entropy of the end-members and calculated as $\Sigma \, x_i \, S^{conf}_i(T_g)$, with $S^{conf}_i(T_g)$ the configurational entropy of $CaSiO_3$ and $MgSiO_3$ glasses (Neuville and Richet 1991). This S^{topo} term corresponds to a mechanical mixing between the two end-members without necessary chemical or physical interaction. S^{topo} can be expressed from the glass structure and varies linearly with composition (Le Losq and Neuville 2017).

S^{mix} corresponds to a chemical mixing term that can be ideal or non-ideal. In the case of Ca/Mg mixing, the simplest hypothesis that can be made is that $S^{mix} = S^{ideal}$, and thus can be written as:

$$S^{ideal} = -n \, R \, \Sigma \, x_i \ln x_i \tag{10}$$

where R is the perfect gas constant and n is the number of atoms exchanged per formula units, 1 in this case, and $x_i = Ca^{2+}/(Ca^{2+}+Mg^{2+})$. From a thermodynamic point of view, the ideal solution implies that end-members components mix randomly in solution.

The excellent agreement (Fig. 7B) between entropies obtained from viscosity measurements and values predicted using Equations (6–10) shows that the ideal mixing hypothesis for Ca/Mg is consistent with the data. It should be noted that the mixing effect is predominant near T_g. The configuration entropy of the intermediate compositions is therefore much higher than that of the end-members, explaining the minimum in viscosity observed upon Ca/Mg mixing. When temperature increases, the entropy of the end-members increases rapidly and S^{mix}, which is constant, eventually becomes small compared to $\Sigma x_i S^{conf}_i(T_g)$ in Equation (9). Therefore, at high temperatures, S^{mix} is negligible and this results in a quasi-linear variations in viscosity, driven by the linear variation of C_{pl} upon Ca/Mg mixing.

These macroscopic conclusions about Ca/Mg mixing are in very good agreement with nanoscale investigations made by ^{17}O NMR (Allwardt and Stebbins 2004) and neutron diffraction and RMC (Cormier and Cuello 2013). From neutron diffraction, the silicate network shows small but significant changes with the Mg/Ca exchange and a significant intermixing of Ca and Mg can be observed (Cormier and Cuello 2013). From another point of view, ^{17}O NMR spectra of the diopside glass $(CaMg)SiO_3$ shows that only one broad non-bridging oxygen atom (NBO) peak is visible, and encompasses the entire range of chemical shifts ranging from Ca-NBO to Mg-NBO. Comparison of the isotropic projections from 3QMAS NMR to 1D spectra predicted using a random model show that Ca/Mg mixing in Ca/Mg-silicate glasses is generally highly disordered (Allwardt and Stebbins 2004), except maybe near the glass transition temperature where a more ordered glass is possible as shown by neutron diffraction made on $CaSiO_3$ glass by Gaskell et al. (1985).

Validity of the viscosimetry approach to determine $S^{conf}(T_g)$. We have just shown that it is possible to determine configurational entropies using viscosity measurements and the AG theory. Are those values of $S^{conf}(T_g)$ truly representative of the glass residual entropy? To answer this question, we can compare $S^{conf}(T_g)$ obtained from calorimetric measurements to those obtained from fitting viscosity data. Data from 11 very different compositions, including B_2O_3, SiO_2 or $KAlSi_3O_8$, show very good agreement between $S^{conf}(T_g)$ values retrieved from viscosity measurements or calorimetric measurements (Fig. 8). This thus validates the use of Equation (6) to determine $S^{conf}(T_g)$ of melts.

In the case of Na/K mixing in silicate melts, only one set of measurements is available from Poole (1948). Using those data, Richet (1984) proposed that an ideal, random Na/K mixing occurs in silicate melts, and viscosity can be modelled using the ideal mixing calculation of S^{conf}; some differences between calculated and measured viscosity in the supercooled silicate melts are visible in Richet (1984) work, however they arise from the way the data and model are represented. The more recent study of Le Losq and Neuville (2017) resolved this, showing that it is clearly possible to model the viscosity of Na–K silicate melts while assuming a random mixing of the alkalis (see the section *Silicate glasses and others network formers*).

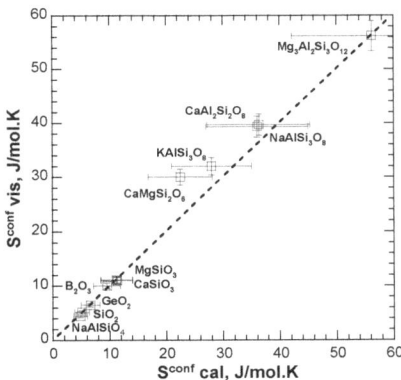

Figure 8. Viscosimetry configurational entropy versus calorimetry configurational entropy for 10 different oxide glasses for which determination of calorimetric measurements is possible. Calorimetric data are from Richet (1984) for SiO_2, $NaAlSi_3O_8$, $KAlSi_3O_8$, de Ligny and Westrum (1996) for GeO_2 and B_2O_3, Richet et al. (1991) for $NaAlSiO_4$, Tequi et al. (1991) for $Mg_3Al_2Si_3O_{12}$, and Richet et al. (1986) for $MgSiO_3$, $CaSiO_3$, $CaMgSi_2O_6$ and $Ca_2Al_2Si_2O_8$. Viscosity data are from Hetherington et al. (1964) and Urbain et al. (1982) for SiO_2, Neuville and Richet (1991) and Le Losq and Neuville (2013) and Le Losq et al. (2017) for alkali and alkaline-earth silicate and aluminosilicate compositions, from Fontana and Plummer (1966) for GeO_2, and from Eppler (1966), Macedo and Litovitz. (1965), Napolitano et al. (1965) for B_2O_3.

Mixing alkali and alkaline-earth elements in silicate glasses

Most industrial and geologic glasses are silicate and aluminosilicate glasses containing a mixture of alkali and/or alkaline-earth elements. Igneous rocks (Morey and Bowen 1925) as well as nearly 90% of industrial glasses contain mixtures of alkali and alkaline-earth elements. We can cite the particular case of soda-lime silica glasses, extensively used for the construction of conventional windows and container glasses. The importance of $M^{2+}O$–M^+_2O–SiO_2 glasses (with M^+ an alkali and M^{2+} an alkaline-earth element) further increased recently because they also are used for the development of bioactive glasses (Hench 1991; Kim et al. 1995; Clupper and Hench 2003; Hill and Brauer 2011; Brauer 2015). Given the importance of alkali and alkaline-earth metal cations in glasses, many studies were performed to understand the $M^{2+}O$–M^+_2O mixing in silicate glasses. They are essentially focused on the glass mechanical properties or devitrification (Morey 1930; Frischat and Sebastian 1985; Koike and Tomozawa 2007; Koike et al. 2007), on glass structure (Buckerman and Müller-Warmuth 1992; Lockyer et al. 1995; Jones et al. 2001; Lee and Stebbins 2003; Lee et al. 2003, Neuville 2005, 2006, Hill and Brauer 2011; Brauer 2015; Moulton and Henderson 2021 and references therein) or thermodynamic properties (Natrup et al. 2005; Neuville 2005, 2006; Richet et al. 2009b; Inaba et al. 2010; Sugawara et al. 2013). Given the above-mentioned importance of $M^{2+}O$–M^+_2O–SiO_2 glasses, we will in this section bring new insights to the effect of $M^{2+}O$–M^+_2O mixture on the properties of silicate glass and melt.

Glass formation domains. The substitution of alkali elements by alkaline-earth elements allows extending the glass forming domain of silica glasses to lower silica contents (Fig. 9), and, more generally, it allows improving many glass properties (Moore and Carey 1951; Imaoka and Yamazaki 1963; Tomozawa 1978, 1999; Gahay and Tomozawa 1989, 1984; Neuville 2005, 2006). For $M^{2+}O$ and M^+O_2 bearing silicate glasses, the vitrification domain starts near 50 mol% silica in most cases. For Mg silicates, it starts at 45.7 mol% (Moore and Carey 1951). It was generally accepted that the presence of Mg–O bonds allows preserving somehow the continuity of the silicate network, with Mg acting to some extent as a network former (Moore and Carey 1951), an idea in agreement with the low coordination number of Mg discussed previously. In the case of CaO–Na_2O–SiO_2 glasses, the first phase diagram was made by Morey and Bowen (1925). It revealed the domain where glasses can be formed, but also an immiscibility domain

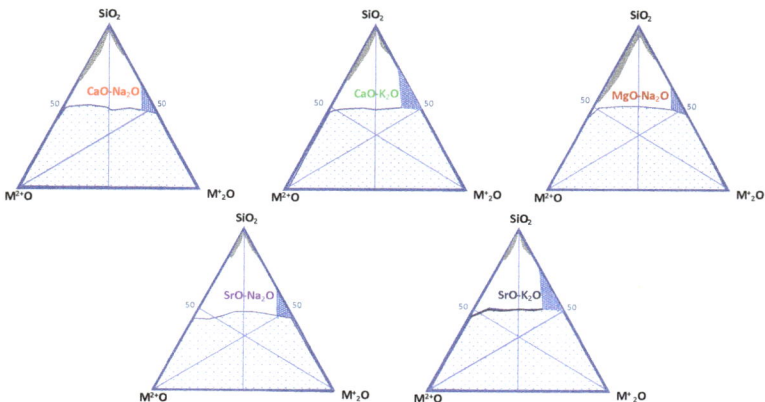

Figure 9. The upper part of the ternary diagram corresponds to the glass forming region of glass between $M^{2+}O$ and M^+_2O, with $M^{2+}=Mg^{2+}$, Ca^{2+}, Sr^{2+} and $M^+=Na^+$, K^+ (redrafted from Imaoka and Yamazaki 1963 and Levin et al. 1964). The **white areas** are the glass making domain, the **light dotted areas** are the crystallization domains, the **dark dotted lines** correspond to the glass making domain, but strongly hygroscopic, and the **gray areas** to unmixed glass.

in which phase separation is observed. This immiscibility domain is very large near the CaO–SiO_2 binary, extending between 62 and 100 mol% of silica. It decreases rapidly with addition of Na_2O in the glass (Morey and Bowen 1925; Greig 1927; Hudon and Baker 2002a,b). In the immiscibility domain, different phases exist: cristobalite, SiO_2, wollastonite $CaSiO_3$, sodium disilicate $Na_2Si_2O_5$, devitrite $Na_2Ca_3Si_6O_{16}$, and a non-defined compound $Na_2Ca_2Si_3O_9$. The proportion of these phases vary as a function of the bulk composition and the cooling rate (Morey and Bowen 1925; Greig 1927; Tomozawa 1978, 1999; Gahay and Tomozawa 1989; Hudon and Baker 2002a,b; see also the Schuller chapter in Neuville et al. 2017).

Similar observations can be made in the other M^+_2O–$M^{2+}O$–SiO_2 systems, with $M^+ = Li^+$, Na^+, K^+ and $M^{2+} = Mg^{2+}$, Ca^{2+}, Sr^{2+}, Ba^{2+}. Generally, a pyrosilicate phase, $M^{2+}SiO_3$, and one or two phases $M^+_yM^{2+}_xSi_zO_u$ crystallize at high silica contents (> 70 mol%, x,y,z,u are different proportions of each crystalline phases), except for $M^{2+}=Ba^{2+}$ where the glass forming domain is large but mostly unconstrained. Bender et al. (2002) achieved the synthesis of Ba silicate glasses up to the BS7 ($BaO/SiO_2=7$ in mol%) composition, and Frantz and Mysen (1995) investigated some of those glass compositions by Raman spectroscopy.

To summarize, in all of the M^+_2O–$M^{2+}O$–SiO_2 ternary diagrams, the glass forming domains are relatively small and essentially exist at silica contents higher than 50 mol%, with exception of barium silicates. In the case of sodium and potassium silicates, a large zone of high hygroscopicity exists below the glass forming domain, typically between 66 and 50 mol% of silica for sodium silicate glasses and between 70 and 50 mol% for potassium silicate glasses (Tomozawa 1978, 1999; Gahay and Tomozawa 1989). In general, one should be careful with potassium silicate glasses that are always highly hygroscopic.

Viscosity variations. Figure 10A shows viscosity measurements of silicate melts for M^{2+}–Na mixing with $M^{2+} = Ca^{2+}$, Sr^{2+}. Ca- and Sr-silicate melts show similar behavior. The addition of those elements produces a large increase in melt viscosity at constant temperature, in agreement with previous measurements (English 1923). The lack of measurement between 10^2 and 10^9 Pa·s results from the very rapid rate of crystallization in this viscosity-temperature range for mixed alkali and alkaline-earth silicate glasses (Meiling and Uhlmann 1977; Mastelaro et al. 2000; Neuville 2005, 2006). In the Figure 10B, we clearly observe that addition of 10 mole% Na_2O in an alkaline-earth silicate melt has a stronger effect on melt viscosity than addition of 10 mole% $M^{2+}O$ to a soda-silicate melt composition. This indicates that a sodium silicate glass network can incorporate alkaline-earth elements more easily than the opposite. A striking difference exists in the behavior of the viscosity between pure Na-silicate melts and pure alkaline-earth silicate melts (CN60.00 or SN60.00). Indeed, at constant viscosity and near T_g (Fig. 10B), i.e., near 1000 K, the viscosity of the end-member compounds differ by 10 orders of magnitude. This difference decreases with temperature: it is less than 0.5 order of magnitude at 1600 K. Furthermore, at 1600 K, the viscosity of the Na_2O–$M^{2+}O$–SiO_2 melts decreases almost linearly with $Na_2O/(Na_2O + M^{2+}O)$. Similar behaviors were already observed by English (1923) in the same system with 75 mole% SiO_2 and 25 mole% Na_2O, and with a substitution of Na_2O by up to 12 mole% CaO.

The viscosity variations observed in Figure 10 can be understood using Equations (6–9). It is possible to determine how configurational entropy varies upon the mixing of Na_2O with M^{2+} in the glasses (Fig. 11A). S^{conf} can be decomposed in S^{topo} and S^{mix} (Eqn. 9). S^{topo} is equal to the sum of endmembers. S^{mix}, in the case of M^+/M^{2+} mixing, has a complex expression that depends on the way the cations mix into the glass structure.

Between the calcium and sodium silicate glasses, $S^{conf}(T_g)$ varies non-linearly with $Na_2O/(Na_2O+CaO)$. It increases rapidly with increasing $Na_2O/(Na_2O+CaO)$ to 0.2, then it slightly decreases with further substitution Ca^{2+} by Na^+. Similar observations are made in the SrO–Na_2O–SiO_2 system (Neuville 2005). The observed variations strongly depart from the

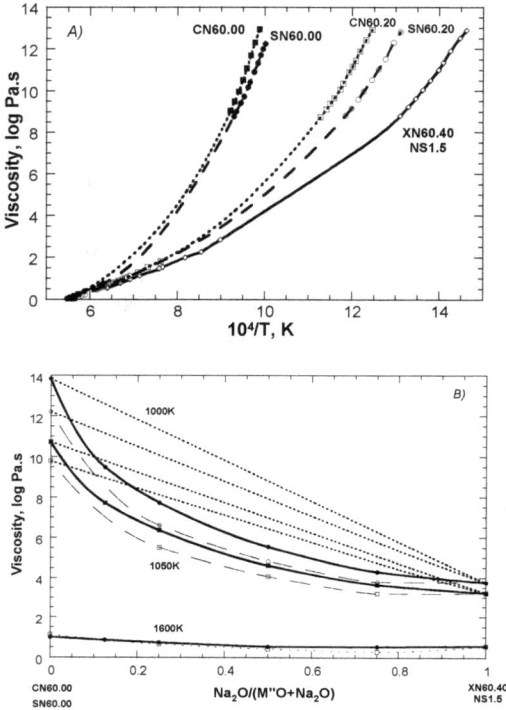

Figure 10. A) Viscosity versus $1/T$ for silicate melts ranging from the pure alkaline-earth silicate melt) to pure soda silicate (XN60.40) composition names are: XN60.Y, X=C or S for the Ca or Sr system, Y corresponds to Na_2O content, 60 to SiO_2 in mole percent, and X=100 − (60+Y) all in mole percent. **B)** Viscosities of Na/M^{2+} silicate melts versus $Na_2O/(Na_2O + M^{2+}O)$ at constant temperature, M^{2+}=Ca, Sr, dashed lines correspond to the linear viscosity variation at 1000 K, 1050 K and 1600 K, **solid and open symbols** correspond to the CaO–Na_2O–SiO_2 and SrO–Na_2O–SiO_2 systems, respectively compiled from Neuville (2005, 2006).

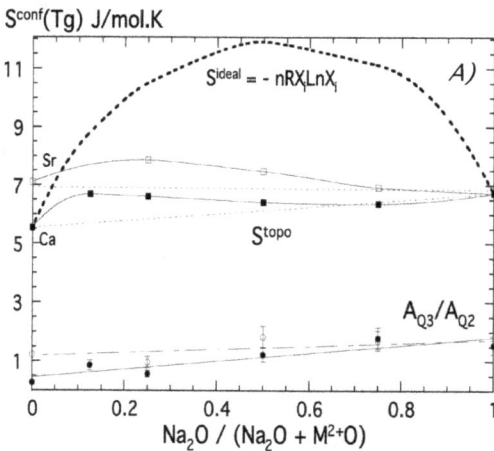

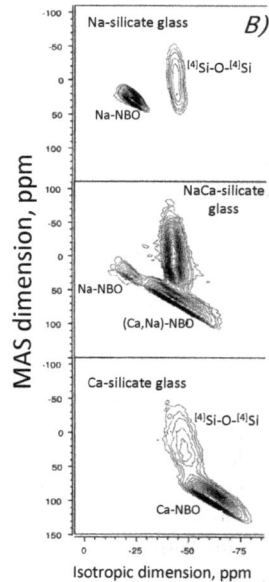

Figure 11. A) $Na_2O/(Na_2O+M^{2+}O)$ dependence of configurational entropy, $S^{conf}(T_g)$, A_{Q3}/A_{Q2} ratio for in **black symbol** CaO and in **open symbol** for SrO (entropy are in J/mol.K and A_{Q3}/A_{Q2} without units and are calculated from the Q^3 and Q^2 areas of the Raman spectra, see Neuville (2005, 2006) for more detail). **B)** ^{17}O 3QMAS MAS spectra for Ca-Na silicate glasses at 9.4 T redrafted from Lee and Stebbins (2003).

predictions made by the ideal model given in Equation (10) (Fig. 11A). In the case of Na^+ and Ca^{2+}, or of Na^+ and Sr^{2+}, it is clear that, as their $S^{conf}(T_g)$ variations are close to linear trends, those atoms do not mix randomly in the silicate melts. The fits of $S^{conf}(T_g)$ presented in Figure 11A thus require the use of a non-ideal mixing model. To model S^{mix} in the case of a non-ideal mixing of alkali and alkaline-earth elements, Neuville (2005, 2006) proposed to use the equation:

$$S^{mix} = W_{NM} \, x_n \, y_c^3 \qquad (11)$$

where $x_n = Na_2O/(Na_2O+M^{2+}O)$ and $y_c = 1-x_n$, and W_{NM} is a constant.

Structure of mixed glasses. Equation (11) implies a significant change in the residual entropy that originates from a re-arrangement of the glass structure as Ca^{2+} is replaced by Na^+. Spectroscopic methods that probe the glass structure allow us to investigate such structural changes. ^{23}Na and ^{17}O Nuclear Magnetic Resonance (NMR) spectroscopy data shown a non-random distribution in the case of Na^+-Ca^{2+} mixing, with in particular a prevalence of Na^+–Ca^{2+} pairs in the system (Fig. 11B), and several partially resolved non-bridging oxygen peaks such as Na^+–NBO, and mixed peaks (Na^+, Ca^{2+})–NBO (Lee and Stebbins 2003). The observed fractions of Na^+–NBO are smaller than those predicted by random distributions of Na^+ and Ca^{2+}, suggesting preference for dissimilar pairs (Lee and Stebbins 2003) which is also in good agreement (Florian et al. 1996; Lee and Stebbins 2002; Lee et al. 2003; Lee 2005). In the previous chapter, Raman spectroscopy soda and lime silicate glasses were shown in Figure 9. Different Q^n contributions are visible in those spectra. From these Q^n contributions, the ratio of the area between Q^3 and Q^2 units, A_{Q3}/A_{Q2} can be calculated. Results of such calculation are plotted in Figure 11A. It shows that the fraction of Q^3 units in the glass increases with increasing the $Na_2O/(Na_2O+M^{2+}O)$ ratio. $M^{2+}O$ silicate glasses are thus richer in Q^2 and Q^4 units than alkali silicate glasses (Frantz and Mysen 1995, Neuville 2005, 2006), because increasing the ionic field strength of metal cation in the glass shifts to the right the equilibrium of the reaction:

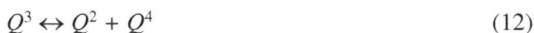

$$Q^3 \leftrightarrow Q^2 + Q^4 \qquad (12)$$

This observation will be used in the section *Silicate glasses and others network formers* for highlighting how one can build models of the structure, entropy and viscosity of silicate melts. Huang et al. (2016) investigated Ca_3SiO_5 composition by using $^{17}O \rightarrow \,^{29}Si$ CP-HETCOR NMR experiment and they proposed that free oxygen can also play a small role structure and properties in the mixed glasses, what is possible but this experiment must be considered with caution, firstly it is a particularly difficult composition to manufacture and secondly, only the nominal compositions are given and not those analyzed.

Network polymerization. The substitution of 30 mol% of CaO by Na_2O in an alkaline-earth silicate glass results in a moderate increase of ~ 0.9 J mol^{-1} K^{-1} of $S^{conf}(T_g)$ (Fig. 11A). This indicates an increase in the structural disorder in the glass, possibly linked to change in the glass network polymerization. Structural changes that may account for the entropy behaviour may be discerned in the Raman spectra of the glasses. In particular, the ratio of the areas of Raman bands assigned to Q^3 and Q^2 units, A_{Q3}/A_{Q2}, varies with $M^+_2O/(M^+_2O+M^{2+}O)$, indicating a change in the Q^n speciation distribution (Neuville 2006). There is a distinct increase of this ratio with adding Na_2O in the alkaline-earth silicate glasses. In the case of Na/Ca and Na/Sr mixing, Raman spectroscopy observations shown in Figure 11A are in good agreement with previous observations by Raman spectroscopy (McMillan et al. 1982; McMillan 1984; Frantz and Mysen 1995) and by ^{17}O NMR (Maekawa et al. 1991; Buckerman and Müller-Warmuth 1992; Jones et al. 2001, Lee and Stebbins 2003; Lee et al. 2003), all showing a decrease in Q^3 fraction upon substitution of alkali by alkaline-earth element in silicate glasses. Moreover, the increase in Q^2 units with adding alkaline-earth elements in soda silicate glasses can be correlated to increasing Boson peak frequency and intensity (Neuville 2006).

Note that below 250 cm^{-1}, there is only a scattering continuum and the Raleigh tail of the exciting line, except at very low frequency where there is the so-called boson peak (Buchenau et al. 1986; Malinovsky and Sokolov 1986). This peak has been ascribed to excitations associated with rotational motions of almost rigid tetrahedra (Buchenau et al. 1986; Helhen et al. 2000, 2002). Helhen et al. (2002) consider that this peak, observed for different silicate glasses, increases in intensity and shifts to higher frequency with higher distortion of the SiO$_4$ tetrahedra. Several others interpretations are proposed: Champagnon and collaborators proposed that Boson peak can be link with structural homogeneity in the glass (Champagnon et al. 2009), and the other hand Chumakov and collaborators proposed that Boson peak can be attributed to some fluctuation of disorder in glass (Chumakov et al. 2014) and finally Schirmacher and collaborators proposed an intermediate interpretation by considering that the Boson peak can corresponds to the spatial fluctuations of elastic constants caused by the structural disorder of the amorphous materials (Schirmacher et al. 2015; Schirmacher and Ruocco 2020). If we consider that molecular dynamic simulations provided by Cormack and coworkers suggest that Si–NBO bond length is about 0.05Å shorter than the Si–BO bond length (Henderson 1995; Cormack and Du 2001; Du and Cormack 2004)[5], we can propose that Q^2 units are more distorted than Q^3 or Q^4 units. This implies that Q^4 and Q^0 units are regular tetrahedra with Q^4 bigger than Q^0 and they are less distorted than the Q^1, Q^2, Q^3 units species (Nakamura et al. 2013). It is probable that Q^2 units are the more distorted tetrahedra with 2 distances T–BO and 2 distances T–NBO which can be in good agreement with the variation of the frequency of boson peak. So, we proposed that the variation on the Boson peak observed as a function of chemical change in simple systems with cation substitution can be attributed to tetrahedra distortion as observed on SiO$_2$ glass with pressure. This variation can also result of structural heterogeneity produce by the fact that soda silicate glasses are richer in Q^3 species than lime-silicate glasses. These structural heterogeneities can also be close the idea of percolation channels proposed by Greaves (1985).

Fragility. Angell (1991) proposed a classification of liquids between strong and fragile: a strong liquid shows a linear variation like SiO$_2$ in the diagram log η versus T_g/T whereas this linear relation is not preserved in a fragile liquid (Fig. 12). The melt fragility, m, is related to its thermodynamic properties via the relationship:

$$m = B_e/S^{conf}(T_g) \times T_g \times (1 + C_p^{conf}(T_g)/S^{conf}(T_g))$$

(13)

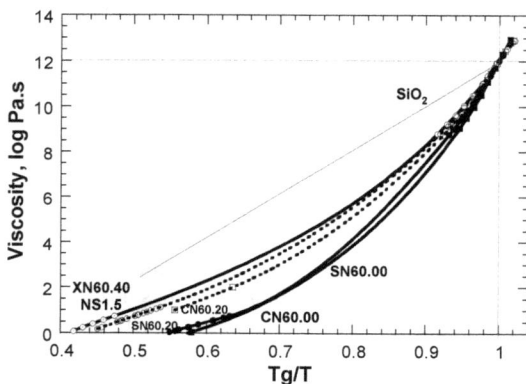

Figure 12. Viscosity of soda–M^{2+}O silicate glasses versus T/T_g. Error bars are smaller than symbol size, from Neuville (2005, 2006).

---

[5] Du and Cormack (2004) proposed from MD simulation that when the alkali content vary between 10 and 50 mole percent the d(Si–BO) vary between 1.625 and 1.629Å and d(Si–NBO) varies between 1.559 and 1.579 Å. Henderson (1995) proposed from EXAFS measurements that d(Si–O) equal 1.61±0.02 Å and 1.58±0.02 Å for NS1.5 glass which is in good agreement with MD.

In the case of the M^+_2O–$M^{2+}O$–SiO_2 system, the viscosities of the alkali- alkaline-earth silicate melts are plotted as a function of T_g/T in Figure 12. The strongest liquid is the pure soda silicate, XN60.40 (NS1.5), whereas the most fragile liquids are the pure alkaline-earth silicate melts. The Ca- and Sr-silicate melts have similar fragility, despite the Ca-silicate melt being more viscous than Sr-silicate melt. Intermediate chemical compositions mixing Na^+ with Ca^{2+} or Sr^{2+} present an intermediate behavior between end-members. The behavior of the melt fragility correlates well with melt C_p^{conf} values, which decrease of ~20% upon replacing M^{2+} with Na ($M^{2+} = Ca^{2+}$ or Sr^{2+}; Neuville 2005, 2006). This observation on the decrease of the C_p^{conf} is in good agreement with the increase of the $\Delta C_p = C_{pl}/C_{pg(Tg)}$ which varies between 1.2 for the Na-silicate glass up to 1.27 and 1.29 for Ca- and Sr-silicate glasses. Note that, Angell (1991) proposed that ΔC_p vary between 1.1 and 1.3 for a strong and fragile liquid respectively.

Melt configurational entropy, fragility, Q^n speciation, mobility and network connectivity. The distribution of cations in the glass network can have important implications for transport properties such as conductivity, viscosity and fragility. In the case of the Na/Ca substitution in silicate melts, Roling and Ingram (2000) have shown that the mobility of Ca^{2+} cations is enhanced upon their replacement by Na^+. This replacement of Ca^{2+} by Na^+ produces a positive coupling effect on the movements of divalent cations that correlates well with the increase in $S^{conf}(T_g)$ (Fig. 11A). Conversely, the activation energies related to movements of Na^+ is slightly changed. The cation mobility strongly depends on the glass structure, and the structural changes taking place during the Na/Ca substitution in the Ca-silicate glass enhance the mobility of Ca^{2+}. The modification in network topology will provide new types of empty sites that could be used for Ca movements (Cormier and Neuville 2004). The mobility of the divalent cation is then assisted by positive coupling with the more mobile monovalent cation (Magnien et al. 2004, 2008).

To summarize the observations made in the case of the mixing between alkali, M^+, and alkaline-earth, M^{2+}, elements in silicate glasses and melts:

- alkali and alkaline-earth cations do not mix randomly,
- increased alkali content increases the formation of Q^3 species,
- diffusivity increase with alkaline-earth element (Roling and Ingram 2000),
- fragility increases with alkaline-earth content,
- the distortion of the tetrahedra suggested by the boson peak variation and the distortion of the network increases with increasing MO content,
- the proportion of Q^2 species increases with $M^{2+}O$ content and in the same way, the intensity and the frequency position of the boson peak increase. The fact that the alkaline-earth silicate glasses around than 60% of silica correspond to a mixture of Q^2 and of a small amount of Q^4 species may explain the phase separation observed in melt containing more than 60% of SiO_2 in the $M^{2+}O$–SiO_2 binary system (see Fig. 9). These two interconnected Q^n species can also be an image of the percolation channels (Greaves et al. 1981; Greaves 1985; Frischat et al. 2001; Le Losq et al. 2017).

Silicate glasses and others network formers

Lead. The replacement of SiO_2 by PbO produces a small decrease in the glass molar volume until the Si/Pb ratio reaches a value of 0.5. Further PbO addition results in a molar volume increase. The PbO amorphous molar volume value can be obtained by extrapolation and is 25.0 cm^3 (Robie et al. 1979; see arrow in Fig. 5) which is similar to the molar volume of litharge, PbO with Pb^{2+} in four-fold coordination. Variations in the molar volume of lead silicate glasses correlate well with structural observations made via Raman spectroscopy (Fig. 9 in Drewitt et al. 2022, this volume, Ben Kacem et al. 2017). For small amounts of lead oxide added to silica glass (<20 mole%), Pb is essentially in six-fold coordination and its introduction into silica results in the formation of non-bridging oxygens, clearly visible in the Q^n species distribution in the Raman spectra (Ben Kacem et al. 2017). This is accompanied by a decrease in melt viscosity and glass transition temperature. At PbO concentrations higher than 20 mol%, a mixture of essentially two lead species, in 4- and 6-fold coordinations, exists. Kohara et al. (2010) and more recently Sampaio et al. (2018) mentioned that Pb can be distributed between 3-, 4-, 5- and 6-fold coordination but recent X-Ray diffraction shows that Pb is essentially distribute in 4- and 6-fold coordination (Drewitt et al. 2022, this volume). Pb in four-fold coordination can be considered as a weak network former (Fayon et al. 1998, 1999; Feller et al. 2010; Ben Kacem et al. 2017) as [4]Pb creates a network of interconnected tetragonal pyramid PbO_4 (Morikava et al. 1982). This is visible in Raman spectra, where an intense band at 140 cm^{-1} can be assigned to Pb–O–Pb covalent bonds (see Drewitt et al. 2022, this volume and Ben Kacem et al. 2017). This band at 140 cm-1 is clearly at higher frequency than that observed in alamosite, $PbSiO_3$, at 110 cm^{-1} and attributed to Pb in 6-fold coordination (Worrell and Henshall 1978; Ben Kacem 2017; Pena et al. 2017). The coexistence of network former PbO_4 and network modifier PbO_6 entities agree with the pair correlation function obtained from the X-ray diffraction data (see Fig. 6B in Drewitt et al. 2022, this volume), which show that the proportion of PbO_4 increases with decreasing SiO_2/PbO. The mixing of the tetragonal pyramid PbO_4 with SiO_4 tetrahedra is probably mechanical in nature, i.e., the two polyhedra cohabit and are interconnected but their mixture does not result in an excess of entropy as indicated by the linear decreases of the glass transition temperature between 30 and 80% of SiO_2 (Fig. 13). The fast decrease in T_g upon addition of ~10 mol% PbO probably relates to the network modifier role of PbO_6 octahedra, containing some NBOs in their first coordination shell.

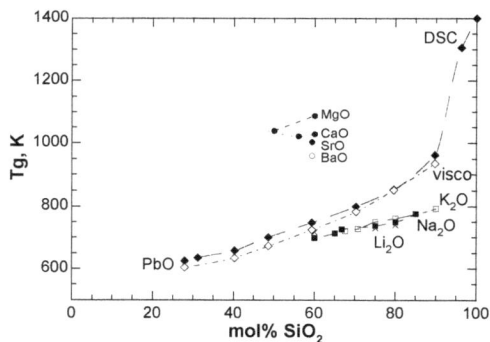

Figure 13. Glass transition temperature as a function of the fraction of SiO_2 in PbO–SiO_2 glasses. Data for Ca^{2+}, Mg^{2+}, Na^+, K^+, Li^+ silicate glasses are original data from Neuville, Ca50 and Mg50 (Neuville and Richet 1991), Ca60 and NS1.5, NS2, NS4 from Neuville (2006), NS3 from Le Losq et al. (2014); Sr60 from Neuville (2005) and Pb-silicate glasses open circle are from viscosity measurements and full circle are from DSC measurements. (Ben Kacem et al. 2017). Note that all T_g are obtained using viscosity measurements to be comparable except those obtained by Ben Kacem with DSC.

The introduction of PbO thus produces the transformation of Q^n units into Q^{n-1} units as PbO_6 are network modifier units. The appearance of these Q^{n-1} species is documented by changes in the high frequency envelope (800—1200cm^{-1}) of the Raman spectra of the glasses (Worrell and Henshall 1978; Ben Kacem et al. 2017). The Raman spectra of PbO-SiO_2 glasses corresponds to the combination of two distinct endmember Raman signals: that of a lead-silicate glass network with Q^{n-1} species, and that of an amorphous lead network characterized by a strong peak at 140 cm^{-1}. The relative abundances of these signals, and of their associated networks, vary as a function of the SiO_2/PbO ratio and this is in good agreement with observation made using ^{17}O NMR by Lee and Kim (2015). Lee and Kim (2015) also observed a small amount of free oxygen as early mention by Sampaio et al. (2018), but this small amount cannot explain the T_g and viscosity variations.

In this sense, lead is an example of an amphoteric[6] element that plays two different roles, network former or network modifier, depending on melt chemical composition. Another example of an element presenting such behavior is Fe^{3+}, which, depending on its coordination, plays different structural roles (Moretti 2005). Indeed, while Fe^{3+} in 4- or 5-fold coordination can be considered as network formers, Fe^{3+} in 6-fold coordination behaves as a network modifier in silicate melts. It can be found in the literature that such elements are "intermediate" and often aluminum is associated with this terminology. However, the story is more complex than this, and Al^{3+} does not really present amphoteric behavior per se as highly coordinated Al species are not surrounded by NBOs such that [4]Al is exclusively a network former element. Its behavior is however complex, as we shall see in the sections below.

Al in silicate glasses. A well-known feature of silicate mineral structures is that Si cations can be replaced isomorphously by aluminum cations, provided that either mono or divalent cations can ensure charge compensation of Al polyhedra to maintain local neutralization of electrical charges in the structure. As such, those mono or divalent cations are called charge compensators. However, it is also possible to obtain alumina-silica glasses without the presence of charge compensator cations. From a research point of view, this is very interesting because $Al_2O_3–SiO_2$ glasses are important materials for the understanding of how network former elements mix. $Al_2O_3–SiO_2$ glasses have a very high liquidus (Sossman 1933), and are difficult to synthesize except by fast quench or sol-gel methods (e.g., see Wang et al. 2020). With those methods, it is possible to obtain $Al_2O_3–SiO_2$ glasses in three different regions, near 15 mol%, 50 mol% or 60 mol% of silica. The molar volume of $xAl_2O_3–(1-x)SiO_2$ glasses calculated from density measurement (Wang et al. 2020) varies linearly between 27.3 up to 31 cm^3 for 60% of Al_2O_3 (Fig. 6).

From the data on Al_2O_3 (corundum), the molar volume of Al in six-fold coordination is 25.57 cm^3 (Robie et al. 1979). The molar volume of Al in five-fold coordination is unknown, but the extrapolation of the $Al_2O_3–SiO_2$ line in Figure 5 yields a value of ~32 cm^3 at 0 mol% of SiO_2. Poe et al. (1992) proposed that Al_2SiO_5 glass is composed by 50% of [5]Al, 40% of [4]Al and 10% of [6]Al.

However, no information on the geometry of the [5]Al oxygen polyhedra is available: it could be a tetragonal pyramid, or a trigonal bi-pyramid. The present data do not allow assessment of this problem, but it is possible to make an assumption: the geometric shape of [5]Al will depend on the nearest charge compensator cation. For example, the smaller the quadrupolar moment, Cq, observed in ^{27}Al NMR is, the more [5]Al will look like a square base pyramid. The role of [4]Al, [5]Al and [6]Al in the glass and melt structure will be discussed later in the section *Aluminosilicate glasses and melts*.

Silicate melts: how can we use existing structural knowledge to model melt properties

With the above information from the traditional glass forming ranges, we understand that we know for silicate melts and glasses:

- the glass has a network of Q^n units, whose concentrations can be determined via NMR or Raman spectroscopy (Dupree et al. 1986; Farnan and Stebbins 1990; Stebbins et al. 1992; Mysen 1990; Maekawa et al. 1991; Mysen and Frantz 1993, 1994; Zhang et al. 1996; Sen et al. 2003; Davis et al. 2010; Ispa et al. 2010; Nesbitt et al. 2018, 2021; O'Shaughnessy et al. 2020)

- glass and melt heat capacities, which can be calculated, and

- their rheological properties.

---

[6] An amphoteric element can react both as an acid or a base; see Moretti (2005) for details on this concept for silicate melts.

It has been a longstanding goal to link those observations via a single model that could predict the properties and structure of glasses and melts. A pioneering start was the work of Mysen (1995), who proposed to calculate the melt configuration heat capacity from partial heat capacity values $C_p^{conf}{}_{Q^n}$ assigned to each Q^n unit with a fraction x_{Q^n} :

$$C_p^{conf} = \Sigma\, x_{Q^n} C_p^{conf}{}_{Q^n} \tag{14}$$

Links between melt thermodynamic parameters, rheology and structure were further established by subsequent publications (Neuville and Mysen 1996; Neuville 2006; Mysen 1990, 1995; Le Losq and Neuville 2013, 2017; Le Losq et al. 2015, 2017; Russell and Giordano 2017). Recently, Le Losq and Neuville (2017) demonstrated how, using the Adam and Gibbs theory and Q^n values from NMR spectroscopy, it is possible to determine with a high precision the viscosity of $Na_2O–K_2O–SiO_2$ melts. In this section, we will show how such a model can be constructed, and what opportunities for extension are envisioned. The application of such modeling for aluminosilicate melts will further be discussed in the section *Aluminosilicate glasses and melts*.

Using a principle similar to that of Equation (14), we may be able to calculate values of $S^{conf}(T_g)$ and B_e in Equations (6–9) from partial values assigned to Q^n units in the glass, i.e., in the melt at the glass transition temperature. This is possible because B_e is temperature-independent (Adam and Gibbs 1965), and $S^{conf}(T)$ can be estimated from $S^{conf}(T_g)$ and C_p^{conf}, which can be easily determined from viscosity data and modeled in silicate melts, respectively. This implies that we can use glass structural data to build a model for melts. However, we need to be able to model precisely how the distribution of Q^n units vary with glass composition. Fortunately, the distribution of Q^n units mostly depends on (i) the glass silica concentration, and (ii) the ionic field strength (IFS) of network modifier metal cations. Numerous ^{29}Si NMR spectroscopy data were acquired on silicate glasses and allow us to know precisely how Q^n units distribute depending on glass composition. They allow, in particular, determining the equilibrium constants of the Q^n dissociation reactions

$$2Q^n = Q^{n-1} + Q^{n+1} \tag{15}$$

which is at play in silicate glasses and melts. This explains the deviations of the Q^n distribution measured compared to silicate minerals, where binary combinations of Q^n units are observed (e.g., $Q^4–Q^3$, $Q^3–Q^2$, etc.). In glasses and melts, randomness implies that the Q^n distribution departs from the binary model. The existing ^{29}Si NMR and Raman data allow one to estimate how the melt composition influences such deviation. In Figure 14, we collected K_{eq} equilibrium constants for the dissociation of Q^3 units and of Q^2 units, and reported them as a function of the ionic field strength. In this Figure, most data are from NMR spectroscopy because data interpretation from ^{29}Si NMR 1D and 2D methods (Edén 2012) is less subject to debate, despite some possible complications that highlight that the determination of Q^n unit distribution always is an arduous task (Pedone et al. 2010; Charpentier et al. 2013). Indeed, Raman data requires the use of Raman cross sections, determined by calibration against ^{29}Si NMR data, to transform the area of Raman peaks assigned to different Q^n units into concentrations of those Q^n units. This is a first barrier as Raman cross-sections are not straightforward to obtain, a second one being the fit of the spectra that can be a topic of debate (Bancroft et al. 2018; Nesbitt et al. 2018, 2021). The very few K_{eq} values from Raman data used in Figure 14 do not come from possibly controversial traditional peak fitting methods, but from the use of linear alternative least square methods to retrieve partial Raman spectra for the different Q^n units in binary silicate glasses, a method developed by Zakaznova et al. (2007) and Malfait et al. (2007, 2009). While it is very powerful for binary silicate glasses, it relies on composition-specific, constant Raman signal shapes for the different Q^n units. This hypothesis is valid in binary systems, but this method is limited because (i) partial Raman spectra for one binary system (e.g., $Na_2O–SiO_2$) cannot be used for others and (ii) this starting hypothesis becomes void for ternary and more complex systems. This is because partial Raman spectra and Raman cross-sections differ depending on the network modifier present into the glass (e.g., Malfait et al. 2009; Woelffel et al. 2015).

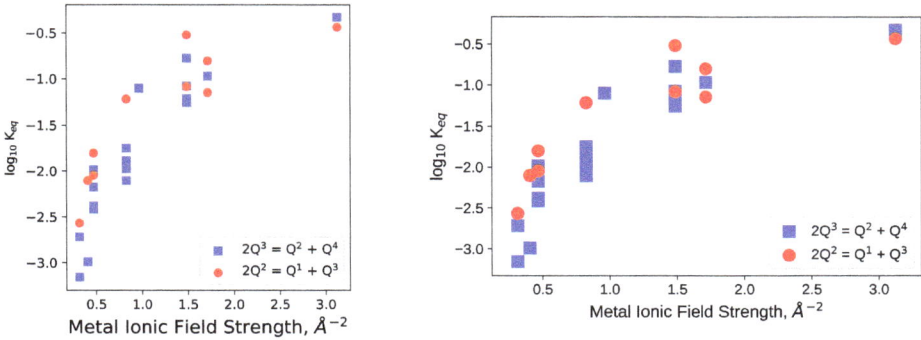

Figure 14. Equilibrium constants of the dissociation of Q^3 and Q^2 units represented as a function of the metal ionic field strength. Metal IFS was calculated assuming a coordination number of 6 and ionic radius from Whittaker and Muntus (1970). Equilibrium constants were calculated from the data of Emerson (1989), Emerson and Bray (1994), Maekawa et al. (1991), Bray et al. (1991), Zhang et al. (1996, 1997), Schneider et al. (2003), Voigt et al. (2005), Larson et al. (2006), Malfait et al. (2007, 2009), Zakaznova-Herzog et al. (2007), Sen et al. (2009), Davis et al. (2010) and Herzog and Zakaznova-Herzog (2011).

An idea that arises when seeing Figure 14 is to parametrize the K_{eq} values as a function of the mean ionic field strength of metals cations in silicate glasses. Indeed, K_{eq} values for Q^3 and Q^2 units change systematically depending on the ionic field strengths of the metal cation in the glass structure. It thus seems possible to calculate K_{eq} values from the mean ionic field strengths of metal cations in the glass. For ternary and more complex silicate glasses, this relies on the fact that mixing different cations in silicate glasses affects linearly the distribution of Q^n units. Fortunately, this is a good approximation. Indeed, from ^{29}Si and ^{23}Na MAS NMR spectroscopy, the Q^n unit distribution upon Rb–Na mixing in alkali trisilicate glasses does not systematically change (Hater et al. 1989). This was also reported during Na and Li mixing in $(Na,Li)_2Si_2O_5$ glasses (Ali et al. 1995; Sen et al. 1996). From ^6Li, ^7Li and ^{29}Si static NMR spectroscopy data, Bray et al. (1991) observed a linear change of the Q^3 units fraction upon mixing Li and K in disilicate glasses. The ^{29}Si static NMR data from Emerson and Bray (1994) further suggest a slight departure from linearity of the Q^3 unit fraction when mixing $Na_2Si_2O_5$ and $Cs_2Si_2O_5$ glasses. Such interpretation may be consistent with ^{17}O Dynamic Angle Spinning NMR data from mixed $Na_2Si_2O_5$–$K_2Si_2O_5$ glasses (Florian et al. 1996), showing slightly non-linear changes of the fractions of Bridging (BO) and Non-Bridging (NBO) oxygen anions with changes in the glasses K/(K + Na) ratios. Such observation agrees with the recent report of Le Losq and Neuville (2017), which showed that the Raman areas of Q^n units in alkali trisilicate glasses vary linearly within error bars.

From the above review, it thus should be possible to calculate K_{eq} values for Q^3 and Q^2 unit dissociation given the mean IFS of metal cations present in the silicate glass. However, despite a good correlation, scattering is visible in Figure 14. Such scattering results from second order effects that may be related to some local peculiarities of the effect of each cation on K_{eq} values as well as our lack of a perfect understanding of the coordination number of each cation. To circumvent this problem, we propose to calculate the mean K_{eq} values in silicate glasses based on the combination of K_{eq} values assigned to each cations (from their calculation in the binary $M^x_{2/x}O$–SiO_2 systems):

$$K_{eq} \text{ glass} = \Sigma \, K_{eq}{}^i \, x_i \tag{16}$$

where x_i is the fraction of a given cation relative to the sum of all cations. Assuming linear variations in the Q^n unit fractions upon mixing different cations, a valid hypothesis given existing data as indicated previously, this will allow obtaining the most precise model for the variations in Q^n units in Al-free silicate glasses (the case of aluminium deserves its own section, such that its effect will be discussed in the section *Aluminosilicate glasses and melts*). K_{eq}^i values determined for each system are provided in Table 1, based on the data reported in Figure 14.

Table 1. Mean values of the K_{eq} coefficients for the dissociation of Q^3 or Q^2 units in M^+O_2–SiO_2 or $M^{2+}O$–SiO_2 glasses, calculated from the data presented in Figure 14. Standard deviations from the mean are provided in parenthesis, when possible.

Metal cation	$K_{eq}\ 2Q^3 = Q^2 + Q^4$	$K_{eq}\ 2Q^2 = Q^1 + Q^3$
Li	0.09(5)	0.2(2)
Na	0.012(4)	0.06(-)
K	0.006(3)	0.012(5)
Rb	0.001(-)	?
Cs	0.001(-)	0.003(-)
Mg	0.464(-)	0.364(-)
Ca	0.106(-)	0.156(-)
Ba	0.078(-)	?

Now that we can calculate the fractions of Q^n units for various silicate compositions, it opens doors to calculate the terms of the Adam and Gibbs equation: B_e and $S^{conf}(T_g)$. Le Losq and Neuville (2017) leveraged such ability for K_2O–Na_2O–SiO_2 melts with 60 to 100 mol% SiO_2, taking into account the ideal mixing of Na and K in silicate melts (Richet 1984) when developing equations for $S^{conf}(T_g)$ and B_e. For $S^{conf}(T_g)$, they assumed that, added to the topological entropy S^{topo}, excess entropies arise from the ideal Si mixing between the different Q^n units S^{mix}_{Si} and random mixing of Na and K into the glass network S^{mix}_{Na-K}, such that:

$$S^{conf}(T_g) = S^{topo} + S^{mix}_{Si} + S^{mix}_{Na-K} \tag{17}$$

with

$$S^{topo} = \sum_{n=2}^{3} x_{Q^n_{Na_{env}}} S^{conf}_{Q^n_{Na_{env}}} + \sum_{n=2}^{3} x_{Q^n_{K_{env}}} S^{conf}_{Q^n_{K_{env}}} + x_{Q^4} S^{conf}_{Q^4} \tag{18}$$

$$S^{mix}_{Si} = \frac{-x_{Si}}{x_O} \times 2R \sum_{n=2}^{4} x_{Q^n} \ln\left(x_{Q^n}\right) \tag{19}$$

and

$$S^{mix}_{Na-K} = \frac{-\left(x_{Na} + x_K\right)}{x_O} \times 2R\left(X_K \ln\left(X_K\right) + \left(1 - X_K\right) \times \ln\left(1 - X_K\right)\right) \tag{20}$$

with x_{Si}, x_O, x_{Na}, and x_K, the atomic fractions of Si, O, Na and K respectively, $X_K = K/(K+Na)$, R the gas constant, and x_{Q^n}, the Q^n fractions. We note that, as Le Losq and Neuville (2017) investigated melts with NBO/T lower than 1.5, equations above were limited to Q^2 to Q^4 units. In addition, in the topological contribution (Eqn. 18), $Q^n_{Na_{env}}$ and $Q^n_{K_{env}}$ units with their associated $S^{conf}_{Q^n_{Na_{env}}}$ and $S^{conf}_{Q^n_{K_{env}}}$ values are distinguished; those are the Q^n units, with $n < 4$, that have NBO bonded preferentially to Na or K, respectively. This is an approximation introduced to take into account the fact that Q^n units with NBO bonded to Na or K may present different S^{conf} values, as shown by the difference of $S^{conf}(T_g)$ between sodic and potassic silicate glasses.

For the calculation of B_e, the link between $S^{conf}(T)$ and B_e needs to be considered. Indeed,

$$B_e = \frac{\Delta \mu s_c}{k_B} \tag{21}$$

and

$$S^{conf}(T_g) = \frac{N_A}{z(T_g)} S_c \tag{22}$$

with $\Delta\mu$ the Gibbs free-energy barrier opposed to the cooperative rearrangements of molecular subunits of size $z(T_g)$ and intrinsic entropy S_c, k_B the Boltzmann constant and N_A the Avogadro constant (Adam and Gibbs 1965; Richet 1984). S_c represents the entropy of the molecular subunits involved in the melt viscous flow and relaxation process. It appears in both Equations (21) and (22), yielding a strong correlation between $S^{conf}(T_g)$ and B_e (Neuville and Richet 1991). From Equations (20) and (22), we can write

$$B_e = \frac{\Delta\mu z(T_g) S^{conf}(T_g)}{R} \tag{23}$$

with R the perfect gas constant ($= N_A k_B$). From Equation (23) and considering that $\Delta\mu$ and S_c are temperature-independent (Richet 1984), B_e should be expressed similarly to $S^{conf}(T_g)$, from the melt structure recorded at T_g in glasses. Considering that, Le Losq and Neuville (2017) assumed that B_e can be expressed as the sum of partial molar B_e terms for the different $Q^n_{M_{env}}$ and Q^4 units, with additional non-linear contributions arising from the role of $S^{conf}(T_g)$ in Equation (23). The latter non-linear contributions are obtained by directly injecting Equations (19) and (20) with K_1 and K_2 scaling terms in the expression of B_e:

$$B_e = \sum_{n=2}^{3} x_{Q^n_{Na_{env}}} B_{eQ^n_{Na_{env}}} + \sum_{n=2}^{3} x_{Q^n_{K_{env}}} B_{eQ^n_{K_{env}}} + x_{Q^4} B_{eQ^4} + K_1 S^{mix}_{Si} + K_2 S^{mix}_{Na-K} \tag{24}$$

Using this model, Le Losq and Neuville (2017) showed that it is possible to calculate the viscosity of the alkali silicate melt with a root mean squared error of 0.2 log Pa·s. This error is much lower than those of parametric models, typical presenting root mean squared errors on their parametrization datasets of ~0.6–0.7 log Pa·s (e.g., Hui and Zhang 2017; Giordano et al. 2008).

Le Losq and Neuville (2017) used cubic splines for interpolation of the Q^n distribution in their model. With the approach presented in this chapter, it is possible to replace this by the K_{eq}-based model for Q^n distribution in silicate melts in order to improve the future deployment and development of such a thermodynamic viscosity model. Preliminary tests yield good predictions with standard errors of 0.3 log Pa·s, a value slightly higher than the initial value of 0.2 log Pa·s from Le Losq and Neuville (2017). This difference arises from the fact that the initial cubic spline modeling of the Q^n distribution in K_2O–Na_2O–SiO_2 melts is slightly more accurate than the global structural model based on the K_{eq} values calculations. Differences are in the range of 1–2%. This actually highlights how sensitive a model can be to melt structure: an error of only 1–2% on the determination of the structure alters measurably the accuracy of the rheological model.

ALUMINOSILICATE GLASSES AND MELTS

Aluminum is a key element in Earth and material sciences because it is present, in various amounts, in most geologic and industrial glasses, and it drastically affects their properties.

Generally, a well-known feature of the structure of silicate minerals is that silicon ions may be replaced isomorphously by aluminum ions provided that either mono or divalent cations are available to maintain local neutralization of electrical charges in the structure (Putnis 1992). This feature is clearly visible in tectosilicates that present an interconnected three-dimensional silica-like tetrahedral structure, in which the mono or divalent cations are accommodated in the framework free volumes. In these minerals, the ratio $R=M^+_2O/Al_2O_3$ or $M^{2+}O/Al_2O_3$ is unity. The metal cations play charge compensator roles, and no non-bridging oxygens are present. This is also generally the case in tectosilicate glasses, albeit some possible minor amounts of NBOs may be present (Stebbins and Xu 1997). However, an additional complexity in aluminosilicate glass is the variable coordination number of Al. Indeed, while Al mostly substitutes for Si in the SiO_4 tetrahedra in silicate glass, it also is found in five and six-fold coordinations, in proportions that depend on melt chemical composition, system pressure and temperature, and glass-forming cooling rate (Lacy 1963; McMillan et al. 1982; Murdoch et al. 1985; Ohtani et al. 1985; Risbud et al. 1987; Merzbacher et al. 1990, 1991; Merzbacher and White 1991; Sato et al. 1991; Coté et al. 1992; Stebbins and Xu 1997; Lee and Stebbins 2000, 2002, 2009; Allwardt et al. 2004, 2007; Stebbins et al. 2000, 2008; Allward and Stebbins 2005a,b, 2007; Neuville et al. 2004a,b, 2006, 2008a,b, 2010; Lee 2005; Henderson et al. 2007; Stebbins 2008; Le Losq et al. 2014). The varying coordination number of Al raises the question of its role in the melt structure depending on its coordination. Indeed, following the terminology proposed by Zachariasen (1932), while Al in four-fold coordination is a network former, Al in six-fold coordination should be a network modifier, and Al in five-fold coordination should probably also be considered as a network modifier. We will review such questions, and more generally the connection between aluminosilicate melt structure and properties in this section.

Molar volume, coordination number and structure

Density and molar volume are properties easy to measure in glasses, and bring important pieces of information regarding their internal structure, and by extension, that of the melt at the glass transition. Variations in the molar volumes of alkali and alkaline-earth silicate and aluminosilicate glasses as a function of silica concentration show similar trends, except for the heaviest elements, like potassium, cesium and rubidium (Fig. 15). From silica glass with $V_m=27.31cm^3$, it is found that generally the molar volume decreases with the decrease in SiO_2, except for K-, Cs-, Rb- silicate glasses. For the same SiO_2 content, the magnitude of the decrease of the molar volume depends on the ionic field strength of the metal cation: it is moderate for Na^+ and very important for Mg^{2+} (for example V_m of Ca silicate $=21.21cm^3$ and $24.14cm^3$ for Na-silicate, both at 60 mol% of SiO_2—note that we compare Ca and Na silicate glass and not Mg-silicate at 60 of silica, because this glass is made with a high cooling rate, and its fictive temperature is probably very different than that for the Na-silicate glass). In the case of potassium, V_m always increases regardless of the glass aluminum content. For other cations, V_m variations in aluminosilicate glasses are more complex. First, the M^{2+}/Al or M^+/Al ratio directly influences the variations of V_m with the glass silica concentration. Secondly, a general concave trend of the V_m versus SiO_2 relationship is observed according to previous work (Seifert et al. 1982; Doweidar 1999; Doweidar et al. 1999). Taking the SiO_2–$CaAl_2O_4$ binary mixture as an example, and starting from silica, V_m decreases with silica decrease until 50 mol% SiO_2 ($V_m=25.81cm^3$ for anorthite glass), and then increases up to a value of ~27 cm³ for the pure $CaAl_2O_4$ end-member for example. Such non-linear V_m trends indicate that V_m modelling in aluminosilicate glasses cannot be obtained from a linear combination of partial V_m values assigned to the different oxides. To circumvent this problem, several papers have modelled the density of glass and melts as a function of their expected Q^n speciation (Doweidar 1999; Doweidar et al. 1999) or other entities like partial volume of oxide components (Lange 1996, 1997; Bottinga and Weill 1970).

One of the most important pieces of information arising from V_m data analysis is the possibility to provide knowledge regarding the coordination numbers of metal cations as well as aluminum. We have mentioned, in the section *Alkali or earth alkaline silicate glasses and*

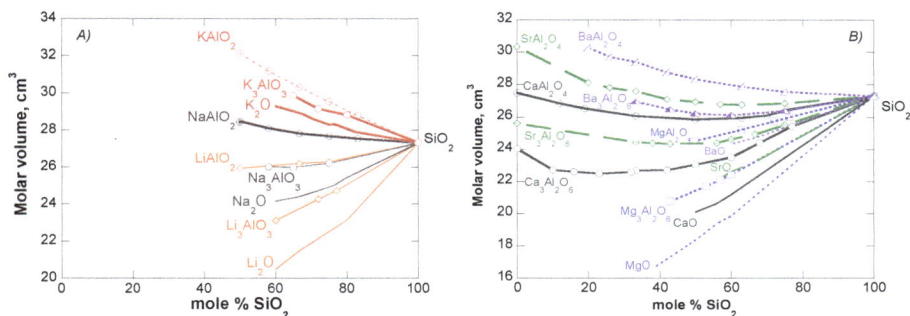

Figure 15. A) molar volume for alkali and **B)** for alkaline-earth silicate and Al-silicate glasses, data for NAS, KAS compositions (Le Losq et al. 2015, 2017; Neuville 1991, 2006), LAS (Strukelj 2008), MAS, CAS system (Neuville 1992), SAS data from Novikov et al. (2017), BAS original data and from Novikov (2017).

melts (Fig. 5), that metal cations' CNs remain almost constant in silicate glasses. It is also well known that the molar volume of the SiO_4^{4-} tetrahedra doesn't change a lot as a function of the Q^n species because the d(Si–BO) or d(Si–NBO) vary less than 0.05 Å (Cormack and Du 2001; Du and Cormack 2004). This variation corresponds to a variation of around 3% on the distance between Si–BO and Si–NBO which causes a variation of the volume of the tetrahedra between 12.4 and 9.3% as a function of the amount of alkali content respectively 10 and 50%, if we used the distances given by Cormack and collaborators. In the case of aluminum, the molar volume of AlO_4^{5-} tetrahedra doesn't change because Al-O distances stay almost constant as a function of the different Q^n species for Al[7], and the most important change in the molar volume of Al units come from the change in the coordination of Al and not of the Q^n species of Al. If we consider that the volumes vary very slightly in the ternary diagrams this implies that the molar volume variation observed between silicates and aluminosilicates glasses is produced essentially by changes in coordination number of alkaline or alkaline-earth. Before discussing this, we will quantify the different proportions of aluminum.

Proportion of Al in five-fold coordination

Given the role of Al coordination on V_m and other melt/glass properties, it is clear that Al coordination should be quantified. Fortunately, through ^{27}Al NMR spectroscopy, several studies allow drawing a picture of Al coordination variations with composition, temperature and pressure.

The presence of [5]Al is systematically observed in the middle of the ternary diagrams of Li aluminosilicate, and alkaline-earths aluminosilicate, including peralkaline glasses in which Al in five-fold coordination is not expected considering the stoichiometry (Neuville et al. 2006, 2007, 2008b, 2010). Moreover, the proportion of [5]Al is always maximum for glasses with about 50 mol% of SiO_2 (Neuville et al 2004, 2006, 2008, Novikov et al. 2017). On the contrary, Na or K tectosilicate and peralumininous glasses have small concentrations of Al in five-fold coordination, which becomes undetectable in peralkaline compositions (Stebbins and Farnan 1992; Allwardt et al. 2005a,b; Lee et al. 2009; Le Losq et al. 2014). The variation of NMR chemical shift values for [5]Al and [6]Al sites with SiO_2 content are less dependent on silica content than [4]Al sites (Figure 13b, from the previous chapter). The NMR chemical shift values are directly related to the inter-tetrahedral angle or the first-neighbor cations (Stebbins and Farnan 1992). This can imply that [5]Al and [6]Al sites are more connected with Al in four fold coordination than with Si in different Q^n species which does not respect the Lowenstein rule's,

---

[7] The distance Al–O increases from 1.76 up to 1.86 Å when the coordination number of Al changes from 4 to 6 (Cormier et al. 2000, 2003; Hennet et al. 2016). This implies that Al–O distance vary smaller when Al change of Q^n species in tetrahedra.

but in the majority of studies this rule is avoided (Neuville et al. 2008a, 2010; Hiet et al. 2009; Lee and Stebbins 2009). 2D NMR ^{27}Al–^{29}Si correlations (Hiet et al. 2009; Lee and Stebbins 2009) show that $^{[5]}$Al is more connected with Si than with Al in lanthanium aluminosilicate glasses (Florian et al. 2007). Highly coordinate Al species, especially $^{[5]}$Al, are likely to have a positive formation enthalpy and their abundance in the melt is expected to increase with increasing temperature as observed by some authors (Neuville et al. 2008b; Stebbins 2008). However, the ^{27}Al NMR spectra at high temperature shifts to lower ^{27}Al chemical shift values with increasing temperature, this can be interpret as resulting from the appearance of $^{[5]}$Al and $^{[6]}$Al contributions located at lower chemical shifts than that of $^{[4]}$Al (Kanehashi and Stebbins 2007; Thompson and Stebbins 2011, 2012, 2013). Signatures of Al in high coordination are also clearly visible using X-ray absorption spectroscopy at the Al K-edge (Neuville et al. 2008b). In the case of CA50.25 (anorthite melts composition), Neuville et al (2008b) show an increase from 8% up to 22% of $^{[5]}$Al between room temperature and 1900 K, in good agreement with high temperature Al NMR (Stebbins 2008). A significantly higher proportion of $^{[5]}$Al is observed in magnesium aluminosilicate glasses than in the others aluminosilicate systems on the tectosilicate join (Neuville et al. 2008b; Novikov et al. 2017; Fig. 23B) which can be explained by the promotion of highly coordinated Al species with the presence of metal cations with relatively high ionic field strength, as shown in boroaluminate by Bunker et al. (1991) and in La-aluminosilicate glass compositions (Florian et al. 2007). Such an idea agrees with the increase in the disorder when the cation field strength decreases (Mysen et al. 1982; Navrotsky et al. 1982, 1985; Murdoch et al. 1985; Sharma et al 1985; Lee and Stebbins 2000, 2002; Neuville et al. 2008a; Wu and Stebbins 2009; Lee et al. 2016, Le Losq et al. 2017).

Several papers on ^{27}Al NMR spectroscopy have shown that, for R = M/Al varying between infinity and 1 (silicate to tectosilicate compositions), Al is essentially in four-fold coordination (Murdoch et al. 1985; Risbud et al. 1987; Merzbacher et al. 1990, 1991; Merzbacher and White 1991; Coté et al. 1992; McMillan and Kirkpatrick 1992; Stebbins and Xu 1997; Stebbins et al. 2000, 2008; Neuville et al. 2004a,b, 2006, 2007, 2008a; Massiot et al. 2008; Richet et al. 2009a; Lee and Stebbins 2009, 2016; Le Losq et al. 2014; Novikov et al. 2017; Florian et al. 2018; Charpentier et al. 2018). In the CAS system, 93% of Al is in 4-fold coordination, and the proportion of $^{[5]}$Al increases slowly up to 7% for the anorthite glass composition on the tectosilicate join (Neuville, et al. 2006). In the MAS system, the proportion of $^{[5]}$Al is a little higher, 12% for Mg-anorthite glass (Neuville et al. 2008a). For all others M^+ or M^{2+} Al-silicate systems, the proportion of $^{[5]}$Al is lower and the amounts decrease with increasing z/r (z : cation charge and r atomic radius, see Neuville et al. 2008a or Stebbins et al. 2008, Novikov et al. 2017). Figures 16 A and B show the total amount of $^{[5]}$Al in mol% in the CAS and MAS system.

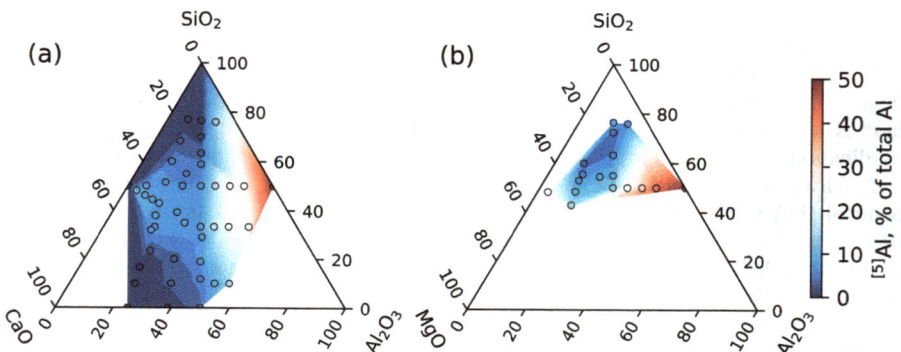

Figure 16. Percentage of $^{[5]}$Al of total Al_2O_3 in the CAS (**left**) and MAS (**right**) systems. Data from Neuville et al. (2004, 2006, 2007, 2008b, 2010), Massiot et al. (2008), Licheron et al. (2011).

Molar volume, AlO_4^{4-} and implications for the coordination number of metal cations

In the previous sections, we saw that the molar volume of Si in the different Q^n species can vary by less than 10% when n varies from 4 to 0 and that the molar volume of Al stays almost constant for Al in four-fold tetrahedra. Despite this, the molar volume V_m of a lime aluminosilicate glass with 50% of SiO_2 can vary from 20 to 25.9 cm^3 between the wollastonite and anorthite compositions, this represents an increase of 25%. This increase is 12% between silicate glasses and the glasses on the R = 3 join, and of 12% again between the join R = 3 and the join R = 1 with $R = MO/Al_2O_3$ with M = Mg, Ca. This huge increases of V_m is a maximum at 50 mol% silica.

In the case of calcium aluminate glasses, the molar volume difference between C3A ($Ca_3Al_2O_6$) and CA ($CaAl_2O_4$) glasses is only 6.8% (Fig. 15B). In Figure 15B, the increase of molar volume between join R = 3 and 1 can also result from a change in the coordination number of calcium because the aluminum stays in four-fold coordination between C3A and CA glasses. In this case, Ca varies between regular octahedra where Ca is in 6-fold coordination to a distorted site in 7–8-fold coordination (Cormier et al. 2001, 2003; Neuville et al. 2008b, 2010; Drewitt et al. 2012, 2017; Jakse et al. 2012; Takahashi et al. 2015; Cicconi et al. 2016; Hennet et al. 2016).

Generally, the coordination number of alkali or alkaline-earth elements is difficult to determine, but changes in the V_m of a glass can inform us about CN changes. For instance, the abovely variation in the CN of Ca is in good agreement with the changes observed via XANES at the Ca K-edge in Ca aluminosilicate glasses (Cormier and Neuville 2004; Neuville et al. 2004b, 2008; Cicconi et al. 2016). In particular, Cicconi et al. (2016) observed a shift to greater energy and a decrease in intensity of the pre-edge in the XANES spectra at the Ca K-edge between wollastonite to anorthite glasses. This shift can also be correlated with the new peak ω_2 observed in the depolarized Raman spectra, VH, (see Fig. 13, Drewitt et al. 2022, this volume). The peak ω_2 decreases when R goes from infinity to 1 (to silicate to tectosilicate glasses)

In the case of the NAS system ($Na_2O–Al_2O_3–SiO_2$), the peak ω_2 is also clearly visible for silicate glasses in the VH Raman spectra and it disappears for $R = Na_2O/Al_2O_3 = 1$ (Hehlen and Neuville 2015). This peak ω_2 exists in all silicate glasses with $M^+ = Na$, K and $M^{2+} = Mg$, Ca, Sr, Ba (Hehlen and Neuville 2015, 2020) and it's intensity decreases when Al replaces M^+ or M^{2+}. In the NAS glass system, a shift is also observed in the chemical shift, δ, of ^{23}Na NMR between NS3 and albite glass (Le Losq et al. 2015) following a trend that can be correlated with variations in XANES at the Na K-edge between the NS3 and albite glasses (Neuville et al. 2004c).

In the case of the SAS ($SrO–Al_2O_3–SiO_2$) system, Novikov et al. (2017) also observed some change in the first XANES oscillation which can also be attributed to a change in CN of strontium.

To summarize, ^{23}Na NMR, XANES at the Na, Sr, Ca K-edge, and depolarized Raman spectra, VH, all show changes that can be correlated with an observed increase in the molar volume between silicate and tectosilicate glasses. V_m varies between 4% and 16% with changing the M^+/Al or M^{2+}/Al ratio at a given silica content and depends clearly on the metal cation size. The V_m changes can be assigned to changes in the coordination of metal cations as their role in the aluminosilicate network varies with the M/Al ratio. Similarly, features observed in NMR, XANES or Raman spectroscopies can be directly correlated with such change in the role of alkali or alkaline-earth elements between network modifier and charge compensator. The concept of network modifier and charge compensator was introduced by Zachariazen but never really demonstrated. Now, with the technical evolution of spectroscopy methods, but with measuring the molar volume determined from density measurements, it is possible to have an idea of the role alkaline elements.

In Figure 17, the molar volume of various glasses is shown as a function of mole percent of Al_2O_3 at constant silica. The glass compositions vary between $MSiO_3$ up to Al_2SiO_5, M = Mg, Ca, Sr, Ba. Four trends are clearly visible for the four alkaline-earth aluminosilicate glass families. In the case of $MgSiO_3$-Al_2SiO_3, the molar volume varies almost linearly with Al_2O_3 (the linear variation corresponds to the dashed line). This can imply:

i.)   that the coordination number of Mg is almost constant in this system and is not really affected by the Mg/Al substitution, an idea in good agreement with XANES at the Mg K-edge (Trcera et al. 2009);

ii)   that the molar volume depends on the proportion of Al in five-fold coordination.

For the Ca, Sr and Ba aluminosilicate glasses the changes observed in the slopes near 25 mol% of Al_2O_3 correspond to the tectosilicate line (R = $M^{2+}O$/Al_2O_3 = 1), characterized by an increase in the proportion of $^{[5]}Al$ in the glass (Neuville et al. 2006, 2008; Novikov 2017; Novikov et al. 2017). This Figure clearly shows two domains, one where V_m is affected by aluminum, the peraluminous domains, and one where V_m depends on an increase in the coordination of the alkaline-earth element, the peralkaline domain (Cicconi et al. 2016; Novikov et al. 2017).

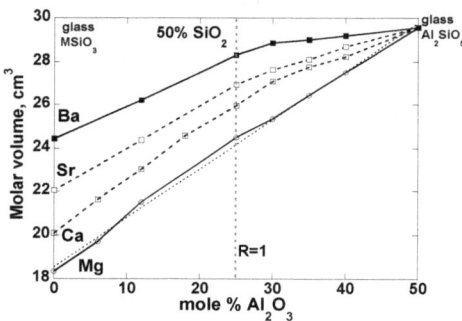

Figure 17. molar volume at constant SiO_2 = 50 mol% for four alkaline-earth aluminosilicate glass compositions. (The **dashed line** between $MgSiO_3$ and Al_2SiO_5 glasses correspond to this equation $18.547+0.22497\times Al_2O_3$). Values are calculated from density measurements from Neuville (1992) for Ca and Mg system per-alkaline compositions, Neuville et al. (2006, 2008) for Ca and Mg Per-aluminate compositions, Novikov et al. (2017) for Sr compositions, Novikov (2017) for Ba compositions and Al_2SiO_5 are from Wang et al. (2020).

Glass transition temperature

In general, a decrease in T_g occurs with decreasing SiO_2 content at all R ratios (R = M^+O_2/Al_2O_3 or $M^{2+}O$/Al_2O_3) (Fig. 18). It is particularly important at silica contents between 100 and ~75 mol%. In detail, this T_g decrease is more significant in Al-free silicate melts than in tectosilicate glasses; the lower R, the lower the effect of changing SiO_2 on T_g. Typically, at $R=1$ in tectosilicate compositions, the decrease in T_g is of ~300 K for $MAl_2Si_2O_8$ glasses, and 200 K and 300 K for respectively $KAlSiO_4$ and $NaAlSiO_4$ glasses. The large contrast between T_g variations in Al-free silicate and tectosilicate glass compositions results from the changing role of metal cations upon aluminum addition into the melt network. Indeed, while in silicate glasses, alkali and alkaline-earth elements are network modifiers that break the 3D tetrahedral network, in aluminosilicate glasses they ensure the charge compensation of AlO_4^- tetrahedral units and thus participate in increasing melt polymerization. This translates to higher T_g along the M^+AlO_2–SiO_2 or $M^{2+}Al_2O_4$–SiO_2 binaries than along the M^+O_2–SiO_2 or $M^{2+}O$–SiO_2 binaries.

Link between structure and properties of aluminosilicate melts: example of the CaO–Al_2O_3–SiO_2 system

To discuss and illustrate the links between glass structure and glass/melt properties, the CaO–Al_2O_3–SiO_2 (CAS) glass system will be taken as an example in the following. The CAS system has the advantage having the largest glass forming domain (Neuville et al. 2006). The observations and interpretations made in this system can be extended to other ternary system.

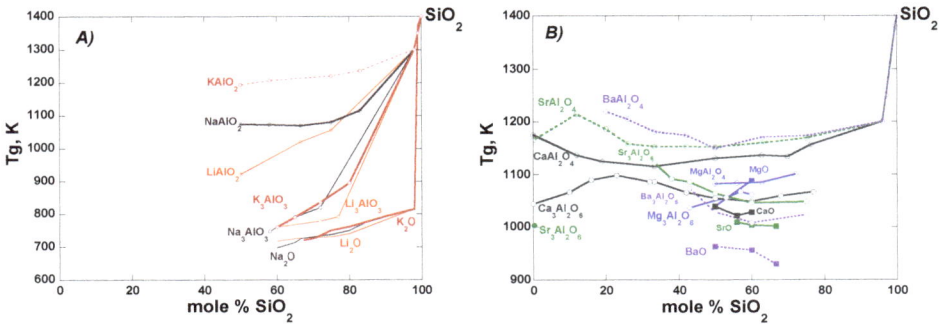

Figure 18. Glass transition temperature T_g, in K, of silicate and aluminosilicate glasses with alkali elements (**A**) and alkaline-earth elements (**B**). The glass transition temperatures were determined from viscosity measurements, i.e., they correspond to the temperature at which log $\eta = 12$ Pas. Data from Neuville (1991) for Li, K silicates; Neuville (2006) and Le Losq et al. (2014) for sodic systems (Neuville 1992) for Ca and Mg, Novikov et al. (2017) for Sr and Novikov (2017) for Ba.

In general, in the CAS system, an important decrease in T_g is observed with decreasing SiO_2 content, which results from network depolymerization, except at R = 1. In the latter case, melt polymerization is almost constant and the decrease in T_g results from the formation of weaker Si–O–Al bonds as Al substitutes for Si in Q^4 units (Navrosky et al. 1985; Cormier et al. 2001, 2005; Neuville et al. 2004a; Hennet et al. 2016). T_g varies by about four hundred degrees between pure SiO_2 ($T_g \sim 1480$ K) and the center of the ternary system ($T_g = 1115$ K for CA33.33, $M^{2+}AX.Y$ where M^{2+}=alkaline-earth element and A=Al_2O_3, X = SiO_2%; i.e., 33.3 mol%, Y = %Al_2O_3 and $M^{2+}O = (100-(X+Y)$ in this case, CaO=33.3%). At R = 1, glasses with less than 30 mol% SiO_2 display a continuous increase in T_g upon decreasing SiO_2 content, and T_g reaches a new local maximum for the $CaAl_2O_4$ composition at 1175 K; the minimum in T_g at R=1 is centered between 40 and 50 mol% SiO_2. The observed behavior in T_g at R = 1 results from the competition between Si and Al as network formers. As indicated by the difficulty of forming glasses at high Al concentrations (e.g., along the SiO_2–Al_2O_3 binary) as well as a lowering of T_g with addition of Al. Al can be considered as a relatively poor network former compared to Si. This is further corroborated by variations in the melt fragility (i.e., the derivative of the melt viscosity as a function of T at T_g): the strongest existing melt is SiO_2, and Al-rich melts have higher fragility, with the most fragile being $CaAl_2O_4$ in the CAS system (Fig. 19). This is also in good agreement with the variation of the C_p^{conf} of SiO_2 and $CaAl_2O_4$, which varies between 8.2 to 13.8 J/mol K. This ΔC_p between liquid and glass at T_g can be expressed in term of C_{p_l}/C_{p_g} ratio as proposed by Angel (1991). This ratio varies between 1.1 for SiO_2, the strongest melt to 1.37 for $CaAl_2O_4$, the more fragile melt in Figure 19.

As mentioned previously, the CAS system presents one of the largest glass-forming domains. Indeed, it is possible to quench glasses even without silica; e.g., like the C12A7 composition (CXAY with C=CaO, A=Al_2O_3, X and Y are their relative proportions), using classic quench rates. With increasing quench rates, e.g., via laser levitation heating, it is possible to make glasses between C3A and CA1.5 (from 75 down to 40 mol% CaO). In this condition, the addition of a small amount of SiO_2 in CaO–Al_2O_3 glasses extends the glass forming ability and lowers the melt liquidus temperature. Other macroscopic properties are also markedly changed with silica introduction in CaO-Al_2O_3 melts/glasses, suggesting extensive structural modifications. For instance, an anomalous behavior of T_g is clearly observed at low silica content, with a maximum in T_g visible at ~20 mol% SiO_2 along the $Ca_3Al_2O_6$–SiO_2 binary (Fig. 18B; Higby et al. 1990; Neuville 1992; Neuville et al. 2004a, 2010; Cormier et al. 2005). Note that this anomalous behavior of T_g results from the change in the polymerization of Al which varies between Q^2 to Q^4 species along the join R=3 between C3A and CA20.40 glasses (Cormier et al. 2001, 2003, 2005; Neuville et al. 2004b).

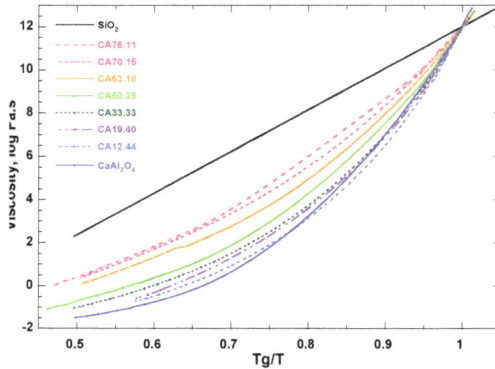

Figure 19. "Angel plot" showing log η versus T_g/T for melts with R = CaO/Al$_2$O$_3$ = 1 in the CAS system.

To explore further the properties of CAS glass compositions, viscosity along the join R = 1 (CaAl$_2$O$_4$–SiO$_2$ binary) is plotted in Figure 20. At high temperatures, viscosity decreases constantly with addition of CaAl$_2$O$_4$ (Fig. 20B). At supercooled temperatures, near T_g, melt viscosity decreases by ~ orders of magnitude with adding 30 mol% CaAl$_2$O$_4$ to silica, then a further addition of 30 mol% CaAl$_2$O$_4$ yields a decrease in viscosity of 1 order of magnitude. Between 40 and 0 mol% SiO$_2$, melt viscosity increases ~5 orders of magnitude. The deviation from a linear variation (Fig. 20B) is 4 orders of magnitude at 1175 K, and less than 2 orders of magnitude at 1900 K. It thus decreases with increasing temperature. These changes thus deviate strongly and negatively from a linear variation, particularly at undercooled conditions. Such behavior may originate from mixing between Si and Al in Q^4 units that occurs in such tectosilicate melts (Seifert et al. 1982), this mixing resulting in excess entropy and leading to non-linear, convex variation of the melt viscosity as a function of the Si/(Si+Al) ratio (e.g., Neuville and Mysen 1996).

In all others glass systems, alkali or alkaline-earth aluminosilicate, viscosity shows similar behavior along the join R = 1, an important decrease between 100 and 70% of SiO$_2$, and for lower silica content, a smaller decrease of the viscosity (Alkaline see Le Losq et al. 2017; Mg, Neuville 1992; Sr, Novikov et al 2017; Ba, Novikov 2017).

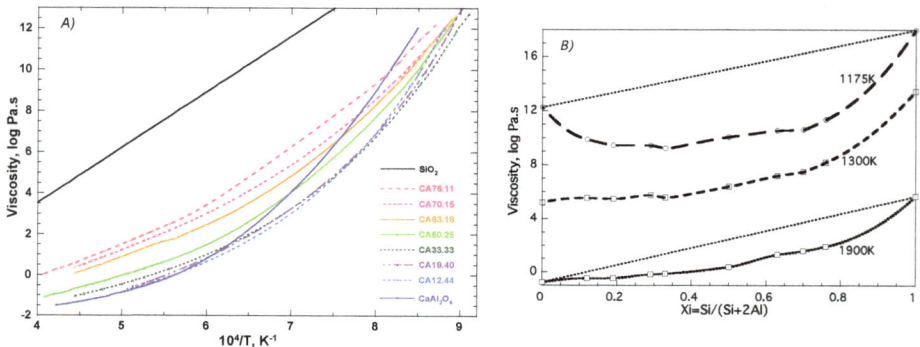

Figure 20. A) Viscosity of melts between silica and CaAl$_2$O$_4$ compositions, CAX.Y mean: X = SiO$_2$%, Y = Al$_2$O$_3$, CaO = 100− (X+Y). **B)** viscosity at constant temperature, 1150, 1200, 1800 K as a function of X_i = Si/(Si+2Al). SiO$_2$ data are from Hetherington et al. (1964) Leko (1979) and Urbain et al. (1982), CaAl$_2$O$_4$ from Urbain (1983), Neuville (1992).

As already mentioned, the configurational heat capacity, C_p^{conf}, is required in order to determine the configurational entropy, $S^{conf}(T)$, from Equations (6) and (7). In the $CaO-Al_2O_3-SiO_2$ system, the configurational heat capacity can be determined from the difference between the heat capacity of the melt at T and that of the glass at T_g (Eqn. 8). Several models exist, like those of Richet and Bottinga (1985) and Richet (1987), and allow one to calculate the configurational heat capacity with precision. The more recent works of Richet and Neuville (1992) and Courtial (1993) and allow refining the heat capacity modelling in the $CaO-Al_2O_3-SiO_2$ system, with using the following partial molar heat capacity equations (J mol^{-1} K^{-1}) for the glass and liquid (the estimated uncertainties is about 1% for the partial molar heat capacities following Courtial and Richet 1993):

$$C_{p_g\ SiO_2} = 127.2 - 0.010777\,T - 431270/T^2 - 1463.9/T^{0.5} \tag{25}$$

$$C_{p_g\ Al_2O_3} = 175.46 - 0.005839\,T - 1347000/T^2 - 1370/T^{0.5} \tag{26}$$

$$C_{p_g\ CaO} = 39.159 + 0.01865\,T - 152300/T^2 \tag{27}$$

$$C_{p_l\ SiO_2} = 81.37 \tag{28}$$

$$C_{p_l\ Al_2O_3} = 85.78 - 130.216T \tag{29}$$

$$C_{p_l\ CaO} = 86.05 \tag{30}$$

With these partial molar heat capacity terms, $C_p^{conf}(T_g)$ can be calculated easily and used for the calculation of $S^{conf}(T_g)$, which are plotted in Figure 21. It is clearly visible that $S^{conf}(T_g)$ of silica glass is almost the same than that of $CaAl_2O_4$, CA glass. Assuming that $S^{conf}(T_g)$ can be decomposed into topological and excess terms, we can model the observed variations as

$$S^{topo} = X_{SiO_2} S^{conf}\left(T_g\right)_{SiO_2} + \left(1 - X_{SiO_2}\right) S^{conf}\left(T_g\right)_{CaAl_2O_4} \tag{31}$$

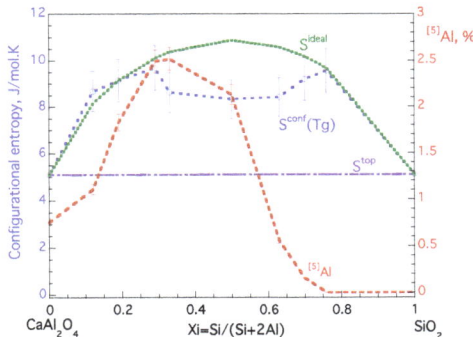

Figure 21. Configurational entropy at the glass transition temperature, $S^{conf}(T_g)$ (Neuville 1992), and real proportion of [5]Al in mole percent in glasses along the $SiO_2-CaAl_2O_4$ binary (Neuville et al. 2004a,b, 2006). Each $S^{conf}(T_g)$ value was calculated from the viscosity data obtained for each composition.

S^{topo} is thus assumed to vary linearly between the two end-members SiO_2 and $CaAl_2O_4$ glasses. This term S^{topo}, is independent of temperature and corresponds to a simple mechanical mixing between two entities. The term S^{ideal}, defined in Equation (10), can be taken as equal to the entropy obtained from an ideal random mixing of two species, in this case Si and Al. At high or low concentration of silica, $S^{conf}(T_g)$ variations seem to follow the ones predicted assuming such an ideal mixing between Si and Al. This suggests that $CaAl_2O_4$ randomly replace Si_2O_4 in the melt/glass structure. However, between 33 and 75 percent of silica, $S^{conf}(T_g)$ variations depart from the predictions of this ideal mixing model. In this silica concentration range, a decrease of ~25% in $S^{conf}(T_g)$ is observed, and there is a maximum difference of ~2.5 J mol^{-1} K^{-1} between $S^{conf}(T_g)$ values and the ideal model at around 50 mol% SiO_2 (Fig. 21).

The observed $S^{conf}(T_g)$ variations (Fig. 21) suggest that :

i. at low or high silica concentrations, Si and [4]Al mix almost randomly in the melt/glass structure,

ii. a deviation from ideal mixing exists at intermediate silica contents. The latter deviation can be rationalized upon considering that Al in five-fold coordination represents around 2.5% of total Al in the CA50.25 glass.

Excess [5]Al tends to decrease $S^{conf}(T_g)$ and in parallel increase the glass transition temperature, as shown by Le Losq et al. (2014) in the $Na_2O–Al_2O_3–SiO_2$ system. A similar effect is potentially observed here, with the presence of [5]Al possibly producing a decrease in $S^{conf}(T_g)$ and also an increase in T_g because [5]Al and [4]Si do not interact and do not mix randomly. One should be careful with extrapolating such behavior because it is only valid close to the glass transition. Indeed, at high temperatures (close or above the liquidus), Neuville et al (2008b) proposed that [5]Al plays a role similar to [5]Si, a transient species ensuring the exchange of O atoms during viscous flow according to Stebbins (1991). Similarly, in polymerized aluminosilicate networks, we can infer that five-fold coordinated network formers (Si or Al) ensure cationic mobility at high temperature. In depolymerized networks, Q^1, Q^2 and Q^3 units ensure network mobility (Farnan and Stebbins 1990, 1994; Stebbins 1991; Neuville et al. 2008b; Le Losq et al. 2014). Such assumptions agree well with the data near the C3A domain (see the section *Aluminate glasses and melts*). From Figure 21, it is further possible to make the assumption that [5]Al plays a different role in silica-rich or alumina-rich glasses and melts. Indeed, in silica-rich networks (>60 mol% SiO_2), [5]Al increases network connectivity and this results in a decrease in $S^{conf}(T_g)$. In an aluminate-rich network, a few percent of [5]Al plays a role similar to [4]Al, which exchanges randomly with Si.

Other systems show behaviors along the tectosilicate join similar to that previously presented for $CaAl_2O_4–Si_2O_4$. In particular, Raman spectra along all tectosilicate joints in the LAS, NAS, KAS, MAS, SAS, BAS systems show similar variations. They all present a linear decrease in the frequency of the bands assigned to T–O symmetric and asymmetric stretching of Q^4 units with decreasing SiO_2 concentration. This correlates very well with increasing ^{27}Al NMR chemical shifts with decreasing silica content in tectosilicate glasses (see Fig. 12, Drewitt et al. 2022, this volume). Those Raman and NMR variations arise from the variation of the T–O distance between Si–O and Al–O, as the Si–O distance is shorter than the Al–O distance.

In the Figure 22, the viscosities of tectosilicate melts with 75 and 50 mol% silica are plotted as a function of T_g/T. The strongest glass in the Angell plot corresponds to the SiO_2 composition, and the more fragile the composition, the greater the curvature. The fragility of melts decreases from Mg, Ca, Sr to Ba for the alkaline-earth elements. Alkali tectosilicate compositions show similar fragilities at 50 mol% silica (Fig. 22B). At 75 mol% (Fig. 22A), Li and K tectosilicate compositions also present a similar fragility while the Na-tectosilicate composition is the strongest one.

From the CAS system to other chemical systems

For tectosilicate glasses and melts, the structure seems to present a random distribution and random substitution between SiO_4^{4-} and AlO_4^{5-} units, in violation of the Loewenstein's aluminum avoidance principle (Loewenstein 1954). The Si/Al substitution is clearly confirmed by the linear variation of the chemical shift of the ^{27}Al NMR and of the frequency of the Raman

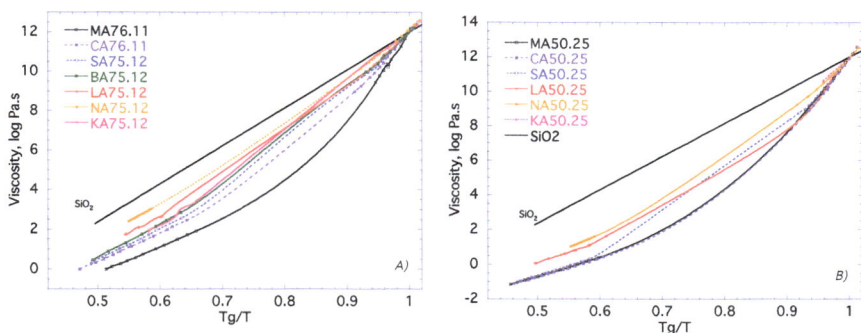

Figure 22. Viscosity of MAX.Y melts versus T_g/T, with M = Li$_2$, Na$_2$, K$_2$, Mg, Ca, Sr, Ba. X = SiO$_2$, Y = Al$_2$O$_3$, and MO = 100 − (SiO$_2$+Al$_2$O$_3$) in mole%, A) for 75 mole% of SiO$_2$, and B for 50 mole% of SiO$_2$. (data are from Le Losq 2012 for LAS, Le Losq et al. 2017, for NAS and KAS, Neuville 1991 for CAS and MAS, Novikov et al.2017, for SAS, and Novikov 2018 for BAS).

band in the 800–1200cm^{-1} domain. We can propose that alkali or alkaline-earth elements are in the free-volumes of the Al$_x$–Si$_{(1-x)}$O$_2$ network, ensuring the charge compensation of AlO$_4$$^{5-}$ units. The Figure 23 shows the proportion of [5]Al in tectosilicate glasses versus SiO$_2$ content. The proportion of [5]Al is a maximum in the middle of the ternary system, in the peraluminous domain for all ternary systems MAS (with S = SiO$_2$, A = Al$_2$O$_3$, M = Li$_2$O, Na$_2$O, K$_2$O, MgO, CaO, SrO, BaO), and increases with the decrease of z/r^2 (Stebbins and Farnan 1992; Lee and Stebbins 1999, 2000; Stebbins et al. 1999, 2008; Allward et al 2003, 2007; Neuville et al. 2004, 2006, 2008, 2010; Stebbins 2008, 2016; Hiet et al. 2009; Lee et al. 2009, 2016; Le Losq et al. 2017; Novikov et al. 2017; Alu et al. 2018). In the NAS and KAS system, [5]Al is not detected in tectosilicate compositions, and it is only present in peraluminous domain (Allwardt, et al. 2003; Le Losq et al. 2014).

In tectosilicate glasses, the NMR chemical shifts of [5]Al and [4]Al decrease linearly with the silica content (Fig. 23A), due to an increase in the mean T–O–T inter-tetrahedral angle as Al substitutes for Si in tetrahedral units (Stebbins and Farnan 1992; Neuville et al. 2006, 2008). At a given silica content, small variations in the ^{27}Al NMR chemical shift are observed, and correlate with the electronegativity and size of the first-neighbor metal cations in the glass structure. These variations arise from small changes in the Si/Al distribution and in the mean inter-tetrahedral T–O–T angle as the metal cation field strength and property changes.

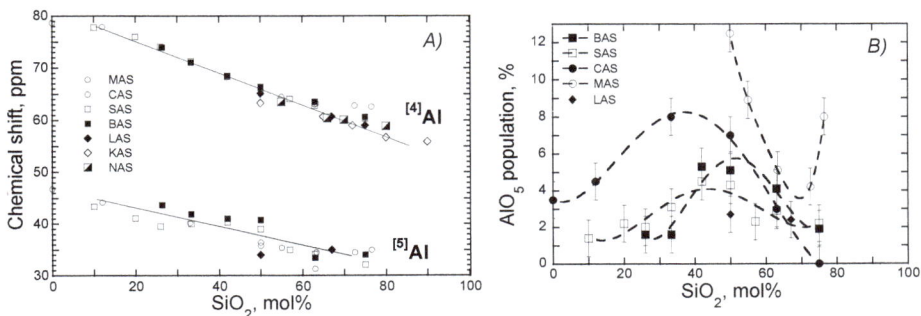

Figure 23. A) NMR chemical shift of [4]Al and [5]Al and **B)** proportion of [5]Al in LAS, MAS, CAS, SAS, BAS system along the join R = 1, the tectosilicate join (note that there is not [5]Al in Na and K tectosilicate glasses). Data from Le Losq (2012), Neuville et al. (2006, 2008), Novikov et al. (2017); Novikov, (2017). Error bars are smaller than symbol size.

In peralkaline glasses, the degree of polymerization plays a small role on the [4]Al environment. Mysen et al. (2003) showed that, in sodium aluminosiicates, Al remains mostly in Q^4 units. Using Raman and viscosity data, Le Losq et al. (2014) corroborated such a view for this system. Other studies suggest that this remains true probably in most aluminosilicate compositions at silica contents higher than 40 mol% (Neuville et al. 2004b, 2008, Novikov et al. 2017); a strong preference of [4]Al for the most polymerized structural units is expected according to experimental (Mysen et al. 1981; Cormier et al. 2000, 2005; Neuville et al. 2004b, 2008; Novikov et al. 2017) and numerical studies (Cormier et al. 2003, Cormier 2019).

Alkaline-earth mixing in aluminosilicate glasses and melts

The substitution of one alkaline-earth element by another has little Effect on glass Raman spectra (Merzbacher and White 1991; Neuville et al. 2008b). The Si/Al substitution actually has a greater influence on the aluminosilicate network than the alkaline-earth element substitution (see also Drewitt et al. 2022, this volume). In the case of Ca–Mg aluminosilicates, this agrees with an ideal mixing of Ca and Mg in aluminosilicates as inferred from macroscopic viscosity measurements (Neuville and Richet 1991). This random distribution was confirmed by ^{17}O NMR data, which show that NBOs are equally associated with Mg or Ca cations (Farnan and Stebbins 1990, 1994; Allwardt and Stebbins 2004). However, the network is more perturbed by the presence of Mg than by Ca because of the higher ionic field strength of Mg (Navrotsky et al. 1982; Roy and Navrotsky 1985) and the Ca/Mg coordination number should play an important role. The coordination number of Mg is well characterized (Fiske and Stebbins 1994; Kroeker and Stebbins 2000; Trcera et al. 2009): it is 4 in silicate glasses, and increases up to 5–6 in aluminosilicate glasses, in agreement with glass molar volumes variations (Fig. 15). Turning to Ca, in silicate glasses it is mainly located in distorted sites with 6–7 oxygen neighbors (Cormier et al. 2003; Neuville et al. 2004b, 2008; Cicconi et al. 2016). Its CN increases with substitution by Al as shown by the variation of the molar volume between $CaSiO_3$ and Al_2SiO_5 glasses (Fig. 17). At high Al_2O_3 content ($R \geq 1$), in Ca/Mg aluminosilicate glasses, Ca is in 7–8 fold coordination (Shimoda et al. 2007a) whereas Mg is in 6-fold coordination (Shimoda et al. 2007b; Trcera et al. 2009), as corroborated by data like molar volume, X-ray diffraction spectra, XANES spectra at the Ca and Mg K-edge, and ^{43}Ca and ^{25}Mg NMR experiments (Fiske and Stebbins 1994; Kroeker and Stebbins 2000; Trcera et al. 2009).

The above discussion raises a paradox: Ca and Mg always have different coordination numbers but seem to mix randomly in silicate and aluminosilicate melts and glasses. First, this may highlight the difficulty in determining alkali or alkaline-earth environments that are loosely held in large sites with fairly irregular coordination (Cormier and Neuville 2004; Neuville et al. 2008). Nevertheless, there is a general consensus for lower coordination for Mg than Ca. The low Mg coordination number and the possibility to have MgO_4 tetrahedra in Mg-silicate glass suggest that Mg will not be available for charge compensation of Al in tetrahedral position. Alternatively, its small size favors its localization in network tetrahedral cavities, yielding important distortions of the aluminosilicate network to accommodate such Mg coordination, which could result in the formation of highly-coordinated Al. This interpretation might justify the modifier role played by Mg despite their low coordination (Shimoda et al. 2007b; Guignard and Cormier 2008; Trcera et al. 2009), in agreement with recent NMR data suggesting that substitution of Na by Mg in aluminosilicates in glass promotes melt depolymerisation (Sreenivasan et al. 2020). Similar observations can be made regarding the role of Zn in silicate or aluminosilicate glasses (Novikov 2017). The fact that the viscosities of the Mg or Zn glass compositions are lower than that of Ca glass compositions for the same amount of SiO_2, confirms the role of network modifier of Mg and Zn while they are preferably in 4-fold coordination (Neuville and Richet 1991; Neuville 1992; Toplis and Dingwell 2004; Neuville et al. 2008).

In mixed CMAS glasses, Raman, NMR and neutron diffraction data (Allwardt et al. 2003, Lee et al. 2003; Lee and Stebbins 2003; Allwardt and Stebbins 2004; Neuville et al.

2006, 2008; Guignard and Cormier 2008; Cormier and Coello 2013; Cormier 2019) are in good agreement with random mixing between Ca and Mg in the melt structure, as initially proposed by Neuville and Richet (1991). This implies that intermediate CMAS melts can be considered as being derived from the random mixing of their CAS and MAS end-members, e.g., $Ca_3Al_2Si_3O_{12}$–$Mg_3Al_2Si_3O_{12}$ for garnet-like glasses and $CaAlSi_2O_8$–$MgAlSi_2O_8$ for anorthite-like compositions (Fig. 24). An interesting observation is that Ca/Mg mixing shows a random distribution and the network polymerization is not really affected by the Ca/Mg substitution, but the proportion of [5]Al increases with the Ca/Mg substitution (Neuville et al. 2008b).

Unfortunately, knowledge regarding the mixing between Sr/Ca, Ba/Mg or more generally between an alkali with an alkaline-earth element in aluminosilicate compositions remain scarce, and it is not possible to propose a general conclusion. It is possible to say that Si and Al are almost randomly distributed in tectosilicate glass compositions, and metal cations are located in cavities of the 3D polyhedral network and ensure the charge compensation of the AlO_4^{5-} tetrahedra. This assumption is in good agreement with the linear variation of the T–O distances from 1.66 up to 1.76 Å between SiO_2 and $CaAl_2O_4$ glasses (Fig. 13b of Drewitt et al. 2022, this volume). The T–O distances are also well correlated with the linear variation of the NMR chemical shift of ^{27}Al and the linear variation of the frequency of the Q^4 species along the tectosilicate joint (compiled in Figs. 13a,b of Drewitt et al. 2022, this volume). They may mix randomly like Ca and Mg, but a non-random distribution with increasing network segregation has also been reported in alkali tectosilicate melts (Le Losq et al. 2017). This highlights the complexity of the structure of aluminosilicate melts and glasses, a problem when considering the modelling of their properties. Other problems are the presence of excess NBOs, [5]Al and tri-coordinated oxygens (Stebbins and Xu 1997; Le Losq et al. 2014). For instance, the presence of [5]Al can be important, reaching concentrations up to 12% of the Al_2O_3 in the middle of the MAS ternary system, and the presence of [5]Al results in increasing the glass transition temperature and decreasing configurational entropy. Metal cations could be located in cavities of the 3D polyhedral network, in a manner consistent with the Compensated Continuous Random Network model (Greaves and Ngai 1995; Greaves and Sen 2007; Le Losq et al. 2017); they ensure the charge compensation of the AlO_4^{5-} tetrahedra. In the potassium aluminosilicate ternary, a non-random distribution of Si and Al between framework units may be accompanied by an increasing segregation of metal cations in clusters or channels. This has been proposed to explain the variations in the structure and properties of potassium tectosilicate melts, for instance (Le Losq and Neuville 2013; Le Losq et al. 2017). This is a problem when considering the modelling of their properties, and a large number of approximation may be necessary as we shall see in the later section.

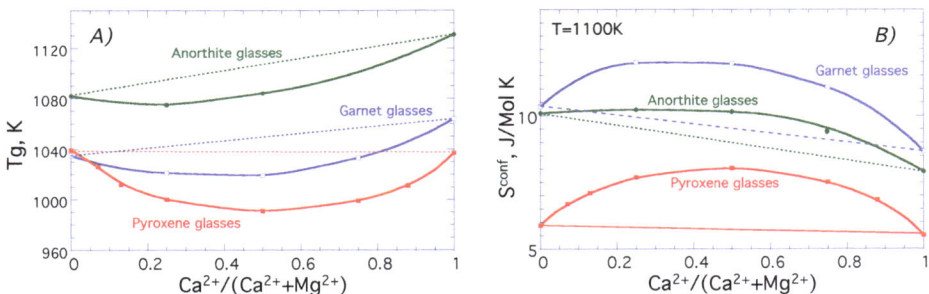

Figure 24. A) glass transition temperature and **B)** configurational entropy at 1100 K upon Ca/Mg mixing in pyroxene ($CaSiO_3$–$MgSiO_3$), garnet ($Ca_3Al_2Si_3O_{12}$–$Mg_3Al_2Si_3O_{12}$) and anorthite ($CaAl_2Si_2O_8$–$MgAl_2Si_2O_8$) glass compositions, with respectively 0, 12.3 and 25.0 mol% Al_2O_3. Data for pyroxene and garnet glasses are from Neuville and Richet (1991) and original data for anorthite glasses.

Models for alkali aluminosilicate melts

The above sections allow us to have a glimpse into the complexity of aluminosilicate melts. Even leaving aside the potential effects of parameters like [5]Al, there is no data for the distribution of Q^n species in aluminosilicate melts because ^{29}Si NMR spectroscopy loses resolution in aluminosilicate compositions due to Si–Al interactions resulting in significant signal broadening and increased complexity of the peaks. Mysen et al. (2003) quantified the fractions of the different Q^n species in Na aluminosilicate glasses by modelling the ^{29}Si NMR data. Their findings allow us to constrain the K_{eq} value for the dissociation $2\,Q^3 = Q^2 + Q^4$ in sodium aluminosilicate glasses (Fig. 25). We observe that Al introduction into the glass network promotes the dissociation of Q^3 units.

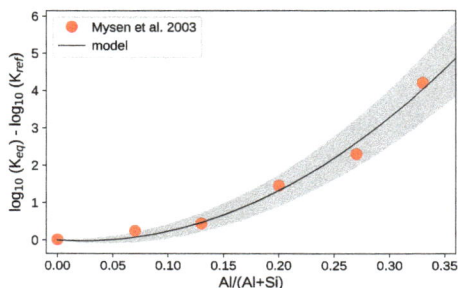

Figure 25. Relative value of the equilibrium constant K_{eq} of the reaction $2\,Q^3 = Q^2 + Q^4$ obtained from Q^n data in Na aluminosilicate melts calculated by Mysen et al. (2003), reported as a function of the Al/(Al+Si) ratio of glasses. K_{ref} is the value of K_{eq} in the glass without Al. The **black line** represents a polynomial modelling of the data, and the **grey areas** are the two-sigma confidence interval of the model predictions. Using this model, it is possible to estimate the Q^n distribution in alkali aluminosilicate glasses, but one must note that such relation remains unknown for other elements like K, Ca or Mg.

This result is the only experimental one to the knowledge of authors that could be used to model the Q^n distribution in aluminosilicates based on the principle of using partial K_{eq} values as highlighted in the section *Silicate melts: how can we use existing structural knowledge to model melt properties*. Other models (e.g., Doweidar 1999), like the associate solution one, are purely theoretical and remain unconstrained by experimental data, as they are not available. This does not prevent one from building models of glass and melt properties. For instance, Doweidar (1999) simply used a binary model for calculating the density of Na_2O–Al_2O_3–SiO_2 glasses from partial molar values of different Na–Q^n units. The structural model employed there is clearly not appropriate, but interestingly, this does not prevent a good prediction of glass densities. A better attempt to link melt composition, structure and viscosity is that of Starodub et al. (2019), who calculated the Q^n distribution in Na_2O–K_2O–Al_2O_3–SiO_2 melts from an associate solution model, and then linked the fractions of Q^n units and the concentrations of cations to the parameters of the Avramov–Milchev equation for calculation of melt viscosity. This is an interesting model, but it suffers from two problems. First, they did not select the viscosity data used for fitting their model, as they used the large SciGlass database, and the results of their model will thus suffer from the propagation of large errors affecting some experimental data. Secondly, the Avramov–Milchev (Avramov and Milchev 1988) equation works well but it is less attractive than the Adam-Gibbs theory because it does not allow one to link the chemical, structural, thermodynamic and dynamic dimension of the problem.

An extension of the model presented in the section *Silicate melts: how can we use existing structural knowledge to model melt properties* could allow one to circumvent such caveats. We will explore if such a model is possible to implement below for the Na_2O–K_2O–Al_2O_3–SiO_2 system. A first step would be to combine the silicate structural model to the relationship presented in Figure 25. This will allow one to approximate the Q^n speciation in aluminosilicate glasses. Then, we need to know the effects of Na–K mixing in aluminosilicate melts and of Al concentration on the glass configurational entropy. Combining viscosity and Raman data with thermodynamic calculations, Le Losq et al. (2014) have shown that Al introduction in NaSi$_2$O$_7$ yields a rapid network polymerisation (Fig. 26A) that is accompanied by a jump in glass configurational entropy and an increase in glass T_g (Fig. 26B). At Al/(Al+Na) higher than 0.5, we see a continuous decrease in $S^{conf}(T_g)$ that is accompanied by increasing T_g.

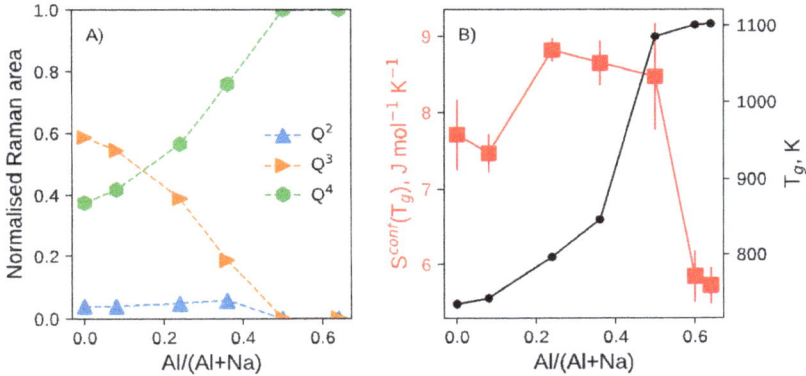

Figure 26. A) Fractions of the Q^n units Raman signal, and **B)** entropy and glass transition temperature of Na aluminosilicate glasses with 75 mol% SiO$_2$ but varying Al/(Al+Na) ratios. Data from Le Losq et al. (2014).

Variations in $S^{conf}(T_g)$ are thus complex functions of the melt Al/(Al+M) ratio. As an approximation, one could assume that Al and Si mix randomly in Q^4 units in the silicate network, as proposed for Na tectosilicate melts by Neuville and Mysen (1996). Note that it is important to remember that unlike the Ca or Mg aluminosilicate system, the ternary system Na$_2$O–Al$_2$O$_3$–SiO$_2$ does not contain Al in 5-fold coordination except in the peraluminous domain (Lee et al. 2003; Le Losq et al. 2014). This hypothesis implies that the Lowenstein Al–O–Al avoidance rule will not be respected, and is thus in contradiction with the interpretation of NMR data of Na aluminosilicate glasses (Lee and Stebbins 1999; Lee et al. 2003). However, this hypothesis allows reproduction of the variations in configurational entropy observed when Al substitutes Si in Na aluminosilicate melts (Neuville and Mysen 1996; Le Losq 2012). One may thus consider it as adequate for the aim of building a model of configurational entropy variations. An additional point is to consider that Q^4_{Al} charged-balanced by Na ($Q^4_{Al,Na_{env}}$) or K ($Q^4_{Al,K_{env}}$) probably present different partial configurational entropies (respectively $S^{conf}_{Q^4_{Al,Na_{env}}}$ and $S^{conf}_{Q^4_{Al,K_{env}}}$). S^{topo} (Eqn. 16) could then be written as:

$$S^{topo} = \sum_{n=2}^{3} x_{Q^n_{Si,Na_{env}}} S^{conf}_{Q^n_{Si,Na_{env}}} + \sum_{n=2}^{3} x_{Q^n_{Si,K_{env}}} S^{conf}_{Q^n_{Si,K_{env}}} + x_{Q^4_{Si}} S^{conf}_{Q^4_{Si}} + x_{Q^4_{Al,Na_{env}}} S^{conf}_{Q^4_{Al,Na_{env}}} + x_{Q^4_{Al,K_{env}}} S^{conf}_{Q^4_{Al,K_{env}}} \quad (32)$$

From this equation, we need to discriminate the fraction of charge-compensator and network modifiers (Na, K), in order to calculate the variation in the fractions of the Q^n units. This could be done easily assuming that the fractions of charge-compensator and network modifiers Na and K are just equal to the weighted average fractions of Na and K. This assumes no preference of Na and K to specific environments, an hypothesis that surely could be challenged.

We see that many hypotheses have been already proposed in order to start building a model to express $S^{conf}(T_g)$. This is without even considering the calculation of any entropy resulting from cationic mixing, S^{mix}. The picture becomes even more complex at this point.

Indeed, while variations of $S^{\text{conf}}(T_g)$ upon mixing Na and K in silicate melts can be explained by a random mixing of those cations (Richet 1984; see the section *Silicate glasses and others network formers*, Eqn. 18): mixing Na and K in the presence of Al results in a different behaviour. Entropy variations upon mixing Na and K in tectosilicate melts ($R = 1$) are far from ideal (Le Losq and Neuville 2013). Viscosity, Raman, NMR and molecular dynamic simulation data all indicate that Na and K actually are present in different environments in tectosilicate melts (Le Losq et al. 2017), with K tending to be present in large percolation channels in the polyhedral Al-Si network. This was recently challenged by a new analysis of viscosity data along the SiO_2–$(Na,K)AlO_2$ binary by Robert et al. (2019). Those authors fitted viscosity data of mixed Na–K tectosilicate melts, leaving the A_e parameter (Eqn. 6) free. This yields results where it is possible to fit viscosity upon Na–K mixing in tectosilicate melts assuming a random Na–K distribution, like in silicate melts. In such cases, most of the variations in viscosity seemed to be explained by variations in A_e. While this is an interesting take on this problem, this is probably not a good representation of the reality. Indeed, many authors demonstrated that the pre-exponential term in viscosity equations tend to converge towards a common value (Persikov 1991; Giordano and Dingwell 2003; Giordano et al. 2008, Giordano and Russell 2018). The A_e parameter is unlikely to vary largely in aluminosilicate melts (e.g., Russell and Giordano 2005, 2017 and references therein). It thus seems improbable that A_e varies largely upon substitution of Na by K at fixed SiO_2 and Al_2O_3 contents. Raman spectroscopy data and molecular dynamic simulations (Le Losq and Neuville 2013; Le Losq et al. 2017, 2019) further indicate that Na tectosilicate and K tectosilicate melts do not have the same structure, such that Na and K do not mix randomly in such systems. The mixing of Na- and K- tectosilicates thus appears to be of a mechanical mixing, and should accordingly yield a pseudo-linear variation of $S^{\text{conf}}(T_g)$. This agrees with conclusions from Le Losq and Neuville (2013) as well as Le Losq et al. (2017), which report pseudo-linear variations in $S^{\text{conf}}(T_g)$ upon Na–K mixing.

Considering that mixing Na and K compensators does not affect entropy, we need to consider three sources for the calculation of S^{mix} : (i) the ideal mixing of Si between different Q^n units, (ii) the ideal mixing of Si and Al in Q^4 units and (iii) the ideal mixing of Na and K network modifiers. We thus have:

$$S_{\text{Si-Al}}^{\text{mix}} = -2\frac{x_{\text{Si}\in Q^4} + x_{\text{Al}}}{x_o}R\left[\frac{\text{Al}}{\text{Al+Si}}\log\frac{\text{Al}}{\text{Al+Si}} + \left(1 - \frac{\text{Al}}{\text{Al+Si}}\right)\log\left(1 - \frac{\text{Al}}{\text{Al+Si}}\right)\right] \quad (33)$$

with $x_{\text{Si}\in Q^4}$ the mole fraction of Si in Q^4 units,

$$S_{\text{Si}}^{\text{mix}} = \frac{-x_{\text{Si}}}{x_O} \times 2R\sum_{n=2}^{4} x_{Q_{\text{Si}}^n} \log\left(x_{Q_{\text{Si}}^n}\right) \quad (34)$$

and assuming a distribution of Na and K as charge-compensator and network modifier that varies linearly with Al/(Al+M), we finally have

$$S_{\text{Na-K modifiers}}^{\text{mix}} = MODS \times \left[-2R\frac{x_{\text{Na}} + x_{\text{K}}}{x_O}\left(x_{\text{K}}\log\left(x_{\text{K}}\right) + \left(1 - x_{\text{K}}\right)\log\left(1 - x_{\text{K}}\right)\right)\right] \quad (35)$$

with MODS the fraction of modifiers contributing to S^{mix}; MODS is equal to $(1 - Al/(Na+K))$ for Al < (Na+K), and is equal to 0 for Al ≥ (Na+K).

The sum of Equations (32–35) allows the calculation of $S^{\text{conf}}(T_g)$. B_e can be calculated following a similar logic to that used for silicate melts (section *Silicate melts: how can we use existing structural knowledge to model melt properties*):

$$B_e = \sum_{n=2}^{3} x_{(Q^n_{(Si,Na_{env})})} B_{e(Q^n_{(Si,Na_{env})})} +$$

$$\sum_{n=2}^{3} x_{(Q^n_{(Si,K_{env})})} B_{e(Q^n_{(Si,K_{env})})} + x_{(Q^4_{Si})} B_{e(Q^4_{Si})} + x_{(Q^4_{Al,Na_{env}})} B_{e(Q^4_{Al,Na_{env}})} + \tag{36}$$

$$x_{(Q^4_{Al,K_{env}})} B_{e(Q^4_{Al,K_{env}})} + K_1 S^{mix}_{Si} + K_2 S^{mix}_{(Na-K \text{ modifiers})} + K_3 S^{mix}_{Si-Al}$$

with K_1, K_2 and K_3 being proportionality constants accounting for the influence of the intrinsic entropy in B_e (section *Silicate glasses and others network formers*; Le Losq and Neuville 2017). A_e is left as a common parameter.

Using the equations described above, and selecting viscosity data from the literature for parameter tuning via least-square (see Fig. 27), we are able to build a model for the viscosity of melts in the $Na_2O–K_2O–SiO_2–Al_2O_3$ system, with compositions from 50 mol% (in aluminosilicates; 60 in silicates) to 100 mol% SiO_2, and NBO/T ranging from 0 to 1.33. Root-mean-square-errors between viscosity predictions and measurements are equal or lower than 0.4 log Pa·s. This demonstrates that building a structural model of the properties of aluminosilicate melts is possible. However, the above model is incomplete because it relies on many hypotheses, and does not take into account (yet) many specifics like the role of $^{[5]}Al$ on $S^{conf}(T_g)$. Despite this, this is an encouraging step toward the construction of structural models that usually achieve a high degree of precision in the predictions they make. New developments in modelling the melt structure and in linking it with melt properties will surely allow improvements of such structural models in the future.

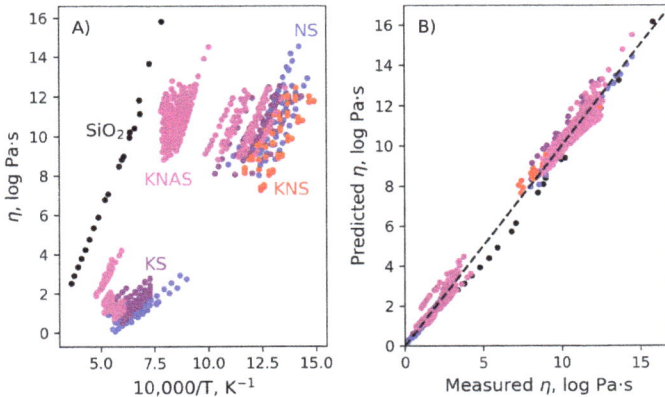

Figure 27. A) Viscosity data used for the model calibration. Data ($n=817$) are those compiled by Le Losq and Neuville (2017) for silicates (NS, KS, KNS and SiO_2), and for aluminosilicates (KNAS) they were compiled from Riebling (1966), Taylor and Rindome (1970), Urbain et al. (1982), Le Losq (2012), Le Losq and Neuville (2013), Le Losq et al. (2014, 2017). The model parameters were optimised by No-U-Turn Markov Chain Monte Carlo sampling via the PyMC3 Python library. The root-mean-squared error between predictions and measurements is of 0.4 log unit.

ALUMINATE GLASSES AND MELTS

Aluminate glasses and melts are of interest from a fundamental structural point of view: they contain only Al as a network former and, hence, allow better understanding of the role of Al in glasses and melts. The study of aluminate materials is also very important to understand the formation and evolution of parental bodies in the universe (Leger et al. 2009; Batalha et al. 2011). Aluminate materials further represent a technological material of interest because of their good IR transmission and ultralow optical losses (Higby et al. 1990) that make them attractive candidates for low loss optical fibers such as infrared waveguides (Lines et al. 1989; King and Shelby 1996). In addition, calcium aluminate glasses have been found to be photosensitive to ultraviolet radiations, leading to potential applications in photometric devices for information storage purposes (Hosono et al. 1985). They also have excellent mechanical properties, such that calcium aluminate glass fibres have been proposed for the reinforcement of cement composites (Wallenberger et al. 2004; El Hayek 2017).

Looking at aluminate glasses allows one to better understand the fundamental role of aluminum in glasses and liquids, a task somehow difficult in aluminosilicate compositions because of the superposition of many complexities (see section *Aluminosilicate glasses and melts*). Numerous studies have probed the structure of calcium aluminate crystals, glasses and melts (e.g., McMillan and Piriou 1983; Hannon and Parker 2000; Benmore et al. 2003; Neuville et al. 2008, 2010; Licheron et al. 2011; Drewitt et al. 2012, 2017). From those, it appears that four compositional "domains" can be distinguished: Al_2O_3, Al_2O_3–CA, CA–C3A and C3A–CaO with CA = $CaAl_2O_4$, C3A = $Ca_3Al_2O_6$.

Al_2O_3

Crystalline Al_2O_3 has corundum structure with Al in 6-fold coordination (Rankin 1915). Several studies, using ^{27}Al NMR at high temperature, show that Al_2O_3 liquid is a mixture of 60% $^{[6]}Al$–40% $^{[4]}Al$. This was confirmed by XANES at the Al K-edge and FDMNES simulation (Neuville et al. 2009). Recently, Shi et al. (2019) proposed, from new X-ray diffraction at high temperature measurements, that Al is essentially in five-fold coordination in the Al_2O_3 liquid; this interpretation is based on the Al–O distance. The ^{27}Al NMR chemical shift is 53 ppm (Couture et al. 1990; Florian et al. 1995; Ansell et al. 1997), a value that can be interpreted as $^{[5]}Al$, or as a mixture of $^{[4]}Al$ and $^{[6]}Al$. When assuming a mixture of $^{[4]}$ Al and $^{[6]}Al$, the high temperature ^{27}Al NMR data allow determining an Al NMR relaxation time in excellent agreement with the relaxation time obtained from viscosity measurements (Urbain et al. 1982). Besides, when comparing XANES data at the Al K-edge measured at high temperature in the liquid state to spectra simulated using the FDMNES code (Joly 2001), the assumption of a mixture of $^{[4]}Al$ and $^{[6]}Al$ provides the best reproduction of the experimental data (Neuville et al. 2009). From all the high temperature NMR and XANES spectroscopies and/or neutron diffraction measurements, it is at the moment preferable to consider that, in liquid alumina, Al is present as a mixture of $^{[4]}Al$ and $^{[6]}Al$.

We have essentially mentioned Al_2O_3 in its crystalline or liquid forms, however, some studies have shown that it is possible to obtain Al_2O_3 in its amorphous form, it is usually obtained by fast quenching and in thin film form (Lee et al. 2010; Kim et al. 2014). In that case, Al is essentially a mix of 4- and 5-fold coordination with a small amount of 6-fold coordination (Lee et al. 2010; Kim et al. 2014). The difference between the Al coordination between glass and liquid state can be easily explained by the high cooling rate.

Al_2O_3–$CaAl_2O_4$ compositions

With the introduction of calcium (or others alkaline-earth elements) in alumina liquid, a first crystalline phase can be found: grossite, $CaAl_4O_7$, called CA2, with AlO_4 tetrahedra only in Q^4 units although one of the O atoms forms an "oxygen tricluster" which links three AlO_4 tetrahedra (Stebbins et al 2001; Iuga et al. 2005; Neuville et al. 2010). With increasing Ca content up to the $CaAl_2O_4$ crystal composition, the amount of tricluster oxygen decreases and Al stays in four fold coordination. The $CaAl_4O_7$ composition always crystallizes while $CaAl_2O_4$ (CA) can be quenched to a glass by fast quench methods. In the CA glass, $^{[4]}Al$ is dominant and $^{[5]}Al$ representing 3.5% of total Al (Neuville et al. 2006). In the CA liquid, $^{[4]}Al$ remains dominant but the proportion of $^{[5]}Al$ increases up to around 40 mol% near 2000 K, as observed from XANES at the Al K-edge and neutron diffraction (Neuville et al. 2008a; Drewitt et al. 2012). Oxygen triclusters also probably are present but remain unquantified during these changes. The coordination of calcium, which is 7-fold in the glasses, increases in the melts as indicated by the evolution of the pre-edge of Ca K-edge XANES spectra between room temperature and superliquidus temperatures (Neuville et al. 2008a).

CA–C3A compositions

For compositions between $50 > Al_2O_3 > 25$ mol%, Al is only in tetrahedral coordination in both the glasses and the crystals, with a number of bridging oxygens varying between 4 to 2 respectively for CA and C3A (McMillan and Piriou 1983; Neuville et al. 2004b, 2006, 2010; Licheron et al. 2011; Drewitt et al. 2017). In C3A crystal or glass ($Ca_3Al_2O_6$), Al can be found in Q^2 units, and, with increasing temperature, Al K-edge XANES show very small changes (Neuville et al. 2008a). The later observation implies that Al stays in Q^2 species with increasing temperature without creation of $^{[5]}Al$. The cationic mobility is only insured by Q^2 species, and not by highly coordinated units as proposed by Stebbins and Farnan (1992) for polymerized melts. Calcium cations in crystalline C3A and glass is in 6-fold coordination, and, near 2000 K, a small pre-edge in Ca K-edge XANES spectra is observed; this implies that the Ca coordination increases with increasing temperature (Neuville et al. 2008a).

C3A–CaO compositions

For compositions with $Al_2O_3 < 25$ mol%, no glass or stable crystalline phases are observed. The knowledge of Al and Ca coordination in such conditions remains very scarce, if not non-existent.

Link with observations in other binary systems

In $SrO–Al_2O_3$ and $BaO–Al_2O_3$ systems, the liquidus temperatures are higher than in the $CaO–Al_2O_3$ system. This observation seems logical since there is agreement with an increase of liquidus temperature with increasing the size of the alkaline-earth elements in alkaline-earth aluminosilicate systems (Méducin et al. 2004; Licheron et al. 2011; Shan et al. 2018). Turning to $MgO–Al_2O_3$, Mg^{2+} forms a very stable spinel with alumina; it is not possible to quench the $MgAl_2O_4$ composition or other magnesium aluminate composition even with using fast-quench methods. In $MgAl_2O_4$ crystals, Mg and Al are in 4 and 6-fold coordinations, respectively, at room temperature and that can change with temperature (Andreozzi et al. 2000; Neuville et al. 2009).

Figure 28 summarizes the evolution of the Al coordination in aluminate glasses and also shows how the molar volume of glass changes with Al_2O_3 content. The glasses CxA or SxA, with x varying between 3 and 1, are made by levitation and fast quench, and all these glasses show Al in 4-fold coordination (Neuville et al. 2006, 2008b, 2010, Licheron et al. 2010). For Ca or Sr aluminates, a linear variation of the molar volume with the alumina content is observed between 25 and 50 mole percent of alumina, and Ca is in 6-fold coordination and Al in 4-fold coordination (Neuville et al. 2008b). If we consider linear extrapolations of these trends, it is possible to propose, considering only Al in four-fold coordination and Ca and Sr in 6-fold coordinations, that the respective partial molar volumes of Al_2O_3, CaO and SrO are respectively of 39.1, 17.0 and 21.0 cm^3 mol^{-1}. Such estimations are in very good agreement with the values proposed by Robie (1978) for the molar volume of CaO and SrO.

To complete the observations made on the molar volume, the glass transition temperatures for CxA and SxA glasses were measured and plotted in Figure 28. The glass transition temperature of the different CxA and SxA glasses were obtained by DSC measurements with a heating rate of 10 °/min. Two linear trends are observed along the binaries $CaO–Al_2O_3$ and $SrO–Al_2O_3$. They converge to similar values when extrapolated to the alumina end-member. From this, we can propose an "apparent" partial glass transition temperature for Al_2O_3 at 1420 ± 10 K (Fig. 29). Following this logic, we also propose partial T_g's for CaO and SrO in 6-fold coordination at 932 ± 10 and 857 ± 10 K. The estimated T_{gi} for Al_2O_3 with Al in four-fold coordination is in very good agreement with calculation of phase diagrams using CALPAHD program (Pitsch, et al. 2021 submitted).

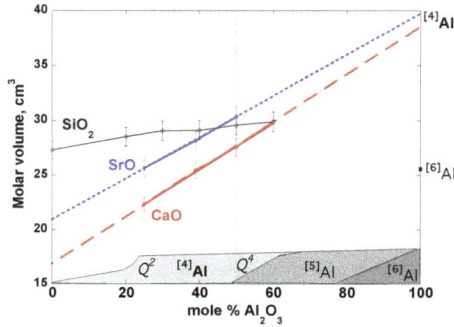

Figure 28. Coordination of Al between MO–Al$_2$O$_3$ and variation of molar volume of aluminate glasses. Molar volumes are calculated from the density measurements for Al$_2$O$_3$–SiO$_2$ (Wang et al. 2020), Al$_2$O$_3$–CaO (Neuville et al. 2006, 2010; Licheron et al. 2011), and Al$_2$O$_3$–SrO (Licheron et al. 2011). **Solid lines** correspond to fits of the data with a line, and **dashed lines** show the extrapolation of those fits to end members.

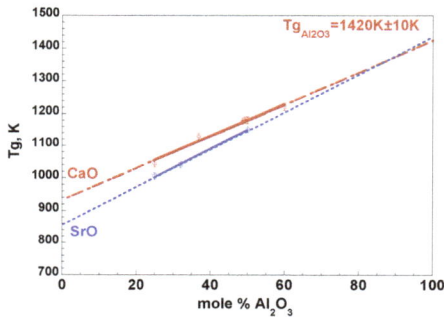

Figure 29. Glass transition temperature obtained from DSC measurements with and heating rate of 10°/min for aluminate glasses. **Full line** corresponds to linear variation fitting datas and **dashed line** corresponds to extrapolation up to end members.

Recently Shan et al (2018) made some calcium-strontium peraluminous glasses at 58% mole of Al$_2$O$_3$ using levitation device and fast cooling rate. They investigated macroscopic properties of these glasses and they showed that an ideal mixing term between Ca and Sr clearly exist to explain Vickers micro-hardness and glass transition temperatures variation between Ca/Sr aluminate glasses.

CONCLUSION AND PERSPECTIVES

Since the 1980s, it has been possible to study the structure of a liquid at high temperature by several spectroscopic methods: Raman, NMR, X-ray absorption and X-ray and neutron diffraction techniques (Neuville et al. 2014b). This allows us to understand the local order and the medium range order directly in the liquid state and thus be able to link the high temperature structure to the measurements of properties directly accessible at high temperature, like heat capacity, viscosity, conductivity, and diffusivity. We have seen that such an approach is very promising because it allows us to have a global view of a glass and a liquid and, thus, to better understand the structural changes that take place in a liquid. This improved understanding of silicate glasses and liquids has great applications in both volcanology and geochemistry where we are better able to understand changes in eruptive dynamism, mass transfers and also chemical balances during crystallization processes for example. But this work allows also a better understanding of industrial processes during melting, glass-forming or even during the processes of glass-ceramic. To summarize the changes that can occur in a cooling liquid, Figure 30, obtained for a composition of $Ca_3Al_2Si_3O_{12}$ liquid, shows both the changes that occur in the liquid state and during its cooling. We can see that at high temperature, above the liquidus, T_m, 1580K, the relaxation times obtained from the ^{27}Al NMR, t_{NMR}, from measurements of conductivity, t_σ, or from viscosities, t_η converge and are similar. It simply means that in the liquid state all cations, whatever their roles in the melts, move in relation to each other. From a structural point of view, it is a liquid state, characterized by the particularly intense pre-edge of the XANES spectra at the Ca K-edge at high temperature. On the other hand, when temperature decreases, there is a decoupling between the various relaxations, this means that the different atoms begin to move differently. The network modifiers, like Ca, continue to move as shown by the relaxation t_σ, while the relaxation calculated from the viscosity measurements, t_η, shows a strong increase, meaning that the tetrahedral network composed of tetrahedral SiO_4 and AlO_4 moves more slowly, which translates into an increase in the viscosity curve. This decoupling between the various families of atoms appears at about 1100 K for this chemical composition, several orders of magnitude separate the relaxation times obtained from the conductivity measurements, t_σ, from those obtained from viscosity measurements, t_η. These differences show, on the one hand, that the aluminosilicate network freezes with a significant increase of t_η and viscosity, and that calcium atoms put themselves in a position close to that they will have in the crystalline phase, as shown by the changes observed in the pre-edge of the XANES spectra at the Ca K-edge. In the end, we find that close to T_g, Ca is probably in a crystallographic site almost identical to the one it will have in a crystal.

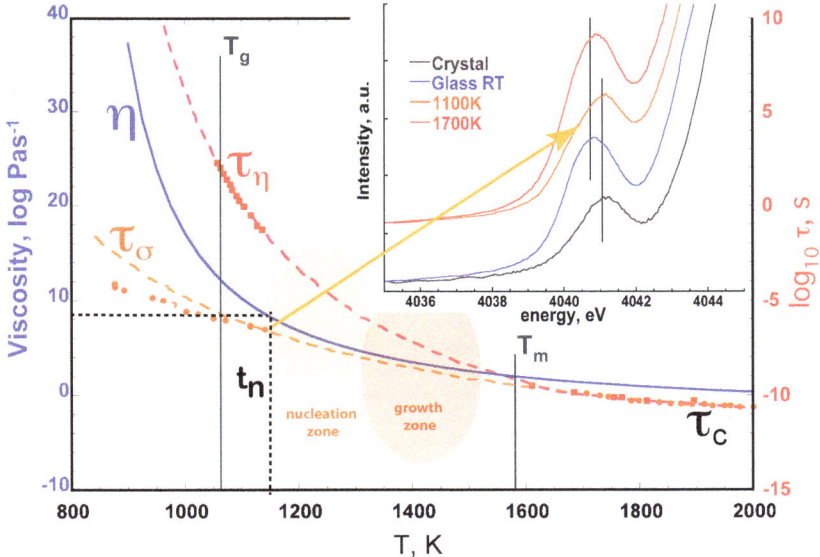

Figure 30. viscosity of $Ca_3Al_2Si_3O_{12}$ liquid from Neuville and Richet (1991), relaxation t_η is calculated from viscosity measurements, t_σ and t_{NMR}, are from Gruener et al. (2001) and pre-edge XANES spectra at the Ca K-edge are from Neuville et al. (2008b)

Figure 30 completes Figure 4, which helped to understand the links between potential movements in a liquid, various entropic states and viscosity, but this latter figure indicates that at high temperature all atoms can move quickly relative to each other and that at lower temperatures, atoms are in different potentials energy states and that changes will take more time.

In 2021, it is therefore possible and necessary to understand the structure and properties of a glass or/and a liquid throughout its history during a natural or industrial process. The current computational techniques discussed in the following chapters can show and predict these different states, and it is easy to imagine that the new techniques based on Machine Learning will allow modeling of the structure and physical properties and be able to predict them globally (Le Losq et al. 2021), this will have great consequences for both earth sciences and materials sciences.

REFERENCES

Adam G, Gibbs JH (1965) On the temperature dependence of cooperative relaxation properties in glass-forming liquids. J Chem Phys 43:139–146

Ali FC, Chadwick AV, Greaves GN, Jermy MC, Ngai KL, Smith ME (1995) Examination of the mixed-alkali effect in (Li,Na) disilicate glasses by nuclear magnetic resonance, conductivity measurements. Solid State Nucl Magn Reson 5:133–143

Allu AR, Gaddam A, Ganisetti S, Balaji S, Siegel R, Mather GC, Fabian M, Pascual MJ, Ditaranto N, Milius W, Senker J (2018) Structure and crystallization of alkaline-earth aluminosilicate glasses: prevention of the alumina-avoidance principle. J Phys Chem B 122:4737–4747

Allwardt JR, Stebbins JF (2004) Ca-Mg and K-Mg mixing around non-bridging O atoms in silicate glasses: An investigation using ^{17}O MAS and 3QMAS NMR. Am Mineral 89:777–784

Allwardt JR, Lee SK, Stebbins JF (2003) Bonding preferences of non-bridging O atoms: evidence from ^{17}O MAS and 3QMAS NMR on calcium aluminate and low-silica Ca-aluminosilicate glasses. Am Mineral 88:949–954

Allwardt JR, Stebbins JF, Schmidt BC, Frost DJ, Withers AC, Hirschmann MM (2005a) Al coordination and density of high-pressure aluminosilicate glasses. Am Mineral 90:1218–1222

Allwardt JR, Poe BT, Stebbins JF (2005b) The effect of fictive temperature on Al-coordination in high-pressure (10 GPa) Na aluminosilicate glasses. Am Mineral 90:1453–1457

Allwardt JR, Stebbins JF, Terasaki H, Du LD, Frost DJ, Withers AC, Hirschmann MM, Suzuki A, Ohtani E (2007) Effect of structural transitions on properties of high-pressure silicate melts: ^{27}Al NMR, glass densities, and melt viscosities. Am Mineral 92:1093–1104

Andreozzi GB, Princivalle F, Skogby H, Della Giusta A (2000) Cation ordering and structural variations with temperature in $MgAl_2O_4$ spinel: an X-ray single-crystal study. Am Mineral 85:1164–1171

Angell CA (1991) Relaxation in liquids, polymers and plastic crystals strong/fragile patterns and problems. J Non-Cryst Solids 131–133:13–31

Ansell S, Krishnan S, Weber J K R, Felten J F, Nordine P C, Beno M A, Price D L, Saboungi M-L (1997) Structure of liquid aluminum oxide. Phys Rev Lett 78:464–466

Avramov I, Milchev A (1988) Effect of disorder on diffusion and viscosity in condensed systems. J Non-Cryst Solids 104:253–260

Batalha NM, Borucki WJ, Bryson ST, Buchhave LA, Caldwell DA, Christensen-Dalsgaard J, Ciardi D, Dunham EW, Fressin F, III TNG, Gilliland RL, Haas MR, Howell SB, Jenkins JM, Kjeldsen H, Koch DG, David W Latham, Lissauer JJ, Marcy GW, Rowe JF, Sasselov DD, Seager S, Steffen JH, Torres G, Basri GS, Brown TM, Charbonneau D, Christiansen J, Clarke B, Cochran WD, Dupree A, Fabrycky DC, Fischer D, Ford EB, Fortney J, Girouard FR, Holman MJ, Johnson J, Isaacson H, Klaus TC, Pavel Machalek, Moorehead AV, Morehead RC, Ragozzine D, Tenenbaum P, Twicken J, Quinn S, VanCleve J, Walkowicz LM, Welsh WF, Devore E, Gould A (2011) Kepler's first rocky planet: Kepler-10b. Astrophys J 729:27

Bacon CR (1977) High-temperature heat content and heat capacity of silicate glasses: experimental determination and a model for calculation. Am J Sci 277:109

Bancroft M, Nesbitt HW, Henderson GS, O'Shaughnessy C, Withers AC, Neuville DR (2018) Lorentzian dominated lineshapes, linewidths for Raman symmetric stretch peaks (800–1200 cm^{-1}) in Q^n ($n = 1$–3) species of alkali silicate glasses/melts. J Non-Cryst Solids 484:72–83

Ben Kacem I, Gautron L, Coillot D, Neuville DR (2017) Structure and properties of lead silicate glasses and melts. Chem Geol 461:104–114

Bender S, Franke R, Hartmann E, Lansmann V, Jansen M, Hormes J (2002) X-ray absorption and photoemission electron spectroscopic investigation of crystalline and amorphous barium silicates. J Non-Cryst Solids 298:99–108

Benmore CJ, Weber JKR, Sampath S, Siewenie J, Urquidi J, Tangeman JA (2003) A neutron, X-ray diffraction study of calcium aluminate glasses. J Phys: Condens Matter 15:S2413–S2423

Bockris JOM, Mackenzie JD, Kitchener E (1955) Viscous flow in silica and binary liquid silcates. Farad Soc Lond Trans 51:1734–1749

Bødker MS, Youngman RE, Mauro J, Smedskjaer MM (2020) Mixed alkali effect in silicate glass structure: viewpoint of ^{29}Si nuclear magnetic resonance and statistical mechanics. J Phys Chem B 124:10292–10299

Bottinga Y, Weill DF (1970) Densities of liquid silicate systems calculated from partial molar volumes of oxide components. Am J Sci 269:169

Bottinga Y, Weill DF (1972) The viscosity of magmatic silicate liquids: a model for calculation. Am J Sci 272:438–475

Brauer DS (2015) Bioactive glasses—structure, properties. Angew Chem Int Edit 54:4160–4181

Bray PJ, Emerson JF, Lee D, Feller SA, Bain DL, Feil DA (1991) NMR, NQR studies of glass structure J Non-Cryst Solids 129 (1991) 240–248

Brückner R (1970) Properties and structure of vitreous silica. II J Non-Cryst Solids 5:177–216

Buckerman W-A, Müller-Warmuth W (1992) A further 29Si NMR study on binary alkali silicate glasses. Glastech Ber 65:18–21

Bunker BC, Kirkpatrick RJ, Brow RK, Turner GL, Nelson C (1991) Local structure of alkaline-eath boroaluminate crystals and glasses. II, ^{11}B and ^{27}Al MAS NMR spectroscopy of alkaline-earth boroaluminate glasses. J Am Ceram Soc 74:1430–1438

Carmichael ISE, Nicholls J, Spera FJ, Wood BJ, Nelson SA (1977) High-temperature of silicate liquids: application to the equilibration and ascent of basic magma. Philos Trans R Soc London A 286:373

Champagnon B, Wondraczek L, Deschamps T (2009) Boson peak, structural inhomogeneity, light scattering and transparency of silicate glasses. J Non-Cryst Solids 355:712–714

Charpentier T, Okhotnikov K, Novikov A, Hennet L, Fischer H, Neuville DR, Florian P (2018) Structure of strontium aluminosilicate glasses from molecular dynamics simulations, neutron diffraction, nuclear magnetic resonance studies. J Phys Chem 122:9567–9583

Chumakov AI, Monaco G, Fontana A, Bosak A, Hermann RP, Bessas D, Wehinger B, Crichton WA, Krisch M, Rüffer R, Baldi G (2014) Role of disorder in the thermodynamics and atomic dynamics of glasses. Phys Rev Lett 112:025502

Cicconi MR, de Ligny D, Gallo T M, Neuville DR (2016) Ca Neighbors from XANES spectroscopy: a tool to investigate structure, redox and nucleation processes in silicate glasses, melts and crystals. Am Mineral 101:1232–1236

Clupper DC, Hench LL (2003) Crystallization kinetics of tape cast bioactive glass 45S5. J Non-Cryst Solids 318:43–48

Cormack AN, Du J (2001) Molecular dynamics simulations of soda-lime silicate glasses. J Non-Cryst Solids 293–295:283–289

Cormier L (2019) Neutron and X-ray diffraction of glass. *In:* Springer Hand Book of Glass, Springer, p 1045–1090

Cormier L, Cuello G (2013) Structural investigation of glasses along the MgSiO$_3$–CaSiO$_3$ join: diffraction studies. Geochim Cosmochim Acta 122:498–510

Cormier L, Neuville DR (2004) Ca and Na environments in Na$_2$O–CaO–Al$_2$O$_3$–SiO$_2$ glasses: influence of cation mixing and cation-network interactions. Chem Geol 213:103–113

Cormier L, Neuville DR, Calas G (2000) Structure and properties of low-silica calcium aluminosilicate glasses. J Non-Crystal Solids 274:110–114

Cormier L, Ghaleb D, Neuville DR, Delaye JM, Calas G (2003) Network polymerization of calcium aluminosilicate glasses: a molecular dynamics and reverse Monte Carlo study. J Non-Crystal Solids 332:255–270

Cormier L, Neuville DR, Calas G (2005) Structure of low-silica calcium aluminosilicate glasses. J Am Ceram Soc 88:2292–2299

Coté B, Massiot D, Taulelle F, Coutures J-P (1992) ^{27}Al NMR spectroscopy of aluminosilicate melts and glasses. Chem Geol 96:367–370

Courtial P (1993) Propriétés thermodynamiques des silicates fondus et des minéraux au voisinage de la fusion. Thèse Université Paris 7

Courtial P, Richet P (1993) Heat capacity of magnesium aluminosilicate melts. Geochim Cosmochim Acta 57:1267–1275

Coutures J-P, Massiot D, Bessada C, Echegut P, Rifflet J-C, Taulelle F (1990) Etude par RMN ^{27}Al d'aluminates liquides dans le domaine 1600–2100°C. CR Acad Sci Paris 310:1041–1045

Davis RO, Jones GO (1953) Thermodynamic and kinetic properties of glasses. Adv Phys 2:370

Davis M, Kaseman D, Parvani S, Sanders K, Grandinetti P, Massiot D, Florian P (2010) $Q(n)$ species distribution in K$_2$O·2SiO$_2$ glass by ^{29}Si magic angle flipping NMR. J Phys Chem A 114:5503–5508

Day DE (1976) Mixed alkali glasses—Their properties and uses. J Non-Cryst Solids 21:343–372

Descamps M (2017) États amorphe et vitreux des composés moléculaires et pharmaceutiques—Propriétés générales. Techniques de l'IngéniEur, PHA2030 v1

Dingwell DB, SL Webb, (1990) Relaxation of silicate glass. Eur J Mineral 2:427

Doremus RH (1973) Glass Science. John Wiley and Sons, New York

Doweidar H (1999) Density-structure correlations in silicate glasses. J Non-Cryst Solids 249:194–200

Doweidar H, Feller S, Affatigato M, Tischendorf B, Ma C, Hammarsten E (1999) Density and molar volume of extremely modified alkali silicate glasses, Phys Chem Glasses 40:339–344

Drewitt JWE L Hennet, Zeidler A, Jahn S PS Salmon, Neuville DR, Fischer HE (2012) Structural transformations on vitrification in the fragile glass forming system CaAl$_2$O$_4$. Phys Rev Lett 109:235501–235506

Drewitt JWE, Barnes AC, Jahn S, Kohn SC, Walter MJ, Novikov A, Neuville DR, Fischer HE, Hennet L (2017) Structure of liquid tri-calcium aluminate. Phys Rev B 95:064203–064214

Drewitt JWE, Hennet L, Neuville DR (2022) From short to medium range order in glasses and melts by diffraction and Raman spectroscopy. Rev Mineral Geochem 87:55–103

Du J, Cormack AM (2004) The medium range structure of sodium silicate glasses: a molecular dynamics simulation. J Non-Cryst Solids 349:66–79

Dulong PL, Petit AT (1819) Recherches sur quelques points importants de la théorie de la chaleur. Annales chimie physique 10:395–413

Dupree R, Holland D, Williams DS (1986) The structure of binary alkali silicate glasses. J Non-Cryst Solids 81:185–200

Dupree R, Ford N, Holland D (1987) Examination of the ^{29}Si environment in the PbO–SiO$_2$ system by magic angle spinning nuclear magnetic resonance. Pt. 1. Glasses. Phys Chem Glasses 28:78–87

Edén M (2012) NMR studies of oxide-based glasses. Annu Rep Prog Chem, Sect C: Phys Chem 108:177–221

Emerson JF, Bray PJ (1994) Nuclear magnetic resonance, transmission electron microscopy studies of mixed-alkali silicate glasses. J Non-Cryst Solids 169:87–95

Emerson JF, Stallworth PE, Bray PJ (1989) High-field ^{29}Si NMR studies of alkali silicate glasses. J Non-Cryst Solids 113:253–259

English S (1923) The effect of composition on the viscosity of glass. Part II J Soc Glass Technol 8:205–251

Eppler RA (1966), Viscosity of molten B$_2$O$_3$. J Am Ceram Soc 49:679–680

Fanderlik I (1990) Silica Glass and Its Application. Elsevier, Amsterdam

Farnan I, Stebbins JF (1990) High-temperature ^{29}Si NMR investigation of solid and molten silicates. J Am Chem Soc 112:32

Farnan I, Stebbins JF (1994) The nature of the glass-transition in a silica-rich oxide melt. Nature 265:1206–1209

Fayon F, Bessada C, Massiot D, Farnan I, Coutures JP (1998) ^{29}Si and ^{207}Pb NMR study of local order in lead silicate glasses. J Non-Cryst Solids 232:403–408

Fayon F, Landron C, Sakurai K, Bessada C, Massiot D (1999) Pb^{2+} environment in lead silicate glasses probed by Pb-LIII edge XAFS and ^{207}Pb NMR. J Non-Cryst Solids 243:39–44

Feller S, Lodden G, Riley A, Edwards T, Croskrey J, Schue A, Liss D, Stentz D, Blair S, Kelley M, Smith G, Singleton S, Affatigato M, Holland D, Smith ME, Kamitsos EL, Varsamis CPE, Loannou E (2010) A multispectroscopic structural study of lead silicate glasses over an extended range of compositions. J Non-Cryst Solids 356:304–313

Fiske PS, Stebbins JF (1994) The structural role of Mg in silicate liquids: A high-temperature ^{25}Mg, ^{23}Na, and ^{29}Si NMR study. Am Mineral 79:848–861

Florian P, Massiot D, Poe B, Farnan I, Couture JP (1995) A time resolved ^{27}Al NMR study of the cooling process of liquid alumina from 2450°C to crystallisation. Solid State Magn Reson 5:233–238

Florian P, Vermillion KE, Grandinetti PJ, Farnan I, Stebbins JF (1996) Cation distribution in mixed alkali disilicate glasses. J Am Chem Soc 118:3493–3497

Florian P, Sadiki N, Massiot D, Coutures JP (2007) ^{27}Al NMR study of the structure of lanthanum- and yttrium-based aluminosilicate glasses and melts. J Phys Chem B 111:9747–9757

Florian P, Novikov A, Drewitt JWE, Hennet L, Sarou-Kanian V, Massiot D, Fischer HE, Neuville DR (2018) Structure and dynamics of high-temperature strontium aluminosilicate melts. Phys Chem Chem Phys 20:27865–27877

Fontana EH, Plummer WA (1966) A study of viscosity-temperature relation-ship in the GeO$_2$ and SiO$_2$ system. Phys Chem Glass 7:139

Frantz JD, Mysen BO (1995) Raman spectra and structure of BaO–SiO$_2$, SrO–SiO$_2$ and CaO–SiO$_2$ melts to 1600°C. Chem Geol 121:155–176

Frischat GH, Sebastian K (1985) Leach resistance of nitrogen-containing Na$_2$O–CaO–SiO$_2$ glasses. J Am Ceram Soc 68:C305–C307

Frischat GH, Poggemann JF, Heide G (2001) Nanostructure and atomic structure of glass seen by atomic force microscopy. J Non-Cryst Solids 345–346:197–202

Fulcher GS (1925) Analysis of recent measurements of the viscosity of glasses. J Am Ceram Soc 8:339

Galeener FL, Leadbetter AJ, Stringfellow MW (1983) Comparison of the neutron, Raman and infrared vibrational spectra of vitreous SiO$_2$, GeO$_2$, BeF$_2$. Phys Rev B 27:1052–1079

Gaskell P, Eckersley MC, Barnes AC, Chieux P (1991) Medium-range order in the cation distribution of a calcium silicate glass. Nature 350:675–677

Gahay VMC, Tomozawa M (1989) The origin of phase separation in silicate melts and glasses. J Non-Cryst Solids 109:27–34

Georges AM, Stebbins JF (1998) Structure and dynamics of magnesium in silicate melts: A high-temperature ^{25}Mg NMR study. Am Mineral 83:1022–1029

Gibbs JH, Dimarzio EA (1958) Nature of the glass transition and the glassy state. J Chem Phys 28:373

Giordano D, Dingwell DB (2003). Non-Arrhenian multicomponent melt viscosity: a model. Earth Planet Sci Lett 208:337–349. Erratum Earth Planet Sci Lett 221:449

Giordano D, Russell JK (2018) Towards a structural model for the viscosity of geological melts. Earth Planet Sci Lett 501:202–212

Giordano D, Russell JK, Dingwell DB (2008) Viscosity of magmatic liquids: A model. Earth Planet Sci Lett 271:123–134

Goldstein M (1969) Viscous liquids and the glass transition: a potential energy barrier picture. J Chem Phys 51:3728

Goldstein M (1976) Viscous liquids and the glass transition. V Sources of the excess specific heat of the liquid. J Chem Phys 64:4767

Goldstein M (2011) On the reality of the residual entropy of glasses and disordered crystals: The entropy of mixing. J Non-Cryst Solids 357:463–465

Greaves GN (1985) EXAFS, the structure of glass. J Non-Cryst Solids 71:203–217

Greaves GN, Fontaine A, Lagarde P, Raoux D, Gurman SJ (1981) Local structure of silicate glasses. Nature 293:611–616

Greig JW (1927) Immiscibility in silicate melts. Am J Sci 13:133–154

Grest GS, Cohen MH (1980) Liquids, glasses, and the glass transition: a free-volume approach. Adv Chem Phys 48:455–525

Gruener G, Odier P, Meneses DD, Florian P, Richet P (2001) Bulk and local dynamics in glass-forming liquids: A viscosity, electrical conductivity, and study of aluminosilicate melts. Phys Rev B 64:024206

Guignard M, Cormier L (2008) Environments of Mg and Al in MgO–Al$_2$O$_3$–SiO$_2$ glasses: a study coupling neutron and X-ray diffraction and Reverse Monte Carlo simulations. Chem Geol 256:111–118

Hannon AC, Parker JM (2000) The structure of aluminate glasses by neutron diffraction. J Non-Cryst Solids 274:102–109

Hater W, Müller-Warmuth W, Meier M, Frischat GH (1989) High-resolution solid-state NMR studies of mixed-alkali silicate glasses. J Non-Cryst Solids 113:210–212

El Hayek R, Ferey F, Florian P, Pisch A, Neuville DR (2017) Structure and properties of lime alumino-borate glasses and melts. Chem Geol 461:75–81

Hehlen B, Neuville DR (2015) Raman response of network modifier cations in alumino-silicate glasses. J Phys Chem B 119:4093–4098

Hehlen B, Neuville DR (2020) Non-network-former cations in oxide glasses spotted by Raman scattering. Phys Chem Chem Phys 22:12724–1273

Hehlen B, Courtens E, Yamanka A, Inoue K (2002) Nature of the Boson peak of silica glasses from hyper-Raman scattering. J Non-Cryst Solids 307:185–190

Hench LL (1991) Bioceramics : from concept ot clinic. J Am Ceram Soc 74:1487–1510

Henderson GS (1995) A Si K-edge EXAFS/XANES study of sodium silicate glasses. J Non-Cryst Solids 183:43–50

Henderson GS, Stebbins JF (2022) The short-range order (SRO) and structure. Rev Mineral Geochem 87:1–53

Henderson GS, Neuville DR, Cormier L (2007) An O K-edge XANES study of calcium aluminates. Can J Chem 85:801–806

Hennet L Drewitt JWE, Neuville DR, Cristiglio V, Kozaily J, Brassamin S, Zanghi D, Fischer HE (2016) Neutron diffraction of calcium aluminosilicate glasses and melts. J Non-Cryst Solids 451:89–93

Herzog F et Zakaznova-Herzog VP (2011) Quantitative Raman spectroscopy: Challenges, shortfalls, and solutions— Application to calcium silicate glasses. Am Mineral 96:914–927

Hetherington G Jack, H, Kennedy JC (1964) The viscosity of vitreous silica. Phys Chem Glass 5:130–137

Hiet J, Deschamps M, Pellerin N, Fayon F, Massiot D (2009) Probing chemical disorder in glasses using silicon-29 NMR spectral editing. Phys Chem Chem Phys 11:6935

Higby PL, Ginther RJ, Aggarwal ID, Friebele EJ (1990) Glass formation and thermal properties of low-silica calcium aluminosilicate glasses. J Non-Cryst Solids 126:209–215

Hill R, Brauer DS (2011) Predicting the bioactivity of glasses using the network connectivity or split network models. J Non-Cryst Solids 357:3884–3887

Hosono H Yamazaki K, Abe Y (1985) Dopant-free ultraviolet-sensitive calcium aluminate glasses. J Am Ceram Soc 68:C-304–C-305

Hudon P, Baker DR (2002a) The nature of phase separation in binary oxide melts and glasses. I Silicate systems. J Non-Cryst Solids 303:299–345

Hudon P, Baker DR (2002b) The nature of phase separation in binary oxide melts, glasses. II Selective solution mechanism. J Non-Cryst Solids 303 (2002) 346–353

Hung I, Gan Z, Gor'kov PL, Kaseman DC, Sen S, LaComb M, Stebbins JF (2016) Detection of "free" oxide ions in low-silica Ca/Mg silicate glasses: Results from ^{17}O → ^{29}Si HETCOR NMR. J Non-Cryst Solids 445–44:1–6

Hui H, Zhang Y (2007) Toward a general viscosity equation for natural anhydrous, hydrous silicate melts. Geochim Cosmochim Acta 71:403–416

Imaoka M, Yamazaki T (1963) Studies of the glass-formation range of silicate systems.investigations on the glass-formation range, 2. J Ceram Assoc Jpn 79:215–232

Inaba S, Fujino S, Sakai K (2010) Non-contact measurement of the viscosity of a soda–lime–silica melt using electric field tweezers. Phys Chem Glasses: Eur J Glass Sci Technol B 51:304–308

Ispa S, Charpentier T, Mauri F, Neuville DR (2010) Structural properties of lithium and sodium tetrasilicate glasses: molecular dynamics simulations vs NMR experimental and first principles data. Solid State Sci 12:183–192

Iuga D, Morais C, Gan Z, Neuville DR, Cormier L, Massiot D (2005) ^{27}Al/^{17}O NMR correlation in crystalline and vitreous materials. J Am Chem Soc 127:11540–11542

Jakse N, Bouhadja M, Kozaily J, Hennet L, Drewitt JWE, Neuville DR, Fischer H E, Cristiglio V, Pasturel A (2012) Interplay between non-bridging oxygen, triclusters, and fivefold Al coordination in low silica content calcium aluminosilicate melts. Appl Phys Lett 101:201903–201908

Joly Y (2001) X-ray absorption near edge structure calculations beyond the muffin-tin approximation. Phys Rev B 63:125120

Jones AR, Winter R, Greaves GN, Smith IH (2001) MAS NMR study of soda-lime silicate glasses with variables degree of polymerisation. J Non-Cryst Solids 293–295:87–92

Kanehashi K, Stebbins JF (2007) In situ high temperature ^{27}Al NMR study of structure and dynamics in a calcium aluminosilicate glass and melt. J Non-Cryst Solids 353:4001–4010

Kashio S, Iguchi Y, Goto T, Nishina Y, Fuwa T (1980) Raman spectroscopic study on the structure of silicate slag. Trans Iron Steel Inst Jpn 20:251–253

Kauzmann W (1948) The nature of the glassy state and the behavior of liquids at low temperatures. Chem Rev 43:219–243

Kim H-M, Miyaji F, Kokubo T (1995) Bioactivity of Na$_2$O–CaO–SiO$_2$ glasses. J Am Ceram Soc 78:2405–2411

Kim N Bassiri R, Fejer MM, Stebbins JF (2014) The structure of ion beamsputtered amorphous alumina films and effects of Zn doping: High-resolution ^{27}Al NMR J Non-Cryst Solids 405:1–6

King W, Shelby J (1996) Strontium calcium aluminate glasses. Phys Chem Glass 37:1–14

Kohara S, Ohno M, Takata T, Usuki L, Morita H, Suzuya S, Akola S, Pusztai L (2010) Lead silicate glasses: Binary network-former glasses with large amounts of free volume. Phys Rev B 82:134209

Koike A, Tomozawa M (2007) IR investigation of density changes of silica glass and soda-lime silicate glass caused by microhardness indentation. J Non-Cryst Solids 353:2318–2327

Koike A, Tomozawa M, Ito S (2007) Sub-critical crack growth rate of soda-lime-silicate glass and less brittle glass as a function of fictive temperature. J Non-Cryst Solids 353:2675–2680

Kroecker S, Stebbins JF (2000) Magnesium coordination environments in glasses and minerals: New insight from high-field magnesium-25 MAS NMR. Am Mineral 85:1459–1464

Kusabiraki K (1986) Infrared spectra of vitreous silica and sodium silicates containing titanium. J Non-Cryst Solids 79:208–212

Kusabiraki K, Shiraishi KT (1981) The infrared spectrum of vitreous fayalite. J Non-Cryst Solids 44:365–368

Lacy ED (1963) Aluminium in glasses and melts. Phys Chem Glasses 4:234–238

Lange RA (1996) Temperature independent thermal expansivities of sodium aluminosilicate melts. Geochim Cosmochim Acta 60:4989–4996

Lange RA (1997) A revised model for the density and thermal expansivity of K_2O–Na_2O–CaO–MgO–Al_2O_3–SiO_2 liquids from 700 to 1900 K: extension to crustal magmatic temperatures. Contrib Mineral Petrol 130:1–11

Lange RA, Carmichael ISE (1987) Densities of Na_2O–K_2O–CaO–MgO–Fe_2O_3–Al_2O_3–TiO–SiO_2 liquids: New-measurements and derived partial molar properties. Geochim Cosmochim Acta 51:2931–2946

Lange RA, Navrotsky A (1992) Heat capacities of Fe_2O_3-bearing silicate liquids. Contrib Mineral Petrol 110:311–320

Larson C Doerr, Affatigato M, Feller S, Holland D, Smith ME (2006) A ^{29}Si MAS NMR study of silicate glasses with a high-lithium content. J Phys: Condens Matter 18:11323–11331

Leko BK (1979) Viscosity of vitreous silica. Fiz Khim Stekla 5:258–278

Leko VK, Gusakova NK, Meshcheryakova EV, Prokhorova TI (1977) The effect of impurity alkali oxides, hydroxyl groups, Al_2O_3, and Ga_2O_3 on the viscosity of vitreous silica. Soy J Glass Phys Chem 3:204–210

Lee SK (2005) Structure, the extent of disorder in quaternary (Ca-Mg and Ca-Na) aluminosilicate glasses and melts. Am Mineral 90:1393–1401

Lee SK, Kim EJ (2015) Probing metal-bridging oxygen and configurational disorder in amorphous lead silicates: Insights from ^{17}O solid-state nuclear magnetic resonance. J Phys Chem C 119:1748–1756

Lee SK, Stebbins JF (1999) The degree of aluminum avoidance in aluminosilicate glasses. Am Mineral 84:937–945

Lee SK, Stebbins JF (2000) Al–O–Al and Si–O–Si sites in framework aluminosilicate glasses with Si/Al=1: quantification of framework disorder. J Non-Cryst Solids 270:260–264

Lee SK, Stebbins JF (2002) The extent of inter-mixing among framework units in silicate glasses and melts. Geochim Cosmochim Acta 66:303–309

Lee SK, Stebbins JF (2003) Nature of cationmixing and ordering in Na-Ca silicate glasses and melts. J Phys Chem B 107:3141–3148

Lee SK, Stebbins J (2009) Effects of the degree of polymerization on the structure of sodium silicate and aluminosilicate glasses and melts: An ^{17}O NMR study. Geochim Cosmochim Acta 73:1109–1119

Lee SK, Mysen BO, Cody GD (2003) Chemical order in mixed-cation silicate glasses and melts. Phys Rev B 68:214206

Lee SK, Park SY, Yi YS, Moon J (2010) Structure and disorder in amorphous alumina thin films: Insights from high-resolution. J Phys Chem C 114:13890–13894

Lee SK, Kim H-I, Kim EJ, Mun KY, Ryu S (2016) Extent of disorder in magnesium aluminosilicate glasses: Insights from ^{27}Al and ^{17}O NMR. J Phys Chem C 120:737–749

Léger A, Rouan D, Schneider J, Barge P, Fridlund M, Samuel B, Ollivier M, Guenther E, Deleuil M, Deeg HJ, Auvergne M, Alonso R, Aigrain S, Alapini A, Almenara JM, Baglin A, Barbieri M, Bruntt H, Bordé P, Bouchy F, Cabrera J, Catala C, Carone L, Carpano S, Csizmadia S, Dvorak R, Erikson A, Ferraz-Mello S, Foing B, Fressin F, Gandolfi D, Gillon M, Gondoin P, Grasset O, Guillot T, Hatzes A, Hébrard G, Jorda L, Lammer H, Llebaria A, Loeillet B, Mayor M, Mazeh T, Moutou C, Pätzold M, Pont F, Queloz D, Rauer H, Renner S, Samadi R, Shporer A, Sotin C, Tingley B, Wuchterl G, Adda M, Agogu P, Appourchaux T, Ballans H, Baron P, Beaufort T, Bellenger R, Berlin R, Bernardi P, Blouin D, Baudin F, Bodin P, Boisnard L, Boit L, Bonneau F, Borzeix S, Briet R, Buey J-T, Butler B, Cailleau D, Cautain R, Chabaud P-Y, Chaintreuil S, Chiavassa F, Costes V, Parrho VC, Fialho FDO, Decaudin M, Defise J-M, Djalal S, Epstein G, Exil G-E, Fauré C, Fenouillet T, Gaboriaud A, Gallic A, Gamet P, Gavalda F, Grolleau E, Gruneisen R, Gueguen L, Guis V, Guivarc'h V, Guterman P, Hallouard D, Hasiba J, Heuripeau F, Huntzinger G, Hustaix H, Imad C, Imbert C, Johlander B, Jouret M, Journoud P, Karioty F, Kerjean L, Lafaille V, Lafond L, Lam-Trong T, Landiech P, Lapeyrere V, Larqué T, Laudet P, Lautier N, Lecann H, Lefevre L, Leruyet B, Levacher P, Magnan A, Mazy E, Mertens F, Mesnager J-M, Meunier J-C, Michel J-P, Monjoin W, Naudet D, Nguyen-Kim K, Orcesi J-L, Ottacher H, Perez R, Peter G, Plasson P, Plesseria J-Y,Pontet B, Pradines A, Quentin C, Reynaud J-L, Rolland G, Rollenhagen F, Romagnan R, Russ N, Schmidt R, Schwartz N, Sebbag I, Sedes G, Smit H, Steller MB, Sunter W, Surace C, Tello M, Tiphène D, Toulouse P, Ulmer B, Vandermarcq O, Vergnault E, Vuillemin A, Zanatta P (2009) Transiting exoplanets from the CoRoT space mission—VIII CoRoT-7b: the first super-Earth with measured radius. Astron Astrophys 506:287–302

Levin EM, Robbins CR, McMurdie HF (1964) Phase Diagrams for Ceramists, 2nd ed. Am Ceram Soc, Columbus, Ohio

Le Losq C, Neuville DR (2013) Effect of K/Na mixing on the structure and rheology of tectosilicate silica-rich melts. Chem Geol 346:57–71

Le Losq Ch, Neuville DR (2017) Molecular structure, configurational entropy and viscosity of silicate melts: link through the Adam and Gibbs theory of viscous flow. J Non-Cryst Solids 463:175–188

Le Losq Ch, Neuville DR, Florian P, Henderson GS, Massiot D (2014) Role of Al^{3+} on rheology and nano-structural changes of sodium silicate and aluminosilicate glasses and melts. Geochim Cosmochim Acta 126:495–517

Le Losq C, Neuville DR, Florian P, Massiot D, Zhou Z, Chen W, Greaves N (2017) Percolation channels: a universal idea to describe the atomic structure of glasses and melts. Sci Rep 7:16490

Le Losq C, Cicconi MR, Greaves GN, Neuville DR (2019) Silicate glasses. *In:* Springer Handbook of Glass, Springer, p 441–488

Le Losq C, Valentine A, Mysen BO, Neuville DR (2021) Deep learning model to predict the structure and properties of aluminosilicate glasses and melts. Geochim Cosmochim Acta 314:27–54

Licheron M, Montouillout V, Millot F, Neuville DR (2011) Raman, ^{27}Al NMR structure investigations of aluminate glasses: $(1-x)Al_2O_3$- xMO, with M=Ca Sr, Ba and $0.5<x<0.75$). J Non-Cryst Solids 257:2796–2801

de Ligny D, Westrum EF (1996) Entropy of calcium and magnesium aluminosilicate glasses. Chem Geol 128:113–128

Lines ME, MacChesney JB, Lyons KB, Bruce AJ, Miller AE, Nassau K (1989) Calcium aluminate glasses as pontential ultralow-loss optical materials at 1.5–1.9 mm. J Non-Cryst Solids 107:251–260

Lockyer MWG, Holland D, Dupree R (1995) NMR investigation of the structure of some bioactive and related glasses. J Non-Cryst Solids 188:207–219

Loewenstein W (1954) The distribution of aluminum in the tetrahedra of silicates and aluminates. Am Mineral 39:92–96

Macedo PB, Litovitz PB (1965), On the relative roles of free volume and activation energy in the viscosity of liquids. J Chem Phys 42:24

McMillan PF (1984) Structural studies of silicate glasses and melts—Applications and limitations of Raman spectroscopy. Am Mineral 69:622–644

McMillan PF, Kirkpatrick RJ (1992) Al coordination in magnesium aluminosilicate glasses. Am Mineral 77:898–900

McMillan P, Piriou B (1983) Raman spectroscopy of calcium aluminate glasses and crystals. J Non-Cryst Solids 55:221–242

McMillan P, Piriou B, Navrotsky A (1982) A Raman spectroscopic study of glasses along the joins silica-calcium aluminate, silica-sodium aluminate, and silica-potassium aluminate. Geochim Cosmochim Acta 46:2021–2037

Maekawa H, Maekawa T, Kawamura K, Yokokawa T (1991) The structural group of alkali silicate glasses determined from ^{29}Si MAS-NMR. J Non-Cryst Solids 127:53–64

Magnien V, Neuville DR, Cormier L, Mysen BO, Richet P (2004) Kinetics of iron oxidation in silicate melts: A preliminary XANES study. Chem Geol 213:253–263

Magnien V, Neuville DR, Cormier L, Roux J, Hazemann J-L, de Ligny D, Pascarelli S, Vickridge I, Pinet O, Richet P (2008) Kinetics and mechanisms of iron redox reactions in silicate melts: The effects of temperature and alkali cations. Geochim Cosmochim Acta 72:2157–2168

Malfait WJ (2009) Quantitative Raman spectroscopy: speciation of cesium silicate glasses. J Raman Spectrosc 40:1895–1901

Malfait WJ, Zakaznova-Herzog VP, Halter WE (2007) Quantitative Raman spectroscopy: high-temperature speciation of potassium silicate melts. J Non-Cryst Solids 353:4029–4042

Massiot D, Fayon F, Montouillout V, Pellerin N, Hiet J, Roiland C, Florian P, Coutures J-P , Cormier L, Neuville DR (2008) Structure and dynamics of oxide melts and glasses: a view from multinuclear and high temperature NMR. J Non-Crystal Solids 354:249–254

Mastelaro V, Zanotto E, Lequeux N, Cortes R (2000) Relationship between short-range order and ease of nucleation in $Na_2Ca_2Si_3O_9$, $CaSiO_3$ and $PbSiO_3$ glasses. J Non-Cryst Solids 262:191–199

Mauro J (2011) Through a glass, darkly: dispelling three common misconceptions in glass science. Int J Appl Glass Sci 2:245–261

Mauro JC, Yue Y, Ellison AJ, Gupta PK, Allan DC (2009) Viscosity of glass-forming liquids. PNAS 106:19780–19784

Méducin F, Redfern SAT, Le Godec Y, Stone HJ, Tucker MG, Dove MT, Marshall WG (2004) Study of cation order-disorder in $MgAl_2O_4$ spinel by in situ neutron diffraction up to 1600K and 3.2 GPa. Am Mineral 89:981–986

Meiling GS, Uhlmann DR (1977) Crystallisation and melting kinetics of soda disilicate. Phys Chem Glass 8:62–68

Merzbacher CI, White WB (1991) The structure of alkaline-earth aluminosilicate glasses as determined by vibrational spectroscopy. J Non-Cryst Solids 130:18–34

Merzbacher CI, Sheriff BL, Hartman JS, White WB (1990) A high-resolution ^{29}Si and ^{27}Al NMR study of alkaline-earth aluminosilicate glasses. J Non-Cryst Solids 124:194–206

Merzbacher CI, McGrath KJ, Highby PL (1991) ^{29}Si NMR and infrared reflectance spectroscopy of low-silica calcium aluminosilicate glasses. J Non-Cryst Solids 136:249–259

Moore C (1951) Limiting compositions of binary glasses of the type $xR_2O \cdot SiO_2$ and ternary glasses of types $yR_2O \cdot yR_2O_3 \cdot SiO_2$ in relation to glass structure. Trans Soc Glass Technol 35:43–57

Moretti R (2005). Polymerisation, basicity, oxidation state and their role in ionic modelling of silicate melts. Ann Geophys 48:583–608

Morey G (1930) The devitrification of soda-lime-silica glasses. J Am Ceram Soc 683–714

Morey GW, Bowen NL (1925) The melting relations of the soda-lime-silica glasses. J Soc Glass Technol 9:226–264

Morikawa H, Takagi Y, Ohno H (1982) Structural analysis of 2PbO·SiO$_2$ glass. J Non-Cryst Solids 53:173–182

Mossa S, La Nave E, Stanley HE, Donati C, Sciortino F, Tartaglia P (2002) Dynamics and configurational entropy in the LW model for supercooled orthoterphenyl. arXiv:cond-mat/0111519v2

Moulton BJA, Henderson GS (2021) Glasses: alkali and alkaline-earth silicates *In:* Encyclopedia of Materials: Technical Ceramics and Glasses. Vol. 2. Pomeroy M (ed) Elsevier, p 462–482

Moynihan CT, Gupta PK (1978) The order parameter model for structural relaxation in glass. J Non-Cryst Solids 29:143

Moynihan CT, Easteal AJ, Wilder J (1974) Dependence of the glass transition temperature on heating and cooling rate. J Phys Chem 78:2873–2879

Moynihan CT, Sasabe H, Tucker J (1976) Kinetics of the glass transition in a calcium-potassium nitrate melt. Proc Electrochem Soc 6:182–194

Moynihan CT, Easteal AJ, Tran DC, Wilder JA, Donovan EP (1976a) Heat capacity and structural relaxation of mixed-alkali glasses. J Phys Chem 78:2673

Moynihan CT, Macedo PB, Montrose CJ, Gupta PK, De Bolt MA, Dill JF, Dom BE, Drake PW, Easteal AJ, Elterman PB, Moeller RP, Sasabe M, Wilder JA (1976b) Structural relaxation in vitreous materials. Ann New-York Acad Sci 279:15

Murdoch JB, Stebbins JF, Carmichael ISE (1985) High-resolution ^{29}Si NMR study of silicate and aluminosilicate glasses: the effect of network modifying cations. Am Mineral 70:332–343

Musgraves JD, Hu J, Calvez L (Eds) Springer Handbook of Glass, Springer

Mysen BO (1990) Role of Al in depolymerized, peralkaline aluminosilicate melts in the systems Li$_2$O–Al$_2$O$_3$–SiO$_2$, Na$_2$O–Al$_2$O$_3$–SiO$_2$, and K$_2$O–Al$_2$O$_3$–SiO$_2$. Am Mineral 75:20–134

Mysen BO (1995) Experimental, in situ, high-temperature studies of properties and structure of silicate melts relevant to magmatic processes. Eur J Mineral 7:745–766

Mysen BO (1999) Structure and properties of magmatic liquids: From haplobasalt to haploandesite. Geochim Cosmochim Acta 63:95–112

Mysen BO, Frantz JD (1993) Structure of silicate melts at high temperature: in-situ measurements in the system BaO–SiO$_2$. Am Mineral 78:699–709

Mysen BO, Frantz JD (1994) Structure of haplobasaltic liquids at magmatic temperatures: in situ, high-temperature study ofmelts on the join Na$_2$Si$_2$O$_5$–Na$_2$(NaAl)$_2$O$_5$. Geochim Cosmochim Acta 58:1711–1733

Mysen BO, Virgo D, Kushiro I (1981) The structural role of aluminum in silicate melt—A Raman spectroscopic study at 1 atmosphere. Am Mineral 66:678–701

Mysen BO, Virgo D, Seifert FA (1982) The structure of silicate melts: Implications for chemical and physical properties of natural magma. Rev Geophys 20:353–383

Mysen BO, Lucier A, Cody GD (2003) The structural behavior of Al^{3+} in peralkaline melts and glasses in the system Na$_2$O–Al$_2$O$_3$–SiO$_2$. Am Mineral 88:1668–1678

Nakamura K, Takahashi Y, Osada M, Fujiwara T (2013) Low-frequency Raman scattering in binary silicate glass: Boson peak frequency and its general expression. J Ceram Soc Japan 121:1012–1014

Napolitano A, Macedo PB, Hawkins EG (1965) Viscosity and density of boron trioxide. J Am Ceram Soc 48:613–616

Navrotsky A, Perandeau P, McMillan PF, Coutures JP (1982) A thermochemical study of glasses and crystals along the joins silica-calcium aluminate and silica-sodium aluminate. Geochim Cosmochim Acta 46:2039–2049

Navrotsky A, Geisinger KP, Gibbs GV (1985) The tetrahedral framework in glasses and melts—inferences from molecular orbital calculations and implications for structure, thermodynamics, and physical properties. Phys Chem Mineral 11:284–298

Natrup FV, Bracht H, Murugavel S, Roling B (2005) Cation diffusion and ionic conductivity in soda-lime silicate glasses. Phys Chem Chem Phys 7:2279–2286

Nernst W (1911) Theoretical Chemistry from the Standpoint of Avogadro's Rule & Thermodynamics, McMillan

Nesbitt HW, O'Shaughnessy C, Bancroft GM, Henderson GS, Neuville DR (2018) Factors affecting line shapes and intensities of Q^3 and Q^4 Raman bands of Cs silicate glasses. Chem Geol 505:1–11

Nesbitt HW, Henderson GS, Bancroft GM, Neuville DR (2021) Spectral resolution and Raman Q^3 and Q^2 cross sections. Chem Geol 562:120040

Neuville DR (1992) Etude des Propriétés Thermodynamiques et Rhéologiques des Silicates Fondus. Thèse de l'Université de Paris VII, spécialité Géochimie Fondamentale

Neuville DR (2005) Structure, properties in (Sr, Na) silicate glasses and melts. Phys Chem Glasses 46:112–119

Neuville DR (2006) Viscosity, structure, mixing in (Ca, Na) silicate melts. Chem Geol 229:28–42.

Neuville DR, Richet P (1990) Viscosité et entropie des silicates fondus. Rivista del la Staz. Sper Vetro 6:213–221

Neuville DR, Richet P (1991) Viscosity, mixing in molten (Ca,Mg) pyroxenes and garnets. Geochim Cosmochim Acta 55:1011–1021

Neuville DR, Courtial P, Dingwell DB, Richet P (1993) Thermodynamic and rheological properties of rhyolite and andesite melts. Contrib Mineral Petrol 113:571–581

Neuville DR, Mysen BO (1996) Role of aluminium in the silicate network: in situ, high-temperature study of glasses and melts on the join SiO$_2$–NaAlO$_2$. Geochim Cosmochim Acta 60:1727–1737

Neuville DR, Cormier LR, Flank AM, Prado RJ, Lagarde P (2004a) Na K-edge XANES spectra of minerals and glasses. Eur J Mineral 16:809–816

Neuville DR, Cormier L, Flank AM, Lagarde P Massiot D (2004b) Aluminum X-ray absorption near edges structure in minerals and glasses. Chem Geol 213:153–163

Neuville DR, Cormier L, Massiot D (2004c) Role of aluminium in peraluminous region in the CAS system. Geochim Cosmochim Acta 68:5071–5079

Neuville DR, Cormier L, Flank AM, Lagarde P (2005) A XANES study at the Na and Al K-edge of soda-lime aluminosilicate glasses. Phys Scripta 115:316–318

Neuville DR, Cormier L, Massiot D (2006) Al speciation in calcium aluminosilicate glasses: A NMR and Raman spectrocospie. Chem Geol 229:173–185

Neuville DR Cormier L, Montouillout V, Massiot D (2007) Local environment of Al in aluminosilicate glasses: a NMR point of view. J Non-cryst Solids 353:180–185

Neuville DR, Cormier L, Montouillout V, Florian P, Millot F, Rifflet JC, Massiot D (2008a) Structure of Mg- and Mg/Ca aluminosilicate glasses: ^{27}Al NMR and Raman spectroscopy investigations. Am Mineral 83:1721–1731

Neuville DR Cormier L, Flank AM, de Ligny D, Roux J, Lagarde P (2008b) Environment around Al, Si and Ca in aluminate and aluminosilicate melts by X-ray absorption spectroscopy at high temperature. Am Mineral 93:228–234

Neuville DR, de Ligny D, Cormier L, Henderson GS, Roux J, Flank AM, Lagarde P (2009) The crystal and melt structure of spinel and alumina at high temperature: an in-situ XANES study at the Al and Mg K-edge. Geochim Cosmochim Acta 73:3410–3422

Neuville DR, Henderson GS, Cormier L, Massiot D (2010) Structure of CaO–Al_2O_3 crystal, glasses and liquids, using X-ray absorption at Al L and K edges and NMR spectroscopy. Am Mineral 95:1580–1589

Neuville DR, Henderson GS, de Ligny D (2014a) Advances in Raman spectroscopy applied to earth, material sciences. Rev Mineral Geochem 78:509–541

Neuville DR, Hennet L, Florian P, de Ligny D (2014b) In situ high temperature experiment. Rev Mineral Geochem 78:779–800

Novikov A (2017) Structure et propriétés des aluminosilicates de Sr, Ba et Zn, Université d'Orléans

Novikov A, Neuville DR, Hennet L, Thiaudière D, Gueguen Y, Florian P (2017) Al and Sr environment in tectosilicate glasses and melts: viscosity, Raman and NMR investigation. Chem Geol 461:115–127

Ohtani E, Taulelle F, Angell CA (1985) Al^{3+} coordination changes in aluminosilicates under pressure. Nature 314:78–82

O'Shaughnessy C, Henderson GS, Nesbitt HW, Bancroft GM, Neuville DR (2017) The structure of caesium silicate glasses and melts: implications for the interpretation of Raman spectra. Chem Geol 461:82–95

O'Shaughnessy C, Henderson G S, Nesbitt HW, Bancroft GM, Neuville DR (2020) The behaviour of modifier cations in alkali silicate glasses. J Am Ceram Soc 103:3991–4001

Parks GS, Huffman HM (1926) Glass as a fourth state of matter. Science 64:364

Parks GS, Huffman HM (1927) Studies on glass. I The transition between the glassy and liquid states in the case of some simple organic compound. J Phys Chem 31:1842

Pena RB, Sampaio DV, Lancelotti RF, Cunha TR, Zanotto ED, Pizani PS (2020) In-situ Raman spectroscopy unveils metastable crystallization in lead metasilicate glass. J Non-Cryst Solids 546:120254

Persikov ES (1991) The viscosity of magmatic liquids : experiment, generalized patterns. A model for calculation and prediction. Applications. Adv Phys Geochem 9:1–40

Petit AT, Dulong PL (1819) Recherches sur quelques points importants de la théorie de la chaleur. Ann chim phys 10:395–413

Poe BT, McMillan PF, Angell CA, Sato RK (1992) Al and Si coordination in SiO_2–Al_2O_3 glasses and liquids: a study by NMR and IR spectroscopy and MD simulations. Chem Geol 96:333–349

Poe BT, Romano C, Zotov N,Cinin G, Marcelli A (2001) Compression mechanism in aluminosilicate melts: Raman and XANES stpectrocopy of glasses quenched from pressures up to 10Gpa. Chem Geol 174:21–31

Poole JP (1948) Viscosité à basse température des verres alcalino-silicatés. Verres Réfrac. 2:222–230

Popov AI (2018) What is glass? J Non-Cryst Solids 502:249–250

Poulain M (1983) Halide glasses. J Non-Cryst Solids 56:1–14

Poulain M, Lucas J (1970) New transition metal fluorozirconates C R Acad Sci 281:2345–2349

Putnis A (1992) An Introduction to Mineral Sciences. Cambridge University Press

Rankin GA (1915) The ternary system CaO–Al_2O_3–SiO_2. Am J Sci 39:1–79

Rekhson SM (1975) Relaxation of glass. Sov J Glass Phys Chem 1:417

Rekhson SL (1980) Viscosity and stress relaxation in commercial glasses in the glass transition region. J Non-Cryst Solids 38–39:457–462

Rekhson SM (1989) Computer modeling of glass behavior. Congress on glass in Leningrad USSR

Richet P (1984) Viscosity and configurational entropy of silicate melts. Geochim Cosmochim Acta 48:471–483

Richet P (1987) Heat capacity of silicate glasses. Chem Geol 62:111–124

Richet P (2009) Residual and configurational entropy: Quantitative checks through applications of Adam-Gibbs theory to the viscosity of silicate melts. J Non-Cryst Solids 355:628–635

Richet P, Bottinga Y (1984) Glass transitions and thermodynamic properties of amorphous SiO_2, $NaAlSi_nO_{2n+2}$ and $KAlSi_3O_8$. Geochim Cosmochim Acta 48:453–470

Richet P, Bottinga Y (1985) Heat capacity of aluminum-free liquid silicates. Geochim Cosmochim Acta 49:471–486

Richet P, Neuville DR (1992) Thermodynamics of silicates melts: Configurational properties. Adv Phys Geochem 10:132–161

Richet P, Robie RA, Hemingway BS (1986) Low temperature heat capacity of diopside glass (CaMgSi$_2$O$_6$): a calorimetric test of the configurational entropy theory applied to the viscosity of liquid silicates. Geochim Cosmochim Acta 50:1521–1533

Richet P, Nidaira A, Neuville DR, Atake T (2009a) Aluminum speciation, vibrational entropy and short-range order in calcium aluminosilicate glasses Geochim Cosmochim Acta 73:3894–3904

Richet P, Nidaira A, Neuville DR, Atake T (2009b) Low-temperature heat capacity and short-range order in alkaline-earth metasilicates. Am Mineral 94:1591–1596

Riebling E F (1966) Structure of sodium aluminosilicate melts containing at least 50 mol% SiO$_2$ at 1500°C. J Chem Phys 44:2857–2865

Risbud SH, Kirkpatrick RJ, Taglialavore AP, Montez B (1987) Solid-sate NMR evidence of 4-, 5- and 6-fold aluminum site in roller-quenched SiO$_2$–Al$_2$O$_3$ glasses. J Am Ceram Soc 70:C10–C12

Ritland HN (1954) Density phenomena in the transformation range of a borosilicate crown glass. J Am Ceram Soc 37:370

Robert G, Smith R A, Whittington A G (2019) Viscosity of melts in the NaAlSiO$_4$–KAlSiO$_4$–SiO$_2$ system: Configurational entropy modelling. J Non-Cryst Solids 524:119635

Robie RA, Hemingway BS, Fisher JR (1979) Thermodynamic properties of mineral and related substance ar 298,15K and 1 bar Pressure and at higher temperature. USGS Bull 1432

Roling B, Ingram MD (2000) Mixed alkaline-earth effects in ion conducting glasses. J Non-Cryst Solids 265:113–119

Rosenhauer M, Scarfe CM, Virgo D (1979) Pressure dependence of the glass transition temperature in glasses of diopside, albite, and sodium trisilicate composition. Carnegie Inst Wash Yearb 78:547

Roy BN, Navrostsky A (1984) Thermochemistry of charge-coupled subsitution in silicate glasses: the system M1/nn+AlO$_2$–SiO$_2$ (M=Li Na, K Rb, Cs Mg, Ca Sr, Ba, Pb) J Am Ceram Soc 67:606–610

Russell JK, Giordano D (2005) A model for silicate melt viscosity in the system CaMgSi$_2$O$_6$–CaAl$_2$Si$_2$O$_8$–NaAlSi$_3$O$_8$, Geochim Cosmochim Acta 69:5333–5349

Russell J K, Giordano D (2017) Modelling configurational entropy of silicate melts. Chem Geol 461:140–151

Ryan M, Blevins S (1987) Viscosity of silicate melts and glasses. USGS 1764

Sampaio DV, Picinin A, Moulton BJ, Rino JP, Pizani PS, Zanotto ED (2018) Raman scattering and molecular dynamics investigation of lead metasilicate glass and supercooled liquid structures. J Non-Cryst Solids 499:300–308

Sato RK, McMillan PF, Dennison P, Dupree R (1991) High-resolution ^{27}Al and ^{29}Si MAS NMR investigation of SiO$_2$–Al$_2$O$_3$ glasses. J Phys Chem 95:4483–4489

Schreiber HD (1986) Redox processes in glass-forming melts. J Non-Cryst Solids 84:129–141

Schmelzer J, Tropin T (2018) Glass transition, crystallization of glass-forming melts, and entropy. Entropy 20:103

Schmelzer JWP, Abyzov AS, Fokin WM (2016) Thermodynamic aspects of pressure-induced crystallization: Kauzmann pressure. Int J Appl Glass Sce 7:474–486

Sakamaki T, Ohtani E (2022) High pressure melts. Rev Mineral Geochem 87:557–574

Sasabe H, De Bolt MA, Macedo PB, Moynihan CT (1977) Structural relaxation in an alkali-lime-silicate glass. Proc XIth International Congress on Glass 1:339

Sharma SK, Virgo D, Mysen BO (1978) Structural of glasses, melts of Na$_2$O$_x$SiO$_2$ (x = 1,2,3) composition from Raman spectroscopy. Carnegie Inst Wash Yearb 77:649

Sharma SK, Philpotts JA, Matson DW (1985) Ring distributions in alkali- and alkaline-earth aluminosilicate framework glasses—A Raman spectroscopic study. J Non-Cryst Solids 71:403–410

Shaw HR (1972) Viscosities of magmatic silicate liquids: an empirical method of prediction. Am J Sci 272 870–893

Shi C, Alderman O, Berman D, Du J, Neuefeind J, Tamalonis A, Weber R, You J, Benmore C (2019) The structure of amorphous and deeply supercooled liquid alumina. Front Mater 6:38

Schneider VR Mastelaro ED Zanotto BA Shakhmatkin NM Vedishcheva AC Wright H Panepucci (2003) Q^n distribution in stoichiometric silicate glasses: thermodynamic calculations and ^{29}Si high resolution NMR measurements. J Non-cryst Solids 325:164–178

Schuller S (2017) Phase separation processes in glass *In:* From Glass to Crystal Neuville et al. 2017 EDP Sciences, p 125–154

Shevyakov AM, Trofinenko AV, Sizonznko AP, Burkov VP, Zhuravlev GI (1978) Study of the structure and crystallisation of melts of the system Li$_2$O–SiO$_2$ by high temperature infrared spectroscopy. Zh Priklad Khim 51:2612

Schairer JF, Bowen NL (1956) The system Na$_2$O–Al$_2$O$_3$–SiO$_2$. Am J Sci 254:129–195

Scherer GW (1984) Use of the Adam–Gibbs equation in the analysis of structural relaxation. J Am Ceram Soc 67:504–511

Seifert F A, Mysen B O, Virgo D (1982). Three-dimensional network structure in the systems SiO$_2$–NaAlO$_2$, SiO$_2$–CaAl$_2$O$_2$ and SiO$_2$–MgAl$_2$O$_2$. Am Mineral 67:696–711

Sen S, Youngman RE (2003) NMR study of Q-speciation and connectivity in K$_2$O–SiO$_2$ glasses with high silica content. J Non-Cryst Solids 331:100–107

Sen S, George AM, Stebbins JF (1996) Ionic conduction and mixed cation effect in silicate glasses and liquids: ^{23}Na and ^{7}Li NMR spin-lattice relaxation and a multiple-barrier model of percolation. J Non-Cryst Solids 197:53–59

Sen S, Maekawa H, Papatheodorou G (2009) Short-range structure of invert glasses along the pseudo-binary join MgSiO$_3$–Mg$_2$SiO$_4$:Results from ^{29}Si and ^{25}Mg MAS NMR Spectroscopy. J Phys Chem B 2009:113 46,

Shan Z, Liu S, Tao H, Yue Y (2018) Mixed alkaline-earth effects on several mechanical and thermophysical properties of aluminate glasses and melts. J Am Ceram Soc 102:1128–1136

Shimoda K, Tobu Y, Shimoikeda Y, Nemoto T, Saito K (2007a) Multiple Ca^{2+} environments in silicate glasses by high-resolution ^{43}Ca MQMAS NMR technique at high, ultra-high (21.8 T) magnetic fields. J Magn Reson 186:114–117

Shimoda K, Tobu Y, Hatakeyma M, Nemoto T, Saito K (2007b) Structural investigation of Mg local environments in silicate glasses by ultra-high field ^{25}Mg 3QMAS NMR spectroscopy. Am Mineral 92:695–698

Schirmacher W, Ruocco G (2020) Heterogeneous elasticity: The tale of the boson peak. arXiv:2009.05970

Schirmacher W, Ruocco G Mazzone V (2015) Heterogeneous viscoelasticity: a combined theory of dynamic and elastic heterogeneity. Phys Rev Lett 115:015901

Simmons JH, Mills SA, Napolitano A (1974) Viscous flow in glass during phase separation. J Am Ceram Soc 57:109

Simmons JH, Napolitano A, Macedo PB (1970) Supercritical viscosity anomaly in oxide mixtures. J Chem Phys 53:170

Sipp A, Neuville DR, Richet P (1997) Viscosity and configurational entropy of borosilicate melts. J Non-Crystal Solids 211:281–293

Sossman DB (1933) The "physical chemistry" of a system of refractory components. J Am Ceram Soc 16:54–57

Sreenivasan H Kinnunen P, Adesanya E, Patanen M, Kantola A (2020) Field strength of network-modifying cation dictates the structure of (Na-Mg) aluminosilicate glasses. Front Mater 7:267

Starodub K, Wu G, Yazhenskikh E, Müller M, Khvan A, Kondratiev A (2019) An Avramov-based viscosity model for the SiO_2–Al_2O_3–Na_2O–K_2O system in a wide temperature range. Ceram Int 45:12169–12181

Stebbins JF (1991) NMR evidence for five-coordinated silicon in silicate glass at atmospheric pressure. Nature 351:638–639

Stebbins JF (2008) Temperature effects on the network structure of oxide melts and their consequences for configurational heat capacity. Chem Geol 256:80–91

Stebbins JF, Farnan I (1992) Effects of high temperature on silicate liquid structure: a multinuclear NMR study. Science 255:586–589

Stebbins JF, Xu Z (1997) NMR evidence for excess non-bridging oxygen in an aluminosilicate glass. Nature 390:60–62

Stebbins JF, Farnan I, Xue X (1992) The structure, dynamics of alkali silicate liquids: a view from NMR spectroscopy, Chem Geol 96:371–385

Stebbins JF, Carmichael ISE, Moret LK (1984) Heat capacities and entropies of silicate liquids and glasses. Contrib Mineral Petrol 86:131–148

Stebbins JF, Lee SK, Oglesby JV (1999) Al–O–Al oxygen sites in crystalline aluminates and aluminosilicate glasses; high-resolution oxygen-17 NMR results. Am Mineral 84:983–986

Stebbins JF, Kroeker S, Lee SK, Kiczenski TJ (2000) Quantification of five- and six-coordinated aluminum ions in aluminosilicate and fluoride-containing glasses by high-field, high-resolution ^{27}Al NMR J Non-Cryst Solids 275:1–6

Stebbins JF, Dubinsky EV, Kanehashi K, Kelsey KE (2008) Temperature effects on non-bridging oxygen and aluminum coordination number in calcium aluminosilicate glasses and melts. Geochim Cosmochim Acta 72:910–925

Strukelj E (2008) Propriété des verres d'aluminosilicate de lithium. M2, Université Paris 6

Sugawara T, Seto M, Kato M, Yoshida S, Matsuoka J, Miura Y (2013) Na_2O activity, thermodynamic mixing properties of SiO_2–Na_2O–CaO melt. J Non-Cryst Solids 371–372 (2013) 58–65

Suzuki A, Ohtani E, Funakoshi K, Terasuki H, Kubo T (2002) Viscosity of albite melt at high-pressure and high-temperature. Phys Chem Mineral 29:159–165

Takahashi S, Neuville DR, Takebe H (2015) Thermal properties, density and structure of percalcic and peraluminus CaO–Al_2O_3–SiO_2 glasses. J Non-Cryst Solids 411:5–12

Tamman G (1925) States of Aggregation, translated by Mehl RF from the second German edition. D. Van Nostrand Co., New York

Tammann G, Hesse W (1926) Die Abhängigkeit der Viskosität von der Temperatur bei unterkühlten Flüssigkeiten. Z Anorg Allg Chem 156:245

Tangemann, J, Lange RA (1998) The effect of Al^{3+}, Fe^{3+}, and Ti^{4+} on the configurational heat capacities of sodium silicate liquids. Phys Chem Minerals 26:83–99

Taylor TD, Rindone GE (1970) Properties of soda aluminosilicate glasses: V. low-temperature viscosities. J Am Ceram Soc 53:692–695

Téqui C, Robie RA, Hemingway BS, Neuville DR, Richet P (1991) Melting, thermodynamic properties of pyrope $(Mg_3Al_2Si_3O_{12})$. Geochim Cosmochim Acta 55:1005–1011

Thompson LM, Stebbins JF (2011) Non-bridging oxygen and high-coordinated aluminum in metaluminous and peraluminous calcium and potassium aluminosilicate glasses: High-resolution ^{17}O and ^{27}Al MAS NMR results. Am Mineral 96:841–853

Thompson LM, Stebbins JF (2012) Non-stoichiometric non-bridging oxygens and five-coordinated aluminum in alkaline-earth aluminosilicate glasses: Effect of modifier cation size. J Non-Cryst Solids 358:1783–1789

Thompson LM, Stebbins JF (2013) Interaction between composition and temperature effects on non-bridging oxygen and high-coordinated aluminum in calcium aluminosilicate glasses. Am Mineral 98:1980–1987

Tomozawa M (1978) Compositional changes as evidence for spinodal decomposition in glass. J Am Ceram Soc 61:444–447

Tomozawa M (1999) A source of the immiscibility controversy of borate and borosilicate glass systems. J Am Ceram Soc 82:206–209

Toplis M, Dingwell DB (2004) Shear viscosities of CaO–Al_2O_3–SiO_2 and MgO–Al_2O_3–SiO_2 liquids: Implications for the structural role of aluminium and the degree of polymerisation of synthetic and natural aluminosilicate melts. Geochim Cosmochim Acta 68:5169–5188

Tool AQ, Eichlin CG (1931) Variations caused in the heating curves of glass by heat treatment. J Am Ceram Soc 14:276

Trcera N, Cabaret D, Rossano S, Farges F, Flank A-M, Lagarde P (2009) Experimental and theoretical study of the structural environment of magnesium in minerals and silicate glasses using X-ray absorption near-edge structure. Phys Chem Minerals 36:241–257

Urbain G (1972) Etude expérimentale de la viscosité de silicoalumineux liquides et essai d'interprétation structurale. Thèse de Doctorat d'Etat es Sciences Physiques, Université Paris VI

Urbain G (1983) Viscosités de liquide du système CaO–Al_2O_3. Revue Internationale des Hautes Températures et Réfractaires 20:135–139

Urbain G, Bottinga Y, Richet P (1982) Viscosity of liquid silica, silicates, alumino-silicates, Geochim Cosmochim Acta 46:1061–1072

Villeneuve N, Neuville DR Boivin P, Bachelery P, Richet P (2008) Magma crystallization, viscosity: A study of molten basalts from the Piton de la Fournaise volcano (La Réunion island). Chem Geol 256:242–251

Vogel H (1921) Das Temperaturabhängigkeitsgesetz der Viskosität von Flüssigkeiten. Phys Z 22:645

Voigt U Lammert H, Eckert H, Heuer A (2005) Cation clustering in lithium silicate glasses: Quantitative description by solid-state NMR and molecular dynamics simulations. Phys Rev B 72:064207

Wallenberger FT, Hicks RJ, Bierhals AT (2004) Design of environmentally friendly fiberglass compositions: ternary eutectic SiO_2–Al_2O_3–CaO compositions, structures and properties. J Non-Cryst Solids 349:377–387

Wang Y, Sakamaki T, Skinner LB, Jing Z, Yu T, Kono Y, Park C, Shen G, Rivers ML, Sutton SR (2014) Atomistic insight into viscosity, density of silicate melts under pressure. Nat Commun 5:3241

Wang Y, Wei S, Cicconi MR, Tsuji Y, Shimizu M, Shimotsuma Y, Miura K, Peng GD, Neuville DR, Poumellec B, Lancry M (2020) Femtosecond laser direct writing in SiO_2–Al_2O_3 binary glasses and thermal stability of Type II permanent modifications. J Am Ceram Soc 103:4286–4294

Whittaker M (1970) Ionic radii for use in geochemistry. Geochim Cosmochim Acta 23:945–956

Winter A (1943) Transformation region of glass. J Am Ceram Soc 26:189–200

Woelffel W, Claireaux C, Toplis MJ, Burov E, Barthel, É., Shukla A, Biscaras J, Chopinet M-H, Gouillart E (2015) Analysis of soda-lime glasses using non-negative matrix factor deconvolution of Raman spectra. J Non-Cryst Solids 428:121–131

Worrell CA, Henshall T (1978) Vibrational spectroscopic studies of some lead silicate glasses. J Non-Cryst Solids 29:283–299

Wong J, Angell AC (1976) Glass Structure. Marcel Dekker. New-York

Wright A (1990) Diffraction studies of glass structure. J Non-Cryst Solids 123 (1990) 129–148

Wu J, Stebbins JF (2009) Effects of cation field strength on the structure of aluminoborosilicate glasses: high-resolution ^{11}B, ^{27}Al and ^{23}Na MAS NMR. J Non-Cryst Solids 355:556–562

Zakaznova-Herzog VP, Malfait WJ, Herzog F, Halter WE (2007) Quantitative Raman spectroscopy: Principles and application to potassium silicate glasses. J Non-Cryst Solids 353:4015–4028

Zanotto ED, Mauro JC (2017) The glassy state of matter: Its definition and ultimate fate. J Non-Cryst Solids 471:490–495

Zachariasen WH (1932) The atomic arrangement in glass. J Am Chem Soc 54:103841–103851

Zhang P, Dunlap C, Florian P, Grandinetti PJ, Farnan I, Stebbins JF (1996) Silicon site distributions in an alkali silicate glass derived by two-dimensional 29Si nuclear magnetic resonance. J Non-Cryst Solids 204:294–300

Zhang P, Grandinetti P, Stebbins JF (1997) anionic species determination in $CaSiO_3$ glass using two-dimensional ^{29}Si NMR J Phys Chem B 101:4004–4008

Reviews in Mineralogy & Geochemistry
Vol. 87 pp. 163-191, 2022
Copyright © Mineralogical Society of America

4

Topology and Rigidity of Silicate Melts and Glasses

Matthieu Micoulaut

Sorbonne Université
Laboratoire de Physique Théorique de la Matière Condensée
Boite 121, 4 place Jussieu
75252 Paris cedex 05
France

mmi@lptmc.jussieu.fr

Mathieu Bauchy

Physics of Amorphous and Inorganic Solids Laboratory (PARISlab)
Department of Civil and Environmental Engineering,
University of California, Los Angeles
420 Westwood Plaza, 5731
Los Angeles, CA 90095
USA

bauchy@ucla.edu

INTRODUCTION

From a fundamental viewpoint, glasses are out of equilibrium systems, so that concepts and techniques from statistical mechanics cannot be applied in a straightforward fashion given the presence of broken ergodicity, which results from an incomplete sampling of the phase space on reasonable time scales. The breakdown of the relationship between spontaneous fluctuations and dissipation leads to time dependent thermodynamic quantities that induce ageing phenomena typical of glasses, i.e., their properties can vary quite substantially with the thermal history of the melt and the waiting time before the experiment/measurement is performed. Given these intrinsic features of the glassy state, atomistic simulations have proven to be extremely useful for the understanding of structure and dynamics or relaxation (Kob and Binder 2005). However, while the available computing power has increased exponentially over the past decades, large enough computing power for direct molecular dynamics (MD) simulations of glass with large system sizes and realistic laboratory time scales is still out of reach. The method appears not necessarily helpful in the context of property optimization or new functionalities ruled by changes in composition, which would require a huge amount of time spent in exploring the compositional phase diagram.

Among promising alternatives, albeit carrying its own limitations, approaches using topology and rigidity focus on the key microscopic physics governing the thermal, mechanical and rheological properties of glass and amorphous materials, while filtering out unnecessary numerical details which ultimately do not affect the overall results. The concept of rigidity in disordered solids such as glasses, amorphous networks or sphere packing actually traces back to the early work of Maxwell (1864) on the stability of trusses and macroscopic structures such as bridges, and to the introduction of mechanical constraints by Lagrange (1788). Inspired by this pioneering work Phillips (1979) proposed at the beginning of the 80's to apply the analysis of trusses to disordered molecular networks such as glasses or amorphous solids which exhibit a lack of periodicity at long range but have some structural order at short range.

1529-6466/22/0087-0004$05.00 (print)
1943-2666/22/0087-0004$05.00 (online)

http://dx.doi.org/10.2138/rmg.2022.87.04

Corresponding materials were then viewed in terms of their topological and rigidity properties, involving in most cases the way atoms connect together, and, more importantly, the way they interact together. The transposition appears to be somewhat obvious: nodes are being replaced by atoms and the bar tensions are replaced by the most relevant interactions at the molecular level which are 2-body radial and 3-body angular interactions resulting in stretching and bending motions. Using the tools provided by structural rigidity which predicts the flexibility of ensembles formed by rigid bodies connected by flexible linkages or hinges, mechanical properties of glassy networks were calculated from the enumeration of constraints at zero temperature. It was, furthermore, highlighted that the notion of mechanical isostaticity (i.e., optimally constrained) is crucial for the promotion of glass-forming tendency in covalent systems. While the approach has been limited to chalcogenide glasses, recent applications of this approach have enabled the prediction of mechanical and thermal properties from a combination of topological methods using empirical models and molecular simulations.

In the present chapter, we review these recent contributions. We first introduce the basic notions of rigidity theory, that is, how an atomic network constrained by interactions can be treated from a mechanical viewpoint as a collection of bars and nodes on which an appropriate constraint enumeration allows identification of a stability point, known as the isostatic condition. Then, we review the application of such ideas to silicate glasses, prior to an extension of the models incorporating a temperature dependence. By combining molecular dynamics simulations and rigidity, one is also able to investigate multicomponent systems under various excitations (temperature, pressure, deformation,...) and this allows one to reduce the complexity of the involved phenomena to a picture that reveals the crucial role played by network topology while also providing clues for property optimization ruled by changes in composition.

ROLE OF NETWORK RIGIDITY

There is an attractive way to analyze and predict various properties in the liquid state (viscosity, diffusivity,...), glass transition-related properties (relaxation, fragility,...) as well as typical glass e.g., mechanical properties using rigidity theory. Using the underlying network structure, this framework provides, indeed, an atomic scale approach to understanding the physico-chemical behavior of network glasses.

Rigidity theory of network glasses

Basic concepts and ideas of mechanical constraints have been introduced in the pioneering contributions of Lagrange and Maxwell (Lagrange 1788; Maxwell 1864). The approach has been extended to disordered atomic networks (Phillips 1979, 1981) and it was recognized that the glass-forming tendency of covalent alloys is optimized for particular compositions in chalcogenide network glasses. Specifically, glasses with some optimal connectivity or mean coordination number $\bar{r}$ satisfying exactly the Maxwell stability criterion of mechanically isostatic structures could be formed with a minimal cooling rate (Azoulay et al. 1975). In the Maxwell approach, this condition obtains when $n_c = n_d$, where n_c is the count of atomic constraint density arising from interactions and n_d the network dimensionality (usually 3).

In glasses and liquids, the dominant interactions are usually near-neighbor bond-stretching (BS) and next-near-neighbor bond-bending (BB) forces and the number of constraints n_c per atom can be exactly computed using a mean-field approach. One obtains:

$$n_c = \frac{\sum_{r \geq 2} n_r \left(\dfrac{r}{2} + 2r - 3 \right)}{\sum_{r \geq 2} n_r} \tag{1}$$

where n_r is the concentration of species being r-fold coordinated. The contribution of the two terms in the numerator arises from BS and BB contributions. Each bond being shared by two neighbors, the density is $r/2$ for bond-stretching (BS) constraints for an r-fold atom. For BB constraints, it has to be remarked that a two-fold (e.g., oxygen) atom involves only one angle, and each additional bond needs the definition of two more angles, leading to the estimate of $(2r - 3)$.

By defining the network mean coordination number $\bar{r}$ of the network $\bar{r} = \sum_{r\geq2}rn_r$ (Eqn. 1) reduce to:

$$n_c = \frac{\bar{r}}{2} + (2\bar{r} - 3) \qquad (2)$$

The application of the Maxwell stability criterion now indicates (i.e., solving Eqn. 2) that isostatic glasses and liquids ($n_c = 3$) are expected to be found at the mean coordination number of $\bar{r} = 2.40$ in 3D (Phillips 1979), corresponding usually to a nonstoichiometric composition where the glass-forming tendency has been found to be optimized experimentally (Bresser et al. 1986).

The physical content and more rigorous picture of this *"heuristic"* stability criterion has been established from a vibrational analysis of bond-depleted random networks (Thorpe 1983) constrained by harmonic BS and BB interactions. By performing an eigenmode analysis of the dynamical matrix, it has been demonstrated that the number f of eigenmodes having a zero frequency vanishes when rigidity percolates in the network which is achieved for $\bar{r} = 2.38$. It has, thus been acknowledged that the Maxwell isostatic condition $n_c = n_d$ defines a mechanical stiffness transition that is driven by the network connectivity $\bar{r}$, which acts as an external parameter like the temperature in a ferromagnetic transition, the order parameter being the density of floppy modes (similar to magnetization in the same analogue of the ferromagnetic transition). For $\bar{r} > \bar{r}_c$, redundant constraints produce internally stressed networks and correspond to a stressed-rigid phase (He and Thorpe 1985; Thorpe 1985). For $n_c < n_d$ however, floppy modes can proliferate, and these lead to a flexible phase where local deformations with a low cost in energy are possible (Kamitakahara et al. 1991), the density being given by: $f = 3 - n_c$ (Thorpe 1983). Different experimental probes of this peculiar transition have been reported in chalcogenides from Raman scattering (Boolchand et al. 1995), stress relaxation (Böhmer and Angell 1992), viscosity measurements (Tatsumisago et al. 1990), vibrational density of states (Kamitakahara et al. 1991), Brillouin scattering (Duquesne and Bellassa 1985), resistivity (Asokan et al. 1989). For a full account of early experimental probes and verification of rigidity percolation as well as the mathematical description using graph theory, one should refer to books devoted to the subject (Thorpe and Duxbury 1999; Phillips and Thorpe 2001; Micoulaut and Popescu 2009).

The situation in silicate glasses

While the application of constraint enumeration has been first performed on chalcogenide glasses, the question of its applicability to silicate glasses has emerged and discussed in the 1990's. It has been realized that for one-fold terminal atoms (as e.g., expected for alkali ions in the neighbourhood of a NBO), a special count has to be achieved as no BB constraints are involved (Boolchand and Thorpe 1994; Zhang and Boolchand 1994). This count leads, indeed, to the correct prediction of the threshold composition in sodium silicates, albeit counter-intuitive given the expected larger coordination number (five) of sodium that might induce some additional stiffening (Zwanziger et al. 1995). However, it has become obvious that for non-directional bondings as those expected for Na–O ionic interactions, some constraints might be ineffective (see below; Micoulaut 2008). This is also revealed by the analysis from molecular dynamics simulations which suggest that alkali–oxygen distances are not all equivalent, the Na–NBO distances being slightly shorter than the Na–BO distances (Cormier et al. 2003). Experimentally, Debye–Waller factors measured in X-ray absorption studies on Ba or Ca binary silicates are also slightly different, suggesting that the O neighbors of Ca are, again, not all equivalent (Taniguchi et al. 1997).

In the initial and mean-field approach, the coordination numbers r of Si, O, and the alkali cation are respectively set to four, two, and one, and depending on the size of the cation a certain number of BB (angular) constraints can be broken, i.e. their angular excursion (as determined from bond angle distributions) can be modified and eventually increased because of size mismatch.

The simplest case is the lithium silicate system $(1-x)SiO_2-xLi_2O$ for which all constraints are considered intact due to the small size of the Li cation (Fig. 1), the ionic radius of Li being of the same order as the one of oxygen so that there is no size mismatch.

(a)   **(b)**

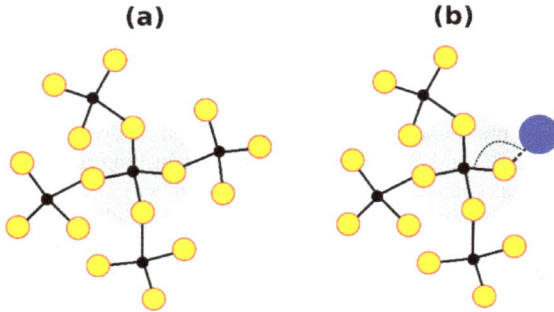

Figure 1. The elemental building blocks of silicate networks: **(a)** $SiO_{4/2}$ tetrahedron (Q^4 species) connected to the rest of the network by four BOs, and **(b)** an $NaSiO_{5/2}$ tetrahedron (Q^3 species) having a sodium atom in the vicinity of an oxygen that is a so-called NBO. The angular excursion around the NBO can be small (in Li silicates) or large (in Na silicates), the latter situation leading to a broken bond-bending constraint for the NBO (see Table 1).

Table 1 provides a survey of the enumeration of constraints of the various alkali and alkaline-earth silicate systems (Micoulaut 2008). Using Table 1 and Equation (1), it is easy to check that for the Li case, the number of floppy modes $f = 3 - n_c$ is equal to:

$$f_{Li} = 3 - \frac{1}{3}\left[11 - 8x\right] \tag{3}$$

which vanishes at the the composition $x = x_c = 0.250$. In a mean-field picture of rigidity, Li silicate glasses in the silica rich domain can therefore be considered as stressed rigid ($n_c > 3$), whereas the increased depolymerization leads to a flexible phase ($n_c < 3$) at higher concentrations.

With increasing metal–O bond distance (from 1.97 Å for Li to 2.36 Å for Na), the cation field strength should be weaker (Mysen and Richet 2019). According to Pauling, the bond strength (i.e., the mechanical constraint) is a systematic function of the bond distance so that an increase of the metal–oxygen bond would induce a reduction of the constraint. This suggests that an increase in cation size will induce additional local degrees of freedom or broken constraints arising from weaker interactions, necessarily a NBO bond-bending, given the reduced BB energy (Kamitakahara et al. 1991). For the Na case, a broken NBO BB constraint shifts the rigidity transition to $x = x_c = 0.20$, the global tendency being that flexibility onsets at lower compositions as an increased number of constraints becomes ineffective. In magnesium and calcium silicates, with the alkaline earth cation having $r = 2$, the stability criterion is found for $x_c = 0.50$ which corresponds to the Wollastonite or Enstatite composition around which stable glasses can be formed (Micoulaut et al. 2005; Bourgel et al. 2009).

Table 1. BS and BB constraint counting in alkali silicate $(1-x)SiO_2-xM_2O$, and alkaline earth silicate glasses $(1-x)SiO_2-xMO$, total number of constraints n_c, and the location of the isostatic composition x_c.

	Si		BO		NBO		M		n_c	x_c
	Si^{BS}	Si^{BB}	BO^{BS}	BO^{BB}	NBO^{BS}	NBO^{BB}	M^{BS}	M^{BB}		
Li	$2(1-x)$	$5(1-x)$	$2-3x$	$2-3x$	$2x$	$2x$	x	–	$11-8x$	0.250
Na	$2(1-x)$	$5(1-x)$	$2-3x$	$2-3x$	$2x$	–	x	–	$11-10x$	0.200
K	$2(1-x)$	$5(1-x)$	$2-3x$	$2-5x$	$2x$	–	x	–	$11-12x$	0.167
Rb	$2(1-x)$	$5(1-x)$	$2-3x$	$2-7x$	$2x$	–	x	–	$11-14x$	0.143
Cs	$2(1-x)$	$5(1-x)$	$2-3x$	$2-9x$	$2x$	–	x	–	$11-16x$	0.125
Mg, Ca	$2(1-x)$	$5(1-x)$	$2-3x$	$2-3x$	$2x$	$2x$	x	x	$11-7x$	0.500
Sr	$2(1-x)$	$5(1-x)$	$2-3x$	$2-3x$	$2x$	$2x$	x	–	$11-8x$	0.400
Ba	$2(1-x)$	$5(1-x)$	$2-3x$	$2-3x$	$2x$	–	x	–	$11-10x$	0.285

Experimental verification. Compositional trends can be quite useful in detecting the signature of rigidity percolation while providing also a simple picture for the understanding of threshold behaviors in physical and chemical properties of silicates. It has been suggested that materials which are optimally constrained (i.e., $n_c = 3$) should form glasses more easily (Phillips 1979) as observed experimentally for the Ge–Se system (Azoulay et al. 1975) or, more recently from the difference temperature $\Delta T = T_x - T_g$ between the glass (T_g) and the crystallization (T_x) (Piarristeguy et al. 2015). The critical cooling rate to avoid crystallization or the vitrification enthalpy as a function of composition appears to be a reasonable indicator for optimal glass formation. Such a critical cooling rate has been measured as a function of composition for Na and K silicates (Fang et al. 1983), and a minimum is obtained at a composition that is located in the interval where optimal glass formation is expected from Maxwell counting (respectively $x = 20$ and 16.7%). It parallels recent findings on the NS4 glass ($x = 20\%$) which has been found to display minimal ageing phenomena (Ruta et al. 2014), in agreement with the fact that enthalpic relaxation has been optimized in the absence of stress. In this respect, the enthalpy of vitrification measured from calorimetry (Tatsumisago et al. 1990) should also display characteristic changes when the system changes from a flexible to a stressed rigid phase. A study of vitrification in $MgO–SiO_2$ melts between the enstatite (50%) and forsterite (66% MgO) compositions has shown that the heat of vitrification increases (Tangeman et al. 2001), an increase that is accompanied by a large change in connectivity (Wilding et al. 2004), the optimal glass composition in Mg silicates being estimated to be at $x = x_c = 50\%$ (Table 1). This composition can be produced by conventional methods, whereas extensions of the glass-forming region toward glasses with high melting temperatures (e.g., forsterite) needs high-temperature containerless synthesis (Kohara et al. 2004).

An indirect indication of glass-forming tendency is also found from the inspection of viscosity with composition at a fixed temperature, as the glass-forming tendency is known to improve for systems that are able to increase their melt viscosity down to lower temperatures (Richet et al. 2006). For such materials able to explore freezing-point depressions (i.e., eutectics), glass forms, indeed, more easily as the liquid is able to reach lower temperatures and higher viscosities which act as a barrier for possible recrystallization. When the viscosity of different (Li, Na, K) silicates are investigated along their liquidus branches, compositions having a minimum in the critical cooling rate correlate very well with the viscosity maximum from the freezing depressions (Richet et al. 2006; Micoulaut 2008).

Raman scattering provides an interesting probe of local and intermediate-range structural changes across the rigidity transition, and in selected cases some modes have been sensitive to the nature of the elastic phases with optical elasticities (i.e., the Raman mode frequency for stretching vibrations) displaying typical features of rigidity, i.e., a power-law variation typical of the elasticity of stressed rigid amorphous networks (Feng et al. 1997). In contrast with chalcogenide network glasses, silicate glasses display a much larger variety of possible Raman modes, most of them being clearly assigned (Zotov 2001). In the context of flexible to rigid transitions, a Raman study of the Calcium and Barium silicate systems has shown that certain modes (e.g., the stressed Q^4 rigid $SiO_{4/2}$ corresponding to a stretching mode) exhibit a marked change in behavior in the line frequency and linewidth at concentrations which are close to the thresholds predicted from the mean-field estimate of Table 1 (Micoulaut et al. 2005; Bourgel et al. 2009).

A strong piece of evidence for flexible to rigid thresholds can be detected from Brillouin scattering (Vaills et al. 2001, 2005; Chaimbault 2004), the measured elastic energy change upon annealing $\Delta\Phi(x)$ between as-quenched (virgin) and annealed samples being almost zero in the stressed rigid phase (e.g., $x < 20\%$ in sodium silicates, Fig. 2) and increases linearly for a modifier content that is larger than the threshold x_c predicted by Maxwell constraint counting (Table 1). The threshold behavior is an indication that changes in the elastic energy occur only in the flexible phase where local deformation modes induce a measurable value for the associated relaxation energy. The shift of the threshold with cation size (from Na to K, see Fig. 2) is also compatible with the possibility that some constraints become ineffective (Table 1). Using Brillouin experiments, it has been found that the spectra between as-quenched (virgin) and annealed samples can be very different and depends substantially on the composition, while also permitting to measure and elastic energy ΔF under annealing which has the property $\Delta F = 0$ for $x < x_c$ and $\Delta F > 0$ for $x > x_c$. It has been found that such an elastic energy is a measure of the floppy mode density of the network, i.e., $\Delta F \propto Ef$ with E the floppy mode energy. A fit using Equation (3) (or Table 1) in the identified flexible phase has led to a floppy mode energy of $E = 0.11$ meV and 0.12 meV for potassium and sodium, respectively (Micoulaut 2008). Using, the same technique, longitudinal $C_{11}(x)$ and shear $C_{44}(x)$ elastic constants exhibit a power-law variation for the Na silicate, with an exponent p of $1.68-1.69$ (Vaills et al. 2005), in fair agreement with the theoretical prediction of Thorpe (1983).

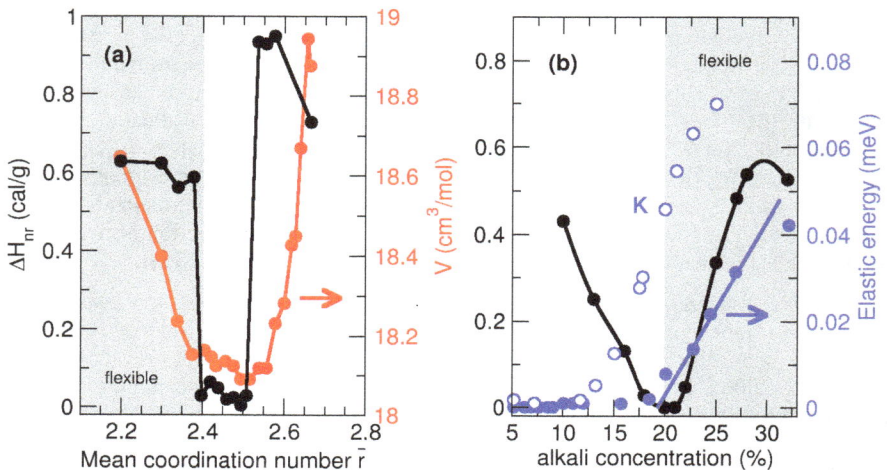

Figure 2. (a) Anomalies (i.e., thresholds or extrema) in physical properties in Ge–Se glasses rescaled as a function of the number of topological constraints (Bhosle et al. 2011): Non reversing heat flow ΔH_{nr} (left axis, black symbols) and molar volume (right axis, red symbols). **(b)** Non-reversing enthalpy ΔH_{nr} in sodium silicates $(1-x)SiO_2–xNa_2O$ (black filled symbols) and Elastic energy released during annealing (right axis) as a function of modifier content in Na (blue filled circles) and K (blue open circles) silicates and. Adapted from (Vaills et al. 2005; Micoulaut 2008).

Lastly, the signature of the loss of rigidity with increasing modifier content is detected in the behavior of the dc ionic conductivity σ, as two obvious conduction regimes are identified over two compositional intervals. A first regime sets in at low modifier concentration where the conductivity is negligibly small because of strong barriers against local deformation due to the stressed rigid nature of the network. Once the system has become flexible, a second regime displays a conductivity trend that is exponential or with a power-law behavior. From a series of investigations on different modified oxides (Novita et al. 2007; Bourgel et al. 2009; Micoulaut et al. 2009), it has become clear that the onset of ionic conductivity in the flexible phase is related to the breakdown of the stressed-rigid network for $x > x_c$.

The special case of silica. The case of vitreous silica desserves some attention. The Si–O–Si bond angle distribution has been extensively studied by various experimental techniques (Mozzi and Warren 1969; Galeener 1985; Pettifer et al. 1988; Poulsen et al. 1995) and molecular simulations (for details, see Yuan and Cormack 2003), the more recent measurements leading to a mean bond angle of about 148–152° with a full-width at half maximum of 17–38°. Because of this broad angular distribution, O can display wide excursions around its average position, with a restoring force that is weak (Smith et al. 1995). On this basis, it has been proposed that bond-bending constraints of O should be broken in vitreous silica (Zhang and Boolchand 1994), and a mean-field constraint count leads indeed to $n_c = 3$, the optimally constrained nature of the network being suggested to be responsible for the relative ease of glass formation of silica.

The addition of alkali modifiers induces a narrowing of the Si–O–Si bond angle distribution and a subsequent lengthening of the Si–O bond distance (Boisen et al. 1990), as demonstrated from a X-ray absorption study (Henderson 1995), and from an NMR study on silica and K silicates (Farnan et al. 1992). The increased presence of Na atoms leads to an increase of the Si–O distance but to a decrease of the Si–Si distance, implying a reduction of the Si–O–Si bond angle distribution that is characterized from a reduced full-width at half maximum for the $K_2Si_4O_9$ glass, the average bond angle remaining essentially the same. The narrowing of the bond angle distribution indicates that the broken O bond-bending constraints of silica are restored in alkali silicates and corresponding BB constraints must be taken into account (Table 1). This view is supported by the potential energy curve determined by an ab inito molecular orbital approach (Newton and Gibbs 1980), which shows that lower angles (i.e., lower than 120°) lead to an increased interaction energy.

Rigidity Hamiltonians and floppy modes

The rigidity approach builds essentially on fully connected networks, that is, at $T = 0$ when neither bonds nor constraints are broken by thermal activation. In fact, an inspection of Equation (1) shows that temperature is absent and this represents a serious drawback. The use of this initial theory is certainly valid as long as one is considering strong covalent bonds or when the viscosity η is very large once the system is trapped in the glassy state, but obviously some elements are missing in order to appropriately treat a high temperature liquid. However, for covalent liquids Nuclear Magnetic Resonance indicate that the low temperature constraint count can be readily extended to the liquid (Bermejo et al. 2008).

Naumis has established a rigidity related Hamiltonian that serves for subsequent modelling in multicomponent silicates (Naumis 2005, 2006), and it builds on the fact that cyclic variables in phase space are identified with floppy modes $f = 3 - n_c$ because when one of these variables is changed, the system will display a change in energy that is close to zero due to the small energetic contribution of such deformation modes (about 5 meV, Kamitakahara et al. 1991). Since it is assumed that floppy modes have a zero frequency, they will not contribute to the total energy of the system so that the sum over coordinates only runs up to $3N(1 - f)$ and the Hamiltonian reads:

$$H = \sum_{j=1}^{3N} \frac{P_j^2}{2m} + \sum_{j=1}^{3N(1-f)} \frac{1}{2} m\omega_j^2 Q_j^2 \qquad (4)$$

where Q_j (position) and P_j (momentum) are the j-th normal mode coordinates in phase space, and w_j is the corresponding eigenfrequency of each normal mode. The corresponding partition function derived from Equation (4) leads to the expression of the free energy and the specific heat which are found to depend on the fraction of floppy modes, that is, on the rigidity status of the network (Naumis 2006). The number of accessible states $\Omega(f)$ can also be calculated in the microcanonical ensemble, and using the Boltzmann relation $S_c(f) = k_B \ln \Omega(f)$, one finds that the configurational entropy S_c provided by an energy landscape altered by the presence of floppy modes is simply given by: $S_c = f N k_B \ln V$, i.e., the floppy mode density is contributing to the configurational entropy and the dynamics of the glass-forming system. There is not a unique relationship for $S_c(f)$ as an alternative class of interaction potentials (i.e., different from the harmonic one shown in Equation (4) has led to a slightly different behavior with S_c scaling as f^3 (Foffi and Sciortino 2006). However, it is essentially the result from Naumis that has been used in the modelling of transport and relaxation properties of silicates, as discussed next.

An elemental viscosity fit. Using those ideas on constraints, in combination with the Adam-Gibbs model for viscosity $\ln \eta = \ln \eta_\infty + A/TS_c$, Mauro and co-workers (Mauro et al. 2009a) have shown that the expression of the configurational entropy S_c as a function of the density of topological degrees of freedom f induces a new form for the viscosity dependence, the Mauro-Yue-Ellison-Gupta-Allan (MYEGA) form given by:

$$\ln \eta = \ln \eta_\infty + \frac{K}{T} \exp\left[\frac{C}{T}\right] \qquad (5)$$

which avoids a divergence at low temperature found in the Vogel–Fulcher–Tamman equation, and has been tested for a variety of systems. Here η_∞ is the viscosity at infinite (high) temperature and C and K are constants which can be related to the fragility and the glass transition temperature.

Temperature dependent constraints

Building on the Naumis relationship between floppy modes or topological degrees of freedom and the configurational entropy Sc (Naumis 2005), Gupta and Mauro have extended the initial $T=0$ rigidity theory and the constraint enumeration to account explicitly for thermal effects (Mauro and Gupta 2009), in an analytical model that uses a two-state thermodynamic step function $q(T)$. This function quantifies the number of rigid BS and BB constraints as a function of temperature, and has the property of a step function with two obvious limits as all relevant constraints can be either intact at low temperature ($q(T)=1$, as in the initial theory) or fully broken at high temperature ($q(\infty)=0$) because temperature is much larger than the constraint energy barriers. At a finite temperature, however, only a fraction of these constraints can become rigid once their associated energy is less than $k_B T$. This modifies Equation (1) which becomes:

$$n_c = \frac{\sum_{r\geq 2} n_r \left[\frac{r}{2} q_r^{BS}(T) + (2r-3) q_r^{BB}(T) \right]}{\sum_{r\geq 2} n_r} \qquad (6)$$

where $q_r^{BS}(T)$ and $q_r^{BS}(T)$ represent the step functions associated with BS and BB interactions, respectively. Different forms can be proposed for such functions, based either on an energy landscape approach (Mauro et al. 2009c) with an onset temperature T_0 or involving a simple exponential activation energy (Angell et al. 1999). A general behavior of $q(T)$ can be computed for any thermodynamic condition from MD simulations (Bauchy and Micoulaut 2013a, see below).

A certain number of thermal and relaxation properties of silicate glasses have been determined from a direct application of Equation (6) in conjunction with structure models.

In order to calculate the glass transition temperature, it is, again, assumed that the Adams–Gibbs model for viscosity $\ln \eta = \ln \eta_\infty + A/TS_c$ holds in the temperature range under consideration, and that the corresponding barrier height A is a slowly varying function with composition. Using the estimation of the configurational entropy from the floppy mode density (Naumis 2005), the general statement of $\eta(T_g) = 10^{12}$ Pa.s is written for two compositions x and x_R, following:

$$\frac{T_g(x)}{T_g(x_R)} = \frac{S_c\left(T_g(x_R), x_R\right)}{S_c\left(T_g(x), x\right)} = \frac{f\left(T_g(x_R), x_R\right)}{f\left(T_g(x), x\right)} = \frac{3 - n_c\left(T_g(x_R), x_R\right)}{3 - n_c\left(T_g(x), x\right)} \tag{7}$$

and $T_g(x)$ can now be determined as a function of composition from the glass transition temperature of a reference composition x_R.

Using the expression for $S_c(T_g(x), x)$ in Equation (7) and the definition of fragility:

$$M = \left[\frac{\partial \ln \eta}{\partial T / T_g}\right]_{T=T_g} \tag{8}$$

one can express the fragility index M as a function of composition:

$$M = M_0\left[1 + \frac{\partial \ln S_c(T, x)}{\partial \ln T}\right] \tag{9}$$

These predictions have been verified for simple chalcogenides (Mauro and Gupta 2009), borates (Mauro et al. 2009b), borosilicates (Smedskjaer et al. 2011) phosphates (Hermansen et al. 2014), or borophosphate glasses (Hermansen et al. 2015). Equation (9) leads to a good reproduction of fragility data with composition (Fig. 3), but requires a certain number of onset temperatures T_0 that act as parameters for the theory. Such fragility predictions permits one to infer which interactions (species dependent bending or stretching interactions) contribute most to the evolution of fragility with composition. For instance, in borosilicates, M is found to be mostly influenced by the bond angle interactions constraining Silicon and Boron bending motion, whereas the BS interactions contribute only at a large silicon to boron ratio.

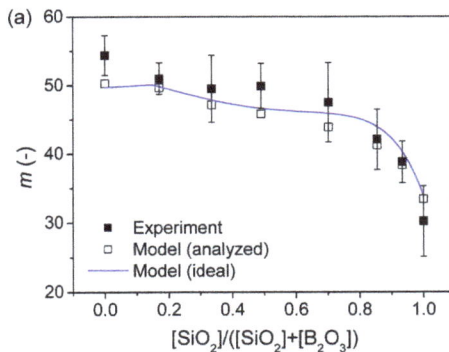

Figure 3. Modeling the composition dependence of the liquid fragility index (m) for borosilicate glasses (Smedskjaer et al. 2011). (a) Prediction of the fragility index (solid line) as a function of composition, compared to experimental data measured from DSC experiments. Reprinted with permission from Smedskjaer MM, Bauchy M, Mauro JC, Rzoska SJ, Bockowski M (2015) Unique effects of thermal and pressure histories on glass hardness: Structural and topological origin. J Chem Phys 143:164505. Copyright (2011) American Chemical Society.

Using the same formalism, the heat capacity change ΔC_p at the glass transition can be readily estimated and compared to measurements accessed from DSC. Using the temperature dependence $q(T)$ for the constraints, it is assumed that the dominant contribution to the heat capacity at the glass transition comes from configurational contributions $C_{p,conf}$ that can be related to S_c, so that one has: $\Delta C_p = C_{pl} - C_{pg} \approx C_{p,conf}$ with C_{pl} and C_{pg} the heat capacity of the liquid and the glass, respectively. The heat capacity change can be written as a function of the configurational enthalpy and S_c:

$$\Delta C_p = \left(\frac{\partial H_{conf}}{\partial T} \right)_p = \left(\frac{\partial H_{conf}}{\partial \ln S_c} \right)_p \left(\frac{\partial \ln S_c}{\partial T} \right)_p \tag{10}$$

and using the Adam–Gibbs expression and Equation (9), one can, furthermore, write the jump of the heat capacity at the glass transition with composition:

$$\Delta C_p \left(x_i, T_g \left(x_i \right) \right) = \frac{1}{T_g \left(x_i \right)} \left(\frac{\partial H_{conf}}{\partial \ln S_c} \right)_{p,T_g} \left(\frac{M \left(x_i \right)}{M} - 1 \right) \tag{11}$$

Equation (11) can be recast in a more compact form given that $S_c(T_g)$ is inversely proportional to T_g, and by assuming that $\partial H_{conf} / \partial S_c$ is, by definition, equal to a configurational temperature at constant pressure (Araujo and Mauro 2010) which is close to the glass transition temperature. This, ultimately, leads to a decomposition-dependent prediction of the heat capacity jump at T_g:

$$\Delta C_p \left(x_i, T_g \left(x_i \right) \right) \approx \frac{\left[A \left(x_i \right) \right]_R}{T_g \left(x_i \right)} \left(\frac{M \left(x_i \right)}{M} - 1 \right) \tag{12}$$

As in the previous examples, the heat capacity jump is evaluated with respect to some reference composition because the parameter $[A(x_i)]_R$, appearing in equation (12) connects the configurational entropy $S_c(T_g)$ with T_g for a reference composition x_R. Applications have been performed on the borosilicate glasses and successfully compared to experimental measurements (Smedskjaer et al. 2011). Again, such predictions have the merit to accurately reproduce experimental data, and to provide some insight into the validity of structural models that can be checked independently from a variety of other experimental (e.g., spectroscopic) probes.

MOLECULAR DYNAMICS AND RIGIDITY

A more general way to enumerate topological constraints for any thermodynamic condition, including under pressure, builds on MD simulations which also permits establishing relationships with thermodynamic and dynamic properties from the ensemble averaged statistics of such atomic scale simulations. In all approaches, classical or first principles (FPMD), Newton's equation of motion is solved for a system of N atoms or ions, representing a glass or a liquid material. Recent applications give a very accurate description of the structural and dynamic properties of most archetypal network-forming systems: SiO_2 (Zeidler et al. 2014), GeO_2 (Micoulaut et al. 2006; Salmon et al. 2012), B_2O_3 (Ferlat et al. 2008; Zeidler et al. 2014) in the glassy or liquid state, and in ambient or densified conditions. Similar agreement holds for the modified oxides such as alkali silicates (Cormack et al. 2002; Horbach and Kob 1999; Bauchy et al. 2013), sodalime silicates (Cormack and Du 2001; Laurent et al. 2014), borosilicates (Cormier et al. 2000; Pedesseau et al. 2015) or aluminosilicates (Winkler et al. 2003).

Topological constraints are extracted from atomic scale trajectories by recording the radial and angular motion of atoms. This connects directly to the enumeration of BS and BB constraining interactions but instead of treating the forces mathematically and querying motion – the standard procedure in MD simulations – one identifies the absence of a restoring force (i.e., a constraint) with an increased atomic radial or angular motion. This strategy is somewhat different from the '*culture of force*' discussed by Wilcek (2004) given that one does not necessarily need to formulate the physical origin of the forces to extract the constraints. As a result, large radial and angular excursions of doublets (bonds) or triplets (of atoms (angles)) will be identified with broken radial or angular constraints. In practice, once the atomic scale trajectories have been accumulated at different thermodynamic conditions from MD, a structural analysis is applied in relationship with the constraint counting (Fig. 4).

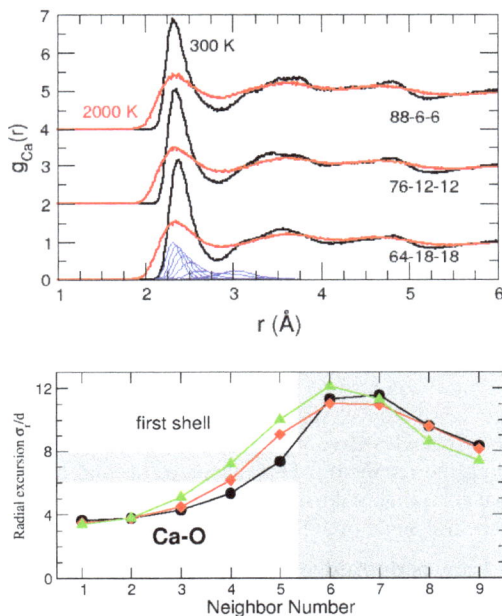

Figure 4. (a) Ca-centered partial pair correlation function for three compositions in soda-lime silicate glasses SiO_2–CaO–Na_2O at two different temperatures (300 K, black; 2,000 K, red). Blue curves correspond to calculated neighbor distributions functions (Laurent et al. 2014). **(b)** Corresponding Ca–O radial standard deviations σ_r/d as a function of neighbor number for three compositions (colored curves). Here d is the peak position of the relevant neighbor distributions. The shaded areas correspond either to the second shell of neighbors, which is not relevant for constraint counting, or to the approximate limit ($7 \pm 1\%$) between intact (low σ_r) or broken (large σ_r) bond-stretching constraints. Reprinted with permission from Laurent L, Mantisi B, Micoulaut M (2014) Structure and topology of soda-lime silicate glasses: implications for window glass. J Phys Chem B 118:12750–12762. Copyright (2014) American Chemical Society.

Bond-bending and bond-stretching

NDFs can be defined by fixing the neighbor number n (first, second, etc) during the bond lifetime, the sum of all NDFs yielding the usual i-centred pair correlation function $g_i(r)$. An integration up to the first minimum gives the coordination numbers r_i, and hence the corresponding number of bond-stretching constraints $r_i/2$ that can be estimated independently from the radial excursions of such bonds defining NDFs.

To determine BB constraints one uses partial bond angle distributions (PBADs) $P(\theta_{ij})$ defined as follows (Micoulaut et al. 2010; Bauchy et al. 2011): for each type of a central atom 0, the N first neighbors i are selected, leading to $N(N-1)/2$ possible angles $i0j$ ($i = 1...N-1, j = 2...N$), i.e., 102, 103, 203, etc. The standard deviation σ_{ι_φ} of each distribution $P(\theta_{ij})$ gives a quantitative estimate of the angular excursion around a bond angle and provides a measure of the bond-bending strength (Fig. 5). Large excursions correspond to a bond-bending weakness giving rise to an ineffective constraint, whereas small values for σ_{ι_φ} correspond to an intact bond-bending constraint which maintains a rigid angle at a fixed value.

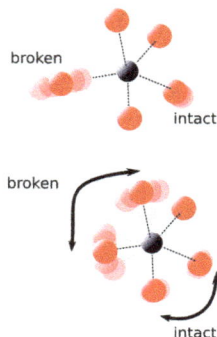

Figure 5. Schematic method of constraint counting from MD generated configurations. Large (small) radial (**top**) or angular (**bottom**) excursions around a mean value are characterized by large (small) standard deviations on radial or angular distributions, representing broken (intact) constraints. Reprinted with permission from Laurent L, Mantisi B, Micoulaut M (2014) Structure and topology of soda-lime silicate glasses: implications for window glass. J Phys Chem B 118:12750–12762. Copyright (2014) American Chemical Society.

Figure 6 shows the PBADs for the silicon atom in a soda-lime silicate glass (Laurent at al. 2014). Broad angular distributions are found in most of the situations, but a certain number of sharper distributions (colored) can also be found, and these are identified with intact angular constraints because these arise from a small motion around an average bond angle. In the displayed example, only six angles have nearly identical and sharp distributions, and these are the six angles defining the tetrahedra. The calculated second moments are shown in the inset and one realizes that six standard deviations are nearly identical, associated with bending motions around the tetrahedral angle of 109°.

Applications have been performed on a variety of systems including chalcogenides and modified oxides, or densified oxide glasses and liquids (see applications below). In the forthcoming, we will apply such methods on silicate network glasses and liquids which are probably the most well documented systems in the context of flexible to rigid transitions.

The enumeration of constraints on realistic models of silicates glasses (Laurent et al. 2014) and liquids (Bauchy and Micoulaut 2011, 2013a) shows that six Si standard deviations have a low value (10°), i.e., four times smaller than all the other angles (Fig. 6). One thus recovers the result found for the stoichiometric oxides (SiO_2, GeO_2), Bauchy et al. 2011).

Behavior in the liquid phase

With such MD simulations, the number of topological constraints can be investigated as a function of temperature and composition in a certain number of glass-forming systems. For all, intact constraints dominate at 300 K (leading to a low standard deviation) but will soften as temperature increases. A limit between a "broken" (or ineffective) and an "intact" constraint can be defined from reasonable limits suggested by for example, the inspection of calculated values in the glassy state (Fig. 6, inset). We note, indeed, that intact constraints arise from strong bonds or angles having a low standard deviation of about 0.02 Å or 5–10 degrees (Bauchy and Micoulaut 2011). On this basis, the constraint density can be evaluated as a function of temperature and pressure and corresponding probabilities indicate that they behave rather similarly to the onset functions $q(T)$ proposed by Mauro and Gupta (Bauchy and Micoulaut 2011, 2013a).

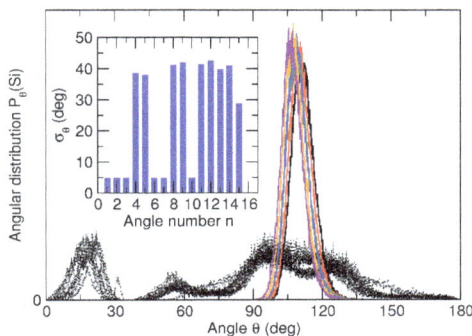

Figure 6. Angles around a silicon atom in a soda-lime silicate glass, defined by the $N=6$ first neighbors leading to 15 possible partial bond angle distributions (Laurent et al. 2014). Colored curves correspond to PBADs having a low standard deviation s_θ (inset, blue bars). Reprinted with permission from Laurent L, Mantisi B, Micoulaut M (2014) Structure and topology of soda-lime silicate glasses: implications for window glass. J Phys Chem B 118:12750–12762. Copyright (2014) American Chemical Society.

ISOSTATIC RELAXATION

The use of MD simulation together with constraint enumeration reveals that isostatic glasses fulfilling merely $n_c = 3$ relax in a very different fashion from other glass-forming liquids. The behavior of transport properties appears to be different too, as recently demonstrated for densified silicates (Bauchy et al. 2013; Bauchy and Micoulaut 2015). These conclusions build on the computation of liquid viscosity using the Green–Kubo (GK) formalism which is based on the calculation of the stress tensor auto-correlation function, given by:

$$\eta = \frac{1}{k_B T V} \int_0^\infty P_{\alpha\beta}(t) P_{\alpha\beta}(0) dt \tag{13}$$

using off-diagonal $(\alpha \neq \beta)$ components $P_{\alpha\beta}(t)$ of the molecular stress tensor defined by:

$$P_{\alpha\beta}(t) = \sum_{i=1}^N m_i v_i^\alpha v_i^\beta + \sum_{i=1}^N \sum_{j>i}^N F_{ij}^\alpha r_{ij}^\beta \tag{14}$$

where the brackets in Equation (13) refer to an average over the whole system. In Equation (14), m_i is the mass of atom i, and F_{ij} is the component α of the force between the atoms i and j, and r_{ij}^β and v_i being the β component of the distance between two atoms i and j, and the velocity of atom i, respectively. Such calculated viscosities are investigated at fixed pressure/density as a function of inverse temperature in an Arrhenius plot, a linear behavior is obtained (Fig. 7) which is found to be compatible with experimental measurements at zero pressure (Bockris et al. 1955; Knoche et al. 1994; Neuville 2006). The corresponding calculated activation energy E_A for viscous flow is found to display a similar trend as the one determined from the calculated diffusivity, and, interestingly, both display a minimum with the calculated number of constraints. This suggests that isostatic networks ($n_c = 3$) will lead to singular relaxation behavior with weaker energy barriers, as also detected for a select number of chalcogenide liquids (Yang et al. 2012; Gunasekera et al. 2013; Ravindren et al. 2014) where the condition of isostaticity ($n_c = 3$) coincides with a network mean coordination number of $\bar{r} = 2.4$. As one also has $M = E_A \ln 10 / k_B T_g$, the combination of a minimum in activation energy and a smooth or constant behavior of T_g usually also leads to a minimum in fragility for isostatic liquids. Such observations have been made for certain chalcogenide and modified oxides melts for which minima in E_A and M coincide and, under certain assumptions regarding structure, the link with the isostatic nature of the network could be established (Chakraborty and Boolchand 2014; Mohanty et al. 2019).

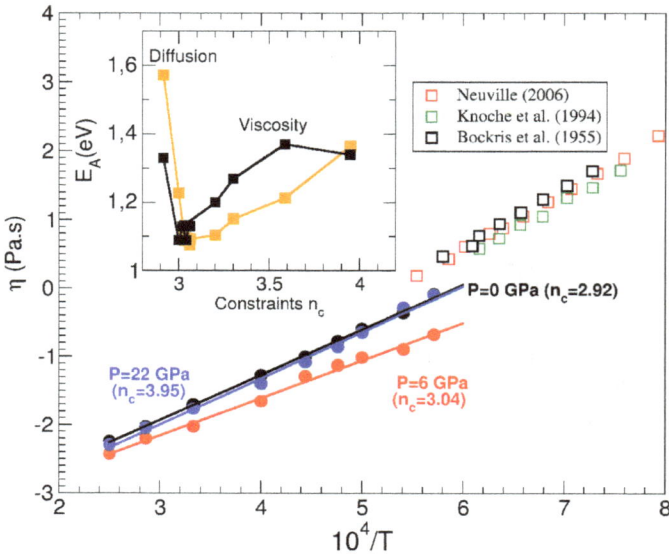

Figure 7. Calculated viscosity using Equation (13) in a densified sodium silicate for different system pressures, compared to experimental data (Bockris et al. 1955; Knoche et al. 1994; Neuville 2006), inset: Calculated activation energies E_A for diffusion (red) and viscosity (black) as a function of the number of constraints n_c, derived from an Arrhenius plot of oxygen diffusivity and Green–Kubo calculated viscosity (Bauchy and Micoulaut 2015).

Such anomalies have been found to also occur in other dynamic properties such as the structural relaxation time determined from the long time behavior of the intermediate scattering function which measures the decay of density–density correlations in Fourier space (Bauchy and Micoulaut 2015; Yildirim et al. 2016). They are also supported from the results of a simplified Kirkwood–Keating model containing only harmonic interactions between atoms. In this model, the glass transitions of isostatic glass-forming liquids involve an activation energy for relaxation time which is minimum (Micoulaut 2010).

Reversibility windows

Another intriguing observation related to isostatic glasses is a minimum in enthalpic relaxation that is, by far, much smaller than in flexible or stressed rigid glasses. Experimentally, such anomalies lead to reversibility windows measured from calorimetry, i.e., there is a tendency for the system to undergo a glass transition with a minimum of thermal changes. This obviously relates to the particular relaxation behavior described above.

Signature from MD simulations. When MD numerical cycles are performed across the glass transition from a high temperature liquid, one finds a hysteresis between the cooling and heating curve. This behavior simply signals the nature of the glassy state that induces slow dynamics at $T < T_g$ with a subsequent volume or enthalpy relaxation as the glass is annealed back to the liquid phase, consistent with experimental observation (Fig. 8). However, it has been found that for selected thermodynamic conditions (pressure, density) and fixed numerical cooling/heating rate the hysteresis curves become minuscule (Bauchy and Micoulaut 2015; Mantisi et al. 2015), and the cooling/heating curves nearly overlap. When the area of the enthalpy/energy (or volume) hysteresis is investigated as a function of pressure or density, a deep minimum is found which reveals a so-called *reversibility window* (RW, Fig. 8 inset).

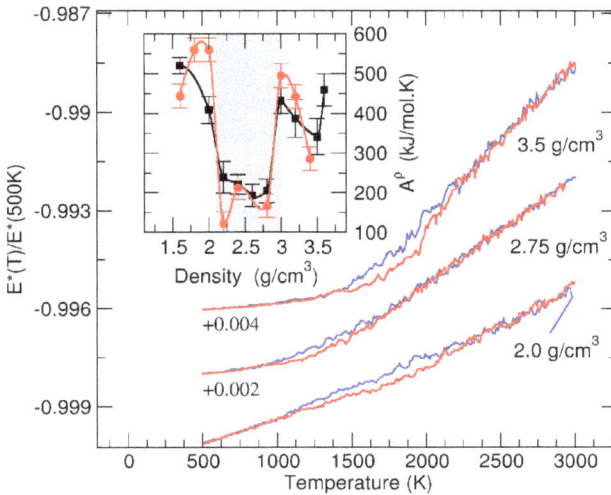

Figure 8. Numerical cooling-heating cycle of a densified sodium silicate for different densities representing a rescaled energy $E^*(T)/E^*(500\,\text{K})$ with temperature. The inset represents the hysteresis area A^ρ for different cooling rates as a function of system density, and defines a reversibility window (Mantisi et al. 2015). Curves are shifted by 0.002 for clarity. Reprinted figure with permission from B. Mantisi, M. Bauchy, and M. Micoulaut, Phys. Rev. B 92:134201 (2015). Copyright (2015)by the American Physical Society.

These thermal anomalies are actually linked with the isostatic nature of the glass-forming liquid, as detected from the MD-based constraint count which shows that the total number of constraints shows a plateau-like behavior around the isostatic value $n_c = 3$ in the temperature range where the glass transition occurs (Fig. 9). The detail shows that angular adaptation drives the mechanical evolution of the liquid under pressure because BS constraints increase due to conversion of the low pressure tetrahedral order ($r = 4$) into octahedral order ($r = 6$) which dominates at elevated pressure (Bauchy and Micoulaut 2015; Mantisi et al. 2015; Yildirim et al. 2016), and which is typical of the short-range order of stishovite. However, for systems beyond a certain threshold pressure/density, further compression leads to a decrease of the number of BB (angular) constraints which indicates that some of the angular interactions have softened (Bauchy and Micoulaut 2015). Results indicate an obvious correlation between the RW thresholds and those obtained from the constraint count, while identifying the isostatic nature of the RW (Fig. 10).

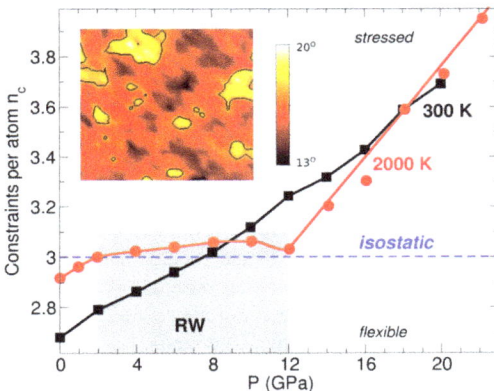

Figure 9. Constraint density n_c as a function of pressure in supercooled (2000 K) and glassy sodium disilicate (NS2). The plateau behavior in the liquid corresponds to pressures (gray zone) for which cooling/heating cycles are minimum (adapted from Bauchy and Micoulaut 2015). The inset shows the BO BB constraint density inside the simulation box, and highlights the heterogeneous character of such constraints (Bauchy and Micoulaut 2013b). Adapted figure with permission from B. Mantisi, M. Bauchy, and M. Micoulaut, Phys. Rev. B 92:134201 (2015). Copyright (2015)by the American Physical Society.

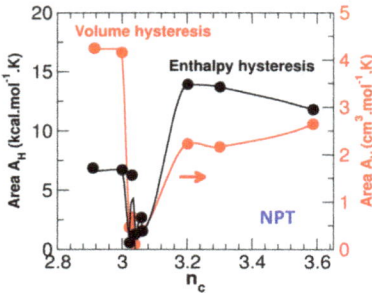

Figure 10. Area of enthalpy and volume hysteresis of densified sodium disilicates in numerical cooling/heating cycles performed in NPT Ensemble as a function of constraint density (Bauchy and Micoulaut 2015). Reprinted figure with permission from B. Mantisi, M. Bauchy, and M. Micoulaut, Phys. Rev. B 92:134201 (2015). Copyright (2015)by the American Physical Society.

In addition, it has been found that the distribution of constraints is not randomly distributed (Fig. 9, inset), and corresponding liquids display a heterogeneous distribution with zones of thermally activated broken constraints that dominate the liquid dynamics (Bauchy and Micoulaut 2013b; Yildirim et al. 2016). In addition, the spatial extent of these flexible regions shows a percolative behavior which is not only connected to flexible–rigid transitions but also to dynamic heterogeneities (Bauchy and Micoulaut 2017). For the particular case of sodium silicate liquids, the temperature at which constraints become homogeneously distributed across the liquid structure is found to depend both on pressure and temperature.

Experimental signature from calorimetry. To support the MD results, a vast body of experimental data has been accumulated during the past fifteen years. One of the most direct signatures of reversibility windows which has a nearly one-to-one correspondence with the result from MD is provided from mDSC measurements (modulated differential scanning calorimetry) which exhibit a deep minimum (Fig. 2) or even a vanishing of a so-called non-reversing enthalpy (ΔH_{nr}) which represents a rough measure of the enthalpy of relaxation during the glass transition. The use of such calorimetric methods to detect the nearly reversible character of the glass transition has not been without controversy, in part, because of the intrinsic measurement of ΔH_{nr} depends on the imposed frequency, albeit frequency corrections are brought to the tempereature scans and these build on a robust methodology (Verdonck et al. 1999).

A large number of network glasses (chalcogenides, oxides) display RWs, and these are summarized in Figure 11. These represent systems which cover various bonding types, ranging from ionic (silicates), iono-covalent, covalent (Ge–Se), or semi-metallic (Ge–Te–In–Ag). For simple covalent network glasses such as Ge_xSe_{1-x}, the experimental boundaries of the RW are found close to the mean-field estimate of the isostatic criterion (Eqn. 1) satisfying $n_c = 3$ because coordination numbers of Ge/Si and S/Se can be safely determined from the 8-$\mathcal{N}$ (octet) rule to yield an estimate of the constraints $n_c = 2 + 5x$ using Equation (1). For such systems, aspects of topology fully control the evolution of rigidity with composition, as there is a weak chemical effect in the case of isovalent Ge/Si or S/Se substitution, and the lower boundary of the RW ($x_c = 20\%$) coincides with the Phillips–Thorpe mean-field rigidity transition $n_c = 3$. However, for most of the systems among which silicates, uncertainties persist regarding the constraint count derived from the local structures (e.g., Q^n species) and geometries as coordination numbers or constraints must be derived from specific structural models.

The presence of relaxation phenomena (Fig. 7) that induce RWs for select compositions leads to various other anomalous behaviors in physical, chemical and mechanical properties (see applications) which manifest by maxima or minima with thermodynamic conditions (temperature, composition, pressure). These signatures provide other alternative and complementary evidence of the RW signature from calorimetric measurements. For instance, it has been suggested that glasses will display an increased tendency towards space-filling because of the isostatic nature of the networks (i.e., absence of stress), which manifests by a minimum in the molar volume (Fig. 2a), a salient feature that has been reported for different silicates, germanates and oxides with composition (Rompicharla et al. 2008; Bourgel et al. 2009) or under pressure (Trachenko et al. 2004). Ionic conductors display an onset of ionic conduction only in compositions belonging to the flexible phase (Novita et al. 2007), i.e., when the network can be more easily deformed at a local level because of the presence of floppy modes which promote mobility. This leads to an exponential increase in the conductivity once $n_c < 3$, as detected from the evolution of ionic conduction in alkali and alkaline earth silicates (Micoulaut 2008).

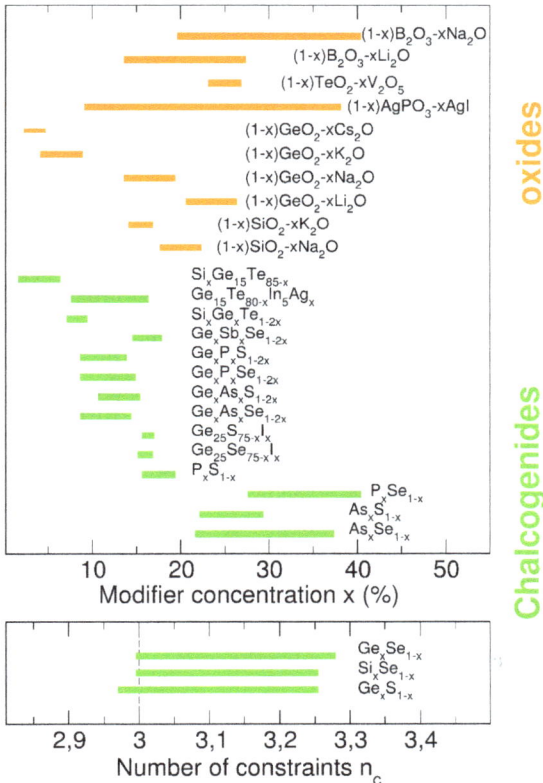

Figure 11. Location of experimental reversibility windows (RW) driven by composition for different chalcogenide and modified oxide glass systems (Mantisi et al. 2015). In the same families of modified oxides (e.g., silicates see Fig. 2 or germanates), there is an effect due to the cation size. Using the octet rule, only a select number of systems can be represented as a function of n_c using the mean-field estimate of Equation (1). Reprinted figure with permission from B. Mantisi, M. Bauchy, and M. Micoulaut, Phys. Rev. B 92:134201 (2015). Copyright (2015)by the American Physical Society.

Nature of reversibility windows from models. While the results of Figures 9 and 10 obviously point to the isostatic nature of the RW, a certain number of other/complementary scenarios have been proposed to describe the observed behaviors, building on the role of fluctuations in the emergence of a double threshold/transition that defines an intermediate phase (IP) between the flexible and the stressed rigid phase (Thorpe et al. 2000; Micoulaut and Phillips 2003; Barré et al. 2005; Brière et al. 2007). Other models inspired by the mean-field treatment of jammed solids neglect the role of coordination fluctuations and propose that the IP is the result of a local distinct configurations that organize due to long-range elastic interactions (Yan and Wyart 2014; Yan 2018).

An alternative path builds on the initial mean-field rigidity treatment and a certain number of models (Thorpe et al. 2000; Brière et al. 2007) propose that at finite temperature, the glass forming under increasing stress induced by the growing cross-linking density, is able to adapt during the cooling in order to reduce the energy, a feature termed as "self-organization" (Thorpe et al. 2000). In detail, it can be shown that such adaptation manifests by an increased angular excursion for the BO atoms in densified liquid silicates which have networks with a growing bond density induced by pressure (Bauchy and Micoulaut 2013a). On simple bar networks or cluster approaches, the IP is defined by a double percolative transition at $\bar{r}_1$ (rigidity) and $\bar{r}_2$ (stress), which can be fully characterized from algortihms taking into account the non-local characteristics of rigidity, and this allows one to locate over-constrained zones in an amorphous network, and to define the size of isostatic rigid and stressed-rigid zones or clusters as a function of the control parameter $\bar{r}$ (Thorpe et al. 2000; Micoulaut and Phillips 2003). In absence of self-organization or network adaptation, both transitions at $\bar{r}_1$ and $\bar{r}_2$ coalesce into the single mean-field transition. Since such approaches or alternatives (Barré et al. 2005) do not explicitly consider temperature effects (and the underlying associated energy barriers), Mousseau and co-workers have considered the possibility of having finite energy barriers to allow for some equilibration (Chubynsky et al. 2006; Brière et al. 2007). The main outcome from these models is that an increased sensitivity for single bond addition or removal exists close to the IP, and this suggests that the system is maintained in a critical state on the rigid–floppy boundary throughout the IP, as also suggested in the liquid phase in densified silicates (Bauchy and Micoulaut 2017).

Using the phenomenology of the elasticity of soft spheres and jamming transitions in combination with a mean-field scenario of rigidity, different authors (Yan and Wyart 2014, Yan 2018) have also shown that temperature considerably affects the way an amorphous network becomes rigid under a coordination number increase, and the existence of an isostatic RW not only depends on T, but also on the relative strength of weak forces present in the system. The related vibrational analysis furthermore suggests that vibrational modes for IP compositions are similar to the anomalous modes observed in the packing of particles near jamming, thus providing an interesting connection with jamming transitions that might also be embedded in the anomalous variation of the molar volume (Fig. 2a). The mean-field scenario for the IP is also used for the problem of rigidity percolation model on a Bethe lattice (Moukarzel 2013) and under certain conditions of bond addition/removal, two discontinuous transitions are found, and the associated IP displays an enhanced isostaticity at the flexible boundary. The main result is that throughout the IP, the network displays a low density of redundant bonds and contains, therefore, low self-stress.

OTHER APPLICATIONS

Diffusivity anomalies

A certain number of densified tetrahedral liquids display diffusivity anomalies, following the well-known example of densified water (Errington and Debenedetti 2001), i.e., D exhibits minima and maxima with pressure/density at fixed temperature in the liquid state. This is also the case for silica (Shell et al. 2002), germania (Pacaud and Micoulaut 2015) and sodium disilicate NS2 (Bauchy et al. 2013). The diffusivity anomalies for the NS2 liquid actually also relate to the elastic nature of the liquid, flexible or stressed, and the correlation becomes obvious when the sodium D_{Na} and oxygen D_O diffusivities are represented, instead of density, as a function of the constraint density (Fig. 12). One now recognizes that such diffusivity anomalies of liquid NS2 connects to the RW, and these are driven by the evolution of constraint density which is the dominant feature controlling the evolution of the transport coefficient in liquids under pressure. In stressed-rigid liquids, both Na and O diffusivities behave the same with an exponential decay of the form $\exp[-An_c]$ and a coefficient A being similar for both species. Close to the isostatic condition however, transport coefficients evolve rapidly and for liquids in the flexible phase (where the $P=0$ belongs to the flexible phase), sodium diffusivity are now markedly larger.

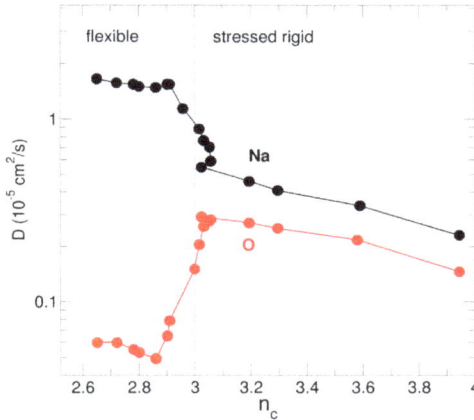

Figure 12. Calculated oxygen and sodium diffusivity in a liquid (2000 K) sodium silicate $2\,SiO_2$–Na_2O as a function of the calculated number of constraints n_c. Transport properties are obviously affected by the elastic nature of the liquid. Adapted from Bauchy et al. (2013) and Bauchy and Micoulaut (2015).

It has been furthermore identified that the diffusivity minima D_{min} (e.g., at $2.1\,g/cm^3$ or $n_c=2.87$ for 2000 K) and maxima D_{max} (at $3.2\ g/cm3$ or $n_c=3.05$) can be related to the boundaries of the RW (Mantisi et al. 2015). In light of these correlations, the locus of D_{min} is related to the boundary for the onset of a rigid but stress-free network-forming liquid, whereas the location of D_{max} is related to the upper boundary of the IP.

Prediction of glass hardness

The hardness of glasses captures their resistance to permanent deformations (Oliver and Pharr 1992; Smedskjaer et al. 2015). Interestingly, although hardness is intrinsically a far-from-equilibrium property, it has been suggested to be governed by an equilibrium structural feature, namely, the number of constraints per atom n_c. Smedskjaer et al. (2010) proposed that glass hardness can be expressed as:

$$H = \left(\frac{\partial H}{\partial n_c} \right) \left[n_c - n_{crit} \right] \qquad (15)$$

This model relies on the following reasoning. First, a critical minimum number of constraints (n_{crit}) is required for the glass to be macroscopically cohesive. The value $n_{crit}=2.5$ (i.e., the minimum number of constraints that are needed to achieve rigidity in two dimensions) was found to be appropriate to describe the hardness of several oxide glasses (Smedskjaer et al. 2010), although this value does not apply to chalcogenide glasses (Bauchy et al. 2015c). Starting from this minimum number of constraints, each additional constraint then contributes to increasing hardness, wherein the contribution of each unit constraint per atom is given by the parameter $\partial H / \partial n_c$. Note that this term has thus far remained a fitting parameter (i.e., with no clear physical foundation) that depends on the indenter geometry, the indentation load, and the specific glass family (although this parameter is composition-independent within a given glass family). This equation has been found to offer realistic predictions of glass hardness for various oxide systems (Smedskjaer et al. 2010; Smedskjaer 2014).

As a recent development, it was suggested that BS and BB constraints may not contribute in the same fashion to increase hardness. In details, hardness has been found to primarily be governed by the number of angular BB constraints, while remaining fairly unaffected by the number of radial BS constraints (see Fig. 13) (Bauchy et al. 2015c). This behavior was suggested to arise from the fact that, during indentation, glass networks deform by following lowest-energy paths—which suggests that deformation occurs via the breakage of weak angular BB constraints rather than stronger radial BS constraints. Note that it has also recently been proposed that hardness is governed by the volumic density (rather than atomic density) of constraints (Zheng et al. 2017).

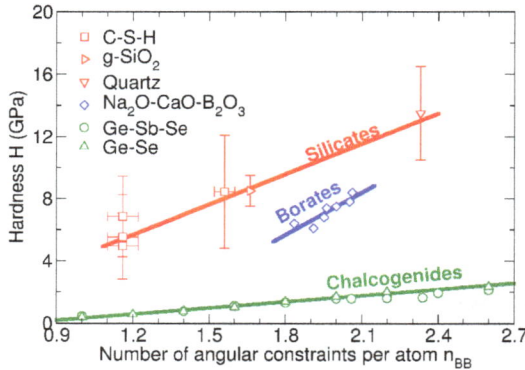

Figure 13. Hardness of various oxide and chalcogenide phases as a function of their number of angular BB constraints per atom (Swiler et al. 1990; Sreeram et al. 1991; Varshneya and Mauro 2007; Smedskjaer et al. 2010; Abdolhosseini Qomi et al. 2013 2014, 2017; Bauchy et al. 2014a,b,c, 2015c; Qomi et al. 2015a,b; Bauchy 2017).

Prediction of glass stiffness

Stiffness characterizes the resistance of glasses to reversible, elastic deformations (Rouxel 2006). In contrast with hardness, stiffness is a near-equilibrium property and, hence, it is less surprising that the macroscopic stiffness of glasses should depend on the density of constraints. Indeed, within the framework of topological constraint theory, glasses can be viewed as networks, wherein atoms are connected to each other via some "small springs" (i.e., the interatomic mechanical constraints). Hence, it is indeed not surprising that the macroscopic stiffness of glasses should be governed by the number of interatomic constraints—in the same fashion as one can calculate the resultant stiffness of a system of several springs in series and/or parallel. In fact, the relationship between stiffness and topological constraints finds its roots in the origins of topological constraint theory (Thorpe 1983). In model lattices, it was shown, that stiffness vanishes in flexible systems ($n_c < 3$) and then scales with n_c once the system is stressed-rigid ($n_c > 3$). A similar relationship was also observed in amorphous semiconductors (Paquette et al. 2017; Bhattarai et al. 2018; King et al. 2018; Su et al. 2018) and chalcogenide glasses (Plucinski and Zwanziger 2015).

More recently, some topological models of glass stiffness applicable to silicate glasses have been proposed (Wilkinson et al. 2019; Yang et al. 2019b). The model proposed by Yang et al. builds on the original model of glass hardness introduced by Smedskjaer et al. (see above) (Yang et al. 2019b). Since the stiffness (e.g., the Young's modulus E) has the dimension of an energy per unit of volume, it was postulated that E can be expressed in terms of the volumetric density of the energy created by each constraint. Further, it was postulated that BS and BB constraints do not contribute with equal weight to increasing the Young's modulus. This arises from the fact that BS and BB constraints exhibit different free energies and that different types of constraints may be activated under different loading conditions (Gupta and Mauro 2009; Bauchy and Micoulaut 2011; Bauchy et al. 2015c). Based on these considerations, the following model was proposed:

$$E = \varepsilon_{BS} n_{BS} + \varepsilon_{BB} n_{BB} \qquad (16)$$

where n_{BS} and n_{BB} are the volumetric density of BS and BB constraints, respectively, and ε_{BS} and ε_{BB} are some fitting parameters that correspond to the typical energies of BS and BB constraints, respectively. This model assumes that the Young's modulus of glasses vanishes when the density of BS and BB constraints becomes zero. As shown in Figure 14, the topological model offers excellent agreement with experimental and molecular dynamics data (Yang et al. 2019a,b) with significant improvement as compared to the traditional Makishima–Mackenzie model (Makishima and Mackenzie 1973).

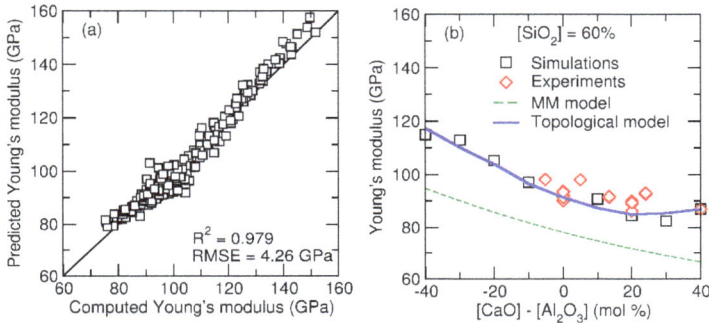

Figure 14. (a) Comparison between the Young's modulus values predicted by the topological model (Eqn. 16) and computed by molecular dynamics simulations. A coefficient of determination $R^2 = 0.979$ and a root mean squared error (RMSE) of 4.26 GPa are obtained. (b) Comparison between the Young's modulus values computed by molecular dynamics simulations, predicted by the topological model, and predicted by the Makishima–Mackenzie (MM) model for a series of $(CaO)_x(Al_2O_3)_{40-x}(SiO_2)_{60}$ glasses (Makishima and Mackenzie 1973). The data are compared with available experimental data (Eagan and Swearekgen 1978; Yamane and Okuyama 1982; Ecolivet and Verdier 1984; Rocherulle et al. 1989; Yasui and Utsuno 1993; Inaba et al. 2000, 2001; Sugimura et al. 2002; Hwa et al. 2003; Gross et al. 2009; Bansal and Doremus 2013; Weigel et al. 2016).

Origin of fracture toughness anomalies

The fracture toughness of glasses captures their resistance to crack propagation and fracture (Bauchy et al. 2015a; Wang et al. 2015; Yu et al. 2015; Januchta et al. 2017a,b; Rouxel and Yoshida 2017; Frederiksen et al. 2018). Interestingly, various glasses have been observed to exhibit maximum fracture toughness at the isostatic threshold (see Fig. 15) (Bauchy et al. 2016; Bauchy 2017). This behavior was explained as follows. On the one hand, flexible glasses ($n_c < 3$) exhibit low cohesion (low surface energy) due to their low connectivity. This tends to limit their resistance to cracking. On the other hand, stressed-rigid glasses ($n_c > 3$) break in a brittle fashion since their fully locked network cannot exhibit any ductile atomic reorganization upon loading. In between these two regimes, isostatic glasses ($n_c = 3$) feature an optimal balance between cohesion and ability to plastically deform (Bauchy et al. 2016).

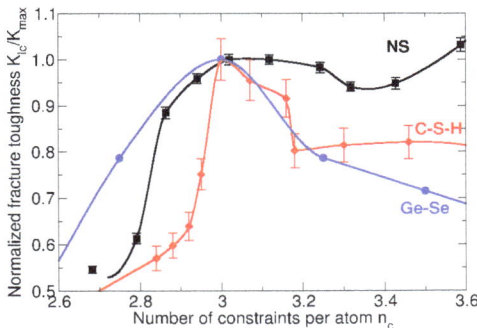

Figure 15. Fracture toughness (normalized by the fracture toughness at $n_c = 3$) of densified sodium silicate glasses (NS), calcium–silicate–hydrates (C–S–H), and Ge–Se glasses as a function of their number of constraints per atom (Guin et al. 2002; Yu et al. 2015; Bauchy et al. 2015a, 2016).

Prediction of dissolution kinetics

Recently, a topological model was introduced to predict the dissolution rate K of oxide glasses in dilute conditions (i.e., forward rate, far from saturation). It was proposed that the dissolution rate can be expressed in terms of the number of constraints per atom as (Pignatelli et al. 2016b):

$$K = K_0 \exp\left(-\frac{n_c E_0}{RT}\right) \qquad (17)$$

where K_0 is a rate constant that depends on the solution phase pH/chemistry (i.e., the barrier-less dissolution rate of a completely depolymerized material for which $n_c = 0$), $E_0 = 20\text{-to-}25\,\text{kJ/mol}$ is the energy barrier that needs to be overcome to break a unit atomic constraint, R is the perfect gas constant, and T is the temperature. Based on this equation, the effective activation energy of dissolution can be expressed as:

$$E_A^{\text{eff}} = n_c E_0 \qquad (18)$$

The physical root of this model can be understood as follows. Starting from a hypothetic fully depolymerized phase ($n_c = 0$), each additional unit constraint per atom tends to increase the dissolution activation energy and, hence, reduce the kinetics of dissolution (Pignatelli et al. 2016b).

This was explained by the fact that the number of constraints per atom n_c captures the local stiffness of the atomic network—so that the stiffer the network, the higher the energy cost associated with local atomic reorganizations. Indeed, independently of its underlying mechanism (e.g., hydrolysis or ion exchange), dissolution involves some atomic reorganizations within the network. For instance, hydrolysis requires the formation of intermediate over-coordinated species (5-fold coordinated Si or three-fold coordinated O), while ion exchange requires some local opening of the network to enable the jump of mobile cations from one pocket to another (Bunker 1994). In both cases, the activation energy associated with these atomic reorganizations depends on the ability of the network to locally adjust its structure, which decreases as its local stiffness increases (Guo et al. 2016). This picture echoes results from density functional theory, which have shown that the activation energy associated with the hydrolysis of inter-tetrahedra bridging oxygen atoms increases with network connectivity and, hence, rigidity (Pelmenschikov et al. 2000). As shown in Fig. 16, this model has been extensively validated over a broad range of oxides and is able to predict the dissolution rate of silicate glasses over four orders of magnitude (Oey et al. 2015, 2017a,b,c; Pignatelli et al. 2016a,c; Hsiao et al. 2017; Mascaraque et al. 2017, 2019).

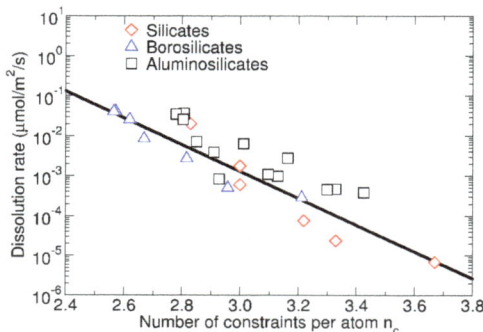

Figure 16. Dissolution rate of various silicate phases as a function of the number of constraints per atom (Oey et al. 2017a,b; Mascaraque et al. 2017, 2019; Pignatelli et al. 2016b).

Other applications and conclusion

Several other glass behaviors and properties have been shown to be largely governed by the topology of their atomic networks (Micoulaut and Yue 2017). Notably, isostatic glasses have been found to feature various unusual and often desirable behaviors, namely, (i) maximum strengthening following ion exchange (Wang and Bauchy 2015; Svenson et al. 2016; Wang et al. 2017a,b), (ii) maximum elastic volume recovery upon loading–unloading cycles (Mauro and Varshneya 2007; Bauchy et al. 2017), (iii) minimum relaxation and aging (Chen et al. 2010; Bauchy and Micoulaut 2015; Mantisi et al. 2015; Bauchy et al. 2017), (iv) minimum propensity for creep (Masoero et al. 2015; Bauchy et al. 2015b, 2017; Pignatelli et al. 2016a), and (v) maximum resistance to irradiation (Krishnan et al. 2017; Wang et al. 2017a). Sub-critical crack growth in oxide glasses exposed to sustained load in humid conditions has also been shown to be controlled by the atomic topology (Smedskjaer and Bauchy 2015). More recently, it has been suggested that topology of the glass surface (rather than that of the bulk) is governing the degree of hydrophilicity/hydrophobicity of silica surfaces (Yu et al. 2018).

Besides oxide glasses, the thermal, mechanical, electrical, and optical properties of amorphous SiC:H thin films were shown to be correlated to the topology of their atomic network (King et al. 2013). Rigidity concepts have also enabled the nanoengineering of new cementitious phases (Bauchy 2017; Liu et al. 2019). The susceptibility of granular systems was found to be maximal at the isostatic threshold (Moukarzel 1998, 2002). Fast-ion conduction was observed to offer a clear signature of the intermediate phase (Novita et al. 2007, 2009; Micoulaut et al. 2009). The performances of phase-change materials were noted to be controlled by their atomic topology (Micoulaut et al. 2010). Finally, it was found that protein folding is controlled by the topology of their molecular architecture (Rader et al. 2002; Phillips 2004, 2009). The large variety of properties and material families for which topological engineering concepts have been successfully applied establish network topology as a generic, fundamental, and promising approach to decode the genome of complex materials.

In this context, further applications to silicate glasses and melts should provide new attractive ways to understand and predict physical and chemical properties as a function of composition and temperature.

REFERENCES

Abdolhosseini Qomi MJ, Bauchy M, Pellenq RJ-M, Ulm F-J (2013) Applying tools from glass science to study calcium–silicate–hydrates. *In:* CONCREEP-9: Mechanics, Physics of Creep, Shrinkage, and Durability of Concrete: A Tribute to Zdenek P. Bazant. Proc 9th International Conference on Creep, Shrinkage, and Durability Mechanics, Cambridge, Mass, Sept 22–25, 2013, Ulm F-J, Hamlin JM, Pellenq RJ-M (eds) ASCE Publications, p 7–85

Abdolhosseini Qomi MJ, Krakowiak KJ, Bauchy M, Stewart KL, Shahsavari R, Brommer DB, Baronnet A, Buehler MJ, Yip S, Ulm FJ, Van Vliet KJ, Pellenq R (2014) Combinatorial molecular optimization of cement hydrates. Nat Commun 5:4960

Abdolhosseini Qomi MJ, Ebrahimi D, Bauchy M, Pellenq R, Ulm FJ (2017) Methodology for estimation of nanoscale hardness via atomistic simulations. J Nanomech Micromech 7:04017011

Araujo RJ, Mauro JC (2010) The thermodynamic significance of order parameters during glass relaxation. J Am Ceram Soc 93: 1026–1031

Angell CA, Richards BE, Velikov V (1999) Simple glass-forming liquids: their definition, fragilities, and landscape excitation profiles. J Phys:Condens Matter 11:A75

Asokan S, Prasad MVN, Parthasarathy G, Gopal ESR (1989) Mechanical and chemical thresholds in IV–VI chalcogenide glasses. Phys Rev Lett 62:808–811

Azoulay R, Thibierge H, Brenac A (1975) Devitrification characteristics of Ge_xSe_{1-x} glasses. J Non-Cryst Solids 18:33–53

Bansal NP, Doremus RH (eds) (2013) Handbook of Glass Properties. Elsevier

Barré J, Bishop AR, Lookman T, Saxena A (2005) Adaptability and "intermediate phase" in randomly connected networks. Phys Rev Lett 94:208701

Bauchy M (2017) Nanoengineering of concrete via topological constraint theory. MRS Bull 42:50–54

Bauchy M, Micoulaut M (2011) Atomic scale foundation of temperature dependent constraints in glasses and liquids. J Non-Cryst Solids 357:2530

Bauchy M, Micoulaut M (2013a) Transport anomalies and adaptative pressure-dependent topological constraints in tetrahedral liquids: evidence for a reversibility window analogue. Phys Rev Lett 110:095501

Bauchy M, Micoulaut M (2013b) Percolative heterogeneous topological constraints and fragility in glass-forming liquids. EPL 104:56002

Bauchy M, Micoulaut M (2015) Densified network glasses and liquids with thermodynamically reversible and structurally adaptive behavior. Nat Commun 6:6398

Bauchy M, Micoulaut M (2017) Evidence for anomalous dynamic heterogeneities in isostatic supercooled liquids. Phys Rev Lett 118:145502

Bauchy M, Micoulaut M, Celino M, Boero M, Le Roux S, Massobrio C (2011) Angular rigidity in tetrahedral network glasses with changing composition. Phys Rev B 83:054201

Bauchy M, Guillot B, Micoulaut M, Sator N (2013) Viscosity and viscosity anomalies in model silicates and magmas: a numerical investigation. Chem Geol 346:47–63

Bauchy M, Abdolhosseini Qomi MJ, Bichara C, Ulm FJ, Pellenq R (2014a) Nanoscale structure of cement:viewpoint of rigidity theory. J Phys Chem C 118:12485–12493

Bauchy M, Abdolhosseini Qomi MJ, Pellenq RJM, Ulm FJ (2014b) Is cement a glassy material? *In:* Computational Modelling of Concrete Structures. 1st edn. Bicanic N, Mang H, Meschke G, de Borst R (eds), CRC Press, p 169

Bauchy M, Abdolhosseini Qomi MJ, Ulm, F-J, Pellenq RJ-M (2014c) Order and disorder in calcium–silicate–hydrate. J Chem Phys 140:214503

Bauchy M, Laubie H, Abdolhosseini Qomi MJ, Hoover CG, Ulm F-J, Pellenq RJ-M (2015a) Fracture toughness of calcium–silicate–hydrate from molecular dynamics simulations. J Non-Cryst Solids 419:58–64

Bauchy M, Masoero E, Ulm F-J, Pellenq R (2015b) Creep of bulk C–S–H: Insights from molecular dynamics simulations. *In:* CONCREEP 10: Proc 10th International Conference on Mechanics and Physics of Creep, Shrinkage, and Durability of Concrete and Concrete Structures, Vienna, Austria, September 21–23, 2015. Hellmich C, Pichler B, Kollegger J (eds) ASCE, p 511–516

Bauchy M, Qomi MJA, Bichara C, Ulm FJ, Pellenq R (2015c) Rigidity transition in materials:hardness is driven by weak atomic constraints. Phys Rev Lett 114:125502

Bauchy M, Wang B, Wang M, Yu Y, Qomi MJA, Smedskjaer MM, Bichara C, Ulm FJ, Pellenq R (2016) Fracture toughness anomalies:Viewpoint of topological constraint theory. Acta Mater 121:234–239

Bauchy M, Wang M, Yu Y, Wang B, Anoop Krishnan, Masoero E, Ulm FJ, Pellenq R (2017) Topological control on the structural relaxation of atomic networks under stress. Phys Rev Lett 119:035502

Bermejo FJ, Cabrillo C, Bychkov E, Fouquet P, Ehlers G, Häussler W, Price DL, Saboungi ML (2008) Tracking the effects of rigidity percolation down to the liquid state: relaxational dynamics of binary chalcogen melts. Phys Rev Lett 100:245902

Bhattarai G, Dhungana S, Nordell BJ, Caruso AN, Paquette MM, Lanford WA, King SW (2018) Underlying role of mechanical rigidity and topological constraints in physical sputtering and reactive ion etching of amorphous materials. Phys Rev Mater 2:055602

Bhosle S, Boolchand P, Micoulaut M, Massobrio C (2011) Meeting experimental challenges to physics of network glasses:assessing the role of sample homogeneity. Solid State Comm 151:1851–1855

Bockris JOM, Mackenzie JD, Kitchener JA (1955). Viscous flow in silica and binary liquid silicates. Trans Faraday Soc 51:1734–1748

Böhmer R, Angell CA (1992) Correlations of the nonexponentiality and state dependence of mechanical relaxations with bond connectivity in Ge–As–Se supercooled liquids. Phys Rev B 45:10091–10094

Boisen MB, Gibbs GV, Downs RT, D'Arco P (1990) The dependence of the Si–O bond length on structural parameters in coesite, the silica polymorphs, and the clathrasils. Am Mineral 75:748–754

Boolchand P, Thorpe MF (1994) Glass-forming tendency, percolation of rigidity, and onefold-coordinated atoms in covalent networks. Phys Rev B 50:10366–10368

Boolchand P, Bresser WJ, Zhang M, Wu Y, Wells J, Enzweiler RN (1995) Lamb–Mössbauer factors as a local probe of floppy modes in network glasses. J Non-Cryst Solids 182:143–154

Bourgel C, Micoulaut M, Malki M, Simon P (2009) Molar volume minimum and adaptative rigid networks in relationship with the intermediate phase in glasses. Phys Rev B 79:024201

Bresser WJ, Suranyi P, Boolchand P (1986) Rigidity percolation and molecular clustering in network glasses. Phys Rev Lett 56:2493–2496

Brière MA, Chubynsky MV, Mousseau N (2007) Self-organized criticality in the intermediate phase of rigidity percolation. Phys Rev E 75:056108

Bunker BC (1994) Molecular mechanisms for corrosion of silica and silicate glasses. J Non-Cryst Solids 179:300–308

Chaimbault F (2004) Transitions de rigidité dans les silicates d'alcalins, PhD Thesis, Université d'Orléans

Chakraborty D, Poolchand P (2014) Topological origin of fragility, network adaptation, and rigidity and stress transitions in especially homogenized nonstoichiometric binary Ge$_x$S$_{100-x}$ glasses. J Phys Chem B 118:2249–2263

Chen P, Boolchand P, Georgiev DG (2010) Long term aging of selenide glasses:evidence of sub-T_g endotherms and pre-T_g exotherms. J Phys-Condens Matter 22:065104

Chubynsky MV, Brière MA, Mousseau N (2006) Self-organization with equilibration: A model for the intermediate phase in rigidity percolation. Phys Rev E 74:016116

Cormack AN, Du J (2001) Molecular dynamics simulations of soda–lime–silicate glasses. J Non-Cryst Solids 293:283–289

Cormack AN, Du J, Zeitler JN (2002) Alkali ion migration mechanisms in silicate glasses probed by molecular dynamics simulations. Phys Chem Chem Phys43:193

Cormier L, Ghaleb D, Delaye JM, Calas G (2000) Competition for charge compensation in borosilicate glasses: Wide-angle X-ray scattering and molecular dynamics calculations. Phys Rev B 61:14495–14499

Cormier L, Ghaleb D, Neuville DR, Delaye JM, Calas G (2003) Chemical dependence of network topology of calcium aluminosilicate glasses: A computer simulation study. J Non-Cryst Solids 332:255–270

Duquesne JY, Bellassa G (1985) Ultrasonic properties of Se–Ge glasses, between 1 K and 100 K. J Phys Colloq 46 C 10:445–448

Eagan RJ, Swearekgen JC (1978) Effect of composition on the mechanical properties of aluminosilicate and borosilicate glasses. J Am Ceram Soc 61:27–30

Ecolivet C, Verdier P (1984) Proprietes elastiques et indices de refraction de verres azotes. Mater Res Bull 19:227–231

Errington JR, Debenedetti PG (2001) Relationship between structural order and the anomalies of liquid water. Nature 409:318–321

Fang CY, Yinnon H, Uhlmann DR (1983) A kinetic treatment of glass formation VIII: Critical cooling rates for Na_2O–SiO_2 and K_2O–SiO_2 glasses. J Non-Cryst Solids 57:465–471

Farnan I, Grandinetti PJ, Baltisberger JH, Stebbins J, Werner U, Eastman MA, Pines A (1992) Quantification of the disorder in network-modified silicate glasses. Nature 358:31–35

Feng X, Bresser WJ, Boolchand P (1997) Direct evidence for stiffness threshold in chalcogenide glasses. Phys Rev Lett 78:4422– 4425

Ferlat G, Charpentier T, Seitsonen AP, Takada A, Lazzeri M, Cormier L, Calas G, Mauri F (2008) Boroxol rings in liquid and vitreous B_2O_3 from first principles. Phys Rev Lett 101:065504

Foffi G, Sciortino F (2006) Extended law of corresponding states in short-range square-wells : A potential energy landscape study. Phys Rev E 74:050401

Frederiksen KF, Januchta K, Mascaraque N, Youngman RE, Bauchy M, Rzoska SJ, Bockowski M, Smedskjaer MM (2018) Structural compromise between high hardness and crack resistance in aluminoborate glasses. J Phys Chem B 122:6287–6295

Galeener FL (1985) A model for the distribution of bond angles in vitreous SiO_2. Philos Mag B 51:L1–L8

Gross TM, Tomozawa M, Koike A (2009) A glass with high crack initiation load: Role of fictive temperature-independent mechanical properties. J Non-Cryst Solids 355:563–568

Guin J-P, Rouxel T, Sanglebœuf J-C, Melscoët I, Lucas J (2002) Hardness, toughness, and scratchability of germanium–selenium chalcogenide glasses. J Am Ceram Soc 85:1545–1552

Gunasekera K, Bhosle S, Boolchand P, Micoulaut M (2013) Topology, super-strong melts and homogeneity of network glass forming liquids. J Chem Phys 139:164511

Guo P, Wang B, Bauchy M, Sant G (2016) Misfit stresses caused by atomic size mismatch:the origin of doping-induced destabilization of dicalcium silicate. Cryst Growth Des 16:3124–3132

Gupta PK, Mauro JC (2009) Composition dependence of glass transition temperature and fragility. I. A topological model incorporating temperature-dependent constraints. J Chem Phys 130:094503

He H, Thorpe MF (1985) Elastic properties of glasses. Phys Rev Lett 54:2107–2110

Henderson GS (1995) A Si K-edge EXAFS/XANES study of sodium silicate glasses. J Non-Cryst Solids 183:43–50

Hermansen C, Mauro JC, Yue Y (2014) A model for phosphate glass topology considering the modifying ion sub-network. J Chem Phys 140:154501

Hermansen C, Youngman RE, Wang J, Yue Y (2015) Structural and topological aspects of borophosphate glasses and their relation to physical properties. J Chem Phys 142:184503

Horbach J, Kob W (1999) Structure and dynamics of sodium disilicate. Philos Mag B 79:1981–1986

Hsiao YH, La Plante EC, Krishnan NA, Le Pape Y, Neithalath N, Bauchy M, Sant G (2017) Effects of irradiation on albite's chemical durability. J Phys Chem A 121:7835–7845

Hwa L-G, Hsieh K-J, Liu L-C (2003) Elastic moduli of low-silica calcium alumino-silicate glasses. Mater Chem Phys 78:105–110

Inaba S, Todaka S, Ohta Y, Morinaga K (2000) Equation for estimating the young’s modulus, shear modulus and Vickers hardness of aluminosilicate glasses. J Jpn Inst Met 64:177–183

Inaba S, Oda S, Morinaga K (2001) Equation for estimating the thermal diffusivity, specific heat and thermal conductivity of oxide glasses. J Jpn Inst Met 65:680–687

Januchta K, Youngman RE, Goel A, Bauchy M, Rzoska SJ, Bockowski M, Smedskjaer MM (2017a) Structural origin of high crack resistance in sodium aluminoborate glasses. J Non-Cryst Solids 460:54–65

Januchta K, Youngman RE, Goel A, Bauchy M, Logunov SL, Rzoska SJ, Bockowski M, Jensen LR, Smedskjaer MM (2017b) Discovery of ultra-crack-resistant oxide glasses with adaptive networks. Chem Mater 29:5865–5876

Kamitakahara W, Cappelletti RL, Boolchand P, Halfpap B, Gompf F, Neumann DA, Mutka H (1991) Vibrational densities of states and network rigidity of chalcogenide glasses. Phys Rev B 44:94–100

King SW, Bielefeld J, Xu G, Lanford WA, Matsuda Y, Dauskardt RH, Kim N, Hondongwa D, Olasov L, Daly B, Stan G (2013) Influence of network bond percolation on the thermal, mechanical, electrical and optical properties of high and low-k a-SiC:H thin films. J Non-Cryst Solids 379:67–79

King SW, Ross L, Lanford WA (2018) Narrowing of the Boolchand intermediate phase window for amorphous hydrogenated silicon carbide. J Non-Cryst Solids 499:252–256

Knoche R, Dingwell DB, Seifert FA, Webb SL (1994) Non-linear properties of supercooled liquids in the system Na_2O–SiO_2. Chem Geol:116:1–16

Kob W, Binder K (2005) Glassy Materials and Disordered Solids. World Scientific, Singapore

Kohara S, Suzuya K, Takeuchi K, Loong CK, Grimsditch M, Weber JKR, Tangeman JA, Key TS (2004) Glass formation at the limit of insufficient network formers. Science 303:1649–1652

Krishnan NMA, Wang B, Sant G, Phillips JC, Bauchy M (2017) Revealing the effect of irradiation on cement hydrates:evidence from a topological self-organization. ACS Appl Mater Interfaces 9:32377–32385

Lagrange JL (1788) Mécanique Analytique (Paris)

Laurent L, Mantisi B, Micoulaut M (2014) Structure and topology of soda-lime silicate glasses: implications for window glass. J Phys Chem B 118:12750–12762

Liu H, Du T, Krishnan NMA, Li L, Bauchy M (2019) Topological optimization of cementitious binders: Advances and challenges. Cem Concr Compos 101:5–14

Makishima A, Mackenzie JD (1973) Direct calculation of Young's modulus of glass. J Non-Cryst Solids 12:35–45

Mantisi B, Bauchy M, Micoulaut M (2015) Cycling through the glass transition: evidence for reversibility windows and dynamic anomalies. Phys Rev B 92:134201

Mascaraque N, Bauchy M, Fierro JLG, Rzoska SJ, Bockowski, Smedskjaer M (2017) Dissolution kinetics of hot compressed oxide glasses. J Phys Chem B 121:9063–9072

Mascaraque N, Januchta K, Frederiksen KF, Youngman RE, Bauchy M, Smedskjaer MM (2019) Structural dependence of chemical durability in modified aluminoborate glasses. J Am Ceram Soc 102:1157–1168

Masoero E, Bauchy M, Del Gado E, Manzano H, Pellenq R, Ulm FJ, Yip S (2015) Kinetic simulations of cement creep:mechanisms from shear deformations of glasses. *In:* CONCREEP 10: Proc 10th International Conference on Mechanics and Physics of Creep, Shrinkage, and Durability of Concrete and Concrete Structures, Vienna, Austria, September 21–23, 2015. Hellmich C, Pichler B, Kollegger J (eds) ASCE, p 555–564

Mauro JC, Gupta PK (2009) Composition dependence of glass transition temperature and fragility. I A topological model incorporating temperature-dependent constraints. J Chem Phys 130:094503

Mauro JC, Varshneya AK (2007) Modeling of rigidity percolation and incipient plasticity in germanium–selenium glasses. J Am Ceram Soc 90:192–198

Mauro JC, Yue Y, Ellison AJ, Gupta PK, Allan DC (2009a) Viscosity of glass-forming liquids. PNAS 106:19780-19784

Mauro JC, Gupta PK, Loucks RJ (2009b) Composition dependence of glass transition temperature and fragility. II. A topological model of alkali borate liquids. J Chem Phys 130:234503

Mauro JC, Loucks RJ, Gupta PK (2009c) Fictive temperature and the glassy state. J Am Ceram Soc 92:75–86

Maxwell JC (1864) On the calculation of the equilibrium and stiffness of frames, Philos Mag 27:294–299

Micoulaut M (2008) Constrained interactions, rigidity, adaptive networks, and their role for the description of silicates. Am Mineral 93:1732–1748

Micoulaut M (2010) Linking rigidity with enthalpic changes at the glass transition and the fragility of glass-forming liquids:insight from a simple oscillator model. J Phys: Condens Matter 22:285101

Micoulaut M, Phillips J (2003) Rings and rigidity transitions in network glasses. Phys Rev B 67:104204

Micoulaut M, Yue Y (2017) Material functionalities from molecular rigidity: Maxwell's modern legacy. MRS Bull 42:18–22

Micoulaut M, Popescu M (eds) (2009) Rigidity, Boolchand Intermediate Phases in Nanomaterials, Series: Optoelectronic Materials and Devices Vol 6, INOE Publishing House, Bucharest

Micoulaut M, Malki M, Simon P, Canizares A (2005) On the rigid to floppy transitions in calcium silicate glasses from Raman scattering and cluster constraint analysis. Philos Mag 85:3357–3378

Micoulaut M, Guissani Y, Guillot B (2006) Simulated structural and thermal properties of glassy and liquid germania. Phys Rev E 73:031504

Micoulaut M, Malki M, Novita DI, Boolchand P (2009), Fast ion conduction and flexibility and rigidity of solid electrolyte glasses. Phys Rev B 80:184205

Micoulaut M, Otjacques C, Raty JY, Bichara C (2010) Understanding phase-change materials from the viewpoint of Maxwell rigidity. Phys Rev B 81:174206

Mohanty C, Mandal A, Gogi VK, Chen P, Novita D, Chbeir R, Bauchy M, Micoulaut M, Boolchand P (2019) Linking melt dynamics with topological phases and molecular structure of sodium phosphate glasses from calorimetry, Raman scattering, and infrared reflectance. Front Mater 6:69

Mozzi RL, Warren BE (1969) The structure of vitreous silica. J Appl Crystallogr 2:164–172

Moukarzel CF (1998) Isostatic phase transition and instability in stiff granular materials. Phys Rev Lett 81:1634–1637

Moukarzel CF (2002) Granular matter instability:a structural rigidity point of view. *In:* Thorpe MF, Duxbury PM (eds) Rigidity Theory and Applications. Springer USA, p 125–142

Moukarzel C (2013) Two rigidity-percolation transitions on binary Bethe networks and the intermediate phase in glass. Phys Rev E 88:062121

Mysen BO, Richet P (2019) Silicate Glasses and Melts, 2nd edn. Elsevier, Amsterdam

Naumis GG (2005) Energy landscape and rigidity. Phys Rev B 71:026114

Naumis GG (2006) Variation of the glass transition temperature with rigidity and chemical composition. Phys Rev B 73:172202

Neuville DR (2006) Viscosity, structure, mixing in (Ca, Na) silicate melts. Chem Geol 299:28–41

Newton MD, Gibbs GV (1980) Ab inito calculated geometries of charge distributions for H_4SiO_4 and $H_6Si_2O_7$ compared with experimental values for silicates and siloxanes. Phys Chem Mineral 6:221–246

Novita DI, Boolchand P, Malki M, Micoulaut M (2007) Fast ion-conduction and flexibility of glassy networks. Phys Rev Lett 98:195501

Novita DI, Boolchand P, Malki M, Micoulaut M (2009) Elastic flexibility, fast-ion conduction, boson and floppy modes in $AgPO_3$–AgI glasses. J Phys: Condens Matter 21:205106

Oey T, Huang C, Worley R, Ho S, Timmons J, Cheung KL, Kumar A, Bauchy M, Sant G (2015) Linking fly ash composition to performance in cementitious systems. Proc World of Coal Ash, Nashville, TN, May 5–7, 2015

Oey T, Hsiao Y-H, Callagon E, Wang B, Pignatelli I, Bauchy M, Sant G (2017a) Rate controls on silicate dissolution in cementitious environments. RILEM Tech Lett 2:67–73

Oey T, Kumar A, Pignatelli I, Yu Y, Neithalath N, Bullard JW, Bauchy M, Sant G (2017b) Topological controls on the dissolution kinetics of glassy aluminosilicates. J Am Ceram Soc 100:5521–5527

Oey T, Timmons J, Stutzman P, Bullard JW, Balonis M, Bauchy M, Sant G (2017c) An improved basis for characterizing the suitability of fly ash as a cement replacement agent. J Am Ceram Soc 100:4785–4800

Oliver WC, Pharr GM (1992) Improved technique for determining hardness and elastic modulus using load and displacement sensing indentation experiments. J Mater Res 7:1564–1583

Pacaud F, Micoulaut M (2015) Thermodynamic precursors, liquid–liquid transitions, dynamic and topological anomalies in densified liquid germania. J Chem Phys 143:064502

Paquette MM, Nordell BJ, Caruso AN, Sato M, Fujiwara H, King SW (2017) Optimization of amorphous semiconductors and low- / high-*k* dielectrics through percolation and topological constraint theory. MRS Bull 42:39–44

Pedesseau L, Ispas S, Kob W (2015) First-principles study of a sodium borosilicate glass-former. I. The liquid state. Phys Rev B 91:134202

Pelmenschikov A, Strandh H, Pettersson LGM, Leszczynski J (2000) Lattice resistance to hydrolysis of Si–O–Si bonds of silicate minerals: ab initio calculations of a single water attack onto the (001), (111) β-cristobalite surfaces. J Phys Chem B 104:5779–5783

Pettifer RF, Dupree R, Farnan I, Sternberg U (1988) NMR determinations of Si–O–Si bond angle distribution in silica. J Non-Cryst Solids 106:408–412

Phillips JC (1979) Topology and covalent non-crystalline solids I: Short-range order in chalcogenide alloys. J Non-Cryst Solids 34:155–181

Phillips JC (1981) Structural model of two-level glass system. Phys Rev B 24:1744–1750

Phillips JC (2004) Constraint theory and hierarchical protein dynamics. J Phys-Condens Matter 16:S5065–S5072

Phillips JC (2009) Scaling and self-organized criticality in proteins: Lysozyme C. Phys Rev E 80:051916

Phillips JC, Thorpe MF (eds) Phase transitions and Self-Organization in Electronic and Molecular Networks. Plenum, New York

Piarristeguy A, Micoulaut M, Escalier R, Jovari P, Kaban I, Van Eijk J, Luckas J, Ravindren S, Boolchand P, Pradel A (2015) Structural singularities in Ge_xTe_{100-x} films. J Chem Phys 143:074502

Pignatelli I, Kumar A, Alizadeh R, Le Pape Y, Bauchy M, Sant G (2016a) A dissolution–precipitation mechanism is at the origin of concrete creep in moist environments. J Chem Phys 145:054701

Pignatelli I, Kumar A, Bauchy M, Sant G (2016b) Topological control on silicates' dissolution kinetics. Langmuir 32:4434–4439

Pignatelli I, Kumar A, Field KG, Wang B, Yu Y, Le Pape Y, Bauchy M, Sant G (2016c) Direct experimental evidence for differing reactivity alterations of minerals following irradiation:the case of calcite and quartz. Sci Rep 6:20155

Plucinski M, Zwanziger JW (2015) Topological constraints and the Makishima–Mackenzie model. J Non-Cryst Solids 429:20–23

Poulsen HF, Neuefeind J, Neumann HB, Schneider JR, Zeidler MD (1995) Amorphous silica studied by high-energy X-ray diffraction. J Non-Cryst Solids 188:63–74

Qomi MJA, Bauchy M, Ulm F-J, Pellenq R (2015a) Polymorphism, its implications on structure–property correlation in calcium-silicate-hydrates. *In:* Nanotechnology in Construction. Sobolev K, Shah SP (eds) Springer International Publishing, p 99–108

Qomi MJA, Masoero E, Bauchy M, Ulm FJ, Del Gado E, Pellenq R (2015b) C-S-H across length scales: from nano to micron. *In:* CONCREEP 10: Proc 10th International Conference on Mechanics and Physics of Creep, Shrinkage, and Durability of Concrete and Concrete Structures, Vienna, Austria, September 21–23, 2015. Hellmich C, Pichler B, Kollegger J (eds) ASCE, p 39–48

Rader AJ, Hespenheide BM, Kuhn LA, Thorpe MF (2002) Protein unfolding: Rigidity lost. PNAS 99:3540–3545

Ravindren S, Gunasekera G, Tucker Z, Diebold A, Boolchand P, Micoulaut M (2014) Crucial effect of melt homogenization on the fragility of non-stoichiometric chalcogenides. J Chem Phys 140:134501

Richet P, Roskosz M, Roux J (2006) Glass formation in silicates:Insights from composition. Chem Geol 225:388–401

Rocherulle J, Ecolivet C, Poulain M, Verdier P, Laurent Y (1989) Elastic moduli of oxynitride glasses: Extension of Makishima and Mackenzie's theory. J Non-Cryst Solids 108:187–193

Rompicharla R, Novita DI, Chen P, Boolchand P, Micoulaut M, Huff W (2008) Abrupt boundaries of intermediate phases and space filling in oxide glasses. J Phys: Condens Matter 20:202101

Rouxel T (2006) Elastic properties of glasses: a multiscale approach. C R Méc 334:743–753

Rouxel T, Yoshida S (2017) The fracture toughness of inorganic glasses. J Am Ceram Soc 100:4374–4396

Ruta B, Baldi G, Chushkin Y, RuffLé B, Cristofolini L, Fontana A, Zanatta M, Nazzani F (2014) Revealing the fast atomic motion of network glasses. Nat Commun 5:3939

Salmon PS, Drewitt JWE, Whittaker DAJ, Zeidler A, Wezka K, Bull CL, Tucker MG, Wilding MC, Guthrie M, Marrocchelli D (2012) Mechanisms of network collapse in GeO_2 glass. J Phys: Condens Matter 24:415102

Shell MS, Debenedetti PG, Panagiotopoulos AZ (2002) Molecular structural order and anomalies in liquid silica. Phys Rev E 66:011202

Smedskjaer MM (2014) Topological model for boroaluminosilicate glass hardness. Front Mater 1:23

Smedskjaer MM, Bauchy M (2015) Sub-critical crack growth in silicate glasses: Role of network topology. Appl Phys Lett 107:141901

Smedskjaer MM, Mauro JC, Yue Y (2010) Prediction of glass hardness using temperature-dependent constraint theory. Phys Rev Lett 105:115503

Smedskjaer M, Mauro JC, Youngman RE, Hogue CL, Potuzak M, Yue Y (2011) Topological principles of borosilicate glass chemistry. J Phys Chem B 115:12930–12946

Smedskjaer MM, Bauchy M, Mauro JC, Rzoska SJ, Bockowski M (2015) Unique effects of thermal and pressure histories on glass hardness: Structural and topological origin. J Chem Phys 143:164505

Smith W, Greaves GN, Gillan MJ (1995) Computer simulations of sodium disilicate liquids. J Chem Phys 103:3091

Sreeram AN, Varshneya AK, Swiler DR (1991) Microhardness and indentation toughness versus average coordination number in isostructural chalcogenide glass systems. J Non-Cryst Solids 130:225–235

Su Q, King S, Li L, Wang T, Gigax J, Shao L, Lanford WA, Nastasi M (2018) Microstructure–mechanical properties correlation in irradiated amorphous SiOC. Scr Mater 146:316–320

Sugimura S, Inaba S, Abe H, Morinaga K (2002) Compositional dependence of mechanical properties in aluminosilicate, borate and phosphate glasses. J Ceram Soc Jpn 110:1103–1106

Svenson MN, Thirion LM, Youngman RE, Mauro JC, Bauchy M, Rzoska SJ, Bockowski M, Smedskjaer MM (2016) Effects of thermal and pressure histories on the chemical strengthening of sodium aluminosilicate Front Mat Glass Sci 3:14

Swiler D, Varshneya AK, Callahan R (1990) Microhardness, surface toughness and average coordination number in chalcogenide glasses. J Non-Cryst Solids 125:250–257

Tangeman J, Phillips BA, Navrotsky A, Weber JK, Hiwson AD, Key TS (2001) Vitreous forsterite (Mg_2SiO_4): Synthesis, structure, and thermochemistry. Geophys Res Lett 28:2517–2520

Taniguchi, T, Okuno M,, Matsumoto T (1997) X-ray diffraction and EXAFS studies of silicate glasses containing Mg, Ca, and Ba atoms. J Non-Cryst Solids 211:56–63

Tatsumisago M, Halfpap BL, Green JL, Lindsay SM,, Angell CA (1990) Fragility of Ge–As–Se glass-forming liquids in relation to rigidity percolation and the Kauzmann paradox. Phys Rev Lett 64:1549–1552

Thorpe MF (1983) Continuous deformations in random networks. J Non-Cryst Solids 57:355–370

Thorpe MF (1985) Rigidity percolation in glassy structures. J Non-Cryst. Solids 76:109–116

Thorpe MF, Duxbury PM (eds) (1999) Rigidity Theory, Applications. Kluwer Academic, New York

Thorpe MF, Phillips JC (eds) (2001) Phase Transitions, Self-organization in Electronic, Molecular Networks. Kluwer Academic, New York

Thorpe MF, Jacobs DJ, Chubynsky MV, Phillips JC (2000) Self-organization in network glasses. J Non-Cryst Solids 266–268:59

Trachenko K, Dove MT, Brazhkin VV, El'kin FS (2004) Network rigidity and properties of SiO_2 and GeO_2 glasses under pressure. Phys Rev Lett 93:135502

Vaills Y, Luspin Y, Hauret G (2001) Annealing effects in SiO_2–Na_2O glasses investigated by Brillouin scattering. J Non-Cryst Solids 286:224–234

Vaills Y, Qu T, Micoulaut M, Chaimbault F, Boolchand P (2005) Direct evidence of rigidity loss and self-organization in silicates. J Phys: Condens Matter 17:4889–4896

Varshneya AK, Mauro DJ (2007) Microhardness, indentation toughness, elasticity, plasticity, and brittleness of Ge–Sb–Se chalcogenide glasses. J Non-Cryst Solids 353:1291–1297

Verdonck E, Schapp K, Thomas LC (1999) A discussion of the principles and applications of modulated DSC. Int J Pharm 192:3–20

Wang M, Bauchy M (2015) Ion-exchange strengthening of glasses:atomic topology matters. ArXiv150507880 Cond-Mat

Wang B, Yu Y, Lee YJ, Bauchy M (2015) Intrinsic nano-ductility of glasses: the critical role of composition. Front Mater 2:11

Wang B, Krishnan NA, Yu Y, Wang M, Le Pape Y, Sant G, Bauchy M (2017a) Irradiation-induced topological transition in SiO₂: Structural signature of networks' rigidity. J Non-Cryst Solids 463:25–30

Wang M, Smedskjaer MM, Mauro JC, Sant G, Bauchy M (2017b) Topological origin of the network dilation anomaly in ion-exchanged glasses. Phys Rev Appl 8:054040

Wang M, Wang B, Krishnan NMA, Yu Y, Smedskjaer MM, Mauro JC, Sant G, Bauchy M (2017c) Ion exchange strengthening and thermal expansion of glasses: Common origin and critical role of network connectivity. J Non-Cryst Solids 455:70–74

Weigel C, Le Losq C, Vialla R, Dupas C, Clement S, Neuville DR, Ruffle B (2016) Elastic moduli of XAlSiO₄ aluminosilicate glasses:effects of charge-balancing cations. J Non-Cryst Solids 447:267–272

Wilcek F (2004) Whence the force of $F = ma$? I: Culture shock. Phys Today 57:11–12

Wilding MC, Benmore CJ, Tangeman JA, Sampath S (2004) Evidence of different structures in magnesium silicate liquids:coordination change in forsterite- to enstatite-composition glasses. Chem Geol 213:281–291

Wilkinson CJ, Zheng Q, Huang L, Mauro JC (2019) Topological constraint model for the elasticity of glass-forming systems. J Non-Cryst Solids X 2:100019

Winkler A, Horbach J, Kob W, Binder K (2003) Structure and diffusion in amorphous aluminum silicate:A molecular dynamics computer simulation. J Chem Phys 120:384

Yamane M, Okuyama M (1982) Coordination number of aluminum ions in alkali-free alumino–silicate glasses. J Non-Cryst Solids 52:217–226

Yan L (2018) Entropy favors heterogeneous structures of networks near the rigidity threshold. Nat Commun 9:1359

Yan L, Wyart M (2014) Evolution of covalent networks under cooling:contrasting the rigidity window and jamming scenarios. Phys Rev Lett 113:215504

Yang G, Gulbiten O, Gueguen Y, Bureau B, Sangleboeuf JCh, Roiland C, King EA, Lucas P (2012) Physical properties of the Ge$_x$Se$_{1-x}$ glasses in the $0 < x < 0.42$ range in correlation with their structure. Phys Rev B 85:144107

Yang K, Xu X, Yang B, Yang B, Cook B, Ramos H, Krishnan NMA, Smedskjaer MM, Hoover C, Bauchy M (2019a) Predicting the Young's modulus of silicate glasses using high-throughput molecular dynamics simulations and machine learning. Sci Rep 9:8739

Yang K, Yang B, Xu X, Hoover C, Smedskjaer MM, Bauchy M (2019b) Prediction of the Young's modulus of silicate glasses by topological constraint theory. J Non-Cryst Solids 514:15–19

Yasui I, Utsuno F (1993) Material design of glasses based on database – INTERGLAD. *In:* Doyama M, Kihara J, Tanaka M, Yamamoto R (eds) Computer Aided Innovation of New Materials II. Elsevier, Oxford, 1539–1544

Yildirim Y, Raty JY, Micoulaut M (2016) Revealing the crucial role of molecular rigidity on the fragility evolution of glass-forming liquids. Nat Commun 7:11086

Yu Y, Wang B, Lee YJ, Bauchy M (2015) Fracture toughness of silicate glasses: insights from molecular dynamics simulations. *In:* Symposium UU—Structure–Property Relations in Amorphous Solids. MRS Proc. CUP

Yu Y, Krishnan NMA, Smedskjaer MM, Sant G, Bauchy M (2018) The hydrophilic-to-hydrophobic transition in glassy silica is driven by the atomic topology of its surface. J Chem Phys 148:074503

Yuan X, Cormack AN (2003) Si–O–Si bond angle and torsion angle distribution in vitreous silica and sodium silicate glasses. J Non-Cryst Solids 319:31–43

Zhang M, Boolchand P (1994) The central role of broken bond bending constraints in promoting glass formation in the oxides. Science 266:1355–1357

Zheng Q, Yue Y, Mauro JC (2017) Density of topological constraints as a metric for predicting glass hardness. Appl Phys Lett 111:011907

Zeidler A, Wezka K, Rowlands RF, Whittaker DAJ, Salmon PS, Polidori A, Drewitt JE, Klotz S, Fischer HE, Wilding MC, Bull CL, Tucker MG, Wilson M (2014) High-pressure transformation of SiO₂ glass from a tetrahedral to an octahedral network:a joint approach using neutron diffraction and molecular dynamics. Phys Rev Lett 113:135501

Zotov N (2001) Effect of composition on the vibrational properties of sodium silicate glasses. J Non-Cryst Solids 287:231–236

Zwanziger JW, Tagg SL, Huffman JC (1995) Comment on broken bond-bending constraints and glass formation in the oxides. Science 268:1510

Reviews in Mineralogy & Geochemistry
Vol. 87 pp. 193-227, 2022
Copyright © Mineralogical Society of America

5

Molecular Simulations of Oxide and Silicate Melts and Glasses

Sandro Jahn

Institute of Geology and Mineralogy
University of Cologne
Zülpicher Straße 49b
50674 Köln
Germany.

s.jahn@uni-koeln.de

INTRODUCTION

The last twenty-five years have seen a tremendous increase in computing power. It is therefore not surprising that computer simulations of melts and glasses have become an indispensable tool to study the structure and properties of geological materials in general and of melts and glasses in particular. Although many of the molecular-scale simulation methods had already been developed at the time of volume 32, *Structure, Dynamics and Properties of Silicate Melts*, of Reviews in Mineralogy (Stebbins et al. 1995), their applications were still quite limited. This is especially true for *ab initio* or *first-principles* simulations that are now widely used to complement experiments or to make predictions of glass or melt structures and properties at extreme conditions. In the meantime two volumes of Reviews in Mineralogy and Geochemistry, volume 42 *Molecular Modeling Theory: Applications in the Geosciences* edited by Cygan and Kubicki (2001) and volume 71 *Theoretical and Computational Methods in Mineral Physics: Geophysical Applications* edited by Wentzcovitch and Stixrude (2010), demonstrated the importance and increasing power of molecular modeling approaches in the Geosciences. Each of these volumes had a chapter dedicated to the simulations of melts and glasses (Poole et al. 1995; Garofalini 2001; Karki 2010). Other review articles included examples from melt and glass simulations in the fields of theoretical spectroscopy (Kubicki 2001; Tossell 2001; Jahn and Kowalski 2014) and diffusion (de Koker and Stixrude 2010).

This chapter is intended to review the progress made in molecular dynamics simulations of oxide and silicate melts and glasses with a focus on the last twenty years. This includes both classical and *ab initio* approaches. As the number of simulation studies performed by far exceeds the scope of this chapter, the examples cited are somewhat selective and certainly incomplete. The methodology is introduced in a rather descriptive way and especially destined for inexperienced readers. More rigorous theory is provided in most of the chapters cited above as well as in text books about molecular modeling methods, see e.g., Allen and Tildesley (1987), Frenkel and Smit (2002), Marx and Hutter (2009). The strengths and weaknesses of the current approaches will be illustrated with a number of examples. Upcoming methods and challenges for future research will be discussed.

SIMULATION METHODS

Atomic-scale simulations of melts and glasses typically start with the setup of a simulation cell that contains a finite number of particles (atoms, ions or molecules) representing the system to be studied. The initial distribution of these particles can either be random with certain

---

1529-6466/22/0087-0005$05.00 (print)
1943-2666/22/0087-0005$05.00 (online) http://dx.doi.org/10.2138/rmg.2022.87.05

constraints on the minimum distance between particles or be derived from previous simulations. In the latter case, elements may have to be exchanged and particles to be added or removed. It is also common to start with a perfect or defect crystal structure that is melted at (very) high temperatures. As melts and glasses are condensed phases, periodic boundary conditions are mostly used to describe the interaction of particles. That means that particles interact across the simulation cell boundaries with particles in neighboring periodic images of the simulation cell. In molecular dynamics simulations, particles moving out on one side of the simulation box enter again from the opposite side. Periodic boundary conditions avoid surface effects that would be considerable for typical systems containing between a few hundred and a few million particles.

Having defined the simulation cell, the total energy of the system is computed using a particle interaction model. This model is the core ingredient of atomic-scale simulations as it determines the level of theory, the accuracy of the model and the accessible time and length scales of the simulation. Most classical interaction models describe the potential energy by analytical functions as a sum of pairwise interactions between the particles. Parameters of these functions need to be parameterized, either by reference to experimental data or to higher level theory. Three- or many-body terms may be added to those potentials either explicitly or implicitly, depending on the type of interaction model. Alternatively, electronic structure methods based on a quantum-mechanical treatment of the particle interactions have become increasingly available for the simulation of melts and glasses in the last 20 years or so. These methods are computationally much more demanding as the total energy is computed from wave equations describing the electronic structure of the material. Besides a number of necessary approximations described below, they are essentially parameter-free and therefore called first-principles or *ab initio* methods. An upcoming approach is the use of machine-learning or neural network potentials. Such potentials have the efficiency of traditional classical potentials but do not depend on a specific analytical function. In the following more details to the three different approaches to describe the particle interactions will be provided.

Classical potentials

In a classical description, all oxide and silicate melts and glasses consist of anions (at least oxygen) and cations (e.g., Si, Al, Na, K, Ca, Mg) that experience electrostatic interactions, of which the Coulomb interaction is by far the strongest. Also, all ions have a finite radius and if two ions approach each other too closely they experience repulsive interactions (this is actually due to the repulsion between the electrons of the the different ions). In addition, rather weak attractive dispersion (van der Waals) interactions arise from fluctuations of the electronic charge densities of atoms or ions. Thus, the simplest pairwise ionic potential between ions i and j separated by a distance r_{ij} used for the systems of interest here is the Born–Mayer potential

$$V\left(r_{ij}\right) = \frac{q_i q_j}{4\pi\varepsilon_0 r_{ij}} + A_{ij}\exp\left(-B_{ij}r_{ij}\right) - \frac{C_{ij}}{r_{ij}^6} \tag{1}$$

where the first term on the right side is the Coulomb potential with q_i and q_j being the charges of ions i and j, and ε_0 is the vacuum permittivity. The potential has three parameters A_{ij}, B_{ij} and C_{ij} for each pair of ions describing the short-range repulsion (second term) and the dispersion (third term) interaction, respectively. Born–Mayer potentials have been used for a long time and for many systems (already outlined by Poole et al. (1995)). More recent parameterizations include potentials for the five oxide system Na_2O–CaO–MgO–Al_2O_3–SiO_2 (Matsui 1994, 1998b) and for the nine oxide system including in addition K_2O, FeO, Fe_2O_3 and TiO_2 (Guillot and Sator 2007a). Recently, the latter set of potentials was extended to include B_2O_3 (Wang et al. 2018). The so-called Teter potential originally used for Na_2O–SiO_2 glasses (Cormack et al. 2002) was further developed to include, e.g., phosphate and fluoride ions (Lusvardi et al. 2008). Most of these potentials use partial charges for the ions as this leads to better melt or glass structures for this type of potential than using formal charges. Partial charges somehow correct for the partly covalent nature of the bonds in absence of an explicit account for many-body interactions or directional bonds. Some of the Born–Mayer type potentials are extensions of an early popular model for SiO_2 (often called BKS potential named after the authors of the paper, van Beest et al. 1990) and zeolites (Kramer et al. 1991) with additional terms to correct instabilities or charge neutrality issues, see e.g., Lacks et al. (2007) and Zhang et al. (2010) for the CaO–MgO–SiO_2 and Horbach et al. (2001) for the Na_2O–SiO_2 system. Ionic potentials that do not include polarization effects, flexible ionic radii or directional bonds are referred to as rigid ion models.

Despite their successful use in simulation studies of certain melt and glass properties, it is well known that due to their simplicity rigid ion models have a number of limitations. For instance, they do not account explicitly for ionic polarization effects and partially covalent bonds that have significant effects on the atomic structure and the vibrational properties not only of crystals but also of glasses and melts (Wilson et al. 1996b). Depending on the property of interest, new fitting methods may lead to an improved performance of rigid ion models, see e.g., Sundararaman et al. (2018, 2019). However, inherent problems of those models such as the missing polarizability cannot be overcome and alternative descriptions of the atomic interactions are indispensible.

Shell models (Dick and Overhauser 1958) describe especially the large anions by a positively charged core and negative shell that are connected by springs. External electric fields, e.g., from other ions, may then induce a dipolar polarization of the ion. Based on an early shell model for SiO_2 (Sanders et al. 1984) Tilocca et al. (2006) parameterized a potential for the system Na_2O–CaO–SiO_2 that includes both a core–shell approach and an additional three-body interaction term to control the O–Si–O bond angle. The breathing shell model is an extension to the shell model and also allows for the dynamic change of the ionic radius Matsui (1998a). This seems to be important if the potential is to be used in various coordination environments or in a wide range of pressures. However, this type of potential has been employed so far mainly to crystalline oxides.

Another type of polarizable ion models is based on a multipole expansion of the electronic charge density of ions and an explicit parameterization of the charge–dipole, dipole–dipole, etc., electrostatic interactions (Stone 1996). Some of these models consider polarization effects up to the quadrupolar level and account for breathing effects similar to the breathing shell model as well as ionic deformation effects. The increasing complexity of these interaction potentials is justified on physical grounds with the expectation to capture all relevant interactions explicitly and hence to improve the accuracy and predictive power of the interaction model. However, compared to the more empirical model approaches the resulting potentials often have a relatively large number of parameters that are not easily constrained by experimental data. Instead most of these models have been optimized by reference to electronic structure calculations (Tangney and Scandolo 2002; Aguado et al. 2003; Salanne et al. 2012). The latter are usually calculations in the framework of density functional theory (DFT, see below) for a number of atomic structures representing crystalline or non-crystalline phases. For each structure, several properties such as interatomic forces, the total energy and the stress tensor of the simulation cell, or induced ionic dipoles or computed. In the fitting procedure the force field parameters are varied until respective properties obtained from the classical potential fit best the DFT values. Initially, such potentials were developed for simple oxides such as SiO_2 (Wilson et al. 1996b; Tangney and Scandolo 2002) or Al_2O_3 (Jahn et al. 2006). More recent models were parameterized for the systems CaO–MgO–Al_2O_3–SiO_2 (Jahn and Madden 2007a), CaO–Al_2O_3–Y_2O_3–La_2O_3–SiO_2 (Haigis et al. 2013; Wagner et al. 2017a) or Na_2O–B_2O_3–SiO_2–La_2O_3 (Pacaud et al. 2017, 2018).

A problem one is faced with when using advanced ionic potentials is the computational cost. Depending on the level of complexity simulations with a polarizable and deformable ion model may easily slow down the calculations by one or two orders of magnitude compared to those using a simple Born–Mayer potential. This limits the accessible system size and simulation time and/or requires more substantial computing resources. On the other hand, the simulations are still significantly more efficient than electronic structure methods (see below), which is a motivation to continue their development.

Additional potential terms have been added to rigid ion models to account for bonded interactions in covalent systems. This includes, e.g., Morse-type functions

$$D_{ij}\left(\left[1-\exp\left(-a_{ij}\left(r_{ij}-r_0\right)\right)\right]^2-1\right) \tag{2}$$

where D_{ij} is the bond dissociation energy, a_{ij} is a measure of the bond strength and r_0 is the equilibrium bond distance. A successful set of potentials was parameterized, e.g., by Pedone et al. (2006) for simulations of multicomponent oxide and silicate glasses. Other potentials also contain three- or four–body angular terms. For instance, Pedone et al. (2008) parameterized a shell model for the SiO_2–H_2O system that includes both Morse–type and angular terms to approximate the complex interactions in hydrous silica. The Guillot and Sator (2007a) potential was extended by bonded interactions to account better for covalent effects in the interactions within the oxides (Dufils et al. 2017, 2020) and to include volatile species such as CO_2 (Guillot and Sator 2011) or H_2O (Dufils et al. 2020). Noble gas components were added to the Guillot and Sator (2007a) potential by Guillot and Sator (2012).

To be able to redistribute electronic charges or to change bonding during a classical MD simulation, reactive force fields have been developed and applied to a number of systems. An example is the coordination-dependent charge transfer potential (Huang and Kieffer 2006), which was used to predict the thermomechanical properties of B_2O_3. Bond-order-based reactive force fields of ReaxFF-type (van Duin et al. 2001) were parameterized, e.g., for the silica-water system (Fogarty et al. 2010) and for pure SiO_2 (Yu et al. 2016). Another type of reactive potential, the Mahadevan–Garofalini force field, MGFF (Mahadevan and Garofalini 2007), was further developed to study hydrous sodium silicate glasses (Mahadevan et al. 2019). At present, reactive force field parameters are not easily transferable between different coordination environments and therefore these potentials have been used mostly for glasses with few components and in a restricted range of temperatures and pressures.

Electronic structure methods

As already mentioned a major advance since the Poole et al. (1995) review paper is the enormous increase in available computing power, which now allows the molecular modeling of melts and glasses using first-principles electronic structure calculations on a routine basis. For condensed matter most simulations are based on density functional theory (DFT), which is also most relevant here. As the methodology is introduced in several excellent textbooks or review articles (e.g., Marx and Hutter 2009; Wentzcovitch and Stixrude 2010) we will concentrate here on summarizing a few practical aspects of DFT-based simulations. The use of *ab initio* calculations for computing vibrational or electronic excitation spectra of Earth materials was reviewed recently by Jahn and Kowalski (2014).

The interactions between particles in the simulation box are computed in a self-consistent calculation, in which the electronic charge distribution is optimized to yield the lowest total energy. According to the basic theorems of DFT there is a unique solution to the problem of finding the ground state electronic structure. Once this ground state is obtained for a set of atomic positions the total energy of the system, forces between the atoms and other derived properties such as the pressure or stress tensor may be derived. The electronic charge density is described in terms of single electron wavefunctions, the so-called Kohn–Sham orbitals, each of them experiencing an effective potential by all the other electrons and by the atomic nuclei. As these Kohn–Sham orbitals are complex functions in three-dimensional space they are expanded in a plane wave or Gaussian basis set (or a combination of both). One important input parameter of a DFT calculation is the energy cutoff that determines how many functions are included in the basis set. The cutoff has to be large enough to provide a converged total energy. Increasing the cutoff also leads to a more expensive calculation.

As only the outer electrons contribute to the chemical bonding, electrons from the inner shells are often combined with the nucleus into so-called pseudo-potentials. They are constructed in a way that outside a certain radius the pseudo-core (i.e., nucleus and inner electrons) reproduces the behavior of a respective all electron reference state whereas inside the pseudo-core the electronic structure is smoothed out, which greatly speeds up the calculation. There are tools to construct pseudo-potentials from scratch but most codes provide a database of pseudo-potentials that may be used after testing that they are suitable for the specific system. At the core of a DFT calculation is the exchange-correlation functional that approximates the quantum mechanical interaction between electrons, which cannot be solved exactly. Many different functional have been proposed over the years but in most studies of melts and glasses those using the local density approximation (LDA) or the generalized gradient approximation (GGA) have been employed thus far. While LDA simulations usually result in stronger bonds and higher density compared to experiment, GGA has the opposite effect. Depending on the specific system it may be useful to include explicitly the weak van der Waals interactions that are not well described by standard DFT. For melts and glasses containing ions with partially filled d- or f-orbitals (e.g., Fe^{2+} or Fe^{3+}), spin-polarized DFT calculations need to be performed. In the latter case, separate charge density calculations are performed for the two different spin directions. In DFT+U calculations an additional correction term, the Hubbard U, is introduced to correct for systematic errors in predicting the electronic band structure within standard DFT, which is especially important for transition metal oxide compounds (Coccoccioni 2010). Hybrid functionals combine a certain amount of exact exchange energy with DFT exchange and correlation. As these approaches have only been used in few cases for the simulation of oxide and silicate glasses and melts we do not provide further technical details here and refer the reader to the literature cited at the beginning of this section.

Molecular dynamics simulations

After having defined a simulation cell and a particle interaction potential the atomic structure of the glass or melt can be studied. The most intuitive approach is to perform a molecular dynamics (MD) simulation. Using this method Newton's equations of motion are integrated numerically on the time scale of molecular and atomic motions. The force acting on each particle in the simulation cell is computed as the negative gradient of the potential energy at the position of the particle. If the force $\mathbf{F}$ is non-zero the particle will be accelerated according Newton's second law $\mathbf{F} = m\mathbf{a}$, where $\mathbf{a}$ is the acceleration and m the mass of the particle. For many particles in the simulation box this leads to a set of coupled differential equations that are solved numerically by propagating the system forward in space and time using a small time step of typically 10^{-16} to 10^{-15} seconds. In each time step, particle positions and velocities are updated. Temperature is defined by the mean particle velocities according to statistical mechanics: $\langle mv^2 / 2 \rangle = 3k_B T / 2$, where v is the velocity, k_B the Boltzmann constant and angular brackets denote an average over particles and time. MD simulations are referred to as classical MD if a classical potential is used and as *ab initio* or first-principles MD if the interaction is described by electronic structure calculations. In both cases, the MD produces trajectories according to classical (Newtonian) mechanics. For some light elements such as H or He, a quantum treatment of the dynamics is sometimes required, which can be approached using path integral MD simulations (Marx and Parrinello 1996). For an isolated system, which is also called microcanonical or *NVE* ensemble, the total energy E is conserved and the simulation box should equilibrate itself at a defined pressure and temperature. In this case, the volume V and the number of particles N are also constant. Often, MD simulations are performed at constant T and/or P by coupling the system to a thermostat and/or a barostat (*NVT* and *NPT* ensembles). A more detailed introduction to the MD method is provided, e.g., in text books by Allen and Tildesley (1987), Frenkel and Smit (2002) or Marx and Hutter (2009).

Thermodynamic equilibrium can only be reached if the temperature is high enough and particles are allowed to sample the configurational space. For oxide and silicate melts this minimum temperature is often well above 2000 K even at atmospheric pressure, which constitutes a major obstacle for modeling of natural melts that often occur at lower temperatures. As shown e.g., by Spiekermann et al. (2016) the lifetime of the strong Si–O bond increases exponentially with decreasing temperature. At 2000 K it reaches a few hundred picoseconds, which especially for *ab initio* MD simulations is on the order of or even succeeds the total simulation time. As a consequence, MD simulations are either performed at very high temperature (which is often the case in studies of silicate and oxide melts) or the systems quickly reach the regime of an undercooled liquid or even of a glass at temperatures well above the experimental glass transition temperature. The simulated glass then may resemble the structure of the liquid as structural changes observed in real glass-forming systems well below 2000 K cannot be captured anymore by MD simulation. Thus, the construction of a good glass model from MD simulations remains a great challenge. On the other hand, MD simulations can easily reach extreme P/T conditions that are very difficult to achieve experimentally.

Monte-Carlo simulations

While in the majority of the molecular modeling studies of melts and glasses of the last 25 years molecular dynamics simulations have been employed, the Monte-Carlo method may constitute a valuable alternative in cases where the dynamics (i.e., time) is not important. This is the case, e.g., for purely structural studies or for the calculation of thermodynamic properties. The idea behind Monte-Carlo simulations is to sample the structure of a melt or glass from random configurations weighted with a certain probability that depends on the respective energy. In practice, new structures are generated from a previous configuration (similar to one snapshot in the molecular dynamics simulation) by displacing atoms in a random way. The new configuration is accepted with a certain probability that depends on the energy change with respect to the previous configuration and on temperature. Often the Metropolis sampling procedure is used. So far, the energies have been computed mostly from classical potentials but *ab initio* approaches can be used as well. Average structural and thermodynamic properties are derived from a large number of Monte-Carlo generated atomic configurations. For details on Monte-Carlo methods the reader to referred to the introductory text books (Allen and Tildesley 1987; Frenkel and Smit 2002) and to the chapter by Poole et al. (1995). For the interpretation of neutron and X-ray diffraction data of liquids and glasses two related Monte-Carlo approaches have been employed, the Reverse Monte-Carlo, RMC (McGreevy 2001), and the Empirical Potential Structure Refinement, EPSR (Soper 2000), methods. Both methods use the experimental data in conjunction with a Monte-Carlo simulation to develop a structural model for the melt or glass. While in RMC the structure is optimized to reproduce the experimental structure factors and radial distribution functions (for their definition see below), in EPSR a classical potential is modified to predict a structure that reproduces the experimental data. In both cases the structural solution is not unique and therefore the obtained structural model should be verified by other computational or experimental methods.

Data analysis

Atomic structure. One of the main motivations for performing molecular dynamics or Monte-Carlo simulations is to develop structural models for melts or glasses. As disordered materials do not possess a long-range atomic order their physical properties are usually isotropic and their structures are described in terms of radial distribution functions $g(r)$. As silicates and oxides consist of more than one chemical element, partial radial distribution functions, $g_{AB}(r) = g_{BA}(r)$ are defined for each distinct pair of atoms A and B

$$g_{AB}(r) = \frac{1}{c_A c_B \rho_0 N} \left\langle \sum_{j=1}^{N_A} \sum_{k=1}^{N_B} \delta\left(\mathbf{r} - \left(\mathbf{r}_j - \mathbf{r}_k\right)\right) \right\rangle \tag{3}$$

These functions describe the mean atomic density of atoms B at a distance r from atoms A relative to the average atomic density of atoms B in the system. ρ_0 is the total atomic density of the system. At long distances r the distribution is random and $g_{AB}(r)$ is equal to one. The first peak of $g_{AB}(r)$ is usually well defined and describes the first coordination shell of atoms B around atoms A (or the other way around). Higher coordination shells are often identified but they increasingly overlap with each other.

The coordination number CN of atom A by atoms B is usually derived from integration over the first peak of $g_{AB}(r)$ up to the first minimum at r_{cut}

$$CN = 4\pi\rho_0 c_B \int_0^{r_{cut}} g_{AB}(r) r^2 dr \tag{4}$$

Due to the disordered nature of the melt and glass structures, both $g_{AB}(r)$ and CN are usually averaged over all atoms in the simulation cell and over the different configurations sampled in the course of the simulation.

The Fourier transformation of $g_{AB}(r)$ yields the partial static structure factor

$$S_{AB}(Q) = \delta_{AB} + \sqrt{c_A c_B}\, \rho_0 \int_0^\infty 4\pi r^2 \left(g_{AB}(r) - 1\right) \frac{\sin(Qr)}{Qr} dr \tag{5}$$

where Q is the magnitude of a wavevector or scattering vector accessible in diffraction experiments. Weighting the $S_{AB}(Q)$ with the concentrations c_A and c_B and the respective scattering factors, e.g., neutron scattering lengths b_A and b_B or X-ray form factors $f_A(Q)$ and $f_B(Q)$ (see Drewitt et al. 2022, this volume), leads to the total static structure factors for neutrons

$$S_N(Q) = \sum_A \sum_B (c_A c_B)^{1/2} \frac{b_A b_B}{\langle b^2 \rangle} S_{AB}(Q) \tag{6}$$

and for X-rays

$$S_X(Q) = \sum_A \sum_B (c_A c_B)^{1/2} \frac{f_A(Q) f_B(Q)}{\langle f(Q)^2 \rangle} S_{AB}(Q) \tag{7}$$

with $\langle b^2 \rangle = \sum_j c_j b_j^2$ and $\langle f(Q)^2 \rangle = \sum_j c_j f_j(Q)^2$. These functions are directly comparable to experimental diffraction data. Structure factors and radial distribution functions contain structural information well beyond the nearest neighbor shell and are thus representatives of the intermediate-range order in the melt or glass.

Vibrational spectra. Raman and infrared (IR) spectroscopy as well as inelastic X-ray and neutron scattering are common experimental methods to study the vibrational properties of materials. Often, these techniques are used to obtain fingerprint information about the molecular building blocks and about chemical bonding in a sample. If properly normalized quantitative information about species concentrations may be extracted (see, e.g., Drewitt et al. 2022, this volume). In melts and glasses, the vibrational bands are often rather broad and strongly overlap, which complicates their assignment.

Vibrational spectra may also be derived from trajectories of MD simulations. The most straightforward approach is to compute the spectrum of the velocity autocorrelation function (Allen and Tildesley 1987; Frenkel and Smit 2002). The latter correlates the velocity of a single particle at time t to the velocity of the same particle at an initial time t_0. During a vibration this autocorrelation oscillates and a Fourier transformation yields the respective frequency.

The obtained power spectra are usually averaged over all atoms of one type. They contain all Fourier components of the atomic motions during the MD simulation (including some noise) and in general show a similar complexity as the experimental spectra. A major advantage of the simulated spectra is that they are element- or even atom-specific. For glasses with relaxed atomic positions, vibrational spectra may also be computed by lattice dynamics approaches (see recent reviews of Gale and Wright (2010) for classical and Baroni et al. (2010); Wentzcovitch et al. (2010) for *ab initio* lattice dynamics).

The calculation of theoretical Raman or IR spectra from MD simulations requires additional information such as the dipole moment (IR) or the polarizability (Raman) of the simulation cell. Due to the selection rules, their time evolution determine whether individual modes are visible and, if so, their intensities. IR spectra have been derived from classical MD simulations by computing the Fourier transformation of the total polarization autocorrelation function, e.g., for amorphous SiO_2 (Wilson et al. 1996b). In this study it was shown that rigid ion models such as the Born–Mayer potential are insufficient to describe the vibrational spectrum of amorphous SiO_2 due to the neglect of ionic polarization effects. *Ab initio* IR spectra are routinely derived on the basis of density functional perturbation theory (Gonze and Lee 1997; Baroni et al. 2001), e.g., for amorphous SiO_2 (Giacomazzi et al. 2009). Raman spectra have been derived from combining lattice dynamics with bond polarizability models, e.g., for amorphous SiO_2 by Zotov et al. (1999) using a classical potential and by Giacomazzi et al. (2009) in a DFT-based study. An alternative approach for computing IR and Raman spectra was introduced by Umari and Pasquarello (2002) by using DFT calculations within a finite electric field. Putrino and Parrinello (2002) computed the autocorrelation function of the polarizability tensor during an *ab initio* MD simulation, which provides a direct but also computationally quite expensive access to the Raman spectrum. A more detailed description of these advanced methods and further references can be found in reviews by Kubicki (2001) and Jahn and Kowalski (2014).

For the interpretation and band assignment it is useful to decompose bulk vibrational spectra into partial contributions from individual molecules or structural units and, even further, into separate vibrational modes that contribute at different frequencies to the bulk vibrational spectrum. Traditionally, such information was obtained by normal mode analysis in the gas phase under the assumption that similar molecular structural units exist in the condensed phase (e.g., Kubicki et al. 1993; Kubicki and Sykes 1993). However, the ability of this approach to account for the variability of structural environments in the condensed phase is limited. Further, high temperature and/or high pressure effects are difficult to consider. A relatively straightforward approach to extract partial vibrational spectra from MD trajectories that requires only little additional computational effort is to project the eigenvectors or particle velocities on normal mode directions (see e.g., Sarnthein et al. 1997; Zotov et al. 1999; Spiekermann et al. 2012). The method is illustrated in Figure 1. The power spectrum of the velocity autocorrelation function introduced above (denoted by full VDOS - vibrational density of states) has a complex shape with several peaks and shoulders. Projections of the particle velocities of a SiO_4 unit on the tetrahedral normal mode directions, e.g., of a symmetric stretching vibration (A_1), and Fourier transformation of the projected velocity autocorrelation functions leads to the so-called quasi-normal mode v_1 shown in Figure 1. The same is done for the other three tetrahedral modes, the asymmetric stretching mode v_3 (F_2), the E-bending mode v_2 (E) and the umbrella bending mode v_4 (F_2). With this procedure well-defined partial contributions to the vibrational spectra are extracted. Some of the quasi-normal modes are not fully decoupled, which leads some 'ghost' spectral density, e.g., of v_4 in the spectral range of v_3 (Spiekermann et al. 2012). Figure 1 also shows a temperature dependence of the mode frequencies.

The connectivity of network-forming units, mostly SiO_4 or AlO_4 tetrahedra, is commonly expressed in terms of Q^n-species, where n denotes the number of bridging oxygens, which ranges between zero (isolated tetrahedron) to four (fully polymerized tetrahedron).

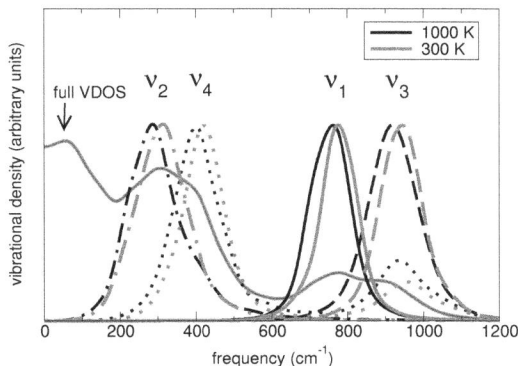

Figure 1. Spectral density of the four tetrahedral quasi-normal modes of a SiO_4 monomer at 300 K and 1000 K in the system SiO_2–H_2O derived from the mode-projected velocity autocorrelation function. The total power spectrum denoted full VDOS is shown for comparison (modified from Spiekermann et al. (2012)).

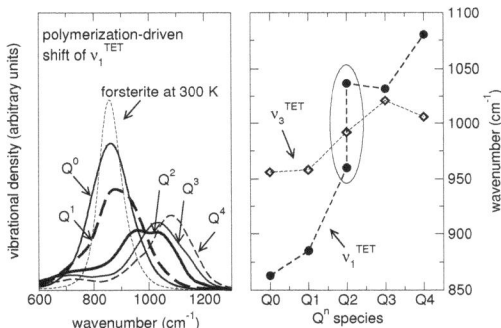

Figure 2. *Left:* symmetric tetrahedral ν_1 modes of MgO–SiO_2 glasses from a mode-projection analysis of *ab initio* MD simulations for different Q^n species. For comparison the corresponding spectrum of a crystal-line phase with forsterite structure is shown with dashed lines. *Right:* evolution of peak frequencies for the symmetric ν_1 and asymmetric ν_3 modes as a function of Q^n (modified from Spiekermann et al. 2013).

The application of the mode-projection method to different Q^n-species shows (Fig. 2), in agreement with experiment, a shift of the tetrahedral ν_1 mode towards higher wavenumbers for increasing degree of polymerization (Spiekermann et al. 2013). The largest shift is observed between Q^1 and Q^3 whereas the vibrational band of Q^2 shows a double-peak structure, which is consistent with previous simulations (Zotov et al. 1999) and a principal component analysis of experimental Raman data (Malfait et al. 2008). Interestingly, the asymmetric tetrahedral ν_3 mode does not show this splitting (Fig. 2). Including a few additional modes such as the symmetric stretch of the Si_2O_7 dimer ($\nu_1{}^{DIM}(Q^1)$) and the asymmetric stretch of the Si–O–Si bridging oxygen ($\nu_3{}^{BO}$) experimental spectra were re-interpreted (Spiekermann et al. 2013). A comparison to the conventional mode assignment is shown in Figure 3.

Unfortunately, the simple mode projection method does not yield information about Raman or IR intensities. One solution for this problem is to couple the mode projection with a bond polarizability model (Zotov et al. 1999). However, parameters for the latter have only been derived for a few simple model systems. Another promising approach is to compute local IR and Raman spectra from *ab initio* MD trajectories using either maximally localized Wannier functions (Thomas et al. 2013) or from Voronoi tesselation of the electron density (Thomas et al. 2015). Both approaches are implemented in the TRAVIS code (Brehm and Kirchner 2011).

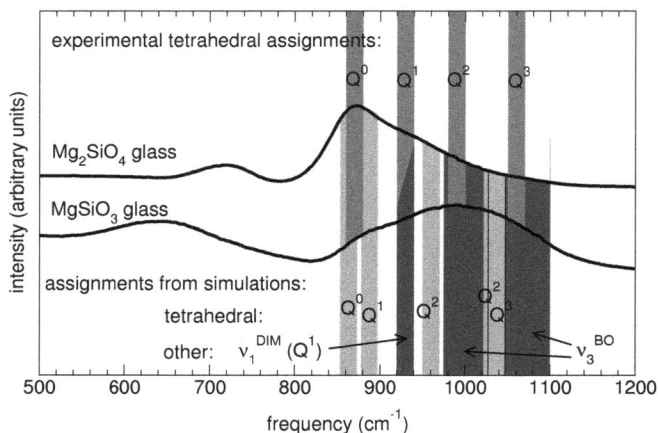

Figure 3. Mode assignment of experimental Raman spectra (Kalampounias et al. 2009) with conventional peak fitting (*top*) and resulting from the mode-projection analysis of *ab initio* MD simulation data (modified from Spiekermann et al. (2013)).

Element-specific spectroscopy. Diffraction experiments and to some extent vibrational spectroscopies provide structural information on the short to intermediate range order in the melt or glass. However, data obtained from these methods are not necessarily element-specific and often only a weighted superposition of partial contributions are measured. Other techniques such as X-ray absorption fine structure (XAFS) or nuclear magnetic resonance (NMR) spectroscopy are suitable approaches to obtain site-specific information about melt and glass structures. Theoretical spectroscopy has become a powerful method to interpret experimental spectra and to test structure models from simulations against experiments. Technically, the calculation of XAFS or NMR spectra requires an input structure model, which may be, e.g., a molecular cluster or a snapshot from an MD simulation. For this atomic configuration an electronic structure calculation is performed to obtain the respective spectrum. The different theoretical approaches have been reviewed recently (Charpentier et al. 2013; Jahn and Kowalski 2014; Pedone 2016) and more details about the different types of spectroscopy are provided in the dedicated chapters of this volume (Henderson and Stebbins 2022, this volume). A notable new development is the use of machine learning approaches to predict NMR parameters in aluminosilicate glasses (Chaker et al. 2019).

Physical and thermodynamic properties. To understand the role of melts and glasses in large-scale geological structures and processes, knowledge of the detailed atomic and electronic structure of the materials is sometimes not required. Instead, physical properties such as the viscosity or diffusion coefficients need to be derived. There are a number of well established methods to extract such properties from an MD simulation trajectory (Allen and Tildesley 1987; Frenkel and Smit 2002). Self-diffusion coefficients of individual atomic species are readily extracted from the linear slopes of the atomic mean square displacements. Viscosities are typically obtained from integration of the stress tensor autocorrelation function. Especially for the studies of melts (and glasses) at high pressures the MD simulation data relating pressure, temperature, volume and total energy of the system have been used to parameterize equations of state or more advanced thermodynamic models that may be further used in larger scale models. For instance, a self-consistent thermodynamic description of silicate melts was developed, e.g., by de Koker and Stixrude (2009). A different model applied to classical MD simulations was presented by Ghiorso et al. (2009).

SIMULATIONS OF OXIDE AND SILICATE MELTS AND GLASSES

Oxide melts and glasses at ambient pressure

SiO_2 and GeO_2 melts and glasses. SiO_2 is a prototype for a strong glass former. Its atomic structure and dynamics has therefore been studied extensively to understand better the different relaxation processes in the (supercooled) liquid that eventually leads to glass formation. Many of these studies used classical MD simulations with the well established BKS potential (van Beest et al. 1990) and simulation times of nanoseconds that are outside the common range of *ab initio* MD simulations. The structure of SiO_2 liquid up to temperatures as high as 4000 K can be described as a random tetrahedral network (e.g., Horbach and Kob 1999a). From a fundamental point the glass-forming process as described, e.g., by mode-coupling theory (Götze and Sjögren 1992) has been addressed in MD simulations by quantifying the structural relaxation processes and the intermediate range structural order in the supercooled liquid (e.g., Horbach and Kob 1999a; Vogel and Glotzer 2004a,b). The self-diffusion coefficients and the viscosity of the melt show a transition from Arrhenius-behavior at lower temperatures to non-Arrhenius behavior at high temperatures, which was interpreted as a fragile-to-strong crossover (Horbach and Kob 1999a; Saksaengwijit et al. 2004; Vogel and Glotzer 2004a). The transition temperature derived from the simulations is at 3330 K (Horbach and Kob 1999a). A similar behavior is also observed for GeO_2 but at much lower temperature (Micoulaut et al. 2006).

Zhang et al. (2004) showed that for BKS silica the transport coefficients strongly depend on system size with variations in the viscosity of more than a factor of two between systems with 180 and 2400 atoms in the simulation box at 3310 K. Carré et al. (2007) performed MD simulations with truncated Coulomb interactions that are computationally much more efficient for large systems than conventional MD simulations with Ewald sums. They demonstrated that this approach yields very good results if the interaction cutoff is chosen large enough (10 Å) to capture essential features of the intermediate range order of the connected SiO_4 tetrahedra. A new method to fit interatomic potentials for SiO_2 by reference to Car–Parrinello-type *ab initio* MD simulations was proposed by Carré et al. (2008). A comparative MD study including both classical potentials and *ab initio* simulations for GeO_2 showed good consistency between the different simulation approaches but also confirmed the strong finite size effects when looking at medium range structural properties such as ring statistics (Hawlitzky et al. 2008). The comparison of vibrational spectra of vitreous SiO_2 and GeO_2 from density-functional theory calculations with experiments support the picture that the structure of these systems can be described as a continuous random network of corner-sharing tetrahedra (Giacomazzi and Pasquarello 2007; Giacomazzi et al. 2009). This network depolymerizes at very high temperature and low density close to the critical point (Green et al. 2018).

Al_2O_3 and other M_2O_3 melts, B_2O_3 melt and glass. Motivated by experimental work on the structure of liquid alumina using aerodynamic levitation in conjunction with X-ray and neutron diffraction (Ansell et al. 1997; Landron et al. 2001), a number of MD simulation studies were performed to investigate the Al coordination and the linkage of AlO_x polyhedra in the melt (San Miguel et al. 1998; Hemmati et al. 1999; Gutierrez et al. 2000; Jahn and Madden 2007b) and in the amorphous (simulation quenched melt) state (Gutierrez and Johansson 2002; Davis and Gutiérrez 2011). During melting of the thermodynamically stable Al_2O_3 polymorph a remarkable change in Al coordination from octahedral to predominantly tetrahedral is observed. As a consequence the density contrast between the melt and the solid is large, about 20% upon melting, and the liquid can be deeply undercooled in a containerless environment before it eventually crystallizes (Krishnan et al. 2005). Al_2O_3 is an especially challenging system for classical interaction potentials as the cubic bixbyite structure is predicted to be the most stable crystalline phase (instead of corundum) for most classical force fields. Quadrupolar polarization effects that are missing from most classical potentials seem to stabilize the corundum structure (Wilson et al. 1996a; Jahn et al. 2006). The MD simulations of alumina melt yield a rather wide

range of average Al coordinations from 4.1 (Gutierrez et al. 2000) to 4.9 (Hemmati et al. 1999). Other studies predict a mixture of 4- and 5-fold coordinated Al with an average close to 4.5 (San Miguel et al. 1998; Jahn and Madden 2007b), which is closer to experimental data. Verma et al. (2011) report an average Al coordination of 5.2 at ambient pressure from *ab initio* MD simulations. Besides differences in the interaction potential a significant second reason for these discrepancies seems to be the quite different densities used to model the liquid at atmospheric pressure. From the results of a high pressure study of liquid Al_2O_3 (Verma et al. 2011) it can be estimated that the density spread corresponds to a pressure range of about 10 GPa, which results in a difference in Al–O coordination of at least 0.5.

Apart from studying the structure of liquid Al_2O_3, Jahn and Madden (2007b, 2008) computed the dynamic structure factor $S(Q,\omega)$ to interpret inelastic X-ray scattering experiments of Sinn et al. (2003). The dynamic structure factors were described by a generalized hydrodynamic model, in which the line widths are related to different relaxation processes and to frequency-dependent transport coefficients. Verma et al. (2011) derived thermodynamic properties such as the thermal expansion coefficient, the bulk modulus and Grüneisen parameter. An interesting phenomenon that has been studied intensively for Al_2O_3–Y_2O_3 system is a liquid–liquid phase transition that leads to a coexistence of two different structures with the same chemical composition in the amorphous state (first observed by Aasland and McMillan (1994)) or in supercooled liquid (Greaves et al. 2008). Already in the end member liquids there are indications of a heterogeneous distribution of 4-fold and 5,6-fold coordinated Al (Jahn and Madden 2007b) or Y (Belonoshko et al. 2001), which leads to dynamic regions with different density on the nanometer scale. Classical MD simulations using a polarizable ion model were performed to support the interpretation of neutron and X-ray diffraction experiments, in which indications of a liquid–liquid phase transition were found at a composition close to 20% Y_2O_3 (McMillan et al. 2003; Wilding et al. 2013). Wilson and McMillan (2004) compared Al_2O_3–Y_2O_3 to Al_2O_3–La_2O_3 melts and found subtle changes in the structures due to the different ionic radii of Y and La. Structural analysis of Al_2O_3–Y_2O_3 glasses from classical MD simulations (Du et al. 2009) showed good agreement with experimental diffraction data and a very diverse structural environment of AlO_x and YO_y polyhedra (with x being mostly equal to 4 or 5 and y to 6, 7 and 8). A higher spectral density in the vibrational density of states is observed in the Raman spectra and by MD simulation for the low-density amorphous form, which suggests that the liquid–liquid phase transition is driven by differences in density and entropy (Wilding et al. 2014). The existence of a liquid–liquid phase transition in Al_2O_3–Y_2O_3 melts has been disputed, e.g., in an experimental study by Barnes et al. (2009) who could not reproduce the results of Greaves et al. (2008). Unfortunately, this controversy cannot be resolved by MD simulations alone as structural heterogeneities on the molecular scale do not necessarily lead to a phase separation in the thermodynamic sense.

B_2O_3 is another important component of many oxide and silicate melts. Pure liquid and glassy B_2O_3 has been of some interest as it is a glass forming system with a low glass transition temperature with a structure quite different from SiO_2 and GeO_2. *Ab initio* MD simulations show that in the liquid state boron atoms are predominantly threefold coordinated to oxygen atoms and oxygen atoms bridge two neighboring boron atoms (Ohmura and Shimojo 2008). In the glassy state planar boroxol rings (three-membered BO_3 units) seem to dominate the structure (Umari and Pasquarello 2005; Ferlat et al. 2008). The structural changes between liquid and glassy state was investigated in detail by Scherer et al. (2019) using both classical and *ab initio* MD approaches. They conclude that due to the complex chemical interaction in this system effective potentials have to include at least a 3-body interaction term, i.e., purely pairwise interaction models are insufficient to capture the essential interactions. Scherer et al. (2019) also provide a comprehensive summary of previous simulation studies of B_2O_3 melt and glass including a comparison of different interaction potentials.

Binary MO–Al$_2$O$_3$ melts and glasses. The technical developments of container-less levitation methods in conjunction with laser heating and X–ray or neutron diffraction also stimulated structural studies of binary MgO–Al$_2$O$_3$ and CaO–Al$_2$O$_3$ melts and glasses. Complementary atomic-scale simulations are even more important for binary than for simple oxides as the partial radial distribution functions strongly overlap. This is especially critical for the first peaks of the anion–cation $g_{AB}(r)$. In this case the peak overlap may prevent a meaningful determination of nearest neighbor distances or coordination numbers from a single diffraction experiment (Hennet et al. 2007; Drewitt et al. 2011). Motivated by X-ray diffraction (Hennet et al. 2007) and inelastic X-ray diffraction (Pozdnyakova et al. 2007) experiments Jahn (2008) performed classical MD simulations of MgO–Al$_2$O$_3$ melts using a polarizable ion model. With increasing MgO content, both the Al and Mg coordination numbers slightly decreased from 4.4 to 4.3 and from 5.2 to 4.9, respectively. While pure Al$_2$O$_3$ melt has a large oxygen deficit to form a tetrahedral network of AlO$_4$ tetrahedra (similar e.g., to SiO$_2$), additional Mg ions act as charge compensators that promote network formation. Although the MgO–Al$_2$O$_3$ system is not glass-forming, a shallow maximum in the viscosity of the melt is observed close to the metaluminous composition (MgAl$_2$O$_4$). More recently, the MgO–Al$_2$O$_3$ melt system was studied by *ab initio* MD simulations (Karki et al. 2019). The structural parameters are largely consistent with the classical MD results. Thermodynamic properties show that the mixture is close to ideal. This behavior is quite different to MgO–SiO$_2$, which shows a miscibility gap on the SiO$_2$-rich side of the phase diagram. Karki et al. (2019) concluded that although alumina and silica are generally considered to play a similar structural role in the polymerization of aluminosilicates, both oxide components behave differently in terms of thermodynamic and transport properties.

The CaO–Al$_2$O$_3$ system is especially interesting as it is glass forming in the absence of any other network-forming cation than Al. Thus, melts and glasses of this binary melt have been studied intensively both by experiments and by MD simulations. Kang et al. (2006) used a Born–Mayer-type potential plus some additional covalent interaction terms to predict a structural model for calcium aluminate glasses. Their computed structure factors and radial distribution functions are in very good agreement with experimental data. For compositions close to the eutectic with 37 mol% of Al$_2$O$_3$ the aluminum is found in a predominantly tetrahedral coordination, whereas the coordination number increases for high Al$_2$O$_3$ contents, which is in agreement with the studies of pure Al$_2$O$_3$ melt discussed above. The mean Ca coordination varied between 5.6 and 5.9, i.e., it is considerably larger than that of Mg due to the larger ionic radius. A similar study with a different pair potential yielded a Ca coordination of 6.2 (Thomas et al. 2006). Of the six oxygen neighbors, three are typically bridging oxygens whereas the other three are non-bridging oxygens. Bridging oxygens are usually connected to two Al and two Ca, whereas non-bridging oxygens have one Al and three Ca neighbors. Cristiglio et al. (2008, 2010) performed *ab initio* MD simulations of CaO–Al$_2$O$_3$ melts and reproduced well respective neutron diffraction data. A combined investigation of calcium aluminate melts by neutron and X-ray diffraction complemented by MD simulations with a polarizable ion model (Drewitt et al. 2011) allowed to extract partial structural information due to the different cross-sections for neutrons and X-rays. Also, the difficulty to extract information about the Ca coordination environment from those experiments alone was demonstrated by comparing simulation-based total and partial radial distribution functions. To solve this problem new neutron diffraction experiments were performed using the method of Ca isotope substitution, in which three measurements with different Ca scattering lengths were made (Drewitt et al. 2012, 2017). By combining the data sets the partial $g_{CaCa}(r)$ can be extracted and directly compared to simulation results (with quite satisfactory agreement). In CaAl$_2$O$_4$, the amount of 5-fold coordinated Al is reduced considerably during fast cooling across the glass transition. Further, the MD simulations predict a change in CaO$_x$ polyhedral connectivity between the melt and the glass (Drewitt et al. 2012). In Ca$_3$Al$_2$O$_6$, which is at the Ca-rich end of the glass forming region, most AlO$_4$ tetrahedra are still part of an infinite network (see Fig. 4), which is not the case anymore for higher CaO concentrations (Drewitt et al. 2017). The structural evolution of CaAl$_2$O$_4$ melts and glasses at high pressures was investigated by Drewitt et al. (2015).

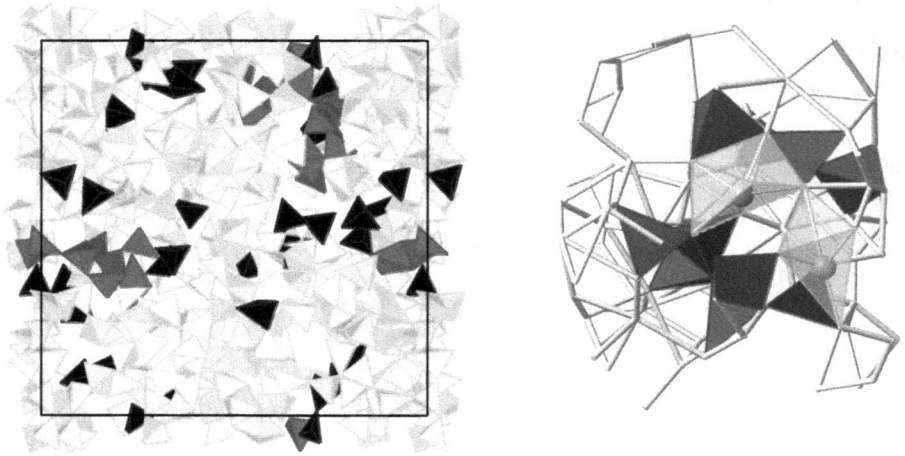

Figure 4. *Left:* AlO_4 network structure in $Ca_3Al_2O_6$ melt. The light tetrahedra are connected to a single infinite network whereas gray tetrahedra represent small clusters of 11 to 12 connected tetrahedra. Black tetrahedra are either AlO_4 monomers or Al_2O_7 dimers. *Right:* representative snapshot from the simulation showing two strongly distorted Ca-centered polyhedra with coordinations 5 and 6 (transparent light gray). The dark polyhedra are neighboring AlO_4 tetrahedra and sticks represent cation–oxygen bonds (figure modified from Drewitt et al. 2017).

Silicate melts and glasses at ambient pressure

Silicate melts. Many of the molecular dynamics simulations of silicate melts at ambient conditions have been focused on the $CaO–MgO–Al_2O_3–SiO_2$ and on the alkali silicate systems. One motivation for these studies was structure prediction and comparison to experimental data. Static structure factors were derived from MD simulations and compared to neutron and X-ray diffraction. Another focus was set on studying the structural and dynamic changes between the melt, the supercooled liquid and the glassy state.

Structural differences between $CaO–Al_2O_3–SiO_2$ and pure SiO_2 melt were subject of an *ab initio* MD simulation study by Benoit et al. (2001). The calcium aluminate melt was found to have more non-bridging oxygen ions than nominally derived form the stoichiometry, which is mainly due to 5-fold coordinated Al and the formation of $(Al,Si)O_3$ triclusters. Further, the Al avoidance rule is violated in the melt. Two-fold rings are observed in both melts as well as a small amount of five-fold coordinated Si. In a classical MD simulation study, Morgan and Spera (2001) investigated structural, physical and thermodynamic properties of $CaAl_2Si_2O_8$ in the liquid state and across the glass transition. The latter was defined as the temperature at which the speciation in the melt was frozen in and for the potential used was identified at 2800 K.

Classical MD simulations were used to complement X-ray and neutron diffraction studies of $CaSiO_3$ melt and glass (Benmore et al. 2010; Skinner et al. 2012). During cooling of the melt the coordination number of Ca increases from about five to about six in the glass. Further, the authors concluded that the fragile viscosity behavior and the glass-forming ability of $CaSiO_3$ can be structurally rationalized by the formation of edge- and face-sharing CaO_x-polyhedra whereas the melt is dominated by corner-sharing Ca polyhedra. This behavior is similar to the $CaAl_2O_4$ system already discussed above (Drewitt et al. 2012). The ring statistics for both $CaSiO_3$ melt and glass is dominated by 5- and 6-membered rings.

The atomic structure and dynamics of sodium silicate and aluminum silicate melts was subject of a number of large scale classical MD simulation studies by Horbach and co-workers

(Horbach and Kob 1999b; Horbach et al. 2001, 2002, 2003; Horbach and Kob 2002; Winkler et al. 2004; Pfleiderer et al. 2006; Voigtmann and Horbach 2006; Binder et al. 2007). Besides a detailed analysis of the atomic-scale structures of the melts they identified the basic diffusion mechanisms for Na and O diffusion and investigated the structural relaxation processes in the viscous liquids by analyzing the intermediate scattering functions (Horbach and Kob 2002; Voigtmann and Horbach 2006). An important finding was that Na diffuses in channels embedded in the SiO_2 matrix (Horbach et al. 2002), which is consistent with experimental data (Meyer et al. 2004). In aluminum silicate melts both Al and Si are found to be primarily in tetrahedral coordination, especially towards lower temperatures (2300 K). However, two-membered rings and triclusters were concentrated close to the Al ions. The authors also found evidence for a heterogeneous distribution of Si and Al and described this observation as an Al-rich network structure percolating through the SiO_2 network. The self-diffusion coefficients of O and Al were two to three times larger than those of Si (Winkler et al. 2004; Pfleiderer et al. 2006). Fast ionic diffusivity was also found for Li in an *ab initio* MD simulation study of lithium disilicate melt and glass (Du and Corrales 2006).

CaO–Al_2O_3–SiO_2 liquids as a model system for molten slags were investigated in a combined neutron diffraction and *ab initio* MD simulation study by Liu et al. (2016). The structure factors from experiments and simulations were found to be in good agreement. The network-modifying role of the Ca cations and the network-forming character of Al were confirmed, the latter considering also five-fold coordinated Al and $(Al,Si)O_3$ triclusters as motifs for a structurally more complex network. A detailed analysis of the modeled structures suggests that tetrahedral order in the melt is a major factor determining the excess entropy and that higher Ca and lower Al concentrations lower the entropy of the liquid due to the greater tetrahedrality of those melts. The atomic structure and dynamics of CaO–Al_2O_3–SiO_2 melts with low silica content predicted from classical and *ab initio* simulations were assessed by Bouhadja et al. (2013). Besides looking at structural parameters, the intermediate scattering functions were analyzed in terms of mode-coupling theory, which provides information about the atomic-scale relaxation processes that determine the behavior of the melt above and across the glass transition. Bouhadja et al. (2014) performed a comprehensive classical MD simulation study of calcium aluminosilicate melts and found that the non-bridging oxygens linked to Si or Al strongly influence both the glass transition temperature and the fragility along different (CaO–Al_2O_3)–(SiO_2) joins. Structure and transport properties of CaO–MgO–Al_2O_3–SiO_2 melt with focus on ionic conductivity were studied using classical MD by Mongalo et al. (2016).

An NMR spectroscopy study of Mg_2SiO_4 glass was complemented by *ab initio* MD simulations of Mg_2SiO_4 melt (Sen and Tangeman 2008). Both experimental spectra and simulations yielded a considerable amount of bridging oxygens, which are absent in the crystalline phases such as forsterite. The authors concluded that this structural difference between the melt/ glass and the crystalline phases could influence the crystallization and segregation behavior of the early terrestrial magma ocean and the chemical differentiation of the Earth up to the present day. A pure simulation study combining classical and *ab initio* approaches was focused on the structure, thermodynamic and transport properties of MgO–SiO_2 liquids (Zhang 2011).

Going beyond the simplified model compositions used in most MD simulation studies of silicate melts, Guillot and Sator (2007a) parameterized a classical force field for a nine component oxide system. With this potential a number of structural, thermodynamic and physical properties of silicate melts with compositions close to natural systems were reasonably well reproduced. Not surprisingly, the transferability of the model is limited for extreme compositions, such as very silica-rich or iron-rich melts. Also, due to the simplicity of the model, Fe^{2+} and Fe^{3+} were parameterized as two independent species and therefore charge transfer is not possible. In a subsequent comparative *ab initio* MD study, Vuilleumier et al. (2009) showed that the structural predictions of the Guillot and Sator (2007a) force field do not differ significantly from the DFT-based calculations but ionic diffusivities and vibrational properties are more sensitive to the details of the interaction potential. The inability of rigid ion models to predict reasonable intensities of the infrared spectra of the liquid or glass due to the lack of polarizability was shown by Wilson et al. (1996b). A more detailed investigation of the electronic structure in silicate melts was performed by Vuilleumier et al. (2011).

The structure and transport properties of a MgO–SiO$_2$ liquid in confinement between two forsterite grains as a model for a melt-wetted grain boundary was studied by Gurmani et al. (2011) using a polarizable ion potential and classical MD simulations. The interfacial layer between mineral and melt of up to two nanometers thickness show distinctly different structures and self-diffusion coefficients than the bulk. The surface energy of the crystal also has an influence on particle mobility at the interface with the slowing down being more pronounced for high energy surfaces. Ultrathin melt films on well-wetted polycrystalline fabrics may also have a reduced electrical conductivity (Gurmani et al. 2011).

Very recently, Xiao and Stixrude (2018) predicted the liquid–vapor phase diagram of MgSiO$_3$ from *ab initio* MD simulations at very high temperatures using a simulation box that contained both the liquid and the vapor phase. While at high T the liquid became increasingly enriched in Mg the vapor phase consisted predominantly of SiO and O$_2$. The liquid–vapor critical point was estimated to be at approximately 6600 K (Xiao and Stixrude 2018).

The effect of melt structure on (trace) element partitioning was explored by classical and *ab initio* MD simulations of silicate melts with different degrees of polymerization containing Y, La or As as minor elements (Haigis et al. 2013; Wagner et al. 2017b). The studies were motivated by experimental trace element partitioning data between titanite and melt (Prowatke and Klemme 2005) and between immiscible melts (Schmidt et al. 2006) that showed a preferential incorporation of the rare earth elements into the less polymerized melts. Structurally, systematic changes in coordination numbers and cation-oxygen distances for Y and La were observed in the simulations when varying the melt polymerization. To make an at least semi-quantitative link to the partitioning experiments, the free energy change of exchange reactions of a minor (Y, La, As) and a major element (Al) between two different melts was estimated using the alchemical transmutation approach of thermodynamic integration. The exchange reaction is decomposed into two partial reactions, in each of which one atom (e.g., Al) is slowly transmuted into the other (e.g., Y). Technically, this is done by performing two independent calculations of the forces, one for the melt containing the Al atom and the other for the Y-bearing melt. The forces for the MD simulation are calculated as a weighted average of the two and the weighting factor is changed in steps from 100% Al to 100% Y. The difference between the free energies of the partial reactions then yields the free energy change of the exchange reaction (Haigis et al. 2013; Wagner et al. 2017b). Using this method the preferential incorporation of Y and La into the less polymerized melt was confirmed. In two *ab initio* studies element partitioning between metal and silicate melt was explored (Zhang and Yin 2012; Künzel et al. 2017). However, significant method development is still needed to make predictions from such simulations more quantitative.

Silicate glasses. Due to the wide-spread technological interest in silicate glasses, their structure and physical properties have been subject to a large number of molecular dynamics simulation studies. A comprehensive review of these investigations is beyond the scope of this paper and the reader is also referred to other review papers on simulations of silicate glasses, e.g., Garofalini (2001); Pedone (2009); Micoulaut and Bauchy 2022, this volume. Many of the MD simulation studies used classical potentials, sometimes in combination with *ab initio* MD, to develop structure models of alkali and earth-alkali silicate and aluminosilicate glasses. The model predictions were assessed by reference to experimental data, e.g., from neutron and X-ray diffraction (Tilocca et al. 2006; Wagner et al. 2017b), XAFS (Trcera et al. 2011; Tamura et al. 2012), NMR spectroscopy (Charpentier et al. 2004; Pedone et al. 2010; Gambuzzi et al. 2014) or vibrational spectroscopy (IR and Raman, Zotov et al. (1999)). Generally, the agreement of the published data is reasonable, which suggests that the simulated structures are useful for a detailed analysis of the coordination environments of various ions and of the intermediate range order as expressed, e.g., by ring structures or Q^n species (e.g., Du and Cormack 2004; Tilocca et al. 2006; Du 2009; Xiang et al. 2013; Charpentier et al. 2018). Having established a structural model for the glass, the goal of some studies has been to establish structure-property relations, e.g., by investigating the evolution of the computed mechanical properties (Young's modulus, shear modulus, bulk modulus, hardness, etc.) as a function of glass composition and glass structure (e.g., Pedone et al. 2006; Du 2009; Xiang et al. 2013). Others related structural changes to chemical durability of the glass (Christie and Tilocca 2010) or to transport coefficients (Tandia et al. 2011).

Bioactive glasses have attracted interest in the molecular modeling community due to their use in biomedicine (Tilocca 2009, 2010a). A number of simulation studies have been performed to develop an atomic-scale understanding of such glasses, especially of phosphosilicate glasses. These glasses are chemically rather complex and commonly contain at least four oxide components (Na_2O–CaO–SiO_2–P_2O_5). In such cases the quality of the obtained structure model may not only depend primarily on the choice of the interaction potential but also on finite simulation time and system size effects. Tilocca (2013) studied the effect of cooling rate and simulation box size on the intermediate range order of the glass and concluded that beyond a certain convergence range (cooling rates in the order of 2 K/ps and system sizes of 3000 atoms) the structures do not change significantly anymore when increasing further the system size or decreasing the cooling rate. The structural role of fluorine on potentially bioactive glasses was investigated by Lusvardi et al. (2008); Christie et al. (2011) and Pedone et al. (2012). The simulations show that fluorine is almost entirely bonded to network-modifying cations such as Na and Ca. The addition of fluorine also seems to strengthen the polymerization of the silicate tetrahedra due to the removal of network modifiers from the silicious matrix. Na migration pathways in a bioactive glass were studied by *ab initio* MD simulations (Tilocca 2010b).

Borosilicate glasses constitute another class of multicomponent glasses with applications, e.g., as bioactive glasses or for the storage of radioactive waste. Due to their favorable mechanical properties and chemical durability they are also widely used in glass industry. Molecular modeling of borosilicates is challenging due to the two different network formers, Si and B, the latter having two different coordination environments, BO_3 and BO_4. The relative abundance of BO_3 and BO_4 changes with chemical composition, temperature and pressure (e.g., Kieu et al. 2011; Connelly et al. 2011). Recently, two new force fields were parameterized for the simulation of borosilicates, a rigid ion model based on the Guillot and Sator (2007a) potential for silicates (Wang et al. 2018) and a polarizable ion potential for sodium borosilicate glasses optimized by reference to DFT calculations only (Pacaud et al. 2017). The latter was extended to include La_2O_3 (Pacaud et al. 2018). Both potentials reproduce well experimental neutron diffraction data. They also make reasonable predictions for the variation of glass density as a function of glass composition (although the polarizable potential underestimates the glass density by up to 7 %) and for the change between BO_3 and BO_4 during cooling (Pacaud et al. 2017) and with increasing SiO_2 content (Wang et al. 2018).

In some cases structural models from classical and *ab initio* MD simulations were compared to validate the classical MD approach (e.g., Tilocca et al. 2006; Wagner et al. 2017b). Comparison of different classical potentials usually reveals that shell models or polarizable ion models outperform rigid ion models (Tilocca et al. 2006; Tilocca 2008), which confirms that the explicit account of ionic polarization effects improves the description of particle interactions in oxides and silicates. By combining diffraction experiments and MD simulations a lot can be learned about such complex glass structures as will be demonstrated in the following by a case study of the structure of $Y_4Al_8Si_{19}O_{56}$ glass (Wagner et al. 2017a). The MD simulations were performed with two different polarizable ion potentials, one of them considering dipolar polarization effects (dippim) and the other comprising polarization effects up to the quadrupolar level as well as deformable ions (quaim). Glass structures were obtained from quenching several snapshots from MD simulations of the melt and averaging over the different quench configurations. Figure 5 shows the derived static structure factors $S(Q)$ from the simulations compared to the respective neutron and X-ray diffraction data of glasses with the same composition.

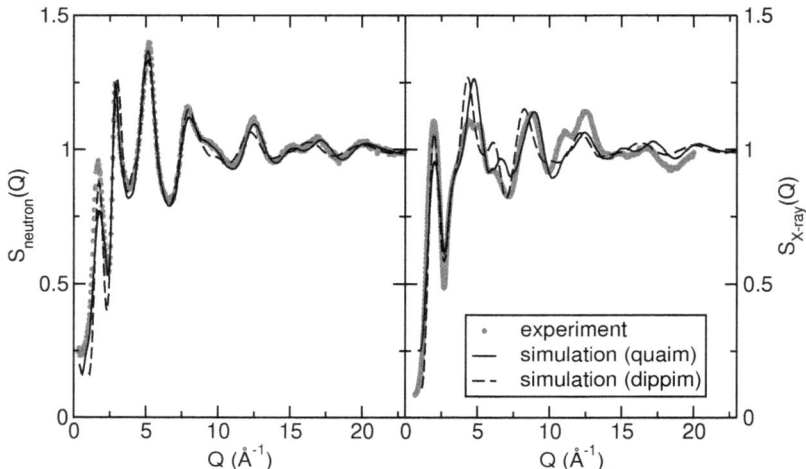

Figure 5. Static structure factors of $Y_4Al_8Si_{19}O_{56}$ glass from neutron (*left*) and X-ray (*right*) diffraction compared to predictions from simulations using two different polarizable ion potentials (modified from Wagner et al. 2017a).

The neutron $S(Q)$ is well reproduced by the simulations, with the quaim showing a somewhat better agreement than the dippim potential. Larger differences are observed for the X-ray $S(Q)$, both in the positions of some peaks and in the absolute intensities. (A better agreement of the simulated $S(Q)$ with the neutron diffraction data was also observed by (Bouhadja et al. 2014) for $CaO-Al_2O_3-SiO_2$ glasses.) These differences may be rationalized by looking at the weighing factors for the partial structure factors $S_{AB}(Q)$, i.e., the product of scattering lengths and concentrations (see methods section above). In neutron diffraction, all atoms show a roughly similar scattering power with coherent scattering lengths b_i ranging between 3.45 fm (Al) and 7.75 fm (Y) (Sears 1992). Thus, the total scattering is dominated by the most abundant atom, oxygen. X-ray scattering power is proportional to the number of electrons (36 for Y^{3+} vs. 10 for O^{2-}, Al^{3+} and Si^{4+}) and thus the Y partials have a much higher weight. This indicates that while the global structure of the glass is relatively well captured in both simulations, the local structure (nearest and next nearest neighbors) between experiment and simulations differs to a larger extent, especially around the Y ions. This is better illustrated by looking at the Fourier transform of the $S(Q)$, i.e., the total radial distribution functions, in Figure 6. Slightly different peak positions in the total $g(r)$ indicate that the quaim model predicts anion-cation and cation–cation nearest neighbor distances that are smaller than in the real glass whereas the oxygen-oxygen distances agree rather well. The dippim model shows the opposite behavior. Further, the first peak in the $g(r)$ is higher and narrower in the simulations, which suggests a better defined nearest neighbor coordination shell (assuming that the experimental $g(r)$ is not artificially broadened during the Fourier transformation using a window function).

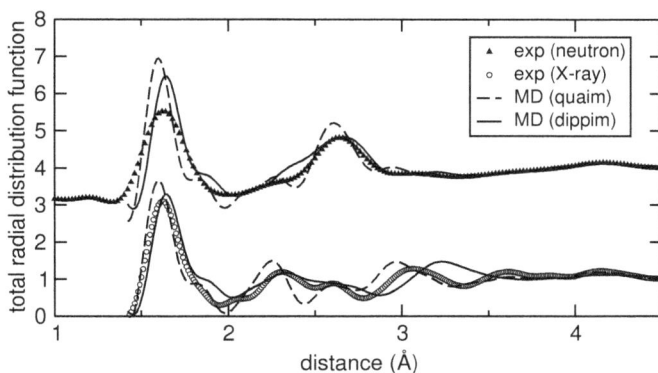

Figure 6. Total radial distribution functions of $Y_4Al_8Si_{19}O_{56}$ glass for neutron (***top***) and X-ray (***bottom***) diffraction obtained from Fourier transformation of the structure factors shown in Fig. 5 (modified from Wagner et al. 2017a).

Looking at the partial $g_{AB}(r)$ provides further insight into the structural differences between the simulation models. In Fig. 7, those functions are shown together with the total X-ray $g(r)$ in the distance range between 2.5 Å and 4.5 Å, which is representative of the nearest neighbor cation–cation distances. The peak in $g(r)$ at approximately 3 Å arises from a superposition of various cation–cation pairs. Only the Y–Y distance is found at significantly larger distance with a first maximum between 3.5 Å and 3.7 Å. The underestimation (quaim) or overestimation (dippim) of the two different models is clearly visible again. However, there is a significant additional difference between the $g_{AB}(r)$ of the two simulation models. The Y–Al, Y–Si and Y–Y radial distribution functions of the quaim potential show a double peak structure whereas the dippim potential predicts single peak structures of the respective partials. Closer inspection reveals a different topology of the glass structures. While in the quaim structure model the Y–O polyhedra are connected to other cation polyhedra by shared corners and edges the dippim model predicts only edge-sharing connectivity. Comparison to the experimental $g(r)$ suggests that the former structure model is more realistic for the measured sample.

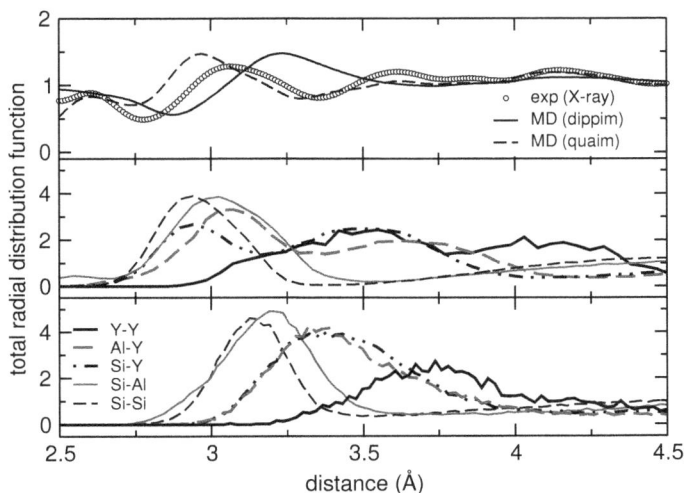

Figure 7. Total X-ray $g(r)$ (***top***) and partial $g_{AB}(r)$ (***middle*** for quaim and ***bottom*** for dippim potential) of $Y_4Al_8Si_{19}O_{56}$ glass in the range of the nearest neighbor cation–cation distances (modified from Wagner et al. (2017a)).

Hydrous silica and silicate melts and glasses. Water has a strong effect on the properties of silicate melts and glasses. In *ab initio* Car–Parrinello MD simulations of hydrous silica melt at 3000 K and 3500 K it was shown that in a system with 30 SiO_2 and 4 H_2O, which corresponds to a water content of 3.84 wt%, almost all the water is dissolved as hydroxyl groups (Pöhlmann et al. 2004). In a follow-up study, Benoit et al. (2008) the electronic structure of defects in quenched glasses with the same composition were investigated in detail. Changing the speciation by lowering the temperature in the simulation becomes increasingly difficult due to the low probability of breaking Si–O bonds. As shown by Spiekermann et al. (2016) the lifetime of Si–O bonds exceeds the typical time scale of *ab initio* MD simulations, i.e., 10^1 to 10^2 ps, already at temperatures well above 2000 K. Without breaking such bonds it is virtually impossible to observe the formation of H_2O molecules by combination of two OH-groups and the formation of a bridging oxygen. Hydrolysis reactions have been artificially imposed in a simulation study of water solubility in calcium aluminosilicate glasses (Bouyer et al. 2010). Alternatively, Monte-Carlo simulations were performed to obtain structural models for hydrous melts and glasses (Anderson et al. 2008a,b; Batyrev et al. 2008; Dupuis et al. 2019). Conventional classical potentials are not able to capture hydrolysis reactions, therefore a number of reactive force fields have been developed and applied (e.g., Cote et al. 2017; Dupuis et al. 2019).

Melts and glasses at high pressure

Classical molecular dynamics simulations. Following the pioneering work by Woodcock et al. (1976) on SiO_2 the behavior of oxide and silicate melts and glasses at high pressure has been subject to many molecular simulation studies. In the 1980s and 1990s structural models were obtained using more or less simple force fields. Major results obtained until 1995 were reviewed by Poole et al. (1995) and will not be discussed here again. The early studies were focused mainly on structural changes in the melts and glasses as expressed, e.g., by the increase in cation coordination numbers with increasing pressure, on pressure-volume equations of state, and on the evolution of transport properties with pressure and temperature. Already Woodcock et al. (1976) found a maximum in the self-diffusion coefficients of SiO_2 liquid under compression, which was interpreted as due to the breakdown of the tetrahedral network structure.

Tangney and Scandolo (2002) compared the structural changes in liquid silica with pressure predicted by the BKS potential (van Beest et al. 1990) and by a new *ab initio* optimized polarizable ion potential with those found in *ab initio* MD simulations. While the transition from tetrahedral to octahedral Si happens at much lower pressure for the BKS potential, the polarizable potential and the *ab initio* data show a consistent behavior. Using the same polarizable potential, Liang et al. (2007) investigated the origin of the minimum in the mechanical strength of SiO_2 glass around 10 GPa and identified the pressure-induced appearance of fivefold defects as a main reason. The maximum compressibility of SiO_2 glass close to 2 GPa was explained in another classical MD simulation study by rigidity arguments (Walker et al. 2007). While the structure is rather rigid at lower and higher pressures, it becomes flexible and softer at intermediate pressures. Horbach (2008) used another interaction potential to study the compression of SiO_2 melt, reproduced the experimentally known maximum in the Si self-diffusion coefficient close to 20 GPa and discussed the relation between this anomalous behavior in the transport properties and the structural changes. Classical MD simulations complemented X-ray and neutron diffraction experiments on the structural behavior of GeO_2 glass under compression where the tetrahedral to octahedral Ge transition occurs between 5 and 15 GPa (Guthrie et al. 2004).

Matsui (1996) studied the structure, bulk moduli and thermal expansion coefficients of $CaO–MgO–Al_2O_3–SiO_2$ melts at pressures up to 20 GPa and temperatures between 1900 K and 2700 K. He found a density crossover between diopside ($CaMgSi_2O_6$) solid and liquid at 11 GPa. Compared to other simulation studies, the temperatures were rather low and the paper contains no information whether all systems are still liquid or already frozen in under the high pressure conditions. Using a different rigid ion potential, Nevins and Spera (1998) and Bryce et al. (1999) investigated the relationship between the structure and self-diffusion coefficients of $CaAl_2Si_2O_8$ and $NaAlO_2$–SiO_2 melts at high pressures, respectively. Here, the temperatures were chosen to be at least 4000 K, which ensured sufficiently fast self-diffusion. Although it may not have been the goal of these studies to connect directly to experiments, the choice of such high temperatures underlines the difficulties of modeling silicate liquids at experimental or geological conditions. The temperature at which the simulated $CaAl_2Si_2O_8$ freezes in ('computer glass transition') was estimated to be at 2800 K (Morgan and Spera 2001). In a later simulation study, the P–T range of the simulations was extended (Spera et al. 2009) to develop an equation of state for $CaAl_2Si_2O_8$ up to 120 GPa and 6000 K (Ghiorso et al. 2009). Comparison with shock wave data suggested that the predicted densities of the simulated liquid were roughly 10% too low at pressures above 20 GPa (Ghiorso et al. 2009). These simulations of $CaAl_2Si_2O_8$ melt show a significant change in the melts structure with increasing pressure, i.e., the transition of 4-fold coordinated Si and Al to 5- and 6-fold coordination as well as a considerable increase of the average Ca coordination from 7 to about 10. Thereby the strongest change is observed between ambient pressure and 20 GPa. Simulation studies in a similar range of P/T conditions were performed by the same research group for Mg_2SiO_4 (Martin et al. 2009), $MgSiO_3$ (Nevis et al. 2009; Spera et al. 2011) and $CaAl_2Si_2O_8$–$CaMgSi_2O_6$ (Martin et al. 2012) melts. More recently, Neilson et al. (2016) investigated $NaAlSi_3O_8$ melt by molecular dynamics simulation in the pressure range up to about 35 GPa and developed a thermodynamic model.

Classical MD simulation studies of the simplified mantle melt system (CaO)–MgO–SiO_2 were also performed by other groups. Lacks et al. (2007) used a modified BKS potential (van Beest et al. 1990) to investigate the systematic change of melt properties along the MgO–SiO_2 join as a function of pressure and composition. They reproduced the maximum in the oxygen self-diffusion coefficient for Si-rich compositions (already observed by Woodcock et al. (1976)) and related minima in the viscosity. At low pressures, their model predicts a negative excess volume that decreases with increasing pressure. While the viscosity and the oxygen self-diffusion coefficient of the melts vary by three to four orders of magnitude at ambient pressure, they converge to similar values within one order of magnitude at pressures above 15 GPa. The compositional space was later extended by Zhang et al. (2010) to include CaO. Similar results were obtained by Guillot and Sator (2007a,b) using a different rigid ion potential. Adjaoud et al. (2008, 2011) used the polarizable ion model of Jahn and Madden (2007a) to investigate the behavior of Mg_2SiO_4 melt up to 35 GPa to understand better the physical state of a magma ocean (Fig. 8). Compared to the rigid ion model MD simulations (Guillot and Sator 2007b; Lacks et al. 2007; Martin et al. 2009) the increase of Si coordination with pressure is much less pronounced whereas the Mg coordination change is very similar. This indicates that the Si–O interaction of the Jahn and Madden (2007a) potential is somewhat too strong. The resulting self-diffusion coefficients are lower and the viscosities are higher by about half an order of magnitude compared to the other simulation studies but seem to fit better extrapolations of experimental data (Adjaoud et al. 2008).

Changes in the structure of silicate glasses at high pressures were investigated by Mead and Mountjoy (2006) for CaO–SiO_2 up to 10 GPa, Shimoda and Okuno (2006) for $CaSiO_3$–$MgSiO_3$ up to 15 GPa, and Drewitt et al. (2015) for CaO–Al_2O_3–SiO_2 up to 30 GPa. Shimoda and Okuno (2006) performed their simulations under cold compression starting from a melt-quenched glass at atmospheric pressure. While the Ca–O and Mg–O coordination number increased with increasing pressure, the Si–O coordination remained almost constant and only a change in the Si–O–Si bond angle was observed. On the contrary, Mead and Mountjoy (2006) and Drewitt et al. (2015) produced the model glass structures by quenching a melt at high pressure. In these studies the increase in coordination numbers of both network-forming and network-modifying cations was reported. The structural changes in $(Na_2O)_{0.3}$–$(SiO_2)_{0.7}$ melts and glasses were modeled by classical MD in a wide range of densities from 2.5 to 5.5 g/cm^3 (Bauchy 2012), which included the transition from 4-fold to 6-fold coordinated Si. Changes in the intermediate range order of this system were discussed.

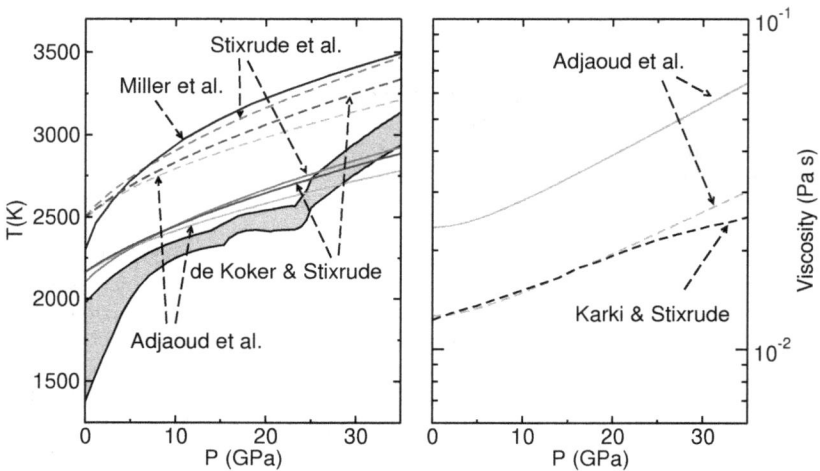

Figure 8. *Left*: adiabatic temperature profiles for a Mg_2SiO_4 magma ocean from two *ab initio* (Stixrude et al. 2009; de Koker and Stixrude 2009) and one classical (Adjaoud et al. 2011) MD studies for two different surface temperatures (full and dashed lines, respectively). Also shown are an experimentally obtained adiabat for a komatiite melt (Miller et al. 1991) and the mantle melting interval (gray shaded area). The simulation models predict that magma ocean crystallization should start at mid-mantle depths. *Right*: viscosity of $MgSiO_3$ (Karki and Stixrude 2010b) and Mg_2SiO_4 (Adjaoud et al. 2011) melts along the adiabats. (figure modified after Adjaoud et al. (2011)).

Going beyond binary or ternary systems, Guillot and Sator (2007a,b) developed and applied a rigid ion model to more complex melt compositions with up to nine components. This allowed to study magmatic liquids in a wide range of geologically relevant compositions including rhyolite, basalt and peridotite. Two separate sets of potentials were optimized for Fe^{2+} and Fe^{3+}. The thermodynamic, structural and transport properties of the melts at ambient pressure (Guillot and Sator 2007a) as well as at high pressures (Guillot and Sator 2007b) are largely consistent with the other rigid ion model simulations described above. In subsequent studies the Guillot and Sator (2007a) potential was extended to investigate the incorporation of CO_2 into silicate melts up to 15 GPa (Guillot and Sator 2011) and the structural environments of noble gases in silicate melts up to 20 GPa (Guillot and Sator 2012). A dedicated investigation of the transport properties of the basaltic melt composition was reported by Bauchy et al. (2013). Motivated by the fact that most of the rigid ion potentials predict self-diffusivities that are too high and viscosities that are too low (Guillot and Sator 2007a; Bauchy et al. 2013), Dufils et al. (2017) developed a new potential including an explicit term for covalent interactions. MD simulations of a basaltic melt using this new model resulted in transport properties much closer to experimental data. Subsequently, the same potential was used to investigate other terrestrial, lunar and planetary melts (Dufils et al. 2018). Very recently, the model was extended to include water and hydrous species (Dufils et al. 2020).

Results of different simulation studies were used to discuss the relation between self-diffusion coefficient and viscosity, which is commonly expressed by either the Stokes–Einstein or the Eyring equation. For depolymerized silicate melts with rather well defined and stable species (e.g., SiO_4^{4-} and Mg^{2+}) the Stokes–Einstein equation, which describes the motion of a Brownian particle, seems to be more appropriate whereas the Eyring equation that assumes jump diffusion seems to provide a better description for polymerized melts and to some extent also for the melts at high pressures (Lacks et al. 2007; Martin et al. 2009; Adjaoud et al. 2011; Ghosh and Karki 2011; Bauchy et al. 2013). However, one has to keep in mind that both relations are phenomenological end member scenarios and a rigorous treatment of the viscosity-diffusion relation for the general case is still missing.

Ab initio molecular dynamics simulations. The possibility to perform *ab initio* MD simulations constituted a major step towards more quantitative and predictive modeling of melts and glasses, especially under experimentally unexplored conditions. Empirical force fields that describe ionic interactions in an approximate way and that were mostly optimized to reproduce certain melt properties at ambient conditions are not expected to predict accurate melt properties at very high pressures and temperatures. This was already demonstrated in the first high-pressure *ab initio* MD simulation study of SiO_2 melt (Trave et al. 2002). Although the increasing Si coordination with pressure is qualitatively predicted by the popular BKS force field (van Beest et al. 1990) the predicted pressure range of this transition is quite different from the *ab initio* prediction (Trave et al. 2002). On the other hand, *ab initio* MD simulations are still restricted to system sizes of a few hundred atoms and simulation times of a few tens to hundreds of picoseconds. Especially for polymerized melts this limits severely the possibility to explore the configurational space on the length scale of intermediate-range order, which includes the formation of ring structures. While the nearest neighbor coordination numbers and physical properties such as equations of state or transport properties are often consistent between different *ab initio* studies, the ring statistics may differ significantly (Trave et al. 2002; Karki et al. 2007). This still existing deficiency of the *ab initio* approach should be born in mind when looking at the otherwise great success story presented in the following.

A pioneering *ab initio* MD simulation study of a mantle melt ($MgSiO_3$) was published by Stixrude and Karki (2005). The simulations demonstrated the Si coordination change from 4 to 6 in the pressure range of the Earth's mantle. Further, a Mie–Grüneisen equation of state was parameterized and the melting curve of $MgSiO_3$ was predicted and compared to the few and scattered experimental data. With the use of a GGA instead of an LDA functional, Wan et al. (2007) obtained similar results for the equation of state of $MgSiO_3$ melt but with a systemic offset compared to the results of Stixrude and Karki (2005), as expected for the two different functionals. Wan et al. (2007) also made *ab initio* predictions for the self-diffusion coefficients and the viscosity of liquid $MgSiO_3$, and discussed a possible density crossover between (Mg,Fe)SiO_3 solid and melt in the lowermost mantle due to the preferential partitioning of Fe into the melt. Both Stixrude and Karki (2005) and Wan et al. (2007) agree in that for the pure $MgSiO_3$ system a density crossover is not expected under Earth's mantle conditions. In a subsequent paper, de Koker et al. (2008) reported structural, thermodynamic and transport properties of Mg_2SiO_4 melt in a wide range of P and T. In this system, a solid–liquid density crossover was predicted at P=13 GPa and T=2550 K. Following up on the *ab initio* MD simulations, thermodynamic models were developed for the $MgO–SiO_2$ system (de Koker and Stixrude 2009; Stixrude et al. 2009; de Koker et al. 2013). This included prediction of melting curves and Hugoniots (i.e., the curves that describe all possible thermodynamic states that can be reached from an initial condition by a single shock), and discussions of solid–liquid density crossovers. An important conclusion from Stixrude et al. (2009) was that due to the increase of the Grüneisen parameter with compression the mantle isentrop is hotter than previously assumed and that a magma ocean would start crystallizing at mid-mantle depths. Karki and Stixrude (2010b) predicted from first principles the evolution of the viscosity of $MgSiO_3$ melt with pressure, temperature and water content. While the isothermal viscosity of the liquid changes by two orders of magnitude between 0 and 150 GPa, the addition of a hydrous component reduces the viscosity substantially. Assuming an adiabatic temperature behavior of a magma ocean in the early history of the Earth, the viscosity increase from the surface to the core–mantle boundary is predicted to be not more than one order of magnitude (see Fig. 8), which has strong implications on the cooling history of the magma ocean. More extensive studies of the transport properties of $MgO–SiO_2$ liquids were made by Ghosh and Karki (2011) and Karki et al. (2013). Boates and Bonev (2013) performed simulations up to very high P and T (600 GPa, 20000 K) and predicted the decomposition of $MgSiO_3$ into MgO solid and SiO_2 melt above the solidus of $MgSiO_3$ and pressures above 150 GPa. Recently, Wolf and Bower (2018) developed a new equation of state for $MgSiO_3$ melt based on simulation results of Spera et al. (2011) and Stixrude et al. (2009). *Ab initio* MD simulations of silicate melts were reviewed by Karki (2010).

A number of *ab initio* MD simulation studies were concerned with the structure, thermodynamics and transport properties of calcium- or sodium-bearing aluminosilicate melts (de Koker 2010; Karki et al. 2011; Ni and de Koker 2011). These studies confirm that observations from classical MD that the coordination number change from four to six happens earlier for Al than for Si, which is apparent already at ambient pressure, where the mean Al coordination is somewhat above 4.0. Similar studies were reported for calcium silicate or calcium magnesium silicate melts (Sun et al. 2011; Verma and Karki 2012; Bajgain et al. 2015b). It is interesting to note that the Ca coordination number in $CaSiO_3$ is substantially higher than in CaO liquid. In the silicate melts it increases from about seven to ten between ambient pressure and 120 GPa, whereas the increase in the oxide melt is between five and eight (Bajgain et al. 2015b). The difference between silicate and oxide melt is less pronounced for the Mg coordination that increases from about five to seven in MgO liquid (Karki et al. 2006) and is only slightly higher in the different Mg silicate melts.

A very recent field of research is the investigation of Fe-bearing silicate and oxide melts at high pressure. Fe is an important component in mantle melts and has a strong influence on melt properties due to its heavy mass (compared to the other elements in mantle melts) and its electronic properties. Due to the inability of standard DFT to predict correctly the electronic band structure of transition metal compounds the FeO or Fe_2O_3 component in silicate and oxide melts had been neglected in *ab initio* MD simulations for a long time. For the crystalline state it has become standard in the computational mineral (and solid state) physics community to perform DFT simulations with a Hubbard U correction that helps to localize the d-orbitals (Cococcioni 2010). In practical terms the *ab initio* MD simulations of such systems are much more demanding than in the conventional case as (1) the DFT calculations have to be performed spin-polarized, i.e., with two separate charge densities for the two spin states, and (2) the convergence of the self-consistent DFT loop is often rather slow. Nevertheless, a few successful studies of Fe-bearing melts have been completed recently. Munoz Ramo and Stixrude (2014) studied the equation of state of Fe_2SiO_4 liquid up to 300 GPa and 6000 K. They found that in contrast to crystals the high-spin to low-spin transition in the melt is very broad and extends to about 200 GPa. Further, in the lowermost mantle, the eutectic melts of the $MgO–FeO–SiO_2$ liquids are predicted to be denser than the coexisting crystals. The broad spin transition range and solid–liquid density crossover was confirmed in two studies of liquid (Mg,Fe)O (Ghosh and Karki 2016; Holmström and Stixrude 2016). Implications of the density crossover in Fe-bearing $MgSiO_3$ for the chemical evolution of the lower mantle were discussed by Karki et al. (2018). The pure $FeSiO_3$ system was studied by Sun et al. (2019). Finally, the electronic conductivity of solid and liquid (Mg,Fe)O in a wide range of pressures and temperatures was predicted by Holmström et al. (2018). These recent studies impressively demonstrate the great potential of DFT-based approaches to model physical properties that are not accessible by classical simulations but that are needed by geophysicists and geodynamicists.

A few *ab initio* MD studies addressed the structure of oxide and silicate glasses at high pressure. A majority of them complemented X-ray diffraction or spectroscopy experiments that often are still not feasible at simultaneous high P and T (i.e., for high pressure melts). Wu et al. (2012) developed a structural model for SiO_2 glass to interpret results from O K-edge X-ray Raman scattering, X-ray diffraction and Brillouin spectroscopy. They were able to assign spectral features in the X-ray Raman data to the increased packing of oxygen around Si atoms and correlated the increase in sound velocity to structural changes in the glass. A coordination change of Si by O from 4 to 6 with a high abundance of intermediate 5-fold coordinated Si in the pressure range of the Earth's mantle was observed. Above 100 GPa a small amount of 7-fold coordinated Si was identified (Wu et al. 2012; Petitgirard et al. 2019) but the average Si coordination remains close to six at least to conditions of the Earth's core-mantle boundary. On the contrary, X-ray diffraction data by Prescher et al. (2017) indicate a mean Si coordination significantly larger than six above 100 GPa. Complementary MD simulations confirm that at even higher pressures the structure of SiO_2 glass evolves from a more or less random close packed structure of oxygen towards a close packing of oxygen and silicon (Prescher et al. 2017). Structural snapshots from these simulations are shown in Figure 9. Average coordination numbers above 6 are predicted for GeO_2 glass already above 50 GPa by *ab initio* MD simulations (Du and Tse 2017) but this is not confirmed by recent experiments up to pressures of 100 GPa (Spiekermann et al. 2019). To resolve these apparently conflicting results further studies are needed. A first step could be to compare the methods coordination numbers are extracted from different experimental approaches and from MD simulations. The Si coordination in two silicate glasses, $MgSiO_3$ (Ghosh et al. 2014) and $CaAl_2Si_2O_8$ (Ghosh and Karki 2018), does not exceed 6.0 in a pressure range up to 150 GPa, regardless of cold or hot compression during the *ab initio* MD simulations.

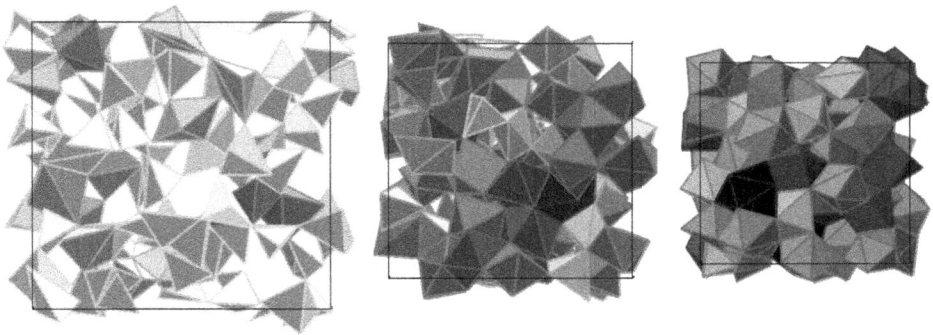

Figure 9. Snapshots from *ab initio* MD simulations of SiO_2 melt/glass at T=4000 K and densities of 2.66 (density of quartz), 4.40 (density of stishovite) and 7.00 g/cm^3 (Prescher et al. 2017). At low pressures (left) the melt is dominated by tetrahedrally coordinated silicon with some transient 3- and 5-fold coordinated Si due to the high temperature. At lower mantle conditions (middle, pressure of approximately 55 GPa (Karki et al. 2007)) the mean coordination is close to six whereas at much higher density (right) 7- and 8-fold coordinated Si become increasingly dominant. The gray scale varies between very light for 3-fold coordination to very dark for 8-fold coordination.

Volatiles play an important role in magmatic systems. Their solubility and speciation in silicate melts strongly depends on pressure, temperature and melt composition. As modeling volatile-bearing silicate melts is still difficult to do using classical MD simulations (although some attempts have been made, see e.g., Guillot and Sator 2011, 2012; Dufils et al. 2020 described above) most studies of hydrous and other volatile-bearing melts at high pressures have used an *ab initio* MD simulation approach. The structural changes and equations of state of hydrous $MgSiO_3$ melt under Earth's mantle conditions were investigated by Mookherjee et al. (2008). The H_2O speciation was found to be dominated by hydroxyl groups and H_2O molecules at low pressures. More complex extended structures were observed and visualized at high pressures (Mookherjee et al. 2008; Karki et al. 2010). The addition of about 8 wt% of water to SiO_2 liquid at 3000 K was shown to increase the diffusivity of O and Si by about one order of magnitude whereas the viscosity decreased by a similar amount (Karki and Stixrude 2010a). Protons are highly mobile in these melts. Bajgain et al. (2015a) concluded that the basaltic melt–water system is close to ideal and fully miscible over most of the mantle conditions, which would allow incorporation of a significant amount of hydrous component into deep mantle melts or an early magma ocean. The effect of water on the structure and physical properties of aluminosilicate melts up to 30 GPa was investigated by Bajgain et al. (2019). As expected the addition of water reduces the density and bulk modulus of the melt as well as the viscosity although this difference seems to decrease with increasing pressure. The hydrous component also has a noticeable influence on the pressure at which the maximum diffusivity and minimum viscosity are observed. These anomalous transport properties are typical for polymerized melts. The ideal mixing behavior of the melt–water system in the mantle is confirmed by simulations of hydrous Fe-rich silicate melts (Du et al. 2019). Despite the density reduction due to the hydrous component such (partial) melts may be gravitationally stable in the lowermost mantle if the Fe content of the melt is sufficiently high. This might also explain the presence of a dense layer at the core–mantle boundary with low seismic velocity.

Besides hydrous melts another focus of recent interest has been on the incorporation of CO_2 into silicate melts. Vuilleumier et al. (2015) investigated the CO_2 speciation in simplified basaltic and kimberlitic melts at 12 GPa and 2073 K. They found a predominance of carbonate groups with minor concentrations of CO_2 and transient species such as $C_2O_5^{2-}$. The carbonate groups modify the structure of the silicate liquid and are observed mainly as free carbonate or non-bridging carbonate. In the kimberlitic melt most of the carbonate groups are linked to the alkaline earth cations. For somewhat different *P/T* conditions and melt compositions *ab initio* MD simulation results were compared to NMR measurements of glasses that were synthesized at high pressure. While the evolution of the Al coordination with pressure was found to be consistent between simulation and experiment, significant differences were observed for the CO_2 speciation. Clearly, further systematic work is needed to understand this discrepancy. Ghosh and Karki (2017) studied transport properties of carbonated $MgSiO_3$ melt at mantle conditions. They reported a decrease in viscosity under carbonation but almost no change in the electrical conductivity of the melt. If carbon is added to the melt in different oxidation states (e.g., as CO or C) carbon atoms tend to cluster in the melt, which may indicate a possible immiscibility (Ghosh et al. 2017). The formation of tetracarbonate species (CO_4 units) and complex oxo-carbon polymers (C_xO_y) was observed in another *ab initio* MD simulation study of pyrolitic melt at lower mantle conditions (Solomatova et al. 2019).

CONCLUSIONS

The last two decades have seen significant progress in our understanding of the molecular structure and the physical properties of oxide and silicate melts and glasses. Molecular dynamics simulations have not only helped to construct more realistic structural models for increasingly complex melts and glasses, but they have also provided new insight into the structure-property relations. This progress has been achieved due to the increased computing power, which allowed *ab initio* MD simulations to be performed on a more or less standard basis. More recently it has become feasible to perform electronic structure calculations of chemically complex systems such as Fe-bearing melts and glasses, which provides access to electronic excitation spectra as well as to transport and thermodynamic properties, which may consider, e.g., electronic spin transitions. As the *ab initio* simulations are still restricted to relatively small system sizes, the development of new classical interatomic potentials is still of great importance. Such potentials are now available for a wide range of chemical compositions. Shortcomings of rigid ion models have been overcome to some extent by the development of polarizable ion models or reactive force fields. However, chemically reactive systems that involve, e.g., hydrolysis reactions or redox processes remain challenging for molecular simulations. Potentially very useful in this respect could be novel machine-learning or neural network potentials that are trained by results from *ab initio* simulations but that do not require a specific functional form for the potential. MD simulations have also been very successful to predict melt properties at high pressures and temperatures, i.e., at conditions of the Earth's and other planetary body's interiors. On the other hand, the modeling of a typical crustal melt close to the liquidus temperature or even below is still largely out of reach for MD simulations due to the finite simulation times and the slow kinetics of Si–O bond formation and dissociation. The same is true for the development of structural models of glasses, especially for those glasses that have a significantly different structure/speciation than the respective melt as the accessible quench rates in the simulations are still many orders of magnitude faster than those of real systems. Advanced sampling methods are highly desired in this respect. Finally, theoretical spectroscopy and free energy modeling approaches have advanced a lot for crystalline materials and have also been successfully applied to melts and glasses. Quantitative predictions of Raman and IR spectra for bulk melts and glasses, or of element partition coefficients and isotope fractionation factors including melt or glass phases remain challenging but should become feasible in the foreseeable future.

REFERENCES

Aasland S, McMillan PF (1994) Density-driven liquid–liquid phase separation in the system $Al_2O_3–Y_2O_3$. Nature 369:633–636

Adjaoud O, Steinle-Neumann G, Jahn S (2008) Mg_2SiO_4 liquid under high pressure from molecular dynamics. Chem Geol 256:185–192

Adjaoud O, Steinle-Neumann G, Jahn S (2011) Transport properties of Mg_2SiO_4 liquid at high pressure: Physical state of a magma ocean. Earth Planet Sci Lett 312:463–470

Aguado A, Bernasconi L, Jahn S, Madden PA (2003) Multipoles and interaction potentials in ionic materials from planewave-DFT calculations. Faraday Discuss 124:171–184

Allen MP, Tildesley DJ (1987) Computer Simulations of Liquids. Oxford University Press

Anderson KE, Grauvilardell LC, Hirschmann MM, Siepmann JI (2008a) Structure and speciation in hydrous silica melts. 2. Pressure effects. J Phys Chem B 112:13015–13021

Anderson KE, Hirschmann MM, Siepmann JI (2008b) Structure and speciation in hydrous silica melts. 1. Temperature and composition effects. J Phys Chem B 112:13005–13014

Ansell S, Krishnan S, Weber, J KR, Felten JJ, Nordinem PC, Beno MA, Price DL, Saboungi, M-L (1997) Structure of liquid aluminum oxide. Phys Rev Lett 78:464–466

Bajgain S, Ghosh DB, Karki BB (2015a) Structure and density of basaltic melts at mantle conditions from first-principles simulations. Nat Commun 6:8578

Bajgain SK, Ghosh DB, Karki BB (2015b) First-principles simulations of CaO and $CaSiO_3$ liquids: structure, thermodynamics and diffusion. Phys Chem Minerals 42:393–404

Bajgain SK, Peng Y, Mookherjee M, Jing Z, Solomon M (2019) Properties of hydrous aluminosilicate melts at high pressure. ACS Earth Space Chem 3:390–402

Barnes AC, Skinner LB, Salmon PS, Bytchkov A, Pozdnyakova I, Farmer TO, Fischer HE (2009) Liquid–liquid phase transition in supercooled yttria–alumina. Phys Rev Lett 103:225702

Baroni S, deGironcoli S, DalCorso A, Giannozzi P (2001) Phonons and related crystal properties from density-functional perturbation theory. Rev Mod Phys 73:515–562

Baroni S, Giannozzi P, Isaev E (2010) Density-functional perturbation theory for quasi-harmonic calculations. Rev Mineral Geochem 71:39–57

Batyrev IG, Tuttle B, Fleetwood DM, Schrimpf RD, Tsetseris L, Pantelides ST (2008) Reactions of water molecules in silica-based network glasses. Phys Rev Lett 100:105503

Bauchy M (2012) Structural, vibrational, and thermal properties of densified silicates: Insights from molecular dynamics. J Chem Phys 137:044510

Bauchy M, Guillot B, Micoulaut M, Sator N (2013) Viscosity and viscosity anomalies of model silicates and magmas: A numerical investigation. Chem Geol 346:47–56

Belonoshko AB, Gutierrez G, Ahuja R, Johansson B (2001) Molecular dynamics simulation of the structure of yttria Y_2O_3 phases using pairwise interactions. Phys Rev B 64:184103

Benmore CJ, Weber, J KR, Wilding MC, Du J, Parise JB (2010) Temperature-dependent structural heterogeneity in calcium silicate liquids. Phys Rev B 82:224202

Benoit M, Ispas S, Tuckerman ME (2001) Structural properties of molten silicates from ab initio molecular-dynamics simulations: Comparison between $CaO–Al_2O_3–SiO_2$ and SiO_2. Phys Rev B 64:224205

Benoit M, Pöhlmann M, Kob W (2008) On the nature of native defects in high OH-content silica glasses: A first-principles study. Europhys Lett 82:57004

Binder K, Horbach J, Knoth H, Pfleiderer P (2007) Computer simulation of molten silica and related glass forming fluids: recent progress. J Phys: Condens Matter 19:205102

Boates B, Bonev S (2013) Demixing instability in dense molten $MgSiO_3$ and the phase diagram of MgO. Phys Rev Lett 110:135504

Bouhadja M, Jakse N, Pasturel A (2013) Structural and dynamic properties of calcium aluminosilicate melts: a molecular dynamics study. J Chem Phys 138:224510

Bouhadja M, Jakse N, Pasturel A (2014) Striking role of non-bridging oxygen on glass transition temperature of calcium aluminosilicate glass-formers. J Chem Phys 140:234507

Bouyer F, Geneste G, Ispas S, Kob W, Ganster P (2010) Water solubility in calcium aluminosilicate glasses investigated by first-principles techniques. J Sol State Chem 183:2786–2796

Brehm M, Kirchner B (2011) Travis—a free analyzer and visualizer for Monte-Carlo and molecular dynamics trajectories. J Chem Inf Model 51:2007–2023

Bryce JG, Spera FJ, Stein DJ (1999) Pressure dependence of self-diffusion in the $NaAlO_2$–SiO_2 system: Compositional effects and mechanisms. Am Mineral 84:345–356

Carré A, Berthier L, Horbach J, Ispas S, Kob W (2007) Amorphous silica modeled with truncated and screened Coulomb interactions: A molecular dynamics simulation study. J Chem Phys 127:114512

Carré A, Horbach J, Ispas S, Kob W (2008) New fitting scheme to obtain effective potential from Car–Parrinello molecular-dynamics simulations: Application to silica. Europhys Lett 82:17001

Chaker Z, Salanne M, Delaye, J-M, Charpentier T (2019) NMR shifts in aluminosilicate glasses via machine learning. Phys Chem Chem Phys 21:21709–21725

Charpentier T, Ispas S, Profeta M, Mauri F, Pickard CJ (2004) First-principles calculation of ^{17}O, ^{29}Si, and ^{23}Na NMR spectra of sodium silicate crystals and glasses. J Phys Chem B 108:4147–4161

Charpentier T, Menziani MC, Pedone A (2013) Computational simulations of solid state NMR spectra: a new era in structure determination of oxide glasses. RSC Adv 3:10550–10578

Charpentier T, Okhotnikov K, Novikov AN, Hennet L, Fischer HE, Neuville DR, Florian P (2018) Structure of strontium aluminosilicate glasses from molecular dynamics simulations, neutron diffraction, and nuclear magnetic resonance studies. J Phys Chem B 122:9567–9583

Christie JK, Tilocca A (2010) Aluminosilicate glasses as yttrium vectors for in situ radiotherapy: Understanding composition-durability effects through molecular dynamics simulations. Chem Mater 22:3725–3734

Christie JK, Pedone A, Menziani MC, Tilocca A (2011) Fluorine environment in bioactive glasses: ab initio molecular dynamics simulations. J Phys Chem B 115:2038–2045

Cococcioni M (2010) Accurate and efficient calculations on strongly correlated minerals with the LDA+U method: Review and perspectives. Rev Mineral Geochem 71:147–167

Connelly AJ, Travis KP, Hand RJ, Hyatt NC, Maddrell E (2011) Composition-structure relationships in simplified nuclear waste glasses: 1. mixed alkali borosilicate glasses. J Am Ceram Soc 94:151–159

Cormack AN, Du J, Zeitler TR (2002) Alkali ion migration mechanisms in silicate glasses probed by molecular dynamics simulations. Phys Chem Chem Phys 4:3193–3197

Cote AS, Cormack AN, Tilocca A (2017) Reactive molecular dynamics: an effective tool for modelling the sol–gel synthesis of bioglasses. J Mater Sci 52:9006–9013

Cristiglio V, Hennet L, Cuello GJ, Pozdnyakova I, Johnson MR, Fischer HE, Zanghi D, Price DL (2008) Local structure of liquid $CaAl_2O_4$ from ab initio molecular dynamics simulations. J Non-Cryst Solids 354:5337–5339

Cristiglio V, Cuello GJ, Hennet L, Pozdnyakova I, Leydier M, Kozaily J, Fischer HE, Johnson MR, Price DL (2010) Neutron diffraction study of molten calcium aluminates. J Non-Cryst Solids 356:2492–2496

Cygan RT, Kubicki JD (Eds.) (2001) Molecular Modeling Theory: Applications in the Geosciences. Reviews in Mineralogy and Geochemistry, Volume 42, GS, MSA, Washington

Davis S, Gutiérrez G (2011) Structural, elastic, vibrational and electronic properties of amorphous Al_2O_3 from ab initio calculations. J Phys: Condens Matter 23:495401

deKoker N (2010) Structure, thermodynamics, and diffusion in $CaAl_2Si_2O_8$ liquid from first-principles molecular dynamics. Geochim Cosmochim Acta 74:5657–5671

deKoker N, Stixrude L (2009) Self-consistent thermodynamic description of silicate liquids, with application to shock melting of MgO periclase and $MgSiO_3$ perovskite. Geophys J Int 178:162–179

deKoker N, Stixrude L (2010) Theoretical computation of diffusion in minerals and melts. Rev Mineral Geochem 72:971–996

deKoker NP, Stixrude L, Karki BB (2008) Thermodynamics, structure, dynamics, and freezing of Mg_2SiO_4 liquid at high pressure. Geochim Cosmochim Acta 72:1427–1441

deKoker N, Karki BB, Stixrude L (2013) Thermodynamics of the MgO–SiO_2 liquid system in Earth's lowermost mantle from first principles. Earth Planet Sci Lett 361:58–63

Dick BG, Overhauser AW (1958) Theory of the dielectric constants of alkali halide crystals. Phys Rev 112:90–103

Drewitt, JWE, Jahn S, Cristiglio V, Bytchkov A, Leydier M, Brassamin S, Fischer HE, Hennet L (2011) The structure of liquid calcium aluminates as investigated by neutron and high-energy X-ray diffraction in combination with molecular dynamics simulation methods. J Phys: Condens Matter 23:155101

Drewitt, JWE, Hennet L, Zeidler A, Jahn S, Salmon PS, Neuville DR, Fischer HE (2012) Structural transformations on vitrification in the fragile glass forming system $CaAl_2O_4$. Phys Rev Lett 109:235501

Drewitt JWE, Jahn S, Sanloup C, deGrouchy C, Garbarino G, Hennet L (2015) Development of chemical and topological structure in aluminosilicate liquids and glasses at high pressure. J Phys: Condens Matter 27:105103

Drewitt JWE, Barnes AC, Jahn S, Kohn SC, Walter MJ, Novikov AN, Neuville DR, Fischer HE, Hennet L (2017) Structure of liquid tri-calcium aluminate. Phys Rev B 95:064203

Drewitt JWE, Hennet L, Neuville DR (2022) From short to medium range order in glasses and melts by diffraction and Raman spectroscopy. Rev Mineral Geochem 87:55-103

Du J (2009) Molecular dynamics simulations of the structure and properties of low silica yttrium aluminosilicate glasses. J Am Ceram Soc 92:87–95

Du J, Cormack AN (2004) The medium range structure of sodium silicate glasses: a molecular dynamics simulation. J Non-Cryst Solids 349:66–79

Du J, Corrales LR (2006) Structure, dynamics, and electronic properties of lithium disilicate melt and glass. J Chem Phys 125:114702

Du X, Tse JS (2017) Oxygen packing fraction and the structure of silicon and germanium oxide glasses. J Phys Chem .B 121:10726–10732

Du J, Benmore CJ, Corrales R, Hart RT, Weber JKR (2009) A molecular dynamics simulation interpretation of neutron and X-ray diffraction measurements on single phase Y_2O_3–Al_2O_3 glasses. J Phys: Condens Matter 21:205102

Du Z, Deng J, Miyazaki Y, Mao H, Karki BB, Lee, KKM (2019) Fate of hydrous Fe-rich silicate melt in Earth's deep mantle. Geophys Res Lett 46:9466–9473

Dufils T, Folliet N, Mantisi B, Sator N, Guillot B (2017) Properties of magmatic liquids by molecular dynamics simulation: the example of a MORB melt. Chem Geol 461:34–46

Dufils T, Sator N, Guillot B (2018) Properties of planetary silicate melts by molecular dynamics simulation. Chem Geol 493:298–315

Dufils T, Sator N, Guillot B (2020) A comprehensive molecular dynamics simulation study of hydrous magmatic liquids. Chem Geol 533:119300

Dupuis R, Beland LK, Pellenq, R J-M (2019) Molecular simulation of silica gels: Formation, dilution, and drying. Phys Rev Mater 3:075603

Ferlat G, Charpentier T, Seitsonen AP, Takada A, Lazzeri M, Cormier L, Calas G, Mauri F (2008) Boroxol rings in liquid and vitreous B_2O_3 from first principles. Phys Rev Lett 101:065504

Fogarty JC, Aktulga HM, Grama AY, van Duin, A CT, Pandit SA (2010) A reactive molecular dynamics simulation of the silica–water interface. J Chem Phys 132:174704

Frenkel D, Smit BJ (2002) Understanding Molecular Simulation: From Algorithms to Applications. Academic Press, San Diego

Gale JD, Wright K (2010) Lattice dynamics from force-fields as a technique for mineral physics. Rev Mineral Geochem 71:391–411

Gambuzzi E, Pedone A, Menziani MC, Angeli F, Caurant D, Charpentier T (2014) Probing silicon and aluminium chemical environments in silicate and aluminosilicate glasses by solid state NMR spectroscopy and accurate first-principles calculations. Geochim Cosmochim Acta 125:170–185

Garofalini SH (2001) Molecular dynamics simulations of silicate glasses and glass surfaces. Rev Mineral Geochem 42:131–168

Ghiorso MS, Nevins D, Cutler I, Spera FJ (2009) Molecular dynamics studies of $CaAl_2Si_2O_8$ liquid. Part II: equation of state and a thermodynamic model. Geochim Cosmochim Acta 73:6937–6951

Ghosh DB, Karki BB (2011) Diffusion and viscosity of Mg_2SiO_4 liquid at high pressure from first-principles simulations. Geochim Cosmochim Acta 75:4591–4600

Ghosh DB, Karki BB (2016) Solid–liquid density and spin crossovers in (Mg,Fe)O system at deep mantle conditions. Sci Rep 6:37269

Ghosh DB, Karki BB (2017) Transport properties of carbonated silicate melt at high pressure. Sci Adv 3:e1701840

Ghosh DB, Karki BB (2018) First-principles molecular dynamics simulations of anorthite ($CaAl_2Si_2O_8$) glass at high pressure. Phys Chem Minerals 45:575–587

Ghosh DB, Karki BB, Stixrude L (2014) First-principles molecular dynamics simulations of $MgSiO_3$ glass: Structure, density, and elasticity at high pressure. Am Mineral 99:1304–1314

Ghosh DB, Bajgain SK, Mookherjee M, Karki BB (2017) Carbon-bearing silicate melt at deep mantle conditions. Sci Rep 7:848

Giacomazzi L, Pasquarello A (2007) Vibrational spectra of vitreous SiO_2 and vitreous GeO_2 from first principles. J Phys: Condens Matter 19:415112

Giacomazzi L, Umari P, Pasquarello A (2009) Medium-range structure of vitreous SiO_2 obtained through first-principles investigation of vibrational spectra. Phys Rev B 79:064202

Gonze X, Lee C (1997) Dynamical matrices, Born effective charges, dielectric permittivity tensors, and interatomic force constants from density-functional perturbation theory. Phys Rev B 55:10355–10368

Götze W, Sjögren L (1992) Relaxation processes in supercooled liquids. Rep Prog. Phys 55:241–376

Greaves GN, Wilding MC, Fearn S, Langstaff D, Kargl F, Cox S, VuVan Q, Majerus O, Benmore CJ, Weber R, Martin CM, Hennet L (2008) Detection of first-order liquid/liquid phase transformations in yttrium oxide–aluminum oxide melts. Science 322:566–570

Green, E CR, Artacho E, Connolly, J AD (2018) Bulk properties and near-critical behaviour of SiO_2 fluid. Earth Planet Sci Lett 491:11–20

Guillot B, Sator N (2007a) A computer simulation study of natural silicate melts. Part I: low pressure properties. Geochim Cosmochim Acta 71:1249–1265

Guillot B, Sator N (2007b) A computer simulation study of natural silicate melts. Part II: high pressure properties. Geochim Cosmochim Acta 71:4538–4556

Guillot B, Sator N (2011) Carbon dioxide in silicate melts: A molecular dynamics simulation study. Geochim Cosmochim Acta 75:1829–1857

Guillot B, Sator N (2012) Noble gases in high-pressure silicate liquids: A computer simulation study. Geochim Cosmochim Acta 80:51–69

Gurmani SF, Jahn S, Brasse H, Schilling FR (2011) Atomic scale view on partially molten rocks: Molecular dynamics simulations of melt-wetted olivine grain boundaries. J Geophys Res 116:B12209

Guthrie M, Tulk CA, Benmore CJ, Xu J, Yarger JL, Klug DD, Tse JS, Mao, H-K, Hemley RJ (2004) Formation and structure of a dense octahedral glass. Phys Rev Lett 93:115502

Gutierrez G, Belonoshko AB, Ahuja R, Johansson B (2000) Structural properties of liquid Al_2O_3: A molecular dynamics study. Phys Rev E 61:2723–2729

Gutierrez G, Johansson B (2002) Molecular dynamics study of structural properties of amorphous Al_2O_3. Phys Rev B 65:104202

Haigis V, Salanne M, Simon S, Wilke M, Jahn S (2013) Molecular dynamics simulations of Y in silicate melts and implications for trace element partitioning. Chem Geol 346:14–21

Hawlitzky M, Horbach J, Ispas S, Krack M, Binder K (2008) Comparative classical and 'ab initio' molecular dynamics study of molten and glassy germanium dioxide. J Phys: Condens Matter 20:285106

Hemmati M, Wilson M, Madden PA (1999) Structure of liquid Al_2O_3 from a computer simulation model. J Phys Chem B 103:4023–4028

Henderson GS, Stebbins JF (2022) The short-range order (SRO) and structure. Rev Mineral Geochem 87:1-53

Hennet L, Pozdnyakova I, Cristiglio V, Cuello GJ, Jahn S, Krishnan S, Saboungi, M-L, Price DL (2007) Short- and intermediate-range order in levitated liquid aluminates. J Phys: Condens Matter 19:455210

Holmström E, Stixrude L (2016) Spin crossover in liquid (Mg,Fe)O at extreme conditions. Phys Rev B 93:195142

Holmström E, Stixrude L, Scipioni R, Foster AS (2018) Electronic conductivity of solid and liquid (Mg,Fe)O computed from first principles. Earth Planet Sci Lett 490:11–19

Horbach J (2008) Molecular dynamics computer simulation of amorphous silica under high pressure. J Phys: Condens Matter 20:244118

Horbach J, Kob W (1999a) Static and dynamic properties of a viscous silica melt. Phys Rev B 60:3169

Horbach J, Kob W (1999b) Structure and dynamics of sodium disilicate. Phil Mag B 79:1981–1986

Horbach J, Kob W (2002) The structural relaxation of molten sodium disilicate. J Phys: Condens Matter 14:9237–9253

Horbach J, Kob W, Binder K (2001) Structural and dynamical properties of sodium silicate melts: an investigation by molecular dynamics computer simulation. Chem Geol 174:87–101

Horbach J, Kob W, Binder K (2002) Dynamics of sodium in sodium disilicate: Channel relaxation and sodium diffusion. Phys Rev Lett 88:125502

Horbach J, Kob W, Binder K (2003) The dynamics of melts containing mobile ions: computer simulations of sodium silicates. J Phys: Condens Matter 15:S903–S908

Huang L, Kieffer J (2006) Thermomechanical anomalies and polyamorphism in B_2O_3 glass: A molecular dynamics simulation study. Phys Rev B 74:224107

Jahn S (2008) Atomic structure and transport properties of MgO–Al_2O_3 melts: A molecular dynamics simulation study. Am Mineral 93:1486–1492

Jahn S, Kowalski PM (2014) Theoretical approaches to structure and spectroscopy of Earth materials. Rev Mineral Geochem 78:691–743

Jahn S, Madden PA (2007a) Modeling Earth materials from crustal to lower mantle conditions: A transferable set of interaction potentials for the CMAS system. Phys Earth Planet Int 162:129–139

Jahn S, Madden PA (2007b) Structure and dynamics in liquid alumina: simulations with an ab initio interaction potential. J Non-Cryst Solids 353:3500–3504

Jahn S, Madden PA (2008) Atomic dynamics of alumina melt: A molecular dynamics simulation study. Condens Matter Phys 11:169–178

Jahn S, Madden PA, Wilson M (2006) Transferable interaction model for Al_2O_3. Phys Rev B 74:024112

Kalampounias AG, Nasikas NK, Papatheodorou GN (2009) Glass formation and structure in the $MgSiO_3$–Mg_2SiO_4 pseudobinary system: From degraded networks to ioniclike glasses. J Chem Phys 131:114513

Kang E, Lee S, Hannon AC (2006) Molecular dynamics simulations of calcium aluminate glasses. J Non-Cryst Solids 352:725–736

Karki BB (2010) First-principles molecular dynamics simulations of silicate melts: Structural and dynamical properties. Rev Mineral Geochem 71:355–389

Karki BB, Stixrude L (2010a) First-principles study of enhancement of transport properties of silica melt by water. Phys Rev Lett 104:215901

Karki BB, Stixrude LP (2010b) Viscosity of $MgSiO_3$ liquid at Earth's mantle conditions: Implication for an early magma ocean. Science 328:740–742

Karki BB, Bhattarai D, Stixrude L (2006) First-principles calculations of the structural, dynamical, and electronic properties of liquid MgO Phys Rev B 73:174208

Karki BB, Bhattarai D, Stixrude L (2007) First-principles simulations of liquid silica: Structural and dynamical behavior at high pressure. Phys Rev B 76:104205

Karki BB, Bhattarai D, Mookherjee M, Stixrude L (2010) Visualization-based analysis of structural and dynamical properties of simulated hydrous silicate melt. Phys Chem Minerals 37:103–117

Karki BB, Bohara B, Stixrude L (2011) First-principles study of diffusion and viscosity of anorthite ($CaAl_2Si_2O_8$) liquid at high pressure. Am Mineral 96:744–751

Karki BB, Zhang J, Stixrude L (2013) First principles viscosity and derived models for $MgO–SiO_2$ melt system at high temperature. Geophys Res Lett 40:94–99

Karki BB, Ghosh DB, Maharjan C, Karato S, Park J (2018) Density-pressure profiles of Fe-bearing $MgSiO_3$ liquid: Effects of valence and spin states, and implications for the chemical evolution of the lower mantle. Geophys Res Lett 45:3959–3966

Karki BB, Maharjan C, Ghosh DB (2019) Thermodynamics, structure, and transport properties of the $MgO–Al_2O_3$ liquid system. Phys Chem Minerals 46:501–512

Kieu, L-H, Delaye, J-M, Cormier L, Stolz C (2011) Development of empirical potentials for sodium borosilicate glass systems. J Non-Cryst Solids 357:3313–3321

Kramer GJ, deMan, A JM, van Santen RA (1991) Zeolites versus aluminosilicate clusters: The validity of a local description. J Am Chem Soc 113:6435–6441

Krishnan S, Hennet L, Jahn S, Key TA, Saboungi, M-L, Madden PA, Price DL (2005) The structure of normal and supercooled liquid aluminum oxide. Chem Mater 17:2662–2666

Kubicki JD (2001) Interpretation of vibrational spectra using molecular orbital theory calculations. Rev Mineral Geochem 42:459–483

Kubicki JD, Sykes D (1993) Molecular orbital calculations of vibrations in three-membered aluminosilicate rings. Phys Chem Minerals 19:381–391

Kubicki JD, Sykes D, Rossman GR (1993) Calculated trends of OH infrared stretching vibrations with composition and structure in aluminosilicate molecules. Phys Chem Minerals 20:425–432

Künzel D, Wagner J, Jahn S (2017) Ni partitioning between metal and silicate melts: An exploratory ab initio molecular dynamics simulation study. Chem Geol 461:47–53

Lacks DJ, Rear DB, van Orman JA (2007) Molecular dynamics investigation of viscosity, chemical diffusivities and partial molar volumes of liquids along the $MgO–SiO_2$ join as functions of pressure. Geochim Cosmochim Acta 71:1312–1323

Landron C, Hennet L, Jenkins TE, Greaves GN, Coutures JP, Soper AK (2001) Liquid alumina: Detailed atomic coordination determined from neutron diffraction data using empirical potential structure refinement. Phys Rev Lett 86:4839–4842

Liang Y, Miranda CR, Scandolo S (2007) Mechanical strength and coordination defects in compressed silica glass: Molecular dynamics simulations. Phys Rev B 75:024205

Liu M, Jacob A, Schmetterer C, Masset PJ, Hennet L, Fischer HE, Kozaily J, Jahn S, Gray-Weale A (2016) From atomic structure to excess entropy: a neutron diffraction and density functional theory study of $CaO–Al_2O_3–SiO_2$ melts. J Phys: Condens Matter 28:135102

Lusvardi G, Malavasi G, Cortada M, Menabue L, Menziani MC, Pedone A, Segre U (2008) Elucidation of the structural role of fluorine in potentially bioactive glasses by experimental and computational investigation. J Phys Chem B 112:12730–12739

Mahadevan TS, Garofalini SH (2007) Dissociative water potential for molecular dynamics simulations. J Phys Chem B 111:8919–8927

Mahadevan TS, Sun W, Du J (2019) Development of water reactive potentials for sodium silicate glasses. J Phys Chem B 123:4452–4461

Malfait WJ, Zakaznova-Herzog VP, Halter WE (2008) Quantitative Raman spectroscopy: Speciation of Na-silicate glasses and melts. Am Mineral 93:1505–1518

Martin GB, Spera FJ, Ghiorso MS, Nevins D (2009) Structure, thermodynamic, and transport properties of molten Mg_2SiO_4: molecular dynamics simulations and model EOS. Am Mineral 94:693–703

Martin GB, Ghiorso M, Spera FJ (2012) Transport properties and equation of state of 1-bar eutectic melt in the system $CaAl_2Si_2O_8–CaMgSi_2O_6$ by molecular dynamics simulation. Am Mineral 97:1155–1164

Marx D, Hutter J (2009) Ab Initio Molecular Dynamics: Basic Theory and Advanced Methods. Cambridge University Press

Marx D, Parrinello M (1996) Ab initio path integral molecular dynamics: Basic ideas. J Chem Phys 104:4077–4082

Matsui M (1994) A transferable interatomic potential model for crystals and melts in the system $CaO–MgO–Al_2O_3–SiO_2$. Mineral Mag 58a:571–572

Matsui M (1996) Molecular dynamics simulation of structures, bulk moduli, and volume thermal expansivities of silicate liquids in the system $CaO–MgO–Al_2O_3–SiO_2$. Geophys Res Lett 23:395–398

Matsui M (1998a) Breathing shell model in molecular dynamics simulation: Application to MgO and CaO J Chem Phys 108:3304–3309

Matsui M (1998b) Computational modelling of crystals and liquids in the system Na_2O–CaO–MgO–Al_2O_3–SiO_2. Properties of Earth and planetary materials at high pressure and temperature. *In:* Properties of Earth and Planetary Materials at High Pressure and Temperature. Manghnani MH, Yagi T (eds) Geophysical Monographs Series Vol 101 AGU, p. 145–151

McGreevy RL (2001) Reverse Monte Carlo modelling. J Phys: Condens Matter 13:R877–R913

McMillan PF, Wilson M, Wilding MC (2003) Polyamorphism in aluminate liquids. J Phys: Condens Matter 15:6105–6121

Mead RN, Mountjoy G (2006) A molecular dynamics study of densification mechanisms in calcium silicate glasses $CaSi_2O_5$ and $CaSiO_3$ at pressures of 5 and 10 GPa. J Chem Phys 125:154501

Meyer A, Horbach J, Kob W, Kargl F, Schober H (2004) Channel formation and intermediate range order in sodium silicate melts and glasses. Phys Rev Lett 93:027801

Micoulaut M, Bauchy M (2022) Topology and rigidity of silicate melts and glasses. Rev Mineral Geochem 87:163–191

Micoulaut M, Guissani Y, Guillot B (2006) Simulated structural and thermal properties of glassy and liquid germania. Phys Rev E 73:031504

Miller GH, Stolper EM, Ahrens TJ (1991) The equation of state of a molten komatiite 2. Application to komatiite petrogenesis and the Hadean mantle. J Geophys Res 96:11849–11864

Mongalo L, Lopis AS, Venter GA (2016) Molecular dynamics simulations of the structural properties and electrical conductivities of CaO–MgO–Al_2O_3–SiO_2 melts. J Non-Cryst Solids 452:194–202

Mookherjee M, Stixrude L, Karki B (2008) Hydrous silicate melt at high pressure. Nature 452:983–986

Morgan NA, Spera FJ (2001) A molecular dynamics study of the glass transition in $CaAl_2Si_2O_8$: thermodynamics and tracer diffusion. Am Mineral 86:915–926

MunozRamo D, Stixrude L (2014) Spin crossover in Fe_2SiO_4 liquid at high pressure. Geophys Res Lett 41:4512–4518

Neilson RT, Spera FJ, Ghiorso MS (2016) Thermodynamics, self-diffusion, and structure of liquid $NaAlSi_3O_8$ to 30 GPa by classical molecular dynamics simulations. Am Mineral 101:2029–2040

Nevins D, Spera FJ (1998) Molecular dynamics simulations of molten $CaAl_2Si_2O_8$: dependence of structure and properties on pressure. Am Mineral 83:1220–1230

Nevis D, Spera FJ, Ghiorso MS (2009) Shear viscosity and diffusion in liquid $MgSiO_3$: Transport properties and implications for terrestrial planet magma oceans. Am Mineral 94:975–980

Ni H, deKoker N (2011) Thermodynamics, diffusion and structure of $NaAlSi_2O_6$ liquid at mantle conditions: A first-principles molecular dynamics investigation. J Geophys Res 116:B09202

Ohmura S, Shimojo F (2008) Mechanism of atomic diffusion in liquid B_2O_3: an ab initio molecular-dynamics study. Phys Rev B 78:224206

Pacaud F, Delaye, J-M, Charpentier T, Cormier L, Salanne M (2017) Structural study of Na_2O–B_2O_3–SiO_2 glasses from molecular simulations using a polarizable force field. J Chem Phys 147:161711

Pacaud F, Salanne M, Charpentier T, Cormier L, Delaye, J-M (2018) Structural study of Na_2O–B_2O_3–SiO_2–La_2O_3 glasses from molecular simulations using a polarizable force field. J Non-Cryst Solids 499:371–379

Pedone A (2009) Properties calculations of silica-based glasses by atomistic simulations techniques: A review. J Phys Chem C 113:20773–20784

Pedone A (2016) Recent advances in solid-state NMR computational spectroscopy: The case of alumino-silicate glasses. Int. J Quantum Chem 116:1520–1531

Pedone A, Malavasi G, Menziani MC, Cormack AN, Segre U (2006) A new self-consistent empirical interatomic potential model for oxides, silicates, and silica-based glasses. J Phys Chem B 110:11780–11795

Pedone A, Malavasi G, Menziani MC, Segre U, Musso F, Corno M, Civalleri B, Ugliengo P (2008) FFSiOH: a new force field for silica polymorphs and their hydroxylated surfaces based on periodic B3LYP calculations. Chem Mater 20:2522–2531

Pedone A, Charpentier T, Menziani MC (2010) Multinuclear NMR of $CaSiO_3$ glass: simulation from first-principles. Phys Chem Chem Phys 12:6054–6066

Pedone A, Charpentier T, Menziani MC (2012) The structure of fluoride-containing bioactive glasses: new insights from first-principles calculations and solid state NMR spectroscopy. J Mater Chem 22:12599–12608

Petitgirard S, Sahle CJ, Weis C, Gilmore K, Spiekermann G, Tse JS, Wilke M, Cavallari C, Cerantola V, Sternemann C (2019) Magma properties at deep Earth's conditions from electronic structure of silica. Geochem Persp Lett 9:32–37

Pfleiderer P, Horbach J, Binder K (2006) Structure and transport properties of amorphous aluminium silicates: Computer simulation studies. Chem Geol 229:186–197

Pöhlmann M, Benoit M, Kob W (2004) First-principles molecular dynamics simulation of a hydrous silica melt: Structural properties and hydrogen diffusion mechanism. Phys Rev B 70:184209

Poole PH, McMillan PF, Wolf GH (1995) Computer simulations of silicate melts. Rev Mineral Geochem 32:563–616

Pozdnyakova I, Hennet L, Brun, J-F, Zanghi D, Brassamin S, Cristiglio V, Price DL, Albergamo F, Bytchkov A, Jahn S, Saboungi, M-L (2007) Longitudinal excitations in Mg–Al–O refractory oxide melts studied by inelastic X-ray scattering. J Chem Phys 126:114505

Prescher C, Prakapenka VB, Stefanski J, Jahn S, Skinner LB, Wang Y (2017) Beyond sixfold coordinated Si in SiO_2 glass at ultrahigh pressures PNAS 114:10041–10046

Prowatke S, Klemme S (2005) Effect of melt composition in the partitioning of trace elements between titanite and silicate melt. Geochim Cosmochim Acta 69:695–709

Putrino A, Parrinello M (2002) Anharmonic raman spectra in high-pressure ice from ab initio simulations. Phys Rev Lett 88:176401

Saksaengwijit A, Reinisch J, Heuer A (2004) Origin of the fragile-to-strong crossover in liquid silica as expressed by its potential-energy landscape. Phys Rev Lett 93:235701

Salanne M, Rotenberg B, Jahn S, Vuilleumier R, Madden PA (2012) Including many-body effects in models for ionic liquids. Theor Chem Acc 131:1143

SanMiguel MA, Sanz JF, Alvarez LJ, Odriozola JA (1998) Molecular-dynamics simulations of liquid aluminum oxide. Phys Rev B 58:2369–2371

Sanders MJ, Leslie M, Catlow CRA (1984) Interatomic potential for SiO_2. J Chem Soc, Chem Commun 1271–1273

Sarnthein J, Pasquarello A, Car R (1997) Origin of the high-frequency doublet in the vibrational spectrum of vitreous SiO_2. Science 275:1925–1927

Scherer C, Schmid F, Letz M, Horbach J (2019) Structure and dynamics of B_2O_3 melts and glasses: From ab initio to classical molecular dynamics simulations. Comput Mater Sci 159:73–85

Schmidt MW, Connolly JAD, Günther D, Bogaerts M (2006) Element partitioning: The role of melt structure and composition. Science 312:1646–1650

Sears VF (1992) Neutron scattering lengths and cross sections. Neutron News 3:26–37

Sen S, Tangeman J (2008) Evidence for anomalously large degree of polymerization in Mg_2SiO_4 glass and melt. Am Mineral 93:946–949

Shimoda K, Okuno M (2006) Molecular dynamics study of $CaSiO_3$–$MgSiO_3$ glasses under high pressure. J Phys: Condens Matter 18:6531–6544

Sinn H, Glorieux B, Hennet L, Alatas A, Hu M, Alp, E E, Bermejo FJ, Price DL, Saboungi, M-L (2003) Microscopic dynamics of liquid aluminium oxide. Science 299:2047–2049

Skinner LB, Benmore, J, Weber JKR, Tumber S, Lazareva L, Neuefeind J, Santodonato L, Du J, Parise JB (2012) Structure of molten $CaSiO_3$: neutron diffraction isotope substitution with aerodynamic levitation and molecular dynamics study. J Phys Chem B 116:13439–13447

Solomatova NV, Caracas R, Manning CE (2019) Carbon sequestration during core formation implied by complex carbon polymerization. Nat Commun 10:789

Soper AK (2000) The radial distribution functions of water and ice from 220 to 673 K and at pressures up to 400 MPa. Chem Phys 258:121–137

Spera FJ, Nevins D, Ghiorso M, Cutler I (2009) Structure, thermodynamic and transport properties of $CaAl_2Si_2O_8$ liquid. Part I: molecular dynamics simulations. Geochim Cosmochim Acta 73:6918–6936

Spera FJ, Ghiorso MS, Nevins D (2011) Structure, thermodynamic and transport properties of liquid $MgSiO_3$: comparison of molecular models and laboratory results. Geochim Cosmochim Acta 75:1272–1296

Spiekermann G, Steele-MacInnis M, Kowalski PM, Schmidt C, Jahn S (2013) Vibrational properties of silica species in MgO–SiO_2 glasses using ab initio molecular dynamics. Chem Geol 346:22–33

Spiekermann G, Steele-MacInnis M, Schmidt C, Jahn S (2012) Vibrational mode frequencies of silica species in SiO_2–H_2O liquids and glasses from ab initio molecular dynamics. J Chem Phys 136:154501

Spiekermann G, Wilke M, Jahn S (2016) Structural and dynamical properties of supercritical H_2O–SiO_2 fluids studied by ab initio molecular dynamics. Chem Geol 426:85–94

Spiekermann G, Harder M, Gilmore K, Zalden P, Sahle CJ, Petitgirard S, Wilke M, Biedermann N, Weis C, Morgenroth W, Tse JS, Kulik E, Nishiyama N, Yavas H, Sternemann C (2019) Persistent octahedral coordination in amorphous GeO_2 up to 100 GPa by Kβ" X-ray emission spectroscopy. Phys Rev X 9:011025

Stebbins JF, McMillan PF, Dingwell DB (Eds.) (1995) Structure, Dynamics and Properties of Silicate Melts. Reviews in Mineralogy, Vol 32, MSA, Washington

Stixrude L, Karki B (2005) Structure and freezing of $MgSiO_3$ liquid in Earth's lower mantle. Science 310:297–299

Stixrude L, deKoker N, Sun N, Mookherjee M, Karki BB (2009) Thermodynamics of silicate liquids in the deep earth. Earth Planet Sci Lett 278:226–232

Ston AJ (1996) The Theory of Intermolecular Forces. Oxford University Press, Oxford

Sun N, Stixrude L, de Koker N, Karki BB (2011) First principles molecular dynamics simulations of diopside $(CaMgSi_2O_6)$ liquid to high pressure. Geochim Cosmochim Acta 75:3792–3802

Sun Y, Zhou H, Yin K, Lu X (2019) First-principles study of thermodynamics and spin transition in $FeSiO_3$ liquid at high pressure. Geophys Res Lett 46:3706–3716

Sundararaman S, Huang L, Ispas S, Kob W (2018) New optimization scheme to obtain interaction potentials for oxide glasses. J Chem Phys 148:194504

Sundararaman S, Huang L, Ispas S, Kob W (2019) New interaction potentials for alkali and alkaline-earth aluminosilicate glasses. J Chem Phys 150:154505

Tamura T, Tanaka S, Kohyama M (2012) Full-PAW calculations of XANES/ELNES spectra of Ti-bearing oxide crystals and TiO–SiO glasses: Relation between pre-edge peaks and Ti coordination. Phys Rev B 85:205210

Tandia A, Timofeev NT, Mauro JC, Vargheese KD (2011) Defect-mediated self-diffusion in calcium aluminosilicate glasses: A molecular modeling study. J Non-Cryst Solids 357:1780–1786

Tangney P, Scandolo S (2002) An ab initio parametrized interatomic force field for silica. J Chem Phys 117:8898–8904

Thomas BWM, Mead RN, Mountjoy G (2006) A molecular dynamics study of the atomic structure of $(CaO)_x$ $(Al_2O_3)_{1-x}$ glass with $x=0.625$ close to the eutectic. J Phys: Condens Matter 18:4697–4708

Thomas M, Brehm M, Fligg R, Vöhringer P, Kirchner B (2013) Computing vibrational spectra from ab initio molecular dynamics. Phys Chem Chem Phys 15:6608–6622

Thomas M, Brehm M, Kirchner B (2015) Voronoi dipole moments for the simulation of bulk phase vibrational spectra. Phys Chem Chem Phys 17:3207–3213

Tilocca A (2008) Short- and medium-range structure of multicomponent bioactive glasses and melts: An assessment of the performances of shell-model and rigid-ion potentials. J Chem Phys 129:084504

Tilocca A (2009) Structural models of bioactive glasses from molecular dynamics simulations. Proc R Soc A 465:1003–1027

Tilocca A (2010a) Model of structure, dynamics and reactivity of bioglasses: a review. J Mater Chem 20:6848–6858

Tilocca A (2010b) Sodium migration pathways in multicomponent silicate glasses: Car–Parrinello molecular dynamics simulations. J Chem Phys 133:014701

Tilocca A (2013) Cooling rate and size effects on the medium-range structure of multicomponent oxide glasses simulated by molecular dynamics. J Chem Phys 139:114501

Tilocca A, deLeeuw NH, Cormack AN (2006) Shell-model molecular dynamics calculations of modified silicate glasses. Phys Rev B 73:104209

Tossell JD (2001) Calculating the NMR properties of minerals, glasses, and aqueous species. Rev Mineral Geochem 42:437–458

Trave A, Tangney P, Scandolo S, Pasquarello A, Car R (2002) Pressure-induced structural changes in liquid SiO_2 from ab initio simulations. Phys Rev Lett 89:245504

Trcera N, Rossano S, Madjer K, Cabaret D (2011) Contribution of molecular dynamics simulations and ab initio calculations to the interpretation of Mg K-edge experimental XANES in K_2O–MgO–$3SiO_2$ glass. J Phys: Condens Matter 23:255401

Umari P, Pasquarello A (2002) Ab initio molecular dynamics in a finite homogeneous electric field. Phys Rev Lett 89:157602

Umari P, Pasquarello A (2005) Fraction of boroxol rings in vitreous boron oxide from first-principles analysis of Raman and NMR spectra. Phys Rev Lett 95:137401

van Beest BWH, Kramer GJ, van Santen RA (1990) Force fields of silicas and aluminophosphates based on ab initio calculations. Phys Rev Lett 64:1955–1958

van Duin ACT, Dasgupta S, Lorant F, Goddard III WA (2001) ReaxFF: a reactive force field for hydrocarbons. J Phys Chem A 105:9396–9409

Verma AK, Karki BB (2012) First-principles study of self-diffusion and viscous flow in diopside $(CaMgSi_2O_6)$ liquid. Am Mineral 97:2049–2055

Verma AK, Modak P, Karki BB (2011) First-principles simulations of thermodynamical and structural properties of liquid Al_2O_3 under pressure. Phys Rev B 84:174116

Vogel M, Glotzer SC (2004a) Spatially heterogeneous dynamics and dynamic facilitation in a model of viscous silica. Phys Rev Lett 92:255901

Vogel M, Glotzer SC (2004b) Temperature dependence of spacially heterogeneous dynamics in a model of viscous silica. Phys Rev E 70:061504

Voigtmann T, Horbach J (2006) Slow dynamics in ion-conducting sodium silicate melts: Simulation and mode-coupling theory. Europhys Lett 74:459–465

Vuilleumier R, Sator N, Guillot B (2009) Computer modeling of natural silicate melts: What can we learn from ab initio simulations. Geochim Cosmochim Acta 73:6313–6339

Vuilleumier R, Sator N, Guillot B (2011) Electronic redistribution around oxygen atoms in silicate melts by ab initio molecular dynamics simulation. J Non-Cryst Solids 357:2555–2561

Vuilleumier R, Seitsonen AP, Sator N, Guillot B (2015) Carbon dioxide in silicate melts at upper mantle conditions: Insights from atomistic simulations. Chem Geol 418:77–88

Wagner J, Haigis V, Leydier M, Bytchkov A, Cristiglio V, Fischer HE, Sadiki N, Zanghi D, Hennet L, Jahn S (2017a) The structure of Y- and La-bearing aluminosilicate glasses and melts: a combined molecular dynamics and diffraction study. Chem Geol 461:23–33

Wagner J, Künzel D, Haigis V, Jahn S (2017b) Trace element partitioning between silicate melts - a molecular dynamics approach. Geochim Cosmochim Acta 205:245–255

Walker AM, Sullivan LA, Trachenko K, Bruin RP, White TOH, Dove MT, Tyer RP, Todorov IT, Wells SA (2007) The origin of the compressibility anomaly in amorphous silica: a molecular dynamics study. J Phys: Condens Matter 19:275210

Wan JTK, Duffy TS, Scandolo S, Car R (2007) First-principles study of density, viscosity, and diffusion coefficients of liquid $MgSiO_3$ at conditions of the Earth's deep mantle. J Geophys Res 112:B03208

Wang M, Krishnan NMA, Wang B, Smedskjaer MM, Mauro JC, Bauchy M (2018) A new transferable interatomic potential for molecular dynamics simulations of borosilicate glasses. J Non-Cryst Solids 498:294–304

Wentzcovitch R, Stixrude L (Eds.) (2010) Theoretical and Computational Methods in Mineral Physics: Geophysical Applications. Reviews in Mineral and Geochemistry Vol 71, MSA GS, Chantilly

Wentzcovitch R, Yu YG, Wu Z (2010) Thermodynamic properties and phase relations in mantle minerals investigated by first principles quasiharmonic theory. Rev Mineral Geochem 71:59–98

Wilding MC, Wilson M, Benmore CJ, Weber JKR, McMillan PF (2013) Structural changes in supercooled Al_2O_3–Y_2O_3 liquids. Phys Chem Chem Phys 15:8589–8605

Wilding MC, Wilson M, McMillan PF, Deschamps T, Champagnon B (2014) Low frequency vibrational dynamics and polyamorphism in Y_2O_3-Al_2O_3 glasses. Phys Chem Chem Phys 16:22083–22096

Wilson M, Exner M, Huang Y-M, Finnis MW (1996a) Transferable model for the atomistic simulation of Al_2O_3. Phys Rev B 54:15683–15689

Wilson M, Madden PA, Hemmati M, Angell CA (1996b) Polarization effects, network dynamics, and the infrared spectrum of amorphous SiO_2. Phys Rev Lett 77:4023–4026

Wilson M, McMillan PF (2004) Interpretation of X-ray and neutron diffraction patterns for liquid and amorphous yttrium and lanthanum aluminum oxides from computer simulation. Phys Rev B 69:054206

Winkler A, Horbach J, Kob W, Binder W (2004) Structure and diffusion in amorphous aluminium silicate: A molecular dynamics computer simulation. J Chem Phys 120:384–393

Wolf AS, Bower DJ (2018) An equation of state for high pressure–temperature liquids (RTpress) with application to $MgSiO_3$ melt. Phys Earth Planet Inter 278:59–74

Woodcock, LV Angell CA, Cheeseman P (1976) Molecular dynamics studies of the vitreous state: simple ionic systems and silica. J Chem Phys 65:1565–1577

Wu M, Liang Y, Jiang, J-Z, Tse JS (2012) Structure and properties of dense silica glass. Sci Rep 2:398

Xiang Y, Du J, Smedskjaer MM, Mauro JC (2013) Structure and properties of sodium aluminosilicate glasses from molecular dynamics simulations. J Chem Phys 139:044507

Xiao B, Stixrude L (2018) Critical vaporization of $MgSiO_3$. PNAS 115:5371–5376

Yu Y, Wang B, Wang M, Sant G, Bauchy M (2016) Revisiting silica with ReaxFF: towards improved predictions of glass structure and properties via reactive molecular dynamics. J Non-Cryst Solids 443:148–154

Zhang L (2011) Thermodynamic properties calculation for MgO–SiO_2 liquids using both empirical and first-principles molecular dynamics. Chem Phys Phys Chem 13:21009–21015

Zhang Y, Yin Q-Z (2012) Carbon and other light element contents in the Earth's core based on first-principles molecular dynamics. PNAS 109:19579–19583

Zhang Y, Guo G, Refson K, Zhao Y (2004) Finite-size effect at both high and low temperatures in molecular dynamics calculations of the self-diffusion coefficient and viscosity of liquid silica. J Phys: Condens Matter 16:9127–9136

Zhang L, VanOrman JA, Lacks DJ (2010) Molecular dynamics investigation of MgO–CaO–SiO_2 liquids: Influence of pressure and composition on density and transport properties. Chem Geol 275:50–57

Zotov N, Ebbsjö I, Timpel D Keppler H (1999) Calculation of Raman spectra and vibrational properties of silicate glasses: Comparison between $Na_2Si_4O_9$ and SiO_2 glass. Phys Rev B 60:6383–6397

Reviews in Mineralogy & Geochemistry
Vol. 87 pp. 229-281, 2022
Copyright © Mineralogical Society of America

6

Mechanical Properties of Oxide Glasses

Bruno Poletto Rodrigues[1], Theany To[2]
Morten M. Smedskjaer[2], Lothar Wondraczek[1,*]

[1] Otto Schott Institute of Materials Research
Friedrich Schiller University Jena,
07743 Jena, Germany
[2] Department of Chemistry and Bioscience
Aalborg University
9220 Aalborg, Denmark

lothar.wondraczek@uni-jena.de

INTRODUCTION

Oxide glasses are among the oldest known types of man-made materials (Harden 1968); they have become ubiquitous in modern life, from windows, containers and tableware to optical fiber. Even before any synthetic glass was ever produced, glasses occurring in nature were already used in the form of tools or jewelry. While for the latter, the peculiar optical properties of transparent variants of natural glasses (such as Libyan desert glass) were of interest, knives or arrowheads made from obsidian benefitted from the characteristic mechanical behavior, which manifests in sharp fracture edges and high stiffness. There are various natural processes that can produce glassy minerals, such as obsidian, pumice, hyaloclastite and sideromelane formed from silicic or basaltic magma; buchite and parabasalt from contact or combustion metamorphism; or fulgurites and impactites from lightning or meteorite impacts (Bentor 1984; Cicconi and Neuville 2019; Okrusch and Frimmel 2020). While the importance of the viscous flow behavior of the parent (magmatic or other) liquid is obvious, there is also an interest in understanding the physical properties of the resulting quenched glasses. This includes the mechanical properties, from elastic deformation to brittle fracture: aside the many practical aspects of mechanical strength, fracture resistance and reliability, a material's response to mechanical load can reveal a surprising level of fundamental insight into its constitution across a broad range of length scales. A widely known example is Mohs' scale of hardness (Mohs 1821), which was devised in the early 1820s as a means to classify minerals on the basis of their relative resistance to scratch damage. This early conjecture evolved into concepts which allowed to relate material hardness to lattice anharmonicity (Plendl and Gielisse 1962) and valence charge density (Liu and Cohen 1989), and have been used to develop design strategies for superhard materials (Haines et al. 2001).

Knowledge about the mechanical properties of glasses (including natural / mineral glasses) mostly comes from comparison to commercial or synthetic materials (e.g., vitreous silica, aluminosilicate and soda-lime silicate glasses). In this paper, the fundamentals of mechanical properties of oxide glasses will be reviewed, covering the breadth from experimental measurement techniques to modeling approaches. As known by its users, oxide glasses are brittle materials. When they are subjected to mechanical loading, stresses are amplified by the presence of defects (primarily surface flaws), which can lead to catastrophic fracture. No permanent deformation on the macroscale occurs prior to failure, as easily observed when piecing together the broken shards. This property is fundamentally related to the interplay of structural disorder and bond characteristics: the absence of lattice periodicity and, at the

1529-6466/22/0087-0006$10.00 (print)
1943-2666/22/0087-0006$10.00 (online)

http://dx.doi.org/10.2138/rmg.2022.87.06

same time, the presence of highly directional ionic or covalent bonds reduce the ability to adapt—by atomic motion—to external stress fields. This has been a major bottleneck in the further use of oxide glasses, as it leads to their notorious fracture behavior (Wondraczek et al. 2011). On the other hand, their distinct mechanical properties have also enabled a variety of applications, starting from the earliest examples of obsidian knives used for the specific stiffness and fracture behaviour of natural glasses, and having evolved into the most modern kinds of scratch resistant cover glasses for electronic devices and displays or high-strength / high specific stiffness reinforcement fiber.

In addition to understanding damage or fracture events, elastic deformation is the initial step in any glass deformation process—and thus also the starting point of this review article. As with their crystalline counterparts, knowledge on the elastic behavior of glasses can provide unique insights into their structure on both short- and medium-range length scales. Furthermore, although no macroscale plastic deformation occurs, glasses may exhibit ductile behaviour on the nanoscale. Observation of such ductility does not only allow for design strategies towards technical glasses with improved mechanical performance, but also provides knowledge on the structural dynamics, pressure response and low-temperature flow behavior. Finally, we will emphasize the roles of structural heterogeneity, environmental effects, and—from a technical viewpoint—strengthening techniques.

ELASTIC CONSTANTS AND POISSON'S RATIO

In its most general definition, material properties are a set of distinct attributes that characterize the materials' response to stimuli (Ashby 2016). Changes in geometry are described by the displacement vector, which is determined uniquely from the difference between the initial and final (or, undeformed and deformed) states, and this vector defines the material's response to strain. On the other hand, stresses are applied stimuli defined from applied vectorial forces and their orientation in relation to a particular plane. Forces normal to a plane result in either *tensile* or *compressive* stresses (by convention tensile stresses are positive and compressive stresses negative), while forces parallel to a plane result in *shear* stresses (Fig. 1). Isostatic, or hydrostatic, stresses (primarily compressive) are applied equally over the whole surface.

The theory of elasticity is a branch of solid mechanics focused on describing the deformations of solids in the limit of small strains, which vanish after the force application is removed (Lurie 2005). The basis of the theory of elasticity was established by Robert Hooke and his observation related to the force and deformation in springs (Hooke first announced his law as an anagram in Latin before publishing the solution "as the extension, so the force" two years later, see Petroski 1996). The Cauchy-generalized Hooke's law is:

$$\sigma_{ij} = C_{ijkl} \cdot \varepsilon_{kl} \tag{1}$$

where σ_{ij} and ε_{kl} are the stress and strain components, respectively, and C_{ijkl} are the elastic constants. The elastic response of materials is fully described by the 21 independent components

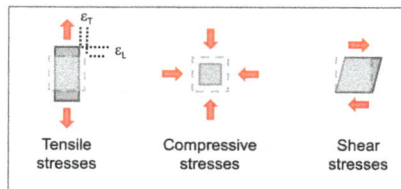

Figure 1. Illustration of elastic deformation under (from left to right) tensile, isostatic compressive, and shear stresses.

(reduced from 36 by the similarity condition $C_{ijkl} = C_{klij}$) of the 4th rank elastic constant tensor. Highly symmetrical crystalline structures like those of the cubic system (simple, body- and face-centered (de Wolff et al. 1985)) reduce the elastic constant tensor to the C_{xxxx}, C_{xxyy} and C_{yzyz} components (also written in Voigt notation as C_{11}, C_{12} and C_{44}; Helnwein 2001), while completely isotropic materials such as polycrystalline bodies or glasses, further reduce it to two independent elastic parameters: the first and second Lamé constants $\lambda = C_{12}$ and $\mu = C_{44}$, with $C_{11} = C_{12} + 2C_{44}$ (Pyke 2002; Lalena and Cleary 2005). For most deformation phenomena, the shear (G) and bulk (K) moduli are used instead,

$$G = \mu = C_{44} = \frac{C_{11} - C_{12}}{2} \tag{2a}$$

$$K = \lambda + \frac{2}{3}\mu = \frac{C_{11} + 2C_{12}}{3} \tag{2b}$$

as they have well-defined physical meaning and describe the materials' resistance to volume-invariant shape deformation (in the case of G) and to shape-invariant volume deformation (in the case of K). The response of glasses (and solid materials in general) to tensile stresses is usually neither volume- nor shape-invariant: extension in the stress direction contracts the material along the perpendicular directions. The material resistance to deformation and the relation between longitudinal (ε_L) and transversal (ε_T) strain are given by Young's modulus E and Poisson's ratio ν, respectively, in terms of K and G (Maceri 2010):

$$E = \frac{9KG}{3K + G} \tag{3a}$$

$$\nu = -\frac{\varepsilon_T}{\varepsilon_L} = \frac{3K - 2G}{2(3K + G)} \tag{3b}$$

For a material to be mechanically stable, both K and G must be positive values. This puts physical limits on the value of Poisson's ratio: for $K \gg G$, $\nu \to \frac{1}{2}$ which is the special case for volume-conserving deformation as observed in rubbers and liquids, and for $G \gg K$, $\nu \to 1$ and the material expands instead of contracting under tensile stress. These latter *auxetic* materials are a result of molecular or microstructural interactions (Evans and Alderson 2000), and no such glasses are currently known. Another special case is when $\nu = 0$ and the material (such as cork) does not laterally deform under elongational stresses.

Alternatively, the problem of deformation can be approached from the point of view of thermodynamics. The amount of work done by external stresses contributes to the free energy density $\Psi = F/V$ as:

$$d\Psi = -\Omega dT + \sigma_{ij}\, d\varepsilon_{ij} \tag{4}$$

where Ω is the entropy density, σ_{ij} and ε_{ij} are the ij components of the stress and strain tensors. The deformed state is defined by the strain tensor that minimizes Ψ, from which result the following general relations between Ψ, ε and the stress and elastic tensors σ and C (Buehler 2008):

$$\sigma_{ij} = \frac{\partial \Psi}{\partial \varepsilon_{ij}} \tag{5}$$

$$C_{ijkl} = \frac{\partial^2 \Psi}{\partial \varepsilon_{ij} \partial \varepsilon_{kl}} \tag{6}$$

For a rigorous derivation, see Refs. (Weiner 2002; Rankin et al. 2013; Hentschke 2017).

The mechanistic and thermodynamic approaches to elasticity give consistent results at different length scales, hinting that a deeper connection can be found between them. One intuitively expects the stiffness of a particular material to be related to the cohesive forces. A physical basis for this intuition is provided by the Cauchy–Born rule, which relates the macroscopic strain directly to the microscopic displacements (Alexander 1998; Weinan and Ming 2007). As such, this intrinsically links the characteristics of interatomic bonding with observable, large-scale mechanical properties. The characteristics of bonding are traditionally used to classify materials in the broad classes of ionic, metallic, covalent and molecular solids (Alemany and Canadell 2014). The ionic bond results from the electrostatic interaction between alternating cations and anions in a regular structure; electrons are fully transferred from the positive to the negative ions, and the charges are homogeneously distributed along the ionic surfaces. This makes the bonding non-directional and ionic solids thus usually crystallize in close-packed, high-coordination-number structures. Non-directionality is also a property of metallic bonding, where the ions form a rigid, periodic lattice around which the valence electrons move freely, like particles in a gas. Due to the degeneracy of the valence energy band in metallic-bonded solids, they can easily form alloys with elements of different valency, and they can also adopt close-packed structures. In the case of covalent bonding, the electron sharing happens pairwise between two neighboring atoms; because of the rearrangement of the outer electronic shells, the covalent bond is strongly directional, therefore covalent solids have lower packing density from the added angular constraints. Finally, molecular solids encompass a wide range of materials where the constituent atoms or molecules are weakly bound via electrodynamic effects such as van der Waals bonds, dipole–dipole interactions and hydrogen bonding (Gilman 2009; Levitin 2014). Naturally this classification is not that clear-cut in real materials; the examples of purely ionic or covalent bonding are rare, with the majority of bonds lying somewhere in the continuum between these extremes depending on their electronegativity (Pauling 1960). For example, this is the case for the bonding in most oxide glasses (Calahoo and Wondraczek 2020). Even the metallic bond, which is close to its ideal description for metals with s or p valence bands, takes an angular-dependent covalent character when the d orbitals of transition metals are considered (Lalena and Cleary 2005).

Since the chemical bonding determines both the strength of the interatomic interactions and the short- and medium-range geometrical constraints dictating the rules of atomic arrangement, in principle, the measurement of the elastic properties provides a window into the short- and medium-range order of any material. This is straightforward for crystalline materials, where periodic boundary conditions apply and the local bond energy density is known or experimentally accessible. In the realm of oxide glasses, relations have been drawn between properties that depend on the *average energy density*, such as Young's modulus and the glass transition temperature (Rouxel 2006), or bulk modulus and the *average dissociation energy* (Pyke 2002; Rouxel 2007): specific structural features are disregarded in these simplistic approaches as they ignore aperiodic heterogeneity, a universal communality among all glass materials. We will discuss such relations in the later parts of this review.

FRACTURE AND BRITTLENESS

As already noted in the introduction, oxide glasses are *the* archetype brittle material. The origin of the field of fracture mechanics is traditionally attributed to publication of the Griffith failure criterion (Gdoutos 1990), but Griffith's work was an extension on the "classic" failure criteria of von Mises, Tresca, and Coulomb-Mohr. These failure criteria were based on the concept of a "critical applied stress"—a well-defined stress limit that is an inherent material property (Prager 1947; Christensen 2013). However, experimental observations showed that the failure strength of any given material was difficult to reproduce, and that materials with high theoretical strength also produced data with large experimental scatter (Lawn 1993). The theoretical strength σ_{th} of a perfectly brittle material can be estimated from the Young's modulus E, the specific surface energy γ_s, and the equilibrium interatomic distance a_0 (Orowan 1934; Petch 1954):

$$\sigma_{th} = \sqrt{\frac{E \cdot \gamma_s}{a_0}} \tag{7}$$

Still, the experimental failure stresses are usually 100 to 1000 times lower than this estimation (Bradt et al. 1974). For example, the theoretical strength of silica glass is estimated to be around 20 GPa, while the experimentally determined failure strength of glass specimens is approximately 100 MPa (Brambilla and Payne 2009; Le Bouhris 2014).

A first insight into this discrepancy was offered by Inglis (1913), considering the stress distribution of an infinite plate containing an elliptical hole with dimensions $2a$ and $2b$ for the major and minor axis, respectively. The internal radius of curvature $\rho = b^2/a$ is minimum at the major axis extremes, where the local stress σ is given as a function of the applied stress σ_∞ and the hole geometry (we note that according to the following equation, the stress amplification depends only on the geometrical shape of the hole, not on its size):

$$\sigma = \sigma_\infty \cdot \left[1 + 2\sqrt{\frac{a}{\rho}} \right] \tag{8}$$

In other terms, the applied stress is amplified at the apex of the hole, as depicted in Figure 2a (Wondraczek 2019).

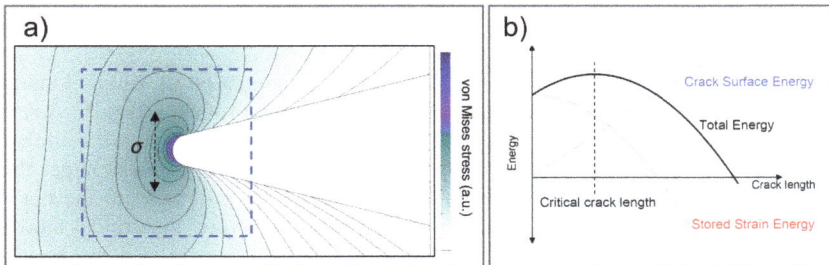

Figure 2. (a) Schematic of stress concentration at the crack tip (the color scale indicates von Mises stress levels). **(b)** Representation of Griffith's energy balance governing crack growth.

For the case of an extremely thin hole, $a \gg b$, the expression above simplifies to $\sigma = 2\sigma_\infty\sqrt{a/\rho}$. This expression can be used to estimate the stress amplification experienced by brittle materials: considering that crack sizes are usually around 10 to 100 mm and that for an atomically sharp crack the crack tip radius of curvature would be similar to the interatomic spacing (0.5 to 1 nm), stress concentration values of 100 to 1000 times the applied stress are predicted, very consistent with the experimental data (Wiederhorn 1984). Inglis' results also suggested that the stresses around the elliptical hole begin to deviate from the applied stress within a distance a from the hole, with the highest gradient of stress concentration being confined to a region of dimension ρ close to the tip of the ellipse. This is an important result, as it explains how the presence of defects and structural irregularities magnifies the applied stresses, causing brittle materials to fail at loads much lower than the theoretical limit. But it could not explain why longer cracks usually propagate more easily than shorter ones or indicate the condition which would lead to material failure under stress (Bradt et al. 1974). If cracking (bond rupture) could be avoided up until the amplified stress field reaches the threshold of shear deformation (e.g., by bond switching reactions), such deformation would lead to an increase in ρ and, hence, reduced amplification. Pronounced plastic deformation observed in thin-layers of amorphous oxides has been related to this effect, whereby flawless geometric confinement was supposed to enable the required levels of homogeneous stress (Frankberg et al. 2019).

Griffith then considered the process of material failure as a problem of energy balance. The stored strain energy U_e in a semi-infinite sheet of volume V and thickness B under tensile stress σ is:

$$U_e = \frac{\sigma^2}{2E'} \cdot V \tag{9}$$

where E' is equal to E if the body is in plane stress, or $E/(1-v^2)$ for plane strain, with v as the Poisson's ratio. However, the introduction of defects changes the elastic behaviour of this semi-infinite sheet, so Griffith applied Inglis' solution to the stress and strain fields for the case of an infinitesimally thin ellipse, now a crack. He then reached the following result for the stored strain energy as a function of crack length (Griffith 1921):

$$U_e = \frac{\sigma^2}{2E'} \cdot V - \frac{\sigma^2}{E'} \cdot \pi B a^2 \tag{10}$$

The other contribution is given by the surface energy U_s of the defect, representing the energy necessary to create the surface of a crack with total length $2a$ in a sheet of thickness B, $U_s = 4\pi B a_s$. Thus, the total energy of this system, U, equals:

$$U(a, \sigma) = U_e + U_s = \frac{\sigma^2}{2E'} \cdot V - \frac{\sigma^2}{E'} \cdot \pi B a^2 + 4aB\gamma_s \tag{11}$$

This expression demonstrates that, for a particular combination of sample loading and elastic moduli, there is a critical crack size a_c at the maximum of total energy U. Any cracks with sizes smaller than the critical size are stable, since $\partial U/\partial a > 0$ for $0 < a < a_c$ and therefore the extension of such cracks is energetically unfavorable. Oppositely, cracks larger than the critical size are unstable, as $\partial U/\partial a < 0$ for $0 > a > a_c$ so any extension to the crack size is energetically favorable and causes catastrophic failure. Therefore, the critical flaw size is the smallest crack which leads to catastrophic failure and is characterized by the unstable equilibrium $\partial U/\partial a = 0$ and $\partial^2 U/\partial a^2 < 0$ (see Fig. 2b). From these conditions, the failure stress σ_f for a material containing a flaw of size a_c is given by Griffith's failure criterion:

$$\sigma_f = \sqrt{\frac{2\gamma_s E'}{\pi a_c}} \tag{12}$$

We refer to Munz and Fett (2015) for a rigorous derivation of Griffith's relation.

Several authors have expanded on Griffith's work, obtaining his failure criterion for a variety of sample and crack geometries. These solutions have the general form of

$$\sigma_f = Y \cdot \sqrt{\frac{\gamma_s E'}{a_c}} \tag{13}$$

where Y is a constant depending on sample and crack geometry, and mode of loading (Bradt et al. 1974). Griffith successfully applied his analysis for the failure stresses of glass fibers and glass tubes, finding good agreement between the measured surface energy and the value estimated from fracture data (Gdoutos 1990; Lawn 1993). While Griffith's criterion showed good agreement with fracture data from glass samples, this material is the closest real equivalence to the ideally isotropic and ideally brittle cracked semi-infinite sheets from his original derivation. For ductile materials, such as steels and aluminum alloys, Griffith's failure criterion gave unreasonably high estimates for the surface energy. This discrepancy was addressed independently by Irwin and Orowan, who proposed that crack extension is not the only "sink" of stored strain energy, as it can also be consumed by plastic deformation in the region close to the crack tip, denominated the "plastic zone", and by other dissipative processes such as heating or particle emission (Dickinson et al. 1981; Langford et al. 1987). Thus, two new measures are introduced: (i) R, which is the material resistance to crack extension and combines the contributions of the specific surface energy γ_s, the plastic deformation γ_p, and other effects γ_o, $R = 2\gamma_s + \gamma_p + \gamma_o$; and (ii) $G(a)$, the energy release rate per unit area of the crack given by:

$$G(a) = -\frac{\partial U_e}{\partial A} = \frac{\sigma^2 \pi a}{E'} \tag{14}$$

Consequently, failure is achieved when the energy release rate reaches a critical value G_c equal to the material resistance to crack extension $G(a) = G_c = R$ (Ibrahim 2015). The failure stress is then given by:

$$\sigma_f = \sqrt{\frac{G_c E'}{\pi a_c}} \tag{15}$$

Additionally, the loci for stable and unstable crack growth are not anymore defined by crack size, but from the relations between the material properties R and G: for $G \leq R$, G_c, the crack growth is stable, while for $G > R$, G_c, the crack growth is unstable and catastrophic. Another related concept is that of the R-curve, which is a plot of R as a function of crack size. In the ideal case of a homogeneous, isotropic material without microstructure (such as in a homogenous glass), R is a material property and the R-curve is flat. However, the tailoring of microstructure in polycrystalline ceramics, phase-separated glasses and glass-ceramics can induce favorable interactions between the advancing crack and the microstructure, generating a rising R-curve and resulting in a "tougher" material, less sensitive to brittle fracture. This is because the rising R-curve "converts" unstably-extending cracks into arrested cracks through the action of the toughening mechanisms, arising from the crack–microstructure interactions (Wiederhorn 1984; Lawn 1993; Becher and Rose 2006; Serbena et al. 2015).

The unification between the energy balance approach of Griffith and the stress amplification from Inglis' analysis was achieved via the works of Westergaard and Irwin. Westergaard's contribution was to find an exact solution for the stress field surrounding a crack (Westergaard 1939), which takes the form

$$\sigma(x) = \frac{\sigma_\infty}{\sqrt{1 - \left(\frac{a}{x}\right)^2}} \tag{16}$$

when considered at the crack plane, with x being the distance from the center of the crack. Note that this equation implies a stress singularity at the crack tips $\sigma_\infty \sqrt{\pi a}$, the existence of which is still an open subject in the field of linear elastic fracture mechanics (Zhu et al. 2019). Irwin's insight was to realize that by changing the coordinate system from Cartesian (x, y) to polar (r, θ), Westergaard's exact solution is well approximated near the crack tip $(r \leq a/10)$ by a set of general equations (Irwin 1957):

$$\sigma_{ij}(r,\theta) = \frac{\sigma_\infty \sqrt{\pi a}}{\sqrt{2\pi r}} \cdot f_{ij}(\theta) + \dots \tag{17}$$

where σ_{ij} are the Cauchy stresses, r is the distance from the crack tip, θ is the angle from the crack plane, and $f_{ij}(\theta)$ are functions dependent on the loading condition and sample/crack geometry. This general form is also independent of the fracture mode being considered: tensile, in-plane shear or out-of-plane shear (Modes I, II, and III, respectively). Irwin recognized that the expression $\sigma_\infty \sqrt{\pi a}$ was common to all general equations and was location-independent, meaning it uniquely describes the stress amplification at the crack tip. He named this quantity the stress intensity factor, K, a constant that determines the crack behavior, and, in the case of brittle materials, provides a complete description of failure. Thus, if the stress intensity factor in any two samples is the same, the crack tip they describe will behave similarly, independently of sample history and external conditions (Fineberg and Marder 1999). Considering the failure criterion in Mode I, the critical stress intensity factor $K_{Ic} = \sigma_f \sqrt{\pi a}$ can be related to the critical energy release rate G_{Ic}:

$$G_{Ic} = \frac{\sigma_f^2 \pi a}{E'} = \frac{K_{Ic}^2}{E'} \tag{18}$$

$$K_{Ic} = \sqrt{G_{Ic} E'} = \sqrt{R E'} \tag{19}$$

Since R and E' are material dependent but sample independent, K_{Ic} is also a material property and defines the fracture toughness (Bradt et al. 1974). Therefore, the failure criterion can be defined as a function of the critical stress intensity factor, with catastrophic crack propagation occurring for $K_I \geq K_{IC}$. It is important to highlight that fracture toughness is not an indication of material strength, but of the materials' sensitivity to the presence of flaws. Both polymers with low yield points and ceramics with high failure strengths have low fracture toughness (generally < 10 MPa m$^{0.5}$), while metals and metallic alloys of varying yield stresses have high fracture toughness (> 40 MPa m$^{0.5}$) (Callister Jr and Rethwisch 2018). As such, the measured strength of polymers and ceramics is determined by the flaw population, which in turn is determined by the processing technique. For metals, the measured strength reflects to a much larger extent the intrinsic material properties.

EXPERIMENTAL METHODS FOR STRENGTH AND TOUGHNESS TESTING

The strength of glass, *intrinsically*, should depend only on the chemical composition and the thermo-mechanical history of the glass. However, as discussed in *Fracture and Brittleness*, flaws and microscopic defects present either on the surface or in the bulk of the glass cause the strength measurement to only provide an extrinsic (*practical*) value, which is usually a few orders of magnitude lower than the intrinsic property (Baker and Preston 1946; Levengood 1958; Naray-Szabo and Ladik 1960; Kurkjian et al. 2010). Besides, environmental fatigue (known as *stress corrosion*) is another reason of the low practical strength of glass (Baker and Preston 1946). Fatigue in glass has also been reported in samples under low loading rates even in inert environments such as in vacuum. Thus, when we compare measured strength values of one glass to another, we should consider the following questions: (i) Are the measured glasses pristine (flawless)? (ii) Are the glasses tested in the same environment? (iii) Are the glasses measured either in the same timescale or with the same rate? For commodity glass products, the most relevant loading situations are bending or biaxial deformation, so most widespread testing protocols focus on these two cases. For uniaxial tests, we will describe two-, three-, and four-point bending, and for biaxial tests we will focus on Ring-on-Ring (RoR) testing as an example (Salem and Jenkins 1999; ASTM 2013). We note that because the strength of glass depends on many parameters, the strength results are reported in terms of a compressed exponential distribution of the probability of fracture, the so-called Weibull distribution (see *Weibull Distribution and Probability Plots*). Finally, we will also describe the methods for fracture toughness testing.

Biaxial test

The biaxial test used in strength measurement of glass has a lower support in the form of a ring or three balls, and an upper support as ball (point load), ring or pressure. Based on their configurations, the methods are called Ball-on-Ring (BoR), Ring-on-Ring (RoR), Pressure-on-Ring (PoR), Ball-on-Ball (BoB), Ring-on-Ball (RoB) and Pressure-on-Ball (PoB) (see **Fig. 3**) (Shetty et al. 1983; With and Wagemans 1989; Salem and Jenkins 1999; Börger et al. 2002; Zhao et al. 2008). A variant of the PoR test is the bulge test, which can be employed to study the mechanical properties of thin glass or other membranes (Meszaros et al. 2012).

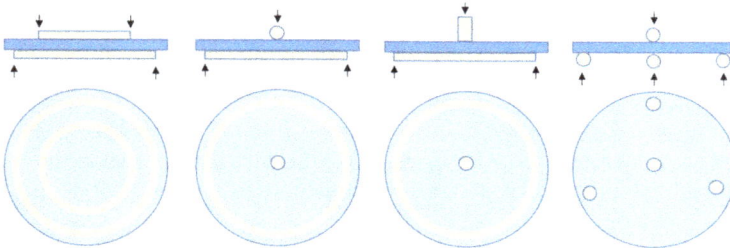

Figure 3. Biaxial test set-ups used to determine the practical fracture strength of glass. From left to right: Ring-on-Ring (RoR), Ball-on-Ring (BoR), Pressure-on-Ring (PoR) and Ball-on-Ball (BoB). The **blue circle** represents the glass samples; **light yellow** features represent the supports (sample holders) and the **black arrows** indicate the directions of the applied stresses.

Among these methods, RoR is the most popular one because a large volume of the test specimen is subjected to stresses, it is convenient to operate, and it is reported to have good agreement with plate theory (Timoshenko and Woinowsky-Krieger 1959). The specimen used in a biaxial experiment can be of circular or square geometry, with the circular one being preferable for its equalized length between the specimen edges to the support ring. The challenge of the RoR, as well as other biaxial tests, is to: (i) keep the specimen as centered as possible between the upper and lower support rings, (ii) minimize the friction and (iii) distribute the load equally on the contact surface between the specimen to the two rings. The first challenge can be solved by connecting the load ring to the load cell and using this load ring to guide the support ring to fix in the center. The two latter challenges can be solved by positioning a compliant layer between each contact surface of the specimen to each ring. During fracture, the crack can initiate from the inside of the load ring or outside of the load ring. The tensile stress can then be calculated accounting for the actual position of the crack origin. If the crack origin is inside the load ring, the radial (s_r) and hoop (s_θ) stress are identical and expressed by:

$$\sigma_r = \sigma_0 = \frac{3F}{2\pi h^2}\left((1-v)\frac{D_s^2 - D_L^2}{2D^2} + (1+v)\ln\frac{D_s}{D_L}\right) \tag{20}$$

If not,

$$\sigma_r = \frac{3F}{2\pi h^2}\left((1-v)\frac{D_s^2 - D_L^2}{2D^2} - (1-v)\frac{d^2 - D_L^2}{2d^2} + (1+v)\ln\frac{D_s}{d}\right) \tag{21a}$$

$$\sigma_\theta = \frac{3F}{2\pi h^2}\left((1-v)\frac{D_s^2 - D_L^2}{2D^2} + (1-v)\frac{d^2 - D_L^2}{2d^2} + (1+v)\ln\frac{D_s}{d}\right) \tag{21b}$$

where F is the load (N) at fracture, D_S and D_L are the diameters of the support and load rings (mm), respectively, D is the test specimen diameter (mm), d is the diameter of the circle on which the crack origin is located to the center of specimen (mm), h is the test specimen thickness (mm), and v is Poisson's ratio. Whenever the crack origin is in between the load ring and the support ring, the hoop stress can be used to estimate the strength since the crack has initiated tangentially. In any situation, bending of the specimen must be avoided during RoR testing, which can be compensated by adapting the employed ring sizes and their ratio relative to the sample thickness. However, for this reason, RoR is typically not applicable to the testing of thin glass sheet, for example, for glass sheet with a thickness far below 1 mm.

Uniaxial test

Tensile tests. The tensile test is another conventional method for measuring the strength of glass materials, usually glass fibers (see Fig. 4a) (Cameron 1966; Ernsberger 1969; Kennedy et al. 1980; Pukh et al. 2005). Unlike the two-point bending test, tensile tests provide direct information on deformation and applied load, which can be simply converted to stress and strain data. One major difficulty of tensile tests is securing the sample prior to the measurement; gripping the sample with enough strength to avoid slipping or drilling the sample both create large defects, which contribute to the reported scatter in the measured strength values (Kurkjian et al. 2003).

Two-point bending. The two-point uniaxial bending technique was first used around the 1950s (Murgatroyd 1944; France et al. 1980). In this test, one support is fixed and another support is moved in most cases. As shown in Figure 4b, a fiber sample is bent in a "U" shape between two supporting faceplates. This bending test is popular for measuring the strength of glass fibers or thin sheet, replacing the difficulty in performing the pure traction method. It is important that the two support plates have the surfaces polished to reduce the risk of contact damage to the fiber. Two-point bending is a simple method and can be used to test freshly-drawn fibers with pristine surface. The fracture stress (σ_f) can be calculated as:

$$\sigma_f = 1.198 \frac{E_g d_g}{D - d} \tag{22}$$

where E_g is the Young's modulus of the glass fiber, d_g the diameter of the glass fiber, D the distance between the two supporters at the fracture of bending fiber and d is the whole dimension of the fiber including any coating material (i.e., $d \geq d_g$). In the case there is no extra coating, $d = d_g$.

Because of its simplicity, the two-point bending test has been used in inert conditions; the obtained inert strength was found to be close to the calculated theoretical strength

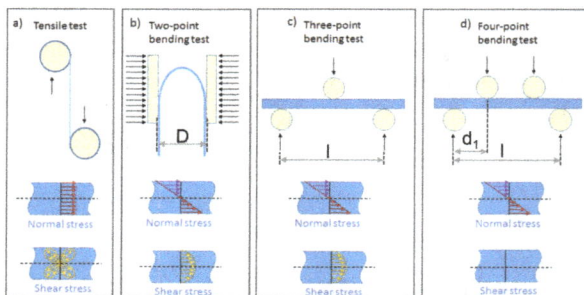

Figure 4. Uniaxial test set-ups used to determine the practical fracture strength of glass. The **blue bar** represents the glass samples; **light yellow** represents the supports, and **black arrows** represents the applied forces. Normal and shear stress conditions in the sample are also sketched, with **purple, red** and **yellow** arrows as tensile, compressive and shear stresses, and the dashed lines show the neutral lines (where stress is zero). Geometrical constants D, l and d_1 are explained in the main text.

(Kurkjian et al. 2003). The fracture stress from this method can be simply written as $\sigma_f = E_g \varepsilon_g$, where ε_g equals to $1.198 d_g/(D-d)$. The method has therefore been used to determine the inert failure strains for glasses such as sodium silicate and sodium aluminosilicate (Lower et al. 2004a,b), and also to study effects of the glass' thermo-mechanical history (Brow et al. 2009). A known drawback of this relatively simple method is that it cannot be used to directly measure the failure stress due to the non-linear elastic modulus (Matthewson et al. 1986).

Three- and four-point bending. Three- and four-point bending tests can also be used to determine the strength of glass fibers, but they are more popular for measurements on either rectangular plates or cylindrical samples (Baratta et al. 1987; Quinn 1988). The flexural strength σ_{fl} of a sample is calculated from the failure load F_f, and sample and experiment geometry:

$$\sigma_{fl,3pb} = \frac{3F_f l}{2bh^2} \tag{23a}$$

$$\sigma_{fl,4pb} = \frac{3F_f d_1}{bh^2} \tag{23b}$$

where l is the support span, d_1 is the distance between the outer support and the inner loading arm, b and h are the samples width and thickness, respectively. For cylindrical samples with diameter D, we have:

$$\sigma_{fl,3pb} = \frac{8F_f l}{\pi D^3} \tag{24a}$$

$$\sigma_{fl,4pb} = \frac{16F_f d_1}{\pi D^3} \tag{24b}$$

Three-point bending has a simpler experimental setup and is less sensitive to sample geometry and alignment errors, while four-point bending has the advantages of a larger probed volume and having a constant value of bending moment with zero shear force in between the two inner loading arms, eliminating the effect of shear forces (Fig. 4d). When using three- or four-point bending methods, the four edges of the rectangular beam need to be polished to ensure that the fracture origin is not from those edges. To identify where the crack initiates, a *post mortem* study is usually performed (see *Fractography*).

Fracture toughness test

Strength tests can also be used to determine fracture toughness, however, in the case of glasses, there is no standard measurement method, mainly due to the difficulty in maintaining the quasistatic equilibrium necessary to correctly estimate K_{Ic} (see *Fracture and Brittleness*); if the crack speed propagation becomes too large (threshold values are approximately 30% of the materials Rayleigh wave speed), several dynamic effects like crack path deflection, crack branching, and crack surface roughening can occur (Fineberg and Marder 1999; Sharon and Fineberg 1999), severely complicating the data analysis. The measurement of fracture toughness started from the flat slab methods such as double torsion and double cantilever beam in the early 1970s (Wiederhorn et al. 1974a,b). Later, point methods (indentation fracture methods, see below) were developed based on calibration, mostly from the flat slab methods (Lawn and Wilshaw 1975; Evans and Charles 1976). After the 1980s, beam methods such as single-edge notched beam, single-edge precracked beam (SEPB), chevron-notched short beam, chevron-notched beam (CNB) and surface crack in flexure were developed and tested on different types of glasses (Shinkai et al. 1981; Nose and Fuji 1988; Reddy et al. 1988; Fett et al. 1991; Dériano et al. 2004; Quinn and Bradt 2007; To et al. 2018). A present consensus is that both SEPB and CNB are self-consistent methods, and therefore recommended for measuring the fracture toughness of glasses (Rouxel and Yoshida 2017).

Indentation fracture toughness. Developed by Evans and Charles in the 1970s (Evans and Charles 1976; Lawn et al. 1980), determining a toughness value from indentation experiments became very popular since it is a fast, straightforward and cheap method, easy to apply on brittle materials. In principle, the only things needed are a polished sample, an indenter and a microscope. In its original formulation, the indentation fracture toughness K_{IFT} was related to the applied indentation load P and the length of the cracks generated at the corners of the imprint, c,

$$K_{IFT} = 0.0752 \cdot \frac{P}{c^{3/2}} \tag{25}$$

Later, several models were developed, most notable, taking into account some material properties (i.e., Young's modulus E and Meyer's hardness H). An exhaustive review of those models (which depend on the indentation crack pattern, e.g., ring, cone, half-penny and Palmqvist cracks), can be found in Ponton and Rawlings (1989a,b).

The experimental procedure consists in the preparation of a polished surface, which is then indented at loads high enough to generate cracks on the residual imprint's corners, but not so high as to generate spalling or edge cracks. Usually "sharp" indenters, such as Vickers, Berkovich or cube-corner geometries, are used as they provide well-defined corner tips from which cracks can easily nucleate and propagate, and allow for cracking to occur at relatively low indentation loads (Januchta and Smedskjaer 2019). The resulting crack geometry is an important parameter for fracture toughness determination; its dimensions and geometrical relations to the residual indent are used in the estimation of the fracture toughness (Ponton and Rawlings 1989a,b). Most glasses and glass-ceramics exhibit half-penny cracks (Yoshida 2019). While extremely convenient, it must be noted that this methodology does not accurately measure fracture toughness, but instead a related quantity, usually called indentation fracture resistance or indentation fracture toughness (not to be confused with the crack initiation resistance, CR, which is also derived from indentation). This property depends strongly on the stress fields around and below the indenter as they evolve during the cycle of loading and unloading, and their interaction with the extending crack (Li et al. 1989; Ponton and Rawlings 1989a,b; Ghosh et al. 1991; Quinn and Bradt 2007). For example, we note that no indentation toughness can be defined if there are no cracks at a given load, although the material of course still behaves according to its underlying toughness at that loading condition. Part of the disagreement between "indentation fracture toughness" and the well-defined critical stress intensity factor for mode I loading ("fracture toughness") can be ascribed to densification. For example, recent work has shown that the disagreement can, at least partially, be overcome by the use of a sharper indenter tip that limits the amount of indentation-induced densification (Gross et al. 2020).

WEIBULL DISTRIBUTION AND PROBABILITY PLOTS

As mentioned previously, the experimental strength data of oxide glasses can vary significantly based on the different sizes and shapes of the pre-existing flaws as well as on the different sizes of the specimens. The simple usage of mean and standard deviation values would be insufficient to represent the strength distribution. Instead, the Weibull distribution is used to describe the statistics of glass strength data. This distribution belongs to the family of extreme value distributions (Coles 2001), and is mathematically defined as a continuous probability distribution with a density function DF:

$$DF(x \mid a,b,c) = \frac{c}{b}\left(\frac{x-a}{b}\right)^{c-1} \exp\left\{-\left(\frac{x-a}{b}\right)^{c}\right\}; x \geq a; a \in R; b, c \in \mathbb{R}^{+} \tag{26}$$

Satisfying the conditions $\mathrm{DF}(x) \geq 0 \forall x$ and $\int_{-\infty}^{+\infty} \mathrm{DF}(x)\mathrm{d}x = 1$. Its cumulative distribution function CDF:

$$\mathrm{CDF}(x \mid a,b,c) = 1 - \exp\left\{-\left(\frac{x-a}{b}\right)^c\right\} \tag{27}$$

satisfies $\lim_{x \to 0} \mathrm{CDF}(x) = 0$ and $\lim_{x \to \infty} \mathrm{CDF}(x) = 1$.

Weibull introduced this distribution while discussing the experimental results from tensile, torsion, and bending tests in stearic acid and plaster-of-Paris rods (Weibull 1939a,b). The distribution can be used to describe the observed mechanical behavior of such brittle materials, where the probability of failure of each sample depends both on the applied load and on specimen size (Brook 1991; Munz and Fett 1999). While the first remark is trivial, the second follows from the fact that, in brittle materials, the specimen's failure strength is directly related to the size of the largest flaw in that specimen. Since big specimens can accommodate larger flaws than small specimens, it follows that the mean strength of a set of small specimens is larger than the mean strength of a set of large samples (Fig. 5a). Weibull's key assumption for representing the behavior of brittle materials was the "weakest link hypothesis", that is, the specimen fails if its weakest volume element fails. In contrast, failure in tough materials first occurs when the whole system cannot anymore accommodate the applied load after a number of "link failures", causing the load to be redistributed to the remaining ones (Jayatilaka and Trustrum 1977; Kittl and Díaz 1990; Danzer 2006).

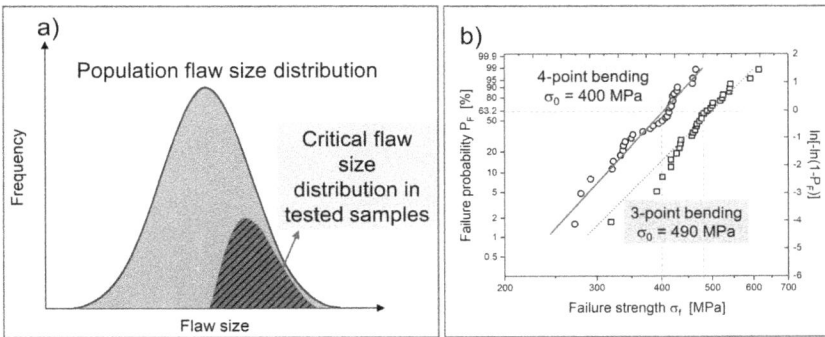

Figure 5. (a) Schematic of the real flaw size distribution in comparison to the flaw sizes probed in mechanical testing. **(b)** Volume effect on sample strength: 3-point-bending stresses a smaller volume than 4-point-bending during mechanical test, resulting in "stronger" samples—Data from Quinn and Quinn (2010).

Consider the failure probability P_F of a volume element V_0 under stress σ is $f(\sigma)$ (also known as the cumulative hazard rate in life testing), then the survival probability P_S of a volume element ΔV is:

$$P_S(\Delta V, \sigma) = 1 - f(\sigma) \cdot \frac{\Delta V}{V_0} \tag{28}$$

For a larger sample with volume $V_0 + \Delta V$, the survival probability is given by:

$$P_S(V_0 + \Delta V, \sigma) = P_S(V_0, \sigma) \cdot P_S(\Delta V, \sigma) = P_S(V_0, \sigma) \cdot \left[1 - f(\sigma) \cdot \frac{\Delta V}{V_0}\right] \tag{29}$$

As ΔV gets infinitesimally small, $\Delta V \to \partial V$ and:

$$P_S\left(V_0 + \partial V, \sigma\right) - P_S\left(V_0, \sigma\right) = -P_S\left(V_0, \sigma\right) \cdot f\left(\sigma\right) \cdot \frac{\partial V}{V_0} \tag{30}$$

Rearranging and integrating yields:

$$\int \frac{\partial P_S\left(V_0, \sigma\right)}{P_S\left(V_0, \sigma\right)} = \int -f\left(\sigma\right) \cdot \frac{\partial V}{V_0} \tag{31a}$$

$$\therefore P_S\left(V_0, \sigma\right) = \exp\left(-\frac{f\left(\sigma\right) \cdot V}{V_0}\right) \tag{31b}$$

In his work, Weibull argues that the simplest material function $f(\sigma)$, which satisfies the conditions of being positive and non-decreasing with increasing σ, is:

$$f\left(\sigma\right) = \left(\frac{\sigma - \sigma_u}{\sigma_0}\right)^m \tag{32}$$

Therefore, the failure probability P_F is given by:

$$P_F = 1 - P_S = 1 - \exp\left[-\frac{V}{V_0} \cdot \left(\frac{\sigma - \sigma_u}{\sigma_0}\right)^m\right] \tag{33}$$

from which the classical three-parameter Weibull CDF (Eqn. 26) is recovered, with:

$$\begin{cases} a = \sigma_u \\ b = \sigma_0 \cdot \left(\dfrac{V_0}{V}\right)^{1/m} \\ c = m \end{cases} \tag{34}$$

It is worth noting that this distribution rightly bears Weibull's name, but it was independently described by Frechet, and Rosin, Rammler and Sperling (Brown and Wohletz 1995; Cook and DelRio 2019). The same probability distribution can also be derived from models other than the weakest link theory, such as the shot-noise model, the hazard rate approach, and the broken-stick model (Rinne 2008). Additionally, the physics of low-probability events has a variety of applications, and consequently the Weibull distribution has been utilized beyond the field of failure mechanics (Mauro and Smedskjaer 2012). For example, the Weibull distribution and the Kohlrausch stretched exponential function for glass relaxation share a common functional form, indicating some shared underlying physics. Just as inhomogeneities in the flaw distribution cause a departure from ideal Weibull statistics, heterogeneous glass samples exhibit departure from ideal stretched exponential relaxation.

In terms of material properties, $a = \sigma_u$ is the threshold stress below which the failure probability is zero (in the case of brittle materials $a = \sigma_u = 0$ and the formula reduces to the 2-parameter Weibull distribution), $b = \sigma_0 \cdot (V_0/V)^{1/m}$ is the scale parameter, characteristic strength or median strength, which depends on the stress configuration and sample geometry, and represents the stress when the failure probability equals $1 - \exp(-1) = 63.2\%$. The mean strength (σ for which the failure probability is 50%) can be calculated as

$\bar{\sigma} = \sigma_u + \left[\sigma_0 \cdot \left(\frac{V_0}{V} \right)^{1/m} \cdot \Gamma \left(\frac{1}{m} + 1 \right) \right]$. Note that both the median and average strength are inversely proportional to the tested volume V; this volume dependence is clearly illustrated when considering the experimental data from same geometry and same material samples tested in 3- and 4-point bending tests, which are shifted along the horizontal axis in a Weibull plot (Fig. 5b) Finally, $c = m$ is the shape parameter or Weibull modulus, defined by the degree of data scatter: the larger the value of m, the more narrow the probability distribution; physically m reflects the relative event rate, with $m < 1$ indicating a relative decrease in event rate with increasing stress, while for $m = 1$ the failure rate is stress-independent, and for $m > 1$ the failure rate increases with stress. Generally, metals and metal alloys have Weibull modulus in the range of 50 to 100, while for "tough" ceramics and polymers that range is between 10 and 40, and for "brittle" ceramics and polymers the values of Weibull modulus are below 10 (Ono 2019). From a glass application perspective, the Weibull modulus should be as high as possible so as to reduce the sample-to-sample variations in observable strength.

The Weibull analysis of experimental data is usually carried out from the Weibull strength distribution graph, prepared via a linearization of the expression for failure probability (for a defined test volume):

$$\ln\left[-\ln\left(1 - P_F \right) \right] = m \cdot \ln\left(\sigma - \sigma_u \right) - m \cdot \ln \sigma_0 \qquad (35)$$

Either Ordinary Least Squares Regression (OLSR) or Maximum Likelihood Estimation (MLE) can be applied to estimate the parameters m and σ_0. While OLSR is commonly employed due to its ease of use and ubiquitous implementation in various software packages, MLE is preferred as the confidence intervals of its estimates are markedly narrower in comparison to the LSR estimates (Quinn and Quinn 2010). For further details on both estimation methods, see (Sen and Srivastava 1990; Yan and Su 2009). As a member of the extreme value distributions family, the quality of the parameter estimation in Weibull analysis is very sensitive to the size of the used dataset, showing increasing sensitivity as the Weibull modulus decreases (Chang et al. 2006). While it is possible to estimate the minimum number of independent measurements needed to achieve a target bound ration for a given significance level, either analytically or through simulations (Guo and Gerokostopoulos 2013; Yang et al. 2019a), reliable estimates are generally achieved for dataset with at least 30 observations. Finally, it is important to choose an unbiased probability estimator to convert the measured failure strength data into failure probability necessary for further calculations. These probability estimators are calculated based on data ranking, after being organized in ascending order (lowest value has rank 1). Both Hazen's estimator $\widehat{P_F} = (i - 0.5)/n$ or the Median Rank estimator $\widehat{P_F} = (i - 0.3)/(n + 0.4)$, with i being the data rank and n the dataset size, are widely used in the Weibull analysis literature (Gorjan and Ambrožič 2012; Nohut 2014; Datsiou and Overend 2018).

STRESS CORROSION AND FATIGUE

The long-term strength of oxide glasses is different from their short-term strength. This fatigue effect arises as the strength is affected by the duration or rate of loading. Since glass strength is controlled by the critical flaw size (see *Fracture and Brittleness*), fatigue is related to crack growth under load as affected by the environment (typically water, so-called stress corrosion) (Gy 2003). Fatigue affects the performance of a wide range of glass applications, from the growth of cracks in windshields due to impact by small gravel to optical fibers that are coated to prevent water entry and thus improve the long-term reliability, to stress corrosion and weathering in naturally occurring glasses. It is thus important to understand the structural mechanism responsible for fatigue to be able to predict and, ideally, limit the growth of cracks. The relation between atomistic structure and stress corrosion cracking remains largely unknown (Rodrigues et al. 2017; Rountree 2017). Such crack growth is termed sub-critical and can be represented by crack growth velocity v vs. stress intensity factor K as shown in Figure 6 (Wiederhorn et al. 1982; Lawn 1993), where $K < K_{Ic}$ (see *Fracture and Brittleness*). Alternatively, it can also be plotted as $v(G)$, where G is the strain energy release rate (Ciccotti 2009). Following the pioneering work of Wiederhorn (Wiederhorn 1967; Wiederhorn and Bolz 1970), the curve involves three or four characteristic growth regions, which we will briefly summarize in this chapter. Finally, we will discuss some of the insights gained into the stress corrosion mechanism.

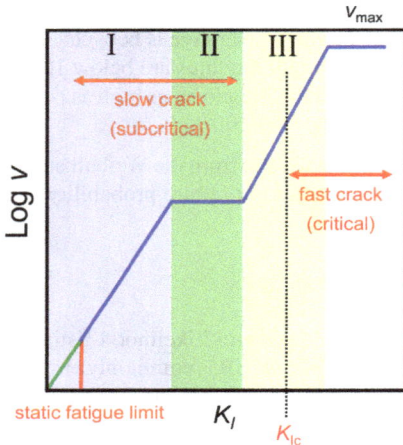

Figure 6. Schematic showing the crack growth velocity v vs. the stress intensity factor K_I (opening mode) in a glass. The shaded regions I–III represent the three regimes of crack propagation from sub-critical (stress corrosion, (**I**) to saturation (**II**) and critical fracture (**III**).

Region 0: environmental limit

Region 0 is only observed for some glasses, for which a threshold behavior (environmental limit) is found at $K = K_0$ (or $G = G_0$). That is, the crack growth velocity decreases dramatically to a very low value when K has decreased to the threshold K_0 value. Alkali-rich glasses generally exhibit the environmental limit and thus a critical stress is required for the crack to propagate (Gehrke et al. 1991). However, no such limit is observed for amorphous silica, thus, a crack will propagate even under minimal stresses. This region is thus important for lifetime analyses of glass materials, but its origin is still subject to debate. Lawn (1993) has proposed that it arises from a shielding zone formed around the crack tip, which dissipates energy. The variation in K_0 with the glass composition has been ascribed to stress-induced diffusion of alkali ions, compressive stress formation in the shielding zone, crack tip blunting, and pH variation at the crack tip (Rountree 2017). Alternatively, following Griffith's theory, a thermodynamic threshold could exist at $G = G_0 = 2\gamma_S$, where γ_S is the energy to create a new surface during bond breaking. However, this would imply crack healing for $G < G_0$, which is not observed experimentally. Nonetheless, it is striking that the typical threshold value of G_0 (0.8 J m^{-2}) is similar to the glass surface tension (Ciccotti 2009). Work by Wiederhorn et al. (2013) has also discussed the phenomenon in terms of water-driven swelling in the crack-tip region. This is because the water diffusion layer results in a compression layer, which lowers the effective stress intensity factor. In any case, additional research is still needed to understand the details of region 0.

Region I: stress corrosion

In this region, the crack growth velocity increases with the stress intensity factor and is limited by the chemical reaction rate at the crack tip. As such, it is termed as the stress corrosion regime. For a given K value, the velocity increases with both the relative humidity and temperature in this region. In addition, the slope of the curve is glass composition dependent. The reaction mechanism between the glass and the environment at the crack tip has been associated with stress-enhanced thermal activation of a dissociative hydrolysis reaction (Fig. 7; Michalske and Freiman 1983).

Following Wiederhorn, v depends on the time for this chemical reaction to occur at the crack tip (Wiederhorn 1967; Wiederhorn and Bolz 1970). Based on reaction-rate theory, Wiederhorn proposed the following equation to fit the data in region I (Wiederhorn 1967),

$$v = A \left(\frac{p_{H_2O}}{p_0} \right)^m \exp\left((-\Delta E_a + bK)/RT \right) \tag{36}$$

where A, m, ΔE_a, and b are adjustable parameters that depend on the glass composition, T is the environmental temperature, R is the gas constant, and p_{H_2O} and p_0 are the partial pressure of the vapor phase in the atmosphere and the total atmospheric pressure, respectively. It is also possible to fit the data in region I with a different exponential equation or with a power law (Ciccotti 2009; Rodrigues et al. 2017), but Equation (36) provides a good connection with the stress reaction, where ΔE_a can be interpreted as the activation energy in the absence of stress and b can be related to the crack tip activation volume and the crack tip radius of curvature. Equation (36) also accounts for the experimentally observed thermal activation.

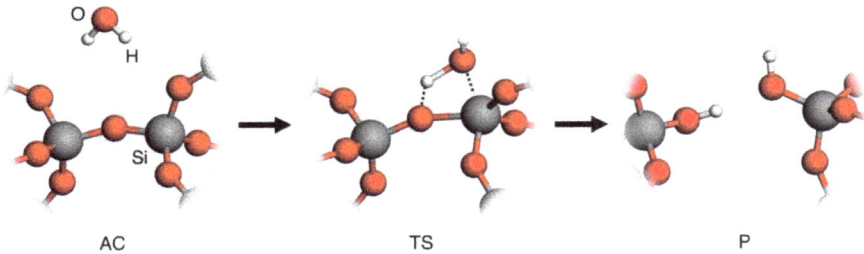

Figure 7. Water-assisted cleavage of the Si–O–Si bond: adsorption complex (**AC**), transition structure (**TS**), product (**P**)—Used by permission of Wiley, from Hühn et al. (2017), Journal of Computation Chemistry, Vol. 38, Fig. 03, p 2351.

Region II: transport kinetics

In this region, the crack growth velocity is limited by the transport kinetics, i.e., the time needed for the reactant to reach the crack front. As such, v is almost independent of the applied stress at the crack tip. However, for a given glass, increasing humidity leads to an increase in the velocity, similarly to the behavior in region I. This increase follows the empirical relation of (Wiederhorn 1967)

$$v = v_0 \cdot p_{H_2O} \cdot D_{H_2O} \tag{37}$$

where D_{H_2O} is the diffusion coefficient of molecular water in air and v_0 is an empirical parameter (Rountree 2017). The transport limitation likely arises from the reduction in diffusivity of water molecules in a vapour or solute due to the decrease in the average free path (Michalske and Bunker 1987). If water molecules permeate the stressed glass network in molecular form, the bulk diffusion rate could also be the limiting factor for region II (Ciccotti 2009).

Region III: inert crack growth

In this region, the crack growth velocity increases again with increasing K (but $K < K_{Ic}$ still applies) and becomes too fast for the chemical reactant to reach the crack front, i.e., the stress–corrosion reactions are suppressed. Consequently, v is independent on the environmental humidity in region III, but remains thermally activated. The velocity reaches very high values in this region, but energy barriers still need to be overcome for the crack to propagate. When the energy barriers cease to exist and mechanical energy is converted to kinetic energy in an unstable manner, we have reached the end of region III and thus $K = K_{Ic}$ (Thomson et al. 1971). In practice, however, it is very challenging to determine this end point and K_{Ic} is typically identified from the position of region III (Ciccotti 2009).

Stress corrosion mechanism

The aim of glass scientists and engineers is to limit subcritical crack growth. To fully avoid it, K at the crack must be less than the environmental limit discussed above. As soon as the crack front begins its propagation, it must ideally be stopped before its velocity reaches Regions II or III, easily leading to dynamic fracture (Rountree 2017). Understanding the stress corrosion mechanism is thus crucial, but also remains debated (Ciccotti and George 2020). Most importantly, there exists a competition between diffusion of water into the crack cavity and within the glass network, corrosion of the network itself, and leaching of mobile alkali ions under the chemical or stress gradient. There has also been discussion about crack tip blunting phenomena. That is, whether the initial atomically sharp crack tip can change its radius during propagation. The crack tip radius of curvature could be a function of glass composition as well as K and eventually become so blunt that the crack tip can stop growing. This is complicated by the fact that the dissolution reactions are also affected by the tip radius and, thus, external stresses (Ito and Tomozawa 1982). The possibilities of plastic deformation (Marsh 1964b) and capillary condensation (Fig. 8) (Wondraczek et al. 2006; Ciccotti et al. 2008) at the crack tip have also been discussed, as well as structural relaxation driven by ion diffusion due to a stress gradient near the crack tip. The latter has received evidence from *in situ* atomic force microscopy analysis of a propagating crack in soda-lime silicate glasses (Célarié et al. 2007). Further progress is likely to come from a better understanding of the glass composition and structure dependence of the stress corrosion phenomena, as most of the previous studies have focused on silica and soda-lime silicate glasses. Advances within direct imaging, structural characterization and reactive molecular dynamics simulations are expected which will further clarify our present understanding.

INDENTATION HARDNESS, SCRATCH RESISTANCE AND CRACKING

On a macroscopic scale, oxide glasses do not undergo any substantial plastic deformation. However, when an oxide glass is penetrated by a sharp object at the microscale, it has long been known that it is possible to avoid brittle fracture when the applied load is sufficiently small (Taylor 1949) and that, hence, plastic deformation occurs. Plastic deformations have also been observed from, e.g., compaction experiments under high hydrostatic pressure (Bridgman and Šimon 1953) and scratch grooves (Gehlhoff and Thomas 1926; Moayedi et al. 2018). These early studies evolved to a broader research field of glass indentation (Peter 1970; Januchta and Smedskjaer 2019; Yoshida 2019), which is using controlled contact loading experiments to understand elasticity, plasticity, yielding and microscopic fracture. From such experiments, it is possible to obtain information about hardness and crack initiation resistance, as well as elasticity if combined with a depth-sensor. To this end, indentation studies offer very attractive techniques, considering the short experimental time, facile sample preparation, ability for local, microscopic testing and combination with spectroscopic analyses (Raman, FTIR), and small overall sample size required: indentation studies may be of particular interest for mineral

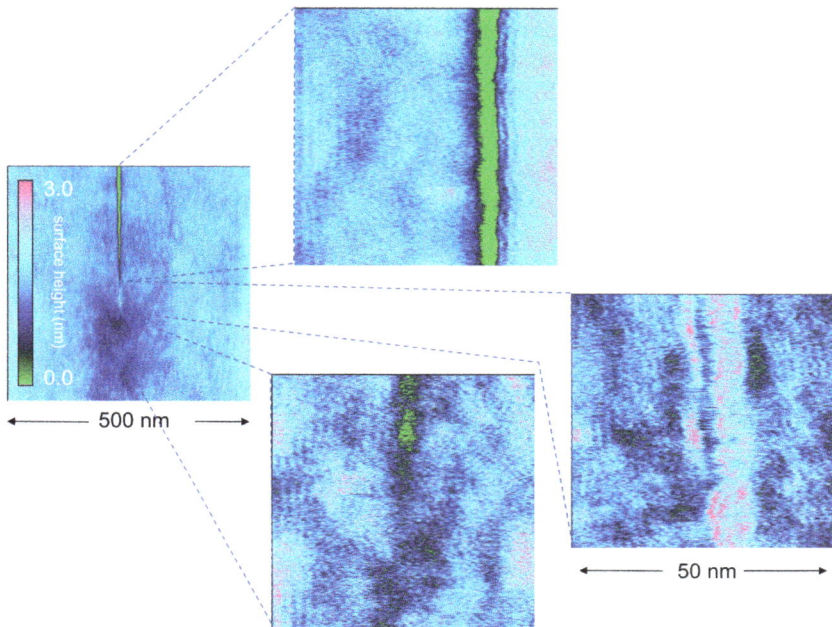

Figure 8. Atomic force microscopy images of an advancing, sub-critical crack in vitreous silica. The magnified images show the discontinuous nature of the crack tip (cavitation), liquid or gel condensation near the crack tip, and the crack opening region.

glasses and glassy inclusions, whose size and shape do not always permit the measurement of mechanical properties through traditional methods. However, typical indentation experiments involve complicated stress fields underneath the indenter, depending on indenter geometry and set-up, loading rates, and studied glass composition. This often makes interpretation of the obtained data difficult. In the following, we will briefly review the state-of-the-art regarding stress fields, deformation mechanism, hardness, strain-rate sensitivity, cracking and scratching phenomena. The focus will be on micro-indentation using a Vickers diamond tip (136° four-sided pyramid), but we note that the tip geometry has a strong influence on the deformation mechanism and ease of cracking (Gross 2012).

Stress fields

Whether a glass exhibits an elastic or elastic–plastic indentation response depends on whether the applied load and tip geometry allow stresses to reach the yield stress criterion. That is, if the load is sufficiently small and/or the tip is sufficiently blunt, the glass deforms purely elastically (Hertz 1881). When the load is higher and/or the tip is sharper, the glass experiences larger stresses and it is permanently deformed through a combination of densification and shear flow. This creates a hemi-spherical zone of compacted material surrounding the indentation cavity. Upon indenter unloading, the elastic deformation attempts to recover, but this is not fully possible as the elastic field acts beyond the range of the compacted hemi-sphere. This puts the elastic zone under residual tensile stress and the densified zone under compressive stress, which can in turn lead to crack initiation at the elastic-plastic boundary (Yoffe 1982).

For conical indentation, Yoffe derived formulas for calculating the stresses acting in the three-dimensional half-space surrounding the indentation cavity (Yoffe 1982). Based on this analysis, two parameters are found to determine whether the stresses are compressive or tensile, namely, the normal load and the blister field strength. Knowledge about the sign of the stress in different directions can be used to predict cracking behavior, since tensile stresses are the driving force for crack initiation. The blister field strength is caused by a strain nucleus built on the two forces acting along the surface and the force acting along the vertical axis (Sellappan et al. 2013). It has been proposed that it is determined by Young's modulus and Poisson's ratio of the glass as well as the deformation (i.e., energy dissipation) mechanism (Sellappan et al. 2013), as discussed in the following.

Deformation mechanism

In addition to elastic deformation, indentation-induced deformation is typically attributed to a combination of shear (plastic) flow and densification. Taylor (1949) made the first observation of shear flow, later supported by the findings of Douglas (1958) and Marsh (1964a). Peter (1970) and Evers (1967) proposed densification as an additional deformation mechanism, as also supported by previous high-pressure experiments (Bridgman and Šimon 1953). Observation of pile-up around the indentation imprint above the surface is a typical indication of shear flow (Fig. 9), but there has also been evidence of shear bands in some oxide glasses, i.e., parallel waves of high local strain moving outward from the point of contact (Gross et al. 2018).

While shear flow conserves the total volume, densification leads to a more tightly packed glass network, for example involving changes in bond angles. The densified contribution to indentation deformation has been found to be recoverable by thermal annealing even at a temperature below the glass transition temperature (T_g) and, thus, before substantial viscous relaxation occurs (Neely and Mackenzie 1968). Such sub-T_g annealing (normally ~$0.9\,T_g$) combined with a measurement of the indent cavity volume before and after relaxation can be used to estimate the relative contribution of densification to indentation deformation in normal (Yoshida et al. 2005) as well as lateral (scratch) geometry loading (Moayedi et al. 2018).

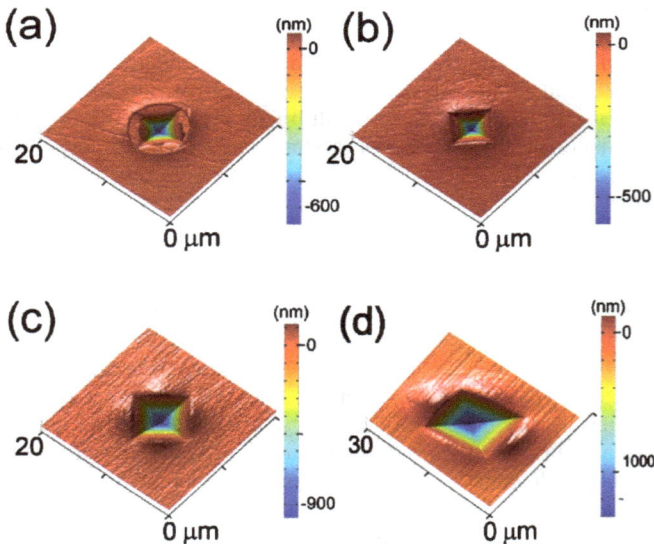

Figure 9. AFM images of Vickers indentation imprints in pristine vitreous SiO_2 (**a**), densified vitreous SiO_2 (**b**), bulk metallic glass (**c**), and metallic platinum (**d**), showing the range of shear behavior from crystalline metals with large pile-up to pristine SiO_2 with minimal pile-up and cracking. Reprinted from Rouxel et al. (2010), *Journal of Applied Physics*, Vol. 107, Fig. 3, p 094309-4 with permission of the AIP Publishing.

Specifically, the so-called volume recovery ratio (V_R) can be determined as a measure of the densification fraction relative to the contribution of shear deformation. It was frequently reported that the glass composition strongly affects the deformation mode, with the general trend that higher modifier content leads to enhanced shear flow relative to densification.

In addition to the relative extent of densification and shear flow, the type of structural changes associated with densification have also been found to be important (Januchta et al. 2017; Benzine et al. 2018). Initially, such studies have been focused on *post mortem* analysis of the zone surrounding the indent, e.g., by using micro-Raman spectroscopy (Winterstein-Beckmann et al. 2014a,b), but recent studies have also pointed-out the importance of (transient) structural changes *in situ* (Yoshida et al. 2019; Fuhrmann et al. 2020; Jaccani et al. 2020). In amorphous silica and many silicate glasses, structural changes are believed to involve a decrease in the inter-tetrahedral Si–O–Si angle and redistribution of ring sizes (Perriot et al. 2006). In borate and boron-containing glasses, an increase in boron coordination number occurs in addition to changes in B–O–B bond angles and ring statistics (Limbach et al. 2015a; Januchta et al. 2017).

Hardness

Hardness is an important property, representing the glass' resistance to permanent deformation by a penetrating object. Hardness (H), or Vickers hardness (H_V) when using a Vickers tip, is defined as the maximum force on the indenter divided by the surface area of the residual imprint. This contrasts with Mohs' original hardness scale, which is based on scratch experiments. The intuitive understanding is that there is a direct relation between the hardness of a material and its scratch resistance, both reflecting the limit of plastic yielding (Tabor 1956). However, more recent studies indicate that for glasses, the ratio between normal hardness and lateral (scratch) hardness is material-dependent (Sawamura and Wondraczek 2018).

The Vickers hardness is calculated from the applied load P and the contact area A_c as $H_V = P/A_c$. The projected area can be estimated from microscopic images of the residual imprint, where d is the average diagonal length of the residual imprint and $A_c = 2\sin(136°/2)/(2d)^2$. The contact area can also be estimated from the contact depth h_c as $A_c = 24.5h_c^2$ (Oliver and Pharr 1992). Despite being an extrinsic mechanical property that depends on various experimental conditions, several semi-empirical models based on, e.g., bond strength (Yamane and Mackenzie 1974) or network rigidity (Smedskjaer et al. 2010a) have been proposed to predict the glass composition dependence of Vickers hardness within a set of fixed measurement conditions.

Indentation hardness is often observed to depend on the applied load. That is, the hardness of most materials (including oxide glasses) typically decreases with increasing indentation load or size. This phenomenon is known as the indentation size effect (ISE) (Bull et al. 1989; Sangwal 2009). Figure 10 shows a schematic illustration of the ISE, with the plateau of hardness at a sufficiently high load. The extent of the ISE can be described by Bernhardt's empirical relation, $P = a_1 \cdot d + a_2 \cdot d^2$, where a_1 is a measure of the extent of the ISE, and a_2 is a measure of the load-independent part of hardness (Bernhardt 1941). The origin of the ISE continues to be a subject of debate, with attributions to, e.g., friction (Li et al. 1993), dislocations (Nix and Gao 1998), densification and fictive temperature change (Gross and Tomozawa 2008), and shear thinning (Kazembeyki et al. 2019).

Strain-rate sensitivity

Not all deformation occurs instantaneously under load (Grau et al. 2005). In metals, time-dependent deformation is a result of the kinetics of atomic diffusion and the movement of dislocations, while in inorganic glasses it is due to structural rearrangements largely controlled by viscous flow (Chu and Li 1977; Guin et al. 2002; Choi et al. 2012). Combining the power-law dependence between stress and deformation (strain) rate $\dot{\varepsilon} = K\sigma^n$ with Tabor's (Tabor 1970) observations relating the surface hardness to the flow stress in metals (Shen et al. 2012), the strain-rate sensitivity is defined as

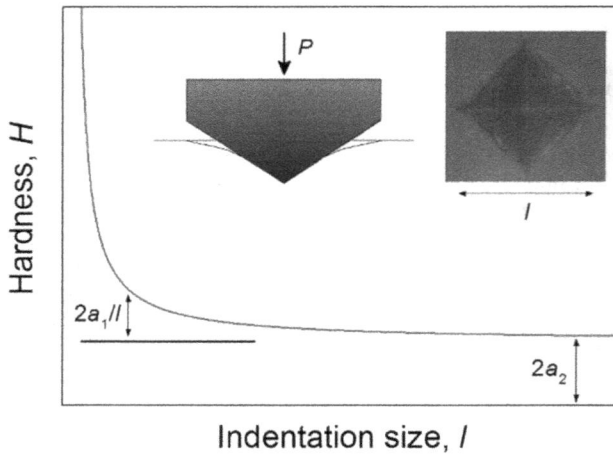

Figure 10. Schematic illustration of the typical dependence of indentation hardness (H) on indentation length (*l*) as produced by varying the applied load (P). This illustrates the so-called indentation size effect (ISE). Reprinted from Smedskjaer (2014) Applied Physics Letters, Vol. 104, Fig. 1, p 251906-1, with permission of AIP Publishing.

$$m = \frac{1}{n} = \frac{\ln H}{\ln \dot{\varepsilon}} \qquad (38)$$

The limiting value of $m = 0$ applies to a rigid and perfectly plastic material, while $m = 1$ applies to a linear viscous material (Bower et al. 1993; Goodall and Clyne 2006). Therefore, the strain-rate sensitivity is also a measure of the localization of plastic deformation under stress, with linear viscous materials showing homogeneously distributed plastic deformation, while rigid and perfectly plastic materials have heterogeneously distributed plastic deformation and the overall structure deforms elastically with any plastic events being highly-localized (Limbach et al. 2017). This time-dependent material response is determined by the interplay between chemical bonding and network topology (see Fig. 11): glasses with high bond strength, high network connectivity and low Poisson's ratio (such as SiO_2, B_2O_3) or high packing density and high Poisson's ratio (such as metallic or ionic glasses) show low values of m; on the other hand, glasses in between these extremes show a large variability in their measured strain-rate sensitivity, which then depends on the specific deformation mechanism (Limbach et al. 2014). The strain-rate sensitivity can also be used to estimate the size of the plastic events, which has recently been shown to correlate with atomic mobility in phosphate glasses (Rodrigues et al. 2019).

Indentation cracking

At sufficiently high loads, cracking patterns develop around the indentation imprint. The type and extent of cracks depend on factors such as the glass composition and processing history, indenter tip geometry, and loading rate (Januchta and Smedskjaer 2019; Yoshida 2019). Although it is not possible to determine fracture toughness accurately using Vickers indentation (see *Experimental Methods for Strength and Toughness Testing*), indentation has proven to be a helpful method for comparing the damage resistance of different glasses (Arora et al. 1979). This is because in many practical situations, the contact with a sharp object is the process that leads to the nucleation of a critical flaw.

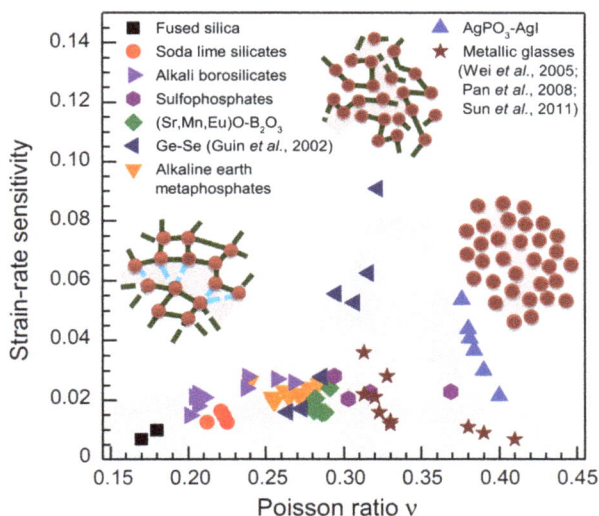

Figure 11. Strain-rate sensitivity as a function of Poisson's ratio for a wide range of glass compositions— Used under CC BY-NC-ND license, from Limbach et al. (2014), Journal of Non-Crystalline Solids, Vol. 404, Fig. 6, p 132.

While amorphous silica forms ring cracks around the Vickers indentation imprint, soda lime silica is prone to form radial cracks from the corners of the indent. This marked difference led to the classification of glass compositions according to their characteristic deformation mechanism and thus cracking pattern, with silica being denoted "anomalous" and soda lime silica being "normal" (Arora et al. 1979). Based on Yoffe's stress field relations and an estimation of the blister field (Yoffe 1982), the driving force for the different crack patterns (such as ring, radial, median, and lateral—see Fig. 12) can be predicted. Hence, the predominant cracking pattern can also be predicted. This analysis clearly shows how the deformation mechanism determines the cracking. That is, compositions with low Poisson's ratio and low modulus-to-hardness ratio are prone to exhibit ring cracks, while glasses with high values of these ratios are prone to exhibit radial and lateral cracks. Median cracks dominate in the intermediate region.

Many glasses within the aluminosilicate and borosilicate families exhibit corner cracking. This allows for a simple quantification of the glasses' resistance toward crack initiation using the approach of Wada et al. (1974). In this method, a series of Vickers imprints are produced at different loads and some time after unloading, the number of corner cracks is counted. The load that corresponds to an average of two cracks per indent (corresponding to 50% initiation probability for a four-sided Vickers imprint) is defined as the crack resistance CR (see Fig. 13).

We note that CR only describes the resistance to corner cracks and depends on extrinsic factors such as loading rate, bluntness of the tip, time between unloading and recording the cracks, and atmospheric conditions (Yoshida et al. 2016; Bechgaard et al. 2018), making comparison of data from different studies challenging. However, for constant conditions, it is clear that the glass composition greatly affects the value of CR (Fig. 13) (Limbach et al. 2015a). Indeed, there has been a great interest in the discovery of new crack-resistant glasses with particularly high CR, from the so-called "less-brittle" glass by Sehgal and Ito (2005) to recent borate-based compositions (Kato et al. 2010; Januchta et al. 2019).

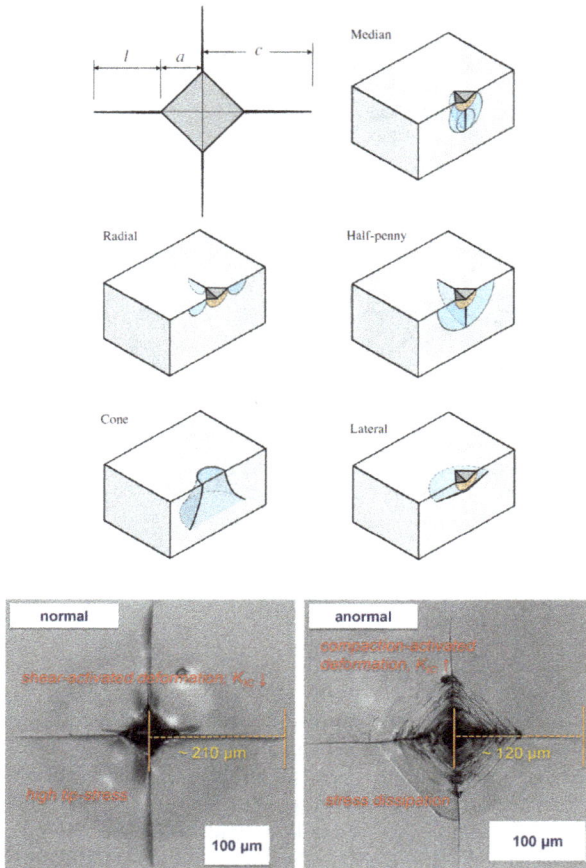

Figure 12. Characteristic cracking systems arising from Vickers indentation of oxide glasses—Used by permission of John Wiley and Sons, from Cook and Pharr (1990), Journal of American Ceramic Society, Vol. 73, Fig. 1, p. 789.

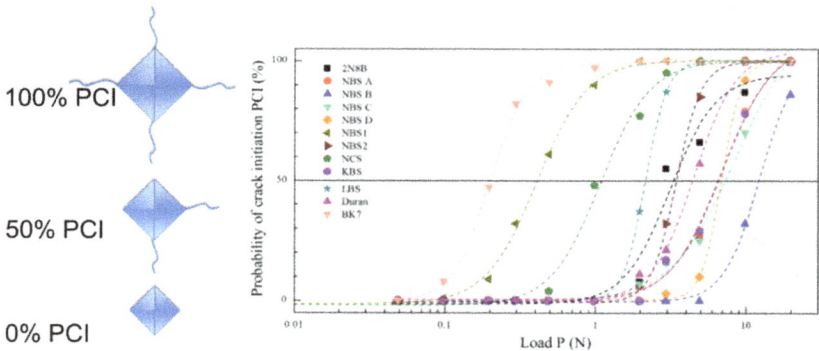

Figure 13. Dependence of the probability of radial crack initiation on the applied Vickers indentation load for a range of different sodium borosilicate glasses. The load corresponding to the probability of 50% corresponds to the crack resistance CR. Right panel used under CC BY-NC-ND license, from Limbach, Winterstein-Beckmann et al. (2015), Journal of Non-Crystalline Solids, Vol. 417-418, Fig. 3, p 21.

Scratch resistance

Since Mohs, studies of hardness and scratch resistance have been closely related, whereby connections between normal (indentation) hardness and scratch behavior are often drawn due to the similarity of experimental observations. However, while normal indentation experiments usually involve quasi-isostatic stress fields (Smedskjaer 2014; Januchta and Smedskjaer 2019; Yoshida 2019), lateral surface deformation is more complex in terms of the variety of overlapping reactions (stick–slip reactions, shear deformation, compaction, plowing deformation, cracking, acoustic interactions, etc.). Direct studies of the surface scratch resistance and abrasion behavior of glasses have therefore been mostly phenomenological (e.g., Le Houérou et al. 2005; Nielsen et al. 2015). When sliding a sharp diamond tip over a glass surface with a gradually increasing normal load F_N, three characteristic regimes of damage are typically observed *after scratching* (Fig. 14) (Le Houérou et al. 2005): a plastic regime in which a scratch groove is created, but no visible cracks are found, a regime of chipping and radial cracking, and a microabrasive regime which is attained after a certain load threshold is reached. Both regimes where reported to depend on the chemical composition of the glass, but also on surface quality and constitution.

In analogy to normal hardness, the scratch hardness of a glass was defined as the resistance of the material surface to lateral deformation (Yoshida et al. 2001),

$$H_S = \frac{\partial W_S}{\partial V_S} \tag{39}$$

where H_s denotes the work W_s needed for generating a persistent scratch groove with the volume V_s during sliding a sharp stylus across the glass surface. The W_s is obtained from the integral of the lateral force F_L observed during displacement of the stylus over the scratch length L_S, $W_s = \int F_L \, dL_s$. Other than in normal indentation experiments, this means that the indenter is displaced not only in normal direction (depth), but also laterally, whereby the lateral force is recorded *in situ* (Sawamura et al. 2018). This approach requires analysis below the thresholds of surface fracture (i.e., within the *elastic–plastic regime;* Sawamura et al. 2019).

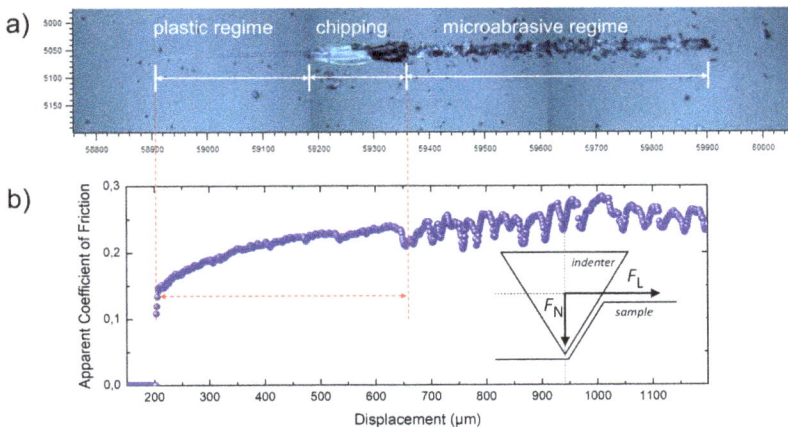

Figure 14. **(a)** Optical micrograph of a scratch generated on vitreous silica with scratching speed of 10 μm/s under increasing normal load (3 mN/s), using a Berkovich diamond stylus in edge-forward configuration. The corresponding apparent friction coefficient recording in situ by monitoring the lateral force F_L is depicted in **(b)**, where the serrations which occur at ~ 650 μm displacement indicate microabrasion. Adopted in modified form and used under CC BY-NC-ND license from Moayedi and Wondraczek (2017), Journal of Non-Crystalline Solids, Vol. 470, Fig. 1, p. 139.

For a broad variety of glasses, the scratch hardness was reported to depend on bond energy density U_0/V_0 and structural cohesion (Fig. 15; Sawamura and Wondraczek 2018), whereby an empirical parameter F was conjectured to describe the divergence between the macroscopic bulk modulus and the mean bond energy density.

$$K = F \frac{U_0}{V_0} \tag{40}$$

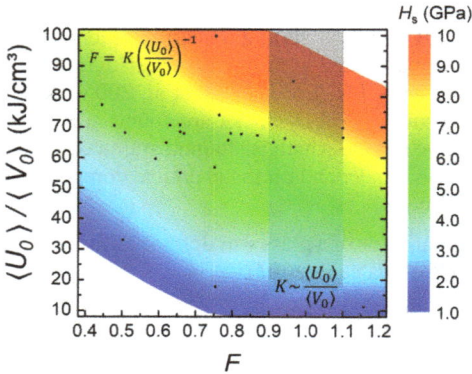

Figure 15. Average mean-field energy density $<U_0/V_0>$, empirical cohesion factor F and scratch hardness H_s of glassy materials. The value of F is a measure of structural heterogeneity. According to Equation (40), it describes the deviation between bulk modulus and the mean-field estimate of bond energy density. Here, lower values of F reflect increasing heterogeneity, and thus H_s is lower than in glasses with more homogeneous spatial distribution of bond energy. Used under CC BY-NC-ND license, from (Sawamura and Wondraczek 2018), Physical Review Materials, Vol. 2, Fig. 3, p 092601-3.

FRACTOGRAPHY

To optimize glass materials for various applications, there is a need to understand why, where and when they start to break. The study on the origin of cracking, crack patterns and their interpretation in terms of materials properties is referred to as *fractography* (Frechette 1990; ASTM 2019). Fractographic analysis can be performed at both macroscopic (from crack pattern) or microscopic (fracture surface) scale, and a fractographer is akin to an investigator or detective who is an expert in microscopy (Zamanzadeh et al. 2007; Quinn 2020). For example, we can illustrate this by considering the case of a glass object found broken after use. The primary and secondary investigations are on macroscopic and microscopic scales, respectively. Initially, some assumptions would be made on the potential reasons for the cracking process. Initial investigations are often to locate the crack origin, i.e., the point where the initial crack started. The investigation is to identify whether single-line or multiple-line cracking occurred and if there is a concentration of those line cracks somewhere on the object (macroscopic study: fracture pattern). Afterwards, a microscopic study (fracture surfaces) needs to be performed, e.g., by using an optical microscope. For example, this is done to investigate whether the crack started from the outside bottom or the from some inside part of the object. Following this example, we will therefore in this chapter first introduce the macroscopic study (so-called crack branching pattern and angle) and then the microscopic study (so-called fracture surfaces).

Crack branching pattern and angle

The broken glass pieces are typically where the investigation is initiated. The reconstruction of the broken pieces would lead to the study of the crack pattern. The general crack pattern can pave a way to the fracture origin, the fracture cause, the fracture energy, and the stress state (Frechette 1990; Quinn 2020). For example, in a biaxial ring-on-ring test (RoR, see *Experimental Methods for Strength and Toughness Testing*), the reconstruction of the broken samples allows us to know whether the crack started from the inside of the loading ring, the support ring, or the sample edge (Fig. 16).

This will lead to the proper analysis of the fracture stress and strain to which the sample was exposed during the test. In fact, only upon conducting this preliminary fractographic analysis, it becomes possible to qualitatively determine the strength of the tested glasses by simply counting the number of the fragments. If there are only four fragments (as seen in Fig. 16-left), the stored energy before fracture is small and so is the strength of the tested samples (Johnson and Holloway 1966; Döll 1975). Stronger samples will have a larger number of fragments. From this sample reconstruction, we can also obtain information about the crack branching angles, which was compared to the stress state for the first time in 1935 (Fig. 17; Preston 1935). The biaxial test that has the fracture origin at the inside of the load ring produces a crack branching angle at a value as high as 180°, while the biaxial test on bottles produces moderate angles around 90°, and a uniaxial test produces crack branching angles as low as 45° (Preston 1935; Bullock and Kaae 1979; Rice 1984; Frechette 1990; Quinn and Wirth 1990; Shinkai 1994).

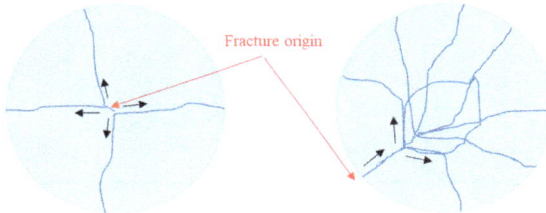

Figure 16. Illustration of two possible branching patterns of ring-on-ring cracking (RoR—see *Experimental Methods for Strength and Toughness Testing*) post mortem. **Left**: the crack starts from inside of the load ring with four fragments. **Right**: the crack starts from the edge of the sample with many fragments. The light blue circle represents the glass specimen, the **dark blue line** the crack pattern, the **red arrow** the fracture origin and the **black arrows** the crack direction.

Fracture surfaces (mirrors, mist and hackle)

As the strength of brittle glasses depends on the pre-existing flaw size and type, specific knowledge about the flaw is as important as the fracture stress in a strength experiment. Indeed, the study of fracture surfaces can reveal many direct and indirect mechanical properties (Rice 1984). Figure 18 shows the optical microscopic image on a fused silica glass rod broken by a two-point bending test, together with the identification of the fracture origin, mirror, mist and hackle.

The mirror region (mostly in form of circle or semicircle) is the smoothest part of the fracture surface, where the crack accelerates from the fracture origin. Mist marks the region where the crack velocity was slowing down. In the hackle region, the surface topography evolves towards a less-grained appearance. Based on the crack velocity role in branching, the relationship between fracture stress as well as fracture toughness can be estimated from these surface features (Levengood 1958; Shand 1959; Johnson and Holloway 1968). For instance, the relationship between fracture stress and hackle size was reported as (Johnson and Holloway 1968),

$$\sigma_f = A_h R_h^{1/2} \tag{41}$$

where σ_f, A_h and R_h are the fracture stress, a proportionality constant and the distance from the fracture origin to the mist/hackle boundary, respectively. Equation (41) is similar to the equations of Griffith or Irwin (Eqns. 12, 15). The stress at fracture strength can be rewritten as (Bansal 1977),

$$\sigma_f = Y^{-1} K_b R_h^{1/2} \tag{42}$$

where Y and K_b are a parameter depending on flaw and specimen geometry and the critical stress-intensity factor for branching (fracture toughness if in mode I), respectively.

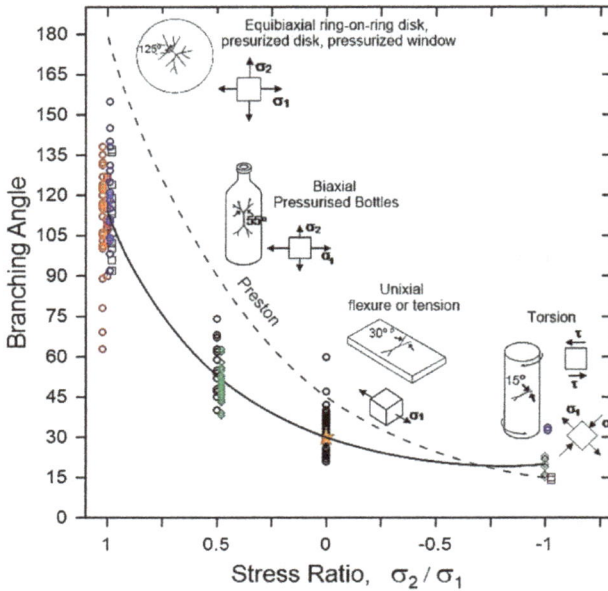

Figure 17. Branching angle versus stress state. Used with permission by the National Institute of Standards and Technology, from Quinn (2020), NIST Special Publication (SP) 960-16, Fig. 4.8, p 4-9.

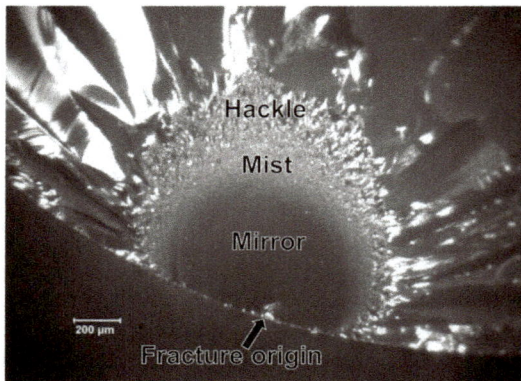

Figure 18. View at a fracture surface on a fused silica glass rod, broken by two-point bending (see *Experimental Methods for Strength and Toughness Testing*), with its fracture origin, mirror, mist and hackle regions indicated. Modified from and used with permission by the NIST, from Quinn (2020), NIST Special Publication (SP) 960-16, Fig. 5.2, p 5-5.

RESIDUAL STRESS AND METHODS FOR ENHANCING THE PRACTICAL STRENGTH OF GLASSES

The practical strength of glass products is typically addressed by *post-processing*, i.e., secondary processing which is applied after glass forming. Post-processing targets persistent surface modification, whereby the surface chemical composition, mass density (free volume, fictive temperature), residual stress or topography are altered to enhance the material's defect resistance, hardness, stiffness, or fracture strength. On the other hand, the fracture toughness is typically not enhanced by the traditional post-processing techniques. In particular, methods by which a residual layer of compressive stress is generated on the glass surface are common for a wide variety of glass applications: surface quenching by rapid heat extraction from above T_g (thermal strengthening) and ion exchange treatments at temperatures below T_g (chemical strengthening). The presence of a surface residual compressive layer counteracts any tensile stresses which could be imposed during handling, and is also assumed to decrease the surface cracking propensity and thus increase the damage resistance. Interestingly, the effects generated from such industrial treatments can also be found in natural materials and minerals, where they might have similar effects on the experimentally observed mechanical properties (Holzhausen and Johnson 1979; Nichols and Varnes 1984).

Thermal strengthening

A popular example of a thermally strengthened glass is the Batavian tear (or Prince Rupert's drop; Aben et al. 2016), which is fabricated by quenching a drop of molten soda lime silicate glass in water. In this process, the cooling rate which is imposed on the drop surface reaches the order of 10^5 K/min (Sajzew and Wondraczek 2021), by far exceeding the rate of conductive heat transport within the drop's interior. Hence, a gradient in cooling rate is generated from the outside (high cooling rate) to the inside (low cooling rate) while undergoing the transition from melt to glass. This translates to a gradient in fictive temperature and, hence, uncompensated strain, imposing a residual stress profile which is readily visible by investigation of the stress optical retardation (Fig. 19).

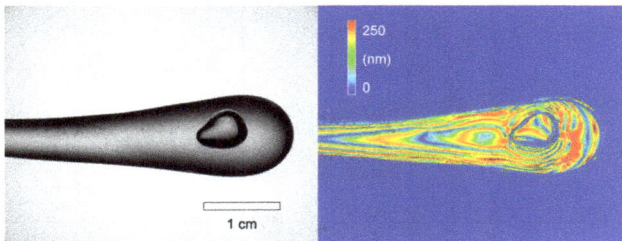

Figure 19. Front-end of a Batavian tear (or Prince Rupert's drop) fabricated by quenching a drop of soda lime silicate glass in water (**left**). The right panel shows the relative stress optical retardation in the same sample. The center object is a bubble embedded in the glass.

The same principle is applied to thermally strengthened technical glasses, whereby glass sheets, tableware or (in some cases) containers are exposed to rapid heat extraction at entrant temperatures of up to 250 K above T_g. Using streams of air or other gases, heat transfer coefficients of 100—500 W/(m².K) are achieved, well-below what would be expected for quenching in water, but still sufficient to generate notable variations in the local cooling rate across the object for a sheet/wall thickness in the range of a few millimeters (Fig. 20a). The question as to whether a thermal gradient can be established within a given sample geometry depends on the interplay of conductive heat transport across the material, and the rate of heat extraction through its surface, embedded in the Biot number Bi,

$$Bi = \frac{hL_c}{\lambda} \tag{43}$$

with the characteristic length L_c of an object with volume V and surface area A, $L_c = V/A$, the thermal conductivity λ and the heat transfer coefficient h. Homogeneous cooling (or heating) can be expected when $Bi < 0.1$; otherwise, thermal gradients will be established, which— in glasses—result in residual stress. For example, for a typical silicate glass with a thermal conductivity of ~ 1 W/(m.K) (Freeman and Anderson 1986; Sørensen et al. 2020) cooled under a flow of air (h ~ 100 W/(m².K)), the achievable L_c for which thermal gradients are avoided is 0.1 cm (corresponding to a bead with a diameter of 6 mm). In larger objects, there would be thermal gradients and, thus, residual stress. For rapid quenching in liquids, mist or fast gas jets, heat transfer coefficients in the range of 10^3–10^4 W/(m².K) may be expected, thus, L_c would be in the range of 10–100 μm.

The residual stress profile as a function of position x along a cross-section $\sigma(x)$ can be estimated from the thermal gradient which is locally established while cooling through T_g, and the linear thermal expansion coefficient α of the glass. For a glass sheet with thickness d, the residual stress is depicted in Figure 20b, and given by,

$$\sigma(x) = -\frac{\alpha E}{1-\nu}\left[\frac{1}{d}\int_{-\frac{d}{2}}^{\frac{d}{2}} T(x,t_c)\,dx - T(x,t_c)\right] \qquad (44)$$

with the critical time t_c being the time at which the center point of the glass object (at $x = 0$) undergoes the glass transition (for simplicity, an instantaneous transition is usually assumed for this). The surface compressive stress σ_s obtained in this way is compensated by an interior tensile stress field reaching about half the magnitude of σ_s.

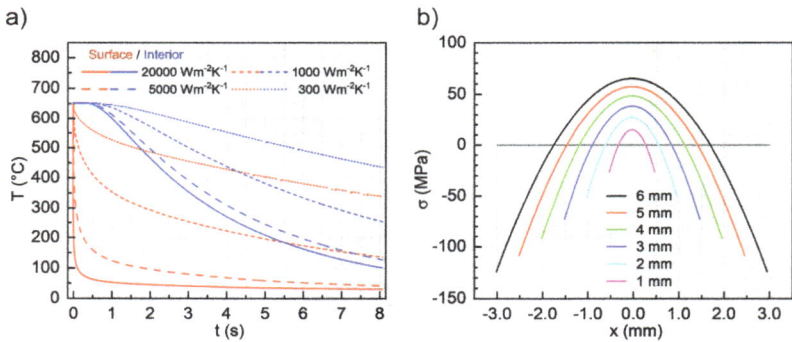

Figure 20. (a) Calculated temperatures at the surface and the center of a 4 mm soda lime silicate glass sheet as a function of time for varying heat transfer coefficients. **(b)** Residual stress profiles in rapidly cooled glass sheet with varying thickness after cooling from 650 °C to 20 °C (h = 300 W/m²K, λ = 1 W/mK, T_g = 550 °C, E = 72 GPa, ν = 0.23 and α = 12·10⁶ K⁻¹). Analytical model adopted from Sajzew and Wondraczek (2021).

Chemical strengthening

Chemical strengthening relies on diffusive surface ion exchange (Karlsson et al. 2010; Varshneya 2010), whereby the temperature is chosen so that high interdiffusion coefficients may be reached, but concurrent stress relaxation is kept as low as possible. A diffusion process is induced by exposing the glass to alkaline salt melts, either through full immersion in a bath of liquid salt, or through deposition of salt layers on the glass surface using pastes or sprays, which are subsequently melted to form a liquid salt layer (Karlsson et al. 2010). The exchange reaction occurs between alkali ions from the salt and from within the glass, caused by a concentration gradient at the glass/liquid interface. Typical exchange temperatures are in the range of 350 to 500 °C, depending on salt bath composition, type of glass and target parameters in terms of the attained diffusion profile and residual stress (magnitude of surface compressive stress CS and depth of the compressive stress layer DOL). For the salt batch, nitrates or mixtures of nitrates, chlorides and sulfates are usually employed.

While soda lime silicate glasses such as those employed for the vast majority of commodity glass products are a very attractive target for chemical strengthening (see Fig. 21), the highest levels of CS (~1 GPa) and DOL (~100 μm) are typically achieved in aluminosilicate glasses adapted for high diffusion rates, network connectivity, and K/Na-for-Na/Li exchange reactions. The diffusion profile $C(z)$ is often (Gy 2008) estimated using

$$C(z) = C_s \cdot \mathrm{erfc}\left(\frac{z}{2\sqrt{Dt}} \right) \tag{45}$$

where C_s is the surface concentration of the invading anion, and D is an effective diffusion coefficient (Karlsson et al. 2017). The resulting stress profile is then obtained through

$$\sigma(z) = \frac{BE}{1-\nu}\left[C(z) - \overline{C} \right] \tag{46}$$

with the base (average) concentration of the invading alkaline species $\overline{C}$ and the linear network dilation coefficient B.

Analogous to the coefficient of thermal expansion for temperature gradients, the parameter B (also called the Cooper coefficient) represents the amount of uncompensated linear elastic strain imposed per unit volume of exchanged alkali species, i.e., the degree of *stuffing* caused by the exchange of smaller ions from within the glass (Li^+, Na^+) through the larger invading species (Na^+, K^+, Cs^+).

$$B = \frac{1}{3V_0} \frac{\partial V_0}{\partial C} \tag{47}$$

Despite some progress which has been made in understanding the structural origin of the residual stress profile, the details of the process remain unclear. Particularly, the so-called network dilation anomaly has been a major puzzle. This anomaly describes the fact that when B is evaluated using Equation (47) from the difference in the molar volume of pure end-member glasses (e.g., as-quenched lithium aluminosilicate and sodium aluminosilicate glasses in the case of Na-for-Li exchange), it is a factor of 2-4 times higher than the B value that would be obtained from stress measurements using Equation (46). Recent simulation studies have proposed that the anomaly arises from a difference in local structural environments of the alkali cations in as-quenched and ion-exchanged glasses (Tandia et al. 2012), with no anomaly found for isostatic glasses (Wang et al. 2017).

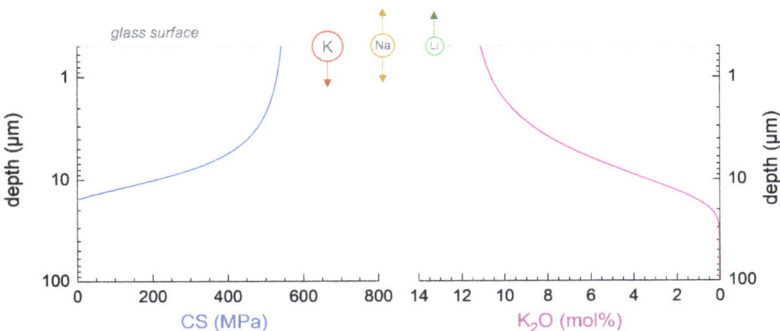

Figure 21. Chemical strengthening, shown by way of example for a soda lime silicate glass after Na^+/K^+-ion exchange for 10 h at 450 °C: (**left**) compressive stress CS as a function of distance from the glass surface and (**right**) variation of the K_2O concentration. Adapted under CC-BY-NC license from Sani et al. (2021), *Journal of the American Ceramic Society*, Vol. 104, Fig. 2, p3175.

Rodrigues et al.

MECHANICAL PROPERTY EXAMPLES

The knowledge of physical properties allows us to classify materials into groups (such as done using the Mohs scale in the context of minerals) or to choose the optimum material for a given application (in technical context). In such considerations, the material properties are usually separated into families, e.g, chemical, physical, and mechanical, although such separation is often ambiguous due to strong (but sometimes non-intuitive) interrelations between the different manifestations of material behaviour. In the following, we will introduce examples of mechanical properties (John 1992). For the class of oxide glasses, the first systematic studies of the mechanical properties and their dependence on glass composition were conducted by Winkelmann and Schott (1894), who adapted a variety of experimental techniques which are still used today. As discussed throughout this review, toughness, which is an estimation of how much energy a material can consume before fracture, is an important property for glasses (Griffith 1921; Launey and Ritchie 2009). In the following, we will therefore compare toughness parameters to two elastic properties. Namely, fracture toughness and fracture energy compared to Young's modulus and Poisson's ratio, respectively. Finally, we discuss various physical properties that are related to Poisson's ratio.

Fracture toughness versus Young's modulus

In many cases of materials selection and design, engineers need to know the material strength. The same is often true in the context of geoscience, although not related to consumer application, but, e.g., to predictive modeling and simulation of geophysical phenomena. Knowledge of the fracture toughness is usually required in addition to a simple strength value (or, as in the case of brittle materials, parameters of the Weibull strength distribution). A metric for fracture toughness that relates to strength is the critical stress intensity factor K_{Ic}, which describes the stress intensity (the amplified stress due to defect size) at which crack propagation becomes critical (Irwin 1957). An important tool in materials selection is the so-called Ashby plot (Ashby and Cebon 1993), which relates the properties of different materials and different classes of materials in a scatter plot. Such plots can be used to select a damage-resistant material for a given application (Ritchie and Dzenis 2008; Demetriou et al. 2011; Ritchie 2011); as an example, an Ashby plot of fracture toughness and Young's modulus for different materials (from foams to metals) is shown in Figure 22. Oxide ceramics and glasses are the good material options if stiffness is the most important criterion for a given application, whereas polymers and foams are preferred if flexibility is the important feature. Besides, if both stiffness and toughness are important, then metals should be considered. The lower limit for K_{Ic} in Figure 22 shows the necessary condition for *critical* fracture. Subcritical fracture occurs below this lower limit, for example, due to stress-corrosion (see *Stress Corrosion and Fatigue*).

Fracture energy versus Poisson's ratio

In general, the fracture toughness of ceramics and glasses is higher than that of polymers, but polymers, especially in the form of composites, are used more frequently than ceramics and glasses in engineering structures (Ashby and Cebon 1993). This is because, unlike polymers, ceramics and glasses fracture in a brittle manner, thus requiring more caution in terms of applications. From the energy-based definition, we have $K_{\mathrm{Ic}} = (EG_{\mathrm{frac}})^{\frac{1}{2}}$, where G is the fracture energy. As shown in Figure 22, some polymers have similar K_{Ic} but much lower E compared to most ceramics and glasses. This is because those polymers need much more energy to fracture when compared to ceramics and glasses. This fracture energy has been shown to have an apparent correlation with elasticity, namely Poisson's ratio, based on a study of metallic glasses (Lewandowski et al. 2005). As a guideline to overcome brittleness, it was suggested also for oxide glasses to seek for sufficiently high Poisson's ratio. To this end, the fracture energy is used to determine whether a material is brittle or ductile by replotting it against Poisson's ratio, shown in Figure 23 (Østergaard et al. 2019). The postulated transition region is at a

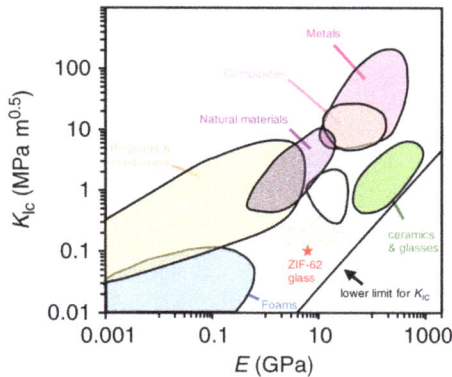

Figure 22. Ashby plot of the relation between fracture toughness (K_{Ic}) and Young's modulus (E) of different classes of materials. Used under CC BY-NC-ND license from To et al. (2020), Nature Communications, Vol. 11, Fig. 3e, p 2593-5.

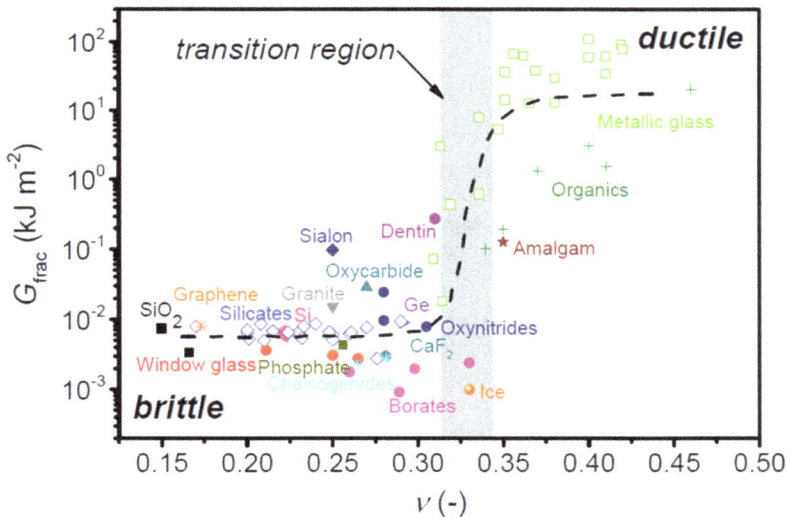

Figure 23. Relation between fracture energy (G_{frac}) and Poisson's ratio (v). Used with permission under CC-BY-NC-ND license, from Østergaard et al. (2019), Materials Vol. 12, Fig. 1, p 2.

Poisson's ratio of around 0.32. Most oxide glasses have the Poisson's ratio below 0.32 and as a result have G_{frac} lower than 0.01 kJ m^{-2} (attributed as being "brittle"). In contrast, metals and metallic glasses have Poisson's ratio and fracture energy above 0.33 and 10 kJ m^{-2}, respectively (attributed as being "ductile"). However, the suggested strategy to reduce the brittleness by enhancing Poisson's ratio still lacks experimental confirmation for the case of oxide glasses.

Properties related to Poisson's ratio

As highlighted above, Poisson's ratio of glasses is an important and interesting material property (Greaves et al. 2011). It is defined as the negative ratio between the transversal and longitudinal strains under uniaxial stress in the elastic regime. Unlike the elastic constants, the Poisson's ratio seems to be more sensitive to structure, with relatively general correlations having been discovered between Poisson's ratio and either atomic packing density or network connectivity, for a wide range of glass compositions (Rouxel 2006; Greaves et al. 2011). That is, Poisson's ratio has been shown to be connected to how a material structure is packed; a trend between Poisson's ratio and atomic packing density has been reported, i.e., the Poisson's ratio apparently increases with the atomic packing density (Greaves et al. 2011). A direct relation between Poisson's ratio and atomic packing density was further proposed for silicate glasses (Makishima and Mackenzie 1975). However, a more recent study (taking into account a broader range of glass compositions) did not confirm a one-to-one relation between these two properties (Østergaard et al. 2019). Poisson's ratio has also been shown to be related to network connectivity, i.e., the average coordination number in the case of chalcogenide glasses, or the number of bridging oxygen in the case of oxide glasses (Rouxel 2007).

Stress optical coefficient and persistent anisotropy

While glasses are usually isotropic materials, they also exhibit photoelasticity, i.e., a reversible change in permittivity as a result of elastic deformation. This induces transient optical anisotropy (birefringence), which can be used to monitor transient or residual stresses. The stress-optical coefficient C relates the induced optical retardation Δ for light polarized in the direction of the principal stress axis relative to the orthogonal one to the principal stress and sample thickness d,

$$C = \frac{\Delta}{\sigma d} \tag{48}$$

Values of C (in unit brewster or 10^{-12} m²/N) are usually measured in uniaxial compression experiments and tabulated in catalogues for technical glasses. To date, the structural origin and predictability of C from glass composition are only semi-empirically known (Smedskjaer et al. 2012). From the viewpoint of technical application, glasses with $C \sim 0$ or $C < 0$ are of particular interest to avoid or compensate stress-induced birefringence in optical devices, for example, using $SnO–P_2O_5$ glasses (Guignard et al. 2007).

Aside photoelasticity, structural anisotropy can also be frozen-in during quenching from the liquid state, leading to a permanently anisotropic glass (Brueckner 1996). This can happen as a result of rapid deformation, when the applied strain rate exceeded the rate of structural (and stress) relaxation so that a texture is induced in the liquid state (for example, by orientation in the direction of flow, denoted flow-anisotropy) and subsequently frozen-in. For example, rapidly drawn or spun glass fiber may exhibit anisotropic behavior (Ya et al. 2008), but also glasses which formed while slowly cooling from above T_g under load (Wu et al. 2009). The excess enthalpy which is induced in this way can be monitored by differential scanning calorimetry (Martin et al. 2005); upon thermal annealing, the induced anisotropy is fully reversible (Inaba et al. 2015).

PERSPECTIVE: TOPOLOGICAL CONSTRAINT THEORY

In the last decades, significant progress has been made in the efforts to correlate mechanical behavior to the chemical composition and structure of oxide glasses (Wondraczek et al. 2011). The current understanding of glass structure has evolved considerably since the seminal works of Goldschmidt (1911, 1928), Lebedev (1921) and Zachariasen (1932). For oxides, their network is comprised of shared oxygens from the first-coordination shells of the cations present in the glass composition (Mysen and Richet 2019). Generally, these oxygen coordination shells are either stable structural units, with well-defined coordination numbers and interatomic distances and angles, in the case of classical "network formers" such as SiO_2,

B_2O_3, P_2O_5 and GeO_2; or less defined coordination shells showing a range of coordination numbers and interatomic distances and angles, as in classical "network modifiers", like alkali and alkali-earth oxides. There are several classification criteria for oxides (Dietzel 1942; Sun 1947; Wright and Thorpe 2013), but all agree that network formers have a stronger covalent bonding characteristic, while network modifiers are strongly ionic. Naturally, reality is not so clear cut, and there are several oxides such as MgO, Al_2O_3 or TiO_2 that lie between formers and modifiers, behaving as both formers and modifiers in different glass compositions (Pedone et al. 2008, 2012; Stebbins et al. 2013). The geometrical correlations between short-range oxygen coordination shells of network formers and modifiers and their medium- and long-range linkages define the glass network topology.

Phillips and Thorpe pioneered the topological approach to understanding how the glass structure informs the properties (Phillips 1979; Phillips and Thorpe 1985), where the glass network is abstracted as a scaffold with the atoms behaving as connecting joints and the chemical bonds between them as joints. This allows for the calculation of Maxwellian rigidity (Maxwell 1864) as a function of the coordination number by counting how many linear and angular constraints each structural joint has. Gupta later considered the glass structure as a network of rigid polytopes connected by their vertices, which is mathematically equivalent to the Phillips–Thorpe approach (Gupta 1993). These approaches were firstly applied to the problem of glass formation in chalcogenide glasses, and it was found that optimal glass formation is achieved when the glass network has an average coordination number $\bar{r} = 2.4$, which is the onset of rigidity percolation (Fig. 24) (He and Thorpe 1985; Seddon 1997; Mitkova and Boolchand 1998), with similar results reported in oxide glasses as well (Avramov et al. 2000; Bocker and Rüssel 2017; Liu et al. 2017). The temperature-dependent bond constraint theory (TDBCT) is an extension of the Phillips–Thorpe approach developed by Gupta and Mauro, which borrows concepts from the potential energy landscape (Sciortino 2005; Mauro et al. 2008) (PEL) description of glasses to add temperature dependence to the network constraints, which in turn are characterized by a constraint onset temperature, characteristic frequency, and observation time (Gupta and Mauro 2009; Mauro et al. 2009). The TDBCT has been successfully applied

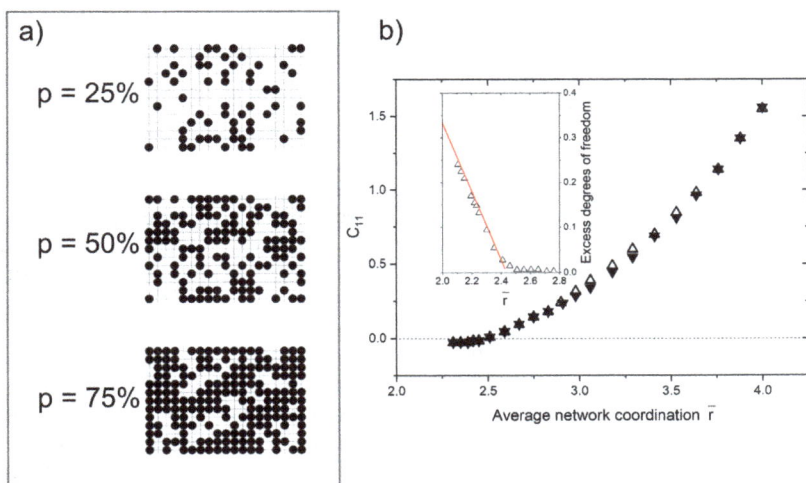

Figure 24. (a) Site percolation on a two-dimensional square grid as a function of the probability of site occupancy p. (b) Simulated elastic modulus C_{11} as a function of the average network coordination, with the inset showing the evolution of excess degrees of freedom f. Data from He and Thorpe (1985) and Zallen (1998).

to predict the compositional dependence (combined with the knowledge of glass structure) of the glass transition temperature and liquid fragility of several glass systems (Smedskjaer et al. 2010b; Fu and Mauro 2013; Jiang et al. 2013; Mauro et al. 2013; Hermansen et al. 2014, 2015; Rodrigues and Wondraczek 2014). Alternatively, a coarse-graining model of the network to a bond lattice of weakest links has also been used to predict the compositional scaling of liquid fragility (Sidebottom and Schnell 2013).

In contrast with thermal properties, which can be directly related to the number of constraints through the PEL and the Adam–Gibbs formalisms, the connection between topological constraints and the mechanical properties of glasses is more tenuous. Early simulation work from He and Thorpe suggests a strong correlation between the average coordination number and the calculated elastic constants, which vanish as $\bar{r}$ approaches the rigidity percolation threshold (He and Thorpe 1985). Traditionally, the compositional dependence of the elastic constants (E, G, K) is estimated with the Makishima–Mackenzie (MM) model, which uses bond dissociation energies and the atomic packing fraction (Makishima and Mackenzie 1973, 1975). However, since the MM model does not consider effects of network connectivity, it consistently underestimates the elastic moduli of several glass systems (Limbach et al. 2015b; Plucinski and Zwanziger 2015; Shi et al. 2020). Recent models attempt to overcome this deficiency by defining a volumetric density of constraints (Yang et al. 2019b; Wilkinson et al. 2019) coupled with the concept of bond constraint strength (Rodrigues and Wondraczek 2014), which become the critical variables to describing both the compositional and the temperature dependence of the elastic moduli. Another mechanical property which has been successfully modelled using bond TDBCT is surface hardness (Smedskjaer et al. 2010a) and this approach has been applied to several oxide glass systems (Smedskjaer et al. 2010c; Zheng et al. 2012, 2017; Zeng et al. 2014; Jiang et al. 2014).

PERSPECTIVE: STRUCTURAL HETEROGENEITY AND NON-AFFINE DEFORMATION

When observed on a macroscopic scale, spatial homogeneity and the absence of microstructure underlie the physical behavior of glassy materials, e.g., their optical transparency, characteristic fracture patterns or distinct mechanisms of charge, heat, and sound transport. In many aspects, the ultimate limit of all these properties is related to the length scale below which material homogeneity is broken; in glasses, this is at the super-structural scale. Since the earliest considerations of glass structure (such as Zachariasen's network hypothesis; Zachariasen 1932) or the free-volume model of Turnbull and Cohen (1961), aperiodic fluctuations in the atomic packing density, in network rigidity and in the distribution of chemical species were acknowledged to constitute the primary difference to a crystalline lattice. Such types of structural heterogeneity have been related to the heterogeneous dynamics of the liquid from which the glass is derived (Angell 1995; Ediger 2000). The manifestations of this structural heterogeneity are manifold, including the limited applicability of linear elasticity and continuum models when observing the mechanical behavior of glasses at short length scales, for example, in the roughness and fractal dimension of fracture surfaces (Daguier et al. 1997; Bonamy et al. 2006), or in the non-affine displacement of atoms and structural units during material deformation (Tanguy et al. 2002; Wittmer et al. 2002). The associated strain localization phenomena and non-affine (inhomogeneous) deformation are common to disordered materials across the broadest range of length scales (Cubuk et al. 2017) and material chemistries (Nicolas et al. 2018), from metallic and oxide glasses to granular media, polymers, dry and wet foams or colloidal suspensions. The extent of the non-affine deformation is on the same order of magnitude as the corresponding affine deformation, and thus cannot be ignored when estimating elastic properties of disordered materials (Alexander 1998; Goldenberg et al. 2007). Structural disorder leads to random spatial fluctuations of the elastic constants with a certain correlation length ξ_G (Fig. 25) (Marruzzo et al. 2013; Mizuno et al. 2013). For two- and three-dimensional Lennard–Jones glasses, correlated particle displacements with a correlation length of 20–30 interparticle spacings were found (Leonforte et al. 2005), and similar observations were made in a computer-generated model of amorphous silica, however, arguing that "*the only practical way to quantify* (elastic heterogeneity) *consists in direct molecular simulations*" (Léonforte et al. 2006). Mizuno et al. studied various methods for the spatial reconstruction of Lennard–Jones glasses from local stress—local strain data, pointing out that heterogeneous non-affine elasticity causes the elastic constants of the bulk material to be smaller than the local average: macroscopic elastic constants such as the shear modulus G_{exp} measured using ultrasonic or other techniques differ from local properties when analyzed within the scale of ξ_G (Mizuno et al. 2013), and are generally lower than the geometric mean value G_0. This is reflected in the non-affinity parameter $n = 1 - G_{exp}/G_0$. In practical terms, $n < 1$ signifies strain localization and the interplay between "*soft*" and "*hard*" regions within a glass lattice. Such classification of soft and hard regions in a glass has also been used in recent studies based on a machine learning classification algorithm to construct a structural metric that is correlated to the mobility of local atomic structures. A non-intuitive structural descriptor (called "softness") was defined that presents an intriguing correlation with the propensity for a given atom to undergo relaxation, although the approach has so far been limited to computer model glasses (Schoenholz et al. 2016; Bapst et al. 2020).

Quantification of elastic heterogeneity and the non-affine contributions to glass deformation is pivotal for understanding the mechanical properties of complex glasses (e.g., hardness, elastic constants), in which the fundamental tool of lattice symmetry is not available (Pan et al. 2021). Heterogeneous network topology was identified to mediate nanoscale ductility in silicate glasses (Wang et al. 2016), and also contributes to microscopic deformation of vitreous silica (Benzine et al. 2018) or thin-film amorphous alumina (Frankberg et al. 2019). Quantitative descriptors of structural disorder might help to construct predictive models beyond brute-force regression of bond energy density (Makishima and Mackenzie 1975; Shi et al. 2020). However, outside of computational simulation, such insight has been elusive, in particular, for chemically complex glass formulations such as multicomponent oxides and silicates. This is despite the significant progress which has been made in analytical imaging methodology (e.g., enabling real-time imaging of atomic rearrangement during material strain in amorphous silica by transmission electron microscopy; Huang et al. 2013). For example, electron correlation microscopy has been employed to map the spatial distribution of shear relaxation times in metallic glasses near T_g (Zhang et al. 2018), but this has remained challenging

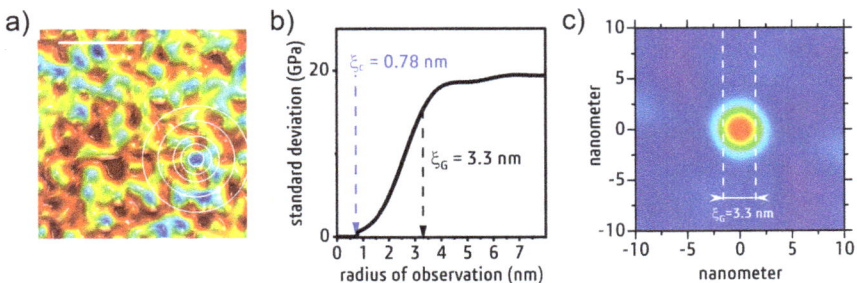

Figure 25. (a) Schematic representation of elastic heterogeneity, assuming a random fluctuation of the shear modulus with a correlation length ξ_G and a Gaussian decay function. Numerical sampling on such a map yields the local standard deviation of shear modulus as a function of sampling radius, shown in **(b)** for the vicinity of a soft region as marked in panel (a). The scale bar in (a) is 20 nm. The schematic corresponds to the typical length-scale of heterogeneity of ~ 3.3 nm which is found in vitreous silica. **(c)** Autocorrelation function of the map shown in (a), visualizing the correlation length ξ_G.

for oxide glasses due to sample stability. Similarly, force microscopy has been applied to probe the spatial heterogeneity of the elastic properties of metallic glasses (Wagner et al. 2011), but has not reached the spatial resolution for mapping on the scale of ξ_G. In the absence of such direct imaging data, theoretical frameworks such as the heterogeneous elasticity theory (Schirmacher 2006) remain untested against the physical nature of heterogeneity (for example, in terms of fluctuating mass density, chemical composition or network rigidity). Therefore, the discovery of material commonalities, universal descriptors for non-affinity and heterogeneous disorder, and quantitative scaling relations remain a major challenge.

PERSPECTIVE: MODELING AND SIMULATION

In brittle oxides, local stresses cannot readily be dissipated by plastic deformation, which can lead to stress concentration and catastrophic fracture. Calculating the distribution of local stresses and resulting fracture patterns in glasses is thus an important task. While experimental measurements of, e.g., surface stresses are possible (via, for example, indentation and nanoindentation (Carlsson and Larsson 2001; Xu and Li 2008), X-ray diffraction for crystalline materials (Mary et al. 2005) or birefringence measurements for transparent materials (Primak 1969), information about the three-dimensional distribution of stress has typically been obtained numerically via the finite element method (FEM). However, FEM is not able to handle stress discontinuities, which is possible with peridynamic simulations. With some differences, both FEM and peridynamic simulations can thus be used to study stress distribution and fracture patterns on the micro- and macrostructural scales, but they both lack the atomic scale information that can be obtained by molecular dynamic (MD) simulations. On the other hand, due to the high computational cost, MD simulations cannot yet be used to study macroscopic crack propagation. In the following section, we review the applications of FEM, peridynamics, and MD simulations to understand glass mechanical properties. Finally, we conclude by considering recent advances in machine learning (ML) algorithms to predict the composition dependence of mechanical properties and eventually design glass microstructures with tailored fracture patterns.

Finite element methods

The basics of FEM can be found elsewhere (Fish and Belytschko 2007; Zienkiewicz et al. 2013); we will focus on its applications to glass mechanics. When performing FEM calculations using an available software, several choices need to be made, for example, considering the mesh refinement, plane or out-of-plane stresses, small or large deflections etc. Material properties such as Young's modulus and Poisson's ratio are needed as input, but often not available in sufficient depth (in particular, transient properties, interdependencies). A dense mesh is needed when high stress gradients are expected, as regions with almost invariant stress can be simulated with fewer elements. As summarized by Belis et al. (2019), the common sources of errors in FEM calculations are modeling error (wrong theory), user error (wrong input), software error (wrong code), discretization error (too coarse mesh), and numerical error (round-off errors).

Modeling the complicated stress field during sharp contact deformation (i.e., indentation) using FEM has received considerable attention. The first notable work was done by Imaoka and Yasui (Imaoka and Yasui 1976; Yasui and Imaoka 1982), who studied the elastic-plastic behavior of glasses with a constant Poisson's ratio of 0.22 by considering the densification contribution to indentation deformation. Since then, there has been focus on developing an accurate constitutive law that describes the glass' response to indentation. Von Mises, Mohr–Coulomb, and Drucker–Prager criteria are some of the models used to understand the indentation deformation (Kermouche et al. 2008; Keryvin et al. 2014; Molnár et al. 2016; Bruns et al. 2020). These approaches are employed to numerically simulate elastic spring-back upon unloading, densification maps, extent of pile-up, and load-displacement curves and compare with the corresponding experimental results. An example of a FEM simulated indentation-induced densification map for amorphous silica is presented in Figure 26, including a comparison with micro-Raman spectroscopy data (Kermouche et al. 2008). As expected, densification is most pronounced just below the tip. Overall, the densification gradient calculated from FEM agrees well with the experimental data, despite some minor differences in absolute values.

It is also possible to study fracture and cracking using FEM. Again considering the case of indentation-induced damage, techniques such as cohesive zone FEM (Bruns et al. 2017) and discrete element modeling (André et al. 2013) have been used. As for other FEM techniques, different material models are compared with respect to their ability to reproduce the experimental results. For example, median/radial cracking has been simulated using cohesive FEM approach via the introduction of cohesive element planes (aligned along the indenter edges perpendicular to the indented surface) to the finite element mesh (see Fig. 27). Glass densification was found to lead to shorter crack lengths and slower crack growth for sharper indenter tip geometries (Bruns et al. 2017). In a recent FEM study (Barthel et al. 2020), it was shown in contrast to previous studies that shear flow stress instead of densification plays the primary role in determining the indentation response of silicate glasses. Crack initiation was proposed to be controlled by the material damage incurred through shear flow.

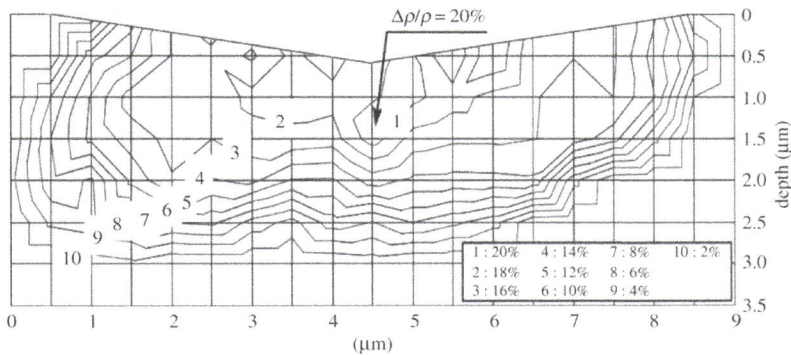

Figure 26. Indentation-induced densification map for amorphous silica. Used with permission by The Royal Society, from Rouxel (2015), Philosophical Transactions of the Royal Society A: Mathematical, Physical and Engineering Sciences, Vol. 343, Fig. 10a, p 2014014-11.

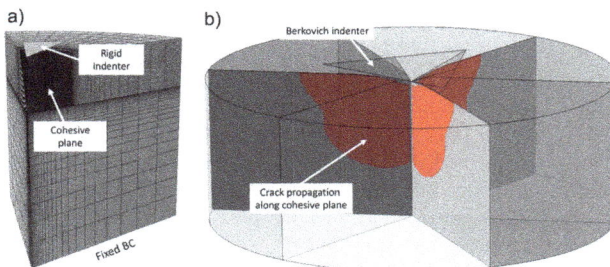

Figure 27. Cohesive zone finite element model of indentation-induced deformation and cracking using a Berkovich tip. The mesh is refined near the indenter contact to model the larger strains and strain gradients in this region. (**A**) Crack plane aligning with the indenter edge. (**B**) Enlarged view of the resulting crack geometry (**red regions**) upon completed unloading. Used with permission by John Wiley and Sons, from Bruns et al. (2017), Journal of the American Ceramic Society, Vol. 100, Fig. 2, p 1931.

Peridynamics

As described above, FEM can be used to study stress fields and cracking in oxide glasses. However, FEM cannot handle stress discontinuities, which may occur around crack tips or at interfaces. This can be made possible by replacing the partial differential equations of solid mechanics with integral equations (Silling 2000; Silling and Lehoucq 2008)—this nonlocal reformulation is the basis for the peridynamic techniques that we consider in the following. It has also been shown that peridynamics better predicts crack branching than FEM techniques (Agwai et al. 2011). An important consequence of being a nonlocal theory is that peridynamics is a meshless scheme and thus scale-invariant, i.e., different length and time scales can be used. In peridynamics, solid particles interact with each other over a characteristic length scale, the so-called horizon. Both state-based (Silling et al. 2007) and bond-based (Silling 2000) formulations have been introduced, with the former being able to account for ductile deformations and varying values of Poisson's ratio. Similar to FEM, material properties such as density, elastic moduli, and yield stress are needed as inputs to the model.

Peridynamics have been used to simulate various fracture phenomena in oxide glasses. For example, Ha and Bobaru (2011) studied dynamic crack propagation including crack branching and crack-path instability in a simulated borosilicate (Duran 50) glass. Their bond-based model was able to replicate the experimentally observed successive branching events and shows that secondary cracks form as a direct consequence of wave propagation and reflection from the boundaries. Hu et al. (2013) used peridynamics to study the impact damage induced by a spherical projectile on a thin glass plate with a thin polycarbonate backing plate. The experimentally observed fracture patterns were captured by the model, providing insights to the early stages of the brittle damage evolution in the glass plate. Recently, Tang et al. (2018) studied the effect of nanoscale phase separation in a simulated glass on the crack propagation mechanism. Figure 28 shows a summary of the observed effects of toughness and stiffness of the nanoscale inclusions on the overall fracture energy of the phase-separated glass. That is, the simulations revealed how phase separation can increase the fracture toughness significantly and also provided insight into the toughening mechanism.

Molecular dynamics

MD simulations can be a highly valuable tool to obtain atomic-scale insight into the structural origin of deformation and fracture mechanisms. The simulated samples can be subjected to, e.g., uniaxial, hydrostatic, or complex indentation-induced strain. The basics of MD simulations of oxide glasses can be found elsewhere (Massobrio et al. 2015), but the most important ingredient is the interaction potential between the atoms. Using an *ab initio* potential, the interactions are calculated from the electronic structure of the material. This provides very accurate results, but it is also very computationally expensive, typically limited to a system size of less than 10^3 atoms and a time scale of less than 10 nanoseconds (Zhang et al. 2020). Unfortunately, this makes it difficult to apply *ab initio* MD simulations to study glass deformation and fracture phenomena, since a larger system is typically needed to avoid problematic size effects. Instead, empirical (classical) or reactive potentials can be used. Although they are not as accurate as *ab initio* potentials, significantly larger systems (possibly 10^6–10^7 atoms) and longer time scales (10s of microseconds) are enabled. Assuming a reliable potential is available for the oxide glass composition of interest (Sundararaman et al. 2018), this allows for mechanical properties to be simulated and the cooling rates needed to form the glasses during melt-quenching will also be closer to experiments as compared to *ab initio* MD. Zhang et al. (2020) have recently compared different interaction potentials to simulate the fracture properties of silica and sodium silicate glasses. While the glass structure does not show a large variation with the potential used, the mechanical properties from failure stress to elastic moduli depend strongly on it. Moreover, they also found that systems of at least 75,000 and 300,000 atoms are needed for silica and sodium silicate systems, respectively, to avoid the finite size effects.

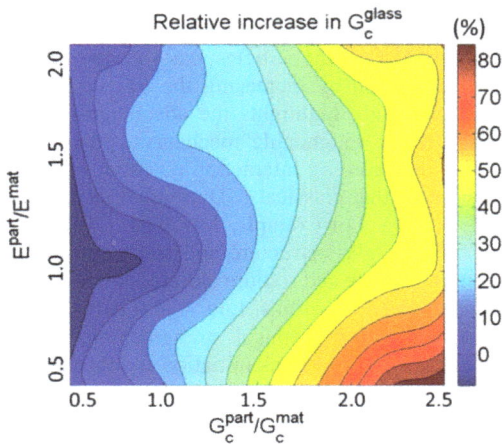

Figure 28. Contour plot showing the effects of stiffness (E) and fracture energy (G_c) on the fracture energy of the phase-separated glass (G_c^{glass}), obtained by peridynamic simulation. "Part" and "mat" refer to the nanoparticles and glass matrix, respectively. The largest increase in fracture energy is obtained when the nanoparticles have low stiffness but high fracture energy, as seen in the lower right corner. Used with permission by the AIP, from Tang et al. (2018), *Physical Review Materials*, Vol. 2, Fig. 5, p 113602-5.

Without going into further detail within the scope of this review, we consider a few typical applications of MD simulations as conducted to understand glass mechanics. Wang et al. (2015) used classical MD simulations to study the fracture behavior of silica, sodium silicate, and calcium aluminosilicate glasses. For the case of calcium aluminosilicate, Figure 29 shows how a crack is first created in the simulated sample with stress concentration at the tip and then the system is stretched. Realistic fracture toughness and critical energy release rate

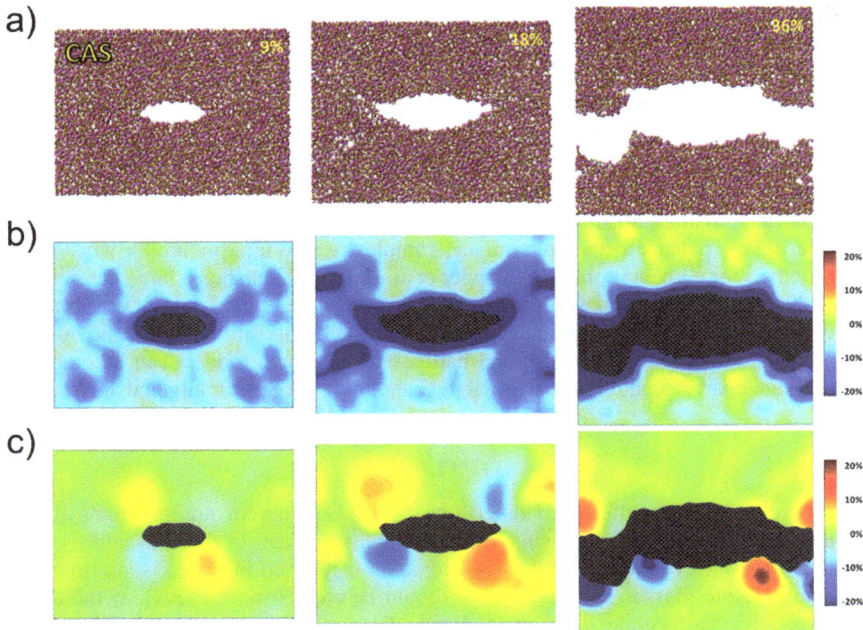

Figure 29. MD-simulated calcium aluminosilicate glass samples with a precrack under strains of 9, 18, and 36%. (**a**) Atomic configurations, with Si, Al, Ca, and O atoms represented in **gold, yellow, magenta**, and **red**, respectively. (**b**) Local density change compared with the bulk density. (**c**) Local shear strain. Used under CC BY-NC-ND license, from Wang et al. (2015), *Frontiers in Materials*, Vol. 2, Fig. 3, p 5.

values could be computed using this approach, showing distinct differences in the degree of nanoscale ductility among the three glasses. In a different study, the fracture behavior of a soda lime silica glass with embedded amorphous Al_2O_3 nanoparticles was simulated (Urata et al. 2018). The nanoparticles limit crack propagation by increasing the fracture surface area and dissipate energy through cracking. This study highlights the advantage of using MD simulations to obtain atomistic insights into the toughening mechanism provided by the nanoparticles, of course with the usual caveats regarding interaction potential reliability and finite size and/or time effects. Finally, we note that classical MD simulations can also be used to simulate indentation-induced deformation. For example, Liu et al. (2020) used simulated nanoindentation tests to compare how the glass structure and stress/strain fields develop underneath the indenter in sodium aluminosilicate and sodium aluminoborate glasses, demonstrating the structural differences between Al and B atoms. Luo et al. (2016) studied the indentation deformation mechanism in glasses after chemical strengthening, thermal tempering, and lamination, highlighting the driving forces for crack initiation and thus guiding the experimental design of post-processing protocols for glasses.

Machine learning

ML is expected to accelerate the physical understanding and discovery of new glass materials (Liu et al. 2019). In general, ML methods use information in the available datasets for interpolation or extrapolation to currently unavailable date regimes. This ML pipeline first involves the collection or generation of consistent data (from experiments or simulations) to build a database of the dependence of, in the present case, mechanical properties on glass composition and processing conditions. ML can then be used to infer non-obvious patterns within the dataset to establish a predictive model. In supervised ML, both inputs (e.g., glass composition) and outputs (e.g., fracture toughness) are known to the algorithm. In unsupervised ML, there is no labeling of the dataset and the output information is thus unknown. For example, the K-mean algorithm can be used to identify clusters of data points that share similar characteristics.

The applications of ML within glass science have mostly focused on supervised regression algorithms to develop composition–property models (Tandia et al. 2019). For example, Yang et al. (2019c) used MD simulations to train four ML algorithms (polynomial regression, LASSO, random forest, and artificial neural network) to predict the composition dependence of Young's modulus in calcium aluminosilicate glasses, see Figure 30. This study points to the importance of considering the balance between accuracy, complexity, and interpretability when choosing the ML method for a given problem. Ravinder et al. (2020) established a large dataset of more than 100,000 oxide glass compositions with up to 37 components. This dataset was used to train high-fidelity deep neural networks and then predict the composition dependence of various properties, including density, moduli, and hardness. Thus, such models can be employed to develop design charts for aiding the rational design of new glasses for targeted applications.

OUTLOOK

The mechanical properties of oxide glasses, as reviewed herein, are important for a range of industrial applications as well as natural phenomena relevant to the geosciences. Most of the studies, in both experiments and simulations, have focused on silica, soda-lime silicate, and aluminosilicate glasses. However, as it should be clear from this review, the details of mechanical deformation under different loading and environmental conditions depend strongly on glass structure and thus the glass composition and processing conditions (regardless as to whether the latter were natural or technical). The proposition that the understanding of glass mechanics is still at a nascent state is substantiated by the fact that a well-established constitutive law for the microductile behavior of glass is still largely missing.

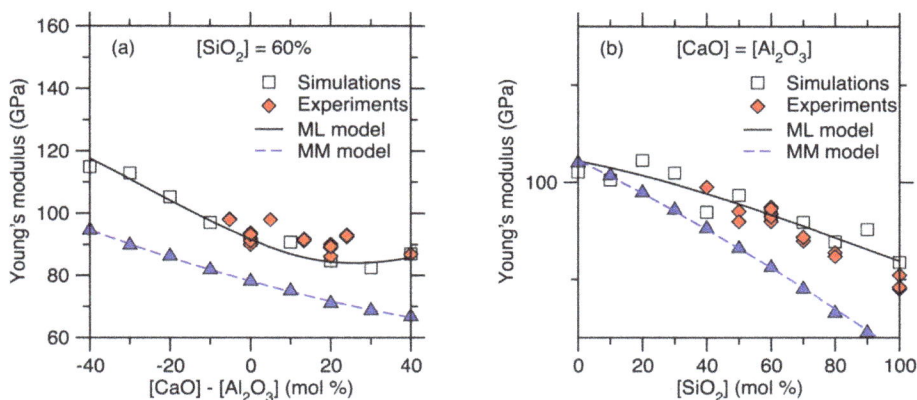

Figure 30. Comparison between experimental values of Young's modulus for calcium alunimosilicate glasses and predictions from molecular dynamic simulations, machine learning algorithms and the Makishima-Mackenzie model. Used under CC BY-NC-ND license, from Yang et al. (2019c), Scientific Reports, Vol. 9, Fig. 9, p 8739-8.

Despite its apparent simplicity, systematic studies on the composition and stress dependence of well-known phenomena and properties such as fatigue, elastic moduli, toughness and transient phenomena are therefore still needed to expand our understanding of glass mechanics (Wondraczek et al. 2022). Ideally, such studies should be coupled with *ex situ* or *in situ* characterization of glass structure as well as stress and strain states. The improved spatial and temporal resolution of X-ray and neutron analytics could help to overcome some of the present limitations, as well as the improved computational power that allows simulations to become more realistic and slowly approach the scale of laboratory experiments. Studies on the relations between glass structure and mechanical properties have hitherto focused mostly on short range, i.e., nearest-neighbor interactions, whereas the roles of medium-range order and heterogeneous elasticity remain insufficiently understood. This is even more relevant considering the many technical and natural glasses which feature nanoscale heterogeneity such as that arising from, phase separation.

In this review, we did not cover temperature and pressure dependent studies and the thermomechanical history of glasses, which present another important parameter affecting mechanical properties (Kapoor et al. 2017; Salmon and Huang 2017).

The interplay between elastic deformation, bond rupture, and chemical reactions with water is a further topic of high interest. As reviewed herein, the temperature dependence of the stress corrosion reactions leads to a temperature dependence of the fatigue resistance. Such detailed understanding is the foundation needed to develop models that can predict the lifetime performance of industrial glass products as well as the continuous evolution of the mechanical properties of natural glasses.

ACKNOWLEDGEMENT

L.W. acknowledges funding from the German Science Foundation through its priority program PP 1594. M.M.S. acknowledges funding from VILLUM FONDEN (grant no. 13253)

REFERENCES

Aben H, Anton J, Õis M, Viswanathan K, Chandrasekar S, Chaudhri MM (2016) On the extraordinary strength of Prince Rupert's drops. Appl Physics Lett 109:231903

Agwai A, Guven I, Madenci E (2011) Predicting crack propagation with peridynamics: a comparative study. Intl J Fract 171:65–78

Alemany P, Canadell E (2014) Chemical bonding in solids. *In:* Frenking G, Shaik S (Eds.), The Chemical Bond: Chemical Bonding Across the Periodic Table .Wiley-VCHVerlag GmbH and Co. KGaA, p 445–476

Alexander S (1998) Amorphous solids: their structure, lattice dynamics, elasticity. Phys Rep 296:65–236

André D, Jebahi M, Iordanoff I, Charles J, Néauport J (2013) Using the discrete element method to simulate brittle fracture in the indentation of a silica glass with a blunt indenter. Comput Methods Appl Mech Eng 265:136–147

Angell CA (1995) Formation of glasses from liquids, biopolymers. Science 267:1924–1935

Arora A, Marshall DB, Lawn BR, Swain MV (1979) Indentation deformation/fracture of normal, anomalous glasses. J Non-Cryst Solids 31:415–428

Ashby MF (2016) Engineering materials, their properties. *In:*Materials Selection in Mechanical Design. Butterworth-Heinemann

Ashby MF, Cebon D (1993) Materials selection in mechanical design. J Phys (paris) IV 3(C7):1–9

ASTM (2013) ASTMC1161-13, Standard test method for flexural strength of advanced ceramics at ambient temperature. ASTM International

ASTM (2019) ASTMC1322-15: Standard practice for fractography, characterization of fracture origins in advanced ceramics. ASTM International

Avramov I, Keding R, Rüssel C (2000) Crystallization kinetics, rigidity percolation in glass-forming melts. J Non-Cryst Solids 272:147–153

Baker TC, Preston FW (1946) The effect of water on the strength of glass. J Appl Phys 17:179–188

Bansal GK (1977) On fracture mirror formation in glass, polycrystalline ceramics. Philos Mag 35:935–944

Bapst V, Keck T, Grabska-Barwińska A, Donner C, Cubuk ED, Schoenholz SS, Obika A, Nelson AWR, Back T, Hassabis D, Kohli P (2020) Unveiling the predictive power of static structure in glassy systems. Nat Phys 16:448–454

Baratta FI, Matthews WT, Quinn GD (1987) Erros associated with flexure testing of brittle materials. Army Lab Command Watertown Ma Material Technology Lab.

Barthel E, Keryvin V, Rosales-Sosa G, Kermouche G (2020) Indentation cracking in silicate glasses is directed by shear flow, not by densification. Acta Mater 194:473–481

Becher PF, Rose F (2006) Toughening mechanisms in ceramic systems. *In*: Materials Science and Technology. Wiley-VCH Verlag GmbH and Co. KGaA

Bechgaard TK, Mauro JC, Smedskjaer MM (2018) Time and humidity dependence of indentation cracking in aluminosilicate glasses. J Non-Cryst Solids 491:64–70

Belis J, Louter C, Nielsen JH, Schneider J (2019) Architectural Glass. *In:* Musgraves JD, Hu J, Calvez L (Eds.) Springer Handbook of Glass, Springer, Cham, p 1781–1819

Bentor YK (1984) Combustion-metamorphic glasses. J Non-Cryst Solids 67:433–448

Benzine O, Bruns S, Pan Z, Durst K, Wondraczek L (2018) Local deformation of glasses is mediated by rigidity fluctuation on nanometer scale. Adv Sci 5:1800916

Bernhardt EO (1941) On microhardness of solids at the limit of Kick's Similarity Law. Z Metallkunde 33:135–144

Bocker C, Rüssel C (2017) Percolation, phase separation, crystallisation. Phys Chem Glasses: Eur J Glass Sci Technol Part B 58:133–141

Bonamy D, Ponson L, Prades S, Bouchaud E, Guillot C (2006) Scaling exponents for fracture surfaces in homogeneous glass, glassy ceramics. Phys Rev Lett 97:135504

Börger A, Supancic P, Danzer R (2002) The ball on three balls test for strength testing of brittle discs: stress distribution in the disc. J Eur Ceram Soc 22 :1425–1436

Bower AF, Fleck NA, Needleman A, Ogbonna N (1993) Indentation of a power law creeping solid. Proc R Soc London, Ser A 441:97–124

Bradt RC, Hasselman DPH, Lange FF (Eds.) (1974) Concepts, Flaws, and Fractography. Springer US

Brambilla G, Payne DN (2009) The ultimate strength of glass silica nanowires. Nano Lett 9:831–835

Bridgman PW, Šimon I (1953) Effects of very high pressures on glass. J Appl Phys 24:405–413

Brook RJ (Ed.) (1991) Concise Encyclopedia of Advanced Ceramic Materials. Elsevier

Brow RK, Lower NP, Kurkjian CR, Li H (2009) The effects of melt history on the failure characteristics of pristine glass fibres. Phys Chem Glasses: Eur J Glass Sci Technol Part B 50:31–33

Brown WK, Wohletz KH (1995) Derivation of the Weibull distribution based on physical principles, its connection to the Rosin–Rammler, lognormal distributions. J Appl Phys 78:2758–2763

Brueckner R (1996) Anisotropic glasses and glass melts: a survey. Glass Sci Technol 69:396–411

Bruns S, Johanns KE, Rehman HUR, Pharr GM, Durst K (2017) Constitutive modeling of indentation cracking in fused silica. J Am Ceram Soc 100:1928–1940

Bruns S, Uesbeck T, Fuhrmann S, Tarragó Aymerich M, Wondraczek L, de Ligny D, Durst K (2020) Indentation densification of fused silica assessed by raman spectroscopy, constitutive finite element analysis. J Am Ceram Soc 103:3076–3088

Buehler MJ (2008) Atomistic Modeling of Materials Failure. Springer US

Bull SJ, Page TF, Yoffe EH (1989) An explanation of the indentation size effect in ceramics. Philos Mag Lett 59:281–288

Bullock RE, Kaae JL (1979) Size effect on the strength of glassy carbon. J Mater Sci 14:920–930

Calahoo C, Wondraczek L (2020) Ionic glasses: Structure, properties and classification. J Non-Cryst Solids X 8:100054

Callister Jr WD, Rethwisch DG (2018) Materials Science, Engineering: An Introduction (10th ed.) Wiley

Cameron NM (1966) Relation between melt treatment, glass fiber strength. J Am Ceram Soc 49:144–148

Carlsson S, Larsson P (2001) On the determination of residual stress and strain fields by sharp indentation testing. part ii: experimental investigation. Acta Mater 49:2193–2203

Célarié F, Ciccotti M, Marlière C (2007) Stress-enhanced ion diffusion at the vicinity of a crack tip as evidenced by atomic force microscopy in silicate glasses. J Non-Cryst Solids 353:51–68

Chang H-J, Huang K-C, Wu C-H (2006) Determination of sample size in using Central Limit Theorem for Weibull distribution. Info Manage Sci 17:31–46

Choi I-C, Yoo B-G, Kim Y-J, Jang J-I (2012) Indentation creep revisited. J Mater Res27:3–11

Christensen RM (2013) The Theory of Materials Failure. OUP Oxford

Chu SNG, Li JCM (1977) Impression creep; a new creep test. J Mater Sci 12:2200–2208

Ciccotti M (2009) Stress-corrosion mechanisms in silicate glasses. J Phys D: Appl Phys 42:214006

Cicconi MR, Neuville DR (2019) Natural glasses. *In:* Musgraves JD, Hu J, Calvez L(Eds.) Springer Handbook of Glass, Springer International Publishing, p 771–812

Ciccotti M, George M, Ranieri V, Wondraczek L, Marlière C (2008) Dynamic condensation of water at crack tips in fused silica glass. J Non-Cryst Solids 354:564–568

Ciccotti M, George M (2020) In situ AFM investigations, fracture mechanics modeling of slow fracture propagation in oxide, polymer glasses. *In:* Handbook of Materials Modeling, Springer International Publishing, p 771–812

Coles S (2001) Classical extreme value theory, models. *In:* An Introduction to Statistical Modeling of Extreme Values, Springer, London, p 45–73

Cook RF, DelRio FW (2019) Material flaw populations, component strength distributions in the context of the weibull function. Exp Mech 59:279–293

Cook RF, Pharr GM (1990) Direct observation, analysis of indentation cracking in glasses, ceramics. J Am Ceram Soc 73:787–817

Cubuk ED, Ivancic RJS, Schoenholz SS, Strickland DJ, Basu A, Davidson ZS, Fontaine J, Hor JL, Huang Y-R, Jiang Y, Keim NC, Koshigan KD, Lefever JA, Liu T, Ma X-G, Magagnosc DJ, Morrow E, Ortiz CP, Rieser JM, Liu AJ (2017) Structure–property relationships from universal signatures of plasticity in disordered solids. Science 358:1033–1037

Daguier P, Nghiem B, Bouchaud E, Creuzet F (1997) Pinning, depinning of crack fronts in heterogeneous materials. Phys Rev Lett 78:1062–1065

Danzer R (2006) Some notes on the correlation between fracture, defect statistics: Are Weibull statistics valid for very small specimens? J Eur Ceram Soc 26:3043–3049

Datsiou KC, Overend M (2018) Weibull parameter estimation and goodness-of-fit for glass strength data. Struct Safety 73:29–41

de Wolff PM, Belov NV, Bertaut EF, Buerger MJ, Donnay JDH, Fischer W, Hahn T, Koptsik VA, Mackay AL, Wondratschek H, Wilson AJC, Abrahams SC (1985) Nomenclature for crystal families, Bravais-lattice types, arithmetic classes. Report of the International Union of Crystallography Ad-Hoc Committee on the Nomenclature of Symmetry. Acta Crystallogr Section A 41:278–280

Demetriou MD, Launey ME, Garrett G, Schramm JP, Hofmann DC, Johnson WL, Ritchie RO (2011) A damage-tolerant glass. Nat Mater 10:123–128

Dériano S, Jarry A, Rouxel T, Sangleboeuf J-C, Hampshire S (2004) The indentation fracture toughness (K_C), its parameters: the case of silica-rich glasses. J Non-Cryst Solids 344:44–50

Dickinson JT, Donaldson EE, Park MK (1981) The emission of electrons, positive ions from fracture of materials. J Mater Sci 16:2897–2908

Dietzel A (1942) Die Kationenfeldstaerken und ihre Beziehung zu Entlastungsvorgaengen, zur Verbindungsbildung und zu den Schmelzpunkten von Silikaten. Z Elektrochem Angew Phys Chemie 48:9–23

Döll W (1975) Investigations of the crack branching energy. Int J Fract 11:184–186

Douglas RW (1958) Comments on indentation tests on glass. J Soc Glass Technol 42:145–57T

Ediger MD (2000) Spatially heterogeneous dynamics in supercooled liquids. Annu Rev Phys Chem 51:99–128

Ernsberger FM (1969) Tensile, compressive strength of pristine glasses by an oblate bubble technique. Phys Chem Glasses 10:240–245

Evans KE, Alderson A (2000) Auxetic materials: functional materials, structures from lateral thinking! Adv Mater 12:617–628

Evans AG, Charles EA (1976) Fracture toughness determinations by indentation. J Am Ceram Soc 59:371–372

Evers M (1967) Plastic deformation of glass with diamond indenters. Glastech Ber 40:41–43

Fett T, Germerdonk K, Grossmueller A, Keller K, Munz D (1991) Subcritical crack growth, threshold in borosilicate glass. J Mater Sci 26:253–257

Fineberg J, Marder M (1999) Instability in dynamic fracture. Phys Rep 313:1–108

Fish J, Belytschko T (2007) A First Course in Finite Elements. Wiley

France PW, Paradine MJ, Reeve MH, Newns GR (1980) Liquid nitrogen strengths of coated optical glass fibres. J Mater Sci 15:825–830

Frankberg EJ, Kalikka J, García Ferré F, Joly-Pottuz L, Salminen T, Hintikka J, Hokka M, Koneti S, Douillard T, Le Saint B, Kreiml P, Cordill MJ, Epicier T, Stauffer D, Vanazzi M, Roiban L, Akola J, Di Fonzo F, Levänen E, Masenelli-Varlot K (2019) Highly ductile amorphous oxide at room temperature, high strain rate. Science 366:864–869

Frechette VD (1990) Failure analysis of brittle materials. *In*: Advances in Ceramics. Wiley

Freeman JJ, Anderson AC (1986) Thermal conductivity of amorphous solids. Phys Rev B 34:5684–5690

Fu AI, Mauro JC (2013) Topology of alkali phosphate glass networks. J Non-Cryst Solids 361:57–62

Fuhrmann S, de Macedo GNBM, Limbach R, Krywka C, Bruns S, Durst K, Wondraczek L (2020) Indentation-induced structural changes in vitreous silica probed by in-situ small-angle X-ray scattering. Front Mater 7:173

Gdoutos EE (1990) Fracture Mechanics Criteria and Applications. Springer Netherlands

Gehlhoff G, Thomas M (1926) Die physikalischen Eigenschaften der Gläser in Abhängigkeit von der Zusammensetzung. Z Tech Physik 6:260–278

Gehrke E, Ullner C, Hähnert M (1991) Fatigue limit, crack arrest in alkali-containing silicate glasses. J Mater Sci 26:5445–5455

Ghosh A, Kobayashi AS, Li Z, Henager CHJ, Bradt RC (1991) Vickers microindentation toughness of a sintered SiC in the median-crack regime. Pacific Northwest Lab., Richland, WA (United States) PNL-SA-16927

Gilman JJ (2009) Chemical Bonding. *In*: Chemistry, Physics of Mechanical Hardness, John Wiley and Sons, p. 27–50

Goldenberg C, Tanguy A, Barrat J-L (2007) Particle displacements in the elastic deformation of amorphous materials: Local fluctuations vs. non-affine field. Europhys Lett 80:16003

Goldschmidt VM (1911) Die Gesetze der Mineralassoziation vom Standpunkt der Phasenregel. Z Anorg Chemie 71:313–322

Goldschmidt VM (1928) Der Kristallbau und die Arten der chemischen Bindung. Z Elektrochem Angew Phys Chemie 34:453–463

Goodall R, Clyne TW (2006) A critical appraisal of the extraction of creep parameters from nanoindentation data obtained at room temperature. Acta Mater 54:5489–5499

Gorjan L, Ambrožič M (2012) Bend strength of alumina ceramics: A comparison of Weibull statistics with other statistics based on very large experimental data set. J Eur Ceram Soc 32:1221–1227

Grau P, Berg G, Meinhard H, Mosch S (2005) Strain rate dependence of the hardness of glass, Meyer's Law. J Am Ceram Soc 81:1557–1564

Greaves GN, Greer AL, Lakes RS, Rouxel T (2011) Poisson's ratio and modern materials. Nat Mater 10: 823–837

Griffith AA (1921) The phenomena of rupture, flow in solids. Philos Trans R Soc London, Ser A 221:163–198

Gross TM (2012) Deformation, cracking behavior of glasses indented with diamond tips of various sharpness. J Non-Cryst Solids 358:3445–3452

Gross TM, Tomozawa M (2008) Fictive temperature-independent density, minimum indentation size effect in calcium aluminosilicate glass. J Appl Phys 104:063529

Gross TM, Wu J, Baker DE, Price JJ, Yongsunthon R (2018) Crack-resistant glass with high shear band density. J Non-Cryst Solids 494:13–20

Gross TM, Liu H, Zhai Y, Huang L, Wu J (2020) The impact of densification on indentation fracture toughness measurements. J Am Ceram Soc 103:3920–3929

Guignard M, Albrecht L, Zwanziger JW (2007) Zero-stress optic glass without lead. Chem Mater 19:286–290

Guin J-P, Rouxel T, Keryvin V, Sanglebœuf J-C, Serre I, Lucas J (2002) Indentation creep of Ge–Se chalcogenide glasses below T_g: elastic recovery, non-Newtonian flow. J Non-Cryst Solids 298:260–269

Guo H, Gerokostopoulos A (2013) Determining the right sample size for your test: Theory and application. 2013 Annual Reliability and Maintainability Symposium

Gupta PK (1993) Rigidity, connectivity, glass-forming ability. J Am Ceram Soc 76:1088–1095

Gupta PK, Mauro JC (2009) Composition dependence of glass transition temperature, fragility. IA Topological model incorporating temperature-dependent constraints. J Chem Phys 130:094503

Gy R (2003) Stress corrosion of silicate glass: a review. J Non-Cryst Solids 316:1–11

Gy R (2008) Ion exchange for glass strengthening. Mater Sci Eng B 149:159–165

Ha YD, Bobaru F (2011) Characteristics of dynamic brittle fracture captured with peridynamics. Eng Fract Mech 78:1156–1168

Haines J, Léger JM, Bocquillon G (2001) Synthesis, design of superhard materials. Annu Rev Mater Res 31:1–23

Harden DB (1968) Ancient glass, I: Pre-Roman. Archaeol J 125:46–72

He H, Thorpe MF (1985) Elastic properties of glasses. Phys Rev Lett 54:2107–2110

Helnwein P (2001) Some remarks on the compressed matrix representation of symmetric second-order, fourth-order tensors. Comput Methods Appl Mech Eng 190:2753–2770

Hentschke R (2017) Basic equations of the theory of elasticity. *In*: Classical Mechanics, Springer International Publishing, p 291–364

Hermansen C, Guo X, Youngman RE, Mauro JC, Smedskjaer MM, Yue Y (2015) Structure–topology–property correlations of sodium phosphosilicate glasses. J Chem Phys 143:064510

Hermansen C, Rodrigues BP, Wondraczek L, Yue Y (2014) An extended topological model for binary phosphate glasses. J Chem Phys 141:244502

Hertz H (1881) Ueber die Beruerung fester elastischer Koerper. J Reine Angew Math 92:156–171

Holzhausen GR, Johnson AM (1979) The concept of residual stress in rock. Tectonophysics 58:237–267

Hu W, Wang Y, Yu J, Yen C-F, Bobaru F (2013) Impact damage on a thin glass plate with a thin polycarbonate backing. Int J Impact Eng 62:152–165

Huang PY, Kurasch S, Alden JS, Shekhawat A, Alemi AA, McEuen PL, Sethna JP, Kaiser U, Muller DA (2013) Imaging atomic rearrangements in two-dimensional silica glass: watching silica's dance. Science 342:224–227

Hühn C, Erlebach A, Mey D, Wondraczek L, Sierka M (2017) Ab initio energetics of Si–O bond cleavage. J Comput Chem 38:2349–2353

Ibrahim RA (2015) Overview of structural life assessment, reliability, Part I: Basic ingredients of fracture mechanics. J Ship Prod Des 31:1–42

Imaoka M, Yasui I (1976) Finite element analysis of indentation on glass. J Non-Cryst Solids 22:315–329

Inaba S, Hosono H, Ito S (2015) Entropic shrinkage of an oxide glass. Nat Mater 14:312–317

Inglis CE (1913) Stresses in a plate due to the presence of cracks and sharp corners. Proc R Inst Naval Archit 55:219–230

Irwin GR (1957) Analysis of stresses and strains near the end of a crack transversing a plate. J Appl Mech 24: 361–364

Ito S, Tomozawa M (1982) Crack blunting of high-silica glass. J Am Ceram Soc 65:368–371

Jaccani SP, Youngman RE, Mauro JC, Huang L (2020) Understanding cracking behavior of glass from its response to hydrostatic compression. Phys Rev Mater 4:063607

Januchta K, Smedskjaer MM (2019) Indentation deformation in oxide glasses: Quantification, structural changes, and relation to cracking. J Non-Cryst Solids X 1:100007

Januchta K, Youngman RE, Goel A, Bauchy M, Logunov SL, Rzoska SJ, Bockowski M, Jensen LR, Smedskjaer MM (2017) Discovery of ultra-crack-resistant oxide glasses with adaptive networks. Chem Mater 29:5865–5876

Januchta K, Stepniewska M, Jensen LR, Zhang Y, Somers MAJ, Bauchy M, Yue Y, Smedskjaer MM (2019) Breaking the limit of micro-ductility in oxide glasses. Adv Sci 6:1901281

Jayatilaka ADS, Trustrum K (1977) Statistical approach to brittle fracture. J Mater Sci 12:1426–1430

Jiang Q, Zeng H, Liu Z, Ren J, Chen G, Wang Z, Sun L, Zhao D (2013) Glass transition temperature, topological constraints of sodium borophosphate glass-forming liquids. J Chem Phys 139:124502

Jiang Q, Zeng H, Li X, Ren J, Chen G, Liu F (2014) Tailoring sodium silicophosphate glasses containing SiO_6-octahedra through structural rules, topological principles. J Chem Phys 141:124506

John V (1992) Introduction to Engineering Materials. Palgrave Macmillan UK

Johnson JW, Holloway DG (1966) On the shape, size of the fracture zones on glass fracture surfaces. Philos Mag A 14:731–743

Johnson JW, Holloway DG (1968) Microstructure of the mist zone on glass fracture surfaces. Philos Mag A 17:899–910

Kapoor S, Wondraczek L, Smedskjaer MM (2017) Pressure-induced densification of oxide glasses at the glass transition. Front Mater 4:1

Karlsson S, Jonson B, Stalhandske C (2010) The technology of chemical glass strenghtening—a review. Glass Technol: Eur J Glass Sci Technol, Part A 51:41–54

Karlsson S, Wondraczek L, Ali S, Jonson B (2017) Trends in effective diffusion coefficients for ion-exchange strengthening of soda-lime-silicate glasses. Front Mater 4:13

Kato Y, Yamazaki H, Yoshida S, Matsuoka J (2010) Effect of densification on crack initiation under Vickers indentation test. J Non-Cryst Solids 356:1768–1773

Kazembeyki M, Bauchy M, Hoover CG (2019) New insights into the indentation size effect in silicate glasses. J Non-Cryst Solids 521:119494

Kennedy CR, Bradt RR, Rindone GE (1980) The strenght of binary alkali silicate glasses. Phys Chem Glasses 21:99–105

Kermouche G, Barthel E, Vandembroucq D, Dubujet P (2008) Mechanical modeling of indentation-induced densification in amorphous silica. Acta Mater 56:3222–3228

Keryvin V, Meng J-X, Gicquel S, Guin J-P, Charleux L, Sanglebœuf J-C, Pilvin P, Rouxel T, Le Quilliec G (2014) Constitutive modeling of the densification process in silica glass under hydrostatic compression. Acta Mater 62:250–257

Kittl P, Díaz G (1990) Size effect on fracture strength in the probabilistic strength of materials. Reliab Eng Syst Sa 28:9–21

Kurkjian CR, Gupta PK, Brow RK, Lower N (2003) The intrinsic strength, fatigue of oxide glasses. J Non-Cryst Solids 316:114–124

Kurkjian CR, Gupta PK, Brow RK (2010) The strength of silicate glasses: what we know, what we do need to know? Int J Appl Glass Sci 1:27–37

Lalena JN, Cleary DA (2005) Mechanical properties. *In*: Principles of Inorganic Materials Design, John Wiley and Sons, Inc, p 299–334

Langford SC, Dickinson JT, Jensen LC (1987) Simultaneous measurements of the electron, photon emission accompanying fracture of single-crystal MgO. J Appl Phys 62:1437–1449

Launey ME, Ritchie RO (2009) On the fracture toughness of advanced materials. Adv Mater 21:2103–2110

Lawn B (1993) Fracture of Brittle Solids. Cambridge University Press

Lawn BR, Wilshaw R (1975) Indentation fracture: principles, applications. J Mater Sci 10:1049–1081

Lawn BR, Evans AG, Marshall DB (1980) Elastic/plastic indentation damage in ceramics: the median/radial crack system. J Am Ceram Soc 63:574–581

Le Bouhris E (2014) Glass: Mechanics and Technology. John Wiley and Sons

Le Houérou V, Sanglebœuf J-C, Rouxel T (2005) Scratchability of soda-lime silica (SLS) glasses: dynamic fracture analysis. Key Eng Mater 290:31–38

Lebedev AA (1921) Ueber Polymorphismus und das Kuehlen von Glas. Arb Staatl Opt Inst Leningrad 2:3

Leonforte F, Boissière R, Tanguy A, Wittmer JP, Barrat J-L (2005) Continuum limit of amorphous elastic bodies. IIIThree-dimensional systems. Phys Rev B 72:224206

Léonforte F, Tanguy A, Wittmer JP, Barrat J-L (2006) Inhomogeneous elastic response of silica glass. Phys Rev Lett 97:055501

Levengood WC (1958) Effect of origin flaw characteristics on glass strength. J Appl Phys 29:820–826

Levitin V (2014) Electrons in crystals, the Bloch waves in crystals. *In:* Interatomic Bonding in Solids: Fundamentals, Simulation, Applications, Wiley-VCH Verlag GmbH and Co. KGaA, p 79–94

Lewandowski JJ, Wang WH, Greer AL (2005) Intrinsic plasticity or brittleness of metallic glasses. Philos Mag Lett 85:77–87

Li Z, Ghosh A, Kobayashi AS, Bradt RC (1989) Indentation fracture toughness of sintered silicon carbide in the palmqvist crack regime. J Am Ceram Soc 72:904–911

Li H, Ghosh A, Han YH, Bradt RC (1993) The frictional component of the indentation size effect in low load microhardness testing. J Mater Res 8:1028–1032

Limbach R, Rodrigues BP, Wondraczek L (2014) Strain-rate sensitivity of glasses. J Non-Cryst Solids, 404, 124–134

Limbach R, Winterstein-Beckmann A, Dellith J, Möncke D, Wondraczek L (2015a) Plasticity, crack initiation and defect resistance in alkali-borosilicate glasses: From normal to anomalous behavior. J Non-Cryst Solids, 417–418:15–27

Limbach R, Rodrigues BP, Möncke D, Wondraczek L (2015b) Elasticity, deformation and fracture of mixed fluoride–phosphate glasses. J Non-Cryst Solids 430:99–107

Limbach R, Kosiba K, Pauly S, Kühn U, Wondraczek L (2017) Serrated flow of CuZr-based bulk metallic glasses probed by nanoindentation: Role of the activation barrier, size and distribution of shear transformation zones. J Non-Cryst Solids 459:130–141

Liu AY, Cohen ML (1989) Prediction of new low compressibility solids. Science 245:841–842

Liu S, Kong Y, Tao H, Sang Y (2017) Crystallization of a highly viscous multicomponent silicate glass: Rigidity percolation, evidence of structural heterogeneity. J Eur Ceram Soc 37:715–720

Liu H, Fu Z, Yang K, Xu X, Bauchy M (2019) Machine learning for glass science and engineering: A review. J Non-Cryst Solids X 4:100036

Liu H, Deng B, Sundararaman S, Shi Y, Huang L (2020) Understanding the response of aluminosilicate, aluminoborate glasses to sharp contact loading using molecular dynamics simulation. J Appl Phys 128:035106

Lower NP, Brow RK, Kurkjian CR (2004a) Inert failure strains of sodium aluminosilicate glass fibers. J Non-Cryst Solids 344:17–21

Lower NP, Brow RK, Kurkjian CR (2004b) Inert failure strain studies of sodium silicate glass fibers. J Non-Cryst Solids 349:168–172

Luo J, Lezzi PJ, Vargheese KD, Tandia A, Harris JT, Gross TM, Mauro JC (2016) Competing indentation deformation mechanisms in glass using different strengthening methods. Front Mater 3:52

Lurie AI (2005) Theory of Elasticity. *In:* Babitsky VI, Wittenburg J (Eds.), Springer Berlin Heidelberg, p 243–407

Maceri A (2010) Theory of Elasticity. Springer Berlin Heidelberg

Makishima A, Mackenzie JD (1973) Direct calculation of Young's modulus of glass. J Non-Cryst Solids 12:35–45

Makishima A, Mackenzie JD (1975) Calculation of bulk modulus, shear modulus, Poisson's ratio of glass. J Non-Cryst Solids 17:147–157

Marruzzo A, Schirmacher W, Fratalocchi A, Ruocco G (2013) Heterogeneous shear elasticity of glasses: the origin of the boson peak. Sci Rep 3:1407

Marsh DM (1964a) Plastic flow in glass. Proc R Soc London, Ser A 279:420–435

Marsh DM (1964b) Plastic flow, fracture of glass. Proc R Soc London, Ser A 282:33–43

Martin B, Wondraczek L, Deubener J, Yue Y (2005) Mechanically induced excess enthalpy in inorganic glasses. Appl Phys Lett 86:121917

Mary N, Vignal V, Oltra R, Coudreuse L (2005) Finite-element, XRD methods for the determination of the residual surface stress field, the elastic–plastic behaviour of duplex steels. Philos Mag 85:1227–1242

Massobrio C, Du J, Bernasconi M, Salmon PS (Eds.) (2015) Molecular Dynamics Simulations of Disordered Materials: From Network Glasses to Phase-Change Memory Alloys. Springer International Publishing

Matthewson MJ, Kurkjian CR, Gulati ST (1986) Strength measurement of optical fibers by bending. J Am Ceram Soc 69:815–821

Mauro JC, Loucks RJ, Varshneya AK, Gupta PK (2008) Enthalpy landscapes, the glass transition. Sci Model Simul 15:241–281

Mauro JC, Gupta PK, Loucks RJ (2009) Composition dependence of glass transition temperature, fragility. IIA topological model of alkali borate liquids. J Chem Phys 130:234503

Mauro JC, Ellison AJ, Allan DC, Smedskjaer MM (2013) Topological model for the viscosity of multicomponent glass-forming liquids. Int J Appl Glass Sci 4:408–413

Maxwell JC (1864) On the calculation of the equilibrium, stiffness of frames. London, Edinburgh, Dublin Philos Mag J Sci 27:294–299

Meszaros R, Merle B, Wild M, Durst K, Göken M, Wondraczek L (2012) Effect of thermal annealing on the mechanical properties of low-emissivity physical vapor deposited multilayer-coatings for architectural applications. Thin Solid Films 520:7130–7135

Michalske TA, Bunker BC (1987) Steric effects in stress corrosion fracture of glass. J Am Ceram Soc 70:780–784

Michalske TA, Freiman SW (1983) A molecular mechanism for stress corrosion in vitreous silica. J Am Ceram Soc 66:284–288

Mitkova M, Boolchand P (1998) Microscopic origin of the glass forming tendency in chalcohalides, constraint theory. J Non-Cryst Solids 240:1–21

Mizuno H, Mossa S, Barrat J-L (2013) Measuring spatial distribution of the local elastic modulus in glasses. Phys Rev E 87:042306

Moayedi E, Wondraczek L (2017) Quantitative analysis of scratch-induced microabrasion on silica glass. J Non-Cryst Solids 470:138–144

Moayedi E, Sawamura S, Hennig J, Gnecco E, Wondraczek L (2018) Relaxation of scratch-induced surface deformation in silicate glasses: Role of densification and shear flow in lateral indentation experiments. J Non-Cryst Solids 500:382–387

Mohs F (1821) Die Charaktere der Klassen, Ordnungen, Geschlechter und Arten oder die Charakteristik des naturhistorischen Mineral-Systemes. Arnold, Dresden

Molnár G, Ganster P, Tanguy A, Barthel E, Kermouche G (2016) Densification dependent yield criteria for sodium silicate glasses—An atomistic simulation approach. Acta Mater 111:129–137

Munz D, Fett T (Eds.) (1999) Ceramics, Volume 36. Springer Berlin Heidelberg

Munz D, Fett T (2015) The Griffith Relation–An Historical Review. KIT Scientific Working Papers 34

Murgatroyd JB (1944) The strength of glass fibers, Part 1: elastic properties. J Soc Glass Technol 28:368–387

Mysen B, Richet P (2019) Structure, property concepts. *In*: Silicate Glasses, Melts, Elsevier, p 109–141

Naray-Szabo I, Ladik J (1960) Strength of silica glass. Nature 188:226–227

Neely JE, Mackenzie JD (1968) Hardness, low-temperature deformation of silica glass. J Mater Sci 3:603–609

Nichols TC, Varnes DJ (1984) Residual stress, rocks. *In:* Finkl CW (Ed.) Applied Geology, Kluwer Academic Publishers, p 461–465

Nicolas A, Ferrero EE, Martens K, Barrat J-L (2018) Deformation, flow of amorphous solids: Insights from elastoplastic models. Rev Modern Phys 90:045006

Nielsen KH, Karlsson S, Limbach R, Wondraczek L (2015) Quantitative image analysis for evaluating the abrasion resistance of nanoporous silica films on glass. Sci Rep 5:17708

Nix WD, Gao H (1998) Indentation size effects in crystalline materials: A law for strain gradient plasticity. J Mech Phys Solids 46:411–425

Nohut S (2014) Influence of sample size on strength distribution of advanced ceramics. Ceram Int 40:4285–4295

Nose T, Fuji T (1988) Evaluation of fracture toughness for ceramic materials by a single-edge-precracked-beam method. J Am Ceram Soc 71:328–333

Okrusch M, Frimmel HE (2020) Igneous rocks. *In*: Mineralogy, Springer Berlin Heidelberg, p 249–274

Oliver WC, Pharr GM (1992) An improved technique for determining hardness, elastic modulus using load, displacement sensing indentation experiments. J Mater Res 7:1564–1583

Ono K (2019) A simple estimation method of Weibull modulus, verification with strength data. Appl Sci 9:1575

Orowan E (1934) Die mechanischen Festigkeitseigenschaften und die Realstruktur der Kristalle. Z Kristall 89:327–343

Østergaard MB, Hansen SR, Januchta K, To T, Rzoska SJ, Bockowski M, Bauchy M, Smedskjaer MM (2019) Revisiting the dependence of poisson's ratio on liquid fragility, atomic packing density in oxide glasses. Materials 12:2439

Pan Z, Benzine O, Sawamura S, Limbach R, Koike A, Bennett T, Wilde G, Schirmacher W, Wondraczek L. Disorder classification of the vibrational spectra of modern glasses. Phys Rev B 104:134106

Pauling L (1960) The Nature of the Chemical Bond and the Structure of Molecules and Crystals: An Introduction to Modern Structural Chemistry. Cornell University Press

Pedone A, Malavasi G, Menziani MC, Segre U, Cormack AN (2008) Role of magnesium in soda-lime glasses: insight into structural, transport, mechanical properties through computer simulations. J Phys Chem C 112:11034–11041

Pedone A, Gambuzzi E, Menziani MC (2012) Unambiguous description of the oxygen environment in multicomponent aluminosilicate glasses from ^{17}O solid state NMR computational spectroscopy. J Phys Chem C 116:14599–14609

Perriot A, Vandembroucq D, Barthel E, Martinez V, Grosvalet L, Martinet C, Champagnon B (2006) Raman microspectroscopic characterization of amorphous silica plastic behavior. J Am Ceram Soc 89:596–601

Petch NJ (1954) The fracture of metals. Prog Metal Phys 5:1–52

Peter KW (1970) Densification, flow phenomena of glass in indentation experiments. J Non-Cryst Solids 5:103–115

Petroski H (1996) Invention by design: How engineers get from thought to thing. Harvard University Press

Phillips JC (1979) Topology of covalent non-crystalline solids I: Short-range order in chalcogenide alloys. J Non-Cryst Solids 34:153–181

Phillips JC, Thorpe MF (1985) Constraint theory, vector percolation, glass formation. Solid State Commun 53:699–702

Plendl JN, Gielisse PJ (1962) Hardness of nonmetallic solids on an atomic basis. Phys Rev 125:828–832

Plucinski M, Zwanziger JW (2015) Topological constraints and the Makishima–Mackenzie model. J Non-Cryst Solids 429:20–23

Ponton CB, Rawlings RD (1989a) Vickers indentation fracture toughness test Part 1 Review of literature, formulation of standardised indentation toughness equations. Mater Sci Technol 5:865–872

Ponton CB, Rawlings RD (1989b) Vickers indentation fracture toughness test Part 2 Application, critical evaluation of standardised indentation toughness equations. Mater Sci Technol 5:961–976

Prager W (1947) An introduction to the mathematical theory of plasticity. J Appl Phys 18:375–383

Preston FW (1935) The angle of forking of glass cracks as an indicator of the stress system. J Am Ceram Soc 18:175–176

Primak W (1969) Determination of small dilatations and surface stress by birefringence measurements. Surf Sci 16:398–427

Pukh VP, Baikova LG, Kireenko MF, Tikhonova LV, Kazannikova TP, Sinani AB (2005) Atomic structure, strength of inorganic glasses. Phys Solid State 47:876

Pyke M (2002) Elastic properties, pressure effects *In*: Rao KJ (Ed.) Structural Chemistry of Glasses, Elsevier

Quinn GD (1988) Fractographic analysis and the army flexure test method. *In*: Advances in Ceramics Volume 22: Fractography of Glasses and Ceramics, p 314–344

Quinn GD (2020) Fractography of Ceramics and Glasses. NIST Special Publication 960-16e2

Quinn GD, Bradt RC (2007) On the Vickers indentation fracture toughness test. J Am Ceram Soc 90:673–680

Quinn JB, Quinn GD (2010) A practical, systematic review of Weibull statistics for reporting strengths of dental materials. Dental Mater 26:135–147

Quinn GD, Wirth G (1990) Multiaxial strength, stress rupture of hot pressed silicon nitride. J Eur Ceram Soc 6:169–177

Rankin DWH, Mitzel N, Morrison C (2013) Theoretical methods. *In*: Structural Methods in Molecular Inorganic Chemistry, John Wiley and Sons, Inc, p 45–78

Ravinder R, Sridhara KH, Bishnoi S, Grover HS, Bauchy M, Jayadeva, Kodamana H, Krishnan NMA (2020) Deep learning aided rational design of oxide glasses. Mater Horiz 7:1819–1827

Reddy KPR, Fontana EH, Helfinstine JD (1988) Fracture toughness measurement of glass, ceramic materials using chevron-notched specimens. J Am Ceram Soc 71:C-310–C-313

Rice RW (1984) Ceramic fracture features, observations, mechanism, uses. *In*: Mecholsky JJ (Ed.) Fractography of Ceramic, Metal Failures. ASTM International, p. 5–103

Rinne H (2008) The Weibull Distribution: A Handbook. CRC Press

Ritchie RO, Dzenis Y (2008) The quest for stronger, tougher materials. Science 320:448

Ritchie RO (2011) The conflicts between strength, toughness. Nat Mater 10:817–822

Rodrigues BP, Wondraczek L (2014) Cationic constraint effects in metaphosphate glasses. J Chem Phys 140:214501

Rodrigues BP, Hühn C, Erlebach A, Mey D, Sierka M, Wondraczek L (2017) Parametrization in models of subcritical glass fracture: activation offset, concerted activation. Front Mater 4:20

Rodrigues BP, Limbach R, de Souza GB, Ebendorff-Heidepriem H, Wondraczek L (2019) Correlation between ionic mobility and plastic flow events in $NaPO_3$–$NaCl$–Na_2SO_4 glasses. Front Mater 6:128

Rountree CL (2017) Recent progress to understand stress corrosion cracking in sodium borosilicate glasses: linking the chemical composition to structural, physical, fracture properties. J Phys D: Appl Phys 50:343002

Rouxel T (2006) Elastic properties of glasses: a multiscale approach. C R Mécanique 334:743–753

Rouxel T (2007) Elastic properties, short-to medium-range order in glasses. J Am Ceram Soc 90:3019–3039

Rouxel T (2015) Driving force for indentation cracking in glass: composition, pressure, temperature dependence. Philos Trans R Society A 373:20140140

Rouxel T, Yoshida S (2017) The fracture toughness of inorganic glasses. J Am Ceram Soc 100:4374–4396

Rouxel T, Ji H, Guin JP, Augereau F, Rufflé B (2010) Indentation deformation mechanism in glass: Densification versus shear flow. J Appl Phys 107:094903

Sajzew R, Wondraczek L (2021) Thermal strengthening of low-expansion glasses, thin-walled glass products by ultrafast heat extraction. J Am Ceram Soc 104:3187–3197

Salem JA, Jenkins MG (1999) A test apparatus for measuring the biaxial strength of brittle materials. Exp Tech 23:19–23

Salmon PS, Huang L (2017) Impact of pressure on the structure of glass, its material properties. MRS Bull 42:734–737

Sangwal K (2009) Review: Indentation size effect, indentation cracks, microhardness measurement of brittle crystalline solids - some basic concepts, trends. Cryst Res Technol 44:1019–1037

Sani G, Limbach R, Dellith J, Sökmen, İ, Wondraczek L (2021) Surface damage resistance, yielding of chemically strengthened silicate glasses: From normal indentation to scratch loading. J Am Ceram Soc 104:3167–3186

Sawamura S, Limbach R, Behrens H, Wondraczek L (2018) Lateral deformation and defect resistance of compacted silica glass: Quantification of the scratching hardness of brittle glasses. J Non-Cryst Solids 481:503–511

Sawamura S, Wondraczek L (2018) Scratch hardness of glass. Phys Rev Mater 2:092601

Sawamura S, Limbach R, Wilhelmy S, Koike A, Wondraczek L (2019) Scratch-induced yielding, ductile fracture in silicate glasses probed by nanoindentation. J Am Ceram Soc 102:7299–7311

Schirmacher W (2006) Thermal conductivity of glassy materials, the "Boson peak". Europhys Lett 73:892–898

Schoenholz SS, Cubuk ED, Sussman DM, Kaxiras E, Liu AJ (2016) A structural approach to relaxation in glassy liquids. Nat Phys 12:469–471

Sciortino F (2005) Potential energy landscape description of supercooled liquids, glasses. J Stat Mech Theor Exp 2005:P05015

Seddon AB (1997) Chalcohalides: glass-forming systems and progress in application of percolation theory. J Non-Cryst Solids 213–214:22–29

Sehgal J, Ito S (2005) A new low-brittleness glass in the soda-lime-silica glass family. J Am Ceram Soc 81:2485–2488

Sellappan P, Rouxel T, Celarie F, Becker E, Houizot P, Conradt R (2013) Composition dependence of indentation deformation, indentation cracking in glass. Acta Mater 61:5949–5965

Sen A, Srivastava M (1990) Regression Analysis: Theory, Methods, and Applications. Springer-Verlag

Serbena FC, Mathias I, Foerster CE, Zanotto ED (2015) Crystallization toughening of a model glass-ceramic. Acta Mater 86:216–228

Shand EB (1959) Breaking stress of glass determined from dimensions of fracture mirrors. J Am Ceram Soc 42:474–477

Sharon E, Fineberg J (1999) Confirming the continuum theory of dynamic brittle fracture for fast cracks. Nature 397:333–335

Shen L, Cheong WCD, Foo YL, Chen Z (2012) Nanoindentation creep of tin and aluminium: A comparative study between constant load and constant strain rate methods. Mater Sc Eng A 532:505–510

Shetty DK, Rosenfield AR, Dukworth WH, Held PR (1983) A biaxial-flexure test for evaluating ceramic strengths. J Am Ceram Soc 66:36–42

Shi Y, Tandia A, Deng B, Elliott SR, Bauchy M (2020) Revisiting the Makishima–Mackenzie model for predicting the young's modulus of oxide glasses. Acta Mater 195:252–262

Shinkai N (1994) The fracture, fractography of flat glass. *In*: Fractography of Glass, Springer US, p 253–297

Shinkai N, Bradt RC, Rindone GE (1981) Fracture toughness of fused SiO_2, float glass at elevated temperatures. J Am Ceram Soc 64:426–430

Sidebottom DL, Schnell SE (2013) Role of intermediate-range order in predicting the fragility of network-forming liquids near the rigidity transition. Phys Rev B 87:054202

Silling SA (2000) Reformulation of elasticity theory for discontinuities, long-range forces. J Mech Phys Solids 48:175–209

Silling SA, Lehoucq RB (2008) Convergence of peridynamics to classical elasticity theory. J Elast 93:13–37

Silling SA, Epton M, Weckner O, Xu J, Askari E (2007) Peridynamic states, constitutive modeling. J Elast 88:151–184

Smedskjaer MM (2014) Indentation size effect, the plastic compressibility of glass. Appl Phys Lett 104:251906

Smedskjaer MM, Mauro JC, Yue Y (2010a) Prediction of glass hardness using temperature-dependent constraint theory. Phys Rev Lett 105:115503

Smedskjaer MM, Mauro JC, Sen S, Yue Y (2010b) Quantitative design of glassy materials using temperature-dependent constraint theory. Chem Mater 22:5358–5365

Smedskjaer MM, Mauro JC, Sen S, Deubener J, Yue Y (2010c) Impact of network topology on cationic diffusion, hardness of borate glass surfaces. J Chem Phys 133:154509

Smedskjaer MM, Saxton SA, Ellison AJ, Mauro JC (2012) Photoelastic response of alkaline earth aluminosilicate glasses. Optics Lett 37:293

Sørensen SS, Pedersen EJ, Paulsen FK, Adamsen IH, Laursen JL, Christensen S, Johra H, Jensen LR, Smedskjaer MM (2020) Heat conduction in oxide glasses: Balancing diffusons, propagons by network rigidity. Appl Phys Lett 117:031901

Stebbins JF, Wu J, Thompson LM (2013) Interactions between network cation coordination and non-bridging oxygen abundance in oxide glasses and melts: Insights from NMR spectroscopy. Chem Geol 346:34–46

Sun K-H (1947) Fundamental condition of glass formation. J Am Ceram Soc 30:277–281

Sundararaman S, Huang L, Ispas S, Kob W (2018) New optimization scheme to obtain interaction potentials for oxide glasses. J Chem Phys 148:194504

Tabor D (1956) The physical meaning of indentation, scratch hardness. Br J Appl Phys 7:159–166

Tabor D (1970) The hardness of solids. Rev Phys Technol 1:145–179

Tandia A, Onbasli MC, Mauro JC (2019) Machine learning for glass modeling. *In*: Musgraves JD, Hu J, Calvez L (Eds.), Springer Handbook of Glass, Springer, Cham, p 1157–1192

Tang, L, Krishnan, NM A, Berjikian, J, Rivera, J, Smedskjaer, MM, Mauro, JC, Zhou, W, Bauchy, M (2018) Effect of nanoscale phase separation on the fracture behavior of glasses: Toward tough, yet transparent glasses. Phys Rev Mater 2:113602

Tanguy A, Wittmer, JP, Leonforte, F, Barrat, J-L (2002) Continuum limit of amorphous elastic bodies: A finite-size study of low-frequency harmonic vibrations. Phys Rev B 66:174205

Taylor WE (1949) Plastic deformation of optical glass. Nature 163:323–323

Thomson, R, Hsieh, C, Rana, V (1971) Lattice trapping of fracture cracks. J Appl Phys 42:3154–3160

Timoshenko S, Woinowsky-Krieger S (1959) Theory of plates and shells. McGraw-Hill

To T, Célarié F, Roux-Langlois C, Bazin A, Gueguen Y, Orain H, Le Fur M, Burgaud V, Rouxel, T (2018) Fracture toughness, fracture energy, slow crack growth of glass as investigated by the Single-Edge Precracked Beam (SEPB), Chevron-Notched Beam (CNB) methods. Acta Mater 146:1–11

To T, Sørensen SS, Stepniewska M, Qiao A, Jensen LR, Bauchy M, Yue Y, Smedskjaer MM (2020) Fracture toughness of a metal-organic framework glass. Nat Commun 11:2593

Turnbull D, Cohen MH (1961) Free-volume model of the amorphous phase: glass transition. J Chem Phys 34:120–125

Urata S, Ando R, Ono M, Hayashi Y (2018) Molecular dynamics study on nano-particles reinforced oxide glass. J Am Ceram Soc 101:2266–2276

Varshneya AK (2010) Chemical strengthening of glass: lessons learned, yet to be learned. Int J Appl Glass Sci 1:131–142

Wada M, Furukawa H, Fujita K (1974) Crack resistance of glass on vickers indentation. Proc Xth Int Congr Glass, 39

Wagner H, Bedorf D, Küchemann S, Schwabe M, Zhang B, Arnold W, Samwer K (2011) Local elastic properties of a metallic glass. Nat Mater 10:439–442

Wang B, Wang H, Lee YJ, Bauchy M (2015) Intrinsic nano-ductility of glasses: the critical role of composition. Front Mater 2:11

Wang B, Yu Y, Wang M, Mauro JC, Bauchy M (2016) Nanoductility in silicate glasses is driven by topological heterogeneity. Phys Rev B 93:064202

Weibull W (1939a) A Statistical Theory of the Strength of Material. Ingenioers Vetenskaps Akademiens Handligar Report 151

Weibull W (1939b) The Phenomenon of Rupture in Solids. Ingenioers Vetenskaps Akademiens Handligar Report 153

Weinan E, Ming P (2007) Cauchy–Born Rule, the stability of crystalline solids: static problems. Arch Ration Mech Anal 183:241–297

Weiner, JH (2002) Corresponding concepts in thermodynamics, statistical mechanics. *In:* Statistical Mechanics of Elasticity (2nd ed.), Dover Publications, p 85–126

Westergaard HM (1939) Bearing pressures and cracks. J Appl Mech 6:A49–A53

Wiederhorn SM (1967) Influence of water vapor on crack propagation in soda-lime glass. J Am Ceram Soc 50:407–414

Wiederhorn SM (1984) Brittle fracture, toughening mechanisms in ceramics. Annu Rev Mater Sci 14:373–403

Wiederhorn SM, Bolz LH (1970) Stress corrosion, static fatigue of glass. J Am Ceram Soc 53:543–548

Wiederhorn SM, Johnson H, Diness AM, Heuer AH (1974a) Fracture of glass in vacuum. J Am Ceram Soc 57:336–341

Wiederhorn SM, Evans AG, Fuller ER, Johnson H (1974b) Application of fracture mechanics to space-shuttle windows. J Am Ceram Soc 57:319–323

Wiederhorn SM, Freiman SW, Fuller ER, Simmons CJ (1982) Effects of water, other dielectrics on crack growth. J Mater Sci 17:3460–3478

Wiederhorn SM, Fett T, Rizzi G, Hoffmann MJ, Guin J-P (2013) The effect of water penetration on crack growth in silica glass. Eng Fract Mech 100:3–16

Wilkinson CJ, Zheng Q, Huang L, Mauro JC (2019) Topological constraint model for the elasticity of glass-forming systems. J Non-Cryst Solids: X 2:100019

Winkelmann A, Schott O (1894) Ueber die Elastizität und über die Zug- und Druckfestigkeit verschiedener neuer Gläser in ihrer Abhängigkeit von der chemischen Zusammensetzung. Ann Phys 287:697–729

Winterstein-Beckmann A, Möncke D, Palles D, Kamitsos EI, Wondraczek L (2014a) Raman spectroscopic study of structural changes induced by micro-indentation in low alkali borosilicate glasses. J Non-Cryst Solids 401:110–114

Winterstein-Beckmann A, Möncke D, Palles D, Kamitsos EI, Wondraczek L (2014b) A Raman-spectroscopic study of indentation-induced structural changes in technical alkali-borosilicate glasses with varying silicate network connectivity. J Non-Cryst Solids 405:196–206

With G, Wagemans HHM (1989) Ball-on-ring test revisited. J Am Ceram Soc 72:1538–1541

Wittmer JP, Tanguy A, Barrat J-L, Lewis L (2002) Vibrations of amorphous, nanometric structures: When does continuum theory apply? Europhys Lett 57:423–429

Wondraczek L (2019) Overcoming glass brittleness. Science 366:804–805

Wondraczek L, Dittmar A, Oelgardt C, Celarie F, Ciccotti M, Marliere C (2006) Real-time observation of a non-equilibrium liquid condensate confined at tensile crack tips in oxide glasses. J Am Ceram Soc 89:746–749

Wondraczek L, Mauro JC, Eckert J, Kühn U, Horbach J, Deubener J, Rouxel T (2011) Towards ultrastrong glasses. Adv Mater 23:4578–4586

Wondraczek L, Bouchbinder E, Ehrlicher A, Mauro JC, Sajzew R, Smedskjaer MM (2022) Advancing the mechanical performance of glasses: Perspectives and challenges. Adv Mater 34:2109029

Wright AC, Thorpe MF (2013) Eighty years of random networks. Phys Status Solidi B)250:931–936

Wu J, Deubener J, Stebbins JF, Grygarova L, Behrens H, Wondraczek L, Yue Y (2009) Structural response of a highly viscous aluminoborosilicate melt to isotropic, anisotropic compressions. J Chem Phys 131:104504

Xu Z-H, Li X (2008) Residual stress determination using nanoindentation technique. *In:* Yang F, Li J (Eds.) Micro, Nano Mechanical Testing of Materials and Devices, Springer Boston, p 139–153

Ya M, Deubener J, Yue Y (2008) Enthalpy, anisotropy relaxation of glass fibers. J Am Ceram Soc 91:745–752

Yamane M, Mackenzie JD (1974) Vicker's hardness of glass. J Non-Cryst Solids 15:153–164

Yan X, Su XG (2009) Linear Regression Analysis. World Scientific

Yang Y, Li W, Tang W, Li B, Zhang D (2019a) Sample sizes based on Weibull distribution, normal distribution for FRP tensile coupon test. Materials 12:126

Yang K, Yang B, Xu X, Hoover C, Smedskjaer MM, Bauchy M (2019b) Prediction of the Young's modulus of silicate glasses by topological constraint theory. J Non-Cryst Solids 514:15–19

Yang K, Xu X, Yang B, Cook B, Ramos H, Krishnan NMA, Smedskjaer MM, Hoover C, Bauchy M (2019c) Predicting the young's modulus of silicate glasses using high-throughput molecular dynamics simulations, machine learning. Sci Rep 9:8739

Yasui I, Imaoka M (1982) Finite element analysis of identation on glass (II). J Non-Cryst Solids 50:219–232

Yoffe EH (1982) Elastic stress fields caused by indenting brittle materials. Philos Mag A 46:617–628

Yoshida S (2019) Indentation deformation and cracking in oxide glass–toward understanding of crack nucleation. J Non-Cryst Solids X 1:100009

Yoshida S, Tanaka H, Hayashi T, Matsuoka J, Soga N (2001) Scratch resistance of sodium borosilicate glass. J Ceram Soc Jpn 109:511–515

Yoshida S, Sanglebœuf J-C, Rouxel T (2005) Quantitative evaluation of indentation-induced densification in glass. J Mater Res 20:3404–3412

Yoshida S, Wada K, Fujimura T, Yamada A, Kato M, Matsuoka J, Soga N (2016) Evaluation of sinking-in and cracking behavior of soda-lime glass under varying angle of trigonal pyramid indenter. Front Mater 3:54

Yoshida S, Nguyen TH, Yamada A, Matsuoka J (2019) In-situ Raman measurements of silicate glasses during Vickers indentation. Mater Trans 60:1428–1432

Zachariasen WH (1932) The atomic arrangement in glass. J Am Chem Soc 54:3841–3851

Zallen R (1998) The percolation model. *In*: The Physics of Amorphous Solids, p 135–204, Wiley-VCH Verlag GmbH

Zamanzadeh M, Larkin ES, Bayer GT, Linhart WJ (2007) Failure analysis and investigation methods for boiler tube failures. Corrosion 2007 Conference and Expo

Zeng H, Jiang Q, Liu Z, Li X, Ren J, Chen G, Liu F, Peng S (2014) Unique sodium phosphosilicate glasses designed through extended topological constraint theory. J Phys Chem B 118:5177–5183

Zhang P, Maldonis JJ, Liu Z, Schroers J, Voyles PM (2018) Spatially heterogeneous dynamics in a metallic glass forming liquid imaged by electron correlation microscopy. Nat Commun 9:1129

Zhang Z, Ispas S, Kob W (2020) The critical role of the interaction potential and simulation protocol for the structural and mechanical properties of sodosilicate glasses. J Non-Cryst Solids 532:119895

Zhao J-H, Tellkamp J, Gupta V, Edwards D (2008) Experimental evaluations of the strength of silicon die by 3-point-bend versus ball-on-ring tests. 2008 11th Intersociety Conference on Thermal and Thermomechanical Phenomena in Electronic Systems, p 687–694

Zheng Q, Potuzak M, Mauro JC, Smedskjaer MM, Youngman RE, Yue Y (2012) Composition–structure–property relationships in boroaluminosilicate glasses. J Non-Cryst Solids 358:993–1002

Zheng Q, Yue Y, Mauro JC (2017) Density of topological constraints as a metric for predicting glass hardness. Appl Phys Lett 111:011907

Zhu F, Ji X, He P, Zheng B, Zhang K (2019) On stress singularity at crack tip in elasticity. Results Phys 13:102210

Zienkiewicz OC, Taylor RL, Zhu JZ (2013) The Finite Element Method: its Basis and Fundamentals. Butterworth-Heinemann

Reviews in Mineralogy & Geochemistry
Vol. 87 pp. 283-337, 2022
Copyright © Mineralogical Society of America

7

Diffusion in Melts and Magmas

Youxue Zhang, Ting Gan

Department of Earth and Environmental Sciences
The University of Michigan
Ann Arbor, MI 48109-1005, USA

youxue@umich.edu; ganting@umich.edu

INTRODUCTION

Diffusion results from random motion of particles and entities. Diffusion in melts and magmas is due to thermally excited random motion of atoms, ions, and clusters, and plays a critical role in magmatic and volcanic processes. In melts and magmas, diffusion is one of the two mechanisms of mass transfer; the other being bulk flow (referred to as convection or advection). When both are present, diffusion refers to the dispersive motion relative to the mean bulk flow in a given reference frame (Richter et al. 1998). Diffusion plays critical roles in controlling magma mixing (Watson 1982; Koyaguchi 1985, 1989; Lesher 1994; Huber et al. 2009; Guo and Zhang 2020), mineral growth and dissolution rates in magmas (e.g., Watson 1982; Harrison and Watson 1983; Zhang et al. 1989; Newcombe et al. 2014; Macris et al. 2018), bubble growth and dissolution rates in magmas (Sparks 1978; Proussevitch and Sahagian 1998; Liu and Zhang 2000; Zhang 2013), and elemental and isotope fractionation during mineral growth and dissolution (Jambon 1980; Richter et al. 1999, 2003; Watson and Muller 2009; Chopra et al. 2012; Watkins et al. 2014, 2017; Holycross and Watson 2016, 2018). As a result, diffusion also plays an essential role in explosive volcanic eruptions and magma crystallization. Furthermore, diffusion has important applications in geospeedometry (Lasaga 1983, 1998; Zhang 1994, 2008; Trail et al. 2016; Zhang and Xu 2016).

Experimental investigation of diffusion in geologically relevant silicate melts began to flourish in the 1970's when micro-analytical measurements of diffusion profiles became available. (A summary of measurement techniques of diffusion profiles can be found in Cherniak et al. 2010.) In addition to the vast number of papers published since then, numerous books and reviews are available for diffusion in silicate melts. Hofmann et al. (1974) edited a book titled "Geochemical Transport and Kinetics" published by Carnegie Institution of Washington. This was the first landmark book summarizing the field. Lasaga and Kirkpatrick (1981) edited a book "Kinetics of Geochemical Processes" as volume 8 of the *Reviews in Mineralogy* (later becoming *Reviews in Mineralogy and Geochemistry*) series. Zhang and Cherniak (2010) edited "*Diffusion in Minerals and Melts*" as volume 72 of *Reviews of Mineralogy and Geochemistry* series, in which one chapter focused on diffusion theory, five chapters on diffusion in silicate melts (Behrens 2010; Lesher 2010; Liang 2010; Zhang and Ni 2010; Zhang et al. 2010), and other chapters were on experimental, analytical, and computational methods, and diffusion in minerals. Several textbooks covered the principles and applications of diffusion theories (Kirkaldy and Young 1987; Shewmon 1989; Cussler 1997; Lasaga 1998; Zhang 2008; Vrentas and Vrentas 2016), and two classic books covered the mathematics of diffusion (Carslaw and Jaeger 1959; Crank 1975). In preparing for this review chapter, we thought carefully about what to cover for this vast field, and decided to briefly go through the fundamentals of diffusion (more complete review can be found in Chakraborty 1995; Zhang 2008, 2010) and solutions to often-encountered diffusion problems, and then focus on post-2010 diffusion studies on silicate melts and magmas. Here, melts refer to (mostly natural) silicate liquid, and magmas refer to crystal-bearing and/or bubble-bearing melts in which the continuous phase is the melt. There is a large body of work on diffusion in glasses, especially in the materials science literature, which is not covered in this review.

1529-6466/22/0087-0007$10.00 (print)
1943-2666/22/0087-0007$10.00 (online)
http://dx.doi.org/10.2138/rmg.2022.87.07

FUNDAMENTALS OF DIFFUSION

Fick's laws

In a compositionally homogeneous phase, diffusion (thermally activated random motion of atoms) is present but would not result in measurable changes in the phase unless the phase is thermodynamically unstable. When there are concentration differences in the phase, diffusion tends to erase these differences and homogenize the composition. The rate at which diffusion proceeds to homogenize a phase is characterized by two Fick's laws. By analogy to Fourier's law that describes the heat flux to be proportional to the temperature gradient, the first Fick's law describes diffusive flux to be proportional to the concentration gradient. In one-dimensional space, it takes the following form:

$$\mathbf{J} = -D\frac{\partial C}{\partial x} \tag{1}$$

where $\mathbf{J}$ is diffusive flux along x direction, D is the diffusion coefficient (or diffusivity) in m^2/s, C is concentration in kg/m^3 or mol/m^3, and $\partial C/\partial x$ is the concentration gradient along x. Symbols are summarized in Table 1. In three dimensions, Fick's first law takes the following form:

$$\mathbf{J} = -D\nabla C \tag{2}$$

where ∇C is the concentration gradient. Melts and magmas considered in this chapter are isotropic media, and hence D does not depend on directions. Therefore, D is a scalar in this chapter (in minerals, D is in general a second-order tensor; Zhang 2010). Values of D in silicate melts are typically of the order 10^{-12} m^2/s, and hence $\mu m^2/s$ ($=10^{-12}$ m^2/s) is often used as the unit of D in this chapter, where it is convenient.

Fick's first law describes the mass flux due to diffusion, and cannot be directly used to calculate how concentrations in a phase would change with time. By incorporating mass conservation into Fick's first law, it is possible to derive Fick's second law. In one-dimensional diffusion, Fick's second law takes the following form:

$$\frac{\partial C}{\partial t} = \frac{\partial}{\partial x}(D\frac{\partial C}{\partial x}) \tag{3a}$$

If D is independent of concentration and distance, the above equation becomes:

$$\frac{\partial C}{\partial t} = D\frac{\partial^2 C}{\partial x^2} \tag{3b}$$

In three dimensions, Fick's second law takes the following form:

$$\frac{\partial C}{\partial t} = \nabla(D\nabla C) \approx D\nabla^2 C \tag{4}$$

Equations (3) and (4) are often referred to as the diffusion equation. Given initial and boundary conditions, Equation (3) or (4) can be solved to determine changes of the concentration in space and time (Carslaw and Jaeger 1959; Crank 1975). Note that even though C in Equations (1) and (2) are in the unit of kg/m^3 or mol/m^3, C in Equations (3) and (4) can also be in other units such as mass fraction, or mass ppm as long as the mass density is roughly constant, or mole fraction if the molar density is roughly constant. To avoid confusion, w rather than C will be used when mass fraction of mass ppm is used as concentration (Table 1).

Table 1. Symbols

$\mathbf{A}$	diagonal matrix in multicomponent diffusion solutions
a	radius, also a parameter for SiO_2 or H_2O_m diffusivity
$C_{i,j}$	concentration (in kg/m^3 or mol/m^3) of component i in phase j
C_{ave}	weighted average concentration in a multi-phase system; $C_{i,\text{ave}} = \phi_1 C_{i,1} + \phi_2 C_{i,2} + \ldots$
D	diffusivity, a scalar in melts, glasses, and magmas containing random crystals
D_0	pre-exponential factor for diffusion in the Arrhenius relation
$D_{w=0}$	diffusivity of a component when its own concentration approaches zero
D_H	diffusivity of the heavy isotope
D_L	diffusivity of the light isotope
$D_{i,j}$	diffusivity of component i in phase j
D_{bulk}	bulk diffusivity in a multiphase media, defined by $\mathbf{J}_{i,\text{bulk}} = -D_{i,\text{bulk}} \, C_{i,\text{ave}}$
D_{eff}	effective diffusivity in crystal-bearing and/or bubble-bearing magmas
$\mathbf{D}$	diffusivity matrix
E	activation energy for diffusion in the Arrhenius relation
$\mathbf{J}$	diffusion flux (a vector)
K	partition coefficient, $K = C_1/C_2 = w_1\rho_1/(w_2\rho_2)$; also equilibrium constant
L	thickness; also dissolution distance
m_H	atomic mass of a heavy isotope
m_L	atomic mass of a light isotope
M_H	molecular mass of a molecule containing the heavy isotope
M_L	molecular mass of a molecule containing the light isotope
$M_{i,j}$	diffusion mobility coefficient of component i in phase j (in ideal systems, $M_{i,j} = D_{i,j}C_{i,j}$)
N	number of components in a system
n	used in multicomponent diffusion in which $n = N-1$
P	pressure
$\mathbf{P}$	eigenvector matrix
$\mathbf{Q}$	diagonal matrix in multicomponent diffusion solutions
R	universal gas constant (8.31447 J mol^{-1} K^{-1})
r	radial coordinate
T	temperature (in K)
t	time
W_i	atomic mass of component i (in kg/mol)
$w_{i,j}$	mass fraction (concentration) of component i in phase j
X_i	mole fraction of component i in the gas phase, also cation mole fraction of i in a melt
x, y, z	spatial coordinate along x-direction, y-direction and z-direction
x_c	characteristic diffusion distance
α	a dimensionless parameter for calculating dissolution distance L
β	an empirical fit parameter to relate diffusivity of heavy and light isotopes to their masses
ΔM	mass gain or loss
ϕ_j	volume fraction of phase j
Λ	diagonal matrix made of eigenvalues
λ_i	the ith eigenvalue
$\mu_{i,j}$	chemical potential (in J/mol) of component i in phase j
ρ_j	density of phase j
σ_j	electric conductivity of phase j. Also standard deviation.

Various kinds of diffusion and diffusivities

Numerous kinds of diffusion have been defined and discussed in the literature, and the definitions are not always consistent. Below is a summary of the many types of diffusion, often encountered in the geological literature.

Based on geometry, diffusion may be classified as one-dimensional, two-dimensional and three-dimensional diffusion. Based on the types of the diffusing material, there can be isotropic (melts, liquids, glasses and magmas and cubic symmetry minerals) or anisotropic diffusion (diffusion in lower-symmetry minerals). Based on the diffusing component or species, diffusion may be classified as follows:

- **Self diffusion.** Strictly speaking, self diffusion means the diffusion of the exact same species in a homogeneous system, not even with isotopic differences. Such self diffusion can only be computationally studied (e.g., De Koker and Stixrude 2010), but cannot be measured analytically. In practice, measured self diffusivity means diffusion of different isotopes in an otherwise chemically homogeneous system (e.g., Liang et al. 1996a; Richter et al. 1999, 2003; Watkins et al. 2014). Self diffusion of a given isotope at constant temperature and pressure can always be well characterized by a constant diffusivity. Note that different isotopes of the same element diffuse at slightly different rate, leading to isotope fractionation (e.g., Richter et al. 1999, 2003; Watkins et al. 2017) to be discussed in a later section.

- **Tracer diffusion.** In mostly early (1970s and 1980s) experimental studies, a tracer (often a radioactive isotope such as ^{86}Rb, Jambon and Carron 1976) is deposited on the surface of a glass of initially uniform composition. The sample is then heated to high temperature to allow the tracer to diffuse into the sample. Such diffusion is termed tracer diffusion. Tracer diffusion can often be characterized by a constant diffusivity.

- **Trace element diffusion without major element concentration gradients.** More recently (1990s and forward), trace element diffusion is often investigated using diffusion couple experiments (e.g., Mungall et al. 1999; Behrens and Hahn 2009; Holycross and Watson 2016, 2018), with the two sides of the diffusion couple having roughly the same chemical composition except for a trace element or multiple trace elements (at < 1000 ppm level) whose diffusivities are probed. These trace element diffusivities are expected to be similar to radioactive tracer diffusivities. To distinguish from trace element diffusion in the presence of major element concentration gradients, this type of trace element diffusion will be referred to as TED1 (trace element diffusion 1).

- **Chemical diffusion.** This category includes all other kinds of diffusion. Chemical diffusion occurs when there are major concentration gradients (or more precisely chemical potential gradients). If there are only two components in the system, the chemical diffusion is *binary diffusion* (also referred to as *mutual diffusion*). Binary diffusivity usually depends on composition. Diffusion in a system of three or more components is referred to as *multicomponent diffusion*. (If there is only one component, then it is self diffusion and cannot be measured.) To quantify multicomponent diffusion, one single diffusion coefficient is not sufficient. Instead, a *multicomponent diffusion matrix* is necessary, in which the on-diagonal terms characterize the effect of a component on its own diffusion, and the off-diagonal terms characterize the effect of other components on its diffusion. In a multicomponent system, if concentration gradients of only two components exist initially, the diffusion of these two components is referred to as *interdiffusion*. The other components can also show diffusion profiles, which are due to effects of multicomponent diffusion. Diffusion of isotopes in a compositionally heterogeneous system is referred to as *isotope diffusion* (it would be self diffusion if chemically homogeneous). For the diffusion of trace elements (at <1000 ppm concentration level) in a multicomponent system with or without major chemical concentration gradients, it is *trace element diffusion*, which is further distinguished as TED1 (in the absence of major chemical concentration gradients) and TED2 (in the presence of major chemical concentration gradients) in this work. TED1 is expected to be similar to tracer diffusion, whereas TED2 displays all the complexity of multicomponent diffusion including nonmonotonic profiles (Zhang et al. 1989). In a binary or multicomponent system, if one component can be present in multiple species and we consider the diffusion of different species, the diffusion of the component is termed *multi-species diffusion*. During multicomponent diffusion, if we consider the diffusion of only one component and treat all other components as one combined "component", then the diffusion is called *effective binary diffusion* (EBD, which may mean either effective binary diffusion, or effective binary diffusivity). EBD has been further classified into first kind and second kind (Zhang et al. 2010). The first kind of effective binary diffusion (*FEBD*) is when all concentration gradients are due to one component, all other components being diluted by the component. FEBD is similar to tracer diffusion or TED1 except the concentration of the diffusing component can be higher in FEBD. The second kind of effective binary diffusion (*SEBD*) includes all other situations. In this work, we reclassify EBD into principally one-concentration-gradient diffusion (POCGD, same as FEBD), interdiffusion (ID), and other types of EBD (OEBD). EBD treatment can only handle concentration profiles that are monotonic. If a component displays mass motion from low to high concentration leading to a nonmonotonic profile, it is called *uphill diffusion*, which cannot be treated by EBD.

Dependence of D on temperature, pressure, and melt composition

The value of D characterizes the diffusion rate. Hence, it is critical to know D under various conditions, and how it varies with other parameters. Based on experimental studies, it is known that D of a component in silicate melts depends strongly on temperature, weakly on pressure, in a complex manner on the melt composition, and sometimes on its own concentration.

The dependence of D on temperature is well characterized by the Arrhenius relation:

$$D = D_0 e^{-E/(RT)} \qquad (5)$$

where R is the universal gas constant (8.31447 J mol^{-1} K^{-1}), T is temperature in K, E is the activation energy (the energy difference between the activated state and normal state), and D_0 is, for lack of a better term, the pre-exponential factor. D_0 is also the hypothetical diffusivity when $T = \infty$. Even though viscosity of melt-glass has often been found to be and successfully modeled as non-Arrhenian (e.g., Hess and Dingwell 1996; Zhang et al. 2003; Hui and Zhang 2007; Giordano et al. 2008), it is difficult to think of a case where D is unambiguously non-Arrhenian.

The dependence of D on pressure is weaker but also more complicated. A relation including both the temperature and pressure dependence is:

$$D = D_0 e^{-(E + P\Delta V)/(RT)} \qquad (6)$$

where P is pressure and ΔV is the activation volume (the volume difference between the activated state and normal state). In this equation, $P\Delta V$ is an energy term and plays a similar role as the activation every E. However, unlike the activation energy, which is always positive for diffusion, ΔV may be either positive (D decreasing with P) or negative (D increasing with P); it may also change signs as pressure varies. For example, Shimizu and Kushiro (1984) reported that oxygen self diffusivity decreases with pressure in diopside melt (positive ΔV) but increases with pressure in jadeite melt (negative ΔV) at $P \leq 2$ GPa, and Tinker and Lesher (2001) showed that Si and O self diffusivity in dacite melt increases with pressure from 1 to 4 GPa (negative ΔV), and then decreases with further increase of pressure to 5.7 GPa (positive ΔV). Experimental data by Chen and Zhang (2008) indicate that the effective binary diffusivity of MgO in basalt melt is roughly independent of pressure from 0.5 to 1.4 GPa.

Diffusivity in silicate melts depends on the major oxide composition of the melts. For example, diffusivity of an element in dry basalt melt is higher than in dry rhyolite melt at the same temperature and pressure, except for He, Li and Na (Behrens 2010; Henderson et al. 1985; Zhang et al. 2010). The dependence of D on melt composition is complicated and there is no theoretical formulation. Many authors tried to develop empirical relations. Mungall (2002) made great effort to model tracer diffusivity of many elements in silicate melts as a function of viscosity and compositional parameters such as ionic radius r, Z^2/r (where Z is valence), $Al/(Na+K+H)$, and M/O ratio where M is the total number of divalent and univalent cations, and O is total number of oxygen. Later studies (e.g., Behrens and Hahn 2009; Zhang et al. 2010; Yu et al. 2019) evaluated the empirical model of Mungall (2002) and concluded the model may be used as an order of magnitude estimate for tracer diffusivities but not accurate enough for practical applications. Fanara et al. (2017) provided fits of diffusivities and obtained $D\eta^{0.7} = 10^{-9.98}$ for trivalent cations, $D\eta^{0.59} = 10^{-9.42}$ for divalent cations, and $D\eta^{0.13}/r^3 = 10^{-1.76}$ for univalent cations, where η is viscosity in Pa·s, D is in m²/s, and r is ionic radius in angstrom. The equation for the univalent cations does not seem to be correct. These equations do not distinguishing diffusivities of different divalent cations (i.e., treating diffusivities of Mg, Ca, Sr and Ba to be the same) or different trivalent cations (treating diffusivities of REE, Al, Cr^{3+} and Ga^{3+} to be the same), and hence, they at the best would provide an order of magnitude estimate of diffusivities. In addition to these general models, other authors have examined how diffusivity of a given component in a specific system depends on composition using simple and empirical composition parameters, often in the form of $\ln D$ being linear to some concentration (mass fraction or mole fraction), such as H_2O (Behrens and Zhang 2001), or SiO_2 (Watson 1982; Lesher and Walker 1986; Koyaguchi 1989; Macris et al. 2018), or $Si+Al$ (Zhang et al. 2010; Zhang and Xu 2016; Yu et al. 2019), or $ASI = Al/(Na+K+2Ca+2Mg)$ (Behrens 2010). Occasionally, a linear dependence of $\ln D$ on the square root of H_2O concentration seems to fit data best (e.g., Zhang et al. 2010, REE diffusion). Nonetheless, the compositional dependence of diffusivity is still not well quantified due to the large number of components that may affect a given diffusivity in natural silicate melts.

The diffusivity of some components in silicate melts may depend on its own concentration, such as SiO_2 (e.g., Watson 1982; Koyaguchi 1989; Macris et al. 2018), and H_2O (Shaw 1974; Zhang et al. 1991a; Zhang and Behrens 2000; Ni and Zhang 2008, 2018). In the former case, SiO_2 is a major component and controls the melt structure (e.g., degree of polymerization). Hence, the dependence of SiO_2 diffusivity on its own concentration is not surprising, and in fact, Yu et al. (2019) showed that it is $Si+Al$ rather than Si that controls Si diffusion. Hence, the dependence on its own concentration in this case is related to the compositional or structural effect. In the latter case, H_2O diffusivity depends on H_2O concentration due to two factors. One is that H_2O dissolves in silicate melts as two species: molecular H_2O (H_2O_m) and hydroxyl (OH) (Stolper 1982a,b). H_2O_m diffuses more rapidly than OH (Doremus 1969; Zhang et al. 1991a; Ni and Zhang 2018), and the proportion of H_2O_m in total H_2O (H_2O_t) increases as H_2O_t concentration increases (Stolper 1982a,b). This leads to a rough linearity between H_2O_t diffusivity and H_2O_t concentration at low H_2O_t concentrations (< 2 wt%). The second factor is that $\ln D_{H_2O_m}$ (as well as $\ln D$ of many other elements) increases linearly with H_2O_t leading to faster than linear increase between $D_{H_2O_t}$ and H_2O_t (Zhang and Behrens 2000; Ni and Zhang 2008, 2018). Hence, part of the dependence of H_2O diffusivity on its own concentration is due to the speciation of H_2O, and part of it is due to compositional dependence. The diffusion of SiO_2 and H_2O will be discussed further in this chapter.

The relation between self or tracer diffusivity and viscosity has been examined extensively and many famous equations (such as the Stokes–Einstein equation, Einstein 1905, and the Eyring equation, Eyring 1936) of inverse proportionality between diffusivity and viscosity have been developed. Some authors have taken these equations for granted.

However, these equations cannot be applied to the diffusion of most components. For example, self and tracer diffusivities may either increase with melt viscosity (for He, Li, and Na; Henderson et al. 1985; Behrens 2010; Zhang et al. 2010), or decrease with melt viscosity (for most other elements). When self or tracer diffusivity decreases with viscosity, the Stokes–Einstein equation and the Eyring equation still often do not work well (Zhang and Ni 2010; Zhang et al. 2010; Ni et al. 2015). For example, for O diffusion in hydrous silicate melts, the error by either of these equations may be many orders of magnitude (Zhang and Ni 2010). The best applications seem to be the Eyring equation for Si or O self diffusivity in anhydrous silicate melts to within a factor of 3 (e.g., Shimizu and Kushiro 1984; Reid et al. 2001; Tinker et al. 2004). Dingwell (1990) and Fanara et al. (2017) discussed the relations between diffusivity of different ions and viscosity. As discussed earlier, Mungall (2002) and Fanara et al. (2017) made effort to quantify relations between diffusivity of different groups of elements and viscosity. We will not examine diffusivity–viscosity relations further.

SOME USEFUL SOLUTIONS TO THE DIFFUSION EQUATION AND EXPERIMENTAL DESIGNS FOR OBTAINING DIFFUSIVITY

Analytical solutions for some often encountered and relatively simple diffusion problems (Fig. 1) are provided in this section without derivations. Readers interested in the associated derivations are referred to textbooks such as Carslaw and Jaeger (1959), Crank (1975), and Zhang (2008). These solutions are often used in experimental studies of diffusion and can sometimes be applied to treat natural diffusion problems by using approximations and simplifications.

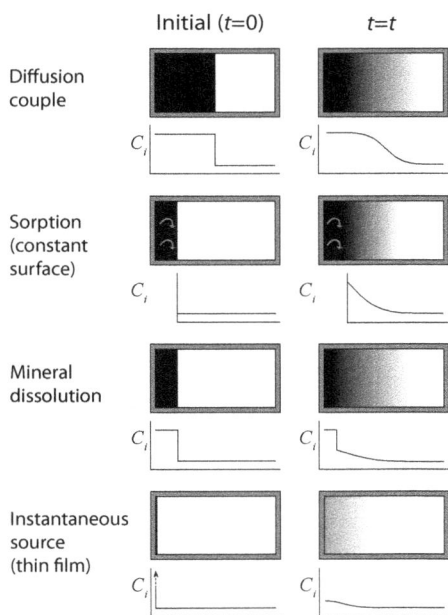

Figure 1. Four diffusion problems that are often encountered in experimental determination of diffusion coefficients and in geological applications. The **left-hand side** shows the initial configuration and the initial concentration profile, and the **right-hand side** shows the effect of diffusion on the distribution of the diffusant and the concentration profile. For the case of diffusion couple setup, the black part means the initial high concentration at $x<0$ (where $x=0$ is the interface). For the case of sorption, the black part means the ambient convecting and uniform gas phase. For the case of mineral dissolution, the black part means the dissolving mineral. For the case of instantaneous source, the initial concentration at the surface (an infinitesimally thin film) is very high as indicated by the arrow. Modified after Watson and Dohmen (2010).

Diffusion couples

When two melts of different compositions (each melt is uniform in composition) are brought into contact in the laboratory or in nature, the diffusion problem is referred to as a diffusion couple (Fig. 1). Define the contact plane to be $x = 0$. Then, one side is at $x < 0$, and the other side is at $x > 0$. Consider the situation when the diffusion distance is small compared to the thickness of the two melts (i.e., diffusion from the interface has not reached the far ends). For self diffusion, binary diffusion with a constant diffusivity, trace or minor element diffusion in a roughly uniform major oxide composition, or for a component in a multicomponent system that can be characterized by a constant effective binary diffusivity, the analytical solution is (Carslaw and Jaeger 1959; Crank 1975):

$$ w = \frac{w_A + w_B}{2} + \frac{w_B - w_A}{2} \mathrm{erf} \frac{x}{\sqrt{4Dt}} \tag{7} $$

where w_A and w_B are the initial mass fraction of the component in melt at $x < 0$ and at $x > 0$, w is the mass fraction of the component at any x and any $t > 0$, and erf is the error function (Carslaw and Jaeger 1959; Crank 1975; Zhang 2008). Equation (7) shows that at a given time t, the diffusion profile (meaning w versus x) is an error function. The diffusion profiles for diffusion couples at $t = 0$ and $t = t$ are shown in Figure. 1. As t increases, the length of the diffusion profile increases. An example of actual experimental data and fit of the data by Equation (7) is shown in Figure. 2.

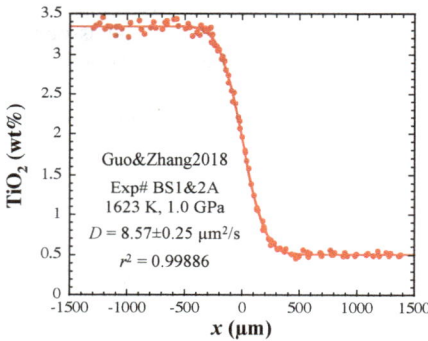

Figure 2. TiO$_2$ diffusion profile from a multicomponent diffusion couple experiment. It is treated as effective binary diffusion with a constant diffusivity. Points are measured data. The flat regions on each side show that diffusion has not reached the far ends. The solid curve is a nonlinear least-squares fit using Equation (7). The fit is excellent, and provides the effective binary diffusivity. Data are from Guo and Zhang (2018).

Sorption or desorption

A gas component may dissolve into or exsolve from a melt or glass that may contain some uniform initial concentration of the component $w_{initial}$. Often the surface concentration of the gas component is fixed by the external gas pressure to be $w_{surface}$ (Sorption in Fig. 1). Define the position of the surface to be $x = 0$. If the diffusivity is constant and diffusion has not reached the far end (if sorption from two parallel surfaces, then diffusion has not reached the center) of the melt or glass, the analytical solution is:

$$ w = w_{surface} + (w_{initial} - w_{surface}) \mathrm{erf} \frac{x}{\sqrt{4Dt}} \tag{8} $$

If the surface concentration is zero (desorption into vacuum), the above equation becomes:

$$ w = w_{initial} \, \mathrm{erf} \frac{x}{\sqrt{4Dt}} \tag{8a} $$

If the initial concentration is zero (sorption), then Equation (8) becomes:

$$ w = w_{surface} \left(1 - \mathrm{erf} \frac{x}{\sqrt{4Dt}}\right) = w_{surface} \, \mathrm{erfc} \frac{x}{\sqrt{4Dt}} \tag{8b} $$

These equations are often used to fit diffusion profiles resulting from sorption or desorption. An example of experimentally generated concentration data with a fit using Equation (8) is shown in Fig. 3. In addition to gas diffusion, isotope diffusion is sometimes accomplished by using an isotopically enriched gas to maintain a constant isotope ratio at the mineral or glass surface and allowing the isotope to diffuse into the solid (e.g., Williams 1965; Ryerson et al. 1989).

In sorption or desorption experiments, sometimes the concentration profile at a given time is not measured due to, e.g., analytical difficulty, but the mass gain or loss of the sample is measured as a function of time. Consider a sample that is a thin plate with uniform thickness L with sorption or desorption from both surfaces. Define ΔM_t and ΔM_∞ to be the amount of the gas component entering (or exiting) the plate at time t and time ∞. When $\Delta M_t/\Delta M_\infty \leq 0.6$, the mass gain or loss can be described by the following equation:

$$\frac{\Delta M_t}{\Delta M_\infty} = \frac{4\sqrt{D}}{L\sqrt{\pi}}\sqrt{t} \tag{9}$$

By plotting $\Delta M_t/\Delta M_\infty$ versus $\sqrt{t}$, one would get a straight line passing through the origin (0,0). Fitting the straight line by a proportionality equation, D can be calculated from the slope.

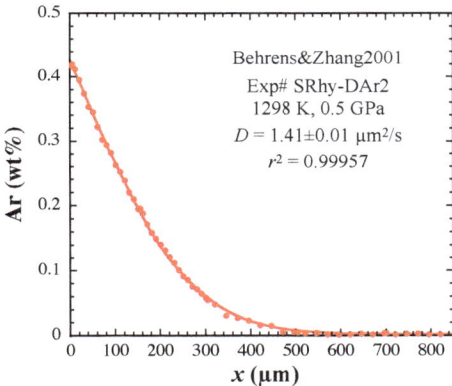

Figure 3. Experimental Ar diffusion profile from an Ar sorption experiment. Points are measured data. The solid curve is a nonlinear least-squares fit using Equation (8). The fit is excellent, and provides the effective binary diffusivity (POCGD). Data are from Behrens and Zhang (2001).

Diffusion in melts during diffusive mineral dissolution

One method to experimentally investigate diffusion in melts is to use crystal dissolution to provide a source for some component (e.g., Harrison and Watson 1983). Often, interface equilibrium between the dissolving crystal and the melt is rapidly reached (Zhang et al. 1989; Liang 2000; Chen and Zhang 2008; Zhang 2008; Yu et al. 2016), meaning that the interface melt composition is fixed, and the dissolving mineral recedes (Mineral dissolution in Fig. 1). Consider the case when convection in the melt can be ignored (e.g., the mineral does not sink in the melt). Assume that the diffusion of a component can be described as by a constant effective binary diffusivity. If the dissolution thickness of the crystal is negligible compared to the diffusion distance, and diffusion has not reached the far end, then the analytical solution for one-dimensional diffusion would be similar to that of the sorption problem Equation (8). If the dissolution thickness is not negligible, the analytical solution for one-dimensional diffusive dissolution is as follows:

$$w = w_{\text{initial}} + (w_{\text{interface}} - w_{\text{initial}})\frac{\text{erfc}\dfrac{(x-L)}{\sqrt{4Dt}}}{\text{erfc}\dfrac{(-L)}{\sqrt{4Dt}}} \tag{10}$$

where $w_{initial}$ and $w_{interface}$ are the initial and interface concentrations in the melt, and L is the growth thickness of the melt, which is related to the dissolution thickness of the crystal (L_c) by $L = L_c(\rho_{crystal}/\rho_{melt})$, where ρ means density, and can be calculated as follows:

$$L = \alpha\sqrt{4Dt} \tag{10a}$$

with α solved from:

$$\frac{(w_{interface} - w_{initial})}{(w_{crystal} - w_{interface})} = \sqrt{\pi}\alpha e^{\alpha^2} \mathrm{erfc}(-\alpha) \tag{10b}$$

where $w_{crystal}$ is the concentration in the crystal. An example of experimental data and a fit to the data is shown in Fig. 4.

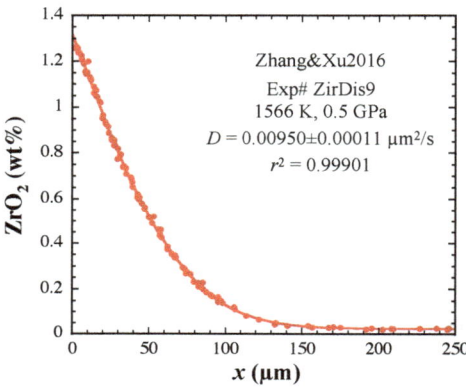

Zhang&Xu2016
Exp# ZirDis9
1566 K, 0.5 GPa
$D = 0.00950 \pm 0.00011\ \mu m^2/s$
$r^2 = 0.99901$

Figure 4. Experimental ZrO_2 diffusion profile from a zircon dissolution experiment. Points are measured data. The solid curve is a nonlinear least-squares fit using Equation (10) in which the melt growth thickness $L = 0.9\ \mu m$ as obtained from experimental data. The fit is excellent, and provides the effective binary diffusivity (POCGD) of Zr. Data are from Zhang and Xu (2016).

Thin-source diffusion

In this method, a fixed (and often undefined) amount of substance (often a radioactive tracer) is deposited on the surface as a thin layer with uniform thickness. Tracer diffusivity is typically constant. If the thin layer (the location is defined as $x = 0$) is sandwiched between two cylinders, then diffusion goes to both directions. Before diffusion reaches the far end, the analytical solution is:

$$C = \frac{M_0}{\sqrt{4\pi Dt}} e^{-x^2/(4Dt)} \tag{11}$$

where C is concentration in kg/m^3 or mol/m^3, and M_0 is deposited mass per unit area (kg/m^2 or mol/m^2). Often M_0 is not known, and the concentration profile is measured at a given time. Hence, the concentration profile would be fit in the following form:

$$C = C_0 e^{-x^2/(4Dt)} \tag{11a}$$

where C_0 is concentration at $x = 0$.

If the thin layer is on the surface of a cylinder and diffusion goes to one direction only (instantaneous source in Fig. 1), then, at a given x, the concentration is two times the concentration given by Equation (11):

$$C = \frac{M_0}{\sqrt{\pi Dt}} e^{-x^2/(4Dt)} = C_0 e^{-x^2/(4Dt)} \tag{12}$$

Measured concentration profiles at a given time t also follow Equation (11a) but C_0 in the case of one-sided diffusion is two times C_0 in the case of two-sided diffusion for a given M_0.

Isotropic diffusion in spheres

Degassing or regassing of a spherical melt or glass belongs to this class of diffusion problems. Melt and glass are isotropic so that D does not vary with diffusion directions. Assume a constant initial concentration (w_{initial}) in the sphere, a constant surface concentration (w_{surface}), and a constant diffusivity D. Then the analytical solution is:

$$\frac{C - C_{\text{surface}}}{C_{\text{surface}} - C_{\text{initial}}} = \frac{2a}{\pi r} \sum_{n=1}^{\infty} \frac{(-1)^n}{n} \sin \frac{n\pi r}{a} e^{-n^2\pi^2 Dt/a^2} \tag{13a}$$

where a is the radius of the sphere, and r is the radial coordinate. The concentration at the center ($r=0$) can be found by

$$\frac{C_{\text{center}} - C_{\text{surface}}}{C_{\text{surface}} - C_{\text{initial}}} = 2\sum_{n=1}^{\infty} (-1)^n e^{-n^2\pi^2 Dt/a^2} \tag{13b}$$

The total amount of mass entering or leaving the sphere is:

$$\frac{\Delta M_t}{\Delta M_\infty} = 1 - \frac{6}{\pi^2} \sum_{n=1}^{\infty} \frac{1}{n^2} e^{-n^2\pi^2 Dt/a^2} \tag{13c}$$

where ΔM_∞ is the final mass gain or loss as t approaches ∞. In other words, ΔM_∞ is the mass gain or loss at equilibrium, and equals $4\pi a^3 (C_{\text{surface}} - C_{\text{initial}})/3$. $\Delta M_t/\Delta M_\infty$ is a measure of how close the system is to equilibrium. If $\Delta M_t/\Delta M_\infty = 0$, then diffusion is just beginning. If $\Delta M_t/\Delta M_\infty = 1$, it means that equilibrium is reached.

Equations (13a–c) converge rapidly for $Dt/a^2 > 0.1$. For smaller Dt/a^2 values, the following three equations may be used for rapid convergence:

$$\frac{C - C_{\text{initial}}}{C_{\text{initial}} - C_{\text{surface}}} = \frac{a}{r} \sum_{n=0}^{\infty} \left[\text{erfc} \frac{(2n+1)a - r}{\sqrt{4Dt}} - \text{erfc} \frac{(2n+1)a + r}{\sqrt{4Dt}} \right] \tag{14a}$$

$$\frac{C_{\text{center}} - C_{\text{initial}}}{C_{\text{initial}} - C_{\text{surface}}} = \frac{2a}{\sqrt{\pi Dt}} \sum_{n=0}^{\infty} e^{-(2n+1)^2 a^2/(4Dt)} \tag{14b}$$

$$\frac{\Delta M_t}{\Delta M_\infty} = \frac{6}{\sqrt{\pi}} \frac{\sqrt{Dt}}{a} \left[1 + 2\sqrt{\pi} \sum_{n=1}^{\infty} \text{ierfc} \frac{na}{\sqrt{Dt}} \right] - 3\frac{Dt}{a^2} \tag{14c}$$

where ierfc is integrated complementary error function. An example of fitting can be found in Zhang (2008, Fig. 3-30a).

Variable diffusivity along a profile

Solutions presented above are all for constant diffusivity along a diffusion profile, which typically happens when the variation in every major oxide concentration is small (e.g., $\Delta w < 4$ wt%). Sometimes, one-dimensional diffusion profiles deviate clearly from error functions and cannot be fit by constant-D solutions. In such cases, there is often no analytical solution. To fit the data, one may guess a relation between D and the composition (e.g., $\ln D$ is linear to concentration, meaning D is an exponential function of the concentration), numerically solve the diffusion problem, and use the numerical solution to fit the experimental diffusion profile (Zhang et al. 1991a; Zhang and Behrens 2000; Yang et al. 2016; Macris et al. 2018; Yu et al. 2019). For example, Fig. 5 shows an SiO_2 diffusion profile during quartz dissolution. Total SiO_2 concentration variation is very large, 50 wt% to about 90 wt%. The effective binary

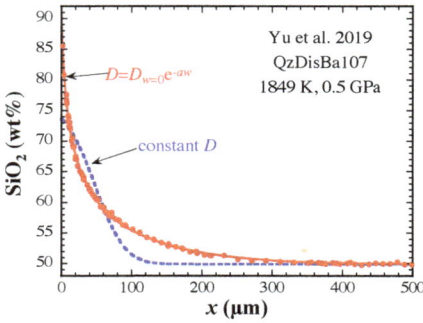

Figure 5. Experimental SiO_2 diffusion profile from a quartz dissolution experiment. **Red points** are measured data. The data indicate very steep slope near the interface ($x = 0$), which descends into a much shallower slope at larger x (e.g., $x = 50$ μm), implying much smaller diffusivity near the interface than in the far-field. The **dashed blue curve** is a nonlinear least-squares fit using constant D (Eqn. 10) in which $L = 34.9$ μm as obtained from experimental data. The fit does not match the data. The **solid red curve** is a nonlinear least squares fit by assuming D_{SiO_2} decreases exponentially as SiO_2 concentration increases. The fit is excellent, and verifies the chosen functional dependence of D_{SiO_2}. Data are from Yu et al. (2019).

diffusivity D_{SiO_2} across the profile is not constant due to such major composition variations. Fitting the concentration profile by a constant D using Equation (10) (blue dashed curve in Fig. 5) does not match the data points well. By assuming that D_{SiO_2} depends exponentially on SiO_2 wt% ($D = D_{w=0}e^{-aw}$, where $D_{w=0}$ and a are two fit parameters and w is wt% of SiO_2), the fit curve (red solid curve) matches the data very well.

If one wishes to examine the relation between D and composition without any bias of a presumed functional form, then Boltzmann analysis may be applied to the diffusion couple problem (Matano 1933: Sauer and Freise 1962), sorption problem, or mineral dissolution problem (Watson 1982; Yu et al. 2019). For a diffusion couple experiment, from the concentration profile $w(x)$ at a given time t, one method to obtain D at a given position x_0 or a given concentration w_{x_0} (w_{x_0} is w at $x = x_0$) is the Boltzmann–Matano method (Matano 1933):

$$D_{x=x_0} = \frac{\int_{w_{x_0}}^{W_\infty} x\,dw}{2t(dw/dx)|_{x=x_0}} \tag{15}$$

where x is distance from the Matano interface, x_0 is the position at which D is calculated, and t is the experimental duration. In using the above equation, it is necessary to first smooth the concentration profile $w(x)$, and also obtain the Matano interface position so that

$$\int_{w_{-\infty}}^{w_\infty} x\,dw = \int_0^\infty (w_\infty - w)dx - \int_{-\infty}^0 (w - w_{-\infty})dx = 0 \tag{16}$$

An alternative Boltzmann method to calculate D at a given position or concentration based on a diffusion couple profile without finding the Matano interface is given by Sauer and Freise (1962):

$$D = \frac{1}{2t(dy/dx)|_{x=x_0}}\left[y_{x_0}\int_{x_0}^\infty (1-y)dx + (1-y_{x_0})\int_{-\infty}^{x_0} y\,dx\right] \tag{17a}$$

where $y = (w-w_{min})/(w_{max}-w_{min})$ so that $y = 0$ at $x = -\infty$ and $y = 1$ at $x = \infty$ (that is, minimum concentration w_{min} is at $x = -\infty$, and maximum concentration w_{max} is at $x = \infty$). If the side of $x > 0$ has lower concentration so that $y = 1$ at $x = -\infty$ and $y = 0$ at $x = \infty$, then the equation becomes:

$$D = \frac{-1}{2t(dy/dx)|_{x=x_0}}\left[(1-y_{x_0})\int_{x_0}^\infty y\,dx + y_{x_0}\int_{-\infty}^{x_0} (1-y)dx\right] \tag{17b}$$

The advantage of the Sauer and Freise (1962) method is that there is no need to find the Matano interface.

For diffusive mineral dissolution experiments, D at a given position x_0 can be calculated using the following equation (Yu et al. 2019):

$$D = \frac{1}{2t(dw/dx)|_{x=x_0}} \left[\int_{w_{x_0}}^{w_\infty} x\,dw + \frac{(w_\infty - w_{x_0})}{(w_c - w_\infty)} \int_{w_0}^{w_\infty} x\,dw \right] \qquad (18)$$

where $w_0 = w|_{x=0}$ is the concentration at the interface melt (note that $x = 0$ is the mineral–melt interface, which is directly measured, rather than calculated as in the case of the Matano interface), and w_c is the concentration of the component in the dissolving crystal.

Diffusion distance and square root of time relation

The analytical solutions (Eqns. 7, 8, 10, and 11) for one-dimensional diffusion typically indicate that concentration depends on $x/(4Dt)^{1/2}$. That is, at a given $x/(4Dt)^{1/2}$, or at $x = 2a(Dt)^{1/2}$ where a is a constant, the concentration is constant regardless of any variation in x and t. Hence, diffusion distance is proportional to $\sqrt{Dt}$. At a given D, the diffusion distance is proportional to square root of time. This is referred to as the square root of time relation, or sometimes the parabolic relation. Often a characteristic distance x_c is roughly defined as

$$x_c \approx \sqrt{Dt} \qquad (19)$$

To be more precise, Zhang (2008) defined the mid-concentration distance to be the distance from the interface at which the concentration is $0.5(w_{interface} + w_{farfield})$. For constant D, the mid-concentration distance x_{mid} for diffusion couple and sorption/desorption can be expressed as (Zhang 2008):

$$x_{mid} = 0.953872\,\sqrt{Dt} \qquad (20)$$

Because diffusion distance is proportional to square root of time, diffusion-controlled processes (such as diffusion-controlled crystal growth, crystal dissolution, oxidation, dehydration, etc.) are often said to follow the parabolic law (t is linear to x^2, e.g., Yu et al. 2016). Conversely, if a process follows the parabolic law, the process is often identified to be diffusion controlled.

MULTICOMPONENT DIFFUSION

Natural silicate melts typically contain 5 to 10 major oxides (≥ 1 wt%) plus minor (0.1 to 1.0 wt%) and trace components (<0.1 wt%). Therefore, diffusion in geological melts is always multicomponent in nature even though usually treated by EBD. The general theory of multicomponent diffusion is well developed. Because the concentration gradient of any one component would affect the diffusive flux of not only itself, but also other components, multicomponent diffusion must be described by a diffusion matrix (De Groot and Mazur 1962; Liang et al. 1997; Zhang 2008, 2010; Liang 2010; Lierenfeld et al. 2019). There are at least two manifestations of multicomponent diffusion compared to binary diffusion. One is uphill diffusion in a stable phase, in which a component diffuses from low concentration to high concentration, resulting in a non-monotonic concentration profile, such as one maximum or minimum during mineral dissolution (Na_2O profile in Fig. 6), or a pair of minimum and maximum in diffusion couples (e.g., see Al_2O_3, FeO, CaO and Na_2O profiles in Fig. 7 later). Applying the effective binary diffusion treatment would fail because the extracted D values would vary from positive to negative, and negative D values are incorrect for stable phases. Another manifestation of multicomponent diffusion is the coordinated motion among many components, resulting in concentration profiles of similar lengths (Fig. 6) for components with widely different self or tracer diffusivities. Coordinated diffusion, with many components showing similar diffusion distances, is often observed when the major concentration gradient is in SiO_2 (Fig. 6). One explanation for coordinated motion of many different components is that a few slowly diffusing major components (such as SiO_2

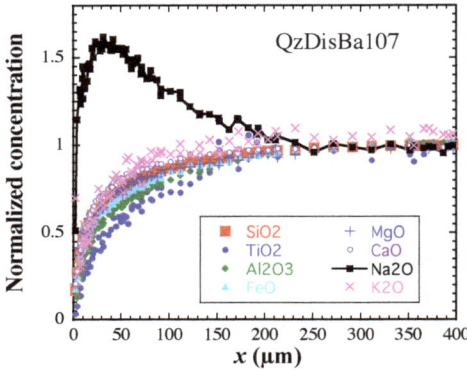

Figure 6. Concentration profiles in the melt during quartz dissolution in basalt (Yu et al. 2019). For easier comparison, the concentration profiles are normalized so that the far-field concentration is 1, and the interface concentration is zero. Na_2O (**black solid squares**) displays obvious uphill diffusion. All other oxides show similar diffusion distance, even though their tracer diffusivity may differ by orders of magnitude. In terms of profile lengths, Ti > Al > Fe ≥ Si ≈ Mg ≥ Ca > K. This sequence is different from the sequence for tracer diffusivities (see Eqn. 28 later).

and Al_2O_3 for aluminosilicate melts) control the chemical potential of other components. The components that diffuse more rapidly redistribute following the chemical potential gradients of the slowly diffusing components, which means more rapidly diffusing components follow the concentration gradients of SiO_2 and Al_2O_3 (Watson 1976, 1982; Zhang 1993), with similar apparent diffusivity. The effect of SiO_2 and Al_2O_3 on the chemical potential and diffusion of other components may be roughly modeled (Zhang 1993). The coordinated motion can still be treated by the effective binary diffusion method even though the extracted EBD can only be applied to diffusion problems with similar concentration gradients and composition.

Liang (2010) provided a thorough review of multicomponent diffusion work. Because the EBD approach is not disappearing anytime soon, especially for minor and trace elements, here we first briefly review and reclassify the effective binary diffusion approach. We then outline the theory of multicomponent diffusion following De Groot and Mazur (1962) and Zhang (2008). Finally, we discuss recent multicomponent diffusion work since the review by Liang (2010).

Effective binary diffusion

Up to a few years ago, the only practical approach in treating diffusion in natural basalt to rhyolite melts, which are multicomponent in nature, is the effective binary diffusion treatment. Cooper (1968) discussed limitations and applications of the effective binary treatment. Although significant progress has been made and we are beginning to use multicomponent diffusion matrix to treat diffusion in basalt (e.g., Guo and Zhang 2018, 2020), our opinion is that we still have a long way to go to treat multicomponent diffusion in numerous natural silicate melts using the diffusion matrix approach. Hence, effective binary diffusion treatment is here to stay in the near future (e.g., next 20 years) in dealing with major element diffusion in natural silicate melts. Furthermore, we are very far from using multicomponent diffusion matrix to treat minor and trace element diffusion. For all these reasons, effective binary diffusion still deserves attention. Rigorously speaking, even tracer diffusion is still in the presence of concentration gradients of other components and hence may be regarded as a kind of effective binary diffusion although the main concentration gradient is in one component (the tracer) only.

When using the effective binary approach, the diffusivity is termed effective binary diffusivity (EBD) or effective binary diffusion coefficient (EBDC). In this approach, the diffusant of interest is treated as one component, and all other components are treated as one combined "component". All solutions to the binary diffusion problems (Eqns. 5–20) are applicable to effective binary diffusion. This treatment can only treat monotonic profiles. For example, Figures 2–5 are all effective binary diffusion profiles. Nonmonotonic profiles, such as Na_2O profile in Figure 6, and Al_2O_3, FeO, CaO, and Na_2O profiles in Figure 8 in a later section, cannot be treated using the effective binary approach. There is a modified effective binary diffusion model (Zhang 1993), which can treat nonmonotonic diffusion profiles, but it has not been much applied.

Because effective binary diffusion covers many different scenarios of diffusion, we suggest that when effective binary diffusivities are mentioned, the type of experiments is included, such as EBD of Zr during zircon dissolution, or EBD of SiO_2 during cassiterite dissolution into a rhyolite melt, etc. Zhang (2010) divided effective binary diffusion into two categories: the first type of effective binary diffusion (abbreviated as FEBD) and the second type of effective binary diffusion (SEBD). In this work, we aim to improve the classification rational, and classify the types of EBD based on how an EBD can be uniquely specified: (i) principally one-concentration-gradient diffusion (POCGD) in multicomponent system, (ii) interdiffusion (ID) in multicomponent system, and (iii) other types of effective binary diffusion (OEBD) in multicomponent system. These are further elucidated below.

POCGD (same as FEBD in Zhang 2010) is the diffusion of a component A into or out of an initially uniform composition (such as sorption, desorption, and thin source diffusion). Other components diffuse mainly in response to the concentration gradient of this component A and their diffusion is typically not considered. POCGD also includes diffusion couples in which the initial concentration gradient is only in a single component A and all other components are the same except for the dilution by component A. When the concentration of the component in POCGD is below 1000 ppm, then it becomes TED1. For example, sorption of Ar into a glass or melt (Carroll 1991; Carroll and Stolper 1991; Behrens and Zhang 2001), hydration or dehydration of a glass or melt or H_2O diffusion couples (Shaw 1974; Zhang et al. 1991a; Zhang and Stolper 1991; Zhang and Behrens 2000; Ni et al. 2013; Ni and Zhang 2018), Zr diffusion in a melt during zircon dissolution into the melt (Harrison and Watson 1983; Zhang and Xu 2016), SiO_2 diffusion in a melt during quartz dissolution into the melt (Watson 1982; Yu et al. 2019), Sn diffusion in a melt during cassiterite dissolution into the melt (Yang et al. 2016), are all examples of POCGD. Diffusivities of POCGD depend only on the bulk composition including the concentration of the diffusing component, but not on other factors (other concentration gradients are all related to the diffusion of the component in consideration). Therefore, when specifying POCGD, one only needs to specify the bulk composition in addition to temperature and pressure. If one is interested in the diffusion of other components (such as Si diffusion in the melt during cassiterite dissolution in a rhyolite), EBD of these other components would be other types of EBD and depend on the major concentration gradients.

Another type of diffusion in the category of effective binary diffusion that is worth special mention is **interdiffusion (ID)**, in which the initial concentration gradients exist only for two compensating components A and B. Because of the motion of other components, effective binary diffusivity of component A may differ from that of B. Components other than A and B typically cannot be treated by effective binary diffusion due to uphill diffusion. To specify an interdiffusivity, it is necessary to include both the bulk composition and the counter-diffusion component, such as interdiffusivity of SiO_2 during SiO_2–K_2O interdiffusion in a basalt, or that of SiO_2 during SiO_2–Al_2O_3 interdiffusion in a basalt. The interdiffusivity of SiO_2 during SiO_2–K_2O interdiffusion in basalt does not necessarily equal to the interdiffusivity of K_2O during SiO_2–K_2O interdiffusion in basalt, or the interdiffusivity of SiO_2 during SiO_2–Al_2O_3 interdiffusion in basalt. For example, interdiffusivity (effective binary diffusivity) of SiO_2 in a haplobasalt2 at 1773 K and 1.0 GPa is 15.7 ± 1.5 μm²/s for SiO_2–Al_2O_3 interdiffusion, and 103 ± 20 μm²/s for SiO_2–K_2O interdiffusion (Guo and Zhang 2016), a variation by a factor of 6. The interdiffusivity of SiO_2 in basalt11a at 1773 K and 1.0 GPa is 6.6 ± 1.6 μm²/s for SiO_2–TiO_2 interdiffusion, and 88 ± 11 μm²/s for SiO_2–K_2O interdiffusion, a variation by a factor of 13 (Guo and Zhang, 2020). The interdiffusivity of CaO in haplobasalt2 is 60 ± 2 μm²/s for SiO_2–CaO interdiffusion, and 116 ± 7 μm²/s for MgO–CaO interdiffusion (Guo and Zhang 2016).

All other types of effective binary diffusivities are more complicated, and are termed, lacking a better term, other types of effective binary diffusion (**OEBD**). Some examples include: SiO_2 diffusion during cassiterite dissolution into a rhyolite melt, Na_2O diffusion during hydration of a melt, Al_2O_3 diffusion during quartz dissolution, diffusion of all components in a basalt–rhyolite diffusion couple or during diopside dissolution into a basalt. Because EBD values depend on directions and relative magnitudes of concentration gradients, specification of the experiments may guide users in choosing the most appropriate EBDs. For example, to model olivine growth in a basalt (Newcombe et al. 2014, 2020), the most appropriate MgO EBD (an OEBD) is that during olivine dissolution in a similar basalt, rather than MgO EBD during diopside dissolution, or MgO EBD in a basalt–rhyolite diffusion couple, or Mg tracer diffusivity or self diffusivity. To model diffusion during mixing of two melts, the most appropriate EBDs are those extracted from diffusion couples made of these two melts.

In terms of applicability, POCGD has the widest applicability. It depends only on the bulk composition (in addition to temperature and pressure). Interdiffusivity depends on both the bulk composition and the counter-diffusion component. Once these are specified, then interdiffusivity is also specified. The other EBDs, or OEBDs, have limited applicability: one must specify the bulk composition as well as concentration gradients to apply. The concentration gradients can be specified in a number of ways, such as MORB–rhyolite diffusion couple, diopside dissolution/growth in a basalt, etc.

Multicomponent diffusion theory

Fick's first law (Eqn. 1) describes diffusive flux in a binary system. In an N-component system ($N \geq 3$), because the summation of concentrations of all components must be 100%, there are $N-1$ independent components. Define the Nth component to be the dependent component, and let $n = N-1$. Because the concentration gradient of any component would contribute to the diffusion of other components, the expanded Fick's law for one-dimensional diffusion takes the following form (the intricacy of the reference frame is not discussed here; interested readers are referred to Brady 1975; Chakraborty 1995; Zhang 2008):

$$\mathbf{J}_1 = -D_{11}^{[N]}\frac{\partial C_1}{\partial x} - D_{12}^{[N]}\frac{\partial C_2}{\partial x} - \cdots - D_{1n}^{[N]}\frac{\partial C_n}{\partial x}$$

$$\mathbf{J}_2 = -D_{21}^{[N]}\frac{\partial C_1}{\partial x} - D_{22}^{[N]}\frac{\partial C_2}{\partial x} - \cdots - D_{2n}^{[N]}\frac{\partial C_n}{\partial x}$$

$$\mathbf{J}_n = -D_{n1}^{[N]}\frac{\partial C_1}{\partial x} - D_{n2}^{[N]}\frac{\partial C_2}{\partial x} - \cdots - D_{nn}^{[N]}\frac{\partial C_n}{\partial x}$$

where $D_{ii}^{[N]}$ characterizes the diffusive flux of component i due to its own concentration gradient $\partial C_i/\partial x$ when the Nth component is used as the dependent component, and $D_{ij}^{[N]}$ ($i \neq j$) characterizes the diffusive flux of component i due to concentration gradient of another component j, $\partial C_j/\partial x$. In other words, $D_{ij}^{[N]}$ ($i \neq j$) describes the cross effect of concentration gradient of component j on the diffusion of component i. In matrix notation, the above set of equations can be written as:

$$\begin{pmatrix} \mathbf{J}_1 \\ \mathbf{J}_2 \\ \vdots \\ \mathbf{J}_n \end{pmatrix} = -\begin{pmatrix} D_{11}^{[N]} & D_{12}^{[N]} & \cdots & D_{1n}^{[N]} \\ D_{21}^{[N]} & D_{22}^{[N]} & \cdots & D_{2n}^{[N]} \\ \vdots & \vdots & \vdots & \vdots \\ D_{n1}^{[N]} & D_{n2}^{[N]} & \cdots & D_{nn}^{[N]} \end{pmatrix}\begin{pmatrix} \partial C_1/\partial x \\ \partial C_2/\partial x \\ \vdots \\ \partial C_n/\partial x \end{pmatrix} = -\mathbf{D}^{[N]}\frac{\partial}{\partial x}\begin{pmatrix} C_1 \\ C_2 \\ \vdots \\ C_n \end{pmatrix} \qquad (21)$$

where $\mathbf{D}^{[N]}$ is referred to as the diffusion matrix, and the superscript $[N]$ means that the Nth component is taken as the dependent component.

Fick's second law in a multicomponent system takes the following form:

$$\frac{\partial}{\partial t}\begin{pmatrix} C_1 \\ C_2 \\ \vdots \\ C_n \end{pmatrix} = \frac{\partial}{\partial x}\left[\mathbf{D}^{[N]} \frac{\partial}{\partial x}\begin{pmatrix} C_1 \\ C_2 \\ \vdots \\ C_n \end{pmatrix}\right] \tag{22}$$

If the **D**-matrix is independent of composition and x, then

$$\frac{\partial}{\partial t}\begin{pmatrix} C_1 \\ C_2 \\ \vdots \\ C_n \end{pmatrix} = \mathbf{D}^{[N]} \frac{\partial^2}{\partial x^2}\begin{pmatrix} C_1 \\ C_2 \\ \vdots \\ C_n \end{pmatrix} \tag{22a}$$

Because melt density is roughly constant, the concentration above may be in either kg/m³, or mass fraction or wt% (w). If a different component k is used as the dependent component, then the concentration vector would be different, $(C_1,..., C_{k-1}, C_{k+1},..., C_N)$, and the **D** matrix would be different. Methods for obtaining $\mathbf{D}^{[k]}$ from $\mathbf{D}^{[N]}$ can be found in Guo and Zhang (2016).

The above diffusion equation can be solved by the diagonalization of **D** using eigenvalues and eigenvectors:

$$\mathbf{D} = \mathbf{P\Lambda P}^{-1} \tag{23}$$

where $\mathbf{\Lambda}$ is a diagonal matrix with each diagonal element λ_i being the eigenvalues, and **P** is the eigenvector matrix, with column j corresponding to eigenvalue λ_j.

A number of analytical solutions have been obtained for the case of constant multicomponent diffusion matrix. For a diffusion couple, before diffusion reaches the far ends, the solution is (Liang, 2010):

$$\mathbf{w} = \frac{\mathbf{w}_A + \mathbf{w}_B}{2} + \mathbf{PQP}^{-1}\frac{\mathbf{w}_B - \mathbf{w}_A}{2} \tag{24}$$

where **w** is a column vector of concentrations, $\mathbf{w}_A$ and $\mathbf{w}_B$ are the initial concentration vectors at $x < 0$ and $x > 0$, **P** is the eigenvector matrix, and **Q** is a diagonal matrix with each diagonal term $Q_{ii} = \mathrm{erf}(x/\sqrt{4\lambda_i t})$ and off-diagonal terms $Q_{ij} = 0$ for $i \neq j$.

For one-dimensional diffusive mineral dissolution before diffusion reaches the far end of the melt, the analytical solution is (Guo and Zhang 2016):

$$\mathbf{w} = \mathbf{w}_{\text{initial}} + \mathbf{PAP}^{-1}(\mathbf{w}_{\text{interface}} - \mathbf{w}_{\text{initial}}) \tag{25}$$

where **P** is the eigenvector matrix, and **A** is a diagonal matrix with $A_{ij} = 0$ if $i \neq j$, and

$$A_{ii} = \frac{\mathrm{erfc}\dfrac{(x-L)}{\sqrt{4\lambda_i t}}}{\mathrm{erfc}\dfrac{(-L)}{\sqrt{4\lambda_i t}}} \tag{26}$$

where L is the melt growth distance (see Eqn. 10). For a discussion of determining $\mathbf{w}_{\text{interface}}$ and L, please refer to Guo and Zhang (2016).

Varshneya and Cooper (1972) used eigenvectors of diffusion matrices to infer exchange mechanisms and also hinted that the eigenvectors might be independent of temperature in ternary SiO_2–SrO–K_2O melts. Chakraborty et al. (1995b) found that diffusion eigenvectors are insensitive to composition in ternary SiO_2–Al_2O_3–K_2O melts and to temperature, and each eigenvalue depends on temperature following the Arrhenius relation and on melt composition. The constancy of eigenvectors and the Arrhenian behavior of eigenvalues would significantly simplify the quantification of multicomponent diffusion. Claireaux et al. (2016, 2019) and Guo and Zhang (2016, 2018, 2020) applied and extended the concepts and approaches to multicomponent diffusion in a quaternary SiO_2–Al_2O_3–CaO–Na_2O melt (NCAS in Table 2), a seven–component haplobasalt (haplobasalt2 in Table 2) and an eight–component basalt (basalt11a in Table 2).

The above summary highlights that the general multicomponent diffusion theory is well developed. The difficulty in applying the theory is in the unavailability of the diffusion matrix. Below we summarize recent efforts to determine the diffusion matrix in aluminosilicate melts.

Recent studies of multicomponent diffusion

There has been major progress in multicomponent diffusion in silicate melts since the review of Liang (2010). Watkins et al. (2014) expanded multicomponent diffusion theory to treat simultaneous isotope diffusion and multicomponent diffusion. Claireaux et al. (2016, 2019) carried out diffusion couple experiments at 1473 to 1633 K to quantify multicomponent diffusion in a quaternary system SiO_2–Al_2O_3–CaO–Na_2O (NCAS in Table 2). Guo and Zhang (2016) studied multicomponent diffusion in a seven component Fe–free haplobasalt SiO_2–TiO_2–Al_2O_3–MgO–CaO–Na_2O–K_2O (haplobasalt2 in Table 2) at 1773 K. Pablo et al. (2017) examined multicomponent diffusion in a ternary sodium borosilicate melt (average composition $68SiO_2$–$18B_2O_3$–$14Na_2O$ by mol%) at 973–1373 K. Guo and Zhang (2018, 2020) investigated multicomponent diffusion in an eight component basalt (basalt11a in Table 2) at 1533 to 1773 K, which has a similar composition to a MORB from Juan de Fuca Ridge except with increased K_2O to resolve the effect of K_2O. The compositions of these silicate melts except for the borosilicate melts are listed in Table 2, and the results from these studies are summarized below.

Table 2. Nominal composition of melts (wt%) investigated for multicomponent diffusion

Melt	SiO_2	TiO_2	Al_2O_3	FeO	MgO	CaO	Na_2O	K_2O	Refs.
NCAS	64.5		11.4			10.8	13.3		1,2
haplobasalt2	50.0	1.50	15.0		10.0	19.0	3.00	1.50	3
basalt11a	51.0	2.00	14.0	11.5	6.5	10.5	3.00	1.50	4,5

References: 1. Claireaux et al. (2016); 2. Claireaux et al. (2019); 3. Guo and Zhang (2016); 4. Guo and Zhang (2018); 5. Guo and Zhang (2020). Effort is made so that the name of each melt is the same or similar to those in Table 1 of Zhang et al. (2010) for easy cross reference. For example, the composition of basalt11a in this Table is similar to that of basalt11 in Table 1 of Zhang et al. (2010).

Multicomponent diffusion in NCAS quaternary system. Claireaux et al. (2016, 2019) investigated multicomponent diffusion in the quaternary system SiO_2–Al_2O_3–CaO–Na_2O (composition NCAS in Table 2) at 1473, 1553 and 1633 K. They obtained the diffusion matrix at each of the three temperatures, and found that the eigenvectors of the three diffusion matrices are similar, and the eigenvalues depend on temperature following the Arrhenius relation, which are consistent with previous studies of multicomponent diffusion in silicate melts of the following compositions: $68\,SiO_2$–$17\,SrO$–$21\,K_2O$ (Varshneya and Cooper 1972), SiO_2–Al_2O_3–CaO (Sugawara et al. 1977; Oishi et al. 1982; Liang et al. 1996b; Liang and Davis 2002), SiO_2–Al_2O_3–K_2O with ~75 wt% SiO_2 (Chakraborty et al. 1995a,b), and SiO_2–$NaAlSi_3O_8$–$KAlSi_3O_8$–H_2O (Mungall et al. 1998). Table 3 lists the three common eigenvectors using SiO_2 as the dependent component, and the Arrhenius equation for calculating the eigenvalues. The eigenvector corresponding to the smallest eigenvalue (λ_1 in Table 3) is mostly the exchange of Al_2O_3 with CaO plus some SiO_2 (the eigenvector component for the Al_2O_3 component is positive, those for CaO and SiO_2 are negative; and 0.06 for Na_2O is considered to be small and negligible here), that to the middle eigenvalue (λ_2 in Table 3) is mostly the exchange of CaO with SiO_2 plus some Al_2O_3, and that to the greatest eigenvalue (λ_3 in Table 3) is mostly the exchange of Na_2O with CaO plus a little SiO_2. These eigenvectors and associated eigenvalues are consistent with expectation that exchange of higher–valence (or network forming) components is slow and that involving lower–valence components is rapid. To calculate the diffusion matrix at a given temperature, one uses Equation (23), in which $\mathbf{P}$ is the three–component eigenvector matrix in Table 3 (by removing the SiO_2 row) and Λ is a diagonal matrix with diagonal elements being λ_1, λ_2 and λ_3.

Table 3. Eigenvectors and eigenvalues for NCAS melt in Table 2 at 1473–1633 K.

Eigenvalues	λ_1	λ_2	λ_3
in m^2/s	$e^{-9.967-29267/T}$	$e^{-6.195-32624/T}$	$e^{-13.697-15541/T}$
Eigenvectors	v_1	v_2	v_3
SiO_2	−0.33	−0.67	−0.10
Al_2O_3	0.83	−0.32	−0.01
CaO	−0.56	0.95	−0.65
Na_2O	0.06	0.04	0.76

Note: Data are from Claireaux et al. (2019). Eigenvalues are arranged by increasing size. SiO_2 is the dependent component. All-component eigenvectors are listed for convenience of examining diffusion exchange mechanisms. The SiO_2 component of each eigenvector is calculated to be the negative sum of all the independent components. The all-component eigenvectors are not unitized. The unitized independent three-component eigenvectors (matrix $\mathbf{P}$ used in Eqns. 23–25) can be obtained by removing the SiO_2 row.

Multicomponent diffusion in haplobasalt2. Guo and Zhang (2016, 2018, 2019a) carried out this study to develop the best strategy for tackling multicomponent diffusion in natural basalt, one of the most common crustal rock types. An Fe-free haplobasalt (haplobasalt2 in Table 2) was chosen. Trial and Spera (1994) suggested that in an N-component system, at least $N-1$ "orthogonal" diffusion couples are required to extract the diffusion coefficient matrix. Hence, for this 7-component system, 6 orthogonal diffusion couples are a minimum. Guo and Zhang (2016) designed the experiments as follows. The haplobasalt2 composition in Table 2 is used as the base composition. Each diffusion couple is made of two halves, in which one half deviates from the base composition by $+1.5$ wt% in component i (often SiO_2) and -1.5 wt% in another component j ($i{\neq}j$), so that the total is 100 wt%, and the other half is opposite, containing 1.5 wt% less in component i, and 1.5 wt% more in component j compared to the base composition. Hence, the number of different glasses with specific compositions that must be prepared is two times the number of diffusion couple experiments. Guo and Zhang (2016) carried out 9 diffusion couple experiments. The first six of them have concentration gradients in SiO_2 and another component, TiO_2, Al_2O_3, MgO, CaO, Na_2O, and K_2O respectively. These six diffusion couples may be regarded as the necessary 6 "orthogonal" couples. Three additional diffusion couple experiments were carried out, with opposing (or interdiffusing) components of TiO_2–MgO, MgO–CaO, CaO–Na_2O. Furthermore, an anorthite dissolution experiment in the base melt composition was carried out. The diffusion matrix is a 6×6 matrix and has been obtained from the first six diffusion couple experiments (which are deemed a minimum) denoted as $\mathbf{D_1}$ matrix (Guo and Zhang 2016), all nine diffusion couple experiments ($\mathbf{D_2}$ matrix; Guo and Zhang 2016), and combined fitting of nine diffusion couple experiments plus one anorthite dissolution experiments ($\mathbf{D_3}$ matrix; Guo and Zhang 2018). With more experiments, the error on the $\mathbf{D}$ matrix is reduced slightly. The mean relative error (here the mean relative error on a matrix is defined to be $\Sigma\sigma_{ij}/\Sigma|D_{ij}|$, summed over all matrix elements) is 7.3% for $\mathbf{D_1}$, 6.3% for $\mathbf{D_2}$, and 5.7% for $\mathbf{D_3}$ (Guo and Zhang 2016, 2018, note that there are corrections; Guo and Zhang 2019a,b). The relative error decreases fairly slowly as the number of experiments increases. A linear extrapolation suggests that 23 experiments at a given temperature would be needed for this 7-component system to reach a mean relative error of $\leq 1\%$. Table 4 shows matrix $\mathbf{D_3}$ (based on 9 diffusion couple experiments and one dissolution experiment) as well as associated eigenvalues and eigenvectors. Figure 7 shows fits to experimental concentration profiles in an experiment (Guo and Zhang 2018).

The diffusion eigenvectors listed in Table 4 are explained as follows. The eigenvector corresponding to the smallest eigenvalue is largely due to Si–Al exchange. That to the second smallest eigenvalue is largely due to Si–Ti exchange (more specifically, exchange of Ti and minor Ca+Mg with Si and minor amount of other components). That to the third smallest eigenvalue is due to divalent cations exchanging with all other components. That to the fourth smallest (also the third largest) eigenvalue is mostly due to Ca exchanging with other components. That to the second largest eigenvalue is due to Ca+K exchanging with all other components. And the eigenvector corresponding to the largest eigenvalue is due to the exchange of Na with all other components. Note that there is no simple Na–K exchange eigenvector. The exchange mechanisms and associated eigenvalues are also consistent with expectation.

Table 4. Diffusion matrix $\mathbf{D^{[Si]}}$ for haplobasalt2 melt at 1773 K.

D ($\mu m^2/s$)	TiO$_2$	Al$_2$O$_3$	MgO	CaO	Na$_2$O	K$_2$O
TiO$_2$	18.78±0.32	−0.81±0.23	−4.20±0.47	−11.10±1.16	−27.13±2.85	−15.54±3.27
Al$_2$O$_3$	−4.72±0.96	8.96±0.43	−17.40±0.96	−36.01±1.95	−60.32±4.74	−80.65±5.33
MgO	−6.77±1.13	0.22±0.58	39.02±1.23	−39.62±2.38	−82.61±5.54	−45.38±7.01
CaO	−11.20±1.30	−4.56±0.62	−27.62±1.29	64.89±2.58	−31.03±5.49	30.37±7.40
Na$_2$O	27.40±1.25	11.66±0.64	48.66±1.28	59.90±1.83	341.56±3.92	98.05±6.13
K$_2$O	5.39±0.50	5.98±0.22	11.67±0.46	15.20±0.93	−0.37±1.88	114.29±2.43
Eigenvalues	λ_1	λ_2	λ_3	λ_4	λ_5	λ_6
in $\mu m^2/s$	13.73±0.26	19.88±0.34	35.59±0.99	80.95±2.26	122.02±3.29	315.33±4.55
Eigenvectors	v_1	v_2	v_3	v_4	v_5	v_6
SiO$_2$	−0.88	−0.95	−0.45	0.07	−0.15	−0.34
TiO$_2$	−0.03±0.02	0.90±0.18	−0.15±0.03	−0.06±0.02	−0.05±0.02	−0.08±0.03
Al$_2$O$_3$	0.99±0.13	−0.20±0.15	−0.35±0.05	−0.09±0.03	−0.37±0.04	−0.15±0.05
MgO	−0.07±0.02	0.18±0.05	0.69±0.06	−0.57±0.14	−0.27±0.10	−0.26±0.08
CaO	0.07±0.01	0.30±0.03	0.58±0.04	0.80±0.20	0.63±0.15	−0.09±0.03
Na$_2$O	−0.02±0.003	−0.12±0.01	−0.14±0.01	−0.01±0.01	−0.33±0.03	0.95±0.51
K$_2$O	−0.06	−0.11	−0.18	−0.14	0.54	−0.03

Note: The composition of the haplobasalt2 melt (Guo and Zhang 2016) is listed in Table 2. Data in the table are mostly from Guo and Zhang (2018) but error estimation of eigenvalues and eigenvectors is from Guo and Zhang (2020). See footnote in Table 3 for all-component eigenvectors (i.e., the calculation of the SiO$_2$ component in an eigenvector).

Multicomponent diffusion in a basalt. Following the study on haplobasalt2 discussed above, Guo and Zhang (2018, 2019b) investigated an eight-component FeO-bearing basalt (basalt11a in Table 2) at 1623 K. The experimental strategy is similar to that in Guo and Zhang (2016). All diffusion couples have initial concentration gradients in only two components. That is, they were interdiffusion experiments. Seven diffusion couple experiments were carried out, with initial concentration gradients in SiO$_2$ and one of the other seven components in turn.

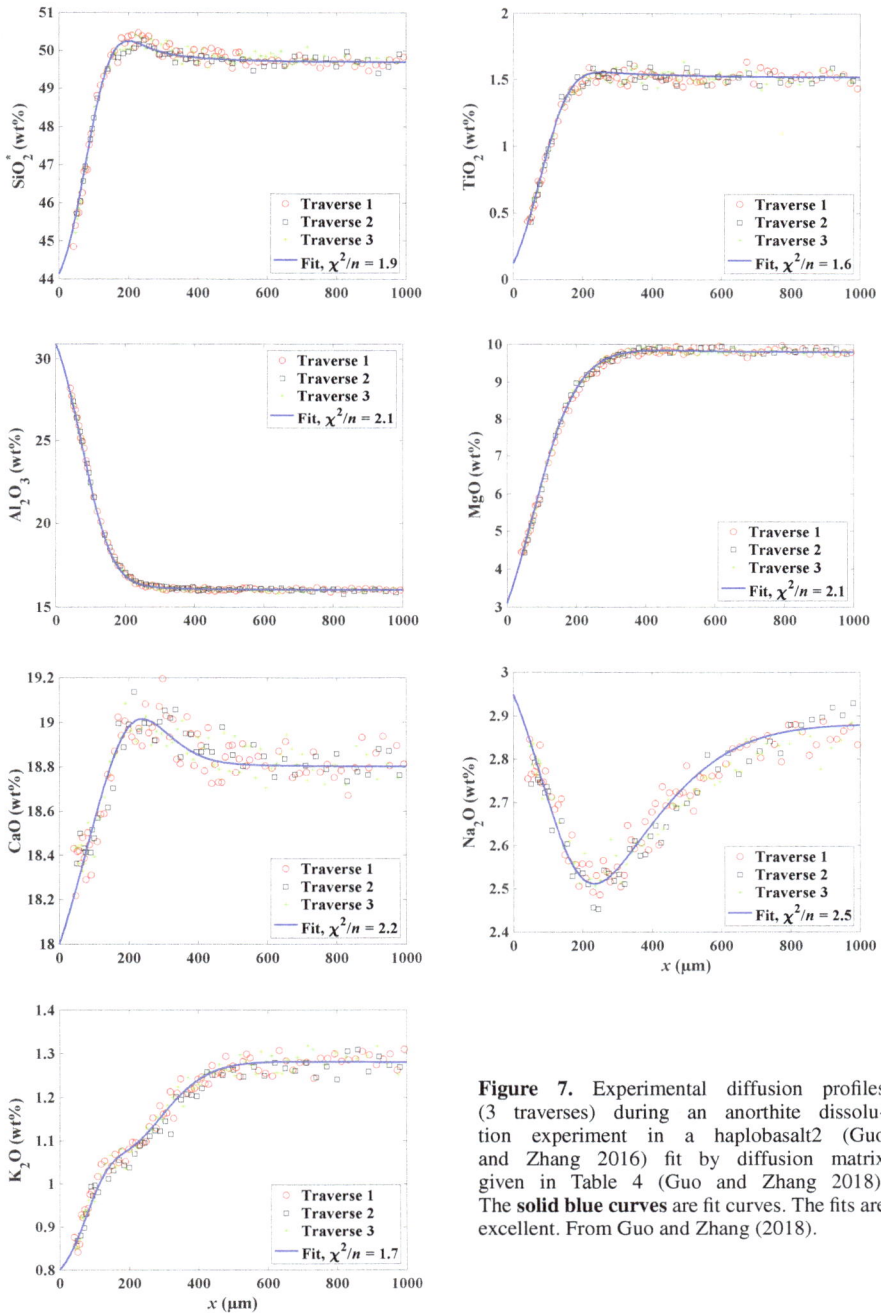

Figure 7. Experimental diffusion profiles (3 traverses) during an anorthite dissolution experiment in a haplobasalt2 (Guo and Zhang 2016) fit by diffusion matrix given in Table 4 (Guo and Zhang 2018). The **solid blue curves** are fit curves. The fits are excellent. From Guo and Zhang (2018).

Two other experiments are Al_2O_3–CaO and MgO–K_2O interdiffusion couples. Diffusion matrix was obtained from nine diffusion couple experiments ($\mathbf{D_1}$), as well as nine diffusion couple experiments plus results of mineral dissolution experiments from literature ($\mathbf{D_2}$). The latter diffusion matrix, which is best constrained, is shown in Table 5, together with eigenvalues and eigenvectors. The diffusion profiles of all oxides in one of the experiments and the fits of the profiles are shown in Fig. 8. It can be seen that all major features are well fit, including the uphill diffusion profiles. Nonetheless, there are still small misfits, and future improvements are necessary.

Table 5. Diffusion matrix $\mathbf{D^{[Si]}}$ for basalt melt at 1623 K.

D ($\mu m^2/s$)	TiO_2	Al_2O_3	FeO	MgO	CaO	Na_2O	K_2O
TiO_2	7.81±0.32	-0.25±0.07	-1.53±0.23	-2.02±0.31	-2.76±0.46	-6.43±1.67	-3.17±0.72
Al_2O_3	-0.81±0.70	5.69±0.14	-7.85±0.43	-6.77±0.55	-14.96±0.82	-29.73±2.90	-20.57±1.43
FeO	-21.66±1.30	-3.70±0.27	23.21±0.91	-31.24±1.08	-38.91±1.57	-72.85±5.73	-46.15±2.48
MgO	-5.52±0.73	1.11±0.16	-7.93±0.55	27.21±0.64	-21.46±0.90	-39.21±3.11	-7.33±1.54
CaO	13.58±1.48	-4.62±0.21	-17.94±0.69	-8.93±0.80	37.88±1.16	-38.15±4.14	15.44±1.66
Na_2O	19.68±1.90	10.28±0.29	28.83±0.90	39.57±0.87	57.21±1.31	243.78±4.65	77.02±1.74
K_2O	5.54±0.33	1.42±0.07	3.24±0.19	4.47±0.28	8.47±0.44	21.37±1.51	39.83±0.73
Eigenvalues							
in $\mu m^2/s$	λ_1	λ_2	λ_3	λ_4	λ_5	λ_6	λ_7
	6.43±0.12	8.18±0.28	14.95±0.27	31.43±0.61	41.68±1.00	58.24±0.77	224.52±4.70
Eigenvectors							
	v_1	v_2	v_3	v_4	v_5	v_6	v_7
SiO_2	-1.06	-1.44	-0.23	-0.45	0.04	0.05	-0.28
TiO_2	0.04±0.02	0.57±0.06	-0.16±0.03	-0.03±0.01	0.01±0.01	-0.01±.003	-0.02±0.01
Al_2O_3	0.99±0.22	0.75±0.27	-0.64±0.14	-0.25±0.03	0.13±0.02	-0.08±0.01	-0.11±0.01
FeO	0.07±0.01	0.30±0.04	0.55±0.10	0.83±0.09	-0.51±0.09	-0.42±0.05	-0.30±0.03
MgO	-0.04±0.01	0.02±0.03	0.37±0.06	-0.24±0.08	0.75±0.19	-0.30±0.04	-0.16±0.02
CaO	0.10±0.02	-0.01±0.03	0.31±0.05	-0.09±0.04	-0.39±0.10	0.84±0.10	-0.14±0.03
Na_2O	-0.06±0.01	-0.08±0.01	-0.15±0.03	-0.17±0.01	0.06±0.01	-0.15±0.01	0.92±0.12
K_2O	-0.04	-0.11	-0.04	0.39	-0.08	0.06	0.09

Note: The composition of the basalt melt is listed in Table 2. Data are mostly from Guo and Zhang (2018) but error estimation of eigenvalues and eigenvectors is from Guo and Zhang (2020). See footnote in Table 3 for all-component eigenvectors.

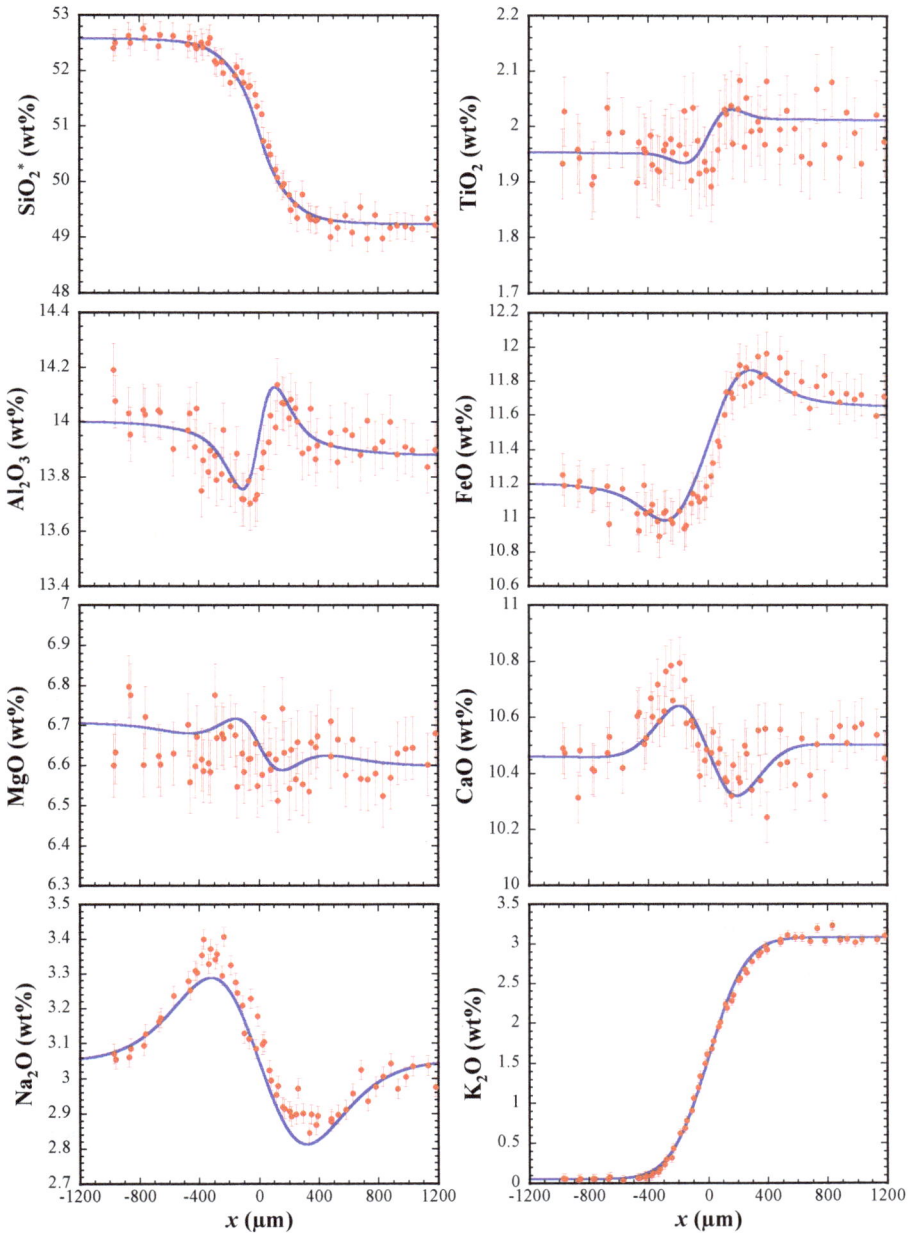

Figure 8. Concentration profiles in a diffusion couple experiment on multicomponent diffusion in a basalt. Solid blue curves are fit curves using [**D**] matrix in Table 5. From Guo and Zhang (2018).

The diffusion eigenvectors listed in Table 5 are explained as follows. The eigenvector corresponding to the smallest eigenvalue is largely due to Si–Al exchange. That to the second smallest eigenvalue is largely due to Si exchange with Al + Ti + Fe. That to the third smallest eigenvalue is due to divalent cations exchanging with all other components. That to the fourth smallest (also the fourth largest) eigenvalue is due to Fe + K exchanging with other components. That to the third largest eigenvalue is due to Fe + Ca exchanging with mostly Mg. That to the second largest eigenvalue is largely due to Ca exchanging with other components. And the eigenvector corresponding to the largest eigenvalue is due to exchange of Na with all other components. These eigenvectors are similar to those in the seven-component haplobasalt2 although the presence of three divalent cations introduces some complexity. Hence, studies of the haplobasalt2 and basalt systems are revealing similar diffusion mechanisms.

Guo and Zhang (2020) continued the study of Guo and Zhang (2018) and examined the temperature dependence of diffusion in basalt11a (Table 2). They reported 18 new diffusion couple experiments, nine each at 1533 K and 1773 K. Diffusion matrices at the two temperatures were determined from the experimental diffusion profiles. These results were combined with those at 1623 K in Guo and Zhang (2018) to examine the temperature dependence of the diffusion matrix, diffusion eigenvectors and eigenvalues. The hypothesis of constant eigenvectors (Varshneya and Cooper 1972; Chakraborty et al. 1995) is roughly but not rigorously verified: the eigenvectors at three different temperatures show similarity but are not identical within error. In addition, they found that some eigenvalues are nearly identical, and defined the phenomenon as near degeneracy of eigenvalues. In mathematical (strict) degeneracy of eigenvalues, eigenvectors are not uniquely defined because any linear combination of two eigenvectors is another eigenvector. In the case of near degeneracy of eigenvalues, eigenvectors are still uniquely defined but more constraints (e.g., more experimental data or higher quality data) are needed to resolve the eigenvectors. This difficulty to resolve the eigenvectors might explain that the extracted eigenvectors at three different temperatures are not identical within error. The occurrence of near degeneracy means that an increase of only one additional component from haplobasalt2 and basalt11a significantly increases the level of difficulty of obtaining accurate eigenvectors. Guo and Zhang (2020) nonetheless made effort to estimate average eigenvectors based on the data at the three temperatures, and redetermined eigenvalues at each temperature using the average eigenvectors. The eigenvalues are shown in Figure 9 in an Arrhenius plot. The average eigenvectors and temperature-dependent eigenvalues are listed Table 6. Diffusion matrix in basalt at a given temperature between 1533 and 1773 K can be estimated using Equation (23), where $\mathbf{P}$ is the temperature-invariant eigenvector matrix (Table 6) and each λ_i is calculated at the given temperature using expressions in Table 6. Using the formulation, a diffusion matrix was calculated at 1673 K and was used to predict diffusion

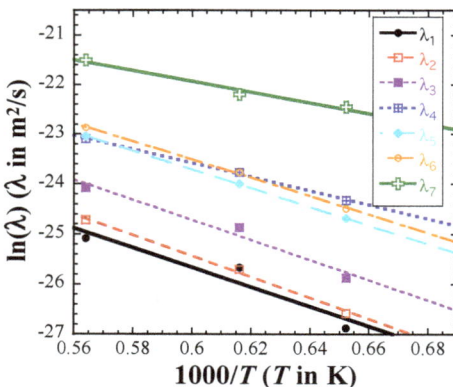

Figure 9. Arrhenius plot of eigenvalues for diffusion in basalt11a. Data are from Guo and Zhang (2020). The lines are least-squares linear fits. The fit equations are provided in Table 6. Eigenvalues λ_1 and λ_2 are nearly degenerate (difference is $\leq$ 0.4 natural logarithm units). λ_4, λ_5 and λ_6 are triply nearly degenerate. From Guo and Zhang (2020).

Table 6. Temperature dependence of eigenvalues [$\lambda(T)$] and the invariant eigenvector matrix **P** for basalt11a in the temperature range from 1533 to 1773 K.

	Eigenvalues (m²/s)						
	λ_1	λ_2	λ_3	λ_4	λ_5	λ_6	λ_7
	$e^{-13.88-19636/T}$	$e^{-12.89-20912/T}$	$e^{-12.73-19987/T}$	$e^{-15.26-13880T}$	$e^{-12.57-18569/T}$	$e^{-12.55-18279/T}$	$e^{-15.45-10808/T}$
	Invariant eigenvectors						
	v_1	v_2	v_3	v_4	v_5	v_6	v_7
TiO$_2$	−0.76	−0.20	−0.18	−0.02	−0.02	−0.02	−0.02
Al$_2$O$_3$	−0.18	0.97	−0.47	−0.15	−0.01	−0.07	−0.10
FeO	−0.51	0.00	0.66	0.86	0.06	−0.41	−0.36
MgO	−0.17	−0.03	0.41	−0.14	−0.71	−0.32	−0.15
CaO	−0.22	0.12	0.33	−0.33	0.70	0.79	−0.08
Na$_2$O	0.17	−0.04	−0.18	−0.12	−0.04	−0.19	0.91
K$_2$O	0.13	−0.02	−0.09	0.32	−0.10	0.25	0.06

Note: T is in K, and λ_i values are in m²/s. Eigenvectors are for independent components with SiO$_2$ as the dependent component

profiles during mineral dissolution with preliminary success except for the K$_2$O diffusion profiles (Guo and Zhang 2020). Magma mixing in the Bushveld Complex at 1473 K is also calculated, revealing possible mixing-generated sulfide ore formation (Guo and Zhang 2020).

In summary, major progresses have been made in recent years on multicomponent diffusion in silicate melts, including natural basalt. Even in an extensively studied basalt, there is still uncertainty in the eigenvectors and eigenvalues, likely due to additional complexity introduced by near degeneracy of eigenvalues. There are still misfits in reproducing experimental diffusion profiles, especially in mineral dissolution experiments. Future work will need to rigorously test whether eigenvectors in natural silicate melts depend on temperature as well as melt composition. If eigenvectors do not depend on temperature or composition, then we would be well on our way to use multicomponent diffusion matrix to treat major oxide diffusion in natural silicate melts in various magmatic processes.

TRACER AND EFFECTIVE BINARY DIFFUSION DATA

In volume 72 (entitled "Diffusion in Minerals and Melts") of Reviews in Mineralogy and Geochemistry published in 2010, five chapters (Behrens 2010; Lesher 2010; Liang 2010; Zhang and Ni 2010; Zhang et al. 2010) thoroughly reviewed diffusion coefficients in silicate melts, covering noble gases (He, Ne, Ar, Kr, Xe, and Rn, Behrens 2010), H, C, and O (Zhang et al. 2010), plus diffusion data on 59 other elements. For most elements, some diffusion data were available. However, no diffusion data were available for N, As, Bi, Se, I, V, Cu, Mo, In, Tm, Ru, Rh, Pd, Ag, Os, Ir, Pt, Au, and Hg (plus most synthetic elements) in natural or nearly natural silicate melts as of 2010. The order of elements/oxides ranked by tracer diffusivity and POCGD from high to low is roughly as follows in rhyolite melt (Behrens 2010; Zhang et al. 2010; Ni et al. 2017; Holycross and Watson 2018):

$$H_2 > He > Li \approx Na > Cu > K > Ne > Ar \approx CO_2 \approx Cl \approx Rb \approx Sb \approx F >$$
$$Ba \approx Cs \approx Sr > Ca > Mg > Be \approx B \approx Ta \approx Nb \approx Y \approx REE > Zr > U \approx \qquad (27)$$
$$Hf \approx Ti \approx Ge \approx Th \approx Si \approx P$$

Rare earth elements have similar diffusivity, but there is consistently slight decrease of diffusivity from D_{La} to D_{Lu}. H$_2$O diffusivity is not included in the sequence because it depends strongly on total H_2O concentration (see discussion below). In other melts, the sequence is similar, although there may be small variations. For example, in basalt melt, the updated sequence is roughly (Behrens 2010; Zhang et al. 2010, and new data):

$$He > Ne > Li > Na \approx Cu > F \approx Cd > Cl \approx Mn \approx Co \approx Ca \approx Sr > Rb \approx$$
$$Br \approx CO_2 \approx Ba > V \approx Tl \approx Cs \approx Pb \approx Y \approx REE > Sc > Te \approx Ti \approx O \approx U \qquad (28)$$
$$\approx Nb > Th \approx Zr \approx Ta > Hf > P \geq Si$$

where the position of Cu, Rb, V, Sc, U, Nb, Th, Zr, Ta, Hf, P and Si are based on new data (Watson et al. 2015; Holycross and Watson 2016; Ni et al. 2017) to be reviewed below. Many empirical fit equations were given in Behrens (2010), Zhang and Ni (2010), and Zhang et al. (2010) for the purpose of estimating elemental diffusivities.

Since the reviews in 2010, new diffusion data and models have been reported for H_2O (Persikov et al. 2010, 2014; Fanara et al. 2013; Ni et al. 2013; Zhang et al. 2017; Ni and Zhang 2018; Kuroda and Tachibana 2019; Newcombe et al. 2019), Li (Holycross et al. 2018), F and Cl (Bohm and Schmidt 2013), Al (Yu et al. 2016), Si (Gonzalez-Garcia et al. 2017; Yu et al. 2019), P (Watson et al. 2015), S (Frischat and Szurman 2011; Lierenfeld et al. 2018), Cl (Yoshimura 2018), Cu (Ni and Zhang 2016; Ni et al. 2017, 2018), Zr (Zhang and Xu 2016), Sn (Yang et al. 2016), Sr and Ba (Fanara et al. 2017), and Mo and W (Zhang et al. 2018). Hence, Cu and Mo no longer belong to the list of elements with no diffusion data. Still, diffusion of 16 nonradioactive elements (N, As, Bi, Se, I, V, In, Tm, Ru, Rh, Pd, Ag, Os, Ir, Pt, Au, and Hg) has not been investigated yet, most of which are chalcophile and siderophile elements. Absence of Tm diffusion data is not expected to be much missed because REE diffusivities are highly consistent and Tm diffusivity can be predicted from diffusivities of other REE's (see Eqns. 45a–c later).

In addition, some papers reported diffusion data on a large number of elements. Holycross and Watson (2016) determined trace element diffusivity (close to TED1) of 25 elements (Sc, V, Rb, Sr, Y, Zr, Nb, Ba, La, Ce, Pr, Nd, Sm, Eu, Gd, Tb, Dy, Ho, Er, Yb, Lu, Hf, Ta, Th and U) in nominally dry basalt melt. Holycross and Watson (2018) measured trace element diffusivity (close to TED1) of 21 elements (Sc, V, Y, Zr, Nb, La, Ce, Pr, Nd, Sm, Eu, Gd, Tb, Dy, Ho, Er, Yb, Lu, Hf, Th and U) in hydrous rhyolite melt. Gonzalez-Garcia et al. (2017, 2018) obtained effective binary diffusivities (OEBD) of 19 major and trace elements (Si, Ti, Fe, Mg, Ca, K, Rb, Cs, Sr, Ba, Co, Sn, Eu, Ta, V, Cr, Hf, Th, U; other elements show uphill diffusion) in shoshonite–rhyolite diffusion couples. Posner et al. (2018) evaluated self diffusivity of O, Si, Mg, and Ca, and interdiffusion of Ni and Co in a peridotite melt at very high pressures of 4–24 GPa and very high temperatures (≥ 2150 K). These heroic efforts greatly expanded the diffusion database.

We review below experimental diffusion data since 2010. The following review will not be nearly as systematic as the several chapters in 2010 (Behrens 2010; Lesher 2010; Liang 2010; Zhang and Ni 2010; Zhang et al. 2010), but will focus on new advances on diffusion in natural or nearly natural melts in recent years. In addition, more emphasis will be on TED1 and POCGD because they only depend on the bulk composition and not on concentration gradients. As it will be seen, the Holy Grail of determining the composition dependence of diffusivity is still elusive, and empirical equations accounting for compositional dependence developed in earlier papers often cannot predict later published data in melts with different compositions.

H_2O diffusion

H_2O diffusion is the best example of multi-species diffusion. Due to the importance of H_2O diffusion in volcanic eruption dynamics, exsolution of hydrothermal fluids, bubble growth as well as the importance of H_2O in controlling magma evolution, and due to the complexity of the H_2O diffusion process, H_2O diffusion has been investigated extensively and is probably the best studied diffusion problem in geology literature (e.g., Shaw 1974; Delaney and Karsten 1981; Karsten et al. 1982; Stanton et al. 1985; Wasserburg 1988; Zhang and Stolper 1991; Zhang et al.

1991a,b, 2017, 2019a,b; Jambon et al. 1992; Nowak and Behrens 1997; Zhang and Behrens 2000; Freda et al. 2003; Behrens et al. 2004, 2007; Liu et al. 2004; Okumura and Nakashima 2004, 2006; Ni and Zhang 2008, 2018; Ni et al. 2009a,b, 2013; Wang et al. 2009; Persikov et al. 2010, 2014; Zhang and Ni 2010; Fanara et al. 2013; Kuroda and Tachibana 2019; Newcombe et al. 2019). Because there is major advancement since 2010, below we briefly summarize the earlier developments and then focus on recent progress since the review of Zhang and Ni (2010). The compositions of silicate melts that have been investigated for H_2O diffusion are listed in Table 7.

Table 7. Chemical composition on dy basis in H_2O diffusion studies in geology literature.

Melt	SiO$_2$ wt%	TiO$_2$ wt%	Al$_2$O$_3$ wt%	FeO wt%	MgO wt%	CaO wt%	Na$_2$O wt%	K$_2$O wt%	X_{Si}	W g/mol	Ref.
rhyolite14a	76.6	0.07	13.2	0.64	0.05	0.57	4.15	4.83	0.711	32.52	1–12
CBS-NSL	75.9	0.20	10.2	4.33	0.0	0.09	5.19	4.61	0.704	33.23	13, 14
GMR-MAC	72.7	0.16	15.2	1.02	0.16	0.76	4.21	4.01	0.686	32.60	11, 13
dacite5	67.5	0.77	15.7	4.28	1.43	4.40	3.58	2.15	0.632	33.49	15
HA2	66.3	0	17.6	0	1.38	2.50	10.45	0	0.592	33.05	12
Ab75Di25	65.8	0	15.1	0	3.17	7.0	8.93	0	0.582	33.53	16
dacite3a	65.4	0.73	15.9	4.44	2.02	4.96	3.88	2.59	0.608	33.84	17–19
andesite7	62.5	0.7	16.7	5.55	2.97	6.48	3.2	1.69	0.583	34.13	12
HA1a	62.3	0	19.8	0.02	2.30	10.2	4.12	1.00	0.570	33.55	20, 21
Ab50Di50	62.2	0	10.5	0	6.79	14.2	6.34	0	0.555	34.45	16
trachyte0b	60.5	0.48	17.8	7.14	0.21	1.72	5.22	7.28	0.553	35.25	22
trachyte0a	59.9	0.39	18.0	3.86	0.89	2.92	4.05	8.35	0.555	34.94	23
phonolite1a	58.9	0.76	19.9	3.61	0.69	3.90	5.96	6.87	0.529	35.04	24
andesite1a	57.2	0.84	17.5	7.58	4.27	7.59	3.31	1.60	0.530	34.98	17
haplobasalt3	52.0	1.06	16.3	0.03	11.2	15.3	2.79	0.89	0.465	35.04	24
basalt11	50.6	1.88	13.9	12.5	6.56	11.4	2.64	0.17	0.475	36.59	25
An36Di64	49.6	0.02	17.5	0.03	9.89	23.8	0.07	0.01	0.448	35.55	26
green glass	48.3	0.39	8.17	15.9	17.4	8.98	0	0	0.450	37.16	27
basalt0	46.1	1.50	16.1	10.8	7.60	13.3	3.56	0.76	0.423	37.15	15
LB2a	43.6	3.46	8.96	21.8	13.1	8.74	0.01	0.00	0.419	38.59	26
yellow glass	43.5	3.11	7.86	21.9	13.2	8.24	0.44	0	0.422	38.69	27

Note: Compositions are listed in decreasing SiO$_2$ order. Similar melt compositions (defined to be ≤ 1.5 wt% difference in every oxide concentrations) are averaged, e.g., rhyolite14a includes many high-silica rhyolites and AOQ(Ab38Or34Qz28). HA: haploandesite. LB, green glass, and yellow glass: lunar basalts. X_{Si} is cation mole fraction of Si on dry basis. W is mass of the melt per mole of oxygen on dry basis (Stolper 1982a,b; Zhang 1999). See footnotes in Table 2 for more explanation of melt names.

References: 1. Shaw (1974); 2. Delaney and Karsten (1981); 3. Karsten et al. (1982); 4. Stanton et al. (1985); 5. Zhang et al. (1991a); 6. Jambon et al. (1992); 7. Nowak and Behrens (1997); 8. Zhang and Behrens (2000); 9. Okumura and Nakashima (2004); 10. Behrens et al. (2007); 11. Ni and Zhang (2008); 12. Persikov et al. (2014); 13. Behrens and Zhang (2009); 14. Wang et al. (2009); 15. Okumura and Nakashima (2006); 16. Persikov et al. (2010); 17. Behrens et al. (2004); 18. Liu et al. (2004); 19. Ni et al. (2009a); 20. Ni et al. (2009b); 21. Ni et al. (2013); 22. Fanara et al. (2013); 23. Freda et al. (2003); 24. Zhang et al. (2017); 25. Zhang and Stolper (1991); 26. Newcombe et al. (2019); 27. Zhang et al. (2019b).

Dissolved H_2O component in silicate melts is present as at least two species, neutral and free H_2O molecules (referred to as H_2O_m), and charged and bonded hydroxyl groups (referred to as OH) (Stolper 1982a,b). The two species interconvert in the melt structure:

$$H_2O_m \text{ (melt)} + O \text{ (melt)} \rightleftharpoons 2OH \text{ (melt)}, \qquad (29)$$

with an equilibrium constant

$$K = [OH]^2/([H_2O_m][O]), \qquad (30)$$

where brackets mean mole fractions, increasing with temperature (Zhang et al. 1995, 1997). Due to the above speciation reaction, OH is the dominant species at low total H_2O content (referred to as H_2O_t hereafter; H_2O refers to the component) such as ≤ 1 wt%, and H_2O_m is the dominant species at high H_2O_t such as ≥ 5 wt%. According to the above reaction, the mole fraction of H_2O_t is expressed as:

$$[H_2O_t] = [H_2O_m] + 0.5[OH]. \qquad (31)$$

On the other hand, the mass fraction of H_2O_t is expressed as:

$$w_{H_2O_t} = w_{H_2O_m} + w_{OH}, \qquad (32)$$

where w_{OH} does not mean the actual OH mass fraction, but by convention it means the mass fraction of H_2O that is present in the melt or glass as OH (Stolper 1982a,b; Zhang 1999). The mole fractions are defined on a single oxygen basis as follows:

$$[H_2O_t] = (w_{H_2O_t}/18.015)/\{w_{H_2O_t}/18.015 + (1-w_{H_2O_t})/W\}, \qquad (33a)$$

$$[H_2O_m] = [H_2O_t]w_{H_2O_m}/w_{H_2O_t}, \qquad (33b)$$

$$[OH] = 2\{[H_2O_t] - [H_2O_m]\}, \qquad (33c)$$

$$[O] = 1 - [H_2O_m] - [OH], \qquad (33d)$$

where 18.015 is the molecular mass of H_2O in g/mol, and W is the mass of the dry melt per mole of oxygen in g/mol. Values of W for investigated melts are listed in Table 7.

Experimental studies of H_2O diffusion before 1990 (Shaw 1974; Delaney and Karsten 1981; Karsten et al. 1982; Stanton et al. 1985) found that H_2O diffusivity depends strongly on H_2O concentration in addition to the temperature dependence. Zhang et al. (1991a) investigated H_2O diffusion in rhyolite14a containing ≤ 1.7 wt% H_2O_t. Based on measured H_2O_m and OH concentration profiles by FTIR, they considered the contribution of both H_2O_m and OH and treated one-dimensional diffusion of H_2O_t using the following multi-species diffusion equation:

$$\frac{\partial[H_2O_t]}{\partial t} = \frac{\partial}{\partial x}\left(D_{H_2O_m}\frac{\partial[H_2O_m]}{\partial x} + D_{OH}\frac{\partial[OH]/2}{\partial x}\right) \qquad (34)$$

where $D_{H_2O_m}$ and D_{OH} are the diffusivity (POCGD) of H_2O_m and OH. Hence, $D_{H_2O_t}$ is related to species diffusivities as follows:

$$D_{H_2O_t} = D_{H_2O_m}\frac{d[H_2O_m]}{d[H_2O_t]} + D_{OH}(1 - \frac{d[H_2O_m]}{d[H_2O_t]}) \qquad (35)$$

The differential in the above equation can be found as (Wang et al. 2009):

$$\frac{d[H_2O_m]}{d[H_2O_t]} = 1 - \frac{(0.5 - X)}{\sqrt{X(1-X)(\frac{4}{K}-1)+0.25}} \qquad (36)$$

where $X = [H_2O_t]$. Zhang et al. (1991a) found that in rhyolite melt and glass, $D_{H_2O_m}$ was roughly constant in their samples (0.1 to 1.7 wt% H_2O_t), and D_{OH} is too small (compared to $D_{H_2O_m}$) to be resolved. That is, OH diffusion is negligible and the diffusion of the H_2O component is accomplished by H_2O_m diffusion and interconversion of OH and H_2O_m. Even when H_2O_t is as low as 0.18 wt%, meaning that more than 90% of H_2O_t is present as OH, contribution of OH diffusion to H_2O_t diffusion is still negligible and unresolvable. The speciation-diffusion model leads to a proportionality between $D_{H_2O_t}$ and H_2O_t content at low H_2O_t (e.g., <2 wt%).

Nowak and Behrens (1997) found that $D_{H_2O_t}$ is no longer proportional to H_2O_t when H_2O_t is > 3 wt%. Zhang and Behrens (2000) extended the multi-species H_2O diffusion model in rhyolite to high H_2O_t, and found that H_2O_m diffusivity ($D_{H_2O_m}$) is no longer a constant, but depends on H_2O_t concentration exponentially:

$$D_{H_2O_m} = D_{X=0}e^{aX},$$ (37)

where $X = [H_2O_t]$, a is a constant depending on T, and $D_{X=0}$ is $D_{H_2O_m}$ at zero H_2O_t. D_{OH} was still not resolved from the experimental data. This formulation has been adopted by subsequent studies until 2013 (Okumura and Nakashima 2004, 2006; Behrens et al. 2004, 2007; Liu et al. 2004; Ni and Zhang 2008; Ni et al. 2009a,b; Wang et al. 2009; Persikov et al. 2010, 2014; Fanara et al. 2013).

Behrens et al. (2004) hinted at OH contribution to H_2O diffusion in diffusion couple experiments in andesite1a melt at 1608–1848 K. Ni et al. (2013) were the first to resolve the noticeable role of OH diffusion contributing to H_2O_t diffusion in a haploandesite melt (HA1a in Table 7) when H_2O_t is low (< 1 wt%) at 1619-1842 K and 1 GPa. They assumed constant OH diffusivity and found $D_{OH}/D_{X=0}$ (note that $D_{X=0}$ is $D_{H_2O_m}$ at zero H_2O_t) ranging from 0.09 to 0.24. Zhang et al. (2017) (note that this Zhang is L. Zhang) investigated H_2O diffusion in haplobasalt3 melt containing 0.03–2.02 wt% H_2O_t and quantified both OH and H_2O_m diffusivities, obtaining $D_{OH}/D_{X=0}$ ranging from 0.10 to 0.17. The success in resolving OH diffusivity in haploandesite (Ni et al. 2013) and haplobasalt (Zhang et al. 2017) confirmed the importance of OH diffusion in depolymerized melt at magmatic temperatures and low H_2O_t.

Ni and Zhang (2018) constructed a general model for H_2O diffusivity in calc-alkaline silicate melts and glasses using literature data. The model did not include trachyte (Freda et al. 2003; Fanara et al. 2013), phonolite (Fanara et al. 2013), or peralkaline rhyolite (Behrens and Zhang 2009; Wang et al. 2009). The model parameterized K, a and $D_{X=0}$, and D_{OH} as a function of the cation mole fraction of Si in dry melt (X_{Si}; values are listed in Table 7), T and P as follows:

$$\ln K = X_{Si}\left(2.6 - \frac{4339}{T}\right)$$ (38a)

$$a = -94.07 + 74.112X_{Si} + \frac{198508 - 166674X_{Si}}{T}$$ (38b)

$$\ln D_{X=0} = 8.02 - 31X_{Si} + 2.348X_{Si}P + \frac{121824X_{Si} - 118323\sqrt{X_{Si}} - (10016X_{Si} - 3648)P}{T}$$ (38c)

$$\ln\frac{D_{OH}}{D_{X=0}} = -56.09 - 115.93X_{Si} + \sqrt{X_{Si}}\left(160.54 - \frac{3970}{T}\right)$$ (38d)

where P is in GPa, T is in K, and $D_{X=0}$ and D_{OH} are in m²/s. Once K, a, $D_{X=0}$ and D_{OH} are calculated from Equations (38a–d), $D_{H_2O_m}$ can be calculated from Equation (37), and then $D_{H_2O_t}$ can be calculated from Equation (35) with the differential from Equation (36). A supplementary excel file is available in Ni and Zhang (2018) for the calculation. Calculations indicate that OH contribution to H_2O diffusion increases with increasing temperature and decreasing SiO_2 concentration. Ni and Zhang (2018) concluded that in rhyolite and dacite glass and melt, contribution of OH to H_2O diffusion is rarely noticeable, whereas in andesite and basalt melt, contribution of OH to H_2O diffusion becomes important at $T \geq 1200$ K.

The synthesis model of H_2O speciation and diffusion by Ni and Zhang (2018) represents a major step forward. Nonetheless, the model is unlikely to be the last word on H_2O diffusion. In their model, data to constrain OH diffusivity are limited. In addition, the compositional coverage by the model does not include peralkaline rhyolite, or phonolite, or trachyte. Newcombe et al. (2019) investigated H_2O diffusion in An36Di64 melt and lunar mare basalt, and found their data on An36Di64 are in reasonable agreement with the model, but those on lunar mare basalt (lower SiO_2 and Al_2O_3 and much higher FeO) are off the model by a factor of 6. Zhang et al. (2019b) determined H_2O_t diffusivity in lunar green glass and yellow glass, and their data are off the model of Ni and Zhang (2018) by a factor of 3 to 8. Future improvement is expected to require more data at low H_2O_t to better resolve OH diffusivity and how it depends on H_2O_t, as well as more compositional coverage (e.g., the role of Al_2O_3, FeO, and alkalis).

Diffusion of alkalis

Zhang et al. (2010) reviewed alkali diffusion data. Li tracer diffusivity does not vary much with composition in dry Ab39Or61, albite, rhyolite, dacite, andesite and basalt melts (Zhang et al. 2010), and Rb tracer diffusivity does not vary much with composition in dry jadeite, albite, rhyolite, haploandesite, trachyte and phonolite. The primary dependence is on temperature. Rb diffusivity increases with H_2O content. Na, K, and Cs tracer diffusivity depends more on melt composition.

Ni (2012) reevaluated existing data on alkali tracer diffusion and developed specific models for each alkali element. The new empirical equation for calculation of D_{Li} is (Ni 2012):

$$\ln D_{Li} = -13.09 - \frac{9722.7 + 1171.4 f_1 + 4943 f_2}{T} \tag{39}$$

where

$$f_1 = \frac{K\#}{K\# + 0.1(1 - K\#)} \tag{39a}$$

and

$$f_2 = \frac{Ca\#}{Ca\# + 3(1 - Ca\#)} \tag{39b}$$

where $K\# = K/(K+Na)$, $Ca\# = Ca/(Ca+K+Na)$, with Na, K and Ca being cation mole fractions. For Na, K, Rb, and Cs, Ni (2012) developed the following empirical equations:

$$\ln D_{Na} = -13.77 - \frac{8815.8 + 1308.2 f_1 + 15164 f_2}{T} \tag{40}$$

$$\ln D_K = -14.81 - \frac{11125 + 1277.6 f_3}{T} \tag{41}$$

$$\ln D_{Rb} = -15.73 - \frac{11376 + 4022.9 f_3}{T} \tag{42}$$

$$\ln D_{Cs} = -11.87 - \frac{5352.8 f_1 + 233.52 F - 30124 Al}{T} \tag{43}$$

$$f_3 = \frac{Na\#}{Na\# + 0.1(1-Na\#)} = \frac{1-K\#}{1-0.9K\#}$$

where Na#=Na/(Na+K), $F = SiO_2 + TiO_2 + Al_2O_3 + P_2O_5$ in wt%, and AI is peralkalinity defined to be the greater of (Na+K–Al)/O (where Na, K, Al and O are atomic fractions) and zero.

Holycross et al. (2018) studied Li trace element diffusivity (TED1) in wet rhyolite (6.0 wt% H_2O). The data are not used to evaluate the model of Ni (2012) because the latter does not contain data from hydrous melts. Gonzalez-Garcia et al. (2018) obtained OEBD of Rb in shoshonite–rhyolite diffusion couple using the Boltzmann–Matano method. These OEBD values are not expected to be similar to tracer diffusivities, and hence are not used to test the model of Ni (2012). Holycross and Watson (2016) reported new Rb trace element diffusion (close to TED1) data in nominally dry basalt melt, which can be used to test Equation (42). The predicted Rb diffusivity using Equation (42) is lower than the experimental data by 0.87 to 1.52 $\ln D$ units, which is not too bad. On the other hand, due to the weak dependence of Rb diffusivity on melt composition, if the Rb Arrhenius equation for rhyolite (Eqn. 13 in Zhang et al. 2010) is used to predict Rb diffusivity, the predicted values are lower than experimental data by only 0.51 to 1.16 $\ln D$ units, better than the predicted values using Equation (42). Hence, except for its ability to reconcile Rb diffusivity in orthoclase melt, the Rb diffusivity model by Ni (2012) does not improve prediction compared to simply assuming no variation from rhyolite to basalt.

Cu diffusion

The absence of Cu diffusion data in natural silicate melts has been remedied by three recent papers (33 data points). Ni and Zhang (2016) investigated Cu diffusion in basalt1a (composition listed in Table 8) melt using diffusion couples. Ni et al. (2017, 2018) reported Cu diffusion data in various rhyolite melts. The diffusion data in the three papers may all be viewed as POCGD, and they are highly consistent (note that Ni et al. 2017 and Ni et al. 2018 are different authors from different laboratories). Figure 10 shows all available Cu diffusion data in natural melts (and comparison with Li, Na and K diffusivity shown as numbered lines). Cu diffusivity is very high in rhyolite to basalt melts, higher than H_2O_t diffusivity at the same H_2O concentration. In dry basalt, Cu diffusivity is similar to Na diffusivity (overlapping in Fig. 10). In dry rhyolite, Cu diffusivity lies between Na and K and closer to K (Ni et al. 2017). These observations can be explained by Cu diffusion as univalent cation Cu^+. For the composition (including H_2O) effect, Ni et al. (2017) found that a single compositional parameter Si+Al–H seems to adequately capture the dependence of D_{Cu} on composition, where Si, Al and H are cation mole fractions on wet basis. The pressure effect was not well resolved by Ni et al. (2017). Their equation without including the pressure effect predicts the later published data in Ni et al. (2018) well (within 0.22 $\ln D$ units) except for the data at low pressures of 0.15 GPa. Following Ni et al. (2017) but including the pressure effect, the following empirical equation is obtained for Cu diffusivity in dry basalt and dry and wet (up to 6 wt% H_2O) rhyolite at 973–1848 K and ≤1.5 GPa:

$$\ln D_{Cu} = -16.68 + 2.872(Si+Al-H) - \frac{5103 + 8259(Si+Al-H) + 411.7P}{T} \tag{44}$$

where P is in GPa and D is in m²/s. The above equation predicts all Cu diffusion data in Ni and Zhang (2016) and Ni et al. (2017, 2018) to within 0.23 $\ln D$ units (1σ error 0.12 $\ln D$ units). This accuracy is among the highest of all empirical predictive equations for diffusivity data across different compositions. We recommend its use to predict Cu^+ diffusion in other natural silicate melts if no experimental data are available.

Table 8. Chemical compositions (on dry basis) for trace element diffusion studies.

	SiO$_2$	TiO$_2$	Al$_2$O$_3$	FeO	MnO	MgO	CaO	Na$_2$O	K$_2$O	Ref.
rhyolite14b	76.8	0.15	13.3	0.78	0.08	0.08	0.62	4.04	3.91	1
rhyolite8a	73.2	0.11	13.8	2.14	0.08	0.18	0.92	4.22	5.31	2
NCO	72.9	0.22	14.2	1.93	0.06	0.18	0.86	4.73	4.24	3
rhyolite3a	70.4	0.29	16.3	1.21	0.00	0.59	1.69	3.93	5.15	4
dacite3a	65.0	0.54	16.5	3.88	0.10	2.23	4.97	4.49	1.52	5
phonolite2	58.5	0.68	19.9	3.53	0.21	0.36	0.74	9.90	5.67	6
shoshonite	53.3	0.69	16.4	8.14	0.21	4.64	8.04	5.46	3.05	2
basalt8a	50.0	1.62	16.0	9.40	0.25	8.50	10.79	3.00	0.20	7, 8
basalt11	49.9	1.83	13.5	12.9	0.22	6.81	10.8	2.65	0.17	3
basalt6	48.5	2.7	13.8	12.7	0.00	7.55	10.9	2.50	0.41	9
basalt1a	46.9	1.65	17.66	10.6	0.00	5.86	10.6	4.43	2.02	10
peridotite	46.1		4.0	8.8		37.5	3.6			11

References: 1. Holycross and Watson (2018); 2. Gonzalez-Garcia et al. (2017, 2018); 3. Yu et al. (2019); 4. Zhang et al. (2018); 5. Lierenfeld et al. (2018); 6. Bohm and Schmidt (2013); 7. Watson et al. (2015); 8. Holycross and Watson (2016); 9. Lesher et al. (1996); 10. Ni and Zhang (2016); 11. Posner et al. (2018). The peridotite composition includes 1 wt% NiO on one side and 1 wt% CoO on the other side. See footnotes in Table 2 for melt names.

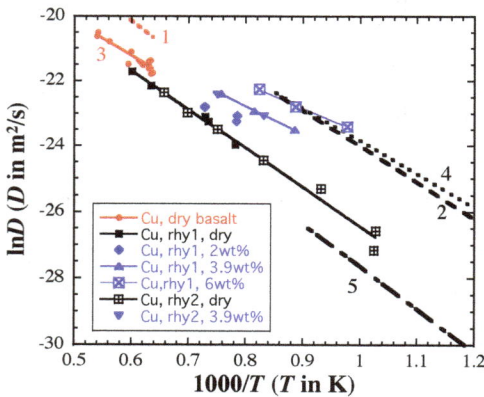

Figure 10. Cu diffusivity in rhyolite to basalt compared with diffusivity of Li, Na, and K. **Red color** for dry basalt. **Black color** for dry rhyolite. **Blue color** for wet rhyolite. **Points with solid lines** are for Cu diffusion data (Ni and Zhang 2016; Ni et al. 2017, 2018). The two lines for Cu diffusivity in dry rhyolite overlap and cannot be seen individually. **Numbered lines** are for diffusion data of Li, Na and K. 1 (**red short-dash line**): Li diffusivity in dry basalt (Lowry et al. 1981); 2 (**black long-dash line**): Li in dry rhyolite (Jambon and Semet 1978); 3 (**red long-dash line**): Na in dry basalt (Lowry et al. 1982). (Line 3 cannot be seen because it overlaps with Cu diffusion line in basalt); 4 (**black short-dash line**): Na in dry rhyolite (Jambon 1982); 5 (**black dot-dash line**): K in dry rhyolite (Jambon 1982).

A note about the calculation of cation mole fractions of Si, Al and H. Often the oxide wt% is given on the dry basis for easy comparison with other melts and then H_2O wt% is separately given. In such cases, calculation of cation mole fractions on dry basis is straightforward. However, for the calculation of cation mole fractions on wet basis (i.e., cation mole fraction of H is also calculated), the non-H_2O oxide wt% must first be calculated by multiplying $(1 - w_{H_2O})$, where w_{H_2O} is the mass fraction of H_2O. Then the reported H_2O wt% and the recalculated wt% of other oxides are used to calculate cation mole fraction. If the oxide wt% is given on wet basis (actual concentrations), then no such conversion of multiplying by $(1 - w_{H_2O})$ is needed.

In addition to the above studies, Von der Gonna and Russel (2000), and Kaufmann and Russel (2008, 2010, 2011) obtained Cu diffusivity in SiO_2–Na_2O, SiO_2–CaO–Na_2O, SiO_2–Al_2O_3–Na_2O, SiO_2–Al_2O_3–CaO–Na_2O melts using square wave voltametry. The Cu diffusivity data determined using the voltametry method are for a mixture of Cu^+ and Cu^{2+} at subequal proportions, and Kaufmann and Russel (2011) derived them to roughly equal to $2D_{Cu^{2+}}$, meaning that these diffusivities are expected to be much smaller than diffusivity of Cu^+ determined by Ni and Zhang (2016) and Ni et al. (2017, 2018). Furthermore, the compositions are very different from natural silicate melts. Equation (44) cannot be applied to predict these diffusivities.

Diffusion of Sc, Y, and REE

Holycross and Watson (2016, 2018) produced high quality trace element diffusion data in both dry basalt8a and wet rhyolite14b containing ~4.1 wt% H_2O and ~6.2 wt% H_2O using diffusion couple experiments. The compositions of basalt8a and rhyolite14b are listed in Table 8. The chemical concentration gradients are only on some 20 trace elements, not on major elements. In principle, the presence of concentration gradients of other trace elements could affect the diffusivity of a given trace element. However, such effect is unlikely to be significant. Hence, the diffusivities are close to TED1. Holycross and Watson (2016, 2018) reported a large number of diffusion data and they are highly self consistent. Diffusion coefficients decrease slightly from La to Lu, by about 20% in dry basalt8a melt (Fig. 11), and slightly more in wet rhyolite14b melt.

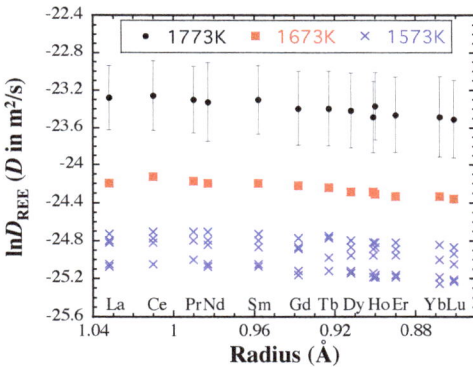

Figure 11. Diffusion coefficients of rare earth elements in basalt8a at three temperatures (Holycross and Watson 2016) as a function of trivalent ionic radius in octahedral sites. Y diffusivity is also shown (almost overlapping with Ho). Error bars at 1σ level are shown at 1773K, and they are similar at other temperatures. Another measure of error is by comparison of the five experiments at 1573 K with different durations. It appears that Ce has the highest diffusivity among the REE, but the difference between Ce and La diffusivity is tiny (~0.04 ln D units) compared to the error (~0.3 ln D units). From La to Lu, ln D decreases by about 0.2.

The activation energy E and pre-exponential factor D_0 based on diffusivities extracted from diffusion profiles for each element are listed in Table 9. From E and D_0 listed, D at a given temperature can be calculated using Equation (5). Holycross and Watson (2016) showed that both E and $\log D_0$ depend roughly linearly on the REE–O bond length (Cicconi et al. 2013). For hydrous rhyolite, the relations with REE elemental sequence shown in Holycross and Watson (2018) are slightly curved. Because REE–O bond lengths are not available for all REE, we use ionic radius to fit all trivalent REE (excluding Eu) diffusion data in the three different melts:

$$\ln D_{REE^{3+}}^{dry\ basalt8a} = -10.03 - \frac{25131 - 1738r}{T} \tag{45a}$$

$$\ln D_{REE^{3+}}^{rhy\,14b+4.1wt\%H_2O} = -9.12 - \frac{24194 + 16516(1.097 - r)^2}{T} \tag{45b}$$

$$\ln D_{REE^{3+}}^{rhy\,14b+6.2wt\%H_2O} = -8.35 - \frac{23250 + 21657(1.069 - r)^2}{T} \tag{45c}$$

where r is trivalent ionic radius of REE in Å in octahedral site from Shannon (1976) (listed in Table 9 for convenience), T is in K, and D is in m²/s. Trial fittings show that adding a dependence of $\ln D_0$ on the ionic radius does not improve the fitting. Equation (45a) reproduces all the diffusivities in Supplementary Table B of Holycross and Watson (2016) to within 0.22 $\ln D$ units, and the REE diffusivities in Supplementary Table A to within 0.24 $\ln D$ units. Equation (45b) reproduced experimental data of trivalent REE diffusivity in rhyolite14b containing ~4.1 wt% H_2O within 0.28 $\ln D$ units. Equation (45c) reproduced experimental data of trivalent REE diffusivity in rhyolite14b containing ~6.2 wt% H_2O within 0.31 $\ln D$ units.

Table 9. Diffusion parameters for some trace elements.

	r (Å)	basalt8a E	$\log D_0$	rhy14b+4.1wt%H_2O E	$\log D_0$	rhy14b+6.2wt%H_2O E	$\log D_0$
Li						39.31	−7.35
Rb		178.33	−4.69				
Sr		161.7	−5.10				
Ba		181.1	−4.67				
V	0.640	203.3	−4.06	185.0	−4.90	222.4	−2.67
Sc	0.745	202.6	−4.14	228.8	−3.42	211.4	−3.24
Y	0.900	195.1	−4.39	188.3	−4.66	165.7	−5.09
La	1.032	191.41	−4.43	188.31	−4.51	203.34	−3.21
Ce	1.01	192.75	−4.41	201.66	−3.97	198.90	−3.33
Pr	0.99	193.18	−4.40	194.63	−4.27	202.07	−3.28
Nd	0.983	193.36	−4.40	203.61	−3.93	200.60	−3.35
Sm	0.958	194.30	−4.37	206.69	−3.84	205.46	−3.18
Eu		188.98	−4.41	166.10	−4.98		
Gd	0.938	194.87	−4.37	209.08	−3.79	193.84	−3.74
Tb	0.923	195.55	−4.35	201.13	−4.14	203.97	−3.32
Dy	0.912	196.42	−4.33	214.66	−3.60	185.72	−4.17
Ho	0.901	196.77	−4.33	210.90	−3.80	190.93	−3.91
Er	0.890	197.42	−4.31	201.01	−4.22	210.29	−3.98
Yb	0.868	198.07	−4.30	218.59	−3.54	171.42	−4.90
Lu	0.861	198.86	−4.29	209.63	−3.93	196.97	−3.82
Zr		219.7	−3.85	182.4	−5.45	155.2	−6.36
Hf		223.8	−3.81	231.1	−3.52		
Th		213.4	−4.02			176.7	−5.10
U		212.0	−3.98	228.8	−3.42	267.8	−1.06
Nb		206.1	−4.18	214.5	−4.03	179.5	−4.91
Ta		218.2	−3.92				
P		147.0	−6.30				

Note: The unit of E is kJ/mol. The unit of D_0 is m²/s. Note $\log D_0$ values rather than $\ln D_0$ are listed following the original authors. Compositions of basalt8a and rhyolite14b (rhy14b) are listed in Table 8. Data are from Holycross and Watson (2016, 2018), except for Li (Holycross et al. 2018) and P (Watson et al. 2015). For REE, E and $\log D_0$ are based on ratio-fitting method in Holycross and Watson (2016, 2018), which have better consistency. Ionic radii are for trivalent cations in octahedral sites from Shannon (1976). A single coordination (octahedral) is used for consistency with no implication on the real coordination number.

Such high accuracy reflects the high self-consistency of the REE diffusion data in Holycross and Watson (2016, 2018). Equations (45a–c) should be able to predict Eu^{3+} (and hence assess the contribution of Eu^{3+} and Eu^{2+} to Eu diffusion) and Tm diffusivity even though no Tm diffusion data were available in the literature. All three equations predict Y diffusion data in Holycross and Watson (2016, 2018) well, within 0.24 ln$D$ units, and Equation (45a) also predicts Sc diffusivity within 0.28 ln$D$ units. Using Equations (45b) and (45c) to predict Sc diffusivity would lead to large errors (1.0 ln$D$ units). Equations (45a–c) mean that there is larger difference in La to Lu diffusivities in wet rhyolite14b melts than in dry basalt8a.

Diffusivities of Li, Rb, Sr, Ba, Sn, V, Zr, Hf, Th, U, Nb and Ta

Holycross and Watson (2016, 2018) also reported high quality trace element diffusion (close to TED1) data for Rb, Sr, Ba, V, Zr, Hf, Th, U, Nb and Ta in dry basalt8a (Table 8), V, Zr, Hf, U, and Nb in rhyolite14b (Table 8) containing 4.1 wt% H_2O, and V, Zr, Th, U and Nb in rhyolite14b containing ~6.2 wt% H_2O. Watson et al. (2015) determined P diffusivity (POCGD) in dry basalt8a using diffusion couple experiments. The activation energies and pre-exponential factors for these elements in dry basalt8a and wet rhyolite14b are listed in Table 9. The diffusivities of tetravalent and pentavalent ions (HFSE) in basalts are shown in Figure 12. They are similar to each other (within ~0.5 ln$D$ units in basalt8a at 1500 to 1600 K), with

$$D_U \approx D_{Nb} > D_{Th} \geq D_{Zr} \geq D_{Ta} > D_{Hf} > D_P. \tag{46}$$

Si diffusivities in basalts with slightly different compositions (self diffusion in basalt6 in Table 8, and POCGD in basalt11 (Juan de Fuca MORB), and interdiffusivity in synthetic basalts) are also shown in Figure 12 for comparison. Due to slightly different compositions, direct comparison of HFSE diffusivities with Si diffusivity is not possible. By correcting to the same composition using Yu et al. (2019), Si diffusivity is equal to or slightly smaller than P diffusivity. The information is used in updating the diffusivity sequence in basalt (Eqn. 28). Figure 12 also shows the high self-consistency in Si self diffusivity and Si POCGD, but high variability in Si interdiffusivity in a given melt.

Holycross et al. (2018) conducted Li diffusion couple experiments in wet rhyolite14b containing 6.0 wt% H_2O at 1063–1148 K and 1.0 GPa, and acquired Li trace element diffusivities (TED1) in addition to Li isotope fractionation profiles. Li diffusivities (TED1) in wet rhyolite at

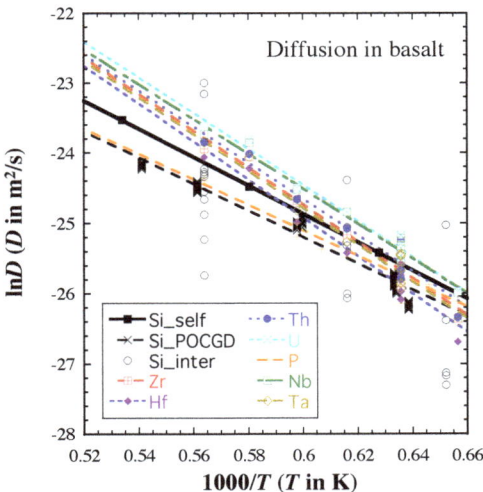

Figure 12. Comparison of HFSE diffusivities. Diffusivities of Zr, Hf, Th, U, P, Nb and Ta are for basalt8a (Table 8) at 1 GPa and are from Holycross and Watson (2016) and Watson et al. (2015). Si self diffusivities are for composition basalt6 (Table 8) at 1 GPa from Lesher et al. (1996). Si POCGD values are for basalt11 (Table 8) from quartz dissolution experiments at 0.5 GPa by Yu et al. (2019). Si interdiffusivity values are for basalt11a and haplobasalt2 (Table 2) at 0.5 to 1 GPa from multicomponent diffusion experiments of Guo and Zhang (2016, 2018, 2020). The large variability of Si interdiffusivity is due to different counter-diffusion component.

1 GPa are higher by 1.8 ln D units than Li tracer diffusion data in a dry rhyolite at 1 atm (Jambon and Semet 1978). The activation energy and pre-exponential factor are listed in Table 9.

Yang et al. (2016) carried out diffusive cassiterite dissolution experiments in various dry and wet rhyolites to determine Sn diffusivity. Sn diffusivity depends on its oxidation state (Sn^{2+} and Sn^{4+}). It was inferred that when graphite capsule is used, Sn in rhyolite melt is mostly divalent. The main concentration gradient is in SnO, and other components are diluted by additional SnO. Hence, the diffusion data are POCGD. Divalent Sn diffusivity in various reduced rhyolites at 1023-1373 K, 0.5 GPa, and 0–5.9 wt% H_2O can be described as follows:

$$\ln D_{SnO}^{silicic\ melts} = -18.194 + 17(0.76 - w_{SiO_2}) - \frac{19418 - 138900 w_{H_2O}}{T} \tag{47}$$

where w_{SiO_2} and w_{H_2O} are mass fraction (not wt%) of SiO_2 and H_2O, T is in K, and D is in m²/s.

Zhang and Xu (2016) carried out diffusive zircon dissolution experiments in various dry and wet rhyolites to determine Zr diffusivity. Even though zircon also contains SiO_2, the dissolution leads to mainly ZrO_2 concentration gradient and the rest are mostly dilution by ZrO_2. Hence, these diffusivities are close to POCGD's. They considered all Zr diffusion data available at the time and came up with the following equation to relate Zr POCGD with T (1270–1890 K), P (0.5–1.5 GPa), and melt composition in various dry and wet rhyolites:

$$\ln D_{Zr}^{rhyolites} = -14.42 - \frac{38784(Si+Al) - 1836P - 3172}{T} \tag{48}$$

where Si + Al is the sum of Si and Al cation mole fractions calculated on wet basis (i.e., H^+ mole fraction is counted), P is in GPa, and T is in K. The effect of H_2O on Zr diffusivity seems to be simply its dilution of the network formers. The above equation reproduces the experimental Zr diffusion data of Zhang and Xu (2016) to within 0.59 ln D units (1σ error 0.29 ln D units). On the Zr diffusion data in rhyolites by Holycross and Watson (2018), the equation predicts six out of seven diffusivity values in the ~6.2 wt% H_2O rhyolite to within 0.27 ln D units, but for the three diffusivities in the ~4.1 wt% rhyolite and the other diffusivity in the ~6.2 wt% H_2O rhyolite, the error ranges from 0.88 to 1.88 ln D units. More effort is needed in the future to derive more accurate general expressions on the compositional dependence of Zr diffusivity.

SiO₂ diffusion

Yu et al. (2019) carried out quartz dissolution experiments in nominally dry rhyolite (NCO listed in Table 8) containing 0.10 wt% H_2O and nominally dry basalt11 (Table 8) containing 0.32 wt% H_2O, and determined effective binary diffusivity of SiO_2. SiO_2 is the major concentration gradient, and gradients of other oxides are largely due to the dilution of SiO_2 (also due to multicomponent diffusion effects, Fig. 6). Hence, technically SiO_2 diffusivity is still POCGD even though the strong SiO_2 concentration gradient causes diffusion of many oxides. Previously, it was thought that SiO_2 diffusivity largely depends on SiO_2 concentration (e.g., Watson 1982; Koyaguchi 1989; Lesher and Walker 1986; Richter et al. 2003; Macris et al. 2018). An SiO_2 concentration profile during any single experiment can indeed be modeled well assuming ln D_{Si} is linear to SiO_2 concentration (Fig. 5), and Boltzmann analysis of SiO_2 concentration profile in every experiment using Equation (18) also shows such a dependence (Yu et al. 2019). However, when D_{Si} values extracted from quartz dissolution in rhyolite are compared to those from quartz dissolution in basalt, it becomes clear that D_{Si} depends on Si + Al rather than SiO_2 alone. As shown in Figure 13, when ln D_{Si} is plotted against SiO_2 concentration, the trends for D_{Si} from quartz dissolution in rhyolite are offset from those for quartz dissolution in basalt. On the other hand, when ln D_{Si} is plotted against Si + Al cation mole fractions, the trends in rhyolite roughly line up with those in basalt.

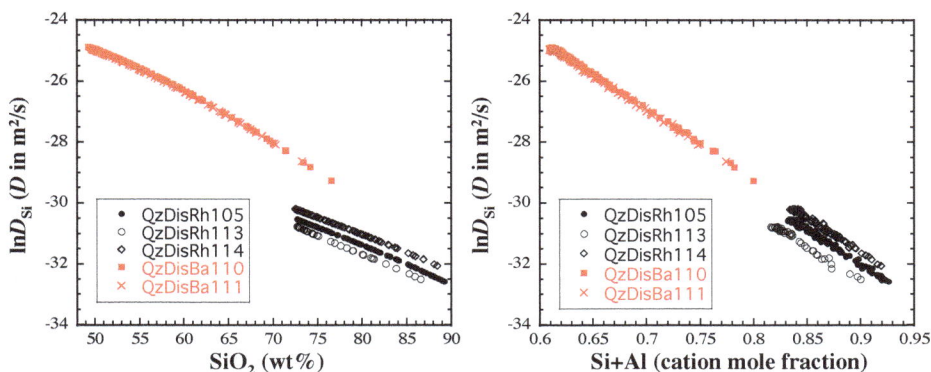

Figure 13. Si diffusivity from functional fitting results (**points**) as a function of SiO_2 concentration (**left**) or Si+Al mole fraction (**right**). **Black points** are from three quartz dissolution experiments in rhyolite (NCO, composition listed in Table 8) and **red points** are from two experiments in basalt (basalt11, composition listed in Table 8) melts at $1404 \pm 10°C$.

Using Si+Al cation mole fraction on wet basis, Yu et al. (2019) obtained the following equation for SiO_2 diffusivity (POCGD) in basalt to rhyolite at 1123–1873 K and 0.5 GPa:

$$\ln D_{Si}^{\text{quartz dissolution}} = -11.41 - 2.758(\text{Si+Al}) - \frac{38829(\text{Si+Al}) - 3826}{T} \tag{49}$$

where T is in K and D is in m^2/s. The above equation reproduces experimental data points in Yu et al. (2019) within $0.95 \ln D$ units (1σ error is $0.32 \ln D$ units). Hence, the accuracy in predicting D_{Si} using the above equation is much worse than that in predicting D_{Cu} using Equation (44), or in predicting REE diffusivities using Equations (45a–c). Some of the inaccuracy is almost certainly due to the dependence of D_{Si} on concentrations of other major oxides, but such dependence cannot be quantified yet. Limited data examined by Yu et al. (2019) seem to indicate that the above equation would work for wet rhyolite too, meaning that the effect of H_2O on reducing D_{Si} is largely due to its dilution of Si+Al cation mole fraction. Equation (49) reproduces Si self diffusivities at 1 GPa (Lesher et al. 1996) within $0.23 \ln D$ units (excellent accuracy). The SiO_2 EBD values during shoshonite–rhyolite diffusion couple experiments (Gonzalez-Garcia et al. 2017) cannot be reproduced well, with maximum deviation of $2.3 \ln D$ units (one order of magnitude) and 1σ error of $0.66 \ln D$ units, reflecting the dependence of OEBD on concentration gradients. Due to much higher pressure (4–24 GPa) and higher MgO contents (37 wt%), the SiO_2 self diffusivity in peridotite at ultrahigh pressures (Posner et al. 2018) also cannot be reproduced well by Equation (49) (which is for 0.5 GPa), with maximum deviation of $3.1 \ln D$ units.

Self diffusion of O, Si, Mg and Ca, and interdiffusivity of Ni and Co in a peridotite melt

Posner et al. (2018) investigated the self diffusion of O, Si, Mg, and Ca, and interdiffusion of Ni and Co in a peridotite melt at 2150–2623 K and 4-24 GPa using diffusion couple experiments. The melt composition is listed in Table 8, with one side of the diffusion couple containing 1 wt% NiO and the other side containing 1 wt% CoO. These are difficult experiments at extreme conditions. The pressure and temperature of the experiments co-varied and hence it is difficult to separate the effects of pressure and temperature on the diffusivities. By fixing the activation energy to some values, the self diffusivities presented by Posner et al. (2018) show a complicated pressure dependence. The diffusivities decrease with increasing pressure from 4 to 8 GPa, then increase with increasing pressure from 8 to 12 GPa, and then decrease again. For Ni and Co interdiffusivity, the pressure dependence is weaker.

Diffusion of Mo and W

Zhang et al. (2018) investigated Mo and W diffusion (close to TED1) in a rhyolite melt (rhyolite3a in Table 8) using both diffusion couple and Mo saturation experiments on both dry and wet melts at 1273–1873 K and 1 GPa. Their work provided the first data (13 points) on Mo diffusion, and was the second investigation (4 points) on W diffusion in aluminosilicate melts. They found that in dry rhyolite3a melt, Mo and W have similar (within 0.16 $\ln D$ units) diffusivities. Adding H_2O increases Mo diffusivity significantly. The Arrhenius relations for Mo and W diffusion are as follows:

$$\ln D_{Mo}^{Dry \ \& \ wet \ Rhyolite2} = -4.47 - 200 w_{H_2O} - \frac{44534 - 532358 w_{H_2O}}{T} \tag{50}$$

$$\ln D_{W}^{Dry \ Rhyolite2} = -2.95 - \frac{47628}{T} \tag{51}$$

where w_{H_2O} is the mass fraction of H_2O, T is in K and D is in m²/s. Equation (50) reproduces Mo diffusion data in Zhang et al. (2018) to within 0.63 $\ln D$ units (1σ uncertainty is 0.31 $\ln D$ units). Equation (51) reproduces W diffusion data in dry rhyolite3a in Zhang et al. (2018) to within 0.14 $\ln D$ units. Mo and W diffusivities in rhyolite3a are compared with those of Nb and Zr in rhyolite14b in Fig. 14. Note that rhyolite14b is much more silicic than rhyolite3a (Table 8). The activation energy for Mo diffusion in rhyolite3a containing ~5 wt% H_2O is smaller than those for Nb and Zr diffusion in rhyolite14b containing ~4.1 and ~6.2 wt% H_2O, leading to $D_{Mo} > D_{Nb}$ at $T < 1200$ K, and $D_{Mo} < D_{Nb}$ at $T > 1400$ K. Nonetheless, Mo and W diffusivities are small and are not very different from other HFSE.

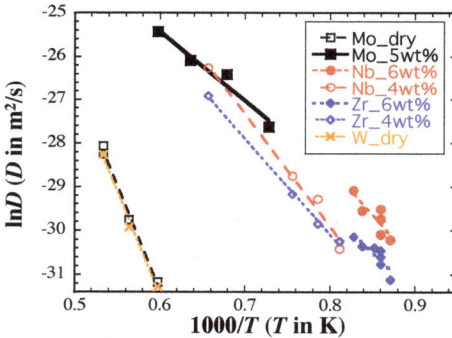

Figure 14. Mo and W trace element diffusivities (TED1) in rhyolite3a (Zhang et al. 2018) compared to Nb and Zr trace element diffusivities (TED1) in rhyolite14b (Holycross and Watson 2018). The composition of rhyolite3a and rhyolite14b are listed in Table 8, and rhyolite14b is more silicic than rhyolite3a. The limited data shows smaller activation energy for Mo diffusion in wet rhyolite3a containing ~5.0 wt% H_2O than Nb and Zr diffusion in wet rhyolite14b containing ~4.1 wt% and ~6.2 wt% H_2O.

Diffusion of F, Cl, and S

Bohm and Schmidt (2013) studied F and Cl diffusion (close to POCGD) in a phonolite2 melt (Table 8) containing ≤ 2.4 wt% H_2O using diffusion couple experiments at 1073–1473 K and 0.1 GPa. In dry phonolite2, F diffusivity is higher than Cl diffusivity by about an order of magnitude. The composition and H_2O concentration range investigated are similar to those by Balcone-Boissard et al. (2009) but the diffusion data do not line up in one trend, indicating either subtle dependence on composition or inter-laboratory inconsistency. Bohm and Schmidt (2013) provided the following Arrhenius equations for F and Cl diffusivities:

$$\ln D_{F}^{dry \ phonolite2} = -18.24 - \frac{12003}{T} \tag{52a}$$

$$\ln D_F^{\text{phonolite2+2.1wt\%H}_2\text{O}} = -17.36 - \frac{11678}{T} \tag{52b}$$

$$\ln D_{Cl}^{\text{dry phonolite2}} = -15.78 - \frac{18413}{T} \tag{53a}$$

$$\ln D_{Cl}^{\text{phonolite2+2.4wt\%H}_2\text{O}} = -13.72 - \frac{18570}{T} \tag{53b}$$

where T is in K and D is in m^2/s.

Yoshimura (2018) examined Cl diffusion in a high-silica rhyolite containing $\leq$ 1.2 wt% H_2O and also reported Ca diffusion data as a byproduct. The composition of the rhyolite is similar to rhyolite14b in Table 8. The Cl diffusion data in dry rhyolite in Yoshimura (2018) are 2–3 orders of magnitude lower than those in Bai and Koster van Groos (1994). Yoshimura (2018) explained this by compromising of the latter data by Na infiltration. After removing the data by Bai and Koster van Groos (1994), Cl diffusivity decreases from basalt to phonolite to rhyolite.

Lierenfeld et al. (2018) examined sulfur diffusion (TED1) in wet dacite melt (4.5–6.0 wt% H_2O) using diffusion couple experiments at 1223–1373 K, 0.20–0.25 GPa, and at logf_{O_2} of FMQ-0.8 (S is dominantly S^{2-}) and FMQ$+2.5$ (S is dominantly S^{6+}). The composition of the dacite is listed as dacite3a in Table 7. The effect of oxidation state on sulfur diffusivity was anticipated but previously unresolved due to data scatter (Behrens and Stelling 2011). With well-designed experiments, Lierenfeld et al. (2018) clearly resolved the effect of logf_{O_2} on S diffusivity, and found that S diffusivities at FMQ-0.8 is about 15 times those at FMQ$+2.5$. The equations to describe sulfur diffusivity in dacite containing 4.5 wt% H_2O at 0.2 GPa are as follows (Lierenfeld et al. 2018):

$$\ln D_{S \text{ at QFM+2.5}}^{\text{Dacite+4.5wt\%H}_2\text{O}} = -13.63 - \frac{16513}{T} \tag{54a}$$

$$\ln D_{S \text{ at QFM-0.8}}^{\text{Dacite+4.5wt\%H}_2\text{O}} = -11.93 - \frac{15118}{T} \tag{54b}$$

where T is in K. Their sulfur diffusion data at 6.0 wt% H_2O are scattered.

Major and trace element diffusion (OEBD) in shoshonite–rhyolite diffusion couple

Gonzalez-Garcia et al. (2017, 2018) carried out shoshonite–rhyolite diffusion couple experiments in both dry and wet ($\leq$2.0 wt% H_2O) conditions at 1473 K and 0.05–0.5 GPa. The compositions of the shoshonite and rhyolite (rhylite8a) are listed in Table 8. Due to the presence of significant concentration gradients in all major oxides, the diffusivities belong to the other types of effective binary diffusivities (OEBD). Numerous elements (e.g., Al, Na, La, Ce, Pr, Nd, Sm, Gd, Tb) show uphill diffusion, and OEBD cannot be extracted for them. For 19 elements with monotonic concentration profiles (Si, Ti, Fe, Mg, Ca, K, Rb, Cs, Sr, Ba, Co, Sn, Eu, V, Cr, Hf, Th, U, Ta), they found that the shapes of the profiles indicate that the diffusivity of each element depends on the bulk composition, as expected since SiO_2 concentration varies from 53 to 73 wt%. They extracted a large number of diffusion coefficients using Boltzmann–Matano analysis. These diffusivities depend on the direction and magnitude of the concentration gradient of all major oxides in addition to the dependence on the bulk composition. Their best applicability is to investigate the kinetics and dynamics of shoshonite–rhyolite mixing. Importantly, Gonzalez-Garcia et al. (2017, 2018) provided data to examine how OEBD values of many elements depend on H_2O concentration, which were previously unavailable. Such dependence might be applicable to the diffusion of these elements under other conditions (such as tracer diffusivity or POCGD).

Data by Gonzalez-Garcia et al. (2017, 2018) show that many components move in coordinated fashion with similar diffusivities, which are in agreement with observations by Watson (1982), Koyaguchi (1989), Richter et al. (2003), Macris et al. (2018), Yu et al. (2019), among others. Gonzalez-Garcia et al. (2017) found that OEDB values increases with increasing H_2O by 0.8–2.3 $\ln D$ units per wt% of H_2O, and decreases with increasing SiO_2 by 0.02–0.12 $\ln D$ units per wt% SiO_2. The latter is roughly consistent with Yu et al. (2019), but predicted D_{SiO_2} using Equation (49) is on average lower than OEBD of SiO_2 (Gonzalez-Garcia et al. 2017) by 0.7 $\ln D$ units with large scatters, revealing the role of different concentration gradients in OEBD, or more generally, multicomponent effects. No general equations were provided by Gonzalez-Garcia et al. (2017, 2018) to relate D with melt composition and pressure.

DIFFUSIVE ELEMENTAL AND ISOTOPE FRACTIONATION DURING MAGMATIC PROCESSES

Diffusion is ubiquitous in magmas. Therefore, it is of interest to understand the possibility and magnitude of diffusive fractionation of isotopes and elements in magmas. Equilibrium fractionation of isotopes and elements is fairly well understood (e.g., Gast 1968; Shaw 1970; Allegre and Minster 1978, and numerous partitioning studies). On the other hand, attention on diffusive fractionation of isotopes and elements is more recent. Jambon (1980) first proposed that isotopes could be diffusively fractionated, which can be recorded by growing crystals from magmas. Richter et al. (1999) were the first to measure diffusive isotope fractionation of $^{48}Ca/^{40}Ca$ in CAS system and $^{76}Ge/^{70}Ge$ in GeO_2 melt using spiked isotopes. Chopra et al. (2012) investigated possible diffusive isotope fractionation in igneous rocks and showed that current instrumental capability can measure such fractionations. If isotope ratios can be fractionated, elemental ratios and patterns can of course also be fractionated by diffusion. Holycross and Watson (2016, 2018) and Watson (2017) discussed diffusive elemental fractionation.

Diffusive fractionation requires different diffusivities. Elemental diffusivities for most elements are available (see reviews by Zhang et al. 2010 and this work) and can be used to discuss elemental fractionation. For isotope diffusion, differences in diffusivities of isotopes usually cannot be resolved by measuring diffusivities of individual isotopes. Instead, the ratio of diffusivities of different isotopes is determined from experimental isotope ratio profiles and related to the mass ratio of the isotopes. If each isotope diffuses freely as individual atoms, diffusivities of heavy and light isotopes can be related by Graham's law (Richter et al. 2003):

$$\frac{D_H}{D_L} = \sqrt{\frac{m_L}{m_H}} \tag{55}$$

where m_H and m_L are the atomic masses, and D_H and D_L are the diffusivities of heavy and light isotopes. If heavy and light isotopes diffuse freely as individual neutral molecules, then

$$\frac{D_H}{D_L} = \sqrt{\frac{M_L}{M_H}} \tag{56}$$

where M_H and M_L are the molecular masses of those containing heavy and light isotopes. If isotopes diffuse as clusters exchanging with other species, then (Richter et al. 2003)

$$\frac{D_H}{D_L} = \sqrt{\frac{M_L(M_H + M)}{M_H(M_L + M)}} \tag{57}$$

where M is the mass of the counter-diffusing species. However, silicate melts are complicated and the diffusion species and mechanisms are complicated (e.g., see multicomponent diffusion eigenvectors) and not accurately known. Hence, an empirical approach is used to characterize the relation between D_H and D_L as follows (Richter et al. 1999):

$$\frac{D_{\mathrm{H}}}{D_{\mathrm{L}}} = \left(\frac{m_{\mathrm{L}}}{m_{\mathrm{H}}}\right)^{\beta} \tag{58}$$

where β is an empirical fit parameter.

Consider isotope fractionation of a given element in a diffusion couple. Suppose the left hand side has a lower concentration and the right hand side has a higher concentration. Treat the diffusion as effective binary diffusion. The element diffuses from the right hand side to the left hand side. The light isotope diffuses more rapidly, and hence is enriched in the LHS. In other words, the LHS is depleted in the heavy isotope, and the RHS is enriched in the heavy isotope. Quantitatively, the concentration profile of each isotopes is an error function (Eqn. 7). Hence, if the effective binary diffusivity is roughly constant, the isotope ratio, using $^{41}K/^{39}K$ as an example, is expressed as follows:

$$\frac{^{41}K}{^{39}K} = \frac{0.5(C_{41,\mathrm{LHS}} + C_{41,\mathrm{RHS}}) + 0.5(C_{41,\mathrm{RHS}} - C_{41,\mathrm{LHS}})\,\mathrm{erf}\dfrac{x - x_0}{\sqrt{4D_{41}t}}}{0.5(C_{39,\mathrm{LHS}} + C_{39,\mathrm{RHS}}) + 0.5(C_{39,\mathrm{RHS}} - C_{39,\mathrm{LHS}})\,\mathrm{erf}\dfrac{x - x_0}{\sqrt{4D_{39}t}}} \tag{59}$$

where x_0 is the interface position, x increases from LHS to RHS, and subscripts 41 and 39 mean ^{41}K and ^{39}K. Converting to the δ-notation and using the initial ratio as standard lead to:

$$\delta\frac{^{41}K}{^{39}K} = \left(\frac{(1 + \dfrac{C_{\mathrm{RHS}}}{C_{\mathrm{LHS}}}) + (\dfrac{C_{\mathrm{RHS}}}{C_{\mathrm{LHS}}} - 1)\,\mathrm{erf}\dfrac{x - x_0}{\sqrt{4D_{39}(m_{39}/m_{41})^{\beta}t}}}{(1 + \dfrac{C_{\mathrm{RHS}}}{C_{\mathrm{LHS}}}) + (\dfrac{C_{\mathrm{RHS}}}{C_{\mathrm{LHS}}} - 1)\,\mathrm{erf}\dfrac{x - x_0}{\sqrt{4D_{39}t}}} - 1\right) 1000\%_0 \tag{60}$$

where $C_{\mathrm{RHS}}/C_{\mathrm{LHS}}$ is the initial concentration ratio of the RHS to the LHD (or concentration contrast). Model calculations (Fig. 15) using Equation (60) show that the magnitude of diffusive isotope fractionation depends on two parameters, one is the β value, and the other is the concentration ratio of the high concentration side to the low concentration side of the diffusion couple. By increasing the β value, or the concentration ratio, the magnitude of isotope fractionation increases. For example, if $\beta = 0.12$ and the concentration ratio is 60, then the total variation of $\delta^{41}K/^{39}K$ would be about 10‰. If $\beta = 0.12$ and the concentration ratio is 2, then the variation of $\delta^{41}K/^{39}K$ would be about 1‰. Both of these fractionations are measurable (Zhang et al. 2019a). One example of real data and fit is shown in Figure 16.

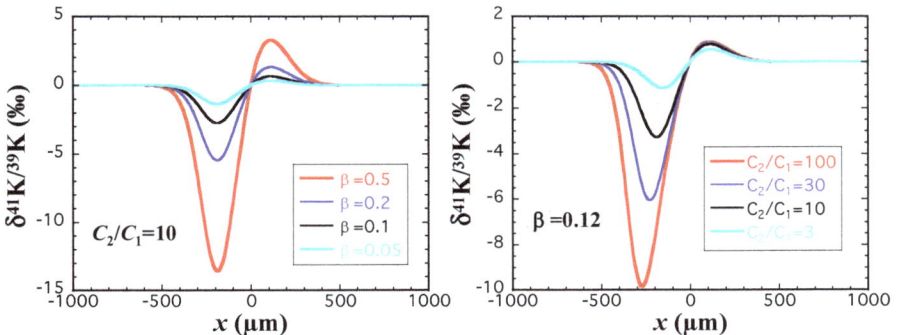

Figure 15. Calculated isotope fractionation in a diffusion couple as a function of β and concentration ratio $C_2/C_1 = C_{\mathrm{RHS}}/C_{\mathrm{LHS}}$. If $\beta = 0$ or $C_2/C_1 = 1$, there would be no isotope fractionation using effective binary treatment.

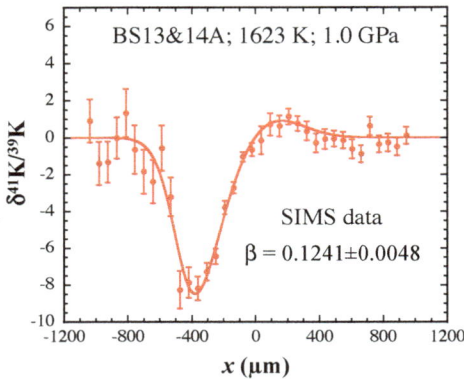

Figure 16. A $^{41}K/^{39}K$ isotope ratio profile (Zhang et al. 2019a) in a multicomponent diffusion couple experiment by Guo and Zhang (2018). The measurement is made by Secondary Ion Mass Spectrometry at Caltech Microanalysis Center. The initial concentration of K_2O is ~0.05 wt% at $x < 0$, and 3.06 wt% at $x > 0$. From Zhang et al. (2019a).

Richter et al. (1999) were the first to determine β values in CAS and GeO_2 melts. Richter et al. (2003, 2008, 2009) experimentally evaluated diffusive fractionation of $^7Li/^6Li$, $^{26}Mg/^{24}Mg$, $^{44}Ca/^{40}Ca$, and $^{56}Fe/^{54}Fe$ in dry basalt–rhyolite diffusion couple and obtained: $\beta_{Li} = 0.215$, $\beta_{Mg} = 0.05$, $\beta_{Ca} = 0.075$, and $\beta_{Fe} = 0.03$. Watkins et al. (2009, 2011, 2014) examined $^{26}Mg/^{24}Mg$ and $^{44}Ca/^{40}Ca$ fractionation in basalt–rhyolite, albite–anorthite, and albite–diopside diffusion couples and SiO_2–CaO–Na_2O system, and found that β_i (where i is an element) increases with D_i/D_{Si}. Watkins et al. (2014) developed the theory to treat isotope diffusion in the context of multicomponent diffusion. More experimentally determined β values and the associated experimental conditions can be found in Table 10.

Table 10. Experimentally determined β values for diffusive isotope fractionation.

Isotopes	β	D_i/D_{Si}	Melt	T (K)	P (GPa)	Ref
$^7Li/^6Li$	0.215 ± 0.005	290	basalt–rhyolite	1623–1723	1.2–1.3	1
$^7Li/^6Li$	0.228	2560	wet rhyolite	1103	1.2	2
$^{26}Mg/^{24}Mg$	0.05 ± 0.01	~1	basalt–rhyolite	1673	1.0–1.2	3
$^{26}Mg/^{24}Mg$	0.10 ± 0.01	1.5	albite–diopside	1723	0.8	4
$^{26}Mg/^{24}Mg$	0.045	~1	basalt–rhyolite	1773	1.45	5
$^{37}Cl/^{35}Cl$	0.09 ± 0.02		dacite	1473–1623	1.0	11
$^{41}K/^{39}K$	~0.12	1.64	basalt	1623	1.0	6
$^{48}Ca/^{40}Ca$	~0.08		CAS	1773	1.0	7
$^{44}Ca/^{40}Ca$	0.075 ± 0.025	1.6	basalt–rhyolite	1623–1723	1.2–1.3	1
$^{44}Ca/^{40}Ca$	0.035 ± 0.005	2.2	basalt–rhyolite	1723	1.0–1.3	8
$^{44}Ca/^{40}Ca$	0.21 ± 0.015	23	albite–anorthite	1723	0.8	4
$^{44}Ca/^{40}Ca$	0.165 ± 0.01	6.3	albite–diopside	1723	0.8	4
$^{44}Ca/^{40}Ca$	0.10		NCS(Ca–Na)	1523	0.8	9
$^{44}Ca/^{40}Ca$	0.035	~1	NCS(Ca–Si)	1523	0.8	9
$^{56}Fe/^{54}Fe$	0.03 ± 0.01	1.3	basalt–rhyolite	1673	1.0–1.2	10
$^{76}Ge/^{70}Ge$	< 0.025		GeO_2	1673	0.5	7

Note: Melt: CAS means CaO–Al_2O_3–SiO_2 system; NCS(Ca–Na) means CaO–Na_2O interdiffusion in Na_2O–CaO–SiO_2 system; NCS(Ca–Si) means CaO–SiO_2 interdiffusion in Na_2O–CaO–SiO_2 system;

References: 1. Richter et al. (2003); 2. Holycross et al. (2018); 3. Richter et al. (2008); 4. Watkins et al. (2011); 5. Chopra et al. (2012); 6. Zhang et al. (2019a); 7. Richter et al. (1999); 8. Watkins et al. (2009); 9. Watkins et al. (2014); 10. Richter et al. (2009); ; 11. Fortin et al. (2017)

Diffusion also leads to elemental fractionation in magmas (Holycross and Watson 2016, 2018; Watson 2017). Holycross and Watson (2016) examined the fractionation of La/Lu ratio in a diffusion couple and found that diffusive fractionation can be significant and measurable. Here, we use REE diffusion coefficients in Equations (45a–c) to model diffusive fractionation of REE patterns in dry basalt to wet rhyolite along a diffusion couple profile. Eu diffusivity is assumed to be 2 times Eu^{3+} diffusivity. To illustrate an extreme (unrealistic) case, we set the initial concentration contrast to be 200: the left hand side initially has the same REE concentration as in chondrites and the right hand side initially contains 200 times chondrite REE concentration. Some calculated REE patterns in basalt and wet rhyolites are shown in Fig. 17. In this extreme case, the REE pattern is fractionated significantly, and more so in rhyolite than in basalt due to larger differences between La and Lu diffusivities in rhyolite. There is also a large Eu anomaly ($Eu/Eu^* \approx 3$). As the initial concentration contrast decreases, the maximum fractionation decreases and the location of the maximum fractionation moves closer to the interface. If the concentration contrast is reduced to a factor of 2, then the REE pattern is much less fractionated, with normalized $(La/Lu)_{CI}$ ratio fractionated by $\leq 2.1\%$ in basalt and $\leq 8.4\%$ in rhyolite $+6.2$ wt% H_2O, and $Eu/Eu^* \leq 1.074$.

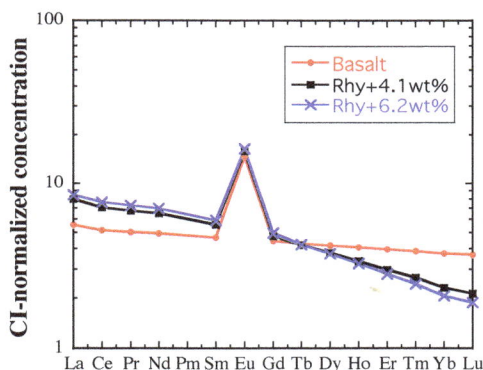

Figure 17. Calculated REE patterns due to diffusive fractionation in a diffusion couple. The initial concentrations in the diffusion couple are: same as chondrite at $x < 0$, and 200 times chondrite at $x > 0$. The diffusivities are calculated from Equations (45a–c), at 1473 K for basalt, 1273 K for rhyolite $+4.1$ wt% H_2O, and 1173 K for rhyolite $+6.2$ wt% H_2O. D_{Eu} is set to be two times the diffusivity of Eu^{3+}. The patterns shown here are at $x = -1.5(Dt)^{1/2}$, where D is the average for all trivalent REE. This is roughly the position where the largest diffusive fractionation occurs.

DIFFUSIVITY IN CRYSTAL-BEARING AND BUBBLE-BEARING MAGMAS

Many diffusion media in geology are heterogeneous media, either due to the presence of multiple phases (such as mantle rocks, or magmas containing phenocrysts and/or bubbles), or the presence of boundaries that show different diffusion properties (e.g., Dohmen and Milke 2010). In such a system, diffusion at a length scale much larger than the heterogeneity (i.e., grain size) may be characterized by a bulk diffusivity or effective diffusivity, which for a given component i may be defined by:

$$\mathbf{J}_{i,\text{bulk}} = -D_{i,\text{bulk}}\nabla C_{i,\text{ave}}, \tag{61}$$

where $\mathbf{J}_{i,\text{bulk}}$ is the bulk flux, $D_{i,\text{bulk}}$ is the bulk diffusivity and $C_{i,\text{ave}}$ is the average concentration (in kg/m^3 or mol/m^3) of component i. $C_{i,\text{ave}}$ is defined as:

$$C_{i,\text{ave}} = \phi_1 C_{i,1} + \phi_2 C_{i,2} + ..., \tag{62}$$

where ϕ_1 and ϕ_2 are volume fractions of phases 1 and 2, and $C_{i,1}$ and $C_{i,2}$ are concentrations (in kg/m^3 or mol/m^3). To treat bulk diffusion in a heterogeneous medium, it is necessary to know how the bulk diffusivity is related to individual-phase diffusivity. To simplify the task, all ϕ_i's are assumed to be constant so that growth and dissolution of crystals and bubbles are not considered in this section. Including growth and dissolution would require another set of kinetic equations to be solved together with diffusion and is beyond the scope of this review.

To the authors' knowledge, Brady (1983) first introduced the treatment of diffusion in heterogeneous media to geology literature and derived relations between bulk diffusivity and individual-phase diffusivities using the similarity between diffusivity and thermal conductivity. Unfortunately, the similarity does not hold perfectly, leading to errors in the derived relations. These errors were also found in other studies and famous books (e.g., Bell and Crank 1974; Crank 1975, Chapter 12; Davis et al. 1975; Cussler 1997, section 6.5.2). Zhang and Liu (2012) identified the error by realizing a key difference between diffusivity and conductivity. During thermal conduction, the heat flux is written to be proportional to temperature gradient. During diffusion, mass flux is normally written to be proportional to the concentration gradient. The difference is that temperature is a continuous function when a phase boundary is crossed, but concentration of a component is not continuous at local equilibrium. The discontinuity means that earlier derived relations for bulk or effective diffusivity in heterogeneous media only apply when the partition coefficient is 1 between the phases. Once this is realized, because chemical potential is continuous across phase boundaries, new analogy equations can be written between diffusion mobility and thermal conductivity where diffusion mobility M is defined as:

$$\mathbf{J}_i = -M_i \nabla \frac{\mu_i}{RT} \tag{63}$$

$\mathbf{J}_i$ is mass flux, μ_i is chemical potential, and M_i is the mobility of component i. Zhang and Liu (2012) showed that in ideal and roughly ideal mixtures (note that Fick's law only applies to ideal and roughly ideal systems),

$$M_i \approx D_i C_i. \tag{64}$$

Hence, in deriving the relation between bulk diffusivity and individual-phase diffusivity, $D_i C_i$ together (rather than D_i alone) should replace thermal conductivity in relating bulk conductivity and individual-phase conductivity. Zhang and Liu (2012) discussed some applications of the new analogy relations. Here we discuss bulk (or effective) diffusivity in crystal-bearing and bubble bearing magmas. We limit our discussion to low percentages of crystals and bubbles so that they do not interact with each other. What exactly is meant by low percentage is not precisely defined, but we expect that the derived relations are applicable at a volume fraction $\phi \leq 0.1$ and possibly at ϕ up to 0.2.

Thermal conductivity and electrical conductivity for heterogeneous media follow similar relations (Kerner 1956; Hashin and Shtrikman 1963). Maxwell (1873, p. 365) derived an expression for electrical resistivity when there are numerous spheres of phase 1 in phase 2, which can be written in terms of electrical conductivity:

$$\sigma_{\text{bulk}} = \frac{2\sigma_2 + \sigma_1 - 2\phi(\sigma_2 - \sigma_1)}{2\sigma_2 + \sigma_1 + \phi(\sigma_2 - \sigma_1)} \sigma_2 \tag{65}$$

where σ means electrical conductivity, and ϕ means the volume fraction of phase 1 in the continuous phase 2. Using Zhang and Liu (2012) analogy between diffusivity and thermal conductivity, we obtain:

$$\frac{D_{\text{bulk}}}{D_2} = \frac{2 + KD_1 / D_2 - 2\phi(1 - KD_1 / D_2)}{2 + KD_1 / D_2 + \phi(1 - KD_1 / D_2)} \frac{C_2}{C_{\text{ave}}} \tag{66}$$

where D_1 and C_1 are diffusivity and concentration (kg/m^3 or mol/m^3) in the dispersed phase 1 (such as crystals and/or bubbles), D_2 and C_2 are diffusivity and concentration (kg/m^3 or mol/m^3) in the continuous phase 2 (the melt in this work), $K = C_1/C_2 = \rho_1 w_1/(\rho_2 w_2)$ is the partition coefficient (taking into consideration the density difference between crystals and melt), and $C_{\text{ave}} = \phi C_1 + (1-\phi)C_2$. The above derivation assumes that local equilibrium is reached between the dispersed spherical particles (all of which are the same phase, defined as phase 1) and the continuous melt (phase 2). That is, the diffusion distance in the dispersed spheres must be about the same as the radius. In addition, the diffusion distance in the continuous phase must be ≥ 10 times the particle diameter.

The bulk diffusivity is further elucidated below. If one plots bulk concentration (Eqn. 62) versus distance, the diffusivity obtained from fitting the profile is the bulk diffusivity. In addition, if one plots the concentration in any individual phase versus distance, the diffusivity obtained from fitting the profile is also the bulk diffusivity. This is illustrated in Figure 18 for the case of melt and one phenocryst phase, with four panels: (a) a hypothetical measured concentration profile as one moves along a line that encounters both phases (this profile has spikes and cannot be fit by a constant D), (b) the average concentration profile, (c) the concentrations measured in the melt, and (d) the concentrations measured in phenocryst grains. Note that the average concentration, the concentration in the melt, and the concentration in phenocryst grains, are all proportional to one another. Hence, the three profiles in Figure 18b,c,d are identical when normalized to the concentration at $x = 0$, and this normalized profile is characterized by a diffusivity equaling D_{bulk}. Therefore, in the case of diffusion in crystal or bubble-bearing magma, assuming local equilibrium between crystals and melt and between bubbles and melt, the diffusivity obtained by measuring concentrations profiles in any single phase (continuous phase, or many discrete grains along a direction) is also the bulk diffusivity or the effective diffusivity.

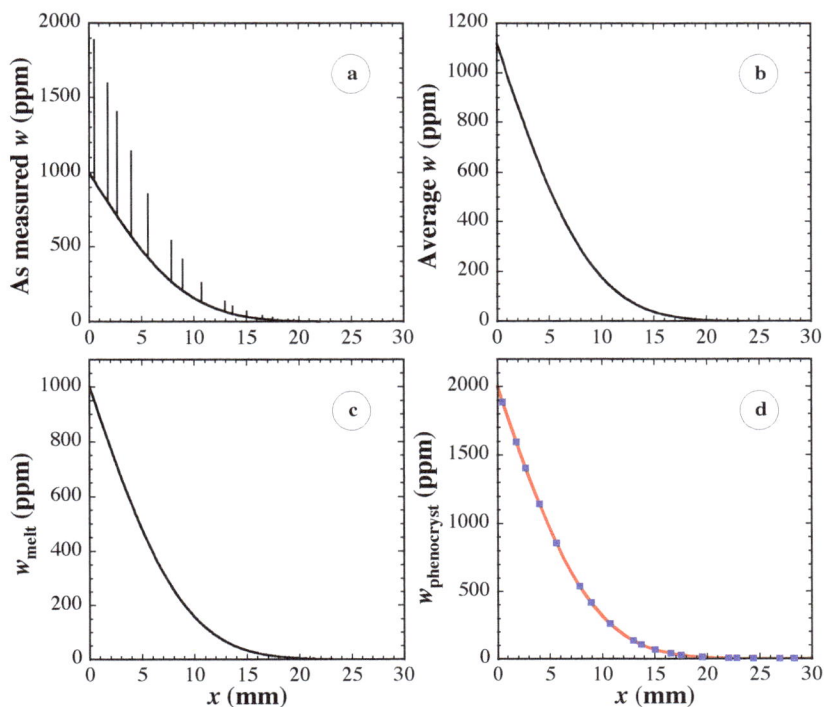

Figure 18. Calculated concentration profiles in a microphenocryst-bearing magma for the case of semi-infinite diffusion from one surface. Equilibrium elemental partition is assumed and the concentration in the microphenocryst is 2 times that in the melt. **(a)** A hypothetical measured profile when the measured points often encounter melt (glass) but occasionally encounter microphenocryst grains (**spikes in the curve**); **(b)** Calculated average concentration profile based on constant microphenocryst fraction; **(c)** Concentration profile by measuring points in the melt; **(d)** concentration profile by measuring points (**blue squares**) in different grains of the microphenocryst and fit to the points.

Crystal-bearing magmas

Here we apply Equation (66) to estimate bulk diffusivity in crystal-bearing magmas assuming equilibrium is reached between crystals and the melt. Let phase 1 be crystals, and phase 2 (continuous phase) be melt in Equation (66). Some limiting cases of interest are discussed below.

At the limit of very small $KD_{crystal}/D_{melt} \ll 1$ (this applies to essentially all components), Equation (66) becomes

$$\frac{D_{bulk}}{D_{melt}} = \frac{2(1-\phi)}{2+\phi} \frac{1}{\phi K + (1-\phi)} \tag{67}$$

The above equation applies when the following conditions are satisfied:

(1) Diffusion distance in the melt is much greater than the diameter of the phenocrysts;

(2) Diffusion in the phenocrysts roughly reached the center of average phenocrysts, so that the phenocrysts are roughly in local equilibrium with the melt.

We temporarily define diffusion reaching the center to mean the center concentration reaching at least 72% equilibrium (which means that the whole phenocrysts reached >91.5% equilibrium), leading to $D_{crystal}t/a^2 \geq 0.2$ where a is radius (Crank 1975, p. 92). For typical diffusivities in minerals and melts, when condition 2 is satisfied, then condition 1 is also satisfied. For example, if the average phenocryst diameter is 0.1 mm ($a = 0.05$ mm), $D_{crystal} = 10^{-15}$ m²/s (Spandler and O'Neill 2010), and $D_{melt} = 5 \times 10^{-12}$ m²/s, then $D_{crystal}t/a^2 \geq 0.2$ means $t \geq 5.79$ days. After this time the diffusion distance in the melt is ≥ 1.58 mm, much greater than the phenocryst diameter. Therefore, condition 1 is also satisfied.

For a highly incompatible element ($K \ll 1$), Equation (66) becomes:

$$\frac{D_{bulk}}{D_{melt}} = \frac{1}{1+0.5\phi} \tag{68}$$

To apply Equation (68), the two conditions listed below Equation (67) as well as $K \ll 1$ must be satisfied.

The variation of D_{bulk}/D_{melt} as a function of ϕ and K is plotted in Figure 19. It can be seen that D_{bulk} is always smaller than D_{melt}. For $K < 1$, the effect of a small fraction of phenocrysts is within uncertainty of diffusion data ($\sim$30%, Zhang et al. 2010) at ≤ 20 vol% of phenocrysts. However, for highly compatible elements, D_{bulk} can be a factor of 3 lower than D_{melt} at 20 vol% of phenocrysts.

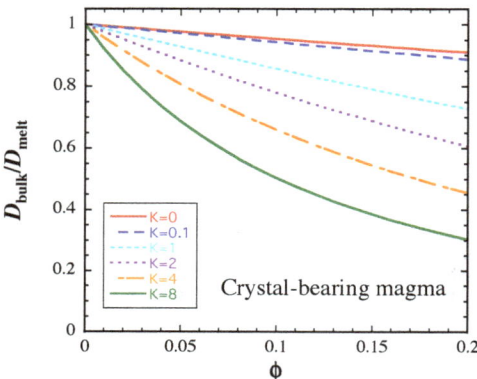

Figure 19. The dependence of D_{bulk}/D_{melt} on the volume fraction of crystals and partition coefficient $K = C_1/C_2 = \rho_1 w_1/(\rho_2 w_2)$ (where subscript 1 means crystal and 2 means melt) at the limit of $KD_1/D_2 \ll 1$.

If there is no rough equilibrium between the phenocrysts and melt, condition 2 below Equation (67) is not satisfied. Then Equation (67) cannot be applied. We consider another limiting case of constant crystal composition during a short-duration experiment. Because the phenocryst particles do not participate in the diffusion, one may just consider diffusion in the melt by ignoring partitioning. In this case, the phenocrysts play the role of small inert (non-active) blocks increasing tortuosity because atoms must diffuse around these particles, but do not participate in the compositional exchange. Mathematically, this may be treated using Equation (66) by adopting $K = 0$. Therefore, the effective diffusivity in the melt is given by Equation (68). That is, Equation (68) also describes how the tortuosity effect reduces the effective melt diffusivity in the limiting case that the crystals do not participate in the diffusion. Note that C_{ave} is irrelevant in the case of nonreactive phenocrysts because concentrations in the phenocrysts may be high or low but they do not participate in diffusion. Hence, effective diffusivity in the melt is a better term than bulk diffusivity to describe diffusion in this case. We temporarily define nonreactive crystals by $D_{crystal}t/a^2 \leq 10^{-4}$. For example, if the average phenocryst diameter is 100 μm ($a = 50$ μm), $D_{crystal} = 10^{-15}$ m²/s $= 10^{-3}$ μm²/s (Spandler and O'Neill 2010), then t must be ≤ 250 s, meaning diffusion distance in the crystal is $(D_{crystal}t)^{1/2} = 0.5$ μm, for Equation (68) to be applicable. In such a case, if $\phi = 0.2$, then $D_{eff}/D_{melt} = 0.91$. Because experimental diffusivities in silicate melts often have a relative error of 30% (e.g., Chen and Zhang 2008, 2009; Zhang et al. 2010), the effect of $\leq 20\%$ nonreactive crystals is within diffusion data/model uncertainty.

For the time regime of $10^{-4} < D_{crystal}t/a^2 < 0.2$, one may use Equations (68) and (67) to find the lower and upper limits of D_{eff} in the melt. For a more precise estimation, a weighted average of the upper and lower limits is taken using the degree of equilibrium of the phenocrysts as weight:

$$\frac{D_{eff}}{D_{melt}} = \frac{2(1-\phi)}{2+\phi}\frac{1}{\phi K+(1-\phi)}(\frac{\Delta M_t}{\Delta M_\infty})+(\frac{1-\phi}{1+0.5\phi})(1-\frac{\Delta M_t}{\Delta M_\infty}) \tag{69}$$

where $\Delta M_t/\Delta M_\infty$ (which may be estimated using Eqn. 13c or 14c) means the degree of equilibrium for diffusion in phenocryst grains.

Bubble-bearing magmas

We now estimate bulk diffusivity in bubble-bearing magmas using Equation (66), with phase 1 being bubbles, and phase 2 being melt. Equation (66) for this specific case can be written as

$$\frac{D_{bulk}}{D_{melt}} = \frac{2+KD_{bubble}/D_{melt}-2\phi(1-KD_{bubble}/D_{melt})}{2+KD_{bubble}/D_{melt}+\phi(1-KD_{bubble}/D_{melt})}\frac{1}{\phi K+(1-\phi)} \tag{70}$$

where $K_i = C_{i,bubble}/C_{i,melt}$, where $C_{i,bubble}$ and $C_{i,melt}$ must be in kg/m³ or mol/m³. There is only one condition for Equation (70) to be applicable: the diffusion distance in the melt must be much greater than the diameter of the bubbles. Adopting the unit of kg/m³ for concentrations in both phase, then $C_{i,bubble} = W_i X_{i,bubble}P/(RT)$ where $X_{i,bubble}$ is mole fraction of component i in the bubble and W_i is the molar mass of i in kg/mol, and $C_{i,melt} = w_{i,melt}\rho_{melt}$, where $w_{i,melt}$ is the mass fraction of i in the melt. Hence,

$$K_i = \frac{C_{i,bubble}}{C_{i,melt}} = (\frac{W_i X_{i,bubble}P}{w_{i,melt}\rho_{melt}RT}) = \frac{W_i}{S_i \rho_{melt}RT} \tag{71}$$

where $Si = w_i,melt/(X_{i,bubble}P)$ is solubility of i in the melt in mass fraction per Pa. The estimated values of K_i for H_2O and CO_2 at some conditions are listed in Table 11. For CO_2 and other gas species with solubility proportional to pressure, K_i is roughly a constant in a given melt. For H_2O, K_i increases as pressure increases. The values of $K_i D_{i,bubble}/D_{i,melt}$ are listed in the last column of Table 10 as Ratio. It can be seen that $K_i D_{i,bubble}/D_{i,melt}$ for gas species is much greater than 1.

Table 11. Estimated values of K_i for H_2O and CO_2.

Melt	Species	P MPa	T K	$C_{i,\text{bubble}}$ kg/m³	$D_{i,\text{bubble}}$ μm²/s	$w_{i,\text{melt}}$	$C_{i,\text{melt}}$ kg/m³	K_i	$D_{i,\text{melt}}$ μm²/s	Ratio
Rhyolite	H_2O	200	1100	394	72,000	0.0603	139	2.84	63.4	3,220
Rhyolite	H_2O	50	1200	90.3	328,000	0.0256	58.8	1.53	16.7	30,100
Rhyolite	H_2O	1	1400	1.55	2.07×10^7	0.00245	5.64	0.27	3.71	1.5×10^6
Basalt	CO_2	200	1500	706	26,400	0.001	2.7	261	6.07	1.1×10^6
Basalt	CO_2	1	1500	3.53	5.28×10^6	0.000005	0.0135	261	7.86	1.8×10^8

Note: H_2O and CO_2 solubilities are calculated based on Zhang et al. (2007). Bubble is assumed to be pure H_2O for the first three cases and pure CO_2 for the last two cases. Melt density is taken as 2300 kg/m³ for rhyolite and 2700 kg/m³ for basalt. $D_{i,\text{bubble}}$ is estimated using elementary theory of diffusion in gases ($D = lv/3$ where l is the mean free path and v is the mean thermal speed). $D_{i,\text{melt}}$ is calculated from Ni and Zhang (2018) for H_2O, and Zhang and Ni (2010) for CO_2. Ratio $= K_i D_{i,\text{bubble}}/D_{i,\text{melt}}$.

For the limiting rare case of $K_i D_{i,\text{bubble}}/D_{i,\text{melt}} \ll 1$ (i.e., for a component that does not go into the bubbles at all), Equation (70) simplifies to:

$$\frac{D_{\text{bulk}}}{D_{\text{melt}}} = \frac{1}{1 + 0.5\phi} \qquad (72)$$

Note that Equation (72) is the same as Equation (68), meaning that the presence of bubbles increases the tortuosity of the diffusion path. Because $D_{i,\text{bubble}}/D_{i,\text{melt}}$ is typically $\gg 1$ (decreasing with increasing pressure in the bubbles) and most components have some solubility in the vapor phase, the applicability of Equation (72) is very limited because K_i would need to be really small (e.g., $<10^{-6}$). For example, a rough estimation for TiO_2 component in pure H_2O fluid phase in equilibrium with a basalt results in a $K_i \approx 0.003$ at 1273 K and 600 MPa (~100 ppm Ti in fluid, Antignano and Manning, 2008; ~2 wt% TiO_2 in hydrous basalt, Ryerson and Watson, 1987) and $K_i D_{i,\text{bubble}}/D_{i,\text{melt}} > 3$ depending on the pressure and how dissolved H_2O in basalt melt increases $D_{\text{Ti,melt}}$. Hence, even for the TiO_2 component that has low solubility in a fluid phase, Equation (72) still does not apply at 1273 K and 600 MPa.

A more widely applicable limiting case is $K_i D_{i,\text{bubble}}/D_{i,\text{melt}} \gg 1$. Then Equation (70) simplifies to:

$$\frac{D_{\text{bulk}}}{D_{\text{melt}}} = \frac{(1 + 2\phi)}{(1 - \phi)} \frac{1}{\phi K + (1 - \phi)} \qquad (73)$$

Equation (73) is expected to apply well to H_2O and CO_2 and most other gas components (some $K_i D_{i,\text{bubble}}/D_{i,\text{melt}}$ values are listed in the last column of Table 11). Note that accurate values of $D_{i,\text{bubble}}$ and $D_{i,\text{melt}}$ are not needed as long as $K_i D_{i,\text{bubble}}/D_{i,\text{melt}} \gg 1$. Fig. 20 displays how $D_{\text{bulk}}/D_{\text{melt}}$ depends on ϕ and K. Note that D_{bulk} may be greater or smaller than D_{melt} depending on the value of K. When K is large (e.g., > 100, for CO_2), $D_{\text{bulk}}/D_{\text{melt}}$ can be much smaller than 1 even at a few percent of bubbles.

In literature studies of H_2O diffusion, often there were a few volume percent of bubbles present in the experimental charges. The authors stated that a few volume percent of bubbles would not affect the extracted diffusion coefficient of H_2O significantly (e.g., Zhang et al. 1991a). Our results in Figure 20 show the presence of 2 vol% of bubbles would increase the bulk diffusivity by less than 10%, and hence validate their statement. On the other hand, the effect of 2 vol% of bubbles could decrease the bulk diffusivity of CO_2 in melt by a factor of 6. The reduction of bulk diffusivity in the melt may sound counterintuitive. The reason is that even at 2 vol% of bubbles, most CO_2 is in bubbles, rather than in the melt. Although

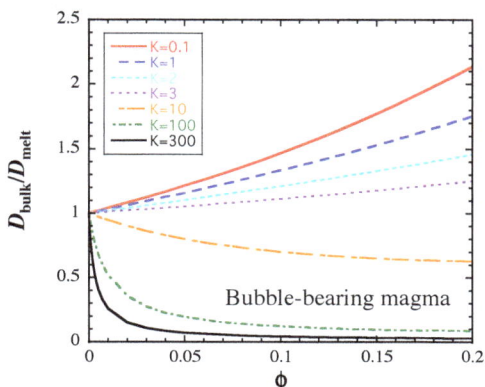

Figure 20. The dependence of D_{bulk}/D_{melt} on the volume fraction of bubbles and partition coefficient K at the limit of $KD_{bubble}/D_{melt} \gg 1$. The relation applies to essentially all gas components. Table 11 gives some estimated K and KD_{bubble}/D_{melt} values. No $K = 0$ limiting curve is shown because it violates the condition that $KD_{bubble}/D_{melt} \gg 1$.

diffusion in bubbles is rapid, bubbles are isolated from one another. Hence, CO_2 transport is by diffusion in the melt, whereas bubbles buffer the CO_2 concentration in the melt to some degree. Hence, as CO_2 diffuses in the melt, CO_2 concentration in the melt does not change so much as in the case of no bubbles, leading to a decrease in the effective CO_2 diffusivity. Because the presence of bubbles may significantly impact on CO_2 diffusivity as well as diffusivity of other gases (such as noble gases) with low solubility in silicate melts, it is critical to prevent bubbles in the experimental charges for diffusion of these gases (e.g., Spickenbom et al. 2010).

CONCLUSIONS

This review on diffusion in silicate melts and magmas mostly covers the progresses since the publication of *Diffusion in Minerals and Melts* as volume 72 of *Reviews in Mineralogy and Geochemistry* in 2010. Major advancement has been made in a number of fields. Multicomponent diffusion studies have made it possible to roughly predict diffusion of all major components in basalt melt during mixing of different basalts or mineral dissolution and growth. Diffusive isotope fractionation in melts has been examined for various elements in different melts, and simultaneous treatment of isotope diffusion and multicomponent diffusion has been developed. Theory has become available to treat diffusion in crystal-bearing or bubble-bearing magmas. In terms of diffusion data, great efforts have been made by some authors to generate a large number of data, which have been applied to model and understand diffusive elemental fractionation and have applications in many other diffusion problems.

ACKNOWLEDGEMENT

We thank an anonymous reviewer for constructive and insightful comments. This work is partially supported by NSF grants EAR-1829822 and EAR-2020603, and NASA grant 80NSSC19K0782.

REFERENCES

Allegre CJ, Minster JF (1978) Quantitative models of trace element behavior in magmatic processes. Earth Planet Sci Lett 38:1–25
Antignano A, Manning CE (2008) Rutile solubility in H_2O, H_2O–SiO_2, and H_2O–$NaAlSi_3O_8$ fluids at 0.7–2.0 GPa and 700–1000°C: implications for mobility of nominally insoluble elements. Chem Geol 255:283–293
Bai TB, Koster van Groos AF (1994) Diffusion of chlorine in granitic melts. Geochim Cosmochim Acta 58:113–123

Balcone-Boissard H, Baker DR, Villemant B, Boudon G (2009) F and Cl diffusion in phonolitic melts: influence of the Na/K ratio. Chem Geol 263:89–98

Behrens H (2010) Noble gas diffusion in silicate glasses and melts. Rev Mineral Geochem 72:227–267

Behrens H, Hahn M (2009) Trace element diffusion and viscous flow in potassium-rich trachytic and phonolitic melts. Chem Geol 259:63–77

Behrens H, Stelling J (2011) Diffusion and redox reactions of sulfur in silicate melts. Rev Mineral Geochem 73:79–111

Behrens H, Zhang Y (2001) Ar diffusion in hydrous silicic melts: implications for volatile diffusion mechanisms and fractionation. Earth Planet Sci Lett 192:363–376

Behrens H, Zhang Y (2009) H_2O diffusion in peralkaline to peraluminous rhyolitic melts. Contrib Mineral Petrol 157:765–780

Behrens H, Zhang Y, Leschik M, Miedenbeck M, Heide G, Frischat GH (2007) Molecular H_2O as carrier for oxygen diffusion in hydrous silicate melts. Earth Planet Sci Lett 254:69–76

Behrens H, Zhang Y, Xu Z (2004) H_2O diffusion in dacitic and andesitic melts. Geochim Cosmochim Acta 68:5139–5150

Bell GE, Crank J (1974) Influence of imbedded particles on steady-state diffusion. J Chem Soc Faraday Trans II 70:1259–1273

Bohm A, Schmidt BC (2013) Fluorine and chlorine diffusion in phonolitic melt. Chem Geol 346:162–171

Brady JB (1975) Reference frames and diffusion coefficients. Am J Sci 275:954–983

Brady JB (1983) Intergranular diffusion in metamorphic rocks. Am J Sci 283-A:181–200

Carroll MR (1991) Diffusion of Ar in rhyolite, orthoclase and albite composition glasses. Earth Planet Sci Lett 103:156–168

Carroll MR, Stolper EM (1991) Argon solubility and diffusion in silica glass: implications for the solution behavior of molecular gases. Geochim Cosmochim Acta 55:211–225

Carslaw HS, Jaeger JC (1959) Conduction of Heat in Solids. Clarendon Press, Oxford

Chakraborty S (1995) Diffusion in silicate melts. Rev Mineral 32:411–503

Chakraborty S, Dingwell DB, Rubie DC (1995a) Multicomponent diffusion in ternary silicate melts in the system $K_2O–Al_2O_3–SiO_2$: I. Experimental measurements. Geochim Cosmochim Acta 59:255–264

Chakraborty S, Dingwell DB, Rubie DC (1995b) Multicomponent diffusion in ternary silicate melts in the system $K_2O–Al_2O_3–SiO_2$: II. mechanisms, systematics, and geological applications. Geochim Cosmochim Acta 59:265–277

Chen Y, Zhang Y (2008) Olivine dissolution in basaltic melt. Geochim Cosmochim Acta 72:4756–4777

Chen Y, Zhang Y (2009) Clinopyroxene dissolution in basaltic melt. Geochim Cosmochim Acta 73:5730–5747

Cherniak DJ, Hervig RL, Koepke J, Zhang Y, Zhao D (2010) Analytical methods in diffusion studies. Rev Mineral Geochem 72:107–170

Chopra R, Richter FM, Watson EB, Scullard CR (2012) Magnesium isotope fractionation by chemical diffusion in natural settings and in laboratory analogues. Geochim Cosmochim Acta 88:1–18

Cicconi MR, Giuli G, Paris E, Courtial P, Dingwell DB (2013) XAS investigation of rare earth elements in sodium disilicate glasses. J Non-Cryst Sol 362:162–168

Claireaux C, Chopinet MH, Burov E, Gouillart E, Roskosz M, Toplis MJ (2016) Atomic mobility in calcium and sodium aluminosilicate melts at 1200°C. Geochim Cosmochim Acta 192:235–247

Claireaux C, Chopinet MH, Burov E, Montigaud H, Roskosz M, Toplis MJ, Gouillart E (2019) Influence of temperature on multicomponent diffusion in calcium and sodium aluminosilicate melts. J Non-Cryst Sol 505:170–180

Cooper AR (1968) The use and limitations of the concept of an effective binary diffusion coefficient for multi-component diffusion. *In*: Mass Transport in Oxides. Vol 296. Wachman JB, Franklin AD, (eds). Nat Bur Stand Spec Publ, p 79–84

Crank J (1975) The Mathematics of Diffusion. Clarendon Press, Oxford

Cussler EL (1997) Diffusion: Mass Transfer in Fluid Systems. Cambridge Univ Press, Cambridge, England

Davis HT, Valencourt LR, Johnson CE (1975) Transport processes in composite media. J Am Ceram Soc 58:446–452

De Groot SR, Mazur P (1962) Non-Equilibrium Thermodynamics. Interscience, New York

De Koker N, Stixrude L (2010) Theoretical computation of diffusion in minerals and melts. Rev Mineral Geochem 72:971–996

Delaney JR, Karsten JL (1981) Ion microprobe studies of water in silicate melts: concentration-dependent water diffusion in obsidian. Earth Planet Sci Lett 52:191–202

Dingwell DB (1990) Effects of structural relaxation on cationic tracer diffusion in silicate melts. Chem Geol 82:209–216

Dohmen R, Milke R (2010) Diffusion in polycrystalline materials: grain boundaries, mathematical models, and experimental data. Rev Mineral Geochem 72:921–970

Doremus RH (1969) The diffusion of water in fused silica. *In*: Reactivity of Solids. Mitchell JW, Devries RC, Roberts RW, Cannon P (eds). Wiley, New York, p 667–673

Einstein A (1905) The motion of small particles suspended in static liquids required by the molecular kinetic theory of heat. Ann Phys 17:549–560

Eyring H (1936) Viscosity, plasticity, and diffusion as examples of absolute reaction rates. J Chem Phys 4:283–291

Fanara S, Behrens H, Zhang Y (2013) Water diffusion in potassium-rich phonolitic and trachytic melts. Chem Geol 346:149–161

Fanara S, Sengupta P, Becker H-W, Rogalla D, Chakraborty S (2017) Diffusion across the glass transition in silicate melts: systematic correlations, new experimental data for Sr and Ba in calcium-aluminosilicate glass and general mechanisms of ionic transport. J Non-Cryst Sol 455:6–16

Fortin M-A, Watson EB, Stern R (2017) The isotope mass effect on chlorine diffusion in dacite melt, with implications for fractionation during bubble growth. Earth Planet Sci Lett 480:15–24

Freda C, Baker DR, Romano C, Scarlato P (2003) Water diffusion in natural potassic melts. Geol Soc Spec Publ 213:53–62

Frischat GH, Szurman M (2011) Role of sulfur and its diffusion in silicate glass melts. Appl Glass Sci 2:47–51

Gast PW (1968) Trace element fractionation and the origin of tholeiitic and alkaline magma types. Geochim Cosmochim Acta 32:1057–1086

Giordano D, Russel JK, Dingwell D (2008) Viscosity of magmatic liquids: a model. Earth Planet Sci Lett 271:123–134

Gonzalez-Garcia D, Behrens H, Petrelli M, Vetere F, Morgavi D, Zhang C, Perugini D (2017) Water-enhanced interdiffusion of major elements between natural shoshonite and high-K rhyolite melts. Chem Geol 466:86–101

Gonzalez-Garcia D, Petrelli M, Behrens H, Vetere F, Fischer LA, Morgavi D, Perugini D (2018) Diffusive exchange of trace elements between alkaline melts: implications for element fractionation and time scale estimations during magma mixing. Geochim Cosmochim Acta 233:95–114

Guo C, Zhang Y (2016) Multicomponent diffusion in silicate melts: SiO_2–TiO_2–Al_2O_3–MgO–CaO–Na_2O–K_2O system. Geochim Cosmochim Acta 195:126–141

Guo C, Zhang Y (2018) Multicomponent diffusion in basaltic melts at 1350°C. Geochim Cosmochim Acta 228:190–204

Guo C, Zhang Y (2019a) Corrigendum to "Multicomponent diffusion in silicate melts: SiO_2–TiO2–Al_2O_3–MgO–CaO–Na_2O–K_2O system" [Geochim. Cosmochim. Acta 195 (2016) 126–141]. Geochim Cosmochim Acta 259:412

Guo C, Zhang Y (2019b) Corrigendum to "Multicomponent diffusion in basaltic melts at 1350 °C" [Geochim. Cosmochim. Acta 228 (2018) 190–204]. Geochim Cosmochim Acta 259:413

Guo C, Zhang Y (2020) Multicomponent diffusion in a basaltic melt: temperature dependence. Chem Geol 549: 119700

Harrison TM, Watson EB (1983) Kinetics of zircon dissolution and zirconium diffusion in granitic melts of variable water content. Contrib Mineral Petrol 84:66–72

Hashin Z, Shtrikman S (1963) Conductivity of polycrystals. Phys Rev 130:129–133

Henderson P, Nolan J, Cunningham GC, Lowry RK (1985) Structural controls and mechanisms of diffusion in natural silicate melts. Contrib Mineral Petrol 89:263–272

Hess K, Dingwell DB (1996) Viscosities of hydrous leucogranitic melts: a non-Arrhenian model. Am Mineral 81:1297–1300

Hofmann AW, Giletti BJ, Yoder HS, Yund RA (eds) (1974) Geochemical Transport and Kinetics. Carnegie Institution of Washington Publ, Washington, DC

Holycross ME, Watson EB (2016) Diffusive fractionation of trace elements in basaltic melt. Contrib Mineral Petrol 171:80

Holycross ME, Watson EB (2018) Trace element diffusion and kinetic fractionation in wet rhyolitic melt. Geochim Cosmochim Acta 232:14–29

Holycross ME, Watson EB, Richter FM, Villeneuve J (2018) Diffusive fractionation of Li isotopes in wet, highly silicic melts. Geochem Persp Lett 6:39–42

Huber C, Bachmann O, Manga M (2009) Homogenization processes in silicic magma chambers by stirring and mushification (latent heat buffering). Earth Planet Sci Lett 283:38–47

Hui H, Zhang Y (2007) Toward a general viscosity equation for natural anhydrous and hydrous silicate melts. Geochim Cosmochim Acta 71:403–416

Jambon A (1980) Isotopic fractionation: a kinetic model for crystals growing from magmatic melts. Geochim Cosmochim Acta 44:1373–1380

Jambon A (1982) Tracer diffusion in granitic melts: experimental results for Na, K, Rb, Cs, Ca, Sr, Ba, Ce, Eu to 1300°C and a model of calculation. J Geophys Res 87:10797–10810

Jambon A, Carron JP (1976) Diffusion of Na, K, Rb and Cs in glasses of albite and orthoclase composition. Geochim Cosmochim Acta 49:897–903

Jambon A, Semet MP (1978) Lithium diffusion in silicate glasses of albite, orthoclase, and obsidian compositions: an ion-microprobe determination. Earth Planet Sci Lett 37:445–450

Jambon A, Zhang Y, Stolper EM (1992) Experimental dehydration of natural obsidian and estimation of D_{H2O} at low water contents. Geochim Cosmochim Acta 56:2931–2935

Karsten JL, Holloway JR, Delaney JR (1982) Ion microprobe studies of water in silicate melts: temperature-dependent water diffusion in obsidian. Earth Planet Sci Lett 59:420–428

Kaufmann J, Russel C (2008) Redox behavior and diffusion of copper in soda-lime-silica melts. J Non-Cryst Sol 354:4614–4619

Kaufmann J, Russel C (2010) Diffusion of copper in soda-silicate and soda-lime-silicate melts. J Non-Cryst Sol 356:1158–1162

Kaufmann J, Russel C (2011) Diffusivity of copper in aluminosilicate melts, studied by square wave voltametry. Phys Chem Glas 52:101–106

Kerner EH (1956) The electrical conductivity of composite media. Proc Phys Soc B 69:802–807

Kirkaldy JS, Young DJ (1987) Diffusion in the Condensed State. The Institute of Metals, London

Koyaguchi T (1985) Magma mixing in a conduit. J Volcanol Geotherm Res 25:365–369

Koyaguchi T (1989) Chemical gradient at diffusive interfaces in magma chambers. Contrib Mineral Petrol 103:143–152

Kuroda M, Tachibana S (2019) Effect of structural dynamical property of melt on water diffusion in rhyolite melt. ACS Earth Space Chem 3:2058–2062

Lasaga AC (1983) Geospeedometry: an extension of geothermometry. *In*: Kinetics and Equilibrium in Mineral Reactions. Saxena SK (ed) Springer-Verlag, New York

Lasaga AC (1998) Kinetic Theory in the Earth Sciences. Princeton University Press, Princeton, NJ

Lasaga AC, Kirkpatrick RJ (eds) (1981) Kinetics of Geochemical Processes. Mineralogical Society of America

Lesher CE (1994) Kinetics of Sr and Nd exchange in silicate liquids: theory, experiments, and applications to uphill diffusion, isotopic equilibrium and irreversible mixing of magmas. J Geophys Res 99:9585–9604

Lesher CE (2010) Self-diffusion in silicate melts: theory, observations and applications to magmatic systems. Rev Mineral Geochem 72:269–309

Lesher CE, Walker D (1986) Solution properties of silicate liquids from thermal diffusion experiments. Geochim Cosmochim Acta 50:1397–1411

Lesher CE, Hervig RL, Tinker D (1996) Self diffusion of network formers (silicon and oxygen) in naturally occurring basaltic liquid. Geochim Cosmochim Acta 60:405–413

Liang Y (2000) Dissolution in molten silicates: effects of solid solution. Geochim Cosmochim Acta 64:1617–1627

Liang Y (2010) Multicomponent diffusion in molten silicates: theory, experiments, and geological applications. Rev Mineral Geochem 72:409–446

Liang Y, Richter FM, Davis AM, Watson EB (1996a) Diffusion in silicate melts, I: self diffusion in CaO–Al_2O_3–SiO_2 at 1500°C and 1 GPa. Geochim Cosmochim Acta 60:4353–4367

Liang Y, Richter FM, Watson EB (1996b) Diffusion in silicate melts, II: multicomponent diffusion in CaO–Al_2O_3–SiO_2 at 1500°C and 1 GPa. Geochim Cosmochim Acta 60:5021–5035

Liang Y, Richter FM, Chamberlin L (1997) Diffusion in silicate melts, III: empirical models for multicomponent diffusion. Geochim Cosmochim Acta 61:5295–5312

Liang Y, Davis AM (2002) Energetics of multicomponent diffusion in molten CaO–Al_2O_3–SiO_2. Geochim Cosmochim Acta 66:635–646

Lierenfeld MB, Zajacz Z, Bachmann O, Ulmer P (2018) Sulfur diffusion in dacitic melt at various oxidation states: implications for volcanic degassing. Geochim Cosmochim Acta 226:50–68

Lierenfeld MB, Zhong X, Reusser E, Kunze K, Putlitz B, Ulmer P (2019) Species diffusion in clinopyroxene solid solution in the diopside-anorthite system. Contrib Mineral Petrol 174:46

Liu Y, Zhang Y (2000) Bubble growth in rhyolitic melt. Earth Planet Sci Lett 181:251–264

Liu Y, Zhang Y, Behrens H (2004) H_2O diffusion in dacitic melt. Chem Geol 209:327–340

Lowry RK, Reed SJB, Nolan J, Henderson P, Long JVP (1981) Lithium tracer-diffusion in an alkali-basaltic melt - an ion-microprobe determination. Earth Planet Sci Lett 53:36–40

Lowry RK, Henderson P, Nolan J (1982) Tracer diffusions of some alkali, alkaline-earth and transition element ions in a basaltic and andesitic melt, and the implications concerning melt structure. Contrib Mineral Petrol 80:254–261

Macris CA, Asimow PD, Badro J, Eiler JM, Zhang Y, Stolper EM (2018) Seconds after impact: insights into the thermal history of impact ejecta from diffusion between lechateliertite and host glass in tektites and experiments. Geochim Cosmochim Acta 241:69–94

Matano C (1933) On the relation between the diffusion coefficient and concentrations of solid metals. Japan J Phys 8:109–113

Maxwell JC (1873) Treatise on Electricity and Magnetism. Oxford Univ Press, London

Mungall JE (2002) Empirical models relating viscosity and tracer diffusion in magmatic silicate melts. Geochim Cosmochim Acta 66:125–143

Mungall JE, Romano C, Dingwell DB (1998) Multicomponent diffusion in the molten system K_2O–Na_2O–Al_2O_3–SiO_2–H_2O. Am Mineral 83:685–699

Mungall JE, Dingwell DB, Chaussidon M (1999) Chemical diffusivities of 18 trace elements in granitoid melts. Geochim Cosmochim Acta 63:2599–2610

Newcombe ME, Fabbrizio A, Zhang Y, Ma C, Le Voyer M, Guan Y, Eiler JM, Saal AE, Stolper EM (2014) Chemical zonation in olivine-hosted melt inclusions. Contrib Mineral Petrol 168:1030

Newcombe ME, Beckett JR, Baker MB, Newman S, Guan Y, Eiler JM, Stolper EM (2019) Effects of pH_2O, pH_2, and fO_2 on the diffusion of H-bearing species in lunar basaltic liquid and an iron-free basaltic analog at 1 atm. Geochim Cosmochim Acta 250:316–343

Newcombe ME, Plank T, Zhang Y, Holycross ME, Barth A, Lloyd AS, Ferguson D, Houghton BF, Hauri E (2020) Magma pressure–temperature–time paths during mafic explosive eruptions. Front Earth Sci 8:531911

Ni H (2012) Compositional dependence of alkali diffusivity in silicate melts: mixed alkali effect and pseudo-alkali effect. Am Mineral 97:70–79

Ni H, Zhang Y (2008) H_2O diffusion models in rhyolitic melt with new high pressure data. Chem Geol 250:68–78

Ni P, Zhang Y (2016) Cu diffusion in a basaltic melt. Am Mineral 101:1474–1482

Ni H, Zhang L (2018) A general model of water diffusivity in calc-alkaline silicate melts and glasses. Chem Geol 478:60–68

Ni H, Behrens H, Zhang Y (2009b) Water diffusion in dacitic melt. Geochim Cosmochim Acta 73:3642–3655
Ni H, Liu Y, Wang L, Zhang Y (2009a) Water speciation and diffusion in haploandesitic melts at 743–873 K and 100 MPa. Geochim Cosmochim Acta 73:3630–3641
Ni H, Xu Z, Zhang Y (2013) Hydroxyl and molecular H_2O diffusivity in a haploandesitic melt. Geochim Cosmochim Acta 103:36–48
Ni H, Hui H, Steinle-Neumann G (2015) Transport properties of silicate melts. Rev Geophys 53:715–744
Ni P, Zhang Y, Simon A, Gagnon J (2017) Cu and Fe diffusion in rhyolitic melts during chalcocite "dissolution": implications for porphyry ore deposits and tektites. Am Mineral 102:1287–1301
Ni H, Shi H, Zhang L, Li W-C, Guo X, Liang T (2018) Cu diffusivity in granitic melts with application to the formation of porphyry Cu deposits. Contrib Mineral Petrol 173:50
Nowak M, Behrens H (1997) An experimental investigation on diffusion of water in haplogranitic melts. Contrib Mineral Petrol 126:365–376
Oishi Y, Nanba M, Pask JA (1982) Analysis of liquid-state interdiffusion in the system $CaO–Al_2O_3–SiO_2$ using multiatomic ion models. J Am Cer Soc 65:247–253
Okumura S, Nakashima S (2004) Water diffusivity in rhyolitic glasses as determined by in situ IR spectroscopy. Phys Chem Minerals 31:183–189
Okumura S, Nakashima S (2006) Water diffusion in basaltic to dacitic glasses. Chem Geol 227:70–82
Pablo H, Schuller S, Toplis MJ, Gouillart E, Mostefaoui S, Charpentier T, Roskosz M (2017) Multicomponent diffusion in sodium borosilicate glasses. J Non-Cryst Sol 478:29–40
Persikov ES, Newman S, Bukhtiyarov PG, Nekrasov AN, Stolper EM (2010) Experimental study of water diffusion in haplobasaltic and haploandesitic melts. Chem Geol 276:241–256
Persikov ES, Bukhtiyarov PG, Nekrasov AN, Bondarenko GV (2014) Concentration dependence of water diffusion in obsidian and dacitic melts at high pressures. Geochem Int 52:365–371
Posner ES, Schmickler B, Rubie DC (2018) Self-diffusion and chemical diffusion in peridotite melt at high pressure and implications for magma ocean viscosities. Chem Geol 502:66–75
Proussevitch AA, Sahagian DL (1998) Dynamics and energetics of bubble growth in magmas: analytical formulation and numerical modeling. J Geophys Res 103:18223–18251
Reid JE, Poe BT, Rubie DC, Zotov N, Wiedenbeck M (2001) The self-diffusion of silicon and oxygen in diopside ($CaMgSi_2O_6$) liquid up to 15 GPa. Chem Geol 174:77–86
Richter F, Liang Y, Minarik WG (1998) Multicomponent diffusion and convection in molten $MgO–Al_2O_3–SiO_2$. Geochim Cosmochim Acta 62:1985–1991
Richter FM, Liang Y, Davis AM (1999) Isotope fractionation by diffusion in molten oxides. Geochim Cosmochim Acta 63:2853–2861
Richter FM, Davis AM, DePaolo DJ, Watson EB (2003) Isotope fractionation by chemical diffusion between molten basalt and rhyolite. Geochim Cosmochim Acta 67:3905–3923
Richter RM, Watson EB, Mendybaev RA, Teng FZ, Janney PE (2008) Magnesium isotope fractionation in silicate melts by chemical and thermal diffusion. Geochim Cosmochim Acta 72:206–220
Richter FM, Watson EB, Mendybaev RA, Dauphas N, Georg B, Watkins JM, Valley JW (2009) Isotopic fractionation of the major elements of molten basalt by chemical and thermal diffusion. Geochim Cosmochim Acta 73:4250–4263
Ryerson FJ, Watson EB (1987) Rutile saturation in magmas: implication for Ti-Nb-Ta depletion in island-arc basalts. Earth Planet Sci Lett 86:225–239
Ryerson FJ, Durham WB, Cherniak DJ, Lanford WA (1989) Oxygen diffusion in olivine: effect of oxygen fugacity and implications for creep. J Geophys Res 94:4105–4118
Sauer VF, Freise V (1962) Diffusion in binaren Gemischen mit Volumenanderung. Z Elektrochem Angew Phys Chem 66:353–363
Shannon RD (1976) Revised effective ionic radii and systematic studies of interatomic distances in halides and chalcogenides. Acta Cryst A32:751–767
Shaw DM (1970) Trace element fractionation during anatexis. Geochim Cosmochim Acta 34:237–243
Shaw HR (1974) Diffusion of H_2O in granitic liquids, I: experimental data; II: mass transfer in magma chambers. *In*: Geochemical Transport and Kinetics. Vol 634. Hofmann AW, Giletti BJ, Yoder HS, Yund RA, (eds). Carnegie Institution of Washington Publ, Washington, DC, p 139–170
Shewmon PG (1989) Diffusion in Solids. Minerals, Metals & Materials Society, Warrendale, PA
Shimizu N, Kushiro I (1984) Diffusivity of oxygen in jadeite and diopside melts at high pressures. Geochim Cosmochim Acta 48:1295–1303
Spandler C, O'Neill HSC (2010) Diffusion and partition coefficients of minor and trace elements in San Carlos olivine at 1300°C with some geochemical implications. Contrib Mineral Petrol 159:791–818
Sparks RSJ (1978) The dynamics of bubble formation and growth in magmas: A review and analysis. J Volcanol Geotherm Res 3:1–37
Spickenbom K, Sierralta M, Nowak M (2010) Carbon dioxide and argon diffusion in silicate melts: insights into the CO_2 speciation in magmas. Geochim Cosmochim Acta 74:6541–6564
Stanton TR, Holloway JR, Hervig RL, Stolper EM (1985) Isotopic effect on water diffusivity in silicic melts. Eos 66:1131

Stolper EM (1982a) Water in silicate glasses: an infrared spectroscopic study. Contrib Mineral Petrol 81:1–17

Stolper EM (1982b) The speciation of water in silicate melts. Geochim Cosmochim Acta 46:2609–2620

Sugawara H, Nagata K, Goto KS (1977) Interdiffusivities matrix of $CaO–Al_2O_3–SiO_2$ melt at 1723 K to 1823 K. Metall Trans 8B:605–612

Tinker D, Lesher CE (2001) Self diffusion of Si and O in dacitic liquid at high pressures. Am Mineral 86:1–13

Tinker D, Lesher CE, Baxter GM, Uchida T, Wang Y (2004) High-pressure viscometry of polymerized silicate melts and limitations of the Eyring equation. Am Mineral 89:1701–1708

Trial AF, Spera FJ (1994) Measuring the multicomponent diffusion matrix: experimental design and data analysis for silicate melts. Geochim Cosmochim Acta 58:3769–3783

Trail D, Cherniak DJ, Watson EB, Harrison TM, Weiss BP, Szumila I (2016) Li zoning in zircon as a potential geospeedometer and peak temperature indicator. Contrib Mineral Petrol 171:25

Varshneya AK, Cooper AR (1972) Diffusion in the system $K_2O–SrO–SiO_2$, III: interdiffusion coefficients. J Am Ceram Soc 55:312–317

Von der Gonna G, Russel C (2000) Diffusivity of various polyvalent elements in a $Na_2O\cdot2SiO_2$ glass melt. J Non–Cryst Solids 261:204–210

Vrentas JS, Vrentas CM (2016) Diffusion and Mass Transfer. CRC Press

Wang H, Xu Z, Behrens H, Zhang Y (2009) Water diffusion in Mount Changbai peralkaline rhyolitic melt. Contrib Mineral Petrol 158:471–484

Wasserburg GJ (1988) Diffusion of water in silicate melts. J Geol 96:363–367

Watkins JM, DePaolo DJ, Huber C, Ryerson FJ (2009) Liquid composition-dependence of calcium isotope fractionation during diffusion in molten silicates. Geochim Cosmochim Acta 73:7341–7359

Watkins JM, DePaolo DJ, Ryerson FJ, Peterson BT (2011) Influence of liquid structure on diffusive isotope separation in molten silicates and aqueous solutions. Geochim Cosmochim Acta 75:3103–3118

Watkins JM, Liang Y, Richter F, Ryerson FJ, DePaolo DJ (2014) Diffusion of multi-isotopic chemical species in molten silicates. Geochim Cosmochim Acta 139:313–326

Watkins JM, DePaolo DJ, Watson EB (2017) Kinetic fractionation of non-traditional stable isotopes by diffusion and crystal growth reactions. Rev Mineral Geochem 82:85–125

Watson EB (1976) Two-liquid partitioning coefficients: experimental data and geochemical implications. Contrib Mineral Petrol 56:119–134

Watson EB (1982) Basalt contamination by continental crust: some experiments and models. Contrib Mineral Petrol 80:73–87

Watson EB (2017) Diffusive fractionation of volatiles and their isotopes during bubble growth in magmas. Contrib Mineral Petrol 172:61

Watson EB, Dohmen R (2010) Non-traditional and emerging methods for characterizing diffusion in minerals and mineral aggregates. Rev Mineral Geochem 72:61–105

Watson EB, Muller T (2009) Non-equilibrium isotopic and elemental fractionation during diffusion-controlled crystal growth under static and dynamic conditions. Chem Geol 267:111–124

Watson EB, Cherniak DJ, Holycross ME (2015) Diffusion of phosphorus in olivine and molten basalt. Am Mineral 100:2053–2065

Williams EL (1965) Diffusion of oxygen in fused silica. J Am Ceram Soc 48:190–194

Yang Y, Zhang Y, Simon A, Ni P (2016) Cassiterite dissolution and Sn diffusion in silicate melts of variable water content. Chem Geol 441:162–176

Yoshimura S (2018) Chlorine diffusion in rhyolite under low-H_2O conditions. Chem Geol 483:619–630

Yu Y, Zhang Y, Chen Y, Xu Z (2016) Kinetics of anorthite dissolution in basaltic melt. Geochim Cosmochim Acta 179:257–274

Yu Y, Zhang Y, Yang Y (2019) Kinetics of quartz dissolution in natural silicate melts and dependence of SiO_2 diffusivity on melt composition. ACS Earth Space Chem 3:599–616

Zhang Y (1993) A modified effective binary diffusion model. J Geophys Res 98:11901–11920

Zhang Y (1994) Reaction kinetics, geospeedometry, and relaxation theory. Earth Planet Sci Lett 122:373–391

Zhang Y (1999) H_2O in rhyolitic glasses and melts: measurement, speciation, solubility, and diffusion. Rev Geophys 37:493–516

Zhang Y (2008) Geochemical Kinetics. Princeton University Press, Princeton, NJ

Zhang Y (2010) Diffusion in minerals and melts: theoretical background. Rev Mineral Geochem 72:5–59

Zhang Y (2013) Kinetics and dynamics of mass-transfer-controlled mineral and bubble dissolution or growth: a review. Eur J Mineral 25:255–266

Zhang Y, Behrens H (2000) H_2O diffusion in rhyolitic melts and glasses. Chem Geol 169:243–262

Zhang Y, Cherniak DJ (eds) (2010) Diffusion in Minerals and Melts. Mineralogical Society of America

Zhang Y, Liu L (2012) On diffusion in heterogeneous media. Am J Sci 312:1028–1047

Zhang Y, Ni H (2010) Diffusion of H, C, and O components in silicate melts. Rev Mineral Geochem 72:171–225

Zhang Y, Stolper EM (1991) Water diffusion in basaltic melts. Nature 351:306–309

Zhang Y, Xu Z (2016) Zircon saturation and Zr diffusion in rhyolitic melts, and zircon growth geospeedometer. Am Mineral 101:1252–1267

Zhang Y, Walker D, Lesher CE (1989) Diffusive crystal dissolution. Contrib Mineral Petrol 102:492–513

Zhang Y, Stolper EM, Wasserburg GJ (1991a) Diffusion of water in rhyolitic glasses. Geochim Cosmochim Acta 55:441–456

Zhang Y, Stolper EM, Wasserburg GJ (1991b) Diffusion of a multi-species component and its role in the diffusion of water and oxygen in silicates. Earth Planet Sci Lett 103:228–240

Zhang Y, Stolper EM, Ihinger PD (1995) Kinetics of reaction $H_2O + O = 2OH$ in rhyolitic glasses: preliminary results. Am Mineral 80:593–612

Zhang Y, Jenkins J, Xu Z (1997) Kinetics of the reaction $H_2O + O = 2OH$ in rhyolitic glasses upon cooling: geospeedometry and comparison with glass transition. Geochim Cosmochim Acta 61:2167–2173

Zhang Y, Xu Z, Liu Y (2003) Viscosity of hydrous rhyolitic melts inferred from kinetic experiments, and a new viscosity model. Am Mineral 88:1741–1752

Zhang Y, Xu Z, Zhu M, Wang H (2007) Silicate melt properties and volcanic eruptions. Rev Geophys 45:RG4004

Zhang Y, Ni H, Chen Y (2010) Diffusion data in silicate melts. Rev Mineral Geochem 72:311–408

Zhang L, Guo X, Wang Q, Ding J, Ni H (2017) Diffusion of hydrous species in model basaltic melt. Geochim Cosmochim Acta 215:377–386

Zhang P, Zhang L, Wang Z, Li WC, Guo X, Ni H (2018) Diffusion of molybdenum and tungsten in anhydrous and hydrous granitic melts. Am Mineral 103:1966–1974

Zhang Y, Gan T, Guan Y (2019a) K isotope fractionation in diffusion couples of molten basalts. Abstract , Goldschmidt Conference

Zhang L, Guo X, Li W-C, Ding J, Zhou D, Zhang L, Ni H (2019b) Reassessment of pre-eruptive water content of lunar volcanic glass based on new data of water diffusivity. Earth Planet Sci Lett 522:40–47

Reviews in Mineralogy & Geochemistry
Vol. 87 pp. 339-403, 2022
Copyright © Mineralogical Society of America

8

Silicate Melt Thermochemistry and the Redox State of Magmas

Roberto Moretti[1,2]

[1]*Université de Paris*
Institut de Physique du Globe de Paris
UMR 7154 CNRS
Paris
France
[2]*Observatoire Volcanologique et Sismologique de Guadeloupe*
Institut de Physique du Globe de Paris
Gourbeyre
France

moretti@ipgp.fr

Giulio Ottonello[3]

[3]*Laboratorio di Geochimica at DISTAV*
University of Genoa
Genova
Italy

INTRODUCTION AND RATIONALE

Redox state is a term widely adopted in geochemistry and petrology. It refers, sometimes very qualitatively, to the ensemble of conditions and variables governing oxidation and reduction reactions fixing mass and energy transfer in magmas, and in particular ruling the speciation of fluids discharged by magmatic systems. We can reasonably say that the redox "state" in all hot environments of the Earth's interior is related to oxygen, which controls exchanges of elements such as iron and sulfur in their different oxidation states (or oxidation numbers). Oxidation state is the (hypothetical or *formal*) charge that redox-sensitive elements would have if all bonds to other atoms (e.g., oxygen) were 100% ionic, without covalent bond fraction: the more electronegative atom in a bond would have possession of the bonding electrons. Obviously, in reality, no bond is 100% ionic and mixture of electrovalent and covalent bonding actually occurs in compounds. However, redox reactions, as taught at the general chemistry level, involve changes in oxidation state expressed as either positive or negative integers. This aspect determines a consequent "syntaxis", which includes the speciation state of studied substances in melts and magmas and in which formal charges are reported with Arabic numbers whereas oxidation numbers are given with Roman numerals (Jorgensen 1969; Ottonello et al. 2001; Moretti 2005, 2021; Moretti and Neuville 2021). This choice avoids confusion between assigned oxidation states on coordinating atoms and the distribution of the effective charge in the reactive entity.

Magmas are the main carrier of mass, hence oxygen, through Earth's interiors up to surface, and their liquid portion, the silicate melt, operates as a high-temperature chemical solvent for metal oxides. Four major features appear when addressing silicate melts and attempting at their phenomenological descriptions:

1529-6466/22/0087-0008$10.00 (print)
1943-2666/22/0087-0008$10.00 (online)

1. compositionally they are a sub-class of oxide melts, and in fact they are typically described as a combination of oxides or mineral like components;

2. they show quasi-lattice properties, with a short-range ordering leading to distinct sites for anions and cations;

3. they are ionic liquids, in which the charge is transported essentially by cations whereas anions are stationary;

4. they may vitrify below some characteristic temperature (T_g, at glass transition) forming metastable disordered glasses.

According to point 1) the thermodynamic description of melts, particularly in the geological literature, was so far mainly based on the mixing of oxides or mineral-like components, which are either expected to form by the ordering of Si, Al units at the liquidus temperature or result from a normative calculation (see also Papale et al. 2022, this volume). Nevertheless, observations 2 to 4 clearly support the idea that silicate melts are made of charged entities, i.e., cations and oxyanions, which may form glasses if organized in large networks or long chains.

The electric charge is primarily transported by cations, but oxygen in its forms (O^{2-} and O-based anionic complexes) does not contribute substantially to the charge transport (Dickson and Dismukes 1962; Dancy and Derge 1966; Cook and Cooper 1990, 2000; Cooper et al. 1996a,b; Magnien et al. 2006, 2008; Cochain et al. 2012, 2013; Le Losq et al. 2020). This is a strong difference when comparing silicate melts to the aqueous solutions, because in the latter the concentration of hydroxide ions or water is always high enough to sustain high current densities (Allanore 2013). Silicate melts are then electrolytes in which, contrary to water, solvated oxides determine a new structural condition of the solvent itself, i.e., the "network".

In such an electrolytic medium the oxygen ligand is considerably affected and oxygen, particularly the oxide ion, O^{2-}, cannot be identified as the solvent, despite its nominal abundance. In fact, different oxygen species originate making the silicate melts a highly interconnected (polymerized) matrix in which solvation units cannot be easily defined and both ionic and covalent bonds rule the reactive entities that make up the melt network. Spectroscopic investigations at all (short- to long-) scales of ordering show that the silicate melt is made by a network of Al,Si polymeric oxyanions, or polyoxyanions (the simplest oxyanion being SiO_4^{4-}), which prevent or hinder the orderly orientation of atoms in the network themselves and also in atoms not associated with the network in the liquid, near the melting point or liquidus temperature (Sun 1947). The melt and glass structures display elements of long-range randomness coexisting with medium- and short-range order (Gaskell et al. 1991), the latter being dominated by the structure of SiO_4 rigid units (Greaves et al. 1981) and its radial limit shrinking with the introduction of alkalis and other so-called network modifiers (Waseda and Suito 1977). Besides, there is no systematic difference between melt and glass structure of the same composition and the effects of thermal expansion are *de facto* negligible on both mean structure and single interionic distances (Brown et al. 1995; Ottonello 2005, and references therein).

These fundamental aspects are in line with the Random Network Model (RNM) of Zachariasen (1932, 1935) and Warren (1933) and comply with the Bernal's (1959) definition of a liquid as an "homogeneous, coherent and essentially irregular assemblage of molecules containing no crystalline regions or holes large enough to admit another molecule" (see also Ottonello 2005). In such a context, the topological part of the problem is then relevant, but the thermodynamic definition of the configurational disorder that arises is far from being fully achieved due to the complexity of phenomena taking place in the inhomogeneity range. In particular, we see that the coordination number (CN) of so-called network modifiers in melts and glasses displays a large scatter (Table 1), which suggests that they adapt themselves to the geometry determined by the much stronger—and essentially covalent—bonds of network

Table 1. Ranges and/or mean M–O coordination number (at 1 bar) of cations in silicate melts and glasses.

M	Acidic	Amphoteric	Basic	Notes
Si	4			
Ge	4[*]			
Al	4–6			
Al	5			Al–rich melts[†]
Fe[III]		4–5[†]		
Ti		4–5		up to 6 in early works[†]
Zr		6–8		
Ni		4–6		
Li			4[*,†]	
Na			7–8	
K			5–6	5–7 (see refs. in [*])
Rb			8[†]	
Cs			8	
Ca			4–6	
Mg			4–6	
Fe[II]			4–6	
Mn			4–6	
Pb			8	

Notes:
[*]Based on the extensive compilation of Brown et al. (1995) and the review by Henderson (2005).
[†]The attribution to acidic, amphotheric and basic oxides as in Ottonello and Moretti (2004) and Ottonello (2005).

formers (Ottonello 2005). This behaviour was interpreted by the Author as due to zeolite-like networks with cation-bearing cages of variable size (Fig. 1). This structural view of both melts and glasses bears many analogies with the evidences of nano-scale percolation channels in which Al charge compensators (e.g., alkali metals) accumulate and migrate (Vessal et al. 1992; Greaves and Ngai 1995; Greaves ad Sen 2007; Le Losq et al. 2017; see also Fig. 1).

This nature of melts greatly contributes to the non-ideality that arises from mixing oxide components and/or postulated molecular structures in liquid. For this reason, the phenomenological formalization of the mixing properties, particularly the thermodynamic mixing free energy of excess, via interaction parameters (such as for regular or sub-regular models; e.g; Papale et al. 2022, this volume) and their polynomial expansion have limited success in decrypting the complex features of inhomogeneity ranges. As noted by Fraser (2005), fitting of coefficients in (sub)regular solution models (e.g., Berman and Brown 1984; Ghiorso and Carmichael 1983; Ghiorso and Sack 1995; Ghiorso et al. 2002; see also Papale et al. 2022, this volume) is based on calorimetric and phase equilibrium data for the compositional ranges in which such data are provided. Therefore, inaccurate models are obtained for those compositions that are not part of the calibration database, particularly alkali-rich ones (e.g., Asimow et al. 2001). These approaches have proved to be geologically useful, but if on one side their success is intrinsically limited by the availability of experimental data, such that some interpolated data are from compounds observed in the solid phase, on the other side a great part of the challenge resides in fact in describing the coulombic nature of the melt,

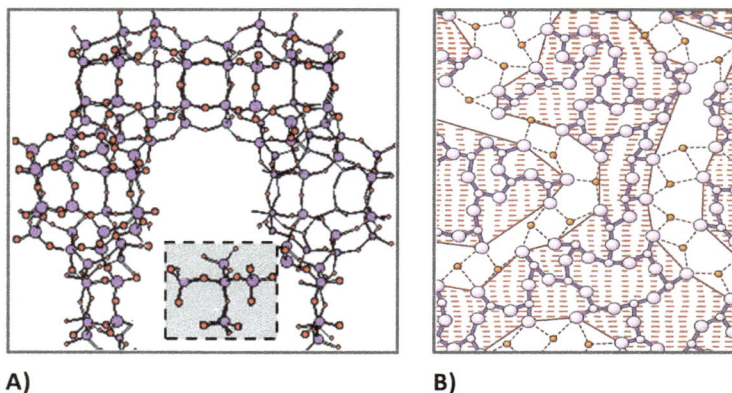

Figure 1: **Panel A**) The (3D) short- and medium-range periodicity of silicate melts and glasses locally re-semble that of all-Si zeolites, with cationic "cages" of variable dimensions, depending on silicate network connectivity. Long-range periodicity is locally interrupted by depolymerization, and the relative arrange-ment of irreducible units (see the **dashed inset**) may contribute to structural instability (and consequent unmixing) through strain energy contributions (modified from Ottonello 2005). **Panel B**) A 2D modified random network (MRN) for an oxide glass. Covalent bonds are shown by **solid lines**, and ionic bonds, by **dotted lines**. **Dark grey bands** are the percolation channels. **Black atoms** denote alkalis. **Small white atoms** represent silicon. **Large white atoms** are for oxygen (redrawn from Greaves 1985).

which exacerbates the differences in the chemical nature of the elements in it (Allanore 2015; Mysen and Richet 2018). This is relevant in the domains of glass science and particularly of extraction metallurgy, whereas descriptions aimed at geological approaches deal with the limited polymerised multi-component region in which the high-energy interaction between silica and "basic" metal oxide (Table 1) is essentially excluded (Fraser 2005).

Basically, we cannot forget that aluminosilicate melts, and more generally all oxide melts, are solvents whose properties shift continuously, and non-linearly, with composition: both covalent and ionic bonds govern the oxyanions that form and their properties. This scenario is completely different, again, to that of aqueous solutions, where solvation shells in which covalent forces are exhausted can be clearly defined: this yields solubility of metals and their compounds very low and allows physico-chemical approaches based the dilute electrolyte concept to work with very good predictive capacity, as visible in redox potential (E) vs pH diagrams (pH = $-\log a_{H^+}$, a_{H^+} being the proton activity). On the contrary, the distinction between solvent and solute in melts is blurred because speciation changes with the marked network de-condensation (depolymerization) from pure silica to metal-oxide rich compositions (Toop and Samis 1962a,b; Hess 1971, 1980; Ottonello 2001; Ottonello and Moretti 2004; Mysen and Richet 2005; Moretti et al. 2014). This poses problems to identify directly, from spectroscopic studies, the structural complexes intervening in relevant acid–base and redox exchanges because the anionic framework of silicate melts makes in fact solute and solvents so intimately related that one cannot identify a solvation shell and then treat as a polarizable continuum the solvent outside such a shell (Moretti et al. 2014).

However, any mixing model implicitly reflects to some extent the interionic interactions involving oxyanions and then how electronic clouds deform and interact in giving to rise to chemical bonds. Therefore, mixing model of melts (and their vitreous counterparts: glasses) succeed when they encompass, either implicitly or explicitly, the changes of the polarization state of ions with composition. Briefly, mixing models of melts and glasses must translate in terms of basic thermodynamic parameters the distortion of the electron cloud of the negatively charged ion by the positively charged ion. Some important parameters and properties relate to this distortion, such as electronegativity, optical basicity, electronic polarizability, all measuring the donor power of oxygen in a ligand (Jorgensen 1969; Duffy and Ingram 1971, 1975, 1976; Duffy 1992; Dimitrov and Komatsu 1999, 2000; Ottonello et al. 2001; Moretti 2005, 2021; Dimitrov et al. 2018). This power is reduced when the "free" oxide ion O^{2-} (the strongest base in oxidic media; Flood and Forland 1947; Duffy and Ingram 1971) is influenced by highly polarizing ions, such as S^{6+}, P^{5+} and Si^{4+}, which build-up oxyanions within which covalent bond is established. These parameters, then, measure the ionic (or covalent) fraction of the chemical bond established between the cations and the surrounding oxygen ligand.

We then see that this leads us straight to the Lewis definition (e.g., Satchell and Satchell 1971) of acid and base: the Lewis base is a species, such as O^{2-}, that donates an electron pair (i.e., a nucleophile) and will have lone-pair electrons; the Lewis acid is a species that accepts an electron pair (i.e., an electrophile) and will have vacant orbitals. The formation of silicate melt oxyanions is then the formation of coordination compounds in which O^{2-}, the Lewis base, is the ligand of a coordination compound with the metal (a network former) acting as the Lewis acid. It could seem that all Lewis acid–base reactions would be redox reactions, but this is strictly true under the even more qualitative acid–base definition of Usanovich, for which little is known but for which any substance that loses electrons is a base while any substance that gains electrons is an acid (Satchell and Satchell 1971).

Here is then a major point: Bronsted chemical reactions are divided between those which exchange electrons (redox reactions) and those which exchange protons (acid–base reactions). The Bronsted sharp distinction implies that any entity reacting with the characteristic acid or base of a solvent (e.g., hydrated H_3O^+ or OH^- in aqueous solutions) does not modify the oxidation state of its individual atoms. This sort of axiom is visible in vertical lines bordering stability fields in predominance diagrams of the type $\log a_i$ vs pH or E vs pH but is contradicted by some mechanisms such as chlorine disproportionation to Cl^{-I} and Cl^{I} (Jorgensen 1969):

$$Cl_{2(g)} + 2OH^- \leftrightarrow Cl^- + ClO^- + H_2O \tag{1}$$

Similar mechanisms, contradicting the Bronsted sharp divide between acid–base and redox reaction schemes exist also in silicate melts, such as in the case of the SO_2 disproportionation in sulfide, S^{2-}, and sulfate, SO_4^{2-}, upon reaction with the O^{2-} base (Moretti 2021):

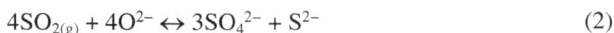

$$4SO_{2(g)} + 4O^{2-} \leftrightarrow 3SO_4^{2-} + S^{2-} \tag{2}$$

This reveals that there are elements of internal convergence between acid–base chemistry and redox chemistry, which are the yin and yang of the chemistry (Cicconi et al. 2020a) and that are not considered by typical operational scales for basicity such as pH. These elements become evident when interpreting in terms of Lewis description the half-reaction:

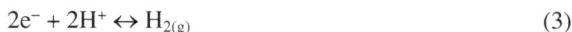

$$2e^- + 2H^+ \leftrightarrow H_{2(g)} \tag{3}$$

which is the link joining redox and acid–base reactions in proton-bearing systems (e.g., aqueous solutions). Similarly, we may anticipate here that following half-reaction

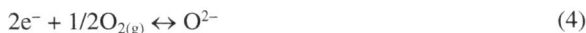

$$2e^- + 1/2O_{2(g)} \leftrightarrow O^{2-} \tag{4}$$

is the link joining redox and acid–base reactions in oxide systems such as silicate melts and glasses (see also Ottonello et al. 2001; Moretti 2005, 2021). Even though we have stated this from the point of view of the Lewis basicity, the analogy is quite straightforward also if we compare the Bronsted and Lux–Flood (Flood and Forland 1947) definitions of acid–

base behavior: just as the measurement of pH is the key to studies of acid–base reactions, those of reactions involving the O^{2-} ion in silicate melts naturally go through the measurement of $-\log a_{O^{2-}}$, (or pO^{2-}), a magnitude similar to pH and playing an identical role. Same as a pH indicator electrode, a pO^{2-} indicator electrode is the essential instrument to study acid–base properties of silicate melts.

The redox/acid–base link in oxide media is much stronger than in aqueous solutions and in fact approaches have been proposed in terms of parameters related to the Lewis acid–base character (network formers and their oxides, such as SiO_2 and Al_2O_3 are acids; network modifiers and their oxides such as MgO, CaO, Na_2O are bases) by using electronegativity and/or optical basicity (Duffy and Ingram 1971, 1975, 1976; Duffy 1992; Dimitrov and Komatsu 1999, 2000; Ottonello et al. 2001; Moretti 2005; Dimitrov et al. 2018).

These approaches allow distinguishing and calculating at least three types of oxygen (bridging, non-bridging and so-called free oxygen, or oxide ion) whose mixing determines the polarization state and the extent of the silicate network (polymerization), hence 1) the reactive properties of the melt mixture as a function of composition (Toop and Samis 1962a,b; Allanore 2013, 2015 and references therein; Moretti 2021 and references therein) and 2) the configurational disorder and the inhomogeneity ranges often observed at high SiO_2 content (e.g., Ottonello 2005, 2021 and references therein). However, silicate melts and glasses still lack a fully developed framework for acid–base properties, which could also be extended to all oxide melts and glasses to 1) formalize the reactive properties of the oxyanionic solvent and of "solvated" species and 2) describe how the speciation of metals affects the silicate network and then its polymerization and properties as a solvent.

The aim of this paper is to provide the basics about the joint redox and acid–base reactivity in melts, too often overlooked in the Earth Sciences literature. Few semi-quantitative approaches are the premise to more complex theories and models (e.g., the ionic–polymeric one) about the role played by bulk composition and melt/glass structure. After an overview of the redox in "magmatology" (Kushiro 2010), we show how to deal with the current knowledge of silicate melt structure to describe redox exchanges of iron and sulfur. To do so, we assume an understanding of the bonding, structure, and properties of individual molecules in melts, and describe how the composition, via melt/glass structure determines its acid–base behavior, or vice versa. Species and ionic complexes participate in reactions where charge (electrons, cations and anions) is transferred from a cation to the surrounding oxygen ligand and in which the oxidation state plays a basic role to "convert" structural findings about iron and sulfur into speciation hypotheses for chemical reactions.

Finally, we will also show that *ab-initio* computations can predict the iron redox behavior via the accurate description of bonding at the atomic level and how this changes as a function of how the solute (iron) "sees" the solvent and thus "feels" non-idealities during dissolution. The description we propose follows from the Tomasi's Polarized Continuum Model (PCM; Floris and Tomasi 1989; Floris et al. 1991; Tomasi and Persico 1994; Tomasi et al. 1999) and was already adopted to silicate melts for noble gas solubility (Ottonello and Richet 2014), silica wet solidus (Ottonello et al. 2015), water solubility and speciation (Ottonello et al. 2018) and forsterite–diopside–silica melting (Ottonello 2021). In this approach there is no need to consider the structure of the silicate framework, because the controlling parameters are given by the dielectric properties of the medium.

Our purpose in this contribution is not to promote a particular type of approach or model rather than another, but to orient the reader among the pros and cons of the available redox formulations, particularly for what concerns the role of composition on melt/glass reactivity and particularly redox reactivity.

THE EMERGENCE AND RISE OF THE CONCEPT OF OXYGEN FUGACITY

Oxygen exchange between metals and their oxides

Silicate melts exhibit a large range of thermal stability, with high temperature conditions that favour fast kinetics of redox exchanges. Because of this and their highly adaptable framework, silicate melts can dissolve important amounts of metals. This implies that nearly all metals can participate in a redox reaction in which oxygen is involved and that determine their partitioning in silicate melts:

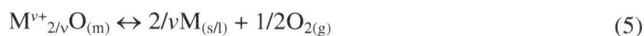

$$M^{v+}_{2/v}O_{(m)} \leftrightarrow 2/vM_{(s/l)} + 1/2O_{2(g)} \tag{5}$$

with v the charge (positive) of the metal cation M in the corresponding oxide and the subscripts s, m, l and g denoting solid, melt, liquid or gas phase (note that henceforth we will distinguish between components relaxed in a silicate melt and those making up a separate oxide liquid essentially non-silicate). From left to right, Reaction (5) is the main target of extractive metallurgy, but also summarizes the ensemble of processes that involved metal segregation from magma and that determined Earth's constitution.

Reaction (5) is perhaps the most effective way to show that melt thermodynamics is intimately connected to redox exchanges involving oxygen. Reaction (5), rewritten in its more general form for component $M^{v+}_{2/v}O$ (either as solid or liquid, including the melt phase) and for one mol of O_2, is the basis for the theory and practice of extractive metallurgy:

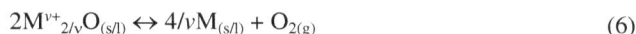

$$2M^{v+}_{2/v}O_{(s/l)} \leftrightarrow 4/vM_{(s/l)} + O_{2(g)} \tag{6}$$

Assessing Reaction (6) has contributed to the systematic thermodynamic catalogue summarized by so-called Ellingham diagrams (Ellingham 1944; Fig. 2), which are used in metallurgy to evaluate the ease of reduction of metal oxides, as well as chlorides, sulfides and sulfates. The diagram shows the variation versus temperature of the Gibbs free energy change of reactions of the type of (6), ΔG^0, which is related to the formation of oxides. The diagram can be used to predict the temperature and estimate the oxygen fugacity under which an ore will be reduced to its metal. The Gibbs energy change of Reaction (6) related to the formation energy of a compound at temperature T, (the Gibbs energy change when one mole of a compound is formed from elements at $P = 1$ bar) is given by:

$$\Delta G^0 = \Delta H^0 - T\Delta S = \left[\Delta H^0_{298} + \int_{298}^{T} C_p dT \right] - T\Delta S_{298} + \int_{298}^{T} \frac{\Delta C_p}{T} dT \tag{7}$$

To compare the relative stabilities of the various oxide components, the Ellingham diagram for oxidation reactions involves one mole of oxygen. For the oxidation of a metal, ΔG^0, represents the chemical affinity of the metal for oxygen. When the magnitude of ΔG^0 is negative, the oxide-bearing phase is stable over the metal and oxygen gas. Furthermore, the more negative the value, the more stable the oxide is. The Ellingham diagram also indicates which element will reduce which metal oxide.

When both M and $M^{v+}_{2/v}O_{(s/l)}$ in Reaction (6) are in their standard states, the equilibrium constant, K_6, corresponding to this reaction is expressed as:

$$K_6 = \frac{\left(a_{M^{2/v}_{s/l}}\right)^{4/v}}{\left(a_{M^{v+}_{2/v}O_{s/l}}\right)^2} \cdot a_{O_2} = \frac{f_{O_2}}{f^0_{O_2}} \tag{8}$$

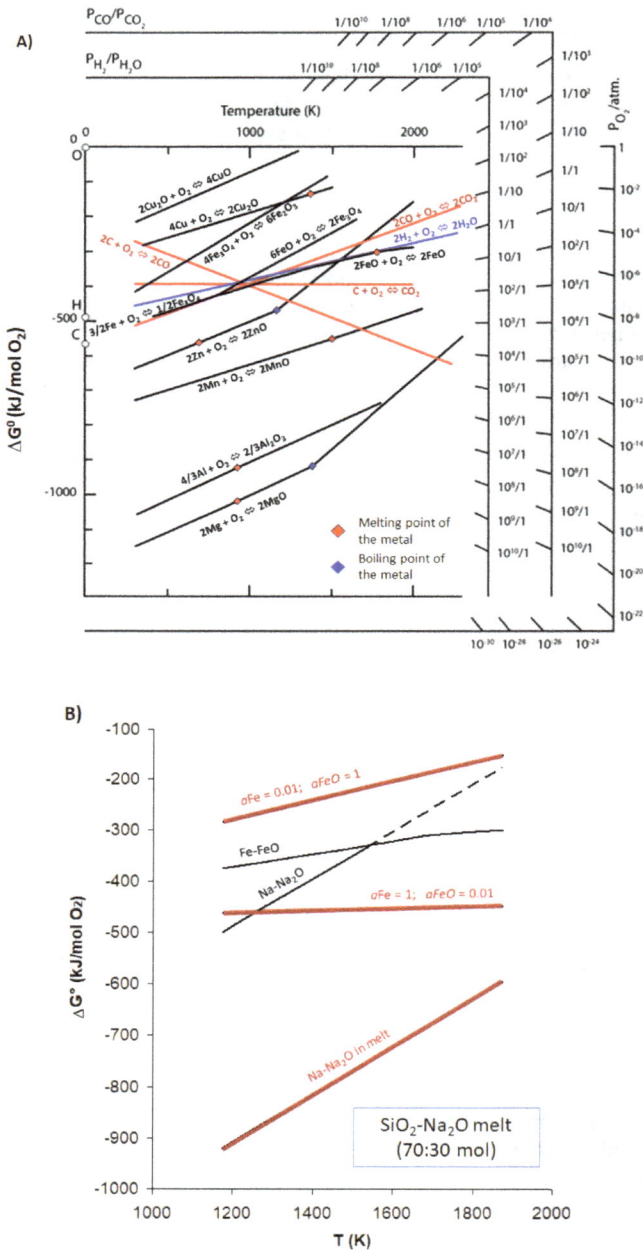

Figure 2. Panel A) Ellingham diagram for some relevant oxides including CO_2 equilibria, water and scales for oxygen fugacity and related quantities via CO/CO_2 and H_2/H_2O ratios at $P_{tot} = 1$ atm. The scale of P_{H_2}/P_{H_2O} ratio is used in exactly the same manner as that of P_{CO}/P_{CO_2} ratio, except that the point H on the ordinate at $T = 0$ K is used instead of point C. Modified from Zhang et al. (2014). **Panel B)** Ellingham diagram restricted to iron and sodium oxides in a Na_2O–SiO_2 melt (30:70 molar) doped in iron. Also shown are energies for pure liquid oxide phases (Fe–FeO and Na–Na_2O lines). Modified after Paul (1990).

where the superscript ^{0} is for the pure gas component gas fugacity at standard state (in this case 1 bar and T of interest). If, at any temperature, the acting oxygen fugacity is greater than the calculated value from Equation (8), spontaneous oxidation of metal M occurs, while oxide $M^{v+}_{2/v}O_{(s/l)}$ decomposes to metal M and gaseous oxygen at the oxygen partial pressure less than the equilibrium value. Therefore, an element is unstable, and its oxide is stable at higher oxygen potentials than its $\Delta G_f^0 - T$ line on the Ellingham diagram. Therefore, the larger negative value of ΔG_f^0 an oxide has, the more stable it is.

In order to illustrate the effect of the fluid phase, refined versions of the Ellingham diagrams included scales designed to read directly the equilibrium oxygen partial pressure or the corresponding P_{H_2}/P_{H_2O} or P_{CO}/P_{CO_2} ratios between metal and its oxide by drawing the line connecting point O for P_{O_2}, C for P_{CO}/P_{CO_2}, or H for P_{H_2}/P_{H_2O} in Figure 2 and the condition of interest (Reaction 6 and Eqn. 8 at T of interest).

The Ellingham diagram of Figure 2a is valid only for pure oxides. As noted by Paul (1990), oxides of transition metals such as iron do not mix ideally with glass-forming melts. Therefore, for melts and glasses activity corrections must be applied to the Ellingham diagram to adopt it profitably to such liquids. Based on the data of Paul (1990), in the Ellingham diagram of Figure 2b we can appreciate the corrections applied to FeO and Na_2O equilibria for a binary Na_2O-SiO_2 melt doped in iron. We can then conclude that the factors determining oxidation–reduction in an oxide melt of a *given composition* (at 1-bar) are 1) the oxygen fugacity in the melt system, 2) the equilibrium constant of Reaction (6), K_6, depending on the standard free energy (Eqn. 7) at temperature of interest, and 3) the activity of the transition metal oxides in the melt.

Reaction (6) also illustrates how pairs of metals and their oxides, both having unitary activity, can be used as so-called redox "buffers", such that f_{O_2} values can be easily fixed at any temperature. Even in presence of a third phase, such as silicate melts or any other liquid, gas–solid assemblages allow a straight application of Equation (8) to impose O_2 activitiy, unless solid phases are not refractory and dissolve other components exchanges with the coexisting liquid. A typical example of one of these gas–solid equilibria fixing f_{O_2} is the metal iron and wustite (here considered ideally stoichiometric) pair (so-called IW buffer):

$$2Fe + O_2 \leftrightarrow 2FeO \ (\text{iron–wustite; IW}) \tag{9}$$

In this framework, geochemistry and petrology benefitted from the continuous exchange of knowledge with metallurgy. To understand high temperature geological systems, geoscientists turned in fact their attention to the transfer of molecular oxygen among molecular components such as oxides or mineral-like macromolecular entities. The practice between geoscientists become to assess criteria for f_{O_2} (or a_{O_2}) estimations disconnected from the formal description of the acid–base character of magmas. In particular techniques were set involving mineral phases coexisting in igneous rock to establish thermodynamic or empiric laws and applicable to quenched glasses via indirect measurements most often of spectroscopic nature. This change of perspective reflects the obvious consideration that geoscientists deal with samples (solidified rocks) made accessible at Earth's surface and which represent the final products at the end of a long thermal and chemical journey, which the *a posteriori* reconstruction of the objective of the geochemical (*lato sensu*) investigation.

The fact that most of the chemical analyses were from techniques in which oxygen was not directly determined but allowed to give oxides has also further favored these approaches. Petrologists then develop techniques to constrain f_{O_2} to investigate melting and sub-solidus conditions of oxides and silicates. Early f_{O_2} control techniques were applied by Darken and Gurry (1945, 1946) via gas mixing in their study of the Fe–O system. Later Eugster developed the double capsule technique, with a Pt capsule containing the starting material, surrounded by a larger gold capsule. A metal–oxide or oxide–oxide pair + H_2O was placed between the two metal containers and reaction involving OH-bearing minerals provided a fixed and known hydrogen fugacity (Eugster 1977), scaled to f_{O_2} via the dissociation of water. A similar procedure is adopted for water solubility studies in melts.

Since Eugster many metal–oxide and oxide–oxide assemblages have been used in experimental petrology that can be called again "redox buffers". Our understanding of the role of f_{O_2} on melt phase equilibria and mineral composition on Earth has greatly benefitted of these advancements, which made possible to collect systematically glasses (quenched melts) to measure ratios of metals in their different oxidation states, particularly Fe^{II}/Fe^{III}, obtained for different P, T, f_{O_2} and melt/glass composition.

Oxygen fugacity and iron-bearing solid phases

Because iron is orders of magnitude more abundant than other transition elements, the proportion of ferric (Fe^{III}) and ferrous (Fe^{II}) iron may control (or buffer) the overall oxygen fugacity (f_{O_2}) of magmas. Therefore, iron redox equilibria and structural conditions in silicate melts and rock mineral assemblages strongly affect magma properties and drive phase equilibria. Reactions such as (9) have are then useful to provide a scale with geological significance for f_{O_2} conditions that is recorded in rock and minerals formed in past and present Earth environments, from the core (in which Fe^0 dominates) up to shallow crust and through all igneous environments, where it exists as Fe^{II} and Fe^{III}. To provide a systematics for the a_{O_2} conditions the following gas–solid reactions, other than Reaction (9), were then assessed in the literature and can be found in many textbooks (Frost, 1991; Fig. 3):

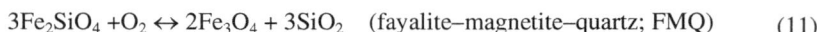

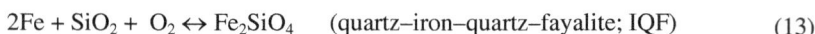

$$4Fe_3O_4 + O_2 \leftrightarrow 6Fe_2O_3 \qquad \text{(magnetite–hematite; MH)} \qquad (10)$$

$$3Fe_2SiO_4 + O_2 \leftrightarrow 2Fe_3O_4 + 3SiO_2 \quad \text{(fayalite–magnetite–quartz; FMQ)} \qquad (11)$$

$$3FeO + O_2 \leftrightarrow Fe_3O_4 \qquad \text{(wustite–magnetite; WM)} \qquad (12)$$

$$2Fe + SiO_2 + O_2 \leftrightarrow Fe_2SiO_4 \quad \text{(quartz–iron–quartz–fayalite; IQF)} \qquad (13)$$

In all these equilibria oxide components display unitary activities, which then appear as pure phases. Therefore, their equilibrium constants simply describe the variations of O_2 activity (a_{O_2}) with temperature $T(K)$:

$$\log f_{O_2} = A/T + B \qquad (14)$$

as well as of O_2 fugacitity (f_{O_2}) with temperature, T (in K), and pressure, P (in bar) :

$$\log f_{O_2} = A/T + B + C(P-1)/T \qquad (15)$$

with A, B and C constants. The pressure term, C, allows computing directly f_{O_2} rather than a_{O_2} at the pressure of interest. Mineral assemblages making up Reactions (9) to (13) are not necessarily occurring in deep Earth and igneous environments, whereas the actual multi-component space of phases fix the via multiple equilibria in which solid solutions and/or the presence of relatively mobile liquid (silicate melts) and/or supercritical fluids play a fundamental role.

However, when one of these mineral "buffers" is selected as reference, the system $\log f_{O_2}$ can be given as a relative value:

$$\Delta \text{buffer} = \log f_{O_2} = \log f_{O_2, \text{buffer}} \qquad (16)$$

A relative f_{O_2} scale embodying temperature effects can thus be set, which allow tracking f_{O_2} variations with respect to a reference. Nevertheless, this has not (or at least not necessarily) a real meaning about log f_{O_2} evolution in igneous systems and related environments (e.g., Moretti and Stefansson 2020). A common misconception is that in a system a given value of Δbuffer (e.g., ΔQFM = 1) may represent some kind of specific, even "genetic" value, which is characteristic of the whole "rock system" throughout its thermal and chemical evolution. However, it is useful to remind that in natural environments oxygen activity (hence f_{O_2}) varies to accommodate the compositional variations and the speciation state of the mineral/melt/fluid phases. In particular, when highly mobile volatile phase and their components are involved, f_{O_2} can thus be fixed by factors that can be considered as external to the system.

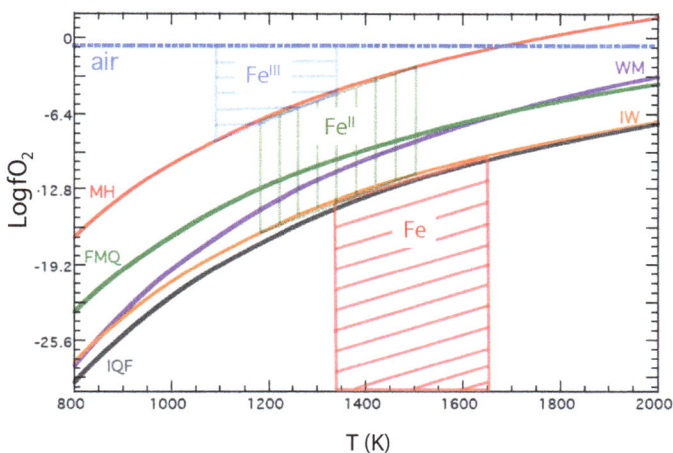

Figure 3. Common solid buffers of f_{O_2} used in petrology, volcanology and geochemistry. Along each line pure phases coexist stably at corresponding T and f_{O_2} values (see Reactions 9 to 13). Oxygen fugacity values were computed for a total pressure of 1 bar (**solid lines**). Also reported is the value of f_{O_2} in air ($P_{O_2} \sim 0.21$ atm). Dashed areas denote typical conditions for iron in its three different oxidation states (see Frost 1991). Temperature interval for each vertical band is only qualitative and does not reflect any Earth's environment associated with a specific occurrence of iron in one or more of its oxidation states.

On the other hand, the evolution of the igneous rock system cannot be approximated by a unique Fe^{II}/Fe^{III} ratio, inherited by the system when it was completely molten, above the liquidus. As noted by Frost (1991), two rock systems that have ideally crystallized along the QFM buffer may have different proportions of fayalite and magnetite because of the crystallization style but different Fe^{II}/Fe^{III} bulk ratios. This does not exclude, however, that the melt may locally fix the redox potential of the system via iron oxidation state, with the Fe^{II}/Fe^{III} ratio approaching unity. This may occur, for instance, in some systems subject to re-melting (e.g., Moretti et al. 2013).

Summarizing, the oxygen fugacity is a thermodynamic parameter used to conveniently report on the redox state of one or more systems, highly useful even if O_2 is not an existing gaseous species that can be detected in a system accessible to measurements. The true redox variables are then oxidation states of iron and other elements in minerals and melts. The common practice is then to measure the concentration ratio of redox couples of multiple valence elements in melts (Fe^{II}/Fe^{III}, but also S^{-II}/S^{VI}, V^{III}/V^V etc) or in gases (e.g., H_2/H_2O, CO/CO_2, H_2S/SO_2) and relate them to f_{O_2} via thermodynamic calculations. With the analytical information extracted from quenched melts, thermodynamic back-calculations of f_{O_2} turn out to be affected by many more uncertainties that are related to the thermodynamic description of melt mixing mixing properties. Although the formulations available in the literature (see next section *Modelling the Redox State of Silicate Melts*) may suggest that it is a simple operation, back-calculating f_{O_2} conditions from melt chips is much less straightforward than in the case of a) volcanic gases, in which governing equilibria are directly solved if the gas analysis is provided (e.g., Giggenbach 1980, 1987; Aiuppa et al. 2011 and references therein) or b) solid–solid equilibria, such as in case of coexisting iron-titanium oxide solid solutions of titanomagnetite (Fe_3O_4–Fe_2TiO_4) and hemo-ilmenite (Fe_2O_3–$FeTiO_3$) (Buddington and Lindsley 1964) or for peridotite assemblages in the mantle (e.g., Mattioli and Wood 1988; Gudmundsson and Wood 1995), for which solid solutions are well assessed, and allow an accurate treatment of component activities from mineral analyses. On the contrary f_{O_2} estimates by metal oxidation states measured in glasses often result by oxybarometric functions in which fitted parameters do not represent thermodynamic parameters and whose validity is then limited to their domain of calibration.

MODELLING THE REDOX STATE OF SILICATE MELTS

Non-reactive species-based approaches to iron redox

The problem of activity–composition relations. The possibility offered by analytical techniques to measure directly the oxidation state of redox-sensitive elements in quenched glasses (e.g., sulfur and iron redox ratios in melt inclusions) has offered the opportunity to develop empirical and theoretical formulations to retrieve the evolution of for the studied system. Nevertheless, different formulations yield different or even contrasting results for FeO/Fe_2O_3 and/or S^{VI}/S^{-II} from the same set of data (e.g., Oppenheimer et al. 2011; Moussallam et al. 2014).

Many of these discrepancies are imputable to how compositional variables are treated. Bulk composition plays a pivotal role in determining the nature of the ligand (i.e., the solvent structure, such as the melt oxygen network), hence the type and amount of reactive species that it forms by complexing central atoms. Because of their polymerized nature, which prevents a precise distinction between solute and solvent like in aqueous solutions, melt composition largely affect the ligand constitution and then the speciation state, hence Fe^{II}/Fe^{III} ratios. Basically, the oxide component reaction:

$$FeO_{1.5(melt)} \leftrightarrow FeO_{(melt)} + 1/4 O_2 \tag{17}$$

does not offer to regular or sub-regular mixing models the possibility to find accurate and internally consistent expressions for the activity coefficients of *oxide components* γ_{FeO} and $\gamma_{FeO_{1.5}}$ (with $FeO_{1.5}$ replacing conveniently Fe_2O_3) over a large compositional range (e.g., from mafic to silicic) that solve the equilibrium constant of Reaction (17) at 1-bar:

$$\log K_{17} = \log \frac{a_{FeO_{melt}}}{a_{FeO_{1.5,melt}}} + \frac{1}{4}\log f_{O_2} = \log \frac{\gamma_{FeO_{melt}}}{\gamma_{FeO_{1.5,melt}}} + \log \frac{X_{Fe^{II}_{melt}}}{X_{Fe^{III}_{melt}}} + \frac{1}{4}\log f_{O_2} \tag{18}$$

with a denoting activities, X molar fractions and γ representing the activity coefficients of the iron oxides

This expression can be tested for a simple binary high-T melt equilibrated in air ($P = 1$ bar; $f_{O_2} = 0.21$), like in glass industry batches. In particular, we can investigate varations of iron oxidation state when the activity of O^{2-} increases, for example by adding alkalies to the batch In particular, it is well known since earlier 1-bar experiences from glass science (e.g., Paul and Douglas 1965; Fig. 4) that alkali addition (i.e., increase of the basic R_2O, R being Li, Na or K) and consequent increase of $a_{O^{2-}}$ favor Fe^{3+} (Fig. 4a)

This behavior cannot be explained by Reaction (17) via canonical expressions of the activity coefficients γ for iron oxides in Equation (18) ($\log K_{17}$), which can be calculated by subtracting the $\log \dfrac{X_{Fe^{2+}_{melt}}}{X_{Fe^{3+}_{melt}}} + \dfrac{1}{4}\log f_{O_2}$ quantity from $\log K_{17}$. Indeed, Figure 4b shows the results of this exercise by using 1) $\log K_{17}$ (Eqn. 18) from Ottonello et al. 2001 for melt-relaxed FeO–FeO$_{1.5}$ components, i.e.:

$$\log K_{17} = 6364.8/T + 2.8792 \qquad (19)$$

which is obtained on the basis of a plethora of experimental data on after standard state homogenization, or 2) alternatively by adopting the JANAF thermochemical tables (Chase 1998) for liquid (not relaxed in a melt) FeO and crystalline Fe$_2$O$_3$, integrated for the Gibbs energy of Fe$_2$O$_3$ melting (at 1 bar) as in Jaysaryua et al. (2004):

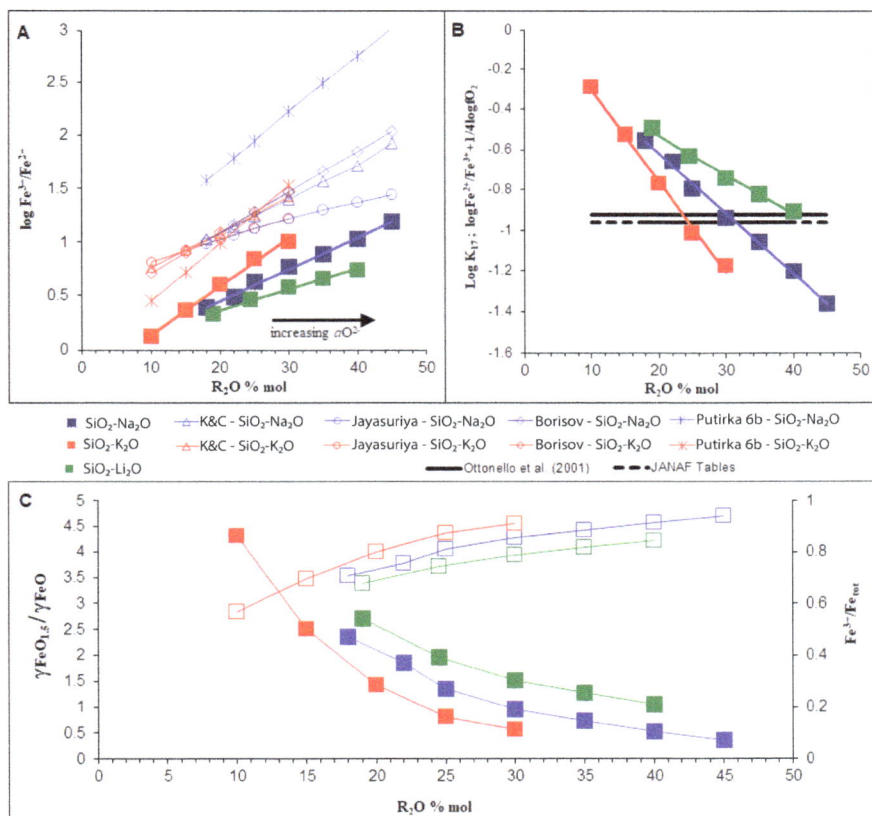

Figure 4. Panel A) Measured (**solid filled squares**) oxidized/reduced iron of for simple SiO$_2$–R$_2$O binaries at 1400 °C equilibrated on air (Paul and Douglas 1965) versus the molar content of alkali oxide (in mol%). Also shown is the performance (**empty symbols and asterisks**) of some empirical/semi-empirical expressions mentioned in the text (Kress and Carmichael 1991; Jayasuriya et al. 2004; Putirka et al. 2016; Borisov et al. 2018), all overestimating oxidized iron. **Panel B)** Equilibrium constant of Reaction (17) at 1400 °C (**black lines; solid:** Ottonello et al. 2001; **semi-dashed**: Chase 1998) and of its compositional term. **Panel C)** Activity coefficient ratios (**filled symbols**) that should reconcile $\log K_{17}$ with the measured oxidized/reduced iron ratio. Also shown is the relative proportion of oxidized iron over total iron (**empty symbols and secondary Y-axis**). Redrawn after Moretti (2021) and expanded.

$$G_{fus}^{\circ 1,T} = 24.5 \cdot \left[T_m - T \right] \left(J / mol; \, T_m = 1838 K \right) \tag{20}$$

with T_m melting temperature of crystalline Fe_2O_3 and giving:

$$\log K_{17} = -10800/T + 5.498 \tag{21}$$

We see that to fit experimental redox data for increasing alkali content, γ increases for FeO and decreases for $FeO_{1.5}$ (Fig. 4c), which is then stabilized. This behavior goes opposite to the predictions of Equation (18) and is counterintuitive because the ratio of activity coefficients should tend to 1 in the low concentration (0.7–0.8 mol %; i.e., diluted conditions) range of total iron in these glass samples.

Jayasuriya et al. (2004) proposed the adoption of a symmetric regular solution mixing model (see Papale et al. 2022, this volume) to solve Equation (18). The excess Gibbs free energy of mixing (G^{ex}) is given by:

$$G^{ex} = \sum_{i=1}^{n-1} \sum_{j>1}^{n} X_i X_j W_{i-j} \tag{22}$$

with W being interaction parameters of enthalpic nature involving pairs of ith and j^{th} melt oxide components (see also Papale et al. 2022, this volume). Therefore:

$$\ln \gamma_i = \frac{\sum_{j \neq i}^{n} X_j W_{i-j}}{RT} - \frac{G^{ex}}{RT} \tag{23}$$

the ratio of activity coefficients of iron species is then expressed by:

$$\ln \gamma \frac{FeO_{1.5}}{FeO} = \frac{\sum_{j}^{n} X_j \left(W_{FeO_{1.5}-j} - W_{FeO-j} \right)}{RT} - \frac{\left(X_{FeO_{1.5}} - X_{FeO} \right) W_{FeO-FeO_{1.5}}}{RT} \tag{24}$$

By regression of available experimental data Jayasuriya et al. (2004) rework Equation (18) and finally obtain:

$$\ln \frac{X_{Fe_{melt}^{III}}}{X_{Fe_{melt}^{II}}} = \frac{a}{T} + b + \sum_i c_i X_i + d \frac{\left(X_{FeO_{1.5}} - X_{FeO} \right)}{T} + \frac{1}{4} \ln f_{O_2} \tag{25}$$

with a, b, c_i and d fitting parameters. The performance of Equation (25) on alkali-rich binary melts can be appreciated in Figure 4a (Jayasuriya data series) and clearly shows that coefficients calibrated on multicomponent melts of geological interest cannot be used. More generally, solid solutions models based on regular or even sub-regular approaches fail in alkali-rich compositions and in SiO_2–MO binaries (see also Fraser 2005; Ottonello 2001, 2005, 2021 and references therein).

The search for one general formulation for all melt compositions of interest in petrology and geochemistry led many authors to formulate empirical expressions based on adjustable parameters, with less formal rigor than the one requested by Equation (18). Among these we can recall the Kress and Carmichael and its antecedent expressions (Kress and Carmichael 1991 and references therein) and the Borisov et al. (2018), which use expression of the type:

$$\ln \frac{X_{Fe_{melt}^{III}}}{X_{Fe_{melt}^{II}}} = \frac{a}{T} + b + \sum_i c_i X_i + d \cdot \ln f_{O_2} \tag{26}$$

or again the Equations (6b,c) in Putirka (2006), which are similar to Equation (26) but combine molar fractions of oxide components or include other quantities such as the ratio of so-called non-bridging oxygens over network forming cations (NBOs), thus recalling an earlier expression proposed by Mysen (1991).

Nevertheless, these fitting procedures do the job of fitting the iron oxidation state in multicomponent systems, within the compositional domain in which they have been calibrated. Instead, they cannot reproduce the effects on iron oxidation state due to the true nature of the silicate melt, that is, the displacement of the effective charge on the oxide-ligand and metal ions in silicate melts and glasses depending on composition (e.g., Duffy 1992; Dimitrov and Komatsu 1999; Ottonello et al. 2001; Moretti 2005). This nature emerges remarkably in SiO_2–MO binaries, particularly silica–alkali oxides (Fig. 4a), whereas multicomponent systems of geological interest tend to average the distribution of the effective charge on metals and surrounding oxygens, which results in a limited region of composition space as well as polymerization and then configurational disorder (Fraser 1975).

It must be noted that the discrepancy from reaction stoichiometry (i.e., $d \neq 0.25$) in Equation (26) implies the loss of internal consistency and the failure of the Gibbs–Duhem relation for all component activities within the same phase (e.g., Lewis and Randall 1961). Furthermore, altering the O_2 stoichiometic coefficient via a fitting procedure implies the violation of mass conservation which characterizes any chemical reaction. Put in a simple way, expressions based on Equation (26) treat terms, resulting on biased calculations of fluid speciation and poor redox constraints on phase equilibria.

Iron redox at P > 1 bar. When increasing pressure, a silicate melt tends to dissolve volatile components; water particularly (see Papale et al. 2022, this volume), whose presence can affect chemical interactions, hence iron oxidation state. This will be addressed in the section *Acid–base, redox properties and the oxygen connection: arguments in support of polymeric models*, whereas here we will review and quantify the effect of pressure on volatile-free melts. First of all, $\log K_{17}$ (Eqn. 18) must be rewritten to include volume terms, i.e.:

$$\log K_{17}(P,T) = \log K_{17}(1,T) + \frac{1}{2.303RT} \int_1^P \Delta \overline{V}_{FeO_{1.5}-FeO} \, dP - \frac{1}{2.303 \cdot 4RT} \int_1^P \overline{V}_{O_{2(g)}} \, dP \tag{27}$$

where $\log K_{17}(1,T)$ can be given by either Equation (19) or (21) and overbar symbol refer to partial molar properties, with $\Delta \overline{V}_{FeO_{1.5}-FeO}$ as the difference between partial molar volumes of $FeO_{1.5}$ and FeO at P and T of interest, and $\overline{V}_{O_{2(g)}}$ as the partial molar volume of oxygen at P and T of interest.

Because the last integral term is already included in the definition of f_{O_2}, we can re-arrange Equation (27) to write:

$$\log\frac{\gamma_{\text{FeO}_{1.5}}}{\gamma_{\text{FeO}}} = \frac{1}{4}\log f_{\text{O}_2} - \log\frac{X_{\text{Fe}^{III}}}{X_{\text{Fe}^{II}}} - \log K_{17}(1,T) - \frac{1}{2.303RT}\int_1^P \Delta\overline{V}_{\text{FeO}_{1.5}-\text{FeO}}\,dP \qquad (28)$$

in which $\gamma_{\text{FeO}_{1.5}}$ and γ_{FeO} are, again, the activity coefficients of iron species but defined at P and T of interest and $\log K_{17}(1,T)$ takes the $a+b/T$ form as in Equations (19) or (21). The volume term in Equation (28) is typically treated as independent of composition, based on the data and relations from Lange and Carmichael (1989), Kress and Carmichael (1991) and Lange (1994). These authors in fact provide a compilation of melt density data based on the assumption that the partial molar volume of each oxide is independent of composition. They state that this assumption is valid within experimental errors that affect the investigated compositional interval (metaluminous and peralkaline melts). The melt volume may then be well reproduced by:

$$V_{\text{melt}}^{P,T} = \sum_{i=1}^{n}\left[\overline{V}_i^{P_r,T_r} + \frac{\partial\overline{V}_i}{\partial T}(T-T_r) + \frac{\partial\overline{V}_i}{\partial P}(P-P_r)\right]x_i \qquad (29)$$

with x_i the molar fraction of the ith oxide component, T_r as the reference temperature ($T_r = 1673\,\text{K}$), and P_r as the reference pressure ($P_r = 1$ bar). For each component, values of the reference volume ($\overline{V}_i^{P_r,T_r}$) partial derivatives used for (isobaric) thermal expansion and (isothermal) compressibility in Equation (29) are tabulated constant terms. It must be noted that errors associated with Equation (29) increase in presence of a) FeO and Fe_2O_3 components, likely because of uncertainties in the iron oxidation state or composition-induced coordination changes of Fe^{3+} (C.N.: IV, V, and VI), as discussed in previous section and as suggested by the partial molar volume value of Fe_2O_3 in alkali silicate melts (Liu and Lange 2006), b) TiO_2, whose molar volume depends on composition in alkali silicate melts, thus being a true *partial* molar volume. Indeed, we should recall that if the partial molar volume of a component in the mixture corresponds to a reference molar volume taken as a function of P and T but not of composition, the excess partial molar volume of that component in the melt is zero. The consequence is that the mixing model is isometric, and W_{ij} mixing terms (Eqns. 22 to 24) and activity coefficients are P-independent (see also Papale et al. 2022, this volume).

Lange and coauthors recommend to use Equation (29) up to 20 kbar, as for higher pressures it becomes important to consider changes of the compressibility coefficient with pressure. For this purpose, we should also note that Equation (29) is a functional form, and does not have a general thermodynamic meaning. In fact, volumetric (isobaric) thermal expansivity (α_v) and (isothermal) volumetric compression coefficient (β_v), or its reciprocal (bulk modulus: k_T), are second derivatives of the Gibbs free energy, and they are respectively given by:

$$\alpha_V = \frac{1}{V}\left(\frac{\partial V}{\partial T}\right)_P = \frac{1}{V}\frac{\partial^2 G}{\partial T\partial P} \qquad (30)$$

$$\beta_V = \frac{1}{k_T} = -\frac{1}{V}\left(\frac{\partial V}{\partial P}\right)_T = -\frac{1}{V}\frac{\partial^2 G}{\partial P^2} \qquad (31)$$

and in this form, they relate each other via:

$$\frac{\alpha_V}{\beta_V} = \left(\frac{\partial S}{\partial V}\right)_T \qquad (32)$$

or

$$c_p = c_V + \frac{TV\alpha_V^2}{\beta_V} \tag{33}$$

where c_p and c_v are the isobaric and isochoric heat capacities respectively and the $\frac{TV\alpha_V^2}{\beta_V}$ term is the anharmonic contribution. The correct expression for the volume of the ith component should then read as:

$$\bar{V}_i^{T,P} = \bar{V}_i^{0,T_r,P_r} \cdot \exp\left[\alpha_{V,FeO}\left(T - T_r\right)\right] \cdot \exp\left[-\beta_{V,FeO}\left(P - P_r\right)\right] \tag{34}$$

Despite these evident limits, Equation (29) does the job and is behind the extension of Equations such as (26) to high-pressure conditions. The final form of the so-called Kress and Carmichael (1991) equation:

$$\ln\frac{X_{Fe_2O_3}}{X_{FeO}} = a \cdot \ln f_{O_2} + \frac{b}{T} + c + \sum_i d_i X_i + e\left[1 - \frac{T_0}{T} - \ln\frac{T}{T_0}\right]$$
$$+ f\frac{P}{T} + g\frac{\left(T - T_0\right)P}{T} + h\frac{P^2}{T} \tag{35}$$

is one of most used equations in the petrological literature, and it is also included in the MELTS code by Ghiorso and Sack (1995), although its adoption is not internally consistent with the regular solution model for silicate melts at the core of the MELTS procedure. It is worth noting that Equation (35) is an operational form not respectful of mass action requirements for reaction $2FeO + 1/2O_2 \rightarrow Fe_2O_3$ (see also Reaction 17). However, this choice shows a better agreement with trends of iron oxidation state for the same composition at fixed temperature (Kress and Camichael 1991).

The Lange and coworkers' partial molar volumes of iron species were also used by O'Neill et al. (2006), Zhang et al. (2017) and Armstrong et al. (2019) to calibrate geophysical equations of state for an andesitic melt (easily quenchable to a glass) at high-pressure (up to 23 GPa; Armstrong et al. 2019) and discuss redox changes upon magma ascent from a mantle source and gradients of iron oxidation in a magma ocean.

The data from O'Neill et al. (2006) and Zhang et al. (2017) on iron oxidation from high-pressure andesites (Figs. 5a,b) of nearly the same composition can be treated to further test the thermodynamic features behind iron oxidation at high pressure. By introducing Equation (19), Equation (28) can be rewritten as:

$$\frac{1}{2.303RT}\int_1^P \Delta\bar{V}_{FeO_{1.5}-FeO}\,dP = \frac{1}{4}\log f_{O_2} - \log\frac{X_{Fe^{III}}}{X_{Fe^{II}}}$$
$$- \log\frac{\gamma_{FeO_{1.5}}}{\gamma_{FeO}} + \frac{6364.8}{T} - 2.8192 \tag{36}$$

Because the composition is nearly the same for all samples in the dataset of Figure 5, mixing of iron component can be considered isometric (regular solution approach), then we can reasonably reduce Equation (24) to:

$$\log \frac{\text{FeO}_{1.5}}{\text{FeO}} = \frac{C}{2.303 \text{R}T} \tag{37}$$

with C a constant factor, independent of P and T, which summarizes the interaction coefficients for the andesitic composition. Its value is directly provided by the 1-bar sample $_P$VF1 in Zhang et al. (2017) and is -10221.52 J/mol. We can then solve Equation (36) for the $\int_1^P \Delta \overline{V}_{\text{FeO}_{1.5}\text{-FeO}} dP$ term, whose values are plotted in Figure 5c. Considering that from Equation (34) we obtain:

$$\int_1^P \Delta \overline{V}_{\text{FeO}_{1.5}\text{-FeO}}\, dP = P \left\{ \begin{array}{l} \overline{V}_{\text{FeO}_{1.5}}^{0,T_r,P_r} \exp\left[\alpha_{V,\text{FeO}_{1.5}}\left(T-T_r\right)\right]\left(1-\dfrac{1}{2}\beta_{V,\text{FeO}_{1.5}}P\right) \\[2mm] -\overline{V}_{\text{FeO}}^{0,T_r,P_r} \exp\left[\alpha_{V,\text{FeO}}\left(T-T_r\right)\right]\left(1-\dfrac{1}{2}\beta_{V,\text{FeO}}P\right) \end{array} \right\} \tag{38}$$

the reference pressure (1 bar) having been ignored in the development of calculations. By approximating $\beta_{V,\text{FeO}_{1.5}} = \beta_{V,\text{FeO}} = \beta_V$, Equation (38) can be re-arranged as follows:

$$\int_1^P \Delta \overline{V}_{\text{FeO}_{1.5}\text{-FeO}}\, dP = \left(P-\frac{1}{2}\beta_V P^2\right)\left\{ \begin{array}{l} \overline{V}_{\text{FeO}_{1.5}}^{0,T_r,P_r} \exp\left[\alpha_{V,\text{FeO}_{1.5}}\left(T-T_r\right)\right] - \\[2mm] \overline{V}_{\text{FeO}}^{0,T_r,P_r} \exp\left[\alpha_{V,\text{FeO}}\left(T-T_r\right)\right] \end{array} \right\} \tag{39}$$

which has the form of a quadratic equation of the type $AP-\dfrac{1}{2}A\beta_V P^2$, with A replacing the term between curly brackets.

The parabolic fit provided in Figure 5c is consistent with Equation (39), and no appreciable improvement of the fit quality is obtained if the intercept is set different from zero. Besides, a linear fit is more than sufficient to explain data, but we keep here the parabolic one for its relevance with respect to Equation (39). Indeed, it returns a (isothermal) compressibility, β_v, with a value of 4.82×10^{-6} bar (valid at $P > 0.4$ GPa; Fig. 5c), which is in line with the estimate of Guo et al. (2014) ($4.72 \pm 0.46 \times 10^{-6}$ bar) for FeO. It would be possible to marginally improve fitting precision by considering the role of the different thermal expansivities of FeO and FeO$_{1.5}$, but this is outside our scopes here. Besides, approximation may reside in the 1-bar Gibbs free energy (Eqn. 19) as well as in the relatively crude approximation to P-independent activity coefficients (Eqn. 37).

By setting $T_r = 1400\,°C$, which is the temperature of the 1 bar point (VF1 in Zhang et al. 2017), we estimate $\Delta \overline{V}_{\text{FeO}_{1.5}\text{-FeO}} = -0.462$ J/bar for $P > 0.4$ GPa. We also see that for $P < 0.4$ GPa the $\Delta \overline{V}_{\text{FeO}_{1.5}\text{-FeO}}$ (at $1400\,°C$) is positive, at a value (0.604 J/bar) in line with the estimate from Lange (1994), which gives 0.742 J/bar at $1400\,°C$ (Fig. 5c).

Assuming the validity of the regular solution approach for a multi-component andesitic melt (Jayasurya et al. 2004), pressure has no effect of the FeO and FeO$_{1.5}$ activity coefficient: the change of sign in the $\overline{V}_{\text{FeO}_{1.5}}^{0,T_r,P_r} \exp\left[\alpha_{V,\text{FeO}_{1.5}}\left(T-T_r\right)\right] - \overline{V}_{\text{FeO}}^{0,T_r,P_r} \exp\left[\alpha_{V,\text{FeO}}\left(T-T_r\right)\right]$ quantity for $P > 0.4$ GPa must be determined by a change in the mean coordination number and then speciation of trivalent iron, which at high pressure acts as network modifier. This behaviour seems then to confirm the early findings by Mysen and Virgo (1985), who suggested Fe^{3+} in six-fold coordination at $P > 10$ kbar.

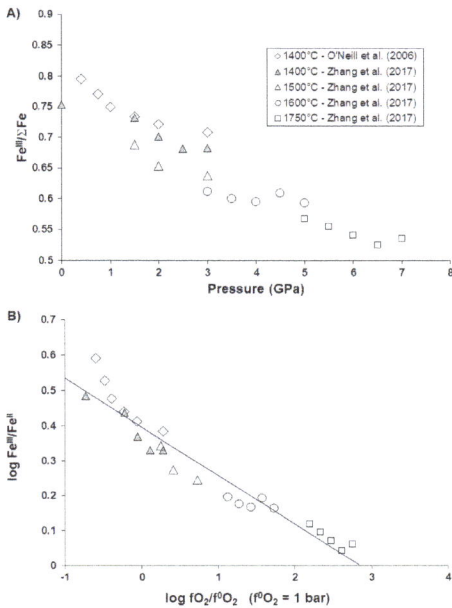

Figure 5. Panel A) Iron oxidation state (as relative proportion of Fe^{III}) of synthetic andesites versus pressure (in GPa for consistency with cited studies). **Panel B)** Iron oxidation state (as trivalent over divalent ratio) versus oxygen fugacity for the same samples; note that the slope of the straight line is -0.13. **Panel C)** $\int_1^P \Delta \overline{V}_{FeO_{1.5}-FeO} \, dP$ -term versus pressure (in bar). See the text for details.

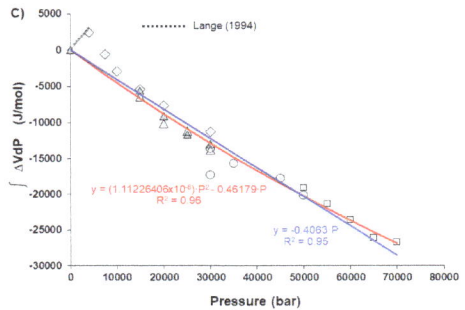

It is worth noting that using Lange and Carmichael data gives $\int_1^P \Delta \overline{V}_{FeO_{1.5}-FeO} \, dP = 0.742 \, [P - 0.0001](J/mol)$, hence 2960 J/mol at 0.4 GPa, which is in line with the experimental determination in Figure 5c, thus suggesting that the cross-over point (i.e, the pressure at which Fe^{III} changes coordination from IV to VI, or more precisely, Fe^{III} in C.N. VI is dominant) occurs not far from 0.4 GPa. If this will be confirmed by further experimental investigation, we should then expect a strong effect on redox, volatile solubility and phase equilibria and of course transport properties of magmas ponding at mid-crustal levels (see also Mysen and Virgo 1985). In particular, exsolution of deep CO_2 (Papale et al. 2022, this volume) and the dynamics of magma sources at the bottom of volcanic plumbing systems may be deeply affected.

The ionic approach and the role of the ligand

The advantages of the ionic "semantics". Rather than using the oxide notation typical of mineral chemistry, the adoption of an ionic notation has to be preferred as it is a straightforward and simpler way to highlight oxidation states and introduce, via charge balance constraints, the stoichiometry of reactions that involve species having a structural significance. Nevertheless, this demands a correct approach to the speciation state of the melt phase, that is, the correct identification of actual chemical reactive entities. These entities (moieties) can be molecules, simple ions or complexes with a formal charge and which intervene in the process of charge transfer (the reaction). The charge transfer involves donor–acceptor pairs, as defined by Lewis: an acid is any substance *acceptor* of an electron pair and a base is any substance *donor* of an electron pair. The speciation state cannot be simply assumed, but it must explain observed data, such as oxidation states and/or solubility behaviors to constrain reaction stoichiometry (e.g., Fincham and Richardson 1954; Moretti et al. 2003; Moretti and Ottonello 2005) and reconcile with spectroscopic observations, particularly those about oxygen coordination around cations (e.g., Le Losq et al. 2020; Neuville et al. 2021 and references therein).

As we already mentioned in the *Introduction*, in oxygen-based, non-protonated substances such as molten oxide and particularly silicate melts, the half-reaction (4), or *oxygen electrode*, represents the main redox exchange and an acid–base exchange in the sense of Lewis (Flood and Forland 1947; Ottonello et al. 2001; Moretti 2005, 2021; Cicconi et al. 2020b). More importantly, equilibrium associated with the reaction clearly relates the molecular oxygen activity a_{O_2}, (then its fugacity, f_{O_2}) to the activity of oxygen ion O^{2-}. The latter can be described as a single, free (unpolarized) oxide ion which is the basic brick that depending on adjacent cations (hence, under different polarization states) builds oxyanion melts and glasses, and particularly the oxygen ligand that frame the silicate network (Flood and Forland 1947; Duffy and Ingram 1971).

The *oxygen electrode* (Eqn. 4) implies that any shift of the (Lewis) acid–base character of the melt is also a redox shift. Variations in composition (i.e., any change of the activity of O^{2-} at constant temperature) will determine a shift of f_{O_2} and vice-versa.

We can already state that acid–base and redox properties are then intimately interrelated and that in a perfectly closed system f_{O_2} is related to the composition. Nevertheless, the link between redox and acid–base features represented by the oxygen electrode Reaction (4), needs a parametrization to define $a_{O^{2-}}$ in an operational way. We need in fact to refer to O^{2-} concentration and its variations during the building an evolution of the silicate network, which follows with the change of mean O^{2-} polarization state (Ottonello et al. 2001; Moretti 2005).

Same as for the electrochemistry of aqueous solution based on the *standard hydrogen electrode* (half-reaction 3), in the melt phase equilibria between two species of redox sensitive elements will result from the sum of two half-reactions: one is the oxygen electrode Equation (4), describing the ligand (or the solvent), and the other involves entities (molecules or ion complexes).

For example, the simple half-reactions such as

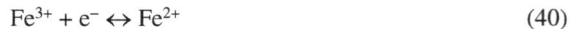

$$Fe^{3+} + e^- \leftrightarrow Fe^{2+} \tag{40}$$

will then combine with half-reaction (4) to give

$$Fe^{3+} + 1/2O^{2-} \leftrightarrow Fe^{2+} + 1/4O_2 \tag{41}$$

whose equilibrium constant is:

$$\log K_{41} = \log \frac{X_{Fe^{2+}_{melt}}}{X_{Fe^{3+}_{melt}}} - \frac{1}{2}\log X_{O^{2-}_{melt}} + \frac{1}{4}\log f_{O_2} \tag{42}$$

and where X denote concentration, of the ionic species of interest (Fe^{2+}, Fe^{3+}, O^{2-}). We use ion concentrations instead of activities under the simple assumption that cations are randomly distributed over a lattice of independent ionic entities. Equation (42) is quite similar to the equilibrium constant that can be computed from the reaction involving macroscopic iron oxide components (Reaction 18), but Equation (42) includes a sort of network contribution via $X_{O^{2-}}$ (or $a_{O^{2-}}$), which is actually a measure of the system basicity. We start seeing that the compositional control on the iron oxidation state can be very strong and also non-linearly related to the melt structure via $X_{O^{2-}}$. Besides, the compositional control can give rise to a dual behavior (particularly relevant for potassium-bearing systems). Such a dual behavior (Fig. 6) cannot be reproduced via classical solution models based on the macroscopic oxide components FeO and Fe_2O_3. A more subtle approach is necessary which accounts for the ionic speciation state via other chemical mechanisms other than Reaction (42).

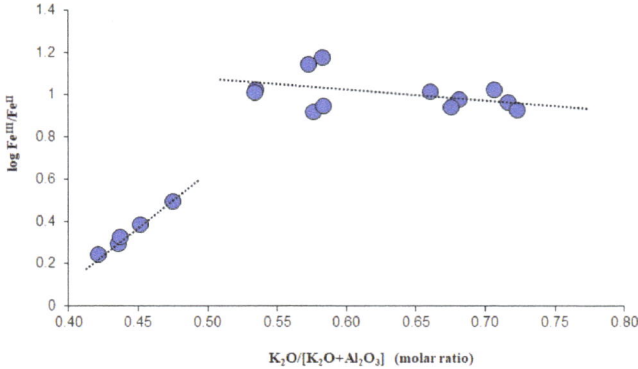

Figure 6. Iron redox ratios from data in Dickenson and Hess (1982): at low-Fe_{tot} contents increasing the $K_2O/(K_2O+Al_2O_3)$ ratio yields a shift from oxidation to reduction.

Ottonello et al. (2001), Moretti and Ottonello (2001, 2003a), Moretti and Papale (2004), Moretti (2005) and more recently Le Losq et al. (2020) have shown that Reaction (41) cannot account for trivalent iron in coordination IV and forming a stable polyhedron, FeO_4^{5-}. This requires extending the iron speciation state in melts and and write half-reaction

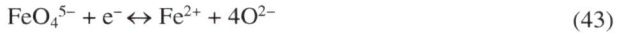

$$FeO_4^{5-} + e^- \leftrightarrow Fe^{2+} + 4O^{2-} \tag{43}$$

which combined with the oxygen electrode (4) yields

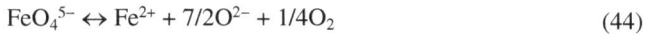

$$FeO_4^{5-} \leftrightarrow Fe^{2+} + 7/2O^{2-} + 1/4O_2 \tag{44}$$

Reactions (43) and (44) remind us that the speciation state of iron cannot be treated only considering simple Fe^{2+} and Fe^{3+} cations. These in fact represent network modifying species that do not give rise to a coordination polyhedron in the sense of Pauling, which is characterized by a strong covalency of Fe–O bonds. On the contrary, in FeO_4^{5-} trivalent iron behaves as a network former.

Because Reaction (44) displays oxygen ions among the products, increasing $a_{O^{2-}}$ (or $X_{O^{2-}}$), at fixed T and f_{O_2}, leads to iron oxidation, in agreement with data in Figures 4a, 6 and as required by its equilibrium constant:

$$\log K_{44} = \log \frac{X_{Fe^{2+}_{melt}}}{X_{FeO_4^{5-}_{melt}}} + \frac{7}{2}\log X_{O^{2-}_{melt}} + \frac{1}{4}\log f_{O_2} \tag{45}$$

In a melt, the bulk iron ratio will then result from the competitive effect of Reactions (41) and (44). By using Roman numerals to identify oxidation numbers, we have:

$$\log \frac{Fe^{III}}{Fe^{II}} = \log \frac{N_{Fe^{3+}_{melt}} + N_{FeO_4^{5-}_{melt}}}{N_{Fe^{2+}_{melt}}} \tag{46}$$

with N indicating number of moles and replacing X, the mol fraction, because molar concentrations and activities, in ionic models should be computed over the appropriate matrix (the anionic population for FeO_4^{5-}, the cationic one for Fe^{3+} and Fe^{2+}) for any fused salt (Temkin 1945), including silicate melts.

Following the guidelines in Fraser (1975), who treated the case of europium (see also Cicconi et al. 2020c) Equation (44) can be reduced to (Ottonello et al. 2001; Moretti 2005):

$$FeO^{2-} \leftrightarrow Fe^{2+} + \frac{3}{2}O^{2-} - \frac{1}{4}O_2 \qquad (47)$$

hence

$$\log K_{47} = \log \frac{X_{Fe_{melt}^{2+}}}{X_{FeO_{2,melt}^-}} + \frac{3}{2}\log X_{O_{melt}^{2-}} + \frac{1}{4}\log f_{O_2} \qquad (48)$$

The use of the FeO_2^- species, instead of FeO_4^{5-}, does not modify the conceptual scheme and allows an even more precise calibration of the equilibrium constant because it limits the propagation of errors related to the estimation of $X_{O^{2-}}$ (or $a_{O^{2-}}$). It is worth stressing that Reaction (47) allows appreciating that O^{2-} is six times more influential than O_2 in affecting the redox state of Fe in silicate melts (Nesbitt et al. 2020).

Based on the ionic notation and the semantics for charge transfer via oxide ions and electron exchange, attempts were made to establish electrochemical series in SiO_2–MO binaries, e.g., Schreiber 1987) but also in ternary joins such as the diopside one (Semkow and Haskin 1985; Colson et al. 1990). The target was an electrochemical potential scale of half-reactions of the type

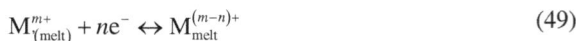

$$M_{(melt)}^{m+} + ne^- \leftrightarrow M_{melt}^{(m-n)+} \qquad (49)$$

which combines to half-reaction (4) to give the following general redox exchange

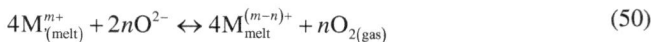

$$4M_{(melt)}^{m+} + 2nO^{2-} \leftrightarrow 4M_{melt}^{(m-n)+} + nO_{2(gas)} \qquad (50)$$

Nevertheless, such series do not consider the effect of the solvent, the melt and its structure, in determining the speciation state (e.g., anionic or cationic), that is, their role of network former or modifier (see also Table 1). This also includes the amphoteric behaviour of some dissolved oxides such as Fe_2O_3 or Eu_2O_3, which can behave either as acids, yielding FeO_2^- (i.e., FeO_4^{5-} tetrahedral units) and EuO_2^- (i.e., EuO_4^{5-}) or bases, yielding Fe^{3+} and Eu^{3+} cations (Fraser 1975; Ottonello et al. 2001; Moretti 2005; Le Losq et al. 2020). The multiple speciation behaviours determined by association or dissociation of O^{2-} and electron transfer can be summarized by the following reaction mechanism (e.g., Moretti 2005; Pinet et al. 2006):

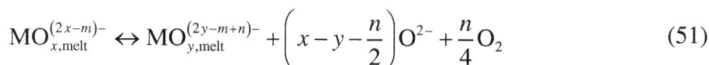

$$MO_{x,melt}^{(2x-m)-} \leftrightarrow MO_{y,melt}^{(2y-m+n)-} + \left(x - y - \frac{n}{2}\right)O^{2-} + \frac{n}{4}O_2 \qquad (51)$$

Acid–base, redox properties and the oxygen connection: arguments in support of polymeric models. Despite the limits summarized above, to assess redox (and acid–base) exchanges in silicate melts we can still calculate the cationic and anionic populations that make up the silicate network and over which concentrations/activities of reactive species can be defined. Consequently, the concentrations of network modifiers (such as Fe^{2+}, or any other network modifier, including the Fe^{3+} cation) will be calculated with reference to the cationic population, whereas the concentration of complexes building the silicate network (*structons*, following the terminology of Fraser 1977) and formed by tetrahedral units coordinated by Si^{4+}, Al^{3+} and also by Fe^{3+} must be calculated over the anionic population. The latter includes a) O^{2-} and b) sulfide and sulfate groups, although these are not part of the network (e.g., Moretti and Ottonello 2003a,b, 2005).

The conceptual basis is the same as the (ideal) Temkin model for fused salt (Temkin 1945), but in silicate melt the complication arises from the connectivity of tetrahedral units sharing apical oxygens (so-called Bridging Oxygens, BOs) and making up a polymerized network without long-range ordering (Mysen and Richet 2018 and references therein). We keep here only the basics behind melt polymerization, without the details on how to assess it, which instead are described in a wide literature (Moretti 2005 and references therein). Polymerization is treated as the mixing of three kinds of oxygens: 1) bridging oxygens (BO), denoted as O^0, 2) non-bridging oxygens (NBO), which are denoted as O^- and represent unshared (covalently unbounded) apical oxygens resulting from the inter-tetrahedral disruption of the network due to network modifiers and 3) the non-network associated or free oxygens, O^{2-} which are the oxygens not bounded to network formers in the formation of a coordination polyhedron (Fincham and Richardson 1954; Toop and Samis 1962a,b; Fraser 1975; Ottonello et al. 2001; Moretti 2005; Mysen and Richet 2018; Nesbitt et al. 2015a,b; Le Losq et al. 2019 and references therein).

Melt polymerization is then the results of the equilibrium reaction of the three oxygen species (Fincham and Richardson 1954; Toop and Samis 1962a,b; Fraser 1977; Ottonello et al. 2001; Ottonello and Moretti 2004; Moretti 2005)

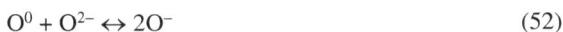

$$O^0 + O^{2-} \leftrightarrow 2O^- \tag{52}$$

Reaction equilibrium $\log K_{52}$ (so-called melt polymerization constant; Toop and Samis 1962a,b) shifts with composition because any silicate network is a solvent *per se* with peculiar dielectric and electrostriction properties, hence with a specific structural framework whose polymerization is the main chemical descriptor. Because of this, the continuous shift of polymerization with composition impacts on redox equilibrium and occurs in a non-linear way via $a_{O^{2-}}$ (see also Moretti 2005 for details). In fact, depending on bulk composition, the polymerization equilibrium $\log K_{52}$ involves a change in the oxidation state of oxygen and hence in the polarization state of the ligand and in network relaxation (Ottonello and Moretti 2001; Moretti 2005). This affects the electron pressure of the ligand (the Lewis donor) surrounding the cations (the acceptors), thus the $a_{O^{2-}}$ term entering Reactions such as (41) and (44 or 47). Variations of the activity of O^{2-} measure the changes in the charge transfer operated by the oxygen network in response to compositional variations. We then see why redox changes are also acid–base changes for the melt system.

Ionic–polymeric approcheas are then a consistent way to deal with the intimate connection between network polymerization and acid–base properties of dissolved oxides. These model give the semantics to write chemical exchanges but without the detail of a true structural description. For example, X-ray Photoelectron Spectroscopic (XPS) results for K-silicate glasses show the presence of at least two types of BO, which can be distinguished by the number of K ions in the moiety (Nesbitt et al. 2017). This yields BO atoms of differing electron density and then affects the charge distribution in a Si tetrahedron and bonding properties. Nevertheless, ionic–polymeric models accomodate these composition-related features by parameterizing the energy shift of the melt solvent with composition rather than by increasing the number of oxygen species in the chemical systems and thus the number of reactions of the type of Equation (52).

Adopted reactive species in ionic–polymeric models may not explain the whole structural complexity revealed by many spectroscopic techniques (e.g., Henderson et al. 2014 and references therein) but have to be consistent with observations on the short-range ordering, in order to discriminate between a) cations in true coordination polyhedra making up the network, such as Fe^{III} in FeO_4^{5-}, b) cations acting as network modifiers, such as Fe^{2+} and Fe^{3+}, and c) anionic groups not associated with the network, such as sulfide and sulfate. In the case of network modifying behavior, the species under consideration are simple cations (Fe^{2+} or Fe^{3+}) because of bonds with surrounding oxygens are mainly electrostatic, rather than covalent (Moretti 2005). For this reason it is not necessary to write network-modifying iron species in octahedral or higher coordinations, (i.e., $Fe^{II}O_6^{10-}$ and $Fe^{III}O_6^{9-}$) as these would not correspond to true polyhedra that build the network and speciate over the anionic matrix. Rather they are simple ionic couplings $Fe^{2+} \cdot 6O^{2-}$ and $Fe^{3+} \cdot 6O^{2-}$. Therefore reactions:

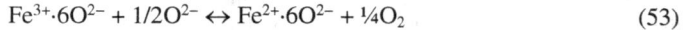

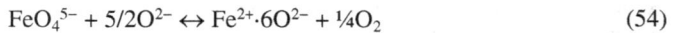

$$Fe^{3+} \cdot 6O^{2-} + 1/2O^{2-} \leftrightarrow Fe^{2+} \cdot 6O^{2-} + \frac{1}{4}O_2 \tag{53}$$

$$FeO_4^{5-} + 5/2O^{2-} \leftrightarrow Fe^{2+} \cdot 6O^{2-} + \frac{1}{4}O_2 \tag{54}$$

reduce to Reactions (41) and (44), respectively. In case of network forming behavior, a true polyhedron (in the sense of Pauling 1960) is formed (FeO_4^{5-} tetrahedron) , with $Fe^{III}-O$ bonds essentially covalent. This behavior is confirmed by structural studies and appear to be the normal condition for trivalent iron (Le Losq et al. 2020 and references therein), in agreement with the well-known fact that a strong Lewis bases such as O^{2-} stabilizes high oxidation states in covalent coordination compounds (e.g., Chiorboli 1980).

If now we want to link the process of ion formation to the dissolution of oxide components, we see that the network-modifying behavior corresponds to dissociation of O^{2-} by the iron macroscopic oxide

$$FeO_{(adc)} \leftrightarrow Fe^{2+}_{(La)} + O^{2-}_{(Lb)} \tag{55}$$

$$Fe_2O_{3(adc)} \leftrightarrow 2Fe^{3+}_{(La)} + 3O^{2-}_{(Lb)} \tag{56}$$

in which iron oxides may be seen as the Lewis adduct complexes (adc as subscript) and their dissolved forms Lewis acids (La; i.e., Fe^{3+} the network modifying cation) and the O^{2-} base (Lb). On the other hand, network-forming behavior of trivalent iron oxide (now the Lewis acid) leads to associate O^{2-} (consumption of the Lewis base) and stabilizes a network structural unit (a coordination polyhedron, that is, the Lewis adduct complex):

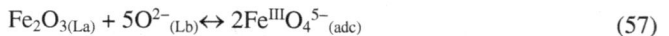

$$Fe_2O_{3(La)} + 5O^{2-}_{(Lb)} \leftrightarrow 2Fe^{III}O_4^{5-}_{(adc)} \tag{57}$$

which can also be expressed as (see Reaction 47)

$$Fe_2O_{3(La)} + O^{2-}_{(Lb)} \leftrightarrow 2Fe^{III}O_2^-_{(adc)} \tag{58}$$

Now, to realize the analogy between $a_{O^{2-}}$ and pH we should consider that the above Reactions (55) to (58) imply the Lux–Flood acid–base formalism (Flood and Forland 1947) for molten oxides and silicate melts:

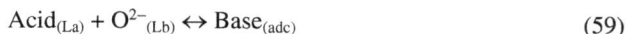

$$Acid_{(La)} + O^{2-}_{(Lb)} \leftrightarrow Base_{(adc)} \tag{59}$$

Reaction (59) defines acid–base exchanges on the basis of the transfer of an oxygen ion and recalls the perhaps more familiar acid–base definition of Bronsted–Lowry for aqueous solutions:

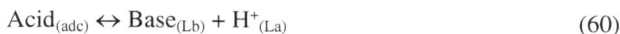

$$\text{Acid}_{(\text{adc})} \leftrightarrow \text{Base}_{(\text{Lb})} + \text{H}^+{}_{(\text{La})} \tag{60}$$

in which the acid–base character is determined by proton transfer.

The behavior of trivalent iron, Fe^{III}, which can be either former or modifier, is then related to the dual or *amphoteric* behavior of Fe_2O_3 (Reactions 56 and 58). Based on Equation (59), cations behaving as network modifiers correspond to Lux–Flood basic oxides, whereas cations behaving as network formers correspond to Lux–Flood acidic oxides. The amphoteric behavior can also be seen in terms of Lewis acid–base behavior.

Because in Reaction (59) i) the Lux–Flood acid is also a Lewis acid, ii) O^{2-} is the Lewis base and iii) the Lux–Flood base is the Lewis adduct complex, we see that there is a considerable overlap between Lux–Flood and Lewis acid–base definitions. Reaction (59) represents then the obvious mechanism to set a basicity indicator in melts and glasses via $a_{O^{2-}}$, same as pH as acidity indicator via Equation (49). On the other hand, Equation (59) represents also the transfer of polarizable oxygen ion to an acidic solute (iron cations in Reactions 55–56; trivalent iron oxide with in Reactions 57–58) leading to formation of an oxyanion species (e.g., iron oxides in Reactions 55–56; the network structural units in Reactions 57–58) characterized by covalent bonds. This overlap implies the possibility of setting a quantitative scale of melt basicity unique for all oxide systems including silicate melts of any composition under the more comprehensive and general Lewis definition. Duffy and Ingram (1971, 1975) and Duffy (1992) have shown that a metrics can be defined on the basis of optical basicity, a parameter experimentally observed with the shift of UV absorption bands of metals in glasses (or other media). The observed shift measures the mean polarization state of the various ligands, such as oxide ions in natural silicate melts and glasses, and their ability to transfer fractional electronic charges to the central metal cation and form covalent bonds. Therefore, optical basicity is directly related to $a_{O^{2-}}$ as defined in the Lux–Flood acid–base formalism (Eqn. 59). The concept of optical basicity arises primarily from the systematic study of the orbital expansion (or "nephelauxetic effect") induced by the increased localized donor pressure on p-block metals (e.g., Tl^I, Pb^{II}, Bi^{III}), which are used at trace levels as probes to measure the shift of UV absorption bands. Optical basicity can also be calculated, depending upon the medium stoichiometry, the oxidation numbers of the cations, and upon a "basicity moderating" parameter related to the Pauling electronegativity (Duffy and Ingram 1975; Duffy 1992; Moretti 2005).

Based on the optical basicity contrast of the anionic and cationic matrixes Ottonello et al. (2001) proposed an assessment of the polymerization constant K_{41}, allowing the computation of the three species of oxygen and related activities, particularly $a_{O^{2-}}$, which measures the Lux–Flood basicity appearing in redox equilibria so far described. This led the Authors to compute the bulk Fe^{III}/Fe^{II} ratio as:

$$\frac{Fe^{III}}{Fe^{II}} = \frac{N_{Fe^{III}}}{N_{Fe^{II}}} = \frac{\left[FeO_2^- \right] \sum \text{anions} + \left[Fe^{3+} \right] \sum \text{cations}}{\left[Fe^{2+} \right] \sum \text{cations}}$$

$$= \frac{f_{O_2}^{1/4}}{K_{17}} \times \frac{K_{58}^{1/2} a_{O^{2-}}^2 \sum \text{anions} + K_{56}^{1/2} \sum \text{cations}}{a_{O^{2-}}^{1/2} K_{55} \sum \text{cations}} \tag{61}$$

Note that with Equation (61) we start using square brackets to define ion activities, meaning that these are computed as the concentration of the ion of interest over the appropriate matrix (either cationic or anionic). By incorporating the amphoteric behavior of trivalent iron, Fe^{III}, Equation (61) can be used to solve counterintuitive behaviors like those shown in Figure 4 (Ottonello et al. 2001; Moretti 2005). Ottonello et al. (2001) have also shown that oxide dissolutions Reactions (55, 56, 58) can be seen as transposition reactions to link thermodynamically macroscopic oxide components and the ionic species that detail the solution behavior of iron oxides. Therefore, their reaction Gibbs energies (or equilibrium constants, i.e., K_{55}, K_{56} and K_{58}) are the energy differences between the standard state for pure molten iron oxides between 1) that at P and T of interest (or 1 bar and T of interest) and 2) the one of completely dissolved ionic substance. The first is the standard state used to compute reaction equilibria (typically from thermochemical tables, such as NIST-JANAF ones; Chase et al. 1998), whereas the second is the standard state in which each ion speciates over the corresponding matrix (fused salts Temkin model). If we take the example of Reaction (55), in the latter standard state the activity of FeO in Reaction (55) should be one (e.g., Lewis and Randall 1970, chapter 22). We then have (Ottonello et al; 2001):

$$\log K_{55} = \frac{\left[Fe^{2+}\right]\left[O^{2-}\right]}{a_{FeO}} \neq 1 = 1.1529 - \frac{1622.4}{T} \tag{62}$$

whose value fills the energy gap between the two standard states. Equilibria of the type of Equation (62), in which oxide components and the ion species derived from their dissociation appear, are the key to reconcile experiments based on the macroscopic oxide notation with a detailed ionic description of the dissociation products (Ottonello et al. 2001). This is an important point, as in many studies oxide components and ionic species are mixed, but without any consideration about their different standard states.

Equation (62) may be easily appreciated in the context of the Temkin model for ion activities on the appropriate (either cationic or anionic) matrix:

$$a_{MO(Temkin)} = a_{M^{2+}} a_{O^{2-}} = x_{M^{2+}} x_{O^{2-}} = \frac{n_{M^{2+}}}{\sum cations} \cdot \frac{n_{O^{2-}}}{\sum anions}$$

$$= \frac{n_{M^{2+}}}{\sum cations} \cdot \frac{n_{O^{2-}}}{\sum structons + \sum "free" anions + n_{O^{2-}}} \tag{63}$$

written for a generic divalent basic oxide MO. Equations of the type (63) measure the strength of the Lux–Flood acid–base pair (Flood and Forland 1947). For example, Fe_2O_3 basic dissociation (Reaction 58) has an equilibrium constant (Ottonello et al. 2001)

$$\log K_{58} = \frac{\left[Fe^{3+}\right]^2 \left[O^{2-}\right]^3}{a_{Fe_2O_3}} \neq 1 = 1.8285 - \frac{4100.2}{T} \tag{64}$$

We then see that at 1400 °C $a_{O^{2-}}$ is higher for Reaction (58) than (55).

A full description of the procedure to set the polymeric model, estimate the amount of structons and compute iron oxidation as a function of compositions is given in Ottonello et al. (2001) and Moretti (2005), whereas some important elements can be found in Papale et al. (2022, this volume). Briefly, the Toop–Samis polymeric scheme is adopted to compute O^0, O^- and O^{2-} via the polymerization constant parameterized on the basis of composition via atomistic parameters, such as the basicity moderating parameters (a measure of the covalence of a cation in the oxygen ligand, then of the solvation character of the cation), for both network modifiers ($\varphi_{M_i^{\nu+}}$) and network formers ($\varphi_{T_j^{n+}}$), through the following equation:

$$K_{52} = \exp\left[4.662 \times \left(\sum_i x_{M_i^{y+}} \varphi_{M_i^{y+}} - \sum_j x_{T_j^{n+}} \varphi_{T_j^{n+}}\right) - 1.1445\right] \qquad (65)$$

where $x_{M_i^{y+}}$ and $x_{T_j^{n+}}$ are atom fractions of network modifiers (including H^+; see Moretti et al. 2014) and network formers, respectively. Equation (65) establishes a formal link between the acid–base properties of the medium (expressed as a contrast between network formers' and network modifiers' basicity in the sense of Lewis; Moretti 2021) and the polymerization constant K_{52}. Equation (65) does not consider the effect of temperature and can be seen as the high-temperature limit for the melt polymerization constant K_{52}. The importance of Equation (65) is that it provides a criterion to combine the contribution of each melt oxide component by defining a basicity contrast between the anionic (network-forming) and the cationic (network-modifying) matrixes. Such a composition-driven contrast is then a measure of the Lewis acid–base character of the oxyanionic substance following Duffy and Ingram (1971), which is then anchored to the Lux–Flood acid–base definition (Eqn. 59) via melt polymerization in the Toop–Samis approach.

Ottonello (2001, 2005) and Ottonello and Moretti (2004) report the further developments of the melt polymeric model about, based on a refined and T-dependent definition of the polymerization constant. These improvements were successfully tested by reproducing activities in 1-bar ternary systems, but they have not been yet used in multicomponent systems or for recalibrating the redox model.

Among the "free" anions appearing at the denominator of the right-most member of Equation (63), we must also count sulphur species (S^{2-} and SO_4^{2-}; see also Papale et al. 2022, this volume) and hydroxyl groups. The latter are obviously related to the amount of water dissolved at pressure in the melt. Water can in fact be an abundant component, given its low molecular weight, and cause important compositional dependencies of the iron oxidation state during melt dehydration and magma differentiation. For an isothermal process, these dependencies will then add up to the eventual redox changes imposed by pressure excursions (e.g., Moretti and Stefansson 2020 and references therein).

Nevertheless, data of Fe^{III}/Fe_{tot} in the literature are discrepant about the role played by water. Following Moore et al. (1995), and also Waters and Lange (2016), water does not affect the ferric to ferrous iron ratio, which is a record of other processes having imposed the oxygen fugacity.

However, according to Baker and Rutherford (1996) and Gaillard et al. (2001) water affects the ferric to ferrous ratio. In some region of the $P–T–f_{O_2}$ space it may cause either a decrease or an increase of oxidation. For example, water-bearing rhyolitic melts have higher ferric to ferrous ratio than anhydrous melts of the same composition (Baker and Rutherford 1996). The same occurs in metaluminous melts, but at higher temperatures ($T > 900\ °C$) and around f_{O_2} values consistent with the coexistence of metal nickel and its oxide (so-called buffer NNO), whereas in peralkaline melts such an increase is observed at high T (Baker and Rutherford 1996). Gaillard et al. (2001) generalise this perspective, observing an increase of the ferric to ferrous ratio of iron in hydrous melts at $\log f_{O_2} < NNO + 1.5$ for all studied compositions, metaluminous and rhyolitic melts and natural peraluminous and peralkaline obsidians. However, they find that above NNO+1.5 water does not affect anymore the ferric to ferrous iron ratio, controlled by the anhydrous composition in agreement with Moore et al. (1995). Wilke et al. (2002) investigated tonalitic melts at 850 °C, whose ferric to ferrous iron ratio showed a marked decrease with respect to the values computed through the Kress and Carmichael (1991) and then based on the anhydrous composition. Nevertheless, this effect is mainly ascribed to the inaccurate calibration of the Kress–Carmichael equation at low T rather than to the water content of melts. Botcharnikov et al (2005) studied hydrous ferribasalts up to 500 MPa and found increasing iron oxidation with water content, well-reproduced by the model of Moretti (2005). Also, Schuessler et al. (2008) highlighted a slight oxidation with increasing water content in hydrous phono-tephritic melts.

Some of the confusion on the effect of dissolved water on the iron redox state of melts may be due to the non-precise assessment of the relative role of pressure and water amount on the polymerization properties. However, Moretti (2005, 2014) have shown that the observed non-linear behaviours of iron oxidation with water may be ascribed to the dual (amphoteric) dissociation of water in melts, which may experience both basic and acidic dissociations. This amphoteric behaviour, which depends on composition, gives rise to T–OH terminal groups, molecular H_2O entities but also "free" OH-groups, and explains many experimental spectroscopic evidences (Moretti et al. 2014 and references therein). Indeed, the water component not only breaks the silicate network chains via:

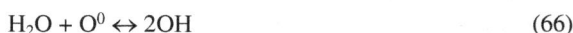

$$H_2O + O^0 \leftrightarrow 2OH \tag{66}$$

as earlier proposed by Fraser (1975) (see also Papale et al. 2022, this volume, their Eqn. 73), but in basic and depolymerized melts it may speciate as a Lux–Flood acid generating "free" hydroxyl:

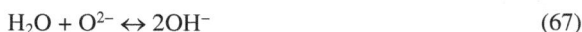

$$H_2O + O^{2-} \leftrightarrow 2OH^- \tag{67}$$

Moretti (2005) and Moretti et al. (2014) modelled these competitive effects by accounting for the amphoteric behavior of the water component by setting a formally consistent set of reactions, in which ionic species intervene. Water speciation was incorporated into a Temkin ionic–polymeric model of molten silicate (Temkin 1945; Ottonello et al. 2001; Moretti 2005) via reaction:

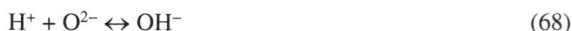

$$H^+ + O^{2-} \leftrightarrow OH^- \tag{68}$$

which results by combining the Lux–Flood acidic water dissociation, Reaction (67), with the Lux–Flood basic dissociation

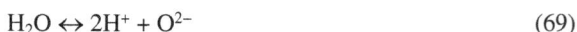

$$H_2O \leftrightarrow 2H^+ + O^{2-} \tag{69}$$

which combines to the Toop–Samis Reaction (52) to give Reaction (66) by formation of terminal T–OH groups. The model results match fairly well the Raman assignments and concentrations reported by Le Losq et al. (2013).

As summarized in Papale et al. (2022, this volume) such an ionic–polymeric model has then been used to parameterize water autoionization in silicate melts by recombining ionic entities defined in the Temkin standard state (Moretti et al. 2014).

In order to model the effect on iron hydrous state in hydrous melts, based on experimental data up to 500 MPa, Moretti (2005) propose 1) how to initialize Equation 63 and compute $\frac{n_{M^{2+}}}{\sum \text{cations}}$ (to which contributes H^+ from Reaction 68) and $\frac{n_{O^{2-}}}{\sum \text{structons} + \sum \text{"free" anions} + n_{O^{2-}}}$, to which contribute O^{2-} and OH^- from Reaction (68), and 2) how to introduce the effect of pressure on the reaction set made by Equations (17, 55, 56, 58).

The Author considered the molar volumes of melt ionic species defined in the polymeric approach as well as melt iron oxides. Molar volumes, isothermal compressibilities and isobaric thermal expansivities for iron oxide components (FeO and Fe_2O_3) were taken by Lange (1994). Volumes of ionic species were calculated on the basis of ionic radii of Shannon (1976) assuming that the 'effective molar volume' of each ionic species equals that of a mole of spherical molecules each characterized by its appropriate Shannon radius. For the FeO_2^- species the molar volume correspond to that of a sphere of radius $r_{O^{2-}} + \frac{1}{2} r_{Fe^{3+}(IV)}$. Since the spherical volume

Table 2. Molar volumes employed for macroscopic and ion iron species involved in Reactions (17, 55, 56, 58). For ion species the adopted ionic radius is reported.

	Molar Volume @ 298.15 K 298.15 K (cc/mol)	Ionic radius (Å)	Reference
FeO	9.64	–	Lange (1994)
Fe_2O_3	29.63	–	Lange (1994)
Fe^{2+}	0.90	0.78	Shannon (1976)
Fe^{3+}	0.51	0.645	Shannon (1976)
O^{2-}	6.92	1.40	Shannon (1976)
FeO_2^-	75.99	3.29	Shannon (1976)

associated to this radius should represent, at a first approximation, the 'effective volume' of the $Fe^{III}O_4^{5-}$ complex, the volume of two oxide ions O^{2-} was simply subtracted in order to obtain the 'effective volume' of the FeO_2^- species. This choice implies that the volume change reaction for the association Reaction (58) is zero. The set of adopted values is summarized in Table 2.

Figure 7 shows the results of model calculation on the same ferribasalts of Botcharnikov et al. (2005). The diagram allows evaluating, for example, the set of conditions for which either $\log f_{O_2}$ or Fe^{III}/Fe^{II} are kept constant. The same oxidation state of iron can be obtained by different couples of $\log f_{O_2}$ and dissolved H_2O. For example, if we move along the $Fe^{III}/Fe^{II} = 0.2$ isoline we can have $\log f_{O_2} = -8.9$ and $H_2O = 6$ wt% as well as $\log f_{O_2} = -7.6$ and $H_2O = 0$. So, the same Fe^{III}/Fe^{II} ratio can be obtained within variations of 1.3–1.2 log units in oxygen

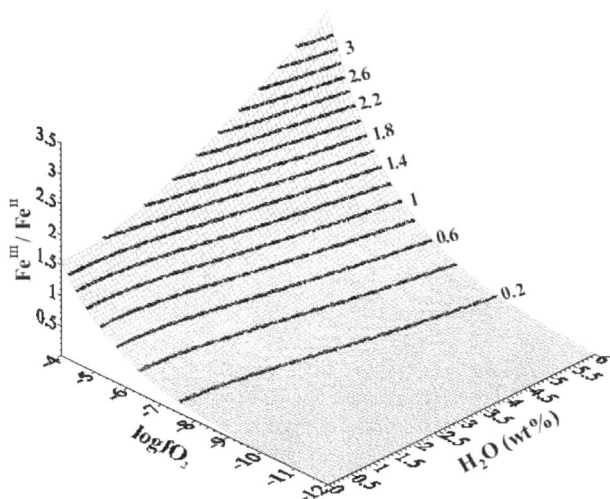

Figure 7. Shape of the iron redox ratio surface in ferribasalts from Botcharnikov et al. (2005), modeled via Moretti (2005) as a function of dissolved water content and oxygen fugacity. Also traced are some isolines of constant Fe^{III}/Fe^{II}.

fugacity but with 6 wt% of difference in the water content. At the same time, $\log f_{O_2}$ can be kept constant upon minor variations of the Fe^{III}/Fe^{II} ratio and large variations in dissolved water. The very likely occurrence of simultaneous variation of similar entity in ascending and degassing magmas may explain the Waters and Lange (2016) assertion that H_2O degassing has no effect on the oxidation state of magmatic liquids.

Oxygen fugacity, sulfur and joint Fe–S redox exchanges

The problem of unsolved compositional behaviors due to speciation and that are not accounted by typical oxide-based approaches to mixtures, is exacerbated when dealing with the mutual exchanges involving iron and another redox sensitive element, such as sulfur.

Sulfur-bearing melt species have a special role since the oxidation of sulfide to sulfate involves eight electrons: for any increment of the Fe^{III}/Fe^{II} redox ratio, there is an eight-fold increment for sulfur species (S^{-II}/S^{VI}; e.g., Douglas and Zaman 1969; Moretti and Ottonello 2003a; Nash et al. 2019; Cicconi et al. 2020b; Moretti and Stefansson 2020). Sulfur in magmas partitions between different phases (gas, solids, such as pyrrhotite and anhydrite and liquid as well, such as immiscible Fe–O–S liquids; Haughton et al. 1974; Moretti and Baker 2008 and references therein). The large electron transfer makes S^{-II}/S^{VI} a highly sensitive indicator to f_{O_2} changes in a narrow range (typically around QFM and NNO buffers in magmatic melts; Moretti 2021 and references therein), whereas its effectiveness as a buffer of the redox potential is limited by the abundance of sulfur in magma, significantly lower than iron. The two oxidation states of sulfur in melts, S^{-II} and S^{VI}, are well known to occur typically in the redox couple involving the simple anion S^{2-} and the anionic complex SO_4^{2-} (Fincham and Richardson 1954; Carroll and Rutherford 1988; Paris et al. 2001; Baker and Moretti 2011 and references therein). Their exchange gives rise to the following half-reaction:

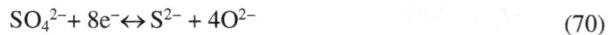

$$SO_4^{2-} + 8e^- \leftrightarrow S^{2-} + 4O^{2-} \tag{70}$$

which combines with the oxygen electrode half-reaction (4) (taken eight times) to give:

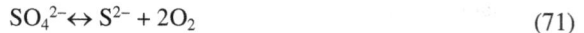

$$SO_4^{2-} \leftrightarrow S^{2-} + 2O_2 \tag{71}$$

Reaction (71) may suggest the occurrence of loose and dispersed sulfide and sulfate anions, both speciating over the melt anionic matrix. This idea is also a consequence of the relatively low abundance of sulfur dissolved in natural melts (generally less than 1 wt%; Wallace and Edmonds 2011 and references therein) and has led to empirical formulations calibrated on compositionally restricted datasets in which the sulfate/sulfide ratio is assumed to be independent of the melt composition (Wallace and Carmichael 1994; Matthews et al. 1999; Jugo et al. 2005, 2010). For example, based on a small dataset of 22 submarine glasses dominated by reduced sulfur, Wallace and Carmichael (1994) extended the results of the earlier Carroll and Rutherford (1988) electron microprobe study and proposed the following equation:

$$\log\frac{X_{SO_4^{2-}}}{X_{S^{2-}}} = 1.02\log f_{O_2} + \frac{25410}{T} - 10 \tag{72}$$

in which the factor multiplying $\log f_{O_2}$ is sensibly lower than 2, the stoichiometric coefficient required by Equation (71). Like for Reaction (17), this low value of the $\log f_{O_2}$ factor cannot be explained by the activity coefficient ratio of sulfide and sulfate deviating from unity. Rather, it must embody compositional variations apparently not considered in Reaction (71) (Moretti and Ottonello 2005).

All these empirical models refuse any role of composition on the oxidation state simply because not made explicit in Reaction (71). However, this reaction results by subtracting each other composition-dependent dissolution equilibria of SO_4^{2-} and S^{2-} from a sulfur gaseous species such as S_2 or SO_2 (see Papale et al. 2022, this volume and references therein):

$$O^{2-} + 1/2S_2 \leftrightarrow S^{2-} + 1/2O_2 \tag{73}$$

$$O^{2-} + 1/2S_2 + 3/2O_2 \leftrightarrow SO_4^{2-} \tag{74}$$

The role of composition in shifting K_{71} is then implicit in the gas–melt dissolution of sulfur in its two species. However, because analytical measurements involve S as S^{-II} in S^{2-} and S as S^{VI} in SO_4^{2-}, these anions being judged to be independent of the melt network/structure, many workers have assumed quite naively that composition plays no role in fixing the sulfate/sulfide equilibrium or that compositional dependencies cancel each other when subtracting Reaction (73) from (74) (Jugo 2005, 2010; Nash et al. 2019; see discussion in Moretti 2021). Nevertheless, Reaction (71) depends on composition (Moretti et al. 2003b, 2005; Moretti 2021) simply because both S^{2-} and SO_4^{2-} anions are not in an unpolarized state, same as O^{2-} (see Moretti 2021). Instead, they are solvated into the melt structure and thus are more or less polarized depending on whole composition and not only on adjacent cations (Moretti and Ottonello 2003b, 2005). Based on this consideration, Moretti and Ottonello (2003b, 2005) modelled sulfur equilibria (73) and (74) in order to provide a description of S dissolution based on f_{O_2} and f_{S_2}. The two competing solubility equilibria were then modelled by weighting for each experimental composition the metal oxide–metal sulfide and metal oxide–metal sulfate pairs contributing to S^{2-} and SO_4^{2-} dissolution (see also Papale et al. 2022, this volume). Weighting factors were established via the Flood–Grjotheim thermochemical cycle by using *electrically equivalent ion fractions* (Flood and Grjotheim 1952; Moretti and Ottonello 2003b, 2005):

$$Y_{M_i^{y+}} = \frac{v_i^+ X_{M_i^{y+}}}{\sum v_j^+ X_{M_j^{y+}}} \tag{75}$$

which enter the Flood–Grjotheim thermochemical cycle in the form (see also Papale et al. 2022, this volume):

$$\ln K'_{O-S} = \sum_{i=1}^{n} Y'_{M_i^{v+}} \ln C_{O-S,M_i}^{anneal.} K_{O-S,M_i} \tag{76}$$

where K'_{O-S} is the bulk exchange reaction (oxide–sulfide or oxide–sulfate) in the melt, K_{O-S,M_i} is the exchange reaction between (pure) *liquid* metal (oxide, sulfide or sulfate) components not relaxed in the melt structure and $C_{O-S,M_i}^{anneal.}$ is the annealing entropy for the single exchange reaction, representing the difference of entropy when passing from the condition of pure *liquid* components to that of *melt* components. Annealing terms have then the meaning of scaling factors for thermodynamic standard state transposition (Moretti and Ottonello 2005).

Assessing the two solubility equilibria for sulfide and sulfate species with respect to a gas phase automatically leads to the assessment of $\log K_{71}$ by subtracting one mechanism from the other. On this basis, Moretti and Ottonello (2005) showed that they could reproduce independent (i.e., not included in the model calibration database) microprobe and gravimetric SO_4^{2-}/S^{2-} data in the $\log f_{O_2}$ range $-2 < QFM < 2$, where electron microprobe measurements of both species were shown to be most reliable. This corroborates the use of the Flood–Grjotheim cycle (corrected by Moretti and Ottonello 2005 with the addition of the annealing entropy) to describe multicomponent interactions in the melt phase, including polymerization and then mixing properties (see also Ottonello 2001; Ottonello and Moretti 2004).

Figure 8a plots S^{VI}/S_{tot} ratios versus $\log f_{O_2}$ (as relative to the fayalite–magnetite–quartz redox buffer, FMQ) from empirical formulas proposed in the literature. All curves show that sulfide to sulfate transition occurs over a narrow f_{O_2} range (about two log units), much narrower than that of iron, reported for comparison in Fig. 8a (red solid line). Note the shift to lower relative f_{O_2} values of Nagashima and Katsura (1973) data, which were questioned and dismissed by Jugo et al. (2005) who stated that departure of these data from existing empirical fits (see Fig. 8a) was due to a wrong f_{O_2} assessment, based on the pretended outdated and inaccurate fluxing technique.

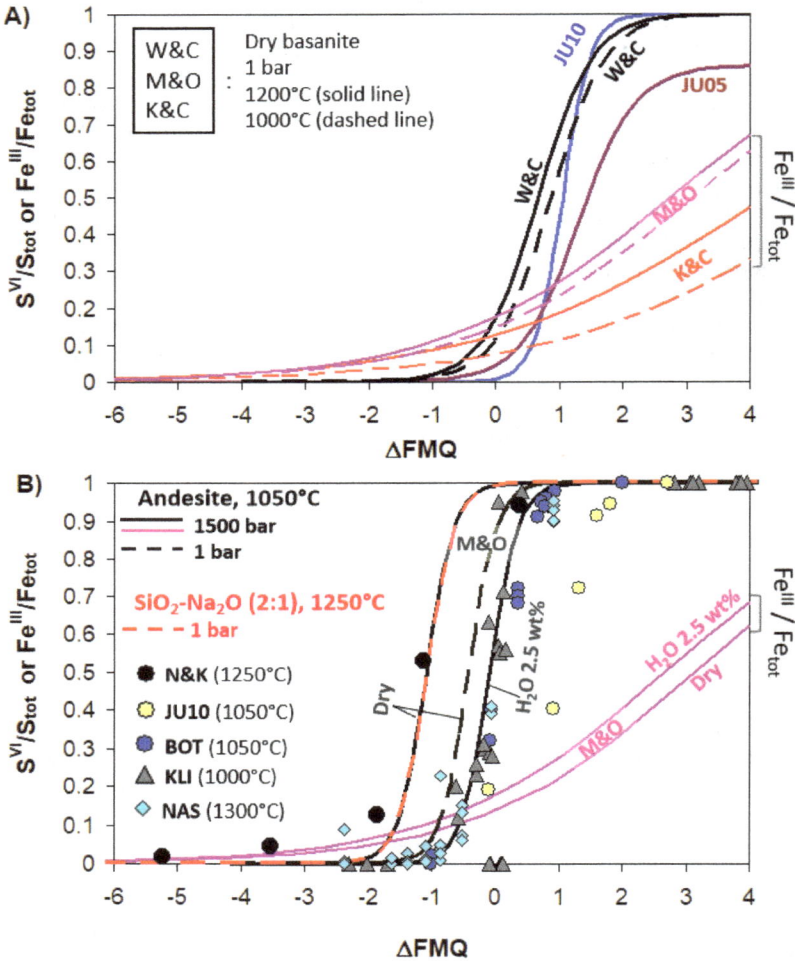

Figure 8. Panel A) Sulfate/sulfide ratio from empirical expressions reported in the literature (JU05: Jugo et al. 2005; JU10: Jugo et al. 2010; W&C: Wallace and Carmichael 1994). The Wallace and Carmichael (1994) equation ("W&C", **black lines**) was drawn for 1000 °C and 1200 °C. Fe^{III}/Fe_{tot} lines refer to 1-bar dry basanite at 1200 °C and 1000 °C via the Kress and Carmichael (1991) equation (K&C) and the Ottonello and Moretti (2001) and Moretti (2005) model (M&O). Modified from Klimm et al. (2012) and Moretti et al (2021). **Panel B)** XANES based S^{VI}/S_{tot} ratios (Nagashima and Katsura 1973: SiO_2–Na_2O melt in 2:1 molar proportion at 1 bar and 1250 °C [N&K]; Jugo et al. 2010: dry and hydrous basalts at 1300 °C, 10 kbar and 1050 °C, 2 kbar, respectively [JU10]; Botcharnikov et al. 2011: hydrous andesite at 1050 °C [BOT]; Kimm et al. 2012: soda lime–silica, potassium–lime–silica, albite and trondhjemite compositions at 1000 °C [KLI]; Nash et al 2019: synthetic Martian basalt, primitive Fe-rich terrestrial basalt, natural dacite and rhyolite, all at at 1300 °C [NAS]). **Solid black lines** have been computed via the CTSFG model of Moretti and Ottonello (2005; see also Papale et al. 2022, this volume) (M&O) on andesite at 1050 °C and 1500 bar, under dry and hydrous conditions (see also Eqn. 79). Note that the M&O model applied to this composition predicts that increasing pressure may increase substantially the sulfur oxidation state at fixed relative f_{O_2}. An opposite role is observed for water addition at fixed relative f_{O_2}. In case of iron oxidation state, adding water promote an increase of the iron oxidation state. The interplay of effects due to composition, temperature and pressure is such that the S^{VI}/S_{tot} vs f_{O_2} profiles of the 1-bar SiO_2–Na_2O melt at 1250 °C and dry andesite at 1500 bar and 1050 °C are indistinguishable. Modified from Moretti (2021).

Figure 8b reports XANES experimental datasets, selected for their accuracy, and shows that the S-redox curve shifts in response to compositional variations. The figure also shows the shift of about one log unit in S^{VI}/S_{tot} predicted by the Moretti and Ottonello (2005) model for a dry and a hydrous (2.5 wt%) andesite (Carmichael 2002) at 1500 bar and 1050 °C. The simulated hydrous andesite compares very well to the andesite investigated at the same P and T by Botcharnikov et al. (2011), to the soda–lime–silica, potassium–lime–silica, albite and trondhjemite compositions in Klimm et al. (2012) at 1000 °C, and with model Martian basalts, primitive Fe-rich terrestrial basalts, natural dacites and rhyolites from Nash et al. (2019) at 1300 °C. The dry andesite curve is shifted to lower f_{O_2} and cannot be distinguished by the model-generated curve running through the experimental data (at 1250 °C) on the SiO_2–Na_2O (2:1) binary (Nagashima and Katsura 1973).

We then see that composition also affects the sulfur oxidation state and among the many compositional variables, concentration of highly soluble water plays a major role. However, this effect can be partially balanced by the effect that pressure has on Equation (76) via volume terms (see Papale et al. 2022, this volume). The intricate interplay of temperature, pressure and composition on Reaction (71) may then produce profiles of sulfur oxidation state that tend to overlap for different sets of data, partly explaining the effort reported in the literature for simple expressions.

In agreement with the strong impact of water on rheological properties and melt phase equilibria (see also previous paragraph), the effect of dissolved water is a major one in driving the shift in acid–base properties hence in the speciation state, that is, in the number, kind and size of chemically reactive structural groups. As an example of the compositional control of water on the sulfur oxidation state of silicate melts, we can recall here the simple expressions given in Marini et al. (2011) for Mt. Mazama rhyodacite at 900 °C:

$$\log\frac{X_{SO_4^{2-}}}{X_{S^{2-}}} = -0.556 C_{H_2O} + 2\log f_{O_2} + 25.372 \tag{77}$$

and for Mt. Etna basalts at 1200 °C:

$$\log\frac{X_{SO_4^{2-}}}{X_{S^{2-}}} = -0.374 C_{H_2O} + 2\log f_{O_2} + 16.158 \tag{78}$$

C_{H_2O} being the water concentration in wt%. A similar expression, over a wide range of T, can be given for the andesite plotted in Figure 9. By using the Moretti and Ottonello (2005) model (M&O in Fig. 8) for different temperatures up to 1300 °C, we obtain (Moretti 2021):

$$\log\frac{X_{SO_4^{2-}}}{X_{S^{2-}}} = 2\log f_{O_2} - C_{H_2O}\left(\frac{4486.4}{T} - 2.512\right) - 23.125 + \frac{59660}{T} \tag{79}$$

If we look at Figure 8b, we see that the sulfate/sulfide ratio decreases, the effect being dramatic for the S^{VI}/S_{tot} ratio between $-2 < \Delta FMQ < 0$. The figure also shows that for a given Fe^{III}/Fe^{II} ratio, water loss increases oxygen fugacity of only 0.2 log-units at most, whereas for a given SO_4^{2-}/S^{2-} ratio water exsolution decreases oxygen fugacity (see also Eqns. 77, 78), up to one log-unit when dehydration is achieved. We can further exploit the plotted behaviors if we combine iron and sulfur redox equilibria and plot the oxidized/reduced log-ratios to appreciate shifts induced by water (Fig. 9). By combining half-reactions (70) and (40) (taken eight times) we obtain an expression for mutual Fe–S interactions in melts:

$$8Fe^{3+} + S^{2-} + 4O^{2-} \leftrightarrow 8Fe^{2+} + SO_4^{\ 2-} \tag{80}$$

where O^{2-} species appear on the left side. Therefore, by increasing the $a_{O^{2-}}$ (and in turn the basicity) by *adding* alkalis such as Na_2O or K_2O to the melt, the equilibrium is shifted rightward, which implies sulfur oxidation and iron reduction.

If we now consider Reaction (43) (always taken eight times) instead of Reaction (40), we obtain:

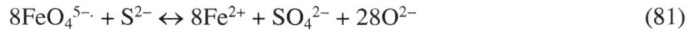

$$8FeO_4^{5-\cdot} + S^{2-} \leftrightarrow 8Fe^{2+} + SO_4^{2-} + 28O^{2-} \tag{81}$$

which can be reduced (see Reaction 47) to

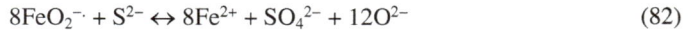

$$8FeO_2^{-\cdot} + S^{2-} \leftrightarrow 8Fe^{2+} + SO_4^{2-} + 12O^{2-} \tag{82}$$

from which we see that adding alkali species reduces sulfur and oxidizes iron. In melts both Reactions (80) and (82) (or alternatively Reaction 81) occur, combined with the sulfur redox equilibrium to obtain the following equilibrium relations

$$\log\frac{X_{SO_4^{2-}}}{X_{S^{2-}}} = 8\log\frac{X_{Fe^{3+}}}{X_{Fe^{2+}}} + 4\log a_{O^{2-}} + \log K_{80} \tag{83}$$

$$\log\frac{X_{SO_4^{2-}}}{X_{S^{2-}}} = 8\log\frac{X_{FeO_2^-}}{X_{Fe^{2+}}} - 12\log a_{O^{2-}} + \log K_{82} \tag{84}$$

The simultaneous occurrence of Equations (83) and (84) allows modeling the mutual interaction of iron and sulfur in melts (Moretti and Ottonello 2003a), although in most cases one of the two equilibria prevail over the other, which is then negligible. This is the case for the andesite plotted in Figures 8b and 9, modelled via Moretti and Ottonello (2005) and reflecting essentially Equation (84). Log-ratios of iron and sulfur redox state define (at constant T) lines of slope 8, constrained by stoichiometry of Reactions (82–84) (Fig. 9a). On the other hand, at a given f_{O_2} and for varying T, the same composition yields a slope lower than the stoichiometric one (Fig. 9b).

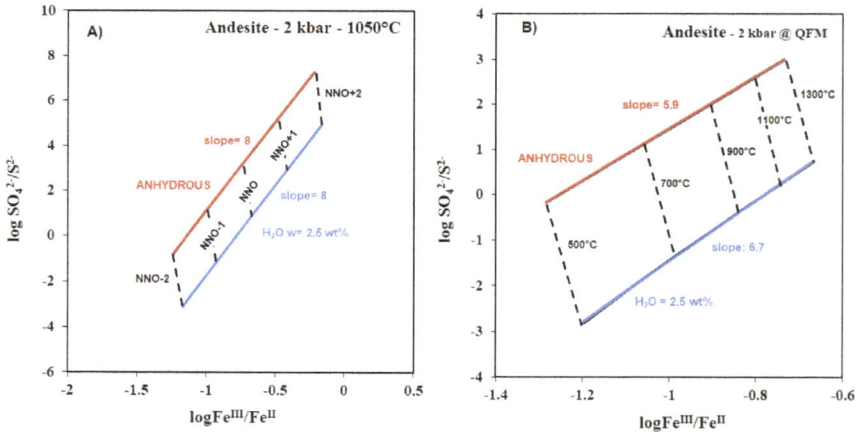

Figure 9. Panel A) Calculation of mutual interaction of iron and sulfur redox couples in dry and hydrous andesites via **Equation (79)** for different f_{O_2} at $T = 1050$ °C and $P = 2$ kbar. Theoretical slopes at 8 are displayed at varying f_{O_2} for each composition. **Panel B)** Same as panel A but for different T along the QFM buffer. Theoretical slopes are not maintained, each composition having a specific slope < 8. Modifed from Moretti (2021).

Fe–S mutual interactions (Eqns. 83 and 84) tell us that only iron fulfills the two basic requirements for an efficient redox couple in natural silicate melts: 1) that a redox-sensitive element must occur in two oxidation states giving rise to species with comparable thermodynamic stabilities, and 2) that these species must be abundant and pervade their chemical environment.

On the contrary, the sulfate–sulfide pair cannot be a good candidate for a buffer, given the generally low abundance of sulfur in melts. Besides, the transition from fully oxidized sulfate to fully reduced sulfide occurs in a relatively narrow range (two log-units) whereas the Fe^{III}/Fe^{II} pair can accommodate much larger f_{O_2} variations (Figs. 8 and 9). Therefore, compared to iron, sulfur has a limited "buffering" redox capacity, but for the same reason it can be a precise indicator of f_{O_2}. Nevertheless, the high sensitivity of the sulfate/sulfide pair with respect to f_{O_2} challenges the calibration of accurate oxythermometers.

In many studies oxide components and ionic species are mixed, but without any consideration about the different standard states. For example, in the case of joint Fe and S redox variations, it is easy to find the following

$$S^{2-} + 4Fe_2O_3 \leftrightarrow SO_4^{2-} + 8FeO \tag{85}$$

Because Equation (85) mixes up iron *oxide components* and sulfur *ionic species,* it cannot have the pretention to be used to constrain thermodynamically the Fe–S interaction from an analytical database about Fe and S redox ratios. In fact, pseudo-reactions such as (85) are solved only empirically by coupling non-thermodynamic expressions of iron and sulfur redox states, in order to constrain possible T–f_{O_2} solutions from measured Fe^{III}/Fe^{II} and S^{VI}/S^{-II} pairs on natural samples, via empirical equations of the type of Kress and Carmichael (1991), Wallace and Carmichael (1994) or Jugo et al. (2010).

In a recent paper, Nash et al. (2019) describe mutual Fe–S relations in melt as due to the following electron exchange:

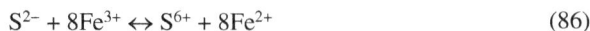

$$S^{2-} + 8Fe^{3+} \leftrightarrow S^{6+} + 8Fe^{2+} \tag{86}$$

which assumes a fully ionized solution that a priori cannot account for the complexity of the melt network, hence the role of composition via the speciation state of iron and sulfur in the melt. Moreover, to constrain the equilibrium constant of Reaction (86), the Authors impose the T-dependence obtained by thermodynamic data on the homogenous *solid phase* reaction:

$$FeS + 8FeO_{1.5} \leftrightarrow 8FeO + FeSO_4 \tag{87}$$

which crudely assumes that the fusion of solid phases plus their relaxation (dissolution) in the melt phase is uniquely determined by entropic terms. Finally, Nash et al. (2019) propose the following equation:

$$\log \frac{X_{SO_4^{2-}}}{X_{S^{2-}}} = 8\log \frac{X_{Fe^{3+}}}{X_{Fe^{2+}}} + \frac{8.7436 \cdot 10^6}{T^2} - \frac{27703}{T} + 20.273 \tag{88}$$

which demands that heat capacities (C_p) of components are not constant in the melt phase. On the basis of their results the Authors state that composition does not affect the sulfate/sulfide ratio. The Fe^{III}/Fe^{II} ratio in Equation (88) can be converted to $\log f_{O_2}$ (e.g., via Kress and Carmichael 1991), to show that the sulfur redox ratio depends only and dramatically on temperature, at least for selected sets of data about the sulfate/sulfide ratio in Fe-bearing melts and glasses. Nash et al. (2019) explain the f_{O_2}-shift between datasets that do not differ substantially each other uniquely on the basis of temperature and discharge compositional effects.

The pseudo-equilibrium constant effects on the component activity coefficients, discharging non-idealities on a T^2 term derived from enthalpy fitting of Fe–S exchange in the solid phase, rather than in pure-liquid or melt.

THE (FULL) *AB-INITIO* PERSPECTIVE: THE CASE OF IRON REDOX

If in aqueous solutions the formation of ions *sensu stricto* is due to the high dielectric constant of the aqueous medium (efficient shielding of charge and formation of hydration spheres), in contrast the dielectric properties of silicate melts are very different, no volumetric electrostriction is parameterized, and covalency forces do not exhaust within a single polyhedron or a combination of them, but distribute over the network (Moretti et al. 2014).

These limits can be circumvented by *ab-initio* computation. In their explicit many-body approach *ab-initio* simulations demand a great computational power making their success restricted to narrow compositional ranges, for which structural predictions matched Nuclear Magnetic Resonance (NMR) or other spectroscopic features (Jahn 2022, this volume). It is however possible to use so-called implicit approaches, such as the quantum-Polarization Continuum model and its extensions to melts (Ottonello and Richet 2014; Ottonello et al. 2015, 2018; Ottonello 2021) and precisely calculate (in our case at 25 °C and 1-bar) the energy of the solvation process (Appendix I) even in absence of an explicit computation visualizing how the melt network responds.

Electron transfer and solute–solvent interactions in melts

Because we are reviewing the nature of redox reactions in the silicate melts, it is almost compulsory, though somewhat tautological, to provide here the *ab-initio* framework of the thermochemistry of electron (e^-) and the physical chemistry of the gaseous diatomic oxygen ($O_{2,g}$) the diatomic molecule in solution ($O_{2,soln}$) and the monoatomic free ion in solution (O^{2-}_{soln}).

The electron. The conventional magnitudes adopted to depict the reactivity of gaseous ions have little to do with the actual magnitudes arising from electron–electron and electron–proton interactions. Basically, we have three possible ways to convey the true magnitudes to the conventional ones: the Electron Convention (EC), the Ion Convention (IC) and the Fermi–Dirac modification of EC (FD).

The EC considers the electron to be equivalent to an element to all extents and therefore its enthalpy of formation (H_f) and entropy of formation (S_f) is defined as equal to zero to all temperatures. In the EC, the enthalpy of the electron at 298.15 K is assumed to correspond to the integrated heat capacity of an ideal gas following Boltzmann statistics ($5/2\,RT$), which gives 6.197 kJ/mol at 298 K. The entropy at T (20.979 J/mol·K at 298 K) is likewise evaluated as that of an ideal gas and is simply translational (plus a degeneracy effect with $G = 2$) and is evaluated likewise, through the Sackur–Tetrode equation:

$$S_{e,EC} = \frac{5}{2}R + R\ln\left[\frac{\left(2\pi m_e k_B \overline{T}\right)^{3/2} k_B G}{P_0 h^3}\right] \tag{89}$$

with R = gas constant; m_e = rest mass of electron; k_B = Boltzmann's constant; h = Planck's constant and P_0 standard state pressure. Bartmess (1994) suggested that the electron must rather obey, at the gaseous state, Fermi–Dirac (FD) statistics and its correct integrated heat capacity should be 0.033 eV (3.146 kJ/mol) instead of 6.197 kJ/mol. In the ion-convention (IC) all magnitudes are set to zero.

Table 3. Thermochemistry of the electron in the three conventions (Bartmess 1994). Data are in kJ/mol (H, G, E) and J/(mol·K) (S, C_P). Parentheses are used for assumed values. E refers to the total (internal) energy (see Mitchell 1928 and Bartmess 1994 for details).

	EC	IC	FD
H_f	0	0	0
G_f	0	(0)	0
$H_{298} - H_0$	6.197	0	3.146
S_{298}	20.979	(0)	22.734
C_P	20.786		17.129
E_{298}	3.730		3.720

Table 3 (from Bartmess 1994) lists the various magnitudes in the three conventions. We will adopt hereafter the EC convention, which is the one adopted in most thermochemical tabulations (Wagman et al. 1982; Robie and Hemingway 1985; Barin 1995; Chase 1998).

Molecular oxygen. Molecular oxygen has unusual properties arising from the fact that its outer electrons are not in a closed shell. Its ground state is $^3\Sigma^-_g$ (i.e., a triply degenerate even multiplet with sigma symmetry) as theoretically and experimentally observed (Meckler 1953; Kotani et al. 1957; Slater 1963). Due to this fact, the oxygen molecule in its ground state has a magnetic moment arising from the spin. This magnetic moments is observed not only in the spectrum of the molecule but also in large-scale properties in the O_2 gas ($O_{2,g}$) and in the O_2 liquid ($O_{2,l}$). If $O_{2,l}$ is poured between the poles of a large magnet, it is oriented in the magnetic field "*in the same way that iron filings would be*" (Slater 1963). $O_{2,g}$ is the standard state adopted in expressing the enthalpy of formation (H_f) and the Gibbs free energy of formation (G_f) from the element at stable state. Hence, these values are conventionally set to zero (as already seen for the electron EC convention; cf. Table 3). The tabulated isobaric heat capacity at 298.15 K and $P = 10^5$ Pa is $C_{P,298.15} = 29.376$ J/(mol·K) in the JANAF tables (Chase 1998) or 29.355 (mol·K) in the NBS tables (Wagman et al. 1982). The entropy is $S_{298.15} = 205.147$ J/(mol·K) (Chase 1998) or $S_{298.15} = 205.138$ J/(mol·K) (Wagman et al. 1982). The molar volume is that of an ideal gas ($V_{298.15} = 2478.97$ J·bar^{-1}). The data at our disposal for $O_{2,sln}$ are scanty and/or contradictory. The NBS tables assign to $O_{2,sln}$ in water $G^0_f = 16.4$ kJ/mol and $H^0_f = -11.7$ kJ/mol (here we pose 0 at apex, for simplicity, to denote P, T room conditions; i.e., $T = 298.15$ K; $P = 10^5$ Pa) and $S^0 = 110.9$ J/(mol·K). Solubility experiments (Battino et al 1983) depict however a solvation Gibbs free energy $\Delta G^0_{solv} = 26.48$ kJ/mol, a solvation enthalpy $\Delta H^0_{solv} = -12.11$ kJ/mol and a solvation entropy $\Delta S^0_{solv} = -129.5$ J/(mol·K), which, when coupled with the EC data for $O_{2,g}$ return $G^0_f = 26.48$ kJ/mol, $H^0_f = -12.11$ kJ/mol, $S^0 = 75.6$ J/(mol·K). The *ab-initio* values obtained in this study are $G^0_f = 26.5$ kJ/mol, $H^0_f = -15.4$ kJ/mol, $S^0 = 64.6$ J/(mol·K) in sufficient agreement with experiments. Gas-phase calculations carried out at the Density Functional Theory (DFT) level with a hybrid functional (DFT/B3LYP) and a sufficiently extended basis set [6-31+G(d,p)] [1] return for $O_{2,g}$ $G^0_f = -1.643$ kJ/mol, $H^0_f = -1.665$ kJ/mol and $S^0 = 205.073$ J/(mol·K). Obviously, both G^0_f and H^0_f should be zero. The origin of the discrepancy in our case stems partly on Basis Set Superposition Error (BSSE) effects, and partly on the fact that we adopted the same computational parameters in computing the total energy of the diatomic molecule and of the isolated gaseous atoms potential [2]. For our purpose this bias is quite acceptable and we will

---

[1] B3LYP is a hybrid functional combining a Becke 3-parameters exchange functional (Becke 1993) with the Lee-Yang-Parr functional (LYP; Lee et al. 1988) expressing the Colle - Salvetti correlation energy density (Colle and Salvetti 1975). It constitutes the most widely adopted DFT procedure in practical calculations. A 6-31+G(d,p) is a split-valence basis set (Ditchfield et al. 1971) with added diffuse (+) and polarization (d,p) terms.

[2] Indeed the valence electrons in the isolated (gaseous) atom should be allowed to occupy wider portions of the space around the nucleus and the adoption of more diffuse functions for the isolated atom could be appropriate. A more consistent practice using localized basis set, is to assign to the isolated atom the total energy (E) that reproduces exactly a null conventional) enthalpy of formation from the elements, for the various elements in

not introduce any correction to the strict theory. Thermophysical properties of $O_{2,sln}$ may be obtained by application of the Integral Equation Formalism of Tomasi's Polarized Continuum Model (Floris and Tomasi 1986; Tomasi and Persico 1994; Barone et al. 1997) coupled with the Pierotti–Reiss Scaled Particle Theory (Pierotti 1963, 1965, 1976; Reiss 1965; Reiss et al. 1959, 1960, 1961; see Appendix I for computational details).

Free oxygen in solution. As far as we know there are no experimental data concerning the thermodynamic properties of O^{2-} in solution and also the gas-phase properties are scanty. Our calculations at B3LYP/6-31G(d,p) theory level assign to O^{2-}_g a strong electron affinity resulting in a quite unstable ion ($G^0_f = 778.6$ kJ/mol). However, when in solution O^{2-}_{sln} is stabilized by a strong electrostatic interaction with the solvent molecules (Table 4). Dispersive and repulsive interactions counterbalance each other and the net result is that the standard potential of the redox couple $\frac{1}{2}O_{2,g} + 2e^- \rightarrow O^{2-}_{sln}$ (Reaction 4) is little affected by the solvent composition (Table 5) [$E^0(V) = -1.8 \pm 1.1$]. Incidentally, we stress that, in computing the standard redox potential one encounters the difficulty of assigning the correct thermodynamic magnitudes to electrons in solution. This problem is circumvented by adopting the procedure of Larson et al. (1968): instead of writing the redox reaction as usual

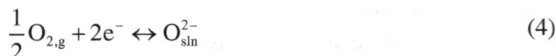

$$\frac{1}{2}O_{2,g} + 2e^- \leftrightarrow O^{2-}_{sln} \tag{4}$$

the redox balance is rather obtained as

$$\frac{1}{2}O_{2,g} + H_{2,g} \leftrightarrow O^{2-}_{sln} + 2H^+_{sln} \tag{90}$$

The identity of the two formulation can be easily understood when subtracting Reaction (4) from (90). In fact, this brings to

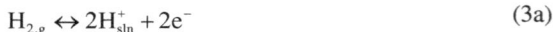

$$H_{2,g} \leftrightarrow 2H^+_{sln} + 2e^- \tag{3a}$$

which is the standard hydrogen redox potential (see Reaction 3), implying a (conventional) zero Gibbs free energy difference. The standard redox potential computed for fictive silicate melt solvents is not dissimilar from what computed for water and is virtually unaffected by composition.

The normal oxygen electrode

Table 4 lists the thermophysical properties of $O_{2,sln}$ and O^{2-}_{sln} in various substances of interest in this study. It is a representative series of silicate melts aiming to cover the compositional range observed in nature. These are the very same substances investigated in the past to parameterize the solvent properties of inert gases in melts (Ottonello and Richet 2014). Water is also listed for simple comparative purposes. Whereas the partial molar Gibbs free energy required to create a cavity in a fluid of hard spheres (G_{cav}) is based on the statistical mechanical theory of fluids and the properties of exact radial distribution functions (Reiss et al. 1959, 1960, 1961; Reiss 1965), its enthalpic (H_{cav}) and entropic (S_{cav}) component, and the various solution energy terms (ΔG_s, ΔH_s, ΔS_s, ΔC_P) as well, are based on the scaled particle theory essentially developed by Pierotti (1963, 1965, 1976). The interaction energy $E_{int} = E_{elec} + E_{disp} + E_{rep}$ is obtained ab initio from the Integral Equation Formalism for the Polarizable Continuum Model (IEFPCM; Cancés et al. 1997) operative in the GAUSSIAN code (Frish et al. 2004)[3].

their stable aggregation state at 298.15 K, 1 bar. In the case of elements that form stable diatomic molecules such as O_2 the calculation is straightforward (subtract the Zero Point Correction from the electronic energy of the diatomic molecule and divide by two).

[3] Because the three forms of energy are purely enthalpic in nature the identification of the corresponding Gibbs free energy is immediate. We also stress here that Pierotti's formulation of the solution energy does not correspond to the solvation energy (G_{solv}) that can be obtained directly by the GAUSSIAN code because the volume

Calculations were carried out at the DFT/B3LYP theory level with a 6-31+G(d,p) basis set. The radii adopted to conform the Solvent Accessible Surface (SAS) of Tomasi's model for the neutral species are the UAHF (United Atom Topological Model). The adopted electrostatic scaling parameter is also the nominal one adopted by the code ($\gamma_{el} = 1.2$).

In Table 5 are listed the standard state thermodynamic properties obtained through thermochemical cycle calculations from the *ab-initio* values. A concise appraisal of the procedure is given in Appendix I[4]. The standard state is that of "*hypothetical one molal solution referred to infinite dilution*". At the risk to appear tedious we want here focus the reader's attention on the beauty of this thermodynamic construction and on the intimate link of thermodynamics with the atomistic approach. Consider a generic strong electrolyte e, dissociating according to the scheme:

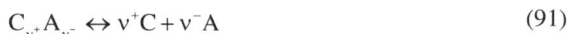

$$C_{v^+}A_{v^-} \leftrightarrow v^+C + v^-A \qquad (91)$$

with C = cation, A = anion, e = $C_{v^+}A_{v^-}$. Its thermodynamic activity is described at high dilution by Henry's law:

$$a_e = K_H m_e^v \qquad (92)$$

with m_e indicating the solute molality of the electrolyte and:

$$v = v^+ + v^- \qquad (93)$$

Let us now imagine plotting the activity of electrolyte e as a function of molality, elevated to stoichiometric factor v (m_e^v), as shown in Figure 10. Graphically in this sort of plot Henry's law K_H represents the slope of Equation (4) (or 90) at infinite dilution. Extending the slope of Henry's law up to value $m_e^v = 1$ (dashed line in Fig. 10) and arbitrarily fixing the ordinate scale so that activity is one at the point defined by the extension of the Henry's law slope at $m_e^v = 1$, we can construct the condition of "*hypothetical one-molal solution referred to infinite dilution*". This condition not only obeys the unitary activity implicit in the standard state definition, but results in an activity coefficient of one at infinite dilution (i.e., $a_e = m_e^v$). The thermodynamic properties of all solutes conform to this construction: basically what it is reproduced is the energy (i.e., enthalpy, entropy, heat capacity, volume) of a moiety at molecular level translated into a macroscopic magnitude valid at the standard state. The term "*hypothetical*" emphasizes the fact that the adopted reference condition does not correspond to the energy properties of the actual solution at the same concentration level (see Fig. 10).

It should be obvious now to anyone that to each solvent pertains a peculiar solute-solvent interaction, hence, we cannot expect to observe identical interaction terms (and corresponding thermodynamic parameters) for solvents differing appreciably in composition and rheology as silicate melts, nor that their properties be identical to what experimentally observed in other solvents such as water.

When plotted against the number density of the solvent (ρ) (Fig. 11a) the discrete values of the various forms of energy exhibit regular trends, which may be eventually replaced by continuous first-order functions as already observed for H_2O_{sln} (Ottonello et al. 2018). The adoption of a continuous function for replacing discrete *ab-initio* calculations is not compulsory, but allows a significant reduction of computational time when operating on an extended database. The O_2 solubility in silicate melts is quite similar to what observed in water. The amount of dissolved O_2 increases regularly with the increase of numeral density of the solvent from ~7 ppm for the less dense to ~26 ppm for the most dense fictive silicate liquid

---

terms are not accounted for in the code and G_{solv} is a simple summation of $G_{cav} + H_i$.

[4] Note that here we adopt the absolute-to-conventional scaling factors arising from the hydration terms of Tissandier et al. (1998), coupled with the gas-phase conventional properties of the gaseous proton (Chase 1998) and corrected for the liberation free energy contribution (Ben Naim 1987).

Table 4. Solution energy terms for O_2 and O_2^- (in italic) in various media at $T = 25$ °C, $P = 1$ bar. The interaction Gibbs free energy ΔG_{int} is the summation of electrostatic + repulsive + dispersive terms computed at B3LYP/6-31G+(d,p) theory level. Data are in kJ/mol (G, H) and J/(mol·K) (S, C_P). The Henry's law constant and the molar amount of dissolved oxygen (ppm) are also listed. at equilibrium with a pure O_2 atmosphere at $T = 298.15$ K and $P = 10^5$ Pa.

Solvent	ΔG_{cav}	H_{cav}	S_{cav}	$\Delta G_i \approx H_{int}$	ΔG_{sln}	ΔH_{sln}	ΔS_{sln}	$\Delta C_P, a$	$\Delta C_P, b \times 10^3$	$\ln K_H$	$X_{O_2,sln} \times 10^6$
andesite[1]	24.6	0.8	−79.8	−12.2	29.5	−13.8	−145.3	1.542	−17.255	11.9	6.9
	11.7	*0.3*	*−37.9*	*−1115.1*	*−1086.3*	*−1117.2*	*−103.4*	*−56.557*	*−7.243*		
leuc–bas[1]	24.7	1.3	−78.5	−12.9	29	−14	−144.4	8.25	−29.715	11	16
	11.7	*0.6*	*−37.3*	*−1167.6*	*−1138.7*	*−1169.5*	*−103.1*	*−94.087*	*−12.467*		
tholeiite[1]	24.7	1.1	−79.1	−12.7	29.2	−14	−144.9	5.716	−24.941	11.1	14.7
	11.7	*0.5*	*−37.6*	*−1069.3*	*−1040.4*	*−1071.2*	*−103.4*	*−74.204*	*−10.462*		
AOB[1]	25.1	1.4	−79.5	−14.4	28.2	−15.4	−146.3	9.095	−31.517	10.7	22.5
	11.8	*0.6*	*−37.6*	*−1185.2*	*−1155.9*	*−1187*	*−104.3*	*−101.388*	*−13.178*		
ugandite[1]	25.3	1.7	−79	−15.1	27.8	−15.8	−146.2	12.936	−39.036	10.5	26.4
	11.9	*0.8*	*−37.3*	*−1097*	*−1067.6*	*−1098.7*	*−104.4*	*−113.454*	*−16.306*		
entatite[2]	25.7	1	−82.7	−16.9	26.6	−18.4	−151	3.887	−22.008	10.7	21.6
	12	*0.4*	*−38.7*	*−1153.9*	*−1124*	*−1155.9*	*−107*	*−73.461*	*−9.145*		
mean comp[3]	25.1	1.3	−80	−14.3	28.3	−15.5	−146.8	7.503	−28.484	11.4	11
	11.8	*0.6*	*−37.8*	*−1089.8*	*−1060.5*	*−1091.7*	*−104.6*	*−85.458*	*−11.907*		
BH–257[4]	25.2	1.2	−80.4	−14.9	28	−16	−147.6	7.269	−28.114	11.3	12.6
	11.9	*0.6*	*−37.9*	*−1167*	*−1137.6*	*−1168.9*	*−105*	*−90.702*	*−11.737*		
BH–266[4]	25.3	1.3	−80.7	−15.4	27.6	−16.5	−148.1	7.645	−28.899	11.1	14.5
	11.9	*0.6*	*−38*	*−1175.3*	*−1145.8*	*−1177.2*	*−105.4*	*−93.816*	*−12.053*		
NS1–G205[4]	24.7	1.5	−77.8	−12.5	29.3	−13.4	−143.5	10.446	−33.815	11.8	7.3
	11.7	*0.7*	*−37*	*−985.4*	*−956.6*	*−987.2*	*−102.7*	*−88.635*	*−14.198*		
NS2–G205[4]	24.6	1.3	−78.3	−12.3	29.4	−13.5	−143.9	7.796	−28.781	11.9	7

Solvent	ΔG_{cav}	H_{cav}	S_{cav}	$\Delta G_i \approx H_{int}$	ΔG_{sln}	ΔH_{sln}	ΔS_{sln}	$\Delta C_P, a$	$\Delta C_P, b \times 10^3$	$\ln K_H$	$X_{O_2,sln} \times 10^6$
NS3–G205[4]	11.7	0.6	-37.2	-991.9	-963.1	-993.8	-102.9	-77.701	-12.085	11.9	6.7
	24.6	1	-79.1	-12.1	29.5	-13.6	-144.6	4.248	-22.175		
CMS1–G212[4]	11.7	0.4	-37.6	-1000.4	-971.7	-1002.4	-103.1	-62.844	-9.312	11	16.3
	25.4	1.2	-81.2	-15.9	27.3	-17	-148.8	7.062	-27.855		
CMS2–G212[4]	11.9	0.5	-38.1	-1207	-1177.3	-1208.8	-105.8	-93.511	-11.606	11	16.7
	25.5	1.1	-81.6	-16	27.3	-17.3	-149.4	5.846	-25.573		
CMS3–G212[4]	11.9	0.5	-38.3	-1085.3	-1055.6	-1087.3	-106.1	-78.313	-10.649	11	16.7
	25.5	1.1	-81.6	-16	27.3	-17.3	-149.4	5.846	-25.573		
NCS1–G205[4]	11.9	0.5	-38.5	-1086.5	-1056.8	-1088.5	-106.4	-72.108	-9.638	11.9	6.8
	24.6	1.1	-78.6	-12.2	29.5	-13.5	-144.1	6.167	-25.721		
NCS2–G205[4]	11.7	0.5	-37.4	-995.5	-966.7	-997.4	-102.9	-70.842	-10.802	11.6	9.3
	24.9	1.3	-79.1	-13.6	28.7	-14.6	-145.4	8.663	-30.576		
NCS3–G211[4]	11.8	0.6	-37.4	-1023.7	-994.5	-1025.5	-103.8	-84.971	-12.805	11.4	11.1
	25.1	1.4	-79.7	-14.4	28.3	-15.4	-146.6	8.633	-30.629		
MS1–G156[4]	11.8	0.6	-37.6	-1047.2	-1017.9	-1049	-104.5	-87.456	-12.803	10.9	19
	25.6	0.9	-82.6	-16.5	26.9	-18	-150.7	3.343	-20.944		
molten silica[5]	12	0.4	-38.7	-1024.7	-994.9	-1026.8	-106.8	-62.712	-8.71	11.4	11.3
	22.8	0	-76.3	-11.5	28.2	-13.9	-141.4	-8.121	-0.327		
	11.6	0	-38.8	-1024.7	-996.2	-1027.2	-104	-9.241	-0.148		
water	25.6	3.9	-72.8	-17.1	26.5	-15.4	-140.6	32.564	-64.147	10.7	22.9[7]
	12.0	1.7	-34.3	-1434.2	-1404.3	-1434.8	-102.1	-341.513	-7.312		

Notes: (1) Lux (1987); (2) Kirsten (1968); (3) Jambon et al. (1986); (4) Hiyagon and Ozima (1986); (5) Shibata et al. (1998); (6) Ottonello and Richet (2014); (7) Battino et al. (1983).

Table 5. Thermodynamic properties of O_2 and O^{2-} (in italic) in various media at $T = 25$ °C, $P = 1$ bar. $E^0(V)$ is the computed standard redox potential of Reaction (4).

Solvent	G^0_f (kJ/mol)	H^0_f (kJ/mol)	S^0 [J/(mol·K)]	V (J/bar)	E^0 (V)
andesite[1]	29.5	–13.8	59.9	2.41	
	451.5	*449.7*	*227.2*	*–0.35*	*–2.34*
leuc–bas[1]	29	–14	60.8	2.41	
	399.1	*397.4*	*227.5*	*–0.34*	*–2.068*
tholeiite[1]	29.2	–14	60.2	2.37	
	497.4	*495.6*	*227.2*	*–0.37*	*–2.578*
AOB[1]	28.2	–15.4	58.8	2.34	
	381.9	*379.8*	*226.3*	*–0.38*	*–1.979*
ugandite[1]	27.8	–15.8	59	2.33	
	470.2	*468.1*	*226.2*	*–0.38*	*–2.437*
entatite[2]	26.6	–18.4	54.1	2.24	
	413.8	*410.9*	*223.6*	*–0.44*	*–2.144*
mean comp[3]	28.3	–15.5	58.4	2.34	
	477.3	*475.2*	*226*	*–0.38*	*–2.473*
BH–257[4]	28	–16	57.6	2.31	
	400.2	*397.9*	*225.6*	*–0.4*	*–2.074*
BH–266[4]	27.6	–16.5	57	2.3	
	392	*389.7*	*225.2*	*–0.4*	*–2.032*
NS1–G205[5]	29.3	–13.4	61.7	2.48	
	581.2	*579.7*	*228*	*–0.3*	*–3.012*
NS2–G205[5]	29.4	–13.5	61.2	2.47	
	574.7	*573*	*227.8*	*–0.31*	*–2.978*
NS3–G205[5]	29.5	–13.6	60.6	2.47	
	566.1	*564.4*	*227.5*	*–0.31*	*–2.934*
CMS1–G212[5]	27.3	–17	56.3	2.29	
	360.5	*358*	*224.8*	*–0.41*	*–1.868*
CMS2–G212[5]	27.3	–17.3	55.8	2.28	
	482.2	*479.6*	*224.5*	*–0.42*	*–2.499*
CMS3–G212[5]	27.1	–17.6	55.2	2.27	
	481	*478.3*	*224.2*	*–0.42*	*–2.493*
NCS1–G205[5]	29.5	–13.5	61	2.47	
	571.1	*569.5*	*227.7*	*–0.31*	*–2.96*
NCS2–G205[5]	28.7	–14.6	59.7	2.4	
	543.3	*541.4*	*226.8*	*–0.35*	*–2.815*
NCS3–G211[5]	28.3	–15.4	58.6	2.36	
	519.9	*517.8*	*226.1*	*–0.37*	*–2.694*
MS1–G156[5]	27	–18	54.5	2.25	
	542.9	*540.1*	*223.8*	*–0.43*	*–2.813*
molten silica[6]	29.8	–13.9	58.4	2.44	
	541.6	*539.6*	*226.7*	*–0.33*	*–2.807*
water	26.5	–15.4	64.6	3.9	
	$(16.4^7–26.5^8)$	$(–11.7^7–11.1^8)$	$(110.9^7–75.7^8)$		
	133.5	*132.1*	*228.5*	*0.68*	*–0.692*

Notes: [1] Lux (1987); [2] Kirsten (1968); [3] Janbon et al. (1986); [4] Hiyagon and Ozima (1986); [5] Shibata et al. (1998); [6] Ottonello and Richet (2014); [7] Wagman et al. (1982); [8] Battino et al. (1983)

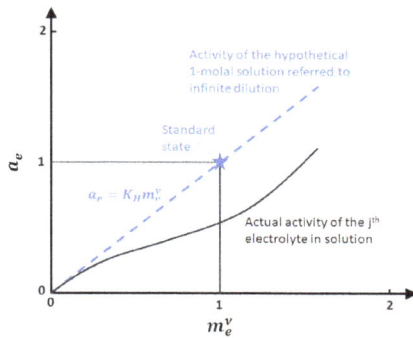

Figure 10. Construction of the standard state of "hypothetical one-molal solution referred to infinite dilution". The standard state is represented by the **blue star** at $m_e^v = 1$ on the tangent to the actual activity curve at very low molalities(modified from Ottonello 1997).

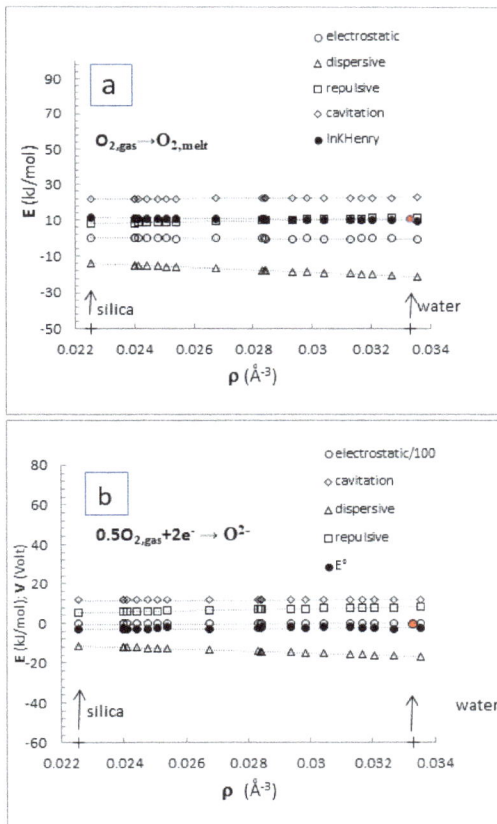

Figure 11. Components of the solute/solvent interaction of $O_{2,sln}$ (**a**), and $O^{2-}{}_{,sln}$ (**b**) resolved for a silicate melts population deemed to be sufficiently representative of the natural compositional realm (Ottonello and Richet 2014) and plotted against the number density of the solvent. The Henry's law constant of the solution process of O_2 is also shown in (**a**) and the standard potential for O^{2-} is shown in (**b**). The O^{2-} standard potential appears to be quite insensitive to melt compositional changes. The symbols $\ln K_H$ (a) and $E°$ (b) for solvent water are filled in **red** to better envisage the difference with silicate melt solvents.

Ab-initio iron redox

Fe^{2+} and Fe^{3+} are here considered to be always in High Spin condition with multiplicity 5 and 6, respectively. The adopted electrostatic scaling parameter is again the nominal one adopted by the GAUSSIAN code ($\gamma_{el}= 1.2$). Based on the few literature estimates existing for the enthalpy of formation of Fe^{2+}_{aq} and Fe^{3+}_{aq} from the elements (Shock and Helgeson 1988; and references therein) it is apparent that also in the case of iron it is appropriate to adopt scaled ionic radii (i.e., $r' = 0.975 + 0.152r^{VI}$; $r'Fe^{2+} = 1.116$ Å; $r'Fe^{3+} = 1.095$ Å). The *ab-initio* thermochemical properties of Fe^{2+}, Fe^{3+} and FeO_2^- ions in solution are listed in Tables 6 and 7.

The data concerning volume must be considered as provisional and adopted with caution (see Appendix to this purpose). In Figure 12a,b,c we see how the standard redox potential of the three redox couples involving iron changes with the rheology of the solvent.

Table 6. Solution energies for Fe^{2+}, Fe^{3+} and FeO_2^- in various media at $T = 25$ °C, $P = 1$ bar. Symbols and references as in Table 2. Solute radii adopted to conform the cavitation energy are 1.116 Å for Fe^{2+}_{sln}, 1.095 Å for Fe^{3+}_{sln} and 2.384 Å for $FeO_2^-_{sln}$.

Solvent	Solute	ΔG_{cav}	H_{cav}	S_{cav}	$\Delta G_i \approx H_{int}$	ΔG_s	ΔH_s	ΔS_s
andesite[1]	Fe^{2+}	9.8	0.3	−32	−1650	−1623.1	−1652.2	−97.5
	Fe^{3+}	9.6	0.3	−31.1	−3711	−3684.4	−3713.2	−96.6
	FeO_2^-	31.7	1.1	−102.6	−253.2	−204.5	−254.6	−168.1
leuc–bas[1]	Fe^{2+}	9.9	0.5	−31.4	−1724.8	−1697.7	−1726.8	−97.3
	Fe^{3+}	9.6	0.5	−30.5	−3878.4	−3851.6	−3880.4	−96.3
	FeO_2^-	31.9	1.9	−100.7	−266.7	−217.6	−267.3	−166.5
tholeiite[1]	Fe^{2+}	9.9	0.4	−31.7	−1584	−1557	−1586	−97.5
	Fe^{3+}	9.7	0.7	−30.3	−3648.6	−3621.2	−3650.3	−97.4
	FeO_2^-	31.9	1.6	−101.6	−242.5	−193.5	−243.4	−167.4
AOB[1]	Fe^{2+}	9.9	0.5	−31.6	−1748.8	−1721.4	−1750.7	−98.4
	Fe^{3+}	9.8	0.4	−31.6	−3827.9	−3800.2	−3829.9	−99.8
	FeO_2^-	32.4	2	−102.1	−272.5	−222.6	−273	−168.8
ugandite[1]	Fe^{2+}	10	0.6	−31.4	−1622.1	−1594.5	−1623.9	−98.5
	Fe^{3+}	9.7	0.5	−30.8	−3626.5	−3599.3	−3628.5	−97.6
	FeO_2^-	32.6	2.5	−101.2	−251.3	−201	−251.2	−168.3
entatite[2]	Fe^{2+}	10.1	0.4	−32.5	−1702.2	−1674.2	−1704.3	−100.8
	Fe^{3+}	9.7	0.5	−30.9	−3873.3	−3846	−3875.3	−98.1
	FeO_2^-	33.2	1.4	−106.7	−266.8	−215.6	−267.8	−175
mean comp[3]	Fe^{2+}	9.9	0.5	−31.8	−1612.2	−1584.8	−1614.2	−98.6
	Fe^{3+}	9.7	0.5	−30.8	−3626.5	−3599.3	−3628.5	−97.6
	FeO_2^-	32.4	1.8	−102.8	−248.9	−198.9	−249.5	−169.6
BH–257[4]	Fe^{2+}	10	0.5	−31.9	−1722.6	−1695	−1724.5	−99
	Fe^{3+}	9.7	0.5	−30.9	−3873.3	−3846	−3875.3	−98.1
	FeO_2^-	32.8	1.8	−103.8	−270.9	−220.5	−271.5	−171.2
BH–266[4]	Fe^{2+}	10	0.5	−31.9	−1734	−1706.3	−1735.9	−99.3
	Fe^{3+}	9.7	0.5	−31	−3898.9	−3871.5	−3900.9	−98.4
	FeO_2^-	32.8	1.8	−103.8	−270.9	−220.5	−271.5	−171.2

Solvent	Solute	ΔG_{cav}	H_{cav}	S_{cav}	$\Delta G_i \approx H_{int}$	ΔG_s	ΔH_s	ΔS_s
NS1–G205[5]	Fe^{2+}	9.8	0.5	−31.2	−1463.3	−1436.3	−1465.1	−96.9
	Fe^{3+}	9.6	0.6	−30.2	−3292.9	−3266.2	−3294.8	−95.9
	FeO_2^-	31.8	2.1	−99.6	−222.5	−173.5	−222.8	−165.3
NS2–G205[5]	Fe^{2+}	9.8	0.5	−31.4	−1472.8	−1445.8	−1474.8	−97
	Fe^{3+}	9.6	0.5	−30.5	−3314.3	−3287.6	−3316.2	−96.1
	FeO_2^-	31.8	1.8	−100.5	−223.9	−175	−224.5	−166.1
NS3–G205[5]	Fe^{2+}	9.8	0.4	−31.7	−1485.2	−1458.2	−1487.2	−97.2
	Fe^{3+}	9.6	0.4	−30.8	−3342	−3315.3	−3344	−96.3
	FeO_2^-	31.7	1.4	−101.6	−225.6	−176.9	−226.7	−167.1
CMS1–G212[5]	Fe^{2+}	10	0.5	−32	−1778.8	−1751.1	−1780.8	−99.7
	Fe^{3+}	9.7	0.5	−31.1	−3999.2	−3971.7	−4001.1	−98.7
	FeO_2^-	32.9	1.8	−104.4	−279.3	−228.7	−280	−172.1
CMS2–G212[5]	Fe^{2+}	10	0.4	−32.2	−1604.6	−1576.8	−1606.6	−100
	Fe^{3+}	9.7	0.4	−31.2	−3609.4	−3581.8	−3611.4	−99
	FeO_2^-	32.9	1.6	−105.1	−249.2	−198.4	−250	−172.9
CMS3–G212[5]	Fe^{2+}	10	0.4	−32.3	−1606.2	−1578.4	−1608.3	−100.2
	Fe^{3+}	9.8	0.4	−31.4	−3613.1	−3585.5	−3615.1	−99.3
	FeO_2^-	33	1.5	−105.8	−249.7	−198.8	−250.6	−173.7
NCS1–G205[5]	Fe^{2+}	9.8	0.4	−31.5	−1478	−1451	−1480	−97.1
	Fe^{3+}	9.6	0.4	−30.6	−3325.8	−3299.1	−3327.8	−96.1
	FeO_2^-	31.7	1.6	−100.9	−224.6	−175.8	−225.4	−166.4
NCS2–G205[5]	Fe^{2+}	9.9	0.5	−31.5	−1517.7	−1490.4	−1519.6	−97.9
	Fe^{3+}	9.6	0.5	−30.6	−3414.9	−3387.9	−3416.7	−96.9
	FeO_2^-	32.2	1.9	−101.5	−232.3	−182.8	−232.8	−167.8
NCS3–G211[5]	Fe^{2+}	9.9	0.5	−31.7	−1551	−1523.5	−1552.9	−98.5
	Fe^{3+}	9.7	0.5	−30.7	−3489.4	−3462.2	−3491.3	−97.5
	FeO_2^-	32.4	1.9	−102.4	−238.6	−188.7	−239.1	−169.2
MS1–G156[5]	Fe^{2+}	10	0.3	−32.5	−1609.8	−1581.9	−1611.8	−100.6
	Fe^{3+}	9.8	0.4	−31.6	−3621	−3593.3	−3623	−99.6
	FeO_2^-	33.1	1.3	−106.6	−250.5	−199.6	−251.7	−174.6
molten silica[6]	Fe^{2+}	9.8	0	−32.8	−1520.8	−1494.1	−1523.3	−97.9
	Fe^{3+}	9.5	0	−31.9	−3421.8	−3395.3	−3424.3	−97.1
	FeO_2^-	31.4	0	−105.2	−230.7	−182.4	−233.2	−170.3
water	Fe^{2+}	10	0.2	−33.2	−2100.4	−2072.5	−2102.7	−101.5
	Fe^{3+}	9.8	0.1	−32.3	−4716.9	−4689.2	−4719.2	−100.6
	FeO_2^-	33.2	0.5	−109.4	−343.3	−292.2	−345.2	−177.8

Notes: [1] Lux (1987); [2] Kirsten (1968); [3] Janbon et al. (1986); [4] Hiyagon and Ozima (1986); [5] Shibata et al. (1998); [6] Ottonello and Richet (2014)

Moretti & Ottonello

Table 7. Thermodynamic properties of Fe^{2+}, Fe^{3+} and FeO_2^- in various media at $T = 25$ °C, $P = 1$ bar in various media. The corresponding standard redox potentials of the ionic solutes are also listed. Symbols and references as in Table 5.

Solvent	Solute	G^0_f	H^0_f	S^0	V^0	E^0
andesite[1]	Fe^{2+}	357	354	−112.4	−3.08	1.85
	Fe^{3+}	966.5	955.7	−203.8	−5.12	3.339
	FeO_2^-	24.1	−2.5	208.8	3.33	0.249
leuc–bas[1]	Fe^{2+}	282.4	279.5	−112.1	−3.08	1.463
	Fe^{3+}	799.2	788.5	−203.6	−5.12	2.761
	FeO_2^-	10.9	−15.3	210.3	3.33	0.113
tholeiite[1]	Fe^{2+}	423.1	420.2	−112.4	−3.1	2.193
	Fe^{3+}	1114.3	1103.5	−203.8	−5.13	3.849
	FeO_2^-	35	8.6	209.4	3.27	0.363
AOB[1]	Fe^{2+}	258.7	255.5	−113.2	−3.11	1.341
	Fe^{3+}	746	735	−204.7	−5.14	2.577
	FeO_2^-	5.9	−21	208	3.24	0.061
ugandite[1]	Fe^{2+}	385.6	382.4	−113.2	−3.11	1.998
	Fe^{3+}	1029.6	1018.6	−204.8	−5.15	3.557
	FeO_2^-	27.5	0.6	208.6	3.23	0.285
entatite[2]	Fe^{2+}	305.9	301.9	−115.7	−3.16	1.585
	Fe^{3+}	850.7	839	−207.1	−5.19	2.939
	FeO_2^-	12.9	−15.8	201.9	3.08	0.133
mean comp[3]	Fe^{2+}	395.3	392	−113.4	−3.11	2.049
	Fe^{3+}	1051.5	1040.4	−204.9	−5.14	3.633
	FeO_2^-	29.6	2.5	207.3	3.24	0.307
BH–257[4]	Fe^{2+}	285.1	281.7	−113.9	−3.13	1.477
	Fe^{3+}	804.8	793.6	−205.3	−5.16	2.78
	FeO_2^-	10.4	−17	206.3	3.19	0.107
BH–266[4]	Fe^{2+}	273.8	270.3	−114.2	−3.13	1.419
	Fe^{3+}	779.3	768	−205.6	−5.16	2.692
	FeO_2^-	8	−19.5	205.7	3.17	0.083
NS1–G205[5]	Fe^{2+}	543.9	541.1	−111.7	−3.05	2.818
	Fe^{3+}	1384.7	1374.1	−203.2	−5.08	4.784
	FeO_2^-	55	29.1	211.6	3.45	0.57
NS2–G205[5]	Fe^{2+}	534.3	531.4	−111.9	−3.05	2.769
	Fe^{3+}	1363.3	1352.7	−203.3	−5.08	4.71
	FeO_2^-	53.5	27.5	210.8	3.44	0.555
NS3–G205[5]	Fe^{2+}	521.9	519	−112.1	−3.05	2.704
	Fe^{3+}	1335.5	1324.9	−203.5	−5.09	4.614
	FeO_2^-	51.6	25.3	209.8	3.43	0.535
CMS1–G212[5]	Fe^{2+}	229	225.4	−114.6	−3.14	1.187
	Fe^{3+}	679.2	667.8	−206	−5.17	2.346
	FeO_2^-	−0.2	−28	204.8	3.15	−0.002

Solvent	Solute	G^0_f	H^0_f	S^0	V^0	E^0
CMS2–G212[5]	Fe^{2+}	403.3	399.6	−114.8	−3.14	2.09
	Fe^{3+}	1069	1057.5	−206.3	−5.17	3.693
	FeO_2^-	30.1	2	204	3.14	0.312
CMS3–G212[5]	Fe^{2+}	401.7	397.9	−115.1	−3.15	2.082
	Fe^{3+}	1065.3	1053.8	−206.5	−5.18	3.68
	FeO_2^-	29.7	1.4	203.2	3.12	0.308
NCS1–G205[5]	Fe^{2+}	529.1	526.2	−111.9	−3.05	2.742
	Fe^{3+}	1351.7	1341.1	−203.4	−5.09	4.67
	FeO_2^-	52.8	26.6	210.4	3.43	0.547
NCS2–G205[5]	Fe^{2+}	489.7	486.6	−112.7	−3.08	2.537
	Fe^{3+}	1263	1252.1	−204.2	−5.12	4.363
	FeO_2^-	45.7	19.1	209.1	3.33	0.474
NCS3–G211[5]	Fe^{2+}	456.6	453.3	−113.3	−3.1	2.366
	Fe^{3+}	1188.6	1177.6	−204.8	−5.14	4.106
	FeO_2^-	39.9	12.9	207.7	3.26	0.413
MS1–G156[5]	Fe^{2+}	398.2	394.3	−115.5	−3.16	2.064
	Fe^{3+}	1057.5	1045.9	−206.8	−5.19	3.653
	FeO_2^-	28.9	0.4	202.2	3.1	0.3
molten silica[6]	Fe^{2+}	486	482.9	−112.9	−3.07	2.519
	Fe^{3+}	1255.5	1244.7	−204.2	−5.1	4.338
	FeO_2^-	46.1	18.9	206.6	3.38	0.478
water	Fe^{2+}	−92.3	−96.6	−116.5	−2.22	0.479
	Fe^{3+}	−38.4	−50.3	−207.8	−4.28	0.133
	FeO_2^-	−63.7	−93.1	199.1	6.18	0.66

Notes: [1] Lux (1987); [2] Kirsten (1968); [3] Janbon et al. (1986); [4] Hiyagon and Ozima (1986); [5] Shibata et al. (1998); [6] Ottonello and Richet (2014)

For Fe^{2+} and Fe^{3+} the dispersive and repulsive interactions are virtually absent and the interaction is purely electrostatic. As already seen for $O_{2,aq}$ and O^{2-}_{sln}, also for the three iron species the effect of composition is quite regular and only for Fe^{3+} one may envisage a certain effect within the limits of the procedure. More important for our purposes is to stress the differences in the electrode potentials exerted in silicate melts with respect to those pertaining to water solvent. This is perhaps more evident in terms of galvanic cell potentials, when the redox potentials of the iron species are coupled with the normal oxygen electrode (Fig. 13). While in solvent water the two redox couples are nearly coincident, they are substantial different and opposite in silicate melt solvents.

We anticipated that a highly polarizable solvent such as water could hinder the rotational motion in solution lowering thus the effective bulk entropy and enhancing, consequently, the Gibbs free energy of formation from the gaseous state. Our calculations however indicate that the *ab-initio* values obtained for the bulk entropy of the solute species are quite consistent with the experimentally derived values (cf. Table A1). Nevertheless, we cannot exclude that this effect could be observed also in silicate melts but this would be unlikely because the melt solvents are less polarizable than water. To confirm our anticipation, it would be enough to measure the *in-situ* amount of dissolved oxygen in solution in a representative set of samples. We emphasize to this purpose that, concerning the water solvent, switching the ΔG^0_f of $O_{2,sln}$ from 16.4 kJ/mol

(Wagman et al. 1982) to 26.5 kJ/mol (Battino et al. 1983) corresponds to decreasing the molar amount of dissolved oxygen from 1339 ppm to roughly 23 ppm for a water equilibrated with a pure oxygen atmosphere at $P = 10^5$ Pa and 298.15 K. Obviously in doing the experiment one should avoid all side-effect linked to eventual disruption of the medium.

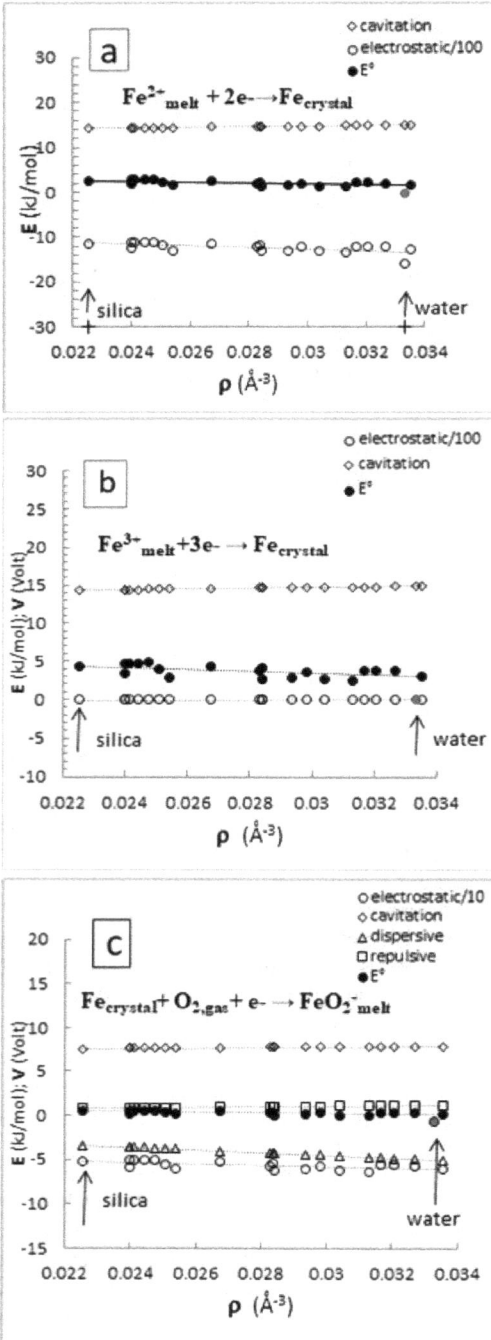

Figure 12. Components of the solute/solvent interaction of Fe^{2+} (**a**), Fe^{3+} (**b**) and FeO_2^- (**c**) resolved for a silicate melts population deemed to be sufficiently representative of the natural compositional realm (Ottonello and Richet 2014) and plotted against the number density of the solvent. The standard redox potentials are also shown. The symbols for $E°$ in solvent water are filled in **red** to better envisage the difference with silicate melt solvents.

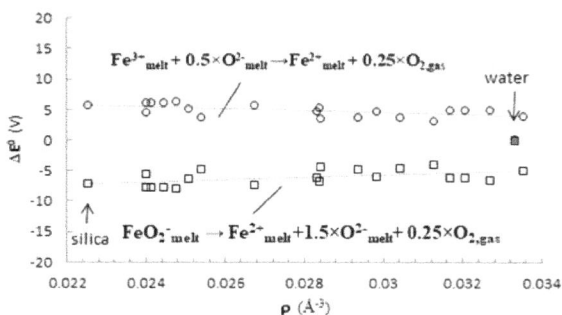

Figure 13. Galvanic potential differences for the two redox couples of iron.

This consideration brings us to the 2nd important question: how much the medium affect the redox state of iron ($f_{O_2,g}$, T and P being held constant) ? If we combine the two redox couples already seen in Figure 13, we see that indeed, when passing from the *"fictive"* melt medium to an aqueous solution we should observe a disproportioning of the reduced vs oxidized species, these last becoming more stable in the new solvent (Fig. 14).

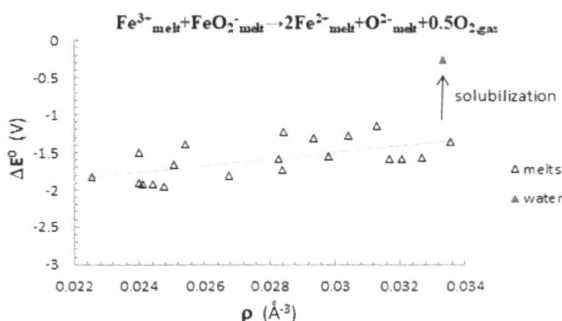

Figure 14. Galvanic potential for the bulk reduction of iron. In aqueous solution the potential is less negative implying a tendency to oxidation (f_{O_2}) T and P being held constant) during solubilization.

It is worth noting that a main implication of this finding is that wet chemistry methods, largely used in old determinations prior to the advent of *in-situ* Mossbauer and XANES techniques, may not provide a reliable information of the iron oxidation state in rocks and glasses.

REMARKS AND PERSPECTIVES

In this review we have stressed on the adoption of the ionic notation to describe redox exchanges of silicate melt via ionic species consistent with the melt acid–base character. The latter can be determined via melt polymerization, accounting for how it changes with melt composition. Therefore, composition determines the "free oxygen" (or oxide ion) activity, $a_{O^{2-}}$ which is like a_{H^+} in waters. "Free" (or more appropriately non-network associated; Nesbitt et al. 2011) oxygen enters redox exchanges and also represents the basicity of the system. In silicate melts, the formal link between redox and acid–base exchanges is represented by the "normal oxygen electrode", which relates $a_{O^{2-}}$ to oxygen fugacity, f_{O_2}, the master variable used to quantify the redox potential and its dependence on bulk composition. *Ab-initio* computation shows that the oxygen electrode is very little dependent on composition.

By means of the ionic notation we can formally evaluate how both iron and sulfur redox state may be affected by compositional effects. The compositionally simpler is the system the larger are these effects, as on the binary SiO_2–MO join, where the coulombic nature of melts exacerbates the difference in the chemical nature of metals in determining the interconnected (polymerized) silicate structure. Silicate melts are in fact a special category of fused salts in which the residual charge distribution resulting from bonding of bridging oxygen to silicon (or other network former) allows oxygen bonding with other cations. This characteristic, however, does not allow a clear distinction between solute and solvent like in aqueous solutions. The speciation state of silicate melt is then much more complicated than in aqueous solutions, due to difficulties in discriminating reactive entities. This speciation changes along with the marked depolymerization of the silicate framework from pure SiO_2 to metal-oxide rich compositions (Toop and Samis 1962a,b; Mysen and Richet 2018).

The case of iron is quite emblematic, as its redox parameterization must include the amphoteric behavior of trivalent iron, which can act as either network former of modifier depending on melt basicity. Although natural melts tend to average the effects of composition, the redox state of sulfur too cannot be approximated by operational forms that depend only on temperature and f_{O_2}. If the problem is how to solve activity coefficients of FeO and Fe_2O_3 in solution, the answer to the problem resides, like for aqueous solutions, in iron speciation in the melt. Again, the main difference with respect to aqueous solutions, is the dependence of speciation on solvent composition

As extensively discussed here and elsewhere, non-idealities are reflected by the shape of the excess Gibbs free energy of mixing, which is correctly reproduced only by ionic–polymeric approaches to silicate melts mixing properties (e.g, Hillert et al. 1985; Ottonello 2001, 2005, 2021; Ottonello et al. 2001; Ottonello and Moretti 2004; Moretti 2005, 2021; Moretti and Ottonello 2003a, 2005; Fraser 2005; Mao et al. 2006). Besides, non-idealities in the mixing properties are associated with counterintuitive redox behaviors that cannot be accounted for by expansions of activity coefficients terms used in regular or sub-regular approaches to Reaction (17). The use of these approaches, as well as the purely empirical calibrations inspired by them, must be used with caution because they are restricted to calibration databases in which a limited polymerized region of composition-space is energetically rather homogenous (Fraser 2005). Typically, this region is that of geologically relevant melts with the exclusion of the alkaline domain. Even in these case, the calibration of interaction parameter must rigorously follow mixing theory for the development and definition of all terms (see Papale et al. 2022, this volume). In this study we have shown how critical is the choice of $\Delta G^0 1, T$, and of volume terms for the definition of Equilibrium (17) at $P > 1$ bar. Clearly, and quite tautologically, big questions in magmatism and planetary evolution cannot be addressed based on thermodynamics which loses consistency because it does not keep tracking, for instance, of standard states and is reduced to a simple fitting of measured masses or related parameters.

Depending on melt composition, alkali addition (i.e., increasing $a_{O^{2-}}$) can either oxidize or reduce iron in the melt. This reflects the change of speciation due to the amphoteric behavior of Fe^{III}, which can behave as either network former or modifier (Ottonello et al. 2001; Moretti 2005, 2020; Le Losq et al. 2020). Models that define the melt (oxo)acidity (Reaction 59) hence $a_{O^{2-}}$ (e.g., polymeric models based on the Toop and Samis mixing of bridging, non-bridging and free oxygens) allows solving speciation and set activity–composition relations of ionic and molecular species, similarly to aqueous solutions and molten salts. These models have shown the ability to solve polymerization redox and sulfur solubility. They can be applied to any other chemical exchange in which composition determines a variation in the melt solvent properties (via polymerization) and owing the flexibility of the Flood–Grjotheim thermochemical cycle (with the addition of annealing terms as in Moretti and Ottonello 2005), in the multicomponent system, the deriving thermodynamic treatment can then be indifferently applied to both simple (i.e., metallurgical slags) and complex silicate melts. As noted by Moretti and Ottonello (2005) *"this is an unconditional (but often overlooked) prerequisite of any thermodynamic parameterization of reactive properties of a given phase"*.

In aqueous solutions pairs of electromotive force (redox potential, E) and pH can be easily measured in laboratory cells and in the field by probe electrodes and directly compared to thermodynamic assessments and diagrammed phase stability. Magma-related samples do not offer the possibility to probe the conditions (temperature, pressure, gas composition and also phase proportion) under which they equilibrated prior to become accessible to our observations. On the other hand, engineering electrodes for analytical control directly in melts at high temperature and as working assembly for anode reaction (inverse of Reaction 4) is a challenging task. Engineering difficulties are enhanced by the limited flux of oxide ions and oxyanions, which is very limited compared to cations (Allanore et al. 2015).

For natural samples, the original equilibrium conditions cannot be restored and it is then necessary to perform equilibrium assumption, such as that oxidation state was preserved in glasses, and make back-calculations. The impossibility to restore and measure the original system like in waters has surely contributed to overlook the role played by acid–base properties and historically favoured the oxide-based redox approaches to melts, thus focusing mainly on the concept of f_{O_2} in petrology and geochemistry. Nevertheless, the aforesaid difficulty in engineering the use of the silicate melt as an electrolytic medium and the important heritage of mineral chemistry also pushed metallurgists to oxide-based approached, as testified by Ellingham's diagrams. The acid–base descriptions was then purely qualitative (e.g., silicic for acidic and mafic for basic) for too long time.

In case of natural samples, it is common to use microbeam spectroscopic techniques (e.g., XANES) to measure masses of elements in their different oxidation state in quenched volcanic glasses and then relate such ratios to f_{O_2}. The modelling of joint Fe and S redox exchanges is a good example and also a major issue for understanding redox variations measured in natural samples and discern whether the primary redox signature inherited from the phase assembly and distinctive mineralogy of the source, and the redox contribution intrinsic to magma evolution on its rise to surface. In this review we have discussed different approaches to determine f_{O_2} when a reliable solid–gas buffer is not available in the system. The oxygen fugacity is a thermodynamic parameter that can be defined also in absence of a fluid phase. Even when representing a virtual quantity, it allows systematically report on the relative oxidation state characterizing different magmatic- and also non magmatic- environments in relation to geodynamics and independently of composition (see for instance Fig. 1 in Moretti and Stefansson 2020).

In a volcanic–magmatic system, tracking the f_{O_2} allows describing fluid evolution from depths of magma source or accumulation to surface, where volcanic gases are discharged and direct estimates of f_{O_2} are either feasible or retrieved from the analysis of discharged gas composition. In these environments, the buffering of the redox potential depends largely on fluid–rock(crystals)–melt interactions. Modeling and observations show that the oxygen fugacity of magmas can be strongly affected by volatile exsolution and degassing during magma ascent. The f_{O_2} values measured in volcanic gases and plumes may only partly due to magma source redox (Moretti and Stefansson 2020 and references therein). Several different examples (Oppenheimer et al. 2011; Moussallam et al. 2014, 2017; Brounce et al. 2017) show that the best way to reconstruct the log f_{O_2} variations of a magmatic–volcanic system is the backtracking of f_{O_2} in natural samples (e.g., CO_2/CO, H_2O/H_2 and SO_2/H_2S in volcanic gases, iron and sulfur redox ratios in glasses, f_{O_2}–T pairs from mineral assemblages). Nevertheless, the performance of f_{O_2} backtracking from measured Fe and S redox states in natural glasses is highly dependent on the choice of adopted oxythermobarometers and may lead to contrasting interpretations (see Moretti and Stefansson 2020) especially because compositional effects are treated very differently by operational forms proposed in the literature. In order to reduced ambiguities in data treatment, compositional dependences cannot be neglected when assessing f_{O_2} from iron and/or sulfur oxidation state in natural glasses.

Besides, it is of fundamental importance to describe accurately the f_{O_2}-f_{S_2} petrogenetic grid for natural silicate melts (e.g., Haughton et al. 1974; Moretti and Baker 2008 and references therein) coexisting with S-bearing liquid and solid phases, and provide the link to late-magmatic and post-magmatic stages, at the transition with deep hydrothermal conditions dominated by highly charged acid solutions (brines and/or superctitical fluids) that transport and deposit metals after scavenging them from the magma body at their bottom.

The approaches shown and mentioned in this study should provide the basis for a well-established coding of mixing, acid–based redox properties of melts, anchored to an oxobasicity scale. There are grounds on which thermodynamically generate pO^{2-} based phase diagrams, such as E–pO^2 and $\log f_{O_2}$–pO^{2-} diagrams, in a way analogous to pH-based diagrams used for aqueous/hydrothermal solutions and corrosion science. The joint description of acid–base properties and redox exchanges via predominance and stability diagrams would reduce ambiguities about the composition-dependent relation with f_{O_2} of metals in their different oxidation states. Besides, this would better constrain the role of redox state on Earth's evolution since its formation and accretion.

For this purpose, *ab-initio* computational experiments are pivotal, as they allow treating reactive solubility and thus speciation solving for the polarization effects in solution. Besides, the bulk solute/solvent interaction may be represented as a continuous function of the number density of the solvent, which allows proposing a model of general validity that can be used to solve for compositional dependencies.

In this study we have applied the Polarizable Continuum Model to compute the Gibbs free energy of solvation of the species involved in iron reactions that were also used to illustrate the ionic–polymeric approaches. The derived *ab-initio* thermochemical properties listed here allow estimating (for the moment only at 25 °C, 1 bar) the iron redox couples discussed throughout this study. Even though quite preliminary, results show that the completely different polarization state acting in silicate melts separates effectively the two main redox couples of iron (Fe^{3+}/Fe^{2+} and FeO_2^-/Fe^{2+}): if these are nearly equal in a solvent with the properties of water they are completely different and tend to be opposite in silicate melts, in line with the contrasting structural role (former vs modifier) reported by spectroscopic analyses and necessary to explain non-idealities and counterintuive iron redox (and acid–base) behaviors that have been observed with along with compositional variations.

CODE AVAILABILITY

The iron-redox code (Ottonello et al. 2001; Moretti 2005) can be accessed from https://github.com/charlesll/iron-magma.

The CTFSFG code (Moretti and Ottonello 2005) can be accessed from https://github.com/charlesll/sulfur-magma.

The code for water amphoteric behavior (Moretti et al. 2014) and speciation can be accessed from https://github.com/charlesll/water-speciation-magma.

At these sites, Fortran source codes are available together with instructions to run them and relevant information on licensing and credit to be given.

APPENDIX

Isolated molecules (gaseous state)

Adopting the integrated heat capacity of a gas following Boltzmann statistics, the enthalpy of formation from the elements at standard state (stable element, $T=298.15$ K, $P=1$ bar) of the gaseous species is:

$$\bar{H}^0_{f,298.15} = H_{corr,298.15} - D_0 + \sum_{i=1}^{n} \nu_i H_{f,A_i,0} - \sum_{i=1}^{n} \nu_i \Delta H_{element_i,0 \to 298.15} + \frac{5}{2}RT\xi_M \qquad (A\text{-}1)$$

The zero-point dissociation energy of the gaseous molecule into gaseous atoms (D_0) is obtained by subtracting from the sum of the electronic energies of the individual gaseous atoms at the a-thermal limit (0 K) the total energy and zero point correction (ZPC) of the gaseous molecule :

$$D_0 = \sum_{i=1}^{n} \nu_i E_{e,A_i} - \left(E_{e,M} + E_{ZPC,M}\right) \qquad (A\text{-}2)$$

The energy terms E_{e,A_i}, $E_{e,M}$ and $E_{ZPC,M}$ in Equation (A-2) are all obtained by experiment, and, as such, available in literature (e.g., the NIST-JANAF or NBS thermo-chemical tables).

Calculations necessary to retrieve absolute properties from *ab-initio* magnitudes are resumed in the thermo-chemical cycle of Figure A-1.

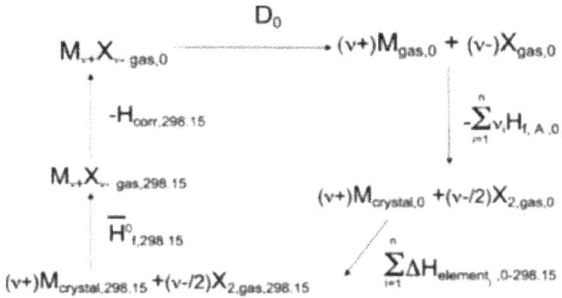

Figure A-1. Thermochemical cycle adopted for the determination of the enthalpy of formation from the elements at $T=298.15$ K, $P=1$ bar. D_0 is the zero-point dissociation energy of the gaseous molecule. $H_{f,Ai,0}$ is the zero-point enthalpy of formation of gaseous atoms from stable elements. $\Delta H_{element,0-298.15}$ is the 0 K to 298.15 K enthalpy difference.

$H_{corr,T}$ in this cycle is the thermal correction to the enthalpy of the isolated gaseous (non-linear) molecule arising from internal energy:

$$H_{corr,T} = Nk_B T^2 \left[\left(\frac{\partial \ln q_{trans}}{\partial T}\right)_V + \left(\frac{\partial \ln q_{rot}}{\partial T}\right)_V + \left(\frac{\partial \ln q_{vib}}{\partial T}\right)_V\right] + PV = $$
$$\frac{3}{2}Nk_B T + \frac{3}{2}Nk_B T + Nk_B \sum_i X_i T\left(\frac{1}{2} + \frac{1}{e^{-X_i}-1}\right) + Nk_B T \qquad (A\text{-}3)$$

where q_{trans}, q_{rot}, q_{vib} are the translational, rotational and vibrational contributions to the microscopic partition function[5,6], h is Planck's constant, k_B is Boltzmann's constant, N is the number of gaseous molecules (i.e., $N=1$ for an isolated gaseous molecule and $N=N_0$ Avogadro's number for one mole of substance), X terms are vibrational temperatures ($X_i = h\varpi_i / k_B T$, with $v_i = i^{th}$ vibrational normal mode), and the last term (PV product) arises from a simple application of the perfect gas law.

The entropy of any isolated molecule can be calculated from first principles by applying (McQuarrie and Simon 1997):

$$S = Nk_B + Nk_B \ln\left(\frac{q(V,T)}{N}\right) + Nk_B T \left(\frac{\partial \ln q(V,T)}{\partial T}\right)_v \tag{A-4}$$

Splitting the microscopic partition function $q(V,T)$ in its translational, rotational, vibrational and electronic components and scaling in molar notation we may write:

$$\bar{S}^0 = S_{trans} + S_{rot} + S_{vib} + S_e = R\left(\ln q_{trans} + \frac{3}{2}\right) +$$

$$R\left(\ln q_{rot} + \frac{3}{2}\right) + R\sum_i \left(\frac{X_i}{e^{X_i}-1} - \ln\left(1-e^{-X_i}\right)\right) + R\ln q_e + R \tag{A-5}$$

In estimating the translational partition function of isolated gaseous molecules, the V dependency is converted into a P dependency by application of perfect gas law (Ochterski 2000):

$$q_{trans} = \left(\frac{2pmk_B T}{h^2}\right)^{32} V = \left(\frac{2pmk_B T}{h^2}\right)^{32} \frac{k_B T}{P} \tag{A-6}$$

The rotational partition function is:

$$q_{rot} = \frac{\pi^{1/2}}{\sigma_{rot}}\left(\frac{T^{3/2}}{(\Theta_{rot,x} \times \Theta_{rot,y} \times \Theta_{rot,z})^{1/2}}\right) \tag{A-7}$$

with $\Theta_{rot} = \dfrac{h^2}{8\pi^2 I k_B}$, I = momentum of inertia and σ_{rot} = rotational symmetry factor.

The Gibbs free energy of formation of the molecule from the stable elements at standard state is readily obtained by applying:

$$\bar{G}^0_{f,298.15} = \bar{H}^0_{f,298.15} - 298.15\left(\bar{S}^0_{298.15} - \sum_{i=1}^n v_i S^0_{A,i}\right) \tag{A-8}$$

where $S^0_{A,i}$ is the standard state entropy of the monatomic stable elements (i.e., $S^0_H \cdot \frac{1}{2} S^0_{H_2,gas}$).

Neutral and charged species in solution

In the Integral Equation Formalism of the Polarized Continuum Model (IEFPCM) the Gibbs free energy of solvation (ΔG^0_{solv}) is conceived as composed of two main terms (coupling work and molecular motion), further subdivided into several components:

---

[5] Assuming the first and higher excited states to be inaccessible at any T, the electronic contribution to the (A-3) partition function (q_e) corresponds to the electronic spin multiplicity of the molecule, which gives no contribution to the internal energy of the molecule.

[6] For a single atom the rotational partition function is identically 1 at all T, so that there are no contributions or rotational terms to internal energy, heat capacity and entropy; for a linear molecule the rotational contribution varies linearly with T so that the partial derivative of the natural logarithm of the rotational term reduces to T^{-1}.

$$\Delta G_{\text{solv}}^0 = W(M/S) + \Delta G_{\text{mm}} =$$

$$\Delta G_{\text{el}} + \Delta G_{\text{cav}} + \Delta G_{\text{disp}} + \Delta G_{\text{rep}} +$$

$$RT \ln\left(\frac{q_{\text{rot,g}} \cdot q_{\text{vib,g}}}{q_{\text{rot,s}} \cdot q_{\text{vib,s}}}\right) - RT \ln\left(\frac{n_{M,g} \cdot \Lambda_{M,g}^3}{n_{M,s} \cdot \Lambda_{M,s}^3}\right) \tag{A-9}$$

where $n_{M,g}$ and $n_{M,S}$ are the numeral densities of the molecule M in the gas and in the solvent and $\Lambda_{M,g}$, $\Lambda_{M,s}$ are the momentum partition functions of M in the two states. The *"coupling work"* terms (first four terms on the right-end part of Eqn. A-9; i.e electrostatic solute–solvent interaction, cavitation, dispersion and repulsive contribution) are related to the work necessary to *"build up"* the molecule M in the solvent S (Tomasi and Persico 1994).

Absolute solvation free energy are usually computed by Boltzmann averaging over all minima (Ben Naim 1992; Barone et al. 1997), i.e.:

$$\exp\left[\frac{-\Delta G_{\text{solv}}}{RT}\right] = \sum_c P_c\left[\frac{-\Delta G_{\text{solv},c}}{RT}\right] \tag{A-10}$$

Assuming that only a single conformer (c) exists in the gas phase is equivalent to stating that its equilibrium mole fraction P_c is equal to 1 in the gas phase and that eventual minor structural modifications taking place during solvation and related to the reaction field are fully accounted for by the variational principle. In the simple cases investigated in this study this assumption is exact.

For the solute molecule, interactions with the solvent perturbs its energy. This is quantum-mechanically expressed as:

$$\left[\hat{H}^0 + \hat{V}_R\right]\Psi = E\Psi \tag{A-11}$$

where $\hat{H}^0$ is the Hamiltonian of the solute *in-vacuo*, $\hat{V}_R$ the electrical potential of the reaction field, E is the energy and Ψ the solute wave function in solution.

The dispersive and repulsive contributions may be expressed as in terms of perturbing operators ($\hat{h}_{\text{rep}}, \hat{h}_{\text{dis}} \cdot \hat{X}_{\text{dis}}$) (Amovilli and Mennucci 1997):

$$\Delta G_{\text{rep}} = \left\langle \Psi \left| \hat{h}_{\text{rep}} \right| \Psi \right\rangle \tag{A-12}$$

$$\Delta G_{\text{dis}} = \left\langle \Psi \left| \hat{h}_{\text{dis}} + \frac{1}{2}\hat{X}_{\text{dis}} \right| \Psi \right\rangle \tag{A-13}$$

Practically, however, the operators are calibrated in such a way as to reproduce the effects obtained with the classical approximation (Floris and Tomasi 1986; Floris et al. 1991). The sum of solute–solvent dispersive and repulsive interactions $\Delta G_{\text{disp}} + \Delta G_{\text{rep}}$ is thus computed as double summations of terms extending to the solute atoms placed within the cavity and solvent atoms outside a cavity limited by a tessellated surface:

$$\Delta G_{\text{dis}} + \Delta G_{\text{rep}} = \overset{\text{solute}}{\sum_a} \overset{\text{solvent}}{\sum_s} \rho_s \overset{\text{tesserae}}{\sum_i} a_i \Gamma_{as} \tag{A-14}$$

where ρ_s is the number density of the solvent in atomic units $\rho_s = N_s/V_s$ (i.e., number of solvent molecules N_s per solvent volume V_s expressed in Å^{-3}), a_i the area of the ith tessera composing the surface and Γ_{as} a set of dispersive and repulsive parameters obtained from crystallographic data (Caillet and Claverie 1978). The electrostatic, dispersive and repulsive contributions to the solvation energy are purely enthalpic in nature, so that we may pose:

$$S_{el} = 0 \rightarrow H_{el} = \Delta G_{el} \qquad (A\text{-}15,1)$$

$$S_{dis} = 0 \rightarrow H_{dis} = \Delta G_{dis} \qquad (A\text{-}15,2)$$

$$S_{rep} = 0 \rightarrow H_{rep} = \Delta G_{rep} \qquad (A\text{-}15,3)$$

The Gibbs free energy of cavitation (ΔG_{cav}) is perhaps the most intriguing form of energy governing the solubility at molecular scale. We adopt here the formulation of Pierotti (1963, 1965, 1976) for the Gibbs free energy of cavitation (ΔG_{cav}) and its enthalpic and entropic components (H_{cav}, S_{cav}, respectively), which are based on the scaled particle theory (SPT) of Reiss and coworkers (Reiss 1965; Reiss et al. 1959, 1960, 1961). Denoting by σ_s the diameter of the solvent we express the reversible work necessary to create in the solvent a cavity of radius $r \le \sigma_s/2$ as:

$$W_0(r,\rho) = k_B T \ln p_{0,i} \qquad (A\text{-}15)$$

where $p_{0,i}$ the complement to 1 of the probability of finding a solute molecule i at the place of a solvent molecular center:

$$p_{0,i} = (1 - p_{0,r}) = \left(1 - \frac{4}{3}\pi r^3 \rho_s\right) \qquad (A\text{-}16)$$

To make room in the solvent for a solute of diameter σ_i the necessary cavity must have a radius $r = \dfrac{\sigma_s + \sigma_i}{2}$.
Denoting by $Y = \dfrac{\pi \rho_s \sigma_s^3}{6}$ the reduced number density and $\varsigma = \sigma_i/\sigma_s$, we write the work done at atomic scale as:

$$\frac{W(r,\rho)}{kT} = -\ln(1-Y) + \left(\frac{3Y}{1-Y}\right)\varsigma + \left[\left(\frac{3Y}{1-Y}\right) + \frac{9}{2}\left(\frac{Y}{1-Y}\right)^2\right]\varsigma^2 + \frac{YP}{\rho_s kT}\varsigma^3 \qquad (A\text{-}17)$$

where P is the (isotropic) pressure. The partial molar Gibbs free energy of cavitation may be assumed to be equal to the reversible work $\overline{G}_{cav} = W(r,\rho)\Delta G_{cav}$. The cavitation enthalpy at molar scale has the following form (Pierotti 1976):

$$\overline{H}_{cav} = Y\alpha_s RT^2 (1-Y)^{-3}\left[(1-Y)^2 + 3(1-Y)\varsigma + 3(1+2Y)\varsigma^2\right] + Y\frac{N_0 P}{\rho}\varsigma^3 \qquad (A\text{-}18)$$

where α_s is the volume thermal expansion of the solvent and N_0 is Avogadro's number, whereas the cavitation entropy may be obtained from the thermodynamic relationship (Pierotti 1976)

$$\overline{S}_{cav} = -\left(\Delta \overline{G}_{cav} - \overline{H}_{cav}\right)/T \qquad (A\text{-}19)$$

We recall now that the bulk Gibbs free energy of solution (ΔG_{sln}) for a gaseous, monatomic species in its stable standard state is simply related to the Henry's Law constant K_H by:

$$\Delta G_{sln} = RT\ln K_H = \Delta \overline{G}_{cav} + \Delta \overline{G}_{el} + \Delta \overline{G}_{disp} + \Delta \overline{G}_{rep} + RT\ln\left(\frac{RT}{\overline{V}_s}\right) \qquad (A\text{-}20)$$

where $\overline{V}_s$ is the molar volume of the solvent. Adoption of this simplified form implies that one assumes the rotational, vibrational and momentum partition functions to be identical in the two states (cf. Eqn. A-9).

The enthalpy of solution is given by the partial derivative of Henry's Law constant (Pierotti 1976):

$$\Delta H_{\text{sln}} = \left(\frac{\partial \ln K_H}{\partial (1/RT)}\right)_P = \bar{H}_{\text{cav}} + \bar{H}_{\text{el}} + \bar{H}_{\text{disp}} + \bar{H}_{\text{rep}} - RT + \alpha_s RT^2 \quad (A\text{-}21)$$

The entropy of solution is (Pierotti 1976):

$$\Delta S_{\text{Sln}} = -\left(\frac{\partial \Delta G_s}{\partial T}\right)_P = \bar{S}_{\text{cav}} + \bar{S}_{\text{el}} + \bar{S}_{\text{disp}} + \bar{S}_{\text{rep}} - R \ln\left(\frac{RT}{\bar{V}_s}\right) + \alpha_s RT \quad (A\text{-}22)$$

because the electronic, dispersive and repulsive contributions to the Gibbs free energy of solution in Equation (A-20) are purely enthalpic (which is implicit in the *ab-initio* assessment). With this proviso, the corresponding entropic terms in Equation (A-22) vanish.

The isobaric heat capacity of solution is (Pierotti 1963, 1965):

$$\Delta C_P, s\ln = \bar{C}_{P_{\text{cav}}} + \bar{C}_{P_{\text{int}}} + 2\alpha_s RT + RT^2 (\partial \alpha_s / \partial T)_P \quad (A\text{-}23)$$

with

$$\bar{C}_{P_{\text{cav}}} = \left[\frac{2}{T} - \alpha_s + \frac{(\partial \alpha_s / \partial T)_P}{\alpha_s}\right] \times \bar{H}'_{\text{cav}} -$$
$$R\left[\frac{Y\alpha_s T}{(1-Y)^2}\right]^2 \times \left[(1-Y)^2 + 6\varsigma(1-Y) + 3\varsigma^2(4Y+5)\right] \quad (A\text{-}24)$$

where $\bar{H}'_{\text{cav}}$ is given by Equation (A-18) without the last term on the right-side and

$$\bar{C}_{P_{\text{int}}} = \bar{C}_{P_{\text{el}}} + \bar{C}_{P_{\text{disp}}} + \bar{C}_{P_{\text{rep}}} = \alpha_s \times \bar{H}_{\text{int}} \quad (A\text{-}25)$$

with

$$\bar{H}_{\text{int}} = \bar{H}_{\text{el}} + \bar{H}_{\text{disp}} + \bar{H}_{\text{rep}} \quad (A\text{-}26)$$

The partial molar volume of the solute at infinite dilution is given by (Pierotti 1976)

$$\bar{V}_s = \bar{V}_{\text{cav}} + \bar{V}_{\text{int}} + \beta RT \quad (A\text{-}27)$$

$$\bar{V}_{\text{cav}} = 82.05 \times \frac{\beta}{\alpha} \times \frac{\bar{H}'_{\text{cav}}}{RT} + \frac{N_o \pi \sigma_i^3}{6} \quad (A\text{-}28)$$

$$\bar{V}_{\text{int}} = \beta_S \bar{H}_{\text{int}} \quad (A\text{-}29)$$

where β_S is the isothermal compressibility of the solvent. Equations (A-25) and (A-29) are strictly valid only for neutral non-polar molecules and must be replaced with estimates based on experiments.

We remand to Ottonello and Richet (2014) and Ottonello et al. (2015, 2018) for practical applications of the theory to natural substances. When applied to water as a solvent, the theory returns the values listed in Table A-1. The adopted solute radii for Fe^{2+} and Fe^{3+} are those returning H^0_f values consistent with the estimates of Shock and Helgeson (1988) while for the O^{2-} aqueous ions it was adopted the crystal radius of Shannon and Prewitt (r_{solute} = 1.26 Å) which is internally consistent with the theoretical estimates of Fumi and Tosi (1964) for F^-, given the assigned difference between the ionic radii of F^- and O^{2-}. As we may see in Figure A-2 the adopted value is not far from the bottom of the potential wells depicting the solute/solvent interaction in water and molten silica. Hence with the adopted value we attain practically the maximum stabilization effect operated by the solvent.

Table A1. Thermodynamic data for stable elements and solute species in water. *Ab-initio* values are compared with literature estimates. The *ab-initio* values for V^0 and C_P at $P = 105$ Pa and $T = 298.15$ K must be considered as provisional.

Species	G^0_f kJ/mol	H^0_f kJ/mol	S^0 J/(mol·K)	C_P J/(mol·K)	V^0 J/bar	Refs
$H_{2,\,g}$	0	0	130.7	28.8	–	1
	–	–	130.3	–	–	2
$O_{2,\,g}$	0	0	205.1	29.4	–	1
	0	0	205.1	29.4	–	3
	–	–	205.1	–	–	2
$O_{2,\,aq}$	16.4	–11.7	110.9	–	–	3
	26.5	–12.1	75.6	–	–	4
	26.5	–15.4	64.6	40.5	3.2	2
$O_{2^-\,g}$	779	778.6	143.3	–	–	2
$O_{2^-\,aq}$	133.5	132.1	228.5	–322.9	0.7	2
$Fe_{crystal}$	0	0	27.3	25.0	–	5
$Fe^{2+}_{\,g}$	–	2749.9	–	–	–	3
	2738.6	2794.4	172.3	–	–	2
$Fe^{2+}_{\,aq}$	–78.9	–89.1	–137.7	–	–	3
	–91.2	–92.5	–107.1			6
	–78.9	–89.1	–138			7
	–96.5	–95.8	–82.8	–33.1	–2.2	8
	–91.5					9
	–92.3	–96.6	–116.5	–40.6	–2.2	2
$Fe^{3+}_{\,g}$	–	5712.8	–	–	–	3
	5788.8	5851.2	173.8	–	–	2
$Fe^{3+}_{\,aq}$	–4.7	–48.5	–315.9	–	–	3
	–4.6	–48.5	–316	–	–	7
	–16.7	–50.2	–280.3	–	–	6
	–17.1	–49.5	–277.4	–34.1	–4.3	8
	–11.4	–	–	–	–	9
	–38.4	–50.3	–207.8	–108.1	–4.3	2
$FeO_2^-_{\,gas}$	–150.9	–142	283.2	–	–	2
$FeO_2^-_{\,aq}$	–63.7	–93.1	199.1	1.5	6.2	2

Notes: (1) Chase (1998); (2) this work; (3) Wagman et al. (1982); (4) Battino et al. (1983); (5) Barin (1995); (6) Larson et al. (1968); (7) Robie et al. (1978); (8) Shock and Helgeson (1988); (9) Nylen and Wigren (1971).

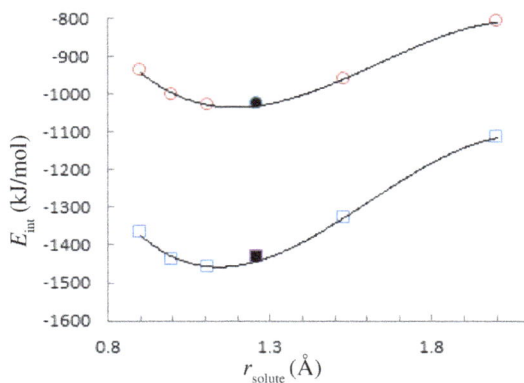

Figure A-2. solute solvent interaction potential well for O^{2-} in molten silica and water. The interaction energy is given by the summation of electrostatic + dispersive + repulsive terms. The cavitation energy is almost purely entropic and does not contribute to the interaction enthalpy. The **filled symbols** denote the adopted value ($r_{solute} = 1.26$ Å).

ACKNOWLEDGEMENTS

The authors thank the volume editors for their continuous support. Sandro Jahn and an anonymous reviewer provided thorough and detailed reviews that improved the manuscript. This study contributes to the IdEx Université de Paris ANR-18-IDEX-0001.

REFERENCES

Aiuppa A, Shinohara H, Tamburello G, Giudice G, Liuzzo M, Moretti R (2011) Hydrogen in the gas plume of an open-vent volcano, Mount Etna, Italy. J Geophys Res: Solid Earth 116(B10):B10204

Allanore A (2013) Electrochemical engineering of anodic oxygen evolution in molten oxides. Electrochim Acta 110:587–592

Allanore A (2015) Features and challenges of molten oxide electrolytes for metal extraction. J Electrochem Soc 162:E13–E22

Amovilli C, Mennucci B (1997) Self-consistent-field calculation of Pauli repulsion and dispersion contributions to the solvation free energy in the polarizable continuum model. J Phys Chem B 101:1051–1057

Armstrong K, Frost DJ, McCammon CA, Rubie DC, Boffa-Ballaran TB (2019) Deep magma ocean formation set the oxidation state of Earth's mantle. Science 365:903–906

Asimow PD, Hirschmann M, Stolper EM (2001) Calculation of peridotite partial melting from thermodynamic models of minerals and melts, IV. Adiabatic decompression and the composition and mean properties of mid-ocean ridge basalts. J Petrol 42:963–998

Baker DR, Moretti R (2011) Modeling the solubility of sulfur in magmas: a 50-year old geochemical challenge. Rev Mineral Geochem 73:167–213

Baker LL, Rutherford MJ (1996) Sulfur diffusion in rhyolite melts. Contrib Mineral Petrol 123:335–344

Barin I (1995) Thermochemical Data of Pure Substances, VCH Verlagsgesellschaft mbH, Weinheim, Germany)

Barone V, Cossi M, Tomasi J (1997) A new definition of cavities for the computation of solvation free energies by the polarizable continuum model. J Chem Phys 107:3210–3221

Bartmess JE (1994) Thermodynamics of the electron and proton. J Phys Chem 98:6420–6424

Battino R, Rettich TR, Tominaga *T* (1983) The solubility of oxygen and ozone in liquids. J Phys Chem Ref Data 12:163–178

Becke AD (1993) A new mixing of Hartree–Fock and local density-functional theories. J Chem Phys 98:1372–1377

Ben Naim A (1987) Solvation Thermodynamics. Plenum Press. New York

Berman RG, Brownn TH (1984) A thermodynamic model for multicomponent melts, with application to the system CaO–Al$_2$O$_3$–SiO$_2$. Geochim Cosmochim Acta 48:661–678

Bernal JD (1959) A geometrical approach to the structure of liquids. Nature 183:141–147

Borisov A, Behrens H, Holtz F (2018) Ferric/ferrous ratio in silicate melts: a new model for 1 atm data with special emphasis on the effects of melt composition. Contrib Mineral Petrol 173:1–15

Botcharnikov RE, Koepke J, Holtz F, McCammon C, Wilke M (2005) The effect of water activity on the oxidation and structural state of Fe in a ferro-basaltic melt. Geochim Cosmochim Acta 69:5071–5085

Botchamikov RE, Linnen RL, Wilke M, Holtz F, Jugo PJ, Berndt J (2011) High gold concentrations in sulphide-bearing magma under oxidizing conditions. Nat Geosci 4:112–115

Brounce M, Stolper E, Eiler J (2017) Redox variations in Mauna Kea lavas, the oxygen fugacity of the Hawaiian plume, and the role of volcanic gases in Earth's oxygenation. PNAS 114:8997–9002

Brown Jr GE, Farges F, Calas G (2018) X-ray scattering and X-ray spectroscopy studies of silicate melts. Rev Min Geochem 32:317–410

Buddington AF, Lindsley DH (1964) Iron-titanium oxide minerals and synthetic equivalents. J Petrol 5:310–357

Caillet J, Claverie P, Pullman B (1978) On the conformational varieties of acetylcholine in the crystals of its halides. Acta Crystallogr B 34:3266–3272

Cancès MT, Mennucci B, Tomasi J (1997) A new integral equation formalism for the polarizable continuum model: Theoretical background and applications to isotropic and anisotropic dielectrics. J Chem Phys 107:3032–3041

Carmichael IS (2002) The andesite aqueduct: perspectives on the evolution of intermediate magmatism in west-central (105–99 W) Mexico. Contrib Mineral Petrol 143:641–663

Carroll MR, Rutherford MJ (1988) Sulfur speciation in hydrous experimental glasses of varying oxidation state; results from measured wavelength shifts of sulfur X-rays. Am Mineral 73:845–849

Chase MW, Jr (1998) NIST-JANAF Thermochemical Tables. Fourth Edition. J Phys Chem Ref Data Monograph 9

Chiorboli P (1980) Fondamenti di Chimica, UTET, Torino

Cicconi MR, Moretti R, Neuville DR (2020a) Earth's electrodes. Elements 16:157–160

Cicconi MR, Le Losq C, Moretti R, Neuville DR (2020b) Magmas are the largest repositories and carriers of earth's redox processes. Elements 16:173–178

Cicconi MR, Le Losq C, Henderson G, Neuville DR (2021) The redox behavior of rare earth elements. *In:* Magma Redox Geochemistry, (eds. Moretti R, Neuville DR) Geophysical Monograph Vol 266, American Geophysical Union. John Wiley & Sons Inc, p 381–398

Cochain B, Neuville DR, Henderson GS, McCammon C, Pinet O, Richet P (2012) Iron content, redox state and structure of sodium borosilicate glasses: A Raman, Mössbauer and boron K-edge XANES spectroscopy study. J Am Ceram Soc 94:1–12

Cochain B, Neuville DR, de Ligny D, Malki M, Testemale D, Pinet O, Richet P (2013) Dynamics of iron-bearing borosilicate melts: Effects of melt structure and composition on viscosity, electrical conductivity and kinetics of redox reactions. J Non-Cryst Solids 373–374:8–27

Colle R, Salvetti D (1975) Approximate calculation of the correlation energy for the closed shells. Theor Chim Acta 37:329–334

Colson RO, Haskin LA, Crane D (1990) Electrochemistry of cations in diopsidic melt: Determining diffusion rates and redox potentials from voltammetric curves. Geochim Cosmochim Acta 54:3353–3367

Cook GB, Cooper RF (2000) Iron concentration and the physical processes of dynamic oxidation in alkaline earth aluminosilicate glass. Am Mineral 85:397–406

Cook GB, Cooper RF, Wu T (1990) Chemical diffusion and crystalline nucleation during oxidation of ferrous ironbearing magnesium aluminosilicate glass. J Non-Crystallogr Solids 120:207–222

Cooper RF, Fanselow JB, Poker DB (1996a) The mechanism of oxidation of a basaltic glass: chemical diffusion of network-modifying cations. Geochim Cosmochim Acta 60:3253–3265

Cooper RF, Fanselow JB, Weber JKR, Merkley DR, Poker DB (1996b) Dynamics of oxidation of a Fe^{2+}-bearing aluminosilicate (basaltic) melt. Science 274:1173–1176

Dancy EA, Derge GJ (1966) Electrical conductivity of FeO_x–CaO slags. Trans Metall Soc AIME 236:1642

Darken L, Gurry RW (1945) The system iron–oxygen. I. The wüstite field and related equilibria. J Am Chem Soc 67:1398–1412

Darken L, Gurry RW (1946) The system iron–oxygen. II. Equilibrium and thermodynamics of liquid oxide and other phases. J Am Chem Soc 68:798–816

Dickenson MP, Hess PC (1982) Redox equilibria and the structural role of iron in alumino-silicate melts. Contrib Mineral Petrol 78:352–357

Dickson WR, Dismukes EB (1962) Electrolysis of FeO–CaO–SiO_2 melts. Trans Metall Soc AIME 224:505

Dimitrov V, Komatsu T (1999) Electronic polarizability, optical basicity and non-linear optical properties of oxide glasses. J Non-Cryst Solids 249:160–179

Dimitrov V, Komatsu T (2000) Interionic interactions, electronic polarizability and optical basicity of oxide glasses. J Ceram Soc Jpn 108:330–338

Dimitrov V, Tasheva T, Komatsu T (2018) Group optical basicity and single bond strength of oxide glasses. J Chem Technol Metall 53:1038–1046

Ditchfield R, Hehre WJ, Pople JA (1971) Self-consistent molecular-orbital methods. IX. An extended Gaussian-type basis for molecular-orbital studies of organic molecules. J Chem Phys 54:724–728

Douglas RW, Zaman MS (1969) Chromophore in iron–sulphur amber glasses. Phys Chem Glasses 10:125

Duffy JA (1992) A review of optical basicity and its applications to oxidic systems. Geochim Cosmochim Acta 57:3961–3970

Duffy JA, Ingram MD (1971) Establishment of an optical scale for Lewis basicity in inorganic oxyacids, molten salts, and glasses. J Am Chem Soc 93:6448–6454

Duffy JA, Ingram MD (1975) Optical basicity–IV: Influence of electronegativity on the Lewis basicity and solvent properties of molten oxyanion salts and glasses. J Inorg Nucl Chem 37:1203–1206

Duffy JA, Ingram MD (1976) An interpretation of glass chemistry in terms of the optical basicity concept. J Non-Cryst Solids 21:373–410

Ellingham HJT (1944) Reducibility of oxides and sulfides in metallurgical processes. J Soc Chem Ind 63:125–133

Eugster HP (1977) Compositions and thermodynamics of metamorphic solutions. *In:* Fraser DG (ed) Thermodynamics in Geology. D Reidel Publishing Company, Dordrecht, p 183–202

Fincham CJB, Richardson FD (1954) The behaviour of sulphur in silicate and aluminate melts. Proc R Soc (London), Ser A 223:40–62

Flood H, Förland T (1947) The acidic and basic properties of oxides. Acta Chem Scand 1:592–604

Flood H, Grjotheim T (1952) Thermodynamic calculation of slag equilibria. J Iron Steel Inst. 171:64–80

Floris FM, Tomasi J (1989) Evaluation of the dispersion contribution to the solvation energy. A simple computational model in the continuum approximation. J Comput Chem 10:616–627

Floris FM, Tomasi J, Pascual-Ahuir JL (1991) Dispersion and repulsion contributions to the solvation energy: Refinements to a simple computational model in the continuum approximation. J Comput Chem 12:784–791

Fraser DG (1975) Activities of trace elements in silicate melts. Geochim Cosmochim Acta 39:1525–1530

Fraser DG (2005) Acid–base properties and structons: towards a structural model for predicting the thermodynamic properties of silicate melts. Ann Geophys 48:549–559

Frisch MJ, Trucks GW, Schlegel HB, Scuseria GE, Robb MA, Cheeseman JR, Montgomery JJA, Vreven T, Kudin KN, Burant JC, Millam JM, Iyengar SS, Tomasi J, Barone V, Mennucci B, Cossi M, Scalmani G, Rega N, Petersson GA, Nakatsuji H, Hada M, Ehara M, Toyota K, Fukuda R, Hasegawa J, Ishida M, Nakajima T, Honda Y, Kitao O, Nakai H, Klene M, Li X, Knox JE, Hratchian HP, Cross JB, Bakken V, Adamo C, Jaramillo J, Gomperts R, Stratmann RE, Yazyev O, Austin AJ, Cammi R, Pomelli C, Ochterski JW, Ayala PY, Morokuma K, Voth GA, Salvador P, Dannenberg JJ, Zakrzewski VG, Dapprich S, Daniels AD, Strain MC, Farkas O, Malick DK, Rabuck AD, Raghavachari K, Foresman JB, Ortiz JV, Cui Q, Baboul AG, Clifford S, Cioslowski J, Stefanov BB, Liu G, Liashenko A, Piskorz P, Komaromi I, Martin RL, Fox DJ, Keith T, Al-Laham MA, Peng CY, Nanayakkara A, Challacombe M, Gill PMW, Johnson B, Chen W, Wong MW, Gonzalez C, Pople JA (2004) Gussian03; Revision B.05 ed.; Gaussian, Inc.: Wallingford, CT

Frost BR (1991) Introduction to oxygen fugacity and its petrologic importance. Rev Mineral Geochem 25:1–9

Fumi FG, Tosi MP (1964) Ionic sizes and Born repulsive parameters in the NaCl-type alkali halides, I: The Huggins-Mayer and Pauling forms. J Phys Chem Solids 25:31–43

Gaillard F, Scaillet B, Pichavant M, Bény JM (2001) The effect of water and f_{O_2} on the ferric-ferrous ratio of silicic melts. Chem Geol 174:255–273

Gaskell PH, Eckersley MC, Barnes AC, Chieux P (1991) Medium-range order in the cation distribution of a calcium silicate glass. Nature 350:675–677

Ghiorso MS, Sack RO (1995) Chemical mass transfer in magmatic processes IV. A revised and internally consistent thermodynamic model for the interpolation and extrapolation of liquid–solid equilibria in magmatic systems at elevated temperatures and pressures. Contrib Mineral Petrol 119:197–212

Ghiorso MS, Carmichael IS, Rivers ML, Sack RO (1983) The Gibbs free energy of mixing of natural silicate liquids; an expanded regular solution approximation for the calculation of magmatic intensive variables. Contrib Mineral Petrol 84:107–145

Ghiorso MS, Hirschmann MM, Reiners PW, Kress VC (2002) The pMELTS: A revision of MELTS for improved calculation of phase relations and major element partitioning related to partial melting of the mantle to 3 GPa. Geochem Geophys Geosystems 3:1–35

Giggenbach WF (1980) Geothermal gas equilibria. Geochim Cosmochim Acta 44:2021–2032

Giggenbach WF (1987) Redox processes governing the chemistry of fumarolic gas discharges from White Island, New Zealand. Appl Geochem 2:143–161

Greaves GN, Ngai KL (1995) Reconciling ionic-transport properties with atomic structure in oxide glasses. Phys Rev B 52:6358–6380

Greaves GN, Sen S (2007) Inorganic glasses, glass-forming liquids and amorphizing solids. Adv. Phys 56! 1–166

Greaves GN, Fontaine A, Lagarde P, Raoux D, Gurman SJ (1981) Local structure of silicate glasses. Nature 293:611–616

Gudmundsson G, Wood BJ (1995) Experimental tests of garnet peridotite oxygen barometry. Contrib Mineral Petrol 119:56–67

Guo X, Lange RA, Ai Y (2014) Density and sound speed measurements on model basalt (An–Di–Hd) liquids at one bar: New constraints on the partial molar volume and compressibility of the FeO component. Earth Planet Sci Lett 388:283–292

Haughton DR, Roeder PL, Skinner BJ (1974) Solubility of sulfur in mafic magmas. Econ Geol 69:451–467

Henderson GS (2005) The structure of silicate melts: a glass perspective. Can Mineral 43:1921–1958

Hess PC (1971) Polymer model of silicate melts. Geochim Cosmochim Acta 35:289–306

Hess PC (1980) Polymerization model for silicate melts. *In:* Physics of Magmatic Processes (Hargraves RB, ed.), Princeton University Press, pp 1–48

Hillert M, Jansson BO, Sundman BO (1985) A two-sublattice model for molten solutions with different tendency for ionization. Metall Trans A 16:261–266

Hiyagon H, Ozima M (1986) Partition of noble gases between olivine and basalt melt. Geochim Cosmochim Acta 50:2045–2057

Kirsten T (1968) Incorporation of rare gases in solidifying enstatite melts. J GeoPhys Res. 73:2807–2810

Klimm K, Kohn SC, Botcharnikov RE (2012) The dissolution mechanism of sulphur in hydrous silicate melts. II: Solubility and speciation of sulphur in hydrous silicate melts as a function of f_{O_2}. Chem Geol 322:250–267

Kotani M, Mizuno Y, Kayama K, Ishigurp E (1957) Electronic structure of simple homonuclear diatomic molecules I. Oxygen molecule. J Chem Phys Soc Jpn 12:707–736

Kress VC, Carmichael ISE (1991) The compressibility of silicate liquids containing Fe_2O_3 and the effect of composition, temperature, oxygen fugacity and pressure on their redox states. Contrib Mineral Petrol 108:82–92

Kushiro I (2010) Toward the Development of "Magmatology". Annu Rev Earth Planet Sci 38:1–16

Jahn S (2022) Molecular simulations of oxide and silicate melts and glasses. Rev Mineral Geochem 87:193-227

Jambon A, Weber H , Braun O (1986) Solubility of He, Ne, Ar, Kr and Xe in a basalt melt in the range 1250–1600 °C. Geochemical implications. Geochim Cosmochim Acta 50:401–408

Jayasuriya KD, O'Neill HSC, Berry AJ, Campbell SJ (2004) A Mossbauer study of the oxidation state of Fe in silicate melts. Am Mineral 89:1597–1609

Jørgensen CK (1969) Oxidation Numbers and Oxidation States. Springer-Verlag, Berlin-Heidelberg-New York

Jugo PJ, Luth RW, Richards JP (2005) An experimental study of the sulfur content in basaltic melts saturated with immiscible sulfide or sulfate liquids at 1300° C and 10 GPa. J Petrol 46:783–798

Jugo PJ, Wilke M, Botcharnikov RE (2010) Sulfur K-edge XANES analysis of natural and synthetic basaltic glasses: Implications for S speciation and S content as function of oxygen fugacity. Geochim Cosmochim Acta 74:5926–5938

Lange RA (1994) The effect of H_2O, CO_2 and F on the density and viscosity of silicate melts. Rev Mineral 30:331–369

Lange RA, Carmichael ISE (1989) Ferric-ferrous equilibria in Na_2O–FeO–Fe_2O_3–SiO_2 melts: Effects of analytical techniques on derived partial molar volumes. Geochim Cosmochim Acta 53:2195–2204

Larson JW, Cerutti P, Garber HK, Hepler LG (1968) Electrode potentials and thermodynamic data for aqueous ions. Copper, zinc, cadmium, iron, cobalt and nickel. J Phys Chem 78:2902–2907

Lee C, Yang W, Parr RG (1988) Development of the Colle–Salvetti correlation-energy formula into a functional of the electron density. Phys Rev B:37:785–789

Le Losq C, Moretti R, Neuville DR (2013) Speciation and amphoteric behaviour of water in aluminosilicate melts and glasses: high-temperature Raman spectroscopy and reaction equilibria. Eur J Mineral 25:777–790

Le Losq C, Neuville DR, Chen W, Florian P, Massiot D, Zhou Z, Greaves GN (2017) Percolation channels: a universal idea to describe the atomic structure and dynamics of glasses and melts. Sci Rep 7:1–12

Le Losq C, Cicconi M, Greaves GN, Neuville DR (2019) Silicate glasses. *In:* Springer Handbook of Glass Springer, Cham, p 441–503

Le Losq C, Moretti R, Oppenheimer C, Baudelet F, Neuville DR (2020) Dynamic *in situ* experimental Fe K-edge XANES study of iron oxidation state and coordination in magmas: application to Erebus phonolite. Contrib Mineral Petrol 175:1–13

Lewis GN, Randall M (1961) Thermodynamics, 2nd Edition (Revised by K Pitzer, L Brewer) McGraw-Hill Book Company

Lias SG, Bartmess JE (2004) Gas-phase Ion Thermochemistry. NIST Chemistry WebBook, NIST Standard Reference Database Number 69, http://webbook.nist.gov/chemistry/

Liu Q, Lange RA (2006) The partial molar volume of Fe_2O_3 in alkali silicate melts: Evidence for an average Fe^{3+} coordination number near five. Am Mineral 91:385–393

Lux G (1987) The behavior of noble gases in silicate liquids: Solution, diffusion, bubbles and surface effects, with applications to natural samples. Geochim. Cosmochim. Acta 51:1549–1560

Magnien V, Neuville DR, Cormier L, Roux J, Hazemann JL, Pinet O, Richet P (2006) Kinetics of iron redox reactions in silicate liquids: a high-temperature X-ray absorption and Raman spectroscopy study. J Nucl Mater 352:190–195

Magnien V, Neuville DR, Cormier L, Roux J, Hazemann JL, de Ligny D, Pascarelli S, Vickridge I, Pinet O, Richet P (2008) Kinetics and mechanisms of iron redox reactions in silicate melts: The effects of temperature and alkali cations. Geochim Cosmochim Acta 72:2157–2168.

Mao H, Hillert M, Selleby M, Sundman B (2006) Thermodynamic assessment of the CaO–Al_2O_3–SiO_2 system. J Am Ceram Soc 89:298–308

Marini L, Moretti R, Accornero M (2011) Sulfur isotopes in magmatic–hydrothermal systems, melts, and magmas. Rev Mineral Geochem 73:423–492

Matthews SJ, Moncrieff DHS, Carroll MR (1999) Empirical calibration of the sulphur valence oxygen barometer from natural and experimental glasses: method and applications. Mineral Mag 63:421–431

Mattioli GS, Wood BJ (1988) Magnetite activities across the $MgAl_2O_4$–Fe_3O_4 spinel join, with application to thermobarometric estimates of upper mantle oxygen fugacity. Contrib Mineral Petrol 98:148–162

McQuarrie DA, Simon JD (1997) Physical Chemistry: A Molecular Approach, University Science Books, Sausalito, CA

Meckler A (1953) Electronic energy levels of molecular oxygen. J Chem Phys 21:1750–1762

Mitchell AC (1928) Entropie Des Elektronengases auf Grund der Fermischen Statistik. Z Phys 50:570–576

Moore G, Righter K, Carmichael ISE (1995) The effect of dissolved water on the oxidation state of iron in natural silicate liquids. Contrib Mineral Petrol 120:170–179

Moretti R (2005) Polymerisation, basicity, oxidation state and their role in ionic modelling of silicate melts. Ann Geophys 48:583–608

Moretti R (2021) Ionic syntax and equilibrium approach to redox exchanges in melts: basic concepts and the case of iron and sulfur in degassing magmas. *In:* Magma Redox Geochemistry (eds. Moretti R, Neuville DR), Geophysical Monograph Vol 266, American Geophysical Union, John Wiley & Sons, Inc., P 117–138

Moretti R, Baker DR (2008) Modeling the interplay of f_{O_2} and f_{S_2} along the FeS–silicate melt equilibrium. Chem Geol 256:286–298

Moretti R, Neuville D (2021) Redox equilibria: from basic concepts to the magmatic realm. *In:* Magma Redox Geochemistry, (eds. Moretti R, Neuville DR) Geophysical Monograph 266, American Geophysical Union. John Wiley & Sons, Inc., P 1–17

Moretti R, Ottonello G (2003a) Polymerization and disproportionation of iron and sulfur in silicate melts: insights from an optical basicity-based approach. J Non-Crystallogr Solids 323:111–119

Moretti R, Ottonello G (2003b) A polymeric approach to the sulfide capacity of silicate slags and melts. Metall Mater Trans B 34:399–410

Moretti R, Ottonello G (2005) Solubility and speciation of sulfur in silicate melts: The Conjugated Toop–Samis–Flood–Grjotheim (CTSFG) model. Geochim Cosmochim Acta 69:801–823

Moretti R, Papale P (2004) On the oxidation state and volatile behavior in multicomponent gas–melt equilibria. Chem Geol 213:265–280

Moretti R, Stefánsson A (2020) Volcanic and geothermal redox engines. Elements 16:179–184

Moretti R, Arienzo I, Orsi G, Civetta L, D'Antonio M (2013) The deep plumbing system of Ischia: a physico-chemical window on the fluid-saturated and CO_2-sustained Neapolitan volcanism (southern Italy) J Petrol 54:951–984

Moretti R, Le Losq C, Neuville DR (2014) The amphoteric behavior of water in silicate melts from the point of view of their ionic–polymeric constitution. Chem Geol 367:23–33

Moussallam Y, Oppenheimer C, Scaillet B, Gaillard F, Kyle P, Peters N, Hartley M, Berlo K, Donovan A (2014) Tracking the changing oxidation state of Erebus magmas, from mantle to surface, driven by magma ascent and degassing. Earth Planet Sci Lett 393:200–209

Moussallam Y, Edmonds M, Scaillet B, Peters N, Gennaro E, Sides I, Oppenheimer C (2016) The impact of degassing on the oxidation state of basaltic magmas: a case study of Kilauea volcano. Earth Planet Sci Lett 450:317–325

Mysen BO (1991) Relations between structure, redox equilibria of iron and properties of magmatic liquids, *In:* Physical Chemistry of Magma (Perchuk LL, Kushiro I, eds) Springer, New York, p 41–98

Mysen BO, Richet P (2018) Silicate Glasses and Melts. Elsevier

Mysen BO, Virgo D (1985) Iron-bearing silicate melts: relations between pressure and redox equilibria. Phys Chem Mineral 12:191–200

Nagashima S, Katsura T (1973) The solubility of sulfur in Na_2O–SiO_2 melts under various oxygen partial pressures at 1100° C, 1250° C, and 1300° C. Bull Chem Soc Jpn 46:3099–3103

Nash WM, Smythe DJ, Wood BJ (2019) Compositional and temperature effects on sulfur speciation and solubility in silicate melts. Earth Planet Sci Lett 507:187–198

Nesbitt HW, Bancroft GM, Henderson GS, Ho R, Dalby KN, Huang Y, Yan Z (2011) Bridging, non-bridging and free (O^{2-}) oxygen in Na_2O–SiO_2 glasses: An X-ray photoelectron spectroscopic (XPS) and nuclear magnetic resonance (NMR) study. J Non-Cryst Solids 357:170–180

Nesbitt HW, Henderson GS, Bancroft GM, Ho R (2015a) Experimental evidence for Na coordination to bridging oxygen in Na-silicate glasses: Implications for spectroscopic studies and for the modified random network model. J Non-Cryst Solids 409:139–148

Nesbitt HW, Bancroft GM, Henderson GS, Sawyer R, Secco RA (2015b) Direct and indirect evidence for free oxygen (O^{2-}) in MO-silicate glasses and melts (M= Mg, Ca, Pb) Am Mineral 100:2566–2578

Nesbitt HW, Henderson GS, Bancroft GM, Sawyer R, Secco RA (2017) Bridging oxygen speciation and free oxygen (O^{2-}) in K-silicate glasses: Implications for spectroscopic studies and glass structure. Chem Geol 461:13–22

Nesbitt HW, Bancroft GM, Henderson GS (2020) Polymerization during melting of ortho- and meta-silicates: Effects on Q species stability, heats of fusion, and redox state of mid-ocean range basalts (MORBs) Am Mineral 105:716–726

Neuville DR, Cicconi MR, Le Losq C (2021) How to measure the oxidation state of multivalent elements in minerals, glasses and melts? *In:* Magma Redox Geochemistry, (eds. Moretti R, Neuville DR) Geophysical Monograph Vol 266, American Geophysical Union. John Wiley & Sons, Inc, p 257–281

Nylen P, Wigren N (1971) Stechiometria CEDAM, Padova

O'Neill HSC, Berry AJ, McCammon CC, Jayasuriya KD, Campbell SJ, Foran G (2006) An experimental determination of the effect of pressure on the $Fe^{3+}/\Sigma Fe$ ratio of an anhydrous silicate melt to 3.0 GPa. Am Mineral 91:404–412

Ochterski JW (2000) Thermochemistry in Gaussian © 2000 Gaussian Inc

Oppenheimer C, Moretti R, Kyle PR, Eschenbacher A, Lowenstern JB, Hervig RL, Dunbar NW (2011) Mantle to surface degassing of alkalic magmas at Erebus volcano, Antarctica. Earth Planet Sci Lett 306:261–271

Ottonello G (1997) Principles of Geochemistry. Columbia University Press, New York

Ottonello G (2001) Thermodynamic constraints arising from the polymeric approach to silicate slags: the system CaO–FeO–SiO$_2$ as an example. J Non-Cryst Solids 282:72–85

Ottonello G (2005) Chemical interactions and configurational disorder in silicate melts. Ann Geophys 48:56–581

Ottonello G (2018) *Ab-initio* reactivity of Earth's materials. Riv Nuovo Cimento 41:225–289

Ottonello G (2021) Thermodynamic models of silicate melts. *In:* Encyclopedia of Glass Science, Technology, History, and Culture (Richet P ed.), Vol. I. The American Ceramic Society, John Wiley & Sons, Inc, p 545–558

Ottonello G, Moretti R (2004) Lux–Flood basicity of binary silicate melts. J Phys Chem Solids 65: 1609–1614

Ottonello G, Richet P (2014) The solvation radius of silicate melts based on the solubility of noble gases and scaled particle theory. J Chem Phys 140:044506-1-9

Ottonello G, Moretti R, Marini L, Zuccolini MV (2001) Oxidation state of iron in silicate glasses and melts: a thermochemical model. Chem Geol 174:157–179

Ottonello G, Richet P, Vetuschi Zuccolini M (2015) The wet solidus of silica: Predictions from the scaled particle theory and polarized continuum model. J Chem Phys 142:054503-1-9

Ottonello G, Richet P, Papale P (2018) Bulk solubility and speciation of H$_2$O in silicate melts. Chem Geol 479:176–187

Papale P, Moretti R, Paonita A (2022) Thermodynamics of multi-component gas–melt equilibrium in magmas: Theory, models, and applications. Rev Mineral Geochem 87:431-556

Paris E, Giuli G, Carroll MR, Davoli I (2001) The valence and speciation of sulfur in glasses by X-ray absorption spectroscopy. Can Mineral 39:331–339

Paul A (1990) Oxidation-reduction equilibrium in glasses. J Non-Cryst Solids 123:354–362

Paul A, Douglas RW (1965) Ferric-ferrous equilibrium in binary alakali silicre glasses. Phys Chem Glasses 6:207–211

Pierotti RA (1963) The solubility of gases in liquids. J Phys Chem 67:1840–1845

Pierotti RA (1965) Aqueous solutions of nonpolar gases J.Phys Chem 69:281–288

Pierotti RA (1976) A scaled particle theory of aqueous and nonaqueous solutions. Chem Rev 76:717–726

Pinet O, Phalippou J, Di Nardo C (2006) Modeling the redox equilibrium of the Ce^{4+}/Ce^{3+} couple in silicate glass by voltammetry. J Non-Cryst Solids 352:5382–5390

Reiss H (1965) Scaled particle methods in the statistical thermodynamics of fluids. Adv Chem Phys 9:1–84

Reiss H, Mayer SW (1961) Theory of the surface tension of molten salts. J Chem Phys 34:2001–2003

Reiss H, Frisch HL, Lebowitz JL (1959) Statistical mechanics of rigid spheres. J Chem Phys 31:369–380

Reiss H, Frisch HL, Helfand E, Lebowitz JL (1960) Aspects of the statistical thermodynamics of real fluids. J Chem Phys 32:119–124

Robie RA, Hemingway SB, Fisher JR (1978) Thermodynamics Properties of Minerals and Related Substances at 298.15 K and 1 bar (105 Pascal) Pressure and at the High Temperatures. US Geol Surv Bull 1452

Satchell DPN, Satchell RS (1971) Quantitative aspects of Lewis acidity. Q Rev Chem Soc 25:171–199

Schuessler JA, Botcharnikov RE, Behrens H, Misiti V, Freda C (2008) Amorphous materials: properties, structure, and durability: oxidation state of iron in hydrous phono-tephritic melts. Am Mineral 93:1493–1504

Schreiber HD (1987) An electrochemical series of redox couples in silicate melts: a review and applications to geochemistry. J Geophys Res: Solid Earth 92(B9): 9225–9232

Semkow KW, Haskin LA (1985) Concentrations and behavior of oxygen and oxide ion in melts of composition CaO·MgO·xSiO$_2$. Geochim Cosmochim Acta 49:1897–1908

Shannon RD (1976) Revised effective ionic radii and systematic studies of interatomic distances in halides and chalcogenides. Acta Crystallogr A32:751–767

Shannon RD, Prewitt CT (1969) Effective ionic radii in oxides and fluorides. Acta Crystallogr B25:925–946

Shibata T, Takahashi E, Matsuda JI (1998) Solubility of neon, argon, krypton, and xenon in binary and ternary silicate systems: a new view on noble gas solubility. Geochim. Cosmochim. Acta 62:1241–1253

Shock EL, Helgeson HC (1988) Calculation of the thermodynamic and transport properties of aqueous species at high pressures and temperatures. Correlation algorithms for ionic species and equation of state predictions to 5 kb and 1000 °C. Geochim. Cosmochim. Acta 52:2009–2036

Slater JC (1929) The theory of complex spectra. Phys Rev 34:1293–1322

Sun KH (1947) Fundamental condition of glass formation. J Am Ceram Soc 30:277–281

Temkin M (1945) Mixtures of fused salts as ionic solutions. Acta Phys Chem USSR 20:411

Tomasi J, Persico M (1994) Molecular interactions in solution: An overview of methods based on continuous distributions of the solvent. Chem Rev 94:2027–2094

Tomasi J, Mennucci B, Cancès E (1999) The IEF version of the PCM solvation method: an overview of a new method addressed to study molecular solutes at the QM ab initio level. J Mol Struct 464:211–226

Toop GW, Samis CS (1962a) Some new ionic concepts of silicate slags. Can Metall Q 1:129–152

Toop GW, Samis CS (1962b) Activities of ions in silicate melts. Trans Met Soc AIME 224:878–887

Vessal B, Greaves GN, Marten PT, Chadwick AV, Mole R, Houde-Walter S(1992) Cation microsegregation and ionic mobility in mixed alkali glasses. Nature 356:504–506

Vetuschi Zuccolini M, Ottonello G, Belmonte D (2011) *Ab-initio* assessment of conventional standard state thermodynamic properties for geochemically relevant gaseous and aqueous species. Comput Geosci 37:646–661

Wagman DD, Evans WH, Parker VB, Schumm RH, Halow I, Bailey SM, Churney KL, Nuttall RL (1982) The NBS tables of chemical thermodynamic properties. Selected values for inorganic and C1 and C2 organic substances in SI units. J Phys Chem Reference Data 11, Supplement N.2, 392 pp

Wallace PJ, Carmichael IS (1994) S speciation in submarine basaltic glasses as determined by measurements of S Kα X-ray wavelength shifts. Am Mineral 79:161–167

Wallace PJ, Edmonds M (2011) The sulfur budget in magmas: evidence from melt inclusions, submarine glasses, and volcanic gas emissions. Rev Mineral Geochem 73:215–246

Warren BE (1933) X-ray diffraction of vitreous silica. Zeitschrift für Kristallographie 86:349–358

Waseda Y, Suito H (1977) The structure of molten alkali metal silicates. Trans Iron Steel Inst Jpn 17:82–91

Waters LE, Lange RA (2016) No effect of H_2O degassing on the oxidation state of magmatic liquids. Earth Planet Sci Lett 447:48–59

Wilke M, Behrens H, Burkhard DMJ, Rossano S (2002) The oxidation state of iron in silicic melt at 500 Mpa water pressure. Chem Geol 189:55–67

Zachariasen WH (1932) The atomic arrangement in glass. J Am Chem Soc 54:3841–3851

Zachariasen WH (1935) The vitreous state. The J Chem Phys 3:162–163

Zhang J, Matsuura H, Tsukihashi F (2014) Processes for recycling. *In:* Treatise on Process Metallurgy, Volume 3, 1507–1561

Zhang HL, Hirschmann MM, Cottrell E, Withers AC (2017) Effect of pressure on $Fe^{3+}/\Sigma Fe$ ratio in a mafic magma and consequences for magma ocean redox gradients. Geochim Cosmochim Acta 204:83–103

Reviews in Mineralogy & Geochemistry
Vol. 87 pp. 405-429, 2022
Copyright © Mineralogical Society of America

9

Nucleation, Growth, and Crystallization in Oxide Glass-formers. A Current Perspective

Maziar Montazerian and Edgar Dutra Zanotto*

Center for Research, Technology and Education in Vitreous Materials (CeRTEV),
Department of Materials Engineering (DEMa)
Federal University of São Carlos (UFSCar)
São Carlos, SP, 13.565-905
Brazil

*dedz@ufscar.br

1. INTRODUCTION

Vitrification from the molten state hinges on averting crystallization during the cooling path. On the other hand, critical natural processes, such as the formation of snow and igneous rocks, like obsidian, and technological operations, for example, solidification of metallic alloys and glass ceramization, illustrate the utmost importance of crystallization in our environment and technology. The structural rearrangements fostered by crystal nucleation and growth cause drastic changes in the macroscopic properties of glass-forming melts and magmas. In this article, we summarize and discuss the applicability of the most accepted models to describe crystal nucleation, growth, and overall crystallization in glass-forming systems. We also focus on the significant progress achieved in the understanding of crystallization over the past few decades through the combined use of theoretical models and experiments. Additionally, we highlight selected open problems and directions for future studies.

When a melt is cooled below its liquidus temperature, it becomes a supercooled liquid (SCL) from which one or more crystalline phases will tend to form. This process takes place in two steps; the formation of crystal clusters (nucleation) and their subsequent evolution to macroscopic crystals (growth). The combination of crystal nucleation and growth leads to the phenomenon called crystallization. Crystallization counteracts vitrification, i.e., the (temporary) freezing of a melt into a glass. Hence, prevention of crystallization upon cooling of any liquid, or upon heating gels, gives rise to a glass. On the other hand, uncontrolled crystallization, devitrification, may occur upon heating a glass (Zanotto and Mauro 2017; Zheng et al. 2019). Eight decades ago, G. W. Morey (1938) stated that "Devitrification is the chief factor which limits the composition range of practical glasses, it is an ever-present danger in all glass manufacturing and working, and takes place promptly with any error in composition or technique." Figure 1 shows a naturally occurring and partially crystallized volcanic glass, called obsidian.

Technological breakthroughs, marked by high-tech industrial processes and devices, require a plethora of novel materials, which include glasses and glass-ceramics with unusual microstructures and enhanced properties, such as high transparency, bioactivity, ionic conductivity, and machinability, sometimes combined with adequate dielectric, magnetic, chemical, mechanical, or thermal shock resistance (Zanotto 2010; Montazerian et al. 2015). To meet this demand, significant efforts have focused on the synthesis of new glasses and

1529-6466/22/0087-0009$05.00 (print)
1943-2666/22/0087-0009$05.00 (online)

http://dx.doi.org/10.2138/rmg.2022.87.09

Figure 1. Partially devitrified obsidian volcanic glass showing (snowflake) cristobalite crystals of ~1cm.

glass-ceramics. In both cases, crystallization control plays such a decisive role that reliable models of crystallization processes are needed. Hence, knowledge about the possible pathways of crystallization allows one to formulate kinetic criteria to answer some questions such as: "Under what conditions can a liquid be supercooled and transform into a glass?" Or equivalently, "under what conditions is crystallization expected to occur on the cooling path?" In attempts to produce new glasses, crystal nucleation and growth must be avoided. Conversely, controlled crystallization can be used to synthesize fully crystallized or semi-crystalline glass-ceramics (Montazerian et al. 2015). Several monographs provide detailed information on these materials (Höland and Beall 2012; Gutzow and Schmelzer 2013; Zanotto 2013; Neuville et al. 2017).

However, these technological aspects represent only one side of the pervasive scientific interest in the kinetics of nucleation and crystallization in glasses. In addition to their practical relevance, the highly viscous glass-forming supercooled liquids serve as remarkable experimental models of metastable systems, in which crystallization processes can be initiated, accelerated or delayed. These processes can thus be studied conveniently under widely different conditions on a laboratory timescale. Such analyses also include the crystallization kinetics versus the thermal history of the sample. For this reason, glass-forming liquids have served as *guinea pigs* for testing crystal nucleation and growth theories, providing a deeper insight into different phase transformation processes. And lastly, controlled crystallization often produces uniquely beautiful (and frequently hidden) nano- and microstructures, as demonstrated by Zanotto (2013), which serves as an additional motivation to pursue research in this endless, albeit highly gratifying quest to unveil the deeply hidden intricacies of glass crystallization and the resulting properties of glass-ceramics.

Preventing or inducing controlled crystal nucleation and growth in a glass (or more correctly in a supercooled liquid, SCL) requires a theoretical understanding of these complex phenomena. Therefore, this article outlines the basic fundamental aspects of the crystallization theory of supercooled oxide glass-forming liquids. Section 2 begins with a description of crystal nucleation kinetics in glass-forming liquids. In Section 3, we provide an overview of the basic modes of crystal growth. Section 4 describes the overall crystallization kinetics, i.e., the evolution of the volume fraction of crystalline phases as a function of time. In this regard, we also dwell on glass-forming ability on the cooling path, and glass stability on heating. In addition to some well-established results, we also discuss open problems and possible approaches to their resolution. In Section 5, a summary of selected results and perspectives for future developments completes this article.

TABLE OF SYMBOLS

ΔG	Gibbs free energy difference	T_m	Melting temperature
R	Ideal gas constant	Δh_m	Heat of melting of one crystal phase particle
A	Nucleus surface area	q_m	Heat of melting
N	Number of particles in the nucleus	N_A	Avogadro's number
$\Delta\mu$	Chemical potential difference	ς	Correction factor that varies in the range 0.4 to 0.6
μ_l	Chemical potential of liquid	f	Fraction of preferred growth sites
μ_{cr}	Chemical potential of crystal	Δs_m	Entropy of melting
σ	Electrical conductivity or surface free energy	C_2, C_3	Parameters determing the time required for the formation of the two-dimensional nucleus
C_α	Particle number density in the crystal cluster	B	Combination of parameters proportional to the effective diffusion coefficient
P	Pressure	$<R>$	Average size of the nuclei
T	Temperature	N	Number of supercritical nuclei
G	Gibbs free energy	dt'	time-interval
R_c	Critical radius	V	Volume
ΔG_c	Change in the Gibbs free energy	V_n	Volume crystallized
n_i	Number of particles in the cluster	ω_n	Shape factor
W	Work	$\alpha_n(t)=(V_n(t)/V)$	Time-dependent crystallized fraction
W_c	Work of critical cluster formation	N_0	Number of supercritical clusters
τ	Relaxation time	n	Number of independent spatial directions
J	Rate of formation of supercritical clusters	a	Activity
J_s	Steady-state nucleation rate	T_b	Stokes-Einstein breakdown temperature
t	Time	T_g	Glass transition temperature
C_1	Parameter	T_K	Kauzmann temperature
n_c	Number of particles in a cluster of critical size	Φ	Catalytic activity factor of a heterogeneous nucleation core
τ_R	Maxwellian relaxation time	CRR	Cooperatively rearranging regions
η	Newtonian viscosity	TKS	kinetic spinodal temperature
G_S	Shear modulus	T_{gr}	Reduced glass transition temperatures
t_{ind}	Induction time	GFA	Glass-forming ability
dN	Change of number of clusters of critical size	GS	Glass stability
dt	Time interval	D_u	Diffusion coefficients from crystal growth rates
k_B	Boltzmann's constant	D_η	Diffusion coefficients from viscosity
D	Diffusion coefficient	T_d	Decoupling temperature
d_0	Diameter	CCR	Critical cooling rate

2. CRYSTAL NUCLEATION AND CLASSICAL NUCLEATION THEORY

Microscopy techniques are commonly used to determine crystal nucleation rates in supercooled liquids. The simplest and most used method is to perform heat treatments for different periods at a given temperature, then develop the (nanosized) crystal clusters at a higher temperature, and finally quench the specimens to room temperature. This is the so-called development or Tammann method. The treated samples are then polished for microscopic analysis, where the crystal numbers and sizes are measured. The so-called Tammann method (1898) is commonly used to determine the number density of supercritical nuclei. The number of crystals per unit volume is called crystal number density, N, which is measured at different nucleation times to envisage the variation of N versus time, t. The nucleation rate is the slope of the N vs. t plots. A broad range of nucleation can be determined by this method. For example, the maximum crystal nucleation rates for oxide glasses vary between 10^1 m$^{-3} \cdot$s^{-1} to 10^{17} m$^{-3} \cdot$s^{-1} (Fokin et al. 2006).

The physical nature of nucleation phenomena, in general, and crystal nucleation in supercooled liquids, in particular, was first described by J. W. Gibbs (1926). His basic idea can be illustrated by the free energy change, ΔG, during crystal cluster formation:

$$\Delta G = n\Delta\mu + \sigma A, \ \Delta\mu = \mu_{cr} - \mu_l, \ A = 4\pi R^2, \ n = \frac{4\pi}{3}C_\alpha R^3 \qquad (1)$$

In formulating Equation (1), it is assumed that spherical crystalline nuclei form in an initially homogeneous liquid. These nuclei are described by their radius, R, surface area, A, and the number, n, of particles (atoms, molecules or the basic structural units of the crystalline phase) they contain. In Equation (1), $\Delta\mu$ (<0) is the difference of the chemical potentials per particle in the liquid (μ_l) and the crystal (μ_{cr}), σ is the interfacial energy, and C_α is the particle number density in the crystal cluster. It is also assumed that the elastic strain energy—due to the density difference between the supercooled liquid and the crystalline solid—immediately decays due to liquid flow and does not interfere with the nucleation process. Within these assumptions, this equation is valid only for the simplest case of crystallization when the liquid and crystal have the same chemical composition. This kind of crystallization is denoted as *polymorphic* or *stoichiometric*. Furthermore, it is assumed that the properties of the crystal clusters are size-independent. Qualitatively, the situation does not change for *incongruent* crystallization when the crystal and liquid phases have different compositions.

According to thermodynamic evolution criteria, at constant pressure, P, and temperature, T, spontaneous macroscopic processes are tied to a decrease in the Gibbs free energy, G, of the system. For this reason, the function $\Delta G = \Delta G(R)$ takes on the shape shown in Figure 2a because the thermodynamic driving force for crystallization, $\Delta\mu$, is negative ($\mu_{cr} < \mu_l$). In this case, cluster formation and growth is accompanied by a decrease in the Gibbs free energy because the surface term in the expression for ΔG is positive. In other words, the tendency for decreasing the Gibbs free energy is counteracted by the surface term, i.e., the surface contribution initially leads to an increase in ΔG with increasing crystal size. Therefore, small crystal clusters formed in the system disappear and only clusters larger than a critical size, R_c, can grow to macroscopic dimensions. As demonstrated in Figure 2a, the critical cluster size is defined by the maximum of $\Delta G(R)$. Systems showing such behavior are denoted as metastable. Metastable states are stable with respect to small fluctuations (generating clusters with sizes $R < R_c$) but unstable with respect to larger fluctuations leading to clusters with sizes $R > R_c$. Thus, viable (supercritical) crystal clusters that are capable of deterministic growth must exceed a certain minimum size. This phenomenon was predicted more than 100 years ago and is now seen directly in molecular dynamics simulations, e.g. Prado et al. (2019). It is this criticality that determines the decisive impact of these embryos (sub-critical) and nuclei (supercritical) on the nucleation processes (Kashchiev 2000; Kelton and Greer 2010; Neuville et al. 2017).

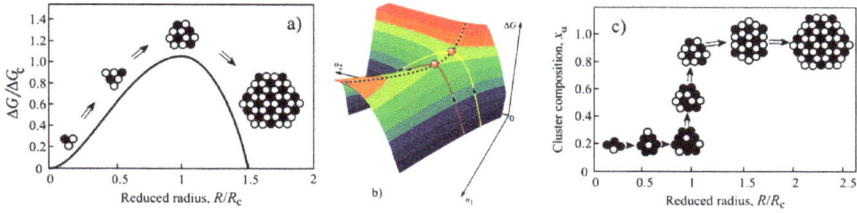

Figure 2. The classical model of nucleation and possible generalizations. **(a)** With only one parameter used to describe the state of the cluster. **(b)** Change of Gibbs free energy in cluster formation when more than one parameter is used. **(c)** An alternative view to the classical scenario of crystallization in multi-component liquids where both the size and composition change (Schmelzer and Schick 2012).

Taking the chemical potential difference and the interfacial energy as size-independent (i.e., employing the so-called "capillarity" approximation), one can derive the critical cluster size and the value of ΔG_c at the critical size from the extreme condition $\Delta G = 0$. These parameters are thus given by:

$$R_c = \frac{2\sigma}{c_\alpha \Delta\mu} \quad \Delta G_c = \frac{1}{3}\sigma A_c = \frac{16\pi}{3}\frac{\sigma^3}{(c_\alpha \Delta\mu)^2} \quad A_c = 4\pi R_c^2 \tag{2}$$

The concepts discussed above are illustrated in Figure 2a within the framework of the classical model of nucleation, whereby the change in the Gibbs free energy of cluster formation reaches a maximum $\Delta G = \Delta G_c$ for the critical cluster size, $R = R_c$. In this model, clusters grow or decay while preserving their properties, so that size is the only parameter specifying their state (Schmelzer and Schick 2012).

A more realistic picture of cluster formation is presented in Figure 2b, where not only the size but also the composition (described by the number of particles, n_i, of two components) of the cluster may change. In this case, the critical cluster corresponds to a saddle point of the Gibbs free energy surface. The evolution to the new phase via the saddle is shown by the red curve. Figure 2c shows an alternative to the classical picture, which is similar to the spinodal decomposition (cf. Gutzow and Schmelzer 2013). In this case, the composition of the crystal cluster changes when a nearly constant size is reached and only after completion of this process are the kinetics governed by the growth of clusters with a roughly constant composition. In several cases in multicomponent systems (Gutzow and Schmelzer 2013), the latter path of evolution (Fig. 2c)—and not the classical picture (Fig. 2a)—may dominate phase transformation.

Critical clusters form by stochastic thermal fluctuations. According to underlying assumptions of statistical physics, the probability of such fluctuations can be expressed as a function of the minimum work of a reversible thermodynamic process. The minimum work to form a critical cluster is $W_c = \Delta G_c$, where ΔG_c is given by Equation (2). This quantity, W_c, the work of critical cluster formation, plays a decisive role in nucleation theory. Then, after a certain time interval, τ (nucleation time-lag), the rate of nucleation, J (the number of supercritical clusters formed per unit time in a unit volume of the liquid), approaches a constant value, the steady-state nucleation rate, J_s. In an early description of this initial period of nucleation by Zeldovich (cf. Gutzow and Schmelzer 2013), the nucleation rate as a function of time, t, was expressed by the simplified relation

$$J(t) = J_s \exp\left(-\frac{\tau}{t}\right) \tag{3}$$

where τ is the (true) nucleation time-lag.

The initial stage of nucleation observed in experiments is often described by the Kashchiev relation (Kaschiev 2000),

$$N(t) = J_s \tau \left[\frac{t - t_0}{\tau} - \frac{\pi^2}{6} - 2 \sum_{m=1}^{\infty} \frac{(-1)^m}{m^2} \exp\left(-m^2 \frac{t - t_0}{\tau} \right) \right], t > t_0 \text{ and } N = 0, t \leq t_0 \qquad (4)$$

where t_0 is the so-called time shift, the observed shift between the N versus time curves experimentally obtained by the double-stage or development method and after single-stage treatments (when a powerful electron microscope is available). This mathematical equation gives a relation for the number, $N(t)$, of supercritical crystallites dependent on time, t. For longer times than the experimental induction time (t_{ind}) sometimes observed in $N(t)$ versus time plots, Equation (4) can be approximated by

$$N(t) \cong J_s \left(t - t_{ind-dev} \right) \qquad t_{ind-dev} = \frac{\pi^2}{6} \tau + t_0 \qquad (5)$$

We emphasize that there is a difference between the $t_{ind-dev}$ (observed after nucleation + development treatments) from the real nucleation induction time, $t_{ind-nucleation}$ that would be obtained from a single-stage treatment. This is exactly what Equations (4) and (5) assume and is illustrated schematically in Figure 3. In this figure, t_1 is the time when the first critical nucleus is formed.

Over a sufficiently long time, Equations (3–5) approach steady-state nucleation conditions, i.e., $(dN/dt) = J_s = \text{constant}$. With $W_c = \Delta G_c$, the steady-state nucleation rate, J_s, can be written as (Gutzow and Schmelzer 2013)

$$J_s = J_0 \exp\left(-\frac{\Delta G_c}{k_B T} \right) = J_0 \exp\left(-\frac{W_c}{k_B T} \right) = J_0 \sqrt{\frac{\sigma}{k_B T}} \left(\frac{D}{d_0^4} \right) \qquad (6)$$

where D is the effective diffusion coefficient controlling nucleation and d_0 is a size parameter or atomic jump distance. Experimental results that illustrate the establishment of a steady-state nucleation rate and its dependence on temperature are shown in Figure 4.

For the case shown in Figure 2a (congruent crystallization, assuming that the cluster properties do not change with size and are the same as those of the new macroscopic phase), D in Equation (6) is the diffusion coefficient of the structural building units in the liquid, and d_0 is their diameter. If several components of the liquid diffuse independently, D must be replaced by an effective diffusion coefficient, which is a combination of the partial diffusion coefficients and the concentrations of the different components in the liquid, and d_0 must be replaced by the average size of these independently moving species (Gutzow and Schmelzer 2013).

In applications of this theory, due to the scarcity of diffusion data for oxide glass-formers, it is often *assumed* that the diffusion coefficient can be replaced by the Newtonian shear viscosity, η, via the Stokes–Einstein–Eyring equation:

$$D \cong \frac{k_B T}{d_0 \eta} \qquad (7)$$

However, its applicability for temperatures around the glass-transition temperature (where homogeneous crystal nucleation is commonly observable) has been questioned even for stoichiometric systems, where decoupling of structural relaxation (expressed by viscosity) and atomic transport (represented by the diffusion coefficient) has often been reported in crystal growth experiments (e.g., Nascimento et al. 2011). However, to the best of our knowledge, **such decoupling has not yet been proved for nucleation processes,** and this is thus a relevant open problem.

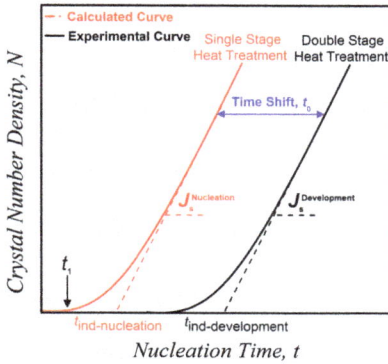

Figure 3. Schematic illustration showing the variation of crystal number density vs. nucleation time. Here, the difference between nucleation induction time in two conditions, viz. double stage heat treatment—laboratory nucleation experimentation—and single-stage heat treatment, which is normally obtained by calculations $J_s^{\text{development}} = J_s^{\text{Nucleation}}$

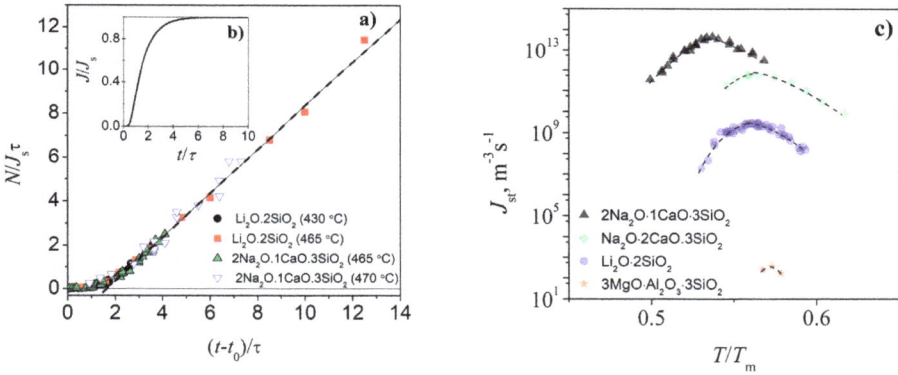

Figure 4. Experimental nucleation rate data for several silicate glasses. **(a)** Reduced crystal number density, $(N(t)/J_s\tau)$, versus reduced nucleation time, $((t-t_0)/\tau)$. The **solid line** is the master curve calculated from Equation (4). **(b)** Reduced nucleation rate versus reduced nucleation time calculated from Equation (4). **(c)** Experimental steady-state nucleation rate, J_s, versus reduced temperature, T/T_m, for four stoichiometric glasses. T_m is the melting temperature. [Reprinted from Fokin VM et al. (2006) Homogeneous crystal nucleation in silicate glasses: A 40 years perspective. J Non-Cryst Solids 352:2681–2714. Copyright (2006), with permission from Elsevier.]

The application of this expression is even more questionable for multicomponent systems. i.e., non-stoichiometric systems in which more than one phase crystallize upon heating (Fokin et al. 2019; Macena et al. 2020). Another issue is related to the case of highly viscous glass-forming melts, for which a non-Newtonian viscosity should be employed to describe viscous flow (Gutzow and Schmelzer 2013). Leaving aside the above-listed reservations, by using the Stokes–Einstein–Eyring relationship, the following expression results for the steady-state nucleation rate:

$$J_s = \frac{\sqrt{\sigma k_B T}}{d_0^5 \eta} \exp\left(-\frac{\Delta G_c}{k_B T}\right) \tag{8}$$

To apply Equation (8) to the interpretation of experimental data, one has to determine the work of critical cluster formation, $W_c = \Delta G_c$, i.e., to specify the thermodynamic driving force, $\Delta\mu$, and the interfacial energy, σ in Equation (2). Assuming that the properties of the crystalline clusters are the same as those of the isochemical macroscopic crystals, one arrives at the simplest approximation by a Taylor expansion of $\Delta\mu(T)$ in the vicinity of the melting temperature:

$$\Delta\mu(T) = \Delta h_\mathrm{m}\left(1 - \frac{T}{T_\mathrm{m}}\right) \tag{9}$$

where Δh_m is the enthalpy of melting per structural unit of the crystal and T_m is the melting temperature. This expression gives an upper bound for $\Delta\mu(T)$ because it neglects the difference in the specific heats of the crystal and SCL.

Since the interfacial energy of the critical nucleus is not directly measurable, it is normally evaluated using the Stefan–Skapski–Turnbull rule (Gutzow and Schmelzer 2013)

$$\sigma = \varsigma\frac{q_\mathrm{m}}{N_A^{1/3} v_\mathrm{m}^{2/3}}\ , \ q_\mathrm{m} = N_A \Delta h_\mathrm{m} \tag{10}$$

In Equation (10), q_m is the molar enthalpy of melting, N_A is Avogadro's number, v_m is the molar volume, and ς is a numerical factor equal to 0.4 to 0.6. By substituting these relations into Equation (8), its temperature dependence can be interpreted. The steady-state nucleation rate J_s is zero at $T = T_\mathrm{m}$, where $\Delta\mu = 0$, cf. Equation (9). Starting from the melting point, the nucleation rate increases with decreasing temperature because of the decrease in the work of critical cluster formation, given by Equation (2), until this trend is overcompensated by the exponential decrease of the diffusivity (increase of viscosity) and results in a maximum. For typical cases of homogeneous nucleation in oxide glasses, the maximum occurs at $T_\mathrm{max} \sim T_\mathrm{g}$, which corresponds to supercoolings of 0.5–0.6 T_m. (Fokin et al. 2006).

Homogeneous nucleation in supercooled glass-forming liquids. In most cases, using viscosity as a proxy to $D(T)$ and a constant, fitted value of σ, this classical approach gives a good temperature dependence, at least for the high-temperature side above T_max, but underestimates the steady-state nucleation rates by **20–55** orders of magnitude, e.g. Fokin et al. (2006). These huge deviations between experiment and theory can be resolved by the introduction of a size or temperature dependence of the interfacial energy, as discussed by Gibbs (1926) and later by others, particularly by Tolman, e.g. Gutzow and Schmelzer (2013). However, this solution does not solve other problems (Fokin et al. 2006), such as the alleged breakdown of the classical nucleation theory (CNT) for temperatures below T_max (Cassar et al. 2020). Another possible solution, resulting from computer simulations and density functional computations, consists of accounting for the size dependence, not only of the surface tension, but also of the other properties of the critical clusters. The internal properties of the clusters generally depend on their sizes. Hence, the surface properties, including the surface tension, must also be size-dependent. Thus, this approach also leads to a size dependence of the surface energy.

On the other hand, recent computer simulations favor the validity of CNT. Rather than using approximations or calculated values for the thermodynamic parameters and diffusivities, the CerTEV, São Carlos group employed parameters directly obtained from molecular dynamic simulations, without any fitting parameter, for different substances and demonstrated that the CNT is indeed a powerful predictor of crystal nucleation rates in some reluctant glass-formers, such as L-J, Ge (Tipeev et al. 2018, 2020) and ZnSe (Separdar et al. 2021). These findings should still be tested with regular glass-formers to generalize CNT's applicability.

With a thermodynamic (generalized Gibbs) approach, which treats the cluster properties as a function of size and degree of supercooling, one concludes that the classical theory—assuming that the clusters have macroscopic properties and employing the capillarity approximation for the interfacial energy—overestimates the work of critical cluster formation, and hence, underestimates the steady-state nucleation rates. Therefore, the classical theory with the capillarity approximation serves as a tool for roughly estimating temperature dependence of nucleation rates, but it must be significantly improved to account for the above-specified effects for a detailed and quantitatively accurate description of the phenomenon (Gutzow and Schmelzer 2013).

Heterogeneous Nucleation. So far, we have considered the case of crystal nuclei that form evenly within a defect-free homogeneous liquid. This mechanism is known as *homogeneous* nucleation. However, nucleation can be readily catalyzed by solid impurities, such as particles embedded in the volume or present on the external surface of glasses. Nucleation originating at such preferential sites is denoted as *heterogeneous* and can be described by the theoretical concepts outlined above if the work of critical cluster formation for homogeneous nucleation, W_c, is replaced by $W_c\Phi$. Here, $\Phi \leq 1$ is the nucleating activity of the heterogeneous nucleation core, and its value depends on the mechanism of catalysis. Heterogeneous nucleation dominates at small supercooling because of the lower work of critical cluster formation than that of homogeneous nucleation. At deep supercoolings, homogeneous nucleation may dominate due to much lower work of critical nucleus formation and the much larger number of sites (all "structural units" of the system) where homogeneous nucleation may proceed (Fokin et al. 2006).

One should note that, in certain cases, the evolution of the crystal phase may not proceed via the red saddle shown in Figure 2b, but via a ridge trajectory indicated by a yellow curve in Figure 2b, if such a trajectory is kinetically favored. This type of behavior may be expected to occur in crystallization occurring at deep supercoolings because of the defective and non-stoichiometric nature of the crystals that might precipitate in the early stages of crystallization (Fokin et al. 2003).

Frequently, several different metastable phases may be formed in the supercooled liquid. As Ostwald suggested many years ago, in such cases the most favorable stable phase is not formed immediately. Instead, the final stable phase frequently crystallizes through several stages in which different metastable phases are formed; this is the so-called Ostwald's rule of stages or Ostwald's step rule.

2.1. Recent findings that warrant further research: Examples of experimental tests

The alleged CNT breakdown at T_{max}. For a variety of oxide glass-forming liquids, the thermodynamic barrier for homogeneous crystal nucleation, W_c, apparently exhibits an unusual increase with decreasing temperatures below the experimental maximum nucleation rate, T_{max} (Fig. 5), ($T_{max} \sim T_g$) which is not compatible with predictions using the CNT. Abyzov et al. (2016) sought possible explanations for the increasing W_c by analyzing whether it could be caused by internal elastic stresses that arise due to density misfits between the crystal and liquid phases. Please recall that this factor was neglected in the derivation of Equation (6).

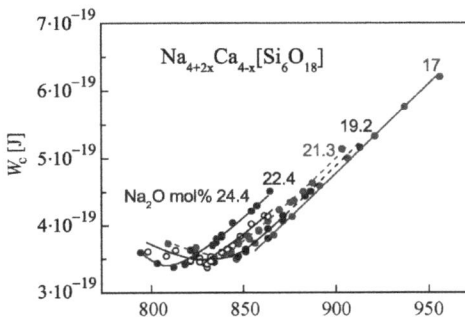

Figure 5. Calculated thermodynamic barrier for nucleation versus temperature for a series of sodium-calcium silicate glasses. W_c exhibits an unusual increase with decreasing temperatures below the maximum nucleation rate, T_{max}. [Reprinted from Abyzov AS et al. (2016) The effect of elastic stresses on the thermodynamic barrier for crystal nucleation. J Non-Cryst Solids 432:325–333. Copyright (2016), with permission from Elsevier.]

For this purpose, the crystal nucleation rate and induction time data for two glasses that display significantly different density misfits between the crystalline and liquid states, lithium and barium disilicates, were employed to determine the work of critical cluster formation, W_c. Quantitative estimates of the effect of the elastic strain energy on W_c were carried out for both glasses. The interplay between stress development and structural relaxation of the SCL was accounted for. Their computations were performed taking into account not only the possibility of precipitation of the most stable crystal phase, but also the fact that different metastable phases might form during the early stages of nucleation. They showed that elastic strain energy indeed reduces the thermodynamic driving force for crystallization, and thus increases the barrier to nucleation. To better illustrate the effect of elastic stress, Figure 6 shows the experimental nucleation rates and the nucleation rates calculated disregarding and accounting for the stresses, and their relaxation for the stable and metastable phases with different melting temperatures, T_m, shown close to the respective curves. In all these cases, the calculated maxima of the nucleation rate are located at temperatures that are lower than the experimental maximum. It is clear that accounting for the elastic strain energy component in the reduction of the thermodynamic driving force decreases the nucleation rates (compare dotted and solid lines). However, as shown by the solid lines and data points, the calculated nucleation rates do not reach the experimental values, and only approach them for lithium disilicate glass (L2S) if a metastable phase having a very low melting point, say 1107 K, appears. Nevertheless, it is important to underline that at 1107 K with any further decrease in temperature, this (invented) metastable phase would be poorly ordered and unstable. Therefore, it seems that the sole effect of elastic strain cannot explain the aforementioned unusual behavior of the thermodynamic barrier at T_{max}. **Hence, a comprehensive explanation for this phenomenon is still lacking** (Abyzov et al. 2016).

In another attempt to explain the break at T_{max}, Fokin et al. (2016) emphasized that the CNT fails to describe crystal nucleation rates in supercooled liquids if one uses a fixed size, d_o, of the "structural units." Some results for silicate glasses support the view that, even for the so-called

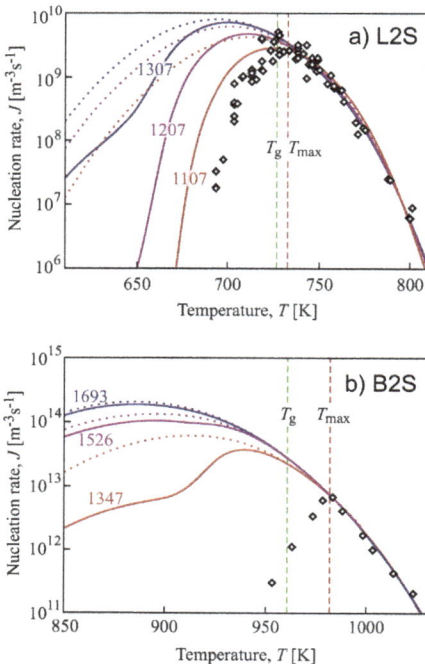

Figure 6. Nucleation rates for lithium disilicate (LS2) **(a)** and barium disilicate (B2S) **(b)** versus temperature. The symbols show measured values, the **dotted lines** show nucleation rates calculated without accounting for stresses. The **solid lines** show the rates taking into account the elastic stresses and their relaxation for the stable macroscopic phase (T_m=1307 °C for LS2 and T_m=1693 °C for B2S) and two metastable phases having different melting temperatures, T_m. [Reprinted from Abyzov AS et al. (2016) The effect of elastic stresses on the thermodynamic barrier for crystal nucleation. J Non-Cryst Solids 432:325–333. Copyright (2016), with permission from Elsevier.]

case of stoichiometric (polymorphic) crystallization, the nucleating phase may have a different composition and/or structure as compared to the parent glass and the evolving macroscopic crystalline phase. This finding perhaps explains the discrepancies between calculated (by CNT) and experimentally observed nucleation rates in deeply undercooled glass-forming liquids (Fokin et al. 2007). Therefore, to reconcile the experimental data and CNT, Fokin et al. (2016) assumed an abnormal increase in the size of the structural units that control nucleation with decreasing temperature for temperatures below T_{max}. This hypothesis was tested for several glass-forming liquids, where crystal formation proceeds by bulk homogeneous nucleation. This study could perhaps explain the temperature dependence of the nucleation rate in the range of $T < T_{max}$, where the description of the nucleation rate by CNT drastically fails. The size of the structural units could be related to the size of the cooperatively rearranging regions (CRR), which are linked to dynamic heterogeneities in glass-forming liquids.

Over the past few decades, a very important discovery in the study of glass-forming liquids was the finding of dynamic heterogeneity referring to the spatiotemporal fluctuations in local dynamics (Ediger 2000). The growth of the dynamic correlation length of the CRR as the temperature decreases toward the glass transition provides a possible approach to understand the dramatic slowdown of dynamics during vitrification (Flenner and Szamel 2010). Thus, more attention has been given to investigating the correlation between structural relaxation with dynamic heterogeneity and crystal nucleation in glass-forming liquids (e.g., Berthier 2011; Henritzi et al. 2015; Gupta et al. 2016). For example, Gupta et al. (2016) referred to the temperature at which the classical critical nucleus size is equal to the average size of the CRR in a supercooled liquid as a "cross-over" temperature. They showed, for the first time, using published nucleation rate, viscosity, and thermo-physical data that the cross-over temperature for the lithium disilicate melt is very close to the temperature corresponding to the maximum in the experimentally observed nucleation rates. They suggested that the abnormal decrease in nucleation rates below the cross-over temperature is most likely because, in this regime, the CRR size controls the critical nucleus size and the nucleation rate. This finding links, for the first time, measured nucleation kinetics to the dynamic heterogeneities in a supercooled liquid (Gupta et al. 2016).

A more straightforward explanation was proposed recently by Cassar et al. (2020). They analyzed literature data for 6 glasses using a rigorous protocol and indicated that the alleged breakdown at T_{max} is apparent only because most researchers are not patient to give long enough heat treatments to reach the steady-state regime. In other words, in most cases, incorrect nucleation rate data have been used to analyze the dynamics below T_{max}. **This problem warrants further investigation.**

I_{max} *versus* T_{gr}. Recently, Abyzov et al. (2018) employed the Classical Nucleation Theory using a characteristic value of the pre-exponential constant and an average (temperature-dependent) interfacial energy and derived an expression to estimate the maximum nucleation rates, I_{max}, as a function of the reduced glass transition temperatures, $T_{gr} \equiv T_g / T_m$ (T_g is the laboratory glass transition temperature and T_m is the melting point or *liquidus* temperature). The theoretical predictions were surprisingly good for 51 out of 54 silicate glass-formers tested and describe the experimental trend well that I_{max} strongly decreases with increasing T_{gr} (Fig. 7). This trend also explains the well-known fact that only silicate glasses having a relatively low T_{gr}, $T_{gr} < 0.6$, show internal homogeneous nucleation in laboratory time/sample-size scales (Abyzov et al. 2018).

Nucleation and the Kauzmann paradox. Zanotto and Cassar (2018) have tried to answer the key question whether any liquid can be cooled down below its melting point to the isentropic (Kauzmann) temperature, T_K, without vitrifying or crystallizing. This long-standing problem concerning the ultimate fate of supercooled liquids is one of the fundamental glitches in materials science. They used thermodynamic and kinetic data and well established theoretical models to estimate the T_{KS} (kinetic spinodal temperature, at which the average

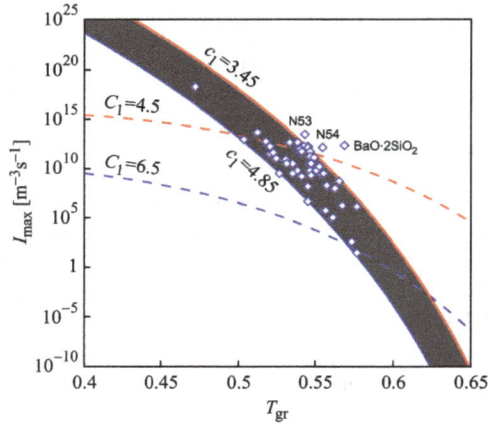

Figure 7. Maximum steady-state nucleation rate, I_{max}, versus reduced glass transition temperature, T_{gr}. The symbols refer to experimental data of 54 silicate glasses. The dashed lines were calculated using **constant** values of $C_1 = c_1 \kappa (T_{gr} - 0.556) = 4.5$ (**top dashed red line**) and $C_1 = 6.5$
(**bottom dashed blue line**). The **solid lines** were recalculated for temperature dependent C_1 us-
ing $T_{max}(T_{gr}) \approx 10^{24} \exp\left\{ -\dfrac{c_1 + \kappa(T_{gr} - 0.556)}{T_{gr}(1 - T_{gr})} \right\}$, where c_1 and κ are the parameters that best fit the current experimental data, e.g. for $c_1 = 3.45$ (**top solid red line**) and $c_1 = 4.85$ (**bottom solid blue line**).
The experimental errors in I_{max} and T_{gr} are of the order of the symbol size. [Reprinted from Abyzov AS et
al. (2018) Predicting homogeneous nucleation rates in silicate glass-formers. J Non-Cryst Solids 500:231–
234. Copyright (2018), with permission from Elsevier.]

time for the first critical crystalline nucleus to appear becomes equal to the average relaxation
time of a supercooled liquid) and the Kauzmann temperature for two substances—a borate
and a silicate glass—which show measurable homogeneous crystal nucleation in laboratory
time scales, as proxies of these families of glass-formers. For both materials, they found that
the T_{KS} are significantly higher than the predicted T_K. Therefore, at ambient pressure, at deep
supercoolings before approaching T_K, crystallization wins the race over structural relaxation.
Hence, the temperature of entropy catastrophe predicted by Kauzmann cannot be reached for
the studied substances; it is averted by incipient crystal nucleation (Zanotto and Cassar 2018).
As several approximations were made for the calculations and they required extrapolations to
very low temperature, this is still considered an **open relevant problem that warrants further
research**. Computer simulations could be particularly relevant to shed light into this problem.

3. BASIC MODELS OF CRYSTAL GROWTH IN SUPERCOOLED LIQUIDS

Microscopy techniques are commonly used to determine crystal growth, $U(T)$, rates in
supercooled liquids. The simplest and most used method is to perform heat treatments for
different periods at a given temperature and quench the material to room temperature. Samples
are then polished for microscopic analysis, where the crystal sizes on the sample surface or in
their interior are measured. Then, the growth rates are calculated from the slopes of the crystal
size radius versus time plots at different temperatures (Fig. 8). Hot stage microscopes can also
be employed, in which case the growing crystals on the sample surface are directly observed *in
situ*. Jiusti et al. (2020) have recently shown (Fig. 9) experimental values of crystal growth rates
for 20 stoichiometric oxide glass formers. They observed that the temperature of maximum
growth rate (T_{max}) lies within the range of 0.90 to 0.98T_L for all the materials investigated,
whereas the maximum crystal growth rate, $U(T_{max})$, varies seven orders of magnitude.
For the sake of simplicity, they used the average value, $T_{max} = 0.94T_L$ in their study for the
derivation of their GFA predictor (see Section 4).

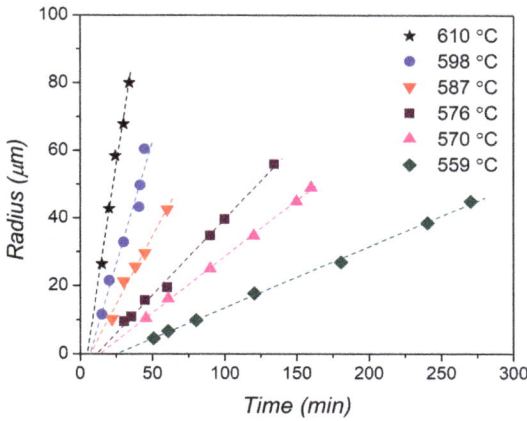

Figure 8. Increase in the crystal radius with time for lithium disilicate glass heat treated at different growth temperatures. [Reprinted from Deubener J et al. (1993) Induction time analysis of nucleation and crystal growth in di- and metasilicate glasses. J Non-Cryst Solids 500:231–234. Copyright (1993), with permission from Elsevier.]

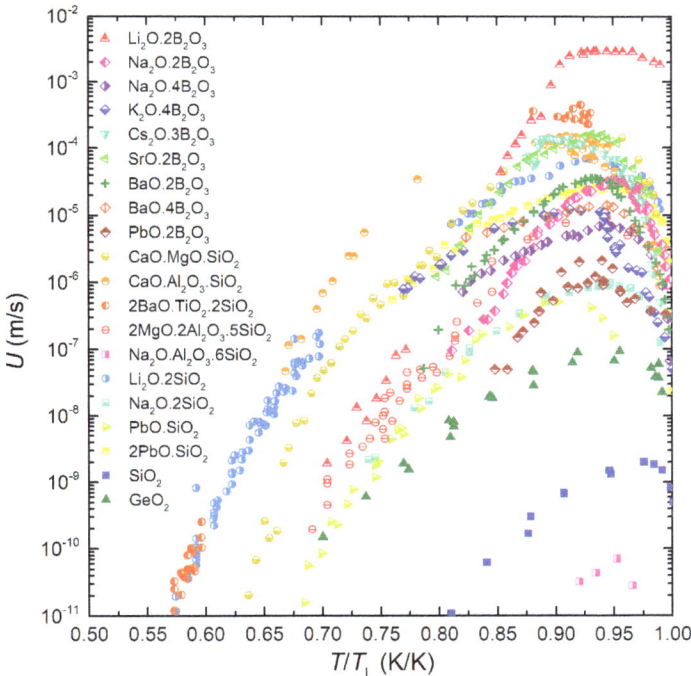

Figure 9. Crystal growth rate versus reduced temperature for 20 stoichiometric oxide glass formers. $U(T_{max})$ always occurs at a temperature between $0.90-0.98T_L$ (**highlighted region**) (Jiusti et al. 2020).

Recently, Reis et al. (2016) proposed and tested a simpler and yet accurate technique capable of determining the crystal growth rate over a fairly wide temperature range by means of a single differential scanning calorimetry (DSC) run. Their method was based on using 50–200 µm thick samples with parallel rough surfaces so that crystal growth is effectively unidirectional and the crystallization fronts have a constant area during the entire crystallization process.

Growth rates are calculated from the expression $U(T) = L \times q \times DSC(T)/A_{peak}$, where $DSC(T)$ is the value of the DSC crystallization curve at each temperature T, A_{peak} is the overall peak area, L is half the sample thickness, and q is the heating rate. This method was tested for different values of L and q for three glasses undergoing predominantly surface nucleation, which have distinctly different crystallization behaviors: stoichiometric lithium disilicate and diopside ($CaO \cdot MgO \cdot 2SiO_2$) and a nonstoichiometric lithium-calcium metasilicate. Growth rates spanning temperature intervals of more than 100 K, including temperature ranges where literature data were scarce due to experimental difficulties, were determined using a single DSC run. The resulting $U(T)$ data were compared with literature data obtained using optical microscopy. The growth rates determined using the proposed method showed excellent agreement with the published data for both stoichiometric glasses and only small errors for the nonstoichiometric glass (Reis et al. 2016; Zheng et al. 2019). Hence, one can use one of these techniques to obtain crystal growth rate curves.

It is known that the properties of the crystal-liquid interface have a decisive influence on the kinetics of crystallization. Theoretical treatments of crystal growth have therefore focused closely on the interfacial structure and its effect on crystallization. With the assumption of congruent polymorphic crystallization, three standard models have been developed for treating crystal growth theoretically (e.g., Uhlmann 1982; Jackson 2004). These models are described briefly below:

(i) *Normal growth.* The interface is pictured as rough at an atomic scale. Growth takes place at step sites, which represent a sizable fraction (0.5–1.0) of the interface. Assuming that this fraction does not change appreciably with temperature, the growth rate, $u(T)$, can be expressed as

$$u = f \frac{D}{4d_0} \left[1 - \exp\left(-\frac{\Delta\mu}{k_B T} \right) \right] \tag{11}$$

where f, fraction of preferred growth sites, is close to unity and $\Delta\mu$ is treated as a positive quantity.

(ii) *Screw dislocation growth.* This model assumes the interface is smooth but imperfect at an atomic scale. Growth takes place at a few step sites provided by screw dislocations that intersect the interface. The growth rate is still given by Equation (11), where f is now the fraction of preferred growth sites (on the dislocation ledges) at the interface. In this case, f is given approximately by $f \approx (T_m - T)/2\pi T_m$ (Nascimento et al. 2011). More generally, according to Jackson (2004), $f = (\Delta s_m/k_B)\xi$ holds, where Δs_m is the entropy of fusion per particle, and ξ is the number of nearest-neighbor sites in a layer parallel to the surface divided by the total number of nearest-neighbor sites. Factor is the largest for the most closely-packed planes of the crystal, for which it is approximately equal to 0.5.

For $f < 2$, the minimum free energy configuration corresponds to half the available sites being filled and represents an atomically rough surface. In contrast, for $f > 2$, the lowest free energy configuration corresponds to a surface where few sites are filled, and a few units are missing from the completed layer, and which represents an atomically smooth interface. Hence, for materials with $\Delta s_m < 2k_B$, the most closely packed interface planes should be rough. For materials with $\Delta s_m < 4k_B$, the most closely-packed surfaces should be smooth, the less tightly packed surfaces rough, and the growth anisotropy rate large.

(iii) *Surface nucleation or two-dimensional growth.* According to this model, the interface is smooth and perfect at an atomic scale, and thus free of intersecting screw dislocations and growth sites. Growth then takes place by the formation and growth of new two-dimensional nuclei at the interface. In this case, the growth rate is expressed by

$$u = C_3 \frac{D}{4d_0^2} \exp\left(-\frac{C_2}{T\Delta T}\right), \tag{12}$$

where C_2 and C_3 are parameters that determine the time required for the formation of the two-dimensional nucleus relative to that required for its propagation across the interface, respectively.

Possible growth modes are illustrated in Figure 10. Similarly to nucleation, the interplay between increasing driving force for crystallization, $\Delta\mu$, and decreasing diffusion coefficient (or increase in viscosity) with decreasing temperature results in a maximum of the crystal growth rates. This maximum is located at higher temperatures than that of the maximum of the steady-state nucleation rate shown in Figure 4c.

Other growth modes exist which are limited not by processes at the liquid-crystal interface but by atomic transport towards the interface. A specific example is a diffusion-limited segregation, which is of particular importance in multicomponent systems. Accounting for size effects on the growth kinetics, the rate for such a growth mode can be expressed as (e.g., Slezov 1999; Jackson 2004).

$$\frac{dR}{dt} = \frac{B}{R}\left(\frac{1}{R_c} - \frac{1}{R}\right) \tag{13}$$

where B is a combination of parameters describing the liquid under consideration, which are proportional to the effective diffusion coefficient governing the rate of supply of the different components to the growing or dissolving cluster.

Equation (13) and its modifications for other growth modes serve as a basis for the theoretical description of the competitive growth of clusters denoted as coarsening or Ostwald ripening. In these late stages of phase formation, larger clusters may continue to grow only when subcritical crystals are dissolved. The theoretical description of this process was first developed by Lifshitz and Slezov (cf. Jackson 2004). Today it is often referred to as the

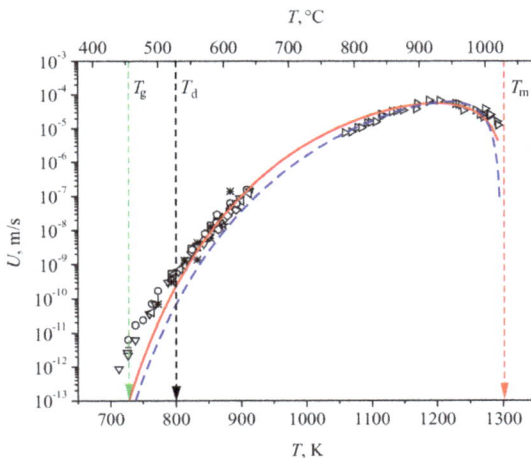

Figure 10. Crystal growth rates for $Li_2O \cdot 2SiO_2$ glasses obtained by different authors. The lines correspond to the screw dislocation mechanism (**full curve**) and two-dimensional surface nucleated growth (**dashed curve**). (T_d: decoupling temperature, T_g: glass transition temperature, T_m: melting temperature). [Reprinted from Nascimento et al. (2011) Dynamic processes in a silicate liquid from above to below the glass transition. J Chem Phys 135:194703, with the permission of AIP Publishing.]

L(ifshitz)S(lezov)W(agner) theory. This theory provides expressions for the average size, R, and the number, N, of supercritical clusters in the system as a function of time. For diffusion-limited growth (Eqn. 13), one obtains

$$R^3 \propto t, N \propto \frac{1}{t} \tag{14}$$

An account of the effect of elastic stresses on coarsening, which leads to qualitative modifications of the coarsening behavior, is reviewed in (Slezov 1999).

3.1 Experimental tests

Nascimento and Zanotto (2006) have analyzed extensive literature data on crystal growth rate and viscosity in the temperature range between $1.1T_g$ (glass transition temperature) and the melting point of silica (SiO_2). They selected $U(T)$ and $\eta(T)$ data for the same silica glass type, having similar impurity contents, and confirmed that the *normal growth* model describes the experimental $U(T)$ data quite well in this wide undercooling range. They then calculated effective diffusion coefficients from crystal growth rate, D_U, and from viscosity, D_η (through the Stokes–Einstein/Eyring equation) and compared these two independent diffusivities with directly measured self-diffusion coefficients of silicon and oxygen in the same silica glass type. Their results showed that silicon (not oxygen) controls the diffusion dynamics involved in both crystal growth and viscous flow in undercooled silica. This study not only unveiled the transport mechanism in this important glass-forming material but also validated the use of (easily measured) viscosity to account for the unknown transport term of the crystal growth expression in a wide range of undercoolings (Nascimento and Zanotto 2006).

Later on, Nascimento and Zanotto (2010) analyzed the kinetic coefficient of crystal growth, $U_{kin} \sim \eta^{-\omega}$, proposed by Ediger (2008), which indicated that the Stokes–Einstein/Eyring (SE/E) equation does not describe the diffusion process controlling crystal growth rates in *fragile* glass-forming liquids. U_{kin} was defined by Ediger (2008), using the *normal* growth model and tested for crystal data for inorganic and organic liquids covering a viscosity range of about 10^4–10^{12} Pa.s. Afterwards, Nascimento and Zanotto 2010 revisited their finding considering two other models: the *screw dislocation (SD)* and the *two-dimensional surface nucleated (2D)* growth models for nine undercooled oxide liquids, in a wider temperature range, from slightly below the melting point down to the glass transition region T_g, thus covering a wider viscosity range: 10^1–10^{13} Pa.s. Then, they normalized the kinetic coefficient (D_U, which scales with $\eta^{-\omega}$, and the exponent ω supposedly depends systematically on the fragility of the liquid: the greater the fragility, the lower the value of ω) for the SD and 2D growth models. These recalculated kinetic coefficients restored the ability of viscosity to describe the transport part of crystal growth rates ($D_U \sim 1/\eta$, $\omega \sim 1$) from low to moderate viscosities ($\eta < 10^6$ Pas), and thus demonstrated that the SE/E equation indeed worked well in this viscosity range for all systems tested. For strong glasses, the SE/E equation described low to high viscosities, from the melting point down to T_g. However, for at least three fragile liquids, diopside (for $T_d = 1.08T_g$, $\eta = 1.6 \times 10^8$ Pas), lead metasilicate (at $1.14T_g$, $\eta = 4.3 \times 10^6$ Pas), and lithium disilicate (at $1.11T_g$, $\eta = 1.6 \times 10^8$ Pas), there were clear signs of a *breakdown* of the SE/E equation at these viscosities. Nascimento and Zanotto demonstrated that viscosity data cannot be used to describe the transport part of the crystal growth (via the SE/E equation) in fragile glasses in the neighborhood of T_g.

In 2015, Schmelzer et al. (2015), derived at a relationship that allows a correlation of the decoupling temperature with the glass transition temperature and the liquid's fragility. The results were confirmed by experimental data. More recently, Cassar et al. (2017) suggested that above the temperature range $1.1T_g$–$1.3T_g$, crystal growth and viscous flow are controlled by the diffusion of silicon and lead in lead metasilicate glass. Below this temperature, crystal growth and viscous flow are more sluggish than the diffusion of silicon and lead. Therefore, T_d marks the temperature where *decoupling* between the (measured) cationic diffusivity and the effective diffusivities calculated from viscosity and crystal growth rates occurs. These authors reasonably proposed that the nature or size of the diffusional entities controlling viscous flow and crystal growth below T_d is quite *different*; the slowest is the one controlling viscous flow, but *both* processes require cooperative movements of some larger structural units rather than jumps of only one or a few isolated atoms (Cassar et al. 2017).

Finally, Cassar et al. (2018a) tested and analyzed 4 different approaches to compute D_U. The classical approach ($D_U \sim \eta^{-1}$) and the fractional viscosity approach of Ediger ($D_U \sim \eta^{-\omega}$) were not able to describe the crystal growth rates near the glass transition temperature of supercooled diopside liquid ($CaMgSi_2O_6$). However, the proposed Arrhenian expression to calculate D_U—gradually changing from a viscosity-controlled to an Arrhenian-controlled process—was able to describe the available data in the whole temperature range and yielded the lowest uncertainty for the adjustable parameters. Their results corroborated the previous finding that viscous flow ceases to control the crystal growth process below the decoupling temperature. Figure 11 shows the overall results of this new approach, i.e., the regression of crystal growth rate data when D_U is calculated following the considerations that D_U gradually changes from viscosity-controlled to Arrhenian-controlled (see Eqn. 15). All available data are well described by the regression. The four adjustable parameters needed for this approach: σ, E_a, D_0, and T_d (see Eqn. 15). The regression yielded $\sigma = 0.223(9)$ J/m², $E_a = 650(50)$ kJ/mol, $\ln(D_0) = 31(6)$, and $T_d = 1100(30)$ K, with RMS = 0.11 (D_0 in m²/s) (Cassar et al. 2018a).

$$D_U = [x]\left[C\frac{T}{\eta_{eq}}\right] + [1-x]\left[D_0 \exp\left(-\frac{E_a}{RT}\right)\right] \tag{15}$$

where $x = \dfrac{1+\tanh(\alpha)}{2}$ (hyperbolic tangent function), $\alpha = \dfrac{T-T_d}{\phi T_d}$ ($\phi = 0.06$). Please note that $C = \dfrac{k_B}{d}$, k_B is the Boltzmann constant and d is the diameter of the moving entity that controls viscous flow, usually assumed to be equal to d_0 (jumping distance of a moving entity).

The above relationships, theory, and fundamentals allow one to describe the growth of crystals with smooth planar or spherical interfaces advancing in the liquid. However, more complex growth patterns do exist, and more complex models of growth are required to properly

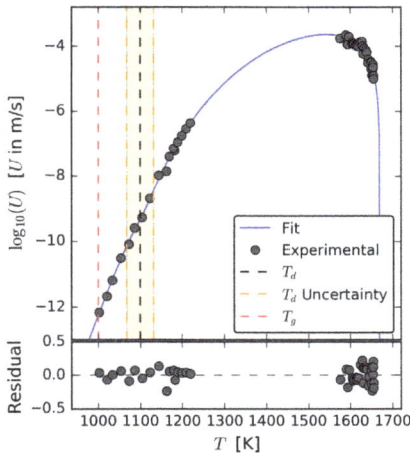

Figure 11. Regression curve of crystal growth rate data with D_U considering Arrhenian-controlled diffusion below T_d and Equation (15) (Cassar et al. 2018a). (T_d: decoupling temperature, T_g: glass transition temperature)

take into account possible interfacial instabilities, surface roughening, or other growth modes such as diffusion-limited aggregation (Jackson 2004). With such complex growth modes, a variety of intricate and beautiful crystal shapes may evolve, some of which are illustrated in Figure 12.

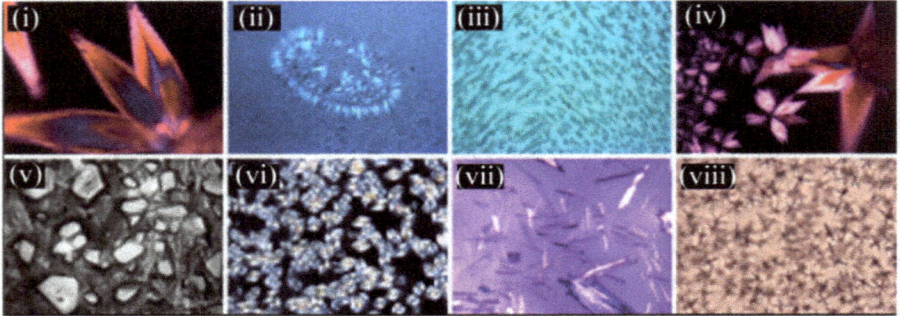

Figure 12. Crystal morphologies formed by nucleation and growth in oxide glass-formers as observed by optical microscopy (crystal sizes from 5 to 100 μm). From top left to bottom right: **(i, ii, iv)** LS crystals nucleated on defects of a $CaO \cdot Li_2O \cdot SiO_2$ glass surface during its preparation via melting–cooling **(iii)** Crystallization propagating from the surface towards the center of a $CaO \cdot Li_2O \cdot SiO_2$ glass specimen; lithium metasilicate crystals nucleated on two perpendicular surfaces and grew towards the sample center. **(v)** Surface of a $CaO \cdot Li_2O \cdot SiO_2$ glass sample after cooling a melt in a DSC furnace; the large-faceted and needle-like crystals are calcium and lithium metasilicates, respectively. **(vi)** Internal crystallization in a Ti-cordierite glass; pure stoichiometric cordierite ($2MgO \cdot 2Al_2O_3 \cdot 5SiO_2$) glass underwent only surface nucleation, but the same glass doped with more than 6 mol% TiO_2 shows internal crystallization of μ-cordierite. **(vii)** Needle-like crystals in $CaO \cdot Li_2O \cdot SiO_2$ eutectic glass formed by internal crystallization in the temperature range between the *solidus* and the *liquidus;* these wollastonite crystals appear on the cooling path. **(viii)** Star-like NaF crystals inside a photo-thermo-refractive (PTR) glass (treatment at a high temperature near the solubility limit).

4. OVERALL CRYSTALLIZATION AND GLASS-FORMING ABILITY: THE JOHNSON–MEHL–AVRAMI–KOLMOGOROV APPROACH

Crystallization of supercooled liquids occurs by a combination of crystal nucleation and growth. The kinetics of such processes is usually described by a theory independently derived between 1937 and 1941 by Johnson, Mehl, Avrami, and Kolmogorov (Kolmogorov 1937; Johnson and Mehl 1939; Avrami 1939, 1940, 1941) denominated (JMAK theory). In this approach, the isothermal evolution of the total amount of the crystalline phase is described as a function of time, accounting simultaneously for nucleation and growth. The basic equations of this approach can be developed as follows.

Let us assume that, in a time interval $dt'(t', t'+dt')$, a number $dN(t') = J(t')[V - V_n(t')]$ of clusters of critical size is formed in the volume $[V - V_n(t')]$. Here, V is the initial volume of the glass-forming melt and $V_n(t')$ the volume already crystallized at time t'. These clusters grow and, at time t, occupy a volume

$$dV_n(t,t') = \omega_n J(t')(V - V_n(t'))dt'\left(\int_{t'}^{t} u(t'')dt''\right)^n \tag{16}$$

where is a shape factor and the integral term describes the growth of the $dN(t')$ clusters formed at t' until time t, i.e., in the time interval $(t - t')$, the exponent n is the number of independent spatial directions of growth. Introducing the ratio, $\alpha_n(t) = (V_n(t)/V)$, between the current volume of the crystalline phase versus the initial volume of the glass-forming melt, one has

$$\alpha_n(t) = 1 - \exp\left(-\frac{\omega_n}{(n+1)} J u^n t^{(n+1)}\right) \tag{17}$$

Integration, i.e., taking the sum over all the time intervals dt' in the range of $(0, t)$, yields

$$\alpha_n(t) = 1 - \exp\left(\omega_n \int_0^t J(t')dt' \left(\int_{t'}^t u(t'')dt''\right)^n\right) \tag{18}$$

Provided the nucleation and growth rates are both constant; one reaches as a special case

$$\alpha_n(t) = 1 - \exp\left(-\frac{\omega_n}{(n+1)} J u^n t^{(n+1)}\right) \tag{19}$$

Conversely, if a number N_0 of supercritical clusters is formed immediately at time $t=0$, growing in n independent spatial directions, one arrives instead at

$$\alpha_n(t) = 1 - \exp\left(-g N_0 u^n t^n\right) \tag{20}$$

The analysis of the time dependence of the α_n-curves thus leads to the indirect determination of nucleation and growth kinetics.

The JMAK theory has been employed in numerous studies to analyze experimental data and determine the degree of crystallinity as a function of time in both isothermal and non-isothermal heat treatments of glasses. Emphasis has usually been given to the determination of the so-called Avrami coefficient $m = n+1$ obtained from the slopes of experimental $\ln[\ln(1-\alpha)^{-1}]$ *versus* $\ln(t)$ plots. An overview of various nucleation and growth mechanisms and the resulting values of the Avrami coefficient are given in Table 1 (Zheng et al. 2019). However, there is some uncertainty in such analyses, because different combinations of nucleation and growth laws may lead to the same Avrami coefficient. For this reason, a separate investigation of the growth kinetics may be required to reach definite conclusions (Johnson and Mehl 1939; Gutzow and Schmelzer 2013).

It is important to underline that the JMAK theory, as given by Equations (19) and (20), does not apply directly to non-isothermal processes. These two equations are derived on the assumption of constant nucleation and growth rates, which does not hold in non-isothermal processes. Therefore, in non-isothermal cases, the general relationships, Equations (17) and (18) must be employed to describe overall crystallization.

Such considerations must also be taken into account when the JMAK formalism is employed to determine whether a liquid will transform into a glass upon cooling or whether it will crystallize. Following Uhlmann (1982), one can consider a supercooled frozen-in liquid a glass if, after vitrification, the volume fraction of the crystal phase does not exceed a certain value of, say, 10^{-6} (the detection limit by microscopy). Using appropriate expressions for nucleation and growth rates, one can then compute (through Eqn. 19 for isothermal conditions) the time required to reach the volume fractions thus defined. In this way, one arrives at the so-called T(ime)T(emperature)T(ransformation)-curves (*TTT*-curves) exemplified in Figure 13 (cf. also Kelton and Greer 2010 and Fig. 10.8 in Gutzow and Schmelzer 2013). These curves give some insight into the characteristic time scales required to prevent measurable crystallization effects. One should keep in mind, however, that these curves overestimate the critical cooling rates for glass formation by about one order of magnitude because, as mentioned earlier, crystallization upon cooling proceeds under non-isothermal conditions.

Table 1. Values of the Avrami exponent (n) for several crystallization mechanisms (Zheng et al. 2019).

Polymorphic change, interface-controlled growth	n	Diffusion-controlled growth	n
Increasing nucleation rate, 3D	> 4	Increasing nucleation rate, 3D	> 2.5
Constant nucleation rate, 3D	4	Constant nucleation rate, 3D	2.5
Decreasing nucleation rate, 3D	3–4	Decreasing nucleation rate, 3D	1.5–2.5
Zero nucleation rate (nucleation site saturation) 3D	3	Constant nucleation rate, 2D	2
Constant nucleation rate, 2-D (plates)	3	Zero nucleation rate, 3D	1.5
Zero nucleation rate, 2-D (plates)	2	Constant nucleation rate, 1D	1.5
Constant nucleation rate (nucleation site saturation), 1D	2	Zero nucleation rate, 2D	1
Zero nucleation rate, 1D (needles)	1	Zero nucleation rate, 1D	0.5

Note: D: growth dimensionality

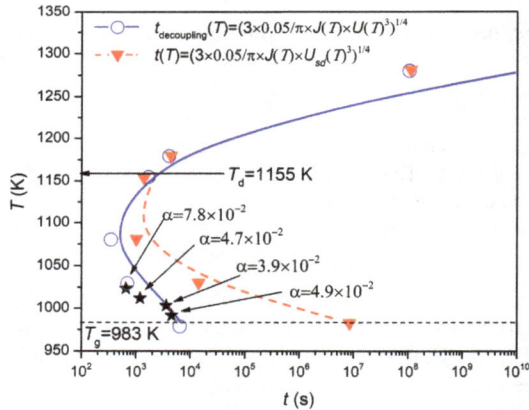

Figure 13. Simulated *TTT*-curves for a BaO·2TiO$_2$·2SiO$_2$ glass with crystallized volume fraction $\alpha = 0.05$ using, in one approach, the screw dislocation growth model both above and below T_d (screw dislocation–**dashed line, triangles**), and in the other the Arrhenius equation below T_d (**solid line, spheres**). Experimental data points (**black stars**) obtained at 993, 1003, 1013, and 1023 K. (T_d: decoupling temperature). [Reprinted from Rodrigues and Zanotto (2012) Evaluation of the guided random parametrization method for critical cooling rate calculations. J Non-Cryst Solids 358:2626–34 Copyright (2012), with permission from Elsevier.]

Using experimental crystal nucleation and growth rate data, Rodrigues and Zanotto (2012) calculated *TTT*-curves for different isothermal and non-isothermal crystallization situations. They also accounted for the breakdown of the Stokes–Einstein–Eyring (SEE) equation at a temperature T_b (somewhat higher than T_g) where the effective diffusion coefficient that controls crystal growth decouples from the value of diffusivity calculated by the SEE equation (Eqn. 7). In Figure 13, we show an example of such a curve for a stoichiometric BaO·2TiO$_2$·2SiO$_2$ glass, which undergoes copious internal homogenous crystal nucleation. The agreement with experimental data (which, in this case, were also obtained in isothermal conditions) is quite impressive, indicating that the JMAK equation is accurate if all the assumptions involved in its derivation are met. From such curves, one can calculate very important properties of supercooled liquids—namely the glass-forming ability, and the glass stability against crystallization on heating.

Glass forming ability (GFA) is the propensity of a melt to vitrify upon cooling. Quantifying glass-forming ability is of utmost importance for the design of new glass compositions, but it is a laborious, time-consuming process. Several methods are available to estimate the GFA (Shelby 2005). The ease of glass formation can be defined by the critical cooling rate (CCR) required to prevent crystallization of a certain sample. The critical cooling rate is a quantitative measure of the ability of a liquid to vitrify and is defined as the slowest rate at which a melt can be cooled from its liquidus temperature (T_m) to below the glass transition temperature (T_g)—which is composition and thermal history dependent—without "detectable" crystallization, i.e., a crystallized volume fraction normally assumed to be in the range of 10^{-6}–10^{-2} (such as α in Fig. 13). The smaller the CCR, the greater the glass forming ability of a liquid. Thus, CCR is a very important characteristic parameter of liquids that should be known to predict the ease or difficulty of glass formation, and, hence, to determine the processing conditions of any glass (Scherer 1991). For instance, the CCR of metallic glasses lie in between 10^6 to 10^1 K/s, whereas commercial oxide glasses, such as window glass, have a CCR of < 10^{-2} K/s.

Figure 14 shows the TTT-curves of two glass-forming melts whose internal (homogeneous) nucleation rates differ by 15 orders of magnitude. Anorthite glass shows a maximum nucleation rate of approximately 10^2 $m^{-3}.s^{-1}$, whereas fresnoite has a maximum of 10^{17} $m^{-3}.s^{-1}$. Even for a small density of surface nucleation sites (N_s), the nose of the TTT curves for heterogeneous surface nucleation leads to shorter times compared to the nose of the homogeneous nucleation TTT curves. The equivalent N_S for a heterogeneous TTT curve to exhibit the same CCR as the homogeneous case would be around 10^{-2} and 10^{-1} sites/m² for anorthite and fresnoite glasses, respectively.

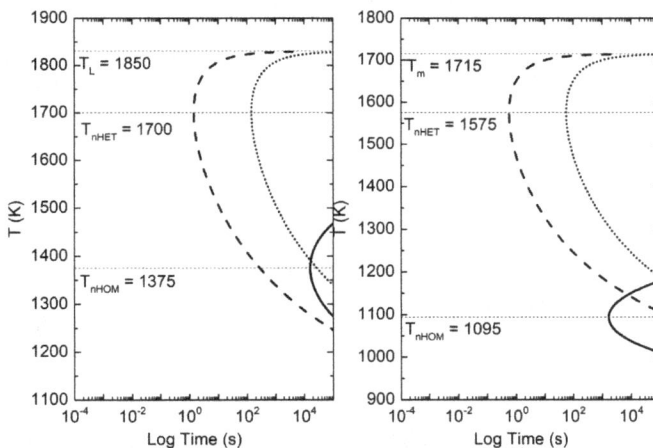

Figure 14. Temperature *versus* time in log scale: TTT curves for anorthite (CaO.Al$_2$O$_3$.2SiO$_2$) (**left**), and fresnoite (2BaO.TiO$_2$.2SiO$_2$) (**right**) considering homogeneous nucleation (——) and heterogeneous surface nucleation for $N_S = 10^5$ (- - -) and $N_S = 10^1$ (·····). The nose times seem to be close to each other, however the time scale is logarithmic. (Jiusti et al. 2020)

Above all, even more important than measuring or estimating the GFA of different substances is the ability to predict the GFA as a function of composition. While this is not yet possible for complex, multicomponent systems, research efforts should and are moving into this direction (Varshneya and Mauro 2019; Zheng et al. 2019). For example, Jiusti et al. (2020) have recently derived a mother parameter, GFA = 1/CCR $\propto [U(T_{max}) \times T_L]^{-1}$, which strongly correlates with the experimental critical cooling rates of oxide glass-formers. A simplified version derived from the mother parameter—which does not need (scarce) crystal growth rate data and only relies on (easily measurable or calculable) viscosity, η, and T_L—GFA $\propto [\eta(T_L)/T_L^2]$—also correlates well with the CCR of several oxide compositions. This new GFA parameter corroborates the widespread concept that substances having high viscosity at T_L, and a low T_L can be easily vitrified, and provides a powerful tool for the quest and design of novel glasses.

4.1 Glass stability against crystallization

While GFA is defined as the resistance to crystallization of a supercooled liquid during cooling, *glass stability* is defined as the resistance to devitrification of a glass or supercoooled liquid during heating. Glass forming ability is most important during processes requiring vitrification, while glass stability (GS) is very important during operations involving thermal treatment of an existing glass, such as annealing, tempering or treatment for ceramization. Although these two properties are not identical, they are frequently confused in the literature and technological practice. It is often (reasonably) assumed that poor glass-forming ability automatically leads to poor glass stability, and vice-versa (Shelby 2005). GS is frequently characterized by the difference between the onset of the glass transformation region (T_g) and the DSC crystallization peak (T_c) or onset of crystallization peak (T_x) for a sample heated at a particular rate. A drawback of this method is that these two temperatures depend not only on the chemical composition but also on glass particle size and heating rate; hence these two experimental parameters should be kept constant for a proper evaluation of glass stability. Some authors argued that the quantity (T_c-T_g) should be normalized by T_g, T_c, T_x, or T_m of the crystalline phase, to compare the behavior of glasses which crystallize in very different temperature ranges. The most well-known parameter is the Hrubý number, which is defined by $H_b = (T_x-T_g)/(T_c-T_g)$. Stable glasses have a $H_b > 0.4$.

However, there is no unanimously accepted criterion for glass stability. As long as samples of different compositions are compared using identical characteristics (particle size and heating rate), several of the proposed parameters (e.g., Hrubý), especially those containing all the three characteristic temperatures, yield similar results. Interested readers are encouraged to refer to a comprehensive review by Zheng et al. (2019), which compares various GS parameters determined by DSC.

5. PERSPECTIVES

Significant advances in the understanding and control of crystal nucleation and growth processes in glass-forming liquids have been achieved over the last five decades. It is now well-established that all materials can vitrify when subjected to sufficiently fast cooling from the liquid state. Thus, novel reluctant glass-forming materials, such as certain metallic and chalcogenide glasses, with unusual properties, have been successfully obtained by very fast quenching. Moreover, controlled, catalyzed internal crystallization of specific glasses has led to a variety of advanced glass-ceramics that are now available on the market. More profound insights into glass crystallization processes, such as precise prediction of nucleation and growth rates, and critical cooling rates for glass formation, based solely on materials properties, will depend critically on new developments in nucleation and growth theories and computer simulations. Artificial Intelligence techniques could be a major player in this context (Cassar et al. 2018b, 2021; Alcobaça et al. 2020; Montazerian et al. 2020).

Furthermore, in their recent comprehensive review, Zheng et al. (2019) elaborated on the versatility and utility of several differential scanning calorimetry (DSC) techniques for examining the dynamics related to nucleation, growth, glass-forming ability and stability.

For example, DSC is very useful for providing estimates of the temperature range where significant nucleation occurs. When properly used, isothermal DSC runs can yield useful information regarding crystallization processes, including the crystal number density, nucleation and growth kinetics, the activation energy for overall crystallization, and the Avrami constant. In addition to DSC, advanced instruments, such as Transmission Electron Microscopy (TEM), Anomalous Small Angle X-ray Scattering (ASAXS), Small-angle Neutron Scattering (SANS), X-ray Absorption Spectroscopy (XAS), Raman Spectroscopy (RS), Nuclear Magnetic Resonance (NMR), Advanced Optical Spectroscopy (OS) and others have been recently employed to study nuclei of critical sizes and medium range order in glasses. They provide critical insight into the complicated and rapidly changing environments in which crystallization happens, helping us to shed light over nucleation and crystallization processes in glass-forming materials. Interested readers are referred to a recent book authored by Neuville and coworkers (Neuville et al. 2017).

Despite the many advances achieved in understanding crystallization, some key problems remain open. Among the most notable, we remark the following:

(i)   Specification of the bulk (structure, composition, density) and surface properties of the critical nuclei as a function of size.

(ii) Description of the temperature dependence of the crystal nucleus–liquid interfacial energy.

(iii) The applicability of the Stokes–Einstein–Eyring (viscosity) relationship in calculating the effective diffusion coefficients that control crystal nucleation.

(iv) Unveiling the cause of the reported breakdown of the CNT in describing the temperature dependence of experimental nucleation rates below T_g.

(v)   A deeper understanding of the relationship between the molecular structure of glass-forming melts and the nucleation mechanism.

(vi) The relation between the sizes of supercritical nuclei vis-à-vis the sizes of co-operatively rearranging regions (CRR) of heterogeneous dynamics (DHD) existing in the structure of viscous liquids

(vii) Comparison of the estimated (by extrapolation) structural relaxation time and the characteristic time for crystallization of glass-forming liquids at the (predicted) Kauzmann temperature, T_K. Such a comparison could resolve the paradox, following Kauzmann's suggestion of the possibility that the putative state of negative entropy may never be reached because at such temperatures, crystallization would always intervene before structural relaxation of any SCL.

All these problems, in addition to several others not mentioned here, such as the development of novel glasses and glass-ceramics, having exotic, unusual compositions and combination of properties, corroborate that glass crystallization is a very dynamic, exciting research topic.

ACKNOWLEDGMENTS

EDZ is indebted to his numerous co-workers and students for the enjoyable and educative joint research on glass crystallization over the past 45 years. Generous and continuous funding by the Brazilian agencies CAPES, CNPq, and São Paulo Research Foundation, FAPESP (CEPID grant #2013/07793-6 and #2015/13314-9) is much appreciated.

REFERENCES

Abyzov AS, Fokin VM, Rodrigues AM, Zanotto ED, Schmelzer JWP (2016) The effect of elastic stresses on the thermodynamic barrier for crystal nucleation. J Non-Cryst Solids 432:325–333

Abyzov AS, Fokin VM, Zanotto ED (2018) Predicting homogeneous nucleation rates in silicate glass-formers. J Non-Cryst Solids 500:231–234

Alcobaça E, Mastelini SM, Botari T, Pimentel BA, Cassar DR, de Carvalho ACPDLF, Zanotto ED (2020) Explainable machine learning algorithms for predicting glass transition temperatures. Acta Mater 188:92–100

Avrami M (1939) Kinetics of phase change. I General theory. J Chem Phys 7:1103–12

Avrami M (1940) Kinetics of phase change. II Transformation time relations for random distribution of nuclei. J Chem Phys 8:212–24

Avrami M (1941) Kinetics of phase change. III Granulation, phase change, and microstructure kinetics of phase change. J Chem Phys 9:177–84

Berthier L (2011) Trend: dynamic heterogeneity in amorphous materials. Physics 4:42

Cassar DR, Lancelotti RF, Nuernberg R, Nascimento MLF, Rodrigues AM, Diz LT, Zanotto ED (2017) Elemental and cooperative diffusion in a liquid, supercooled liquid and glass resolved. J Chem Phys 147:014501

Cassar DR, Rodrigues AM, Nascimento MLF, Zanotto ED (2018a) The diffusion coefficient controlling crystal growth in a silicate glass-former. Inter J Appl Glass Sci 9:373–382

Cassar DR, de Carvalho ACPLF, Zanotto ED (2018b) Predicting glass transition temperatures using neural networks. Acta Mater 159:249–256

Cassar DR, Serra AH, Peitl O, Zanotto ED (2020) Critical assessment of the alleged failure of the Classical Nucleation Theory at low temperatures. J Non-Cryst Solids 547:120297

Cassar DR, Santos GG, Zanotto ED (2021) Designing optical glasses by machine learning coupled with a genetic algorithm. Ceram Inter 47: 10555–10564

Deubener J, Brückner R, Sternitzke M (1993) Induction time analysis of nucleation and crystal growth in di- and metasilicate glasses. J Non-Cryst Solids 163:1–12

Ediger MD (2000) Spatially heterogeneous dynamics in supercooled liquids. Ann Rev Phys Chem 51:99–128

Ediger MD (2008) Crystal growth kinetics exhibit a fragility-dependent decoupling from viscosity. J Chem Phys 128:034709

Flenner E, Szamel G (2010) Dynamic heterogeneity in a glass forming fluid: Susceptibility, structure factor, and correlation length. Phys Rev Lett 105:217801

Fokin VM, Potapov OV, Zanotto ED, Spiandorello FM, Ugolkov VL, Pevzner BZ (2003) Mutant crystals in $Na_2O \cdot 2CaO \cdot 3SiO_2$ glasses. J Non-Cryst Solids 331:240–253

Fokin VM, Zanotto ED, Yuritsyn NS, Schmelzer JWP (2006) Homogeneous crystal nucleation in silicate glasses: A 40 years perspective. J Non-Cryst Solids 352:2681–2714

Fokin VM, Schmelzer JWP, Nascimento MLF, Zanotto ED (2007) Diffusion coefficients for crystal nucleation and growth in deeply undercooled glass-forming liquids. J Chem Phys 126:234507

Fokin VM, Abyzov AS, Zanotto ED, Cassar DR, Rodrigues AM, Schmelzer JWP (2016) Crystal nucleation in glass-forming liquids: Variation of the size of the "structural units" with temperature. J Non-Cryst Solids 447:35–44

Fokin VM, Abyzov AS, Rodrigues AM, Pompermayer RZ, Macena GS, Zanotto ED, Ferreira EB (2019) Effect of non-stoichiometry on the crystal nucleation and growth in oxide glasses. Acta Mater 180:317–328

Gibbs JW (1926) Collected Works, vol. 1, Thermodynamics. Longman, New York

Gupta PK, Cassar DR, Zanotto ED (2016) Role of dynamic heterogeneities in crystal nucleation kinetics in an oxide supercooled liquid. J Chem Phys 145:211920

Gutzow IS, Schmelzer JWP (2013) The Vitreous State: Thermodynamics, Structure, Rheology, and Crystallization. Springer, 2nd enlarged ed. Heidelberg

Henritzi P, Bormuth A, Klameth F, Vogel M (2015) A molecular dynamics simulations study on the relations between dynamical heterogeneity, structural relaxation, and self-diffusion in viscous liquids. J Chem Phys 143:164502

Höland W, Beall GH (2013) Glass-Ceramic Technology. Wiley, Hoboken, New Jersey

Jackson KA (2004) Kinetic Processes. Wiley-VCH, Weinheim

Jiusti J, Zanotto ED, Cassar DR, Andreeta MRB (2020) Viscosity and liquidus based predictor of glass-forming ability of oxide glasses. J Am Ceram Soc 103:921–932

Johari GP, Schmelzer JWP (2014) Crystal nucleation and growth in glass-forming systems: Some new results and open problems. In: Glass: Selected Properties and Crystallization. Schmelzer JWP (ed) de Gruyter, Berlin p 531–590

Johnson WA, Mehl R (1939) Reaction kinetics in processes of nucleation and growth. Trans AIME 135:416–58

Kashchiev D (2000) Nucleation : Basic Theory with Applications. Butterworth Heinemann, Oxford

Kauzmann W (1948) The nature of the glassy state and the behavior of liquids at low temperatures. Chem Rev 43:219–256

Kelton KF, Greer AL (2010) Nucleation in Condensed Matter: Applications in Materials and Biology. Elsevier, Amsterdam

Kolmogorov AN (1937) On the statistical theory of crystallization of metals. Izv Akad Nauk SSSR 3:355–59

Macena GS, Abyzov AS, Fokin VM, Zanotto ED, Ferreira EB (2020) Off-stoichiometry effects on crystal nucleation and growth kinetics in soda-lime-silicate glasses. The combeite (Na_2O•2CaO•$3SiO_2$)—devitrite (Na_2O•3CaO•$6SiO_2$) joint. Ac ta Mater 196:191–199

Montazerian M, Singh SP, Zanotto ED (2015) An analysis of glass-ceramic research and commercialization. Am Ceram Soc Bull 94:30–35

Montazerian M, Zanotto ED, Mauro JC (2020) Model-driven design of bioactive glasses: from molecular dynamics through machine learning. Inter Mater Rev 65(5): 297–321

Morey GW (1954) The Properties of Glass. Reinhold Publishers, New York

Nascimento MLF, Zanotto ED (2006) Mechanisms and dynamics of crystal growth, viscous flow, and self-diffusion in silica glass. Phys Rev B—Condens Matter and Mater Phys 73:024209

Nascimento MLF, Zanotto ED (2010) Does viscosity describe the kinetic barrier for crystal growth from the liquidus to the glass transition? J Chem Phys 133:174701

Nascimento MLF, Fokin VM, Zanotto ED, Abyzov AS (2011) Dynamic processes in a silicate liquid from above to below the glass transition. J Chem Phys 135:194703

Neuville D, Cornier L, Caurant D, Montagne L (eds) (2017) From Glass to Crystal: Nucleation, Growth and Phase Separation, from Research to Applications. EDP Science, London

Prado SCC, Rino JP, Zanotto ED (2019) Successful test of the classical nucleation theory by molecular dynamic simulations of BaS. Comp Mater Sci 161:99–106

Reis RMCV, Fokin VM, Zanotto ED, Lucas P (2016) Determination of crystal growth rates in glasses over a temperature range using a single DSC run. J Am Ceram Soc 99:2001–2008

Rodrigues BP, Zanotto ED (2012) Evaluation of the guided random parametrization method for critical cooling rate calculations. J Non-Cryst Solids 358:2626–34

Scherer GW (1991) Chapter 3: Glass and amorphous materials. *In:* Materials Science and Technology. Vol 9. Cahn RW, Haasen P, Kramer EJ (eds), VCH Publications, New York

Schmelzer JWP, Schick C (2012) Dependence of crystallization processes of glass-forming melts on melt history: A theoretical approach to a quantitative treatment. Phys Chem Glass Eur J Glas Sci Technol Part B 53:99–106

Schmelzer JWP, Abyzov AS, Fokin VM, Schick C, Zanotto ED (2015) Crystallization in glass-forming liquids: Effects of decoupling of diffusion and viscosity on crystal growth. J Non-Cryst Solids 429:45–53

Separdar L, Rino JP, Zanotto ED (2021) Molecular dynamics simulations of spontaneous and seeded nucleation and theoretical calculations for zinc selenide. Comp Mater Sci 187:110124

Shelby JE (2005) Introduction to Glass Science and Technology. The Royal Society of Chemistry, Lonton

Slezov VV (1999) Kinetics of First-Order Phase Transitions. Wiley-VCH, Weinheim

Tammann G (1898) Über die Abhängigkeit der Zahl der Kerne. Z Phys Chem 25:441–479

Tipeev AO, Zanotto ED, Rino JP (2018) Diffusivity, interfacial free energy, and crystal nucleation in a supercooled Lennard–Jones liquid. J Phys Chem C 122:28884–28894

Tipeev AO, Zanotto ED, Rino JP (2020) Crystal nucleation kinetics in supercooled germanium: MD simulations versus experimental data. J Phys Chem B 124:7979–7988

Uhlmann DR (1982) Crystal growth in glass-forming liquids: A ten-year perspective. *In:* Advances in Ceramics. Vol 4. Simmons JH, Uhlmann DR, Beall GH (eds), American Ceramic Society, Columbus, p 80–124

Varshneya AK, Mauro JC (2019) Fundamentals of Inorganic Glasses. 3rd ed. Elsevier, Amsterdam

Volmer M (1939) Kinetik der Phasenbildung. Th. Steinkopff, Dresden

Zanotto ED (2010) A bright future for glass-ceramics. Am Ceram Soc Bull 89:19–27

Zanotto ED (2013) Crystals in Glass: A Hidden Beauty. Wiley, Hoboken New Jersey

Zanotto ED, Cassar DR (2018) The race within supercooled liquids—Relaxation versus crystallization. J Chem Phys 149:02450

Zanotto ED, Mauro JC (2017) The glassy state of matter: Its definition and ultimate fate. J Non-Cryst Solids 471:490–495

Zanotto ED, Tsuchida JE, Schneider JF, Eckert H (2015) Thirty-year quest for structure-nucleation relationships in oxide glasses. Inter Mater Rev 60:376–391

Zheng Q, Zhang Y, Montazerian M, Gulbiten O, Mauro JC, Zanotto ED, Yue Y (2019) Understanding glass through differential scanning calorimetry. Chem Rev 119:7848–7939

Reviews in Mineralogy & Geochemistry
Vol. 87 pp. 431-556, 2022
Copyright © Mineralogical Society of America

10

Thermodynamics of Multi-component Gas–Melt Equilibrium in Magmas: Theory, Models, and Applications

Paolo Papale[1*], Roberto Moretti[2,3], Antonio Paonita[1]**

[1]Istituto Nazionale di Geofisica e Vulcanologia
Italy
[2]Université de Paris, Institut de Physique du Globe de Paris
UMR 7154 CNRS, 1 rue Jussieu
F-75238 Paris
France
[3]Observatoire Volcanologique et Sismologique de Guadeloupe
Institut de Physique du Globe de Paris, Le Houelmont
F-97113 Gourbeyre
France
[*]Sezione di Pisa, via Cesare Battisti 53
56125 Pisa
Italy
[**]Sezione di Palermo, via Ugo la Malfa 153
90146 Palermo
Italy

INTRODUCTION

Volatiles are the fuel and fundamental control for major processes involving magmas and volcanoes, including magma ascent, magma convection and mixing, magma differentiation, and volcanic eruptions. It is not an exaggeration to say that without volatiles in magmas the world would be a profoundly different place, and likely, life would have not flourished. Without volatiles in magmas, you would not be reading these pages. In fact, it was Earth degassing, mostly during volcanic eruptions, that provided the raw materials for the formation of the oceans and the atmosphere. Life on our planet owes much to volatiles transported with magmas and dispersed on the surface during volcanic eruptions.

Magmatic volatiles are not just transported by magmas. They do cause, to a large extent, magma transfer from deep regions to the surface, and do determine the associated dynamics. Accordingly, we strictly need a deep understanding of the behavior of volatiles in magmas, without which we would never penetrate the complex processes of magma ascent, evolution, and eruption. Among the key factors behind such processes, two major ones are: i) the decrease with decreasing pressure of the solubility of volatiles dissolved in the magmatic melt, and ii) the large compressibility of the gas phase causing substantial expansion upon pressure decrease. Other factors can be important; for example, chemical reactions between magmas and wall rocks can result in additional gas phase formation, e.g., by crystallization-driven secondary boiling in magma (Woods and Cardoso 1997) or by decarbonation of the host rocks (Deegan et al. 2010). Such processes can substantially alter the gas output budget at active volcanoes (Chu et al. 2019) as well as their eruption potential and dynamics (Freda et al. 2011). However, no other factor or process appears so fundamental in controlling mass transfer from depth, and ultimately in shaping the same evolution of the Earth, as the two ones above directly involving the relationships between pressure and volatiles.

1529-6466/22/0087-0010$15.00 (print)
1943-2666/22/0087-0010$15.00 (online)

http://dx.doi.org/10.2138/rmg.2022.87.10

Because of their crucial role in Earth history as well as in petrogenesis and volcanic eruption dynamics, magmatic volatiles have been the subject of countless studies. A simple search in the Web of Science with "magma" and "water" as topics, water being the most abundant volatile component in natural magmas, produces about 5500 entries. The vast majority of those studies (and of many others involving other volatile components such as carbon dioxide, sulfur, noble gases, and halogens) aim to reproduce subsurface conditions at various depths below the Earth's surface. These studies directly measure, under lab-controlled conditions, several quantities relevant for the behavior of volatiles in magmas. These include the pure or combined saturation contents of volatile components, their diffusion through the melt and related kinetic delays, the viscous delay to bubble expansion, the effects of volatiles on phase equilibria and on virtually any major magmatic property, etc. Thanks to such a tremendous experimental effort, we know a lot of the behavior of magmatic volatiles over a substantial range of $P-T-X$ conditions (X standing for "composition", generally referred to both volatile and non-volatile components in magmas) (for brief reviews including experimental techniques for magmatic volatiles, see Wallace and Anderson 1999; Moore 2008). Other studies involving magmatic volatiles measure and interpret data on degassing from volcanoes and volcanic areas, mostly with the purpose of constraining magmatic processes at depth and contributing to forecasts of volcanic eruptions.

Understanding volatile–melt thermodynamics is a priority and a milestone for igneous petrology, ever since when petrologists first added water to their experiments. As noted by Burnham (1994), the role of H_2O in granite petrogenesis and its strong impact in lowering melt liquidus temperature even when dissolved in small amounts represented the ultimate evidence ruling out a metasomatic origin for granites. Those discoveries started a new era of understanding, when volatiles ceased to be perceived as the Maxwell demon cited by Bowen, doing what petrologists wished them to do. Since the early experiments of Goranson (1931) on albite and sanidine melts, it became clear that water could not be treated as a simple oxide with effects proportional to its dissolved amount, and that the influence on phase equilibria of even small amounts of dissolved water could not be assessed by simple dilution mechanisms (for a review and analysis of the structure and properties of silicate melts see Moretti and Ottonello 2022, this volume). When the importance of even small amounts of water in the origin of granites was finally understood (Bowen and Tuttle 1950; Tuttle and Bowen 1958), the time was ripe to start penetrating the mechanisms and the interactions responsible of water dissolution in silicate melts, and to develop a robust theoretical framework for predictions and interpretations.

Today petrologists and volcanologists employ 1) thermodynamics to describe the equilibrium conditions between volatiles, melts, and crystalline phases, 2) chemical kinetics to describe the departure from the equilibrium conditions, and 3) thermo-fluid dynamics to describe the motion of gas-bearing magmas as well as the differential motion of the gas phase with respect to the melt (or melt + crystals) phase. A complete treatment of the behavior of volatiles in magmas would require a book by itself, and it is largely beyond the scopes of this contribution. Here we focus on the thermodynamics of multi-component gas–melt equilibrium (gas stands here for the light phase made of volatile components, although at deep magmatic conditions a more appropriate term would be that of supercritical phase). Non-equilibrium conditions due to finite time to achieve equilibrium, not treated here, increase in importance for increasing rates of change of the physico-chemical quantities describing the system, and for increasing viscosity. In fact, the latter increases the time necessary to achieve equilibrium, and the former decreases the time available to achieve equilibrium. Fortunately, equilibrium thermodynamics is a satisfactory approximation in many real situations. When kinetic delays cannot be neglected, equilibrium thermodynamics still provides the direction towards which the conditions are expected to evolve, and the distance from equilibrium provides the force driving system evolution.

This chapter is organized as it follows.

- First, we introduce the fundamental equations of equilibrium thermodynamics starting from basic principles. The only required background knowledge consists in the first and second principles of thermodynamics. That leads naturally to the equilibrium equations, as well as to clear understanding of the meaning of real versus ideal modeling.

- We then concentrate on fully real, multi-component thermodynamics, and on models based on regular solution approaches allowing the extreme compositional variability of silicate melts to be fully accounted for. That section includes the SOLWCAD (Papale et al. 2006) and MagmaSat (Ghiorso and Gualda 2015) models for single component and two-component $H_2O + CO_2$ volatile–melt equilibrium.

- After that, we discuss models that are based on speciation equations that treat solubility as the result of chemical reactions between the volatiles and the melt. We start by reviewing the seminal work by Burnham and co-workers, and that by Stolper and co-workers which led to the model by Dixon et al. (1995) and to the VolatileCalc code for $H_2O + CO_2$ volatile–melt equilibrium by Newman and Lowenstern (2002). Then, we examine the work done on the third major volatile in magmas represented by sulfur, whose dissolution mechanisms are largely controlled by redox conditions; that brings in the concept and the roles of oxygen fugacity in controlling sulfur–melt thermodynamics, and a number of additional basic concepts such as melt polymerization and melt optical basicity in determining the dissolution of volatiles in silicate melts. In this part we present and discuss the works by Moretti and co-workers leading to the CTZFG model for sulfur–melt thermodynamics (Moretti and Ottonello 2005), as well as models for mixed saturation in the C–O–H–S system such as SolEx (Witham et al. 2012), D-Compress (Burgisser et al. 2015), and COHS (Moretti et al. 2003). That section is concluded with a brief discussion on the solubility of halogen species, and with hints on ab-initio and molecular dynamics approaches to model volatile saturation in silicate melts.

- The following section extends the equilibrium equations to minor components represented by noble gases, the study of which complements and enriches the understanding of the deep volcanic processes. Such species carry precious and unique information on the deep thermodynamic conditions and degassing processes, thanks to their largely inert nature which prevents significant modifications by secondary processes in hydrothermal and low-temperature systems. We present and discuss models for pure noble gas as well as for mixtures of noble gases and major volatile components based on classical thermodynamic as well as on statistical mechanics approaches.

- We conclude by showing and discussing applications of volatile–melt thermodynamic equilibrium modeling to real problems in igneous petrology and volcanology. This part is divided into three sections dealing with use of models to i) constrain conditions and processes inside magmatic reservoirs, ii) interpret data from volcano degassing, and iii) model transport processes and describe magma and eruption dynamics.

Finally, this review does not account for the solubility of minor, redox controlled, reactive species such as H_2, CO and CH_4, for which extensive models of saturation conditions in silicate melts have not been produced yet. These species can be important at highly reduced conditions, and can be relevant in the study of the early Earth and its deep interior. Nevertheless, their effectiveness in tracking magma degassing is limited by i) their very low abundance at magmatic temperatures and ii) their high sensitivity to secondary processes, making them good indicators, together with sulfur species, of post-magmatic processes, particularly those in hydrothermal environments (Moretti and Stefansson 2020 and references therein).

There are still many papers where polynomial or other relatively simple expressions are employed to fit experimental data over restricted P–T–X ranges. They have little relevance beyond the limited range of the corresponding experiments, and we do not discuss them.

Table 1. Nomenclature. As a general reference, total quantities are expressed as capital letters, specific quantities are in lower case, while lower case symbols with horizontal bar on top indicate partial molar quantities of the component at the subscript. Components are reported as subscripts, and phases as superscripts. Molar fractions are indicated with y when they refer to the gas phase, and with x when they refer to the liquid phase, both with no additional reference to the phase (no superscript). Locally defined variables are not reported in this Table.

Physical parameters	
F	Helmholtz energy
a	activity
C_P	specific heat at constant pressure
C_{S_2}	sulfide capacity
$C_{S^{6+}}$	sulfate capacity
$D_i^{G/M}$	partition coefficient of component i between gas and melt
f	fugacity
$G, g, \overline{g}_i$	Gibbs energy (total, specific, partial molar for component i)
$H, h, \overline{h}_i$	enthalpy (total, specific, partial molar for component i)
k	Henry's law constant
K_n	Equilibrium constant for reaction n
M	molar weight
N	total number of moles
n_i	number of moles of component i
n (index)	number of components
P	pressure
R	gas constant
S, s	entropy (total, specific)
T	temperature
U	internal energy
$V, v, \overline{v}_i$	volume (total, specific, partial molar for component i)
w_i	mass fraction of component i
$w_{ij}, w_{ij}^{(0)}, w_{ij}^{(1)}$	coefficient of interaction between components i, j
x	mole fraction in the liquid phase
x'	mole fraction in the liquid phase, rescaled to volatile-free
$x_{M_i^{y+}}$	atom fraction of network modifier i
$x_{T_j^{q+}}$	atom fraction of network former j
y	mole fraction in the gas phase
Greek symbols	
γ	activity coefficient
$\gamma_{M_i^{y+}}$	basicity moderating parameter for network formers
$\gamma_{T_j^{q+}}$	basicity moderating parameter for network modifiers
λ	ratio of gas molar mass to melt molar mass

μ	chemical potential
ϕ	fugacity coefficient
Subscripts	
$i, j, \ldots$	indexes for components
MIX	referred to the multiphase mixture
$P, T, \ldots$	computed at constant $P, T, \ldots$
Superscripts	
+	ideal state
0	reference or standard state
$\alpha, \beta, \gamma, \ldots$	indexes for generic phases
E	excess
G	gas (or supercritical) phase
L	liquid phase
M	melt phase
T	total

FUNDAMENTAL EQUATIONS OF EQUILIBRIUM THERMODYNAMICS

Fundaments of thermodynamic equilibrium

Our initial treatment follows that of Prausnitz et al. (1985). Over this entire section, thermodynamic relations are model-independent; whatever relates to use of the theory to construct models of volatile–melt thermodynamics is treated starting from the next section. By exploring the theory of fluid phase equilibria we derive equilibrium equations and highlight the meaning of ideal assumptions that are adopted in some gas–liquid equilibrium models. Knowledge of the first and second principles of thermodynamics is the only prerequisite to fully understand the arguments and equations in this section. We treat each volatile component without regards to chemical reactions that can occur within each phase (e.g., water dissociation into hydroxyl groups). Dealing with macroscopic quantities, thermodynamics allows such an approach, provided that the various quantities appearing in the equilibrium equations are then consistently modeled. Reaction-based approaches to volatile dissolution in silicate melts are treated separately in a dedicated section.

The classical problem of phase equilibrium takes into account the three fundamental processes of i) heat transfer between the phases, ii) displacement of phase boundaries, and iii) mass transfer across phase boundaries. By definition, all of these processes are zeroed at equilibrium (more precisely, their average effects are zeroed; natural equilibria are always dynamic, meaning that heat and mass always flow across phase boundaries, and the phase boundaries always shift. At equilibrium, however, such processes are equally efficient in all directions, so that on average they cancel out). Thermodynamics treats the condition of equilibrium with reference to some thermal (heat flow), mechanical (momentum flow, or boundary displacement), and chemical (mass transfer) potential, which must be the same for all phases in order to achieve equilibrium. The thermal potential is represented by temperature (T), the mechanical potential is represented by pressure (P), and the chemical potential is represented by the quantity with the same name of chemical potential (μ). The condition of thermodynamic equilibrium is therefore that the temperature and pressure of all phases must be the same, and that the chemical potential of all components in all phases must also be the same:

$$T^{(1)} = T^{(2)} = \ldots = T^{(\pi)}$$

$$P^{(1)} = P^{(2)} = \ldots = P^{(\pi)}$$

$$\mu_1^{(1)} = \mu_1^{(2)} = \ldots = \mu_1^{(\pi)}$$

$$\mu_2^{(1)} = \mu_2^{(2)} = \ldots = \mu_2^{(\pi)} \tag{1}$$

$$\vdots \qquad \vdots \qquad \qquad \vdots$$

$$\mu_m^{(1)} = \mu_m^{(2)} = \ldots = \mu_m^{(\pi)}$$

where the equilibrium is extended to m components distributed in π phases.

While pressure and temperature are intuitive quantities that we use in our daily life, the chemical potential is less straightforward. To understand its meaning we start from the following expression representing a combination of the first and second laws of thermodynamics for a homogeneous, closed system:

$$dU \le TdS - PdV \tag{2}$$

with the inequality applying to an irreversible process, and the equality to a reversible process. In the above equation, U is the internal energy, S is entropy, and V is volume. From Equation (2) it follows that

$$dU_{S,V} \le 0 \tag{3}$$

Through Legendre's transformations, and with the same reasoning:

$$dH \le TdS + VdP, \ dH_{S,P} \le 0 \tag{4}$$

$$dF \le -SdT - PdV, \ dF_{T,V} \le 0 \tag{5}$$

$$dG \le -SdT + VdP, \ dG_{T,P} \le 0 \tag{6}$$

where H is enthalpy, F is Helmholtz energy, and G is Gibbs energy (also called Gibbs free energy).

Equations (2–6) tell us that the energy of a homogeneous closed system evolves towards a minimum during the irreversible processes through which equilibrium is achieved, then it remains at that minimum when the equilibrium is achieved (the slow, zero-averaging processes taking place at equilibrium are in fact reversible processes). That explains why many routines employed to determine the equilibrium conditions of complex systems search for the direction that minimizes the energy of the system (whatever energy can be chosen among those at Equations (2–6)).

Equations (2–6) are valid for homogeneous closed systems. However, when dealing with mass transfer between the phases, which is the classical problem of chemical equilibria, each phase is open by definition. In such a case, the reasoning above can be repeated by noticing that each expression for energy must also include a dependence on the amount of each component present in the system (e.g., a given phase). Accordingly, the variation of the internal energy U must be described also in terms of variations of such amounts:

$$U = U(S, V, n_1, n_2, \ldots n_m) \tag{7}$$

$$dU = \left(\frac{\partial U}{\partial S}\right)_{V,n_i} dS + \left(\frac{\partial U}{\partial V}\right)_{S,n_i} dV + \sum_i \left(\frac{\partial U}{\partial n_i}\right)_{S,V,n_{j\ne i}} dn_i \tag{8}$$

where n_i is the number of moles of the ith component. Again, through Legendre's transformations the equilibrium conditions follow:

$$dU \leq TdS - PdV + \sum_i \mu_i dn_i \tag{9}$$

$$dH \leq TdS + VdP + \sum_i \mu_i dn_i \tag{10}$$

$$dF \leq -SdT - PdV + \sum_i \mu_i dn_i \tag{11}$$

$$dG \leq -SdT + VdP + \sum_i \mu_i dn_i \tag{12}$$

where the chemical potential referred to the ith component, μ_i, is given by

$$\mu_i = \left(\frac{dU}{dn_i}\right)_{S,V,n_{j \neq i}} = \left(\frac{dH}{dn_i}\right)_{S,P,n_{j \neq i}} = \left(\frac{dF}{dn_i}\right)_{T,V,n_{j \neq i}} = \left(\frac{dG}{dn_i}\right)_{T,P,n_{j \neq i}} \tag{13}$$

Each of the Equations (9–12) is equally valid in representing the evolution of energy for an open system, and each of the derivatives at Equation (13) is an equally valid definition of the chemical potential. However, Equation (12) expressing the Gibbs energy is of more direct use in most practical applications, since it expresses system energy in terms of changes of pressure and temperature, these being the quantities that are usually kept under control in laboratory experiments, and the quantities that are more directly related to our experience. Accordingly, we define partial molar quantities in terms of P and T:

$$\overline{x}_i = \left(\frac{dX}{dn_i}\right)_{T,P,n_{j \neq i}} \tag{14}$$

with X (capital) being any extensive quantity. A partial molar quantity for a given component in a mixture describes, therefore, the change in the extensive quantity it refers to, when small amounts of that component are added or subtracted at constant P, T and without changing the amount of any other component in the mixture. By comparing Equations (13) and (14), it follows that the chemical potential of the component i in a mixture corresponds to the partial molar Gibbs energy of that component in the mixture:

$$\mu_i = \overline{g}_i \tag{15}$$

Thus, for a pure component:

$$\mu = g, \ G = n\mu \tag{16}$$

The above equations clarify the relevance of the chemical potential in thermodynamic equilibrium. In summary, the equilibrium condition is conveniently expressed in terms of minimization of the Gibbs energy, a condition that for any given P, T pair is realized when the partial molar Gibbs energy, or chemical potential, of any component in the mixture does not evolve further, or in other words, when the potential for further chemical evolution of the system is (on average) zeroed.

We now introduce the concept of fugacity, and show that equivalence of the fugacity of all components in all phases is an equilibrium criterion as general as equivalence of chemical potentials, additionally offering a simpler way of expressing heterogeneous chemical equilibria involving fluid phases.

Let's start by referring to a pure component made of one single mole along a reversible path. From Equations (6) and (16):

$$d\mu = -sdT + vdP \tag{17}$$

From Equation (17) it follows that

$$\left(\frac{d\mu}{dP} \right)_T = v \tag{18}$$

Therefore, by referring to an ideal gas and integrating at constant T:

$$\mu - \mu^0 = RT \ln \frac{P}{P^0} \tag{19}$$

where the superscript 0 refers to a reference state for which the chemical potential is known. Because Equation (18) has been integrated at constant T, the reference state must be at the same temperature as that of interest.

Equation (19) provides a workable expression for the chemical potential, but it is limited to ideal gases. A similar expression that works for any component at any system conditions is given by the Lewis extension of Equation (19):

$$\mu_i - \mu_i^0 = RT \ln \frac{f_i}{f_i^0} \tag{20}$$

where f is fugacity. Fugacity represents therefore a sort of modified pressure which embeds the deviation from ideal behavior of a component in any physical state. Equation (20) shows that μ_i^0 and f_i^0 are not independent: once one is given, the other one is fixed (or in other words, once the reference state is established, μ_i^0 and f_i^0 are both determined).

In order to show that equality of fugacities is an equilibrium criterion as general as equality of chemical potentials, consider a component i in two phases α and β at equilibrium. Equations (1) and (20) tell us that

$$\mu_i^\alpha = \mu_i^{0,\alpha} + RT \ln \frac{f_i^\alpha}{f_i^{0,\alpha}} = \mu_i^\beta = \mu_i^{0,\beta} + RT \ln \frac{f_i^\beta}{f_i^{0,\beta}} \tag{21}$$

There are two options for the reference states related to the two phases: either they are the same, or they are at the same temperature (corresponding to system temperature, see above) but different pressure and/or composition. In the first case $\mu_i^{0,\alpha} = \mu_i^{0,\beta}$, from which it immediately follows that $f_i^\alpha = f_i^\beta$. In the second case an expression of the transformation between the two reference states can be written:

$$\mu_i^{0,\alpha} = \mu_i^{0,\beta} + RT \ln \frac{f_i^{0,\alpha}}{f_i^{0,\beta}} \tag{22}$$

By substituting Equation (22) into Equation (21) and simplifying, the same result follows: $f_i^\alpha = f_i^\beta$. Thus, for any possible case, that is, for any possible choice of the reference state of the component in the different phases, equality of fugacities is the same as equality of chemical potentials. The former can be therefore employed at Equation (1), in substitution of the equality of chemical potentials, as a fully equivalent criterion for chemical equilibrium. With reference to a liquid–gas system, and presupposing equality of temperature and pressure, the new equilibrium equation becomes

$$f_i^G = f_i^L \tag{23}$$

Fugacity has been introduced at Equation (20) as a "modified pressure" embedding deviation from ideal behavior. Accordingly, for a pure, ideal gas, fugacity is equivalent to pressure. For a component in an ideal gas mixture, $f_i = y_i P$, where y_i is the molar fraction of the component in the mixture (and $y_i P$ is called the partial pressure of component i). It follows that for a real gas mixture

$$\frac{f_i}{y_i P} = \phi_i \neq 1 \rightarrow f_i = y_i \phi_i P \tag{24}$$

where the fugacity coefficient ϕ_i expresses the deviation from ideal behavior. Equation (24) is a convenient expression for the fugacity of a component in a gaseous mixture. For liquid mixtures it is usually more convenient to refer fugacity to some quantity that can be immediately related to the energy of the system, as that offers more direct opportunity for appropriate modeling. That is done through the definitions of activity and activity coefficient. The activity of a component in a mixture, a_i, is defined as the ratio between its fugacity at system conditions and at some convenient reference state. The activity coefficient γ_i is instead the ratio between activity and mole fraction:

$$a_i = \frac{f_i}{f_i^0}, \gamma_i = \frac{a_i}{x_i} \rightarrow f_i = \gamma_i x_i f_i^0 \tag{25}$$

where x_i is mole fraction of the component i (note that we indicate mole fractions with the letter y for the gas phase, and x for the liquid phase, see the Nomenclature at Table 1).

Equations (24) and (25) provide two equivalent expressions for the fugacity of a component in a real mixture. The former (Eqn. 24) is usually preferred for the gas phase, and the latter (Eqn. 25) for the liquid phase. The real gas–liquid chemical equilibrium equation (23) can now be written as

$$y_i \phi_i P = \gamma_i x_i f_i^{0L} \tag{26}$$

where the superscript L has been added to f_i^0 to indicate that the reference fugacity corresponds to some state of reference for the component dissolved in the liquid (or melt) phase. Note that for ideal mixtures $\phi_i = \gamma_i = 1$ (an explanation of the second equality is provided below), therefore, the ideal gas–liquid chemical equilibrium equation becomes $y_i P = x_i f_i^{0L}$, or $x_i = (f_i^{0L})^{-1} p_i$ where p_i is the partial pressure of component i. The latter expression corresponds to the largely used (and sometimes abused) Henry's law, which states that at constant P–T the amount of solute (x_i) is proportional to its partial pressure in the gas phase (p_i). The constant of proportionality is the inverse of the reference fugacity of the component in the liquid phase, which is a constant for fixed P–T conditions. Limits to the use of Henry's law in modeling water and carbon dioxide mixed saturation in silicate melts are illustrated in the following section, and for noble gases in the dedicated section.

Let's write Equation (26) with reference to the most abundant volatile component in magmas, water, and assume the gas phase coexisting with the magmatic melt is made of only gaseous water:

$$\phi_{H_2O} P = \gamma_{H_2O} x_{H_2O} f_{H_2O}^{0L} \tag{27}$$

Provided that we know how to express ϕ_{H_2O}, γ_{H_2O}, and $f_{H_2O}^{0L}$, Equation (27) has only one unknown corresponding to x_{H_2O}, the amount of water dissolved in the melt at given P–T conditions and at equilibrium with a pure H_2O gas. That amount is called the solubility of water in that specific melt at the given P–T conditions. Consider now the presence in the system of two volatiles, e.g., consider the second most abundant volatile component in magmas, carbon dioxide, together with water. We can write two expressions of Equation (26), one for H_2O and the other one for CO_2:

$$y_{H_2O}\phi_{H_2O}P = \gamma_{H_2O}x_{H_2O}f_{H_2O}^{0L} \tag{28}$$

$$y_{CO_2}\phi_{CO_2}P = \gamma_{CO_2}x_{CO_2}f_{CO_2}^{0L} \tag{29}$$

The system formed by Equations (28–29) has four unknown, represented by the dissolved amounts of H_2O and CO_2, and the concentrations of H_2O and CO_2 in the gas phase. In order to be solved, two other equations are needed. These are given by two mass conservation equations, the first one stating that the gas phase is made of only H_2O and CO_2, the second one stating that whatever the partitioning of H_2O and CO_2 between the melt and the gas phase, their total mass in the system must be conserved (Papale 1999a):

$$y_{H_2O} + y_{CO_2} = 1 \tag{30}$$

$$\frac{x_{H_2O}^T - x_{H_2O}}{y_{H_2O} - x_{H_2O}} = \frac{x_{CO_2}^T - x_{CO_2}}{y_{CO_2} - x_{CO_2}} \tag{31}$$

where the superscript T refers to the total amount of each volatile component in the system, or in other words, its abundance in the gas–melt system as a whole.

Equations (28–31) form a system of four equations in four unknown, that can be solved. However, the presence of the total amounts of each volatile component in the system introduces a substantial difference with respect to the case of one single volatile considered above. In fact, now the amount of each volatile dissolved in the melt, as well as the composition of the coexisting gas phase, depend on the abundance of each volatile component in the system. In other words, for any given volatile-free melt composition the dissolved amounts do not depend anymore on just thermodynamic quantities (P, T) and they are not a thermodynamic quantity—solubility—themselves; the dissolved amounts define the saturation conditions, and are therefore called the saturation contents, distinct from solubility which is a property of a volatile in a given system and it refers to equilibrium with the pure-component gas phase.

Terms in the equilibrium equations

Solution of Equation (27) for the solubility of a volatile component (in the case of Eqn. 27, H_2O), or of the system of Equations (28–31) for two-component ($H_2O + CO_2$) volatiles, requires that we express in some way the fugacity coefficient in the gas phase ϕ_i, the activity coefficient in the liquid phase γ_i, and the fugacity at reference state for the component in the liquid phase f_i^{0L} (i being each volatile component under consideration). In general, such expressions require a number of assumptions or approximations, pertaining therefore to the field of modeling. The equations presented in the following express theoretical relationships between such quantities and other thermodynamic quantities making their modeling straightforward, and pertain therefore to thermodynamic theory. In other words, the equations below relate the fugacity coefficient, the reference fugacity, and the activity coefficient, to other quantities that can be either measured or estimated from experiments.

We start with the fugacity coefficient in the gas phase, ϕ_i. From the definition of fugacity coefficient given above, and making use of fundamental thermodynamic relationships, we can write:

$$RT\ln\phi_i = RT\ln\frac{f_i}{y_iP} = \int_0^P \left(\bar{v}_i - \frac{RT}{P}\right)dP \tag{32}$$

The integral term gives a clear idea of the meaning of ϕ_i as the deviation from ideal behavior. In fact, for an ideal gas component the integral term is zero (and ϕ_i is 1). Equation (32) shows that the fugacity coefficient of a component in a gas mixture can be determined at any

P–T if we know its partial molar volume as a function of *P–T*, and in particular, if we know it from zero pressure (ideal conditions) to system pressure. Volume (and less directly, partial molar volume) is a quantity that we can treat much better than the abstract fugacity coefficient; in particular, we can measure it, and derive a model—an equation of state, or EOS—from the measurements. Because in a real system the partial molar volume depends on composition, the corresponding EOS must be accurate over the range of possible compositions of interest, besides the range of *P–T* conditions of interest. There are many EOS in the literature that describe the behavior of pure gases as well as gas mixtures, from relatively simple Redlich–Kwong models to ab-initio models that require substantial computational resources (a description of some EOSs employed in gas–melt thermodynamics is provided when dealing with noble gas solubility). In general, the choice of a particular EOS is made on a balance between accuracy and affordable computational time.

We have just seen that the fugacity coefficient of a component in the gas phase can be obtained from volume data only. The same holds for the reference fugacity of a component in the liquid (or melt) state, f_i^{0L}: if an EOS for the volume of the component in such a reference state is known, we can derive an expression for f_i^{0L} that holds for any *P–T* condition covered by the EOS. This expression was derived in Papale (1997), and the following shows how that is done.

Let's start by expressing f_i^{0L} as a function of *P–T*:

$$f_i^{0L} = f_i^{0L}(P,T) \tag{33}$$

By introducing the natural logarithm, and differentiating:

$$d\ln f_i^{0L} = \left(\frac{\partial \ln f_i^{0L}}{\partial T}\right)_P dT + \left(\frac{\partial \ln f_i^{0L}}{\partial P}\right)_T dP \tag{34}$$

Since

$$\left(\frac{\partial \ln f_i^{0L}}{\partial T}\right)_P = -\frac{h_i - h_i^+}{RT^2}, \left(\frac{\partial \ln f_i^{0L}}{\partial P}\right)_T = \frac{v_i}{RT} \tag{35}$$

where h_i is the molar enthalpy in the reference state, h_i^+ is the molar enthalpy in the ideal state, v_i is molar volume in the reference state, and R is the gas constant, it follows

$$\ln f_i^{0L} = \ln f_i^{0L}\left(P^0,T^0\right) + \int_{P^0}^{P} \frac{v_i}{RT} dP - \int_{T^0}^{T} \frac{h_i - h_i^+}{RT^2} dT \tag{36}$$

where P^0 and T^0 refer to conditions at which $f_i^{0L}(P^0,T^0)$ is known. Consider now the second integral at Equation (36). The enthalpy difference at the numerator can be written in terms of volume:

$$h_i - h_i^+ = \int_{P^+}^{P} \left[v_i - T\left(\frac{\partial v_i}{\partial T}\right)_P\right] dP \tag{37}$$

where P^+ refers to a pressure at which the component in its reference state (and system *T*) behaves as ideal. By choosing $P^0=P^+$, and substituting Equation (37) into Equation (36):

$$\ln f_i^{0L} = \ln f_i^{0L}\left(P^0,T^0\right) + \int_{P^0}^{P} \frac{v_i}{RT} dP - \int_{T^0}^{T} \frac{1}{RT^2} \int_{P^0}^{P} \left[v_i - T\left(\frac{\partial v_i}{\partial T}\right)_P\right] dP dT \tag{38}$$

which is the desired expression. Equation (38) allows to determine f_i^{0L} for any P–T condition, provided we have an EOS for v_i, the molar volume of the component in its reference state pertaining to the liquid phase, and a condition defined by $P^0 \cdot T^0$ at which f_i^{0L} is known.

We have seen above that the fugacity coefficient in the gas phase and the reference fugacity in the liquid phase can be computed on the basis of volume data. For the former, those volume data are obtained from measurements on real gases, either pure component or mixtures. For the latter, we need first of all to define a reference state for the component dissolved in the liquid phase, for which volume data (or an EOS allowing volume computations) are needed. The choice of such a reference state is a matter of modeling; different choices are possible, and selecting one or the other is a matter of convenience and correspondence between model predictions and experimental observations. This is considered in the following section dealing with regular solution modeling approaches.

In order to be able to solve the equilibrium Equation (27) (for a pure component gas phase) or (28–31) (for two-component gas phase), we still have to find an expression for the activity coefficient γ_i. Note that until now, none of the terms that have been considered in detail includes any direct dependence on the composition of the liquid (such a dependence may appear in the expression for the reference fugacity of the component in the liquid phase f_i^{0L}, if the molar volume in the reference state, v_i at Equation (38), is found to depend on liquid composition). In fact, that dependence is embedded in the activity coefficient of the component dissolved in the liquid phase. The following shows the relationships between activity coefficient and energetic state of the liquid mixture, which strictly depends on liquid composition.

Equation (15) tells us that the chemical potential of a component corresponds to the partial molar Gibbs energy of that component in the mixture, and Equation (14) provides an expression for that:

$$\mu_i = \left(\frac{dG}{dn_i} \right)_{T,P,n_{j\neq i}} \tag{39}$$

While Equation (20) establishes a link between chemical potential and fugacity, Equation (39) relates chemical potential to Gibbs energy. The Gibbs energy is in turn related to volume and entropy, Equation (6). When a mixture is formed from individual components, the resulting Gibbs energy depends on the volume and entropy of the mixture resulting from the individual components. Let's consider volume first. We have two options: a) the components mix in a non-reactive way, meaning that the volume of the mixture is simply given by the sum of the volumes of the single components (isometric mixing); b) the components react upon mixing, eventually forming new bonds that result in a deviation from isometric mixing (non-isometric mixing). We can express the above through a mixing equation for volume:

$$V_{MIX} = \Sigma V_i + V^E \tag{40}$$

Equation (40) says that, in general, the volume of a mixture is equal to the sum of the volumes of the individual components (V_i) plus an excess term (V^E) that embeds all the deviations from simple isometric mixing.

The situation for entropy is similar, with the addition that the entropy of a mixture always differs from the sum of the entropies of the individual components by at least a quantity which is called "ideal mixing entropy". In fact, entropy can be seen as a measure of the number of states that are accessible to a system; simply putting different 'things' together represents an increase in entropy with respect to the sum of the individual entropies, as it increases the possible configurations of the system thus the number of accessible states. That can be visualized by considering a simple example: imagine you have two cavities in the ground, one besides the other, where you can place two balls that you have in your hands. If you only have red balls, there is only one final possibility: one red ball in each cavity. If, instead, you have a red and a blue ball, the number of possible configurations becomes two: the red ball in the cavity on the right, or the red ball in the cavity on the left. The system formed by two different balls offers more options (or it has more configurations available) with respect to the system with two equal balls. Analogously, if instead of balls we have components, by simply putting them together in a mixture results in an increase of entropy with respect to the sum of the individual entropies, independent from any interaction between the components (affecting further the mixture entropy). In summary, by mixing individual components there will be one additional term in the mixture entropy, due to ideal mixing, which sums up with the excess term that accounts for real behavior involving interactions between the components:

$$S_{\text{MIX}} = \Sigma S_i + S^{\overset{\text{IDEAL}}{\text{MIXING}}} + S^{\text{E}} \tag{41}$$

Because the Gibbs energy depends on volume and entropy, we can expect that it will also be given by the sum of three terms: a first term due to the sum of the Gibbs energies of the individual components, a second term deriving from entropy increase upon ideal mixing, and a third term describing real behavior and embedding the interactions between the individual components:

$$G_{\text{MIX}} = \Sigma G_i + G^{\overset{\text{IDEAL}}{\text{MIXING}}} + G^{\text{E}} \tag{42}$$

Since from Equation (12):

$$\left(\frac{\partial G}{\partial P}\right)_{T,n} = V, \left(\frac{\partial G}{\partial T}\right)_{P,n} = -S \tag{43}$$

it follows that the ideal mixing term in the expression of G_{MIX} at Equation (42), which derives exclusively by an entropy contribution, must not depend on pressure. In fact, Equation (42) can be written as

$$G_{\text{MIX}} = \Sigma n_i \mu_i^0 + NRT \Sigma x_i \ln x_i + G^{\text{E}} \tag{44}$$

where n_i is number of moles of component i, N is total number of moles, $x_i = n_i/N$ is the molar fraction of i, and μ_i^0 is the chemical potential of the individual component i (equal to the molar Gibbs energy of that component, Eqn. 16). Equation (44) says that the Gibbs energy of a mixture is given by a contribution due to sum of the Gibbs energies of the individual components, a contribution due to ideal mixing and related to ideal entropy of mixing, and an excess contribution that derives from chemical interactions between the components in the mixture. From Equation (39):

$$\mu_i = \left(\frac{dG}{dn_i}\right)_{T,P,n_{j\neq i}} = \mu_i^0 + RT \ln x_i + \left(\frac{dG^{\text{E}}}{dn_i}\right)_{T,P,n_{j\neq i}} \tag{45}$$

From Equation (20) and the definition of activity:

$$\mu_i = \mu_i^0 + RT \ln a_i \tag{46}$$

and by comparing Equations (45) and (46):

$$RT \ln a_i = RT \ln x_i + \left(\frac{dG^{\text{E}}}{dn_i}\right)_{T,P,n_{j\neq i}} \tag{47}$$

We can now apply the definition of activity coefficient given above ($a_i = x_i\gamma_i$) to obtain an expression that relates the activity coefficient to the Gibbs energy of the mixture:

$$RT \ln \gamma_i = \left(\frac{\mathrm{d}G^{\mathrm{E}}}{\mathrm{d}n_i} \right)_{T,P,n_{j \neq i}} = \overline{g}_i^{\mathrm{E}} \tag{48}$$

Equation (48) shows that the activity coefficient of a component in a mixture, γ_i, directly relates to the partial molar excess Gibbs energy of that component in the mixture. The relevance of the activity coefficient in describing the real behavior of a system is now fully evident: if the formation of a mixture from its single components does not involve any excess contribution to its final energetic state—that is, if mixture formation is not dissimilar from putting together inert objects so that the final mixture energy is simply the sum of the initial energies of the components (plus an ideal mixing term due to ideal entropy of mixing), then the right-hand side of Equation (48) is equal to zero therefore the activity coefficient of each component is equal to 1, and the activity of each component corresponds to its mole fraction. This is what is called an ideal mixture. For real mixtures, instead, the activity coefficient of each component describes the contribute of that component to the energetic state of the mixture, Equation (48), beyond ideal mixing, that is, when the components are not simple inert objects mechanically put together.

Equation (48) allows us to determine the activity coefficient γ_i for use in the chemical equilibrium equations (27) or (28–29), from an expression for the excess Gibbs energy of the liquid mixture. Such an expression is a matter of modeling, and the following section describes the (generic) regular solution approach that is of widespread use in multi-component phase equilibria including volatile–melt equilibrium. That leads directly to introduce the first class of volatile saturation models described in this chapter.

REGULAR SOLUTION MODELING APPROACH TO VOLATILE–MELT THERMODYNAMICS

Excess Gibbs energy

As it is shown above, the equilibrium equations (27) (pure component gas phase, corresponding to volatile solubility) or (28–31) (two-component gas phase, or saturation conditions) are closed by appropriate expressions for the fugacity coefficient in the gas phase (Eqn. 32), reference fugacity in the liquid phase (Eqn. 38), and activity coefficient in the liquid phase (Eqn. 48). In turn, that requires an EOS for the gas phase ($\overline{v}_i$ at Eqn. 32), an EOS for the reference state of the component in the liquid phase (v_i at Eqn. 38), and an expression for the excess Gibbs energy of the liquid (G^{E} at Eqn. 48). Defining such quantities and their dependence on the P–T–X conditions is usually a matter of the specific problem under consideration, as well as of the desired accuracy in the calculations. In any case, it is relevant to understand that the choice of an EOS for the reference state in the liquid, and the form of the excess Gibbs energy, are not independent. This aspect will be considered in more detail below.

Equation (48) highlights the relevance of a proper expression for the excess Gibbs energy of the liquid mixture formed by the silicate melt plus dissolved volatiles: once such an expression is given, the activity coefficient of each dissolved component (including the dissolved volatile components) is defined, as the activity coefficient of each component is nothing but a measure of the contribution of that component on the excess Gibbs energy (or better said, a measure of the partial molar excess Gibbs energy of that component in the mixture). The excess Gibbs energy, and the activity coefficients associated to the mixture components, describe the deviation of the mixture from ideal behavior, or its real behavior (as shown above, for an ideal mixture $G^{\mathrm{E}} = 0$ and $\gamma = 1$). One of the most common ways to express G^{E} is that of employing an expansion into coefficients describing the interactions between the components making up the mixture (e.g., Gokcen 1996). Such interactions can be of various order, e.g., at the first order we have the interactions between any pair of components, at the second order the interactions between triplets of components (or between each component and all other pairs of components), etc. Obviously, higher orders provide in principle higher accuracy, but require more experimental data to constrain the increased number of interaction coefficients, and may easily result in data overfitting. Therefore, only binary interactions are employed in most practical applications, and symmetric interactions are usually assumed, meaning that the coefficient describing the interaction of component a with component b is assumed to be the same as that describing the interaction of b with a. Additionally, the interaction of a component with itself is assumed to be zero. Such an approach is especially useful when the number of components in the mixture is large; e.g., if only major oxides plus dissolved H_2O and CO_2 were assumed to describe natural magmas, then the number of components would be 12 (10 major oxides plus two volatiles), and the number of symmetric binary interaction coefficients to be calibrated from experimental data would be 60 (given by $12^2/2-12$). The excess Gibbs energy is therefore written as

$$G^E = N\sum_{i=1}^{n-1}\sum_{j=i+1}^{n} x_i x_j w_{ij} \tag{49}$$

where the w_{ij}'s are symmetric binary interaction coefficients:

$$w_{ij} = w_{ji}, \; w_{ii} = 0 \tag{50}$$

A form of the excess Gibbs energy like the one at Equation (49) is usually referred to as a "regular solution" model. Actually, strictly speaking a regular solution is one for which the excess Gibbs energy does not depend on temperature. If G^E is independent from both pressure and temperature, then the model is of the "Scatchard–Hildebrand" type, also termed "strict regular solution" model. In practice, P–T dependency can be embedded in the w_{ij} terms (which would therefore not be "coefficients" anymore), resulting in non-regular (T-dependent) or non-isometric (P-dependent) mixture models. In less strict terms, and according to a large literature, all of these models are termed here "regular solution" models.

In order to understand the meaning and consequences of the above different models for G^E, consider the first of Equations (43), the expression of G at Equation (44), and the definition of partial molar quantity at Equation (14). From these equations we can write an expression for the partial molar volume of each volatile in the liquid mixture:

$$\bar{v}_i = \left(\frac{\partial V}{\partial n_i}\right)_{P,T,n_{j\neq i}} = \frac{\partial}{\partial n_i}\left[\left(\frac{\partial G}{\partial P}\right)_{T,n}\right]_{P,T,n_{j\neq i}} = v_i^0 + \frac{\partial}{\partial n_i}\left[\left(\frac{\partial G^E}{\partial P}\right)_{T,n}\right]_{P,T,n_{j\neq i}} \tag{51}$$

In this equation, v_i^0 is the molar volume of the component i (a dissolved volatile) in its reference state, thus, it coincides with v_i at Equation (38), as the two quantities have the same definition. It is clear, therefore, that the functional form of v_i for use within Equation (38), and that of G^E, are linked, as they contribute to define the partial molar volume of the volatile component dissolved in the liquid. A similar argument leads to define a link between the partial molar entropy and partial molar enthalpy of each dissolved volatile component, and the temperature dependency of the excess Gibbs energy. From the above it immediately follows that any choice of a specific functional form for G^E immediately translates into relevant assumptions on the actual behavior of the mixture and of the dissolved volatile components, impacting the definition of a reference state for the dissolved volatiles; e.g., if the w_{ij} terms involving a given volatile component do not depend on pressure, it follows (Eqn. 51) that the excess partial molar volume of the dissolved volatile component is zero, thus the reference molar volume of the volatile at Equation (38) corresponds to the partial molar volume of that volatile in the mixture. If the reference molar volume is taken—as it is usual, see below—as a function of P and T but not of composition, the consequence of an isometric mixing model (that is, P-independent w_{ij} terms involving volatiles) is a model that assumes that the partial molar volume of the volatile dissolved in the melt does not depend on melt composition. Similarly, T-independent w_{ij} terms involving a volatile would imply composition-independent partial molar entropy of that dissolved volatile component (and only linear dependence on T of the w_{ij} terms involving the volatile would imply composition-independent partial molar enthalpy, see Eqn. 59 below).

The discussion above shows that the choice of an EOS for the reference molar volume of the dissolved volatile at Equation (38) and that for the form of the excess Gibbs energy are not independent, as anticipated above; rather, they concur to provide a description of the real behavior of the volatile dissolved in the melt. It is easily understood that although the equations introduced so far do not deal with dissolution reactions and volatile speciation in the melt (which are instead the basis of treatment in the next section), different functional forms for the w_{ij} terms involving volatile components can mimic some of the consequences of volatile speciation; e.g., causing the partial molar volume of the dissolved volatile to depend in a more or less complex way on melt composition (besides P and T) as if melt composition was controlling the effective volume occupied by the dissolved volatile. That is known to be the case for dissolved carbon dioxide, which is dissolved about entirely in molecular form in acidic melts, and about entirely in the form of carbonate ion (CO_3^{2-}) in mafic melts (see below).

In light of the above discussion, it is necessary to keep in mind that any model is only functional to its purpose (and it should be evaluated only for its performance in relation to that purpose). That may appear as an obvious statement, but it is in fact more relevant than it appears. Translated in practical terms, if the aim of a model is that of computing with a certain accuracy the volatile saturation surface, then it should not be expected to also model with similar accuracy other quantities that appear in the model, e.g., the partial molar volume or partial molar entropy of the dissolved volatiles. In fact, the accuracy required for those quantities in the model is only the one which is necessary to satisfactorily compute the saturation surface, which is the model objective.

Early models

Spera (1974) employed an expansion in binary interaction coefficients similar to Equation (49), but differing from it for the fact that the w_{ij} quantities were assumed to be asymmetric (meaning that in general, $w_{ij} \neq w_{ji}$). In addition, the w_{ij} terms were taken as pressure dependent (thus, according to Equation (50), involving a non-zero excess partial molar volume). The model aimed at describing H_2O solubility in three specific silicate melts of basaltic, andesitic and pegmatitic composition, for which experimental H_2O solubility had been measured for constant temperature and at pressure up to 700 MPa (Burnham and Jahns 1962; Hamilton et al. 1964). The asymmetric assumption allows calculation of both H_2O solubility in the melt phase, and melt solubility in the gas phase, leading to a description of the phase diagram of the melt–H_2O system which included domains of immiscibility of the two components. On the other hand, the asymmetric assumption requires calibration of a double number of model parameters, thus requiring either a sufficiently large number of experimental data or a low number of model components. For this reason Spera used as components only H_2O and a second component with the same composition as the anhydrous melt.

Nicholls (1980) used a formalism based on Equation (49) in the theoretical description of his model for H_2O solubility in silicate melts, but did not derive numerical values for the binary interaction coefficients. Instead, the role of melt composition was embedded in only three adjustable parameters generically defining the melt under consideration (e.g., "basalt", "rhyolite", etc.).

A far more comprehensive approach was followed by Ghiorso et al. (1983), who presented the first model of H_2O solubility in silicate melts of virtually any composition. The model employed equality of chemical potentials between water distributed in the melt phase and in a pure-water gas phase, and constrained the symmetric binary coefficients in Equation (49) through comparison with experimental water solubility data referring to different melt compositions. Although not dealing directly with water speciation in the melt, the model of Ghiorso et al. (1983) (and that of Ghiorso and Gualda 2015, described below in more details) introduce a modification in the relationship between activity coefficient (or chemical potential) of dissolved water, and melt excess Gibbs energy (Eqn. 57 below), to account for their assumption of complete dissociation of dissolved water into hydroxyl groups. That modification assumes therefore that the activity of dissolved water, now referred to hydroxyl groups, is proportional to the square of the concentration of dissolved water on a molecular basis. Although allowing calculation of composition-dependent water solubility in silicate melts, the model in Ghiorso et al. (1983) was derived in a context that was not exclusively nor principally aimed at volatile–melt thermodynamics. As a consequence, the data employed for calibration were only limited, and no extensive comparison between model predictions and experimental data was provided to evaluate model performance. These limitations were largely overcome years later by Papale and co-workers and then by Ghiorso and co-workers, leading to the SOLWCAD and MagmSat models, respectively, for the volatile saturation surface of $H_2O–CO_2$–melt systems of virtually any composition. These two models are described in details in the following.

The SOLWCAD model (Papale et al. 2006)

Papale (1997) calibrated a solubility model for H_2O and CO_2 as pure gas components, which was then refined and extended in Papale (1999a) to two-component $H_2O + CO_2$ saturation for real gas–melt mixtures of virtually any composition from synthetic two-component to natural magmas over a wide P–T range covering up to some tens of km depth below the Earth surface. The model was then re-calibrated in Papale et al. (2006), which is the currently updated reference for the SOLWCAD model.

SOLWCAD solves the equilibrium Equations (28–31), together with a modified Redlich–Kwong EOS from Kerrick and Jacobs (1981) for $H_2O–CO_2$ gas mixtures providing for each P–T–X condition the partial molar volumes of H_2O and CO_2, thus the fugacity coefficients in the gas phase via Equation (32). As it is shown above, solution of the equilibrium equations requires the definition of a reference volume for each dissolved volatile in the liquid phase as a function of P–T for use in Equation (38), and of a proper form for the excess Gibbs energy in order to obtain by its derivation (Eqn. 48) the activity coefficient of each dissolved volatile. For water, a 10-parameter polynomial expression from Burnham and Davis (1971), describing the partial molar volume of water in melts with albitic composition as a function of pressure and temperature, was employed as the reference molar volume of dissolved water. For carbon dioxide no similar equation was available. By analogy with water, a similar 10-parameters equation was initially employed, which was then reduced in Papale et al. (2006) to a 3-parameter expression (linear in P and T). For the excess Gibbs energy, a form similar to Equation (49) was employed, but without any a-priori assumption on the w_{ij} parameters involving each dissolved volatile; instead, many different functional forms were employed, from constant values (strictly regular solution) to various P–T dependencies (non-regular and/ or non-isometric solutions), and the performance deriving from each different choice was compared. The w_{ij} parameters not involving volatiles were taken from Ghiorso et al. (1983) in order to minimize the number of total parameters to calibrate. Accordingly, the modified liquid components used in Ghiorso et al. (1983) are also employed in SOLWCAD. All of the available data pertaining to H_2O solubility, CO_2 solubility, and mixed $H_2O + CO_2$ saturation in silicate melts from 1-component plus volatile(s) H_2O and/or CO_2 to natural compositions were employed in the regression procedure, summing up to >1200 lines in the regression matrix. A robust linear regression technique was designed that automatically identifies outliers (points lying far from the majority of others) and excludes them from the regression. Model performance was evaluated by considering the balance between minimization of the sum of squared residuals and number of model coefficients constrained through regression.

The best forms for the w_{ij} parameters resulting from any subsequent calibration of SOLWCAD (corresponding to the model parameters published in Papale 1997, 1999a, and Papale et al. 2006) turns out to be different for H_2O and CO_2:

$$w_{H_2Oi} = w_{H_2Oi}^{(0)} \tag{52}$$

$$w_{CO_2i} = w_{CO_2i}^{(0)} + w_{CO_2i}^{(1)} \ln \frac{P}{P^0} \tag{53}$$

As discussed above, Equation (52) implies that in order to describe H_2O solubility and mixed H_2O–CO_2 saturation, there is no need to add any non-regular, non-isometric behavior to the liquid mixture as the result of H_2O dissolution; or in other words, the saturation surface of H_2O can be computed by assuming no excess volume or entropy as a result of H_2O dissolution in the melt. On the contrary, Equation (53) implies that the same does not hold for CO_2: when isometric mixture behavior (zero excess volume) is assumed for CO_2 dissolution, the model performance is sensibly worsened, requiring a non-isometric mixing term (the pressure-dependent term in Eqn. 53) the costs of which—doubling the number of model coefficients—are more than compensated by the increase in accuracy. Note the wording used here, and remember (see above) the key concept that *any model is only functional to its purposes*: in no way do we mean that water dissolution in silicate melts *is* a regular, isometric process. Rather, we mean that in order to represent water solubility/saturation, non-regular, non-isometric effects, if existing (as they likely do), can be neglected without substantial loss of accuracy; whereas for carbon dioxide, neglecting non-isometric effects would be too rough an approximation, and the costs of including those effects are more than rewarded in terms of required accuracy.

We should wonder why modeling water and carbon dioxide dissolution in silicate melts requires such markedly different descriptions as from Equations (52–53). One possible answer might be that the CO_2 data are overall less internally consistent than for H_2O. As a matter of fact, in the case of water the original calibration presented in Papale (1997) is substantially the same as that in Papale (1999a) and Papale et al. (2006) in spite of an about 75% expansion of the regression dataset. That is as to say that within model uncertainty (for water, less than 10% in most conditions, comparable with most experimental uncertainties), H_2O solubility data produced during nearly ten years from the first to the last model calibration were anticipated by the model predictions. The situation for CO_2 is radically different; quite often new data contradicted previous ones, and new model calibrations produced therefore significantly different values of the fitting parameters. Under such conditions, the larger number of model parameters required to satisfactorily fit CO_2 saturation contents as from Equation (53) may reflect internal data inconsistencies rather than some physical aspect of CO_2 dissolution. However, the different forms for H_2O and CO_2 at Equations (52–53), not established a-priori but determined a-posteriori from the data themselves, actually embed a substantial, first order difference between H_2O and CO_2 dissolution in silicate melts. This is discussed in the following.

Equation (51), with G^E given by Equation (49) with interaction parameters involving volatiles at Equations (52–53), leads to the following expressions for the partial molar volume of H_2O and CO_2 dissolved in the melt:

$$\bar{v}_{H_2O} = v_{H_2O}^0 + \bar{v}_{H_2O}^E = \bar{v}_{H_2O,\text{albite}}\left(P,T\right) - \tag{54}$$

$$x_{CO_2}\left(1 - x_{H_2O} - x_{CO_2}\right)\frac{1}{P}\sum_{i \neq H_2O=1}^{n} x_i' w_{CO_2i}^{(1)} + \left(1 - x_{H_2O}\right)x_{CO_2}\left(\frac{\partial w_{H_2OCO_2}}{\partial P}\right)_{T,n}$$

$$\bar{v}_{CO_2} = v^0_{CO_2} + \bar{v}^E_{CO_2} = v^0_{CO_2}(P,T) + \tag{55}$$

$$\left(1 - x_{H_2O} - x_{CO_2}\right)\frac{1}{P}\sum_{i \neq H_2O=1}^{n} x'_i w^{(1)}_{CO_2 i} + x_{H_2O}\left(1 - x_{CO_2}\right)\left(\frac{\partial w_{H_2OCO_2}}{\partial P}\right)_{T,n}$$

In Equations (54) and (55) the volatile-free melt composition is expressed in terms of x'_i quantities, which are mole fractions corrected so to be independent from the amount of dissolved volatiles:

$$x'_{i \neq H_2O,CO_2} = \frac{x_i}{1 - x_{H_2O} - x_{CO_2}} \tag{56}$$

The two expressions above for partial molar volumes present a substantial difference deriving from the different forms of Equations (52–53) for H_2O and CO_2: in fact, Equation (54) shows that in the absence of CO_2, the partial molar volume of water is equal to its reference partial molar volume and it is therefore independent from melt composition. On the contrary, the partial molar volume of carbon dioxide always carries an excess contribution, and it always depends on melt composition. That difference has a counterpart in the different dissolution mechanisms of H_2O and CO_2: for the former, the distribution between hydroxyl ions OH^- and molecular water is a strong function of the total amount of dissolved water (i.e., a strong function of pressure) and a much weaker function of melt composition (Stolper 1982a), whereas for the latter, the distribution between carbonatic ions CO_3^{2-} and molecular carbon dioxide is largely controlled by melt composition, with carbonate ions dominating in mafic melts, and molecular CO_2 dominating in more silicic melts (Fine and Stolper 1985, 1986, and many subsequent studies by them and others). Because the effect on mixture volume of dissociated and molecular species is largely different, it is not surprising that a compositional dependence seems to be required to accurately describe CO_2 solubility, while it is unimportant for H_2O solubility. Note, again, that the excess part of the partial molar volume of dissolved H_2O in Equation (54) is just an indirect effect, entirely due to the presence in the mixture of a component, CO_2, that mixes non-isometrically, thus causing itself a composition-dependent excess term in the partial molar volume of all other components. In light of Equations (54–55) and knowledge of the modes of dissolution of water and carbon dioxide in silicate melts, Equations (52–53) take a more profound meaning: the strong control of melt composition in determining the dissolution reactions of carbon dioxide implies a strong dependence of the partial molar volume of dissolved CO_2 on melt composition, thus pressure-dependent interaction coefficients involving CO_2 consistent with non-zero excess volume of CO_2-bearing melts. For water, instead, its partial molar volume is much more a function of pressure than melt composition, and that is accounted for by pressure-dependent reference molar volume of dissolved water with no need of compositional terms thus no need of non-isometric mixing terms. In other words, while for both volatiles their partial molar volumes carry a dependence on melt composition, that dependence turns out to be sufficiently weak for water to be neglected with the purpose of determining H_2O solubility, whereas it appears too strong for carbon dioxide to be neglected without losing too much accuracy in the determination of its solubility.

The discussion above embeds one of the most effective aspects of thermodynamics: being based on macroscopic quantities like volume, energy, entropy, etc., thermodynamics builds a bridge with the microscopic processes concerning chemical reactions taking place at the molecular scale, and even if not taking explicitly account for such reactions, it is able to implicitly and consistently account for them. If we did not know anything about the modes of dissolution of water and carbon dioxide in silicate melts, from solubility data only—which do not include any direct information about chemical reactions—and robust thermodynamic modeling we would be justified in hypothesizing a major difference between H_2O and CO_2 dissolution, such to imply a much stronger compositional control for the latter.

With the w_{ij} parameters involving the dissolved volatiles, Equations (52–53), we have a complete expression for the excess Gibbs energy of the liquid (or melt) phase, Equation (50), that can be derived according to Equation (48) to obtain the activity coefficients of dissolved water and carbon dioxide:

$$RT \ln \gamma_{H_2O} = \left(1 - x_{H_2O}\right)x_{CO_2} w_{H_2OCO_2} + \left(1 - x_{H_2O}\right)\left(1 - x_{H_2O} - x_{CO_2}\right) \times \tag{57}$$

$$\sum_{i \neq CO_2 = 1}^{n} x_i' w_{H_2Oi}^{(0)} - x_{CO_2}\left(1 - x_{H_2O} - x_{CO_2}\right) \times$$

$$\left[\sum_{i \neq H_2O = 1}^{n} x_i' w_{CO_2i}^{(0)} + \ln \frac{P}{P^0} \sum_{i \neq H_2O = 1}^{n} x_i' w_{CO_2i}^{(1)}\right] - \left(1 - x_{H_2O} - x_{CO_2}\right)^2 \sum_{i \neq H_2O, CO_2 = 1}^{n-1} \sum_{j \neq H_2O, CO_2 = i+1}^{n} x_i' x_j' w_{ij}$$

$$RT \ln \gamma_{CO_2} = \left(1 - x_{CO_2}\right)x_{H_2O} w_{H_2OCO_2} - x_{H_2O}\left(1 - x_{H_2O} - x_{CO_2}\right) \times \tag{58}$$

$$\sum_{i \neq CO_2 = 1}^{n} x_i' w_{H_2Oi}^{(0)} + \left(1 - x_{CO_2}\right)\left(1 - x_{H_2O} - x_{CO_2}\right) \times$$

$$\left[\sum_{i \neq H_2O = 1}^{n} x_i' w_{CO_2i}^{(0)} + \ln \frac{P}{P^0} \sum_{i \neq H_2O = 1}^{n} x_i' w_{CO_2i}^{(1)}\right] - \left(1 - x_{H_2O} - x_{CO_2}\right)^2 \sum_{i \neq H_2O, CO_2 = 1}^{n-1} \sum_{j \neq H_2O, CO_2 = i+1}^{n} x_i' x_j' w_{ij}$$

Regression of model coefficients has further shown that the H_2O–CO_2 interaction term, $w_{H_2OCO_2}$, can be neglected without loss of accuracy; and that due to their general low abundance in silicate melts, and corresponding negligible abundance in the experimental melts, the coefficients describing the interaction between dissolved volatiles and Ti and Mn oxides (or more precisely, the pseudo-components in Ghiorso et al. 1983, also employed in this model, involving Ti and Mn) can also be neglected. The resulting model coefficients calibrated by regression of the experimental data is therefore 9 for water (8 H_2O–oxide interaction coefficients, plus the $f_{H_2O}^{0L}(P^0, T^0)$ term in Eqn. 38), and 20 for carbon dioxide (16 CO_2-oxide interaction coefficients, plus the $f_{CO_2}^{0L}(P^0, T^0)$ term in Equation (38), plus three a parameters in the equation $v_{CO_2} = a_1 + a_2 T + a_3 P$ for use in Equation (38), also corresponding to $v_{CO_2}^0$ in Eqn. 55). P^0 in Equations (57–58) is taken equal to 0.1 MPa. The calibrated model parameters and their standard deviations are reported at Table 2 in Papale et al. (2006). The SOLWCAD code can be freely accessed from http://www.pi.ingv.it/progetti/eurovolc/#VA, where online computations can be performed, and from where both a stand-alone application and the open source Fortran routine can be downloaded.

As it is abundantly shown in Papale (1997, 1999a) and Papale et al. (2006), and further illustrated below, SOLWCAD is capable of tracing the effects of even tiny compositional variations. Among the many experimental observations that are satisfactorily modeled by SOLWCAD, we remind here the change from retrograde to prograde solubility of water in albitic melts with increasing pressure (Paillat et al. 1992); the so-called "alkali effect" (Holtz et al. 1995) consisting in a progressive decrease of water solubility with decreasing Na_2O along the albite–K-feldspar join; the so-called "plagioclase minimum" (Paillat et al. 1992) consisting of a minimum in water solubility at about half distance along the albite–anorthite join; the progressive increase in carbon dioxide solubility with increasing degree of silica undersaturation (Blank and Brooker 1994); the high affinity of carbon dioxide with calcium resulting in the capability of carbonatite melts to dissolve up to nearly 30 wt% CO_2 at atmospheric pressure (Moussallam et al. 2015); and many others, either revealed by to-date experimental data or not yet illustrated experimentally.

Duan (2014) set a model of H_2O–CO_2 saturation in silicate melts which is very similar to the SOLWCAD model of Papale et al. (2006), but which employs i) a different equation of state for the gas phase (Duan and Zhang 2006), ii) constant (no pressure dependence) $w_{CO_{2i}}$ terms, and iii) empirical relations for the P–T dependence of the chemical potential of dissolved volatiles in their reference state, thus with no use of Equation (38). The Duan (2014) model is available online at http://calc.geochem-model.org/Test/services/WebForm1.aspx.

The MagmaSat model (Ghiorso and Gualda 2015)

Ghiorso and Gualda (2015) developed a mixed H_2O–CO_2 saturation model which builds upon the previous H_2O solubility model in Ghiorso et al. (1983), briefly introduced above. Ghiorso et al. (1983) adopted oxide components for the melt phase deriving from the calibration of the SILMIN thermodynamic model for melt–crystal equilibrium (the main subject presented in Ghiorso et al. 1983). The new formulation in Ghiorso and Gualda (2015) takes into account the progress in modeling magma–crystal thermodynamics, and adopts oxide components and corresponding binary interaction coefficients from the MELTS model in Ghiorso and Sack (1995) and Gualda et al. (2012) . The most relevant differences with SOLWCAD are the following:

- use of oxide pseudo-components, and corresponding binary interaction coefficients, consistent with the MELTS formulation, allowing computations of H_2O–CO_2 saturation to be performed within the frame of more comprehensive thermodynamic modeling including liquid–solid–gas equilibrium relationships;

- adoption of a more recent formulation for $H_2O + CO_2$ gas phase thermodynamics from Duan and Zhang (2006);

- assumption of P–T independent excess Gibbs energy (strictly regular solution model), implying a-priori neglect of any excess term, thus compositional independence of the partial molar volume and entropy of the dissolved volatiles;

- assumption of a linear relationship between activity and square of the mole fraction of dissolved water, motivated by assumed complete dissolution of water into hydroxyl groups;

- adoption of two different carbon species in the melt, namely molecular CO_2 and calcium carbonate $CaCO_3$.

The total number of model parameters calibrated upon regression of the experimental data is 40 (versus 29 parameters in SOLWCAD), 15 for water and 25 for carbon dioxide. Online calculations with MagmaSat can be done at http://melts.ofm-research.org/CORBA_CTserver/GG-H2O-CO2.html.

Because of the last two points above, the model of Ghiorso and Gualda (2015) includes elements that are more deeply considered in the next section dedicated to reactive species-based approaches. The linear relationship between activity and square of the molar fraction for dissolved H_2O, already introduced above, takes its motivation from experimental data dating back to Burnham and Davis (1971), who found such a linear relationship in the albite–water system for $x_{H_2O} < 0.5$ (and positive deviations above it). Such a relationship is similar to a Henry's law for strong binary electrolytes in aqueous solutions, and for that reason it was termed a "Henry's law analogue" by Burnham (1979). It was interpreted as due to a solution mechanism whereby molecular water in the gaseous phase reacts with oxygens in the silicate melt to form two OH^- groups, or one OH^- plus one OM^- (where M is an exchangeable cation, such as Na in albitic melts). That model, subsequently generalized to magmas of any composition in Burnham (1975, 1994) and Burnham and Nekvasil (1986), is described in details in the following section, and represented for many years the reference for predicting water saturation in silicate melts (e.g., Holloway and Blank 1994). Subsequent approaches, discussed in the next section and based on the seminal work by Stolper (1982a,b), developed further those ideas by considering two equilibria, a heterogeneous one between molecular water in the gas and melt phases, and a homogeneous one within the melt phase between dissolved water in molecular and hydroxyl (OH^-) form. All of those models imply a linear relationship between activity and squared mole fraction of dissolved water, which is then assumed in MagmaSat to take place for any amount of dissolved water. Ghiorso and Gualda (2015) therefore added a water-dependent term to the expression of the excess Gibbs energy at Equation (49):

$$G^E = NRT \left[x_{H_2O} \ln x_{H_2O} + \left(1 - x_{H_2O}\right) \ln\left(1 - x_{H_2O}\right) \right] + N \sum_{i=1}^{n-1} \sum_{j=i+1}^{n} x_i x_j w_{ij} \tag{59}$$

Performance of the SOLWCAD and MagmaSat models

In general terms, the two SOLWCAD and MagmaSat models both describe H_2O and CO_2 solubility, and mixed H_2O–CO_2 saturation, in silicate melts of virtually any composition from simple synthetic melts to natural ones, over an ample *P–T* range encompassing conditions of interest for most problems in crustal petrology and volcanology. Differences in the calculations reflect the different details of the model, and particularly i) the use of different binary interaction coefficients not involving volatiles, ii) the adoption of isometric vs. non-isometric mixing of carbon dioxide in MagmaSat and SOLWCAD, respectively, and iii) the use in MagmaSat of speciation models for the dissolved volatiles.

Figure 1 illustrates examples of calculations for H_2O and CO_2 solubility in a variety of melts:

- Panel a) refers to H_2O, and to several different compositions in the haplogranitic system. For each *P–T* pair in the figure, the compositions on the horizontal axis are arranged with decreasing silica content from left to right. Each ascending ramp in the data as well as the model calculations reflects progressive substitution of potassium with sodium oxide ("alkali effect", Holtz et al. 1995).

- Panel b) refers to water solubility in melts with basaltic, rhyolitic and phonolitic composition. Both the SOLWCAD and MagmaSat models correctly describe the overall experimental trends, and concur with the data to suggest increasing H_2O solubility when moving from basalt to rhyolite to phonolite.

- Panel c) compares experimental data and calculations for CO_2 solubility in a rhyolitic melt. Panel d) shows the case for CO_2 solubility and a tholeiitic melt for various temperatures and three pressure values from 100 MPa to 1.5 GPa.

- Panels e) and f) illustrate the dependence on temperature of CO_2 solubility for two different melts corresponding to a haplo-phonolite (e) and a Ca-rich leucitite (f), for pressures in the range 0.5–2.5 GPa. Both models agree with the data in showing, for both melts, retrograde solubility (i.e., decrease with increasing *T*). In the haplo-phonolite case, however, the MagmaSat model largely underestimates the increase in CO_2 solubility with pressure. In the Ca-rich-leucitite case, the two models tend to either underestimate (SOLWCAD) or overestimate (MagmaSat) CO_2 solubility for any *P–T* in the considered range.

The calculation examples in Figure 1 illustrate the capability of the regular solution modeling approach to trace the complex features of the saturation surface of H_2O and CO_2 in the hyperspace made of *P*, *T*, and *n* compositional axes, with n equal to the number of oxide components (or their combinations) forming natural as well as synthetic silicate melts. The two analyzed models are similar in their general approach, but differ for a number of aspects discussed above. In general, the significantly larger number of regression parameters (40 vs. 29) of the MagmaSat model does not appear to translate into improved performance; whereas non-isometric CO_2 mixing in SOLWCAD may contribute to explain a different behavior with increasing pressure for some of the cases in Figure 1.

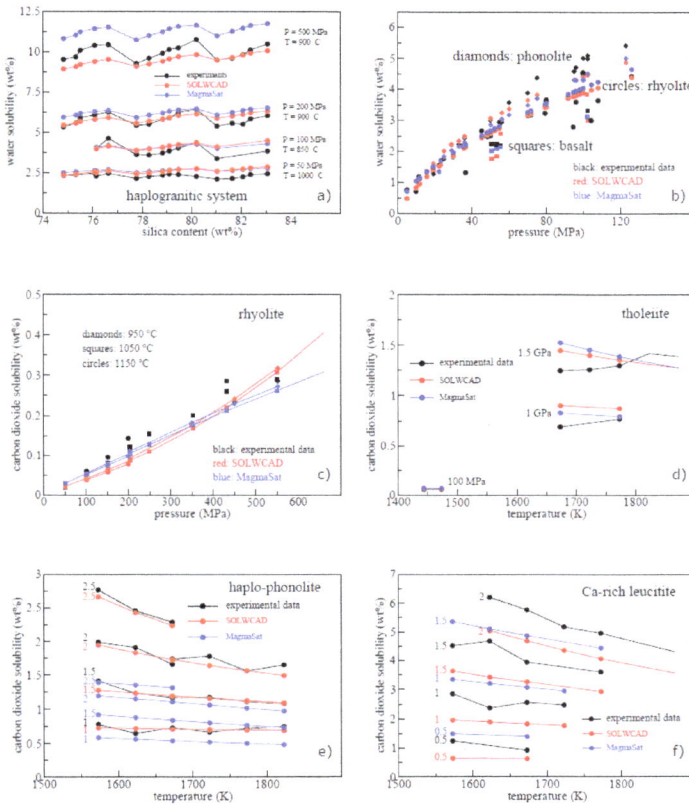

Figure 1. Comparison between experimental data and model calculations with the regular solution models SOLWCAD and MagmaSat, for a variety of silicate melt compositions. **a)** Solubility of water in melts with different compositions in the haplogranitic system (data from Holtz et al. 1995). Every ascending section corresponds to a progressive increase in the Na/K ratio. **b)** Solubility of water in three melts with basaltic (data from Berndt et al. 2002), rhyolitic (data from Silver et al. 1990), and phonolitic (data from Carroll and Blank 1997) composition. **c)** Solubility of carbon dioxide in rhyolite (data from Fogel and Rutherford 1990). **d)** Solubility of carbon dioxide in tholeiite (data from Pan et al. 1991). **e)** Solubility of carbon dioxide in haplo-phonolite (data from Morizet et al. 2002). Numbers are pressure in GPa. **f)** Solubility of carbon dioxide in Ca-rich leucitite (data from Thibault and Holloway 1994). Numbers are pressure in GPa.

One relevant aspect of non-ideal thermodynamic models is that they allow placing limits on the use of simple ideal relationships and assumptions. Figure 2 shows an example referring to Henrian behavior of water and carbon dioxide in equilibrium with natural magmas. Henry's law has been defined above, immediately after the derivation of the fundamental equilibrium Equation (26), where it has been shown that it derives directly from Equation (26) in the ideal limit case. Henry's law applied to magmas states that for multi-component volatiles at constant pressure, temperature, and volatile-free melt composition, the concentration of each volatile in the melt is proportional to its partial pressure in the gas phase (the constant of proportionality being thus the Henry's constant). Therefore, Henry's law allows in principle calculation of isobaric multi-component saturation from solubility data only (that is, from data on one-component gas phase), and for that reason, it looks very attractive as a quick means of computation. However, use of full non-ideal thermodynamic modeling in Figure 2 shows that the region where Henry's law can be applied to natural magmas with reasonable approximation

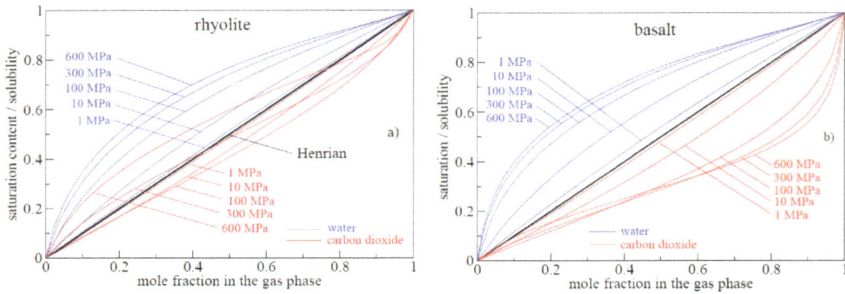

Figure 2. Real vs. ideal (Henrian) behavior of mixed H_2O–CO_2 fluids in silicate melts. The figure refers to two different melts of **a)** rhyolitic (from Taupo volcano, Yamashita 1999, $T = 850$ °C) and **b)** basaltic (from Mount Etna, Spilliaert et al. 2006, $T = 1000$ °C) composition. The vertical axis is normalized saturation content, that is, saturation content for each volatile divided by its solubility at the same P–T–X conditions. In such a plot, Henrian behavior corresponds to a straight line with slope equal to 1, indicated in the two panels by the **thick black line**. Water (**blue lines**) and carbon dioxide (**red lines**) mixed equilibrium conditions are simulated by the fully non-ideal SOLWCAD code (Papale et al. 2006), at the pressure values reported in the figure.

is limited to very low pressure of order 1 to a few MPa. At 10 MPa the deviations from Henry's law are already substantial for both rhyolitic and basaltic melts (and from other compositions not reported in the figure), and they become increasingly large at higher pressure. Water displays positive deviations from Henry's law (that is, more dissolved H_2O than predicted by Henry's law) for all conditions in Figure 2, while carbon dioxide shows instead dominantly negative deviations, but it can display positive deviations at large pressure and water-rich gas phase conditions, as visible in Figure 2a, for rhyolitic compositions.

The example in Figure 2 illustrates the benefits of non-ideal modeling beyond simple calculations, in this case by providing strict boundaries to the adoption of simple relationships (like Henry's law) deriving from ideal simplifications. Even more relevant than that, fully non-ideal modeling can be employed as a further guide in the interpretation of experimental results. We provide one example below.

One of the most noteworthy relationships that has been found to characterize dissolution of water, initially in albite–H_2O systems and then for a number of silicate melt compositions, was demonstrated by Burnham and Davis (1971, 1974), who found close-to-linear trends of activity of dissolved water when plotted against the square of its mole fraction. That is the "Henry's law analogue" relationship introduced above and interpreted in terms of a dissolution reaction whereby each initial water molecule reacts to form two hydroxyl groups (Burnham 1979). No similar relationship enters the formulation of SOLWCAD: from its equations, it can be easily seen that for a given temperature and volatile-free melt composition the activity of dissolved water is a complex function of its mole fraction: $a_{H_2O}^L = x_{H_2O} \exp\left[\left(1 - x_{H_2O}\right)^2 \cdot \text{const} \right]$ where the non-dimensional constant embeds the compositional and temperature terms. However, when the plot is done with SOLWCAD-computed quantities, close-to-linear trends do emerge for a variety of melt compositions and temperatures (Fig. 3). The trends in Figure 3 do not reflect any particular dissolution mechanism, as that is not part of the model formulation; rather, they result from the interplay between the quantities that appear in the above expression for activity, which reflects fundamental thermodynamics with the only addition that the excess Gibbs energy can be expressed as an expansion of binary interaction terms (Eqn. 49) with values calibrated against experimental solubility and saturation data.

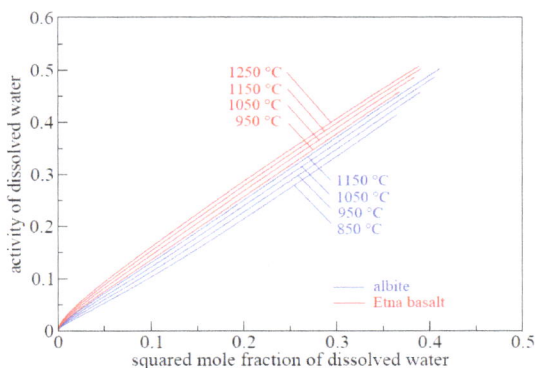

Figure 3. Computed activity–dissolved mole fraction relationships. The composition of the Etna basalt comes from Spilliaert et al. (2006). For clarity, no other compositions have been included in the figure. Although not deriving from any explicit linear relationship in the thermodynamic formulation, the trends above closely approximate linear relationships over most of their length. Calculations from the SOLWCAD model (Papale et al. 2006).

The trends in Figure 3 and the above discussion do not imply that water hydroxylation does not take place. On the contrary, that is known to be one major mechanism of water dissolution in silicate melts (as it is largely discussed in the following section). What Figure 3 provides is an understanding of the contributions that fully non-ideal thermodynamic modeling can bring to the interpretation of experimental data, whereby observed simple trends may be the result of more complex relationships rooted in fundamental aspects of thermodynamics; as well as a warning to the inclusion of artificial dependences between specific quantities in thermodynamic formulations, as that may obscure the actual relationships and result in a deviation from, rather than better approximation of, the observed trends.

Several examples of calculations of the mixed $H_2O + CO_2$ saturation surface can be found in Papale (1999a), Papale et al. (2006), and Ghiorso and Gualda (2015), as well as in the last section below dedicated to applications of the models described here. Other volatile components are considered in the following, including sulfur and noble gases, and calculations of mixed saturation contents are provided. A further section provides a number of examples of how computations of mixed volatile saturation conditions can be employed to solve practical issues in petrology, geochemistry, and volcanology.

REACTIVE SPECIES-BASED APPROACHES TO VOLATILE–MELT EQUILIBRIA

Making solubility models (nearly) ideal: the Burnham model for water solubility

We start this section on reactive species-based approaches to the determination of volatile solubility by presenting the Burnham model, anticipated above when discussing the modification introduced by Ghiorso and Gualda (2015) in describing the excess Gibbs energy of water-bearing melts. Strictly speaking, the Burnham model does not make use of chemical reactions, and it could equally be described elsewhere. However, the experimental findings that are at the basis of the model have been interpreted in terms of H_2O dissociation reactions, which provided a justification to the model formulation; and their impact has been such that other models (e.g., MagmaSat) include ad hoc modifications to account for water dissociation reactions as they were hypothesized and described by Burnham and co-workers.

As we have seen in previous sections, the problem of volatile solubility is a facet of modeling mixing properties in silicate melts (see Moretti and Ottonello 2021). Motivated by the need of understanding the magmatic origin of granites, Burnham and Davis (1974) studied the solubility of H_2O in albite ($NaAlSi_3O_8$) melts at different pressures and temperatures. From Equation (20) they related, for a constant T process, the change of the chemical potential to the change of fugacity (note that in the equations throughout this section we refer to the liquid phase as the "melt" phase, indicated by the superscript "M", to emphasize the relevance of the internal complex topological structure of silicate melts):

$$d\mu_{H_2O}^{M} = RTd\left(\ln f_{H_2O}^{M}\right) \tag{60}$$

and found by experiments that $f_{H_2O}^{M}$ goes to zero as the dissolved water in melts, $x_{H_2O}^{M}$, also goes to zero, but varies as $\left(x_{H_2O}^{M}\right)^2$ up to $x_{H_2O}^{M} \approx 0.5$. The same quadratic relationship applies if activity, $a_{H_2O}^{M}$, is adopted in place of $f_{H_2O}^{M}$ (the activity is in fact equal to fugacity divided by a reference fugacity, Equation (25), with the latter that can be chosen so to be a constant value). These results confirmed previous experimental findings by Hamilton et al. (1964) on basaltic and andesitic melts, and did not represented a true surprise to Burnham.

As anticipated, the fact that water solubility was found to increase with the square root of water fugacity was interpreted as reflecting the fact that dissolution of one mole of H_2O generates two moles of independent species. Burnham (1975) then proposed for hydrous albitic melts the following reaction:

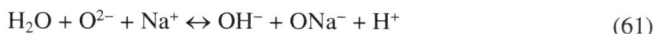

$$H_2O + O^{2-} + Na^+ \leftrightarrow OH^- + ONa^- + H^+ \tag{61}$$

Nevertheless, Burnham and Davis (1974) did not come out with a model of water speciation (see next section) and considered this reaction functional for water solubility but unlikely to occur, suggesting that a free proton cannot exist and that this is probably attached to ONa^-. The activity of H_2O for $x_{H_2O}^{M} \leq 0.5$ was then defined by a Henry's law analogue:

$$a_{H_2O}^{M} = k_{H_2O}^{M,feld}\left(x_{H_2O}^{M}\right)^2 \tag{62}$$

where $k_{H_2O}^{M,feld}$ is analogous to a Henry's law constant for melts of feldspar composition (albite, anorthite, orthoclase) and it was expressed as

$$\begin{aligned}
\ln k_{H_2O}^{M,feld} = 5 + &\left(4.481\times10^{-8}T^2 - 1.51\times10^{-4}T - 1.137\right)\ln P + \\
&\left(1.831\times10^{-8}T^2 - T + 4.656\times10^{-2}\right)\left(\ln P\right)^2 + \\
&7.8\times14.882\times10^{-5}\left(\ln P\right)^3 - 5.012\times10^{-4}\left(\ln P\right)^4 + \\
&5.754\times10^{-3}T - 1.621\times10^{-6}T^2
\end{aligned} \tag{63}$$

with P in bar and T in K.

For $x_{H_2O}^{M} > 0.5$, Equation (63) no longer holds. Extension to large dissolved water contents was accomplished with a relationship proposed in Burnham (1975):

$$a_{H_2O}^{M} = 0.25 k_{H_2O}^{M,feld}\exp\left(\frac{6.52 - 2667/T}{x_{H_2O}^{M} - 0.5}\right) \tag{64}$$

The above formulation was generalized and applied to a wide range of melt compositions, from basalt to granitic pegmatite (Burnahm 1975, 1994; Burnham and Nekvasil 1986), by expressing melt composition on the basis of 8 oxygens, by analogy with albite; e.g., SiO_2 becomes Si_4O_8, and diopside ($CaMgSi_2O_6$) becomes $Ca_{1.33}Mg_{1.33}SI_{2.67}O_8$. When normalized on an 8-oxygen basis, the solubility of many natural melts turned out to be close to that of albite. As outlined by Wood and Fraser (1975), 8-oxygen melt components should mix ideally or very close to ideality, and can represent a judicious choice of components.

Burnham (1981) then showed that in quartz melts (Si_4O_8) water solubility is well reproduced if the quantity 0.47 is added to $\ln k_{H_2O}^{M,feld}$ (so that the first term on the right side of Equation (63) takes the value 5.47 instead of 5). The same correction was found to apply for diopsidic melts, despite the different structural condition. Both compositions show in fact lower water solubility than for albitic melts (Fig. 4). This suggested that the offset value 0.47 between silica and diopside on one side, and feldspar on the other, may result from the additional interaction of water with Al-bearing complexes in feldspar melt compositions. Following this reasoning, nepheline melts ($Na_2Al_2Si_2O_8$) should have $\ln k_{H_2O}^{M,feld}+0.47$, given that there are 2 moles of Al per 8 oxygens. On this basis, Burnham (1994 and references therein) claimed that he could reproduce the much higher solubility of H_2O in nepheline melts, as had been described experimentally by Boettcher and Wyllie (1969). Using the 8 moles convention, Burnham (1975, 1979, 1981) proposed to perform a CIPW-like normative calculation of mineral-like melt components and to apply to $\ln k_{H_2O}^{M,feld}$ a correction factor accounting for the number of hydroxyl–aluminate complexes, N_{Al}:

$$\ln k_{H_2O}^{M} = \ln k_{H_2O}^{M,feld} + 0.47\left(1 - N_{Al}\right) \tag{65}$$

which can be generalized as (e.g., Holloway and Blank 1994):

$$\ln k_{H_2O}^{M} = \ln k_{H_2O}^{M,feld} + 0.47\left(1 - X_{Ab} - X_{Or} - X_{An} - 2X_{Ne}\right) \tag{66}$$

with X_{Ab}, X_{Or}, X_{An}, X_{Ne} being the molar fraction, respectively, of albite, orthoclase, anorthite and nepheline normative components on an 8-oxygen basis (Fig. 4).

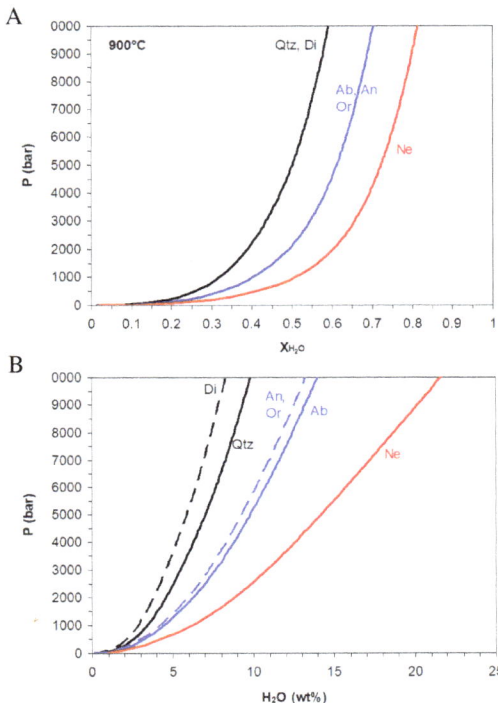

Figure 4. Calculated (Eqns. 63–66) water solubility ($a_{H_2O}= 1$) in feldspar (albite: Ab, anorthite: An; orthoclase: Or), nephelinitic (Ne), diopsidic (Di) and silica (Qtz) melts at 900°C. **a)** Molar fraction of dissolved water; **b)** weight percent of dissolved water. Note that the Burnham model returns the same solubility for Qtz–Di and Ab–Or–An melts when the dissolved mole fractions are computed (panel a). Compositional differences emerge when the different molar weights of the melt are used (panel b) to calculate mass fractions.

The contributions by Burnham and his co-workers have been invaluable. The thermodynamic properties of H_2O at conditions relevant for the Earth's crust were assessed for the first time (Burnham et al 1969), and the determination of partial molar properties of water in melts led to the first quantitative evaluation of the mechanical energy release upon water exsolution from magmas, with relevant implications for explosive volcanic eruption dynamics and igneous breccia formation (Burnham 1985). Today more advanced models, described here and in the previous section, provide more accurate descriptions of water solubility in silicate melts over wide P–T and compositional ranges. Still, the work by Burnham stands as the first comprehensive quantitative description of water–silicate melt thermodynamics, and one that continues to exert a substantial impact on subsequent investigation and models.

General aspects of the reactive species-based approaches to volatile–melt equilibria

Equation (61) suggests that a rather intuitive approach to volatile dissolution in melts is based on chemical reactions involving the volatile species and the silicate matrix. The same equation shows that these reactions cannot be simply expressed by neutral molecules, rather, they require the use of charged species to account for the chemical complexity of melts and their reactivity that modifies the molecular structure of both the solvent (the melt) and the solute (the volatile component). Major volatile solubility mechanisms are not simply mechanical, contrary to noble gases, but imply chemical reactions and a change in the speciation state of the dissolving component. As an example, the reactive solubility of carbon dioxide in silicate melts can be expressed by the heterogeneous (i.e., involving multiple phases) equilibrium

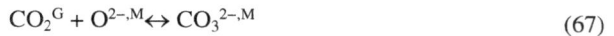

$$CO_2^G + O^{2-,M} \leftrightarrow CO_3^{2-,M} \tag{67}$$

which can be seen as the sum of

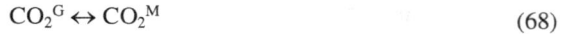

$$CO_2^G \leftrightarrow CO_2^M \tag{68}$$

which is an example of mechanical solubility, and

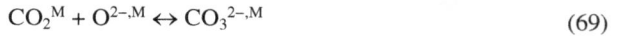

$$CO_2^M + O^{2-,M} \leftrightarrow CO_3^{2-,M} \tag{69}$$

The homogenous melt equilibrium Reaction (69) modifies the melt structure and properties depending on the speciation of dissolved gaseous CO_2 into dissolved molecular CO_2 and carbonate groups CO_3^{2-} (Fine and Stolper 1985, 1986). Reactive homogenous equilibria, such as the one in Reaction (69), imply additional deviations from ideality limiting further Henrian behaviors that may better characterize, to some extent, mechanical solubility mechanisms (e.g. Wilhelm et al. 1977). Therefore, deviations from either Raoult's or Henry's law are largely ascribed to the concentration of ion species such as CO_3^{2-} at Reaction (69), which enter the solubility mechanisms via ponderal factors given by the reaction stoichiometry.

In general, we can state that this class of solubility models is characterized by the appearance of 'collateral' ionic species (such as O^{2-} in Reaction (67)) that illustrate the role played by the composition of the solvent, i.e. the silicate melt, upon formation of reactive complexes. In the example of Reaction (67), the production of carbonate ion is favored by increasing O^{2-}, which we anticipate here to be a measure of the so-called melt basicity. This confirms the same trends observed with the regular solution models (SOLWCAD and MagmaSat) described above: the more abundant the basic oxides (e.g., MgO, CaO, K_2O, Na_2O), the larger the CO_2 solubility. We then start to see that such trends reflect the acid–base character of the melt and its controls on dissolution reactions. Like for any other multiphase system in chemistry, the solution behavior is determined by the acid–base character of the solvent, albeit different definitions and measures of that character (e.g., pH for aqueous solutions).

From the above discussion it follows that reactive volatile solubility models belong to the categories of ionic–polymeric models described in Moretti and Ottonello (2021).

In this section we describe the major features of such models applied to magmatic volatiles, and show how to develop a consistent thermochemical treatment in which species and components have different meanings and roles (see also Moretti and Ottonello 2021).

The models described in this section are intimately constrained by the stoichiometry of reactions on which they rely, and involve species that have distinct controlling effects on melt structure. Developing such a modeling approach strictly requires the identification of the chemical reactive entities, which can be pictured as quasi-chemical species which represent the components in this approach and whose bonding properties determine the solution behavior either as an acid or a base . The speciation state behind these models cannot be simply assumed, rather, it must be supported by spectroscopic observations, besides reproducing or explaining the solubility (and other) data. It should not be expected, however, that a reactive species-based approach includes or describes any single subtleties arising from spectroscopic studies. As for other approaches described above, the requirement is to include the minimum information necessary to achieve satisfactory model performance with reference to the model objective, which is that of computing volatile–melt equilibria. An analogy with thermodynamic models for aqueous solutions is useful. Those models are based on the description of the nature of solute–solvent and solute–solute interactions compatible with the bulk of information from the observed dissociation of aqueous species. These models assess the acid–base character of the solution, expressed in terms of pH, and how that character changes with bulk composition, with very limited knowledge of the structure of the solvent (i.e., liquid water).

H_2O models revisited: the role of speciation in ionic polymeric-approaches

Advances in the polymeric modeling of silicate melts have allowed a far better understanding of the reactivity of silicate melts. As it is discussed above, regular solution models for volatile solubility and mixed volatile saturation do not aim at accurately reproducing all of the thermodynamic properties of silicate melt–volatile systems. Rather, they search for the (relatively) simplest model that satisfactorily represents the thermodynamic properties at the minimum sophistication level required to model volatile solubility. An example comes from Equation (51): although the SOLWCAD (and MagmaSat) model implies a description of partial molar volumes (and partial molar entropies) of dissolved volatiles, those descriptions should only be regarded as the minimum level necessary to achieve satisfactory model performance. In fact, if one makes the derivation at Equation (51) with the binary interaction coefficients involving carbon dioxide at Equation (53), a compositional-dependent partial molar volume of dissolved CO_2 can be computed, but that quantity diverges at pressures approaching atmospheric pressure. That is because for the sake of computing solubility, the $w_{CO_2,i}^{(1)}$ coefficients at Equation (53) are only relevant at high pressure, whereas upon derivation as from Equation (51) they turn out to dominantly define the partial molar volume of CO_2 at low pressure. Clearly, Equation (53) can be adequate to account for the pressure dependence of CO_2 solubility, without necessarily describing CO_2 partial molar volume beyond the level required for the purpose of computing CO_2 solubility. Similarly, the neglect of any T-dependence in the excess Gibbs energy implies that the entropy of mixing is purely ideal, which is not the case when mixing melts with different silica content: large enthalpies of mixing develop, implying T-dependence of the excess Gibbs energy above linear (e.g., Wood and Fraser 1977).

The foundations of the ionic polymeric approach described here are profoundly different from those of regular solution models. Here the structures at ionic scale constitute the basis for calculations of melt properties, in principle including those defining fluid rheology. Because the macroscopic behavior emerges as a consequence of the microscopic arrangement, the dissolution reactions and volatile speciation are or primary relevance. It is worth noting that such models allow describing melts in terms of their acid–base properties (see also Moretti and Ottonello 2021). This provides a conceptual framework similar to that of aqueous solutions, and that allows treating any multiphase equilibrium in which melt reactive entities participate to achieve neutralization, e.g., mineral precipitation, thus determining element partitioning.

Within such a comprehensive approach, the ability to find more convenient ways to describe non-ideal behaviors deriving from constituent interactions upon mixing is fundamental to define a manageable model. Toop and Samis (1962a,b) made the wise choice of ascribing non-idealities to a recurrent reaction mechanism forming a silicate chain (polymer) via formation and disruption of highly reactive oxo-bridges. Based on the work of Fincham and Richardson (1954), Toop and Samis developed a scheme in which the reactivity of oxo-bridges of a silicate oxyanion (chain, ring, tridimensional structures etc…) is independent of the size of the oxyanion. Basically, polymerization proceeds independently of the length of the polymeric chain (principle of equal reactivity of co-condensing groups, see Moretti and Ottonello 2021 and reference therein) and can be described by anionic quasi-chemical species exchanging three kinds of oxygen (from here on, the superscript M for the melt phase is neglected for simplicity; otherwise differently indicated, the chemical species are intended to form part of the melt structure):

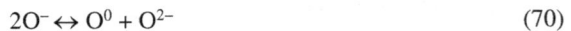

$$2O^- \leftrightarrow O^0 + O^{2-} \tag{70}$$

where O^{2-} are free oxide ions (also called "free oxygens" or better "non-network-associated oxygens", Nesbitt et al. 2011); O^0 are bridging oxygens; and O^- are non-bridging (single-bounded) oxygens. Quasi chemical species are thus represented by the oxygens that exist with three different formal charges $(0, -1, -2)$ depending on the covalent bonds they exchange with the surrounding in the process of silicate network formation or disruption.

Reaction (70) represents the mechanism of acid–base exchanges occurring in melts, being directly anchored to the Lux–Flood acid–base formalism (Flood and Forland 1947) for oxyanionic media:

$$\text{Base} \leftrightarrow \text{Acid} + O^{2-} \tag{71}$$

which, as anticipated above, makes the concentration of free oxide ions a measure of melt basicity in the same way as pH is a measure of acidity in aqueous solutions (Flood and Forland 1947; Duffy and Ingram 1971; Ottonello et al. 2001; Moretti 2005, 2020; Moretti and Ottonello 2021 and references therein).

The equilibrium Reaction (70) has been extensively discussed by Toop and Samis (1962a,b), Hess (1971), Fraser (1975, 1977), Wood and Fraser (1977), Ottonello (1997), Ottonello et al. (2001), Ottonello and Moretti (2004), Moretti (2005), Moretti and Ottonello (2003a, 2005, 2021), showing that its equilibrium constant involves the ideal mixing of three quasi-chemical oxygen species:

$$K_{(70)} = \frac{x_{O^{2-}} \cdot x_{O^0}}{x_{O^-}^2} \tag{72}$$

where x is molar concentration, and the subscript (70) refers to Reaction (70).

Based on the Toop–Samis model, Fraser (1977) first explained the relationship between water solubility and square root of water fugacity, Equation (62), as related to melt depolymerization. The argument is further supported by the well-known marked decrease of melt viscosity with increasing H_2O concentration in the melt (e.g., Richet et al. 1996; Dingwell et al. 1996; and many others). Based on the fundamental hypothesis of the Toop–Samis model that oxygen reactivity is independent from molecular size, Fraser (1977) suggested that

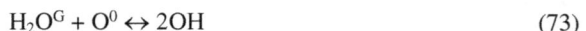

$$H_2O^G + O^0 \leftrightarrow 2OH \tag{73}$$

with OH representing a terminal OH group attached to a silicate polymer, i.e. –Si–OH. As reported in Ottonello (1997) and Moretti et al. (2014), Reaction (73) actually results from three reactions:

1) basic dissociation of H_2O (following the Lux–Flood definition at Reaction (71))

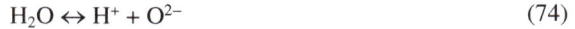

$$H_2O \leftrightarrow H^+ + O^{2-} \qquad (74)$$

2) the Toop–Samis exchange between quasi-chemical oxygen species, Reaction (70);

3) the association of highly reactive protons and non-bridging oxygens (taken two times) to form OH:

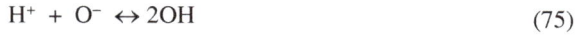

$$H^+ + O^- \leftrightarrow 2OH \qquad (75)$$

Because the (quasi-chemical) oxygen species mix ideally, Fraser (1977) proposed that

$$K_{(73)} = \frac{f^0_{H_2O} \cdot X^2_{OH}}{f_{H_2O} \cdot X_{O^0}} \qquad (76)$$

where f_{H_2O} is expressed in bar, and it is implicitly assumed that the reference fugacity of molecular water in the melt is equal to 1 bar. Reaction (73) implies that $x_{OH} = 2x_{H_2O}$; together with Equation (76), that provided an explanation to the well-known relationship between fugacity and squared concentration of water, first assessed by Burnham as extensively discussed above (Fig. 5; see also Fig. 3 and the related discussion).

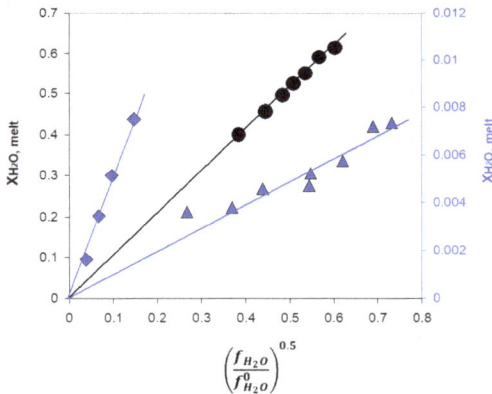

Figure 5. Linear dependence of water solubility (as molar content) versus the square root of water activity (*x*-axis), based on data from Burnham and Davis (1971, 1974) and Burham (1975). **Blue symbols** correspond to the right axis. Modified after Wood and Fraser (1977).

It is worth noting that Equation (76) implies the choice of standard states for all components corresponding to pure component at system *P–T*. If instead a typical standard state corresponding to 1 bar and system T is adopted, then a pressure-dependent exponential term must be added for OH and O^0, related to the variation of the Gibbs energy with pressure along the polybaric water–melt saturation curve. More in general, disregarding pressure-volume terms in Equation (76) seriously affects the validity of a water dissolution model in which only OH groups are formed. It is in fact rather intuitive that OH formation via Reaction (73) is highly favored at very low H_2O fugacity, when a polymerized, dry melt is prone to depolymerize and split in smaller structural units. Once again, that behavior has a parallel in the sharp decrease of viscosity when adding small amounts of water to initially dry melts, an effect the magnitude of which is much higher for more polymerized melts (Schulze et al. 1996; Romano et al. 2001, 2003).

An important new input to reactive water dissolution modeling came from NMR and infra-red spectroscopy on hydrous silicate melts. By studying quenched glasses that were supposed to preserve the chemical features of the parental melts, Stolper (1982a,b) reported evidence of the coexistence of two species of dissolved water, namely hydroxyl groups and molecular water, and showed that the concentration of hydroxyl groups can be lower than that required by the Burnham model to exhaust the available exchangeable cations. Stolper then reworked Reaction (73) as resulting from the sum of two equilibria, one heterogeneous and involving only water in molecular form:

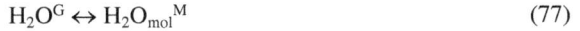

$$H_2O^G \leftrightarrow H_2O_{mol}{}^M \tag{77}$$

and the other one homogenous and involving reaction of molecular water with melt oxygen:

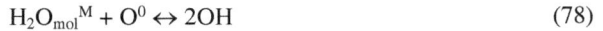

$$H_2O_{mol}{}^M + O^0 \leftrightarrow 2OH \tag{78}$$

In Reactions (77–78) the subscript mol refers to molecular water (species), distinct from the H_2O component that is partitioned between molecular water and OH groups.

Stolper (1982b) further assumed ideal mixing of the three species involved in Reaction (78), neglecting the distinction between the three species of oxygens in the Toop–Samis model (Reaction 70). These choices imply that the activity of each of the three species in Reaction (78) is equal to the corresponding proportion (Stolper 1982b; Silver and Stolper 1989). By indicating with n the number of moles of the species, with N_w and N_s the number of moles of water and water-free silicate melt involved in the mixing process, and with r the number of moles of oxygen atoms in the silicate melt (e.g., $r = 8, 2, 6$ for $NaAlSi_3O_8$, SiO_2, $CaMgSi_2O_6$, respectively), and noticing (Reaction 78) that for every two moles of OH groups in the melt one mole of O^0 is consumed, it follows

$$n_{O^0} = rN_s - 0.5n_{OH} \tag{79}$$

$$n_{H_2O_{mol}} = N_w - 0.5n_{OH} \tag{80}$$

from which it emerges that the activities of OH, O^0 and H_2O_{mol} can be computed by their relative mole fractions. It should be noted that such an ideal mixing of OH, H_2O_{mol} and O^0 species corresponds to non-ideal mixing in terms of (anhydrous) silicate melt and water components (Stolper 1982b), because the species (O and H_2O_{mol}) corresponding to these components exhibit negative deviations from ideality due to dilution by the OH species. This implies that the model fails at high total water concentrations as it cannot treat critical phenomena, such as vapor–melt miscibility gaps that require positive deviations of activities from ideal mixing.

Based on the above relations, the equilibrium constant of Reaction (78) is given by:

$$K_{(78)} = \frac{n_{OH}^2}{n_{O^0} \cdot n_{H_2O_{mol}}} = \frac{n_{OH}^2}{\left(rN_s - 0.5n_{OH}\right)\left(N_w - 0.5n_{OH}\right)} \tag{81}$$

Equation (81) can be solved for n_{OH} for given (experimentally determined) values of $K_{(78)}$ and bulk composition of the system in terms of N_w, N_s and r. On this basis, Stolper (1982b) showed that the weight proportions of H_2O_{mol} and OH vary non-linearly with the total amount of H_2O (Fig. 6). In particular, when the total amount of dissolved water exceeds ~5 wt%, molecular water dominates. The model results were found to reproduce well the spectroscopically observed speciation in albitic and rhyolitic melts, as well as in a few basaltic glasses (Stolper 1982a). At lower total dissolved water contents, hydroxyl groups dominate over molecular water (Fig. 6). Because all oxygens mix ideally (Eqn. 81), the maximum hydroxyl concentration occurs when the proportion of water oxygens represents half of the total oxygen content, i.e., for $N_w/(N_w + rN_s) = 0.5$. The value of $K_{(78)}$ determines the maximum in the OH curve and the curvature of the O^0 and H_2O_{mol} trends in Figure 6, hence the point where the concentration of OH overcomes that of H_2O_{mol}. It is worth noting that complete conversion of dissolved water into hydroxyl groups (Burnham model) implies $K_{(78)} \gg 1$, which is inconsistent with the spectroscopic data provided by Stolper (1982a).

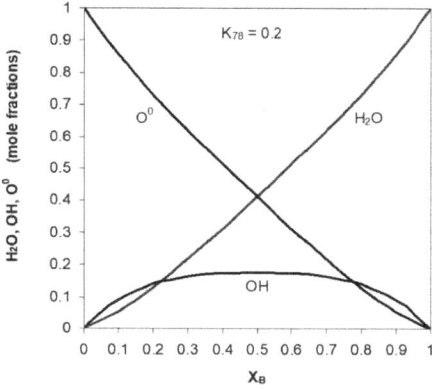

Figure 6. Calculated mole fractions of melt-oxygen species (molecular H_2O, OH groups, oxygen atoms) in melts. X_B is the fraction of oxygens contributed by dissolved water divided by the total number of oxygens in the glass/melt. The computations in the figure employ $K_{(78)}=0.2$, which for low concentrations of dissolved species fits the data for all considered melt compositions. The distribution of oxygen species is symmetric with respect to $X_B=0.5$. Redrawn after Stolper (1982b).

The fact that the model does not distinguish oxygen atoms not associated with water species (i.e., no distinction between bridging, non-bridging and free oxygens, all considered equally reactive) can be justified on a first approximation for melt compositions such as albite, silica, and to some extent, rhyolite. These melts are in fact fully (or highly) polymerized, therefore the distinction between different oxygens becomes less relevant (note however that even for such melts not all bridging oxygens, O^0, are energetically equivalent).

With the introduction of the homogeneous + heterogeneous Reactions (77) and (78), the equilibrium constant for the bulk dissolution Reaction (73) is given by

$$K_{(73)} = K_{(77)} \cdot K_{(78)} = \frac{a_{OH}^2}{a_{H_2O}^G a_{O^0}} \tag{82}$$

If N_w moles of water and N_s moles of melts are mixed (r oxygens per mole) and for $n_{OH} \gg n_{H2O_{mol}}$, that is, $n_{OH} \approx 2N_w$, Equation (82) can be rearranged as:

$$\left[\frac{K_{(77)}K_{(78)}}{4 f_{H_2O}^0} \cdot \frac{\left(rN_s\right)^2 - N_w^2}{\left(N_w + N_s\right)^2} \right] = \frac{X_w^2}{f_{H_2O}} \tag{83}$$

where x_w is the mole fraction of bulk dissolved water. Equation (83) can be used to fit $K_{(77)}$ vs. P for different temperatures and by taking typical values of $K_{(78)}$ between 0.1 and 0.2. This exercise is shown in Figure 7, giving to the quantity $K_{(77)}^{-1}$ the meaning of a Henry's law constant for the equilibrium between water vapor and molecular water in the melt.

Equation (83) implies that for low values of x_w, $rN_s \gg N_w$ and the relationship between x_w^2 and f_{H_2O} becomes close to linear. Therefore, the observation of such a linear relationship does not imply dissolution of water uniquely as hydroxyl groups, but at least for low dissolved water content is an approximation of a more complex relationship involving coexistence of OH groups and molecular water in the melt. According to the original observation by Burnham (1975a), the model of Stolper predicts non-linear behavior at high bulk dissolved water contents ($x_w > 0.5$). In other words, the continuous variation in the proportions of hydroxyl and molecular groups (Reaction 78) explains observations that were originally attributed to a discontinuous change in the nature of the water–melt interaction occurring for $x_w > 0.5$.

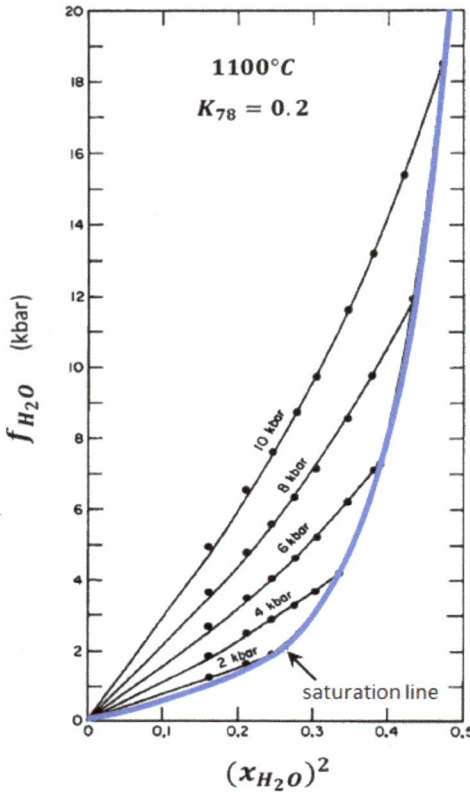

Figure 7. Fugacity of water versus the square of the mole fraction of water in hydrous melts at pressure up to 10 kbar ($T = 1100\ °C$). Black circles refer to the albitic melt from Burnham and Davis (1974). The mole fraction of water is defined as in Burnham and Davis (1974) based on 8 oxygens per formula unit of anhydrous silicate, and it differs from the X_B parameter in Figure 6. Each isobar is drawn by fixing K_{77} for any given pressure and temperature. The concentration of hydroxyl groups comes from Reaction (78), with $K_{78} = 0.2$. Total dissolved water is then the sum of molecular water and water present as hydroxyl groups. At each pressure, K_{77} is such that the calculated curve matches the experimental data from Burnham and Davis (1974) for $x_{H_2O} = 0.25$. The gas-saturation curve enveloping all the isobars on the right side is defined as the point (from Burnham and Davis 1974) at which water activity is equal to 1 at the corresponding P and T. Modified after Stolper et al. (1982b).

The model by Stolper elegantly and simply reconciles spectroscopic evidence with solubility data. An interesting aftermath of this model is that at high pressure (> 1 GPa) water solubility becomes nearly independent from melt composition (Hodges 1974), as water dissolution is dominated by molecular water, whereas the OH concentration, which is highly dependent on melt composition, becomes too little to affect the solubility properties at conditions where the amount of dissolved water is > 15 wt%. Still, some approximations characterized the original formulation of the model: ideal mixing, no temperature dependence, and implicit assumption of equal partial molar volumes of dissolved H_2O_{mol}, OH groups and oxygens (no change in volume accompanying Reaction 78).

The original model by Stolper was expanded in Silver and Stolper (1985) by introducing a P-dependent partial molar volume of H_2O_{mol} ($\bar{V}_{H_2O_{mol}}$) to better account for pressure effects on the reaction constant K_{77}. By using the Murnhagan equation of state (e.g., Anderson 1995) they obtained:

$$\ln K_{77}(P,T) = \ln K_{77}(P^0,T^0) + \ln\left[\frac{f^0_{H_2O}(P,T)}{f^0_{H_2O}(P^0,T^0)}\right] -$$

$$\frac{2\cdot10^5}{3}\bar{V}_{H_2O_{mol}(1bar)}\cdot\frac{\left\{\left[1+2\cdot10^{-5}P\right]^{3/4}-\left[1+2\cdot10^{-5}P^0\right]^{3/4}\right\}}{RT}$$

(84)

where $\bar{v}_{H_2O_{mol}}$ is given in cc/mol, and P^0 and T^0 are reference pressure and temperature (in bar and K, respectively), chosen as a point on the wet solidus on the melt–water phase diagram for a given melt composition (Albite–H_2O: $P^0 = 5000$ bar, $T^0 = 1018$ K, $\bar{v}_{H_2O_{mol}\,(1\,bar)} = 16.4$ cc/mol; Diopside–H_2O: $P^0 = 6000$ bar, $T^0 = 1538$ K, $\bar{v}_{H_2O_{mol}\,(1\,bar)} = 17.8$ cc/mol; Silica–H_2O: $P^0 = 6000$ bar; $T^0 = 1343$ K, $\bar{v}_{H_2O_{mol}\,(1\,bar)} = 17$ cc/mol; Silver and Stolper 1985). This expansion led the authors to conclude that ideal mixing of undistinguished oxygen species with molecular water and hydroxyl groups is consistent with phase equilibrium (solubility and freezing point depression) and spectroscopic data on water speciation in melts. Their approach provides an indirect way to determine $\bar{v}_{H_2O_{mol}}$ from solubility data by re-arranging and differentiating Equation (84):

$$\frac{\partial}{\partial p}\left(\frac{\ln K_{77}}{f_{H_2O}^0(P,T)}\right)_T = -\frac{\bar{v}_{H_2O_{mol}}}{RT} \tag{85}$$

which can be used to infer the partial molar volume of molecular water from plots of $\dfrac{\ln K_{77}}{f_{H_2O}^0(P,T)}$ vs. P.

The Silver and Stolper (1985) expansion of the original Stolper model implies a slight dependence of solubility on temperature, which counterintuitively results from the assumptions of T-independent $\bar{v}_{H_2O_{mol}}$ and K_{78} (Eqn. 81). This slight dependence was found to compare well with the experimental data. Additionally, the Silver and Stolper (1985) model includes a compositional dependence of K_{77} (Eqn. 84), consistent with the principle that structurally different melt oxygens cannot be treated as equally reactive.

In a subsequent work, Silver and Stolper (1990) operated a change of notation, replacing the melt oxygen, O^0, with a generic and undistinguished O^{2-} (not having the meaning of free oxygen as in Reaction 70) therefore denoting hydroxyls as OH^-. Reaction (78) was then written as:

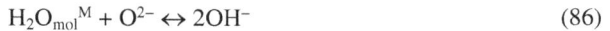

$$H_2O_{mol}^M + O^{2-} \leftrightarrow 2OH^- \tag{86}$$

The authors found that the ideal mixing model of O^{2-}, OH^- and H_2O_{mol} developed in Stolper et al. (1982b) and Silver and Stolper (1985) did not match with the new data for water speciation in hydrous albitic melts at bulk dissolved H_2O contents > 2 wt% (Fig. 8). As a consequence, they modeled the non-ideal interaction of the above three species by alternatively using 1) the formalism for a strictly regular ternary solution, 2) a simplified polymer mixing model using Flory's approximation for the entropy of mixing of chain polymer units, each consisting of l undistinguished oxygen atoms (hence $l-1$ tetrahedra units) in the anhydrous melt split by interaction with H_2O, and 3) mixing models with several types of distinguishable oxygens and hydroxyls, which allow a direct extension of the original ideal mixing model in Stolper (1982b). Models of category 2) were dismissed because the introduction of structural complexities was felt unwarranted, given the substantial lack of structural data on hydrous melts. Models of category 3) were also abandoned because the tendency to include more and more species in compositionally different melts implies a significant risk of overfitting (and violation of the Gibbs rule of phases). Because of its relative simplicity, the authors proposed a regular solution model to account for the newly observed dependence of water speciation on temperature, with T-dependent increase in the proportion of hydroxyl groups (Silver and Stolper 1989; Stolper 1989; Silver et al. 1990). The proposed regular solution model adopts three binary interaction parameters, $w_{H_2O_{mol}-O}$, $w_{H_2O_{mol}-OH}$, and w_{O-OH}, which allow the computation of activities for Reaction (78), hence of K_{78}:

$$a_{H_2O_{mol}}^M = x_{H_2O_{mol}}^M \exp\left\{\frac{1}{RT}\cdot\left[\begin{array}{l} x_O^M x_{OH}^M\left(w_{H_2O_{mol}\text{-}O}-w_{O\text{-}OH}+w_{H_2O_{mol}\text{-}OH}\right)+\\ \left(x_O^M\right)^2\cdot w_{H_2O_{mol}\text{-}O}+\left(x_{OH}^M\right)^2\cdot w_{H_2O_{mol}\text{-}OH}\end{array}\right]\right\} \tag{87}$$

$$a_{OH}^M = x_{OH}^M \exp\left\{\frac{1}{RT}\cdot\left[\begin{array}{l} x_{H_2O_{mol}}^M x_O^M\left(w_{H_2O_{mol}\text{-}OH}-w_{H_2O_{mol}\text{-}O}+w_{O\text{-}OH}\right)+\\ \left(x_{H_2O_{mol}}^M\right)^2\cdot w_{H_2O_{mol}\text{-}OH}+\left(x_O^M\right)^2\cdot w_{O\text{-}OH}\end{array}\right]\right\} \tag{88}$$

$$a_O^M = x_O^M \exp\left\{\frac{1}{RT}\cdot\left[\begin{array}{l} x_{H_2O_{mol}}^M x_{OH}^M\left(w_{O\text{-}OH}-w_{H_2O_{mol}\text{-}OH}+w_{H_2O_{mol}\text{-}O}\right)+\\ \left(x_{H_2O_{mol}}^M\right)^2\cdot w_{H_2O_{mol}\text{-}O}+\left(x_{OH}^M\right)^2\cdot w_{O\text{-}OH}\end{array}\right]\right\} \tag{89}$$

and finally:

$$\ln K_{78} = \frac{w_{H_2O_{mol}\text{-}O}}{RT}\left(x_{OH}^M-1\right)+\frac{2w_{O\text{-}OH}}{RT}\left(1-x_{H_2O_{mol}}^M-\frac{3}{2}x_{OH}^M\right)+$$
$$\frac{2w_{H_2O_{mol}\text{-}OH}}{RT}\left(x_{H_2O_{mol}}^M-\frac{1}{2}x_{OH}^M\right)+\ln\frac{\left(x_{OH}^M\right)^2}{x_{H_2O_{mol}}^M x_{OH}^M} \tag{90}$$

in which no pressure dependence is included, and the last term on the right side corresponds to the ideal mixing term in Equation (81). Equation (90) was then rearranged to give (Silver 1989; Silver and Stolper 1990):

$$-\ln\frac{\left(x_{OH}^M\right)^2}{x_{H_2O_{mol}}^M\left(1-x_{OH}^M-x_{H_2O_{mol}}^M\right)} = A'+B'x_{OH}^M+C'x_{H_2O_{mol}}^M \tag{91}$$

with A', B' and C' being P–T-independent terms determined by regression on a large dataset and defined as (Silver et al. 1990):

$$A' = -\ln K_{78}+\frac{2w_{O\text{-}OH}-w_{H_2O_{mol}\text{-}O}}{RT}=1.093 \tag{92}$$

$$B' = \frac{w_{H_2O_{mol}\text{-}O}-3w_{O\text{-}OH}-w_{H_2O_{mol}\text{-}OH}}{RT}=16.858 \tag{93}$$

$$C' = \frac{w_{H_2O_{mol}\text{-}OH}-w_{O\text{-}OH}}{RT}=7.892 \tag{94}$$

with values referring to rhyolitic melts (Silver et al. 1990). Equation (91) can be solved iteratively to give x_{OH} as a function of $x_{H_2O_{mol}}$, with their sum corresponding to the solubility of water in the melt. Alternatively, by introducing x_B, the fraction of oxygen contributed by water divided by the total number of oxygens, Equations (92–94) can be solved for x_{OH}:

$$-\ln\frac{x_{OH}^2}{\left(x_B-0.5x_{OH}\right)\left(1-x_B-0.5x_{OH}\right)} = A'+\left(B'-0.5C'\right)x_{OH}+C'x_B \tag{95}$$

Assuming arbitrarily that $w_{H_2O_{mol}\text{-}O}=0$, so that $a_{H_2O_{mol}}^M \to x_{H_2O_{mol}}$ when $x_B \to 0$, calibration of Equations (92–94) gives $w_{H_2O_{mol}\text{-}OH}=11.7$ kJ/mol, $w_{O\text{-}OH}=48.5$ kJ/mol and $\ln K_{78}$ (850°C) $=-11.5$ (850 °C being the reference temperature corresponding to the temperature of water saturation of the rhyolitic melt investigated by Silver et al. 1990). Different values, summarizing the compositional dependence, were calculated for Ca–Al–silica, orthoclasic and albitic melts

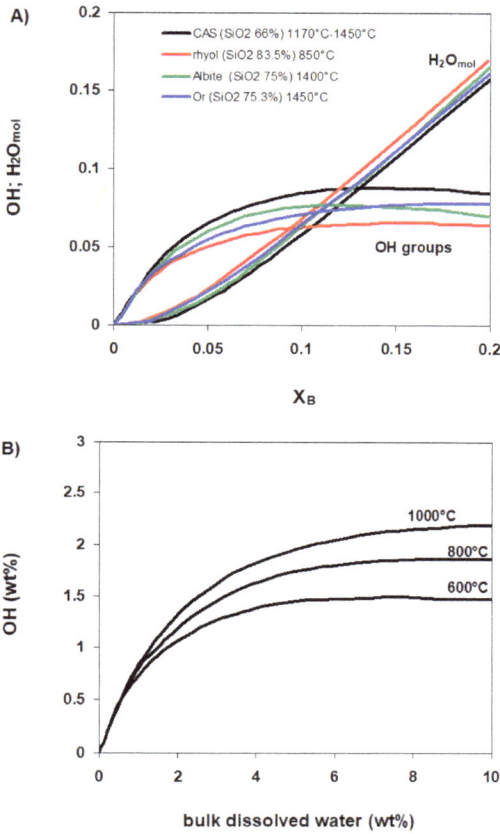

Figure 8. a) Best fits of spectroscopically determined proportions of molecular H_2O and OH groups as a function of x_B for a calcium–aluminum–silica (CAS; 1170–1450 °C), orthoclasic (Or; 1450 °C), rhyolitic (rhyol; 850 °C), and albitic (Ab; 1400 °C) glasses. The data reveal a control of melt composition on water speciation not accounted for in the original Stolper (1982b) model. **b)** Concentration (wt%) of OH groups versus bulk dissolved water (wt%) at three different temperatures using the regular solution model described in Silver and Stolper (1989) and Silver et al. (1990). Modified after Stolper (1989) and Silver et al. (1990).

(Silver et al. 1990). Stolper (1989) and Silver and Stolper (1989) also showed that by using Equation (90) it is possible to evaluate the inherent T-dependence of K_{78} (Fig. 8). By fitting their own spectroscopic data on rhyolite (Silver et al. 1990), they obtained:

$$\ln K_{78} = \frac{-12881}{T(\mathrm{K})} - 0.0261 \tag{96}$$

that at 850 °C gives $\ln K_{78} = -11.5$ as expected.

To complete the model calibration, Silver and Stolper (1989) used the solubility data in the albite–H_2O system from Burnham and Davis (1971) within their regular solution model (Eqns. 87–94) with the assumption that the activity of dissolved bulk and molecular water coincide, and found that $x^{\mathrm{M}}_{H_2O_{mol}} \propto a^{\mathrm{M}}_{H_2O_{mol}} = a^{\mathrm{M}}_{H_2O}$, confirming the earlier use of this approximation by Stolper (1982b). This proportionality is also called in many papers and textbooks a "Henrian behavior", different however from the definition introduced above and deriving from the equilibrium Equation (26) when ideal behavior is assumed, so that the fugacity and activity coefficients (ϕ and γ in Eqn. 26) are both equal to 1. In fact, such an ideal behavior implies that the concentration is equal to the activity, while here the concentration is only proportional to the activity, meaning that $\gamma \neq 1$ but still a constant value. That constant value has the meaning of a shift between standard states, which can be understood by comparing Henry's law defined above: $x_i = \left(f_i^0 \right)^{-1} P_i$ with the one from the present definition of Henrian behavior:

$x_i = \left(\gamma_i f_i^0\right)^{-1} P_i$. For both definitions, highly dilute conditions are required to justify the ideal assumption at their basis.

The basis for the computation of the solubility of water in molecular form by Silver and Stolper (1989) and Silver et al. (1990) is represented by the following Equation (97):

$$a_{H_2O}^M(P_2,T_2) = a_{H_2O}^M(P_1,T_1) \frac{f_{H_2O}^G(P_2,T_2)}{f_{H_2O}^G(P_1,T_1)} \cdot \exp\left\{-\int_{P_1}^{P_2} \frac{V_{H_2O}^{0,M}(P,T_2)}{RT_2} dP + \int_{T_1}^{T_2} \frac{\Delta H_{H_2O}^0(P_1,T)}{RT^2} dT\right\} \quad (97)$$

with $f_{H_2O}^G$ being the fugacity of pure water gas, $V_{H_2O}^{0,M}$ the molar volume of water in the melt, $\Delta H_{H_2O}^0$ the enthalpy change due to Reaction (77), and the subscripts 1 and 2 representing two different P–T conditions. Equation (97) is obtained by equating two expressions of the equilibrium constant for Reaction (77), one in terms of activities, and the other one in terms of reaction energetics. From Equation (97) if the activity of water in a gas-saturated melt is known at given P–T conditions, and with known expressions for the P–T dependence of $V_{H_2O}^{0,M}$ and $\Delta H_{H_2O}^0$, the activity can be computed for any other P–T pair given an equation of state for pure water gas returning $f_{H_2O}^G$. It is interesting to compare this equation to Equation (36), as they appear very similar on a first sight, but are in fact different in their functional form reflecting a fundamentally different meaning: while Equation (36) (and its further development into Equation (38) which is employed in the SOLWCAD code) describes the change in fugacity along any P–T path in a given phase, Equation (97) describes the change in activity (which relates immediately to fugacity) along a P–T path corresponding to the saturation surface for a non-reactive component (like molecular water). It is worth noting that for constant P–T conditions and variable gas composition the volume and enthalpy terms in Equation (97) disappear, and the activity of dissolved molecular water, therefore the concentration of dissolved molecular water, becomes directly proportional to its fugacity in the gas phase, returning the original definition of Henry's law.

Silver et al. (1990) generalized the approach by using measured solubilities in melts of various compositions to fit enthalpy, volume and $x_{H_2O_{mol}}$ in the low-P range. Using the proportionality between concentration and activity discussed above, the pressure integral in Equation (97) can be determined by plotting $\ln\left(x_{H_2O_{mol}}/f_{H_2O}^G\right)$ vs P. Silver et al. (1990) provide a table of values to enter Equation(97), with constant enthalpy and volume terms (except for albitic melts, for which the volume is given as depending on T). As noted by Holloway and Blank (1994), molar volume determinations need high-P solubility data, because low-P data are typically too scattered. As a consequence, molar volumes constrained in this way are well applicable to high concentrations of H_2O in melts, but not to low concentrations (see also Lange 1994).

Dixon et al. (1995a,b) and Dixon (1997) drop the enthalpy term in Equation (97) for basaltic (sensu lato) compositions, thus producing negligible variations of water solubility in temperature intervals typical of basaltic magmas. Dixon (1997) introduces a simple linear extrapolation for the compositional dependence of H_2O solubility as a function of SiO_2, to be used in Equation (97) together with (ideal) substitution of $a_{H_2O}^M$ with x_{H_2O}. The proposed relationship is:

$$x_{H_2O}(P_1,T_1) = -3.04 \cdot 10^{-5} + 1.29 \cdot 10^{-6} SiO_2 \text{ (wt\%)} \quad (98)$$

Once the concentration of dissolved molecular water is computed from Equation (97), the concentration of hydroxyl groups (x_{OH}) can be computed from Equation (91). The total mole fraction of H_2O in the melt is then given by $x_{H_2O} = x_{H_2O_{mol}} + 0.5x_{OH}$ and by using the quantity x_B introduced at Equation (95), the weight fraction w_{H_2O} is:

$$w_{H_2O} = \frac{18.02 x_b}{18.02 x_b + (1 - x_b) M} \tag{99}$$

where M is the molar weight of the anhydrous melt on a single oxygen basis (Silver et al. 1990).

Speciation-based CO_2 models

Early infrared measurements showed that in mafic melts CO_2 dissolves as carbonate, whereas in feldspar melts (albite–anorthite) it dissolves in molecular form, suggesting a dependence on melt basicity (Mysen et al. 1976) which can be summarized by Reaction (67). Similarly, Spera and Bergman (1980) suggested that a mechanism of CO_2 dissolution related to poly-condensation can be envisaged in which carbon dioxide reacts with non-bridging oxygens to polymerize the melt:

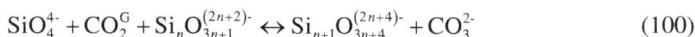

$$SiO_4^{4-} + CO_2^G + Si_n O_{3n+1}^{(2n+2)-} \leftrightarrow Si_{n+1} O_{3n+4}^{(2n+4)-} + CO_3^{2-} \tag{100}$$

where n is a positive integer. Note that Reaction (100) has the same meaning as (Fine and Stolper 1985):

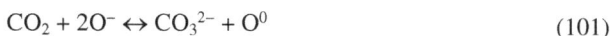

$$CO_2 + 2O^- \leftrightarrow CO_3^{2-} + O^0 \tag{101}$$

which is obtained by summing up Reactions (67) and (70) (neglecting superscripts for phase specification). Reaction (101) applies to a variety of reactive melt entities, and not just polymer chain segments like for Reaction (100). Spera and Bergman (1980) showed that their proposed Reaction (100), here generalized into Reaction (101), agrees with the available data on free energies for carbonation reactions of the type

$$A_2SiO_3{}^M + CO_2{}^G \leftrightarrow SiO_2 + MCO_3{}^M \tag{102}$$

$$M_2SiO_4{}^S + CO_2{}^G \leftrightarrow MSiO_3{}^S + MCO_3{}^S \tag{103}$$

with A an alkali metal atom and M a divalent metal atom, and with the superscripts G, M, S indicating the gas, melt, and solid phase, respectively. The free energy of CO_2 dissolution increases, and the degree of melt polymerization decreases, with increasing experimental CO_2 solubility, in the order Fe $\approx$ Mg $\ll$ Li $<$ Na $<$ K $<$ Ca (Spera and Bergman 1980, and references therein). Based on such a mechanism for molecular level dissolution of CO_2 in silicate melts, Spera and Bergman (1980) then approached Reaction (68) with no reference to the precise form in which CO_2 enters the silicate melt. The equilibrium constant for Reaction (68) was thus written as:

$$K_{68} = \frac{f_{CO_2}^G}{a_{CO_2}^M} \tag{104}$$

which presupposes $f_{CO_2}^{0G} = 1$ (e.g., $f_{CO_2}^G$ in bars and $f_{CO_2}^{0G} = 1$ bar). By expressing K_{68} in energetic terms, and by assuming that i) the dissolved CO_2 behaves ideally, ii) the partial molar volume of CO_2 in the melt, $\bar{v}_{CO_2}^M$, is P-independent, and that iii) the ΔC_P (change of isobaric heat capacity) of reaction is zero (so that the reaction enthalpy and entropy are constant at any T), Equation (104) leads to:

$$R \ln \frac{f_{CO_2}^G}{a_{CO_2}^M} = -\frac{\Delta H_{68(1bar, T_r)}^0}{T} + \Delta S_{68(1bar, T_r)}^0 + \frac{\bar{v}_{CO_2}^M}{T}(P-1) \tag{105}$$

with P in bars, and with T_r indicating reference temperature.

Plotting experimental values of $R \ln \dfrac{f_{CO_2}^G}{a_{CO_2}^M}$ vs. P at constant T, a straight line is then

obtained, with the slope determining $\bar{v}_{CO_2}^M$ and the intercept at 1 bar giving $\dfrac{-\Delta G_{68(1bar,T_r)}^0}{T}$, which

is then fitted for $\Delta H_{68(1bar,T_r)}^0$ and $\Delta S_{68(1bar,T_r)}^0$.

On this basis, Spera and Bergman (1980) fitted reaction enthalpies and entropies and the molar volume of dissolved CO_2 for melt compositions corresponding to andesite, tholeiite, and olivine melilite, and showed that at equal P–T, CO_2 dissolution s favored by increasing molar ratio of network modifiers over network formers, i.e., by increasing melt basicity. The authors also provided robust confirmation to the experimental observation of a positive dependence of CO_2 solubility on pressure (so that CO_2 is inevitably released upon decompression); and because of negative $\Delta H_{68(1bar,T_r)}^0$, they suggested small heating effects due to CO_2 exsolution from the melt.

A major advance in understanding CO_2 dissolution in silicate melts is due to the work by Fine and Stolper (1985, 1986). By using infrared spectroscopy, they found coexistence of molecular CO_2 and carbonate groups in quenched glasses in the system Na_2O–Al_2O_3–SiO_2–CO_2. They thus proposed an approach similar to that in Stolper (1982b) for water, based on Reaction (69) and its equilibrium constant:

$$K_{69} = \frac{a_{CO_3^{2-}}}{a_{O^{2-}} \cdot a_{CO_2,mol}} \tag{106}$$

From Reaction (69), CO_2 dissolution results from mixing of molecular CO_2, carbonates groups, and a sub-category of highly reactive oxygens (not necessarily coinciding with the free oxygens in Reaction 70) which control the interaction. This marks a difference with the case of water dissolution modelling, for which all oxygens are equally reactive. Since O^{2-} is a sub-group of the undistinguished oxygens in the anhydrous, carbon-free melt, it follows that $a_{O^{2-}} < 1$. The three species involved in Reaction (69) are likely to mix far from ideal behavior because of substantial differences in their size, therefore, equating their activities to mole fractions, as it is common with water, may not be justified. However, as a first evaluation of Reaction (69) as the basis for CO_2 dissolution modelling, Fine and Stolper (1985) assumed such an equality between activity and mole fraction for the species involved in the reaction. With this assumption, mole fractions are computed in relation to the number of moles of the species involved in the reaction: if a melt consists of 1 mole of $CO_{2,mol}$, 3 moles of CO_3^{2-}, and 1 mole of $NaAlSi_2O_6$ (6 moles of oxygen), $x_{CO_{2mol}} = 0.1$ and $x_{CO_3}^{2-} = 0.3$.

In particular, given the approximately linear relationship between the measured concentrations of molecular CO_2 and carbonate up to a bulk dissolved CO_2 content of ~2 wt%, they propose that

$$\frac{x_{CO_3^{2-}}}{x_{CO_2,mol}} = K_{69} \cdot a_{O^{2-}} = const \tag{107}$$

Equations (106) and (107) evidence the role of melt basicity: assuming in the first instance that K_{69} does not vary with melt composition, a larger melt basicity (e.g., higher Na/Si ratio) means increased $a_{O^{2-}}$ therefore higher $x_{CO_3}^{2-} / x_{CO_{2mol}}$ ratio. In other words, melts with high basicity are expected to dissolve carbon mostly as carbonate groups, according to the observations. One may see a parallel, already discussed above, between the substantial differences highlighted here between the mechanisms of dissolution of water and carbon dioxide, and the different type of mixing of the two volatiles with silicate melts (isometric for water, non-isometric for carbon dioxide) found to better match, in the regular solution approach of the SOLWCAD model, the observed solubilities in melts over a wide compositional range.

In their subsequent experimental study, Fine and Stolper (1986) showed that Ca–Mg–silicate glasses carry no molecular CO_2 but only carbonate groups, suggesting that $a_{O^{2-}}$ is likely not the only control on the $x_{CO_3^{2-}}/x_{CO_{2mol}}$ ratio. Accordingly, they suggested that some cations, such as Ca, may favour carbonate stabilization more than others, such as Na.

Stolper et al. (1987) then proposed a solubility–speciation model for CO_2 in albitic melts in which solubility is assessed as the sum of dissolved carbonate and dissolved molecular CO_2 (Fig. 9). Based on their extensive experimental data, they fit CO_2 solubility and speciation in a way similar to that for water. Accordingly, they wrote Reaction (68) for molecular CO_2 in the gas and the melt, and referred both of them to the same standard state. On this basis they wrote the analogue of Equation (97) for carbon dioxide:

$$a_{CO_2}^M(P,T) = a_{CO_2}^M(P^0,T^0) \frac{f_{CO_2}^G(P,T)}{f_{CO_2}^G(P^0,T^0)} \cdot \exp\left\{-\frac{\overline{v}_{CO_2}^M(P-P^0)}{RT} - \frac{\Delta H_{(68)}^0(P^0)}{R}\left[\frac{1}{T} - \frac{1}{T^0}\right]\right\} \quad (108)$$

where the symbols have the same meaning, this time for gaseous CO_2 or dissolved molecular CO_2, as for Equation (97), and the integrations are done by assuming P-independent partial molar volume of dissolved molecular CO_2 $\overline{v}_{CO_2}^M$ and T-independent enthalpy of reaction at reference conditions $\Delta H_{(68)}^0(P^0)$. This latter term is allowed to change with pressure according to $\Delta H_{(68)}^0(P) = \Delta H_{(68)}^0(P^0) + \overline{v}_{CO_2}^M(P-P^0)$ (Stolper et al. 1987).

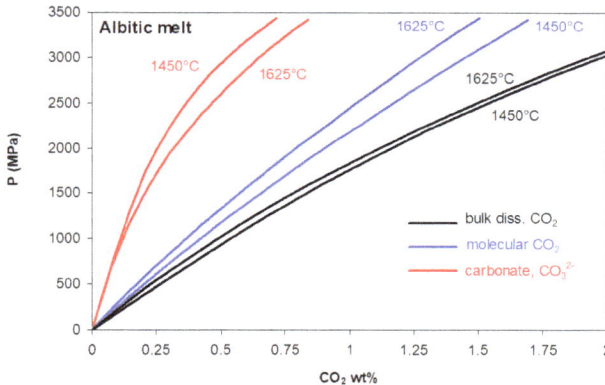

Figure 9. Weight concentration of bulk carbon dioxide, molecular CO_2 and carbonate groups, in albitic melts. Temperature is found to have a negligible effect on the bulk dissolution of CO_2, given its contrasting effects on the separate solubilities of carbonate groups and molecular CO_2. Modified after Stolper et al. (1987).

The ideal mixing assumption allows substitution of mole fractions to activities in Equation (108), and to write the equilibrium constant for Reaction (69), expressing the homogeneous equilibrium between the molecular and carbonate forms of dissolved carbon dioxide, as:

$$K_{69} = \frac{x_{CO_3^{2-}}}{x_{O^{2-}} x_{CO_{2,mol}}} \quad (109)$$

Considering ideal mixing of ionic species at reference conditions P^0, T^0, Equation (109) leads to:

$$K_{69}(P,T) = K_{69}(P^0,T^0) \cdot \exp\left\{-\frac{\overline{v}_{(69)}^M(P-P^0)}{RT} - \frac{\Delta H_{(69)}^0(P^0)}{R}\left[\frac{1}{T} - \frac{1}{T^0}\right]\right\} \qquad (110)$$

where P-T-independent reaction volume and enthalpy changes are considered. Stolper et al. (1987) modelled CO_2 dissolution in albitic melts with Equations (108–109), and found a negligible to slightly negative dependence of bulk dissolved CO_2 on temperature (Fig. 9). They also found that the variations in the $x_{CO_3^{2-}}/x_{CO_{2mol}}$ ratio with P-T are small and gradual, contradicting a previous claim of a sharp variation at $P = 1.5$–3 GPa due to large and sudden change in melt polymerization (Mysen et al. 1976). On the contrary, the bulk CO_2 solubility increase in that pressure range was found to be due essentially to dissolution as molecular carbon dioxide driven by the $V\Delta P$ term in Equation (108). Finally, the authors did not find any enhancement of bulk CO_2 solubility at low water contents, which was found in some experiments and interpreted as increased carbonate stabilization under hydrous conditions; and that is also found, under some conditions, from the regular solution models described above.

Based on newly collected data and the approach by Stolper and co-workers described above, Holloway and Blank (1994) proposed parameters for CO_2 solubility in melts with rhyolitic, tholeiitic, basanitic, and leucititic composition (parameters for MOR basalts can be found in Dixon et al. 1995a). For the mafic melts, for which spectroscopy shows dissolution as only carbonate, the formalism of Equation (110) was employed, referred to Reaction (67) (heterogeneous equilibrium) instead of Reaction (69). Implicit in their treatment, the equilibrium constant of Reaction (67) adopts an ideal assumption and it is written as:

$$K_{67} = \frac{f_{CO_2}^{0G} x_{CO_3^{2-}}}{f_{CO_2}^G x_{O^{2-}}} \qquad (111)$$

which then leads to (Dixon et al. 1995a,b; Dixon 1997):

$$x_{CO_3^{2-}}(P,T) = x_{CO_3^{2-}}(P^0,T^0) \frac{f_{CO_2}^G(P,T)}{f_{CO_2}^G(P^0,T^0)} \times$$
$$\exp\left\{-\frac{V_{(CO_3^{2-}-O^{2-})}^0(P-P^0)}{RT} - \frac{\Delta H_{(67)}^0(P^0)}{R}\left[\frac{1}{T} - \frac{1}{T^0}\right]\right\} \qquad (112)$$

in which $V_{(CO_3^{2-}-O^{2-})}^0 = V_{CO_3^{2-}}^0 - V_{O^{2-}}^0$ and the terms for O^{2-} ($x_{O^{2-}}$) are assumed to not be affected by P-T given the comparably small amount of CO_3^{2-}. As for water, Dixon et al. (1995a,b) and Dixon (1997) determined by fitting the parameters for generic basaltic compositions, dropping the enthalpy term; and as for water, they obtained negligible variations of carbon dioxide solubility in the temperature interval typical of basaltic magmas. Dixon (1997) considered two main factors controlling CO_2 solubility in melts, namely, the role of polymerization (roughly, the amount of silica and alumina, whose decrease favours CO_2 solubility) and the affinity scale for metal-carbonate dissolution (Spera and Bergman; 1980). On that basis they defined a composite empirical parameter that re-groups in two competing terms $Si + Al$ and $Ca + K + Na + Mg + Fe$. The correlation between CO_2 solubility and composition is then integrated in the following relationship:

$$x_{CO_3^{2-}}(P,T) = 8.7 \cdot 10^{-6} - 1.7 \cdot 10^{-7} SiO_2 (wt\%) \qquad (113)$$

which is analogous to Equation (98) for water.

For mafic melts in which CO_2 dissolves only as carbonate, rather than dismissing $x_{O^{2-}}$ Holloway and Blank (1994) considered that $x_{O^{2-}} = 1 - x_{CO_3^{2-}}$, which can be substituted into Equation (111) giving:

$$x_{CO_3^{2-}} = \frac{f_{CO_2}^G}{f_{CO_2}^{0G}} K_{67} \left[1 - \frac{f_{CO_2}^G}{f_{CO_2}^{0G}} K_{67} \right]^{-1} \tag{114}$$

Holloway and Blank (1994) adopted Equations (109–110) (written for K_{67}) by replacing the reaction volume with $V_{(CO_3^{2-} - O^{2-})}^0$. From tabulated values of K_{67}, $f_{CO_2}^{0G}$, and volume and enthalpy terms, they calculated the solubility of CO_2 as carbonate groups only. For rhyolitic melts, for which only molecular CO_2 is found from spectroscopy, Equation (109) is still adopted, but with K_{68} replacing K_{69} as in the Spera and Bergman (1980) model. In this case:

$$x_{CO_2 mol} = \frac{f_{CO_2}^G}{f_{CO_2}^{0G}} K_{68} \tag{115}$$

Weight fractions of dissolved CO_2 are then computed from:

$$w_{CO_2} = \frac{44.01 x_{CO_2}}{44.01 x_{CO_2} + (1 - X_{CO_2}) M} \tag{116}$$

with M being the molar weight of the anhydrous melt on a single oxygen basis.

The VolatileCalc model

Holloway and Blank (1994) employed the models above for water and carbon dioxide solubility to calculate mixed $H_2O + CO_2$ saturation in melts with given specific composition (that is, for melts for which calibrated quantities for use in Eqns. 97 and 108 exist), by coupling the two volatiles through their reciprocal dilution in the gas phase (that is, with real EOS for pure gas components but ideal gas mixing), with no effects of the two dissolved volatiles on each other's saturation content.

The code VolatileCalc (Newman and Lowenstern 2002) formalizes the procedures in Holloway and Blank (1994). In particular, it adopts:

- for water in generic rhyolite, Equation (97) for $x_{H_2O mol}$, and Equation (91) for x_{OH};
- for water in generic basalt, Equation (98) for $x_{H_2O mol}$, and Equation (91) for x_{OH};
- for carbon dioxide in generic rhyolite, Equation (108) for $x_{CO_2 mol}$;
- for carbon dioxide in generic basalt, Equation (113) for $x_{CO_3^{2-}}$.

Because of large use of ideal approximations both in the gas and melt phase, Newman and Lowenstern (2002) recommend the use of VolatileCalc only at low pressure. Calculated saturation curves for generic basalt and rhyolite melts, allowed by VolatileCalc, agree in their general features with those from compositionally dependent non-ideal regular solution models over similar compositions only in the low pressure range below 100 MPa, above which large deviations from ideal behavior are found (Papale 1999a; see also Fig. 2). Besides such limitations, VolatileCalc has been the first multicomponent volatile saturation model to be easily accessible to a wide audience of researchers, boosting further investigation on several aspects related to volatiles in igneous petrology and volcanology.

VolatileCalc is accessible at https://volcanoes.usgs.gov/observatories/yvo/jlowenstern/ other/software_jbl.html.

Further implementations of speciation-based models

Like any other model, the Stolper class of models (including the application made in VolatileCalc) can be further implemented based on the availability of new data. For melts with alkalic basalt composition, Lesne et al. (2011a) proposed to implement water speciation modelling by introducing the following T-dependences for the terms in Equations (91–94):

$$\ln K_{78} = \frac{-2704.4}{RT} + 0.641 \tag{117}$$

$$A' = \ln K_{78} + \frac{49016}{RT} \tag{118}$$

$$B' = \frac{-2153326.51}{RT} \tag{119}$$

$$C' = \frac{1.965495217}{RT} \tag{120}$$

Lesne et al. (2011b) used the Holloway and Blank (1994) approach to CO_2 solubility and determined for basalts (*sensu lato*) $V^0_{(CO_3^{2-} \cdot O^{2-})}$ and $\ln\left(K_{67} \cdot f^{0G}_{CO_2}\right) = 0.893\Pi - 15.247$, with Π defined as (Dixon 1997): $\Pi = -6.50(Si+Al) + 20.17\left(Ca + 0.8K + 0.7Na + 0.4Mg + 0.4Fe^{2+}\right)$ where the components represent atomic fractions.

A more recent parameterization of Π led Shishkina et al.(2014) to propose an empirical expression for CO_2 solubility in mafic melts: $\ln w_{CO_2} = 1.167 \ln P + 0.671\,\Pi$, with w_{CO_2} being carbon dioxide concentration in wt%, and P expressed in MPa (Fig. 10). For the same generic class of melts, the same authors proposed the simple relationship for H_2O solubility: $\ln w_{H_2O} = a(Na+K) + b$, with w_{H_2O} being water concentration in wt%, Na and K expressed as atomic fractions, and a, b being polynomial expressions in pressure. Clearly, these parameterizations account only partially for melt composition variations within the generic class of "mafic melts", and neglect solubility differences due to temperature differences.

Iacono-Marziano et al. (2012) presented a semi-empirical parameterization for H_2O–CO_2 saturation in mafic melts ranging from andesitic to picritic compositions. Their model is based on speciation reactions like those seen above, but it then develops into a semi-empirical parameterization not very dissimilar from other parameterizations not discussed in this review. Carbon dioxide solubility in mafic melts is assessed based on Reaction (67):

$$K_{67} = \frac{a_{CO_3^{2-}}}{f^G_{CO_2} a_{O^{2-}}} = \frac{\gamma_{CO_3^{2-}} x_{CO_3^{2-}}}{f^G_{CO_2} a_{O^{2-}}} \tag{121}$$

Note that lack of the $f^{0G}_{CO_2}$ term and consequent inconsistent units in Equation (121) require the implicit use of $f^{0G}_{CO_2} = 1$ (e.g., $f^{0G}_{CO_2}$ in bars and $f^{0G}_{CO_2} = 1$ bar). Equation (121) is then expanded in log-form considering H, S and V contributions at varying P–T and re-arranged as:

$$\ln x_{CO_3^{2-}} = -\ln \gamma_{CO_3^{2-}} + \ln f_{CO_2} + \ln a_{O^{2-}} - \frac{\Delta H^0}{RT} + \frac{\Delta S^0}{T} + \frac{P\Delta V^0}{RT} \tag{122}$$

with ΔH^0, ΔS^0 and ΔV^0 are reaction enthalpy, entropy and volume change of Reaction (67) at equilibrium.

$x_{CO_3^{2-}}$ is then rescaled to ppm, $(CO_2)^{ppm}$, and the activity coefficient term is taken as:

$$-\ln \gamma_{CO_3^{2-}} = \sum_i x_i d_i \tag{123}$$

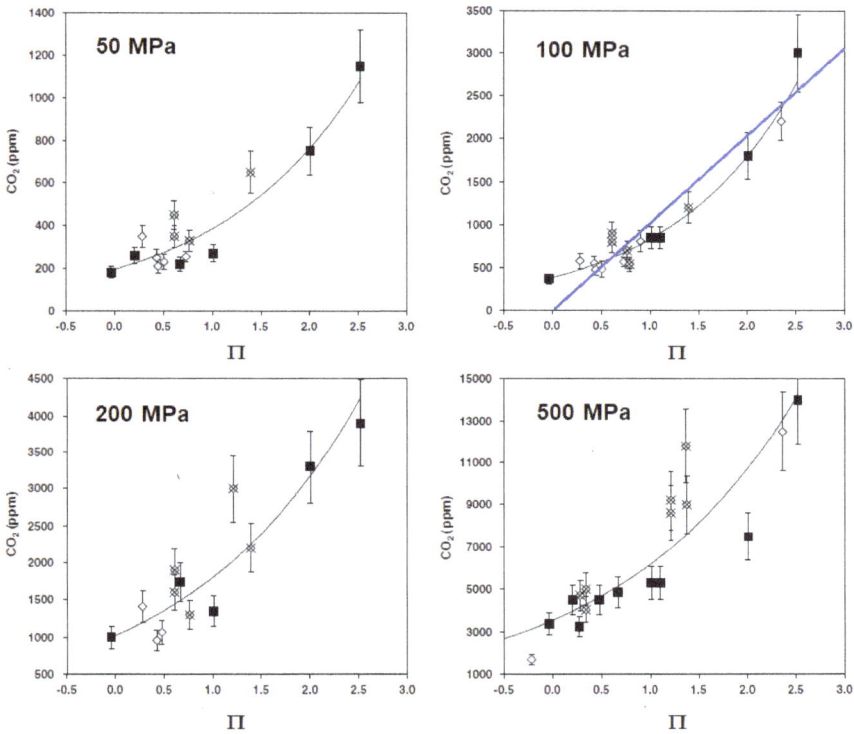

Figure 10. Pressure dependence of CO_2 solubility as a function of the parameter Π defined in Dixon (1997) (the straight **blue line** at 100 MPa corresponds to the relationship by Dixon 1997; see also text)). **Black squares**: compositions studied in Shishkina et al. (2014); **white diamonds**: data used for the exponential fits (**black solid lines**); **crossed symbols**: data points not used for the exponential fits. Modified after Shishkina et al. (2014).

where the d_i terms are meant to express the chemical control of selected oxides or ratio of oxides (x_i) on CO_2 solubility, thus following an approach similar to that of Wallace and Carmichael (1992) for sulfur, although inspired by the conventional Margules mixing formalism. The Authors also assume a functional form for $a_{O^{2-}}$ such that

$$\ln a_{O^{2-}} = b \cdot \frac{\text{NBO}}{\text{O}} \qquad (124)$$

where the term NBO/O indicates the ratio between non-bridging to total oxygens, justified by spectroscopic measurements and theoretical investigations that suggest a correlation between this quantity and CO_2 solubility (Brooker et al. 2001; Guillot and Sator 2011). To assess the number of NBOs, Iacono-Marziano et al. (2012) consider two extreme cases represented by anhydrous melt (Marrocchi and Toplis 2005) and hydrous one, in which all dissolved H_2O is assumed to have converted into OH groups, all of which produce NBOs (see the original paper for the details of the computation).

By further assuming constant enthalpy, entropy and volume terms, Equation (122) is then rewritten as:

$$\ln\left(w_{CO_2}, \text{ppm}\right) = \sum_i x_i d_i + a \cdot \ln P_{CO_2} + b \cdot \frac{\text{NBO}}{\text{O}} + \frac{A}{T} + B + \frac{C \cdot P}{T} \qquad (125)$$

where the CO_2 partial pressure is used instead of fugacity (ideal gas) and d_i, a ,b, A, B and C are fitting parameters.

In the case of water, Iacono-Marziano et al. (2012) assume complete speciation into OH^- groups as for Reaction (73), that they write as:

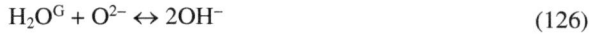

$$H_2O^G + O^{2-} \leftrightarrow 2OH^- \tag{126}$$

and then rework its equilibrium constant to obtain for dissolved water (in wt%) the same functional form as for carbon dioxide in Equation (125):

$$\ln\left(w_{H_2O}, wt\%\right) = \sum_i x_i d_i' + a' \cdot \ln P_{H_2O} + b' \cdot \frac{NBO}{O} + \frac{A'}{T} + B' + \frac{C' \cdot P}{T} \tag{127}$$

After calibration on a dataset of H_2O, CO_2 and $H_2O + CO_2$ saturation conditions for mafic melts, the final parameterization takes the following form:

$$\ln\left(w_{CO_2}, ppm\right) = x_{H_2O} d_{H_2O} + x_{Al} d_{Al} + x_{FeO+MgO} d_{FeO+MgO} + x_{Na_2O+K_2O} d_{Na_2O+K_2O} + \\ a \cdot \ln P_{CO_2} + b \cdot \frac{NBO}{O} + B + \frac{C \cdot P}{T} \tag{128}$$

$$\ln\left(w_{H_2O}, wt\%\right) = a' \cdot \ln P_{H_2O} + b' \cdot \frac{NBO}{O} + B' + \frac{C' \cdot P}{T} \tag{129}$$

which excludes enthalpy terms, therefore limiting the effects of temperature on volatile saturation contents (as in Dixon 1995a,b, 1997), consistent with the limited T range of the calibration database which includes experiments mostly equilibrated between 1200 and 1300 °C, and not exceeding 1100–1400 °C. An application of this model with caveats about its extrapolation outside the calibration domain can be found in Shishkina et al. (2014).

Following the philosophy of Iacono-Marziano et al. (2012), Eguchi and Dasugpta (2018) used a relation similar to Equation (128) for CO_2 solubility in natural silicate melts at graphite/diamond saturation. A purely empirical formulation for CO_2 solubility in low-silica kimberlite melts is given by Moussallam et al. (2015) which takes the form $\ln(wCO_2, wt\%) = A(SiO_2 + Al_2O_3)^3 + B(SiO_2 + Al_2O_3)^2 + C(SiO_2 + Al_2O_3) + D$, where D is a constant and A, B and C are flexible power-law functions of pressure.

Acid–base compositional control on volatile speciation

In the speciation models described so far, the role of the melt in determining the speciation of dissolved water and carbon dioxide is either limited or practically absent. The Burnham model defines but does not implement speciation mechanisms that allow a judicious choice of macroscopic components removing non-ideal Henry's-like behaviour. The suite of Stolper's (and co-workers) models assess water speciation but avoid further complexities related to the different reactive properties of melt oxygens (O^0, O^- and O^{2-}) as from the Lux–Flood formalism (Reaction 70–71) and as it is directly accounted for in the Toop and Samis (1962a,b) approach. Spera and Bergman (1980) and Stolper-like models based on Henrian behaviour for volatile solubility/saturation are not anchored to the melt reactive complexity, and simply employ enthalpy and volume terms for different melt types. The Dixon et al. (1997) extension adopts silica as a control variable for dissolved molecular water and carbonate groups in basaltic (sensu-lato) compositions at low-pressure. Finally, structural parameters (e.g. NBO/O in Iacono-Marziano et al. 2012) are used as semi-empirical parameters anchored to composition but without a theoretical framework for melt reactivity.

Although not fully explicit or accounted for, melt acid–base properties lie on the bottom line of the above approaches. Let's recall that the acid–base character of a solution, hence the property of its species to behave as either a base or an acid (or both) determine, as for aqueous solutions, the reactivity of the solvent (the melt) which controls volatile speciation hence its saturation (see Fraser 1977; Ottonello and Moretti 2004; Moretti 2005; Cicconi et al. 2020a; Moretti and Ottonello 2021). Accordingly, reactive-species models can provide a quantitative description of dissolution and speciation, provided a basicity scale.

In the case of water, Fraser (1977) already noted that *"in very basic melts H_2O may dissolve as an acidic oxide"*, which is something not accounted for by the models seen so far. This fact and the associated implications may perhaps not be immediately evident to the reader, but two examples extracted from the arguments in the previous pages should help:

(1) Dissolution of CO_2 as CO_3^{2-}. Following Reaction (67), increasing carbonate species results in increased melt polymerization, as qualitatively discussed by Spera and Bergman (1980), Fine and Stolper (1985, 1986) and Stolper et al. (1987). Because increased melt polymerization is also a consequence of increased silica concentration (and decreased O^{2-}), the larger the silica content the lower the CO_2 solubility as carbonate in response to the Mass Action Law. This behavior is fully consistent with the notation of Reaction (67), which basically says that CO_2 behaves as an acidic component according to the Lux–Flood definition at Reaction (71).

(2) Dissolution of H_2O as hydroxyl groups. Following Reaction (73), hydroxyl groups cause melt depolymerization, as long structural units of the type $-O-Si-O-Si-O-$ are disrupted into shorter $-O-Si-OH$ units. Accordingly, silica-rich (O^0-rich) polymerized melts dissolve water quite easily based on the mechanism at Reaction (73). If Reaction (75) is subtracted from Reaction (73) while taking into account Reaction (70) between three kinds of reactive oxygen, we obtain Reaction (74) (with H_2O^G instead of H_2O), which says that water is a basic oxide following the Lux–Flood formalism of Reaction (71).

The above examples show that dissolution of CO_2 and H_2O can have opposite effects on melt structure, an aspect which does not emerge from the above speciation-based models, possibly because the competition between polymerization and de-polymerization at low pressure is poorly effective as a consequence of far lower amounts of dissolved CO_2 with respect to H_2O. An example of such a competing behavior, increasing in relevance as pressure increases, is the observed increase in CO_2 saturation content when small amounts of H_2O are dissolved in an initially H_2O-free melt. Such a non-ideal behavior was first observed in pre-1980 experiments by Mysen and co-workers (Mysen 1976; Mysen et al. 1976), then confirmed by Jacobsson (1997), King and Holloway (2002), and Botcharnikov et al. (2005). In light of the examples above, that effect may be a consequence of decreased melt polymerization when adding H_2O, thus increasing CO_2 dissolution. It is relevant to note that such a non-ideal behavior is captured by the regular solution models described above (see Fig. 13 in Papale 1999a, and Fig. 14 in Ghiorso and Gualda 2015), although these models do not account explicitly for melt structural effects.

A first step towards the assessment of the response of the melt solvent in terms of its acid–base properties has been the assessment of the amphoteric behavior of dissolved water, that is, the dual character (acid and base) of H_2O in a melt (as for pure water and water solutions). Several lines of evidences, and particularly the Raman and NMR detection of OH^- (i.e., M–OH groups; Xue and Kanzaki 2004; Le Losq et al. 2013, 2015) suggest such a dual behavior of water and its dependence on melt composition (Xue and Kanzaki 2004; Moretti 2005; Mysen and Cody 2005; Moretti et al. 2014). In particular, in depolymerized melt systems water tends to stabilize the silicate network because it forms complexes with metals subtracting them from their role of network modifiers. This behaviour is similar to that of CO_3^{2-}, and points to water self-ionization (or autoprotolysis); that is, a reaction through which water molecules dissociate into solvated protons H^+ and hydroxide ions, OH^- (Moretti 2005; Moretti et al. 2014; Dufils et al. 2020):

$$H_2O_{mol} \leftrightarrow H^+ + OH^- \tag{130}$$

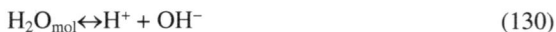

This reaction stands as the basis of the pH scale of acidity and has been one of the most fruitful concepts in general chemistry, used to describe solvent–solute interactions without reference to the structural state of water. Reaction (130) suggests that the concept of dissolved molecular water developed by Stolper and co-workers is actually related to water auto-ionization, and that the species actually involved are, again, H^+ and OH^-. In terms of ionic semantics compatible with the Lux–Flood formalism (Reaction 71), H^+ is the result of the basic dissociation of bulk water (macroscopic component) via Reaction (74), then combining with NBOs to give terminal OH groups, which determine silicate network depolymerization as demanded by the Toop–Samis Reaction (70). The above quotation by Fraser (1977) on the capability of water to dissolve as an acidic oxide in very basic melts can be rendered by the reaction:

$$H_2O^G + O^{2-} \leftrightarrow 2OH^- \tag{131}$$

It is important to note that Reaction (131) has a different meaning with respect to Reaction (86) from Silver and Stolper (1990) and subsequent papers, beyond the reference to molecular water in the gas or the melt phase. In fact, in Reaction (131) O^{2-} refers to oxygens that are not part of the polymerization structure of the melt, or free (non-structural or non-network associated) oxygens as from Reaction (70), while in Reaction (86) they do not necessarily take the same meaning. Similarly, OH^- in Reaction (131) refers to free or non-network associated hydroxyls that consume free oxygens as a consequence of water dissolution. As a side argument, we stress here the relevance of using appropriate notations with clear meanings. As a matter of fact, differently charged ions participating in exchanges with water (or other dissolved volatiles) have radically different meanings in terms of structural and acid–base implications for the melt.

Polymeric models based on Toop–Samis oxygen exchange at Reaction (70) treat simultaneously the basic and acidic dissociation of water, including hydrogen proton H^+ among silicate chain-breaking network modifiers. As for a salt in a mixture, the Toop–Samis exchange of oxygens implies adoption of the Temkin's assumption that the activity of any oxide (including water) in the melt can be expressed in terms of ionic fractions only and that there are no undissociated molecules. This corresponds to a standard state of complete dissociation in which the cations of the mixture are statistically distributed at random over the cation positions regardless of their electric charge, and the same holds for the anions. Therefore, the anions and cations in the Temkin standard state are assumed to occupy positions in their own respective semi-lattices. Reconciling the Temkin model to measured species requires therefore recombining at equilibrium dissociated entities such as H^+ and O^- (Reaction 75) to produce T–OH groups (T being a network former such as Si or Al), whose relative amount measures the strong depolymerizing effect on the melt. OH^- is one among the free anions, including O^{2-}, that complex metals such as Na, K, Mg, Fe^{2+}, and Ca. In doing so, OH^- groups subtract metals from their role of network modifiers, limiting melt depolymerization in metal-rich melts such as basalts. In fact, water acidic dissociation is much more pronounced in mafic than in silicic melts (Fig. 11). This has a macroscopic counterpart in much lower extent of viscosity decrease associated with initial water dissolution in mafic with respect to silicic melts (Mysen and Cody 2005; Moretti et al. 2014; see also Fig. 49c).

Hydrogen protons H^+ can also complex hydroxyl ions OH^- and recombine to form water molecules H_2O_{mol} (Reaction 130), detected in silicate melts by vibrational spectroscopy.

According to the Toop–Samis model for reactive oxygen exchange, water dissolving in silicate melts should then be treated in terms of completely dissociated (i.e., Temkin model) ion products H^+ and OH^-. This is coherent with the well-established Lux–Flood acid–base behavior based on O^{2-} exchange (Reaction 70), which demands dissociation of all macroscopic

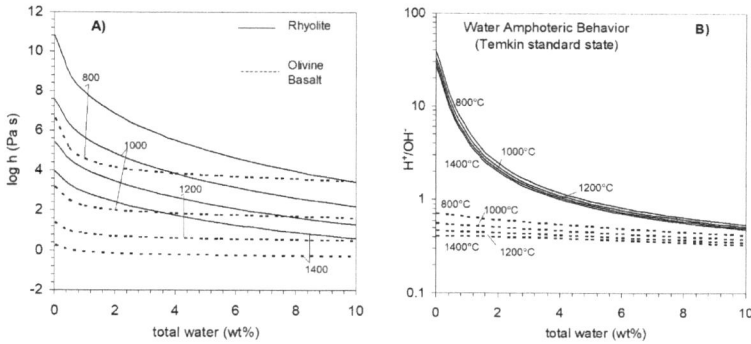

Figure 11. Patterns of melt viscosity and water dissociation quantitities as a function of water content in the melt. **a)** Viscosity of hydrous rhyolitic and basaltic melts (calculated as in Giordano et al. 2008). **b)** Water dissociation from Reaction (139) consistent with the Temkin's model (see also Moretti and Ottonello 2021). For low SiO_2 melts (basalts) the amphoteric behavior of dissolved water is such that the extent of depolymerization and the decrease of viscosity are limited upon further addition of water.

oxide components in ionic species according to acid–base reaction mechanisms (for water, Reaction 74 or 130). This formalism leads to quantifying the proportions of of O^0, O^- and O^{2-} in the melt, whereby $x_{O^{2-}}$ is a measure of melt basicity (see Reaction 71) and its logarithm has practically the same meaning for silicate melts as the pH value in aqueous solutions (Flood and Forland 1947; Trémillon 1967; Duffy and Ingram 1971; Moretti 2005, 2021; Moretti et al. 2014; Cicconi et al. 2020b; Moretti and Ottonello 2021). Within the frame of the Toop–Samis model, the compositional dependence of Reaction (70) reflects the oxygen-based variability of the melt network, which results in a composition-dependent mean polarization state of the silicate medium and thus composition-dependent proportions of the three oxygen species (O^0, O^- and O^{2-}). The compositional shift of the network mean polarization state determines the acid–base properties of the melt (Ottonello and Moretti 2001; Moretti 2005, 2021; Moretti and Ottonello 2021), which are therefore strictly related to the relative proportions of the three kinds of oxygens, thus to melt polymerization. Based on the Toop–Samis formalism, a consistent set of chemical components involved in volatile–melt reactions can be identified so that charge is transferred between reactive chemical entities such as network-modifying cations, network-forming polyanions and free anions not contributing to the network structure, such as O^{2-} and OH^-. The syntax employed to write chemical exchanges relates therefore to aspects of the melt structure, although not being itself a complete structural description of the melt phase.

Moretti et al. (2014) solved water speciation and its amphoteric behavior on the basis of a polymeric framework in which Equation (72) describing the equilibrium of Reaction (70) is solved for its compositional dependence. For a thorough discussion about the implementation of ionic–polymeric models of silicate melts, the reader is referred to Moretti and Ottonello (2021). Here we summarize basic features that are relevant for volatile speciation and that also form the basis for sulfur saturation modeling treated below.

Following Ottonello et al. (2001), K_{70} in Equation (72) is parameterized on the basis of atomistic properties, such as the basicity moderating parameters (a measure of the covalence of a cation in the oxygen ligand, then of the solvation character of the cation; see also Moretti and Ottonello 2021), for both network modifiers ($\gamma_{M_i^{v+}}$) and network formers ($\gamma_{T_j^{n+}}$), through the following equation:

$$K_{70} = \exp\left[4.662 \times \left(\sum_i x_{M_i^{v+}}\gamma_{M_{\boxtimes}^{v+}} - \sum_j x_{T_j^{n+}}\gamma_{T_j^{n+}}\right) - 1.1445\right] \tag{132}$$

where $x_{M_i^{v+}}$ and $x_{T_j^{n+}}$ are atom fractions of network modifiers (including H^+) and network formers, respectively. Equation (132) establishes a formal link between the acid–base properties of the medium (expressed as a balance between network formers' and network modifiers' basicity in the sense of Lewis; Moretti 2021) and the polymerization constant K_{70}. The Toop–Samis model adopts complete dissociation of constituting basic oxides, which actually corresponds to the Temkin (1945) model of fused salts. This ascribes the activity of a molten component to the product of the activities of ionic fractions over the cation and anion matrixes. For a generic divalent oxide that corresponds to:

$$a_{MO(Temkin)} = a_{M^{2+}} a_{O^{2-}} = x_{M^{2+}} x_{O^{2-}} = \frac{n_{M^{2+}}}{\sum cations} \cdot \frac{n_{O^{2-}}}{\sum anions} \tag{133}$$

which defines a standard state corresponding to completely dissociated oxide components, so that for a generic dissolution reaction for a generic divalent oxide (Lux–Flood formalism of Eqn. 71) :

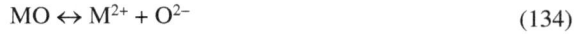

$$MO \leftrightarrow M^{2+} + O^{2-} \tag{134}$$

we get

$$K_{134} = \frac{x_{M^{2+}} x_{O^{2-}}}{a_{MO}} \neq K_{134(Temkin)} = 1 \tag{135}$$

where 1 is the value for the complete dissociation in the Temkin standard state, implying a standard state shift that must be tracked. Given the Temkin-like complete dissociation, the number of moles of bridging oxygens and of free oxygens per mole of melt is:

$$\left(O^\circ\right) = \frac{4N_T - \left(O^-\right)}{2}, \; \left(O^{2-}\right) = \left(1 - N_Y\right) - \frac{\left(O^-\right)}{2} \tag{136}$$

where N_T is the number of moles of tetrahedra in a mole of melt, $1 - N_Y$ represents the number of moles of basic oxide per mole of melt, and the parentheses indicate moles per mole of melt.

Combining Equations (72) and (136) gives:

$$K_{70} = \frac{4N_T - \left(O^-\right)\left[2 - 2N_T - \left(O^-\right)\right]}{4(O^-)^2} \tag{137}$$

that can be solved for (O^-).

In this conceptual framework, the activities of dissociated oxide components correspond to those of Temkin's fused salt notation in its standard state (Temkin 1945), which for polymeric melts are generalized as (Moretti and Ottonello 2005):

$$a_{M_\chi O_\psi} = \left[\frac{\left(M^{v+}\right)}{\sum cations}\right]^\chi \times \left[\frac{\left(O^{2-}\right)}{\left[\left(O^{2-}\right) + \left(OH^-\right) + \sum structons\right]}\right]^\psi \tag{138}$$

Ottonello et al. (2001), Moretti (2005) and Moretti and Ottonello (2021) report the procedure to compute the number of polyanionic units (or structons, in the sense of Fraser 1975) that appear in Equation (138).

In focusing on water speciation, Moretti et al. (2014) subtract Reaction (130) from Reaction (74), that is, they remove the macroscopic oxide water component and equilibrate its dissociation products in the Temkin standard state of complete dissociation, obtaining:

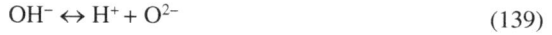

$$OH^- \leftrightarrow H^+ + O^{2-} \tag{139}$$

with equilibrium constant derived in the Temkin standard state as:

$$\log K_{139} = \frac{[H^+][O^{2-}]}{[OH^-]} \tag{140}$$

with square brackets denoting concentrations assumed to correspond to activities.

Moretti et al. (2014) then consider infrared data using the following mass balance:

$$n_{H^{TOT}} = 2n_{H_2O}^{TOT} = n_{OH^{FTIR}} + 2n_{H_2O_{mol}} = n_{T-OH} + n_{OH^-} + 2n_{H_2O_{mol}} \tag{141}$$

where infrared spectroscopy allows determination of OH^{FTIR} given by T–OH plus OH^- (i.e., M–OH) along with H_2O_{mol}. The last step is then reconciling infrared observations with the polymeric model based on complete dissociation of water (Reactions 130 and 74, or their difference given by Reaction 139). Based on existing infrared data of water speciation, the authors showed that the amount of detected T–OH groups equals the amount of H^+ which is residual after recombination to give H_2O_{mol}. This number corresponds to:

$$n_{H^+} = n_{T-OH} = \frac{\Sigma \text{cations}\left(\dfrac{n_{OH^{FTIR}}}{2} - n_{H_2O}^{TOT}\right) + K_{139}n_{O^{2-}}\left(n_{H_2O}^{TOT} + \dfrac{n_{OH^{FTIR}}}{2}\right)}{K_{139}n_{O^{2-}} + \Sigma \text{cations}} \tag{142}$$

Equation (142) is the necessary link to anchor the infrared observational database to the requirements of a polymeric model based on full oxide dissociation typical of the Temkin standard state in ionic–polymeric models. It allows calibrating K_{139} (Eqn. 140) and solve Equation (141) for n_{T-OH}, n_{OH^-} and $2n_{H_2O_{mol}}$. Moretti et al. (2014) showed that infrared speciation is inherited at glass transition temperature, very likely due to hydrogen bonding that favors H_2O_{mol} (Le Losq et al., 2014). Their melt speciation based on Reaction (139) and subsequent recombination to give H_2O_{mol} and T–OH other than OH^- (or M–OH) is in good agreement with i) NMR results by Xue and Kanzaki (2004, 2008) (Fig. 12) reporting low but significant concentration of M–OH groups in mafic compositions, and ii) *in-situ* Raman measurements by Le Losq et al. (2014), and subsequent IR, Raman and NMR measurements by Le Losq et al. (2015). Moretti et al. (2014) could then derive water auto-ionization in melts:

$$\log K_{130} = \log \frac{[H^+][OH^-]}{[H_2O_{mol}]} = -1.416 + \frac{3568.68}{T} \tag{143}$$

which represents a fundamental property of water dissolving in silicate melts (see also Dufils et al. 2020). The value of K_{130} at Equation (143) is in line with that obtained by Raman, determinations in Le Losq et al. (2013), yielding log $K_{130} = 1863.1/T - 1.9585$ in their investigated T-range. The code for water amphoteric behavior and speciation can be accessed from https://github.com/charlesll/water-speciation-magma

Although not done by the previous authors, water solubility can be directly computed by replacing Reaction (139) with Reactions (74) (basic dissociation of water) and (130) (acidic dissociation of water) from which it was derived.

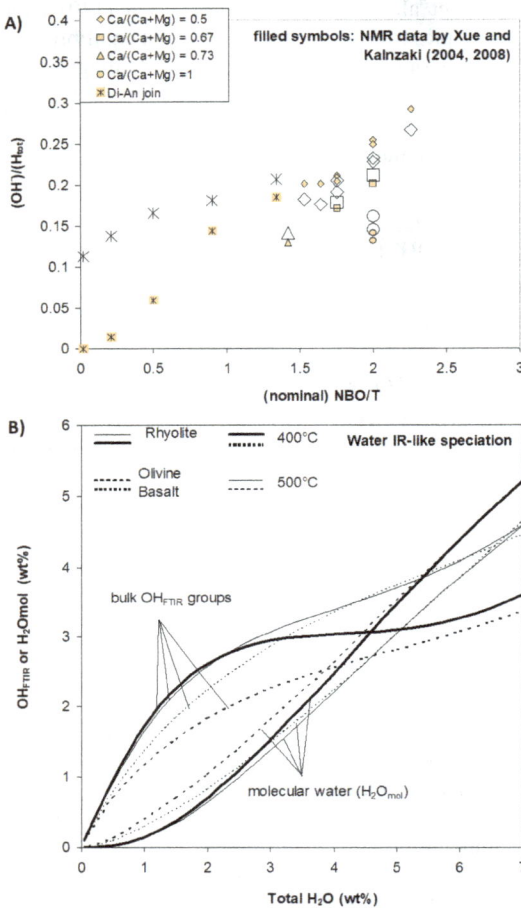

Figure 12. a) NMR data (**filled symbols**) of water speciation (Xue and Kanzaki 2004, 2008) compared to the predictions of the Moretti et al. (2014) model for water amphoteric behavior in melts (**empty symbols**). **b)** Concentration of FTIR-like water species predicted by Moretti et al. (2014) for hydrous rhyolites and basalts at 400 and 500 °C. These two temperatures are typical glass transition temperatures of hydrous melts. They have been computed as the temperature corresponding to a viscosity of 10^{12} Pa s, computed from via the Giordano et al. (2008) model, and refer to rhyolite with 2 wt.% H_2O and basalt with 3.5 wt.% H_2O (500 °C) and rhyolite with 4 wt.% H_2O (400 °C). Modified after Moretti et al. (2014).

The example above shows that the ionic–polymeric treatment of dissolution reactions provides a unified, consistent description of acid–base reactivity determined by the oxides constituting the melt. These arguments are further expanded in the following section, which deals with the solubility of sulfur, the amphoteric volatile par excellence.

Sulfur solubility

Saturation of sulfur in melts under conditions frequent in nature involves the formation of immiscible FeOS liquid as well as solid phases like pyrrhotite and anhydrite. Here we focus on only melt–gas equilibrium, while for the saturation conditions with respect to the above liquid and solid phases we address the reader to the vast existing literature (e.g., Smythe et al. 2017; Baker and Moretti 2011; Moretti and Baker 2008, and references therein).

Since early studies in metallurgy (Fincham and Richardson 1954), sulfur is known to be characterized by a speciation state strongly dependent on oxygen fugacity. Two solubility mechanisms were defined:

$$O^{2-} + \frac{1}{2}S_2 \leftrightarrow S^{2-} + {}^1/_2 O_2 \qquad (144)$$

$$O^{2-} + \frac{1}{2}S_2 + \frac{3}{2}O_2 \leftrightarrow SO_4^{2-} \tag{145}$$

That can be re-written in terms of SO_2 (or any other S-bearing gaseous species):

$$O^{2-} + SO_2 \leftrightarrow S^{2-} + \frac{3}{2}O_2 \tag{146}$$

$$O^{2-} + SO_2 + \frac{1}{2}O_2 \leftrightarrow SO_4^{2-} \tag{147}$$

As for reactions involving water illustrated above, melt basicity, corresponding to the O^{2-} term, appears on the left-side in Reactions (144–145). Assuming ideal mixing of the species involved in Reactions (146) and (147) (therefore assuming equivalence between activity and concentration), and employing the corresponding reaction constants, the roles of melt basicity, oxygen fugacity, and SO_2 fugacity in determining the equilibrium are highlighted (remind that fugacity is equal to activity times a reference fugacity, Eqn. 25):

$$\log\left[S^{2-}\right] = -\frac{3}{2}\log a_{O_2} + \log a_{SO_2} + \log K_{146} + \log\left[O^{2-}\right] \tag{148}$$

$$\log\left[SO_4^{2-}\right] = \frac{1}{2}\log a_{O_2} + \log a_{SO_2} + \log K_{147} + \log\left[O^{2-}\right] \tag{149}$$

where square brackets denote concentration. Equations (148) and (149) illustrate some fundamental aspects of sulfur solubility in silicate melts: i) the bulk solubility of sulfur is given by the sum of dissolved sulfide (Eqn. 148) and sulfate (Eqn. 149), and ii) for given *P*, *T* conditions and melt composition (i.e., for constant $[O^{2-}]$) and fixed fugacity of SO_2, dissolved sulfide and sulfate respectively decrease (Eqn. 148) and increase (Eqn. 149) with increasing oxygen fugacity, following on a log–log scale simple linear relations constrained by the O_2 stoichiometric coefficient in Reactions (146) and (147). Such a dual behavior depicts a solubility minimum in total sulfur as a function of oxygen fugacity, which represents an important theoretical constraint for the correct interpretation of sulfur solubility data, particularly near the QFM buffer (Carroll and Rutherford 1988; Moretti and Ottonello 2005; Moretti 2021). At the same time, it is worth noting that the theoretical constant slopes of log [S] vs. $\log a_{O_2}$ (or $\log f_{O_2}$) are only attained for conditions corresponding to constant sulfur dioxide and free oxygen activity (Moretti et al. 2003a; Moretti and Ottonello 2005; Cicconi et al. 2020). Under such conditions, Equations (148) and (149) lead to:

$$\log w_S = -\frac{3}{2}\log f_{O_2} + const \quad \text{(reduced region)} \tag{150}$$

$$\log w_S = \frac{1}{2}\log f_{O_2} + const \quad \text{(oxidized region)} \tag{151}$$

where w_S is mass fraction of dissolved sulfur. The entire procedure could be re-written starting from Reactions (144) and (145) and fixing f_{S_2} instead of f_{SO_2}, in which case the stoichiometric coefficients at Equations (150–151) would be $-1/2$ for the reduced region, and $3/2$ for the oxidized one.

Matching the theoretical slopes at Equations (150–151) is a condition always achieved in equilibrium experiments on sulfur solubility where fixed stoichiometry is imposed to the gaseous mixture. This is however not the natural case, since rarely is the gas phase sufficiently abundant to buffer the system. While the theoretical slopes are useful to test if sulfur is present entirely as either sulfide or sulfate (e.g, Fincham and Richardson 1954; Moretti and Ottonello 2005; Carroll and Webster 1994; Baker and Moretti 2011, and references therein), directly extending the results of buffered experiments to real system data can lead to serious misinterpretation. Similarly, the interpretation of experimental results requires particular care. An example of the latter is provided in Figure 13, which shows sulfur solubility data for a hydrous $Ab_{35}Di_{10}An_{55}$ melt (Gorbachev and Bezmen 2002). From a log S vs. log f_{SO_2} plot (panel a) one may conclude that the increase of sulfur is related to its enhanced dissolution as sulfate. However, panel b reveals that all points align on the sulfide side where the theoretical slope is −1.5 (Eqn. 150) (the fact that the points do not strictly obey such an alignment is simply due to different pressures and water contents at saturation, which perturb the value of other quantities at Equations (148–149)—or equivalently, which make the last term on the right side at Equations (150–151) to not be strictly constant). Panels c and d in the same figure show the theoretical slopes for an Etna basalt investigated by Lesne et al. (2015), whereas panels e and f show a trachyandesite from

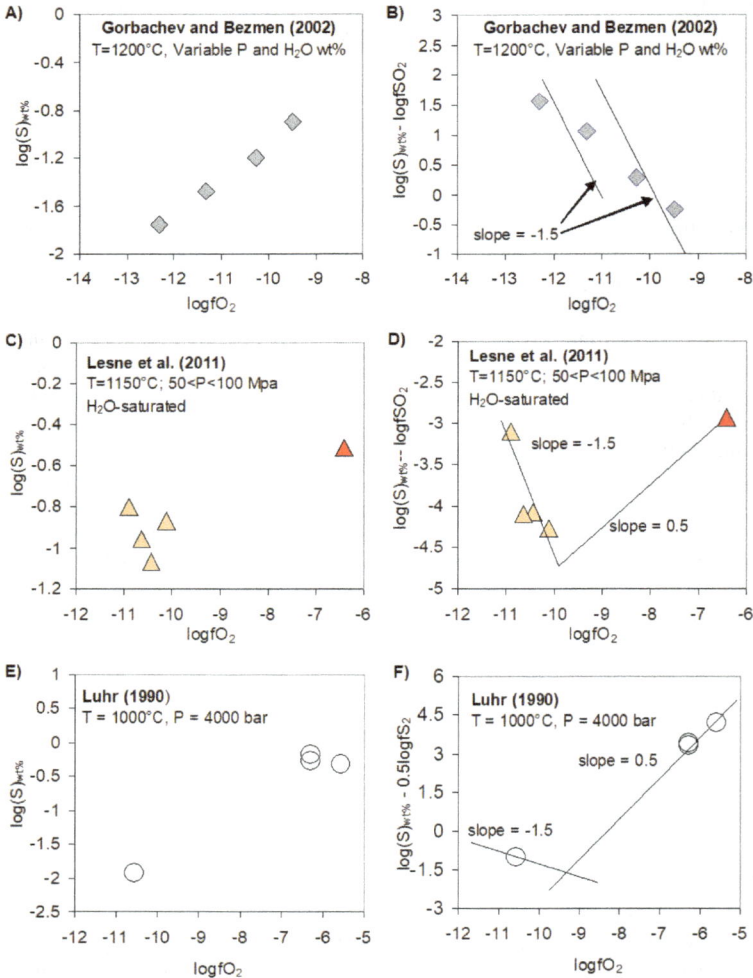

Figure 13. Bulk sulfur solubility and relative sulfide-sulfate solubilities. **a)** and **b)** Hydrous $Ab_{35}Di_{10}An_{55}$ melt ($T = 1200$ °C; Gorbachev and Bezmen 2002). **c)** and **d)** Hydrous Etna composition at 1050 °C (1200 °C for the **red triangle**) and $P < 100$ MPa. **e)** and **f)** Re-melted El-Chichon trachyandesite ($T = 1000$ °C; $P = 2–4$ kbar; Luhr 1990). Modified and extended after Moretti and Ottonello (2005).

El Chichon volcano in Mexico (Luhr 1990) (note that in this latter case f_{S_2} instead of f_{SO_2} has been used). Data points falling close to theoretical lines with negative slope in panels d and f d can be attributed to samples containing sulfide sulfur, whereas those falling close to theoretical lines with positive slope can be attributed to samples containing sulfate sulfur. The position of the solubility minima can be roughly identified.

In summary, while on a log–log scale theoretical sulfide and sulfate solubilities are linearly related to oxygen fugacity, real data can display substantial non-linear trends reflecting the variability of other quantities, particularly f_{O_2} and f_{SO_2} (or f_{S_2}) which enter the sulfur-related Reactions (144–147) (see also the related discussion in Moretti et al. 2003, and the discussion in the section below dedicated to combined C–H–O–S models).

It is worth noting that the subordinate presence of other dissolved sulfur species like S^0 or SO_3^{2-}, in which S appears as S^{4+}, has been object of some conjecture and discussion (e.g., Matjushkin et al. 2016). However, from published data the reduced sulfur species at equilibrium is definitely S^{2-}, and oxidized sulfur in the form of sulfite is likely an analytical artifact related to irradiation with an electron beam during electron microprobe analyses or an intense focused X-ray beam during synchrotron analysis (Fleet et al. 2005; Wilke et al. 2008). It is then justified to model sulfur solubility in melts as the result of redox-dependent sulfur dissolution as sulfide and sulfate. A full discussion of the role of redox features can be found in Moretti and Ottonello (2021) and references therein.

Modeling sulfur solubility in silicate melts: the CTSFG model

Below we introduce general concepts that have been fundamental for modeling sulfur solubility in silicate melts, and build on those concepts to develop first the ideas and equations at the basis of the CTSFG (Conjugated–Toop–Samis–Flood–Grjotheim) model developed in a series of papers by Moretti and co-workers. In the subsequent section we treat other modeling approaches that have been proposed to calculate sulfur solubility in silicate melts, then we describes the principles over which the COSH code is developed, joining the CTSFG and SOLWCAD codes to compute the compositional and redox-state -dependent saturation surface of fluids in the C–O–S–H system.

Based on Reaction (144), which dominates sulfur dissolution at reduced conditions typical of the metallurgical industry, early works in metallurgy have defined the quantity C_{S_2} called sulfide capacity (Fincham and Richardson 1954):

$$C_{S_2} = w_S \left(\frac{f_{O_2}}{f_{S_2}} \right)^{\!\!1/2} \tag{152}$$

which takes a specific value for each different melt composition, and that embodies the dependence of sulfur solubility on both oxygen and sulfur fugacity, their ratio being called the "sulfurizing potential" (Moretti and Ottonello 2005 and references therein). This quantity has been of critical relevance for refining steel and alloys from sulfur. A very large database of values of C_{S_2} at 1 bar has been set for slags, mostly contained within the SiO_2–Al_2O_3–MgO–CaO–FeO–MnO system. Fincham and Richardson (1954) demonstrated that within the limits of their measurements, C_{S_2} is an empirically predictable property. It should be stressed here that the sulfide capacity is a pseudo-equilibrium constant and does not, in any way, locate the sulfur solubility at saturation with other S-bearing liquid or solid phases. Its nature is similar to that of an equilibrium constant as long as C_{S_2} is a constant for each composition (T and P being also constant), a condition which implies sufficiently small amounts of dissolved sulfur so that the activities of all oxide components in the melt are about the same as those observed in the sulfur-free melt. O'Neill and Mavrogenes (2002) fitted experimental sulfide capacities for different silicate melt compositions at 1400 °C with an equation of the type:

$$\ln C_{S_2} = \sum_{i,\text{modifier}} x_i A_i + B_{\text{Fe-Ti}} x_{\text{Fe}} x_{\text{Ti}} + const \tag{153}$$

where the A_i terms relate to the difference in the free energies of formation of sulfide and oxide components, and $B_{\text{Fe-Ti}}$ is a sort of interaction parameter for Fe-Ti bearing melts. Other empirical parameterizations have been proposed in Sosinsky and Sommerville (1986), Young et al. (1992), and Beckett (2002), all based on the concept of optical basicity (introduced below).

Because of the lower interest of oxidized conditions for metallurgy, the related literature is much less abundant than for reduced conditions. In principle, a sulfate capacity analogous to the sulfide capacity at Equation (152) can be defined based on Reaction (145):

$$C_{\text{SO}_4^{2-}} = w_S f_{O_2}^{-3/2} f_{S_2}^{-1/2} \tag{154}$$

However, treating sulfur solubility based on Reactions (144–145) (or (146–147)), or by using sulfide and sulfate capacities defined at Equations (152) and (154), requires full assessment of the oxygen structure in the melt, thus full assessment of melt polymerization in a way similar to that for water speciation (see also Moretti and Ottonello 2021). As for the case of water, mutual interactions of the constituting oxides can be described in the framework of Temkin's fused salt approach (Temkin 1945), implying that ionic species are assumed to follow Raoult's behavior over their respective matrixes. The polymeric nature of silicate melts is thus described in terms of reactivity between charged species and functional groups (cations, free anions and polymeric units or "structons", according to Fraser 1975a,b, 1977), which in turn determine the mixing properties of the phase. In the case of sulfur, its "altervalent" nature (that is, the strong control exerted by oxygen fugacity in determining its partitioning between oxidized and reduced sulfur species as described by Reactions (144–145)) makes the role of melt polymerization, thus of the acid–base properties of the constituting oxides, even more fundamental in determining the overall solubility (Ottonello et al. 2001; Moretti and Ottonello 2003a,b, 2005, 2021; Moretti and Papale 2004; Moretti 2005, 2021; Baker and Moretti 2011; Cicconi et al. 2020; Le Losq et al 2020; Moretti and Stefansson 2020).

Moretti and Ottonello (2003a, 2005) have shown that the compositional dependence of Reactions (144–145) reflects the fact that both S^{2-} and SO_4^{2-} anions are not in a free (or "unpolarized") state, as for the O^{2-} or the OH^- groups seen above (see also Moretti 2021; Moretti and Ottonello 2021). Such anions are solvated into the melt structure and thus their polarization depends on overall melt composition, which is described by the overall oxygen-based silicate network and not by just the adjacent cations. Based on this consideration, Moretti and Ottonello (2003a, 2005) modelled the competing sulfur equilibria pairs (144–145) as:

$$w_S = w_{S,\text{sulfide}} + w_{S,\text{sulfate}} = C_{S^{6+}} f_{O_2}^{3/2} f_{S_2}^{1/2} + C_{S^{2-}} \frac{f_{S_2}^{1/2}}{f_{O_2}^{1/2}} \tag{155}$$

and showed that the capacities in Equation (155) can be conveniently combined by using Flood–Grjotheim electrically equivalent ion fractions defined for a generic cation M of charge v^+ (Flood and Grjotheim 1952; Moretti and Ottonello 2005):

$$x'_{M_i^{v+}} = \frac{v_i^+ n_{M_i^{v+}}}{\sum v_i^+ n_{M_i^{v+}}} \tag{156}$$

The $x'_{M_i^{v+}}$ values represent weighting factors for the contribution of each metal oxide–metal sulfide and metal oxide–metal sulfate pair to S^{2-} and SO_4^{2-} dissolution. In practice, the solubility mechanisms described by Reactions (144–145) can be solved by combining oxide–sulfide and oxide–sulfate exchanges (on a one-oxygen basis) of the type:

$$M_{2/\upsilon}O^M + \frac{1}{2}S_2^G \leftrightarrow M_{2/\upsilon}S^M + \frac{1}{2}O_2^G \tag{157}$$

$$M_{2/\upsilon}O^M + \frac{1}{2}S_2^G + \frac{3}{2}O_2^G \leftrightarrow M_{2/\upsilon}SO_4^M \tag{158}$$

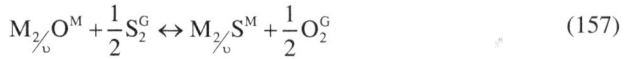

with equilibrium constants:

$$K_{O\text{-}S,M} = \frac{a_{M_{2/\upsilon}S}}{a_{M_{2/\upsilon}O}}\left(\frac{a_{O_2}}{a_{S_2}}\right)^{1/2} \tag{159}$$

$$K_{O\text{-}SO_4,M} = \frac{a_{M_{2/\upsilon}SO_4}}{a_{M_{2/\upsilon}O}}\left(a_{O_2}^{-3/2}a_{S_2}^{-1/2}\right) \tag{160}$$

Each one of the above exchanges includes specific dissociations, and related constants, for oxides and sulfides, or oxides and sulfates. For example, in the case of Reactions (157) and (158) the components undergo basic dissociation following (on a one S atom basis):

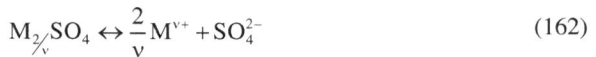

$$M_{2/\upsilon}S \leftrightarrow \frac{2}{\upsilon}M^{\upsilon+} + S^{2-} \tag{161}$$

$$M_{2/\upsilon}SO_4 \leftrightarrow \frac{2}{\upsilon}M^{\upsilon+} + SO_4^{2-} \tag{162}$$

The same applies to the corresponding basic molten oxides, for which (on a one O atom basis):

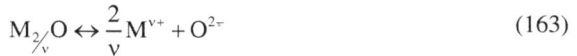

$$M_{2/\upsilon}O \leftrightarrow \frac{2}{\upsilon}M^{\upsilon+} + O^{2-} \tag{163}$$

Reaction (157–158) and (161–163) represent a change of notation with respect to Reactions (144–145) through specification of oxide and sulfide or sulfate components. As a consequence, they imply a change of standard state as they link oxides/sulfides/sulfates (components, with standard state defined at given P and system T, see the discussion immediately following Eqn. 19) and their ion products (species, with standard state in the Temkin model corresponding to the state of complete dissociation, see the section above on acid–base controls on volatile speciation). It follows that the equilibria at Equations (159) and (160) embody energetic transpositions from one standard state to the other (see also Ottonello et al. 2001; Moretti and Ottonello 2003a,b, 2005, 2021; Baker and Moretti 2011; Moretti 2021) that must be accounted for in the solution of the equations

Equations (159) and (160) weighted by electrically equivalent ion fractions defined at Equation (156) can be combined within a thermochemical cycle first proposed by Flood and Grjotheim (1952) and fully implemented by Moretti and Ottonello (2003a, 2005) (see Moretti and Ottonello 2021). This cycle can be equally applied to simple (e.g., metallurgical slags) as well as complex (e.g., natural magmas) silicate melts, providing the formalism to account for the composition-dependent melt solvation properties and polarization state of sulfide and sulfate groups; thus for composition-dependent sulfur solubility in silicate melts. The same formalism allows generalization of the equilibria (157) and (158) referring to some specific oxide-forming cation M with charge $\upsilon+$, to obtain expressions describing the compositional control on oxide–sulfide and oxide–sulfate disproportionation in multicomponent slags or melts:

$$\ln K'_{O\text{-}S} = \sum_{i=1}^{n} x'_{M_i^{\upsilon+}} \ln C^{anneal.}_{O\text{-}S,M_i} K_{O\text{-}S,M_i} \tag{164}$$

$$\ln K'_{O-SO_4} = \sum_{i=1}^{n} x'_{M_i^{\nu+}} \ln C^{anneal.}_{O-SO_4,M_i} K_{O-SO_4,M_i} \tag{165}$$

where $C^{anneal.}_{O-S,M_i}$ and $C^{anneal.}_{O-SO_4,M_i}$ are constants (entropies of annealing) accounting for energy gaps related to the aforementioned differences in standard states, and the multiplying electrically equivalent ion fractions (see Eqn. 156) are computed with reference to the appropriate matrix, either cationic or anionic. In the case of the anionic matrix, the charges of central cations (network formers) are considered.

The determination of the bulk oxide–sulfide and oxide–sulfate disproportionation constants (terms in Eqns. 164–165)) requires knowledge of the individual Gibbs energy of reaction for all components (Eqns. 159–160) as a function of temperature. Unfortunately, the available thermodynamic data are insufficient for such a goal. To overcome this issue, Moretti and Ottonello (2003a), Moretti et al. (2003), and Moretti and Ottonello (2005) treated the oxide–sulfide and oxide–sulfate thermodynamic constants as independent variables in a non-linear minimization routine that compares, through the appropriate stoichiometric conversions, the computed sulfide and sulfate solubility with "measured" sulfide and sulfate capacities. The sulfur disproportionation equilibria were then treated in the context of the Toop–Samis polymeric approach (see above), as it is described in the following, separately, for oxide–sulfide and oxide–sulfate equilibria.

Individual component oxide–sulfide equilibria. As it is described above for water speciation, melt chemistry is expressed in terms of network formers and network modifiers as in Ottonello et al. (2001): Si^{4+}, P^{5+}, Al^{3+}, Fe^{3+} are network formers and speciate over the anionic matrix as long as they are counterbalanced by network modifiers in originating complexes of the type $\left[\dfrac{4-\upsilon}{\chi} M^{\chi+} T^{\upsilon+} \right] O_4^{t-}$ with M indicating the network modifiers (Na^+, K^+, Mg+, Ca^{2+}, Mn^{2+}, Fe^{2+}, H^+, and Fe^{3+} in octahedral coordination). For both regression purposes when not specified, and in direct calculations, the ferric/ferrous ratio is best computed through the thermobarometric function described in Ottonello et al. (2001) and Moretti (2005) ensuring full internal model consistency.

Oxide–sulfide disproportionation equilibrium constants (see Eqn. 157) were then allocated for all oxides of interest as described in Moretti and Ottonello (2003a), and the ensuing equilibrium constants were expanded as an Arrhenian dependence:

$$\ln K_{O-S,M_i} = A_{O-S,M_i} + \frac{B_{O-S,M_i}}{T} \tag{166}$$

where the K_{O-S,M_i} terms express, for each oxide, the equilibrium for Reaction (157) and contribute to the right-hand side of Equation (164). Slope and intercept coefficients A_{O-S,M_i}, B_{O-S,M_i} can be first obtained from the examination of Gibbs energies of reaction between pure liquid $M_{2/\nu}O$ and $M_{2/\nu}S$ and the gaseous phase at different T conditions (Eqn. 161 rewritten for liquid components instead of molten ones). This implies that relaxation effects in the annealing of pure liquids in the melt phase are neglected, and it is equivalent to assuming that differences in the standard states of pure liquid and pure melt components reduce to zero. Transposition of energetic levels between the two standard states, corresponding to the $C^{anneal.}_{O-S,M_i}$ term in Equation (164), are then added as described below. Gibbs energies of reaction were estimated (Moretti and Ottonello 2003a, 2005) for a generic metastable (supercooled and/or superheated) "liquid" $M_{2/\nu}S$ component from salt data in liquid state available in thermochemical compilations from the temperature of fusion to some higher temperature (Barin and Knacke 1973; Robie et al. 1978; Ghiorso et al. 1983), together with isobaric heat capacity of pure liquid $M_{2/\nu}S$:

$$H_{M_{2/v}S}^{L} = H_{M_{2/v}S}^{0L} + \int_{T_f, M_{2/v}S}^{T} C_{P,M_{2/v}S}^{L} dT \tag{167}$$

$$S_{M_{2/v}S}^{L} = S_{M_{2/v}S}^{0L} + \int_{T_f, M_{2/v}S}^{T} \frac{C_{P,M_{2/v}S}^{L}}{T} dT \tag{168}$$

$$G_{M_{2/v}S}^{L} = H_{M_{2/v}S}^{L} - T S_{M_{2/v}S}^{L} \tag{169}$$

As an example, the individual exchange reaction involving silica:

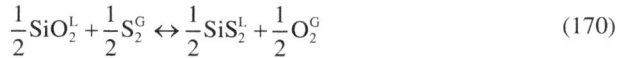

$$\frac{1}{2} SiO_2^{L} + \frac{1}{2} S_2^{G} \leftrightarrow \frac{1}{2} SiS_2^{L} + \frac{1}{2} O_2^{G} \tag{170}$$

results in

$$\ln K_{\text{O-S,Si}} = 2.0492 - \frac{36418}{T} \tag{171}$$

referred to the pure liquid component standard state. Based on a large dataset (>1200 anhydrous compositions) Moretti and Ottonello (2005) adopted non-linear minimization techniques to evaluate annealing entropies ($C_{\text{O-S,M}_i}^{\text{anneal.}}$ terms in Eqn. 164) representing standard state transposition between "pure liquid component" to "pure melt component". Slope coefficients found from the procedure above were let unchanged (that means assuming no enthalpic contribution to standard state transposition) and the optimization was conducted only on the entropic (intercept) terms (see Table 1 in Moretti and Ottonello 2005). A special case regarded assessment of the H_2O–H_2S exchange, for which an enthalpy constraint was taken from the gas phase homogeneous equilibrium:

$$H_2O^{G} + \frac{1}{2} S_2 \leftrightarrow H_2S^{G} + \frac{1}{2} O_2 \tag{172}$$

Individual component oxide–sulfate equilibria. The conceptual framework is similar to oxide–sulfide disproportionation, but involves two main limitations, both of them affecting the accuracy of the Moretti and Ottonello (2005) model for oxide–sulfate equilibria: i) due to much poorer industrial interest for the oxidized size of the equilibria involving sulfur, very few thermodynamic data exist for pure liquid components $M_{2/v}SO_4$, so that Gibbs energies of reaction between pure liquid $M_{2/v}O$ and $M_{2/v}S$ could be directly estimated only for Mg^{2+}, Na^+ and K^+; ii) the database on sulfate solubility is also significantly less extended (about 1/5) than that of sulfide. As an example, maintaining the same Arrhenian T-dependence adopted for oxide–sulfide disproportionation (Eqn. 166), silicon-related A and B parameters were unconstrained as a result of the lack of thermodynamic data for the component $Si(SO_4)_2$. Equation (173) illustrates the strategy that was adopted by Moretti and Ottonello (2005) to overcome the database limitations:

$$B_{\text{MO-MSO}_4} - B_{\text{MO-MS}} =$$
$$\frac{B_{\text{Si}_{0.5}\text{O-Si}_{0.5}\text{SO}_4} + B_{\text{Si}_{0.5}\text{O-Si}_{0.5}\text{SO}_4} + B_{\text{Na}_2\text{O-Na}_2\text{SO}_4} + B_{\text{K}_2\text{O-K}_2\text{SO}_4}}{4} -$$
$$\frac{B_{\text{Si}_{0.5}\text{O-Si}_{0.5}\text{S}} + B_{\text{MgO-MgS}} + B_{\text{Na}_2\text{O-Na}_2\text{S}} + B_{\text{K}_2\text{O-K}_2\text{S}}}{4} = const \tag{173}$$

Equation (173) simply says that the enthalpy of the various (isocoulombic) sulfide–sulfate exchange reactions ($M_{2/v}S + 2O_2 \leftrightarrow M_{2/v}SO_4$) can be approximated to a constant value (which turned out to be 114 kJ/mol), independent from the coordinating metal cation involved in the reaction, and in line with previous observations for solid compounds (Jacob and Jyengar 1982). Moretti and Ottonello (2005) fixed in this way the B parameter for important oxide–sulfate couples, such as those involving aluminum, calcium, titanium, manganese and ferrous iron. For trivalent iron the same B value as for aluminum was assigned based on the similarities in their structural role in melts. Phosphorous and chromium were not treated due to lack of data.

As for sulfide, the slope and intercept terms (A and B) for water were derived from the homogeneous gas-phase reaction at 1 bar:

$$H_2O^G + \frac{3}{2}O_2 + \frac{1}{2}S_2 \leftrightarrow H_2SO_4^G \tag{174}$$

and the B terms for water and other oxide components were let unchanged, whereas the A terms were corrected (as for sulfide reactions) for the annealing term resulting from standard state transposition.

Extension to high pressure conditions. Extension of the model for sulfide and sulfate capacities (Eqns. 159–160) at $P > 0.1$ MPa requires volume terms in the expressions for individual oxide–sulfide or oxide–sulfate equilibria, Equation (166) (referred to sulfide Reactions 157, and its analogous referring to sulfate Reactions 158):

$$\ln K_{O-S,M}^{P-T} = A_{O-S,M} + \frac{B_{O-S,M}}{T} - \frac{1}{RT}\int_{P^0}^{P}\left(\Delta V_m + \Delta V_g\right)dP \tag{175}$$

where P^0 corresponds to 0.1 MPa. The term $-\dfrac{1}{RT}\displaystyle\int_{P^0}^{P}\Delta V_m dP$ describes the effect of pressure on the equilibrium constant by transposing the activities of the annealed liquid components (i.e., melt-phase components instead of pure liquid components) $M_{2/v}S$ and $M_{2/v}SO_4$ from the standard state of "pure liquid component at 0.1 MPa and system T" to the standard state of "pure melt component at system $P–T$". The ΔV_g term (related to gaseous O_2 and S_2 volumes) are instead removed from the integral and are embodied into the experimental/calculated values for fugacities of gaseous O_2 and S_2.

For molten oxides, Moretti and Ottonello (2005) adopted the volume contributions from Lange and Carmichael (1990) and Lange (1994). Based on those values, the molar volumes of water and sulfur trioxide were computed through a 2nd order regression of oxide molar volumes (on a single-oxygen basis) against optical basicity, which corresponds to the reciprocal of the basicity moderating parameters (see Eqn. 132) (Moretti and Baker 2008; Fig. 14). A complete list of optical basicity values can be found in Moretti (2005) and Moretti and Ottonello (2005). That procedure resulted in $V_{SO_3} = 57.71$ cm³/mol and $V_{H_2O} = 16.44$ cm³/mol at 1400 °C. Volumes of pure molten sulfates were calculated isometrically as:

$$V_{M_{2/v^+}SO_4} = V_{M_{2/v^+}O} + V_{SO_3} \tag{176}$$

Molar volumes of sulfide components were obtained from the following relation:

$$\Delta V_{SO_4-S_2} = V_{M_{2/v^+}SO_4} - V_{M_{2/v^+}S} = 41\,\text{cm}^3/\text{mol} \tag{177}$$

The value of 41 cm³/mol in Equation (177) is a constant fitting parameter of the model, calibrated on observed sulfide capacities for water-bearing compositions over a $P–T$ range of conditions.

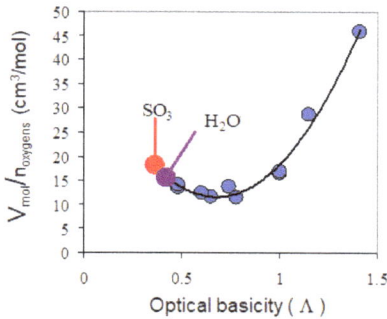

Figure 14. Molar volumes at 1400°C on a single oxygen–oxygen basis versus optical basicity (non-dimensional). The quadratic fit to the data allows estimation of the molar volume for SO_3. Modified after Moretti and Baker (2008).

CTSFG model calculations. The CTSFG sulfur solubility model, described above, has been applied to calculate the sulfide and sulfate capacity of metallurgical (Fig. 15) and geological melts either at 1 bar (Fig. 16) or with compositions constrained for any P–T along the melt–water saturation boundary through the SOLWCAD model (Fig. 17). Exploration of sulfur saturation in a range of P–T–X conditions reveals a number of complexities, mostly regarding the compositional dependence of both sulfide and sulfate capacities, illustrated in the following.

Figure 18 shows a direct relationship between sulfide capacity and temperature for a large compositional range from basalt to rhyolite. Water has a logarithmic effect on sulfide capacity, which decreases with increasing water content. The log C_{S_2} surfaces have a similar shape for all compositions considered in the figure. While for basic melts (panels a and c) high pressure isobars have an almost vertical trend, for acidic melts (panels e and g) the slope of isobars largely depends on the water content. Such trends are very different when the sulfate capacity is considered. Figure 18 (right panels) shows in fact that the sulfate capacity is inversely

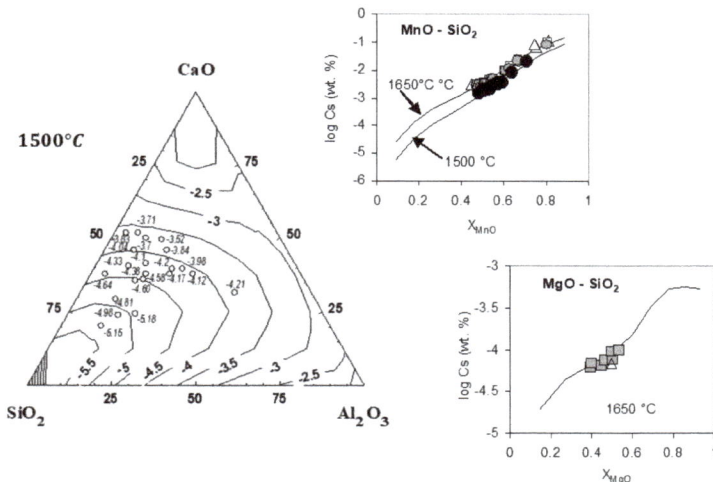

Figure 15. Compositional dependence of sulfide capacities in simple binary or ternary melts (slags) of metallurgical interest. Modified after Moretti and Ottonello (2003a).

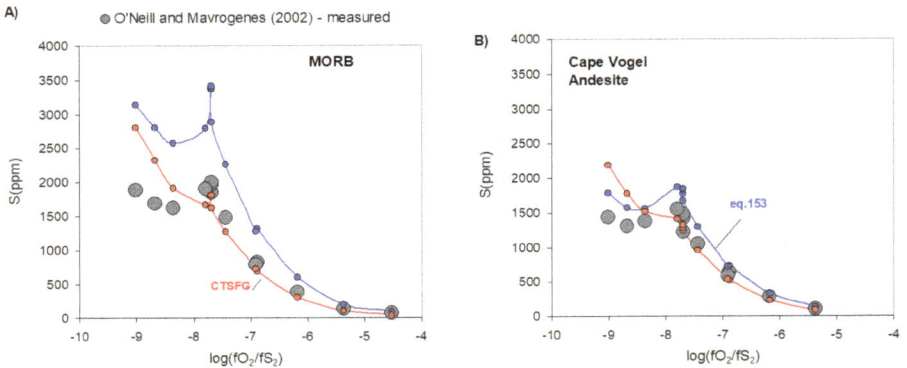

Figure 16. Sulfide solubility in synthetic melt analogues experimentally investigated by O'Neill and Mavrogenes (2002). **Large grey circles**: experimental data; **small circles**: model(s) results. The CTSFG model performance (**red**) is compared with *ad hoc* fit (**blue**, Eqn. 153). Lines are drawn as just a guide for the eyes. Modified after Moretti and Ottonello (2005).

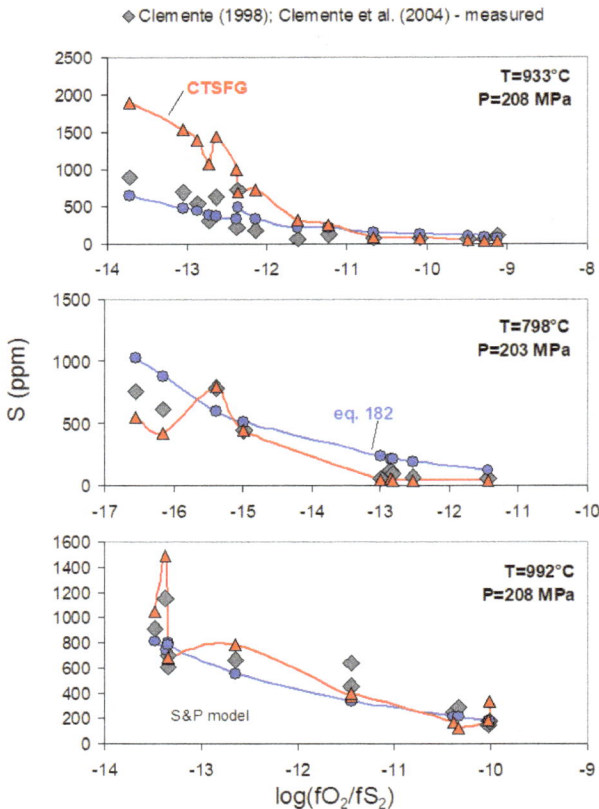

Figure 17. Sulfide solubility in a synthetic hydrous rhyolite. **Large grey symbols**: experimental data from Clemente (1998) and Clemente et al. (2004). The CTSFG model performance (**red**) is compared with *ad hoc* fit (**blue**, Eqn. 182). Lines are drawn as just a guide for the eyes. Modified after Moretti and Ottonello (2005).

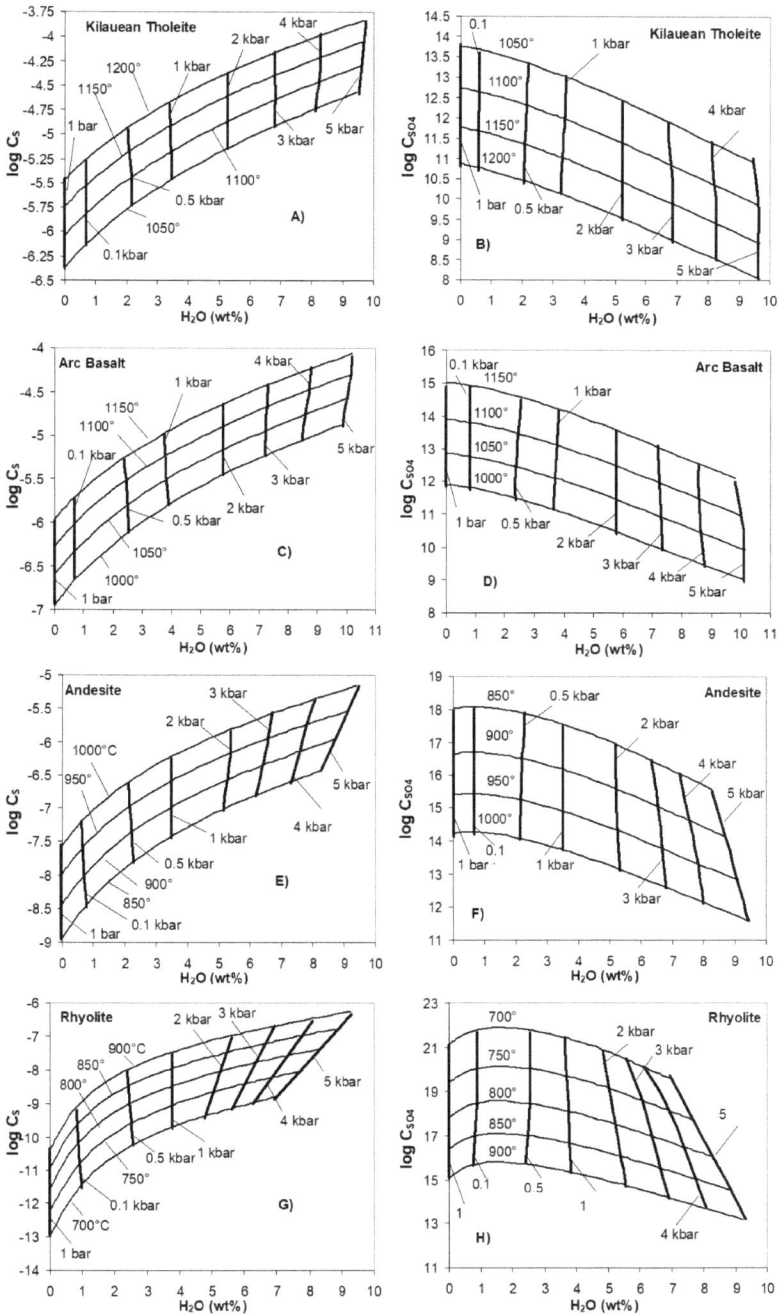

Figure 18. Sulfide (**left panels**) and sulfate (**right panels**) capacities along the melt–water saturation curve for four different melt compositions. Water contents are calculated through the SOLWCAD model. With sulfide an sulfate capacities as from these panels, all possible combination of dissolved sulfur, f_{O_2} and f_{S_2} can be computed via Equation (155). Modified after Moretti and Ottonello (2005).

related to temperature, a behavior which is opposite with respect to that for sulfide capacity. As for sulfide capacity, the addition of water generally results in a decrease of the sulfate capacity, but with more complex relationships with melt composition than for sulfide capacity. In fact, the andesitic and rhyolitic compositions employed in the figure show a maximum in log $C_{SO_4^{2-}}$ for water contents between 1 and 1.6 wt%, which appears to be implicit in the chemical interaction of the $H_2O–SO_4^{2-}$ melt couple. As for sulfide capacity, sulfate capacity isobars display close to vertical trends for basic compositions.

The sulfide and sulfate capacity calculations in Fig. 18, together with the corresponding values of f_{O_2} and f_{S_2}, provide the quantities for determining sulfur solubility via Equation (155).

An important application of the CTSFG model regards the speciation state of sulfur. This is typically described through the following reaction, obtained by subtracting Reaction (144) from Reaction (145) (or 146 from 147) :

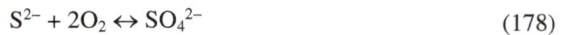

$$S^{2-} + 2O_2 \leftrightarrow SO_4^{2-} \tag{178}$$

with equilibrium constant for any melt composition obtained by subtracting Equation (164) from (165). The speciation between sulfate and sulfide, and thus the redox state of sulfur in melts (Eqn. 178) is implicitly solved in the CTSFG model when calculating the solubility of sulfide and sulfate groups for any melt composition (Fig. 19). Therefore, the equilibrium constant of Reaction (178), K_{78}, is an effective oxybarometer that allows computing oxygen fugacity, f_{O_2}, for any melt composition in which sulfate/sulfide ratios are available. The computation of f_{O_2} is then independent from sulfur solubility (Cicconi et al. 2003b; Moretti and Ottonello 2003b, 2005; Moretti and Stefansson 2020; Moretti 2021) but takes into account the role of melt composition on the sulfate–sulfide speciation. Details on the treatment of redox features, particularly within polymeric modeling, can be found in Moretti and Ottonello (2021).

The CTFSG code can be accessed from https://github.com/charlesll/sulfur-magma.

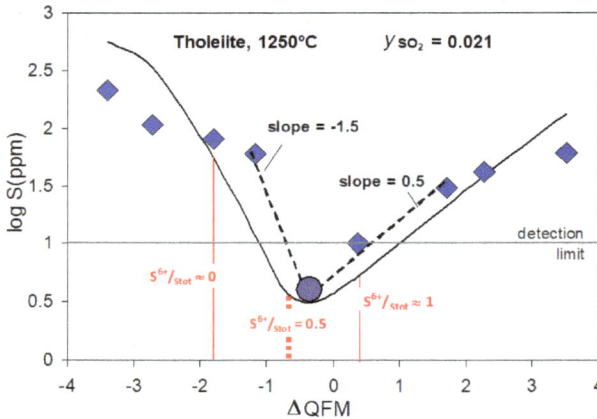

Figure 19. Experimental sulfur solubility from Katsura and Nagashima (1974) compared to CTSFG model results (**solid black line**). The **big circle** shows the position of the expected minimum based on theoretical slopes. Note that 10 ppm represents the detection limit in the experimental investigation. Also reported (**in red**) are relative proportions of sulfide and sulfate as from the CTSFG model. Comparable amounts of sulfide and sulfate occur in proximity of the minimum. Modified after Moretti and Ottonello (2005).

Additional models for sulfur solubility in silicate melts

Other models or parameterizations, including the one of O'Neill and Mavrogenes (2002) introduced above for sulfide capacity (Eqn. 153), attempt to determine the f_{S_2}–f_{O_2}–composition relations to return sulfur solubility. Once again, this review does not include the vast field of conditions related to saturation by solid (FeS, $CaSO_4$) or immiscible sulfide liquid ($FeOS$). Those conditions whereby sulfur is dissolved at either sulfide (Mavrogenes and O'Neill 1999) or anhydrite (Baker and Moretti 2011, and references therein) saturation have great relevance in petrology but they are beyond the scopes of this review, which is instead centered on gas–melt equilibria.

For basic compositions, Wallace and Carmichael (1992) proposed that the most important mechanism by which sulfide dissolves in melts is given by Reaction (162) for iron:

$$FeO^M + \frac{1}{2}S_2^G \leftrightarrow FeS^M + \frac{1}{2}O_2^G \tag{179}$$

with equilibrium constant given by:

$$K_{179} = \frac{\gamma_{FeS} x_{FeS}}{\gamma_{FeO} x_{FeO}} \left(\frac{f_{O_2}}{f_{S_2}} \right)^{\frac{1}{2}} \tag{180}$$

in which the $f_{O_2}^0$ and $f_{S_2}^0$ terms are implicitly assumed to cancel out. Rather than exploring mixing models for the melt phase, the authors assessed empirically the following expression:

$$\ln x_S = a \ln f_{S_2} - b \ln f_{O_2} + c \ln x_{FeO} + \frac{d}{T} + e + \sum_i x_i f_i \tag{181}$$

in which x_i is the mole fraction of the ith oxide component, and the f_i terms are compositional parameters calibrated by data regression along with the a, b, c, d and e terms. Equation (181) seems however inconsistent with Equation (180) and Reaction (179) unless the a and b parameters take a value of 0.5 (which is not the case), that value corresponding to the stoichiometric coefficient of O_2 and S_2 in Reaction (179) (see also Moretti 2002; Moretti and Ottonello 2021).

Scaillet and Pichavant (2005) set a purely empirical model (see Fig. 17) which relates the melt sulfur content to the sulfur fugacity for a variety of melt compositions including hydrous basalts, and proposed that:

$$\log w_S = aP - bT + c\Delta NNO^3 + d\Delta NNO^2 + e\Delta NNO\Delta FFS + f\Delta FFS + \sum_i g_i w_i \tag{182}$$

where ΔNNO and ΔFFS are the referenced $f_{O_2}^0$ and $f_{S_2}^0$ against the Ni–NiO and Fe–FeS solid buffers, respectively, w_i is the mass concentration of component i and a, b, c, d, e, f and g_i are fitting parameters.

A number of authors have approached sulfur solubility modeling by introducing a partition coefficient between the gas and melt phases, $D_S^{G/M}$, defined as:

$$D_S^{G/M} = \frac{C_S^G}{C_S^M} \tag{183}$$

where C_S is the mass concentration of sulfur in the gas or melt phase. For example, Zajacz et al. (2012) equilibrated andesitic melts with an H–O–S bearing volatile phase and found that a Henrian behavior explained the relationships between the dominant S-bearing species in the melt and H_2S in the gas phase. Thus, by considering the reaction:

$$FeO^M + H_2S^G \leftrightarrow FeS^M + H_2O^G \tag{184}$$

they reworked the equilibrium constant (with ideal behavior of FeO^M) to justify $D_S^{G/M} {f_{H_2O}^G}/{f_{H_2O}^{0G}} \propto {1}/{x_{FeO}}$. In a subsequent study, Zajacz (2015) studied the effect of melt polymerization on the quantity $D_S^{G/M} = {x_{SO_3}^M}/{x_{SO_3}^G}$.

A more complex formulation for $D_S^{G/M}$ was provided by Masotta et al. (2016), with reference to oxidized arc magmas:

$$\ln D_{S,mol}^{G/M} = 9.2 - 31.4\frac{NBO}{N_T} - 1.8ASI - 29.5Al\# + 4.2Ca\#$$ (185)

where *ASI* is the alumina saturation index $\left(ASI = \dfrac{x_{Al_2O_3}}{x_{CaO} + x_{Na_2O} + x_{K_2O}}\right)$, and *Al#* and *Ca#* are two empirical compositional parameters $\left(Al\# = \dfrac{x_{Al_2O_3}}{x_{SiO_2} + x_{TiO_2} + x_{Al_2O_3}}, Ca\# = \dfrac{x_{CaO}}{x_{Na_2O} + x_{K_2O}}\right)$.

Lesne et al. (2015) proposed a model in which the gas phase is characterized by the presence of H_2S and/or SO_2 such that the two following reactions can be defined:

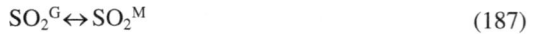

$$H_2S^G \leftrightarrow H_2S^M$$ (186)

$$SO_2^G \leftrightarrow SO_2^M$$ (187)

with equilibrium constants being, respectively:

$$K_{186} = \frac{f_{H_2S}}{x_{H_2S}}$$ (188)

$$K_{187} = \frac{f_{SO_2}}{x_{SO_2}}$$ (189)

in which ideal behavior for the melt species is assumed, and the reference states for each sulfur component in the gas and melt phase coincide. The authors found that in hydrous basalts, contrary to rhyolite (Clemente et al. 2004), the amount of dissolved sulfur does not tend to zero when f_{H_2S} tends to zero. Rather, they found a residual amount of dissolved sulfur ($x_{S,res}$) that increases linearly with the FeO content of the melt according to:

$$x_{S,res} = 0.019339 x_{FeO}$$ (190)

such that $x_{H_2S} = x_{S^{2-}} = x_S - x_{SO_4^{2-}} - x_{S,res}$. In practice, the authors introduced an additional species based on observations and reflecting the affinity of sulfur for iron at low f_{O_2}. This is a major aspect of the very low pressure behavior of the highly soluble sulfur component with respect to other volatile components such as water, carbon dioxide, and noble gases.

To account for the effects of *P* and *T* on K_{186}, Lesne et al. (2015) used an expression of the type of Equation (110), which after some algebra gives:

$$x_{H_2S}(P,T) = \frac{f_{H_2S}^G(P,T)}{K_{186}(P^0,T^0)} \cdot \exp\left\{-\frac{V_{H_2S}^0(P-P^0)}{RT} - \frac{\Delta H_{H_2S}^0(P^0)}{R}\left[\frac{1}{T} - \frac{1}{T^0}\right]\right\}$$ (191)

which was then fitted for P^0, T^0, the reference pressure and temperature, and $\Delta H_{H_2S}^0(P^0)$, the heat of dissolution of H_2S at P^0, while $V_{H_2S}^0$, the partial molar volume of H_2S as a melt species, was set equal to zero. For the SiO_2 range 48–55 wt%, Lesne et al. (2015) parameterized K_{186} as:

$$K_{186} = -1323947.1948 x_{FeO} + 158611.19322$$ (192)

$$\Delta H_{H_2S}^0 (J \cdot mol^{-1}) = -2590400.8 x_{FeO} + 75571.8$$ (193)

Figure 20 shows much higher $f_{S_2}^G$ values for given dissolved sulfur content predicted from the model of Lesne et al. (2015) compared to the CTSFG model, likely reflecting large approximations in extrapolating $f_{S_2}^G$ for samples at equilibrium with a FeS-bearing phase.

The metallurgical literature has also developed thermodynamic models for sulfur saturation in silicate melts (slugs) which apply at 0.1 MPa and anhydrous reduced conditions, those being the common conditions of industrial interest. Among them, the Lehmann and Gaye (1992) and Kang and Pelton (2009) models describe short-range ordering in ionic solutions and relate sulfide capacity to the Gibbs energy of mixing.

A final caveat is on the compositional dependence of the oxidation state of sulfur. While the large amount of oxygen required to oxidize sulfide species (e.g., Nash et al. 2019) may suggest poor relevance of melt composition in determining sulfide–sulfate partitioning in the melt, modeling Reaction (178) as being composition-independent appears totally unjustified. In fact, the experimental evidence and the models developed so far concur in highlighting a strong compositional dependence of sulfur solubility, hence of both sulfide and sulfate capacities. That is particularly true for pressure above atmospheric under hydrous conditions (Moretti and Baker 2008; Beaudry and Grove 2018; Moretti and Stefansson 2020; Moretti 2021). Accordingly, the models described above either include melt composition among the model variables or explicitly refer to some well-defined, limited compositional range. A correct appraisal of the role of composition on sulfide–sulfate speciation is also an essential ingredient to model S-isotope fractionation, particularly during magmatic degassing (Marini et al. 2011).

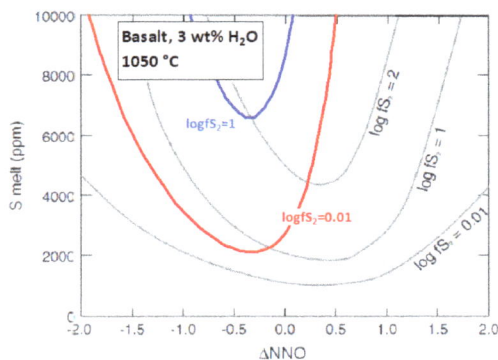

Figure 20. Calculations from the Lesne et al. (2015) model (**grey lines**, modified after Lesne et al. 2015) and comparison with the CTSFG model (**blue and red lines**) for a Stromboli-type basalt.

Combined C–H–O–S–(±Cl) saturation models

This sub-section briefly describes three different approaches that have resulted in three codes—SolEx, D-Compress, and COHS—for the saturation of compositionally rich fluid phases in the C–H–O–S–(±Cl) system. Because such fluids usually constitute more than 99% of natural magmatic emissions, these models are often employed to interpret magma degassing and derive constraints on the deep magmatic processes.

Witham et al. (2012) set the SolEx procedure for mixed C–O–H–S–Cl saturation in generic basaltic melts. For CO_2–H_2O they adopted the model in Dixon (1997), thus providing an implementation similar to the VolatileCalc code described above. Water and carbon dioxide are assumed to volumetrically dominate the gas phase thus their saturation is not influenced by the other volatiles, constituting the background over which the partition of sulfur and chlorine between the H_2O+CO_2-bearing melt and the H_2O+CO_2-dominated gas phase is computed.

The partition coefficients $D_{S,Cl}^{G/M}$ relate to a generic dissolution mechanism for S and Cl of the type $S^G \leftrightarrow S^M$ and $Cl^G \leftrightarrow Cl^M$. The equilibrium constants are written as:

$$K_{S,Cl} = \frac{f_{S,Cl}}{a_{S,Cl}^M} = \frac{\phi_{S,Cl} P \lambda D_{S,Cl}^{G/M}}{\gamma_{S,Cl}} \qquad (194)$$

where λ is the ratio of gas molar mass to melt molar mass, ϕ denotes the fugacity coefficient in the gas phase, γ is the activity coefficient in the melt phase (taken equal to 1), and the partition coefficients for S and Cl are determined by best-fitting the data in Lesne et al. (2011b):

$$\log K_S = 431.69 P^{-0.0074946} - 10.49 + 2.7426 \cdot 10^{-3} P - 1.4891 \cdot 10^{-7} P^2 \qquad (195)$$

with P in bars, while K_{Cl} is assumed to be constant and equal to 0.31.

From Equation (195) it follows that for the generic basaltic compositions for which the model is developed, K_S is assumed to exclusively depend on pressure. The model of Churakov and Gottschalk (2003) is employed to calculate the fugacity coefficients in the gas phase. The SolEx method applies in the 0.5–400 MPa range and for oxidation states $\Delta NNO > 0.5$, implying that the S-bearing gas species is SO_2.

Burgisser et al (2015) propose a method (D-Compress) that returns the composition of volatiles coexisting with three generic melts of basaltic, phonolitic and rhyolitic composition, for temperatures between 790 and 1400 °C and pressure up to 300 MPa. D-Compress represents an evolution of the redox/degassing model of Burgisser and Scaillet (2007), and employs very simple solubility laws, with parameters that can be replaced to match new data. The core of the method is constituted by the calculations for the gas phase, which can be a mixture of basic elements in the C–O–H–S system, with exchange of oxygen between the gas and melt phase regulating the redox state and melt iron ratio. D-Compress is provided as compiled software and source code as electronic supplementary material in Burgisser et al. (2015). The determinations for the gas phase combine mass balances and equilibrium constants involving the following reactions:

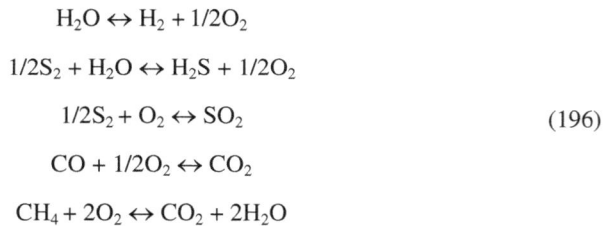

$$H_2O \leftrightarrow H_2 + 1/2 O_2$$

$$1/2 S_2 + H_2O \leftrightarrow H_2S + 1/2 O_2$$

$$1/2 S_2 + O_2 \leftrightarrow SO_2 \qquad (196)$$

$$CO + 1/2 O_2 \leftrightarrow CO_2$$

$$CH_4 + 2 O_2 \leftrightarrow CO_2 + 2 H_2O$$

The fugacity coefficients are assumed to follow the Lewis–Randall rule, meaning that for each component they are taken to coincide with that of the pure component at the same P–T conditions. While the gas phase is made of the nine species appearing in Reactions (196), the dissolved components are H_2S, SO_2, H_2O and CO_2, for which simple dissolution laws are assumed:

$$x_i = a_i f_i^{b_i} \qquad (197)$$

with a and b being fitting parameters (or polynomials in T) for each compositional type (basalt, phonolite, rhyolite). It can be noted that Equation (197) mimics simple power law relationships of the type $w = s P^n$, which were popular in the seventies and eighties to model the effect of pressure on the solubility of single component fluids, particularly water (e.g., Sparks 1978). Equation (197) can be compared to Henry's law, which refers instead to constant P–T condition and variable gas phase composition: $x_i = const. \, P_i$, where P_i is the partial pressure of component i. Because fugacity is partial pressure corrected for non-ideal effects, the power law relationship at Equation (197) implies that Henrian behavior is never approached, not even at very low pressure, unless b_i is close to 1 (generally not the case).

The model does not allow computation of the saturation conditions for temperature larger than the melting temperature of FeS (Moretti and Baker 2008) or for f_{S_2} larger than that at FeS saturation (Liu et al. 2007). Redox equilibrium between ferric and ferrous iron, and consequent O_2 exchange with the gas phase, is computed via the empirical formulation of Kress and Carmichael (1991).

From the description above, it should be clear that a relevant aspect of D-Compress is that of conserving the mass of oxygen throughout its calculations; e.g., along a decompressive path. Under such a constraint, oxygen fugacity is found to vary in highly non-linear ways during decompression accompanying magma ascent. Assuming constant mass of oxygen is a strong constraint for magma thermodynamics, and one which has been highly debated. Ghiorso and Kelemen (1987) defined oxygen as a perfectly mobile component which is continuously added or removed from the system, and supported the necessity of modelling phase equilibria in magmatic systems open to oxygen exchange. Moretti and Stefansson (2020) noted that closed conditions with respect to oxygen exchange may be effective only in the limited regions of high magma acceleration leading to fragmentation of silicic magmas (see Fig. 50 below), while in practically any other condition such an assumption appears unrealistic.

Dealing with compositionally complex systems, the models above adopt quite strong simplifications in their thermodynamic treatment. A different approach was followed by Moretti et al. (2003) and Moretti and Papale (2004) in setting up the COHS model for the system CO_2–H_2O–H_2S–SO_2–silicate melts of virtually any composition. COHS embeds the full SOLWCAD and CTSFG models, fully described above, for H_2O+CO_2 saturation and S solubility and speciation, respectively, closed by mass balance equations analogous to Equations (30) and (31) for H_2O–CO_2–melt systems:

$$\frac{w_{CO_2}^T - w_{CO_2}^G}{w_{CO_2}^L - w_{CO_2}^G} = \frac{k_1\left(w_{H_2O}^T - w_{H_2O}^G\right) - k_2 w_{H_2S}^G}{k_1\left(w_{H_2O}^L - w_{H_2O}^G\right) - k_2 w_{H_2S}^G} \tag{198}$$

$$\frac{w_{CO_2}^T - w_{CO_2}^G}{w_{CO_2}^L - w_{CO_2}^G} = \frac{w_S^T - k_3 w_{SO_2}^G - k_4 w_{H_2S}^G}{w_S^L - k_3 w_{SO_2}^G - k_4 w_{H_2S}^G} \tag{199}$$

$$w_{H_2O}^G + w_{CO_2}^G + w_{SO_2}^G + w_{H_2S}^G = 1 \tag{200}$$

where k_{1-4} are stoichiometric coefficients with value 1/9, 1/17, 1/2, and 16/17, respectively; and by homogeneous equilibria in the gas phase:

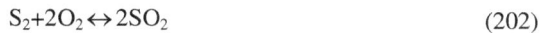

$$2H_2O + 2SO_2 \leftrightarrow 3O_2 + 2H_2S \tag{201}$$

$$S_2 + 2O_2 \leftrightarrow 2SO_2 \tag{202}$$

with equilibrium constants from thermodynamic data in Barin and Knacke (1973) and Barin et al. (1977):

$$\log K_{201} = 8.223 - \frac{54209}{T} \tag{203}$$

$$\log K_{202} = 7.64 - \frac{37794}{T} \tag{204}$$

where T is absolute temperature.

Note that if sulfur is not present in the system, Equation (199) drops together with Equations (201–204), and Equations (198, 200) reduce to Equations (30, 31) for H_2O–CO_2–melt systems, guaranteeing consistency of calculations in the limit $S \rightarrow 0$. For the complete conditions in the C–O–H–S–melt system, COHS allows the determination of the composition of the gas phase in terms of the four major gaseous components above, and that of the melt phase in terms of dissolved H_2O, CO_2, and S, as well as sulfur speciation in S^{2-} and SO_4^{2-}, by accounting for the mutual effects of the dissolved volatiles in the melt phase according to the following: i) the dissolved H_2O has a direct (chemical) effect on S saturation in the melt phase; ii) the same holds for the relationships between H_2O and CO_2 in the melt phase; iii) the dissolved sulfur modifies the activity coefficients of dissolved H_2O and CO_2 by dilution in the melt phase; ii) the same holds for the effect of dissolved CO_2 on dissolved sulfur.

Gaseous molecular sulfur and oxygen, which appear in Reactions (201–202), do not contribute to the mass balance Equations (198–200). For both of them, their abundance in the gas phase for common magmatic conditions is several orders of magnitude less than that of the major species involved in mass balances. For oxygen, that also reflects the choice of treating it as the perfectly mobile component as defined in Ghiorso and Kelemen (1987). S_2 and O_2 fugacities are however necessary to solve the multicomponent volatile saturation equations in COHS. Fugacities of components H_2O, CO_2, SO_2, H_2S in the gas phase are computed from SUPERFLUID (Belonoshko et al. 1992), which accounts for complex compositions in a wide P–T range encompassing that of interest for common magmatic conditions. For any given oxygen fugacity (see below), the fugacity of molecular sulfur in the gas phase is computed from Equations (203–204), and the polymeric approach of CTSFG ensures consistency with ferric/ferrous redox equilibrium in the melt phase, also passed to SOLWCAD, depending on oxygen fugacity and melt composition. Details on the redox treatment applied within COHS modeling can be found in Moretti and Ottonello (2021).

The method workflow is summarized in Figure 21.

In order to apply to a range of possible conditions relevant for natural magmas, COHS allows computations in which the redox constraint is given by one of the following: i) fix f_{O_2}, then determine the ferric/ferrous ratio in the melt; ii) fix the ferric/ferrous ratio in the melt, then determine f_{O_2}; iii) fix the H_2S/SO_2 ratio in the gas phase, then determine f_{O_2} from Reaction (201) and finally the ferric/ferrous ratio in the melt. The model predicts the non-linear behavior of volatile dissolution, in particular sulfur dissolution as sulfide and sulfate depending on oxygen fugacity and on H_2O and CO_2 contents dissolved in the melt (themselves depending on S content as detailed above). Figure 22 shows that while the theoretical trends intrinsic in Reactions (146–147) are embedded in the calculations (panel b), the actual sulfur content variation with the oxidation state is less straightforward, as a result of the mutual interactions between the different volatiles in the melt and gas phases. Figure 23 shows how the redox state (for a given total volatile content assemblage) affects CO_2 and S saturation, respectively, for two different Stromboli-type (panels a, c) and rhyolitic (panels b, d) melts, giving rise to non-linear behaviors that are revealed by the complex modeling approach in COHS. Such non-linear behaviors largely affect the gas phase (Fig. 24) and can lead to serious misinterpretations when measuring the composition of volcanic emissions and their evolution with time. This is further considered in the section below dedicated to the use of multi-component volatile saturation models for the interpretation of data from real volcanic emissions.

Moretti et al. (2018) further implemented COHS by adding chlorine in external calculations (that is, by determining Cl saturation contents for the conditions established by COHS, with Cl not affecting those conditions). By employing the Cl partition coefficients by Alletti et al. (2009), the authors modeled H_2O–CO_2–S–Cl contents in Etnean melts. The CO_2–H_2O–SO_2–H_2S–melt saturation and degassing code (Moretti et al. 2003; Moretti and Papale 2004) is publicly available for download at https://github.com/charlesll//chosetto (DOI: 10.5281/zenodo.5554941).

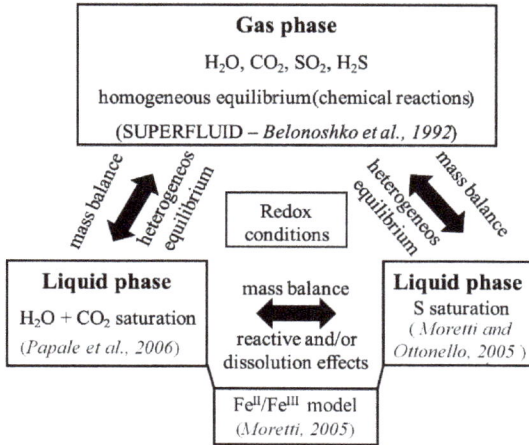

Figure 21. Simplified workflow of the multicomponent CO_2–H_2O–SO_2–H_2S–melt equilibrium COHS model. Modified after Moretti and Papale (2004).

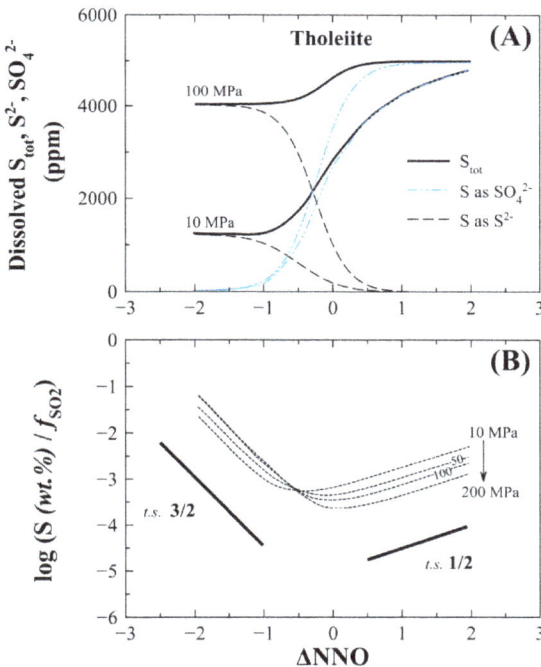

Figure 22. Sulfur saturation computed with COHS (Moretti et al. 2003). **a)** Sulfur saturation in a tholeiitic melt equilibrated at 1400 K and two different pressures of 10 and 100 MPa, as a function of oxygen fugacity. The computation refers to a gas–melt system with $w_{H_2O}^T = 3\,wt\%, w_{CO_2}^T = 1\,wt\%, w_S^T = 0.5\,wt\%$. **b)** Theoretical slopes from Reactions (146–147) are embedded in the calculations, and they do emerge when sulfur saturation is plotted as the normalized quantity on the y-axis. Note the shift to higher oxygen fugacity of the sulfur minimum with increasing pressures, hence increasing dissolved water content. Modified after Moretti et al. (2003).

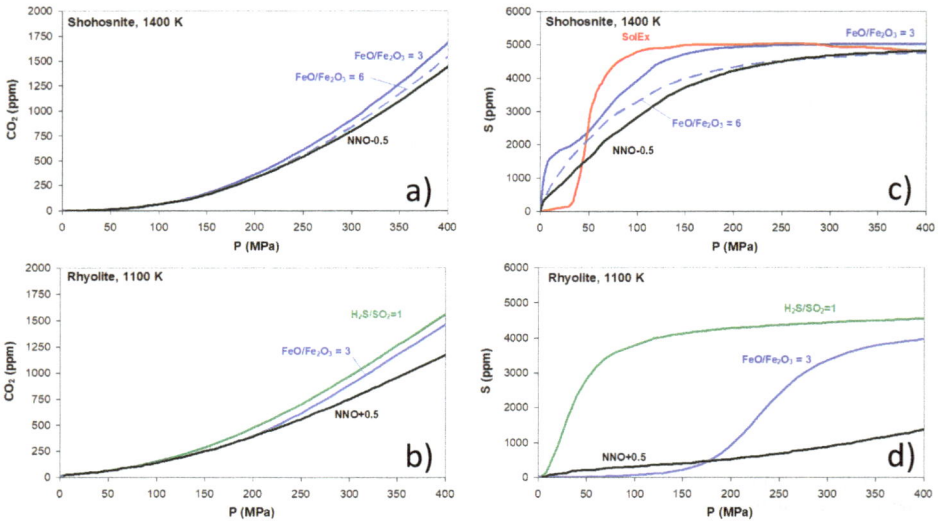

Figure 23. Computed carbon dioxide and sulfur saturation contents for two different melts of shoshonitic (Stromboli-like) and rhyolitic composition at temperature of 1400 and 1100 K, respectively, and three different redox buffers reported on the corresponding lines (iron ratios in the melt phase are in mass, while ratios of S species in the gas phase are in moles). For the shoshonitic case, total (gas–melt) volatile contents are: $w_{H_2O}^T = 4\,wt\%, w_{CO_2}^T = 1\,wt\%, w_S^T = 0.5\,wt\%$, and for the rhyolitic case they are: $w_{H_2O}^T = 6\,wt\%, w_{CO_2}^T = 1\,wt\%, w_S^T = 0.5\,wt\%$. For both melts, more reduced conditions decrease the dissolved contents of both CO_2 and S, with that decrease being larger for the rhyolite. **Panel c** also shows the results of a computation with the SolEx code (Witham et al. 2012). Modified after Moretti and Papale (2004).

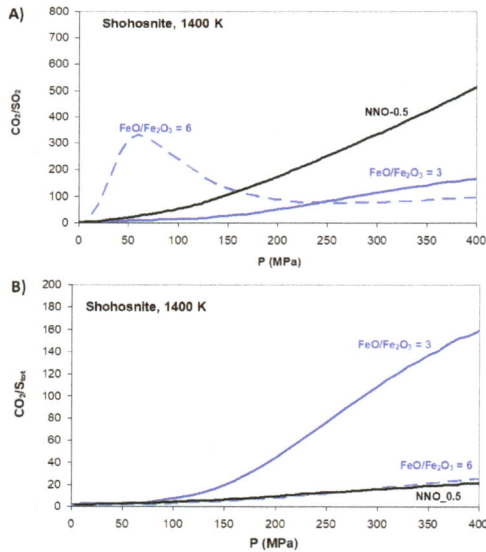

Figure 24. Predicted evolution of the gas phase composition upon closed system decompression, for the same computations reported in Fig. 23a,c. **a)** CO_2/SO_2 molar ratio. **b)** CO_2/S_{tot} molar ratio ($S_{tot} = SO_2+H_2S$). Iron ratios refer to mass. Note the large control on gas phase composition by the redox state, and the maximum in CO_2/SO_2 molar ratio at about 50 MPa for the reduced conditions corresponding to $FeO/Fe_2O_3 = 6$.

Halogen solubility

Understanding and modeling halogen saturation in silicate melts is by far less developed than for major volatile species H_2O, CO_2, and S, and for noble gases considered below. That is likely the consequence of an unfavorable combination; in fact, halogens are much less abundant than major volatiles, thus less crucial for understanding and modeling magmatic degassing, and at the same time they involve substantial theoretical and experimental complexities making their treatment much more difficult than for basically non-reactive noble gases. It is not our purpose to review here the bulk of experimental assessments on halogen solubility. Rather, we limit this section to just a summary of some major aspects that should be accounted for in the modelling of halogen dissolution in silicate melts, and particularly of chlorine.

Shinohara et al. (1984, 1989) provided an experimental summary of halogen partition between felsic melts and co-existing fluids, showing the formation of an important subcritical region giving rise to a chlorine saturation limit linked to the presence of a hydrosaline solution. Metrich and Rutherford (1992) showed the relevance of such findings for real rhyolites. Lowenstern (1994) analysed further the saturation of hydrosaline fluids in peralkaline rhyolites, extending its relevance to other silicic magmas. The existence of a miscibility gap between water-rich gas and Cl-rich brine is therefore a major non-ideal mixing element requiring an appropriate modeling strategy (e.g., Shmulovich and Graham 2004). Such a complex behavior has also been evidenced in mixed saturation experiments with C–O–H–Cl (Webster 1997; Webster and Rebbert 1998) and S–O–H–Cl (Botcharnikov et al. 2004) fluids in rhyolitic melts, where addition of CO_2 is found to expand the miscibility gap resulting in (small) increase in Cl saturation content. Similar non-ideal behaviors have been reported for andesitic (Botcharnikov et al. 2007) as well as phonolitic and trachytic (Webster et al. 2009) melts. Pressure appears to be a major control: at high pressure conditions the dissolved chlorine (as chloride) decreases with pressure, a behavior which has been attributed to the large negative partial molar volume of chloride in the aqueous phase (Shinohara et al. 1989). The negative pressure dependence of Cl solubility implies that the Cl contents of melts may actually increase during magma decompression if the magma coexists with aqueous fluid and Cl-rich brine (Carroll 2005). The physics of chlorine degassing is therefore quite intricate, on one side because of the role played by melt composition (e.g., Metrich and Rutherford 1992), on the other side because several pressure dependencies take place: HCl solubility appears to display a positive pressure dependence while NaCl shows largely negative pressure dependence. Furthermore, the HCl–NaCl exchange reaction between melt and aqueos fluid, while favoring HCl dissolution in the aqueous fluid at lower pressure, largely depends on alkali and aluminum concentrations (Shinohara and Fujimoto 1994; Shinohara 2009). Finally, pressure and temperature are expected to have comparable and opposite effects on the HCl–NaCl exchange (Shinohara and Fujimoto 1994).

From the standpoint of system composition, as mentioned above it appears that aluminum and alkali concentrations play important roles on chlorine behavior in hydrous silicic melts (e.g., Metrich and Rutherford 1992). With fluids of similar Cl molality, higher Cl concentrations are observed in peralkaline phonolitic melts compared to peraluminous phonolitic melts, and Cl concentrations in phonolitic and trachytic melts are approximately twice those found in calc-alkaline rhyolitic melts under similar conditions (Carroll 2005 and references therein).

No miscibility gap and brine formation has been recorded in experiments involving equilibration of H–O–Cl fluids with basaltic melts, for Cl concentrations < 2–3 wt% (Webster et al. 1999, 2004; Stelling et al. 2008), this amount being far above typical Cl concentrations in natural magmas. Therefore, at least for basalts, for which melt–gas (C–H–O–Cl) equilibrium appears to not involve the formation of a coexisting brine, simple modeling approaches based on the definition of (isothermal) gas–melt partition coefficients for Cl have been proposed. Calibration of such coefficients shows a weak dependence on pressure and an increase of Cl solubility for more mafic melts (Webster et al. 1999; Stelling et al. 2008; Alletti et al. 2009). However, such calibrations can be made complex by the presence of several chloride species such as HCl, NaCl and KCl. At high pressure, the behavior of chlorine can be dominated by gas–melt NaCl-based exchanges (Burnham 1979), while the Na and K contents of the melt largely affect the stability of also other chloride species (Shinohara 2009). In andesitic melts equilibrated with H–O–S gas, K and Fe are found to partition more strongly into Cl-bearing volatiles than Na, suggesting that the major chloride components in the gas phase coexisting with intermediate to mafic magmas at upper crustal conditions are NaCl, KCl, $FeCl_2$ and HCl (Zajacz et al. 2012). Beermann et al. (2015) equilibrated Etna basalts with H–O–S fluids and found non-linear partitioning of Cl between water-rich gas and melt that the authors described via a quadratic power functions of the type $y_{Cl} \propto x_{Cl}^2$ for Cl concentrations up to 70–80% of the value corresponding to pure Cl solubility in the melt. The same authors suggest that the formation of Cl complexes with network modifying cations and hydrogen ions may bond to non-bridging oxygens forming hydroxyl groups in the melt.

Fluorine dissolution in melts of geologic interest has been much less studied than chlorine. As noted by Carroll and Webster (1994), experiments on F-bearing melts are highly challenging and may easily alter melt composition and structure, with formation of multiple F-complexes (Mysen et al. 2004), including the formation of Si–F bonds (Dalou et al. 2015). This can be avoided by adding F to starting glasses, for example by substituting 2 moles of NaF to 1 mole of Na_2O (i.e., by adopting the exchange operator F_2O_{-1}) and by using small fluid/melt ratios, which however affect precise and accurate measurements of partitioning. Fluorine partitioning data are rare also as a consequence of difficult measurements of fluorine via electron microprobe. Carroll and Webster (1994) and Baker and Alletti (2012) provide reviews of the partitioning behavior of fluorine. Still, understanding and quantifying the solubility of F is a challenging task in light of high F reactivity with the components in both the melt and gas phases. For this reason, Dalou et al. (2015) refer to partition coefficients for the Si–F complex.

Baker and Alletti (2012) suggest partition coefficients of fluorine in basalts significantly higher than 1 (gas over melt) based on available experimental studies (Alletti 2008; Chevychelov et al. 2008b). That contrasts with the common consideration of fluorine as a compatible element in fluid-saturated melts (e.g., Dolejš and Baker 2007), which would imply minimal loss of F into the volatile phase until very low pressures (e.g., Aiuppa et al. 2009a).

Bromine and iodine partition coefficients were found to be larger than those for chlorine. Bureau et al. (2000) show that $\log D_i^{G/M}$ increases linearly in the order Cl < Br < I. Precipitation of a Br-rich brine between 100 and 200 MPa is observed for Br contents of order thousands of ppm, this content increasing from haplogranitic (around 3000 ppm) to pantelleritic melts (> 1 wt%) (Bureau and Métrich 2003). Bureau et al. (2010) found that the Br partition coefficient between a low-density fluid and haplogranitic melts decreases tending to 1 with increasing temperature and, more markedly, with increasing pressure, with some experimental difficulty related to change of Br concentrations upon experimental sample quenching. Louvel et al. (2020) found similar decrease of $D_{Br}^{G/M}$ with increasing pressure and temperature, ascribing it to increasing structural similarity between the high pressure hydrous melts and the local structure of Br in the gas phase hosting abundant alkali-silica at high P–T conditions. The authors suggest to extend their results to chlorine and iodine.

Despite the scarcity of experimental data and poor developed modeling of fluorine partition, Villemant and Boudon (1998, 1999) and Villemant et al. (2003) have proposed degassing models for Cl, F and Br mainly based upon measurements from natural volcanic (mainly andesitic) products.

First principles and molecular dynamics approaches

The thermodynamic models described so far largely differ in their objectives and simplifying assumptions, ranging from fully non-ideal models describing multi-component volatile saturation as a function of system composition, to ideal models that can be applied to more limited P–T–compositional ranges, to simple parameterizations. All of them, however, have in common the fact that they rest on calibration procedures based on measured or measurement-derived thermodynamic quantities. A radically different approach characterizes *ab initio* or first-principles methods, which probe directly the properties of matter at the microscopic scale to provide estimates of macroscopic observables. These methods have a predictive power that derives from the quantum-mechanical description of interacting atoms and electrons. Density-functional theory (DFT) (e.g., Parr and Yang 1989), provides reasonable scaling with system size and generally good accuracy in reproducing most ground state properties, and is adopted in most applications. In molecular dynamics (MD) approaches, instead, the microscopic path of each individual atom in the system is simulated by solving Newton's equations of motion. Under some circumstances, the computed system evolution can then be related to thermodynamic properties of the system (e.g., Rapaport 1996). Classical MD generally considers the system as composed of massive, point-like nuclei or hard spheres, with forces acting between them. These forces are derived from empirical effective potentials. The accuracy of classical MD depends on the quality of the parameterization of interatomic potentials, which depends in turn on the accuracy of the reference data. The ab initio DFT approach to the electronic-structure can be combined with classical molecular dynamics to provide a more accurate description of thermodynamic properties as well as of chemical reactions and phase stability.

Ottonello et al. (2015, 2018) developed a DFT description of the silicate melt (the solvent) as a continuum dielectric medium surrounding quantum-mechanical water solute. In such a scheme there is no necessity to simulate molecular clusters of the solvent around the solute (explicit solvation), which on the other hand would be too small (i.e. much less than one mole of atoms) due to excessive computational costs, to be really representative of water–melt interactions. In the implicit solvation approach of Ottonello et al. (2015, 2018) the dielectric medium fills the space around and outside a cavity where the solute is confined, and where it relaxes to equilibrate (i.e., dissolve) into the solvent. In the context of continuum models the interaction between the dielectric medium (defined by its static and optical dielectric constants) and the charge distribution of the solute provides the electrostatic part of the solvation free energy. Solvation effects beyond electrostatic screening include cavitation, dispersive and repulsive terms. In principle, the application of continuum models does not require the definition of specific interactions between the solvent and the solute molecules: the structure of the melt is not needed at all, as the truly controlling parameters are the dielectric properties of the solvent medium.

Ottonello et al. (2015) reproduced the observed behavior of H_2O–SiO_2 systems, from pure molten silica to pure water over a wide range of pressure and temperature. They found that the solution energy is dominated by cavitation terms that are mainly entropic in nature. Cavitation energy determines a large negative solution entropy and a consequent marked increase of gas phase fugacity with increasing temperature. The solution enthalpy is negative and dominated by electrostatic terms which display a minimum at about 6 mol% dissolved water. The speciation behavior is interpreted in Ottonello et al. (2018) as due to the formation of energetically efficient hydrogen bonding when OH groups reach an appropriate amount and then a relative positioning with respect to H_2O molecules (see also Stolper 1982b), which in turn do not take a correct position with respect to NBOs to delocalize protons and create additional T–OH groups. The Gibbs energy of hydrogen bonding is of order only a few kJ/mol, but although representing a subordinate contribution to the whole Gibbs energy of solution, it is critical to match experimental bulk H_2O solubility. Because the bulk solute–solvent (water–melt) interaction can be represented as a continuous function of the number density of the solvent (given by the number of solvent molecules per solvent volume), Ottonello et al. (2018) proposed a model of general validity working for any melt composition. The model was tested against the SOLWCAD calibration database (about 950 data for water over a wide P–T range for compositions from synthetic 2-component to natural magmas from ultramafic to rhyolitic and ultra-alkaline), showing good performance. It is worth noting that Ottonello et al. (2018) have simplified the whole approach by excluding the presence of OH⁻ groups derived by water autoprotolysis (Reaction 130), for which there is poor consensus. A description of the DFT technique coupled to an implicit solvation approach can be found in Moretti and Ottonello (2021), with application to iron redox state in silicate melts.

Dufils et al. (2020) investigated the solubility of water in melts and the properties of the hydrous melt through MD simulations at fixed P–T conditions. Following the procedure in Guillot and Sator (2011), they simulated phase coexistence by putting in contact a silicate melt with supercritical H_2O. The solubility is then given by the final counting of hydrogen-species in the melt at equilibrium. They found that the solubility of water changes very little when the melt composition changes from rhyolitic to andesitic to basaltic, but it is strongly enhanced in ultramafic melts. Water speciation at magmatic temperature reveals that the crossover in the concentration of OH groups and molecular water occurs for a water content of about 15 wt%, much higher than the 3–4 wt% observed in glasses (Stolper 1982a; Silver et al. 1990), suggesting a much larger proportion of OH groups at temperature significantly higher than the glass transition temperature. The simulations were carried out by accounting for self-ionization (or autoprotolysys, Reaction 130) in the definition of the water force field, showing that hydroxyl groups are more preferentially linked to metal cations than to network formers, whereas H_2O molecules and H^+ (or better, H_3O^+) are almost exclusively linked to metal cations. Melt polymerization decreases gradually with increasing water content in andesitic and basaltic melts, but it remains almost constant in peridotitic melts, which are by themselves highly depolymerized because of very low silica content. Finally, Dufils et al. (2020) found for all considered compositions a substantially similar linear variation of melt molar volume with water content, implying substantially constant partial molar volume of water largely independent from melt composition (ideal volume of mixing, or zero excess volume). In turn, those findings would imply that the partial molar volume of water does not depend on either total water content or water speciation.

Mookerjee et al. (2008) performed *ab initio* MD simulations (atom nuclei are treated as classical particles but the forces acting on them are quantum mechanical and derived from electronic-structure calculation) of hydrous $MgSIO_3H$ up to 136 GPa to cover the pressure regime of the Earth's mantle. Pressure was found to have a large influence on water speciation, determining an increase with pressure in the proportion of Si–O–H–O–Si polyhedral linkages, –O–H–O–H– chains and O–H–O edge decoration of SiO_6 octahedra. Basically, at very high pressure the whole mixture has remnants of the structure of the pure water component, and a molar volume approximating the partial molar volume of water in the melt. The simulations performed allow the authors to conclude that the increase with pressure of water solubility is essentially unlimited for all mantle P–T conditions.

Carbon dioxide has been also the object of extensive computational simulations. Guillot and Sator (2011) carried out MD simulations on CO_2 solubility in silicate melts. They equilibrated supercritical CO_2 with three different melts (rhyolite, MORB and kimberlite) up to 15 GP and 2273 K, and found that CO_2 solubility is inversely related with temperature and weakly dependent on composition up to 10 GPa, where CO_2 dissolution starts occurring in the order rhyolite < MORB < kimberlite. The solubility of CO_2 is found to increase markedly with pressure (2 wt% CO_2 at 2 GPa and > 25 wt% at 10 GPa). In terms of speciation, the proportion of carbonate groups increases with pressure and decreases with temperature, it is higher in CO_2-undersaturated melts than in saturated ones, and decreases with melt polymerization.

However, the fraction of molecular CO_2 is still relevant in CO_2-saturated basic and ultrabasic melts, in contrast with FTIR measurements showing only carbonate ions in basaltic melts (Fine and Stolper 1985, 1986). Low-temperature extrapolation of simulation results for MORB predicts that the proportion of CO_2 might be negligible in the glass at room temperature. Guillot and Sator (2011) concluded that carbonate species are preferentially associated with NBOs and, subordinately, to BOs, thus providing an explanation of why the concentration of carbonate groups increases with melt depolymerization. Furthermore, modifier cations are not randomly distributed around carbonate groups, at least at the high simulated temperatures. Their results about the carbonate environment seem however to partially contradict the *ab initio* MD results of Villeumier et al. (2015) and Moussallam et al. (2016), showing that in low-SiO_2 (e.g., kimberlitic) melts, the dissolved CO_2 forms free carbonate groups associated with alkaline earth cations but disconnected from the melt network.

Regarding high-pressure (mantle conditions) CO_2 speciation in melts, *ab initio* MD studies of carbonated $MgSiO_3$ melts (Ghosh et al. 2017; Ghosh and Karki 2017) reveal that with increasing temperature the proportion of molecular CO_2 increases, and that with increasing pressure other dissolved carbon species appear in the melt, all polymerized with the silicate framework: polyhedral bound carbonates, tetrahedrally coordinated CO_4, and polymerized di-carbonates. Similar results were found for carbonated pyrolytic and forsteritic melts simulated via ab initio MD by Solomatova et al. (2019, 2020), who also found evidence for Si–C bonds, in agreement with low-pressure studies on highly-reduce ferrobasalts (Kadik 2004) and Si–Li–O–C melts (Sen et al. 2013). The authors also suggested that the CO_2/CO_3^{2-} ratio of classical MD simulations by Guillot and Sator (2011) may be severely overestimated.

MD simulations have been employed also to investigate the behavior of noble gases in silicate melts, but these are presented in the section below, entirely dedicated to noble gases. Besides, an exhaustive review of MD approaches to silicate melts can be found in Jahn (2022, this volume)

The *ab initio* and/or MD simulations briefly summarized above offer a harvest of results that can be compared with those from experimental studies and classical thermodynamic modeling described in this review, significantly complementing our understanding of volatile–melt equilibrium and capabilities to model it. Such an effort, which requires substantial space, is not included here, but we envisage that a detailed analysis of ab initio / MD simulation results may contribute to shed light on several aspects related to different interpretations of experimental data as well as different modeling strategies as they are depicted throughout this chapter.

NOBLE GASES

General aspects of noble gas solubility in magmas

Since pioneering research in the 1960s, the topic of noble gas solubility in silicate melts has continuously grown in relevance for its scientific as well as commercial implications. That growing interest is testified by the continuously increasing experimental investigation on melts and glasses having importance as either commercial materials or for understanding relevant geological processes, translating into large coverage of compositional, pressure and temperature ranges from industrial to deep crust conditions (Doremus 1966; Shackelford et al. 1972; Shelby 1976; Hayatsu and Waboso 1985; Jambon et al. 1986; Lux 1987; White et al. 1989; Nakayama and Shackelford 1990; Carroll and Stolper 1991, 1993a,b; Broadhurst et al. 1992; Roselieb et al. 1992, 1995; Draper and Carroll 1995; Chamorro-Perez et al. 1996, 1998; Shibata et al. 1996, 1998; Shackelford 1999; Paonita et al. 2000; Walter et al. 2000; Schmidt and Keppler 2002; Miyazaki et al. 2004; Marrocchi and Toplis 2005; Bouhifd et al. 2006, 2008; Tournour and Shelby 2008a,b; Iacono-Marziano et al. 2010; Paonita et al. 2012;

Niwa et al. 2013; Fabbrizio et al. 2017; Leroy et al. 2019). The bulk of these studies has clearly shown that noble gas solubility strongly depends on the composition of the solvent (or the silicate melt) and the atomic radius of the specific noble gas (see Paonita 2005, for a review). In general, substantial solubility increase is observed for larger concentration of SiO_2 in the melt (by almost two orders of magnitude for Ar when moving from basalt to rhyolite), and for smaller noble gas atomic size (by almost two orders of magnitude from Xe to He). These trends have suggested that the atoms of dissolved noble gases occupy holes and free spaces within the silicate melt structure, rather than reacting with the melt oxides as for other volatile components examined so far. That is fully consistent with the poorly or non-interactive nature of noble gases, and depicts a so-called mechanical (as opposed to chemical) mechanism of dissolution (Doremus 1966; Studt et al. 1970; Shackelford et al. 1972). For heavier noble gases, recent studies suggest however a certain degree of interaction with the molecular structure of the solvent (Crépisson et al. 2018; Leroy et al. 2018).

Experimental dissolved concentrations of noble gases display near linear trends with pressure up to several hundred MPa. The trends at higher pressure are still debated. Some experiments suggest the occurrence of a drop in Ar solubility at large pressure of order 4–5 GPa for SiO_2 and olivine melts, and larger than that value for anorthite, chondrite, and sanidine melts (Chamorro-Perez et al. 1996, 1998; Bouhifd et al. 2006, 2008). Other studies suggest instead that after a threshold concentration is reached, further pressure increase has no effect on Ar solubility (Schmidt and Keppler 2002; Niwa et al. 2013). A similar behavior has been observed for Xe at about 3.5 GPa, with the exact threshold value depending on melt composition (Leroy et al. 2019). Temperature changes seem to modestly affect noble gas solubility, and the observations are often within experimental uncertainty (see references in Paonita 2005).

A few studies during last 20 years (Paonita et al. 2000, 2012; Fabbrizio et al. 2017) have concerned the solubility of noble gases in presence of H_2O and CO_2, with the idea that such major volatiles may influence noble gas solubility considering that noble gases are just trace components in an H_2O–CO_2 rich gas phase. Paonita et al. (2000) measured He solubility in basaltic and rhyolitic melts at pressure in the range 100–200 MPa and in presence of an H_2O–CO_2 rich gas phase with ≈ 0.1 mol% He (0 to 50 mol% CO_2). The solubility was found to increase by about three times when the melt contained 3 wt% dissolved H_2O, while further dissolved H_2O increase up to 6 wt%, or dissolved CO_2 increase up to 0.05 wt%, had negligible effects. Similar conditions (pressure up to 215 MPa, $P \approx P_{H_2O} + P_{CO_2} \gg P_{noble\ gas}$) were employed to measure Ar solubility, which turned out to be four times higher in a 5 wt% H_2O-bearing basalt melt with respect to the dry melt. The above results are coherent with more recent investigation by Fabbrizio et al. (2017), which however employed different conditions corresponding to $P_{Ar} > P_{H_2O} > P_{CO_2}$. They observed that an increase of dissolved H_2O up to about 2 wt.% (at constant dissolved CO_2 content) did not produce any clear pattern in the Ar solubility in a basalt at 1 GPa, while a further increase above 2 and up to 5 wt.% was associated with a 2 to 3-fold increase in Ar solubility. A higher pressure of 3–5 GPa, or increasing H_2O concentration up to 5.3 wt.% and CO_2 concentration up to 0.5 wt%, did not produce systematic effects on Ar solubility, while a decrease was observed for CO_2 contents > 0.5 wt.%.

It is not among the scope of this chapter to provide a discussion of experimental techniques to measure volatile solubility and saturation contents. A review of the techniques employed for noble gases can be found in Paonita (2005).

Basic thermodynamics of noble gas solubility in silicate melts

The chemically inert nature of noble gases makes them particularly suited to be modeled with a non-reactive approach. The equilibrium partitioning between gas and silicate melt can be therefore described by the reaction:

$$i^G \leftrightarrow i^L \qquad (205)$$

where i refers to each specific noble gas. The equilibrium equation (26) can be written as:

$$f_i / a_i^L = f_i^{0,L} \tag{206}$$

with the same notation introduced above, thus $f_i^{0,L}$ indicates the fugacity of the noble gas in a suitably selected reference state of the pure solute component at system P–T. While Equation(206) has a general validity, commonly very low equilibrium contents of noble gases in silicate melts, usually of the order of ppm, often allow the adoption of the ideal approximation for the melt phase $a_i^L \approx x_i$. In this case (theoretical infinite dilution), a Henrian approximation for any P–T condition is justified. For the gas phase, the ideal approximation $f_i \approx P_i$ is also employed for sufficiently low pressure, leading to two equilibrium equations that have been largely used as a basis for modeling noble gas solubility:

$$\text{high pressure: } f_i / x_i = K_h(P,T), \quad \text{low pressure: } P_i / x_i = K_h(P,T) \tag{207}$$

with $K_h(P, T)$ being a Henrian-like term describing the variation of the Henry's constant (or more precisely, the inverse of the Henry's constant) with pressure and temperature (Fig. 25).

The general (non-ideal) Equation (206) allows computation of gas–melt partitioning provided the availability of i) an EOS describing the non-ideal behavior of the gas phase, ii) an expression for the P–T-dependent reference state fugacity in the melt phase, and iii) an activity–composition relationship for the melt phase. The following illustrates the various modeling approaches employed to provide solutions to Equation (206).

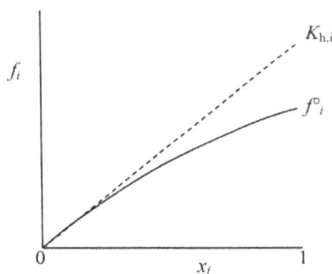

Figure 25. Henrian versus real behavior in a fugacity vs. concentration plot. We show the conceptual comparison between Equation (206) (**solid line**) and Equations (207) (**dotted line**), with their related reference states K_h and f^0. The concentration range where the two curves match is where the Henrian approximation works.

Pure noble gases and noble gas mixtures

We have discussed in the general thermodynamics section at the beginning of this chapter, that computing the fugacity of a component in a real gas mixture (f_i in Eqn. 206, being $f_i^G = f_i^L = f_i$ at equilibrium) requires knowledge of the P–T dependence of the partial molar volume of that component in the gas mixture at system temperature and from zero pressure (ideal state) to system pressure (Eqn. 32). That knowledge is embedded in the EOS (Equation Of State) for the real mixture. An EOS that has been commonly employed for gas mixtures involving noble gases is a modified Redlich–Kwong (MRK) equation (Holloway 1977):

$$P = \frac{RT}{v^G - b} - \frac{a}{\sqrt{T} v^G (v^G + b)} \tag{208}$$

where v^G is the molar volume of the gas phase (pure or mixed), and a and b are parameters that reflect the attractive and repulsive interactions in the system, respectively. For each pure non-polar species (like noble gases) a and b can be obtained from the critical constants of the gas. In the case of noble gas mixtures, the following mixing rules applies:

$$a = \sum_{i=1}^{n}\sum_{j=1}^{n} y_i y_j a_{ij}, \quad b = \sum_{i=1}^{n} y_i b_i \tag{209}$$

where $a_{ij} = (a_i a_j)^{0.5}$ are the cross interaction parameters between 'i' and 'j' species and $y_{i,j}$ is molar fraction in the gas phase. Flowers and Helgeson (1983) obtained an expression for the fugacity coefficient ϕ_i for a fluid described by a MRK EOS and for which the above mixing rules apply:

$$\ln \phi_i = \ln \frac{v^G}{v^G - b} + \frac{b_i}{v^G - b} - \frac{2\sum_{j=1}^{n} x_j a_{ij}}{RT^{3/2} b} \ln \frac{v^G + b}{v^G} +$$
$$\frac{ab_i}{RT^{3/2}b^2}\left[\ln \frac{v^G + b}{v^G} - \frac{b}{v^G + b}\right] - \ln \frac{Pv}{RT} \tag{210}$$

Equation (210) allows to convert partial pressure into fugacity via Equation (24).

A more recent EOS has been developed by Churakov and Gottschalk (2003a,b) based on perturbation theory (Gray and Gubbins 1984), and can be applied to predict thermodynamic properties of most geological fluids. This EOS includes all noble gases, both as pure and mixed components. However, in our knowledge it has not been used yet in modeling noble gas solubility in magmas.

Reference state and activity–composition relationships

In modeling noble gas solubility, the choice of a reference state and activity–composition relationship for use within Equation (206) has been strictly linked to the available dataset for parameter calibration and to the main independent variables to be included in the models. In modeling the effects of P and T on noble gas solubility in a melt with given composition, two terms for the reference molar volume and solution enthalpy can be incorporated in Equation (206) so as to achieve:

$$\ln \frac{f_i}{a_i^L} = \ln f^0 \left(P^0, T^0\right) + \frac{\Delta H_{P^0,T}^0}{R}\left(\frac{1}{T} - \frac{1}{T^0}\right) + \frac{v_{P,T}^0}{RT}\left(P - P^0\right) \tag{211}$$

where P^0 and T^0 define the standard state; $v_{P,T}^0$ and $\Delta H_{P^0,T}^0$ are the reference volume and solution enthalpy of the dissolved noble gas, respectively, and R is the gas constant. The activity–composition relationship depends on the theoretical model hypothesized for the dissolution process. If we assume ideal mixing between the silicate melt and noble gas, then:

$$\ln \frac{f_i}{x_i} = \ln K_h^0 \left(P^0, T^0\right) + \frac{\Delta H_{P^0,T}^0}{R}\left(\frac{1}{T} - \frac{1}{T^0}\right) + \frac{v_{P,T}^0}{RT}\left(P - P^0\right) \tag{212}$$

where the two parameters $v_{P,T}^0$ and $\Delta H_{P^0,T}^0$ define the P–T dependence of the Henry's-like term K_h in Equation (207). The parameters $K_{h\ (P^0,\ T^0)}$, $v_{P,T}^0$ and $\Delta H_{P^0,T}^0$ can be then calibrated for each noble gas and each given melt composition by using experimental data over a range of pressure and temperature. This approach has been employed to successfully reproduce the experimental data on selected melt compositions in a large P–T range (Lux 1987; White et al. 1989; Carroll and Stolper 1991, 1993; Draper and Carroll 1995; Shibata et al. 1998; Schmidt and Keppler 2002; Tornaru and Shelby 2008a,b; Iacono-Marziano et al. 2010) (Fig. 26). The constrained partial molar volumes of heavy noble gases (Ar, Xe) turned out to be comparable to their respective co-volumes (the b parameter) in the gas mixture computed from the MRK equations (208), which is a proxy for the atomic size (White et al. 1989; Schmidt and Keppler 2002). The estimated v^0 values of He, Ar and Xe in glasses, slightly lower than co-volumes (Carroll and Stolper 1991, 1993; Draper and Carroll 1995), suggest that dissolution of noble gases creates just a small amount of new space in the melt, and that the atoms of gas occupy

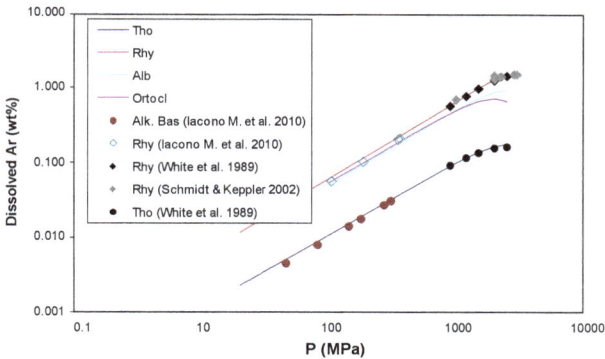

Figure 26. Ar solubility as a function of pressure up to 3 GPa, as predicted by Equation (212). **Symbols** represent experimental data at 1200 °C (Iacono-Marziano et al. 2010) and 1500 °C (White et al. 1989; Schmidt and Keppler 2002). **Lines** represent predictions by Equation (212) at 1200 and 1500 °C for different melt compositions, with parameters calibrated by using the Iacono-Marziano et al. (2010) model. Note i) the much higher Ar solubility in acidic melts with respect to mafic ones, ii) the linear relation between Ar concentration and pressure at least up 1 GPa, and iii) the threshold solubility above 2 GPa.

pre-existing structural holes according to the mechanism of mechanical solubility discussed above. Small v^0 values support Henrian-like behavior of dissolved noble gases for pressure up to at least some hundreds MPa. The Ar molar volume shows negligible dependence on melt composition from tholeiites to rhyolites (White et al. 1989; Schmidt and Keppler 2002; Iacono-Marziano et al. 2010), suggesting that a similar environment surrounds the dissolved atoms in spite of notable structural differences in the silicate structure.

The solution enthalpy ΔH^0 is small in both glasses and melts, supporting the observation of very low dependence of noble gas solubility on temperature (Fig. 27). Negative ΔH^0 values, implying retrograde solubility (solubility decrease with increasing temperature), have been estimated in acidic glasses with composition from albite and rhyolite to pure SiO_2 (Shackelford et al. 1972; Shelby 1972a,b; Carroll and Stolper 1991; Draper and Carroll 1995; Tournour and Shelby 2008a,b). Increasing Na_2O and K_2O concentrations to above 20 mol% seem to invert this solubility–temperature relation and be associated to prograde solubility (Shelby 1973 and 1974). On the other hand, positive enthalpy of solution (prograde solubility) has been commonly measured in basaltic melts (Hayatsu and Waboso 1985; Jambon et al. 1986; Lux 1987) (Fig. 27), as well as in $Na_2O–SiO_2$, $Na_2O–CaO–SiO_2$, and $Na_2O–MgO–SiO_2$ melts (Shibata et al. 1988; White et al. 1989). Within melts having the same composition, heavier noble gases are associated to either more positive or more negative enthalpy than light gases.

In order to explain the above results, the dissolution of an atom of noble gas in a silicate melt can be conveniently seen as a two-step process, with each step having its own enthalpy change: i) the creation of a hole in the melt structure to accommodate the gas atom, and ii) the transfer of the atom into the hole. The former step needs energy, therefore the associated enthalpy change is positive; instead, moving an atom from a gas into a more condensed phase involves heat production thus negative enthalpy change. The whole enthalpy of solution derives from the sum of the two terms, therefore its sign depends on which contribution dominates. Glasses and silica-rich melts have open structures, so we can expect that only little energy is spent to create further space. In such cases most of the energy is spent to move atoms into the melt, resulting in overall negative enthalpy change thus retrograde solubility behavior. For less polymerized, alkali-rich melts, for which substantial energy is spent to create space for noble gas accommodation into the melt structure, the balance can be reversed and result in overall positive enthalpy change thus prograde solubility behavior.

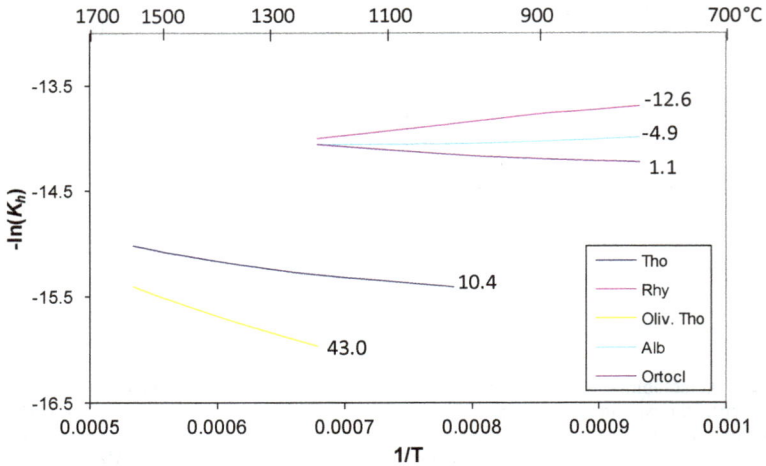

Figure 27. Theoretical effects of temperature on Ar solubility (as the inverse of Henry constant). Shown curves are computed by Equation (212) for mafic to acidic melts at 0.1 MPa, with the solution enthalpy derived from Iacono-Marziano et al. (2010) model. Numbers close to lines are the computed values of solution enthalpy (in kJ/mol).

An alternative approach that has been adopted to model the activity–composition relationship for noble gas solubility in silicate melts. The inert nature of noble gases and their trace abundance in silicate melts justify the following assumptions: i) noble gases dissolve into a fixed population of holes in the silicate melt network; ii) the available sites for a given noble gas are identical; iii) their number is P-T-independent; iiii) gas atoms do not influence each other. With these assumptions, the activity of a dissolved noble gas can be seen as resulting from a balance between the occupied and available sites in the silicate melt network:

$$a_i^L = \frac{n_i^L}{N_s - n_i^L} \tag{213}$$

where n_i^L is the number of dissolved atoms and N_s is the number of available sites (both per unit volume of melt). In Equation (213) the nominator identifies the actual condition compared to a reference at the denominator, which can help understand the relationship between Equation (213) and the usual expression of activity as the ratio between fugacity (actual conditions) and reference fugacity, Equation (25). At constant temperature, and with the further assumption that noble gas dissolution does not imply any change in the space of dissolution sites available within the melt network (or in other words, that the partial molar volume of the dissolved noble gas is zero, or $v^0 = 0$ in Equations (211–212), Equations (211) and (213) can be combined resulting in the well-known formalism of the Langmuir adsorption isotherm:

$$K_L(T) = \frac{n_i^L}{f_i(N_s - n_i^L)} \tag{214}$$

where $K_L(T)$ is the temperature-dependent equilibrium constant.

Use of Equation (214) requires that the available number of sites be calibrated together with the equilibrium constant (Shelby 1976; Roselieb et al. 1992; Carroll and Stolper 1991; Walter et al. 2000). Schmidt and Keppler (2002) employed Equation (214) to model their experiments showing that a threshold Ar concentration is reached at high pressure in silicic and tholeiitic melts (Fig. 28), as previously discussed. They interpreted such a threshold Ar concentration in terms of saturation of the available sites for Ar dissolution (N_s in Eqn. 214), that for rhyolite they constrained to 1.38×10^{21} cm^{-3}. That number represents only 60% of the interstitial holes determined by approximating the structure of the rhyolitic melt as a tridymite-like network, suggesting that only a fraction of the available holes is suited to accommodate Ar atoms. The concentration threshold, thus the estimated value of N_s for Ar dissolution in tholeiite, was found to be nearly one order of magnitude lower than for rhyolite (Fig. 28), in agreement with more compact structure of mafic with respect to silicic melts (Schmidt and Keppler 2002). Finally, estimates of N_s for Xe provided values significant lower than for Ar, consistent with significantly larger size of the former.

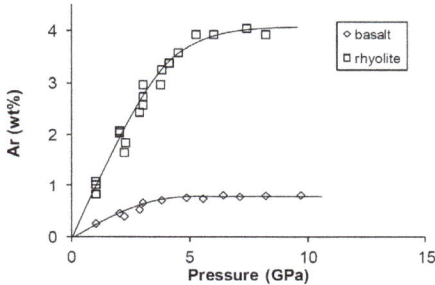

Figure 28. Predicted Ar solubility as a function of pressure from Equation (214). The calculations for both basalt and rhyolite melts employ the parameters given by Schmidt and Keppler (2002). The **symbols** are experimental data from Schmidt and Keppler (2002) for the same melts in the range 1500–2000 °C. Note the threshold concentration in dissolved Ar at about 5 GPa (modified after Paonita 2005)

Additional high-pressure experiments performed by Niwa et al. (2013) confirmed the threshold behavior for Ar solubility in SiO$_2$ melts (Fig. 29). However, as anticipated above, such a behavior contrasts markedly with previous experiments by Chamorro-Perez et al. (1996) and Bouhifd et al. (2006, 2008), who observed in a variety of compositionally different melts a dramatic drop in Ar solubility above a large pressure depending on melt composition. In Al-free melts, such as pure SiO$_2$ and molten olivine, such a threshold pressure was found to be close to 5 GPa, increasing with Al in Al-bearing melts from 10 GPa in chondrite to 14 GPa in sanidine to 17 GPa in anorthite. Bouhifd et al. (2008) related this trend to the well-known change in Al coordination with pressure (Lee et al. 2006 and references therein), whereby progressive increase of highly coordinated Al results in a porosity minimum below which Ar atoms are no longer hosted within the melt network. By contrast, Niwa et al. (2013) hypothesized that the observed drop in Ar solubility was a seeming effect due to presence of crystalline phases in the experimental capsules.

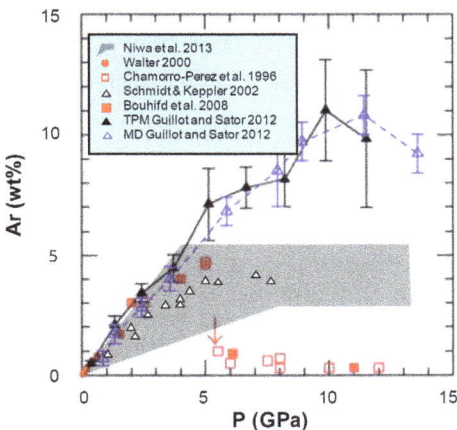

Figure 29. Ar solubility in liquid SiO$_2$ up to very high pressure. TPM and MD are Guillot and Sator (2012)'s computed solubility by means of test particle method and molecular dynamics, respectively (modified from Guillot and Sator 2012).

While the Langmuir adsorption isotherm model at Equation (214) appears to work well for a variety of melt compositions and various noble gases up to very high pressure, other studies including recent ones suggest that heavy noble gases may not conform to the fundamental assumption that the gas atoms do not affect the structure of the solvent (or in other words, $v^0 = 0$ may not hold). In fact, *in-situ* high pressure X-ray absorption spectroscopy studies on the structural environment of dissolved krypton in vitreous silica and sanidine glass show that Kr atoms not only can modify the structural holes where they accommodate (Wulf et al. 1999), but they also create bonds with oxygen inside cages formed by the largest aluminosilicate rings (Crépisson et al. 2018). Bonding of Xe to O at high pressure is also found in haplogranitic magmas by means of *in situ* X-ray diffraction (Leroy et al. 2018). In the case of a light noble gas like He, a high-pressure X-ray diffraction and Raman study in an SiO_2 melt confirms instead a pure interstitial solubility mechanism with no interaction with the silicate network (Shen et al. 2011).

Models based on statistical mechanics

Statistical mechanics has been employed to fit noble gas solubility data and determine a pressure-independent Henrian-like constant. These models assume mechanical solubility and treat the dissolved gas atoms as harmonic oscillators in a structural cavity (or interstitial hole). Starting from the ideal Equation (207), and employing n_i^L defined above as a measure of the dissolved amount of the noble gas under inspection, Doremus (1966), Studt et al. (1970) and Shackelford et al. (1972) obtained the following equation:

$$\frac{n_i^L}{P_i} = K(T) = \left(\frac{h^2}{2\pi m_i kT}\right)^{\frac{3}{2}} (kT)^{-1} N_s \left(\frac{e^{-\frac{\theta}{2T}}}{1 - e^{-\frac{\theta}{T}}}\right)^3 e^{-\frac{E(0)}{RT}} \tag{215}$$

where $K(T)$ is a T-dependent Henrian-like constant (note that it is here the inverse of the similar quantity in Eqn. 207), h and k are the Plank and Boltzmann constants, respectively, m_i is atomic mass, $\theta = hv/k$ is the characteristic temperature of vibration of the atom in that hole type (with v being the frequency of vibration), $E(0)$ is the binding energy, that is, the atom energy at rest relative to the similar rest state in free gas, and N_s is defined at Equation (213).

Equation (215) requires the determination of the three fitting parameters N_s, θ, and $E(0)$. In silica glass, these parameters depend upon the atomic size of the noble gas; for larger noble gases we achieve lower number of sites and frequency, and more negative binding energy (Shelby 1976; Shackelford and Brown 1980; Nakayama and Shackelford 1990). A sharp increase in the binding energy and decrease in the vibrational frequency were obtained by regression along the join Na_2O-SiO_2 when moving from the silica-rich to the sodium-rich edge of the miscibility gap (Shelby 1973). This is in agreement with both the stronger structural constraints and the necessity to spend energy in creating adequate holes in a less polymerized solvent.

Sarda and Guillot (2005) and Guillot and Sarda (2006) wrote the vapor–melt chemical equilibrium equation of noble gases in the form:

$$\frac{n_i^L}{n_i^G} = e^{-\left(\mu_i^{E,L} - \mu_i^{E,G}\right)/kT} = \frac{\delta_i^L}{\delta_i^G} \tag{216}$$

where μ_i^E and δ_i refer to the excess chemical potential (equal to $RT\ln a_i$, Eqns. 20–25) and the "solubility parameter" of the noble gas in the specified phase, k is the Boltzmann constant, and n_i is the noble gas atoms per unit volume (i.e., the number density) in the specified phase. The authors described the gas and melt phases in the framework of the hard sphere fluid model, which is the reference model in liquid state theory. They fitted the Carnahan–Starling EOS for a hard sphere fluid by using P–T data on noble gas–bearing vapor phases, as well as volumetric data of melts. For the noble gas vapor they used the activity-composition relationship embedded in the Carnahan–Starling EOS to calculate $\mu_i^{E,G}$, while for the silicate melt $\mu_i^{E,L}$ was evaluated according to scaled particle theory developed for hard spheres, adding an energetic contribution of dissolution usually treated as a fitting parameter. Equation (216) was then employed to determine the solubility parameter ratio (far right hand side of Eqn. 216) by fitting noble gas solubility data, and then used to compute noble gas solubility. The authors reported solubility calculations for all noble gases, and compared pure Ar solubility in silica, haplogranite, tholeiite and olivine melts for the available experimental data. The model predictions were accurate up to very high pressure excluding the drop in solubility above 5 GPa previously claimed by by Chamorro-Perez et al. (1996) and Bouhifd et al. (2006, 2008) and largely discussed above.

Guillot and Sator (2012) employed the test particle method (TPM) in conjunction with molecular dynamics (MD) simulations to calculate the excess chemical potentials in the melt and gas phase by statistical mechanics relationships, for use in Equation (216). Noble gas solubility was computed for all noble gases in rhyolite, MORB, olivine, enstatite and SiO_2 melts up to 20 GPa pressure and 1400–2000°C temperature. They found a decrease in solubility with increasing size of the noble gas and decreasing silica content of the melt, and an increase in solubility with increasing temperature (that is, positive enthalpy of dissolution) for all noble gases and compositions considered. In light of the discussion in the previous section, positive enthalpy of dissolution implies that the energetic cost for cavity formation in the silicate network is higher than the solvation energy. However, due to the very high explored temperatures their solubility vs. temperature relationship can not easily be compared with the available experimental data, normally achieved below 1500 °C where they suggest instead very modest thermal effects on solubility. The computations by Guillot and Sator (2012) do not reveal any large drop of noble gas solubility above a certain pressure (Fig. 29), confirming therefore the results for Ar solubility by Niwa et al. (2013) with respect to those by Chamorro-Perez et al. (1996) and Bouhifd et al. (2006, 2008), as discussed above.

Models accounting for melt composition

All experiments have confirmed a strong dependence of noble gas solubility on melt composition, justifying substantial efforts in finding an expression for the Henry's like term K_h in Equation (207) (with or without the ideal assumption $f_i = P_i$) which also accounts for the compositional variability of silicate melts. The aspects of noble gas dissolution seen above, whereby mechanical dissolution largely dominates (at least for lighter noble gases), suggest that the silicate network architecture, and some quantity embedding its major properties, may provide an effective compositional parameterization for noble gas solubility. It is useful here to briefly recall some major aspects of silicate melt network architecture.

Silicate melts and glasses basically consist of five structural units, represented by SiO_2 (three-dimensional network), $Si_2O_5^{2-}$ (sheet), $Si_2O_6^{4-}$ (chain), $Si_2O_7^{6-}$ (dimer), and SiO_4^{4-} (monomer) (Virgo et al. 1980). Silicon and other tetrahedrally coordinated cations (Al, Fe^{3+}, Ti^{4+}) act as network-formers, whereas Na^+, K^+, Mg^{2+}, Ca^{2+}, Fe^{2+} are called network-modifiers as they break the silicate polymers into smaller units. It is easily understood that a fully polymerized, three-dimensional SiO_2 structure is associated to a large number of holes where gas atoms can be accommodated, whereas less polymerized structures dominated by 2D or 1D structural units offer far less free space. Accordingly, highly polymerized, SiO_2-rich compositions are associated with higher noble gas solubility (Shibata et al. 1998) (Fig. 30).

A quantitative expression of the degree of polymerization is given by the ratio NBO/T, that is, the number of Non-Bridging Oxygens per atom of a Tetrahedrally coordinated cation (Brawer and White 1975; Mysen et al. 1985). In a fully developed three-dimensional network all oxygens are bridging oxygens and therefore NBO/T = 0. The value of NBO/T increases when other structural units are present, up to NBO/T = 4 when only TO_4^{4-} monomers are present in the melt. On this basis, Shibata et al. (1998) used the following modification of the equilibrium in Equation (207) (without assuming equality of fugacity and partial pressure) to express the role of melt composition on noble gas solubility:

$$\ln \frac{f_i}{x_i} = \ln K_h (P,T) = x_{BO} \ln k_{BO} + x_{NBO} \ln k_{NBO} \tag{217}$$

where x_{BO} and x_{NBO} are the fraction of units with bridging oxygens and with both bridging and non-bridging oxygens, respectively, and k_{BO} and k_{NBO} are the corresponding constants which are different for each different noble gas. Equation (217) has been calibrated for noble gases from Ne to Xe, but a relationship between solubility and the NBO/T ratio exists for helium, too (Fig. 30). Significant negative deviations from the solubility obtained by means of Equation (217) are found in melts with high Al (or other network-formers different from Si) content.

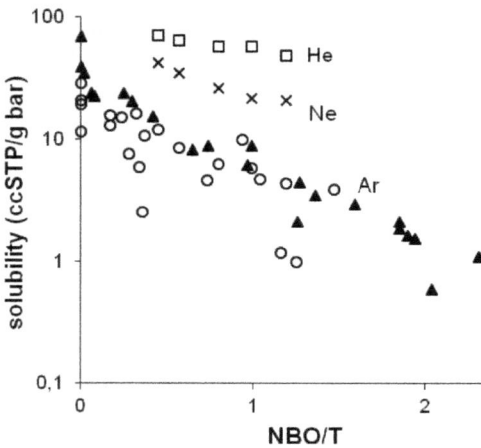

Figure 30. Solubility of light noble gases versus melt polymerization (expressed as NBO/T, see text). Data sources: Shibata et al. (1998) for Ar (**circles**: T/(T+Si) > 0.1, **triangles**: T/(T+Si) < 0.1, where T is the number of tetrahedrally coordinated cations other than silicon); Jambon et al. (1986) and Lux (1987) for He and Ne, respectively. Modified after Shibata et al. (1998).

Jambon (1987) and Chennaoui-Aoudjehane and Jambon (1990) used simple oxide components to calculate the solubility of Ar at 0.1 MPa:

$$\ln x_{Ar} = \sum_{j}^{n} x_j k_j \tag{218}$$

where x_{Ar} is the molar fraction of dissolved Ar at 0.1 MPa and the k_j quantities express the contribution of the individual oxide components in determining Ar solubility. Marrocchi and Toplis (2005) used a similar approach, but they used the tetrahedral units SiO_2, $Si_2O_5^{2-}$ (sheet) and $Si_2O_6^{4-}$ (chain like units) rather than oxides as melt components. For Al-bearing melts, they observed linear correlations between solubility and composition along the tectosilicate joins SiO_2–$NaAlO_2$, SiO_2–$Ca_{0.5}AlO_2$, and SiO_2–$Mg_{0.5}AlO_2$, leading them to employ $NaAlO_2$, $Ca_{0.5}AlO_2$ and $Mg_{0.5}AlO_2$ as components in addition to the above tetrahedral units. The model was calibrated against solubility data at 1600 °C and successfully accounted for reported Ar solubility in simple Al-free and Al-bearing systems as well as in a few natural liquids (Fig. 31), suggesting that Ar incorporation mostly relates to large sites in the melt structure (e.g., holes within rings of n-tetrahedra).

Based on the mechanism of interstitial dissolution in silicate networks, noble gas solubility has been described in terms of relationships with melt properties that are themselves related to free space in the silicate network, and that provide therefore a measure of that space.

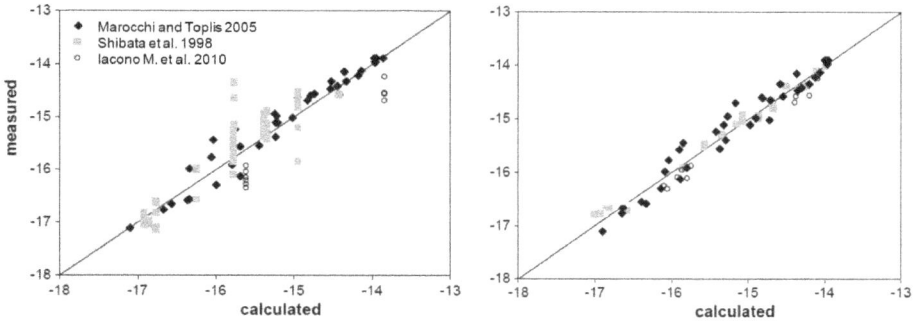

Figure 31. Measured vs. calculated Henry's constants for Ar by using the structural models of Marrocchi and Toplis (2005) (**left**) and Iacono-Marziano et al. (2010) (**right**) (modified after Iacono-Marziano et al. 2010). Data source: synthetic compositions at 1600 °C and 0.1 MPa (Marrocchi and Toplis 2005); synthetic compositions (NS, NCS and CMS) at 1200–1600 °C and 187–198 MPa (Shibata et al. 1998); basalt and rhyolite at 1200 °C and 49–375 MPa (Iacono-Marziano et al. 2010).

This approach has involved use of melt density (Lux 1987; White et al. 1989), melt molar volume (Broadhurst et al. 1992) and melt ionic porosity (Carroll and Stolper 1993a). In particular, the latter authors showed that ionic porosity works better than the other employed properties (density, molar volume) in predicting noble gas solubility. Ionic porosity IP is defined as:

$$IP = 100\left(1 - \frac{v_{ca}}{v^L}\right) \tag{219}$$

where v_{ca} is the molar volume of cations plus anions in one mole of melt (equal to the weighted sum of each ion molar volume) and v^L is melt molar volume. On this basis, Carroll and Stolper (1993a) wrote the first of equilibrium Equations (207) as:

$$\ln\frac{x_i}{P_i} = -\ln K_h(P,T) = \alpha_i IP + \beta_i \tag{220}$$

where α_i and β_i are fitting parameters, different for each ith noble gas, calibrated over experimentally measured solubility in various melt compositions. Carroll and Stolper (1993a) estimated these parameters for all stable noble gases, by using experimental data at 1000–1400 °C and 0.1 MPa. In spite of its general accomplishments, ionic porosity failed to predict solubility in CaO–MgO–Al$_2$O$_3$–SiO$_2$ (CMAS) melts, as well as in simple binary silicate melts (CaO–SiO$_2$, Na$_2$O–SiO$_2$ MgO–SiO$_2$). In fact, while ionic porosity provides a measure of the overall free volume in melt, it does not discriminate between situations characterized by a few large or many small cavities, which would translate into largely different noble gas solubility. As an example, the incorporation of network-modifier cations is likely to reduce the size of interstitial sites such that they can no longer accommodate noble gas atoms (Shibata et al. 1998).

In order to address the above effects, the ionic porosity model has been modified by Iacono-Marziano et al. (2010). They re-defined ionic porosity by considering the effects of different cations on melt porosity. Those effects were taken as being only partially linked to ionic radii, and to reflect also a P–T-dependent individual cation's capability to modify free space by forming, distorting or breaking the tetrahedral network. Accordingly, the term v_{ca} in Equation (219) was replaced by the weighted sum of partial molar-like filled volume terms referred to each jth oxide:

$$v_j(P,T) = v_{ca,j}^0 + v_{s,j}^0 + \lambda_j\left(\frac{1}{T} - \frac{1}{T^0}\right) + \kappa_j\left(P - P^0\right) \tag{221}$$

where $v_{ca,j}^0$ is the same individual component term providing the value of v_{ca} in Equation (219), $v_{s,j}^0$ is the contribution by each individual oxide to large-scale structural effects, and κ_j and λ_j express the dependence on pressure and temperature, respectively, of the quantity v_j referred to each oxide. Equation (220) was then modified to:

$$\ln\frac{x_i}{f_i} = -\ln K_{\rm h}\left(P,T\right) = -_i\left[100 - \frac{100}{v^{\rm L}\left(P,T\right)}\left(\sum_j^n v_j\left(P,T\right)x_j + v_{q,{\rm Na_2O}}\left(P,T\right)x_{\rm Na_2O}^2\right)\right] + \beta_i \quad (222)$$

with $v_{q,{\rm Na_2O}}\left(P,T\right) = v_{q,{\rm Na_2O}}^0 + \lambda_{q,{\rm Na_2O}}\left(1/T - 1/T^0\right) + \kappa_{q,{\rm Na_2O}}\left(P - P^0\right)$ being an additional quadratic compositional term. Iacono-Marziano et al. (2010) calibrated the model given by Equations (221–222) by regression on more than 400 noble gas solubility measurements in natural and synthetic melts. The model satisfactorily reproduces the data over a wide range of temperature (800–1600 °C) and pressure (up to 3 GPa), and it represents the most comprehensive tool to-date for estimating He, Ne and Ar solubility in silicate melts of virtually any composition (Figs. 31 and 32). The model reproduces well the relationships between Ar solubility and temperature for mafic and acidic liquids, and the linear increase in Ar concentration with pressure (Fig. 26). At high pressure (2–3 GPa) the calculated concentrations in the melt display a trend towards a saturation threshold, in agreement with the data by Schmidt and Keppler (2002) that did not form part of the calibration dataset (Fig. 26).

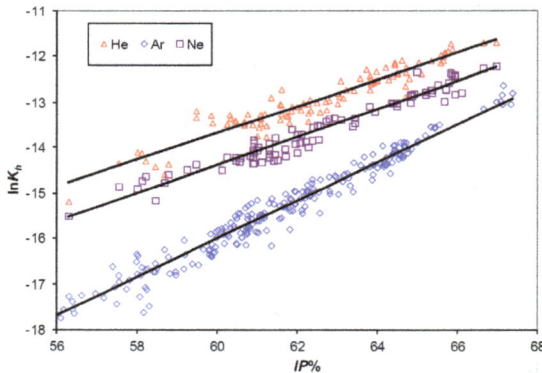

Figure 32. Relationship between Henry's constant for noble gases and ionic porosity of silicate melts. **Lines** are predictions for He, Ne and Ar with the Iacono-Marziano et al. (2010) model, while symbols derive from measured solubilities (data sources: Ar from Hayatsu and Waboso 1958; Jambon et al. 1986; Lux 1987; White et al. 1989; Carroll and Stolper 1991, 1993; Broadhurst et al. 1992; Roselieb et al. 1992; Shibata et al. 1998; Miyazaki et al. 2004; Marrocchi and Toplis 2005; Iacono-Marziano et al. 2010; Ne from Shackelford et al. 1972; Hayatsu and Waboso 1985; Jambon et al. 1986; Lux 1987; Broadhurst et al. 1992; Roselieb et al. 1992; Shibata et al. 1998; Miyazaki et al. 2004; Heber et al. 2007; Tournour and Shelby 2008b; Iacono-Marziano et al. 2010; He from Shackelford et al. 1972; Mulfinger et al. 1972; Jambon et al. 1986; Lux 1987; Shibata et al. 1998; Paonita et al. 2000; Mesko et al. 2000; Mesko and Shelby 2002; Tournour and Shelby 2008a). Ionic porosities are computed according to partial molar volumes by Iacono-Marziano et al. (2010).

Modeling mixed H_2O–CO_2–noble gases

In terrestrial magmas, noble gases can be regarded as trace elements (Moreira and Kurz 2013). The gas phase in natural magmas is normally H_2O–CO_2 dominated (Giggenbach 1996) and even at total pressure of some GPa, the partial pressures of noble gases is usually less than 1 MPa (Carroll and Webster 1994). As a consequence, evaluating possible effects of major volatiles on noble gas solubility is of major relevance, as the direct use of solubility data and models related to pure noble gas or noble gas mixtures in real situations of geologic relevance may lead to misinterpretations. In particular, the thermodynamic properties of noble gases dissolved in H_2O–CO_2-bearing silicate melts is still a challenging issue.

Nuccio and Paonita (2000) modeled the fugacity and activity of noble gases in mixed H_2O–CO_2–noble gas conditions. For the gas phase, they included H_2O and CO_2 in Equations (208–210) by using the cross-coefficient for H_2O–CO_2 interaction in De Santis et al. (1974), and the a and b parameters for pure H_2O and pure CO_2 by Holloway (1977). For the melt phase, they computed the *IP* value (Eqn. 219) by including the partial molar volumes of H_2O and CO_2 in computing v_L, and H and C ion volumes in computing v_{ca}. The approach employed to calculate noble gas partitioning between a gas with given H_2O and CO_2 concentration and a melt with given composition and *P–T* conditions requires: 1) to determine the dissolved H_2O and CO_2 saturation contents with the reasonable assumption that the comparably vanishing amounts of noble gases do not affect them; 2) to calculate the *IP* value for the H_2O–CO_2 bearing melt; 3) to calculate K_h by Equation (220) with α_i and β_i assumed to coincide with the same values for dry melt. From the above, the fugacity of the noble gas at equilibrium is computed from Equation (210), and the dissolved concentration of the noble gas is calculated from Equation (220). Nuccio and Paonita (2000) employed the model in Papale (1999a) (a previously calibrated version of the SOLWCAD model) for H_2O–CO_2 saturation, and the model in Lange and Carmichael (1987) for melt molar volume. Their Extended Ionic Porosity (EIP) model predicts a positive dependence of noble gas solubility on the amount of H_2O dissolved in the melt for melts from basalt to rhyolite, up to about 3 wt% dissolved water above which the effects of further increase in water concentration on noble gas solubility become about negligible (Fig. 33). On the contrary, noble gas solubility is found to slightly decrease when the dissolved CO_2 increases (Fig. 33). The above dependences of noble gas solubility on major volatile contents grow exponentially with increasing noble gas atomic radius. Predictions on H_2O effects are in very good agreement with the experimental measurements in Paonita et al. (2000) showing an about three-fold increase in He solubility when adding 3 wt% H_2O to the melt, and very little effects with further H_2O addition up to 6 wt% (Fig. 33). The measured effect of CO_2 addition (up to 0.05 wt%) was a slight solubility decrease, again in agreement with the theoretical calculations (Fig. 33). The analysis confirmed that the assumption of unchanged α_{He} and β_{He} for hydrated and dry melts is a reliable one. The model was subsequently tested with Ar solubility data in H_2O–CO_2 bearing basalt (Paonita et al. 2012). While it correctly predicted an increase of Ar solubility with increasing dissolved H_2O content, the measured increase was by about a factor 4 for 5 wt% H_2O addition, rather than the predicted increase by about 30 times. As a consequence, Paonita et al. (2012) proposed

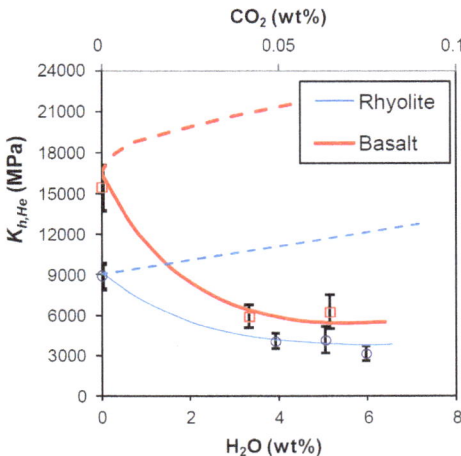

Figure 33. Effect of dissolved H_2O and CO_2 on He Henry's constant. Model curves from Nuccio and Paonita (2000) were obtained for H_2O–He (**solid**) or CO_2–He (**dashed**) with ~0.1 mol% He in vapor and up to 200 MPa of pressure (Henry's constant calculated on 8-oxygen basis melt). Experimental data are for H_2O–He vapor (He ~ 0.1 mol%) from Paonita et al. (2000).

a new calibration of the EIP model for Ar dissolution in the hydrous basalt employed in their experiments, in which they presented a $\ln K_{h,Ar}$ vs. *IP* relationship for that given melt with *IP* depending solely on the dissolved H_2O content. It is clear, however, that many more data are required before a robust, comprehensive model for Ar solubility in H_2O–CO_2 bearing melts is produced.

Sarda and Guillot (2005) and Guillot and Sarda (2006) employed Equation (216) and the approach described thereby to model noble gas partitioning between a CO_2-dominated gas and silicate melts. On such a basis, they determined $\mu_i^{E,L}$ (and then δ_i^L) for a noble gas *i* in a silicate melt and $\mu_i^{E,G}$ (and then δ_i^G) in the CO_2 gas. On this basis, Aubry et al. (2013) employed the equations of state obtained for the two coexisting phases (silicate melt and gaseous CO_2) together with a CO_2 solubility curve deduced from MD simulations (Guillot and Sator 2011 and 2012), and implemented the test particle method (as in Guillot and Sator 2012) for studying the case of noble gas partitioning between gaseous CO_2 and CO_2-saturated melt. They performed a series of MD simulations to calculate the solubility parameters of a noble gas in gaseous CO_2 and in basalt melt (δ_i^G and δ_i^L in Equation 216). The computed ratio between the solubility parameters in Equation (216) was then used to calculate noble gas partition coefficients between the two coexisting phases as a function of pressure. The calculations were extended to all noble gases (Fig. 34) and the corresponding solubility was found to be very similar to that obtained by Guillot and Sator (2012) in their study of the solubility of pure noble gases in dry silicate melts at high pressure, confirming small or negligible effects of CO_2 on noble gas solubility. This result suggests that a pressure-dependent solubility constant for pure noble gas can be used to estimate the amount of noble gas dissolved in a silicate melt coexisting with a gaseous CO_2 phase, once the partial pressure of the noble gas in the gas phase is known.

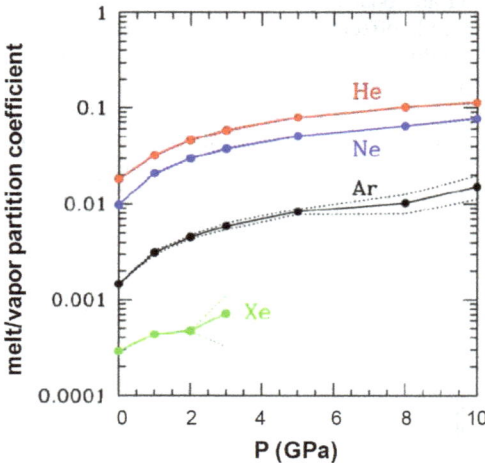

Figure 34. Noble gases partition coefficients between CO_2 gas and basaltic melt up to very high pressure (modified from Aubry et al. 2013). Calculated curves were obtained by TPM coupled to MD simulations of the coexisting vapor and melt at 1600 °C.

More recently, Fabbrizio et al. (2017) applied Equation (212) to deal with the chemical equilibrium of Ar between Ar–H_2O–CO_2 mixed vapor and a basalt melt. The authors performed a best fit calibration of Equation (212) at isothermal conditions ($\Delta H^0 = 0$) by employing both H_2O–CO_2-poor and H_2O–CO_2-rich data. The Ar fugacity in the gas phase was computed by the MRK-EOS at Equation (210) extended to H_2O–CO_2–noble gas systems as described above. A value of v^0 for Ar in Ar–H_2O–CO_2-bearing basalt at 1500 °C and 1 to 5 GPa pressure was then calibrated, corresponding to 24.3 ± 0.3 cm^3 mol^{-1}, which matches with the same value for volatile-free basaltic melts (although it is worth noting that no dependence of v^0 on the amount of dissolved H_2O and CO_2 was included in the model). That suggested that the effect of pressure in the range 1–5 GPa is dominant in controlling Ar solubility compared to the effects of dissolved H_2O and CO_2. As already observed in previous experiments, pressure was found to have a positive effect on argon incorporation in H_2O–CO_2 bearing basalts. In agreement with their experimental data, the model of Fabbrizio et al. (2017) predicts an Ar solubility threshold at 3 GPa, with Ar concentration of ~0.38 wt.% which is of the same order of the amount found in H_2O–CO_2-free basaltic melts. It should be noted here that a comparison of the EIP model described above with the experimental data in Fabbrizio et al. (2017) for Ar–H_2O–CO_2-bearing basalt is not appropriate, because the experimental data were obtained at 1–5 GPa and with $P_{Ar} > P_{H_2O} > P_{CO_2}$, while the EIP model is designed to work at < 1 GPa and for $P \approx P_{H_2O} + P_{CO_2} \gg P_{Ar}$. In fact, the high dissolved Ar (thousands of ppm, close to the Ar saturation threshold) would violate the EIP model condition of infinite Ar dilution.

Ar solubility as a proxy for N_2 solubility

Noble gas solubility has been used in comparison to solubility of molecular nitrogen to achieve information on its solution mechanisms, based on the well-established mechanical dissolution mechanism of noble gases. The solubility of nitrogen in silicate melts depends strongly on oxygen fugacity, as the latter controls the speciation of nitrogen. The few experimental data of nitrogen solubility in melts with composition matching common magmas (andesite and basalt) demonstrate that it dissolves as molecular species N_2 at redox conditions typical of lithospheric magmas (Carroll and Webster 1994, and references therein; Marty et al. 1995; Libourel et al. 2003; Miyazaki et al. 2004). At more reducing conditions (IW buffer or below), the solubility of nitrogen progressively increases with decreasing f_{O_2}, suggesting the formation of chemical bonds with the silicate network (possibly as nitride groups; Libourel et al. 2003). More recent solubility data in simple synthetic melts suggest that even at NNO redox conditions, a fraction of nitrogen can dissolve as NH_2^+ and NH_3 complexes in H_2O–bearing melts at pressures higher than 1 GPa, causing modest increase in solubility (Mysen et al. 2008). Such dissolution mechanism becomes dominant at low f_{O_2} (Mysen et al. 2008; Mysen and Fogel 2010). Under the conditions in which nitrogen dissolution is molecular (as N_2), its solubility practically matches that of Ar (Libourel et al. 2003; Miyazaki et al. 2004; Roskoz et al. 2006). This observation, together with similar size of N_2 molecules and Ar atoms, is a strong support for a mechanical mechanism of solution of N_2 with no or poor interaction with the silicate melt structure (see also Mysen 2013 for a review). As a consequence, Ar is frequently used as a proxy of N_2 in modeling nitrogen solubility in magmas (e.g., as in Fig. 46 below), taking advantage by the wide spectrum of conditions over which their similar behavior is observed, covering most conditions of interest for volcano degassing. Under these f_{O_2} conditions, knowledge of the P–T and melt composition effects on Ar solubility, as well as the related quantitative models, can be applied to predict N_2 solubility in petrological and volcanological studies. As an example, Ar solubility in basalts, computed by using the updated IP model, matches very well with N_2 solubility achieved under oxidizing conditions (>IW redox buffer) by using the recent model of Bernadou et al. (2021), calibrated for basaltic melts and up to high N_2 fugacities. Similarly, measured N_2 solubilities in haplogranites at 200 MPa and NNO+1 (Li et al. 2015) are reproduced by the IP model very well.

Applications

As discussed in the Introduction session, volatile–melt equilibria are a central element for our capability to model the evolution of magmas as well as the dynamics of volcanic eruptions. Volatiles crucially affect practically any magma property and any aspect of their dynamics from depth to final emplacement after the eruption. Models for volatile–melt equilibrium are therefore critical to our understanding of igneous petrology and for volcanology.

Much of the investigation which makes large use of volatile–melt equilibria relates to one or more of the following:

• determination of magma storage conditions;

• interpretation of data from volcanic emissions;

• constraining and modeling magma and eruption dynamics.

In this section we provide, for each one of the above applications, examples showing the variety of relevant insights on volcanic systems and their dynamics, afforded by thermodynamic models of multi-component volatile–melt equilibria.

Determination of magma storage conditions

The theoretical foundation that makes multi-component volatile–melt equilibrium effective and powerful as a means of constraining magma storage condition, is that at equilibrium the number of degrees of freedom of the system is reduced according to phase rule. For the gas–liquid equilibrium between H_2O, CO_2 and a melt with specified volatile-free composition (3 components, C, and 2 phases, P), the phase rule says that the number of degrees of freedom $F = C—P + 2$ is equal to 3. Therefore, once two among equilibrium pressure, temperature and composition are given, the third one is also given. In applications to real magmatic systems the melt composition (including the dissolved volatiles) is usually retrieved by studying melt inclusions, or pockets of melt entrapped within growing crystals. Under favorable conditions, and in particular depending on the characteristics of the lattice arrangement of the host crystal, the entrapped melt can essentially retain its original composition including the dissolved volatiles until analyzed in the lab (there are countless contributions on melt inclusion studies, from the pioneering work by Roedder (e.g., 1976, 1979). Early reviews are those by Johnson et al. (1994) and Lowenstern (1995). The *Reviews in Mineralogy and Geochemistry* volume 69: *Minerals, Inclusions And Volcanic Processes*, edited by Keith D. Putirka and Frank J Tepley III (2008), and in particular the chapters by Kent (2008), by Moore (2008), and by Métrich and Wallace (2008), provides ample review and analysis of melt inclusion methods, experimental procedures, and results, including potentials and limits of the technique). Similarly, temperature is usually constrained by well-known solid–liquid equilibria (e.g., Anderson et al. 2008; Hansteen and Klügel 2008). Therefore, pairs of H_2O and CO_2 dissolved in the original melt and measured in melt inclusions can be employed to constrain the remaining relevant intensive quantities characterizing the equilibrium. In mathematical terms, once the dissolved amounts of the two volatiles are determined from lab analysis, and for given magma temperature, Equations (28–30) constitute a system of three equations in the three unknowns pressure and concentration of each volatile component in the gas phase. Therefore, if melt inclusions can be assumed to reflect the original gas–melt thermodynamic equilibrium in magma, the equilibrium equations allow the determination of the entrapment pressure as well as the composition of the coexisting gas phase. That can be graphically illustrated in diagrams like the one in Figure 35, first introduced by Anderson et al. (1989) and now widely used for melt inclusion studies, thanks to tools that allow their easy and quick production, such as VolatileCalc (Newmann and Lowenstern 2002), SOLWCAD (Papale et al. 2006), and MagmaSat (Ghiorso and Gualda 2014). The diagram in Figure 35 shows that at thermodynamic equilibrium, for any observed pair of H_2O and CO_2 dissolved in a given melt and at given temperature there is one and only one corresponding equilibrium pressure and gas phase composition (note that throughout this section all of the $H_2O–CO_2$–melt equilibrium calculations employ the SOLWCAD model from Papale et al. 2006, described above).

A dissolved CO_2–H_2O diagram like the one in Figure 35 provides information that largely transcends the determination of entrapment pressure and coexisting gas phase, these being

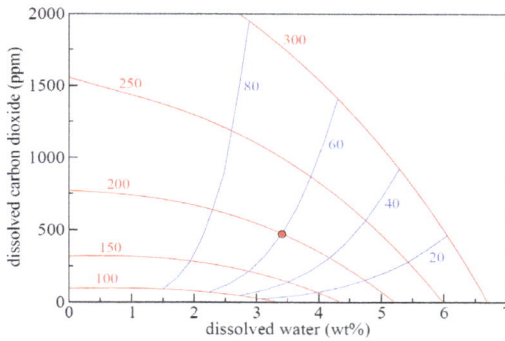

Figure 35. Dissolved CO_2–H_2O diagram for a basaltic melt (composition from Lesne et al. 2011a) at 1200 °C. The descending **red lines** are isobars (numbers are pressure in MPa), while the ascending **blue lines** connect points having same gas phase composition (numbers are CO_2 concentration in wt%). The example point (**red**) represents a pair of dissolved H_2O and CO_2 from a hypothetical melt inclusion. The diagram shows that assuming equilibrium conditions, that melt inclusion originated at 200 MPa from a melt coexisting with a 60 wt% CO_2 gas phase. Calculations made with SOLWCAD.

quantities of fundamental relevance. In fact, series of melt inclusion data from individual eruptions or from multiple eruptions from a given volcano carry information on the structure of the plumbing system, e.g., showing pressure intervals—thus depth intervals—where crystallization is more likely to happen. These intervals are commonly interpreted in terms of regions of magma residence where limited changes in intensive variables allow close-to-equilibrium crystallization; or in other words, in terms of location and vertical extension of magma chambers (Fig. 36). Melt inclusions also carry information on several processes that can affect underground magmatic bodies; e.g., the occurrence at specific depth intervals of closed vs. open system degassing. In the latter case, part of the gas phase is lost by the magma undergoing decompression. Because CO_2 is much less soluble than H_2O in silicate melts, gas loss invariably results in relative H_2O enrichment with respect to CO_2, visible as a shift to the right in a dissolved CO_2—H_2O diagram (Fig. 37) (*cfr.* Newman et al. 1988; Anderson et al. 1989). On the other extreme, relatively shallow magma bodies can be constantly or periodically fluxed by a gas phase of deeper provenance, e.g., released by a deeper magma undergoing open system degassing (Spilliaert et al. 2006; Moretti et al. 2013a,b; Caricchi et al. 2018). Due to low CO_2 solubility with respect to H_2O, a gas phase originating at larger depth tends to be enriched in CO_2 with respect to the gas formed at shallow levels (that also depends, obviously, on other factors, e.g., total abundance of H_2O and CO_2 and melt composition; however, given a magmatic system, the deeper gas is practically always enriched in CO_2 with respect to shallower ones). Fluxing of magma by deeper CO_2-rich gas results in progressive increase of the total CO_2/H_2O ratio, which at isobaric conditions results in a progressive shift towards the CO_2 axis of the dissolved CO_2–H_2O pairs (Fig. 37). In more extreme conditions, the mass flow-rate of the gas fluxing the shallower magmatic system can be sufficiently large to buffer the gas phase and control the gas–melt equilibrium. In such a case, pairs of dissolved H_2O and CO_2 trapped within melt inclusions formed along a decompression path distribute along the corresponding gas phase isopleth (Fig. 37). Comparison with the distribution of melt inclusions in Figure 36a suggests that gas-buffered conditions may be effective at Stromboli (Fig. 36b), which is in fact well-known for a striking disproportion between gas emissions and erupted lava or pyroclasts (Allard et al. 1994; Pino et al. 2011). Aiuppa et al. (2010b) and Pino et al. (2011) have highlighted non-linear relationships between the saturation contents of volatiles in the C–O–H–S system under fluxing by CO_2-rich gas, such to determine an increase in the sulfur content at saturation (Fig. 38) adding complexity to the interpretation of gas monitoring data.

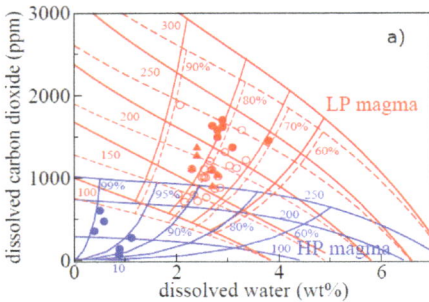

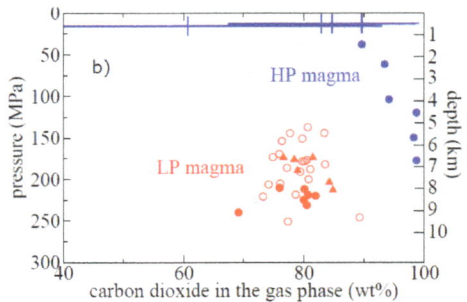

Figure 36. Melt inclusion data from Stromboli volcano, Aeolian Islands, Italy. **a)** Dissolved CO_2–H_2O diagram. The **red lines and symbols** refer to the so-called LP magma which is occasionally discharged during major eruptions and paroxysms, while the **blue lines and symbols** refer to the HP magma which feeds the normal Strombolian activity. **Full and dashed red lines** highlight the effects of compositional variability in LP melt inclusions. The significant compositional differences between the LP and HP melts cause largely different isobars, mostly due to largely reduced CO_2 solubility in the HP melt. Data from Métrich et al. 2001 (**solid triangles**), Bertagnini et al. 2003 (**open circles**), and Métrich et al. 2010 (**solid circles**). **b)** Same data as in a) in a pressure vs. gas phase composition diagram, evidencing that the LP magma resides in a portion of the plumbing system deeper than 5 km (130 MPa) and extending down to > 9 km (250 MPa). As expected, the HP magma occupies the shallower portion of the volcanic system, and its gas phase is highly enriched in carbon dioxide. The topmost inclusions of the HP magma, without symbol and not reported in panel a, have CO_2 contents below the FTIR detection limits (40–50 ppm) thus they have large error bars reflecting large uncertainty in the composition of the coexisting gas phase (but only relatively small uncertainty regarding entrapment pressure/depth). These inclusions are found to originate within the upper 1 km below the volcanic crater. Highly CO_2-enriched melt inclusions (extreme right of the diagram at panel b) suggest that at least occasionally, the HP magma can exist down to 6–7 km. Calculations made with SOLWCAD.

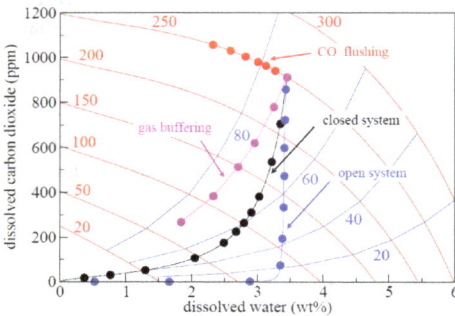

Figure 37. Example curves on a dissolved CO_2–H_2O diagram. The examples show a series of hypothetical melt inclusion data along paths corresponding to closed system decompression (**black**), open system decompression (**blue**, loss of 5 wt% of the gas phase assumed at each subsequent computational step along the decompression path), isobaric CO_2 fluxing (**red**), and decompression buffered by an about 75 wt% CO_2 gas phase (**magenta**). All of the example paths and the isobaric lines refer to a rhyolitic melt from Blank et al. (1993) at 800 °C, and start from total H_2O and CO_2 contents of 4 and 2 wt%, respectively, at a pressure of 250 MPa. Calculations made with SOLWCAD.

The simulated, relatively simple trends in Figure 37 illustrate the rich information that can be retrieved by processing melt inclusion data with multicomponent gas–melt equilibrium modeling. Data from real volcanoes (e.g., those in Fig. 36) can result from a combination of processes like those exemplified in Figure 37, and others (e.g., magma crystallization, convection, mixing, etc.) that can be equally important, especially for the behavior of highly soluble volatile components such as sulfur and chlorine. In fact, concentration in the melt of such highly soluble components at late stages of crystallization can result in cloudy distributions of data points on binary diagrams involving volatile pairs. For example, Moretti et al. (2013b) have shown that the bulk of H_2O–CO_2–S concentrations in melt inclusions from Ischia and Procida islands (Phlegrean Volcanic District, Southern Italy) can be explained by polybaric crystallization under different gas buffering conditions due to CO_2 upstreaming (fluxing) from the mantle. CO_2-dominated gas buffers also shift the oxidation state resulting in melt inclusions that are S-depleted at Procida (prevalently $Fe^{3+}/Fe_{tot} < 1$) and S-enriched at Ischia (prevalently $Fe^{3+}/Fe_{tot} > 1$).

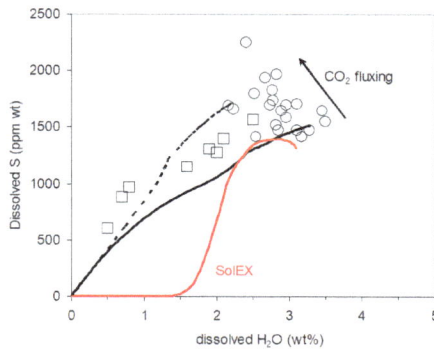

Figure 38. Water and sulfur contents in melt inclusions retrieved in magmas erupted during the 2003 paroxysm at Stromboli. Symbols are data from Bertagnini et al. (2003) (**open circles**) and Métrich et al. (2005) (**open squares**). The **black lines** refers to computations with COHS for decompression under closed system conditions and $w_{H_2O}^T = 3.67 wt\%, w_{CO_2}^T = 2.44 wt\%, w_S^T = 1500 ppm$ (**solid line**), and same amounts with $w_{CO_2}^T$ equal to 17 wt% (**dashed line**) in response to deep CO_2 addition (fluxing). Under such conditions, fluxing is seen to decrease the dissolved H_2O content (as in Fig. 37) and increase the dissolved S content. Also plotted is the H_2O–S closed system evolution computed for the same conditions from the SolEX model. Modified after Pino et al. (2011) and Witham et al. (2012).

By using the COHS model extended to chlorine (described above), Moretti et al. (2018) reconstructed the S–H_2O and Cl–H_2O degassing patterns of several Etna eruptions (Fig. 39), including the 122 BCE basaltic Plinian eruption (Coltelli et al. 1998). By jointly studying CO_2–H_2O–S–Cl dissolution, the authors suggested that the variations in oxidation state (involving also products from the same eruption) emerging from sulfur–water covariation (Fig. 39a) is an apparent effect due to CO_2-fluxing generating batches of magma with heterogeneous total volatile proportions (Fig. 39b). Different dissolved Cl–H_2O relationships (and Na/K melt ratios) between the 122 BC and recent magmas (Fig. 39c) were also interpreted to be a consequence of CO_2-rich gas fluxing.

The above examples demonstrate that melt inclusion studies transcend the definition of magma storage conditions entering the domain of complex dynamic processes, that can be recognized and quantified by processing melt inclusions through modeling of volatile–melt equilibrium thermodynamics. Melt inclusions can also be employed to determine the total abundance of volatiles in magmas (or the mass of the dissolved and exsolved volatiles divided by the mass of the melt plus the gas phase). This is a major and non-trivial application: in fact, it is the total abundance of each volatile component in magma that determines the gas–melt multicomponent equilibria (Eqns. 28–31), the abundance of the gas phase, the density relationships governing the dynamics of convection and mixing between different magmas (Papale et al. 2017), the acceleration and overall dynamics of magma ascent upon eruption on the surface (Papale et al. 1998; Papale and Polacci 1999), and the convective vs. collapsing fate of the volcanic jet above the volcanic crater, and associated far different impacts on the surroundings (Neri et al. 1998; Esposti Ongaro et al. 2006).

As seen above, the dissolved CO_2—H_2O diagram in Figure 35 shows that one H_2O–CO_2 pair (one point in the diagram) is enough to constrain, for a given temperature, the equilibrium pressure and gas phase composition. However, there are infinite pairs of total H_2O and CO_2 corresponding to that equilibrium, or in other words, corresponding to the same set of intensive quantities pressure, concentration of volatiles in the melt, and gas phase composition. Papale (2005) demonstrated that those infinite pairs lie on a straight line, that he termed the Total Volatiles Line (TVL), with each point along the TVL corresponding to same value of any quantity except the relative mass of gas in the gas–melt system. In other words, one single pair of dissolved H_2O and CO_2 does not constrain their total abundance in the magma, but it determines a relationship between such total abundances, represented by the corresponding TVL. In a CO_2—H_2O diagram, TVL's originate from the corresponding melt inclusion; in fact, that would be the case for an under-saturated magma exactly reaching the saturation condition. Therefore, if melt inclusions formed at different pressure correspond to the same total H_2O and CO_2 contents (closed system conditions), the corresponding TVL's will cross each other at the only point along each straight line which is consistent with the equilibrium represented by each melt inclusion. A graphical example is provided in Figure 40. Before analyzing it, it is useful to introduce the TVL equations (Papale 2005):

$$w_{CO_2}^{T} = q + m w_{H_2O}^{T} \tag{223}$$

$$m = \frac{w_{CO_2}^{G} - w_{CO_2}^{L}}{w_{H_2O}^{G} - w_{H_2O}^{L}} \tag{224}$$

$$q = w_{CO_2}^{L} - m w_{H_2O}^{L} \tag{225}$$

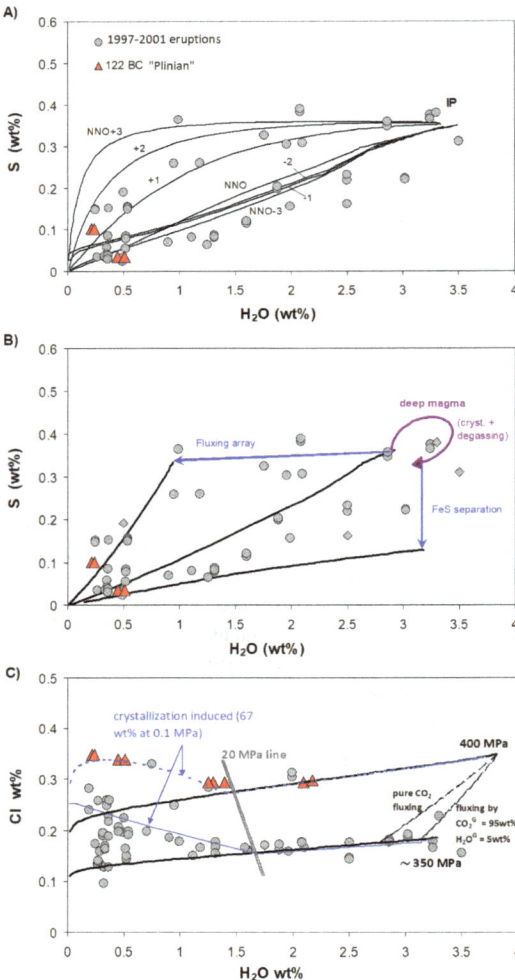

Figure 39. Melt inclusion data and model calculations for Etna magmas with multicomponent C–O–H–S–Cl gas phase. **a)** Dissolved H_2O and S (symbols), and calculations from the COHS(+Cl) model for $-3 \leq NNO \leq 3$ and fixed total volatile content corresponding to $w_{H_2O}^{T} = 3.5$ wt%, $w_{CO_2}^{T} = 2$ wt%, $w_{S}^{T} = 3200$ ppm. Note the relevant changes associated to different assumed redox conditions. **b)** Same as in panel a for NNO redox conditions, with addition of i) CO_2-fluxing (and consequent dehydration) and ii) FeS separation (yielding S depletion). The **curved violet arrow** embraces conditions for the deep crystallization of Etnean primitive melts (see Moretti et al. (2018). **c)** Dissolved H_2O and Cl. Significant crystallization and associated Cl concentration in the melt can explain the trends observed in dehydrated samples below 20 MPa. CO_2 fluxing determines chlorine depletion in the melt phase, and may explain a shift from Na-rich (122 BC) to K-rich (recent products) magmas. Two fluxing paths are shown (CO_2-pure and CO_2 at 95 (wt%). Low pressure data for the 122 BC eruption are best fitted by including crystallization under closed-system conditions for the gas phase, whereas low pressure Cl contents in recent products are best modelled by crystallization occurring under gas phase open-system conditions, with P-dependent Cl partition coefficient from Alletti et al. (2009). Modified after Moretti et al. (2018).

As it is clear from Equations (223–225), determining the TVL only requires dissolved amount of H_2O and CO_2, that are directly measured in melt inclusions, and the composition of the coexisting gas phase, that for any pair of dissolved H_2O and CO_2 is univocally determined (see Fig. 35) and can be computed with a mixed $H_2O + CO_2$ saturation model.

Let's now consider the example in Figure 40. Assume that we have measured the dissolved amount of H_2O and CO_2 as in panel a from a series of melt inclusions, and that we know (from independent information) the magmatic temperature. For simplicity, this example assumes no crystallization therefore the same volatile-free composition (a rhyolite from Blank et al. 1993) for all melt inclusions. Note however that crystallization, causing further volatile concentration with respect to the melt–gas system as well as a melt compositional shift, can be readily accounted for through a modification of Equations (224–225) (Papale 2005). As we have seen above, and as it is also visible in panel a of Figure 40, each melt inclusion constrains the corresponding entrapment pressure as well as the composition of the coexisting gas phase (note that if the melt composition from the different inclusion was not the same, e.g. as a consequence of crystallization, the computation would still be possible but a simple representation like the one in Figure 40a would not be, because each data point would have its own isobars and isopleths). We can use that information to determine with Equations (223–225) the TVL for each pair of dissolved volatiles (or equivalently, for each melt inclusion). Panel b of Figure 40 shows the TVL's determined in this way. The TVL's corresponding to the four high pressure melt inclusions cross each other at the same point in the diagram, meaning that over that pressure range the magma evolved under closed system conditions with respect to the gas phase. The point at which the TVL's cross each other corresponds to the total (i.e., dissolved + exsolved) H_2O and CO_2 contents (in the example, 6 wt% H_2O and 2 wt% CO_2). Different from the above, the remaining five melt inclusions formed at lower pressure show TVL's that do not cross each other at a specific point, rather, they depict a fan-shaped frame which is the typical signature of open system degassing (Papale 2005). The envelope of TVL's corresponding to an open system path constraints the evolution of total H_2O and CO_2 in the system upon degassing, as it is shown by the two additional points in Figure 40b with reported total volatile contents.

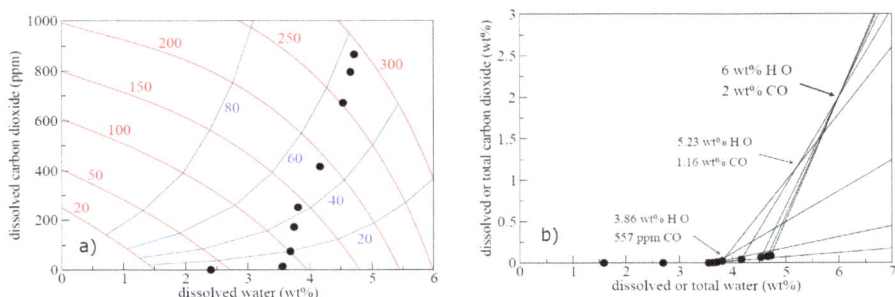

Figure 40. Example of application of the TVL method (Papale 2005) to determine total H_2O and CO_2 contents from melt inclusion data. **a)** Hypothetical melt inclusion data (**black circles**) measured in a melt with composition corresponding to the rhyolitic melt in Blank et al. (1993) (same as in Fig. 37). **b)** Total Volatile Lines (TVL) for the melt inclusions in panel a (also reported in panel b). The computations assume a temperature of 800 °C. The TVL's for high pressure melt inclusions cross each other at a point corresponding to total (in the liquid and gas phases) H_2O and CO_2 in the magma from which the inclusions were formed. Progressive loss of gas under open system conditions (low-*P* inclusions) is revealed as a typical fan-shaped geometry of TVL's. The TVL's for the data lying on the *x* axis (no dissolved CO_2) coincide with the *x* axis. Calculations made with SOLWCAD.

Once the total volatile contents are determined, the corresponding gas fraction can be computed from simple mass balances:

$$w^G = \frac{w^T_{H_2O} + w^T_{CO_2} - w^L_{H_2O} - w^L_{CO_2}}{1 - w^L_{H_2O} - w^L_{CO_2}} \tag{226}$$

$$\varepsilon^G = \frac{w^G}{\rho^G}\left[\frac{w^G}{\rho^G} + \frac{1-w^G}{\rho^L}\right]^{-1} \tag{227}$$

where w^G and ε^G are the weight and volume fraction of the gas phase, respectively, in the gas + liquid system, and ρ is density (corrections to account for crystal content can be easily added). By using the Lange and Carmichael (1987) model for the density of water-bearing silicate melts, the SOLWCAD model for H_2O–CO_2–melt equilibrium (which embeds the Kerrick and Jacobs (1981) EOS for $H_2O + CO_2$ gas mixtures), and neglecting the effect on melt density of small amounts of dissolved carbon dioxide, the upper four melt inclusions in Figure 40 turn out to have formed under closed system conditions in the pressure range 291–189 MPa from a magma that contained 6 wt% total H2O and 2 wt% total CO_2, and that included a 3.36–3.96 wt%, 11.2–16.7 vol% gas phase (both increasing with decreasing pressure) with composition corresponding to 57–49 wt% CO_2 (decreasing with decreasing pressure); all of this information being retrieved by melt inclusion data only, processed with a multicomponent thermodynamic equilibrium model.

The example above, as well as the application of the TVL method to a number of real melt inclusion data in magmas from all around the world (Papale 2005), clearly shows a substantial difference between H_2O and CO_2. While maximum H_2O contents found in ensembles of melt inclusion data can be taken in many cases as a (minimum) approximation of the effective total H_2O content in the magma, total CO_2 contents are commonly orders of magnitude higher than the dissolved amounts, as a consequence of the very low solubility of CO_2 in natural magmas.

Estimations of total, pre-degassing content of volatiles has also taken advantage from the application of equilibrium degassing models to fluid inclusions entrapped in igneous products. Different from melt inclusions, which record the amount of volatiles dissolved in the melt at the time of entrapment (and that are often accompanied by gas bubbles generated by shrinking of the melt or glass after entrapment, Frezzotti 2001), fluid inclusions are interpreted as original gas bubbles that were entrapped in growing minerals or quenched glass, or in other words as relicts of the original volatile exsolution process. Aubry et al. (2013) estimated a range of initial CO_2 in MORB by computing vesicularity and CO_2 content of a MORB ascending to the seafloor and comparing the model results to the same observed quantities in MOR pillow lavas. According to the authors, the observed scattered distribution of vesicularity in MORB samples would be the reflection of equilibrium CO_2 exsolution from various depths in the mantle according to variable initial CO_2 contents. Episodes of gas loss during ascent would further explain samples with particularly low vesiclularity and CO_2 content (Aubry et al. 2013). On this ground, the inventory of noble gases in these glasses, interpreted by means of volatile equilibrium modeling, put key constraints on the complex decompression paths of MORB magmas. The variability of the He/Ar ratios in MORB glasses (see Fig. 41), encompassing a range up to some orders of magnitude higher than the expected initial ratio in the undegassed MORB (estimated from production and accumulation of radiogenic [4]He and [40]Ar nuclides in the mantle), has in fact suggested open system degassing processes in which the poorly soluble Ar is preferenially partitioned into the gas phase with respect to He (Jambon et al. 1985, 1986; Marty and Tolstikhin 1998; Marty and Zimmermann 1999; Moreira and Sarda 2000; Burnard 2001; Burnard et al. 2002, 2003, 2004; Sarda and Moireira 2002; Yamamoto and Burnard 2005). Lower abundances of He and Ar at higher He/Ar ratios and large scattering of data have suggested the occurrence of multistep degassing processes marked by discrete episodes of gas loss (Guillot and Sarda

2006; Aubry et al. 2013; see Fig. 41), suggesting that plumbing systems under mid-ocean ridges may be constituted by multiple zones of magma storage and accumulation. For the sake of completeness, the investigation of noble gas and carbon ratios (He/Ne, He/Ar, He/Xe, He/CO_2) during the last fifteen years has shown that gas–melt equilibrium thermodynamics do not explain by itself the data on MORB fluid inclusions (Paonita and Martelli 2006, 2007; Gonnermann and Mukhopadhyay 2007; Furi et al. 2010; Weston et al. 2015; Tucker et al. 2018). Although having different degree of complexity, all of these models point to to non-equilibrium diffusion-controlled degassing paths during decompressive bubble growth punctuated by episodes of gas loss as a more likely explanation of the data.

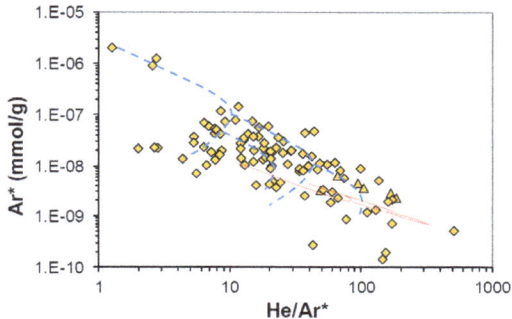

Figure 41. Computed He/Ar ratio (*x*-axis) as a function of Ar abundance in MORB at seafloor eruption conditions. Ar* is radiogenic Ar, or Ar corrected by air addition. Episodes of bubble loss have been simulated at 0.4 and 0.2 GPa starting from a CO_2-saturated magma at 1 GPa and initial He/Ar equal to 1.5. Noble gas solubilities were computed with the model of Nuccio and Paonita (2000). **Dashed and dotted lines** refer to vesicle and melt composition, respectively. Data from the compilation by Paonita and Martelli (2007).

Interpretation of volcano degassing data

Degassing from magmatic bodies stored at depth or ascending toward the surface is the major source of volcanic gas emissions from crater plumes, fumaroles, soil degassing, bubbling pools etc. As volcanic fluids carry information on first order magmatic processes occurring from large depths to near the surface, measuring volcanic gas emissions and interpreting their variations is essential to understand volcano dynamics and produce short term hazard forecasts. Because volcanic emissions are affected to largely different extents by secondary processes (after gas release from magma), the information from volcanic gas studies can be different depending on the investigated (or available) emission sources. On one extreme, crater plumes at open system volcanoes are usually taken as highly representative of the magmatic gas. On the other extreme, fumaroles, soil degassing, and other emissions at volcanoes characterized by the presence of important geothermal circulation systems (as volcanic calderas usually are) largely reflect the complex physico-chemical interactions between deep (magmatic) gases and shallow waters, brines, and rocks. In such cases, discriminating between a magmatic or geothermal origin for the observed variations can be difficult and requires additional measurements (e.g., isotopic ratios) and multi-parametric constraints (e.g., Chiodini et al. 2017; Gresse et al. 2018).

Melt inclusions considered above yield insights into volatile–melt equilibrium taking place at specific *P–T* conditions prior to past eruptions. In contrast, volcanic emissions analyzed here provide an integrated representation of degassing taking place over a range of depths and conditions, combined with many other processes at deep as well as shallow levels, including decarbonation and other reactions involving the surrounding rocks, mixing and reactions with geothermal and meteoric sources, etc. Because of such complexities, more robust interpretations are obtained by combining gas chemistry with melt inclusion data and multicomponent gas–melt equilibrium modeling.

Monitoring of volcanic emissions has greatly progressed thanks to recent technological developments leading to the diffusion and evolution of increasingly more precise, portable, multi-purpose instruments for continuous remote measurements. Describing experimental approaches and techniques is not among the purposes of this chapter, and the reader is referred to an abundant literature on the subject (e.g., Platt 1994; Symonds et al. 2001; Edmonds et al. 2002; Oppenheimer and McGonigle 2004; Shinohara 2005; Aiuppa et al. 2007; Platt and Stutz 2008; Liotta et al. 2010; Roberts et al. 2017; and many others). We only mention here that major benefits for measurements in volcanic plumes came from developments in optical methods leading to OP-FTIR techniques for the concentration and ratios of HCl, HF, SO_2, CO_2, H_2O, CO and OCS (Edmonds et al. 2002), and to DOAS techniques for the total flux of these species (Platt 1994; Platt and Stutz 2008). A further key improvement came from the development of the Multicomponent gas analyzer system (Multi-Gas; Shinohara 2005; Aiuppa et al. 2007), that is a multi-sensor assembly developed to measure concentrations of CO_2 and H_2O by means of non-dispersive IR spectrometers, and SO_2 by electrochemical sensor, allowing therefore the determination of major volatile ratios in volcanic plumes.

One major concept that is routinely applied in the interpretation of volcanic emissions is that of fractionation of volatiles controlled by their differential solubility, whereby gases reflecting higher pressure equilibrium conditions are invariably enriched in the least soluble volatile components. That explains the relevance of measuring major volatile ratios such as CO_2/H_2O and CO_2/SO_2, (or C/S) as well as ratios involving minor volatile components such as He/CO_2, He/Ar and He/Ne. In all such quantities, the concentration of volatiles with markedly different solubility are compared. If, for example, a degassing magma moves towards the surface, then the associated decompression will cause variations which are amplified in the example ratios above, and we will "see" the rise of magma in a progressive or sudden decrease of, e.g., the CO_2/SO_2 ratio (CO_2 being much less soluble in common natural magmas than SO_2). Conversely, if a deeper magmatic body starts releasing gases, those gases will be seen at the surface as an increase in the ratio of less soluble to more soluble components, e.g., an increase in the CO_2/SO_2 ratio, reflecting the deeper conditions at which the gases have been released by the magma. Obviously, the simple patterns described here can be jeopardized by several other factors which affect magma degassing as well as the composition of fluids reaching the surface. Accordingly, the interpretations usually employ several indicators and constraints from multi-parametric monitoring systems.

In general, the magmatic gases derive from a level of gas–melt separation which can reasonably be a foam layer at the roof of a magma batch or a sill-like body (Jaupart and Vergniolle 1989; Menand and Philips 2007), so that the gas ratios can be representative of the pressure at this level. If instead the magmatic gases are released from magma distributed over substantially elongated geometries, then the measured gas ratios are representative of integrated conditions along the magma column (Edmonds 2008). Horizontally elongated structures can represent traps where magma ascent is inhibited and bubbly suspensions can develop at the top of the reservoirs by accumulation of gas bubbles (Menand and Philips 2007), and this type of structures can develop more and more times along a composite magma plumbing system. As a consequence, the modeling and interpretation of measured quantities requires multiple steps to account for repeated decoupling in the bubble–melt system (Edmonds et al. 2010; Paonita et al. 2012, 2013). Additional mechanisms for gas escape can involve the development of permeable fault structures at the conduit walls (Jaupart and Allegre 1991; Kozono and Koyaguchi 2012) and, in highly viscous silicic systems, the formation of a bubble network with hydraulic continuity in a magma column (Eichelberger et al. 1986; Burton et al. 2007; La Spina et al. 2017).

The number of applications of the above concepts and models on volcano degassing is impressive, and has provided enormous information on magmatic processes constraining degassing pressures and depths, composition and temperature of the degassing magmas, total volatile contents in magmas and in degassing mantle sources, and many others. Observations carried out on fluid emissions from many volcanoes all over the world suggest that the CO_2/SO_2 ratio is a more useful indicator for volcanic early warning systems than the $CO_2/(SO_2+H_2S)$ ratio. The latter, instead, can reflect the specific geodynamic context over which the volcanism develops (e.g., Aiuppa et al. 2014). As anticipated above, the CO_2/SO_2 ratio before an eruption may decrease as a consequence of magma rise and associated shallower magma degassing, or increase due to increased contribution to the volcanic emissions by magmatic gas from a deep magma source (sulfur scrubbing may represent a further complication, especially for calderas where hysrothermal systems can be highly developed). More likely, the observed variations at the surface are a combination of these (and possibly others, e.g., *en masse* microlite crystallization upon magma ascent) processes, and robust interpretations require evaluations of multi-parametric observations and data. In various cases, however, compositional shifts towards increased CO_2/SO_2 ratios have been observed at open system volcanoes including Villarrica, Chile (Aiuppa et al. 2017), Masaya, Nicaragua (Aiuppa et al. 2018), Redoubt, Alaska (Werner et al. 2013), Etna and Stromboli, Italy (Aiuppa et al. 2007, 2008, 2009b, 2010a,b) (Fig. 42). Current interpretations, also based on measured H_2O/CO_2 ratios, suggest that volcanic plumes emitted at the surface result from mixing of deep (CO_2-rich) and shallow (CO_2-poor) gas phase contributions. This emerges clearly at Stromboli, where measured plume compositions distribute along the computed curve of mixing between gases exsolved at high and low pressure below the volcano (Fig. 42b). An increased contribution from deep magma degassing, hinted at by increased CO_2/SO_2 ratio, is often observed shortly before the occurrence of a new eruption, and shortly before major explosive events at Stromboli (Fig. 42a) as well as other volcanic systems (e.g., at Villarica, Chile, Fig. 43). Conversely, phreatic and phreatomagmatic eruptions in 2014, 2016 and 2017 at Poàs, Costa Rica, were accompanied or preceded by large decreases in the CO_2/SO_2 ratio (de Moor et al. 2019).

At Etna, CO_2/SO_2 and H_2O/CO_2 ratios measured in the gas plume above the crater in the period 2005–2006 (Aiuppa et al. 2007) revealed consistent changes (when the former increases the latter decreases, according to the solubility relationships between the three volatile components). Application of the COHS model allowed the identification of periods during which volcano degassing was mainly fed by slowly convecting magma close to atmospheric pressure, and periods of resumption of fresh magma supply that anticipated by a few days the occurrence of a summit eruption. In some other situations the CO_2/SO_2 ratio is found to evolved over months or years, e.g. at Redoubt and Kilauea (Werner et al. 2019), interpreted to reflect decompression of magma during its slow ascent to shallow crust levels beneath Redoubt (Werner et al. 2012), and mantle-driven surge in magma supply to Kilauea during 2003–2007 (Poland et al. 2012).

While the CO_2/SO_2 ratio seems particularly relevant as a messenger of magma ascent and increased potential for a new eruption, the H_2S/SO_2 ratios can also be important, especially in those many situations where a shallow hydrothermal system controls the volcanic emissions. The fumarole emissions at Campi Flegrei display practically only H_2S as sulfur component (Chiodini et al. 2010), as a consequence of chemical reactions occurring in the highly reduced hydrothermal environment below the caldera floor. Accordingly, one of the most relevant signs of increased hazard from an impending eruption is believed to be the appearance of SO_2 in the fumarole gases (e.g., Selva et al. 2012), as that would suggest an increased magmatic gas input into the hydrothermal system beyond its reducing capabilities. At the Merapi volcano, Indonesia, increase of both CO_2/SO_2 and H_2S/SO_2 ratios in fumaroles during the months preceding the 2010 explosive eruption were interpreted as an increased contribution from a deep, more reduced source (likely mafic magma) with respect to the shallower source feeding the fumaroles during "normal" time (Surono et al. 2012; Pritchard et al. 2019).

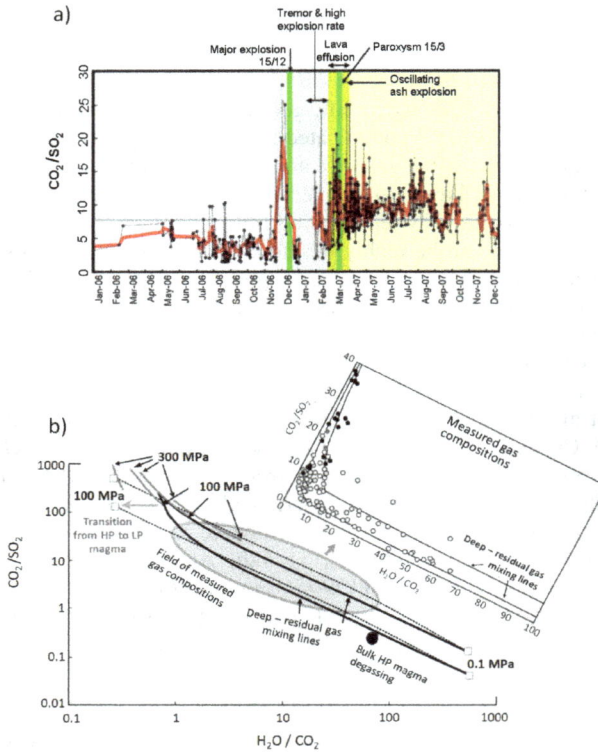

Figure 42. a) Measured CO_2/SO_2 ratio in the volcanic plume above the Stromboli crater in the period 2006–2007, encompassing the major explosion on December 15, 2006, and the paroxysm on March 15, 2007. From Aiuppa et al. (2009b). **b)** Modeled CO_2/SO_2 and H_2O/CO_2 covariations. The **elliptic gray area**, expanded in the upper-right diagram, indicates the range of measured values. **Grey solid lines** refer to modelled gas compositions in the 300–100 MPa range for crystal-poor (low-porphiricity, LP) magma feeding major explosions and paroxysms at Stromboli. **Thin black dashed** lines refer to model results for a magmatic gas formed by decompression (100 MPa to 0.1 MPa) of the degassed, crystal-rich (high-porphiricity, HP) magma occupying the upper volcanic domain. Also traced are lines of mixing line between deep CO_2-rich gas bubbles and shallow gas from the largely degasses upper magma (**thick black lines**). "Bulk HP magma degassing" (**black filled circle**) refers to the possible composition of the gas phase from the shallow magma at 0.1 MPa. In the top-right enlargement, the syn-explosive (**black circles**) and quiescent (**open circles**) gas data show distinct (although partially overlapping) composition. Modified after Aiuppa et al. (2010b).

A general situation for gas emissions at many volcanoes can be that they are fed from both residual degassing of shallow, partially degassed, partially crystallized magma (low CO_2/SO_2 and H_2S/SO_2, high H_2O/CO_2) and deep, volatile-rich, often less chemically evolved magma (high CO_2/SO_2 H_2S/SO_2, low H_2O/CO_2). As it is shown in Fig. 42, changes in measured ratios can reflect changes in the relative contributions, with the further complications that i) decompression during ascent of the deep magmatic component results in the production of less CO_2-rich gas, and ii) the evolution of the redox conditions during magma decompression, thus the evolution of the H_2S/SO_2 ratio in the gas phase accompanying a rising magma, can be very complex. Multi-component, composition-dependent, volatile–melt saturation models can be used to link redox variations observed in gases and magmas down to the magma source, leading to backtrack magmatic degassing from source to surface. Relevant examples are

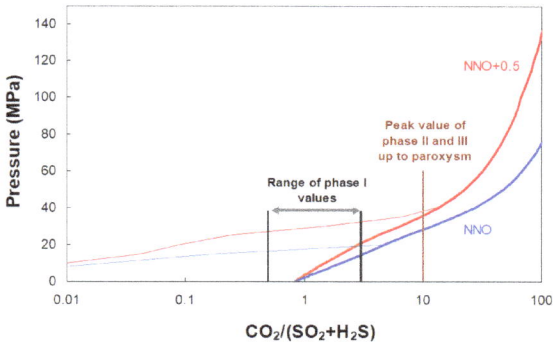

Figure 43. Modelled evolution of the CO_2/S_{tot} ($S_{tot} = SO_2 + H_2S$) ratio in the magmatic gas coexisting with a Villarrica-like melt. Closed-system simulations (**thick blue line**: NNO; **thick red line**: NNO+0.5) of magma depressurization (200 to 0.1 MPa), and open-system simulations (**thin lines**, from 40 MPa for the case NNO+0.5, and from 20 MPa for the case NNO), compared to measured values. Closed-system degassing paths were drawn based on melt inclusion data. According to these data, the two redox conditions define limiting C/S_{tot} values at each pressure step. Measured paroxysmal C/S_{tot} values may then represent single-step-degassing taking place along closed-system patterns. Open system degassing paths were chosen to bracket measured values at the crater during quiet degassing but also to intercept paroxysmal C/S_{tot} values. Background (Phase I) degassing ($CO_2/S_{tot} = 0.8$–3) compares well with residual open-system degassing. Phase II and particularly phase III peak values up to 10 are anticipatory of eruptive activity and show that the CO_2-enriched gas was likely caused by separation of gas originating at 20–35 MPa. Modified after Aiuppa et al. (2017).

provided in Aiuppa et al. (2007, 2010b, 2011), Oppenheimer et al. (2011), and Moussallam et al. (2014), showing that other processes can affect the chemistry of sulfur-bearing volcanic gases, such as precipitation of solid sulfide/sulfate phases as well as immiscible FeOS liquid. Despite not accounted for in current procedures such as COHS, SolEx, or D-COMPRESS, described above, precipitation of S-bearing phases is likely to be a major factor in many natural situations, besides being an important constraint on oxygen fugacity for multi-component gas–melt equilibrium thermodynamics.

Using noble gases with major volatiles

Decompression-induced fractionation among volatile components having different solubility obviously involves minor components, like noble gases, together with the major components considered above. Accordingly, poorly soluble, heavy noble gases (Ar, Kr, Xe) are exsolved from the melt earlier than more soluble He and Ne. Accordingly, the relative abundance of noble gases in volcanic emissions is as much powerful as that of major components for constraining the underground dynamics.

For closed system degassing, fractionation between two noble gases i and j can be written as (Jambon et al. 1986):

$$\frac{x_i}{x_j} = \frac{x_i^T}{x_j^T} \frac{S_i}{S_j} \left[\frac{V^* + \rho^L S_j T_e / T^0}{V^* + \rho^L S_i T_e / T^0} \right] \qquad (228)$$

where $x_{i,j}$ and $x_{i,j}^T$ are noble gas concentration in the melt and total system (i.e., the undersaturated melt), respectively, V^* is the volume of exsolved gas phase per unit volume of melt, representing a proxy of exsolution and degassing, T_e is the temperature of gas–melt equilibrium, T^0 is a reference temperature, ρ^L is melt density and $S_{i,j}$ is the solubility constant $= x_{i,j}/P_{i,j}$, corresponding to the inverse of the Henry's constant K_h defined at Equation (207).

Equation (228) assumes ideal behavior of the gas phase, therefore it is only adequate for sufficiently low pressure conditions. For an infinitely efficient open (Rayleigh) degassing system (for which each infinitesimal parcel of gas is immediately removed from near the melt upon exsolution), noble gas fractionation becomes:

$$\frac{x_i}{x_j} = \frac{x_i^T}{x_j^T}\left(\frac{x_i}{x_i^T}\right)^{1-S_i/S_j} \tag{229}$$

For both ideal closed (Eqn. 228) and open (Eqn. 229) conditions, concentrations in the gas phase (either coexisting with the melt or leaving the melt), $y_{i,j}$, can be computed from:

$$\frac{y_i}{y_j} = \frac{x_i}{x_j}\frac{S_j}{S_i} \tag{230}$$

Computations of noble gas fractionation from Equations (228–230) are reported in Figure 44, showing much higher efficiency of open system degassing in fractionating volatile components having different solubility.

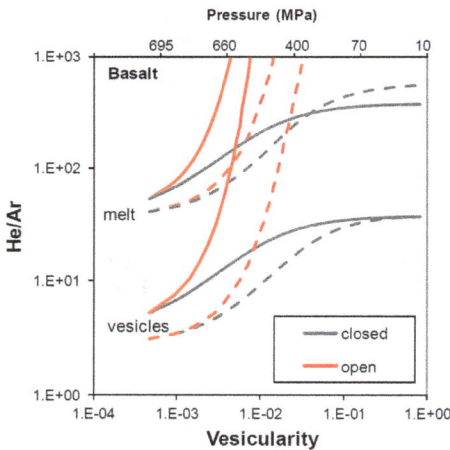

Figure 44. Closed and open system degassing paths for He/Ar for a basaltic melt at 1200 °C. In the case of open system degassing, vesicularity refers to the cumulative volume of gas exsolved (higher volumes approaching unity correspond to high values of He/Ar out of the vertical axis scale). **Solid curves** are computed by Equations (228) (**closed**) and (229) (**open**) with a He/Ar solubility ratio of 10. **Dashed curves** are computations with the Nuccio and Paonita (2001) model with correction for Ar solubility as in Paonita et al. (2012) (model parameters: initial pressure 700 MPa, 3 wt% total H_2O, 0.65 wt% total CO_2). Pressure reported on figure top refers to only the dashed model curves. The mismatch between **dashed and solid curves** is due to dependence of noble gas solubility on the degassing pressure and H_2O–CO_2 contents and degassing (see text).

The value of V^* in Equation (228) is itself a complex function of the degree of magma decompression below saturation pressure, therefore, it is poorly adequate to represent an independent proxy of volatile exsolution (Guillot and Sarda 2006). More refined calculations take into account the evolution of the relevant compositional and thermodynamic quantities with decompression (Nuccio and Paonita 2001; Guillot and Sarda 2006). For closed system conditions, and considering the two major volatiles H_2O and CO_2, the fractionation between two noble gases can be written (Nuccio and Paonita 2001):

$$\frac{y_i}{y_j} = \frac{x_i^T}{x_j^T}\frac{K_{h,i}}{K_{h,j}}\frac{\left(x_{H_2O}^T - x_{H_2O} + x_{CO_2}^T - x_{CO_2}\right)K_{h,j} - \phi_j P\left(x_{H_2O}^T + x_{CO_2}^T - 1\right)}{\left(x_{H_2O}^T - x_{H_2O} + x_{CO_2}^T - x_{CO_2}\right)K_{h,i} - \phi_i P\left(x_{H_2O}^T + x_{CO_2}^T - 1\right)} \tag{231}$$

where the thermodynamic parameters $K_{h,i,j}$ and $\phi_{j,j}$ change with the P–T conditions according to the new H_2O and CO_2 contents in melt and vapor. For an open system, the fractionation among noble gases as a function of magma depressurization can still be computed by step-wise solutions of Equation (231) with removal of the degassed (i.e., lost from the system) amount of volatiles at each computational step (Fig. 44). These calculations require coupling Equation (231) to a model for the saturation of major volatiles and noble gases (see above).

The application of the above concepts and equations to noble gas abundance in volcanic emissions from long time series of data provides relevant insights into the plumbing system and the deep processes at active volcanoes. At Mt. Etna (Italy) long datasets from sampling of low-temperature soil gas emissions and fumaroles along the rim of the Voragine summit crater constitute a unique dataset of chemical compositions and of isotope and elemental abundances of Ar, Ne and He (Caracausi et al. 2003a,b; Rizzo et al. 2006; Paonita 2010; Paonita et al. 2012, 2016). Due to interaction with shallow thermal aquifers and ground water, fumarolic gases and even more, soil gas emissions must be evaluated and filtered to retrieve the magmatic gas component (Liotta et al. 2010; Paonita et al. 2012). Paonita et al. (2012) calculated the decompressive degassing path of Etnean trachybasalts in the system H_2O–CO_2–He–Ar by coupling the SOLWCAD model (Papale et al. 2006) for H_2O–CO_2 saturation and the Nuccio and Paonita (2000) model for noble gases, the latter implemented with updated equations for Ar solubility in H_2O–CO_2-bearing basalts. Olivine-hosted melt inclusion data from the most primitive Etnean magmas, processed as shown above, show entrapment pressures up to 450 MPa at 1150 °C (this temperature being estimated from solid–melt equilibria) and further suggest that the earliest exsolved gas phase contains 75 mol% CO_2. Data from fluid inclusions in olivine from the same primitive magmas allow estimation of initial noble gas and major volatile ratios. With these constraints, isothermal decompressive degassing paths can be computed assuming either closed or open system conditions with respect to the gas phase. Figure 45 shows the results of such calculations on a typical He/Ar vs Ar/CO_2 graph compared with gas emission measurements, suggesting that open system degassing is consistent with the observed one order of magnitude variability in the sampled gases. Summit crater fumaroles align on the same degassing path as for the peripheral emissions, with the former corresponding to shallower degassing levels. Crater fumarole compositions were found to be consistent with mixing between a deep gas released at 200–400 MPa and a shallow gas released at about 130 MPa. Accordingly, measured abundance ratios, like He/Ar ratios, do not depend solely on exsolution and degassing pressure, but mostly reflect variable relative contributions from the deep and shallow magma reservoirs supplying gas at different times.

Fumaroles from the Solfatara crater at Campi Flegrei, Italy, provide a 30 year-long time series of N_2, CO_2, He, and Ar data (Caliro et al. 2007, 2014; Chiodini et al. 2010, 2012). The data show a decrease during last 10 years of the N_2/CO_2, and Ar/CO_2 ratios, and increase of the He/CO_2 and He/Ar* ratios. Caliro et al. (2014) used the above open-system degassing model together with noble gas solubility data and modeling from Iacono-Marziano et al. (2010) and Paonita et al. (2012) to simulate decompressive H_2O–CO_2–He–Ar–N_2 magma degassing, with initial values of the N_2/CO_2, Ar/CO_2 and He/CO_2 ratios as fitting parameters. Comparison with solubility data from the literature indicated a trachybasaltic composition for the degassing magma, while the degassing pressure and initial CO_2 concentration in the gas phase remained unconstrained (Fig. 46). Inclusion of observed volatile output rates among the model constraints led to the conclusion that the degassing source is located at a pressure of about 300 MPa, roughly corresponding to 12 km depth. That conclusion would exclude that a shallow magma at 3–4 km depth, thought to have been emplaced during the 1982–84 bradiseismic crisis, can be the major source of magmatic volatiles at Campi Flegrei. A scenario involving mixing at shallow pressure (< 100 MPa) between gases released at different depths in the Campi Flegrei plumbing system more consistently explains the observations (Fig. 46). The picture proposed by Caliro et al. (2014) is that of a sill-like reservoir at a depth of 3–4 km as the shallow source of volatiles, fluxed by gas coming from a deep magmatic source providing the main contribution to the surface emissions. Periodic inflation of the caldera floor (Chiodini et al. 2010) would then correspond to periodic increase in the gas flux from the deep reservoir.

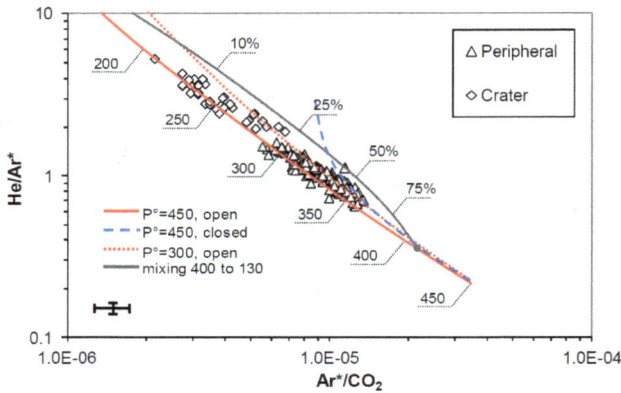

Figure 45. He/Ar vs. Ar/CO₂ decompressive or mixing degassing paths for Etnean basalt at 1150 °C, as computed by Paonita et al. (2012). Ar* is radiogenic Ar, or Ar corrected by air addition. **Lines** named "P°=450, open" and "P°=450, closed" refer to open and closed system degassing, starting from an initial pressure of 450 MPa (labeled numbers are pressure in MPa). The **line** "P°=350, open" is open system degassing from an initial pressure of 350 MPa. The line "mixing 400 to 130" refers to binary mixing between two gas phases exsolved at 400 and 130 MPa **pressure**, respectively (labeled percentages refer to the high pressure end-member). **Symbols** are measured compositional ratios in peripheral emissions and summit crater fumaroles at Mt. Etna.

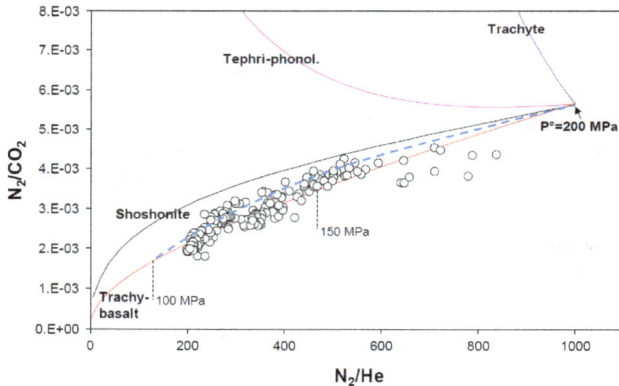

Figure 46. N₂/CO₂ vs. N₂/He decompressive or mixing degassing paths for Campi Flegrei magmas, as computed by Caliro et al. (2014). All decompressive paths (**solid curves**) were performed starting from 60 mol% CO₂ in an initial gas phase at 200 MPa. Ar solubility is used as a proxy for N₂ solubility. The **dashed line** refers to mixing between the assumed initial gas phase and a gas phase exsolved at 100 MPa (33.5 mol% CO₂), both computed from equilibrium with an ascending trachy-basalt. **Circles** are data from fumaroles of La Solfatara crater at Campi Flegrei. Only a trachybasaltic magma composition fits the observed data, while decompressive and mixing paths are poorly distinguishable. Hydrothermal contamination of the magmatic gas ratios is negligible (Caliro et al. 2014).

The examples above illustrate the capability of multi-component gas–melt equilibrium modeling to provide a robust framework for the interpretation of data from volcanic fluid emissions, by constraining major quantities characterizing the volcanic system as well as suggesting major processes occurring from deep regions of the volcanic plumbing system to the surface. Conceptual as well as physico-mathematical models describing a given volcanic system must comply with such constraints, and with many others coming from other disciplines and employing other parameters. In general, multi-component gas–melt equilibrium thermodynamics provides major constraints on total volatile abundances as well as dissolved and exsolved amounts at specific *P–T*–composition conditions, location (in *P* terms) of regions of magma storage and crystallization and intensive quantities characterizing them, extent and characterization of open vs. closed system conditions (relative to the gas phase) across vertically extended plumbing systems, vertical movements of degassing magmas, and others. Accordingly, multi-component gas–melt equilibrium thermodynamics constitutes one major tool to investigate volcanic systems, penetrate their deeper features, recognize ongoing deep and shallow processes undergoing, and support volcanic early warning systems.

Constraining and modeling magma and eruption dynamics

The model applications above illustrate the variety of insights into the complexities of volcanic systems that can be acquired through the study of volatiles, either dissolved in the melt or exsolved in a gas phase. To complete the panorama of the potential applications of volatile saturation models, this section explores instead magma and eruption dynamics, illustrating some of the major roles of magmatic volatiles that have been unveiled by coupling volatile saturation modeling with models of magma and volcano thermo-fluid dynamics.

As stated in the Introduction section, magmatic volatiles are the fuel triggering magma motion and volcanic eruptions. In fact, magmatic systems can be visualized as a reactor forcing magma to be transferred from crustal levels to the surface. The energy comes from the strong dependence of volatile saturation on pressure. Imagine we have a magmatic body somewhere in the crust, that occupies a certain region with some vertical extension. The upper part of the magmatic system is subject to a lower pressure with respect to its deeper portions, as a consequence of increasing magma-static pressure when moving downwards. Because of the presence of poorly soluble volatile components, and particularly carbon dioxide—the second most abundant volatile component after water in most natural magmas—the depth at which saturation conditions are reached is usually very large (Fig. 47). Thus, most natural magmas are over-saturated at crustal level, and the proportion of the gas phase increases with decreasing depth. Gas expansion upon decompression adds to the increase of gas volume and decrease of magma bulk density. In principle, gas bubbles in magmas tend to move upwards under the buoyancy force; however, that force is counteracted by viscous forces, which are more efficient for smaller gas bubbles. In fact, viscous forces depend on surface, which varies with the square of the radius, whereas buoyancy forces depend on volume, thus on the cubic power of the radius. As a consequence, an increase in bubble radius always favors buoyancy over friction. Accordingly, differential movement of gas bubbles is more effective at lower pressure where the gas volume fraction and the size of gas bubbles are larger. While shallow magmatic bodies provide much of the surface volcanic emissions, deep sources can also be effective, adding gas enriched in deep-released, poorly soluble components like carbon dioxide and heavy noble gases, as largely discussed in the previous sections. Loss of gas at shallow level in turn implies local density increase, which can destabilize the magmatic column forcing batches of shallow, degassed, dense magma to sink into the underlying magma, and triggering a convection mechanism whereby new volatile-rich, lighter magma rises up through the magmatic column feeding new magma degassing and repeating the mechanism again and again (Fig. 48). Similar dynamics can be observed at volcanic lakes, where the magma column reaches the surface: batches of gas-rich magma and large gas bubbles periodically reach the lake surface, accompanied by rapid sink of pre-existing, degassed, partially cooled, denser magma forming the previous lake surface. The new magma, now degassed, contributes to the new lake surface, until it sinks back into the magmatic column upon arrival of new gas-rich batches, continuously rejuvenating the lake surface (e.g., Birnbaum et al. 2020).

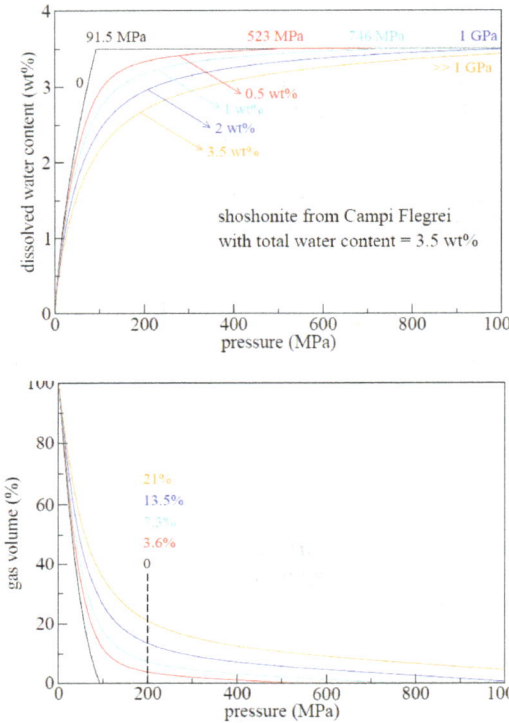

Figure 47. Calculated saturation conditions for a shoshonite from Campi Flegrei at T =1400 K (melt composition from Mangiacapra et al. 2008). **a)** Water saturation content as a function of pressure, for assumed total water content of 3.5 wt% and different total carbon dioxide contents from 0 to 3.5 wt% (**colored numbers corresponding to same color lines**). Also reported are the exsolution pressure corresponding to each line, with corresponding colors. The inclusion of only 0.5 wt% total carbon dioxide results in a nearly 6-fold increase of the saturation pressure, corresponding to a depth increase of about 17.5 km. **b)** Gas volume as a function of pressure. Shoshonitic magma is thought to exist in a large reservoir below Campi Flegrei at a pressure close to 200 MPa (about 8 km depth) (Mangiacapra et al. 2008). The shoshonite within such a reservoir would be largely under-saturated for the employed conditions, while it would contain more than 20 vol% gas for a total carbon dioxide content equal to the total water content. Calculations made with SOLWCAD.

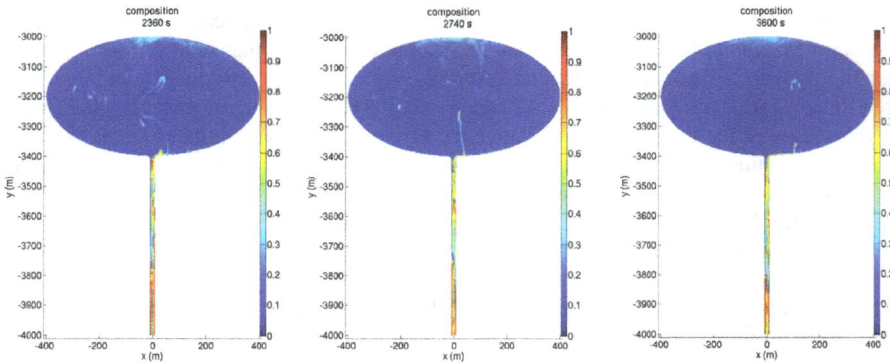

Figure 48. Buoyancy-induced magma chamber replenishment by volatile-rich magma. The three panels refer to different times (indicated in the panels) after initial encounter between a volatile-rich shoshonite (**red**) initially filling the dyke, and a partly degassed phonolite (**blue**) initially occupying the magma chamber. The **color scales** refer to composition (1 = pure shoshonite, 0 = pure phonolite). Discrete plumes of shoshonite-rich magma rise through the chamber and mix with the phonolitic magma giving rise to complex convective patterns. The density difference due to different volatile contents (H_2O and CO_2) provides all of the buoyancy force driving the observed dynamics. Coupling of multicomponent volatile saturation modeling (SOLWCAD) with thermo-fluid dynamic modeling allows computation of the amount and composition of dissolved and exsolved volatiles at any point in space and any instant of time, for any combination of the two magma types, providing therefore with a robust picture of the density and viscosity relationships controlling the convection and mixing dynamics. After Papale et al. (2017).

The convective processes described above, triggered and controlled by magmatic volatiles, explain a number of major direct observations and reconstructions at active and ancient volcanoes, among which: gas discharge often far greater than it can be accommodated by the mass of the erupted magma (e.g., Allard et al. 1994; Gerlach and McGee 1994); periodic injection of new magma batches inside magmatic reservoirs (e.g., Bachmann et al. 2002; Bain et al. 2013); generation of alternate rims of growth and dissolution in crystals sensitive to the amount and activity of water and/or other volatiles (e.g., Wark et al. 2007; Pizarro et al. 2019); CO_2 over-enrichments in melt inclusions (e.g., Barsanti et al. 2009; Caricchi et al. 2018); and in general, efficient mixing of compositionally different magmas erupted at many volcanoes worldwide (e.g., Nakamura 1995; Griffin et al. 2002; Hodge and Jellinek 2020). As shown in the previous section, such processes involving magma degassing and associated magma convection dynamics can also explain complex evolutions in the composition of volcanic gases released at the surface, which are in fact commonly interpreted in terms of magmatic movements and employed within early warning systems.

While volatile exsolution and magma degassing take place, virtually all major magmatic properties evolve. Dissolved water strongly affects magma density and viscosity which play major roles in magma and eruption dynamics. The discussion above illustrates some aspects of the major role played by volatiles on magma density. Magma ascent along a volcanic conduit is a relevant example to illustrate the consequences of degassing-induced viscosity changes on eruption dynamics.

Figure 49 shows the control of dissolved water on melt viscosity, for a variety of natural magmas from mafic to silicic ones. In all cases, a large (one–two orders of magnitude) increase in melt viscosity correlates with water exsolution from the melt, especially below about 50 MPa (roughly 2 km depth). The overall viscosity difference between a fully degassed and water-rich melt largely depends on melt composition, and is much higher for more silicic (more polymerized) melts (Stevenson et al. 1998; Romano et al. 2003; Giordano et al. 2004a,b).

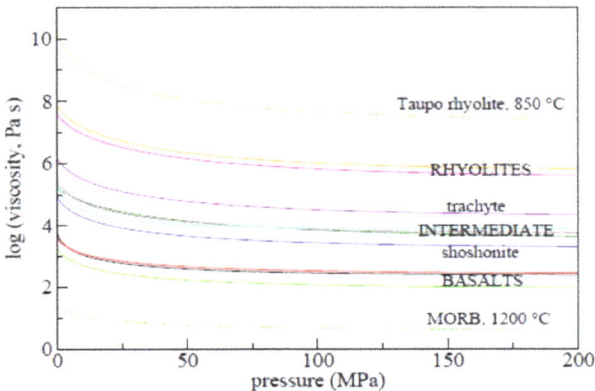

Figure 49. Calculated viscosity of various melts as a function of pressure, by using the Giordano et al. (2008) model for melt viscosity together with SOLWCAD for $H_2O + CO_2$ saturation. In order to highlight the role of compositional differences, all calculations consider the same total volatile contents corresponding to 4 wt% H_2O, 1 wt% CO_2, and same temperature equal to 1100 °C. The viscosity curves related to the most and least viscous compositions among those considered here (the Taupo rhyolite and MORB, respectively), are also shown at a temperature closer to the corresponding natural conditions for the two magma types, expanding significantly the range of natural melt viscosities. Melt compositions from: Etna basalt: Spillaert et al. (2006), Kilauea basalt: Hauri (2002); MORB: Mortimer et al. (2012); shoshonite: Mangiacapra et al. (2008); latite: Di Matteo et al. (2006); andesite: Baker (1982); phonolite: Carroll and Blank (1997); trachyte: Mangiacapra et al. (2008); Pantellerite: Landi and Rotolo (2015); Krafla rhyolite: Masotta et al. (2018); Taupo rhyolite: Yamashita (1999).

Such relationships can be appreciated by taking in mind the depolymerization effects of water dissolved in silicate melts, largely discussed above. The relevance of melt viscosity increase upon water exsolution can be appreciated by simply considering the outstanding role played by magma viscosity on the dynamics of volcanic eruptions, which correlates with a tendency to produce explosive eruptions. Other factors can play important roles in determining the style of an eruption, but among them, viscosity tends to be a dominant one (e.g., Polacci et al. 2004). That is illustrated in the following, where multi-component volatile saturation is used in conjunction with magma ascent dynamics for explosive volcanic eruptions.

Figure 50 shows the calculated 1D multiphase flow dynamics of magma ascending along a volcanic conduit during steady phases of volcanic eruptions. Steady flow refers to a kind of idealized dynamics whereby flow variables and properties change in space but not in time. That is a condition commonly assumed to describe the dynamics during well-developed phases of volcanic eruptions, for which relevant changes in the overall dynamics are observed on a time scale sufficiently long (significantly longer than the transit time of the fluid across the simulated domain). The depicted dynamics are entirely driven by volatile exsolution, without which there would not be any velocity increase along the conduit (panel b), in turn causing high rates of strain and forcing magma to cross the boundary of rate-limited glass transition (Dingwell 1996) resulting in magma fragmentation characterizing all large explosive eruptions (Papale 1999b).

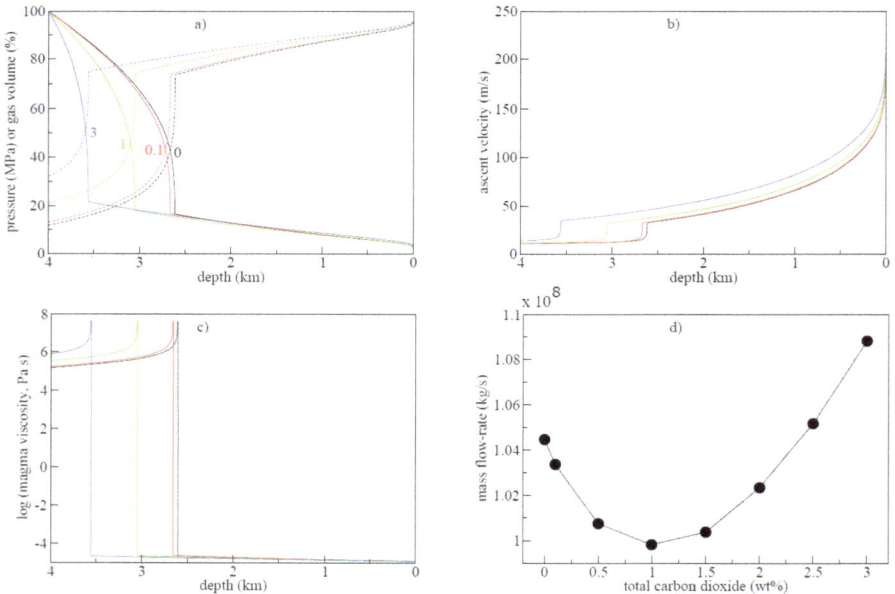

Figure 50. Volatile control on the dynamics of magma ascent. The simulations refer to a rhyolitic melt at 830 °C that flows along a cylindrical conduit with an 80 m diameter from a magma chamber at 100 MPa and 4 km depth. The total (i.e., in the melt+gas system) water content is fixed at 5 wt% in all simulations, while the total carbon dioxide content is varied from zero to 3 wt% (numbers on the curves with corresponding color in panel a). **a)** The **solid lines** are pressure, and the **dashed lines** are gas volume. For the curves in this panel, as well as in panels b) and c), the **singular point with sudden slope** change marks the occurrence of magma fragmentation. **b)** melt ascent velocity. **c)** Magma viscosity (including the effect of non-deformable gas bubbles). Note that the magma viscosity drops by > 12 orders of magnitude across fragmentation, as a consequence of rupture of the highly viscous melt continuum and replacement with hot gas continuum. **d)** Mass flow-rates resulting from the simulations displayed in panels a–c) (plus others not displayed) as a function of total carbon dioxide in magma. Simulations done with the CONDUIT4 model and code (Papale 2001), which embeds the SOLWCAD code for mixed $H_2O + CO_2$ saturation.

The dynamics illustrated in Figure 50, the relationships between flow variables and properties, and the critical role played by volatile exsolution are epitomized in the flow diagram in Figure 51. In order to understand the diagram, it must be noted that steady flow dynamics implies the same mass flow-rate at any level along the volcanic conduit. However, although the arguments below strictly apply to steady flow, the described relationships among flow variables and properties provide a general framework to understand the control by magmatic volatiles on magma ascent and eruption dynamics.

As soon as the magma starts to move upwards, its pressure decreases, and the associated volatile exsolution gives rise to a series of cascading events augmented by the very effective feedback mechanism depicted in Figure 51. Volatile exsolution has two major consequences: 1) it decreases magma bulk density, thus by continuity, increasing magma velocity to sustain the same mass flow-rate, and 2) it increases melt viscosity, Figure 49c. The bulk viscosity, or the viscosity of the gas–melt (and crystals) mixture is in Figure 50c, and increases even more than melt viscosity as long as the capillary number (Ca) associated with gas bubbles is smaller than 1, a situation which holds for most silicic magmas under a large range of conditions (Colucci et al. 2017, present non-Newtonian flow simulations with $Ca \gg 1$ causing local bulk viscosity decrease). In turn, viscosity increase means an increase of the magnitude of frictional forces that dissipate the mechanical energy of the flowing fluid, resulting in further depressurization. That is the feedback mechanism evidenced in Figure 51: pressure decrease upon magma ascent causes a series of changes that end up with further pressure decrease, so that at a certain level, controlled by the pressure–volatile saturation relationships, the rate of pressure decrease becomes very large (in absolute value) (Fig. 50a), and with it, the rate of variation associated with any quantity related to dissolved and exsolved volatiles also largely increases (Fig. 50). Figure 51 illustrates further the associated flow evolution: because velocity increase is tied to the above feedback mechanism, and its rate of change is subject to the same increase as for all other flow variables, the rate of strain, equal to the spatial derivative of velocity, also increases. Dynamic models of magma fragmentation based on either ductile–brittle transition according to Maxwell's theory of visco-elastic fluids (Dingwell 1996; Papale 1999b) or to disequilibrium gas bubble pressure evolution (Melnik 2000) predict the occurrence of magma fragmentation when the product of viscosity and rate of strain overcomes a critical value (that depends on the elastic properties of magma, and for the bubble disequilibrium mechanism, also on kinetic factors). Figure 51 shows that both ingredients for magma fragmentation are there, and they are both related to volatile exsolution and linked together by the depicted feedback mechanism, which is itself a consequence of the pressure–volatile saturation relationships.

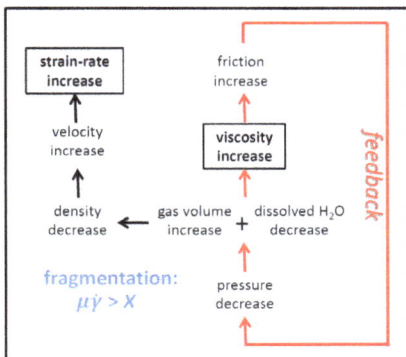

Figure 51. Major processes, and their relationships, during magma ascent towards the surface. A major feedback effect, depicted with red arrows and lines, causes increasingly large rates of change, as from the trends in Figure 50a-c. The criterion for fragmentation, reported in the figure, requires that the product of viscosity μ times strain-rate γ overcomes a critical value defining a rate-limited glass transition (Dingwell 1996). Both viscosity and strain-rate, framed and bold in the figure, increase faster and faster, eventually culminating into magma fragmentation. See text for further explanation.

The above dynamics and the relationships between flow variables and properties, illustrated in Figures 49–51, elucidate the fundamental control exerted by the magmatic volatiles on the dynamics of magma ascent and of volcanic eruptions. The feedback mechanism depicted in Figure 51 stands at the core of such dynamics, and it allows understanding of the major roles played by various quantities. As already discussed above, a critical role is played by magma viscosity, as it is easily understood from its variations spanning nearly ten orders of magnitude for natural magmas (Fig. 49c) and its central role in the overall magma dynamics (Fig. 51). For magmas with very low viscosity, the feedback mechanism depicted in Figure 51 is ineffective and magma fragmentation is not achieved (Papale 1999b). That explains the first order difference observed between basaltic (low viscosity) and rhyolitic (high viscosity) eruptions, with the former mostly characterized by relatively quiet flows of non-fragmented (liquid continuum) lava, and the latter instead largely consisting of explosive ejection of fragmented (gas continuum) pyroclasts.

Figure 50 shows what happens after magma fragmentation: the bulk viscosity drop associated with transition from the liquid to the gas continuum is dramatic, or about 12 orders of magnitude. While the dynamics of the bubbly magma below fragmentation, illustrated in Figure 51, are largely controlled by frictional forces, the same forces become (nearly) negligible above fragmentation, and the flow of magma becomes momentum-controlled. Accordingly, the rates of change of all flow variables suddenly decrease (Fig. 50) as the feedback mechanism in Figure 51 is disrupted. Large rates are restored only close to the conduit exit, and especially for magma ascent velocity, as a consequence of large gas phase expansion when approaching atmospheric pressure (note that the exit pressure in the illustrated simulations is larger than atmospheric, reflecting the achievement of sonic flow conditions at the exit plane associated with choked flow, e.g., Dobran 1992).

Because volatiles are central in the above mechanisms controlling magma ascent dynamics, their abundance and composition is one major factor determining the dynamics of volcanic eruptions. Papale et al. (1998) and Neri et al. (1998) have shown that an increase in the abundance of water results in larger mass flow-rate of the eruption and conditions favoring the sustainment of the volcanic cloud, which is instead progressively destabilized with decreasing water content, initially resulting in partial column collapses and associated pyroclastic flows, then producing total collapse of the volcanic jet above the volcanic vent with large pyroclastic flows when the water content is sufficiently low. Different volatile species, however, can have different effects on the eruption dynamics. Figure 50 shows the effects of adding carbon dioxide to the rising magma. The poorly soluble carbon dioxide component causes an earlier release of water (Fig. 47), with substantial consequences on the overall dynamics. In all simulations in Fig. 50 (as well as those in Papale and Polacci 1999), more carbon dioxide translates in substantial deepening of the fragmentation level, as a result of shifted pressure—gas volume relationships (Fig. 47b). The eruption mass flow-rate is found to vary in a highly non-linear way (Fig. 50d). In fact, on one side more carbon dioxide means less dissolved water thus higher viscosity for a given pressure, resulting in larger friction forces responsible for less efficient flow and deepening of the fragmentation level according to the mechanisms in Figure 51. On the other hand, deepening of the fragmentation level implies a longer stretch of conduit where friction forces are negligible and thus the flow is more efficient. The interplay between these two contrasting effects determine the non-linear trend of mass flow-rate with carbon dioxide abundance in magma, shown in Figure 50d. The resulting sub-aerial dynamics of the mixture of gas and pyroclasts leaving the vent, however, is scarcely affected by the presence of carbon dioxide, and mostly controlled by water, over a range of explored water and carbon dioxide abundances for eruptions at Campi Flegrei (Esposti Ongaro et al. 2006).

CONCLUDING REMARKS

This chapter reviews much of our to-date understanding of multi-component gas–melt equilibrium in silicate systems, exploring the fundamental theoretical aspects and the developed modeling approaches, and providing examples of applications that highlight the great importance of volatiles in magmas. None of the several models illustrated, from those based on regular solution approaches allowing full consideration of non-ideal thermodynamics and compositional complexities of the silicate melts, to those based on reactive species approaches that have similarly evolved to high levels of sophistication, is as accurate and complete as we would like it to be. In thermodynamic terms, no model is fully consistent with the Gibbs–Duhem equation stating the conditions for internal thermodynamic consistency. Still, the progress during the last 30 years has been enormous, and forcedly, this review only partially accounts for any study, model or investigation involving magmatic volatiles. However, there is in the literature no other review paper, in our knowledge, that presents the fundamental theoretical aspects and the modeling approaches for multi-component volatile–melt equilibrium in silicate systems, with a systematic approach and at a level comparable with the one that is found here. Our intention has been that of providing the young generations with the resource that we would have mostly appreciated when we were young scientists ourselves, and our wish is that this chapter will help them exploit their enthusiasm and ideas and achieve far more advanced levels in modeling and understanding volatiles in magmas.

REFERENCES

Aiuppa A, Moretti R, Federico C, Giudice G, Gurrieri S, Liuzzo M, Papale P, Shinohara H, Valenza M (2007) Forecasting Etna eruptions by real-time observation of volcanic gas composition. Geology 35:1115–1118

Aiuppa A, Giudice G, Gurrieri S, Liuzzo M, Burton M, Caltabiano T, McGonigle AJS, Salerno G, Shinohara H, Valenza M (2008) Total volatile flux from Mount Etna. Geophys Res Lett 35:L24302

Aiuppa A, Baker DR, Webster, JD (2009a) Halogens in volcanic systems. Chem Geol 263:1–18

Aiuppa A, Federico C, Giudice G, Giuffrida G, Guida R, Gurrieri S, Liuzzo M, Moretti R, Papale P (2009b) The 2007 eruption of Stromboli volcano: insights from real-time measurements of the volcanic gas plume CO_2/SO_2 ratio. J Volcanol Geotherm Res 182:221–230

Aiuppa A, Burton M, Caltabiano T, Giudice G, Gurrieri S, Liuzzo M, Mure F, Salerno G (2010a) Unusually large magmatic CO_2 gas emissions prior to a basaltic paroxysm. Geophys Res Lett 37:L17303

Aiuppa A, Bertagnini A, Métrich N, Moretti R, Di Muro A, Liuzzo M, Tamburello GA (2010b) Model of degassing for Stromboli volcano. Earth Planet Sci Lett 295:195–204

Aiuppa A, Shinohara H, Tamburello G, Giudice G, Liuzzo M, Moretti R (2011) Hydrogen in the gas plume of an open-vent volcano, Mount Etna, Italy. J Geophys Res: Solid Earth 116:B10204

Aiuppa A, Robidoux P, Tamburello G, Conde V, Galle B, Avard G, Bagnato E, De moor JM, Martinez M, Munoz A (2014) Gas measurements from the Costa Rica–Nicaragua volcanic segment suggest possible along-arc variations in volcanic gas chemistry. Earth Planet Sci Lett 407:134–147

Aiuppa A, Bitetto M, Francofonte V, Velasquez G, Parra CB, Giudice G, Liuzzo M, Moretti R, Moussallam Y, Peters N, Tamburello G, Valderrama OA, Curtis A (2017) A CO_2-gas precursor to the March 2015 Villarica volcano eruption. Geochem Geophys Geosystems 18:2120–2132

Aiuppa A, de Moor JM, Arellano S, Coppola D, Francofonte V, Galle B, Giudice G, Liuzzo M, Mendoza E, Saballos A, Tamburello G, Battaglia A, Bitetto M, Gurrieri S, Laiolo M, Mastrolia A, Moretti R (2018) Tracking formation of a lava lake from ground and space: masaya volcano (Nicaragua) 2014–2017. Geochem Geophys Geosystems 19:496–515

Allard P, Carbonnelle J, Métrich N, Loyer H, Zettwoog P (1994) Sulphur output and magma degassing budget at Stromboli volcano. Nature 368:326–330

Alletti M (2008) Experimental investigation of halogen diffusivity and solubility in Etnean basaltic melts. PhD thesis, University of Palermo

Alletti, M, Baker DR, Scaillet B, Aiuppa A, Moretti R, Ottolini L (2009) Chlorine partitioning between a basaltic melt and $H_2O–CO_2$ fluids at Mount Etna. Chem Geol 263:37–50

Anderson AT Jr, Newman S, Williams SN, Druitt TH, Skirius C, Stolper E (1989) H_2O, CO_2, Cl and gas in Plinian and ash-flow Bishop rhyolite. Geology 17:221–225

Anderson OL (1995) Equations of State of Solids for Geophysics and Ceramic Science. Oxford University Press

Anderson L, Barth AP, Wooden JL, Mazdab F (2008) Thermometers and thermobarometers in granitic systems. Rev Mineral Geochem 69:121–142

Aubry GJ, Sator N, Guillot B (2013) Vesicularity, bubble formation and noble gas fractionation during MORB degassing. Chem Geol 343:85–98

Bachmann, O, Dungan MA, Lipman PW (2002) The Fish Canyon magma body, San Juan volcanic field, Colorado: Rejuvenation and eruption of an upper-crustal batholith. J Petrol 43:1469–1503

Bain AA, Jellinek AM, Wiebe RA (2013) Quantitative field constraints on the dynamics of silicic magma chamber rejuvenation and overturn. Contrib Mineral Petrol 165:1275–1294

Baker PE (1982) Evolution and classification of orogenic volcanic rocks. *In*: Andesites. Thrope RS (Ed) Wiley, New York, p 11–23

Baker DR, Alletti M (2012) Fluid saturation and volatile partitioning between melts and hydrous fluids in crustal magmatic systems: The contribution of experimental measurements and solubility models. Earth Sci Rev 114:298–324

Baker DR, Moretti R (2011) Modeling the solubility of sulfur in magmas: a 50-year old geochemical challenge. Rev Mineral Geochem 73:167–213

Barin I, Knacke O (1973) Thermochemical Properties of Inorganic Substances. Springer-Verlag, Berlin-Heidelberg-New York

Barin I, Knacke O, Kubaschewsky O (1977) Thermochemical Properties of Inorganic Substances. Supplement. Springer-Verlag, Berlin-Heidelberg-New York

Barsanti M, Papale P, Barbato D, Moretti R, Boschi E, Hauri E, Longo A (2009) Heterogeneous large total CO_2 abundance in the shallow magmatic system of Kilauea volcano, Hawaii. J Geophys Res 114:B12201

Beckett JR (2002) Role of basicity and tetrahedral speciation in controlling the thermodynamic properties of silicate liquids, part 1: the system $CaO–MgO–Al_2O_3–SiO_2$. Geochim Cosmochim Acta 66:93–107

Belonoshko A, Saxena SK (1992) A unified equation of state for fluids of C–H–O–N–S–Ar composition and their mixtures up to very high temperatures and pressures. Geochim Cosmochim Acta 56:3611–3626

Beermann O, Botcharnikov RE, Nowak M (2015) Partitioning of sulfur and chlorine between aqueous fluid and basaltic melt at 1050°C, 100 and 200 MPa. Chem Geol 418:132–157

Bernadou F, Gaillard F, Füri E, Marrocchi Y, Slodczyk A (2021) Nitrogen solubility in basaltic silicate melt - Implications for degassing processes, Chem Geol 573:120192

Berndt J, Liebske C, Holtz F, Freise M, Nowak M, Ziegenbein D, Hurkuk W, Koepke J (2002) A combined rapid-quench and H2-membrane setup for internally heated pressure vessels: description and application for water solubility in basaltic glasses. Am Mineral 87:1717–1720

Bertagnini A, Métrich N, Landi P, Rosi M (2003) Stromboli volcano (Aeolian Archipelago, Italy): An open window on the deep-feeding system of a steady state basaltic volcano, J Geophys Res 108 B7:2336

Birnbaum J, Keller T, Suckale J, Lev E (2020) Periodic outgassing as a result of unsteady convection in Ray lava lake, Mount Erebus, Antarctica. Earth Planet Sci Lett 530:115903

Blank JG, Stolper EM, Carroll MR (1993) Solubilities of carbon-dioxide and water in rhyolitic melt at 850 °C and 750 bars. Earth Planet Sci Lett 119:27–36

Blank JG, Brooker RA (1994) Experimental studies of carbon dioxide in silicate melts: solubility, speciation, and stable carbon isotope behavior. Rev Mineral 30:157–186

Boettcher AL, Wyllie PJ (1969) Phase relationships in the system $NaAlSiO_4–SiO_2–H_2O$ to 35 kilobars pressure. Am J Sci 267:875–909

Boettcher SL, Guo Q, Montana A (1989) A simple device for loading gases in high-pressure experiments. Am Mineral 74:1383–1384

Botcharnikov RE, Behrens H, Holtz F, Koepke J, Sato H (2004) Sulfur and chlorine solubility in Mt. Unzen rhyodacitic melt at 850 C and 200 MPa. Chem Geol 213:207–225

Botcharnikov RE, Freise H, Holtz F, Behrens H (2005) Solubility of C–O–H mixtures in natural melts: new experimental data and application range of recent models. Ann Geophys 48:633–646

Botcharnikov RE, Holtz F, Behrens H (2007) The effect of CO_2 on the solubility of $H_2O–Cl$ fluids in andesitic melt. Eur J Mineral 19:671–680

Bouhifd MA, Jephcoat AP (2006) Aluminium control of argon solubility in silicate melts under pressure. Nature 439:961–964

Bouhifd MA, Jephcoat AP, Kelley SP (2008) Argon solubility drop in silicate melts at high pressures: a review of recent experiments. Chem Geol 256:252–258

Bottinga Y, Javoy M (1990) MORB degassing: Bubble growth and ascent. Chem Geol 8:255–270

Bowen FL, Tuttle OF (1950) The system $NaAlSi_3O_8–KAlSi_3O_8–H_2O$. J Geol 58:489–511

Brawer SA, White WB (1975) Raman spectroscopy investigation of the structure of silicate glasses, I, the binary silicate glasses. J Chem Phys 63:2421–2432

Broadhurst CL, Drake MJ, Hagee BE, Bernatowicz TJ (1992) Solubility and partitioning of Ne, Ar, Kr, and Xe in minerals and synthetic basalt melts. Geochim Cosmochim Acta 56:709–723

Brooker RA, Wartho J-A, Carroll MR, Kelley SP, Draper DS (1998) Preliminary UVLAMP determinations of argon partition coefficients for olivine and clinopyroxene grown from silicate melts. Chem Geol 147:185–200

Bureau H, Métrich N (2003) An experimental study of bromine behaviour in water-saturated silicic melts. Geochim Cosmochim Acta 67:1689–1697

Bureau H, Keppler H, Métrich N (2000) Volcanic degassing of bromine and iodine: experimental fluid/melt partitioning data and applications to stratospheric chemistry. Earth Planet Sci Lett 183:51–60

Bureau H, Foy E, Raepsaet C, Somogyi A, Munsch P, Simon G, Kubsky S (2010) Bromine cycle in subduction zones through *in situ* Br monitoring in diamond anvil cells. Geochim Cosmochim Acta 74:3839–3850

Burgisser A, Scaillet B (2007) Redox evolution of a degassing magma rising to the surface. Nature 445:194–197

Burgisser A, Alletti M, Scaillet B (2015) Simulating the behavior of volatiles belonging to the C–O–H–S system in silicate melts under magmatic conditions with the software D-Compress. Computers Geosci 79:1–14

Burnard P (1999) The bubble-by-bubble volatile evolution of two mid-ocean ridge basalts. Earth Planet Sci Lett 174:199–211

Burnard P (2001) Correction for volatile fractionation in ascending magmas: noble gas abundances in primary mantle melts. Geochim Cosmochim Acta 65:2605–2614

Burnard P, Graham DW, Farley KA (2002) Mechanisms of magmatic gas loss along the Southeast Indian ridge and the Amsterdam-St. Paul Plateau. Earth Planet Sci Lett 203:131–148

Burnard P, Harrison D, Turner G, Nesbitt R (2003) Degassing and contamination of noble gases in Mid-Atlantic Ridge basalts. Geochem Geophys Geosystems 4:1002

Burnard P, Graham DW, Farley KA (2004) Fractionation of noble gases (He, Ar) during MORB mantle melting: a case study on the Southeast Indian Ridge. Earth Planet Sci Lett 227:457–472

Burnham CW (1975) Water and magmas: a mixing model. Geochim Cosmochim Acta 39:1077–1084

Burnham CW (1979) The importance of volatile constituents. *In*: The Evolution of the Igneous Rocks, Princeton University Press, p 1077–1084

Burnham CW (1981) The nature of multicomponent aluminosilicate melts, Phys Chem Earth 13:197–229

Burnham CW (1985) Energy release in subvolcanic environments. Econ Geol 80:1515–1522

Burnham CW (1994) Development of the Burnham Model for prediction of H_2O solubility in magmas. Rev Mineral 30:123–130

Burnham CW, Jahns RH (1962) A method for determining the solubility of water in silicate melts, Am J Sci 260:721–745

Burnham CW, Davis NF (1971) The role of H_2O in silicate melts. 1. P–V–T relations in the system $NaAlSi_3O_8$–H_2O to 10 kilobars and 1000 °C. Am J Sci 270:54–79

Burnham CW, Davis NF (1974) The role of H_2O in silicate melts: II Thermodynamics and phase relations in the system $NaAlSi_3O_8$–H_2O to 10 kilobars, 700 °C—1100 °C. Am J Sci 274:902–940

Burnham CW, Nekvasil H (1986) Equilibrium properties of granitic magmas. Am Mineral 71:239–263

Burnham CW, Holloway JR, Davis NF (1969) Thermodynamic properties of water to 1000°C and 10000 bars. Geol Soc Am Spec Paper 132:1–96

Burton MR, Mader HM, Polacci M (2007) The role of gas percolation in quiescent degassing of persistently active basaltic volcanoes. Earth Planet Sci Lett 264:46–60

Caliro S, Chiodini G, Moretti R, Avino R, Granieri D, Russo M and Fiebig J (2007) The origin of the fumaroles of La Solfatara (Campi Flegrei, South Italy). Geochim Cosmochim Acta 71:3040–3055

Caliro S, Chiodini G, Paonita A (2014) Geochemical evidences of magma dynamics at Campi Flegrei (Italy). Geochim Cosmochim Acta 132:1–15

Caracausi A, Favara R, Giammanco S, Nuccio PM, Paonita A, Pecoraino G and Rizzo A (2003a): Mount Etna: Geochemical signals of magma ascent and unusually extensive plumbing system. Geophys Res Lett 30:1057–1060

Caracausi A, Italiano F, Nuccio PM, Paonita A, Rizzo A (2003b) Evidence of deep magma degassing and ascent by geochemistry of peripheral gas emissions at Mt. Etna (Italy): assessment of the magmatic reservoir pressure. J Geophys Res 108 B10: 2463–2484

Caricchi L, Sheldrake T, Blundy J (2018) Modulation of magmatic processes by CO_2 flushing. Earth Planet Sci Lett 491:160–171

Carroll MR (2005) Chlorine solubility in evolved alkaline magmas. Ann Geophys 48:619–631

Carroll MR, Rutherford MJ (1988) Sulfur speciation in hydrous experimental glasses of varying oxidation state; results from measured wavelength shifts of sulfur X-rays. Am Mineral 73:845–849

Carroll MR, Stolper EM (1991) Argon solubility and diffusion in silica glass: implications for the solution behavior of molecular gases, Geochim Cosmochim Acta 55:211–225

Carroll MR, Stolper EM (1993a) Noble gas solubilities in silicate melts and glasses: new experimental results for Ar and the relationship between solubility and ionic porosity. Geochim Cosmochim Acta 57:5039–5051

Carroll MR, Sutton SR, Rivers ML, Woolum DS (1993b) A experimental study of krypton diffusion and solubility in silicic glasses. Chem Geol 109:9–28

Carroll MR, Webster JD (1994) Solubilities of sulfur, noble gases, nitrogen, chlorine, and fluorine in magmas. Rev Mineral 30:231–279

Carroll MR, Blank JG (1997) The solubility of H_2O in phonolitic melts. Am Mineral 82:549–556

Chamorro-Perez E, Gillet P, Jambon A (1996) Argon solubility in silicate melts at very high pressures Experimental set-up and preliminary results for silica and anorthite melts. Earth Planet Sci Lett 145:97–107

Chamorro-Perez E, Gillet P, Jambon A, Bardo J, McMillan P (1998) Low argon solubility in silicate melts at high pressure. Nature 393:352–355

Chamorro EM, Brooker RA, Wartho JA, Wood BJ, Kelley SP, Blundy JD (2002) Ar and K partitioning between clinopyroxene-silicate at 8 GPa. Geochim Cosmochim Acta 66:507–519

Chan SL, Elliot SR (1991) Theoretical study of the interstice statistics of the oxygen sublattice in vitreous SiO_2. Phys Rev B 43:4423–4432

Chennaoui-Aoudjeane H, Jambon A (1990) He solubility in silicate glasses at 250°C: a model for calculation. Eur J Mineral 2:539–545

Chevychelov VY, Bocharnikov RE, Holtz F (2008) Experimental study of chlorine and fluorine partitioning between fluid and subalkaline basaltic melt. Doklady Earth Sci 422:1089

Chiodini G, Caliro S, Cardellini C, Granieri D, Avino R, Baldini A, Donnini M, Minopoli C (2010) Long-term variations of the Campi Flegrei, Italy, volcanic system as revealed by the monitoring of hydrothermal activity. J Geophys. Res 15:B03205

Chiodini G, Caliro S, De Martino P, Avino R, Gherardi F (2012) Early signals of new volcanic unrest at Campi Flegrei caldera? Insights from geochemical data and physical simulations. Geology 40:943–946

Chiodini G, Giudicepietro F, Vandemeulebrouck J, Aiuppa A, Caliro S, De Cesare W, Tamburello G, Avino R, Orazi M, D'Auria L (2017) Fumarolic tremor and geochemical signals during a volcanic unrest. Geology 45:1131–1134

Chu X, Lee C-TA, Dasgupta R, Cao W (2019) The contribution to exogenic CO_2 by contact metamorphism at continental arcs: A coupled model of fluid flux and metamorphic decarbonation. Am J Sci: 631–657

Churakov SV, Gottschalk M (2003a) Perturbation theory based equation of state for polar molecular fluids. I Pure fluids. Geochim Cosmochim Acta 67:2397–2414

Churakov SV, Gottschalk M (2003b) Perturbation theory based equation of state for polar molecular fluids: II fluid mixtures. Geochim Cosmochim Acta 67:2415–2425

Cicconi MR, Le Losq C, Moretti R, Neuville DR (2020a) Magmas are the largest repositories and carriers of earth's redox processes. Elements 16:173–178

Cicconi MR, Moretti R, Neuville DR (2020b) Earth's electrodes. Elements 16:157–160

Clemente B (1998) Etude expérimentale et modélisation de la solubilité du soufre dans les liquides magmatiques. Ph. D thesis, Université de Orléans

Clemente B, Scaillet B, Pichavant M (2004) The solubility of sulfur in hydrous rhyolitic melts. J Petrol 45:2171–2196

Colucci S, Papale P, Montagna CP (2017) Non-Newtonian flow of bubbly magma in volcanic conduits. J Geophys Res 122:1789–1804

Coltelli, M, Del Carlo, P, Vezzoli, L (1998) Discovery of a Plinian basaltic eruption of Roman age at Etna volcano, Italy. Geology 26:1095–1098

Crépisson C, Sanloup C, Cormier L, Blanchard M, Hudspeth J, Rosa AD, Mathon O, Irifune T (2018) Kr environment in feldspathic glass and melt: A high pressure, high temperature X-ray absorption study. Chem Geol 493:525–531

Dalou C, Le Losq C, Mysen BO, Cody GD (2015) Solubility and solution mechanisms of chlorine and fluorine in aluminosilicate melts at high pressure and high temperature. Am Mineral 100:2272–2283

Deegan FM, Troll VR, Freda C, Misiti V, Chadwick JP, McLeod CL, Davisdon JP (2010) Magma–carbonate interaction processes and associated CO_2 release at Merapi Volcano, Indonesia: Insights from experimental petrology. J Petrol 51:1027–1051

De Moor JM, Stix J, Avard G, Muller C, Corrales E, Diaz JA, Alan A, Brenes J, Pacheco J, Aiuppa A, Fischer TP (2019) Insights on hydrothermal—magmatic interactions and eruptive processes at Poas volcano (Costa Rica) from high—frequency gas monitoring and drone measurements. Geophys Res Lett 46:1293–1302

De Santis R, Breedveld GJF, Prausnitz JM (1974) Thermodynamic properties of aqueous gas mixtures at advanced pressures. Ind Eng Chem Process Des Develop 13:374–377

Di Matteo V, Mangiacapra A, Dingwell DB, Orsi G (2006) Water solubility and speciation in shoshonitic and latitic melt composition from Campi Flegrei Caldera (Italy). Chem Geol 229:113–124

Dingwell DB (1996) Volcanic dilemma: Flow or blow? Science 273:1054–1055

Dingwell DB, Romano C, Hess KU (1996) The effect of water on the viscosity of a haplogranitic melt under *PTX* conditions relevant to silicic volcanism. Contrib Mineral Petrol 124:19–28

Dixon JE (1997) Degassing of alkalic basalts. Am Mineral 82:368–378

Dixon JE, Stolper EM, Holloway JR (1995) An experimental study of water and carbon dioxide solubilities in Mid-Ocean Ridge basaltic liquids. Part I: Calibration and solubility models. J Petrol 36:1607–1631

Dixon JE, Stolper EM, Holloway JR (1995b) An experimental study of water and carbon dioxide solubilities in mid-ocean ridge basaltic liquids. Part I: calibration and solubility models. J Petrol 36:1607–1631

Dobran F (1992) Nonequilibrium flow in volcanic conduits and application to the eruption of Mt. St. Helens on May 18, 1980, and Vesuvius in AD 79. J Volcanol Geotherm Res 49:285–311

Dolejš D, Baker DR (2007) Liquidus equilibria in the system $K_2O–Na_2O–Al_2O_3–SiO_2–F_2O_{-1}–H_2O$ to 100 MPa: II Differentiation paths of fluorosilicic magmas in hydrous systems. J Petrol 48:807–828

Doremus H (1966) Physical solubility of gases in fused silica. J Am Ceram Soc 49:461–462

Draper DS, Carroll MR (1995) Argon diffusion and solubility in silicic glasses exposed to an Ar–He mixtures. Earth Planet Sci Lett 132:15–24

Duan Z, Zhang Z (2006) Equation of state of the H_2O, CO_2, and $H_2O–CO_2$ systems up to 10 GPa and 2573.15 K: molecular dynamics simulations with ab initio potential surface. Geochim Cosmochim Acta 70:2311–2324

Duffy JA, Ingram MD (1971) Establishment of an optical scale for Lewis basicity in inorganic oxyacids, molten salts, and glasses. J Am Chem Soc 93:6448–6454

Dufils T, Sator N, Guillot BA (2020) Comprehensive molecular dynamics simulation study of hydrous magmatic liquids. Chem Geol 533:119300

Edmonds M (2008) New geochemical insights into volcanic degassing. Phil Trans R Soc A 366:4559–4579

Edmonds M, Pyle DM, Oppenheimer C (2002) HCl emissions at Soufriere Hills Volcano, Montserrat, West Indies, during a second phase of dome building, November 1999 to September 2000. Bull Volcanol 64:21–30

Edmonds M, Aiuppa A, Humphreys M, Moretti R, Giudice G, Martin RS, Herd RA, Christopher T (2010) Excess volatiles supplied by mingling of mafic magma at an andesite arc volcano. Geochem Geophys Geosystems 11:Q04005

Eichelberger JC, Carrigan CR, Westrich HR, Price RH (1986) Non-explosive silicic volcanism. Nature 323:598–602

Esposti Ongaro T, Papale P, Neri A, Del Seppia D (2006) Influence of carbon dioxide on the large-scale dynamics of magmatic eruptions at Phlegrean Fields (Italy). Geophys Res Lett 33:L06318

Fabbrizio A, Bouhifd MA, Andrault D, Bolfan-Casanova N, Manthilake G, Laporte D (2017) Argon behavior in basaltic melts in presence of a mixed H_2O–CO_2 fluid at upper mantle conditions. Chem Geol 448:100–109

Fincham CJB, Richardson FD (1954) The behaviour of sulphur in silicate and aluminate melts. Proc R Soc London A223:40–62

Fine G, Stolper EM (1985) The speciation of carbon dioxide in sodium aluminosilicate glasses. Contrib Mineral Petrol 91:105–121

Fine G, Stolper EM (1986) Dissolved carbon dioxide in basaltic glasses: concentrations and speciation. Earth Planet Sci Lett 76:263–278

Fisher DE (1970) Heavy rare gas in a Pacific seamount. Earth Planet Sci Lett 9:331–335

Fisher DE (1997) Helium, argon and xenon in crushed and melted MORB. Geochim Cosmochim Acta 61:3003–3012

Fleet ME, Liu X, Harmer SL, King PL (2005) Sulfur K-edge XANES spectroscopy: Chemical state and content of sulfur in silicate glasses. Can Mineral 43:1605–1618

Flood H, Förland T (1947) The acidic and basic properties of oxides. Acta Chem Scand 1:592–604

Flood H, Grjotheim T (1952) Thermodynamic calculation of slag equilibria. J Iron Steel Inst 171: 64–80

Flowers GC, Helgeson HC (1983) Equilibrium and mass transfer during progressive metamorphism of siliceous dolomites. Am J Sci 283:230–286

Fogel RA, Rutherford MJ (1990) The solubility of carbon dioxide in rhyolitic melts: a quantitative FTIR study. Am Mineral 75:311–1326

Frank RC, Swets DE, Lee RW (1961) Diffusion of neon isotopes in fused quartz. J Chem Phys 35:1451–1459

Fraser DG (1975) An investigation of some long-chain oxi-acid systems. PhD Thesis, University of Oxford

Fraser DG (1977) Thermodynamic properties of silicate melts. *In*: Thermodynamics in Geology (Fraser DG Ed) D Reidel Pub Co, Dortrecht-Holland

Freda C, Gaeta M, Giaccio B, Marra F, Palladino DM, Scarlato P, Sottili G (2011) CO_2-driven large mafic explosive eruptions: the Pozzolane Rosse case study from the Colli Albani Volcanic District (Italy). Bull Volcanol 73:241–256

Frezzotti ML (2001) Silicate-melt inclusions in magmatic rocks: applications to petrology. Lithos 55:273–299

Füri E, Hilton DR, Halldórsson SA, Barry PH, Hahm D, Fischer TP, Grönvold K (2010) Apparent decoupling of the He and Ne isotope systematics of the Icelandic mantle: The role of He depletion, melt mixing, degassing fractionation and air interaction. Geochim Cosmochim Acta 74:3307–3332

Gardner JE, Hilton M, Carroll MR (1999) Experimental constrains on degassing of magma: isothermal bubble growth during continuous decompression from high pressure. Earth Planet Sci Lett 168:201–218

Gardner JE, Hilton M, Carroll MR (2000) Bubble growth in highly viscous silicate melts during continuous decompression from high pressure. Geochim Cosmochim Acta 64:1473–1483

Gerlach TM, Nordilie BE (1975) The C–O–H–S gaseous system, part I: composition limits and trends in basaltic glasses. Am J Sci 275:353–376

Gerlach TM, McGee KA (1994) Total sulfur dioxide emissions and pre-eruption vapor-saturated magma at Mount St. Helens 1980–88. Geophys Res Lett 21:2833–2836

Ghiorso MS, Kelemen PB (1987) Evaluating reaction stoichiometry in magmatic systems evolving under generalized thermodynamic constraints: examples comparing isothermal and isenthalpic assimilation. Magmatic Processes: Physicochemical Principles 1:319–336

Ghiorso MS, Sack RO (1995) Chemical mass transfer in magmatic processes IV A revised and internally consistent thermodynamic model for the interpolation and extrapolation of liquid–solid equilibria in magmatic systems at elevated temperatures and pressures. Contrib Mineral Petrol 119:197–212

Ghiorso MS, Carmichael ISE, Rivers ML, Sack RO (1983) The Gibbs free energy of mixing of natural silicate liquids; an expanded regular solution approximation for the calculation of magmatic intensive variables. Contrib Mineral Petrol 84:107–145

Ghiorso MS, Gualda GAR (2015) An H_2O–CO_2 mixed fluid saturation model compatible with rhyolite-MELTS. Contrib Mineral Petrol 169:53

Ghosh DB, Karki BB (2017) Transport properties of carbonated silicate melt at high pressure. Sci Adv 3:e1701840

Ghosh DB, Bajgain SK, Mookherjee M, Karki BB (2017) Carbon-bearing silicate melt at deep mantle conditions. Sci Rep 7:1–8

Giggenbach WF (1996) Chemical composition of volcanic gases, *In*: Monitoring and Mitigation of Volcanic Hazards, Scarpa & Tilling (Eds) Springer, p 221–256

Giordano D, Romano C, Papale P, Dingwell DB (2004a) The viscosity of trachytes, and comparison with basalts, phonolites and rhyolites. Chem Geol 213:49–61

Giordano D, Romano C, Dingwell DB, Poe B, Behrens H (2004b) The combined effects of water and fluorine on the viscosity of silicic magmas. Geochim Cosmochim Acta 68:5159–5168

Giordano D, Russell JK, Dingwell DB (2008) Viscosity of magmatic liquids: A model. Earth Planet Sci Lett 271:123–134

Gokcen NA (1996) Gibbs-Duhem-Margules Laws. J Phase Equil 17 :50–51

Gonnermann H, Mukhopadhyay S (2007) Non-equilibrium degassing and a primordial source for helium in ocean-island volcanism. Nature 449:1037–1040

Goranson RW (1931) The solubility of water in granitic magmas. Am J Sci 22:481–502

Gorbachev P, Bezmen N (2002) Solubility of sulfur in water-saturated An–Di melts under various conditions. EMPG IX, Zurich, J Conf. 7:40 (abstr.)

Gray CG, Gubbins KE (1984) Theory of Molecular Fluids. I Fundamentals. Clarendon Press, Oxford, UK

Gresse M, Vandemeulebrouck J, Byrdina S, Chiodini G, Roux P, Rinaldi AP, Wathelet M, Ricci T, Letort J, Petrillo Z, Tuccimei P, Lucchetti C, Sciarra A (2018) Anatomy of a fumarolic system inferred from a multiphysics approach. Sci Rep 8:7580

Griffin WL, Wang X, Jackson SE, Pearson NJ, O-Reilly SY, Xu X, Zhou X (2002) Zircon chemistry and magma mixing, SE China: *In-situ* analysis of Hf isotopes, Tonglu and Pingtan igneous complexes. Lithos 61:237–269

Gualda GAR, Ghiorso MS, Lemons RV, Carley TL (2012) Rhyolite-MELTS: a modified calibration of MELTS optimized for silica-rich, fluid-bearing magmatic systems. J Petrol 53:875–890

Guillot B, Sarda P (2006) The effect of compression on noble gas solubility in silicate melts and consequences for degassin at mid-ocean ridges. Geochim Cosmochim Acta 70:1215–1230

Guillot B, Sator N (2011) Carbon dioxide in silicate melts: a molecular dynamics simulation study. Geochim Cosmochim Acta 75:1829–1857

Guillot B, Sator N (2012) Noble gases in high-pressure silicate liquids: a computer simulation study. Geochim Cosmochim Acta 80:51–69

Hamilton DL, Burnham CW, Osborn EF (1964) The solubility of water and effects of oxygen fugacity and water content on crystallization in mafic magmas. J Petrol 5:21–39

Hansteen TH, Klügel A (2008) Fluid inclusion thermobarometry as a tracer for magmatic processes. Rev Mineral Geochem 69:143–178

Hauri E (2002) SIMS analysis of volatiles in silicate glasses, 2: Isotopes and abundances in Hawaiian melt inclusions. Chem Geol 183:115–141

Hayatsu A, Waboso CE (1985) The solubility of rare gases in silicate melts and implications for K–Ar dating. Chem Geol 52:97–102

Heber VS, Brooker RH, Kelley SP, Wood BJ (2007) Crystal–melt partitioning of noble gases (helium, neon, argon, krypton, and xenon) for olivine and clinopyroxene. Geochim Cosmochim Acta 71:1041–1061

Hess PC (1971) Polymer model of silicate melts. Geochim Cosmochim Acta 35:289–306

Hodge KF, Jellinek AM (2020) The influence of magma mixing on the composition of andesite magmas and silicic eruption style, Geophys Res Lett 47:e2020GL087439

Hodges FN (1974) The solubility of H_2O in silicate melts. Year Book Carnegie Inst Washington 73:251–255

Holloway JR (1977) Fugacity and activity of molecular species in supercritical fluids, *In*: Thermodynamics in Geology. Fraser DG (Ed) Reidel Pub Co, Dordrecht-Holland, Boston

Holloway JR, Blank JG (1994) Application of experimental results to C–O–H species in natural melts. Rev Mineral 30:187–230

Holtz F, Behrens H, Dingwell DB, Johannes W (1995) H_2O solubility in haplogranitic melts: compositional, pressure, and temperature dependence. Am Mineral 80:94–108

Honda M, Patterson DB (1999) Systematic elemental fractionation of mantle-derived helium, neon and argon in mid-ocean ridge glasses. Geochim Cosmochim Acta 63:2863–2874

Iacono-Marziano G, Paonita A, Rizzo A, Scaillet B, Gaillard F (2010) Noble gas solubilities in silicate melts: New experimental results and a comprehensive model of the effects of liquid composition, temperature and pressure. Chem Geol 279:145–157

Iacono-Marziano G, Morizet Y, Le Trong E, Gaillard F (2012) New experimental data and semi-empirical parameterization of H_2O–CO_2 solubility in mafic melts. Geochim Cosmochim Acta 97:1–23

Ihinger PD, Zhang Y, Stolper EM (1999) The speciation of dissolved water in rhyolitic melt. Geochim Cosmochim Acta 63:3567–3578

Jacob KT, Jyengar GNK (1982) Oxidation of alkaline earth sulfides to sulfates: thermodynamic aspects. Metall Trans B 17B:387–390

Jahn S (2022) Molecular simulations of oxide and silicate melts and glasses. Rev Mineral Geochem 87:193-227

Jakobsson S (1997) Solubility of water and carbon dioxide in an icelandite at 1400 °C and 10 kilobars. Contrib Mineral Petrol 127:129–135

Jambon A (1987) He solubility in silicate melts: a tentative model of calculation. Chem Geol 62:131–136

Jambon A, Weber HW, Begeman F (1985) Helium and argon from an Atlantic MORB glass: concentration, distribution and isotopic composition. Earth Planet Sci Lett 73:255–267

Jambon A, Weber HW, Braun O (1986) Solubility of He, Ne, Ar, Kr, Xe in a basalt melt in the range of 1250–1600°C: Geochemical implications. Geochim Cosmochim Acta 50:401–408

Jaupart C, Allegre CJ (1991) Gas content, eruption rate and instabilities of eruption regime in silicic volcanoes. Earth Planet Sci Lett 102:413–429

Jaupart C, Vergniolle S (1989) The generation and collapse of a foam layer at the roof of a basaltic magma chamber. J Fluid Mech 203:347–380

Johnson MC, Anderson AT, Jr, Rutherford MJ (1994) Pre-eruptive volatile contrnts of magmas. Rev Mineral 30:281–330

Jugo PJ, Wilke M, Botcharnikov RE (2010) Sulfur K-edge XANES analysis of natural and synthetic basaltic glasses: Implications for S speciation and S content as function of oxygen fugacity. Geochim Cosmochim Acta 74:5926–5938

Kadik A, Pineau F, Litvin Y, Jendrzejewski N, Martinez I, Javoy M (2004) Formation of carbon and hydrogen species in magmas at low oxygen fugacity. J Petrol 45:1297–1310

Katsura T, Nagashima S (1974) Solubility of sulfur in some magmas at 1 atmosphere. Geochim Cosmochim Acta 38:517–531

Kelley SP, Arnaud NO, Turner G (1994) High spatial resolution [40]Ar/[39]Ar investigation of quartz, biotite and plagioclase using a new ultra-violet laser probe extraction technique. Geochim Cosmochim Acta 58:3519–3525

Kent AJR (2008) Melt inclusions in basaltic and related volcanic rocks. Rev Mineral Geochem 69:273–332

Kerrick DM, Jacobs GK (1981) A modified Redlich–Kwong equation for H_2O, CO_2, and $H_2O–CO_2$ mixtures at elevated pressures and temperatures, Am J Sci 281:735–767

King PL, Holloway JR (2002) CO_2 solubility and speciation in intermediate (andesitic) melts: the role of H_2O and composition. Geochim Cosmochim Acta 66:1627–1640

Kirsten T (1968) Incorporation of rare gas in solidifying enstatite melts. J Geoph Res 73:2807–2810

Kohn SC, Dupree R, Smith ME (1989) A multinuclear magnetic resonance study of the structure of hydrous albite glasses. Geochim Cosmochim Acta 53:2925–2935

Kohn SC, Smith ME, Dirken PJ, van Eck ERH, Kentgens APM, Dupree R (1998) Sodium environment in hydrous and dry albite glasses: improved ^{23}Na solid state NMR data and their implication for water dissolution mechanisms. Geochim Cosmochim Acta 62:79–87

Kozono T, Koyaguchi T (2012) Effects of gas escape and crystallization on the complexity of conduit flow dynamics during lava dome eruptions. J Geophys Res 117:B08204

Landi P, Rotolo SG (2015) Coolong and crystallization recorded in trachytic enclaves hosted in pantelleritic magmas (Pantelleria, Italy): implications for pantellerite petrogenesis. J Volcanol Geotherm Res 301:169–179

Lange RA (1994) The effect of H_2O, CO_2 and F on the density and viscosity of silicate melts. Rev Mineral 30:331–369

Lange RA, Carmichael ISE (1987) Densities of $Na_2O–K_2O–MgO–FeO–Fe_2O_3–Al_2O_3–TiO_2–SiO_2$ liquids—new measurements and derived partial molar properties. Geochim Cosmochim Acta 51:2931–2946

La Spina G, Polacci M, Burton M, de' Michieli Vitturi M (2017) Numerical investigation of permeability models for low viscosity magmas: Application to the 2007 Stromboli effusive eruption. Earth Planet Sci Lett 473:273–290

Le Losq C, Moretti R, Neuville DR (2013) Speciation and amphoteric behaviour of water in aluminosilicate melts and glasses: high-temperature Raman spectroscopy and reaction equilibria. Eur J Mineral 25:777–790

Le Losq C, Mysen BO, Cody GD (2015) Water and magmas: insights about the water solution mechanisms in alkali silicate melts from infrared, Raman, and ^{29}Si solid-state NMR spectroscopies. Prog Earth Planet Sci 2:22

Le Losq C, Moretti R, Oppenheimer C, Baudelet F, Neuville DR (2020) *In situ* XANES study of the influence of varying temperature and oxygen fugacity on iron oxidation state and coordination in a phonolitic melt. Contrib Mineral Petrol 175:1–13

Lee SK, Cody GD, Fei Y, Mysen BO (2006) The effect of Na/Si on the structure of sodium silicate and aluminosilicate glasses quenched from melts at high pressure: a multi-nuclear (Al-27, Na-23, O-17) 1D and 2D solid-state NMR study. Chem Geol 229:162–172

Leroy C, Sanloup C, Bureau H, Schmidt BC, Konopkova Z, Raepsaet C (2018) Bonding of xenon to oxygen in magmas at depth. Earth Planet Sci Lett 484:103–110

Leroy C, Bureaua H, Seanloup C, Raepsaet C, Glazirind K, Munsche P, Harmanda M, Prouteau G, Khodja H (2019) Xenon and iodine behaviour in magmas. Earth Planet Sci Lett 522:144–154

Lange RM, Charmichael ISE (1987) Densities of $Na_2O–K_2O–CaO–MgO–FeO–Fe_2O_3–Al_2O_3–TiO_2–SiO_2$ liquids: new measurements and derived partial molar properties. Geochim Cosmochim Acta 51:2931–2946

Lesne P, Scaillet B, Pichavant M, Iacono-Marziano G, Bey J-M (2011a) The H_2O solubility of alkali basaltic melts: an experimental study. Contrib Mineral Petrol 162:133–151

Lesne P, Scaillet B, Pichavant M, Beny JM (2011b) The carbon dioxide solubility in alkali basalts: an experimental study. Contrib Mineral Petrol 162:153–168

Lesne P, Scaillet B, Pichavant M (2015) The solubility of sulfur in hydrous basaltic melts. Chem Geol 418:104–116

Li Y, Huang R, Wiedenbeck M, Keppler H (2015) Nitrogen distribution between aqueous fluids and silicate melts. Earth Planet Sci Lett 411:218–228

Libourel G, Marty B, Humbert F (2003) Nitrogen solubility in basaltic melt. Part I Effect of oxygen fugacity. Geochim Cosmochim Acta 67:4123–4135

Liotta M, Paonita A, Caracausi A, Martelli M, Rizzo A, Favara R (2010) Hydrothermal processes governing the geochemistry of the crater fumaroles at Mount Etna volcano (Italy). Chem Geol 278:92–104

Liotta M, Rizzo, Paonita A, Caracausi A, Martelli M (2012) Sulfur isotopic compositions of fumarolic and plume gases at Mount Etna (Italy) and inferences on their magmatic source, Geochem Geophys Geosystems 13:Q05015

Liu Y, Samaha NT, Baker DR (2007) Sulfur concentration at sulfide saturation (SCSS) in magmatic silicate melts. Geochim Cosmochim Acta 71:1783–1799

Louvel M, Sanchez-Valle C, Malfait WJ, Pokrovski GS, Borca CN, Grolimund D (2020) Bromine speciation and partitioning in slab-derived aqueous fluids and silicate melts and implications for halogen transfer in subduction zones. Solid Earth 11:1145–1161

Lowenstern JB (1994) Chlorine, fluid immiscibility, and degassing in peralkaline magmas from Pantelleria, Italy. Am Mineral 79:353–369

Lowenstern JB (1995) Application of silicate–melt inclusions to the study of magmatic volatiles. *In*: Magmas, Fluids, and Ore Deposits. Thompson JFH (Ed). Mineral Assoc Canada Short Course Ser 23, p 71–99

Luhr J (1990) Experimental phase relations of water and sulfur-saturated arc magmas and the 1982 eruption of El-Chichon volcano. J Petrol 31:1071–1114

Lux G (1987) The behavior of noble gases in silicate liquids: solution, diffusion, bubbles, and surface effects, with applications to natural samples. Geochim Cosmochim Acta 51:1549–1560

Mangiacapra A, Moretti R, Rutherford M, Civetta L, Orsi G, Papale P, The deep magmatic system of the Campi Flegrei caldera (Italy). Geophy Res Lett 35 (2008) L21304

Marty B (1995) Nitrogen content of the mantle inferred from N_2–Ar correlation in oceanic basalts. Nature 377:326–329

Marty B, Tolstikhin IN (1998) CO_2 fluxes from mid-ocean ridges, arcs and plumes. Chem Geol 145:233–248

Marty B, Zimmermann L (1999) Volatiles (He, C, N, Ar) in mid-ocean ridge basalts: assessment of shallow-level fractionation and characterization of source composition. Geochim Cosmochim Acta 63:3619–3633

Marty B, Lenoble M, Vassard N (1995) itrogen, helium and argon in basalt: a static mass spectrometric study. Chem Geol 120:183–195

Marrocchi Y, Toplis MJ (2005) Experimental determination of argon solubility in silicate melts: an assessment of the effects of liquid composition and temperature. Geochim Cosmochim Acta 69:5765–5776

Masotta M, Keppler H, Chaudhari A (2016) Fluid–melt partitioning of sulfur in differentiated arc magmas and the sulfur yield of explosive volcanic eruptions. Geochim Cosmochim Acta 176:26–43

Masotta M, Mollo S, Nazzari M, Tecchiato V, Scarlato P, Papale P, Bachmann O (2018) Crystallization and partial melting of rhyolite and felsite rocks at Krafla volcano: A comparative approach based on mineral and glass chemistry of natural and experimental products, Chem Geol 483:603–618

Matsuda J, Marty B (1995) The $^{40}Ar/^{36}Ar$ ratio of the undepleted mantle: a reevaluation. Geophys Res Lett 22:1937–1940

Matjuschkin V, Blundy JD, Brooker RA (2016) The effect of pressure on sulphur speciation in mid- to deep-crustal arc magmas and implications for the formation of porphyry copper deposits. Contrib Mineral Petrol 171: 66

Mavrogenes JA, O'Neill HSC (1999) The relative effects of pressure, temperature and oxygen fugacity on the solubility of sulfide in mafic magmas. Geochim Cosmochim Acta 63:1173–1180

McMillan PF (1994) Water solubility and speciation models. Rev Mineral 30:131–156

Melnik O (2000) Dynamics of two-phase conduit flow of high-viscosity gas-saturated magma: large variations of sustained explosive eruption intensity. Bull Volcanol 62:153–170

Menand T, Philips JC (2007) Gas segregation in dykes and sills. J Volcanol Geothermal Res 159:393–408

Mesko MG, Shelby JE (2002) Helium solubility in ternary soda-lime-silica glasses and melts. Phys. Chem. Glasses 43:91–96

Mesko MG, Newton K, Shelby JE (2000) Helium solubility in sodium silicate glasses and melts. Phys. Chem. Glasses 41:111–116

Metrich N, Rutherford MJ (1992) Experimental study of chlorine behavior in hydrous silicic melts. Geochim Cosmochim Acta 56:607–616

Métrich N, Wallace PJ (2008) Volatile abundances in basaltic magmas and their degassing paths tracked by melt inclusions. Rev Mineral Geochem 69:363–402

Métrich N, Bertagnini A, Landi P, Rosi M (2001) Crystallization driven by decompression and water loss at Stromboli volcano (Aeolian Islands, Italy). J Petrol 42:1471–1490

Métrich N, Bertagnini A, Landi P, Rosi M, Belhadj O (2005) Triggering mechanism at the origin of paroxysms at Stromboli(Aeolian Archipelago, Italy): The 5 April 2003 eruption. Geophys Res Lett 32:L10305

Métrich N, Bertagnini A, Di Muro A (2010) Conditions of magma storage, degassing and ascent at Stromboli: new insights into the volcano plumbing system with inferences on the eruptive dynamics. J Petrol 51:603–626

Miyazaki A, Hiyagon H, Sugiura N, Hirose K, Takahashi E (2004) Solubilities of nitrogen and noble gases in silicate melts under various oxygen fugacities: implications for the origin and degassing history of nitrogen and noble gases in the Earth. Geochim Cosmochim Acta 68:387–401

Mookerjee M, Stixrude L, Karki B (2008) Hydrous silicate melt at high pressure. Nature 452:983–986

Montana A, Guo Q, Boettcher SL, White BS, Brearley M (1993) Xe and Ar in high-pressure silicate liquids. Am Mineral 78:1135–1142

Moore G (2008) Interpreting H_2O and CO_2 contents in melt inclusions: Constraints from solubility experiments and modeling. Rev Mineral Geochem 69:333–362

Moreira MA, Kurz MD (2013) Noble gases as tracers of mantle processes and magmatic degassing. *In:* Noble Gases As Geochemical Tracers. Burnard P (Ed) Springer Heidelberg, p 371–391

Moreira M, Sarda P (2000) Noble gas constraints on degassing processes. Earth Planet Sci Lett 176:375–386

Moretti R (2005) Polymerisation, basicity, oxidation state and their role in ionic modelling of silicate melts. Ann Geophys 48:583–608

Moretti R (2021) Ionic syntax and equilibrium approach to redox exchanges in melts: basic concepts and the case of iron and sulfur in degassing magmas. Chapter 6. *In:* Magma Redox Geochemistry. Moretti R, Neuville DR (Eds) Am Geophys Union Geophysical Monograph 266, John Wiley & Sons, Inc, p 117–138

Moretti R, Baker DR (2008) Modeling of the interplay of fO_2 and fS_2 along the FeS–silicate melt equilibrium. Chem Geol 256:286–298

Moretti R, Neuville D (2021) Redox equilibria: from basic concepts to the magmatic realm. Chapter 1 *In:* Magma Redox Geochemistry. Moretti R, Neuville DR (Eds) Am Geophys Union Geophysical Monograph 266, John Wiley & Sons, Inc, p 1–17

Moretti R, Ottonello GA (2003a) Polymeric approach to the sulfide capacity of silicate slags and melts. Metall Mater Trans B 34B:399–410

Moretti R, Ottonello G (2003b) Polymerization and disproportionation of iron and sulfur in silicate melts: insights from an optical basicity-based approach. J Non-Cryst Solids 323:111–119

Moretti R, Ottonello G (2005) Solubility and speciation of sulfur in silicate melts: the Conjugated–Toop–Samis–Flood–Grjotheim (CTSFG) model. Geochim Cosmochim Acta 69:801–823

Moretti R, Ottonello G (2022) Silicate melt thermochemistry and the redox state of magmas. Rev Mineral Geochem 87:339–403

Moretti R, Papale P (2004) On the oxidation state and volatile behavior in multicomponent gas–melt equilibria. Chem Geol 213:265–280

Moretti R, Stefánsson A (2020) Volcanic and geothermal redox engines. Elements 16:179–184

Moretti R, Papale P, Ottonello G (2003) A model for the saturation of C–O–H–S fluids in silicate melts. *In:* Volcanic Degassing, Spec Publ Geol Soc London 213:81–102

Moretti R, Arienzo I, Civetta L, Orsi G, Papale P (2013a) Multiple magma degassing sources at an explosive volcano. Earth Planet Sci Lett 367:95–104

Moretti R, Arienzo I, Orsi G, Civetta L, D'Antonio M (2013b) The deep plumbing system of Ischia: a physico-chemical window on the fluid-saturated and CO_2-sustained Neapolitan volcanism (southern Italy). J Petrol 54:951–984

Moretti R, Le Losq C, Neuville DR (2014) The amphoteric behavior of water in silicate melts from the point of view of their ionic-polymeric constitution. Chem Geol 367:23–33

Moretti R, Métrich N, Arienzo I, Di Renzo V, Aiuppa A, Allard P (2018) Degassing vs. eruptive styles at Mt. Etna volcano (Sicily, Italy). Part I: Volatile stocking, gas fluxing, and the shift from low-energy to highly explosive basaltic eruptions. Chem Geol 482:1–17

Morizet Y, Brooker RA, Kohn SC (2002) CO_2 in haplo-phonolite melt: solubility, speciation and carbonate complexation. Geochim Cosmochim Acta 66:1809–1820

Mortimer N, Gans PB, Hauff F, Barker DHN (2012) Paleocene MORB and OIB from the Resolution Ridge, Tasman Sea, Austral. J Earth Sci 59:953–964

Moussallam Y, Oppenheimer C, Scaillet B, Gaillard F, Kyle P, Peters N, Hartley M, Berlo K, Donovan A (2014) Tracking the changing oxidation state of Erebus magmas, from mantle to surface, driven by magma ascent and degassing. Earth Planet Sci Lett 393:200–209

Moussallam Y, Florian P, Corradini D, Morizet Y, Sator N, Vuilleumier R, Guillot B, Iacono-Marziano G, Schmidt BC, Gaillard F (2016) The molecular structure of melts along the carbonatite–kimberlite–basalt compositional joint: CO_2 and polymerisation. Earth Planet Sci Lett 434:129–140

Moussallam Y, Morizet Y, Massuyeau M, Laumonier M, Gaillard F (2018) CO_2 solubility in kimberlite melts. Chem Geol 418:198–205

Mysen BO (1976) The role of volatiles in silicate melts: solubility of carbon dioxide and water in feldspars, pyroxene, and feldspathoid melts to 30 kb and 1625 °C. Am J Sci 276:969–996

Mysen BO (2013) Structure–property relationships of COHN-saturated silicate melt coexisting with COHN fluid: A review of *in-situ*, high-temperature, high-pressure experiments. Chem Geol 346:113–124

Mysen BO, Cody GD (2005) Solution mechanisms of H_2O in depolymerized peralkaline melts. Geochim Cosmochim Acta 69:5557–5566

Mysen BO, Fogel ML (2010) Nitrogen and hydrogen isotope compositions and solubility in silicate melts in equilibrium with reduced (N + H)-bearing fluids at high pressure and temperature: effects of melt structure. Am Mineral 95:987–999

Mysen BO, Eggler DH, Seitz MG, Holloway JR (1976) Carbon dioxide solubilities in silicate melts and crystals: Part I Solubility measurements. Am J Sci 276:455–479

Mysen BO, Virgo D, Seifert FA (1985) Relationships between properties and structure of aluminosilicate melts. Am Mineral 70:88–105

Mysen BO, Cody GD, Smith A (2004) Solubility mechanisms of fluorine in peralkaline and meta-aluminous silicate glasses and in melts to magmatic temperatures. Geochim Cosmochim Acta 68:2745–2769

Mysen BO, Yamashita S, Chertkova N (2008) Solubility and solution mechanisms of NOH volatiles in silicate melts at high pressure and temperature–amine groups and hydrogen fugacity. Am Mineral 93:1760–1770

Nakamura M (1995) Continuous mixing of crystal mush and replenished magma in the ongoing Unzen eruption. Geology 23:807–810

Nakayama GS, Shackelford JF (1990) Solubility and diffusivity of argon in vitreous silica. J Non-Cryst Solids 126:249–260

Neri A, Papale P, Macedonio G (1998) The role of magma composition and water content in explosive eruptions. 2. Pyroclastic dispersion dynamics. J Volcanol Geotherm Res 87:95–115

Nesbitt HW, Bancroft GM, Henderson GS, Ho R, Dalby KN, Huang Y, Yan Z (2011) Bridging, non-bridging and free (O^{2-}) oxygen in Na_2O–SiO_2 glasses: An X-ray Photoelectron Spectroscopic (XPS) and Nuclear Magnetic Resonance (NMR) study. J Non-Cryst Solids 357:170–180

Newman S, Lowenstern JB (2002) VolatileCalc: a silicate melt–H_2O–CO_2 solution model written in Visual Basic for Excel. Computers Geosci 28:597–604

Newman S, Epstein S, Stolper E (1988) Water, carbon dioxide, and hydrogen isotopes in glasses from the ca. 1340 AD eruption of the Mono Craters, California: Constraints on degassing phenomena and initial volatile contents. J Volcanol Geotherm Res 35:75–96

Nicholls J (1980) A simple thermodynamic model for estimating the solubility of H_2O in magmas. Contrib Mineral Petrol 74:211–220

Niwa K, Miyakawa C, Yagi T, Matsuda J-C (2013) Argon solubility in SiO_2 melt under high pressures: A new experimental result using laser-heated diamond anvil cell. Earth Planet Sci Lett 363:1–8

Nowak M, Behrens H (1995) The speciation of water in haplogranitic glasses and melts determined by *in-situ* near-infrared spectroscopy. Geochim Cosmochim Acta 59:3445–3450

Nowak M, Behrens H (2001) Water in rhyolitic magmas: getting a grip on a slippery problem. Earth Planet Sci Lett 184:515–522

Nuccio PM, Paonita A (2000) Investigation of the noble gas solubility in H_2O–CO_2 bearing silicate liquids at moderate pressure II: the Extended Ionic Porosity (EIP) model. Earth Planet Sci Lett:183:499–512

Nuccio PM, Paonita A (2001) Magmatic degassing of multicomponent vapors and assessment of magma depth: application to Vulcano Island (Italy). Earth Planet Sci Lett 193:467–481

Nuccio PM, Valenza M (1998) Magma degassing and geochemical detection of its ascent, *In*: Water–Rock Interaction Arehart & Hulston (Eds.) Balkema, p 475–478

O'Neill H St.C, Mavrogenes JA (2002) The sulfide capacity and the sulfur content at sulfide saturation of silicate melts at 1400°C and 1 bar. J Petrol 43:1049–1087

Oglesby JV, Kroeker S, Stebbins JF (2001) Potassium hydrogen disilicate: a possible model compound for ^{17}O NMR spectra of hydrous silicate glasses. Am Mineral 86:341–347

Oppenheimer C, McGonigle AJS (2004) Exploiting ground-based optical sensing technologies for volcanic gas surveillance. Ann Geophys 47:1455–1470

Oppenheimer C, Moretti R, Kyle PR, Eschenbacher A, Lowenstern JB, Hervig RL, Dunbar NW (2011) Mantle to surface degassing of alkalic magmas at Erebus volcano, Antarctica. Earth Planet Sci Lett 306:261–271

Ottonello G (1997) Principles of Geochemistry. Colombia University Press

Ottonello G, Moretti R (2004) Lux–Flood basicity of binary silicate melts. J Phys Chem Solids 65:1609–1614

Ottonello G, Moretti R, Marini L (2001) Vetuschi Zuccolini M On the oxidation state of iron in silicate melts and glasses: a thermochemical model. Chem Geol 174:157–179

Ottonello G, Richet P, Vetuschi Zuccolini M (2015) The wet solidus of silica: Predictions from the scaled particle theory and polarized continuum model. J Chem Phys 142:054503

Ottonello G, Richet P, Papale P (2018) Bulk solubility and speciation of H_2O in silicate melts. Chem Geol 479:176–187

Paillat O, Elphick SC, Brown WL (1992) The solubility of water in $NaAlSi_3O_8$ melts: a re-examination of Ab–H_2O phase relationships and critical behavior at high pressures. Contrib Mineral Petrol 112:490–500

Pan V, Holloway JR, Hervig RL (1991) The pressure and temperature dependence of carbon dioxide solubility in tholeiitic basalt melts. Geochim Cosmochim Acta 55:1587–1595

Paonita A (2005) Noble gas solubility in silicate melts: a review of experimentation and theory and implications regarding magma degassing processes. Ann Geophys 48:647–669

Paonita A (2010) Long-range correlation and nonlinearity in geochemical time series of gas discharges from Mt. Etna, and changes with 2001 and 2002–03 eruptions. Nonlinear Processes Geophys 17:733–751

Paonita A, Martelli M (2006) Magma dynamics at mid-ocean ridges by noble gas kinetic fractionation: assessment of magmatic ascent rates. Earth Planet Sci Lett 241:138–158

Paonita A, Martelli M (2007) A new view of the He–Ar–CO_2 degassing at mid-ocean ridges: homogeneous composition of magmas from the upper mantle. Geochim Cosmochim Acta 71:1747–1763

Paonita A, Gigli G, Gozzi D, Nuccio PM, Trigila R (1999) Investigation of He solubility in H_2O–CO_2 bearing silicate liquids at moderate pressure: an experimental method. Earth Planet Sci Lett 181:595–604

Paonita A, Caracausi A, Iacono-Marziano G, Martelli M, Rizzo A (2012) Geochemical evidence for mixing between fluids exsolved at different depths in the magmatic system of Mt Etna (Italy). Geochim Cosmochim Acta 84:380–394

Paonita A, Federico C, Bonfanti P, Capasso G, Inguaggiato S, Italiano F, Madonia P, Pecoraino G, Sortino F (2013) The episodic and abrupt geochemical changes at La Fossa fumaroles (Vulcano Island, Italy) and related constraints on the dynamics, structure, and compositions of the magmatic system. Geochim Cosmochim Acta 120:158–178

Paonita A, Caracausi A, Martelli M, Rizzo A (2016) Temporal variations of helium isotopes in volcanic gases quantify pre-eruptive refill and pressurization in magma reservoirs: The Mount Etna case. Geology 44:499–502

Papale P (1997) Modeling of the solubility of a one-component H_2O or CO_2 fluid in silicate liquids. Contrib Mineral Petrol 126:237–251

Papale P (1999a) Modeling of the solubility of a two component H_2O+CO_2 fluid in silicate liquids. Am Mineral 84:477–492

Papale P (1999b) Strain-induced magma fragmentation in explosive eruptions. Nature 397:425–428

Papale P (2001) Dynamics of magma flow in volcanic conduits with variable fragmentation efficiency and nonequilibrium pumice degassing. J Geophys Res 106:11043–11065

Papale P (2005) Determination of total H_2O and CO_2 budgets in evolving magmas from melt inclusion data. J Geophys Res 110:B03208

Papale P, Polacci M (1999) Role of carbon dioxide in the dynamics of magma ascent in explosive eruptions. Bull Volcanol 60:583–594

Papale P, Neri A, Macedonio G (1998) The role of magma composition and water content in explosive eruptions. 1. Conduit ascent dynamics. J Volcanol Geotherm Res 87:75–93

Papale P, Moretti R, Barbato D (2006) The compositional dependence of the saturation surface of H_2O+CO_2 fluids in silicate melts. Chem Geol 229:78–95

Papale P, Montagna CP, Longo A (2017) Pressure evolution in shallow magma chambers upon buoyancy-driven replenishment. Geochem Geophys Geosystems 18:1214–1224

Parr RG, Yang W (1989) Density-Functional Theory of Atoms and Molecules. Oxford University Press, New York

Pino NA, Moretti R, Allard P, Boschi E (2011) Seismic precursors of a basaltic paroxysmal explosion track deep gas accumulation and slug upraise. J Geophys Res: Solid Earth 116:B02312

Pizarro C, Parada MA, Contreras C, Morgado E (2019) Cryptic magma recharge associated with the most voluminous 20th century eruptions (1921, 1948 and 1971) at Villarica Volcano. J Volcanol Geotherm Res 384:48–63

Platt U (1994) Differential optical absorption spectroscopy (DOAS). Chem Anal Ser 127:27–83

Platt U, Stutz J (2008) Differential Optical Absorption Spectroscopy - Principles and Applications. Springer, Berlin Heidelberg New York

Polacci M, Papale P, Del Seppia D, Giordano D, Romano C (2004) Dynamics of magma ascent and fragmentation in trachytic versus rhyolitic eruptions. J Volcanol Geotherm Res 131:93–108

Poland MP, Miklius A, Sutton AJ, Thornber CR (2012) A mantle-driven surge in magma supply to Kilauea Volcano during 2003–2007. Nat Geosci 5:295–300

Prausnitz JM, Lichtenthaler RN, Gomes de Azevedo E (1985) Molecular Thermodynamics of Fluid-Phase Equilibria, Prentice-Hall

Pritchard ME, Mather TA, McNutt SR, Delgado FJ, Reath K (2019) Thoughts on the criteria to determine the origin of volcanic unrest as magmatic or non-magmatic. Philos Trans R Soc A 377:20180008

Proussevitch AA, Sahagian DL (1996) Dynamics of coupled diffusive and decompressive bubble growth in magmatic systems. J Geophys Res 101:17447–17456

Proussevitch AA, Sahagian DL (1998) Dynamics and energetics of bubble growth in magmas: analytical formulation and numerical modeling. J Geophys Res 103:18223–18251

Rapaport DC (1996) The Art of Molecular Dynamics Simulation. Cambridge University Press

Richet P, Lejeune AM, Holtz F, Roux J (1996) Water and the viscosity of andesite melts. Chem Geol 128:185–197

Rizzo A, Caracausi A, Favara R, Martelli M, Paonita A, Paternoster M, Nuccio PM, Rosciglione A (2006) New Insights into magma dynamics during last two eruptions of Mount Etna as inferred by geochemical monitoring from 2002 to 2005. Geochem Geophys Geosystems 7:Q06008

Roberts TJ, Lurton T, Giudice G, Liuzzo M, Aiuppa A, Coltelli M, Vignelles D, Salerno G, Couté B, Chartier M, Baron R, Saffell JR, Scaillet B (2017) Validation of a novel multi-gas sensor for volcanic HCl alongside H_2S and SO_2 at Mt. Etna. Bull Volcanol 79:36

Robie RA, Hemingway BS, Fisher JR (1978) Thermodynamic properties of minerals and related substances at 298.15 K and 1 bar (105 pascal) pressure and at higher temperatures. US Geol Surv Bull 1452

Roedder E (1976) Petrologic data from experimental studies on crystallized silicate melt and other inclusions in lunar and Hawaiian olivine. Am Mineral 61:684–690

Roedder E (1979) Origin and significance of magmatic inclusions. Bull Mineral 102:487–510

Romano C, Giordano D, Papale P, Mincione V, Dingwell DB, Rosi M (2003) The dry and hydrous viscosities of alkaline melts from Vesuvius and Phlegrean Fields. Chem Geol 202:23–38

Romano C, Poe B, Mincione V, Hess KU, Dingwell DB (2011) The viscosities of dry and hydrous XAlSi$_3$O$_8$ (X= Li, Na, K, Ca$_{0.5}$, Mg$_{0.5}$) melts. Chem Geol 174:115–132

Roselieb K, Rammensee W, Buttner H, Rosenhauer M (1992) Solubility and diffusion of noble gases in vitreous albite. Chem Geol 96:241–266

Roselieb K, Rammensee W, Buttner H, Rosenhauer M (1995) Diffusion of noble gases in melts of the system SiO$_2$–NaAlSi$_2$O$_6$. Chem Geol 120:1–13

Roskosz M, Mysen B O and Cody G D Dual speciation of nitrogen in silicate melts at high pressure and temperature: an experimental study. Geochim Cosmochim Acta 70 (2006) 2902–2918

Sarda P, Graham D (1990) Mid-oceanic ridge popping rocks: implications for degassing at ridge crests. Earth Planet Sci Lett 97:268–289

Sarda P, Guillot B (2005) Breaking of Henry's law for noble gas and CO$_2$ solubility in silicate melt under pressure. Nature 436:95–98

Sarda P, Morieira M (2002) Vesiculation and vesicle loss in mid-ocean ridge basalt glasses: He, Ne, Ar elemental fractionation and pressure influence. Geochim Cosmochim Acta 66:1449–1458

Scaillet B, Pichavant M (2005) A model of sulphur solubility for hydrous mafic melts: application to the determination of magmatic fluid compositions of Italian volcanoes. Ann Geophys 48:671–698

Schmidt BC, Keppler H (2002) Experimental evidence for high noble gas solubilities in silicate melts under pressures. Earth Planet Sci Lett 195:277–290

Schmidt BC, Riemer T, Kohn SC, Behrens H, Dupree R (2000) Different water solubility mechanisms in hydrous glasses along the quartz–albite join. Evidence from NMR spectroscopy. Geochim Cosmochim Acta 64:513–526

Schmidt BC, Riemer T, Kohn SC, Holtz F, Dupree R (2001) Structural implications of water dissolution in haplogranitic glasses from NMR spectroscopy: influence of total water content and mixed alkali effect. Geochim Cosmochim Acta 65:2949–2964

Schulze F, Behrens H, Holtz F, Roux J, Johannes W (1996) The influence of H$_2$O on the viscosity of a haplogranitic melt. Am Mineral 81:1155–1165

Selva J, Marzocchi W, Papale P, Sandri L (2012) Operational eruption forecasting at high-risk volcanoes: the case of Campi Flegrei, Naples. J Appl Volcanol 1:5

Sen S, Widgeon SJ, Navrotsky A, Mera G, Tavakoli A, Ionescu E, Riedel R (2013) Carbon substitution for oxygen in silicates in planetary interiors. PNAS 110:15904–15907

Shackelford JF (1982) A gas probe analysis of structure in bulk and surface layers of vitreous silica. J Non-Cryst Solids 49:299–307

Shackelford JF (1999) Gas solubility in glasses—principles and structural implications. J Non-Cryst. Solids 253:231–241

Shackelford JF, Brown BD (1980) A gas probe analysis of structure in silicate glasses. J Am Ceram Soc 63:562–565

Shackelford JF, Masaryk JS (1978) The interstitial structure of vitreous silica. J Non-Cryst Solids 30:127–139

Shackelford JF, Studt PL, Fulrath RM (1972) Solubility of gases in glass II He, Ne, and H$_2$ in fused silica. J Appl Phys 43:1619–1626

Shannon RD, Prewitt CT (1969) Effective ionic radii in oxides and fluorides. Acta Cryst 25:925–946

Shelby JE (1972a) Helium migration in natural and synthetic vitreous silica. J Am Ceram Soc 55:61–64

Shelby JE (1972b) Neon migration in vitreous silica. J Am Ceram Soc 55:167–170

Shelby JE (1973) Effects of phase separation on helium migration in sodium silicate glasses. J Am Ceram Soc 56:263–266

Shelby JE (1974) Helium diffusion and solubility in K$_2$O–SiO$_2$ glasses. J Am Ceram Soc 57:236–263

Shelby JE (1976) Pressure dependence of helium and neon in vitreous silica. J Appl Phys 47:135–139

Shen A, Keppler H (1995) Infrared spectroscopy of hydrous silicate melts to 1000°C and 10 kbar: Direct observation of H$_2$O speciation in a diamond-anvil cell. Am Mineral 80:1335–1338

Shen G, Mei Q, Prakapenka VB, Lazor P, Sinogeikin S, Meng Y, Park C (2011) Effect of helium on structure and compression behavior of SiO$_2$ glass. PNAS 108:6004–6007

Shibata T, Takahashi E, Matsuda J (1996) Noble gas solubility in binary CaO–SiO$_2$ system. Geophys Res Lett 23:3139–3142

Shibata T, Takahashi E, Matsuda J (1998) Solubility of neon, argon, krypton and xenon in binary and ternary silicate system: a new view on noble gas solubility. Geochim Cosmochim Acta 62:1241–1253

Shinohara H (2005) A new technique to estimate volcanic gas composition: plume measurements with a portable multi-sensor system. J Volcanol. Geothermal Res 143:319–333

Shinohara H (2009) A missing link between volcanic degassing and experimental studies on chloride partitioning. Chem Geol 263:51–59

Shinohara H, Iiyama JT, Matsuo S (1984) Comportement du chlore dans le système magma granitique-eau. C R Acad Sci Sér 2 Mec Phys Chim Sci Univers Sci Terre 298:741–743

Shinohara H, Iiyama JT, Matsuo S (1989) Partition of chlorine compounds between silicate melt and hydrothermal solutions: I Partition of NaCl–KCl. Geochim Cosmochim Acta 53:2617–2630

Shinohara H, Fujimoto K (1994) Experimental study in the system albite–andalusite–quartz–NaCl–HCl–H$_2$O at 600° C and 400 to 2000 bars. Geochim Cosmochim Acta 58:4857–4866

Shishkina T, Botcharnikov R, Holtz F, Almeev RR, Jazwa AM, Jakubiak AA (2014) Compositional and pressure effects on the solubility of H_2O and CO_2 in mafic melts. Chem Geol 388: 112–129

Shmulovich KI, Graham CM (2004) An experimental study of phase equilibria in the systems H_2O–CO_2–$CaCl_2$ and H_2O–CO_2–NaCl at high pressures and temperatures (500–800 C, 0.5–0.9 GPa): geological and geophysical applications. Contrib Mineral Petrol 146:450–462

Silver L, Stolper E (1985) A thermodynamic model for hydrous silicate melts. J Geol 93:161–177

Silver L, Stolper E (1989) Water in albitic glasses. J Petrol 30:667–709

Silver LA, Ihinger PD, Stolper EM (1990) The influence of bulk composition on the speciation of water in silicate glasses. Contrib Mineral Petrol 104:142–162

Smythe DJ, Wood BJ, Kiseeva ES (2017) The S content of silicate melts at sulfide saturation: new experiments and a model incorporating the effects of sulfide composition. Am Mineral 102:795–803

Solomatova NV, Caracas R, Manning ME (2019) Carbon sequestration during core formation implied by complex carbon polymerization. Nat Commun 10:789

Solomatova N, Caracas R, Cohen R (2020) Carbon speciation and solubility in silicate melts. *In:* Carbon in Earth's Interior. Manning CE, Lin J-F, Mao WL (Eds) AGU Geophysical Monograph 249, p 179–194

Sosinsky DJ, Sommerville ID (1986) The composition and temperature dependence of the sulfide capacity of metallurgical slags. Metall Trans B 17B:331–337

Sparks RSJ (1978) The dynamics of bubble formation and growth in magmas: a review and analysis. J Volcanol Geotherm Res 3:1–37

Spera FJ (1974) A thermodynamic basis for predicting water solubilities in silicate melts and implications for the low velocity zone. Contrib Mineral Petrol 45:175–186

Spera FJ, Bergman SC (1980) Carbon dioxide in igneous petrogenesis: I Contrib Mineral Petrol 74:55–66

Spilliaert N, Allard P, Métrich N, Sobolev AV (2006) Melt inclusion record of the conditions of ascent, degassing, and extrusion of volatile-rich alkali basalt during the powerful 2002 flank eruption of Mount Etna (Italy) J Geophys Res 111:B04203

Stelling J, Botcharnikov RE, Beermann O, Nowak M (2008) Solubility of H_2O- and chlorine-bearing fluids in basaltic melt of Mount Etna at T= 1050–1250 C and P= 200 MPa. Chem Geol 256:102–110

Stevenson RJ, Bagdassarov NS, Dingwell DB, Romano C (1998) The influence of trace amounts of water on the viscosity of rhyolites. Bull Volcanol 60:89–97

Stolper EM (1982a) Water in silicate glasses: an infrared spectroscopic study. Contrib Mineral Petrol 81:1–17

Stolper EM (1982b) The speciation of water in silicate melts. Geochim Cosmochim Acta 46:2609–2620

Stolper EM (1989) Temperature dependence of the speciation of water in rhyolitic melts and glasses. Am Mineral 74:1247–1257

Stolper EM, Fine G, Johnson T, Newman S (1987) Solubility of carbon dioxide in albitic melt. Am Mineral 72:1071–1085

Studt PL, Shackelford JF, Fulrath RM (1970) Solubility of gases in glass—a monoatomic model. J Appl Phys 44: 2777–2780

Surono, Jousset P, Pallister J, Boichu MF, Budisantoso A, Costa F, Andreastuti S, Prata F, Schneider D, Clarisse L, Humaida H, Sumarti S, Bignami C, Griswold J, Carn S, Oppenheimer C, Lavigne F (2012) The 2010 explosive eruption of Java's Merapi volcano- a '100-year' event. J Volcanol Geotherm Res 241:121–135

Sykes D, Kubicki J (1993) A model for H_2O solubility mechanisms in albite melts from infrared spectroscopy and molecular orbital calculations. Geochim Cosmochim Acta 57:1039–1052

Symonds RB, Gerlach TM, Reed MH (2001) Magmatic gas scrubbing: implications for volcano monitoring. J Volcanol Geotherm Res 108:303–341

Taylor BE (1986) Magmatic volatiles: isotopic variations of C, H, and S. Rev Mineral 16:185–225

Temkin M (1945) Mixtures of fused salts as ionic solutions. Acta Phys Chim URSS 20:411–420

Thibault Y, Holloway JR (1994) Solubility of CO_2 in a Ca-rich leucitite: effects of pressure, temperature and oxygen fugacity. Contrib Mineral Petrol 116:216–224

Toop GW, Samis CS (1962a) Some new ionic concepts of silicate slags. Can Metall Q 1:129–152

Toop GW, Samis CS (1962b) Activities of ions in silicate melts. Trans Metall Soc AIME 224:878–887

Tournour CC, Shelby JE (2008a) Helium solubility in alkali silicate glasses and melts. Phys Chem Glasses Eur J Glass Sci Technol B49:207–215

Tournour CC, Shelby JE (2008b) Neon solubility in silicate glasses and melts. Phys Chem Glasses Eur J Glass Sci Technol B49:2237–44

Trémillon B (1971) La chimie en solvants non-aqueux. Presses Universitaires de France, Paris

Tucker JM, Mukhopadhyayc S, Gonnermann HM (2018) Reconstructing mantle carbon and noble gas contents from degassed mid-ocean ridge basalts. Earth Planet Sci Lett 496:108–119

Tuttle OF, Bowen NL (1958) Origin of granite in the light of experimental studies in the system $NaAlSi_3O_8$–$KAlSi_3O_8$–SiO_2–H_2O. Geol Soc Am Mem 74:153

Villemant B, Boudon G (1998) Transition from dome-forming to plinian eruptive styles controlled by H_2O and Cl degassing. Nature 392:65–69

Villemant B, Boudon G (1999) H_2O and halogen (F, Cl, Br) behavior during shallow magma degassing processes. Earth Planet Sci Lett 168:271–286

Villemant B, Boudon G, Nougrigat S, Poteaux S, Michel A (2003) Water and halogens in volcanic clasts: tracers of degassing processes during plinian and dome building eruptions. *In:* Volcanic Degassing. Oppenheimer, C, Pyle DM, Barclay, J (Eds.) Geol Soc London Spec Publ 213, p 63–79

Vuilleumier R, Seitsonen AP, Sator N, Guillot B (2015) Carbon dioxide in silicate melts at upper mantle conditions: Insights from atomistic simulations. Chem Geol 418:77–88

Virgo D, Mysen BO, Kushiro I (1980) Anionic constitution of 1-atmosphere silicate melts: Implications for the structure of igneous melts. Science 208:1371–1373

Wallace P, Carmichael IS (1992) Sulfur in basaltic magmas. Geochim Cosmochim Acta 56:1863–1874

Wallace P, Anderson AT, Jr (1999) Volatiles in Magmas. *In:* Sigurdsson H (Ed.) Encyclopedia of Volcanoes, Elsevier, p 149–170

Walter H, Roselieb K, Buttner H, Rosenhauer M (2000) Pressure dependence of the solubility of Ar and Kr in melts of the system SiO_2–$NaAlSi_2O_6$. Am Mineral 85:1117–1127

Wark DA, Hildreth W, Spear FS, Cherniak DJ, Watson EB (2007) Pre-eruption recharge of the Bishop magma system. Geology 35:235–238

White BS, Brearley M, Montana A (1989) Solubility of argon in silicate liquids at high pressures. Am Mineral 74:513–529

Webster JD (1997) Chloride solubility in felsic melts and the role of chloride in magmatic degassing. J Petrol 38:1793–1807

Webster JD, Rebbert CR (1998) Experimental investigation of H_2O and Cl^- solubilities in F-enriched silicate liquids; implications for volatile saturation of topaz rhyolite magmas. Contrib Mineral Petrol 132:198–207

Webster JD, Sintoni MF, De Vivo B (2009) The partitioning behavior of Cl, S, and H_2O in aqueous vapor–± saline–liquid saturated phonolitic and trachytic melts at 200 MPa. Chem Geol 263:19–36

Werner C, Evans WC, Kelly PJ, McGimsey R, Pfeffer M, Doukas M, Neal C (2012) Deep magmatic degassing versus scrubbing: Elevated CO_2 emissions and C/S in the lead-up to the 2009 eruption of Redoubt Volcano, Alaska. Geochem Geophys Geosystems 13:Q03015

Werner C, Kelly PJ, Doukas M, Lopez T, Pfeffer M, McGimsey R, Neal C (2013) Degassing of CO_2, SO_2, and H_2S associated with the 2009 eruption of Redoubt volcano, Alaska. J Volcanol Geotherm Res 259:270–284

Werner C, Fischer TP, Aiuppa A, Edmonds M, Cardellini C, Carn S, Chiodini G, Cottrell E, Burton M, Shinohara H, Allard P (2019) Carbon dioxide emissions from subaerial volcanic regions: Two decades in review. *In:* Deep Carbon Past to Present. Orcutt BN, Daniel I, Dasgupta R (Eds.) Cambridge University Press, p 188–236

Weston B, Burgess R, Ballentine D (2015) Disequilibrium degassing model determination of the ^3He concentration and $^3He/^{22}Ne$ of the MORB and OIB mantle sources. Earth Planet Sci Lett 410:128–139

Wilhelm E, Battino R, Wilcock RJ (1977) Low-pressure solubility of gases in liquid water. Chem Rev 77:219–262

Wilke M, Jugo PJ, Klimm K, Susini J, Botcharnikov R, Kohn SC, Janousch M (2008) The origin of S^{4+} detected in silicate glasses by XANES. Am Mineral 93:235–240

Witham F, Blundy J, Kohn SC, Lesne P, Dixon J, Churakov SV, Botcharnikov R (2012) SolEx: A model for mixed COHSCl-volatile solubilities and exsolved gas compositions in basalt. Computers Geosci 45:87–97

Withers AC, Zhang Y, Behrens H (1999) Reconciliation of experimental results on H_2O speciation in rhyolitic glass using *in-situ* and quenching techniques. Earth Planet Sci Lett 173:343–349

Wood BJ, Fraser DG (1977) Elementary Thermodynamics for Geologists: Oxford University Press

Woods A, Cardoso S (1997) Triggering basaltic volcanic eruptions by bubble-melt separation. Nature 385:518–520

Wulf R, Calas G, Ramos A, Buttner H, Roselieb K, Rosenhauer M (1999) Structural environment of krypton dissolved in vitreous silica. Am Mineral 84:1461–1463

Xue X, Kanzaki M (2004) Dissolution mechanisms of water in depolymerized silicate melts: Constraints from ^1H and ^{29}Si NMR spectroscopy and ab initio calculations. Geochim Cosmochim Acta 68:5027–5057

Xue X, Kanzaki M (2008) Structure of hydrous aluminosilicate glasses along the diopside–anorthite join: A comprehensive one-and two-dimensional ^1H and ^{27}Al NMR study. Geochim Cosmochim Acta 72:2331–2348

Yamamoto J, Burnard P (2005) Solubility controlled noble gas fractionation during magmatic degassing: implications for noble gas compositions of primary melts OIB and MORB. Geochim Cosmochim Acta 69:727–734

Yamashita S (1999) Experimental study of the effect of temperature on water solubility in natural rhyolite melt to 100 MPa. J Petrol 40:1497–1507

Young RW, Duffy JA, Hassall GJ, Xu Z (1992) Use of optical basicity concept for determining phosphorous and sulphur slag-metal partitions. Ironmaking Steelmaking 19:201–219

Zeng Q, Nekvasil H, Grey CP (1999) Proton environments in hydrous aluminosilicate glasses: a ^1H MAS, $^1H/^{27}Al$, and $^1H/^{23}Na$ TRAPDOR NMR study. J Phys Chem B 103:7406–7415

Zeng Q, Nekvasil H, Grey CP (2000) In support of a depolymeryzation model for water in sodium aluminosilicate glasses: information from NMR spectroscopy. Geochim Cosmochim Acta 64:883–896

Zajacz Z (2015) The effect of melt composition on the partitioning of oxidized sulfur between silicate melts and magmatic volatiles. Geochim Cosmochim Acta 158:223–244

Zajacz Z, Candela PA, Piccoli PM, Sanchez-Valle C (2012) The partitioning of sulfur and chlorine between andesite melts and magmatic volatiles and the exchange coefficients of major cations. Geochim Cosmochim Acta 89:81–101

Reviews in Mineralogy & Geochemistry
Vol. 87 pp. 557-574, 2022
Copyright © Mineralogical Society of America

11

High Pressure Melts

Tatsuya Sakamaki and Eiji Ohtani

Department of Earth Science
Graduate School of Science
Tohoku University
Sendai 980-8578
Japan

sakamaki@tohoku.ac.jp

INTRODUCTION

Magma has been deeply related to the evolution and dynamics of the Earth from the magma ocean in the early Earth to present magmatism. Better understanding of properties of melts at high pressure is critical to evaluate the nature and the fate of melts generated in the deep mantle. In the early Earth, melt properties controlled the crystallization of the magma ocean, which induced chemical segregation and evolution of the interior of the Earth. In addition, the magmatism of the current Earth may occur not only below volcanoes but also in the deep mantle according to the recent geophysical observations. Partial melting in the deep Earth's mantle controls mantle dynamics.

Here, we review the state of knowledge on the high-pressure properties of melts. Firstly, based on geophysical observations, we address the existence of magmas in the interior of the Earth, which yields constraints on the region of mantle melting. Secondly, we illustrate the physical properties of high pressure melts and discuss the migration of magmas in the interior of the Earth. Thirdly, we describe the properties of H_2O-bearing and CO_2-bearing melts for understanding the effect of volatile components on melts. The existence of these volatiles causes decreases in mantle melting temperatures and they are preferentially partitioned into the melts. Therefore, the melts generated in the deep mantle include a large amount of H_2O and/or CO_2 components. Finally, we discuss the terrestrial magma ocean—one of the biggest melting phenomena in the Earth's history—and its crystallization.

EXISTENCE OF MAGMA IN THE INTERIOR OF THE EARTH

Evidence for mantle melting

Partial melting of the Earth's mantle occurred in the uppermost mantle due to the decompression melting beneath mid-oceanic ridges and hot spots, and mantle wedges associated with slab subduction. However, several geophysical observations, such as seismological and magnetotelluric data, indicate that magmas can exist not only in shallower regions but also in the deeper mantle, such as lithosphere-asthenosphere boundary (LAB) (e.g., Schmerr 2012; Naif et al. 2013), the base of upper mantle (e.g., Revenaugh and Sipkin 1994; Tauzin et al. 2010; Toffelmier and Tyburczy 2007) and the base of lower mantle (e.g., Lay et al. 1998). Because the mantle geotherm (e.g., Katsura et al. 2010) is low compared to the melting curve of mantle rocks (e.g., Zhang and Herzberg 1994), deep mantle melting can be promoted by a supply of volatile components (e.g., H_2O and CO_2) which cause a strong depression of melting temperature of mantle rocks (e.g., Hirschmann et al. 1999).

1529-6466/22/0087-0011$05.00 (print)
1943-2666/22/0087-0011$05.00 (online)

http://dx.doi.org/10.2138/rmg.2022.87.11

H_2O in Earth's mantle

H_2O is the most important volatile in the Earth. The flux of H_2O on Earth has been estimated by several previous studies. For example, Peacock (1990) has suggested that the amount of H_2O degassed to the surface through magmatism is 2.0×10^{11} kg/year, and the H_2O flux returned to the mantle by subducting slabs is $\sim 8.7 \times 10^{11}$ kg/year. Thus, 6.7×10^{11} kg/year of H_2O move to the deep interior associated with slab subduction. According to Wallace (2005), there might be a balance in the flux of H_2O between the input through subducting slabs and the output through degassing through arc volcanism to the surface; both fluxes are 3×10^{11} kg/year. On the other hand, van Keken et al. (2011) estimated that one third of the bound H_2O subducted globally in slabs reaches 240 km depth (2.2–3.4×10^{11} kg/year), whereas two thirds of this H_2O is degassed through dehydration of the slabs during subduction (4.8–6.6×10^{11} kg/year). In spite of uncertainties, it is significant to specify the H_2O reservoirs in the mantle because a tiny amount of H_2O can change the properties of mantle materials.

Mantle transition zone

The mantle transition zone is considered to be a H_2O reservoir of the Earth's mantle due to a high water solubility of wadsleyite and ringwoodite (Inoue et al. 1995; Kohlstedt et al. 1996). Supporting evidence has been provided by a discovery of hydrous ringwoodite with 1.5 wt.% H_2O (Pearson et al. 2014) and phase Egg (Wirth et al. 2007) in the diamond inclusions of mantle xenoliths. Ringwoodite is a main component of the lower part of mantle transition zone, and phase Egg is one of the most significant hydrous aluminosilicate minerals (Ono et al. 1999; Sano et al. 2004; Abe et al. 2019). Since their stability field is consistent with a condition of the mantle transition zone, these observations strongly support the existence of a hydrous mantle transition zone. In addition, experimental studies indicate a large contrast in water storage capacity between the mantle transition zone and the upper and lower mantles. The water solubility of wadsleyite and ringwoodite is about 1–3 wt.% (Inoue et al. 1995; Kohlstedt et al. 1996), while those of olivine and bridgmanite are both less than 0.1 wt.% (Kohlstedt et al. 1996; Bolfan-Casanova et al. 2003).

The H_2O-rich mantle transition zone plays an important role as a trigger of mantle melting. In other words, owing to the large difference in H_2O solubility between minerals of mantle transition zone and those of upper and lower mantle, dehydration melting can occur at the boundaries (Bercovici and Karato 2003; Ohtani et al. 2004; Schmandt et al. 2014). For example, several geophysical observations strongly support the existence of melt at the base of the upper mantle (Revenaugh and Sipkin 1994; Song et al. 2004; Toffelmier and Tyburczy 2007). The gravitational stability of hydrous magmas at the base of upper mantle was also experimentally confirmed based on the melt density determinations (Matsukage et al. 2005; Sakamaki et al. 2006). In addition to the boundary between upper mantle and transition zone, a low-velocity seismic anomaly has also been detected at the top of the lower mantle (Schmandt et al. 2014; Liu et al. 2016 2018). This means H_2O-induced mantle melting or aqueous fluid generation occurs not only above but also below the mantle transition zone.

Ultralow velocity zone at the base of lower mantle

Seismological studies have indicated the large low-shear velocity provinces (LLSVP) under the African continent and in the Pacific basin (e.g., Helmberger et al. 2005; Garnero and McNamara 2008; McNamara et al. 2010; Yu and Garnero 2018). Although the low-shear velocity is observed, the LLSVP may not be caused by partial melting of the mantle but rather may be caused by compositional anomalies (e.g., Garnero and McNamara 2008; Deschamps et al. 2012; Ballmer et al. 2016; McNamara and Zhong 2005). On the other hand, the core–mantle boundary (CMB) region has been also examined by seismic analyses (e.g., Garnero et al. 1993; Garnero and Helmberger 1998; Wysession 1996), and an ultralow-velocity zone (ULVZ), which may be partially molten, has been observed (e.g., Williams and Garnero 1996). The distribution of

ULVZ is localized at the CMB and not laterally continuous. The ULVZ shows strong reductions in seismic velocities (e.g., Wen and Helmberger 1998; Garnero and Helmberger 1998; Garnero et al. 1993) and is denser than the surrounding mantle (Rost et al. 2006).

The anomalous seismic features are discontinuously distributed on the CMB, and continuous melting may occur at certain regions by following reasons. Seismic tomography studies have demonstrated that subducted slabs extended to the base of the lower mantle (e.g., van der Hilst et al. 1997; van der Hilst and Karason 1999). Thus, the subducting slab may play an important role in melting at the base of lower mantle. For example, the slab presumably transports H_2O to the deep mantle. The newly discovered hydrous phase, phase H (e.g., Nishi et al. 2014), has a stability field that can be expanded to the lowermost mantle by dissolution of the hydrous AlOOH component (e.g., Ohtani et al. 2014; Ohira et al. 2014). If the stagnant slab with hydrous phase is heated up at the bottom region of the lower mantle, H_2O will be released from the hydrous phase and should influence the formation of ULVZ. In addition, the presence of dense melt at the base of the lower mantle has been suggested by some authors based on experimental studies and *ab-initio* calculations (Ohtani 1983; Stixtude and Karki 2005; Mosenfelder et al. 2007). There are some triggers for mantle melting except for the supply of H_2O. For example, the basaltic component can also decrease the melting temperature of mantle rocks because the solidus temperature of basalt under the CMB conditions is significantly lower than the melting of the average mantle (Andrault et al. 2014). On the other hand, a large amount of FeO could also lower the melting temperature of the mantle rocks (Mao et al. 2005). Thus, the melting of mantle may occur due to the enrichment of FeO, which may be caused by the reaction between liquid outer core and lower mantle minerals (e.g., Knittle and Jeanloz 1991).

PHYSICAL PROPERTY OF HIGH PRESSURE MELTS

Density of melt at high pressure

The density of silicate melts is a central property in influencing not only volcanic activity but also the differentiation of the Earth. Density measurements have been conducted using several methods, including shock compression (e.g., Rigden et al. 1984 1989), sink-float technique in a large-volume press (e.g., Agee and Walker 1988; Ohtani et al. 1993; Suzuki et al. 1995; Agee 1998; Sakamaki et al. 2006), and synchrotron X-ray absorption measurement combining with large-volume presses (e.g., Sakamaki et al. 2009 2010a,b 2011 2013; Sakamaki 2017a; van Kan Parker et al. 2012; Seifert et al. 2013; Malfait et al. 2014a,b; Crépisson et al. 2014).

The dynamic responses of the melts to pressure are very different from those of solids. For example, the quantitative comparison in compressibility between melt and crystal of diopside and anorthite composition has been reported by Ai and Lange (2008), who demonstrated that melt and solid of the same composition have strikingly different compressibilities. The enhanced compressibility of silicate melts at low pressures relative to the crystalline form presumably reflects the response of intermediate-range order structures, due to the evolution of the topology of tetrahedral connectivity towards a more compact network structure. In addition to the higher compressibility of melts, an amount of iron is an important factor controlling the density of magma because iron is heavier component and is generally partitioned into the silicate melts compared to coexisting mantle minerals. The large contrast in the compressibility between melts and crystals together with partitioning of iron into the melts causes density crossover between dense silicate melt and mantle. This allows for melts to stagnate in the mantle due to their gravitational stability.

The density of melts at higher pressure conditions has been studied using first-principles molecular dynamics simulations. Silicate melts are highly compressible at low pressure, which can be attributed to large pressure-induced increases in all cation–anion coordination types

(Bajgain et al. 2015). The changes in the coordination number with increasing pressure become gradual as pressure increases because the distances of cation–anion bonds decrease systematically with pressure. In essence, the melts become much more incompressible ("stiffer") at high pressure.

A melt–crystal density crossover is thus made possible in the deep mantle because of the high compressibility and Fe enrichment of the liquid phase. For example, the basaltic melt density can actually exceed the mantle density at one or more depths (e.g., Ohtani and Maeda 2001; Sanloup et al. 2013b; Bajgain et al. 2015). Compression curves of basaltic melts are shown in Figure 1. This result implies that dense melt could be gravitationally stable at those depths, thereby providing a plausible explanation for seismic low-velocity anomalies. Although there are differences in density in three studies under lower mantle conditions in Figure 1, the densities of basaltic melt are larger than Preliminary reference Earth model (PREM: Dziewonski and Anderson 1981) in the lowermost mantle. This result implies that an existence of stagnant melts at the base of lower mantle where seismic low velocity anomalies have been observed.

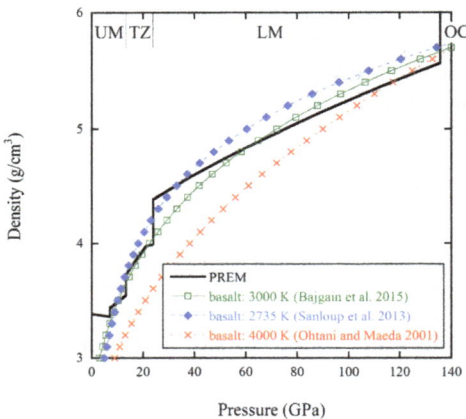

Figure 1. Compression curves of basaltic magmas up to the base of the lower mantle. UM = upper mantle; TZ = transition zone; LM = lower mantle; OC = outer core. The density profile of Preliminary reference Earth model (PREM) is also shown for comparison. Compression curves of Bajgain et al. (2015) and Sanloup et al. (2013b) are fitted by the fourth order Birch–Murnaghan equation of state, that of Ohtani and Maeda (2001) is represented by Vinet equation of state. There is a density crossover between the basaltic magma and Earth's mantle at several depths.

Viscosity of melt at high pressure

The viscosity of silicate melts controls their mobility, and gives us important knowledge about the timescale of volcanic activities and eruption styles in the present Earth. In addition, viscosity plays a central role in the dynamics of a terrestrial magma ocean, core formation, and chemical differentiation of the mantle in the Earth's early history. For silicate melts, viscosity measurements at high pressure have been performed by a falling sphere method using a large volume press (Kushiro 1976, 1978a; Brearley et al. 1986; Mori et al. 2000). The falling sphere method has been significantly advanced by combination with synchrotron-based X-ray radiography, which enables us to determine the melt viscosity more accurately (Suzuki et al. 2002, 2005, 2011; Reid et al. 2003; Liebske et al. 2005; Sakamaki et al. 2013). Because the viscosity of silicate melts is considered to be closely related to the network structure, simultaneous measurements of viscosity and structure at high pressure and high temperature significantly have enhanced our understanding of the relationship between viscosity and structure (e.g., Sakamaki et al. 2013; Kono et al. 2013, 2014a,b 2015).

The ratio between the number of non-bridging oxygen anions (NBO) and that of tetrahedrally coordinated cations (T), NBO/T, is widely used as an indicator of the network structure of silicate melts. The degree of polymerization, which increases with decreasing NBO/T, has long been recognized as a fundamental structural factor influencing physical properties (e.g., Kushiro 1986). Also, the viscosity of silicate melts is characterized by the degree of polymerization. Polymerized melts show higher viscosity at ambient pressure due to their strong three-dimensional network featuring oxygen bridges between tetrahedral cations. Depolymerized melts have lower tetrahedral connectivity and lower viscosity (e.g., Scarfe and Cronin 1986). The contrasting behavior between polymerized and depolymerized melts is also shown in the pressure dependence of viscosity. Although free-volume theory predicts that viscosity increases with pressure (Cohen and Turnbull 1959), isothermal viscosity of polymerized melts (e.g., basaltic melt) decreases with pressure up to 3–5 GPa, above which it turns over to positive pressure dependence (Sakamaki et al. 2013).

The viscosity of several silicate melts as a function of pressure is shown in Figure 2. Typical fully polymerized melts, such as $NaAlSi_3O_8$ and $NaAlSi_2O_6$ melts (Kushiro 1978a; Mori et al. 2000; Suzuki et al. 2011), show a rapid decrease in viscosity with pressure, while the viscosity of depolymerized melts (e.g., $CaSiO_3$, $MgSiO_3$, Fe_2SiO_4 melts) is less sensitive to pressure. In the case of $NaAlSi_2O_6$ melt, its melt structure is also examined under high pressure conditions using X-ray diffraction technique (Sakamaki et al. 2012). Combining viscosity data with structural analysis, a large decrease in the viscosity with pressure is caused by structural changes: the T–O–T angle decreases and the bonds were broken and reformed. Moreover, these structure and viscosity studies of $NaAlSi_2O_6$ melts are consistent with density measurements of the melt (Sakamaki 2017a). A densification of $NaAlSi_2O_6$ melt at low pressure is likely to be dominated by the topological rearrangements. Summarizing the previous studies for polymerized melts, the negative pressure dependence of viscosity in polymerized melts is related to the shrinkage of the intermediate-range order network structure, which causes the narrowing of the T–O–T angle and weakening of the T–O–T network. Experimental data demonstrate that polymerized melts generally possess much higher values of viscosity than those of depolymerized ones and show higher sensitivity to pressure. In sharp contrast, the structure of the $CaMgSi_2O_6$ melt (depolymerized melt: $NBO/T = 2$) is more closely packed and shows little change with increasing pressure (Wang et al. 2014), resulting in an increase in viscosity (Taniguchi 1992; Reid et al. 2003).

Recently, the viscosity measurement of depolymerized melts using an ultrahigh-speed camera led to a new insight (Spice et al. 2015; Cochain et al. 2017). These melts showed small negative pressure dependence of viscosity, although previous works expected an increase in viscosity with pressure. Even fully depolymerized Fe_2SiO_4 melts ($NBO/T = 4$: Spice et al. 2015) denoted a slight decrease in viscosity with pressure, and this may relate to the increase in Fe–O coordination number, which has been identified by structural measurement using X-ray diffraction technique (Sanloup et al. 2013a). $MgSiO_3$ and $CaSiO_3$ melts ($NBO/T = 2$: Cochain et al. 2017) have higher viscosity than Fe_2SiO_4 melt. The viscosity feature in three melts shown in Figure 2 may reflect the difference in degree of polymerization between Fe_2SiO_4 melt and $MgSiO_3/CaSiO_3$ melts.

As mentioned above, the pressure dependence of viscosity for silicate melts is unique and is deeply related to the degree of polymerization. Combing the experimental results with computational simulations, Wang et al. (2014) proposed the following model for better understanding of melt viscosity under high pressure conditions. In polymerized melts, the pressure of the viscosity turnover corresponds to the tetrahedral packing limit, below which the structure is compressed through tightening of the intertetrahedral bond angle, resulting in continual breakup of tetrahedral connectivity, reduction in the ratio of bonding oxygen, and viscosity decrease. Above the turnover pressure, the coordination number of Si and/or Al increases to allow further packing and formation of triclusters, with increasing viscosity. Wang et al. (2014) also mentioned the viscosity change of depolymerized melt, which has a less breakable TO_4 network because some corners are already unconnected. Therefore, the network structure rearranges without breaking significant amounts of BO bonds, and the viscosity becomes less sensitive to pressure.

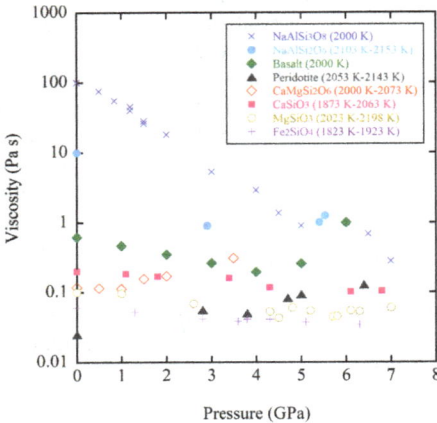

Figure 2. Viscosity of silicate melts as a function of pressure. Following data are used: $NaAlSi_3O_8$ at 2000 K (Kushiro 1978a; Mori et al. 2000), $NaAlSi_2O_6$ at 2103–2153 K (Suzuki et al. 2011), basalt at 2000 K (Sakamaki et al. 2013), peridotite at 2053–2143 K (Liebske et al. 2005), $CaMgSi_2O_6$ at 2000–2073 K (Taniguchi 1992; Reid et al. 2003), $CaSiO_3$ at 1873–2063 K and $MgSiO_3$ at 2023–2198 K (Cochain et al. 2017), Fe_2SiO_4 at 1823–1923 K (Spice et al. 2015). Polymerized silicate melts show higher viscosity and their pressure dependency is clearly negative, whereas the viscosity of depolymerized melts is less sensitive to pressure. Apparent reversal in pressure dependence is observed in basaltic melt.

Although the NBO fraction in silicate glass decreases gradually with pressure (e.g., Lee 2011), NBO/T under ambient conditions seems to be a good indicator to understand the behavior of silicate melts. The situation becomes especially difficult at high pressures, as NBO/T is strongly pressure-dependent and becomes meaningless when the Si, Al, and Fe cations become octahedrally coordinated. This model can complement previous thermodynamic treatments on the pressure dependence of the viscosity of silicate melts (Bottinga and Richet 1995). Pressure-induced changes in the amounts of BO and NBO in the network structure may provide a physical basis for the contribution of polymerization to configurational entropy in the Adam–Gibbs viscosity model (Bottinga and Richet 1995; Lesher 2010). The contrasting behavior between nominally polymerized and depolymerized melts diminishes with pressure in response to the smaller structural distinction between these melts. As shown in Figure 2, in the case of ambient pressure, the viscosity of fully polymerized melts is different by about four orders of magnitude from that of depolymerized melts. On the other hand, the viscosity difference becomes one order of magnitude around 7 GPa. Thus, above 7 GPa, NBO/T is no longer relevant for defining the structure, and viscosities of silicate melts fall into a narrower range.

First-principles molecular dynamics simulations provide valuable knowledge about viscosity and diffusion of silicate melts at much higher pressure. The calculated viscosity of melts at high pressure is sensitive to temperature (e.g., Karki et al. 2011, 2018; Verma and Karki 2012). In the case of basaltic melt (Karki et al. 2018), the viscosity increases monotonically with pressure at high temperature (> 4000 K). In contrast, the viscosity at lower temperature (< 3000 K) decreases initially and then increases rapidly with pressure. The calculated activation volume shows slightly negative value at low pressure and temperature, implying an anomalous dynamic regime, and the volume becomes positive under much higher pressure. The anomalous behavior of the melt viscosity with pressure at low pressure is consistent with experimental studies (e.g., Sakamaki et al. 2013). Computational and experimental results propose that the low-pressure dynamic anomaly is a universal behavior for silicate melts.

H_2O-BEARING MELTS

Generation of hydrous magma in the interior of the Earth

The Earth is the water planet, and the majority of its surface is covered by oceans, representing a large amount of H_2O. The existence of H_2O strongly influences Earth's dynamics, and H_2O plays an important role in the magmatism of the interior of the Earth.

It is well known that small amount of H_2O decreases the melting temperature of mantle rocks. If free H_2O exists in the deep mantle, the H_2O causes a decrease in melting temperature of surrounding rocks and generates hydrous magmas.

Subducting slabs containing hydrous minerals provide H_2O into the mantle transition zone, and H_2O is stored as hydrous phases such as wadsleyite and ringwoodite. Since these minerals have comparatively high H_2O capacity (Inoue et al 1995; Kohlstedt et al. 1996), the mantle transition zone can act as a H_2O reservoir and source as described in an earlier section. Therefore, hydrous conditions can be created at the bottom of the upper mantle and the top of the lower mantle which have lower capacity of H_2O than mantle transition zone. For example, voluminous continental flood basalts can be produced by H_2O-bearing melts derived from local upwelling of hydrous transition zone rocks (e.g., Wang et al. 2015). Also, geophysical studies proposed an existence of melt layer above the 410 km discontinuity (e.g., Revenaugh and Sipkin 1994; Toffelmier and Tyburczy 2007; Tauzin et al. 2010), and dehydration melting may occur at the boundary between upper mantle and mantle transition zone. Similar phenomena have been suggested in the region below 660 km discontinuity (e.g., Schmandt et al. 2014).

Although the solubility of H_2O in silicate melts is low at lower pressure, it rises dramatically with increasing pressure (e.g., Khitarov and Kadik 1973). Thus, silicate melts are important reservoir for H_2O and play a crucial role in water circulation in the interior of the Earth (Hirschmann 2006; Hirschmann and Kohlstedt 2012). The addition of H_2O also fundamentally changes physical properties of melt, such as density and viscosity, which strongly influence mobility of magmas.

H_2O effect on the melt density

The density of silicate melt can be calculated using the partial molar volumes of component oxides and their pressure- and temperature- dependences (Bottinga and Weill 1970; Lange and Carmichael 1987 1989). The addition of H_2O decreases the density of a silicate melt due to the large partial molar volume of H_2O in silicate melts. The density of hydrous magmas and the partial molar volume of H_2O have been measured over wide ranges of temperature, pressure, and composition using several techniques. Experimental studies indicate that the partial molar volume of H_2O in hydrous silicate melts is lower than in free H_2O, and independent of melt composition and structure (Orlova 1964; Burnham and Davis 1971; Ochs and Lange 1997 1999; Richet and Polian 1998; Richet et al. 2000). Important insights have been obtained based on studies of hydrous glasses quenched from high pressure and temperature, and the compositional independence of the partial molar volume of H_2O has been demonstrated (Ochs and Lange 1997, 1999). The use of sink–float techniques using a large volume apparatus has enabled us to expand the experimental condition of density measurements and a wide range of data for hydrous melt has been collected (Matsukage et al. 2005; Sakamaki et al. 2006; Agee 2008; Jing and Karato 2012). In addition, synchrotron X-ray absorption method is also powerful tool for precise density measurement of hydrous silicate melts (e.g., Sakamaki et al. 2009; Malfait et al. 2014a,b). The experimental data have been incorporated into an equation of state for the partial molar volume of H_2O in silicate melt and the density of hydrous silicate melts can be calculated at high pressure and temperature up to the top of lower mantle conditions (Sakamaki 2017b). The equation of state could reproduce most experimental data as shown in Figure 3. A rapid decrease in the partial molar volume of H_2O in silicate melt can be found at lower pressure, and the change in the volume becomes small with pressure. The pressure effect on the partial molar volume of H_2O in melt is divided into three pressure regions: 1) high compressibility region below 0.5 GPa; 2) intermediate region between 0.5 GPa and 5 GPa; 3) low compressibility region above 5 GPa. The pressure effect on the partial molar volume of H_2O in melts is about 10 times larger than other oxides, such as SiO_2, Al_2O_3, FeO, MgO and CaO, below 0.5 GPa. However, the pressure-induced decrease in the partial molar volume of H_2O in melts becomes smaller with pressure, especially above 5 GPa. Thus the compressibility of hydrous magmas is much higher than that of dry melts only under lower pressure condition (< 5 GPa) and the compressibility contrast between hydrous and dry melts is greatly reduced with pressure. Figure 3 also indicates that the temperature effect on the partial molar volume of H_2O in melt decreases with increasing pressure. At ambient pressure, the difference in partial molar volume of H_2O in melt at temperatures between 2473 K and 1273 K is 19.0 cm^3/mol. The differences at 10 GPa and 20 GPa are 5.1 cm^3/mol and 3.1 cm^3/mol, respectively.

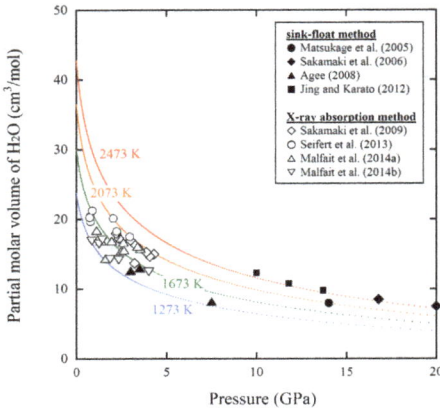

Figure 3. Isothermal compression curves of the partial molar volume of H_2O in silicate melt. The curves are drawn by the Vinet equation of state with parameters obtained by Sakamaki (2017b). Experimental results of the partial molar volume of H_2O are also plotted for comparison. The results of sink-float method and X-ray absorption method are shown by close and open symbols, respectively. The pressure dependence of the partial molar volume of H_2O is divided into following three regions: 1) high compressibility region below 0.5 GPa; 2) intermediate region between 0.5 and 5 GPa; 3) low compressibility region above 5 GPa.

The behavior of hydrous magmas above and below the mantle transition zone

Understanding the pressure- and temperature-dependence of density of hydrous silicate liquids is essential for discussing the ponding of hydrous silicate melts in the interior of the Earth, where geophysical anomalies have been observed. Assuming that melt compositions formed at the base of the upper mantle are similar to partial melts formed by melting of dry peridotite (IT8720: Ito and Takahashi 1987), the density of hydrous melt with 8 wt% H_2O is estimated to be equal to that of PREM (Dziewonski and Anderson 1981) at the base of upper mantle conditions. Compression curves of magmas are shown in Figure 4, which compares PREM with densities of dry and hydrous melts. The driving force for ascending magma is the density contrast between the magma and the surrounding mantle rocks. Because the density of hydrous magma is close to that of the mantle at a depth of 410-km seismic discontinuity, the magma does not move upward and stagnate around the 410-km discontinuity. In other words, the hydrous magma may be gravitationally stable in the case that the H_2O content of magma is less than 8 wt%, and may be responsible for the seismic anomaly at the base of lower mantle (e.g., Revenaugh and Sipkin 1994; Tauzin et al. 2010).

The dehydration melting has been suggested not only at the base of the upper mantle but also at the top of the lower mantle (Schmandt et al. 2014). In the case of melting at the top of lower mantle under hydrous condition, the compositions of partial melts are MgO-rich and SiO_2-poor (Nakajima et al. 2019). Since the melt density is small compared to the density of mantle below the 660-km seismic discontinuity, the magma can float at the top of lower mantle. The residual mantle rocks separate from the melt due to the large density difference at the top of lower mantle. The downward moving mantle rock is poor in melt components, that is, the lower mantle composition below the melt layer can be MgO-poor and SiO_2-rich. The melt at the top of lower mantle acts as a filter, and MgO-poor and SiO_2-rich residual rocks sinks to lower mantle. These results can provide a possible mechanism for formation of the chemical heterogeneity at the top of the lower mantle.

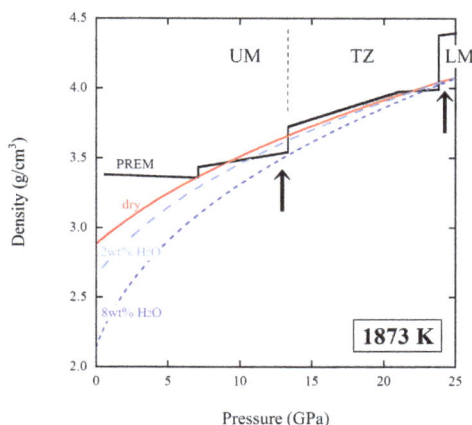

Figure 4. Comparison of density between magmas and PREM. UM = upper mantle; TZ = transition zone; LM = lower mantle. The curves are drawn by the Birch–Murnaghan equation of state with parameters at 1873 K obtained by Sakamaki (2017b). Hydrous peridotitic magma formed by dehydration melting can be gravitational stable at the base of upper mantle and the top of lower mantle which are indicated by arrows.

The important point is that hydrous silicate melts can be denser at the base of upper mantle and less dense at the top of lower mantle compared to surrounding mantle rocks as shown by arrows in Figure 4 (Nakajima et al. 2019; Ohtani et al. 2018). These ponded melts would explain the geophysical anomaly and some H_2O-rich melts may go back to the mantle transition zone.

H_2O effect on the melt viscosity

Although the viscosity of melts is affected by pressure, temperature and compositions, H_2O has a greater effect on viscosity than any other oxide components (e.g., Dingwell 2015). The principal effect of H_2O is to depolymerize the silicate network and thus weakening of the silicate melt, which results in the strong decrease in melt viscosity. The structural role of H_2O in silicate melts is different from that of the other network modifiers (Giordano et al. 2008). The greatest effect on reduction of melt viscosity is for the first few wt% of dissolved H_2O (Dingwell et al. 1996; Richet et al. 1996; Whittington et al. 2001). For example, the influence of H_2O on viscosity of rhyolite melt is strong at low H_2O contents: 1 wt.% H_2O corresponding to as much as 4 log-units in viscosity (e.g., Ardia et al. 2008). In the case of high H_2O-contents (>10 wt.%), the decrease in melt viscosity is only an additional 1 log-unit.

Hui et al. (2009) inferred viscosity of hydrous rhyolitic melts containing 0.8–4 wt% H_2O at pressures up to 2.83 GPa in the high viscosity range. This represents the first viscosity data at low temperature (high viscosity) and pressures above 1 GPa. Combining the data of this study with literature data (Shaw 1963; Schulze et al. 1996; Ardia et al. 2008), they constructed a general viscosity model of rhyolitic melts as a function of temperature, pressure and H_2O-content from low to high viscosity range. The model can be utilized for interpolation within the range of available data. Based on the model, although the strongest effect of pressure on the viscosity of rhyolitic melts is observed at low H_2O contents, the pressure effect on the melt viscosity is hidden at high H_2O contents.

Dissolved H_2O in silicate melt causes a more mobile melt structure, which results from the depolymerization of the melt by forming hydroxyl groups. In other words, the addition of H_2O may change the pressure dependence of melt viscosity. For example, viscosity of dry andesite melt decreases with pressure (Kushiro et al. 1976), while that of hydrous andesite melt with 3 wt% H_2O increases with pressure (Kushiro 1978b). This means that the negative pressure dependence of viscosity of dry polymerized melt may change to positive pressure dependence after adding a certain amount of H_2O. In the case of basaltic melts, which are intermediate polymerized melts, the viscosity considerably decreases with H_2O dissolution at moderate pressure and remains unchanged at high pressures (Persikov 1991). The unique feature of dry basaltic melts is viscosity minimum around 5 GPa (Sakamaki et al. 2013), and the hydrous basaltic melt retains the negative pressure dependence of the viscosity and the minimum viscosity at about 5 GPa. In contrast, H_2O dissolution in kimberlite melts, which are highly depolymerized melts, has almost no significant effect on the change in the viscosity at high pressure (Persikov et al. 2017) because there are less networks broken by H_2O. Thus, the viscosity of hydrous melts and its pressure dependence are related to the initial degree of polymerization; polymerized melts are more sensitive to the addition of H_2O than depolymerized melts.

CO_2-BEARING MELTS

CO_2-rich magma in the interior of the Earth

Carbon dioxide is almost as significant as water in terms of volatile component in the Earth's interior. Carbon-rich magmas, such as kimberlite, are generated in the deep mantle (e.g., Keshav et al. 2005). CO_2-induced decreases in the melting temperature of mantle peridotite have been well known (e.g., Ghosh et al. 2014). Since CO_2 prefers to be partitioned into silicate melts relative to mantle minerals in the interior of the Earth, carbon-rich magma may exist in the interior of the Earth (e.g., Dasgupta and Hirschmann 2010; Keshav et al. 2011). For example, the presence of carbonatitic melt in the asthenosphere has been supported experimentally based on the electrical conductivity measurement (Gaillard et al. 2008).

The oxygen fugacity decreases significantly with depth and the lower mantle is likely to be under strongly reducing conditions (Frost and McCammon 2008). An upwelling mantle could undergo redox melting across the discontinuity between the lower mantle and mantle transition zone (e.g., Rohrbach and Schmidt 2011). On the other hand, Stagno and Frost (2010) argued that diamond is stable in the deep upper mantle in the normal mantle oxygen fugacity, and the CO_2-bearing melt can be formed only due to the redox melting of uprising plumes of diamond-bearing peridotites. The process results in formation of CO_2-rich melt pockets. Since carbon plays an important role in the Earth's interior, the physicochemical property of carbon–bearing melts is an interesting research target (e.g., Dasgupta 2013).

CO_2 effect on the melt density

Because CO_2 has very high solubility in molten silicates compared to silicate minerals (Shcheka et al. 2006; Keppler et al. 2003), most of CO_2 are partitioned into silicate melts (Dasgupta and Hischmann 2010; Ni and Keppler 2013; Dasgupta et al. 2013). The dissolution of CO_2 in silicate melts reduces density (e.g., Liu and Lange 2003). At low pressures, this effect will be small due to the low solubility of CO_2 in the melts (e.g., Khitarov and Kadik 1973). However, it may become very significant under deep mantle conditions because the CO_2 solubility into the melts increases with pressure. Based on density measurements at high pressure, the partial molar volume of CO_2 in silicate melts was determined experimentally (e.g., Ghosh et al. 2007; Sakamaki et al. 2011). As pressure increases, the partial molar volume of CO_2 decreases rapidly initially and then gradually. Since the partial molar volume of CO_2 in silicate melts is less compressible than that of H_2O as shown in Figure 5, CO_2 is more effective in reducing the density of the melts at high pressure (Sakamaki et al. 2011). First principles molecular dynamics simulations also calculated the partial molar volume of CO_2 in melt up to lower mantle conditions (Ghosh et al. 2017), the partial molar volume of CO_2 for all of previous works converges to 10 cm^3/mol above 100 GPa. Compared to this value, the partial molar volumes of H_2O is systematically smaller (< 10 cm^3/mol) at higher pressure (Mookherjee et al. 2008).

CO_2-induced partial melting has been proposed to occur at the top of the mantle transition zone and in the uppermost parts of the lower mantle either owing to the changes in the oxidation state (Rohrbach and Schmidt 2011; Stagno et al. 2013; Dasgupta et al. 2013) and/ or due to the lowering of solidus temperatures so that the solidus intersects the mantle adiabat

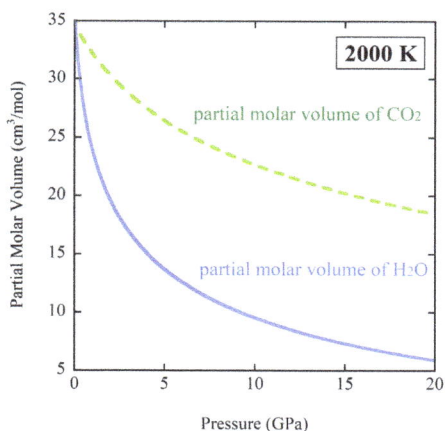

Figure 5. Comparison of compression curve of the partial molar volume of CO_2 in melt with that of H_2O in melt. The curves indicate the partial molar volume at 2000 K expressed by the Vinet equation of state using parameters for CO_2 reported by Ghosh et al. (2007) and Sakamaki et al. (2011) and H_2O reported by Sakamaki et al. (2006) and Sakamaki (2017b), respectively. Due to less compressibility of the partial molar volume of CO_2 in silicate melts, the volume is enough larger than that of other components.

(Litasov et al. 2013; Thompson et al. 2016). The CO_2-bearing silicate melts provide us a viable mechanism to explain geophysical anomalies of the deep mantle (e.g., Revenaugh and Sipkin 1994; Schmandt et al. 2014). Based on sink-float experiments and the first principles molecular dynamics simulations, the addition of 5-8 wt.% of CO_2 can produce a neutrally buoyant silicate melt at the top of the mantle transition zone (Ghosh et al. 2007; Ghosh et al. 2017). The calculated density of the silicate melt containing 5 wt.% CO_2 at 2200 K is greater than the mantle density at the bottom of the mantle transition zone and smaller than the density at the top of lower mantle (Ghosh et al. 2017). Similarly, the melt density with 5 wt.% CO_2 at 4000 K may exceed the density of lowermost mantle. Although CO_2 lowers the density of silicate melts, the effect is modest under lower mantle conditions. Hence, silicate melt atop the core–mantle boundary could keep large amounts of CO_2 efficiently without gravitational instability.

CO_2 effect on the melt viscosity

High-pressure falling-sphere experiments demonstrated that dissolved CO_2 induced a decrease in viscosity of $NaAlSi_3O_8$ melt at high pressure, while no detectable change in viscosity was observed in $CaNaAlSi_2O_7$ melt with the addition of 2 wt% CO_2 (Brearley and Montana 1989). In addition, the viscosity of $KAlSi_3O_8$ melt (White and Montana 1990) and $NaAlSi_2O_6$ melt (Suzuki 2018) decreased with addition of CO_2. The viscosity of CO_2–bearing $NaAlSi_2O_6$ melt was one to two orders of magnitude lower than that of the pure jadeite melt. Focusing on the linear relationship between viscosity and reciprocal temperature, the temperature dependence of the viscosity of $NaAlSi_2O_6$ + 0.5 wt% CO_2 melt (Suzuki 2018) was similar to that of $KAlSi_3O_8$ + 0.5 wt% CO_2 melt (White and Montana 1990), which may imply a similar effect on the viscosity by dissolving CO_2. The slope of linearity between viscosity and reciprocal temperature becomes steeper with increasing dissolved CO_2 amount in the melt (Suzuki 2018). The changes in slope may relate to a structural modification of silicate melts by the addition of CO_2. Classifying melts based on the composition, $NaAlSi_3O_8$, $KAlSi_3O_8$ and $NaAlSi_2O_6$ melts are fully polymerized, while the $CaNaAlSi_2O_7$ melt has a relatively depolymerized network structure. Because the viscosity of $CaNaAlSi_2O_7$ melts is less sensitive to the dissolving CO_2, there is a possibility that the difference in CO_2 effect on the viscosity is caused by the difference in degree of polymerization of these melts.

Viscosity of carbonate melt

In the early stages of in-situ measurement, viscosities of carbonate melts ($K_2Mg(CO_3)_2$ and $K_2Ca(CO_3)_2$ compositions) have been examined at high pressure by Dobson et al. (1996). A recent study succeeded in measuring the viscosity of $CaCO_3$ and dolomite, $(Mg_{0.40}Fe_{0.09}Ca_{0.51})$

CO_3 melts by using the ultrafast X-ray imaging, and the melts showed ultralow viscosity (Kono et al. 2014a). There is almost no compositional dependence in the viscosity between $CaCO_3$ and $(Mg_{0.40}Fe_{0.09}Ca_{0.51})CO_3$ melts and no pressure effect on the viscosity was observed under experimental conditions up to 6.2 GPa. Moreover, the viscosity of Na_2CO_3 melts was also measured using the same technique (Stagno et al. 2018). Based on recent viscosity measurements of the carbonate melts, the melts have negligible pressure dependence, whereas they have a small compositional dependence among alkali metal-bearing and alkaline earth metal-bearing carbonate melts. These low viscosities of carbonate melts are orders of magnitude lower than those of silicate melts as shown in Figure 6, where viscosities of volatile-bearing silicate melts are also plotted for comparison. Although the dissolution of volatile components in the silicate melts decreases the melt viscosity efficiently, the carbonate melts show much less viscosity.

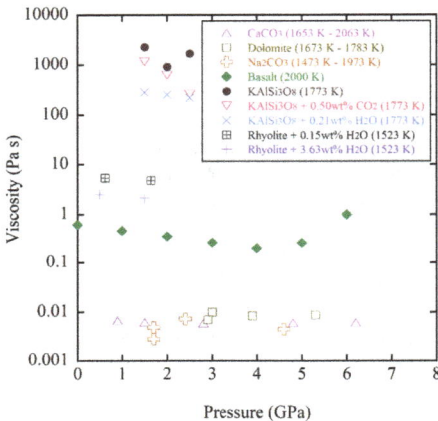

Figure 6. Viscosity of carbonate melts as a function of pressure. The viscosities of $CaCO_3$ (1653–2063 K) and dolomite (1673–1783 K) melts were reported by Kono et al. (2014a) and those of Na_2CO_3 melt (1473–1973 K) were measured by Stagno et al. (2018), respectively. For comparison, the viscosity of dry basaltic melt at 2000 K (Sakamaki et al. 2013), dry and CO_2/H_2O bearing $KAlSi_3O_8$ melts at 1773 K (White and Montana 1990), and hydrous rhyolite melts at 1523 K (Ardia et al. 2008) are also shown. The viscosity of carbonate melts is much smaller than silicate melts.

MAGMA OCEAN IN THE EARLY EARTH

Melting in the early Earth's mantle

Important melting of the early Earth was caused by the accretion of the Earth in primordial or steam atmosphere (Kaula 1979; Hayashi et al. 1979; Abe and Matsui 1986; Olsen and Sharp 2019) and also by the giant impact event (e.g., Nakajima and Stevenson 2015), which heated the interior of the Earth by the release of energy induced by the collision and subsequent gravitational segregation of the core (Herzberg et al. 2010; Rubie et al. 2015). The CMB condition might have led to temperatures higher (Davies et al. 2015; Labrosse 2015; Nakagawa and Tackley 2010) than the mantle solidus (e.g., Ohtani 1983; Andrault et al. 2011; Fiquet et al. 2010). Therefore, the accretional and/or giant impact event could have caused melting of the entire mantle.

Dynamics of the magma ocean crystallization

The crystallization of the magma ocean is the primary event in the early Earth for the formation of a chemical stratification (e.g. Ohtani 1985; Boyet and Carlson 2005; Caro et al. 2005). In terms of the crystallization depth of magma ocean, the difference in temperature between mantle liquidus (e.g., Ohtani 2009; Fiquet et al. 2010; Andrault et al. 2011) and the adiabatic profile of magma ocean is significant. In other words, the crystallization of magma ocean might start at the bottom or in the intermediate depth of the lower mantle where the temperature crossover between magma ocean and mantle liquidus would have occurred. If bridgmanite, which is a liquidus phase, segregated from the crystallizing magma ocean, the melt composition should shift toward higher Fe content because Fe enters preferentially into the melt (e.g. Boukare et al. 2015). Although early formed crystals sank, the crystals reached neutral buoyancy after some amount of bridgmanite removal from magma ocean (Andrault et al. 2017). The Fe-condensed magmas became higher density and sank to the bottom. This effect could have contributed to produce an early basal magma ocean and some melt layers might have survived to the present (Labrosse et al. 2007).

REFERENCES

Abe Y, Matsui T (1986) Early evolution of the Earth: accretion, atmosphere formation and thermal history. J Geophys Res 91:E291–E302

Abe R, Shibazaki Y, Ozawa S, Ohira I, Tobe H, Suzuki A (2018) In situ X–ray diffraction studies of hydrous aluminosilicate at high pressure and temperature. J Mineral Petrol Sci 113:106–111

Agee CB (1998) Crystal–liquid density inversions in terrestrial and lunar magmas. Phys Earth Planet Inter 107:63–74

Agee CB (2008) Static compression of hydrous silicate melt and the effect of water on planetary differentiation. Earth Planet Sci Lett 265:641–654

Agee CB, Walker D (1988) Static compression and olivine flotation in ultrabasic silicate liquid. J Geophys Res 93:3437–3449

Ai Y, Lange RA (2008) New acoustic velocity measurements on CaO–MgO–Al$_2$O$_3$–SiO$_2$ liquids: Reevaluation of the volume and compressibility of CaMgSi$_2$O$_6$–CaAl$_2$Si$_2$O$_8$ liquids to 25 GPa. J Geophys Res 113:B04203

Andrault D, Bolfan-Casanova N, Lo Nigro G, Bouhifd MA, Garbarino G, Mezouar M (2011) Melting curve of the deep mantle applied to properties of early magma ocean and actual core–mantle boundary. Earth Planet Sci Lett 304:251–259

Andrault D, Pesce G, Bouhifd MA, Bolfan-Casanova N, Henot JM, Mezouar M (2014) Melting of subducted basalt at the core–mantle boundary. Science 344:892–895

Andrault D, Bolfan-Casanova N, Bouhifd MA, Boujibar A, Garbarino G, Manthilake G, Mezouar M, Monteux J, Parisiades P, Pesce G (2017) Toward a coherent model for the melting behavior of the deep Earth's mantle. Phys Earth Planet Inter 265:67–81

Ardia P, Giordano D, Schmidt MW (2008) A model for the viscosity of rhyolite as a function of H$_2$O-content and pressure: a calibration based on centrifuge piston cylinder experiments. Geochim Cosmochim Acta 72:6103–6123

Bajgain S, Ghosh DB, Karki BB (2015) Structure and density of basaltic melts at mantle conditions from first-principles simulations. Nat Commun 6:8578

Ballmer MD, Schumacher L, Lekic V, Thomas C, Ito G (2016) Compositional layering within the large low shear-wave velocity provinces in the lower mantle. Geochem Geophys Geosyst 17:5056–5077

Bercovici D, Karato S (2003) Whole-mantle convection and the transition-zone water filter. Nature 425:39–44

Bolfan-Casanova N, Keppler H, Rubie DC (2003) Water partitioning at 660 km depth and evidence for very low water solubility in magnesium silicate perovskite. Geophys Res Lett 30, doi: 10.1029/2003GL017182

Bottinga Y, Richet P (1995) Silicate melts: The "anomalous" pressure dependence of the viscosity. Geochim Cosmochim Acta 59:2725–2731

Bottinga Y, Weill DF (1970) Densities of silicate liquid systems calculated from partial molar volumes of oxide components. Am J Sci 269:169–182

Boukare CE, Ricard Y, Fiquet G (2015) Thermodynamics of the MgO–FeO–SiO$_2$ system up to 140 GPa: application to the crystallization of Earth's magma ocean. J Geophys Res: Solid Earth 120:6085–6101

Boyet M, Carlson RW (2005) Nd-142 evidence for early (>4.53 Ga) global differentiation of the silicate Earth. Science 309:576–581

Brearley M, Montana A (1989) The effect of CO$_2$ on the viscosity of silicate liquids at high pressure. Geochim Cosmochim Acta 53:2609–2616

Brearley M, Dickinson Jr JE, Scarfe CM (1986) Pressure dependence of melt viscosities on the join diopside–albite. Geochim Cosmochim Acta 50:2563–2570

Burnham CW, Davis NF (1971) The role of H$_2$O in silicate melts: I. *P–V–T* relations in the system NaAlSi$_3$O$_8$–H$_2$O to 10 kilobars and 1000 °C Am J Sci 270:54–79

Caro G, Bourdon B, Wood BJ, Corgne A (2005) Trace-element fractionation in Hadean mantle generated by melt segregation from a magma ocean. Nature 436:246–249

Cochain B, Sanloup C, Leroy C, Kono Y (2017) Viscosity of mafic magmas at high pressures. Geophys Res Lett 44:818–826

Cohen MH, Turnbull D (1959) Molecular transport in liquids and glasses. J Chem Phys 31:1164–1169

Crépisson C, Morard G, Bureau H, Prouteau G, Morizet Y, Petitgirrd S, Sanloup C (2014) Magmas trapped at the continental lithosphere-asthenosphere boundary. Earth Planet Sci Lett 393:105–112

Dasgupta R (2013) Ingassing, storage, and outgassing of terrestrial carbon through geologic time. Rev Mineral Geochem 75:183–229

Dasgupta R, Hischmann MM (2010) The deep carbon cycle and melting in Earth's interior. Earth Planet Sci Lett 298:1–13

Dasgupta R, Mallik A, Tsuno K, Withers AC, Hirth G, Hirschmann MM (2013) Carbon-dioxide-rich silicate melt in the Earth's upper mantle. Nature 493:211–215

Davies C, Pozzo M, Gubbins D, Alfe D (2015) Constraints from material properties on the dynamics and evolution of Earth's core. Nat Geosci 8:678–685

Deschamps F, Cobden L, Tackley PJ (2012) The primitive nature of large low shear-wave velocity provinces. Earth Planet Sci Lett 349:198–208

Dingwell DB (2015) Properties of rocks and minerals-diffusion, viscosity, and flow of melts. *In:* Treatise on Geophysics. Schubert G (ed) Elsevier, Amsterdam, p 473–486

Dingwell DB, Romano C, Hess K-U (1996) The effect of water on the viscosity of a haplogranitic melt under *P–T–X* conditions relevant to silicic volcanism. Contrib Mineral Petrol 124:19–28

Dobson DP, Jones AP, Rabe R, Sekine T, Kurita K, Taniguchi T, Kondo T, Kato T, Shimomura O, Urakawa S (1996) In-situ measurement of viscosity and density of carbonate melts at high pressure. Earth Planet Sci Lett 143:207–215

Dziewonski AM, Anderson DL (1981) Preliminary reference Earth model. Phys Earth Planet Inter 25:297–356

Fiquet G, Auzende AL, Siebert J, Corgne A, Bureau H, Ozawa H, Garbarino G (2010) Melting of peridotite to 140 Gigapascals. Science 329:1516–1518

Frost DJ, McCammon CA (2008) The redox state of the Earth's mantle. Ann Rev Earth Planet Sci 36:389–420

Gaillard F, Malki M, Iacono-Marziano G, Pichavant M, Scaillet B (2008) Carbonatite melts and electrical conductivity in the asthenosphere. Science 322:1363–1365

Garnero EJ, Helmberger DV (1998) Further structural constraints and uncertainties of a thin laterally varying ultralow-velocity layer at the base of the mantle. J Geophys Res Solid Earth 103:12495–12509

Garnero EJ, McNamara AK (2008) Structure and dynamics of Earth's lower mantle. Science 320:626–628

Garnero EJ, Grand SP, Helmberger DV (1993) Low P-wave velocity at the base of the mantle. Geophys Res Lett 20:1843–1846

Ghosh S, Ohtani E, Litasov KD, Suzuki A, Sakamaki T (2007) Stability of carbonated magmas at the base of the Earth's upper mantle. Geophys Res Lett 34:L22312

Ghosh S, Litasov KD, Ohtani E (2014) Phase relations and melting of carbonated peridotite between 10 and 20 GPa: a proxy for alkali- and CO_2-rich silicate melts in the deep mantle. Contrib Mineral Petrol 167:964

Ghosh DB, Bajgain SK, Mookherjee M, Karki BB (2017) Carbon-bearing silicate melt at deep mantle conditions. Sci Rep 7:848

Giordano D, Russell JK, Dingwell DB (2008) Viscosity of magmatic liquids: a model. Earth Planet Sci Lett 271:123–134

Hayashi C, Nakazawa K, Mizuno H (1979) Earth's melting due to the blanketing effect of the primordial dense atmosphere. Earth Planet Sci Lett 43:22–28

Helmberger D, Lay T, Ni S, Gurnis M (2005) Deep mantle structure and the post-perovskite phase transition. Proc Natl Acad Sci USA 102:17257–17263

Herzberg C, Condie K, Korenaga J (2010) Thermal history of the Earth and its petrological expression. Earth Planet Sci Lett 292:79–88

Hirschmann MM (2006) Water, melting, and the deep Earth H_2O cycle. Annu Rev Earth Planet Sci 34:629–653

Hirschmann MM, Kohlstedt DL (2012) Water in Earth's mantle. Phys Today 65:40–45

Hirschmann MM, Asimow PD, Ghiorso MS, Stolper EM (1999) Calculation of peridotite partial melting from thermodynamic models of minerals and melts. III. Controls on isobaric melt production and the effect of water on melt production. J Petrol 40:831–851

Inoue T, Yurimoto H, Kudoh Y (1995) Hydrous modified spinel, $Mg_{1.75}SiH_{0.5}O_4$: a new water reservoir in the mantle transition region. Geophys Res Lett 22:117–120

Hui H, Zhang Y, Xu Z, Del Gaudio P, Behrens H (2009) Pressure dependence of viscosity of rhyolitic melts. Geochim Cosmochim Acta 73:3680–3693

Ito E, Takahashi E (1987) Melting of peridotite at uppermost lower-mantle conditions. Nature 328:514–517

Jing Z, Karato S (2012) Effect of H_2O on the density of silicate melts at high pressures: Static experiments and the application of a modified hard-sphere model of equation of state. Geochim Cosmochim Acta 85: 357–372

Karki BB, Bohara B, Stixrude L (2011) First-principles study of diffusion and viscosity of anorthite ($CaAl_2Si_2O_8$) liquid at high pressure. Am Mineral 96:744–751

Karki BB, Ghosh DB, Bajgain SK (2018) Simulation of silicate melts under pressure. *In:* Magmas Under Pressure. Kono Y, Sanloup C (ed) Elsevier, Amsterdam, p.419–453

Katsura T, Yoneda A, Yamazaki D, Yoshino T, Ito E (2010) Adiabatic temperature profile in the mantle. Phys Earth Planet Inter 183:212–218

Kaula WM (1979) Thermal evolution of the Earth and Moon growing by planetesimal impacts. J Geophys Res 84:999–1008

Keppler H, Wiedenbeck M, Shcheka SS (2003) Carbon solubility in olivine and the mode of carbon storage in the Earth's mantle. Nature 424:414–416

Keshav S, Corgne A, Gudfinnisson GH, Bizimis M, McDonough WF, Fei Y (2005) Kimberlite petrogenesis: insights from clinopyroxene–melt partitioning experiments at 6 GPa in the $CaO–MgO–Al_2O_3–SiO_2–CO_2$ system. Geochim Cosmochim Acta 69:2829–2845

Keshav S, Gudfinnsson GH, Presnall DC (2011) Melting phase relations of simplified carbonated peridotite at 12–26 GPa in the systems CaO–MgO–SiO$_2$–CO$_2$ and CaO–MgO–Al$_2$O$_3$–SiO$_2$–CO$_2$: highly calcic magmas in the transition zone of the Earth. J Petrol 52:2265–2291

Khitarov NI, Kadik AA (1973) Water and carbon dioxide in magmatic melts and peculiarities of the melting process. Contrib Mineral Petrol 41:205–215

Kohlstedt DL, Keppler H, Rubie DC (1996) Solubility of water in the α, β and γ phases of (Mg,Fe)$_2$SiO$_4$. Contrib Mineral Petrol 123:345–357

Kono Y, Kenney-Benson C, Park C, Shen G, Wang Y (2013) Anomaly in the viscosity of liquid KCl at high pressures. Phys Rev B 87:024302

Kono Y, Kenney-Benson C, Hummer D, Ohfuji H, Park C, Shen G, Wang Y, Kavner A, Manning CE (2014a) Ultralow viscosity of carbonate melts at high pressures. Nat Commun 5:5091

Kono Y, Park C, Kenney-Benson C, Shen G, Wang Y (2014b) Toward comprehensive studies of liquids at high pressures and high temperatures: Combined structure, elastic wave velocity, and viscosity measurements in the Paris-Edinburgh cell. Phys Earth Planet Inter 183:196–211

Kono Y, Kenney-Benson C, Shibazaki Y, Park C, Shen G, Wang Y (2015) High-pressure viscosity of liquid Fe and FeS revisited by falling sphere viscometry using ultrafast X-ray imaging. Phys Earth Planet Inter 241:57–64

Knittle E, Jeanloz R (1991) Earth's core–mantle boundary: Results of experiments at high pressures and temperatures. Science 251:1438–1443

Kushiro I (1976) Change in viscosity and structure of melt of NaAlSi$_2$O$_6$ composition at high pressures. J Geophys Res 81:6347–6350

Kushito I (1978a) Viscosity and structure of albite (NaAlSi$_3$O$_8$) melt at high pressures. Earth Planet Sci Lett 41:87–90

Kushiro I (1978b) Density and viscosity of hydrous calc-alkalic andesite magma at high pressures. Carnegie Inst Washington Year Book 77:675–677

Kushiro I (1986) Viscosity of partial melts in the upper mantle. J Geophys Res 91:9343–9350

Kushiro I, Yoder JHS, Mysen BO (1976) Viscosities of basalt and andesite melts at high pressures. J Geophys Res 81:6351–6356

Labrosse S (2015) Thermal evolution of the core with a high thermal conductivity. Phys Earth Planet Inter 247:36–55

Labrosse S, Hernlund JW, Coltice N (2007) A crystallizing dense magma ocean at the base of the Earth's mantle. Nature 450:866–869

Lange RA, Carmichael ISE (1987) Densities of Na$_2$O–K$_2$O–CaO–MgO–FeO–Fe$_2$O$_3$–Al$_2$O$_3$–TiO$_2$–SiO$_2$ liquids: new measurements and derived partial molar properties. Geochim Cosmochim Acta 51:2931–2946

Lange RA, Carmichael ISE (1989) Ferric–ferrous equilibria in Na$_2$O–FeO–Fe$_2$O$_3$–SiO$_2$ melts: effects of analytical techniques on derived partial molar volumes. Geochim Cosmochim Acta 53:2195–2204

Lay T, Williams Q, Garnero EJ (1998) The core–mantle boundary layer and deep Earth dynamics. Nature 392:461–468

Lee SK (2011) Simplicity in melt densification in multi-component magmatic reservoirs in Earth's interior revealed by multinuclear magnetic resonance. PNAS 108:6847–6852

Lesher CE (2010) Self-diffusion in silicate melts: theory, observations and applications to magmatic systems. Rev Mineral Geochem 72:269–309

Liebske C, Schmickler B, Terasaki H, Poe BT, Suzuki A, Funakoshi K, Ando R, Rubie DC (2005) Viscosity of peridotite liquid up to 13 GPa: Implications for magma ocean viscosities. Earth Planet Sci Lett 240:589–604

Litasov KD, Shatskiy A, Ohtani E, Yaxley GM (2013) Solidus of alkaline carbonatite in the deep mantle. Geology 41:79–82

Liu Q, Lange RA (2003) New density measurements on carbonate liquids and the partial molar volume of the CaCO$_3$ component. Contrib Mineral Petrol 146:370–381

Liu Z, Park J, Karato S (2016) Seismological detection of low-velocity anomalies surrounding the mantle transition zone in Japan subduction zone. Geophys Res Lett 43:2480–2487

Liu Z, Park J, Karato S (2018) Seismic evidence for water transport out of the mantle transition zone beneath the European Alps. Earth Sci Lett 482:93–104

Malfait WJ, Seifert R, Perirgirard S, Perrillat JP, Mezouar M, Ota T, Nakamura E, Lerch P, Sachez-Valle C (2014a) Supervolcano eruptions driven by melt buoyancy in large silicic magma chambers. Nat Geo 7:122–125

Malfait WJ, Seifert R, Petitgirard S, Mezouar M, Sanchez-Valle C (2014b) The density of andesitic melts and the compressibility of dissolved water in silicate melts at crustal and upper mantle conditions. Earth Planet Sci Lett 393:31–38

Mao WL, Meng Y, Shen G, Prakapenka VB, Campbell AJ, Heinz DL, Shu J, Caracas R, Cohen RE, Fei Y, Hemley RJ, Mao HK (2005) Iron-rich silicate in the Earth's D″ layer. PNAS 102:9751–9753

Matsukage KN, Jing Z, Karato S (2005) Density of hydrous silicate melt at the conditions of Earth's deep upper mantle. Nature 437:488–491

McNamara AK, Zhong S (2005) Thermochemical structures beneath Africa and the Pacific Ocean. Nature 437:1136–1139

McNamara AK, Garnero EJ, Rost S (2010) Tracking deep mantle reservoirs with ultra-low velocity zones. Earth Planet Sci Lett 299:1–9

Mookherjee M, Stixrude L, Karki BB (2008) Hydrous silicate melt at high pressure. Nature 452:983–986

Mori S, Ohtani E, Suzuki A (2000) Viscosity of the albite melt up 7 GPa at 2000 K. Earth Planet Sci Lett 175:87–92

Mosenfelder JL, Asimow PD, Ahrens TJ (2007) Thermodynamic properties of Mg_2SiO_4 from shock measurements to 200 GPa on forsterite and wadsleyite. J Geophys Res 112:B06208

Naif S, Key K, Constable S, Evans RL (2013) Melt-rich channel observed at the lithosphere-asthenosphere boundary. Nature 495:356–359

Nakagawa T, Tackley PJ (2010) Influence of initial CMB temperature and other parameters on the thermal evolution of Earth's core resulting from thermochemical spherical mantle convection. Geochem Geophys Geosyst 11:Q06001

Nakajima A, Sakamaki T, Kawazoe T, Suzuki A (2019) Hydrous magnesium-rich magma genesis at the top of the lower mantle. Sci Rep 9:7420

Nakajima M, Stevenson DJ (2015) Melting and mixing states of the Earth's mantle after the moon-forming impact. Earth Planet Sci Lett 427:286–295

Ni H, Keppler H (2013) Carbon in silicate melts. Rev Mineral Geochem 75:251–287

Nishi M, Irifune T, Tsuchiya J, Tange Y, Nishihara Y, Fujino K, Higo Y (2014) Stability of hydrous silicate at high pressures and water transport to the deep lower mantle. Nat. Geosci. 7:224–227

Ochs FA, Lange RA (1997) The partial molar volume, thermal expansivity, and compressibility of H_2O in $NaAlSi_3O_8$ liquid: new measurements and an internally consistent model. Contrib Mineral Petrol 142:235–243

Ochs FA, Lange RA (1999) The density of hydrous magmatic liquids. Science 283:1314–1317

Ohira I, Ohtani E, Sakai T, Miyahara M, Hirao N, Ohishi Y, Nishijima M (2014) Stability of a hydrous δ-phase, AlOOH–$MgSiO_2(OH)_2$, and a mechanism for water transport into the base of lower mantle. Earth Planet Sci Lett 401:12–17

Ohtani E (1983) Melting temperature distribution and fractionation in the lower mantle. Phys Earth Planet Inter 33:12–25

Ohtani E (1985) The primordial terrestrial magma ocean and its implication for the stratification of the mantle. Phys Earth Planet Inter 38:70–80

Ohtani E (2009) Melting relations and the equation of state of magmas at high pressure: Application to geodynamics. Chem Geol 265:279–288

Ohtani E, Suzuki A, Kato T (1993) Flotation of olivine in the peridotite melt at high pressure. Proc Japan Acad 69:23–28

Ohtani E, Maeda M (2001) Density of basaltic melt at high pressure and stability of the melt at the base of the lower mantle. Earth Planet Sci Lett 193:69–75

Ohtani E, Litasov K, Hosoya T, Kubo T, Kondo T (2004) Water transport into the deep mantle and formation of a hydrous transition zone. Phys Earth Planet Inter 143–144:255–269

Ohtani E, Amaike Y, Kamada S, Sakamaki T, Hirao N (2014) Stability of hydrous phase H $MgSiO_4H_2$ under lower mantle conditions. Geophys Res Lett 41:8283–8287

Ohtani E, Yuan L, Ohira I, Shatskiy A, Litasov K (2018) Fate of water transported into the deep mantle by slab subduction. J Asian Earth Sci 167:2–10

Olsen PL, Sharp ZD (2019) Nebular atmosphere to magma ocean: A model for volatile capture during Earth accretion. Phys Earth Planet Inter 294:106294

Ono S (1999) High temperature stability limit of phase egg, $AlSiO_3(OH)$. Contrib Mineral Petrol 137:83–89

Orlova G (1964) Solubility of water in albite melts-under pressure. Int Geol Rev 6:254–258

Peacock SM (1990) Fluid processes in subduction zones. Science 248:329–337

Pearson DG, Brenker FE, Nestola F, McNeill J, Nasdala L, Hutchison MT, Matveev S, Mather K, Silversmit G, Schmitz S, Vekemans B, Vincze L (2014) Hydrous mantle transition zone indicated by ringwoodite included within diamond. Nature 507:221–224

Persikov ES (1991) The viscosity of magmatic liquids: experiment, generalized patterns; a model for calculation and prediction: application. *In:* Physical Chemistry of Magmas. Perchuk LL, Kushiro I (ed) Springer-Verlag, New York, p 1–40

Persikov ES, Bukhtiyarov PG, Sokol AG (2017) Viscosity of hydrous kimberlite and basaltic melts at high pressures. Russian Geol Geophys 58:1093–1100

Reid JE, Suzuki A, Funakoshi K, Terasaki H, Poe BT, Rubie DC, Ohtani E (2003) The viscosity of $CaMgSi_2O_6$ liquid at pressures up to 13 GPa. Phys Earth Planet Inter 139:45–54

Revenaugh J, Sipkin SA (1994) Seismic evidence for silicate melt atop the 410-km mantle discontinuity. Nature 369:474–476

Richet P, Polian A (1998) Water as a dense icelike component in silicate glasses. Science 28:396–398

Richet P, Lejeune A-M, Holtz F, Roux J (1996) Water and the viscosity of andesite melts. Chem Geol 128:185–197

Richet P, Whittington A, Holtz F, Behrens H, Ohlhorst S, Wilke M (2000) Water and the density of silicate glasses. Contrib Mineral Petrol 138:337–347

Rigden SM, Ahrens TJ, Stolper EM (1984) Densities of liquid silicates at high pressures. Science 226:1071–1074

Rigden SM, Ahrens TJ, Stolper EM (1989) High-pressure equation of state of molten anorthite and diopside. J Geophys Res 94:9508–9522

Rohrbach A, Schmidt MW (2011) Redox freezing and melting in the Earth's deep mantle resulting from carbon iron redox coupling. Nature 472:209–212

Rost S, Garnero EJ, Williams Q (2006) Fine-scale ultralow-velocity zone structure from high-frequency seismic array data. J Geophys Res Solid Earth 111:B09310

Rubie DC, Nimmo HJ, Melosh HJ (2015) Formation of the Earth's core. *In:* Treatise on Geophysics. Schubert G (ed) Elsevier, Amsterdam, p 43–79

Sakamaki T (2017a) Density of jadeite melts under high pressure and high temperature conditions. J Mineral Petrol Sci 112:300–307

Sakamaki T (2017b) Density of hydrous magma. Chem Geol 475:135–139

Sakamaki T, Suzuki A, Ohtani E (2006) Stability of hydrous melt at the base of the Earth's upper mantle. Nature 439:192–194

Sakamaki T, Ohtani E, Urakawa S, Suzuki A, Katayama Y (2009) Measurement of hydrous peridotite magma density at high pressure using the X-ray absorption method. Earth Planet Sci Lett 287:293–297

Sakamaki T, Ohtani E, Urakawa S, Suzuki A, Katayama Y (2010a) Density of dry peridotite magma at high pressure using an X-ray absorption method. Am Mineral 95:144–147

Sakamaki T, Ohtani E, Urakawa S, Suzuki A, Katayama Y, Zhao D (2010b) Density of high-Ti basalt magma at high pressure and origin of heterogeneities in the lunar mantle. Earth Planet Sci Lett 299:285–289

Sakamaki T, Ohtani E, Urakawa S, Terasaki H, Katayama Y (2011) Density of carbonated peridotite magma at high pressure using an X-ray absorption method. Am Mineral 96:553–557

Sakamaki T, Wang Y, Park C, Yu T, Shen G (2012) Structure of jadeite melt at high pressures up to 4.9 GPa. J Appl Phys 111:112623

Sakamaki T, Suzuki A, Ohtani E, Terasaki H, Urakawa S, Katayama Y, Funakoshi K, Wang Y, Hernlund JW, Ballmer MD (2013) Ponded melt at the boundary between the lithosphere and asthenosphere. Nat Geosci 6:1041–1044

Sanloup C, Drewitt JWE, Crépisson C, Kono Y, Park C, McCammon C, Hennet L, Brassamin S, Bytchkov A (2013a) Structure and density of molten fayalite at high pressure. Geochim Cosmochim Acta 118:118–128

Sanloup C, Drewitt JWE, Konôpková Z, Dalladay-Simpson P, Morton DM, Rai N, van Westrenen W, Morgenroth W (2013b) Structural change in molten basalt at deep mantle conditions. Nature 105:104–107

Sano A, Ohtani E, Kubo T, Funakoshi K (2004) In situ X-ray observation of decomposition of hydrous aluminum silicate AlSiO$_3$OH and aluminum oxide hydroxide δ-AlOOH at high pressure and temperature. J Phys Chem Solids 65:1547–1554

Scarfe CM, Cronin DJ (1986) Viscosity-temperature relationships of melts at 1 atm in the system diopside-albite. Am Mineral 71:767–771

Schmandt B, Jacobsen SD, Becker TW, Liu Z, Dueker KG (2014) Dehydration melting at the top of the lower mantle. Science 344:1265–1268

Schmerr N (2012) The Gutenberg discontinuity: Melt at the lithosphere–asthenosphere boundary. Nature 335:1480–1483

Seifert R, Malfait WJ, Petitgirard S, Sanchez-Valle C (2013) Density of phonolitic magmas and time scales of crystal fraction in magma chambers. Earth Planet Sci Lett 381:12–20

Schulze F, Behrens H, Holtz F, Roux J, Johannes W (1996) The influence of H$_2$O on the viscosity of a haplogranitic melt. Am Mineral 81:1155–1165

Shaw HR (1963) Obsidian-H$_2$O viscosities at 1000 and 2000 bars in the temperature range 700 to 900 C°. J Geophys Res 68:6337–6343

Shcheka SS, Wiedenbeck M, Frost DJ, Keppler H (2006) Carbon solubility in mantle minerals. Earth Planet Sci Lett 245:730–742

Song TRA, Helmberger DV, Grand SP (2004) Low-velocity zone atop the 410-km seismic discontinuity in the northwestern United States. Nature 427:530–533

Spice H, Sanloup C, Cochain B, De Grouchy C, Kono Y (2015) Viscosity of liquid fayalite up to 9 GPa. Geochim Cosmochim Acta 148:219–227

Stagno V, Frost DJ (2010) Carbon speciation in the asthenosphere: Experimental measurements of the redox conditions at which carbonate-bearing melts coexist with graphite or diamond in peridotite assemblages. Earth Planet Sci Lett 300:72–84

Stagno V, Ojwang DO, McCammon CA, Frost DJ (2013) The oxidation state of the mantle and the extraction of carbon from Earth's interior. Nature 493:84–88

Stagno V, Stopponi V, Kono Y, Manning CE, Irifune T (2018) Experimental determination of the viscosity of Na$_2$CO$_3$ melt between 1.7 and 4.6 GPa at 1200–1700 °C: Implications for the rheology of carbonatite magmas in the Earth's upper mantle. Chem Geol 501:19–25

Stixrude L, Karki BB (2005) Structure and freezing of MgSiO$_3$ liquid in Earth's lower mantle. Science 310:297–299

Suzuki A (2018) Effect of carbon dioxide on the viscosity of a melt of jadeite composition at high pressure. J Mineral Petrol Sci 113:47–50

Suzuki A, Ohtani E, Kato T (1995) Flotation of diamond in mantle melt at high pressure. Science 269:216–218

Suzuki A, Ohtani E, Funakoshi K, Terasaki H, Kubo T (2002) Viscosity of albite melt at high pressure and high temperature. Phys Chem Minerals 29:159–165

Suzuki A, Ohtani E, Terasaki H, Funakoshi K (2005) Viscosity of silicate melts in CaMgSi$_2$O$_6$–NaAlSi$_2$O$_6$ system at high pressure. Phys Chem Minerals 32:140–145

Suzuki A, Ohtani E, Terasaki H, Nishida K, Hayashi H, Sakamaki T, Shibazaki Y, Kikegawa T (2011) Pressure and temperature dependence of the viscosity of a NaAlSi$_2$O$_6$ melt. Phys Chem Minerals 38:59–64

Taniguchi H (1992) Entropy dependence of viscosity and the glass transition temperature of melts in the system diopside-anorthite. Contrib Mineral Petrol 109:295–303

Tauzin B, Bebayle E, Wittlinger G (2010) Seismic evidence for a global low-velocity layer within the Earth's upper mantle. Nat Geosci 3:718–721

Thompson AR, Walter MJ, Kohn SC, Brooker RA (2016) Slab melting as a barrier to deep carbon subduction. Nature 529:76–79

Toffelmier DA, Tyburczy JA (2007) Electromagnetic detection of a 410-km-deep melt layer in the southwestern United States. Nature 447:991–994

van der Hilst RD, Widiyantoro S, Engdahl ER (1997) Evidence for deep mantle circulation from global tomography. Nature 386:578–584

van der Hilst RD, Karason H (1999) Compositional heterogeneity in the bottom 1000 kilometers of Earth's mantle: toward a hybrid convection model. Science 283:1885–1888

van Kan Parker M, Sanloup C, Sator N, Guillot B, Tronche EJ, Perrillat JP, Mezouar M, Rai N, van Westrenen W (2012) Neutral buoyancy of titanium-rich melts in the deep lunar interior. Nat Geosci 5:186–189

van Keken PE, Hacker BR, Syrause EM, Abers GA (2011) Subduction factory: 4 Depth-dependent flux of H_2O from subducting slabs worldwide. J Geophys Res 116:B01401

Verma AK, Karki BB (2012) First-principles study of self-diffusion and viscous flow in diopside ($CaMgSi_2O_6$) liquid. Am Mineral 97:2049–2055

Wallace PJ (2005) Volatiles in subduction zone magmas: concentrations and fluxes based on melt inclusion and volcanic gas data. J Volcanol Geotherm Res 140:217–240

Wang Y, Sakamaki T, Skinner LB, Jing Z, Yu T, Kono Y, Park C, Shen G, Rivers ML, Sutton SR (2014) Atomistic insight into viscosity and density of silicate melts under pressure. Nat Commun 5:3241

Wang X-C, Wilde SA, Li Q-L, Yang Y-N (2015) Continental flood basalts derived from the hydrous mantle transition zone. Nat Commun 6:7700

Wen LX, Helmberger DV (1998) A two-dimensional P–SV hybrid method and its application to modeling localized structures near the core–mantle boundary. J Geophys Res Solid Earth 103:17901–17918

White BS, Montana A (1990) The effect of H_2O and CO_2 on the viscosity of sanidine liquid at high pressures. J Geophys Res Solid Earth 95:15683–15693

Whittington A, Richet P, Linard Y, Holtz F (2001) The viscosity of hydrous phonolites and trachytes. Chem Geol 174:209–223

Williams Q, Garnero EJ (1996) Seismic evidence for partial melt at the base of earth's mantle. Science 273:1528–1530

Wirth R, Vollmer C, Brenker F, Matsyuk S, Kaminsky F (2007) Inclusions of nanocrystalline hydrous aluminium silicate "Phase Egg" in superdeep diamonds from Juina (Mato Grosso State, Brazil). Earth Planet Sci Lett 259:384–399

Wysession ME (1996) Large-scale structure at the core–mantle boundary from diffracted waves. Nature 382:244–248

Yu S, Garnero EJ (2018) A global assessment of ultra-low velocity zones. Geochem Geophys Geosys 19:396–414

Zhang J, Herzberg C (1994) Melting experiments on anhydrous peridotite KLB-1 from 5.0 to 22.5 GPa. J Geophys Res 99:17729–17742

Reviews in Mineralogy & Geochemistry
Vol. 87 pp. 575-606, 2022
Copyright © Mineralogical Society of America

Volatile-bearing Partial Melts in the Lithospheric and Sub-Lithospheric Mantle on Earth and Other Rocky Planets

Rajdeep Dasgupta[1*], Proteek Chowdhury[1,2], James Eguchi[1,2], Chenguang Sun[1,3], Sriparna Saha[1,4]

[1]*Department of Earth, Environmental and Planetary Sciences, Rice University*
6100 Main Street, MS 126, Houston, TX 77005, USA
[2]*Current address: Department of Earth and Planetary Sciences*
University of California at Riverside, 900 University Avenue, Riverside, CA 92521, USA
[3]*Department of Geological Sciences, Jackson School of Geosciences*
The University of Texas at Austin, Austin, TX 78712, USA
[4]*Current address: School of Teacher Education and School of Earth and Environment*
University of Canterbury, Private Bag 4800, Christchurch 8140, New Zealand

rajdeep.dasgupta@rice.edu; proteekc@ucr.edu; james.eguchi@ucr.edu;
csun@jsg.utexas.edu; sriparna.saha@pg.canterbury.ac.nz

INTRODUCTION

Generation and extraction of partial melts from the interior of rocky planets are essential steps for supplying heat and life-essential nutrients to the surface environments. Mantle-derived partial melts dissolve key volatile elements such as carbon, hydrogen, and sulfur and when brought to the surface or near-surface conditions, liberate them as gases such as carbon dioxide (CO_2), water vapor (H_2O), and sulfur dioxide (SO_2) or hydrogen sulfide (H_2S). These gases influence the composition of the atmosphere–ocean systems, silicate-weathering feedback, (Walker et al. 1981) and/or ocean chemistry, thereby affecting the climate on many timescales (e.g., Robock 2000). Indeed, a number of well-documented events that affected atmospheric conditions and life forms on the surface of Earth have been linked to emissions of gases such as CO_2, which are released through magmatic activities and volcanism (e.g., Payne et al. 2010; Schaller et al. 2011; Eguchi et al. 2020). Many of these exsolved gases modulate the surface temperature of the planets by absorbing incoming stellar radiation. Therefore, for Earth-like rocky planets in the habitable zone, partial melting and volatile degassing processes also directly allow stability of liquid water on the planet's surface. In addition to influencing physical and chemical environment, the magmatically released gases serve as ingredients that are essential for carbon-based life—the only form of life known thus far. For volcanically–tectonically active planets such as Earth, the process of magmatism or partial melting is known to have operated over the entire geologic history; even for other Solar System bodies such as Mars, this process has operated over much of the planet's history. Hence, it is of fundamental importance, for assessing the evolution of habitable conditions of rocky planets, to understand where, how much, and what compositions of volatile-bearing partial melts are generated at the subsurface of Earth and other rocky planets through space and time.

On the modern Earth, partial melts on the one hand act as a carrier of dissolved volatiles species (H, C, S, N, and halogens), while on the other hand major volatile species such as CO_2 and H_2O affect the thermodynamic stability (Brey and Green 1977; Dasgupta et al. 2007b; Médard and Grove 2008) and dynamic behavior of partial melts (Lange 1994). The chief volatile species that affect the stability of mantle melts varies as a function of tectonic setting. For example, mantle wedge decompression partial melting is known to be mostly overprinted by a hydrous flux (e.g., Ulmer 2001; Grove et al. 2006), whereas initiation of deep melting at any other mantle domains that are nominally anhydrous is argued to be influenced mainly by CO_2 or a mixed CO_2–H_2O system(e.g., Dasgupta et al. 2007b; Hirschmann 2010; Dasgupta 2018). We will summarize in this synthesis the current understanding of partial melt stability in the Earth's sub-lithospheric and lithospheric mantle and how such stability is affected by dissolved volatiles. We will give specific attention to how, as a function of major element bulk composition of the mantle rocks as well as their H_2O and CO_2 contents, the storage of C and H at subsolidus conditions may differ and how that in turn influences the melting behavior of the Earth's mantle. Partial melting is directly affected by CO_2–H_2O, however, such volatile-induced melting also can extract other major volatile element such as sulfur. We will also combine our latest understanding of sulfur dissolution in mantle melts and discuss how the melt compositions is expected to vary from oceanic to continental mantle in terms of relative budgets of C, S, and H.

For most of the geologic history, Earth's shallow mantle where partial melting commences is thought to have remained at redox state such that the oxidized form of major volatile elements (CO_2, carbonates, H_2O or hydroxyl) are of relevance. However, mantle domains in deeper portion of the continental lithosphere or deeper oceanic mantle domains can be more reduced, with carbon stored in reduced form, as graphite/diamond and/or in Fe–Ni–S–C alloy (Frost and McCammon 2008; Luth and Stachel 2014; Tsuno and Dasgupta 2015; Zhang et al. 2019). Similar storage of carbon in reduced forms is also suggested or expected for other rocky planetary mantles such as those of the Moon, Mars, and Mercury(e.g., Sato et al. 1973; Sato 1979; Righter et al. 2008; Nicholis and Rutherford 2009; Li et al. 2017). On the other hand, sulfur storage may mostly be in the form of sulfides in all planetary mantles in our Solar System (Zolotov et al. 2013; Zhang et al. 2018; Brenan et al. 2019). The composition of residual sulfide is, however, expected to vary substantially as the mantle gets extremely reduced (e.g. Namur et al. 2016a). A critical question is whether redox state of planetary mantles affect the efficiency of extraction of volatiles through magmatic processes. Therefore, we will also consider in this paper what concentrations of dissolved C, S, and H may be expected in partial melts generated from reduced mantles. Importantly, we will discuss how volatile concentrations in such reduced partial melts compare with volatile concentrations expected for oxidized melts generated at the shallow upper mantle of Earth.

THE EFFECTS OF MAJOR VOLATILES
ON MELTING OF THE EARTH'S MODERN MANTLE

The two major volatile components that expand the *P–T* stability of partial melts are H_2O and CO_2, i.e., the oxidized form of carbon and hydrogen. The role of H_2O has been widely recognized and discussed in the literature for causing magma generation in convergent boundaries on Earth. Indeed, the subduction zone magmas contain distinctly higher abundances of dissolved H_2O compared to all other natural melts from intraplate and divergent plate boundaries(e.g., Sobolev and Chaussidon 1996; Plank et al. 2013), suggesting that the primary melts generated in mantle depths in this setting are also much more H_2O-rich. The origin of primary H_2O-rich partial melts in convergent margins is largely due to the process of subduction, which brings hydrous minerals formed at the sea-floor to mantle depths (e.g., Schmidt and Poli 1998). Because of the breakdown of the hydrous silicate minerals in the downgoing plate, rocks at the base of the mantle wedge can potentially experience a flux of an aqueous fluid phase. This can lead to initiation of melting being caused chiefly by aqueous fluids (Ulmer 2001; Grove et al. 2006). While the roles of slab-derived H_2O-rich fluids (Till et al. 2012; Grove and Till 2019) or hydrous partial melts (Mallik et al. 2015, 2016; Saha et al. 2018) in initiating mantle wedge melting is obvious from the geochemistry of subduction zone magmas, the role of volatile-induced melting in other tectonic settings is more subtle (Dasgupta et al. 2007b, 2013; Mierdel et al. 2007; Hirschmann 2010; Sifré et al. 2014; Massuyeau et al. 2015; Dasgupta 2018). However, decades of work combining natural observations of the chemistry of intraplate magmas from both oceanic and continental settings with data from high $P-T$ experiments suggest that incipient melting caused by CO_2, H_2O, or CO_2+H_2O are pervasive at many other geodynamic settings as well. The following sections will summarize the constraints that exist on the stability and volatile contents in partial melts at different lithospheric and sub-lithospheric depths in different settings.

Solidus of CO_2, H_2O, and CO_2+H_2O-bearing peridotite

In order to constrain the stability of volatile-bearing partial melts, their proportions, and their compositions, one must first determine the solidi of mantle lithologies in the presence of volatile components. The influence of H_2O or CO_2 components on the mantle solidus depends on the storage mechanism of these chemical species, which in turn depends on their concentrations and the $P-T$ conditions (Dasgupta 2018). At shallow depths, the storage of H_2O in nominally anhydrous minerals is limited, and only a few tens of ppm by weight can be dissolved in major mantle minerals at sub-solidus conditions (Kohlstedt et al. 1996; Hirschmann et al. 2005; Hirschmann 2006). Similarly, carbon storage in nominally carbon-free mantle silicates is thought to be minimal (Keppler et al. 2003; Shcheka et al. 2006), and with carbonates only becoming stable at pressures greater than ~2 GPa, there are no other solid accessory phases of oxidized carbon stable at shallow domains. Hence, at shallow depths, the chief repository of water and carbon are in the form of aqueous fluid and CO_2-rich fluid. The effect of H_2O-rich fluid saturation on the solidus of silicate rocks such as peridotite at shallow depth increases as a function of pressure because the solubility of H_2O in near-solidus silicate melt increases with pressure. Therefore, from ambient pressure to ~3 GPa, the aqueous fluid saturated solidus has a negative Clapeyron slope (Fig. 1); only at pressures above 3–4 GPa does

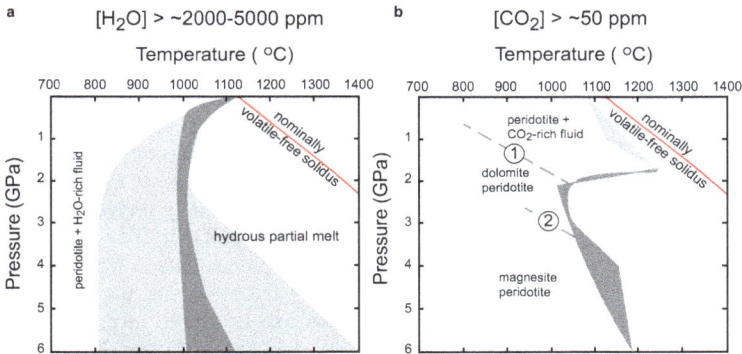

Figure 1. Solidus of Earth's mantle peridotite saturated in or in the presence of H_2O-fluid (**a**) and CO_2-fluid or carbonate minerals (**b**) modified after Dasgupta (2018). H_2O-fluid-present solidus of peridotite becomes relevant for melting depths in the Earth's mantle with concentration typically at or above 2000–5000 ppm as below this bulk H_2O content, H_2O may be present chiefly in hydrous silicate mineral such as amphibole and/or phlogopite or in nominally anhydrous minerals (NAMs). In contrast, CO_2-fluid-present or carbonate mineral-present solidus of peridotite is the relevant solidus of peridotite, with as little as 50 ppm CO_2 in the system. Another key difference between the solidus topology in (**a**) versus (**b**) is that at high concentration, H_2O-fluid is the stable phase at the subsolidus conditions at all pressures whereas CO_2-fluid reacts with peridotite to form dolomite solid solution at ~2 GPa and then magnesite solid solution at ~3 GPa (the carbonation reaction boundaries are marked by dashed lines). Reaction marked '1': olivine + Ca-rich pyroxene + CO_2 = Ca-poor pyroxene + dolomite and '2': Ca-poor pyroxene + CO_2 = Ca-rich pyroxene + magnesite proceed to the right with increasing pressure and are taken from Brey et al. (1983). The **light grey shaded field** in (**a**) marks the full range of aqueous fluid-present peridotite solidus constrained in experimental studies, with the lowest solidus based on Grove et al. (2006), Till et al. (2012), and Mysen and Boetcher (1975) and the highest solidus bound given by Green et al. (2011). The **dark grey shaded region** in (**a**) marks the locus of the H_2O-fluid-saturated solidus by most studies (e.g. Kushiro et al. 1968; Green 1973; Millhollen et al. 1974; Kawamoto and Holloway 1997). The **light shaded band** in (**b**) is the range of CO_2-fluid-present solidus of peridotite (Falloon and Green 1989; Dasgupta 2013, 2018) and the **dark grey band** in (**b**) is the carbonated peridotite solidus (Falloon and Green 1989; Dasgupta 2013). Also shown for reference, in both panels, is the nominally volatile-free solidus of peridotite from Hirschmann (2000).

the aqueous fluid-saturated solidus attain a positive slope. Because dissolved water increases the entropy of the silicate melts, thereby stabilizing it at lower temperatures, the extent of solidus depression at any given pressure is directly proportional to the dissolved H_2O content in the silicate melt. Dissolved H_2O contents in peridotite partial melts increase with increasing pressure until reaching a maximum at ~3 GPa, which explains why the effect of dissolved water on the solidus depression increases with pressure from ambient pressure to ~3 GPa.

In contrast to the effect of H_2O-rich fluid on peridotite solidus, the effect of CO_2-rich fluid on peridotite solidus at shallow depths ($\lesssim 2$ GPa for natural peridotite compositions) is limited. This is because that CO_2 solubility in silicate melt at shallow depths is only few thousands of ppm by weight (e.g., Dixon et al. 1995; Dixon 1997; Brooker et al. 2001; Papale et al. 2006; Ghiorso and Gualda 2015; Eguchi and Dasgupta 2018a) in contrast to water solubility in silicate melt, which reaches several weight percent, even at low pressures (e.g., Stolper 1982; Ghiorso and Gualda 2015). At pressures greater than ~2 GPa, CO_2-rich fluids react with subsolidus peridotite, leading to the formation of equilibrium carbonate minerals, such as dolomite and/or magnesite, depending on the peridotite composition (Brey et al. 1983; Falloon and Green 1989; Saha et al. 2018; Saha and Dasgupta 2019). The stabilization of carbonate minerals as accessory phases has immense influence on the solidus location of peridotite. Formation of carbonate minerals leads to near-isobaric drop in the solidus of the peridotite at ~2 GPa (Fig. 1). This is the pressure at which the carbonated peridotite solidus intersects the carbonation reaction. Hence at pressures >2 GPa, the solidus of peridotite is a carbonated peridotite solidus and its temperature drops several hundred degrees below the nominally CO_2-free peridotite solidus (Fig. 1).

The effects of CO_2 and H_2O contents and major element bulk composition on the solidus

Other than the fundamental difference in the location and topology of the H_2O-fluid-saturated versus the CO_2-fluid/carbonate-saturated peridotite solidus, another key difference between the effect of H_2O and CO_2 on peridotite solidus is the abundance at which H_2O and CO_2 affect the peridotite solidus (Fig. 1). H_2O-fluid saturated solidus becomes relevant if the bulk H_2O content of the system exceeds the storage capacity of solid silicate minerals at subsolidus conditions. The minerals that store H_2O in peridotite are nominally anhydrous silicates, with low concentrations of bulk H_2O, i.e., 50–1000 ppm by weight of H_2O (Hirschmann et al. 2005; O'Leary et al. 2010; Dasgupta 2018). For bulk H_2O contents between 0.005–0.1 wt% and 0.2–0.5 wt%, H_2O in peridotite mostly resides in hydrous minerals such as amphibole and phlogopite. The exact limit of H_2O content that can be stored in nominally anhydrous mantle minerals versus in hydrous silicate minerals varies depending on the bulk composition of peridotite. The following section will address some variations in the loci of melting that may be relevant as a function of compositional variation in peridotite. For an average fertile peridotite composition, the H_2O-rich fluid saturated solidus shown in Figure 1 is only relevant for systems with bulk H_2O contents greater than 0.2–0.5 wt% H_2O. This contrasts with the concentration range of CO_2 for which the CO_2-fluid saturated or carbonated peridotite solidus is relevant. Given that carbon dissolution in nominally carbon-free silicates is much more limited (few ppm by weight; Keppler et al. 2003; Shcheka et al. 2006) compared to the dissolution of hydrogen, the carbonated peridotite solidus becomes the relevant solidus of peridotite once CO_2 contents exceed just a few tens of ppm of CO_2, at depths greater >70 km. This carbonated peridotite solidus has not been documented to be dependent on bulk CO_2 content in the ppm range. Therefore, for nominally volatile-free systems, with only 100s of ppm of volatile species, CO_2 has a greater influence in establishing the peridotite solidus compared to the effects of H_2O.

Unlike the effects of CO_2, the effects of H_2O on the solidus of peridotite vary depending on the bulk H_2O contents (e.g., Ulmer 2001). With water contents of up to ~1000 ppm, shallow mantle peridotite may accommodate all the water in nominally anhydrous minerals (Kohlstedt et al. 1996; Hirschmann et al. 2005). With such a water storage mechanism, the solidus depression of peridotite is directly proportional to bulk water content (Fig. 2) and with up to ~500 ppm H_2O, the isobaric solidus depression is < 100 °C (Hirschmann et al. 2009; O'Leary et al. 2010). However, if the water content in the system exceeds ~1000 ppm, hydrous silicate minerals such as amphibole, phlogopite, and chlorite could become stable in peridotitic systems. In such scenarios, the isobaric solidus, for the depth range over which one of these hydrous phases is stable at subsolidus conditions, is the dehydration solidus (Wyllie and Wolf 1993; Green et al. 2014). For K-depleted peridotite compositions that are thought to be common in the Earth's upper mantle, the most relevant hydrous mineral phase in the shallow mantle where melting is most prevalent is pargasitic amphibole (Green et al. 2011; Saha and Dasgupta 2019; Lara and Dasgupta 2020). Hence between ~0.5–1.0 GPa and ~3–4 GPa, the aqueous fluid saturated solidus is replaced by the dehydration solidus of amphibole-peridotite (Fig. 2). For peridotite compositions richer in K, laboratory experiments show phlogopite along with amphibole or phlogopite alone may be the subsolidus phase that would be stable in peridotitic bulk composition (Fig. 3) (Saha et al. 2018, 2021). The stability of amphibole versus phlogopite depends on Na_2O/K_2O ratio of the peridotite composition in question along with water content (Saha and Dasgupta 2019), as shown in Figure 3. In bulk compositions where phlogopitic mica as opposed to amphibole is the more relevant phase, the dehydration solidus (of phlogopite-peridotite) tends to be higher as phlogopite can be stable to much higher temperatures than amphibole (Mallik et al. 2015; Saha et al. 2018). The exact temperature of the phlogopite-out boundary and thus the dehydration solidus at a

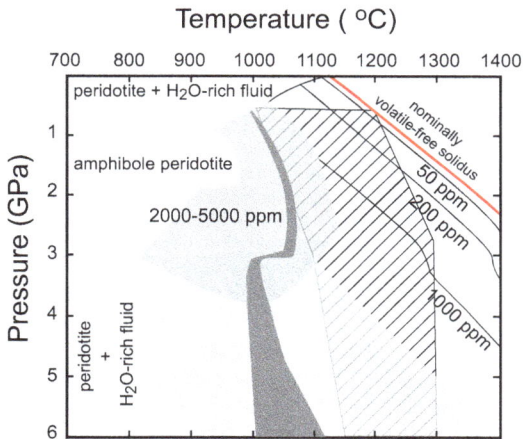

Figure 2. Solidus of fertile mantle peridotite with bulk H_2O content varying between 50 and ~5000 ppm. For bulk H_2O content varying between 0 and ~2000 ppm, the mantle peridotite is fluid-absent and the entire water inventory being contained in NAMs at subsolidus conditions. At this concentration range, the isobaric solidus depression is directly proportional to water content. At H_2O content ≳ 2000 ppm and ≲ 5000 ppm for fertile peridotite bulk compositions, water chiefly resides in amphibole at the subsolidus conditions and the hydrous peridotite solidus between ~0.5–1.0 GPa and 2–4 GPa becomes the dehydration solidus of amphibole peridotite (**intermediate grey shaded** region that bounds the entire range of possible amphibole breakdown conditions: e.g., (Wallace and Green 1991; Mandler and Grove 2016; Saha et al. 2018); the preferred dehydration solidus of amphibole peridotite is marked with a narrower band with **darker shades of grey**). Above ~2–4 GPa, i.e., above the stability of pargasitic amphibole, even for 2000–5000 ppm H_2O, fluid-present solidus of peridotite (**light grey** and **dark grey bands** taken from Fig. 1) takes effect. The nominally H_2O-free solidi with H_2O content of 50–1000 ppm are based on the bulk water partition coefficient and freezing point depression calculation of O'Leary et al. (2010). Also, shown for reference are the loci of phlogopite-out dehydration solidi of peridotite as compiled in Saha et al. (2018; **hatched field**), which can only be relevant for a K-rich local mantle domains. The loci of the of hatched field shows that for K-rich and moderately H_2O-rich mantle domains, the aqueous fluid-present solidus may not be relevant even to pressures up to 5–6 GPa or even as high as 7 GPa where phlogopite may be stable (Konzett and Ulmer 1999).

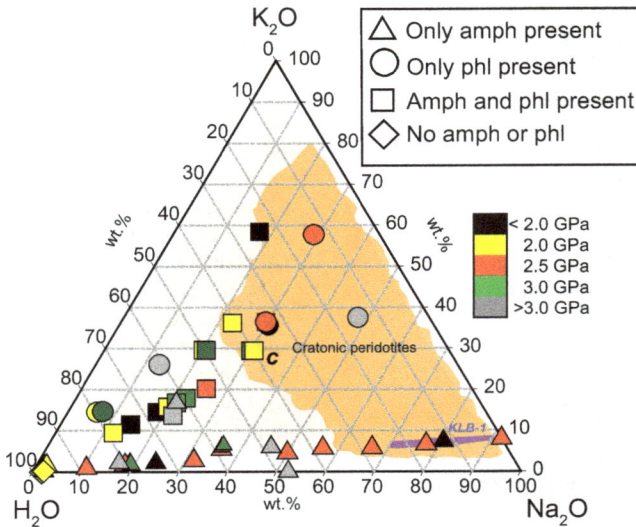

Figure 3. Na$_2$O–K$_2$O–H$_2$O pseudo-ternary space in wt% showing the experimental peridotite bulk compositions at $T < 1100\,°C$ and $P \sim 1$–4 GPa that stabilize only amphibole [**circles**: (Wallace and Green 1991; Niida and Green 1999; Green et al. 2011,2014; Mandler and Grove 2016)], only phlogopite [triangles: (Thibault et al. 1992; Conceição and Green 2004; Saha and Dasgupta 2019)], both amphibole and phlogopite [**squares**: (Mengel and Green 1989; Fumagalli et al. 2009; Green et al. 2011; Tumiati et al. 2013; Condamine and Médard 2014; Saha et al. 2018)], and neither amphibole nor phlogopite [**diamonds**: (Grove et al. 2006; Till et al. 2012)]. Symbol colors mark experimental pressures. Also shown for comparison are natural fertile peridotite KLB-1 (Davis et al. 2009) and field of cratonic peridotites (**darker orange** – most dominant compositions; **lighter orange** – the entire range of compositions) assuming that the natural peridotite may contain between 100 and 1000 ppm bulk H$_2$O. The figure suggests that the fertile peridotite in the Earth's mantle likely would contain pargasitic amphibole as the stable hydrous phase in water-rich, subsolidus conditions whereas depleted but variably metasomatized peridotite in the cratonic lithospheric mantle could contain either amphibole or phlogopite-rich mica or both depending on bulk compositions.

given pressure depends on the bulk water content of the system. For example, fluid saturated systems (i.e., K$_2$O/H$_2$O weight ratio <~2.2) with higher bulk water destabilizes phlogopite completely at lower temperatures than fluid saturated systems with lower bulk H$_2$O (Mallik et al. 2016; Saha and Dasgupta 2019). In addition, stability of phlogopite is also enhanced by incorporation of fluorine and chlorine (Foley et al. 1986). The phlogopite-out boundary and therefore the dehydration solidus of phlogopite-peridotite observed in various experimental studies were shown in Figure 12 of Mallik et al. (2015) and Figure 6 of Saha et al. (2018). Figure 2 reproduces the phlogopite-out boundaries as summarized in Saha et al. (2018). Compared to the amphibole–peridotite dehydration solidi, the phlogopite-out boundaries can be as much as 100–250 °C higher at any given pressure. Furthermore, phlogopite is stable to pressures distinctly higher than amphibole. Hence for K-rich domains in the mantle, the fluid absent, dehydration solidus can be similar to those expected for nominally anhydrous solidus with 100–1000 ppm H$_2$O. Importantly, for aqueous fluid-present solidus to be relevant, bulk water in excess of the storage capacity of phlogopite-peridotite would be needed to pressures of 5–6 GPa and as high as 7 GPa (Konzett and Ulmer 1999). Beyond the stability field of phlogopite, at pressures as high as 9 GPa, K-rich systems may also observe the stability of K-richterite (Konzett et al. 1997; Konzett and Ulmer 1999). Therefore, for K-rich peridotitic systems, K-richterite dehydration solidus may be applicable at pressures above the stability of phlogopite.

Carbonated peridotite solidus during open system processes

Solidus determination of $CO_2 \pm H_2O$-bearing peridotitic systems thus far focused mostly on obtaining a robust constraint on the solidus for a peridotite composition that is widely accepted to be most relevant (e.g., peridotite xenolith KLB-1) and for a volatile-poor condition (e.g., only hundreds of ppm CO_2). Yet, in the Earth's mantle, low-degree carbonated melts generated in deeper domains can act as a flux in shallower domains. Similarly, carbonated melts derived from subducting or recycled oceanic crust or sediments can act as a flux to overlying mantle. In both of these cases local mantle domains may have major and minor element concentrations and CO_2 contents that deviate significantly from KLB-1 + ppm level CO_2 composition. Hence a key question becomes, as CO_2-rich melt flux impregnates a mantle domain and locally modifies the bulk composition beyond what is expected for the ambient Earth's mantle, how does the equilibrated phase assemblage differ from nominally CO_2-free peridotite systems? The challenge in answering the question is that in open systems with variable melt flux, not only does the bulk CO_2 of the system change but also the concentration of many other key elements, including alkalis. Sun and Dasgupta (2019), using literature phase equilibria data of peridotite + $CO_2 \pm H_2O$, closely related to natural compositions, developed an empirical model to predict phase proportions of carbonated ± hydrous peridotite systems. The model of Sun and Dasgupta (2019) can be used to predict local equilibrium assemblages (of nominally anhydrous minerals and carbonate minerals or carbonated melts) for bulk compositions that maybe realized owing to metasomatism or autometasomatism, i.e., open system processes. Figure 4 shows predictions of conditions at which carbonated melt would completely freeze in through reactive crystallization processes in peridotitic bulk compositions. Sun and Dasgupta (2019) described this as the carbonation freezing front, which occurs during the process of percolation of carbonated melt through peridotite matrix. Although the Sun and Dasgupta (2019) model closely reproduces the solidi of carbonated peridotites determined experimentally by Dasgupta and Hirschmann (2006) (0.29 wt% Na_2O and 2.51 wt% CO_2) and Ghosh et al. (2009) (0.50 wt% Na_2O and 5.0 wt% CO_2), this model still needs to be verified via careful, targeted experiments using peridotites of variable composition and CO_2 contents. However, the current model shows that the carbonation freezing front or the solidus of carbonated peridotite can vary significantly as a function of bulk compositions. In particular, for a given bulk CO_2 content, enhancement of alkali such as Na_2O in the bulk peridotite can cause lowering of the carbonated peridotite solidus (Dasgupta and Hirschmann 2007a; Sun and Dasgupta 2019). Similarly, peridotitic bulk compositions

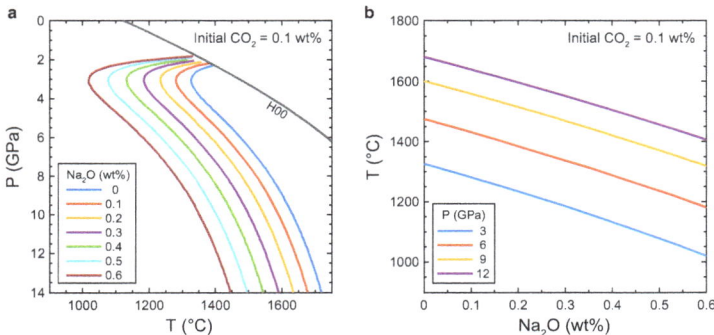

Figure 4. Carbonation freezing fronts (CFFs) of peridotite with 1000 ppm CO_2 and variable Na_2O contents, based on the model of Sun and Dasgupta (2019), which allows calculation of P–T condition at which carbonated melt in equilibrium with mantle peridotite undergoes complete reactive crystallization. In other words, the model lines in **(a)** mark the calculated solidus of carbonated peridotite of different bulk compositions, i.e., variable Na_2O/CO_2 ratio. **(b)** Carbonation freezing fronts for a peridotite with 1000 ppm CO_2 as function of variable bulk Na_2O and grouped into different pressure.

that are depleted in alkali (e.g., cratonic lithosphere mantle) may have carbonated solidi that are elevated compared to what is thought to be appropriate for a model carbonated peridotite KLB-1. The empirical model of Sun and Dasgupta (2019) captures several aspects of the phase equilibria of carbonated peridotite systems, such as melting reactions. The model can be used with confidence for making predictions of equilibrium phase assemblages for peridotite+CO_2±H_2O systems with locally high CO_2 contents. However, accurate melt fraction predictions for trace-bulk CO_2 systems remain highly uncertain even with this model because the model was calibrated empirically for systems with 0.9–17.1 wt% CO_2.

The composition of H_2O–CO_2-bearing partial melts from nominally volatile-free mantle—beneath oceans and continents

The previous sections established that as a function of CO_2 and H_2O contents, the topology, location, and melting reaction across the solidus vary significantly, with H_2O likely having a larger effect in freezing point depression than CO_2 when present as a fluid phase at subsolidus conditions. When both volatiles are present, the presence of dissolved CO_2 in the fluid lowers the activity of H_2O (a_{H_2O}) in the fluid phase. This results in an increase in the H_2O-fluid-present solidus—a feature that has been observed for H_2O-rich fluid-present solidus of pelites (Tsuno and Dasgupta 2012; Mann and Schmidt 2015) and has also been confirmed for a peridotite bulk composition (Saha and Dasgupta 2019). However, at pressures ≳2 GPa, and with concentrations of just tens to hundreds of ppm, which are relevant for most of the Earth's mantle, trace carbonate has a far greater effect than that of H_2O on the peridotite solidus (Fig. 1). Across this carbonated peridotite solidus, the near-solidus melt is carbonatitic, i.e., nearly pure carbonate melts (melt CO_2 content, $C_L^{CO_2}$ = ~40 wt% and SiO_2 content typically < 5–10 wt%). Because the CO_2 content of mantle peridotite partial melt at dolomite/magnesite saturation is 37–44 wt% (Dalton and Presnall 1998; Dasgupta and Hirschmann 2007b), the residual carbonate mineral gets exhausted at extremely low-degrees of melting (Dasgupta and Hirschmann 2006). With mantle CO_2 contents in the range of 50–200 ppm, similar to those constrained for MORB source mantle, the carbonatitic melt fraction across the solidus is estimated to be 0.01–0.05 wt% and at this extent of melting, the residual mineral carbonate phase gets exhausted within ~10 °C of the solidus.

The change of melt composition and the extent of melting above the carbonated peridotite solidus have been constrained in nominally H_2O-free, carbonated peridotite systems (e.g., Dasgupta et al. 2007a,b, 2013; Ghosh et al. 2014) and summarized recently in Dasgupta (2018). These studies on natural carbonated peridotite confirmed some previous observations in simplified CMAS+CO_2 peridotite system (e.g., Gudfinnsson and Presnall 2005). It is observed that a transition from carbonatitic melt to carbonate-rich silicate melt takes place between the carbonated peridotite solidus and the volatile-free peridotite solidus. However, importantly, this transition occurs at a relatively lower temperature with respect to the volatile-free solidus with increasing pressures (Gudfinnsson and Presnall 2005; Dasgupta et al. 2007a,b, 2013). One important difference between the natural systems and the simplified CMAS systems, however, is that in the former the onset of carbonated silicate melting occurs at distinctly lower temperature at any given pressure. Figure 5a, modified from Dasgupta (2018), captures this behavior, where natural peridotite melt CO_2 contents are shown between the nominally dry, carbonated and nominally volatile-free fertile peridotite solidus in *P–T* space. The melt CO_2 content as a function of *P–T* should change if trace quantities of H_2O are added to the system, even for the same peridotite major element bulk composition. This is because, H_2O being an incompatible species during mantle melting, partitions into the low-degree melts, enhances the melt fraction, and lowers the CO_2 content (at a given *P–T*) that would otherwise be stable for nominally anhydrous systems. Several previous studies calculated this effect (e.g. Dasgupta et al. 2007b, 2013; Hirschmann 2010; Dasgupta 2018), by applying the estimated $D_{H_2O}^{peridotite/carb.\ silicate\ melt}$, which constrains the melt H_2O content for a given melt fraction, *F*, and the cryoscopic equation (Eqn. 1).

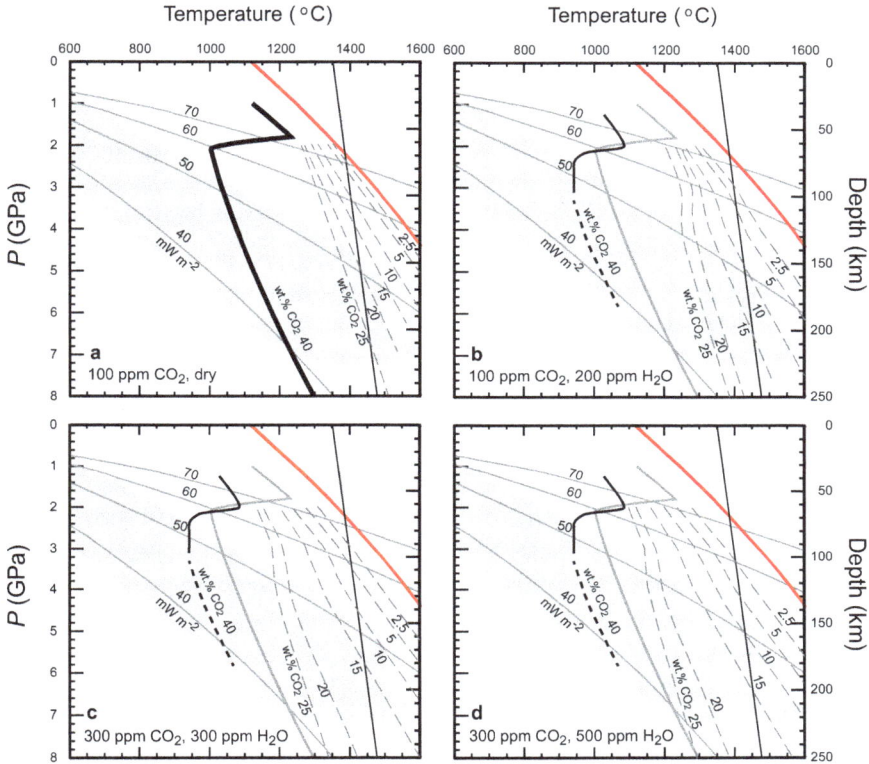

Figure 5. The conditions of near-solidus melting and carbonated silicate melt generation in **(a)** dry carbonated fertile peridotite system (Dasgupta et al. 2013) and **(b–d)** carbonated fertile peridotite system with hundreds of ppm water in NAMs (Dasgupta et al. 2013; Dasgupta, 2018) and comparison of such with depth–temperature conditions beneath oceanic ridges and continents. The **light dashed grey curves** represent melt CO_2 isopleths, with CO_2 contents in weight percent marked against each curve. The **solid grey curves** represent conductive geotherms below the continents (Lee et al. 2011) with corresponding surface heat flux in mWm^{-2} marked against each curve and the **black curve** marks the solid mantle adiabat for corresponding mantle potential temperature of 1350 °C. Also plotted for reference, the nominally volatile-free peridotite solidus from Hirschmann (2000). In **(b)**, **(c)**, and **(d)**, the H_2O content of the carbonated silicate melt along each melt-CO_2 isopleths vary according to the variation of $D_{H_2O}^{peridotite/carb.\ silicate\ melt}$ and the melt fraction of the carbonated silicate melt. **(b)**, **(c)**, and **(d)** also show nominally dry carbonated peridotite solidus (thick grey curve from Dasgupta 2013) and the hydrous carbonated peridotite solidus, in **black** (Wallace and Green 1988).

$$T = \frac{T_{peridotite}^{carb\ silicate\ melting(F)}}{(1 - (R/\Delta S_{peridotite}^{fusion})\ln(1 - X_{CO_3^{2-}}^{melt} - X_{OH^-}^{melt}))} \tag{1}$$

where T is the temperature of hydrous carbonated melt stabilization, $T_{peridotite}^{carb\ silicate\ melting(F)}$ is the temperature of stabilization of a dry carbonated silicate melt of a given melt fraction, F. $X_{OH^-}^{melt}$ and $X_{CO_3^{2-}}^{melt}$ are the mole fraction of H_2O and CO_2 in the melt, respectively and $\Delta S_{peridotite}^{fusion}$ is the entropy of near-solidus fusion of peridotite. This simplified approach, which ignores the combined effects of CO_2 and H_2O on melt compositions, non-ideal interactions between H_2O and CO_2 components and all other major element components in the melt, provides a guide as to what the H_2O contents of carbonated melts could be. This approach also suggests what the combined effects of CO_2 and H_2O could be in stabilizing hydrous, carbonated melts of a given melt fraction at temperatures below the volatile-free peridotite solidus. The results of this approach are reproduced from Dasgupta (2018) in Figure 5b–d, using a depth dependent $D_{H_2O}^{\text{peridotite/carb. silicate melt}}$ given in Dasgupta (2018). Figure 5 suggests that partitioning of water into carbonated silicate melt enhances the stability of such melt. Therefore, higher H_2O in the bulk system would facilitate the stability of carbonated melts at lower temperatures.

Following continental geotherms with equivalent surface heat flux of 40–50 mW/m², production of a carbonated silicate melt with ~25 wt% CO_2 can occur anywhere between ~8.5 and 3.5 GPa for CO_2 and H_2O content of bulk peridotite varying between 100 ppm CO_2, 200 ppm H_2O and 300 ppm CO_2, 500 ppm H_2O (Fig. 5). With bulk H_2O content of peridotite varying from 200 ppm (with bulk CO_2 of 100 ppm) to 500 ppm (with bulk CO_2 of 300 ppm), the water content of such carbonated silicate melt varies from ~5.9 to 13.5 wt%. Figure 6 plots the CO_2/H_2O ratio of hydrous carbonated melts along two representative cratonic mantle geotherms. Figure 6 shows that within the cratonic mantle lithosphere, CO_2/H_2O ratios of carbonated silicate melts decrease with increasing depth. Figure 6 also shows that low-degree carbonated melts generated along cooler cratonic mantle geotherms have higher CO_2/H_2O ratios compared to those melts generated along hotter geotherms. The estimation of the H_2O contents of carbonated melts comes entirely from the application of water partitioning between NAMs and equilibrium carbonated melts and does not take into account what roles melt compositions could play in altering the CO_2 and H_2O content for mixed volatile systems.

The carbonated melts also exhibit some distinct characteristics in terms of other major elements in the melt. The most pronounced chararacteristic is the silica-poor nature of these

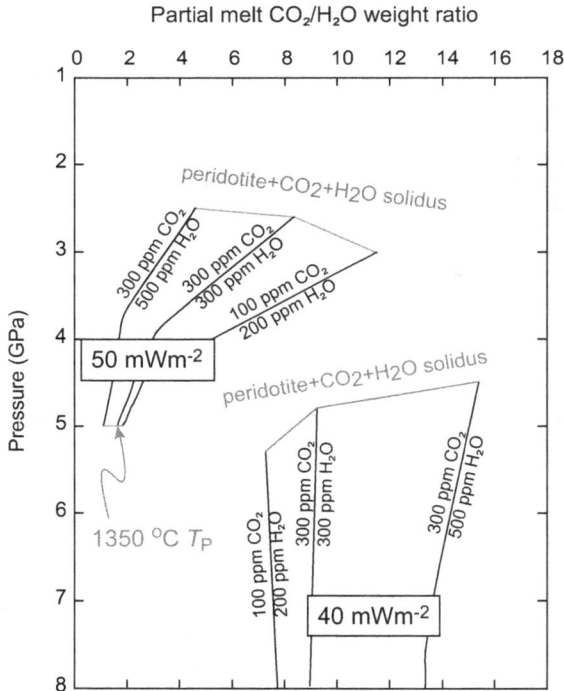

Figure 6. CO_2/H_2O weight ratio of incipient partial melts generated in the cratonic lithospheric mantle along two different geotherms (corresponding surface heat flux of 40 and 50 mWm⁻²) for mantle bulk composition with three different volatile budgets (100 ppm CO_2 + 200 ppm H_2O; 300 ppm CO_2 + 300 ppm H_2O, and 300 ppm CO_2 + 500 ppm H_2O). The plotted curves reflect the intersection of the melt compositional isopleths plotted for different bulk compositions in Figure 5 and the cratonic lithosphere geotherms.

melts, where the degree of silica undersaturation increases as a function of increasing dissolved CO_2 contents. Figure 7a shows the relation between dissolved CO_2 contents of carbonated peridotite partial melts and their silica contents from experiments at nominally anhydrous conditions. It can be seen in Figure 7a that there is a perfect linear negative correlation between CO_2 and SiO_2 contents of peridotite partial melts. While this observation of strongly silica-poor character of partial melts being caused by dissolved CO_2 (as CO_3^{2-}) has been shown before (Dasgupta et al. 2007a, 2013; Ghosh et al. 2014), it remained unclear what the effects of mixed CO_2–H_2O volatiles would be. In other words, does the CO_2–SiO_2 trend hold for systems where the melts are also variably hydrous? To address this, in Figure 7a we also plot recent experimental partial melt compositions from the studies of Stamm and Schmidt (2017) and Dvir and Kessel (2017) where the melts were variably hydrous, with H_2O contents of CO_2-bearing melts being estimated by mass balance calculations. Interestingly it can be seen that the CO_2–SiO_2 relationship is unmodified even if the melts are rich in H_2O with estimated concentration varying between ~1 and 5–7 wt%. In particular, with H_2O content up to ~5 wt%, strongly carbonated melts can still be generated with strong silica depletion. It is also worth pointing out that the CO_2–SiO_2 relationship observed in mantle-derived partial melts shows no pressure dependence for melts generated at pressures $\gtrsim 2$ GPa. This suggests that silica activity in the melt based barometer (e.g., Albarede 1992; Lee et al. 2009) cannot be applied to deep melts derived from the garnet stability field that are likely to be variably carbonated.

Another key aspect of low-degree CO_2-bearing melts is the strong enrichment in CaO, which increases with increasing CO_2 contents based on partial melting experiments in peridotite and mixed peridotite-eclogite systems (Fig. 6b; Dasgupta et al. 2007a, 2013; Mallik and Dasgupta 2013, 2014; Ghosh et al. 2014; Sun and Dasgupta 2019). Increased CaO contents of CO_2-bearing mantle melts can be understood through the strong affinity of CO_3^{2-} for Ca^{2+} as observed in many CO_2 solubility models (e.g., Ghiorso and Gualda 2015; Eguchi and Dasgupta 2018). This is also consistent with the Gibbs free energies of carbonation reactions (Spera and Bergman 1980), which suggests CO_3^{2-} should have the strongest affinity for Ca^{2+} (Dixon 1997). Again, we also check in Fig. 7b whether the enrichment in CaO content of carbonated partial melts is affected by dissolved H_2O. It is observed that with

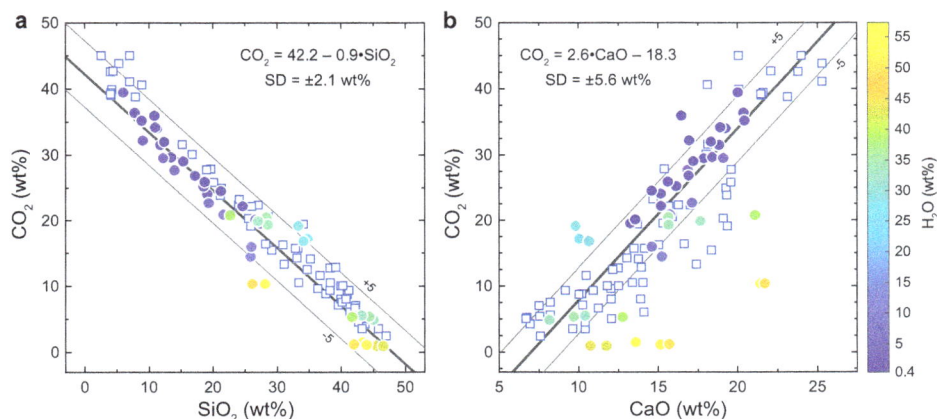

Figure 7. Compositions of volatile-rich partial melts generated from partial melting of CO_2-bearing and CO_2+H_2O-bearing peridotite systems in **(a)** CO_2–SiO_2 and **(b)** CO_2–CaO space. Low-degree partial melts of CO_2 and H_2O-poor fertile peridotite, i.e., similar to those of the Earth's ambient upper mantle are CO_2- and CaO-rich and SiO_2-poor. Lower the degree of melting, for a given bulk composition, the lower the SiO_2 content and higher the CaO. Nominally anhydrous carbonated peridotite partial melt compositions are from Dasgupta and Hirschmann (2007a,b), Dasgupta et al. (2013) and Sun and Dasgupta (2019) and the peridotite + CO_2 + H_2O system data are from Stamm and Schmidt (2017) and Dvir and Kessel (2017). The melt compositions are color coded with the estimated water contents.

dissolved water content ≤~5–6 wt%, the CO_2–CaO relation is unmodified. However, with much greater concentration of water, CaO contents of partial melts with mixed CO_2–H_2O volatiles show greater scatter. Figure 7, therefore, supports the argument that volatile-rich melts from the Earth's oxidized upper mantle are SiO_2-poor and CaO-rich, not dissimilar from what is obtained in nominally dry carbonated mantle melting experiments. If the water content of these melts does not exceed 5–10 wt%, the overall systematics shown in Figure 7 would hold. The low-silica, low-alumina, and high-Ca, carbonated and hydrous partial melts generated in the laboratory high pressure experiments are similar to the systematics observed in natural kimberlites (e.g., Price et al. 2000; Becker and Roex 2005) and other silica-poor magmas such as carbonatites, melilitites, nephelinites, and basanites. It is worth noting that following the affinity for major alkaline earth metal, i.e., Ca^{2+}, composition-dependent CO_2 solubility models also show enhancement of CO_2 concentration with alkali metals Na and K (e.g., Iacono-Marziano et al. 2012; Ghiorso and Gualda 2015; Eguchi and Dasgupta 2018), although the relative importance of Na and K remain debated with some studies finding Na to be far more important in enhancing CO_2 solubility compared to K (Vetere et al. 2014; Muth et al. 2020) whereas at least one study arguing K to be more important in enhancing CO_3^{2-} dissolution (Morizet et al. 2014). Furthermore, unlike the strong affinity for Ca^{2+}, CO_3^{2-} anion does not show similarly strong affinity for Mg^{2+} in silicate melts and CO_2 solubility increases much more if the increase of NBO/T is caused by an increase of Ca rather than by an increase of Mg (e.g., Duncan et al. 2017; Eguchi and Dasgupta 2018). In fact, with increasing Mg/Ca ratio in the silicate melt, CO_2 solubility diminishes (Morizet et al. 2017, 2019).

THE EFFECTS OF VOLATILE STORAGE IN ACCESSORY PHASES VERSUS MAJOR MANTLE MINERALS

Subsolidus mantles of rocky planets including Earth can primarily store the major volatiles either in the nominally volatile-free, dominant mantle minerals or in accessory phases. The preceding sections described how as a function of depth and concentration, the subsolidus storage of water can be chiefly in the form of fluid, hydrous silicate minerals, or in nominally anhydrous minerals. However, such changes in the water-storing phases is applicable only in scenarios where the bulk water varies over a large concentration range, from ≳1 wt% to ≲500 ppm (0.05 wt%). In volatile-poor, sub-lithospheric mantles of rocky planets, all H_2O components are dissolved in nominally anhydrous minerals (NAMs). In contrast, storage of other volatile elements, even at low concentrations varies from element to element and if stored in an accessory phase, the phase of importance can vary as a function of oxygen fugacity. Consequently, the fate of a given volatile element and how efficiently it is extracted via mantle melting from planetary interiors to surface environments may vary significantly.

Table 1 summarizes our current understanding of the storage of hydrogen, carbon, and sulfur at the onset of melting and how they are expected to vary as a function of oxygen fugacity. As can be seen from Table 1, at the low concentrations (tens to hundreds of ppm by weight) expected for most of the volatile-depleted rocky planetary mantles, hydrogen is known to be stored in NAMs, whereas carbon and sulfur are stored mostly in accessory phases. The accessory phases for carbon storage at the shallow melting depths of Earth are carbonate minerals such as dolomite and magnesite solid solutions. For reduced planetary mantles, such as those of Mars, the Moon, or even the extremely reduced mantle of Mercury, it is expected to be graphite/diamond. Significant storage of carbon at these reduced conditions may be expected in stable metal sulfide phases as well (e.g., Tsuno and Dasgupta 2015; Zhang et al. 2019). Sulfur storage in the Earth's mantle is thought to be minimal in nominally sulfur-free silicate minerals, with the highest partitioning into clinopyroxene (Callegaro et al. 2014). Only in relatively oxidized mantle domains such as those above subducting slabs or in the subducted crustal assemblages, sulfur storage in more oxidized accessory phases such as anhydrite

Table 1. Storage of major volatiles in the solid mantle as a function of oxygen fugacity.

fO_2 (ΔIW)	Example of planetary mantles	C storage	H storage	S storage
+4.5	Earth–subduction zones	CO_2 fluid/dolomite/magnesite	in NAMs	Anhydrite
~+3.5	Earth	CO_2 fluid/dolomite/magnesite	in NAMs	Fe–Ni sulfide
−1	Mars and the Moon	graphite/diamond	in NAMs	Fe–Ni sulfide
<−4	Mercury, Aubrite parent body	graphite/diamond	in NAMs	Ca–Mg–Fe sulfide

Footnotes:
Chief storage of C, H and S at the onset of melting for different planetary bodies in the Solar System and how such storage mechanisms vary as a function of fO_2 mantle.
NAMs - nominally anhydrous minerals.

feasible (Table 1) (Chowdhury and Dasgupta 2019; Zajacz and Tsay 2019). In all other planetary mantles that are similarly, or more reduced than the Earth's sub-ridge oceanic mantle or sub continental lithospheric mantle, sulfur is expected to be in solid or molten metal sulfides. In the mantles of Earth, the Moon, and Mars, sulfur is expected to be hosted in Fe–Ni-sulfides (Lorand 1991; Brenan et al. 2019), whereas in the extremely reduced mantle of Mercury, sulfur may reside in Ca–Mg–Fe-sulfides (Malavergne et al. 2014; Namur et al. 2016a).

The subsolidus storage mechanism of major volatiles imparts a key control on both the location of the solidus (the effect of H_2O and CO_2) and the composition of the near-solidus melt, including its volatile content. If the volatile species of interest is chiefly stored in an accessory phase, high concentrations of the volatile species in the melt at saturation (C_{sat}) of the accessory phase would mean that at a relatively low extent of melting (F), the accessory phase would melt out. In contrast, if the saturation concentration of a volatile species is low, the volatile-bearing accessory phase would remain in the residue over a large extent of mantle melting (F). For volatile species that are stored in accessory volatile-rich phases (e.g., carbonate for C, sulfide for S), at the onset of melting their concentrations in the melt are constrained by the saturation values. Only after the exhaustion of the accessory phase, can the volatile element act as a highly incompatible element. Specifically, if

$$C_{sat} < C_0/[D + F(1-D)] \qquad (2)$$

the volatile concentration is set by C_{sat} whereas if

$$C_{sat} > C_0/[D + F(1-D)] \qquad (3)$$

the accessory phase hosting the volatile species (graphite/diamond for C and sulfide or anhydrite for S; Table 1) is exhausted and the volatile content of the melt is set by extent of melting (F), the bulk concentration of volatiles (C_0), and the bulk partition coefficient of the volatile species between the silicate mineral assemblage and the partial melt (D).

The following section will present the current understanding of the fate of C, H, and S during mantle melting. It will address how as a function of varying volatile storage at different conditions of planetary mantles, the volatile contents of mantle-derived melts may vary.

Mobilization of carbon versus sulfur-bearing accessory phase during melting of the Earth's mantle

Easily fusible accessory phase such as dolomite or magnesite solid solution directly influences the solidus location, by stabilizing a melt phase that is carbonate-rich. This leads to efficient melting of the mineral carbonate phase at the onset of melting in the Earth's oxidized upper mantle, leading carbon to behave as a highly incompatible element. Consequently, very low-degree melt generated from the Earth's oxidized upper mantle are extremely CO_2-rich (~37–45 wt% CO_2 or ~12 wt% C); e.g., Dalton and Presnall 1998; Dasgupta and Hirschmann 2007b.

In contrast to accessory carbonate minerals, sulfur-bearing accessory phase mono-sulfide solid solutions or mono-sulfide liquid solutions impart only a limited control on the near-solidus melting behavior of Earth's silicate mantle. Although sulfide solid solution itself undergoes melting below the peridotite solidus (Bockrath et al. 2004; Zhang and Hirschmann 2016), given the limited mobility of such sulfide melt in the mantle matrix (e.g., Minarik et al. 1996; Shannon and Agee 1996), molten sulfide acts as a residual phase during mantle melting. Furthermore, because sulfide melts and silicate melts are non-conjugate immiscible melts, sulfide melt does not aid in silicate dissolution and no continuous transition exists between sulfide-rich melt and silicate melt. Therefore, even though sulfide melting occurs at temperatures lower than the peridotite solidus, the extraction of sulfur from the Earth's mantle is dependent on the sulfur content at sulfide saturation of mantle partial melts.

Similar to sulfide, carbon may also act as a refractory phase during mantle melting, if stored in graphite or diamond, i.e., in reduced mantle domains such as deep continental lithospheric mantle (e.g., Woodland and Koch 2003; Frost and McCammon 2008; Stagno et al. 2013). If silicate melts could form in such reduced mantle domains on Earth, then carbon extraction would also depend on carbon concentration of silicate melt at graphite/diamond saturation. However, except for special circumstances such as generation of silicate partial melt from more easily fusible subducted crustal lithologies (Eguchi and Dasgupta 2017, 2018b) or generation of hydrous silicate melts atop the mantle transition zone (e.g., Bercovici and Karato 2003), silicate melt generation under deep (> 5–6 GPa), reduced, upper mantle conditions of Earth is not common. Hence, we will not discuss further extraction of carbon from Earth's graphite/diamond bearing mantle. However, such processes become applicable for other reduced planetary mantles and will be discussed in a forthcoming section.

Sulfur mobilization by incipient melt in the Earth's mantle

The sulfur concentration of basaltic melts at sulfide saturation in the Earth's upper mantle is only ~1000–2000 ppm (e.g., Mavrogenes and O'Neill 1999; O'Neill and Mavrogenes 2002; Fortin et al. 2015; Ding and Dasgupta 2017, 2018; Smythe et al. 2017). Using coupled thermodynamic mantle melting and sulfur solubility models, Ding and Dasgupta (2017, 2018) considered in detail how as a function of mantle melting and mantle potential temperatures, exhaustion of sulfide during mantle melting is likely to take place. These authors showed that depending on the initial sulfur abundance in the mantle and mantle potential temperature, sulfide exhaustion could take place between 10 and 15% melting for MORB source mantle or even lower for hotter OIB source mantle. Thus, at melting degrees lower than 10–15 wt%, mantle partial melts are likely to have S contents set by sulfide saturation (sulfur concentration at sulfide saturation: SCSS) values, and only above melting degrees of 10–15% S would behave as an incompatible element. However, Ding and Dasgupta (2017, 2018), as well as many other geochemical studies, did not consider sulfur mobility via incipient volatile-rich melts. Hence, it remained unclear what type of geochemical fractionation between sulfur and other key volatiles such as carbon and water could take place during low degree melting in the Earth's mantle.

The key questions are (1) how efficiently mantle-derived melts extract sulfur from sulfide-bearing mantle lithologies and (2) whether $CO_2 \pm H_2O$-rich mantle melts can efficiently mobilize sulfur from the Earth's interior. Both questions have implications for whether mantle-derived primary melts are sulfide-saturated and whether mantle sulfides are present in all mantle domains (Fig. 8). Until recently, SCSS was only constrained experimentally for nominally volatile-free silicate melts or hydrous silicate melts (e.g., Mavrogenes and O'Neill 1999; O'Neill and Mavrogenes 2002; Liu et al. 2007; Li and Ripley 2009; Ding et al. 2014, 2018; Fortin et al. 2015; Smythe et al. 2017). Recent studies of Woodland et al. (2019) and Chowdhury and Dasgupta (2020) provided the first data for SCSS of carbonated melts at mantle conditions. These studies showed that in comparison to basaltic melts, SCSS of carbonated melts are distinctly lower. In particular, Chowdhury and Dasgupta (2020) showed that for CO_2-rich melts, SCSS diminishes

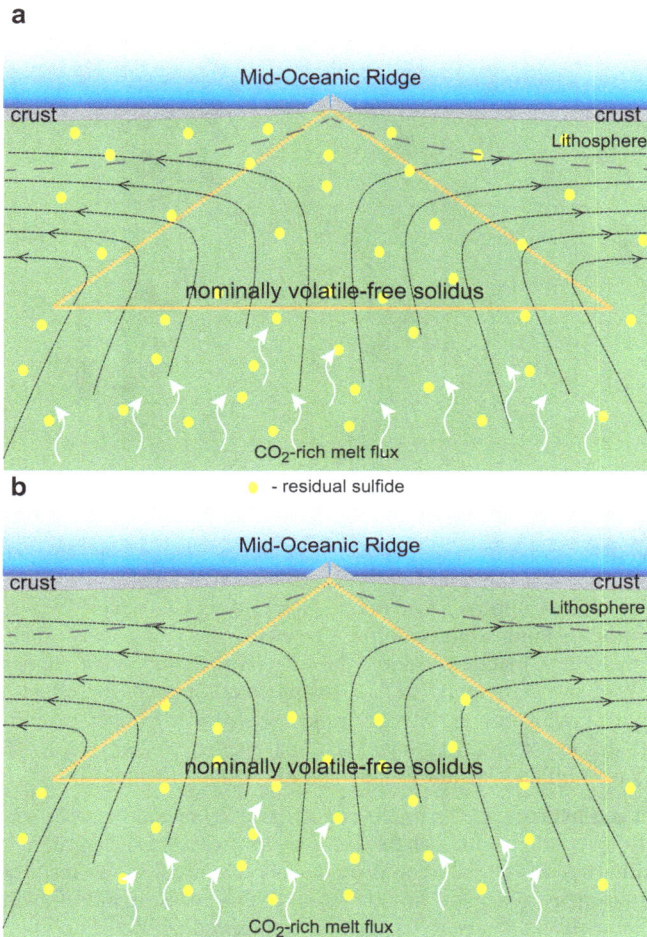

Figure 8. Sulfide extraction framework for Earth's oceanic mantle beneath mid-oceanic ridges. **(a)** Scenario in which incipient volatile-rich melt and dominant basaltic melting do not consume sulfide from the melting triangle owing either to relatively low SCSS, or high enough sulfide abundance in the mantle, or not high enough extent of melting or all of these factors acting in tandem. **(b)** Scenario in which incipient volatile-rich melt and larger volume basaltic melt consume sulfide in the melting triangle owing to high SCSS, or high enough extent of melting, or low abundance of sulfide abundance or all of these factors acting together.

with decreasing silica, i.e., as melt compositions evolve from silica-poor basaltic to carbonatitic through carbonated silicate compositions of kimberlitic or melilititic or nephelinitic affinity, SCSS drops (Fig. 9). Chowdhury and Dasgupta (2020) provide a parameterization to determine SCSS of partial melts as the melt composition evolves from near-solidus, low-temperature carbonatitic melt to higher temperature basaltic melt near and across the volatile-free peridotite solidus. Application of these new SCSS data and model suggests that the CO_2-rich and H_2O-bearing incipient melt beneath ocean ridges has a limited effect in mobilizing sulfur from sulfide-bearing mantle. Therefore, deep carbonated melting does not consume much sulfide at the onset of silicate melting, and only a minimal drop in the residual sulfide budget is expected via extraction of deep carbonated silicate melt (Chowdhury and Dasgupta 2020).

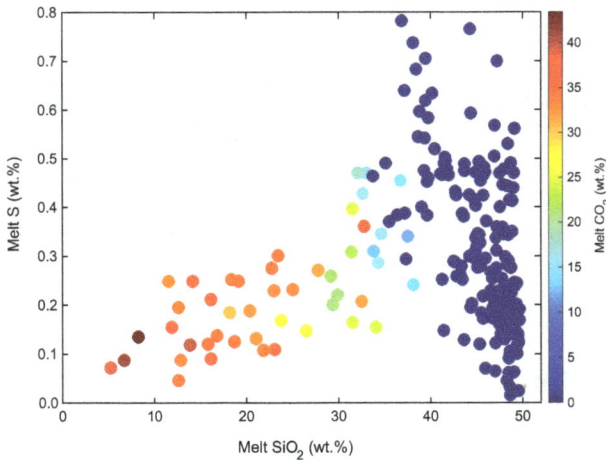

Figure 9. A plot of sulfur versus melt silica content showing the sulfur concentrations at sulfide saturation for carbonatitic to basaltic melts from high pressure-temperature experimental studies (P = 1 bar-10.5 GPa and T = 1150–1800 °C). The symbols are color coded for estimated melt CO_2 contents. The compilation of experimental SCSS for carbonated and nominally CO_2-free mantle melts is from Chowdhury and Dasgupta (2020).

Sulfur extraction, as opposed to carbon, therefore, occurs chiefly via silicate melting across the volatile-free solidus of peridotite. The exact SCSS of basaltic melt generated via decompression melting of mantle peridotite depends on P, T, and major element composition of such melts. Ding and Dasgupta (2017, 2018) modeled in detail the effects of each of the intensive variables, taking into account possible variations in mantle potential temperature during adiabatic decompression melting. Such analyses suggested that 10–20% melting is necessary to exhaust sulfide if mantle S abundance is ~100–200 ppm.

Carbonated and hydrous silicate melts can act as chief agents in mobilizing sulfur at the subcontinental lithospheric mantle, where owing to cooler temperatures along the conductive geotherms, volatile-free melt generation is precluded. Figure 10 shows the modeled SCSS of $CO_2 \pm H_2O$-bearing melts generated in the cratonic lithospheric mantle for different surface heat fluxes. These melts are generated across the carbonated peridotite solidus, for nominally H_2O-free mantle, with all water being stored in nominally anhydrous silicates at the onset of carbonated melting. Owing to lower temperatures, SCSS of carbonated melts at continental mantle conditions is generally lower compared to SCSS of similar melts along the oceanic adiabat. Therefore, carbonated melting in the cratonic mantle generates a melt composition that is high in C but very low in S, resulting in a high C/S ratio.

It is also possible that in the metasomatized cratonic mantle, a hydrous silicate melt is generated across the dehydration solidus of amphibole peridotite. Figure 10 compares the SCSS of such hydrous melts with those of carbonated melts along various shield geotherms in the cratonic mantle. Owing to lower temperature and despite the strongly H_2O-rich character of these melts, the sulfur dissolution capacity at sulfide saturation for these melts appears to be even lower compared to the carbonated melts. This suggests that although CO_2-rich melts with variable water content or H_2O-rich melts from metasomatized regions may be generated in the lithosphere, such melts do not mobilize sulfur efficiently. Irrespective of the cause of melting, i.e., either CO_2 or H_2O, the melt flux from the cold lithospheric mantle has high C/S (Fig. 11) and H/S ratios. This contrasts with melt composition generated via shallow melting beneath oceanic ridges or in intraplate settings where melts are generally CO_2 and H_2O-poor but may contain S content in excess of 1000 ppm. In such scenarios and C/S and the H/S ratios are distinctly lower (Fig. 11).

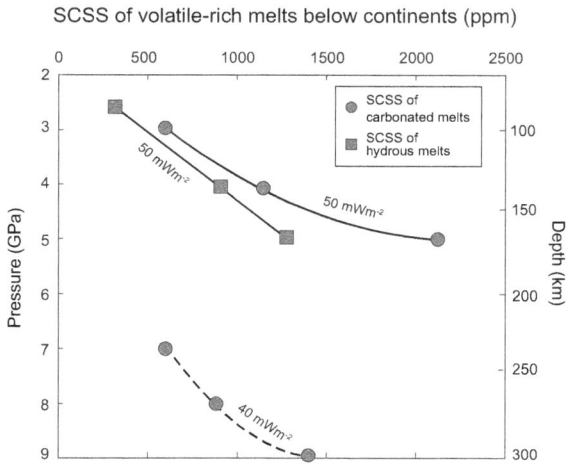

Figure 10. SCSS as a function of pressure (GPa) or depth (km) along continental geotherms with surface heat flux of 50 mW/m^2 (solid line) and 40 mW/m^2 (dashed line). The **square symbols** are SCSS of hydrous peridotite partial melts from Green et al. (2014) generated across the amphibole–peridotite dehydration solidus, calculated using the SCSS model of Fortin et al. (2015) and the **circles** are SCSS of carbonated peridotite partial melts calculated using Chowdhury and Dasgupta (2020) model. Along both geotherms, the shallower melt is at the carbonated peridotite solidus with 100 ppm CO_2 (Dasgupta 2018), which has low SCSS compared to the melt at greater depth (convective mantle adiabat of potential temperature, $T_P \sim 1350\,^\circ C$). For hydrous partial melts, the shallower melt is at the solidus of pargasite-bearing lherzolite with subsolidus water stored in pargasite and NAMs, has slightly lower SCSS compared to melt at greater depth (convective mantle adiabat of $T_P \sim 1350\,^\circ C$).

Extraction of C–S–H volatiles from other rocky mantles in the Solar System

The oxidized shallow mantle of the modern Earth allows C–O–H volatiles to be present in the form of oxidized species such as carbonates or H_2O-rich fluid. Volatiles in these oxidized forms lower the melting temperatures of mantle silicates and causes generation of partial melts from a larger volume of the mantle (Fig. 12). Plus, as discussed earlier, the composition of mantle partial melts can be rich in CO_2 and/or H_2O. If present in the form of more refractory phase such as graphite/diamond, carbon does not impart much control in modifying the solidus of the mantle. Hence for a volatile-poor, reduced mantle, a much smaller volume of the mantle would get processed through the melting region per unit time. However, even in this scenario, a basaltic silicate melt generated across the nominally volatile-free solidus can dissolve a finite concentration of volatile elements such as carbon and sulfur. For sulfur extraction, this process is not dissimilar to how it gets extracted from most of Earth's convective mantle. Although if the mantle domains are more reduced than is thought to be the case for most of Earth's history, dissolved SO_4^{2-} cannot contribute to sulfur release from sulfide saturated mantle. Furthermore, under extremely reduced conditions, the saturation values of graphite or sulfide equilibrated mantle-derived melts may be very different.

Conditions and compositions of the mantles of rocky planets may vary due to differences in planetary building blocks and/or differences in the conditions of early differentiation and core-mantle separation. For example, compared to the Earth's mantle, the mantle of Mars is thought to be much more FeO-rich and have a lower Mg# (100 × molar MgO/(MgO+FeO)). On the other hand, the mantle of Mercury is thought to be FeO-poor. These compositional differences indeed have effects on the conditions of melting and the generated compositions of basaltic melts (Bertka and Holloway 1994a; Hirschmann 2000; Namur et al. 2016b; Duncan et al. 2018; Ding et al. 2020). Such differences in the compositions of primary basaltic melts and the

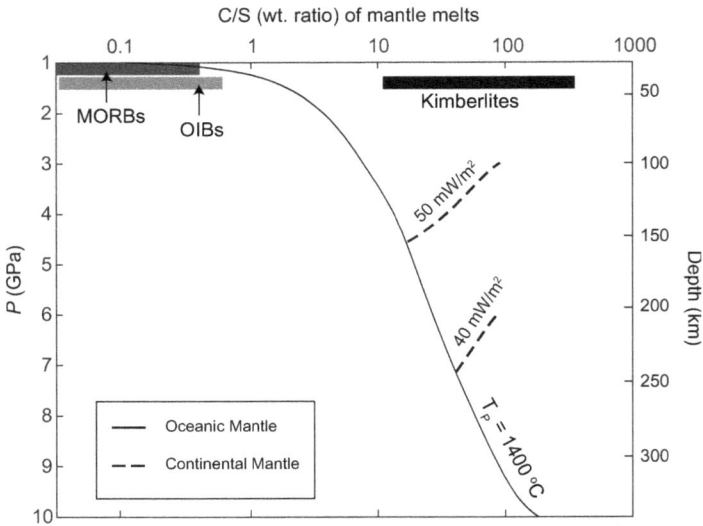

Figure 11. Estimated C/S weight ratio of Earth's mantle-derived partial melts as a function of pressure and depth along continental geotherms (dashed curves) and mid-oceanic ridge adiabat (solid curve). The two continental shield geotherms correspond to surface heat flux of 40 and 50 mW/m² and the convective mantle adiabat corresponds to mantle potential temperature of 1400 °C. C content of partial melts along the continental geotherms are from Dasgupta (2018) for a dry peridotite bulk composition with 100 ppm CO_2 and the corresponding S contents are calculated using the model of Chowdhury and Dasgupta (2020). For the oceanic mantle, partial melt C contents are from Dasgupta et al. (2013), again for a dry peridotite with 100 ppm bulk CO_2; the corresponding melt S contents are calculated using the SCSS model of Chowdhury and Dasgupta (2020). Also plotted for comparison are the range of C/S ratios measured or estimated for different terrestrial magmas that are thought to be undegassed or minimally degassed – mid-ocean ridge basalts [MORBs; (Saal et al. 2002; Miller et al. 2019)], ocean island basalts [OIBs; (Moussallam et al. 2016; Moussallam et al. 2019)], and kimberlites (Howarth and Büttner 2019). For MORBs and OIBs, the C/S ratios are calculated based on measured concentrations of C and S. For kimberlites, the S contents are measured whereas the C contents are estimated based on the reported silica contents and the CO_2–SiO_2 relationship shown in Fig. 7a.

conditions of their stability of course affect the volatile contents of such melts. However, the compositional differences between the mantles of Earth, Mars, the Moon, and Mercury is less significant compared to the differences in the redox state among these bodies. Among these four planetary bodies, Earth's mantle is the most oxidized (Christie et al. 1986; Bezos and Humler 2005; Cottrell and Kelley 2011), followed by that of Mars and the Moon (Wadhwa 2001; Wadhwa 2008), whereas Mercury and Aubrite parent body are thought to be the most reduced (e.g., Casanova et al. 1993; Wadhwa 2008; Namur et al. 2016a). The difference between the oxygen fugacity (fO_2) of the Earth's mantle and those of the Moon and Mars are ~4 ± 1 log units, and the difference between the Earth and that of the Mercurian mantle is as much as 9 ± 2 log units. Such large differences in the fO_2 conditions of different planetary mantles has huge implications for the volatile storage and degassing efficiency of individual volatile species. Among the bodies being considered, Earth's shallow mantle is the only reservoir where carbon is expected to exist as CO_2-rich fluid or mineral carbonate. In all other cases, graphite is likely the stable carbon-bearing phase, except in very reducing mantles, where a certain fraction of carbon may be dissolved in a Fe–Ni alloy phase. For sulfur, sulfide is thought to be the chief repository for all planetary mantles although the composition of sulfide and its metal/sulfur ratio is expected to vary significantly with decreasing fO_2. Therefore, extraction of carbon and sulfur from a large swath of planetary mantles relies heavily on the carbon content at graphite saturation (CCGS) and sulfur content at sulfide saturation (SCSS) of basaltic melts generated at mantle conditions and especially how they vary as a function of fO_2.

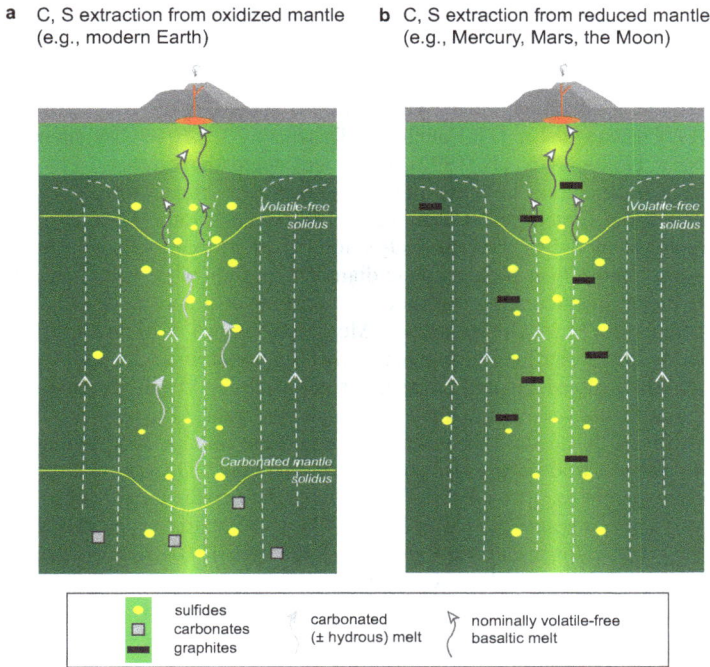

Figure 12. Cartoon showing difference in the framework of extraction of carbon and sulfur from planetary mantles as a function of varying oxygen fugacity, fO_2. (**a**) Scenario of melting and volatile extraction from volatile-poor yet relatively oxidized mantle such as that of the modern Earth. (**b**) Scenario of partial melting and volatile extraction from volatile-poor but relatively reduced mantle such as those of the Moon, Mars, and Mercury. In (**a**), carbon and sulfur storage in the subsolidus mantle is assumed to be chiefly in the form of mineral carbonate and sulfide (molten or solid), respectively. Whereas in (**b**), carbon and sulfur storage in the subsolidus mantle is assumed to be chiefly in the form of graphite/diamond and sulfide (molten or solid), respectively. In (**a**), deeper onset of carbonated melting leads to efficient extraction of C and H (stored in NAMs) from a larger mantle volume compared to that in (**b**), where the onset of melting is controlled by the nominally volatile-free solidus. Extraction of sulfur between (**a**) and (**b**) differs mainly by lower overall SCSS for more oxidized, basaltic and carbonated melts in (**a**) and higher SCSS, which increases with decreasing fO_2, for reduced basaltic melts in (**b**).

Carbon contents of graphite-saturated mantle melts—application to planetary mantles

In natural silicate melts, CO_2 can dissolve both as molecular CO_2 and CO_3^{2-}, with CO_3^{2-} being more important in mafic melts (e.g. Fine and Stolper 1986; Holloway and Blank 1994; Dixon 1997; Duncan et al. 2017) and molecular CO_2 being more important in felsic melts (e.g. Stolper et al. 1987; Fogel and Rutherford 1990; Blank et al. 1993; Behrens et al. 2004; Duncan and Dasgupta 2014, 2015, 2017). The dissolution reactions for molecular CO_2 and CO_3^{2-} are:

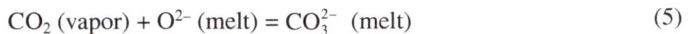

$$CO_2 \text{ (vapor)} = CO_2 \text{ (melt)} \tag{4}$$

$$CO_2 \text{ (vapor)} + O^{2-} \text{ (melt)} = CO_3^{2-} \text{ (melt)} \tag{5}$$

However, at fO_2 conditions that fall distinctly below the CCO buffer ($C + 0.5 O_2 = CO$; $C + O_2 = CO_2$), no vapor phase can be present in equilibrium with silicate melt and an additional graphite/diamond oxidation reaction as given below needs to be considered to fully understand CO_2 dissolution in silicate melt:

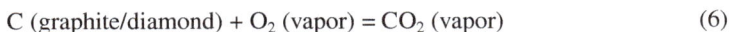

$$C \text{ (graphite/diamond)} + O_2 \text{ (vapor)} = CO_2 \text{ (vapor)} \tag{6}$$

Equation (6) demonstrates that at graphite saturation, fCO_2 is a function of fO_2. With lower fO_2 resulting in lower fCO_2, which translates to lower carbon content, dissolved as CO_2, in the melt, following equations (4) and (5).

Several studies (Hirschmann and Withers 2008; Stanley et al. 2011, 2012; Duncan and Dasgupta 2017; Duncan et al. 2017; Eguchi and Dasgupta 2017,2018a,b; Dasgupta and Grewal 2019), following Holloway et al. (1992), have considered the extent of CO_2 dissolution in silicate melt of a given composition as a function of pressure, temperature, and oxygen fugacity at graphite saturation. Eguchi and Dasgupta (2018) extended these efforts and developed an internally consistent model to compute CO_2 solubility in silicate melts as a function of P, T, melt composition, and fO_2 at both graphite/diamond saturation as well fluid saturation. The results of this model in calculating carbon concentrations of graphite- or CO_2 vapor-saturated basaltic melt for Earth, Mars, the Moon, and Mercury is shown in Figure 13a. As can be seen from Figure 13a, only under the Earth's mantle melting conditions, can a CO_2-vapor–basaltic-melt equilibrium be realized, resulting in as much as ~1000–3000 ppm dissolved C, which may be carried by mantle-derived basaltic melt to the crustal depths. Of course as discussed before, mantle melting on Earth can also take place in the carbonate stability field, and hence at fO_2 near the graphite/diamond-carbonated phase equilibrium, strongly CO_2-rich melt may be generated from the Earth's mantle (e.g., Stagno et al. 2013). Therefore, the C content of carbonatitic to low-CO_2 alkalic melts are also plotted in Figure 13a. In comparison, the predicted C contents dissolved as CO_2 of the lunar and Martian basalts are in the 1–100 ppm range and those from Mercurian mantle are in the 0.001–0.01 ppm range. Therefore, Figure 13a shows that with all other factors being equal, with the control of fO_2 alone, nominally anhydrous basalts from different planetary mantles may extract orders of magnitude different carbon contents.

The model of Eguchi and Dasgupta (2018) only considered dissolution of carbon as CO_2 or CO_3^{2-}. However, depending on the water content of different planetary basalts and the fO_2 conditions of their generation, fH_2 may also be significant. Equation (7) shows that for a given activity of H_2O, hydrogen fugacity, fH_2 would increase with decreasing fO_2.

$$H_2O \text{ (vapor)} = H_2 \text{ (vapor)} + 0.5O_2 \qquad (7)$$

Therefore, for hydrous basalts generated at graphite-saturated, reduced conditions, carbon may dissolve in basaltic melt also as hydrogenated species, which would be favored at higher fH_2 conditions, following a reaction such as:

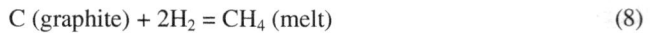

$$C \text{ (graphite)} + 2H_2 = CH_4 \text{ (melt)} \qquad (8)$$

Li et al. (2017) developed empirical models of total C dissolution in silicate melt as a function of melt composition, fO_2, P, T, and melt H_2O content for two distinct fO_2 ranges. Given the model of Li et al. (2017) was calibrated with total carbon analyzed using secondary ionization mass spectrometry (SIMS), it can potentially account for the possible presence of carbon dissolved as species other than CO_3^{2-}. The results of this model are also plotted in Figure 13a assuming total water content amounting to a mole fraction of water (X_{H_2O}) of 0.001. The model results of Li et al. (2017) show that at higher fO_2 planetary mantles, such as those of the Earth, Mars, and the Moon, the total dissolved C in basaltic melt at graphite saturation largely follows the behavior captured via dissolution of CO_3^{2-}. This suggests that under these conditions, even for partial melts that are mildly hydrous, CO_3^{2-} is the only species that is critical to consider. However, some studies argued that dissolution of carbon monoxide, CO could be more important than previously thought (Yoshioka et al. 2019). The model of Li et al. (2017) suggests at more reducing conditions, other species of carbon such as hydrogenated species such as methyl groups or other C–H species may become important and the total C solubility at graphite saturation may deviate more and more from carbonate dissolution alone with falling fO_2. If this is relevant, C

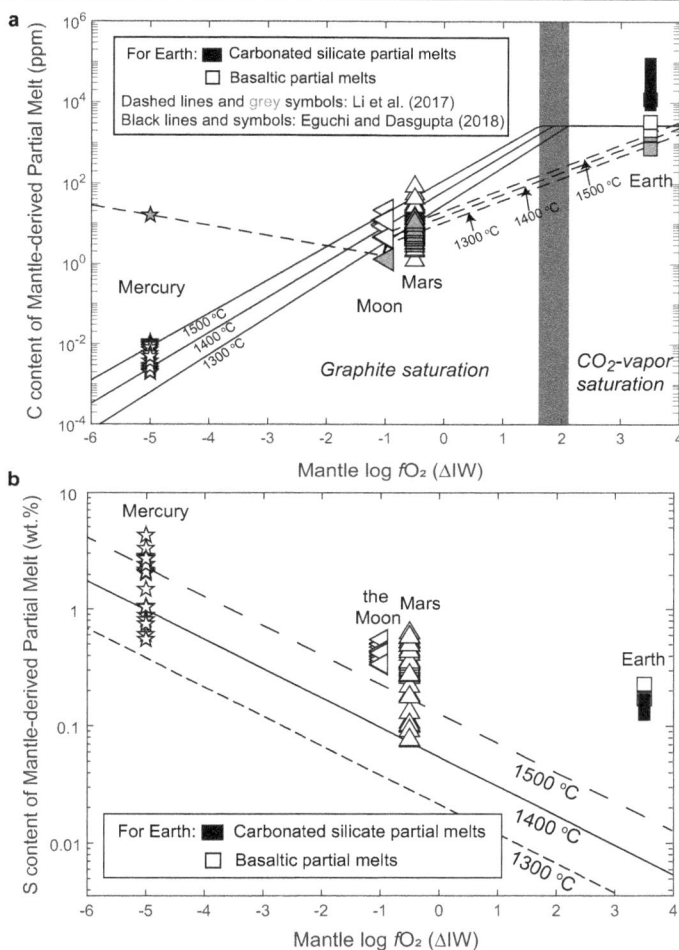

Figure 13. Log-log plot showing the **(a)** carbon and **(b)** sulfur concentration at graphite and sulfide saturation (CCGS and SCSS) of mantle partial melts from Earth (**squares**), Mars (**upright triangles**), the Moon (**triangles pointing to the left**), and Mercury (**stars**) as a function of oxygen fugacity (fO_2) thought to be prevailing in these planetary mantles over most of their geologic history. The mantle partial melt compositions for Earth are taken from experiments of Baker et al. (1995) and Walter (1998) for basaltic silicate melts and Hirose (1997), Dasgupta et al. (2007a, 2013), Stamm and Schmidt (2017), and Sun and Dasgupta (2019) for carbonated melts. For Mars, the mantle partial melt compositions are from Bertka and Holloway (1994b), Agee and Draper (2004), Collinet et al. (2015), and Filiberto and Dasgupta (2015). The lunar partial melt compositions are those used in Ding et al. (2018) and the compositions of partial melts for Mercury are from Namur et al. (2016a). In **(a)**, the **open symbols** and the **solid lines** reflect modeled CO_2 concentration at graphite saturation based on the model of Eguchi and Dasgupta (2018a). The **dashed lines** and the **grey symbols** are total dissolved carbon content based on the empirical calibration of Li et al. (2017). The **solid and dashed black lines** are CCGS values of a peridotite partial melt composition (Baker et al. 1995) calculated using Eguchi and Dasgupta (2018a) and Li et al. (2017) models, respectively at 1300 °C, 1400 °C and 1500 °C. In **(b)**, sulfur concentration at sulfide saturation (SCSS as wt%) of various partial melts are calculated using the exact compositions and conditions as in (a). The **black lines** show how the SCSS of a peridotite partial melt composition used in **(a)** (Baker et al. 1995) increases with decreasing fO_2, following the model of Namur et al. (2016a) at 1 GPa and 1300 °C (**short-dashed**), 1400 °C (**solid**), and 1500 °C (**dashed**). SCSS for different planet-specific basaltic melt compositions are calculated using Smythe et al. (2017)—Earth, Ding et al. (2014)—Mars, Ding et al. (2018)—the Moon, and Namur et al. (2016a)—Mercury. SCSS of carbonated partial melts for Earth is based on the model of Chowdhury and Dasgupta (2020). The **vertical grey band** separates fO_2 conditions for CO_2-rich vapor stability (to the right) versus graphite stability (to the left).

extraction of similarly reduced mantles as that of Mercury may not be as inefficient and there could be as much as 10s of ppm C in mantle-derived melts from Mercury-type reduced mantles.

Taken together, the calculations presented above highlights the importance of oxygen fugacity being a controlling variable in dictating the efficiency of carbon extraction at graphite saturation. With more reducing conditions, C dissolution diminishes in planetary basaltic melts.

Sulfur concentrations at sulfide saturation—application to planetary mantles

Similar to extraction of carbon, sulfur extraction via dissolution of sulfide into mantle-derived silicate melt also is thought to vary significantly as a function of oxygen fugacity. At reducing conditions ($\log fO_2 < IW + 3$, where IW represents the fO_2 of the iron–wustite buffer and is given by the reaction: $Fe + 0.5O_2 = FeO$), sulfur (S^{2-}) replaces oxygen (O^{2-}) in the anion sublattice of the silicate melt following the expression (O'Neill and Mavrogenes 2002):

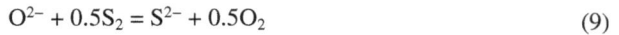

$$O^{2-} + 0.5S_2 = S^{2-} + 0.5O_2 \qquad (9)$$

Equation (9) suggests that as fO_2 decreases, sulfur dissolution should be enhanced, where fugacity of S_2, fS_2 is set by the equilibrium between silicate melt and sulfide melt following the equilibrium

$$MO^{Sil} + 0.5S_2 = MS^{Sul} + 0.5O_2 \qquad (10)$$

In Equation (10), the cation M can be either Fe, which is the case for sulfide melt-silicate melt equilibrium under conditions expected for Earth, Mars, and the Moon, or some combination of Fe, Ca, and Mg under the extremely reducing conditions prevailing in planetary mantles such as Mercury's.

Figure 13b plots the calculated SCSS of the same experimental partial melts used for CCGS calculations in Figure 13a. It can be observed that the sulfur carrying capacity of mafic magmas shows an opposite trend as a function of fO_2 compared to the CCGS trend. Sulfur carrying capacity at sulfide saturation increases slightly going from relatively FeO-poor mantle melts of Earth to more FeO rich partial melts of the Moon and Mars. However, for all these bodies the SCSS of basalts lie in the range of few thousands of ppm. For Earth, carbonated mantle melts of course mostly have lower SCSS as presented earlier, and are also plotted in Figure 13b for completeness. However, with further reduction, and fO_2 conditions of approximately $IW - 5$, SCSS of mafic melts reaches as high as several weight percent for planetary mantles such as that of Mercury's (e.g., Namur et al. 2016a). The SCSS model of Namur et al. (2016a) captures the trend of SCSS increase with decreasing fO_2. However, at the higher end of fO_2, i.e., between IW and $IW + 5$ it underpredicts the SCSS of mantle-derived mafic melts such as those of the Moon, Mars, and Earth as revealed from the application of other SCSS models, which were designed specifically to constrain SCSS of these planetary basalts (Ding et al. 2014, 2018; Smythe et al. 2017).

Fractionation of carbon and sulfur during mantle melting at varying oxygen fugacity

The calculations of CCGS and SCSS presented above leads to an important prediction of carbon-sulfur fractionation during magma generation from planetary mantles that differ in their oxygen fugacity by several log units. Figure 14 demonstrates this by plotting the C/S weight ratio of planetary mantle partial melts for different Solar System bodies that vary significantly in the redox state of their interiors. For Earth, the C/S ratio of the primary basaltic melt can vary from ~0.1 to 10 and if carbonated magmas are considered, can even approach ~100. These high C/S ratios of magmatic flux for Earth are in stark contrast with the ones that are modeled to be derived from more reduced, graphite-saturated mantles. For magmatic fluxes from the lunar and Martian mantles, the C/S ratio is approximately between ~0.01 and 0.001. For Mercurian-type reduced mantles, the C/S ratio can be as low as between ~10^{-7} and 10^{-6}, if C dissolution is

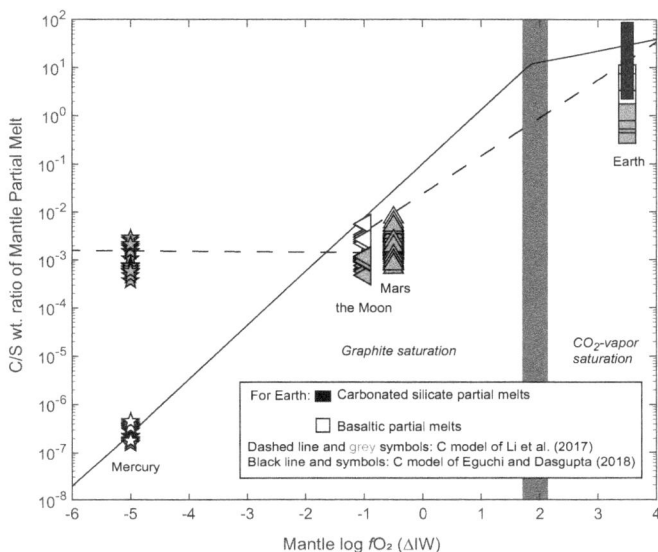

Figure 14. A log–log plot showing the carbon/sulfur (C/S) ratio of mantle-derived partial melts plotted as a function of oxygen fugacity (fO_2) for Earth, Mars, the Moon and Mercury. The **solid black line** is calculated using Eguchi and Dasgupta (2018a) model for C and Namur et al. (2016a) model for S. The **dashed line** is calculated using Li et al. (2017) model for C and Namur et al. (2016a) model for S. The **grey and open** symbols are derived using the same solubility models for C and S as used in Figure 13. The vertical **grey band** separates fO_2 conditions for CO_2-rich vapor stability (to the **right**) versus graphite stability (to the **left**).

entirely in the form of CO_3^{2-}. On the other hand, if increased fugacity of other gas species, such as hydrogen is relevant, C dissolution may be somewhat higher and the C/S ratio of Mercurian mantle-derived melts may not be much lower than those from the Moon and Mars.

Carbon, sulfur and water as a function of melting degree and mantle redox

The discussion in the preceding section assumed that graphite and sulfide remained in the residue and controlled the carbon and sulfur contents of mantle-derived melts via saturation. However, depending on the degree of mantle melting and the initial abundance of volatile-bearing accessory phases, the melting process may consume such phases. For example, Figure 15 shows the coupled effects of bulk mantle S contents, degree of melting, and SCSS values for various rocky planets in the Solar System on the expected abundance of sulfur in mantle partial melts. For simplicity, a fixed SCSS value is used for Earth, Mars, and the Moon. In detail, the SCSS of mantle-derived partial melts for the Moon and Mars are somewhat higher than the basaltic melts on Earth (Fig. 13b). Furthermore, in Figure 15 a fixed SCSS value is used for each planetary body as a function of the extent of melting. However, in natural systems, the SCSS of mantle melts also changes with the degree of melting as the P, T, and melt composition changes with the degree of melting. These effects have been modeled in detail for Earth and Mars by Ding and Dasgupta (2017, 2018) and Ding et al. (2014), respectively. The focus of Figure 15, however, is to highlight the distinct difference in SCSS of extremely reduced Mercurian basalts and basalts for other rocky planets in the Solar System. Figure 15 shows that for a sulfur-poor mantle with only 100–300 ppm S, terrestrial, lunar, and Martian mantles may remain sulfide saturated with large extents of melting, but an extremely reduced mantle similar to Mercury's would be sulfide undersaturated. Therefore, high sulfur abundances in lavas from the different geochemical provinces of Mercury as estimated using data from the MESSENGER mission (Peplowski et al. 2015; Weider et al. 2015; Namur et al. 2016a), suggest that the Mercurian mantle must be much more S-rich compared to the terrestrial mantle.

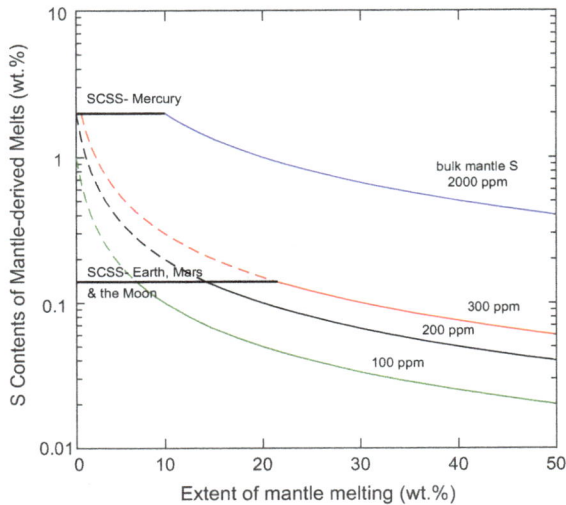

Figure 15. Sulfur contents of mantle melts for various rocky bodies in our Solar System plotted as a function of the degree of mantle melting. SCSS at Mercurian mantle conditions is distinctly higher that the SCSS of other planetary basalts. As a consequence, for the same sulfur content in the mantle source (e.g., 200–300 ppm S), the mantle melts from Earth, Mars, and the Moon are likely to be sulfide saturated over most realistic extent of partial melting and thus the sulfur contents being dictated by SCSS whereas Mercurian mantle melt. However, with the same extent of melting (e.g., 5–20 wt%), Mercurian basalts would be sulfide undersaturated with S contents decreasing with the degree of melting.

Information on the volatile abundance of planetary interiors are often lacking, especially for distant bodies in the Solar System and beyond. Therefore, an accurate prediction of what a mantle-derived melt flux would bring to the surface environments in terms of volatile concentrations is difficult. However, important insights can be gained, by constraining what the expected fractionation patterns between the major volatiles, C, S, and H would be, controlled entirely by mantle redox. For this analysis, one can at first order, assume that all planetary mantles started with the same volatile inventory. Figure 16 presents calculations to showcase such volatile element evolution as a function of melting degree for three distinct planetary mantle fO_2s. One key underlying piece of information for such modeling is whether fractionation of C and S is argued to be heavily affected by carbonated melting or by CCGS and SCSS along with melting degree. The extraction of water is expected to take place by partitioning of water between NAMs and partial melts under all oxygen fugacity conditions (e.g., Withers et al. 2011). However, further studies, will be needed to confirm the behavior of H during melting under extremely reducing conditions.

Figure 16 shows that for Earth's mantle, C acts as a highly incompatible element for effectively the entire melting interval whereas H is less incompatible. Sulfur contents for a large fraction of melting interval is set by SCSS. As a consequence C/H and C/S ratios start at > 1 for low extent of melting and decreases with increasing melting degree. At 5–20 wt% melting, depending on the SCSS and initial sulfide abundance in the mantle, sulfide may get exhausted from the Earth's mantle (e.g., Ding and Dasgupta 2017, 2018) and thus with melting extent greater than that, the C/S ratio does not change.

The C/H and C/S systematics of mantle partial melts from the reduced mantles are distinctly different compared to those from the Earth's oxidized mantle. Irrespective of the exact fO_2 of the reduced mantle, the C/S and C/H ratios of low-degree melts start out low and increase with increasing degrees of melting. The C/S ratio of reduced mantle melts can be

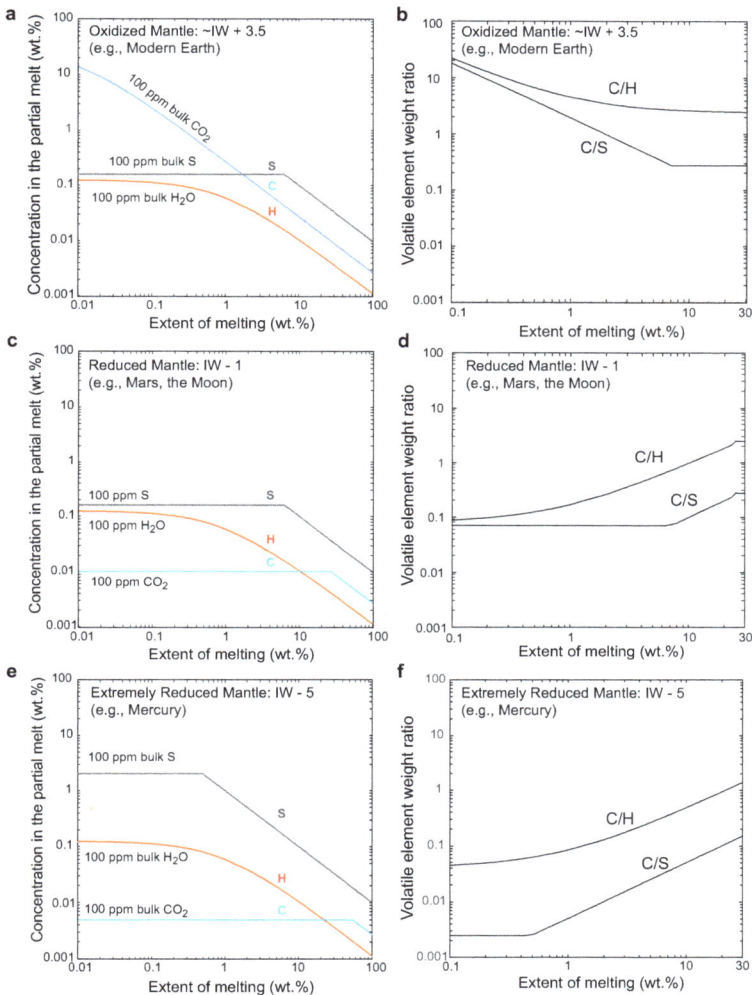

Figure 16. Modeled variation of carbon, water (plotted as hydrogen), and sulfur concentration (**a, c, and e**) and C/H and C/S ratio (**b, d, f**) in the mantle-derived partial melts as a function of degree of melting for three distinct planetary mantle redox state. (**a, b**) oxidized mantle similar to that of the modern Earth at $\log fO_2 \sim IW + 3.5$, (**c, d**) reduced mantle similar to those of the Moon and Mars at $\log fO_2 \sim IW - 1$, and (**e, f**) extremely reduced mantle similar to that of the mantle of Mercury at $\log fO_2 \sim IW - 5$. The calculations are for rocky mantles with initial H_2O, CO_2, and S contents of 100 ppm each, i.e., ~11 ppm H, ~27 ppm C, and 100 ppm S, similar to the depleted upper mantle of Earth processed by mid-oceanic ridge magmatism. With these concentrations of volatiles at the onset of melting, all H is thought to be stored in NAMs, all C is in the form of graphite or carbonate, and all S is in sulfide phases. If sulfide, carbonate, and graphite are present, the partial melt compositions in terms of C and S contents are those set by carbonate saturation, graphite saturation, and sulfide saturation, respectively. Owing to high CO_2 content of near solidus partial melt of peridotite at the carbonate stability field, mineral carbonate melts out at vanishingly small amount of melting and for the entire melting range shown, carbon acts as a highly incompatible element. For simplicity, fixed CCGS, SCSS values were used for each of the planetary mantle. SCSS of Earth, Mars, and the Moon have been taken as a fixed value of 1400 ppm (Mavrogenes and O'Neill 1999; Ding et al. 2014, 2018; Smythe et al. 2017) and that for Mercury as 2 wt% (Namur et al. 2016a). Once the accessory, volatile-bearing residual phase is melted out, the bulk peridotite–melt partition coefficient used for carbon, sulfur, and hydrogen are 0.0001 (Rosenthal et al. 2015), 0.0001 (assumed), and 0.009 (e.g. O'Leary et al. 2010), respectively.

constant across a large melting interval, i.e., in the presence of residual graphite and sulfide. However, the exact C/S ratio would vary depending on absolute mantle fO_2, which controls SCSS and CCGS values. For example, Figure 16 shows the C/S ratio for the lunar and Martian mantles melts are close to 0.1 and those from the Mercurian mantle are in the range 0.001– 0.01. Only if the sulfur abundance were as low as in the Earth's mantle, would the C/S ratio increase, owing to consumption of the residual sulfide phase at some finite extent of melting.

CONCLUDING REMARKS

Oxidized forms of carbon and hydrogen in the mantle have significant influence in stabilizing mantle partial melts at depths well below the volatile-free mantle solidus. This leads to much larger volumes of the mantle being processed per unit planetary evolution timescale (Dasgupta and Hirschmann 2006; Sandu et al. 2011; Fuentes et al. 2019), resulting in efficient extraction of highly incompatible elements including life-essential elements, noble gases, and heat producing elements. Oxidized species of carbon and hydrogen allow incipient, volatile-rich melts to form even within the thermal boundary layer and thus influence lithospheric properties (e.g., Dasgupta 2018; Saha and Dasgupta 2019) such as electrical conductivity, seismic shear wave velocity, and rheology. If present as a free vapor phase, H_2O-rich vapor has a larger influence in depressing the solidus of mantle rocks than the influence of CO_2-rich vapor. However, at the volatile-depleted conditions relevant for most regions of the Earth, a free, aqueous vapor phase is unexpected. In such scenarios oxidized carbon, present as carbonate minerals, imparts a larger control in stabilizing trace amounts of carbonated melts well below the volatile-free mantle solidus. Therefore, for a similar abundance of H_2O and CO_2, the cratonic lithospheric mantle generates an incipient melt flux that has CO_2/H_2O (C/H) ratio >1 and as high as 10–15. At lithospheric and sub-lithospheric depths, these carbonated melts display a strong negative correlation between SiO_2 and CO_2 and a positive correlation between CaO and CO_2. Up to 5–10 wt% dissolved water does not affect the major element compositional attributes of these silica undersaturated melts. CO_2-rich melts do not mobilize significant sulfur from the sulfide-saturated upper mantle of Earth and as a consequence, the oxidized mantle of Earth, over its entire melting range, releases carbon much more efficiently compared to sulfur. Shallow basaltic melts, which constitute the greatest volume of crust-forming magma have a C/S ratio ~1, whereas low-degree CO_2-rich melts may have C/S ratios as high as 80. The shallowest equilibrium melts beneath continents are the ones with the highest C/S ratios; however, similarly high C/S ratios can only be obtained in the oceanic mantle at great depths via carbonated or redox melting.

Whereas oxidized C–O–H vapor imparts a strong control on partial melt stability in rocky mantles, storage in reduced phases has no significant effects on partial melt stability. Therefore, for reduced planetary mantles such as those of Mars and Mercury, a much smaller volume of mantle gets processed through the melting zone per unit time. While Earth's oxidized upper mantle allows generation of partial melts with different C/S and C/H ratios with melts varying between basalts to carbonatitic melts, the reduced planetary mantles supply melt inputs to the crustal reservoirs that are distinctly different and are expected to be much less variable. Using a few other rocky bodies in our Solar System with available estimates on the fO_2 state of their mantle domains, we compared and contrasted how the major volatile contents and major volatile ratios of the mantle-derived melts could vary as a function of the mantle redox state. A major controller of the volatile abundance of mantle-derived basaltic melts is how the mineralogy and composition of the volatile-bearing host phase changes as a function of redox, which plays a key role in controlling their dissolution in silicate. The C/S ratio is expected to decrease as the fO_2 of the planetary mantle decreases. This is because the CCGS and SCSS values decrease and increase with decreasing fO_2, respectively. The C/H ratio of mantle melts is also expected to be much lower for a more reduced, nominally ahydrous mantle.

Extraction of C, S, and water from the planetary mantles is a critical step in setting and maintaining the chemistry of the ocean-atmosphere system for rocky planets, which in turn modulates the surface temperature of planets. The synthesis in this chapter provides a framework for computing the primary magmatic flux of major life-essential volatiles from planetary mantles. An Earth-like planet with a relatively oxidized upper mantle seems unique in terms of extracting volatiles more efficiently due to the active roles of fluids in stabilizing partial melts at deeper depths. Furthermore, the relative abundances of carbon to sulfur or carbon to water in the primary magmas are expected to change significantly depending on redox state and extent of melting, where the latter is largely controlled by the thermal state of the planet. Future studies aimed at constraining mantle degassing for establishing secondary atmospheres of rocky planets should therefore consider how the volatile inventory in the mantle-derived melt is expected to change as a function of the interior oxidation state.

ACKNOWLEDGMENTS

We gratefully acknowledge useful review and editorial comments by Zoltan Zajacz and Grant Henderson, respectively. This work received support from NASA grant 80NSSC18K0828 and NSF grant EAR-1763226 to RD.

REFERENCES

Agee CB, Draper DS (2004) Experimental constraints on the origin of Martian meteorites and the composition of the Martian mantle. Earth Planet Sci Lett 224:415–429

Albarede F (1992) How deep do common basaltic magmas form and differentiate? J Geophys Res: Solid Earth 97:10997–11009

Baker MB, Hirschmann MM, Ghiorso MS, Stolper EM (1995) Compositions of near-solidus peridotite melts from experiments and thermodynamic calculations. Nature 375:308–311

Becker M, Roex APL (2005) Geochemistry of South African on- and off-craton, Group I and Group II kimberlites: petrogenesis and source region evolution. J Petrol 47:673–703

Behrens H, Ohlhorst S, Holtz F, Champenois M (2004) CO_2 solubility in dacitic melts equilibrated with H_2O–CO_2 fluids: Implications for modeling the solubility of CO_2 in silicic melts. Geochim Cosmochim Acta 68:4687–4703

Bercovici D, Karato S-i (2003) Whole-mantle convection and the transition-zone water filter. Nature 425:39–44

Bertka CM, Holloway JR (1994a) Anhydrous partial melting of an iron-rich mantle I: subsolidus phase assemblages and partial melting phase relations at 10 to 30 kbar. Contrib Mineral Petrol 115:313–322

Bertka CM, Holloway JR (1994b) Anhydrous partial melting of an iron-rich mantle II: primary melt compositions at 15 kbar. Contrib Mineral Petrol 115:323–338

Bezos A, Humler E (2005) The Fe3+/ΣFe ratios of MORB glasses and their implications for mantle melting. Geochim Cosmochim Acta 69:711–725

Blank JG, Stolper EM, Carroll MR (1993) Solubilities of carbon dioxide, water in rhyolitic melt at 850°C and 750 bars. Earth Planet Sci Lett 119:27–36

Bockrath C, Ballhaus C, Holzheid A (2004) Fractionation of the platinum-group elements during mantle melting. Science 305:1951–1953

Brenan JM, Mungall JE, Bennett NR (2019) Abundance of highly siderophile elements in lunar basalts controlled by iron sulfide melt. Nat Geosci 12:701–706

Brey G, Green DH (1977) Systematic study of liquidus phase relations in olivine melilitite+H_2O+CO_2 at high pressures and petrogenesis of an olivine melilitite magma. Contrib Mineral Petrol 61:141–162

Brey G, Brice WR, Ellis DJ, Green DH, Harris KL, Ryabchikov ID (1983) Pyroxene–carbonate reactions in the upper mantle. Earth Planet Sci Lett 62:63–74

Brooker RA, Kohn SC, Holloway JR, McMillan PF (2001) Structural controls on the solubility of CO_2 in silicate melts Part I: bulk solubility data. Chem Geol 174:225–239

Casanova I, Keil K, Newsom HE (1993) Composition of metal in aubrites: Constraints on core formation. Geochim Cosmochim Acta 57:675–682

Callegaro S, Baker DR, De Min A, Marzoli A, Geraki K, Bertrand H, Viti C, Nestola F (2014) Microanalyses link sulfur from large igneous provinces and Mesozoic mass extinctions. Geology 42:895–898

Chowdhury P, Dasgupta R (2019) Effect of sulfate on the basaltic liquidus, Sulfur Concentration at Anhydrite Saturation (SCAS) of hydrous basalts – Implications for sulfur cycle in subduction zones. Chem Geol 522:162–174

Chowdhury P, Dasgupta R (2020) Sulfur extraction via carbonated melts from sulfide-bearing mantle lithologies— Implications for deep sulfur cycle and mantle redox. Geochim Cosmochim Acta 269:376–397

Christie DM, Carmichael ISE, Langmuir CH (1986) Oxidation states of mid-ocean ridge basalt glasses. Earth Planet Sci Lett 79:397–411

Collinet M, Médard E, Charlier B, Vander Auwera J, Grove TL (2015) Melting of the primitive martian mantle at 0.5–2.2 GPa and the origin of basalts and alkaline rocks on Mars. Earth Planet Sci Lett 427:83–94

Conceição RV, Green DH (2004) Derivation of potassic (shoshonitic) magmas by decompression melting of phlogopite+pargasite lherzolite. Lithos 72:209–229

Condamine P, Médard E (2014) Experimental melting of phlogopite-bearing mantle at 1 GPa: Implications for potassic magmatism. Earth Planet Sci Lett 397:80–92

Cottrell E, Kelley KA (2011) The oxidation state of Fe in MORB glasses and the oxygen fugacity of the upper mantle. Earth Planet Sci Lett 305:270–282

Dalton JA, Presnall DC (1998) Carbonatitic melts along the solidus of model lherzolite in the system CaO–MgO–Al_2O_3–SiO_2–CO_2 from 3 to 7 GPa. Contrib Mineral Petrol 131:123–135

Dasgupta R (2013) Ingassing, storage, and outgassing of terrestrial carbon through geologic time. Rev Min Geochem 75:183–229

Dasgupta R (2018) Volatile bearing partial melts beneath oceans and continents—where, how much, and of what compositions? Am J Sci 318:141–165

Dasgupta R, Grewal DS (2019) Origin, early differentiation of carbon and associated life-essential volatile elements on Earth. *In:* Orcutt B, Daniel I, Dasgupta R (Ed.), Deep Carbon: Past to Present. Cambridge University Press, Cambridge, p 4–39

Dasgupta R, Hirschmann MM (2006) Melting in the Earth's deep upper mantle caused by carbon dioxide. Nature 440:659–662

Dasgupta R, Hirschmann MM (2007a) Effect of variable carbonate concentration on the solidus of mantle peridotite. Am. Mineral 92:370–379

Dasgupta R, Hirschmann MM (2007b) A modified iterative sandwich method for determination of near-solidus partial melt compositions. II. Application to determination of near-solidus melt compositions of carbonated peridotite. Contrib Mineral Petrol 154:647–661

Dasgupta R, Hirschmann MM, Smith ND (2007a) Partial melting experiments of peridotite+CO_2 at 3 GPa and genesis of alkalic ocean island basalts. J Petrol 48:2093–2124

Dasgupta R, Hirschmann MM, Smith ND (2007b) Water follows carbon: CO_2 incites deep silicate melting and dehydration beneath mid-ocean ridges. Geology 35:135–138

Dasgupta R, Mallik A, Tsuno K, Withers AC, Hirth G, Hirschmann MM (2013) Carbon-dioxide-rich silicate melt in the Earth's upper mantle. Nature 493:211–215

Davis FA, Tangeman JA, Tenner TJ, Hirschmann MM (2009) The composition of KLB-1 peridotite. Am Mineral 94:176–180

Ding S, Dasgupta R (2017) The fate of sulfide during decompression melting of peridotite—implications for sulfur inventory of the MORB-source depleted upper mantle. Earth Planet Sci Lett 459:183–195

Ding S, Dasgupta R (2018) Sulfur inventory of ocean island basalt source regions constrained by modeling the fate of sulfide during decompression melting of a heterogenous mantle. J Petrol 59:1281–1308

Ding S, Dasgupta R, Tsuno K (2014) Sulfur concentration of martian basalts at sulfide saturation at high pressures and temperatures – implications for deep sulfur cycle on Mars. Geochim Cosmochim Acta 131:227–246

Ding S, Hough T, Dasgupta R (2018) New high pressure experiments on sulfide saturation of high-FeO* basalts with variable TiO_2 contents—Implications for the sulfur inventory of the lunar interior. Geochim Cosmochim Acta 222:319–339

Ding S, Dasgupta R, Tsuno K (2020) The solidus and melt productivity of nominally anhydrous Martian mantle constrained by new high pressure–temperature experiments—Implications for crustal production and mantle source evolution. J Geophys Res Planets 123:e2019JE006078

Dixon JE (1997) Degassing of alkalic basalts. Am Mineral 82:368–378

Dixon JE, Stolper EM, Holloway JR (1995) An experimental study of water and carbon dioxide solubilities in mid-ocean ridge basaltic liquids. Part I: Calibrations and solubility models. J Petrol 36:1607–1631

Duncan MS, Dasgupta R (2014) CO_2 solubility and speciation in rhyolitic sediment partial melts at 1.5–3 GPa—Implications for carbon flux in subduction zones. Geochim Cosmochim Acta 124:328–347

Duncan MS, Dasgupta R (2015) Pressure and temperature dependence of CO_2 solubility in hydrous rhyolitic melt—Implications for carbon transfer to mantle source of volcanic arcs via partial melt of subducting crustal lithologies. Contrib Mineral Petrol 169:1–19

Duncan MS, Dasgupta R (2017) Rise of Earth's atmospheric oxygen controlled by efficient subduction of organic carbon. Nat Geosci 10:387–392

Duncan MS, Dasgupta R, Tsuno K (2017) Experimental determination of CO_2 content at graphite saturation along a natural basalt-peridotite melt join: Implications for the fate of carbon in terrestrial magma oceans. Earth Planet Sci Lett 466:115–128

Duncan MS, Schmerr NC, Bertka CM, Fei Y (2018) Extending the solidus for a model iron-rich martian mantle composition to 25 GPa. Geophys Res Lett 45:10,211–10,220

Dvir O, Kessel R (2017) The effect of CO_2 on the water-saturated solidus of K-poor peridotite between 4 and 6 GPa. Geochim Cosmochim Acta 206:184–200

Eguchi J, Dasgupta R (2017) CO_2 content of andesitic melts at graphite saturated upper mantle conditions with implications for redox state of oceanic basalt source regions and remobilization of reduced carbon from subducted eclogite. Contrib Mineral Petrol 172:12

Eguchi J, Dasgupta R (2018a) A CO_2 solubility model for silicate melts from fluid saturation to graphite or diamond saturation. Chem Geol 487:23–38

Eguchi J, Dasgupta R (2018b) Redox state of the convective mantle from CO_2-trace element systematics of oceanic basalts. Geochem Persp Lett 8:17–21

Eguchi J, Seales J, Dasgupta R (2020) Great Oxidation and Lomagundi events linked by deep cycling and enhanced degassing of carbon. Nature Geosci 13:71–76

Falloon TJ, Green DH (1989) The solidus of carbonated, fertile peridotite. Earth Planet Sci Lett 94:364–370

Filiberto J, Dasgupta R (2015) Constraints on the depth and thermal vigor of melting in the Martian mantle. J Geophys Res Planets 120:109–122

Fine G, Stolper E (1986) Dissolved carbon dioxide in basaltic glasses: concentrations, speciation. Earth Planet Sci Lett 76:263–278

Fogel RA, Rutherford MJ (1990) The solubility of carbon dioxide in rhyolitic melts; a quantitative FTIR study. Am Mineral 75:1311–1326

Foley SF, Taylor WR, Green DH (1986) The effect of fluorine on phase relationships in the system $KAlSiO_4$–Mg_2SiO_4–SiO_2 at 28 kbar and the solution mechanism of fluorine in silicate melts. Contrib Mineral Petrol 93:46–55

Fortin M-A, Riddle J, Desjardins-Langlais Y, Baker DR (2015) The effect of water on the sulfur concentration at sulfide saturation (SCSS) in natural melts. Geochim Cosmochim Acta 160:100–116

Frost DJ, McCammon CA (2008) The redox state of earth's mantle. annual. Rev. Earth Planet Sci 36:389–420

Fuentes JJ, Crowley JW, Dasgupta R, Mitrovica JX (2019) The influence of plate tectonic style on melt production and CO_2 outgassing flux at mid-ocean ridges. Earth Planet Sci Lett 511:154–163

Fumagalli P, Zanchetta S, Poli S (2009) Alkali in phlogopite and amphibole and their effects on phase relations in metasomatized peridotites: a high-pressure study. Contrib Mineral Petrol 158:723–737

Ghiorso MS, Gualda GAR (2015) An H_2O–CO_2 mixed fluid saturation model compatible with rhyolite–MELTS. Contrib Mineral Petrol 169:53

Ghosh S, Ohtani E, Litasov KD and Terasaki H (2009) Solidus of carbonated peridotite from 10 to 20 GPa and origin of magnesiocarbonatite melt in the Earth's deep mantle. Chem Geol 262:17–28

Ghosh S, Litasov K, Ohtani E (2014) Phase relations and melting of carbonated peridotite between 10 and 20 GPa: a proxy for alkali- and CO_2-rich silicate melts in the deep mantle. Contrib Mineral Petrol 167:1–23

Green DH (1973) Experimental melting studies on a model upper mantle composition at high pressure under water-saturated, water-undersaturated conditions. Earth Planet Sci Lett 19:37–53

Green DH, Hibberson WO, Kovacs I, Rosenthal A (2011) Water and its influence on the lithosphere–asthenosphere boundary. Nature 472:504–504

Green DH, Hibberson WO, Rosenthal A, Kovács I, Yaxley GM, Falloon TJ, Brink F (2014) Experimental Study of the Influence of Water on Melting and Phase Assemblages in the Upper Mantle. J Petrol 55:2067–2096

Grove TL, Till CB (2019) H_2O-rich mantle melting near the slab–wedge interface. Contrib Mineral Petrol 174:80

Grove TL, Chatterjee N, Parman SW, Médard E (2006) The influence of H_2O on mantle wedge melting. Earth Planet Sci Lett 249:74–89

Gudfinnsson G, Presnall DC (2005) Continuous gradations among primary carbonatitic, kimberlitic, melilititic, basaltic, picritic, and komatiitic melts in equilibrium with garnet lherzolite at 3–8 GPa. J Petrol 46:1645–1659

Hirose K (1997) Partial melt compositions of carbonated peridotite at 3 GPa and role of CO2 in alkali-basalt magma generation. Geophys Res Lett 24:2837–2840

Hirschmann MM (2000) The mantle solidus: experimental constraints and the effect of peridotite composition. Geochem Geophys Geosys 1:2000GC000070

Hirschmann MM (2006) Water, melting, and the deep Earth H_2O cycle. Ann Rev Earth Planet Sci 34:629–653

Hirschmann MM (2010) Partial melt in the oceanic low velocity zone. Phys Earth Planet Int 179:60–71

Hirschmann MM, Withers AC (2008) Ventilation of CO_2 from a reduced mantle and consequences for the early Martian greenhouse. Earth Planet Sci Lett 270:147–155

Hirschmann MM, Aubaud C, Withers AC (2005) Storage capacity of H_2O in nominally anhydrous minerals in the upper mantle. Earth Planet Sci Lett 236:167–181

Hirschmann MM, Tenner T, Aubaud C, Withers AC (2009) Dehydration melting of nominally anhydrous mantle: The primacy of partitioning. Phys Earth Planet Int 176:54–68

Holloway JR, Blank JG (1994) Application of experimental results to C–O–H species in natural melts. Rev Mineral 30:186–230

Holloway JR, Pan V, Gudmundsson G (1992) High-pressure fluid-absent melting experiments in the presence of graphite; oxygen fugacity, ferric/ferrous ratio and dissolved CO_2. Eur J Mineral 4:105–114

Howarth GH, Büttner SH (2019) New constraints on archetypal South African kimberlite petrogenesis from quenched glass-rich melt inclusions in olivine megacrysts. Gond Res 68:116–126

Iacono-Marziano G, Morizet Y, Le Trong E, Gaillard F (2012) New experimental data and semi-empirical parameterization of H_2O–CO_2 solubility in mafic melts. Geochim Cosmochim Acta 97:1–23

Kawamoto T, Holloway JR (1997) Melting temperature and partial melt chemistry of H_2O-saturated mantle peridotite to 11 Gigapascals. Science 276:240–243

Keppler H, Wiedenbeck M, Shcheka SS (2003) Carbon solubility in olivine and the mode of carbon storage in the Earth's mantle. Nature 424:414–416

Kohlstedt DL, Keppler H, Rubie DC (1996) Solubility of water in the α, β, γ phases of $(Mg,Fe)_2SiO_4$. Contrib Mineral Petrol 123:345–357

Konzett J, Ulmer P (1999) The stability of hydrous potassic phases in lherzolitic mantle—an experimental study to 9.5 GPa in simplified and natural bulk compositions. J Petrol 40:629–652

Konzett J, Sweeney RJ, Thompson AB, Ulmer P (1997) Potassium amphibole stability in the upper mantle: an experimental study in a peralkaline KNCMASH system to 8.5 GPa. J Petrol 38:537–568

Kushiro I, Syono Y, Akimoto S-I (1968) Melting of a peridotite nodule at high pressures and high water pressures. J Geophys Res 73:6023–6029

Lange RA (1994) The effect of H_2O, CO_2, and F on the density and viscosity of silicate melts. Rev Mineral 30: 331–369

Lara M, Dasgupta R (2020) Partial melting of a depleted peridotite metasomatized by a MORB-derived hydrous silicate melt - Implications for subduction zone magmatism. Geochim Cosmochim Acta 290:137–161

Lee C-TA, Luffi P, Plank T, Dalton H, Leeman WP (2009) Constraints on the depths and temperatures of basaltic magma generation on Earth and other terrestrial planets. Earth Planet Sci Lett 279:20–33

Lee C-TA, Luffi P, Chin EJ (2011) Building and destroying continental mantle. Ann Rev Earth Planet Sci 39:59–90

Li C, Ripley EM (2009) Sulfur contents at sulfide-liquid or anhydrite saturation in silicate melts: empirical equations and example applications. Econ Geol 104:405–412

Li Y, Dasgupta R, Tsuno K (2017) Carbon contents in reduced basalts at graphite saturation: Implications for the degassing of Mars, Mercury, and the Moon. J Geophys Res Planets 122 :300–1320

Liu Y, Samaha N-T, Baker DR (2007) Sulfur concentration at sulfide saturation (SCSS) in magmatic silicate melts. Geochim Cosmochim Acta 71:1783–1799

Lorand J-P (1991) Sulphide petrology, sulphur geochemistry of orogenic lherzolites: A comparative study of the Pyrenean bodies (France), the Lanzo massif (Italy). J Petrol Special Volume 2:77–95

Luth R, Stachel T (2014) The buffering capacity of lithospheric mantle: implications for diamond formation. Contrib Mineral Petrol 168:1–12

Malavergne V, Cordier P, Righter K, Brunet F, Zanda B, Addad A, Smith T, Bureau H, Surblé S, Raepsaet C, Charon E, Hewins RH (2014) How Mercury can be the most reduced terrestrial planet and still store iron in its mantle. Earth Planet Sci Lett 394:186–197

Mallik A, Dasgupta R (2013) Reactive infiltration of MORB-eclogite-derived carbonated silicate melt into fertile peridotite at 3 GPa and genesis of alkalic magmas. J Petrol 54:2267-2300.

Mallik A, Dasgupta R (2014) Effect of variable CO_2 on eclogite-derived andesite–lherzolite reaction at 3 GPa— Implications for mantle source characteristics of alkalic ocean island basalts. Geochem Geophys Geosyst 15:1533–1557

Mallik A, Nelson J, Dasgupta R (2015) Partial melting of fertile peridotite fluxed by hydrous rhyolitic melt at 2–3 GPa: implications for mantle wedge hybridization by sediment melt and generation of ultrapotassic magmas in convergent margins. Contrib Mineral Petrol 169:1–24

Mallik A, Dasgupta R, Tsuno K, Nelson J (2016) Effects of water, depth and temperature on partial melting of mantle-wedge fluxed by hydrous sediment-melt in subduction zones. Geochim Cosmochim Acta 195:226–243

Mandler BE, Grove TL (2016) Controls on the stability and composition of amphibole in the Earth's mantle. Contrib Mineral Petrol 171:1–20

Mann U, Schmidt MW (2015) Melting of pelitic sediments at subarc depths: 1. Flux vs. fluid-absent melting and a parameterization of melt productivity. Chem Geol 404:150–167

Massuyeau M, Gardés E, Morizet Y, Gaillard F (2015) A model for the activity of silica along the carbonatite–kimberlite–mellilitite–basanite melt compositional joint. Chem Geol 418:206–216

Mavrogenes JA, O'Neill HSC (1999) The relative effects of pressure, temperature, oxygen fugacity on the solubility of sulfide in mafic magmas. Geochim Cosmochim Acta 63:1173–1180

Médard E, Grove T (2008) The effect of H_2O on the olivine liquidus of basaltic melts: experiments and thermodynamic models. Contrib Mineral Petrol 155:417–432

Mengel K, Green DH (1989) Stability of amphibole and phlogopite in metasomatized peridotite under water-saturated and water-undersaturated conditions. *In:* Kimberlites and Related Rocks, Proc 4th International Kimberlite Conference Perth, Australia, 1986, Geol Soc Aust Spec Pub 14, Blackwell Scientific, p 571–581

Mierdel K, Keppler H, Smyth JR, Langenhorst F (2007) Water solubility in aluminous orthopyroxene and the origin of Earth's asthenosphere. Science 315:364–368

Miller WGR, Maclennan J, Shorttle O, Gaetani GA, Le Roux V, Klein F (2019) Estimating the carbon content of the deep mantle with Icelandic melt inclusions. Earth Planet Sci Lett 523:115699

Millhollen GL, Irving AJ, Wyllie PJ (1974) Melting interval of peridotite with 5.7 per cent water to 30 kilobars. The J Geol 82:575–587

Minarik WG, Ryerson FJ, Watson EB (1996) Textural entrapment of core-forming melts. Science 272:530–533

Morizet Y, Paris M, Gaillard F, Scaillet B (2014) Carbon dioxide in silica-undersaturated melt. Part I: The effect of mixed alkalis (K and Na) on CO_2 solubility and speciation. Geochim Cosmochim Acta 141:45–61

Morizet Y, Paris M, Sifré D, Di Carlo I, Gaillard F (2017) The effect of Mg concentration in silicate glasses on CO_2 solubility and solution mechanism: Implication for natural magmatic systems. Geochim Cosmochim Acta 198:115–130

Morizet Y, Trcera N, Larre C, Rivoal M, Le Menn E, Vantelon D, Gaillard F (2019) X-ray absorption spectroscopic investigation of the Ca and Mg environments in CO_2-bearing silicate glasses. Chem Geol 510:91–102

Moussallam Y, Edmonds M, Scaillet B, Peters N, Gennaro E, Sides I, Oppenheimer C (2016) The impact of degassing on the oxidation state of basaltic magmas: A case study of Kīlauea volcano. Earth Planet Sci Lett 450:317–325

Moussallam Y, Longpré M-A, McCammon C, Gomez-Ulla A, Rose-Koga EF, Scaillet B, Peters N, Gennaro E, Paris R, Oppenheimer C (2019) Mantle plumes are oxidised. Earth Planet Sci Lett 527:115798

Muth M, Duncan MS, Dasgupta R (2020) The effect of variable Na/K on the CO_2 content of slab-derived rhyolitic melts. I Manning CE, Lin J-F and Mao WL Eds.), Carbon in Earth's Interior, Geophysical Monograph 249. John Wiley & Sons, Inc., p 195–208

Mysen BO, Boettcher AL (1975) Melting of a hydrous mantle: I. Phase relations of natural peridotite at high pressures and temperatures with controlled activities of water, carbon dioxide, and hydrogen. J Petrol 16:520–548

Namur O, Charlier B, Holtz F, Cartier C, McCammon C (2016a) Sulfur solubility in reduced mafic silicate melts: Implications for the speciation and distribution of sulfur on Mercury. Earth Planet Sci Lett 448:102–114

Namur O, Collinet M, Charlier B, Grove TL, Holtz F, McCammon C (2016b) Melting processes and mantle sources of lavas on Mercury. Earth Planet Sci Lett 439:117–128

Nicholis MG, Rutherford MJ (2009) Graphite oxidation in the Apollo 17 orange glass magma: Implications for the generation of a lunar volcanic gas phase. Geochim Cosmochim Acta 73:5905–5917

Niida K, Green HD (1999) Stability and chemical composition of pargasitic amphibole in MORB pyrolite under upper mantle conditions. Contrib Mineral Petrol 135:18–40

O'Leary JA, Gaetani GA, Hauri EH (2010) The effect of tetrahedral Al^{3+} on the partitioning of water between clinopyroxene and silicate melt. Earth Planet Sci Lett 297:111–120

O'Neill HSC, Mavrogenes JA (2002) The sulfide capacity and the sulfur content at sulfide saturation of silicate melts at 1400°C and 1 bar. J Petrol 43:1049–1087

Papale P, Moretti R, Barbato D (2006) The compositional dependence of the saturation surface of $H_2O + CO_2$ fluids in silicate melts. Chem Geol 229:78–95

Payne JL, Turchyn AV, Paytan A, DePaolo DJ, Lehrmann DJ, Yu M, Wei J (2010) Calcium isotope constraints on the end-Permian mass extinction. PNAS 107:8543–8548

Peplowski PN, Lawrence DJ, Feldman WC, Goldsten JO, Bazell D, Evans LG, Head JW, Nittler LR, Solomon SC, Weider SZ (2015) Geochemical terranes of Mercury's northern hemisphere as revealed by MESSENGER neutron measurements. Icarus 253:346–363

Plank T, Kelley KA, Zimmer MM, Hauri EH, Wallace PJ (2013) Why do mafic arc magmas contain ~4 wt% water on average? Earth Planet Sci Lett 364:168–179

Price SE, Russell JK, Kopylova MG (2000) Primitive magma from the Jericho pipe, NWT, Canada: Constraints on primary kimberlite melt chemistry. J Petrol 41:789–808

Righter K, Yang H, Costin G, Downs RT (2008) Oxygen fugacity in the Martian mantle controlled by carbon: New constraints from the nakhlite MIL 03346. Meteor. Planet Sci 43:1709–1723

Robock A (2000) Volcanic eruptions and climate. Rev Geophys 38:191–219

Rosenthal A, Hauri EH, Hirschmann MM (2015) Experimental determination of C, F, and H partitioning between mantle minerals and carbonated basalt, CO_2/Ba and CO_2/Nb systematics of partial melting, and the CO_2 contents of basaltic source regions. Earth Planet Sci Lett 412:77–87

Saal AE, Hauri E, Langmuir CH, Perfit MR (2002) Vapour undersaturation in primitive mid-ocean-ridge basalt and the volatile content of Earth's upper mantle. Nature 419:451–455

Saha S, Dasgupta R (2019) Phase relations of a depleted peridotite fluxed by a CO_2–H_2O fluid– implications for the stability of partial melts versus volatile-bearing mineral phases in the cratonic mantle. J Geophys Res: Solid Earth 124:10089–10106

Saha S, Dasgupta R, Tsuno K (2018) High pressure phase relations of a depleted peridotite fluxed by CO_2–H_2O bearing siliceous melts and the origin of mid-lithospheric discontinuity. Geochem Geophys Geosyst 19:595–620

Saha S, Peng, Y, Dasgupta R, Mookherjee M, Fischer K (2021) Assessing the presence of volatile-bearing mineral phases in the cratonic mantle as a possible cause of mid-lithospheric discontinuities. Earth Planet Sci Lett 553:116602

Sandu C, Lenardic A, McGovern P (2011) The effects of deep water cycling on planetary thermal evolution. J Geophys Res: Solid Earth 116: B12404

Sato M (1979) The driving mechanism of lunar pyroclstic eruptions inferred from the oxygen fugacity behavior of Apollo 17 orange glass. Proc Lunar Planet Sci Conf 10:311–325

Sato M, Hickling NL, McLane JE (1973) Oxygen fugacity values of Apollo 12, 14, and 15 lunar samples and reduced state of lunar magmas. Proc Lunar Planet Sci Conf 4:1061–1079

Schaller MF, Wright JD, Kent DV (2011) Atmospheric P_{CO_2} perturbations associated with the Central Atlantic Magmatic Province. Science 331:1404–1409

Schmidt MW, Poli S (1998) Experimentally based water budgets for dehydrating slabs and consequences for arc magma generation. Earth Planet Sci Lett 163:361–379

Shannon MC, Agee CB (1996) High pressure constraints on percolative core formation. Geophys Res Lett 23:2717–2720

Shcheka SS, Wiedenbeck M, Frost DJ, Keppler H (2006) Carbon solubility in mantle minerals. Earth Planet Sci Lett 245:730–742

Sifré D, Gardes E, Massuyeau M, Hashim L, Hier-Majumder S, Gaillard F (2014) Electrical conductivity during incipient melting in the oceanic low-velocity zone. Nature 509:81–85

Smythe DJ, Wood BJ, Kiseeva ES (2017) The S content of silicate melts at sulfide saturation: New experiments and a model incorporating the effects of sulfide composition. Am Mineral 102:795–803

Sobolev AV, Chaussidon M (1996) H_2O concentrations in primary melts from supra-subduction zones and mid-ocean ridges: Implications for H_2O storage and recycling in the mantle. Earth Planet Sci Lett 137:45–55

Spera FJ, Bergman SC (1980) Carbon dioxide in igneous petrogenesis: I. Contrib Mineral Petrol 74:55–66

Stagno V, Ojwang DO, McCammon CA, Frost DJ (2013) The oxidation state of the mantle and the extraction of carbon from Earth's interior. Nature 493:84–88

Stamm N, Schmidt MW (2017) Asthenospheric kimberlites: Volatile contents and bulk compositions at 7 GPa. Earth Planet Sci Lett 474:309–321

Stanley BD, Hirschmann MM, Withers AC (2011) CO_2 solubility in Martian basalts and Martian atmospheric evolution. Geochim Cosmochim Acta 75:5987–6003

Stanley BD, Schaub DR, Hirschmann MM (2012) CO_2 solubility in primitive martian basalts similar to Yamato 980459, the effect of composition on CO_2 solubility of basalts, and the evolution of the martian atmosphere. Am Mineral 97:1841–1848

Stolper E (1982) Water in silicate glasses: An infrared spectroscopic study. Contrib Mineral Petrol 81:1–17

Stolper E, Fine G, Johnson T, Newman S (1987) Solubility of carbon dioxide in albitic melt. Am Mineral 72:1071–1085

Sun C, Dasgupta R (2019) Slab-mantle interaction, carbon transport, and kimberlite generation in the deep upper mantle. Earth Planet Sci Lett 506:38–52

Thibault Y, Edgar AD, Lloyd FE (1992) Experimental investigation of melts from a carbonated phlogopite lherzolite: Implications for metasomatism in the continental lithospheric mantle. Am Mineral 77:784–794

Till C, Grove T, Withers A (2012) The beginnings of hydrous mantle wedge melting. Contrib Mineral Petrol 163:669–688

Tsuno K, Dasgupta R (2012) The effect of carbonates on near-solidus melting of pelite at 3 GPa: Relative efficiency of H_2O and CO_2 subduction. Earth Planet Sci Lett 319–320:185–196

Tsuno K, Dasgupta R (2015) Fe–Ni–Cu–C–S phase relations at high pressures and temperatures – The role of sulfur in carbon storage and diamond stability at mid- to deep-upper mantle. Earth Planet Sci Lett 412:132–142

Tumiati S, Fumagalli P, Tiraboschi C, Poli S (2013) An experimental study on COH-bearing peridotite up to 3·2 GPa and implications for crust–mantle recycling. J Petrol 54:453–479

Ulmer P (2001) Partial melting in the mantle wedge—the role of H_2O in the genesis of mantle-derived 'arc-related' magmas. Phys Earth Planet Int. 127:215–232

Vetere F, Holtz F, Behrens H, Botcharnikov RE, Fanara S (2014) The effect of alkalis and polymerization on the solubility of H_2O and CO_2 in alkali-rich silicate melts. Contrib Mineral Petrol 167:1014

Wadhwa M (2001) Redox state of Mars' upper mantle and crust from Eu anomalies in shergottite pyroxenes. Science 291:1527–1530

Wadhwa M (2008) Redox conditions on small bodies, the Moon and Mars. Rev Min Geochem 68:493–510

Walker JCG, Hays PB, Kasting JF (1981) A negative feedback mechanism for the long-term stabilization of Earth's surface temperature. J Geophys Res. 86:9776–9782

Wallace ME, Green DH (1988) An experimental determination of primary carbonatite magma composition. Nature 335:343–346

Wallace M, Green DH (1991) The effect of bulk rock composition on the stability of amphibole in the upper mantle: Implications for solidus positions and mantle metasomatism. Min Petrol 44:1–19

Walter MJ (1998) Melting of garnet peridotite and the origin of komatiite and depleted lithosphere. J Petrol 39:29–60

Weider SZ, Nittler LR, Starr RD, Crapster-Pregont EJ, Peplowski PN, Denevi BW, Head JW, Byrne PK, Hauck SA, Ebel DS, Solomon SC (2015) Evidence for geochemical terranes on Mercury: Global mapping of major elements with MESSENGER's X-Ray Spectrometer. Earth Planet Sci Lett 416:109–120

Withers AC, Hirschmann MM, Tenner TJ (2011) The effect of Fe on olivine H_2O storage capacity: Consequences for H_2O in the martian mantle. Am Mineral 96:1039–1053

Woodland AB, Girnis AV, Bulatov VK, Brey GP, Höfer HE (2019) Experimental study of sulfur solubility in silicate–carbonate melts at 5–10.5 GPa. Chem Geol 505:12–22

Woodland AB, Koch M (2003) Variation in oxygen fugacity with depth in the upper mantle beneath the Kaapvaal craton, South Africa. Earth Planet Sci Lett 214:295–310

Wyllie PJ, Wolf MB (1993) Amphibolite dehydration-melting: sorting out the solidus. Geol Soc Lon, Spl Pubs 76:405–416

Yoshioka T, Nakashima D, Nakamura T, Shcheka S, Keppler H (2019) Carbon solubility in silicate melts in equilibrium with a CO–CO_2 gas phase and graphite. Geochim Cosmochim Acta 259:129–143

Zajacz Z, Tsay A (2019) An accurate model to predict sulfur concentration at anhydrite saturation in silicate melts. Geochim Cosmochim Acta 261:288–304

Zhang Z, Hirschmann MM (2016) Experimental constraints on mantle sulfide melting up to 8 GPa. Am Mineral 101:181–192

Zhang Z, von der Handt A, Hirschmann MM (2018) An experimental study of Fe-Ni exchange between sulfide melt and olivine at upper mantle conditions: implications for mantle sulfide compositions and phase equilibria. Contrib Mineral Petrol 173:19

Zhang Z, Qin T, Pommier A, Hirschmann MM (2019) Carbon storage in Fe–Ni–S liquids in the deep upper mantle and its relation to diamond and Fe-Ni alloy precipitation. Earth Planet Sci Lett 520:164–174

Zolotov MY, Sprague AL, Hauck II SA, Nittler LR, Solomon SC, Weider SZ (2013) The redox state, FeO content, and origin of sulfur-rich magmas on Mercury. J Geophys Res: Planets 118:138–146

Reviews in Mineralogy & Geochemistry
Vol. 87 pp. 607-638, 2022
Copyright © Mineralogical Society of America

13

Decrypting Magma Mixing in Igneous Systems

Daniele Morgavi[1*], Mickael Laumonier[2], Maurizio Petrelli[1], Donald B. Dingwell[3]

[1] Dept. of Physics and Geology
University of Perugia
Piazza Università
06100 Perugia
Italy
[2] Université Clermont Auvergne
CNRS, IRD, OPGC
Laboratoire Magmas et Volcans
F-63000 Clermont-Ferrand
France
[3] Dept. Earth and Environmental Sciences
Ludwig-Maximilian-University (LMU)
Theresienstrasse 41/III
80333 München
Germany

daniele.morgavi@unipg.it

1. INTRODUCTION

This chapter is intended to give an overview on the fundamental concepts behind the processes of magma mixing, broadly considering interactions between different magmas. The effort of the authors has been to select, from the ample literature, the key concepts to provide the best understanding of the physical and chemical processes occurring during magma mixing. In particular, the review is structured in: 1) an historical perspective, recounting the discovery and evolution of our understanding of magma mixing through time; 2) definitions of mixing and mingling and their major geological evidence from the field; 3) scaling rules and numerical modelling, with a deep overview on the kinematics of magma mixing; 4) a synopsis of experimental investigations of the complexity of magma mixing processes; 5) the implications of magma mixing in volcanic systems together with highlights, outstanding issues and possible future developments.

2. HISTORICAL PERSPECTIVE (1851 TO MODERN TIMES)

To the best of our knowledge, the first investigation devoted to magma mixing recorded in the literature is the work of Robert W. Bunsen (Bunsen 1851), a chemist from Heidelberg University who reported and interpreted the chemical variation of igneous rocks from the western region of Iceland. Through chemical analyses Bunsen highlighted that the linear correlation between pairs of chemical elements in binary plots in those Icelandic rocks is the consequence of "simple" binary mixing between two magmas with different chemical compositions. The use of mixing to explain magmatic evolution (especially by a chemist) was highly contested with in particular the strongest opposition coming from Wolfgang Sartorius Freiherr von Waltershausen (1853, esp. p. 413–421). Von Waltershausen, an expert on the volcanism of Iceland and Etna

1529-6466/22/0087-0013$05.00 (print)
1943-2666/22/0087-0013$05.00 (online)

http://dx.doi.org/10.2138/rmg.2022.87.13

at that time, argued against the method used by Bunsen of averaging rock analyses to calculate the starting end-members that eventually took part in the mixing process. Von Walterhausen criticized not only the arbitrary choice of the end-members but also disliked Bunsen's notion of an extensive layering of felsic/mafic rocks and magmas beneath Iceland. Such opposition inhibited the attention of geologists being paid to magma mixing processes for some time (see Wilcox 1999 for a deeper historical review on magma mixing).

Decades later, the experimental and thermodynamic work of Norman L. Bowen (e.g. (Bowen 1928) stymied discussions of magma mixing processes in its profound emphasis on internal igneous differentiation; whereby the conceptual model of fractional crystallization was firmly established as the most fundamental petrological process for generating the diversity of igneous rocks. Although Bowen did not explicitly deny the possibility of magma mixing, he reinterpreted field evidence of magma mixing as immiscibility of liquids (e.g. Bowen 1928). In 1944, Wilcox published a work on the Gardner River complex (Yellowstone, USA; Wilcox 1944) that is now considered to be a milestone for magma mixing recognition and remained one of the few papers on the topic. Mixing received more and more attention in the 1950's and started to be really accepted by the geological community from the 1960's. For example, Williams (1952) noticed glass shards with different colours and refractive indexes in tuffs collected in Costa Rica, and conclude that they were formed during the eruption of an heterogeneous magma. A year later, Wager and Bailey demonstrated that the commonly observed crystal haloes around mafic enclaves is a texture resulting from the quench of a mafic liquid magma juxtaposed against a felsic, colder one (Wager and Bailey 1953). This latter publication was the first of a long series reporting textures of magma mingling and mixing from the 1950s to the 1960s (Bailey and McCallien 1957; Elwell 1958; Elwell et al. 1960, 1962; Gibson and Walker 1963; King 1964, 1965; Blake et al. 1965; Wager et al. 1965; Blake 1966; Walker and Skelhorn 1966; Thompson 1968; Didier and Lameyre 1969; Kranck 1969).

In the 1970s, facing the plethora of unequivocal evidence in both plutonic and volcanic rocks, geoscientists began to investigate magma mixing more deeply (see for better details (Wilcox 1999). For example, Didier discussed the implications of fine-grained ("microgranular") mafic enclaves as indicators of magma mixing in "Granites and Their Enclaves" (Didier 1973) and Sparks and Sigurdsson proposed magma mixing as a mechanism for triggering felsic explosive eruptions (Sparks and Sigurdsson 1977) wherein the replenishment of a felsic, partly-crystallized reservoir by a mafic, hotter one leads to an unstable system where the base of the felsic magma becomes less dense, generating convection and mingling. Eichelberger made a step forward in the understanding of magma interaction by interpreting andesitic rocks as a mixture of melts from the upper mantle and lower crust from petrological evidence (Eichelberger 1980, 1978). This point of view has subsequently been nuanced: andesitic volcanism likely involves both the addition of new material to the crust and the fractionation of pre-existing crust (Annen et al. 2006; Blatter et al. 2013; Laumonier et al. 2017). Investigations about magma mixing kept increasing to a peak in the late 1980s (Sparks et al. 1980; Dickin and Exley 1981; Hibbard 1981; Sigurdsson and Sparks 1981; Huppert et al. 1982a, 1983, 1984, 1985; Grove and Baker 1983; Reid et al. 1983; Marshall and Sparks 1984; Sakuyama 1984; Blake and Ivey 1986; Sparks and Marshall 1986; Huppert and Sparks 1988; Koyaguchi and Blake 1989; Oldenburg et al. 1989). Magma mixing was finally recognized as a widespread process involved in many contexts with consequences for petrogenesis and eruption triggers (Blundy and Sparks 1992; Wiebe 1994; Nakamura 1995; Snyder and Tait 1996; Kress 1997; Jellinek and Kerr 1999). From the 2000s to the present day, more studies employing the development of new methods and techniques, have started to unravel systematically the complexity of magma mixing, and to further link it to eruptive events and related timescales (Bergantz 2000; Blake and Fink 2000; Coombs et al. 2000; Couch et al. 2001; Harford and Sparks 2001; Poli and Perugini 2002; Bachmann and Bergantz 2003; Perugini and Poli 2004; Petrelli et al. 2005, 2011, 2018;

Perugini et al. 2006a, 2010, 2013, 2015; Vigneresse 2007; Sparks et al. 2008; Kratzmann et al. 2009; Ruprecht and Bachmann 2010; De Campos et al. 2011; Zellmer et al. 2012; Morgavi et al. 2013a,b,c, 2015, 2016; Ruprecht and Plank 2013; Chamberlain et al. 2014; El Omari et al. 2015; Rossi et al. 2017, 2019; Laeger et al. 2017, 2019; Astbury et al. 2018; Fig. 1).

Since the very first hypotheses about the origin of mixed igneous rocks (e.g. Bunsen 1851), a plethora of evidence of magma mixing processes have accumulated in the literature (cf. Fig. 2), in discussions of many tectonic environments and throughout geological time. In the following, we explain magma mixing and mingling definition and the major geological evidence from the field.

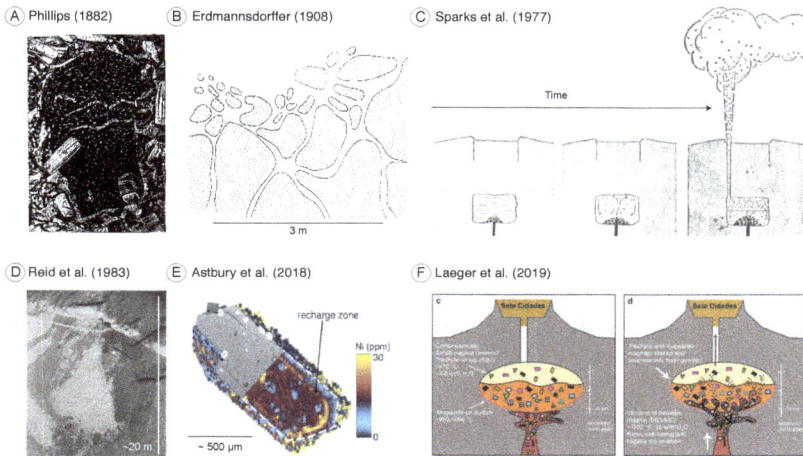

Figure 1. A) mafic enclaves of the Shap granite host (England; Phillips 1882). **B)** Drawing of contact between mafic and granitic rocks in the Brocken granitic massif (Germany; Erdmannsdorffer 1908). **C)** Sketch of volcanic plumbing system evolution during a magma mixing triggered eruption (Sparks and Sigurdsson 1977). **D)** Assimilation and magma mixing evidence at the "southern tip of Baja California" (Reid et al. 1998). **E)** Recharge zone in a cpx from Campi Flegrei recent activity highlighted by LA-ICP-MS crystal mapping (Astbury et al. 2018). **F)** Volcanic plumbing system dynamic at the Sete Cidades Volcano (São Miguel, Azores; Laeger et al. 2019).

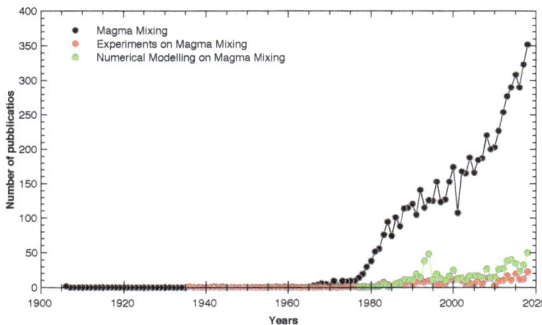

Figure 2. Plot showing the number of articles published on magma mixing processes from the year 1900 to 2018. We searched in the published scientific record (source data: Scopus, August 2018), the words: magma mixing (**black dots**), experiments on magma mixing (**red dots**) and modeling on magma mixing (**green dots**).

3. DEFINITIONS AND FIELD EVIDENCE OF MAGMA MIXING

3.1 Definitions of mixing and mingling

Anderson (1976) provided a definition of magma mixing reporting that the process occurs when two or more compositionally distinct magmas are mixed together and the melt of each is blended into a compositionally uniform magma. The petrological-volcanological community often splits magma mixing into two separate physico-chemical processes: i) mechanical mixing (also referred to as "magma mingling"), by which two or more batches of magma mingle without chemical exchanges in between them, and ii) a chemical mixing (also referred to as "magma mixing") consisting of chemical exchange between the interacting magmas in which elements diffuse from one magma due to compositional gradients continuously generated by the mechanical dispersion of the two magmas (e.g. Flinders and Clemens 1996). Magma mingling is mainly controlled by the rheology of the two interacting magmas. As an example, decreasing the viscosity contrast results in progressively more efficient mingling dynamics (e.g. Sparks and Marshall 1986; Grasset and Albarède 1994; Bateman 1995; Poli et al. 1996; Perugini and Poli 2005; Laumonier et al. 2014b). Magma mixing" is driven by the mobility of chemical elements in the melt fractions of the two magmas, and it is mainly matter of the dynamics occurring within the system and time (e.g. Baker 1990; Lesher 1990; Perugini et al. 2006b; Morgavi et al. 2013b). Linear variations in inter-elemental plots for a set of rock samples have long been considered as the sole evidence for the occurrence of magma mixing (e.g. Fourcade and Allegre 1981). The adoption of the above conceptual models led to the common practice of applying the term magma mingling to indicate the process acting to physically disperse (no chemical exchange is involved) two or more magmas, whereas the term magma mixing indicates that the mingling process is also accompanied by chemical exchange. Such a conceptual approach may allow us simplifying the complexity of the magma mixing processes and make it more tractable from the petrological point of view, but this terminology is unfortunately not consistently used in the literature, causing some confusion. Even if it is not always easy to clearly discriminate between the two processes, mingling with no mixing should be and appears to be quite a rare process in nature as physical dispersion and chemical exchange typically occur in tandem (e.g. Morgavi et al. 2016, 2019; Perugini and Poli 2012).

3.2 Mixing and mingling structures

The most common evidence for magma mixing in igneous rocks is the occurrence of textural heterogeneities, extensively discussed in the last decades (e.g. Eichelberger 1975; Anderson 1976; Bacon 1986; Didier and Barbarin 1991; Wada 1995; De Rosa et al. 1996, 2002; Ventura 1998; Morgavi et al. 2016; Smith 2000; Snyder 2000; Perugini et al. 2002, 2007; Perugini and Poli 2005, 2012; Pritchard et al. 2013). In order to classify magma mixing structures, the evidence of mechanical mixing in igneous rocks can be roughly divided into four groups: a) flow structures, b) enclaves, c) fragmented dykes and septa, and d) physico-chemical disequilibria in melts and crystals (e.g. Walker and Skelhorn 1966; Hibbard 1981; Didier and Barbarin 1991; Flinders and Clemens 1996; Perugini et al. 2002; Streck 2008). These structures are detailed below.

Here flow structures refer to portions of a magma dispersed within a host that show a high spatial continuity (e.g. Wada 1995; De Rosa et al. 1996). Generally, these structures can be easily recognized in field outcrops as they show alternating light and dark coloured bands composed of magmas with different compositions (Fig. 3). Fluid structures highlight, within the rock, the flow fields associated with the dynamics that caused the magmas to interact, generating structures that wrap around each other, giving the rock an extremely dynamic appearance. An example from a Grizzly Lake outcrop in Yellowstone National Park (USA); (Pritchard et al. 2013; Morgavi et al. 2016) is shown in Figure 3B: flow bands of rhyolitic magma (white) intrude into a basaltic/hybrid magma (red to dark grey).

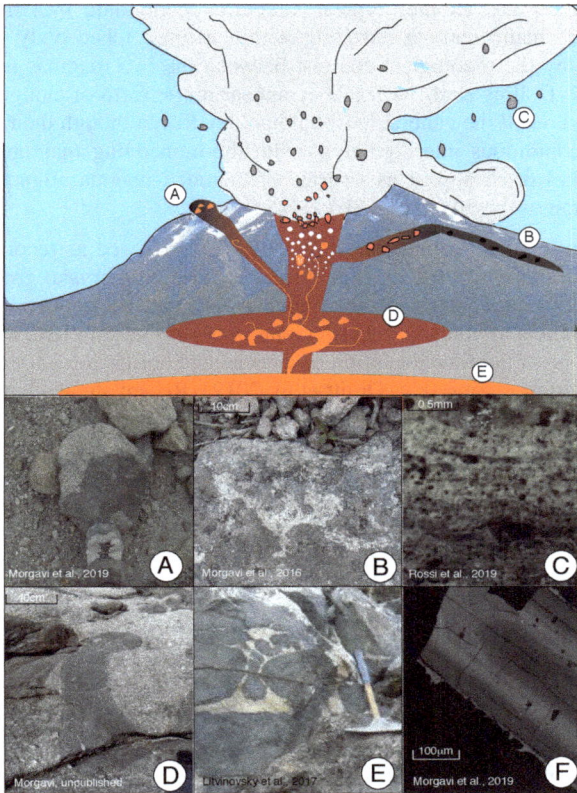

Figure 3. Sketch showing the occurrence of magma mixing processes in various igneous settings from magma chambers to volcanic conduits. Pictures below from A to F illustrate the different lithologies present in the above sketch. **A)** mafic enclave in andesitic rocks (Soufriere hills volcano, Montserrat, Caribbean territory). **B)** Filament-like structures of rhyolite and basaltic interacting magmas (Grizzly lake complex, Yellowstone Volcano USA, Morgavi et al. 2016). **C)** Bunded pumice (Aeolian Islands, Italy; Rossi et al. 2019). **D)** Monte Capanne pluton showing evidence of magma mixing (Island of Elba, Tuscan Magmatic Province, Italy). **E)** Composite dikes in granitoid suites (Transbaikalia, Russia, Litvinovsky et al. 2017). **F)** Zoned crystal from the 4.67 ± 0.09 ka Agnano Monte-Spina eruption (Phlegrean Fields), has a darker (i.e. Ba-poorer) resorbed core and resorption events (Morgavi et al. 2019).

Magmatic enclaves are probably the structural evidence that are widely believed to characterize mostly magma mixing processes. The term magmatic enclave is used to identify a discrete portion of a magma occurring within its host magma with a contrasted composition (e.g. Bacon 1986; Didier and Barbarin 1991). Generally, enclaves display quite sharp contacts with the host rock, although it is not rare to observe that some enclaves display engulfments and disruptions of their boundaries due to infiltration of the host magma. Some examples of enclaves found in the volcanic rocks from Soufrière Hills are shown in Figure 3A,B.

In plutonic systems, enclaves are one of the most apparent indications of magmatic interaction processes (Fig. 3D). They are commonly named Microgranular Mafic Enclaves (MME) based on their texture showing their rapid cooling in a colder host (e.g. Frost and Mahood 1987; Didier and Barbarin 1991; Poli and Tommasini 1991). This definition refers to the fact that in most cases, mafic enclaves show a much finer grain size and less-evolved mineralogical and geochemical characteristics than that of the host rock.

Fragmented dikes are another typical structure of plutonic systems. Such features generally consist of mafic magma intruding within a more felsic body whose dispersion has been inhibited by the rheological contrast between the two magmas (e.g. Takada 1988; Tobisch et al. 1997; Collins et al. 1999). They appear in the form of elongated bodies inside the host magma and recall the general morphology of a dike, although their spatial continuity is interrupted by continuous interdigitations with the embedding magma. For this reason, the fragmented dikes often appear as swarms of included magma aligned along the main propagation direction of the dikes (Fig. 3E).

Disequilibrium textures in minerals (Fig. 3F) can be viewed as records of the thermal and compositional variation of the magmatic system during the crystal growth or resorption due to magma mixing processes. As the zoning pattern can be well-preserved in minerals from both the plutonic and volcanic rocks, crystal populations from both environments have been used to reconstruct the time evolution of thermal and compositional exchanges between the two magmas during mixing (e.g. Druitt et al. 2012). Recent studies have highlighted the importance of detailed investigations of crystal compositional variability, not only for the reconstruction of the fluid-dynamic regime governing the evolution of the magmatic body, but also for understanding the length-scale of the compositional variability induced by the mixing process; whereby the latter may be considered as a proxy to estimate the residence time of magmas in sub-volcanic reservoirs prior to eruption (e.g. Costa and Chakraborty 2004; Martin et al. 2008; Druitt et al. 2012; Chamberlain et al. 2014; Perugini et al. 2015a).

4. SCALING RULES AND NUMERICAL MODELLING

4.1 Stretching and folding plus diffusion: kinematic description of magma mixing

From the fluid-dynamical point of view, the process of magma mixing involves the length scale reduction of magma heterogeneities by mechanical dispersion (i.e., stretching and folding) and chemical diffusion. As a consequence, magma homogenization or hybridization (i.e. a complete mixing), results from the stretching and folding of chemically different (felsic and mafic, for example) magmas, termed "end-members", producing thin filaments, where the fluxes resulting from chemical diffusion continue to act until end-members can no longer be distinguished (Perugini and Poli 2012). It should be noted that crystal and volatile exchanges between end-members are often involved during magma deformation resulting from mixing (Huber et al. 2009; Wiesmaier et al. 2015).

During the mixing processes, the degree of magma homogenization depends on many parameters, such as the geometry of the system (e.g. magma chamber or conduit shape and size, the volumes of end-members involved), magma properties, e.g., viscosity and density, the driving forces and time (Oldenburg and Spera 1991; Bergantz 2000; Perugini et al. 2006b, 2008, 2015b; Spera et al. 2016). The interplay of these parameters affects the spatial-temporal evolution of the system leading to the production of various magma mingling and mixing structures (filaments, enclaves, mineral disequilibria, geochemical variability) and eventually a hybrid magma. To elucidate the effects of stretching and folding processes in magmas, Figure 4A illustrates a simple fluid dynamic experiment: a black fluid, representing a mafic magma, interacts with a white one (a more felsic magma) by progressive stretching and folding, finally generating a complex lamellar pattern of flow structures (Perugini and Poli 2012). Note that magma mixing structures can range over many length scales producing fractal (scale-invariant) structures (Perugini and Poli 2012). The presence of fractal structures is a typical consequence of the development of a chaotic dynamics within the system (Perugini and Poli 2012; El Omari et al. 2015; Morgavi et al. 2016; Petrelli et al. 2016).

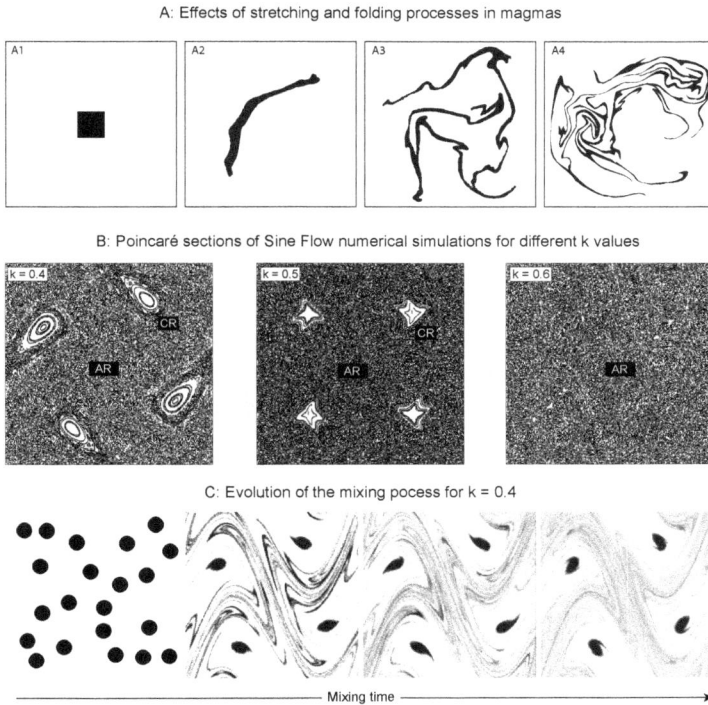

Figure 4. A) Effects of stretching and folding during chaotic magma mixing (modified from De Campos et al. 2010). B) Poincaré sections showing the evolution of the Sine Flow scheme for different *k* values (modified from Perugini et al. 2004). C) Evolution of the sine flow model for *k* = 0.4 (modified from Perugini et al. 2004).

The fractal nature of magma mixing has been extensively reported in the literature (Perugini et al. 2004, 2006a, 2008, 2015), and widely studied by numerical experiments based on iterative protocols of stretching and folding, plus diffusion (e.g. the Sine Flow; Perugini et al. 2004, 2006a; Petrelli et al. 2005). As an example, Figure 4B describes the flow regimes associated with the Sine Flow protocol, a numerical scheme that has been widely utilized as a proxy for magma mixing (e.g. Perugini et al. 2003, 2004). The sine flow is defined by two motions (Liu et al. 1994):

$$x_{t+1} = x_t + \frac{k_s}{2} \cdot \sin\left(2\pi \cdot y_t\right)\left[mod1\right] \qquad (1)$$

$$y_{t+1} = y_t + \frac{k_s}{2} \cdot \sin\left(2\pi \cdot x_{t+1}\right)\left[mod1\right] \qquad (2)$$

where (x_t, y_t) and (x_{t+1}, y_{t+1}) are the coordinate of each fluid particle at the time t and $t+1$, respectively. Also, the k_s parameter regulates the relative proportions of filament-like regions and coherent regions in the mixing system (see below). In particular, an increase in k leads to an increase of the stretching and folding efficiency. Finally, the $[mod1]$ [mod notation implies that the space domain of the system is periodic between zero and one; i.e. $0 \le (x,y) \le 1$. The flow is defined on a 2D torus meaning that whenever a particle exits the unit square, it re-enters the box through the opposite side (Liu et al. 1994; Perugini et al. 2004).

Despite the simplicity of the sine flow, it is able to capture the main relationships between the dynamics occurring during the interaction of different magmas and the textures produced by magma mixing, widely observed in nature (e.g. Perugini et al. 2003a, 2004; Petrelli et al. 2006). In detail, the Figure 4B reports the Poincaré section for the sine-flow scheme, highlighting the main characteristics of the systems (see Perugini et al. 2003a for more details). A Poincaré section is a technique used for analysis of dynamical systems. It replaces the flow of an nth-order continuous-time system with an $(n - 1)^{th}$-order discrete-time system called the Poincaré map (Parker and Chua 1989). The Poincaré map's usefulness lies in the reduction of order and the capability to highlight the essential dynamical structures of complex systems (Parker and Chua 1989). Figure 4B shows regular regions, consisting of closed trajectories (elliptic regions), coexisting with irregular regions where the elements are iterated in a more irregular way (dotted regions). The coexistence of different fluid-dynamic regimes is typical of fluid mixing systems, such as magma mixing (Petrelli et al. 2005; Perugini and Poli 2012; El Omari et al. 2015). The main consequence is that irregular and regular regions of the map (Fig. 4B: $k = 0.4$) will represent different portions of the system where magmas will be well mixed (Active Regions, AR) and poorly mixed (Coherent Regions, CR), respectively. Increasing the k value (Fig. 4B: $k = 0.5$ and 0.6), CR disappear progressively in favor of AR leading to the development of fully developed chaos (Perugini et al. 2003).

To facilitate visualization of these phenomena, Figure 4C reports the initial configuration where blobs of an intruding magma (black circles) are placed inside a hosting magma (white colored square). This initial configuration is a convenient selection aimed at the reduction of computation time, which does not significantly affect the results of the simulation. Figure 4C highlights that stretching and folding dynamics, actively working in active regions (AR), will rapidly disperse the intruding magma in the system, increasing the interface between the interacting magmas and allowing for an efficient homogenization by chemical diffusion. On the contrary, blobs of the intruding magma that are placed inside coherent regions (CR), will poorly deform, limiting the interface size between both components and inhibiting chemical homogenization at short timescales.

Examples of the application of chaotic flow fields to magma mixing have been reported by Perugini et al. (2004), and Petrelli et al. (2005). In detail, Perugini et al. (2004) investigated the morphological characteristics of mixing structures of pumice samples from the 13 ka Upper Pollara eruption (Salina Island, southern Italy). Chemical variations generated by numerical simulations have been quantitatively compared with digital images extracted from natural samples (Perugini et al. 2004). Results showed that the pyroclastic rocks of the Upper Pollara sequence were produced by mixing ~25–40 vol% of andesite and ~75–50 vol% of rhyolite. Also, Perugini et al. (2004) estimated the Reynolds number (Re; see below) occurring during the mixing process. The results reported by Perugini et al. (2004) showed that Re was between ca. 500 and 7000, with a positive correlation between the initial percentage of mafic magma. Also, the results suggested that the magma-mixing process mainly occurred in the conduit.

Petrelli et al. (2005) used the sine flow protocol to study the process of magma mixing and assess the time-scales of enclave homogenization. Results indicated that the homogenization time of enclaves in active regions (AR) is several orders of magnitude faster than coherent regions (CR). As an example, an enclave with a diameter of 100 cm can be homogenized in the CR in ~380 years, assuming an advection velocity of 10 cm/year. On the contrary, in CR, the same enclave would require 6.5×10^5 years for a complete homogenization.

The journal-bearing flow is an alternative kinematic protocol that has been successfully used to study the process of magma mixing (de Campos et al. 2011; Morgavi et al. 2013a, 2015). The basic geometry of a journal-bearing flow simulation is reported in Figure 5 and it consists of an outer cylinder, filled with the liquids of interest, and an inner cylinder, which is located off-center. The geometry of the journal-bearing flow is characterized by two parameters:

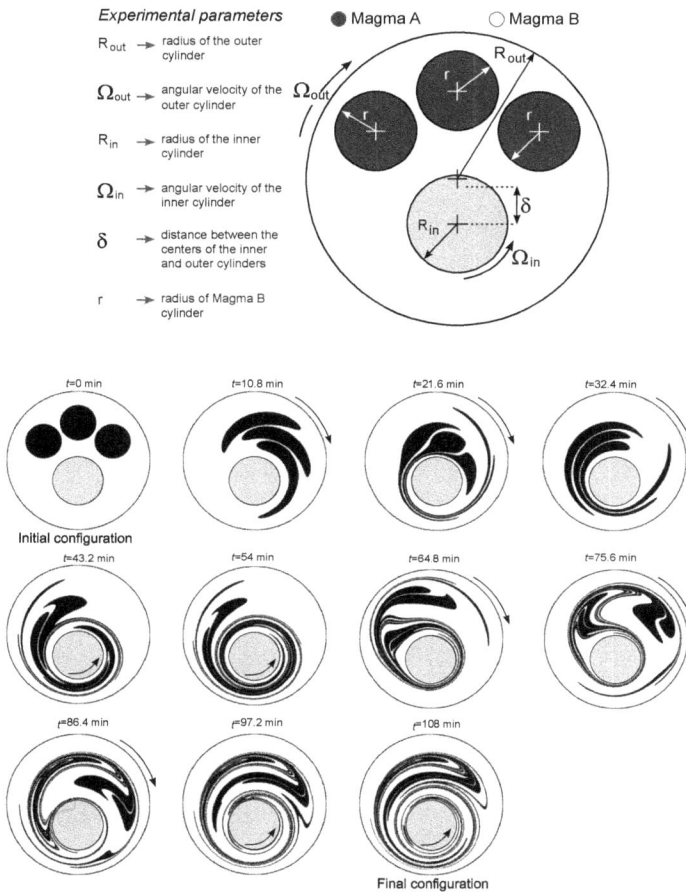

Figure 5. Initial configuration and time evolution of the Journal Bearing Flow system utilized for magma mixing experiments. More details in De Campos et al. (2011) and Perugini et al. (2012).

the ratio $r = R_{in}/R_{out}$ (i.e. the ration of the radii of the two cylinders), and the eccentricity ratio $\varepsilon = \delta/R_{out}$, where δ is the distance between the centers of the inner and outer cylinders (R_{in} and R_{out}; Fig. 5). The alternating rotation of the two cylinders in opposite directions triggers the mixing process (Ottino 1989). Figure 5 reports the outcome of the numerical simulation starting at t=0, and for the entire duration of the mixing protocol reported by Perugini et al. (2012). Both melts are intimately dispersed by the flow field producing a complex pattern of filaments after only a few stretching and folding cycles (Fig. 5). The results reported by De Campos et al. (2011) strongly support the hypothesis that chaotic flow fields represent powerful dynamics for the blending of magmas.

4.2 Complete fluid dynamic description of magma mixing

A complete fluid-dynamic numerical modelling of magmatic systems undergoing the interaction between different magmas involves the solution of flow balance in mass, momentum, and total energy, expressed as partial differential equations (Oldenburg et al. 1989; Bergantz 2000; Petrelli et al. 2011).

In particular, the continuum formulation (i.e. Eulerian approach) replaces the fundamental molecular structure of the interacting magmas with a limited set of properties, defined at each point in the fluid and tracked over the time evolution of the system. In detail, each magma property is considered to be a continuous function of position and time (Oldenburg et al. 1989). This approach provides an accurate description of the mixing process as the initial assumption of magmas as a 'continuum' fluid can be assumed being valid, i.e. when crystal– or/and gas–melt separations and, in particular, their relative movements and interactions can be neglected (Oldenburg et al. 1989).

To understand the basics of magma mixing simulations in the continuum, consider its formulation in the simplest case of melt-melt interaction above the liquidus temperature, at pressures greater than the saturation threshold (i.e. liquid–liquid interaction only; Fig. 6A), and where the thermal diffusivity can be neglected over the kinematic viscosity (i.e. infinite Prandtl number; Oldenburg et al. 1989; Petrelli et al. 2011).

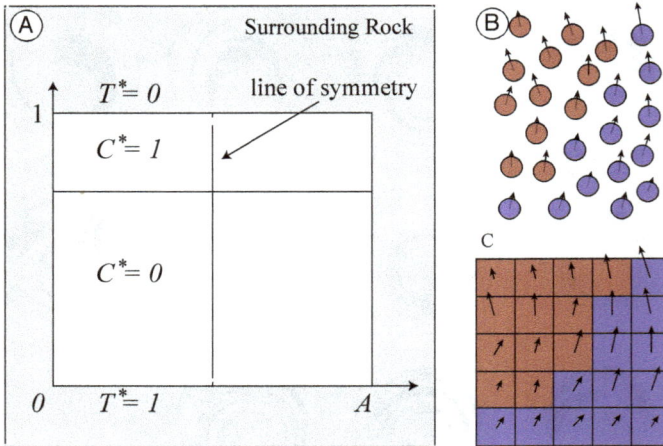

Figure 6. Schematic exemplifications of the Lagrangian (A) and Eulerian (B) approaches for the simulation of magma mixing. (C) Initial configuration utilized by Oldenburg et al. (1989) and Petrelli et al. (2011) for the simulation of the mixing process in a magma chamber.

A linearized equation of states is often employed to account for density variation as a function of temperature (T) and concentration (C):

$$\rho = \rho_{bf}\left[1 - \alpha(T - T_b) + \beta(C - C_f)\right] \tag{3}$$

where ρ [kg/m^3] is the fluid density, α [K^{-1}] and β [wt.%$^{-1}$] are the thermal and chemical expansivity (assumed to be independent of pressure, temperature and melt composition), T [K] is the temperature, C [wt.%] is the concentration (SiO$_2$ is typically utilized as the reference chemical specie), ρ_{bf} is the reference density of the magma at a concentration C_f and temperature T_b. As further simplifications, it is often assumed that there are no internal heat sources in the system domain and the temperature is fixed at the top (T_t) and bottom (T_b) boundaries of the magma chamber (with $T_b > T_t$). Also, vertical boundaries of the chamber are considered as adiabatic (i.e. no heat is transferred in or out the magma chamber during the mixing process; Fig. 6A).

The temperature (T) and the concentration (C) can be converted to dimensionless values utilizing the following relations (Oldenburg et al. 1989; Petrelli et al. 2011):

$$T^* = \frac{(T - T_f)\zeta}{dq_b} \qquad (4)$$

$$C^* = \frac{C - C_m}{C_f - C_m} \qquad (5)$$

where T_f [K], ζ [W m^{-1} K^{-1}], q_b [W/m^2], d [m], C_m [wt.%], and C_f [wt.%] are the temperature of the felsic magma, the thermal conductivity, the heat flux at the floor of the chamber, the height of the chamber, the concentration of the mafic magma and the concentration of the felsic magma, respectively. The time t [s] can be made dimensionless using the following relation:

$$t^* = \frac{k}{d^2} \cdot t \qquad (6)$$

where k [m^2/s] is the thermal diffusivity.

As a consequence, the dimensionless formulation of the problem can be described using the following equations (Oldenburg et al. 1989):

$$\nabla \cdot V = 0 \qquad (7)$$

$$\frac{\partial V}{\partial t^*} + (V \cdot \nabla)V = -\nabla P + \nabla \cdot \left(\eta^* \left((\nabla V) + (\nabla V)' \right) \right) + \mathrm{Ra}\left[T^* + \mathrm{R}_\rho \left(C^* - 1 \right) \right] \qquad (8)$$

$$\frac{\partial T^*}{\partial t^*} + \nabla \left(T^* V \right) = \nabla^2 T^* \qquad (9)$$

$$\frac{\partial C^*}{\partial t^*} + \nabla \left(C^* V \right) = \frac{1}{\mathrm{Le}} \nabla^2 C^* \qquad (10)$$

where V is the velocity field. The equations reported above are governed by three main parameters (Oldenburg et al. 1989): (1) the Rayleigh number Ra, based on heat flux imposed along the base or "floor" of the chamber; (2) the buoyancy ratio R_ρ, i.e. the ratio of chemical over thermal buoyancy forces, and (3) Lewis number Le:

$$\mathrm{Ra} = \frac{g\alpha q_b d^4}{k\zeta\eta} \qquad (11)$$

$$R_\rho = \frac{\beta\zeta(C_f - C_m)}{\alpha dq_b} \qquad (12)$$

$$\mathrm{Le} = \frac{k}{D} \qquad (13)$$

The two other main parameters governing the evolution of the system are the viscosity ratio (r_η) and the aspect ratio (i.e, $A = w/d$) of the chamber, respectively.

A useful metric that can be utilized to measure the goodness of mixing is the relaxation of concentration variance decay i.e. intensity of segregation; (Oldenburg et al. 1989; Morgavi et al. 2013; Perugini et al. 2013). For any chemical component (C_i) of the magmatic system, the relaxation of concentration variance decay, $\sigma_n^2(C_i)_t$, is defined as the variance at any time, $\sigma^2(C_i)_t$, divided by its initial value $\sigma^2(C_i)_{t=0}$ (Oldenburg et al. 1989; Morgavi et al. 2013; Perugini et al. 2015):

$$\sigma_n^2(C_i)_t = \frac{\sigma^2(C_i)_t}{\sigma^2(C_i)_{t=0}} \qquad (14)$$

In any system, the initial value of $\sigma_n^2(C_i)_{t=0}$ is equal to 1. Also, the relation $\sigma_n^2(C_i)_{t=\infty}=0$ is valid for any chemical species characterized by nonzero chemical diffusion coefficient.

The above equations have been numerically solved by Oldenburg et al. (1989) showing that a complex evolution characterizes the magma mixing process, with numerous flow reversals and local unmixing events. Under the initial and boundary conditions investigated by Oldenburg et al. (1989), the homogenization time (t_{mix}), defined as the time required to attain the $\sigma_n^2(C_i)_t$ threshold of 0.0025, is roughly proportional to $r_\eta^{1/2}Le^{1/2}Ra^{-1}R_\rho^2$ (Fig. 7). When translated to r_η, La, Ra, and R_ρ with suitable values for natural systems, this relationship provides a range of mixing dimensional times. The homogenization process ranges from about 1/10 to 10 times d^2/k (Oldenburg et al. 1989). Mixing times are at a minimum for square geometry (i.e. a large-volume, thermally well-connected basaltic body). On the contrary, the process of magma mixing is inhibited by the formation of multiple cells of different composition in sill like bodies (Oldenburg et al. 1989).

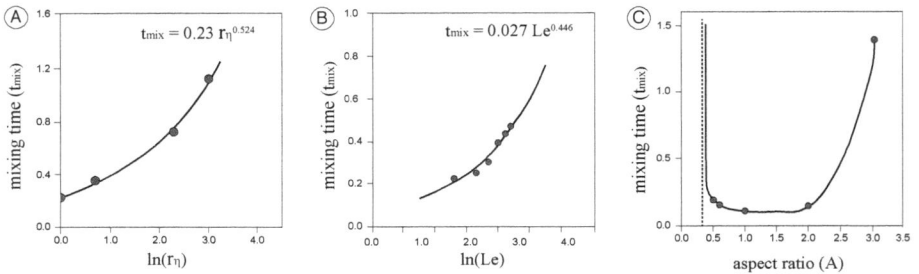

Figure 7. Mixing time (t_{mix}) as function of **(A)** the viscosity ratio between the interacting magmas (r_η), **(B)** the Lewis numner (Le) and **(C)** the aspect ratio of the chamber (A). (Modified from Oldenburg et al. 1989).

Using a similar configuration to the one investigated by Oldenburg et al. (1989), Petrelli et al. (2011) estimated the hybridization time-scales of magmas during chaotic mixing. To estimate the 'degree of chaoticity' within the system, Petrelli et al. (2011) utilized the 'finite time Lyapunov exponent' (Ottino 1989; Farnetani and Samuel 2003). In detail, Petrelli et al. (2011) passively tracked the trajectories of pairs of fluid particles placed at an infinitesimal distance δ_0 and subsequently applied the following equation (Ottino 1989; Farnetani and Samuel 2003):

$$\lambda = \lim_{\substack{t^*\to\infty \\ \delta_0\to 0}}\left[\frac{1}{t^*}\ln\left(\frac{\delta}{\delta_0}\right)\right]\approx\frac{1}{t^*}\ln\left(\frac{\delta}{\delta_0}\right) \tag{15}$$

Positive Lyapunov exponents characterize magmatic system subjected to "sensitivity upon initial conditions" (Ottino 1989) that is typical of chaotic systems (Perugini et al. 2003, 2004, 2006; Perugini and Poli 2012). As an example, the results reported by Petrelli et al. (2011) suggested that magmatic structures (e.g. enclaves) characterized by lateral dimension of the order of ~1000 m will be stretched to a filament-like structure with a thickness of 15 m after 320 years and down to less than 4.0 mm after 1900 years (Fig. 8). To note, a 4.0 mm-thick filament of magma can be homogenized by diffusion in ~10 years if we used a chemical diffusion coefficient of 10^{-14} m²/s. Such homogenization time-scales are orders of magnitude lower than typical life-times estimated for magma reservoirs, in particular for volcanic environments, indicating that fast hybridization of magmas can be rapidly achieved when chaotic dynamics govern the evolution of the magmatic system (Petrelli et al. 2011).

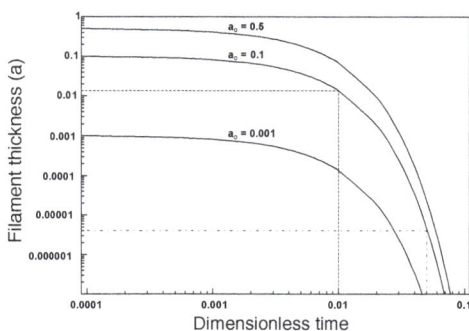

Figure 8. Evolution of the produced magma filament thickness (a) when starting from spherical volumes of a mafic enclave with initial diameter (a_0) located in a chaotic region. The filament thickness (a) decays quickly with time as the magma is stretched, generating filaments up to 4 orders of magnitude minimum thinner than the initial diameter (a_0). As an example, the thickness of a filament generated from a blob with an initial diameter $a_0 = 0.1$ is 0.015 after a dimensionless time of 10^{-2} (**dashed lines**), and it decays to 4×10^{-6} when a_0 is equal to 0.05 (**dot-dashed lines**). In dimensional terms this means that a mafic mass with an initial diameter of 100 m within a magma chamber with $d = 1000$ m is stretched to a filament-like structure with a thickness of 15 m after a time $t = 320$ years and to 4.0 mm after $t = 1900$ years (modified from Petrelli et al. 2011).

The approach proposed by Oldenburg et al. (1989) and Petrelli et al. (2011) is appealing in its simplicity, its application is limited to natural systems where, for example, the gas pressure is above the saturation threshold (i.e. volatile rich, or shallow systems) and the magma rheology is non-Newtonian (e.g. crystallinity above 30 %; Caricchi et al. 2007).

For some specific cases, improved approaches have been proposed but still remain in the continuum description. As an example, Bergantz (2000) extended the above formulation to multiphase systems utilizing an Eulerian–Eulerian two-phase model (Fig. 9A–C). It is a two-phase simulation based on the assumption that all the space can be occupied by both phases (Bergantz 2000). It further assumes that the local phase volume fraction is a continuous variable, representing the existence probability of the phase in any spatial volume. Within this volume, each phase is characterized by a single and distinct velocity. The study reported by Bergantz (2000), focused on the process of magma mixing on multiphase systems characterized by melt and crystal phases. It investigated three distinct regimes characterized by different fluid-dynamical regimes. These regimes are characterized by different Reynolds numbers (Re) defined as (Bergantz 2000):

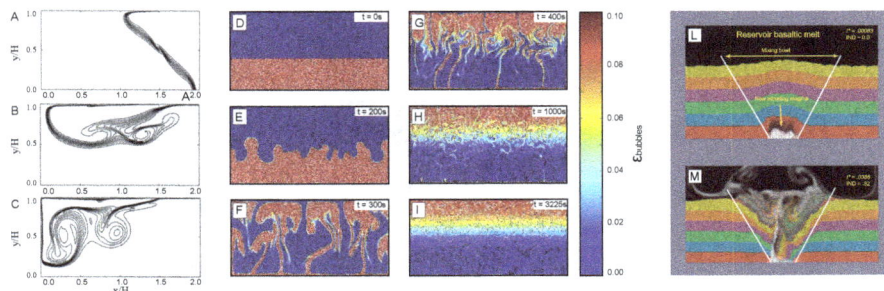

Figure 9. The use of mixed Euleran–Lagrangian approaches. **A–F**) Different stages in the overturn of a lower, bubble-rich layer (modified from Ruprecht et al. 2008). **G–I**) Collapse of the intrusive contact for a Reynolds number of 1000 and an aspect ratio of 0.5. **L**) Intrusion of a basaltic magma from below, colored in white, in an olivine mush with about 40% porosity (modified from Bergantz 2000). The resident basaltic melt is colored in black. The intrusion induces a viscoplastic response generating a mixing bowl (**L**) and a subsequent overturn (**M**), generating mixing among crystals and melts (modified from Schleicher et al. 2016).

$$\mathrm{Re} \approx \frac{H}{\eta}\sqrt{gHA_t} = Gr^{\frac{1}{2}} \tag{16}$$

where H, η, g, A_t, and Gr are the height of unstable interface between the intruding and the resident magma, the kinematic viscosity of the system, the acceleration of gravity, and the Atwood ratio, respectively. The Atwood ratio is defined as (Bergantz 2000):

$$A_t = \frac{\rho_r - \rho_i}{\rho_r + \rho_i} \tag{17}$$

with ρ_r and ρ_i being the density of the resident and the intruding magmas, respectively.

The results reported by Bergantz (2000) highlight that for Reynolds numbers larger than 10^2, the intrusive contact will collapse generating chaotic dynamics. For Reynolds numbers from 10 to 10^2, unmixed material will survive. Finally, for Reynolds numbers approaching 1, internal slumping and folding can occur.

Further examples where the continuum approach is extended to multiphase numerical experiments are provided by Montagna et al. (2015) and Garg et al. (2019) (Fig. 10). In detail, Garg et al. (2019) extended the Eluerian formulation to multiphase systems by treating the magma as a mixture of two components, each one characterized by a melt with given composition and volatile content (i.e. H_2O and CO_2). The results reported by Garg et al. (2019) showed that the fluid-dynamical evolution of magma mixing in shallow systems can result in time persistent (i.e. for long time) configurations characterized by the coexistence of magmas from nearly pure to variably mixed end-member compositions (Fig. 10). Further, short mixing time scales may relate

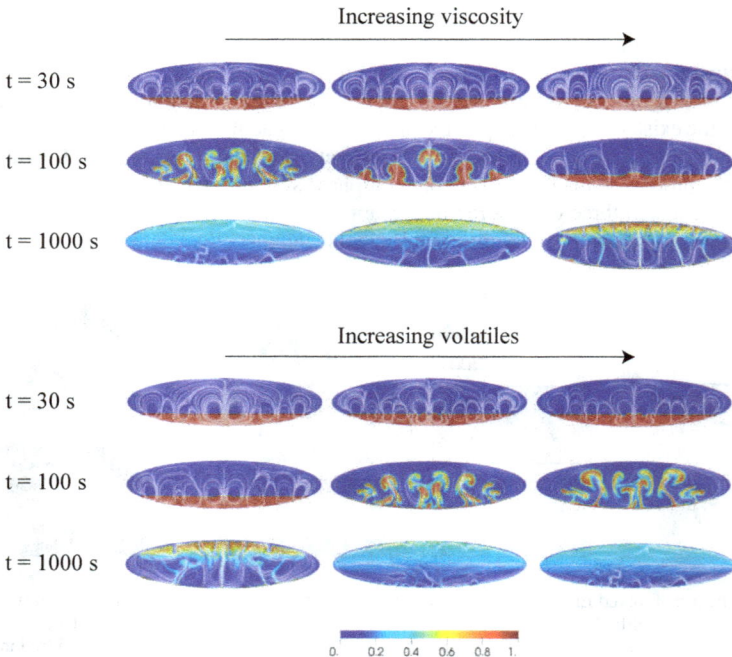

Figure 10. Evolution of the mixing process with time as function of the viscosity (**top panel**) and volatiles (**bottom panel**). Modified from Garg et al. (2019).

to syn-eruptive processes (Cioni et al. 1995; Garg et al. 2019). Finally, Garg et al. (2019) provided evidence that chemical heterogeneities that are typically found in the eruptive products are not necessarily representative of the new magma refilling just before the eruption. The numerical experiments by Longo et al. (2012) investigated magma convection and mixing accompanied by periodic refilling. They showed that the presence of CO_2 in the refilling magma strongly influences the evolution of the system. The increase of CO_2 produces a shift from simple plume rising and spreading near the chamber top, to complex behaviors and flow evolution of the system, including large scale vorticity and extensive mixing. In addition, lower chamber depths and lower magma viscosities significantly enhance the efficiency of magma mixing (Longo et al. 2012).

Using a similar approach to the one proposed by Longo et al. (2012), Montagna et al. (2015) investigated the arrival of magma from depth into shallow reservoirs, potentially triggering an eruption. In detail, Montagna et al. (2015) highlighted that over few hours, magmas appear to mingle throughout the reservoir, leading to hard-to-identify convective patterns. Given the short timescales of magma mixing processes and eruption triggering, the authors suggested that it would be advantageous to routinely detect geophysical signals of magma refilling for eruption forecasting and mitigation actions (Montagna et al. 2015).

To improve the modelling of multiphase systems (i.e. melt plus crystals and/or bubbles) hybrid approaches based on both the continuum (i.e. Eulerian) and the modelling of discrete particles (i.e. Lagrangian) have been successfully proposed (Ruprecht et al. 2008; Bergantz et al. 2015; Schleicher et al. 2016). Typically, the melt phase has been still described in the continuum, whereas crystals and/or bubble have been modeled as discrete particles (Fig. 9).

As an example Ruprecht et al. (2008) approached the multiphase character of natural magmas by computing a bubble-free magma in the continuum (i.e. melt plus "background crystals"), whereas bubbles are modeled as rigid spheres (Ruprecht et al. 2008) (Fig. 9D–I). The simulation domain was fixed to a 200 m wide and 100 m high rectangle. The main aim of Ruprecht et al. (2008) was to investigate the transport and zoning of crystals associated with a gas-driven mixing event. All simulations highlighted chaotic flow dynamics with fast overturn timescales of minutes to hours (Ruprecht et al. 2008). Also, Ruprecht et al. (2008) showed that the gathering and dispersal of crystals is strongest and most uniform for high Re numbers. On the contrary, at low Re numbers, crystal populations are characterized by less gathering of crystals that originated from distal portions of the magma body. During an overturn, the crystals pass through environments of changing chemical potential experiencing complex dynamics of growth and dissolution. Results highlight that crystals most likely record their initial as well as their final chemical environment only, given crystallization times are slower than overturn timescales. On the contrary, dissolution and advection timescales are of same order of magnitude, suggesting that dissolution could significantly occur during the overturn.

Bergantz et al. (2015) approached the multiphase character of magmas by employing the discrete element method-computational fluid dynamics (DEM-CFD; Fig. 9L–M). It consists of a Lagrangian–Eulerian approach for the solid and fluid phases, respectively. In particular, the solid phases are explicitly resolved, whereas the fluid phase is treated as a continuum. The advantage on this hybrid Lagrangian–Eulerian method over the pure continuum approach consists in the explicit modelling of collisions, sustained frictional contact, buoyancy, fluid drag and interphase momentum transport between phases (Bergantz et al. 2015; Fig. 9L–M). The discrete-element numerical model has been utilized by Bergantz et al. (2015) to simulate the refilling of magma mushes by new magma batches. Results highlighted that the injection of new magma into a mushy reservoir causes a viscoplastic response of the resident magma (Bergantz et al. 2015). On the basis of the injection rates, Bergantz et al. (2015) pointed out three distinct dynamic regimes. If the magma injection rate is slow, the intruded magma penetrates and spreads through the crystal mush (Bergantz et al. 2015). The propagation of granular fluid stresses associated with a new intrusion generates localized conjugate failure modes or soft faults, and these delimit a fluidized region called the mixing bowl (Bergantz et al. 2015; Schleicher et al. 2016). With increasing velocity, the injected magma generates a stable cavity of fluidized magma, isolated from the rest of the system (Bergantz et al. 2015). Further increasing the intruding rates, the fluidized magma incorporates crystals from the walls of the system, bringing together crystals from different parts of the system and leaving little melt unmixed (Bergantz et al. 2015).

The DEM-CFD is described in detail by Schleicher et al. (2016). To trigger magma mixing in a crystal-rich mush, the intruding melt must fluidize settled crystals by locally decreasing their packing (Schleicher et al. 2016; Figure 11). This fluidization process requires the intrusion of injected melt within the crystal mush at a sufficient rate to overcome the weight of the crystals in the bed. This rate is known as the minimum fluidization velocity (U_{mf}) and it can be calculated by solving a modified formulation of the Ergun equation accounting for the new melt intruding only a portion of the base of the lateral extent of the mush (Bergantz et al. 2015; Schleicher et al. 2016):

$$\frac{1}{2}\theta\beta I_w^2 \ln\left(\frac{2H_0}{I_w}\right)U_{mf}^2 + \theta\alpha I_w\left(H_0 - \frac{I_w}{2}\right)U_{mf} = \left(I_w + H_0\tan\theta\right)H_0\left(\rho_p - \rho_f\right)g\left(1 - \varepsilon_f\right) \quad (18)$$

$$\alpha = 150\frac{\left(1-\varepsilon_f\right)^2}{\varepsilon_f^3}\frac{\mu_f}{d_p^2} \quad (19)$$

$$\beta = 1.75\frac{\left(1-\varepsilon_f\right)}{\varepsilon_f^3}\frac{\rho_f}{d_p} \quad (20)$$

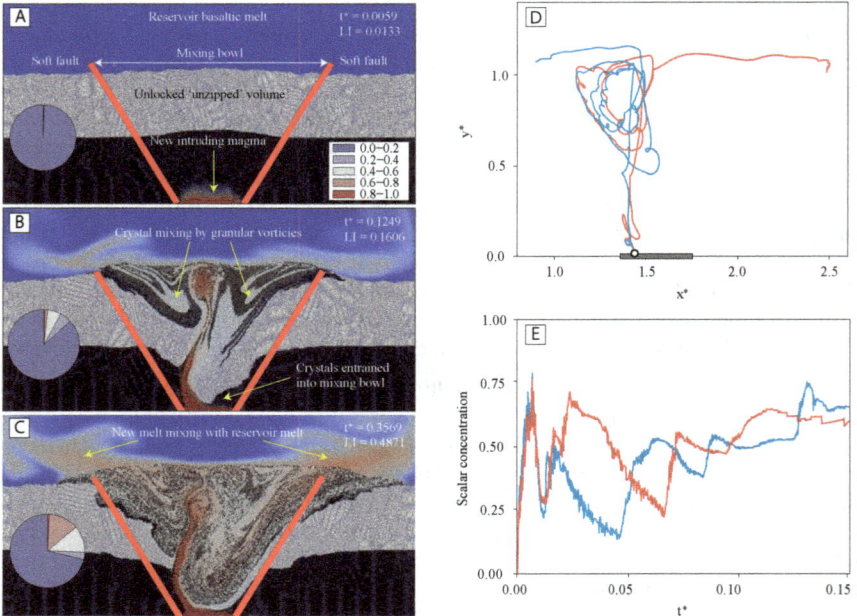

Figure 11. Open-system dynamics and magma mixing in crystal mushes. **A)** Generation of the mixing bowl by viscoplastic failure; **B)** the magma input unlocks the mixing bowl; **C)** Circulation and mixing. **D)** Trajectories of two crystals highlighting the sensitive dependence on initial conditions (i.e. developing of chaotic dynamics). **E)** Concentration of the melt that interacts with the two crystals during their transport (modified from Bergantz et al. 2015).

Theta (θ) is the angle from the vertical at the edges of the fluidized region (mixing bowl), I_w is the width of the intrusion inlet, H_0 is the height of particle bed affected by pressure drop, ε_f is the fluid fraction (porosity), μ_f is the dynamic viscosity of the fluid, ρ_f is the fluid density, d_p is the particle diameter, ρ_p is the particle density, and g is the magnitude of gravitational acceleration (Schleicher et al. 2016). The main results reported by Schleicher et al. (2016) highlight that magma mixing of a basaltic crystal mush can occur within days with geologically inferred magma intrusion rates (Fig. 11).

To the best of our knowledge, a purely Lagrangian description of the magma mixing process, i.e. where the smallest identifiable element or fundamental entity is a single melt molecule, bubble or crystal particle allowing a full multiphase description of the system, has not been attempted to date. The advantage of this captivating and challenging approach is that it may yield a very precise description of each state of magma mixing at any temperature, pressure, and composition of the interacting magmas. The major disadvantage, however, is that we must track an enormously large number of molecules, hardly feasible in most circumstances.

5. EXPERIMENTAL STUDIES FOR DECIPHERING THE COMPLEXITY OF MAGMA MIXING

Experimental studies have been set up to decipher the nature of magma mixing, leading to the reproduction of interactions between different magmas through textural, chemical and rheological features. A portion of the experimental investigations has employed analogue materials (e.g. gels or aqueous solutions) highlighting the effects of viscosity and density on mixing (e.g. McBirney 1980; Huppert et al. 1984; Campbell and Turner 1985, 1986; Blake and Campbell 1986; Blake and Ivey 1986; Turner and Campbell 1986; Sparks and Marshall 1986; Koyaguchi and Blake 1989; Thomas et al. 1993; Snyder and Tait 1996; Jellinek and Kerr 1999; Jellinek et al. 1999; Sato and Sato 2009; Woods and Cowan 2009; Spina et al. 2019). Other experiments have been conducted at magmatic conditions enabling the investigation of real magmas at physical and chemical conditions close to natural systems (e.g. Kouchi and Sunagawa 1983, 1985; Johnston and Wyllie 1988; Carroll and Wyllie 1989; Wyllie et al. 1989; Van Der Laan and Wyllie 1993; De Campos et al. 2004, 2008; Morgavi et al. 2013a; Laumonier et al. 2014a,b, 2015; Pistone et al. 2016).

5.1 Reproducing the textural evidences of magma mixing

The typical sequence of physical interaction between different magmas (i.e. mingling) corresponds to the disruption of the (mafic) magma into a (felsic) reservoir. The different steps of such sequence were observed in deformation experiments involving dry magmas (Laumonier et al. 2014a). As detailed below, the disruption of the mafic magma mostly depends on the amount of strain and on the viscosity of each end-members through their crystal fraction (Morgavi et al. 2013b,c; Laumonier et al. 2014a) (Fig. 12). First, deformation generates a wavy interface between end-members, leading to the formation of blobs, fingers, and other irregularities (Perugini and Poli 2005; Laumonier et al. 2014a) (Fig. 12A). Then, parcels of the mafic term separate from the main body, thus creating enclaves. Further deformation of the system leads to the stretching/disaggregation of enclaves, the isolation of crystals from the end-members, the formation of schlieren and the generation of banded magmas (De Campos et al. 2011 ; Morgavi et al. 2013b; Laumonier et al. 2014a; Fig. 12B).

Kouchi and Sunagawa (1983, 1985) produced pioneering experiments reproducing many of the mixing and mingling textures found in nature. Their experimental protocol included the juxtaposition and shearing by rotation of contrasted magmas (basalt and dacite) at high temperature and atmospheric pressure, producing a homogeneous liquid, intermediate in composition, where chemically zoned dacitic enclaves persist (Fig. 12C). Similar textures

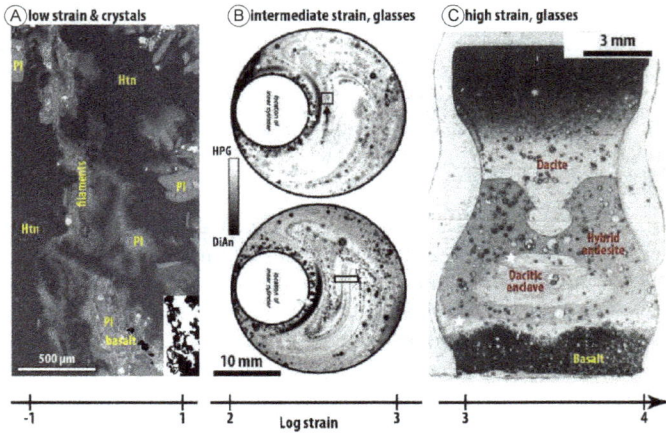

Figure 12. Mingling and mixing textures obtained experimentally at various strain rates and conditions, and between different magmas. **A)** Scanning Electron Microscope image showing isolated crystals (mostly plagioclase Pl), enclaves and filaments produced between a glassy haplotonalite (Htn) and a crystal-bearing basalt under shearing (the shear direction is vertical) (Laumonier 2013). **B)** Filaments and layering obtained under chaotic conditions between a haplogranite (HPG) and a haplobasalt (DiAn) (de Campos et al. 2011). **C)** enclaves and hybrid produced during high strain, shear experiment (the shear direction is horizontal) between a basalt and dacite (Kouchi and Sunagawa 1982).

were obtained by Zimanowski et al. (2004) by juxtaposing and deforming dry basalts and rhyolites at 1325 °C: the produced texture were millimeter-scale droplets of mafic, partially crystallized magma dispersed in the host with a composition varying from felsic to intermediate according to the run duration. To note, the size of the droplets depends on the amount of deformation, thus resembling natural observations (e.g. De Rosa et al. 1996). In the studies reported above, the amount of strain was relatively high ($\gamma > 10^3$). However, experiments conducted at magmatic conditions and moderate strain (i.e. $\gamma < 4$) produced enclave disruption into small parcels of mafic to intermediate magmas or isolated crystals as long as the viscosity contrast between end-members was low (Laumonier et al. 2014a,b). As a consequence, mafic enclaves can be "quickly" disaggregated at shear rates expected in magmatic reservoirs (Spera et al. 1988). On the contrary, enclave preservation can provide insights on the occurrence of a minimum viscosity contrast (0.5 log unit) between juxtaposed magmas during the interaction (Laumonier et al. 2014a,b). Also, the transition from enclave preservation to enclave disaggregation occurs in a narrow temperature interval (<10 °C) and it is related to the crystal fraction and crystal aspect ratios, affecting the rheological behavior of the system and ultimately producing a rigid network (Philpotts et al. 1998; Martin et al. 2006; Champallier et al. 2008; Picard et al. 2013). The same chemical systems but under hydrous conditions gives a transition over a larger interval (~50 °C), likely due to the similar viscosities between the magmas involved in the system (Laumonier et al. 2015). Note that in the case of a water-saturated magmatic system, bubbles enhance mixing by generating plume-like filament when crossing the interface between end-members (see Wiesmaier et al. 2015; Fig. 13 A–D).

The origin of schlieren was described as the result of the relatively intense elongation of enclaves under high amounts of shear strain such as in conduit conditions (Seaman et al. 1995). This has been verified in studies where magmas are poorly crystallized. As an example, parcels of intermediate magma after repeated stretching and folding of rhyolitic and basaltic melts were produced under chaotic conditions (De Campos et al. 2011; Morgavi et al. 2013c; Rossi et al. 2017; Fig. 12B). In order to preserve chemically contrasted layers, magmas should be quenched relatively quickly, since the development of those bands greatly increase the surface

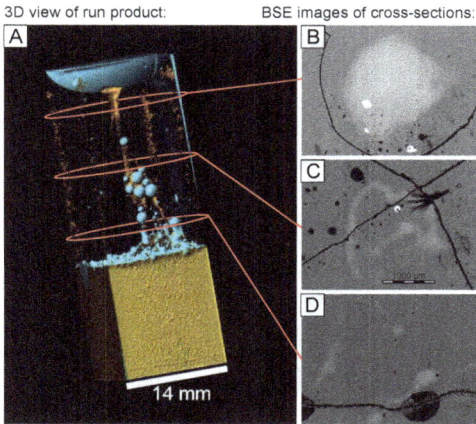

3D view of run product: BSE images of cross-sections:

Figure 13. Plume-like filaments generated by the entrainment of basalt into rhyolite during the raise if bubbles (Wiesmaier et al. 2015). **D)** Block 3D with bubbles (**blue**), basalt (**orange**) and rhyolite (**transparent**) where 2D sections are localized and shown via SEM images at different locations (**A to D**).

of the interface between end-members, and consequently enhances chemical exchanges and hybridization (Morgavi et al. 2013a). A possibility is the preferential formation of bands in lava flows or in the conduit, rather than in reservoirs where interdiffusion between chemically contrasted layers has time to homogenize at magmatic temperatures. It is worth noting that when produced, filaments are cystal-free (Laumonier et al. 2014a), suggesting that schlieren come from crystal-free magmas or from melt extracted out of a mush or enclaves.

5.2 Rheological constraints

Many studies have focused on the rheological control on magma mixing and mingling for providing the key concepts for the decrypting of such complex petrological process (e.g. Hibbard and Watters 1985; Bebien et al. 1987; Didier and Barbarin 1991; Fernandez and Gasquet 1994; Perugini and Poli 2005). In detail, several studies showed that magma mixing is the most efficient between end-members with similar viscosities (Huppert et al. 1982b, 1984, 1986; Kouchi and Sunagawa 1983, 1985; Blake and Ivey 1986; Campbell and Turner 1986; Sparks and Marshall 1986; Koyaguchi and Blake 1989; Turner and Foden 1996; Jellinek and Kerr 1999; Laumonier et al. 2014b). For instance, Huppert et al. (1982b, 1984) highlighted mixing is produced when the viscosity difference between end-members is low, although mingling can be produced at higher viscosity contrast. However, magma mixing was also successfully demonstrated even between magmas with large contrasts in viscosity (Freundt and Tait 1986; Koyaguchi and Blake 1989; Jellinek et al. 1999; de Campos et al. 2011; Perugini et al. 2012; Morgavi et al. 2013c). Such contrast reaches up to 4 orders of magnitude between a mafic and a haplogranite co-deformed under chaotic conditions as expected in conduit, lava flows, or in the magma reservoir experiencing convection (de Campos et al. 2011; Perugini et al. 2012; Morgavi et al. 2013c). Similarly to the conclusions after chaotic deformation experiments, a viscosity contrast does not necessarily prevent mixing if magmas incur deformation such as repeated intrusion into reservoirs (Koyaguchi and Blake 1989; Sato and Sato 2009).

Whatever the viscosity contrast, the absolute viscosity has been argued to be a key factor in magma mixing, with a maximum viscosity value of 10^5 Pa.s (Woods and Cowan 2009). In their pioneering work, Sparks and Marshall (1986) insisted on the importance of the rheological conditions of liquid parcels for mixing and not the whole magma because complete mixing occurs only if magmas behave as liquids. Since thermal diffusion is typically orders of magnitude faster than chemical diffusion, mingling magmas can be expected to thermally equilibrate before significant mixing occurs. Thus, capacities of mixing depend on the physical and chemical properties of magmas after their thermal re-equilibration (Sparks and Marshall 1986).

Such conditions are represented by three distinct regions in a graph plotting the fraction of mafic magma injected in a differentiated reservoir vs. the MgO concentration of the mafic magma (Fig. 14). In the region corresponding to a low fraction of mafic magma with a relatively high MgO content, magma mixing may not be possible because the mafic magma quickly cools down, crystallizes and becomes hardly mobile before chemical homogenization (solid field in Fig. 14). In the region representing the injection of a low fraction of mafic magma, but with a low MgO content (Viscous fluid), magma mixing is possible although limited; both end-members have similar viscosities. Complete homogenization can occur when both magmas behaves as fluids (Fluid field, Fig. 14). Sparks and Marshall (1986) calculated that the viscosity of such scenario should not exceed 10^6 Pa.s, thus slightly above the viscosity conditions defined by Woods and Cowan (2009). The numerical predictions of Sparks and Marshall (1986) were verified at magmatic conditions and low strain rates by Laumonier et al. (2014b). Although the authors did not produce a complete hybrid, mingling and mixing occurs when the viscosity of the system is below 10^7 Pa.s and when end-members (basalt, dacite and haplotonalite) deform together, i.e. when they have similar apparent viscosities, within a half order of magnitude. In all cases, the ratio of injected magma must be relatively high in order to prevent the rapid cooling of the mafic magma (Sparks and Marshall 1986; Frost and Mahood 1987; Laumonier et al. 2014b). The higher the volume of injected mafic magmas, the more mafic the resulting hybrid.

Generally speaking, the injection of a less viscous magma into a more viscous one creates a Saffman–Taylor instability at the interface resulting in wavy contacts that develops into so called viscous fingering (Didier and Barbarin 1991; Perugini and Poli 2005; Fig. 15A–H). The produced interface morphologies depend on the viscosity ratio between the interacting magmas as noted above. Furthermore, these morphologies have been successfully characterized by empirical relationships relating their fractal dimension (i.e. a parameter able to quantify the complexity of a geometrical pattern; Mandelbrot 1982) and the viscosity ratio between the interacting magmas (Perugini and Poli 2005). According to the amount of strain induced in the system and the rheological contrasts with the hosting magma, pieces of mafic magma may detach and form enclaves (Perugini and Poli 2012). If the host magma is characterized by a relatively high viscosity (e.g. highly crystallized), the injection of the mafic magma preferentially occurs in fractures, creating dykes (Hibbard and Watters 1985). Also, interface instabilities between the interacting magmas may cause dyke dislocations, leading to enclaves swarm (Bebien et al. 1987; Hibbard and Watters 1985). Fernandez and Gasquet (1994) provided a detailed investigation of the rheological evolution of coexisting felsic and mafic magmas accounting for the effect of cooling, increase of silica content and increase of crystallinity. Fernandez and Gasquet (1994) identified two main rheological thresholds at about 35% and 65% and pointed out three major steps in the rheological behaviour of magmas during cooling (Fig. 15I–L): 1) A Newtonian state at low volume fraction of crystals (for crystallinity values lower than about 35%), 2) shear thinning with yield stress, or Bingham behaviour (for crystallinity values between about 35% and 65%), and 3) "Solid-like" behaviour (for crystallinity values larger than about 65%). The inversion temperature (T) threshold was identified, at which viscosities of both magmas are the same (Fernandez and Gasquet 1994). This specific threshold has major petrologic significance because the viscosity ratio between the interacting magmas inverts during cooling when the temperature of magma falls below the inversion temperature (Fernandez and Gasquet 1994; Fig.16I–L).

Further rheological constraints come from the study of the interaction between rhyolite and latite rocks in Lipari Islands (Italy) reported by Davì et al. (2009). Davì et al. (2009) tracked in detail the evolution of latitic melt viscosities as function of water content and temperature changes. The authors concluded that mingling occurs when the mafic end-member is more viscous than the felsic one, while the opposite case produces mixing.

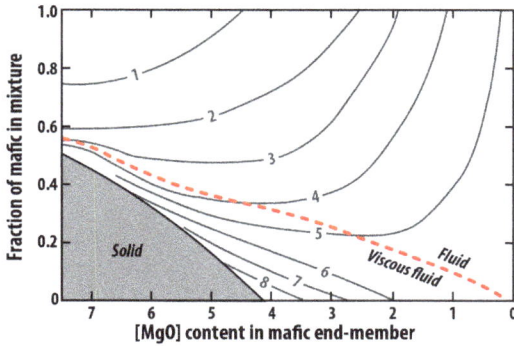

Figure 14. "composition-mixing diagram" after (Sparks and Marshall 1986) representing the physical conditions of the mafic magma after injection in a felsic reservoir.

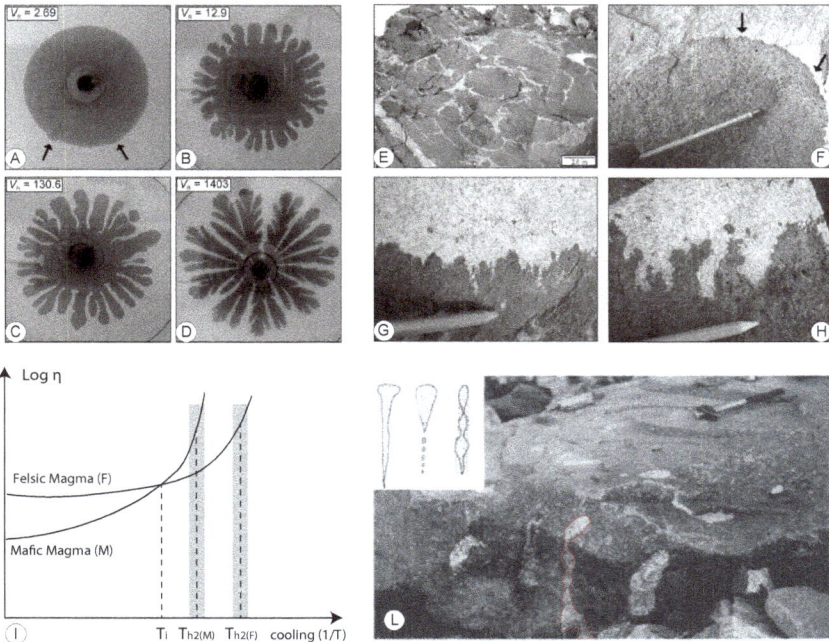

Figure 15. A–D) Optical images after viscous fingering experiments using glycerin and water solutions showing various morphologies of interfaces revealing Saffman-taylor instability (**black arrows in A**) (after Perugini and Poli 2005). **E to H)** Natural examples of magic–felsic interactions from the Terra Nova intrusive complex (Antarctica) with magic pillows (**E**) and small-scale interface with digitations (**F to H**) (after Perugini and Poli 2005). **I)** Apparent viscosities of felsic and mafic magmas during the cooling of the system. T_i is the inversion temperature whereas T_{h2} is the second rheological threshold (modified from (Fernandez and Gasquet 1994). **L)** Blobs of felsic magma rising in the dioritic host (modified from (Fernandez and Gasquet 1994).

5.3 Flow regimes in magma chambers and mixing

Turner and Campbell (1986) reviewed the application of analogue fluid dynamics magma chambers investigating the processes producing diversity in igneous rocks, such as fractional crystallization, assimilation and magma mixing. Turner and Campbell (1986) highlighted system geometry, viscosity contrasts and density variations as the main parameters governing the dynamical behaviour of magma chambers and the composition of the resulting rocks.

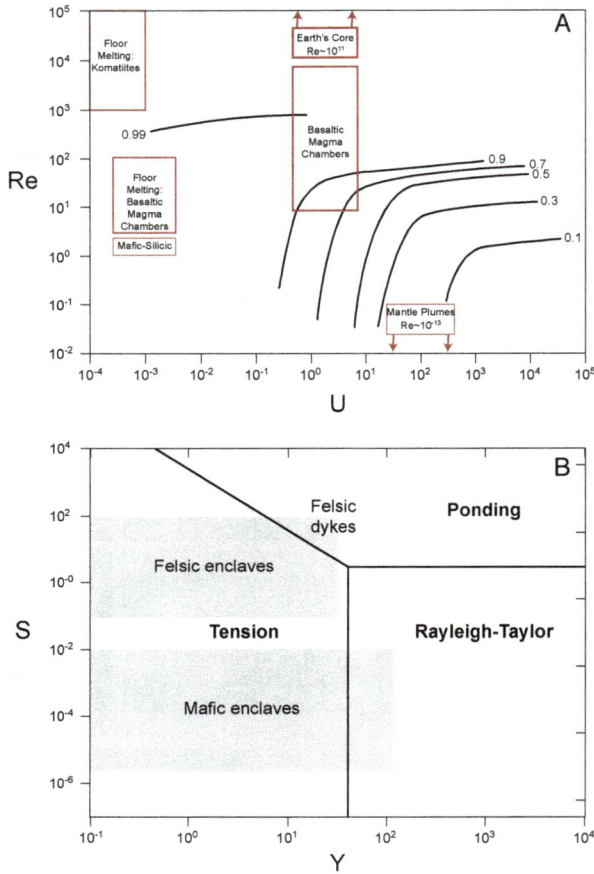

Figure 16. A) Diagram of the mixing efficiency (curves from 0.1–low efficiency to 0.99–high efficiency) over a large region of Re–U parameter space, after (Jellinek and Kerr 1999). Several geological applications are shown such as floor melting komatiites, Earth Core, basaltic magma chambers, composite magma chambers and mantle plumes. **B)** Diagram showing the behavior of the system as function of the dimensionless numbers Y and S. Please refer to the text for further details. The transitions between the different regimes (i.e. Tension, Rayleigh-Taylor, and Ponding) are shown by the solid lines at $Y_c \sim 50$ and $S_c \sim 3$ (modified from Hodge et al. 2012).

The experiments reviewed by Turner and Campbell (1986) highlighted the increase in mixing intensity of a resident magma with a replenishing one as a function of the injection rate and on the physical (i.e. densities) and rheological (i.e. absolute viscosities and viscosity contrasts) parameters governing the system. In detail, the slow injection of a relatively dense magma is expected to produce the spread of the injected magma on the floor of the chamber, resulting in a low degree of mixing with the resident magma, even if it is of comparable viscosity (Turner and Campbell 1986). The same volume of magma, injected with high upward momentum, will trigger turbulent dynamics within the system and efficient mixing between the injected and the resident magmas (Turner and Campbell 1986). The resulting convective dynamics are a function of the geometry and dimension of the system, the rheology of the interacting magmas, and by the processes (mainly thermal and compositional) affecting density changes in the magmas (Turner and Campbell 1986).

From a series of experiments conducted on analogue solutions, Jellinek et al. (1999) measured the fraction of interface between end-members and intermediate liquids, thus proposing a quantification of the mixing efficiency or intensity (Fig. 16A). The results reported by Jellinek et al. (1999) highlight that the mixing process is favoured when the viscosity ratio between the host and the intruding magma is low (i.e. $r_n \leq 10^0$) and the Reynolds number is large (Fig. 16A). The efficiency of magma mixing, i.e. up to hybridization, is also favoured by volatiles decreasing the crystal content in magmas, the solidus temperature, and increasing element diffusivity, high heat input (added by hot magma replenishment) and transport dynamics (De Campos et al. 2011; Melekhova et al. 2013; Laumonier et al. 2014b, 2015).

In addition, Hodge et al. (2012a) developed the $Y–S$ regime diagram to unravel the history of deformation and disaggregation of a more mafic magmas injected into actively convecting silicic magma chambers through dikes (Fig. 16B). In the $Y–S$ regime diagram, the parameter S is the ratio between the timescale (t_{rt}) necessary for the development of a gravitational Rayleigh–Taylor-type (R–T) instability on an injected plume, tilted with respect to gravity (Skilbeck and Whitehead 1978), and the timescale (t_s) for an injection to rise or sink through the magma chamber (Hodge et al. 2012):

$$S = \frac{t_{rt}}{t_s} \tag{21}$$

The Y parameter is the ratio between the timescale for shearing of a rheologically complex dike by flow in the magma chamber (t_f) to the timescale for lateral disaggregation or yielding of the dike (t_y) (Hodge et al. 2012):

$$Y = \frac{t_f}{t_y} \tag{22}$$

The $Y–S$ regime diagram, reported in Figure 16B, applies to enclaves found in lava flows and accounts for the chamber conditions (geometry and flow dynamics), the viscosity ratio between enclaves and host magmas, and the yield strength of the injected magma (Hodge et al. 2012).

5.4 Experiments involving crystal disequilibrium during magma mixing

Crystal zoning during magma mixing has been extensively studied, but the mechanisms of its formation are poorly explained in the literature. Zoning results in crystallization under new conditions (mostly pressure, temperature) or from a liquid with a different composition, such as potentially after replenishment of the reservoir (Hibbard 1991; Couch et al. 2001). For example, a wide variety in chemical composition was measured in crystals moved from a basalt into a felsic melt (Laumonier et al. 2014a): plagioclase crystals displayed anorthite contents ranging between those of both end-members ($An_{30}–An_{80}$) due to resorption and recrystallization in the intermediate liquid. In the most deformed regions of the same experiments, crystals of clinopyroxene are immersed in felsic liquid, after the replacement of the mafic melt by felsic one, or by extraction of crystals from the mafic term (Laumonier et al. 2014a). Such out-of-equilibrium crystals did not have enough time to re-equilibrate with their host. Similarly, the juxtaposition of dacitic and andesitic hydrous magmas creates a layer where crystals grow towards the mafic end-member as a comb layer (Pistone et al. 2013).

6. CHEMICAL INVESTIGATIONS: FROM A SIMPLE LINEAR RELATION TO A COMPLEX INTERPLAY WITH SYSTEM DYNAMICS

Considering bulk rock geochemical data on binary plots (element vs. element), mixing is typically identified when all the elements follow a linear trend (e.g. Fourcade and Allegre 1981) resulting in a hybrid chemical composition C:

$$C = XC_m + (1-X)C_f \tag{23}$$

where C_m and C_f are the composition of the mafic and the felsic end members respectively, X is the relative amount of the mafic magma in the system. Before the occurrence of the mixing process, the system is defined by two single compositional values (i.e. C_m and C_f) characterized by X values equal to 1 and 0, respectively. In the case of a complete mixing (i.e. homogeneous system), the equation reduces to a single point: the hybrid composition. Extending the observations to elemental or isotopic ratios, the mixing equation assumes a more general hyperbolic formulation (Langmuir et al. 1978):

$$Ax + Bxy + Cy + D = 0 \tag{24}$$

where x and y are general variables along the abscissa and ordinate, respectively; and A, B, and C are coefficients of the variables x and y (Langmuir et al. 1978).

Despite the simplicity of the above mentioned equations, they were successfully used to identify mixing process in many magmatic systems (Langmuir et al. 1978; Fourcade and Allegre 1981; Perugini and Poli 2012). However, several studies on the mineralogical and geochemical features of mixed rocks (e.g. Hibbard 1981; Wallace and Bergantz 2002; Costa and Chakraborty 2004; Perugini et al. 2005; Słaby et al. 2008; Masotta et al. 2018), as well as those focused on quantitative analyses of morphologies related to textural heterogeneity (e.g. Wada 1995; De Rosa et al. 2002; Perugini et al. 2002, 2004; Poli and Perugini 2002), have highlighted the importance of system dynamics in producing the substantial complexity of geochemical variations and textural patterns found in the resultant rocks (e.g. Flinders and Clemens 1996; De Campos et al. 2011; Morgavi et al. 2013a,b,c, 2016). A major feature revealed by the results of chaotic mixing experiments is the diffusive fractionation process due to the development in time of chaotic mixing dynamics (Perugini et al. 2006, 2008). This process explains the strong deviation of element concentration in intermediate products from the expected linear distribution in inter-elemental plots, according to typical conceptual models describing magma mixing processes (e.g. Fourcade and Allegre 1981; Perugini and Poli 2012 and references therein). The time evolution of compositional exchanges between magmas from the experiments can be effectively modelled, leading to the prospect that the record of magma mixing processes may serve as chronometers to estimate the time interval between mixing and eruption (Perugini et al. 2010, 2015).

7. IMPLICATION OF MAGMA MIXING AND FUTURE DEVELOPMENTS

Volcanic systems can be very dangerous and are amongst the most complex geological environments to be investigated. Many volcanic centers are notoriously variable in their eruptive behaviour with each eruption being potentially very distinct in style materials and magnitude from the previous ones.. Most of the highly explosive volcanic eruptions on Earth, have been triggered by refilling of a magma chamber by a hotter and less evolved magma, which causes mixing between the host and the intruding magma. Famous examples include, Grizzly lake, Yellowstone, (Morgavi et al. 2016), Vesuvius 79 AD (Cioni et al. 1995), Krakatau 1883 (Self 1992), and Pinatubo 1991 (Kress 1997). Recent studies have shown that relatively small-scale eruptions, such as the 2010 eruption of Eyjafjallajokull volcano, can also be triggered by a mixing event (Sigmundsson et al. 2010; Laeger et al. 2017). As documented in the previous sections, injection of a more mafic magma commonly triggers convection dynamics and widespread mixing that in turn causes inflation of the magma chamber, superheating of the felsic magma and "fluidization" of the magmatic mass (e.g. Sparks and Sigurdsson 1977). Vesiculation induced by convection increases magma pressure and can fracture the volcanic edifice (crater, plug, flanks) triggering an explosive eruption. Primitive magma injection and subsequent mixing may be accompanied by local earthquakes, gravity changes and ground deformation, including ground oscillations with ultra-long-periods, that can now be accurately detected (e.g. Williams-Jones and Rymer 2002; Longo et al. 2012). It will be clear that none of the currently available monitoring methods (e.g. gas emissions, gravity changes, ground deformation or local earthquakes) is able to recognize the time remaining before a volcanic eruption. The reason is that we have no information regarding the time elapsed between the onset of magma chamber instability and the initiation of an eruption. The knowledge of the time elapsing between the beginning of mixing (and associated geophysical signals) and eruption is, thus, of paramount importance in order to attempt eruption forecasting. In detail, the estimation of unlocking timescales (i.e. the time needed to reactivate a magmatic system by a single or repeated intrusions by hotter magma) and characterization of magmatic systems before an eruption, is one of the main challenges of current petrological and volcanological investigations (Bachmann and Bergantz 2003; Bachmann et al. 2005; Till et al. 2015; Cooper et al. 2016; Sliwinski et al. 2017; Spera and Bohrson 2018). As reported above, composite investigations (i.e. thermal, fluid-dynamical, and thermodynamic modelling) are the most effective approach (e.g. Petrelli et al. 2018). For example, Spera and Bohrson (2018) performed heat transfer modelling of felsic magma mushes, heated by the recharge of mafic magmas. Also, fluid dynamics investigations on the same topic were performed by other researchers (e.g. Huber et al. 2010; Bachmann and Huber 2016). The results pointed out a wide range of unlocking timescales, ranging from $\sim 10^{-2}$ to $\sim 10^{6}$ years, mainly depending on the dimension of the system and its dynamical evolution, posing a large problem of interpretation of seemingly inactive volcanoes (e.g. Laumonier et al. 2019). Also, the modelling of chemical diffusion in crystals was widely applied to infer the timescales of eruptions involving magma mixing (Costa and Chakraborty 2004; Turner and Costa 2007; Costa et al. 2008; Costa and Morgan 2010; Druitt et al. 2012; Petrone et al. 2016). Results points to a broad range of timescales, ranging from minutes to kilo-years (Petrelli and Zellmer 2020), and further highlighting the complex interplay among many process and dynamics affecting a volcanic plumbing system before an eruption. Finally, mixing to eruption timescales estimated using the concentration variance decay point to very short timing occurring between the beginning of the last mixing event and an eruption (Morgavi et al. 2013; Perugini et al. 2015).

In conclusion, the many studies related to magma mixing since Bunsen's work (Bunsen 1851) (Fig. 1) reveal a widespread occurrence and variegated nature of the mixing process through the wide documentation of field evidence and geochemical behavior. However, we are still far from understanding how all mixing related processes interact with each other. In particular, we have barely begun to exploit the potential of laboratory studies and computer simulations for decrypting the above processes. Future work should 1) tackle the consequences of pressure variation (decompression upon magma ascent) on mixing through new experimental devices, 2) decipher the link between mixing and geophysical signals in active volcanoes with the development of innovative techniques and experimental to field apparatuses, 3) develop numerical modelling merging mingling and mixing evidence altogether such as field observation, geochemical data on rock sample, experimental on natural and analogue material. All of the above-mentioned future works should finally be correlated to the geophysical signals from volcanic data set. The exploration of those new fields of research will provide crucial information about additional processes that can play a fundamental role in mixing and thus provide vital information to better mitigate volcanic hazards.

ACKNOWLEDGEMENTS

The authors acknowledge Grant S. Henderson for the detailed and constructive comments that greatly helped the authors in improving the manuscript. ML acknowledges the Laboratory of Excellence ClerVolc (contribution n°374), DBD acknowledges the support of the 2018 ERC Advanced Grant EAVESDROP (834225). DM acknowledges the MIUR project n. PRIN2017-2017LMNLAW "Connect4Carbon".

REFERENCES

Anderson AT (1976) Magma mixing: petrological process and volcanological tool. J Volcanol Geotherm Res 1:3–33

Andújar J, Scaillet B (2012) Relationships between pre-eruptive conditions and eruptive styles of phonolite-trachyte magmas. Lithos 152:122–131

Annen C, Blundy JD, Sparks RSJ (2006) The genesis of intermediate and silicic magmas in deep crustal hot zones. J Petrol 47:505–539

Astbury RL, Petrelli M, Ubide T, Stock MJ, Arienzo I, D'Antonio M, Perugini D (2018) Tracking plumbing system dynamics at the Campi Flegrei caldera, Italy: High-resolution trace element mapping of the Astroni crystal cargo. Lithos 318–319:464–477

Bachmann O, Bergantz GW (2003) Rejuvenation of the Fish Canyon magma body: A window into the evolution of large-volume silicic magma systems. Geology 31:789–792

Bachmann O, Huber C (2016) Silicic magma reservoirs in the Earth's crust. Am Mineral 101:2377–2404

Bachmann O, Dungan MA, Bussy F (2005) Insights into shallow magmatic processes in large silicic magma bodies: the trace element record in the Fish Canyon magma body, Colorado. Contrib Mineral Petrol 149:338–349

Bacon CR (1986) Magmatic inclusions in silicic and intermediate volcanic rocks. J Geophys Res 91:6091

Bailey EB, McCallien WJ (1957) Composite minor intrusions, and the Slieve Gullion complex, Ireland. Geol J 1:466–501

Baker DR (1990) Chemical interdiffusion of dacite and rhyolite: anhydrous measurements at 1 atm and 10 kbar, application of transition state theory, and diffusion in zoned magma chambers. Contrib Mineral Petrol 104:407–423

Bateman R (1995) The interplay between crystallization, replenishment and hybridization in large felsic magma chambers. Earth Sci Rev 39:91–106

Bebien J, Gagny C, Tanani SS (1987) Are the associations of acid and basic magmas fractal objects? C R Acad Sci Ser II 305:277–280

Bergantz GW (2000) On the dynamics of magma mixing by reintrusion: Implications for pluton assembly processes. J Struct Geol 22:1297–1309

Bergantz GW, Schleicher JM, Burgisser A (2015) Open-system dynamics and mixing in magma mushes. Nat GeoSci 8:793–796

Blake DH (1966) The net-veined complex of the Austurhorn intrusion, southeastern Iceland. J Geol 74:891–907

Blake S, Campbell IH (1986) The dynamics of magma-mixing during flow in volcanic conduits. Contrib Mineral Petrol 94:72–81

Blake S, Fink J (2000) On the deformation and freezing of enclaves during magma mixing. J Volcanol Geotherm Res 95:1–8

Blake S, Ivey GN (1986) Magma-mixing and the dynamics of withdrawal from stratified reservoirs. J Volcanol Geotherm Res 27:153–178

Blake DH, Elwell RWD, Gibson IL, Skelhorn RR, Walker GPL (1965) Some relationships resulting from the intimate association of acid and basic magmas. Q. J Geol Soc 121:31–49

Blatter DL, Sisson TW, Hankins, WB (2013) Crystallization of oxidized, moderately hydrous arc basalt at mid- to lower-crustal pressures: Implications for andesite genesis. Contrib Mineral Petrol 166:861–886

Blundy JD, Sparks RSJ (1992) Petrogenesis of mafic inclusions in granitoids of the Adamello Massif, Italy. J Petrol 33:1039–1104

Bowen NL (1928) The Evolution of the Igneous Rocks. Princeton University Press, Princeton, N.J.

Bunsen R (1851) Ueber die Processe der vulkanischen Gesteinsbildungen Islands. Ann Phys 159:197–272

Campbell IH, Turner JS (1985) Turbulent mixing between fluids with different viscosities. Nature 313:39

Campbell IH, Turner JS (1986) The influence of viscosity on fountains in magma chambers. J Petrol 27:1–30

Caricchi L, Burlini L, Ulmer P, Gerya T, Vassalli M, Papale P (2007) Non-Newtonian rheology of crystal-bearing magmas and implications for magma ascent dynamics. Earth Planet Sci Lett 264:402–419

Carroll MR, Wyllie PJ (1989) Granite melt convecting in an experimental micro-magma chamber at 1 050 °C, 15 kbar. Eur J Mineral 1:249–260

Chamberlain KJ, Morgan DJ, Wilson CJN (2014) Timescales of mixing and mobilisation in the Bishop Tuff magma body: perspectives from diffusion chronometry. Contrib Mineral Petrol 168:1034

Champallier R, Bystricky M, Arbaret L (2008) Experimental investigation of magma rheology at 300 MPa: From pure hydrous melt to 76 vol.% of crystals. Earth Planet Sci Lett 267:571–583

Cioni R, Civetta L, Marianelli P, Metrich N, Santacroce R, Sbrana A (1995) Compositional layering and syn-eruptive mixing of a periodically refilled shallow magma chamber: the AD 79 Plinian eruption of Vesuvius. J Petrol 36:739–776

Collins WJ, Richards SR, Healy BE, Ellison PI (2001) Origin of heterogeneous mafic enclaves by two-stage hybridisation in magma conduits (dykes) below and in magma chambers. Trans R Soc Edinburgh 91:27 – 45

Coombs ML, Eichelberger JC, Rutherford MJ (2000) Magma storage and mixing conditions for the 1953–1974 eruptions of Southwest Trident volcano, Katmai National Park, Alaska. Contrib Mineral Petrol 140:99–118

Cooper GF, Wilson CJN, Millet MA, Baker JA (2016) Generation and rejuvenation of a supervolcanic magmatic system: A case study from Mangakino volcanic centre, New Zealand. J Petrol 57:1135–1170

Costa F, Chakraborty S (2004) Decadal time gaps between mafic intrusion and silicic eruption obtained from chemical zoning patterns in olivine. Earth Planet Sci Lett 227:517–530

Costa F, Morgan D (2010) Timescales of Magmatic Processes, Timescales of Magmatic Processes: From Core to Atmosphere. John Wiley & Sons, Ltd, Chichester, UK

Costa F, Dohmen R, Chakraborty S (2008) Time scales of magmatic processes from modelling the zoning patterns of crystals. Rev Mineral Geochem 69:545–594

Couch S, Sparks RSJ, Carroll MR (2001) Mineral disequilibrium in lavas explained by convective self-mixing in open magma chambers. Nature 411:1037–1039

Davì M, Behrens H, Vetere F, De Rosa R (2009) The viscosity of latitic melts from Lipari (Aeolian Islands, Italy): Inference on mixing-mingling processes in magmas. Chem. Geol 259:89–97

De Campos CP, Dingwell DB, Fehr KT (2004) Decoupled convection cells from mixing experiments with alkaline melts from Phlegrean Fields. Chem. Geol 213:227–251

De Campos CP, Dingwell DB, Perugini D, Civetta L, Fehr TK (2008) Heterogeneities in magma chambers: Insights from the behavior of major and minor elements during mixing experiments with natural alkaline melts. Chem. Geol 256:130–144

De Campos CP, Perugini D, Ertel-Ingrisch W, Dingwell DB, Poli G (2011) Enhancement of magma mixing efficiency by chaotic dynamics: An experimental study. Contrib Mineral Petrol 161:863–881

De Rosa R, Mazzuoli R, Ventura G (1996) Relationships between deformation and mixing processes in lava flows: A case study from Salina (Aeolian Islands, Tyrrhenian Sea). Bull Volcanol 58:286–297

De Rosa R, Donato P, Ventura G (2002) Relationships of mingled/mixed magmas: An example from the Upper Pollara eruption (Salina Island, southern TyrrhenianSea, Italy). Lithos 65:299–311

Dickin AP, Exley RA (1981) Isotopic and geochemical evidence for magma mixing in the petrogenesis of the Coire Uaigneich Granophyre, Isle of Skye, N.W. Scotland. Contrib Mineral Petrol 76:98–108

Didier J (1973) Granites and Their Enclaves. Elsevier Scientific Publishing Co

Didier J, Barbarin B (eds) (1991) Enclaves and Granite Petrology. Developments in Petrology 13, Elsevier Science

Didier J, Lameyre J (1969) Association de granites et de diorites quartziques au Peyron, près de Burzet (Ardèche, Massif Central français). C R Acad Sci Paris 268:368–371

Druitt TH, Costa F, Deloule E, Dungan M, Scaillet B (2012) Decadal to monthly timescales of magma transfer and reservoir growth at a caldera volcano. Nature 482:77–80

Eichelberger JC (1975) Origin of andesite and dacite: Evidence of mixing at Glass Mountain in California and at other circum-Pacific volcanoes. Bull Geol Soc Am 86:1381–1391

Eichelberger JC (1978) Andesitic volcanism and crustal evolution. Nature 275:21–27

Eichelberger JC (1980) Vesiculation of mafic magma during replenishment of silicic magma reservoirs. Nature 288:446–450

El Omari K, Le Guer Y, Perugini D, Petrelli M (2015) Cooling of a magmatic system under thermal chaotic mixing. Pure Appl Geophys 172:1835–1849

Elwell RWD (1958) Granophyre and hybrid pipes in a dolerite layer of Slieve Gullion. J Geol 66:57–71

Elwell RWD, Skelhorn RR, Drysdall AR (1960) Inclined granitic pipes in the diorites of Guernsey. Geol Mag. 97:89–105

Elwell RWD, Skelhorn RR, Drysdall AR (1962) Net-veining in the diorite of northeast Guernsey, Channel islands. J Geol 70:215–226

Erdmannsdörffer OH (1908) Über Bau und Bildingsweise des Brockenmassivs. Königlich Preussischen Geologischen Landesanstalt und Bergakademie, Jahrbuch für 1905 26:379–405

Farnetani CG, Samuel H (2003) Lagrangian structures and stirring in the Earth's mantle. Earth Planet Sci Lett 206:335–348

Fernandez AN, Gasquet DR (1994) Relative rheological evolution of chemically contrasted coeval magmas: example of the Tichka plutonic complex (Morocco). Contrib Mineral Petrol 116:316–326

Flinders J, Clemens JD (1996) Non-linear dynamics, chaos, complexity and enclaves in granitoid magmas. Trans R Soc Edinburgh, Earth Sci 87:217–223

Fourcade S, Allegre CJ (1981) Trace elements behavior in granite genesis: A case study The calc-alkaline plutonic association from the Querigut complex (France). Contrib Mineral Petrol 76:177–195

Freundt A, Tait SR (1986) The entrainment of high-viscosity magma into low-viscosity magma in eruption conduits. Bull Volcanol 48:325–339

Frost TP, Mahood G a (1987) Field, chemical, and physical constraints on mafic-felsic magma interaction in the Lamarck Granodiorite, Sierra Nevada, California. Geol Soc Am Bull 99:272

Garg D, Papale P, Colucci S, Longo A (2019) Long-lived compositional heterogeneities in magma chambers, and implications for volcanic hazard Sci Rep 9:1–13

Gibson IL, Walker GPL (1963) Some composite rhyolite/basalt lavas and related composite dykes in eastern Iceland. Proc Geol Assoc 74, 301–IN3

Grasset O, Albarède F (1994) Hybridization of mingling magmas with different densities. Earth Planet Sci Lett 121:327–332

Grove TL, Baker MB (1983) Effects of melt density on magma mixing in calc-alkaline series lavas. Nature 305:416–418

Harford CL, Sparks RSJ (2001) Recent remobilisation of shallow-level intrusions on Montserrat revealed by hydrogen isotope composition of amphiboles. Earth Planet Sci Lett 185:285–297

Hibbard MJ (1981) The magma mixing origin of mantled feldspars. Contrib Mineral Petrol 76:158–170

Hibbard MJ (1991) Textural anatomy of twelve magma-mixed granitoid systems. *In:* Enclaves and Granite Petrology. Developments in Petrology 13, Didier J Barbarin B (Eds.), Elsevier Science, p 431–444

Hibbard MJ, Watters RJ (1985) Fracturing and diking in incompletely crystallized granitic plutons. Lithos 18:1–12

Hodge KF, Carazzo G, Jellinek AM (2012) Experimental constraints on the deformation and breakup of injected magma. Earth Planet Sci Lett 325–326:52–62

Huber C, Bachmann O, Manga M (2009) Homogenization processes in silicic magma chambers by stirring and mushification (latent heat buffering). Earth Planet Sci Lett 283:38–47

Huber C, Bachmann O, Dufek J (2010) The limitations of melting on the reactivation of silicic mushes. J Volcanol Geotherm Res 195:97–105

Huppert HE, Sparks RSJ (1985) Cooling and contamination of mafic and ultramafic magmas during ascent through continental crust. Earth Planet Sci Lett 74:371–386

Huppert HE, Sparks RSJ, Turner JS (1982a) Effects of volatiles on mixing in calc-alkaline magma systems. Nature 297:554–557

Huppert HE, Stewart Turner J, Sparks RSJ (1982b) Replenished magma chambers: effects of compositional zonation and input rates. Earth Planet Sci Lett 57:345–357

Huppert HE, Sparks RSJ, Turner JS (1983) Laboratory investigations of viscous effects in replenished magma chambers. Earth Planet Sci Lett 65:377–381

Huppert HE, Sparks RSJ, Turner JS (1984) Some effects of viscosity on the dynamics of replenished magma chambers. J Geophys Res 89(B8):6857–6877

Huppert HE, Sparks RSJ, Wilson JR, Hallworth MA, Leitch AM (1986) Laboratory experiments with aqueous solutions modelling magma chamber processes II. Cooling and crystallization along inclined planes. *In:* Orig. igneous layering, Springer, Dordrecht, p 539–568

Huppert HE, Sparks RSJ (1988) The generation of granitic magmas by intrusion of basalt into continental crust. J Petrol 29:599–624

Jellinek AM, Kerr RC (1999) Mixing and compositional stratification produced by natural convection: 2. Applications to the differentiation of basaltic and silicic magma chambers and komatiite lava flows. J Geophys Res Solid Earth 104:7203–7218

Jellinek AM, Kerr RC, Griffiths RW (1999) Mixing and compositional stratification produced by natural convection 1. Experiments and their application to Earth's core and mantle. J Geophys Res Solid Earth 104:7183–7201

Johnston AD, Wyllie PJ (1988) Interaction of granitic and basic magmas: experimental observations on contamination processes at 10 kbar with H_2O. Contrib Mineral Petrol 98:352–362

King BC (1964) The nature of basic igneous rocks and their relations with associated acid rocks. Part IV. Sci Prog 282–292

King B (1965) The nature of basic igneous rocks and their relations with associated acid rocks. VI. Sci Prog 53:437–446

Kouchi A, Sunagawa I (1982) Experimental study of mixing of basaltic and dacitic magmas. Sci Rep Tohoku Univ Ser 3:163–175

Kouchi A, Sunagawa I (1983) Mixing basaltic and dacitic magmas by forced convection. Nature 304:527

Kouchi A, Sunagawa I (1985) A model for mixing basaltic and dacitic magmas as deduced from experimental data. Contrib Mineral Petrol 89:17–23

Koyaguchi T, Blake S (1989) The dynamics of magma mixing in a rising magma batch. Bull Volcanol 52:127–137

Kranck EH (1969) Anorthosites and rapakivi, magmas from the lower crust *In:* Origin of anorthosites and related rocks. Isachsen YV (ed.) NY State Mus Sci Serv Mem 18: 93–97

Kratzmann DJ, Carey S, Scasso R, Naranjo JA (2009) Compositional variations and magma mixing in the 1991 eruptions of Hudson volcano, Chile 71:419–439

Kress V (1997) Magma mixing as a source for Pinatubo sulphur. Nature 389:591–593

Laeger K, Petrelli M, Andronico D, Misiti V, Scarlato P, Cimarelli C, Taddeucci J, Del Bello E, Perugini D (2017) High-resolution geochemistry of volcanic ash highlights complex magma dynamics during the Eyjafjallajökull 2010 eruption. Am Mineral 102:1173–1186

Laeger K, Petrelli M, Morgavi D, Lustrino M, Pimentel A, Paredes-Mariño J, Astbury RL, Kueppers U, Porreca M, Perugini D (2019) Pre-eruptive conditions and triggering mechanism of the ~16 ka Santa Bárbara explosive eruption of Sete Cidades Volcano (São Miguel, Azores). Contrib Mineral Petrol 174:11

Langmuir CH, Vocke RD, Hanson GN, Hart SR (1978) A general mixing equation with applications to Icelandic basalts. Earth Planet Sci Lett 37:380–392

Laumonier M (2013) Mélange de magmas à HP–HT: contraintes expérimentales et application au magmatisme d'arc. PhD dissertation, Université d'Orléans

Laumonier M, Gaillard F, Muir D, Blundy J, Unsworth M (2017) Giant magmatic water reservoirs at mid-crustal depth inferred from electrical conductivity and the growth of the continental crust. Earth Planet Sci Lett 457:173–180

Laumonier M, Scaillet B, Arbaret L, Champallier R (2014a) Experimental simulation of magma mixing at high pressure. Lithos 196:281–300

Laumonier M, Scaillet B, Pichavant M, Champallier R, Andujar J, Arbaret L (2014b) On the conditions of magma mixing and its bearing on andesite production in the crust. Nat Commun 5:5607

Laumonier M, Scaillet B, Arbaret L, Andújar J, Champallier R (2015) Experimental mixing of hydrous magmas. Chem Geol 418:158–170

Laumonier M, Karakas O, Bachmann O, Gaillard F, Lukács R, Seghedi I, Menand T, Harangi S (2019) Evidence for a persistent magma reservoir with large melt content beneath an apparently extinct volcano. Earth Planet Sci Lett 521:79–90

Lesher CE (1990) Decoupling of chemical and isotopic exchange during magma mixing. Nature 344:235–237

Liu M, Muzzio FJ, Peskin RL (1994) Quantification of mixing in aperiodic chaotic flows. Chaos, Solitons Fractals 4:869–893

Longo A, Papale P, Vassalli M, Saccorotti G, Montagna CP, Cassioli A, Giudice S, Boschi E (2012) Magma convection and mixing dynamics as a source of Ultra-Long-Period oscillations. Bull Volcanol 74:873–880

Mandelbrot BB (1982) The Fractal Geometry of Nature. W. H. Freeman and Co

Marshall LA, Sparks RSJ (1984) Origin of some mixed-magma and net-veined ring intrusions. J Geol Soc London 141:171–182

Martin VM, Holness MB, Pyle DM (2006) Textural analysis of magmatic enclaves from the Kameni Islands, Santorini, Greece. J Volcanol Geotherm Res 154:89–102

Martin VM, Morgan DJ, Jerram DA, Caddick MJ, Prior DJ, Davidson JP (2008) Bang! Month-scale eruption triggering at Santorini volcano. Science 321:1178

Masotta M, Laumonier M, McCammon C (2018) Transport of melt and volatiles in magmas inferred from kinetic experiments on the partial melting of granitic rocks. Lithos 318–319:434–447

McBirney AR (1980) Mixing and unmixing of magmas. J Volcanol Geotherm Res 7:357–371

Melekhova E, Annen C, Blundy J (2013) Compositional gaps in igneous rock suites controlled by magma system heat and water content. Nat Geosci 6:385–390

Montagna CPP, Papale P, Longo A (2015) Timescales of mingling in shallow magmatic reservoirs. *In:* Chemical, Physical and Temporal Evolution of Magmatic Systems. Caricchi L, Blundy JD (eds) Geol Soc Spec Publ 422:131–140

Morgavi D, Arzilli F, Pritchard C, Perugini D, Mancini L, Larson P, Dingwell DB (2016) The Grizzly Lake complex (Yellowstone Volcano, USA): Mixing between basalt and rhyolite unraveled by microanalysis and X-ray microtomography. Lithos 260:457–474

Morgavi D, Perugini D, de Campos CP, Ertel-Ingrisch W, Dingwell DB (2013a) Time evolution of chemical exchanges during mixing of rhyolitic and basaltic melts. Contrib Mineral Petrol 166:615–638

Morgavi D, Perugini D, De Campos CP, Ertel-Ingrisch W, Dingwell DB (2013b). Morphochemistry of patterns produced by mixing of rhyolitic and basaltic melts. J Volcanol Geotherm Res 253:87–96

Morgavi D, Perugini D, De Campos CP, Ertl-Ingrisch W, Lavallée Y, Morgan L, Dingwell DB (2013c) Interactions between rhyolitic and basaltic melts unraveled by chaotic mixing experiments. Chem Geol 346:119–212

Morgavi D, Petrelli M, Vetere FP, González-García D, Perugini D (2015) High-temperature apparatus for chaotic mixing of natural silicate melts. Rev Sci Instrum 86:105108

Morgavi D, Arienzo I, Montagna C, Perugini D, Dingwell DB (2019) Magma mixing: history and dynamics of an eruption trigger. *In:* Gottsmann J Neuberg J Scheu B (eds) Volcanic Unrest. Advances in Volcanology. Springer, p 123

Nakamura M (1995) Continuous mixing of crystal mush and replenished magma in the ongoing Unzen eruption. Geology 23:807–810

Oldenburg CM, Spera FJ (1991) Numerical modeling of solidification and convection in a viscous pure binary eutectic system. Int. J Heat Mass Transf. 34:2107–2121

Oldenburg CM, Spera FJ, Yuen DA, Sewell G (1989) Dynamic mixing in magma bodies: theory, simulations, and implications. J Geophys Res 94:9215–9236

Ottino JMM (1989) The Kinematics of Mixing: Stretching, Chaos, and Transport. Cambridge, U.K. Cambridge Univ. Press

Parker TS, Chua LO (1989) Poincaré Maps. *In:* Practical Numerical Algorithms for Chaotic Systems. Springer New York, New York, NY, p 31–56

Perugini D, Poli G (2004) Analysis and numerical simulation of chaotic advection and chemical diffusion during magma mixing: Petrological implications. Lithos 78:43–66

Perugini D, Poli G (2005) Viscous fingering during replenishment of felsic magma chambers by continuous inputs of mafic magmas: Field evidence and fluid-mechanics experiments. Geology 33:5–8

Perugini D, Poli G (2012) The mixing of magmas in plutonic and volcanic environments: Analogies and differences. Lithos 153:261–277

Perugini D, Poli G, Gatta GD (2002) Analysis and simulation of magma mixing processes in 3D. Lithos 65:313–330

Perugini D, Poli G, Mazzuoli R (2003) Chaotic advection, fractals and diffusion during mixing of magmas: Evidence from lava flows. J Volcanol Geotherm Res 124:255–279

Perugini D, Ventura G, Petrelli M, Poli G (2004) Kinematic significance of morphological structures generated by mixing of magmas: A case study from Salina Island (southern Italy). Earth Planet Sci Lett 222:1051–1066

Perugini D, Poli G, Valentini L (2005) Strange attractors in plagioclase oscillatory zoning: petrological implications. Contrib Mineral Petrol 149:482–497

Perugini D, Petrelli M, Poli G (2006a) Analysis of concentration patterns in volcanic rocks: Insights into dynamics of highly explosive volcanic eruptions. Phys A Stat Mech Appl 370:741–746

Perugini D, Petrelli M, Poli G (2006b) Diffusive fractionation of trace elements by chaotic mixing of magmas. Earth Planet Sci Lett 243:669–680

Perugini D, Valentini L, Poli G (2007) Insights into magma chamber processes from the analysis of size distribution of enclaves in lava flows: A case study from Vulcano Island (Southern Italy). J Volcanol Geotherm Res 166:193–203

Perugini D, De Campos CP, Dingwell DB, Petrelli M, Poli G (2008) Trace element mobility during magma mixing: Preliminary experimental results. Chem. Geol 256:145–156

Perugini D, Poli G, Petrelli M, De Campos CP, Dingwell DB (2010) Time-scales of recent Phlegrean Fields eruptions inferred from the application of a 'diffusive fractionation' model of trace elements. Bull Volcanol 72:431–447

Perugini D, De Campos CP, Ertel-Ingrisch W, Dingwell DB (2012) The space and time complexity of chaotic mixing of silicate melts: Implications for igneous petrology. Lithos 155:326–340

Perugini D, De Campos CP, Dingwell DB, Dorfman A (2013) Relaxation of concentration variance: A new tool to measure chemical element mobility during mixing of magmas. Chem Geol 335:8–23

Perugini D, De Campos CP, Petrelli M, Dingwell DB (2015a) Concentration variance decay during magma mixing: A volcanic chronometer. Sci Rep 5:1–10

Perugini D, De Campos CP, Petrelli M, Morgavi D, Vetere FP, Dingwell DB (2015b) Quantifying magma mixing with the Shannon entropy: Application to simulations and experiments. Lithos 236–237:299–310

Petrelli M, Perugini D, Poli G (2005) Time-scales of hybridisation of magmatic enclaves in regular and chaotic flow fields: Petrologic and volcanologic implications. Bull Volcanol 68:285–293

Petrelli M, Perugini D, Poli G (2011) Transition to chaos and implications for time-scales of magma hybridization during mixing processes in magma chambers. Lithos 125:211–220

Petrelli M, El Omari K, Le Guer Y, Perugini D (2016) Effects of chaotic advection on the timescales of cooling and crystallization of magma bodies at mid crustal levels. Geochem Geophys Geosystems 17:425–441

Petrelli M, El Omari K, Spina L, Le Guer Y, La Spina G, Perugini D (2018) Timescales of water accumulation in magmas and implications for short warning times of explosive eruptions. Nat Commun 9:770

Petrelli M, Zellmer GF (2020) Rates and timescales of magma transfer, storage, emplacement, and eruption. *In:* Dynamic Magma Evolution, F. Vetere (Ed.), AGU, Chapter 1, p 1–41

Petrone CM, Bugatti G, Braschi E, Tommasini S (2016) Pre-eruptive magmatic processes re-timed using a non-isothermal approach to magma chamber dynamics. Nat Commun 7:1–11

Philpotts AR, Shi J, Brustman C (1998) Role of plagioclase crystal chains in the differentiation of partly crystallized basaltic magma. Nature 395:343–346

Phillips JA (1880) On concretionary patches and fragments of other rocks contained in granite. Q J Geol Soc Lond 36:1–22

Picard D, Arbaret L, Pichavant M, Champallier R, Launeau P (2013) The rheological transition in plagioclase-bearing magmas. J Geophys Res Solid Earth 118:1363–1377

Pistone M, Caricchi L, Ulmer P, Reusser E, Ardia P (2013) Rheology of volatile-bearing crystal mushes: Mobilization vs. viscous death. Chem Geol 345:16–39

Pistone M, Blundy JD, Brooker RA, EIMF (2016) Textural and chemical consequences of interaction between hydrous mafic and felsic magmas: an experimental study. Contrib Mineral Petrol 171

Poli G, Perugini D (2002) Strange attractors in magmas: evidence from lava flows. Lithos 65:287–297

Poli GE, Tommasini S (1991) Model for the origin and significance of microgranular enclaves in calc-alkaline granitoids. J Petrol 32:657–666

Poli G, Tommasini S, Halliday AN (1996) Trace element and isotopic exchange during acid–basic magma interaction processes. Trans R Soc Edinburgh, Earth Sci 87:225–232

Pritchard CJ, Larson PB, Spell TL, Tarbert KD (2013) Eruption-triggered mixing of extra-caldera basalt and rhyolite complexes along the East Gallatin–Washburn fault zone, Yellowstone National Park, WY, USA. Lithos 175–176:163–177

Reid JB, Evans OC, Fates DG, Reid Jr JB, Evans OC, Fates DG (1983) Magma mixing in granitic rocks of the central Sierra Nevada, California. Earth Planet Sci Lett 66:243–261

Rossi S, Petrelli M, Morgavi D, González-García D, Fischer LA, Vetere FP, Perugini D (2017) Exponential decay of concentration variance during magma mixing: Robustness of a volcanic chronometer and implications for the homogenization of chemical heterogeneities in magmatic systems. Lithos 286–287:396–407

Rossi S, Petrelli M, Morgavi D, Vetere FP, Almeev RR, Astbury RL, Perugini D (2019) Role of magma mixing in the pre-eruptive dynamics of the Aeolian Islands volcanoes (Southern Tyrrhenian Sea, Italy). Lithos 324–325:165–179

Ruprecht P, Bachmann O (2010) Pre-eruptive reheating during magma mixing at Quizapu volcano and the implications for the explosiveness of silicic arc volcanoes. Geology 38:919–922

Ruprecht P, Bergantz GW, Dufek J (2008) Modeling of gas-driven magmatic overturn: Tracking of phenocryst dispersal and gathering during magma mixing. Geochem Geophys Geosystems 9:2008Q07017

Ruprecht P, Plank T (2013) Feeding andesitic eruptions with a high-speed connection from the mantle. Nature 500:68–72

Sartorius von Waltershausen W (1853) Ueber die vulkanischen Gesteine in Sicilien und Island und ihre submarine Umbildung. Göttingen, Dietrichschen Buchhandlung, 532 p

Sakuyama M (1984) Magma mixing and magma plumbing systems in island arcs. Bull Volcanol 47:685–703

Sato E, Sato H (2009) Study of effect of magma pocket on mixing of two magmas with different viscosities and densities by analogue experiments. J Volcanol Geotherm Res 181:115–123

Schleicher JM, Bergantz GW, Breidenthal RE, Burgisser A (2016) Time scales of crystal mixing in magma mushes. Geophys Res Lett 43:1543–1550

Seaman SJ, Scherer EE, Standish JJ (1995) Multistage magma mingling and the origin of flow banding in the Aliso lava dome, Tumacacori Mountains, southern Arizona. J Geophys Res Solid Earth 100:8381–8398

Self S (1992) Krakatau revisited: The course of events and interpretation of the 1883 eruption. GeoJournal 28:109–121

Sigmundsson F, Hreinsdóttir S, Hooper A, Árnadóttir T, Pedersen R, Roberts MJ, Óskarsson N, Auriac A, Decriem J, Einarsson P, Geirsson H, Hensch M, Ófeigsson BG, Sturkell E, Sveinbjörnsson H, Feigl KL (2010) Intrusion triggering of the 2010 Eyjafjallajökull explosive eruption. Nature 468:426–432

Sigurdsson H, Sparks RSJ (1981) Petrology of rhyolitic and mixed magma ejecta from the 1875 eruption of Askja, Iceland. J Petrol 22:41–84

Słaby E, Götze J, Wörner G, Simon K, Wrzalik R, Śmigielski M (2008) K-feldspar phenocrysts in microgranular magmatic enclaves: A cathodoluminescence and geochemical study of crystal growth as a marker of magma mingling dynamics. Lithos 105:85–97

Sliwinski JT, Bachmann O, Dungan MA, Huber C, Deering CD, Lipman PW, Martin LHJ, Liebske C (2017) Rapid pre-eruptive thermal rejuvenation in a large silicic magma body: the case of the Masonic Park Tuff, Southern Rocky Mountain volcanic field, CO, USA. Contrib Mineral Petrol 172:30

Smith JV (2000) Structures on interfaces of mingled magmas, Stewart Island, New Zealand. J Struct Geol 22:123–133

Snyder D (2000) Thermal effects of the intrusion of basaltic magma into a more silicic magma chamber and implications for eruption triggering. Earth Planet Sci Lett 175:257–273

Snyder D, Tait S (1996) Magma mixing by convective entrainment. Nature 379:529–531

Sparks RSJ, Folkes CB, Humphreys MCS, Barfod DN, Clavero J, Sunagua MC, McNutt SR, Pritchard ME (2008) Uturuncu volcano, Bolivia: Volcanic unrest due to mid-crustal magma intrusion. Am J Sci 308:727–769

Sparks SRJ, Sigurdsson H, Wilson L (1977) Magma mixing: a mechanism for triggering acid explosive eruptions. Nature 267:315–318

Sparks RSJ, Meyer P, Sigurdsson H (1980) Density variation amongst mid-ocean ridge basalts: Implications for magma mixing and the scarcity of primitive lavas. Earth Planet Sci Lett 46:419–430

Sparks RSJ, Marshall LA (1986) Thermal and mechanical constraints on mixing between mafic and silicic magmas. J Volcanol Geotherm Res 29:99–124

Spera FJ, Bohrson WA (2018) Rejuvenation of crustal magma mush: A tale of multiply nested processes and timescales. Am J Sci 318:90–140

Spera FJ, Borgia A, Strimple J, Feigenson M (1988) Rheology of melts and magmatic suspensions. 1. Design and calibration of concentric cylinder viscometer with application to rhyolitic magma. J Geophys Res 93:10273–10294

Spera FJ, Schmidt JS, Bohrson WA, Brown GA (2016) Dynamics and thermodynamics of magma mixing: Insights from a simple exploratory model. Am Mineral 101:627–643

Streck MJ (2008) Mineral textures and zoning as evidence for open system processes. Rev Mineral Geochem 69:595–622

Takada A (1988) Subvolcanic structure of the central dike swarm associated with the ring complexes in the Shitara district, central Japan. Bull Volcanol 50:106–118

Takeuchi S (2011) Preeruptive magma viscosity: An important measure of magma eruptibility. J Geophys Res Solid Earth 116:B10201

Thomas N, Tait S, Koyaguchi T (1993) Mixing of stratified liquids by the motion of gas bubbles: application to magma mixing. Earth Planet Sci Lett 115:161–175

Thompson RN (1968) Tertiary granites and associated rocks of the Marsco area, Isle of Skye. Q J Geol Soc London 124:349–380

Till CB, Vazquez JA, Boyce JW (2015) Months between rejuvenation and volcanic eruption at Yellowstone caldera, Wyoming. Geology 43:695–698

Tobisch OT, McNulty BA, Vernon RH(1997) Microgranitoid enclave swarms in granitic plutons, central Sierra Nevada, California, Lithos, Volume 40, Issues 2–4,

Turner JS, Campbell IH (1986) Convection and mixing in magma chambers. Earth Sci Rev 23:255–352

Turner S, Costa F (2007) Measuring timescales of magmatic evolution. Elements 3:267–272

Turner S, Foden J (1996) Magma mingling in late-Delamerian A-type granites at Mannum, South Australia. Mineral Petrol 56:147–169

van der Laan SR, Wyllie PJ (1993) Experimental interaction of granitic and basaltic magmas and implications for mafic enclaves. J Petrol 34:491–517

Ventura G (1998) Kinematic significance of mingling-rolling structures in lava flows: A case study from Porri Volcano (Salina, Southern Tyrrhenian Sea). Bull Volcanol 59:394–403

Vigneresse JL (2007) The role of discontinuous magma inputs in felsic magma and ore generation. Ore Geol Rev 30:181–216

Wada K (1995) Fractal structure of heterogeneous ejecta from the Me-akan volcano, eastern Hokkaido, Japan: implications for mixing mechanism in a volcanic conduit. J Volcanol Geotherm Res 66:69–79

Wager L, Vincent E, Brown G, Bell J (1965) Marscoite and related rocks of the Western Red Hills Complex Isle of Skye. Philos. Trans R Soc A Math Phys Eng Sci 257:273–307

Wager LR, Bailey EB (1953) Basic magma chilled against acid magma. Nature 172:68–69

Walker GPL, Skelhorn RR (1966) Some associations of acid and basic igneous rocks. Earth Sci Rev 2:93–109

Wallace GS, Bergantz GW (2002) Wavelet-based correlation (WBC) of zoned crystal populations and magma mixing. Earth Planet Sci Lett 202:133–145

Wiebe RA (1994) Silicic magma chambers as traps for basaltic magmas: The Cadillac Mountain Intrusive Complex, Mount Desert Island, Maine. J Geol 102:423–437

Wiesmaier S, Morgavi D, Renggli CJ, Perugini D, De Campos CP, Hess K (2015) Magma mixing enhanced by bubble segregation 1007–1023

Wilcox RE (1944) Rhyolite-basalt complex on Gardiner River, Yellowstone Park, Wyoming. Bull Geol Soc Am 55:1081–1096

Wilcox RE (1999) The idea of magma mixing: history of a struggle for acceptance. J Geol 107:421–432

Williams H (1952) Volcanic history of the Meseta Central Occidental, Costa Rica. Univ Calif Publ Geol Sci 29:145–180

Williams-Jones G, Rymer H (2002) Detecting volcanic eruption precursors: A new method using gravity and deformation measurements. J Volcanol Geotherm Res 113:379–389

Woods AW, Cowan A (2009) Magma mixing triggered during volcanic eruptions. Earth Planet Sci Lett 288:132–137

Wyllie PJ, Carroll MR, Johnston AD, Rutter MJ, Sekine T, Van Der Laan SR (1989) Interactions among magmas and rocks in subduction zone regions: experimental studies from slab to mantle to crust. Eur. J Mineral 1:165–179

Zellmer GF, Sheth HC, Iizuka Y, Lai Y-J (2012) Remobilization of granitoid rocks through mafic recharge: Evidence from basalt–trachyte mingling and hybridization in the Manori–Gorai area, Mumbai, Deccan Traps. Bull Volcanol 74:47–66

Zimanowski B, Buttner R, Koopmann A (2004) Experiments on magma mixing. Geophys Res Lett 31:1–3

Reviews in Mineralogy & Geochemistry
Vol. 87 pp. 639–720, 2022
Copyright © Mineralogical Society of America

14

Magma / Suspension Rheology

Stephan Kolzenburg,[1,2,3] Magdalena O. Chevrel,[4] Donald B. Dingwell[2]

[1]Department of Geology
University at Buffalo
126 Cooke Hall
Buffalo, NY 14260-4130, USA

[2]Department of Earth and Environmental Sciences
Ludwig-Maximilians-University Munich
Theresienstr. 41 / III
80333, Munich, Germany

[3]Department of Earth and Planetary Sciences
McGill University
3450 University Street
Montreal, H3A 0E8, Quebec, Canada

[4]Laboratoire Magmas et Volcans,
Université Clermont Auvergne, CNRS, IRD, OPGC,
6 Avenue Blaise Pascal
f-63000, Clermont-Ferrand, France

stephank@buffalo.edu; oryaelle.chevrel@ird.fr; dingwell@lmu.de

1. THEORETICAL CONSIDERATIONS

When considering the rheology of a suspension the fundamental physical aspect to be quantified is the redistribution of strain between the deformable matrix fraction of the physical mixture and the suspended phase or phases. For the case of solid particle suspensions in magma and lava the suspended crystalline phases will generally be devoid of significant internal strain during flow. With some exceptions (Cordonnier et al. 2009; Kendrick et al. 2017), the stresses driving the magmatic deformation and flow, are generally not high enough to induce significant crystal deformation (see however below). Thus, the primary task becomes one of defining the geometry of the suspended phases and their potential redistribution and reorientation during viscous flow. For the case of significantly deformable suspended phases (e.g., gas-filled vesicles, immiscible oxide or sulfide melts as well as enclaves suspended during magma mixing) the distribution of the strain between the matrix liquid and the vesicular gas or suspended fluid inclusions must be dealt with using the concept of the deformability of the bubble or inclusion shape (Taylor 1932; Stein and Spera 1992). Taken together, the rheology of a multiphase liquid-supported lava or magma can then be seen as the trivial but not simple task of quantifying strain partitioning between the vesicles and melt together with the quantification of the distribution of all suspended phases and the shape consequences of the vesicle deformation.

Generally, in particle (i.e., crystal) suspension rheology the distribution of the non-deformable space associated with the suspended crystalline phase is parameterized in terms of crystal fraction, size, shape, orientation, and their respective distributions. Beyond this more or less classical treatment of suspension rheology lie however further considerations which are specific to the case of a silicate magma or lava from which the crystalline phases are growing during flow.

1529-6466/22/0087-0014$10.00 (print)
1943-2666/22/0087-0014$10.00 (online)

http://dx.doi.org/10.2138/rmg.2022.87.14

Firstly, in any protracted crystallization scenario between liquidus and solidus, the composition of the liquid matrix must evolve with crystallization. Given the sensitivity of silicate melt viscosity to composition and its growing importance with decreasing temperature, a very precise control on the matrix liquid composition is required in any study of silicate melt suspension rheology in order to distinguish the chemical effects from the physical effects of crystallization on rheology. Fortunately, for systems which quench naturally or which can be quenched in experiments to a glassy matrix, the shift in viscosity due to the shift in chemistry of the melt phase can be easily determined via scanning calorimetry using the principle of a shift factor between melt viscosity and melt calorimetric glass transition temperature. The groundwork for this technique has been well laid (Gottsmann et al. 2002; Giordano et al. 2005) and it has been applied in previous studies of suspension rheology involving microlites (Stevenson et al. 1996, 2001).

Somewhat more challenging is the case of potential chemical melt gradients. If crystal growth is driven by diffusion-controlled processes, then crystal growth will lead to chemical compositional gradients in the immediate vicinity of the growing crystal–melt interfaces. For crystal growth we may expect that the shift in SiO_2 content due to the generally anticipated non-eutectic crystal growth may lead to significant viscosity gradients around growing crystals. For the case of growing vesicles, it is the water content that may vary in cases where the vesicle growth is diffusion controlled by water (Hurwitz and Navon 1994; Humphreys et al. 2008; McIntosh et al. 2014). These water gradients should generate very substantial viscosity gradients and in general this effect should become more extreme with decreasing pressure and thus total water contents of the system.

Finally, we must not forget volume relaxation effects (Dingwell and Webb 1989; Bagdassarov and Dingwell 1993a,b) or non-Newtonian melt viscosity effects (Li and Uhlmann 1970; Dingwell and Webb 1989). For vesicle-free suspensions, volume relaxation will only be relevant in frequency domain rheology measurements where the total strain on the system is very low. For vesicle-rich systems, the rheological description of the suspension must take into account the compressibility and/or decompressive growth of vesicles and the resulting volume deformation contributions.

For melt rheology (i.e., pure liquids), prediction of the occurrence of non-Newtonian behavior is straightforward if the stress distribution is known and the strain rate can be observed. The general rule of 2.5 log units distant from the calculated stress relaxation time of the Newtonian liquid via the Maxwell law is a good predictor. The challenge may however lie in the determination of stress distribution in multiphase samples where the stress may accumulate in the liquid phase in local restrictions and therefore drive non-Newtonian response in a spatially localized manner (Caricchi et al. 2007; Lavallée et al. 2007; Ishibashi 2009; Deubelbeiss et al. 2011); see also Lavallée and Kendrick (2022).

2. CONVENTIONAL DESCRIPTIONS OF RHEOLOGICAL DATA

The shear viscosity of pure silicate melts has been described over a large compositional and temperature range (see Russell et al. 2022, this volume). It has been documented that silicate melts behave as Newtonian fluids as long as the inverse strain rate remains smaller than the melt's structural relaxation timescale τ, which is described by the Maxwell relationship; $\tau = \eta_0/G_\infty$, where η_0 is the shear viscosity at zero frequency and G_∞ is the melt's shear modulus at infinite frequency. Thus, τ describes the timescale at which the melt's microstructure can respond to accommodate strain and identifies the transition between relaxed flow and unrelaxed fracture. Dingwell and Webb (1989) and Webb and Dingwell (1990b) describe that the onset of non-Newtonian flow occurs when:

$$\tau = \frac{10^{-3} G_\infty}{\eta_0} \qquad (2.1)$$

with the assumption that $G_\infty = 10^{10}$ Pa for silicate melts, which has proven to accurately reproduce the rheological behavior of most silicate melts (Dingwell and Webb 1989). All data and models reviewed below concern strain-rate regimes where the melt phase of the suspension remains in the Newtonian regime and any non-Newtonian effects derive from the presence of the suspended particles (crystals) and/or bubbles.

Parameterization of the rheology of any suspension requires establishing the relationship between the imposed shear stress ($\dot{\varepsilon}$) and the resulting deformation rate (i.e., shear strain-rate), or *vice versa*. To assess the material's flow behavior, the imposed stress is plotted against strain rate, producing flow curves (Lenk 1967). Examples of common flow curves are shown in Figure 2.1 and all variables used in this chapter are defined in Table 1.1. The flow regime is defined as Newtonian when shear stress (σ) is proportional to strain rate ($\dot{\varepsilon}$):

$$\sigma = \eta\dot{\varepsilon} \tag{2.2}$$

where η defines the Newtonian viscosity. Relaxed silicate melts display Newtonian viscosity that is independent of strain-rate but usually strongly dependent on other variables (e.g., temperature, composition, redox—see Russell et al. 2022, this volume). However, the presence of a solid or a gas phase can cause the viscous flow behavior of a suspension to deviate from that of its liquid component as well as introduce non-Newtonian behavior and thus requires more complex parameterization strategies. The flow regime is non-Newtonian (see Fig. 2.1) when shear stress is not proportional to strain rate (shear thinning or thickening behavior), when a minimum stress is required before viscous flow initiates (Bingham regime) or when both behaviors are combined (Herschel–Bulkley model), the latter of which is expressed as:

$$\sigma = \tau_y + K\dot{\varepsilon}^n \tag{2.3}$$

where τ_y is the yield stress that is to be overcome in order to initiate flow; K is the flow consistency and n the flow index; describing the degree of non-Newtonian behavior. For Newtonian materials $\tau_y = 0$, $n = 1$ and $K = \eta$; for shear-thickening ($n > 1$) and for shear-thinning materials ($n < 1$) (Herschel and Bulkley 1926).

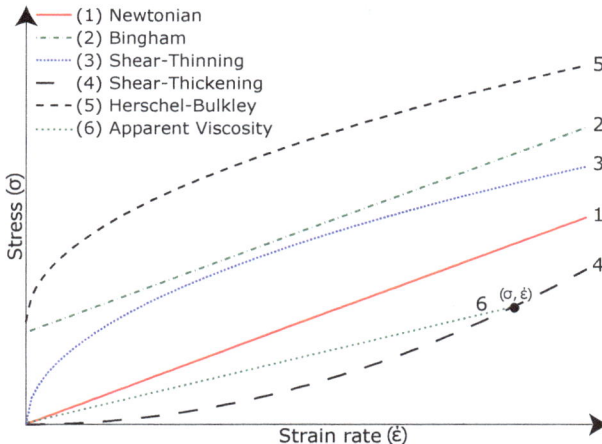

Figure 2.1. Flow curve examples: **1.** Newtonian $\sigma = \eta\dot{\varepsilon}$. **2.** Bingham $\sigma = \eta_B\dot{\varepsilon} + \tau_y$ with τ_y being the yield stress and η_B the Bingham viscosity. **3.** Shear thinning $\sigma = K\dot{\varepsilon}^n$ with n, the flow index, < 1 and K the consistency. **4.** Shear thickening $\sigma = K\dot{\varepsilon}^n$ with $n > 1$. **5.** Herschel–Bulkley $\sigma = \tau_y + K\dot{\varepsilon}^n$ plotted as shear thinning (i.e., $n > 1$). **6.** Apparent viscosity $\eta_{app} = \sigma/\dot{\varepsilon}$, is the slope of a line from the origin to a specific point along any flow curve in σ–$\dot{\varepsilon}$ space (i.e., the momentary observation at a set of experimental conditions). Every flow curve is constructed from multiple such measurements of apparent viscosity.

Table 1.1. Table of symbols.

Parameter	Unit	Description	Equation
		Section 2: Conventional Decriptions of Rheological Data	
τ	s	Timescale at which the melt's microstructure can respond to accommodate strain	2.1
η_0	Pa s	Shear viscosity at zero frequency	2.1
G_∞	10^{10} Pa	Melt's shear modulus at infinite frequency	2.1
σ	Pa	Stress	2.2
$\dot{\varepsilon}$	s^{-1}	Strain rate	2.2
η	Pa s	Newtonian viscosity of a pure silicate liquid	2.2
τ_y	Pa	Yield stress	2.3
K	Pa s	Consistency	2.3
n	–	Flow index	2.3
η_B	Pa s	Bingham viscosity	Fig. 2.1
η_r	–	Relative viscosity	2.5
η_{app}	Pa s	Apparent viscosity	2.4
η_{eff}	Pa s	Effective viscosity	2.5
η_{bulk}	Pa s	Bulk Viscosity	2.6
η_v	Pa s	Volume Viscosity	2.6
η_s	Pa s	Shear Viscosity	2.6
		Section 3: Analogue Experiments	
Re_p	–	Particle Reynolds number	3.1
St	–	Stokes number	3.2
Pe	–	Péclet number	3.3
T	K	Temperature	3.3
Sc	–	Schmidt number	3.4
η_l	Pa s	Viscosity of the liquid	3.1
ρ_l	kg/m^3	Density of the liquid	3.1
ρ_p	kg/m^3	Density of the particle	3.2
a_p	m	Particle radius	3.1
a_b	m	Bubble radius	3.1
κ	–	Characteristic length scale of the particle	3.2
k	J K^{-1}	Boltzmann constant	3.3
Ca	–	Capillary number	3.5
Γ	N m^{-1}	Bubble surface tension	3.5
		Section 4: High Temperature Experiments	
Ω	rad/s	Angular velocity	4.1
R_o	m	Outer radius, i.e., radius of the crucible	4.1
R_i	m	Inner radius, i.e., radius of the spindle	4.1

Parameter	Unit	Description	Equation
M	N m	Torque	4.2
l	m	Effective immersed length of the spindle	4.2
$\dot{\varepsilon}_e$	–	Engineering strain rate	4.3
F	N	Force	4.4
h	m	Sample height	4.4
Δ_h	m	Sample height variation	4.4
Δ_t	s	Time variation	4.4
V	m^3	Sample volume	4.4
S_{eff}	m^2	Effective surface area	4.6
Φ_b	Fraction (0–1)	Bubble volume fraction	4.8
S_{abs}	m^2	Absolute surface area	4.8
V_0	m^3	Initial sample volume	4.9
t	s	Time	4.9
k	–	Coefficient to track the porosity loss	4.9
ϕ_i	Fraction (0–1)	Initial sample porosity	4.9
ϕ_f	Fraction (0–1)	Final sample porosity	4.9
t_f	s	Final time	4.9
ε_T	–	Observed strain	4.11
α	–	Fitting constant	4.11
d	m	Sample diameter	4.14
E	Pa	Young's modulus	4.11
η_{0x}	Pa s	Viscosity of the deposit at zero porosity	4.11
G^*	Pa	Complex shear modulus	4.12
ε_r	–	Strain at any radius r	4.14
O	rad	Angle of deflection	4.14
j	$\sqrt{-1}$	Constant	4.12
σ_0	Pa	Stress amplitude	4.12
ε_0	s^{-1}	Strain amplitude	4.12
ℓ	s	Phase lag	4.12
η^*	Pa s	Complex viscosity	4.13
Ȯ	rad/s	Twist rate	4.13
ω	Hz	Angular frequency	4.13
$\dot{\varepsilon}_r$	–	Strain-rate at any given radius	4.15
ω	rev/s	Twist rate	4.15
r	m	Radius of the sample	4.18

Parameter	Unit	Description	Equation
		Section 5: Field Rheology	
F	N	Force of penetration (viscous drag)	5.1
u	m/s	Speed of penetration	5.1
R_{eff}	m	Effective radius of the vane	5.1
		Section 6: Parameterization Strategies	
Φ	Fraction (0–1)	Particle volume fraction	6.1
B	–	Einstein coefficient	6.1
Φ_m	Fraction (0–1)	Maximum packing fraction	6.5
β	–	Fitting coefficient	6.11
Φ_*	Fraction (0–1)	Critical particle fraction at the onset of the exponential increase of viscosity	6.14
δ, γ, ξ	–	Empirical parameters that vary with strain-rate and particles shape	6.14
$\eta_r(max)$	–	Relative viscosity of the suspension at complete solidification i.e., $\Phi = 1$	6.15
k, n	–	Fitting parameters shaping of the sigmoidal function.	6.15
η_f	Pa s	Viscosity of a suspension with fine particle	6.16
η_c	Pa s	Viscosity of a suspension with coarse particle	6.16
$\Phi_c, \gamma_c, \xi_c, \Phi_{*c}$	–	Fitting parameters for coarse particles	6.17
$\Phi_f, \gamma_f, \xi_f, \Phi_{*f}$	–	Fitting parameters for fine particles	6.17
r_p	–	Particle aspect ratio	6.20
Φ_{ml}	Fraction (0–1)	Maximum packing fraction for equant particles and given as 0.656 for smooth and 0.55 for rough particles	6.20
b	–	Fitting parameter: 1.08 for smooth and 1 for rough particles	6.20
Φ_{m0}	–	Φ_{m0} represents Φ_m for a monomodal suspension derived from Equation 6.19	6.22
ϱ	Fraction (0–1)	Polydispersity (tends to 0 for polydisperse size distribution and tends to 1 for monodisperse suspension)	6.22
S_p	–	Ratio of the specific surface area of a polydisperse system	6.23
S_m	–	Ratio of the specific surface area of a monodisperse system	6.23
r	m	Particle radius	6.23
K_r	–	Relative consistency	6.24
n_{min}	–	Flow index at $\Phi = \Phi_m$	6.30
Φ_C	Fraction (0–1)	Critical particle fraction for yield stress that is determined experimentally	6.34
D_p	feet	Particle diameter	6.35
δ		Shape factor (ratio of the surface area of a sphere of equivalent volume to the surface area of the particle)	6.35
σ_g		Geometric standard deviation for particle diameter; σ_g 2.02	6.35
A, Φ_C, m		Fitting parameters = 0.848; 0.1978, 0.8364, respectively	6.35
τ_c	Pa	Characteristic stress (or inter-particle cohesion)	6.37

Parameter	Unit	Description	Equation
τ^*	Pa	Yield stress coefficient; a fitting parameter that tends to increase with particle aspect ratio	6.39
λ_b	s	Bubble relaxation timescale	6.40
K_b	–	Characteristic stress (or inter-particle cohesion)	6.40
Cd	–	Dynamic capillary number	6.41
$\ddot{\varepsilon}$	s	Change in strain-rate	6.41
η_{r0}	–	Relative viscosity of the suspension at the low Ca limit	6.51
$\eta_{r\infty}$	–	Relative viscosity of the suspension at high Ca limit	6.51
C_X	–	Capillarity which captures the combined effect of shear and flow steadiness	6.52
Φ_t	–	Total suspended phases volume	6.54
Φ_b	–	Volume fraction occupied by bubbles	6. 54
Φ_c	–	Volume fraction occupied by particles (crystals)	6. 54
Φ_l	–	Volume fraction occupied by liquid phase	6. 54
$\eta_b(\Phi_b)$	Pa s	Viscosity of the bubble suspension (bubble + liquid)	6.59
$\eta_{r,b}(\Phi_b)$	–	Relative viscosity of bubbles suspension	6.59
$\eta_c(\Phi)$	Pa s	Viscosity of the crystal suspension (crystal + liquid)	6.62
$\eta_{r,c}(\Phi)$	–	Relative viscosity of crystals suspension	6.63

Magmatic suspensions, once exceeding the dilute regime (particle fractions $\lesssim 30$ vol%), in most cases have a strain-rate dependent viscosity and their flow behavior is best described by a flow curve rather than a single Newtonian viscosity value. The common practice, when preforming rheological experiments, is to construct these flow curves via systematic measurement of a set of apparent viscosity (η_{app}) values, which are the momentary σ–$\dot{\varepsilon}$ observations (i.e., the ratio of a given stress to a given strain-rate; 6 in Fig. 2.1) at varying experimental conditions:

$$\eta_{app} = \frac{\sigma}{\dot{\varepsilon}} \qquad (2.4)$$

This apparent viscosity is, however, different from the Newtonian viscosity of a fluid even though both have the same dimension. The apparent viscosity is only valid for a specific subset of deformation conditions and varies with strain-rate. Hence it only represents a singular point on the whole flow curve but does not provide an exhaustive description of the fluid's rheology. Yet, it is useful in many cases to discuss rheological data in terms of apparent viscosity as long as it is clearly identified as such.

The presence of a suspended phase can cause the viscous flow behavior of a suspension to deviate from that of its liquid component as well as introduce non-Newtonian behavior. To describe the rheological behavior of a suspension, the term relative viscosity, η_r, is frequently employed. The relative viscosity is the ratio of the viscosity of the suspension, also called the effective viscosity (Petford 2009), to the Newtonian viscosity, η, of the suspending liquid:

$$\eta_r = \frac{\eta_{eff}}{\eta} \qquad (2.5)$$

Parameterization strategies for magmatic suspensions thus aim to describe either the relative viscosity or the stress–strain-rate response of a suspension in the form $\eta_r = f(\Phi)$ or $\sigma = f(\varepsilon)$, where Φ is the volume fraction of the suspended phase (particles (crystals) or bubbles).

A further consideration to be made is that liquids and multiphase suspensions may deform under a variety of stress environments. In magmatic suspensions, shear deformation is by far the most common deformation environment, but tensile, compressive and oscillatory stresses may also become relevant for example during unloading in a volcanic flank collapse scenario or when seismic waves pass through a magma storage system. Tensile or compressive stresses may cause effects of volume viscosity to become important in addition to shear viscosity. In such cases, a bulk (or longitudinal viscosity) (η_{bulk}) is defined that is the product of volume (η_v) and shear viscosity (η_s):

$$\eta_{bulk} = \eta_v + \frac{4}{3}\eta_s \tag{2.6}$$

For the sake of brevity, the term viscosity refers to shear viscosity in this chapter unless otherwise specified. Bulk and volume viscosities become important when studying compressible materials such as bubble bearing suspensions, as it will become apparent later in this chapter for the case of parallel plate and oscillatory viscometry on bubbly melts. As an example: in parallel plate measurements, at high porosity, vesicular melts may behave in pure uniaxial compression (i.e., no bulging or sample translation) and in this case, the measured viscosity is equivalent to the samples' longitudinal viscosity. Therefore, while simple parallel plate rheometry is operationally attractive, separate determination of either shear or volume viscosity is required to completely resolve the samples' rheology, a caveat that rotational and torsional measurements avoid.

3. EXPERIMENTS ON ANALOGUE MATERIALS

Understanding the rheological behavior of magma is critical to determining magma ascent rates, force balances, flow structures as well as lava flow velocities and runout distances and even magma failure. Flow rates and internal force balances, in turn, are core parameters driving gas exsolution and crystallization. During crystallisation, the growth of solid crystals as well as the evolution of the interstitial liquid causes the transfer of components from the liquid into the crystal population. Similar transfer processes occur during degassing, i.e., bubbles form as volatiles are exsolving from the liquid into the bubble. Thus, there is an intimate feedback between a magmas textural and chemical evolution and its rheology. The technical challenges presented by measurements under natural conditions and high temperature experimentation, combined with the motivation to systematically explore texture dependent variations in the rheology of two- and three-phase suspensions have inspired a wealth of experiments on analogue materials.

The largest part of analogue studies investigates two phase suspensions of either liquid and particles (simulating crystal bearing magma) or liquid and bubbles (simulating exsolved volatiles). The central goal of these studies is to mimic the flow of natural magmas and lavas and to derive constitutive equations describing their rheology. Analogue experiments are advantageous because they allow simplification of the rheological characterization of multiphase suspensions since they do not involve the need for high temperature or pressure equipment. The sample texture (i.e., solid and/or bubble content, size and distribution) can be precisely controlled because no (or only previously anticipated) chemical interactions occur. As a result, analogue experiments are often more reproducible than experiments with natural silicate liquids and single process parameters can readily be isolated and investigated. Oftentimes, transparent materials can be used, allowing assessment of the 3D and 4D (i.e., 3D plus time) systematics of the process. Heterogeneities of natural systems can be removed, allowing an investigation of the fundamental physical behavior of suspensions. Questions arising from field relations can be simplified and geometric relations reproduced to identify and map the process guiding parameters. As a result, these types of experiments can often reveal processes or process-systematics that may not be readily evident in natural scenarios. However, there are also a number of caveats when employing analogue materials to derive a physical understanding of magma rheology. The complexity of natural processes (geometries, dimensions, pressurization, crystallization and degassing) is difficult to recreate, requiring simplification of the process for experimentation. The necessary simplifications made during experiment design and monitoring frequently exclude the important chemical and physical feedback mechanisms that arise in natural systems, outlined above. To date, few analogue experiments have attempted to reproduce transient processes occurring in natural multiphase suspensions (e.g., crystallization and degassing).

While any experimental dataset allows to derive models describing its physical behavior, validation of the derived models cannot simply rest on further experimentation but must involve expansion and testing of the model in flow situations relevant to natural scenarios. However, application of the derived models for natural scenarios requires careful scaling of experiment size and process parameters (e.g., energy and/or forces) and validation of the derived constitutive equations on field examples can be very difficult. One of the most important considerations when carrying out analogue experimentation is therefore to precisely constrain what the experiment is set out to model. This is almost exclusively done via characterization of natural materials and/or field observations in concert with dimensional analysis. Doing so is critical since it is the only way that the derived results and constitutive equations have relevance to the natural environment.

Analogue experiments that aim to understand natural phenomena must ensure that the physical relationships of the simulated interactions compare to those of the natural system. The choice of the employed analogue material is of paramount importance in order to accurately mimic the physical behavior of natural material. Dimensional analysis is routinely employed to ensure accurate scaling of the investigated process. Below, we review the most relevant nondimensional parameters for analogue modeling of particle and bubble suspensions and present a brief overview of common experimental materials and strategies.

Particle suspension analogues

The core nondimensional parameters for analogue crystal suspensions of magmas are:

1) The particle Reynolds number, which describes the ratio of inertial forces to viscous forces of a particle within a fluid:

$$Re_p = \frac{\rho_l a_p^2 \dot{\varepsilon}}{\eta_l} \tag{3.1}$$

where ρ_l and η_l are the density and the viscosity of the liquid phase, respectively and a_p is the particle mean radius. This ratio describes whether the particles' behavior falls in the non-inertial and laminar regime ($Re_p \ll 1$) or in the inertial and turbulent regime ($Re_p \gg 1$). Magmatic flow is commonly restricted to low particle Reynolds numbers (i.e., the laminar regime).

2) The Stokes number describes the coupling between the solid particles and the liquid phase as the ratio of the characteristic time of the motion of a particle subject to viscous drag and inertia, to the characteristic time of flow of the suspension:

$$St = \frac{\rho_p a_p^3 \dot{\varepsilon}}{\kappa \eta_l} \tag{3.2}$$

where κ is a the characteristic length scale of the particle, which is proportional to its radius (Coussot and Ancey 1999) but further depends on its shape, size and orientation and ρ_p is the mean particle density. This ratio allows distinguishing whether the fluid–particle coupling is weak for St > 1; i.e., solid and fluid phases move independently and particles behave separated from the fluid; or whether the fluid coupling is strong for St ≪ 1; i.e., solid and fluid phase do not move independently and the particles behave as part of the fluid mixture.

3) The Péclet number, which relates the effects of diffusion (Brownian motion; i.e., the random motion of particles suspended in a fluid resulting from their collision with the thermal movement of molecules in the fluid) to advection (flow forces within the liquid):

$$\text{Pe} = \frac{6\pi\eta_l a_p^3 \dot\varepsilon}{kT} \tag{3.3}$$

where k is the Boltzmann constant (1.38×10^{-23} J K^{-1}) and T is the temperature (in Kelvin). Brownian effects dominate at Pe < 10^3 (Stickel and Powell 2005), disturbing the particle alignment and thus enhancing viscous dissipation.

In the space of these three nondimensional parameters, the regime relevant to magmatic flow is hydrodynamic at (Pe ≫ 10^3), viscous (Re$_p$ ≪ 10^{-3}) and strongly-coupled (St ≪ 1). If the particle Reynolds number is small (Re$_p$ < 10^{-3}) and the Péclet number is large (Pe > 10^3), both Pe and Re$_p$ can be neglected, and the viscosity is a unique value at every particle concentration. The suspension thus behaves Newtonian, for a specific set of conditions of shear rates defined by values of a_p, η_l, and ρ_l. The size of this "window" scales according to the Schmidt number:

$$\text{Sc} = \frac{\text{Pe}}{\text{Re}_p} = \frac{6\pi\eta_l^2 a_p}{\rho_l kT} \tag{3.4}$$

A suspension may behave Newtonian for greater ranges of shear rate as particle size and fluid viscosity increase, such that Sc ≫ 1. Thus, the Schmidt number may be used to anticipate the regime in which non-Newtonian effects may become important.

Bubble suspension analogues

The capillary number is the main non-dimensional parameter relevant for analogue bubble suspensions in flow scenarios where stress and strain rate are constant. It reflects the balance between the deforming, viscous, force acting on a bubble and the restoring force of the bubble's surface tension:

$$\text{Ca} = \frac{\eta_l a_b \dot\varepsilon}{\Gamma} \tag{3.5}$$

where a_b is the radius of the spherical bubble and Γ is the surface tension at the bubble–melt interface.

While the melt viscosity is predominantly controlled by temperature and composition (Giordano et al. 2008a; Russell et al. 2022, this volume), the dependence of surface tension on temperature and composition of the melt is much weaker (Walker and Mullins 1981; Bagdassarov et al. 2000).However, the surface tension of silicate melts can vary strongly as a function of melt water content, varying from ~ 0.05 Nm^{-1} for hydrous melts (Mangan and Sisson 2005) to around 0.3 Nm^{-1} for dry melts (Bagdassarov et al. 2000).

The effect of increasing bubble volume fraction on suspension viscosity depends strongly on Ca. When surface tension dominates deformation, Ca is small and bubble shapes are spherical, whereas when the deforming stress dominates, Ca is high, and the bubbles are highly elongated. This changepoint is important because the contribution of a bubble or bubble-population to the viscosity of the bulk suspension drastically varies with shape. The effect of increasing bubble volume fraction on the suspension viscosity thus depends strongly on Ca. At low capillary number, bubbles act to increase suspension viscosity since "rigid" spherical bubbles represent an obstacle to the flow field. At constant bubble volume, this effect decreases with increasing bubble elongation. At high capillary number, bubbles are highly elongate and flow line distortion is small. Thus, bubbles at high Ca act to decrease suspension viscosity by introducing free slip surfaces in the liquid (Rust and Manga 2002). Considerable research effort has been devoted to the parameterization of the deformation behavior of bubbles in steady and unsteady flow conditions. These are reviewed in the *"Parameterization strategies"* section in this chapter.

Experimental materials and measurement strategies

Depending on the application and level of complexity, a variety of analogue materials have been used to investigate multiphase rheology. Early works used materials such as sugar solutions in water (Einstein 1906), aqueous solutions of lead nitrate and glycerol (Ward and Whitmore 1950), latex (Maron and Levy-Pascal 1955), xylene, bromonaphthalene, or glycerine (Gay et al. 1969; Wildemuth and Williams 1985). The various liquids were largely chosen either for their viscosity at measurement conditions or for varying density at constant viscosity. However, many of these proved difficult to work with because their properties are hard to scale for application to magmatic and volcanic flows. There is a range of suitable liquids that have been used to simulate specific scenarios or rheological behaviors, which are not reviewed in detail here. Instead, the reader is referred to the work of Kavanagh et al. (2018) where the authors present a comprehensive overview over methods and properties of materials used in analogue modeling of volcanic processes. A further valuable resource is the online compilation of analogue materials for volcanology compiled by A. Rust (https://sites.google.com/site/volcanologyanalogues/home). This database provides information on a wide range of materials that have been used in volcanology. The online catalogue describes their properties and applications, while also discussing their respective limitations. The most commonly employed suspending liquids over the past decades are corn- or golden-syrup (Llewellin et al. 2002b; Rust and Manga 2002; Bagdassarov and Pinkerton 2004; Soule and Cashman 2005; Castruccio et al. 2010; Mueller et al. 2011; Jones and Llewellin 2021) and silicon oils (Sumita and Manga 2008; Mueller et al. 2010; Cimarelli et al. 2011; Del Gaudio et al. 2013, 2014; Moitra and Gonnermann 2015; Spina et al. 2016; Klein et al. 2018). The advantage of corn- or golden-syrup is that it can be mixed with water in any desired proportion. Increasing the amount of water reduces the viscosity of the suspending liquid, thus allowing to simulate, for example, varying composition and/or temperature of a magma. This does, however, make them susceptible to drying out over long experimental timescales or to take up water in very humid climates. Silicon oils on the other hand are advantageous because they exist over a very wide range of viscosities ($\sim$0.1–10000 Pa s), are generally inert and do not take up or loose humidity.

Using these liquids as suspending phase, analogue suspensions for rheological investigations are commonly prepared as two-phase suspensions: bubbles in liquid (such as in Rust and Manga 2002, Llewellin et al. 2002b), or particles in liquid (such as in Mueller et al. 2010, Moitra and Gonnermann 2015); and rarely as three-phase suspensions (bubbles and particles in a liquid); see Truby et al. (2015) for an example. This is largely owing to the complexity of interactions in three phase suspensions, which require detailed characterization of two-phase flow prior to attempting parameterization of three-phase suspensions. The most common material used to simulate spherical particles are glass beads of varying sizes, whereas glass and mineral fibres are commonly used as prolate particles (aspect ratio > 1) and particles such as mineral and glass flakes or glitter are used as oblate particles (aspect ratio < 1). Preparation of these suspensions can be challenging and time consuming because, for example, air is frequently entrained during the preparation of particle suspensions and can be difficult to remove. Further, once a homogenized particle suspension is prepared, there is a finite amount of time for which it remains homogenous since the particles begin to settle (or float) through the suspending liquid, a process that is accelerated when aiming to remove bubbles using a centrifuge. The most common gasses to introduce bubbles in the suspending liquid are air or inert gasses such as nitrogen and argon. Preparation of bubbly suspensions can be equally challenging, since bubbles tend to rise rapidly through the suspending liquid, and it is difficult to attain a homogenous bubble size distribution. Thus, introducing bubbles in a controlled fashion requires elaborate experimental devices such as pre-measurement frothing machines used in Llewellin et al. (2002b) or *in situ* bubble supply as used in Rust and Manga (2002).

Rheometry on these analogue suspensions is for the most part performed in one of three types of rotational viscometers 1) concentric cylinder, 2) cone and plate or 3) rotating parallel plate. These techniques allow for essentially infinite strain and are thus optimal for experiments aimed at systematically mapping the rheological response of a suspension. In order to accurately measure a samples' stress–strain-rate response at varying shear rates or shear stresses it is crucial to allow for the sample to reach textural and mechanical equilibrium. It has been noted that achieving an equilibrium state requires some "pre-shear treatment" (i.e., deformation to large strains prior to measurement) because particles that are, initially, randomly oriented require a finite amount of strain before reaching equilibrium orientation distributions. This pre-shear treatment reportedly requires total strains of ~50–200 until the measurement reaches a steady value; see Figure 7a in Mueller et al. (2010). The data and parameterization strategies derived from analogue experimentation are reviewed, together with those of high temperature experimentation on natural materials in the section "*Parameterization Strategies*" later in this chapter.

Published measurements on the rheology of analogue suspensions are, to date, almost exclusively performed on non-reactive materials and in textural equilibrium. These have been critical to developing a basic understanding of the effect of crystals and bubbles on rheological properties of natural melts and magmas at steady conditions. Their application to natural environments is, however, limited since most magmatic and volcanic processes are reactive and/or operate under disequilibrium conditions, where variations in the deformation- and cooling-rates can affect both crystallization and degassing processes. These, in turn, alter the texture and rheology of the magma or lava. Recently, however, Cimarelli et al. (2011) used a slowly solidifying epoxy resin to simulate matrix crystallization of a particle suspension and Spina et al. (2016) exploited the solubility of Ar gas in silicon oil to study the vesiculation dynamics of analogue suspensions of varying particle concentrations. These experiments highlight the vast potential that still remains to be explored in studies on analogue material suspensions. Nonetheless, due to the complexity and potentially unforeseen dynamics of natural systems, combination of analogue and high temperature experimentation on natural materials is critical to generate a holistic understanding of magma transport processes.

4. EXPERIMENTS ON HIGH TEMPERATURE SILICATE MELT SUSPENSIONS

Over the past decades, considerable effort was devoted to studying the rheology of magmatic suspensions at temperatures relevant to natural processes. Due to the complexity of the problem and the technical challenges associated with high temperature experimentation, advances in the study of natural magmatic suspensions or silicate melt analogues are closely coupled to the development of new experimental machinery or methods. Most experiments to date consider the rheology of two-phase suspensions (either silicate melt and crystals or silicate melt and bubbles). Measurements on three-phase suspensions (melt–crystal–bubble) at magmatic conditions are only recently starting to become more numerous. Experimental methods for high temperature suspension rheology vary drastically depending on the nature of the investigated sample and the measurement conditions (temperature, pressure, redox conditions). This is because the viscosity of magmatic suspensions can span over 12 orders of magnitude and may involve both Newtonian and non-Newtonian behavior. A complete rheological characterization of a magmatic suspension therefore requires a combination of several experimental methods. Owing to these challenges, most studies present data for a specific sub-set of conditions that are accessible by the employed apparatus. Further, *in situ* textural observations (as available for some suspension rheology measurements on analogue materials) are not yet possible during high temperature experimentation because the samples are housed in furnaces and often require containment in containers that are opaque to visible light and X-rays. However, some promising tomographic methods are beginning to become available (Dobson et al. 2020) and will likely fill this gap in the coming years (see the section "*Technological Advances*" in this chapter).

The most frequently used methods for high temperature suspension rheometry are concentric cylinder, rotational viscometry (Dingwell and Virgo 1987; Spera et al. 1988), parallel plate, uniaxial compression viscometry (Bagdassarov and Dingwell 1992; Hess et al. 2007) and torsion viscometry in Paterson-type devices (Paterson and Olgaard 2000; Rutter et al. 2006; Caricchi et al. 2007). Schematics of the devices and geometries are shown in Figure 4.1.

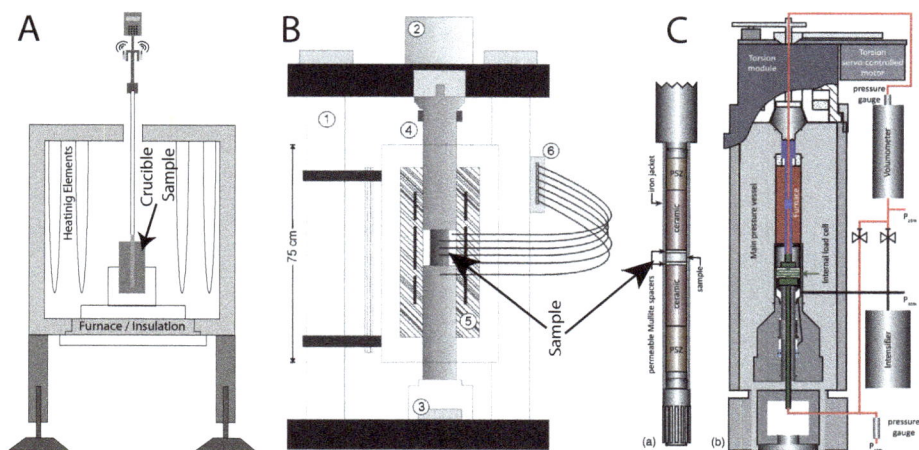

Figure 4.1 Schematics of the most common experimental devices for rheometry of magmatic suspensions. **A)** Concentric cylinder device for rotational viscometry; re-drawn from Kolzenburg et al. (2016); see also Dingwell (1986); **B)** Parallel plate device for uniaxial compression viscometry; reproduced from Hess et al. (2007), with the permission of AIP publishing; **C)** Paterson apparatus (Paterson and Olgaard 2000) for torsion viscometry; reproduced from Kushnir et al. (2017), with permission of Elsevier.

Limitations in the experimental parameters accessible by the respective devices (i.e., temperature, stress, strain-rate and oxygen fugacity) broadly divide the available data into two categories:

1. Low viscosity suspensions (commonly below $\sim10^4$ Pa s) measured via concentric cylinder viscometry and at low degrees of undercooling (i.e., near the liquidus). The bulk compositions used in this approach usually span silica contents between ~35 to 55 wt. % SiO_2, i.e., Basalts and Foidites to Andesites (Fig. 4.2). This is because of their relatively low viscosity and comparatively fast crystallization kinetics, which allow steady state conditions in texture and suspension viscosity to be reached on the timescales of hours to days and permit quenching of the samples for textural characterization. The fast kinetics also allow for experimentation at near natural conditions (i.e., in thermal, textural and chemical disequilibrium).

2. High viscosity suspensions (commonly above $\sim10^6$ Pa s) measured via parallel plate viscometry in uniaxial compression or via torsion viscometry. The bulk

compositions used in this approach are usually > 60 wt. % SiO_2 (i.e., Andesites to Rhyolites) with only a few examples of measurements at lower silica contents (Fig. 4.2). These experiments are typically performed just above the melt's glass transition temperature (i.e., in the supercooled liquid state). This is because these high viscosity systems have comparatively slow crystallization and vesiculation kinetics, which allows negligible textural and or chemical change of the sample over the course of the experiment and to maintain a steady state suspension viscosity over the measurement timescales (minutes to hours).

Other rheometric methods such as fibre elongation (Li and Uhlmann 1970; Webb and Dingwell 1990a), falling body viscometry (Mackenzie 1956; Kushiro et al. 1976) or three point bending (McBirney and Murase 1984), that have proven useful to investigate for example the influence of pressure on melt viscosity or non-Newtonian effects in unrelaxed, high viscosity melts have largely been abandoned for suspension rheology. The same is true for centrifuge experiments that were initially attempted for two and three phase rheometry by Roeder and Dixon (1977). Micro-penetration, that simulates the geometry of a falling sphere experiment (Hess and Dingwell 1996) has proven inadequate for determination of suspension rheology in the laboratory since it probes a comparatively small volume of melt adjacent to the indenter rather than the bulk of the sample. However, the penetration method is well suited when scaled for field measurements, where the penetrating rod is much larger than the average crystal (see "*Field Rheology*" section later in this chapter). Since both method and apparatus place tight constraints on measurement conditions, we review the advances in magma suspension rheometry grouped by the three most common experimental methods 1) concentric cylinder, 2) parallel plate and 3) torsion.

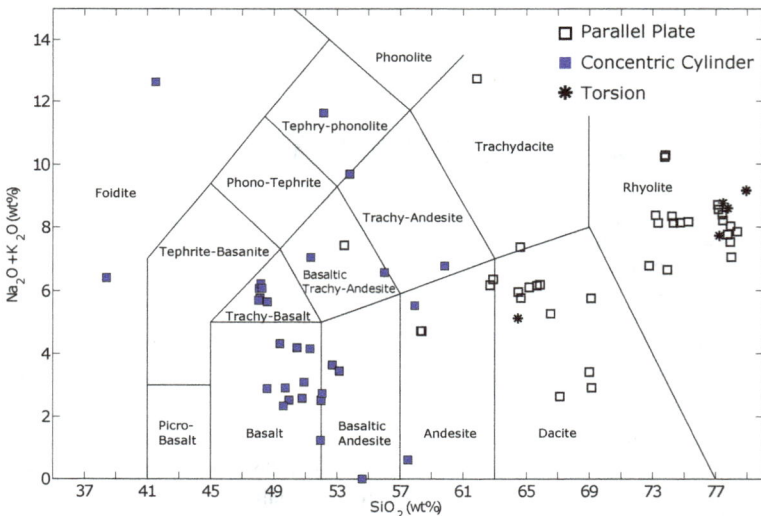

Figure 4.2. Total alkali versus silica diagram showing the compositions of published multiphase rheology measurements on lavas. Measurements of low viscosity suspensions (Foidite to Basaltic Andesite) are performed exclusively via concentric cylinder methods, whereas high viscosity magmas (Andesite to Rhyolite) are almost exclusively performed via parallel plate and torsional viscometry.

Concentric cylinder experiments

Concentric cylinder viscometry has long been a go-to technique for measuring the viscosity of silicate melts at super-liquidus temperatures (Shaw 1969; Cukierman et al. 1972;

Dingwell and Virgo 1987; Spera et al. 1988). It is commonly employed in combination with viscosity measurements near the glass transition temperature, e.g., via micro-penetration viscometry (Hess and Dingwell 1996) or estimation of the melt viscosity by application of shift factors to calorimetric data (Stevenson et al. 1995; Gottsmann et al. 2002; Giordano et al. 2005, 2008b) to interpolate the theoretical temperature-dependent viscosity of the pure melt across the crystallization interval (Tammann and Hesse 1926). This theoretical curve of the temperature–viscosity relationship of the crystal free melt is required to quantify the contribution of crystals and/or bubbles on the rheology of the bulk sample. The temperature range covered by concentric cylinder viscometry spans from 800 to 1700 °C. Theoretically, concentric cylinder viscometry allows imposing a vast range of strain rates but, in practice, the range of accessible strain rates is largely limited by the rheometer used for measurement and the method of sample containment. In current experimental geometries, the strain-rate limit is defined by either the torque limit of the rheometer or by the strength of the coupling between sample container and its holder, as the sample container may start slipping at high torque. Strain rates imposed in published data range from ~0.005 s^{-1} to ~9 s^{-1}. (Dingwell and Virgo 1988; Stein and Spera 1998; Vona and Romano 2013; Kolzenburg et al. 2018b, 2020)

In most experiments, temperature is not measured directly in the suspension because insertion of a thermocouple would disturb flow within the sample and thus affect the rheological measurement. Instead, the furnace control temperature is calibrated to the measured temperature of a non-crystalizing melt that is measured directly via insertion of a thermocouple (commonly platinum–rhodium alloys; type S or B). Concentric cylinder measurements are usually performed at atmospheric pressure and in air but experimentation under controlled atmospheres (e.g., more reducing conditions) is possible (Dingwell and Virgo 1987; Chevrel et al. 2014; Kolzenburg et al. 2018c, 2020). The importance of measurements at varying oxygen fugacity is increasingly recognized due its effect on melt viscosity (for example by reduction of Fe^{3+} to Fe^{2+}) and on the onset of crystallization as well as phase equilibria. To date, no apparatus exists that allows for concentric cylinder viscometry under pressure, a key component affecting crystal phase assembly and bubble nucleation and growth dynamics in magmatic systems but there are active efforts to expand concentric cylinder measurement capacity in that direction (see section *"Technological Advances"* later in this chapter).

For concentric cylinder experiments, the sample is melted in a cylindrical container and housed in a box or tube furnace. A cylindrical spindle is then inserted into the melt, generating the concentric cylinder measurement geometry (Fig. 4.1A). All parts in contact with the melt are commonly made of Pt–Rh alloys to withstand the high experimental temperatures and to avoid reaction between the melt and the container or spindle. More cost-effective materials, such as alumina ceramics, Pt–Rh alloys sheathed alumina ceramics or graphite, that allow for extraction of the entire sample without disturbing sample texture have been used with limited success. Alumina ceramics have proven inadequate for experiments involving low viscosity melts and high temperatures since they are soluble in the melt and induce contamination. Such contamination changes the melt composition and, with that, its viscosity, phase relations and crystallization kinetics. Nonetheless, ceramics are promising candidates for experimentation with high viscosity melts and at relatively low temperatures, which would result in negligible contamination during experimentation for durations of several days due to the much lower diffusivities in high viscosity systems and at low temperature. Graphite containers and spindles require low oxygen fugacity to avoid combustion during the experiment, which would in turn induce changes in the measurement geometry or, at worst, result in crucible failure and leakage into the furnace.

Concentric cylinder measurements quantify the torque exerted by the magmatic liquid or suspension on the rotating spindle immersed into the sample. This torque is proportional to the apparent viscosity of the sample at the imposed experimental conditions.

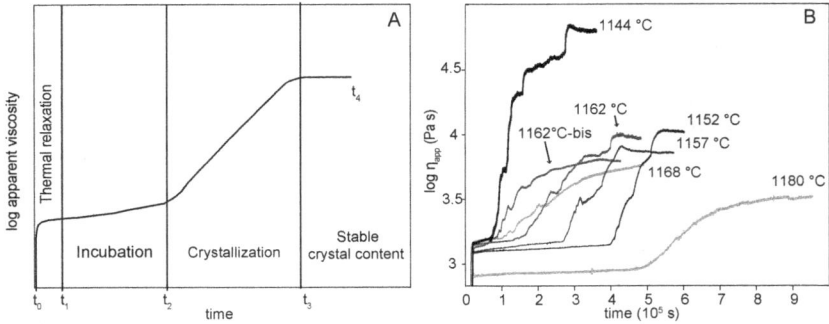

Figure 4.3 A) Schematic viscosity-time path of crystallization experiments, after Vona and Romano (2013) and Chevrel et al. (2015). As a general trend, four regions can be recognized with increasing time (t): 1) t_0–t_1: viscosity increase due to thermal equilibration of the melt to the experimental temperature; 2) t_1–t_2: "incubation time" time-invariant viscosity of the metastable liquid before crystallization. 3) t_2–t_3: viscosity increase due to crystallization; 4) t_3–t_4: time-invariant viscosity of the crystal–melt suspension with a constant crystal content (thermodynamic and textural equilibrium). **B)** Sample Dataset of the time and temperature dependence of the evolution of apparent viscosity during experiments at constant sub-liquidus temperature and constant shear rate; from Chevrel et al. (2015).

Experiments can be performed at either constant strain rate (i.e., torque is measured while the spindle is rotated at a constant rate) or constant stress (i.e., rotation rate is measured while the spindle is rotated at a constant torque). Standard concentric cylinder viscometry aims to achieve a linear velocity profile across the liquid (i.e., between inner and outer cylinder). However, linear velocity profiles are only achieved when the gap between the two cylinders is narrow, a geometry which does not lend itself to the study of particle suspensions. Therefore, most experiments on natural silicate melts are performed in wide gap geometries, where the flow velocity field is non-linear. This method can be employed to study silicate melt suspensions since it allows sufficient space to accommodate crystals and/or bubbles in the gap between cylinder and spindle while returning accurate viscosity measurements. Further, the concentric cylinder geometry allows for continuous viscosity measurement as a function of varying experimental conditions or textural changes in the sample, and infinite strain. This makes it more flexible than other methods that have limitations in total achievable strain (e.g., parallel plate viscometry). Following Couette theory, the shear stress (σ) and shear strain-rate ($\dot{\varepsilon}$) measured at the inner cylinder are described by:

$$\dot{\varepsilon} = \frac{2\Omega}{n\left(1-\left(\dfrac{R_i}{R_o}\right)^{\frac{2}{n}}\right)} \tag{4.1}$$

$$\sigma = \frac{M}{2\pi r R_i^2 l} \tag{4.2}$$

where Ω is the angular velocity, M the measured torque, R_i and R_o the radii of the inner and outer cylinder, respectively, l is the effective immersed length of the spindle and n is the flow index. The value of the flow index (n) is determined as the slope of M vs. Ω in log-log space. From Equations (4.1) and (4.2) the sample's flow curve (σ vs. $\dot{\varepsilon}$; Fig. 2.1) can be constructed to describe the rheological behavior of the measured suspension. As apparent from Equation (4.1), the wide gap setup has the disadvantage that the shear strain-rate depends on both the measurement geometry and the flow index value (n).

Alternatively, the torque-viscosity relationship can be calibrated against materials of known viscosity over the desired range of torque and rotation rate, as routinely done for viscometry of pure silicate melts. The most frequently used standards for calibration are glasses from either the National Institute of Standards and Technology (NIST; formerly National Bureau of Standards, NBS) or from the Deutsche Glas Gesellschaft (DGG) for which the temperature viscosity relationships are accurately known. Using linear fits to the calibration data, a torque measurement at a certain rotation rate corresponds to the apparent viscosity of an unknown suspension.

Tying sample texture to the rheological response of a suspension fundamentally relies on accurate quantification of textural parameters including crystal or bubble content, shape (aspect ratio) and size-distribution, and particle surface roughness (see also the "*Parameterization Strategies*"section later in this chapter). Recovering these data in combination with rheological measurements lies at the heart of the motivation for most available measurements of concentric cylinder suspension rheometry.

The vast majority of studies address the effect of crystals on suspension rheology and only few concentric cylinder experiments measure bubble bearing melts. This is because concentric cylinder measurements are commonly performed near the melt's liquidus temperature (i.e., at low viscosities), where the large density differential between bubbles and melt allow for bubble percolation (i.e., buoyant separation of the exsolved gas phase) over the course of the experiment and, thus, thermomechanical equilibrium cannot be achieved for most magma compositions. The much lower density differential between common crystals and silicate melts allows maintaining textural homogeneity (i.e., constant crystal contents and shapes) over timescales sufficiently long for experimentation (hours to days). Hence, concentric cylinder viscometry is an ideal method to study magmatic low viscosity melt–particle suspensions. In the following, we review measurements of melt + crystals suspensions under equilibrium and disequilibrium conditions and their use for mapping of the full rheological behavior of crystallising lava. We also include a review of the few measurements of melt + bubble suspensions employing this experimental approach.

Crystal-bearing suspensions at equilibrium conditions. For experimentation at equilibrium conditions, the sample is first molten at super liquidus temperatures to determine the viscosity of the crystal free melt. Subsequently, the sample is cooled to sub-liquidus temperatures at which crystallization is expected to occur. The environmental parameters (temperature, oxygen fugacity and shear-rate) are then maintained constant and the rheological response of the sample is monitored as a fuction of time. Some of the first experiments of this type were inspired by and combined with field measurements, which are reviewed later in this chapter. These were performed on lavas from Hawaii (Shaw 1969) and Etna (Gauthier et al. 1973). Both studies note profound sub-liquidus deviations from the liquid viscosity trend as well as from Newtonian behavior at deformation rates relevant to lava flows. Subsequent studies, such as Ryerson et al. (1988) and Pinkerton and Stevenson (1992), began to focus on systematic laboratory experimentation, and the combination with field measurements was discontinued after the work of Pinkerton and Norton (1995); see the "*Field Rheology*" section later in this chapter. While the aforementioned studies attempted connecting measurements of crystal contents to the rheological data using complementary petrological experiments, none of them present systematic textural analyses of the experimental samples themselves. This combination was introduced by Sato (2005) who highlighted the need for more detailed experimentation by documenting a profound discrepancy between the rheology of elongate plagioclase bearing suspensions with respect to suspensions of spherical particles (Marsh 1981) (see section "*Parameterization strategies*" later in this chapter for more details). Efforts to recover textural data and crystallization kinetics of the experimental samples were developed only later (Ishibashi and Sato 2007, 2010; Ishibashi 2009). These studies provided unprecedented detail on the non-Newtonian effects of magmatic suspensions including thixotropy, shear thinning and apparent yield stress. Some other studies (Sonder et al. 2006; Hobiger et al. 2011) present systematic mapping of shear rate effects on magmatic suspensions at various temperatures but omit textural data, thus impeding parameterization.

Continuous temporal monitoring of the viscosity evolution from super liquidus to steady state sub-liquidus conditions, which allows commenting on the process kinetics, combined with systematic mapping of shear rate effects and analysis of textural features of the sample were first presented about a decade ago (Vona et al. 2011) but are starting to become more numerous in recent years (Vetere et al. 2013, 2017, 2019; Vona and Romano 2013; Sehlke et al. 2014; Chevrel et al. 2015; Sehlke and Whittington 2015; Campagnola et al. 2016; Soldati et al. 2016, 2017; Liu et al. 2017; Morrison et al. 2020). Figure 4.3 shows the schematic temporal evolution of viscosity in this type of experiment. The data show a profound time dependence and follow a characteristic four stage evolution:

1.   During the initial temperature drop from super-liquidus temperatures to the experimental sub-liquidus temperature, the measured viscosity increases rapidly due to the temperature dependence of melt viscosity (thermal relaxation).

2.   A period of constant, or very slowly increasing, viscosity, commonly termed "incubation time", which has been shown to range from minutes to days. The duration is a function of both the degree of undercooling (i.e., temperature difference between sample temperature and liquidus temperature) and melt-viscosity. The detailed kinetics of this incubation period remain poorly understood and incubation times are poorly reproducible in experiment. Nonetheless, systematic changes are documented and, generally speaking, increasing undercooling decreases incubation time, whereas increasing viscosity increases incubation time.

3.   A relatively sharp increase in viscosity that is a result of the effect of crystal growth throughout the sample. It has been shown that crystallisation is often not homogeneous but propagates through the crucible from the walls of both the crucible and the spindle toward the sample center, a result of increased nucleation efficiency at both surfaces (Chevrel et al. 2015; Kolzenburg et al. 2018). This increase of viscosity gradually decelerates until reaching thermodynamic equilibrium (i.e., constant crystal content) and mechanical equilibrium (i.e., homogenous crystal alignment within the flow field).

4.   A plateau in the measured viscosity, which represents the apparent viscosity of the suspension after reaching thermodynamic and textural equilibrium (i.e., sample temperature, crystal content and torque measurement are constant). When reaching this plateau, it is common to vary the shear rate to investigate the non-Newtonian behavior of the suspension. At the end, the measurement is stopped, and the sample is quenched for textural analyses.

Attaining this equilibrium state (t_4 onward) prior to any further rheological investigation of the sample (such as variations in shear rate) or quenching of the sample for textural analyses is important because the sample temperature is not measured directly. Constant thermal conditions as well as cessation of crystal growth are assumed to be achieved once a steady torque reading is reached. The time required to reach this steady state depends strongly on the sample composition and the melt viscosity since both affect the crystallization kinetics. In this type of experiment, the equilibrium state can be reached within few hours for low viscosity samples (Vona et al. 2011), while it may take several days for higher viscosity samples (Chevrel et al. 2015). Detailed studies on this incubation time are few and mostly provide measurements of nucleation delay in the absence of deformation (Swanson 1977; Couch et al. 2003; Hammer 2004). These studies also provide quantitative descriptions of nucleation and growth rates and note that the incubation time varies systematically as a function of proximity to the melt liquidus (i.e., degree of undercooling), where a lower degree of undercooling results in longer incubation times. Systematic understanding of the nucleation delay requires extensive experimental efforts coupled with thermodynamic modeling. Recently, Rusiecka et al. (2020) have performed such work and developed a model predicting the nucleation delay of olivine, plagioclase, and clinopyroxene in basaltic melts, and alkali feldspar and quartz in felsic melts that is based on classical nucleation theory and benchmarked with experimental data from experiments performed in the absence of deformation in a piston cylinder apparatus.

Since the sample reaches a quantifiable textural equilibrium in these experiments, the data allow establishing the relationships between texture and rheology as it is commonly done in experiments on analogue materials (see "*Analogue Materials*" and "*Parameterization Strategies*" sections later in this chapter). Systematic experimentation has, however, shown that the sample texture attained in the steady state is heavily dependent on the thermal- and deformation-history of the sample in the stages before reaching thermomechanical equilibrium (Kouchi et al. 1986; Vona and Romano 2013; Kolzenburg et al. 2018b). As such, the recovered experimental data are pertinent to making quantitative ties between texture and rheology but data from the rheological evolution of the sample on the path to the equilibrium state, are tightly restricted to the specific experimental conditions. Further, strain accommodation in the sample may vary from being homogenously distributed to being heavily localized. Therefore, a complete analysis requires the textures to remain undisturbed during quench. In an effort to do so, Chevrel et al. (2015) employed a new kind of spindle, where an alumina ceramic is wrapped in thin Pt-foil, allowing to maintain the entire sample undisturbed during quench without sacrificing much precious metal lab ware. Their results show strong textural organization and crystal alignment during the crystallisation stage (*t*3).

All available concentric cylinder measurements at equilibrium conditions are performed on relatively low silica content and low viscosity melts (Fig. 4.2), where the fast crystallization kinetics allow experimentation over manageable timeframes. Further, they are dominantly measured in air and at atmospheric pressures. Therefore, most data are acquired and interpreted with respect to the emplacement of lava flows. With decreasing temperature (i.e., increasing degree of undercooling) all studies observe increasing effective viscosity as a result of increasing crystal fraction. The onset of non-Newtonian, shear thinning, behavior is documented at crystal contents above ~5 vol% and becomes more pronounced as undercooling, and therewith crystal content, increases. While some studies comment on the potential existence of thixotropy (Sato 2005; Ishibashi and Sato 2007, 2010; Campagnola et al. 2016) in the measured magmatic suspensions, conclusive evidence is not available to date. This is likely because strain-dependent changes in the textural configuration of the experimental sample impede reproducing previous textural states that would be required to ascertain thixotropic behavior (Barnes 1997).

Overall, the available data highlight that detailed knowledge of magma and lava undercooling, as well as strain and strain-rate dependent effects on the crystallization kinetics, is required to assess the resulting changes in rheology and, therewith, flow velocity of magmas and lavas. This outlines the necessity of incorporating the complex feedback mechanisms between flow environment (i.e., slope for lavas and pressure differential for magma plumbing), flow velocity and lava rheology into transport models of magmatic suspensions.

Crystal-bearing suspensions at disequilibrium conditions. The experimental efforts reviewed above are dedicated to understanding the multiphase rheology of lava at constant environmental and textural conditions. This is because this type of experiment provides the data required to derive empirical rheologic laws from the experimental data (see section "*Parameterization Strategies*" for details later in this chapter). However, the conditions of subsurface magma migration and flow of lava on the surface of Earth and other Planets are inherently dynamic and induce disequilibrium. Measured and modelled cooling rates of basaltic lavas during flow and ascent range from 0.01 to 20 °C/min (Huppert et al. 1984; Flynn and Mouginis-Mark 1992; Hon et al. 1994; Cashman et al. 1999; Witter and Harris 2007; La Spina et al. 2015, 2016; Kolzenburg et al. 2017). These values are mostly representative of conduit wall contacts and lava flow crusts and can, therefore, be taken as maximum values that are expected to be lower in the interior of well insulated systems. Decompression during ascent is another factor promoting crystal growth and influencing its kinetics (Hammer and Rutherford 2002; Blundy et al. 2006; Arzilli and Carroll 2013). In addition, shear-rates during viscous transport in volcanic plumbing systems can range from ~70 s^{-1} in Plinian eruptions (Papale 1999) to as low as 10^{-9} s^{-1} in slowly convecting magma chambers (Nicolas and Ildefonse 1996) and effusion-rates for basaltic eruptions range between 1–1000 m^{3}s^{-1} (Harris and Rowland 2009; Coppola et al. 2017). For common lava flow geometries (thickness between 2 and 10 m; widths between 200–1000 m) this translates to shear-rates between ~0.001–2.5 s^{-1}. Further, magmas are generated at low oxygen fugacities (fO_2) and then transported and erupted on Earth's surface, moving towards increasingly oxidizing environments. The effect of oxygen fugacity on the viscosity and structure of silicate melts relevant to natural compositions has been investigated for a range of compositions (Hamilton et al. 1964; Mysen and Virgo 1978; Mysen et al. 1984; Dingwell and Virgo 1987; Dingwell 1991; Herd 2003; Liebske et al. 2003; Sato 2005; Vetere et al. 2008; Chevrel et al. 2013a; Kolzenburg et al. 2018). Oxygen fugacity also strongly affects the stability of Fe-bearing phases, the onset of crystallization and degassing, as well as the melts crystallization-path and -kinetics and glass transition temperature (T_g) under both static (i.e., constant T and P) and dynamic (decreasing P and T) conditions (Hamilton et al. 1964; Sato 1978; Toplis and Carroll 1995; Bouhifd et al. 2004; Markl et al. 2010; Arzilli and Carroll 2013; La Spina et al. 2016; Kolzenburg et al. 2020). Therefore, evaluating the influence of the evolving environmental parameters on the transport and emplacement dynamics of magmatic suspensions requires systematic characterization of their rheological properties at non-isothermal and non-equilibrium conditions.

The importance of disequilibrium effects on crystal growth has been recognized for decades (Walker et al. 1976; Pinkerton and Sparks 1978; Coish and Taylor 1979; Gamble and Taylor 1980; Lofgren 1980; Long and Wood 1986; Hammer 2006; Arzilli and Carroll 2013; Vetere et al. 2015; Arzilli et al. 2018; Kolzenburg et al. 2020) and has inspired experimental studies investigating the cooling- and shear-rate dependence of the dynamic rheology of crystallizing silicate melts at conditions near those of natural emplacement scenarios (Kouchi et al. 1986; Ryerson et al. 1988; Giordano et al. 2007; Vona and Romano 2013; Kolzenburg et al. 2020). The number of published rheological studies at disequilibrium conditions is, however, low and systematic mapping of the effects of cooling rate (Giordano et al. 2007; Kolzenburg et al. 2016, 2017, 2019; Vetere et al. 2019), shear rate (Kolzenburg et al. 2018b) and oxygen fugacity (Kolzenburg et al. 2018c, 2020) as well as their interdependence is only beginning in recent years. This is largely because disequilibrium experimentation does, to date, not allow for textural characterization of the sample during experimentation, which would be necessary for standard rheological parameterization (see the "*Parameterization Strategies*" section later in this chapter). Diffusion and crystal growth are very rapid at the high degrees of undercooling reached in constant cooling disequilibrium experiments and it is therefore not possible to extract and quench the experimental charges sufficiently fast to investigate their textures. Quantification of textures during disequilibrium experiments would require data at high spatial and temporal resolution, such as *in situ* tomographic data, as it is starting to become available for analogue materials (Dobson et al. 2020). Further, variations in the thermal inertia of the experimental apparatus induce characteristic thermal paths in the investigated melts which require *in situ* thermal monitoring of the sample. This was hindered in concentric cylinder viscometry, as hard-wired data transmission compromises the highly sensitive torque measurements for accurate viscosity determination. In temperature-stepping experiments this can be overcome by calibrating the sample against the furnace temperature (Dingwell 1986). At disequilibrium, however, the release of latent heat of crystallization (Settle 1979; Lange et al. 1994; Blundy et al. 2006), redox foaming (i.e., liberation of oxygen gas bubbles during reduction of Fe_2O_3 to FeO; see also Dingwell and Virgo (1987)), heat advection and changing heat capacity may also influence the samples' thermal state. Further, temperature calibration without deformation, cannot assess viscous-heating effects potentially acting at high viscosities and/or high shear rates (Hess et al. 2008; Cordonnier et al. 2012a). A new experimental device that allows for *in situ* thermal characterization of rheological measurements was presented in Kolzenburg et al. (2016) and may serve to address the above questions in the future (see for example Kolzenburg et al. 2020).

The available disequilibrium data document that, at the conditions charted to date, cooling rate is the governing parameter determining the rheological evolution of magma and shear rate and oxygen fugacity play smaller roles. All presented experimental data describe a systematic sub-liquidus rheological evolution (Fig. 4.4). The measured apparent viscosity initially follows the pure liquid curve and once crystallization occurs, the apparent viscosity of the suspension (also sometimes called the effective viscosity; Petford, 2009) deviates from this curve towards higher viscosity values. Over the course of the experiment, the apparent viscosity of the magmatic suspension gradually increases with increasing undercooling until reaching a point where the apparent viscosity of the suspension rises steeply, terminating its capacity to flow at the "rheological cut off temperature" (T_{cutoff}); see Kolzenburg et al. (2017). This point is commonly assigned as the stopping condition for cooling limited lava flow behavior (Wilson and Head 1994; Harris and Rowland 2009) but systematic experimental determination of this temperature only started over the past few years. The available data show that increasing cooling rate delays the crystallisation onset and hence decreases the temperature at which the initial departure from the liquid curve occurs as well as reducing T_{cutoff}. Decreasing oxygen fugacity decreases the temperature of the crystallisation onset and hence decreases the temperature at which the suspension viscosity departs from liquid viscosity and the T_{cutoff} (Kolzenburg et al. 2018c, 2020). Conversely, increasing shear-rate promotes crystallization (Kouchi et al. 1986; Vona and Romano 2013; Kolzenburg et al. 2018b; Tripoli et al. 2019) and therefore acts to increase both the temperature of the departure from liquid viscosity and T_{cutoff}. This is the case for any investigated cooling rate and composition. Currently available data (see Fig. 4.4. as example) suggest that, while inducing important changes, shear-rate does not out-scale the effects of cooling-rate or oxygen fugacity (Kolzenburg et al. 2018b). While standard rheological parameterization of disequilibrium data is not possible due to the lack of quantitative textural data, T_{cutoff} measurements present a promising approach to develop an empirical understanding of stopping criteria that can be implemented in magma and lava transport models at low computational cost (Harris and Rowland 2001; Chevrel et al. 2018).

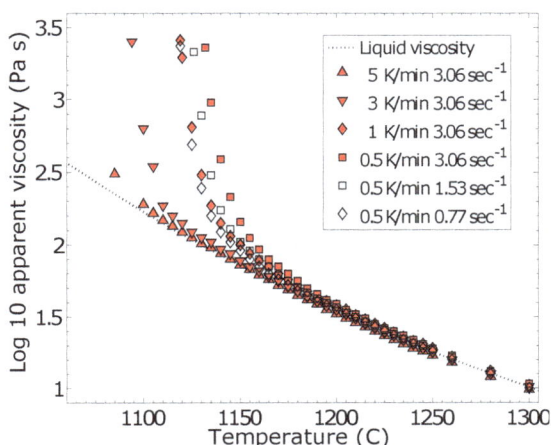

Figure 4.4. Evolution of apparent viscosity under constant cooling at various cooling and shear rates; data from Kolzenburg et al. (2017). All data describe a similar trend, where during constant cooling from super-liquidus to sub-liquidus temperatures the apparent viscosity initially follows the pure liquid curve and once crystallization occurs, the apparent viscosity deviates from this curve towards higher viscosity values.

Rheological mapping of crystal-bearing suspensions. To describe the full rheological evolution of a magma batch over the course of its journey to and on the surface requires rheological characterization over a wide range of sub-liquidus temperatures and shear-rate conditions. This can be done by combining data from equilibrium and disequilibrium experiments as well as from other techniques (e.g., parallel plate). An example of such a rheological map, representing the physical behavior of a given starting bulk rock composition from super- to sub-liquidus (crystal-bearing suspension) conditions, is given in Figure 4.5. The melt compositions of Vona et al. (2011) and Kolzenburg et al. (2018c), albeit stemming from different eruptions, are within analytical error for all major oxides and hence all variations in the rheology of the crystallizing basalt are controlled by variations in the volumetric fractions of crystals and bubbles rather than melt composition. The megacryst-bearing lava measured in Vona et al. (2017) is slightly more silicic ($\sim +4$ wt.% SiO_2) and hence somewhat more viscous. Comparison of the pure liquid viscosity of the re-melted bulk rock and the separated groundmass presented in Vona et al. (2017) document that, for basaltic melts, crystallization induced variations in melt composition result in small changes (<0.2 log units) in the liquid viscosity of the evolving suspension. Note that this is due to the compositional similarity of the melt and the crystallizing phases. This effect will be larger if the crystal growth induces significant changes in the melt composition, which is not the case in this example.

The non-isothermal viscosity data from Kolzenburg et al. (2018b) document that both cooling-rate and shear rate exert a modulating effect on the disequilibrium rheology of the Etna melt. However, the concentric cylinder measurements are mechanically restricted to relatively low viscosities. Measurements beyond the mechanical limit of concentric cylinder viscometry were presented in (Vona et al. 2017) who employed parallel plate viscometry to measure the viscosity of three phase magmatic suspensions (this method is reviewed in detail in the following section). Combined, the data from both methods form a continuing trend with respect to the concentric cylinder viscometry measurements (Kolzenburg et al. 2018b) (Fig. 4.5) but document lower lava

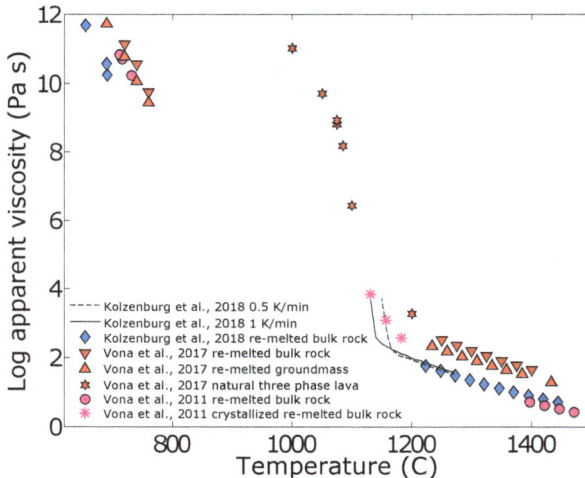

Figure 4.5. Summary plot of apparent viscosity vs. temperature for published melt and crystal-suspension viscosity data on re-melted lava samples from Mt Etna, Italy. Viscosity of the re-melted bulk rocks and groundmass (megacrysts removed) were performed at 1) super-liquidus temperatures (i.e., pure melt at >1200 °C) via concentric cylinder, 2) low temperature (<800 °C) via micro penetration and differential scanning calorimetry; 3) at sub liquidus conditions via concentric cylinder viscometry at constant temperature (**snowflakes**) or at varying cooling- and shear-rates (**continuous and dashed lines**); Viscosity of the natural rock (i.e., the lava including groundmass, bubble and crystal) was measured at subliquidus temperature (between 1200 and 1000°C) via parallel plate viscometry using unconfined uniaxial deformation (**stars**).

viscosities than extrapolation from the two-phase measurements would suggest. This is likely a result of the differences in sample texture, where all concentric cylinder data are restricted to bubble free two-phase suspensions of crystals and melt, whereas the parallel plate viscometry data present measurements on three phase (i.e., crystal and bubble bearing) suspensions. In summary, the rheological evolution of lava at sub-liquidus conditions can be reconstructed neatly by combining datasets from differing sources. This is also shown in (Pinkerton and Norton 1995; Sehlke et al. 2014; Kolzenburg et al. 2019), where laboratory and field estimates of lava rheology at emplacement conditions are compared. The respective data fall within a close range, highlighting the potential of data correlation from various sources and the need to expand the available experimental database in order to generate a holistic view of the dynamics of magma and lava transport. Nonetheless, to date no study has shown direct correlation of field and laboratory measurements that proves beyond reasonable doubt that a one to one comparison is possible. This is largely due to the respective technical limitations of both field and laboratory rheometry.

Additionally, increasing volatile contents push the onset of crystallization to lower temperatures and increasing pressure may change the crystallisation sequence e.g., (Hamilton et al. 1964; Gualda et al. 2012; Arzilli and Carroll 2013). To date, no experimental infrastructure exists to investigate either of these effects during rheological measurements and as a result, the effects of pressure and volatile content remain uncharted. Further, the available data on disequilibrium rheology do not account for the presence of bubbles, which is largely due to the unfavorable balance between machine constraints (low torque limits inhibit measurements at sufficiently high melt viscosities to retain bubbles) and melt crystallization kinetics (high viscosity melts crystalize slowly, requiring extremely slow cooling rates). The only apparatus capable of such measurements was presented by Stein and Spera (1998) but has since been put out of comission. A promising approach for direct measurements of three phase mixtures at natural disequilibrium conditions is field rheology, e.g., Chevrel et al. (2019a), as reviewed later in this chapter.

Data from concentric cylinder measurements at both thermal equilibrium and disequilibrium enable us to produce time–temperature–transformation (TTT) diagrams. These diagrams map the time and temperature dependence of the crystallization process and, thus, qualitatively reconstruct the process kinetics. This kind of analysis of rheological data was first presented in Chevrel et al. (2015) on the example of an andesitic melt from Tungurahua, Ecuador. Figure 4.6 shows a comparison of TTT diagrams of experiments at constant undercooling by Chevrel et al. (2015) and Vona et al. (2011) with those from experiments at disequilibrium presented in Kolzenburg et al. (2017) and Kolzenburg et al. (2019). Since the location of these transitions in time–temperature space are deduced from change points in the rheological data, they can only be tied qualitatively to classical, texture derived, time–temperature–transformation data from crystallization experiments as reviewed in for example Hammer (2008). A more quantitative connection between these two types of datasets is, however, possible, when the rheological data are combined with textural analyses and models for particle suspension rheology (Kolzenburg et al. 2020).

At thermal equilibrium (Fig. 4.6A), the onset of crystallization occurs slightly earlier with decreasing temperature, reflecting the increased supersaturation of the crystallizing phases. Note that all these experiments reach a constant crystal content but remain melt-particle suspensions (i.e., they do not solidify completely). Due to the higher melt viscosity, the andesitic sample of Chevrel et al. (2015) has slower phase dynamics than the basaltic composition measured in Vona et al. (2011) and thus, both onset and equilibrium state are shifted to longer times. The two datasets at disequilibrium conditions (Fig. 4.6B) describe broadly similar trends, where the onset of crystallization (open symbols) and the point of rheological solidification (black filled symbols) shift to higher temperatures and longer times with decreasing cooling rate. The difference between the two disequilibrium datasets can be explained by the variation in composition, where the lower viscosity lava from Piton de la

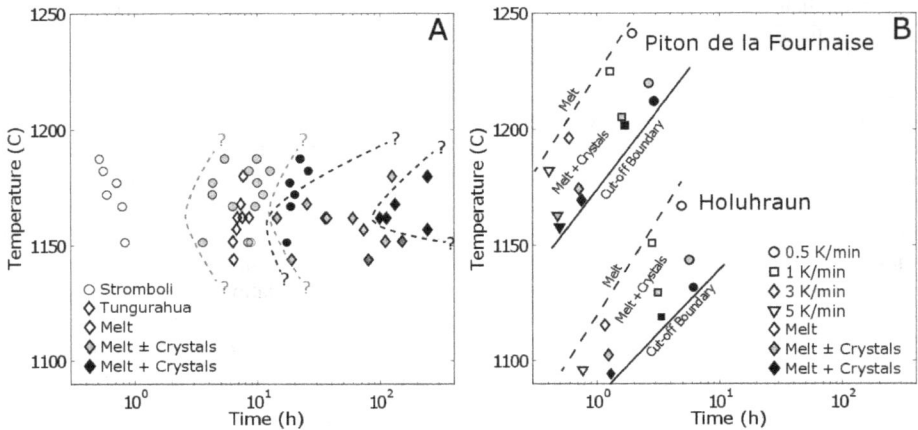

Figure 4.6. Time–Temperature–Transformation (TTT) diagrams reconstructed from viscosity measurements at atmospheric conditions (i.e., pressure and oxygen fugacity). The data plotted in **A** are reconstructed from the viscosity–time paths of thermal equilibrium measurements as shown in Figure 4.3 and were extracted from Vona et al. (2011) (**circles**) and Chevrel et al. (2015) (**diamonds**). **Open symbols** are the pure liquid phase (t_1 in Fig. 4.3); **gray symbols** describe the time of the crystallization onset (t_2 in Fig. 4.3); **black symbols** correspond to the thermodynamic equilibrium – constant crystal content (t_3 onward). **Dashed black and gray lines** delineate the time of the crystallization onset (low temperature) and thermodynamic equilibrium (high temperatures), also called the "crystallization nose", for the Vona et al. (2011) and Chevrel et al. (2015) datasets, respectively. The data plotted in **B** are reconstructed from the viscosity–time paths of measurements under constant cooling – thermal disequilibrium – (Kolzenburg et al. 2017, 2019). The individual data points for each cooling rate represent the times at which 1) the relative viscosity departs significantly from the liquid viscosity (**open symbols**; $\eta_r > 0.2$; highest temperature) 2) the relative viscosity starts to increase exponentially (**gray symbols**; $\eta_r > 2$; intermediate temperature) 3) the measurement limit at which the sample exceeds the torque limit of the rheometer and its relative viscosity tents towards infinity i.e., solidification (**black symbols**). Dotted and bold lines represent the crystallization onset and the rheological cut off, respectively.

Fournaise (La Réunion) has higher liquidus temperatures (hence the higher temperatures of the crystallization onset and solidification point) and faster crystallization kinetics (hence the shorter times required to reach both the crystallization onset and solidification point) than the Holuhraun lavas (Iceland). A prominent feature when comparing data from equilibrium and disequilibrium experiments is that at thermal equilibrium (Fig. 4.6A), the crystallization process shows a dominant time dependence (solidification occurs from left to right in the diagram), whereas at thermal disequilibrium (Fig. 4.6B), the crystallization process shows a dominant temperature dependence (solidification occurs from top to bottom in the diagram for any given cooling rate). This is because the available disequilibrium data are restricted to rather high cooling rates (> 0.5 K/min) and it is expected that at lower cooling rates (approaching equilibrium conditions), the dependence of temperature would decrease while the time dependence would increase thus merging the two experimental datasets. The measurements on the lava from Piton de la Fournaise suggest an increase in the time dependence of the process between the highest and lowest cooling rate but, to date, no coherent dataset exists that investigates this transition for a single composition and a broad range of cooling rates.

Bubble bearing suspensions

Concentric cylinder rheometry data on bubble bearing suspensions are scarce. This is largely due to the thermal and mechanical constraints of the method. Retaining bubbles in a melt at high temperatures for the duration of rheology experiments (several hours to days) requires relatively high melt viscosities ($>10^5$ Pa s) to minimize bubble percolation (i.e., volatile exsolution and loss upwards through the melt). This, in turn, requires a device design that permits high torques. Only few datasets on concentric cylinder measurements of bubble bearing suspensions are published (Stein and Spera 1992, 2012; Bagdassarov and Dingwell 1993a) and few studies describe devices that could access such experimental conditions in the laboratory (Stein and Spera 1998; Morgavi et al. 2015; Kolzenburg et al. 2016). Alternatively, it was suggested that field rheometry could be used for measuring the rheology of bubble-bearing low viscosity suspension (Chevrel et al. 2018), but to date no systematic study has been published.

Parallel plate experiments

Parallel plate suspension viscometry is largely dedicated to understanding the multiphase rheology of lava and magma at constant environmental and textural conditions. The compositions used in studies employing this approach vary from Andesites to Rhyolites with few examples of Basalts (Fig. 4.2). This method allows for simple experimentation on multiphase suspensions of synthetic or natural samples, since the samples can be readily recovered by core drilling from any glass bearing rock sample. This permits detailed characterization of sample texture, an important factor for parameterization of the rheological data (see the "*Parameterization Strategies*" section later in this chapter). For accurate measurement it is crucial to ensure parallel faces of the sample surfaces in contact with the compacting pistons since uneven faces result in anomalously high strain-rates due to the reduced initial contact area and thus underestimation of the samples viscosity. The cylindrical cores are loaded in a uniaxial press (Fig. 4.1B) and heated to the desired experimental temperature. Sample temperature can be monitored by insertion of thermocouples or by calibration to reference samples. Experiments are performed near the melt's glass transition (i.e., in the supercooled liquid state), where the viscosity is sufficiently high to neglect crystallization and vesiculation timescales relative to the duration of the experiment. This minimises textural and chemical change of the sample over the course of the experiment and maintains steady state suspension viscosity over the measurement timescales (minutes to hours).

Parallel plate viscometry uses cylindrical samples that are deformed in uniaxial compression either at constant load or at constant strain-rate. Constant strain-rate experiments record stress-time relationships sensed via a load cell. Constant load experiments record strain-time relationships measured by dilatometry. Although the experiments involve relatively small total strains it is still possible to run stress or strain-rate stepping experiments (Fig. 4.7A). The simplest approach to calculate the apparent viscosity of the sample in uniaxial compression parallel plate experiments is to derive it directly from the ratio of the recorded stress (σ, i.e., load over surface area in contact with the piston) and engineering-strain-rate ($\dot{\varepsilon}_e$; i.e., variation of the sample length ($\Delta l/l$) per unit time) (Quane and Russell 2005; Avard and Whittington 2012; Heap et al. 2014; Ryan et al. 2019a):

$$\eta_{app} = \frac{\sigma}{\dot{\varepsilon}_e} \qquad (4.3)$$

However, at the imposed strains (up to ~30–40 %), the sample geometry changes, and the sample strain cannot simply be described as $\Delta l/l$. Thus, Recovering the samples shear viscosity from parallel-plate viscometry requires that the geometrical change is accounted for. This approach rests on the mathematical description of the flow process by two differential equations, the Navier–Stokes equation and the continuity equation, following the Wallace plastimeter methodology (Rowlatt 1956). For isothermal conditions and with some simplifying assumptions, an analytical solution to these was obtained by Gent (1960). Based on experiments on coal tar and cross correlation with concentric cylinder viscometry, Gent (1960) validated the proposed, geometry independent (isovolumetric), theoretical formulation to recover shear viscosity from parallel plate experiments. Following this approach, calculation of the samples' shear viscosity (η_s) from the experimental data rests on one of two geometrical assumptions for the deformation mechanism.

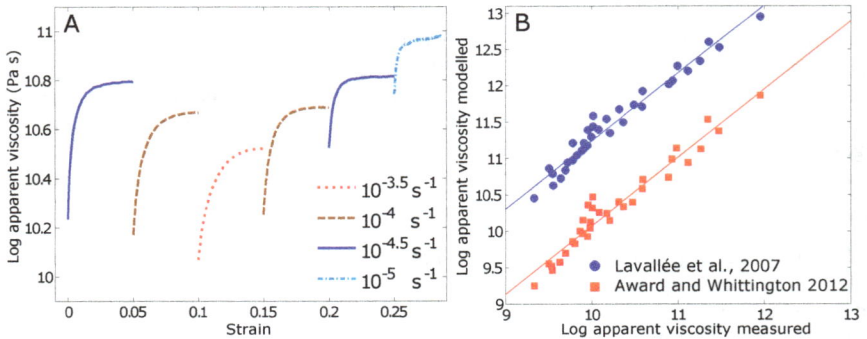

Figure 4.7. (A) Plot of apparent viscosity vs. strain for a strain rate stepping experiment on Chaos Crags (USA) dome lava (unpublished data by Kolzenburg et al.). For a given strain rate, the recorded load (and thus also the measured viscosity) initially increases rapidly until reaching a constant value equivalent to the relaxed rate of continuous sample deformation. The data describe clear shear thinning behavior (i.e., increasing strain-rate decreases apparent viscosity). **(B)** Comparison between the non-Newtonian models of Lavallée et al. (2007) and Avard and Whittington (2012) derived from uniaxial parallel plate experiments (plot modified after Avard and Whittington 2012 Fig, 12a). Note that this figure compares the model of Lavallée et al. (2007) to the one presented in Avard and Whittington (2012). As such, this is not an assessment of overall quality of one model over the other but a demonstration of the fact that that neither of these non-Newtonian models can comprehensively describe the rheology of dome lavas but each one is applicable only within the framework it was developed in. This highlights the need for further and systematic rheometry in order to derive a holistic rheological model.

1. The "no slip" condition, where the contact surface area between the sample and the parallel plates remains constant and the cylinder accommodates all deformation by bulging/barrelling. In this case the shear viscosity is calculated as follows:

$$\eta_s = \frac{2\pi F h^5}{3V \dfrac{\Delta h}{\Delta t}\left(2\pi h^3 + V\right)} \tag{4.4}$$

    where F the applied force, h the sample height, V the sample volume and t is time. Δt and ΔV are time and volume variation, respectively

2. The "perfect slip" condition, where the contact surface area between the sample and the plates increases with deformation while the cylinder does not bulge (i.e., retains vertical sides). In this case the shear viscosity is calculated as follows:

$$\eta_s = \frac{h^2 F}{3V \dfrac{\Delta h}{\Delta t}} \tag{4.5}$$

Both equations are valid only for experimental geometries where the sample diameter is smaller than the diameter of the parallel plates (i.e., no sample extrusion) and the sample is assmed to be both incompressible and stay cylindrically symmetrical (i.e., not sagging under its own weight).

Further, at high porosity, vesicular melts may behave in pure uniaxial compression, meaning that neither bulging nor sample translation occurs. It is important to note here, that the deformation of bubbles contributes a significant volume viscosity to the macroscopic flow of vesicular materials and due to the compressibility of pore fluids or gases, the sample volume does, in most cases, not remain constant during experimentation. In this case, the measured viscosity is equivalent to the samples' bulk (or longitudinal) viscosity (Eqn. 2.6) and therefore additional determination of shear or volume viscosity is required to completely resolve the samples' rheology.

The main limitation of parallel plate viscometry for application to natural magmatic suspensions has long been the relatively small sample size. Most common parallel plate viscometers are restricted to cylindrical sample diameters below 10-15 mm, largely because of the balance between the high stresses required for deformation of high viscosity samples near the glass transition and the limitations in compressive strength of the materials used as pistons driving the parallel plates (ceramics, silica glass or high temperature metal alloys). For a sample to be representative of the whole rock texture, however, its diameter sample should be ~8–10 times that of the largest crystal or bubble present (Ulusay and Hudson 2006). Seeing as bubble and crystal sizes in igneous rocks commonly range from hundreds of microns up to a few centimetres, rheological studies were initially limited to microlite bearing melts with crystal sizes below ~350 micron (Lejeune and Richet 1995; Stevenson et al. 1996) and bubble sizes below 1 mm (Bagdassarov and Dingwell 1992). Although the available experimental data are few, only a limited number of studies were presented following the initial qualitative descriptions in the aforementioned contributions and there is significant potential for expansion and improvement.

The first magmatic suspensions studied with parallel plate viscometry concerned vesicular rhyolite and were presented in Bagdassarov and Dingwell (1992). However, as outlined above, the equations provided by Gent (1960) are valid only for isovolumetric systems. On first glance this makes them inapplicable to bubble bearing systems, since bubbles contain compressible gas. Several approaches have been presented to account for this issue but all of them remain to be validated by cross correlation with other rheometric methods.

Bagdassarov and Dingwell (1992) modify the perfect slip equation provided in Gent (1960) by expanding sample volume to:

$$V = h \times S_{eff} \tag{4.6}$$

And then simplifying Equation (4.5) to:

$$\eta_s = \frac{hF}{3S_{eff} \frac{\Delta h}{\Delta t}} \tag{4.7}$$

Note that they introduce S_{eff}, the effective surface area of the sample, which is corrected for the sample's surface porosity in contact with the plates:

$$S_{eff} = S_{abs} \times (1 - \Phi_b) \tag{4.8}$$

where S_{abs} being the total sample surface area in contact with the plates and Φ_b the bubble volume fraction.

Vona et al. (2016) also adapted the equation for perfect slip conditions and expanded it to account for variations in sample volume over the course of the experiment. They achieve this by measuring sample volume before and after the viscosity measurement and then distributing the volume reduction linearly over the course of the experiment:

$$\eta_{app} = \frac{h^2 F}{3[V_0(1+kt)] \frac{\Delta h}{\Delta t}} \tag{4.9}$$

where V_0 is the initial sample volume and k is a coefficient tracking the porosity loss. It is defined as $k = 0.01(\phi_f - \phi_i) / t_f$, where ϕ_i is the initial porosity, ϕ_f is the final porosity, and t_f is the final time. This linear distribution of volume reduction, however, does not account for the compressibility of the gas phase in the bubble, which induces a non-linear transition from simple core shortening during the initial deformation phase to the perfect slip condition reached at the end of the experiment. As such this approach is only applicable for low strain experiments.

Bubbly silicate melts or foams are essentially impermeable unless they undergo bubble wall fragmentation (Taisne and Jaupart 2008; Takeuchi et al. 2008, 2009; Caricchi et al. 2011; Shields et al. 2014; Von Aulock et al. 2017; Ryan et al. 2019a,b). Hence they can act as gas springs due to the compressibility of the pore gas (Jellinek and Bercovici 2011). It would therefore be favorable to treat the sample as one volumetric unit and use the entire sample (melt + bubbles) surface area when converting measurement data to viscosity values (see Eqn. 4.10).

This approach could, in principle, be developed to also account for the non-linear compressibility of gases for the study of suspensions of pressurized bubbles. Ultimately, the simplest and most favorable way to recover shear viscosity from parallel plate experiments is a modified version of the method proposed by Bagdassarov and Dingwell (1992) that uses the absolute sample surface area (rather than a porosity correction) in the form of:

$$\eta_{\text{app}} = \frac{hF}{3S_{\text{abs}} \dfrac{\Delta h}{\Delta t}} \tag{4.10}$$

Due to the range of proposed data reduction approaches reviewed above, rigorous comparison of published date is not possible and hence systematic parameterization of bubble suspension and three phase suspension rheology has been impeded to date. In the following we present a review of published experimental work on magmatic suspensions, focusing on the employed apparatuses, textural variations and core findings of the respective studies.

The first experiments on bubble-free particle suspension obtained with parallel plate viscometry were performed with unimodal enstatite spherules suspended in an enstatite melt (Lejeune and Richet (1995). This study describes Newtonian behavior up to crystal volume fractions of $\Phi = 0.4$. Beyond the $\Phi = 0.4$ threshold Stevenson et al. (1996) document non-Newtonian shear thinning effects and the development of an apparent yield stress. At $\Phi > 0.7$ the deformation behavior is reported to evolve towards non-uniform distribution of crystals and melt with the onset of brittle processes (solid-like behavior). The authors state that the data support the simple Einstein–Roscoe model (see *"Parameterization Strategies"* section later in this chapter for details) for suspensions of spherical particles but also noted that crystal size distribution and shape may be of importance to fully describe textures relevant to natural samples. The latter point was demonstrated by Stevenson et al. (1996) who performed measurements on obsidian containing prismatic to needle shaped microlites and demonstrated the Einstein–Roscoe model may drastically underestimate the effective suspension viscosity once particle shapes deviate from spherical.

The sample size limitation was addressed by Quane et al. (2004) who introduced a new device capable of high temperature (up to 1100 °C) experimentation on samples of up to ~7 cm diameter and loads of up to ~1100 kg. This high-temperature, low-load apparatus was employed in a number of studies, dominantly aimed at quantifying the rheological behavior of welding and compacting volcaniclastic deposits, on both natural and analogue materials at atmospheric conditions (Quane and Russell 2005, 2006; Quane et al. 2009; Heap et al. 2014) and was later modified to allow for experimentation at elevated H_2O pore fluid pressures (Robert et al. 2008). Based on some of these data, Russell and Quane (2005) developed an empirical rheological model for welding and compacting deposits that has found wide application in numerical studies, e.g., Kolzenburg et al. (2019). The observed strain (ε_T) is ascribed to a combination of a time-dependent viscous compaction (ε_v) and a time-independent mechanical compaction (ε_m) described by:

$$\varepsilon_T = \frac{(1-\phi_0)}{\alpha} \ln\left\{ 1 + \frac{\alpha\sigma\Delta t}{\eta_{0x}(-1\phi_0)} \exp^{\frac{\alpha\phi_0}{1-\phi_0}} \right\} + \frac{\sigma}{E(1-\phi_0)} \tag{4.11}$$

where ϕ_0 is the original porosity, σ is the stress acting on the sample, and η_{0x} and E are the viscosity and Young's modulus of the deposit at zero porosity, respectively. In this model, the effective viscosity of the crystal-bearing melt without bubbles (i.e., η_{0x} where $\phi_0 = 0$) is determined as a function of temperature. The α value is used to predict the bulk viscosity of the mixture as a function of porosity, and the increase in relative viscosity with decreasing porosity (Quane et al. 2009). Published values of α range from ~0.7 to ~5.5 (Ducamp and Raj 1989; Quane and Russell 2005; Quane et al. 2009; Heap et al. 2014) and vary as a consequence of differences in sample microstructure (i.e., porosity, crystal content, pore size and shape, particle size and shape, and pore and particle size distribution, amongst others). See Wadsworth et al. (2022, this volume) for further details on sintering and welding.

A similar device to that introduced in Quane et al. (2004) was also employed by Vona et al. (2017) to study the multiphase rheology of megacryst-rich magmas. This study documents the lava's complex rheological response related to a non-homogenous deformation of the natural sample (e.g., viscous and/or brittle shear localization), favored by the presence of bubbles. The authors argue that the obtained flow parameters can be considered as representative of the bulk rheology of natural magmas, characterized by similar non-homogeneous deformation styles. The optimal scaling relation of sample size $\gg$ largest crystal or bubble could however not be reached by this study due to machine limitations.

To date there are only few parallel plate devices that are capable of performing high load high temperature experiments on large samples. The design for a high load, high temperature deformation apparatus was introduced in Hess et al. (2007) and allows measurements on geologically relevant sample dimensions (up to 100 mm in both length and diameter). This resulted in several seminal contributions advancing the understanding of non-Newtonian effects such as viscous heating in pure melts (Hess et al. 2008) as well as strain-rate dependent rheology and viscous limit of flow during dome building eruptions (Lavallée et al. 2007, 2008, 2012; Cordonnier et al. 2009, 2012b); see also Lavallée and Kendrick (2022, this volume) . The large sample size and high load capacity directed use of this device towards investigating magma failure and shear localisation (Lavallée et al. 2008, 2012, 2013; Coats et al. 2018); mapping the boundary between viscous and elastic deformation mechanisms and exploring suspensions of high crystal volume fractions $\Phi = 0.5$–0.6. The first published datasets suggested that highly crystalline suspensions of varying composition and texture can be described by a single non-Newtonian rheological law (Lavallée et al. 2007). Successive experiments, however, showed that their behavior is more complex, involving decreases in viscosity with increasing total strain and strain-rate, resulting from changes in sample texture (i.e., breaking of crystals and textural reorganization rather than true non-Newtonian flow behavior). Therefore, further systematic characterization is needed in order to derive holistic rheological laws for magmatic three phase suspensions at eruptive conditions (Cordonnier et al. 2009; Avard and Whittington 2012).

Some interesting non-isothermal parallel plate viscometry experiments were presented in Bouhifd et al. (2004) and Villeneuve et al. (2008), in which samples of glassy basalt were heated from below their glass transition temperature to 1250 °C and 1300 °C, respectively. This induced crystallization of the glassy basalt sample while passing from the supercooled liquid state through the crystallization window to super-liquidus temperatures. The results highlight the profound effect that crystallization has to increase the effective viscosity of basaltic melts and also indicate that decreasing oxygen fugacity may inhibit crystallization in the supercooled liquid field. Unfortunately, neither of the studies provided accurate determinations of the viscosity–crystal fraction relationships or textural characterization, rendering the data not suitable for parameterization and derivation of rheological flow-laws.

Dynamic experiments tracking rheological changes during vesiculation of synthetic three phase magmas were presented in Pistone et al. (2017). The experiments document that during foaming (i.e., nucleation and growth of gas-pressurised bubbles) and inflation, the rheological lubrication of the system is dictated by the initial crystallinity. At $\Phi < 0.6$, gas bubbles form and coalesce during expansion and viscous deformation, favoring strain localisation and gas permeability within shear bands, reducing the samples overall apparent viscosity. At $\Phi = 0.6$–0.7, gas exsolution generates pressurized pores that remain trapped in the solid crystal clusters and promote the formation of microscopic fractures within both melt and crystals, driving the system to brittle behavior. At higher crystallinity ($\Phi > 0.8$), vesiculation leads to large overpressures, triggering extensive brittle fragmentation. This novel approach to using parallel plate viscometry as a method to probe dynamic changes in flow behavior of three phase suspensions magma is very promising and presents a potentially very fruitful research path to be explored for natural magmas in the years to come.

Torsion experiments

Oscillatory torsion measurements. The first high temperature viscometry experiments on magmatic suspensions performed in torsion used a custom built forced sinusoidal torsion deformation device that induced very small oscillatory strain ($\varepsilon < 10^{-3}$ rad) in a cylindrical sample while monitoring the torque as a function of deflection angle (Bagdassarov and Dingwell 1993b; Bagdassarov et al. 1994). In oscillatory experiments, the total stress comprises two components: an elastic component in-phase with the strain and a viscous component in-phase with the strain rate. Together, these signals generate a sinusoidal signal that, in viscous and viscoelastic materials, is phase-shifted with respect to the applied stress. Measurements are performed over a range of oscillatory frequencies to determine the material's frequency dependent viscosity function. The resulting stress–strain wave forms are described in terms of complex variables having both real and imaginary parts. The complex shear modulus, G^*, is then defined as follows

$$G^* = \frac{\text{complex stress amplitude}}{\text{complex strain amplitude}} = \frac{\sigma_0}{\varepsilon_0}\cos\ell + \frac{\sigma_0}{\varepsilon_0}j\sin\ell \qquad (4.12)$$

where ε_0 and σ_0 are the stress- and strain-amplitude, respectively, ℓ is the phase lag (i.e., loss angle) and $j = \sqrt{-1}$.

The complex viscosity is then defined as:

$$\eta^* = \frac{\text{complex stress amplitude}}{\text{complex strain rate amplitude}} = \frac{\sigma_0}{j\varepsilon_0\varpi}e^{j\ell} = \frac{G^*}{j\varpi} \qquad (4.13)$$

with $\varpi = 2\pi f$ being the angular frequency and f the frequency.

Bagdassarov and Dingwell (1993b) and Bagdassarov et al. (1994) resolve the frequency dependent shear viscosity as a real and imaginary component in a frequency range of 0.005–10 Hz and temperature range of 600–900 °C. The authors describe that the recovered shear viscosities of magmatic suspensions at low frequencies and high temperatures compare well with data from parallel plate viscometry. They find the relaxed shear viscosity of vesicular rhyolites to decrease progressively with increasing bubble content. As documented previously by Stein and Spera (1992), magma viscosity can either increase or decrease with bubble content, depending on the rate and state (steady vs. unsteady) of strain during magmatic flow. Oscillatory torsion viscometry at high temperatures was largely abandoned after these studies with exception of one dataset on natural three phase suspensions from Etna, Hawaii and Vesuvius published in James et al. (2004). They document non-Newtonian, shear-thinning rheology in all samples for temperatures between 200 and 1150 °C but do not rigorously correlate the data to sample texture.

It is important to note that there is a fundamental difference between rheometry using oscillation experiments and all other experimental methods reviewed in this chapter. The strain induced in the sample is extremely small in most forced oscillation experiments and hence, the texture of the sample undergoes negligible change. Measurements under forced oscillation, thus resolve the rheological regime near zero strain. This allows measuring the viscous response of a sample without inducing strain dependent effects such as bubble or crystal deformation and/or re-orientation. Sumita and Manga (2008) present a study of the oscillatory rheology of analogue suspensions of spherical particles at volume fractions of $0.2 < \Phi < 0.6$. They present data for the response of the elastic and viscous moduli as a function of oscillation frequency and find a strain dependent rheological evolution where the suspension displays three rheological regimes. The response is initially linear viscoelastic, becomes shear-thinning with increasing strain and transitions to shear-thickening at high strain amplitudes. While these data are of great value to understanding for example the viscous dissipation of seismic energy in magma bodies, they are not directly applicable to processes where large strains are expected (e.g., magma migration and lava flow). Correlation of frequency dependent data to data from methods probing higher strain regimes remains challenging, largely due to the paucity of data since torsion experimentation following the above reviewed studies focused chiefly on deformation in simple shear at much higher total strains.

Simple shear torsion measurements. Simple shear torsion viscometry is predominantly performed in Paterson type high temperature, moderate pressure apparatuses that were originally designed for rock deformation experiments. The Paterson apparatus (Paterson and Olgaard 2000) allows for sample deformation in compression and/or torsion at confining pressures of up to 500 MPa (~ equivalent to the pressure at 15 km depth) and at temperatures of up to ~1325 °C. This makes them well suited to study the effect of moderate pressures and, with that, variations in volatile contents on the rheology of magmatic suspensions. Cylindrical samples of 8–15 mm diameter and between 4 and 15 mm length are loaded in custom built load-column assemblies made of alumina and zirconia pistons enclosed in a copper or iron jacket, which is housed in an internally heated gas pressure vessel; see Figure 4.1C. Determination of sample viscosity is based on torque and rotation rate data recorded during measurement, where both the effective stress and strain rate need to be calculated following the procedure presented in Paterson and Olgaard (2000) to account for the experimental geometry and non-linear distribution of both parameters within the sample.

One inherent issue in torsion experiments, is that the strain rate of a cylindrical sample is non uniform with the highest rate at the outside of the cylinder and zero strain-rate at the center (Paterson and Olgaard 2000). However, because of the constraints of the geometry, planes of material normal to the deformation axis will remain planar, and radial lines of material normal to the deformation axis will remain linear. Any element along a radius of the cylinder undergoes simple shear in a plane normal to the radius (r), the displacement being in the direction normal to the cylinder axis.

The shear strain ε_r at any radius r is thus given by:

$$\varepsilon_r = \frac{O}{h} \tag{4.14}$$

where h is the length of the sample and O being the angle of deflection and so the strain-rate at any given radius is given by:

$$\dot{\varepsilon}_r = \frac{r\dot{O}}{h} \times s^{-1} \tag{4.15}$$

Quantification of the experimental data commonly rests purely on the maximum shear strain ε, at the surface of the cylinder, which is given by:

$$\varepsilon = \frac{dO}{2h} \tag{4.16}$$

where d is the sample diameter and the corresponding maximum strain rate $\dot{\varepsilon}$ is given by:

$$\dot{\varepsilon} = \frac{d\dot{O}}{2h} = \frac{\pi d\omega}{h} \qquad (4.17)$$

where $\dot{O}$ and ω are the twist rate in radians per second and in revolutions per second, respectively. The torque (M) measured in torsion represents the integrated shear stress (σ_s) over the entire cylinder and is expressed as follows:

$$M = 2\pi \int_0^{\frac{d}{2}} \sigma_s r^{2} dr \qquad 4.18$$

Hence, for Newtonian liquids it related to viscosity as follows:

$$M = \frac{\pi d^3}{12} \eta \dot{\varepsilon} \qquad 4.19$$

whereas when assuming a Herschel–Bulkley flow law it is described as follows:

$$M = \frac{\pi \tau_y d^3}{12} + \frac{\pi d^3}{4(n+3)} K \dot{\varepsilon}^n \qquad 4.20$$

This highlights that, while strain-rate dependent non-Newtonian effects can be investigated qualitatively in a Paterson apparatus using the standard sample assembly geometry (Fig. 4.1C), rigorous strain-rate dependent rheological quantification is not possible for cylindrical samples since the strain rate varies drastically along the sample radius within the specimen. One way to overcome this issue is to use hollow cylinders as proposed by (Paterson and Olgaard 2000) who filled the hollow cylinders with a material of similar compressibility but lower internal friction. Sample strain is then confined to a narrow annulus for which it can be treated as near constant. The torque resulting from this inner plug can be calibrated and thus both shear stress and shear strain rate can be recovered, allowing for more accurate rheological measurements. To date this approach has not been employed for measurements concerning the multiphase rheology of magmatic suspensions due to a combination of the challenges of sample preparation as well as the limitations in the sample-to-crystal size ratio.

Caricchi et al. (2007), Arbaret et al. (2007) and Champelier et al. (2008) provided the first systematic viscosity measurements and textural analyses using this apparatus for experiments on synthetic suspensions. Caricchi et al. (2007) present measurements of equant quartz grains of $\Phi = 0.5$–0.8 suspended in a hydrous haplogranitic melt, whereas Arbaret et al. (2007) present measurements of $\Phi = 0$–0.76 of equant corundum grains suspended in a hydrous haplogranitic melt. The experimental data document increasing relative viscosity with increasing crystal fraction, shear thinning, and strain dependent viscosity decreases through textural reorganization. They present a comprehensive rheological dataset and derived an empirical set of equations by combining the experimental data with literature values on dilute suspensions of spherical particles. The resulting rheological model describes the interdependent effects of crystal volume fraction and strain-rate on the relative viscosity of two-phase suspensions of silicate melt and crystals. These span crystal volume fractions of $\Phi = 0$–0.8 and strain rates from $\sim 10^{-3}$ to 10^{-6} s^{-1}. They use the model to discuss the implications of non-Newtonian rheology for magma ascent, fragmentation and degassing. An example dataset from Caricchi et al. (2007) is shown in Figure 4.8. Expansion of the investigated crystal volume fractions to a range between $\Phi = 0.16$–0.76 of equant particles presented in Champallier et al. (2008), who echoed the rheological findings of Caricchi et al. (2007) and also documented the absence of measurable yield stress and thixotropic behavior. They further stress that a complete understanding of magmatic multiphase rheology requires experiments with different shapes, sizes and aspect ratios of crystals.

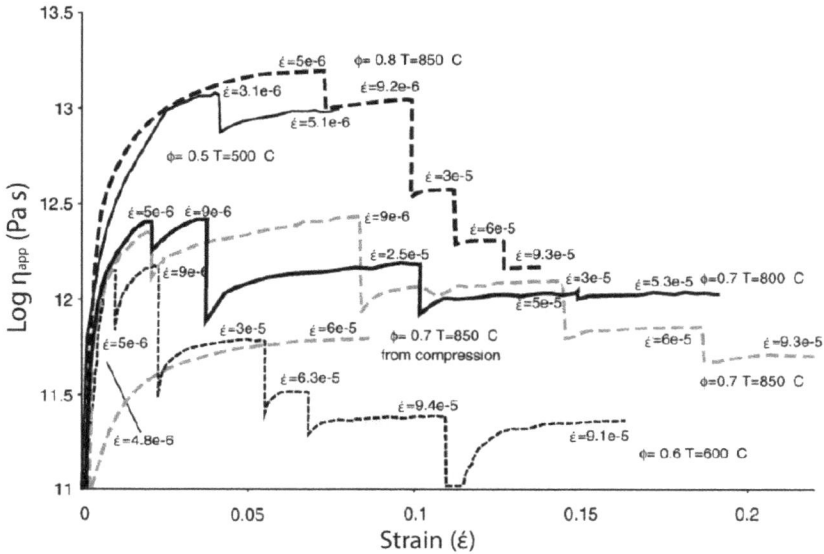

Figure 4.8. Apparent viscosities obtained from individual strain rates ($\eta_{app} = \sigma/\dot{\varepsilon}$) plotted versus strain; modified from Caricchi et al. (2007). Increasing the strain rate for any given crystal fraction (Φ) and temperature (T) induces a decrease of viscosity, which is interpreted as shear thinning behavior in Caricchi et al. (2007).

Pichavant et al. (2016) expanded the range of compositions and crystal aspect ratio for two phase melt–particle suspensions through measurements of plagioclase crystals suspended in a basaltic melt. Their results highlight the interplay between rheology and the maturation of microstructures with increasing absolute strain and the importance of crystal shape on suspension rheology. The study concludes that magmatic suspensions of plagioclase have viscosities approximately five orders of magnitude higher than suspensions of equivalent crystallinities made of isometric particles such as quartz that were presented in Caricchi et al. (2007). This discrepancy highlights the need for more systematic experimental mapping of all relevant rheological parameters to derive theoretical models describing multiphase magma rheology as a function of textural parameters (dominantly crystal-content, -shape, size distributions, pressure, and volatile content; see also the *"Parameterization Strategies"* section later in this chapter).

The first systematic investigations of three phase (crystals, bubbles and hydrous silicate melt) suspensions were spearheaded in experimental efforts using a Paterson type apparatus. This is because it allows containment of pressurized bubbles at simulated magmatic conditions. These studies highlighted significant changes in rheology between two and three phase suspensions. Pistone et al. (2012, 2013) present a series of measurements on three-phase, hydrous (2.26–2.52 wt.% H_2O), haplogranitic magmas, composed of quartz crystals ($\Phi = 0.24$–0.65), pressurized CO_2-rich gas bubbles ($\Phi_b = 0.09$–0.12) and melt in different proportions. The results show that three-phase magmas are characterized by a rheological behavior that is substantially different with respect to suspensions containing only crystals or only bubbles. Three-phase suspension rheology is found to be strongly strain-rate dependent (i.e., non-Newtonian). Both shear thinning and shear thickening non-Newtonian regimes were observed. Apparent shear thinning occurs in crystal-rich magmas ($\Phi = 0.55$–0.65; $\Phi_b = 0.09$–0.12) as a result of crystal size reduction and shear localization. Note however that viscosity reduction from crystal breakage constitutes a change in sample componentry rather than being a rheological phenomenon from textural re-organization (Cordonnier et al. 2009). Pistone et al. (2013) show that when bubbles are retained in the suspension, the presence of limited amount of gas bubbles results in a prominent decrease in viscosity compared to the rheology of bubble-free, crystal-bearing systems, e.g., at $\Phi = 0.7$, a decrease of about 4 orders of magnitude in relative viscosity is caused by adding 10 vol% of bubbles. At intermediate crystallinity ($\Phi = 0.44$; $\Phi_b = 0.12$) both shear thickening and thinning occur. Contrary to previous data on two phase suspensions, the authors report that shear thickening prevails in dilute suspensions ($\Phi = 0.24$; $\Phi_b = 0.12$), when bubble coalescence and outgassing dominate. This highlights that crystallization and outgassing of magmatic bodies can lead to a substantial increase of viscosity inducing their "viscous death" and transition to elastic behavior. On the other hand, the significant viscosity decrease associated with limited volumes of gas can promote re-mobilization of large plutonic magma bodies and the generation of large explosive eruptions. Pistone et al. (2016) presented some further measurements on the material synthesized and characterized in Pistone et al. (2012). Based on this database they derived empirical flow laws for multiphase magmatic suspensions. Their applicability remains to be tested both experimentally as well as via numerical and field studies. The insight that degassing may profoundly affect suspension viscosity inspired several contributions that employ torsion experiments to study the viscous to brittle transition and constrain the parameters required to develop shear fracturing of particle suspensions (Cordonnier et al. 2012b) and/or bubble coalescence (Kushnir et al. 2017). Both processes act to generate permeability and facilitate outgassing (Caricchi et al. 2011; Shields et al. 2014; Kushnir et al. 2017). The available data highlight that generation of a permeable network heavily relies on brittle deformation mechanisms (i.e., melt fracturing) to generate the connectivity required for permeability development; see also (Ryan et al. 2019b). For further details please also see Wadsworth et al. (2022, this volume) and Lavallée and Kendrick (2022, this volume).

Okumura et al. (2016) performed the first high temperature viscometry experiments on natural volcanic rocks with $\Phi = 0.16-0.45$; $\Phi_b < 0.04$. They highlight that in natural samples, the effective viscosity of the magma may increase with respect to previous experimental results due to textural heterogeneity. These heterogeneities introduce local stiff regions within the sample owing to crystal interaction. This highlights that similar to small scale parallel plate viscometry, sample size is one of the main limitations of viscometry in a Paterson type apparatus.

5. FIELD RHEOLOGY

Laboratory viscometry of silicate melt or analogue material suspensions, as reviewed in the earlier sections of this chapter, provides unique information for understanding the flow dynamics of magmas and lavas. However, it has proven difficult, if not impossible, to reproduce natural conditions in the laboratory, in particular in terms of redox state, crystal- and bubble-contents, sizes, shapes and their respective distributions as well as pressure (see Section 4. *Experiments on high temperature silicate melt suspensions*). In order for the laboratory data to become relevant for the interpretation or forecasting of flow emplacement they need to be directly linked to natural settings and conditions. Numerous studies have estimated lava viscosity from the dynamics of the lava flow as a whole, using either the velocities of lava flowing within channels or at flow fronts (Nichols 1939; Krauskopf 1948; Rose 1973; Walker et al. 1973; Harris et al. 2004; James et al. 2007) or from solidified flows, using their final dimensions, e.g., (Hulme 1974; Fink and Zimbelman 1986; Moore 1987; Kilburn and Lopes 1991) However, these methods are only able to provide spatially and temporally integrated values of the flow as a whole, averaging viscosity gradients across the flow (e.g., effect of cooler surface, base and sides) and viscosity variations with time (Kolzenburg et al. 2018). Other studies have tried to provide a parameterization to link the bulk chemical composition to the rheological properties of the whole lava flow (Hulme 1974; Lev et al. 2012), which, while operationally favorable due to its simplicity, does not allow tying the lava viscosity to the texture and temperature of the suspension itself. This has inspired several research groups to attempt direct viscosity measurements in the field, on active lava flows. These measurements of lava viscosity use specialized instrumentation and aim to capture the rheological flow curve of the lava in its natural state, including all intrinsic and extrinsic dynamics such as crystal and bubble content, temperature, oxygen fugacity etc. Such rheological field data, when associated with temperature measurement, detailed textural analyses of the samples, and dedicated experimentation in the laboratory, can provide a holistic assessment and parameterization of three-phase suspension rheology at natural conditions. However, due to the challenges associated with the development and deployment of this specialized instrumentation and the intrinsic hazards of work on active volcanoes, field rheology data remain very scarce; see Chevrel et al. (2019a) for a detailed review.

The success of field rheology campaigns is fundamentally controlled by the eruptive dynamics of the volcano, accessibility of the flow as well as external conditions (e.g., weather). While an eruption might be in progress and the infrastructure may be available, field measurements are not always possible, as direct approach to and contact with active lava are required. All the aforementioned conditions must be favorable to perform accurate and reproducible field rheometry. A further complication arises from the fact that, working in a dynamic environment like active lava implies that the lava is continuously changing (e.g., cooling, crystallization, degassing) and hence stable conditions are rarely encountered.

Current field rheology data is restricted to low silica content lava (<55 wt. %) and low crystal contents $\Phi < 0.45$. This is largely due to their relatively low viscosity at eruption temperatures ($<10^6$ Pa s). However, an important limitation even in low viscosity systems is the rapid cooling of the lava surface. The time of the measurement therefore needs to be shorter than crust formation. Lavas of higher viscosity have not been measured to date because of the extremely challenging conditions of approach. High viscosity lava flows are usually 'a'ā to block types with an outer, fragmented, surface that makes it extremely difficult to access the molten interior. Further, the time required to measure high viscosities may expose the operator to risks from falling blocks. Lastly, there are currently no instruments available for measuring at sufficiently high torque. Due to this mechanical limitation, the devices employed in the studies reviewed below were so far only able to access a narrow range of flow types (i.e., pahoehoe lobes and stable small lava channels).

In over 60 years (1948 to 2019), only eleven studies on field lava rheology measurements have been published (Einarsson 1949; Shaw et al. 1968; Gauthier et al. 1973; Pinkerton and Sparks 1978; Panov et al. 1988; Pinkerton et al. 1995a,b; Pinkerton and Norton 1995; Belousov and Belousova 2018; Chevrel et al. 2018). Two types of instruments have been used; both are derived from methodologies applied in laboratory rheometry: 1) penetrometers, where a penetrating body is pushed into or onto the lava by applying an axial force; this method is commonly used for relatively high viscosity lavas ($>10^3$ Pa s) and 2) rotational viscometers, analogous to the concentric cylinder method, where a measurement body is pushed into the lava and then rotated; this method is commonly used for lower viscosity lavas ($<10^4$ Pa s). The penetrometer method is either based on calculation of the viscous drag following Stokes' law (falling sphere theory) for penetrometers with semi-spherical heads (Panov et al. 1988; Belousov and Belousova 2018) or viscosity is obtained from calibration of the relationship between the penetrating force, the speed of penetration and fluid viscosity of a known liquid (Einarsson 1949, 1966; Gauthier et al. 1973; Pinkerton and Sparks 1978). Penetration rheometry has also been employed to measure the apparent yield stress of lava by recording the minimum force required to initiate movement (Pinkerton and Sparks 1978). Rotational viscometers, like in the laboratory, involve a rotating spindle immersed into the lava and viscosity is obtained from the wide-gap concentric cylinder theory (see the *Experiments on high temperature silicate melt suspensions* section earlier in this chapter for details).

Low carbon stainless steel alloys have proven most suitable for construction of both types of field rheometers because they are highly resistant to both mechanical stress and thermal exposure, readily available, and reasonably cost-effective. Although the use of steel may potentially cause iron contamination of the lava, the degree of contamination is considered to be insignificant due to the much shorter timescale of the measurements with respect to the diffusion speed of iron in silicate melts. Initial attempts to measure the viscosity of lava used crude instruments (such as forcing a rod by hand into flowing lava), and even the latest instruments (motor-driven rotational viscometer) are significantly less refined than those one would encounter in a well-equipped laboratory. This highlights that advances in instrumentation are still required to advance field rheology to a point where it can become a routine technique for parameterization of natural magmatic suspensions.

Falling sphere

To date, the falling sphere method was employed only once by Shaw et al. (1968) for measurements in the Makaopuhi lava lake. A stainless-steel sphere (2.5 cm diameter) was attached to a fine stainless-steel wire that was passed through a drill hole in the thick, solidified surface of the lava lake. The sphere was released into the lava in its hottest part and the movement of the wire (velocity and distance of penetration) behind the descending sphere was measured to calculate the viscosity via Stokes law. The strain rate during measurement was as low as 0.004 s^{-1} given the settling velocity of 3.5×10^{-3} cm/s. The viscosity is reported as 10^3–$10^{3.25}$ Pa s. This method could only be employed on a stable lava lake of sufficient depth, and with a suitably thick crust to support the measurement system.

Penetrometers

Penetrometers have been employed from the first published measurement by Einarsson (1949) until recently Belousov and Belousova (2018). This type of instrument is advantageous because it is light, easily transportable to the field, easy to build (the simplest versions consist of a rod equipped with a force gauge) and permits quick measurements over a wide range of viscosity (10^3–10^7 Pa s). The three kinds of penetrometers used to date are: 1) the simple penetrometer consisting of a pole pushed into the lava by hand, 2) the ballistic penetrometer that is a spear shot with a crossbow into the lava, and 3) the dynamic penetrometer that operates by propelling a piston once it has been placed inside the molten lava.

Simple penetrometers, where a rod is thrust by hand into the lava, allow the operator to qualitatively assess the viscosity during the measurements from to the amount of force and the time needed to penetrate the lava. In low viscosity lava, the rod may sink under its own weight while for higher viscosity the operator needs to apply the entire body weight to push the rod into the lava. Quantification of viscosity has been established via a relationship between viscosity and velocity of penetration from repeated measurements of fluid with known viscosity (Einarsson 1949; Pinkerton and Sparks 1978). More advanced penetrometers are equipped with a force gauge to record the force applied on the rod and designed with a semi-spherical head (Fig. 5.1B). Neglecting the potential effect of lava sticking to the rod, the viscosity of the fluid equates to the viscous drag on a half sphere that can be obtained via Stokes' law, given the force acting on the penetrating rod advancing through a viscous medium is known (Panov et al. 1988; Belousov and Belousova 2018):

$$\eta_{\text{app}} = \frac{F}{3\pi u R_{\text{eff}}} \tag{5.1}$$

where F is the force of penetration (viscous drag), u is the speed of penetration, and R_{eff} is the effective radius of the rod.

The ballistic penetrometer was used only once and involves shooting a spear at high-velocity perpendicularly into the lava and measuring its penetration depth (Gauthier 1971; Gauthier et al. 1973). The viscosity measurement relies entirely on previous laboratory

Figure 5.1. Field penetrometers: **A)** Photo and sketch of the dynamic penetrometer employed by Pinkerton and Sparks (1978); **B)** simple penetrometer with dynamometric gauge (Belousov and Belousova 2018).

calibration of the same spear geometry on liquids of different viscosities. The high initial penetration velocity prevents lava advance rates from influencing the measurement and limits cooling of the lava around the spear during penetration.

The disadvantage of simple and ballistic penetrometers is that the viscosity is an integrated value of the lava properties from the cooler surface of the flow to the hotter core. The measurements are therefore biased by the higher force required to penetrate the more viscous outer layer. Thus, this type of penetrometer tends to provide a semi-quantitative measurement of the rheology of the cooler exterior of a flow, and little indication of the rheological characteristics of the hot interior. To overcome this issue, Pinkerton and Sparks (1978) introduced the dynamic penetrometer (Fig. 5.1). It consists of a piston that is protected from the cooler crust by an outer stainless-steel tube. Once it is passed through the lava crust and has reached thermal equilibrium with the surrounding lava, penetration is activated (i.e., only the tip of the penetrometer is moving forward). This penetrometer was equipped with a pre-compressed spring to provide a controlled force during penetration which is recorded together with the piston advance rate. These data can then be converted into shear stress and strain-rate measurements (Pinkerton and Sparks 1978).

Another approach was proposed by Belousov and Belousova (2018) who measured the viscosity profile across small pāhoehoe lobes (Fig. 5.1B). To do so, they recorded the speed of penetration under a constant load from the surface to the base of the lobe. This enabled them to document the higher viscosity at the lobe surface, which quickly decreases directly below the skin, reaching a minimum in the interior of the lobe and then increases again toward the base.

Rotational viscometers

Two types of rotational viscometers have been employed in the field: a fixed rig installed on the frozen surface of a lava lake with a rotating spindle lowered into the molten core and controlled rotation rate, Fig. 5.2A; Shaw et al. (1968) and a portable instrument with a strain-rate controlled spindle inserted into the lava (Figs. 5.2B and 5.3); (Pinkerton and Sparks 1978; Pinkerton 1994; Pinkerton et al. 1995a; Chevrel et al. 2018). Both types of rotational viscometers rely on wide-gap concentric cylinder theory, where the torque is converted into shear stress and the rotational velocity into strain-rate using Couette theory and the spindle geometry (Dingwell 1986; Spera et al. 1988); see also Section 4. Equations (4.1) and (4.2). Unlike most laboratory experiments where the immersed spindle is cylindrical, vane geometry (i.e., four orthogonal blades; see Fig. 5.2B) is favorable for use in the field due to 1) lower weight, 2) ease of penetration, 3) reduced disturbance of lava during insertion, 4) reduced slippage between the edge of the vane and the lava (Shaw et al. 1968). The material between the vanes is trapped, forming a virtual cylinder of sample material whose equivalent diameter is used for viscosity calculation (Eqns. 4.1 and 4.2).

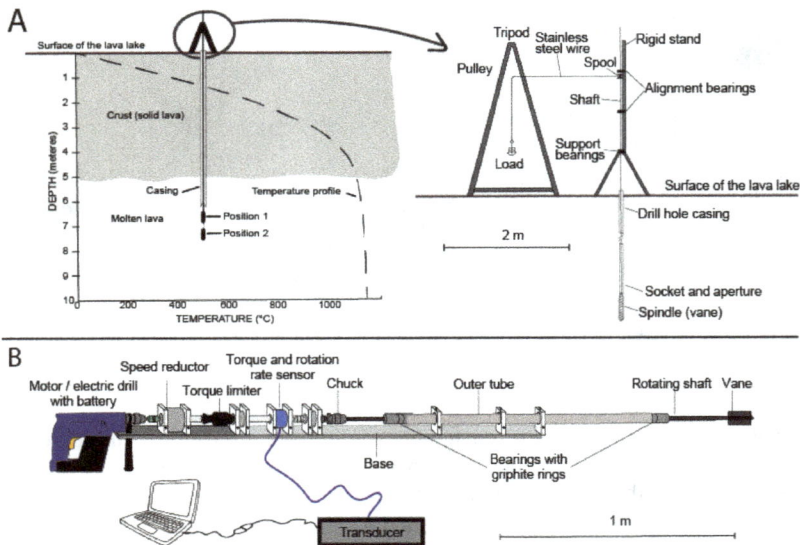

Figure 5.2. Sketch of a fixed rotational viscometer (**A**) ; modified from Shaw et al. (1968) and a motor-driven rotational viscometer (**B**) employed in Pinkerton et al. (1995b) and Chevrel et al. (2018).

The fixed rotational viscometer can only be used in the unique setting where a thick, frozen, stable surface is available for installation (Fig. 5.2A); (Shaw et al. 1968). The experimental setup consists of a fixed stand with a rotating shaft and vane attached to its lower end that is lowered vertically into the lava (Fig. 5.2A). A wire is spooled to the shaft, passed through a pulley mounted on a tripod and attached to a load. This permits to pull the wire and to rotate the shaft. By changing the load weight, different rotational torques can be applied to the vane. Flow curves are then obtained by measuring the resulting rotational speed.

Portable rotational viscometers have been used several times and the design has evolved over the years from manually activated to motor-driven (Fig. 5.3). The manually activated shear vane consists of a stainless steel vane attached to a torque wrench (Fig. 5.3A), which allows yield stress to be measured by applying torque slowly until the shear vane begins to rotate (Pinkerton and Sparks 1978). This system was also used to measure viscosity over a range of strain rate by monitoring the rotation speeds via an optical tachometer (Pinkerton and Norton 1995). The first motor-driven rotational viscometer consisted of a rotating steel vane attached to a drill hammer (Fig. 5.3B). The torques measured at different rotation rates (monitored with optical tachometer) are recorded using a torque meter, mounted coaxially between the drive train and the shear-vane. This device had a relatively low torque limit and was employed on natrocarbonatite lavas at Oldoinyo Lengai by (Pinkerton et al. 1995a). The second generation of motor-driven rotational viscometers was modified for silicate lavas (that have higher temperature and higher viscosity than natrocarbonatite lava). The motor is connected to a reduction gearbox, equipped with a torque limiter (to not break the instrument) and a new, combined, torque-rotation rate sensor (Figs. 5.2b, 5.3c,d). This viscometer has the capacity of varying the rotational speed; and thus to apply a large range of strain rates (0.2–3 s^{-1}) (Pinkerton 1994; Pinkerton and Norton 1995). The range of applicable strain rates is limited by the machine's torque range (< 2 N) and vane geometry and results in a range of stresses of 100–1000 Pa. By changing the torque sensor to a higher torque range of the same instrument, Chevrel et al. (2018) were able to reach higher stresses (up to 2500 Pa) and strain rates (up to 6 s^{-1}) but the lower strain-rate limit was consequently increased to 1 s^{-1} given the lava viscosity and the vane geometry.

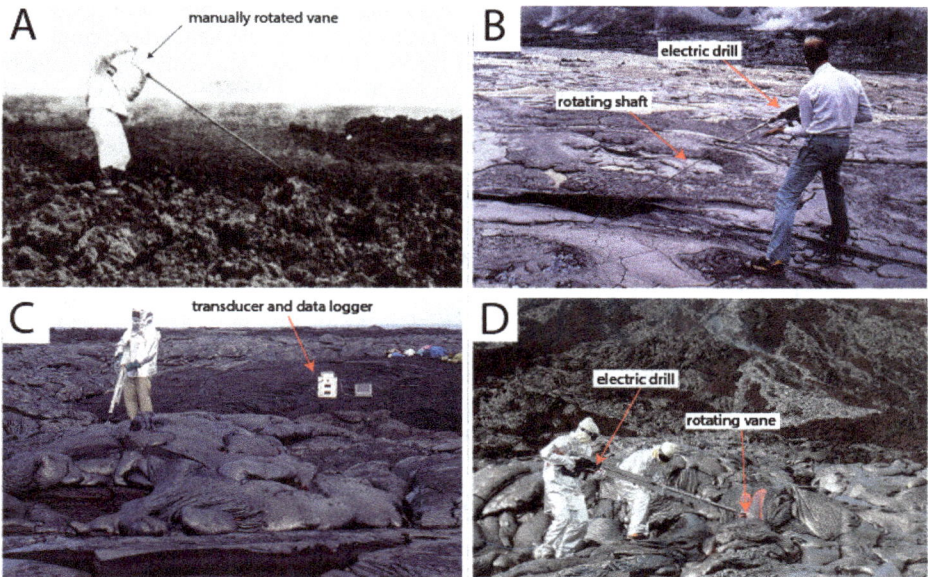

Figure 5.3. Photographs of all generations of portable rotational viscometers; **A)** manual rotational viscometer; photo from Chester et al. (2012); **B)** first motor-driven rotational viscometer employed on natrocarbonatite lavas employed by Pinkerton et al. (1995a); **C)** first measurements of the motor-driven rotational viscometer on pāhoehoe lavas by Pinkerton (1994) **D)** latest measurements using the motor-driven rotational viscometer on pāhoehoe lavas by Chevrel et al. (2018). Figure modified from Chevrel et al. (2019a).

Toward parameterization: requirements for future field viscometry

Field viscometry must always be performed in combination with temperature measurements and lava sampling. This is because the subsequent textural and petrographic analyses of these samples and data are key to understanding how crystal and bubble content affect rheology during lava emplacement. The molten lava must be sampled and quenched rapidly to conserve the texture at the location of measurement. In order to use field rheometry to its full potential for mapping of lava rheology in 4D throughout an entire lava flow, future field campaigns should combine measurements of rheological properties, temperature and lava texture as a function of distance from the vent to the front and across the flow. This likely also requires multiple field rheometers to be deployed synchronously in order to resolve the lava flows' rheology in 4D.

Recently Chevrel et al. (2018) showed that field viscosity measurements could not be readily compared with sub-liquidus laboratory measurements on natural lavas because the shear stresses acting during lava flow emplacement are, so far, mechanically inaccessible in the laboratory. Further, experiments performed in the laboratory tend to overestimate viscosity at a given temperature because of the shift of crystallisation toward higher temperature due to lack of volatiles in the melt, experimentation at ambient pressure and oxygen fugacity, as well as the dynamic thermal and chemical disequilibrium present during lava flow on the surface. Facilitated by the combined measurement of viscosity, temperature and petrology, Chevrel et al. (2018) further show that volatile and bubble content must be considered if the viscosity of active lava should be estimated via combined rheological and petrological models. They state, that to reproduce the field viscosity measurement using available models (see section "*Parameterization Strategies*"), three-phase parameterisation with 50% bubbles and 15% crystals is needed. Good agreement between field measurement and model-based viscosity estimation is obtained when considering elongated crystals (aspect ratio of 2.4) and deformed bubbles (Ca ≫ 1) (see section *Parameterization Strategies*" for details).

Reducing errors associated with field measurements requires accurate sensors as well as meticulous setup and calibration, which is difficult to achieve in the field where conditions are more dynamic and less controlled than in the lab. Field measurements will always be constrained by the balance between measurement machinery and the lava's thermodynamics. To reduce the effects of crust formation during measurements the instruments need to be pre-heated and inserted into fresh, molten lava through emerging breaches in the crust, or at the breaking point of pāhoehoe lobes where little-to-no crust is present. Further, the measurement timescale needs to be shorter than the timescale of cooling and crust formation. Both issues may be minimized by employing a dynamic penetrometer (Pinkerton and Sparks 1978), which allows triggering the sensor once the lavas isothermal core is reached. These are, however, limited to a specific range of stress–strain-rate conditions (See Fig. 5.4).

Figure 5.4 further highlights the current gaps in achievable measurement conditions. The development of the next generation of field rheometers thus needs to address several core issues. They need to be robust, light enough to be carried over rough ground and to remote locations, easily mounted and easy to handle (ideally by one person). Additionally, in order to capture the full rheological behavior of lava, field rheometers need to be able to apply a range of stress–strain-rate combinations. Current limitations include low strain rates ($< 1\,s^{-1}$) and low shear stresses (< 200 Pa). The aforementioned issues place high demands on the dimensions of the shear vane or spindle, torque sensor capability and motor power. The technological advances since the early measurements (e.g., electronically controlled motors and sensors) make the development of a new generation of rotational viscometer the most suitable approach for future measurements on active lavas. Combination of several methods is the most promising approach to recover a more complete rheological map and to describe the full flow curves for natural lavas. An attempt to measure the variation of viscosity over a large range of strain rate for Hawaiian lavas, obtained via field viscometry, is presented (Chevrel et al. 2018) and revealed a shear-thinning behavior.

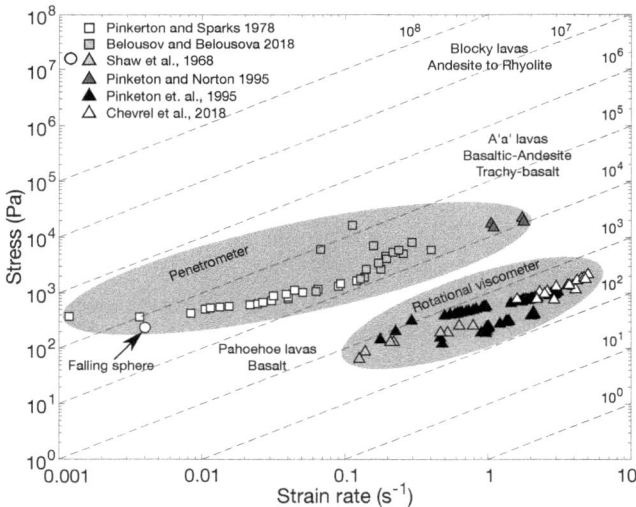

Figure 5.4. Regime plot of stress vs strain rate, mapping the measurement capacities of field rheometers. The plot shows the domains where lava viscosity has been measured with current hardware (**shaded areas**). It is important to state that neither method (rotational or penetration) is superior to the other but instead, they are to be regarded as complementary and employed depending on the viscosity regime that is encountered in the field. **Dashed lines** are iso-viscous conditions, **squares** and **triangles** represent data acquired with penetrometers and rotational viscometers, respectively. The **highlighted circle** represents the only data acquired by falling sphere method.

6. PARAMETERIZATION STRATEGIES

A central goal of most of experimental rheology is to generate predictive models that describe the rheology of multiphase suspensions as a function of extrinsic (temperature, shear rate, etc.) and intrinsic (melt- and crystal-composition, texture etc.) parameters. After reviewing the broad range of experimental approaches on both analogue and natural materials we now review the strategies for parameterization of the derived data that aim to generate a systematic understanding of multiphase magma and lava rheology. Parameterization of the data recovered from analogue, high-temperature and field experimentation has largely advanced in parallel with the increasing availability of experimental data. Thus, the derived models are intimately tied to the experimental approaches and have evolved predominantly as parameterization strategies for two phase suspensions of either particle- or bubble-suspensions and few attempts of the description of three-phase flow have been presented.

This review of parameterization strategies rides on the nomenclature introduced in the overview of common ways to describe rheological data that was presented in the first two sections of this chapter. Here we first presented parameterization efforts for particle suspensions, followed by those for bubble suspensions and finally those for three-phase suspensions. These models and the underlying physics are dominantly developed on the basis of data derived from analogue experiments, since these have produced the most complete and well constrained datasets for all relevant variables. The strategies reviewed here are focused on parameterization of shear viscosity. Multiphase suspensions may, however, also deform under tensile or compressive stresses, where the effects of volume viscosity become important. In such cases, longitudinal or bulk viscosities are defined that are products of volume and shear viscosities (see the "*Conventional descriptions of rheological data*" section). Also, note that due to the extremely narrow range of applicability of empirical, petrological approaches that relate relative viscosity to temperature as they were presented for example in Shaw (1969) or Dragoni and Tallarico (1994) we do not include them in this review.

Particle suspensions

In crystalline suspensions, the crystal phase or phases act as non-deformable particles which increase the suspension viscosity via both hydrodynamic effects (the melt has to flow around and between the particles) at low particles volume fraction (Φ) and mechanical interaction among particles at intermediate to high Φ. For low particle fractions (i.e., in the dilute regime) it is recognized that suspensions maintain a Newtonian character (i.e., a linear stress–strain-rate relationship). Within increasing particle fraction, once moving beyond a certain Φ threshold (that varies between 0.05 and 0.25, depending on the particle aspect ratio and surface roughness), suspensions becomes non-Newtonian, commonly exhibiting shear thinning behavior. Once Φ is sufficiently large (semi-dilute to concentrated regime), particles start to interact with each other and may align themselves with the flow direction or form force chains, resulting in a stronger shear thinning behavior and/or the development of an apparent yield stress (at $0.25 \lesssim \Phi \lesssim 0.6$). When the maximum packing fraction (Φ_m) is reached at even higher particle concentrations (concentrated regime) the particles may develop a solid interlocking network, which causes a drastic increase in the effective viscosity of the suspension and initiates the transition from suspension to creep rheology (Kohlstedt and Zimmerman 1996; Petford 2003; Lavallée et al. 2007).

Parameterization of the relative suspension viscosity η_r

The increase of viscosity and the evolution from Newtonian to non-Newtonian behavior during crystallisation depend fundamentally on the crystal content, shape, surface texture, and size distribution as well as the imposed strain rates. The development and nature of non-Newtonian behavior then result from the nature of particle–particle interaction during flow. Early empirical models consider the dilute regime (i.e., $\Phi \lesssim 0.25$) where particle interactions are limited. They are able to reproduce the measured relative viscosities of a broad range of suspensions with varying particle shapes, size distributions and surface properties very accurately and prove impressively robust within their Φ limits. These models are founded on the work of Einstein (1906) who first described the relative viscosity of suspensions of solid spheres (Eqn. 6.1, Table 6.1). The goodness of fit of this model was later improved for a larger dataset by Guth and Gold (1938) using a second degree polynomial function (Eqn. 6.2, Table 6.1) and by Vand (1948) using an exponential term (Eqn. 6.3, Table 6.1). However, these models break down once the particle volume fraction exceeds the dilute regime, as they do not account for the effects of particle–particle interaction (Fig. 6.1). The first model aiming to capture the onset of the concentrated regime was presented by Roscoe (1952) (Eqn. 6.4, Table 6.1). This model, often called the Einstein–Roscoe equation, was frequently used for the study of magmas and lavas over the last decades (Shaw 1965; Marsh 1981; Murase et al. 1985; Ryerson et al. 1988; Pinkerton and Stevenson 1992). While these pioneer models are based on data for suspensions of solid spheres, their fitting parameters can be modified to capture variations in particle shapes and size dispersions as well.

Subsequent parameterization strategies aimed to capture the effects that arise in the semi-dilute regime ($\Phi < \Phi_m$). These describe particle suspension viscosity as a function of the relation between the volume fraction of suspended particles and a maximum particle volume fraction. The assumption is that when Φ_m is reached, there is no longer enough "free space" for particles to move past each other, causing the particle population to become "jammed", generating a rigid framework and thus the suspension becomes un-deformable with an infinite viscosity, Φ_m. These models are largely based on the work of Eilers (1941) and predict infinite viscosity at $\Phi = \Phi_m$ (Eqn. 6.5, Table 6.1). The model of Eilers (1941) was later simplified by Maron and Pierce (1956) (Eqn. 6.6, Table 6.1) and expanded by Krieger and Dougherty (1959) to include a shape dependent Einstein coefficient B (Eqn. 6.7, Table 6.1). The latter two equations serve as the base for a number of other proposed models (Eqns. 6.8–6.13, Table 6.1) (Gay et al. 1969; Chong et al. 1971; Wildemuth and Williams 1984; Shapiro and Probstein 1992; Faroughi and Huber 2015; Moitra and Gonnermann 2015). The fitting parameters in these models are dependent on particle shape and surface roughness, which are usually well defined in analogue experiments (Mueller et al. 2010; Mader et al. 2013; Klein et al. 2018) but difficult to quantify for natural magmatic suspensions. The model complexity varies largely as a function of the experimental data that the authors aim to describe.

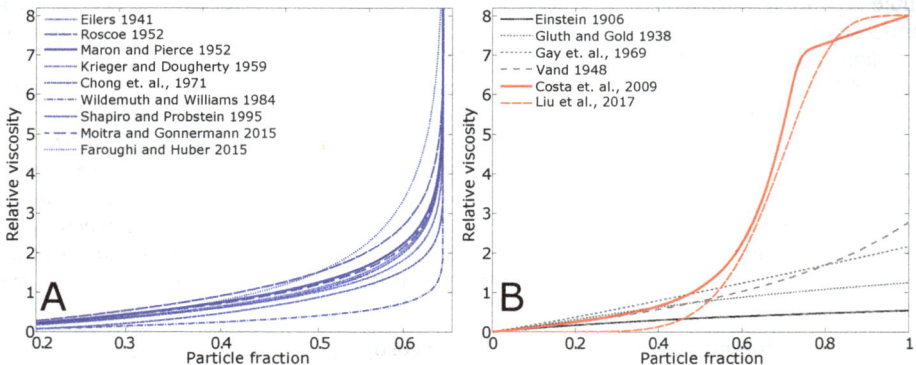

Figure 6.1. Examples of models of particle suspension rheology. **A)** models predicting a complete rheological lock-up at $\Phi = \Phi_m$ plotted for $\Phi_m = 0.65$ using the equations given in Table . Note that for Roscoe (1952) (Eqn. 6.4) this means $R=1.58$. **B)** models without rheological lock-up using the equations given in table 6.1. For Costa et al. (2009) (Eqn. 6.14) we plot the model with $B = 2.5$, $\Phi_* = 0.65$, $\alpha = 5$, $\delta = 13 - \alpha$ and $\xi = 10^{-4}$) and for Liu et al. (2017) we plot the model for $k=10$, $n=7$ and $\eta_{r(max)} = 8$.

publications are reported with the reference.

Equation	Comment	Reference	Eqn.
$\eta_r = 1 + B\Phi$	For dilute particle content ($\Phi<0.25$) with $B=2.5$ for spheres	Einstein (1906; p.300)	(6.1)
$\eta_r = 1 + B\Phi + B_1 \times \Phi^2$	For dilute particle content ($\Phi<0.25$) with $B=2.5$ and $B_1=14.1$ for spheres	Guth and Gold (1938)	(6.2)
$\eta_r = e^{\frac{B\Phi}{1-\alpha\Phi}}$	With $\alpha = 0.60937$	Vand (1948; Eqn. 6.8)	(6.3)
$\eta_r = 1 - R\Phi^{-2.5}$	With $R = \dfrac{1}{\Phi_m} = 1.35$ for spheres	Rosecoe (1952; Eqn. 3)	(6.4)
$\eta_r = \left(1 + \dfrac{1.25\Phi}{1-\Phi/\Phi_m}\right)^2$		Eilers (1941; pp.321)	(6.5)
$\eta_r = \left(1 - \dfrac{\Phi}{\Phi_m}\right)^{-2}$		Maron and Pierce (1956; Eqn. 30)	(6.6)
$\eta_r = \left(1 - \dfrac{\Phi}{\Phi_m}\right)^{-B\Phi_m}$	With $B = 2.5$ for spheres but can be fitted for other particle shapes	Krieger and Dougherty (1959; Eqn. 27)	(6.7)
$\eta_r = \exp\left\{\left[2.5 + \dfrac{\Phi}{\Phi_m - \Phi}\right]^\alpha \times \dfrac{\Phi}{\Phi_m}\right\}$	With $\alpha = 0.48$	Gay et al. (1969; Eqn. 19)	(6.8)
$\eta_r = \left(1 + \alpha\dfrac{\frac{\Phi}{\Phi_m}}{1-\frac{\Phi}{\Phi_m}}\right)^2$	With $\alpha = 0.75$	Chong et al. (1971; Eqn. 1)	(6.9)

Equation	Comment	Reference	Eqn.
$$\eta_r = \cfrac{1}{\left(\cfrac{\frac{\Phi}{\Phi_m}}{1-\frac{\Phi}{\Phi_m}}\right)}$$		Wildemuth and Williams (1984; Eqn. 1)	(6.10)
$$\eta_r = 1 + \frac{3\pi}{8}\frac{\beta}{\beta+1}\left(\frac{3+4.5\beta+\beta^2}{\beta+1} - 3\left(1+\frac{1}{\beta}\right)\ln(\beta+1)\right)$$	With $\beta = \cfrac{\left(\frac{\Phi}{\Phi_m}\right)^{\frac{1}{3}}}{1-\left(\frac{\Phi}{\Phi_m}\right)^{\frac{1}{3}}}$	Shapiro and Probstein (1992; Eqn. 1)	(6.11)
$$\eta_r = \left(1 - \frac{\Phi}{\Phi_m}\right)^{-\alpha}$$	With $\alpha = 1.92$	Moitra and Gonnermann (2015; Eqn. 13)	(6.12)
$$\eta_r = \left(1 - \frac{\Phi_m - \Phi}{\Phi_m(1-\Phi)}\right)^{\frac{2.5\Phi_m}{1-\Phi_m}}$$		Faroughi and Huber (2015; Eqn. 46)	(6.13)
$$\eta_r = \frac{1+\varphi^\delta}{\left[1-\Phi_*\right]^{\beta\Phi_*}}$$	With: $\Phi_* = (1-\xi)\times\mathrm{erf}\left[\dfrac{\sqrt{\pi}}{2(1-\xi)}\varphi(1+\varphi^\gamma)\right]$ $\varphi = \Phi/\Phi_*$, and ε, γ, δ, Φ_* are fitting parameters	Costa et al. (2009; Eqn. 1 + 2 + 6)	(6.14)
$$\eta_r = \eta_{r(max)}\frac{1-\exp(-k\Phi^n)}{1-\exp(-k)}$$	$\eta_{r(max)}$ is the relative viscosity of the suspension at complete solidification i.e., $\Phi = 1$; k and n are fitting parameters	Liu et al. (2017; Eqn. 6)	(6.15)

All these models produce satisfactory fits to most experimental data in the dilute and semi-dilute regime but are not appropriate for concentrated magmatic suspensions because they cannot assess viscosity for particle concentrations above Φ_m, where they tend to infinity. However, it is documented that materials are able to flow even when $\Phi > \Phi_m$, because Φ_m increases as particles align during flow, or when the particle size distribution is polydisperse (space between large particles can be filled with smaller particles). In such cases (high particle fraction suspensions; Φ = 0.5–0.8) the rheological response is primarily dictated by the crystal phase and phase-assembly while the importance of the interstitial melt is reduced. This observation implies that models for dilute suspensions will break down for suspensions of $\Phi \gtrsim 0.6$ (Lejeune and Richet 1995; Lavallée et al. 2007) because the materials are still able to flow at $\Phi > \Phi_m$ and hence their viscosity cannot be infinity. Further, experimental data document that even solid materials are able to deform via creep when Φ = 1. Therefore, a more complete rheological model has to encompass rheological transitions from a low relative viscosity regime, where the rheology is determined by the suspending liquid, to a high viscosity regime where the effect of particles dominates.

Such a parameterization approach was first introduced for magmatic suspensions in Costa (2005). On the basis of this model, Caricchi et al. (2007) developed an empirical characterization describing the non-Newtonian, strain-rate-dependent rheological effects of particles in the range of Φ = 0–0.8 that is based on experimental data obtained via torsion experiments on synthetic silicate melt suspensions (see the "*High temperature experiments*" section). They provide a 3D equation for the Φ–$\dot{\epsilon}$ dependence of the viscosity of partially crystallized magmas by fitting their experimental data to the model of Costa (2005). The Costa (2005) model was subsequently improved and generalised into a more comprehensive model in Costa et al. (2009) (Eqn. 6.14, Table 6.1), which has found frequent application in a range of studies concerning the rheology of geomaterials (Jamieson et al. 2011; Mandler and Elkins–Tanton 2013; Bachmann and Huber 2016; Cashman et al. 2017). The complex nature of the model stems from its goal to describe the entire range of Φ = 0–1 and deserves some more detailed consideration. The Costal et al. (2009) model is calibrated on analogue suspensions of particle fractions, between Φ = 0.1–0.8 and describes the relative viscosity increase of two-phase mixtures as a sigmoidal curve that initially increases exponentially until reaching a critical particle fraction, beyond which the exponential increase decays (Fig. 6.1B). This model is the first to cover the transition from the regime where the deformation behavior is controlled by melt viscosity up to the beginning of the regime where the deformation behavior is controlled by a solid framework of interlocking particles and is the most complete description of particle suspension rheology to date. Nonetheless, this model still requires characterization of fitting parameters as a function of particle size and shape as well as their respective distributions (Eqn. 6.14, Table 6.1). One of these parameters is defined as the critical solid fraction at the onset of the acceleration of viscosity (Φ_*). Another parameter (γ) describes the rate of relative viscosity increase with particle volume fraction, as $\Phi \rightarrow \Phi_*$; and the increase of η at $\Phi > \Phi_*$ is then expressed in terms of $\delta = A - \gamma$ with A = 13 as an empirical constant. ξ is an empirical fitting parameter commonly $\ll 1$ that allows optimization of the model for a given dataset. For large Φ, the second term in the numerator is a minute correction when $\Phi < \Phi_*$, while it becomes important when $\Phi > \Phi_*$. At $\Phi = \Phi_*$, the relative viscosity depends on Φ and Φ_*, only. With decreasing Φ at $\Phi < \Phi_*$, the model (Eqn. 6.14, Table 6.1) tends to the Krieger and Dougherty (1959) equation (Eqn. 6.7, Table 6.1) and when $\Phi \rightarrow 0$, it recovers the Einstein (1906) equation (Eqn. 6.1, Table 6.1).

Recently, Liu et al. (2017) proposed a simpler parameterization approach aimed at covering the range of Φ = 0–1. The model is based on a sigmoidal function and only requires three fitting parameters (Eqn. 6.15, Table 6.1). While this model is attractive due to its flexibility, simplicity and low number of fitting parameters, it does not lend itself for easy expansion to incorporate the details of variations in particle aspect ratio or polydispersity.

Using the Costa et al. (2009) model, this could be done by combining the Φ/Φ_* term with the empirical data of for example Mueller et al. (2010) or Klein et al. (2018). Nonetheless, both models of Costa et al. (2009) and Liu et al. (2017) remain to be validated by a complete rheological dataset covering the entire range of $\Phi = 0$–1.

Capturing polydispersity in relative viscosity models. An important issue that remains to be addressed in the models reviewed above is the fact that natural magmatic suspensions most frequently contain microlites in association with larger phenocrysts. This polydisperse distribution can act to moderate the onset of non-Newtonian effects because they are directly tied to the size and shape of the suspended particles and their respective distribution. Thus, a reasonable prediction of the relative viscosity of particle suspensions requires the consideration of particle volume fraction as well as their shape- and size-distribution.

Early work on bidisperse suspensions by Farris (1968) proposed that polydisperse suspensions may be treated as an incremental system where the coarse solid fraction acts as a monodisperse suspension in a separate, monodisperse fluid containing the finer particles. Farris (1968) considered a bidisperse system, where coarser and finer fractions have the same shape and behave independently of each other and proposed the following treatment:

$$\eta_r = \left(\frac{\eta_f}{\eta_l}\right)\left(\frac{\eta_c}{\eta_f}\right) \tag{6.16}$$

where η_l, η_f and η_c are the liquid viscosity, fine particle suspension viscosity, and coarse particle suspension viscosity, respectively (i.e., the finer grained suspension is treated as the effective medium; see the end of this section for details).

Cimarelli et al. (2011) used analogue experiments to measure the rheology of suspensions with bimodal particle size and shape distributions and show that increasing the relative proportion of prolate to equant particles at constant Φ can increase the relative viscosity by up to three orders of magnitude. Based on their data, they expand the modeling approach of Costa et al. (2009) for a variety of bidisperse suspensions. Their parameterisation rides on a variable that, instead of representing the monodisperse particle content (Φ), considers the ratio of fine (Φ_f) to coarse (Φ_c) particles $x = \dfrac{\Phi_f}{\Phi_f + \Phi_c}$: to modulate the fitting parameters of Costa et al. (2009). They report thefollowing parameterization to Equation (6.14):

$$\Phi_* = \Phi_{*f} x^\alpha + \Phi_{*c}\left(1 - x\right)^\alpha \text{ with } \alpha = 1.34 \tag{6.17}$$

$$\gamma = \gamma_f x \gamma_c \left(1 - x\right) \tag{6.18}$$

$$\xi = \xi_f x \xi_c \left(1 - x\right) \tag{6.19}$$

The symbols are explained in Table 1.1.

Recently, Klein et al. (2018) presented a different parameterization using the Maron and Pierce (1956) model (Eqn. 6.6, Table 6.1). They substitute the shape parameter of the Mueller et al. (2010) model for Φ_m in Equation (6.25) to derive a model for polydisperse suspensions with non-spherical particles, which they then implement into the Maron–Pierce model (Eqn.6.6, Table 6.1):

$$\eta_r = \left(1 - \Phi\left[1 - \left(\left(1 - \left\{\Phi_{m1}\exp\left[-\frac{\left(\log_{10} r_p\right)^2}{2b^2}\right]\right\}\right)\gamma^\alpha\right]\right)^{-1}\right)^{-2} \tag{6.20}$$

The aspect ratio r_p is defined as $r_p = \dfrac{l_a}{l_b}$, and l_a is the particle's axis of rotational symmetry and l_b is its maximum length perpendicular to that and non-spherical particles are approximated as prolate/oblate spheroids or cylinders (Mueller et al. 2010).

Parameterization of the maximum packing fraction. The maximum packing fraction (Φ_m) reflects the particle concentration at which the viscosity of a suspension of specific particle sizes, shapes and distributions tends towards infinity (i.e., a solid framework of interlocking particles develops and induces a transition from viscous flow to elastic or creep deformation). Φ_m is commonly determined empirically but can also be determined geometrically or numerically (Evans and Gibson 1986; Gan et al. 2004). Experimental values of Φ_m are often found to be lower than predicted from the geometrical analyses due to random heterogeneities in packing density as well as variations in packing efficiency resulting from particle roughness (Mader et al. 2013). Φ_m was found to strongly depend on particle shapes and size distributions (Cimarelli et al. 2011; Mueller et al. 2011; Klein et al. 2017)

Systematic studies have shown that for monodisperse suspensions, Φ_m decreases with particle shape anisotropy (i.e., elongate or prolate), surface roughness and decreasing particle alignment, while it increases for equant particles and polydisperse mixtures (Chong et al. 1971; Lejeune and Richet 1995; Saar et al. 2001; Caricchi et al. 2007; Vona et al. 2011; Mader et al. 2013). Often, Φ_m is determined by fitting the experimental data to rheological models like the Maron–Pierce equation (Eqn. 6.6. Table 6.1). Systematic measurement of values for Φ_m allowed derivation of empirical models that describe its dependence on aspect ratio and particle roughness (Mueller et al. 2011); Figure 6.2:

$$\Phi_m = \Phi_{m1} \exp\left[-\frac{\left(\log_{10} r_p\right)^2}{2b^2} \right] \qquad (6.21)$$

where Φ_{m1} is the maximum packing fraction for equant particles and given as 0.656 and 0.55 for smooth and rough particles, respectively (Mader et al. 2013); b is a fitting parameter equal to 1.08 and 1 for smooth and rough particles, respectively. Klein et al. (2018) demonstrate that the only parameter in Eqn. 6.20 that is directly affected by particle shape is the monomodal Φ_m.

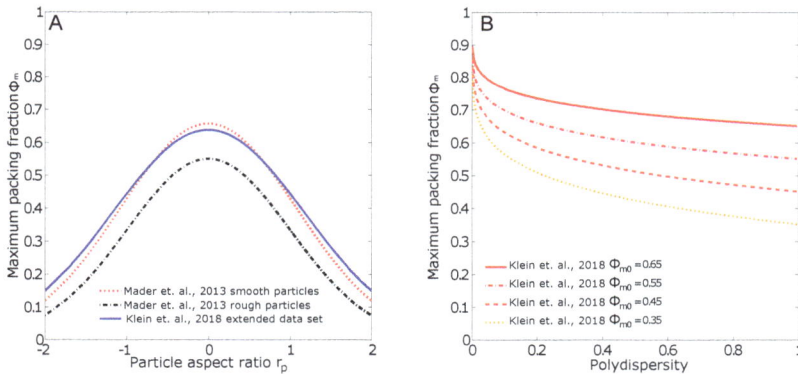

Figure 6.2. A) Summary of models of maximum packing fraction as a function of particle aspect ratio. All models are based on the equation of Mueller et al. (2010) for which Mader et al. (2013) highlighted the separation of smooth and rough particles and Klein et al. (2018) presented an improved fit based on an extended dataset. Note that these are established for monodisperse particle suspensions. **B)** The dependence of maximum packing fraction on particle polydispersity after Klein et al. (2018), plotted here for Φ_{m0} of 0.65–0.35. Note that $\varrho = 1$ is the monodisperse limit (i.e., polydispersity increases with decreasing ϱ) and Φ_{m0} is the maximum packing fraction of a monodisperse suspension and is derived from subplot A.

To address the issue of suspension with polydisperse particle size populations, highlighted in the previous section, Klein et al. (2017) show that the polydispersity (Φ) of solid particles is the main parameter that influences Φ_m and hence the viscosity of a given suspension. Using the Maron–Pierce equation (Eqn. 6.6, Table 1.1), they derived an empirical model that relates polydispersity of a particle size population with the maximum packing fraction Φ_m of a suspension and takes the shape of:

$$\Phi_m = 1 - ((1 - \Phi_{m0})\, \varrho^\alpha) \qquad (6.22)$$

where Φ_{m0} represents Φ_m for a monomodal suspension derived from Equation (6.21) (solid curve in Fig. 6.2A), $\alpha = 0.173$ is an empirical constant and ϱ is the polydispersity of the suspension. The polydispersity represents the ratio of the specific surface area of a polydisperse system S_p to that of a monodisperse system S_m at the same volume fraction with radius r (Torquato 2013). Thus, increasing polydispersity decreases the surface area ratio (ϱ) and $\varrho = 1$ represents the monodisperse limit (i.e., lower values indicate higher degrees of polydispersity):

$$\varrho = \frac{S_p}{S_m} = \frac{r\langle r\rangle}{\langle r^3\rangle} = \frac{\langle r\rangle\langle r^2\rangle}{\langle r^3\rangle} \qquad (6.23)$$

with $\langle r^n\rangle$ being the n-th moment of a given size distribution.

This approach was further expanded for application to polydisperse suspensions of varying aspect ratios in Klein et al. (2018), where the authors present a re-parameterization of the Mueller et al. (2010) data for smooth particles, based on an experimental dataset expanded to lower r_p and they recover a best fit for $\Phi_{m1} = 0.637$ and $b = 1.171$. Using this updated fitting approach and data sets of polydisperse suspensions with mean aspect ratios of ~7 and ~0.09, Klein et al. (2018) showed that Equation (6.22) is a robust tool to also estimate Φ_m for polydisperse distributions of suspensions with particles of aspect ratios above and below 1. Their analyses show that Φ_m systematically decreases, while maintaining the general relation of decreasing Φ_m with increasing polydispersity ϱ (Fig. 6.2B).

In summary, Φ_m decreases as particle aspect ratios deviate from unity and as particle surface roughness increases. Increasing polydispersity (i.e., decreasing ϱ) on the other hand can drastically increase Φ_m by allowing the interparticle space to be filled with smaller particles. It is important to note, however, that the models plotted in Figure 6.2A apply to monodisperse suspensions only but that these can be expanded for application to polydisperse suspensions as shown in Figure 6.2B. However, if both parameters are not constant (e.g., in a suspension of particles of varying size and aspect ratio), Φ_m cannot be predicted with great confidence. No constitutive model for polydisperse suspensions of particles with varying aspect ratio has been presented to date. The data presented by Moitra and Gonnermann (2015) show that varying the proportions of particles of differing aspect ratio can affect Φ_m by up to 25%. Nonetheless, a first order estimate can still be deduced from the work of Klein et al. (2018) by using the mean particle aspect ratio of a suspension of interest.

Parameterization of suspension consistency and flow index. Non-Newtonian effects cannot be described completely by a single, strain-rate-independent viscosity and thus the concept of relative viscosity, which considers the effective viscosity of a suspension at specific stress–strain-rate condition, is, at times, insufficient. A full rheological description requires establishing non-linear flow curves. The most frequently used approach to do so is by employing the three-parameter Herschel–Bulkley model (Eqn. 2.3), accounting for the yield stress, the flow consistency and the flow index of the suspension. As mentioned earlier, the non-Newtonian character increases with increasing Φ, resulting dominantly in shear-thinning behavior. It is therefore necessary to parameterise the consistency and the flow index as a function of Φ and strain rate.

Wildemuth and Williams (1984) showed that the relative viscosities of a large range of suspensions of varying particle shapes collapse on a single curve when plotted against Φ/Φ_m. This finding was echoed in Mueller et al. (2010), who evaluated the relative consistency:

$$K_r = \frac{K}{\eta_0} \qquad 6.24$$

of their experimental data by fitting a modified Maron–Pierce equation in the form of:

$$K_r = \left(1 - \frac{\Phi}{\Phi_m}\right)^{-2} \qquad 6.25$$

The results of their analysis show that the relative consistency of a suspension depends strongly on particle aspect ratio, thus demonstrating that the dependence of K on particle aspect ratio is caused by the dependence of the maximum packing fraction Φ_m on aspect ratio. This correlation was validated for suspensions of varying size and shape modalities by Moitra and Gonnermann (2015); see Figure 6.3.

The flow index, n, characterizes the nature and intensity of the shear rate dependence of viscous flow (i.e., curvature and slope in K–Φ space). The more the flow index differs from 1, the more non-Newtonian the fluid becomes. For non-Newtonian materials without yield stress ($\tau_y = 0$), the flow curve function is $\sigma = K\dot{\varepsilon}^n$ and characterization of the nature and intensity of shear rate dependent changes in rheology are provided by models of the variation of the flow index n as a function of Φ. As the apparent viscosity of the suspension is $\eta_{app} = \dfrac{\sigma}{\dot{\varepsilon}}$, the relative viscosity is defined as:

$$\eta_r = \left(\frac{K}{\eta_l}\right)\dot{\varepsilon}^{n-1} \qquad (6.26)$$

There are several studies presenting parameterization of natural magmatic suspensions or their analogues that show a decrease in effective viscosity with increasing shear rate at any given Φ (i.e., shear thinning behavior; $n < 1$) (Mader et al. 2013). These all describe similar trends where, all other parameters being constant, n decreases with increasing Φ. Since the available

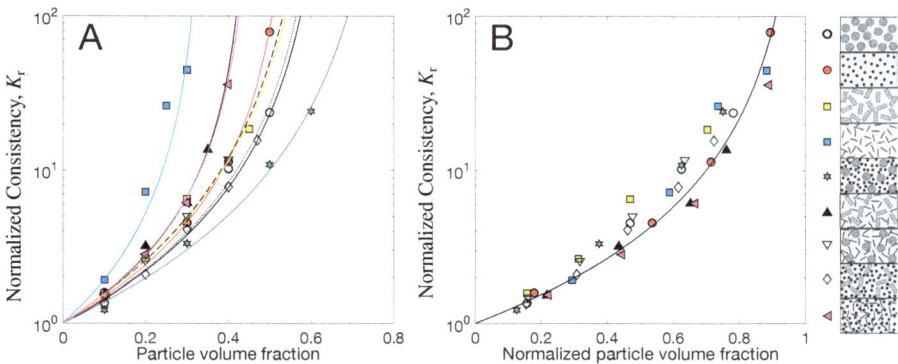

Figure 6.3. Relative consistency of monomodal and polymodal suspensions of varying particle aspect ratios plotted against particle volume fraction (A) and normalized particle volume fraction (Φ/Φ_m) (B), modified after Moitra and Gonnermann (2015). Lines in (A) are the best-fits of a Maron–Pierce type model (Eqn. 6.25) where the exponent (−2) has been adapted to fit each dataset and the line in (B) is the best for all dataset of a Maron–Pierce type model (Eqn. 6.25) where the exponent is −1.92 (see Moitra and Gonnermann (2015) for further details).

models describe the flow index as a function of Φ/Φ_m, it is important to view them in the light of the parameterization of Φ_m presented earlier in this section. This is highlighted in Mueller et al. (2010) and Moitra and Gonnermann (2015), who show that the onset of shear-thinning behavior commences at lower Φ for higher aspect ratio particles and that flow index values do not collapse onto a single curve by normalising the Φ/Φ_m (Fig. 6.3). Based on the measured systematic dependence of n (Φ/Φ_m) on particle aspect ratio, Mueller et al. (2010) derive an empirical relationship (Eqn. 6.28, Table 6.2). Their results show that non-Newtonian behavior is primarily controlled by particle aspect ratio rather than consistency or yield stress. Moitra and Gonnermann (2015) echo these findings and show that across different suspension types, at constant Φ, n decreases with increasing particle aspect ratio, and/or with decreasing particle size heterogeneity. However, instead of incorporating particle aspect ratio into the model, they introduce n_{min} that is the corresponding value of n at Φ_m (Eqn. 6.30, Table 6.2). Both Φ_m and n_{min} have to be estimated or determined experimentally but can also, alternatively, be treated as fitting parameters for data of unknown Φ_m and n_{min}. Truby et al. (2015) present measurements on three phase suspensions and derive the first, empirical, parameterization of the flow index for bubble-bearing suspensions (Eqn. 6.31, Table 6.2) and for multiphase suspensions, accounting for the volume fraction of particles and bubbles (Eqn. 6.32, Table 6.2).

Due to the higher technical difficulties and uncertainties of high temperature experimentation compared to experiments on analogue materials and the increased complexity of the sample textures as well as their quantification, the number of high temperature datasets parameterizing n, K and Φ_m remains scarce compared to those of analogue suspensions. To date, only two parameterizations of the flow index as a function of Φ/Φ_m have been presented that derive from experimentation on natural magmatic suspensions, namely Ishibashi (2009) (Eqn. 6.27, Table 6.2) and Vona et al. (2011) (Eqn. 6.29, Table 6.2).

Table 6.2. Equations for parameterization of the flow index.

Equation	Reference	Comment	Eqn. #
$n = 1 - 2\alpha\left[\ln\left(1 - \dfrac{\Phi}{\Phi_m}\right)\right]^2$	Ishibashi (2009) Eqn. 9b	From HT experiments, with $\alpha = 0.118$	(6.27)
$n = 1 - 0.2r_p\left(\dfrac{\Phi}{\Phi_m}\right)^4$	Mueller et al. (2010) Eqn. 5.2	From analogue experiments	(6.28)
$n = 1 + 2\alpha\log\left(1 - \dfrac{\Phi}{\Phi_m}\right)$	Vona et al. (2011) Eqn. 20	From HT experiments, with $\alpha = 0.118$	(6.29)
$n = 1 - \left(1 - n_{min}\right)\left(\dfrac{\Phi}{\Phi_m}\right)^{2.3}$	Moitra and Gonnermann (2015) Eqn. 14	From analogue experiments	(6.30)
$n = 1 - 0.334\Phi_b$	Truby et al. (2015) Eqn. 6.1	From analogue experiments on bubbles	(6.31)
$n = 1 - 0.2\left(\dfrac{\Phi}{\Phi_m}\right)^4 - 0.334\Phi_b$	Truby et al. (2015) Eqn. 6.2	From analogue experiments on multiphase mixtures	(6.32)

All presented models in Figure 6.4 show reasonable agreement up to $\Phi/\Phi_m = 0.5$ but they deviate significantly at higher Φ/Φ_m. This is likely a result of the variation in the textural makeup of the suspensions measured in the different studies (i.e., particle size, particle shape as well as orientation and distribution), which end up influencing the parameterization. While the relative consistency/viscosity of the investigated suspensions can be described satisfactorily by Φ_m alone, which is only dependent on particle shape and polydispersity (see section on parameterization of Φ_m above), the non-Newtonian character of particle suspensions has a more complex relation to its textural features. Thus, it is likely owing to the paucity of existing experimental data on polydisperse suspensions that a general functional description of the effect of textural parameters on the non-Newtonian character of suspensions has not been presented so far.

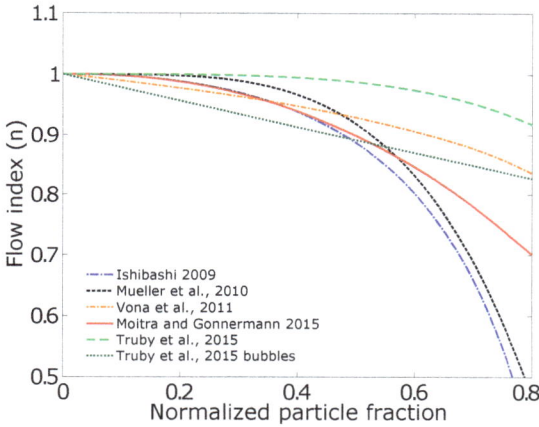

Figure 6.4. Models of the variation of the flow index as function of particle fraction, normalised for $\Phi_m = 0.65$. All models are summarized in Table 6.2. The Ishibashi (2009) model is fitted for Vona et al. (2011) data with $\alpha = 0.118$. Mueller et al. (2010) is plotted for the mean particle aspect ratio, $r_p = 6.5$. Moitra and Gonnermann (2015) is plotted for $n_{min} = 0.5$. Truby et al. (2015) is plotted for $\Phi_b = 0$, and Truby et al. (2015) bubbles for the effect of bubbles on a particle free melt.

Parameterization of the yield stress τ_y

To complete the description of non-Newtonian flow laws, using the Bingham equation ($\sigma = \eta_B \dot{\varepsilon} + \tau_y$) or the three parameter Herschel–Bulkley equation (Eqn. 2.3), for suspensions with an apparent yield stress, a parameterization of the yield stress (τ_y) as a function of Φ is required. Yield stress is defined as the stress under which a suspension first begins to show liquid-like behavior (i.e., continuous deformation) as stress is increased from zero or, similarly, the point where, when decreasing the applied stress, solid-like behavior (i.e., no continuous deformation) is first noticed. In particle suspensions, yield stress is thought to develop when a network of interacting particles is generated (Philpotts and Carroll 1996). At low shear stresses, this network is deformed elastically and ultimately broken up once the yield stress is reached, allowing the suspension to flow (Heymann et al. 2002). The first parameterization of yield stress in magmas was given by (Ryerson et al. 1988) and is based on experiments on natural lava at high temperature (Eqn. 6.33, Table 6.3). More recently, based on channelized flows of analogue material, Castruccio et al. (2010) proposes a different equation (Eqn. 6.34, Table 6.3) where a critical particle fraction, Φ_C, for development of a yield stress, is given (and reported as $\Phi_C = 0.27$).

However, these do not include any parameterization of the efficiency of particle–particle interaction as a function of textural parameters such as aspect ratio or packing density.

Table 6.3. Equations for parameterization of the yield stress.

Equation	Reference	Comment	Eqn. #
$\tau_y = 6500\Phi^{2.85}$	Ryerson et al. (1988) Eqn. 17		(6.33)
$\tau_y = 5\times10^6\left(\Phi - \Phi_C\right)^8$	Castruccio et al. (2010) Eqn. 18	With $\Phi_C = 0.27$	(6.34)
$\tau_y = 200\left(\dfrac{D_p}{\Phi_m - \Phi}\right)\left(\dfrac{\Phi_m}{1-\Phi_m}\right)^2\left(\dfrac{1}{\eth^{1.5}\sigma_g^2}\right)$	Gay et al. (1969) Eqn. 27	Where for spheres $\eth = 0.118$. σ_g is the geometric standard deviation for particle diameter, reported as 2.02	(6.35)
$\tau_y = \left[A\left(\dfrac{\dfrac{\Phi}{\Phi_C}-1}{1-\dfrac{\Phi}{\Phi_m}}\right)\right]^{\frac{1}{m}}$	Wildemuth and Williams (1985) Eqn. 2b	A, Φ_C and m are constants reported as 0.848, 0.1978 and 0.8364,	(6.36)
$\tau_y = \left(\dfrac{\left(\dfrac{\Phi}{\Phi_C}-1\right)}{\left(1-\dfrac{\Phi}{\Phi_m}\right)}\right)^{\frac{1}{p}}\tau_c$	Zhou et al. (1995) Eqn. 8	With $\Phi_C = 0.2$ and $p = 1.3$. τ_c is the inter-particle cohesion, expressed as: $\tau_c = \dfrac{1}{6C\pi a^3}$ where a is the particle radius; C is a fit parameter	(6.37)
$\tau_y = \left(1-\dfrac{\Phi}{\Phi_m}\right)^{-2} - 1$	Heymann et al. (2002) Eqn. 9		(6.38)
$\tau_y = \tau^*\left[\left(1-\dfrac{\Phi}{\Phi_m}\right)^{-2} - \left(1-\dfrac{\Phi_C}{\Phi_m}\right)^{-2}\right]$	Moitra and Gonnermann (2015) Eqn. 15	τ^* is a fitting parameter.	(6.39)

It is intuitive that the efficiency of formation of interacting particle networks varies also with particle concentration, size and shape. Thus, a model capturing these complexities needs to be able to account for their variation. Some experimental datasets document that yield stress arises at lower particle concentrations for higher particle aspect ratios due to more efficient particle–particle interaction. As a result, more robust parameterization approaches of τ_y as a function of Φ frequently include its relation to Φ_m allowing to capture the fact that the development of yield stress is strongly dependent on particle packing efficiency (Eqn. 6.33 to Eqn.6.39, Table 6.3) (Gay et al. 1969; Wildemuth and Williams 1985; Ryerson et al. 1988; Zhou et al. 1995; Heymann et al. 2002; Castruccio et al. 2010; Moitra and Gonnermann 2015).

Most datasets that serve as a basis to develop these relationships report that yield stress is small or negligible for $\Phi \ll \Phi_m$ but rapidly becomes appreciable when $\Phi/\Phi_m \geq 0.8$ (Mueller et al. 2010). Further, yield stress tends to be larger for smaller particles, as these can achieve high packing densities and thus generate stronger networks (Heymann et al. 2002). Published models for the relationship between τ_y and Φ predict vastly different values (see Fig. 6.5), which is

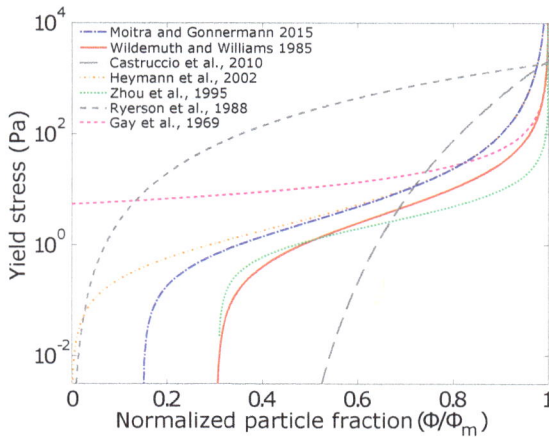

Figure 6.5. Examples of models for yield stress as a function of particle content, plotted for Φ/Φ_m with $\Phi_m = 0.65$. Models are summarized in Table 6.3. Gay et al. (1969) is plotted for Dp = 0.000833333 (i.e., 250 micron particle size). Moitra and Gonnermann (2015) is plotted for $\tau^* = 1$ and $\Phi_C = 0.0975$.

likely due to the fact that they were individually derived from, or optimized for, a narrow range of rheological data and suspension textures. However, with exception of the models presented by Ryerson et al. (1988) and Castruccio et al. (2010), that are derived from high temperature and analogue flow experiments, respectively, the models show broad agreement that yield stress rapidly increases at $\Phi \rightarrow \Phi_m$. As evident from the variation in the predicted values, the available experimental data prove difficult to model and no robust relationship between Φ, Φ_m and τ_y is available. Heymann et al. (2002) suggest that smaller particle sizes produce higher yield stresses as a result of increased particle–particle interactions and Mader et al. (2013) discuss the importance of aspect ratio for the development of yield stress. However, the conditions under which an apparent yield stress is measured is fundamentally linked to the measurement limitations of the rheometric apparatus (predominantly the lower shear rate and -stress limit).

The inconsistencies in the determination of yield stress likely root in the fact that the yield stress is best described as a transition zone between two regimes of drastically varying deformation timescales (Barnes 1999), rather than being a rheological constant, which it is commonly interpreted to be. For a detailed discussion of the nature of the yield stress, its relevance and application to natural multiphase suspensions please see Figure 6.9 in the later section "*The yield stress dilemma*" below.

Unsteady flow in particle suspensions. While unsteady flow in bubble suspensions is relatively well understood (see below), this is not the case for particle suspensions. Much of the data reviewed above consider steady (i.e., continuous) flow. Data from oscillatory measurements on particle suspensions, which can measure unsteady flow (aiming for example to describe the rheological effect of seismic waves propagating through a suspension) are scarce. Sumita and Manga (2008) present a study of the oscillatory rheology of analogue suspensions of spherical particles at volume fractions of $0.2 < \Phi < 0.6$. They present data for the response of the elastic and viscous moduli as a function of oscillation frequency and find a strain dependent rheological evolution where the suspension displays three rheological regimes. The response is initially linear viscoelastic, becomes shear-thinning with increasing strain amplitude and transitions to shear-thickening at high strain amplitudes. They suggest the suspensions may be best described by combining both relaxation and retardation phenomena (i.e., a Burgers model; see Findley et al. (1977). However, experimental data remain scarce and thus no systematic description of suspension rheology in unsteady flow has been presented to date.

Mader et al. (2013) evaluate oscillatory viscometry data for particle volume fractions in the concentrated regime $\Phi > 0.25$ and report that the complex viscosity measured under forced oscillation is consistently lower than the effective viscosity measured in steady shear. While the physical origin of this behavior remains unclear, Mader et al. (2013) provide a convincing hypothesis. They suggest that the lower viscosity values measured in oscillation results from variations in the nature of particle–particle interactions between the two flow regimes. They postulate, that particles within suspensions undergoing constant shear (i.e., large total strains) experience numerous and repeated interactions, hence increasing the suspension's viscosity. Particles within suspensions subject to small amplitude oscillations may not move far enough to interact strongly with one another, hence their effect on viscosity is less pronounced. They provide a comparison of previously unpublished data and highlight that this observation appears to be independent of particle aspect ratio.

Bubble suspensions

While the importance of gas bubbles for the rheology of magmas had been evident for several decades (Einarsson 1949), experimental measurements on natural magmatic bubbly melts remained largely absent until the late 80's and early 90's (Spera et al. 1988; Bagdassarov and Dingwell 1992; Stein and Spera 1992; Bagdassarov and Dingwell 1993a; Lejeune et al. 1999). These were paralleled by systematic experimentation on bubble-bearing analogue materials (e.g., Kraynik 1988 and references therein) and theoretical approaches (Taylor 1932; Schowalter 1978; Manga et al. 1998; Manga and Loewenberg 2001). Together, the numerical and experimental evidence from both analogue and natural bubble suspensions document that the presence of bubbles can both increase and decrease the effective viscosity of a suspension depending on the force balance around the suspended bubble. The theoretical framework defining the differing flow regimes in magmatic bubble suspensions was first rigorously defined in Rust and Manga (2002) and Llewellin et al. (2002b). These studies document that under steady flow conditions the force balance around bubbles is well described by the capillary number (Eqn. 3.5).

At low capillary number, bubbles act to increase suspension viscosity since the bubbles remain spherical and thus represent an obstacle to the flow field. At constant bubble volume and increasing capillary number, this effect decreases with increasing bubble elongation. At large capillary numbers, bubbles are highly elongate and flow line distortion is small, thus bubbles act to decrease suspension viscosity as they introduce free slip surfaces in the liquid (Rust and Manga 2002). The effect is plotted for a range of Ca in Figure 6.6A, Equation 6.51.

Under changing flow conditions, the capillary number has to be described as the ratio of the bubble relaxation time λ_b to the fluid deformation timescale $(1/\dot{\varepsilon})$, thus Ca$= \lambda_b \dot{\varepsilon}$. The bubble relaxation time describes the timescale on which a deformed bubble returns to spherical shape under the action of surface tension in the absence of shear or, similarly, the timescale over which a bubble can respond to a change in its shear environment. For a single bubble in a Newtonian liquid, the bubble relaxation time is:

$$\lambda_b = K_b \frac{\eta_0 a}{\Gamma} \tag{6.40}$$

where K_b is a dimensionless constant that is a function of the bubble volume fraction. For dilute systems, $K_b = 1$, i.e., the bubble does not interact with other bubbles or particles. In non-dilute suspensions, $K_b > 1$; the bubbles interact with each other and their deformation is affected by neighboring bubbles. Theoretical and experimental investigations in this regime have found that in such systems λ_b increases with increasing gas volume fraction (Oldroyd 1953; Oosterbroek and Mellema 1981; Loewenberg and Hinch 1996; Llewellin et al. 2002b; Rust and Manga 2002).

For flow to be steady, the shear conditions must either remain constant or change slowly with respect to λ_b. A further complication arises when flow is unsteady (i.e., strain-rate varies as a

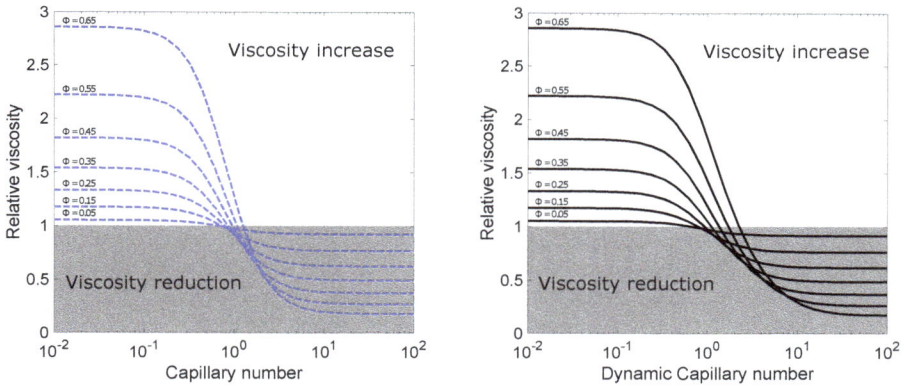

Figure 6.6. The effect of bubble relaxation on relative viscosity under **A)** steady flow conditions, where Ca is the relevant dimensionless parameter for bubble shape and **B)** unsteady flow conditions, where Cd is the relevant dimensionless parameter for bubble shape. Note that the difference in flow condition is not of drastic effect to the relative viscosity. An approximate solution for conditions where both parameters vary was proposed in Mader et al. (2013) (see Eqn. 6.52)

function of time, where the rate- of change in strain rate may vary as well). This is often the case for volcanic flows when a magma batch moves through the conduit to the surface. Unsteady flow introduces an effect of bubble elasticity resulting from the bubble surface tension and thus, in an unsteady flow, energy can be stored and released during changes in the shear conditions. The rheology thus becomes viscoelastic and requires quantification of both the viscous and elastic moduli (Bagdassarov and Dingwell 1993b; Llewellin et al. 2002a,b; Rust and Manga 2002)

Rust and Manga (2002) and Llewellin et al. (2002b) consider a characteristic timescale for the rate of change of the shear environment (λ_b) that describes the time required for the change in strain-rate ($\ddot{\varepsilon}$) to be of the same magnitude as the strain-rate itself, i.e., $\lambda_b = \frac{\dot{\varepsilon}}{\ddot{\varepsilon}}$. If $\lambda_b < \frac{\dot{\varepsilon}}{\ddot{\varepsilon}}$, the bubbles can adjust their shape sufficiently fast to maintain an equilibrium shape (corresponding to the specific strain-rate environment) and flow can be considered steady. On the other hand, if $\lambda_b > \frac{\dot{\varepsilon}}{\ddot{\varepsilon}}$, the bubbles do not manage to reach an equilibrium shape on the timescale of change in the strain rate environment and the flow is unsteady. To describe this flow steadiness Llewellin et al. (2002b) introduce the dynamic capillary number:

$$Cd = \lambda_b \frac{\ddot{\varepsilon}}{\dot{\varepsilon}} \tag{6.41}$$

when Cd ≪ 1 bubble relaxation is fast with respect to the timescale of change in strain rate; the flow is approximately steady, and the viscosity of the suspension can be described by the capillary number Ca. In this regime the viscosity of the bubble suspension can increase or decrease with Φ_b depending on the bubble shape. When Cd ≫ 1, the bubbles do not relax fast enough to reach equilibrium shapes; Ca is then undefined, and the bubbles deform with the flow. Thus flow-line distortion past the bubble is reduced, inducing free-slip surfaces. The strain will then be accommodated via changes in the bubble shape and, since their viscosity is negligible with respect to the viscosity of the melt, the effective viscosity of a bubble suspension in unsteady flow is lower than the viscosity of the bubble free melt (i.e., $\eta_r < 1$). This results in a decrease in the suspension viscosity as Φ_b increases. The effect is plotted for a range of Cd in Figure 6.6B, Equation (6.52).

Early models on bubble suspension rheology, describing the relative suspension viscosity as a function of bubble volume fraction Φ_b did not incorporate this variation in deformation regime and are thus, very restricted in their applicability. The first model of this kind was presented by Taylor (1932) who extended Einstein's theory to emulsions of two liquids and derived Equation (6.42) (Table 6.4) for a suspended phase of negligible viscosity. Later experiments on bitumen emulsions by Eilers (1941) and Eilers (1943) found a more pronounced increase in relative viscosity at higher Φ_b and they presented a model with two different fitting parameters (Eqn. 6.43, Table 6.4). Both these approaches assume steady deformation (Cd $\ll$ 1) and that the bubbles' deformation from spherical is negligible (Ca $\ll$ 1). Thus, due to the distortion of flow lines around the bubbles, the relative viscosity increases with increasing Φ_b (Fig. 6.7A). Mackenzie (1950) presented a theoretical study on the mechanics of a solid containing spherical holes that takes the form of Equation 6.44, Table 6.4. Although this theory can in principle be applied to describe bubble suspensions, it neglects surface tension and thus only holds for flow conditions where the bubbles maintain their spherical shape (Ca $\ll$ 1). Later models are based on experiments within the steady regime (Cd $\ll$ 1) but at varying degrees of bubble deformation (Ca > 1). Thus, in these models, due to the reduction in distortion of flow lines around the bubbles with elongation, the relative viscosity decreases with increasing Φ_b (Fig. 6.7B).

Hashin and Shtrikman (1963) presented a modified theoretical approach for a suspension of randomly positioned spherical bubbles, describing the viscosity reducing bubble deformation regime (Eqn. 6.45, Table 6.4). Subsequently, a number of studies were presented that measured and empirically parameterized the effect of porosity on the relative viscosity of suspensions within natural and synthetic glass powder analogues during compaction (Eqns. 6.46–6.48, Table 6.4). Note that we use ϕ for models considering pore space between granular materials whereas we use Φ_b for models considering bubbles in a liquid, the important difference between these two textures is discussed in detail below.

The first parameterization of experiments on bubbly volcanic melts were presented by Bagdassarov and Dingwell (1992), who describe a marked decrease in relative viscosity with increasing bubble volume fraction (Eqn. 6.49, Table 6.4). More recently, Vona et al. (2016) performed uniaxial compression experiments on foamed rhyolites at bubble fractions up to 0.65 and documented a similarly strong decrease in relative viscosity, which they fit using a modified version of the model proposed by Ducamp and Raj (1989) that allows more flexibility in the model shape at low Φ_b and takes the form of (Eqn. 6.50, Table 6.4).

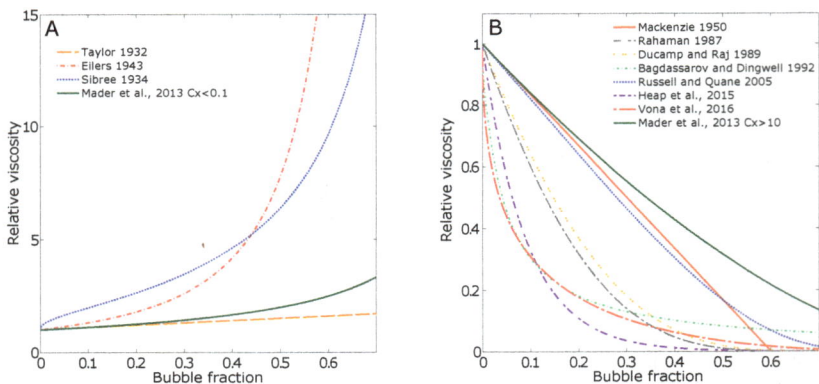

Figure 6.7. Models of relative viscosity vs.= bubble volume fraction for steady deformation (Cd $\ll$ 1) for **A)** bubble deformation from spherical is negligible (Ca $\ll$ 1) using the equations given in Table 6.4. We plot Mader et al. (2013) for Cx < 0.1 and **B)** varying degrees of bubble deformation (Ca > 1). We plot Mader et al. (2013) for Cx < 10.

The striking variation in the predicted relative viscosities plotted in Figure 6.7A,B highlight the textural complexity of the problem. It is worth noting that models derived from compaction of particulate materials presented by Rahaman et al. (1987), Ducamp and Raj (1989), Quane and Russell (2005), Quane et al. (2009) and Heap et al. (2014) predict a much more pronounced reduction on relative viscosity with increasing Φ_b than the theoretical or semi theoretical models of Mader et al. (2013) and Mackenzie (1950), which consider bubbly melts and unconnected pore space, respectively. Only the model presented in Quane and Russell (2005) shows reasonable agreement with the theoretical or semi theoretical models (Fig. 6.7B). This discrepancy between data describing bubbly liquids and those describing compacting granular material is likely because in compacting granular material the pore space is both non-spherical and highly connected. Because the pore space is highly connected, it forbids treating the pores as separate entities since gas can readily be displaced within the pore space (i.e., move within

Table 6.4. Equation for parameterization the flow of bubbly or porous melts.

Equation	Reference	Comment	Eqn.
$\eta_r = 1 + \Phi_b$	Taylor (1932) Eqn. 19		(6.42)
$\eta_r = \left(1 + \dfrac{1.25\Phi_b}{1 - b\Phi_b}\right)^2$	Eilers (1943) Eqn. 3	With $b = 1.28$–1.30.	(6.43)
$\eta_r = 1 - \dfrac{5}{3}\Phi_b$	Mackenzie (1950) Eqn. 24		(6.44)
$\eta_r = 1 - \dfrac{5\Phi_b}{3 + 2\Phi_b}$	Hashin and Shtrikman (1963) Eqn. 3.17		(6.45)
$\eta_r = e^{-b\phi}$	Rahaman et al. (1987) Eqn. 35	With $b = 11.2$	(6.46)
$\eta_r = e^{-b\left(\frac{\phi}{1-\phi}\right)}$	Ducamp and Raj (1989) Eqn. 8	With $2.5 < b < 4$	(6.47)
$\eta_r = 10^{\frac{\alpha\phi}{1-\phi}}$	Russell and Quane (2005) Quane et al. (2009) Heap et al. (2014) Eqn. 7	Where α is a material-dependent fitting coefficient: $\alpha = 5.3$ for glass beads, $\alpha = 0.78$ for Bandelier Tuff ash. $\alpha = 2$ for crystal-bearing melts from Mount Meager.	(6.48)
$\eta_r = \dfrac{1}{1 + b\Phi_b}$	Bagdassarov and Dingwell (1992) Eqn. 5	With $b = 22.4$	(6.49)
$\eta_r = e^{-\alpha\left(\frac{\Phi_b}{1-\Phi_b}\right)^\beta}$	Vona et al. (2016) Eqn. 4	With $\alpha = 1.47$ and $\beta = 0.48$.	(6.50)
$\eta_r = \eta_{r\infty} + \dfrac{\eta_{r0} - \eta_{r\infty}}{1 + (KCa)^m}$	Rust and Manga (2002) Eqn. 9 Llewellin et al. (2002b)	With $K = 6/5$ and $m = 2$ η_{r0} and $\eta_{r\infty}$ represent the relative viscosity at the low and high Ca limit	(6.51)
$\eta_r = \eta_{r\infty} + \dfrac{\eta_{r0} - \eta_{r\infty}}{1 + Cx^m}$	Mader et al. (2013) Eqn. 32	With $Cx = \sqrt{Ca^2 Cd^2}$, $\eta_{r0} = (1 - \Phi_b)^{-1}$, $\eta_{r\infty} = (1 - \Phi_b)^{\frac{5}{3}}$ For details please see the text in the following paragraphs	(6.52)
$\eta_r = \dfrac{1}{1 - (b\Phi_b)^{\frac{1}{3}}}$	Sibree (1934) Eqn. 1 optimizing Hatschek (1911)	With $b = 1.2$	(6.53)

or escape from the system). The flow process in such compacting materials, where gas escapes through a compacting particulate framework is very different to that of a bubbly liquid that flows as a unit and thus flow lines are affected by the presence and nature of bubbles. Nonetheless, both the data (Fig. 6.8) and derived models (Fig. 6.7) presented by Bagdassarov and Dingwell (1992), Bagdassarov and Dingwell (1993b) and Vona et al. (2016), that were developed on the basis of experiments on bubbly melts, also predict a more pronounced reduction on relative viscosity with increasing Φ_b than the theoretical or semi theoretical models of for example Mader et al. (2013), Manga and Loewenberg (2001) and Mackenzie (1950), see Figure 6.8.

Re-evaluation of the data of Vona et al. (2016); (Vona pers. comm., and Sicola et al. 2021) reveal that the measurements likely underestimate the effective viscosity of the sample. This is because the samples developed high connectivity between the bubbles due to bubble wall breakage during post foaming quench. This alteration of sample texture resulted in deformation behavior more akin to collapsing foam or compacting particle mixtures and, hence, drastically reduced effective viscosity. New experimental data by Sicola et al. (2021) that were measured immediately after foaming and without any thermal or mechanical stressing between foaming and viscosity measurement document a less pronounced viscosity reduction and better agreement with the models derived from analogue suspensions. Further, re-evaluation of the data reduction strategy of Bagdassarov and Dingwell (1992); (for details please refer to *"Experiments on high temperature silicate melt suspensions"* section) indicates that these measurements also underestimate the effective viscosity and that corrected values of relative viscosity are higher than

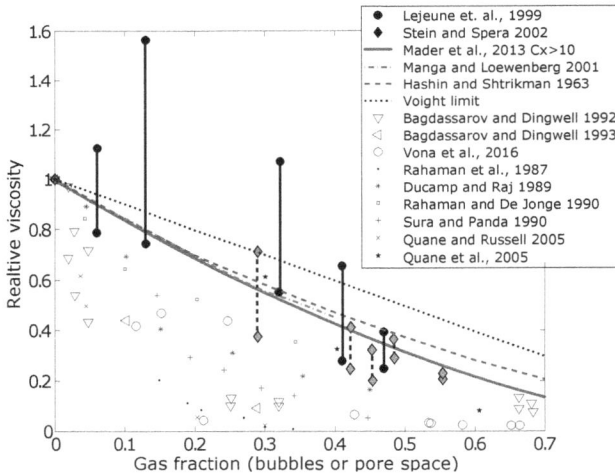

Figure 6.8. Available experimental data on porous melts and bubble suspensions. Maximum and minimum values of validated rheological data on bubble suspensions are plotted as large filled symbols and the spread at each bubble volume fraction is highlighted by the connecting bar. Note that the data from Lejeune et al. (1999) were measured at varying temperatures and thus there may be an error in bubble volume fraction due to bubble expansion/contraction when varying experimental conditions, which is the likely cause for the spread in the measured relative viscosities. Data from rheological measurements for which under-estimation of the relative viscosity is likely are plotted as **large open symbols**, see main text for details. Data from experiments on compaction of a melt framework with communicating pore space are plotted as **small symbols**. Note that while these experimental data are valid for sintering and compaction of fragmental volcaniclastic materials, the presence of an interconnected pore network is not representative of bubbly magma. This compilation highlights that validated measurements on natural bubbly melts agree reasonably well with theoretical and empirical models derived from bubble suspensions of analogue materials, potentially resolving the highly debated spread in measured values and consoling measurements on natural materials with those on analogue suspensions. **Lines** represent different theoretical and semi theoretical models on bubble suspensions that are validated by measurements on analogue materials.

predicted by the respective models (Fig. 6.8). Re-processing of the data following the improved data reduction mechanism proposed in Section 4 (Eqn. 4.10) might allow to unite these data with those of analogue experiments, theoretical models and the experiments of Lejeune et al. (1999) and Stein and Spera (2002).

Nonetheless, after accounting for the parallel plate viscometry data deviating from the models derived from analogue suspensions, the low effective viscosities reported from the in oscillatory measurements presented in Bagdassarov and Dingwell (1993b) remain at odds with the remaining dataset. This is likely due to the fact that the strain in oscillatory measurements is extremely small. These measurements are advantageous for probing sample viscosity without imposing any relevant textural deformation or re-orientation, as it is the case during for example the viscous dissipation of seismic energy in magmatic bodies. However, these measurements are not directly applicable to magma migration and lava flow since the bubbles do not deform significantly or reach an equilibrium shape and, thus, their contribution to the effective shear viscosity during flow is not fully expressed. Similar strain dependent effects in oscillatory measurements have been documented for particle suspensions by Sumita and Manga (2008), who present a study of the rheology of analogue suspensions of spherical particles. They document a strain dependent rheological evolution during oscillation, where the suspension displays three rheological regimes. At low strain amplitude, the response is linear viscoelastic, becomes shear-thinning with increasing strain amplitude and transitions to shear-thickening at high strain amplitudes. Oscillatory measurements at low strain amplitude are nonetheless highly valuable for systematic development of a strain-dependent understanding of the effect of bubbles on suspension rheology, because the measurements represent the low strain endmember. Development of a complete, strain dependent, dataset would require correlation of frequency dependent data to data from methods probing higher strain regimes. To date, this remains challenging, largely due to the paucity of experimental data and because torsion experimentation following in the above reviewed studies focused chiefly on deformation in simple shear and much higher total strains.

The variation in flow process and the re-evaluation of the data of Vona et al. (2016), Bagdassarov and Dingwell (1992) and Bagdassarov and Dingwell (1993b) potentially allows to address the large spread in the models predicting a reduction on relative viscosity in Figure 6.7B, narrowing the spread of models applicable to bubbly melts. However, it does not allow reconciliation of the apparent contradiction between the theoretical and experimental studies that describe η_r increasing with Φ_b (Fig. 6.7A) and those that describe the opposite (Fig. 6.7B). Doing so requires a model that also incorporates the different flow regimes and bubble deformation timescales and thus can describe both the increase in relative viscosity for Ca < 1 and the decrease of the relative viscosity in the case of Ca > 1 as well as the dynamic changes of steady to unsteady flow regimes. To capture this broad range, both Llewellin et al. (2002b) and Rust and Manga (2002) performed experiments across a wide range of Ca and Cd. Both studies confirm that bubbles may increase or decrease the viscosity of a suspension depending on the conditions of shear.

Based on the work of Frankel and Acrivos (1970), who present a complete physical analyses of emulsion rheology and derive a constitutive equation that is restricted to dilute emulsions of high viscosity contrast and where the bubbles remain approximately spherical, Llewellin et al. (2002b) and Rust and Manga (2002) propose an expression for steady, simple-shear flow (i.e., Cd $\ll$ 1; Ca variable) in the form of a Cross model (Cross 1965):

$$\eta_r = \eta_{r\infty} + \frac{\eta_{r0} - \eta_{r\infty}}{1 + \left(K Ca\right)^m} \tag{6.51}$$

where $K = 6/5$ and $m = 2$, and η_{r0} and $\eta_{r\infty}$ represent the relative viscosity of the suspension at the low and high Ca limit, respectively

On the basis of the work of Pal (2003) and Llewellin and Manga (2005); Mader et al. (2013) generalize the low (η_{r0}; Ca < 0.1) and high ($\eta_{r\infty}$; Ca > 10) capillary number limits for application to non-dilute suspensions, proposing:

$$\eta_{r0} = \left(1 - \Phi_b\right)^{-1} \tag{6.52a}$$

$$\eta_{r\infty} = \left(1 - \Phi_b\right)^{\frac{5}{3}} \tag{6.52b}$$

Further, Mader et al. (2013) extend the work presented in (Llewellin et al. 2002a) to unsteady flow in simple-shear encompassing large changes in strain and strain rate (i.e., both Ca and Cd may vary simultaneously and independently). They develop a complex general expression for η_r (Ca, Cd), based on which they propose the following approximation that shows good agreement with the more complex theoretical relationships:

$$\eta_r = \eta_{r\infty} + \frac{\eta_{r0} - \eta_{r\infty}}{1 + Cx^m} \tag{6.52}$$

where $m = 2$ for a monodisperse suspension and Cx is the capillarity, defined as:

$$Cx = \sqrt{Ca^2 Cd^2} \tag{6.52c}$$

The two end member cases for the low (η_{r0}; Ca < 0.1) and high ($\eta_{r\infty}$; Ca > 10) capillary number limits are plotted as dotted lines in Figure 6.7A and 6.7B, respectively. Note that the parameterization proposed in Mader et al. (2013) covers the whole range of Ca and Cd and thus represents a completely new analysis that demonstrates that the dependence of viscosity on Ca and on Cd is practically indistinguishable and can be reduced to the newly introduced term Cx, the capillarity, which captures the combined effect of shear and flow steadiness, effectively reducing the dimensionality of the problem. However, as evident from the summary plot in Figures 6.7B and 6.8, the model does not reproduce the drastic decrease in relative viscosity depicted in some experimental data. This is likely owing to the fact that most experimental data derive from measurements of polydisperse suspensions, which are not described by the model as well as the variations in sample texture and pore space connectivity outlined earlier.

Figure 6.7A also shows the model deduced from high Ca experiments on bubbles in a fungicide liquid stabilised with an organic solvent by Sibree (1934) which is based on the model for foam rheology presented in Hatschek (1911):

$$\eta_r = \frac{1}{1 - \left(b\Phi_b\right)^{\frac{1}{3}}} \tag{6.53}$$

with $b = 1.2$. These experiments show an increase in relative viscosity albeit being performed at high capillary numbers. This is because the textural framework of the suspension transitions from a bubbly liquid to a polyhedral foam at $\Phi_b \gtrsim 0.5$–0.7 (depending on the nature of the bubbly liquid i.e., the maximum packing fraction for the bubble population). The rheology of a bubbly melt can change drastically across this transition due to self-stabilization of the bubble walls as they transition from curved bubble walls to more straight melt films that intersect at plateau boarders in sets of three. This structure is not stable at high strain and can thus result in rapid breakdown of the foam through bubble wall failure and gas escape (Ryan et al. 2019a). A constitutive description of flow in this regime has, however, not been published to date.

In summary, comparing available models for the relative viscosity of bubble suspensions (Fig. 6.7) with those for particle suspensions (Fig. 6.1) it is evident, that increasing the bubble volume fraction affects the relative viscosity of a suspension much less than adding the same volume fraction of solid particles. While many parameterization strategies have been proposed for bubble bearing melts, there is no systematic agreement between them over any region of Φ_b and no single functional shape seems able to fully describe the complexity of the process. This is largely rooted in two reasons: 1) the fact that bubbles have drastically varying effects on suspension viscosity depending on the deformation regime they encounter (described by the bubble's capillary number which itself is a function of the bubble size and the properties of the suspending liquid), and 2) at elevated bubble fractions ($\Phi_b \gtrsim 0.5–0.7$) the textural framework of the suspension transitions from a bubbly liquid to a foam and its rheology can change drastically across this transition including the development of yield stress via self-stabilization of bubble walls.

The re-evaluation and of existing models presented in this section reveals that to date, the theoretical or semi theoretical models of Mader et al. (2013) and Mackenzie (1950) (black lines in Figure 6.7B) represent the most robust parameterization of the rheology of bubbly melts (with the Mader et al. (2013) model also allowing to parameterize varying Ca and Cd). Based on available measurements on magmatic suspensions it seems that for most experimentally investigated cases, flow occurs in the regime of Ca $\gg$ 1 and Cd $\gg$ 1 and thus bubbles in most flow scenarios relevant to magma emplacement or lava flow on the surface, will act to decrease the effective viscosity of the magmatic suspension. This review, however, also highlights that available experimental data and field measurements are scarce. Significant effort and careful and systematic experimentation on bubble suspensions of a wide range of textural makeup is needed for validation and / or expansion of these models and to make them usable to assess deformation in magmatic and volcanic environments. Further, the discrepancy between experimental data on analogue materials and natural and synthetic silicate melts remains to be addressed.

Unsteady flow in bubble suspensions. In oscillatory measurements the imposed strain is commonly small and thus, the suspended bubbles remain approximately spherical for most experimental conditions. However, the bubble relaxation timescale remains an important contributor. This was studied in Llewellin et al. (2002b), who provide a comprehensive dataset on unsteady flow in bubbly suspensions, which served as the basis to define the dynamic capillary number Cd (Fig. 6.6B). At low Cd and Ca, the bubble relaxation timescale is faster than the deformation and thus the bubbles remain spherical and act to increase the effective viscosity of the suspension. With increasing oscillation frequency (i.e., increasing Cd) bubbles remain spherical (due to the low imposed strain) but deformation is faster than the bubble relaxation timescale. Thus, the bubbles accommodate deformation through internal strain and act to decrease the effective viscosity of the suspension. Further, the specific effect of the suspended phase (for every investigated Ca and Cd regime) becomes more important with increasing strain and increasing bubble volume fraction. Combined with the existing understanding of bubble–melt–interface tension, the effect of bubbles in unsteady flow is described well by the parameterization framework reviewed earlier in this chapter.

Three-phase suspensions (crystal + bubble + melt)

Few experimental data on magmatic three phase suspensions are available (see the section on high temperature experiments). The available studies commonly observe pseudo-plastic behavior and a strong shear thinning component for all the investigated suspensions and provide empirical equations describing the effective viscosity (η_{eff}) or stress as a function of temperature and strain rate for multiphase magmas (Russell and Quane 2005; Lavallée et al. 2007; Avard and Whittington 2012; Pistone 2012; Pistone et al. 2013; Heap et al. 2014). The differences between these empirical models are small and, nonetheless, a new empirical description was proposed with almost every new experimental dataset. This highlights that, while empirical models are operationally attractive due to their simplicity and are seemingly robust in the light of the respective dataset, a single empirically derived equation for the relative viscosity of a specific lava as a function of temperature is insufficient to capture the complexity of lava dome rheology (See also Fig. 4.7). A more robust parameterization thus requires detailed characterization of the sample texture and its composition to allow application of the parameterization approaches reviewed above.

While few experimental studies have aimed to parameterize the rheology of crystal and bubble-bearing magmas in this way, several theoretical or semi theoretical approaches have employed effective medium theory to do so. Effective medium theory treats multiphase and polydisperse suspensions as incremental products of suspensions of the phase in question within a medium that is characterised by the viscosity of a suspension made of the other phase. For example, Farris (1968) proposed that bidisperse suspensions may be treated as a system where the coarse solid fraction acts as a monodisperse suspension in a separate, monodisperse fluid containing the finer particles (Eqn. 6.16). In order to apply this approach, it is crucial to clearly define the volume fractions of the respective phases in the suspension, as these are the core input parameter to any model employed in effective medium theory.

Volume fractions in three phase suspensions. Particle (crystal)- and gas volume fraction are strictly defined for any suspension when considering the entirety of the suspension. In this case, the concentration of the total suspended phases Φ_t is defined as:

$$\Phi_t = \frac{\Phi_b + \Phi_c}{\Phi_l + \Phi_b + \Phi_c} \tag{6.54}$$

where Φ_l, Φ_c and Φ_b are the volume fraction of liquid, particles and bubbles, respectively.

This strict definition, however, does not apply directly when employing effective medium theory for multi-phase suspensions and varies depending on whether one considers a bubble suspension or a particle suspension as the effective medium. Thus, it is important to clearly define what process is to be investigated when employing effective medium theory. When the bubble suspension is treated as the effective medium, the appropriate definitions are:

$$\Phi_b = \frac{\Phi_b}{\Phi_l + \Phi_b} \tag{6.55}$$

$$\Phi_c = \frac{\Phi_c}{\Phi_l + \Phi_b + \Phi_c} \tag{6.56}$$

This approach is favorable for example when investigating the crystallization of a bubbly magma at constant bubble volume fraction. Whereas if the particle (crystal) suspension is treated as the effective medium, the appropriate definitions are:

$$\Phi_b = \frac{\Phi_b}{\Phi_l + \Phi_b + \Phi_c} \tag{6.57}$$

$$\Phi_c = \frac{\Phi_c}{\Phi_l + \Phi_c} \tag{6.58}$$

This approach is favorable for example when investigating the vesiculation of a magma at constant crystal volume fraction.

Applying effective medium theory. Having defined the volume fractions relevant to the scenario of interest, effective medium theory is applied as follows:

When the bubble suspension is treated as the effective medium, the viscosity of the three-phase suspension (η_{eff}) is the product of the viscosity of the effective medium ($\eta_b\Phi_b$) and the relative viscosity of crystals ($\eta_{r,c}\Phi_c$):

$$\eta_{eff} = \eta_b(\Phi_b) \times \eta_{r,c}(\Phi_c) \tag{6.59}$$

with:

$$\eta_b\left(\Phi_b\right) = \eta_l \times \eta_{r,b}\left(\Phi_b\right) \tag{6.60}$$

where η_l is the viscosity of the liquid phase and $(\eta_{r,b}\Phi_b)$ is the relative viscosity of bubble suspension and Φ_b and Φ_c are calculated with Equations (6.55) and (6.56).

This approach was applied in Truby et al. (2015), where the effect of particles $(\eta_{r,c}\Phi_c)$ is calculated with the Maron and Pierce (1956) equation (Eqn. 6.6), whereas the effect of bubbles is calculated following the approach presented in Mader et al. (2013). Under their experimental conditions, bubbles form the effective medium and deform under low Ca. On this basis, Truby et al. (2015) provide an empirical parameterisation for the flow index :

$$n = 1 - 0.2\left(\frac{\Phi_c}{\Phi_m}\right)^4 - 0.334\Phi_b \tag{6.61}$$

Furthermore, this treatment allowed Truby et al. (2015) to explicitly include variations in Φ_m, including the effects of particle aspect ratio as well as surface roughness. Note that under the condition of $\Phi \sim \Phi_m$ or $\Phi_m = 1$; Equation (6.65) reduces to the Phan-Thien and Pham (1997) model (see below). This approach was applied for example in Kilgour et al. (2016) and Beckett et al. (2014) to reconstruct the rheology of multiphase magmas for the examples of Mt. Ruapehu and Stromboli, respectively.

When the particle suspension is treated as the effective medium, the viscosity of the three-phase suspension (η_{eff}) is the product of the viscosity of the effective medium ($\eta_c\Phi_c$) and the relative viscosity of bubbles ($\eta_{r,b}\Phi_b$):

$$\eta_{eff} = \eta_c\left(\Phi_c\right) \times \eta_{r,b}\left(\Phi_b\right) \tag{6.62}$$

with:

$$\eta_c\left(\Phi_c\right) = \eta_l \times \eta_{r,c}\left(\Phi_c\right) \tag{6.63}$$

where η_l is the viscosity of the liquid phase and $\eta_{r,c}(\Phi_c)$ is the relative viscosity of the crystal suspension and Φ_b and Φ_c are calculated with Equations (6.57) and (6.58). This approach was applied to estimate the viscosity of basaltic lava flows, using the crystal suspension as the effective medium (Rhéty et al. 2017; Dietterich et al. 2018; Chevrel et al. 2019b). Recently, Chevrel et al. (2018), considering the effect of deformable bubbles (Eqn. 6.62) on a crystal suspension effective medium (Eqn. 6.63), provide a robust reconstruction the viscosity of an active pāhoehoe lava that was measured independently using a field viscometer.

An earlier theoretical parameterization was proposed by Phan-Thien and Pham (1997) and later applied by Harris and Allen (2008) to study of basaltic magmas from Mauna Loa and Mount Etna. Depending on the relative size of particles and vesicles, Phan-Thien and Pham (1997) propose three separate equations to capture the multiphase flow behavior:

1. For crystals smaller than vesicles:

$$\eta_{eff} = \eta_l\left(1 - \frac{\Phi_c}{1-\Phi_b}\right)^{-\frac{5}{2}}\left(1-\Phi_b\right)^{-1} \tag{6.64}$$

2. For crystals and vesicles of equal size:

$$\eta_{eff} = \eta_l\left[1 - \Phi_c - \Phi_b\right]^{\frac{-(5\Phi_c + 2\Phi_b)}{2(\Phi_c + \Phi_b)}} \tag{6.65}$$

3. For crystals larger than vesicles:

$$\eta_{\text{eff}} = \eta_1 \left(1 - \frac{\Phi_b}{1-\Phi_c}\right)^{-1} (1-\Phi_c)^{-\frac{5}{2}} \tag{6.66}$$

In these equations, Φ_c and Φ_b are the volume fraction of crystals and bubble, respectively, relative to the liquid faction (Φ_1) and $\Phi_c + \Phi_b + \Phi_1 = 1$.

Note, that the treatment of Phan-Thien and Pham (1997) does not account for textural variations and is only applicable to spherical particles. Further, it does not account for strain-rate dependent effects or allow for substitution of different models for the rheology of either component of the suspension.

In summary, while effective medium theory is a promising approach, especially in the dilute regime, this method exclusively uses models for two-phase suspensions and assumes that bubbles remain unaffected by the crystal framework or *vice versa*. The resulting effective viscosity of the multiphase suspension is then simply defined as:

$$\eta_{\text{eff}} = \eta_1 \times \eta_{r,c} \times \eta_{r,b} \tag{6.67}$$

where relative viscosity of crystals and bubbles can be estimated using any of the models presented in Table 6.1 (Fig. 6.1) and in Table 6.4 (Fig. 6.7), respectively, but considering the phase fraction depending on the considered effective medium, as described above.

Bubble–crystal interactions that occur in a natural magma can therefore not be captured by this simple series of two separate models. Nonetheless, until a full three-phase rheological model is developed, effective medium treatment of multiphase suspensions can be considered the most reasonable approach, for investigation of dynamic changes in suspension rheology for example in modeling applications or for the reconstruction of the rheology of a sample recovered from the field. It is however important to be aware of the assumptions and limitations tied to the selected approach and the simplifications and generalizations that need to be made when selecting and applying the appropriate equations because the individual models for multiphase rheology are very sensitive to the required input parameters.

The yield stress dilemma

The existence of a yield stress resulting from particle–particle interactions has broadly been accepted in lava and magma rheology. The formation of such interacting particle networks varies with particle concentration, size and shape. Yield stress is observed at lower particle concentrations for higher particle aspect ratios due to the more efficient particle–particle interaction. Further, yield stress tends to be larger for smaller particles as these may easily achieve high packing densities and thus generate stronger networks (Heymann et al. 2002). A numerical attempt to define the critical particle volume fraction at which touching particle networks may form in non-sheared suspension was presented by Saar et al. (2001).

However, the existence of a yield stress *sensu stricto* has been disproven for a broad variety of materials including solids, soft solids, liquids and suspensions; for details see Barnes (1999) and references therein. It has been shown that, although many materials display a dramatic change in mechanical properties over a small range of stress (i.e., an apparent yield stress), these materials, after an initial linear elastic response to the applied stress, display steady, slow deformation (i.e., Newtonian behavior) when stressed for long times below this apparent yield stress. Barnes (1999) presents a comprehensive review on this issue and highlights that this transition from very high (creep) viscosity ($>10^6$ Pa s) to mobile liquid (<0.1 Pa s) may take place over a single order of magnitude in applied stress. He highlighted that, when viewed on a linear plot, this extreme behavior may easily lead to the interpretation that the material possesses a yield stress. In fact, in many cases the flow curves appear to be adequately described by a simple Bingham equation. He noted, however, that if viewed on a logarithmic basis (Fig. 6.9), a pattern of Newtonian / power-law / Newtonian flow behavior at low / transitional / and high stresses, respectively, emerges, leading him to cite the Heraclitan expression πάντα ῥεῖ (panta rhei) "everything flows".

Although Barnes (1999) shows that yield stresses do not exist as a physical property (i.e., as a critical stress below which no flow takes place), the concept of a yield stress has proved very useful with application to magma migration and lava flow emplacement (Hulme 1974). This is largely because the radical transition in flow response rate across the apparent yield stress change zone pushes the magmatic suspension in a regime where the timescales of other processes such as heat loss and crystallization act on much faster timescales than the deformation timescale and hence the material can be treated as effectively solid under those conditions. Apparent yield stress in magmas and lavas is reported in several studies (Shaw 1969; Murase and Mc Birney 1973; Pinkerton and Stevenson 1992; Pinkerton and Norton 1995; Zhou et al. 1995; Philpotts and Carroll 1996; Cashman et al. 1999; Hoover et al. 2001; Heymann et al. 2002) but no link to experimental measurements is made. Attempts to join morphology derived yield stress estimates and model-based yield stress estimates (based on the textural properties of lavas) were presented in Chevrel et al. (2013b) and Castruccio et al. (2014). Nonetheless, care has to be taken when applying the concepts derived from experiments aimed at investigating yield stress effects (Fink and Griffiths 1992; Griffiths and Fink 1992) to natural scenarios. It should be considered, that once a lava flow crust is sufficiently solidified to hinder further advance, the flow as a whole it is still capable of deformation but behaves immobile on the time scale of observation. This point (i.e., the apparent yield stress) is therefore very valuable as limiting parameter in models considering the advance of lava flows (Dragoni et al. 1986; Harris and Rowland 2001). A further complication arises when deriving apparent yield stresses from geometrical measurements of lava flows using the approaches presented in for example Hulme (1974) or Moore et al. (1978). For these approaches to be employed accurately it is crucial to constrain the lava emplacement history, as late stage inflation (a common process in lava flows) may drastically increase the recovered apparent yield stresses (Kolzenburg et al. 2018a).

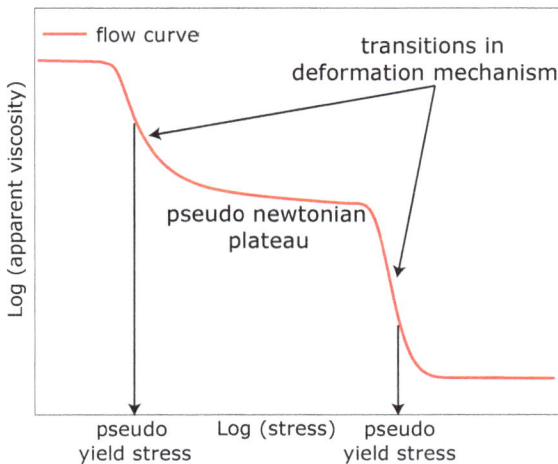

Figure 6.9. Schematic representation of an apparent viscosity measurement as a function of stress (i.e., flow curve) showing the succession of pseudo-Newtonian behavior (plateau) and transitions in deformation mechanism (increase in apparent viscosity with decreasing stress) that are frequently interpreted as pseudo yield stresses (Barnes 1999). One example would be the transition from viscous flow to crystal creep during crystallization of a sample. Since these transitions can be on the scale of up to several orders of magnitude, they frequently exceed the measurement range of a specific measurement device or geometry and are thus easily mistaken for material property limits and thus interpreted as yield strengths.

7. TECHNOLOGICAL ADVANCES

While a number of experimental methods are available for rheometry of homogenous silicate liquids, the selection becomes much more narrow for rheometry of multiphase suspensions. This is largely due to the small sample volumes or small strains that methods like fibre elongation, micro-penetration or falling body experiments, can achieve. The main technological advances for measurement of two- and three-phase suspensions have been made by enhancing the measurement capacity and accuracy of existing methods as well as by combining these measurement approaches with *in situ* imaging or sensing techniques. All these advances are rooted in the desire to better constrain the evolution of textural and thermal properties during flow and, therewith, improve the parameterization efforts reviewed earlier in this chapter.

Advances in experimental devices and methods over the past decades include:

1. High torque and high temperature concentric cylinder rheometers such as the one employed in Stein and Spera (2002). These devices allow to systematically measure the rheology of magmatic suspensions and emulsions over a wide range of viscosity and strain rate, thus generating deformation maps over a wide range of compositions and emplacement scenarios. This reduces the need for piecing together data sets from multiple methods or samples, increasing the accuracy of the results. Further, such devices permit experimentation at high total strain, which is important when considering that strain may affect the rheological response of magmatic suspensions and emulsions (Mueller et al. 2011; Ryan et al. 2019a).

2. High load and high temperature parallel plate rheometers such as the one presented in Hess et al. (2007). These devices are required for very high viscosity samples and to overcome issues of sample scaling (mainly the ratio of sample size with respect to the largest crystal or bubble). This allows for investigation of crystal sizes and size-distributions relevant to natural magmatic systems (Lavallée et al. 2007).

3. In situ thermal measurements during high temperature concentric cylinder rheometry such as developed in Kolzenburg et al. (2016). This allows simultaneous monitoring of both the rheological and thermal evolution of a crystallizing melt, expanding the capacity for measurements in thermal disequilibrium and analysis of process kinetics (Kolzenburg et al. 2018b).

4. Improved instrument capacity for in situ rheological measurements in the field (Chevrel et al. 2019a). While safe access to active lava remains difficult, enhanced (or novel) versions of field rheometers will enable increased measurement precision during three-phase rheometry of natural lavas that is required for benchmarking of analogue and high temperature laboratory experiments.

Advances in in situ imaging techniques over the past decades include high resolution X-ray computed tomography for both 3D and 4D (3D plus time) analysis of magmatic suspensions. 3D characterization has become almost a standard technique for detailed textural analyses of geologic samples, see for example Polacci et al. (2010) for a review. It has also been used for characterization of experimental samples pre- and post-deformation (Okumura et al. 2010; Shields et al. 2014; Ryan et al. 2015; Dobson et al. 2020). These have allowed investigating deformation processes recorded in natural samples at much greater detail than classic 2D petrographic methods. Wright and Weinberg (2009) for example highlight the textural heterogeneity of foamed magmas around solid particles and note profound strain localization within natural samples. They argue that strain localization may produce plug flow within ascending magmas, resulting in localized, strain-rate regime dependent fragmentation in the lead up to wholesale fragmentation of foamed magmas. 4D *in situ* textural characterization during high temperature crystallization, foaming or deformation experiments, facilitated by synchrotron tomography is a rather recent and extremely promising advancement since it allows real time monitoring of the textural evolution the investigated samples during crystallization, foaming and deformation (Song et al. 2001; Polacci et al. 2010, 2018; Madonna et al. 2013; Arzilli et al. 2015; Pistone et al. 2015; Pleše et al. 2018; Dobson et al. 2020). Synchrotron tomography becomes especially valuable for characterization of multiphase flow since all participating components can be monitored in great spatial and temporal detail. 4D tomography is starting to be coupled with devices for rheometry, which may in the future allow for *in situ* measurement of both the crystallization kinetics and the rheological response of evolving natural systems (Dobson et al. 2020). Ohtani et al. (2005) and Raterron and Merkel (2009) presented early viscosity and density measurements of melts and glasses at high pressure and temperature in a synchrotron multi anvil apparatus. An application to natural multiphase magmatic suspensions was published by Okumura et al. (2013) who developed a new apparatus for torsion experiments at high temperature in combination with synchrotron tomography. Haboub et al. (2014) developed a new apparatus for contemporaneous synchrotron radiation tomography and high temperature deformation in compression and tension. This device was recently employed by Tripoli et al. (2019) to study the deformation dependent crystallization kinetics of basaltic melts but, unfortunately, rheological data could not be recovered since the sample was too low in viscosity. Further, a number of abstracts have been presented on 4D rheometry measurements of both natural and analogue multiphase suspensions (Dobson et al. 2016; Coats et al. 2017; Lavallée et al. 2019) but comprehensive 4D-rheological datasets have not been published to date.

8. OUTSTANDING CHALLENGES

At the end of a comprehensive review of magma rheology, McBirney and Murase (1984) noted that little is known about rates and mechanisms of flow at very low stress and over long duration and that, while the importance of thermal feedback mechanisms were identified as very important (Shaw 1969), non-isothermal data are absent. Interestingly, over 30 years later, while much has been learned about magma rheology (especially for silicate liquids) sub-liquidus rheology and the effects of disequilibrium and low stress environments remain largely uncharted. Here we can only re-iterate the concluding statement of McBirney and Murase (1984), that few fields of geological research hold greater potential for enhancing our understanding of basic magmatic and tectonic processes, an insight that was recently highlighted also with respect to the formation and dynamics of trans-crustal magma reservoirs (Sparks et al. 2019). In this chapter, we have summarized the major advances in multiphase suspension rheology of lavas, magmas and their analogues that were achieved over the past decades. Combined with advances in rheological measurement capacities, these produce a constantly growing rheological database that can be used for validation and improvement of empirical and theoretical models. However, as highlighted in the respective sections, there are a number of outstanding challenges that remain to be addressed in order to generate a holistic understanding of magma and lava flow properties. In this section we want to highlight the most notable ones.

Models for multiphase rheology

A constitutive model for bubble-bearing melts (Eqns. 6.51 and 6.52) has been developed and has been validated on a broad range of analogue suspensions. The advantage of this model is that it is based on physical theory, validated by experiments and describes both steady and unsteady flow and thus, the model parameters are directly related to physical processes. Data for magmatic bubble-bearing silicate melts are scarce. The available data have proven problematic for cross correlation with analogue measurements due to the varying nature of deformation in flowing bubble suspensions and compacting or sintering porous particulate material as well as measurement uncertainties (Fig. 6.8) and the much greater viscosity contrast between suspending liquid and the suspended gas phase. To date, few data are available that could be used to cross validate these models (specifically for magmatic suspensions) and for developing adequate experimental strategies. Consequently, providing such data remains an open challenge.

The available models for particle-bearing suspensions are almost exclusively restricted to steady flow conditions and vary widely in accuracy and applicability. Further, they are commonly restricted to narrow textural and deformational (strain and strain-rate) regimes. This is largely because the models are predominantly empirical in nature and no physical theory has been presented that links the model parameters to the physics of flow. The situation becomes worse for multiphase (bubble + crystal or polydisperse crystal-bearing) suspensions where with increasing volume fraction of bubbles or particles, the interaction between the different components becomes more important. While available models provide decent descriptions in the dilute regime (i.e., total fraction of suspended phases $\Phi_t < 0.5$), they fail at high bubble or particle fractions. For example, at $\Phi_b > {\sim}0.5{-}0.7$, bubble suspensions become foam-like, forming polyhedral cells of gas separated by thin films of liquid, which drastically changes their rheology. Equivalently, at particle contents $\Phi > \Phi_m$, the suspension transitions from viscous dominated to a range of shear localization and/or plastic deformation phenomena, which are not captured by most published models (except Costa et al. (2009) and derivate models); for details on these topics see Lavallée and Kendrick (2022, this volume). Thus, while the two-phase end-members (bubbles and particles) of suspension rheology are reasonably well-described individually, the critical future research required to form a holistic description of natural magmas and lavas needs to address the physics and disequilibrium dynamics involved in three-phase rheology, including polydisperse size-distribution and the interaction between bubbles and particles.

Reactive flow and phase dynamics

Most current parameterization efforts assume that the solid and/or bubble fraction has no physical or chemical relation to the liquid in which it is suspended. This is very clearly not the case for magmas and lavas, that are reactive materials, where both crystals and bubbles nucleate and grow during cooling and decompression. These reactions change the textural state as well as the chemical composition and the rheological behavior of the liquid. The magnitude of this effect heavily depends on the initial melt composition and the nature and phase dynamics of the growing crystals, all of which vary as a function of the environmental conditions. The viscosity changes induced by the chemical evolution of the melt may be as small as < 0.1 log units in basaltic lavas, where initially, crystallization does not increase the residual silica content by much but can be much greater than the effect of the suspended crystals once the residual melt becomes more evolved and silica rich. While the fundamental importance of disequilibrium has been recognized early on, data availability remains scarce and significant experimental effort in this field is still required to cover the most relevant compositions and to experimentally map the range of environmental parameters pertinent to flow of natural silicate melts under disequilibrium. Such a database would then allow deducing the underlying process systematics and expanding these into a theoretical description of the flow behavior of magma and lava that includes crystallization kinetics and diffusion of elements between the melt and the suspended solid and gas phases. Such a complete viscosity model of a crystallizing magma would need to include multidimensional rheologic maps (i.e., and expansion of Figs. 4.5 and 4.6 as a 4 or 5 dimensional TTT–rheology diagram that includes variations in stress and strain-rate).

Additionally, a parameter that remains entirely unexplored for natural silicate melt suspension rheology and is only scarcely touched upon for the rheology of analogue suspensions is the effect of bubble dynamics under elevated and changing pressure conditions. As outlined in this chapter, pressure changes modulate the phase dynamics of both gas and crystal phases and are thus critical to the understanding of the rheology of a crystallizing magma during its journey from source to surface. Systematic exploration of the effect of pressure will require rheological measurements during decompression to reproduce the natural dynamics of bubble and crystal nucleation and growth. This goal will, in future research, benefit greatly from the advances in tomographic methods, as these allow tracking the physical phase evolution of magma in 4D.

Filling data gaps

Analogue suspension experiments have covered the largest part of suspension textures encountered in the field. For natural magmatic suspensions at high temperature, however, the data are much more limited. Measurements on natural magmatic particle suspensions are becoming more numerous but they predominantly cover basaltic compositions and rather low viscosities (with $\eta \lesssim 10^4$ Pas). Measurements on high viscosity systems (with $\eta \gtrsim 10^8$ Pas) and bubble-bearing melts are only slowly becoming available and there is a marked gap in compositions and textures in the intermediate field (see Fig. 4.2). This limited coverage of the intermediate range largely reflects the mechanical limitations of the available rheometers and highlights that the development of advanced rheometers, able to cover a wider range of torque, is critical to cover the gap between low torque–high strain concentric cylinder and high load–low strain parallel plate measurements. The current inconsistency between these two methods is not least due to the limitations in total strain that can be imposed in uniaxial compression or in torsion.

Another major limitation to generating a complete understanding of multiphase magma rheology is that research to date largely focused on two-phase suspensions. The advances in two phase suspension rheology that were made over the past decades have now reached a point at which there is sufficiently detailed knowledge to tackle the next greater challenge, that is, the rheology of three-phase magmas (i.e., with suspended bubbles and crystals). Doing so will require an expanded experimental capacity to build a dataset capturing the additional complexities of the physics involved in the interaction between bubbles and particles. Several groups have started to tackle these challenges (Harris and Allen 2008; Caricchi et al. 2011; Pistone et al. 2012; Chevrel et al. 2013b; Shields et al. 2014; Truby et al. 2015; Vasseur et al. 2015; Pistone et al. 2016; Vona et al. 2017) but there is a vast open field of environmental parameters that remains to be explored.

Further, in order to connect the theoretical and experimental insights on multiphase magma rheology to natural flow scenarios it is important that the distinct lack of *in situ* field measurements of lava rheology is addressed. These are currently limited largely due to the logistical and technical difficulties as well as the hazards associated with collecting these data. Expansion of this slim dataset and the narrow range of measurable conditions would be a significant advancement, as these allow for combined rheological, textural and petrographic analysis. These measurements are key to understanding how crystal and bubble content and their temporal and spatial variation affect rheology during lava emplacement. Future field campaigns should, thus, focus on measuring the relevant lava properties (e.g., temperature, rheology and lava texture) as a function of distance from the vent to the front and across the flow, in order to map lava rheology in 4D through the flow. Field measurements will always be constrained by the balance between measurement machinery (Fig. 5.4) and the lava's thermal dynamics. Thus, it is necessary to develop advanced equipment for field measurement that is readily deployable while also improving the measurement uncertainty. The most suitable way to achieving this is likely a combination of rotational and penetration rheometers that allow covering the broad range of viscosities and thermal regimes encountered in nature.

Connecting magma rheology and rock mechanics

Most rheological models for multiphase suspensions break down beyond the semi-dilute regime where, at geologically relevant stresses and strain rates, a transition from the viscous to the elastic field is documented. Costa et al. (2009) and Liu et al. (2017) proposed the first models aiming to account for this transition zone but experimental data for the validation of these models across the entire range of suspended particle factions is unavailable to date. The above review of recent field and laboratory studies of the rheological properties of natural silicate melts highlights that, at present, viscosity and yield stress are treated as separate properties. However, as outlined in the section "the yield stress dilemma", these two properties are closely related and represent two components of the much broader phenomenon of deformation of igneous rocks and melts at high temperature. An upcoming challenge therefore is to generate cross disciplinary datasets that will be able to connect the fields of rheology, rock mechanics, volcanology and numerical modeling.

Exploiting multidisciplinary datasets

While theoretical and more complex and complete descriptions of multiphase rheology are continuously being developed, their application fundamentally relies on parameters that are non-quantifiable prior to or during eruption (e.g., particle size, shape, aspect ratio, surface roughness) and that constantly evolve during transport of the magmatic suspension beneath or at the Earth's surface. Hence, their applicability to hazard forecasting or the prediction of eruption dynamics during or in preparation for volcanic crises is extremely limited. One of the large outstanding challenges is to develop rheological descriptions that robustly describe the flow of magma and lava across geologically relevant parameter space and that can be tied to data that can readily be measured during volcanic activity. For example, Coppola et al. (2013) introduced an empirical approach tying satellite derived radiant density measurements to the bulk rheology of lava. When applied at sufficiently high resolution and with high acquisition frequency (minutes to hours), this is a promising approach to inform emplacement models in near real time (Harris et al. 2019). However, to date, few attempts have been made to cross-correlate satellite, field- and laboratory measurements (Harris et al. 2019). Such a cross correlation would be very valuable in order to render the satellite derived data applicable to near real time hazard assessment and forecasting.

Further, combination of the recent laboratory developments and the growing availability of complementary datasets from a range of sub-disciplines (e.g., tomography, remote-sensing, drone technology, high-speed computation facilities) may lead to a more in depth understanding of magma and lava transport. These are beginning to be employed and tested in concert on long lasting (e.g., Chaiten, 2008–2010; Holuhraun, 2014–2015; Hawaii, 2018), or frequent, short duration (Piton de la Fournaise) lava flow eruptions, which are proving to be spectacular opportunities to perform interdisciplinary studies. Merging these interdisciplinary datasets promises to stimulate the development of new and enhanced tools for near real-time monitoring and forecasting of volcanic activity.

Characterizing nanoscale processes

In this review particle size was only mentioned when presenting the effect of polydisperse crystal size distributions. However, it has become apparent that nanolites (crystals at scales smaller than about 100 nanometers) can impact volcanic processes by affecting the viscosity of magmatic suspensions as well as by modulating gas exsolution. (Sharp et al. 1996) first identified nanometer scale crystallinity in natural volcanic glasses using transmission electron microscopy (TEM) observations. They coined the phrase "nanolites" for this previously unnoticed component of the grain size distribution in eruptive products and observed that it was a grain size distribution distinct from the commonly observed microlite size distribution. Nanolite bearing glasses frequently appear to be homogenous glasses to the naked eye and to the most common micro analysis techniques, such as scanning electron microscopes (SEM) and electron probes (EPMA). In recent years the presence of nanolites has been confirmed in a wide range of eruptive and experimental products using TEM or other nanoscale observations such as Raman spectroscopy (Barone et al. 2016; Burgess et al. 2016; Di Genova et al. 2018; Mujin and Nakamura 2014). Nanolites have been shown to affect important volcanic processes such as bubble nucleation (Cáceres et al. 2020; Di Genova et al. 2020; Kolzenburg et al. 2020) and, with that, changes in volcanic explosivity and rheological transitions due to vesiculation and/or crystal growth (Bouhifd et al. 2004; Di Genova et al. 2020; Liebske et al. 2003; Mujin et al. 2017; Villeneuve et al. 2008). While it had been suggested previously that very fine particles can have a stronger effect on suspension rheology than an equal amount of larger particles for both geological suspensions (Del Gaudio et al. 2013) and synthetic suspensions (Rudyak 2013; Taylor et al. 2013), recent research suggests that this effect may be even stronger for nano scale particles either due to the effect of crystal growth on the melt structure (Di Genova et al. 2017) or due to physical / geometrical effects, where aggregates of a large number of fine particles act to immobilize a significantly larger volume of melt in their interstitial space than an equal volume fraction of larger particles (Di Genova et al. 2020). Little is known about the exact processes that cause these size dependent effects and we will likely see an emerging body of literature on this topic concerned with geological melts forming over the coming decade.

ACKNOWLEDGMENTS

We thank Satoshi Okumura, Grant Henderson and an anonymous reviewer for constructive comments that helped improve this chapter. We are also grateful for discussions with many friends and colleagues on the topic of magma and lava rheology that helped generate a broad overview over this topic. SK acknowledges the support of a H2020 Marie Skłodowska-Curie fellowship DYNAVOLC – No.795044. MOC acknowledges the Agence National de la Recherche through the project LAVA (Program: DS0902 2016; Project: ANR-16 CE39-0009); this is ANR-LAVA contribution no. 17 and it is Laboratory of Excellence ClerVolc contribution number 479. DBD acknowledges the support of ERC 2018 Advanced Grant 834225 (EAVESDROP).

REFERENCES

Arbaret L, Bystricky M, Champallier R (2007) Microstructures and rheology of hydrous synthetic magmatic suspensions deformed in torsion at high pressure. J Geophys Res: Solid Earth 112:B10208
Arzilli F, Carroll MR (2013) Crystallization kinetics of alkali feldspars in cooling and decompression-induced crystallization experiments in trachytic melt. Contrib Mineral Petrol 166:1011–1027
Arzilli F, Mancini L, Voltolini M, Cicconi MR, Mohammadi S, Giuli G, Mainprice D, Paris E, Barou F, Carroll MR (2015) Near-liquidus growth of feldspar spherulites in trachytic melts: 3D morphologies and implications in crystallization mechanisms. Lithos 216–217:93–105
Arzilli F, Polacci M, La Spina G, Le Gall N, Cai B, Hartley M, Di Genova D, Vo N, Bamber E, Nonni S (2018) Rapid growth of plagioclase: implications for basaltic Plinian eruption. EGU GeneralAssembly 13431
Avard G, Whittington AG (2012) Rheology of arc dacite lavas: experimental determination at low strain rates. Bull Volcanol 74:1039–1056
Bachmann O, Huber C (2016) Silicic magma reservoirs in the Earth's crust. Am Mineral 101:2377–2404
Bagdassarov N, Dingwell DB (1992) A rheological investigation of vesicular rhyolite. J Volcanol Geotherm Res:307–322
Bagdassarov N, Pinkerton H (2004) A review and investigation of the Non-Newtonian properties of lavas based on laboratory experiments with analogue materials. J Volcanol Geotherm Res 132:115–136
Bagdassarov N, Dorfman A, Dingwell DB (2000) Effect of alkalis, phosphorus, and water on the surface tension of haplogranite melt. Am Mineral 85:33–40
Bagdassarov NS, Dingwell DB (1993a) Deformation of foamed rhyolites under internal and external stresses: an experimental investigation. Bull Volcanol 55:147–154
Bagdassarov NS, Dingwell DB (1993b) Frequency dependent rheology of vesicular rhyolite. J Geophys Res: Solid Earth 98:6477–6487
Bagdassarov NS, Dingwell DB, Webb SL (1994) Viscoelasticity of crystal- and bubble-bearing rhyolite melts. Phys Earth Planet Inter 83:83–99
Barnes HA (1997) Thixotropy—a review. J Non-Newtonian Fluid Mech 70:1–33
Barnes HA (1999) The yield stress—a review or 'παντα ρει'—everything flows? J Non-Newtonian Fluid Mech 81:133–178
Barone G, Mazzoleni P, Corsaro RA, Costagliola P, Di Benedetto F, Ciliberto E, Gimeno D, Bongiorno C, Spinella C (2016) Nanoscale surface modification of Mt. Etna volcanic ashes. Geochim Cosmochim Acta 174:70–84
Beckett F, Burton M, Mader H, Phillips J, Polacci M, Rust A, Witham F (2014) Conduit convection driving persistent degassing at basaltic volcanoes. J Volcanol Geotherm Res 283:19–35
Belousov A, Belousova M (2018) Dynamics and viscosity of 'a'a and pahoehoe lava flows of the 2012–2013 eruption of Tolbachik volcano, Kamchatka (Russia). Bull Volcanol 80:6
Blundy J, Cashman K, Humphreys M (2006) Magma heating by decompression-driven crystallization beneath andesite volcanoes. Nature 443:76

Bouhifd MA, Richet P, Besson P, Roskosz M, Ingrin J (2004) Redox state, microstructure and viscosity of a partially crystallized basalt melt. Earth Planet Sci Lett 218:31–44

Burgess KD, Stroud RM, Dyar MD, McCanta MC (2016) Submicrometer-scale spatial heterogeneity in silicate glasses using aberration-corrected scanning transmission electron microscopy. Am Mineral 101:2677–2688

Cáceres F, Wadsworth FB, Scheu B, Colombier M, Madonna C, Cimarelli C, Hess K-U, Kaliwoda M, Ruthensteiner B, Dingwell DB (2020) Can nanolites enhance eruption explosivity? Geology 48:997–1001

Campagnola S, Vona A, Romano C, Giordano G (2016) Crystallization kinetics and rheology of leucite-bearing tephriphonolite magmas from the Colli Albani volcano (Italy). Chem Geol 424:12–29

Caricchi L, Burlini L, Ulmer P, Gerya T, Vassalli M, Papale P (2007) Non-Newtonian rheology of crystal-bearing magmas and implications for magma ascent dynamics. Earth Planet Sci Lett 264:402–419

Caricchi L, Pommier A, Pistone M, Castro J, Burgisser A, Perugini D (2011) Strain-induced magma degassing: insights from simple-shear experiments on bubble bearing melts. Bull Volcanol 73:1245–1257

Cashman KV, Thornber C, Kauahikaua JP (1999) Cooling and crystallization of lava in open channels, and the transition of Pāhoehoe Lava to 'a'ā. Bull Volcanol 61:306–323

Cashman KV, Sparks RSJ, Blundy JD (2017) Vertically extensive and unstable magmatic systems: A unified view of igneous processes. Science 355

Castruccio A, Rust AC, Sparks RSJ (2010) Rheology and flow of crystal-bearing lavas: Insights from analogue gravity currents. Earth Planet Sci Lett 297:471–480

Castruccio A, Rust A, Sparks R (2014) Assessing lava flow evolution from post-eruption field data using Herschel–Bulkley rheology. J Volcanol Geotherm Res 275:71–84

Champallier R, Bystricky M, Arbaret L (2008) Experimental investigation of magma rheology at 300 MPa: From pure hydrous melt to 76 vol.% of crystals. Earth Planet Sci Lett 267:571–583

Chester DK, Duncan AM, Guest JE, Kilburn C (2012) Mount Etna: the Anatomy of a Volcano. Springer Science and Business Media

Chevrel MO, Pinkerton H, Harris AJL (2019a) Measuring the viscosity of lava in the field: A review. Earth Sci Rev 196:102852

Chevrel MO, Baratoux D, Hess K-U, Dingwell DB (2014) Viscous flow behavior of tholeiitic and alkaline Fe-rich martian basalts. Geochim Cosmochim Acta 124:348–365

Chevrel MO, Giordano D, Potuzak M, Courtial P, Dingwell DB (2013a) Physical properties of $CaAl_2Si_2O_8$–$CaMgSi_2O_6$–FeO–Fe_2O_3 melts: Analogues for extra-terrestrial basalt. Chem Geol 346:93–105

Chevrel MO, Platz T, Hauber E, Baratoux D, Lavallée Y, Dingwell DB (2013b) Lava flow rheology: A comparison of morphological and petrological methods. Earth Planet Sci Lett 384:109–120

Chevrel MO, Harris AJ, James MR, Calabrò L, Gurioli L, Pinkerton H (2018) The viscosity of pāhoehoe lava: in situ syn-eruptive measurements from Kilauea, Hawaii 2. Earth Planet Sci Lett

Chevrel MO, Harris A, Ajas A, Biren J, Gurioli L, Calabrò L (2019b) Investigating physical and thermal interactions between lava and trees: the case of Kīlauea's July 1974 flow. Bull Volcanol 81:6

Chevrel MO, Cimarelli C, deBiasi L, Hanson JB, Lavallée Y, Arzilli F, Dingwell DB (2015) Viscosity measurements of crystallizing andesite from Tungurahua volcano (Ecuador). Geochem Geophys Geosystems 16:870–889

Chong J, Christiansen E, Baer A (1971) Rheology of concentrated suspensions. J Appl Polym Sci 15:2007–2021

Cimarelli C, Costa A, Mueller S, Mader HM (2011) Rheology of magmas with bimodal crystal size and shape distributions: Insights from analog experiments. Geomchem Geophys Geosyst 12:Q07024

Coats R, Kendrick JE, Wallace PA, Miwa T, Hornby AJ, Ashworth JD, Matsushima T, Lavallée Y (2018) Failure criteria for porous dome rocks and lavas: a study of Mt. Unzen, Japan. Solid Earth 9:1299–1328

Coats R, Cai B, Kendrick J, Wallace P, Hornby A, Miwa T, von Aulock F, Ashworth J, Godinho J, Atwood R (2017) Understanding the rheology of two and three-phase magmas. AGU Fall Meeting Abstracts

Coish R, Taylor LA (1979) The effects of cooling rate on texture and pyroxene chemistry in DSDP Leg 34 basalt: a microprobe study. Earth Planet Sci Lett 42:389–398

Coppola D, Laiolo M, Piscopo D, Cigolini C (2013) Rheological control on the radiant density of active lava flows and domes. J Volcanol Geotherm Res 249:39–48

Coppola D, Ripepe M, Laiolo M, Cigolini C (2017) Modelling satellite-derived magma discharge to explain caldera collapse. Geology 45:523–526

Cordonnier B, Hess KU, Lavallée Y, Dingwell DB (2009) Rheological properties of dome lavas: Case study of Unzen volcano. Earth Planet Sci Lett 279:263–272

Cordonnier B, Schmalholz S, Hess KU, Dingwell D (2012a) Viscous heating in silicate melts: An experimental and numerical comparison. J Geophys Res: Solid Earth (1978–2012) 117:B02203

Cordonnier B, Caricchi L, Pistone M, Castro J, Hess K-U, Gottschaller S, Manga M, Dingwell D, Burlini L (2012b) The viscous-brittle transition of crystal-bearing silicic melt: Direct observation of magma rupture and healing. Geology 40:611–614

Costa A (2005) Viscosity of high crystal content melts: dependence on solid fraction. Geophys Res Lett 32:L22308

Costa A, Caricchi L, Bagdassarov N (2009) A model for the rheology of particle-bearing suspensions and partially molten rocks. Geomchem Geophys Geosyst 10:Q03010

Couch S, Sparks R, Carroll M (2003) The kinetics of degassing-induced crystallization at Soufriere Hills Volcano, Montserrat. J Petrol 44:1477–1502

Coussot P, Ancey C (1999) Rheophysical classification of concentrated suspensions and granular pastes. Phys Rev E 59:4445–4457

Cross MM (1965) Rheology of non-Newtonian fluids: a new flow equation for pseudoplastic systems. J Colloid Sci 20:417–437

Cukierman M, Tutts P, Uhlmann D (1972) Viscous flow behavior of lunar compositions 14259 and 14310. Lunar Planet Sci Conf Proc 3:2619

Del Gaudio P (2014) Rheology of bimodal crystals suspensions: Results from analogue experiments and implications for magma ascent. Geomchem Geophys Geosyst 15:284–291

Del Gaudio P, Ventura G, Taddeucci J (2013) The effect of particle size on the rheology of liquid–solid mixtures with application to lava flows: Results from analogue experiments. Geomchem Geophys Geosyst 14:2661–2669

Deubelbeiss Y, Kaus BJ, Connolly JA, Caricchi L (2011) Potential causes for the non-Newtonian rheology of crystal-bearing magmas. Geomchem Geophys Geosyst 12:Q05007

Di Genova D, Kolzenburg S, Wiesmaier S, Dallanave E, Neuville DR, Hess KU, Dingwell DB (2017) A compositional tipping point governing the mobilization and eruption style of rhyolitic magma. Nature 552:235

Di Genova D, Caracciolo A and Kolzenburg S (2018) Measuring the degree of "nanotilization" of volcanic glasses: Understanding syn-eruptive processes recorded in melt inclusions. Lithos 318–319:209–218

Di Genova D, Brooker RA, Mader HM, Drewitt JWE, Longo A, Deubener J, Neuville DR, Fanara S, Shebanova O, Anzellini S, Arzilli F, Bamber EC, Hennet L, La Spina G, Miyajima N (2020) In situ observation of nanolite growth in volcanic melt: A driving force for explosive eruptions. Sci Adv 6: eabb0413

Dieterich HR, Downs DT, Stelten ME, Zahran H (2018) Reconstructing lava flow emplacement histories with rheological and morphological analyses: the Harrat Rahat volcanic field, Kingdom of Saudi Arabia. Bull Volcanol 80:85

Dingwell D (1991) Redox viscometry of some Fe-bearing silicate melts. Am Mineral 76:1560–1562

Dingwell DB (1986) Viscosity-temperature relationships in the system $Na_2Si_2O_5$–$Na_4Al_2O_5$. Geochim Cosmochim Acta 50:1261–1265

Dingwell DB, Virgo D (1987) The effect of oxidation state on the viscosity of melts in the system Na_2O–FeO–Fe_2O_3–SiO_2. Geochim Cosmochim Acta 51:195–205

Dingwell DB, Virgo D (1988) Viscosities of melts in the Na_2O–FeO–Fe_2O_3–SiO_2 system and factors controlling relative viscosities of fully polymerized silicate melts. Geochim Cosmochim Acta 52:395–403

Dingwell DB, Webb SL (1989) Structural relaxation in silicate melts and non-Newtonian melts rheology in geologic processes. Phys Chem Mineral 16:508–516

Dobson K, Wadsworth F, Di Genova D, Kolzenburg S, Vasseur J, Marone F, Dingwell D (2016) Magmas on the move: in situ 4d experimental investigation into the rheology and mobility of three-phase magmas using ultra fast X-ray tomography. AGU Fall Meeting Abstracts

Dobson KJ, Allabar A, Bretagne E, Coumans J, Cassidy M, Cimarelli C, Coats R, Connolley T, Courtois L, Dingwell DB, Di Genova D (2020) Quantifying microstructural evolution in moving magma. Front Earth Sci 8:287

Dragoni M, Tallarico A (1994) The effect of crystallization on the rheology and dynamics of lava flows. J Volcanol Geotherm Res 59:241–252

Dragoni M, Bonafede M, Boschi E (1986) Downslope flow models of a Bingham liquid: implications for lava flows. J Volcanol Geotherm Res 30:305–325

Ducamp VC, Raj R (1989) Shear and densification of glass powder compacts. J Am Ceram Soc 72:798–804

Eilers H (1943) Die viskositäts-konzentrationsabhängigkeit kolloider Systeme in organischen Lösungsmitteln. Kolloid-Z 102:154–169

Eilers vH (1941) Die viskosität von Emulsionen hochviskoser Stoffe als Funktion der Konzentration. Kolloid Z 97:313–321

Einarsson T (1949) Studies of the Pleistocene in Eyjafjörður, Middle Northern Iceland. Prentsmiðjan Leiftur

Einarsson T (1966) Studies of temperature, viscosity, density and some types of materials produced in the Surtsey eruption. Surtsey Res Progr Rep 1:163–179

Einstein A (1906) Eine neue Bestimmung der Moleküldimensionen. Ann Phys 324:289–306

Evans K, Gibson A (1986) Prediction of the maximum packing fraction achievable in randomly oriented short-fibre composites. Compos Sci Technol 25:149–162

Faroughi SA, Huber C (2015) A generalized equation for rheology of emulsions and suspensions of deformable particles subjected to simple shear at low Reynolds number. Rheol Acta 54:85–108

Farris R (1968) Prediction of the viscosity of multimodal suspensions from unimodal viscosity data. Trans Soc Rheol 12:281–301

Findley WN, Lai JS, Onaran K, Christensen R (1977) Creep and Relaxation of nonlinear viscoelastic materials with an introduction to linear viscoelasticity. American Society of Mechanical Engineers Digital Collection

Fink JH, Zimbelman JR (1986) Rheology of the 1983 Royal Gardens basalt flows, Kilauea volcano, Hawaii. Bull Volcanol 48:87–96

Fink JH, Griffiths RW (1992) A laboratory analog study of the surface morphology of lava flows extruded from point and line sources. J Volcanol Geotherm Res 54:19–32

Flynn LP, Mouginis-Mark PJ (1992) Cooling rate of an active Hawaiian lava flow from nighttime spectroradiometer measurements. Geophys Res Lett 19:1783–1786

Frankel N, Acrivos A (1970) The constitutive equation for a dilute emulsion. J Fluid Mech 44:65–78

Gamble RP, Taylor LA (1980) Crystal/liquid partitioning in augite: effects of cooling rate. Earth Planet Sci Lett 47:21–33

Gan M, Gopinathan N, Jia X, Williams RA (2004) Predicting packing characteristics of particles of arbitrary shapes. KONA Powder Part J 22:82–93

Gauthier F (1971) Etude comparative des caractéristiques rhéologiques de laves basaltiques en laboratoire et dur le terrain (Doctoral dissertation).

Gauthier F, Guest JE, Skelhorn RR (1973) Mount Etna and the 1971 eruption - Field and laboratory studies of the rheology of Mount Etna lava. Phil Trans R Soc London Ser A 274:83–98

Gay E, Nelson P, Armstrong W (1969) Flow properties of suspensions with high solids concentration. AIChE J 15:815–822

Gent A (1960) Theory of the parallel plate viscometer. Br J Appl Phys 11:85

Giordano D, Nichols ARL, Dingwell DB (2005) Glass transition temperatures of natural hydrous melts: a relationship with shear viscosity and implications for the welding process. J Volcanol Geotherm Res 142:105–118

Giordano D, Russell JK, Dingwell DB (2008a) Viscosity of magmatic liquids: A model. Earth Planet Sci Lett 271:123–134

Giordano D, Potuzak M, Romano C, Dingwell DB, Nowak M (2008b) Viscosity and glass transition temperature of hydrous melts in the system CaAl2Si2O8–CaMgSi2O6. Chem Geol 256:203–215

Giordano D, Polacci M, Longo A, Papale P, Dingwell DB, Boschi E, Kasereka M (2007) Thermo-rheological magma control on the impact of highly fluid lava flows at Mt. Nyiragongo. Geophys Res Lett 34:L06301

Gottsmann J, Giordano D, Dingwell DB (2002) Predicting shear viscosity during volcanic processes at the glass transition: a calorimetric calibration. Earth Planet Sci Lett 198:417–427

Griffiths RW, Fink JH (1992) The morphology of lava flows in planetary environments: Predictions from analog experiments. J Geophys Res 97(B13):19739–19748

Gualda GA, Ghiorso MS, Lemons RV, Carley TL (2012) Rhyolite-MELTS: a modified calibration of MELTS optimized for silica-rich, fluid-bearing magmatic systems. J Petrol 53:875–890

Guth E, Gold O (1938) Viscosity and electroviscous effect of the AgI sol. II. Influence of the concentration of AgI and of electrolyte on the viscosity. Phys Rev 53:322

Haboub A, Bale HA, Nasiatka JR, Cox BN, Marshall DB, Ritchie RO, MacDowell AA (2014) Tensile testing of materials at high temperatures above 1700 °C with in situ synchrotron X-ray micro-tomography. Rev Sci Instrum 85:083702

Hamilton D, Burnham CW, Osborn E (1964) The solubility of water and effects of oxygen fugacity and water content on crystallization in mafic magmas. J Petrol 5:21–39

Hammer JE (2004) Crystal nucleation in hydrous rhyolite: Experimental data applied to classical theory. Am Mineral 89:1673–1679

Hammer JE (2006) Influence of fO_2 and cooling rate on the kinetics and energetics of Fe-rich basalt crystallization. Earth Planet Sci Lett 248:618–637

Hammer JE (2008) Experimental studies of the kinetics and energetics of magma crystallization. Rev Mineral Geochem 69:9–59

Hammer JE, Rutherford MJ (2002) An experimental study of the kinetics of decompression-induced crystallization in silicic melt. J Geophys Res: Solid Earth 107:ECV 8-1:8–24

Harris A, Rowland S (2009) Effusion rate controls on lava flow length and the role of heat loss: a review. Studies in volcanology: the legacy of George Walker Special Publications of IAVCEI 2:33–51

Harris AJ, Rowland S (2001) FLOWGO: a kinematic thermo-rheological model for lava flowing in a channel. Bull Volcanol 63:20–44

Harris AJ, Chevrel MO, Coppola D, Ramsey M, Hrysiewicz A, Thivet S, Villeneuve N, Favalli M, Peltier A, Kowalski P (2019) Validation of an integrated satellite-data-driven response to an effusive crisis: the April–May 2018 eruption of Piton de la Fournaise. Ann Geophys, Istituto Nazionale di Geofisica e Vulcanologia 61: 10.4401/ag-7972

Harris AJL, Flynn LP, Matias O, Rose WI, Cornejo J (2004) The evolution of an active silicic lava flow field: an ETM+ perspective. J Volcanol GeothermRes 135:147–168

Hashin Z, Shtrikman S (1963) A variational approach to the theory of the elastic behaviour of multiphase materials. J Mech Phys Solids 11:127–140

Hatschek E (1911) Die Viskosität der Dispersoide. Colloid Polym Sci 8:34–39

Heap MJ, Kolzenburg S, Russell JK, Campbell ME, Welles J, Farquharson JI, Ryan A (2014) Conditions and timescales for welding block-and-ash flow deposits. J Volcanol Geotherm Res 289:202–209

Herd CD (2003) The oxygen fugacity of olivine-phyric martian basalts and the components within the mantle and crust of Mars. Meteorit Planet Sci 38:1793–1805

Herschel W, Bulkley R (1926) Measurement of consistency as applied to rubber-benzene solutions. *In:* Measurement of Consistency as Applied to Rubber–Benzene Solutions. Vol 26, p 621–633

Hess K, Dingwell D (1996) Viscosities of hydrous leucogranitic melts: A non-Arrhenian model. Am Mineral 81:1297–1300

Hess K-U, Cordonnier B, Lavallée Y, Dingwell DB (2007) High-load, high-temperature deformation apparatus for synthetic and natural silicate melts. Rev Sci Instrum 78:075102.

Hess K-U, Cordonnier B, Lavallée Y, Dingwell DB (2008) Viscous heating in rhyolite: An in situ experimental determination. Earth Planet Sci Lett 275:121–126

Heymann L, Peukert S, Aksel N (2002) On the solid–liquid transition of concentrated suspensions in transient shear flow. Rheol Acta 41:307–315

Hobiger M, Sonder I, Büttner R, Zimanowski B (2011) Viscosity characteristics of selected volcanic rock melts. J Volcanol Geotherm Res 200:27–34

Hon K, Kauahikaua J, Denlinger R, Mackay K (1994) Emplacement and inflation of pahoehoe sheet flows: Observations and measurements of active lava flows on Kilauea Volcano, Hawaii. Geol Soc Am Bull 106:351–370

Hoover SR, Cashman KV, Manga M (2001) The yield strength of subliquidus basalts—experimental results. J Volcanol Geotherm Res 107:1–18

Hulme G (1974) The Interpretation of Lava Flow Morphology. Geophy J Inter 39:361–383

Humphreys MCS, Menand T, Blundy JD, Klimm K (2008) Magma ascent rates in explosive eruptions: Constraints from H_2O diffusion in melt inclusions. Earth Planet Sci Lett 270:25–40

Huppert HE, Sparks RSJ, Turner JS, Arndt NT (1984) Emplacement and cooling of komatiite lavas. Nature 309:19–22

Hurwitz S, Navon O (1994) Bubble nucleation in rhyolitic melts: Experiments at high pressure, temperature, and water content. Earth Planet Sci Lett 122:267–280

Ishibashi H (2009) Non-Newtonian behavior of plagioclase-bearing basaltic magma: Subliquidus viscosity measurement of the 1707 basalt of Fuji volcano, Japan. J Volcanol Geotherm Res 181:78–88

Ishibashi H, Sato H (2007) Viscosity measurements of subliquidus magmas: Alkali olivine basalt from the Higashi-Matsuura district, Southwest Japan. J Volcanol Geotherm Res 160:223–238

Ishibashi H, Sato H (2010) Bingham fluid behavior of plagioclase-bearing basaltic magma: Reanalyses of laboratory viscosity measurements for Fuji 1707 basalt. J Mineral Petrol Sci 105:334–339

James MR, Pinkerton H, Robson S (2007) Image-based measurement of flux variation in distal regions of active lava flows. Geomchem Geophys Geosyst 8:Q03006

James MR, Bagdassarov N, Müller K, Pinkerton H (2004) Viscoelastic behaviour of basaltic lavas. J Volcanol Geotherm Res 132:99–113

Jamieson RA, Unsworth MJ, Harris NB, Rosenberg CL, Schulmann K (2011) Crustal melting and the flow of mountains. Elements 7:253–260

Jellinek AM, Bercovici D (2011) Seismic tremors and magma wagging during explosive volcanism. Nature 470:522–525

Jones TJ, Llewellin EW (2021) Convective tipping point initiates localization of basaltic fissure eruptions. Earth Planet Sci Lett 553:116637

Kavanagh JL, Engwell SL, Martin SA (2018) A review of laboratory and numerical modelling in volcanology. Solid Earth 9:531–571

Kendrick J, Lavallée Y, Mariani E, Dingwell D, Wheeler J, Varley N (2017) Crystal plasticity as an indicator of the viscous–brittle transition in magmas. Nature Commun 8:1–12

Kilburn CR, Lopes RM (1991) General patterns of flow field growth: Aa and blocky lavas. J Geophys Res: Solid Earth 96:19721–19732

Kilgour G, Mader H, Blundy J, Brooker R (2016) Rheological controls on the eruption potential and style of an andesite volcano: A case study from Mt. Ruapehu, New Zealand. J Volcanol Geotherm Res 327:273–287

Klein J, Mueller SP, Castro JM (2017) The influence of crystal size distributions on the rheology of magmas: New insights from analog experiments. Geomchem Geophys Geosyst 18:4055–4073

Klein J, Mueller SP, Helo C, Schweitzer S, Gurioli L, Castro JM (2018) An expanded model and application of the combined effect of crystal-size distribution and crystal shape on the relative viscosity of magmas. J Volcanol Geotherm Res 357:128–133

Kohlstedt DL, Zimmerman ME (1996) Rheology of partially molten mantle rocks. Ann Rev Earth Planet Sci 24:41–62

Kolzenburg S, Giordano D, Cimarelli C, Dingwell DB (2016) In Situ thermal characterization of cooling/crystallizing lavas during rheology measurements and implications for lava flow emplacement. Geochim Cosmochim Acta:244–258

Kolzenburg S, Jaenicke J, Münzer U, Dingwell DB (2018a) The effect of inflation on the morphology-derived rheological parameters of lava flows and its implications for interpreting remote sensing data —A case study on the 2014/2015 eruption at Holuhraun, Iceland. J Volcanol Geotherm Res 357:200–212

Kolzenburg S, Giodano D, Hess KU, Dingwell DB (2018b) Shear rate-dependent disequilibrium rheology and dynamics of basalt solidification. Geophys Res Lett 45:6466–6475

Kolzenburg S, Giordano D, Di Muro A, Dingwell DB (2019) Equilibrium viscosity and disequilibrium rheology of a high magnesium basalt from piton De La Fournaise volcano, La Reunion, Indian Ocean, France. Ann Geophys 62:218

Kolzenburg S, Hess K-U, Berlo K, Dingwell DB (2020) Disequilibrium rheology and crystallization kinetics of basalts and implications for the Phlegrean volcanic district. Front Earth Sci 8:187

Kolzenburg S, Berlo K, Dingweil DB (2020) Vesiculation kinetics of variably crystalline rhyolites, Goldschmidt. Geochemical Society of America, Hawaii

Kolzenburg S, Giordano D, Thordarson T, Höskuldsson A, Dingwell DB (2017) The rheological evolution of the 2014/2015 eruption at Holuhraun, central Iceland. Bull Volcanol 79:45

Kolzenburg S, Di Genova D, Giordano D, Hess KU, Dingwell DB (2018c) The effect of oxygen fugacity on the rheological evolution of crystallizing basaltic melts. Earth Planet Sci Lett 487:21–32

Kouchi A, Tsuchiyama A, Sunagawa I (1986) Effect of stirring on crystallization kinetics of basalt: texture and element partitioning. Contrib Mineral Petrol 93:429–438

Krauskopf KB (1948) Lava movement at Paricutin volcano, Mexico. Geol Soc Am Bull 59:1267–1284

Kraynik AM (1988) Foam flows. Annu Rev Fluid Mec 20:325–357

Krieger IM, Dougherty TJ (1959) A mechanism for non-Newtonian flow in suspensions of rigid spheres. Trans Soc Rheol 3:137–152

Kushiro I, Yoder Jr H, Mysen B (1976) Viscosities of basalt and andesite melts at high pressures. J Geophys Res 81:6351–6356

Kushnir ARL, Martel C, Champallier R, Arbaret L (2017) In situ confirmation of permeability development in shearing bubble-bearing melts and implications for volcanic outgassing. Earth Planet Sci Lett 458:315–326

La Spina, Burton M, Vitturi MdM (2015) Temperature evolution during magma ascent in basaltic effusive eruptions: A numerical application to Stromboli volcano. Earth Planet Sci Lett 426:89–100

La Spina, Burton M, Vitturi MdM, Arzilli F (2016) Role of syn-eruptive plagioclase disequilibrium crystallization in basaltic magma ascent dynamics. Nat Commun 7:13402

Lange RA, Cashman KV, Navrotsky A (1994) Direct measurements of latent heat during crystallization and melting of a ugandite and an olivine basalt. Contrib Mineral Petrol 118:169–181

Lavallée Y, Hess K-U, Cordonnier B, Bruce Dingwell D (2007) Non-Newtonian rheological law for highly crystalline dome lavas. Geology 35:843–846

Lavallée Y, Meredith P, Dingwell D, Hess K-U, Wassermann J, Cordonnier B, Gerik A, Kruhl J (2008) Seismogenic lavas and explosive eruption forecasting. Nature 453:507–510

Lavallée Y, Benson PM, Heap MJ, Hess K-U, Flaws A, Schillinger B, Meredith PG, Dingwell DB (2013) Reconstructing magma failure and the degassing network of dome-building eruptions. Geology 41:515–518

Lavallée Y, Varley N, Alatorre-Ibargüengoitia M, Hess K-U, Kueppers U, Mueller S, Richard D, Scheu B, Spieler O, Dingwell D (2012) Magmatic architecture of dome-building eruptions at Volcán de Colima, Mexico. Bull Volcanol 74:249–260

Lavallée Y, Cai B, Kendrick JE, Dobson K, Von Aulock FW, Kaus B, Godinho J, Atwood R, Courtois L, Azeem M, Holness M, Lee DP (2019) Illuminating shear-induced vesiculation in magma via synchrotron imaging. IUGG 2019, Montreal.

Lavallée Y, Kendrick JE (2022) Strain localization in magmas. Rev Mineral Geochem 87:721-765

Lejeune A, Bottinga Y, Trull T, Richet P (1999) Rheology of bubble-bearing magmas. Earth Planet Sci Lett 166:71–84

Lejeune AM, Richet P (1995) Rheology of crystal-bearing silicate melts: An experimental study at high viscosities. J Geophys Res: Solid Earth 100:4215–4229

Lenk R (1967) A generalized flow theory. J Appl Polym Sci 11:1033–1042

Lev E, Spiegelman M, Wysocki RJ, Karson JA (2012) Investigating lava flow rheology using video analysis and numerical flow models. J Volcanol Geotherm Res 247–248:62–73

Li J, Uhlmann DR (1970) The flow of glass at high stress levels: I. Non-Newtonian behavior of homogeneous 0.08 $Rb_2O \cdot 0.92\ SiO_2$ glasses. J Non-Cryst Solids 3:127–147

Liebske C, Behrens H, Holtz F, Lange RA (2003) The influence of pressure and composition on the viscosity of andesitic melts. Geochim Cosmochim Acta 67:473–485

Liu Z, Pandelaers L, Blanpain B, Guo M (2017) Viscosity of heterogeneous silicate melts: assessment of the measured data and modeling. ISIJ Inter 57:1895–1901

Llewellin E, Manga M (2005) Bubble suspension rheology and implications for conduit flow. J Volcanol Geotherm Res 143:205–217

Llewellin E, Mader H, Wilson S (2002a) The constitutive equation and flow dynamics of bubbly magmas. Geophys Res Lett 29:23-21–23-24

Llewellin E, Mader H, Wilson S (2002b) The rheology of a bubbly liquid. Proc R Soc London. Ser A 458:987–1016

Loewenberg M, Hinch E (1996) Numerical simulation of a concentrated emulsion in shear flow. J Fluid Mech 321:395–419

Lofgren G (1980) Experimental studies on the dynamic crystallization of silicate melts. Phys Magmatic Process 487:551

Long PE, Wood BJ (1986) Structures, textures, and cooling histories of Columbia River basalt flows. Geol Soc Am Bull 97:1144–1155

Mackenzie J (1950) The elastic constants of a solid containing spherical holes. Proc Phys Soc Sec B 63:2–11

Mackenzie J (1956) Simultaneous measurements of density, viscosity, and electric conductivity of melts. Rev Sci Instrum 27:297–299

Mader HM, Llewellin EW, Mueller SP (2013) The rheology of two-phase magmas: A review and analysis. J Volcanol Geotherm Res 257:135–158

Madonna C, Quintal B, Frehner M, Almqvist BSG, Tisato N, Pistone M, Marone F, Saenger EH (2013) Synchrotron-based X-ray tomographic microscopy for rock physics investigations: Synchrotron-based rock images. Geophysics 78:D53-D64

Mandler BE, Elkins–Tanton LT (2013) The origin of eucrites, diogenites, and olivine diogenites: Magma ocean crystallization and shallow magma chamber processes on Vesta. Meteorit Planet Sci 48:2333–2349

Manga M, Loewenberg M (2001) Viscosity of magmas containing highly deformable bubbles. J Volcanol Geotherm Res 105:19–24

Manga M, Castro J, Cashman KV, Loewenberg M (1998) Rheology of bubble-bearing magmas. J Volcanol Geotherm Res 87:15–28

Mangan M, Sisson T (2005) Evolution of melt–vapor surface tension in silicic volcanic systems: Experiments with hydrous melts. J Geophys Res: Solid Earth 110:B01202

Markl G, Marks MA, Frost BR (2010) On the controls of oxygen fugacity in the generation and crystallization of peralkaline melts. J Petrol 51:1831–1847

Maron SH, Levy-Pascal AE (1955) Rheology of synthetic latex: VI. The flow behavior of neoprene latex. J Colloid Sci 10:494–503

Maron SH, Pierce PE (1956) Application of Ree–Eyring generalized flow theory to suspensions of spherical particles. J Colloid Sci 11:80–95

Marsh B (1981) On the crystallinity, probability of occurrence, and rheology of lava and magma. Contrib Mineral Petrol 78:85–98

McBirney AR, Murase T (1984) Rheological properties of magmas. Ann Rev Earth Planet Sci 12:337–357

McIntosh IM, Llewellin EW, Humphreys MCS, Nichols ARL, Burgisser A, Schipper CI, Larsen JF (2014) Distribution of dissolved water in magmatic glass records growth and resorption of bubbles. Earth Planet Sci Lett 401:1–11

Moitra P, Gonnermann HM (2015) Effects of crystal shape- and size-modality on magma rheology. Geomchem Geophys Geosyst16:1–26

Moore H (1987) Preliminary estimates of the rheological properties of 1984 Mauna Loa lava. US Geol Surv Prof Pap 1350:1569–1588

Moore H, Arthur D, Schaber G (1978) Yield strengths of flows on the Earth, Mars, and Moon. Lunar Planet Sci Conf 9:3351–3378

Morgavi D, Petrelli M, Vetere F, González-García D, Perugini D (2015) High-temperature apparatus for chaotic mixing of natural silicate melts. Rev Sci Instrum 86:105108

Morrison AA, Whittington A, Smets B, Kervyn M, Sehlke A (2020) The Rheology of Crystallizing basaltic lavas from Nyiragongo and Nyamuragira volcanoes, DRC. Volcanica 3:1–28

Mueller S, Llewellin EW, Mader HM (2010) The rheology of suspensions of solid particles. Proceedings of the Royal Society A: Mathematical, Phys Eng Sci 466:1201–1228

Mueller S, Llewellin EW, Mader HM (2011) The effect of particle shape on suspension viscosity and implications for magmatic flows. Geophys Res Lett 38:L13316

Mujin M, Nakamura M (2014) A nanolite record of eruption style transition. Geology 42:611–614

Mujin M, Nakamura M, Miyake A (2017) Eruption style and crystal size distributions: Crystallization of groundmass nanolites in the 2011 Shinmoedake eruption. Am Mineral 102:2367–2380

Murase T, Mc Birney AR (1973) Properties of some common igneous rocks and their melts at high temperatures. Geol Soc Am Bull 84:3563–3592

Murase T, McBirney AR, Melson WG (1985) Viscosity of the dome of Mount St. Helens. J Volcanol Geotherm Res 24:193–204

Mysen BO, Virgo D (1978) Influence of pressure, temperature, and bulk composition on melt structures in the system $NaAlSi_2O_6-NaFe^{3+}Si_2O_6$. Am J Sci 278:1307–1322

Mysen BO, Virgo D, Seifert FA (1984) Redox equilibria of iron in alkaline earth silicate melts; relationships between melt structure, oxygen fugacity, temperature and properties of iron-bearing silicate liquids. Am Mineral 69:834–847

Nichols RL (1939) Viscosity of lava. J Geol 47:290–302

Nicolas A, Ildefonse B (1996) Flow mechanism and viscosity in basaltic magma chambers. Geophys Res Lett 23:2013–2016

Ohtani E, Suzuki A, Ando R, Urakawa S, Funakoshi K, Katayama Y (2005) Viscosity and density measurements of melts and glasses at high pressure and temperature by using the multi-anvil apparatus and synchrotron X-ray radiation. *In*: Advances in High-Pressure Technology for Geophysical Applications. Elsevier, p 195–209

Okumura S, Nakamura M, Nakano T, Uesugi K, Tsuchiyama A (2010) Shear deformation experiments on vesicular rhyolite: Implications for brittle fracturing, degassing, and compaction of magmas in volcanic conduits. J Geophys Res: Solid Earth 115:B06201

Okumura S, Nakamura M, Uesugi K, Nakano T, Fujioka T (2013) Coupled effect of magma degassing and rheology on silicic volcanism. Earth Planet Sci Lett 362:163–170

Okumura S, Kushnir ARL, Martel C, Champallier R, Thibault Q, Takeuchi S (2016) Rheology of crystal-bearing natural magmas: Torsional deformation experiments at 800 °C and 100MPa. J Volcanol Geotherm Res 328:237–246

Oldroyd J (1953) The elastic and viscous properties of emulsions and suspensions. Proc R Soc London Ser A Math Phys Sci 218:122–132

Oosterbroek M, Mellema J (1981) Linear viscoelasticity of emulsions: I. The effect of an interfacial film on the dynamic viscosity of nondilute emulsions. J Colloid Interfac Sci 84:14–26

Pal R (2003) Rheological behavior of bubble-bearing magmas. Earth Planet Sci Lett 207:165–179

Panov VK, Slezin YB, Storcheus AV (1988) Mechanical properties of lava extruded in the 1983 Predskazanny eruption (Klyuchevskoi volcano). J Volcanol Seismol 7:25–37

Papale P (1999) Strain-induced magma fragmentation in explosive eruptions. Nature 397:425–428

Paterson MS, Olgaard DL (2000) Rock deformation tests to large shear strains in torsion. J Structur Geol 22:1341–1358

Petford N (2003) Rheology of granitic magmas during ascent and emplacement. Annu Rev Earth Planet Sci 31:399–427

Petford N (2009) Which effective viscosity? Mineral Mag 73:167–191

Phan-Thien N, Pham D (1997) Differential multiphase models for polydispersed suspensions and particulate solids. J Non-Newtonian Fluid Mech 72:305–318

Philpotts AR, Carroll M (1996) Physical properties of partly melted tholeiitic basalt. Geology 24:1029–1032

Pichavant M, Brugier Y, Di Muro A (2016) Petrological and experimental constraints on the evolution of Piton de la Fournaise magmas. *In*: Active Volcanoes of the Southwest Indian Ocean. Springer, p 171–184

Pinkerton H (1994) Rheological and related properties of lavas. Etna: Magma and lava flow modeling and volcanic system definition aimed at hazard assessment:76–89

Pinkerton H, Sparks RSJ (1978) Field measurements of the rheology of lava. Nature 276:383–385

Pinkerton H, Stevenson RJ (1992) Methods of determining the rheological properties of magmas at sub-liquidus temperatures. J Volcanol Geotherm Res 53:47–66

Pinkerton H, Norton G (1995) Rheological properties of basaltic lavas at sub-liquidus temperatures: laboratory and field measurements on lavas from Mount Etna. J Volcanol Geotherm Res 68:307–323

Pinkerton H, Norton G, Dawson J, Pyle D (1995a) Field observations and measurements of the physical properties of Oldoinyo Lengai alkali carbonatite lavas, November 1988. *In:* Carbonatite Volcanism. Springer, p 23–36

Pinkerton H, Herd R, Kent R, Wilson L (1995b) Field measurements of the rheological properties of basaltic lavas. Lunar Planet Sci Conf Vol 26

Pistone M (2012) Physical Properties of Crystal- and Bubble -Bearing Magmas. PhD ETH Zuerich, Zuerich

Pistone M, Cordonnier B, Ulmer P, Caricchi L (2016) Rheological flow laws for multiphase magmas: An empirical approach. J Volcanol Geotherm Res 321:158–170

Pistone M, Whittington AG, Andrews B, Cottrell E (2017) Crystal-rich lava dome extrusion during vesiculation: An experimental study. J Volcanol Geotherm Res 347:1–4

Pistone M, Caricchi L, Ulmer P, Reusser E, Ardia P (2013) Rheology of volatile-bearing crystal mushes: mobilization vs. viscous death. Chem Geol 345:16–39

Pistone M, Caricchi L, Ulmer P, Burlini L, Ardia P, Reusser E, Marone F, Arbaret L (2012) Deformation experiments of bubble- and crystal-bearing magmas: Rheological and microstructural analysis. J Geophys Res: Solid Earth (1978–2012) 117:B05208

Pistone M, Arzilli F, Dobson KJ, Cordonnier B, Reusser E, Ulmer P, Marone F, Whittington AG, Mancini L, Fife JL (2015) Gas-driven filter pressing in magmas: Insights into in-situ melt segregation from crystal mushes. Geology 43:699–702

Pleše P, Higgins M, Mancini L, Lanzafame G, Brun F, Fife J, Casselman J, Baker D (2018) Dynamic observations of vesiculation reveal the role of silicate crystals in bubble nucleation and growth in andesitic magmas. Lithos 296:532–546

Polacci M, Mancini L, Baker DR (2010) The contribution of synchrotron X-ray computed microtomography to understanding volcanic processes. J Synchrotron Radiat 17:215–221

Polacci M, Arzilli F, La Spina G, et al. (2018) Crystallisation in basaltic magmas revealed via in situ 4D synchrotron X-ray microtomography. Sci Rep 8:8377

Quane SL, Russell JK (2005) Welding; insights from high-temperature analogue experiments. J Volcanol Geotherm Res 142:67–87

Quane SL, Russell JK (2006) Bulk and particle strain analysis in high-temperature deformation experiments. J Volcanol Geotherm Res 154:63–73

Quane SL, Russell JKR, Kennedy L (2004) A low-load, high-temperature deformation apparatus for volcanological studies. Am Mineral 89:873–877

Quane SL, Russell JK, Friedlander EA (2009) Time scales of compaction in volcanic systems. Geology 37:471–474

Rahaman MN, De Jonghe LC, Scherer GW, Brook RJ (1987) Creep and densification during sintering of glass powder compacts. J Am Ceram Soc 70:766–774

Raterron P, Merkel S (2009) In situ rheological measurements at extreme pressure and temperature using synchrotron X-ray diffraction and radiography. J Synchrotron Radiat 16:748–756

Rhéty M, Harris A, Villeneuve N, Gurioli L, Médard E, Chevrel O, Bachélery P (2017) A comparison of cooling- and volume-limited flow systems: Examples from channels in the Piton de la Fournaise April 2007 lava flow field. Geochem Geophys Geosyst 18:3270–3291

Robert G, Russell JK, Giordano D, Romano C (2008) High-temperature deformation of volcanic materials in the presence of water. Am Min 93:74–80

Roeder PL, Dixon JM (1977) A centrifuge furnace for separating phases at high temperature in experimental petrology. Can J Earth Sci 14:1077–1084

Roscoe R (1952) The viscosity of suspensions of rigid spheres. Br J Appl Phys 3:267

Rose WI (1973) Pattern and mechanism of volcanic activity at the Santiaguito volcanic dome, Guatemala. Bull Volcanol 37:73

Rowlatt MA (1956) Compression plastimeter. United States patent US 2,754,675

Rudyak VY (2013) Viscosity of nanofluids. Why it is not described by the classical theories. Adv Nanoparticles 2:266

Rusiecka MK, Bilodeau M, Baker DR (2020) Quantification of nucleation delay in magmatic systems: experimental and theoretical approach. Contrib Mineral Petrol 175:47

Russell JK, Quane SL (2005) Rheology of welding: inversion of field constraints. J Volcanol Geotherm Res 142:173–191

Russell JK, Hess K-U, Dingwell DB (2022) Models for viscosity of geological melts. Rev Mineral Geochem 87:841–885

Rust AC, Manga M (2002) Effects of bubble deformation on the viscosity of dilute suspensions. J Non-Newtonian Fluid Mech 104:53–63

Rutter EH, Brodie KH, Irving DH (2006) Flow of synthetic, wet, partially molten "granite" under undrained conditions: An experimental study. J Geophys Res: Solid Earth 111

Ryan A, Russell J, Heap M, Kolzenburg S, Vona A, Kushnir A (2019a) Strain-dependent rheology of silicate melt foams: importance for outgassing of silicic lavas. J Geophys Res: Solid Earth 124:8167–8186

Ryan AG, Russell JK, Nichols AR, Hess K-U, Porritt LA (2015) Experiments and models on H_2O retrograde solubility in volcanic systems. Am Mineral 100:774–786

Ryan AG, Kolzenburg S, Vona A, Heap MJ, Russell JK, Badger S (2019b) A proxy for magmatic foams: FOAMGLAS®, a closed-cell glass insulation. J Non-Cryst Solids: X 1:100001

Ryerson F, Weed H, Piwinskii A (1988) Rheology of subliquidus magmas: 1. Picritic compositions. J Geophys Res: Solid Earth 93:3421–3436

Saar MO, Manga M, Cashman KV, Fremouw S (2001) Numerical models of the onset of yield strength in crystal–melt suspensions. Earth Planet Sci Lett 187:367–379

Sato H (2005) Viscosity measurement of subliquidus magmas:1707 basalt of Fuji volcano. J Mineral Petrol Sci 100:133–142

Sato M (1978) Oxygen fugacity of basaltic magmas and the role of gas-forming elements. Geophys Res Lett 5:447–449

Schowalter WR (1978) Mechanics of Non-Newtonian Fluid. Pergamon

Sehlke A, Whittington AG (2015) Rheology of lava flows on Mercury: An analog experimental study. J Geophys Res: Planets 120:1924–1955

Sehlke A, Whittington A, Robert B, Harris A, Gurioli L, Médard E (2014) Pahoehoe to `a`a transition of Hawaiian lavas: an experimental study. Bull Volcanol 76:1–20

Settle M (1979) Lava rheology: Thermal buffering produced by the latent heat of crystallization. Lunar Planet Sci Conf 10:1107–1109

Shapiro AP, Probstein RF (1992) Random packings of spheres and fluidity limits of monodisperse and bidisperse suspensions. Phys Rev Lett 68:1422

Sharp TG, Stevenson RJ, Dingwell DB (1996) Microlites and" nanolites" in rhyolitic glass: Microstructural and chemical characterization. Bull Volcanol 57:631–640

Shaw H, Wright T, Peck D, Okamura R (1968) The viscosity of basaltic magma; an analysis of field measurements in Makaopuhi lava lake, Hawaii. Am J Sci 266:225–264

Shaw HR (1965) Comments on viscosity, crystal settling, and convection in granitic magmas. Am J Sci 263:120–152

Shaw HR (1969) Rheology of basalt in the melting range. J Petrol 10:510–535

Shields JK, Mader HM, Pistone M, Caricchi L, Floess D, Putlitz B (2014) Strain-induced outgassing of three-phase magmas during simple shear. J Geophys Res: Solid Earth 119:6936–6957

Sibree J (1934) The viscosity of froth. Trans Faraday Soc 30:325–331

Sicola S, Vona A, Ryan AG, Russell JK, Romano C (2021) The effect of pores (fluid-filled vs. drained) on magma rheology. Chem Geol 569:120147

Soldati A, Sehlke A, Chigna G, Whittington A (2016) Field and experimental constraints on the rheology of arc basaltic lavas: the January 2014 Eruption of Pacaya (Guatemala). Bull Volcanol 78:1–19

Soldati A, Beem J, Gomez F, Huntley JW, Robertson T, Whittington A (2017) Emplacement dynamics and timescale of a Holocene flow from the Cima Volcanic Field (CA): Insights from rheology and morphology. J Volcanol Geotherm Res 347:91–111

Sonder I, Zimanowski B, Büttner R (2006) Non-newtonian viscosity of basaltic magma. Geophys Res Lett 33

Song S-R, Jones KW, Lindquist BW, Dowd BA, Sahagian DL (2001) Synchrotron X-ray computed microtomography: studies on vesiculated basaltic rocks. Bull Volcanol 63:252–263

Soule SA, Cashman KV (2005) Shear rate dependence of the pāhoehoe-to-'a 'ā transition: analog experiments. Geology 33:361–364

Sparks RSJ, Annen C, Blundy JD, Cashman KV, Rust AC, Jackson MD (2019) Formation and dynamics of magma reservoirs. Philos Trans R Soc A 377:20180019

Spera FJ, Borgia A, Strimple J, Feigenson M (1988) Rheology of melts and magmatic suspensions: 1. Design and calibration of concentric cylinder viscometer with application to rhyolitic magma. J Geophys Res: Solid Earth 93:10273–10294

Spina L, Cimarelli C, Scheu B, Di Genova D, Dingwell DB (2016) On the slow decompressive response of volatile- and crystal-bearing magmas: An analogue experimental investigation. Earth Planet Sci Lett 433:44–53

Stein DJ, Spera FJ (1992) Rheology and microstructure of magmatic emulsions: theory and experiments. J Volcanol Geotherm Res 49:157–174

Stein D, Spera F (1998) New high-temperature rotational rheometer for silicate melts, magmatic suspensions, and emulsions. Rev Sci Instrum 69:3398–3402

Stein DJ, Spera FJ (2002) Shear viscosity of rhyolite-vapor emulsions at magmatic temperatures by concentric cylinder rheometry. J Volcanol Geotherm Res 113:243–258

Stevenson RJ, Dingwell DB, Webb S, Bagdassarov N (1995) The equivalence of enthalpy and shear stress relaxation in rhyolitic obsidians and quantification of the liquid-glass transition in volcanic processes. J Volcanol GeothermRes 68:297–306

Stevenson RJ, Dingwell DB, Webb SL, Sharp TG (1996) Viscosity of microlite-bearing rhyolitic obsidians: an experimental study. Bull Volcanol 58:298–309

Stevenson RJ, Dingwell DB, Bagdassarov NS, Manley CR (2001) Measurement and implication of "effective" viscosity for rhyolite flow emplacement. Bull Volcanol 63:227–237

Stickel JJ, Powell RL (2005) Fluid mechanics and rheology of dense suspensions. Annu Rev Fluid Mech 37:129–149

Sumita I, Manga M (2008) Suspension rheology under oscillatory shear and its geophysical implications. Earth Planet Sci Lett 269:468–477

Swanson SE (1977) Relation of nucleation and crystal-growth rate to the development of granitic textures. Am Mineral 62:966–978

Taisne B, Jaupart C (2008) Magma degassing and intermittent lava dome growth. Geophys Res Lett 35

Takeuchi S, Nakashima S, Tomiya A (2008) Permeability measurements of natural and experimental volcanic materials with a simple permeameter: toward an understanding of magmatic degassing processes. J Volcanol Geotherm Res 177:329–339

Takeuchi S, Tomiya A, Shinohara H (2009) Degassing conditions for permeable silicic magmas: Implications from decompression experiments with constant rates. Earth Planet Sci Lett 283:101–110

Tammann G, Hesse W (1926) Die Abhängigkeit der Viscosität von der Temperatur bie unterkühlten Flüssigkeiten. Z Anorg Allg Chem 156:245–257

Taylor GI (1932) The viscosity of a fluid containing small drops of another fluid. Proc R Soc London Ser A 138:41–48

Taylor R, Coulombe S, Otanicar T, Phelan P, Gunawan A, Lv W, Rosengarten G, Prasher R, Tyagi H, (2013) Small particles, big impacts: A review of the diverse applications of nanofluids. J Appl Physics 113:1

Toplis M, Carroll M (1995) An experimental study of the influence of oxygen fugacity on Fe-Ti oxide stability, phase relations, and mineral—melt equilibria in ferro-basaltic systems. J Petrol 36:1137–1170

Torquato S (2013) Random Heterogeneous Materials: Microstructure and Macroscopic Properties. Springer Science and Business Media

Tripoli B, Manga M, Mayeux J, Barnard H (2019) The effects of deformation on the early crystallization kinetics of basaltic magmas. Front Earth Sci 7:250

Truby J, Mueller S, Llewellin E, Mader H (2015) The rheology of three-phase suspensions at low bubble capillary number. Proc R Soc A 471:20140557

Ulusay R, Hudson JA (2006) The complete ISRM suggested methods for rock characterization, testing and monitoring:1974–2006. The International Society for Rock Mechanics and Rock Engineering, Ankara, Turkey

Vand V (1948) Viscosity of solutions and suspensions. I. Theory. J Phys Chem 52:277–299

Vasseur J, Wadsworth FB, Lavallée Y, Bell AF, Main IG, Dingwell DB (2015) Heterogeneity: the key to failure forecasting. Sci Rep 5:13259

Vetere F, Sato H, Ishibashi H, De Rosa R, Donato P (2013) Viscosity changes during crystallization of a shoshonitic magma: new insights on lava flow emplacement. J Mineral Petrol Sci 108:144–160

Vetere F, Behrens H, Schuessler JA, Holtz F, Misiti V, Borchers L (2008) Viscosity of andesite melts and its implication for magma mixing prior to Unzen 1991–1995 eruption. J Volcanol Geotherm Res 175:208–217

Vetere F, Murri M, Alvaro M, Domeneghetti MC, Rossi S, Pisello A, Perugini D, Holtz F (2019) Viscosity of pyroxenite melt and its evolution during cooling. J Geophys Res: Planets 124:1451–1469

Vetere F, Iezzi G, Behrens H, Holtz F, Ventura G, Misiti V, Cavallo A, Mollo S, Dietrich M (2015) Glass forming ability and crystallisation behaviour of sub-alkaline silicate melts. Earth Sci Rev 150:25–44

Vetere F, Rossi S, Namur O, Morgavi D, Misiti V, Mancinelli P, Petrelli M, Pauselli C, Perugini D (2017) Experimental constraints on the rheology, eruption and emplacement dynamics of analog lavas comparable to Mercury's northern volcanic plains. J Geophys Res: Planets 122:1522–1538

Villeneuve N, Neuville DR, Boivin P, Bachèlery P, Richet P (2008) Magma crystallization and viscosity: a study of molten basalts from the Piton de la Fournaise volcano (La Réunion island). Chem Geol 256:242–251

Von Aulock FW, Kennedy BM, Maksimenko A, Wadsworth FB, Lavallée Y (2017) Outgassing from Open and Closed Magma Foams. Front Earth Sci 5:46

Vona A, Romano C (2013) The effects of undercooling and deformation rates on the crystallization kinetics of Stromboli and Etna basalts. Contrib Mineral Petrol 166:491–509

Vona A, Romano C, Dingwell DB, Giordano D (2011) The rheology of crystal-bearing basaltic magmas from Stromboli and Etna. Geochim Cosmochim Acta 75:3214–3236

Vona A, Ryan AG, Russell JK, Romano C (2016) Models for viscosity and shear localization in bubble-rich magmas. Earth Planet Sci Lett 449:26–38

Vona A, Di Piazza A, Nicotra E, Romano C, Viccaro M, Giordano G (2017) The complex rheology of megacryst-rich magmas: The case of the mugearitic "cicirara" lavas of Mt. Etna volcano. Chem Geol 458:48–67

Wadsworth FB, Vasseur J, Llewellin EW, Dingwell DB (2022) Hot sintering of melts, glasses and magmas. Rev Mineral Geochem 87:801-840

Walker D, Mullins O (1981) Surface tension of natural silicate melts from 1,200–1,500 C and implications for melt structure. Contrib Mineral Petrol 76:455–462

Walker D, Kirkpatrick R, Longhi J, Hays J (1976) Crystallization history of lunar picritic basalt sample 12002: phase-equilibria and cooling-rate studies. Geol Soc Am Bull 87:646–656

Walker G, Huntingdon A, Sanders A, Dinsdale J (1973) Lengths of lava flows [and discussion]. Phil Trans R Soc London A 74:107–118

Ward S, Whitmore R (1950) Studies of the viscosity and sedimentation of suspensions Part 1. The viscosity of suspension of spherical particles. Br J Appl Phys 1:286–290

Webb SL, Dingwell DB (1990a) Non-Newtonian rheology of igneous melts at high stresses and strain rates: experimental results for rhyolite, andesite, basalt, and nephelinite. J Geophys Res 95:15695–15701

Webb SL, Dingwell DB (1990b) The onset of non-Newtonian rheology of silicate melts. Phys Chem Mineral 17:125–132

Wildemuth C, Williams M (1984) Viscosity of suspensions modeled with a shear-dependent maximum packing fraction. Rheol Acta 23:627–635

Wildemuth C, Williams M (1985) A new interpretation of viscosity and yield stress in dense slurries: coal and other irregular particles. Rheol Acta 24:75–91

Wilson L, Head JW (1994) Mars: Review and analysis of volcanic eruption theory and relationships to observed landforms. Rev Geophys 32:221–263

Witter JB, Harris AJ (2007) Field measurements of heat loss from skylights and lava tube systems. J Geophys Res: Solid Earth (1978–2012) 112:B01203

Wright HMN, Weinberg RF (2009) Strain localization in vesicular magma: Implications for rheology and fragmentation. Geology 37:1023–1026

Zhou JZ, Uhlherr PH, Luo FT (1995) Yield stress and maximum packing fraction of concentrated suspensions. Rheol Acta 34:544–561

Reviews in Mineralogy & Geochemistry
Vol. 87 pp. 721-765, 2022
Copyright © Mineralogical Society of America

15

Strain Localization in Magmas

Yan Lavallée

Department of Earth, Ocean and Ecological Science
University of Liverpool
Liverpool, L69 3GP
United Kingdom

Jackie E. Kendrick

School of Geosciences
University of Edinburgh
James Hutton Road
Edinburgh EH9 3FE
United Kingdom

INTRODUCTION

One of the most intriguing and (from the point of view of modeling) poorly understood aspects of geomaterial mechanics is strain localization. Strain localization is a common feature of viscous, elastic and/ or plastic materials undergoing non-homogeneous deformation. During the deformation of rocks or magmatic suspensions, strain may be variably partitioned (spatially and in magnitude) between phases or multi-scalar heterogeneities of variable strengths, which may promote the development of shear zones and/ or shear bands (e.g., Wright and Weinberg 2009); as such, strain localization is a scale-dependent phenomenon that may range between discrete and pervasive deformation. Strain localization takes place in a variety of geologic settings ranging from mantle convection, to tectonic plate dynamics, faulting, earthquakes and magma transport. One of its greatest consequences can be the onset of catastrophic material failure (although not invariably, nor a prerequisite). The ability of geomaterials to localize deformation may thus be viewed as a measure of *"the fragility of the Earth"*—a threshold for the occurrence of a wide range of geological hazards. Magma transport is itself a form of strain localization in the Earth and at specific volcanoes, the remarkable, unpredictable and alarming occurrence of eruptions, switching from low-risk effusive to high-risk explosive behavior is a direct consequence of strain localization in magma.

Magmas comprise a wide spectrum of attributes. Central to their definition is the presence of melt (sometimes melts) and invariably, magmas contain different fractions of exsolved volatile bubbles, crystals and rock fragments. They are widespread in the Earth's crust and mantle (arguably down to the core–mantle boundary; Sakamaki and Ohtani 2022, this volume), and they are sometimes referred to as crystal mushes or migmatites (following extensive crystallization or partial melting, respectively) and upon extrusion, as lavas. Magmas have often been categorized according to their attributes: single-phase melts, crystal-poor suspensions, crystal-rich suspensions, bubbly suspensions, and multiphase suspensions (see Kolzenburg et al. 2022, this volume, for a comprehensive review). Understanding the rheology of such diverse suspensions is amongst the grand challenges of geosciences as shear viscosity may span more than some 25 orders of magnitude, thus its study requires judicious

1529-6466/22/0087-0015$05.00 (print)
1943-2666/22/0087-0015$05.00 (online)

http://dx.doi.org/10.2138/rmg.2022.87.15

approaches. Importantly, during deformation of magmatic suspensions—as well as granular materials (recognizing that partially crystalline suspensions exhibit key characteristics of cohesionless granular materials; e.g., Gourlay and Dahle 2007)—strain can partition between phases so that wholesale flow properties (i.e., viscosity, consistency or friction coefficient) can evolve as strain localizes. The resultant modification of the suspensions may further impact their permeable network (transporting migrating melt and/ or exsolved volatiles) as well as the development of fractures contributing to magmatic fragmentation or rupture and fault slip. In most magmatic and volcanic scenarios, strain localization may (for most rheological considerations) be considered the norm.

Strain localization has been studied using field descriptions, numerical simulations and laboratory constraints. Here we briefly introduce field observations and some key aspects of numerical models before reviewing the results of laboratory experiments which have helped constrain the evolution and localization of strain in multiphase magmas; here defined by the presence of a melt phase, plus vesicles and/or crystals. In particular, we review how strain localization in magmas is prompted by their multiphase nature, which promotes stress concentration and thus strain partitioning between phases, each of which may obey contrasting deformation mechanisms. Silicate melts flow in a relaxed state by diffusion-related processes and break when they undergo the glass transition. In contrast, crystals and bubbles may variably deform (via crystal plasticity and shape deformation, respectively). As a result, melt–crystal–bubble mixtures can adopt variable configurations depending on the fraction of these phases and the local temperature, stress conditions and deformation history. Thus, the interplay between different deformation mechanisms and the resultant macroscopic deformation mode (ductile vs brittle) during strain localization carries implications for transport, seismicity, construction of the permeable network, and eruption style.

MATERIAL DEFORMATION AND STRAIN LOCALIZATION

The deformation of geomaterials is complex. Whereas in some cases strain may be homogenous, strain is generally not uniform; it is most commonly heterogeneous in multiphase systems and in extreme cases, strain may be discontinuous if material undergoes rupture.

Evidence for strain localization in magmas: the geologic record

Exposed intrusive and extrusive igneous rocks (including volcaniclastics), exhibit evidence of strain localization. This is commonly evidenced through the alignment and/ or stretching of enclaves, lithics, crystals, bubbles, melt and fractures due to shear, both pure and simple. These can be observed at various scales and in different parts of igneous bodies. Strain primarily localizes near the margins of intrusions (e.g., along dyke margins; Coward 1980) and eruptive products (e.g., flow levees), but may also localize in the core of intrusions (e.g., flows bands from convective currents) or extrusive bodies (e.g., fractures within lava spines; tuffisites within volcanic conduits; and compressional ridges in lava flows). Figure 1 shows some examples of strain localization in magmas and lavas. At large scale, we conventionally envisage shear zones developed through simple shear to epitomize strain localization, but strain localization can also take place via pure shear such as in compactional areas of a collapsing foam in obsidian lava domes. At smaller scale, strain localization is commonly observed in the development of flow bands, crystallographic preferred orientation, tiling, bubble elongation and fractures.

Importantly, what these examples display is that strain localization is not simply a geometric pattern nor is it necessarily the end game; i.e., flow banding or rupture are not points of no return. Strain localization is a kinetic process; thus in the lifetime of any magmatic system, the dominance of different deformation mechanisms (hence degree of strain localization) may switch repeatedly. This is best exemplified at volcanoes with pulsatory magma ascent (such

Figure 1. Examples of strain localization in magmas and lavas. **a)** The near-vertical magma spine erupted in 1902 at Mont Pelée, Martinique. Photograph from Alfred Lacroix. **b)** Fault-controlled gas-and-ash emissions at Santiaguito, Guatemala, pictured in 2012. **c)** Pronounced levees and crease structures in the andesitic Kalama age lava flows at Mount St. Helens, USA. **d)** Breccia cross–cutting flow bands in an andesite block from Volcán de Colima, Mexico. **e)** Internal structure of a tuffisite vein exposed in a fractured andesite dome block from Volcán de Colima, Mexico (Kendrick et al. 2016); the tuffisite consist of variably sintered fragments, inhomogenously distributed due to venting, sintering and post-emplacement deformation. **f)** Banded rhyolitic obsidian in Panum Crater, Long Valley, USA. **g)** Banded, faulted and healed obsidian in a block at Obsidian Dome, Long Valley, USA. **h)** Flow bands in granite showing biotite-rich (darker) and biotite-poor (pink) layers, from St. Peter Suite at Point Brown, Australia. Photograph courtesy of Roberto Weinberg. **i)** Mafic enclaves indicating shear in proximity to a granitic protrusion at Point Brown, Australia; elongate, elliptic enclaves at the margins indicate higher strains. Photograph courtesy of Roberto Weinberg. **j)** Post-emplacement transverse offset of a mafic dyke (intruded along a localized fault) due to ductile deformation of the mafic mush host. Rock in a collection held at Harvard University.

as during spine extrusion; Fig 1a), cyclicity in gas emissions and explosions (Fig. 1b), and in eruptive products, with the development of flow levees (Fig. 1c), brecciation (Fig. 1d), local fragmentation followed by generation and injection of tuffisite veins (Fig. 1e) and cross-cutting relationships between shear bands, faults and fractures (Fig. 1f,g). Such relative kinetic markers are also present in magmatic flow banding fabrics (Fig. 1h), in mingled magmas (Fig. 1i) and during dyke injection along fractures inside magma mushes (Fig. 1j).

Strain regimes

There are two types of shear strain—pure and simple—depending on the symmetry of the components of the local strain tensor. [Here, we note that the principal surfaces of stress and strain may not always exist for evolving materials with three dimensionally varying states of stress or strain (Treagus and Lisle 1997), but we refer to these descriptions as they help resolve macroscopic observations.] Pure shear is induced by a normal, orthogonal stress and involves only compression (in the direction of applied stress) and transversal extension (or vice versa); there is no rotation in the process and the finite strain ellipsoid is coaxial (as the strain axes of infinitesimal strain increments remain parallel). This may result in constructional strain (if a body elongates parallel to the principal applied stress direction), flattening strain (if a body compacts parallel to the principal applied stress direction) or plain strain (if the material undergoes zero strain in the direction normal to the principal applied stress direction; see Flinn 1962 and "The deformation of early linear structures in areas of repeated folding" in Ramsay 1967 for reviews). On the other hand, simple shear is induced by a tangential stress and involves rotation, defined by the angular shear, and the finite strain ellipsoid exhibits no symmetry (i.e., strain is non-coaxial as the angles of the strain axes of the infinitesimal strain ellipsoids are non-parallel, evolving following each strain increment). In both cases, deformation should involve no volume change (i.e., isovolumetric.) Yet, as we will see in this chapter, the deformation of most geological materials, and certainly magmas (with the exception of pure silicate melts), results in volume changes. Moreover, geomaterials are not simply exposed to one of these two strain regimes; it is most common for both pure and simple shear to act together, though generally to different extents. For instance, phase rotation may take place in multiphase materials due to non-uniform strain even under seemingly pure shear conditions. A such, pure and simple shear are not so simple in multiphase magmas.

Stress and strain regimes in magmatic environments

The state of stress may vary significantly in magmatic bodies; as a result, pure and simple shear may dominate in different areas. For instance, in quasi-stationary magmatic bodies, the overburden of the roof rock (during stoping or caldera subsidence), may promote pure shear (Fig. 2a), whereas in more dynamic systems in which magma convects, propagates into dykes, is replenished, or is intercepted by faults, currents may favor the local or wholesale development of simple shear (Fig. 2b).

In volcanic conduits, flow vectors are commonly sketched to illustrate various flow types and distributions of dominant strain regimes (Fig. 2c). Flow in a conduit results in lateral velocity variations, which promotes the development of simple shear near the margins (Fig. 2c). In Newtonian bodies for which viscosity is rate independent, strain is pervasive across the conduits, albeit at different strain rates, and the margins are not expected to develop slip at the boundaries. This is often referred to as a Hagen–Poiseuille flow. Yet, even in this relatively simple rheological case, the reconstruction of strain history has been deemed challenging if the conduit is inclined, since non-orthogonal stresses modify the velocity vectors (Mukherjee 2012). In non-Newtonian bodies, for which the viscosity is rate and strain dependent, strain preferentially localizes near conduit margins, ultimately resulting in plug-like flow (where simple shear flanks a relatively undeformed core). In extreme cases, shear may induce rupture and a switch to fault-slip controlled ascent. Irrespective of rheology, axial variations in ascent

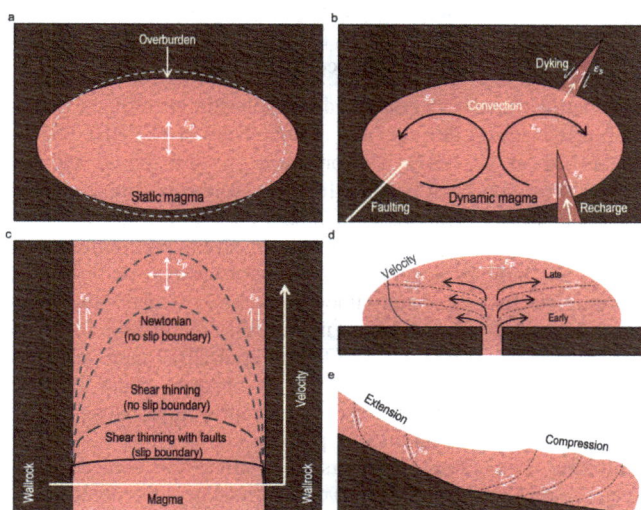

Figure 2. Schematic representations of strain localization in different intrusive and extrusive settings. **a)** Pure shear strain (ε_p) is active in magma chambers, driven by overburden. **b)** A magma chamber can also experience simple shear strain (ε_s) due to convection, magma recharge events, propagation of dykes and tectonic faulting. **c)** Velocity profiles across a magmatic conduit for Newtonian and shear thinning fluid flow, indicating the prevalence of pure shear in the center, with progressive prevalence of simple shear towards the margins. **d)** The extrusion of lava onto horizontal substrata is exemplified by pure shear at the apex of the flow and near-horizontal simple shear planes in the lateral part of the flow. **e)** Lava extruded onto tilted ground experiences simple shear in both extension and compression along its length.

velocities may promote pure shear in the core of the magmatic column: i.e., constriction if the ascent rate locally increases, and flattening if the ascent rate locally decreases (Fig. 2c).

When magma extrudes at the Earth's surface, for example in a lava flow or lava dome emplaced on a horizontal substratum, the apex may be in a pure shear regime, whilst lateral regions may spread via simple shear (Fig. 2d). The latter is commonly seen by the ropy texture and clinker or blocky textures on the surface of pahoehoe and a'ā lavas, respectively, as well as by slickensides at the bases of fractured slabs rafting onto lava near flow fronts. As lava flows downslope, variable rates of simple shear may lead to the development of extensional and compressive stresses, which may cause listric faults and ridges, respectively (Fig. 2e).

Evolution of properties and conditions during magma transport

Magma transport is generally driven by pressure gradients, resulting in decompression, which commonly triggers vesiculation and crystallization (Martel and Schmidt 2003; Castro and Dingwell 2009; Lindoo et al. 2016). Upon transport, temperature may decrease due to contact with cooler rocks or water, or increase due to magma recharge or shear heating (both viscous energy dissipation and frictional heating; see section *Frictional melting and thermal vesiculation*); cooling tends to promote crystallization (Cashman 1993), vitrification (Gottsmann and Dingwell 2001) and volatile resorption (McIntosh et al. 2014), whilst heating can promote crystal breakdown (De Angelis et al. 2015), melting (Kendrick et al. 2014a) and volatile exsolution and vesiculation (Lavallée et al. 2015a). Phase transformations are associated with volume changes (Richet and Bottinga 1986), which locally stress and strain magma. During cooling and crystallization, localized convection cells develop (Mattsson et al. 2011), due to contraction-driven jointing which promotes advective fluid ingress in the late stage straining of magma (Lamur et al. 2018). Similarly, the evolution of conduit breccias or tuffisites involves magma rupture, followed by incursion of gas and pyroclasts, and finally venting and sintering

(Wadsworth et al. 2022, this volume), which are all evidence of strain localization due to locally transient material properties. Thus, even in an exceptional case where magma or lava remains seemingly stationary, evolving conditions induce strain localization (Caricchi et al. 2014).

Numerical simulations have been employed to resolve the rheological behavior and physical evolution of multiphase magmas during shear (Deubelbeiss et al. 2010, 2011) or densification via sintering (Wadsworth et al. 2021). The presence of solid, liquid and gaseous phases in magmas concentrates stress on each phase when the system is subjected to deformation. As each phase entertains contrasting rheology and "strength" (gas < melt < crystals), they variably accumulate and relax the applied stress, resulting in strain partitioning between magmatic phases. For example, gas bubbles generally (though not always) deform more than the melt, whilst crystals accumulate stress and may undergo some crystal plasticity. Local fluctuations in dynamic pressures in a stressed melt phase may further modify the local chemical potential (cf. Wheeler 2014), triggering crystallization (Tripoli et al. 2019) and vesiculation (Lavallée et al. 2017). Consequently, stress and strain partitioning promote strain localization. Given that strain localization arises from locally contrasting relaxation timescales, it need not necessarily be stimulated by the presence of additional phases, i.e., crystals or bubbles. In magmatic environments exhibiting temperature gradients (e.g., along the cooler margins of pahoehoe flows or where magmas develop shear heating) or chemical heterogeneities (e.g., when magmas mingle or segregate), local viscosity contrasts may equally favor strain localization.

Deformation modes: ductile vs brittle

Before delving into a description of ductile and brittle deformation, we take the opportunity to recall a recent statement by Sparks et al. (2019) in their comprehensive review on the *Formation and dynamics of magma reservoirs*: "Terminology provides an essential pre-requisite for scientific discourse, but can present challenges, for example, when the terminology has genetic connotations; where meaning is implicit rather than explicit; if terms are ambiguous; or if meanings are not fully agreed within a scientific community or change with time. These difficulties are widespread in the igneous and volcanic literature." This remark is especially apt when discussing ductile and brittle deformation, terms which are frequently employed with different connotations. Below we summarize our usage of these terms, such that the reader may accurately interpret the text.

Rock deformation has long been characterized as brittle or ductile. As such, it is only natural that we employ similar terminology in volcanology to describe the deformation of magmas, which may flow or rupture. In early structural studies, the brittle–ductile transition was simply defined via strain markers, where brittle referred to ruptured materials along localized faults, ductile to pervasive strain, and transitional behavior to a combination of the two (Fig. 3a; e.g., Ramsay 1967). In experimental rock physics, brittle and ductile deformation equally relate to the macroscopic behavior of rocks and their occurrence is characterized by changes in shear stress and at times, volume. In the brittle field, rocks succumb to failure along localized (sets of) dilatant fractures, without significant prior deformation; glasses are a prime example of brittle materials which break abruptly (Célarié et al. 2003). In the ductile field, rock deformation results in more distributed, pervasive strain (except for the possible development of variably localized compaction bands and/ or shear bands at relatively high angles to the applied stress; Wong et al. 2001), which generally causes a net compaction of the rock (Rutter 1986; Heap et al. 2015a, 2017). Mechanically, brittle deformation may be accompanied by an initial phase of strain hardening as a rock yields beyond its elastic stress limit, followed by a stress drop as the material catastrophically ruptures (dark blue line, Fig. 3b); beyond this point, fault friction controls slip along the fracture. In contrast, ductile deformation may be accompanied by sustained strain hardening during compaction (green line, Fig. 3b; see Heap and Violay 2021, for a review of deformation modes in volcanic rocks). Under transitional conditions, rocks may rupture, or tear, to differing extents,

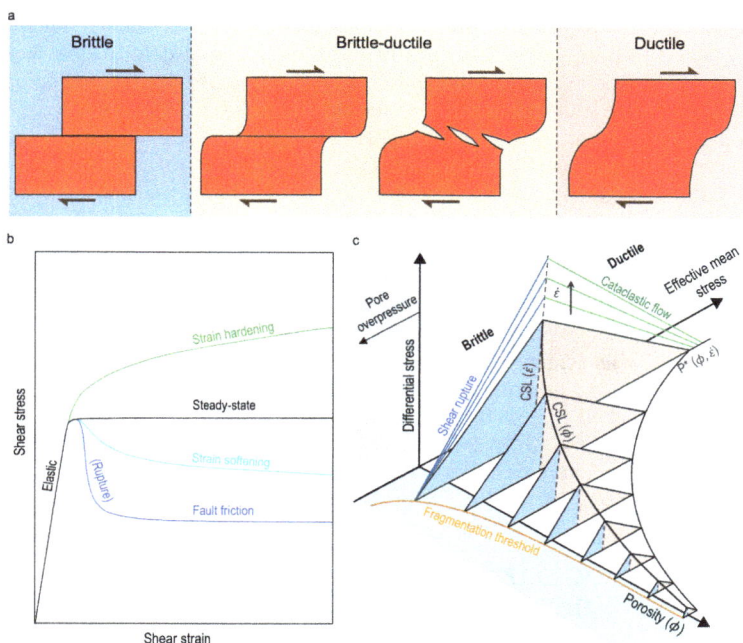

Figure 3. a) Shear strain localization in the ductile, brittle and transitional (brittle–ductile) regime. In the brittle regime, shear strain is localized on a single plane; as the component of ductility increases, shear strain becomes increasingly distributed, and transverse tears may occur. In the ductile regime strain is distributed. After Ramsay (1967). **b)** Representation of stress–strain curves and associated terminology during rock and magma deformation, with ductile behaviour in **green**, and brittle in **blue**. **c)** Schematic of the yield cap for vesicular volcanic rocks and magmas showing the brittle and ductile fields as a function of effective mean stress (equal to the difference between the average principal stresses and pore pressures). In the brittle field, strength increases with effective pressure, applied strain rate or with reducing porosity; in cases of pore overpressure (i.e., at negative effective pressures) magma ruptures at the fragmentation threshold. In the ductile field, strength decreases with effective mean stress, porosity and/or a reduction in applied strain rate. Modified from Lavallée and Kendrick (2020).

causing a period of strain softening (green line, Fig. 3b). This distinction in macroscopic behavior refers to deformation modes and not deformation mechanisms, yet it remains that brittle can also be used to define the deformation mechanism of discrete fracture events. Thus, one needs to be explicit about their inferences. So, in the above definition, compaction of porous materials or deformation of wet sand are regarded as ductile deformation modes, irrespective of the deformation mechanisms at work (which may be fracturing, grain rotation, slip, etc.), and brittle deformation nearly always results in dilation. Yield caps are commonly employed to constrain the conditions for brittle and ductile deformation (Fig. 3c); they constrain the shear strength of rock as a function of effective mean stress (equal to the average of the three primary stresses minus the pore pressure). In the brittle field the shear strength increases as a function of effective mean stress and in the ductile field the opposite is observed down to P^*, the point at which a porous rock compacts due to pore collapse and exhibits no shear strength, becoming more compliant (Loaiza et al. 2012). The transition between the ductile and brittle regimes for a given material is termed the critical-state line (CSL; Barton 1976). The strength of rocks and as a result, the size of their yield caps non-linearly decreases as a function porosity (Fig. 3c). In conditions of pore overpressure (where pore pressure exceeds the effective mean stress), the yield cap may extend to meet the conditions for fragmentation, commonly referred to as the fragmentation threshold (Fig. 3c).

In volcanology, this definition of brittle is applicable to magmas undergoing rupture or fragmentation, yet with only rare distinctions between microscopic deformation mechanism and bulk rupture. As for ductile, it is commonly used synonymous with viscous deformation; yet, it is not necessarily given that magma in the ductile regime is always wholly viscous. In melts, viscous flow ought to be considered as a deformation mechanism; but melts in magmatic suspensions may locally experience other deformation mechanisms (e.g., brittle rupture) even during macroscopically ductile deformation (e.g., during cataclastic flow; Fig. 3c). Ductile deformation of magma would be best described by rock physics' definition of ductility, whereby deformation is isovolumetric or compactant; though we must acknowledge that localized dilatancy may also play a part, especially in suspensions subject to component reorganization. Experimental studies have shown that whilst strain may be pervasive in porous magmatic suspensions, they experience increasing degrees of strain localization as they cool below the glass transition and enter the brittle regime (Heap et al. 2017). Recent efforts have been made to correctly identify and describe both the viscous–brittle and ductile–brittle transitions of multiphase magma (Lavallée et al. 2008, 2013; Cordonnier et al. 2012b; Kendrick et al. 2013, 2017; Coats et al. 2018; Wadsworth et al. 2018). Here we will abide to these definitions and will challenge interpretations where necessary, based on the observations made herein (see section *Brittle versus ductile deformation modes in rocks and magmas*).

STRAIN LOCALIZATION: AN INTERPLAY BETWEEN DEFORMATION MECHANISMS

The development of strain localization results from a complex interplay of deformation mechanisms, extant at different conditions. Here we review the occurrence of (i) viscous flow and energy dissipation, (ii) bubble deformation, coalescence and alignment, (iii) crystal alignment and deformation, (iv) rupture, and (v) frictional processes, as deformation mechanisms central to strain localization in magmas, lavas and fragmental (granular) materials.

Viscous flow and energy dissipation

Silicate melts are viscoelastic bodies (Dingwell and Webb 1989, 1990). Thus, they exhibit both viscous and elastic characteristics upon deformation, commonly illustrated using a dashpot and spring in series. When subjected to an applied stress, melt initially undergoes instantaneous elastic deformation, followed by a period of stress relaxation and viscous flow, characterized by strain evolving linearly with time. When the applied stress is released, the melt instantaneously recovers the strain from the initial elastic deformation and then slowly recovers the strain imparted during stress relaxation, so that only the strain associated with viscous flow remains. Mechanistically, elasticity results from the stretching of bonds whereas viscous flow results from diffusion of atoms and molecules. Upon viscous deformation, some energy is dissipated via conversion to heat, resulting in hysteresis in the mechanical data (i.e., visible in stress–strain curves).

As silicate melts are viscoelastic bodies, they can be well-described by Maxwell's concept of relaxation, so that the timescale of relaxation (τ) is proportional to the ratio between the shear viscosity (η_s) and the rigidity modulus at infinite frequency (G_∞):

$$\tau = \frac{\eta_s}{G_\infty} \tag{1}$$

This timescale is commonly referred to as the glass transition. The rheology of silicate melts has been the subject of extensive experimental studies with the result that we have obtained robust models of the chemical and temperature dependence of viscosity (Giordano et al. 2008). The viscosity of silicate melts can vary by some fifteen orders of magnitude over magmatically relevant conditions. In contrast, the elastic modulus of silicate melt has been shown to vary much less as a function of chemistry and temperature and can be approximated at 10^{10} Pa (Dingwell and Webb 1989), simplifying its application to natural scenarios. As silicate liquids exhibit Newtonian rheology, their viscosity is found to be rate independent. Exceptions arise when the applied strain rate approaches the relaxation rate of the melt structure (i.e., at $1/\tau$), which manifests as a viscosity decrease as the melt nears the kinetic conditions for rupture. It has been suggested that structural breakdown of the melt is the cause for this apparent non-Newtonian behavior near the glass transition interval (Webb and Dingwell 1990a,b; Bottinga 1994), whereas later studies have challenged the origin of this behavior (Hess et al. 2008; Cordonnier et al. 2012a; Wadsworth et al. 2018).

Viscous energy dissipation, commonly referred to as shear heating or viscous heating, is the conversion of energy to heat during flow. This heat causes temperature to rise, which results in a reduction of the melt's viscosity (assuming it is not lost). The magnitude of heat exchange can be assessed via the Brinkman dimensionless number which is the ratio between viscous energy dissipation versus loss by conduction. Experimental studies have measured viscous heating by inserting thermocouples in deforming melts (Hess et al. 2008; Cordonnier et al. 2012a) and suspensions (Lavallée et al. 2007, 2008). In silicate melts, viscous heating can be readily measured and was found to be proportional to viscosity and applied strain rate (Hess et al. 2008); even a few seconds of deformation at eruptive strain rates can generate temperature increases on the order of tens of degrees. Indeed, viscous heating leading to viscosity reduction, was found to be significant at strain rates approaching the glass transition of highly viscous melts (Cordonnier et al. 2012a; Wadsworth et al. 2018), providing a viable potential origin for the non-Newtonian behavior observed as melts approach their elastic limit (e.g., Webb and Dingwell 1990a,b); they found that there may be different regimes in which structural breakdown or viscous heating are the dominant contributors to the observed non-Newtonian behavior. Any development of temperature gradients due to heat gain and loss will promote strain localization. In magmatic suspensions, viscous heating is difficult to appraise numerically as the stress concentrated in the melt phase is poorly estimated. It is also difficult to measure as it is challenging to ensure thermocouples are inserted in the melt phase and because crystals act as heat sinks. Lavallée et al. (2008) showed that suspensions can heat by some degrees in a few seconds of deformation leading to rupture. Such thermally-driven viscosity reduction is locally limited by thermal diffusivity and therefore would accentuate strain partitioning (cf., Holtzman et al. 2005). The segregation of melt into anastomosed bands, separate from crystal-rich regions (see section *Development of crystal fabrics*) may be especially prone to develop viscous energy dissipation, depending on the volume fraction of melt, the angle of the band with respect to the principal shear stress and the amplification of strain rate in the melt band. Thus, further work is necessary in order to constrain viscous energy dissipation in magmatic suspensions and develop more robust rheological models, as it may encourage strain localization and promote apparent non-Newtonian behavior.

Bubble deformation and alignment

Exposed shallow conduits (Tuffen and Dingwell 2005), lava flows (Shields et al. 2016), lava domes (Eichelberger et al. 1986; Castro et al. 2002; Rust et al. 2003) and pyroclasts (Martí et al. 1999; Rust et al. 2003; Dingwell et al. 2016) portray the complex strain history of porous and/or vesicular (bubbly) magmas during volcanic eruptions. These igneous structures reveal the contrasting fate of isolated and connected bubbles which control the development of permeability that dictates gas emissions and volcanic eruption style (Eichelberger et al. 1986).

Isolated bubbles. Gas bubbles present in magmatic suspensions may be subjected to shear, which may be substantial during flow (Bagdassarov and Dingwell 1992; Manga et al. 1998; Lejeune et al. 1999). The deformation of bubbles can be evaluated via the Capillary number (*Ca*) which is the ratio between viscous and capillary forces (Taylor 1934; Llewellin et al. 2002; Rust and Manga 2002). It is commonly expressed as:

$$Ca = \frac{\eta_s a \dot{\varepsilon}}{\Gamma} \qquad (2)$$

which considers the bubbles' equivalent spherical radius (a), the strain rate ($\dot{\varepsilon}$), and the liquid–gas surface tension (Γ). Surface tension is generated by cohesive forces which creates internal pressure and coerces melt surfaces to contract to a minimum area, achieved in a sphere. So, the surface tension timescale (λ), also sometimes termed the characteristic relaxation timescale or bubble relaxation timescale, can be calculated through:

$$\lambda = \frac{\eta_s a}{\Gamma} \qquad (3)$$

Thus, the geometry of a bubble depends on the product of the surface relaxation time and the strain rate:

$$Ca = \lambda \dot{\varepsilon} \qquad (4)$$

Therefore, for small bubbles, high melt viscosities or low strain rates, flow is dominated by capillary forces and surface tension may help bubbles retain spherical shapes; otherwise, bubbles stretch due to viscous drag. The presence of bubbles thus modifies the rheology of lavas: bubbles increase suspension viscosity at low Ca and decrease suspension viscosity at high Ca—a transition centered at $Ca = 1$ (Llewellin and Manga 2005; Mader et al. 2013). [We direct the reader to Mader et al. (2013) for a comprehensive review of two-phase magma rheology.]

Importantly for the problem of strain localization, we must consider (i) the alignment of (individual) bubbles and (ii) the interaction and configuration of a vesicular network in a flow field. First, the capillarity of the suspension controls the orientation of bubbles with respect to the flow field (θ). Rust et al. (2003) compiled theoretical and empirical relationships between capillarity, bubble geometry and their relative orientation θ in different strain regimes (Fig. 4a,b; Table 1). These equations have proved useful in quantifying shear in aphyric lavas and pyroclasts (Rust et al. 2003; Dingwell et al. 2016), whose vesicle geometry and alignment may vary significantly (Fig. 4c–e). For instance, tube pumices exhibit extensive strain and offer unique strain markers of ductile and brittle processes enacted at fragmentation (Martí et al. 1999). Their origin has remained long debated: are they the result of pure (Mader et al. 1996) or simple (Polacci 2005) shear (Fig. 4f)? Reconstruction of vesicles using X-ray or neutron computed tomography now provides a means to apply capillarity to constrain shear regimes extant upon fragmentation (Fig. 4g). Yet, to do so for pyroclasts, one must consider bubble shape relaxation induced by surface tension once the principal shear stress is released upon fragmentation. This can be done by considering the surface tension timescale (Toramaru 1988) as well as knowledge of temperature to resolve the evolution of viscosity, and thus bubble shape evolution expected during cooling down to the glass transition (Fig. 4h). For ballistic bombs and large lapilli, this may simply be done by letting a given clast cool whilst monitoring the temperature

Table 1. Bubble deformation and orientation for pure and simple shear scenarios, as compiled by Rust et al (2003).

Geometrical conditions	Pure shear	Simple shear
Bubble deformation: $Ca \ll 1$	$D = 2Ca$	$D = Ca$
Bubble deformation: $l/a \gg 1$	$l/a = 16Ca^{2*}$	$l/a = 3.45Ca^{1/2}$
Bubble orientation: $Ca \ll 1$	$\theta = 0$	$\theta = \pi/4 - 0.6Ca$
Bubble orientation: $l/a \gg 1$	$\theta = 0$	$\theta = \tan^{-1}(0.359Ca^{-3/4})$

Note: *We refer the reader to Acrivos and Lo (1978) for further detail about the constant of proportionality in the bubble deformation equation.

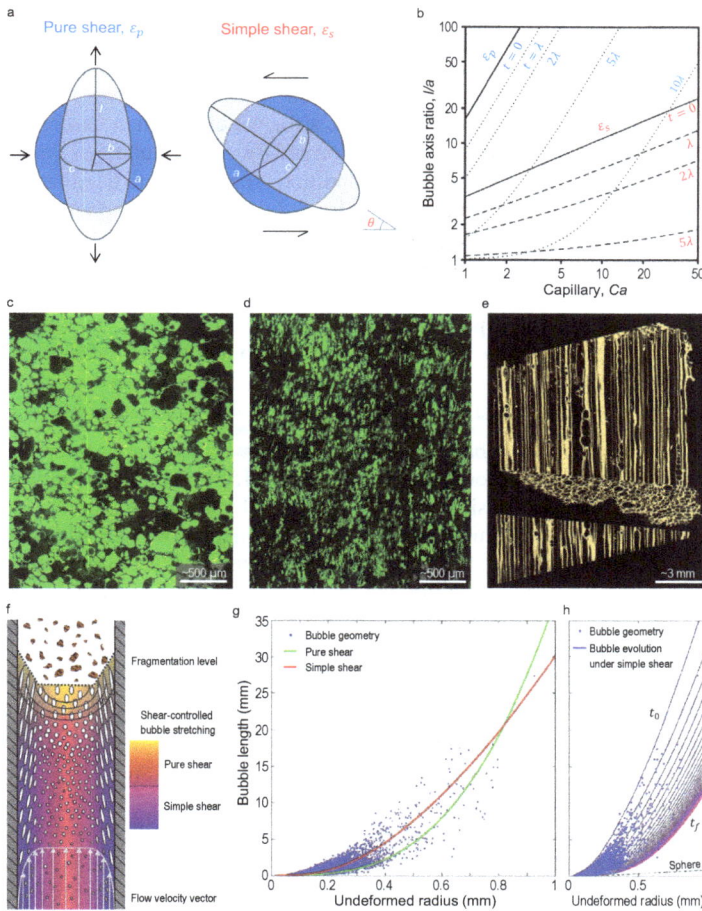

Figure 4. Isolated bubbles as indicators of strain. **a)** The strain ellipsoids for bubbles deformed in pure and simple shear, showing the initial radius (a) of the undeformed bubble (**dark blue**), the modified orientation (θ) of the half length of the long axis (l), and minimum and maximum radius (b and c) which can be used to quantify the 2D ellipticity ($1-c/b$). **b)** Bubble axis ratio (l/a) versus capillary number for bubbles deformed in pure (ε_p) and simple shear (ε_s), with contours for time (t) cast as a function of bubble surface timescale or characteristic relaxation timescale (λ), showing decreasing aspect ratio with time in both pure and simple shear. **c)** Relatively isotropic bubbles and **d)** sheared bubbles in rhyolitic pumices from Tarawera, New Zealand, imaged using UV light on thin sections impregnated by dyed epoxy. **e)** Tomographic reconstruction of a clast of tube pumice from Ramadas Volcanic Centre, Argentina, showing high aspect ratio pores and narrow bubble walls. Image courtesy of Katherine Dobson. **f)** Schematic representation of shear regime within a conduit, reconstructed using bubble shape analysis in **g–h**. **g)** The geometry of the bubble population in the erupted pyroclasts at Ramadas Caldera (Marti et al. 1999) is explored as length versus undeformed radius, compared to the best fit for the case of pure and simple shear, and in h) compared to the fitting curves for simple shear evolution using cooling rate of $10^{-4.9}$ s^{-1} and strain rate of 10^{-2} s^{-1}, since pure shear provides an infeasible solution the bubble textures are attributed to simple shear. Panels **e–h** modified after Dingwell et al. (2016).

(e.g., Dingwell et al. 2016); but for other pyroclasts which may have been insulated in ash clouds or pyroclastic density currents, this may be more difficult to ascertain. In the case of tube pumice, the method has helped demonstrate that simple shear near conduit margins may be responsible for the significant generation of elongate or tubular vesicles in large explosive eruptions.

Second, the deformation conditions, and relaxation state of melts, influence the interaction and configuration of vesicles in magmatic suspensions and foams (Ryan et al. 2019). Using synthetic FOAMGLAS (as analogues to bubbly magmas), Ryan et al. (2019) assessed the strain dependence of the rheological and physical evolution of two-phase suspensions. They observed that foams experiencing conditions for which the melt can relax the applied stress harden during deformation due to compaction and loss in vesicularity. Foams subjected to higher strain rates or cooler, higher-viscosity conditions (i.e., closer to the glass transition) for which the melt cannot easily relax the applied stress show complex responses to deformation (both strain weakening and hardening). In either case, they found that deformation does not generally result in the creation of permeable paths (as bubbles remain isolated). In highly strained, unrelaxed foams however, strain localization can promote bubble coalescence along thin shear bands, oriented perpendicular to the principal stress, resulting in localized outgassing pathways. So, strain localization may be key to modifying the configuration of the vesicular network in magma, prompting permeable flow in otherwise persistently impermeable magmatic foams characterized by isolated bubbles.

The above descriptions of bubbly suspensions are however not appropriate if crystals are abundant, as in many magmatic scenarios. Truby et al. (2015) suggested that in some scenarios, the rheology of multiphase suspensions may be simplified by assuming the presence of crystals in an interstitial suspension containing small bubbles. This may prove helpful to constrain some magmatic environments, yet, it cannot explain some experimental observations such as the recognition that gas may readily flow through crystals (Laumonier et al. 2011; Lavallée et al. 2017), and bubble shapes may evolve non systematically with respect to the macroscopic strain (Dobson et al. 2020), owing to local effects. Our models await further experimental constraints, which may now be possible with the advent of *in operando* testing during X-ray imaging (Baker et al. 2012; Lavallée et al. 2017; Coats 2019; Wadsworth et al. 2019; Dobson et al. 2020).

Connected bubbles and pores. At high vesicularity (e.g., $\gtrsim 30\%$; Klug and Cashman 1996) or upon shear (Okumura et al. 2006) bubbles may coalesce, as commonly observed in eruptive pyroclasts and lavas (Eichelberger et al. 1986; Wright et al. 2006; Colombier et al. 2017; Bain et al. 2019), or they may be connected via fractures (Smith et al. 2001; Mueller et al. 2005; Farquharson et al. 2015; Kendrick et al. 2021). Similarly, in fragmental materials (e.g., breccias and tuffisites), intergranular pores and cracks may provide a variably well connected, porous network (Tuffen and Dingwell 2005; Kendrick et al. 2016). Connected porous networks exhibit non-minimized surface area (i.e., non-spherical), which allows venting and encourages geometrical evolution. Even for restricted eruptive regimes in which surface tension is the dominant acting force, evolution of the porous network (by progressive isolation) promotes strain localization. Recently, several experimental studies have helped constrain the kinetics of densification (down to ca. 3% vesicles) resulting from surface tension (Vasseur et al. 2013; Wadsworth et al. 2014, 2017, 2019). As Equation (3) indicates that the radius of bubbles (or the curvature of the gas–melt interface) controls the surface relaxation timescale, the geometry of connected bubbles (or melt droplets in the process of agglutination) dictates the development of local stresses and the relative anisotropy of the densifying system, so that strain may localize during the process of densification. In cases where crystals are present, they locally impart a rigidity that impacts the gas–melt interface curvature, thus influencing densification rate and extent (Pascual et al. 2002; Heap et al. 2015b; Kendrick et al. 2016). Such work has helped bridge rheological models for the sintering of porous aggregates and flow of bubbly liquids (Vasseur et al. 2013; Wadsworth et al. 2014, 2017, 2019).

For larger suspensions or those at depth (where overburden impacts densification), viscous forces imposed by gravity exceed surface tension. Quane and Russell (2005) experimentally studied the compaction of molten silica bead aggregates characterized by an initially fully connected network of vesicles using uniaxial compressive loading (i.e., pure shear), showing that connectivity of vesicles eases their destruction. They found that strain (ε) evolves nonlinearly following:

$$\varepsilon = \frac{\phi_0 - \phi_t}{1 - \phi_t} \qquad (5)$$

which requires knowledge of the initial vesicularity ϕ_0 and the vesicularity at any given time during deformation (ϕ_t). As a result of this strain hardening, the viscosity of suspensions with connected vesicles generally increases nonlinearly at a reducing rate upon densification (Quane et al. 2009). Kendrick et al. (2013) deformed crystal-rich suspensions with different initial fractions of connected vesicles, using uniaxial compressive tests at variable loads; the results indicated that the applied stress and strain dictate the magnitude of densification, strain hardening, and degree of strain localization. In a follow-up study, Ashwell et al. (2015) deformed crystal-rich and crystal-poor lavas, containing up to ~60% connected vesicles, showing that an abundance of crystals can prevent compaction to vesicularity below 18–20% in the absence of confinement. Under confinement however, the fate of pores may vary depending on the shear regime: torsion experiments have shown that extensive simple shear may serve to enhance connectivity (Okumura et al. 2010); however, during pure shear connected vesicles can be readily squeezed out (Heap et al. 2017). Thus, during flow of multiphase magmas, lavas or granular suspensions, strain is likely to localize preferentially in areas containing large (voluminous) connected vesicles (Dobson et al. 2020), which may promote localized viscous heating, further enhancing strain localization (e.g., Wright and Weinberg 2009).

Permeable flow in porous materials commonly localizes along limited pathways (van der Linden et al. 2019). As such, we anticipate that the development of pore pressure during fluid flow will vary locally, leaving certain areas of vesicular magma (with low gas pressure) to be more vulnerable to external stresses, as is the case for fluid flow in compacting granular materials (van der Linden et al. 2019) and rocks (Eggertsson et al. 2016; Shilko et al. 2018), or in the presence of dykes acting as strain localization buffers to regional deformation (MacDonald et al. 2017). Thus, we advance that strain will localize in such situations, causing local compaction, whilst the vesicular regions sustaining fluid flow, and importantly fluid pressure, may remain porous as commonly observed by the presence of vesicular domains surrounded by dense lava in obsidian flows (Fig. 5a). Similarly, strain localization takes place throughout the formation of tuffisites, which would be rheologically weaker than their host magma, and where pressurized fluids can counteract densification during sintering (Fig. 1e). Finally, strain is inhomogenously distributed during the agglutination and associated shear and compaction of pyroclasts in cinder cones or pyroclastic density currents (Fig. 5b), owing to the contrasting rheologies of pyroclasts and variable development of the porous network within and surrounding them.

Figure 5. Strain in connected porous networks. **a)** Localized densification of porous zones into dense obsidian, leading to compartmentalization of the porous network at Panum crater, Long Valley, USA. **b)** Eutaxitic texture showing compacted pores and ash particles (brownish patches) resulting from vertical compaction and lateral elongation in the Road Kill ignimbrite, USA; modified from Lavallée et al. (2015).

Crystal alignment and deformation

Development of crystal fabrics. The configuration of suspended crystals is commonly modified during shear (Holtzman et al. 2003b; Caricchi et al. 2008; Picard et al. 2011), because crystals may organize and align themselves as observed in migmatites (Blumenfeld and Bouchez 1988), cumulates (Bertolett et al. 2019), sheet-complexes (Kruhl and Vernon 2005), dykes (Shelley 1985; Wada 1992), spines (Kendrick et al. 2012; Pallister et al. 2013; Wallace et al. 2019b), lava domes (Vernon 1987; Walter et al. 2013), lavas flows (Ventura et al. 1996; Castro et al. 2002; Walter et al. 2013) and viscously remobilized wallrocks (Schauroth et al. 2016). The crystals present in magma act as obstacles that hinder viscous flow and so increase magma viscosity (Lejeune and Richet 1995; Caricchi et al. 2007). Experiments on suspensions have shown that strain partitions into regions of elevated melt fraction, because the effective viscosity is lower than in crystal-rich regions (Ventura et al. 1996). Picard et al. (2013) experimentally constrained that the preponderance of strain partitioning fabrics increases with crystal fraction. In crystal-poor suspensions, crystals are (generally) free to rotate and align in the local flow-velocity vector. Vachon et al. (2021) states that "simple shear flow causes preferred crystal orientations that are approximately parallel to the boundary. Where pure shear deformation dominates, there is a tendency for crystals to orient themselves in the direction of the greatest tensile strain rate". The rotation and alignment of crystals locally disturbs the flow field, causing melt to circumvent crystals during shear; as such when micro-layering is present in interstitial melt responding to non-coaxial strain, it may generate micro-folds in an asymmetrical arrangement that reflects the sense of phenocryst rotation (Fig. 6a; Vernon 1987). Where crystals can rotate freely, Mueller et al. (2009) defined two end-member cases: if the long axis of an elongate crystal is aligned parallel to the vorticity vector, the particle rotates constantly about that axis; however if the elongate crystal's long axis is perpendicular to the vorticity vector, the particle tumbles end over end, according to the paper. Ultimately, flow promotes crystal alignment and development of fabrics (Fig. 6b). The aspect ratio of particles (i.e., length: width) controls the evolution of particle orientation with respect to the shear plane as a function of strain (Blumenfeld and Bouchez 1988), thus influencing the development of crystal fabrics, flow bands and tiling (i.e., imbrication structures; Fig. 6c).

In suspensions with crystallinity greater than the critical crystal volume fraction at which a crystal network forms (i.e., a critical fraction dependent on crystal shape; Saar et al. 2001), deformation results in crystal interactions and development of variably crystalline flow bands (Fig. 6d); the development of fabrics is often deemed essential to seemingly unlock crystal frameworks and enable deformation (Bergantz et al. 2017). [Note that although crystal-rich magmas exhibit no yield strength (Caricchi et al. 2007; Lavallée et al. 2007), their relaxation timescale can be relatively long (compared to single phase melts), so the configurational modification of magma during the development of fabrics can facilitate deformation.] Torsion experiments have shown that simple shear promotes the segregation of melt in anastomosing channels surrounding crystal-rich bands (sometimes referred to as non-bands; see Fig. 6e), lowering the viscosity until an equilibrium configuration is reached (Holtzman et al. 2003a; Holtzman and Kohlstedt 2007; Picard et al. 2011). Holtzman et al. (2012) constrained that a higher degree of melt segregation generally enhances strain rates. In their experiments using equant crystals in suspension, Holtzman and Kohlstedt (2007) observed bimodality of melt band angles to the shear stress direction, with thick melt bands generally at 10–25° (though most commonly 15–20°), connected by thin melt bands at low angles of 0–10° (Fig. 6e). They proposed that the same set of processes dictate both the large and small bands: because of strain partitioning, the stress field in the crystal-rich regions between large bands is back-rotated relative to the sense of shear in the crystal-rich non-bands, so that thin melt bands develop at the same angle to their neighboring large melt bands, as the large bands do to the applied shear direction; thus, thin melt bands end approximately parallel to the shear plane.

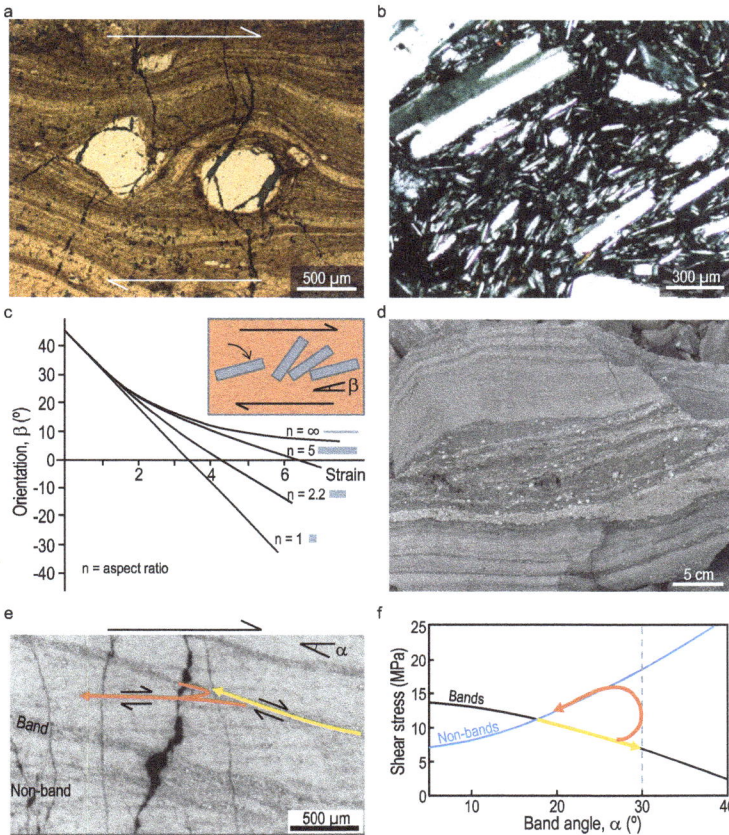

Figure 6. a) Flow bands around crystals in rhyolite from the Nez Percé obsidian dome, Yellowstone, USA **b)** Photomicrograph in plane polarized light of an andesite from Volcán de Colima showing microlite alignment in the groundmass. **c)** Evolution of crystal orientation with respect to shear plane as a function of strain. Crystal orientation is dependent on aspect ratio (length: width). Modified from Blumenfeld and Bouchez (1988). The inset shows crystal alignment and tiling when crystals physically interact during rotation. **d)** Crystal-rich and crystal-poor vesicular bands in a rhyolitic block at Glass Creek dome, Long Valley, USA. **e)** Depiction of the generation of two populations of crystal-poor flow bands during shear of crystal-rich suspensions: the first, comprised of wider "bands" inclined at ~10–25° from shear direction, and second, finer "non-bands" which form by the intermittent locking of grains that cause a jump in shear stress in the wide bands (**yellow arrow**) that releases some of the crystal cargo into fine layers (**red arrow**), approximately parallel to the sense of shear. Such bands accrue and destruct perpetually, so shear textures are ubiquitous in flow scenarios, with angle, band thickness and spacing dictated by crystal cargo and strain rate. **f)** A graphical representation of the flow band formation process in terms of band angle and shear stress for bands and non-bands (with a total segregation factor (S_1) of 0.6). Panels e–f modified from Holtzman et al. (2005).

As such, strain partitioning results in a mechanical affinity for the development of shear at 15–20° angles. As previously stated, strain localization in such melt-rich bands may promote viscous energy dissipation, enabling further localization. Holtzman et al. (2005) observed that the evolution of strain partitioning controls the evolution of energy dissipation and attainment of a steady-state rheology in melt segregated magmas. For steady-state to be achieved, balance must be attained with respect to temperature gain versus loss and melt reorganization versus strain partitioning (in melt- and crystal-rich bands). Therefore, melt must continuously move with respect to the crystals (dependent on the crystal-framework permeability), which they hypothesize may be possible via, what they refer to as, the pumping cycle (Fig. 6e, f). Melt migrates from low (e.g., ~15°) to high (e.g., 30°) angle bands owing to reduced shear stress (and mean pressure). At some critical angle, the melt pressure may increase to near that of the crystal-rich bands, causing a shift in the pressure gradient direction and squeezing of melt towards areas of reduced pressure in low angle-bands; then the process repeats itself under steady-state. If so, then we argue here that this may ensure ubiquity of textural overprints (such as melting and crystallization) imposed by shear (i.e., causing local temperature increases and pressure fluctuations). In a follow-up study Holtzman and Kohlstedt (2007) detailed that the spacing of the segregated melt bands decreases with increasing applied stress, which they relate to the compaction length decreasing with decreasing viscosity at higher stresses. In these experiments they noted no net volume change and argued that any local dilation expected from shear (Reynolds 1885; Gourlay and Dahle 2007) would be counterbalanced by melt redistribution and ingress in low pressure areas of strain localization. Thus shear may also promote physico-chemical differentiation of crystal-rich dome lavas or mushes, which may be important for the generation of magma in the mantle (McKenzie 1985; Kohlstedt and Holtzman 2002; Holtzman et al. 2003b) as well as aphyric magma eruptions. Such filter pressing may be further enhanced by the presence of exsolved volatiles (Pistone et al. 2015).

In crystal-rich suspensions, crystal rotation and alignment contributes to reducing the suspension viscosity, a strain weakening effect which is more significant if crystals are elongate, such as late-stage microlites in extrusive products, phenocrysts in alkaline lavas (such as trachytes and phonolites) and in cumulates produced during mush compaction (Bertolett et al. 2019). Caricchi et al. (2008) deformed trachytic lavas in torsion to map the evolution of viscosity and associated crystal fabrics using shape preferred orientation (SPO), noting flow bands similar to those described by Holtzman et al. (2005). However, in these suspensions of elongate plagioclase crystals melt segregation was not the dominant observation; instead, discrete areas of strain localization developed whereby crystals aligned at ~20–40° from the shear plane (in areas referred to as bands by Holtzman et al. 2005). In the non-band areas, Caricchi et al. (2008) optically saw no development of fabrics from the elongate crystals (unlike quantification in later work by Picard et al. (2011); see below). They noted that in suspensions undergoing strain weakening an additional discrete, low-angle (~10°) band of elongate crystals overprinted the aforementioned fabric.

In suspensions, where crystal alignment is difficult to identify (e.g., in the presence of tabular and/ or equant crystals), one may map the crystallographic preferred orientation (CPO) in three-dimensions (3D), using neutron texture diffraction (Walter et al. 2013) or electron backscattered diffraction (EBSD) during scanning electron microscope imaging (Prior et al. 1999). Neutron texture diffraction to reconstruct CPO has occasionally been employed on natural lavas (Walter et al. 2013; Zucali et al. 2014), showing the alignment of plagioclase and sanidine in the flow field; as it reconstructs a full 3D volume, Walter et al. (2013) were able to show the occurrence of rotation (centered along the b axis) as both the a and c crystallographic axes of these oblate crystals aligned in the flow field; yet, to our knowledge, such textural characterization ought to be more commonly employed on experimentally deformed lavas with known strain histories. Magmatic fabrics, mapped by EBSD, have classically focused on the development of CPO of quartz, olivine, and biotite (Romeo et al. 2007; Zak et al. 2008; Beane and Wiebe 2012; Graeter et al. 2015), but recently the literature has been increasingly populated by studies on plagioclase (Picard et al. 2011; Satsukawa et al. 2013; Ji et al. 2014; Fiedrich et al. 2017; Holness et al. 2017; Kendrick et al. 2017; Bertolett et al. 2019; Wallace et al. 2019b), owing to its importance in magmatic and volcanic systems. Picard et al. (2011) employed EBSD to map CPO of lavas experimentally deformed in torsion; they showed that in shear bands, CPO indeed developed with the alignment of the long crystal axis {100} roughly parallel to the shear plane; yet in non-bands, they constrained a concentration of {100} poles at ~45° (albeit with wide scatter), anti-clockwise from the principal shear plane with girdles of {001} and {100} poles at ~60° and ~50° clockwise, respectively. In a complementary study, Qi et al. (2018) deformed ultramafic magma mushes and observed the development of CPO of olivine crystals with a slightly stronger alignment of [001] than [100] axes in the shear direction; yet, showing little change in either strength or distribution with increasing stress or strain. Comparing their results to CPO developed during rock deformation, Qi et al. (2018) suggest that the presence of melt reduces the CPO.

So, in summary, whereas suspensions with low crystallinity (<50%) tend to develop simple fabrics from crystal rotation, alignment and imbrication (or tiling), crystal-rich suspensions partition strain, resulting in complex anastomosed bands with intricate fabrics: crystal-poor regions develop at relatively low angle from the shear plane and with crystals aligned nearly parallel to the shear plane, and crystal-rich regions develop with crystal alignment up to ~45° anti-clockwise from the principal shear plane, akin to S/C'-like fabrics. Thus, beyond its rheological impact, the development of fabrics (i.e., tiling vs S/C' fabrics) may be used as an indicator of shear, paleo-flow direction, and therefore, areas of strain localization.

Crystal plasticity. In suspensions with high interstitial melt viscosity, high crystal content or low vesicularity, large stresses may accumulate in the crystalline phase (Deubelbeiss et al. 2010, 2011), which may result in crystal plasticity (Kendrick et al. 2017; Wallace et al. 2019b; Wieser et al. 2020) and rupture of phenocrysts in plutonic and eruptive products. It is rather common to witness undulose extinction of quartz and micas (during optical observation of minerals under cross-polarized light), as these minerals are generally deemed weak. But all minerals may be subject to crystal plasticity if the conditions are suitable. Crystal plasticity may occur via different mechanisms: (i) dislocation creep, involving the movement of dislocations (i.e., edge or screw dislocations) in crystal lattices, migrating from the crystal edge inward, causing lattice defects, and (ii) diffusion creep, occurring via diffusion of vacancies through crystal lattices (i.e., Nabarro–Herring creep) or along grain boundaries (i.e., Coble creep when dry or pressure solution when wet). Whereas dislocation creep can result in brittle failure, diffusion creep generally results in plastic deformation; indeed, diffusion creep is particularly sensitive to temperature and operates in crystals at homologous temperatures within ~10% of their melting temperature. [Note that the homologous temperature, defined as the ratio of observation to melting temperatures, is commonly invoked to cast transitions in deformation mechanisms (Murrel 1990).] High-temperature deformation experiments on partially molten olivine aggregates (i.e., magma mush with <12% melt) have shown that diffusion creep generally dominates over dislocation creep at pressures greater than ~100 MPa, depending on melt fraction, water concentration and applied stress (Mei and Kohlstedt 2000; Mei et al. 2002); they found that strain rate increased by a factor of 20 and 40 in the diffusion and dislocation creep regimes, respectively, when the rock underwent 12% partial melting, thus impacting strain localization. So, whereas diffusion creep may be common in deep magmas deforming slowing over long timescales, dislocation creep may be more common in shallow magma mushes (Vinet and Higgins 2010; Bertolett et al. 2019; Wieser et al. 2020) and erupting lavas (Kendrick et al. 2017; Wallace et al. 2019b).

Recent evidence suggests that even under very low confinement, the crystals present in dome lavas may deform plastically prior to failure (Fig. 7a; Kendrick et al. 2017; Wallace et al. 2019b). EBSD mapping of experimentally deformed porous crystal-rich dome lavas by Kendrick et al. (2017) showed that phenocrysts can fracture, whilst microlites may systematically deform via crystal plasticity (Fig. 7b), concluding that crystal plasticity plays a key role in the deformation of crystal-rich lavas. Under unconfined conditions, as experienced in active lava domes, crystals underwent dislocation creep and the misorientation of the crystal lattices within the microlites increased as a function of stress and strain, until rupture (Fig. 7c). Grain size reduction was seen synchronous to deformation, and fragments of broken microlites recorded the highest misorientations (deformation); thus, crystals exhibit a plastic limit during

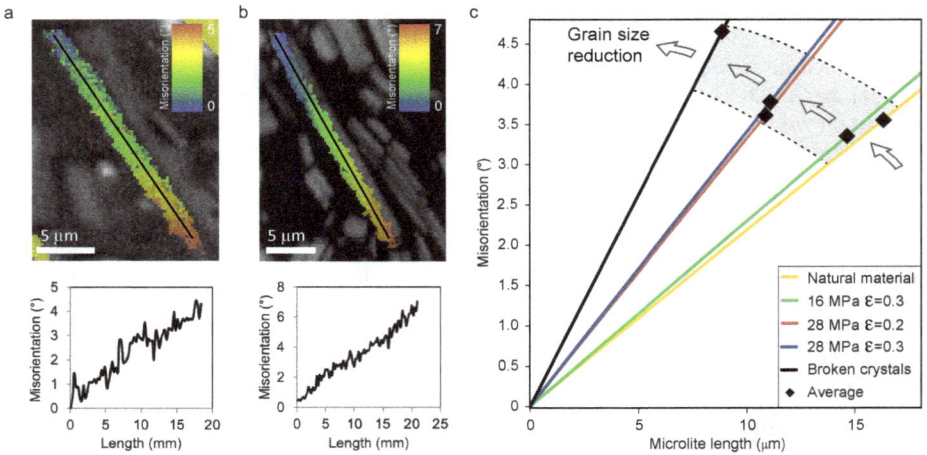

Figure 7. The development of crystal plasticity in plagioclase microlites identified and quantified using electron backscatter diffraction (EBSD). **a)** An example of a plastically deformed microlite occurring naturally in the groundmass of porphyritic andesite from Volcán de Colima, the color scale represents the misorientation angle of the crystal lattice from a reference point (the end of the **black transect line** at misorientation 0, shown by the blue color), with the crystal plastic distortion along the marked transect represented graphically below. **b)** An example of crystal plasticity in the same andesite, this time a cylinder of magma was deformed in a uniaxial press at 940 °C under 28 MPa axial stress to 0.2 strain. **c)** The transects were analyzed for more than 50 microlites for the as-collected lavas and each of a suite of experimentally deformed samples, and the gradient of each defined a misorientation per micron value which defines the gradient of the lines in **c**. Crystal plasticity is seen to increase as a function of applied stress and strain, and additionally the mean grain size decreases (**black diamonds**). When considering only plasticity within fragments of broken microlites, the plasticity is highest (**black line**), suggestive of microlites breaking upon reaching a plastic limit. Data from Kendrick et al. (2017).

dislocation creep. In a follow-up study, Wallace et al. (2019b) mapped crystal-plasticity across a shear zone developed along the spine erupted at Unzen in 1994–1995, finding that the degree of crystal lattice misorientation in plagioclase microlites increased systematically across the shear zone towards the marginal shear plane, and also that weak phenocrysts such as mica suffered substantial crystal plasticity.

The absence of crystal plasticity in several deformed igneous bodies has previously been argued to indicate the presence of melt taking up the strain during deformation (Rosenberg 2004); Lavallée and Kendrick (2020) advanced that this may be the case if the melt viscosity is very low, as even erupted fragments of shallow mush exhibit crystal plasticity (Wieser et al. 2020), and added the caveat that the absence of crystal plasticity may also indicate post-deformation crystallization or longer deformation timescales (i.e., slower strain rates) which limit stress accumulation in the crystalline phase. Note that similarly, the presence of twinned crystals in igneous rocks has often been inferred to result from shear stresses during crystal-crystal interaction (Buerger 1945; Smith and Brown 1988), although Deubelbeiss et al. (2011) showed that stress can accumulate in crystals without the requirement for crystal-crystal interaction due to build-up of stress in the interstitial melt phase. Recently, Brugger and Hammer (2015) demonstrated, by mapping the products of crystallization experiments using EBSD, that twinning may be induced by crystallization (at increasing proportion as a function of undercooling), due to growth defects introduced during the incipient stages of crystallization; thus twinning may not necessarily arise from deformation (Brugger and Hammer 2015), nor should it be used, alone, as an indicator of strain localization. To date, insufficient data exists to define the plastic limit of crystals, necessary to resolve the origin of non-Newtonian rheology of suspensions and the stress–strain history of magmas and lavas. But it is likely that our rheological models for multiphase suspensions (e.g., Costa et al. 2009) may require consideration of crystal-plastic deformation (Ancey 2007) and a complete reconstruction of phase evolution during strain, now possible via *in operando* testing during X-ray imaging (e.g., Polacci et al. 2018; Coats 2019; Dobson et al. 2020).

Multiphase magma rupture

The strength of multiphase magma. Material rupture results from an excess of potential energy, distributed internally by heterogeneities at the atomic level (Kilburn 2012). The development of fractures during material rupture is dependent on the density of the material's constituents (hence on chemistry), on temperature, effective pressure (i.e., confining pressure minus pore pressure), and deviatoric stress (or applied strain rates). During rupture, most of the energy is consumed by surface area creation and the remaining energy converts into heat (Fig. 8) and kinetic energy that displaces fragments, causing fault slip or the ejection of pyroclasts, which may cause an explosive eruption (Alatorre-Ibargüengoitia et al. 2010) or formation of tuffisites (Kolzenburg et al. 2012, 2019; Kendrick et al. 2016). Material rupture generates fragments of all sizes (Kueppers et al. 2006; Kennedy and Russell 2012). Thus, volcanic ash may be produced by rupture due to simple shear as well as during magmatic fragmentation due to pore overpressure (Hornby et al. 2019b).

Although liquids do not have a shear strength, silicate melts can exhibit strength when forced to deform in the glassy, solid state. Silicate melt has been demonstrated to rupture at conditions where the applied strain rate cannot be relaxed and stress accumulates elastically (Dingwell and Webb 1989, 1990). If sufficient stress accumulates (~100–1000 MPa), the melt structure breaks down and the melt macroscopically ruptures (Webb and Dingwell 1990a, b). As melts can repeatedly fracture and heal in volcanic environments (e.g., Tuffen et al. 2003; Wadsworth et al. 2018), Lamur et al. (2019) investigated the time dependence of tensile strength recovery along healing silicate melt interfaces; they observed that healing initiates after a timescale proportional to the relaxation timescale, and therefore melt viscosity. Initially, strength was gained rapidly by the wetting of (and associated obliteration of pores along) the interface due to viscous deformation; then, the experiments showed slower strengthening congruent with diffusional processes. As bubbles may get trapped during the processes, the original "strength of the melt" was never fully recovered, so subsequent ruptures recurred along the same plane; however, if these bubbles can be removed (e.g., by volatile resorption, due to shear, or if healing occurs in low-viscosity melts) then all evidence of fracture-healing may be erased from the melt (e.g., Taddeucci et al. 2021).

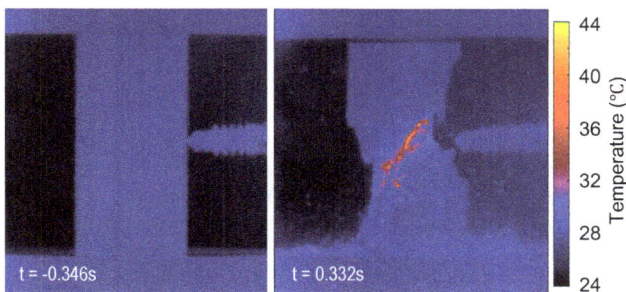

Figure 8. Thermal output generated by failure recorded using a thermographic camera, with images immediately prior to and after wholesale failure (at $t=0$) of a basaltic sample deformed at room temperature under uniaxial compression at a strain rate of 10^{-5} s^{-1}, showing a temperature increase of >14 °C.

The variable abundances of melt, crystals and bubbles in magmas imply that a wide range of shear conditions may lead to rupture. The strength of magmatic suspensions generally decreases as a function of vesicularity and crystallinity. This has been demonstrated for: aphyric suspensions in compression (Vasseur et al. 2013) and tension (Martel et al. 2000); dense, variably-crystalline suspensions in compression (Cordonnier et al. 2012b), but not in tension; and, variably-vesicular, crystal-bearing suspensions in compression (Coats et al. 2018) and tension (Martel et al. 2001; Hornby et al. 2019a). Zhu et al. (2010) provide an analytical approximation of the pore-emanated crack model of Sammis and Ashby (1986), predicting uniaxial compressive shear strength (σ_p) as a function of vesicularity (ϕ) and vesicle radius (a):

$$\sigma_p = \frac{1.325 K_{IC}}{\phi^{0.414} \sqrt{\pi a}} \qquad (6)$$

where K_{IC} is the critical stress intensity factor or "fracture toughness"—a scale-invariant material property describing resistance to tensile failure (Balme et al. 2004). The model predicts that with increased stress applied on a porous medium, cracks propagate from the pores (parallel to the direction of the applied stress) when the stress at the crack tip reaches the critical stress intensity factor. Vasseur et al. (2013) demonstrated that $\frac{K_{IC}}{\sqrt{\pi a}}$ can predict the nonlinear reduction in strength as a function of vesicularity of aphyric magmatic suspensions. In contrast, for crystal-bearing porous suspensions, where strength also decreases as a function of vesicularity but with more variability than aphyric suspensions, the pore-emanated crack model, alone, falls short in resolving the porosity dependence of multiphase suspensions' strength (Coats et al. 2018). Instead, the presence of crystals appears to warrant the consideration of micro-cracks, common in crystalline materials. Coats et al. (2018) invoked the wing-crack model (Ashby and Sammis 1990), which relates tensile stress concentration at the tips of sliding, inclined (45°) pre-existing cracks, to the principal applied stress. Frictional shear resistance (σ_s) of cracks must be overcome to propagate wing cracks (parallel to the principal stress direction) if the local stress exceeds K_{IC}. The unconfined compressive strength can be estimated using the analytical solution (after Baud et al. 2014):

$$\sigma_p = \frac{1.346 K_{IC}}{\sqrt{1+\sigma_s^2} - \sigma_s \sqrt{\pi c}} D_0^{-0.256} \qquad (7)$$

where D_0 is the initial damage, $D_0 = \pi (c \cos \gamma)^2 n_A$, which is a function of the fracture length ($2c$), the number of sliding cracks per unit area, n_A, and the orientation of original microcracks from which wing cracks are generated, $\gamma = 0.5 \tan^{-1}(1/\sigma_s)$ This model however does not consider material porosity so is again insufficient alone to resolve the strength of suspensions. Instead, Coats et al. (2018) propose that future solutions should include consideration of both models—suggestions equally advocated for by the description of volcanic rock strength (Heap et al. 2014; Zhu et al. 2016).

The transience of phases in magmas during transport and eruption implies that strength evolves through time and varies spatially. This is particularly important as spatial variations in magma heterogeneity may favor strain localization. For example the formation of tuffisites (Fig. 1e) may be triggered by locally exceeding magmas' tensile strength, with rupture followed by incursion of gas and ash, and finally venting and sintering (Tuffen and Dingwell 2005; Kendrick et al. 2016; Wadsworth et al. 2021); so, we anticipate variations in strain and rheology (viscosity and strength) depending on the extent of sintering, venting and associated densification. Compressive strength tests on intact and tuffisite-bearing andesites showed that sintering of pyroclasts may heal tuffisite-bearing magmas (Kolzenburg et al. 2012), due to densification (Vasseur et al. 2013), similarly to strength recovery on planar fractures (Lamur et al. 2019). So, heterogeneous, porous domains may temporarily weaken magma until they densify and reach the configurations of the surrounding host magmas. [We direct the reader to Wadsworth et al. (2022, this volume) for a comprehensive review of sintering.]

Rupture development. The architecture of ruptured materials varies widely in nature, from localized fractures or faults to discrete fault arrangements (e.g., en-échelon) and complex fracture networks in shear zones and breccias (Figs. 1, 10, 12, 14, 15; e.g., Sibson 1996). The spectrum of fracture patterns reflects the physical attributes (i.e., the presence of flaws or heterogeneities) and rheological properties of materials, and the variable stress conditions they experienced before, during and after rupture formation. As magmas rupture by behaving brittlely, they tend to break in similar ways to glasses or volcanic rocks; but in cases where the rheological behavior is transitional (e.g., if the melt is locally able to flow viscously, whilst other components or areas rupture), the rupture may proceed differently, and moreover, may subsequently heal. Since the stress conditions control the development of rupture, here we assess magma rupture in extension, compression and torsion; for each configuration, we detail the contributions of material heterogeneities (e.g., the concentration and length of flaws such as porosity, pre-existing cracks, crystals, and grain boundaries), which are central to fracture theory and micro-mechanical models (Griffith 1968; Sammis and Ashby 1986; Ashby and Sammis 1990), and which impacts the rupture of multiphase magmas.

In tension, laboratory samples generally break by the development of a single through-going crack which grows from pre-exiting flaws that concentrate stress in materials. Experiments in which glass was pulled in tension whist imaging the surface topology through atomic force microscopy revealed that rupture develops via the progressive creation and coalescence of nano-cavities, as low-density damage areas discretely localize ahead of a propagating crack tip (Célarié et al. 2003). Early studies demonstrated the time-dependence of glass strength, whereby at lower loads, longer time is required for the fracture to grow and cause wholesale rupture than at higher loads (Gurney 1948). High-temperature, tensional experiments are often conducted to resolve the temperature- and strain rate-dependence of silicate melts' rheology or strength (Webb and Dingwell 1990b; Lamur et al. 2019), and as such, the fracture development in melts is often not texturally reported, although the materials are deemed to fail like glasses. Okumura et al. (2020) conducted high-temperature fiber elongation tests whilst measuring changes in X-ray diffraction patterns and found that melt rupture was not preceded by observable changes in the melt structure at the atomic scale. So, the general view remains that the rupture of single-phase melts and glasses is likely to occur in highly discrete areas at the nanoscale, and upon macroscopic propagation, these fractures tend to adopt conchoidal forms.

In contrast, multiphase magmas tend to rupture in tension via the development of damage at the microscopic scale (Hornby et al. 2019a); that is, at larger scale than single-phase glasses or melts, due to abundance of flaws (pores and/or crystals) to focus stress inside samples. The rupture of multiphase magmas in tension is temperature and rate dependent (Hornby et al. 2019a). Hornby et al. (2019a) demonstrated that the degree of strain localization increases, and strain to failure decreases as a function of deformation rate: at high temperature and/ or low strain rate, when the rheological behavior is transitional, the authors noted slow tearing and pervasive damage as viscous flow simultaneously ensued in the construction of blunt fractures with undulate edges. Conversely, at lower temperature and/ or higher strain rate, when the rheological behavior was fully brittle, the suspension failed via a single, localized microfracture with little prior deformation. In cases where magmas are subjected to rapid decompression, experiments have shown that the offloading waves triggers layer-by-layer fragmentation of laboratory samples as the pressure drop creates a stress gradient through the medium, causing macroscopic tensile fractures at regular intervals (Scheu and Dingwell 2022, this volume; Alidibirov and Dingwell 1996, 2000). In contrast, where the pore pressure exceeds the effective mean stress and tensile strength of magmas, magmatic fragmentation (or pore embrittlement) proceeds by the rupture of bubble walls, extending to differing lengths and in different directions, generating pyroclasts of all sizes and shapes (Kueppers et al. 2006); the fragments may remain together as breccia or the fine and/ or low-density fragments may migrate if the remaining energy (kinetic) is sufficient (Alatorre-Ibargüengoitia et al. 2010), creating tuffisites and/or causing explosive volcanism. The surface area of fragments/ pyroclasts so generated correlate to the applied energy (Kueppers et al. 2006). As such, fracture architecture may more broadly be considered as a record of the deformation history, however unravelling this history is complex and challenging as the spectrum of fracture geometries could represent formation in different discrete events, or evolving conditions, or simultaneous but contrasting manifestation of strain localization in physically and rheologically heterogeneous materials. Therefore locally contrasting fracture geometries are common in extrusive lavas and pyroclasts, and also in supracrustal intrusions.

In compression, strain is generally more widespread than in tension. The rupture of geomaterials in compression has been subject of extensive experimentation [see Paterson and Wong (2005) for a comprehensive review]. Geomaterial rupture occurs via the nucleation of tensile (Mode I) fractures, principally from flaws or heterogeneities in materials (vesicles and crystals), which propagate stably, parallel with the maximum stress direction (σ_1), and which open in the direction of the minimum stress (σ_3); upon increasing applied stress, the fractures grow and coalescence in the build-up to system-scale failure (Fig. 9a; Kilburn 2003, 2012; Paterson and Wong 2005). In this manner, rupture in heterogeneous geomaterials is initially distributed as cracks propagate small distances between flaws (Fig. 9b,c) and strain energy can be readily dissipated elastically; then in its final stage in the approach to failure, fracture coalescence results in strain localization (Wong and Baud 2012). Thus, material rupture (i.e., macroscopically brittle behavior) occurs with little prior strain and generally leads to dilation. This is generally the case under compressional stresses, but in exceptional cases where the stress is sufficient to propagate supershear faults at rates exceeding seismic shear wave velocity (Bouchon and Vallee 2003; Xia et al. 2004; Ben-David et al. 2010), then rupture may proceed via "true" shear faults without the need for the prior development of tensile fractures.

Under compressive stresses, single-phase melts rupture via the development of conchoidal fractures, but the fractures tend to form a broad, complex network (Tuffen et al. 2008), as equally observed in glasses at room temperature (e.g., Kolzenburg et al. 2013). The development of ruptures in compression (including strain to failure and architecture of fracture network) is influenced by the viscosity of the melt (and thus chemistry and temperature) and the strain rate (Dingwell and Webb 1989; 1990); rupture occurs abruptly via localized fractures in the brittle regime (i.e., at high strain rate and melt viscosity), whereas in transitional rheological regimes (Fig. 10), the melt may partially deform viscously whilst some areas undergo progressive rupture (Wadsworth et al. 2018). Experiments in which a silicate melt (encased in a solid conduit-like shell) underwent rupture during very rapid stress fluctuations showed that localized supershear faults may propagate, flanked by a dendritic network of tensile fractures, organized in a Mach cone geometry (Benson et al. 2012; Lavallée et al. 2012a). So the loading conditions are a key control on the development of rupture and the architecture of fault damage in silicate melts.

The rupture of multiphase magmas under compressive stresses is equally temperature, rate (Fig. 11; Lavallée et al. 2008, 2013; Kushnir et al. 2017b; Coats et al. 2018) and pressure dependent (e.g., Laumonier et al. 2011; Cordonnier et al. 2012b). Vasseur et al. (2015) constrained the rupture of porous melts, finding a greater contribution of strain localization associated with rupture of the densest samples. Vasseur et al. (2015) constrained the rupture of variably porous melts, finding that in the most porous samples, fracture damage is initially distributed, and localizes on the approach to failure, whereas denser materials have a greater contribution of strain localization throughout, due to fewer nucleation sites. During cataclastic flow of vesicular magmas, which would prevail at high confining pressure or low pore pressure (that is, at low effective pressure), the compaction of the vesicular network would occur via distributed viscous flow and/ or brittle fractures of bubble walls as the suspension compacts (Heap et al. 2017; Kushnir et al. 2017a); which may lead to the development of compaction bands or evolve into shear-enhanced compaction bands developing at high angle to the maximum applied stress.

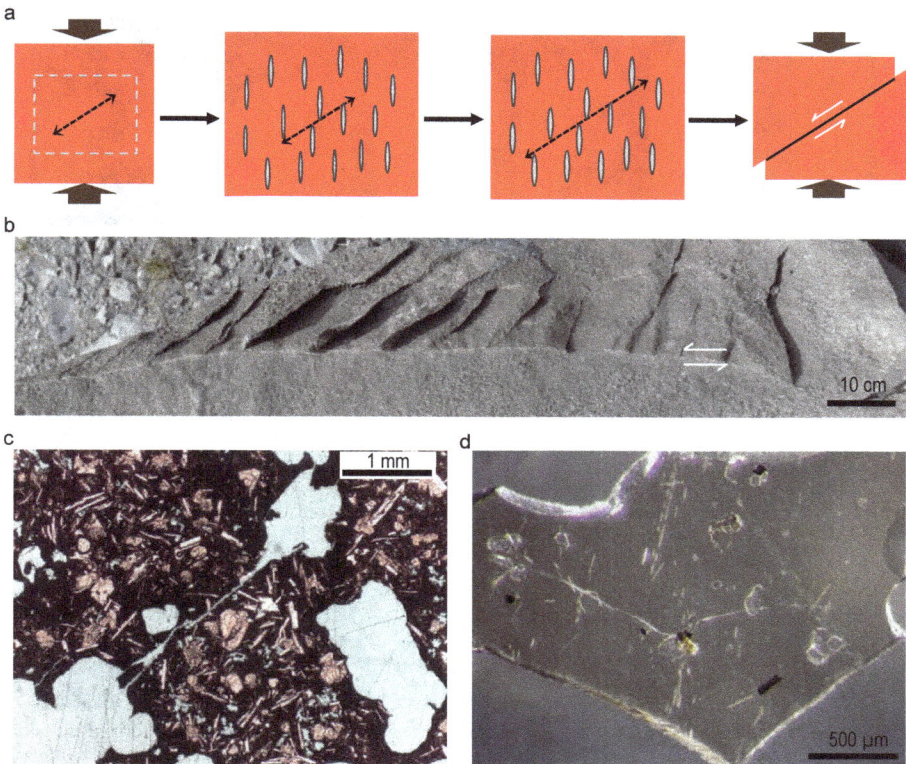

Figure 9. Fracture dynamics in heterogeneous materials. **a)** Schematic representation of damage accumulation in a material under compression. Modified from Kilburn (2003). **b)** Dome lava block exposing multiple mode I fractures torn during late-stage shear faulting at Volcán de Colima, Mexico. **c)** A fracture connecting two pores in basaltic lava from the 2014 Holuhraun fissure eruption, Iceland. The fracture manifests strain localization at a weak point in the heterogeneous lava. **d)** Fractures connecting microlites in rhyolitic obsidian from the IDDP-1 drill-site in Krafla, Iceland, that intersected a magma body at a depth of ~2.1 km during drilling.

Figure 10. Rhyolitic melt deformed under a constant uniaxial stress of 146 MPa at high temperature (745 °C); sample from the Rocche Rosse obsidian lava flow, Lipary, Italy. **a)** A localized fracture has a slightly undulating fracture surface, with conchoidal geometry at its tip. Along the main fracture smaller fractures are also noted. **b)** The deformed sample (with initial aspect ratio of 2:1) shows barreling, with macroscopically ductile behavior driven by viscous flow, yet, at high strain ruptures propagate inward from the sample margin, causing local dilation.

Complementarily, Lavallée et al. (2013) tested the impact of compressive stresses on the architecture of fractures leading to system-size rupture of poorly vesicular, crystal-rich lavas. They observed that at low differential stress, lava flowed primarily viscously, crystals failed, and small tensional tears developed on the periphery of the sample (Fig. 11a), yet no large-scale shear

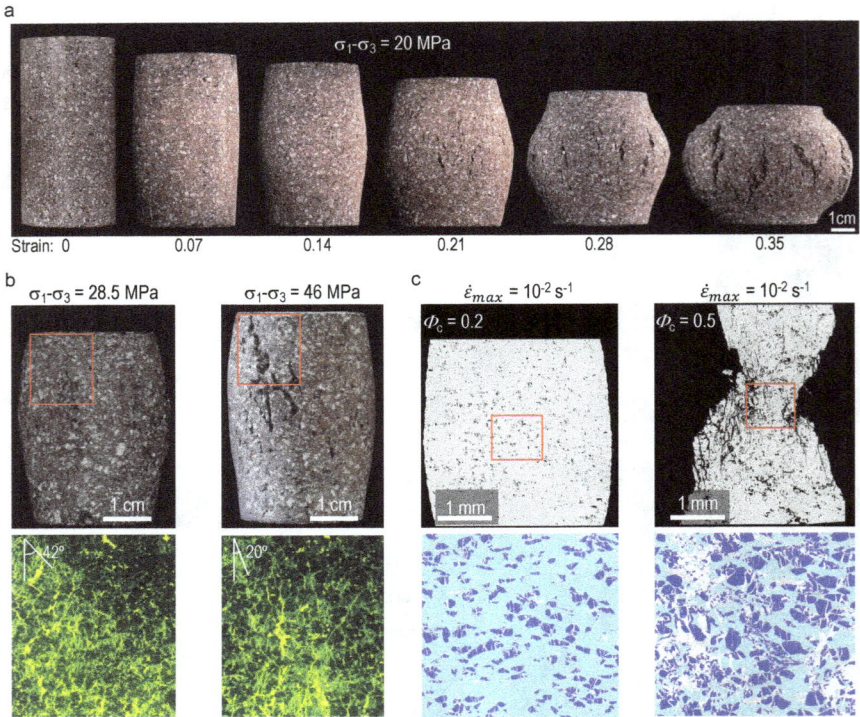

Figure 11. Constrained damage accumulation in multiphase suspensions. **a)** A series of deformed andesitic lavas depicting the evolution of strain localization with increasing strain during constant uniaxial compressive stress at high temperature. After Lavallée et al. (2012). **b)** Fracture damage in andesitic lavas deformed at high temperature under different applied stresses. The barreling and shortening of the sample indicate lower strain to failure at higher normal stress and fluorescent dyed epoxy thin sections show more brittle, localized damage at a steeper angle, closer to the principal applied stress, at higher normal stress. After Lavallée et al. (2012). **c)** Suspensions with different crystallinity (F_c) shown by QEMSCAN (Quantitative Evaluation of Minerals by SCANning electron microscopy), deformed under compression at up to 10^{-2} s^{-1} at high temperature. Above, solids (crystals and glass) are in grey and pore space in black, the more crystalline suspension on the right behaves in a brittle manner, whilst in the lower crystallinity sample damage is distributed. The zoomed-in segments below show fracture damage accumulates in the crystals (dark blue) and glass (pale blue) with pore space (white). Damage first accumulates in the crystals and is restricted to this phase at low crystal content. At higher crystal content fractures also nucleate at crystal margins and propagate between crystals in the suspension. Modified from Coats et al. (2019).

fractures developed inside the sample; thus, crystals (i.e., which cannot easily yield) break first, whilst the melt and gas bubbles may relax the applied stress. In nature, we commonly observe fractured crystals, lithics or enclaves (with contrasting rheology to the host) in otherwise intact lava (Fig. 12a-d); the rupture of crystals can be associated with rupture of the surrounding melt which subsequently relaxed and healed to obscure the evidence (Taddeucci et al. 2021), although this is not a prerequisite as the same texture can result simply from the contrasting rheology of the phases. Lavallée et al. (2013) found that (for the magmatic suspensions tested) application of differential stresses exceeding 20 MPa provoked sample failure, and that rupture associated with applied stresses just sufficient to achieve failure resulted from the rather pervasive distribution of tensile fractures aligned (sub) parallel with the applied stress. These fractures coalesced with hourglass shaped shear zones developed at a ~42° angle from the applied principal stress (Fig. 11b). With increasing applied stress, rupture still ensued from the development of tensile fractures (sub) parallel with the applied stress; yet, the shear zone became more localized and importantly, the angle of the resultant shear structures with respect to the applied principal stress decreased (down to ~20 °) and strain to failure was reduced (Fig. 11b). Such localized shear faults, flanked by tensile fractures are also observed in erupted lava (Fig. 12e; Smith et al. 2001; Lavallée et al. 2021). At higher temperature, this ductile–brittle transition will occur at higher strain rates, or higher stresses (Lavallée et al. 2007, 2008; Ichihara and Rubin 2010; Wadsworth et al. 2018), or for a given set of conditions the likelihood for development of magma rupture increases with crystallinity (Fig. 11c; Coats 2019). Coats (2019) deformed crystal-bearing suspensions under compression whilst imaging the internal structure via X-ray computed tomography at the synchrotron. The 3D reconstructions showed a systematic development of dilatant fractures, nucleating at crystal-melt interfaces and causing crystal rupture or local tearing of the crystal-melt interface. Using compression at different strain rates ($\dot{\varepsilon}$) on low vesicularity (φ_{pi}), suspensions with different crystallinities (φ_x), they defined the creation of fracture damage (φ_p) as:

$$\varphi_p = \left\{ \left[0.36 \log\left(\dot{\varepsilon}\right) + 2.99 \right] \varphi_x \right\} \varepsilon + \varphi_{pi} \tag{8}$$

The contribution from fracture propagation to the net volume increase of deforming crystal-bearing magmas is significant and suggestive of the complex development of local stresses in the vicinity of crystals, promoting pore space in pressure shadows, as commonly seen in nature (Fig. 12a). Rheologically, dilation is argued to be a likely contributor to the non-Newtonian behavior of crystal-bearing suspensions (Fig. 11c; Coats 2019).

Torsional experiments have commonly been employed to evaluate material rupture as they offer the application of extensive simple shear strain. Early on, Gurney (1948) noted that the time dependence of glass rupture (and strength) is -equivalent in torsional and in tensional configurations, as the development of rupture (via propagation of tensile fractures) remains the same despite the contrasting imposed stress fields. Okumura et al. (2010) conducted torsional experiments on aphyric, vesiculated rhyolitic melts and observed that the degree of localization and the extent of fractures developed increase as the suspension viscosity increases (with decreasing temperature). They noted that at the lowest tested temperature (i.e., the most brittle) the fractures generated en-échelon patterns that localized further strain and so compartmentalized vesicular regions (consisting of isolated vesicles) on either side of fractures, suggesting that multiple fracturing events are necessary to create wider fault damage zones and connect vesicles; at higher temperature (i.e., lower viscosity) where the behavior was transitional, they observed that fractures were not as penetrative due to viscous relaxation in adjacent areas. Kushnir et al. (2017a) tested the development of fractures (by monitoring fracture permeability *in operando*) in multiphase magmas subjected to simple shear. They also observed that at moderate strain rates, fracture permeability was established in magma following repeated tensile (Mode I) fracture events upon substantial strain (>3); at higher strain rates, where stresses cannot be relaxed, en-échelon, tensile fractures developed readily and prompted permeable flow shortly after the onset of inelastic deformation. They advance that strain does not immediately localize on Mode I fractures, making them long-lived outgassing pathways. Laumonier et al. (2011) employed torsion experiments on poorly vesicular crystal-bearing suspensions at high pressure and demonstrated that the development of crystal fabrics during strain localization resulted in the redistribution of dispersed vesicles into larger elongate vesicles in Riedel arrangements, as material locally tore. As previously noted, Holtzman and Kohlstedt (2007) observed no dilatant fractures or vesicles in their experiments and suggested that dilation must have been counteracted by melt redistribution. Here, we point out that the vesicularity of the starting material is notably different in the two studies, so the answer may revolve around the vesicularity and resultant degree of saturation of volatiles in the melt phase at the tested conditions.

Figure 12. Fracturing and damage accumulation in magmas. **a)** Pressure shadows around phenocrysts in a crystal-rich lava from Deadman Creek dome, Long Valley, USA. **b)** Large, fractured feldspar phenocryst, exemplifying strain partitioning in crystals during deformation of the crystal-rich lava at Deadman Creek dome, Long Valley, USA. **c)** Photomicrograph of a stepwise fractured biotite in dacitic lava from Mount Unzen, Japan. The smeared nature of the phenocryst alludes to the sense of shear, a neighboring plagioclase is also crushed into angular fragments, whilst the intact nature of the groundmass suggests deformation occurred during magmatic flow. **d)** A banded obsidian from Glass Creek dome (Long Valley, USA), showing brittle deformation of the stiffer, microcrystalline pale layers, whilst glassy layers viscously accommodate strain. **e)** Dilatant shear fracture in the center of the Mount Unzen lava spine erupted in 1994–1995, showing strain partitioning during flow of highly viscous suspensions. The site was explored in detail in Smith et al. (2001).

The above experiments demonstrate that in both pure and simple shear configurations the geometry of shear zones and the magnitude of strain sustained denoted a progressive transition from macroscopically ductile to brittle behavior under increasing loads (or strain rates), increasing viscosity, or if the abundance of flaws changes (i.e., if vesicularity decreases or crystallinity increases). So the natural ranges of stress conditions and magma properties can cause a spectrum of rupture development and architectures. It remains that, in all cases, brittle rupture generally promotes local dilation, especially where strain is localized. Yet, the precise nature of the viscous–brittle, viscous–cataclastic and brittle–cataclastic transitions of magmas (Fig. 3c), which impact their wholesale rheology and rupture, require further attention. The importance of understanding and mapping the architecture of fractures is that their construction regulates permeable pathways for outgassing (See section *Construction versus destruction of permeability*) and is accompanied by a characteristic seismic signatures that help us resolve the state of magmatic systems in (near) real time.

Seismogenicity and failure forecast. Laboratory experiments seeking to resolve the development of magma ruptures have employed acoustic emission (AE) monitoring (Lavallée et al. 2008, 2012a, 2013; Tuffen et al. 2008; Benson et al. 2012; Ashwell et al. 2015; Vasseur et al. 2015, 2017; Ryan et al. 2019). Fracturing events are associated with local stress drops which release elastic waves that can be monitored by sensors in the acoustic spectrum. [Note that here we use the term 'seismic' to describe these acoustic or micro-seismic events.] Lavallée et al. (2008) demonstrated that single phase magmas deformed in the viscous regime are aseismic. At low applied stresses, multiphase magma deformation generates little seismicity, but deformation becomes increasingly seismic when deformed at strain rates approaching the viscous–brittle transition. They found that the overall rate of AE generated during magma rupture increases nonlinearly with the stress (or strain rate) applied (Fig. 13a), indicating an increased dominance of brittle deformation mechanisms with increased strain rate across the ductile–brittle transition of magmas.

As magma failure is accompanied by the complex development of fractures of different sizes, it generates acoustic emissions with variable amplitude–frequency distributions, similar to the magnitude-frequency distribution of seismicity during earthquakes (e.g., Aki 1965) and volcanic unrest (Roberts et al. 2015; Carter et al. 2020). The negative gradient of the slope of amplitude-frequency distribution, which we equally term *b*-value in laboratories, has commonly been analyzed to describe material rupture. The *b*-value accompanying the rupture of magmas increases overall as a function of vesicularity in aphyric suspensions (Vasseur et al. 2015) as the propagation of fractures becomes more pervasive, owing to increased nucleation loci. Yet, in detail, data have shown that fracture propagation becomes increasingly localized (during one test) and the *b*-value decreases on approach to failure (Vasseur et al. 2015). Similarly, in multiphase suspensions the *b*-value decreases as a function of applied stress or strain rate leading to rupture (Lavallée et al. 2008), reflecting the localization of damage at high strain rates, as time is insufficient to allow pervasive fracture generation. Thus, rupture is generally more localized and the *b*-value lower in dense magmas (than in vesicular magmas, for a given strain rate), during deformation at higher strain rates, or at high strains approaching strain to failure, revealing an increased affinity for macroscopically brittle behavior (Fig. 13b).

The acceleration in released AE rate and acoustic energy that occurs during magma deformation provides an excellent scale-dependent proxy to forecast magma rupture (Fig. 13c; Lavallée et al. 2008). Material failure forecast methods (FFMs) have been developed by compiling acoustic emission data associated with rock failure testing, first by Voight (1988, 1989) and later utilized and adapted by many others (Lavallée et al. 2008, 2013; Smith and Kilburn 2010; Bell et al. 2011b; Vasseur et al. 2015, 2017; Hao et al. 2017). FFMs have been applied to volcanic systems to forecast the onset of eruptions based on acceleration of seismicity resulting from rock failure during creation of a volcanic conduit (Voight 1989; Cornelius and Voight 1995; De la Cruz-Reyna et al. 2008; Smith and Kilburn 2010; Bell et al. 2011a, 2018) or transition from effusive to explosive eruption, based on seismicity generated in magma (De la Cruz-Reyna and Reyes-Davila 2001). To resolve the factors controlling the development of seismicity leading to magma failure, Vasseur et al. (2015) evaluated the ability to forecast failure of variably vesicular, aphyric melts. They observed that the abundance of vesicularity, which they cast in terms of a heterogeneity index (i.e., the degree of disorder in suspensions, here, porosity), governs the rate of acceleration of acoustic emissions leading to magma rupture: the failure of dense magma is accompanied by exponential acceleration in AE, whereas the failure of porous magma is characterized by power law acceleration of AE (Fig. 13d). These variable mathematical descriptions impact the duration and extent of

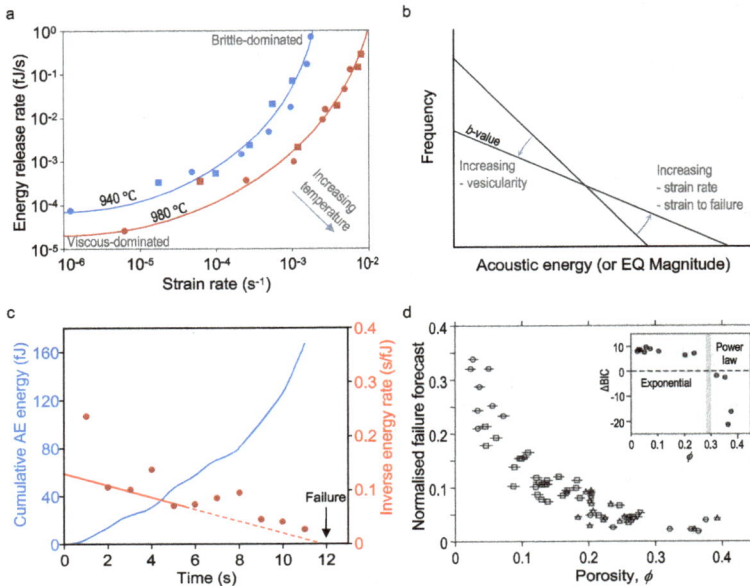

Figure 13. Acoustic emissions during magma deformation. **a)** Energy release rate for a series of experiments at different constant normal stress (therefore strain rate) at two temperatures for lavas from Volcán de Colima, Mexico (**squares**) and Bezymianny, Russia (**circles**). Both lavas behave similarly, Deformation at lower temperature is more seismogenic, releasing more AEs, suggestive of brittle-dominated deformation. As temperature is increased, interstitial melt viscosity is decreased and deformation becomes viscous-dominated (i.e., less seismogenic). Data from Lavallée et al. (2008). **b)** Schematic representation of seismic *b*-value, which characterizes the slope of the frequency-magnitude (or acoustic emission energy) distribution during deformation. The *b*-value may be lowered with increasing vesicularity (heterogeneity) or heightened by increasing strain rate or strain to failure. **c)** Cumulative acoustic emission energy through time during a single deformation test of andesitic lava at magmatic temperature. The inverse AE rate (in 1s intervals) can be used to predict failure time by extrapolation of the minima (**solid to dashed red line**) to the intercept with the *x*-axis, via the failure forecast method (FFM). Data from Lavallée et al. (2008). **d)** Failure forecast time, normalized to observed failure time, casting the accuracy of the failure forecast method given as a function of magma porosity, since more porous samples are more seismogenic, their failure may be more accurately predicted. **Inset**: Acceleration in AE rate can be best described by exponential trends at low porosity ($\lesssim 0.3$) and power law above, as described by the change in the Bayesian information criterion (BIC) parameter. After Vasseur et al. (2017).

seismicity leading to rupture and thus control the accuracy of failure forecasts (Fig. 13d; Vasseur et al. 2015). In a follow-up study, Vasseur et al. (2017) coupled the acoustic signature of porous magma failure with micro-mechanical modeling, based on the pore-emanated crack model of Sammis and Ashby (1986), to resolve the underlying mechanisms for exponential and power-law accelerations, and present geometrical corrections for the scale of heterogeneities, which if carefully considered, increase the accuracy of magma failure forecasts. The contrasting affinities for strain localization in dense versus porous materials influences the development of precursory signals (Vasseur et al. 2017; Kendrick et al. 2021) and thus similar constraints presumably impact our ability to forecast volcanic eruptions.

Fault processes

Frictional slip and shear resistance. Strain localization leading to rupture is ubiquitous across scales (Fig. 14a, b). In fractured volcanic systems (as well as in fragmental, pyroclastic materials), frictional slip may become the dominant deformation mechanism (Kendrick et al. 2012; Kennedy and Russell 2012; Pallister et al. 2013; Lavallée et al. 2014; Hornby et al. 2015; Wallace et al. 2019b). Kendrick and Lavallée (2022, this volume) to our knowledge, are the first to compare and contrast the frictional properties of lavas versus those of rocks. Whilst it demonstrates that strain localizes more in rocks than in lavas (i.e., strain localizes more in brittle materials), much can be learned about strain localization in magma from frictional experiments on rocks, as outlined below (for more detail on fault processes in magmas, see Kendrick and Lavallée 2022, this volume).

The frictional properties of rocks have long been cast via the frictional law introduced by Byerlee (1978). The compiled mechanical data of fractured rocks constrain that the shear resistance (σ_s) along a fault is proportional to the applied normal stress (σ_n) via the frictional coefficient (μ):

$$\sigma_s = \mu\sigma_n \tag{9}$$

But the presence of fluids in fault zones may lead to the development of pore pressure (P_p), which counteracts normal stress (Violay et al. 2014), so that Equation (9) may be rewritten as:

$$\sigma_s = \mu(\sigma_n - \alpha P_p) \tag{10}$$

where α is a geometrical factor (i.e., Biot–Willis coefficient) that describes fluid distribution by assessing the macroscopic fault area (A) and the real contact area (A_r) ($\alpha = 1 - A_r/A$). Byerlee (1978) observed that below 200 MPa normal stress (i.e., in the upper ~8 km of Earth's crust) the friction coefficient can be approximated at 0.85 and rocks essentially exhibit no cohesion. Yet, frictional properties are both time and rate dependent (Rice 1993; Dieterich 1994; Marone 1998); if slip is rate strengthening then slip is deemed stable, whereas if slip is rate weakening, as commonly observed for volcanic rocks (Hornby et al. 2015) and ash gouge (Lavallée et al. 2014), sliding is generally deemed unstable. Unstable slip may result in stick–slip motion (Scholz 1998), producing periodic seismic activity, as for instance observed during spine extrusion (Kendrick et al. 2014b; Lamb et al. 2015). Rotary shear experiments on gouge have shown that an increase in slip rate results in an increased localization of strain, producing characteristic P–Y fabrics (Fig. 14c), as observed in natural volcanic fault zones (Fig. 14d; Kendrick et al. 2012; Pallister et al. 2013; Lavallée et al. 2014). The porosity as well as degree of saturation impact the orientation of shear bands in deforming porous or granular materials (Zhang and VanderHeyden 2002; Zhang et al. 2002); similarly, during fault slip in porous materials, friction may prompt wear and comminution modifying the distribution of strain across slip surfaces (Hughes et al. 2020).

Frictional melting and thermal vesiculation. Friction generates a substantial amount of heat, as theoretically detailed by Carslaw and Jaeger (1959). Frictional heat has been demonstrated to challenge the stability of the mineralogical assemblage in shallow magmas: it may cause melting and vesiculation. At slow slip rates, and thus heating rates, moderate temperature increase may prompt melting at near eutectic conditions, but at fast slip and heating rates, substantial temperature increase may prompt selective melting (individual melting of minerals based on their melting temperature; Shimamoto and Lin 1994; Hirose and Shimamoto 2005; Spray 2010; Lavallée et al. 2012b; Kendrick et al. 2014a; Wallace et al. 2019a), prompting changes in the rheological controls on slip as the frictional melt chemically evolves and suspends crystal restites (Kendrick et al. 2014b; Hornby et al. 2015; Wallace et al. 2019a). Extensive experimental studies have shown that the mineralogy of material imposes important controls on the development and associated shear resistance of frictional melts, and thus on strain localization: frictional melts in basaltic materials tend to lubricate fault zones by imposing lower effective friction coefficients than that predicted from Byerlee's law (Violay et al. 2014), promoting strain localization. In contrast, in intermediate and felsic rocks they tend to act as viscous brakes, especially at low normal stresses ($\lesssim 5$ MPa), causing high effective friction coefficients (Kendrick et al. 2014a,b; Hornby et al. 2015), which may act to

Figure 14. Strain localization during fault slip. **a)** Block of the 1994–1995 lava spine at Mount Unzen, Japan. The spine shows the interplay of brittle–ductile mechanisms and is mantled by fault breccia which has an undulating contact to the dense marginal shear zone. Within the spine shear is distributed across micro- and meso-scopic shear bands with parallel C–S fabrics; shear further localized during the generation of Riedel fractures, developed from the marginal shear zone. **b)** Comminuted and sheared plagioclase phenocrysts grading to euhedral crystal across a distance of ~10 mm, shown in a photomicrograph of the margin of a dacitic block from Mount Unzen, Japan. Flow textures in the groundmass indicate shear occurred in the viscous regime, with brittle fracture partitioned into the crystals. **c)** Grain size reduction in volcanic ash gouge deformed in a rotary shear apparatus under simple shear, showing partitioning of strain into fine- and coarse-grained layers. Sample from Lavallée et al. (2014). **d)** Cataclasite from a lava spine erupted at Mount St. Helens, USA, in 2006, showing lenses of coarse (dark) and fine (pale) comminuted material, and en-échelon brecciation of phenocrysts in the lower edge becoming incorporated into the shear zone. Sample from Lavallée et al. (2014).

distribute shear across wider regions (Lavallée et al. 2012b; Violay et al. 2014; Hornby et al. 2015; Wallace et al. 2019b), especially when surroundings are partially molten (Kendrick and Lavallée 2022, this volume). Such frictional processes may lead to brittle failure in several scenarios, for example, if slip rate is sufficiently high, or as viscosity increases during waning of slip, the frictional melt may be unable to relax imposed stresses and can fail (Kendrick et al. 2014b; Lavallée et al. 2015b). Additionally, frictional heating can locally lower the solubility of water in melts and prompt thermal vesiculation, as observed in experimental products and volcanic ash—a process which may be pivotal in switching eruptive style towards momentary explosive conditions at active lava domes (Lavallée et al. 2015a).

CONSEQUENCES OF STRAIN LOCALIZATION IN MAGMAS

Brittle versus ductile deformation modes in rocks and magmas

Material rheology finds application in a wide range of science and engineering disciplines. Each discipline has defined its own terminology, appropriate for their problems and materials, and at times, this may become confusing. Here we align our nomenclature of magma rheology with that of rock deformation in the hope that it will help unify these closely related fields. As previously stated, the ductile and brittle deformation modes are defined by pervasive or localized shear, resulting in compaction or dilation, respectively (Rutter 1986). Thus, in this definition single phase melts can only behave in a ductile manner if they undergo no volume change, as they do during viscous flow (as silicate melts can only compact if they undergo a change from tetrahedral to octagonal coordination at very high pressure; e.g., Sakamaki and Ohtani 2022, this volume), but they cannot compact and exhibit cataclastic flow; upon brittle rupture, they dilate. Dense crystal-bearing magma cannot compact either, but only deform isovolumetrically or dilatantly. Instead, they exhibit a non-Newtonian rheology, commonly punctuated by distributed dilatation (Coats 2019), crystal rotation (Caricchi et al. 2008), crystal-melt segregation (Holtzman et al. 2003a) and crystal plasticity (Kendrick et al. 2017), even when viscous deformation dominates; this, therefore, would classify the deformation mode as ductile even though the magma locally dilates, shifting to transitional deformation mode at increased shear stresses as strain increasingly localizes to macroscopic fractures. Porous lavas show a wider range of behavior depending on vesicularity and stress conditions. Brittle rupture ensues due to pore overpressure (Spieler et al. 2004) or at high applied shear stresses (Coats et al. 2018), causing dilation (Fig. 3c); but at higher effective mean stresses, magma may compact viscously or via cataclastic flow, both of which are ductile modes of deformation. The issue with these behavioral definitions in terms of magmas is that the deformation of geomaterials varies as a function of dominant deformation mechanism and importantly, scale (as contrasting deformation mechanisms may prevail locally). In the brittle field the occurrence of strain localization is generally attributed to faults or shear zones developing at low angle ($\lesssim 45°$) to the primary applied stress, whereas in the ductile field flow is generally pervasive but compaction bands or shear-enhanced compaction bands may develop at different angles to the primary stress ($> 45°$) depending on the complexity of the porous structure (Heap et al. 2015a; Ryan et al. 2019). Without contextual information to support any interpretation, such strain localization features may be misinterpreted as evidence of brittle deformation, if a microstructural analysis is overlooked. So it may be that the intrinsic heterogeneous nature of multiphase magma ensures that strain commonly localizes during their deformation, albeit at varying scales and angles within the acting stress field, depending on the mechanisms in operation.

Deformation is generally scale dependent and strain localization is always scale dependent. For instance, whereas a porous rock may compact homogeneously at small scale, large porous rock masses more likely comprise areas of strain localization. So, the behavior asserted in small-scale laboratory testing may not (always or simply) upscale. Hence, here we encourage the development of apparatuses for larger sample testing and the search for more powerful scaling solutions involving theoretical and empirical approaches. Despite challenges in reconciling nomenclature with observed behavior, here we conclude that the evolution of volume and the orientation of strain localization features (e.g., faults and shear zones) with respect to the principal shear stress remain good proxies to define the brittle and ductile behavior of rocks and magmas.

Construction versus destruction of permeability

The development of fluid permeability in magma is commonly controlled by vesiculation and shear, and is central to the release of volatiles from high-viscosity magma in which bubbles may not necessarily freely ascend (i.e., low Stokes number of the gas phase; Sparks 2003b), controlling the likelihood of explosive eruption (Eichelberger et al. 1986; Degruyter et al. 2012). Studies have shown that permeability initiates once vesicles coalesce, defined as the percolation threshold. Beyond the percolation threshold, permeability nonlinearly increases as a function of vesicularity and vesicle throat size (Johnson et al. 1987; Klug and Cashman 1996; Saar and Manga 1999; Mueller et al. 2005; Wright et al. 2006; Degruyter et al. 2010; Bai et al. 2011). Permeable flow in magma can only exist whilst bubbles remain connected, which is (in most cases) only possible if pore pressure is sustained; otherwise, local stresses acting on magma will serve to suppress connected bubbles (Heap et al. 2015b; Kushnir et al. 2017a; Ryan et al. 2019) and surface tension will isolate bubbles, thus shutting permeability (Wadsworth et al. 2016, 2017).

During transport, shear may strain bubbles in a way which retains or increases their connectivity, enhancing the resultant permeability (Okumura et al. 2006, 2008; Wright et al. 2006), especially if crystals are present (Laumonier et al. 2011; Okumura et al. 2012; Kendrick et al. 2013; Lavallée et al. 2013; Shields et al. 2014; Kushnir et al. 2017b); shear may however, compact a suspension, shutting permeability (Kendrick et al. 2013; Ashwell et al. 2015). The development of fabrics associated with shear invariably promotes the construction of an anisotropic permeable porous network (Wright et al. 2006; Kendrick et al. 2013; Gaunt et al. 2014; Ashwell et al. 2015; Farquharson et al. 2016a,b; Farquharson and Wadsworth 2018), which is likely a key control on gas emission at volcanoes (Edmonds et al. 2003; Edmonds and Herd 2007), especially if magma ruptures (Okumura et al. 2009, 2010; Lavallée et al. 2013; Gaunt et al. 2014; Kushnir et al. 2017b). Kushnir et al. (2017b) found that the amount of strain required to generate permeable fracture pathways decreases with increasing strain rate. Complementarily, Lavallée et al. (2013) observed that an increase in strain rate also promotes larger, more connected fracture pathways; as magmas are forced across the ductile–brittle transition the fracture geometry evolves, shifting from distributed to localized outgassing fracture channels with increased degrees of anisotropy. Mapping of permeability across the shear zone of the lava spine at Mount Unzen revealed a ~3-m wide region whereby compaction reduced the permeability (and induced permeability anisotropy) near the conduit margin; within ~1 m of the fault plane, the compactant region is overprinted by dilatant bands, containing abundant microfractures, that locally enhance permeability (Fig. 15; Lavallée et al. 2021). This overprint represents a shift from ductile to increasingly brittle deformation as magma ascended to shallower depth and was likely formed under lower effective pressure and variations in strain rate during pulsatory magma ascent. Experiments have shown that upon fracture closure (if the pore pressure is insufficient to counteract local normal stress applied on a fracture) permeability can be reduced almost to the value of the intact (pre-fractured) material (e.g., Eggertsson et al. 2016; Lamur et al. 2017). Furthermore, healing (occurring via wetting and diffusion) of interfaces may ensue at a timescale which depends upon the melt viscosity (Okumura and Sasaki 2014; Lamur et al. 2019), hindered by the presence of bubbles, crystals and lithics (which, to date, remain unconstrained), or by shear which may break bonds along the fault plane, and may be enhanced by frictional heating (Lavallée et al. 2015b). Therefore, the spatially and temporally variable and transient nature of the permeable porous network in magmatic columns promotes periodicity in fluid fluxes (Michaut et al. 2013), as observed by cyclic gas emissions (Edmonds et al. 2003), which may be at the origin of sporadic transitions in eruptive style (Edmonds and Herd 2007). Note that this seemingly good description of magma permeability results primarily from *ex situ* measurements (i.e., by measuring permeability before and after deformation); the only direct determinations for magmas are for material densification due to sintering (Wadsworth et al. 2017, 2021) and rare measurements *in operando* at high temperature (Gaunt et al. 2016; Kushnir et al. 2017a,b).As these challenging measurements accrue, we anticipate numerous changes in our view of shearing magma permeability, permeability anisotropy and as a consequence volcanic outgassing.

Implications for magma seismicity and tilt at volcanoes

The signs of volcanic unrest can vary greatly (Sparks 2003a; Gottsmann et al. 2011). Ground deformation and seismicity remain the most common and reliable sources of information to assess magma transport (Neuberg 2000; Chouet and Matoza 2013; Sigmundsson et al. 2015), the timing of extrusion (Bell et al. 2013) or of a transition in eruptive style (De la Cruz-Reyna and Reyes-Davila 2001). In general, we find that the extent of signals originating from magma

Figure 15. Structure, porosity and permeability across a transect of a sheared block (detached from the towering 1994–1995 spine at Mount Unzen), representing strain localization at the conduit margin. The exhumed block is mantled by fault gouge cross-cutting the marginal high-shear region (see inset). The high-shear zone consists of a compacted cataclasite with parallel C–S fabrics, overprinted by dilatant Riedel fractures, causing permeability anisotropy; shear and anisotropy decrease towards the undeformed core of the cooled magmatic column. Data from Lavallée et al. (2021).

ascent increases as a function of magma viscosity, owing to increased viscous drag (that magnifies the resultant deformation) and brittleness of the material involved (that enhances rupture events). [Moreover, the geologic record does not preserve syn-eruptive porous textures very well in mafic systems, as bubbles and cracks readily relax and compact, whereas it does in intermediate and felsic systems. As a result, our view of magma ascent processes is biased towards eruptions of relatively high viscosity magmas.]

At active open vent systems, such as lava domes, the periodic occurrence of low-frequency earthquakes and their close association with tilt signals commonly depict cyclic behavior (Neuberg et al. 2006; Johnson et al. 2014), and offer key measures of eruption processes (Neuberg 2000) for use in explosion forecasting (Chouet 1996). Low-frequency earthquakes have been postulated to originate from resonance within fluid-filled cracks (Chouet 1996) or magma faulting (Neuberg 2000), likely instigated in zones of strain localization near conduit margins (Tuffen and Dingwell 2005; Okumura et al. 2010, 2012; Lavallée et al. 2013). The observation that low-frequency earthquakes commonly recur from a fixed (non-destructive) source with a repeating triggering mechanism has been used to infer that ascending magma undergoes simple shear and ruptures once it reaches a certain depth according to critical stress and rheological conditions (Iverson et al. 2006; Neuberg et al. 2006; Lavallée et al. 2008; De Angelis and Henton 2011; Kendrick et al. 2014b); and that beyond this point ascent is controlled by fault friction (Lavallée et al. 2012b, 2014; Kendrick et al. 2014b). The extrusion of stiff lava spines has been characterized to be pulsatory, with stick–slip motion along marginal faults leading to drumbeat seismicity, as observed in 2004–2008 at Mount St. Helens, USA (Iverson et al. 2006; Kendrick et al. 2014b), and in 1994–1995 at Unzen volcano (Lamb et al. 2015). Slip on such localized planes may generate substantial frictional heat, even resulting in frictional melts (Kendrick et al. 2012, 2014a; Plail et al. 2014), the rheology of which modifies the localization of strain and resultant fault slip (see Kendrick and Lavallée 2022, this volume).

High resolution geophysical monitoring in combination with petrological investigation of eruptive products can reveal further insights into seismogenic conduit processes. In a recent study on Santiaguito volcano, Guatemala, Lavallée et al. (2015a) showed that volcanic ash produced during periodic gas-and-ash explosions may be accompanied by melting as well as thermally-induced vesiculation. Continuous monitoring of such activity during a five-day period indicated that the lava dome underwent 222 inflation/ deflation cycles at regular intervals of ~26 minutes (Johnson et al. 2014), with similar magnitude and timescales of inflation and deflation. Of these, 154 cycles resulted in the simple release of gas at the tilt apex, whilst 68 released gas and ash explosively. Close examination of the signals showed subtle and consistent contrasts between gas emissions versus explosive events; inflation/ deflation cycles which resulted in the release of gas and ash showed a strikingly more rapid tilt signal in the final stage prior to explosion, accompanied by very long period (VLP) signals (Fig. 16; Lavallée et al. 2015a). In a previous study, Johnson et al. (2008) had shown that during these explosions, the dome may rapidly uplift and subside in a piston-like fashion by 20–50 cm in one second (i.e., <1m/s) along active faults. In tandem, these observations may imply the occurrence of greater mechanical work from strain localization and hence, thermal output, which may be at the origin of the frictional melting and thermal vesiculation observed in the volcanic ash of these explosions (Lavallée et al. 2015a). Although responsible for both significant gas release and ash generation, these frequent explosions at Santiaguito never disrupt the whole dome; this is a common occurrence at active lava domes. How is this "damage control" scenario possible? How is this paradox achieved? It is almost as though the volcanic system has engineered a regime of degassing and stress release which, despite involving local fragmentation (destruction) of lava, maintains the overall integrity of the dome. We posit that periodic shear localization or "sacrificial fragmentation" events may, in some cases, regulate the structural stability of active lava domes over prolonged periods of time. Thus, deep magma shearing combined with shallow fault slip imposes traction along conduit margins that produce characteristic deformation and seismic signals (Neuberg et al. 2018). As such, the close relationships of seismic and tilt signals with respect to magma shearing, faulting and volcanic eruptions, demand a careful assessment of physico-chemical processes during strain localization, which may hold key information about stress distribution (its dissipation by viscous processes and heating), fracture generation, and the thermal budget of magma, influencing eruption style.

Transport in volcanic conduits and during volcanic eruptions: a model

A picture emerges (from integration of laboratory findings, field studies and numerical simulations) that strain localization in magma controls transport and volcanic eruption styles. Figure 17 attempts to depict strain localization and associated deformation mechanisms during magma ascent and eruption. We advance that strain localization during magma transport is enhanced by crystallization and vesiculation as well as magma mixing and lateral temperature variations. The resultant heterogeneity in flow properties of these phases modifies the

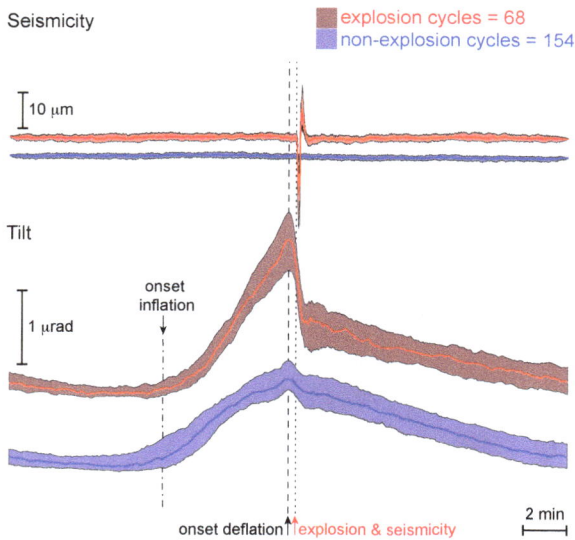

Figure 16. Seismicity and tilt measurements of 68 gas-and-ash explosion cycles and 154 non-explosion cycles recorded at the lava dome of Santiaguito volcano, Guatemala, during 2012. Non-explosive cycles involved 5–6 min of gradual inflation, culminating in gas emission and followed by gradual deflation, with no accompanying seismicity. On the other hand, cycles accompanied by explosions were characterized by more substantial inflation (rate and magnitude), followed by rapid deflation, a gas-and-ash explosion and the release of VLP (very long period) seismic signals, after which deflation followed a similar trend to non-explosive cycles. Data from Lavallée et al. (2015a).

rheology, partitioning stress and strain between melt (sometimes melts), crystals and bubbles, which results in an interplay of deformation mechanisms in each magmatic phase: melt may flow viscously and generate viscous energy dissipation, whilst bubbles and crystals may align in flow bands and deform (via viscous deformation or crystal plasticity, respectively). Shear enhances anisotropy. Whilst shear may locally increase permeability by coalescence of bubbles or dilatancy caused by fracturing, shear-enhanced compaction may lower the permeability, thus affecting outgassing via the porous network. At shallow depth, magma may undergo a shift from macroscopically ductile to macroscopically brittle deformation, so that deep compactant shear zones may be overprinted by local dilatant faults which further modify the permeable porous network.

Rupture entails specific requirements and triggers characteristic seismicity. Upon decompression the melt degasses, causing viscosity to increase and pressure to build in vesicles which, together with the strain rate experienced, challenges the structural stability of magma. If sufficient pressure is accumulated so that critical conditions are met, magma may rupture. From this point the system has two options: 1) if sufficient energy remains fragmental particles, pyroclasts, may be transported in gas mixtures (to erupt explosively or become trapped as tuffisites), or 2) fragments may shear via fault slip or cataclastic flow, and frictional heating may melt crystals and trigger vesiculation (which may enhance the likelihood of sacrificial fragmentation). Planar fractures or faults may heal via wetting and diffusion. Shear and repeated fracturing may subsequently, cyclically recur, causing stick–slip motion or periodic, fracture-controlled explosions as magma effuses. It results that the competition between these deformation mechanisms promotes pulsatory ascent (i.e., rheologically controlled variation in discharge rates). Thus, in this model, strain localization takes place in broad, marginal shear zones undergoing viscous deformation (more common during slow to moderate ascent rate with

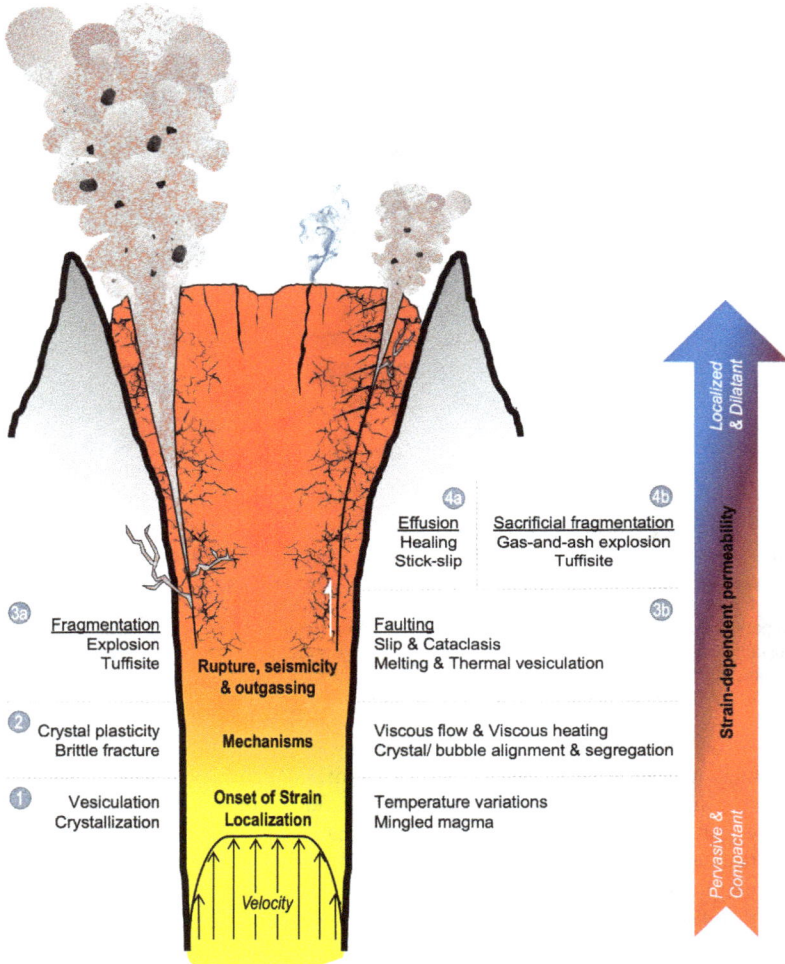

Figure 17. A conceptual model for strain localization in ascending magma in a volcanic conduit. The evolution is depicted through stages 1–4, which describe the contributors to the evolution of deformation mechanisms, from dominantly ductile, pervasive and compactional strain at depth, to brittle, localized and dilatant mechanisms in the shallow conduit, all of which contribute to permeability evolution that acts as a feedback to the ongoing eruptive activity. Such mechanisms are important in all scenarios in which magma is on the move, highlighted here as such for visualization purposes. As magma ascends it may: **1**, comprise multiple batches of melts that become mingled, the melt vesiculates and crystallizes, and is subject to local temperature variations. The rheological evolution will dictate the velocity profile across the conduit as drag at the margins contributes to shear localization. **2**, As strain localizes magmas may suffer a number of mechanisms from viscous flow (including viscous heating) to bubble and crystal alignment and segregation, crystal plasticity and brittle fracture. This can lead to rupture, seismicity and outgassing that may take the form of **3a** in which fragmentation by dilational rupture and decompression that leads to an explosion and deposition of tuffisites or **3b**, faulting, whereby strain is localized on a slip plane that may be subject to cataclasis, frictional heating, crystal melting and thermal vesiculation. The result of the latter (**3b**) can be either **4a**, effusive eruption, if slip planes heal and stick–slip motion may occur, or **4b**, shallow gas-and-ash explosions and tuffisites injection resulting from sacrificial fragmentation of material bounding faults where strain is locally accommodated. Thus, strain localization encompasses a wide spectrum of deformation mechanisms which dictates magma transport and eruption.

respect to the structural relaxation kinetics) and along localized fault zones (during rapid ascent). As such, shear may take place everywhere in conduits, but to differing magnitudes. So, we surmise that magma transport and volcanic eruptions are the manifestation of a complex interplay of deformation mechanisms due to strain localization in magmas.

ACKNOWLEDGEMENTS

We are first and foremost eternally grateful to the European Research Council who has funded much of the work included in this review, via a Starting Grant on "Strain Localisation in Magmas" (SLiM, no 306488). We thank the team of researchers who have participated in this project, both in our laboratory (Adrian Hornby, Amelia Bain, Amy Hughes, Anthony Lamur, Biao Cai, Claire Harnett, Fabian Wadsworth, Felix von Aulock, Fiona Iddon, Guðjón Eggertsson, James Ashworth, Jenny Schauroth, Jérémie Vasseur, Jose Godinho, Joshua Weaver, Kai-Uwe Hess, Katherine Dobson, Lauren Schaefer, Oliver Lamb, Paul Wallace, Rebecca Coats, Richard Wall, Sarah Henton De Angelis, Takahiro Miwa, and many more) during field campaigns (Andreas Rietbrock, Armando Pineda, Benjamin Andrews, Benewan Kennedy, Corrado Cimarelli, Emma Rhodes, Gustavo Chigna, Hugh Tuffen, Jeffrey Johnson, Jonathan Castro, Nick Varley, Silvio De Angelis, Takeshi Matsushima, and many more), and via collaboration (Elisabetta Mariani, Giulio Di Toro, Jamie Farquharson, John Wheeler, Kelly Russell, Loic Courtois, Mike Heap, Peter Lee, Philip Benson, Philip Meredith, Takehiro Hirose, Tom Mitchell, and many more). We also acknowledge support from an Early Career Fellowship (ECF–2016–325) and a Research Fellowship (RF–2019–526\4) of the Leverhulme Trust, awarded to J.E.K. and Y.L., respectively. Finally, we wish to express our most sincere and utmost gratitude to Donald Dingwell for his continued, kind and generous support during and following our doctoral studies.

REFERENCES

Aki K (1965) Maximum likelihood estimate of *b* in the formula log $N = a - bm$ and its confidence limits. Bull Earthquake Res Inst, Tokyo University 43:237–239

Alatorre-Ibargüengoitia MA, Scheu B, Dingwell DB, Delgado-Granados H, Taddeucci J (2010) Energy consumption by magmatic fragmentation and pyroclast ejection during Vulcanian eruptions. Earth Planet Sci Lett 291:60–69

Alidibirov M, Dingwell DB (1996) Magma fragmentation by rapid decompression. Nature 380:146–148

Alidibirov M, Dingwell DB (2000) Three fragmentation mechanisms for highly viscous magma under rapid decompression. J Volcanol Geotherm Res 100:413–421

Ancey C (2007) Plasticity and geophysical flows: A review. J Non-Newtonian Fluid Mech 142:4–35

Ashby MF, Sammis CG (1990) The damage mechanics of brittle solids in compression. Pure Appl Geophys 133:489–521

Ashwell PA, Kendrick JE, Lavallée Y, Kennedy BM, Hess KU, von Aulock FW, Wadsworth FB, Vasseur J, Dingwell DB (2015) Permeability of compacting porous lavas. J Geophys Res-Solid Earth 120:1605–1622

Bagdassarov NS, Dingwell DB (1992) A rheological investigation of vesicular rhyolite. J Volcanol Geotherm Res 50:307–322

Bai LP, Baker DR, Hill RJ (2011) Permeability of vesicular Stromboli basaltic glass: Lattice Boltzmann simulations and laboratory measurements. J Geophys Res-Solid Earth 115:B07201

Bain AA, Lamur A, Kendrick JE, Lavallée Y, Calder ES, Cortes JA, Butler IB, Cortes GP (2019) Constraints on the porosity, permeability and porous micro-structure of highly-crystalline andesitic magma during plug formation. J Volcanol Geotherm Res 379:72–89

Baker DR, Brun F, O'Shaughnessy C, Mancini L, Fife JL, Rivers M (2012) A four-dimensional X-ray tomographic microscopy study of bubble growth in basaltic foam. Nat Commun 3:1135

Balme MR, Rocchi V, Jones C, Sammonds PR, Meredith PG, Boon S (2004) Fracture toughness measurements on igneous rocks using a high-pressure, high-temperature rock fracture mechanics cell. J Volcanol Geotherm Res 132:159–172

Barton N (1976) The shear strength of rock and rock joints. Int J Rock Mech Min Sci Abstracts 13:255–279

Baud P, Wong TF, Zhu W (2014) Effects of porosity and crack density on the compressive strength of rocks. Int J Rock Mech Min Sci 67:202–211

Beane R, Wiebe RA (2012) Origin of quartz clusters in Vinalhaven granite and porphyry, coastal Maine. Contrib Mineral Petrol 163:1069–1082

Bell AF, Naylor M, Main IG (2013) The limits of predictability of volcanic eruptions from accelerating rates of earthquakes. Geophys J Int 194:1541–1553

Bell AF, Naylor M, Heap MJ, Main IG (2011a) Forecasting volcanic eruptions and other material failure phenomena: An evaluation of the failure forecast method. Geophys Res Lett 38: L15304

Bell AF, Greenhough J, Heap MJ, Main IG (2011b) Challenges for forecasting based on accelerating rates of earthquakes at volcanoes and laboratory analogues. Geophys J Int 185:718–723

Bell AF, Naylor M, Hernandez S, Main IG, Gaunt HE, Mothes P, Ruiz M (2018) Volcanic eruption forecasts from accelerating rates of drumbeat long-period earthquakes. Geophys Res Lett 45:1339–1348

Ben-David O, Cohen G, Fineberg J (2010) The dynamics of the onset of frictional slip. Science 330:211–214

Benson PM, Heap MJ, Lavallée Y, Flaws A, Hess KU, Selvadurai APS, Dingwell DB, Schillinger B (2012) Laboratory simulations of tensile fracture development in a volcanic conduit via cyclic magma pressurisation. Earth Planet Sci Lett 349:231–239

Bergantz GW, Schleicher JM, Burgisser A (2017) On the kinematics and dynamics of crystal-rich systems. J Geophys Res-Solid Earth 122:6131–6159

Bertolett EM, Prior DJ, Grayley DM, Hampton SJ, Kennedy BM (2019) Compacted cumulates revealed by electron backscatter diffraction analysis of plutonic lithics. Geology 47:445–448

Blumenfeld P, Bouchez JL (1988) Shear criteria in granite and migmatite deformed in hte magmatic and solid states. J Struct Geol 10:361–372

Bottinga Y (1994) Non-Newtonian rheology of homogeneous silicate melts. Phys Chem Mineral 20:454–459

Bouchon M, Vallee M (2003) Observation of long supershear rupture during the magnitude 8.1 Kunlunshan earthquake. Science 301:824–826

Brugger CR, Hammer JE (2015) Prevalence of growth twins among anhedral plagioclase microlites. Am Mineral 100:385–395

Buerger MJ (1945) The genesis of twin crystals. Am Mineral 30:469–482

Byerlee J (1978) Friction of rocks. Pure Appl Geophys 116:615–626

Caricchi L, Burlini L, Ulmer P, Gerya T, Vassalli M, Papale P (2007) Non-Newtonian rheology of crystal-bearing magmas and implications for magma ascent dynamics. Earth Planet Sci Lett 264:402–419

Caricchi L, Giordano D, Burlini L, Ulmer P, Romano C (2008) Rheological properties of magma from the 1538 eruption of Monte Nuovo (Phlegrean Fields, Italy): An experimental study. Chem Geol 256:158–171

Caricchi L, Biggs J, Annen C, Ebmeier SK (2014) The influence of cooling, crystallisation and re-melting on the interpretation of geodetic signals in volcanic systems. Earth Planet Sci Lett 388:166–174

Carslaw HS, Jaeger JC (1959) Conduction of Heat in Solids. Oxford University Press, Oxford

Carter W, Rietbrock A, Lavallée Y, Gottschammer E, Moreno AD, Kendrick JE, Lamb OD, Wallace PA, Chigna G, De Angelis S (2020) Statistical evidence of transitioning open-vent activity towards a paroxysmal period at Volcan Santiaguito (Guatemala) during 2014–2018. J Volcanol Geotherm Res 398:106891

Cashman KV (1993) Relationship between plagioclase crystallization and cooling rate in basaltic melts. Contrib Mineral Petrol 113:126–142

Castro J, Manga M, Cashman K (2002) Dynamics of obsidian flows inferred from microstructures: insights from microlite preferred orientations. Earth Planet Sci Lett 199:211–226

Castro JM, Dingwell DB (2009) Rapid ascent of rhyolitic magma at Chaiten volcano, Chile. Nature 461:780–784

Célarié F, Prades S, Bonamy D, Ferrero L, Bouchaud E, Guillot C, Marliere C (2003) Glass breaks like metal, but at the nanometer scale. Phys Rev Lett 90:075504

Chouet BA (1996) Long-period volcano seismicity: Its source and use in eruption forecasting. Nature 380:309–316

Chouet BA, Matoza RS (2013) A multi-decadal view of seismic methods for detecting precursors of magma movement and eruption. J Volcanol Geotherm Res 252:108–175

Coats R (2019) The Rheology of Pore and Crystal Bearing Magmas. Doctor of Philosophy, University of Liverpool, Liverpool, UK

Coats R, Kendrick JE, Wallace PA, Miwa T, Hornby AJ, Ashworth JD, Matsushima T, Lavallée Y (2018) Failure criteria for porous dome rocks and lavas: a study of Mt. Unzen, Japan. Solid Earth 9:1299–1328

Colombier M, Wadsworth FB, Gurioli L, Scheu B, Kueppers U, Di Muro A, Dingwell DB (2017) The evolution of pore connectivity in volcanic rocks. Earth Planet Sci Lett 462:99–109

Cordonnier B, Schmalholz SM, Hess KU, Dingwell DB (2012a) Viscous heating in silicate melts: an experimental and numerical comparison. J Geophys Res 117:B02203

Cordonnier B, Caricchi L, Pistone M, Castro J, Hess KU, Gottschaller S, Manga M, Dingwell DB, Burlini L (2012b) The viscous–brittle transition of crystal-bearing silicic melt: Direct observation of magma rupture and healing. Geology 40:611–614

Cornelius RR, Voight B (1995) Graphical and PC-software analysis of volcano eruption precursors according to the materials failure forecast method (FFM). J Volcanol Geotherm Res 64:295–320

Costa A, Caricchi L, Bagdassarov N (2009) A model for the rheology of particle-bearing suspensions and partially molten rocks. Geochem Geophys Geosyst 10:Q03010

Coward MP (1980) The analysis of flow profiles in a basaltic dyke using strained vesicles. J Geol Soc 137:605–615

De Angelis S, Henton SM (2011) On the feasibility of magma fracture within volcanic conduits: Constraints from earthquake data and empirical modelling of magma viscosity. Geophys Res Lett 38:L19310

De Angelis SH, Larsen J, Coombs M, Dunn A, Hayden L (2015) Amphibole reaction rims as a record of pre-eruptive magmatic heating: An experimental approach. Earth Planet Sci Lett 426:235–245

De la Cruz-Reyna S, Reyes-Davila GA (2001) A model to describe precursory material-failure phenomena: applications to short-term forecasting at Colima volcano, Mexico. Bull Volcanol 63:297–308

De la Cruz-Reyna S, Yokoyama I, Martinez-Bringas A, Ramos E (2008) Precursory seismicity of the 1994 eruption of Popocatepetl Volcano, Central Mexico. Bull Volcanol 70:753–767

Degruyter W, Bachmann O, Burgisser A (2010) Controls on magma permeability in the volcanic conduit during the climactic phase of the Kos Plateau Tuff eruption (Aegean Arc). Bull Volcanol 72:63–74

Degruyter W, Bachmann O, Burgisser A, Manga M (2012) The effects of outgassing on the transition between effusive and explosive silicic eruptions. Earth Planet Sci Lett 349:161–170

Deubelbeiss Y, Kaus BJP, Connolly JAD (2010) Direct numerical simulation of two-phase flow: Effective rheology and flow patterns of particle suspensions. Earth Planet Sci Lett 290:1–12

Deubelbeiss Y, Kaus BJP, Connolly JAD, Caricchi L (2011) Potential causes for the non-Newtonian rheology of crystal-bearing magmas. Geochem Geophys Geosyst 12:Q05007

Dieterich J (1994) A constitutive law for rate and state earthquake production and its application to earthquake clustering. J Geophys Res-Solid Earth 99:2601–2618

Dingwell DB, Webb SL (1989) Structural relaxation in silicate melts and non-Newtonian melt rheology in geologic processes. Phys Chem Mineral 16:508–516

Dingwell DB, Webb SL (1990) Relaxation in silicate melts. Euro J Mineral 2:427–449

Dingwell DB, Lavallée Y, Hess KU, Flaws A, Martí J, Nichols ARL, Gilg HA, Schillinger B (2016) Eruptive shearing of tube pumice: pure and simple. Solid Earth 7:1383–1393

Dobson KJ, Allabar A, Bretagne E, Coumans J, Cassidy M, Cimarelli C, Coats R, Connolley T, Courtois L, Dingwell DB, Di Genova D (2020) Quantifying microstructural evolution in moving magma. Front Earth Sci 8:287

Edmonds M, Herd RA (2007) A volcanic degassing event at the explosive-effusive transition. Geophys Res Lett 34: L21310

Edmonds M, Oppenheimer C, Pyle DM, Herd RA, Thompson G (2003) SO_2 emissions from Soufriere Hills Volcano and their relationship to conduit permeability, hydrothermal interaction and degassing regime. J Volcanol Geotherm Res 124:23–43

Eggertsson GH, Lavallée Y, Kendrick JE, Lamur A, Markússon SH (2016) Enhancing permeability by multiple fracturesin the Krafla geothermal reservoir, Iceland. *In:* European Geothermal Congress 2016, 19–23 September, Strasbourg, France

Eichelberger JC, Carrigan CR, Westrich HR, Price RH (1986) Non-explosive silicic volcanism. Nature 323:598–602

Farquharson JI, Wadsworth FB (2018) Upscaling permeability in anisotropic volcanic systems. J Volcanol Geotherm Res 364:35–47

Farquharson J, Heap MJ, Varley NR, Baud P, Reuschle T (2015) Permeability and porosity relationships of edifice-forming andesites: A combined field and laboratory study. J Volcanol Geotherm Res 297:52–68

Farquharson JI, Heap MJ, Baud P (2016a) Strain-induced permeability increase in volcanic rock. Geophys Res Lett 43:11603–11610

Farquharson JI, Heap MJ, Lavallée Y, Varley NR, Baud P (2016b) Evidence for the development of permeability anisotropy in lava domes and volcanic conduits. J Volcanol Geotherm Res 323:163–185

Fiedrich AM, Bachmamann O, Ulmer P, Deering CD, Kunze K, Leuthold J (2017) Mineralogical, geochemical, and textural indicators of crystal accumulation in the Adamello Batholith (Northern Italy). Am Mineral 102:2467–2483

Flinn D (1962) On folding during three dimensional progressive deformation. Geol Soc London 118:385–433

Gaunt HE, Sammonds PR, Meredith PG, Smith R, Pallister JS (2014) Pathways for degassing during the lava dome eruption of Mount St. Helens 2004–2008. Geology 42:947–950

Gaunt HE, Sammonds PR, Meredith PG, Chadderton A (2016) Effect of temperature on the permeability of lava dome rocks from the 2004–2008 eruption of Mount St. Helens. Bull Volcanol 78:30

Giordano D, Russell JK, Dingwell DB (2008) Viscosity of magmatic liquids: A model. Earth Planet Sci Lett 271:123–134

Gottsmann J, Dingwell DB (2001) Cooling dynamics of spatter-fed phonolite obsidian flows on Tenerife, Canary Islands. J Volcanol Geotherm Res 105:323–342

Gottsmann J, De Angelis S, Fournier N, Van Camp M, Sacks S, Linde A, Ripepe M (2011) On the geophysical fingerprint of Vulcanian explosions. Earth Planet Sci Lett 306:98–104

Gourlay CM, Dahle AK (2007) Dilatant shear bands in solidifying metals. Nature 445:70–73

Graeter KA, Beane RJ, Deering CD, Gravley D, Bachmann O (2015) Formation of rhyolite at the Okataina Volcanic Complex, New Zealand: New insights from analysis of quartz clusters in plutonic lithics. Am Mineral 100:1778–1789

Griffith AA (1968) Phenomena of rupture and flow in solids. ASM Trans Q 61:871–906

Gurney C (1948) Delayed fracture of glass under tension, torsion and radial pressure. Nature 161:729–730

Hao S, Yang H, Elsworth D (2017) An accelerating precursor to predict "time-to-failure" in creep and volcanic eruptions. J Volcanol Geotherm Res 343:252–262

Heap MJ, Violay MES (2021) The mechanical behaviour and failure modes of volcanic rocks: a review. Bull Volcanol 83:1–47

Heap MJ, Lavallée Y, Petrakova L, Baud P, Reuschle T, Varley NR, Dingwell DB (2014) Microstructural controls on the physical and mechanical properties of edifice-forming andesites at Volcan de Colima, Mexico. J Geophys Res-Solid Earth 119:2925–2963

Heap MJ, Farquharson JI, Baud P, Lavallée Y, Reuschle T (2015a) Fracture and compaction of andesite in a volcanic edifice. Bull Volcanol 77:1–9

Heap MJ, Farquharson JI, Wadsworth FB, Kolzenburg S, Russell JK (2015b) Timescales for permeability reduction and strength recovery in densifying magma. Earth Planet Sci Lett 429:223–233

Heap MJ, Violay M, Wadsworth FB, Vasseur J (2017) From rock to magma and back again: The evolution of temperature and deformation mechanism in conduit margin zones. Earth Planet Sci Lett 463:92–100

Hess KU, Cordonnier B, Lavallée Y, Dingwell DB (2008) Viscous heating in rhyolite: an in situ determination. Earth Planet Sci Lett 275:121–126

Hirose T, Shimamoto T (2005) Growth of molten zone as a mechanism of slip weakening of simulated faults in gabbro during frictional melting. J Geophys Res-Solid Earth 110:B05202

Holness MB, Vukmanovic Z, Mariani E (2017) Assessing the role of compaction in the formation of adcumulates: a microstructural perspective. J Petrol 58:643–673

Holtzman BK, Kohlstedt DL (2007) Stress-driven melt segregation and strain partitioning in partially molten rocks: Effects of stress and strain. J Petrol 48:2379–2406

Holtzman BK, Groebner NJ, Zimmerman ME, Ginsberg SB, Kohlstedt DL (2003a) Stress-driven melt segregation in partially molten rocks. Geochem Geophys Geosyst 4:8607

Holtzman BK, Kohlstedt DL, Zimmerman ME, Heidelbach F, Hiraga T, Hustoft J (2003b) Melt segregation and strain partitioning: Implications for seismic anisotropy and mantle flow. Science 301:1227–1230

Holtzman BK, Kohlstedt DL, Morgan JP (2005) Viscous energy dissipation and strain partitioning in partially molten rocks. J Petrol 46:2569–2592

Holtzman BK, King DSH, Kohlstedt DL (2012) Effects of stress-driven melt segregation on the viscosity of rocks. Earth Planet Sci Lett 359:184–193

Hornby AJ, Kendrick JE, Lamb OD, Hirose T, De Angelis S, von Aulock FW, Umakoshi K, Miwa T, Henton De Angelis S, Wadsworth FB, Hess KU (2015) Spine growth and seismogenic faulting at Mt. Unzen, Japan. J Geophys Res: Solid Earth 120:2169–9356

Hornby AJ, Lavallée Y, Kendrick JE, De Angelis S, Lamur A, Rietbrock A, Chigna G (2019a) Brittle–ductile deformation and tensile rupture of dome lava during inflation at Santiaguito, Guatemala. J Geophys Res 124:10107–10131

Hornby AJ, Lavallée Y, Kendrick JE, Rollinson G, Butcher AR, Clesham S, Kueppers U, Cimarelli C, Chigna G (2019b) Phase partitioning during fragmentation revealed by QEMSCAN particle mineralogical analysis of volcanic ash. Sci Rep 9:126

Hughes A, Kendrick JE, Lamur A, Wadsworth FB, Wallace PA, Di Toro G, Lavallée Y (2020) Frictional behaviour, wear and comminution of synthetic porous geomaterials. Front Earth Sci 8:562548

Ichihara M, Rubin MB (2010) Brittleness of fracture in flowing magma. J Geophys Res-Solid Earth 115:B12202

Iverson RM, Dzurisin D, Gardner CA, *et al.* (2006) Dynamics of seismogenic volcanic extrusion at Mount St Helens in 2004–05. Nature 444:439–443

Ji S, Shao T, Salisbury MH, Sun S, Michibayashi K, Zhao W, Long C, Liang F, Satsukawa T (2014) Plagioclase preferred orientation and induced seismic anisotropy in mafic igneous rocks. J Geophys Res-Solid Earth 119:8064–8088

Johnson DL, Koplik J, Dashen R (1987) Theory of dynamic permeability and tortuosity in fluid-saturated porous media. J Fluid Mech 176:379–402

Johnson JB, Lees JM, Gerst A, Sahagian D, Varley N (2008) Long-period earthquakes and co-eruptive dome inflation seen with particle image velocimetry. Nature 456:377–381

Johnson JB, Lyons JJ, Andrews BJ, Lees JM (2014) Explosive dome eruptions modulated by periodic gas-driven inflation. Geophys Res Lett 41:6689–6697

Kendrick JE, Lavallée Y (2022) Frictional melting in magma and lava. Rev Min Geochem 87:919–963

Kendrick JE, Lavallée Y, Ferk A, Perugini D, Leonhardt R, Dingwell DB (2012) Extreme frictional processes in the volcanic conduit of Mount St. Helens (USA) during the 2004–2008 eruption. J Struct Geol 38:61–76

Kendrick JE, Lavallée Y, Hess KU, Heap MJ, Gaunt HE, Meredith PG, Dingwell DB (2013) Tracking the permeable porous network during strain-dependent magmatic flow. J Volcanol Geotherm Res 260:117–126

Kendrick JE, Lavallée Y, Hess KU, De Angelis S, Ferk A, Gaunt HE, Meredith PG, Dingwell DB, Leonhardt R (2014a) Seismogenic frictional melting in the magmatic column. Solid Earth 5:199–208

Kendrick JE, Lavallée Y, Hirose T, Di Toro G, Hornby AJ, De Angelis S, Dingwell DB (2014b) Volcanic drumbeat seismicity caused by stick–slip motion and magmatic frictional melting. Nat Geosci 7:438–442

Kendrick JE, Lavallée Y, Varley NR, Wadsworth FB, Lamb OD, Vasseur J (2016) Blowing off steam: Tuffisite formation as a regulator for lava dome eruptions. Front Earth Sci 4:41

Kendrick JE, Lavallée Y, Mariani E, Dingwell DB, Wheeler J, Varley NR (2017) Crystal plasticity as an indicator of the viscous–brittle transition in magmas. Nat Commun 8:1926

Kendrick JE, Schaefer LN, Schauroth J, Bell AF, Lamb OD, Lamur A, Miwa T, Coats R, Lavallée Y, Kennedy BM (2021) Physical and mechanical rock properties of a heterogeneous volcano: the case of Mount Unzen, Japan. Solid Earth 12:633–664

Kennedy LA, Russell JK (2012) Cataclastic production of volcanic ash at Mount Saint Helens. Phys Chem Earth 45–46:40–49

Kilburn CRJ (2003) Multiscale fracturing as a key to forecasting volcanic eruptions. J Volcanol Geotherm Res 125:271–289

Kilburn CRJ (2012) Precursory deformation and fracture before brittle rock failure and potential application to volcanic unrest. J Geophys Res-Solid Earth 117:B02211

Klug C, Cashman KV (1996) Permeability development in vesiculating magmas: Implications for fragmentation. Bull Volcanol 58:87–100

Kohlstedt DL, Holtzman BK (2002) Shearing melt out of the mantle. Geochim Cosmochim Acta 66:A410-A410

Kolzenburg S, Heap MJ, Lavallée Y, Russell JK, Meredith PG, Dingwell DB (2012) Strength and permeability recovery of tuffisite-bearing andesite. Solid Earth 3:191–198

Kolzenburg S, Russell JK, Kennedy LA (2013) Energetics of glass fragmentation: Experiments on synthetic and natural glasses. Geochem Geophys Geosyst 14:4936–4951

Kolzenburg S, Ryan AG, Russell JK (2019) Permeability evolution during non-isothermal compaction in volcanic conduits and tuffisite veins: Implications for pressure monitoring of volcanic edifices. Earth Planet Sci Lett 527

Kolzenburg S, Chevrel MO, Dingwell DB (2022) Magma/suspension rheology. Rev Mineral Geochem 87:639-719

Kruhl JH, Vernon RH (2005) Syndeformational emplacement of a tonalitic sheet-complex in a late-Variscan thrust regime: Fabrics and mechanism of intrusion, Monte'e Series, northeastern Sardinia, Italy. Can Mineral 43:387–407

Kueppers U, Perugini D, Dingwell DB (2006) "Explosive energy" during volcanic eruptions from fractal analysis of pyroclasts. Earth Planet Sci Lett 248:800–807

Kushnir ARL, Martel C, Champallier R, Wadsworth FB (2017a) Permeability evolution in variably glassy basaltic andesites measured under magmatic conditions. Geophys Res Lett 44:10262–10271

Kushnir ARL, Martel C, Champallier R, Arbaret L (2017b) In situ confirmation of permeability development in shearing bubble-bearing melts and implications for volcanic outgassing. Earth Planet Sci Lett 458:315–326

Lamb OD, De Angelis S, Umakoshi K, Hornby AJ, Kendrick JE, Lavallée Y (2015) Repetitive fracturing during spine extrusion at Unzen volcano, Japan. Solid Earth 6:1277–1293

Lamur A, Kendrick JE, Eggertsson GH, Wall RJ, Ashworth JD, Lavallée Y (2017) The permeability of fractured rocks in pressurised volcanic and geothermal systems. Sci Rep 7:6173

Lamur A, Lavallée Y, Iddon F, Hornby AJ, Kendrick JE, von Aulock FW, Wadsworth FB (2018) Disclosing the temperature of columnar jointing in lavas. Nat Commun 9:1432

Lamur A, Kendrick JE, Wadsworth FB, Lavallée Y (2019) Fracture healing and strength recovery in magmatic liquids. Geology 47:195–198

Laumonier M, Arbaret L, Burgisser A, Champallier R (2011) Porosity redistribution enhanced by strain localization in crystal-rich magmas. Geology 39:715–718

Lavallée Y, Kendrick JE (2020) A review of the physical and mechanical properties of volcanic rocks and magmas in the brittle and ductile regimes. *In*: Forecasting and Planning for Volcanic Hazards, Risks, and Disasters Vol 2. Papale P (ed) Elsevier

Lavallée Y, Hess K-U, Cordonnier B, Dingwell DB (2007) Non-Newtonian rheological law for highly crystalline dome lavas. Geology 35:843–846

Lavallée Y, Meredith PG, Dingwell DB, Hess KU, Wassermann J, Cordonnier B, Gerik A, Kruhl JH (2008) Seismogenic lavas and explosive eruption forecasting. Nature 453:507–510

Lavallée Y, Benson PM, Heap MJ, Flaws A, Hess KU, Dingwell DB (2012a) Volcanic conduit failure as a trigger to magma fragmentation. Bull Volcanol 74:11–13

Lavallée Y, Mitchell TM, Heap MJ, Vasseur J, Hess K-U, Hirose T, Dingwell DB (2012b) Experimental generation of volcanic pseudotachylytes: Constraining rheology. J Struct Geol 38:222–233

Lavallée Y, Benson PM, Heap MJ, Hess K-U, Flaws A, Schillinger B, Meredith PG, Dingwell DB (2013) Reconstructing magma failure and the degassing network of dome-building eruptions. Geology 41:515–518

Lavallée Y, Hirose T, Kendrick JE, De Angelis S, Petrakova L, Hornby AJ, Dingwell DB (2014) A frictional law for volcanic ash gouge. Earth Planet Sci Lett 400:177–183

Lavallée Y, Hirose T, Kendrick JE, Hess K-U, Dingwell DB (2015b) Fault rheology beyond frictional melting. PNAS 112:9276–9280

Lavallée Y, Dingwell DB, Johnson JB, Cimarelli C, Hornby AJ, Kendrick JE, Von Aulock FW, Kennedy BM, Andrews BJ, Wadsworth FB, Rhodes E (2015a) Thermal vesiculation during volcanic eruptions. Nature 528:544–547

Lavallée Y, Cai B, Coats R, Kendrick JE, von Aulock FW, Wallace PA, Le Gall N, Godinho J, Dobson K, Atwood R, Holness M (2017) Illuminating magma shearing processes via synchrotron imaging. 2017 European Geoscience Union Annual Assembly, Vienna, Austria, p 17616

Lavallée Y, Miwa T, Ashworth JD, Wallace PA, Kendrick JE, Coats R, Lamur A, Hornby A, Hess KU, Matsushima T, Nakada S (2021) Transient conduit permeability controlled by a shift between compactant shear and seismogenic dilatant rupture at Unzen volcano (Japan). Solid Earth Discussions 1–39 [preprint], https://doi.org/10.5194/se-2021-127

Lejeune AM, Richet P (1995) Rheology of crystal-bearing silicate melts—an experimental-study at high viscosities. J Geophys Res-Solid Earth 100:4215–4229

Lejeune AM, Bottinga Y, Trull TW, Richet P (1999) Rheology of bubble-bearing magmas. Earth Planet Sci Lett 166:71–84

Lindoo A, Larsen JF, Cashman KV, Dunn AL, Neill OK (2016) An experimental study of permeability development as a function of crystal-free melt viscosity. Earth Planet Sci Lett 435:45–54

Llewellin EW, Manga A (2005) Bubble suspension rheology and implications for conduit flow. J Volcanol Geotherm Res 143:205–217

Llewellin EW, Mader HM, Wilson SDR (2002) The rheology of a bubbly liquid. Proc R Soc London Ser A 458:987–1016

Loaiza S, Fortin J, Schubnel A, Gueguen Y, Vinciguerra S, Moreira M (2012) Mechanical behavior and localized failure modes in a porous basalt from the Azores. Geophys Res Lett 39:L19304

MacDonald JM, Magee C, Goodenough KM (2017) Dykes as physical buffers to metamorphic overprinting: an example from the Archaean–Palaeoproterozoic Lewisian Gneiss Complex of NW Scotland. Scottish J Geol 53:41–52

Mader HM, Phillips JC, Sparks RSJ, Sturtevant B (1996) Dynamics of explosive degassing of magma: Observations of fragmenting two-phase flows. J Geophys Res-Solid Earth 101:5547–5560

Mader HM, Llewellin EW, Mueller SP (2013) The rheology of two-phase magmas: A review and analysis. J Volcanol Geotherm Res 257:135–158

Manga M, Castro J, Cashman KV, Loewenberg M (1998) Rheology of bubble-bearing magmas. J Volcanol Geotherm Res 87:15–28

Marone C (1998) Laboratory-derived friction laws and their application to seismic faulting. Annu Rev Earth Planet Sci 26:643–696

Martel C, Schmidt BC (2003) Decompression experiments as an insight into ascent rates of silicic magmas. Contrib Mineral Petrol 144:397–415

Martel C, Dingwell DB, Spieler O, Pichavant M, Wilke M (2000) Fragmentation of foamed silicic melts: an experimental study. Earth Planet Sci Lett 178:47–58

Martel C, Dingwell DB, Spieler O, Pichavant M, Wilke M (2001) Experimental fragmentation of crystal- and vesicle-bearing silicic melts. Bull Volcanol 63:398–405

Martí J, Soriano C, Dingwell DB (1999) Tube pumices as strain markers of the ductile–brittle transition during magma fragmentation. Nature 402:650–653

Mattsson HB, Caricchi L, Almqvist BSG, Caddick MJ, Bosshard SA, Hetenyi G, Hirt AM (2011) Melt migration in basalt columns driven by crystallization-induced pressure gradients. Nat Commun 2:299

McIntosh IM, Llewellin EW, Humphreys MCS, Nichols ARL, Burgisser A, Schipper CI, Larsen JF (2014) Distribution of dissolved water in magmatic glass records growth and resorption of bubbles. Earth Planet Sci Lett 401:1–11

McKenzie D (1985) The extraction of magma from the crust and mantle. Earth Planet Sci Lett 74:81–91

Mei S, Kohlstedt DL (2000) Influence of water on plastic deformation of olivine aggregates 2. Dislocation creep regime. J Geophys Res-Solid Earth 105:21471–21481

Mei S, Bai W, Hiraga T, Kohlstedt DL (2002) Influence of melt on the creep behavior of olivine–basalt aggregates under hydrous conditions. Earth Planet Sci Lett 201:491–507

Michaut C, Ricard Y, Bercovici D, Sparks RSJ (2013) Eruption cyclicity at silicic volcanoes potentially caused by magmatic gas waves. Nat Geosci 6:856–860

Mueller S, Melnik O, Spieler O, Scheu B, Dingwell DB (2005) Permeability and degassing of dome lavas undergoing rapid decompression: An experimental determination. Bull Volcanol 67:526–538

Mukherjee S (2012) Simple shear is not so simple! Kinematics and shear senses in Newtonian viscous simple shear zones. Geol Mag 149:819–826

Murrel SAF (1990) Brittle-to-ductile transitions in polycrystalline non-metallic materials. *In*: Deformation Processes in Minerals, Ceramics and Rocks. Vol 18. Barber DJ, Meredith PG, (eds). Springer, p 424

Neuberg J (2000) Characteristics and causes of shallow seismicity in andesite volcanoes. Philos Trans R Soc London Ser A 358:1533–1546

Neuberg JW, Tuffen H, Collier L, Green D, Powell T, Dingwell D (2006) The trigger mechanism of low-frequency earthquakes on Montserrat. J Volcanol Geotherm Res 153:37–50

Neuberg JW, Collinson ASD, Mothes PA, Ruiz MC, Mather TA (2018) Understanding cyclic seismicity and ground deformation patterns at volcanoes: intriguing lessons from Tungurahua volcano, Ecuador. Earth Planet Sci Lett 482:193–200

Okumura S, Sasaki O (2014) Permeability reduction of fractured rhyolite in volcanic conduits and its control on eruption cyclicity. Geology 42:843–846

Okumura S, Nakamura M, Tsuchiyama A (2006) Shear-induced bubble coalescence in rhyolitic melts with low vesicularity. Geophys Res Lett 33:L20316

Okumura S, Nakamura M, Tsuchiyama A, Nakano T, Uesugi K (2008) Evolution of bubble microstructure in sheared rhyolite: Formation of a channel-like bubble network. J Geophys Res-Solid Earth 113:B07208

Okumura S, Nakamura M, Takeuchi S, Tsuchiyama A, Nakano T, Uesugi K (2009) Magma deformation may induce non-explosive volcanism via degassing through bubble networks. Earth Planet Sci Lett 281:267–274

Okumura S, Nakamura M, Nakano T, Uesugi K, Tsuchiyama A (2010) Shear deformation experiments on vesicular rhyolite: Implications for brittle fracturing, degassing, and compaction of magmas in volcanic conduits. J Geophys Res-Solid Earth 115: B06201

Okumura S, Nakamura M, Nakano T, Uesugi K, Tsuchiyama A (2012) Experimental constraints on permeable gas transport in crystalline silicic magmas. Contrib Mineral Petrol 164:493–504

Okumura S, Uesugi K, Sakamaki T, Goto A, Uesugi M, Takeuchi A (2020) An experimental system for time-resolved X-ray diffraction of deforming silicate melt at high temperature. Rev Sci Instrum 91: 095113.

Pallister JS, Cashman KV, Hagstrum JT, Beeler NM, Moran SC, Denlinger RP (2013) Faulting within the Mount St. Helens conduit and implications for volcanic earthquakes. Geol Soc Am Bull 125:359–376

Pascual M, Duran A, Pascual L (2002) Sintering behaviour of composite materials borosilicate glass–ZrO$_2$ fibre composite materials. J Euro Ceram Soc 22:1513–1524

Paterson MS, Wong T-F (2005) Experimental Rock Deformation—The Brittle Field. Springer Verlag, Berlin

Picard D, Arbaret L, Pichavant M, Champallier R, Launeau P (2011) Rheology and microstructure of experimentally deformed plagioclase suspensions. Geology 39:747–750

Picard D, Arbaret L, Pichavant M, Champallier R, Launeau P (2013) The rheological transition in plagioclase-bearing magmas. J Geophys Res-Solid Earth 118:1363–1377

Pistone M, Arzilli F, Dobson KJ, Cordonnier B, Reusser E, Ulmer P, Marone F, Whittington AG, Mancini L, Fife JL, Blundy JD (2015) Gas-driven filter pressing in magmas: Insights into in-situ melt segregation from crystal mushes. Geology 43:699–702

Plail M, Edmonds M, Humphreys MCS, Barclay J, Herd RA (2014) Geochemical evidence for relict degassing pathways preserved in andesite. Earth Planet Sci Lett 386:21–33

Polacci M (2005) Constraining the dynamics of volcanic eruptions by characterization of pumice textures. Ann Geophys 48:731–738

Polacci M, Arzilli F, La Spina G, Le Gall N, Cai B, Hartley ME, Di Genova D, Vo NT, Nonni S, Atwood RC, Llewellin EW (2018) Crystallisation in basaltic magmas revealed via in situ 4D synchrotron X-ray microtomography. Sci Rep 8:8377

Prior DJ, Boyle AP, Brenker F, Cheadle MC, Day A, Lopez G, Peruzzi L, Potts G, Reddy S, Spiess R, Timms NE (1999) The application of electron backscatter diffraction and orientation contrast imaging in the SEM to textural problems in rocks. Am Mineral 84:1741–1759

Qi C, Hansen LN, Wallis D, Holtzman BK, Kohlstedt DL (2018) Crystallographic preferred orientation of olivine in sheared partially molten rocks: the source of the "a-c Switch". Geochem Geophys Geosyst 19:316–336

Quane SL, Russell JK (2005) Welding: insights from high-temperature analogue experiments. J Volcanol Geotherm Res 142:67–87

Quane SL, Russell JK, Friedlander EA (2009) Time scales of compaction in volcanic systems. Geology 37:471–474

Ramsay JG (1967) Folding and Fracturing Rocks. McGraw-Hill

Reynolds O (1885) On the dilatancy of media composed of rigid particles in contact. Phil Mag 20:469–481

Rice JR (1993) Spatiotemporal complexity of slip on a fault. J Geophys Res-Solid Earth 98:9885–9907

Richet P, Bottinga Y (1986) Thermochemical properties of silicate glasses and liquids: a review. Rev Geophys 24:1–25

Roberts NS, Bell AF, Main IG (2015) Are volcanic seismic *b*-values high, and if so when? J Volcanol Geotherm Res 308:127–141

Romeo I, Capote R, Lunar R, Cayzer N (2007) Polymineralic orientation analysis of magmatic rocks using electron back-scatter diffraction: Implications for igneous fabric origin and evolution. Tectonophysics 444:45–62

Rosenberg CL (2004) Shear zones and magma ascent: A model based on a review of the Tertiary magmatism in the Alps. Tectonics 23:TC3002

Rust AC, Manga M (2002) Effects of bubble deformation on the viscosity of dilute suspensions. J Non-Newtonian Fluid Mech 104:53–63

Rust AC, Manga M, Cashman KV (2003) Determining flow type, shear rate and shear stress in magmas from bubble shapes and orientations. J Volcanol Geotherm Res 122:111–132

Rutter EH (1986) On the nomenclature of mode of failure transitions in rocks. Tectonophysics 122:381–387

Ryan AG, Russell JK, Heap MJ, Kolzenburg S, Vona A, Kushnir ARL (2019) Strain-dependent rheology of silicate melt foams: importance for outgassing of silicic lavas. J Geophys Res-Solid Earth 124:8167–8186

Saar MO, Manga M (1999) Permeability–porosity relationship in vesicular basalts. Geophys Res Lett 26:111–114

Sakamaki T, Ohtani E (2022) High pressure melts. Rev Mineral Geochem 87:557-574

Sammis CG, Ashby MF (1986) The failure of brittle porous solids under compressive stress states. Acta Metall 34:511–526

Satsukawa T, Ildefonse B, Mainprice D, Morales LFG, Michibayashi K, Barou F (2013) A database of plagioclase crystal preferred orientations (CPO) and microstructures—implications for CPO origin, strength, symmetry and seismic anisotropy in gabbroic rocks. Solid Earth 4:511–542

Schauroth J, Wadsworth FB, Kennedy B, von Aulock FW, Lavallée Y, Damby DE, Vasseur J, Scheu B, Dingwell DB (2016) Conduit margin heating and deformation during the AD 1886 basaltic Plinian eruption at Tarawera volcano, New Zealand. Bull Volcanol 78:12

Scheu B, Dingwell DB (2022) Magma fragmentation. Rev Mineral Geochem 87:767-800

Scholz CH (1998) Earthquakes and friction laws. Nature 391:37–42

Shelley D (1985) Determining paleo-flow directions from groundmass fabrics in the lyttleton radial dykes, New Zealand. J Volcanol Geotherm Res 25:69–79

Shields JK, Mader HM, Pistone M, Caricchi L, Floess D, Putlitz B (2014) Strain-induced outgassing of three-phase magmas during simple shear. J Geophys Res-Solid Earth 119:6936–6957

Shields JK, Mader HM, Caricchi L, Tuffen H, Mueller S, Pistone M, Baumgartner L (2016) Unravelling textural heterogeneity in obsidian: Shear-induced outgassing in the Rocche Rosse flow. J Volcanol Geotherm Res 310:137–158

Shilko EV, Dimaki AV, Psakhie SG (2018) Strength of shear bands in fluid-saturated rocks: a nonlinear effect of competition between dilation and fluid flow. Sci Rep 8:1428

Shimamoto T, Lin AM (1994) Is frictional melting equilibrium or non-equilibrium melting? Structural Geology (J Tectonic Res Group Japan) 39:79–84

Lavallée & Kendrick

Sibson RH (1996) Structural permeability of fluid-driven fault-fracture meshes. J Struct Geol 18:1031–1042

Sigmundsson F, Hooper A, Hreinsdóttir S, Vogfjörd KS, Ófeigsson BG, Heimisson ER, Dumont S, Parks M, Spaans K, Gudmundsson GB, Drouin V (2015) Segmented lateral dyke growth in a rifting event at Bardarbunga volcanic system, Iceland. Nature 517:191–195

Smith JV, Brown WL (1988) Feldspar Minerals: Volume 1. Crystal Structures, Physical, Chemical, and Microtextural Properties. Springer Verlag, New York

Smith R, Kilburn CRJ (2010) Forecasting eruptions after long repose intervals from accelerating rates of rock fracture: The June 1991 eruption of Mount Pinatubo, Philippines. J Volcanol Geotherm Res 191:129–136

Smith JV, Miyake Y, Oikawa T (2001) Interpretation of porosity in dacite lava domes as ductile–brittle failure textures. J Volcanol Geotherm Res 112:25–35

Sparks RSJ (2003a) Forecasting volcanic eruptions. Earth Planet Sci Lett 210:1–15

Sparks RSJ (2003b) Dynamics of magma degassing. In: Volcanic Degassing. Vol 213. Oppenheimer C, Pyle DM, Barclay J, (eds). p 5–22

Sparks RSJ, Annen CJ, Blundy J, Cashman K, Rust A, Jackson MD (2019) Formation and dynamics of magma reservoirs. Philos Trans R Soc A 377:20180019

Spieler O, Kennedy B, Kueppers U, Dingwell DB, Scheu B, Taddeucci J (2004) The fragmentation threshold of pyroclastic rocks. Earth Planet Sci Lett 226:139–148

Spray JG (2010) Frictional melting processes in planetary materials: from hypervelocity impact to earthquakes. Annu Rev Earth Planet Sci 38:221–254

Taddeucci J, Cimarelli C, Alatorre-Ibargueengoitia MA, Delgado Granados H, Andronico D, Dl Bello E, Scarlato P, Di Stefano F (2021) Fracturing and healing of basaltic magmas during explosive volcanic eruptions. Nat Geosci 14:248–254

Taylor GI (1934) The formation of emulsions in definable fields of flow. Proc R Soc London Ser A 146:0501–0523

Toramaru A (1988) Formation of propagation pattern in a two-phase flow systems with application to volcanic eruptions. Geophys J Int 95:613–623

Treagus SH, Lisle RJ (1997) Do principal surfaces of stress and strain always exist? J Struct Geol 19:997–1010

Tripoli B, Manga M, Mayeux J, Barnard H (2019) The effects of deformation on the early kinetics of basaltic magmas. Front Earth Sci 7:250

Truby JM, Mueller SP, Llewellin EW, Mader HM (2015) The rheology of three-phase suspensions at low bubble capillary number. Proc R Soc London Ser A 471:20140557

Tuffen H, Dingwell DB (2005) Fault textures in volcanic conduits: evidence for seismic trigger mechanisms during silicic eruptions. Bull Volcanol 67:370–387

Tuffen H, Dingwell DB, Pinkerton H (2003) Repeated fracture and healing of silicic magma generate flow banding and earthquakes? Geology 31:1089–1092

Tuffen H, Smith R, Sammonds PR (2008) Evidence for seismogenic fracture of silicic magma. Nature 453:511–514

Vachon R, Bazargan M, Hieronymus DF, Ronchin E, Almqvist BSG (2021) Crystal rotations and alignment in spatially varying magma flows: 2-D examples of common subvolcanic flow geometries. Geophys J Int 226:709–727

van der Linden JH, Tordesillas A, Narsilio GA (2019) Preferential flow pathways in a deforming granular material: self-organization into functional groups for optimized global transport. Sci Rep 9:18231–18231

Vasseur J, Wadsworth FB, Lavallée Y, Hess K-U, Dingwell DB (2013) Volcanic sintering: Timescales of viscous densification and strength recovery. Geophys Res Lett 40:5658–5664

Vasseur J, Wadsworth FB, Lavallée Y, Bell AF, Main IG, Dingwell DB (2015) Heterogeneity: The key to failure forecasting. Sci Rep 5:13259

Vasseur J, Wadsworth FB, Heap MJ, Main IG, Lavallée Y, Dingwell DB (2017) Does an inter-flaw length control the accuracy of rupture forecasting in geological materials? Earth Planet Sci Lett 475:181–189

Ventura G, DeRosa R, Colletta E, Mazzuoli R (1996) Deformation patterns in a high-viscosity lava flow inferred from the crystal preferred orientation and imbrication structures: An example from Salina (Aeolian Islands, southern Tyrrhenian Sea, Italy). Bull Volcanol 57:555–562

Vernon RH (1987) A microstructural indicator of shear sense in volcanic rocks and its relationship to porphyroblast rotation in metamorphic rocks. J Geol 95:127–133

Vinet N, Higgins MD (2010) Magma solidification processes beneath Kilauea Volcano, Hawaii: a quantitative textural and geochemical study of the 1969–1974 Mauna Ulu lavas. J Petrol 51:1297–1332

Violay M, Di Toro G, Gibert B, Nielsen S, Spagnuolo E, Del Gaudio P, Azais P, Scarlato PG (2014) Effect of glass on the frictional behavior of basalts at seismic slip rates. Geophys Res Lett 41:348–355

Voight B (1988) A method for prediction of volcanic-eruptions. Nature 332:125–130

Voight B (1989) A relation to describe rate-dependent material failure. Science 243:200–203

Wada Y (1992) Magma flow directions inferred from preferred orientations of phenocryst in a composite feeder dyke, Miyake-Jima, Japan. J Volcanol Geotherm Res 49:119–126

Wadsworth FB, Vasseur J, von Aulock FW, Hess K-U, Scheu B, Lavallée Y, Dingwell DB (2014) Nonisothermal viscous sintering of volcanic ash. J Geophys Res-Solid Earth 119:8792–8804

Wadsworth FB, Vasseur J, Scheu B, Kendrick JE, Lavallée Y, Dingwell DB (2016) Universal scaling of fluid permeability during volcanic welding and sediment diagenesis. Geology 44:219–222

Wadsworth FB, Vasseur J, Llewellin EW, Dobson KJ, Colombier M, Von Aulock FW, Fife JL, Wiesmaier S, Hess KU, Scheu B, Lavallée Y (2017) Topological inversions in coalescing granular media control fluid-flow regimes. Phys Rev E 96:033113

Wadsworth FB, Witcher T, Vossen CEJ, Hess K-U, Unwin HE, Scheu B, Castro JM, Dingwell DB (2018) Combined effusive–explosive silicic volcanism straddles the multiphase viscous-to-brittle transition. Nat Commun 9:4696

Wadsworth FB, Vasseur J, Schauroth J, Llewellin EW, Dobson KJ, Havard T, Scheu B, von Aulock FW, Gardner JE, Dingwell DB, Hess KU (2019) A general model for welding of ash particles in volcanic systems validated using in situ X-ray tomography. Earth Planet Sci Lett 525:115726

Wadsworth FB, Vasseur J, Llewellin EW, Dingwell DB (2022) Hot sintering of melts, glasses and magmas. Rev Mineral Geochem 87:801-840

Wadsworth FB, Vasseur J, Llewellin EW, Brown RJ, Tuffen H, Gardner JE, Kendrick JE, Lavallée Y, Dobson KJ, Heap MJ, Dingwell DB (2021) A model for permeability evolution during volcanic welding. J Volcanol Geotherm Res 409:107118

Wallace PA, De Angelis SH, Hornby AJ, Kendrick JE, Clesham S, von Aulock FW, Hughes A, Utley JE, Hirose T, Dingwell DB, Lavallée Y (2019a) Frictional melt homogenisation during fault slip: Geochemical, textural and rheological fingerprints. Geochim Cosmochim Acta 255:265–288

Wallace PA, Kendrick JE, Miwa T, Ashworth JD, Coats R, Utley JE, Henton De Angelis S, Mariani E, Biggin A, Kendrick R, Nakada S (2019b) Petrological architecture of a magmatic shear zone: A multidisciplinary investigation of strain localisation during magma ascent at Unzen Volcano, Japan. J Petrol 60:791–826

Walter JM, Iezzi G, Albertini G, Gunter ME, Piochi M, Ventura G, Jansen E, Fiori F (2013) Enhanced crystal fabric analysis of a lava flow sample by neutron texture diffraction: A case study from the Castello d'Ischia dome. Geochem Geophys Geosyst 14:179–196

Webb SL, Dingwell DB (1990a) The Onset of Non-Newtonian Rheology of Silicate Melts—a Fiber Elongation Study. Phys Chem Mineral 17:125–132

Webb SL, Dingwell DB (1990b) Non-Newtonian rheology of igneous melts at high stresses and strain rate—Experimental results for rhyolite, andesite, basalt, and nepheline. J Geophys Res-Solid Earth Planets 95:15695–15701

Wheeler J (2014) Dramatic effects of stress on metamorphic reactions. Geology 42:647–650

Wieser PE, Edmonds M, Maclennan J, Wheeler J (2020) Microstructural constraints on magmatic mushes under Kilauea Volcano, Hawai'i. Nat Commun 11:14

Wong T-F, Baud P (2012) The brittle–ductile transition in porous rock: A review. J Struct Geol 44:25–53

Wong TF, Baud P, Klein E (2001) Localized failure modes in a compactant porous rock. Geophys Res Lett 28:2521–2524

Wright HMN, Weinberg RF (2009) Strain localization in vesicular magma: Implications for rheology and fragmentation. Geology 37:1023–1026

Wright HMN, Roberts JJ, Cashman KV (2006) Permeability of anisotropic tube pumice: Model calculations and measurements. Geophys Res Lett 33:L17316

Xia KW, Rosakis AJ, Kanamori H (2004) Laboratory earthquakes: The sub-Rayleigh-to-supershear rupture transition. Science 303:1859–1861

Zak J, Verner K, Tycova P (2008) Grain-scale processes in actively deforming magma mushes: New insights from electron backscatter diffraction (EBSD) analysis of biotite schlieren in the Jizera granite, Bohemian Massif. Lithos 106:309–322

Zhang DZ, VanderHeyden WB (2002) The effects of mesoscale structures on the macroscopic momentum equations for two-phase flows. Int J Multiphase Flow 28:805–822

Zhang YQ, Hao H, Yu MH (2002) Effect of porosity on the properties of strain localization in porous media under undrained conditions. Int J Solids Struct 39:1817–1831

Zhu W, Baud P, Wong TF (2010) Micromechanics of cataclastic pore collapse in limestone. J Geophys Res-Solid Earth 115:B04405

Zhu W, Baud P, Vinciguerra S, Wong TF (2016) Micromechanics of brittle faulting and cataclasticflow in Mount Etna basalt. J Geophys Res: Solid Earth 121:4268–4289

Zucali M, Fontana E, Panseri M, Tartarotti P, Capelli S, Ouladdiaf B (2014) Submarine lava flow direction revealed by neutron diffraction analysis in mineral lattice orientation. Geochem Geophys Geosyst 15:765–780

Reviews in Mineralogy & Geochemistry
Vol. 87 pp. 767-800, 2022
Copyright © Mineralogical Society of America

16

Magma Fragmentation

Bettina Scheu and Donald B. Dingwell

Earth and Environmental Sciences
Ludwig-Maximilians-Universität München
Theresienstrasse 41
80333 Munich
Germany

b.scheu@lmu.de, dingwell@lmu.de

INTRODUCTION

The scientific investigation of magma fragmentation and the development of experiments to constrain and quantify its mechanisms, dynamics, efficiency and products was ushered in over a quarter century ago by the realisation that magmas behave like solids during explosive volcanism. Beginning with the initial design and construction of a high temperature, high pressure fragmentation device capable of decompressing dome lavas at magmatic conditions, a major new element of experimental geosciences was enabled. In the ensuing decades the fragmentation process has been systematically disassembled into its constituent parts in experiments designed to fully characterise the central mechanism of explosive volcanism. Here, following a brief preamble on the historical approaches to the problem, we review the understanding of the fragmentation process that has arisen from this multidecadal campaign, including the concepts and quantification of the fragmentation threshold, the speed of the fragmentation front, the fragmentation efficiency and the resultant grain size distribution, the fundamental role of permeability, the energy balance of fragmentation, the specific role of external water and steam in fragmentation and the fresh look at volcanic ash generated by these studies. We conclude with a brief outlook on unsolved problems.

Fragmentation is the process or processes whereby intact magma is transformed into pyroclastic materials (bombs, pumice, ash, etc.). As such, it is magma fragmentation that is the defining attribute of explosive volcanism. The diversity of textures and states of magma, the clear range of dynamics of volcanic explosions, and the variability of the nature of fragmented materials generated in explosive eruptions are all clear testimony to the range of volcanic scenarios that can yield magma fragmentation. Despite this variability the fundamental criteria governing fragmentation; its efficiency, speed and extent, are limited in number. These criteria have been mapped with increasing sophistication and control for the past 25 years. Generally, this places volcanologists in the situation of envisaging scenarios which might explain explosive volcanic behavior and which, importantly, can then be quantified, constrained and modelled with respect to clearly parameterized magma properties. The advent of experimental volcanology has nowhere been felt more strongly than in the mechanistic quantification of explosive volcanism.

For the purposes of quantifying and explaining the phenomenology of processes surrounding explosive eruptions we can break the problem down into three parts: firstly, the state of the magmatic system at storage and the initiation of pre-eruptive ascent; secondly, the scenario for the achievement of conditions whereby fragmentation is initiated and sustained; and thirdly, the fate of fragmented volcanic ejecta during conduit ascent, column ascent, dispersion, deposition and, potentially, welding.

1529-6466/22/0087-0016$05.00 (print)
1943-2666/22/0087-0016$05.00 (online)
http://dx.doi.org/10.2138/rmg.2021.87.16

In this chapter we deal solely with the second stage described above, namely that of the fragmentation process itself. Moreover, we deal with magmatic fragmentation as a mechanistically parameterized process closely linked to the magma properties and we do not delve into fragmentation scenarios. Magmatic fragmentation is sometimes also termed 'dry' magmatic fragmentation referring to the absence of interaction with external water. (Inasmuch as water is often the most important volatile in magmatic systems this is perhaps unfortunate.) Magmatic fragmentation is driven only by the magmatic volatiles. If magma or magmatic fragmentation interact with external water resulting in, or intensifying an ongoing explosive eruption, this eruption is termed hydromagmatic or phreatomagmatic. In some cases, explosive eruptions are driven solely by the violent expansion of volatiles (primarily H_2O, CO_2), heated by rising magma or a hydrothermal system. These eruptions are termed phreatic or hydrothermal eruptions and are considered to not eject juvenile ash, lapilli and bombs. Magma typically fragments in a volcanic conduit at relatively shallow depths in both subaerial and submarine environments. The fragments generated during this so-called 'primary' fragmentation, are then accelerated upwards in the conduit and in most cases ejected out of the vent forming a volcanic plume, ballistic ejecta and/or a pyroclastic density current. During the above, various processes take place that cause further fragment break-up altering the primary fragment size distribution. All these processes may be subsumed under the term 'secondary' fragmentation. The final distribution of fragments ejected and deposited is therefore a sum of various processes.

Below we provide a brief overview of some key elements of fragmentation mechanisms and their associated regimes. We then discuss experimental approaches and theoretical models designed to explore magma fragmentation, followed by insights into the physical nature of the fragmentation behavior, such as threshold, speed and further factors affecting the fragmentation behavior.

SILICATE LIQUIDS—THE BASIS OF MAGMA.

Magma is arguably the most complex of all geomaterials. It can be defined as a multiphase geomaterial (e.g., Kolzenburg et al. 2022, this volume) where the supporting phase is a silicate melt—and it is the presence of a melt phase which is the single necessary material criterion for the definition of magma. As such it is worth briefly reviewing the role of the silicate melt in the fragmentation of magma. The propensity of silicate magma to fragment in explosive volcanism is conditioned on a few key elements of the earth system. One of these is the chemistry of silicate melts that are generated by the melting of the Earth´s interior. The earth (and other terrestrial planets to a less well-known extent) is well-populated by silicate melts with relatively high silica contents (up to ca. 75% SiO_2). This chemical peculiarity of evolved terrestrial planets leads to a fluid (i.e., magma) component within their lithosphere which has, by any measure, peculiar properties. Nowhere are the effects more dramatic that in the composition dependence of the viscosity of silicate melts. The two greatest compositional influences on viscosity of natural silicate melts are silica content and water content (e.g., Dingwell 1996; Giordano et al. 2008). In fact, pure SiO_2 melt has the highest viscosity at all temperatures of any liquid known to man. Thus, foremost amongst the consequences of having present within the lithosphere, melts with silica contents higher than basaltic (i.e., ca. 45 wt.% SiO_2), is the potential for a substantially enhanced viscosity of these melts under eruptive conditions. Magmatic differentiation typically (although not exclusively) drives the melt phase of silicate magmas towards progressively higher silica content, yielding the potential for higher viscosities (Bowen 1928, 1934). Thus, as magmas achieve higher levels in a progressively cooler crust, they will tend to have higher silica contents and higher viscosities. The second great variable of magma chemistry influencing melt viscosity is the water content of the melt. The mobility of water as a volatile phase is enhanced as the magma arrives at higher levels in the lithosphere. It becomes a dominant physical phenomenon accompanying magma ascent immediately prior to and during volcanic eruption. The loss of water from the silicate melt phase has the potential to result in drastically increased melt viscosity (Fig. 1; Hess and Dingwell 1996). The effect accelerates with magma ascent and, as the example of Earth provides, surface conditions of 1 bar or less in pressure combined with the fact that all volcanic glasses on earth are oversaturated in volatiles clearly indicates that disequilibrium degassing during final ascent is ubiquitous. To put it succinctly, were it not for the extreme viscosities of high silica, water-poor liquids, the story of volcanism on Earth would be a very different one indeed.

The general outline of these viscosity trends (the silica effect) and the water effect have been well-known for over 50 years (Friedman et al. 1963; Burnham 1964; Shaw 1972) but their combined influences in multicomponent silicate melts have only been comprehensively parameterized about a decade ago (Giordano et al. 2008; Russell et al. 2022, this volume).

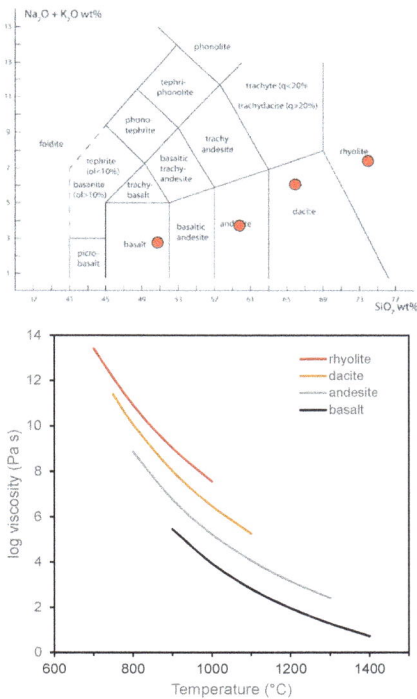

Figure 1. Overview of the viscosity of silicate melts (after Giordano et al. 2008) The **bottom left** plot shows the dependence of melt viscosity on temperature; all melts are shown in a range typical for their respective eruptive temperatures. The basalt contains 52 wt.% SiO_2, the andesite 59 wt.% SiO_2, the dacite 66 wt.% SiO_2, and the rhyolite 74 wt.% SiO_2, all compositions are indicated in the TAS diagram (**left**). The viscosity of a silicate melt is sensitive to the presence and abundance of volatiles (**bottom right**) At constant temperature the melt viscosity decreases with increasing water content, this effect is especially pronounced for low water contents.

FRAGMENTATION—A MATERIALS TRIGGER

Magma fragmentation is the key process of explosive eruptions, distinguishing it from its effusive counterparts. However, the expression of magma fragmentation will inevitably be highly variable in detail, depending as it must on melt and magma properties, conduit dynamics, and eruption environment.

For very low viscous magmas the fragmentation is often rather "fluidal" and described as "fragmentation by fluid dynamic-induced break-up" (e.g., Jones et al. 2019). Therein the disruption of a string of extending magma depends on the interplay between viscous forces and inertial and liquid surface tension forces causing different types of fluid dynamic instabilities.

When liquid surface tension and inertial forces dominate, fragmentation is characterized by the capillary instability timescale τ_{cap}, of the melt string:

$$\tau_{cap} = \sqrt{\frac{\rho d^3}{\Gamma}}$$

with melt density ρ, surface tension Γ the diameter d of the melt string. However, if viscous forces dominate, the melt break-up is governed by the viscous instability timescale τ_{vis}:

$$\tau_{vis} = \frac{\eta d}{\Gamma}$$

where η is the melt viscosity. The dimensionless Ohnesorge number (Oh) can be used to determine which of these forces and associated timescale is dominating this fluid dynamic induced break-up:

$$Oh = \frac{\tau_{vis}}{\tau_{cap}}$$

For $Oh > 1$, the viscous forces dominate and break-up occurs on the viscous instability timescale, whereas for $Oh < 1$, break-up occurs on the capillary instability timescale as the inertial and surface tension forces dominate (Villermaux 2012; Jones et al. 2019).

In contrast, silicic, highly viscous magma will in most cases be disrupted in a regime dominated by localized brittle processes. Highly viscous magma is driven by stresses out of its relaxed state causing elastic strain energy to accumulate in the magma until failure conditions are met. This departure from the relaxed (thermodynamically stable or metastable) state lies in the approaching equivalence of the strain rates of 1) structural relaxation of the liquid state and 2) the deformation sustained by the Newtonian response to driving stresses. For the common case of solid and/or fluid suspended phases in magma the distribution of stress and the localization of strain become more complex as the magma begins to exhibit the elements of a multiphase rheology (e.g., Kolzenburg et al. 2022, this volume).

In the purely viscous state, the magmatic liquid will by definition relax stresses more quickly than the timescale implicit in the strain rate of liquid deformation. As a result, the storage of stress in the liquid does not rise beyond that driving the viscous flow under Newtonian conditions. The quantification of the regimes of fully viscous versus fully elastic stress–strain response is given by the Maxwell relaxation time τ_r (exemplified here for shear deformation)

$$\tau_r = \frac{\eta}{G}$$

where η is the liquid (Newtonian) viscosity and G ($\sim 10^{10}$ Pa) is the shear modulus of the glassy state (e.g., Dingwell and Webb 1989, 1990; Dingwell 1997). An instantaneously applied stress is dissipated over a time proportional to τ_r. This timescale of relaxation can be compared to timescales of other processes, as the shear strain rate γ_s or a characteristic decompression rate (Note: for an elongational strain and associated strain rate G will be replaced by the elastic modulus E). The dimensionless Deborah number, De, is often used in rheology to compare such characteristic timescales where

$$De = \dot{\gamma}_s \tau_r = \dot{\gamma}_s \frac{\eta}{G}$$

De can be used to discriminate between different rheological states of a melt or magma. If $De \ll 1$, the melt is in a relaxed, thermodynamically (meta)stable state and stress is viscous-ly dissipated, following the Newtonian rheology. The viscous dissipation results in structural rearrangement of the melt pervasive to the entire body subjected to the stress (i.e., ductile deformation). If $De \gg 1$, the melt is unrelaxed and stress will accumulate, as the deformation timescale (the inverse of the strain rate) is smaller than the structural relaxation time. This may ultimately result in brittle failure of the melt or magma at localized failure planes. It should be noted that also in this state a melt may experience a certain degree of viscous dissipation (i.e., ductile deformation), however the reaction of the melt to the applied stress is dominated by localized brittle failure. The transition between these two states with respect to relaxation is the glass transition (Fig. 2) and it is typically preceded by a non-Newtonian, shear thinning rheology (Webb and Dingwell 1990a,b).

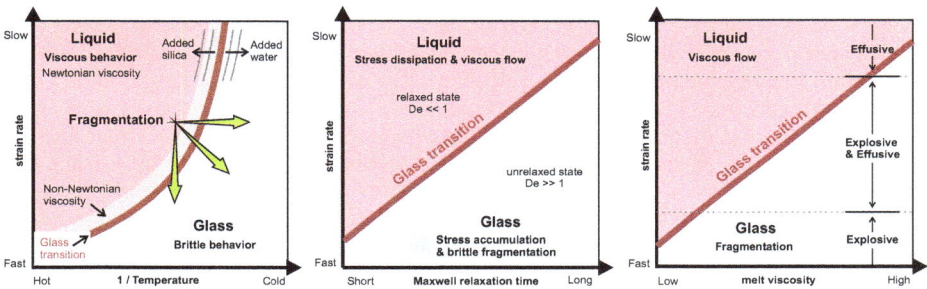

Figure 2. Illustration of the glass transition concept in different parameter spaces: Strain rate–temperature space (**left**), strain rate–relaxation time space (**middle**), strain rate–melt viscosity space (**right**). The glass transition is a kinetic barrier dividing the behavior of silicate melts into two states: viscous liquid and brittle glass (elastic solid). The liquid field refers to the relaxed state of the melt and viscous response to deformation characterized by $De \ll 1$. The glassy field refers to the elastic solid state, where the deformation is faster than structural relaxation time of the melt and stress is accumulated ($De \gg 1$) resulting in elastic response up to brittle failure. The glass transition is very sensitive to even small variations in H_2O and SiO_2 content of the melt, highlighting the important role of degassing and crystallization during magma ascent and eruption. Fragmentation occurs by crossing the glass transition, either by increasing the rate of deformation (strain rate), cooling of the melt, or a combination of both (**green arrows**). The strain rate–melt viscosity space is suited to schematically depict the relation to eruption styles. Low strain rates result predominantly in effusive behavior where minor, often low energetic fragmentation may occur (e.g., fissure eruptions). High strain rates cause predominantly explosive eruptions, which may as a minor component contain a viscous deformation. At intermediate strain rates the eruption behavior often straddles the glass transition and thus cycles between explosive and effusive behavior. Figures modified from Dingwell (1996), Gonnermann and Manga (2003), Cashman and Scheu (2015).

As noted above, due to the enormous range of Newtonian viscosity versus the relative insensitivity of the shear modulus in temperature and composition, this relaxation time is, to a very good approximation, directly proportional to the Newtonian viscosity. Thus the Deborah number at a given strain rate can be accessed via Maxwell´s law at a given viscosity. Melt viscosity is discussed in depth in Russell et al. (2022, this volume); here we would like to note that in magmatic and especially in volcanic processes the viscosity of the melt can vary over several orders of magnitude depending on silica content, the ratio of network formers to modifiers, and especially the melt water content and the temperature (Fig. 1; Hess and Dingwell 1996; Mysen and Richet 2005). In general, magmatic systems drive through differentiation to higher silica contents and higher water contents and lower temperatures. At constant temperature, the specific evolution of viscosity with composition will depend largely on water and silica content with some complexity being added via the agpaitic index, whereby peralkaline differentiation trends will lead to lower viscosities as both peralkalinity and water contents increase (Giordano et al. 2008). In addition to these purely chemical effects, the temperature regime of magmatism drifts with differentiation to lower temperatures whereas the temperature regime of erupting magma drops catastrophically during and immediately following the event. Thus the playing field of the variation of the glass transition temperature with time in magmatic and volcanic systems is rich in complexity.

As the glass transition marks the transition between liquid and solid behavior (Fig. 2) it is not only observed in the rheological response of the system but in terms of all other thermodynamic and several kinetic properties as well. Changes in thermodynamic properties such as heat capacity, expansivity and elasticity are very well documented through experiments (e.g., Dingwell 1995).

In summary, an enrichment in silica content increases the melt viscosity and thus widens the brittle field towards higher temperature. Similarly, the loss of volatiles (e.g water) shifts the glass transition to higher temperatures. With the concepts stated above, it should be apparent that the glass transition can be crossed either by cooling (or heating) or by an increase (or decrease) in the strain rate, or a mixture of both (Fig. 2). The glass transition can be crossed starting from both fields (liquid, glass) and multiple crossings during a volcanic eruption and emplacement of volcanic deposits, are frequent, resulting from changes in deformation timescale caused by accelerating magma, or magma experiencing rapid decompression, or during the ejection of fragments and their ground impact as well as after their emplacement. Welding of pyroclastic deposits—from spatter to ignimbrite—is one excellent example of this complexity whereby initially fragmented materials which re-weld have physically crossed the glass transition at least three times before they cool to ambient temperature (Wilding et al. 1996, 2000). For all these processes the interplay between the structural relaxation of the melt or magma and the strain rate associated to that process defines the type of response between the end-members of ductile-like melt/magma fragmentation by fluid dynamic-induced break-up (Jones et al. 2019) when viscous stresses can be relaxed and disruption is caused by fluid instabilities, and brittle-like fragmentation if the tensile strength is exceeded (Wadsworth et al. 2017, 2018).

For magmas the presence of crystals and bubbles imply further complex interactions. To evaluate the structural relaxation timescale and *De* numbers, we have to consider how the magma viscosity η_m deviates from the melt viscosity. Therefore η_m is often non-dimensionalized by the melt viscosity resulting in the relative viscosity $\eta_r = \eta_m / \eta$. Suspended crystals generally cause an increase in relative viscosity and enhance the strain-rate dependent non-Newtonian rheology, their role in suspension rheology has been parameterized with increasing data and accuracy many times (e.g., Caricchi et al. 2007; Lavallée et al. 2007; Cimarelli et al. 2011; Mueller et al. 2011; Cordonnier et al. 2012; Kolzenburg et al. 2022, this volume). At low strain rates, in the Newtonian field the relative viscosity increases more or less pronouncedly depending on shape and amount of crystals. After transitioning into the non-Newtonian regime at high strain rates, the relative viscosity decreases again as a result of a geometrical redistribution of the crystals. The degree of this shear thinning behavior again depends on crystal shape and load.

The effect of bubbles on magma viscosity is more ambiguous, depending on the shapes of the bubbles which are a result of the balance between surface and viscous forces (e.g., Lejeune et al. 1999; Pal 2003). Surface tension Γ will act trying to return a deformed bubble to its energetically favored spherical shape on the capillary time

$$\tau_c = \frac{\eta r}{\Gamma}$$

with r being the radius of the spherical bubble, or the radius of curvature for the case of a deformed bubble. Opposing this natural tendency to sphericity driven by the surface tension, the deviatoric stresses generating magma flow act to maintain a degree of bubble non-sphericity. This timescale is given by the strain rate of the shape deformation (e.g., by shear or elongational forces):

$$\tau_s = \frac{1}{\dot{\gamma}}$$

The dimensionless Capillary number Ca relates the relative effects of viscous drag forces versus the surface forces acting across an interface between a liquid and a gas (or in other words, the capillary and deformation time scales):

$$Ca = \frac{\tau_c}{\tau_s} = \frac{\eta r \dot{\gamma}}{\Gamma}$$

For $Ca \ll 1$ surface tension forces dominate and the bubbles remain spherical. Due to their effective rigidity (nondeformability) these spherical bubbles exert a volumetric effect comparable to that of crystals on magma viscosity causing its increase (Dingwell 1998; Pal 2003; Llewellin and Manga 2005; Kolzenburg et al. 2022, this volume). For $Ca \gg 1$ viscous forces dominate, resulting in deformed, elongated bubbles, whose aspect ratio increases with Ca. These deformed bubbles decrease the magma viscosity by the action of their deformation from sphericity), indeed for high Ca the relative viscosity η_r drops well below 1 (e.g., Rust and Manga 2002; Llewellin and Manga 2005; Kolzenburg et al. 2022, this volume). A good example of the prodigious further complexity that can be expected in the deformation of highly vesicular magma is provided by tube pumice pyroclasts which exhibit polygonal terminations and parallel "kink" bands (Martí et al. 1999). Due to the variations in bubble sizes and aspect ratios in these tube pumices substantial kinematic analyses of their stress–strain histories have been attempted leading to the conclusion that they record intermittent uniaxial compressive events as well as the tensile fragmentation events leading to their disruption (Martí et al. 1999; Dingwell et al. 2016). The relative degree of pure shear versus simple shear deformation in magmatic vesicles has been investigated using X-ray tomography of such tube pumices on very large vesicle populations with preliminary results indicating a dominance of simple shear (Dingwell et al. 2016).

Once we begin to consider the existence of solid phases included in magmas combined with a highly foamed state we enter a realm where the distributions of the three essential phases—crystals, bubbles and melt—might play a substantial role in conditioning the strength of the failing walls of bubbles during fragmentation (Mungall et al. 1996; Romano et al. 1996). If the included solid phases are extant in sizes down to the microlite and even nanolite levels then the surface energy of very thin bubble walls will potentially sense the presence of these suspended phases. Interaction of micro- to nanolite phases with the bubble walls at these very low lengthscales may impact the eruption style (Platz et al. 2007; Cáceres et al. 2020). It is conceivable therefore that the strength of such complex materials is a property which needs to be expressed in terms of the length-scale of the heterogeneities.

EXPERIMENTAL VOLCANOLOGY OF MAGMATIC FRAGMENTATION

Experimental approaches to understand the mechanisms causative to explosive volcanism reach back nearly five decades. The history and evolution of these experiments can be seen either chronologically or by dividing them into those using natural materials and those that use non-magmatic materials, so-called magma analogues. This duality has to be taken into account in order to appreciate the strengths and limitations of the individual experimental approaches. Natural materials can be extremely complex in composition and texture, thus it may be difficult to obtain identical samples and precisely repeatable starting states. Such problems may be circumvented by the use of analogue materials. The disadvantage of magma analogues is that they are not magma—it is possible to match some properties of magma, however there are always other properties that are not matched, yet might alter the results or limit the exploitation of the experimental results. Therefore, the choice of sample material depends on the research question to be answered. The study of magmatic fragmentation provides a good example of the way dynamic experiments using both analogue and natural materials can interface with hypotheses and numerical models to lead to a better understanding of highly complex processes inaccessible to direct observation.

Early experimental approaches to the fragmentation of magmas inspired a series of laboratory apparatus, each of which was designed for a somewhat different purpose and for the optimization of different observations. The whole field had several origins whose parallel development eventually enriched the discussion of fragmentation substantially.

Early experimental approaches

Some of the earliest estimates of the tensile strength of silicate melts come from the experimental studies of the onset of non-Newtonian rheology in geological melts by Webb and Dingwell (1990a,b). In those studies dilatometric experiments using the fiber elongation technique were successfully manipulated with stepwise increasing driving loads such that the melts were driven into shear thinning rheology. That experimental design led to runaway melt deformation which was ultimately only arrested by either 1) the total runout of the dilatometer or 2) fiber breakage. Due to the meticulous control on load, melt initial dimensions and accumulated strain, the geometry and load on the fiber could be combined to estimate the failure stress for the melt. This led to values in the range of ca. 100 MPa which were at the time the first direct quantitative observation of melt strength at magmatic temperatures.

A next step was accomplished by Romano et al. (1996) who performed a wide-ranging set of experiments on fluid inclusions trapped in vesicular volatile-saturated silicic melts. These strikingly novel experiments took the form of decrepitation temperature determinations on two-phase approximately pure water-bearing fluid inclusions that were previously employed in several methodologically novel studies surrounding the determination of glass transition temperatures in these quenched systems (Romano et al. 1994, 1995). Knowledge of the P–T path of the two phase (liquid-gas) coexistence line of water, together with the temperature data for decrepitation, allowed the calculation of the pressure of decrepitation for these vesicular systems. The results varied from the previously documented range of 100 MPa that had been observed in the fiber failure data of Webb and Dingwell (1990a,b) for the case of sparingly soluble fluids such as CO_2 and Xe; to the much lower strengths of ca. 1 MPa for the case of water-saturated vesicular silicic glasses (Romano et al. 1996). Subsequent TEM exploration of the weak, water-saturated systems revealed the presence of initially radial cracks leading from the vesicle surface into the glass that systematically rotated into concentric fractures within a few microns of depth in the glass (Romano et al. 1996). The real strengths of the water-saturated glasses were subsequently interpreted in terms of the presence of cracking in water-rich glasses undergoing dehydration into vesicles during the quench process (Romano et al. 1996).

To explore magma–water interaction an experimental procedure was designed for nuclear safety analysis; whereby water was injected into a high temperature molten silicate and generated a fragmentation of liquid at the low viscosity end of the spectrum of volcanic magmas by a process termed molten fuel coolant interaction (Wohletz and McQueen 1984; Zimanowski et al. 1991). Typical liquids investigated were carbonates and melilites and the experiments clearly demonstrated that the expansive force of superheated water was capable of massive and widespread fragmentation and ejection of magmatic liquids (Zimanowski et al. 1986, 1997). These experiments were the first to demonstrate the feasibility of high temperature melt-water interactions resulting in explosive phreatomagmatic volcanism.

Figure 3A illustrates early experiments designed by Kieffer and Sturtevant (1984) to investigate volcanic jets by erupting pure gases and later accelerating two phase flow and fragmentation (Mader et al. 1994). A slight modification of this design is used to study the dynamics of accelerating two phase flows in H_2O–CO_2 polymer systems (Zhang et al. 1997).

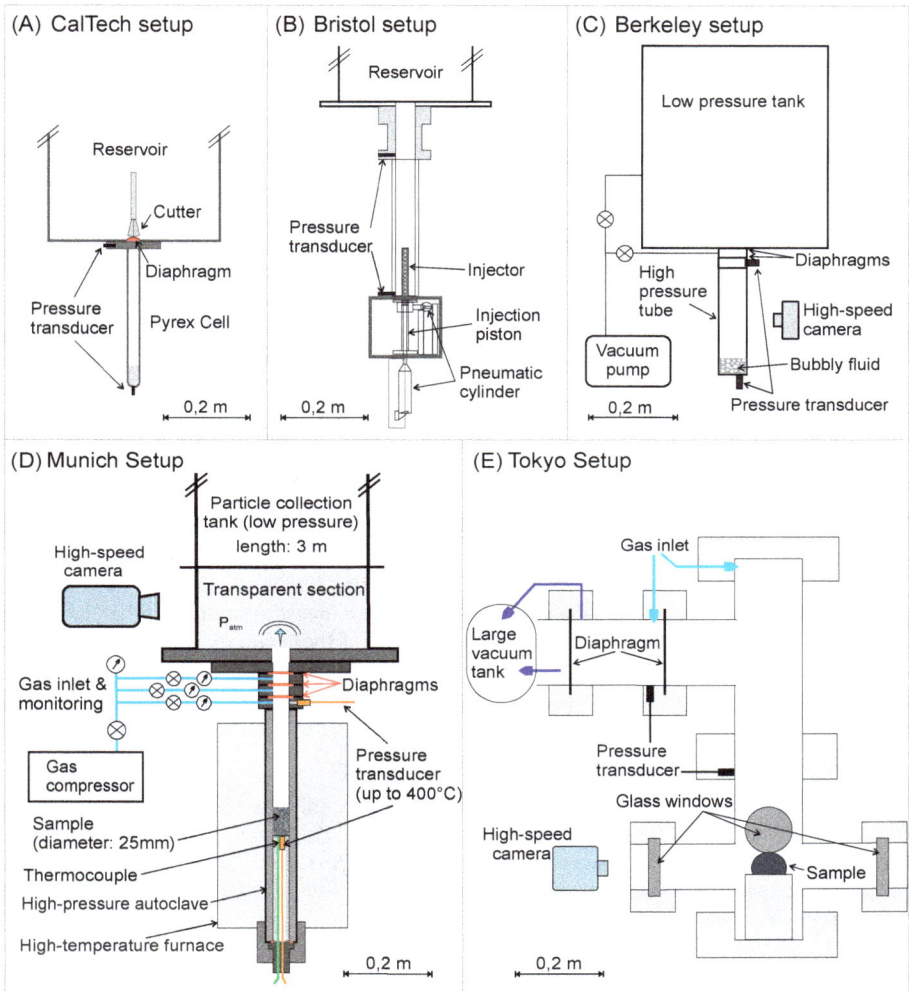

Figure 3. Overview of experimental devices to study magma fragmentation. Devices A–C all operate against an evacuated low pressure side, device D decompresses to ambient pressure, and device E can be operated in both ways. Device D is the only device operating with rocks and viscous magmas under in-situ conditions. Setup (**A**) is used to study accelerating two phase flow and fragmentation, a diaphragm separates the pressurized Pyrex cell from a reservoir, as sample serve for instance gases with a high molecular weight or water saturated with CO_2 (Caltech, modified after Kieffer and Sturtevant 1984, Mader et al. 1994); setup (**B**) represents a modification of (A) in the sense that here rapid acceleration of a two-phase flow is triggered by the vigorous chemical mixing of K_2CO_3 and HCl (Bristol, modified after Mader et al. 1994); the Berkeley setup (**C**) is also similar to (A), but uses a multiple diaphragm system, the dimension of the transparent high pressure section is enlarged, as samples serve xanthan gum water mixtures where bubbles were added prior to decompression by stirring (modified after Namiki and Manga 2005); the Munich setup (**D**) was the first device subjecting natural material as volcanic rocks and viscous magma to rapid decompression as it can be operated up to 900°C, later also silicon oils were used as magma analogue (at ambient temperature); this setup introduced the a system of up to three diaphragms to precisely adjust pressure and triggering, the device was persistently adapted to varying research question and sample sizes (modified after Alidibirov and Dingwell 1996a; Kueppers et al. 2006); setup (**E**) is also equipped with a system of two diaphragms, it operates with maltose–water mixtures; the peculiarity of this device lies in the sample not being confined in the high pressure section (Tokyo, modified after Kameda et al. 2008).

Mader et al. (1994) further modified these experiments by using the vigorous chemical mixing of K_2CO_3 solution and HCl whereby CO_2 gas is generated and an expanding, accelerating two phase flow results (Fig. 3B). These two designs both sought to understand how the two-phase flow led up to and precipitated fragmentation (by using a boiling system). These experiments were accompanied by the application of high speed photography for the optical quantification of kinematic data. Although each of the above experiments unquestionably possessed its own advantages for the understanding of the processes leading to fragmentation, they nevertheless all shared a common disadvantage in the study of the products of fragmentation. The preservation of the state of the fragmented materials at the point of fragmentation, which has become a vast source of information on the process, was impossible in these experiments. This type of setup was revived in mid-2000s (Fig. 3C) to study the decompression of bubbly liquids (Namiki and Manga 2005, 2008).

It was into this field of experimentation that the idea of decompressive fragmentation of viscous magmas was launched in the mid-1990s. Alidibirov and Dingwell (1996a,b) designed and constructed an experimental apparatus (Fig. 3D) to test whether simple decompression of a highly viscous vesicular magma could result in magmatic fragmentation (without external water). Further, the experimentally generated pyroclasts could be compared with natural ash and lapilli. The first successful decompression fragmentation results, published in 1996, demonstrated unequivocally that it was decompression alone which provided the key to explaining the magmatic fragmentation in highly viscous magmas, which generate the largest and most common explosive eruptions in nature. The nature of this device, together with the state and properties of the dome lavas that were investigated, meant that the unaltered kinematic data contained in the pyroclastic products of these fragmentation events could be fully quantified and used to constrain and infer the conditions of fragmentation. This work was central to what has become a quarter century of experimental investigations into the defining process of explosive volcanism.

Fragmentation experiments using magma analogues

One advantage of analogue samples is that they enable the investigation of the role of viscosity of magmatic systems at ambient temperature. Dynamic processes can more easily be observed by high-speed cameras. Fragmentation experiments, sometimes combined with degassing experiments, have been performed on analogue materials such as K_2CO_3–HCl or gum–rosin–acetone mixtures, H_2O–CO_2 polymers and corn syrup (Mader et al. 1994; Zhang et al. 1997; Mourtada-Bonnefoi and Mader 2004; Namiki and Manga 2005). All these experiments use shock-tube devices where the samples are decompressed against a vacuum chamber; decompression is initiated by rupturing of a diaphragm by various means (Fig. 3A–C). More recent experimental devices operate with pressures of up to 10 MPa on the sample side and decompress via a system of several diaphragms to ambient conditions or vacuum (Kameda et al. 2008; Spina et al. 2016). In most devices the sample is confined by sample container or the test cell itself allowing expansion in only one direction; the setup of Kameda et al. (2008) forms an exception, as the sample is unconfined (Fig. 3E). Magma analogues such as xanthan gum–water mixtures (Namiki and Manga 2005, 2008), maltose–water mixtures (Kameda et al. 2008, 2013) and silicone oils (Spina et al. 2016, 2019) enable varying of the viscosity range of analogue material over several log units.

For several studies the starting material is a bubbly liquid where bubbles are introduced into the sample prior to the experiment by a chemical reaction (e.g., Kameda et al. 2008) or mechanical stirring (e.g., Namiki and Manga 2005). In other studies the formation of a bubbly liquid is part of the experiment; either 1) bubbles are the result of the mixing the analogue components leading to explosive boiling independent of decompression (e.g., Mader et al. 1994) or 2) bubbles nucleate and grow as a result of increasing supersaturation of a fluid phase in the sample due to decompression (e.g., Zhang et al. 1997; Ichihara et al. 2002; Spina et al. 2016, 2019).

Fragmentation experiments using silicate melts and volcanic rocks

At noted above, experiments on natural magma using a shock-tube apparatus were carried out first by Alidibirov and Dingwell (1996a,b). In the meantime, a variety of setups have been designed for this apparatus (Fig. 4), each dedicated to explore different aspects of magma fragmentation and factors influencing it (Martel et al. 2000; Spieler et al. 2004; Mueller et al. 2005, 2008; Kueppers et al. 2006; Scheu et al. 2006b; Alatorre-Ibargüengoitia et al. 2010; Kremers et al. 2010; Cimarelli et al. 2014; Mayer et al. 2015; Montanaro et al. 2016b; Paredes-Mariño et al. 2019). The apparatus consists of two main parts; a large low-pressure section (tank; 3 m long, 0.4 m wide) at ambient pressure and temperature to collect the fragmented and ejected clasts, and a cold seal pressure vessel (autoclave) containing the sample. The autoclave can be pressurized to up to 40 MPa using Argon gas, and heated by an external split-tube furnace to 850–900 °C, covering the eruptive conditions typical for silica-rich magma (Scheu et al. 2008; Castro et al. 2009). Samples are heated at a rate of ~ 15 °C/min, followed by a dwell time (depending on sample size and final temperature) to ensure thermal equilibration.

Cylindrical samples with variable sizes from 17–60 mm diameter can be used, depending on setup and sample characteristics. Large pores, crystals or sample heterogeneities require large sample sizes; a ratio of 1:8 between the characteristic dimension of pores or crystals and the sample diameter is aimed for to assess the response of the entire sample to decompression and obtain reproducible results (Kremers et al. 2010; Montanaro et al. 2016a,b).

A system of two to three diaphragms separates the tank from the autoclave (Fig. 3 D; Fig. 4). Metal plates (Cu, Fe, Al) of different thicknesses, imprinted with a circle and a cross to tune their burst pressure and opening style serve as diaphragms. In this way, burst pressures of 0.5–25 MPa for a single diaphragm can be obtained. The use of several diaphragms allows for a gradual pressure built-up between tank and autoclave with the advantage of increased reproducibility of experimental conditions, long and safe standing times prior to a precise triggering of the decompression. Once triggered, a shock wave is generated and travels into the low pressure tank, while a rarefaction wave propagates down into the autoclave towards the sample.

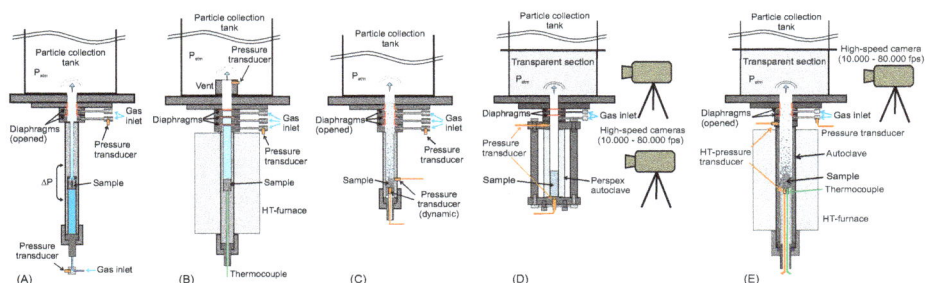

Figure 4. Details and modifications of the Munich fragmentation set-up to serve different research questions. Dedicated autoclaves of stainless steel, acrylic glass or NIMONIC, transparent inlets for visualization, and variable assembly of sensors and ultrahigh-speed cameras are used in these setups to explore: (**A**) gas permeability, (**B**) fragmentation threshold, grain size distribution and volcanic lightning at magmatic temperature, (**C**) fragmentation speed and threshold, grain size distribution and volcanic lightning at ambient conditions, (**D**) visualization of the fragmentation process and ejection behavior of rocks and magma analogues at different decompression rates, (**E**) steam-driven eruptions under elevated temperatures covering a vast variety of hydrothermal conditions. ***Note***: the transparent section just below the particle collections tank can be added to all set-up modification and allows the analysis of ejection behavior. Similarly a vent depicted in (B) only can be added to all set-up modifications. (Modified and expanded from: Mueller et al. 2008; Cashman and Scheu 2015; Montanaro et al. 2016a; Spina et al. 2016.)

For all experiments, the pressure evolution in the autoclave is monitored via a static piezoelectric pressure transducer situated at gas inlet below the undermost diaphragm. Importantly, up to 400 °C, additional dynamic piezoelectric pressure transducers can be added to the autoclave to quantify the speed of the fragmentation.

THEORETICAL MODELS AND CRITERIA FOR MAGMA FRAGMENTATION

During the last decades, magmatic fragmentation and its mechanisms have been widely discussed in theoretical models describing conduit and eruption processes and proposing various criteria for fragmentation to occur (e.g., Sparks 1978; Alidibirov 1994; Papale 1999; Zhang 1999; Alidibirov and Dingwell 2000; Koyaguchi et al. 2008; Fowler et al. 2010; Gonnermann 2015). The criteria have been based on investigations of deposits of explosive eruptions, theoretical models and laboratory experiments. Early models have incorporated magma disruption by bubble coalescence as the main mechanism (Verhoogen 1951). However, McBirney and Murase (1970) and Sparks (1978) demonstrated that coalescence is unlikely to be the leading fragmentation mechanism for highly viscous magmas, although possibly applicable for eruptions of low viscosity magmas. Two main groups of processes leading to fragmentation can be distinguished:

1. **Fragmentation due to rapidly accelerating two phase flow leading to fluid dynamic-induced break-up.** Vesiculation and bubble growth form the driving mechanisms behind expansion and acceleration of the magma during ascent. High strain rates within the magma ultimately result in magma break-up due to fluid dynamic instabilities (Villermaux 2012; Jones et al. 2019). Low viscosity basaltic magma is much more likely to fragment by fluid dynamic-induced break-up compared to its highly viscous silicic counterpart, as the relaxation time scale in low-viscosity magma is very short and the magma can remain in a relaxed state at even high strain rates. In this system the conditions for either viscous or capillary fluid dynamic instabilities are met earlier than elastic strain can accumulate in the magma—a prerequisite for a brittle component of fragmentation. (Dingwell and Webb 1989; Papale 1999; Cashman and Scheu 2015; Gonnermann 2015; Jones et al. 2019). Magma fragmentation by fluid dynamic-induced break-up is a pervasive process, sometimes also termed ductile fragmentation. Typical manifestations of magma fragmenting by this process encompass Hawaiian fire fountains, bursting of bubbles at the surface of lava lakes as well as Strombolian slug bursts (Porritt et al. 2012; Cashman and Scheu 2015; Gonnermann 2015). Fragmentation by brittle fracturing due to high strain rates is likely to be common in Plinian eruptions of silica-rich higher viscosity magma (Dingwell and Webb 1989; Papale 1999; Cashman and Scheu 2015). However also the relatively rare basaltic Plinian eruptions are thought to encompass a brittle component as a consequence of viscosity increases due to efficient crystallization of the basaltic magma (Sable et al. 2006; Moitra et al. 2018; Arzilli et al. 2019).

2. **Fragmentation of vesicular magma due to rapid decompression.** In most cases silicic, highly viscous magma will disrupt dominated by localized brittle processes. Highly viscous magma will be pushed out of its relaxed state much earlier, i.e., by processes with significantly lower deformation rates, than its low viscous counterpart, causing elastic strain to accumulate in the magma until failure conditions are met.

One common endmember-case of this type of magma fragmentation is the fragmentation of vesicular magma due to rapid decompression events, such as dome collapse or the removal of a plug in a Vulcanian-style eruption. As this type of magma fragmentation is a rather rapid physical process, subjecting vesicular magma to high deformation rates, the fracture process is likely to occur by brittle failure.

This type of magma fragmentation has to date received the most comprehensive investigation. It is envisaged for transient eruptions as well as for sustained eruptions, where silicic vesicular magma rapidly ascends into conduit regions exposing it to steep pressure gradients (Melnik 2000; Cashman and Scheu 2015; Gonnermann 2015). However, to cover the full breadth of magma fragmentation styles including the dynamic stress evolution in the magma prior to fragmentation, brittle processes might need to be blended to a certain degree with ductile components accounting for instance for rapid ascent and ongoing vesiculation of highly vesicular magma in Plinian eruptions (Papale 1999, 2001; Koyaguchi et al. 2008). First steps towards this combination have been made using magma analogues (e.g., Kameda et al. 2013; Spina et al. 2019)

Several physical models which have been proposed for brittle magma fragmentation account for the stress distribution caused by pressurized gas within the vesicles of a magma body. McBirney and Murase (1970) used solely elasticity theory and Griffith theory for fracture propagation in an elastic medium. Later models are based on the stress distribution in thin-walled isolated spheres (Alidibirov 1994) or thick walled isolated spheres (Zhang 1999). Based on shock-tube experiments of volcanic rocks under magmatic conditions, Spieler et al. (2004) proposed an empirical fragmentation criterion linking porosity, overpressure and the bulk strength of the magma. Koyaguchi et al. (2008) developed a refined fragmentation model based on the model of thick-walled spheres and Griffith theory. This model is the first theoretical model that satisfactorily describes the fragmentation behavior observed in shock tube experiments (Koyaguchi et al. 2008; Fowler et al. 2010). It should be noted that all the theoretical models mentioned so far are so-called cell models, which consider the stress evolution in a magma in relation to the gas pressure in an isolated cell. These models have likely precipitated the common assumption that only isolated and closed vesicles contribute to pressurization of a magma and are thereby relevant for magma fragmentation. While this might be true for some foam-like structures, it is in sharp contrast to that fraction of explosive products, bearing a high degree of vesicle connectivity, exhibiting a large range of porosities and which can reasonably be inferred to have preserved the physical state at fragmentation (e.g., Eichelberger et al. 1986; Klug and Cashman 1996; Mueller et al. 2005, 2008; Farquharson et al. 2016; Colombier et al. 2017).

Several fracturing processes have been proposed to act during magma fragmentation, including fracturing by elastic unloading and violent gas filtration flow (Alidibirov and Dingwell 2000); however the process of layer-by-layer fragmentation is now widely accepted as being the predominant fracturing process caused by rapid decompression (Alidibirov and Dingwell 2000; Ichihara et al. 2002; Scheu et al. 2008; Fowler et al. 2010; McGuinness et al. 2012; Gonnermann 2015; Fowler and Scheu 2016). Its importance was further strengthened by its visual observation in high-speed recordings of the fragmentation of volcanic rocks (Fig. 5; Fowler et al. 2010; McGuinness et al. 2012; Cashman and Scheu 2015). The latter form the basis of another mathematical model, taking a different approach on fragmentation than the

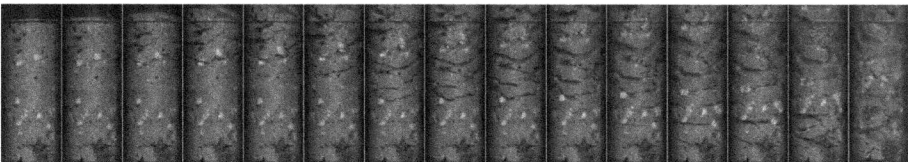

Figure 5. Sequence of a fragmenting dacite sample with 48 % porosity at an applied pressure of 6 MPa, resulting in an energy density of 2.9 MJ/m^3. The video was recorded at 10^4 fps, the time interval between first and last image shown here is about 2×10^{-3} s. The sequence shows the gradual propagation of the fragmentation through the sample; it clearly shows the repetition of the fragmentation process behind the downward migration fragmentation front. These events are termed 'intra-block' fragmentation and are a sign of a medium to high energetic fragmentation process causing several fragmentation generations.

cell model. Fowler et al. (2010) describe the fragmentation process by coupling two processes of different timescales into a force balance: (1) the fast unloading of the solid treated as an elastic body and (2) the slow depressurization of the connected pore space due to permeable outgassing. Also in this model the fracture criterion depends on the local effective tensile stress in the rock overcoming its tensile strength. As the gas depressurization front propagates slowly, the effective tensile stress meets failure conditions at a finite depth in the sample, and fracturing occurs. As the fragments are ejected the gas pressure in the crack rapidly decreases which can lead to continued fragmentation of the ejected fragments following the same force balance and failure criterion as for the first fragmentation event due to continued slow depressurization of the connected pore space in the fragments. The fragmentation events caused by the same mechanism are grouped into fracture or fragmentation families, each time the failure criterion is met again inside an already fragmented particle a new family of fractures ensues (Figs. 5, 9); implications of this will be discuss further in the section on energetic considerations. The number of fracture generations (families) depends on the applied pressure and the sample characteristics such as vesicle size. The experimental generation of fractures (in time and space) is very well reproduced by the numerical solution of this model (McGuinness et al. 2012).

The Fowler et al. (2010) model of magma fragmentation was the first theoretical model explicitly incorporating permeability, providing the possibility to assess the timescales of permeable gas loss against decompression.

MAGMA FRAGMENTATION BEHAVIOR

Fragmentation threshold

The fragmentation threshold as defined by Spieler et al. (2004) describes the minimum applied overpressure needed to fully fragment a sample. It has been explored by rapid decompression experiments of porous volcanic rocks of various chemical compositions and states at 850 °C as well as at ambient temperature and at confining pressures from 0.6–40 MPa (Fig. 6; Alidibirov and Dingwell 1996b; Martel et al. 2000, 2001; Spieler et al. 2004; Kueppers et al. 2006; Mueller et al. 2008; Scheu et al. 2008). In these experiments the samples fragment if the stress exerted due to the decompression front propagating into the sample (see section above for detailed description) exceeds the bulk tensile strength of the rock sample σ_m. Spieler el al. (2004) formulated an empirical relation for the fragmentation threshold ΔP_{fr} and the porosity Φ of the samples to be

$$\Delta P_{fr} = \frac{\sigma_m}{\Phi}$$

where $\sigma_m \approx 1$ MPa and Φ is given as volume fraction.

The cell models discussed above all predict a dependence of ΔP_{fr} on the porosity, however these pre-2004 models all show weaknesses in reproducing the experimental results of Spieler et al. (2004). Only the theoretical model by Koyaguchi et al. (2008), based on experimental data, describes the fragmentation threshold well over a wide range of porosity; therein the fragmentation threshold is given by

$$\Delta P_{fr} = \frac{2\sigma_m \cdot (1-\Phi)}{3\Phi \cdot \sqrt{\Phi^{-\frac{1}{3}}-1}}$$

where σ_m was fitted to be 2.18 MPa.

Experiments on a broad variety of samples with different origin and composition showed that the fragmentation threshold is largely independent of sample composition and state

(Spieler et al. 2004); it is a physical material property governed by the bulk tensile strength of the magma or rock. Material properties such as vesicle sizes and distribution as well as the size, type and distribution of crystals might exert a second order effect on the fragmentation threshold and the fragmentation behavior (Jones et al. 2016).

In these experiments only a minor influence of temperature on the fragmentation behavior could be observed (Fig. 6). Threshold values obtained at ambient temperatures seem to be slightly higher compared to values at 850 °C, however the scatter of both data sets largely overlap. One explanation for this apparent heightening of the threshold could be in a possibly reduced effective tensile strength at magmatic temperatures. This may be mirrored in the lower strength value in the relation by Spieler et al. (2004) describing high-temperature experiments whereas Koyaguchi et al. (2008) used ambient temperature experiments and obtained a slightly higher value for the best fit of the tensile strength. The largely overlapping results at ambient and high temperature clearly demonstrate the overriding dominance of the brittle fragmentation process, even if the samples high-temperature experiments are above their conventional (e.g., calorimetric) glass transition temperatures. In other words, for all rapid decompression experiments the decompression timescale τ_p is much smaller than the timescale for structural relaxation: $\tau_p \ll \tau_r$ (Dingwell 1996; Gonnermann 2015).

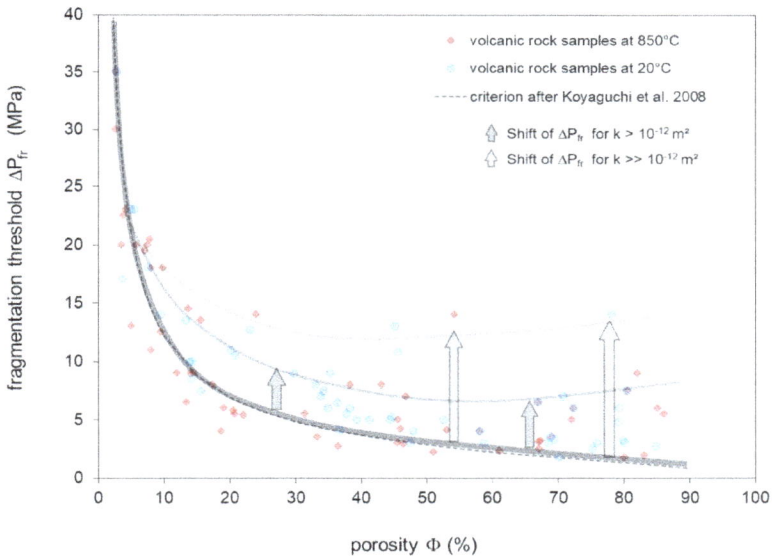

Figure 6. The fragmentation threshold (ΔP_{fr}) describes the applied pressure needed to fully fragment a sample with a given porosity. **Red diamonds** and **blue dots** depict experiments conducted at magmatic temperature (850 °C) and ambient conditions, respectively. The **black dashed line** shows the fragmentation criterion by Koyaguchi et al. (2008), **thick grey line** represents the fragmentation threshold curve based on this criterion. No significant distinction of ΔP_{fr} can be observed between high-temperature and ambient temperature experiments, however a slight tendency towards higher ΔP_{fr} at ambient conditions might be inferred. The samples tested comprise the full range of typical magma composition from basaltic to rhyolitic and revealed that the ΔP_{fr} is large insensitive to chemistry. It is striking that a significant amount of samples over the entire porosity-range and irrespective of experimental, show a deviation of ΔP_{fr} values from the ΔP_{fr} defined by the Koyaguchi criterion to higher values. Common to all these samples is a permeability $k > 10^{-12}$ m²; such permeable samples experience a non-negligible depressurization on the timescale of decompression resulting in a shift to higher ΔP_{fr} values. The higher the permeability, the more pronounced is this shift; the **thin grey lines** depict such ΔP_{fr} shifts. Note that the permeability-shift does affect all porosities, not only very high porosities where it might be most pronounced. (Modified from Mueller et al. 2008).

Influence of permeability on magma fragmentation

During magma ascent bubbles nucleate, grow and eventually start to interact, typically yielding a connected network. The transition from an isolated to a connected and thus permeable bubble network occurs at the so called percolation threshold. The porosity value associated to this threshold has been shown to vary considerably, depending on bubble size distribution, groundmass crystallinity, degree of shear deformation and further processes active during magma ascent and emplacement (Rust and Cashman 2011; Burgisser et al. 2017; Colombier et al. 2017; Kushnir et al. 2017; Lindoo et al. 2017; Giachetti et al. 2019).

The transition from a 1) physically isolated system to a 2) connected, permeable system of vesicles bears important implications for the eruptive behavior of a magma (Gonnermann and Manga 2003; Cashman and Scheu 2015; Colombier et al. 2017). Volatiles exsolved from the magma into the bubble network can be outgassed, which leads to a dissipation of overpressure and chemical shift of the magma due to degassing. In open-conduit scenarios the efficiency of this process is dominated by the magma permeability, whereas in closed-conduit scenarios outgassing is governed by the permeability of the conduit walls and the plug clogging the conduit, which usually exhibit low permeability. Below this dense plug the magma may possess higher permeability and porosity. In such a scenario the magma permeability will result in an even distribution of overpressure over the entire connected pore space. Alternatively, in the absence of pore connectivity a scenario envisaging foamed flow may also yield pressure homogenization via the viscous flow process itself. Hence permeable outgassing can lower the explosivity of volcanic eruptions, or even inhibit them by turning them into effusive eruptions (Gonnermann and Manga 2003; Degruyter et al. 2012; Cashman and Scheu 2015; Colombier et al. 2020). The impact of outgassing on a magmatic system however is again a matter of timescales and may vary for different processes; for the evolution of overpressure in an ascending magma outgassing needs to be balanced against volatile exsolution (degassing) which again depends on pressure. Also, magma fragmentation can be affected by outgassing. Here we discuss the effect on dominantly brittle fragmentation where permeable gas loss is evaluated on the timescales of rapid decompression events. Laboratory experiments revealed that below a critical permeability of 10^{-12} m^2 the onset of magma fragmentation remained unaffected by changes in permeability, whereas above this value the fragmentation threshold increased strongly with increasing permeability (Mueller et al. 2008; Fig. 6). For magma or volcanic rocks with $k < 10^{-12}$ m^2 the timescale of stress buildup in response to rapid decompression (decompression timescale) is much smaller than the timescale for overpressure dissipation by permeable outgassing i.e., $\tau_p \ll \tau_k$. Such magma or volcanic rocks can be treated as effectively impermeable on the timescale of rapid decompression, which enables the use of models based on isolated cells to model their fragmentation behavior (Koyaguchi et al. 2008). However, for highly permeable magma /volcanic rocks these models deviate significantly from the experimental observations, limiting their applicability. Here the theoretical model of fragmentation by Fowler et al. (2010) and McGuinness et al. (2010) shows its strength as it does account for permeability and covers high- as well as low-permeability magma adequately. Mueller et al. (2008) extended the empirical relationship by Spieler et al. (2004) for the fragmentation threshold ΔP_{fr} and the porosity Φ by adding a term for the permeability:

$$\Delta P_{fr} = \frac{\dfrac{a}{\sqrt{k}} + \sigma_m}{\Phi}$$

where a and σ_m are fitting constants with $a = 8.2 \times 10^5$ MPa/m, $\sigma_m = 1.5$ MPa and Φ is given as volume fraction. We wish to stress that although the effect of permeability on magma fragmentation is felt most heavily in pumiceous samples, it is present at all porosities.

Speed of magma fragmentation

The speed of fragmentation describes the propagation of the fragmentation front through a magmatic body. It needs to be carefully discriminated from the speed of crack propagation of individual fractures during fragmentation. The direction of the fragmentation process follows the pressure gradient and is perpendicular to the decompression front and fragmentation surface, whereas the cracks/fractures generally follow the tensile failure mode and thus are oriented normal to the gradient responsible for the geometric expression of the fragmentation surface.

Experimentally the fragmentation speed is obtained from the time delay of the pressure drop over the entire sample and the sample length (Fig. 4C; e.g., Scheu et al. 2006b; Richard et al. 2013). Alternatively the fragmentation speed can be obtained from high-speed video recordings tracing the layer-by-layer fragmentation through the sample. Good agreement is observed between the two methods. Typical values of the fragmentation speed range between less than 5 m/s to above 150 m/s (Kennedy et al. 2005; Scheu et al. 2008; McGuinness et al. 2012; Richard et al. 2013). The speed of fragmentation increases with the logarithm of the energy available for fragmentation (Fig. 7), the latter being provided by the pressurized gas in the (connected) pore space of the sample. In Figure 7A a velocity "0" indicates the initiation of fragmentation (FI). However, at FI the energy is still not sufficient to fully fragment the sample and thus stable fragmentation is not yet reached. At slightly higher energies the fragmentation threshold (ΔP_{fr}) is reached, where the sample is fully fragmented; here the first value of a fragmentation speed can be attributed. Experimental analyses revealed, that the values of FI range very close to those of ΔP_{fr}, indeed their ranges largely overlap given the natural heterogeneity of volcanic rocks and magma. Thus FI and ΔP_{fr} are commonly merged into one range and described as ΔP_{fr}, with the distinction becoming useful to grasp the meaning of a fragmentation speed of "0".

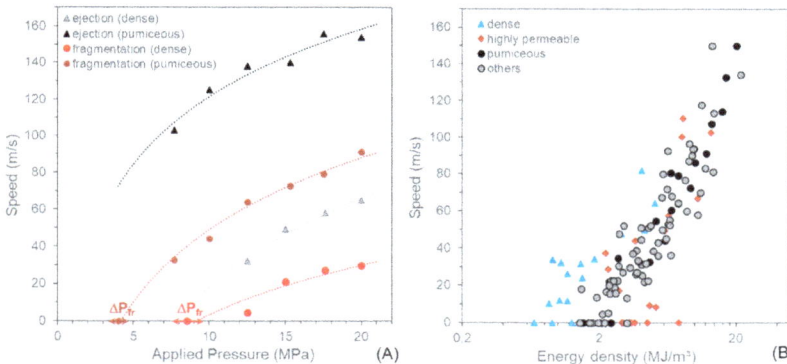

Figure 7 (A). Speed of fragmentation and speed of the ejected particles, exemplary for of dense (15–17% porosity) and pumiceous (> 60% porosity) samples (modified after Alatorre-Ibargüengoitia et al. 2011). The fragmentation speed of "0" corresponds strictly speaking to the initiation of fragmentation (FI); this value is slightly below the threshold ΔP_{fr}, however caused by sample heterogeneities FI and ΔP_{fr} overlap largely and for convenience only ΔP_{fr} is used (and depicted as **arrow**). With increasing applied pressure, fragmentation and ejection speed increase logarithmically, pointing out that the largest variation occurs at values around and slightly above ΔP_{fr}. **(B)** Compilation of fragmentation speed data in relation to the energy density. Dense (<15% porosity), pumiceous (> 60% porosity) as well as highly permeable samples ($k > 4 \times 10^{-12}$ m²) are indicated. In general the fragmentation speed data collapse on a common trend as a function of energy density. Dense samples deviate to lower energy density values, likely due to a significant contribution of elastic energy as result of the high applied pressures needed to fragment these samples (elastic energy is not considered in this plot). For highly permeable samples ΔP_{fr} is shifted to higher values as energy is "drained" out of the samples by permeable outgassing on the timescale of fragmentation, this effect can further be seen at low speed values corresponding to experiments slightly above ΔP_{fr}. (Data compiled from: Kennedy et al. 2005; Mueller et al. 2005, 2008; Scheu et al. 2006b, 2008; Alatorre-Ibargüengoitia et al. 2011; Richard et al. 2013, and including some unpublished data.)

Whereas Figure 7A is best suited to appreciate the role of applied pressure and available pore space on the fragmentation speed, these two variables are collapsed and normalized by the sample volume resulting in energy density values in Figure 7B. A qualitative value of the energy density ρ_E is provided by: $\rho_E = E/V \sim P\Phi$, with porosity given as fraction (detailed in section on energetic considerations). The comparison between energy density and fragmentation speed reveals several vital insights:

Firstly, all data collapse on a common trend that positively correlates fragmentation speed with increasing energy density, whereas the fragmentation threshold collapses to a value of roughly 2 MJ/m³. The speed of fragmentation is largely unaffected by the provenance and composition of fragmented material and mainly influenced by the available energy. The scatter in speed data is widest at very low energy values just above the threshold. Here the fragmentation process is highly susceptible to material heterogeneities, preexisting cracks or tension stored in the rock or magma.

Secondly, there are two groups of samples, dense and highly permeable ones, deviating from the common trend (Fig. 7B). Dense samples (blue diamonds) all have porosities < 15% and deviate from the trend towards lower energy density values with the largest deviations (i.e., lowest values of ρ_E) being caused by samples with only 3–5% porosity. Due to the very low connected porosity of <5%, high pressure values between 25–40 MPa are needed to fragment these samples. In many cases fragments were not ejected due to a lack of available gas inside the pore space and thus energy to propel them out of the autoclave. For such dense samples the fragmentation seems to be at least partially caused by the elastic unloading of the sample in response to decompression (Alidibirov and Dingwell 2000; Anderson 2005). However, elastic energy is not considered in Figure 7B, which accounts for the deviations to lower energy density values in that diagram. Indeed we speculate that the deviation to the mean energy density value might be a measure of the elastic energy contributing to fragmentation for these dense samples. Rapid decompression of dense, purely glassy samples resulted in fragmentation initiation at 50 MPa, which is in good agreement with the tensile strength of silicate melts and corroborates the notion of a mixture of energy sources presented here.

The effect of permeability on the fragmentation speed is rather complex (Richard et al. 2013). For highly permeable samples with $k > 4 \times 10^{-12}$ m² the initiation of fragmentation as well as the fragmentation threshold are clearly shifted to higher energies (Fig. 7B, red samples); this behavior was already described and depicted in Figure 6. For even higher energies, once stable fragmentation is established, two trends have been observed. In most cases the speed increases faster with increasing energies compared to samples not affected by outgassing. However for a few, mostly pumiceous, samples fragmentation speeds remains low. These opposing effects have been attributed to inertial effects in relation to the textural properties of the fragmenting rocks (Richard et al. 2013).

Finally, moderate (> 15 % porosity) to highly porous samples all follow the same trend governed by the energy of the pressurized gas in the connected pore space; elastic energy seems to have only second order effects. Here the fragmentation process is dominated by layer-by-layer fragmentation due to vesicle bursting, which is currently widely accepted to be a main mechanism causing magma fragmentation (Alidibirov and Dingwell 2000; Ichihara et al. 2002; Scheu et al. 2008; Fowler et al. 2010; McGuinness et al. 2012; Gonnermann 2015; Fowler and Scheu 2016).

We note that while in many experiments on magma fragmentation the fragmentation surface forms an uneven but on average straight plane (Fig. 5), the fragmentation surface in volcanic conduits might not. We therefore speculate that the shape of the fragmentation surface is governed by the magma properties, especially porosity and permeability, which usually are considered to vary in proximity to the conduit margins, for instance due to strain localization resulting in high shear and frictional heating or densification (Gonnermann and Manga 2003; Wright and Weinberg 2009; Lavallée et al. 2013; Farquharson et al. 2016; Schauroth et al. 2016). Consequently, the fragmentation surface might be curved in proximity to the conduit walls.

Timescales of magma ascent and fragmentation—Implications for eruption styles

From the above observations we conclude that the speed of propagation of the fragmentation front through magma in the conduit will be very much lower than the acoustic velocities that have been obtained for such materials (under unrelaxed conditions) (Scheu et al. 2006a). In fact, the fragmentation speeds are so low that they may compete with magmatic ascent rates. The ascent of magma has been the focus of a wide range of research studies and publications whereby values associated with explosive eruptions typically reach from 0.01 m/s to >1 m/s, with values being able to reach 10s of m/s especially with regard to the uppermost conduit, accelerated by vesiculation (Papale et al. 1998; Sable et al. 2006; Castro and Dingwell 2009; Cassidy et al. 2018). Thus, we might envisage a highly dynamic scenario that could incorporate a balance between ascent (supply) rate of magma and fragmentation (consumption) rate of the magma. In a simple view of the process one might divide the system up into three stages and associate typical eruption styles, based on the ratio δ of their fragmentation and ascent velocities (Fig. 8):

$$\delta = \frac{v_{\text{frag}}}{v_{\text{ascent}}}$$

1. δ ≫ 1: The fragmentation rate is higher that the magma ascent rate; once fragmentation is initiated the fragmentation surface deepens into the conduit (Fig. 8). The speed of fragmentation might wane and finally cease once encountering non-eruptible magma (e.g too low porosity to meet the fragmentation criterion). Changing magma properties such as porosity or permeability, incomplete ejection of pyroclasts and fall back of material from plume and crater rim, conduit wall collapse, as well as complex feedback mechanism between these factors are likely causes for the waning and ultimately the cessation of fragmentation (Clarke et al. 2002; Kennedy et al. 2005). The regime of δ ≫ 1 describes transient eruptions and typically corresponds to the general case of Vulcanian events. Transient fragmentation events as for instance the Vulcanian eruptions at Soufrière Hills volcano in Montserrat might last for several seconds to minutes only; within that time the fragmentation level can reach down to a few km in depth, 1–2 km in the case of the 1997 Montserrat eruptions (Formenti et al. 2003; Clarke et al. 2002). Vulcanian eruptions may exhibit a cyclicity in the time domain of minutes to days (Druitt et al. 2002; Odbert et al. 2014) with cyclic welding and fragmentation events being common on those timescales (Kolzenburg and Russell 2014; Kolzenburg et al. 2019).

2. δ ≪ 1: The average ascent rate of magma is higher than the average fragmentation rate (Fig. 8). Also in this regime no sustained fragmentation occurs. The regime of δ ≪ 1 might correspond to dome forming eruptions, interspersed with minor fragmentation events. In analogy to the above described fragmentation initiation, short-lived or even local fragmentation events might occur, triggered by local instabilities or decompressive events as landslides or (partial) collapse events—in this scenario the criteria for stable fragmentation is not (yet) met and a triggered fragmentation ceases due to insufficient energy. If subjected to a sufficiently large decompressive event—a major dome collapse or landslide for instance—this regime can tip into the regime of δ ≫ 1 and stable (not sustained) fragmentation ensues (Woods and Koyaguchi 1994; Scheu et al. 2006b; Ogburn et al. 2015; Preece et al. 2016).

3. $\delta \approx 1$: Magma ascent and fragmentation rates are approximately equal (Fig. 8). This regime might pertain to the case of sustained Plinian eruptions, where a stable fragmentation surface is a prerequisite for sustained explosive events (Woods and Koyaguchi 1994; Papale 1999, 2001; Koyaguchi et al. 2008). In this regime a self-stabilizing mechanism is inherent to fragmentation: if the fragmentation level deepens it reaches less vesiculated magma, fragmentation is slowed down and carried back upwards to equilibrium level. The opposite accounts if ascent rates increase and fragmentation is dragged upwards out of equilibrium. As a result fragmentation may oscillate around an equilibrium fragmentation surface; these oscillations might be considered one of several mechanisms causing pulsatory activity in a sustained Plinian eruption (Mason et al. 2006; Cioni et al. 2015).

The considerations above are a first experimental attempt to shed light on the complex interaction and control which the properties of silicate melts and magmas exert on the dynamics of explosive eruptions. In the shallow conduit, vesiculation lowers the volatile content of the melt causing an increase of melt viscosity that is highly non-linear once it drops below 1 wt% (Fig. 1; Hess and Dingwell 1996). Consequently magma viscosity will increase as well. The role of bubbles in particular on magma viscosity may be critical as a viscosity switch between deformable and non-deformable conditions. Once we further consider the viscosity gradient around a bubble resulting from the volatile diffusion into and consequently volatile depletion around a bubble (Mungall et al. 1996; Lensky et al. 2001), the effect on the *Ca* number and the propensity for increasing magma viscosity, the scenario appears increasingly fragile.

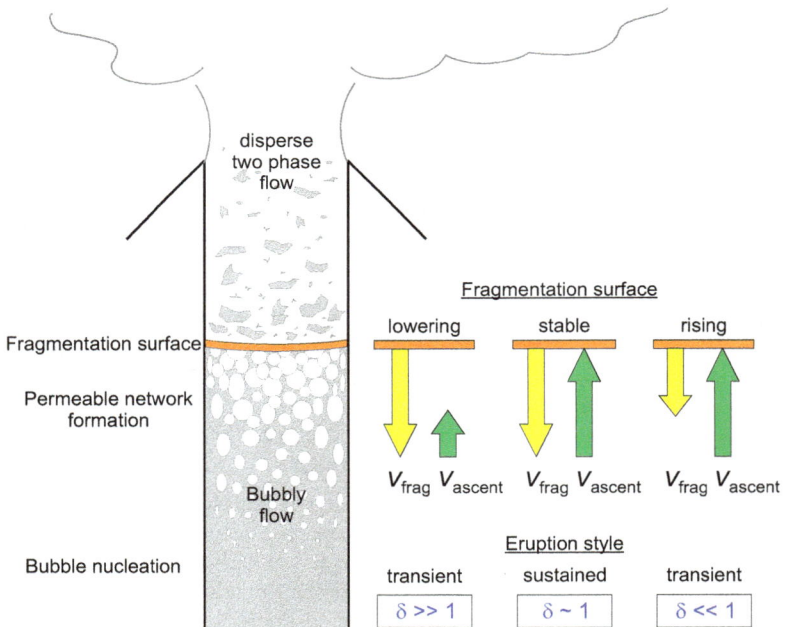

Figure 8. Conceptualized view on timescales of magma fragmentation and magma ascent in the upper conduit. Depending on the ratio, three regimes can be identified and associated to eruption styles. (1) If $\delta \gg 1$, the $\delta = v_{frag}/v_{ascent}$ fragmentation surface lowers into the conduit; a transient. Vulcanian-style eruption results with fragmentation waning and ceasing encountering conditions less prone to fragmentation with depth (2) If $\delta \ll 1$, the fragmentation surface migrates upwards resulting as well in a transient eruption, here eventually ceasing most likely as results of insufficient energy; likely eruption style might encompass dome forming eruptions with low energetic intermittent fragmentation events, (3) If $\delta \approx 1$, the fragmentation surface remains stable during eruption facilitating sustained events as Plinian eruptions.

PRODUCTS OF MAGMA FRAGMENTATION

Energetic considerations of magma fragmentation

During magma fragmentation the potential thermodynamic energy E stored within a volume of magma is used to fracture the magma and create new surfaces in the form of pyroclasts, the surplus energy is mainly converted into kinetic energy of the ejected pyroclasts; a small and thus often neglected part of the energy is converted to friction, heat and elastic energy (Alidibirov 1994; Alatorre-Ibargüengoitia et al. 2010; Arciniega-Ceballos et al. 2015). This view is geared towards predominantly brittle magma fragmentation; it conceptually corresponds to the energy balance suggested by Grady (1982) for a magma undergoing rapid expansion, relating the kinetic energy density of a magma prior to fragmentation to kinetic energy density and surface energy density after fragmentation.

The energy E which drives the fragmentation process is largely provided by the expansion of pressurized gas in the pore space of the samples, as this expansion is quasi instantaneous, we can treat it adiabatically

$$E = \frac{P_0 V_{\text{gas}}}{\gamma - 1} \left[1 - \left(\frac{P_{\text{atm}}}{P_0} \right)^{\frac{\gamma-1}{\gamma}} \right]$$

where γ is the adiabatic coefficient (e.g., $\gamma = 1.67$ for argon), P_{atm} the atmospheric pressure and P_0 the applied overpressure. A volume normalization allows to compare experimental results of for samples different sizes and is expressed as the energy density ρ_E

$$\rho_E = \frac{E}{V}$$

A qualitative value of ρ_E is obtained by the simple multiplication of applied pressure and porosity (as fraction), ignoring the compressibility of gas: $\rho_E \sim P\Phi$

A clear dependence of the fragmentation behavior on the energy available for fragmentation can be established. Above, we described the relationship between fragmentation threshold and speed, next we cover the ejection properties, the pyroclasts generated and the fragmentation mechanism.

The ejection velocities of clasts and gas correlate positively with the energy available for fragmentation; ejection velocities are generally higher than the fragmentation speed of a sample. Fig. 7A shows examples of this behavior for a dense and a pumiceous sample, with peak ejection velocities of 30–60 m/s and 100–160 m/s respectively (Alatorre-Ibargüengoitia et al. 2010, 2011). In other experiments maximum ejection velocities of the clast–gas mixture of up to ~250–300 m/s were recorded at early stages and observed to decay with time according to either power or exponential laws (Alatorre-Ibargüengoitia et al. 2010, 2011; Cigala et al. 2017).

It has been estimated that about 10–15% of the total energy available for fragmentation is consumed to create new surfaces during fragmentation; this energy fraction correlates inversely with porosity, which is intuitive as in highly vesicular magma bubble walls are thinner and solid fractions between bubbles are less prominent i.e., the surface was created prior to fragmentation (Alatorre-Ibargüengoitia et al. 2010; Montanaro et al. 2016a).

The process of the fragmentation is shaped by the total energy available. As briefly noted above, based on theoretical considerations and early concepts, the fragmentation process was perceived as a single and defining event that separates the bubbly flow of magma from the disperse two-phase flow of gas and pyroclasts. i.e., the fragmentation is the discontinuity through which coherent magma transitions to a disperse two-phase flow of gas and particles

(Melnik 2000; Papale 2001; Koyaguchi et al. 2008). Implicit to this concept was the understanding that only one fragmentation event is migrating through a sample or magma body in a layer-by-layer fashion (Alidibirov and Dingwell 2000; Cashman and Scheu 2015). Changes of the fragmentation process with increasing energy were quantified in terms of higher speeds and changes in the grain size distribution of generated pyroclasts (Spieler et al. 2003; Scheu et al. 2006b; Kueppers et al. 2006); yet still under the assumption of a single layer-by-layer fragment event with thinner layers reflecting the smaller grain sizes generated. While the general concept of fragmentation as a discontinuity holds, our view on the fragmentation process has been refined by insights from high-speed recordings as well as congruent theoretical models discussed previously (Fowler et al. 2010; McGuinness et al. 2012; Fowler and Scheu 2016). Our new understanding can be summarized as follows (Fig. 9): for low energies (just above ΔP_{fr}) the fragmentation process indeed consists of one layer-by-layer event migrating through a body, resulting in a grain size distribution heavily skewed towards and with a dominant peak at coarse grain sizes. For medium and high energies, the fragmentation process starts in exactly the same way with layers of the same thickness L_{frag}. The layer thickness L_{frag} results from the failure criterion, the material properties and depressurization profile (McGuinness et al. 2012). For medium to higher energies the first fragmentation occurs while the magma is still sufficiently over-pressurized for the failure criterion to be met again in the already fragmented clasts during continuing decompression. This process is described as intra-block fragmentation (Fowler et al. 2010). In that way several fragmentation generations (fragmentation families) will continue to fracture the sample or magma causing a succession of fragmentation. All of these fragmentation families are considered "primary" in distinction to the later described "secondary" fragmentation processes. As a result the grain size distribution becomes less skewed, the maximum peak shifts successively towards finer grain sizes and the height of the maximum peak decreases. Based on the failure criterion we suggest that there is a minimum thickness/characteristic length scale d for this fragmentation process, which is intrinsic to the magma.

Grain size distribution of fragmentation products

Initially, grain size distributions (GSDs) were fitted empirically using log-normal, Rosin–Rammler and Weibull distributions (Brown and Wohletz 1995; Spieler et al. 2003; Mackaman-Lofland et al. 2014). The sequential fragmentation theory (SFT; Brown 1989; Wohletz et al. 1989) and the application of fractal theory to size distributions of rock fragments, soils as well as natural and experimental pyroclasts distributions (Turcotte 1986; Perfect 1997; Perugini and Kueppers 2012) attempted to overcome this empiricism by providing a more physical basis for these distributions. Both theories rely on an at least partially scale-invariant and thus self-similar random fragmentation process. The foundation is laid already by Kolmogorov (1941), describing how the successive fragmentation of a rock sample can result in a lognormal distribution of its clast sizes.

One of the most striking concepts of GSDs associated with fragmentation families as shown in Fig. 9 is that they can be interpreted as the time evolution of a distribution. As described above, the GSDs in Fig. 9 decay and move towards finer grain sizes with increasing fragmentation energy and thus increasing fragmentation families. The dynamic time evolution of a GSD is, together with the choice of fragmentation kernels, a major distinction between the 'reductive fragmentation theory (RFT)' by Fowler and Scheu (2016) and the SFT. SFT provides a rather static view of a GSD, consisting of several individual GSDs that constitute the final GSD and remain constant over time / during fragmentation. Fowler and Scheu (2016) interpret the GSDs of experiments at different fragmentation energy to represent proxy time sequences of a single evolving GSD, with a final GSD once the minimum thickness d is reached for all fragmentation layers. Their RFT relies on stochastic process theory describing the evolving GSDs by the sum of two stochastic gamma distributions, one for the self-similar coarse fraction and a second describing the non-similar production of fines. The mechanistic

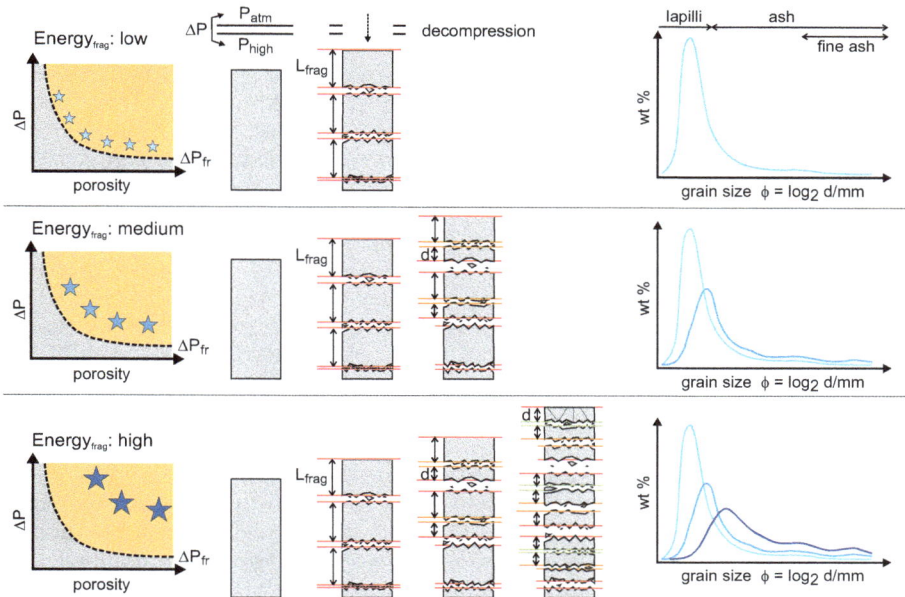

Figure 9. Schematic fragmentation process and resulting grain size distribution (GSD) in relation to energy available for fragmentation. **(Top)** For low energies that are just above the fragmentation threshold (ΔP_{fr}) only one family of fracture is generated during fragmentation, indicated by **red lines** marking the fracture surfaces. The created blocks show a uniform thickness of L_{frag}. The resulting GSD resembles a positively skewed log-normal distribution with a very strong peak (mode) roughly corresponding in size to L_{frag}. *Note*: Already at this stage the entire size range down to fine ash is present **(Middle)** If more energy is available for fragmentation, a second family for fractures (indicated by orange lines) is generated that continue to disintegrate already fragmented blocks, a process known as intra-block fragmentation. The resulting GSD is also a positively skewed log-normal distribution, however the peak is less pronounced and shifted towards finer grain size; the amount of fines and very fines is increased and possible secondary peaks (e.g., caused by crystal fractions) become visible. **(Bottom)** In the case of high fragmentation energy, the process of intra-block fragmentation continues, possibly generating further fracture families (indicated by **orange and green lines**) until the entire sample is disintegrated in rather evenly thin discs of thickness d. Note that the first generation of fractures and their spacing remain the same, irrespective of the energy available. If sufficient energy is available further fracture generations are initiated until the entire sample is fragmented evenly into layers of thickness d. The peak of the resulting log-normal GSD is again lowered, shifted towards finer grain sizes and less skewed. Whereas the wt% of its tail at the fine and very fines increases significantly and possible secondary peaks become more prominent. (*Note*: L_{frag} and d are characteristic to the sample and appears to depend on factors such as dominant vesicle size and scale-dependent permeability).

understanding behind this is that the fragmentation of a coarse clast results in both coarse as well as fine particles, whereas only coarse clasts (presumably $> 2d$) continue to fragment as the fines are too small to maintain sufficient gas pressure to re-approach the fragmentation criterion; this leads to a depletion of coarser clasts and an enrichment of fines during successive fragmentation events. This yields several important implications:

Firstly, the grain size of the coarse peak in an evolved GSD should relate to material properties of the magma and be roughly in the order of $< 2d$. This GSD might be seen as a kind of equilibrium GSD and thus a suitable candidate to evaluate the amount of secondary fragmentation (see later section) in distinction to it. The discrimination of the effect of primary and secondary fragmentation on the GSD of a natural volcanic deposits is not trivial if only two processes contribute and nearly impossible in case of even more (Jones and Russell 2017). Here experimentally generated GSDs might help in closing a gap as they do not reflect significant transport-related sorting, and thus represent pure primary GSDs.

Secondly, during each fracturing as part of a progressing fragmentation event coarse and fine clasts are generated. Despite the common assumption that increasing fragmentation energy should result in finer clasts, several experimental studies demonstrate that fine clasts in the range of ash (< 2 mm) and fine ash (< 63 μm) are commonly generated from the first fragmentation event onwards, regardless of the fragmentation energy (Spieler et al. 2003; Kueppers et al. 2006a; Scheu et al. 2008; Fowler and Scheu 2016). However with increasing energy (equivalent to increasing overpressure at a given porosity), the amount of fines does increase (i.e., the GSD shifts towards smaller grain sizes). Often the smallest grain sizes generated during fragmentation are attributed to bubble wall shards (Heiken and Wohletz 1986; Rust and Cashman 2011). We highlight that, while bubble wall shards certainly contribute to the fines in fragmentation products, there are further primary mechanisms forming them. In fact, fines (< 63 μm) are formed during every brittle fracturing event as a consequence of stress accumulation at the crack tip during fracture propagation (Anderson 2005; Fowler and Scheu 2016; Iravani et al. 2018).

Several measures have been proposed and are used to evaluate the efficiency of fragmentation. Kueppers et al. (2006) used the weight fraction of ash-sized clasts, the increase in surface area, or the shift of the mean diameter in products of fragmentation experiments, to describe the efficiency of fragmentation in relation to the energy available for fragmentation. However, this apparently simple task becomes complicated as soon as we allow for more energy sources, e.g., adding external water. Zimanowski et al. (2003) describe the fragmentation resulting from thermal contraction of the quenched magma as very efficient, as indicated by the high amount of fines generated (see below). However the energy liberated by the near-instantaneous vaporisation of the water can be up to several orders of magnitude higher than the energy by gas expansion and thus would results in a rather low efficiency if treated following the same approach as proposed by Kueppers et al. (2006) for brittle fragmentation by gas expansion.

The considerations above account for fragmentation dominated by brittle processes and are geared towards but not limited to highly viscous magmas. During eruptive activity several energy sources might by accessible, causing a modulation of the GSD. Liu et al. (2015) proposed that the fragmentation efficiency of the 2011 Grímsvötn eruption was increased by residual thermal stresses in glass quenched by glacial water as evidenced by a higher fraction of fines in the resultant deposits. Similar observations are reported by Moreland et al. (2019) for deposits of the 10th century Eldgjá eruption; here the GSDs of its explosive phases revealed a contribution of external water in subglacial phases, likely through the action of thermal granulation (this process is detailed below) on a magmatically fragmenting highly vesicular system.

SECONDARY FRAGMENTATION

Pyroclasts generated by magma fragmentation often experience further size-reducing processes as a result of thermal stresses, abrasion or mechanical interactions of clasts such as high- or low-energy collisions (Freundt and Schmincke 1992; Kaminski and Jaupart 1998). All these processes are subsumed under the term secondary fragmentation in order to discriminate them from the primary fragmentation which turns a coherent magma into a gas–particle flow. Secondary fragmentation can occur within the volcanic conduit, plume or in pyroclastic density currents (PDCs). While all secondary fragmentation processes lead by definition to the size reduction of particles, their effects on the shapes of clasts may differ. Abrasion and low-energy impacts lead to significant rounding of clasts commonly observed in the deposits of PDCs. These clast-rounding processes are associated with an efficient production of fine particles.

Several experimental studies have focused on secondary fragmentation processes predominantly occurring in pyroclastic density currents using impact and abrasion experiments (e.g., Dufek and Manga 2008; Manga et al. 2011; Kueppers et al. 2012; Jones et al. 2016; Hornby et al. 2020). A few experimental studies have explored intra conduit processes using pumice collision (Dufek et al. 2012), attrition (Jones and Russell 2017) or decompression experiments (Paredes-Mariño et al. 2019). Paredes-Mariño et al. (2019) quantified the effect of conduit geometry and clast properties such as size and porosity/density on secondary fragmentation processes. In that study, loose particles of 2–4 mm size were exposed to rapid decompression. During acceleration and movement in the conduit abrasion and particle collisions—either with other particles or resulting from a conduit constriction—efficiently reduced particle size and produced fine ash. However, few pumiceous particles were observed to have undergone primary fragmentation events, demonstrating that for intra-conduit processes, primary and secondary fragmentation events can overlap (Fig. 10). This finding is underpinned by the insights into the primary fragmentation mechanism (Fig. 9), revealing that several fragmentation generations can occur down to a characteristic particle size assuming sufficient energy available (Fowler et al. 2010; Fowler and Scheu 2016).

Due to its efficiency in ash and fine ash production, secondary fragmentation often results in bimodal grain size distributions (Kueppers et al. 2012; Jones and Russell 2017; Paredes-Mariño et al. 2019). The abundance of ash and especially fine ash plays an important role in various processes such as the dynamics of a volcanic plume (greater heights due to more efficient heat transfer) or the mobility of PDCs. Further, the additional fine ash produced by secondary fragmentation increases the hazards to health (Horwell 2007) and aviation (Prata and Tupper 2009; Song et al. 2019).

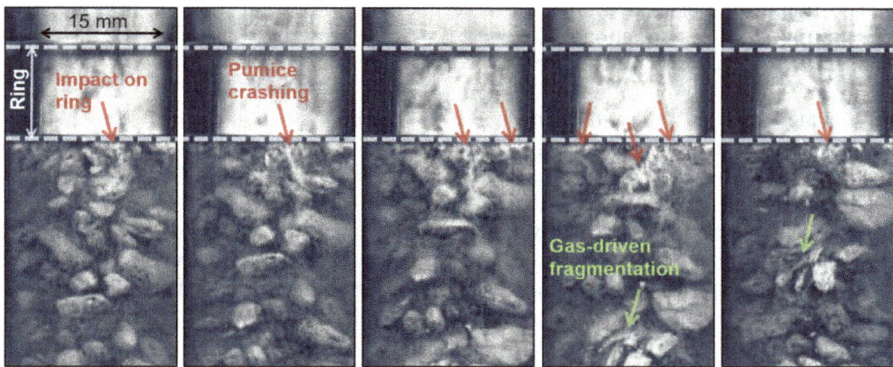

Figure 10. Image sequence of ultrahigh-speed video recording of pumiceous clasts being rapidly decompressed in a conduit with a constriction (**ring**) with the aim to induce secondary fragmentation. After the decompression front propagates into the autoclave, pumice clasts are lofted and accelerated by the expanding gas-phase upwards in the autoclave. Clasts crash against the conduit constriction (**red arrows**), at the same time gas-driven explosion of large pumiceous clasts (**green arrows**) can be observed (modified from Paredes-Mariño et al. 2019).

EXPLOSIVE AND NON-EXPLOSIVE MAGMA WATER INTERACTION

The majority of the interactions of magma with water are non-explosive, taking place at the mid-oceanic ridges worldwide. The basaltic magma is rapidly quenched by the surrounding water and forms pillow-lavas overlying columnar jointed lava flows. Due to the high quench rates, the rims of the pillows remains glassy and are often fractured as a result of thermal stresses. Between pillows glassy fragments are found, termed hyaloclastites, which have been formed by thermal stresses exceeding the strength of the basaltic magma (e.g., van Otterloo et al. 2015). A very similar process takes places in subglacial settings where magma is interacting with ice; the formation of tuyas and tindars in Iceland serves as good example (Smellie 2000, and references therein). Thermal granulation plays also an important role in

Surtseyan eruptions that are shallow to emergent subaqueous eruptions; granulation caused by thermal stresses is an efficient secondary disruption process to create ash and fine ash, which either remain in place, coating their parental clasts or are liberated into the plume (Mastin 2007; Colombier et al. 2019). Further magma–water interaction during Surtseyan eruptions affects the eruption dynamics itself, the vesiculation of magma and magma fragmentation, the uptake of salts by volcanic ash and, in some cases, the initiation of the eruption (Kokelaar 1986; Schipper and White 2016; Colombier et al. 2018, 2019).

Energetic, yet non-explosive magma–water interaction can also contribute to the formation of Pele's hair and tears, long thin glassy fibers and small elongated drops. An experimental study showed that they can form if a string of melt dropping into agitated water is efficiently sheared which induces an extensional strain along the thinning string, ultimately disrupting it (Mastin et al. 2009).

Hydromagmatic or phreatomagmatic eruptions are the manifestation of explosive interaction of magma with variable amounts of external water, either as surface water or ground water. These eruptions draw their energy from the rapid heat transfer through the magma–water interface, causing the water to flash to steam and the magma to quench. The result is an often vigorous eruption accompanied by ejection of large amounts of fine ash and dense blocky and platy grains and clasts (Colgate and Sigurgeirsson 1973; Self and Sparks 1978; Heiken and Wohletz 1985) that tend to differ drastically from the cuspate shapes of disrupted bubbly melt.

The fragmentation in hydromagmatic eruptions is considered to occur via a repeated cycle starting with the formation of a vapor film at the magma–water interface, next by the expansion of the vapor film, followed by condensation and collapse inducing instabilities that fragment the magma at the solid-liquid interface, and finally high heat transfer at the newly created surface. Which mechanisms drive the fragmentation process itself is to date not fully solved; several mechanisms are proposed as for example stress waves due to the collapsing vapor film, thermal stresses in the quenched magma, or hydrodynamic instabilities of the magma–water interface (Sheridan and Wohletz 1983). Experimental studies exploring various aspects of explosive magma water interaction such as the effect of premixing style, the effect of the magma–water ratio or the effect of increasing newly generated surface on explosivity commenced several decades ago and are ongoing (e.g., Wohletz and McQueen 1984; Zimanowski et al. 1991; Büttner et al. 2006; Sonder et al. 2018).

Explosive eruptions may include magmatic as well as hydromagmatic components, with proportions fluctuating over the course of an eruptive event (Self and Sparks 1978; Carey et al. 2010; Moreland et al. 2019).

NON-MAGMATIC FRAGMENTATION: STEAM-DRIVEN ERUPTIONS

Some explosive eruptions erupt exclusively non-juvenile material. These eruptions are driven by the thermal energy of hot fluids, which are driven by rapid expansion of liquids flashed to steam, gases or supercritical fluids steam (Mastin 1995; Browne and Lawless 2001). They are termed either phreatic or hydrothermal eruptions, or simply 'steam-driven eruptions' (Browne and Lawless 2001; Montanaro et al. 2016a,b; Stix and Moor 2018).

Steam-driven eruptions have been largely disregarded in the past as they are typically considered short-lived and rather small. Also deposits of steam-driven eruptions are sparse in the geological record. Yet they are often precursors to larger magmatic eruptions. Steam-driven eruptions itself however very often lack clear precursory signals, resulting in sudden eruptions with little or no warning (Montanaro et al. 2016a; Dempsey et al. 2020). Only in the last decade these complex and highly variable eruptions have gain increased attention by the scientific community (e.g., Thiéry and Mercury 2009; Stix and Moor 2018), largely driven by several violent and/or tragic events such as the 2012 Tongariro eruption in New Zealand (Breard et al. 2014), the 2014 Ontake eruption in Japan (Yamaoka et al. 2016, and refs therein) causing at least 58 fatalities amongst hikers, or the 2019 Whakaari (also known as White Island) off the coast of New Zealand's Northern Island killing or fatally injuring 21 of the 47 persons visiting the island as the eruption occurred (Dempsey et al. 2020).

The fragmentation mechanism operative in steam-driven eruptions exhibits many similarities to that observed in rapid decompression of vesicular magma (Rager et al. 2014; Mayer et al. 2015; Montanaro et al. 2016a,b; Rott et al. 2019; Gallagher et al. 2020). Steam-driven eruptions are very highly variable and their explosivity depends on the thermodynamic state of the fluid being rapidly decompressed. If already present in the gas phase, the fragmentation behavior in terms of fragmentation threshold, fragmentation and ejection speed as well as resulting grain size distribution is comparable to their magmatic counterparts (argon gas expansion). With increasing fraction of a fluid phase the eruption intensity increases as the fluid flashes to steam during decompression. Fragmentation speed and ejection velocities increase, ejection velocities can even double and the grain size distributions shift significantly towards finer grain sizes (Mayer et al. 2016; Montanaro et al. 2016b).

Hydrothermal alteration causes changes not only of the mineralogy but also of the petrophysical properties such as porosity, permeability and strength. Increases and decreases in these properties are possible and often occur coexist at spatial separation due to leaching, precipitation and recrystallization mechanisms. These dynamic changes add to the complexity of steam-driven eruptions as thy directly affect the propensity towards explosive failure as well as the fragmentation behavior once failed (Mayer et al. 2016; Gallagher et al. 2020) Analysis of hydrothermally altered rocks from Campi Flegrei revealed that secondary mineralization can exert a major control on the liberation of significant amount of ultra-fines in the "thoracic" ($< 10 \ \mu m$) or even the respirable ($< 4 \ \mu m$) fraction in eruptions of low energy (Mayer et al. 2016). In a similar way the precipitation of high-temperature silica polymorphs in the pore space of magma by outgassing prior to eruption can enhance the generation of ultra-fines, which pose a threat to human health (e.g., Horwell et al. 2003).

CONCLUDING REMARKS AND FUTURE PERSPECTIVES

Magma fragmentation is the distinctive mechanism dividing effusive from explosive components of eruptive behavior, and thus a defining process in volcanology worthy of intense scrutiny. It is intimately interwoven with multi-phase rheology and the properties of silicate melts; we highlighted several of the complex and often non-linear interactions between them and put them into perspective concerning eruption behavior. Despite our growing understanding of magma fragmentation, multi-phase rheology and the properties of silicate and other melts much remains to be discovered. Despite the theoretical and experimental studies of the last decade the complexity of their interactions are not yet fully revealed. There exists a substantial need for further innovative and novel approaches.

A very recent example is the effect that nanometer-sized crystals (nanolites) might exert on magma rheology and its implications for modulating the eruption style and triggering of explosive eruptions (Sharp et al. 1996; Cáceres et al. 2020; Di Genova et al. 2020). A further example is provided by the application of spinodal decomposition in the analysis of bubble nucleation processes in silicates melts and magmas with direct implications for magma fragmentation and eruption styles (e.g., Allabar and Nowak 2018; Sahagian and Carley 2020).

Potential future research directions include the following: (1) a thorough evaluation of the conditions leading to the formation of nanolites, and exploration of the role different types of nanolites pose for shallow conduit processes as well as the processes and products of magma fragmentation; (2) an in-depth evaluation of the dynamically evolving grain size distributions resulting from magma fragmentation, and their potential dependence on textural and material properties of the fragmenting magma. This likely enables the attribution of GSD fractions to primary and secondary processes; (3) a clarification of the concept of fragmentation efficiency, with the prospect of enabling quantitative tests of hypotheses of fragmentation scenarios; (4) broadening the dynamic range of rapid decompression experiments, enabling the experimental coupling of vesiculation and fragmentation at *in situ* conditions; (5) experiments combining the dynamics of two-phase bubbly flow with magma fragmentation, (6) the expansion of our knowledge on mixed fluids (gas–liquid, H_2O–CO_2, ..) as a source for fragmentation energy and quantification of their impact on the propensity for explosive eruptions.

In this chapter, we have discussed magma fragmentation predominantly in highly viscous, silica-rich melts and magmas as these host the most violent explosive eruptions. We have emphasized several fragmentation mechanisms and explored quantitative links between the fragmentation behavior, process and products in relation to the energy available for fragmentation. Finally, we have visited the interaction of juvenile (magma) and non-juvenile (rock) materials with external water and steam and explored the associated eruption characteristics. A common thread throughout is novel and highly unconventional experimental studies, flanked with theories and concepts 1) tested by or 2) developed based on, experimental volcanology.

ACKNOWLEDGEMENTS

We wish to thank our present and past colleagues as well as affiliates of LMU Volcanology group for numerous discussions on individual aspects on the topic discussed in this chapter. We are grateful to Paolo Papale and Stephan Kolzenburg for constructive review comments that helped to improve the original manuscript. We acknowledge funding from the German Research Foundation (DFG) through the projects number 628578 as well as project number 364653263—TRR 235. DBD acknowledges the support of ERC 2018 ADV Grant 834225 (EAVESDROP).

REFERENCES

Anderson TL (2005) Fracture mechanics: fundamentals and applications. 3rd edition. CRC press

Alatorre-Ibargüengoitia MA, Scheu B, Dingwell DB, Delgado-Granados H, Taddeucci J (2010) Energy consumption by magmatic fragmentation and pyroclast ejection during Vulcanian eruptions, Earth Planet Sci Lett 291:60–69

Alatorre-Ibargüengoitia MA, Scheu B, Dingwell DB (2011) Influence of the fragmentation process on the dynamics of Vulcanian eruptions: an experimental approach, Earth Planet Sci Lett 302:51–59

Alidibirov M (1994) A model for viscous magma fragmentation during volcanic blasts. Bull Volcanol 56:459–65

Alidibirov M, Dingwell DB (1996a) An experimental facility for the investigation of magma fragmentation by rapid decompression, Bull Volcanol 58:411–416

Alidibirov M, Dingwell DB (1996b) Magma fragmentation by rapid decompression. Nature 380:146–48

Alidibirov M, Dingwell DB (2000) Three fragmentation mechanisms for highly viscous magma under rapid decompression: J Volcanol Geotherm Res 100:413–421

Allabar A, Nowak M (2018) Message in a bottle: Spontaneous phase separation of hydrous Vesuvius melt even at low decompression rates. Earth Planet Sci Lett 501:192–201

Arciniega-Ceballos A, Alatorre-Ibargüengoitia M, Scheu B, Dingwell DB (2015) Analysis of source characteristics of experimental gas burst and fragmentation explosions generated by rapid decompression of volcanic rocks. J Geophys Res: Solid Earth 120:5104–5116

Arzilli F, La Spina G, Burton MR, Polacci M, Le Gall N, Hartley ME, Di Genova D, Cai B, Vo NT, Bamber EC, Nonni S (2019) Magma fragmentation in highly explosive basaltic eruptions induced by rapid crystallization. Nat Geosci 12:1023–1028

Bowen NL (1928) The Evolution of the Igneous Rocks: Princeton, New Jersey, Princeton University Press

Bowen NL (1934) Viscosity data for silicate melts transactions. American Geophysical Union 15:249–255

Breard ECP, Lube G, Cronin SJ, Fitzgerald R, Kennedy B, Scheu B, Montanaro C, White JDL, Tost M, Procter JN, Moebis A (2014) Using the spatial distribution and lithology of ballistic blocks to interpret eruption sequence and dynamics: August 6 2012 Upper Te Maari eruption, New Zealand. J Volcanol Geotherm Res 286:373–386

Brown WK (1989) A theory of sequential fragmentation and its astronomical applications. J Astrophys Astron 10:89–112

Browne PRL, Lawless JV (2001) Characteristics of hydrothermal eruptions, with examples from New Zealand and elsewhere. Earth Sci Rev 52:299–331

Brown WK, Wohletz KH (1995) Derivation of the Weibull distribution based on physical principles and its connection to the Rosin–Rammler and lognormal distributions. J Appl Phys 78:2758–2763

Burgisser A, Chevalier L, Gardner JE, Castro JM (2017) The percolation threshold and permeability evolution of ascending magmas. Earth Planet Sci Lett 470:37–47

Burnham CW (1964) Viscosity of a water-rich pegmatite melt at high pressures. Geol Soc Am Spec Pap 76:26

Büttner R, Dellino P, Raue H, Sonder I, Zimanowski B (2006) Stress-induced brittle fragmentation of magmatic melts: theory and experiments. J Geophys Res 111:B08204

Cáceres F, Wadsworth FB, Scheu B, Colombier M, Madonna C, Cimarelli C, Hess K-U, Kaliwoda M, Ruthensteiner B, Dingwell DB (2020) Can nanolites enhance eruption explosivity? Geology 48:997–1001

Carey RJ, Houghton BF, Thordarson T (2010) Tephra dispersal and eruption dynamics of wet and dry phases of the 1875 eruption of Askja Volcano, Iceland. Bull Volcanol 72:259–278

Caricchi L, Burlini L, Ulmer P, Gerya T, Vassalli M, Papale P (2007) Non-Newtonian rheology of crystal-bearing magmas and implications for magma ascent dynamics. Earth Planet Sci Lett 264:402–419

Cashman KV, Scheu B (2015) Magmatic fragmentation. *In:* Sigurdsson H, Houghton B, Rymer H, Stix J, McNutt S (Eds.), The Encyclopedia of Volcanoes, p 459–471

Cassidy M, Manga M, Cashman K, Bachmann O (2018) Controls on explosive-effusive volcanic eruption styles. Nat Commun 9:1–16

Castro JM, Dingwell DB (2009) Rapid ascent of rhyolitic magma at Chaitén volcano, Chile. Nature 461:780–783

Cigala V, Kueppers U, Peña Fernández JJ, Taddeucci J, Sesterhenn J, Dingwell DB (2017) The dynamics of volcanic jets: Temporal evolution of particles exit velocity from shock-tube experiments. J Geophys Res: Solid Earth 122:6031–6045

Cimarelli C, Costa A, Mueller S, Mader HM (2011) Rheology of magmas with bimodal crystal size and shape distributions: Insights from analog experiments. Geochem Geophys Geosyst 12:Q07024

Cimarelli C, Alatorre-Ibargüengoitia M, Kueppers U, Scheu B, Dingwell DB (2014) Experimental generation of volcanic lightning. Geology 12:79–82

Cioni R, Pistolesi M, Rosi M (2015) Plinian and subplinian eruptions. *In:* The Encyclopedia of Volcanoes. Academic Press, p 519–535

Clarke AB, Neri A, Voight B, Macedonio G, Druitt TH (2002) Computational modelling of the transient dynamics of the August 1997 Vulcanian explosions at Soufriere Hills Volcano, Montserrat: influence of initial conduit conditions on near-vent pyroclastic dispersal. Mem Geol Soc London 21, 319–348

Colgate SA, Sigurgeirsson T (1973) Dynamic mixing of water and lava. Nature 244:552–555

Colombier M, Wadsworth FB, Gurioli L, Scheu B, Kueppers U, di Muro A, Dingwell DB (2017) The evolution of pore connectivity in volcanic rocks. Earth Planet Sci Lett 462:99–109

Colombier M, Scheu B, Wadsworth FB, Cronin SJ, Vasseur J, Dobson KJ, Hess K-U, Tost M, Yilmaz T, Cimarelli C, Brenna M, Ruthensteiner B, Dingwell DB (2018) Vesiculation and quenching during Surtseyan eruptions at Hunga Tonga-Hunga Ha'apai volcano, Tonga. J Geophys Res: Solid Earth 123:3762–3779

Colombier M, Scheu B, Kueppers U, Cronin SJ, Mueller SB, Hess K-U, Wadsworth FB, Tost M, Dobson KJ, Ruthensteiner B, Dingwell DB (2019) In situ granulation by thermal stress during subaqueous volcanic eruptions. Geology 47:179–182

Colombier M, Wadsworth FB, Scheu B, Vasseur J, Dobson KJ, Cáceres F, Allabar A, Marone F, Schlepütz CM, Dingwell DB (2020) In situ observation of the percolation threshold in multiphase magma analogues. Bull Volcanol 82:1–15

Cordonnier B, Caricchi L, Pistone M, Castro J, Hess KU, Gottschaller S, Manga M, Dingwell DB, Burlini L (2012) The viscous-brittle transition of crystal-bearing silicic melt: direct observation of magma rupture and healing. Geology 40:611–614

Degruyter W, Bachmann O, Burgisser A, Manga M (2012) The effects of outgassing on the transition between effusive and explosive silicic eruptions. Earth Planet Sci Lett 349:161–170

Dempsey DE, Cronin SJ, Mei S, Kempa-Liehr AW (2020) Automatic precursor recognition and real-time forecasting of sudden explosive volcanic eruptions at Whakaari, New Zealand. Nat Commun 11:1–8

Di Genova D, Brooker RA, Mader HM, Drewitt JW, Longo A, Deubener J, Neuville DR, Fanara S, Shebanova O, Anzellini S, Arzilli F (2020) In situ observation of nanolite growth in volcanic melt: A driving force for explosive eruptions. Sci Adv 6:eabb0413

Dingwell DB (1995) Relaxation in silicate melts: some applications. Rev Mineral 32:21–66

Dingwell DB (1996) Volcanic dilemma: flow or blow? Science 273:1054–55

Dingwell DB (1997) The brittle-ductile transition in high-level granitic magmas: material constraints. J Petrol 38:1635–1644

Dingwell DB (1998) A physical description of magma relevant to explosive silicic volcanism. *In:* Physics of Explosive Volcanic Eruptions. Geol Soc London Spec Publ 145, p 9–26

Dingwell DB, Webb SL (1989) Structural relaxation in silicate melts and non-Newtonian melt rheology in geological processes. Phys Chem Mater 16:508–16

Dingwell DB, Webb SL (1990) Relaxation in silicate melts. Europ. J Mineral 2:427–449

Dingwell DB, Lavallée Y, Hess K-U, Flaws A, Marti J, Nichols ARL, Gilg HA, Schillinger B (2016) Eruptive shearing of tube pumice: pure and simple. Solid Earth 7:1383–1393

Druitt TH, Young SR, Baptie B, Bonadonna C, Calder ES, Clarke AB, Cole PD, Harford CL, Herd RA, Luckett R, Ryan G (2002) Episodes of cyclic Vulcanian explosive activity with fountain collapse at Soufrière Hills Volcano, Montserrat. Mem Geol Soc London 21:281–306

Dufek J, Manga M (2008) In situ production of ash in pyroclastic flows. J Geophys Res Solid Earth 113:B09207

Dufek J, Manga M, Patel A (2012) Granular disruption during explosive volcanic eruptions. Nat Geosci 5:561–564

Eichelberger JC, Carrigan CR, Westrich HR, Price RH (1986) Nonexplosive silicic volcanism. Nature 323:598–602

Farquharson JI, Heap MJ, Lavallée Y, Varley NR, Baud P (2016) Evidence for the development of permeability anisotropy in lava domes and volcanic conduits. J Volcanol Geotherm Res 323:163–185

Fowler AC, Scheu B (2016) The evolution of grain size distribution in explosive rock fragmentation. Proc R Soc A 472:20150843

Fowler AC, Scheu B, Lee WT, McGuinness MJ (2010) A theoretical model of the explosive fragmentation of vesicular magma. Proc R Soc A 466:731–752

Formenti Y, Druitt TH, Kelfoun K (2003) Characterisation of the 1997 Vulcanian explosions of Soufrière Hills Volcano, Montserrat, by video analysis. Bull Volcanol 65:587–605

Freundt A, Schmincke H-U (1992) Abrasion in pyroclastic flows. Geol Rundsch 81:383–389

Friedman I, Long W, Smith RL (1963) Viscosity and water content of rhyolite glass. J Geophys Res 68:6523–6535

Gallagher A, Montanaro C, Cronin S, Scott B, Dingwell DB, Scheu B (2020) Hydrothermal eruption dynamics reflecting vertical variations in host rock geology and geothermal alteration, Champagne Pool, Wai-o-tapu, New Zealand. Bull Volcanol 82:1–19

Giachetti T, Gonnermann HM, Gardner JE, Burgisser A, Hajimirza S, Early TC, Truong N, Toledo P (2019) Bubble coalescence and percolation threshold in expanding rhyolitic magma. Geochem Geophys Geosyst 20:1054–1074

Giordano D, Russell JK, Dingwell DB (2008) Viscosity of magmatic liquids: A model. Earth Planet Sci Lett 271:123–134

Gonnermann HM (2015) Magma fragmentation. Annu Rev Earth Planet Sci 43:431–458

Gonnermann HM, Manga M (2003) Explosive volcanism may not be an inevitable consequence of magma fragmentation. Nature 426:432–35

Grady DE (1982) Local inertial effects in dynamic fragmentation. J Applied Physics 53:322–325

Heiken G, Wohletz K (1985) Volcanic Ash. Berkeley/Los Angeles: University of California Press

Hess K-U, Dingwell DB (1996) Viscosities of hydrous leucogranitic melts: A non-Arrhenian model. Am Mineral 81:1297–1300

Hornby A, Kueppers U, Maurer B, Poetsch C, Dingwell D (2020) Experimental constraints on volcanic ash generation and clast morphometrics in pyroclastic density currents and granular flows. Volcanica 3:263–283

Horwell CJ, Sparks RSJ, Brewer TS, Llewellin EW, Williamson BJ (2003) Characterization of respirable volcanic ash from the Soufrière Hills volcano, Montserrat, with implications for human health hazards. Bull Volcanol 65:346–362

Horwell CJ (2007) Grain-size analysis of volcanic ash for the rapid assessment of respiratory health hazard. J Environ Monitor 9:1107–1115

Ichihara M, Rittel D, Sturtevant B (2002) Fragmentation of a porous viscoelastic material: implications to magma fragmentation. J Geophys Res 107 B10:2229

Iravani A, Åström JA, Ouchterlony F (2018) Physical origin of the fine-particle problem in blasting fragmentation. Phys Rev Appl 10:034001

Jones TJ, Russell JK (2017) Ash production by attrition in volcanic conduits and plumes. Sci Rep 7:5538

Jones TJ, McNamara K, Eychenne J, Rust AC, Cashman KV, Scheu B, Edwards R (2016) Primary and secondary fragmentation of crystal bearing intermediate magma. J Volcanol Geotherm Res 327:70–83

Jones TJ, Reynolds CD, Boothroyd SC (2019) Fluid dynamic induced break-up during volcanic eruptions. Nat Commun 10:3828

Kameda M, Kuribara H, Ichihara M (2008) Dominant time scale for brittle fragmentation of vesicular magma by decompression. Geophys Res Lett 35:L14302

Kameda M, Ichihara M, Shimanuki S, Okabe W, Shida T (2013) Delayed brittle-like fragmentation of vesicular magma analogue by decompression. J Volcanol Geotherm Res 258:113–125

Kaminski É., Jaupart C (1998) The size distribution of pyroclasts and the frag-mentation sequence in explosive volcanic eruptions. J Geophys Res 103:29759–29779

Kennedy B, Spieler O, Scheu B, Kueppers U, Taddeucci J, Dingwell DB (2005) Conduit implosion during Vulcanian eruptions. Geology 33:581–584

Kieffer SW, Sturtevant B (1984) Laboratory studies of volcanic jets. J Geophys Res: Solid Earth 89:8253–8268

Klug C, Cashman KV (1996) Permeability development in vesiculating magmas: implications for fragmentation. Bull Volcanol 58:87–100

Kokelaar P (1986) Magma–water interactions in subaqueous and emergent basaltic volcanism: Bull Volcanol 48:275–289

Kolmogorov AN (1941) Über das logarithmisch normale Verteilungsgesetz der Dimensionen der Teilchen bei Zerstückelung. Dokl. Akad. Nauk. SSSR 31:99–101. (Translated as 'The logarithmically normal law of distribution of dimensions of particles when broken into small parts', NASA technical translation NASATTF–12,287, NASA,Washington DC, June 1969)

Koyaguchi T, Scheu B, Mitani NK, Melnik O (2008) A fragmentation criterion for highly viscous bubbly magmas estimated from shock tube experiments. J Volcanol Geotherm Res 178:58–71

Kolzenburg S, Russell JK (2014) Welding of pyroclastic conduit infill: A mechanism for cyclical explosive eruptions. J Geophys Res: Solid Earth 119:5305–5323

Kolzenburg S, Russell JK (2019) Permeability evolution during non-isothermal compaction in volcanic conduits and tuffisite veins: Implications for pressure monitoring of volcanic edifices. Earth Planet Sci Lett 527:115783

Kolzenburg S, Chevrel MO, Dingwell DB (2022) Magma/suspension rheology. Rev Mineral Geochem 87:639-719

Kremers S, Scheu B, Cordonnier B, Spieler O, Dingwell DB (2010) Influence of decompression rate on fragmentation processes: An experimental study, J Volcanol Geotherm Res 193:182–188

Kueppers U, Scheu B, Spieler O, Dingwell DB (2006) Fragmentation efficiency of explosive volcanic eruptions: a study of experimentally generated pyroclasts, J Volcanol Geotherm Res 153:125–135

Kueppers U, Putz C, Spieler O, Dingwell DB (2012) Abrasion in pyroclastic density currents: insights from tumbling experiments. Phys Chem Earth 45–46:33–39

Kushnir ARL, Martel C, Champallier R, Arbaret L (2017) In situ confirmation of permeability development in shearing bubble-bearing melts and implications for volcanic outgassing bubble-bearing melts and implications for volcanic outgassing. Earth Planet Sci Lett 458:315–326

Lavallée Y, Hess K-U, Cordonnier B, Dingwell DB (2007) A non-Newtonian rheological flow law for highly-crystalline dome lavas. Geology 35:843–846

Lavallée Y, Benson PM, Heap MJ, Hess KU, Flaws A, Schillinger B, Meredith PG, Dingwell DB (2013) Reconstructing magma failure and the degassing network of dome-building eruptions. Geology 41:515–518

Lejeune AM, Bottinga Y, Trull TW, Richet P (1999) Rheology of bubble-bearing magmas. Earth Planet Sci Lett 166:71–84

Lensky NG, Lyakhovsky V, Navon O (2001) Radial variations of melt viscosity around growing bubbles and gas overpressure in vesiculating magmas. Earth Planet Sci Lett 186:1–6

Llewellin EW Manga M (2005) Bubble suspension rheology and implications for conduit flow. J Volcanol Geotherm Res 143:205–217

Lindoo A, Larsen JF, Cashman KV, Oppenheimer J (2017) Crystal controls on permeability development and degassing in basaltic andesite magma. Geol Soc Am 45:831–834

Liu EJ, Cashman KV, Rust AC, Gislason SR (2015) The role of bubbles in generating fine ash during hydromagmatic eruptions. Geology 43:239–242

Mackaman-Lofland C, Brand BD, Taddeucci J, Wohletz K (2014) Sequential fragmentation/transport theory, pyroclast size-density relationships, and the emplacement dynamics of pyroclastic density currents–a case study on the Mt. St. Helens (USA) 1980 eruption. J Volcanol Geotherm Res 275:1–13

Mader HM, Zhang Y, Phillips JC, Sparks RSJ, Sturtevant B, Stolper E (1994) Experimental simulations of explosive degassing of magma. Nature 372:85–88

Manga M, Patel A, Dufek J (2011) Rounding of pumice clasts during transport: field measurements and laboratory studies. Bull Volcanol 73:321–333

Martel C, Dingwell DB, Spieler O, Pichavant M, Wilke M (2000) Fragmentation of foamed silicic melts: an experimental study. Earth Planet Sci Lett 178:47–58

Martel C, Dingwell DB, Spieler O, Pichavant M, Wilke M (2001) Experimental fragmentation of crystal- and vesicle-bearing silicic melts. Bull Volcanol 63:398–405

Martí J, Soriano C, Dingwell DB (1999) Tube pumices as strain markers of the ductile–brittle transition during magma fragmentation. Nature 402:650–653

Mason RM, Starostin AB, Melnik OE, Sparks RSJ (2006) From Vulcanian explosions to sustained explosive eruptions: the role of diffusive mass transfer in conduit flow dynamics. J Volcanol Geotherm Res 153:148–165

Mastin LG (1995) Thermodynamics of gas and steam-blast eruptions, Bull Volcanol 57:85–98

Mastin LG (2007) Generation of fine hydromagmatic ash by growth and disintegration of glassy rinds: J Geophys Res 112:B02203

Mastin LG, Spieler O, Downey WS (2009) An experimental study of hydromagmatic fragmentation through energetic, non-explosive magma–water mixing. J Volcanol Geotherm Res 180:161–70

Mayer K, Scheu B, Gilg HA, Heap MJ, Kennedy BM, Lavallée Y, Letham-Brake M, Dingwell DB (2015) Experimental constraints on phreatic eruption processes at Whakaari (White Island volcano) J Volcanol Geotherm Res 302:150–162

Mayer K, Scheu B, Montanaro C, Yilmaz TI, Isaia R, Aßbichler D, Dingwell DB (2016) Influence of hydrothermal alteration on phreatic eruption processes in Solfatara (Campi Flegrei) J Volcanol Geotherm Res 320:128–143

McBirney,AR, Murase T (1970) Factors governing the formation of pyroclastic rocks. Bull Volcanol 34:372–384

McGuinness MJ, Scheu B, Fowler AC (2012) Explosive fragmentation criteria and velocities for vesicular magma. J Volcanol Geotherm Res 237:81–96

Melnik O (2000) Dynamics of two-phase conduit flow of high-viscosity gas-saturated magma: large variations of sustained explosive eruption intensity. Bull Volcanol 62:153–170

Moitra P, Gonnermann HM, Houghton BF, Tiwary CS (2018) Fragmentation and Plinian eruption of crystallizing basaltic magma. Earth Planet Sci Lett 500:97–104

Montanaro C, Scheu B, Cronin SJ, Breard ECP, Lube G, Dingwell DB (2016a) Experimental estimates of the energy budget of hydrothermal eruptions; application to 2012 Upper Te Maari eruption, New Zealand. Earth Planet Sci Lett 452:281–294

Montanaro C, Scheu B, Mayer K, Orsi G, Moretti R, Isaia R, Dingwell DB (2016b) Experimental investigation of the explosivity of steam-driven eruptions: case study from Solfatara volcano, Campi Flegrei. J Geophys Res: Solid Earth, 121: 7996–8014

Moreland WM, Thordarson T, Houghton BF, Larsen G (2019) Driving mechanisms of subaerial and subglacial explosive episodes during the 10th century Eldgjá fissure eruption, southern Iceland. Volcanica 2:129–150

Mourtada-Bonnefoi CC, Mader HM (2004) Experimental observations of the effect of crystals and pre-existing bubbles on the dynamics and fragmentation of vesiculating flows. J Volcanol Geotherm Res 129:83–97

Mueller S, Spieler O, Scheu B, Dingwell DB (2005) Permeability and degassing of dome lavas undergoing rapid decompression: an experimental determination, Bull Volcanol 67:526–538

Mueller S, Scheu B, Spieler O, Dingwell DB (2008) Permeability control on magma fragmentation. Geology 36:339–402

Mueller S, Llewellin EW, Mader HM (2011) The effect of particle shape on suspension viscosity and implications for magmatic flows. Geophys Res Lett 38:L13316

Mungall JE, Bagdassarov NS, Romano C, Dingwell DB (1996) Numerical modelling of stress generation and microfracturing of vesicle walls in glassy rocks. J Volcanol Geotherm Res 73:33–46

Mysen BO, Richet P (2005) Silicate Glasses and Melts: Properties and Structure, Elsevier, Amsterdam

Namiki A, Manga M (2005) Response of a bubble bearing viscoelastic fluid to rapid decompression: implications for explosive volcanic eruptions. Earth Planet Sci Lett 236:269–84

Namiki A, Manga M (2008) Transition between fragmentation and permeable outgassing of low viscosity magmas. J Volcanol Geotherm Res 169:48–60

Odbert HM, Stewart RC, Wadge G (2014) Cyclic phenomena at the Soufrière Hills volcano, Montserrat. Geological Society, London, Memoirs 39:41–60

Ogburn SE, Loughlin SC, Calder ES (2015) The association of lava dome growth with major explosive activity (VEI ≥ 4): DomeHaz, a global dataset. Bull Volcanol 77:40

Pal R (2003) Rheological behavior of bubble-bearing magmas. Earth Planet Sci Lett 207:165–179

Papale P (1999) Strain-induced magma fragmentation in explosive eruptions. Nature 397:425–428

Papale P (2001) Dynamics of magma flow in volcanic conduits with variable fragmentation efficiency and nonequilibrium pumice degassing. J Geophys Res 106:11043–65

Papale P, Neri A, Macedonio G (1998) The role of magma composition and water content in explosive eruptions: 1. Conduit ascent dynamics. J Volcanol Geotherm Res 87:75–93

Paredes-Mariño J, Scheu B, Montanaro C, Arciniega A, Dingwell DB, Perugini D (2019) Volcanic ash generation: effects of componentry, particles size and conduit geometry on size-reduction processes. Earth Planet Sci Lett 514:13–27

Perfect E (1997) Fractal models for the fragmentation of soils and rocks: a review. Eng Geol 48:185–198

Perugini D, Kueppers U (2012) Fractal analysis of experimentally generated pyroclasts: a tool for volcanic hazard assessment. Acta Geophys 60:682–698

Platz T, Cronin SJ, Cashman KV, Stewart RB, Smith IE (2007) Transition from effusive to explosive phases in andesite eruptions—A case-study from the AD1655 eruption of Mt. Taranaki, New Zealand. J Volcanol Geotherm Res 161:15–34

Porritt LA, Russell JK, Quane SL (2012) Pele's tears and spheres: Examples from Kilauea Iki. Earth Planet Sci Lett 333:171–180

Prata AJ, Tupper A (2009) Aviation hazards from volcanoes: the state of the science. Nat Hazards 51:239–244

Preece K, Gertisser R, Barclay J, Charbonnier SJ, Komorowski JC, Herd RA (2016) Transitions between explosive and effusive phases during the cataclysmic 2010 eruption of Merapi volcano, Java, Indonesia. Bull Volcanol 78:1–16

Rager AH, Smith EI, Scheu B, Dingwell DB (2014) The effects of water vaporization on rock fragmentation during rapid decompression: Implications for the formation of fluidized ejecta on Mars, Earth Planet Sci Lett 385:68–78

Richard D, Scheu B, Mueller SP, Spieler O, Dingwell DB (2013) Outgassing: influence on speed of magma fragmentation, J Geophys Res, Solid Earth 118:862–877

Romano C, Dingwell DB, Sterner SM (1994) Kinetics of quenching of hydrous feldspathic melts: quantification using synthetic fluid inclusions. Am. Mineral 79, 1125–1134

Romano C, Dingwell DB, Behrens H (1995) The temperature-dependence of the speciation of water in $NaAlSi_3O_8$–$KAlSi_3O_8$ melts: an application of fictive temperatures derived from synthetic fluid inclusions. Contrib Mineral Petrol 122:1–10

Romano C, Mungall J, Sharp T, Dingwell DB (1996) Tensile strengths of hydrous vesicular glasses: an experimental study. Am Mineral 81:1148–1154

Rott S, Scheu B, Montanaro CMayer K, Joseph E, Dingwell DB (2019) Hydrothermal eruptions at unstable crater lakes: insights from the Boiling Lake, Dominica, Lesser Antilles. J Volcanol Geotherm Res 381:101–118

Russell JK, Hess K-U, Dingwell DB (2022) Models for viscosity of geological melts. Rev Mineral Geochem 87:841-885

Rust AC, Manga M (2002) Effects of bubble deformation on the viscosity of dilute suspensions. J Non-Newtonian Fluid Mech 104:53–63

Rust AC, Cashman KV (2011) Permeability controls on expansion and size distributions of pyroclasts. J Geophys Res 116:B11202

Sable JE, Houghton BF, Del Carlo P, Coltelli M (2006) Changing conditions of magma ascent and fragmentation during the Etna 122 BC basaltic Plinian eruption: Evidence from clast microtextures. J Volcanol Geotherm Res 158:333–354

Sahagian D, Carley TL (2020) Explosive volcanic eruptions and spinodal decomposition: a different approach to deciphering the tiny bubble paradox. Geochem Geophys Geosyst 21:e2019GC008898

Schauroth J, Wadsworth FB, Kennedy B, von Aulock FW, Lavallée Y, Damby DE, Vasseur J, Scheu B, Dingwell DB (2016) Conduit margin heating and deformation during the AD 1886 basaltic Plinian eruption at Tarawera volcano, New Zealand. Bull Volcanol 78:12

Scheu B, Kern H, Spieler O, Dingwell DB (2006a) Temperature dependence of elastic P-and S-wave velocities in porous Mt. Unzen dacite. J Volcanol Geotherm Res 153:136–147

Scheu B, Spieler O, Dingwell DB (2006b) Dynamics of explosive volcanism at Unzen: an experimental Contribution, Bull Volcanol 69:175–187

Scheu B, Kueppers U, Mueller S, Spieler O, Dingwell DB (2008) Experimental volcanology on eruptive products of Unzen volcano. J Volcanol Geotherm Res 175:110–119

Schipper CI, White JDL (2016) Magma–slurry interaction in Surtseyan eruptions. Geology 44:195–198

Self S, Sparks RSJ (1978) Characteristics of widespread pyroclastic deposits formed by the interaction of silicic magma and water. Bull Volcanol 41:196–212

Sharp TG, Stevenson RJ, Dingwell DB (1996) Microlites and "nanolites" in rhyolitic glass: Microstructural and chemical characterization. Bull Volcanol 57:631–640

Shaw HR (1972) Viscosities of magmatic silicate liquids; an empirical method of prediction. Am J Sci 272:870–893

Sheridan MF, Wohletz KH (1983) Hydrovolcanism: basic consideration and review. J Volcanol Geotherm Res 17:1–29

Smellie JL (2000) Subglacial eruptions. In: Sigurdsson H (ed) Encyclopedia of Volcanoes. Academic Press, San Diego, pp 403–418

Sonder I, Harp AG, Graettinger AH, Moitra P, Valentine GA, Büttner R, Zimanowski B (2018) Meter-scale experiments on magma–water interaction. J Geophys Res:Solid Earth 123:10–597

Song W, Yang S, Fukumoto M, Lavallée Y, Lokachari S, Guo H, You Y, Dingwell DB (2019) Impact interaction of in-flight high-energy molten volcanic ash droplets with jet engines. Acta Materialia 171:119–131

Sparks RSJ (1978) The dynamics of bubble formation and growth in magmas: a review and analysis. J Volcanol Geotherm Res 3:1–37

Spieler O, Alidibirov M, Dingwell DB (2003) Grain-size characteristics of experimental pyroclasts of 1980 Mount St. Helens cryptodome dacite: effects of pressure drop and temperature. Bull Volcanol 65:90–104

Spieler O, Kennedy B, Kueppers U, Dingwell DB, Scheu B, Taddeucci J (2004) A fragmentation threshold for the initiation and cessation of explosive eruptions. Earth Planet Sci Lett 226:139–148

Spina L, Cimarelli C, Scheu B, Di Genova D, Dingwell DB (2016) On the slow decompressive response of volatile- and crystal-bearing magmas: An analogue experimental investigation. Earth Planet Sci Lett 433:44–53

Spina L, Morgavi D, Costa A, Scheu B, Dingwell DB, Perugini D (2019) Gas mobility in rheologically-layered volcanic conduits: the role of decompression rate and crystal content on the ascent dynamics of magmas. Earth Planet Sci Lett 524:115732

Stix J, Moor JMD (2018) Understanding and forecasting phreatic eruptions driven by magmatic degassing. Earth, Planets Space 70:19

Thiéry R, Mercury L (2009) Explosive properties of water in volcanic and hydrothermal systems J Geophys Res 114:B05205

Turcotte DL (1986) Fractals and fragmentation. J Geophys Res 91:1921–1926

van Otterloo J, Cas RAF, Scutter CR (2015) The fracture behaviour of volcanic glass and relevance to quench fragmentation during formation of hyaloclastite and phreatomagmatism. Earth-Sci Rev 151:79–116

Yamaoka K, Geshi N, Hashimoto T, Ingebritsen SE, Oikawa T (2016) Special issue "The phreatic eruption of Mt. Ontake volcano in 2014". Earth Planet Space 68:1–8

Verhoogen J (1951) Mechanics of ash formation. Am J Sci 249:729–739

Villermaux E (2012) The formation of filamentary structures from molten silicates: Pele's hair, angel hair, and blown clinker. C R Mécanique 340:555–564

Wadsworth FB, Witcher T, Vasseur J, Dingwell DB, Scheu B (2017) When does magma break? In: Advances in Volcanology. Springer, Berlin, Heidelberg

Wadsworth FB, Witcher T, Vossen CE, Hess KU, Unwin HE, Scheu B, Castro JM, Dingwell D (2018) Combined effusive-explosive silicic volcanism straddles the multiphase viscous-to-brittle transition. Nat Commun 9:4696

Webb SL, Dingwell DB (1990a) Non-Newtonian rheology of igneous melts at high stresses and strain rates: experimental results for rhyolite, andesite, basalt and nephelinite. J Geophys Res 95:15695–15701

Webb SL, Dingwell DB (1990b) The onset of non-Newtonian rheology of silicate melts: a fiber elongation study. Phys Chem Mineral 17:125–32

Wilding M, Webb SL, Dingwell DB, Ablay G, Marti J (1996) The variation of cooling rates within volcanic facies from Tenerife, Canary Islands. Contrib Mineral Petrol 125:151–160

Wilding M, Dingwell DB, Batiza R, Wilson L (2000) The cooling rates of hyaloclastites: applications of relaxation geospeedometry to undersea volcanics. Bull Volcanol 61:527–536

Wohletz KH, McQueen RG (1984) Experimental studies of hydromagmatic volcanism, explosive volcanism: inception, evolution, and hazards. Studies in Geophysics. National Academy Press, Washington DC, p 158–169

Wohletz KH, Sheridan MF, Brown WK (1989) Particle size distributions and the sequential fragmentation/transport theory applied to volcanic ash. J Geophys Res 94:15703–15721

Wright HM, Weinberg RF (2009) Strain localization in vesicular magma: Implications for rheology and fragmentation. Geology 37:1023–1026

Woods AW, Koyaguchi T (1994) Transitions between explosive and effusive eruptions of silicic magmas. Nature 370:641–644

Zhang YX (1999) A criterion for the fragmentation of bubbly magma based on brittle failure theory. Nature 402:648–50

Zhang Y, Sturtevant B, Stolper EM (1997) Dynamics of gas-driven eruptions: Experimental simulations using CO_2–H_2O–polymer system. J Geophys Res 102:3077–3096

Zimanowski B, Lorenz V, Frohlich G (1986) Experiments on phreatomagmatic explosions with silicate and carbonatitic melts. J Volcanol Geotherm Res 30:149–153

Zimanowski B, Fröhlich G, Lorenz V (1991) Quantitative experiments on phreatomagmatic explosions. J Volcanol Geotherm Res 48:341–348

Zimanowski B, Büttner R, Nestler J (1997) Brite reaction of a high-temperature ion melt. Europhys Lett 38:285

Zimanowski B, Wohletz K, Dellino P, Büttner R (2003) The volcanic ash problem. J Volcanol Geotherm Res 122:1–5

Reviews in Mineralogy & Geochemistry
Vol. 87 pp. 801-840, 2022
Copyright © Mineralogical Society of America

17

Hot Sintering of Melts, Glasses and Magmas

Fabian B. Wadsworth[1], Jérémie Vasseur[2],
Edward W. Llewellin[1], Donald B. Dingwell[2]

[1]*Department of Earth Sciences, Durham University, Durham, United Kingdom*
[2]*Department of Earth and Environmental Sciences*
Ludwig-Maximilians-Universität, Munich, Germany

fabian.b.wadsworth@durham.ac.uk; j.vasseur@lmu.de
ed.llewellin@durham.ac.uk; dingwell@lmu.de

1. INTRODUCTION TO SINTERING IN VOLCANIC ENVIRONMENTS

Magma may fragment as it ascends through the volcanic plumbing system to the Earth surface, forming volcanic ash and pyroclasts (Scheu and Dingwell 2022, this issue). If the pyroclastic products of fragmentation are deposited hot, they can stick, coalesce, and weld together, re-amalgamating into coherent magma. Subaerial deposits include welded, lava-like, and rheomorphic ignimbrites (Branney et al. 1992; Andrews and Branney 2011), and welded fall deposits proximal to an eruptive vent (Houghton and Carey 2015). Welding occurs in the sub-surface at conduit margins (Gardner et al. 2017; Wadsworth et al. 2020), and in tuffisite veins produced when gas–particle dispersions hydro-frack country rock or overlying magma (Tuffen et al. 2003; Tuffen and Dingwell 2005; Heap et al. 2019). Welding also occurs during transport in eruption plumes, as an undesirable consequence of the accumulation of molten silicate particles in the hot zone in jet aircraft (Song et al. 2014; Giehl et al. 2016), and when volcanic ash is struck by volcanic lightning in air or on the ground (Cimarelli et al. 2017; Mueller et al. 2018). In Figure 1, we give a visual tour of some of the volcanic environments in which welding occurs.

Welding of molten magmatic particles to coherent magma effectively reverses the fragmentation process. This reversal can be so total that any evidence for fragmentation is completely over-printed, leading to, for example, lava-like ignimbrites (Branney et al. 1992), so-named because they can be mistakenly interpreted as effusive lavas, rather than the products of explosive volcanism. Wadsworth et al. (2020) proposed that most subaerial rhyolitic lavas may be the product of in-conduit welding, and may not be truly effusive lavas at all. There is growing evidence for this welding origin of rhyolite lavas (Schipper et al. 2021). Therefore, developing theories for welding in magmatic environments is crucial to unpicking volcanic processes that can disguise the product of a vigorous explosive eruption as the product of more gentle lava effusion. Similarly, it has been proposed that welding of material that falls back into volcanic vent areas, or that fills tuffisite veins, can plug the volcanic system, allowing gas pressure to build up, and possibly leading to further explosive activity (Quane et al. 2009; Farquharson et al. 2017; Kolzenburg et al. 2019; Wadsworth et al. 2020).

The physics of welding in volcanic environments is broadly synonymous with *viscous sintering* processes as they are understood in physics and material science research. Sintering (and welding) involves the amalgamation of many individual, loose particles into a single, coherent body. The particles involved may be solid or molten, crystalline or amorphous, and may possess a variety of shapes, sizes, and internal texture. The amalgamation of the particles may be driven by particle–surface forces (e.g., surface tension), inertial forces, or externally applied bulk forces. This diversity of materials and process offers a family of problems of enormous potential complexity. In all cases, sintering of packs of many particles involves the reduction of pore space volume and a resultant bulk density increase (e.g., Prado et al. 2003;

1529-6466/22/0087-0017$05.00 (print)
1943-2666/22/0087-0017$05.00 (online)

http://dx.doi.org/10.2138/rmg.2022.87.17

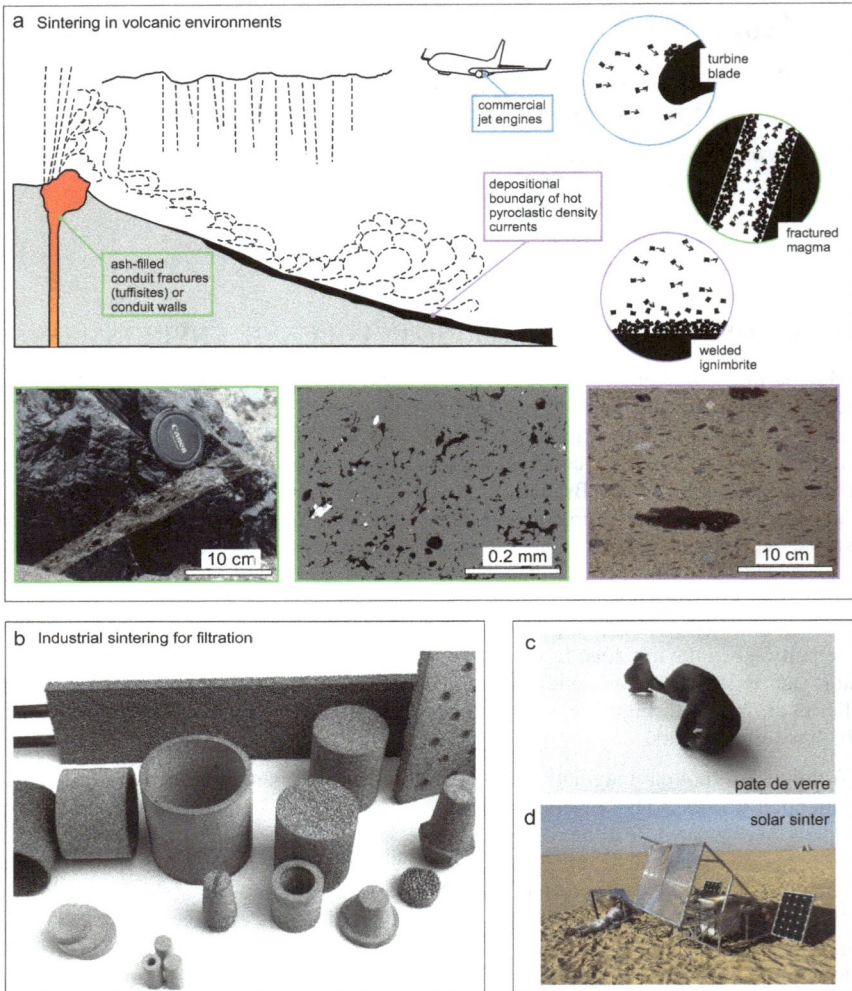

Figure 1. Examples of systems for which sintering is applicable. (**a**) The natural volcanic system in which sintering can occur when volcanic particles are ingested into jet engines and impact the hot moving parts (e.g., turbine blades) or accumulate in ducts and cooling vents (Shinozaki and Roberts 2013; Giehl et al. 2016; Song et al. 2017), during transport and deposition of pyroclastic density currents (Branney et al. 1992; Branney and Kokelaar 2002; Quane and Russell 2005), and within volcanic conduits either during large explosive eruptions (Gardner et al. 2017, 2019) or during hybrid activity forming tuffisite veins (Tuffen and Dingwell 2005; Farquharson et al. 2017; Gardner et al. 2018). The 3 *inset* images in panel (a) are (from left to right) a partially welded tuffisite, a scanning electron microscope image of a partially welded tuffisite (both from Heap et al. 2019), and a welded ignimbrite (from Wadsworth et al. 2021). (**b**) An example of different shapes of industrial sintered filters (image reproduced with permission from Carbis Filtration™). (**c**) An example of pâte de verre mould casting in which sintering is used to create 3-dimensional forms from glass particles (this example is a piece by Alastair Mackie; see Wadsworth et al. 2018). (**d**) An experimental process called 'solar sinter' (www.kayserworks.com) by Markus Kayser in which desert sand is sintered using a Fresnel lens to melt the sand particles to melt droplets, and a motorized sintering container to produce precision 3-dimensional forms; analogous to 3D printing.

Vasseur et al. 2013; Wadsworth et al. 2016), reduction in permeability to fluid in the pore space between the sintering particles (Wadsworth et al. 2017a; Heap et al. 2019), together with increases in both bulk strength (Vasseur et al. 2013) and elastic wave velocities (Vasseur et al. 2016). However, natural volcanic welding processes involve additional factors that render the problem more complex than is typical of controlled, industrial viscous sintering. Such complexities include volatile disequilibrium leading to the degassing or resorption of volatiles during welding (Sparks et al. 1999; Gardner et al. 2018; Wadsworth et al. 2019, 2021), textural heterogeneities such as semi-crystalline or porous particles (Quane and Russell 2005; Heap et al. 2015; Kendrick et al. 2016), and non-isothermal conditions—particularly dynamic cooling during welding (Branney and Kokelaar 2002; Wadsworth et al. 2014, 2019, 2021; Kolzenburg et al. 2019). In this Chapter, we use the term *sintering* to refer to this family of process, but note that we draw no distinction with the process of *welding* as used in volcanology.

Viscous sintering is a fundamental phenomenon, important in a vast array of processes beyond volcanology, from the initial stages of hot planetesimal formation (Sirono 2011), to the production of many types of porous ceramics (Rahaman 2007), glasses (Prado et al. 2003) and filters (Krasnyi et al. 2005). Just as magmatic melts and magmas are usually silicate systems, most industrial applications of viscous sintering also involve glass-forming silicates. Therefore, welding in volcanology is directly analogous to viscous sintering processes in many other contexts. In this Chapter, we explore the dynamics of hot sintering of silicate melts and glasses from phenomenological and theoretical perspectives. We lay out the families of sintering phenomena and summarize existing mathematical theories for sintering, presenting a suite of models that can be used to predict the time-dependence of the physical properties of these systems. We have tried to make these models easy-to-use and analytical, and where possible, theoretical and grounded in physical principles, without empirical adjustment. We test these models against a large collated database of experimental results, and arrive at a preferred, flexible modelling framework that is consistent with all available data, and accounts for the evolution of porosity for both monodisperse and polydisperse systems over a wide range of temperatures and pressures. This modelling framework involves a theory for predicting the distribution of pore sizes in an evolving sintering system, from granular to non-granular.

We explore the application of the modelling framework to volcanic scenarios, and show that it can be used to predict sintering in environments where volatiles, such as water, diffuse into or out of the particles, under non-isothermal conditions of arbitrary thermal pathway, and in situations where the gas pressure and the squeezing pressure on the particles differ. The model is couched in terms of dimensionless parameters, and we explore regimes of behaviour via maps in dimensionless parameter space, to show the regimes in which model assumptions are valid, and to identify the limits of applicability beyond which new dynamics emerge. This general modelling framework is then extended also to predict the increase in material strength and elastic moduli, and the decrease in hydraulic permeability of the pore space during sintering. Finally, we explore what are the outstanding questions in sintering theory and how they may be relevant for future research.

1.1 Families of sintering phenomena

The word 'sintering' has been used to describe different processes. Amongst these, one can distinguish between end-member cases of sintering that occurs in the solid state—termed 'solid-state sintering'- and sintering that occurs in the liquid state—termed 'viscous sintering' or sometimes 'liquid-state sintering'. Viscous sintering may still occur when rigid, solid particles are mixed into the sintering pack of particles such that the particles in the sintering system may be a mixture of solid particles (not undergoing sintering) and viscous particles (or droplets). Similarly, solid-state sintering can occur with a minor viscous component present in the particle mixture. However, in general the physics of mixed systems has received less attention than the physics of either purely viscous or purely solid-state sintering. In this Chapter we focus on viscous sintering, allowing that the viscously sintering pack may include solid particles, but do not consider solid-state sintering. The third phase that is present in sintering systems is the inter-particle phase, which is usually a gas. Thus, in general, sintering occurs in systems of mixtures of gas, liquid, and solid phases. As sintering involves the reduction of the pore spaces between the particles, the gas volume decreases as sintering progresses, such that the relative proportions of each phase are a function of time.

In Figure 2 we show a ternary diagram for which the apexes are defined as the volume fractions of the three phases. The upper part of this phase diagram comprises 'loose' systems that includes dispersions of particles in a gas, where the gas volume fraction is sufficiently high that the particles are unlikely to be in contact with one another and therefore, sintering phenomena are unlikely. We note that Scherer (1977) does explore the sintering of 'low density' glasses, which would fall in our *loose* regime, but we do not consider this region further. The dark grey shaded band in the center of Figure 2 shows typical initial gas volume fractions for stable random packs of particles. The bounds on the dark grey shaded region are given as general end-member packing values for different polydispersivity and angularity of particles (e.g., Wadsworth et al. 2017b). From starting conditions in the dark grey box, systems of any given mixture viscous liquid particles and solid particles will progress away from the gas-phase apex during sintering, as the gas phase is expelled (we note that systems in which the melt phase crystallizes during sintering will have a more complex trajectory in this space). The left edge of the ternary plot represents pure viscous sintering (zero solid particles), while the right edge of the ternary plot represents pure solid-state sintering. This chapter focusses on viscous sintering processes that occur in the red quadrilateral in Figure 2; that is, systems of viscous particles that may include solid particles up to the random close pack of approximately 0.585 solid volume fraction (Boyer et al. 2011) or a maximal random pack at 0.63 (Mueller et al. 2010). At solid volume fractions higher than these values, most or all of the solid particles are in contact with one another, and dominate over viscous particle contact points.

The physics of sintering in the region to the right of the red quadrilateral in Figure 2 is not covered here. While sintering along the right-hand edge of the ternary diagram (solid-state sintering) is well explored in ceramics science (Rahaman 2007) and is known to operate during compaction of crystalline volcanic materials (Ryan et al. 2020), the region to the left of that far-right ternary edge, but to the right of the red quadrilateral, is broadly unexplored. We do not explore this region in this Chapter. We also note that while we do consider mixed solid–viscous sintering within the red quadrilateral in Figure 2, we only focus on situations where the solid phase is not undergoing solid-state sintering, and only represents a solid obstructing phase, modifying purely viscous sintering. This situation is typical of volcanic welding where the solid-state sintering of crystals occurs over longer timescales than viscous sintering of melt particles, such that if the sintering system is a mixture of melt and solid particles, the viscous component will dominate the kinetics.

1.2 Phenomenology and internal texture

In a laboratory setting, sintering systems start as granular powders of glass that can be heated and softened to viscous droplets. In volcanic environments, the system can be emplaced as hot droplets directly. In this Chapter, we use the terms *droplet* or *viscous particle*, and *solid particle* to disambiguate the state of the sintering components. Usually, the particles are initially in 'packs' meaning that they share contacts with other particles around them (i.e., within the grey region in Figure 2). The particle pack can then either be heated (typical in the laboratory or in industrial settings), or it can arrive in the packed state already hot (typical in nature such as volcanic settings). When hot, the particles melt or relax into silicate droplets, at which point the contact points between particles will evolve into widening necks (Frenkel 1945; Eshelby 1949; Eggers et al. 1999; Kang 2004). The material feeding the widening necks is transported by viscous flow from the particle itself, resulting in a mass transport toward the

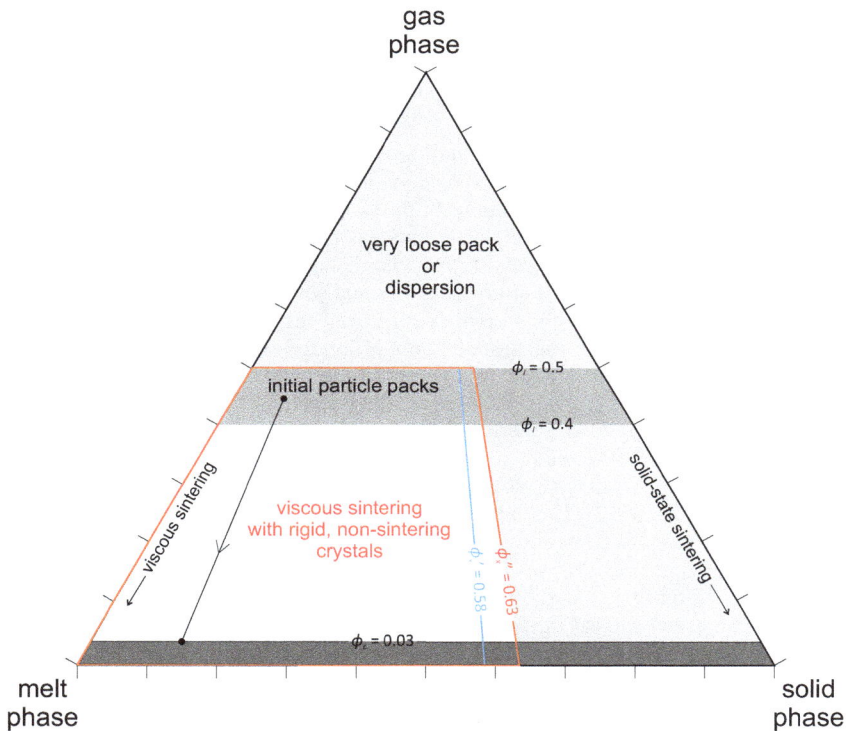

Figure 2. Sintering systems can cover a range of multiphase space. Common to all sintering systems is a gas phase interstitial to particles or droplets of either solids or melt or a mixture of both. The gas phase starts from a volume fraction ϕ_i which is usually $0.4 \lesssim \phi_i \lesssim 0.5$(**shaded region** marked 'initial particle packs') for typical random packs of particles of a range of polydispersivity in particle sizes (Wadsworth et al. 2017b). For any $\phi > \phi_i$, the system is a very loose pack or a dispersion of particles in a gas phase and therefore particle-particle interactions are transient, and sintering is not efficient (although this regime has been studied; Scherer 1977; Scherer and Bachman 1977). From the initial packed state ϕ_i, sintering involves the ejection of gas from the system, resulting in $\phi(t) < \phi_i$. The **black arrowed line** shows a typical sintering trajectory as the gas phase is ejected (this example is for a solid volume fraction of $\phi_x = 0.13$). The final gas volume fraction ϕ_c is given by the percolation threshold, which for spherical particles sintered in the absence of applied pressure is $\phi_c \approx 0.03$ (Wadsworth et al. 2016a). When all of the particles are glass (or melt droplets at high temperature), the sintering follows the left-hand edge of this ternary diagram and is usually termed 'viscous sintering'. When all of the particle are rigid solids, the sintering follows the right-hand edge of this ternary diagram and is termed 'solid-state sintering,' which is usually a slower process than viscous sintering (Ryan et al. 2020). In this Chapter we deal specifically with the system encapsulated by the **red polygon** labelled 'viscous sintering with rigid, non-sintering crystals.' In this Chapter we only deal with mixed viscous–solid systems for which the solid crystals are not themselves undergoing sintering, and instead represent inert resistors to the purely viscous sintering process. We define the right-hand limit of the red polygon as $\phi_x'' = 0.63$ representing the maximum random pack of solid particles that could exist in a final sintered system. Above ϕ_x'', the solids form a rigid framework and sintering is dominated by solid-solid contact points between particles rather than by melt–melt contact points between particles. Also shown is the jamming fraction $\phi_x' = 0.58$ above which the solid particles are jammed (Torquato et al. 2000; Boyer et al. 2011). In this ternary diagram and throughout this Chapter, gas volume fractions ϕ are given as fractions of the bulk system, while solid volume fractions ϕ_x are given as fractions of the solid–melt system only (i.e., on a gas-free basis; see text).

particle–particle contact points (Frenkel 1945). Because the mass in the particles or droplets of these systems is conserved, and because the material is flowing into neck regions that were previously occupied with the interstitial gas phase, the gas phase is displaced and permeably expelled from the sintering system. As liquid flows and reduces the gas volume fraction in the system, the bulk density goes up and the total volume decreases (if we assume the expelled gas is lost from the system and not conserved such that the system is 'open'). We show typical 3D and 2D views of the internal pore geometry during sintering in Figure 3.

What starts out as a system of packed but loose particles—a granular medium—evolves toward a state where the droplet liquid phase is continuous and interconnected, as is the gas phase interstitial to the particles. That is, most of the sintering process is defined as having the internal geometry of a complex and often tortuous gas-liquid manifold, which is constantly in motion. As the gas volume decreases further, the distance between any particular region of liquid and any other region of liquid decreases and, eventually, pore throats close-off. This process toward the end of sintering is only poorly understood in detail, but presumably the closure of pore throats occurs in cascades, pinching off regions of the gas phase. Once pinch-off points develop, gas is no longer fully connected to the outside of the system (Fig. 3) and cannot therefore be permeably expelled any longer. The final state of sintering is then dispersed, isolated regions of gas that relax to spherical bubbles. What began as disconnected particles of liquid in a continuous gas phase, ends as disconnected gas bubbles in a liquid phase (Fig. 3). The average of all of the internal curvature of the sintering system therefore inverts (Wadsworth et al. 2017a; Okuma et al. 2017). The final gas volume fraction that is at equilibrium following these pinch-off events depends on how broad was the distribution of particles in the starting pack (Wadsworth et al. 2017b). However, as a rule of thumb, this final gas volume fraction is around 0.03 (labelled on Figs. 2 and 3), which is similar to the rigorously-determined critical point at which the space exterior to identically overlapping spheres in computer simulations become disconnected (Kertesz 1981; Elam et al. 1984). Simulations have shown a weak dependence of that value on the initial sphere size distribution (Rintoul 2000).

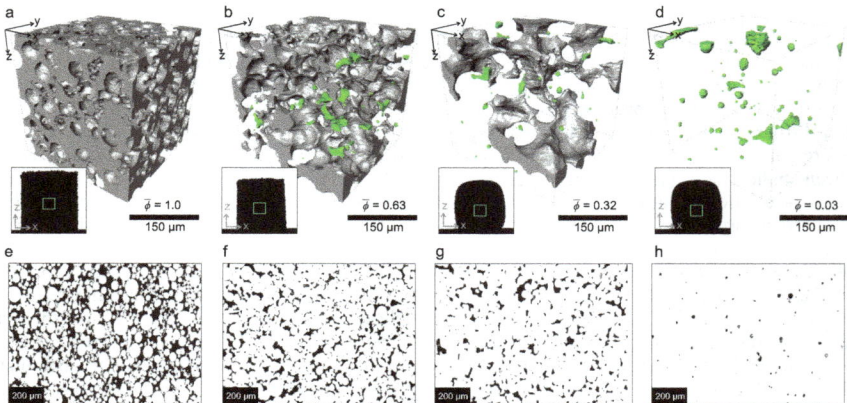

Figure 3. Internal textures of sintering systems captured in 3-dimensions using *in situ* X-ray tomography at high temperature (**a–d**) or in 2-dimensions using scanning electron microscopy at room temperature on quenched and polished samples (**e–h**; Wadsworth et al. 2016b, 2017a). In (a–d) the gas-volume fraction ϕ is normalized to the initial value $\phi_i = 0.48$, to give $\bar{\phi}$ labelled in the images. The **grey volume** is the connected gas volume, the **green volume** is the isolated gas volume, and the high-temperature molten glass particles (melt phase) has been rendered **transparent** for clarity. The images *inset* to (a–d) are 2-dimensional silhouettes of the cross section of a sintering sample at the same porosity as labelled in the main image. The **green square** represents the 2-dimensional field of view in the z–x plane rendered in the main images. (e–h) Scanning electron microscopy images of glass bead samples sintered to different gas volume fractions where the **black phase** represents pores and the **white phase** represents glass. The gas volume fraction in each of (e–h) match the corresponding image above in (a–d).

The texture of a sintering system remains isotropic and homogeneous over scales much larger than the particle sizes (Prado et al. 2001; Wadsworth et al. 2017a). When sintering is performed under an applied uni-directional (or uniaxial) load, the particles are sometimes flattened against one another (Quane and Russell 2005), producing anisotropy. And presumably such anisotropy is developed further if sintering occurs under large shearing stresses. This difference between isotropic and anisotropic resulting textures will influence the physical properties of the system such that, for example, the permeability might also be expected to develop an anisotropy. This Chapter identifies anisotropy as a frontier topic and necessarily deals exclusively with the isotropic case.

1.3 Conceptual approaches

Understanding sintering phenomena has involved an array of approaches. The most fundamental distinction is between 'continuum' approaches and 'micro-mechanical' approaches. The former involves the pursuit of a predictive relationship between the macroscopic changes in a sintering system, such as its bulk volume, and the *bulk* properties of the whole system, such as its bulk viscosity or compressibility, or the bulk stresses acting over the whole volume. In the continuum approach, therefore, the particle or droplet scale processes are neglected except to the extent to which they manifest as some bulk phenomenon. An example of this is where the bulk strain of a sintering sample is related to its bulk viscosity and a bulk driving stress (Scherer 1986; Rahaman et al. 1987; Olevsky 1998; Grunder et al. 2005; Quane and Russell 2005; Olevsky et al. 2006). The continuum approach has the advantage that no micro-structural geometry has to be assumed, and the behaviour can be explained using a few controlling parameters. However, the potential disadvantage is that, in most cases, the bulk properties (such as the bulk viscosity) must be determined empirically from experimental data and cannot be predicted *a priori* for any arbitrary system, such that the constitutive relationships that result can only reliably be used with the range of experimental conditions tested.

The alternative to the continuum approach is the micro-mechanical approach, in which the attempt is made to predict the bulk behaviour from first-principle models for the behaviour at the particle or droplet scales (Frenkel 1945; Mackenzie and Shuttleworth 1949; Olevsky 1998; Prado et al. 2003; Wadsworth et al. 2016a). Insights into the behaviour of the particles or droplets are then scaled to the system size using some assumptions about the number density or coordination of those particles or droplets. The advantage of this approach is that it often results in a first-principles model with no empirical adjustments, such that it can be used confidently outside of the region in which it was experimentally verified. However, these models do often rely on idealized assumptions of geometry (e.g., spherical particles, or cubic particle lattice arrangements), which are not strictly valid in most applied cases.

A third type of model is explicit numerical simulation of the droplet scale dynamics without idealized assumptions (e.g., Kirchhof et al. 2009). However, such simulations are computationally expensive and usually restricted to small numbers of particles or droplets.

Here, we start from the assumption that the desirable result is a micro-mechanical model built from first principles. While we acknowledge that continuum empirical approaches can be expedient and valid, we propose that more insight into sintering phenomena can be gained when the physics are understood from the smallest scales up to the larger scales. In this Chapter, we aim to use these micro-mechanical approaches and to validate them at each step using a compilation of published experimental data. Where possible, we then use this approach to give some physical sense to the other continuum, empirical models that have been proposed.

2. THEORETICAL MODELS FOR SINTERING UNDER NO EXTERNAL LOAD

This Chapter deals with molten systems of droplets, which includes high viscosity softened silicate glass particles. Therefore, we can draw on dimensional analysis for droplet systems in general to consider what dynamic regimes silicate droplets may occupy. We confine this analysis to droplets that are at rest, and we do not consider impacts between droplets and other objects. First, the role played by gravity acting on small droplets is governed by the Eötvös number Eo (or sometimes called the Bond number) which is

$$\text{Eo} = \frac{\rho g R^2}{\Gamma} \tag{1}$$

where ρ is the droplet density, g is the gravitational acceleration, R is the droplet radius, and Γ is the interfacial tension between the droplet and the surrounding fluid. When Eo $\ll 1$, the interfacial tension dominates and the droplet is expected to remain round, while for Eo $\gg 1$, the gravitational force acting on the droplet overcomes the interfacial tension force and the droplet may deform to non-sphericity under its own weight. Second, the Ohnesorge number defines when a droplet undergoing interfacial-tension-driven motion is dominated by viscous forces or inertial forces whereby

$$\text{Oh} = \frac{\mu}{\sqrt{\rho \Gamma R}} \tag{2}$$

when Oh $\gg 1$, viscosity μ dominates, and when Oh $\ll 1$ inertia dominates.

In volcanic environments, most viscous sintering processes occur with ash-sized particles (i.e., $R < 1$mm), at magmatic temperatures where $\mu > 100$ Pa.s, for melt particles with densities around 2000 kg.m^{-3}, and interfacial tension values of 0.3 N.m^{-1} (values are discussed and justified later, in Section 6.1). Given these values, Eo $\ll 1$ and Oh $\gg 1$ are the typical regimes for most volcanic welding conditions. In this Chapter, we concentrate on these regimes (Eo $\ll 1$ and Oh $\gg 1$), while noting that there are some relevant welding scenarios for which gravitational forces become important on the droplet scale (larger particles), and for which inertial forces can play a role (lower viscosities).

For Eo $\ll 1$ and Oh $\gg 1$, and in the absence of an external confining stress on the sintering pack, we can assume that capillary (or surface-tension) forces dictate the dominant time- and pressure-scales. We define these as λ and P_L, respectively

$$\lambda = \frac{\mu L}{\Gamma}; P_L = \frac{2\Gamma}{L} \tag{3}$$

for which L is a lengthscale. We can use these to generalize theoretical sintering models. In order to compare data across a wide range of experimental or natural conditions, we can normalize time by the capillary timescale such that $\bar{t} = t/\lambda = t\,\Gamma/(\mu L)$. However, in many systems, conditions of sintering are not constant. For instance, temperature may vary over a significant interval, affecting μ. For this case of strongly non-isothermal sintering or any situation where μ evolves, Wadsworth et al. (2016) proposed an integral form for the dimensionless time.

$$\bar{t} = \frac{t}{\lambda} = \frac{\Gamma}{L} \int_{t_i}^{t} \frac{1}{\mu} \mathrm{d}t \tag{4}$$

where t_i is an initial time when sintering starts. In this Chapter we will identify two broad families of models for sintering in which Equation (4) will become useful. The fundamental difference between them arises via the definition of the lengthscale L; in one model system, $L \equiv R_i$ where R_i is the initial droplet radius, and in the other model system $L \equiv a_i$, where a_i is an initial pore radius between packed particles in a sintering system. In order to draw the important distinction in these normalizations, in the former model we rename the dimensionless time $\bar{t}$ as $\bar{t}_n$, while in the latter we rename it $\bar{t}_b$ (note that sub-script n is used to mean 'neck' and subscript b is used to mean 'bubble', relating to models that follow).

2.1 Sintering of two droplets: Coalescence

Sintering of two viscous droplets is sometimes also called coalescence when the sintering mechanism is viscous flow. For two-droplet systems, it is the width of the neck between the droplets that is the relevant metric both to measure, and to model. Frenkel (1945) was the first to note that it is the surface tension at the droplet interfaces that drives viscous flow in the interior of the droplets, feeding material into the growing neck, and derived a solution for the neck radius R_n as a function of time t for the case of two droplets of initial equal radius R_i in an inviscid exterior fluid (e.g., air)

$$\frac{R_n}{R_i} \approx \left(C \frac{t\Gamma}{\mu R_i} \right)^{1/2} \tag{5}$$

where C is a constant. Frenkel (1945) found that $C = 3/2$, whereas Eshelby (1949) found $C = 1$. In either case, this solution leads to the proportionality $R_n \propto t^{1/2}$.

Eggers et al. (1999) found a more general model for the viscous coalescence of two droplets surrounded by either a viscous or an inviscid fluid. Here, we focus on the case of an inviscid outer fluid surrounding the two droplets, for which an analytical asymptotic approximation valid up to $R_n/R_i < 0.03$ is

$$\frac{R_n}{R_i} \approx -\frac{1}{\pi} \frac{t\Gamma}{\mu R_i} \ln \left(2^{-1/3} \frac{t\Gamma}{\mu R_i} \right) \tag{6}$$

which suggests that the scaling proportionality is $R_n \propto t \ln t$, rather than $R_n \propto t^{1/2}$.

A problem with the Frenkel (1945) model (Eqn. 5) and the Eggers et al. (1999) asymptotic solution (Eqn. 6), is that neither have the property that R_n approaches a finite value as $t \to \infty$. To address this, Eggers et al. (1999) provide a more rigorous solution for the viscous dynamics of coalescence, which requires a numerical solution and is

$$\frac{1}{R_i} \frac{dR_n}{dt} = -\frac{1}{\pi} \frac{\Gamma}{\mu R_i} \ln \left(2^{-1/3} \frac{R_n}{R_i} \right). \tag{7}$$

Another approach is taken by Hopper (1984) who proposed a solution to the Navier–Stokes equations for the case of viscous flow driven by interfacial tension in the coalescence of two cylinders: that is, a 2D analogue of the 3D sphere-coalescence problem dealt with here. The Hopper (1984) model is

$$\frac{R_n}{R_i} = 2^{1/3} (1-\alpha)(1+\alpha^2)^{-1/2}$$

$$\frac{\Gamma t}{\mu R_i} = 2^{1/3} \frac{\pi}{4} \int_{\alpha^2}^{1} \left[\beta(1+\beta)^{1/2} \kappa(\beta) \right]^{-1} d\beta \tag{8}$$

$$\kappa(\beta) = \int_0^1 \left[(1-\gamma^2)(1-\beta\gamma^2) \right]^{-1/2} d\gamma.$$

where the parts of Equation (8) are solved together and, with use of the numerical quadrature to compute the definite integrals, can be analytical. Similar to the Eggers et al. (1999) model (Eqn. 7), the Hopper (1984) solution has the desirable result that it asymptotes to $R_n/R_i = 2^{1/3}$ as $t \to \infty$. The variable α is the control parameter here and $0 \leq \alpha < 1$ where $\alpha = 1$ corresponds to the initial state and $\alpha = 0$ to large times.

Normalizing the neck radius as $\overline{R}_n = R_n/R_i$ and the time using Equation (4) to give $\overline{t}_n$, we can cast these models (Eqns. 5–7) in dimensionless form, which for brevity, we give in the Supplementary Information. The relative accuracy of each of these approaches to two-droplet coalescence will be assessed later against collated experimental datasets (see Section 6).

2.2 'Free sintering' of many particles or sintering under zero applied load

The two-droplet problem described above has a range of possible solutions, all of which are appealingly simple (e.g., Eqns. 5–8). However, Kirchhof et al. (2009) showed that even adding one more droplet to the system (e.g., a 3-droplet coalescence problem) renders the problem far less amenable to simple analysis or scaling. Finite element models can be used to assess the evolution of systems of 3 or a few droplets in various coordination in 3D space (Kirchhof et al. 2009). However, once a system goes beyond a few droplets, to a many-droplet problem such as is relevant in most practical applications, the evolution of complex geometry is no longer explicitly computable. Therefore, approximations must be made and approximate models must be developed. In this section we summarize the main models. As with the two-droplet problem, here we focus on the case where capillary forces dominate, which is generally referred to as 'free sintering' or 'capillary sintering'.

Upscaling the Frenkel (1945) model. The simplest approach to the problem of many droplets has been to 'upscale' models for two-droplet systems. To give the most simple example, we take the Frenkel (1945) model (recognizing the input of Eshelby (1949), this is sometimes called the Frenkel–Eshelby model), which is given in Equation (5) and predicts the radius of the neck between two coalescing droplets R_n. Kang (2004) proposed a scaling for the radius of curvature on the outside of the growing neck (sometimes called the neck's 'meniscus') that is $h \approx R_n^2/(4R_i)$. Using this approximate h with Equation (5), Frenkel (1945) cast the evolution of the length of a chain of droplets ΔH relative to the initial length of the chain H_i as

$$\frac{\Delta H}{H_i} \approx \frac{h}{R_i} = \frac{3\Gamma t}{8\mu R_i}. \tag{9}$$

The result in Equation (9) can then be used to find the evolution of the bulk gas volume fraction in a pack of cubically packed particles (Prado et al. 2003; Soares et al. 2012), assuming that Equation (9) is valid at every particle–particle contact point and pair

$$\phi = 1 + (\phi_i - 1)\left(1 - \frac{3\Gamma}{8\mu R_i}t\right)^{-3} \tag{10}$$

where ϕ is the volume fraction of gas (i.e., the porosity) between the droplets and ϕ_i is the value of ϕ at t_i. Introducing $\overline{\phi} = \phi/\phi_i$ and using Equation (4) to give the dimensionless capillary neck-growth timescale $\overline{t}_n$, we can render Equation (10) dimensionless

$$\overline{\phi} = \frac{1}{\phi_i} + \left(1 - \frac{1}{\phi_i}\right)\left(1 - \frac{3}{8}\overline{t}_n\right)^{-3}. \tag{11}$$

The Mackenzie–Shuttleworth (1949) model. The limitations—described above—of approaches that simply upscale a two-droplet model to a many-droplet model, highlight the complexity in large sintering systems. An altogether different approach was taken by Mackenzie and Shuttleworth (1949) who envisage that sintering of a pack of particles can be approximated by shrinking of the pores between the particles. Each pore is idealized to be a spherical bubble of radius a, sitting in a shell of melt of radius S (measured from the bubble center; Fig. 4).

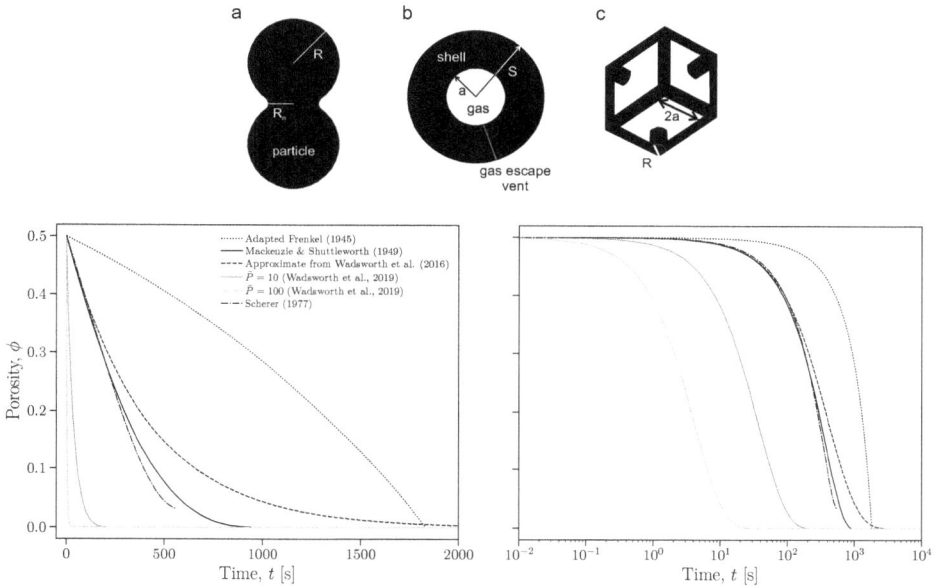

Figure 4. Model geometries (**a–c**) and model predictions (**d**). (**a**) The 'Frenkel' model (Frenkel 1945) for sintering is based on a unit-cell of two particles coalescing, then scaled up to predict bulk porosity. (**b**) The 'Mackenzie–Shuttleworth' model (Mackenzie and Shuttleworth 1949; Wadsworth et al. 2016a) is based on a unit cell of idealized gas bubbles in viscous shells where the gas pressure is kept constant by hypothetical permeable gas escape vents. (**c**) The 'Scherer' model (Scherer 1977) is explicitly formulated for 'low density' systems or dispersions (see Fig. 2) and is based on a unit cell of large open cubic arrays of pores bordered by viscous cylinders. In (**d**) we show an example dimensional solution for the evolution of $\phi(t)$ for each model for the case of $R_i = 10^{-5}$ m, which results in $a_i = 1.8410^{-6}$ m, (see text; Torquato 2013), and $\mu = 10^8$ Pa.s. We also show the pressure-sintering solution for pressures 10 and 100 times the surface tension pressure P_L (Wadsworth et al. 2019).

This model starts from a modified form of the Rayleigh–Plesset equation, which is used to model the dynamics of spherical bubbles in liquids, and which can be cast for bubbles in shells of finite size (rather than bubbles in an infinite liquid; Prousevitch et al. 1993)

$$P_{\mathrm{g}} - P_{\mathrm{l}} = \underbrace{\frac{2\Gamma}{a}}_{\substack{\text{surface} \\ \text{tension} \\ \text{stress}}} + \underbrace{4\mu \frac{da}{dt}\left(\frac{1}{a} - \frac{a^2}{S^3}\right)}_{\substack{\text{Newtonian} \\ \text{viscous stress}}} \tag{12}$$

where P_{g} and P_{l} are, respectively the pressure in the gas phase, and the pressure acting on the outside of the liquid shell, and where inertial terms are neglected because Oh $\gg$ 1 (Eqn. 2). In a real sintering system, the pores interstitial to the particles are interconnected, such that, in reality, these 'bubbles' are vented and open (Wadsworth et al. 2017a), which means that the amount of gas in the bubbles is not constant, that the gas pressure can equilibrate with the liquid pressure, and that $P_{\mathrm{g}} = P_{\mathrm{l}}$, such that the left-hand side of Equation (12) evaluates to 0.

The result is an ordinary differential equation describing the evolution of the gas volume fraction with time (Mackenzie and Shuttleworth 1949)

$$\frac{d\phi}{dt} = -\frac{3\Gamma}{2\mu}\left(\frac{4\pi}{3}N\right)^{1/3} \phi^{2/3}\left(1-\phi\right)^{1/3} \tag{13}$$

where N is the number density of bubbles per unit volume of melt or glass (in units m^{-3}). If we use the relationship between N and ϕ noted by Chiang et al. (1997), $4\pi a_i^3 N/3 = \phi_i/(1-\phi_i)$, where a_i is the bubble radius at time t_i (and at ϕ_i), then we can re-cast the Mackenzie and Shuttleworth (1949) to be a function of the initial conditions only

$$\frac{d\phi}{dt} = -\frac{3\Gamma}{2\mu a_i}\left(\frac{\phi_i}{1-\phi_i}\right)^{1/3}\phi^{2/3}(1-\phi)^{1/3} \tag{14}$$

for which we are making the assumption that N is constant. As we did with Equation (10), we can introduce the normalizations ϕ/ϕ_i and using Equation (4), we can normalize time by assuming here that $L=a_i$ to give $\bar{t}_b$ and to find

$$\frac{d\bar{\phi}}{d\bar{t}_b} = -\frac{3}{2}\left(\frac{1-\phi_i\bar{\phi}}{1-\phi_i}\right)^{1/3}\bar{\phi}^{2/3} \tag{15}$$

This model is often referred to as being valid for the end-stages of sintering (e.g., Scherer 1986; Bellehumeur et al. 1996). This assertion is based on the fact that the model geometry appears at first glance to be composed of isolated pores (or bubbles) in a closed-shell liquid medium (Fig. 4). However, this model is more sophisticated than that, and implicit in the approach encapsulated here is that the closed liquid shells are interconnected by hypothetical vents (Fig. 4), otherwise the amount of gas in the pores would remain constant, and the bubble volume would be stable under any isothermal condition. For these reasons, the Mackenzie and Shuttleworth (1949) model has been shown to be successful at predicting ϕ at intermediate and final (Prado et al. 2001) or all (Wadsworth et al. 2016a, 2017b, a) stages of sintering.

The Scherer (1977) model. Scherer (1977) noted that the description of the pore space between particles as spherical bubbles in liquid shells (see Section 2.2) is less valid for the very high porosities associated with very loose packs of particles. Instead, they imagine that the sintering particles are arranged in straight chains connected in a cubic lattice of chains. They cast the geometry as being cylinders of melt, which represent the chains of particles, with radius equivalent to the particle radii R and length equivalent to the pore size $2a$. This model geometry results in large, cubic pore spaces that are interconnected by definition (as opposed to the Mackenzie and Shuttleworth (1949) model in which the pores are interconnected via hypothetical vents that do not enter explicitly in the modelled dynamics). In this case, the porosity is

$$\phi = 1 - \frac{V_1}{8a^3} = 1 - \left(\frac{3}{4}\pi x^2 - \sqrt{2}x^3\right) \tag{16}$$

where $V_1 = 6\pi R^2 a - 8\sqrt{2}R^3$ is the volume of the liquid phase and $x = R/a$.

Then Scherer (1977) derives the rate of sintering as the cylindrical liquid bridges defining the lattice shrink under interfacial tension. They equate the energy dissipated in viscous flow to the energy change resulting from the reduction in internal surface area. The final relationship between ϕ and t is

$$\frac{d\phi}{dt} = \frac{\Gamma}{2\mu a_i}\frac{d\phi}{dx}\left(\frac{1-\phi}{1-\phi_i}\right)^{1/3}. \tag{17}$$

Normalizing ϕ by ϕ_i and using Equation (4) to define $\bar{t}_b$, we have

$$\frac{d\bar{\phi}}{d\bar{t}_b} = \frac{1}{2}\frac{d\bar{\phi}}{dx}\left(\frac{1-\phi_i\bar{\phi}}{1-\phi_i}\right)^{1/3} \tag{18}$$

The approximate exponential model (Wadsworth et al. 2016a). Both the Mackenzie and Shuttleworth (1949) and Scherer (1977) models require a numerical solution. An analytical solution for the Mackenzie and Shuttleworth (1949) model was found by Wadsworth et al. (2016a) for the case where the pore radius is known as a function of time

$$\phi = \phi_i \exp\left(-\frac{3\Gamma}{2\mu a(t)}t\right) \tag{19}$$

for which the dimensionless solution is

$$\bar{\phi} = \exp\left(-\frac{3}{2}\bar{t}_b^*\right) \tag{20}$$

where $\bar{t}_b^*$ represents Equation (4) but where $L=a(t)$. Equation (20) is of little practical use because $a(t)$ is rarely known *a priori*. However, it has been shown that an exponential function of the type given in Equation (20) captures the broad kinetics of free sintering when the approximation is made that $a(t) = a_i$ for all times, such that $\bar{t}_b^* = \bar{t}_b$ in Equation (20) (Prado et al. 2001, 2003; Wadsworth et al. 2016a).

2.3 Extending sintering models to polydisperse systems

The models described so far can be used with monodisperse values of R_i or a_i in $\bar{t}_n$ or $\bar{t}_b$ respectively. But industrial powders or natural systems are rarely monodisperse. Therefore, it's important to develop a general method for predicting sintering rates, while accounting for polydispersivity.

Here we give a single example using the Mackenzie and Shuttleworth (1949) model, but we note that the same approach could be used for other models. Assuming we have a rigorous relationship between a distribution of initial particle sizes and the distribution of pore sizes between those particles in the initial pack (this is defined and discussed in Section 4.1), then we can apply a convolution method (Wadsworth et al. 2017b). If we define a continuous probability density function $F(\zeta)$, where $\zeta = a/R$, we can solve the dimensional form of the Mackenzie and Shuttleworth (1949) model (Eqn. 14) for every contribution of ζ and therefore every contribution of a_i and integrate to find the evolution of the weighted average of the normalized porosity, which we term $\bar{\bar{\phi}}$

$$\bar{\bar{\phi}}(t) = R_i \int_0^\infty \bar{\phi}(t) F(\zeta) \mathrm{d}\zeta. \tag{21}$$

Procedurally we use Equation (14) to give $\phi(t)$, which is then normalized by ϕ_i to find $\bar{\phi}(t)$ for each class of ζ before using Equation (21) to find $\bar{\bar{\phi}}(t)$.

3. 'PRESSURE SINTERING' OR SINTERING UNDER EXTERNAL LOAD

Section 2 focusses on 'free sintering' or capillary sintering. However, many applications of industrial relevance involve 'pressure sintering', in which the weight of the particles pressurizes the system, or external pressures are applied during sintering (Scherer 1986). Similarly, pressure sintering may be important in many natural applications such as in the welding of thick ignimbrites deposited rapidly (Sparks et al. 1999; Wadsworth et al. 2019). Here we explore available models for pressure sintering, in which the pressure acting on the particles or droplets is large compared with the capillary pressure P_L.

3.1 The extended Mackenzie–Shuttleworth model (Wadsworth et al. 2019)

Returning to Equation (12) as a starting point, we can acknowledge that the pressure acting on the droplets P_1 may be greater than the gas pressure P_g. Following the same derivation as we did to find Equation (13), but without requiring $P_1 = P_g$, we arrive at (Wadsworth et al. 2019, 2021)

$$\frac{d\phi}{dt} = -\frac{3}{4}\frac{\Delta P}{\mu}\phi - \frac{3\Gamma}{2\mu a_i}\left(\frac{\phi_i}{1-\phi_i}\right)^{1/3}\phi^{2/3}\left(1-\phi\right)^{1/3} \tag{22}$$

where $\Delta P = P_1 - P_g$. We can define a pressure scale $\overline{P} = \Delta P / P_L$ using Equation (3) with $P_L = 2\Gamma/a_i$ and the normalizations used previously to render Equation (22) dimensionless

$$\frac{d\overline{\phi}}{d\overline{t}_b} = -\frac{3}{2}\left[\overline{P}\overline{\phi} + \left(\frac{1-\phi_i\overline{\phi}}{1-\phi_i}\right)^{1/3}\overline{\phi}^{2/3}\right]. \tag{23}$$

Here, ΔP represents an additional isotropic pressure term, separate from the capillary term, which arises from the weight of the particle pack or an applied external pressure.

3.2 The extended Scherer model (Scherer 1986)

Scherer (1986) acknowledged that the pressure on the particles or droplets may be acting in one direction and may not be isotropic as assumed in Equation (23). They derived a solution for a uniaxial pressure σ and neglected the gas pressure altogether. This has the implication that their solution is strictly valid only in the case where the gas pressure is negligible compared with the applied uniaxial pressure. In the Supplementary Information we give the full derivation adapted from Scherer (1986), which results in

$$\frac{d\phi}{dt} = \frac{\Gamma}{\mu a_i}\left[\frac{1}{2}\frac{d\phi}{dx}\left(\frac{1-\phi}{1-\phi_i}\right)^{1/3} + \frac{2}{3}(1-\phi)\left(\delta^{1/2}-\delta\right)\overline{\sigma}\exp\left(-2\varepsilon_r\right)\right] \tag{24}$$

where $\delta = [3-2(1-\phi)]/(1-\phi)$, $\overline{\sigma} = \sigma/P_L$, and is therefore analogous to $\overline{P}$ but for which P is replaced by σ. ε_r is the radial strain of the whole sintering system. A dimensionless version of Equation (24) is as follows

$$\frac{d\overline{\phi}}{d\overline{t}_b} = \frac{1}{2}\frac{d\overline{\phi}}{dx}\left(\frac{1-\phi_i\overline{\phi}}{1-\phi_i}\right)^{1/3} + \frac{2}{3}(1-\phi_i\overline{\phi})\left(\delta^{1/2}-\delta\right)\overline{\sigma}\exp\left(-2\varepsilon_r\right) \tag{25}$$

such that the first term on the right-hand side represents the right-hand side of Equation (18), and the second term captures the effect of the uniaxial pressure. The constitutive equations for ε_r and a solution method to Equations (24, 25) are given in the Supplementary Information.

3.3 The Quane and Russell (2005) model.

Quane and Russell (2005) present a simple empirical model for the porosity evolution during volcanic sintering, which they validate using experiments on natural volcanic ash packs as well as glass beads. The model is given by

$$\phi = \frac{A}{A-B}$$
$$A = \ln\left[\frac{B\sigma}{\mu(1-\phi_i)}t + \exp\left(-\frac{B\phi_i}{1-\phi_i}\right)\right] \tag{26}$$

where B is a constant that is empirically determined for each particle type tested. For example, for volcanic ash, $B = 0.78$, while for spherical silicate glass particles, $B = 5.3$ (Quane and Russell 2005;

Quane et al. 2009). The Quane and Russell (2005) model is analytical and easy-to-use, and has therefore been used widely to calculate compaction timescales for rubble-filled volcanic conduits (Quane et al. 2009), the welding time of high-temperature ignimbrites (Russell and Quane 2005; Lavallée et al. 2015), sintering in tuffisite veins (Farquharson et al. 2017; Kolzenburg et al. 2019), and in combination with constitutive equations for physical properties, to compute the permeability and strength during densification of magmas (Heap et al. 2015). We present this model for completeness, but do not analyze it further simply because the purely empirical nature of Equation (26) inevitably limits its applicability and use to the range of conditions of viscosity, uniaxial pressure and particle shape/type under which it was tested.

4. PORE SIZES, SURFACE AREA AND INTER-PARTICLE DISTANCES IN SINTERING SYSTEMS

The models described here can be grouped into those that rely on knowledge of the particle size only (Frenkel 1945; Eshelby 1949) and those that rely on some constraint of the pore sizes between the particles (Mackenzie and Shuttleworth 1949; Wadsworth et al. 2016a, 2019). There are simple methods to measure the particle size distribution of glass particles used in sintering systems (e.g., laser particle size analysis). However, pore size determination is less straightforward. Indeed, because the pores between particles are not discrete objects, it is not immediately obvious what measurement is valid in terms of constraint of a single size or distribution of sizes in continuous pore spaces. Given that the size constraints placed on convolute pore spaces is a foundational quantity underpinning many of the models for sintering, here we present a theoretical approach to predicting characteristics of the pore space (Wadsworth et al. 2016a, 2017b).

These theoretical constraints are based on a large body of work in the characterization of random heterogeneous media (summarized in Torquato 2013). The theoretical approaches can be divided into those relevant for the pores between "hard particles", such as classical solid granular media, or for the pores between "overlapping particles". As discussed, sintering systems begin as packs of particles such that the starting state of a sintering system is in the "hard particles" category. But as sintering progresses, the internal geometry changes. While the particles do not move into one another or overlap in the strict sense, we note that the particle centers do move toward one another as mass flows from the particles into the particle–particle contact point necks. Therefore, the pores approximately evolve according to the theory for the pores between "overlapping particle" packs. This assumption has been borne out by comparison of predictions based on this theory for pore sizes between overlapping particles, and the measured properties of sintered media, such as the fluid permeability (discussed later; Wadsworth et al. 2016b).

In what follows, we present a suite of methods to predict the pore-scale characteristics of hard or overlapping granular media, relevant for sintering systems. Specifically, we provide solutions for the pore sizes, the specific surface area, and the inter-particle distances, all of which are required within this Chapter.

4.1 Pore sizes

The radii of initially spherical glass particles are trivial to constrain using a variety of techniques, which provides constraint of the lengthscale R_i for use with the Frenkel (1945) model. However, the lengthscale a_i that appears in the Mackenzie and Shuttleworth (1949) or the Scherer (1977) models and extensions thereof (Scherer 1986; Wadsworth et al. 2019) is a less easy-to-constrain parameter. Torquato (2013) and Torquato and Avellaneda (1991) provide a rigorous expression for a mean pore size a occurring between particles in arbitrary packing. Their scheme can be cast for a packing of completely impenetrable 'hard' spheres. This is given in the form of a cumulative probability density $F(\zeta)$ of the pore size distribution for which $\zeta = 1/x = a/R$, as described earlier but defined formally here:

$$F(\zeta) = \frac{E_V(\zeta)}{\phi}, \tag{27}$$

where $E_V(\zeta)$ is a pore nearest-neighbour exclusion probability function. To solve for E_V is a non-trivial problem that has received significant attention. A validated expression for E_V as a function of R is given by Torquato (2013) based on Torquato and Avellaneda (1991), which we cast in terms of the gas volume fraction ϕ

$$E_V(\zeta) = \phi \exp\left(-(1-\phi)\left[y_0(1+\zeta)^3 + 3y_1(1+\zeta)^2 + 12y_2(1+\zeta) + y_3\right]\right). \tag{28}$$

Equation (28) is valid for $\zeta \geq 0$ and contains coefficients y_n which are given by

$$y_0 = \frac{2 - \phi + (1-\phi)^2 - (1-\phi)^3}{\phi^3},$$

$$y_1 = \frac{(1-\phi)\left(3(1-\phi)^2 + 4\phi - 7\right)}{2\phi^3}, \tag{29}$$

$$y_2 = \frac{(1-\phi)^2(1+\phi)}{2\phi^3},$$

$$y_3 = -(y_0 + 3y_1 + 12y_2).$$

The nth moment of the probability density function of ζ, termed $\langle \zeta^n \rangle$, is then related to the cumulative probability density function $F(\zeta)$ in Equation (27) by integrating as follows

$$\langle \zeta^n \rangle = n \int_0^\infty \zeta^{n-1} F(\zeta) d\zeta, \tag{30}$$

hence the mean (i.e., $n = 1$) value of a is $\langle a \rangle = \langle \zeta \rangle / \langle R \rangle$.

Equations (27–30) can be used to find $\langle a \rangle$ in the monodisperse limit of R. Torquato (2013) describes the solution of Lu and Torquato (1992) which further constrains a polydisperse solution which is again validated against data. In this form, the pore nearest neighbour exclusion probability function is the polydisperse $e_V(\zeta)$ and is

$$e_V(\zeta) = \phi \exp\left(-2S(1-\phi)\left[\frac{z_0}{8}(1+\zeta)^3 + \frac{z_1}{4}(1+\zeta)^2 + \frac{z_2}{2}(1+\zeta)\right]\right), \tag{31}$$

for which S is the ratio of the specific surface of the polydisperse system to that of the monodisperse system at the same value of ϕ. S is given by

$$S = \frac{\langle R^2 \rangle}{\langle R^3 \rangle}\langle R \rangle, \tag{32}$$

where again $\langle R^n \rangle$ is the nth moment of the probability density function of the particle radius distribution. As before, the coefficients z_n are given by specific solutions, here with a dependence on S and $\langle R^n \rangle$,

$$z_0 = \frac{4\phi\langle R\rangle^2\left[\phi + 3S(1-\phi)\right] + 8\langle R^2\rangle\left[S(1-\phi)\right]^2}{\phi^3\langle R^2\rangle},$$

$$z_1 = \frac{6\phi\langle R\rangle^2 + 9S\langle R^2\rangle(1-\phi)}{\phi^2\langle R^2\rangle},$$

$$z_2 = \frac{3}{\phi}.$$

(33)

To arrive at the polydisperse solution for $\langle a\rangle$ as a function of $\langle R\rangle$, the same method as the monodisperse limit is applied but where $E_V(\zeta)$ in Equation (27) is replaced by $e_V(\zeta)$ and Equation (30) remains unchanged.

The methods given to find characteristic pore sizes between packed particles (Eqns. 27–33) is motivated by the necessity to place constraint on a single value of pore size $\langle a_i\rangle$ for use with the sintering models given in this Chapter. However, we note that, in reality, the system of inter-particle or inter-droplet distances is not equivalent to an ideal spherical pore in a melt shell. Initially and in the monodisperse limit, the system is clearly a pack of particles with a complex interconnected gas pore space (e.g., Fig. 5), which is not well approximated by the model geometries (Fig. 4). Wadsworth et al. (2017b) noted that it may be reasonable to assume that, if the system of initial droplets or particles is highly polydisperse, the approximation of the pore space as spheres becomes more tenable and realistic. Indeed, they noted that experimental data using highly polydisperse populations of particles resulted in a better match to sintering models than did data using monodisperse populations of particles. Therefore, they used a wide range of experimental data to propose an empirical adjustment to $\langle a_i\rangle$ to give an adjusted parameter $\hat{a}_i$ for monodisperse systems. This adjustment takes the form

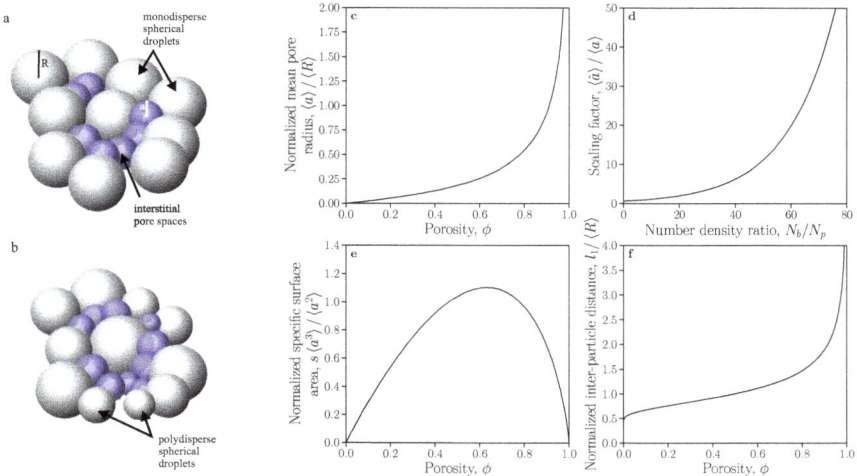

Figure 5. The pore size functions used herein to predict the pore characteristics of sintering systems. (**a–b**) The monodisperse (**a**) and polydisperse (**b**) pictures of a starting point before sintering begins showing the glass particles or droplets in **grey** and the imagined segmentation of the pores into spheres in **purple**. (**c**) The pore radius as a function of porosity from Equations (27–33) (Lu and Torquato 1992; Torquato 2013). (**d**) The empirical scaling factor (Eqn. 34) for use with the Mackenzie and Shuttleworth (1949) model in order to render it applicable to polydisperse systems (Wadsworth et al. 2017b). (**e**) The specific surface area for overlapping sphere systems (Eqns. 35–36). And (**f**) The inter-particle distances from particle centre to particle centre (Eqn. 37).

$$\frac{\hat{a}_i}{\langle a_i \rangle} \approx 0.76 \exp\left(0.05 \frac{N_b}{N_p}\right)$$

(34)

$$N_b = \frac{3\phi_i}{4\pi\langle a_i \rangle^3}; N_p = \frac{3(1-\phi)}{4\pi\langle R_i \rangle^3}$$

where N_b and N_p are the number density of pores and particles, respectively, and we note that they are number density per unit total volume here (as opposed to per unit volume of melt or glass as was the case for N introduced earlier).

4.2 Specific surface area

The specific surface area s is the total surface area of the pore space within a unit volume. This property has been shown to be essential for predicting a range of transport properties of sintering systems (Torquato and Lu 1990; Martys et al. 1994; Torquato 2013; Wadsworth et al. 2016a,b, 2017a). There is a range of methods for computing the specific surface area of porous media. The simplest is for a pack of monodisperse particles, where $s = 3(1-\phi)/R$, or for polydisperse particles $s = 3(1-\phi) \langle R^2 \rangle / \langle R^3 \rangle$. When the particles begin to overlap or interact during sintering, and cease to be a simple pack, the solution becomes non-trivial. For the monodisperse case, this relates to the pore size as

$$s(a) = -\frac{3(1-\phi)\ln(1-\phi)}{\langle a \rangle},$$

(35)

and for the polydisperse case, $\langle a \rangle$ is replaced by $\langle a^3 \rangle / \langle a^2 \rangle$ so that

$$s(a) = -3(1-\phi)\ln(1-\phi)\frac{\langle a^2 \rangle}{\langle a^3 \rangle}.$$

(36)

4.3 Inter-particle distance

The inter-particle distance l_1 is a metric subtly different from the pore size and has been shown to be the controlling lengthscale in the uniaxial compressive strength of sintered glass samples (Vasseur et al. 2017). For monodisperse particles, this is given by

$$l_1 = \frac{G(4/3)}{(-\ln\phi)^{1/3}}\langle R \rangle$$

(37)

where G is the gamma function.

5. EXPERIMENTAL APPROACHES

In Sections 2, 3 and 4, we present purely theoretical constraints on sintering phenomena, each representing a different approach. These can be tested against data from experiments. There is a wide range of experimental techniques and approaches used to assess sintering processes. Here we summarize the principal experimental approaches and, importantly, the approaches used to collect the data that are collated and analyzed herein. In Figure 6 we show the experimental set ups described here in schematic and simplified form; in Table 1, we summarize these approaches and identify the sources of the data that will be analyzed later. We do not specifically discuss so-called flash sintering or microwave sintering, but acknowledge that they have important industrial applications.

Table 1. A guide to empirical data used in this Chapter.

...opic	Relevant theory[+]	Reference for empirical data	Material	Technique	Data analysis
...oalescence of two ...rticles/droplets[*]	Sect. 2.1. Eqn. 5 (Frenkel 1945; Eshelby 1949); Eqns. 6, 7 (Eggers et al. 1999); Eqn. 8 (Hooper 1984).	Kingery and Berg (1955)	Spherical SLS glass beads	Hot microscopy	Fig. 8
		Rosenzweig and Narkis (1981)	Spherical PMMA particles	Hot microscopy	Fig. 8
		Bellehumeur et al. (1996)	Spherical HDPE and LDPE particles	Hot microscopy	Fig. 8
...ntering of many ...rticles/droplets in ...e capillary sintering ...gime	Sect. 2.2. Eqn. 11 (Frenkel 1949; Prado et al. 2001); Eqn. 15 (Mackenzie and Shuttleworth 1949; Wadsworth et al. 2016b); Eqn. 18 (Scherer 1977); Eqn. 20 (Wadsworth et al. 2016b).	Jagota et al. (1990); Wadsworth et al. (2016a, 2017a,b); Vasseur et al. (2016)	Monodisperse and polydisperse spherical SLS glass beads	Optical dilatometry (Fig. 6a)	Fig. 10
		Prado et al. (2001); Vasseur et al. (2013); Wadsworth et al. (2014)	Polydisperse angular SLS and BS glass particles	Optical dilatometry (Fig. 6a) or furnace experiments	Fig. 10
		Lara et al. (2004)	Polydisperse ABS glass	Optical dilatometry (Fig. 6a)	Fig. 10
		Reis et al. (2018)	Polydisperse spherical diopside glass beads	Optical dilatometry (Fig. 6a)	Fig. 10
...ntering of many ...rticles/droplets in ...e pressure-sintering ...gime	Sect. 3. Eqn. 22 (Wadsworth et al. 2019); Eqn. 24 (Scherer et al. 1986); Eqn. 25 (Quane and Russell 2005).	Friedman et al. (1963)	Angular Pyrex™glass particles of unknown distribution	Compaction rig (Fig. 6d)	Fig. 11
		Rahaman et al. (1990)	Monodisperse spherical BS glass beads	Vertical dilatometer with applied load	Fig. 11
...ntering of many ...rticles/droplets with ...ystals	Sect. 7.1. Eqns. 38, 39 (this study).	Eberstein et al. (2009)	Polydisperse aluminum-bearing BS glass particles mixed with ceramic rigid particles	Optical dilatometry (Fig. 6a)	Fig. 12
...ntering of many ...rticles/droplets in the ...pillary regime and ...th diffusive mass ...nsfer	Sect. 7.2. Eqn. 40 (Wadsworth et al. 2019).	Gardner et al. (2018, 2019)	Polydisperse angular rhyolite glass particles with dissolved water	Cold-seal pressure vessel (Fig. 6c)	Fig. 13
		Wadsworth et al. (2019)	Polydisperse angular Krafla rhyolite glass particles with dissolved water	Optical dilatometry (Fig. 6c) and in situ x-ray tomography (Fig. 6b)	Fig. 13

...tes:
...or these studies we only select data for which two particles of equal size are coalescing.
...ere we refer to the dimensionless equations where possible and direct the reader to the relevant section for the ...nensional equivalents.
...S = soda-lime-silica glass. BS = borosilicate glass. PMMA = poly(methyl methacrylate).
...PE and LDPE = high- and low-density polyethylene.

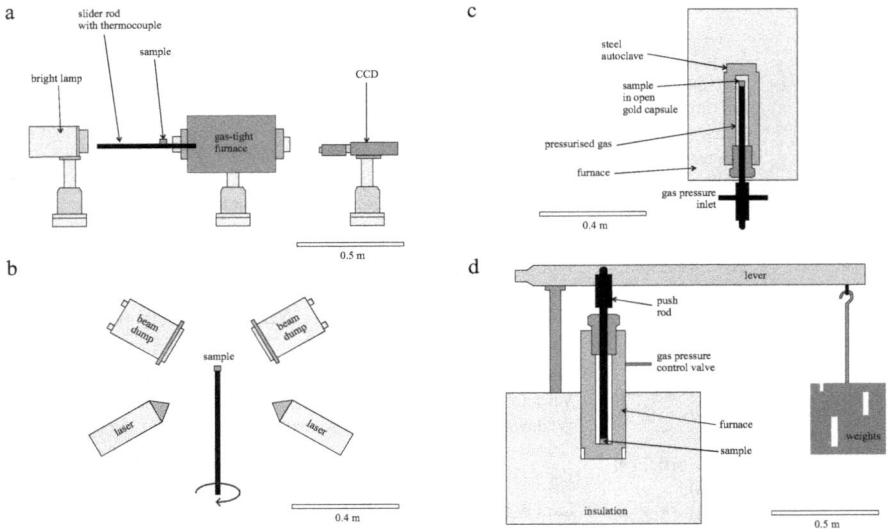

Figure 6. Experimental approaches used to examine sintering phenomena. **(a)** The optical dilatometer. **(b)** The *in situ* sintering set up at the Swiss Light Source synchrotron for use with the X-ray beamline TOMCAT. **(c)** The cold-seal pressure vessel. **(d)** A compaction rig adapted from Friedman et al. (1963).

5.1 Sintering under equilibrium pressures

Sintering is amenable to laboratory analysis using a variety of techniques. Perhaps most simply, powders formed of any crushed glass particle size can be poured into a crucible and heated to a target temperature for a known period of time. The result will be a partially or wholly sintered mass which could be drilled or broken out of the crucible for inspection or further analysis (this technique was used by Vasseur et al. 2013, 2016). However, when sintering is constrained within a crucible (sometimes referred to as 'constrained sintering'), there can be edge effects as the glass particles soften and stick to the walls of the crucible, and this can lead to potential resistance to the sintering stress, depending on the wetting properties of the glass–crucible interfaces. To avoid this, 'unconstrained' or 'free sintering' is often preferable.

Free sintering at low pressure. Free sintering involves the formation of a free-standing pack of particles that can be heated without falling apart (used by Lara et al. 2004; Wadsworth et al. 2016). Free sintering can be performed using simple piles of particles in a heat-resistant dish (Kueppers et al. 2014), but this can make it difficult to form a reproducible initial packing density of the particles. A preferred method is to use a dye press and a push rod to load particles of a known total mass into a known volume, and if the particles are sufficiently small then electrostatic forces between particles will allow this particle pack to be extruded from the dye as an intact, free-standing cylinder (Lara et al. 2004; Song et al. 2014; Wadsworth et al. 2016a). To make it possible to use larger particles, a small droplet of water can be added to the powder. The free-standing packed cylinder of particles can then be heated and the post-experimental run product will not require drilling from a crucible, making further analysis easy.

Optical dilatometry is a technique that allows free-standing cylinders (or any shape) to be monitored visually and continuously through a window during heating in a small precision furnace (Boccaccini and Hamann 1999) that can be programmed to perform any temperature–time pathway at 1 bar pressure (Fig. 6a). Previous work (Wadsworth et al. 2014, 2016a, 2017b, 2019) has used a Hesse Instruments (GmbH) optical dilatometer, which uses in-house software to monitor, in real time, various characteristics of the sample including (but not limited to) the height above the base (in pixels), a maximum sample width (in pixels), and the cross-sectional area of the sample (in pixels squared). The software also records images of the evolving sample captured using the CCD camera. Wadsworth et al. (2016) applied a solid-of-revolution (or solid-of-rotation) method to these images to convert the distance of the sample edge from a mid-line to a volume (in voxels) assuming the sample is axi-symmetric. If the position of the mid-line is at $j = 0$ and j and y are a horizontal and vertical coordinate, respectively, then the volume is $V = \int_0^H \pi j^2 dy$. This can result in a continuous measure of the sample volume with time during any thermal pathway programmed in the furnace, and is therefore a versatile tool for testing sintering dynamics.

As discussed, the porosity ϕ is the typical metric for sintering studies. When the glass particle phase can be assumed to be at constant volume (e.g., no crystallization and no significant volume change associated with volatile movement into or out of the particles), V can then be continuously converted to porosity using $\phi(t) = 1 - (V_f/V)(1 - \phi_f)$. This requires knowledge of the final porosity ϕ_f and the final volume V_f, both of which can be measured after the experiment. This method has the advantage that the measured volumes V and V_f can remain in units of voxels and do not require calibration, except via ϕ_f. Examples of 2D silhouettes captured during optical dilatometry, which are the images used directly in the calculation of $V(t)$, are given in Figure 3 (insets to Figs. 3a–d).

In situ sintering at X-ray beamlines using synchrotron radiation. Using optical dilatometry (Section 5.1), the volume of a sample is observed, but the internal pore structure and the partitioning of the pore space between connected and isolated porosity, as well as length scales of pores or surface area information, cannot be determined. These data can be valuable in constraining the physics of sintering and the properties of sintered materials, and can be reconstructed from 3-dimensional tomographic data. Such data can be captured continuously via in situ, synchrotron-source X-ray tomography of systems that are sintering at high temperature. Wadsworth et al. (2017a) performed sintering tests at the TOMCAT X-ray imaging beamline at the Swiss Light Source synchrotron (Fig. 6b). Heat was controlled using a dual laser heating system (Fife et al. 2012) and monitored using a pyrometer (and calibrated using a variety of in-house techniques). A sequence of radiographs collected while the sample is rotated under laser heating are collected and those are reconstructed into 3-dimensional tomographic data using in-house reconstruction algorithms. A 3-dimensional image processing software, such as Avizo™, can be used to segment the data into pore space volumes and glass volumes (or droplet volumes). Example results are shown in Figures 3a–d, where information about connectivity of pore space across the domain can be extracted. These 3-dimensional data are invaluable for further high-level analysis, such as simulating fluid flow through the pore spaces using techniques such as lattice-Boltzmann simulations (Llewellin 2010), which can, in turn, be used to determine the permeability of sintering materials continuously (Wadsworth et al. 2017a). While X-ray techniques can be performed on quenched post-experimental samples with similar results (Okuma et al. 2017), the continuous nature of *in situ* techniques allows the sintering continuum to be captured in a single experiment.

Sintering at elevated gas pressure. Cold-seal pressure vessels are a mainstay of petrology and geochemistry research, and have the advantage that the temperature and pressure of a small controlled sample can be modified with high precision (e.g., Gardner and Ketcham 2011). Samples are typically sealed in a capsule which is welded shut, and then loaded into a steel autoclave, which is in turn loaded into a furnace (Fig. 6c). The pressure on the sample is controlled by a squeezing gas pressure acting on the sample capsule exterior via a gas pressure in the autoclave. A less widely used variant of this technique is to leave the capsule open so that, for sintering systems, the pressurizing gas medium in the autoclave is also the gas pressure between the particles (Gardner et al. 2018, 2019). This has the advantage that the system can be in the capillary sintering regime, and not in the pressure sintering regime, even at interstitial gas pressures far above atmospheric. As a specific case applicable to volcanic processes, this is relevant because high gas pressures can allow experimental access to sintering of volatile-rich droplets, and droplets with disequilibrium dissolved volatiles concentrations, which would not be readily possible at atmospheric pressure.

5.2 Sintering under differential pressures

Sintering under a uniaxial load is a common way to accelerate sintering in industrial settings (Scherer 1986; Rahaman and De Jonghe 1990) and has been used to simulate compaction of obsidian particles (Friedman et al. 1963) and sintering phenomena in welding ignimbrites after deposition (Quane and Russell 2005; Russell and Quane 2005) or in welding block-and-ash flows (Heap et al. 2014). The procedure varies depending on the exact device or set up being used, but an example is shown in Figure 6d adapted from Friedman et al. (1963) in which the load on the sample is kept constant via a weight assembly. In this specific example, the sample is pre-sintered to give it coherence so that it can be free standing in the furnace under the push rod, but in other experimental set-ups the sample is loaded as a loose powder and pressed in a cup (Heap et al. 2014). The load force is applied in one direction giving rise to the pressure we term σ (Section 3.2). Usually, the unsupported sides are at a fixed gas pressure that is less than σ, with the result that the sample will flow and bulge sideways while also sintering (Quane and Russell 2005). While this set up may not be exactly relevant to any particular natural scenario, its utility in industrial sintering has made it a worthwhile geometry to understand (Scherer 1986; Rahaman et al. 1987; Rahaman and De Jonghe 1990).

6. EMPIRICAL DATA AND ANALYSIS

In this Section we describe collated data from the literature (see Table 1 for a summary). Each dataset used was collected using one of the techniques described in Section 5. Data are identified as relating to free (or capillary) sintering, or pressure sintering, and are analyzed using the theoretical models given in Section 2 or Section 3, respectively. We have collated data from both isothermal and non-isothermal sintering, and across a variety of glass compositions and particle sizes. The analysis we present is cast in dimensionless space, such that the thermal history is accounted for and all data should collapse to a single description, which has been termed the 'master sintering curve' (Uhlmann et al. 1975; Wadsworth et al. 2014).

6.1 Fundamental quantities specific to the data used here

The first step in analyzing experimental data is to constrain the physical properties of the glass used to collect each dataset. From Sections 2, 3 and 4, it's clear that the dominant properties we need to consider are the viscosity of the glass/melt μ, the particle size R or the moments of the distribution of particle sizes $\langle R \rangle$, $\langle R^2 \rangle$ and $\langle R^3 \rangle$ (which are used to compute the pore sizes), the interfacial tension Γ, and the pressures acting on the system P_l or σ, if any.

The viscosity of silicate glass and melt has been tightly constrained for a wide array of compositions over all relevant temperatures of practical interest, leading to some global viscosity models for which the glass composition and temperature are the only inputs required (Fluegel 2007; Giordano et al. 2008). These generally take the form of a Vogel–Fulcher–Tammann equation (Russell et al. 2022, this volume)

$$\log_{10}\left(\mu\right) = W_1 + \frac{W_2}{T - W_3} \tag{38}$$

for which W_1, W_2, and W_3 are coefficients that depend on composition in a manner specific to each model. For the literature data analysed herein, we use Equation (38) with inputs for W_1, W_2, and W_3 given by the authors of the study from which the data comes. In Figure 7a we show the relationship between μ and T for all glass used in this Chapter.

The particle size used in published studies of sintering is generally in the range $10^{-7} < R < 10^{-4}$ m. Importantly, at the conditions of sintering addressed here, these sizes are sufficiently small that Eo $\ll$ 1 computed using Equation (1), such that gravitational forces are negligible on the particle scales. The specific distributions of particle sizes used in studies that are collated herein are given in Figure 7b.

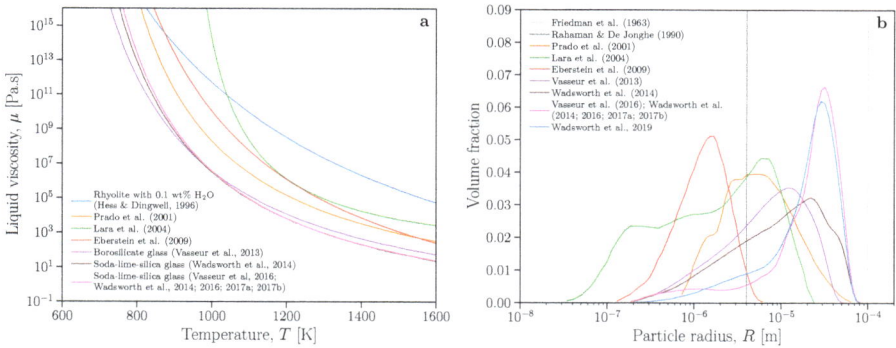

Figure 7. The properties of glass particle populations analyzed herein. **(a)** The melt viscosity as a function of temperature for all glass systems that are used to validate sintering models in this chapter (Hess and Dingwell 1996; Prado et al. 2001; Lara et al. 2004; Eberstein et al. 2009; Vasseur et al. 2013, 2016; Wadsworth et al. 2014, 2016a, 2017a), where in each case, the curve is defined by a Vogel–Fulcher–Tammann functional form fit to data from experimental viscosity determinations on the same glass. **(b)** The particle size distributions used in the studies represented in this chapter covering the range from $3{\times}10^{-8} \lesssim R \lesssim 1{\times}10^{-4}$ m and polydispersivity $0.4 \lesssim S \lesssim 1$. The vertical lines are for the two studies for which an indicative single particle size is given (Friedman et al. 1963; Rahaman and De Jonghe 1990).

For silicate melts and glasses in contact with air or specific gases of laboratory utility, there is also reasonably good constraint for the interfacial tension Γ (Parikh 1958; Kraxner et al. 2009). While it is clear that this property should be dependent on composition and temperature (e.g., Walker and Mullins Jr 1981), there exists no general model for that dependence. We note that the published evidence is that, for glass without large concentrations of dissolved volatiles, such as water, and over relatively small temperature changes, the differences in Γ are small (Walker and Mullins Jr 1981). Therefore, unless otherwise stated in the published sources from which we extract data, we use a general value for dry silicate melts of $\Gamma \approx 0.3\,\mathrm{Nm}^{-1}$ (Vasseur et al. 2013; Wadsworth et al. 2016a).

In Section 6.2, we collate data for non-silicate systems of droplet-droplet pairs sintering. The materials used there are poly(methyl methacrylate) or PMMA (Rosenzweig and Narkis 1981), and high- or low-density polyethylene or HDPE and LDPE (Bellehumeur et al. 1996). The viscosity, and interfacial tension properties of these materials are given in the references originating the experimental data.

Taken together, these constraints of material properties are all that are required to compute λ and P_L using Equation (3), hence, these are the constraints required to normalize time in all experiments using Equation (4). This presents an opportunity to analyze all data in a given regime (e.g., capillary sintering regime) together regardless of the specific experimental conditions— glass composition, particle size, temperature program—under which the sintering occurred.

6.2 Sintering of two droplets or particles

In Figure 8a we show raw data for the evolution of the dimensional neck radius R_n as a function of time. These data include those for the sintering of two equal-sized droplets or glass particles of compositions that include soda-lime-silica glass beads (Kingery and Berg 1955), droplets of poly(methyl methacrylate), or PMMA (Rosenzweig and Narkis 1981), and droplets of both high- and low-density polyethylene, or HDPE and LDPE (Bellehumeur et al. 1996). In Figure 8b we show these same data but normalized as R_n/R_i and where time is normalized to give $\overline{t}_n$ using Equation (4) where $L = R_i$. This procedure of normalization collapses the scattered datasets in Figure 8a to a single monotonically increasing trend in Figure 8b.

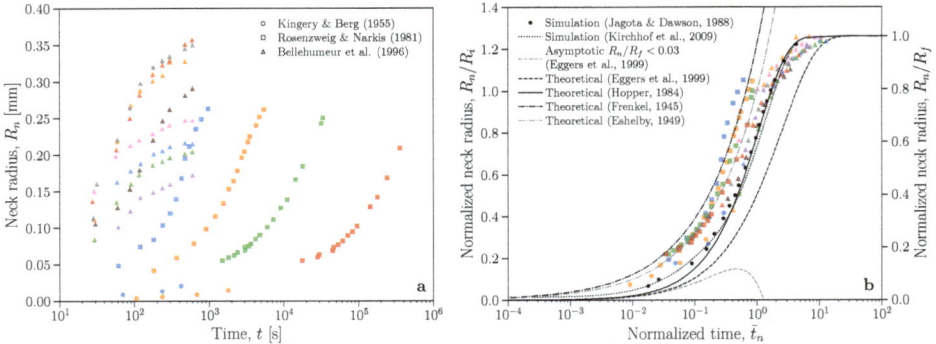

Figure 8. The sintering (or coalescence) of two equal-sized melt droplets. In panel (**a**), we show collated experimental data for the growth of the neck radius R_n with time t for spherical soda-lime-silica glass beads (Kingery and Berg 1955), droplets of poly(methyl methacrylate), or PMMA (Rosenzweig and Narkis 1981), and droplets of both high- and low-density polyethylene, or HDPE and LDPE (Bellehumeur et al. 1996). In panel (**b**) we render these data dimensionless by normalizing the neck radius to the initial drop radius R_n/R_i, and the time by the particle capillary timescale $\bar{t}_n = t\Gamma/(\mu\ R_i)$. In this panel we additionally show the results of finite element simulations (Jagota et al.; Kirchhof et al. 2009) and compare all data with models described in the main text (Frenkel 1945; Eshelby 1949; Hopper 1984; Eggers et al. 1999).

In Section 2.1 we present models for sintering of two droplets in the form $R_n / R_i = f(T)$ (Frenkel 1945; Eshelby 1949; Hopper 1984; Eggers et al. 1999); we present these models in Figure 8b so that their performance can be assessed against the data. We find that, although the Frenkel (1945) and Eshelby (1949) models perform well against the data at early and intermediate times, they fail to capture the late-stage sintering because they do not asymptotically approach an final radius R_f. As such, it is clear that the Hopper (1984) and Eggers et al. (1999) models provide a more complete description of the data in this dimensionless space. In Figure 8b we also show the results of finite element simulations (Jagota and Dawson 1988; Kirchhof et al. 2009). In all cases, the normalization by R_i can be replaced by R_f so that the models $R_n/R_f \to 1$ as $t \to \infty$, and the conversion is $R_n/R_i = 2^{1/3}R_n/R_f$ (in Fig. 8b we give both possibilities cast as two y-axis labels).

6.3 Sintering of large systems of many droplets

For systems of many droplets, the progress of sintering is most usefully quantified through the evolution of porosity with time. We note that, in many studies, the relative density ρ/ρ_0 is measured instead, where ρ_0 is the density of the glass and ρ is the bulk density of the system. Porosity and relative density are related by $\phi = 1 - \rho/\rho_0$.

As discussed in Section 1, during sintering of many particles, the evolution of porosity ϕ in the pack as a function of time is non-linear and decreases monotonically from an initial value ϕ_i to a final trapped porosity ϕ_c, at which point the gas phase becomes fully isolated (Fig. 3). In Figure 9 we show the general trend mapped out by sintering data for $\phi(t)$ in isothermal (Fig. 9a) and non-isothermal (Fig. 9b) conditions. For a given glass composition and particle size distribution, sintering proceeds more rapidly at higher temperature and at higher heating rates (from a common starting point).

Capillary sintering regime. By normalizing time to give $\bar{t}_n$, in the case where $L = R_i$ in the upscaled Frenkel–Eshelby models, or $\bar{t}_b$ in the case where $L = a_i$ in the bubble-based models (Section 2), and by normalizing ϕ to give $\bar{\phi}$, we can arrive at universal 'master' sintering plots. In this space $\bar{\phi}(\bar{t}_n)$ or $\bar{\phi}(\bar{t}_b)$, all data and models for the capillary sintering regime can be plotted together in a single plot regardless of composition or particle size.

In Figure 10 we show the dimensionless solution to the upscaled Frenkel (1945) model (Eqn. 11 with Eqn. 4 where $L = R_i$), the monodisperse solution to the Mackenzie and

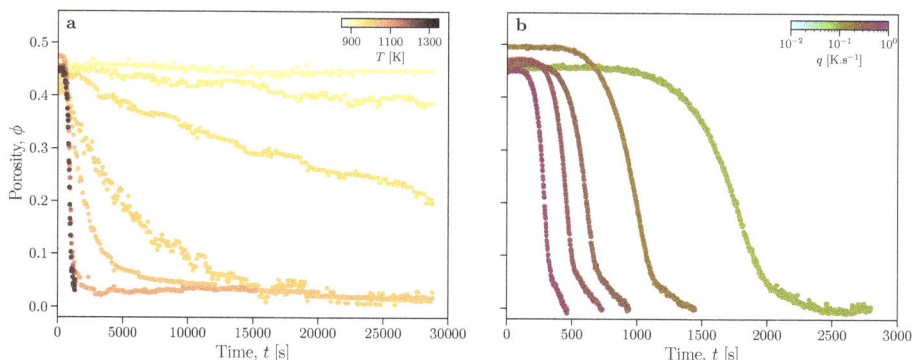

Figure 9. Indicative examples for the evolution of porosity with time during hot sintering of free-standing packs of glass particles observed using optical dilatometry in **(a)** isothermal conditions during sintering of spherical glass beads (data from Wadsworth et al. 2016) and **(b)** conditions of linear heating during sintering of angular glass particles (data from Wadsworth et al. 2014).

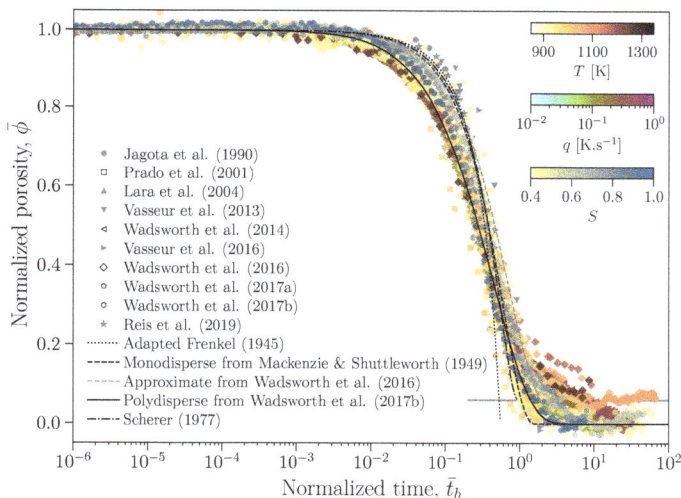

Figure 10. The capillary-sintering regime for packed hot glass particles. The data shown in this plot are compiled from a range of sources and include angular particles of soda-lime-silica glass and borosilicate glass (Prado et al. 2001; Vasseur et al. 2013; Wadsworth et al. 2014), spherical particles of soda-lime-silica glass (Jagota et al. 1990; Vasseur et al. 2016; Wadsworth et al. 2016a, 2017a,b), angular glass particles of aluminium-barium-silicate (Lara et al. 2004) and spherical diopside glass beads (Reis et al. 2018). In order to compare all data performed using this wide range of compositions, at a wide range of temperatures (900–1300 K) or heating rates (0.08–1 K.s^{-1}), and across a wide range of polydispersivity (from $S = 0.4$ to 1), we use the normalizations $\bar{\phi} = \phi/\phi_i$ and $\bar{t}_b = t\Gamma/(\mu \hat{a}_i)$ (valid in the case of isothermal conditions and monodisperse particles; see text for variants thereof used for non-isothermal conditions or polydisperse particles). These compiled data are compared with continuum models for sintering (Frenkel 1945; Mackenzie and Shuttleworth 1949; Scherer 1977; Wadsworth et al. 2016a, 2017b). In order to plot the Frenkel (1945) model in this space, we use $\bar{t}_n$ in place of $\bar{t}_b$. The **horizontal grey line** represents an approximate percolation threshold porosity ϕ_c/ϕ_i at which sintering stops, where for an example case we take $\phi = 0.03$ and $\phi_i = 0.5$.

Shuttleworth (1949) model (Eqn. 15 with Eqn. 4 where $L = \langle a_i \rangle$), the polydisperse adaptation of the Mackenzie and Shuttleworth (1949) model by Wadsworth et al. (2017b) (Eqn. 15 with Eqn. 4 where $L = \langle \hat{a}_i \rangle$ via Eqn. 33), the analytical approximate Wadsworth et al. (2016a) model (Eqn. 20) and the Scherer (1977) model (Eqn. 18 with Eqn. 4 where $L = a_i$). These model solutions are compared with a large compiled dataset for sintering of angular particles of soda-lime-silica and borosilicate glass (Prado et al. 2001; Vasseur et al. 2013; Wadsworth et al. 2014), spherical particles of soda-lime-silica glass (Jagota et al. 1990; Vasseur et al. 2016; Wadsworth et al. 2016a, 2017a,b), angular glass particles of barium-aluminosilicate (Lara et al. 2004) and spherical diopside glass beads (Reis et al. 2018).

It is clear that all models are similar in their performance against the data. Following Wadsworth et al. (2016a) we favor the Mackenzie and Shuttleworth (1949) model for several reasons. Firstly, it captures the late stages of sintering better than the Frenkel (1945) model, which diverges from the data as sintering goes to completion. Secondly, both the Frenkel (1945) and Scherer (1977) models are based on assumptions of cubic arrays of pores, whereas the Mackenzie and Shuttleworth (1949) model and adaptations thereof (Wadsworth et al. 2016a, 2017b) is instead generalized for any random heterogeneous arrangement of pore structure via Equations (27–33) (Torquato 2013). Finally, the starting point for the Mackenzie and Shuttleworth (1949) model (Eqn. 12) includes the separation of P_l and P_g, making it easy to adapt to pressure sintering regimes (see below).

Pressure sintering regime. Most experimental data have been collected on systems in the capillary sintering regime (Section 6.3) using equipment such as an optical dilatometer (Fig. 6a). However, some data have been collected under conditions of applied load σ using equipment such as a uniaxial press (Fig. 6d). For these datasets, we use the Trouton ratio (Trouton 1906) to convert σ to an applied average liquid pressure P_l and assume that pressure sintering is more relevant than capillary sintering when $(P_l - P_g) > P_L$ (discussed later).

In Figure 11 we show data for dimensionless porosity $\bar{\phi}(\bar{t}_b)$ for systems in which pressure is acting on a pack of sintering glass particles at either $\bar{P} = 110$ for angular borosilicate glass particles (Friedman et al. 1963) or $\bar{P} = 1.3$ for spherical borosilicate particles (Rahaman and De Jonghe 1990). Also plotted are the pressure-sintering adaptation of the Mackenzie and Shuttleworth (1949) sintering model (Eqn. 23) and the model from Scherer (1986) (Eqn. 25). Data from Figure 10 for $\bar{P} = 0$ are shown in pale grey for reference. We note that, while the Scherer (1977) model performed well in the capillary sintering regime, the pressure-sintering extension (Scherer, 1986) appears less effective at high $\bar{P}$. The Mackenzie and Shuttleworth (1949) model performs better, with a relatively small overprediction of $\bar{\phi}$ at short $\bar{t}_b$ which is likely to arise because the papers originating the data do not quote a full particle size distribution, such that we are limited to using the monodisperse model for an approximate or mean particle size.

7. SINTERING OF MORE COMPLEX SYSTEMS

In Section 6, we analyzed the two principal sintering regimes relevant for industrial sintering of glass particles/droplets: the capillary sintering regime; and the pressure sintering regime. Here we extend those results to more complex systems. We identify two complexities that provide additional relevance to both industrial and natural scenarios: (1) sintering of mixed glass-crystal systems (see Fig. 2); and (2) sintering of glass particles that are super- or under-saturated in a water relative to the atmosphere in which they are sintering. Understanding the former process (sintering with crystals) is relevant to understand sintering in crystal-bearing magmas (e.g., Kendrick et al. 2016; Farquharson et al. 2017). Understanding the latter process is of less direct relevance in industrial sintering scenarios, but is of central importance in volcanology where almost all magmatic particles or droplets at the Earth's surface are not in chemical equilibrium with their surrounding (Wadsworth et al. 2019, 2021).

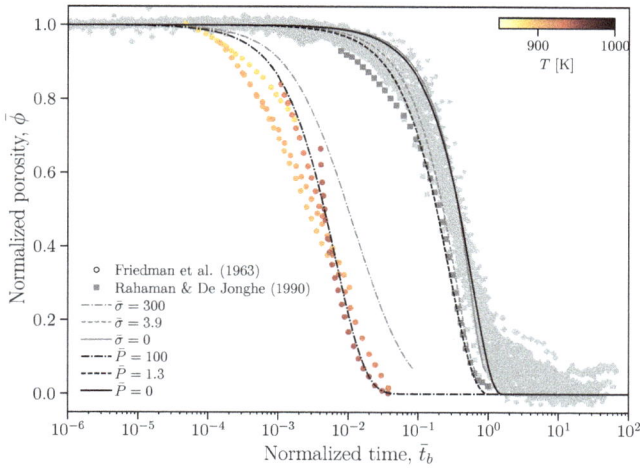

Figure 11. The pressure-sintering regime. Sintering in the case where the difference between the liquid pressure P_l and the pressure of the interstitial gas phase P_g is larger than the Laplace pressure P_L, which equates to $\overline{P} > 1$. Shown here are data in which the pressure acting on a pack of sintering glass particles was elevated by applying a constant load to the pack at either $\overline{P} = 110$ on angular borosilicate glass particles (Friedman et al. 1963) or $\overline{P} = 1.3$ on spherical borosilicate particles (Rahaman and De Jonghe 1990). The model shown in **black** is the pressure-sintering adaptation (see Wadsworth et al. 2019) of the monodisperse sintering model for different $\overline{P}$ (Mackenzie and Shuttleworth 1949; Wadsworth et al. 2017b) and the model from Scherer et al. (1986) is shown in **grey line** styles. Data from Figure 10 for $\overline{P} = 0$ are shown in **pale grey** for reference.

7.1 The effect of rigid crystals

We define the volume fraction of solid particles ϕ_x relative to the volume of viscous droplets, such that $\phi_x = 0$ when the system contains no rigid particles, and $\phi_x = 1$ when all the particles are solid and no viscous component is present. In Figure 2, we show the continuum of sintering systems between those composed of glass particles only (i.e., $\phi_x = 0$) and the solid-state sintering end member at $\phi_x = 1$. In mixed systems that occupy $0 < \phi_x < \phi'_x$, the dynamics of sintering are still dominated by viscous reduction of bulk surface area driven by flow in the glass particle component of the system, but when $\phi_x > 0$, this flow is impeded by the presence of rigid crystals (Eberstein et al. 2009). In Figure 12 we compile a comprehensive dataset in which ϕ_x was varied systematically (Pascual et al. 2005; Eberstein et al. 2009). At low ϕ_x, the sintering collapses to the viscous capillary sintering result (here we only show the Mackenzie and Shuttleworth (1949) sintering model and polydisperse variant; see Fig. 9). However, at high ϕ_x, the sintering appears to be retarded by the presence of crystals.

To our knowledge, there remains no rigorous theoretical treatment to account for this case of sintering with mixed viscous–rigid systems (Pascual et al. 2005; Eberstein et al. 2009). As a starting point, one might assume that the simplest way to account for the way the crystals would obstruct the flow would be via their effect on the glass viscosity. This approach is like assuming that the crystals are small compared with the glass particles and that the crystals are embedded in the glass such that the glass–crystal particles act like a continuum fluid, and that sintering is otherwise the same. This requires using a viscosity model for the effect of rigid crystals, an example of which is the Maron and Pierce (1956) result Kolzenburg et al. 2022, this volume)

$$\eta = \mu \left(1 - \frac{\phi_x}{\phi'_x} \right)^{-2} \tag{39}$$

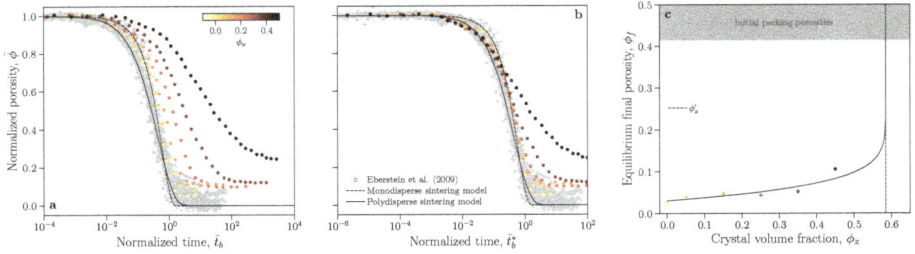

Figure 12. Sintering of glass particles mixed with solid crystalline particles. The data shown are a dataset (Eberstein et al. 2009) in which borosilicate glass particles were mixed with non-reactive alumina ceramic particles α-Al$_2$O$_3$, where the mixture volume fraction is given as ϕ_x. **(a)** Data normalized in the same way as shown in Figure 12, showing that rigid particles retard sintering rates. Data from Figure 10 are shown in pale grey for reference. **(b)** Data from panel (a) but where μ in $\bar{t}_b$ is replaced by a mixture viscosity η using $\eta = \mu(1-\phi_x/\phi_x')^{-2}$ (Maron and Pierce 1956) with a jamming fraction for spherical particles of $\phi_x' = 0.585$ (Boyer et al. 2011). **(c)** The final porosity ϕ_f from panel (a) as a function of ϕ_x. Shown is an empirical function of the form $\phi_f \sim 1 - \phi_x' + (\phi_c - (1-\phi_x'))(1-\phi_x/\phi_x')^c$ with $c=0.1$.

which has been validated widely (Mueller et al. 2010). Using $\phi_x'=0.585$ for the random jamming fraction (Boyer et al. 2011) in Equation (39) and then replacing μ with η in Equation (4) to normalize collated experimental data, we show that this approach is successful in explaining the effect of crystals at short times (i.e., during the early part of sintering using data from Eberstein et al. 2009; Fig. 12). In reality, the crystals are of similar size to the glass particles and are randomly distributed between the glass particles rather than being embedded in a glass continuum. Therefore, this expedient scaling approach may not be valid for all ratios of glass-crystal sizes or conditions. Nevertheless, it appears to be somewhat successful for the data shown here at low to moderate ϕ_x.

The second effect appears to be on the divergence of the porosity toward a final equilibrium value that is $\phi > \phi_c$ for $\phi_x > 0$. This effect appears to be to 'hold open' the pore network and reduce the efficiency with which the glass particles flow into the pore spaces. Using the data, we extract this final porosity ϕ_f for each value of ϕ_x. A purely empirical law for the final porosity should have the quality that $\phi_f \rightarrow \phi_c$ as $\phi_x \rightarrow 0$ and that $\phi_f \rightarrow \phi_i$ as $\phi_x \rightarrow \phi_x'$. This latter statement ignores the potential for solid-state sintering, which presumably operates on timescales far longer than these experiments (e.g., Rahaman 2007). One such empirical law is

$$\phi_f \sim 1 - \phi_x' + \left[\phi_c - \left(1 - \phi_x\right)\right]\left(1 - \frac{\phi_x}{\phi_x'}\right)^c \tag{40}$$

where $c=0.1$ is a fitting parameter (see Fig. 12c). Further work is needed to better understand this effect of rigid particles on the final pore geometries and porosity in sintering of mixed systems.

7.2 Sintering with diffusive hydration or dehydration

Sintering is a key applied problem in volcanology because glassy volcanic ash is often deposited hot such that the particulate deposit undergoes a transition to a non-particulate, sintered (or 'welded') state (Branney and Kokelaar 1992). However, a complicating process is that volcanic ash particles are rarely in chemical equilibrium upon deposition and may be supersaturated in volatiles due to incomplete degassing during eruption. Because these same volatiles influence the viscosity of the glass in the ash, it is important to account for syn-sintering degassing. To do this, we take an example problem of water diffusing out of particles during sintering, which we capture using Fick's second law for diffusive mass transport (Zhang and Gan 2022, this issue), cast in spherical coordinates (Crank 1975)

$$\frac{\partial C_w}{\partial t} = \frac{1}{r^2}\frac{\partial}{\partial r}\left(r^2 D \frac{\partial C_w}{\partial r}\right) \tag{41}$$

where C_w is the water concentration in the glass locally, r is the radial position away from the particle centre, and D is the diffusivity of water in the glass. We can solve Equation (41) subject to the initial condition that C_w is at an initial value for all r at $t = 0$, and boundary conditions that $\partial C_w / \partial r = 0$ at $r = 0$ and $C_w = C_e$ at $r = R$, where C_e is the equilibrium concentration (or solubility) at the particle surface.

The approach outlined here results in a continuous evolution of $C_w(r)$ profiles with time. These can be converted into a single nominal average value of C_w in the particle, which is taken to be the 'controlling' C_w, via $\langle C_w \rangle = \int_0^1 C_w d\bar{r}$, where $\bar{r} = r / R$. Then finally, the averaged $\langle C_w \rangle$ can be converted to an average viscosity for the particle $\langle \mu \rangle$ via a constitutive law for the effect of C_w on μ (Hess and Dingwell 1996; Giordano et al. 2008). The sintering component of the problem is then solved by exchanging μ for time-dependent $\langle \mu \rangle$ in the Mackenzie and Shuttleworth (1949) sintering model and polydisperse variants thereof, or any of the other sintering models considered in this Chapter.

Using the approach outlined here, Wadsworth et al. (2019) analyzed sintering experiments using particles of natural obsidian supersaturated with water (see Fig. 13). This shows that, across a wide range of temperatures or heating rates, the effect of degassing on retarding sintering rates can be accounted for, opening up a wide range of possible volcanological applications in which welding of volcanic ash is important. Details of the constitutive laws for $\mu(T, C_w)$, $D(T)$ and $C_e(T, P_l)$ for the obsidian used are all given in the Supplementary Information or can be found in Wadsworth et al. (2019).

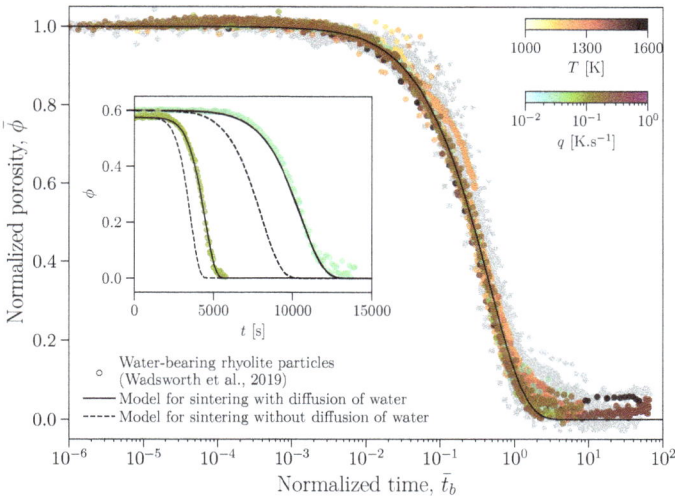

Figure 13. Sintering of natural hot rhyolitic glass particles supersaturated in dissolved water in the low-Peclet number regime Pc < 1 such that they are degassing during sintering (Gardner et al. 2018, 2019; Wadsworth et al. 2019). In the main panel we show data from Wadsworth et al. (2019) collected in isothermal and non-isothermal conditions using optical dilatometry, and isothermal conditions using *in situ* X-ray tomography. In order to model these data, we define $\bar{t}_b = \Gamma t / (\langle \mu \rangle \hat{a}_i)$ in place of $\bar{t}_b = \Gamma t / (\mu \hat{a}_i)$ and the non-isothermal equivalent thereof (see main text). *Inset:* 2 examples from the complete dataset, where the full sintering-diffusion model is shown along with the sintering model assuming no diffusion occurs and the water concentration does not deviate from the initial value. Data from Figure 10 are shown in **pale grey** for reference.

8. PHYSICAL PROPERTIES OF SINTERED SYSTEMS

Understanding how porosity evolves in a sintering system is key to understanding the evolution of other properties that depend on porosity directly. Here we explore the extent to which the experimentally-validated theory for the time-dependence of porosity developed in previous sections can be coupled with constitutive laws for the relationship between porosity and properties, to arrive at a description of the time-dependence of the physical properties of sintering packs. We focus on hydraulic properties (e.g., permeability) and elastic properties (e.g., elastic moduli and material strength).

8.2 The sintered filter: permeability during sintering (Wadsworth et al. 2021)

In the preceding sections, we have focused on the time-dependence of the porosity in sintering systems. The porosity decrease occurs so long as the system is permeable. However, the permeability decreases monotonically as sintering progresses. Being able to predict the permeability for a given porosity during sintering is appealing because sintered glass materials are commonly used as solid filtration devices for fluids (e.g., Krasnyi et al. 2005). To understand how the permeability drops during sintering processes, we need to constrain the permeability as a function of porosity. While the permeability of granular packs of particles has received a lot of attention (Sangani and Acrivos 1982; Carman 1997), sintering systems of particles or droplets that can reach very low porosities are less well studied. Martys et al. (1994) proposed that the permeability in systems of spheres that can interact or overlap is given by

$$k_r = \frac{2\left(1-\left(\phi-\phi_c\right)\right)}{s^2} \tag{42}$$

$$\bar{k} = \frac{k}{k_r} = \left(\phi-\phi_c\right)^b \tag{43}$$

where k_r is the 'reference' permeability and $\bar{k}$ is the permeability of the system normalized by this reference. Here we draw on the constraint of the specific surface area s (Eqns. 35, 36) and b is the percolation exponent. The value of b has been predicted theoretically and from first principles to be $b = 4.4$ (Feng et al. 1987). Therefore, every component of Equation (43) is constrained.

In Figure 14, we show the permeability of sintered materials. The 3D topology of sintering samples was determined using X-ray computed tomography (Wadsworth et al. 2017a) via techniques shown in Figure 6b and permeability was determined via lattice-Boltzmann simulations of fluid flow through each segmented 3-dimensional dataset of the pore volume, using LBflow (Llewellin 2010). These data follow Equation (43) (labelled as 'percolation theory' in Fig. 14a) up to a critical porosity. Above a critical porosity, the data smoothly deviate toward models for granular packs of particles—here we show the model from Sangani and Acrivos (1982). Alongside data from Wadsworth et al. (2017a), we plot data for sintered metallic beads from AmesPore® (using both bronze and stainless steel beads), clean quartz sandstones (Bourbie and Zinszner 1985), sintering dry rhyolite particles (Okumura and Sasaki 2014), variably welded block-and-ash flow deposits (Heap et al. 2015), simulated domains of overlapping spheres (Vasseur and Wadsworth 2017), and simulated domains of body-centered-cubic packs of high porosity granular media (Llewellin 2010).

All data shown in Figure 14a appear consistent with the percolation model (Eqn. 43), which gives us confidence to apply this to understand how permeability may evolve with time during sintering. In Figure 14b, we can map the permeability determined from the sintering data from Wadsworth et al. (2017a) to time directly, as they were collected *in situ* and continuously while the sample was hot and sintering. Here we use the polydisperse sintering model (Mackenzie and Shuttleworth 1949; Wadsworth et al. 2016a, 2017b) and simply convert $\phi(\bar{t}_b)$ to $k(\bar{t}_b)$ via

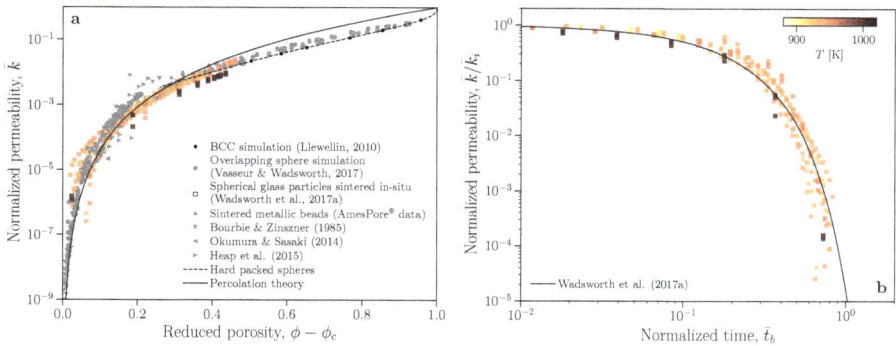

Figure 14. The permeability of sintered materials. **(a)** The relationship between permeability k and porosity ϕ for sintered materials cast as the normalized permeability $\bar{k} = k/k_r$ (see text) and the reduced porosity $\phi - \phi_c$, where ϕ_c is the percolation threshold. Data shown include the permeability of spherical glass bead packs sintered *in situ* using high-temperature X-ray computed tomography (Wadsworth et al. 2017a), sintered metallic beads from AmesPore® (using both bronze and stainless steel beads), clean quartz sandstones (Bourbie and Zinszner 1985), sintering dry rhyolite particles (Okumura and Sasaki 2014), variably welded block-and-ash flow deposits (Heap et al. 2015), simulated domains of overlapping spheres (Vasseur and Wadsworth 2017), and simulated domains of body-centred-cubic (BCC) packs of high porosity granular media (Llewellin 2010). These data are compared against a percolation model for sintering or overlapping granular media $\bar{k} = (\phi - \phi_c)^b$ (derived from first principles (Martys et al. 1994; Wadsworth et al. 2016b) where $b = 4.4$ from percolation theory (Feng et al. 1987), and a high-porosity model for hard packed spheres (Sangani and Acrivos 1982). **(b)** The time-dependence of permeability during sintering toward the percolation threshold (Wadsworth et al. 2017a). Here $\bar{k}$ is normalized a second time to its initial value $\bar{k}_i$ and the relationship given in panel (a) for $\bar{k}(\phi)$ is used in conjuction with the polydisperse sintering model for $\phi(\bar{t}_b)$ shown in Figures 12-15 (Mackenzie and Shuttleworth 1949; Wadsworth et al. 2016a, 2017b).

Equation (43). This represents a continuous model for permeability evolution during sintering, which could be used to predict the permeability that would result from fabricating a sintered glass material at known conditions of manufacture.

8.2 Elastic properties

Sintered materials are often used structurally or in situations where knowledge of their elastic properties is key. The most readily measured properties are the velocities of ultrasonic waves. The compressional wave velocity v_p can be used to constrain the P-wave modulus M via $M = \rho v_p^2$. Importantly, M can be used to constrain other elastic properties such as the dynamic bulk modulus K or the dynamic shear modulus G_s via $K = M(1+\nu)/[3(1-\nu)]$ and $G_s = M(1-2\nu)/[2(1-\nu)]$, where $\nu = 0.238$ is the Poisson's ratio of glass (Berge et al. 1995).

Vasseur et al. (2016) measured v_p for sintered glass materials across a range of ϕ and converted these results to $M(\phi)$, which are reproduced in Figure 15. These data can then be used to validate models for M in order to predict the elastic properties of any sintered material. One of the most well-known bounds on M is the Reuss (1929) lower bound, which states $M \geq [(1-\phi)/M_0 + \phi/M_1]^{-1}$ where M_0 is the P-wave modulus of the glass (i.e., M at $\phi = 0$) and M_1 is the P-wave modulus of the pore fluid (i.e., M at $\phi = 1$). In Figure 15b we show that this bound is reasonable at high ϕ. At lower porosities, we instead compare the data to a cluster expansion result for the P-wave modulus of a dilute suspension of spherical cavities in a solid (Torquato 2013; Vasseur et al. 2016). This is given by

$$ M = M_0 - \phi \left[K_0 \left(1 - \tau_1^{-1}\right) + \frac{4}{3} G_0 \left(1 + \tau_2^{-1}\right) \right] \tag{44} $$

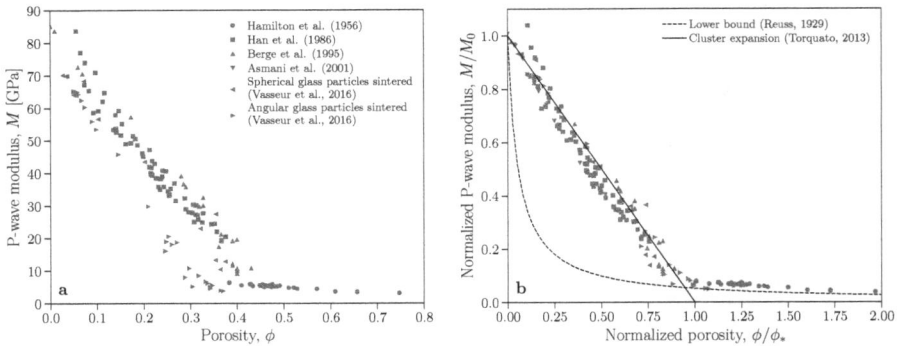

Figure 15. The compressional wave velocity in sintered materials. (**a**) The so-called P-wave modulus M as a function of the total porosity ϕ. Data include sintered angular glass fragments and spherical glass beads (Berge et al. 1995; Vasseur et al. 2016), sandstones (Han et al. 1986), high porosity granular sediments (Hamilton 1956), and sintered alumina ceramics (Asmani et al. 2001). (**b**) The value of M normalized by the P-wave modulus at $\phi = 0$, termed M_0, as a function of ϕ normalized by the pre-sintered packed value of ϕ, termed ϕ^*. Also shown in panel (b) are models for the P-wave modulus based on the cluster expansion method (Torquato 2013) and the well-known granular lower-bound (Reuss 1929).

where $\tau_1 = 4G_0/(3K_0)$ and $\tau_2 = (9K_0 + 8G_0)/(6K_0 + 12G_0)$ and subscript 0 represents the value of that property at $\phi = 0$ (e.g., the value of K at $\phi = 0$ is K_0). Given that K_0 and G_0 are well-known properties of glass and other solids (Berge et al. 1995) and that M_0 can be extrapolated from data such as those given in Figure 15, Equation (44) represents a useful tool for finding $M(\phi)$. The intercept of this model at $M = 0$ occurs at a critical ϕ, which we call ϕ_* and is: $\phi_* = 4M_0/[K_0(1 + \tau_1) + G_0(1 + \tau_2^{-1})/3]$. Normalizing ϕ by this value ϕ_* allows us to compare $M(\phi)$ over a wide range of compositions and grainsize and shape characteristics via the resultant, simple expression

$$\frac{M}{M_0} = 1 - \frac{\phi}{\phi_*}. \tag{45}$$

We find that Equation (45) is a good description of data for sintering glass materials for which $0 < \phi < \phi_*$.

8.3 Compressive strength of sintered materials

The other elastic property that is critical to understand and predict is the material strength of sintered solids. Using data for the uniaxial compressive strength σ_c of sintered solids, in Figure 16 we show that the uniaxial strength is a non-linear decreasing function of porosity (Vasseur et al. 2013, 2017). Given that these solids are not crack-dominated prior to deformation, the most widely used model to render this trend predictive is the pore-emanating crack model (Sammis and Ashby 1986). A full solution to this model in 3-dimensions requires solving several equations simultaneously. However, Zhu et al. (2011) used the 2-dimensional solution to the pore-emanating crack model to find an empirical analytical fit that reproduces the full solution with excellent accuracy across all conditions. Therefore, we follow the same functional form used by Zhu et al. (2011) to fit against the full solution in 3-dimensions and find a general solution

$$\sigma_c = \frac{1.5691}{\phi^{0.43}} \frac{K_{Ic}}{\sqrt{\pi l_1}} \tag{46}$$

where K_{Ic} is the fracture toughness of the solid matrix. For glass, an estimate of $K_{Ic} = 0.7\,\text{MPa.m}^{1/2}$ (Wiederhorn 1969) is used. In solving Equation (46), we are using $l_1(\phi)$ continuously in Figure 16

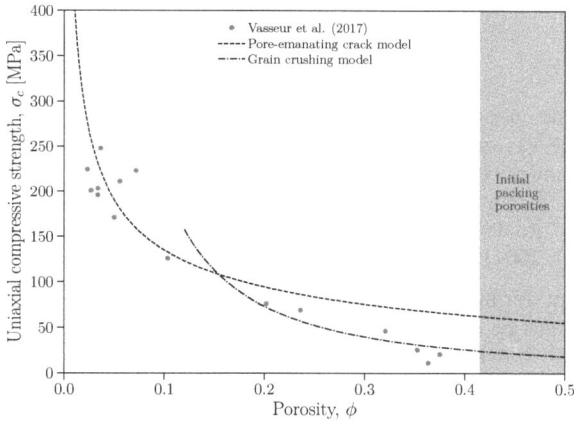

Figure 16. The uniaxial compressive strength of sintered materials as a function of porosity. Here we show data for the strength of sintered glass bead packs (Vasseur et al. 2017), compared with two models—the 'pore-emanating crack model' that predicts the strength as a function of bubble-geometry pores (Sammis and Ashby 1986), and the 'grain crushing model' that predicts the strength as a function of particle–particle contacts (Zhang et al. 1990).

and Equation (37) for the inter-particle distances. This model appears to be valid at low porosities, which is consistent with the fact that the topology of the pore space approaches the geometry of spherical pores, on which the model is based, as sintering progresses to low porosities.

At high porosities, the system is closer to a granular material for which applied uniaxial stresses induce hoop stresses in the particles at particle–particle contact points. For granular packs, the general model is the so-called 'grain crushing' model (Zhang et al. 1990) which is

$$\sigma_c = \frac{2.2\left(1-v^2\right)^2}{E^2\left(1-2v\right)^3}\left(\frac{K_{1c}}{\sqrt{\alpha\phi\langle R\rangle}}\right)^3. \tag{47}$$

where E is the Young's modulus and α is the ratio of the proto crack length to the particle size and is taken to be a constant at $\alpha = 10^{-3}$. Consistent with the granular nature of higher porosity, poorly sintered materials, Equation (47) outperforms Equation (46) when compared with data for which all the input parameters to both models are known (Fig. 16).

9. SINTERING DYNAMICS MAPS

In this Chapter we have identified the dominant regimes that can occur during sintering of glass particles or melt droplets where inertia is negligible. As discussed in Section 3, the balance between external pressures $P_1 - P_g$ and the Laplace pressure P_L dictates whether sintering occurs in the capillary sintering regime or the pressure sintering regime. We defined $\overline{P} = \Delta P / P_L = (P_1 - P_g)\langle R\rangle/(2\Gamma)$ as the dimensionless pressure ratio (Section 3.1), so that capillary sintering occurs for $0 < \overline{P} \ll 1$ while pressure sintering occurs for $\overline{P} \gg 1$.

For sintering to proceed, the gas between the particles or droplets must escape through percolative transport. Over sufficient lengths, this percolative transport of the gas through the system may be slow compared with the sintering itself, in which case sintering could induce an increase in gas pressure and render the problem more complicated than given in Sections 2 or 3. Therefore, we define a dimensionless length based on the balance between percolative transport of the gas and the viscous transport of the melt, which we term $\overline{L}$

$$\overline{L} = L\left(k_\mathrm{r}\frac{\mu}{\mu_g}\right)^{-1/2} \tag{48}$$

where k_r is a reference permeability and μ_g is the gas viscosity. Here, if $\overline{L} \ll 1$ then sintering is not limited by the gas escape pathways, whereas if $\overline{L} \gg 1$ then gas escape may influence sintering processes and a coupled gas percolation and sintering problem must be cast and solved. We term this latter situation of $\overline{L} \gg 1$ the 'compaction' regime, because cognate problems named 'compaction' problems have been addressed for other porous viscous media applications (e.g., Bercovici et al. 2001).

We propose that, for non-inertial, high-Oh sintering, $\overline{L}$ and $\overline{P}$ are sufficient to determine the dynamics that must be accounted for, hence which resultant models may be appropriate. Wadsworth et al. (2019) mapped sintering dynamics on a plot of $\overline{L}$ and $\overline{P}$, shown in Figure 17. Following Wadsworth et al. (2019), we use this sintering map to predict where natural processes sit and add a broad-brush estimate of where industrial processes may sit. We find that only the largest sintering applications in industrial settings fall in the $\overline{L} \gg 1$, which may explain why that regime is under-explored. However, natural and industrial sintering processes straddle the $\overline{P} = 1$ divide, such that the models given in Section 2 should be used for capillary sintering at $\overline{P} < 1$ and the models given in Section 3 should be used for pressure sintering at $\overline{P} > 1$.

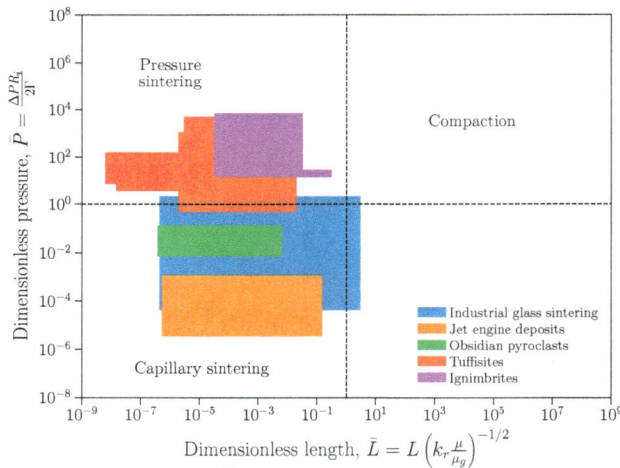

Figure 17. A sintering regime map reproduced following Wadsworth et al. (2019) for high-Oh conditions, where inertia is negligible. The dimensionless pressure separates capillary sintering at low-$\overline{P}$ from pressure sintering at high-$\overline{P}$; the dimensionless length of the system separates small systems in which outgassing of the interstitial gas phase can escape freely at low-$\overline{L}$ from larger systems where permeability limits outgassing at high-$\overline{L}$. The high-$\overline{P}$, high-$\overline{L}$ regime is termed the compaction regime. Most natural magmatic systems and industrial glass sintering systems are initially at low-$\overline{L}$ and span the low-to-high $\overline{P}$ regimes.

10. RECIPES FOR SINTERING AND APPLICATIONS

In Sections 2 and 3, we introduce a variety of theoretical and empirical models for coalescence and sintering. In Sections 6–7 we compare these models with experimental data in the capillary sintering regime (Fig. 10) and the pressure-sintering regime (Fig. 11) and conclude that the pore-based approach of Mackenzie and Shuttleworth 1949 and variants thereof (Wadsworth et al. 2016a, 2017b) is the most versatile and the most accurate predictor of data. The most important features of this model, which justify this choice, are: (1) the microstructural arrangement of pore spaces is not prescribed, such that there is no assumption of microstructure *per se* (c.f. the cubic pore arrangement prescribed in Scherer, 1977); and (2) the gas pressure and liquid pressure can be set to different values, allowing dynamic cases of evolving gas pressure to be solved (the simplest case is for constant, but finite gas pressure: the pressure-sintering regime).

We provide a flow chart for application of this family of models, such that a user can follow a defined work-flow based on what inputs they know/have (Fig. 18). We propose that this diagram represents an effective 'recipe' for modelling many isothermal or non-isothermal sintering processes.

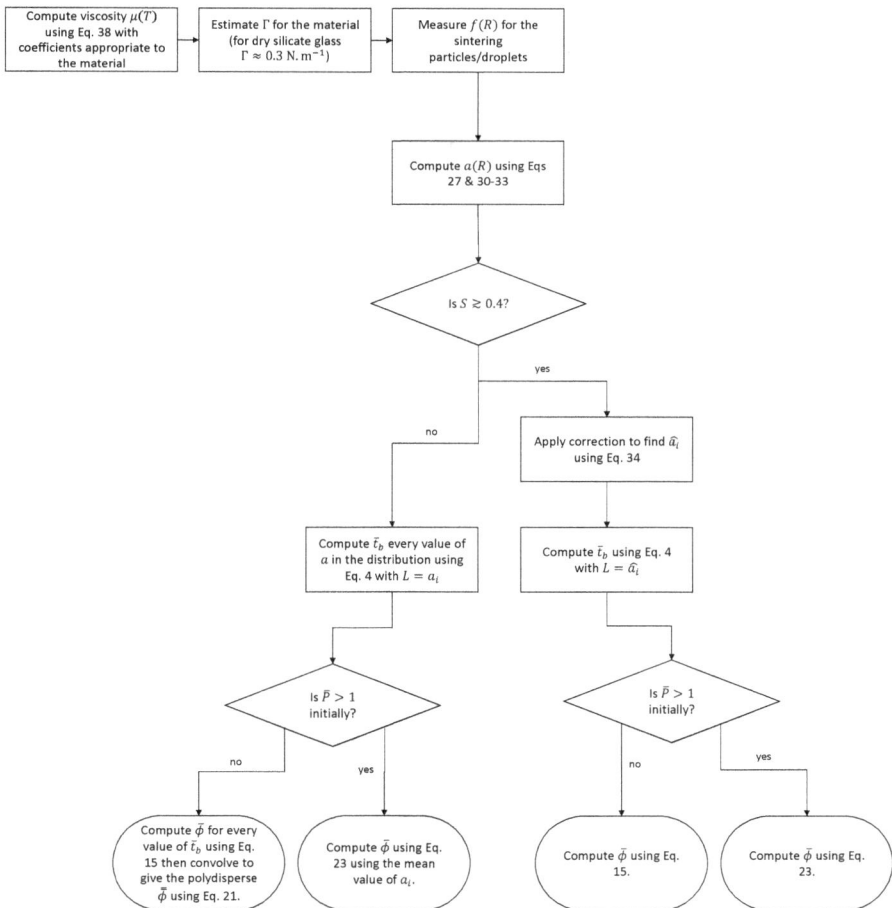

Figure 18. Flow chart summarizing how to use this chapter to compute sintering rates. This map is limited to the Mackenzie and Shuttleworth (1949) model and variants thereof (Wadsworth et al. 2016a, 2019) and does not include explicit mention of how to compute the other physical properties of sintering systems (see text for details of those calculations).

11. OUTLOOK

There are sintering phenomena that are relevant to volcanic scenarios, but which remain to be explored. We propose that the reason these regimes have not been studied, or have received less attention, may be that sintering has typically been a subject of research with application to material fabrication in mind. It may be that other applications, such as natural sintering on Earth (Sparks et al. 1999; Wadsworth et al. 2014, 2019; Gardner et al. 2017, 2018, 2019), or on other planetary bodies (Uhlmann et al. 1975; Sirono 2011), may expose new regimes of general interest.

First, an outstanding issue is the case where shear stresses act on the bulk sintering system, imposing a shearing motion that may accelerate the decrease in porosity with time compared with pressure sintering. This is known to be critically important in the emplacement of ignimbrites, especially those that record large shear strain (Andrews and Branney 2011), and yet has not been widely tested experimentally or theoretically. This would represent a regime analogous to $\bar{P} \gg 1$, but the pressure driving the system is a shear stress and not an isotropic pressure. The deformation of the droplets or particles from spherical would be likely to induce anisotropy of microstructure, and therefore of the resultant permeability and elastic properties (Quane and Russell 2006).

Second, is the case of the inertial sintering regime, which also remains unstudied. Some low viscosity magmas on Earth may agglutinate in the inertial regime, and some very high temperature sintering phenomena applicable to industry may not be purely viscous. In the inertial regime, Eggers et al. (1999) noted that the lengthscale controlling the motion of sintering of two droplets is initially R_n (rather than R_i), and that R_n is initially singular at zero. This implies that Oh is arbitrarily large (if we use R_n in Oh in place of R_i), implying that the initial stage of sintering of any liquids is always viscous rather than inertial. We can rearrange Oh to see that this is true for $R_n < R_c$, where $R_c = \mu^2/(\rho\Gamma)$. For droplets that are larger than R_c, the condition where the neck width is sufficiently large to induce inertial effects will be met. Alternatively, rapid heating to very large temperatures, where glass viscosities are sufficiently low, could push the system into the regime where Oh $\ll$ 1.

Third, non-linear dynamics at high-$\bar{L}$ (i.e., where sintering is limited by percolative gas transport) have received little attention for sintering systems. However, some of the largest ignimbrites on Earth are thought to deposit sufficiently rapidly that compaction may be the dominant densification process (Riehle 1973; Ashwell et al. 2015). Similarly, as industrial processing pushes to produce larger sintered products in shorter spaces of time, the sintering system could be pushed into the high-$\bar{L}$ regime. In this case, as discussed in Section 9, there can be a non-linear feedback between sintering and gas pressurization that can result in porosity waves and inhomogeneous sintering.

Finally, recent work has shown that solid-state sintering in the absence of a melt phase (see Fig. 2), may be important in volcanic settings (Ryan et al. 2018, 2020). While solid-state sintering has been explored extensively for ceramics production processes (Kang 2004; Rahaman 2007), it has only very recently been considered for crystal compositions or composites of crystals relevant in nature. The lack of a general solid-state sintering model that can predict sintering rates for any mineral composite starting pack of powder, means that more work is needed before natural applications can be addressed.

ACKNOWLEDGMENTS

This work is based on a body of published work by the authors that also involved Kate Dobson, Jenny Schauroth, Bettina Scheu, Yan Lavallée, Jackie Kendrick, Kai-Uwe Hess, Felix von Aulock, Michael Heap, Hugh Tuffen, Julie Fife, Mathieu Colombier, Alex Kushnir, James Gardner, and Sebastian Wiesmaier and others. We have benefited from discussions with those authors and also with Oscar Prado, Steve Sparks, Amy Ryan, Stephan Kolzenburg, Jamie Farquharson, and Kelly Russell.

REFERENCES

Andrews GDM, Branney MJ (2011) Emplacement and rheomorphic deformation of a large, lava-like rhyolitic ignimbrite: Grey's Landing, southern Idaho. Geol Soc Am Bull 123:725–743

Ashwell PA, Kendrick JE, Lavallée Y, Kennedy BM, Hess KU, von Aulock FW, Wadsworth FB, Vasseur J, Dingwell DB (2015) Permeability of compacting porous lavas. J Geophys Res 120:1605–1622

Asmani M, Kermel C, Leriche A, Ourak M (2001) Influence of porosity on Youngs modulus and Poisson's ratio in alumina ceramics. J Eur Ceram Soc 21:1081–1086

Bellehumeur CT, Bisaria MK, Vlachopoulos J (1996) An experimental study and model assessment of polymer sintering. Polym Eng Sci 36:2198–2207

Bercovici D, Ricard Y, Schubert G (2001) A two-phase model for compaction and damage: 1. General theory. J Geophys Res Solid Earth 106:8887–8906

Berge PA, Bonner BP, Berryman JG (1995) Ultrasonic velocity–porosity relationships for sandstone analogs made from fused glass beads. Geophysics 60:108–119

Boccaccini AR, Hamann B (1999) Review in situ high-temperature optical microscopy. J Mater Sci 34:5419–5436

Bourbie T, Zinszner B (1985) Hydraulic and acoustic properties as a function of porosity in Fontainebleau sandstone. J Geophys Res Solid Earth 90:11524–11532

Boyer F, Guazzelli É, Pouliquen O (2011) Unifying suspension and granular rheology. Phys Rev Lett 107:188301

Branney MJ, Kokelaar P (1992) A reappraisal of ignimbrite emplacement: progressive aggradation and changes from particulate to non-particulate flow during emplacement of high-grade ignimbrite. Bull Volcanol 54:504–520

Branney MJ, Kokelaar BP (2002) Pyroclastic Density Currents and the Sedimentation of Ignimbrites. Geological Society Memoir Number 27. Geol Soc London

Branney MJ, Kokelaar BP, McConnell B (1992) The Bad Step Tuff: a lava-like rheomorphic ignimbrite in a calc-alkaline piecemeal caldera, English Lake District. Bull Volcanol 54:187–199

Carman PC (1997) Fluid flow through granular beds. Process Saf Environ Prot Trans Inst Chem Eng Part B 75:

Chiang Y-M, Birnie DP, Kingery WD (1997) Physical Ceramics. J. Wiley, NY

Cimarelli C, Yilmaz T, Colombier M, Villanova J, Höfer L, Hess KU, Ruthensteiner B, Dingwell D (2017) Micro-and nano-CT textural analysis of an experimental volcanic fulgurite. EGU Gen Assem Conf Abstr 19:17982

Crank J (1975) The Mathematics of Diffusion. Clarendon Pess, Oxford.

Eberstein M, Reinsch S, Müller R, Deubener J, Schiller WA (2009) Sintering of glass matrix composites with small rigid inclusions. J Eur Ceram Soc 29:2469–2479

Eggers J, Lister J, Stone H (1999) Coalescence of liquid drops. J Fluid Mech 401:293–310

Elam WT, Kerstein AR, Rehr JJ (1984) Critical properties of the void percolation problem for spheres. Phys Rev Lett 52:1516

Eshelby J (1949) Discussion of seminar on the kinetics of sintering. Metall Trans 185:796–813

Farquharson JI, Wadsworth FB, Heap MJ, Baud P (2017) Time-dependent permeability evolution in compacting volcanic fracture systems and implications for gas overpressure. J Volcanol Geotherm Res 339:81–97

Feng S, Halperin B, Sen P (1987) Transport properties of continuum systems near the percolation threshold. Phys Rev B 35:197–214

Fife JL, Rappaz M, Pistone M, Celcer T, Mikuljan G, Stampanoni M (2012) Development of a laser-based heating system for in situ synchrotron-based X-ray tomographic microscopy. J Synchrotron Radiat 19:352–358

Fluegel A (2007) Glass viscosity calculation based on a global statistical modelling approach. Glas Technol J Glas Sci Technol Part A 48:13–30

Frenkel J (1945) Viscous flow of crystalline bodies under the action of surface tension. J Phys 9:385–391

Friedman I, Long W, Smith RL (1963) Viscosity and water content of rhyolite glass. J Geophys Res 68:6523–6535

Gardner JE, Ketcham RA (2011) Bubble nucleation in rhyolite and dacite melts: temperature dependence of surface tension. Contrib Mineral Petrol 162:929–943

Gardner JE, Llewellin EW, Watkins JM, Befus KS (2017) Formation of obsidian pyroclasts by sintering of ash particles in the volcanic conduit. Earth Planet Sci Lett 459:252–263

Gardner JE, Wadsworth FB, Llewellin EW, Watkins JM, Coumans JP (2018) Experimental sintering of ash at conduit conditions and implications for the longevity of tuffisites. Bull Volcanol 80:23

Gardner JE, Wadsworth FB, Llewellin EW, Watkins JM, Coumans JP (2019) Experimental constraints on the textures and origin of obsidian pyroclasts. Bull Volcanol 81:22

Giehl C, Brooker R, Marxer H, Nowak M (2016) An experimental simulation of volcanic ash deposition in gas turbines and implications for jet engine safety. Chem Geol 461:160–170

Giordano D, Russell JK, Dingwell DB (2008) Viscosity of magmatic liquids: a model. Earth Planet Sci Lett 271:123–134

Grunder AL, Laporte D, Druitt TH (2005) Experimental and textural investigation of welding: effects of compaction, sintering, and vapor-phase crystallization in the rhyolitic Rattlesnake Tuff. J Volcanol Geotherm Res 142:89–104

Hamilton EL (1956) Low sound velocities in high-porosity sediments. J Acoust Soc Am 28:16–19

Han D, Nur A, Morgan D (1986) Effects of porosity and clay content on wave velocities in sandstones. Geophysics 51:2093–2107

Heap MJ, Farquharson JI, Wadsworth FB, Kolzenburg S, Russell JK (2015) Timescales for permeability reduction and strength recovery in densifying magma. Earth Planet Sci Lett 429:223–233

Heap MJ, Kolzenburg S, Russell JK, Campbell ME, Welles J, Farquharson JI, Ryan A (2014) Conditions and timescales for welding block-and-ash flow deposits. J Volcanol Geotherm Res 289:202–209

Heap MJ, Tuffen H, Wadsworth FB, Reuschlé T, Castro JM, Schipper CI (2019) The permeability evolution of tuffisites and implications for outgassing through dense rhyolitic magma. J Geophys Res Solid Earth 124:8281–8299

Hess K-UU, Dingwell DB (1996) Viscosities of hydrous leucogranitic melts: A non-Arrhenian model. Am Mineral 81:1297–1300

Hopper RW (1984) Coalescence of two equal cylinders: exact results for creeping viscous plane flow driven by capillarity. J Am Ceram Soc 67:C262–C264

Houghton B, Carey RJ (2015) Pyroclastic fall deposits. Chapter 34 In: The Encyclopedia of Volcanoes, 2nd. ed., Sigurdsson H (ed) Academic Press, Amsterdam, p 599–616

Jagota A, Dawson PR (1988) Micromechanical modeling of powder compacts—I. Unit problems for sintering and traction induced deformation. Acta Metall 36:2551–2561

Jagota A, Mikeska KR, Bordia RK (1990) Isotropic constitutive model for sintering particle packings. J Am Ceram Soc 73:2266–2273

Kang S-JL (2004) Sintering: Densification, Grain Growth and Microstructure. Elsevier Butterworth-Heinemann, Amsterdam

Kendrick JE, Lavallée Y, Varley NR, Wadsworth FB, Lamb OD, Vasseur J (2016) Blowing off steam: Tuffisite formation as a regulator for lava dome eruptions. Front Earth Sci 4:41

Kertesz J (1981) Percolation of holes between overlapping spheres: Monte Carlo calculation of the critical volume fraction. J Phys Lettres 42:393–395

Kingery WD, Berg M (1955) Study of the initial stages of sintering solids by viscous flow, evaporation-condensation, and self-diffusion. J Appl Phys 26:1205–1212

Kirchhof MJ, Schmid HJ, Peukert W (2009) Three-dimensional simulation of viscous-flow agglomerate sintering. Phys Rev E 80:026319

Kolzenburg S, Ryan AG, Russell JK (2019) Permeability evolution during non-isothermal compaction in volcanic conduits and tuffisite veins: Implications for pressure monitoring of volcanic edifices. Earth Planet Sci Lett 527:115783

Kolzenburg S, Chevrel MO, Dingwell DB (2022) Magma/suspension rheology. Rev Mineral Geochem 87:639-719

Krasnyi BL, Tarasovskii VP, Val'dberg AY, Kaznacheeva TO (2005) Porous permeable ceramics for filter elements cleaning hot gases from dust. Glas Ceram (English Transl Steklo i Keramika) 62:134–138

Kraxner J, Liška M, Klement R, Chromčíková M (2009) Surface tension of borosilicate melts with the composition close to the E-glass. Ceramics-Silikáty 53:141–143

Kueppers U, Cimarelli C, Hess KU, Taddeucci J, Wadsworth FB, Dingwell DB (2014) The thermal stability of Eyjafjallajökull ash versus turbine ingestion test sands. J Appl Volcanol 3:4

Lara C, Pascual MJ, Prado MO, Duran A (2004) Sintering of glasses in the system RO–Al$_2$O$_3$–BaO–SiO$_2$ (R=Ca, Mg, Zn) studied by hot-stage microscopy. Solid State Ionics 170:201–208

Lavallée Y, Wadsworth FB, Vasseur J, Russell JK, Andrews GD, Hess KU, von Aulock FW, Kendrick JE, Tuffen H, Biggin AJ, Dingwell DB (2015) Eruption and emplacement timescales of ignimbrite super-eruptions from thermo-kinetics of glass shards. Front Earth Sci 3:2

Llewellin E (2010) LBflow: An extensible lattice Boltzmann framework for the simulation of geophysical flows. Part II: usage and validation. Comput Geosci 36:115–122

Lu B, Torquato S (1992) Nearest-surface distribution functions for polydispersed particle systems. Phys Rev A 45:5530

Mackenzie JK, Shuttleworth R (1949) A phenomenological theory of sintering. Proc Phys Soc Sect B 62:833

Maron S, Pierce P (1956) Application of Ree–Eyring generalized flow theory to suspensions of spherical particles. J Colloid Sci 11:80–95

Martys NS, Torquato S, Bentz DP (1994) Universal scaling of fluid permeability for sphere packings. Phys Rev E 50:403

Mueller S, Llewellin EW, Mader HM (2010) The rheology of suspensions of solid particles. Proc R Soc A Math Phys Eng Sci 466:1201–1228

Mueller SP, Helo C, Keller F, Taddeucci J, Castro JM (2018) First experimental observations on melting and chemical modification of volcanic ash during lightning interaction. Sci Rep 8:1389

Okuma G, Kadowaki D, Hondo T, Tanaka S, Wakai F (2017) Interface topology for distinguishing stages of sintering. Sci Rep 7:11106

Okumura S, Sasaki O (2014) Permeability reduction of fractured rhyolite in volcanic conduits and its control on eruption cyclicity. Geology 42:843–846

Olevsky EA (1998) Theory of sintering: from discrete to continuum. Mater Sci Eng R Reports 23:41–100

Olevsky EA, Tikare V, Garino T (2006) Multi-scale study of sintering: a review. J Am Ceram Soc 89:1914–1922

Parikh NM (1958) Effect of atmosphere on surface tension of glass. J Am Ceram Soc 41:18–22

Pascual MJ, Durán A, Prado MO, Zanotto ED (2005) Model for sintering devitrifying glass particles with embedded rigid fibers. J Am Ceram Soc 88:1427–1434

Prado M, Dutra Zanotto E, Müller R (2001) Model for sintering polydispersed glass particles. J Non Cryst Solids 279:169–178

Prado MO, Zanotto ED, Fredericci C (2003) Sintering polydispersed spherical glass particles. J Mater Res 18:1347–1354

Prousevitch AA, Sahagian DL, Anderson AT (1993) Dynamics of diffusive bubble growth in magmas: Isothermal case. J Geophys Res Solid Earth 98:22283–22307

Quane SL, Russell JK (2005) Welding: insights from high-temperature analogue experiments. J Volcanol Geotherm Res 142:67–87

Quane SL, Russell JK (2006) Bulk and particle strain analysis in high-temperature deformation experiments. J Volcanol Geotherm Res 154:63–73

Quane SL, Russell JK, Friedlander EA (2009) Time scales of compaction in volcanic systems. Geology 37:471–474

Rahaman MN (2007) Sintering of Ceramics. CRC Press, Boca Raton

Rahaman MN, De Jonghe LC (1990) Sintering of spherical glass powder under a uniaxial stress. J Am Ceram Soc 73:707–712

Rahaman MN, De Jonghe LC, Scherer GW, Brook RJ (1987) Creep and densification during sintering of glass powder compacts. J Am Ceram Soc 70:766–774

Reis RM, Barbosa AJ, Ghussn L, Ferreira EB, Prado MO, Zanotto ED (2018) Sintering and rounding kinetics of irregular glass particles. J Am Ceram Soc 102:845–854

Reuss A (1929) Berechnung der Fließgrenze von Mischkristallen auf Grund der Plastizitätsbedingung für Einkristalle. ZAMM-J Appl Math Mech für Angew Math und Mech 9:49–58

Riehle JR (1973) Calculated compaction profiles of rhyolitic ash-flow tuffs. Bull Geol Soc Am 84:2193–2216

Rintoul MD (2000) Precise determination of the void percolation threshold for two distributions of overlapping spheres. Phys Rev E 62:68

Rosenzweig N, Narkis M (1981) Sintering rheology of amorphous polymers. Polym Eng Sci 21:1167–1170

Russell JK, Quane SL (2005) Rheology of welding: inversion of field constraints. J Volcanol Geotherm Res 142:173–191

Russell JK, Hess K-U, Dingwell DB (2022) Models for viscosity of geological melts. Rev Mineral Geochem 87:841-885

Ryan AG, Friedlander EA, Russell JK, Heap MJ, Kennedy LA (2018) Hot pressing in conduit faults during lava dome extrusion: Insights from Mount St. Helens 2004–2008. Earth Planet Sci Lett 482:171–180

Ryan AG, Russell JK, Heap MJ, Zimmerman ME, Wadsworth FB (2020) Timescales of porosity and permeability loss by solid-state sintering. Earth Planet Sci Lett 549:116533

Sammis CG, Ashby MF (1986) The failure of brittle porous solids under compressive stress states. Acta Metall 34:511–526

Sangani AS, Acrivos A (1982) Slow flow through a periodic array of spheres. Int J Multiph Flow 8:343–360

Scherer GW (1977) Sintering of low-density glasses: I, Theory. J Am Ceram Soc 60:236–239

Scherer GW (1986a) Viscous sintering under a uniaxial load. J Am Ceram Soc 69:C206–C207

Scherer GW, Bachman DL (1977) Sintering of low-density glasses: II, Experimental study. J Am Ceram Soc 60:239–243

Scheu B, Dingwell DB (2022) Magma fragmentation. Rev Mineral Geochem 87:767-800

Schipper CI, Castro JM, Kennedy BM, Tuffen H, Whattam J, Wadsworth FB, Paisley R, Fitzgerald RH, Rhodes E, Schaefer LN, Ashwell PA (2021) Silicic conduits as supersized tuffisites: Clastogenic influences on shifting eruption styles at Cordón Caulle volcano (Chile). Bull Volcanol 83:11

Shinozaki M, Roberts K (2013) Deposition of ingested volcanic ash on surfaces in the turbine of a small jet engine. Adv Eng Mater 15:986–994

Sirono S (2011) Planetesimal formation induced by sintering. Astrophys J Lett 733:41

Soares VO, Reis RC, Zanotto ED, Pascual MJ, Durán A (2012) Non-isothermal sinter-crystallization of jagged $Li_2O–Al_2O_3–SiO_2$ glass and simulation using a modified form of the clusters model. J Non Cryst Solids 358:3234

Song W, Hess KU, Damby DE, Wadsworth FB, Lavallée Y, Cimarelli C, Dingwell DB (2014) Fusion characteristics of volcanic ash relevant to aviation hazards. Geophys Res Lett 41:2326–2333

Song W, Lavallée Y, Wadsworth FB, Hess KU, Dingwell DB (2017) Wetting and spreading of molten volcanic ash in jet engines. J Phys Chem Lett 8:1878–1884

Sparks RSJ, Tait SR, Yanev Y (1999) Dense welding caused by volatile resorption. J Geol Soc London 156:217–225

Torquato S (2013) Random Heterogeneous Materials: Microstructure and Macroscopic Properties. Springer Science & Business Media

Torquato S, Avellaneda M (1991) Diffusion and reaction in heterogeneous media: Pore size distribution, relaxation times, and mean survival time. J Chem Phys 95:6477–6489

Torquato S, Lu B (1990) Rigorous bounds on the fluid permeability: effect of polydispersivity in grain size. Phys Fluids A Fluid Dyn 2:487–490

Torquato S, Truskett TM, Debenedetti PG (2000) Is random close packing of spheres well defined? Phys Rev Lett 84:2064–2067

Trouton FT (1906) On the coefficient of viscous traction and its relation to that of viscosity. Proc R Soc London Ser A 77:426–440

Tuffen H, Dingwell DB (2005) Fault textures in volcanic conduits: evidence for seismic trigger mechanisms during silicic eruptions. Bull Volcanol 67:370–387

Tuffen H, Dingwell DB, Pinkerton H (2003) Repeated fracture and healing of silicic magma generate flow banding and earthquakes? Geology 31:1089–1092

Uhlmann DR, Klein L, Hopper RW (1975) Sintering, crystallization, and breccia formation. Moon 13:277–284

Vasseur J, Wadsworth FB (2017) Sphere models for pore geometry and fluid permeability in heterogeneous magmas. Bull Volcanol 79:77

Vasseur J, Wadsworth FB, Lavallée Y, Hess KU, Dingwell DB (2013) Volcanic sintering: Timescales of viscous densification and strength recovery. Geophys Res Lett 40:5658–5664

Vasseur J, Wadsworth FB, Lavallée Y, Dingwell DB (2016) Dynamic elastic moduli during isotropic densification of initially granular media. Geophys J Int 204:1721–1728

Vasseur J, Wadsworth FB, Heap MJ, Main IG, Lavallée Y, Dingwell DB (2017) Does an inter-flaw length control the accuracy of rupture forecasting in geological materials? Earth Planet

Wadsworth FB, Vasseur J, von Aulock FW, Hess KU, Scheu B, Lavallée Y, Dingwell DB (2014) Nonisothermal viscous sintering of volcanic ash. J Geophys Res Solid Earth 119:8792–8804

Wadsworth FB, Vasseur J, Llewellin EW, Dingwell DB (2016a) Sintering of viscous droplets under surface tension. Proc R Soc A Math Phys Eng Sci 472:20150780

Wadsworth FB, Vasseur J, Scheu B, Kendrick JE, Lavallée Y, Dingwell DB (2016b) Universal scaling of fluid permeability during volcanic welding and sediment diagenesis. Geology 44:219–222

Wadsworth FB, Vasseur J, Llewellin EW, Dobson KJ, Colombier M, Von Aulock FW, Fife JL, Wiesmaier S, Hess KU, Scheu B, Lavallée Y (2017a) Topological inversions in coalescing granular media control fluid-flow regimes. Phys Rev E 96:033113

Wadsworth FB, Vasseur J, Llewellin EW, Dingwell DB (2017b) Sintering of polydisperse viscous droplets. Phys Rev E 95:033114

Wadsworth F, Llewellin E, Rennie C, Watkinson C (2018) In Vulcan's forge. Nat Geosci 12:2–3

Wadsworth FB, Vasseur J, Schauroth J, Llewellin EW, Dobson KJ, Havard T, Scheu B, von Aulock FW, Gardner JE, Dingwell DB, Hess KU (2019) A general model for welding of ash particles in volcanic systems validated using in situ X-ray tomography. Earth Planet Sci Lett 525:115726

Wadsworth FB, Llewellin EW, Vasseur J, Gardner JE, Tuffen H (2020) Explosive–effusive volcanic eruption transitions caused by sintering. Sci Adv 6:eaba7940

Wadsworth FB, Vasseur J, Llewellin EW, Brown RJ, Tuffen H, Gardner JE, Kendrick JE, Lavallée Y, Dobson KJ, Heap MJ, Dingwell DB (2021) A model for permeability evolution during volcanic welding. J Volcanol Geotherm Res 409:107118

Walker D, Mullins O, Jr (1981) Surface tension of natural silicate melts from 1,200–1,500 °C and implications for melt structure. Contrib Mineral Petrol 76:455–462

Wiederhorn SM (1969) Fracture surface energy of glass. J Am Ceram Soc 52:99–105

Zhang J, Teng-Fong Wong, Davis DM (1990) Micromechanics of pressure-induced grain crushing in porous rocks. J Geophys Res 95:341–352

Zhang Y, Gan T (2022) Diffusion in melts and magmas. Rev Mineral Geochem 87:283-337

Zhu W, Baud P, Vinciguerra S, Wong T (2011) Micromechanics of brittle faulting and cataclastic flow in Alban Hills tuff. J Geophys Res Solid Earth 116:B06209

Reviews in Mineralogy & Geochemistry
Vol. 87 pp. 841-885, 2022
Copyright © Mineralogical Society of America

18

Models for Viscosity of Geological Melts

James K. Russell[1], Kai-Uwe Hess[2] Donald B. Dingwell[2]

*[1] Earth, Ocean and Atmospheric Sciences
The University of British Columbia
Vancouver, V7W-2S1, British Columbia, Canada*

krussell@eoas.ubc.ca

*[2] Department of Earth and Environmental Sciences
Ludwig-Maximilians-University Munich
Theresienstr. 41 / III
80333, Munich, Germany*

1. INTRODUCTION

Viscosity is the single most important physical property governing the formation, transport and eruption of naturally occurring silicate melts and magmas. Silicate melt viscosity is of fundamental interest to earth scientists, and in particular petrologists, volcanologists and geophysicists, because of the role it plays in governing melt production, magma ascent processes, styles of volcanic eruption, as well as, rates of physicochemical processes (e.g., degassing, crystallization). The viscosity of natural silicate magmas is a complex function of the temperature–pressure–composition dependence of the melt viscosity plus the non-linear effects due to the presence of included phases such as crystals and gas/fluid-filled bubbles (see Kolzenberg et al. 2022, this volume). It is, in particular, the great variation of melt viscosity with temperature and chemical composition which provides the backdrop for the dependence of magma viscosity on suspended crystal and bubble contents and changes in strain rate (Stevenson et al. 1996; Lavallée et al. 2007; Caricchi et al. 2007; Vona et al. 2011, 2013; Okumura et al. 2013; Kolzenburg et al. 2017, 2020; Ryan et al. 2019).

The ability to parameterize and predict the viscosity of geological melts has been of fundamental interest to Earth science for over a century (Becker 1897; Bowen 1915, 1934; Kani and Howaska 1936; Shaw 1963, 1972; Bottinga and Weill 1972) and has been addressed in three main ways.

Firstly, field studies of active lavas have provided estimates of magma rheology. Chevrel et al. (2019) provide a summary and critical analysis of 11 field-based studies published since 1940. Most commonly, average or bulk rheological properties are calculated from measured surface velocities in active lava channels and estimates of the channel's geometry and slope (e.g., Nichols 1939; Lev and James 2014). These calculations assume a rheology (i.e., Newtonian vs. Bingham) and adopt the corresponding equation for the velocity profile in a channel-confined fluid to inform on the average flow rheology (i.e., viscosity and yield strength). In rare instances, measurements of super-elevations of higher velocity lavas measured at corners of active channels are also used to constrain effective viscosity (Chevrel et al. 2019). There have also been efforts to directly measure the lava viscosity using a viscometer inserted into magma flowing in an active channel. The choice of viscometer is critical depending on the expected range of viscosity values and the style and properties of the lava (Chevrel et al. 2019). Performed correctly, these challenging measurements have the potential to track the evolution in rheology driven by changes in temperature, crystallization and degassing (e.g., Kolzenburg et al. 2017, 2018, 2022, this volume).

1529-6466/22/0087-0018$05.00 (print)
1943-2666/22/0087-0018$05.00 (online)

http://dx.doi.org/10.2138/rmg.2022.87.18

Secondly, new, diverse and innovative experimental methods have led to an exponential increase in measurements of viscosity for geological melts. Direct measurements of viscosity for geological melts serve several purposes. These data have filled fundamental gaps in our knowledge of melt physical properties and have elucidated and quantified the viscous properties of most terrestrial melts, including hydrous melts (Fig. 1). Geoscientists concerned with the formation, transport and eruption of silicate magmas can use the experimental data directly where they have similar melt compositions. Collectively, these experimental campaigns have shown naturally-occurring silicate melts to span more than 15 orders of magnitude (10^{-1} to 10^{14} Pa s). These ranges in melt viscosity are primarily in response to variations in temperature (T) and melt composition, especially dissolved volatiles which can generate large (~10^5 Pa s) nonlinear changes in viscosity (Dingwell et al. 1996; Hess and Dingwell 1996; Giordano et al. 2004).

Thirdly, and most importantly, the expansion in published viscosity data has supported the development of new conceptual and predictive models for the viscosities of geological melts at geological conditions (Table 1). The citation records for published viscosity models (cf. Table 1; Fig. 2) provide some indication of the impact of these tools. These predictive models are an especially critical component of computational models designed to simulate magmatic and volcanic processes.

In this chapter, we pursue two main objectives:

- First, we provide a comprehensive review of viscosity models for geological melts (Table 1) including the datasets which form the basis of the models, the mathematical formations used to model the data, and the capacity of each model for extrapolation.

- Second, we provide critical analysis of the strengths and weaknesses of the most widely-used geological models for melt viscosity.

We conclude with a brief discussion of "best practices" and suggestions for future experimentation: what experiments are needed; what is needed to ensure the highest quality data; and what have been and are the potential pitfalls in experimentation. These considerations are intended to help guide the form and attributes of the next-generation models for the viscosity of geological melts.

2. EXPERIMENTAL DETERMINATION OF MELT VISCOSITY

The shear viscosity (η) of silicate melts is a ratio of shear stress to the corresponding shear strain rate and describes the resistance to flow. All geological melts have been shown experimentally to be Newtonian except under shear strain rates approaching those of structural relaxation (Dingwell and Webb 1989; Webb and Dingwell 1990a,b). At shear strain rates approaching the inverse of their Maxwell relaxation time, silicate melts can show non-Newtonian behaviour (i.e., shear thinning) and may ultimately transition to brittle behaviour. The critical conditions for this viscous-brittle transition have been explored experimentally (e.g., fiber elongation: Webb and Dingwell 1990a; uniaxial compression: Cordonnier et al. 2012) because of its relevance to volcanic processes involving fragmentation of the melt (i.e., explosive vs. effusive volcanic eruptions; Dingwell 1996; Papale 1999; diGenova et al. 2017; Wadsworth et al. 2018). Distant from this singular non-Newtonian effect lies a vast range of strain rates over which geological melts behave in a Newtonian fashion. Viscosity determinations made in that Newtonian regime dominate the literature and their Newtonian expression means that results obtained from a wide range of experimental methods can be compared and synthesised into one consistent data set for the composition-, temperature- and pressure-dependence of melt viscosity (Fig. 3; see below).

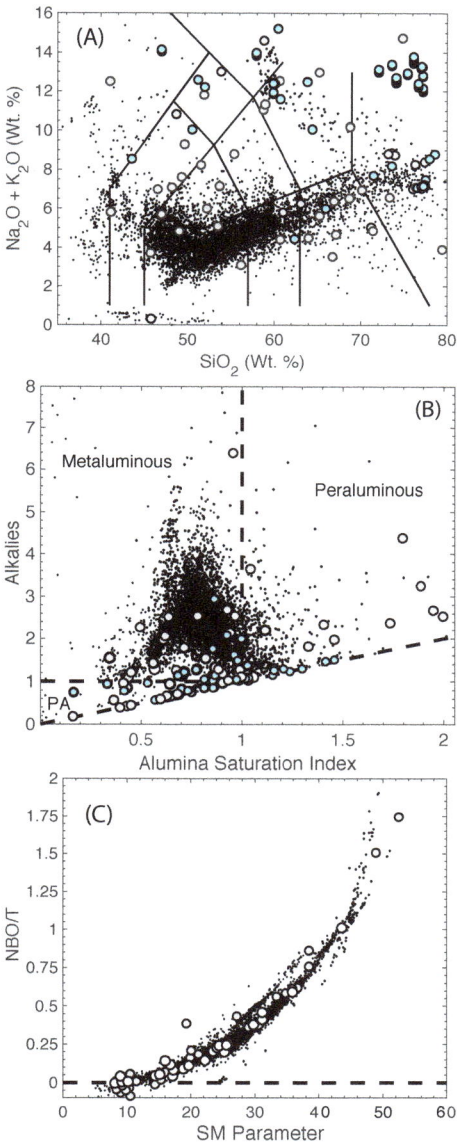

Figure 1. Comparison of the compositional range of natural igneous rocks (**crosses**) to a database of high-temperature viscosity measurements for anhydrous (**white circles**) and hydrous (**blue circles**) geological silicate melts (sources in Giordano et al. 2008). (**A**) Total alkalies vs. silica; (**B**) Alumina saturation index ($Al_2O_3/[CaO + Na_2O + K_2O]$) vs. alkalinity (inverse Agpaitic) index ($Al_2O_3/[Na_2O + K_2O]$); and (**C**) NBO/T vs. SM parameter (Giordano and Dingwell 2003).

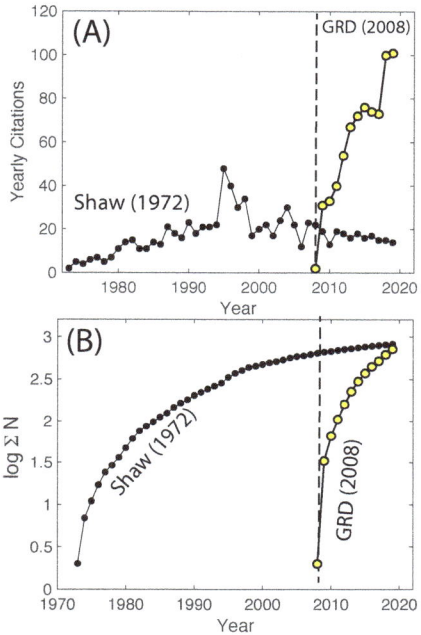

Figure 2. Scientific citations from Web-of-Science© for silicate melt viscosity models of Shaw (1972) and Giordano et al. (2008; GRD) plotted as logarithm of cumulative citations (N) for the period 1972-2020 (see Table 1).

Measurement techniques

A variety of different experimental methods are required to measure melt viscosity over the full range of temperatures and compositions found in natural systems including, concentric cylinder, micropenetration, parallel-plate, beam bending, falling-sphere, and fibre elongation. At temperatures where the viscosity is between 10^{-1} and 10^5, melt viscosity is most commonly measured by concentric cylinder methods. Concentric cylinder measurements involve the so-called Couette geometry of a coaxial arrangement of a cylindrical container and an inner immersed spindle. The technique involves a controlled strain rate (of rotation) and a monitored stress (Dingwell 1986).Its advantages include easy control and variation of the rotation rate whereas its challenges include an inherent radially varying strain rate which may lead to the redistribution of phases in the case of suspension studies.

Table 1. Summary of chemical models for geological melts.

Model Source	Melts[1]	Model[2]	Other	N[3]	ΔT (°C)	ΔP (MPa)	NP[4]	Citation[5]
Bottinga and Weill (1972)	M	A	–	2440		–	285	649
Shaw (1972)	M	A	H_2O	2440		–	7	856
Persikov[6]	M	A	H_2O, CO_2, F, Cl			< 12 Gpa	13–19	(–) 14
Baker (1996)	Granite	AG	H_2O	34	500–900	0.1–1000	7	60
Hess and Dingwell (1996)	Leucogranite	VFT	H_2O	111	611–1643	–	6	454
Ardia et al (2008)	Haplogranite	VFT	H_2O	141	580–1640	500–2500	6	42
Whittington et al. (2009)	Dacite	VFT	H_2O	78	400–1592	–	6	54
Romine and Whittington (2015)	Rhyolite	VFT	H_2O	211	523–1501	0.1–2000	9	14
Romano et al. (2003)	M (2)	VFT	H_2O	116	400–1500	–	12	67
Misiti et al. (2011)	M (2)	–	H_2O	95	560–1600	0.1–500	12	14
Duan (2014)	M (20)	VFT	H_2O, fO_2	220	460–1600	0.1–1500	11	4
Sehlke and Whittington (2016)	Basalt (20)	AG	–	496	650–1600	–	13	19
Giordano and Dingwell (2003)	M (19)	–	–	350	700–1600	–	10–12	157
Giordano et al. (2006)	M (44)	–	–	800	700–1650	–	12–15	77
Hui and Zhang (2007)	M (109)	–	H_2O	1451	300–1705	–	37	187
Giordano et al. (2008)	M (178)	VFT	H_2O, F	1770	245–1705	–	17	874

Notes:

[1] Composition of melts used; M denotes a diverse group of multicomponent melt compositions (number of melts in brackets).
[2] Abbreviations include: Arrhenian (A), Non-Arrhenian (NA), Pressure (P), Adam–Gibbs (AG), Vogel–Fulcher–Tammann (VFT).
[3] Number of experimental data used in calibration.
[4] Number of adjustable parameters in model.
[5] Citations from Web-of-Science© as of March 2021.
[6] Persikov model has sources Persikov (1991; no. of citations not available) and Persikov and Bukhtiyarov (2009).

Where melt viscosity exceeds ~10^7 Pa s, the most common measurement techniques are micropenetration and cylinder-compression viscometry (the latter is commonly referred to as parallel plate viscometry; Gent 1960). These methods exploit the relative differences between times scales for structural relaxation of the melt and rates of crystallization or vesiculation at temperatures immediately above glass transition temperatures (T_g). Operationally, the melt is initially stabilised at superliquidus conditions and then cooled rapidly enough to suppress crystallization or vesiculation, yielding a glass. The glassy sample is then heated up in a temperature "excursion" to slightly above T_g where the experiment can be performed in a supercooled metastable state whilst avoiding the crystallization/foaming window (see Fig. 3).

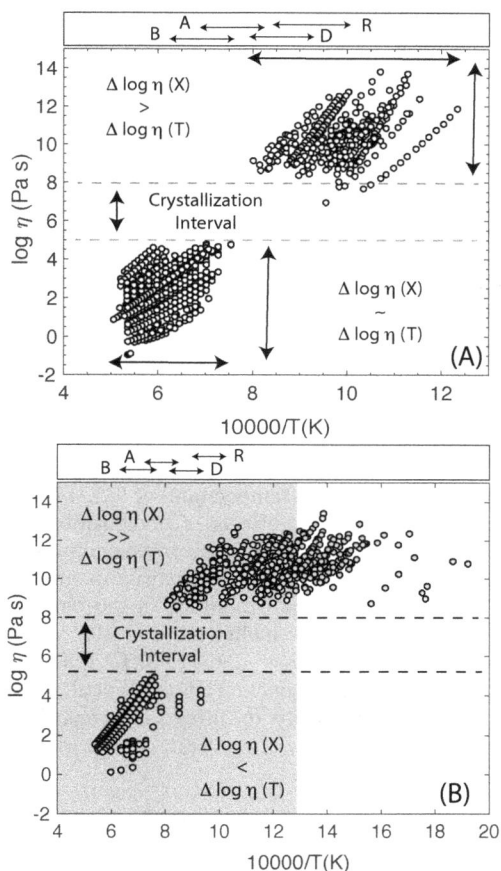

Figure 3. Compilation of modern measurements of viscosity for a range of geological melts (sources in Giordano et al. 2008). (**A**) Viscosity measurements ($N=945$) on 59 anhydrous silicate melts at controlled temperature and atmospheric pressure. (**B**) Viscosity values ($N=843$) of volatile-bearing (H_2O, F) silicate melts. Note change in scale of x-axis; **grey shaded field** denotes the log η–$T(K)$ space corresponding to measured values for anhydrous melts. Crystallization on the time-scales of the experiments creates a data gap between low and high viscosity measurements. The **top bar** on each panel denotes the range of formation and eruption temperatures for anhydrous to hydrous natural basalts (**B**), andesite (**A**), dacite (**D**) and rhyolite (**R**).

The micropenetration and parallel plate techniques have been widely applied to the determination of the viscosity of supercooled metastable melts, both dry and hydrous. The hydrous melts are synthesized at high-T and high-P superliquidus conditions using a conventional piston–cylinder or hydrothermal bomb (Lejeune et al. 1994, Giordano et al. 2004; Vetere et al. 2006). At low-T, the viscosity measurement can often be accomplished in less time than the time required for significant water loss via diffusion and this is a prerequisite for the determination of the viscosity of hydrous melts in a metastable state. At higher temperatures and lower viscosities, phase separation may ensue in the form of foaming, crystallization or liquid–liquid unmixing (depending on melt composition). The common gap in data between 10^5 Pa s and 10^8 Pa s (Fig. 3) is an experimental artifact resulting from this competition between the timescale of melt relaxation (i.e., minimum measurement timescale) and timescales of processes that may modify the material properties (i.e., crystallization, vesiculation or liquid-unmixing).

On occasion, innovative methods have been developed and used to address specific questions that cannot be constrained by a combination of the more common low-viscosity (e.g., concentric-cylinder) and high viscosity (e.g., micropenetration) measurements. Notably, Dorfman et al. (1996) used a centrifuge-assisted falling-sphere technique to measure melt viscosity at intermediate temperatures that are commonly experimentally inaccessible due to enhanced crystallization rates. A similar methodology was used by Ardia et al. (2008) to measure the viscosity of hydrous silicic melts under pressure. In both cases, the centrifuge-assisted method ensured the timescale of viscometry was shorter than for crystallization or foaming in these peralkaline and hydrous (respectively), rhyolitic melts. For some melt compositions, these unique datasets can be critical for more accurately constraining the nature of the melt's temperature dependence (e.g., Arrhenian vs. non-Arrhenian; Russell et al. 2002).

Viscosity data for natural silicate melts

Figure 3 illustrates the range and quality of melt viscosity data available as a result of modern experimentation (sources in Giordano et al. 2008). The dataset includes a wide range of melt compositions that cover the compositional range of most common natural melts (Fig. 1A, B). The parameter NBO/T (ratio of nonbridging oxygens to tetrahedrally coordinated cations) and the SM parameter ($CaO + MgO + MnO + 0.5 FeO_{Total} + Na_2O + K_2O$ mol.%; Giordano and Dingwell 2003), both calculated from melt compositions, serve as chemical proxies for the structural organization of the melt and degree of polymerization (Mysen et al. 1982; Mysen 1988). The melt viscosity dataset (Fig. 3) spans and parallels the range of NBO/T and SM values of the common natural melts indicating the experimental data include melts of a wide range polymerisation (Fig. 1C).

The anhydrous dataset of 952 measurements (Fig. 3A) spans temperatures of 535–1705 °C and viscosities of 10^{-1} to 10^{14} Pa s. At high temperature, the effects of composition and temperature on the range of viscosity values are similar (Fig. 3A). The corresponding data for volatile-rich melts comprise 843 measurements on 143 different compositions and are dominated by variations in H_2O and F contents (Fig. 3B). The magnitude of the range of viscosity values is more or less the same as for the anhydrous melts ($10^{-0.1}$ to $10^{13.4}$ Pa s) but the corresponding temperature range is to much lower values (245–1580 °C) reflecting the marked decrease in melt viscosity with increasing volatile content (Dingwell et al. 1996; Baker 1996; Hess and Dingwell 1996). At low values of viscosity, the effects of composition are subordinate to temperature whereas at high viscosities (low temperature) compositional (i.e., volatiles) effects dominate.

The relative effects of melt composition and temperature on viscosity can be summarized using calculated isokom temperatures (i.e., constant viscosity) for a variety of melt compositions (Fig. 4). The melt compositions range from polymerized (low NBO/T and SM) to highly depolymerized (high NBO/T and SM) (data from Giordano and Russell 2018). At low viscosities (10^4 to 10^6 Pa s) isokom temperatures decrease markedly with increasing SM and NBO/T. At higher viscosities approaching 10^{12} Pa s the decrease in isokom temperature with composition is less apparent. Two calculated isokoms for equivalent melts with 1 wt.% dissolved H_2O illustrate the strong effect of H_2O on melt viscosity (cf. Hess and Dingwell 1996; Dingwell 1998; Giordano et al. 2004). Again the effect of H_2O is most pronounced in more polymerized melts with low NBO/T and SM.

3. MODEL CONSIDERATIONS

Scientific models have the main purpose of representing ideas, replicating data or observations, and making pertinent predictions and reliable extrapolations. Greenwood (1989) eloquently summarized the minimum requirements for a model to be deemed useful and scientifically acceptable. He suggested, *"we should reserve the word 'model' for a well constrained, logical construct (not necessarily mathematical), that has necessary predictable and testable consequences."* To reproduce a set of data does not suffice. A model having no predicted and testable consequences is at risk of being unverifiable, sterile, and of limited use (Popper 1968). The more rigorous the model, the more consequences it implies—each of which is a test of that model.

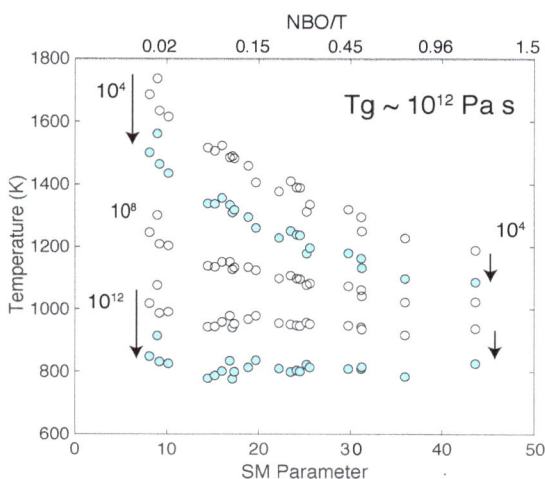

Figure 4. Isokom temperatures for viscosities of 10^4, 10^8, and 10^{12} Pa s plotted as $T(K)$ for 23 melt compositions spanning a wide range of compositions and degrees of polymerization (i.e., NBO/T and SM). The 10^{12} Pa s isokom (**solid circles**) marks the glass transition temperature (T_g). **Blue circles** show depression of isokom temperatures for melts having 1 wt.% dissolved water.

The main goal of viscosity models for geological melts (silicate and otherwise) is to reliably predict melt viscosity at the conditions found in natural systems. Conceptually, good models incorporate abundant and accurate experimental observations and return a reliable tool that reproduces the data and can be extrapolated to domains where data are absent or scant. Our need for models that accurately predict the viscosity of geological melts at conditions relevant to melt production, storage and transport continues to grow as the compositional range of geological melts on Earth and other planets expands (Fig. 1). These models are also required to support the ever more sophisticated numerical models we use for simulating igneous and volcanic processes. Melts in geological systems are susceptible to continuous and dramatic changes in temperature (T) and chemical composition (X) driven by crystallization, mixing with other melts, assimilation of foreign material, changes in redox state, or exsolution or re-dissolution of volatiles. Whilst experiments cannot be expected to cover all of these T-X-driven variations in melt viscosity, models can.

Lastly, it is worth recalling that although terrestrial melts have restricted and characteristic formation and eruption temperatures (Spera 2000; Lesher and Spera 2015), laboratory experiments cover a substantially greater $T(K)$–log η space (cf. Fig. 3A, B). However, models that can be extrapolated over an even wider range of melt compositions (or temperatures and fO_2 conditions) can accommodate potential, unforeseen, extreme compositions arising from protracted differentiation (as stated above), partial melting of crustal rocks, melts produced by frictional heating, lightning strikes or meteor impacts, as well as, new melts discovered on other planetary bodies.

The ideal viscosity model for geological melts will, in addition to reproducing the original data, have the following attributes. It should span all of the compositional range (including volatiles) found in naturally-occurring volcanic rocks and it should be computationally continuous across the entire compositional and temperature spectrum of the database. Viscosity experiments performed at lower temperatures and higher viscosities have shown geological melts to vary in their T-dependence from near Arrhenian (i.e., more polymerized melts with low NBO/T or SM values) to markedly non-Arrhenian (i.e., depolymerized melts having high NBO/T or SM values) (Fig. 3). Thus, an effective model for viscosity of geological melts must also be able to accommodate both strong near-Arrhenian and non-Arrhenian T-dependent behaviour of silicate melts (e.g., Angell 1985).

As noted above, an ideal model should also have predictable and testable consequences (e.g., Popper 1968; Greenwood 1989). For example, models for the viscosity of geological melts should support the calculation of derivative transport properties that can be tested independently, such as melt fragility (m) and glass transition temperature (T_g). Angell (1985) recognized two extreme behaviours of glass-forming liquids depending on their behaviour as they cool to their T_g: strong vs. fragile. Melt fragility is a measure of the degree to which the T-dependence of viscosity deviates from Arrhenian behaviour. In his early work, Angell suggested that strong liquids are near-Arrhenian and characterized by small changes in heat capacity as they cross the glass transition temperature. Conversely, fragile melts are associated with larger changes in heat capacity at the transition and are more strongly non-Arrhenian in their T-dependence.

The degree of fragility is commonly tracked by the model-independent "steepness index" (m; Angell 1985; Plazek and Ngai 1991), which characterizes the slope of the viscosity and the associated average structural relaxation time of the melt with temperature as it approaches T_g. Essentially, it is the slope of the viscosity curve evaluated a temperatures approaching T_g:

$$m = \frac{d\left(\log_{10} \eta\right)}{d\left(T_g\big/T\right)} \text{ where } T \to T_g \tag{1}$$

where T_g is defined as the temperature at which the melt viscosity ~10^{12} Pa s. Lastly, models should be internally consistent with the original tenets or assumptions adopted in the model (i.e., assumed T_g values, constant high-T limit, etc.).

4. TEMPERATURE DEPENDENCE OF MELT VISCOSITY

Melts of different composition can have very different viscosities at equal temperatures (Fig. 4) and can have different temperature dependences (Fig. 3). The simplest temperature dependence shown by silicate melt viscosity is Arrhenian:

$$\eta = Ae^{-E_a/RT} \tag{2}$$

where E_a is the activation energy for viscous flow and the pre-exponential term A represents the viscosity at infinite temperature (η_∞). The values of the two adjustable parameters A and E_a are commonly taken as characteristic for individual melt compositions although theory and practice suggest A is a constant independent of melt composition (see below; Bottinga and Weill 1972; Shaw 1972; Persikov 1991; Persikov and Bukhtiyarov 2009; Russell et al. 2002, 2003).

In general, at high-temperatures activation energy (i.e., slope of log η vs. $1/T$) is highest in high viscosity liquids and decreases with decreasing viscosity. Early experimental measurements of melt viscosity were limited to super-liquidus temperatures where many silicate melts show near-Arrhenian behaviour over the observed temperature range. Ultimately, additional experimental strategies were developed that extended the range of measurements down to temperatures near T_g confirming that many geological melts can exhibit non-Arrhenian T-dependence (i.e., see Richet and Bottinga 1986, 1995; Dingwell et al. 1993).

Non-Arrhenian functions for melt viscosity

Non-Arrhenian melts require higher-order functions to describe the non-linear T-dependence in their viscosity. Numerous functions have been developed to capture the non-Arrhenian temperature dependence and most involve three adjustable parameters (for simplicity referred to here as A, B, C). The reader is referred to Sturm (1980), Bottinga et al. (1995), Richet and Bottinga (1995), Russell et al. (2002, 2003), Russell and Giordano (2005), Mauro et al. (2009), and Kondratiev and Khvan (2016) for discussion of the merits of each of these. Here we briefly review four of these functions.

The Vogel–Fulcher–Tammann (VFT) equation is an empirical means of accommodating the non-Arrhenian temperature dependence of viscosity (Angell 1991; Richet and Bottinga 1995; Bottinga et al. 1995; Rössler et al. 1998):

$$\log \eta = A_{VFT} + \frac{B_{VFT}}{T - C_{VFT}} \tag{3}$$

where A_{VFT}, B_{VFT}, C_{VFT} are adjustable parameters specific to individual melts. The parameter A_{VFT} is the value of log η (Pa·s) at infinite temperature (log η_∞). The parameter B_{VFT} corresponds to the pseudo-activation energy associated with viscous flow and is thought to represent a potential energy barrier obstructing the structural rearrangement of the melt. The C_{VFT} parameter is the temperature (K) at which viscosity becomes infinite. The VFT function is an effective descriptor of the viscosity over the compositional range of most geochemically important melts.

The Adam–Gibbs (AG) functions are based on configurational entropy theory and provide a connection between the transport properties, relaxation timescales of melts, and their thermochemical properties (Adam and Gibbs 1965; Richet 1984; Bottinga et al. 1995; Richet and Bottinga 1995; Bottinga and Richet 1996; Toplis et al. 1997; Russell and Giordano 2017). The T-dependence of melt viscosity (η) is described by:

$$\log \eta = A + \frac{B}{T S_c (T)} \tag{4}$$

where S_c is the configurational entropy of the melt at temperature (T), B is proportional to the activation energy for viscous flow, and A is again the high temperature limiting viscosity (Adam and Gibbs 1965; Richet 1984). The AG function can be expanded by introducing the concept of the glass transition temperature:

$$\log \eta = A + \frac{B}{T \left[S_c (T_g) + \int_{T_g}^{T} \frac{Cp(T)}{T} dt \right]} \tag{5}$$

where $S_c(T_g)$ is the residual configurational entropy of the investigated sample at T_g. Assuming that configurational heat capacity of the melt is constant above T_g and that $S_c \sim 0$ at T_g (Richet 1984; Angell 1991; Toplis et al. 1997) simplifies the AG function to three adjustable parameters:

$$\log \eta = A_{AG} + \frac{B_{AG}}{T \left[\log \frac{T}{C_{AG}} \right]} \tag{6}$$

where C_{AG} is the temperature of the glass transition. Under these conditions, the VTF and AG equations are operationally equivalent (Angell 1991; Bottinga et al. 1995; Richet and Bottinga 1995; Bottinga and Richet 1996; Toplis et al. 1997).

The Avramov and Milchev (1988) model determines the temperature dependence of the average jump frequency of moving entities (in their wording "molecules") and, through it, the viscosity. The main assumption is that energy barriers of different heights appear, with a distribution depending on entropy. The viscosity depends therefore on total entropy, not on configurational entropy alone. Adopting a reference (glass transition) temperature corresponding to a melt viscosity $10^{12.5}$ Pa s and further assuming that the heat capacity of the liquid is independent of temperature they advocated a function (AV) of the form (i.e., Avramov 1998):

$$\log \eta = A_{AV} + \left(\frac{B_{AV}}{T} \right)^{C_{AV}} \tag{7}$$

Recently, Mauro et al. (2009) has explored the attributes of another three-parameter equation (MYEGA: MY) for the temperature dependence of melt viscosity that was first described by Waterton (1932):

$$\log \eta = A_{MY} + \frac{B_{MY}}{T} e^{\frac{C_{MY}}{T}} \tag{8}$$

Their derivation starts by linking the configurational entropy from the Adam–Gibbs approach to a topologically determined number of degrees of freedom per atom. C_{MY} is a normalised activation energy describing the intact degrees of freedom per atom with rising temperature. B_{MY} is essentially the normalised B_{AG} of the Adam–Gibbs equation (Fotheringham 2019).

These four functions (Eqn. 3, 6–8) have been fitted to viscosity data sets for three simple (synthetic) silicate melts (Fig. 5): $CaMgSi_2O_6$ (Diopside, Dp), $CaAl_2Si_2O_8$ (Anorthite, An), and $NaAlSi_3O_8$ (Albite, Ab). One attribute of these melts is the abundance of measurements spanning a large range of temperature (685 to 2175 °C) and viscosity ($10^{-1.2}$ to $10^{-14.5}$ Pa s) (sources in Li et al. 2020). Furthermore, two of the melts have NBO/T values of 0 and one has an NBO/T of 2, whereas two melts are non-Arrhenian and one melt is near-Arrhenian in their T-dependence. The optimal adjustable parameters A, B and C (Table 2) for each melt differ between the four model equations (Eqn. 3, 6–8) because of differences in their form but each function reproduces the data well (Fig. 5A). The greatest discordance occurs at high-T and low viscosity where the AV function shows the largest misfits.

The parameter A in each of the four model equations represents the value of melt viscosity at infinite temperature (log η_∞).

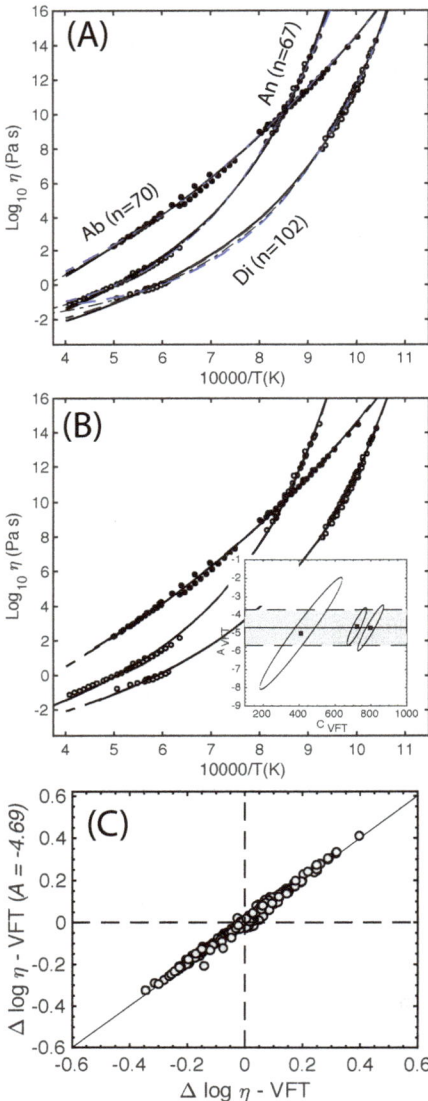

Figure 5. Models for the temperature dependence of melt viscosity fitted to data for melts (Di, An, Ab): **(A)** Comparison of model curves based on VFT (**solid line**), Adam–Gibbs (AG, **dash line**), Avramov (AV, **dash-dot line**), and MYEGA (MY, **blue dash line**) functions. The four functions reproduce the data well and predict similar values of T_g and m (Table 2). **(B)** The VFT function fitted to the Di, An, and Ab database assuming a common value of A (log η_∞) (**solid line**) and compared to VFT function for individual melts (**dashed lines**). Inset shows 99% confidence limits on values of A and C (**solid squares**) from the individual fits; **horizontal line** denotes the optimal value for a common A (−4.69 ± 0.99; Table 4). **(C)** Misfits to data from fitting individual datasets (Di, An, Ab) vs. fitting with a common A.

Interestingly, each function defines very different restricted ranges of A for the three melts (Table 2, A_{VFT}: -5 to -4.6, A_{AG}: -3.7 to -3.6, A_{AV}: -2.1 to -1.4, A_{MY}: -4.2 to -2.3) (Russell et al. 2002; Kozmidis–Petrovic 2014). The optimal values of A may suggest which functions are best for geological melts. For example, the values of A_{AV} are only slightly lower than the lowest measured values of log η (-1.29) whereas model values of A_{MY} span 2 orders of magnitude for these three melts (see below).

Each of these four functions (Eqn. 3, 6–8) allows for the calculation of the derivative properties. Table 3 lists the expressions for calculating values of T_g, taken as the temperature where $\eta \sim 10^{12}$ Pa·s, and melt fragility m from the optimal parameters A, B, and C. This provides an additional means of illustrating the differences between these non-Arrhenian parameterisations by comparing the derivative properties implied by the optimal parameters A, B, and C. Values of T_g and m for the three melts are reported in Table 3 based on the four model equations. All four models predict very similar values of T_g even though the functional forms are very different. Fragility values (m) predicted with the VFT, AG and AV model equations also agree well; notably, the MY function predicts a substantially lower value for m.

Table 2. Formulas to calculate derivative properties (T_g and m) from model values[1] of A, B, C.

Model	T_g (K)[2]	Fragility (m)[3]
VFT	$T_{gVFT} = \dfrac{B_{VFT}}{12 - A_{VFT}} + C_{VFT}$	$m_{VFT} = \dfrac{B_{VFT}}{T_{gVFT}\left[1 - \dfrac{C_{VFT}}{T_{gVFT}}\right]^2}$
AG	$\text{roots} \dfrac{B_{AG}}{12 - A_{AG}} - T_{gAG} \ln\dfrac{T_{gAG}}{C_{AG}}$	$m_{AG} = \dfrac{(12 - A_{AG})\left[\ln\dfrac{T_{gAG}}{C_{AG}} + 1\right]}{\ln\dfrac{T_{gAG}}{C_{AG}}}$
AV[2]	$T_{gAV} = \dfrac{B_{AV}}{[12.5 - A_{AV}]^{1/C_{AV}}}$	$m_{AV} = C_{AV}\left(\dfrac{B_{AV}}{T_{gAV}}\right)^{C_{AV}}$
MYEGA	$\text{roots}(12 - A_{MY})T_{gMY} - B_{MY}e^{C_{MY}/T_{gMY}}$	$m_{MY} = \dfrac{B_{MY}C_{MY}}{T_{gMY}^2}e^{C_{MY}/T_{gMY}}$

Notes:
[1]Values of A, B, C are adjustable parameters for each function (see text) fitted to viscosity data.
[2]T_g is taken as T(K) at which melt has a viscosity of 10^{12} Pa s except AV model which adopts $10^{12.5}$ Pa s
[3]Fragility (m) is taken as the steepness index (see text).

5. THE HIGH-TEMPERATURE LIMIT TO MELT VISCOSITY (A)

The parameter A in the four functions discussed above (Eqn. 3, 6–8) is the value of log η_∞ and, conceptually, represents the high-temperature limit to silicate melt viscosity. Russell et al. (2002, 2003) reviewed the concept of A as a constant for all geological melts (i.e., independent of composition) implying that at high-T all melts ultimately converge to a single, common value of viscosity. This assertion is supported practically and theoretically (Myuller 1955; Eyring et al. 1982; Angell 1985; Russell et al. 2002, 2003).

Table 3. Parameterization of compiled datasets[1] of T(K)–log η for diopside (Dp), anorthite (An), and albite (Ab) melts using VTF, AG, AV, and MYEGA (MY) empirical equations (see text).

Melt Composition		Dp	An	Ab
No. Experiments		102	67	70
NBO/T		2	0	0
SM Parameter		50	25	12.5
VFT	A_{VFT}	−4.63	−4.71	−5.01
	B_{VFT}	4,520	5,536	11,513
	C_{VFT}	721.1	797.9	410.7
	T_g (K)	993	1129	1087
	m	60.7	57.0	27.3
AG	A_{AG}	−3.62	−3.68	−3.57
	B_{AG}	2,350	2,988	9,887
	C_{AG}	700.5	766.0	283.5
	T_g (K)	993	1130	1088
	m	60.4	56.1	27.1
AV	A_{AV}	−1.39	−1.62	−2.13
	B_{AV}	1,801	2,242	4,359
	C_{AV}	4.36	3.81	1.91
	T_g (K)	993	1130	1090
	m	58.4	51.9	27.0
MYEGA	A_{MY}	−2.36	−2.65	−4.21
	B_{MY}	631.4	1127.4	8938.39
	C_{MY}	3096	3037	738.8
	T_g (K)	993	1130	1088
	m	44.8	39.4	11.0

Note: [1]Data sources as in Li et al. (2020).

Practical application of a high-T limit

We demonstrate the practical aspects of the "*common A concept*" using two separate datasets: i) the 3 synthetic silicate melts An–Ab–Di (Fig. 5); and ii) 23 multicomponent natural melts (sources in Giordano and Russell 2018).

Simple System: Di–An–Ab. VFT functions have been fit simultaneously to the An–Ab–Di dataset assuming that all three melts share a common but unknown high–temperature limiting value of viscosity (i.e., $A = \log \eta_\infty$) whilst allowing for individual values of B_{VFT} and C_{VFT} (Table 4). The optimal solution for A_{VFT} is −4.69 (± 0.99) and the experimental data are reproduced exceptionally well (Fig. 5C) despite using two fewer adjustable parameters (i.e., single value for A rather than three). Furthermore, the model functions exactly overlap the individually fitted functions despite the differences in values of A, B, and C (cf. Table 3 vs. Table 4).

As discussed by Russell et al. (2002), the adjustable parameters (i.e., A, B, and C) for individual melts are strongly correlated and the nature of the covariation mainly reflects the distribution (in temperature) of the data. Confidence envelopes on the values for A and C (solid squares) obtained from fitting the three datasets individually (Fig. 5B; Inset) emphasize several important facts. Firstly, they show the strong, model-induced covariance between adjustable parameters inherent in the VFT function, which will be also true for any of the other parametric equations (Eqns. 3–7). This is a caution against attributing "apparent" correlations between parameters (e.g., A, B, and C) solely to variations in melt composition. Secondly, they delineate the range of model values (i.e., A and C) that, when combined in a non-arbitrary way, will reproduce the original data accurately (e.g., Russell et al. 2002). In the case of albite melts the much larger A–C confidence envelope results from applying the 3-parameter VFT function to a near-Arrhenian dataset (linear). The optimal value for the common A (−4.69; Fig. 5B Inset) is well within the confidence limits for the individual model fits.

Table 4. Model parameters from simultaneously fitting the VFT function to the T(K)–log η data for the melts diopside (Dp), anorthite (An), and albite (Ab) (See Fig. 5). The model assumes the three melts share a common high-temperature limiting value (A) and individual values for B and C. Values in brackets are the 99% confidence limits on the optimal model parameters.

Melt Composition		Dp	An	Ab
No. Experiments		102	67	70
VFT	A_{VFT}		−4.69 (0.99)	
	B_{VFT}	4,582 (1156)	5,507 (1352)	10,907 (2074)
	C_{VFT}	718 (57)	799 (63)	434 (99)
	T_g (K)	992	1129	1087
	m	60.2	57.0	27.8
χ^2_{min}		6.61	5.56	6.27

Application to Natural melts. The practicality and mathematical viability of a common A are equally applicable to natural multicomponent melts. We have fit VFT functions to a set of 413 viscosity experiments performed on 23 (n) different anhydrous silicate melt compositions (Giordano and Russell 2018). The experimental data span much of the compositional range found in natural systems and include both near-Arrhenian and strongly non-Arrhenian melts.

Fitting the VFT function to the 23 melts individually requires 69 parameters (Table 5) whilst a VFT optimization based on a common single value of A uses 47 ($2n + 1$) parameters: B and C values for individual melts and a single value for A_{VFT} (−4.48). The original data are reproduced well (±0.5 log units) with fewer parameters (Fig. 6A). Values of B and C for the 23 melt compositions vary between 4,400 and 12,100, and between 245 and 670, respectively (Table 5; Fig. 6C). In contrast, the individual VFT fits generate a much larger range of parameter values (Fig. 6A–C) some of which are unrealistic: A (−11.38 to −3.63), B (4100 to 34,000), and C (−530 to 690). For example, the values of A are excessively low for polymerized melts (i.e., low SM values; Fig. 6B) which tend to have more Arrhenian T-dependence.

Values of fragility calculated from the individual VFT functions versus values associated with a common value of A are compared in Figure 6D. The agreement between the two sets of values indicates that even with ~30% fewer adjustable parameters, the fundamental properties of the original data (i.e., departure from Arrhenian) are preserved. The single discordant data point is a near-Arrhenian (low fragility) rhyolite melt which is essentially overfitted by the 3-parameter VFT function resulting in an infeasible negative value for C (−531) and an unrealistically high value of B (34,000).

Implications of a common A

The modelling based on a single unknown value of A generates results that are indistinguishable from those achieved by fitting data sets individually (Fig. 6). Mathematically, the concept of a common high-T limit to silicate melt viscosity is valid regardless of the model adopted (VTF vs.

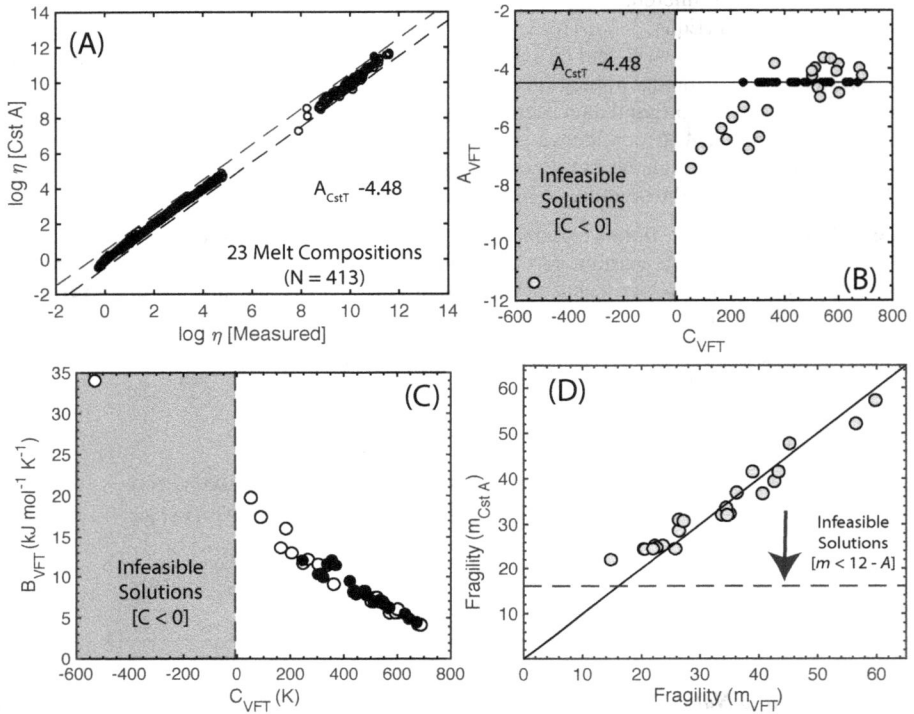

Figure 6. VFT functions fitted to viscosity data ($N=413$) for 23 multicomponent silicate melts (Table 5). (A) Comparison of measured log η to VFT model values constrained to have a common A (log $\eta_\infty = -4.48$). (B) Values of A and C for 23 individual melts fitted independently to VFT functions (**grey symbols**). Values of C less than 0 are infeasible. **Black symbols** denote values of C assuming a common A value (-4.48; dashed line). (C) Comparison of values of B and C for VFT functions fitted to individual datasets (**open circles**) vs. values assuming a common A (**solid circles**). (D) Values of fragility (m) calculated from the individual VFT functions vs. the VFT functions with a common A. **Dashed line** indicates limit to fragility defined by $12 - A_{VFT}$ where A is a constant.

AG vs. MY vs. AV, etc); however, the optimal value of A will be model dependent (Russell et al. 2003). There are (at least) two very real tangible benefits of a constant A for modelling the viscosity of silicate melts. Firstly, it suggests that all effects of melt composition, including water content, can be ascribed to only two (B and C) of the three parameters.

Additionally, the common A constraint reduces the overall uncertainties on all of the adjustable parameters as illustrated by Russell et al. (2002, 2003). The non-linear character of the non-Arrhenian models ensures that there are strong covariances between model parameters which allow for a wide ranges of acceptable values (i.e., A, B, C) for individual melts. In particular, the fitted confidence limits on A for a single melt can easily span 5 log units (i.e., -1 to -7). This is true even where the data are numerous, well-measured, and span a wide range of temperatures and viscosities. Stated another way, there is a substantial range of model values which, when combined in a non-arbitrary way, can accurately reproduce the experimental data. However, when we invoke a common A, the main result is that the acceptable range of A becomes very tight and the "permissive" range of B and C values is greatly reduced. Further details on the numerical justification for a parameterization assuming a common A are provided in Russell et al. (2002, 2003).

Table 5. VFT parameters for fitting T(K)–log η datasets (N=413) for 23 multicomponent silicate (Giordano and Russell 2018) individually and assuming a common value for A. Also reported are chemical proxies for melt structure (NBO/T, SM) and values of T_g and m.

Label	SM	NBO/T	No. Data	VFT Individual Fits					VFT Fits with Common A (−4.48)			
				A	B	C	T_g (K)	m	B	C	T_g (K)	m
Rattlesnake Tuff	8.12	0.00	11	−7.43	19766	52.9	1070	20.4	12096	355.2	1089	24.5
Lipari_RR	9.22	0.01	13	−6.75	17422	90.8	1020	20.6	11593	337.2	1041	24.4
MDV_snt	8.98	0.05	17	−6.43	16039	184.0	1054	22.3	11421	370.1	1063	25.3
Mercato1600	17.15	0.05	23	−5.32	11667	247.6	921	23.7	9927	322.8	925	25.3
PVC	14.47	0.06	25	−5.68	13004	205.4	941	22.6	10393	316.5	947	24.8
MNV	15.23	0.07	19	−6.05	13654	165.0	922	22.0	10305	303.3	929	24.5
AMS_B1	17.35	0.10	11	−3.82	9056	362.2	935	25.8	10309	305.8	931	24.5
Newberry	10.17	0.02	16	−11.38	34019	−530.9	924	14.8	12057	245.7	977	22.0
NYT_lm*13*	18.86	0.12	24	−3.97	7390	514.1	977	33.7	8245	474.2	974	32.1
UNZ	16.86	0.14	20	−3.63	6879	545.1	985	35.0	8247	479.3	980	32.3
MST	19.69	0.15	21	−4.25	7308	503.0	953	34.4	7695	484.9	952	33.6
CI_OFI04	16.01	0.16	20	−5.44	11387	336.0	989	26.4	9452	421.8	995	28.6
FR_a	22.22	0.19	27	−4.66	7437	523.9	970	36.2	7151	537.0	971	36.9
Pompei TR	23.49	0.23	17	−4.08	7014	501.8	938	34.6	7830	446.1	921	32.0
MRP	25.23	0.26	22	−3.84	5636	600.8	957	42.6	6579	554.2	953	39.4
Ves_W	24.15	0.26	14	−6.76	12183	265.8	915	26.4	8101	434.1	926	31.0
Ves_G	24.50	0.28	14	−6.34	11559	304.8	935	27.2	8320	436.7	942	30.7
Min_2b	25.58	0.30	25	−3.66	5629	572.1	932	40.6	6861	510.7	927	36.7
Pollena GM	29.77	0.44	15	−4.09	5648	593.2	944	43.3	6167	568.8	943	41.5
ETN	31.22	0.51	10	−4.84	6019	602.4	960	45.2	5477	629.0	961	47.7
Ves_Gt	31.15	0.53	16	−4.98	6987	532.0	944	38.9	6202	570.2	947	41.5
NYI	35.92	0.73	23	−3.97	4257	677.5	944	56.5	4911	643.5	941	52.1
EIF	43.61	1.16	10	−4.24	4171	687.9	945	59.8	4476	671.8	943	57.2

The concept of a high-T limit to silicate melt viscosity cannot be tested directly because it requires observations at extreme temperatures well outside the range of conventional experimental methods. We can state, however, that the value of A (constant or not) must be less than any of our physical measurements of melt viscosity (e.g., $\leq 10^{-1}$ Pa s for peridotitic melt; Dingwell et al. 2004). Persikov (1991), Russell et al. (2002) and others have argued that at super-liquidus temperatures, all silicate melts become highly dissociated liquids regardless of their structural arrangement at lower temperatures and will converge to a lower limiting value of viscosity. Indeed, there is no direct evidence to suggest that "fragile", "intermediate" and, even, "strong" silicate melts maintain different rheological properties at temperatures well above their respective liquidus temperatures. The implication is that natural melts, as diverse as basalt and rhyolite, should converge to a common viscosity at high-T.

The concept of a constant A is consistent with studies of low-T glass-forming systems including polymer melts, organic liquids or liquid elements (e.g., Angell 1991, Scopigno et al. 2003). In these lower temperature systems, experiments performed at $T \gg T_g$ show that both strong and fragile melts converge to a fixed viscosity of $\sim10^{-5}$ Pa s (e.g., Eyring et al. 1982; Angell 1991; Persikov 1991; Russell et al. 2003; Scopigno et al. 2003). The Maxwell relationship ($\tau = \eta_0/G_\infty$), which informs on the time scales of relaxation processes in melts (Eyring et al. 1982; Angell 1991; Richet and Bottinga 1995; Toplis 1998; Russell et al. 2003; Scopigno et al. 2003 and references therein), provides some constraints on the lower limit to melt viscosity (η_0). The quasilattice vibration period ($\sim10^{-14}$ s) representing the time between successive assaults on the energy barriers to melt rearrangement limits the relaxation time scales (τ) of melt. Assuming an average value of $\sim10^{10}$ Pa for the bulk shear modulus (G_∞) of the melt at infinite high frequency suggests a lower limiting value to viscosity (η_0) of $\sim10^{-4 \pm 2}$ Pa s (e.g., Dingwell and Webb 1989; Toplis 1998). This accords well with the values returned by our optimization based on the VFT function assuming a common unknown A (i.e., $10^{-4.7 \text{ to} -4.5}$).

6. EARLY MODELS AND MODELLING OF GEOLOGICAL MELTS

Here, we review published models for predicting the viscosity of geological melts at geological conditions. The models are diverse in that they adopt different functional forms and span specific and different ranges of melt composition, volatile contents, temperature, and pressure (Table 1). Our analysis clearly illustrates the increasing sophistication and the overall advancement we are making as a scientific community and is also intended to point the way forward.

Pioneering (Arrhenian) models

It is ca. 50 years since Shaw (1972) and Bottinga and Weill (1972) published empirical methods for predicting silicate melt viscosities as a function of temperature (T) and composition. These models are still used to provide estimates of melt viscosity in studies concerning magmatism, volcanism, and planetary differentiation although the citation record shows a substantial decline as more comprehensive models have become available (Fig. 2).

These models were built on an experimental database that was limited in three main ways. Firstly, the database of viscosity measurements on natural silicate melts was sparse. Secondly, the available experimental data did not completely overlap or span the range of natural melt compositions. Thirdly, the majority of data derived from super-liquidus experiments. On this basis, both Shaw (1972) and Bottinga and Weill (1972) adopted an Arrhenian temperature dependence for melt viscosity and the seminal contribution made by these two papers was to provide the first tools for predicting the viscosity of geological melts as a function of composition and temperature.

Bottinga and Weill (1972). The Bottinga and Weill (1972) model (BW) for melt viscosity was calibrated on 2440 log η–T(K) data points derived from 31 synthetic, anhydrous silicate chemical systems spanning 35–91 mol% SiO_2 over a temperature range of 1100–1800 °C. Their model predicts viscosity as a function of melt composition and temperature as:

$$\ln\eta = \sum_i^n x_i D_i \tag{9}$$

where x_i and D_i are the mole fraction of the ith component and an empirical constant for that component, respectively. The components relevant to natural systems in the BW model include: SiO_2, TiO_2, FeO $_{(total)}$, MgO, CaO, Na_2O, K_2O, and alumina was distributed across the components $KAlO_2$, $NaAlO_2$, $CaAl_2O_4$, $MgAl_2O_4$ and $MnAl_2O_4$. Unique values of D_i for each of these 12 oxide components were determined for 5 restricted ranges of SiO_2 contents: 35–45, 45–55, 55–65, 65–75 and 75–81 mol%. Furthermore, for fixed melt compositions, the D_i values varied every 50 °C over the temperature interval 1200–1800 °C.

This bootstrap model reproduced the original datasets well over the range of viscosities 10^{-1} to 10^5 Pa s where most melts show a near-Arrhenian temperature dependence. The BW model can be applied to natural systems (Fig. 7A) but the calculations tend to be laborious because a lookup table prescribes the values of D_i depending on the composition and, then, the temperature of the melt. A more important consideration is the lack of continuous functions describing D_i values as a function of composition and temperature. This is exemplified in the original figure of Carmichael et al. (1974, Fig. 4.7) where they used the BW model to calculate viscosity for 6 liquids at temperatures of 1050–1500 °C (reproduced here as Fig. 7A). Careful examination of the model points they used to define each line (see original) shows slight discontinuities arising from jumping from one set of D_i values to another. This is also apparent in the original paper of Bottinga and Weill (1972, see their Fig. 7) where they computed viscosity for 7 melts over the temperature interval 1200–1800 °C. This issue would only be further exacerbated when modelling viscosities of evolving melts as a function of, both, temperature and composition. Ideally, a predictive model should be continuous across the variable space of interest.

Shaw (1972). Shaw (1972), based on the compilation and calculations of Bottinga and Weill (1972), developed an alternative predictive model (SH) for geological melts viscosity. He took a different approach where he expanded the Arrhenian function (Eqn. 1) to the form:

$$\ln\eta = C_\eta - C_T s + s\left(\frac{10^4}{T(K)}\right) \tag{10}$$

where s is a characteristic slope calculated from melt compositions and C_η and C_T are model constants: -6.4 and 1.5, respectively. The innovation in the SH model is that the compositional effects on viscosity are encapsulated in s which is calculated from the oxides with only 5 model parameters. The model is continuous in temperature and composition space and includes the effects of H_2O. Figure 7B is a recalculation of viscosity curves for hydrous (0 to 10 wt.% H_2O) rhyolite using the Shaw (1972) model as explored by Carmichael et al. (1974, Fig. 4.6). Superimposed on those curves are the inferred solidus and liquidus of felsic intrusive magmas which effectively delineate (reduce) the portions of the model log η–T(°C) curves that are relevant to magmatic processes. The SH model rapidly became, and remained for over 30 years, the primary method of predicting melt viscosity in geological systems (Fig. 2).

Giordano et al. (2008) illustrated the application of the Shaw model to a database of modern viscosities for anhydrous and hydrous natural multicomponent melts (reproduced here in Fig. 7). The SH model is Arrhenian and therefore systematically underestimates anhydrous melt viscosity at lower temperatures (Fig. 7C). This results from an Arrhenian extrapolation to lower temperatures combined with the fact that viscosity curves for all non-Arrhenian silicate

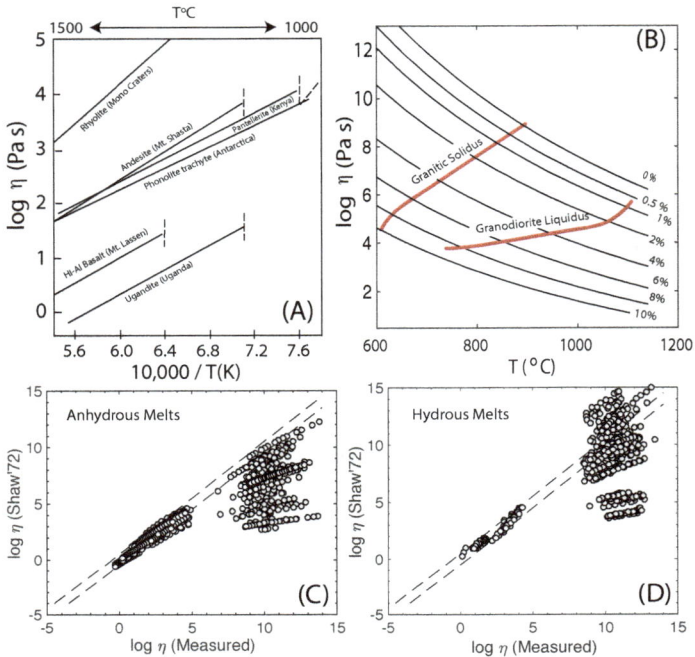

Figure 7. Petrological applications of early non-Arrhenian models for viscosity of geological melts. **(A)** Melt viscosity as a function of temperature for a range of anhydrous volcanic rock compositions calculated with the Bottinga and Weill (1972) model (cf. Carmichael et al. 1974, Fig. 4-7). All calculations are performed at temperatures in excess of the inferred liquidus temperatures (**dashed vertical lines**). **(B)** Viscosity curves for hydrous rhyolite melts calculated with the Shaw (1972) model and plotted vs. $T(°C)$ (labels are wt.% H_2O). Also shown are the approximate granite solidus and granodiorite liquidus (cf. Carmichael et al. 1974, Fig. 4-6). **(C, D)** Viscosity values predicted by model of Shaw (1972) for dataset shown in Figure 3: **(C)** anhydrous melts and **(D)** hydrous melts.

melts are concave up (Fig. 3) and, thus, steepen with decreasing temperature (increasing viscosity). However, the model does a remarkably good job of reproducing melt viscosity at high temperatures (Fig. 7C, D), especially, considering the sparseness of the data that was available to Shaw (1972). The model replicates high-T viscosity data well because most silicate melts show only minor deviations from Arrhenian behaviour above liquidus temperatures. However, it is also testimony to the cleverness of Shaw's parameterization for composition based on 5 model coefficients and 2 constants.

The Shaw model also predicts the high-temperature viscosity of hydrous melts well (Fig. 7D). Hydrous melts of basic and intermediate chemistry tend to be more Arrhenian-like than their anhydrous counterparts and, consequently, the Shaw model probably can be applied to hydrous melts over larger ranges of temperatures. However, at sub-liquidus temperatures, the Shaw model results in much scatter and, both, overshoots and undershoots measurements of viscosity drastically (Fig. 7D).

A final attribute of the Shaw model deserving comment is the high-temperature behaviour of melt viscosity that is implicit to the model's parameterization. In Equation (10) the implied high-temperature limit to melt viscosity (i.e., A (Pa s) = log η_∞) is expressed as:

$$\log\eta_\infty = \frac{(C_\eta - C_T s)}{2.303} - 1 \tag{11}$$

In the Shaw model, there is a compositional dependence to log η_∞ (i.e., A_{Shaw}) embedded in the term C_Ts but it is slight. Values of A_{Shaw} calculated for this same database are restricted to between −6.6 and −4.8 for, both, anhydrous and hydrous melts. Notably, this range is very close to the model A_{VFT} value (−4.48) obtained earlier where we fit a common A to data for 23 anhydrous melts (i.e., Fig. 6 and Table 5, Giordano and Russell 2018). The SH model estimate of a high-T limit to melt viscosity is significantly more restricted and more sensible than the range of values one gets by fitting melts for individual values of A (i.e., Table 5, $A = -11.4$ to −3.6).

Persikov (1980's to 2000's). The work of Persikov (1984, 1991) has resulted in another important set of Arrhenian models for predicting magmatic melt viscosity. He and his co-authors (Persikov and Bukhtiyarov 2009) offer an alternative approach to modelling (PB) melt viscosity at temperatures above and near liquidus conditions as a function of composition including volatiles, temperature, and, both, total and fluid pressure. Their Arrhenian model predicts melt viscosity at temperature and pressure (η_T^P) based on the equation:

$$\eta_T^P = \eta_\infty e^{\left(E_x^P / RT(\text{K})\right)} \tag{12}$$

where the pre-exponential constant η_∞ is the high-temperature limit to melt viscosity. Based on theory, Persikov (1991) assigned *a priori* a constant value (independent of composition and pressure) of $10^{-3.5}$ dPa ($10^{-4.5}$ Pa s) to η_∞ (Persikov and Bukhtiyarov 2009). The effects of variable composition on melt viscosity are accounted for by the variable E_x^P representing the activation energy for viscous flow. Operationally, model values of viscosity (dPa s) are calculated from:

$$\log \eta_T^P = \frac{E_x^P}{4.576T(\text{K})} - 3.5 \tag{13}$$

The activation energy term, E_x^P, depends on a variable K which is proportional to the NBO/T of the melt ($K = 100$ NBO/T) as calculated from the oxide composition in a manner prescribed by Persikov (1984). At low NBO/T, K is adjusted to account potentially for pronounced structural reorganization in the melt based on ratios of Al/(Al + Si). Values of K are also adjusted to account for dissolved H_2O (Table 6). The main attribute of the Persikov and Bukhtiyarov (2009) viscosity model is its capacity to accommodate the effects of composition, including a wide range of dissolved volatile contents, and pressure. Its main limitation is the Arrhenian basis which limits its application to near liquidus conditions or strong liquids.

Table 6. Empirical functions and coefficients from Persikov and Bukhtiyarov (2009) for calculating E_x^P for anhydrous and hydrous melts at 1 atm pressure.

$K^{(1)} = 100$(NBO/T)	$\varepsilon = $Al^{iv}/(Al^{iv} + Si^{iv})	E_x^1 (for $P = 1$ atm)
$K < 17$	$\varepsilon < 0.25$	$E_\varepsilon = 98 - 158\,\varepsilon + 152\,\varepsilon^2$ $E_{X1}^1 = (K/17)\,(51.6 - E_\varepsilon) + E_\varepsilon$
	$\varepsilon > 0.25$	$E_{X1}^1 = 68 - 0.965\,K$
$17 < K < 100$	—	$E_{X2}^1 = 54.4 - 0.165\,K^{(2)}$
$100 < K < 200$	—	$E_{X3}^1 = 42.4 - 0.045\,K$
$200 < K < 400$	—	$E_{X4}^1 = 34.4 - 0.005\,K$

Notes:
[1]Values of K for water content: $K_{H_2O} = K_{Dry} - (K_{Dry} - 2) \times H_2O$ (wt.%)/100
[2]Author supplied spreadsheet (2019) uses: $E_{X2}^1 = 52.4 - 0.165\,K$

Figure 8 compares the measured values of viscosity for anhydrous and hydrous natural multicomponent melts (i.e., data from Giordano et al. 2008) to values predicted by the PB model. Here, we are using the PB model under a restricted range of conditions (i.e., 1 atm) relative to its full capability. The PB model is best at reproducing melt viscosity at high temperatures (Fig. 8A–B) where silicate melts are near-Arrhenian. As with the Shaw model, the PB model cannot capture departures from Arrhenian behaviour that are most apparent in fragile melts and at lower temperatures and systematically underestimates melt viscosity at lower temperatures for the anhydrous dataset (Fig. 8A). For hydrous melts at subliquidus conditions temperatures, some viscosities are predicted well by the PB model whilst others are scattered and, both, overshoot and undershoot measured values (Fig. 8B).

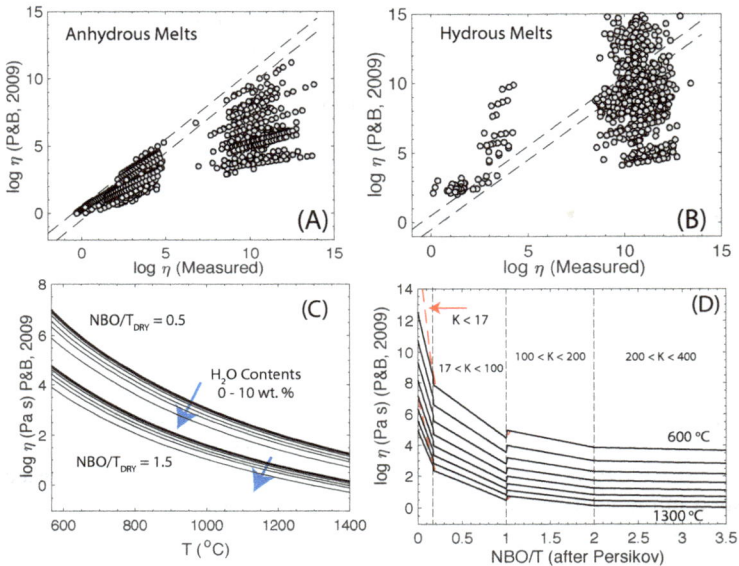

Figure 8. Prediction of melt viscosity using model of Persikov (1991) and Persikov and Bukhtiyarov (2009) (PB). Measured values of log η (as in Fig. 3) vs. values calculated with PB model for (**A**) anhydrous and (**B**) hydrous melts. (**C**) The PB model effects of H_2O (0.5, 1, 2, 4, 6 and 10 wt.%) on dry melts (**heavy line**) having NBO/T values of 0.5 and 1.5 corresponding to a strong, near-Arrhenian and fragile, non-Arrhenian melt, respectively. (**D**) PB viscosities as a function of NBO/T (cf. Persikov 1984) and contoured for temperature. The PB algorithm uses a composition-dependent parameter K (= 100 NBO/T) which changes discontinuously at critical values of K. At low values of K (i.e., < 17), there are two trends (**red dashed lines**) for melts with low (< 0.25) and normal ratios of Al/(Al+Si).

We have used the PB model to explore the model effects of H_2O on melt viscosity. Specifically, we calculate the T-dependant viscosity as a function of H_2O content for a more (NBO/T ~ 0.5; strong) and less (NBO/T ~ 1.5; moderately fragile) polymerized melt using NBO/T as a proxy for melt structure. The PB model suggests that < 2% water has little effect (< 1 order of magnitude) on viscosity and that the effects of water are only slightly more pronounced in strong melts over moderately fragile (Fig. 8C–D).

The core element to the PB model is the calculation of E_x^P from values of K (Table 6) which are, themselves, a function of NBO/T. There are two main issues with this approach. Firstly, by modelling melt viscosity as a function of the parameter NBO/T the specific controls of composition (i.e., individual oxides) on melt viscosity are lost. The countervailing benefit, of course, is that if you only know the NBO/T of the melt you can calculate the viscosity directly. This leads to the second problem which is illustrated in Figure 8D. PB model values of viscosity are calculated as a function of NBO/T values over the temperature range of 600–1300°C. NBO/T values dictate the values of K and, based on the values of K, one of six specific expressions is used to compute the critical parameter E_x^p (Table 6). This creates discontinuities in the compositional dependence of viscosity that are most pronounced at low values of K and low temperatures. As with the BW model, the mathematical rigour of the PB model is reduced because it is not continuous and the discontinuities can become important when considering systems where temperature and composition are evolving. Ideally, a predictive model should be continuous across the variable space of interest.

7. MODELS FOR HYDROUS SILICIC MELTS

Hydrous low-P models for silicic melts

Several models have been developed to predict melt viscosity in geologically-relevant hydrous silicic melts (see Table 1) reflecting the interest and importance of silicic volcanism (explosive eruption of rhyolite and dacite) and silicic magmatism (granitic crustal melts). The models of Baker (1996) and Hess and Dingwell (1996) represent another pair of synchronistic publications heralding a new level of sophistication in modelling of melt viscosity. Both present models developed to address a specific scientific question—the viscosity of hydrous silicic melts. The models are calibrated on simple silicic melt compositions having variable H_2O contents (~1–12 wt.%).

Baker (1996) measured melt viscosity for hydrous granitic melts and used the data to calibrate two empirical equations for predicting melt viscosity at lower ($x_{H_2O} < 0.25$):

$$\log \eta = \left[0.0292649T\,(\mathrm{K}) - 53.198903 \right] x_{H_2O} - 0.0129207T\,(\mathrm{K}) + 24.2973 \qquad (14)$$

and higher x_{H_2O} contents:

$$\log \eta = -2.2977826 x_{H_2O} - 0.0095110T\,(\mathrm{K}) + 15.6293 \qquad (15)$$

where x_{H_2O} is the mole fraction of H_2O based on 8 oxygen atoms.

The Baker model is linear in its T-dependence and water content (Fig. 9A) but, relative to the Shaw (1972) model, has the disadvantage that the two functions (Eqn. 14, 15) create a discontinuity so that viscosity is not continuous across H_2O content (at $x_{H_2O} = 0.25$; Fig. 9A). The magnitude of this discontinuity is temperature-dependent (Fig. 9A, Inset). The model implies substantially different activation energies above and below $x_{H_2O} = 0.25$. Furthermore, there is no high-T limit to viscosity; increasing T causes a linear decrease in $\log \eta_\infty$ which also decreases with increasing H_2O contents. Model values of T_g ($T(\mathrm{K}) \sim 10^{12}$ Pa s) as a function of H_2O content are also complicated by the use of two functions (Fig. 9C). At water contents < 0.25, the T_g values drop off rapidly from 950 K for the dry melt to negative values as x_{H_2O} approaches 0.25. Conversely, at $x_{H_2O} > 0.25$, T_g is nearly constant and shows only a slight decrease with increasing H_2O content.

Hess and Dingwell (1996) published the first non-Arrhenian model (HD) based on the VFT function for predicting melt viscosity in hydrous haplogranitic melts. Their six parameter model has the form:

$$\log \eta = -3.545 + 0.833 \ln \left(w_{H_2O} \right) + \frac{9601 - 2368 \ln \left(w_{H_2O} \right)}{T\,(\mathrm{K}) - \left(195.7 + 32.25 \ln \left(w_{H_2O} \right) \right)} \qquad (16)$$

where w_{H_2O} is the wt.% of dissolved H_2O in the melt. The HD model does a better job than the Baker (1996) model and other Arrhenian models in reproducing melt viscosity in hydrous silicic systems (see Hess and Dingwell 1996). One inconvenience is that the term $\ln w_{H_2O}$ is undefined at zero water content, however, the authors suggest that this condition is never met in Nature. A water content of 1 wt.% reduces the function to:

$$\log \eta = -3.545 + \frac{9601}{T(K) - 195.7} \qquad (17)$$

A calibration based on the term $\ln(1 + w_{H_2O})$ would circumvent that issue.

In addition to reproducing the experimental data well, the main attributes of the HD model are that it is continuous across water content and temperature (Fig. 9B) and predicts reasonable and experimentally validated decreases in T_g with increasing H_2O content. The Shaw (1972) model, although purely Arrhenian, compares reasonably well with the refined HD non-Arrhenian model in, both, predicting melt viscosity (Fig. 9B) and T_g (Fig. 9C) as a function of H_2O content. To a large extent, this reflects the fact that these melts are strong geological melts and their departures from Arrhenian behaviour are moderate.

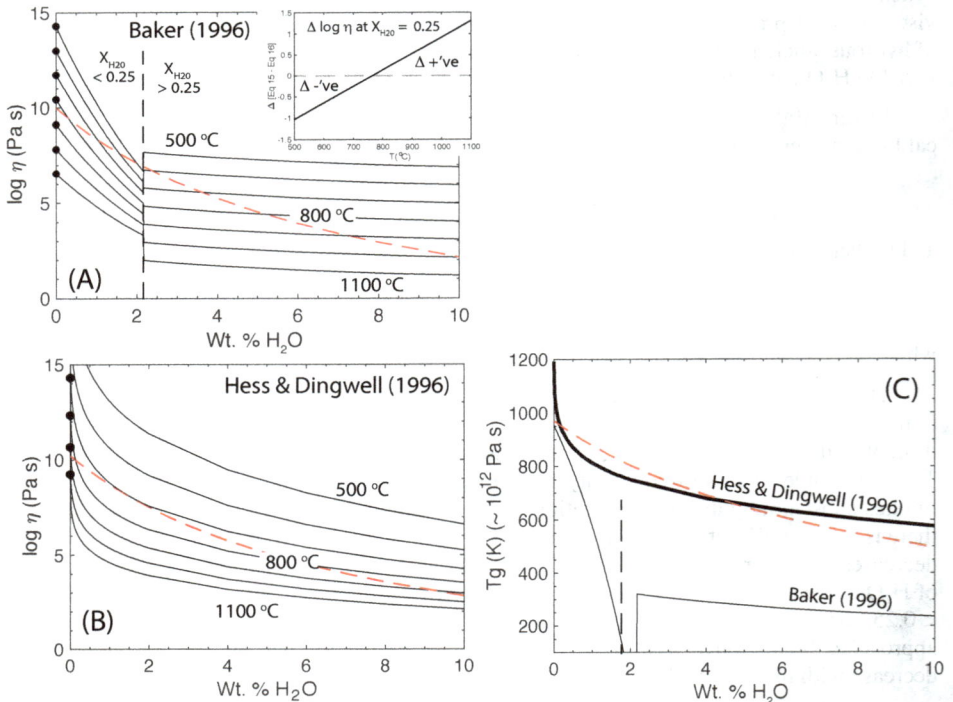

Figure 9. Models for viscosity of hydrous silicic melts. **(A)** The viscosity of granitic melts calculated with Arrhenian model of Baker (1996) for H_2O contents of 0–10 wt.% and over the temperature range 500–1100 °C. **Vertical dashed line** denotes a discontinuous transition in empirical equations used for x_{H_2O} contents below and above 0.25 (see text). Inset shows the magnitude of discontinuity (i.e., $\Delta \log \eta$) as a function of temperature. **Red dashed line** is model of Shaw (1972) for same melt composition at 800 °C. **(B)** Values of viscosity calculated with the non-Arrhenian model of Hess and Dingwell (1996) over the same H_2O contents and temperature as in (A). **Red dashed line** as in (A). **(C)** Values of T_g (K) predicted by the two models and calculated for a range of H_2O contents (see text). Also shown are T_g values implied by the Shaw (1972) model (**red dashed line**).

Hydrous high-*P* models for silicic melts

Ardia et al. (2008) performed a series of innovative centrifuge-aided experiments to measure the viscosity of hydrous (0.15–5.24 wt.%) rhyolite melts as a function of temperature (580–1640 °C) and pressure (5–24 kbar) (Table 1). Using these new data and data from the literature (e.g., Schulze et al. 1996), they parameterized the B_{VFT} and C_{VFT} in the VFT function (i.e., Eqn. 3) to account for mol. % water content ($m_{\mathrm{H_2O}}$) and pressure (*P*) in kilobars as:

$$B_{\mathrm{VFT}} = b_1 + b_2 m_{\mathrm{H_2O}} + b_3 P \qquad (18)$$

and

$$C_{\mathrm{VFT}} = c_1 + c_2 \ln(1 + m_{\mathrm{H_2O}}). \qquad (19)$$

They assumed A_{VFT} was a constant (−4.28) and did not account for compositional variations other than H_2O contents. Allowing *A* to have a dependence on H_2O content did not improve the fit.

This is one of the first generalized models for predicting the viscosity of hydrous (H_2O-undersaturated) silicic melts at elevated temperature and pressures. Their model predicts a strong decrease in melt viscosity with temperature and H_2O content and pressure to have a variable and minor effect (Fig. 10A). The greatest effect of pressure is at low pressures, low *T* and low H_2O contents. Model values of T_g decrease with increasing water contents and are lower than predicted by HD; increasing pressure causes a slight further reduction in T_g (Fig. 10B) Melt fragility decreases slightly with H_2O content but is insensitive to pressure (Fig. 10B, inset).

Their model reproduces the original experimental data well, yet has several limitations. Firstly, the model values of B_{VFT} and C_{VFT} calculated for a range of H_2O contents become unrealistic at H_2O contents > 1.54 wt.%. Calculated values of C_{VFT} become negative and aphysical requiring a substantial corresponding decrease in B_{VFT} (Fig. 10C). The anhydrous melt has B_{VFT} and C_{VFT} values of 11660 and 289K, respectively, whereas at 10 wt.% H_2O the values are 9168 and −228K. Furthermore, at higher water contents the model curves of log η vs. $10000/T(\mathrm{K})$ are concave down which is unrealistic and may undermine their efforts to calculate model values for activation energy associated with viscous flow of these melts (cf. Ardia et al. 2008).

Hui et al. (2009) followed the work of Ardia et al. (2008) to develop a similar model for the viscosity of hydrous rhyolite at pressure based on their new data and compiled literature data. They adopted the Zhang et al. (2003) low-pressure model and fit for additional parameters to account for pressures of 3 GPa over water contents of 0–8 wt.% H_2O. Their model uses a pair of exponential expressions in *P* and *T*. One of the exponential functions is multiplied by a term involving water content and having a total of 10 adjustable parameters. The model reproduces the compiled dataset well. Interestingly, the Hui et al. (2009) model agrees well with the Hess and Dingwell (1996) model in terms of the 1 atm. 800 °C isopleth (Fig. 10A) and in terms of calculated values of T_g (Fig. 10B). The Hui et al. (2009) model also shows pressure to have a subordinate effect on viscosity relative to temperature and H_2O content. One disadvantage of the Hui et al. (2009) model is that it is difficult to explicitly extract derivative properties such as T_g and fragility. A second concern is that values of *A* are independent of H_2O content but are dependent on pressure (Fig. 10D) and unreasonable. Over a pressure range of ~3.5 GPa, *A* increases from −8 to −5.8 Pa s.

Two other models concerning the viscosity of hydrous silicic (dacite to rhyolite) melts are Whittington et al. (2009) and Romine and Whittington (2015). Whittington et al. (2009) contributed experiments on 6 dacite liquids having water contents (*w*) up to 5 wt.% and then parameterized the VFT function for water content ($\log_{10} w + 0.26$). Their model adopts a common

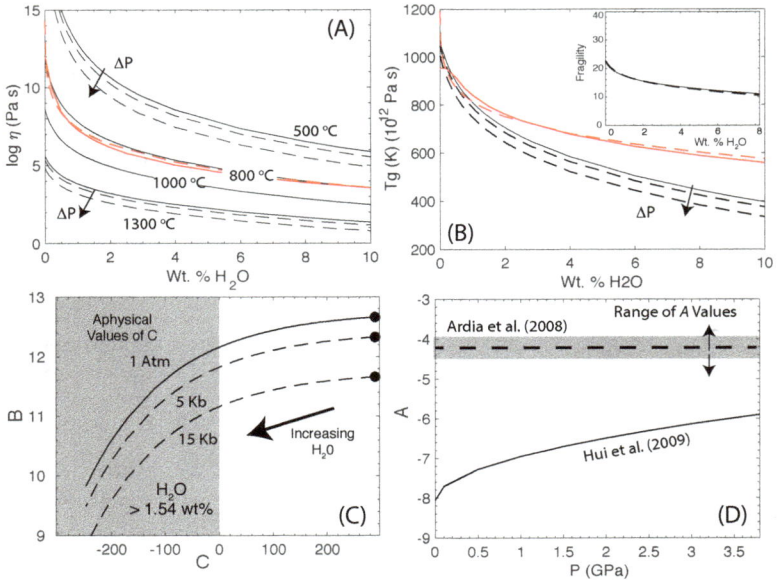

Figure 10. Models for viscosity of silicic melts as a function of temperature, H$_2$O content, and pressure including Ardia et al. (2008) and Hui et al. (2009). (**A**) Ardia et al. (2008) model values of log η vs. wt.% H$_2$O at 1 atm. and contoured for temperature (500, 800, 1000, and 1300 °C); **dashed lines** denote the effects of increasing pressure (5 and 15 Kb) at 500 °C and 1300 °C. Also shown are models of Hess and Dingwell (1996, **dashed red line**) and Hui et al. (2009) (**solid red line**) for melts at 800 °C and 1 atm. (**B**) Values of T_g as a function of H$_2$O content from Ardia et al. (2008) for pressures of 1 atm, 5 Kb, and 15 Kb (**black lines**); inset shows model fragility values at 1 atm., 5 and 15 Kb. T_g values (1 atm) from Hess and Dingwell (1996) and Hui et al. (2009) are shown as in A. (**C**) Ardia et al. (2008) parameters B and C calculated for hydrous silicic melts at 1 atm, 5 and 15 Kb. B and C values are highest for anhydrous end-member melts (**black circles**) and decrease with increasing H$_2$O content. At H$_2$O contents greater than 1.54 wt.% values of C are negative and aphysical. (**D**) Comparison of Ardia et al. value of A (i.e., log $\eta_\infty \sim$ -4.25) to values implicit to Hui et al. (2009) model as function of pressure (see text).

value for A of −4.43 (following Russell et al. 2003). Relative to the 800°C (0.1 MPa) viscosity isopleth of other melt viscosity models (e.g., Hess and Dingwell 1996; Ardia et al. 2008; Hui et al 2009; Romine and Whittington 2015), the Whittington et al. (2009) model predicts lower viscosities (> 1 log unit) at low H$_2$O contents (Fig. 11A). However, T_g–H$_2$O values (Fig. 11B) predicted by their model agree with all other models except Ardia et al. (2008). This suggests, again, that the Ardia et al. (2008) model may not predict derivative properties well. One anomaly of the Whittington et al. (2009) model concerns the model effects of H$_2$O content on B_{VFT} and, thus, C_{VFT} (Fig. 11B). The model values of B_{VFT} and C_{VFT} are positively and linearly correlated. For a single melt composition with variable H$_2$O content, this is not necessarily surprising as H$_2$O is expected to decrease both activation energy (~α to B_{VFT}) and T_g (~α to C_{VFT}). However, across a wide range of H$_2$O contents (0–12 wt.%) B_{VFT} values are near-constant varying by less than 0.5% (7600–7629). A second issue involves the term (log$_{10}$ w+0.26) which goes to zero at H$_2$O contents of 0.74 wt.% (i.e., rather than at 0 wt.%) and defines a base value for B_{VFT} of 7618. Water contents less than 0.74 cause an increase in B_{VFT} and H$_2$O contents > 0.74 wt.% cause a decrease in B_{VFT} values. Regardless of this idiosyncrasy, the overall pattern of B_{VFT} is as expected and shows a non-linear decrease, albeit very minor, with H$_2$O content (Fig. 11B).

Romine and Whittington (2015) produced new viscosity experiments (N=211) on hydrous rhyolite melts at 1 atm. Then using these data and low and high-pressure data compiled (N = 480) from the literature they created an empirical model for silicic melt viscosity that accounts for temperature, H_2O content and pressure. Their model uses a VFT function where the B_{VFT} and C_{VFT} terms are explicitly expanded to accommodate H_2O contents (wt.%) and an additional term is added that accounts for both pressure (P, MPa) and water content. They optimize

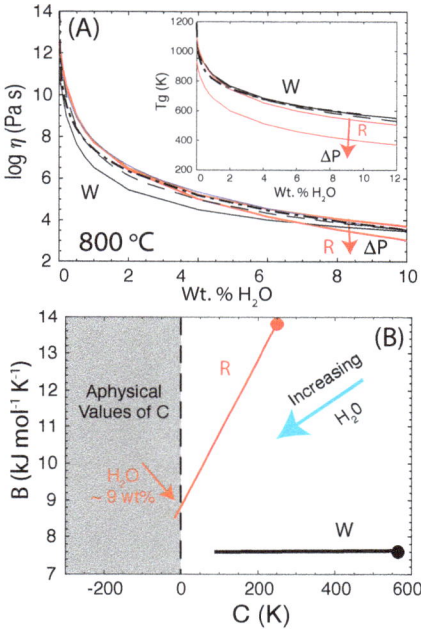

Figure 11. Viscosity models for silicic melts of Whittington et al. (2009, **solid black lines**, W) and Romine and Whittington (2015, **red lines**, R at 1 atm and 15 Kb). (**A**) Model values of log η vs. wt.% H_2O at 800 °C and 0.1 MPa; arrow denotes effects of increasing pressure for Romine and Whittington (2015) model. Also shown are models Hess and Dingwell (1996) and Hui et al. (2009) (**dashed black lines**). Inset shows corresponding T_g values including 1 atm and 15 kb values from Ardia et al. (2008) (**solid blue lines**). (**B**) The values of B and C from models of Whittington et al. (2009) and Romine and Whittington (2015). B and C values are highest for anhydrous end-member melt and decrease with increasing H_2O content. The **grey shaded field** denotes an aphysical region defined by $C < 0$; the Romine and Whittington (2015) model reaches this space at $H_2O > 9$ wt.%. Whittington et al. (2009) model has a near constant value of B over a wide range of H_2O contents.

for a common value for A (−4.40) approximating a high-temperature limit to melt viscosity. The model reproduces the 800 °C (1 atm) viscosity isopleth of Hess and Dingwell (1996) to < 1 log unit and shows pressure to have a small but significant effect on viscosity at high H_2O contents (Fig. 11A). Their model matches the T_g–H_2O curves of the other silicic melt models to within 50 °C (Fig. 11A, inset) which improves on the Ardia et al (2008) model (Fig. 10B). The model predicts a significant decrease in T_g with pressure, as suggested by Ardia et al. (2008), that becomes more pronounced at high H_2O contents (Fig. 11A, Inset). The values of B_{VFT} and C_{VFT} for this model also covary strongly, positively and linearly and both decrease with H_2O in a non-linear (but correlated) manner. The highest values of B_{VFT} and C_{VFT} are associated with the anhydrous melt and the lowest with the highest water contents. One irregularity is that at H_2O contents ≥ 9 wt.%, model values of C are negative and physically unreasonable. Lastly, the value of A (−4.40) in their model does not represent the high-temperature limits to melt viscosity. Rather, that is given by:

$$\log \eta_\infty = A - P[0.00082 + 0.000051w] \tag{20}$$

At 0.1 MPa pressure and high temperatures, the model viscosity converges to a fixed value close to −4.40 but with a very weak, linear dependence on H_2O content. However, at high pressures (≥ 1000 MPa), the model converges to very different lower values of viscosity. These high-temperature limits also show a strong linear dependence on H_2O content: at 1500 MPa the model converges to log $\eta_\infty \sim 10^{-5.6}$ (dry) and $10^{-6.6}$ (12 wt.% H_2O).

8. OTHER MODELS FOR RESTRICTED COMPOSITIONAL RANGES

Several other important non-Arrhenian models have been developed for restricted ranges of melt composition that address specific geological issues or questions. They have the potential attribute of predicting very accurately the viscosities of silicate melts that are relevant to the particular problem. One of the countervailing limitations is that they cannot be extrapolated past their original calibration datasets and the adjustable parameters are commonly less meaningful or even aphysical—a concession made to improve the reproducibility of the data of primary concern.

Models for melts from the Phlegrean Field

Romano et al. (2003) produced new viscosity data on two volcanic materials from Vesuvius (V1631W/G) and the Phlegrean fields (AMS B1/D1) deemed critical for understanding the potential for explosive volcanism in the highly populated region of Naples. They measured the temperature-dependent (400 to 1500 °C) viscosity for two dry melts and the same melts with variable dissolved water contents (0 to 3.8 wt.%). Viscosity measurements for each of these two melts spanned 10^2 to 10^{12} Pa s. They also compiled viscosity data on four other hydrous melts (i.e., trachyte, phonolite and haplogranitic) from the literature (Dingwell et al. 1996, HPG8; Whittington et al. 2001, W Tr, W Ph; Giordano et al. 2000, T Ph). Each dataset for the individual melts, including the anhydrous end-member and their hydrous counterparts, was fit to individual VFT functions modified to accommodate H_2O content:

$$\log\eta = a_1 + a_2 \ln w + \frac{b_1 + b_2 w}{T(\text{K}) - (c_1 + c_2 \ln w)} \tag{21}$$

where w is the wt.% H_2O and a_i, b_i, and c_i are adjustable parameters unique to each of the six melts considered here. Thus, there are 6 adjustable parameters for each anhydrous melt composition (i.e., a total of 18 parameters). The form of Equation (21) is very similar to that proposed by Hess and Dingwell (1996) except that the numerator term involves w (i.e., wt.% H_2O) rather than $\ln w$. Romano et al. (2003) used this modification to allow for zero water content (i.e., undefined condition $\ln w$ at $w=0$).

The model does a good job of reproducing the original data and provides a solid means of tracking the effects of H_2O content on the behaviour and values of melt viscosity for these anhydrous or hydrous melts at geological temperatures (Fig. 12A). Increasing water content causes a significant decrease in viscosity (Fig. 12A). Their study also provides preliminary insight into how H_2O content can be accommodated in models for melt viscosity (i.e., Eqn. 21).

However, there are some issues with the model. Firstly, the model is mathematically undefined at water contents of zero although virtually all geological melts contain some dissolved H_2O. This inconvenience could be remedied, for example, by replacing the term $[c_1 + c_2 \ln w]$ in Equation (21) by $[c \ln(1 + w)]$ using one less adjustable parameter. A second issue is that the model correctly predicts a decrease in fragility with increasing H_2O content (e.g., Giordano et al. 2008) but, at higher H_2O contents, fragility increases in four of the six melts (Fig. 12B). This is a case where the model cannot be extrapolated past the limits of the calibration data (i.e., H_2O > 4 wt.%). The model functions for all six melts show a pronounced and probably unrealistic decrease in Tg with increasing H_2O content from anhydrous values of 900–1175 K to < 500 K at 10 wt.% H_2O (Fig. 12C). Furthermore, the functional form of these models (Eqn. 21) ascribes a strong unrealistic compositional (H_2O content) dependence to the high-T viscosity limit (i.e., A): dry melts have values of −6.7 to −4.2 whilst H_2O-rich melts have values of −7 to −1. Three melts show a pronounced increase in A with H_2O content and three show a strong decrease.

Misiti et al. (2011) produced new data and a predictive model for two other alkaline rock compositions associated with the Campi Flegrei caldera complex. They performed 95 measurements of viscosity on anhydrous and hydrous (0.02 to 3.3 wt.%) equivalent melts of

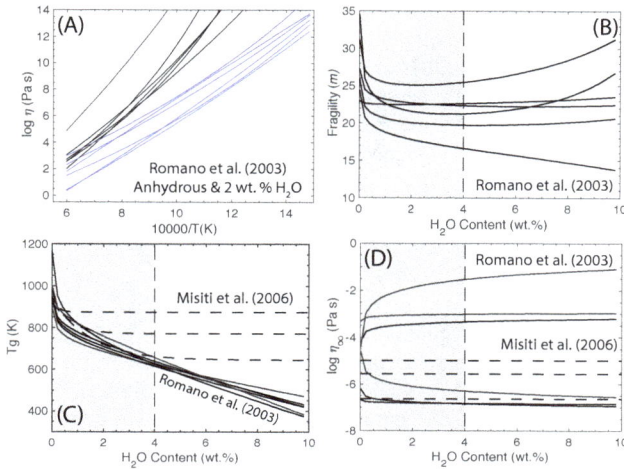

Figure 12. Models for viscosity of hydrous alkaline melts (Romano et al. 2003, **solid lines**; Misiti et al. 2006, **dashed lines**). (**A**) Values of log η vs. 10000/T(K) predicted by model of Romano et al. (2003) for six anhydrous (**black**) silicate melts and for the same melts having 2 wt.% H_2O (**blue**). (**B**) Model values of fragility for six melts of Romano et al. (2003) as a function of H_2O content. (**C**) Calculated values of T_g as a function of H_2O content. (**D**) Values of log η_∞ implied by the two models as a function of H_2O.

shoshonitic (Min) and latitic (FR) composition. Melt viscosity was measured at temperatures of 840 to 1870 K and 1 atm and under a pressure of 500 MPa; viscosity values ranged from $10^{1.5}$ to 10^{10} Pa s. They fit the anhydrous and hydrous data for each melt to an equation of the form:

$$\log \eta = -a + \frac{b}{T(\mathrm{K})-c} + \frac{d}{T(\mathrm{K})-e} e^{\left[g\frac{w}{T(\mathrm{K})}\right]} \tag{22}$$

where w is the wt.% H_2O and a–g are adjustable parameters unique to each melt. Their analysis included previously published data (Misiti et al. 2006) on a trachytic melt composition (dry and hydrous) from Agnano Monte Spina (AMS). The form of the model is an improvement in that it is mathematically defined at zero H_2O. A complication is that model values of T_g and fragility cannot be calculated explicitly but need to be solved for numerically. Calculated values of T_g for the three melts initially decrease appropriately with increasing H_2O contents before remaining constant at $H_2O > \sim 1\%$ (Fig. 12C). The model allows for H_2O-content independent high-temperature limit (i.e., log η_∞), however, the implication of fitting Equation (22) to the individual datasets for each melt composition is that each melt has a different high-temperature limit (log $\eta_\infty = -5$ to -6.6; Fig. 12D). Lastly, we have also calculated viscosity isotherms for the three melts (not plotted); viscosity decreases with increasing H_2O content. The nature of decrease, however, is very strange for the shoshonitic and latitic melts in that the isothermal melt viscosity is constant at water contents > 1%. The model isothermal curves for the trachytic melt (Misiti et al. 2006) are more reasonable suggesting there may be problems with some of the Misiti et al. (2011) datasets.

A model for natural Fe-bearing silicate melts

Most natural silicate melts contain some amount of iron and some terrestrial melts can have 15–18 wt.% FeO(T) (e.g., Williams-Jones et al. 2020). The iron occurs in two main valence states depending on the redox state of the melt: ferrous (Fe^{2+}) and ferric (Fe^{3+}). There is clear evidence on the importance of iron oxide speciation (i.e., FeO vs. Fe_2O_3) for affecting melt viscosity (Dingwell and Virgo 1987; Dingwell 1991; Chevrel et al. 2013, 2014;

Kolzenburg et al. 2018). Duan (2014) developed a melt viscosity model, based on data from the literature, that accounted for melt composition (including H_2O), temperature, pressure and redox state of the iron. The Duan (2014) is calibrated on a compilation of ~200 experiments on 19 Fe-bearing melts spanning a temperature range of 733–1873 K. The model is calibrated for anhydrous to hydrous (<12 wt.%) and for pressure (<1500 MPa). The 19 melt compositions used in calibration include basalt (14), andesite (2) and rhyolite (3).

The functional form of the model is the VFT equation (Eqn. 3) modified as follows:

$$\log \eta = A_D + \frac{b_1 + b_2 \ln\left(JB\right)}{c_1 + c_2 \ln\left(JC\right)} \tag{23}$$

where JB and JC are complex functions of composition, H_2O content and pressure and each contains another three adjustable parameters (b_{3-5} and c_{3-5}, respectively) for a total of 11. All of the anhydrous compositional dependencies are ascribed to a single term (VSM) which is a modified form of the SM parameter (Giordano and Dingwell 2003) and includes measured mole fractions of Fe^{2+} and Fe^{3+}. Where Fe^{2+} and Fe^{3+} are not known, the method of Moretti (2005) is used to calculate the proportions of Fe^{2+} and Fe^{3+} in the silicate melt at the experimental conditions (i.e., $T–P–fO_2$). The VSM parameter is then calculated as [CaO + MgO + MnO + FeO + Na_2O + K_2O]/[SiO_2 + TiO_2 + Al_2O_3 + Fe_2O_3 + P_2O_5] (mol % oxides).

The main attributes of this effort are to create a model for silicate melt viscosity that accounts for H_2O content, pressure, and iron oxidation state. There are several issues, however, that reflect the relatively small dataset used for calibration and the complex form and interdependence of the functions *JB* and *JC*. The model reproduces the calibration dataset very well and reproduces select datasets that were not used for calibration purposes (see Duan 2014). However, when applied to a larger and more diverse dataset the model fails (Fig. 13A) and the magnitude of misfit between model and data increases with decreasing iron content (Fig. 13B).

The form of the functions used to expand the VFT parameters *B* and *C* for the effects of composition induce a strong numerical correlation. This is illustrated in Figure 13C which shows the computed values for diverse anhydrous and hydrous melts (data as in Fig. 3). The anhydrous melts show a very strong positive induced correlation between *B* and *C* which is somewhat counterintuitive. More importantly, the range of model values of *B* and *C* for both these anhydrous and hydrous melts is unrealistically small ($B < 600$ J mol^{-1}, $C \sim 280$ K, Fig. 13C).

The same issues are found in the calculated derivative properties T_g and *m* (Fig. 13D). For anhydrous melts, T_g and *m* show a strong negative correlation and hydrous melts show an unrealistic, strong positive correlation. At least part of the strong covariation in these properties reflects a numerically induced correlation caused by the form of the functions for JB and JC. Another anomaly is that the anhydrous and hydrous melts define two separate trends even though the hydrous melts span a wide range of H_2O contents including anhydrous counterparts. Lastly, the range of fragility values predicted by the Duan (2014) model is substantially smaller (21–30) than found by fitting the original data for the individual melts (10–60).

A model for extraterrestrial melts

Sehlke and Whittington (2016) developed a new model for predicting the viscosity of extra-terrestrial melts as a function of temperature and composition. To achieve this, they produced new high- and low-temperature measurements of viscosity for synthetic multicomponent silicate melts intended to capture the diversity of melts from the extraterrestrial planetary bodies (+ some earth melts). Their dataset includes 496 measurements of viscosity at 1 atm on 20 anhydrous planetary basaltic melts (mainly tholeiitic) estimated to be relevant to Mars (2), Mercury (7), Earth's Moon (4), Io (1) and Vesta (1) and Earth (4).

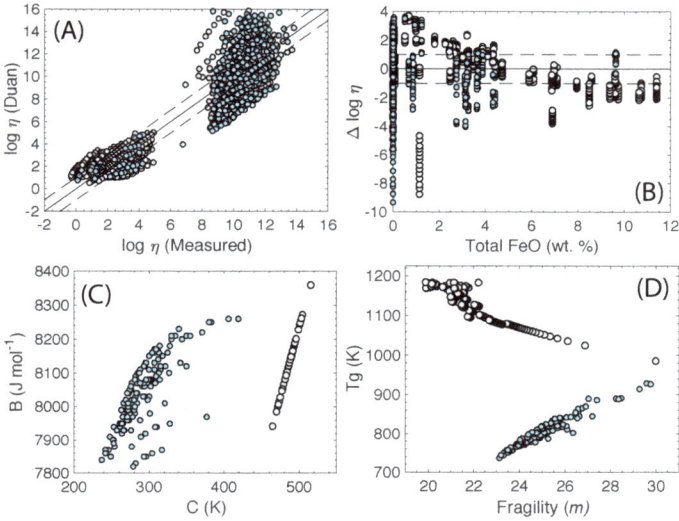

Figure 13. Duan (2014) multicomponent model for silicate melts. **(A)** Comparison of Duan (2014) model predictions applied to anhydrous (**grey**) and hydrous (**blue**) melts in database of Giordano et al. (2008). **(B)** $\Delta \log \eta$ (Observed – Predicted) for data shown in (A) vs. FeO(T) content (wt.%). **(C)** Model values of B_{VFT} and C_{VFT} calculated for anhydrous and hydrous melts (as in A; see text). **(D)** Model values of T_g and melt fragility calculated for same melts as in (A).

Their predictive model for the temperature dependence of viscosity of these melts is based on the Adam–Gibbs formulation (Eqn. 6) expanded as:

$$\log \eta = A_{AG} + \frac{B_{AG}}{T(K)\left[S^c_{T_g} + \Delta Cp_{T_g} \ln\left(\frac{T(K)}{T_g}\right)\right]} \tag{24}$$

where A_{AG} is the high-temperature limit to viscosity and ΔCp_{T_g} is the change in configurational heat capacity taken at the glass transition temperature (T_g). Their model assumes that all melts converge to a common A_{AG} and their optimized value is −3.34 which is similar to values postulated by Russell et al. (2003, $A_{AG} \sim -3.15$) and Russell and Giordano (2017, $A_{AG} \sim -3.51$). Values of ΔCp_{T_g} ($Cp_{Melt,T_g} - Cp_{Glass,T_g}$) are calculated *a priori* using published models for the compositional and temperature dependence of heat capacity for melts and glasses. The model of Stebbins et al. (1984) is used to compute a temperature-independent value of Cp_{Melt,T_g} from the melt compositions. The values of Cp_{Glass,T_g} were calculated for individual melt compositions using the heat capacity model for silicate glasses of Richet (1987). The operational T_g for Cp_{Glass,T_g} and their model is the temperature at which the heat capacity of the glass approximates the Dulong–Petit limit of 3R where R is the ideal gas constant (see Sehlke and Whittington 2016 for details). The model parameters B_{AG} and $S^c_{T_g}$ are treated as unknowns and are dependent on melt composition. B_{AG} is a linear combination of 7 terms of $b_i X_i$ where b_i are adjustable parameters and X_i are oxide concentrations (mol %) or linear combinations of various oxides. Similarly, $S^c_{T_g}$ is a linear combination of 5 terms of $s_i W_i$; the variables s_i are the adjustable parameters and W_i are complex combinations of oxides and values of NBO/T and SM. In summary, the 13 model parameters (A_{AG}, b_i, s_i) are determined by regressing Equation (24) against the 496 measurements of T(K):log η. Note that there is an important errata (Sehlke and Whittington 2017).

The main attributes of this work are three-fold. Firstly, it contributes new experimental data thereby filling a gap in the geochemical inventory. Secondly, they show clearly that published models for the viscosity of multicomponent melts (e.g., Shaw 1972; Giordano et al. 2008) cannot predict the temperature-dependent viscosity for these exotic melt compositions. On that basis, they presented a 13-parameter predictive model (Eqn. 24) which accurately reproduces their original data, as well as, ancillary data not used in the original calibration. It is also the first multi-component model for predicting silicate melt viscosity using the Adam–Gibbs function.

There are, however, some important limitations and weaknesses to the approach used. Firstly, their model, which is driven by physical measurements of viscosity, is bootstrapped to two other models (rather than data) which have their inherent uncertainties. They use these two models for the calorimetric properties of melts and glasses to calculate and treat as knowns the term ΔCp_{Tg} for each melt composition. To create models based on mixed datasets (e.g., rheological vs. calorimetric) can be a challenge because of differences in the observational timescales of the techniques relative to the timescales of melt relaxation (i.e., Deborah number). To mix models can be even more problematic and this may explain why they were forced to adjust the value used for the Dulong–Petit limit. Figure 14 (inset) compares T_g values (i.e., $T(K)$ at $\eta = 10^{12}$ Pa s) implied by the SW model to model values of T_g calculated by fitting the viscosity data for each melt individually (Fig. 14, inset) and shows the SW model to reproduce the ideal VFT-based T_g values for most melts.

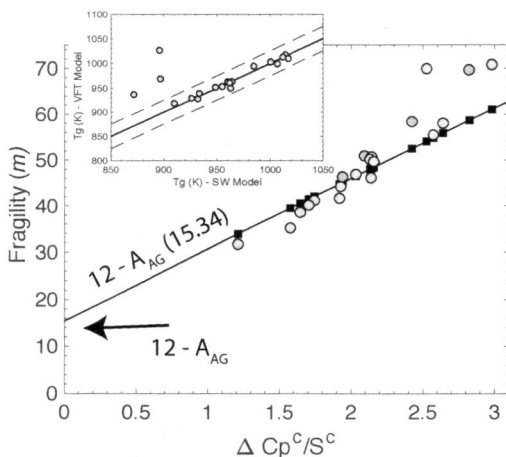

Figure 14. Model results of Sehlke and Whittington (2016) comparing ratio $\Delta Cp_{Tg}/Sc$ to values of fragility; the assumption of a common high-T limit to silicate melt viscosity (i.e., constant A) implies that fragility is linearly dependent on $\Delta Cp_{Tg}/S_c$ where the slope and intercept have the same value ($12 - A$). The data are scattered indicating internal inconsistencies within the model (see Russell and Giordano 2017). **Small black symbols** are where data would ideally plot on theoretical line. Inset compares values of T_g calculated iteratively for the SW model (i.e., $T(K)$ at $\eta = 10^{12}$ Pa s) to values calculated by fitting the original viscosity data for each melt to the VFT function (see text). Most values agree to within ± 25 K (**dashed lines**).

However, the form of the functions adopted for capturing the compositional dependences of B_{AG} and S_{Tg}^c involve 7 and 5 adjustable parameters, respectively. The form of each function (i.e., for B_{AG} and S_{Tg}^c) mixes mol % oxides with NBO/T and SM which are themselves complex functions of oxide concentrations. This obscures and makes it difficult to isolate the effects of individual oxides. Two of the terms contributing to B_{AG} are statistically equal to zero suggesting that the model may be overfitted.

Another indicator of the model's limitation is shown in Figure 14. Russell and Giordano (2017) suggested a highly useful plot of fragility vs. the ratio $\Delta Cp_{T_g} / S^c_{T_g}$ for checking the internal consistency of Adam–Gibbs based models that invoke a constant value of A_{AG}. Such models require values m and $\Delta Cp_{T_g} / S^c_{T_g}$ to define a single linear trend where the slope and intercept have the same value: $[12 - A]$. This specific linear relationship is a consequence of adopting a common (unknown) high-temperature limit to melt viscosity (i.e., A) and using the viscosity data set to solve for an optimal value of A. In the case of the Selhke and Whittington (2016) model, the data are scattered rather than plotting on the theoretical line with slope and intercept of 15.34 (Fig. 14). The scatter here indicates an internal inconsistency perhaps induced by the method used to fix values of ΔCp_{T_g} (i.e., model vs. observed) or from mixing data with external models.

9. MULTICOMPONENT MELT MODELS FOR GEOLOGICAL SYSTEMS

Non-Arrhenian multicomponent melt models (anhydrous)

The ultimate goal is to produce a predictive model for the viscosity of geological melts that spans the range of natural melt compositions found on Earth and, ultimately, other planetary bodies over the range of temperatures and pressures pertinent to magma formation, transport and eruption. The early models of Shaw (1972), Bottinga and Weill (1972) and Persikov (1991) represent major steps forward but as discussed were limited to high-temperature (super liquidus) or strong melts because of the models' Arrhenian temperature dependence.

One of the first models for predicting the viscosity of multicomponent silicate melts with a non-Arrhenian temperature dependence was developed by Giordano and Dingwell (2003 2004). They presented new data on the viscosity of 19 natural melt compositions spanning basanite to phonolite and trachyte to dacite in composition and 10^0–10^{12} Pa s in viscosity (Table 1). They used these 350 experimental measurements to create an empirical model for predicting the viscosity of anhydrous melts as a function of NBO/T values calculated from the melt compositions (mol% oxides). They also defined a new "network modifiers" parameter SM computed from the mol% sums of $Na_2O + K_2O + CaO + MgO + MnO + 0.5$ FeO_{Total}). Their work showed that isothermal viscosity values decreased strongly and regularly with increasing values of, both, NBO/T and SM. On that basis they produced empirical equations for predicting melt viscosity as a function of T(°C) and NBO/T:

$$\log \eta = a_1 \ln\left[\frac{NBO}{T} - a_2 \right] + a_3 \tag{25}$$

or SM:

$$\log \eta = c_1 + \frac{c_2 c_3}{c_3 + SM} \tag{26}$$

The a_i and c_i terms are themselves complicated, non-linear functions of T(°C) involving a total of 12 and 10 adjustable parameters, respectively, optimized against the measured values of melt viscosity (Note: An errata was published to correct both the form of Equation (25) as well as some of the coefficients for the a_i and c_i terms; Giordano and Dingwell 2004).

Their two models (i.e., NBO/T vs. SM) reproduce the original dataset reasonably well but the complexity of the non-linear functions for the a_i and c_i functions makes it difficult to extract some important derivative properties. As noted by the authors, the model cannot be extrapolated outside the temperature range (700–1600°C) of the original observations (Fig. 15A–B). It is also difficult to explore the specific effects of melt chemistry on viscosity because the roles of individual oxides are obscured by lumping the chemistry into a single parameter (NBO/T or SM; Fig. 15B). Calculated values of T_g implicit to the two (NBO/T or SM) models are similar and decrease markedly from ~1300–1400 K to ~ 925–950 K with increasing depolymerization (i.e., increasing SM; Fig. 15C). Their model based on NBO/T implies a slight compositional dependence of log η_∞ which diminishes at NBO/T $\geq$ 0.75 and converges to values of ~ $10^{-4.5}$ to $10^{-5.5}$. Conversely, the SM-based model implies a constant log η_∞ of ~10^{-8} (Fig. 15C, Inset).

Giordano et al. (2006) introduced 144 new measurements of melt viscosity and expanded the range of melt compositions in support of a compiled database of 800 experiments on 44 multicomponent melt compositions (Table 1). The database spanned a temperature range of 700–1650 °C and viscosity range of 10^{-1} to 10^{12} Pa s. They then modified and recalibrated the viscosity model developed by Giordano and Dingwell (2003). Their approach was again to fit the VFT function (Eqn. 3) to the viscosity data for each melt for optimal values of A_{VFT}, B_{VFT} and C_{VFT} but adopting the concept of a single common A_{VFT} for all melts (Russell et al. 2003). Their regression yielded A_{VFT} ~ −4.07 which accords well with previous values from Russell et al. (2002). They then used the VFT functions to calculate model isothermal values of viscosity for each ($N=44$) melt and plotted these values against the corresponding SM parameter. The isothermal values of log η were calculated for 630 °C and at 100 °C increments from 700 to 2000 °C. These model isothermal datasets were then regressed against the SM parameter as in Equation (26) where the coefficients (c_i) are complicated non-linear functions of temperature (Note: there is a typo in their Equation (5) required to compute the coefficient a_3 (+45.5755 instead of −45.5755).

This recalibration of the Giordano and Dingwell (2003) model reproduces most of the viscosity database well. As with the previous model, this formulation cannot be used at temperatures < 650 °C although it improves slightly the behaviour of the lower temperature isothermal viscosity curves (Fig. 15A, B). The model values of T_g are very similar to values from Giordano and Dingwell (2003) (Fig. 15C). A major reservation with their approach, which has undesirable consequences, is that parameterisation is against values of log η

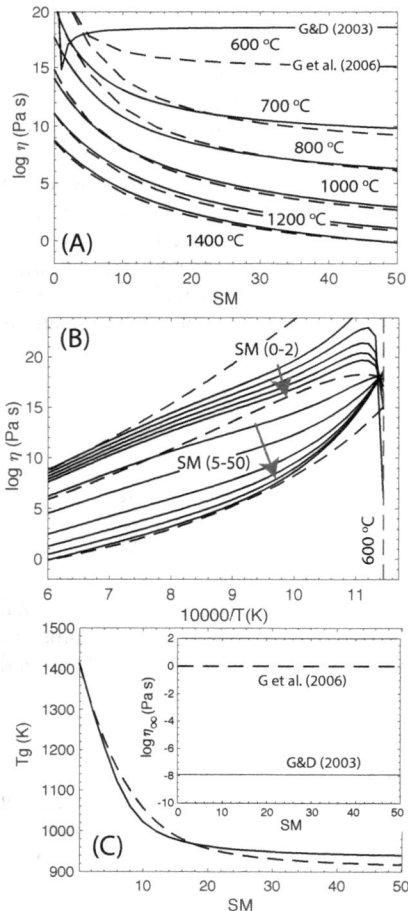

Figure 15. Non-Arrhenian viscosity models for anhydrous multicomponent silicate melts (Giordano and Dingwell 2003, **solid lines**; Giordano et al 2006, **dashed lines**). (**A**) Isothermal viscosity curves as a function of SM parameter showing decrease in viscosity with increasing depolymerization. The model cannot be extrapolated to lower temperatures (< 650 °C). (**B**) Values of log η predicted by Giordano and Dingwell (2003) vs. $10000/T$(K) for melts having SM values of 0–2 and 5–50. **Dashed lines** are for SM values of 0, 5 and 50 using the Giordano et al. (2006) model (**C**) Calculated values of T_g (K) from the two models are similar and decrease markedly from ~1300–1400 K to ~925–950 K with increasing depolymerization (i.e., increasing SM). Inset shows the implied high-T limit to viscosity for the two models; both are compositionally independent but converge to unreasonable values of log η_∞ ~ −8 or 0.

predicted from VFT functions rather than the actual original viscosity data. The consequence is that, although the VFT functions were fit subject to a constant high-temperature limit to viscosity ($A_{VFT} \sim -4.07$), the multicomponent model converges at high-temperature to an unrealistic value of log $\eta_\infty \sim 0$ (Fig. 15C, inset).

One of the main contributions of the Giordano and Dingwell (2003, 2004) and Giordano et al (2006) models is to highlight the complex effects of chemical composition on melt viscosity. The model represented by Equation (26) uses the SM parameter calculated from chemical composition to predict the temperature dependence of melt viscosity. The model reproduces viscosity data for metaluminous melts very well but is less effective for peraluminous and peralkaline melt compositions. This highlights the fact that even melts having the same SM (or NBO/T) values can have very different temperature-dependent viscosities. The Giordano et al. (2006) model (i.e., Eqn. 26) predicts viscosities for peralkaline melts that are too high at lower temperatures and too low viscosities for peraluminous melts at low temperatures (see Fig. 8 in Giordano et al. 2006). As a preliminary fix to this issue, they added a 4th temperature-dependent coefficient which accounts for the effects of excess alumina or alkalies.

Non-Arrhenian multicomponent melt models (hydrous)

Within a single year, two models were published that took modelling silicate melt viscosity significantly further (Hui and Zhang 2007; Giordano et al. 2008). Hui and Zhang (2007) published a non-Arrhenian model (HZ) for hydrous silicate melts based on an empirical equation:

$$\log \eta = A_{HZ} + \frac{B_{HZ}}{T(K)} + e^{\left[C_{HZ} + D_{HZ} / T(K) \right]} \tag{27}$$

where A_{HZ}, B_{HZ}, C_{HZ} and D_{HZ} are linear functions of the mole fractions (x_i) of the oxides in the melt. Each of the parameters A_{HZ}, B_{HZ}, and C_{HZ} is also dependent on the mole fraction of H_2O and D_{HZ} includes an additional H_2O-dependent parameter Z (= $x_{H_2O}^{[1/(1+185.797/T(K))]}$). The model is calibrated against 1451 experimental measurements of melt viscosity on 109 melt compositions, spanning temperatures of 573–1978 K, viscosity values of 10^{-1} to 10^{15} Pa s, and for water contents of < 12.3 wt.% for rhyolite and < 5% for other melts. The model has 37 coefficients and, as one would expect with that many adjustable parameters, reproduces the original data extremely-well to within 0.61 log units at the 95% confidence level.

The HZ model does a very good job of reproducing the original dataset and predicting values of viscosity for melt compositions bounded by the original data. However, there are serious negative consequences that arise from the choice of equation used to capture temperature dependence and from overfitting the data. The number of parameters (37) represents ~34% of the total number of melt compositions in their database. Furthermore, the form of their model (Eqn. 27) implies a strong compositional dependence on the high-temperature limits on melt viscosity (i.e., log η_∞) which is simply not valid. A trait of the HZ model is the very wide range (~10 orders of magnitude) of values of A_{HZ} for melt compositions in their database spans: −9 to −1.3 for anhydrous melts and −9 to 0.04 for hydrous melts (Fig. 16A). These represent aphysical values and a significant flaw in the model limiting its potential use outside of the range of temperatures and melt compositions used to calibrate the 37 parameter model. By way of contrast, Shaw's (1972) simple Arrhenian model has a much narrower range of A values (−6.5 to −4.9; Fig. 16A) and modern practice (Giordano et al. 2008, $A=-4.6\pm0.2$; Persikov and Bukhtiyarov 2009, $A=-4.5$) and theory would advocate for a single value for A that transcends melt composition of between −5 and −4.5 (Fig. 16A).

Another disadvantage of the HZ model is that the form of their base equation precludes a direct connection of the viscosity model to intrinsic melt properties such as glass transition temperature or fragility. An example of this is shown in Figure 16B which plots the model

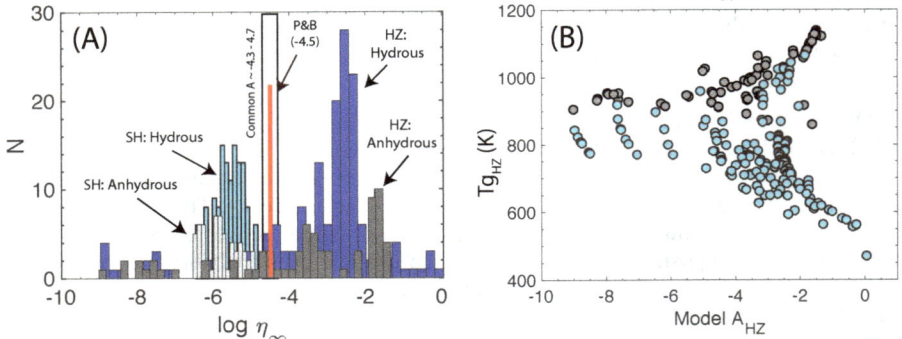

Figure 16. Hui and Zhang (2007) model for viscosity of multicomponent hydrous silicate melts. (**A**) Values of *A* representing the high-*T* limits to silicate melt viscosity (i.e., log η_∞) implicit in HZ model (Hui and Zhang 2007) compared to *A* values predicted by SH model (Shaw 1972), the GRD model (**red bar** at -4.6±0.2; Giordano et al. 2008) or adopted *a priori* by Persikov and Bukhtiyarov (2009; see text for discussion). (**B**) Model values A_{HZ} and T_g (K) predicted by the HZ model for anhydrous (**grey symbols**) and hydrous (**blue symbols**) melts in the Hui and Zhang (2007) database. The strong covariation is numerically induced and emphasizes the negative consequences of ascribing a compositional dependence to *A* (i.e., log η_∞) and overfitting the data (i.e., 37 adjustable parameters).

values of A_{HZ} for melts in HZ database against the predicted value of T_g (K) taken as the temperature at which melt viscosity ~10^{12} Pa s. It is immediately apparent that the predicted values of T_g are numerically driven by the value of A_{HZ} for each melt. For anhydrous melts, T_g increases with increasing A_{HZ} which makes no sense. T_g values decrease for hydrous melts relative to their anhydrous counterparts but again the trends are strongly and negatively correlated to unrealistic values of A_{HZ} (i.e., log η_∞). In summary, the HZ model is excellent for reproducing the original datasets and predicting viscosity for melts close to the calibration dataset but has limited capacity for extrapolation and for predicting other melt properties.

10. GRD MODEL (2008)

The GRD model represents the latest multicomponent chemical model for predicting the viscosity of natural volatile-bearing silicate melts (Giordano et al. 2008, Table 1). The original model was calibrated on > 1750 measurements of viscosity on multicomponent anhydrous and volatile-rich silicate melts spanning 10^{-1} to 10^{14} Pa s. The melt database was representative of a wide range of common silicate volcanic rocks (cf. Fig. 1). Specifically, they compiled viscosity data on 58 anhydrous melt compositions (N ~ 930) spanning a temperature range of 535–1705 °C and on volatile-enriched (H_2O and F) equivalents to the anhydrous melts (N ~ 840) making for >100 different hydrous melt compositions. The latter dataset included viscosity values measured at significantly lower experimental temperatures: 245–1580 °C. The melt database has calculated NBO/T values that span 0 to 1.8 and the range of oxide contents (wt.%) is: SiO_2 (41–79), TiO_2 (0–3), Al_2O_3 (0–23), FeO (0–12), MnO (0–0.3), MgO (0–32), CaO (0–26), Na_2O (0–11), K_2O (0.3–9), P_2O_5 (0–1.2), H_2O (0–8), and F (0–4).

Their model is built on the VFT equation (Eqn. 3) and they assumed *A* to be a single constant (to be solved for) and ascribed all compositional dependencies of viscosity to the terms B_{GRD} and C_{GRD}. They represented B_{GRD} and C_{GRD} as linear combinations of oxide components and a subordinate number of combined oxide cross-terms. The melt viscosity model was calibrated against the viscosity–*T*(K)–melt composition dataset and reduced to 18 adjustable parameters including: i) *A* = −4.55±0.21, ii) 10 coefficients to capture the compositional controls on B_{GRD}, and iii) 7 coefficients for C_{GRD}. Operationally, the temperature dependence of melts is calculated from the mole fractions of oxides (m_i, n_i) and oxide combinations (*M1$_j$*, *M2$_j$*, *N1*, *N2*) as:

$$\log \eta = -4.55 + \frac{\sum_{i=1}^{7} b_i m_i + \sum_{j=1}^{3} b_{1,j} M 1_{1,j} M 2_{1,j}}{T(\text{K}) - \left[\sum_{i=1}^{6} c_i n_i + c_{1,1} N 1_{1,1} N 2_{1,1} \right]} \tag{28}$$

where b_i and c_i are the optimized coefficients (see Table 7).

The model describes the original data well (RMSE ~ 0.40 log units) and the average misfit is between ±0.25 log units for the anhydrous melts and ±0.35 for hydrous melts (Fig. 17). They suggest that the model should be accurate to within ~ 5% relative error; the largest deviations are for low-T, high viscosity (> 10^8 Pa s) data (Fig. 17B).

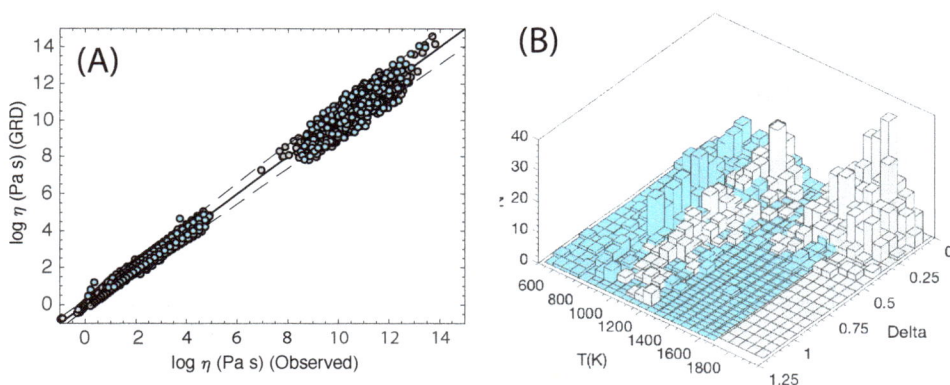

Figure 17. Quality of fit of Giordano et al. (2008; GRD) model. **(A)** Crossplot comparing measured values of log η to predicted values for anhydrous (**grey**) and volatile-rich (**blue**) melt compositions. **(B)** Histogram of absolute values of Δlog η organized by experimental temperature. Most deviations are less than 0.5 log units; greater values of Δlog η correspond to the lower temperature (higher viscosity) measurements for both anhydrous and volatile-rich datasets.

Attributes

The GRD model is continuous in composition and temperature space, predicts the viscosity of natural volatile-bearing silicate melts (including H_2O and F), and can predict other transport properties including glass transition temperatures and melt fragility. Model values of B_{GRD} and C_{GRD} for their melt compositions vary from 4450 to 12,000 J mol^{-1}, and 0 to 668 K, respectively (Fig. 18A). Values B_{GRD} and C_{GRD} for their anhydrous melts show a strong, negative, linear covariation reflecting the effects of depolymerization on both parameters (Fig. 18A, C, D). The values of B_{GRD} are highest where NBO/T is ~0 whereas C_{GRD} values are highest for melts having the highest NBO/T (Fig. 18C, D). The pattern between B_{GRD} and C_{GRD} values changes dramatically but coherently for volatile-rich melts (Fig. 18). Dissolved H_2O causes a marked decrease in values of C_{GRD} (Fig. 18A, D) whereas the effects of H_2O on B_{GRD} are varied and less apparent (Fig. 18A, C).

The GRD model can also be used to calculate derivative transport properties including the glass transition temperature (T_g) and the fragility m of silicate melts. Calculated values of T_g and m for all the melts in the database (Fig. 18B) show a wide range in fragility (m) indicating the presence of both strong and moderately fragile melts. Values of T_g for the anhydrous melts vary slightly (Fig. 18B) and are not strongly correlated with fragility. Values of, both, T_g and m are strongly affected and depressed by volatile contents (Fig. 18B). The major decrease in T_g with increased volatile content accords well with experimental studies (Hess and Dingwell 1996).

Table 7. GRD coefficients for calculating VFT parameters B and C; $A = -4.55 \pm 0.21$ and an example calculation[a].

	Oxides	b_i^b		Oxides	c_i^b	Oxide	Wt. %	B's	C's		
b_1	SiO$_2$+TiO$_2$	159.6 (7)	c_1	SiO$_2$	2.75 (0.4)	SiO$_2$	73.57	11411.65	196.558	A_{VFT}	-4.55
b_2	Al$_2$O$_3$	-173.3 (22)	c_2	TAd	15.7 (1.6)	TiO$_2$	0.06	-1107.53	101.130	B_{VFT}	10264.0
b_3	FeO(T)+MnO+P$_2$O$_5$	72.1 (14)	c_3	FMe	8.3 (0.5)	Al$_2$O$_3$	11.16	60.98	11.252	C_{VFT}	79.50
b_4	MgO	75.7 (13)	c_4	CaO	10.2 (0.7)	FeO	0.99	38.37	33.232	T_g(K)	700
b_5	CaO	-39.0 (9)	c_5	NKf	-12.3 (1.3)	MnO	0.05	-127.00	-61.397	m	18.7
b_6	Na$_2$O + Vc	-84.1 (13)	c_6	ln(1+V)	-99.5 (4)	MgO	0.35	-1235.83	-258.961		
b_7	V + ln(1+H$_2$O)	141.5 (19)	c_{11}	(Al$_2$O$_3$+FM+CaO)	0.30 (0.04)	CaO	3.13	2135.35	57.685	log η^g	4.05
b_{11}	(SiO$_2$ +TiO$_2$)*(FM)	-2.43 (0.3)		$-$ P$_2$O$_5$)×(NK+V)		Na$_2$O	2.35	-235.04			
b_{12}	(SiO$_2$+TA+P$_2$O$_5$)*NK+H$_2$O)	-0.91 (0.3)				K$_2$O	4.49	-1239.33			
b_{13}	(Al$_2$O$_3$)*(NK)	17.6 (1.8)				P$_2$O$_5$	0.00	562.42			
						H$_2$O	3.85				

Notes: [a]Rhyolite melt with 4 wt.% H$_2$O. [b]Numbers in brackets indicate 95% confidence limits on values of model coefficients. [c]H$_2$O+F$_2$O$_{-1}$; [d]TiO$_2$ + Al$_2$O$_3$; [e]FeO(T) + MnO + MgO; [f]Na$_2$O + K$_2$O; [g]log η (Pa s) $= A + B/(T(K) - C)$ and predicted value for this melt at 1273 K is 4.05 Pa s.

The GRD model predicts a near-linear decrease in fragility with increased volatile content; for their database m is 20–58 for anhydrous melts vs. 12–49 for volatile-rich equivalents (Fig. 18B). The implication is that volatiles cause melts that are moderately depolymerized to become less fragile (e.g., stronger as defined by the m parameter) such that dry non-Arrhenian melts tend to become more Arrhenian with increasing volatile content. Strong melts that are moderately polymerized become even more Arrhenian with increased H$_2$O content (Fig. 18B).

We have illustrated the utility of the GRD by calculating transport properties of melts within a ternary defined by end-members basanite, rhyolite and hydrous rhyolite (4 wt.% H$_2$O). The end-members have significantly different compositions and distinct temperature-dependent viscosities as illustrated by the calculated fragilities (18.7–57.7; Fig. 19). Melts resulting from mixing the end-members have intermediate viscosity values and are bounded by basanite and rhyolite at high temperature and rhyolite and hydrous rhyolite at lower temperatures (Fig. 19A). Values of T_g and melt fragility are also bounded by the end members (Fig. 19B) and vary nearly linearly on the two joins involving hydrous rhyolite but distinctly non-linear between basanite and rhyolite.

As stated above, one of the main attributes of the GRD model is its potential for calculating derivative melt properties that are not directly measured (e.g., T_g and m). Activation energy (E_a) represents the energy barrier for viscous flow and incorporates the temperature-dependent structural relaxation times for molecular rearrangement. The original definition of activation energy (E_a) derives from the Arrhenian equation (Glasstone et al. 1941) and is taken as:

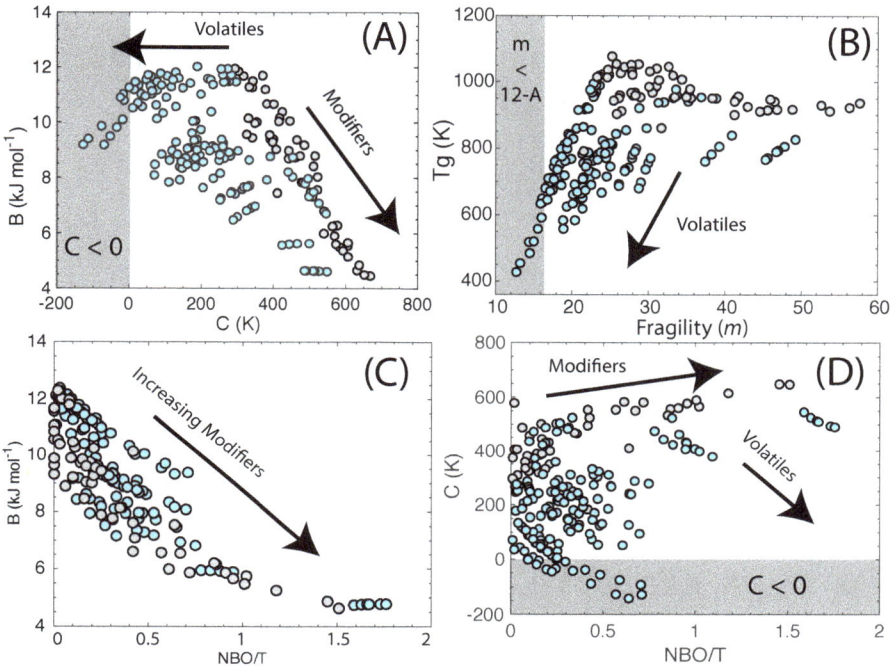

Figure 18. Results from GRD model. (**A**) The model values of B (kJ mol^{-1}) and C (K) plotted against each other for anhydrous (**grey**) and volatile-rich (**blue**) melt compositions. (**B**) Calculated values of T_g and melt fragility for the full suite of melts used to calibrate the GRD model. (**C, D**) Model values of B (kJ mol^{-1}) and C (K) plotted against NBO/T for the same data. (**D**) The **grey shaded boxes** denote fields where the GRD model predicts aphysical values of C or m (see text).

$$\frac{d\ln\eta}{d\left(\frac{1}{T(\text{K})}\right)} = \frac{E_a}{R} \tag{29}$$

where R is the gas constant (8.314 J mol^{-1} K^{-1}). Applying Equation (29) to the VFT function used to model the temperature dependence of silicate melt viscosity (i.e., log η) yields:

$$E_a = \frac{2.303B}{\left(\dfrac{C}{T(\text{K})} - 1\right)^2} \tag{30}$$

The GRD model provides the means to calculate the composition-dependent terms B and C for silicate melts and, thereby, calculate the corresponding values of E_a as a function of melt composition and temperature.

The slopes of the VFT functions for the three end-member melts, and thus values of E_a, vary with temperature (Fig. 19A). Qualitative examination of the VFT functions at temperatures of 700, 1000 and 1400 °C suggest that E_a values for rhyolite and hydrous rhyolite change only slightly across that range of temperatures (i.e., near Arrhenian melts). In contrast, the VFT function for the basanite melt suggests a pronounced change in E_a values with temperature.

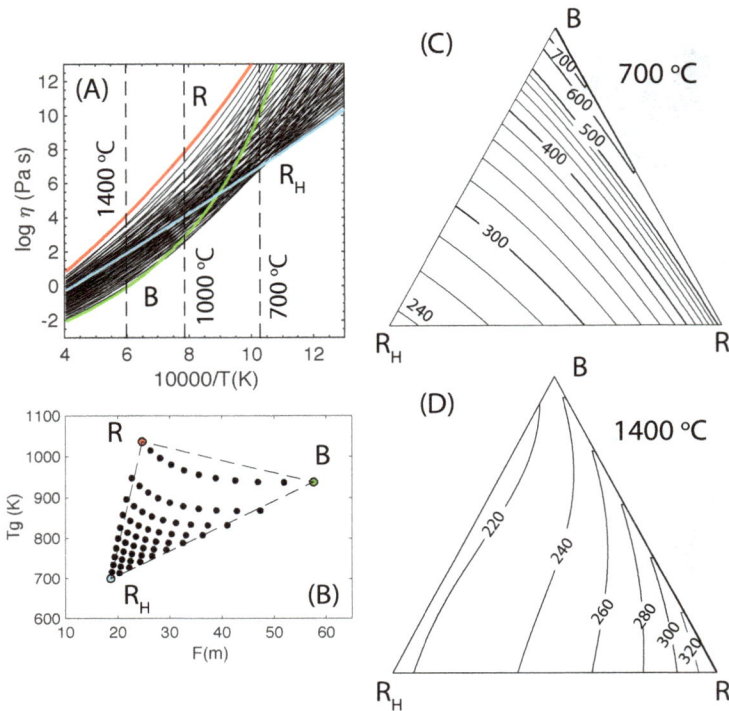

Figure 19. Transport properties for ternary system Basanite (B)–Rhyolite(R)–Hydrous Rhyolite (R_H) calculated with GRD model (**A**) Temperature dependent viscosity curves for melts bounded by B (green curve), R (**red curve**), and R_H (**blue curve**) sampled at 10 wt.% increments. **Vertical lines** denote temperatures of 700 °C, 1000 °C, and 1400 °C. (**B**) Model values of T_g and melt fragility (*m*) for the same melts; dashed lines are linear interpolations between end-members (B (**green**), R (**red**), and R_H (**blue**)). (**C, D**) Contoured values of activation energy (kJ mol^{-1}) calculated for the same melts at 700 °C and 1400 °C, respectively.

These results are illustrated quantitatively for the entire system via ternary diagrams contoured for E_a (kJ mol^{-1}) for temperatures of 700 °C and 1400 °C (Fig. 19C,D). At lower temperatures, E_a values vary by a factor of 3 and contours of E_a are subparallel to the basanite–rhyolite join. The lowest values are for hydrous rhyolite (<240 kJ mol^{-1}) and the highest belong to basanite (>750 kJ mol^{-1}). Rhyolite melt (i.e., dry) has a value of ~500 kJ mol^{-1} indicating the effects of dissolved H_2O are to cause melts to become stronger and more Arrhenian-like. At higher temperatures E_a values decrease for all melts and the differences in E_a across the ternary dissipate; E_a values vary by ~100 kJ mol^{-1}. Contours of E_a are now subparallel to the join between basanite and hydrous rhyolite melts. Basanite shows the greatest decrease ($\Delta \sim 550$ kJ mol^{-1}) whilst hydrous rhyolite changes least ($\Delta \sim 20$ kJ mol^{-1}) consistent with having the lowest initial fragility (*m* = 18.7; Fig. 19B).

Weaknesses

The GRD model replicates its original calibration data set well as shown in Figure 17 where the dashed lines denote 5% relative uncertainty on the measured values. The authors also showed the model to have the capacity for extrapolation and accurate prediction of a small number of viscosity values not used for calibration (see Appendix of Giordano et al 2008). The greatest deviations between their data and model occur at lower temperatures corresponding to where the anhydrous and volatile-enriched melts have high (>10^8 Pa s) viscosities (Fig. 17B). This applies to both strong (near Arrhenian) and more fragile (non-Arrhenian) melts as evidenced by the lack of correlation between $\Delta\log\eta$ values and the NBO/T values computed from the melt compositions (i.e., a proxy for polymerization and fragility).

Although the GRD model has several strengths, a variety of weaknesses are emerging as more experimental data become available, the number and range of users have increased, and the scope of Earth Science research has grown (e.g., planetary exploration; volcanic risk mitigation). In particular, new experimental measurements of melt viscosity provide a continuous test of the GRD model (Fig. 20). Measurements of viscosity for, both anhydrous and hydrous, metaluminous melts not used in the original calibration continue to be well predicted by the GRD model (Fig. 20A). Probably this reflects the fact that the original database was dominated by metaluminous melt compositions. In contrast, corresponding new datasets for peraluminous and peralkaline melts present a problem. Anhydrous peraluminous melts are well predicted at high temperatures but the GRD model underestimates viscosity at lower temperatures approaching T_g (Fig. 20B). One consequence is that model values of T_g corresponding to $\eta=10^{12}$ Pa s will be significantly underestimated. Interestingly hydrous peraluminous melts are better predicted by the GRD model at all temperatures perhaps reflecting the fact that hydrous melts tend to be more Arrhenian than their anhydrous counterparts.

The original GRD database comprised a limited number of peralkaline melt compositions and consequently does not reproduce new peralkaline measurements well (Fig. 20C). As for most melts, the GRD model does a better job of prediction at high temperature but

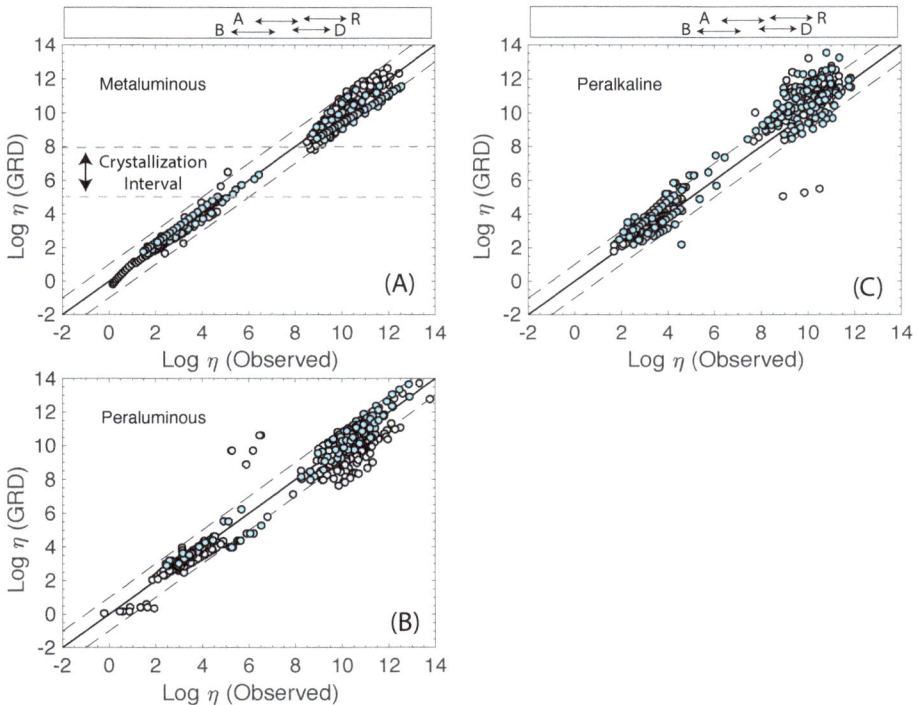

Figure 20. GRD predicted values of viscosity not used in the original model calibration for: (**A**) metaluminous melts, (**B**) peraluminous melts, and (**C**) peralkaline melts. Data include anhydrous (**grey symbols**) and volatile-rich (H_2O, F; **blue symbols**) melts (see text). Dashed lines indicate $\Delta\log\eta$ values of ± 1.

the discrepancies increase dramatically at lower temperatures where $\eta \geq 10^7$ Pa s (Fig. 20C). Lastly, the GRD model fails to reproduce a range of extreme (e.g., high FeO or MgO) melt compositions expected to be present on other planetary models (i.e., Sehlke and Whittington 2016). This limits the overall utility of the GRD model.

To some extent, these failures of the GRD model simply reflect the absence of experimental data for these compositions in the calibration dataset. In this case, a recalibration and reconfiguration of the model with new (expanded) experimental data could be a partial remedy. However, the problem may also be structural indicating that the oxide basis (species) used to model the compositional dependence of B_{GRD} and C_{GRD} needs to be reconsidered. There are, for example, indications that at some level there are structural problems in the GRD model which are summarized in Figure 18. The grey shaded fields (Fig. 18A, B, D) denote inaccessible parameter spaces; any values within these regions are aphysical for silicate melts. A small number of volatile-rich melts, corresponding to moderately strong ($m \sim 21\text{--}24$) anhydrous counterparts, fall into these fields (i.e., C_{GRD} values < 0; Fig. 18A, D). Values of C_{GRD} represent the temperature where the viscosity becomes singular and is related to the Kauzmann temperature, where the entropy difference between solid and liquid is zero. In principle, values of C_{GRD} for silicate melts must always be greater than 0 K. The other inaccessible field is defined by values of melt fragility (Fig. 18B); there is a lower limit to fragility defined by $m \geq 12 - A_{GRD}$ which for the GRD model is a value of ~16.5 (Russell and Giordano 2017). The same small number of hydrous melts which have aphysical values of C_{GRD} also have unrealistic fragility values < ~16.5.

11. QUO VADIMUS

Despite the strengths and diversity of the models designed to predict the viscosity of geological melts, and reviewed here, there remains substantial room for improvement. The most comprehensive (i.e., multicomponent) models for the viscosity of geological melts incorporate the effects of major and minor oxides, as well as, H_2O (and F). However, these models do not account for other volatiles that may be important in some magmatic systems including CO_2, sulphur, chlorine. New experimental data that address this gap are becoming available, in particular for CO_2 (Bourgue and Richet 2001; Di Genova et al. 2016), which should support more comprehensive models.

Commonly, iron has been treated as a single species (e.g., FeO_{total}) whereas silicate melts contain both ferric and ferrous iron and their proportions affect melt structure substantially (Mysen 1988; Dingwell 1991). Previously, there were insufficient experimental data to model the effect of iron redox on melt viscosity partly because the ferric/ferrous ratio was not measured independently. Recent experimental studies have produced data that suggest Fe redox can affect the viscosity of some melts significantly (i.e., peralkaline melts; Dingwell and Virgo 1987; Di Genova et al. 2017) reflecting the fact that Fe^{2+} and Fe^{3+} contribute differently to melt structural organization (e.g., Chevrel et al. 2014; Stabile et al. 2016). The most significant effects are likely to be at highly oxidizing conditions. Incorporating oxidation state considerations may yield significantly better fits, as well as conceptually stronger models.

Planetary exploration and scientific enquiry will become an ever greater part of Earth Sciences. The compositions of magmatic rocks discovered on extra-terrestrial planetary bodies are guaranteed to include compositions not found on Earth (e.g., Chevrel et al. 2013, 2014; Sehlke and Whittington 2016). Furthermore, there remain extreme melt compositions on Earth for which the transport properties are poorly known (e.g., peridotite, komatiite, carbonatite; Dingwell et al. 2004; Di Genova et al. 2016; Sehlke and Whittington 2016). These are areas where future models will depend on new experimental measurements of viscosity for a wider range of melt compositions. Some of these measurements are extremely challenging where viscosities are $< 10^{-2}$ Pa s or for liquids that are highly susceptible to crystallization or unmixing.

Few models for multicomponent geological melts incorporate the effects of pressure. The pressure dependence of viscosity, although minor compared to the effects of varying temperature and composition, is critical if we are to model melt transport within the deeper crust and mantle. To a large extent, T–X–P models for melt viscosity are hindered by a paucity of measurements of melt viscosity at higher pressures. The experiments are laborious, time-consuming, challenging, and have significantly greater analytical uncertainties. More and better measurements of melt viscosity at higher pressure are needed to assess its importance in different geological environments (i.e., volcanic, crustal, mantle), over a variety of geological melts (i.e., strong and fragile), and across the full range of temperatures. Future models for the viscosity of geological melts, that should include the effects of pressure, require these data; producing such data represents the major challenge to experimentalists for the next decade.

Lastly, all models are essentially empirical, in that, the chemical components chosen (Table 1) have no explicit or independent relationship to the structure or speciation of the silicate melt. Ultimately, future models may benefit immensely from the use of a component basis that is, at least in part, a reflection of melt speciation (e.g., Mysen et al. 1982, Mysen 1988; Le Losq and Neuville 2017; Lee et al. 2020). For example, Le Losq and Neuville (2017) developed a model to predict melt viscosity for simple system glasses (i.e., SiO_2–Na_2O–K_2O) based on Q^n-species abundance derived from Raman spectroscopy. Conversely, can a robust viscosity model be used to inform, at least indirectly, on the molecular structural organization of high-T melts (Russell and Giordano 2017; Giordano and Russell 2018)?

12. CLOSING THOUGHTS

Future models for melt viscosity can be improved by creating more data to increase the range of melt compositions studied, to accommodate additional volatile species, including Fe redox, and to constrain the effects of pressure. However, the quality of experimental datasets is of equal or greater importance. Reliable predictive models for melt viscosity benefit greatly from datasets that include: i) accurate experimental temperature measurements, ii) reliable measurements of melt viscosity, and iii) accurate and comprehensive measurements of melt composition.

A consequence of a larger number of laboratories publishing experimental measurements of viscosity is the inherent variability in the quality of the datasets. Poor or erroneous datasets are a real and significant impediment to developing better predictive models because, in disciplines such as ours (Earth Sciences), data remain relatively sparse. For example, we rarely reward (i.e., allow for publication) repeated measurements of melt viscosity on the same melt compositions over the same range of temperatures. Erroneous or flawed data take great effort to identify, verify, and carefully reject when there is little interlaboratory repetition.

Temperature measurements are precise and have little uncertainty. Most well-established laboratories use internal standards to make precise and accurate experimental measurements of viscosity. However, there are several areas where published datasets show significant variations in quality. If we have any hope of modelling the effects of composition on melt viscosity the compositions of the experimental material must be well characterised. The compositions of the melts should be measured directly rather than accepting a nominal pre-melting composition. Ideally, there should be post-experiment verification of the composition to ensure no compositional drift during experimentation. This is of greatest concern where small variations in composition are associated with significant changes in melt structure (i.e., metaluminous melts such as most calc-alkaline rhyolites).

Lastly, many measurements of viscosity are a challenge where the timescale of the experiment approaches the timescales for processes of oxidation/reduction, crystallization or exsolution (i.e., mafic and ultramafic melts; Dingwell et al. 2004). Thorough microscopic, spectroscopic, and analytical study of the run products serves the community well by verifying that the measured viscosity corresponds to the putative melt—rather than a modified material.

ACKNOWLEDGEMENTS

This research was supported by the Natural Sciences and Engineering Research Council of Canada (NSERC; JKR) and an ERC 2018 Advanced Grant 834225 (EAVESDROP; DBD). We acknowledge and thank Stephan Kolzenburg, Amy Ryan, and Tom Jones for helpful reviews of a preliminary manuscript. Subsequently, the manuscript greatly benefitted from critical reviews by Alex Sehlke and Fabian Wadsworth whose comments and suggestions have helped us clarify our message.

REFERENCES

Adam G, Gibbs JH (1965) On the temperature dependence of cooperative relaxation properties in glass-forming liquids. J Chem Phys 43:139–146
Angell CA (1985) Strong and fragile liquids. *In*: Relaxations in Complex Systems. Ngai KL, Wright GB (eds) U.S. Department of Commerce National Technical Information Service, Springfield, Virginia, p 3–11
Angell CA (1991) Relaxation in liquids, polymers and plastic crystals—strong/fragile patterns and problems. J Non-Cryst Solids 131–133:13–31
Ardia P, Giordano D, Schmidt MW (2008) A model for the viscosity of rhyolite as a function of H_2O-content and pressure: a calibration based on centrifuge piston-cylinder experiments. Geochem Cosmochim Acta 72:6103–6123
Avramov I (1998) Viscosity of glass-forming melts. J Non-Cryst Solids 238:6–10
Avramov I, Milchev A (1988) Effect of disorder on diffusion and viscosity in condensed systems. J Non-Cryst Solids 104:253–260
Baker DR (1996) Granitic melt viscosities: Empirical and configurational entropy models for their calculation. Am Mineral 81:126–134
Becker GF (1897) Some queries on rock differentiation. Am J Sci 3:21–40
Bottinga Y, Richet P (1996) Silicate melt structural relaxation; rheology, kinetics, and Adam Gibbs theory. Chem Geol 128:129–141
Bottinga Y, Weill D (1972) The viscosity of magmatic silicate liquids: a model for calculation. Am J Sci 272:438–475
Bottinga Y, Richet P, Sipp A (1995) Viscosity regimes of homogeneous silicate melts. Am Mineral 80:305–319
Bourgue E, Richet P (2001) The effects of dissolved CO_2 on density and viscosity of silicate melts: a preliminary study. Earth Planet Sci Lett 193:57–68
Bowen NL (1915) The crystallization of haplobasaltic, haplodioritic, and related magmas. Am J Sci 236:161–185
Bowen NL (1934) Viscosity data for silicate melts. EOS Trans, Am Geophys Union 15:249–255
Caricchi L, Burlini L, Ulmer P, Gerya T, Vassalli M, Papale P (2007) Non-Newtonian rheology of crystal-bearing magmas and implications for magma ascent dynamics. Earth Planet Sci Lett 264:402–419
Carmichael ISE, Turner FJ, Verhoogen J (1974) Igneous Petrology. McGraw-Hill, New York, USA
Chevrel MO, Giordano D, Potuzak M, Courtial P, Dingwell DB (2013) Physical properties of $CaAl_2Si_2O_8$–$CaMgSi_2O_6$–FeO–Fe_2O_3 melts: Analogues for extra-terrestrial basalt. Chem Geol 346:93–105
Chevrel MO, Baratoux D, Hess K-U, Dingwell DB (2014) Viscous flow behavior of tholeiitic and alkaline Fe-rich martian basalts. Geochem Cosmochim Acta 124:348–365
Chevrel MO, Pinkerton H, Harris AJL (2019) Measuring the viscosity of lava in the field: A review. Earth Sci Rev 196:102852
Cordonnier B, Schmalholz S, Hess KU, Dingwell D (2012) Viscous heating in silicate melts: An experimental and numerical comparison. J Geophys Res 117:B02203
Di Genova D, Cimarelli C, Hess KU, Dingwell DB (2016) An advanced rotational rheometer system for extremely fluid liquids up to 1273 K and applications to alkali carbonate melts. Am Mineral 101:953–959
Di Genova D, Vasseur J, Hess KU, Dingwell DB (2017) Effect of oxygen fugacity on the glass transition, viscosity and structure of silica- and iron-rich magmatic melts. J Non-Cryst Solids 470:78–85
Dingwell DB (1986) Viscosity-temperature relationships in the system $Na_2Si_2O_5$–$Na_4Al_2O_5$. Geochim Cosmochim Acta 50:1261–1265
Dingwell DB (1991) Redox viscometry of some Fe-bearing silicate liquids. Am Mineral 76:1560–1562

Dingwell DB (1996) Volcanic dilemma—flow or blow? Science 273:1054–1055

Dingwell DB (1998) The glass transition in hydrous granitic melts. Phys Earth Planet Int 107:1–8

Dingwell DB, Virgo D (1987) The effect of oxidation state on the viscosity of melts in the system Na_2O–FeO–Fe_2O_3–SiO_2. Geochim Cosmochim Acta 51:195–205

Dingwell DB, Webb SL (1989) Structural relaxation in silicate melts and non-Newtonian melt rheology in geologic processes. Phys Chem Miner 16:508–516

Dingwell DB Bagdassarov NS Bussod GY Webb SL (1993) Magma rheology. *In*: Experiments at High Pressures and Application to the Earth's Mantle. Luth RW (ed) Mineral Assoc Can, Short Course Handbook 21:131–196

Dingwell DB, Romano C, Hess K (1996) The effect of water on the viscosity of a haplogranitic melt under *P*–*T*–*X* conditions relevant to silicic volcanism. Contrib Mineral Petrol 124:19–28

Dingwell DB, Courtial P, Giordano D, Nichols ARL (2004) Viscosity of peridotite liquid. Earth Planet Sci Lett 226:127–138

Dorfman A, Hess KU, Dingwell DB (1996). Centrifuge-assisted falling-sphere viscometry. Eur J Mineral 8:507–514

Duan X (2014) A model for calculating the viscosity of natural iron-bearing silicate melts over a wide range of temperatures, pressures, oxygen fugacites, and compositions. Am Mineral 99:2378–2388

Eyring H, Henderson D, Stover BJ, Eyring EM (1982) Statistical Mechanics and Dynamics, Second edition. John Wiley, New York

Fotheringham U (2019) Viscosity of glass and glass-forming melts. *In:* Springer Handbook of Glass, Musgraves JD, Hu J, Calvez L (eds) Springer Handbooks. Springer, Cham

Gent AN (1960) Theory of the parallel plate viscometer. Brit J Appl Phys 11(2):85–88

Giordano D, Dingwell DB (2003) Non-Arrhenian multicomponent melt viscosity: A model. Earth Planet Sci Lett 208:337–349

Giordano D, Dingwell DB (2004) Erratum to 'Non Arrhenian multicomponent melt viscosity: a model': [Earth and Planetary Science Letters 208 2003. 337–349], Earth Planet Sci Lett 221:49

Giordano D, Russell JK (2018) Towards a structural model for the viscosity of geological melts. Earth Planet Sci Lett 501:202–212

Giordano D, Dingwell DB, Romano C (2000) Viscosity of a Teide phonolite in the welding interval. J Volc Geotherm Res 103:239–245

Giordano D, Romano C, Poe B, Dingwell DB, Behrens H (2004) The combined effects of water and fluorine on the viscosity of silicic magmas. Geochim Cosmochim Acta 68:5159–5168

Giordano D, Mangicapra A, Potuzak M, Russell JK, Romano C, Dingwell DB, Di Muro A (2006) An expanded non-Arrhenian model for silicate melt viscosity: A treatment for metaluminous, peraluminous and peralkaline melts. Chem Geol 229:42–56

Giordano D, Russell JK, Dingwell DB (2008) Viscosity of magmatic liquids: a model. Earth Planet Sci Lett 271:123–134

Glasstone S, Laidler K J, Eyring H (1941). The Theory of Rate Processes. McGraw-Hill, New York

Greenwood HJ (1989) On models and modelling. Can Mineral 27:1–14

Hess K-U, Dingwell DB (1996) Viscosities of hydrous leucogranitic melts: A non-Arrhenian model. Am Mineral 81:1297–1300

Hui H, Zhang Y (2007) Toward a general viscosity equation for natural anhydrous and hydrous silicate melts. Geochim Cosmochim Acta 71:403–416

Hui H, Zhang Y, Xu Z, Del Gaudio P, Behrens H (2009) Pressure dependence of viscosity of rhyolitic melts. Geochim Cosmochim Acta 73:3680–3693

Kani K, Hosokawa K (1936) On the viscosity of silicate rock-forming minerals and igneous rocks. Rev Electrotechn Lab 391:1–105

Kolzenburg S, Giordano D, Thordarson T, Höskuldsson A, Dingwell DB (2017) The rheological evolution of the 2014/2015 eruption at Holuhraun, central Iceland. Bull Volc 79:45

Kolzenburg S, Giordano D, Hess KU, Dingwell DB (2018) Shear rate-dependent disequilibrium rheology and dynamics of basalt solidification. Geophys Res Lett 45:6466–6475

Kolzenburg S, Hess KU, Berlo K, Dingwell DB (2020) Disequilibrium rheology and crystallization kinetics of basalts from Campi Flegrei. Front Earth Sci 8:187

Kolzenburg S, Chevrel MO, Dingwell DB (2022) Magma/suspension rheology. Rev Mineral Geochem 87:639-719

Kondratiev A, Khvan AV (2016) Analysis of viscosity equations relevant to silicate melts and glasses. J Non-Cryst Solids 432:366–383

Kozmidis-Petrovic AF (2014) Equations of viscous flow of silicate liquids with different approaches for universality of high temperature viscosity limit. Process Appl Ceram 8:59–68

Lavallée Y, Hess KU, Cordonnier B, Dingwell DB (2007) Non-Newtonian rheological law for highly crystalline dome lavas. Geology 35:843–846

Lee SK, Mosenfelder JL, Park SY, Lee AC, Asimow PD (2020) Configurational entropy of basaltic melts in Earth's mantle. PNAS17:21938–21944

Lejeune AM, Holtz F, Roux J, Richet P (1994) Rheology of an hydrous andesite: an experimental study at high viscosities. EOS Trans, Am Geophys Union 75:724

Le Losq C, Neuville DR (2017) Molecular structure, configurational entropy and viscosity of silicate melts: Link through the Adam and Gibbs theory of viscous flow. J Non-Cryst Solids 463:175–188

Lesher CE, Spera FJ (2015) Thermodynamic and transport properties of silicate melts and magma. *In*: Encyclopedia of Volcanoes. Sigurdsson H, Houghton B, Rymer H (eds) Academic Press, Heidelberg, Amsterdam, Boston, p 113–141

Lev E, James MR (2014) The influence of cross-sectional channel geometry on rheology and flux estimates for active lava flows. Bull Volc 76:829

Li M, Russell JK, Giordano D (2020) Temperature–pressure–composition model for melt viscosity in the Dp–An–Ab system. Chem Geol 560

Mauro JC, Yue Y, Ellison AJ, Gupta PK, Allan DC (2009) Viscosity of glass-forming liquids. PNAS 106:19780–19784

Moretti R (2005) Polymerisation, basicity, oxidation state and their role in ionic modelling of silicate melts. Annals Geophys 48:583–608

Misiti V, Vetere F, Freda C, Scarlato P, Behrens H, Mangiacapra A, Dingwell DB (2011) A general viscosity model of Campi Flegrei (Italy) melts. Chem Geol 290:50–59

Misiti V, Freda C, Taddeucci J, Romano C, Scarlato P, Longo A, Papale P, Poe BT (2006) The effect of H₂O on the viscosity of K-trachytic melts at magmatic temperatures. Chem Geol 235:124–137

Mysen BO, Virgo D, Seifert FA (1982). The structure of silicate melts: implications for chemical and physical properties of natural magma. Rev Geophys 20:353–383

Mysen BO (1988) Structure and Properties of Silicate Melts. Elsevier, Amsterdam

Myuller RL (1955) A valence theory of viscosity and fluidity for high-melting glass-forming materials in the critical temperature range. Zhurnal Prikladnoi Khimii 28:1077–1087

Nichols RL (1939) Viscosity of Lava. J Geol 47:290–302

Okumura S, Nakamura M, Uesugi K, Nakano T, Fujioka T (2013) Coupled effect of magma degassing and rheology on silicic volcanism. Earth Planet Sci Lett 362:163–170

Papale P (1999) Strain-induced magma fragmentation in explosive eruptions. Nature 397:425–428

Persikov ES (1984) The Viscosity of Magmatic Melts. Nauka, 160 pp. (in Russian).

Persikov ES (1991) The viscosity of magmatic liquids: Experiment generalized patterns, a model for calculation and prediction, applications. *In*: Physical Chemistry of Magmas, Advances in Physical Chemistry. Perchuk LL, Kushiro I (eds) Springer, Berlin, 1–40

Persikov ES, Bukhtiyarov PG (2009) Interrelated structural chemical model to predict and calculate viscosity of magmatic melts and water diffusion in a wide range of compositions and *T–P* parameters of the Earth's crust and upper mantle. Russ Geol Geophys 50:1079–1090

Plazek DJ, Ngai KL (1991) Correlation of polymer segmental chain dynamics with temperature-dependent time-scale shifts. Macromolecules 24:1222–1224

Popper KR (1968) The Logic of Scientific Discovery. Harper and Row, New York

Richet P (1984) Viscosity and configurational entropy of silicate melts. Geochim Cosmochim Acta 48:471–483

Richet P (1987) Heat capacity of silicate glasses. Chem Geol 62:111–124

Richet P, Bottinga Y (1986) Thermochemical properties of silicate glasses and liquids: A review. Rev Geophys 24:1–25

Richet P, Bottinga Y (1995) Rheology and configurational entropy of silicate melts. *In* Structure, Dynamics and Properties of Silicate Melts, Stebbins JF, McMillan PF, Dingwell DB (Eds) Rev Mineral 32:67–94

Romano C, Giordano D, Papale P, Mincione V, Dingwell DB, Rosi M (2003) The dry and hydrous viscosities of alkaline melts from Vesuvius and Phlegrean Fields. Chem Geol 202:23–38

Romine WL, Whittington AG (2015) A simple model for the viscosity of rhyolites as a function of temperature, pressure and water content. Geochim Cosmochim Acta 170:281–300

Rössler E, Hess K-U, Novikov VN (1998) Universal representation of viscosity in glass forming liquids. J Non-Cryst Solids 223:207–222

Russell JK, Giordano D (2005) A model for silicate melt viscosity in the system CaMgSi₂O₆–CaAl₂Si₂O₈–NaAlSi₃O₈. Geochim Cosmochim Acta 69:5333–5349

Russell JK, Giordano D (2017) Modelling the configurational entropy of silicate melts. Chem Geol 461:140–151

Russell JK, Giordano D, Dingwell DB, Hess K-U (2002) Modelling the non-Arrhenian rheology of silicate melts: Numerical considerations. Eur J Mineral 14:417–427

Russell JK, Giordano D, Dingwell D (2003) High-temperature limits on viscosity of non-Arrhenian silicate melts. Am Mineral 88:1390–1394

Ryan AG, Russell JK, Heap MJ, Kolzenburg S, Vona A, Kushnir ARL (2019) Strain-dependent rheology of silicate melt foams: importance for outgassing of silicic lavas. J Geophys Res 124:8167–8186

Schulze F, Behrens H, Holtz F, Roux J, Johannes W (1996) The influence of H₂O on the viscosity of a haplogranitic melt. Am Mineral 81:1155–1165

Scopigno T, Rocco G, Sette F, Monaco G (2003) Is the fragility of a liquid embedded in the properties of its glass? Science 302:849–852

Sehlke A, Whittington AG (2016) The viscosity of planetary tholeiitic melts: A configurational entropy model. Geochim Cosmochim Acta 191:277–299

Sehlke A, Whittington AG (2017) Corrigendum to" The viscosity of planetary tholeiitic melts: A configurational entropy model"[Geochim Cosmochim Acta 191 (2016) 277–299] Geochim Cosmochim Acta 197:474–475

Shaw HR (1963) Obsidian–H_2O viscosities at 1000 and 2000 bars in the temperature range 700° to 900°C. J Geophys Res 68:6337–6343

Shaw HR (1972) Viscosities of magmatic silicate liquids; an empirical method of prediction. Am J Sci 272:870–893

Spera FJ (2000) Physical properties of magma. *In*: Encyclopedia on Volcanoes, Sigurdsson H, Houghton B, Rymer H, Stix J, McNutt S (eds) Academic Press, San Diego, USA, 171–190

Stabile P, Webb S, Knipping JK, Behrens H, Paris E, Giuli G (2016) Viscosity of pantelleritic and alkali-silicate melts: Effect of Fe redox state and Na/(Na+ K) ratio. Chem Geol 442:73–82

Stebbins JF, Carmichael ISE, Moret LK (1984) Heat capacities and entropies of silicate liquids and glasses. Contrib Mineral Petrol 86:131–148

Stevenson RJ, Dingwell DB, Webb SL, Sharp TG (1996) Viscosity of microlite-bearing rhyolitic obsidians: an experimental study. Bull Volc 58:298–309

Sturm KG (1980) Zur Temperaturabhängigkeit der Viskosität von Flüssigkeiten. Glastechn Ber 53:63–76

Toplis MJ (1998) Energy barriers to viscous flow and the prediction of glass transition temperatures of molten silicates. Am Mineral 83:480–490

Toplis MJ, Dingwell DB, Hess KU, Lenci T (1997) Viscosity, fragility and configurational entropy of melts along the join SiO_2–$NaAlSiO_4$. Am Mineral 82:979–990

Vetere F, Behrens H, Holtz F, Neuville DR (2006) Viscosity of andesitic melts—new experimental data and a revised calculation model. Chem Geol 228:233–245

Vona A, Romano C, Dingwell DB, Giordano D (2011) The rheology of crystal-bearing basaltic magmas from Stromboli and Etna. Geochim Cosmochim Acta 75:3214–3236

Vona A, Romano C, Giordano D, Russell JK (2013) The multiphase rheology of magmas from Monte Nuovo (Campi Flegrei, Italy). Chem Geol 346:213–227

Waterton SC (1932). The viscosity-temperature relationship and some inferences on the nature of molten and of plastic glass. J Soc Glass Technol 16:244–249

Webb SL, Dingwell DB (1990a) The onset of non-Newtonian rheology of silicate melts—a fiber elongation study. Phys Chem Miner 17:125–132

Webb SL, Dingwell DB (1990b). Non-Newtonian rheology of igneous melts at high stresses and strain rates: Experimental results for rhyolite, andesite, basalt, and nephelinite. J Geophys Res 95(B10):15695–15701

Wadsworth FB, Witcher T, Vossen CE, Hess K-U, Unwin HE, Scheu B, Castro JM, Dingwell DB (2018) Combined effusive-explosive silicic volcanism straddles the multiphase viscous-to-brittle transition. Nat Comm 9:1–8

Whittington A, Richet P, Linard Y, Holtz F (2001) The viscosity of hydrous phonolites and trachytes. Chem Geol 174:209–223

Whittington AG, Bouhifd MA, Richet P (2009) The viscosity of hydrous $NaAlSi_3O_8$ and granitic melts: Configurational entropy models. Am Mineral 94:1–16

Williams-Jones G, Barendregt R, Russell JK, LeMoigne Y, Gallo R (2020) The age of the Tseax volcanic eruption, British Columbia, Canada. Can J Earth Sci 57:1238–1253

Zhang Y, Xu Z, Liu Y (2003) Viscosity of hydrous rhyolitic melts inferred from kinetic experiments, and a new viscosity model. Am Mineral 88:1741–1752

Reviews in Mineralogy & Geochemistry
Vol. 87 pp. 887-918, 2022
Copyright © Mineralogical Society of America

19

Non-terrestrial Melts, Magmas and Glasses

Guy Libourel[1,2], Pierre Beck[3] and Jean-Alix Barrat[4]

[1]*Université Côte d'Azur, Observatoire de la Côte d'Azur*
CNRS, Laboratoire Lagrange
Boulevard de l'Observatoire
CS 34229, 06304 Nice Cedex 4
France
[2]*Hawai'i Institute of Geophysics and Planetology*
School of Ocean, Earth Science and Technology
University of Hawai'i at Mānoa
Honolulu, Hawai'i 96821
USA
[3]*Université Grenoble Alpes, CNRS, IPAG*
Institut de Planétologie et d'Astrophysique de Grenoble (IPAG) UMR 5274
Grenoble, F-38041
France
[4]*Université de Bretagne Occidentale, CNRS, IRD*
Ifremer, LEMAR UMR 6539, F-29280 Plouzané
France

INTRODUCTION

The classical model for the formation of the Solar System invokes the gravitational collapse of a cold molecular cloud of interstellar gas, its heating, and the formation of a central star (Lewis 1995; Pfalzner et al. 2015). A flattened spinning protoplanetary disk of dust and gas surrounding the protosun extends for 100 AU (1AU $\approx$ Earth to Sun distance $\approx 150 \times 10^6$ km). Initially hot, this disk later cools down (T-Tauri stage). Since the heat flux is greatest close to the sun and at places where the cloud is the densest, the disk is hotter near the center and cooler farther away, suggesting, in the early solar nebula, a temperature decrease with increasing heliocentric distance from the proto-sun. Due to the viscous expansion of the disk, temperature drops rapidly from a few thousand degrees within 1 AU to a few hundred farther out. As the protoplanetary disk cooled down, the gas temperatures dropped below the condensation temperatures of refractory materials, like aluminates, silicates and metals, which formed in the inner Solar System at less than 1AU. Water ice (snow line) formed at around 3–4 AU while other volatiles (ammonia, methane, carbon dioxide, nitrogen) formed farther out. Clumping of these rock/metal condensed materials into planetesimals (in the inner part) and chunks of ice (in the outer part) would eventually yield the actual structure of the Solar System with, in particular, its division between inner terrestrial planets and outer gas giant planets (Lewis 1995).

From its birth 4.6 billion years ago to the present day, the Solar System has thus hosted many environments in which various types of molten silicates, magmas and glasses have been formed. A question arises then about the differences between terrestrial and extraterrestrial igneous materials. On Earth, if basalts show a considerable petrological variety, their chemical and isotopic signatures have been inherited from the same terrestrial mantle melted under a limited range of pressure-temperature conditions. In contrast, the chemical and isotopic compositions of meteorites or samples returned on Earth by space missions indicate at least many tens of distinct parent bodies.

1529-6466/22/0087-0019$05.00 (print)
1943-2666/22/0087-0019$05.00 (online)

http://dx.doi.org/10.2138/rmg.2022.87.19

Asteroids are small, lack atmospheres, and those which accreted fast were heated quickly by short-lived radioisotopes ([26]Al, [60]Fe, etc.) instead of the much longer-lived isotopes of K, Th and U that have mainly powered the heating of planet-embryos and planets (Dauphas and Chaussidon 2011). Relative to the Earth, their magmatic activity had to be very different. From asteroids to planetesimals and planet embryos, there are indeed large differences in the depth (pressure) range over which melts form, the amount of melting that takes place in any given time interval, the rates at which melts migrate, the amounts and compositions of their volatiles, and the way in which they erupt when reaching the surface or crystallize when they do so (Wilson and Keil 2012).

Both celestial ballistic and sample return missions offer us direct access to these extraterrestrial bodies. Meteorites can roughly be divided into chondrites and achondrites. Chondrites are pre-planetary rocks formed from the proto-planetary disk and give insights on the first solid materials in our solar system. Achondrites on the other hand are pieces of a differentiated planetary bodies, like the Moon, Mars or our Earth. From those rocks we can learn about planetary formation, evolution, and differentiation, meaning how planets in general form.

The diversity of achondrites (i.e., basaltic or igneous) and iron meteorites suggests more specifically that a significant population of large asteroid-sized bodies underwent internal melting and differentiation during the first several million years of the Solar System (Caporzen et al. 2011; Elkins-Tanton et al. 2011; Weiss and Elkins-Tanton 2013; Kruijer et al. 2014). In this scenario, early accreting planetesimals retained sufficient short-lived radiogenic isotopes to melt partially or fully the interior except for a conductively cooled crust whose thickness was between a few and several tens of km. Material near the surface of this crust would have undergone minimal metamorphic heating and have been considered as the potential source region for non-metamorphosed chondritic meteorites. Deeper material from the crust would have experienced increasing degrees of metamorphism and eventually underwent partial melting and differentiation. In this model (Fu and Elkins-Tanton 2014), the preservation of either a chondritic or an achondritic crust on a differentiated body would thus depend on the positive or negative buoyancy of silicate melts produced at depth with respect to the overlying crust.

Another peculiarity of extraterrestrial igneous materials resides in the distinctiveness of their environments of formation, i.e., microgravity, high vacuum, high carbon/oxygen ratios, hypervelocity impacts, flash heating, etc (Michel et al. 2015). Extraterrestrial igneous materials are therefore unique samples for inventorying the various celestial bodies, understanding their internal dynamics, and specifying their conditions of formation, which all together provides clues for understanding the evolution of our Solar System since its birth.

Because any concern for completeness would be here illusory, we have focused on three examples of heating processes producing different types of igneous rocks: i) by transient heating events in the protoplanetary disk: chondrules; ii) by [26]Al decay: differentiated bodies, and iii) by impact melting: high-pressure shock melt veins and tektites. Our selected igneous extraterrestrial materials plotted in a total alkali-silica (TAS) diagram (Fig. 1) show us, if it were still necessary, that similar chemical compositions do not imply similar formation processes.

CHONDRULES: THE EARLIEST MOLTEN DROPLETS OF THE SOLAR SYSTEM

Chondrules, millimeter-sized igneous droplets, are one of the major constituents of chondritic meteorites (chondrites), which represent fragments of parent bodies populating the main asteroid belt. Chondrule formation appears to have started contemporaneously with calcium-aluminum-rich refractory inclusions (CAIs) dated at 4567.3 ± 0.16 Ma (Amelin et al. 2010; Connelly et al. 2012) and may have lasted over the entire lifetime of the accretionary disk,

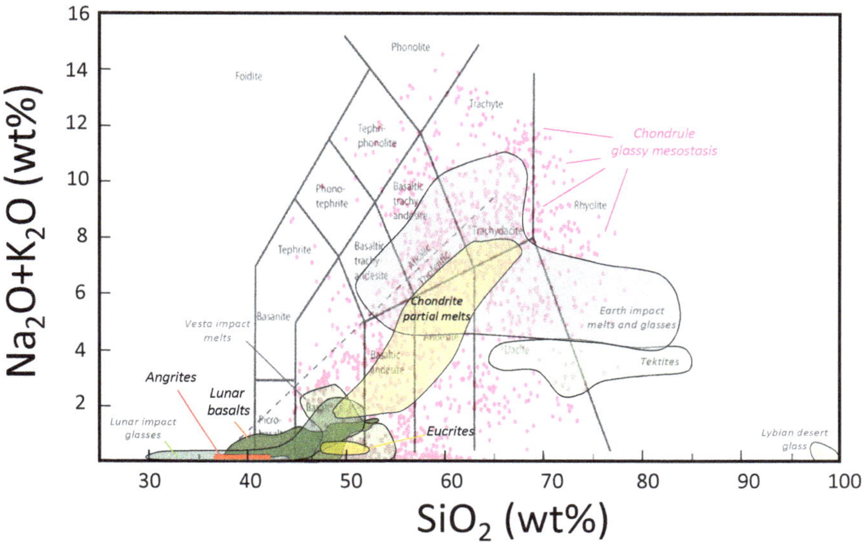

Figure 1. Total Alkali Silica (TAS) diagram of representative extraterrestrial igneous objects depicted in this chapter. **Pink dots**: glass composition of chondrules from ordinary and carbonaceous chondrites; **Reddish** and **yellow areas**: igneous differentiated bodies; **Greenish areas**: impact glasses and tektites; See text for explanation.

up to ~4 Ma (Villeneuve et al. 2009; Connelly et al. 2012; Bollard et al. 2014; Pape et al. 2019). Chondrules may thus provide some of the most powerful constraints on conditions in the solar protoplanetary disk, if the processes that led to their heating, melting and recrystallization can be understood. However, despite being recognized for over 200 years their origins remain enigmatic. Chondrules are believed to have formed by melting of solid precursors, including refractory inclusions (Krot et al. 2018 ; Krot 2019; Marrocchi et al. 2019), in dust-rich regions of the protoplanetary disk during transient heating events of still unknown nature; both nebular and planetary mechanisms are being debated (Alexander and Ebel 2012).

Extraterrestrial igneous droplets

Since chondrules are solidified droplets of molten material, their textures are indicative of the crystallization process and are described using the terms of igneous petrology. Chondrules exhibit a range of textures (Fig. 2), which in order of increasing grain size, and decreasing aspect ratio, are described as cryptocrystalline (CC), radiating (R), barred or dendritic (B), and porphyritic (P). The dominant silicate mineralogy of chondrules consists of olivine, low-Ca pyroxene, commonly overgrown with narrow rims of high-Ca pyroxene (e.g., Jones et al. 2018). FeNi metal, clinopyroxene, plagioclase, oxide and sulfide are the others most frequently associated phases, often bathed in the glassy mesostases. Mineralogical zoned chondrules are the dominant chondrule type with a core enriched in olivine crystals surrounded by a rim of low-Ca pyroxene laths hosting poikilitic olivines (Friend et al. 2016, Barosch et al. 2019). Some chondrules are rounded as they were once entirely molten but many are irregular in shape because they were only partly melted or because they accreted other particles as they solidified (Jacquet 2021).

Barred textures are usually composed of bars of olivine (barred olivine, or BO chondrules), and radiating textures are usually composed of needles of pyroxene (radiating pyroxene, or RP chondrules). Porphyritic textures, the dominant type in chondrites, can include any ratio of olivine to pyroxene (PO = porphyritic olivine with >90% olivine, PP = porphyritic pyroxene with >90% pyroxene, POP = porphyritic olivine and pyroxene). A further chondrule classification refers to the initial (unequilibrated) FeO content of olivine and pyroxene (Jones et al. 2018): Chondrules which have Fa and Fs <10 mol %, are designated as type I, and those with Fa and Fs >10 mol %, are designated as type II (Fa = atomic Fe/(Fe+Mg) in olivine; Fs = atomic Fe/(Fe+Mg+Ca) in pyroxene). Most type I chondrules, particularly in carbonaceous and enstatite chondrites, have Fa < 2 mole %. Type I and type II chondrules differ in several important petrologic respects. For example, type I chondrules contain abundant Fe,Ni metal blebs suggesting reducing conditions of formation, i.e., with log fO_2 (oxygen fugacity) well below the Fe/FeO (IW) buffer curve. Olivine and pyroxene grains in type II chondrules (Fig. 2) are often larger than those in their type I counterparts, and commonly show significant Fe–Mg core to rim zoning arising from disequilibrium growth.

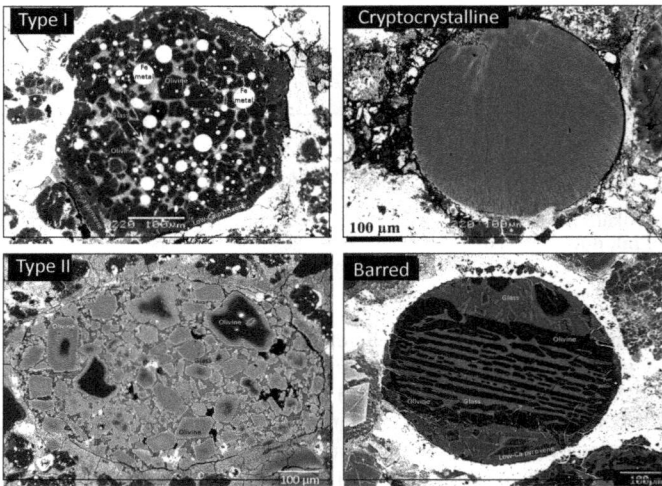

Figure 2. Typical textures of natural chondrules from a variety of chondrites (backscattered electron images). Nonporphyritic chondrules include cryptocrystalline (CC), barred (or dendritic) and radiated (not shown) textures. Radiating textures are commonly pyroxene (P)-rich (RP); barred and dendritic textures are commonly olivine (O)-rich (BO). Porphyritic (P) textures include olivine-rich (PO, or type A), olivine- and pyroxene-rich (POP, or type AB), and pyroxene-rich (PP, or type B). Chondrules with MgO-rich (Mg# >90) primary olivine and pyroxene compositions are designated type I, and chondrules with more FeO-rich (Mg# <90) primary olivine and pyroxene compositions are designated type II (Mg# = atomic Mg/(Mg + Fe)). Hence, for example, a chondrule with MgO-rich olivine and pyroxene is a type IAB, POP chondrule.

Material interstitial to olivine and pyroxene is described as chondrule mesostasis. In unequilibrated chondrites, mesostasis in many chondrules is glassy. There are significant inter- and intra-chondrule variations in chemical compositions of chondrule glasses, including mesostases and melt inclusions : (in wt%) 40 < SiO_2 < 80, 0 < CaO < 25, 10 < Al_2O_3 <30, 0 < MgO < 8, 0.2 < TiO_2 < 2, 0 < Na_2O < 10 (Table 1; Figs. 1 and 3). In spite of the observed diversity of chondrule textures, mineralogy, and bulk chemistry, chondrule glasses depict a simple compositional array: silica content is negatively correlated with CaO, Al_2O_3, MgO and TiO_2 and positively correlated with Na_2O and K_2O (Jones and Scott 1992; Libourel et al. 2006). The glass represents the final liquid present during cooling, and commonly contains fine-grained elongate microcrystallites of various phases (e.g., olivine, low- or high Ca-pyroxene, oxides, etc) which are interpreted as quench crystals. Mesostasis in unequilibrated chondrites can also be predominantly crystalline, typically consisting of an intergrowth of plagioclase and high-Ca pyroxene with minor amounts of glass.

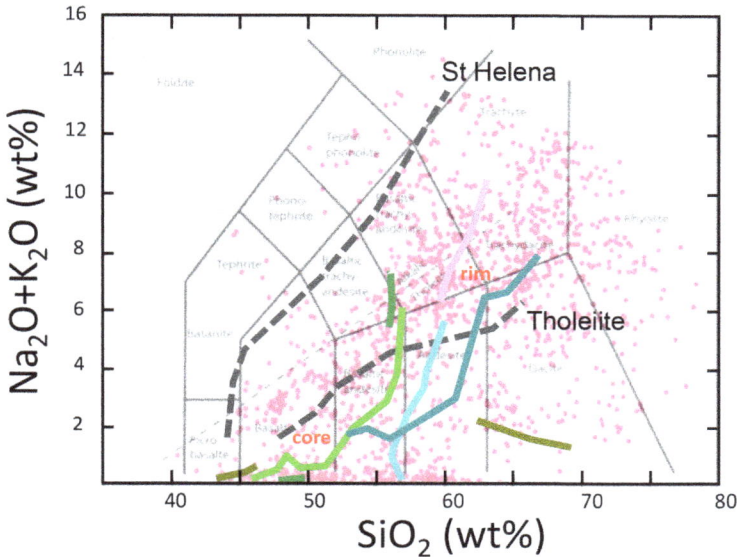

Figure 3. TAS diagram for chondrule melt in chondrites. **Pink dots**: glass composition of chondrule meso-stases and melt inclusions from ordinary (OC) and carbonaceous (CC) chondrites. **Colored lines**: core to rim traverse for representative OC and CC chondrules (Grossman et al. 2002; Krot et al. 2004; Libourel et al. 2006). **Dark grey dashed lines**: Differentiation trends for representative basanatic (St Helena; Kawabata et al. 2011) and tholeiitic basalts (Th; Villiger et al. 2004). Notice the significant chemical core to rim zoning of mesostases in mm-sized chondrules.

Chondrules also contain an iron-nickel-sulfur component, containing minerals such as troilite, pyrrhotite, pentlandite, Fe,Ni metal, and magnetite. Minor chromite is common in type II chondrules, and spinel occurs rarely in type I chondrules. SiO_2 polymorphs occurs commonly in chondrules from enstatite (E) and carbonaceous Renazzo type (CR) chondrites, and rarely in ordinary (O) and other carbonaceous (C) chondrite groups. In addition to chondrules dominated by olivine and pyroxene, there are well-defined aluminium-rich chondrules (ARC), which include plagioclase-rich chondrules (PRC) and glass-rich chondrules.

Chondrules are present in the different chondrite groups with specific characteritics (Krot and Scott 2005; Rubin 2010; Jones 2012). This includes differences in chondrule mineralogy, size and abundance, as well as the distribution of different textural types of chondrules. They are generally attributed to differences in the chemical compositions of chondrule precursor assemblages, the local dust/gas ratio, and/or their individual thermal histories. Viable astrophysical models of chondrule formation must be able to account for these property differences (Jones 2012; Jones et al. 2018).

Chondrule thermal history

Dynamic crystallization (cooling rates) experiments have successfully reproduced chondrule textures and therefore, until recently, provided the main constraints on chondrule thermal history, including peak temperatures and cooling rates of chondrule melts (Hewins et al. 2005; Desch and Connolly 2002; Desch et al. 2012; Jones et al. 2018). Nucleation and growth theory states that spontaneous crystallization from a melt will not start at saturation but only when the solution becomes supersaturated. Homogeneous nucleation poses a large energy barrier, which is easier to overcome at higher level of supersaturation. This is why significant degrees of undercooling (see Jones et al. 2018 for review) are generally invoked for olivine crystallization in chondrule melts (Fig. 4). By increasing the degree of undercooling (e.g., cooling rate), it was shown experimentally that the olivine habit evolves following the sequence: polyhedral, skeletal and dendritic (Jones et al. 2018), matching well with porphyritic, barred and radial textures depicted in chondrules.

In this scenario, chondrule precursors consist of fine-grained mineral grains, and the time spent at peak temperatures is relatively short (less than two hours). The liquidus temperature is a strong function of chondrule composition, and varies from around 1400–1700 °C (e.g., Radomsky and Hewins 1990). In general, porphyritic textures are produced when the peak temperature is below the liquidus, thus preserving multiple nucleation sites. Barred textures are produced when the peak temperature is slightly above the liquidus temperature, such that most, but not all, nucleation embryos are destroyed. Radial textures are produced from temperatures that clearly exceed the liquidus temperature, and in which few or no nucleation sites are available. These general statements are based on numerous experimental studies that have been described in detail in previous reviews (Lofgren 1996; Desch and Connolly 2002; Hewins et al. 2005; Desch et al. 2012). These experimental data give our best estimates of cooling rates: 1-3000 °C.h^{-1} for chondrule textures.

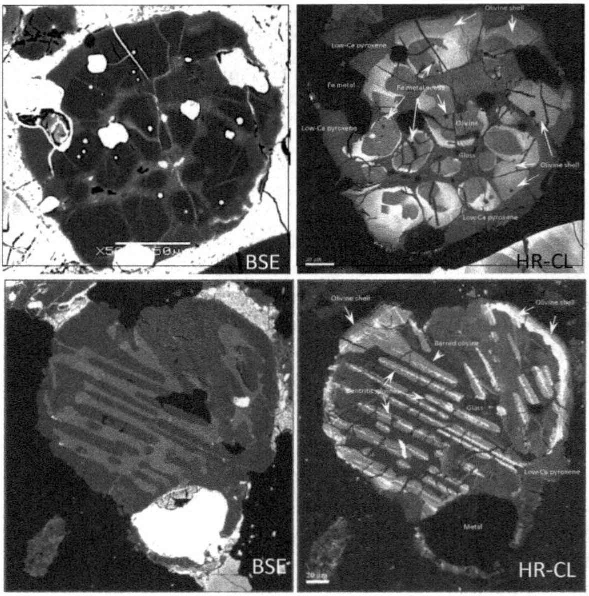

Figure 4. Back scattrered electron (BSE) versis high-resolution CL (HR-CL) images of representative Type I chondrules. Up left and right: Yamato 81020, Carbonaceous Ornans-type CO3.05, Ch#30b. This porphyritic olivine pyroxene chondrule shows the complex CL figures in each olivine with a core and euhedral outer edges. Olivine cores always contain metal and glassy inclusions. This porphyritic olivine pyroxene chondrule shows an example of Mg-rich olivine core with asymmetric overgrowth toward the periphery of the chondrule, forming a more or less continuous shell of Mg-rich olivine at the outer edge, formed of successive olivine overgrowth layers. Low-Ca pyroxene postdates the olivine growth. Down left and right: Yamato 81020, CO3.05, Ch#9. Barred olivine chondrule. CL images allow resolving the first dendritic growth of each olivine bar in this chondrule. Notice the notable absence of FeNi-metal blebs in olivine bars. The olivine rim that partly covers the chondrule surface corresponds to a multi-layer Mg-rich olivine shell, similar to those observed at the outer edge of porphyritic chondrule. The unconformity between the shell and the olivine bars and their common crystallographic orientation suggest that the Mg-rich olivine shell formed after the olivine bars by an epitaxial growth mechanism.

Table 1. Representative compositions (in wt.%) of glassy mesostases and melt inclusions in olivine in chemically-zoned type I porphyritic chondrules from ordinary (Semarkona, flashy green in Fig. 3) and carbonaceous (PCA91082) chondrites.

| | *chondrule 5, PCA91082* | | | | | | | | |
	melt inclusions		core				$\longrightarrow$		rim
SiO_2	45.3	42.4	53.5	55.0	55.6	56.6	64.9	67.3	79.6
TiO_2	1.2	1.3	0.31	0.35	0.41	0.32	0.32	0.27	0.15
Al_2O_3	26.1	27.5	24.5	22.3	21.5	22.9	17.2	17.3	8.7
Cr_2O_3	0.46	0.27	0.24	0.53	0.42	0.44	0.29	0.16	0.11
FeO	0.76	0.62	0.45	0.87	1.1	0.40	0.39	0.43	2.4
MnO	< 0.07	<0.07	0.46	0.56	1.4	0.99	1.4	1.2	0.38
MgO	4.7	4.0	3.7	5.5	4.4	3.3	1.8	0.86	1.4
CaO	19.7	22.3	13.5	13.1	11	11.1	7.2	5.0	0.98
Na_2O	<0.06	<0.06	2.1	1.7	3.1	3.3	3.9	6.6	2.1
K_2O	<0.04	<0.04	0.07	0.07	0.11	0.16	0.82	0.57	1.7
Total	98.3	98.4	98.9	99.9	99.1	99.4	98.2	99.7	97.6

| | *chondrule 7, Semarkona* | | | | | | |
	melt inclusions		core			$\longrightarrow$	rim
SiO_2	56.7	40.5	46.0	48.5	52.4	56.0	56.8
TiO_2	0.57	1.2	1.3	1.1	0.76	0.54	0.32
Al_2O_3	15.6	28.3	25.7	22.4	20.7	21.2	22.1
Cr_2O_3	0.18	0.22	0.24	0.34	0.36	0.12	0.18
FeO	0.15	0.34	0.19	1.7	0.38	0.27	0.39
MnO	<0.07	<0.07	<0.07	<0.07	0.18	0.27	0.26
MgO	4.7	4.1	5.4	5.1	4.6	3.5	3.5
CaO	20.9	22.6	19.2	18.0	18.0	13.3	10.9
Na_2O	0.09	0.75	0.23	0.85	1.3	3.3	5.4
K_2O	<0.04	<0.04	<0.04	0.05	<0.04	0.06	0.27
Total	98.9	98.0	98.3	98.0	98.7	98.4	100.2

By analogy with terrestrial petrology rules, the interpretation of chondrule textures in these experiments largely rest on the degree of undercooling and the availability of nucleation sites for crystal growth as the melt cools (Lofgren and Russell 1986 ; Radomsky and Hewins 1990; Lofgren 1996). Because of the importance of nucleation considerations, the relationship between texture and chondrule cooling rate is, to a certain extent, ambiguous. It is possible to produce different textures at the same cooling rate if the initial availability of nucleation sites is variable (Lofgren and Russell 1986; Lofgren 1996). Since availability of nucleation sites is a function of several variables, including precursor grain size, peak temperature, and time, textures alone do not provide unique thermal histories for specific textural types of chondrules.

Thermal history of chondrules has also been constrained from the mineral chemistry and zoning of olivine and pyroxene (see Jones et al. 2018 for review). Fe–Mg chemical zoning profiles of olivine (considering diffusional modification of zoning profiles as crystals grow by fractional crystallization from a chondrule melt) or diffusion profiles of Fe–Mg between Mg-rich olivine relict and its Fe-rich overgrowth have been extensively used (Jones and Lofgren 1993; Greeney and Ruzicka 2004; Hewins et al. 2009; Miyamoto et al. 2009; Villeneuve et al. 2015; Stockdale 2020). By taking advantage of the most recent study on type II porphyritic chondrules (Stockdale 2020), determined chondrule cooling rates for chondrules range from 5 to 8000 °C.h^{-1} and non-linear cooling histories had to be considered to produce good model fits to many of the observed diffusion profiles. On average, these inferred cooling rates are similar to those determined by dynamic crystallisation experiments. The range however includes much more rapid cooling rates and suggests porphyritic chondrules can form at much more rapid cooling rates than suggested by dynamic crystallisation experiments.

Since the cooling rates chondrules experienced during their formation are key constraints on chondrule formation mechanisms, new studies both conventional and innovative, e.g., trace element profile in FeNi metal, mineral defect and dislocation, exsolution and microstructure in pyroxene, calorimetric properties of the glass across the glass transition, are clearly needed in order to characterize peak temperatures and thermal histories as recorded by the various phases in a given chondrule, from chondrule to chondrule in a given chondrite, from chondrite to chondrite and from different types of chondrites, i.e., carbonaceous and non-carbonaceous chondrites.

Evidence for gas–melt interaction during chondrule formation

There has been a longstanding debate about whether chondrules behaved as open chemical systems, that is, gaining or losing material by exchange with surrounding vapor during cooling (Sears et al. 1996; Connolly and Jones 2016). Alternatively, chondrule chemical and isotopic compositions may primarily record the compositions of their precursors, and they remained essentially closed systems upon melting and solidification. A preponderance of recent experimental and observational evidence indicates that chondrules did behave as open systems while molten or partially molten in the nebular setting (Ebel et al. 2018).

The chemical composition of chondrule mesostases in all chondrite groups (Table 1) show a strong negative correlation between SiO_2 and Al_2O_3, CaO, or TiO_2, which is opposite to trends expected based on closed-system crystallization of olivine and low-Ca pyroxene from chondrule melts (Libourel et al. 2006). An open-system behavior of chondrule melts with extensive chemical exchange with the surrounding nebular gas seems instead to prevail. Similarly in type I chondrules (Fig. 2), the frequently observed low-Ca pyroxene shell at chondrule edges and the correlative core to edge silica activity gradient have also been considered as resulting from gas–melt interaction processes. Tissandier et al. (2002) demonstrated experimentally that partially molten olivine bearing chondrule-like samples exposed to gaseous SiO incorporate silica, which induces the peripheral low-Ca pyroxene saturation of the melt and crystallization of low-Ca pyroxene. The presence of low-Ca pyroxene in the periphery of many chondrules or the partial resorption of olivine, e.g., olivine poikilitically enclosed by low-Ca pyroxene

(Soulie et al. 2017) are very likely resulting from such gas–melt interactions, which can be conveniently modeled by a reaction of the type:

$$Mg_2SiO_4 \text{ (ol)} + SiO \text{ (gas)} + \tfrac{1}{2} O_2 \text{ (gas)} \leftrightarrow Mg_2SiO_4 \text{ (ol)} + SiO_2 \text{ (melt)} \leftrightarrow Mg_2Si_2O_6 \text{ (low-Ca px)}$$

Accordingly, low-Ca pyroxene at chondrule edges have been interpreted as the results of gas–melt interaction during which their crystallization results from the partial dissolution of olivine due to the elevated silica activity in the peripheral melt, in response to the high partial pressure of $PSiO_{(gas)}$ of the surrounding gas (see also Friend et al. 2016; Marrocchi et al. 2019). In CR carbonaceous chondrites, direct $SiO_{(gas)}$ condensation into chondrule melts is also inferred for the formation of silica-rich igneous rims around magnesian type I POP chondrules, consisting of igneously-zoned low-Ca and high-Ca pyroxenes, glassy mesostasis, Fe, Ni-metal nodules, and a nearly pure crystalline SiO_2 phase (Krot et al. 2004). In such a scenario, the nature of the type I chondrules, i.e., PO, POP, PP is controlled by Henry law behavior of the $PSiO_{(gas)}$ and/or the duration of such gas–melt interactions.

The preponderance of gas–melt interaction during chondrule formation is further supported by glassy mesostases that are concentrically zoned in Na and K, with enrichments near the outer margins (Matsunami et al. 1993; Grossman et al. 2002; Libourel et al. 2003; Nagahara et al. 2008; Hewins et al. 2012). Figure 3 depicts some examples of the core to-rim chemical zoning measured in the mesostasis of mm-sized type I PO and POP chondrules. From both carbonaceous and ordinary chondrites, the prominent enrichment in alkalis and silica measured in a given chondrule is striking, roughly equaling the overall range of silica variation of terrestrial magmatism for some chondrules (Fig. 3).

However, for low or intermediate cooling rates (see above), low oxygen fugacities (log $fO_2 \ll$ IW; Grossman et al. 2008) and low pressures (total pressure $\approx 10^{-6}$ to 10^{-3} bars) of the solar protoplanetary disk (nebula), experiments, natural analogs, and theoretical calculations all show that there should be extensive evaporation of major and minor elements, in the order S > Na, K > Fe > Si > Mg for FeNi metal-bearing type I chondrules (Alexander et al. 2008 and reference therein). If elemental fractionations in chondrules are a function of volatility, it comes that the more volatile elements (such as S, and alkalis) should be entirely absent, which they are not. In addition, clinopyroxene–glass partitioning demonstrates that the Na contents of the glass in the central regions of chondrules were established before solidification (Libourel et al. 2003). Furthermore, the fractionated elements should exhibit large and systematic isotopic fractionations, which they do not (Alexander et al. 2008; Ebel et al. 2018).

Sodium diffusion in molten silicates being four orders magnitude faster than diffusion of Si leaves open the possibility of Na influx into chondrules at high temperature prior parent body accretion. Taking into account that Na solubility from an alkali-rich vapor is primarily controlled by the silica content of the melt (Mathieu et al. 2008, 2011), the enrichment/ heterogeneity in Na_2O content in chondrule mesostases could reflect variations of the $PSiO_{(g)}$ of the gaseous surroundings, which in turn provide a sound explanation for the preferential enrichment of alkali and silica in low-Ca pyroxene-rich chondrules (Fig. 3).

Further evidences of gas–melt interactions have been recently posited from a high-resolution cathodoluminescence survey of FeO-poor chondrules from various chondrite groups (Libourel and Portail 2018). Changes of cathodoluminescence activator concentrations of magnesium-rich olivines reveal overlooked internal zoning structures and multi-layered shells at chondrule margins (Fig. 4), which provide evidence for high temperature gas-assisted near-equilibrium epitaxial growth of olivines during chondrule formation. It is shown that this interaction with the surrounding gas (i.e., the partial pressures reigning in the gaseous environment and the duration of the gas–melt interaction) rather than various cooling histories, defined chondrule composition and texture. High partial pressures of gaseous Mg and SiO are required to maintain olivine saturation in chondrules in their solar protoplanetary disk-forming region (Nagahara et al. 2008). Libourel and Portail (2018) also infer that porphyritic or barred textures of chondrules (Fig. 4) have the same thermal history. Their difference is simply controlled by the presence or the absence of metal grains that acted as seeding agents at high temperature above 1500–1600 °C (to overcome the nucleation barrier) during near equilibrium conditions crystallization of chondrule melts. Porphyritic and barred chondrules are inferred to terminate their formation by fast cooling/quench on the order of 10^3–10^4 °C.h^{-1}, in order to prevent further crystallization and elemental diffusion. Fast cooling is also needed to quench the silica-poor depolymerized liquids bathing chondrule crystals into glasses. Despite some clear advances in the field (see Russel et al. 2018 and references therein), more work needs to be done for understanding the chondrule formation processes and in particular how to disentangle thermal from chemical effects during open system melt crystallization, i.e., undercooling/cooling rates versus gas–melt interaction.

Astrophysical implications

Many melting mechanisms (e.g., gravitational shock waves, current sheets, planetary embryo bow shocks, impact jets, etc.) have been put forward over the last decades to explain the igneous features of chondrules (see review in Russell et al. 2018) and their location of formation in the disk (e.g., Kruijer et al. 2020; Williams et al. 2020). However no single process is able yet to satisfy all of the characteristics of chondrules/chondrites highlighted by decades of research (Russell et al. 2018). Does this mean that a mechanism involving multiple processes interacting with each other may be required to address the intricate problem posed by chondrule formation? Without being able to decide on this issue yet, disruptive and vaporizing collisions between planetesimals as proposed recently by Stewart et al. (2019) for chondrule and chondrite formation are going in this direction. For instance, the expansion of impact vapor plumes by driving high-temperature bow shocks in the solar nebula and the subsequent hydrodynamic collapse of the plume by favoring mixing between nebular gas and dust with materials from the original planetesimals offer a range of processes and physicochemical conditions that deserve to be explored in more details, some of which resembling those highlighted above, e.g., gas–melt interaction, moderately volatile element behavior, non-linear cooling rates, etc. Nevertheless the fact remains that understanding chondrule formation is not at hand yet, and the key question to know if chondrules are a by-product of planet formation rather than representing an important stage in the planet-building process is still pending.

MAGMATIC MELTS FROM PROTOPLANETS

The first protoplanets accreted very early and were the place of intense magmatic activity due to the heat generated by the disintegration of ^{26}Al (e.g., Kruijer et al. 2014). The composition and diversity of the magmas generated by these objects is a widely open question. Elements of answers can be provided by the study of meteorites. More than 60,000 meteorites are now inventoried. The vast majority of them are chondrites and correspond to rocks that were primitive, brecciated or metamorphosed to varying degrees, but which have not undergone partial melting. Only less than 4000 meteorites, the achondrites, derive from the mantle or crust of differentiated or partially differentiated bodies. This number may seem large, and could suggest that we have a good picture of the diversity of magmas produced by the first objects in the solar system. This is not the case. Chemical and isotopic data show that these rocks come from no more than 40 distinct parent bodies (Greenwood et al. 2018). The meteorite sampling appears even less representative when we realize that about 95% of the achondrites derive from only two parent bodies: about 75% of these meteorites, the howardites–eucrites–diogenites or HEDs, originate from the crust of the 4-Vesta asteroid (e.g., McSween et al. 2011; Russell et al. 2012), and about 20% from the mantle of a fairly large object (at least 700 km in diameter, and maybe more) now disrupted (e.g., Downes et al. 2008; Warren 2012).

Although all these observations demonstrate that the sampling of the melts produced by the magmatic activity of the asteroids or early protoplanets through meteorites is not representative of their diversity, a very large amount of data has been obtained on these rocks, and allows us to discuss the genesis of the primordial crusts and the compositions of the melts that built them.

The vast majority of the protoplanetary lava samples at our disposal are basaltic rocks with very low alkali contents (Fig. 5 and Table 2). In addition to these, there are a handful of rocks richer in silica of andesitic or trachyandesitic compositions. All these rocks are obviously not representative of all the magmas generated by the first protoplanets. We also have in the meteorite collections numerous pyroxenic cumulates (diogenites, aubrites), or melting residues (ureilites, brachinites, lodranites...) for which the associated liquids are still unknown. We will review in this section the data we have on protoplanetary magma samples available in the meteorite collections.

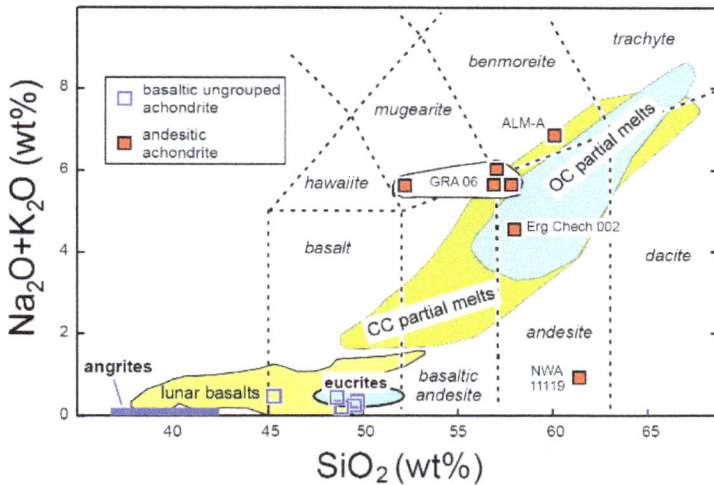

Figure 5. TAS diagram for protoplanetary melts. Lunar basalts and melts produced by partial melting of of carbonaceous (CC) and ordinary (OC) chondritic systems (Collinet and Grove 2020) are shown for comparison.

Alkali-depleted protoplanetary basalts.

Eucrites from the asteroid 4-Vesta make up the bulk of the sampling, to which are added other very similar lavas whose isotopic characteristics show that they have formed on other parent bodies, and angrites, particular lavas which are ultra-depleted in Na and K.

Eucrites. With about 1500 known meteorites, eucrites are by far the best-known extraterrestrial magmatic rocks (Mittlefehldt 2015). They are mainly composed of calcic plagioclases (in the range of bytownite to anorthite), pyroxenes, with minor amounts of silica (cristobalite, tridymite or quartz), chromite, iron sulphides (troilite), phosphates, and metal. While the mineralogy is fairly simple, the textures show an extremely complex history. First, most eucrites are breccias containing debris from one or more lithologies. The larger fragments often show doleritic textures suggesting cooling in thick flows or shallow intrusions (Fig. 6). Pyroxenes, originally pigeonites, are generally composed of augitic exsolutions contained in a low-Ca pyroxene. The composition of these pyroxenes and the thicknesses of the exsolutions indicate equilibrium temperatures in the order of 700–900°C during a prolonged thermal

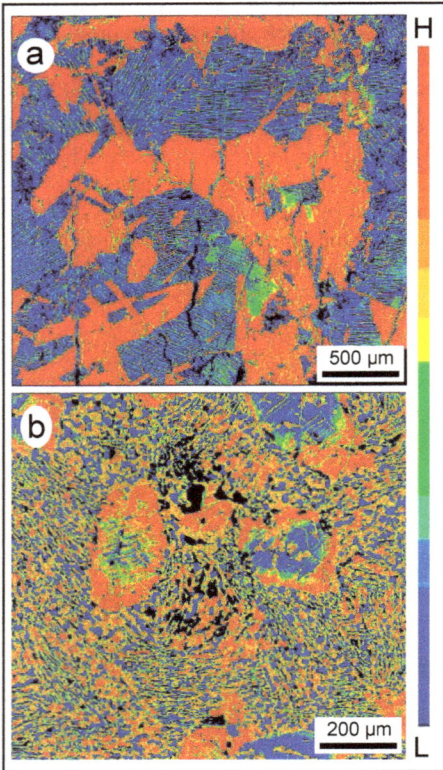

Figure 6. Calcium map of two clasts with different textures from eucrites. **a)** ophitic clast in Millbillil-lie displaying large plagioclase (**red**) and low-Ca pyroxenes (**blue**) with fine exsolutions of augite (**green**); **b)** fine-grained clast in Northwest 4523, displaying zoned pyroxene crystals with low-Ca cores (**blue or green**) and augitic rims (**red**) set in a groundmass composed mainly of pyroxene (**blue, green**) and plagioclase (**red**). Notice the exsolutions in the pyroxenes.

metamorphism which has erased the chemical zoning of the pyroxenes (e.g., Yamaguchi et al. 1996). However, rare eucrites escaped this thermal metamorphism and still show pyroxenes with magmatic zonings. In addition to these metamorphic transformations, some eucrites show evidence of late hydrothermal circulation, resulting in Fe enrichments in the fractures of the pyroxenes and veinlets made of Fe-rich olivines (e.g., Barrat et al. 2011; Warren et al. 2014).

In spite of the transformations they have undergone, the chemical composition of eucrites makes it possible to reconstitute an important part of the differentiation of their parent body. These rocks are basalts rich in Fe (FeO*=17.5-21 wt%) and very poor in alkali (e.g., Mittlefehldt 2015, Table 2). The Na_2O contents are at most 0.5–0.6 wt%. Mg# numbers (=100 Mg/ (Mg+Fe), atomic) range from about 40 to 30. These compositions are quite unusual for basalts but resemble those of some lunar basalts (Fig. 5). The major elements showed early on that two main trends seemed to control the chemical evolution from a main group (e.g., Stolper 1977): the Nuevo Laredo trend and the Stannern trend named after two meteorites. Following the experimental work of Stolper (1977) and the geochemical models by Warren and Jerde (1987), the Nuevo Laredo trend, a decrease in Mg# number coupled with an increase in the concentrations of incompatible elements, is explained by fractional crystallisation. The Stannern trend, an increase of incompatible element concentrations without changing the Mg# number, has been the subject of much more discussions. It was first proposed that partial melting of a homogeneous source could explain this trend, the eucrites of the Stannern trend, richer in incompatible elements (e.g., Ti), resulted from lower melting rates than the eucrites of the main group (e.g., Consolmagno and Drake 1977). This hypothesis, which was popular

Table 2. Major element abundances in basaltic and andesitic melts from protoplanets (in wt%).

	SiO$_2$	TiO$_2$	Al$_2$O$_3$	Cr$_2$O$_3$	FeO*	MnO	MgO	CaO	Na$_2$O	K$_2$O	P$_2$O$_5$	Total	Reference
Angrites													
LEW 87051	40.4	0.73	9.19	0.17	19	0.24	19.4	10.8	0.0234	0.04	0.08	100.03	Mittlefehldt, Lindstrom (1990)
D'Orbigny	38.4	0.89	12.4	0.04	24.7	0.28	6.49	15	0.0172		0.16	98.38	Mittlefehldt et al. (2002)
Basaltic eucrites													
Juvinas	49.34	0.638	13	0.34	18.82	0.56	7.27	10.38	0.47	0.04	0.092	100.95	McCarthy et al (1973)
Nuevo Laredo	49.46	0.95	11.78	0.29	20.27	0.56	5.46	10.4	0.57	0.05	0.11	99.9	Duke and Silver (1967)
Stannern	49.7	0.98	12.33	0.34	17.78	0.525	6.97	10.67	0.62	0.07	0.102	100.08	McCarthy et al (1973)
Ungrouped basaltic achondrites													
Pasamonte	48.59	0.65	12.7	0.33	19.64	0.56	6.77	10.25	0.45	0.05	0.1	100.09	Duke and Silver (1967)
Bunburra	49.56	0.9	11.97	0.33	17.65	0.58	5.87	9.4	0.33	0.05	0.05	96.7	Benedix et al. (2018)
NWA 011	45.63	0.92	13.12	0.24	21.3	0.4	6.66	11.11	0.45	0.03	<0.02	99.86	Yamaguchi et al. (2002)
NWA 1240	49.4	0.31	9.99	0.71	20.19	0.57	10.36	8.12	0.25	<0.03	0.09	99.99	Barrat et al. (2003)
Ibitira	48.81	0.9	12.97	0.38	18.19	0.47	7.28	10.99	0.21	0.03	0.1	100.33	Jarosewich (1990)
Andesitic achondrites													
GRA 06	55.98	0.1275	16.95	0.1	9.12	0.09	2.695	6.49	5.51	0.23	2.48	99.76	Day et al. (2009)
ALM-A	60.07	0.67	14.66	0.28	5.57	0.27	4.81	7.29	6.59	0.29	0.52	101.02	Bischoff et al. (2014)
NWA 11119	61.37	0.18	19.05	0.17	1.49	0.22	4.52	12.08	0.89	0.04		100	Srinivasan et al. (2018)
EC 002	58.01	0.36	9.12	0.42	11.22	0.47	7.06	8.31	4.2	0.34	0.06	99.58	Barrat et al. (2021)

for a long time, only explained the trace element contents imperfectly. Today, the eucrites of the Stannern trend are explained by crustal contamination. Models of trace element abundance (Barrat et al. 2007), corroborated by partial melting experiments of eucrites (Yamaguchi et al. 2013; Crossley et al. 2018), indicate that these magmas would result from the assimilation of partial-melting products of the eucritic crust by main group eucrites. The discovery of restitic eucrites reinforces this model (Yamaguchi et al. 2009).

Eucrites are also characterized by distinct oxygen isotopic compositions, and have very homogeneous $\Delta^{17}O$ values close to -0.24 ‰ (Greenwood et al. 2005). The differentiation of their parent body was certainly controlled by a magma ocean. These lavas are moreover seen by many authors as evolved liquids resulting from the crystallization of a magma ocean, where various cumulates have also crystallized (Righter and Drake 1997; Ruzicka et al. 1997; Warren 1997). Diogenites which are orthopyroxenitic cumulates displaying the same $\Delta^{17}O$ as eucrites were often assumed formed in the magma ocean (e.g., Warren 1997). However, the trace element geochemistry indicates that the parent liquids of these cumulates were different from those of the eucrites (e.g., Mittlefehldt 1994; Barrat et al. 2008), which rules out this hypothesis and implies a more complex magmatic history. Alternatively, these cumulates could form from magmas produced by the melting of magma ocean cumulates. They would then be the result of a different and later magmatic step in the history of the parent body (e.g., Yamaguchi et al. 2011). The parent magmas of the diogenites are unknown.

Angrites. These meteorites are extremely rare since only about thirty stones are known (Keil 2012). They are unbrecciated rocks mainly composed of Al–Ti-bearing diopside–hedenbergite (fassaite), calcic olivine and anorthite. Notable accessory phases are kirschteinite, spinel, troilite, ilmenite, titanomagnetite, titanite, celsian, and carbonates. Their textures make it possible to distinguish between volcanic and plutonic rocks whose cooling rate estimates made by different authors give conflicting results. In any case, these meteorites derive from a single parent body, characterized by very homogeneous O isotopic signatures ($\Delta^{17}O = -0.07$ ‰; Greenwood et al. 2005). As illustrated by their very particular mineralogy, these rocks have very unusual chemical compositions (Table 2). They are poor in SiO_2, and contain practically no Na and K. They also have high Ca/Al ratios (CaO (wt%)/Al_2O_3 (wt%)>1 and reaching 1.2 for many of them), which contrast with those of other basalts, extraterrestrial or not (CaO/Al_2O_3< 1). The origin of this feature is not understood, and could reflect the involvement of spinel (e.g., Mittlefehldt and Lindstrom 1990), or an excess of calcium attested by the presence of magmatic carbonate (e.g., Jambon et al. 2005). The angrites have extremely old crystallization ages, of the order of 4,564 Ga (Schiller et al. 2015), and are among the oldest known magmatic rocks. The composition, differentiation and chronology of the parent body of these rocks is a subject of intense discussion.

Ungrouped basaltic meteorites. Detailed petrological and isotopic studies have shown that some meteorites, initially assumed to be eucritic, originated from different parent bodies. The first of these reclassified rocks was North West Africa (NWA) 011 (and its pairings). This rock is a recrystallized breccia mainly composed of calcic plagioclases and pyroxenes with augitic exsolutions, with the same accessory minerals as the eucrites. Yamaguchi et al (2002) showed that its oxygen isotopic composition is clearly not eucritic ($\Delta^{17}O = -1.8$ ‰). Later, it was shown that other basaltic meteorites such as NWA 1240 (Fig. 7), Pasamonte or Bunburra Rockhole, although mineralogically or geochemically similar to eucrites, have O isotopic compositions which distinguish them from eucrites (Scott et al. 2009; Benedix et al. 2017). Thus, Vesta and the parent body of the angrites were not exceptions, and probably many other bodies generated basaltic lavas very poor in alkaline elements. It is obviously difficult to discuss the history of these parent bodies when there are no constraints on their bulk compositions, the origin of the Na and K depletions, and when we only have one or at best two samples derived from them.

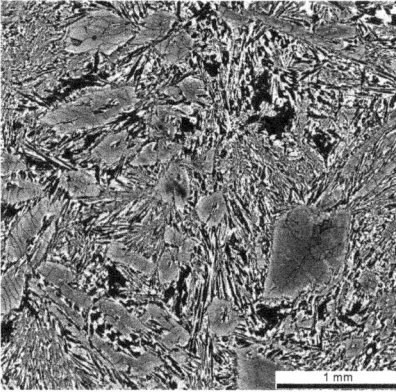

Figure 7. Backscattered-electron image of a polished section of the ungrouped Northwest Africa 1240 achondrite displaying low-Ca pyroxene phenocrysts (**gray**) set in a groundmass made of pyroxene (**gray to light gray**) and plagioclase (**black**).

Andesitic or trachyandesitic achondrites

Until about ten years ago, the prevailing opinion was that the differentiation of the first protoplanets essentially generated basaltic lava like those sampled by the basaltic achondrites. The discovery in 2006 of two paired meteorites Graves Nunatak (GRA) 06128 and 06129 (thereafter GRA 06) with silica and alkali contents comparable to those of terrestrial andesites or trachyandesites, has allowed this point of view to evolve (Day et al. 2009; Shearer et al. 2010). Since then, a few other andesitic or trachyandesitic achondrites were discovered such as Almahata Sitta trachyandesite (nicknamed ALM-A), NWA 11119, or more recently Erg Chech (EC) 002 which is currently being actively studied (Bischoff et al. 2014; Srinivasan et al. 2018; Barrat et al. 2021). None of these rocks originate from the same parent body, suggesting that the formation of andesitic melts may have been quite frequent in protoplanets. Subsequently, melting experiments (Usui et al. 2015; Lunning et al. 2017; Collinet and Grove 2020) have shown that the partial melting of chondrites could perfectly generate andesitic or trachyandesitic magmas rich in silica, alumina and alkalis (Fig. 5). These experimental results are very important and allow us to imagine that the primordial crusts of many protoplanets were in fact andesitic and not necessarily basaltic.

The few andesitic achondrites are very different from each other both texturally and mineralogically. GRA 006 is a medium-grained, foliated rock made of an assemblage of sodic plagioclases, two pyroxenes, olivine, phosphates and sulphides. The chemical composition of the phases is quite homogeneous and suggests a prolonged metamorphism. ALM-A, NWA 11119 and EC 002 have volcanic textures indicating much shorter cooling rates (Fig. 8). Of course, the composition of their minerals are different, but all three have crystallized plagioclases, high and low-Ca pyroxenes, and accessory phosphates, chromite, ilmenite and metal. NWA 11119 is remarkable for its porphyritic texture and tridymite crystals.

From a chemical point of view, these rocks are also very different. While their silica content (55–62 wt%) does of course distinguish them from previous basaltic rocks, it is noticeable that they are not systematically rich in Na and K. NWA 11119 is rather poor in alkaline elements and probably derives from a parent body poor in volatiles like those of the previous basaltic achondrites. The other andesitic achondrites derive from parent bodies much richer in volatiles. A link has been proposed between GRA 06 and brachinites (e.g., Day et al. 2009), and ALM-A undoubtedly originates from the parent body of ureilites. The parent body of EC 002 was a protoplanet of non-carbonaceous chondritic composition rich in Na and K (Barrat et al. 2021). This meteorite is the oldest lava known at present and is thought to have crystallised 2.25 Myr only after the Ca-Al rich inclusions. As with other achondrites, these meteorites here only provide fragmentary information about their parent bodies and the history of their differentiation.

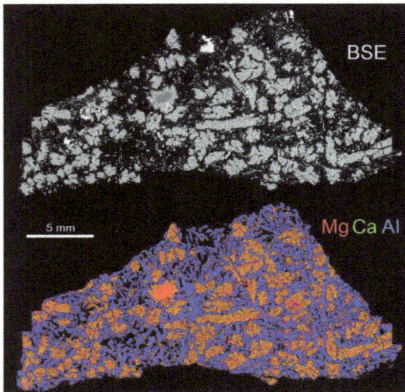

Figure 8. Backscattered electron (BSE) image and false-colored X-ray map of a polished section of EC 002 (courtesy of Akira Yamaguchi). Pyroxenes are **orange to red**, and feldspars are **blue** in the false-colored X-ray.

IMPACT-RELATED MELTS AND GLASSES

Because of the elevated relative velocities, collisions between solar system objects are highly energetic. Today, the typical impact velocity of asteroidal bodies on Earth is around 20 km/s while relative velocities between asteroids in the main-belt are of the order of 5 km/s (Steel et al. 1998). When the collision occurs, the kinetic energy of the impactor is transferred to the target in the form of momentum and heat. Given their supersonic impact velocities, these collisions generate shock waves, which have the potential to produce a significant dynamic compression of the target under high strain rates, and then deposit a residual heat after the passage of the shock (i.e. irreversible adiabatic compression). As noted by Osinski et al. (2012), a specificity of impact melting when compared to endogenic melting processes is the timescale of melting and the fact that important superheating can occur, up to total vaporization. Details on impact related melting can also be found in Cicconi et al. (2022, this volume).

Because impacts are a major geological process at the surface of Solar System bodies, and one of the fingerprints of impacts is the production of amorphous phases and silicate melts, we will review in this section impact related amorphization and melting of planetary materials.

Impact induced melting and vaporisation, lessons from terrestrial impactites

At the time of the writing, the Earth hosts about 190 confirmed impact crater sites (http://passc.net/EarthImpactDatabase/), which offer the opportunity to study the impact process "post-mortem". Impact-related rocks are referred to as impactites (Stoeffler and Grieve 2007), and include a number of melt-bearing or melt-related rocks (an extensive review on terrestrial impact melts and glasses can be found in Dressler and Reimold 2001).

Tektites and micro-tektites are dark glassy objects encountered as distal deposits, which are found over a specific geographical location (strewn field). The presence or association to relict or shock metamorphic phases confirms the impact origin of tektites (Koeberl 1992), as well as geochemical signatures showing the incorporation of minute amounts of extra-terrestrial materials in tektites (Koeberl and Shirey 1993; Barrat et al. 1997). Spherules-rich beds produced by impact events have also been found in the sedimentary records, and sometimes can be associated to known impact craters (Simonson and Glass 2004). Overall, the major and trace element composition of tektites suggest an origin by melting of the upper continental crust and the presence of [10]Be anomalies in tektites has been taken as evidence that the protolith was a very superficial horizon (Pal et al. 1982). Tektites are depleted in the most volatile elements (H_2O, Beran and Koeberl 1997; Cu, Zn Moynier et al. 2010, Jiang et al. 2019, K for micro-tektites Humayun and Koeberl 2004) with sometimes-associated isotopic fractionation. The depletion of volatile species in tektite compared to the expected protolith has been explained by evaporative loss of the silicate liquid (Moynier et al. 2010), or bubble-generation and migration in the melt (Melosh and Artemevia 2004). Another specificity of tektites is the reduced nature of the glass, with very low ferric/ferrous iron ratio, when compared to usually encountered terrestrial volcanic rocks (Rossano et al. 1999).

Tektites and microtektites represent only a minor fraction of the melt volume produced during an impact. The largest volume of impact melt lithologies is found in proximal impactites (within a few tens of crater diameter). Often-recrystallized silicate melt occur as allochtonous melt sheets, as fragments in polymict impact breccias, or as dykes and veins within basements lithologies or displaced rocks (Stöffler and Grieve 2007). The volume of impact melt appears to scale with the transient cavity diameter at least for impacts on crystalline targets (Grieve and Cintala 1992). In the case of the Manicouagan crater with a rim diameter of around 100 km, the total volume of melt estimated is around 1200 km^3 (Grieve and Cintala 1992), roughly equivalent to 40 000 years of Kilauea activity. The largest melt sheet that has been mapped within the crater is 55 km in diameter, with a thickness up to 230 m (Floran et al. 1978).

In Figure 9, the composition of various impact related glasses and melts are plotted in the total alkali–silica diagram (TAS). This diagram reveals that without a few exceptions, there is an overlap between major element chemistry of terrestrial impact-melt and impact glasses. Tektites appear to have on average higher SiO_2 abundance than Earth impact glasses as well as a lower alkali abundance. Lybian desert glasses lie at an extremity of this diagram, since their composition is >98 wt. % SiO_2, a specificity explained by an impact on quartz arenites (Barrat et al. 1997).

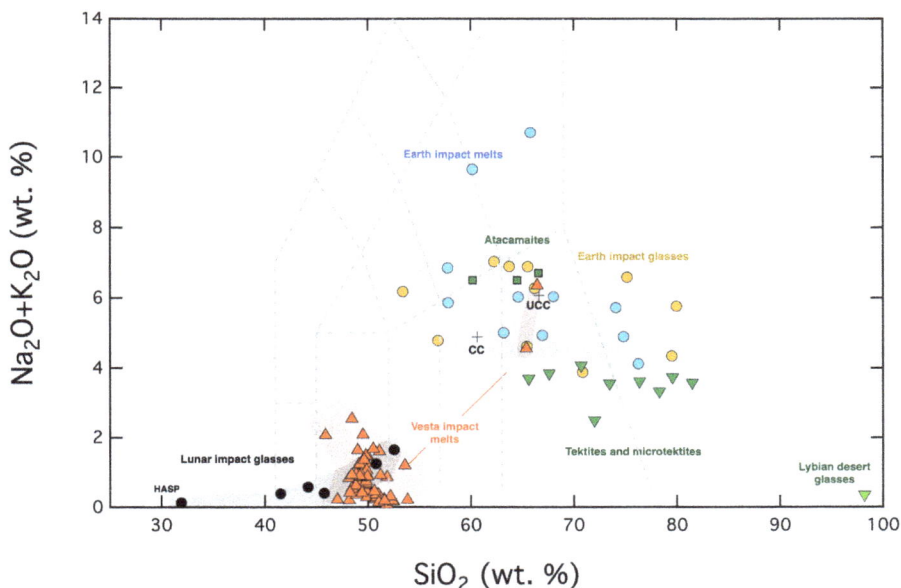

Figure 9. TAS diagram for various impact related melts and glass on Earth, the Moon and Vesta. See text and table for references to the dataset used in this diagram. CC: Continental Crust; UCC: Upper Continental Crust.

Impact-related amorphization of minerals: dense glass or not

Two different types of glasses can form in response to shock compression. A first type of glass is produced by elevated shock pressures and post-shock temperatures, which lead to almost complete melting of the target upon decompression. In that case, the formed glass solidifies under low-pressure and is structurally identical to glasses formed by quenching of a melt under ambient conditions. A very common example of these "low-density" glasses is *lechatelierite*, an amorphous form of SiO_2, which is recognizable in impactite thin section by the presence of flow texture, vesicles and a refractive index in the range (1.458–1.460) (Stöffler and Langenhorst 1994). If the shock-pressure is not too high, the heat deposited post-shock will be lower than required for total melting, and another type of glass may be produced: diaplectic glass (Engelhardt and Berscht 1969). By definition a diaplectic glass is a glass pseudo-morphosing a pre-existing crystal that lacks evidence of melting (absence of flow texture and vesicles). Diaplectic glasses of quartz and feldspar have been described in impactites and shock–recovery experiments. Structurally, diaplectic glasses are different from a glass of similar composition formed by quenching of a melt. They are denser and present a higher degree of long-range structural order. Diaplectic silica glass has a typical refractive index of 1.461–1.468 and a density higher that lechatelierite (2.206–2.276; Stoffler and Langenhorst 1994). Shock-wave amorphisation and densification has been reproduced in shock-recovery experiments (Stoffler et al. 1991; Okuno et al. 1999; Reynard et al. 1999). Shock-recovered anorthite composition glass (24 GPa) was found to be denser by 2.2 % than at ambient conditions (Reynard et al. 1999) with a refractive index increase of 0.8 %.

Pressure-induced amorphisation has been documented in high-pressure diamond-anvil cell experiments initially on water ice (Mishima 1984) followed by quartz (Hemley et al. 1988) and anorthite (Williams and Jeanloz 1989). In these static experiments, pressure induced amorphization can occur during cold compressions when high-pressure polymorphs formation are hindered by sluggish kinetic, and when a glass form becomes denser than the compressed low-pressure polymorphs. While a similar scenario has been somehow proposed for the formation of diaplectic glasses in meteorites (Williams and Jeanloz 1989), the impact community appears to favor a formation of diaplectic glass by quenching of a melt during decompression for quartz (Langenhorst 1994) as well as for feldspar (Chen and El Goresy 2000; El Goresy et al. 2013).

Diaplectic silica glass is often found in terrestrial impactite. Diaplectic feldspar glass is more common in meteorites and referred to as maskelynite (quartz is very rare in meteorites), a terminology proposed by Gustav Tshermak (1872) following observations of a non-birefringent form of feldspar in the Shergotty meteorite. Maskelynite has been found in several classes of extra-terrestrial samples and there seems to be a correlation between the frequency of maskelynite and the size of the parent body (Rubin 2015). Maskelynite is particularly abundant in Martian meteorites, is sometimes found in lunar samples, and is rare in HED meteorites.

High-pressure melting, and high-pressure phase bearing melt.

Several meteorite families display local evidence of impact melting in the form of shock melt veins (Fig. 10) and melt pockets (Fig. 11). Macroscopically, shock melt veins are black linear features a few tens of microns to a few mms thick, sometimes branching out. Melt pockets are tens of μm to a few mm-sized areas that were molten upon passage of the shock wave. The presence of shock melt veins is an indicator of a high-degree of shock compression according to the classification of Stöffler et al. (1991) of ordinary chondrites (shock stage 6/7). Shock melt veins and melt pockets have been found in ordinary and enstatite chondrites, eucrites, Lunar meteorites, Martian meteorites and iron meteorites. They are dark in macroscopic aspect, and often show microscopic evidence of melting by the presence of round sulfides grains (troilite) explained by the immiscibility of sulfide and silicate melts (Fig. 10). Numerous workers have studied shock melt veins because they host a menagerie of high-pressure minerals (Binns et al. 1969; Sharp and De Carli 2006; Gillet et al. 2007) including several polymorphs of Mg_2SiO_4 and $MgSiO_3$ (Fig. 10). A possible mechanism for generating shock melt veins is frictional heating (Stoffler et al. 1991; Langenhorst and Poirier 2000), implying that the temperature increase is local: shock effects recorded in melt veins are different from those observed in

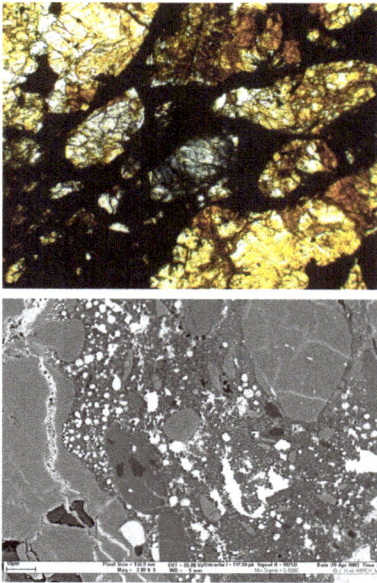

Figure 10. Top) Transmitted light image a shock vein in the Tenham meteorite (L6). The width of the scene is around 500 μm. Translucent silicate clasts are embedded in a dark opaque matrix, made of a melt quenched at high-pressure. Most of the silicate clasts remain unaffected by the melting process, except the blue crystal in the center of the vein. The color is due to the transformation of olivine to ringwoodite; the phase change is associated to a change in the Fe^{2+} crystal-field transition band, which produces the blue color of ringwoodite. **Bottom**) SEM image of another shock vein in the Tenham meteorite. Note the presence of rounded sulfurs and metal grains diagnostic of melting. The fine-grained matrix contains entrained fragments of olivine (often transformed to ringwoodite and wadsleite), pyroxene (often transformed to akimotoite) and plagioclase (often transformed to maskeynite or hollandite). The oval-shaped olivine grain in the top right corner shows variables shades of gray due to partial phase transition of olivine to denser polymorphs.

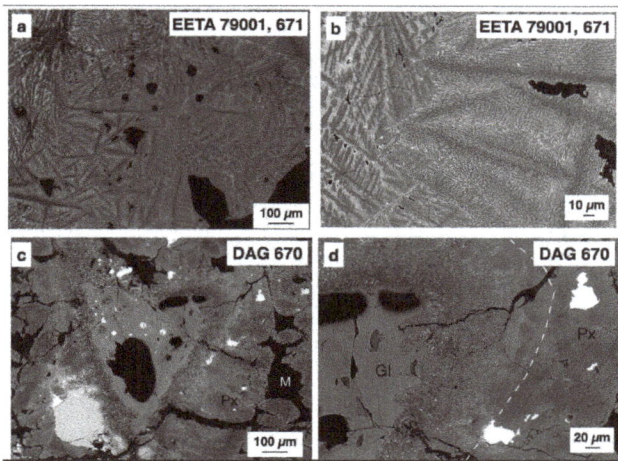

Figure 11. Textures of impact-melt pockets in martian meteorites EETA 79001 **(a,b)** and DAG 670 **(c,d)**. Melt-pockets are often encountered as mm-sized area in martian meteorites, and up to cm-size in the case of EETA 79001. The large melt pockets in EETA 79001 are relatively homogeneous compared to melt pockets from other shergotties, and present zoned skeletal crystals of pyroxene. Micrometer-sized sulfide spheres also occur. The melt pocket in DAG 670 appears more heterogeneous, with areas enriched in maskelynite and the presence of large sulfide globules. Note also the evolution in the (d) image (from left to right), from a glassy area, a zone with dendritic crystals, and the host rock.

the bulk rock, in particular those kinetically activated such as reconstructive phase transitions (Beck et al. 2005). In some cases, the silicate liquid from shock veins may have crystallized under high-pressure, as indicated by the presence of high-pressure liquidus assemblage (Chen et al. 1996). Cooling models have been formulated for shock melt veins than can explain a more frequent survival of high-pressure mineral in thin (<1 mm) rather than in larger (mm) veins (Langenhorst and Poirier 2000; Shaw and Walton 2013; Hu and Sharp 2017).

Shock related melt pockets have been found among chondrites and basaltic achondrites. They are particularly abundant in Martian meteorites where they are host to high-pressure minerals (Beck et al. 2004; Fig. 11). A special case is the Antarctic meteorite Elephant Moraine 79001, a Martian basalt containing unusually large melt pockets that have been described as a specific lithology (Fig. 11). The concentration of rare gases in this lithology, in abundance similar to Martian atmosphere, has been a strong argument in favor of a Martian origin for the SNC meteorites (Bogard and Johnson 1983). The formation of melt pockets has been explained by energy dissipation when the shock wave encounters impedance contrasts, at the interface between distinct minerals (Stoffler et al. 1991) or if voids were present (Beck et al. 2007). Defocused microprobe analysis of melt pockets in the NWA 856 shergottite revealed a composition intermediate between maskelynite and pyroxene (Jambon et al. 2002) indicated that both mineral can melt during the formation of melt pockets in shergottite.

Transported glass fragments and glass beads in extra-terrestrial regolith and regolith breccias

On planetary bodies, impact breccias have both a deep and vast "drainage basin" and can be seen as high priority target to sample the compositional diversity of surface rocks. The Apollo missions only sampled a few locations on the moon and detailed attention was given to soils and impact breccia. They often contained glass spherules or fragments originating from distant area of the Moon, and may sample original lithologies from the lunar surface. The study of lunar impact-melt and glass-bearing breccias also provides chronological clues to the lunar cratering record (Fernandes et al. 2013; Zellner et al. 2019). Within the lunar regolith glasses are present in the form of agglutinates (local melting of the regolith due to micrometeorites impact, Cintala 1992), coatings on grain (post-impact vapor deposit), or larger fragments deposited ballistically (μm–mm). Glassy agglutinates and surface coatings have been particularly investigated to understand the darkening of the lunar surface with age (so called space weathering, Hapke 2001, Pieters et al. 1993).

Ballistically implemented glass fragment are abundant in the lunar regolith and come in many different colors and composition (Chao et al. 1970; Fig. 12). Their composition in major and trace elements has been used to fingerprint their source lithologies. A major challenge (and debate) in the study of lunar glasses is to be able to distinguish impact-related glasses from those of volcanic glasses, referred to as pristine glasses. Pristine glasses are defined by the combination of several chemical and petrographical properties including (non exhaustively) the absence of schlieren and exotic clast/inclusions, intrasample chemical homogeneity and intersample homogeneity or crystal/liquid fractionation trends. For a full description of pristine glasses, the reader is referred to Delano (1986).

Impact related glasses are the dominant fraction of Apollo 16 soils (95 %) and were classified in six groups by Ziegler et al. (2006). Major element compositions of such glasses are provided in Table 3. They are generally mafic and alkali poor (Fig. 9). A peculiar type of glass was also identified by Naney et al. (1976) in Apollo 16 samples, as unusually rich in Al (31–36 wt. %) and depleted in Si (30–34 wt. %) the so-called HASP glasses (High-Alumina Silica Poor, see Fig. 9 and Table 3). Similarities between HASP major element composition and Ca-Al-rich inclusions from meteorites were initially noted (Naney et al. 1976), but the preferred model for producing these peculiar glasses is fractional vaporization of a basaltic liquid, a process likely at stake in lunar impact glass formation in general (Naney et al. 1976; Delano et al. 1981).

The howardite-eucrite-diogenite meteorite suite is a large family of extra-terrestrial rocks originating from a large differentiated asteroid (4-Vesta). Among those rocks, howardite are breccias that often contain implanted solar wind gases testifying of a regolithic nature (40 %, Bischoff 2006). Howardites (and less frequently polymict eucrite) contains impact melt clasts and impact glass reminiscent of observations of the lunar surface (Noonan 1974; Labotka

Table 3: Composition of various impact related melts.

	n	SiO2	TiO2	Al2O3	Cr2O3	Fe2O3	FeO	FeOTot	MnO	NiO	MgO	CaO	Na2O	K2O	P2O5	LOI	H2O	S	Total	Ref
HED Impact Spherules and Glasses																				
Barrat et al. (2009a)																				
Kapoeta. spherules (S) or glassy fragments (F)																				
BR3. S		50.36	0.43	9.13	0.67		18.32		0.51		14.21	7.25	0.36	0.03	0.04				101.31	[1]
DC21. F		51.57	0.35	6.7	0.65		17.47		0.52		17.83	5.66	0.22	0.01	0.01				100.99	[1]
Bununu. spherules (S) or glassy fragments (F)																				
B2. S		50.83	0.44	8.91	0.61		17.46		0.59		14.43	7.13	0.25	0.04	0.04				100.73	[1]
B3. F		49.73	0.74	12.58	0.26		18.84		0.62		7.42	10.86	0.39	0.08	0.05				101.57	[1]
Y-7308. spherules (S) or glassy fragments (F)																				
85-4. 1 F		52.35	0.17	4.4	1.21		15.58		0.5		21.9	3.49	0.07	0.12					99.79	[1]
104-2. 1 S		53.83	0.21	4.5	1.24		15.04		0.56		20.7	3.68	0.13	0.08					99.97	[1]
NWA 1664. fusion crust. spherules (S) or glassy fragment (F)																				
1664P. 5 F		48.69	0.86	12.48	0.28		19.82		0.6		6.04	11.12	0.25	0.35	0.1				100.59	[1]
1664B1. 1 F		49.68	0.58	11.78	0.44		17.74		0.53		9.23	9.97	0.21	0.44	0.04				100.64	[1]
1664B5.3S		48.43158	0.56724	9.70466	0.59598		17.7351		0.54866		13.17774	8.40882	0.0807	0.46146	0.00586				99.7178	[1]
1664B6.1S		53.57	0.4	10.1	0.47		11.46		0.4		16.89	6.18	0.14	1.06	0.01				100.68	[1]
NWA 1664. composite spherule (1664B1.1S)																				
mafic glasses																				
1664B1. 1 S. core		51.1	0.73	13.38	0.45		13.17		0.47		8.34	10.91	0.2	1.41	0.01				100.17	[1]
1664B1. 1 S. rim		49.82	0.72	13.17	0.43		15.52		0.54		7.72	10.82	1.21	0.06	0.07				100.08	[1]
felsic glass																				
high K	11	66.41	1.04	17.87	0.26		1.97		0.08		0.21	4.99	0.56	5.8	0.01				99.2	[1]
low K	2	65.36	1.11	18.96	0.49		1.96		0.05		0.19	7.04	0.61	3.95	0.00				99.72	[1]

	n	SiO$_2$	TiO$_2$	Al$_2$O$_3$	Cr$_2$O$_3$	Fe$_2$O$_3$	FeO	FeOTot	MnO	NiO	MgO	CaO	Na$_2$O	K$_2$O	P$_2$O$_5$	LOI	H$_2$O	S	Total	Ref
NWA 1769, spherule with an orthopyroxene nucleus (1769C, 2S)																				
glass		49.62	0.62	11.72	0.54		16.75		0.61		9.53	9.14	0.16	1.21	0.05				99.95	[1]
Opx		54.68	0.06	0.73	0.61		15.02		0.5		27.13	1.14	0.01	0.00	0.02				99.9	[1]
Lunar impact glasses																				
Zeigler et al. (2006). Naney et al. (1976)																				
Low-Ti basaltic glass	41	45.77	4.06	10.06	0.3		20.23		0.26		8.69	9.57	0.28	0.13	0.07				99.42	[2]
High-Ti basaltic glass	17	41.53	8.58	10.49	0.44		17.88		0.25		8.98	10.9	0.34	0.05	0.09				99.53	[2]
Basaltic-andesite glass	24	52.53	3.58	12.9	0.13		13.28		0.18		4.82	9.1	0.97	0.68	0.51				98.68	[2]
High-Al basaltic glass	27	44.2	3.89	14.6	0.35		14.64		0.19		10.22	11	0.49	0.09	0.11				99.78	[2]
"KREEPy"		50.79	2.47	15.47	0.18		11.24		0.15		7.87	9.89	0.73	0.52	0.18				99.49	[2]
High alumina silica poor (HASP)	53	31.91	0.82	34.16	0.03		4.98				8.49	19.52	0.13						100.04	[3]
Earth Impact glasses																				
Compilation by Dressler and Reimold (2001)																				
Ries crater		63.71	0.8	14.98		1.37	3.02		0.1		2.21	3.74	3.45	3.44	0.33		2.7		99.85	[4]
Haughton crater		64.35	0.02	19.27	0.01		0.14		0.02	0.01	0.04	0.05	1.15	11.59	0.07				96.72	[5]
Popigai crater		65.41	0.63	13.11				5.76			2.77	2.6	1.88	2.73	0.07				94.96	[6]
Mistatin crater	58	53.39	0.97	21.56	0.04			5.94	0.06		1.4	8.31	5.05	1.13					97.85	[7]

	n	SiO₂	TiO₂	Al₂O₃	Cr₂O₃	Fe₂O₃	FeO	FeOTot	MnO	NiO	MgO	CaO	Na₂O	K₂O	P₂O₅	LOI	H₂O	S	Total	Ref
Sudbury. Onaping formation																				
Gray	610	62.23	0.84	13.72	0.03		4.98		0.16		3.39	4.89	4.45	2.58	0.07				97.34	[8]
Black	80	56.81	0.84	9.26	0.04		11.12		0.36		8.1	6.92	2.46	2.32	0.16				98.39	[9]
Wanapitei crater																				
		79.5	0.21	9.76	0.01			1.62	0.01		0.81	0.46	1.88	2.46	0.07				96.79	[10]
		66.18	0.35	16.41	0.02			3.4	0.02		1.57	0.86	3.11	3.17	0.09				95.18	[10]
		75.15	0.42	11.38				2.09	0.04		1.11	0.39	2.67	3.92	0.05				97.22	[10]
		79.95	0.08	9.3				0.84			0.57	0.43	2.62	3.14					96.93	[10]
Earth Impact melts																				
Compilation by Dressler and Reimold (2001)																				
Popigai crater																				
	163	63.17	0.73	14.54				6.74			3.38	3.7	2.29	2.71	0.08				97.34	[6]
Mistatin crater																				
	40	57.8	0.98	19.59		6.42			0.11		1.52	7.75	3.86	2	0.31				100.34	[11]
Vredefort crater																				
	28	66.92	0.5	12.65		7.26			0.14		3.56	3.89	2.68	2.25	0.11	0.07			100.03	[12]
Morokweng crater																				
	45	64.58	0.49	13.44		5.9			0.06		3.73	3.41	3.88	2.15	0.12	2.01			99.77	[13]
Manicouagan crater																				
	24	57.75	0.77	16.51		3.98	2.29		0.11		3.5	5.92	3.82	3.03	0.22		1.73	0.01	99.64	[14]
Sudbury. Onaping formation																				
	79	68.01	0.55	13.28			4.63	0.1	2.56		2.02	3.66	2.38	0.05	0.22		1.65	0.06	99.17	[8]
Wanapitei crater																				
		74.05	0.19	12.32		0.57	1.25		0.02		0.97	0.83	2.92	2.8					95.92	[15]
		74.8	0.2	11.62				2.3			1.62	1.3	2.62	2.27					96.73	[15]
		76.25	0.2	10.82		0.62	1.8		0.03		1.83	1.31	1.77	2.35					96.98	[15]

n	SiO$_2$	TiO$_2$	Al$_2$O$_3$	Cr$_2$O$_3$	Fe$_2$O$_3$	FeO	FeOTot	MnO	NiO	MgO	CaO	Na$_2$O	K$_2$O	P$_2$O5	LOI	H$_2$O	S	Total	Ref
Tektites and Lybian desert glasses and Atacamaites																			
Compilation by Dressler and Reimold (2001) and from Fudali (1981) and Gattacceca et al. (2021)																			
Australasian																			
Australites	73.45	0.69	11.53				4.69			2.05	3.5	1.28	2.28					99.47	[16]
Microtektite	72	0.8	13.6				4.9			2.4	3.6	0.6	1.9					99.8	[16]
Muon Nong	78.3	0.63	10.18				3.75	0.08		1.43	1.21	0.92	2.41					98.91	[17]
North American																			
Bediasites	76.37	0.76	13.78				3.98			0.63	0.65	1.54	2.08					99.79	[18]
Georgiaites	81.5	0.49	10.71				2.5			0.55	0.51	1.19	2.39					99.84	[18]
Microtektites	70.7	0.8	15.4				5			1.77	1.61	1.05	3.02					99.35	[16]
Ivory Coast																			
Tektites	67.58	0.56	16.74				6.16	0.06		3.46	1.38	1.9	1.95					99.79	[19]
Microtektites	65.6	0.7	15.6				6.9			4.6	1.4	1.9	1.8					98.5	[16]
Central European																			
Moldavites	79.57	0.34	9.55				1.72	0.07		1.74	2.7	0.38	3.36					99.43	[20]
Lybian desert glasses																			
	98.2	0.23	0.7		0.53	0.24			0.02	0.01	0.3	0.33	0.02					100.58	[21]
Atacamaites																			
	64.3	0.52	12.72			9.09		0.08		1.91	4.45	3.58	2.91	0.1				99.66	[22]
	66.4	0.53	12.99			6.88		0.08		1.84	4.32	3.6	3.06	0.09				99.79	
	60.1	0.54	13.28			11.89		0.08		1.45	5.57	3.54	2.94	0.07				99.46	

Notes: [1] Barrat et al. (2009a) [2] Ziegler et al. (2006) [3] Naney et al. (1976) [4] Engelhardt and Hörz (1965) and Engelhardt (1967) [5] Dressler and Sharpton (1998) [6] Masaitis 1994 [7] Grieve (1975) [8] Dressler et al. (1996) [9] Muir and Peredery (1984) [10] Dressler et al. (1997) [11] Marchand and Crocket (1977) [12] Therriault et al. (1997) [13] Dressler and Reimold (2001) [14] Grieve and Floran (1978) [15] Grieve and Ber (1994) [16] Koeberl (1990) [17] Koeberl (1992) [18] Koeberl and Glass (1988) [19] Koeberl et al. (1998) [20] Engelhardt et al. (1987) [21] Fudali (1981) [22] Gattacceca et al. (2021).

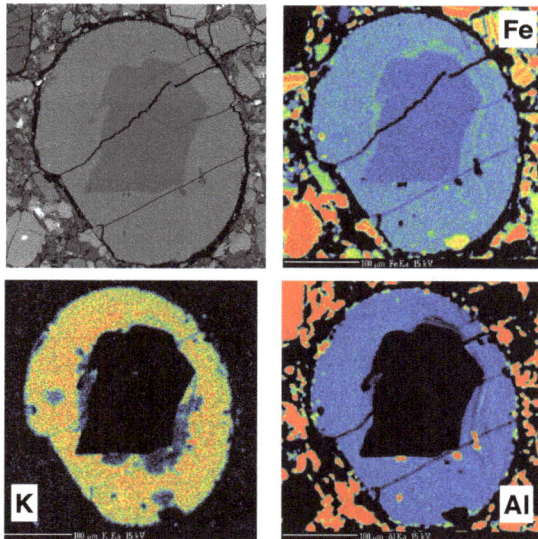

Figure 12. Impact spherule in the howardite Northwest Africa 1769. It is often not obvious to distinguish "pristine glass" of volcanic origins, from impact related glasses. One of the criteria to identify impact related spherule is the presence of clasts with the spherule, as can be seen in these SEM and composition maps. The impact spherule hosts a large grain of pyroxene, that can be identified based on the presence of iron and absence of K, Al. Small, Al-rich grains are also present within the spherule that are small feldspar grains present in the spherule as well. The Fe-rich aureole around the pyroxene grains reflects chemical interaction between the pyroxene crystal and the silicate liquid.

and Papike 1980; Fuhrman and Papike 1981; Mittlefehldt and Lindstrom 1997; Barrat et al. 2009a,b). The major element composition of these glasses reveals the presence of two groups. The first group is similar in major element composition to howardite and eucrite (i.e., low-K, Mg # 41-72) and probably originates from melting of "classical" HED. The second group if however rich in K (K_2O up to 2 wt.%) and does not have equivalent rock composition in the HED meteorite collection. This emphasizes the interest of studying glass fragments in meteoritic breccias, which seems to present a larger diversity in the nature of the lithologies that can be sampled when compared to larger rocks.

Prospects

Our knowledge of the cratering processes is somehow limited by the lack of a direct observation of a large impact. Still, knowledge will be gained in the next decades by new observations of impact structures (detailed geological work, drilling) and identification of new impact structures (for instance the recent identification of the Hiawatha crater in Greenland; Kjaer et al. 2016). In addition, applications of new geochemical techniques to impactites, and the development of original numerical and laboratory experiments will continue to improve our understanding of impact melts and glass formation.

In the case of larger impacts (crater diameter >100 m), the timescale of the pressure pulse is significantly longer than what can be achieved today through dynamic compression experiments. The high strain rates and the very fast increase in pressure experienced by compressed material upon passage of the shock wave may lead to specific transformation pathway of minerals (i.e., the comparison to static experiments is complex). The design of new experiments trying to reproduce both the strain rate and the duration of the pressure pulse (10^{-3}–1 s) may help understanding the mechanism at stake during the formation of high-pressure minerals. A major novelty in the field of dynamic compression is the development of time-resolved shock wave experiment. While such experiment does not reproduce the duration of the pressure pulse in a large impact, they are remarkable and exciting. By using pulsed X-ray source of high-brightness (synchrotron or free-electron lasers), it is now possible to observe the temporal evolution of matter during a dynamic compression (Radousky et al. 2021). Such experiments may help understand high-pressure phase formation, phase-transition kinetics, melting mechanisms, and dense glass formation upon shock compression.

Impact melts and glasses are extremely precious since they can be used to date impact crater structure (and to connect tektite formation to a specific impact crater). Also, more recently, efforts have been made to connect these impact ages to the formation of asteroids families, and as such to trace the source region of the impactors among asteroids (Bottke et al 2007; Masiero et al. 2011). Today, ages of asteroid family are somehow model dependent and have some uncertainty. The ongoing GAIA mission (ESA) is currently providing asteroid orbits with a high accuracy for an immense set of objects. This mission will unable us to identify and date new asteroids families, and to define more accurately existing asteroid families. Such an approach may help understand the collisional history of the asteroid belt, and then the cratering record of terrestrial planets. In parallel, we should pursue approaches that enable to date the shock event experienced by meteorites (Li and Hsu 2018; Amsellem et al. 2020), to provide further constrains on the dynamical evolution of the Solar System.

ACKNOWLEDGEMENTS

The authors thank the editors of this issue for giving them the opportunity to show that extraterrestrial materials and objects offer a wide variety of igneous processes and intellectual challenges which, it is hoped, new generations will not be afraid to tackle.

REFERENCES

Alexander CMOD, Grossman JN, Ebel DS, Ciesla FJ (2008) The formation conditions of chondrules and chondrites. Science 320:1617–1619

Alexander CMOD, Ebel DS (2012) Questions, questions: can the contradictions between the petrologic, isotopic, thermodynamic and astrophysical constraints on chondrule formation be resolved? Meteorit Planet Sci 47:1157–1175

Amelin Y, Kaltenbach A, Iizuka T, Stirling CH, Ireland TR, Petaev M, Jacobsen SB (2010) U–Pb chronology of the solar system's oldest solids with variable $^{238}U/^{235}U$ Earth Planet Sci Lett 300:343–350

Amsellem E, Moynier F, Mahan B, Beck P (2020) Timing of thermal metamorphism in CM chondrites: Implications for Ryugu and Bennu future sample return. Icarus 339:113593

Barosch J, Hezel DC, Ebel DS, Friend P (2019) Mineralogically zoned chondrules in ordinary chondrites as evidence for open system chondrule behaviour. Geochim Cosmochim Acta 249:1–16

Barrat JA, Jahn BM, Amossé J, Rocchia R, Keller F, Poupeau GR, Diemer E (1997) Geochemistry and origin of Libyan Desert glasses. Geochim Cosmochim Acta 61:1953

Barrat JA, Jambon A, Bohn M, Blichert-Toft J, Sautter V, Gopel C, Gillet P, Boudouma O, Keller F (2003) Petrology and geochemistry of the unbrecciated achondrite Northwest Africa 1240 (NWA 1240): an HED parent body impact melt. Geochim Cosmochim Acta 67:3959–3970

Barrat JA, Yamaguchi A, Greenwood RC, Bohn M, Cotten J, Benoit M, Franchi IA (2007) The Stannern trend eucrites: Contamination of main group eucritic magmas by crustal partial melts. Geochim Cosmochim Acta 71:4108–4124

Barrat JA, Yamaguchi A, Greenwood RC, Benoit M, Cotten J, Bohn M, Franchi IA (2008) Geochemistry of diogenites: still more diversity in their parental melts. Meteorit Planet Sci 43:1759–1775

Barrat J-A, Yamaguchi A, Greenwood RC, Bollinger C, Bohn M, Franchi IA (2009a) Trace element geochemistry of K-rich impact spherules from howardites. Geochim Cosmochim Acta 73:5944

Barrat JA, Bohn M, Gillet P, Yamaguchi A (2009b) Evidence for K-rich terranes on Vesta from impact spherules. Meteorit Planet Sci 44:359

Barrat JA, Yamaguchi A, Bunch TE, Bohn M, Bollinger C, Ceuleneer G (2011) Possible fluid–rock interactions on differentiated asteroids recorded in eucritic meteorites. Geochim Cosmochim Acta 75:3839–3852

Barrat JA, Chaussidon M, Yamaguchi A, Beck P, Villeneuve J, Byrne DJ, Broadley MW, Marty B (2021) A 4665 Myr old andesite from an extinct chondritic protoplanet. PNAS 118: e2026129118

Beck P, Gillet P, Gautron L, Daniel I, El Goresy A (2004) A new natural high-pressure (Na,Ca)-hexaluminosilicate $[(Ca_xNa_{1-x})Al^{3+}_xSi_{3-x}O_{11}]$ in shocked Martian meteorites. Earth Planet Sci Lett 219:1–12

Beck P, Ferroir T, Gillet P (2007) Shock-induced compaction, melting, and entrapment of atmospheric gases in Martian meteorites. Geophys Res Lett 34:L01203

Benedix GK, Bland PA, Friedrich JM, Mittlefehldt DW, Sanborn ME, Yin QZ, Greenwood RC, Franchi IA, Bevan AW, Towner MC, Perrotta GC (2017) Bunburra Rockhole: Exploring the geology of a new differentiated asteroid. Geochim Cosmochim Acta 208:145–159

Beran A, Koeberl C (1997) Water in tektites and impact glasses by FTIR spectrometry. Meteorit Planet Sci 32:211

Binns RA (1969) Ringwoodite, natural $(Mg,Fe)_2SiO_4$ spinel in the tenham meteorite. Nature 221:943

Bischoff A, Scott ERD, Metzler K, Goodrich CA (2006) Nature and origins of meteoritic breccias. *In:* Meteorites and the Early Solar System II. Lauretta DS, McSween HY (Eds), Univ Arizona Press, p 679–712

Bischoff A, Horstmann M, Barrat JA, Chaussidon M, Pack A, Herwartz D, Ward D, Vollmer C, Decker S (2014) Trachyandesitic volcanism in the early Solar System. PNAS 111:12689–12692

Bogard DD, Johnson P (1983) Martian gases in an antarctic meteorite? Science 221:651

Bollard J, Connelly JN, Bizzarro M (2015) Pb-Pb dating of individual chondrules from the CBa chondrite Gujba: assessment of the impact plume formation model. Meteorit Planet Sci 50:1197–1216

Bottke WF, Vokrouhlický D, Nesvorný D (2007) An asteroid breakup 160 Myr ago as the probable source of the K/T impactor. Nature 449:48

Carporzen L, Weiss BP, Elkins-Tanton LT, Shuster DL, Ebel D, Gattacceca J (2011) Magnetic evidence for a partially differentiated carbonaceous chondrite parent body. PNAS 108:6386–6389

Chen M, El Goresy A (1999) The nature of "maskelymite" in shocked meteorites: not diaplectic glass but glass quenched from shock-induced dense melt at high pressures. Meteorit Planet Sci Suppl 34:A24

Cicconi MR, McCloy JS, Neuville DR (2022) Non-magmatic glasses. Rev Mineral Geochem 87:965–1014

Cintala MJ (1992) Impact-induced thermal effects in the lunar and mercurian regoliths. J Geophys Res 97:947

Collinet M, Grove TL (2020) Widespread production of silica- and alkali-rich melts at the onset of planetesimal melting. Geochim Cosmochim Acta 277:334–357

Connelly JN, Bizzarro M, Krot AN, Nordlund Å, Wielandt D, Ivanova MA (2012) The absolute chronology and thermal processing of solids in the solar protoplanetary disk. Science 338:651–655

Connolly HC, Jones RH (2016) Chondrules: The canonical and noncanonical views: A review of chondrule formation. J Geophys Res: Planets 121:1885–1899

Consolmagno GJ, Drake MJ (1977) Composition and evolution of the eucrite parent body; evidence from rare earth elements. Geochim Cosmochim Acta 41:1271–1282

Crossley SD, Lunning NG, Mayne RG, McCoy TJ, Yang S, Humayun M, Ash RD, Sunshine JM, Greenwood RC, Franchi IA (2018) Experimental insights into Stannern-trend eucrite petrogenesis. Meteorit Planet Sci 53:2122–2137

Dauphas N, Chaussidon M (2011) A perspective from extinct radionuclides on a young stellar object: the sun and its accretion disk. Annu Rev Earth Planet Sci 39:351–386

Day JM, Ash RD, Liu Y, Bellucci JJ, Rumble III D, McDonough WF, Walker RJ, Taylor LA (2009) Early formation of evolved asteroidal crust. Nature 457:179–182

Delano JW (1986) Pristine lunar glasses: criteria, data, and implications. J Geophys Res 91:D201

Desch SJ, Connolly, Jr. HC (2002) A model of the thermal processing of particles in solar nebula shocks: Application to the cooling rate of chondrules. Meteorit Planet Sci 37:183–207

Desch SJ, Morris MA, Connolly Jr. HC, Boss AP (2012) The importance of experiments: Constraints on chondrule formation models. Meteorit Planet Sci 47:1139–1156

Downes H, Mittlefehldt DW, Kita NT, Valley JW (2008) Evidence from polymict ureilite meteorites for a disrupted and re-accreted single ureilite parent asteroid gardened by several distinct impactors. Geochim Cosmochim Acta 72:4825–4844

Dressler BO, Reimold WU (2001) Terrestrial impact melt rocks and glasses. Earth Science Reviews 56 (205)

Dressler BO, Weiser T, Brockmeyer P (1996) Recrystallized impact glasses of the Onaping Formation and the Sudbury Igneous Complex, Sudbury Structure, Ontario, Canada. Geochim Cosmochim Acta 60:2019

Dressler BO, Crabtree D, Schuraytz DC (1997) Incipient melt formation and devitrification at the Wanapitei impact structure, Ontario, Canada. Meteorit Planet Sci 32:249

Dressler BO, Sharpton VL (1998) Coexisting pseudotachylite and rock glasses at the Haughton Impact Crater, Canada. Lunar Planet Sci Conf, p 1384

Duke MB, Silver LT (1967) Petrology of eucrites, howardites and mesosiderites. Geochim Cosmochim Acta 31:1637–1665

Ebel DS, Alexander CMO, Libourel G (2018) Vapor–melt exchange: constraints on chondrite formation conditions and processes. *In:* Chondrules: Records of Protoplanetary Disk Processes Cambridge Planetary Science. Cambridge University Press, p 151–174

El Goresy A, Gillet P, Miyahara M, Ohtani E, Ozawa S, Beck P, Montagnac G (2013) Shock-induced deformation of Shergottites: Shock-pressures and perturbations of magmatic ages on Mars. Geochim Cosmochim Acta 101:233

Elkins-Tanton LT, Weiss BP, Zuber MT (2011) Chondrites as samples of differentiated planetesimals. Earth Planet Sci Lett 305:1–10

Engelhardt WV, Bertsch W (1969) Shock induced planar deformation structures in quartz from the Ries crater, Germany. Contrib Mineral Petrol 20 (203)

Engelhardt WV, Hörz F (1965) Riesgläser und Moldavite. Geochim Cosmochim Acta 29:609

Engelhardt WV, Arndt J, Stöffler D, Müller WF, Jeziorkowski H, Gubser RA (1967) Diaplektische Gläser in den Breccien des Ries von Nördlingen als Anzeichen für Stoßwellenmetamorphose. Contrib Mineral Petrol 15:93

Engelhardt WV, Luft E, Arndt J, Schock H, Weiskirchner W (1987) Origin of moldavites. Geochim Cosmochim Acta 51:1425

Fernandes VA, Fritz J, Weiss BP, Garrick-Bethell I, Shuster DL (2013) The bombardment history of the Moon as recorded by $^{40}Ar–^{39}Ar$ chronology. Meteorit Planet Sci 48:241

Floran RJ, Grieve RAF, Phinney WC, Warner JL, Simonds CH, Blanchard DP, Dence MR (1978) Manicouagan impact melt, Quebec. 1. Stratigraphy, petrology, and chemistry. J Geophys Res 83:2737

Friend P, Hezel DC, Mucerschi D (2016) The conditions of chondrule formation, part II: open system. Geochim Cosmochim Acta 173:198–209

Fu RR, Elkins-Tanton LT (2014) The fate of magmas in planetesimals and the retention of primitive chondritic crusts. Earth Planet Sci Lett 390:128–137

Fuhrman M, Papike JJ (1981) Howardites: samples of the regolith of the eucrite parent-body petrology of Bholgati, Bununu, and Kapoeta. Lunar Planet Sci Conf, p 309

Gattacceca J, Devouard B, Barrat J-A, Rochette P, Balestrieri ML, Bigazzi G, Ménard G, Moustard F, Dos Santos E, Scorzelli R, Valenzuela M, Quesnel Y, Gounelle M, Debaille V, Beck P, Bonal L, Reynard B, Warner M (2021) A 650 km². Miocene strewnfield of splash-form impact glasses in the Atacama Desert, Chile. Earth Planet Sci Lett 569:117049

Gillet P, El Goresy A, Beck P, Chen M (2007) High-pressure mineral assemblages in shocked meteorites and shocked terrestrial rocks: Mechanisms of phase transformations and constraints to pressure and temperature histories. *In:* Advances in High-Pressure Mineralogy, Ohtani E (Ed) Geol Soc Am Spec Pap 421, p 57–82

Greeney S, Ruzicka A (2004) Relict forsterite in chondrules: implications for cooling rates. Lunar Planet Sci XXXV, #1426

Greenwood RC, Franchi IA, Jambon A, Buchanan PC (2005) Widespread magma oceans on asteroidal bodies in the early Solar System. Nature 435:916–918

Greenwood RC, Burbine TH, Miller MF, Franchi IA (2017). Melting and differentiation of early-formed asteroids: The perspective from high precision oxygen isotope studies. Chemie der Erde 77:1–43

Grieve RAF (1975) Petrology and chemistry of the impact melt at Mistastin Lake crater, Labrador. Geol Soc Am Bull 86:1617

Grieve RAF, Cintala MJ (1992) An analysis of differential impact melt–crater scaling and implications for the terrestrial impact record. Meteoritics 27:526

Grossman JN, Alexander CMO, Wang JH, Brearley AJ (2002) Zoned chondrules in Semarkona: evidence for high and low-temperature processing. Meteorit Planet Sci 37:49–73

Grossman L, Beckett JR, Fedkin AV, Simon SB, Ciesla FJ (2008) Redox conditions in the solar nebula: observational, experimental, and theoretical constraints. Rev Mineral Geochem 68:93–140

Hapke B (2001) Space weathering from Mercury to the asteroid belt. J Geophys Res 106:10039

Hemley RJ, Jephcoat AP, Mao HK, Ming LC, Manghnani MH (1988) Pressure-induced amorphization of crystalline silica. Nature 334:52

Hewins RH, Connolly HC Jr., Lofgren GE, Libourel G (2005) Experimental constraints on chondrule formation. *In:* Chondrites and the Protoplanetary Disk. AN Krot ER D Scott and B Reipurth (eds) Astron Soc Pacific Conf Ser 341, p 286–316

Hewins RH, Ganguly J, Mariani E (2009) Diffusion modeling of cooling rates of relict olivine in Semarkona chondrules 40th Lunar Planet Sci Conf, #1513

Hewins RH, Zanda B, Bendersky C (2012) Evaporation and recondensation of sodium in Semarkona Type II chondrules. Geochim Cosmochim Acta 78:1–17

Hu J, Sharp TG (2017) Back-transformation of high-pressure minerals in shocked chondrites: Low-pressure mineral evidence for strong shock. Geochim Cosmochim Acta 215:277

Humayun M, Koeberl C (2004) Potassium isotopic composition of Australasian tektites. Meteorit Planet Sci 39:1509

Jacquet E (2021) Collisions and compositional variability in chondrule-forming events. Geochim Cosmochim Acta 296:18–37

Jambon A, Barrat JA, Sautter V, Gillet P, Göpel C, Javoy M, Joron JL, Lesourd M (2002) The basaltic shergottite Northwest Africa 856 (NWA 856): Petrology and chemistry. Meteorit Planet Sci 37:1147

Jambon A, Barrat JA, Boudouma O, Fonteilles M, Badia D, Gopel C, Bohn M (2005) Mineralogy and petrology of the angrite Northwest Africa 1296. Meteorit Planet Sci 40:361–375

Jarosewich E (1990) Chemical analyses of meteorites : a compilation of stony and iron meteorites. Meteoritics 25:323–337

Jiang Y, Chen H, Fegley B, Lodders K, Hsu W, Jacobsen SB, Wang K (2019) Implications of K, Cu and Zn isotopes for the formation of tektites. Geochim Cosmochim Acta 259:170

Jones RH (2012) Petrographic on the diversity of chondrule reservoirs in the protoplanetary disk. Meteorit Planet Sci 47:1176–1190

Jones RH, Lofgren GE (1993) A comparison of FeO-rich, porphyritic olivine chondrules in unequilibrated chondrites and experimental analogues. Meteoritics 28:213–221

Jones RH, Villeneuve J, Libourel G (2018) Thermal histories of chondrules: Petrologic observations and experimental constraints. *In:* Chondrules and the Protoplanetary Disk SS. Russell HC Connolly Jr. AN Krot (Eds) Cambridge University Press, p 57–90

Kawabata H, Hanyu T, Chang Q, Kimura JI, Nichols ARL, Tatsumi Y (2011) The petrology and geochemistry of St. Helena alkali basalts: evaluation of the oceanic crust-recycling model for HIMU OIB J Petrol 52:791–838

Keil K (2012) Angrites, a small but diverse suite of ancient, silica-undersaturated volcanic-plutonic mafic meteorites, and the history of their parent asteroid. Chemie der Erde 72:191–218

Kjær KH, Larsen NK, Binder T, Bjørk AA, Eisen O, Fahnestock MA, Funder S, Garde AA, Haack H, Helm V, Houmark-Nielsen M (2018) A large impact crater beneath Hiawatha Glacier in northwest Greenland. Sci Adv 4:eaar8173

Koeberl C (1990) The geochemistry of tektites: an overview. Tectonophysics 171:405

Koeberl C (1992) Tektite origin by hypervelocity asteroidal or cometary impact: The quest for the source craters. *In*: Large Meteorite Impacts and Planetary Evolution. Dressier BO, Grieve RAF, Sharpton VL (eds) Geol Soc Am Spec Pap 293 p. 42

Koeberl C, Glass BP (1988) Chemical composition of North American microtektites and tektite fragments from Barbados and DSDP Site 612 on the continental slope off New Jersey. Earth Planet Sci Lett 87:286

Koeberl C, Shirey SB (1993) Detection of a meteoritic component in Ivory Coast tektites with rhenium–osmium isotopes. Science 261:595

Koeberl C, Reimold WU, Blum JD, Chamberlain CP (1998) Petrology and geochemistry of target rocks from the Bosumtwi impact structure, Ghana, and comparison with Ivory Coast tektites. Geochim Cosmochim Acta 62:2179

Krot AN (2019) Refractory inclusions in carbonaceous chondrites: Records of early solar system processes. Meteorit Planet Sci 39:1931–1955

Krot AN, Libourel G, Goodrich CA, Petaev MI (2004) Silica-rich igneous rims around magnesian chondrules in CR carbonaceous chondrites: Evidence for condensation origin from fractionated nebular gas. Meteorit Planet Sci 39:1931–1955

Krot AN, Scott ERD, Reipurth B (2005) Chondrites and the protoplanetary disk. *In:* Chondrites and the Protoplanetary Disk. Astron Soc Pacific vol 341

Krot AN, Nagashima K, Libourel G, Miller KE (2018) Multiple mechanisms of transient heating events in the protoplanetary disk: Evidence from precursors of chondrules and igneous Ca, Al-rich inclusions. *In:* Chondrules and the protoplanetary disk. Russell SS, Connolly HC Jr., Krot AN (Eds) Cambridge: Cambridge University Press, p 11–56

Kruijer TS, Touboul M, Fischer-Gödde M, Bermingham KR, Walker RJ, Kleine T (2014) Protracted core formation and rapid accretion of protoplanets. Science 344:1150–1154

Kruijer TS, Burkhardt C, Budde G, Kleine T (2017) Age of Jupiter inferred from the distinct genetics and formation times of meteorites. PNAS 114:6712–6716

Kruijer TS, Kleine T, Borg LE (2020) The great isotopic dichotomy of the early Solar System. Nat Astron 4:32–40

Labotka TC, Papike JJ (1980) Regolith of the eucrite parent-body petrology of the howardite meteorites, *In:* Lunar Planet Sci Conf, p 593

Langenhorst F (1994) Shock experiments on pre-heated α- and β-quartz: II X-ray and TEM investigations. Earth Planet Sci Lett 128:683

Langenhorst F, Poirier J-P (2000) Anatomy of black veins in Zagami: clues to the formation of high-pressure phases. Earth Planet Sci Lett 184:37

Lewis JS (1995) Physics and Chemistry of the Solar System. Academic Press

Li S, Hsu W (2018) Dating phosphates of the strongly shocked Suizhou chondrite. Am Mineral 103:1789

Libourel G, Portail M (2018) Chondrules as direct thermochemical sensors of solar protoplanetary disk gas. Sci Adv 4:eaar3321

Libourel G, Krot AN, Tissandier L (2003) Evidence for high temperature condensation of moderately volatile elements during chondrule formation. Lunar Planet Sci Conf 2003, #1558

Libourel G, Krot AN, Tissandier L (2006) Role of gas–melt interaction during chondrule formation. Earth Planet Sci Lett 251:232–240

Lofgren GE (1996) A dynamic crystallization model for chondrule melts. *In:* Chondrules and the protoplanetary disk. Hewins RH, Jones RH, Scott ERD (Eds) Cambridge University Press, p 187–196

Lofgren GE, Russell WJ (1986) Dynamic crystallization of chondrule melts of porphyritic and radial pyroxene composition. Geochim Cosmochim Acta 50:1715–1726

Lunning NG, Gardner-Vandy KG, Sosa ES, McCoy TJ, Bullock ES, Corrigan CM (2017) Partial melting of oxidized planetesimals: An experimental study to test the formation of oligoclase-rich achondrites Graves Nunataks 06128 and 06129. Geochim Cosmochim Acta 214:73–85

Marchand M, Crocket JH (1977) Sr isotopes and trace element geochemistry of the impact melt and target rocks at the Mistastin Lake crater, Labrador. Geochim Cosmochim Acta 41:1487

Marrocchi Y, Villeneuve J, Batanova V, Piani L, Jacquet E (2018) Oxygen isotopic diversity of chondrule precursors and the nebular origin of chondrules. Earth Planet Sci Lett 496:132–141

Marrocchi Y, Euverte R, Villeneuve J, Batanova V, Welsch B, Ferrière L, Jacquet E (2019) Formation of CV chondrules by recycling of amoeboid olivine aggregate-like precursors. Geochim Cosmochim Acta 247:121–141

Masaitis VL (1994) Impactites from Popigai crater. *In:* Large Meteorite Impacts and Planetary Evolution. Dressier BO, Grieve RAF, Sharpton VL (eds) Geol Soc Am Spec Pap 293, p 153–162

Masiero JRAK, Mainzer T, Grav JM, Bauer RM, Cutri J, Dailey PRM, Eisenhardt RS, McMillan TB, Spahr MF, Skrutskie D, Tholen RG, Walker EL, Wright E, DeBaun D, Elsbury T, Gomillion GS, Wilkins A (2011) Main Belt Asteroids with WISE/NEOWISE I Preliminary Albedos and Diameters. Astrophys J 741:68

Mathieu R, Khedim H, Libourel G, Podor R, Tissandier L, Deloule E, Faure F, Rapin C, Vilasi M (2008) Control of alkali-metal oxide activity in molten silicates. J Non-Cryst. Solids 354:5079–5083

Mathieu R, Libourel G, Deloule E, Tissandier L, Rapin C, Podor R (2011) Na_2O solubility in $CaO–MgO–SiO_2$ melts. Geochim Cosmochim Acta 75:608–628

Matsunami S, Ninagawa K, Nishimura S, Kubono N, Yamamoto I, Kohata M, Wada T, Yamashita Y, Lu J, Sears DWG, Nishimura H (1993) Thermoluminescence and compositional zoning in the mesostasis of a Semarkona A1 chondrule and new insights into the chondrule-forming process. Geochim Cosmochim Acta 57:2101–2110

McCarthy TS, Erlank AJ, Willis JP (1973) On the origin of eucrites and diogenites. Earth Planet Sci Lett 18:433–442

McSween HY, Mittlefehldt DW, Beck AW, Mayne RG, McCoy TJ (2011) HED meteorites and their relationship to the geology of Vesta and the Dawn mission. Space Sci Rev 63:141–174

Melosh HJ, Artemieva N (2004) How does tektite glass lose its water? *In:* Lunar Planet Sci Conf, p 1723

Michel P, DeMeo FE, Bottke WF (2015) Asteroids IV Tucson: University of Arizona Press

Mittlefehldt DW (1994) The genesis of diogenites and HED parent body petrogenesis. Geochim Cosmochim Acta 58:1537–1552

Mittlefehldt DW (2015) Asteroid (4) Vesta: I The howardite–eucrite–diogenite (HED) clan of meteorites. Chemie der Erde 75:155–183

Mittlefehldt DW, Lindstrom MM (1990) Geochemistry and genesis of the angrites. Geochim Cosmochim Acta 54:3209–3218

Mittlefehldt DW, Lindstrom MM (1997) Magnesian basalt clasts from the EET 92014 and Kapoeta howardites and a discussion of alleged primary magnesian HED basalts. Geochim Cosmochim Acta 61:453

Mittlefehldt DW, Killgore M, Lee MT (2002) Petrology and geochemistry of D'Orbigny, geochemistry of Sahara 99555, and the origin of angrites. Meteorit Planet Sci 37:345–369

Miyamoto M, Mikouchi T, Jones RH (2009) Cooling rates of porphyritic olivine chondrules in the Semarkona (LL3.00) ordinary chondrite: A model for diffusional equilibration of olivine during fractional crystallization. Meteorit Planet Sci 44:521–530

Moynier F, Koeberl C, Beck P, Jourdan F, Telouk P (2010) Isotopic fractionation of Cu in tektites. Geochim Cosmochim Acta 74:799

Muer TL, Peredery WV (1984) The Onaping Formation *In:* The Geology and Ore Deposits of the Sudbury Structure, p 139–210

Nagahara H, Kita NT, Ozawa K, Morishita Y (2008) Condensation of major elements during chondrule formation and its implication to the origin of chondrules. Geochim Cosmochim Acta 72:1442–1465

Naney MT, Crowl DM, Papike JJ (1976) The Apollo 16 drill core: statistical analysis of glass chemistry and the characterization of a high alumina-silica poor (HASP) glass. Lunar Planet Sci Conf Proceedings 1:155

Noonan AF, Rajan RS, Chodos AA (1974) Microprobe analyses of glassy particles from howardites. Meteoritics 9:385

Okuno M, Reynard B, Shimada Y, Syono Y, Willaime C (1999) A Raman spectroscopic study of shock-wave densification of vitreous silica. Phys Chem Mineral 26:304

Osinski GR, Grieve RA, Marion C, Chanou A (2012) Impact melting. *In:* Impact Cratering: Processes and Products: London, Blackwell Publishing, p 125–145

Pal DK, Moniot RK, Kruse TH, Herzog GF, Tuniz C (1982) Beryllium-10 in Australasian tektites - Evidence for a sedimentary precursor. Science 218:787

Pape J, Mezger K, Bouvier AS, Baumgartner LP (2019) Time and duration of chondrule formation: Constraints from $^{26}Al–^{26}Mg$ ages of individual chondrules. Geochim Cosmochim Acta 244:416–436

Pieters CM, Fischer EM, Rode O, Basu A (1993) Optical effects of space weathering: The role of the finest fraction. J Geophys Res 98:20817

Pfalzner S, Davies MB, Gounelle M, Johansen A, Münker C, Lacerda P, Veras D (2015) The formation of the solar system. Physica Scripta 90:068001

Radomsky PM, Hewins RH (1990) Formation conditions of pyroxene–olivine and magnesian olivine chondrules. Geochim Cosmochim Acta 54:3475–3490

Radousky HB, Armstrong MR, Goldman N (2021) Time resolved x-ray diffraction in shock compressed systems. J Appl Phys 129:040901

Reynard B, Okuno M, Shimada Y, Syono Y, Willaime C (1999) A Raman spectroscopic study of shock-wave densification of anorthite $CaAl_2Si_2O_8$ glass. Phys Chem Mineral 26:432

Righter K, Drake MJ (1997) A magma ocean on Vesta: core formation and petro-genesis of eucrites and diogenites. Meteorit Planet Sci 32:929–944

Rossano S, Balan E, Morin G, Bauer J-P, Calas G, Brouder C (1999) ^{57}Fe Mössbauer spectroscopy of tektites. Phys Chem Mineral 26:530

Rubin AE (2010) Physical properties of chondrules in different chondrite groups: Implications for multiple melting events in dusty environments. Geochim Cosmochim Acta 74:4807–4828

Rubin AE (2015) Maskelynite in asteroidal, lunar and planetary basaltic meteorites: An indicator of shock pressure during impact ejection from their parent bodies. Icarus 257:221

Russell CT, Raymond CA, Coradini A, McSween HY, Zuber MT, Nathues A, De Sanctis MC, Jaumann R, Konopliv AS, Preusker F, Asmar SW (2012) Dawn at Vesta : testing the protoplanetary paradigm. Science 336:684–686

Russell SS, Connolly Jr HC, Krot AN (2018) Chondrules: Records of Protoplanetary Disk Processes. Cambridge University Press

Ruzicka A, Snyder GA, Taylor LA (1997) Vesta as the howardite, eucrite and diogenite parent body: implications for the size of a core and for large-scale differentiation. Meteorit Planet Sci 32:825–840

Schiller M, Connelly JN, Glad AC, Mikouchi T, Bizzarro M (2015) Early accretion of protoplanets inferred from a reduced inner Solar System ^{26}Al inventory. Earth Planet Sci Lett 15:45–54

Scott ERD, Greenwood RC, Franchi IA, Sanders IS (2009) Oxygen isotopic con-straints on the origin and parent bodies of eucrites, diogenites, and howardites. Geochim Cosmochim Acta 73:5835–5853

Sears DWG, Huang S, Benoit PH (1996) Open-system behavior during chondrule formation *In*: RH Hewins RH Jones ERD Scott (Eds.), Chondrules and the Protoplanetary Disk, Cambridge University Press, Cambridge 1996, p 221–232

Sharp TG, de Carli PS (2006) Shock effects in meteorites. *In:* Meteorites and the Early Solar System II. Lauretta DS, McSween HY (Eds), Univ Arizona Press, p 653–677

Shaw CSJ, Walton E (2013) Thermal modeling of shock melts in Martian meteorites: Implications for preserving Martian atmospheric signatures and crystallization of high-pressure minerals from shock melts. Meteorit Planet Sci 48:758

Shearer CK, Burger PV, Neal C, Sharp Z, Spivak-Birndorf L, Borg L, Fernandes VA, Papike JJ, Karner J, Wadhwa M, Gaffney A (2010) Non-basaltic asteroidal melting during the earliest stages of solar system evolution. A view from Antarctic achondrites Graves Nunatak 06128 and 06129. Geochim Cosmochim Acta 74:1172–1199

Simonson BM, Glass BB (2004) Spherule layers records of ancient impacts. Annu Rev Earth Planet Sci 32:329

Soulié C, Libourel G, Tissandier L (2017) Olivine dissolution in molten silicates: An experimental study with application to chondrule formation. Meteorit Planet Sci 52:225–250

Srinivasan P, Dunlap DR, Agee CB, Wadhwa M, Coleff D, Ziegler K, Zeigler R, McCubbin FM (2018) Silica-rich volcanism in the early solar system dated at 4.565 Ga. Nat Commun 9:3036

Steel D (1998) Distributions and moments of asteroid and comet impact speeds upon the Earth and Mars. Planet Space Sci 46:473

Stewart ST, Carter PJ, Davies EJ, Lock SJ, Kraus RG, Root S, Petaev MI, Jacobsen SB (2019) Collapsing impact vapor plume model for chondrule and chondrite formation. 50th Lunar Planet Sci Conf 2019, # 2132

Stockdale SC (2020) Constraining the Cooling Rates of Chondrules. PhD thesis. The Open University

Stoeffler D, Keil K, Scott ERD (1991) Shock metamorphism of ordinary chondrites. Geochim Cosmochim Acta 55:3845

Stoffler D, Grieve RAF (2007) Impactites. Chapter 2.11 *In:* Metamorphic Rocks: A Classification and Glossary of Terms, Recommendations of the International Union of Geological Sciences. Fettes, D, Desmons J (Eds) Cambridge University Press, Cambridge, UK

Stoffler D, Langenhorst F (1994) Shock metamorphism of quartz in nature and experiment: I Basic observation and theory. Meteoritics 29:155

Stolper E (1977) Experimental petrology of eucritic meteorites. Geochim Cosmochim Acta 41:587–611

Tissandier L, Libourel G, Robert F (2002) Gas–melt interactions and their bearing on chondrule formation. Meteorit Planet Sci 37:1377–1389

Tschermak G (1872) Die Meteoriten von Schergotty und Gopal-. pur. Sitzber. Akad Wiss Wien Math Naturwiss Kl Abt 122–146

Usui T, Jones JH, Mittlefehldt DW (2015) A partial melting study of an ordinary (H) chondrite composition with application to the unique achondrite Graves Nunataks 06128 and 06129. Meteorit Planet Sci 50:759–781

Villeneuve J, Chaussidon M, Libourel G (2009) Homogeneous distribution of ^{26}Al in the solar system from the mg isotopic composition of chondrules. Science 325: 985–988

Villeneuve J, Libourel G, Soulié C (2015) Relationships between type I and type II chondrules: Implications on chondrule formation processes. Geochim Cosmochim Acta 160:277–305

Villiger S, Ulmer P, Muntener O, Thompson AB (2004) The liquid line of descent of anhydrous, mantle-derived, tholeiitic liquids by fractional and equilibrium crystallization-an experimental study at 1.0 GPa. J Petrol 45:2369.2388

Warren PH (1997) Magnesium oxide-iron oxide mass balance constraints and a more detailed model for the relationship between eucrites and diogenites. Meteorit Planet Sci 32:945–963

Warren PH (2012) Parent body depth–pressure–temperature relationships and the style of the ureilite anatexis. Meteorit Planet Sci 47:209–227

Warren PH, Jerde EA (1987) Composition and origin of Nuevo Laredo Trendeucrites. Geochim Cosmochim Acta 51:713–725

Warren PH, Rubin AE, Isa J, Gessler N, Ahn I, Choi BG (2014) Northwest Africa 5738: Multistage fluid-driven secondary alteration in an extraordinarily evolved eucrite. Geochim Cosmochim Acta 141:199–227

Weiss BP, Elkins-Tanton LT (2013) Differentiated planetesimals and the parent bodies of chondrites. Annu Rev Earth Planet Sci 41:529–560

Williams Q, Jeanloz R (1989) Static amorphization of anorthite at 300 K and comparison with diaplectic glass. Nature 338:413

Williams CD, Sanborn ME, Defouilloy C, Yin Q-Z, Kita NT, Ebel DS, Yamakawa A, Yamashita K (2020) Chondrules reveal large-scale outward transport of inner Solar System materials in the protoplanetary disk. PNAS 117:23426–23435

Wilson L, Keil K (2012) Volcanic activity on differentiated asteroids: a review and analysis. Chemie der Erde 72:289–322

Yamaguchi A, Taylor GJ, Keil K (1996) Global crustal metamorphism of the eucrite parent body. Icarus 124:97–112

Yamaguchi A, Clayton RN, Mayeda TK, Ebihara M, Oura Y, Miura YN, Haramura H, Misawa K, Kojima H, Nagao K (2002) A new source of basaltic meteorites inferred from Northwest Africa 011. Science 296:334–336

Yamaguchi A, Barrat JA, Greenwood RC, Shirai N, Okamoto C, Setoyanagi,T, Ebihara M, Franchi IA, Bohn M (2009) Crustal partial melting on Vesta: evidence from highly metamorphosed eucrites. Geochim Cosmochim Acta 73:7162–7182

Yamaguchi A, Barrat JA, Ito M, Bohn M (2011) Post eucritic magmatism on Vesta: Evidence from the petrology and thermal history of diogenites. J Geophysical Res-Planets 116:E08009

Yamaguchi A, Mikouchi T, Ito M, Shirai N, Barrat JA, Messenger S, Ebihara M (2013) Experimental evidence of fast transport of trace elements in planetary basaltic crusts by high temperature metamorphism. Earth Planet Sci Lett 368:101–109

Zeigler RA, Korotev RL, Jolliff BL, Haskin LA, Floss C (2006) The geochemistry and provenance of Apollo 16 mafic glasses. Geochim Cosmochim Acta 70:6050

Zellner NEB (2019) Lunar impact glasses: probing the moon's surface and constraining its impact history. J Geophys Res (Planets) 124:2686

Reviews in Mineralogy & Geochemistry
Vol. 87 pp. 919-963, 2022
Copyright © Mineralogical Society of America

Frictional Melting in Magma and Lava

Jackie E. Kendrick[1] and Yan Lavallée[2]

[1]School of Geosciences
University of Edinburgh
James Hutton Road
Edinburgh EH9 3FE
United Kingdom

[2]Department of Earth, Ocean and Ecological Sciences
University of Liverpool
4 Brownlow Street
Liverpool L69 3GP
United Kingdom

INTRODUCTION

The product of frictional melting of geomaterials is termed "pseudotachylyte". The name, first coined by Shand (1916), represents the visual similarity to the lava "tachylyte", being a dark aphanitic rock with a glassy appearance. Pseudotachylytes have been referred to by many names since their first identification, including trap-shotten gneiss (Holland 1900), hyalomylonite (Masch et al. 1985) and frictionite (Maddock 1986), the latter of which is still occasionally used. Controversy remains as to the precise defining characteristics of pseudotachylytes (Magloughlin and Spray 1992; Rowe et al. 2005; Spray 2010) and even to an extent, the mechanisms by which they form (e.g., Lin 2007). Here we adopt the definition of Spray (2010), i.e., that pseudotachylytes originate by frictional melting on a slip plane, comprising a formerly liquid matrix that may or may not host "survivor" mineral and/or lithic clasts.

Pseudotachylyte has now been identified in numerous settings, from impact craters on Earth (Reimold 1995; Spray 1998; Melosh 2005) and the moon (Christie et al. 1973), to tectonic faults (Sibson 1975; Swanson 1992; Curewitz and Karson 1999; Di Toro et al. 2009), basal slip surfaces of gravity slides (Hacker et al. 2014; Biek et al. 2019), landslides (Masch et al. 1985; Maddock 1986; Lin et al. 2001; De Blasio and Elverhøi 2008) and most recently, in active volcanic environments in avalanches (Bernard and van Wyk de Vries 2017), sector collapses (Legros et al. 2000; Hughes et al. 2020), caldera superfaults (Kokelaar 2007), the deposits of tumbling blocks in pyroclastic flows (Grunewald et al. 2000; Schwarzkopf et al. 2001) and in primary extrusive (Kendrick et al. 2012, 2014b) and explosive (Lavallée et al. 2015b) volcanic products. Despite the simultaneous occurrence of large stress-strain fluctuations as well as intense events of mass transfer (e.g., magma ascent, sector collapse, caldera collapse, block-and-ash flows) pseudotachylytes have rarely been noted in volcanic materials and more seldom still in the eruptive products of volcanic eruptions. However, recently there has been increased recognition that fault-like processes in the volcanic conduit contribute towards controlling eruption style (Cassidy et al. 2018) and as such, the causes and effects of frictional melting in volcanic rocks and magmas represent a research area in its infancy. The frequent and, as yet, unpredictable transition from effusive to explosive behavior is common to active composite volcanoes, yet our understanding of the processes which control this evolution is poor. The key to this catastrophic transition rests in the rheological response of the magma. Magmas are complex suspensions of crystals and gas bubbles in an amorphous silicate melt.

During ascent in volcanic conduits, magma evolves constantly due to changes in pressure and temperature (e.g., Sparks 1997) that drive gas exsolution (i.e., degassing) and crystallization (Cashman and Blundy 2000; Melnik and Sparks 2005). In particular, more silicic magmas evolve from a regime controlled by viscous flow to that where they may fracture and fragment (Papale 2001; Gonnermann and Manga 2013; Okumura and Kozono 2017), owing to increasing viscosity of the degassing, crystallizing magmatic suspension and the elastic response of the melt when subjected to stress variations shorter than the structural relaxation timescale (Dingwell and Webb 1989; Papale 1999). Such transient behavior dictates the volcano's ability to build pressure via the creation and destruction of permeable pathways: On one hand, when the viscosity of the melt is low enough to allow gas bubbles to rise buoyantly through the magma, and to accommodate the strain experienced during deformation, gas can escape passively, alleviating the potential explosiveness of an eruption (e.g., Burton et al. 2007; Gonnermann and Manga 2013); On the other hand, when the melt viscosity is too high and permeability is too low to accommodate the rate of gas exsolution or deformation, stress accumulates (e.g., Costa and Macedonio 2002; Edmonds 2008). Accumulated gas may rupture magma (Alidibirov and Dingwell 2000; Scheu et al. 2008; Arciniega-Ceballos et al. 2015), dissipating the stored energy (Alatorre-Ibargüengoitia et al. 2010). If sufficient energy remains to propel the particles from the conduit, failure may be accompanied by an explosive eruption (e.g., Gonnermann and Manga 2013).

The tendency for non-Newtonian suspensions to localize strain ensures that in the upper conduit of many composite volcanoes, plug flow prevails (e.g., Lensky et al. 2008; Calder et al. 2015; Okumura and Kozono 2017). High-viscosity magmas may evolve to form a magma plug, which may heighten the ability for gas pressure to build below the dense magma mass by closed-system degassing, enhancing the possibility of wholesale fragmentation and an explosive eruption (Diller et al. 2006; de' Michieli Vitturi et al. 2008; Wright and Weinberg 2009; Preece et al. 2016). Alternatively, if the conditions for magma failure are met over a protracted temporal or spatial scale, a shear zone or damage halo may flank the magma column (Tuffen and Dingwell 2005; Edmonds and Herd 2007; Kendrick et al. 2012; Lavallée et al. 2012b; Pallister et al. 2013; Wallace et al. 2019a), channeling outgassing in this region via anisotropic fracture networks within the marginal fault zone (Rust et al. 2004; Castro et al. 2012; Lavallée et al. 2013; Gaunt et al. 2014; Kendrick et al. 2014b; Plail et al. 2014). Changes in magma viscosity (Hale and Wadge 2008) or magma flux (Okumura and Kozono 2017) will perturb the stability of the system in such a scenario, and can result in a shift in eruption style (e.g., Denlinger and Hoblitt 1999; Diller et al. 2006).

Subtle changes in seismicity (Green and Neuberg 2006; Umakoshi et al. 2008; Thelen et al. 2011; Lamb et al. 2014, 2015), tilt and ground deformation (Chadwick Jr et al. 1988; Voight et al. 1998; Albino et al. 2011; Neuberg et al. 2018) and gas flux measurements (Edmonds 2008; Chiodini et al. 2016) are used to interpret the state of magma during an eruption. Magma failure (rupture) and slip releases characteristic low-frequency (LF) or long-period (LP) seismicity (Neuberg 2000; McNutt 2005; Chouet and Matoza 2013), a component of a spectrum of seismic signals released during unrest (McNutt 1996; Nakada et al. 1999). The repetitive nature of seismicity in the upper conduit has been directly correlated to the step-wise ascent of magma via stick-slip motion at a number of dome-building volcanoes (e.g., Voight et al. 1999; Moran et al. 2005; Lensky et al. 2008; Umakoshi et al. 2008; Costa et al. 2012; Lamb et al. 2015). Slip and traction along these conduit margin shear zones has been posited as the cause of cyclic inflation-deflation signals observed at differing timescales during volcanic eruptions (e.g., Green et al. 2006; Hale and Muhlhaus 2007; Lavallée et al. 2015b; Neuberg et al. 2018).

Examination of fault products in lava domes, exhumed conduits and lava spines attests to such fault-controlled ascent dynamics (Kendrick et al. 2012, 2014b; Kennedy and Russell 2012; Miwa et al. 2013; Pallister et al. 2013; Calder et al. 2015; Wallace et al. 2019a).

Deposits suggest the upper hundreds of meters of ascent may be controlled by conduit margin shear zones hosting a range of faulting-generated products, from gouge formed via comminution (Kennedy and Russell 2012; Pallister et al. 2013), injection veins of granular tuffisites (Tuffen et al. 2003; Castro et al. 2012; Berlo et al. 2013), to fault breccias (Rust et al. 2004; Castro et al. 2012; Hornby et al. 2015), cataclasites and unltracataclasites (Tuffen and Dingwell 2005; Cashman et al. 2008; Pallister et al. 2013), areas of viscous remobilization due to shear heating (Wallace et al. 2019a) and pseudotachylytes generated by frictional melting (Kendrick et al. 2012). In volcanoes, the possibility for frictional melting during fault slip may be particularly high as volcanoes offer uniquely favorable conditions for it, as: (1) the distinct elevated temperature conditions are "above" those of the Earth's geotherm; (2) the driving force of magma ascent is buoyancy, thus slip rate at conduit margins is critically controlled by external forcing; and (3) the presence of glass, an amorphous solid which undergoes a (rheological) thermo-kinetic transition (according to temperature and strain-rate) permits viscous remobilization at temperatures lower than the melting point of the mineral constituents of magmas.

In this chapter we build upon the recent paradigm that viscous magma ascent may be in-part controlled by fault processes along the conduit margin. We explore the occurrence of frictional melting in magmas and lavas during eruptive activity and other strain localization scenarios and provide details of frictional heating and melting during fault friction from the literature. These topics will be covered from the standpoint of experimental techniques, advances and results that have shaped our knowledge of frictional melting. Finally, we will reflect upon the implications of frictional melting in volcanic environments and consider the challenges and future directions in this line of study.

FAULT FRICTION

Identifying pseudotachylytes

Fault-generated pseudotachylyte has been described in a broad range of rock types, though the vast majority are found in crystalline plutonic or metamorphic rocks (>95%; Sibson and Toy 2006). Pseudotachylytes tend to be underrepresented in the rock record in comparison to the frequency of seismogenic fault slip events (Kirkpatrick and Rowe 2013). For many years the lack of pseudotachylytes was interpreted as proof that an elusive set of circumstances must be met to generate frictional melt (e.g., Rempel and Rice 2006; Sibson and Toy 2006), despite early theoretical propositions (e.g., McKenzie and Brune 1972) and experimental evidence that rock friction readily results in melting (e.g., Spray 1987; Tsutsumi and Shimamoto 1997; Di Toro et al. 2006). Whilst other dynamic fault weakening effects such as thermal pressurization of a liquid phase (water) may be invoked to explain the scarcity of pseudotachylytes (e.g., Rice 2006), recent works suggest that it is rather the post-formation fragility of pseudotachylytes that result in their underrepresentation (Kirkpatrick and Rowe 2013). In natural fault rocks, conditions such as temperature, strain rate and fluids likely fluctuate throughout formation and so cataclasites, mylonites, gouge and pseudotachylytes may repeatedly form and deconstruct in unison or in subsequent slip cycles (Magloughlin 1989; Swanson 1992; McNulty 1995; Di Toro and Pennacchioni 2005; Rowe et al. 2005; Lin 2007; Kim et al. 2010; Pittarello et al. 2012), and may also be altered, recrystallized, or hydrated (e.g., Camacho et al. 1995; Kirkpatrick and Rowe 2013). In volcanic settings pseudotachylytes may further be affected by viscous processes if they form in magmas that remain in the viscous field, i.e., with an interstitial melt phase that remains at temperatures above the glass transition (Kendrick et al. 2014b). In addition, faulted magmas will endure many hundreds of meters more slip during ascent to the surface, and the progressive shallowing of the slip surfaces at the conduit margin may change the dominant deformation mechanism from compactional to dilational (e.g., Lamb et al. 2015; Lavallée and Kendrick 2020), such that pseudotachylytes may be overprinted by fault gouge (Kendrick et al. 2012).

As pseudotachylytes form in a wide variety of host rocks there is no single defining feature that can be applied to classify them, and as such, identification of pseudotachylytes requires a combination of approaches. Whilst the classic view that pseudotachylyte forms glassy, black bands of millimeters to 10's of centimeters thickness and lateral extent of up to many 10's of meters is occasionally manifest, more often than not, such distinctive linear features have been erased or overprinted, and identifying pseudotachylyte is fraught with challenges (Kirkpatrick and Rowe 2013). At outcrop scale, pseudotachylytes can also be seen in lenses and irregularly shaped pockets that host substantial volumes of melt, and interstitial to angular-clast breccias that often occupy fault intersections or topological embayments (Sibson 1975). Pseudotachylytes may also be preserved as injection features (Kirkpatrick et al. 2009), even when the primary fault surface has been erased. In other instances, pseudotachylyte may only survive as fragments in reworked cataclasites, where clasts are susceptible to brittle damage, rounding, hydrothermal alteration and recrystallisation and thus are likely to be difficult to identify (Kirkpatrick and Rowe 2013).

In most shallow crustal settings frictional melting is followed by relatively fast cooling (10's of seconds; Mitchell et al. 2016), with the melt quenching to glass in rare instances (Philpotts 1964; Lin 1994b; Obata and Karato 1995; Goodwin 1999; Rowe et al. 2005) or rapidly crystallizing to form distinctive dark, aphanitic textures (Lin 1994a; Otsuki et al. 2003; Di Toro and Pennacchioni 2005). In rock types where the mineral assemblage is complex, and constituent minerals have contrasting melting temperatures, then the pseudotachylyte might be a result of selective melting of certain minerals (Magloughlin and Spray 1992; Spray 1992, 1993; Lin and Shimamoto 1998). Selective melting can also be enhanced by different strengths of component minerals, which likely plays a part in the early stages of frictional melting (Spray 2010). The differentiation imparted by selective melting can form pseudotachylytes which have distinct bulk chemical composition compared to their host rock (Philpotts 1964; Maddock 1992; O'Hara 1992; Lavallée et al. 2015b; Ishikawa and Ujiie 2019), thus geochemical techniques such as X-ray fluorescence may aid in their identification. Specifically, frictional melts may be enriched in iron compared to their host rocks (Petrík et al. 2003; Zhang et al. 2018) and several studies have revealed high magnetic susceptibility or remanence (Fukuchi 2003; Ferré et al. 2005, 2012) or distinct anisotropy of magnetic susceptibility (Molina Garza et al. 2009) of pseudotachylytes. As such, domain state analysis magnetic tests may also be important in identifying the occurrence of frictional melting and revealing the thermal history resulting from coseismic slip (Ferré et al. 2005, 2012; Freund et al. 2007; Pei et al. 2014; Yang et al. 2020).

Even in the case of complete melting of the mineral assemblage, frictional melts have been consistently shown to be enriched in certain trace elements (e.g., Maddock 1992; Killick 1994), which may be used to identify the source rock as well as the conditions of formation (e.g., O'Hara 1992). During cooling of pseudotachylytes, different mineral phases crystallize due to different pressure and/or temperature conditions from the formation of the original host rock (Camacho et al. 1995; Curewitz and Karson 1999; Barker 2005; Kendrick et al. 2012, 2014b; Pittarello et al. 2012; Price et al. 2012; Deseta et al. 2014). The mineral assemblage of recrystallized pseudotachylytes is commonly dominated by small, euhedral crystals or may be cryptocrystalline (Sibson and Toy 2006; Kirkpatrick and Rowe 2013). In relatively pristine, well-preserved pseudotachylytes, flow bands may still be evident and both dendritic crystals (Magloughlin 1989) and spherulites (Di Toro and Pennacchioni 2005) have been suggested to represent particularly rapid quenching. Thermochronology of the crystallized pseudotachylytes can be used to constrain the host rock temperature and thermal history of the melt and, by comparison to typical geotherms, can provide information about the formation depth (Kirkpatrick et al. 2012; Yamada et al. 2012). As well as recrystallisation from frictional melting, survivor clasts or restite crystals (with higher melting temperatures) that were suspended in the frictional melt may be preserved; these clasts are often rounded and can aid in identifying the occurrence of frictional melting as well as discerning the temperature and timescale of melting (Lin 1999; Ray 2004; Kendrick et al. 2012; Behera et al. 2017; Sarkar et al. 2019). Pseudotachylytes may also contain small, near-spherical immiscible sulfide droplets that further indicate frictional melting (Magloughlin 2005; Plail et al. 2014).

Pseudotachylytes have been found to contain vesicles (Scott and Drever 1954; Maddock et al. 1987; Lin 1994b; Lavallée et al. 2015b) and bubble collapse structures (Magloughlin 2011) which may infill to form amygdules post-emplacement (Rowe et al. 2005; Kirkpatrick and Rowe 2013). Vesicles or bubbles likely result from either incorporation of aqueous fluids hosted in neighboring rocks or melting of hydrous minerals (e.g., amphibole, gypsum, zeolite, clays, phyllosilicates) releasing water into the frictional melt directly (Magloughlin 2011). Diffusional growth which occurs as temperature increases or pressure decreases will also contribute to the vesicularity of the pseudotachylyte (Maddock et al. 1987), resulting in highly vesicular frictional melts in shallow faults (Maddock et al. 1987) and injection-veins (Takagi et al. 2007). In their work on volcanic ash, Lavallée et al. (2015b) postulated that thermal vesiculation, i.e., bubble growth in the melt phase of igneous rocks by exsolution of water due to solubility lowering during frictional heating, impacts frictional melt viscosity and can induce fragmentation of lavas undergoing frictional slip. Such a process, i.e., thermal vesiculation, may be relevant to the occurrence of pseudotachylytes in a range of volcanic settings, where host rocks have variably hydrated interstitial melt or glass (Kendrick et al. 2014b; Bernard and van Wyk de Vries 2017; Hughes et al. 2020). For example, incorporation of primitive, vesicular pseudotachylyte fragments into cataclasites bounding an intact, dense pseudotachylyte layer at the base of a volcanic debris avalanche suggests that thermal vesiculation may play an important role in the stability of frictional melt zones, which can pivot between brittle and viscous processes (Hughes et al. 2020).

Other textural distinctions between fault rocks and their host rock may allude to the process of formation, for example pseudotachylyte can form anomalously dense layers or bands (Ozawa and Takizawa 2007; Kendrick et al. 2014b; Plail et al. 2014), which are especially noticeable in porous rocks. Frictional melting may then result in a stronger pseudotachylyte than the surrounding rocks (Mitchell et al. 2016; Proctor and Lockner 2016), forcing any future slip events to be accommodated by rupture of the adjacent materials, forming layered or banded shear zones with cataclasites and pseudotachylytes, which have been observed in volcanic rocks (e.g., Kendrick et al. 2014b). Typically, porosity contrast between adjacent layers or porosity anisotropy results in the channeling of fluid, with many shear zones consisting of a low-permeability fault core and high-permeability damage zone (e.g., Evans et al. 1997). The flushing of fluids may overprint many of the characteristic features of a pseudotachylyte and may contribute to their scarcity in the rock record (Kirkpatrick and Rowe 2013; Mittempergher et al. 2014). Yet in volcanic settings, overprints caused by magmatic degassing may help identify pseudotachylytes due to their unique geochemical signature (e.g., Plail et al. 2014). Where such overprints do not facilitate identification, by employing multiple approaches briefly outlined here to investigate fault rocks, pseudotachylytes may still be identified with confidence, and indeed documented examples even in active volcanic settings continue to accrue.

Fault slip in tectonic and volcanic environments

Structural, seismological and experimental studies continue to vastly improve our understanding of the differing conditions which result in faulting and the post-development influence of faulted rocks on slip behavior (Sibson 1977; Cowie and Scholz 1992; Linker and Dieterich 1992; Wong et al. 1992; Scholz et al. 1993; Wells and Coppersmith 1994; Chester and Chester 1998; O'Hara 2001; Abe et al. 2002; Ben-Zion and Sammis 2003; Monzawa and Otsuki 2003; Rempel 2006; Di Toro et al. 2008; Lyakhovsky and Ben-Zion 2008; Marone and Richardson 2010; Faulkner et al. 2011; Niemeijer et al. 2012).

Fault slip occurs in rocks as the accommodation of the different relative motions of two component parts. Yet, a degree of controversy surrounds the laws which describe slip behavior (e.g., Scholz 1998) and several aspects of the mechanics controlling slip remain poorly understood. Specifically, questions remain with regard to: (1) the stress build-up required to trigger slip, often termed fault "strength" and whether this accumulation of shear stress is released by the stress drop experienced during an earthquake or whether some residual stresses are stored, maintaining the system in a stressed state (e.g., McGarr 1999; Copley 2018); and (2) the temporal and spatial variation in seismic versus aseismic slip, and thus the controls which define whether slip manifests as an earthquake or a prolonged period of creep (Dieterich 1979).

In the context of volcanic complexes, such unresolved questions may be viewed in a different light, since fault slip in volcanoes is driven by external forcing in most scenarios, and as such, fault friction serves as a feedback controlling slip resistance, strain localization, permeable architecture, etc., rather than defining whether slip will or will not occur. For example, frictional contacts at the borders of ascending viscous magmas are generated by the buoyancy-driven ascent of the magma, and frictional surfaces at the base of debris avalanches are formed due to the mass movement following gravitational collapse (Fig. 1).

In magmatic conduits magma chamber overpressure and magma buoyancy control ascent, ascent rate controls degassing and crystallization and hence conditions experienced by the magma, including shear stress (de' Michieli Vitturi et al. 2008; Bain et al. 2021). The presence of a localized shear zone at the conduit margins not only dictates the resistance imposed at the margin (to control fault slip), but can further contribute to the dynamics of magma ascent (Costa et al. 2012; Kendrick et al. 2014a; Cassidy et al. 2018). A characteristic of the non-Newtonian, shear thinning rheology of multiphase magmas is that viscosity is reduced as the rate of deformation increases (Costa and Macedonio 2002) thus the propensity of magmatic suspensions to localize strain (see Lavallée and Kendrick 2022, this volume) can create a runaway effect via strain weakening (Deubelbeiss et al. 2011; Kendrick et al. 2013), further reducing viscosity in localized areas whilst the bulk of the magma may be relatively unaffected (Wright and Weinberg 2009; Calder et al. 2015).

Volcanic conduits further differ from traditionally considered faulting environments due to their substantially elevated ambient temperature conditions; magma typically ascends in the range of 700-1200 °C, depending on composition of the system. The temperature profile within volcanic conduits is relatively poorly constrained, relying primarily upon geothermobarometry based on the equilibrium of crystal assemblages and on melt inclusions (e.g., Blundy et al. 2006). Moreover, universal descriptions for thermal inputs due to latent heat of crystallization (e.g., Blundy et al. 2006), shear heating (Costa and Macedonio 2005; Hale et al. 2007) and fault-friction (Kendrick et al. 2012, 2014a,b; Lavallée et al. 2012a, 2015a,b; Okumura et al. 2015) remain to be devised and as such local fluctuations and short temporal perturbations of temperature in particular are difficult to resolve. In other frictional melt generating faulting scenarios in volcanic complexes, such as sector collapse or avalanche events (Legros et al. 2000; Hacker et al. 2014; Hughes et al. 2020), or the tumbling of large blocks from lava dome eruptions (Grunewald et al. 2000; Schwarzkopf et al. 2001), thermal evolution during slip may be more akin to traditional contexts of fault slip, in which opposing surfaces begin in temperature equilibrium and return rapidly to that condition post-slip (Fig. 1).

In volcanic settings it is also important to consider the presence of fluids and gases. When fractures occur, particularly dilatant fractures in shallow settings, fluids may preferentially flush into them (Brace et al. 1966), increasing local fluid pressure which counteracts the normal stress acting on the fault, thus decreasing the shear resistance, which could favor slip (Beeler et al. 2016; Zhu et al. 2020). However, the gas may be able to readily escape if a permeable fracture network exists, so the timescale and extent to which fluids impact fault slip depends on the local drainage conditions (e.g., Brace and Martin 1968; Rempe et al. 2020).

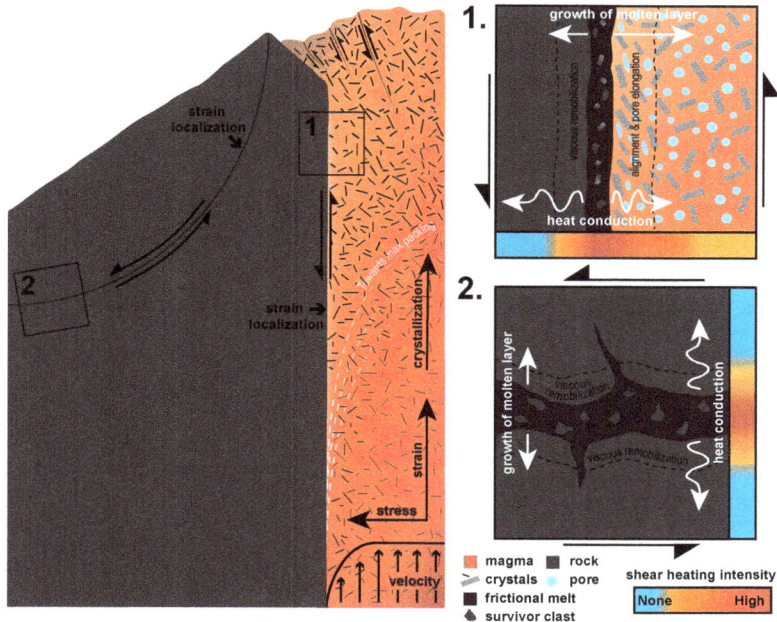

Figure 1. Schematic diagram of the scenarios of fault slip in an active volcanic complex. A cross section of the conduit and volcanic edifice shows magma ascending. As magma ascends degassing and crystallization increase its viscosity, and the flow transitions towards plug-like ascent, in which strain is localized near the conduit margin, and shear zones develop in the erupting magma (after Kendrick et al. 2017; see also Lavallée and Kendrick 2022, this volume). Within the edifice, pre-existing weaknesses or new instabilities may be triggered by dynamic forcing such as rainfall or seismicity, which can cause flank instability and gravitational collapses in which strain is localized on basal contacts (Acocella 2021). **Boxes 1 and 2** mark the positions of insets 1 and 2 which show the processes acting on the shear zones in more detail. In both insets the intensity of shear heating is indicated by the color bars running perpendicular to the shear planes. **Inset 1** shows how, once frictional melt is formed at a conduit margin, shear heating within the frictional melt sustains melt production and causes conduction of heat into both the country rock and adjacent magma, which gradually widens the molten zone. The additional heat contribution from the slip zone into the already molten magma would serve to reduce viscosity, which itself increases the contribution of shear heating, and enhances crystal rotation and alignment, and pore elongation (e.g., Wallace et al. 2019a). Such shear zones serve to increase permeability anisotropy, which can channel fluids from the degassing magma (e.g., Lavallée and Kendrick 2022, this volume). Both lithic "survivor" clasts and mineral restites may become suspended in the frictional melt layer. If interstitial glass is present in the bounding country rock, viscous remobilization may occur in the areas above T_g (defined by **dashed lines**), causing either dilation or compaction (e.g., Lavallée and Kendrick 2020). **Inset 2** shows how a frictional melt layer within volcanic rock (e.g., in the case of sector collapse) sustains heat generation via shear heating, how heat conducts away from the slip zone into the under- and overlying rock, and how the melt interface migrates (melt zone widens) into these units according to the melting temperature of the constituent phases of the rock (after Hirose and Shimamoto 2005b). Suspended lithic "survivor" clasts are common due to the transient (~disequilibrium) nature of melting, and melt may be extracted from the slip surface by injection into the adjacent rock.

Moreover, in magmas the creation of fractures and the associated decompression can enhance volatile exsolution, alleviating the overpressure which drives ascent (e.g., Gonnermann and Manga 2013), and inducing crystallization (Tait et al. 1989; Blundy et al. 2006) that could serve to increase viscosity (though crystallization also releases heat, which could reduce viscosity of the surviving melt phase) and slow ascent (e.g., Melnik and Sparks 2002; Burgisser and Degruyter 2015). As such, there remains much to be understood with regard to the coupled processes of deformation/fault slip, heating and fluid flow in active volcanic environments, where poor preservation of evidence (due to high temperatures, remobilization, alteration) hampers interpretations.

Dynamics of frictional sliding

Friction occurs whenever two bodies slide past one another, accommodating the difference in relative motion (Kennedy 2000), the study of which is termed tribology. Bowden and Tabor (1950) proposed one of the most widely accepted models for frictional sliding, whereby interfacial friction is attributed to two contributing factors acting on microcontacts: (1) formation and shearing of contact bridges between surface asperities, i.e., adhesive, and (2) plastic deformation of the asperities via ploughing. The real contact surface or area (A_r) is comprised of these microcontacts, which is smaller than the geometric contact surface area (A). When two surfaces are loaded together, A_r increases with time via growth of existing contacts and initiation of new contacts, until the area is sufficient to elastically support the contact load. As such, A_r also increases with increasing normal stress. Rabinowicz (1951) proposed a distinction between static and kinetic friction that had been first posited in the 1800's. In his depiction, static friction was subject to the time-dependence as described above (A_r increases over time and static friction increases). Kinetic or dynamic friction represents the friction during sliding. Dynamic friction is always lower than static friction (e.g., Scholz and Engelder 1976) due to the contact being transient, with the implication that A_r decreases as slip velocity increases. Later, Rabinowicz (1958) invoked these mechanisms to describe stick-slip behavior in a number of scenarios, before Brace and Byerlee (1966) proposed stick-slip as a mechanism for crustal earthquakes.

Our current understanding of frictional sliding in rocks is shaped by Byerlee (1978) who proposed a frictional law which built upon the Mohr–Coulomb failure criterion. Mohr–Coulomb theory assumes that for isotropic materials, failure occurs at specific combinations of the maximum and minimum principal stresses, disregarding the effect of intermediate stresses or the state of strain. Byerlee extended this criterion, by analysis of a compilation of experimental data obtained on the rupture of intact rocks as well as on rock-on-rock frictional sliding, stating that shear stress (τ) is proportional to the applied normal stress (σ_n) via the friction coefficient (μ), irrespective of whether the rock is intact or already ruptured:

$$\tau = \mu\sigma_n \tag{1}$$

At low normal stresses (<200 MPa) $\mu = 0.85$, whilst at higher normal stresses (>200 MPa) $\mu = 0.6$ and cohesion comes into effect (nominally $+50$ MPa), as suggested in the Mohr–Coulomb failure criterion. This friction coefficient holds true for sliding on planes of variable roughness, and the failure of intact rocks.

In the presence of pore fluids, which may saturate rocks, Terzaghi (1923) demonstrated that pore fluid pressure (P_p) serves to counteract σ_n. As such, Biot and Willis (1957) showed that the normal stress in Equation (1) may be substituted for the effective normal stress ($\sigma_{eff} = \sigma_n - \alpha P_p$) incorporating a geometrical factor referred to as the Biot–Willis coefficient, or effective stress coefficient ($\alpha = 1 - A_r/A$) which weights geometric/ macroscopic fault area (A) and the real contact area (A_r). As such, the shear resistance along a fault hosting pressurized pore fluid may be expressed as:

$$\tau = \mu(\sigma_n - \alpha P_p) \tag{2}$$

such that the weighted friction coefficient becomes:

$$\mu = \frac{\tau}{\sigma_n - \alpha P_p} \tag{3}$$

In shallow faulting scenarios, A_r is often negligible and it is thus assumed that the effective stress coefficient approaches unity (i.e., $\alpha \approx 1$; Rempe et al. 2020). This unweighted scenario is referred to as the Terzaghi stress, in which pore pressure counteracts normal stress across the whole surface area (Terzaghi 1923), an inference which depends on the drainage state, but which in most laboratory experiments this is likely to be true (Beeler et al. 2016).

Rock interfaces exhibit rate-and-state dependent behavior during frictional sliding (Dieterich 1979, 1992; Ruina 1983; Nielsen et al. 2000). This observation in many ways builds upon the early work by Rabinowicz (1951). The most common form relates the friction coefficient to the deformation rate (V) and state (θ) via:

$$\mu\left(V,\theta\right) = \mu^* + a\ln\left(\frac{V}{V^*}\right) + b\ln\left(\frac{V\theta}{D_c}\right) \qquad (4)$$

where parameters a and b are proportionality constants defining instant (rate) and time-dependent (state) evolution, D_c is the characteristic slip distance over which evolution occurs and μ^* is the friction coefficient measured at velocity V^*. There are two empirical equations (Ruina 1983) which describe the evolution (i.e., the time derivative, $\dot{\theta}$) of the state parameter θ, defined as the ageing law (named as such because at $V=0$, θ increases through time):

$$\dot{\theta} = 1 - \left(\frac{V\theta}{D_c}\right) \qquad (5)$$

and the slip law:

$$\dot{\theta} = -\left(\frac{V\theta}{D_c}\right)\ln\left(\frac{V\theta}{D_c}\right) \qquad (6)$$

so called because when $V=0$ then $\dot{\theta}=0$, thus the evolution of θ is only dependent upon slip. Therefore, in either case at steady-state slip conditions, the state parameter θ can essentially be reduced to D_c/V (e.g., van den Ende et al. 2018) so that the equation can be rewritten in the form:

$$\mu_{ss}\left(V\right) = \mu^* + (a - b)\ln\left(\frac{V}{V^*}\right) \qquad (7)$$

where the expression $a - b$ (which may be positive or negative) is commonly used as a metric to describe the velocity-dependence of the friction coefficient at steady state (μ_{ss}). Most laboratory data of fault slip are framed as such. Frictional sliding may be velocity strengthening (if resistance to sliding increases with increasing velocity) which promotes stable aseismic slip, and is characterized by $a > b$ (i.e., positive $a - b$), or velocity weakening where $a < b$ (resistance to sliding decreases with increasing velocity) i.e., likely to produce earthquake instabilities (Dieterich 1979; Dieterich and Kilgore 1994). In the latter case, weakening mechanisms depend on lithology, and yet, despite the importance of rock type and environmental conditions (such as stress state and fluid saturation), the near-universal occurrence of dynamic weakening (i.e., negative $a - b$) has been inferred during coseismic slip (Di Toro et al. 2011).

In simple terms, dynamic weakening may essentially be attributed to frictional heating. During frictional sliding, mechanical energy is transformed into internal energy or heat, concentrated at the contact area between the solid bodies. The real contact area involved in the generation of heat is a topic of ongoing investigations; some research suggests plastic deformation processes in the bulk solid are responsible for heating (Rigney and Hirth 1979) while others invoke only the top atomic layers (Landman et al. 1996). Given the spectrum of componentry of geomaterials, and the complexity of their deformation mode (e.g., Byerlee 1978), such subtleties of the microscopic process are unlikely to be individually addressed in the definition of constitutive laws for frictional sliding in rocks. Yet, we find that these processes are active across the spectrum of lithologies and despite several unknowns, there is near-universal agreement that almost all the energy generated during sliding is converted to heat via frictional heating (Kennedy 2000), and that this heat is then conducted through the solid away from the contact area (e.g., Carslaw and Jaeger 1959). The total heat input at a contact (Q_c) is defined as:

$$Q_c = \mu \sigma_n V A_c t \tag{8}$$

where A_c is the area of contact and t is the duration over which slip occurs (recall that $\mu = \tau/\sigma_n$). The heat so generated may conduct away from the contact zone at slow slip speeds, whereas at faster, coseismic slip rates (>0.1 m.s^{-1}) contact temperature increases rapidly. Assuming an infinitesimally thin slip zone over which heating is active, and that heat can only be lost perpendicular to that slip surface, then this relationship is described by the diffusion equation of Carslaw and Jaeger (1959), and the temperature change (ΔT) caused by a given slip event of time t_x can be computed via:

$$\Delta T = \frac{\tau V_e}{\rho C_p} \sqrt{\frac{t_x}{\pi k}} \tag{9}$$

where ρ is mass density, C_p is heat capacity and k is diffusivity, and when the frictional work rate or power density (Ω_p) is considered constant (after Nielsen et al. 2010b), given that:

$$\Omega_p = \tau V_e \tag{10}$$

Temperature increase can result in dynamic weakening via a number of mechanisms, including: thermal pressurization of pore fluid (Sibson 1973); elastohydrodynamic lubrication (Brodsky and Kanamori 2001); chemical decomposition (relevant to rocks containing volatile-bearing minerals; e.g., Han et al. 2007; Hirose and Bystricky 2007); mineral breakdown (Noda et al. 2011); silica gel formation (Di Toro et al. 2004; Kirkpatrick et al. 2013); flash heating at asperity contacts, resulting in thermal (Rice 2006; Beeler et al. 2008) and mechanical weakening (Weber et al. 2019) and frictional melting (Spray 1993; Hirose and Shimamoto 2005b; Di Toro et al. 2006). In the case of ascending magma at elevated temperatures in a conduit, such assertions are vital to understand the occurrence and impact of fault slip. Indeed, melting of the crystal cargo in magma (and of the conduit wall rock) may be an inevitable consequence of faulting in hot magmas and lavas, and as such here we will focus on the mechanisms which result in frictional melting to produce pseudotachylyte, and specifically those advances elucidated by experimentation.

FRICTIONAL MELTING

The history of experimental approaches

Laboratory experimentation allows for the probing of contributing factors towards an observed phenomenon. In the case of frictional melting, experimentation has been conducted over the last several decades in order to unravel the physical conditions that lead to and follow frictional melting on a slip plane, as well as the chemical processes taking place during melting. In the case of frictional sliding, the motivation for this research is to unravel the effect of the established mechanics on ongoing slip processes. Historically, many shear configurations have been implemented to study fault friction, including direct shear (Wang et al. 1975), double shear (Dieterich 1972), biaxial (Scholz and Engelder 1976) and triaxial (Byerlee 1967) set-ups, with good agreement of findings across each method and apparatus (Byerlee 1978). Subsequently many such approaches have been used to study frictional melting, for example: Spray (1987) used a frictional welding machine to reproduce pseudotachylytes, providing an excellent description of the melting process, though lacking the constraints of the causal conditions;

Serendipitously, Killick (1990) noted the production of pseudotachylyte whilst drilling Proterozoic lavas at 2.4 km depth; Lockner et al. (2017) amongst others developed frictional melts over short slip distances in triaxial cells at simulated deep crustal confining pressures, yet the fault geometry in such tests prohibits decoupling of shear and normal stresses. To date, none of these approaches offer the dynamism of the high-velocity rotary shear apparatus first designed by Shimamoto in the late 1980's (Shimamoto and Tsutsumi 1994), and since modified to tackle different aspects and conditions of extreme fault friction by several research groups around the world.

The rotary shear apparatus sees two cylindrical (rock) samples placed end-on-end to simulate a slip surface and allows the collection of torque (used to calculate shear stress) and axial shortening data at an applied rotation rate and axial stress over large total displacements (e.g., Ma et al. 2014). The only disadvantage of the cylindrical set-up is the variation in slip rate across the slip surface, which is minimized by using hollow cores. Following Shimamoto and Tsutsumi (1994) a nominal equivalent rotation velocity (V_e) can then be calculated from the rotation rate (R) and the outer and inner radii of the hollow cylinders (r_0 and r_i respectively) via:

$$V_e = \frac{4\pi R\left(r_o^2 + r_i r_o + r_i^2\right)}{3(r_o + r_i)} \tag{11}$$

Assuming that the shear stress is constant over the sliding surface area (S) in this geometry:

$$S = \pi(r_i^2 - r_0^2) \tag{12}$$

Then, the rate of frictional work (W) may also be given by:

$$W = \tau V_e S \tag{13}$$

Kitajima et al. (2010) explored this assumption in more detail, but here as in other studies (e.g., Lin 2007; Boulton et al. 2017) we abide by the assumption that shear stress acts uniformly over the sliding surface. Because geometrical constraints of the apparatus mean that normal stress often cannot exceed ~10 MPa, several authors have posited that the advantage of using W (or by extension frictional work rate or power density) to compare response to slip, is that larger displacements can substitute for unattainable higher normal stresses (Lin 2007; Lockner et al. 2017); although, during melting and fault slip in the presence of a frictional melt layer, such substitutions may not suffice.

Mechanical response to melting

As many natural pseudotachylytes have been observed in intrusive igneous rocks the majority of work on frictional melting has been conducted on them. A number of studies have recreated the geometry and properties of frictional melt using rotary shear experiments (Fig. 2), that also allow us to interpret the conditions leading to their formation (Shimamoto and Tsutsumi 1994; Tsutsumi and Shimamoto 1997; Lin and Shimamoto 1998; Hirose and Shimamoto 2005a,b; Di Toro et al. 2006; Nielsen et al. 2008, 2010b; Del Gaudio et al. 2009; Lavallée et al. 2012a, 2015a; Kendrick et al. 2014a; Violay et al. 2014; Hornby et al. 2015). Such tests can be challenging due to decomposition of phases and differential thermal expansion during such rapid heating rates, which can result in sample rupture. Rocks containing quartz (such as granites) present a particular challenge due to the increased rate of expansivity across the α–β transition resulting in fracturing (Ohtomo and Shimamoto 1994). Glass-rich materials may experience a similar outcome when they undergo viscous remobilization (when crossing the glass transition), which can make the experimental study of volcanic materials undergoing frictional slip difficult (Lavallée et al. 2015).

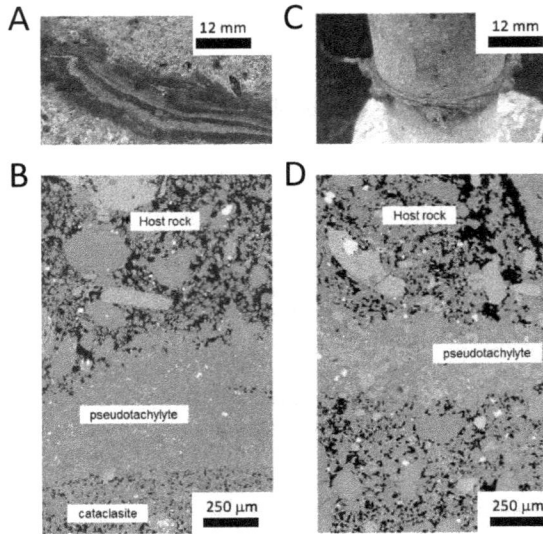

Figure 2. Pseudotachylyte in volcanic rocks: (**A**) Hand-specimen from an andesitic block at Soufriere Hills volcano (Montserrat), showing multiple generations of interlayered pseudotachylyte and cataclasite which have been viscously remobilized in the partially molten magma prior to eruption, so that they laterally pinch in and out. (**B**) The same material as in A in a backscattered electron image of a thin section, showing the coarse phenocrysts and finer glassy groundmass of the porous host andesite, with grain size reduction from the host material to the cataclasite, and densified pseudotachylyte comprising rounded surviving phenocrysts and an equant, interlocking mineral assemblage. (**C**) An example of a pseudotachylyte generated in a dacite from Mount St. Helens (USA) using a high velocity rotary shear test at 2 MPa of normal stress and 1 m.s^{-1} slip rate. (**D**) The same material as in C in a backscattered electron image of a thin section, showing large phenocrysts and glass-bearing porous groundmass in the host rock, with a dense pseudotachylyte layer of comparable thickness to the natural example in B, with a number of surviving plagioclase crystal fragments held in suspension in an interstitial glass (i.e., the pseudotachylyte). Textural differences may result from the duration of melting (unconstrained in the natural example) and the more rapid cooling of pseudotachylyte generated experimentally.

During fault slip at constant normal stress and at velocities that result in melting, shear stress on the slip plane evolves (Fig. 3A). This evolution was examined by Hirose and Shimamoto (2005b) who dissected gabbros subjected to different slip distances to propose a model for the progression of fault slip. At the onset of sliding an initial phase of grain plucking and comminution results in an early peak shear stress (hence, peak friction; P_1 in Fig. 3A), which is followed by an initial weakening phase corresponding to flash heating and smoothing of asperities, reaching a preliminary period of steady-state sliding (stable shear stress and friction coefficient; SS_1). A second peak in shear stress then initiates, which is interpreted as the production of isolated melt patches as frictional melting ensues (Hirose and Shimamoto 2005b). As the melt patches grow, shear stress increases further until a second peak (P_2) at which point a continuous molten layer exists on the slip plane. Shear stress then decreases over a characteristic distance (D_c; Fig. 3B), which corresponds to the melt thickening, until a second steady-state period of sliding is achieved (SS_2 in Fig. 3A).

Numerous studies have shown that the onset of melting strengthens the slip zone, i.e., shear resistance increases to a higher second peak than that achieved during rock–rock sliding (Hirose and Shimamoto 2005b). This is attributed to a lubricating cataclastic gouge layer on the contact surface during the initial period of slip (P_1 to SS_1 in Fig. 3) which comminutes and allows minor shear-weakening within the first few meters of slip (Hetzel et al. 1996; Reches and Lockner

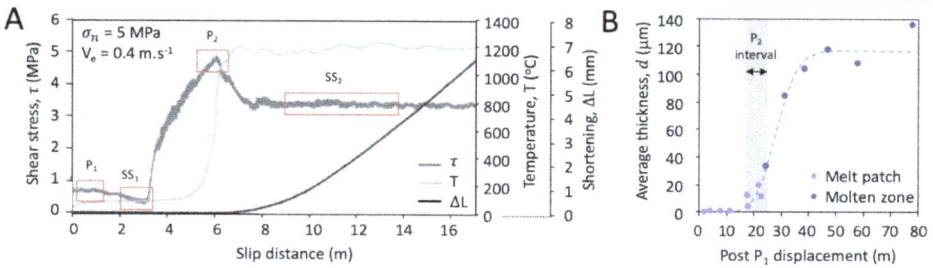

Figure 3. (A) Friction experiment on a dacite (after Kendrick et al. 2014) showing how frictional sliding produces an early peak (P_1) as slip ensues, followed by slip weakening to an initial steady-state shear stress (SS_1) due to flash heating and lubricating gouge, followed by a rapid increase in shear stress as melt patches begin to form on the slip interface, the melt layer coalesces at a second peak (P_2), shear stress then decreases as the melt zone widens before attaining steady state slip (SS_2) controlled by the viscosity of the melt. Shortening begins as the melt layer coalesces at P_2 and accelerates until also reaching a steady state, where melt is produced and expelled at a constant rate. Temperature monitored on the sample surface reaches an initial plateau due to flash heating and then increases as melt becomes visible at the sample surface just before P_2. **(B)** Area-averaged thickness of melt patches and a frictional melt layer plotted against the displacement after P_1 during friction experiments at 1.2–1.5 MPa normal stress in gabbros, the dashed line is the least squares fit (after Hirose and Shimamoto 2005). Melt thickness initially increases slowly as patches appear, then accelerates quickly to a peak, and then stabilizes.

2010; Spagnuolo et al. 2016), during this stage the temperature monitored on the sample surface is relatively stable. The ultra-fine crystal fragments produced during this stage then begin to preferentially melt due to their larger surface area (Spray 1992), generating heterogeneously distributed melt patches on the slip surface (Tsutsumi and Shimamoto 1997; Hirose and Shimamoto 2005b; Chen et al. 2017). The small melt patches act as a viscous brake and rapidly increase shear resistance (Tsutsumi and Shimamoto 1997; Fialko and Khazan 2005; Hirose and Shimamoto 2005b) and correspondingly locally raise temperatures further (Fig. 3A). As the temperature increases, melt patches coalesce resulting in a molten layer with abundant crystal fragments (e.g., Fig. 4), as seen at the onset of melting (e.g., Wallace et al. 2019b). With further melting the volume of melt to crystals increases until a critical melt fraction is reached (Fialko and Khazan 2005; Rosenberg and Handy 2005; Chen et al. 2017) and temperature stabilizes.

Once melt is present, the shear resistance during slip is strongly influenced by the melt's rheological properties; melt viscosity, suspended pores and particles and the strain-rate (Spray 1993; Hirose and Shimamoto 2005a,b; Lavallée et al. 2012a; Hornby et al. 2015). The heat source may essentially be considered as viscous shear, and heat is similarly depleted by

Figure 4. A frictional melt zone (FMZ) generated in a dacite (from Mount Unzen, Japan) during a rotary shear experiment with a slip rate of 1.45 m.s^{-1}, normal stress of 3 MPa and total displacement of 19.6 m. The sample exhibits embayment of the FMZ into adjacent hydrous amphibole (Am) crystals as they preferentially melt, with one showing a reaction front (black band) along the melt contact. Also shows viscous remobilization in the adjacent glassy groundmass identified by rotation of the microlites in the direction of shear in the host rock, and suspension of plagioclase crystals in the melt as they have higher melting temperatures (modified from Hornby et al. 2015).

conduction, latent heat and removal of melt. The apparent viscosity (η_{app}) of the melt plus suspended crystals on the slip plane can be calculated from the monitored shear stress and shear strain rate ($\dot{\varepsilon}$) via:

$$\eta_{app} = \frac{\tau}{\dot{\varepsilon}} \qquad (14)$$

where $\dot{\varepsilon}$ is:

$$\dot{\varepsilon} = \frac{V_e}{d} \qquad (15)$$

For a given melt layer thickness (d) and equivalent velocity (V_e). As melt thickness cannot be accurately characterized during slip, thickness must be measured from dissected experimental specimens. Thickness of the melt layer increases during slip due to further melting and may stabilize during steady-state sliding (e.g., Ferrand et al. 2021). Thickening generally results in a reduction in shear stress (e.g., Tsutsumi and Shimamoto 1997), from P_2 to SS_2 (Fig. 3), as the distance between the two opposing interfaces increases, referred to as the Stefan problem (Hirose and Shimamoto 2005a). The Stefan problem describes the migration of the melt–solid interface (the melting surface) into the bulk rock under non-linear conditions (Nielsen et al. 2010b), i.e., at conditions prior to achieving steady state under a constant slip rate (Hirose and Shimamoto 2005a; Nielsen et al. 2008). The melting surface migration is controlled by the melting of the rocks' constituent phases (see section *Selective melting*), which depending upon the temperature gradient around the slip zone, also dictates the roughness of this interface (Nielsen et al. 2010a).

Once steady state slip is achieved in the presence of melt, a simple diffusion equation yields a relatively accurate thermal profile and provides a solution for shear stress, melt viscosity, and thickness (Nielsen et al. 2008). Analysis has shown that preserved pseudotachylyte thickness is proportional to displacement across scales from the laboratory to the field for a wide range of lithologies (Ferrand et al. 2021). For a given rock type, melt thickness varies as a function slip velocity and normal stress (Nielsen et al. 2008), but can be highly variable within a single slip zone: A study by Hornby et al. (2015) measured slip zone thicknesses in experimentally generated pseudotachylytes in dacites, which showed variability of >100 μm within single slip zone (Fig. 4). In volcanic rocks, a potential contributing factor in addition to the impact of the mineralogical assemblage is the presence of volcanic glass in the host rock, which may facilitate viscous remobilization of areas adjacent to frictional melts (see Fig. 1), widening the slip zone and contributing to the evolution of the frictional melt zone (Lavallée et al. 2012a; Kendrick et al. 2014a; Violay et al. 2014; Hornby et al. 2015). Solutions for apparent viscosity based on mechanical data provide good approximations and are useful to compare the relative behavior of different materials where the use of the friction coefficient may not be entirely appropriate (i.e., in the presence of melt). Apparent viscosity measurements may be validated using careful geochemical analysis of frictional melts and modeling of the non-Arrhenian melt viscosity plus the effect of suspended survivor clasts. For a description of fault slip in the presence of a melt layer see the section *Frictional melt rheology*.

Selective melting

The onset of melting and growth of a molten zone during fault slip are controlled by the mineralogy of the host rocks (e.g., Spray 1992). To accurately constrain the processes of frictional heating during fault slip and understand the progression of frictional melting, quantification of temperature during experiments is vital. In natural pseudotachylytes temperature has commonly been estimated from its mineral assemblage, assuming recrystallisation occurred under equilibrium (e.g., Magloughlin and Spray 1992). Yet, frictional melting is a rapid, dynamic and chemically chaotic process (Sibson 1975) that results from conversion of thermomechanical energy on a slip plane to heat at strain-rates $>10^{-2}$ s^{-1} and slip velocities >0.1 m s^{-1}. Thermal monitoring of experiments has recorded heating rates of >1000 °C.s^{-1} (e.g., Lavallée et al. 2012a, 2015a; Kendrick et al. 2014a), and the geochemical signatures of natural and experimentally-derived pseudotachylytes show that frictional melts are formed by selective melting of their host rocks (e.g., Magloughlin and Spray 1992; O'Hara 1992; Lavallée et al. 2015b; Wallace et al. 2019b). The constituent rock-forming minerals thus control the progression of frictional melting (Fig. 5), where selective melting occurs in the order of their solidus temperatures (Spray 1992). As such, frictional melting can be defined as a non-equilibrium process (Sibson 1975; Spray 1992; Lin and Shimamoto 1998) and the compositional evolution of frictional melts may be used to constrain the conditions of formation of natural pseudotachylytes, especially duration of slip (e.g., Jiang et al. 2015).

A framework for the progressive melting of constituent mineral phases (under non-isothermal conditions) is given in Spray (2010), whereby minerals with the lowest strength and breakdown temperatures are preferentially comminuted and melted to create a polyphase suspension comprising mineral and rock fragments enclosed within a liquid. Comminution is the process of crushing and grinding a solid material to form smaller particles, which often results in a power-law particle-size distribution represented by a large number of small fragments, an intermediate number of mid-size fragments, and a small number of large fragments (e.g., Sammis et al. 1987; Guo and Morgan 2006; Spray 2010; Kennedy and Russell 2012). von Rittinger (1867) proposed that the energy consumed in the size reduction of solids is proportional to the new surface area produced, a theory which appears to be supported by comminution products which show that the normally scale-invariant power-law distribution falls apart at very fine grain size (which has been termed the comminution limit) because fracturing smaller particles requires disproportionately more energy than fracturing larger ones (e.g., Lowrison 1974). Moreover, comminution is relatively inefficient; compared to the mechanical energy input only 1–2% contributes to grain size reduction (Tromans 2008), with the bulk of remaining energy converted to heat. Since melting then ensues in the comminuted gouge, it follows that the progression of melting is highly dependent upon yield strength and fracture toughness of the material which controls this comminution, as well as the breakdown or melting temperature of the constituent minerals involved (Fig. 5).

Amongst rock-forming minerals the types of bonds control their fracture toughness, covalent-bonded minerals exhibit the highest fracture toughness (e.g., diamond), followed by ionic-bonded (e.g., metal oxides), then ionocovalent (e.g., silicates) phases and finally ionocovalent or van der Waals bonded minerals (i.e., many phyllosilicates). This trend is approximately comparable with other metrics of hardness and toughness including yield strength (Fig. 5); for a comprehensive description of the variability of yield strength, fracture toughness, Mohs number and indentation hardness for rock-forming minerals, see Spray (2010). As comminution and heating progress, thermal conductivity also becomes important. The contrasting thermal conductivity of minerals means some (i.e., poor conductors) are more prone to undergo thermally induced fracturing. Decrepitation (fracture by differential thermal expansion) can be triggered, contributing to grain size reduction (Spray 2010). This secondary phase of fragmentation mechanisms occurs in addition to the grain size reduction by comminution and may be more effective in different (less conductive) phases. This contributes to the non-linearity of the melting process (e.g., Fig. 3), since the Gibbs–Thomson equation dictates that the smaller the fragment of a mineral, the lower the melting temperature becomes (e.g., Lee et al. 2017), and thus melting onset can occur during frictional sliding at much lower temperatures than their equilibrium melting temperature.

Here we will focus exclusively on the progression of frictional melting in magmas and lavas. Differences in mineral assemblages are expected both between volcanic systems, and within a single system. In a given magmatic center (especially intermediate volcanic systems) mineralogy and crystallinity can be naturally variable over short to long duration, either

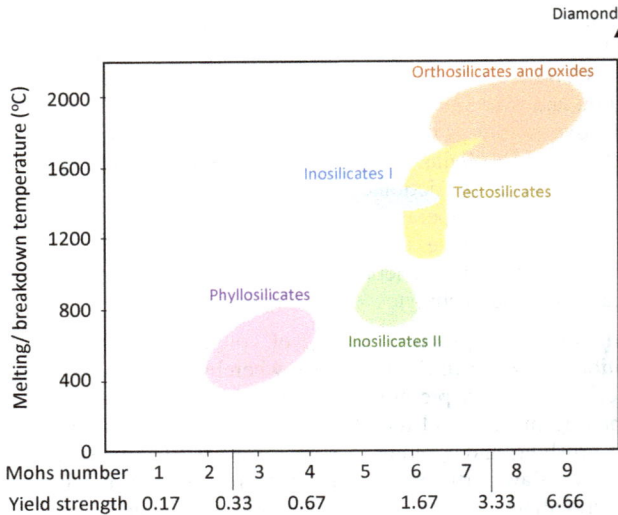

Figure 5. Mineral stability at temperature. Approximate melting or breakdown temperature is shown versus Mohs number and yield strength (in GPa) for groups of rock-forming minerals (after Spray 2010). Following the nomenclature of Spray et al. (2010) inosilicates I refers to single-chain inosilicates such as ortho- or clinopyroxene, and inosilicates II refers to double-chain inosilicates such as amphiboles. Orthosilicates and oxides include rutile, zircon and corundum; tectosilicates include quartz, plagioclase and orthoclase; and phyllosilicates include mica and serpentine.

reflecting magma recharge or changes in storage or ascent conditions (Smith and Leeman 1987; Scott et al. 2013). For example, previous studies have reported that elevated magma-water content in a plumbing system which introduces hydrous phases, often coincides with an increase in explosivity (Luhr and Carmichael 1990). In terms of frictional behavior, the presence of hydrous phases (e.g., amphibole, gypsum, zeolite, clays, phyllosilicates) are likely to bring forward melting onset and impact the rheology of the frictional melts produced due to their typically lower melting points and the associated release of water (e.g., Allen 1979). Thus, it is important to ascertain the effect of mineralogical changes on frictional melting, and the feedback this will have on magma ascent dynamics.

Experiments on volcanic rocks demonstrate the propensity for frictional melting after a few cm's slip at upper conduit conditions (Kendrick et al. 2014a; Wallace et al. 2019a). Selective melting of minerals in different volcanic materials have been shown to produce heterogenous melts (Lavallée et al. 2012a; Hornby et al. 2015; Wallace et al. 2019b). This results from chemically distinct "protomelts" or "schlieren", the products of melting of individual minerals or partial mixing of these protomelts (Masch et al. 1985; Grunewald et al. 2000; Hornby et al. 2015; Wallace et al. 2019b). Protomelts represent minerals with the lowest melting temperatures (e.g., Shimamoto and Lin 1994; Hirose and Shimamoto 2005a,b; Lavallée et al. 2012a; Hornby et al. 2015). Selective melting occurs both within the comminuted gouge formed during initiation of slip, and once initiated, it is also the process by which the melt layer grows, melting the adjacent host rock by contact heating, and melting minerals with low melting temperatures further away by conduction or diffusion of heat from the slip interface. Thus, selective melting also enables the suspension of survivor clasts (that is un-melted restite lithic or crystal fragments) and contributes to embayments in the otherwise planar frictional melt boundary as certain minerals are preferentially consumed (Fig. 4). Here the Gibbs–Thomson effect may once again come into play, where the local melting temperature of a mineral's external surface (in contact with the melt zone) is reduced along the irregular surface topology (Hirose and Shimamoto 2003; Nielsen et al. 2010a; Lee et al. 2017; Wallace et al. 2019b).

It therefore follows that the melting surface (i.e., the melt–solid interface) topology is sensitive to the rate of melting; with faster rates leading to lower roughness (Nielsen et al. 2010a).

Recently Wallace et al. (2019b) undertook a systematic study host-rock mineralogy on compositional and textural homogenization in different lavas. Specifically, the effect of the presence of hydrous amphibole was considered by comparing an amphibole-bearing andesite from Soufrière Hills Volcano (Montserrat) to an amphibole-free andesite from Volcán de Colima (Mexico). Results showed that the amphibole-bearing andesite began melting after a shorter slip distance at the same velocity and normal stress, corresponding to the lower melting temperature of the amphibole (Fig. 5). The amphibole-bearing lava also reached a lower peak and steady state shear stress (Wallace et al. 2019b). The frictional melt zone progressively homogenized during fault slip, as revealed by textural examination and chemical transects at different slip distances beyond the attainment of steady-state slip (Fig. 6; Wallace et al. 2019b). At all stages of slip, survivor clasts were suspended in the melt. The proportion of suspended crystals remained relatively constant throughout slip (at steady state), though the size and rounding of the suspended clasts appeared to increase as the melt zone matured (Fig. 6). The Gibbs–Thomson effect states that local melting temperature of a mineral in contact with melt is lowered in proportion to its curvature (Nielsen et al. 2010a), an effect which both enhances the rounding of suspended particles, and increases the complexity of the melt–rock interface during the sustained presence of frictional melt (Hirose and Shimamoto 2003).

Careful chemical timeseries of frictional melting products can elucidate the mineral contributions to the frictional melt (Wallace et al. 2019b). For example, for amphibole-bearing and amphibole-free andesites (Fig. 7), variability in major element oxide compositions were

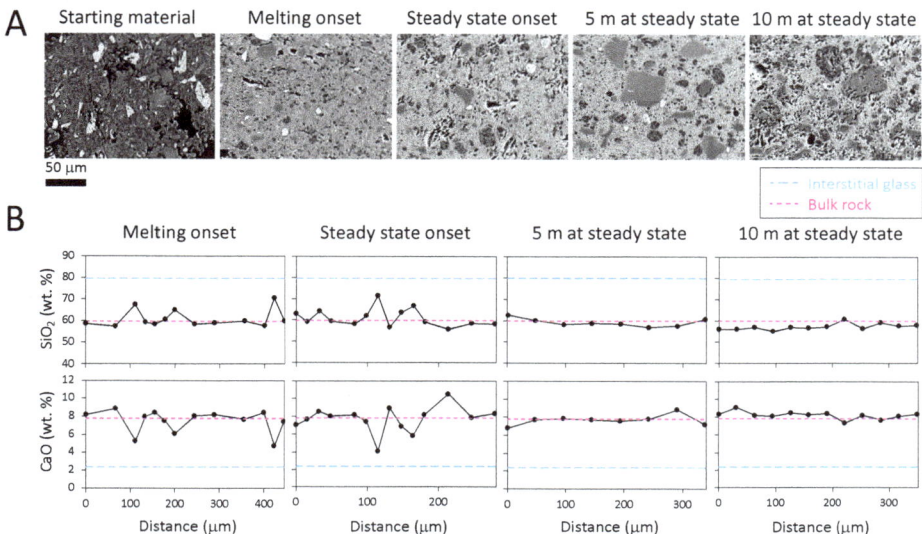

Figure 6. Physical and chemical homogenization of a frictional melt in an andesite from Soufrière Hills (Montserrat) containing plagioclase, hornblende, clinopyroxene and orthopyroxene, iron-titanium oxides and rhyolitic interstitial glass deformed in rotary shear at $1\,m.s^{-1}$ and normal stress of 1 MPa. (A) A time series of frictional melt textures generated at melting onset, at the attainment of steady state sliding plus 5 m and 10 m slip at steady state, compared to the host groundmass, showing increasing size of suspended clasts as the melt zone thickens towards steady-state sliding. (B) Example profiles of major element oxide compositions across the frictional melts at the same slip distances in A, showing variability in the melts around the bulk rock composition, with variability decreasing, indicating homogenization with increasing slip distance (data from Wallace et al. 2019b).

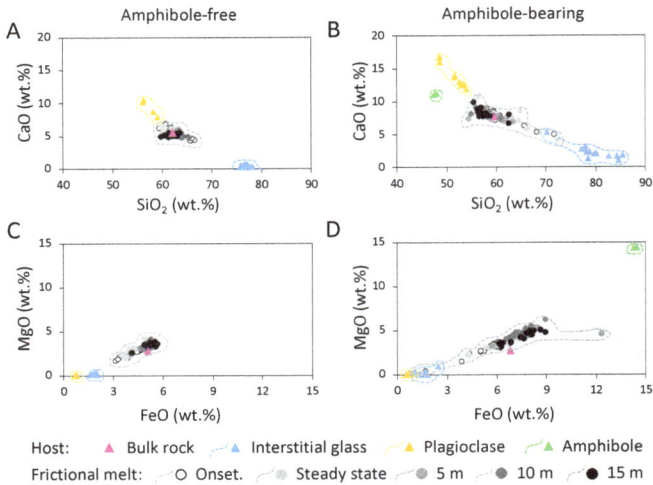

Figure 7. Binary plots showing the evolution of frictional melt composition for representative major element oxides (SiO_2, CaO, FeO and MgO) in an amphibole-free andesite from Volcán de Colima (Mexico) and an amphibole-bearing andesite from Soufrière Hills (Montserrat) with increasing slip displacement from the onset of melting until 15 m sliding beyond the attainment of steady-state slip. For comparison, the bulk rock composition, host-rock plagioclase, interstitial glass, and amphibole (where present) are plotted. In the presence of amphibole, the frictional melt composition is much more variable, and slowly converges towards a more mafic composition than the bulk, whereas in the amphibole-free case, the frictional melt has little variability from the outset, and homogenizes towards the bulk composition with increasing slip distance "(data from Wallace et al. 2019b).

considerably higher in the amphibole-bearing andesite, suggesting that greater heterogeneity results from slip in rocks which have mineral compositions with disparate melting points. Throughout slip, the chemical composition of the frictional melts trended towards that of the bulk rock for amphibole-free lava, whereas the amphibole-bearing sample melt became progressively more mafic than the host (Wallace et al. 2019b), as has been noted in natural pseudotachylytes (Camacho et al. 1995). More broadly, the melting of polyphase rocks by selective melting may produce frictional melt that is more ferromagnesian and basic than the bulk rock, forming pseudotachylytes of distinct composition to the host materials (Spray 2010). Detailed examination of the onset of melting and the progression of homogenization, characterized experimentally, may help unravel the timeframes of fault slip to create natural pseudotachylytes.

An examination of the slip distance after which melting may be achieved in different lithologies reveals that the mineralogical assemblage of each lava composition systematically controls melting onset distance (Fig. 8A). In the lithologies examined, we note that for a given slip rate and normal stress condition there is a clear tendency for basalts to melt most rapidly, followed by andesites, with dacites taking the longest slip distance to melt (note that distance to melt here is identified by the peak shear stress caused by melt, P_2 in Fig. 3, not the inflection from SS_1 to P_2 caused by the first appearance of melt patches, which may follow a different sequence). For a given slip velocity, increasing normal stress non-linearly shortens the distance required to melt to as little as a few cm's slip at >5 MPa at slip velocities >1 m.s^{-1} (e.g., Kendrick et al. 2014a; Violay et al. 2014). At a given normal stress, slip velocity controls the distance over which melting is achieved, showing the interplay between how rapidly heat is produced versus dissipated/ conducted away from the slip zone (as described by Eqn. 9).

Hirose and Shimamoto (2005b) posited that the slip weakening distance (D_c) would be directly controlled by the rate of melting of the mineral assemblage as the melt–solid interface migrates into the host rock (as described previously). In their further examination of the problem, they proposed that both slip weakening and the thickening of the molten layer are characterized by exponential changes with comparable characteristic distances, and thus they proposed that D_c can be explained by the thickening of the melt zone in both experimental and natural faults (Hirose and Shimamoto 2005a). Slip rate also has an important control on the slip weakening distance (Fig. 8B); D_c is reduced at faster slip rates (e.g., Niemeijer et al. 2011) likely due to the interplay between heat production and loss described by the Stefan problem (Nielsen et al. 2010b). The data shown in Figure 8B are for a narrow range of normal stresses, 1.2–1.5 MPa, yet, previous work has found that slip weakening distance decreases inversely with normal stress for a given rock type (Nielsen et al. 2008; Niemeijer et al. 2011); although a systematic study remains to be conducted. Following weakening, once steady-state slip is achieved, the rheology of the melt, plus any suspended particles, controls fault slip and the temperature generated by shear heating (Fig. 1).

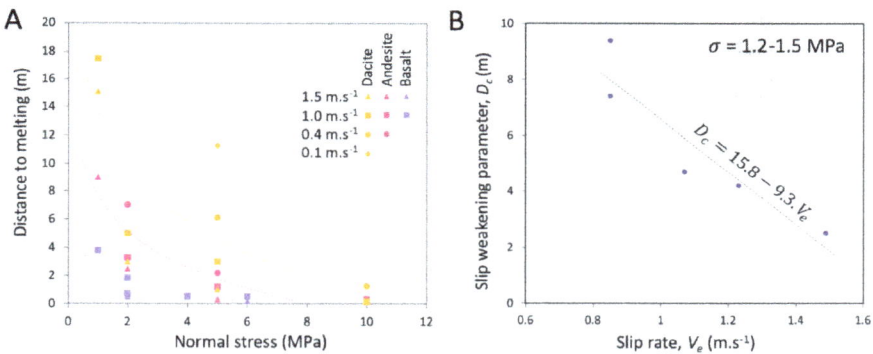

Figure 8. Characteristics of frictional melting dynamics. **(A)** Distance to melting, defined as the second peak in shear stress (P_2 in Fig. 3), as a function of applied normal stress for basalt, andesite and dacite samples at a range of slip rates. Data shows that the slip distance required for melting decreases non-linearly with increasing applied normal stress, and at a given normal stress faster slip rate results in melting over shorter distances (e.g., note the dacite's distance to melting decreasing from >11 m to ~1 m as slip velocity is increased from 0.1 to 1.5 m.s^{-1} at 5 MPa normal stress). A clear tendency for basalts to melt most rapidly, followed by andesites, with dacites taking the longest slip distance to melt is seen. Dashed lines are illustrative least squares fits through the dataset available and do not capture the slip rate variation. Data for the andesite and dacite are from Kendrick et al. (2014a), data for the basalt is unpublished. **(B)** Once a frictional melt layer covers the slip surface, shear stress begins to decrease from a peak (P_2 in Fig. 3) to its steady state condition (SS_2) as the melt thickens. The distance over which weakening occurs is described by the slip weakening parameter, D_c, which is approximately linearly dependent on slip rate, as demonstrated with gabbro samples from Hirose and Shimamoto (2005a).

Frictional melt rheology

Shear stress. Pseudotachylyte generation greatly alters the dynamics of fault slip (e.g., Otsuki et al. 2003; Di Toro et al. 2006; Nielsen et al. 2010b) such that melt rheology controls the slip zone properties (Tsutsumi and Shimamoto 1997; Hirose and Shimamoto 2005a,b; Nielsen et al. 2008; Niemeijer et al. 2011; Lavallée et al. 2012a, 2015a; Kendrick et al. 2014a; Hornby et al. 2015). Selective melting controls melt onset and homogenization timescale as the melt chemically evolves and lithic and crystal restites become suspended (Hornby et al. 2015; Wallace et al. 2019b). The final pseudotachylyte (including melt and suspended clasts) may approximate the bulk chemistry of the host rock in natural (Kendrick et al. 2012, 2014b) and experimental samples (Hornby et al. 2015), or may be more mafic than the host rock, especially where hydrous phases are present, again, as demonstrated by natural (Lin 1994b; Camacho et al. 1995; Andersen et al. 2008) and experimental (Wallace et al.

2019b) pseudotachylytes. Whilst friction coefficient is not, in principle, appropriate to describe fault slip in the presence of melt, the relationship of monitored shear stress to imposed normal stress and slip rate conditions still elucidates the shear resistance imposed on the fault, and is occasionally termed the effective friction coefficient.

In the presence of a molten layer the rheological properties (i.e., strain-rate dependent melt viscosity, suspended particles and bubbles) dictate materials' ongoing response to sliding (Spray 1993; Hirose and Shimamoto 2005a; Di Toro et al. 2006; Nielsen et al. 2008, 2010b; Niemeijer et al. 2011; Lavallée et al. 2012a, 2015a; Kendrick et al. 2014a; Violay et al. 2014; Hornby et al. 2015; Wallace et al. 2019b). Importantly, the composition of frictional melts, controlled by the composition and mineralogical componentry of the rock, dictates the shear resistance imposed by them (Fig. 9). Frictional melts in more mafic rocks, such as basalts, tend to lubricate slip zones (e.g., Violay et al. 2014), imposing lower effective friction coefficients than that predicted from Byerlee's law (Eqn. 1). In felsic and intermediate rocks, including andesite and dacite, frictional melts tend to act as viscous brakes, causing high shear stresses, and effective friction coefficients higher than the Byerlee frictional envelope, especially at low normal stresses relevant shallow faulting environments including volcanic centers (Fig. 9; Lavallée et al. 2012a; Kendrick et al. 2014b; Hornby et al. 2015; Wallace et al. 2019b). This observation is matched by modeling results, which further suggest that viscous braking becomes less effective at higher normal stresses (Fialko and Khazan 2005), for example in subduction zone faulting. At these conditions frictional melt appears to lubricate slip, seemingly irrespective of melt composition (e.g., Di Toro et al. 2006). In all compositions, waning slip velocity may result in the viscous brake effect coming into play (e.g., Spray 2005), as a decrease in shear rates increases the apparent viscosity of shear thinning suspensions and decreases thermal input, thus promoting cooling which increases the melt viscosity (Lavallée et al. 2012a; Kendrick et al. 2014a).

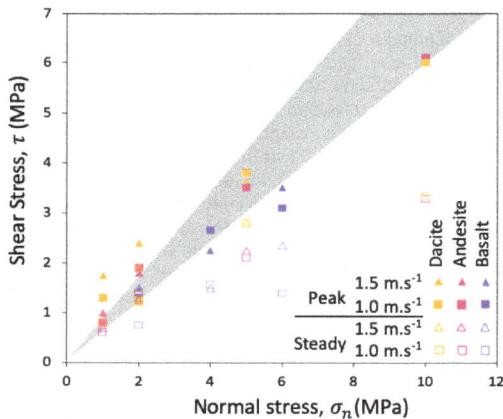

Figure 9. Shear resistance of basalt, andesite and dacite rocks. The data were collected from a suite of experiments performed at variable normal stresses and two slip rates, showing peak shear stress (P_2) and steady-state shear stress (SS_2) response of the rocks undergoing frictional melting. Typically, at low normal stresses the peak shear stress is higher than the shear resistance anticipated with Byerlee's law (**grey area**), and in more silicic compositions (dacite) the shear stress remains elevated at steady-state sliding conditions. Thus, at the onset of melting shear resistance may be higher than would be in an intact rock and thus the material acts as a viscous brake. As normal stress is increased the shear stress response increases at lower rate, dropping below the zone given for frictional sliding by Byerlee, and thus the melt may be lubricating with respect to intact rock. Data for the basalt is unpublished and data for andesite and dacite is from Kendrick et al. (2014a).

An aspect of frictional melting which is commonly overlooked is the high shear resistance imposed during the onset of melting. The onset of frictional melting always imposes marked strengthening of faults (Tsutsumi and Shimamoto 1997; Hirose and Shimamoto 2005b; Chen and Rempel 2014). This transitional period in which shear resistance ramps from an early stable sliding period of the rock–rock contact (SS_1 in Fig. 3) to a second peak shear resistance (P_2 in Fig. 3) can last several meters, especially at modest slip rates or normal stresses (Tsutsumi and Shimamoto 1997; Lavallée et al. 2012a; Kendrick et al. 2014b; Hornby et al. 2015; Wallace et al. 2019b). The peak shear stress exceeds the shear stress of steady-state sliding in the presence of a thickened melt layer (SS_2 in Fig. 3), and often significantly exceeds the frictional envelope of Byerlee (solid symbols, Fig. 9). This may be highly relevant for frictional melts observed in certain short-lived slip events (e.g., Sibson 1975). As such there is a discrepancy in the way that melt is discussed as a lubricator or viscous brake, as all melts when first generated serve to temporarily increase resistance to slip. As such, situations where flash heating plays a role but melting can be avoided may have the lowest friction coefficients of all (e.g., Tsutsumi and Shimamoto 1997; Rice 2006), whereas those situations in which melting is able to begin may hinder or even halt slip. After the fault zone passes from hosting individual melt patches to hosting a molten layer (e.g., Hirose and Shimamoto 2005b), and the melt zone thickens, melt can be a brake or lubricator controlled by its rheology. Tsutsumi and Shimamoto (1997) stress the importance of the inclusion of such non-linear dynamics into modeling of earthquake initiation processes, and such considerations are equally relevant to faulting in volcanic environments.

Theoretical solutions. Several studies have attempted to describe the effect of slip in a thin melt layer. A suite of constitutive relations and differential equations describe the thickness, temperature and viscosity during viscous fluid flow on a slip plane (Fialko and Khazan 2005; Nielsen et al. 2008, 2010b). Under steady-state conditions, the diffusion equation (which yields a simple thermal profile) further allows constraint of the shear stress (e.g., Nielsen et al. 2008). As the primary aims of these studies are to interpret and mimic experimental observations such that the model may be extended or extrapolated to the study of earthquake source parameters at depth, the models mathematically capture the essential characteristics of the dynamics of fault slip in the presence of melt (e.g., Nielsen et al. 2010b). The summation is that a theoretical solution describing the coupling of shear heating, thermal diffusion, and melt expulsion can be obtained to describe shear stress at steady state slip conditions, via:

$$\tau \approx \sigma_n^{\frac{1}{4}} \left(\frac{A}{\sqrt{r}} \right) \sqrt{\frac{\log\left(\frac{2V_e}{V_c}\right)}{\left(\frac{V_e}{V_c}\right)}} \tag{16}$$

after Nielsen et al. (2010b), where is r the radius of the contact area, A is a dimensional normalizing factor (incorporating rock and melt parameters and another geometrical factor) and V_c is a characteristic rate (that may be considered a critical slip velocity where a behavior transitions from a velocity-hardening to velocity-weakening). Such solutions offer a reasonable fit to laboratory data and are invaluable in the upscaling of laboratory work to various crustal scenarios, and yet the subtleties of slip dynamics are not captured. For example, equivalent viscosity of the melt only depends on slip rate and a series of fixed constitutive parameters; and as a consequence, for fixed velocity, the shear stress will vary only because of changes in thickness of the melt layer. For systems where direct observation is possible (e.g., the laboratory), it is thus advisable to also take a direct approach to the quantification of viscosity.

Melt viscosity. Numerous experimental investigations have noted that slip is velocity weakening (~rate weakening) in the presence of a molten layer (Fig. 10A; Tsutsumi and Shimamoto 1997; Kendrick et al. 2014a; Violay et al. 2014). This means that for a given normal stress, an increase of slip rate results in a decrease in monitored shear stress (and hence decrease in effective friction coefficient). This relationship follows a non-linear reduction in shear stress

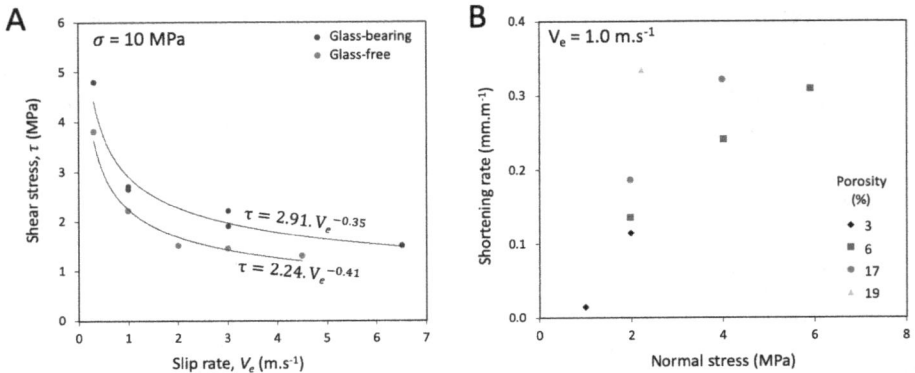

Figure 10. Mechanics of frictional melt. (**A**) The steady-state shear stress in the presence of a frictional melt (SS$_2$) is velocity weakening, with shear stress decreasing as slip rate increases at a given normal stress. Here, glass-bearing and glass-free basalts (from Violay et al 2014) during frictional sliding at 10 MPa normal stress illustrate the effects of shear thinning rheology on shear stress during slip, and highlight that the presence of glass may serve to increase shear stress at a given slip condition, potentially due to widening of the slip zone by viscous remobilization of the adjacent host rock at temperatures above the glass transition. (**B**) Steady-state sliding ensues once a melt layer is produced and has achieved a stable thickness, where melt is expelled at an equivalent rate to its production. Melt production and expulsion is monitored via shortening, shown here as a function of normal stress. Experiments on variably porous basalts (unpublished) demonstrate that shortening rate (measured in millimeters per meter of slip) is faster at higher normal stress, and for a given normal stress is higher for more porous samples.

as a function of slip rate, with the degree of reduction controlled by the specific frictional melt viscosity (Fig. 10A). Tsutsumi and Shimamoto (1997) attribute this to a thicker melt layer at a higher velocity caused by a higher production rate of melt. Whilst such transient phenomena likely contribute to the velocity weakening of melt-bearing faults, experiments by Kendrick et al. (2014a) where slip rate was fluctuated showed the near-instantaneous response of shear stress, more rapidly acclimatizing to the new slip rate than the predicted equilibration to a new melt zone thickness in both the case of velocity increase and decrease. Such observations are key in unravelling the rheology and viscosity of frictional melt suspensions.

As described in the section *Mechanical response to melting*, apparent viscosity of a slip zone can be retrieved from the mechanical data during a friction experiment using Equation (13). This solution relies on constraint of melt thickness, which is assumed to be constant throughout steady state slip due to achieving a balance between melt production rate (controlled by slip rate and normal stress) and melt expulsion (Nielsen et al. 2008; Niemeijer et al. 2011). During experiments shortening is monitored, and as shortening rate achieves a steady state then the melt zone is interpreted to have a stable thickness. In the presence of melt, shortening rate is faster at higher normal stress and faster at higher slip rate, but structural heterogeneities in the host rock, such as grain size or porosity variation can also impact melt thickness. Using experiments on variably porous basalts with the same bulk chemistry and mineralogical assemblages, Figure 10B shows how steady state shortening rate (due to equilibrated melting and melt expulsion) varies as a function of host rock porosity during direct shear experiments: Higher shortening rates are achieved in more porous samples, which is interpreted to be due to the increased ability for melt to infiltrate the host rock at higher porosity, enhancing melting (Hirose and Shimamoto 2005b; Nielsen et al. 2010a). This suggests that higher porosity host rocks could result in thicker melt zones, and that local porosity heterogeneities may impact melt thickness, which has been relatively overlooked to date. This is likely to be important in volcanic rocks which have anisotropic porous networks and porosities spanning 0–98% (e.g., Gonnermann and Manga 2007; Cashman and Scheu 2015).

An approach which takes the chemical composition of the frictional melt, the monitored temperature, thickness and volume of suspended particles to accurately model the apparent viscosity of the frictional melt zone has been utilized by a number of studies (e.g., Lavallée et al. 2012a; Hornby et al. 2015; Hughes et al. 2020). Silicate melts are viscoelastic bodies (Dingwell and Webb 1989). Work on the viscosities of silicate melts has resulted in a statistically robust dataset over a wide compositional range, all displaying strong non-Arrhenian temperature-dependence (Hess and Dingwell 1996; Giordano et al. 2006). A substantial body of work over the last 4–5 decades has dissected through the chemical composition, temperature, pressure and strain rate dependence of silicate melts' viscosity (see this volume). The temperature dependence of melt viscosity is well constrained for a range of compositions, presented as an interactive calculator by Giordano et al. (2008). Their viscosity calculator uses 10 major and minor oxide components and the volatile phases H_2O and F, calibrated against a large dataset that spans the range of most frictional melts which could be formed by selective or wholesale melting of many rocks. Thus, by taking the major and trace element oxide composition of the frictional melt, analyzed by electron probe micro analysis (EPMA), one can estimate the temperature-dependence of melts' viscosity; the range of viscosities of dry rhyolitic, andesitic and basaltic melts are shown in Figure 11A. Hornby et al. (2015) showed how the more mafic early stage melts (protomelts) exhibited lower viscosities than the more evolved melts produced following longer slip durations, thus frictional melts may transiently move through the compositional, temperature and hence viscosity space represented by Figure 11A during their development.

Water may be incorporated into the slip zone during frictional melting, either by the release of volatiles from the melting of hydrous phases, from hydrated interstitial glass in the host rock or from liquid water in pore space (Magloughlin 2011; Lavallée et al. 2015b). The presence of dissolved water in silicate melts serves to decrease their viscosity (Fig. 11B; Hess and Dingwell 1996); the effect is non-linear for a given temperature, such that the first 1% addition of H_2O reduces viscosity more significantly than any subsequent fraction. Water solubility is temperature dependent (see section *Thermal vesiculation*), so in situations where shear heating increases the temperature of frictional melts, the solubility of water decreases (Liu et al. 2005), triggering volatile exsolution which would cause a net increase in the melt viscosity (Hess and Dingwell 1996); yet, this chemical effect would be accompanied by the physical presence of vesicles, which also modify the flow field and the resultant apparent viscosity of the suspension (e.g., Manga et al. 1998). Pseudotachylytes, especially protomelts from selective melting, and those formed in shallow crustal settings are often characterized by vesicular textures (Maddock et al. 1987; Takagi et al. 2007; Lavallée et al. 2015b; De Blasio and Medici 2017; Hughes et al. 2020).

Suspension viscosity. Pseudotachylytes frequently preserve evidence of both vesicles and crystals in suspension. Selective melting during frictional melt production means that minerals with higher melting points may survive as suspended clasts within frictional melt, and vesicles are formed by the incorporation of fluids from mineral breakdown, interstitial glass/melt devolatization or trapping of groundwater.

A substantial body of work has demonstrated the influence of suspended crystals or bubbles on non-Newtonian viscosity and numerous parameterizations have been generated, yet few have tackled their combined impact (Lejeune and Richet 1995; Manga et al. 1998; Llewellin et al. 2002; Caricchi et al. 2007; Lavallée et al. 2007; Costa et al. 2009; Cimarelli et al. 2011; Mueller et al. 2011; Mader et al. 2013; Lesher and Spera 2015; Truby et al. 2015). As an illustrative example of the impact of suspended crystals on viscosity, a range of particle concentrations are plotted in Figure 11C. As particle fraction increases, melt viscosity non-linearly increases (Fig. 11D), following the relationship commonly referred to as the Einstein-Roscoe equation, whereby relative viscosity (η_{rel}) is related to crystallinity (φ) and the maximum packing fraction (φ_0) via:

$$\eta_{rel} = \frac{\eta_{app}}{\eta_m} = \left(1 - \frac{\varphi}{\varphi_0}\right)^{-\frac{5}{2}} \tag{17}$$

where η_{app} is the viscosity of the suspension (apparent viscosity) and η_m is the melt viscosity (see Lesher and Spera (2015) for a review of suspensions' apparent viscosity). For simplicity here, we represent the relative viscosity of a suspension in Figure 11C assuming spherical particles (due to the rounding of suspended particles in a melt following Gibbs–Thomson theory), and a maximum packing fraction of 64% (based on monodisperse spheres). Spherical particles may reasonably approximate equant crystals, but in the presence of elongate crystals, the maximum packing fraction would decrease. Changing aspect ratio or polydispersity of the suspended crystal cargo would change the amount of relative viscosity increase as a function of crystallinity (Fig. 11D), and hence change the absolute viscosities of the suspension at given crystal fractions (Fig. 11C). What Equation (11) fails to capture (Fig. 11C), and which has been demonstrated by a number of studies (e.g., Caricchi et al. 2007), is the reduction in relative viscosity as a function of increasing strain rate (Fig. 11D) owing to the shear thinning nature of magmatic suspensions. This may be caused by the concentration of stress and strain around the bubbles and particles, which has been shown to reduce the maximum shear strain rate that can be relaxed by ~2 orders of magnitude (e.g., Gottsmann et al. 2009; Cordonnier et al. 2012; Coats et al. 2018). Particle size, aspect ratio, dispersity (of size and shape) and maximum packing dictate the magnitude of the reduction in relative viscosity as a function of strain rate (for an example see Fig. 11E). A further impact of suspensions (as opposed to pure melts) during deformation, is that they can evolve with strain, exhibiting strain hardening or strain weakening behavior, for example resulting from pore collapse, or crystal alignment, respectively (Lavallée and Kendrick 2020) adding further to the complexity of suspensions' response to evolving conditions of fault slip.

A few authors have implemented complex non-Newtonian suspension rheology solutions to frictional melts (Fig. 10F; Lavallée et al. 2012a; Hornby et al. 2015; Wallace et al. 2019b; Hughes et al. 2020). In these examples the apparent viscosity of the frictional melt suspension was estimated using the strain rate-dependent rheological model for magmatic suspensions of Costa et al. (2009) using fitting parameters from the experimental data of e.g., Caricchi et al. (2007), and maximum packing fraction estimates with the model of Mueller et al. (2011) for the restite clast population aspect ratio distribution. The temperature monitored during the experiment (either by thermal camera or by thermocouples adjacent to the slip surface) was then mirrored by the viscosity evolution as a function of slip (Fig. 11F), though assumptions must still be made as to strain rate over an evolving thickness of frictional melt.

Modeled values of apparent viscosity can be compared to the viscosity experimentally determined from the mechanical data using Equation (14). The comparison yields the observation that apparent viscosity from mechanical data is lower than that which is modeled, suggesting a number of contributing factors: (1) The temperature monitored on the surface of the sample is likely a minimum value, and on the slip zone may be higher, which means viscosity would be lower than that modeled due to the viscosity of the melt phase itself being lower; (2) Both methods rely upon melt thickness, which is either estimated or based upon dissected samples; in the mechanical case this directly impacts the strain rate and accordingly the apparent viscosity (via Eqns. 13 and 14, respectively) and in the modeled case melt thickness impacts the calculated strain rate (Eqn. 13), which dictates the degree of shear thinning of the suspension; (3) The geometry of the experiment means that slip rate varies across the annular contact zone, and because the suspension is non-Newtonian, this means shear resistance is not equal at all distances from the center of rotation.

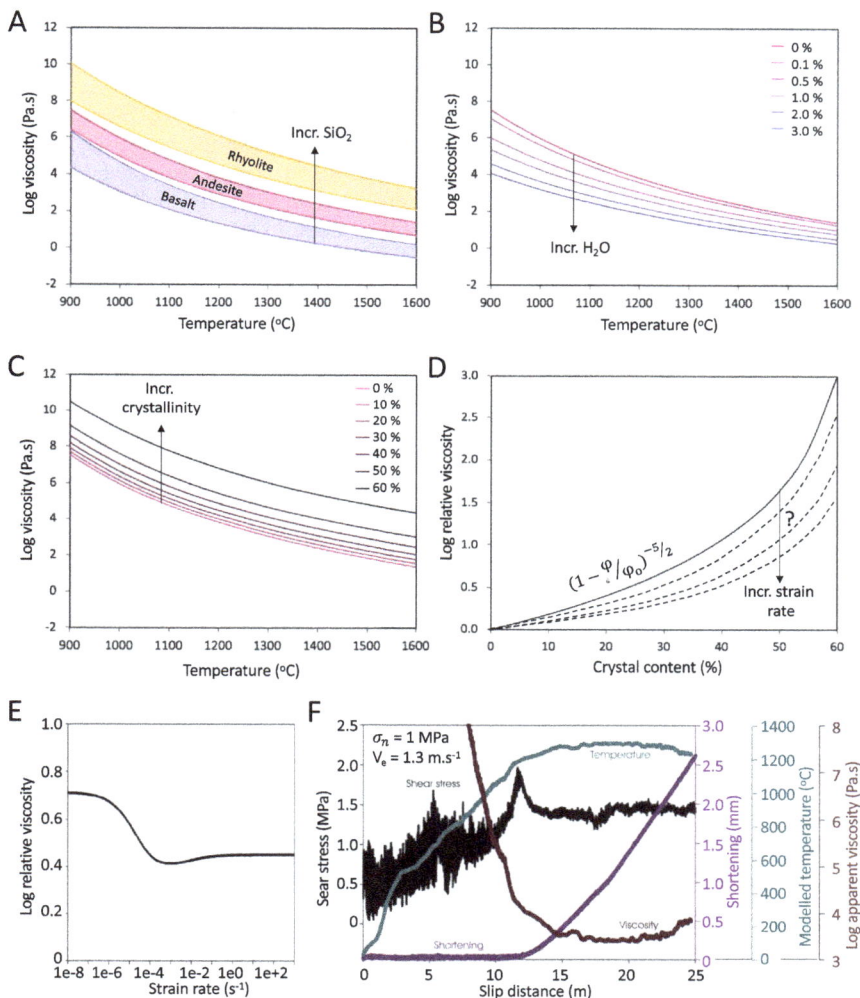

Figure 11. The rheology of silicate melts: (**A**) Effect of chemical composition on melt viscosity. Illustrative examples are given for the range of basalt, andesite and rhyolite compositions (dry endmembers). The effect of increasing SiO_2 is shown, though other major and trace element oxides also exert a control. (**B**) Effect of dissolved water concentration on viscosity of an andesitic melt across a range of temperatures. (**C**) Effect of crystallinity (up to 60 vol.%) on the viscosity of a suspension containing an andesitic melt (the same as in B) across a range of temperatures, modeled using spherical particles (similar to rounded survivor clasts' geometry) and a maximum packing fraction of 64%, using the method Einstein–Roscoe equation, Eqn. 17 (see Lesher and Spera 2015). (**D**) The impact of crystallinity in C on the relative viscosity (i.e., the suspension viscosity divided by the melt viscosity) of suspensions for static conditions; with increased strain rate the relative viscosity of the suspension decreases. (**E**) Plot of the strain rate dependence of relative viscosity for a melt with 30 vol.% crystallinity computed with the Costa et al. (2009) viscosity model (after Lavallée et al. 2012); the data suggests a rate-independent viscosity regime at very low strain rate, although this remains poorly constrained. (**F**) Evolution of apparent viscosity of a frictional melt zone (using the suspension modeled in E based on corrected temperature (monitored by thermocouples near the slip zone) during a rotary shear experiment (using averaged strain rate), with shear stress, shortening and temperature shown. Viscosity decreases rapidly as the melt patches coalesce (during strengthening) and as the melt zone widens (during weakening).

The brittle field. As frictional melts are viscoelastic bodies, they exhibit components of both viscous and elastic materials, in different proportions (e.g., Dingwell and Webb 1989; Webb and Dingwell 1990). The point at which the dominant deformation mechanism switches from one (e.g., viscous) to the other (e.g., elastic) is referred to as the glass transition, T_g, which divides the liquid state from the solid state. In the liquid state, the melt can relax a stress and the body behaves as a Newtonian liquid. In the glass transition interval, the melt struggles to relax the applied stress and may suffer structural breakdown, promoting a non-Newtonian rheology. In the solid state, the melt behaves like a glass and cannot relax the applied stress, so if sufficient stress is accumulated, the material may rupture (e.g., Dingwell and Webb 1989; Webb and Dingwell 1990). The glass transition is a thermo-kinetic barrier, requiring both consideration of temperature and timescale; and thus, strain rates and heating/ cooling rates. The glass transition reflects the relaxation timescale or conversely, the relaxation rate; i.e., if the relaxation rate exceeds the strain rate experienced, then the melt flows viscously, but if it is too slow, the melt may rupture. For convenience, the glass transition is often stated to occur when the melt viscosity exceeds 10^{12} Pa.s, but in detail, the strain rate and heating/cooling rate control the temperature and thus the viscosity at which T_g is crossed. Thus during slip of a melt-bearing fault, the frictional melt may be forced to succumb to the glass transition via several scenarios (e.g., Lavallée et al. 2015a). For example: (1) Intuitively, as slip velocity wanes and thermal input by shear heating decreases, the melt may cool, causing an increase in frictional melt viscosity, and a decrease in relaxation rate; this may result in quenching to a glass or if the strain rate (or applied stress) is great enough it may lead to fragmentation; (2) An increase in slip rate that cannot be equilibrated by thickening of the melt zone resulting in increased strain rate; (3) Local narrowing of the slip plane (e.g., between resistant clasts or minerals) that locally elevates strain rate; (4) A change in the crystal cargo or vesicularity, since suspended clasts or bubbles increases relaxation timescale compared to their pure melt equivalent, which means that the same conditions are more likely to result in brittle failure (Lavallée and Kendrick 2020). Such transient behavior suggests that slip in the presence of frictional melt is not controlled merely by viscosity, but also by an interplay of viscoelastic forces, centered around the glass transition, involving an element of brittle fracturing of the molten phase. This may provide a mechanism for the concurrent formation of pseudotachylytes and cataclasites, and may force us to reassess the active modes of deformation during fault slip, both in volcanic settings and traditionally considered tectonic occurrences.

Viscous remobilization

The presence of glass in many volcanic rocks, and as such the provision of the glass transition, demands for a broadening of the definition of frictional melts, and pseudotachylytes, during faulting in volcanic rocks. Since volcanic rocks often contain a fraction of interstitial groundmass glass the area adjacent to the slip zone may be subject to viscous remobilization during frictional heating (e.g., Lavallée et al. 2012a; Hornby et al. 2015; Wallace et al. 2019b). As previously stated, the temperature of the glass transition is dictated by rates (whether strain rate or heating/ cooling rate) and the silicate melt composition. Faster heating rates or faster shear rates push T_g to higher temperatures (Webb and Dingwell 1990; Gottsmann and Dingwell 2001). Neglecting consideration of rates (and simplifying T_g to a viscosity of 10^{12} Pa.s) T_g ranges from ~600 °C for basaltic glasses to up to ~800 °C for rhyolitic compositions (e.g., Giordano et al. 2008). It is important to note that interstitial glass is more silicic than the bulk composition of volcanic rocks, being the product of fractional crystallization, and that glass is rare in more mafic rocks whose kinetics favor crystallization. During heating in the approach to T_g the structure of a glass begins to seek a state of local equilibrium (with respect to pressure and temperature conditions) via configurational changes in its structure. As T_g is reached, it relaxes any predisposed stress, manifest by a rapid increase in the volumetric expansion rate (Dingwell and Webb 1989). Above T_g the glass behaves as a liquid, able to dissipate applied stresses by relaxation. This characteristic means that adjacent to frictional melts, e.g., that may exceed 1100 °C (Lavallée et al. 2012a, 2015b; Kendrick et al. 2014a; Hornby et al. 2015), the interstitial glass in the wall rock is hot and able to viscously flow. For example Hornby et al. (2015) recognized, via the alignment of microlites, bands of viscous remobilization more than 200 μm wide on either side of experimentally generated frictional melts in dacite lavas with rhyolitic interstitial glass (Fig. 4). The presence of glass may thus widen shear zones, which could lead to the incorporation of more suspended clasts into the shear zone (e.g., Violay et al. 2014) as minerals with higher melting temperature become entrained as the melt/ solid boundary migrates into the rock. Viscous remobilization may also contribute towards the strengthening phase of fault evolution, as melt patches begin to form on the slip surface, potentially serving to increase the magnitude of strengthening at melting onset. Moreover, viscous remobilization could account for the higher shear stress observed in glass-bearing compared to glass free materials (Fig. 10A; Violay et al. 2014). These areas will serve to dissipate strain, reducing the proportion of shear stress exerted on frictional melt (thus promoting an increase in apparent viscosity). In a natural fault slip environment, the presence of glass-bearing wall rocks could facilitate viscous coupling to the frictional melt, increasing the shear resistance during faulting events and potentially hindering slip (Lavallée et al. 2012a; Kendrick et al. 2014a,b; Violay et al. 2014; Hornby et al. 2015; Wallace et al. 2019a). The remobilized areas will have long structural relaxation timescales, and exhibit relatively high viscosities (compared to the frictional melt zone) meaning they may still experience brittle failure if the strain rate exceeds this timescale or shear stress exceeds the strength (Lavallée et al. 2015a). Using single-phase amorphous silicate glasses Lavallée et al. (2015a) showed how the heating rate at which T_g was met controlled the response to frictional sliding, whereby the relaxation timescale, controlled by the melt viscosity according to Maxwell's law of viscoelasticity, enables slip to persist in the solid state until sufficient heat is generated to reduce the viscosity and allow remobilization in the liquid state. Complementarily, in their shear experiments on crushed obsidian at high temperature Okumura et al. (2015) noted cataclastic processes in rapidly sheared shards above T_g, highlighting that comminution may still occur in hot magmas, depending on strain rate, as has been evidenced in several magmatic shear zones (e.g., Cashman et al. 2008; Kendrick et al. 2014b; Wallace et al. 2019a).

Thermal vesiculation

The volatiles dissolved (i.e., primarily water), in frictional melt and in the interstitial melt or glass phase of adjacent magma or wall rock, respectively, can exsolve to form vesicles. Water solubility in melt (Fig. 12A) is dependent on pressure and temperature conditions (Liu et al. 2005); water solubility is retrograde at low pressure, but prograde above ~500 MPa (Paillat et al. 1992). Thus, in volcanic settings and shallow tectonic contexts, the exsolution of water may be triggered by either decompression or by increasing temperature. The latter case may thus be prompted by frictional heating on a slip plane within glass-bearing rocks (Lavallée et al. 2015b). Recently, careful examination of experimental products and volcanic ash generated along active faults during lava dome eruptions showed evidence for thermal vesiculation by volatile release from the interstitial glass of the host andesite (Lavallée et al. 2015b). But the same mechanism may be true for both interstitial glass in the host rock, and for the melt produced by frictional melting. Once melt is present, water can be released, or conversely resorbed, by perturbations of pressure or temperature (Fig. 12). The amount of water released due to fault slip will depend on the magnitude of the heating event, but also on the local pressure conditions which dictates the initial volatile solubility and hence water content of the melt (Fig. 12). Devolatilisation also results in an increase in melt viscosity (see Fig. 11B), and the degree to which water loss increases viscosity depends upon its initial concentration and so, saturation level, the consequence of which is that a devolatilisation of the same amount will more drastically increase the viscosity of a shallow-dwelling melt than a deeper one.

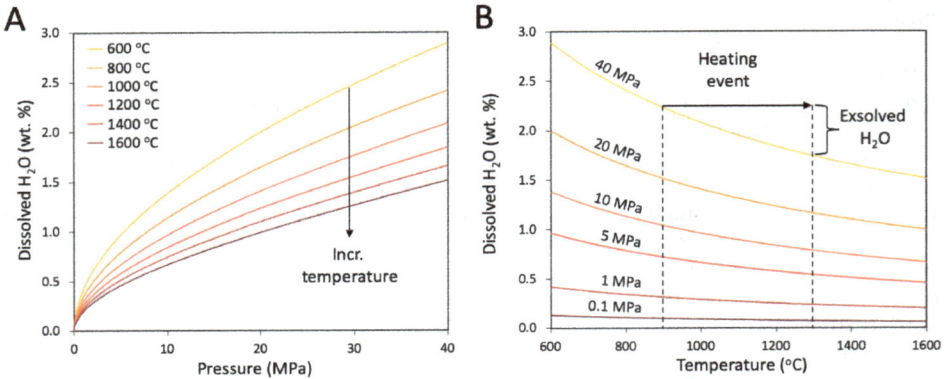

Figure 12. Water solubility in melt: (**A**) Water solubility shown as dissolved H_2O as a function of pressure for a range of temperatures, calculated using the relationship presented in Liu et al. (2005) for a rhyolitic melt composition. (**B**) Thermobarometric limits on water concentration from Liu et al. (2005) show how a heating event such as frictional heating induces exsolution of water, and the magnitude of the exsolution for a given heating event depends on pressure conditions. Calculated using a rhyolitic frictional melts, after Lavallée et al. (2015).

The vesiculation triggered by rapid heating may be sudden, and could even induce fragmentation (due to rapid strain rates, cf. Dingwell 1996), thanks to the increasing viscosity of dehydrating melt (Hess and Dingwell 1996), and lower strength of vesicular materials. This could manifest in a number of ways in conduit-derived volcanic pseudotachylytes. For example, thermal vesiculation may: (1) Be the cause of small sacrificial fragmentation events, whereby localized areas of a lava dome and magmatic conduit experience intense frictional work producing heat, which then vesiculates and fragments in response, producing a small gas-and-ash explosion, but which allows the remainder of the magma to go unperturbed, prolonging its structural stability (Lavallée et al. 2015b; De Angelis et al. 2016); (2) Explain the very steep temperature profiles observed during frictional slip in volcanic rocks deformed experimentally (e.g., Fig. 3A), by causing grain size reduction by fragmentation, increasing surface area and decrepitation, hence lowering melting temperatures via the Gibbs–Thomson equation; (3) Act as a feedback mechanism, with rapid straining allowing melt to remain in the brittle regime, explaining the diachronous occurrence of cataclasites and pseudotachylytes (e.g., Kendrick et al. 2014b), as described in the section on *Brittle failure*.

In other shallow settings such as shallow faults, sector collapses, debris avalanches and gravity slides, vesicular pseudotachylytes have also been noted (Maddock et al. 1987; Takagi et al. 2007). It may be possible to discern if thermal vesiculation played a part in their formation, although distinguishing thermal vesiculation from the incorporation of water from breakdown of hydrous phases during melting of minerals or from saturated rocks is challenging. Identifying thermal vesiculation is further complicated by preservation of pseudotachylytes which is hindered by numerous overprinting mechanisms (Kirkpatrick and Rowe 2013) and the tendency for water to resorb in melt during cooling (Fig. 12), such that porous structures and volatile contents of pseudotachylytes are unlikely to represent their formation characteristics. Where rapid quenching is possible, for example in injection veins or fragments of vesicular pseudotachylytes that are preserved in cataclasite, it may be possible to interpret porous structures for formation conditions (e.g., Maddock et al. 1987). Current calculations of pseudotachylyte formation conditions tend to rest on thermochronology (e.g., Kirkpatrick et al. 2012), though volcanic environments with their high ambient temperatures are not ideal settings for such approaches.

Elevated ambient temperature

We have little data on the influence of elevated starting temperature (such as seen in volcanoes) on fault slip, as such this section has been kept separate in order to briefly examine preliminary experimental observations and speculate as to the cause and consequence of these results in light of our observations of natural systems. The technological challenge of performing experiments at high slip rates over long slip distances at elevated starting temperatures means there is a scarcity of data in this regime. Noda et al. (2011) performed rotary shear friction experiments on dolerite at temperatures up to 1,000 °C, yet they used a slip rate of 0.010 m.s^{-1} at 1 MPa normal stress to isolate the effect of temperature from other weakening mechanisms under stable slip conditions, suggesting that rate weakening observed in the absence of melt may be due in part to temperature increase. Here we will explore the rheological response, temperature profile and textures of a glass-bearing dacite from Mount Unzen (Japan) deformed in a rotary shear apparatus at an ambient temperature of 600 °C (from here-on termed HT) at high velocity (1 m.s^{-1}), a normal stress of 2 MPa over a slip distance of 17 m, generated in Wallace et al. (2019a) in comparison to a room-temperature (RT) equivalent generated in Hornby et al. (2015), neither of which have been previously examined in this light.

Figure 13 shows the shear stress response of both experiments. The RT experiment progressed via the typical stages of fault slip (depicted in Fig. 3), with an initial peak (P$_1$) during rock–rock contact, a weakening phase to lubricated sliding by comminution and flash heating (SS$_1$), a strengthening phase where melt patches initiate, increasing in shear stress to a second peak (P$_2$) that is followed by weakening as the melt zone thickens and stable sliding (SS$_2$) as the melt generation and melt expulsion rate equilibrate. The HT experiment however showed no initial peak (P$_1$), initial weakening, or stable sliding (SS$_1$) of the rock–rock contact, and instead moved directly into strengthening initiated by molten patches forming and potentially viscous remobilization of the interstitial glass of the lava. This strengthening period is followed by a broad and unstable peak (P$_2$) and subsequently a less pronounced weakening phase than the RT equivalent. Stable sliding (SS$_2$) ensues for a brief time, but after ~8 m of slip shear stress begins to increase and becomes markedly higher than in the RT experiment. Until this point the shortening curves are relatively similar despite the differences in shear stress, but at this point they diverge as shortening rate accelerates in the sample at HT. The temperature profiles (dashed lines, Fig. 13) are however disparate from the start. The RT experiment follows a non-linear heating profile (similar to that in Fig. 3), which depicts the progression from comminution and grain size reduction, flash heating and decrepitation that further enhances grain size reduction (Spray 2010), to the initiating of melting of small grains and asperities and their rapid coalescence to form a frictional melt layer (e.g., Hirose and Shimamoto 2005b). In the HT example, heating is near linear. This may reflect the diminished effect of flash heating at elevated starting temperature (e.g., Passelègue et al. 2014), and also the less effective grain size reduction by comminution and by thermal fracturing, since decrepitation can be attributed in part to differential thermal expansion of minerals (Spray 2010), which here would already be equilibrated to the ambient condition of 600 °C. An absence of smaller grains, which preferentially melt (e.g., Lee et al. 2017), would mean that melting progresses more gradually (though, beginning at conditions closer to melting).

A closer examination of the evolution of temperature across the shear zones in the RT and HT experiments (Fig. 14) reveals that not only does peak temperature increase throughout slip, but the width of the high temperature zone increases, meaning that increasing portions of the rock are at temperatures above the glass transition and above the onset of melting (T_{mo}) of the mineral assemblage (In Fig. 14 denoted as d$_g$ for width above T_g and d_{mo} for width above T_{mo}). In the RT example d_g and d_{mo} widen but achieve relatively stable width after around 8 m (8 s) of slip, which corresponds to the attainment of steady state slip (SS$_2$) and a steady shortening rate. On the other hand, in the HT experiment d_g and d_{mo} continue to increase throughout slip (to

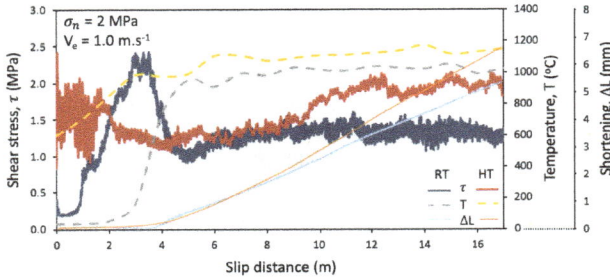

Figure 13. Friction experiments performed on dacite samples at room temperature (RT) and an elevated ambient starting temperature of 600 °C (HT) using a rotary shear apparatus. The results show that the RT experiment follows the conventional progression from P_1 to SS_1, to P_2 and eventually SS_2 similar to the example in Figure 3, and that shortening accelerates from the creation of the melt layer until it achieves steady state, beyond which it follows a linear relationship. However, the HT experiment exhibits no discernible P_1 and SS_1 phases, instead moving directly to melting, shown by the broad peak in shear stress from 0–2 m slip distance; it is followed by a less pronounced weakening stage as the melt zone thickens, a shortening rate that continues to accelerate throughout slip, and a shear stress that also gradually increases throughout the nominal SS_2 phase. These observations are complemented by temperature measurements (showing the maximum temperature on a representative transect perpendicular to the slip surface) of the sample surface which indicate a steady increase from the onset of sliding for the HT sample, compared to a more typical heating profile of the RT sample where temperature ramps rapidly after a few m's slip as melt patches begin to form. In the HT example, melting onset was likely near-instantaneous, and increasing shear stress through SS_2 may represent viscous remobilization of the glass phase and partial melting of mineral phases in the host rock. RT data from Hornby et al. (2015) and HT data from Wallace et al. (2019a).

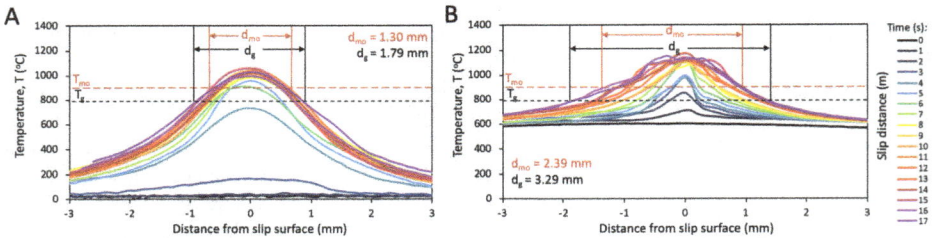

Figure 14. Sample surface temperature perpendicular to the slip surface over a 6 mm wide thermal profile measured every 1m slip in dacite samples using a thermographic camera, for the mechanical data shown in Figure 13 (at 1 m.s⁻¹ and normal stress of 2 MPa). The glass transition temperature $T_g \cong 790$ °C and the melting onset temperature $T_{mo} \cong 900$ °C are marked against the temperature profiles for **(A)** the experiment starting at room-temperature (RT) and **(B)** an experiment starting at 600 °C (HT). Both show how the peak temperature increases through time and eventually stabilizes (after ~7 m at RT and ~6 m at HT); the data show that the hot zone thickens to a relatively stable width in the RT example, but continues to thicken throughout slip (17m) in the HT example. We further characterize d_g as the thickness of the area able to viscously remobilize as the interstitial glass exceeds T_g, and d_{mo}, the thickness of the area where partial melting can occur above T_{mo}, both of which are nearly doubled in width in the HT sample. Reprocessing of RT data from Hornby et al. (2015) and HT data from Wallace et al. (2019a).

17 m), which may account for the increasing shear stress during what ought to be steady-state sliding (SS_2) and to the continuously accelerating shortening rate observed during slip (Fig. 13). In the HT experiment d_g (the portion above T_g) widens near-linearly throughout the experiment, whereas d_{mo} (the portion above T_{mo}) potentially approaches stabilization beyond 15 m slip. Thermal conductivity theory dictates that temperatures should eventually stabilize, but evidence suggests this may be after a longer slip distance at elevated starting temperature. This observation perhaps indicates that slip in volcanic settings, particularly in magmas, would

struggle to achieve steady-state sliding, and that shear stress may continue to increase during slip until the resistance is eventually so high that the viscous brake comes into effect, which could halt slip or cause failure by localizing strain elsewhere.

The absolute values of d_g (width above T_g) are almost twice as large in the HT example, indicating that a much larger area is able to viscously mobilize (Fig. 14). The wider zone of viscous remobilization could explain the higher shear stress in the HT as compared to RT experiment (Fig. 13) as viscous remobilization is postulated to be the cause of higher steady-state shear stress during slip in glass-bearing lavas (cf. Violay et al. 2014). In nature, large areas of viscous remobilization caused by shear localization and heating were invoked to explain the rotation and deformation of phenocrysts and microlites in the upper few 100 m's of the conduit at Mount Unzen (Fig. 1; Wallace et al. 2019a). The zone of viscous remobilization would also have a thermally insulating effect in nature and the experiments, impacting conduction of heat away from the slip zone, and accordingly, the growth of the molten zone (Fig. 1) described by the Stefan problem.

We see evidence of this in the thickness of the melt zone itself, d_{mo} is also almost twice as wide in the HT compared to the RT example (Fig. 14), suggesting that mineral breakdown and partial melting can occur over a wider area in initially-HT shear zones. In the aforementioned natural example at Mount Unzen, the shear zone revealed a systematic variation in thermally-induced breakdown textures in amphiboles which were observed as much as 2 m from the primary slip surface (Wallace et al. 2019a). The width of the areas that exceeded T_g and T_{mo} (Fig. 14) are compared to the dissected HT sample in thin section in Fig. 15. The >250 mm thick melt layer in the HT experiment compares to a ~150 mm layer in the RT experiment (Hornby et al. 2015). The visible impact of viscous remobilization is seen in the microlite alignment in the zone adjacent to the frictional melt (Fig. 15A,C) as well as the compaction of pore space in this region (Fig. 15B). Viscous remobilization has been shown to result in more effective pore closure than brittle deformation (Heap et al. 2017), resulting in densification and anisotropy development, which in natural shear zones would impact the percolation of fluids. In the experimental sample, we see that suspended survivor clasts are primarily higher-melting temperature plagioclase crystals, with some Fe–Ti oxides (Fig. 15C). Decrepitation of minerals adjacent to the slip zone is evident (Fig. 15D), and these directly feed into the melt zone as suspended particles. The onset of mineral breakdown is seen in the rims of plagioclase phenocrysts adjacent to the slip zone (which exceeded T_{mo}), whilst those further away have none (Fig. 15A). Biotite, containing water in its structure, shows vesicular reaction rims adjacent to the melt (Fig. 15E), and microlites adjacent to the slip zone are extensively molten, contributing to heterogeneous schlieren and melts that infiltrate fractures in phenocrysts (Fig. 15D–F).

Many of the textures seen in the HT experimental sample reflect those seen in nature, especially those within volcanic pseudotachylytes (Kendrick et al. 2012, 2014b; Lavallée et al. 2015b). Finally, the onset of crystallization is captured in Fig. 15E. Typically, experimentally generated pseudotachylytes cool very rapidly when slip is halted and the thermal input drops due to cool ambient conditions, but in the HT example the sample was left to cool within the furnace (at ~10 °C.min^{-1}), which provided time for crystallization. Such cooling timescales are rapid compared to that envisaged for pseudotachylytes that form in active volcanic complexes (Kendrick et al. 2014b). This brings an important consideration for the evidence of frictional melting in volcanic environments, as potentially the presence of elevated temperatures after the formation of a pseudotachylyte enables complete crystallization of frictional melt layers and obscures their occurrence. Even if the frictional melt represented only selective melting, and was more mafic than the host rock, then recrystallisation of mafic minerals such as hornblende, pyroxene or magnetite (e.g., Lin 2008) would still surround equant survivor clasts of plagioclase and other minerals with high melting temperatures. Thus, finer grained or more mafic bands may represent historic fault slip and frictional melting in volcanic materials at elevated temperature (Kendrick et al. 2012, 2014b; Plail et al. 2014).

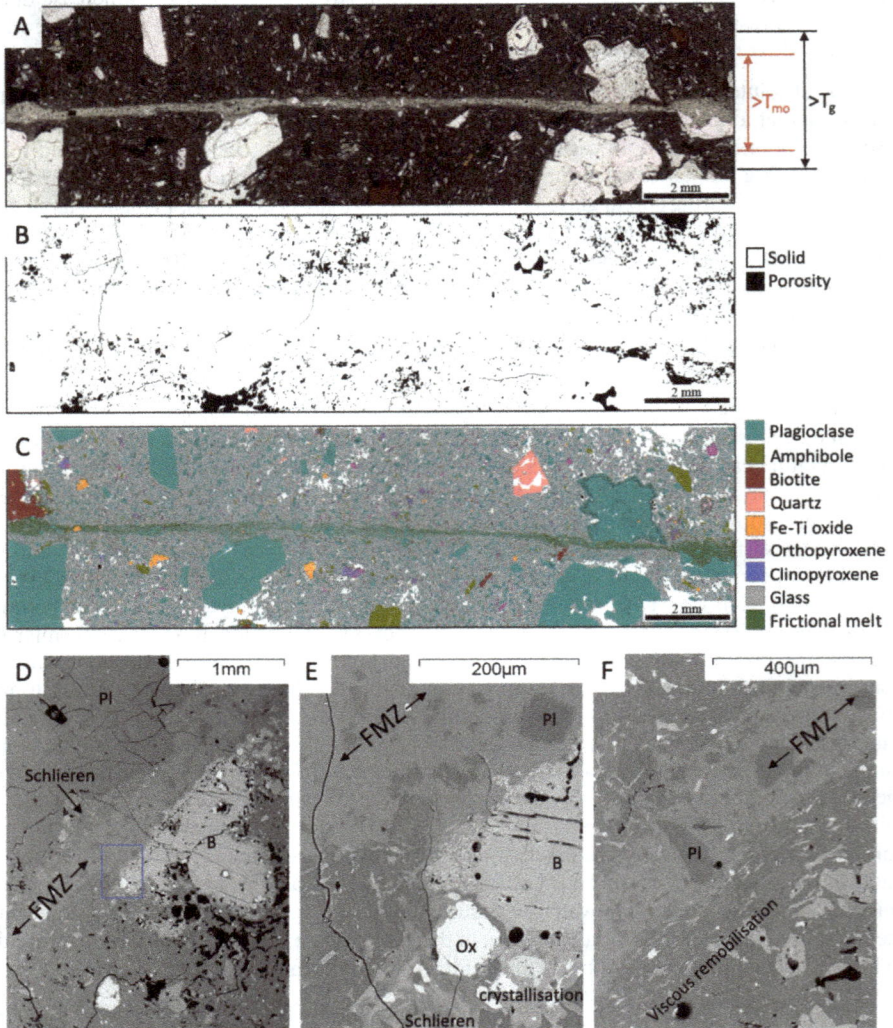

Figure 15. Examination of a frictional melt zone in a dacitic lava, formed during a rotary shear experiment at 600 °C (mechanical data shown in Fig. 13 and temperature evolution in Fig. 14b). The frictional melt zone is shown in (**A**) plane-polarized light using an optical microscope, (**B**) as solid (**white**) and porosity (**black**) binary images generated from (**C**) mineralogical distribution map resolved by QEMSCAN, highlighting embayments of the frictional melt into neighboring phenocrysts (especially biotite), densification of the host rock in areas neighboring the melt, potentially due to partial melting at $>T_{mo}$ and viscous remobilization of interstitial glass $>T_g$. Textures are examined in detail in backscattered electron images of the frictional melt zone (FMZ) in D–F. (**D**) Frictional melt zone sandwiched between a plagioclase [Pl] and biotite [B] phenocryst showing the interaction with both; the plagioclase is fractured, and small fragments are suspended in the frictional melt. Angular embayments are filled by filaments (or schlieren) of protomelts, whereas the biotite has an undulating contact with the melt and has a rim of vesiculation adjacent to the FMZ; the area in the blue box is shown in greater detail in (**E**) where the biotite is directly bordered by protomelts which crystallized during post-experiment cooling (which takes place more slowly following HT experiments due to elevated ambient temperature conditions); note an oxide [Ox] next to the biotite is shown to be stable at the conditions locally experienced. (**F**) Area of viscous remobilization, showing elongate schlierens of partial melt, observed in the host adjacent to the FMZ. The partial melts occupy fractures in phenocrysts of plagioclase to the left of the slip zone and variably rounded survivor clasts of plagioclase are suspended in the melt. Sample from Wallace et al. (2019a).

Fault healing and cyclic rupture

One of the largest unknowns in the study of fault rupture concerns fault healing through time (e.g., Copley 2018). In tectonic settings, sealing, healing and strength recovery of faults has often been considered as time-dependent, occurring over timescales of decades to centuries (e.g., Scholz 2019). In active volcanoes the processes driving healing may differ to tectonic faulting due to their distinct temperature conditions above those typical of the Earth's geotherm. Even in faulting conditions starting from ambient temperatures, recent experimental work has shown that fault strengthening can be a near-instantaneous process upon frictional melting (e.g., Hirose and Shimamoto 2005b; Mitchell et al. 2016; Proctor and Lockner 2016; Hayward and Cox 2017).

Field examination of eruptive products reveals that magma failure in the conduit is a common process (Tuffen et al. 2003; Cashman et al. 2008; Kendrick et al. 2012; Pallister et al. 2013; Lavallée et al. 2015b; Wallace et al. 2019a), influencing the explosivity of ongoing activity (Tuffen and Dingwell 2005; Lensky et al. 2008; De Angelis and Henton 2011; Costa et al. 2012; Okumura and Kozono 2017). Frequently observed cyclic/ repetitive seismic signals at many volcanoes worldwide have been attributed to repeated magma failure or slip on a fault plane at the conduit margin at shallow depth (Fig. 1; Denlinger and Hoblitt 1999; Neuberg et al. 2006; Lensky et al. 2008; Costa et al. 2012; Hornby et al. 2015; Lamb et al. 2015). Evidence for comminution, brecciation (Rust et al. 2004; Goto et al. 2008; Castro et al. 2012), cataclasis (Tuffen and Dingwell 2005; Kennedy et al. 2009; Pallister et al. 2013), frictional melting (Kendrick et al. 2012, 2014b; Plail et al. 2014), and enhanced viscous remobilization of magma near the conduit margin (Wallace et al. 2019a) supports the view that, in viscous magmatic systems, the bulk of the magma is able to ascend as a plug, due to strain localization at the conduit margin (Fig. 1; e.g., Costa et al. 2007b). The consequences of this strain localization at the conduit margin are complex and numerous.

Temporal and spatial evolution of fault zone products results from fluctuations in temperature (e.g., Heap et al. 2017) and variation of local stresses, caused by e.g., changing ascent rate, decreasing normal stress during shallowing, changes in conduit width or geometry or influx of fluids. Once formed, frictional melt can be lubricating, or can act as a viscous brake, and can dynamically shift from one to another. How fault slip occurs will impact how the shear zone evolves, how shear heating is developed, how heat is dissipated (Fig. 1) and how subsequent episodes of slip proceed. These changes act as feedbacks to the ongoing magma ascent, enhancing instability and contributing to the cyclicity of fault slip (e.g., Kendrick et al. 2014a). Fault strengthening occurs at the onset of melting (e.g., Tsutsumi and Shimamoto 1997; Hirose and Shimamoto 2005b; Kendrick et al. 2014a). Where the rocks or magma contain interstitial glass or melt, frictional melt can viscously heal to the host rock on a rapid timescale (~seconds, depending on viscosity), as indicated by the healing timescale of melt (Yoshimura and Nakamura 2010; Lamur et al. 2019). Even without interstitial glass, welding of the frictional melt can activate almost immediately, even during ongoing fault displacement (Hayward and Cox 2017). Depending on the timescale of healing, this could serve to either: (1) if inefficient, allow strain to remain localized on a defined slip plane, requiring overcoming of resistance imposed by the healing surfaces for further slip (Lockner et al. 2017), which in a conduit could accentuate stick slip cycles (Kendrick et al. 2014a); or; (2) if efficient, result in slip zones that are stronger than the rock or magma from which they formed (e.g., Mitchell et al. 2016; Proctor and Lockner 2016; Lee et al. 2020), which would dissipate damage during slip and encourage subsequent fractures to occur in the neighboring areas (e.g., Hayward and Cox 2017), generating multiple slip planes of pseudotachylyte (Kendrick et al. 2014b). Scenario 1 may be most relevant to high viscosity magmatic systems, for example more silicic melts or magmas with higher crystal contents, or cooler settings, which heals comparatively slowly (cf. Lamur et al. 2019), whereas scenario 2 may be most applicable to lower viscosity melts, and highly vesicular magmas in which compaction is efficient (Shields et al. 2016) or porous rocks subjected to melting.

Fault slip and healing will also impact permeability of the fault and adjacent areas, which will impact the drainage conditions of the fault plane (Beeler et al. 2016), which in turn impacts the stress conditions experienced (see Eqn. 2). In viscous systems, e.g., in the presence of melt, this evolution may be particularly dynamic as fracture zones shut, seal and heal, reducing the permeability of the system (Tuffen et al. 2003; Yoshimura and Nakamura 2010; Cabrera et al. 2011; Okumura and Sasaki 2014; Lamur et al. 2019). The effective stress coefficient α (introduced in Eqns. 2–3, which weights the geometric fault area and the real contact area) may shift rapidly due to wetting of the fault interface (e.g., Lamur et al. 2019), thus the area over which pressurized fluids counteract normal stress may be transient in melt-bearing fault zones. The area accessible to fluids defines the ability for the volcano to outgas and alleviate pressure. If the shear zone developed is sufficiently permeable, outgassing may release excess pressure, favoring effusive activity, however, if permeability is lower and pressure can accumulate, it may drive an explosive episode (e.g., Lavallée and Kendrick 2020). In combination with seismicity and ground deformation measurements, gas flux measurements during volcanic crises allude to the percolation of gases through the permeable subsurface (e.g., Edmonds 2008), which can provide clues as to the state of unrest of a volcanic system, though most inferences of conduit processes originate from the examination and experimental interrogation of eruptive products.

Fracture system longevity controls both strength recovery, and the extent to which the magma can outgas before the pathways close and heal, allowing the system to pressurize again in a quasi-continuous cycle (Okumura and Sasaki 2014). A full understanding of the timescale of strength recovery, and thus the ability to build and release stress during cyclic magma ascent remains elusive, yet paramount, for the development of a comprehensive model addressing shifts in eruptive behavior.

SUMMARY AND CONCLUDING REMARKS

Whilst this chapter is by no means a comprehensive overview of frictional melting, it should serve as a starting point, briefly overviewing the occurrence of frictional melting, distinguishing features and approaches used to interrogate pseudotachylyte formation, before moving to focus exclusively on frictional melting in volcanic materials, specifically during the ascent of viscous magma. The ensuing summary briefly encapsulates the observations that have shaped our view of frictional melting in volcanic conduits, rather than attempting to re-cover ground covered earlier in the chapter.

In volcanic environments, the transition from endogenous to exogenous growth can be attributed to a shift in magma rheology into the brittle regime (Hale and Wadge 2008), and thus the ascent of high-viscosity magma can form discrete shear zones along conduit margins. The conduit margin may then be considered analogous to a fault zone, owing to the elastic response of a highly viscous magma subjected to stress variations shorter than the structural relaxation timescale. It has become increasingly recognized, via theoretical models (Costa et al. 2007a; Hale and Muhlhaus 2007; Gonnermann and Manga 2013), experimentation (Lavallée et al. 2012a, 2015a) and detailed petrological constraints (Tuffen and Dingwell 2005; Wright and Weinberg 2009; Wallace et al. 2019a), that strain localization may locally raise magmatic temperatures. The plug-like flow that characterizes high viscosity magma ascent enhances these systems susceptibility to frictional heating, contributing to shifts in rheology that in-part dictate eruption style.

To our benefit a plethora of work has explored the occurrence of pseudotachylytes (e.g., Sibson 1975), their distinguishing features (e.g., Kirkpatrick and Rowe 2013) and the process of formation (e.g., Spray 1992) in a wide range of materials. Rotary shear experiments, were first introduced to the geological community in the early 1990's (Shimamoto and Tsutsumi 1994), spurring an influx of experimental data on frictional melting (Tsutsumi and Shimamoto 1997; Lin and Shimamoto 1998; Nakamura et al. 2002; Hirose and Shimamoto 2005a; Niemeijer et al. 2011). Experiments on volcanic rocks have shown that fault friction localized on narrow planes can generate temperature rises of >1000 °C per second, resulting in melting (Lavallée et al. 2012a, 2015a; Kendrick et al. 2014a,b; Hornby et al. 2015). Frictional melting can alter the physical and chemical properties of the magma: driving mineral reactions; melting crystalline phases; triggering devolatilisation and vesiculation; inducing fragmentation; lowering interstitial melt viscosity; altering magnetic properties; redistributing or closing-off porosity that alters degassing pathways and efficiently healing fractures. Volcanic rocks demonstrate a propensity for frictional melting in as little as a few centimeters of slip, and fewer from high ambient volcanic temperatures (the distinct mineralogy, presence of interstitial glass that allows viscous remobilization and larger surface area afforded by high porosities facilitates rapid melting as compared to many other lithologies), and show that even from room temperature, enough heat is generated to induce frictional melting at as little as 0.1 m.s^{-1}—a rate commonly achieved during magma ascent. Frictional melts have been proposed as one of the major dynamic weakening mechanisms that reduce shear resistance during co-seismic slip, and at high normal stresses melt has a lubricating effect on fault slip. However, during the onset of melting and throughout slip at low normal stresses, particularly in more silicic melts, melt-bearing slip surfaces show exceptionally high shear stresses (compared with a solid rock–rock contact or gouge-hosting shear zone), revealing frictional melt's role as a viscous brake in many scenarios.

While recent experimental efforts have helped define the fault slip processes occurring in volcanic material (Moore et al. 2008; Kennedy et al. 2009; Lavallée et al. 2012a; Kendrick et al. 2014a,b; Hornby et al. 2015; Wallace et al. 2019a,b) the new model of fault-controlled viscous magma ascent lacks a solid mechanical description of dynamic slip processes at the high temperature conditions extant in volcanic conduits. Preliminary results indicate that: (1) comminution and grain size reduction may be less effective at elevated starting temperature; (2) flash heating becomes less effective with increasing ambient temperature; (3) heating may be more linear due to efficiency of heat dissipation from the slip zone and lack of smaller particles; (4) shear zones are likely to be wider at high ambient temperatures (due to thermal insulation and the presence of interstitial glass that allows strain to be distributed over a wider area); (5) shear resistance is likely to be higher in slip zones at higher temperature (potentially due to adjacent viscous remobilization, and wider, more thermally insulated shear zones; Fig. 1); (6) the attainment of steady-state slip will take longer slip distances due to sluggish temperature equilibration; (7) post-slip healing efficiency may be enhanced. All of which sums to point to enhanced instability of fault slip at elevated temperatures. Understanding the conduit margin processes that regulate magma ascent is vital to understanding the style and timescale of an eruption, as well as establishing links to monitored geophysical signals. These recent developments have compelled the volcanological community to begin reassessing the driving forces of eruption dynamics, marking the beginning of a new chapter in volcanological investigations.

ACKNOWLEDGEMENTS

J.E. Kendrick acknowledges the support of an Early Career Fellowship of the Leverhulme Trust. Yan Lavallée was supported by the Natural Environment Research Council (grant no. NE/T007796/1) and a Research Fellowship of the Leverhulme Trust. Both authors were additionally supported by a Starting Grant of the European Research Council (grant no. 306488).

REFERENCES

Abe S, Dieterich JH, Mora P, Place D (2002) Simulation of the influence of rate- and state-dependent friction on the macroscopic behavior of complex fault zones with the lattice solid model. Pure Appl Geophys 159:1967–1983

Acocella V (2021) Volcano flank instability and collapse. *In*: Volcano-Tectonic Processes. Acocella V, (ed) Springer International Publishing, Cham, p 205–244

Alatorre-Ibargüengoitia MA, Scheu B, Dingwell DB, Delgado-Granados H, Taddeucci J (2010) Energy consumption by magmatic fragmentation and pyroclast ejection during Vulcanian eruptions. Earth Planet Sci Lett 291:60–69

Albino F, Pinel V, Massol H, Collombet M (2011) Conditions for detection of ground deformation induced by conduit flow and evolution. J Geophys Res: Solid Earth 116

Alidibirov M, Dingwell DB (2000) Three fragmentation mechanisms for highly viscous magma under rapid decompression. J Volcanol Geotherm Res 100:413–421

Allen AR (1979) Mechanism of frictional fusion in fault zones. J Struct Geol 1:231–243

Andersen TB, Mair K, Austrheim H, Podladchikov YY, Vrijmoed JC (2008) Stress release in exhumed intermediate and deep earthquakes determined from ultramafic pseudotachylyte. Geology 36:995–998

Arciniega-Ceballos A, Alatorre-Ibargüengoitia M, Scheu B, Dingwell D (2015) Analysis of source characteristics of experimental gas burst and fragmentation explosions generated by rapid decompression of volcanic rocks. J Geophys Res: Solid Earth 120:5104–5116

Bain AA, Kendrick JE, Lamur A, Lavallée Y, Calder ES, Cortés JA, Cortés GP, Gómez Martinez D, Torres RA (2021) Micro-textural controls on magma rheology and vulcanian explosion cyclicity. Front Earth Sci 8

Barker SLL (2005) Pseudotachylyte-generating faults in Central Otago, New Zealand. Tectonophysics 397:211–223

Beeler NM, Tullis TE, Goldsby DL (2008) Constitutive relationships and physical basis of fault strength due to flash heating. J Geophys Res: Solid Earth 113:B01401

Beeler NM, Hirth G, Thomas A, Bürgmann R (2016) Effective stress, friction, and deep crustal faulting. J Geophys Res: Solid Earth 121:1040–1059

Behera BM, Thirukumaran V, Soni A, Mishra PK, Biswal TK (2017) Size distribution and roundness of clasts within pseudotachylytes of the Gangavalli Shear Zone, Salem, Tamil Nadu: An insight into its origin and tectonic significance. J Earth Syst Sci 126:46

Ben-Zion Y, Sammis CG (2003) Characterization of Fault Zones. Pure Appl Geophys 160:677–715

Berlo K, Tuffen H, Smith VC, Castro JM, Pyle DM, Mather TA, Geraki K (2013) Element variations in rhyolitic magma resulting from gas transport. Geochim Cosmochim Acta 121:436–451

Bernard K, van Wyk de Vries B (2017) Volcanic avalanche fault zone with pseudotachylite and gouge in French Massif Central. J Volcanol Geotherm Res 347:112–135

Biek RF, Rowley PD, Hacker DB (2019) The Gigantic Markagunt and Sevier Gravity Slides resulting from Mid-Cenozoic catastrophic mega-scale failure of the Marysvale Volcanic Field, Utah, USA. GSA Field Guide 56, Geological Society of America

Biot MA, Willis DG (1957) The elastic coefficients of the theory of consolidation. J Appl Mech 24:594–601

Blundy J, Cashman K, Humphreys M (2006) Magma heating by decompression-driven crystallization beneath andesite volcanoes. Nature 443:76–80

Boulton C, Yao L, Faulkner DR, Townend J, Toy VG, Sutherland R, Ma S, Shimamoto T (2017) High-velocity frictional properties of Alpine Fault rocks: Mechanical data, microstructural analysis, and implications for rupture propagation. J Struct Geol 97:71–92

Bowden FP, Tabor D (1950) The Friction and Lubrication of Solids, Pt 2. Oxford University Press

Brace WF, Byerlee JD (1966) Stick-slip as a mechanism for earthquakes. Science 153:990–992

Brace WF, Martin RJ (1968) A test of the law of effective stress for crystalline rocks of low porosity. Int J Rock Mech Min Sci Geomech Abstr 5:415–426

Brace WF, Paulding B, Scholz C (1966) Dilatancy in the fracture of crystalline rocks. J Geophys Res 71:3939–3953

Brodsky EE, Kanamori H (2001) Elastohydrodynamic lubrication of faults. J Geophys Res 106:16357–16374

Burgisser A, Degruyter W (2015) Magma ascent and degassing at shallow levels. Chapter 11 *In*: The Encyclopedia of Volcanoes (Second Edition). Sigurdsson H (ed) Academic Press, Amsterdam, p 225–236

Burton MR, Mader HM, Polacci M (2007) The role of gas percolation in quiescent degassing of persistently active basaltic volcanoes. Earth Planet Sci Lett 264:46–60

Byerlee JD (1967) Frictional characteristics of granite under high confining pressure. J Geophys Res 72:3639–3648

Byerlee J (1978) Friction of rocks. Pure Appl Geophys 116:615–626

Cabrera A, Weinberg RF, Wright HMN, Zlotnik S, Cas RAF (2011) Melt fracturing and healing: A mechanism for degassing and origin of silicic obsidian. Geology 39:67–70

Calder ES, Lavallée Y, Kendrick JE, Bernstein M (2015) Lava dome eruptions. Chapter 18 *In*: The Encyclopedia of Volcanoes (Second Edition). Sigurdsson H, (ed) Academic Press, Amsterdam, p 343–362

Camacho A, Vernon RH, Fitz Gerald JD (1995) Large volumes of anhydrous pseudotachylyte in the Woodroffe Thrust, eastern Musgrave Ranges, Australia. J Struct Geol 17:371–383

Caricchi L, Burlini L, Ulmer P, Gerya T, Vassalli M, Papale P (2007) Non-Newtonian rheology of crystal-bearing magmas and implications for magma ascent dynamics. Earth Planet Sci Lett 264:402–419

Carslaw HS, Jaeger JC (1959) Conduction of Heat in Solids. Oxford University Press

Cashman K, Blundy J (2000) Degassing and crystallization of ascending andesite and dacite. Philos Trans R Soc London 358:1487–1513

Cashman C, Scheu B (2015) Magmatic fragmentation. Chaper 25 *In*: Encyclopedia of Volcanoes. 2nd Edition. Sigurdsson H, Houghton B, Rymer H, Stix J, McNutt S (eds). Elsevier, p 459–471

Cashman KV, Thornber CR, Pallister JS (2008) From dome to dust: shallow crystallization and fragmentation of conduit magma during the 2004–2006 dome extrusion of Mount St. Helens, Washington. *In*: A Volcano Rekindled: The Renewed Eruption of Mount St Helens, 2004–2006. Sherrod DR, Scott WE, Stauffer PH, (eds). U.S. Geological Survey Professionnal Paper 1750, p 387–413

Cassidy M, Manga M, Cashman K, Bachmann O (2018) Controls on explosive–effusive volcanic eruption styles. Nat Commun 9:2839

Castro JM, Cordonnier B, Tuffen H, Tobin MJ, Puskar L, Martin MC, Bechtel HA (2012) The role of melt–fracture degassing in defusing explosive rhyolite eruptions at volcán Chaitén. Earth Planet Sci Lett 333–334:63–69

Chadwick Jr WW, Archuleta RJ, Swanson DA (1988) The mechanics of ground deformation precursory to dome-building extrusions at Mount St. Helens 1981–1982. J Geophys Res: Solid Earth 93:4351–4366

Chen J, Rempel AW (2014) Progressive flash heating and the evolution of high-velocity rock friction. J Geophys Res: Solid Earth 119:3182–3200

Chen X, Elwood Madden AS, Reches Ze (2017) Friction evolution of granitic faults: heating controlled transition from powder lubrication to frictional melt. J Geophys Res: Solid Earth 122:9275–9289

Chester FM, Chester JS (1998) Ultracataclasite structure and friction processes of the Punchbowl fault, San Andreas system, California. Tectonophysics 295:199–221

Chiodini G, Paonita A, Aiuppa A, Costa A, Caliro S, De Martino P, Acocella V, Vandemeulebrouck J (2016) Magmas near the critical degassing pressure drive volcanic unrest towards a critical state. Nat Commun 7:13712

Chouet BA, Matoza RS (2013) A multi-decadal view of seismic methods for detecting precursors of magma movement and eruption. J Volcanol Geotherm Res 252:108–175

Christie JM, Griggs DT, Heuer AH, Nord GL, Jr., Radcliffe SV, Lally JS, Fisher RM (1973) Electron petrography of Apollo 14 and 15 breccias and shock-produced analogs. Lunar Planet Sci Confer Proc 4:365

Cimarelli C, Costa A, Mueller S, Mader HM (2011) Rheology of magmas with bimodal crystal size and shape distributions: Insights from analog experiments. Geochem Geophys Geosyst 12:Q07024

Coats R, Kendrick JE, Wallace PA, Miwa T, Hornby AJ, Ashworth JD, Matsushima T, Lavallée Y (2018) Failure criteria for porous dome rocks and lavas: a study of Mt. Unzen, Japan. Solid Earth Discuss 2018:1–42

Copley A (2018) The strength of earthquake-generating faults. J Geol Soc 175:1–12

Cordonnier B, Caricchi L, Pistone M, Castro J, Hess K-U, Gottschaller S, Manga M, Dingwell DB, Burlini L (2012) The viscous–brittle transition of crystal-bearing silicic melt: Direct observation of magma rupture and healing. Geology 40:611–614

Costa A, Macedonio G (2002) Nonlinear phenomena in fluids with temperature-dependent viscosity: An hysteresis model for magma flow in conduits. Geophys Res Lett 29:1402

Costa A, Macedonio G (2005) Viscous heating effects in fluids with temperature-dependent viscosity: triggering of secondary flows. J Fluid Mech 540:21–38

Cowie PA, Scholz CH (1992) Displacement-length scaling relationship for faults: data synthesis and discussion. J Struct Geol 14:1149–1156

Costa A, Melnik O, Vedeneeva E (2007a) Thermal effects during magma ascent in conduits. J Geophys Res: Solid Earth 112:B12205

Costa A, Melnik O, Sparks RSJ, Voight B (2007b) Control of magma flow in dykes on cyclic lava dome extrusion. Geophys Res Lett 34: L02303

Costa A, Caricchi L, Bagdassarov N (2009) A model for the rheology of particle-bearing suspensions and partially molten rocks. Geochemistry, Geophysics, Geosystems 10:Q03010

Costa A, Wadge G, Melnik O (2012) Cyclic extrusion of a lava dome based on a stick-slip mechanism. Earth Planet Sci Lett 337–338:39–46

Curewitz D, Karson JA (1999) Ultracataclasis, sintering, and frictional melting in pseudotachylytes from East Greenland. J Struct Geol 21:1693–1713

De Angelis S, Henton SM (2011) On the feasibility of magma fracture within volcanic conduits: Constraints from earthquake data and empirical modelling of magma viscosity. Geophys Res Lett 38:L19310

De Angelis S, Lamb OD, Lamur A, Hornby AJ, von Aulock F, Chigna G, Lavallée Y, Rietbrock A (2016) Characterization of moderate ash-and-gas explosions at Santiaguito volcano, Guatemala, from infrasound waveform inversion and thermal infrared measurements. Geophys Res Lett 43:6220–6227

De Blasio FV, Elverhøi A (2008) A model for frictional melt production beneath large rock avalanches. J Geophys Res: Earth Surface 113:F02014

De Blasio FV, Medici L (2017) Microscopic model of rock melting beneath landslides calibrated on the mineralogical analysis of the Köfels frictionite. Landslides 14:337–350

de' Michieli Vitturi M, Clarke AB, Neri A, Voight B (2008) Effects of conduit geometry on magma ascent dynamics in dome-forming eruptions. Earth Planet Sci Lett 272:567–578

Del Gaudio P, Di Toro G, Han R, Hirose T, Nielsen S, Shimamoto T, Cavallo A (2009) Frictional melting of peridotite and seismic slip. J Geophys Res: Solid Earth 114:B06306

Denlinger RP, Hoblitt RP (1999) Cyclic eruptive behavior of silicic volcanoes. Geology 27:459–462

Deseta N, Andersen TB, Ashwal LD (2014) A weakening mechanism for intermediate-depth seismicity? Detailed petrographic and microtextural observations from blueschist facies pseudotachylytes, Cape Corse, Corsica. Tectonophysics 610:138–149

Deubelbeiss Y, Kaus BJP, Connolly JAD, Caricchi L (2011) Potential causes for the non-Newtonian rheology of crystal-bearing magmas. Geochem Geophys Geosyst 12:Q05007

Di Toro G, Pennacchioni G (2005) Fault plane processes and mesoscopic structure of a strong-type seismogenic fault in tonalites (Adamello batholith, Southern Alps). Tectonophysics 402:55–80

Di Toro G, Goldsby DL, Tullis TE (2004) Friction falls towards zero in quartz rock as slip velocity approaches seismic rates. Nature 427:436–439

Di Toro G, Hirose T, Nielsen S, Pennacchioni G, Shimamoto T (2006) Natural and experimental evidence of melt lubrication of faults during earthquakes. Science 311:647–649

Di Toro G, Pennacchioni G, Nielsen S (2008) Pseudotachylytes and earthquake source mechanics. Chapter 5 *In:* International Geophysics. Vol. 94 (Eiichi F ed.) Academic Press, p 87–133

Di Toro G, Han R, Hirose T, De Paola N, Nielsen S, Mizoguchi K, Ferri F, Cocco M, Shimamoto T (2011) Fault lubrication during earthquakes. Nature 471:494–498

Dieterich JH (1972) Time-dependent friction in rocks. J Geophys Res (1896–1977) 77:3690–3697

Dieterich JH (1979) Modeling of rock friction: 1. Experimental results and constitutive equations. J Geophys Res: Solid Earth 84:2161–2168

Dieterich JH (1992) Earthquake nucleation on faults with rate-and state-dependent strength. Tectonophysics 211:115–134

Dieterich JH, Kilgore BD (1994) Direct observation of frictional contacts: New insights for state-dependent properties. Pure Appl Geophys 143:283–302

Diller K, Clarke AB, Voight B, Neri A (2006) Mechanisms of conduit plug formation: Implications for vulcanian explosions. Geophys Res Lett 33:L20302

Dingwell DB (1996) Volcanic dilemma: Flow or blow? Science 273:1054–1055

Dingwell D, Webb SL (1989) Structural relaxation in silicate melts and Non-Newtonian melt rheology in geologic processes. Phys Chem Mineral 16:508–516

Edmonds M (2008) New geochemical insights into volcanic degassing. Philos Trans R Soc A 366:4559–4579

Edmonds M, Herd RA (2007) A volcanic degassing event at the explosive–effusive transition. Geophys Res Lett 34:L21310

Evans JP, Forster CB, Goddard JV (1997) Permeability of fault-related rocks, and implications for hydraulic structure of fault zones. J Struct Geol 19:1393–1404

Faulkner DR, Mitchell TM, Jensen E, Cembrano J (2011) Scaling of fault damage zones with displacement and the implications for fault growth processes. J Geophys Res 116:B05403

Ferrand TP, Nielsen S, Labrousse L, Schubnel A (2021) Scaling seismic fault thickness from the laboratory to the field. J Geophys Res: Solid Earth 126:e2020JB020694

Ferré EC, Zechmeister MS, Geissman JW, MathanaSekaran N, Kocak K (2005) The origin of high magnetic remanence in fault pseudotachylites: Theoretical considerations and implication for coseismic electrical currents. Tectonophysics 402:125–139

Ferré EC, Geissman JW, Zechmeister MS (2012) Magnetic properties of fault pseudotachylytes in granites. J Geophys Res: Solid Earth 117:B01106

Fialko Y, Khazan Y (2005) Fusion by earthquake fault friction: Stick or slip? J Geophys Res: Solid Earth 110:B12407

Freund F, Salgueiro da Silva MA, Lau BWS, Takeuchi A, Jones HH (2007) Electric currents along earthquake faults and the magnetization of pseudotachylite veins. Tectonophysics 431:131–141

Fukuchi T (2003) Strong ferrimagnetic resonance signal and magnetic susceptibility of the Nojima pseudotachylyte in Japan and their implication for coseismic electromagnetic changes. J Geophys Res: Solid Earth 108: 2312

Gaunt HE, Sammonds PR, Meredith PG, Smith R, Pallister JS (2014) Pathways for degassing during the lava dome eruption of Mount St. Helens 2004–2008. Geology 42:947–950

Giordano D, Mangiacapra A, Potuzak M, Russell JK, Romano C, Dingwell DB, Di Muro A (2006) An expanded non-Arrhenian model for silicate melt viscosity: A treatment for metaluminous, peraluminous and peralkaline liquids. Chem Geol 229:42–56

Giordano D, Russell JK, Dingwell DB (2008) Viscosity of magmatic liquids: A model. Earth Planet Sci Lett 271:123–134

Gonnermann HM, Manga M (2007) The fluid mechanics inside a volcano. Annu Rev Fluid Mech 39:321–356

Gonnermann HM, Manga M (2013) Dynamics of magma ascent in the volcanic conduit. *In:* Modeling Volcanic Processes: The Physics and Mathematics of Volcanism. Lopes RMC, Fagents SA, Gregg TKP, (eds). Cambridge University Press, Cambridge, p 55–84

Goodwin LB (1999) Controls on pseudotachylyte formation during tectonic exhumation in the South Mountains metamorphic core complex, Arizona. Geol Soc London Spec Publ 154:325–342

Goto Y, Nakada S, Kurokawa M, Shimano T, Sugimoto T, Sakuma S, Hoshizumi H, Yoshimoto M, Uto K (2008) Character and origin of lithofacies in the conduit of Unzen volcano, Japan. J Volcanol Geotherm Res 175:45–59

Gottsmann J, Dingwell DB (2001) Cooling dynamics of spatter-fed phonolite obsidian flows on Tenerife, Canary Islands. J Volcanol Geotherm Res 105:323–342

Gottsmann J, Lavallée Y, Martí J, Aguirre-Díaz G (2009) Magma–tectonic interaction and the eruption of silicic batholiths. Earth Planet Sci Lett 284:426–434

Green DN, Neuberg J (2006) Waveform classification of volcanic low-frequency earthquake swarms and its implication at Soufrière Hills Volcano, Montserrat. J Volcanol Geotherm Res 153:51–63

Green DN, Neuberg J, Cayol V (2006) Shear stress along the conduit wall as a plausible source of tilt at Soufriere Hills volcano, Montserrat. Geophys Res Lett 33:L10306

Grunewald U, Sparks RSJ, Kearns S, Komorowski JC (2000) Friction marks on blocks from pyroclastic flows at the Soufriere Hills volcano, Montserrat: Implications for flow mechanisms. Geology 28:827–830

Guo Y, Morgan JK (2006) The frictional and micromechanical effects of grain comminution in fault gouge from distinct element simulations. J Geophys Res: Solid Earth 111:B12406

Hacker DB, Biek RF, Rowley PD (2014) Catastrophic emplacement of the gigantic Markagunt gravity slide, southwest Utah (USA): Implications for hazards associated with sector collapse of volcanic fields. Geology 42:943–946

Hale AJ, Muhlhaus HB (2007) Modelling shear bands in a volcanic conduit: Implications for over-pressures and extrusion-rates. Earth Planet Sci Lett 263:74–87

Hale AJ, Wadge G (2008) The transition from endogenous to exogenous growth of lava domes with the development of shear bands. J Volcanol Geotherm Res 171:237–257

Hale AJ, Wadge G, Muhlhaus HB (2007) The influence of viscous and latent heating on crystal-rich magma flow in a conduit. Geophys J Int 171:1406–1429

Han R, Shimamoto T, Hirose T, Ree J-H, Ando J-i (2007) Ultralow friction of carbonate faults caused by thermal decomposition. Science 316:878–881

Hayward KS, Cox SF (2017) Melt welding and its role in fault reactivation and localization of fracture damage in seismically active faults. J Geophys Res: Solid Earth 122:9689–9713

Heap MJ, Violay M, Wadsworth FB, Vasseur J (2017) From rock to magma and back again: The evolution of temperature and deformation mechanism in conduit margin zones. Earth Planet Sci Lett 463:92–100

Hess KU, Dingwell DB (1996) Viscosities of hydrous leucogranitic melts: A non-Arrhenian model. Am Mineral 81:1297–1300

Hetzel R, Altenberger U, Strecker MR (1996) Structural and chemical evolution of pseudotachylytes during seismic events. Mineral Petrol 58:33–50

Hirose T, Bystricky M (2007) Extreme dynamic weakening of faults during dehydration by coseismic shear heating. Geophys Res Lett 34:L14311

Hirose T, Shimamoto T (2003) Fractal dimension of molten surfaces as a possible parameter to infer the slip-weakening distance of faults from natural pseudotachylytes. J Struct Geol 25:1569–1574

Hirose T, Shimamoto T (2005a) Slip-weakening distance of faults during frictional melting as inferred from experimental and natural pseudotachylytes. Bull Seismol Soc Am 95:1666–1673

Hirose T, Shimamoto T (2005b) Growth of molten zone as a mechanism of slip weakening of simulated faults in gabbro during frictional melting. J Geophys Res: Solid Earth 110:B05202

Holland TH (1900) The Charnockite Series, A group of Archean hypersthenic rocks in peninsular India. Mem Geol Surv India 28:119–249

Hornby AJ, Kendrick JE, Lamb OD, Hirose T, De Angelis S, von Aulock FW, Umakoshi K, Miwa T, Henton De Angelis S, Wadsworth FB, Hess KU (2015) Spine growth and seismogenic faulting at Mt. Unzen, Japan. J Geophys Res: Solid Earth 20:4034–4054

Hughes A, Kendrick JE, Salas G, Wallace PA, Legros F, Di Toro G, Lavallée Y (2020) Shear localisation, strain partitioning and frictional melting in a debris avalanche generated by volcanic flank collapse. J Struct Geol 140:104132

Ishikawa T, Ujiie K (2019) Geochemical analysis unveils frictional melting processes in a subduction zone fault. Geology 47:343–346

Jiang H, Lee C-TA, Morgan JK, Ross CH (2015) Geochemistry and thermodynamics of an earthquake: A case study of pseudotachylites within mylonitic granitoid. Earth Planet Sci Lett 430:235–248

Kendrick JE, Lavallée Y, Ferk A, Perugini D, Leonhardt R, Dingwell DB (2012) Extreme frictional processes in the volcanic conduit of Mount St. Helens (USA) during the 2004–2008 eruption. J Struct Geol 38:61–76

Kendrick JE, Lavallée Y, Hess KU, Heap MJ, Gaunt HE, Meredith PG, Dingwell DB (2013) Tracking the permeable porous network during strain-dependent magmatic flow. J Volcanol Geotherm Res 260:117–126

Kendrick JE, Lavallée Y, Hirose T, Di Toro G, Hornby AJ, De Angelis S, Dingwell DB (2014a) Volcanic drumbeat seismicity caused by stick-slip motion and magmatic frictional melting. Nature Geosci 7:438–442

Kendrick JE, Lavallée Y, Hess KU, De Angelis S, Ferk A, Gaunt HE, Meredith PG, Dingwell DB, Leonhardt R (2014b) Seismogenic frictional melting in the magmatic column. Solid Earth 5:199–208

Kendrick JE, Lavallée Y, Mariani E, Dingwell DB, Wheeler J, Varley NR (2017) Crystal plasticity as an indicator of the viscous-brittle transition in magmas. Nat Commun 8:1926

Kennedy FE (2000) Frictional heating and contact temperatures. *In*: Modern Tribology Handbook. Bhushan B (ed) CRC Press

Kennedy LA, Russell JK (2012) Cataclastic production of volcanic ash at Mount Saint Helens. Phys Chem Earth, Parts A/B/C 45–46:40–49

Kennedy LA, Russell JK, Nelles E (2009) Origins of Mount St. Helens cataclasites: Experimental insights. Am Mineral 94:995–1004

Killick AM (1990) Pseudotachylite generated as a result of a drilling "burn-in". Tectonophysics 171:221–227

Killick AM (1994) The geochemistry of pseudotachylyte and its host rocks from the West Rand Goldfield, Witwatersrand Basin, South Africa: implications for pseudotachylyte genesis. Lithos 32:193–205

Kim JW, Ree JH, Han R, Shimamoto T (2010) Experimental evidence for the simultaneous formation of pseudotachylyte and mylonite in the brittle regime. Geology 38:1143–1146

Kirkpatrick JD, Rowe CD (2013) Disappearing ink: How pseudotachylytes are lost from the rock record. J Struct Geol 52:183–198

Kirkpatrick JD, Shipton ZK, Persano C (2009) Pseudotachylytes: rarely generated, rarely preserved, or rarely reported? Bull Seismol Soc Am 99:382–388

Kirkpatrick JD, Dobson KJ, Mark DF, Shipton ZK, Brodsky EE, Stuart FM (2012) The depth of pseudotachylyte formation from detailed thermochronology and constraints on coseismic stress drop variability. J Geophys Res: Solid Earth 117:B06406

Kirkpatrick JD, Rowe CD, White JC, Brodsky EE (2013) Silica gel formation during fault slip: Evidence from the rock record. Geology 41:1015–1018

Kitajima H, Chester JS, Chester FM, Shimamoto T (2010) High-speed friction of disaggregated ultracataclasite in rotary shear: Characterization of frictional heating, mechanical behavior, and microstructure evolution. J Geophys Res: Solid Earth 115:B08408

Kokelaar P (2007) Friction melting, catastrophic dilation and breccia formation along caldera superfaults. J Geol Soc 164:751–754

Lamb OD, Varley NR, Mather TA, Pyle DM, Smith PJ, Liu EJ (2014) Multiple timescales of cyclical behaviour observed at two dome-forming eruptions. J Volcanol Geotherm Res 284:106–121

Lamb OD, De Angelis S, Umakoshi K, Hornby AJ, Kendrick JE, Lavallée Y (2015) Repetitive fracturing during spine extrusion at Unzen volcano, Japan. Solid Earth 6:1277–1293

Lamur A, Kendrick JE, Wadsworth FB, Lavallée Y (2019) Fracture healing and strength recovery in magmatic liquids. Geology 47:195–198

Landman U, Luedtke WD, Gao J (1996) Atomic-scale issues in tribology: interfacial junctions and nano-elastohydrodynamics. Langmuir 12:4514–4528

Lavallée Y, Kendrick JE (2020) A review of the physical and mechanical properties of volcanic rocks and magmas in the brittle and ductile fields. In: Forecasting and planning for volcanic hazards, risks, and disasters. Papale P, (ed) Elsevier

Lavallée Y, Kendrick JE (2022) Strain localization in magmas. Rev Mineral Geochem 87:721-765

Lavallée Y, Hess KU, Cordonnier B, Dingwell DB (2007) Non-Newtonian rheological law for highly crystalline dome lavas. Geology 35:843–846

Lavallée Y, Mitchell TM, Heap MJ, Vasseur J, Hess K-U, Hirose T, Dingwell DB (2012a) Experimental generation of volcanic pseudotachylytes: Constraining rheology. J Struct Geol 38:222–233

Lavallée Y, Varley N, Alatorre-Ibargüengoitia M, Hess KU, Kueppers U, Mueller S, Richard D, Scheu B, Spieler O, Dingwell D (2012b) Magmatic architecture of dome-building eruptions at Volcán de Colima, Mexico. Bull Volcanol 74:249–260

Lavallée Y, Benson PM, Heap MJ, Hess KU, Flaws A, Schillinger B, Meredith PG, Dingwell DB (2013) Reconstructing magma failure and the degassing network of dome-building eruptions. Geology 41:515–518

Lavallée Y, Hirose T, Kendrick JE, Hess K-U, Dingwell DB (2015a) Fault rheology beyond frictional melting. PNAS 112:9276

Lavallée Y, Dingwell DB, Johnson JB, Cimarelli C, Hornby AJ, Kendrick JE, Von Aulock FW, Kennedy BM, Andrews BJ, Wadsworth FB, Rhodes E (2015b) Thermal vesiculation during volcanic eruptions. Nature 528:544–547

Lee SK, Han R, Kim EJ, Jeong GY, Khim H, Hirose T (2017) Quasi-equilibrium melting of quartzite upon extreme friction. Nat Geosci 10:436–441

Lee AL, Lloyd GE, Torvela T, Walker AM (2020) Evolution of a shear zone before, during and after melting. J Geol Soc 177:738–751

Legros F, Cantagrel JM, Devouard B (2000) Pseudotachylyte (frictionite) at the base of the Arequipa volcanic landslide deposit (Peru): Implications for emplacement mechanisms. J Geol 108:601–611

Lejeune AM, Richet P (1995) Rheology of crystal-bearing silicate melts—an experimental study at high viscosities. J Geophys Res 100:4215–4229

Lensky NG, Sparks RSJ, Navon O, Lyakhovsky V (2008) Cyclic activity at Soufrière Hills Volcano, Montserrat: degassing-induced pressurization and stick-slip extrusion. Geol Soc London Spec Publ 307:169–188

Lesher CE, Spera FJ (2015) Thermodynamic and transport properties of silicate melts and magma. Chapter 5 In: The Encyclopedia of Volcanoes (Second Edition). Sigurdsson H, (ed) Academic Press, Amsterdam, p 113–141

Lin A (1994a) Microlite morphology and chemistry in pseudotachylite, from the Fuyun Fault Zone, China. J Geol 102:317–329

Lin A (1994b) Glassy pseudotachylyte veins from the Fuyun fault zone, northwest China. J Struct Geol 16:71–83

Lin AM (1999) Roundness of clasts in pseudotachylytes and cataclastic rocks as an indicator of frictional melting. J Struct Geol 21:473–478

Lin A (2007) Fossil earthquakes: The Formation and Preservation of Pseudotachylytes. Lecture Notes in Earth Sciences. Springer Berlin Heidelberg, Berlin, Heidelberg

Lin A (2008) Chemical composition and melting processes of pseudotachylyte. In: Fossil Earthquakes: The Formation and Preservation of Pseudotachylytes. Lin A (ed) Springer Berlin Heidelberg, Berlin, Heidelberg, p 159–176

Lin A, Shimamoto T (1998) Selective melting processes as inferred from experimentally generated pseudotachylytes. J Asian Earth Sci 16:533–545

Lin A, Chen A, Liau C-F, Lee C-T, Lin C-C, Lin P-S, Wen S-C, Ouchi T (2001) Frictional fusion due to coseismic landsliding during the 1999 Chi-Chi (Taiwan) ML 7.3 Earthquake. Geophys Res Lett 28:4011–4014

Linker MF, Dieterich JH (1992) Effects of variable normal stress on rock friction: observations and constitutive equations. J Geophys Res 97:4923–4940

Liu Y, Zhang Y, Behrens H (2005) Solubility of H_2O in rhyolitic melts at low pressures and a new empirical model for mixed H_2O–CO_2 solubility in rhyolitic melts. J Volcanol Geotherm Res 143:219–235

Llewellin EW, Mader HM, Wilson SDR (2002) The constitutive equation and flow dynamics of bubbly magmas. Geophys Res Lett 29:2170

Lockner DA, Kilgore BD, Beeler NM, Moore DE (2017) The transition from frictional sliding to shear melting in laboratory stick-slip experiments. *In*: Fault Zone Dynamic Processes: Evolution of Fault Properties during Seismic Rupture. Wiley, p 105–130

Luhr JF, Carmichael ISE (1990) Petrological monitoring of cyclical eruptive activity at Volcán Colima, Mexico. J Volcanol Geotherm Res 42:235–260

Lyakhovsky V, Ben-Zion Y (2008) Scaling relations of earthquakes and aseismic deformation in a damage rheology model. Geophys J Int 172:651–662

Ma S, Shimamoto T, Yao L, Togo T, Kitajima H (2014) A rotary-shear low to high-velocity friction apparatus in Beijing to study rock friction at plate to seismic slip rates. Earthquake Sci 27:469–497

Maddock RH (1986) Frictional melting in landslide-generated frictionites (hyalomylonites) and fault-generated pseudotachylytes-discussion. Tectonophysics 128:151–153

Maddock RH (1992) Effects of lithology, cataclasis and melting on the composition of fault-generated pseudotachylytes in Lewisian gneiss, Scotland. Tectonophysics 204:261–278

Maddock RH, Grocott J, Van Nes M (1987) Vesicles, amygdales and similar structures in fault-generated pseudotachylytes. Lithos 20:419–432

Mader H, Llewellin E, Mueller S (2013) The rheology of two-phase magmas: A review and analysis. J Volcanol Geotherm Res 257:135–158

Magloughlin JF (1989) The nature and significance of pseudotachylite from the Nason terrane, North Cascade Mountains, Washington. J Struct Geol 11:907–917

Magloughlin JF (2005) Immiscible sulfide droplets in pseudotachylyte: Evidence for high temperature (>1200 °C) melts. Tectonophysics 402:81–91

Magloughlin JF (2011) Bubble collapse structure: a microstructural record of fluids, bubble formation and collapse, and mineralization in pseudotachylyte. J Geol 119:351–371

Magloughlin JF, Spray JG (1992) Frictional melting processes and products in geological materials. Tectonophysics 204:197–204

Manga M, Castro J, Cashman KV, Loewenberg M (1998) Rheology of bubble-bearing magmas. J Volcanol Geotherm Res 87:15–28

Marone C, Richardson E (2010) Learning to read fault-slip behavior from fault-zone structure. Geology 38:767–768

Masch L, Wenk HR, Preuss E (1985) Electron microscopy study of hyalomylonites—evidence for frictional melting in landslides. Tectonophysics 115:131–160

McGarr A (1999) On relating apparent stress to the stress causing earthquake fault slip. J Geophys Res: Solid Earth 104:3003–3011

McKenzie D, Brune JN (1972) Melting on fault planes during large earthquakes. Geophys J R Astronom Soc 29:65–78

McNulty BA (1995) Pseudotachylyte generated in the semi-brittle and brittle regimes, Bench Canyon shear zone, central Sierra Nevada. J Struct Geol 17:1507–1521

McNutt SR (1996) Seismic monitoring and eruption forecasting of volcanoes: a review of the state-of-the-art and case histories. *In*: Monitoring and Mitigation of Volcano Hazards. Scarpa R, Tilling RI, (eds). Springer-Verlag, Berlin, p 99–146

McNutt SR (2005) Volcanic seismology. Annu Rev Earth Planet Sci 33:461–491

Melnik O, Sparks RSJ (2002) Dynamics of magma ascent and lava extrusion at Soufrière Hills Volcano, Montserrat. Geol Soc London Memoirs 21:153–171

Melnik OE, Sparks RSJ (2005) Controls on conduit magma flow dynamics during lava dome building eruptions. J Geophys Res 110:B02209

Melosh HJ (2005) The mechanics of pseudotachylite formation in impact events. *In*: Impact Tectonics. Koeberl C, Henkel H (eds). Springer Berlin Heidelberg, Berlin, Heidelberg, p 55–80

Mitchell TM, Toy V, Di Toro G, Renner J, Sibson RH (2016) Fault welding by pseudotachylyte formation. Geology 44:1059–1062

Mittempergher S, Dallai L, Pennacchioni G, Renard F, Di Toro G (2014) Origin of hydrous fluids at seismogenic depth: Constraints from natural and experimental fault rocks. Earth Planet Sci Lett 385:97–109

Miwa T, Okumura S, Matsushima T, Shimizu H (2013) Asymmetric deformation structure of lava spine in Unzen Volcano Japan. In Book Asymmetric deformation structure of lava spine in Unzen Volcano Japan. 2013 Fall Meeting, AGU, San Francisco, Abst. V14A-03

Molina Garza RS, Geissman J, Wawrzyniec T, Weber B, Martínez ML, Aranda-Gómez J (2009) An integrated magnetic and geological study of cataclasite-dominated pseudotachylytes in the Chiapas Massif, Mexico: a snapshot of stress orientation following slip. Geophys J Int 177:891–912

Monzawa N, Otsuki K (2003) Comminution and fluidization of granular fault materials: implications for fault slip behavior. Tectonophysics 367:127–143

Moore PL, Iverson NR, Iverson RM (2008) Frictional properties of the Mount St. Helens gouge. *In*: A Volcano Rekindled: The Renewed Eruption of Mount St Helens, 2004–2006. Sherrod DR, Scott WE, Stauffer PH (eds). U.S. Geological Survey Professionnal Paper 1750, p 415–424

Moran S, Malone S, Qamar A, Thelen W, Waite G, Horton S, LaHusen R, Major JJ (2005) Overview of seismicity associated with the 2004–2005 eruption of Mount St. Helens. AGU Fall Meeting Abstr: V52B-02

Mueller S, Llewellin EW, Mader HM (2011) The effect of particle shape on suspension viscosity and implications for magmatic flows. Geophys Res Lett 38:L13316

Nakada S, Shimizu H, Ohta K (1999) Overview of the 1990–1995 eruption at Unzen Volcano. J Volcanol Geotherm Res 89:1–22

Nakamura N, Hirose T, Borradaile GJ (2002) Laboratory verification of submicron magnetite production in pseudotachylytes: relevance for paleointensity studies. Earth Planet Sci Lett 201:13–18

Neuberg J (2000) Characteristics and causes of shallow seismicity in andesite volcanoes. Philos Transactions R Soc London 358:533–1546

Neuberg JW, Tuffen H, Collier L, Green D, Powell T, Dingwell D (2006) The trigger mechanism of low-frequency earthquakes on Montserrat. J Volcanol Geotherm Res 153:37–50

Neuberg JW, Collinson ASD, Mothes PA, Ruiz CM, Aguaiza S (2018) Understanding cyclic seismicity and ground deformation patterns at volcanoes: Intriguing lessons from Tungurahua volcano, Ecuador. Earth Planet Sci Lett 482:193–200

Nielsen SB, Carlson JM, Olsen KB (2000) Influence of friction and fault geometry on earthquake rupture. J Geophys Res 105:6069–6088

Nielsen S, Di Toro G, Hirose T, Shimamoto T (2008) Frictional melt and seismic slip. J Geophys Res: Solid Earth 113:B01308

Nielsen S, Toro GD, Griffith WA (2010a) Friction and roughness of a melting rock surface. Geophys J Int 182:299–310

Nielsen S, Mosca P, Giberti G, Di Toro G, Hirose T, Shimamoto T (2010b) On the transient behavior of frictional melt during seismic slip. J Geophys Res-Solid Earth 115:B10301

Niemeijer AR, Di Toro G, Nielsen S, Di Felice F (2011) Frictional melting of gabbro under extreme experimental conditions of normal stress, acceleration and sliding velocity. J Geophys Res: Solid Earth 116:B07404

Niemeijer A, Di Toro G, Griffith WA, Bistacchi A, Smith SAF, Nielsen S (2012) Inferring earthquake physics and chemistry using an integrated field and laboratory approach. J Struct Geol 39:2–36

Noda H, Kanagawa K, Hirose T, Inoue A (2011) Frictional experiments of dolerite at intermediate slip rates with controlled temperature: Rate weakening or temperature weakening? J Geophys Res: Solid Earth 116:B07306

O'Hara K (1992) Major- and trace-element constraints on the petrogenesis of a fault-related pseudotachylyte, western Blue Ridge province, North Carolina. Tectonophysics 204:279–288

O'Hara KD (2001) A pseudotachylyte geothermometer. J Struct Geol 23:1345–1357

Obata M, Karato S-i (1995) Ultramafic pseudotachylite from the Balmuccia peridotite, Ivrea-Verbano zone, northern Italy. Tectonophysics 242:313–328

Ohtomo Y, Shimamoto T (1994) Significance of thermal fracturing in the generation of fault gouge during rapid fault motion: An experimental verification. Struct Geol J Tectonic Res Group Jpn 39:135–144

Okumura S, Kozono T (2017) Silicic lava effusion controlled by the transition from viscous magma flow to friction controlled flow. Geophys Res Lett 44:3608–3614

Okumura S, Sasaki O (2014) Permeability reduction of fractured rhyolite in volcanic conduits and its control on eruption cyclicity. Geology 42:843–846

Okumura S, Uesugi K, Nakamura M, Sasaki O (2015) Rheological transitions in high-temperature volcanic fault zones. J Geophys Res: Solid Earth 120:2974–2987

Otsuki K, Monzawa N, Nagase T (2003) Fluidization and melting of fault gouge during seismic slip: Identification in the Nojima fault zone and implications for focal earthquake mechanisms. J Geophys Res-Solid Earth 108:2192

Ozawa K, Takizawa S (2007) Amorphous material formed by the mechanochemical effect in natural pseudotachylyte of crushing origin: A case study of the Iida-Matsukawa Fault, Nagano Prefecture, Central Japan. J Struct Geol 29:1855–1869

Paillat O, Elphick SC, Brown WL (1992) The solubility of water in $NaAlSi_3O_8$ melts: a re-examination of $Ab–H_2O$ phase relationships and critical behaviour at high pressures. Contrib Mineral Petrol 112:490–500

Pallister JS, Cashman KV, Hagstrum JT, Beeler NM, Moran SC, Denlinger RP (2013) Faulting within the Mount St. Helens conduit and implications for volcanic earthquakes. GSA Bull 125:359–376

Papale P (1999) Strain-induced magma fragmentation in explosive eruptions. Nature 397:425–428

Papale P (2001) Dynamics of magma flow in volcanic conduits with variable fragmentation efficiency and nonequilibrium pumice degassing. J Geophys Res: Solid Earth 106:11043–11065

Passelègue FX, Goldsby DL, Fabbri O (2014) The influence of ambient fault temperature on flash-heating phenomena. Geophys Res Lett 41:828–835

Pei J, Zhou Z, Dong S, Tang L (2014) Magnetic evidence revealing frictional heating from fault rocks in granites. Tectonophysics 637:207–217

Petrík I, Nabelek PI, Janák M, Plašienka D (2003) Conditions of formation and crystallization kinetics of highly oxidized pseudotachylytes from the High Tatras (Slovakia). J Petrol 44:901–927

Philpotts AR (1964) Origin of pseudotachylites. Am J Sci 262:1008–1035

Pittarello L, Pennacchioni G, Di Toro G (2012) Amphibolite-facies pseudotachylytes in Premosello metagabbro and felsic mylonites (Ivrea Zone, Italy). Tectonophysics 580:43–57

Plail M, Edmonds M, Humphreys MCS, Barclay J, Herd RA (2014) Geochemical evidence for relict degassing pathways preserved in andesite. Earth Planet Sci Lett 386:21–33

Preece K, Gertisser R, Barclay J, Charbonnier SJ, Komorowski J-C, Herd RA (2016) Transitions between explosive and effusive phases during the cataclysmic 2010 eruption of Merapi volcano, Java, Indonesia. Bull Volcanol 78:54

Price NA, Johnson SE, Gerbi CC, West DP (2012) Identifying deformed pseudotachylyte and its influence on the strength and evolution of a crustal shear zone at the base of the seismogenic zone. Tectonophysics 518–521:63–83

Proctor B, Lockner DA (2016) Pseudotachylyte increases the post-slip strength of faults. Geology 44:1003–1006

Rabinowicz E (1951) The nature of the static and kinetic coefficients of friction. J Appl Physics 22:1373–1379

Rabinowicz E (1958) The intrinsic variables affecting the stick-slip process. Proc Phys Soc 71:668–675

Ray SK (2004) Melt–clast interaction and power-law size distribution of clasts in pseudotachylytes. J Struct Geol 26:1831–1843

Reches Z, Lockner DA (2010) Fault weakening and earthquake instability by powder lubrication. Nature 467:452-U102

Reimold WU (1995) Pseudotachylite in impact structures—generation by friction melting and shock brecciation?: A review and discussion. Earth Sci Rev 39:247–265

Rempe M, Di Toro G, Mitchell TM, Smith SAF, Hirose T, Renner J (2020) Influence of effective stress and pore fluid pressure on fault strength and slip localization in carbonate slip zones. J Geophys Res: Solid Earth 125:e2020JB019805

Rempel AW (2006) The effects of flash-weakening and damage on the evolution of fault strength and temperature. *In*: Earthquakes: Radiated Energy and the Physics of Faulting. Vol 170. Abercrombie R, McGarr A, DiToro G, Kanamori H, (eds). Amer Geophysical Union, Washington, p 263–270

Rempel AW, Rice JR (2006) Thermal pressurization and onset of melting in fault zones. J Geophys Res: Solid Earth 111

Rice JR (2006) Heating and weakening of faults during earthquake slip. J Geophys Res 111:B05311

Rigney DA, Hirth JP (1979) Plastic deformation and sliding friction of metals. Wear 53:345–370

Rittinger V (1867) Lehrbuch der Aufbersitungskunde. Berlin

Rosenberg CL, Handy MR (2005) Experimental deformation of partially melted granite revisited: implications for the continental crust. J Metamorph Petrol 23:19–28

Rowe CD, Moore JC, Meneghini F, McKeirnan AW (2005) Large-scale pseudotachylytes and fluidized cataclasites from an ancient subduction thrust fault. Geology 33:937–940

Ruina A (1983) Slip instability and state variable friction laws. J Geophys Res 88:10359–10370

Rust AC, Cashman KV, Wallace PJ (2004) Magma degassing buffered by vapor flow through brecciated conduit margins. Geology 32:349–352

Sammis C, King G, Biegel R (1987) The kinematics of gouge deformation. Pure Appl Geophys 125:777–812

Sarkar A, Chattopadhyay A, Singh T (2019) Roundness of survivor clasts as a discriminator for melting and crushing origin of fault rocks: A reappraisal. J Earth System Sci 128:51

Scheu B, Ichihara M, Spieler O, Dingwell DB (2008) A closer look at magmatic fragmentation. Scheu B, Ichihara M, Spieler O, Dingwell DB. A closer look at magmatic fragmentation. AGU Fall Meeting Abstr V43C-1813

Scholz CH (1998) Earthquakes and friction laws. Nature 391:37–42

Scholz CH (2019) The Mechanics of Earthquakes and Faulting. Cambridge University Press, Cambridge

Scholz CH, Engelder JT (1976) The role of asperity indentation and ploughing in rock friction — I: Asperity creep and stick-slip. Int J Rock Mech Min Sci Geomech Abstracts 13:149–154

Scholz CH, Dawers NH, Yu JZ, Anders MH (1993) Fault growth and fault scaling laws—preliminary results. J Geophys Res-Solid Earth 98:21951–21961

Schwarzkopf, Schmincke, Troll (2001) Pseudotachylite on impact marks of block surfaces in block-and-ash flows at Merapi volcano, Central Java, Indonesia. Int J Earth Sci 90:769–775

Scott JS, Drever HI (1954) X. Frictional fusion along a Himalayan Thrust. Proc R Soc Edinburgh Sect B Biology 65:121–142

Scott JAJ, Pyle DM, Mather TA, Rose WI (2013) Geochemistry and evolution of the Santiaguito volcanic dome complex, Guatemala. J Volcanol Geotherm Res 252:92–107

Shand SJ (1916) The pseudotachylyte of Parijs (Orange Free State), and its relation to 'Trap-Shotten Gneiss' and 'Flinty Crush-Rock'. Q J Geol Soc 72:198–221

Shields JK, Mader HM, Caricchi L, Tuffen H, Mueller S, Pistone M, Baumgartner L (2016) Unravelling textural heterogeneity in obsidian: Shear-induced outgassing in the Rocche Rosse flow. J Volcanol Geotherm Res 310:137–158

Shimamoto T, Lin AM (1994) Is frictional melting equilibrium or non-equilibrium melting? Struct Geol (J Tectonic Res Group Japan) 39:79–84

Shimamoto T, Tsutsumi A (1994) A new rotary-shear high-velocity frictional testing machine: Its basic design and scope of research. Struct Geol 39:65–78

Sibson RH (1973) Interactions between temperature and pore-fluid pressure during earthquake faulting and a mechanism for partial or total stress relief. Nat Phys Sci 243:66–68

Sibson RH (1975) Generation of pseudotachylyte by ancient seismic faulting. R Astronom Soc London J 43:775–794

Sibson RH (1977) Fault rocks and fault mechanisms. J Geol Soc 133:191–213

Sibson RH, Toy VG (2006) The habitat of fault-generated pseudotachylyte: presence vs. absence of friction-melt. *In*: Earthquakes: Radiated Energy and the Physics of Faulting. Abercrombie R, McGarr A, Di Toro G, Kanamori H (eds) AGU Geophysical Monograph 170:153–166

Smith DR, Leeman WP (1987) Petrogenesis of Mount St. Helens dacitic magmas. J Geophys Res: Solid Earth 92:10313–10334

Spagnuolo E, Nielsen S, Violay M, Di Toro G (2016) An empirically based steady state friction law and implications for fault stability. Geophys Res Lett 43:3263–3271

Sparks RSJ (1997) Causes and consequences of pressurisation in lava dome eruptions. Earth Planet Sci Lett 150:177–189

Spray JG (1987) Artificial generation of pseudotachylyte using friction welding apparatus: simulation of melting on a fault plane. J Struct Geol 9:49–60

Spray JG (1992) A physical basis for the frictional melting of some rock-forming minerals. Tectonophysics 204:205–221

Spray JG (1993) Viscosity determinations of some frictionally generated silicate melts: Implications for fault zone rheology at high strain rates. J Geophys Res: Solid Earth 98:8053–8068

Spray JG (1998) Localized shock- and friction-induced melting in response to hypervelocity impact. Geol Soc London, Spec Publ 140:195–204

Spray JG (2005) Evidence for melt lubrication during large earthquakes. Geophys Res Lett 32: L07301

Spray JG (2010) Frictional melting processes in planetary materials: from hypervelocity impact to earthquakes. Annu Rev Earth Planet Sci 38:221–254

Swanson MT (1992) Fault structure, wear mechanisms and rupture processes in pseudotachylyte generation. Tectonophysics 204:223–242

Tait S, Jaupart C, Vergniolle S (1989) Pressure, gas content and eruption periodicity of a shallow, crystallising magma chamber. Earth Planet Sci Lett 92:107–123

Takagi H, Arita K, Danhara T, Iwano H (2007) Timing of the Tsergo Ri landslide, Langtang Himal, determined by fission-track dating of pseudotachylyte. J Asian Earth Sci 29:466–472

Terzaghi K (1923) Die Berechnung der Durchlassigkeitsziffer des Tones aus Dem Verlauf der Hidrodynamichen. Span-Nungserscheinungen Akademie der Wissenschaften in Wien; Mathematish-Naturwissen-SchaftilicheKlasse: Mainz, Germany, p 125–138

Thelen W, Malone S, West M (2011) Multiplets: Their behavior and utility at dacitic and andesitic volcanic centers. J Geophys Res:Solid Earth 116:B08210

Tromans D (2008) Mineral comminution: Energy efficiency considerations. Miner Eng 21:613–620

Truby JM, Mueller SP, Llewellin EW, Mader HM (2015) The rheology of three-phase suspensions at low bubble capillary number. Proc R Soc London A 471:20140557

Tsutsumi A, Shimamoto T (1997) High-velocity frictional properties of gabbro. Geophys Res Lett 24:699–702

Tuffen H, Dingwell D (2005) Fault textures in volcanic conduits: evidence for seismic trigger mechanisms during silicic eruptions. Bull Volcanol 67:370–387

Tuffen H, Dingwell DB, Pinkerton H (2003) Repeated fracture and healing of silicic magma generate flow banding and earthquakes? Geology 31:1089–1092

Umakoshi K, Takamura N, Shinzato N, Uchida K, Matsuwo N, Shimizu H (2008) Seismicity associated with the 1991–1995 dome growth at Unzen Volcano, Japan. J Volcanol Geotherm Res 175:91–99

van den Ende MPA, Chen J, Ampuero JP, Niemeijer AR (2018) A comparison between rate-and-state friction and microphysical models, based on numerical simulations of fault slip. Tectonophysics 733:273–295

Violay M, Di Toro G, Gibert B, Nielsen S, Spagnuolo E, Del Gaudio P, Azais P, Scarlato PG (2014) Effect of glass on the frictional behavior of basalts at seismic slip rates. Geophys Res Lett 41:348–355

Voight B, Hoblitt RP, Clarke AB, Lockhart AB, Miller AD, Lynch L, McMahon J (1998) Remarkable cyclic ground deformation monitored in real-time on Montserrat, and its use in eruption forecasting. Geophys Res Lett 25:3405–3408

Voight B, Sparks RS, Miller AD, Stewart RC, Hoblitt RP, Clarke A, Ewart J, Aspinall WP, Baptie B, Calder ES, Cole P (1999) Magma flow instability and cyclic activity at Soufriere Hills Volcano, Montserrat, British West Indies. Science 283:1138–1142

Wallace PA, Kendrick JE, Miwa T, Ashworth JD, Coats R, Utley JE, Henton De Angelis S, Mariani E, Biggin A, Kendrick R, Nakada S (2019a) Petrological architecture of a magmatic shear zone: a multidisciplinary investigation of strain localisation during magma ascent at Unzen Volcano, Japan. J Petrol 60:791–826

Wallace PA, De Angelis SH, Hornby AJ, Kendrick JE, Clesham S, von Aulock FW, Hughes A, Utley JE, Hirose T, Dingwell DB, Lavallée Y (2019b) Frictional melt homogenisation during fault slip: Geochemical, textural and rheological fingerprints. Geochim Cosmochim Acta 255:265–288

Wang C-y, Goodman RE, Sundaram PN (1975) Variations of VP and VS in granite premonitory to shear rupture and stick-slip sliding: Application to earthquake prediction. Geophys Res Lett 2:309–311

Webb SL, Dingwell DB (1990) Non-Newtonian rheology of igneous melts at high stresses and strain rates: experimental results for rhyolite, andesite, basalt and nephelinite. J Geophys Res 95:15695–15701

Weber B, Suhina T, Brouwer AM, Bonn D (2019) Frictional weakening of slip interfaces. Sci Adv 5:eaav7603

Wells DL, Coppersmith KJ (1994) New empirical relationships among magnitude, rupture length, rupture width, rupture area, and surface displacement. Bull Seismol Soc Am 84:974–1002

Wong T-f, Gu Y, Yanagidani T, Zhao Y (1992) Chapter 5 Stabilization of Faulting by Cumulative Slip. *In*: International Geophysics. Vol Volume 51. Brian E, Teng-fong W, (eds). Academic Press, p 119–143

Wright HMN, Weinberg RF (2009) Strain localization in vesicular magma: Implications for rheology and fragmentation. Geology 37:1023–1026

Yamada K, Hanamuro T, Tagami T, Shimada K, Takagi H, Yamada R, Umeda K (2012) The first (U–Th)/He thermochronology of pseudotachylyte from the Median Tectonic Line, southwest Japan. J Asian Earth Sci 45:17–23

Yang T, Chou Y-M, Ferré EC, Dekkers MJ, Chen J, Yeh E-C, Tanikawa W (2020) Faulting Processes Unveiled by Magnetic Properties of Fault Rocks. Rev Geophys 58:e2019RG000690

Yoshimura S, Nakamura M (2010) Fracture healing in a magma: An experimental approach and implications for volcanic seismicity and degassing. J Geophys Res-Solid Earth 115:B09209

Zhang S, Tullis TE, Scruggs VJ (1999) Permeability anisotropy and pressure dependency of permeability in experimentally sheared gouge materials. J Struct Geol 21:795–806

Zhang L, Li H, Sun Z, Chou Y-M, Cao Y, Wang H (2018) Metallic iron formed by melting: A new mechanism for magnetic highs in pseudotachylyte. Geology 46:779–782

Zhu W, Allison KL, Dunham EM, Yang Y (2020) Fault valving and pore pressure evolution in simulations of earthquake sequences and aseismic slip. Nat Commun 11:4833

Reviews in Mineralogy & Geochemistry
Vol. 87 pp. 965-1014, 2022
Copyright © Mineralogical Society of America

21

Non-Magmatic Glasses

Maria Rita Cicconi[1,*], John S. McCloy[2], Daniel R. Neuville[1]

*[1]Université de Paris
Insitut de Physique du Globe de Paris
Géomatériaux, CNRS
1 rue Jussieu
F-75005 Paris
France*

maria.rita.cicconi@fau.de

neuville@ipgp.fr

*[2]School of Mechanical and Materials Engineering
Washington State University
PO Box 642920
Pullman, WA 99164–2920
USA*

john.mccloy@wsu.edu

*[*]Correspondence address:
WW3—Friedrich-Alexander-Universität Erlangen-Nürnberg
D-91058 Erlangen
Germany*

OVERVIEW

On Earth, natural glasses are typically produced by rapid cooling of melts, and as in the case of minerals and rocks, natural glasses can provide key information on the evolution of the Earth. However, natural glasses are products not solely terrestrial, and different formation mechanisms give rise to a variety of natural amorphous materials. In this chapter, we provide an overview of the different natural glasses of non-magmatic origin and on their formation mechanisms. We focus on natural glasses formed by mechanisms other than magmatic activity and included are metamorphic glasses and glasses produced from highly energetic events (shock metamorphism). The study of these materials has strong repercussions on planetary surface processes, paleogeography/paleoecology, and even on the origin of life.

ACRONYMS AND GLOSSARY

Hypervelocity impacts	Impacts, involving impacting bodies that are traveling at speed (generally greater than a few km/s) higher enough to generate shock waves upon impact.
K–Pg	Cretaceous–Paleogene (K–Pg) boundary (~66 million years ago)
KT	Cretaceous–Tertiary boundary (former name for K–Pg)
Lechatelierite	shock-fused SiO_2 glass
LDG	Libyan Desert Glass
m-Tek or m-T	microtektites; Small distal ejecta with diameter less than 0.1 cm.

1529-6466/22/0087-0021$05.00 (print)
1943-2666/22/0087-0021$05.00 (online)
http://dx.doi.org/10.2138/rmg.2022.87.21

MN	Muong Nong-type tektites
PDFs	Planar deformation features; microscopic parallel, isotropic features in minerals that originate from elevated shock metamorphism.
SF	tektite/mictotektites strewn fields
TAS	Total-Alkalis versus Silica diagram
Tektites	mm- to cm-scale, glassy particles of ballistically transported impact melt, formed by the impact of an extraterrestrial projectile.
T_g	Glass transition temperature
YD	Younger Dryas is a geological period from ~12,900 to ~11,700 BP
YDB	Younger Dryas boundary
XAS	X-ray Absorption Spectroscopy

INTRODUCTION

This chapter aims to review natural glasses formed by mechanisms other than magmatic activity. After a brief description of the most common natural amorphous materials and of the different mechanisms of formation, we report some examples of metamorphic glasses and glasses from highly energetic events. In the end, we provide an overview of the properties of these natural amorphous materials.

In the last decades, several reviews, for specific natural glasses, have been published (e.g., Glass 1990; Koeberl 1997; Eby et al. 2010 ; Pasek et al. 2012; Glass and Simonson 2013), with more recent general reviews on different natural glasses provided in Heide and Heide (2011), Glass (2016), Cicconi and Neuville (2019), McCloy (2019).

The composition and origin of natural glasses

Natural glasses have different origins and chemical compositions, though their chemical variability matches the differentiation found in many common types of volcanic rocks. An overview of the enormous variability of natural glasses occurring on Earth and Lunar soil can be appreciated in the TAS diagram (total alkalis vs. silica; Fig. 1). These amorphous materials have SiO_2 contents between ~33–99 wt.%, and alkali contents ranging between 0 and 15 wt.%, thus covering many common types of volcanic rocks. Despite glasses having differentiated chemically in similar ways to many common volcanic rocks, most of the time, non-magmatic glasses have experienced extreme conditions of formation, far from those of common igneous materials. Indeed, most of the glasses considered in this chapter have cooling rates orders of magnitude higher than volcanic terrestrial (or lunar) glasses, and/or have been subject to extremely elevated peak pressures. For instance, cooling rates may range from extremely high values, as in fulgurite formation (~10^{10}°C/min) and submarine basaltic eruptions (10^7°C/min), to moderate values in tektites 10^4–10^{-2}°C/min, to very slow rates in massive obsidian flows 10^{-2}–10^{-4}°C /min (Switzer and Melson 1972; Weeks et al. 1984; Wilding et al. 1996a,b; Rietmeijer et al. 1999; Potuzak et al. 2008). By comparison, water, which is considered to be a very weak glass former, requires cooling rates on the order of 10^6°C/s to form an amorphous solid (Debenedetti and Stanley 2003).

Before describing non-magmatic glasses, we would like to provide an overview of all amorphous materials found on Earth (and on other terrestrial planets), and of the various mechanisms of formation. The most well-known natural amorphous materials are of magmatic origin and include basaltic glasses and obsidians. Basaltic glasses have an average composition of about (wt.%) 50–54% SiO_2, 12–17% Al_2O_3, 8–12% FeO_{tot}, 2–4% alkali ($K_2O + Na_2O$), 15–20% alkaline-earth ($CaO + MgO$) (Cicconi and Neuville 2019 and references therein) and their low viscosity favors crystallization (devitrification). Volcanic glasses produced upon rapid cooling of basaltic melts are called sideromelane, but they also occur as volcanic ash, fibers, and teardrops (i.e., Pele's Hair and Pele's Tears) and more rarely form solidified foam—reticulite.

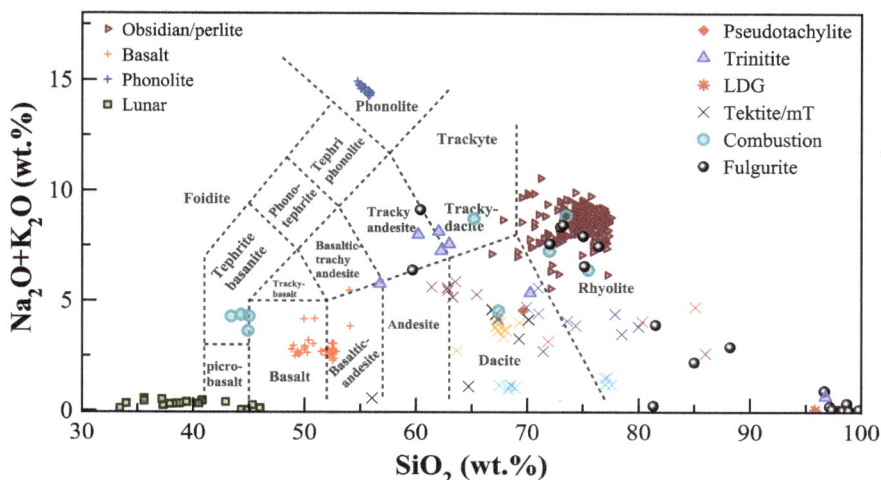

Figure 1. SiO$_2$ vs. total alkali (wt.%) diagram for several natural non-magmatic glasses described in this chapter. Glasses deriving from magmatic processes are reported for comparison. Data compilation from Cicconi and Neuville (2019). Combustion glasses from Table 1. LDG = Libyan Desert Glass; mT = microtektites.

Obsidian glass was first described in Pliny The Elder's Natural History with the name of "*obsiana*", so-called because of its similarity to a very dark stone found in Ethiopia by Obsius. This glass has accompanied and influenced human evolution since prehistoric times, and nowadays it still enters the popular culture, even if with more fanciful names (i.e., dragonglass, after the fantasy drama television series Game of Thrones). Obsidians are generally subalkalic rhyolitic with an average composition of about (wt.%) 72–77% SiO$_2$, 10–15% Al$_2$O$_3$, 1–2% FeO$_{tot}$, 7–10% alkali (K$_2$O + Na$_2$O), 0.5–2% alkaline-earth (CaO + MgO) (Cicconi and Neuville 2019 and references therein). The glass-forming processes of obsidian melts are strongly influenced by the content, size, and shape of microlites, and by the contents of volatile components (such as water, fluorine, and chlorine, sulphur and carbon oxides), and small variations in volatile contents can cause important changes in the flow dynamics of obsidian melts (Carmichael 1979; Castro et al. 2002; Heide and Heide 2011 and references therein).

Other natural amorphous materials are formed by metamorphic processes. For instance, the so-called pyrometamorphic glasses form due to burning of fossil fuels such as coal and natural gas or other organic material (McCloy 2019 and references therein). The term pyrometamorphism, which defines a type of contact metamorphism, was originally proposed in 1912 by Brauns (Brauns 1912; Grapes 2010 and references therein) to describe a high-temperature/low-pressure metamorphism observed in schist xenoliths in trachyte and phonolite magma of the East Eifel area (Germany). The term buchite is used to define those partially or completely melted materials as a consequence of pyrometamorphism.

Glasses from highly energetic events are formed in a completely different way, and with a completely different timescale than magmatic and metamorphic ones. For instance, impact glass formation is related to the collision of an extraterrestrial body on the surface of the Earth. Thus, they derive from shock metamorphism of existing silicate rocks and sediments. Impacts or airbursts (shock melting caused by a cosmic object exploding in the atmosphere) can be either natural or artificial, and both provide shock markers due to the extreme high temperatures and pressures. Among natural glasses that experienced extremely high temperatures in a very short time, there are also the fulgurites, formed following lightning strikes.

Finally, biomineralization processes can produce amorphous materials. These are considered eco-friendly, and thus they have captured the attention of organic-/inorganic chemists and materials scientists (Cicconi and Neuville 2019 and references therein). Several siliceous marine organisms exhibit discontinuous, three-dimensional frameworks of short chains of SiO_4 tetrahedra, bonded with apical hydroxyls. The low-temperature hydrated variety of silica, opal ($SiO_2 \cdot nH_2O$), is a biomineral and displays different arrangements of amorphous SiO_2, water, and cristobalite, and/or tridymite (see Cicconi and Neuville 2019 and references therein). Depending on the arrangements, it is possible to distinguish three opal types: i) opal-C (cristobalite); ii) opal-CT (cristobalite and tridymite); iii) opal-A (X-ray-amorphous opal). The latter can be further divided in opal-AN (e.g., hyalite) and opal-AG with an amorphous silica gel structure. In a "maturation process" (Ostwald ripening) opals are transformed as follows: Opal-AG→Opal-CT→Opal-C→ microcrystalline quartz (Skinner and Jahren 2003 and references therein).

METAMORPHIC GLASSES

Pyrometamorphic glasses

General features of pyrometamorphism. Pyrometamorphic glasses are an important class of amorphous materials summarized by Grapes (2010). The term 'pyrometamorphism' was originally used by Brauns (1912) to describe changes in contact zones (i.e., aureoles) between magma intrusions and country rock, where high temperature and atmospheric pressure result in particular mineralogical and morphological changes. These high temperature changes are often observed in xenoliths within later igneous rocks, near shallow magmatic intrusions, and within tuffs and breccias. Brauns assumed that pyrometamorphism must create melting in the heat-affected rocks and vitrification on cooling. Other terms have been introduced but have fallen out of favor, such as optalic metamorphism, emphasizing the transient, quickly dissipated heat such as used when 'baking a brick.' Pyrometamorphized rocks tend to lose all volatiles, and melts recrystallize with anhydrous mineral assemblages, mostly oxides and simple silicates, but at least 60 different minerals have been reported from this prograde metamorphic process (Sokol et al. 2007). Some characteristic evidence of such processes include bleaching of carbonate rocks, reddening of iron-containing rocks, fusion of mineral grains, hardening, and similar morphological and chemical changes to those observed in earthenware ceramic and clay brick-making processes. Heat for pyrometamorphic transformations need not come from contact magmatic heat, but also from other sources, such as combustion of coal beds and organic-rich sediments or even lightning. The Subcommission on the Systematics of Metamorphic Rocks (SCMR) of the International Union of Geological Sciences (IUGS) categorizes these effects as subvariants of contact metamorphism, namely 'burning/combustion metamorphism' and 'lightning metamorphism,' i.e., fulgurites (Callegari and Pertsev 2007).

While contact and lightning metamorphism are highly localized, high-temperature changes in rock due to coal bed combustion can be regionally extensive. Well documented examples in the USA occur in the Grimes Canyon area, Monterey Formation, California (Bentor et al. 1981); the Powder River Basin, Wyoming (Herring and Modreski 1986; Cosca et al. 1989; Clark and Peacor 1992); and various coal-bearing areas of Montana and Colorado (Rogers 1918). The phenomenon of combustion metamorphism has been found to be quite common (Bentor et al. 1981; Sokol et al. 2007; Grapes 2010), including examples in China (de Boer et al. 2001), Mongolia (Peretyazhko et al. 2018), Indonesia, India, Russia (e.g., Chelyabinsk brown coal basin) (Sokol et al. 1998), Iran, Iraq, Jordan (Khoury et al. 2015), Israel e.g., Hatrurim basin (Ron and Kolodny 1992; Vapnik et al. 2007; Sokol et al. 2014), Czechia, Germany, England, Italy (Melluso et al. 2004), Mali (Svensen et al. 2003), Canada (Mathews and Bustin 1984; Canil et al. 2018), Venezuela, Colombia, New Zealand (Tulloch and Campbell 1993), and Australia (Rattigan 1967).

Some terms are commonly used for pyrometamorphic rocks as follows. Hornfels forms where clay-rich rocks contact a hot igneous body, and partially or entirely recrystallize after in-situ melting, resulting in baked and hardened silicate + oxide systems. Clinker originally referred to altered or burned coal, but now refers generally to burnt brick-like rock which 'rings' ('clinks') when struck. Note that clinker is also the high-temperature processed calcium aluminosilicate used as the reactive precursor for Portland cement. These same 'clinker' minerals, e.g., belite sulfoaluminate, have been found in Israel in natural pyrometamorphic contexts (Sokol et al. 2014). Buchite, in contrast to hornfels, is largely vitrified from high heat applied to sandstones or pelites, resulting in hard, fused, and glassy material; the term is not restricted to coal fire lithologies, so in that case, should be referred to as 'buchite clinker.' Finally, paralava resembles artificial slag or basalt, is generally vesicular and aphanitic, sometimes shows evidence of flow, and formed from burning and melting of sedimentary rocks (shale, sandstone, marl) by proximal combusting coal seams. The distinction among these is based on both protolith and peak temperature, which can be ~400 to >1600 °C locally, resulting in a continuum from baked/burnt rock to fused-grained rock (clinker) to partially melted buchite, to wholly melted and partially devitrified rock (paralava), often all in the same area. Terminology is not uniformly employed, and in certain localities clinker is still called 'scoria' despite the different volcanic origin assumed from that term. Use of the terms 'buchite' and 'paralava' imply the observable presence of glass resulting from a quenched rock melt (Grapes 2010). Table 1 lists the chemical composition of different combustion glasses.

Coal fires and spontaneous combustion. Coals in the Western USA, known to spontaneously combust, are typically lignite or sub-bituminous grade (Rogers 1918; Heffern et al. 2007). Spontaneous combustion occurs due to a complex process involving absorption of oxygen, such as through cracks, and oxidation of unsaturated hydrocarbons (Stracher 2007). The main exothermic reaction is that of carbon reacting with oxygen gas to form CO_2 (Grapes 2010). These early reactions take place at temperatures as low as 80 °C, but this oxidation further generates heat until about 200 °C, where autogenous oxidation occurs, followed by ignition at 350–400 °C. Fine dust lignite, however, which has typically absorbed a large amount of oxygen, can ignite at temperatures as low as 150 °C (Rogers 1918). Moisture is said to exacerbate this effect, with escaping hydrogen-containing gases also playing a role. Other factors influencing the combustion reaction include coal factors such as rank and pyrite content, reaction factors such as particle size and temperature, and macroscopic factors such as air flow, overbedding rock type, and thermal conduction (Grapes 2010). Often areas can burn in the absence of additional oxygen, producing highly reducing conditions and the creation of coal ash with evidence of metallic iron (de Boer et al. 2001; Grapes 2010).

Combustion of coal beds has been observed in modern times on the coast of Dorset, England; in the 'Smoking Hills' of Canada (Mathews and Bustin 1984); and at the 'Burning Mountain' (Mt. Wingen) in New South Wales, Australia (Bentor et al. 1981). The Smoking Hills are thought to have been burning for at least 150 years, while the Burning Mountain has been combusting for at least 15,000 years. The Posidonia shales in Germany, though not currently combusting, are thought to have been active c. 1000–1500 CE (Bentor et al. 1981). Many other examples are known (Kuenzer and Stracher 2012).

Examples of creation of pyrometamorphic materials in the historic past are known from burning of a wide range of fossil fuels, from spontaneous combustion but also other natural or anthropogenic ignition (Kuenzer and Stracher 2012). Combustion in coal mining 'spoil heaps' has resulted in recent pyrometamorphic processes in Russia (Chelyabinsk) (Sokol et al. 1998) and Italy (Ricetto) (Capitanio et al. 2004; Stoppa et al. 2005) (Fig. 2). One of the interesting features of the assemblage in Russia is the presence of pure glassy carbon, known as shungite (Beyssac et al. 2002; Melezhik et al. 2004 ; Kwiecinska et al. 2007; Golubev et al. 2016). Additionally, minerals have been observed which were crystallized from the gas phase in cracks, including pure oxide, sulfide, silicate, and carbide single crystals.

Table 1. Chemical composition (wt.%) of different combustion glasses.

	Glass in Paralava		Paralava					Glass								Clinker	Biomass slag
	Ricetto, Italy	Khazakstan	Russia, Chelyabinsk	USA, PRB Wyom.	China	Silicate paralava Canada	Khazakstan	Mongolia (1)	Italy, Ricetto	USA, Grimes, Calif.	USA, PRB WY	Mongolia (2)	Canada, Shrimpton	New Zealand	Canada, Tranquille	USA, PRB WY.	Botswana
SiO$_2$	44.40	60.97	44.86	49.50	51.47	58.88	58.73	69.63	67.42	74.00	73.86	46.48	59.76	65.495	53.17	68.46	66.19
TiO$_2$	0.94	0.38	1.00	0.66	0.61	1.71	0.34	0.91	0.60	0.35	0.43	0.60	0.74	17.65	1.10	0.74	0.71
Al$_2$O$_3$	20.05	7.78	17.59	16.54	18.35	20.43	7.91	15.07	16.65	12.20	13.13	25.22	15.31	0.35	17.28	16.14	2.96
Fe$_2$O$_3$ tot	7.97	2.62	14.36	22.51	13.71	11.40	3.03	2.71	4.75	2.32	1.47	3.76	7.13	2.83	7.75	3.41	3.09
MnO	0.15	0.14	0.24	0.26	0.45	0.12	0.09	0.14	0.08	0.04			0.43		0.16	0.04	0.16
MgO	1.67	3.08	5.15	1.96	4.59	2.23	4.78	0.31	1.48	0.18	0.32	0.21	2.04	0.85	1.06	2.56	2.53
CaO	21.40	8.87	15.13	7.57	9.62	1.27	12.76	1.07	3.60	1.44	1.81	3.62	4.29	2.64	3.68	4.78	4.52
Na$_2$O	0.58	8.96	0.10	0.01	0.02	0.10	7.45	2.66	2.31	2.39	0.41	3.24	4.25	7.15	4.91	0.30	0.64
K$_2$O	3.56	3.88	1.43	1.48	1.00	2.59	3.08	5.06	4.55	4.34	7.97	9.80	4.89	1.58	9.11	2.47	14.27
P$_2$O$_5$	0.37	1.48		0.35	0.31	0.31	0.83			0.10	0.19					0.14	1.71
BaO												0.28					
SrO										0.60		0.16					
SO$_3$		0.42															0.07
TOT	101.08	98.59	99.83	100.82	100.12	99.01	98.99	97.54	101.4	97.96	99.57	93.36	98.83	98.55	98.22	99.02	96.85
Ref.	[1]	[2]	[3]	[4]	[5]	[6]	[1]	[7]	[8]	[9]	[4]	[7]	[10]	[11]	[10]	[4]	[12]

Footnotes: [1] Melluso et al. 2004; [2] Grapes et al. 2013; [3] Sokol et al. 1998; [4] Cosca et al. 1989; [5] Grapes et al. 2009; [6] Piepjohn et al. 2007; [7] Peretyazhko et al. 2018; [8] Capitanio et al. 2004; [9] Bentor 1984; [10] Canil et al. 2018; [11] Coombs et al. 2008; [12] Thy et al. 1995.

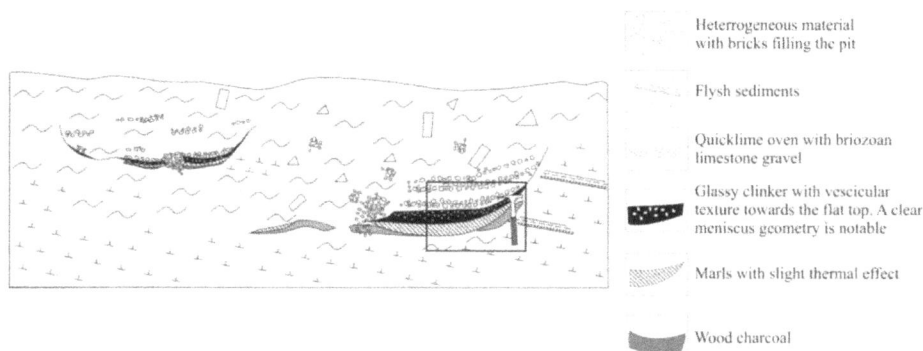

Figure 2. Pyrometamorphic material at Ricetto (Italy). Modified from Figure 1, Stoppa et al. (2005). MSA copyright.

Other processes fueling pyrometamorphism. It should be noted that pyrometamorphic rocks and glasses can be produced by high temperature burning using any fuel, not just coal. Many examples have been described where the fuel was biomass or organic material covered by sediment. This organic material can be deposited naturally, and subsequently be covered to ignite and burn in the subsurface under dry conditions, such as examples in Chile (Roperch et al. 2017) and Africa (Melson and Potts 2002; Svensen et al. 2003). Alternatively, organic deposits can be a result of human activity, such as archaeologically observed middens (Thy et al. 1995) or modern-day haystacks (Baker and Baker 1964), the burning of which produces glassy material.

Organic gases can be fuels for rock burning as well. One of the most studied areas of pyrometamorphic rocks, the Dead Sea region in Israel, has been interpreted as having a pyrometamorphic origin, from a mud volcano and associated methane and hydrocarbon gas combustion (Sokol et al. 2007, 2010). Gas seeps and their associated 'eternal flames' from spontaneous combustion have played important roles in human history, especially in religious and mythological traditions. For example, the methane seeps known as Chimera in southwest Turkey was the site of the first Olympic fire in the ancient Greek world, and the site of a temple of Hephaestus (Vulcan) the god of fire and metallurgy (Hosgormez et al. 2008; Etiope 2015), and it still burns today. All kerogens of various grades, from bitumen to oil, are potential sources of fuel for creation of pyrometamorphic rocks and glass.

Pyrometamorphic melts are also known to be produced from oilfield fire. Characteristic glasses, known as tengizites after the Tengiz oildfield in Kazakhstan, formed in this inferno (Kokh et al. 2016). Tengizites, essentially a type of paralava/slag, contain 59–69 wt.% SiO_2, 7.3–9.7 wt.% Al_2O_3, 12.8–17.9 wt.% CaO, 2.0–3.7 wt.% MgO, 2.0–3.0 wt.% Na_2O, 1.3–1.9 wt.% K_2O, <0.3 wt.% H_2O and <0.4 wt.% total volatiles, as a result of melting sand and clayey silt along with biogenic calcite (Kokh et al. 2016). The associated buchite at the site is known as mesolite, consisting of partially melted and highly altered sediment, and compared to tengizite has lower SiO_2, higher Al_2O_3, MgO, K_2O, TiO_2, and total volatiles, but lower FeO/Fe_2O_3 ratio.

One detailed example. Compositions of pyrometamorphic glasses vary drastically depending on the local chemical environment derived from the protolith, extent of melting, mixing of the melt, and subsequent crystallization. For example, most Western USA coal-bearing regions are interbedded with predominantly shale and sandstone (Rogers 1918). The famous example Grimes Canyon in California consists of protoliths of mudstone containing clays, detrital quartz and feldspar, opal, apatite, biogenic carbonates (calcite and dolomite), and gypsum, while also being bituminous, containing more than 10% carbon (Bentor et al. 1981). Some diatomites, shales, and phosphorites also were locally metamorphized; phosphorite contains predominantly detrital apatite, which melts at 1650 °C (Bentor 1984). Some constituents elevated in the metamorphized rocks such as sulfate, and in other cases chloride, likely came from groundwater introduction during the combustion process (Rogers 1918; Bentor et al. 1981).

Bentor (1984) distinguishes between combustion 'glasses' and 'crystalline slags' for these materials, depending on the peak melting temperature achieved. 'Glasses' result from low temperature melts ≤1000 °C, are enriched relative to the starting mudstone in Na_2O (avg. 2.39 wt.%), K_2O (avg. 4.34 wt.%), Al_2O_3 (avg. 12.2 wt.%), and SiO_2 (avg. 74.0 wt.%) and maintain high viscosity in the small volume seams between unmelted rock, thus preventing crystallization (Bentor 1984). 'Crystalline slags' melted at ≥1100 °C, dissolving even apatite crystals and other refractory minerals, resulting in a lower viscosity, more basalt-like melt (i.e., avg. wt.%, 55.6 SiO_2, 9.0 CaO, 8.6 Al_2O_3, 2.8 Fe_2O_3, 1.9 SO_3, 1.8 MgO, 1.6 P_2O_5), which underwent extensive crystallization on cooling (hence 'slag'). These 'slags' are essentially synonymous with paralavas. There is strong evidence of silicate/phosphate liquid immiscibility from the lower temperature silicate melt and higher temperature phosphorite melt in these cases (Bentor et al. 1981; Bentor 1984; Capitanio 2005; Peretyazhko et al. 2018).

Two samples of Grimes Canyon were investigated by X-ray diffraction. Not all phases could be accurately identified by XRD, but diffraction indicated the majority was amorphous glass (50–75%), with large fractions of plagioclase feldspar (12–25%), F-apatite (6–7%), calcite (3–4%), pyroxene (~augite) (1–9%), and tridymite (1–4%). The lighter portions of the banded material contained more carbonate and phosphate crystalline phases (i.e., calcite and apatite). These phases are broadly consistent with those observed in Bentor et al. (1981). In Wyoming Powder River Basin paralavas, phenocrysts of olivine, cordierite, and spinel (magnetite-hercynite-ulvospinel) are also common, along with many unusual minerals and mineral associations (Cosca et al. 1989). In the Apennine region in Italy, pyrometamorphic outcrops include wollastonite and melilite, likely from protoliths of mixed marly sediments of carbonates and shales (Melluso et al. 2004).

In summary, pyrometamorphic glasses and rocks are highly variable due to the varying temperatures, mineral protoliths, and local environmental conditions, resulting in equally variable disequilibrium melting, volatilization, mixing, and crystallization (Sokol et al. 1998; Canil et al. 2018; Peretyazhko et al. 2018).

Pseudotachylite or frictionites

Pseudotachylite is a generic name for friction melts, and usually, in the geological terminology, pseudotachylites are classified as fault rocks (Killick 2003) even if the genesis could be quite broad. In agreement with Heide and Heide (2011), we classify these glasses as metamorphic ones, and in turn, among those glasses that originated from pre-existing rocks and display marked modifications due to changes in the pressure–temperature conditions.

What we now know as friction melts (and especially those with a pumice texture) were originally associated exclusively with volcanic origin or with impact events. The term itself was introduced to describe the abundant dark glassy veins in Vrederfort (South Africa), the largest impact structure on Earth (Reimold and Gibson 2005). Nowadays, pseudotachylites are associated with large impact structures, earthquake-generated layers (veins), and very large rock avalanches (e.g. Reimold and Gibson 2005; Spray 2010). Indeed, the exclusive volcanic or impact-associated origin hypotheses have been both ruled out from the study of pumiceous rocks from Ötz Valley (Köfels landslide, see Erismann et al. 1977; De Blasio and Medici 2017). Since then, several studies have focused on fault-related friction melts that form by co-seismic friction (fault pseudotachylite) or avalanches (e.g., Ferré et al. 2005 ; Weidinger and Korup 2009; Weidinger et al. 2014).

The typical generic definition provided for pseudotachylite is: "dense rock produced in the compression and shear associated with intense fault movements, involving extreme mylonitization or partial melting" (Bates and Jackson 1987). Thus, the term "pseudotachylite" should be used to describe rocks that present unambiguous evidence of high temperatures. Because sometimes this name is used regardless of the origin of the glass, it is recommended to point out the differences between impact-related and fault-related pseudotachylites, such as the thickness of the material, or their formation history (Kenkmann et al. 2000; Reimold and Gibson 2005). The mechanisms of formation of (impact)-pseudotachylites in large impact structures are described in Reimold and Gibson (2005). In the following, we will briefly describe pseudotachylites associated with very large landslides or formed along a fault plane during seismic deformation.

Fault pseudotachylites usually are veins of a few millimeters to a few centimeters in width that are related to large magnitude seismic events and located mostly in the upper crust (Ferré et al. 2005). Large seismic events can induce a local increase of the temperature and melting of the host rocks along the fault plane, resulting in the creation of amorphous (or cryptocrystalline) veins in response to frictional fusion. Depending on the maximum temperature achieved during seismic slip, the mineral phases present in the host rocks will be either melted or preserved as relict grains in the glassy matrix (Spray 1992). The chemical composition of the fault pseudotachylites is often slightly more mafic than the host rock, but the glass chemistry is dependent on bulk composition, mineral assemblage, and texture of the host rocks (Maddock 1992; Di Toro et al. 2005). The study of the microlites and of the composition of the glass matrix can be used to determine the temperature regimes associated with seismic events. For example, artificial pseudotachylites have been produced by direct high-speed friction experiments on granite and gabbro compositions (e.g., Lin and Shimamoto 1998) and the estimated melt temperatures of natural and experimental pseudotachylites are in the range of 750–1400 °C, and locally higher than 1700 °C (Lin and Shimamoto 1998 and references therein). Moreover, these friction experiments confirmed that the amorphous phase formed is more mafic (has a lower SiO_2 content) and hydrous than the host rock, and that fault-pseudotachylites can form by frictional heating at shallow depth. Heide and Heide (2011) also report melting temperatures of 1700 °C because of the presence of lechatelierite inclusions in frictionite melts. Nestola et al. (2010) report the occurrence of hexagonal $CaAl_2Si_2O_8$ (dmisteinbergite) in a fault pseudotachylite from the Gole Larghe Fault (Adamello, Italy), a polymorph that is formed at high temperatures (1200–1400 °C) by rapid cooling. Constraint of the temperature regimes during seismic events is considered essential in order to better understand fault processes and to infer earthquake source parameters.

Friction melts can be generated as well during very large landslides (Erismann et al. 1977; Legros et al. 2000; Weidinger et al. 2014). Well-studied landslides-derived pseudotachylites are in Köfels (Austria) and Himalaya (Nepal). The latter is a giant rockslide (dislocation of ~170 m) that caused the formation of a homogeneous glassy layer with a thickness between 1–3 cm (Masch and Preuss 1977; Heide and Heide 2011). Masch and Preuss (1977) report a detailed study of both events, and they observed that the glass matrix is chemically heterogeneous with schlieren, bubbles, and relicts from the parent rock materials (partial to almost complete melting of host rocks of granitic to granodioritic composition). Moreover, they report the occurrence of glasses with pure quartz, plagioclase, and alkali-feldspar compositions. Weidinger et al. (2014) report a review of several basal deposits of giant rockslides and defined that very short (<10 s) giant moving rockslides could produce friction-induced partial melting with basal temperatures >1500 °C (in anhydrous conditions).

GLASSES FROM HIGHLY ENERGETIC EVENTS

We will describe amorphous materials deriving from highly energetic events, either natural or artificial. First of all, glasses produced by natural events, such as tektites and impact spherules, will be taken into account. Then, we are going to describe some enigmatic natural glasses that are almost certainly of impact origin (accepted or assumed) but that are not found in stratigraphic contexts. Notable examples of such glasses are Libyan Desert Glass, Darwin Glass, and Dakhleh Glass. Finally, we are going to describe other glasses, resulting from a rapid "shock" event, the fulgurites. At the end, we will also consider a glass, not strictly natural, but that can be considered as an analogue of impact glasses: the atomic glass—trinitite.

Impactites

Impact processes are able to greatly influence the evolution of planets. Looking at the Moon, we can recognize with the bare eye that the surface is heavily affected by impact cratering. Besides our satellite, planets such as Mercury, Mars, and many satellites in the Solar System have circular depressions that testify the occurrence of numerous impacts, with only a few exceptions, including Earth, where the topography has been remodeled by active geological processes. The first impact crater directly witnessed on Earth in modern times (the so-called Carancas event) is very recent. It occurred near the southern shore of Lake Titicaca in Peru on September 2007, where the impact of a stony meteorite produced a 13.5 m crater (Tancredi et al. 2009).

The recognition of impact cratering as a fundamental process in the evolution of a planet has occurred rather recently as well, with the discovery of small spherules distributed worldwide in the Cretaceous- Paleogene K–Pg layer (formally known as KT, Cretaceous–Tertiary), first detected in Gubbio, Italy by Alvarez and coauthors (1980). These spherules and the associated layer contain chemical and physical evidence for a major impact that caused (supposedly) the mass extinction event at the K–Pg boundary. In 1991, a 170–180 km diameter crater was identified on the Yucatan peninsula as the impact site (Chicxulub crater) of a 10 to 15 km asteroid that induced the cataclysmic event responsible for the formation of the K–Pg worldwide ejecta horizons and caused the end-Cretaceous mass extinction, around 65 Ma ago (Hildebrand et al. 1991).

As might be expected, impacts have caused catastrophic effects on climate and the landscape; however, they also produce beneficial effects, such as the economic significance of impact deposits (i.e., Sudbury, Canada; Vredefort, South Africa), or the associated hydrothermal systems created (Osinski and Pierazzo 2012; Pierazzo and Melosh 2012).

The impact of a large object (>50 m for a stony object), moving at a great speed (>11 km s^{-1}; French 1998), with the surface will create an impact crater after a very short sequence of complex events, which can be summarized into three main stages: contact and compression, excavation, and modifications (Melosh 2011). The generic term impactite refers to the large variety of materials (rocks, melts) formed by the hypervelocity impact(s) of a large extraterrestrial body. Distinct materials are generated by the forces of a hypervelocity impact, starting with a shock wave (compression stage) and followed by decompression from peak shock pressures, with associated heat generation and material transport (excavation stage) (Reimold and Jourdan 2012). The post-shock heat causes the solid-state deformation and/or melting of the target crustal rock and even its vaporization, and the original material is entirely altered by the extreme heat and pressure regimes. Simulations of impact processes have established that the timescale of crater formation is extremely short, compared to many other geological processes, and is in the order of seconds to minutes for craters ranging from 1 to 100 km. Thus impactites will present very distinctive characteristics (French 1998; Stöffler and Grieve 2007; Stöffler et al. 2018). Impactite is the commonly used (and misused) word for all shocked, melted or vaporized materials as a consequence of a hypervelocity impact, and because of the confusion, a study group proposed a classification and a nomenclature of these materials, based on geological setting, texture, composition, and degree of shock metamorphism (Stöffler and Grieve 2007; Stöffler et al. 2018). The first grouping is based on the location of the materials, and "proximal impactites" is used for materials found in the proximity of the impact crater, whereas the term "distal impactites" refers to materials that do not occur directly in or around a source crater (i.e., distal ejecta). Further sub-classifications have been applied to proximal impactites, and according to Stöffler and Grieve 2007, proximal can be further divided into three subgroups: shocked rocks, impact melt-rocks and impact breccias. Depending on the peak pressure and the post-shock temperatures, both minerals and original rocks texture will present shock effects that allow identifying progressive stages of shock metamorphism. For instance, by increasing shock pressure, there are i) Planar Fractures (PFs) and Planar Deformation Features (PDFs) in minerals such as quartz, zircon and feldspars; ii) high-pressure quartz polymorphs (coesite, stishovite); iii) diaplectic mineral glasses (produced without fusion); iv) fused mineral glasses (produced with fusion), and v) melts (see Fig. 3; Reimold and Jourdan 2012). For example, quartz grains exposed to shock compression develop planar microstructures depending on the pressure range: PFs for pressure range of 5–8 GPa, and PDFs over the pressure range of 5–10 GPa to ~35 GPa. Shock-induced deformation in zircon grains have been reported as well, indicating shock pressures higher than 20 GPa (Ferrière and Osinski 2012). Impact diamonds have been reported as well in many impact sites. The transition from carbon (graphite) to diamond and lonsdaleite is around 13–15 GPa, with temperature in the range of 1300–2000 K (French 1998; Pratesi 2009).

Figure 3 clearly shows that the pressure conditions for normal crustal metamorphism (regional, contact) differs completely from shock metamorphism, resulting in the production of characteristic features and materials. In normal crustal metamorphism, the range of temperature and pressure is around ≤1000 °C and 1–3 GPa, respectively, and rocks and mineral transformations occur in very long times (>10^5 years), thus approaching equilibrium (French 1998). On the contrary, shock metamorphism is a relatively instantaneous process, with peak pressures that reach >100 GPa, and temperatures much higher than 2000 °C (French 1998). Because of the rapid process, particularly of cooling, quenched amorphous and crystalline metastable phases are characteristic. For a detailed description of all shocked materials, readers are referred to the comprehensive work of Stöffler and Langenhorst (1994) and to Osinski et al. (2012).

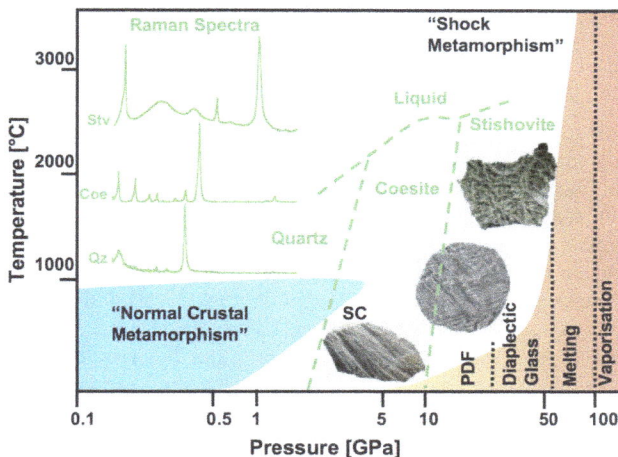

Figure 3. Pressure (*P*)–temperature (*T*) diagram for normal (crustal) and shock metamorphisms. Ranges for silica polymorphs and shock markers are reported. Figure from Reimold and Jourdan (2012). Copyright MSA.

In the following, we will mainly discuss the most homogeneous glasses created by hypervelocity impact events, i.e., the distal ejecta tektites, but we will also provide a brief overview of some other distal ejecta and proximal glasses. To distinguish proximal impact glasses, volcanic glasses, and distal ejecta tektites, Koeberl (2013) provides some characteristics that are distinctive of the distal ejecta and enable their recognition. Tektites are:

    i) amorphous and fairly homogeneous (no crystallites);

    ii) contain lechatelierite (amorphous SiO_2);

    iii) occur within definite areas, called strewn-fields (SF) and are associated with a single source impact crater;

    iv) do not occur directly in or around a source crater (they are distal ejecta).

Moreover, tektites are depleted in water (0.002 to 0.02 wt.%, at least an order of magnitude lower than the H_2O content of volcanic glasses) and are highly reduced (almost all iron occurs as Fe^{2+}) (Fudali et al. 1987; Koeberl 1994, 2013; Beran and Koeberl 1997; Rossano et al. 1999; Melosh and Artemieva 2004; Giuli et al. 2010a, 2013a, 2014a; Giuli 2017).

Distal Ejecta. As a consequence of a hypervelocity impact, the uppermost (200 m) surficial target rock is melted and ejected. A small portion of this molten ejected material will fly far away from the source crater, and the materials are called distal ejecta (according to Montanari and Koeberl 2000, ~90 % of the ejecta are proximal i.e., deposited within five crater radii from the impact site). The flying melts are quickly quenched to form fairly homogeneous amorphous materials with typical shapes (e.g., splash forms) like spheres, teardrops, disc-shaped form, and are known as tektites (Fig. 4). A partial crystallization of primary microlites/crystallites (e.g., clinopyroxene or Ni-rich spinel) upon cooling could occur, and the material formed is called microkrystite (Glass and Simonson 2013 and references therein).

Distal ejecta deposits (or air-fall beds) are deposited far away from the source crater and can be divided into two types: i) tektites and microtektites, and ii) spherule beds (Osinski et al. 2012). A more enigmatic kind of distal ejecta found on Earth consists of glassy materials, not volcanic in origin, but not having spherical shapes and not classified as tektites. Examples of these enigmatic glasses are Libyan Desert Glass, and Darwin glass (Glass and Simonson 2013). The reasons behind the formation of spherules and tektites remain poorly understood, and no widely accepted model or evidence has been found to explain why just a few impacts form large amounts of high-speed ejected glasses and why only a few provide tektites (Howard 2011). Thus, tektite formation must require specific impact conditions (see below).

On Earth, impact craters are not easily found (190 confirmed impact structure as of July 2019; Earth Impact Database 2019) because their morphologies have been modified by hydrothermal and chemical alteration and by tectonic processes. Thus, most of the impact craters which occurred on Earth have been subsequently destroyed or covered. For instance, the Chicxulub impact structure is located beneath ~1 km of sediment and half offshore of the Yucatan Peninsula. Until 1991, the year of the discovery, the only proof of a connection between end-Cretaceous mass extinction (K–Pg boundary layer) and a possible cataclysmic impact was the worldwide occurrence of distal ejecta horizons. Hence, sometimes, distal ejecta are the only remaining witnesses of large impact events, and they provide essential information regarding planetary processes when found in the stratigraphic record. The importance of air-fall beds has been compared to the importance of volcaniclastic layers (Glass 2016).

When dealing with distal ejecta, or in general, with glass spherules, the most critical point is to gather enough evidence to rule out the volcanic origin. These pieces of evidence are usually called 'impact markers' and include chemical, isotopic, and mineralogical marks that indicate the involvement of a cosmic body. Examples of markers that indicate an extraterrestrial contribution

Figure 4. Australasian tektites with various shapes; dumbbell, teardrop, and fragments. Dumbbells are formed by the rapid rotation of melt droplets ejected into the atmosphere. When the rotation is high enough, the dumbbell is broken to form teardrops.

are the elevated content of siderophile elements (Ir, platinum group elements, PGE), the occurrence of shocked minerals and/or Ni-rich spinels and/or impact diamonds, and atypical isotope ratios. The complete list of the well-established or new 'impact markers' is beyond the scope of this chapter, and the readers are referred to the exhaustive contribution of Goderis et al. (2012).

Tektites and microtektites

Tektites are small, typically dark, glassy objects that have been transported through the atmosphere hundreds to thousands of kilometers from the impact site, and are found only in some regions of the Earth's surface, called tektite strewn fields (SF). Tektites are generally chemically homogeneous, Si-rich glasses of various sizes (usually >1 cm), with typical aerodynamic shapes and very characteristic surface features (Fig. 4). Microtektites are microscopic tektites, with a diameter <0.1 cm, primarily found in deep-sea sediments.

Tektites can be further subdivided into three types, depending on their forms: Muong Nong-Type (or layered), ablated/flanged tektites (or aerodynamically shaped tektites), and the most common ones, the splash forms (or normal tektites). The morphology of the most common tektites varies between spherical, dumbbell-shaped, and teardrops as a result of rapid rotation of the molten flying material (Fig. 4).

Muong Nong-Type tektites (MN—named after a region in Laos) are unusually large tektites (up to several tens of centimeters in size) with layered structures. Koeberl (1992a) observed that Muong Nong-Type tektites are enriched in volatile elements (the halogens, Cu, Zn, Ga, As, Se, Pb), and have higher water contents compared to splash-forms and ablated tektites. Moreover, MN tektites present chemical heterogeneity (darker and lighter layers) and may contain relict mineral grains (e.g., corundum, quartz, chromite, and cristobalite) and very large bubbles (Glass and Simonson 2013). Despite the differences with other tektites, MN-Type tektites have an average chemical composition (major element chemistry) and age matching those of common tektites. Moreover, MN-Type tektites strongly differ from volcanic glasses because of the presence of shocked mineral inclusions, differences both in major and trace element contents (e.g., REE patterns), lower water content, and highly reduced iron (Koeberl 1992a). The presence of relict mineral grains and the relatively higher water content of Muong Nong-Type tektites suggest that these glasses have experienced the lowest formation temperatures of all tektites (Koeberl 1992a), as does the presence of volatiles (halogens, Zn) and the vesicularity of lechatelierite particles (Montanari and Koeber 2000). It is assumed that Muong Nong-type tektites have been deposited closer to the source crater, and based on [10]Be data, they should derive from a greater depth in the target deposits than most other tektites (Ma et al. 2004). The layered structure of these tektites, with their chemical heterogeneities, and the presence of relict mineral grains, may serve to better constrain tektite conditions of formation because they supposedly represent a major link between target rocks and tektites (Montanari and Koeber 2000).

To date, of the four known tektite strewn-fields, all but one have been linked to source craters, based on geographic location, geochemical evidence, and composition (Koeberl 2013 and references therein) (Fig. 5). The oldest strewn field known is the North American (NA) one of 35.3 Ma age (±0.1) associated with the 85 km diameter Chesapeake Bay (USA) impact structure and includes Bediasites, Georgianites, Barbados and Cuba tektites. The Central European (CE) or moldavite strewn field of 14.8 Ma age (±0.2; Schmieder et al. 2018) is associated with the Nördlinger Ries impact structure of about 24 km in diameter (Nördlinger Ries, Bavaria, Germany). There is another impact crater, the Steinheim crater, ~3.8 km in diameter, located about 42 kilometers west-southwest from the center of Nördlinger Ries. These two craters are believed to have formed nearly simultaneously by the impact of a binary asteroid (Stöffler et al. 2002). Moldavites are found mainly within a 200–450 km range from the Ries crater, in the Czech Republic (Bohemia, Moravia), Austria, and Germany. Recently, a sub-strewn field in Poland has been reported as well (Brachaniec et al. 2014), circa 485 km from the Ries crater. Moldavites have a unique green color and have strong surface sculpture derived from erosion. Stöffler et al. (2002), and more recently Artemieva et al. (2013) modeled the formation of the Central European strewn field, suggesting a 1.1–1.5 km stony meteorite impactor hitting the target surface with an angle of 30°, at 15–18 km/s.

The Ivory Coast (IC) tektite strewn field is associated with the 1.07 Ma old Bosumtwi crater (10.5–11 km diameter, Ghana, Africa) (Koeberl et al. 2007). The Bosumtwi crater is the youngest well-preserved impact structure known, and it is almost entirely filled by a lake (Lake Bosumtwi). The youngest strewn-field, of 803 ± 3 ka (Lee and Wei 2000), is the Australasian one, for which no source crater has been identified so far. Tektites of the Australasian strewn field (AA) include australites, thailandites, indochinites, philippinites, and javanites. The AA strewn field is very extensive, and findings of microtektites on land and in oceans (microtektite-bearing deep-sea cores) have improved the determination of the AA geographic distribution, which is spread from the southeastern region of Asia down to Australia and Tasmania (Fig. 5). Recently, smaller tektites (microtektites) that were found in four different sites in northern Victoria Land Transantarctic Mountains (TAM, Antarctica) have been related to AA, and there is clear evidence (Folco et al. 2008, 2009, 2010, 2011; Giuli et al. 2014a) that these samples represent a major southeastward extension of the Australasian strewn field.

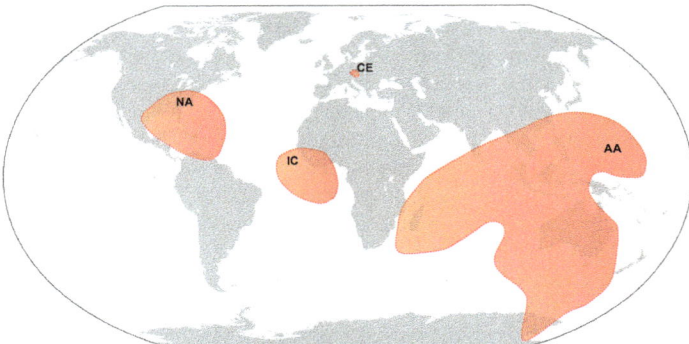

Figure 5. Approximate location and extension of the four strewn fields: NA (North American), IC (Ivory Coast), CE (Central Europe), and AA (Australasian). The known source craters are Chesapeake Bay (NA), Ries (CE), and Bosumtwi crater (IC). Redrawn after Folco et al. (2011, 2016); Glass (2016).

Thus, the AA strewn field covers more than 10% of the Earth's surface, and even if no source crater has been identified yet, several authors suggest its location somewhere in the Indochina region (Folco et al. 2016). Besides these four well-known tektite strewn fields, recently, two others have been proposed, one from central American (Belize glasses), and the second from Uruguay (uruguaites) (Senftle et al. 2000; Povenmire et al. 2011; Giuli et al. 2014b; Povenmire and Cornec 2015; Ferrière et al. 2017).

The chemical composition of tektites is fairly homogeneous within the same strewn-field, and also among the four strewn-fields, the chemistry of all distal ejecta shows that these glasses are aluminosilicates with high SiO_2 contents and depleted in alkalis (see Table 2).

Tektites are usually black to dark green, except moldavites that have a characteristic lighter green color and do not contain microlites or relict minerals, but contain vesicles, mostly spherical, or elongated as evidence of glass flow. Figure 6 shows the light transmitted optical image of an Australasian tektite. Flow lines and small vesicles are clearly visible. Bubbles are very frequent in tektites, and many studies have been performed in order to determine the composition, the content, and the pressure of the trapped gases. These data are very important because they provide information on the nature of the atmosphere during formation (quenching). Oxygen, C-based gases (CO, CO_2, CH_4), and noble gases have been identified in different proportions, and sometimes contradictory results have been presented, mainly because of the extraction methods used. Early studies reported the presence of CO_2, CO, H_2, SO_2, CH_4, N_2, and O_2 with total gas contents ranging between 0.23 and 0.82 cm^3/g (Ottemann 1966 and references therein). Gentner and Zähringer (1959) and Zähringer and Gentner (1963) studied noble gases in several tektites and found heavy noble gas (Ar: Kr: Xe) ratios very similar to that of the Earth's atmosphere, thus confirming the terrestrial origin of these specimens. Moreover, the established contents of argon (^{40}Ar) was fundamental to determine the age of tektites across the AA strewn fields and to confirm that the different specimens were of the same age, and in turn, of the same origin. A recent review by Žák et al. (2012) lists all studies concerning C-based gases in tektites (excluding noble-gases). More recent studies on gas ratios and noble gas isotopic ratios confirmed that the gases have a terrestrial atmosphere origin, with a typical enrichment of lighter noble gases (He, Ne) compared to the heavy ones (Ar, Kr, Xe) (e.g., O'Keefe et al. 1962; Müller and Gentner 1968; Jessberger and Gentner 1972; Matsubara and Matsuda 1991; Matsuda et al. 1996). The estimated bubble internal pressures are very low ($\ll 1$atm), thus suggesting that vesicle formation must result from internal gas pressure during tektite cooling, and this happened in an atmosphere under low pressures (at high altitude). For instance, Matsuda et al. (1996) reported a bubble pressure of $\sim 10^{-4}$ atm, and a solidification of the molten ejecta that should have occurred high in the stratosphere at altitudes above 20–40 km, in agreement with earlier observations (Rost 1964).

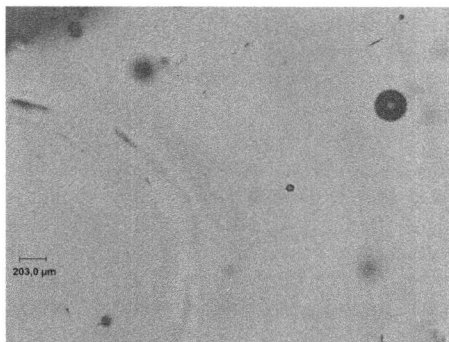

Figure 6. Optical microscope image (transmitted light) of an Australasian tektite slice. Several bubbles of different dimensions and flow lines are visible.

Table 2. Average major oxide compositions of tektite and microtektites (wt.%).

	NA[§]	NA micro[‡]	CE[#]	IC[§]	IC micro[‡]	AA[$]	MN-type[$]	TAM micro[*]	AA-HMg micro[⊕]
SiO$_2$	81.80	73.50	80.30	67.58	67.37	73.70	78.00	72.27	54.73
TiO$_2$	0.51	0.68	0.32	0.56	0.54	0.78	0.62	0.88	0.92
Al$_2$O$_3$	11.20	14.08	10.10	16.74	16.68	12.30	10.60	15.07	17.41
FeO	2.64	3.99	1.69	6.16	6.46	4.35	3.81	3.77	6.02
MnO			0.10	0.06		0.10	0.08		
CaO	0.45	1.59	1.35	1.38	1.57	2.41	1.51	3.49	4.82
MgO	0.61	1.81	1.69	3.46	3.72	2.21	1.74	2.87	15.36
Na$_2$O	0.94	1.87	0.55	1.90	1.89	1.37	1.33	0.25	0.64
K$_2$O	2.44	2.24	3.48	1.95	1.77	2.78	2.18	0.93	0.11

Footnotes: § Glass (2016); ‡ Giuli et al. (2013a); # Žák et al. (2012); $ Mizera et al. (2016); * Giuli et al. (2014a) (TAM: microtektites from Victoria Land Transantarctic Mountains); ⊕ Glass et al. (2004). For the original data source and error, see the cited references

Three of the four tektite strewn-fields (SF) so far known also present examples of microtektites (diameter usually <0.1 cm): the North American, the Ivory Coast, and the Australasian SF. These spherules have been found in deep-sea deposits (see, e.g., Glass 1967, 1972, 1978, 1990) and are very important for defining the extension of the strewn fields (e.g., Glass and Zwart 1979), for constraining the stratigraphic age of tektites, and for providing an indication regarding the location of possible source craters (e.g., Glass and Pizzuto 1994). Ivory Coast microtektites are found as far as >2,000 km away from their source, whereas Australasian microtektites have also been found at extraordinary distances (>10,000 km) from the hypothetical source crater. Microtektites, like tektites, show various morphologies, with oblate/prolate spheres, dumbbell- and teardrop-shaped, and may contain lechatelierite and vesicles. However, microtektites display a larger range of colors than tektites, with the majority being transparent, colorless, or greenish/yellowish. Trace elements and isotopic abundances confirm that microtektites are genetically related to tektites in the associated strewn field (Frey 1977). However, microtektites usually show a wider compositional range than tektites, even if generally they are similar to the tektites from the same strewn-field (Table 2). For example, it is possible to recognize IC and AA microtektites based on their alkaline earth ratios (MgO/CaO).

There is a particular sub-group of AA microtektites that presents notably high Mg contents (denoted HMg; up to 24 wt.% MgO) and lower SiO$_2$ contents than "normal" AA-microtektites, with a bottle-green color and highly eroded state (Glass and Simonson 2013). A detailed study of major and trace elements of >100 microtektites from the different strewn fields has been performed by Glass et al. (2004), who found, besides the HMg microtektites, some glasses with high Ni contents (denoted HNi; up to ~470 ppm). HMg microtektites have been found also in the IC strewn-field. The study of the trace elements in microtektites also confirmed that these small distal ejecta derive from the upper continental crust.

Everything seemed to point to microtektites being just smaller tektites. However, lately, some differences have been found, namely on the Fe oxidation state of North American microtektites (Giuli et al. 2013a). These authors have shown that some North American microtektites present higher Fe^{3+}/Fe^{2+} ratio (up to 0.61), compared to the respective tektites, implying that different formation mechanisms are involved for such small objects. Interestingly, for these microtektites, there is a positive correlation with the distance from the known source crater, and more oxidized conditions are reported for longer distances. This seems to be in contrast with the data available on other microtektites recovered at much further distances (i.e., AA microtektites). Because of the relatively limited information available on tektite/

microtektite formation mechanisms (see below), it is difficult to provide an interpretation of those data. Do microtektites register another "path" of the ejecta, or perhaps a different timeframe? Do they have different cooling rates? Further experimental studies, along with modeling and further ejecta discoveries, will help to answer these questions, and probably the broader question of "how do distal ejecta form?".

Other distal ejecta. Besides tektites and microtektites, it is worth mentioning the occurrence of many other spherule-associated layers across the world. For example, a clinopyroxene-bearing spherule layer (cpx-spherules) has been traced in many locations on Earth. The cpx-spherules associated layer is enriched in Ir and is ~10–20 ka older than the North American microtektite one (Glass et al. 2004 and references therein). These spherules have been associated with the Popigai complex crater (Siberia, Russia), an extremely large structure (~100 km diameter) of 35.7 ± 0.2 Ma (Melosh 2011).

As mentioned before, the most notorious and studied distal ejecta deposits are the K–Pg (formerly KT) spherule ones, associated with the 170–180 km Chicxulub crater (Yucatan, Mexico). The impact event responsible for this crater had enough energy to distribute ejecta worldwide. The small K–Pg spherules (100–500 μm), that resemble microtektites, were first detected in the Cretaceous–Paleogene K–Pg layer in Gubbio (Italy) (Alvarez et al. 1980). K–Pg distal impact ejecta horizons are associated with Ir enrichments, siderophile element anomalies, shocked minerals, and high-pressure polymorphs (shocked quartz grains, coesite and stishovite). K–Pg spherule layers have a global geographical extension with more than 350 sites identified, but because of the poor preservation of the claystone at the K–Pg boundary, in the early 80s there were some debates as to the origin of these spherules, with some authors supporting an impact hypothesis (e.g., Smit and Klaver 1981), and others attributing an authigenic origin for the spherules (e.g., Izett 1987). A few years later, many authors, by studying many K–Pg spherules, and in particular the Si-rich spherules preserved at the K–Pg layer at Beloc (Haiti), provided clear evidences of an impact origin, based on geochemical data, enrichment in platinum group elements, the presence of lechatelierite and shocked quartz grains (Bohor 1990; Sigurdsson et al. 1991; Koeberl 1992b; Koeberl and Sigurdsson 1992). Despite the alteration of some deposits, according to Morgan et al. (2006) there is a correlation between shock markers and paleodistances from the impact site, and in particular the number and size of the spherules and shocked minerals are inversely proportional to their distance from Chicxulub. This evidence has been used to confirm the occurrence of a single highly energetic impact event related to the formation of the K–Pg spherule horizons, and to provide insights on the obliquity of the projectile.

Among the proposed impact-related Cenozoic distal ejecta layers, there is one horizon related to a hypothesized impact event, that according to Firestone et al. (2007), may have occurred at the beginning of the Younger Dryas (YD ~12.8 ka). In this case, the impact event is not strictly related to a collision with the Earth's surface, but (allegedly) to an airburst, that is a shock wave caused by the explosion of a cosmic object in the atmosphere (such as the well-known 1908 Tunguska airburst, Siberia; e.g. Melosh 1989 and references therein). The Younger Dryas boundary is a <5 cm-thick sediment layer dated ~12.8 ka in several sites in North America, Belgium, and Syria (Firestone et al. 2007; Bunch et al. 2012; Wittke et al. 2013), that contains impact-related markers, such as Ir enriched grains, magnetic micro-spherules (10–250 μm), vesicular carbon spherules, glass-like carbon, shock-fused vesicular lechatelierite, impact nano-diamonds, and fullerenes containing extraterrestrial concentrations of ^{3}He (helium-3 is extremely rare in terrestrial crustal rocks; Koeberl 2013). Still, the YD impact remains just a hypothesis because most of the reported "impact-related" markers have not been confirmed in independent studies or are not considered unambiguous impact proxies. For instance, fullerene with anomalous ^{3}He contents has not been found in all sites, and no shocked minerals have been reported. Moreover, Paquay et al. (2009) measured both PGE concentrations and Os isotope ratio (^{187}Os/^{188}Os) in the bulk sediments without finding evidence of extraterrestrial components.

In a review, Pinter et al. (2011) report that none of the evidence described by Firestone et al. (2007) were incontrovertible, and more recently, van Hoesel et al. (2014), in another review on the YD impact markers, concluded that no unambiguous markers have been found so far, and that the data available still cannot support the claim that there was a Younger Dryas impact event.

Enigmatic impact glasses

Glass and Simonson (2013), in their contribution on distal impact ejecta, describe some glasses, almost certainly of impact origin, but not found in stratigraphic contexts. Most of the time, no source craters could be associated with these materials; hence the melting has been often attributed to radiation from airburst. These natural glasses have been found in several locations on Earth and cannot be classified as tektites or spherules. Examples of such glasses are: glass fragments from South Ural (Russia) called urengoites (Deutsch et al. 1997), glass fragments from Egypt (Dakhleh glass; Osinski et al. 2007), the well-known Libyan Desert Glass, and Darwin Glass (with a putative associated crater; Meisel et al. 1990).

In an area of ~6500 km^2 in southwest Egypt, close to the border with Libya, natural Si-rich glass fragments, known as Libyan Desert Glass (LDG) are found (Rocchia et al. 1996). Fission track dating provides an average age of ~29 Ma (28.5±2.3 Ma to 29.4±0.5 Ma (Storzer and Wagner 1977), and 28.5±0.8 Ma (Bigazzi and Michele 1996). LDG occurs as centimeter- to decimeter-sized, irregularly shaped, and strongly wind-eroded glass pieces (Fig. 7), and one of the most famous specimens is the one used as the yellowish scarab found in Tutankhamen's armor. Libyan Desert Glasses are very silica-rich (~96.5–99 wt.% SiO$_2$; see Fig. 1; Table 3) and show a limited variation in major and trace element abundances (the remaining few wt.% are oxides of aluminum, iron, calcium, and magnesium). They are homogeneous with a characteristic pale yellow/yellow-green color and characterized by the presence of schlieren (Glass and Simonson 2013). Some minor elongated vesicles and spherulitic inclusions of cristobalite have been reported as well, and in some cases baddeleyite (as a high-temperature breakdown product of zircon), rutile, staurolite, cordierite, kyanite, reidite and mullite have been detected (Storzer and Koeberl 1991; Rocchia et al. 1996; Greshake et al. 2010, 2018; Gomez-Nubla et al. 2017; Cavosie and Koeberl 2019).

Since its discovery, early in the 20th century (Clayton 1934) the origin of the Libyan Desert Glass still represents an unanswered enigma to all scientists and researchers. However, an impact-related origin seems the most plausible mechanism. In fact, LDG fragments are thought to be the remains of a glassy surface layer, resulting from the high temperature melting of sandstones/desert sand, caused either by a meteorite impact or—to explain the absence of an impact crater—by airburst (Seebaugh and Strauss 1984; Aboud 2009). The rare earth element (REE) pattern of LDG is typical of upper continental crust, but the analysis of desert sand and sandstones found in the Libyan Desert Glass strewn-field do not perfectly match, thus no rock precursors and crater could be definitively identified (Storzer and Koeberl 1991). Evidence for an impact origin includes the presence of schlieren and flow textures, detectable amount of Ir, lechatelierite, and baddeleyite and, in general, high pressure–temperature phases (Storzer and Koeberl 1991; Barrat et al. 1997; Swaenen et al. 2010). Moreover, the high concentration of PGE, the presence of graphite-rich inclusions (ribbons), and reduced Fe species are additional evidence for an impact origin (Barrat et al. 1997 ; Pratesi et al. 2002; Giuli et al. 2003). Some specimens of LDG present marked layering, with brownish or grayish-bluish streaks (in general, both are referred to as dark streaks). The brownish streaks are enriched in siderophile elements, Mg, Fe and Ni, and the Os isotopic data indicates the extraterrestrial component (Barrat et al. 1997; Pratesi et al. 2002). Observations by Transmission Electron Microscopy (TEM) revealed the occurrence of Al-, Fe- and Mg-enriched amorphous nano-spherules (80–100 nm) in the LDG dark streaks, as a result of silicate–silicate liquid immiscibility (Pratesi et al. 2002). Similarly, other impact glasses from Zhamanshin crater (blue zhamanshinites; Kazakhstan) contain spherical nano-inclusions (~100 nm) of a second glass enriched in Fe, Ca, Mg and P (Zolensky and Koeberl 1991).

Table 3. Average major oxide compositions of impact glasses (wt.%).

	LDG[§]	LDG dark streak[⊕]	Wabar dark[#]	Wabar light[#]	Aouelloul[#]	Suevite glass—Ries[#]	Darwin[‡]	Darwin*[§]	Atacamaite[§]
SiO$_2$	98.4	95.85	87.45	92.88	86.51	66.19	76.47–93.85	82.74	68.74
TiO$_2$	0.12	0.18	0.15	0.12	0.55	0.82	0.22–0.8	0.609	0.43
Al$_2$O$_3$	1.19	1.48	1.77	2.64	5.76	14.74	3.14–11.45	7.89	10.66
FeO	0.12	0.98	5.77	0.53	1.59	3.86	0.84–5.87	2.39	5.40
MnO		0.02						0.01	0.06
CaO	0.01	0.08	1.90	1.46	0.73	3.96	0.02–0.25	0.04	3.59
MgO	0.01	1.38	0.60	0.47	0.91	2.60	0.24–4.0	0.87	1.52
Na$_2$O	0.005	0.02	0.39	0.42	0.14	3.62	0.0–0.21	0.09	2.77
K$_2$O	0.009	0.01	0.58	1.61	2.05	3.26	0.75–2.71	2.05	2.61

Footnote: Data from [§]Koeberl (1997); [⊕]Giuli et al. (2003); [‡]Ottemann (1966); [*]Howard (2008); [§]unpublished data. *average chemical composition of the specimen shown in Figure 8

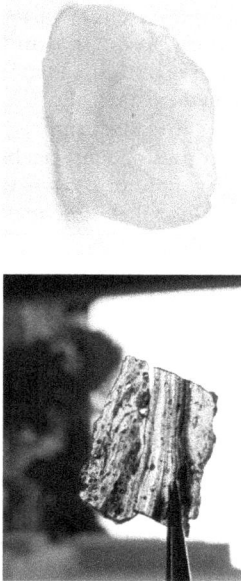

Figure 7. Left side: LDG with the typical aeolian erosion aspect. **Right side:** slice of Darwin glass shows pronounced layering and flow structures, and elongated vesicles.

Based on the amorphous and mineral phases occurring in the different LDG specimens, the range of temperature and pressure have been constrained to T >1900 K and pressure from 10 to >30 GPa (Pratesi et al. 2002; Gomez-Nubla et al., 2017; Cavoisie and Koeberl 2019). However, while high pressure occurred, most probably during melting and ejection, the quenching of the molten material happened quickly at atmospheric pressure (Greshake et al. 2018).

The occurrence of a meteoritic component suggests the impact of an extraterrestrial object and not an airburst. This seems to be confirmed by recent work by Koeberl and Ferrière (2019) that reports the discovery of shock markers (e.g. PFs and PDFs on quartz grains) in bedrock samples recovered in the Libyan Desert Glass strewn-field, and the suggestion of a deeply eroded impact crater in the area. Moreover, the detection of reidite, a high pressure (>30 GPa) polymorph of zircon ZrSiO$_4$, indicates that the pressure–temperature regimes are in line with a crater-forming impact (Cavoisie and Koeberl 2019).

Another impact glass that cannot be referred to as tektite is found in a strewn-field of ~400 km² in western Tasmania (Australia): the Darwin glass. The age of this glass, estimated by Ar–Ar methods, is 816 ± 7 ka (Lo et al. 2002), thus very close to the Australasian tektites, although their major and trace element compositions are very different, indicating that the temporal association is coincidental. The Darwin crater, a small (~1.2 km) simple impact crater formed in sedimentary target rocks, was discovered in 1972 and proposed as the source of Darwin glasses (Ford 1972; Fudali and Ford 1979). Howard and Haines (2007) carried out a detailed petrographic study of the crater-filling samples, but no conclusive shock markers have been found (e.g., PDFs in quartz grains). Nevertheless, the geochemistry of the target rocks and glasses, the location of the crater, and the glass distribution (Darwin strewn-field) all point to the Darwin crater as the source of the ejecta. Darwin glasses occur in an area larger than 400 km², with distances of at least 20 km (Howard 2011) from the putative Darwin crater, thus belonging to the distal ejecta group. Glasses generally occur as irregular centimeter-sized fragments, or masses, even if small splash-forms (spheres and teardrops <5 mm) can be found across the Darwin strewn-field. The color ranges from white, gray, light or dark green, dark brown, to black, and the glasses are generally vesicular and often exhibit flow structure marked by bands of elliptical vesicles (see Figs. 7 and 8) (Glass and Simonson 2013).

Interestingly, the proportion of white glasses is highest in the proximity of the crater, and the proportion of darker glasses, as well as the proportion of splashform shapes, increase with distance from the putative crater (Howard 2008, 2009; Gomez-Nubla et al. 2015). Darwin glasses are SiO_2 rich and extremely depleted in Na_2O and CaO. The chemistry suggests the presence of (at least) two main glass groups: the first one, which represents ~80% of the samples, is enriched in SiO_2 (average ~87 wt.%), whereas the second one has a lower SiO_2 content (average ~81 wt.%), higher contents of MgO and FeO, and is enriched in Ni, Co and Cr (Howard 2008) (see Table 3). Optical microscope images of a Darwin glass show the contrast between layers of a lighter colored glass (light brown) and a darker one (dark brown); furthermore, turbulent flow structures and bubbles of different dimensions and shapes are present (Fig. 8). All textural features visible in this specimen are very similar to those observed by See et al. (1998) in an impact glass bomb from the Ries crater. Raman spectroscopy data associated with the glass specimen shown in Figures 7 and 8 are reported in the section *Glass Properties*

Analyses by Raman spectroscopy allowed the identification of small inclusions of α-cristobalite, and iron or iron/nickel oxides (Gomez-Nubla et al. 2015). Recently, Raman spectroscopy has also been employed to detect and identify organic matter in both LDG and Darwin glasses (Gómez-Nubla et al. 2018). Based on the results, and the identification of e.g., fullerene-type compounds,

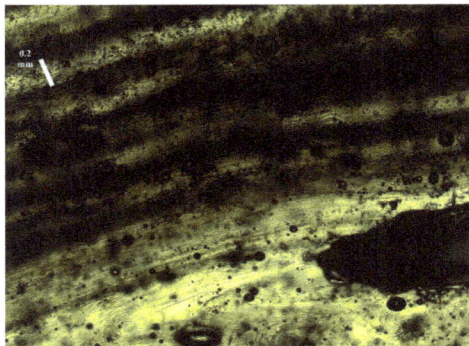

Figure 8. Optical microscope view (transmitted light) of the Darwin glass slice shown in Figure 7. Darker and lighter layering, flow structures, vesicles, and voids are clearly visible.

the authors constrained a pressure of ~15 GPa and temperatures between 670 and 1900 K. The preservation of organic material in inclusions of LDG and Darwin glasses has been explained by the trapping of the organic matter within the impact melt, and a fast quench that avoided the total decomposition of the material (Gómez-Nubla et al. 2018; Howard et al. 2013).

Formation of distal ejecta. The origin and the formation mechanisms of tektites, microtektites, and spherules have significantly been debated. Indeed, starting in the 50s, there was a great dispute on the terrestrial or extraterrestrial (lunar) origin of these ejecta (e.g., Urey 1955, 1963; Barnes 1958; O'Keefe and Shute 1963; O'Keefe 1969). Even after 1969, and the availability of lunar samples, the debate was on, and Schnetzler (1970) published a provocative obituary with the famous statement "The lunar origin of tektites [...] died on July 20, 1969. The cause of death has been diagnosed as a massive overdose of lunar data". The dispute has lasted until recently with O'Keefe (1994) still supporting the theory that tektites were ejected from volcanoes on the Moon, but, in the end, the terrestrial origin of tektites has been finally accepted, thanks to all geochemical and isotopic data available.

If the origin has been settled, the formation mechanisms are not completely clear yet. New distal ejecta findings, improved experimental data and numerical modeling have strongly enhanced the understanding of the physical–chemical conditions occurring during a hypervelocity impact (Melosh 1989, 2013). Nevertheless, the reasons why only a few impacts have produced significant amounts of distal ejecta is uncertain. Moreover, spherule formation from the plume is not fully constrained as well, because physical and mathematical models are hampered by the limited information available on how materials respond to extremely high stresses in highly non-equilibrium conditions. One of the most critical factors for distal ejecta formation is thought to be the obliquity of the impact (the angle between the projectile and target impact), which determine the maximum shock pressure and temperature regimes. Moreover, the presence of water in the target rocks has also been pointed out as a key factor as well (Artemieva 2002; Stöffler et al. 2002; Howard 2011).

Distal ejecta are believed to be formed during the very initial stage of the excavation (cratering process) when the projectile and the target materials (with a volume comparable to that of the impactor) are highly shocked, melted and vaporized. The highly pressurized mixture of vaporized rock, molten material, small solid fragments, and hot atmospheric gases constitute the so-called vapor plume (aka impact plume or fireball) (Melosh 2013), that expands away from the forming crater. The vapor plume pushes away the ambient atmosphere in its path and this allows ejected melt to follow long ballistic trajectories (cf. Howard 2011; see also Melosh 1989), and for large enough impacts the vapor plume might expand and burst the top of the atmosphere allowing materials to be dispersed at great distances, even on a global scale. Numerical modeling on spherule (or tektite) nucleation and condensation from the vapor plume have been performed by e.g., Johnson and Melosh (2012, 2014); Melosh (1989); Pierazzo et al. (1998). Johnson and Melosh (2012, 2014) differentiate two origins for spherules, based on their thermodynamic histories: i) vapor condensed droplets and ii) melt droplets, in agreement with early observations on other ejecta (Engelhardt et al. 1987). Because of the extremely high temperatures expected in the plume, the melt will have a low viscosity (7×10^{-6} Pa s, according to Melosh and Artemieva 2004), and volatile-bearing bubbles will rapidly travel through the melt and burst at the liquid surface, resulting in a liquid phase quickly depleted in water and other volatile elements, such as Na and K (Melosh and Artemieva 2004). The presence of a superheated melt in the vapor plum might explain not only the occurrence of alkali depletion, and the low water content in the ejecta, but also the oxygen volatilization and, in turn, the highly reducing conditions.

Proximal impactite. Glasses deposited within five crater radii from the impact site are referred to as proximal. These amorphous materials can resemble tektites, but they show higher chemical heterogeneity, relatively higher water contents (0.02–0.06 wt.% H_2O), and contaminants from the impactor (McPherson et al. 1984; Devouard et al. 2014; Koeberl et al. 2019). Example of glasses associated with impact sites are Irghizites (Kazakhstan), Aouelloul glasses (Mauritania), Lonar Crater Glass (India), Waber Glass (Saudi Arabia) and Atacamaites (Chile).

Associated with the Zhamanshin impact structure (Kazakhstan), there are a variety of impact glasses. The main groups are i) Irghizites, glasses with aerodynamic flight-shapes that resemble tektites, but with higher water contents, and with an average SiO_2 content of ~74 wt.%; ii) Zhamanshinites that can be found closer to the source crater and are much larger than Irghizites. Zhamanshinites are very heterogeneous, with layered structures, flow structures and based on the chemistry they are further differentiated in different subgroups (Fredriksson et al. 1977; Koeber and Fredriksson 1986), with glasses Si-enriched (>70 wt.% SiO_2), and glasses strongly depleted in silica (<55 wt.%), called basic impactites. An additional amorphous material is referred to as blue Zhamanshinite. It has high SiO_2 and CaO contents and presents evidence of liquid immiscibility with the occurrence of amorphous inclusions that provide the characteristic turquoise to very dark blue color (Koeber and Fredriksson 1986; Koeberl 1988; Zolensky and Koeberl 1991). Nano-sized Ca–Fe–Mg–P-rich silicate immiscible liquid inclusions scatter the light (Rayleigh scattering), resulting in a blue coloration in the glass.

Africa has 19 confirmed impact structures (Reimold and Koeberl 2014), and among them, there is the Aouelloul structure in Mauritania, a relatively young (3.1 ± 0.3 Ma) crater of ~390 m diameter. The enrichment in siderophile elements and the occurrence of lechatelierite, baddeleyite, Ni-rich Fe-spherules are the impact markers identified. The glasses associated (Aouelloul glasses) spread in the crater's proximity, and similar to the Darwin glasses, are very heterogeneous. They are water-depleted silicate glasses, and the impact nature of these glasses is also supported by the abundant schlieren texture and different bulk chemistries, and the presence of unambiguous meteoritic component.

Wabar crater is a very young impact structure in Saudi Arabia (290 ± 38 years). The associated glasses are both large fragments and small (≤1 cm) glassy aerodynamic flight-shaped and spheres (Prescott et al. 2004; Hamann et al. 2013). Wabar glasses are known to have several chemical heterogeneities, with flow structures, emulsions of Fe-rich ultrabasic silicate glasses, FeNi globules and areas marked with light- and dark-brown glasses, respectively depleted or enriched in Fe (Ottemann 1966; Hörz et al. 1989; Hamann et al. 2013; see Table 3). See et al. (1998) by studying the microstructure of the glasses, highlighted the textural and chemical heterogeneity that are present in these impact glasses and the occurrence of interstitial enrichments of Fe and Ni, most probably resulting from the impactor.

From the Atacama Desert (Chile), thousands of aerodynamically shaped black glasses have been recovered and named atacamaites. These glasses have minor vesicularity, inclusions of an extremely silica-rich glass (lechatelierite), turbulent flow structures, substantial chemical variations (e.g., FeO contents vary from 5 wt.% to 15 wt.%), and extremely high contents of Ni, Co, Re, Ir and Pt (orders of magnitude higher than the average upper continental crust). Moreover, water contents are higher (~130 ppm) than those associated with tektites. All these characteristics point to the impact nature of these glasses (Devouard et al. 2014; Dos Santos et al. 2015; Koeberl et al. 2019).

For the sake of comprehensiveness, we will briefly describe an example of a proximal impact breccia, the suevite, from the Ries crater type locality. An exhaustive review on Ries crater and suevite is reported by Stöffler et al. (2013; the modeling part is in Artemieva et al. 2013). The term suevite was first used to describe an impact breccia formed at the Ries impact crater (Bavaria, Germany); nowadays the term is used more in general to describe polymict (consisting of heterogeneous fragments) impact breccia containing clastic matrix/groundmass from the target rocks, shocked minerals, and glass. At the Ries impact site, suevite can be found both in the crater and outside the rim (outer or fallout suevite), associated with the different (five) shock stages. Because of the occurrence of a full set of rocks subjected to different degrees of shock metamorphism (and distal ejecta as well), the Ries crater is one of the most studied and has been the type locality for discovering and classification of all shocked materials (Stöffler 1971; Stöffler et al. 2013). Figure 9 reports a photo of a glass-rich suevitic impact breccia (fallout suevite from Otting quarry; Ries crater, Germany). The heterogeneous nature of the sample is clear, with lithics, relict mineral grains, and abundant glass portions. From this locality was first reported the occurrence of coesite (Shoemaker and Chao 1961). According to Stöffler et al. (2013) the fallout suevite (also called outer suevite) is almost completely made up of lithic clasts derived from the target crystalline basement rocks, with a few percents of sedimentary origin (<6% sedimentary lithic) and contains rocks of all stages of shock metamorphism: from the unshocked materials of stage 0 (0–10 GPa), to moderate shocked (stages I and II: 10–60 GPa), up to stage IV (>60GPa). There is also abundant glass material.

Figure 9. Fallout suevite from Otting quarry (Ries crater, Germany). The polymictic sample has lithics (brighter spots), relict mineral grains and abundant glass (darker portions).

Oberdorfer (1905) observed the presence of spherical dark-brown amorphous bodies surrounded by a light brown glass, in suevite deposits in the west side of the Ries crater. Textural features were indicative of silicate–silicate melt immiscibility. Other textural evidence of liquid immiscibility were reported by Graup (1999), but in this case, for an immiscibility between silicate and carbonate melts. Crystalline calcite is a common constituent of suevite, and it was believed to derive from post-impact hydrothermal activities (Engelhardt 1967, 1972). However, because of the presence of gas vesicles in carbonate flow lines, or carbonate globules embedded in silicate glass (and vice versa), and the presence of deformed and coalescing spheres of carbonate within the silicate melt, the carbonate component is likely a result of impact-shock melting, and not of secondary origin (Graup 1999).

As observed above, impact glasses (including the enigmatic ones) are more heterogeneous than distal ejecta (i.e., tektites). For example, Irghizites, Wabar, and Darwin glasses present at least two major chemical bulk compositions. Table 3 reports the average major oxide composition for some of the impact glasses discussed here. The chemical variability among the same impact glasses reflects the differences in the mechanisms of formation of these glasses compared to tektites, especially those related to different time-scales.

Fulgurite

Fulgurites are formed as a result of fusion and modifications of rocks (s.l.) by lightning, and the name derives from the Latin word "*fulgur*" (lightning). Usually they consist of irregularly shaped tubes ranging from approximately 1 cm in diameter to 1 mm, even if they

Cicconi et al.

may extend laterally or vertically for up to 10 m (Essene and Fisher 1986). Famous fulgurites specimens are those coming from the Saharan desert; however, fulgurites can occur on any soil, either formed naturally or triggered in the proximity to high voltage power lines.

In this section, we will introduce the lightning process, the fulgurite morphology and mineralogy, and we will try to arouse interest on the multidisciplinary aspects related to fulgurites. Indeed, the study of these specimens and their mechanisms of formation has started to become more and more significant. Starting from the curiosity of understanding more in detail these fascinating objects, researchers from several disciplines (physics, atmospheric sciences, geochemistry, materials science, and impact-related sciences) have gathered and compared much information. The results obtained are well beyond pure scientific curiosity, with key implications in biogeochemistry and the formation of life (see below).

Lightning is a very common phenomenon but is not uniformly distributed across the Earth and equatorial conditions (warm, moist) greatly favor the occurrence of lightning. For reference, the place on Earth with the highest number of lightning flashes for year is Lake Maracaibo (Venezuela) with an average of ~232 flashes (per km^2 per year).

This common phenomenon is extremely energetic, and a lightning flash may dissipate ~10^9 J (Borucki and Chameides 1984; Rakov and Uman 2003), raising, instantaneously (<milliseconds) the surrounding air temperatures to a range of 10^5 K (Uman 1964; Uman and Krider 1989). Thus, a really high amount of energy strikes the surface target material, providing a local temperature rise higher than 2000 K, and with even greater temperatures expected on the "current" path (Uman and Krider 1989). Such an amount of energy is able to melt/vaporize the inorganic target materials, incinerating the organic ones along the current path, and inducing the formation of voids and vesicles from the release of the volatilized components. The heating rates have been constrained to be in the order of $\geq 10^3$ K/s (Uman 1964; Krider et al. 1968; Grapes and Müller-Sigmund 2010; Pasek et al. 2012; Sengupta 2017), with the high temperatures lasting microseconds. The ultrafast cooling (Grapes and Müller-Sigmund 2010) leads to the formation of very homogeneous glass layers and the abundant occurrence of lechatelierite. The result of such a fast event is a cylinder (often presenting lateral branches) with an inner smooth glassy portion surrounded by an irregular outer surface that can be composed of melted, partially melted, and unmelted mineral grains and rock fragments, and with many spherical and elongated vesicles occurring in all portions (Fig. 10).

Lightning strikes are, therefore, rapid high energetic phenomena able to induce strong physical and chemical modifications in the target materials. The morphology of the fulgurite specimens formed will be strongly dependent on energy and timescale of the strike, on target composition and heterogeneity, etc. The bulk composition of fulgurites is quite broad when considering the range of sample: a reflection of the different target materials (Fig. 1). However, the chemistry of the glassy part is much narrower and reveals a volatilization of most of the major elements, resulting in an amorphous phase enriched in silica, and depleted in alkalis. Most of the studies report the

Figure 10. Fulgurite specimens showing different morphologies.

occurrence of glass heterogeneities, with lechatelierite glass (98–100 wt.% SiO_2) occurring in different percentages, and secondary glass phases having relatively lower amounts of SiO_2, and higher proportions of Al_2O_3 and FeO (Frenzel and Stahle 1984; Essene and Fisher 1986; Pasek et al. 2012). Pasek et al. (2012) classified fulgurites based on morphology, mineralogy, and petrology, providing five main types depending on the target rocks (sand, soil, rocks):

- Fulgurites formed in quartz sand are classified as type I, and usually exhibit thin glassy walls; type I can contain one or two melts consisting prevalently of lechatelierite, and sometimes, also a SiO_2-rich melt with higher concentrations of Al_2O_3 and/or FeO. In the groundmass has also been reported enrichments in Zr oxide- and Fe–Ti oxide-rich glass.

- Fulgurites type II have a thicker glass portion compared to type I. The glassy portion is more compositionally varied (<50% lechatelierite) because they formed in different environments than silica sand, such as unconsolidated soils containing clays minerals, small rocks, and so on.

- Type III consists of lechatelierite and feldspar glasses, and a calcite-rich matrix. Fulgurites III are mostly found in calcite-rich soils and are the densest (average density[1] 2.1 ± 0.5 g/cm^3).

- Type IV fulgurites are heterogeneous melts, with the inner portion rarely consisting of lechatelierite, and the outer portion consisting of unmelted (or partially melted) rocks and mineral grains. These fulgurites usually form in bulk rocks and have densities similar to those of the target rocks.

- Type V are droplet fulgurites that have a homogeneous glass matrix. The two main oxides contained in type V droplets are SiO_2 and K_2O, which are enriched relative to the originating fulgurite, whereas other oxides are depleted.

An updated classification (Pasek and Pasek 2018) includes also specimens artificially formed, either from high voltage power lines, or triggered by high-voltage electrical arcing.

Since fulgurites result from a quick high energetic event that also produces a shock wave, in many specimens, shock markers typical of impact glasses have been identified. For instance, shock lamellae in quartz and kink bands in plagioclase have been reported, with estimated shock pressures >10 GPa (Gieré et al. 2015). Other studies (either experimental or based on modeling) reported even higher shock peak pressure above 20 GPa (Carter et al. 2010; Ende et al. 2012). Artificial produced (triggered) fulgurites confirmed that temperature and pressure regimes needed to produce shocked materials could be achieved by lightning strikes (Chen et al. 2017). The occurrence of such shock features in fulgurites, until then considered diagnostic of impactites, raised questions about samples of uncertain origin, such as irregular glassy objects having shocked quartz grains, but unlikely related to any impact event (Melosh 2017).

Besides the occurrence of shock markers, in many fulgurites O-free phases and metals indicative of extremely reducing conditions have been observed, such as those typical of meteorite phases. These findings have significant implications, especially concerning the origin of life (see below). The occurrence of metallic phases has been reported for decades (Anderson 1925; Essene and Fisher 1986; Wasserman and Melosh 2001). Essene and Fisher (1986) explained via thermodynamic calculations that the occurrence of metallic globules rich in silicon, spheroids of silicon-bearing metals (99.5 at.% of metallic silicon phase with minor amounts of titanium, iron, and phosphorus), and Fe-rich spherules, required extremely high temperatures (>2000 K) and reducing conditions close to the SiO_2–Si buffer. However, the authors could not decipher (and rule out) the role of other agents, such as the presence of carbonaceous materials, or the degassing of oxygen or the formation of nitrogen oxide gases.

---

[1] The density value reported is related to the density of the material as approximated by a whole fulgurite cylinder Pasek et al. (2012)..

Other metallic phases, such as Fe–Si–Al–Ti metallic droplets, or phosphide phases usually common in meteorites (e.g., Fe_2Si, Fe_3P) have been reported as well (Pasek et al. 2012 and references therein). Rowan and Ahrens (1994) report mineral reduction in MORB (mid-oceanic ridge basalt) glasses by shock experiments. Shock pressures between ~0.8 and ~6 GPa induced a change of the glass chemistry (enrichment in SiO_2) and the formation of Fe-, Si- and Mo-rich metallic microspheres embedded in the shocked glass. Based on the experimental conditions and the occurrence of iron silicide alloys, the authors estimated that an oxygen fugacity ~8 log units below the initial one was needed in order to form such metallic phases. Jones et al. (2005) artificially (triggered-lightning) produced specimens composed of 99.9% pure binary oxides of manganese and nickel in order to study the reduction mechanisms, and while they observed the formation of nickel oxide particles, the manganese oxide fulgurite showed no metallic phase formation. Thus, the thermodynamic stability of the oxides involved has a key role in the reduction mechanisms. Based on their experiments, Jones et al. (2005) suggest different possible mechanisms for the reduction associated with lightning, and they highlighted that the presence of carbon is not required for oxide reduction during fulgurite formation.

The presence of reduced phosphorus species in fulgurites is particularly important. A relative enrichment in phosphorus is not unusual in fulgurites (Martin Crespo et al. 2009; Pasek and Block 2009; Pasek et al. 2012), but the occurrence of phosphides has different implications. In iron meteorites, pallasites, and enstatite chondrites, phosphorus commonly occurs as phosphide (e.g., schreibersite $(Fe,Ni)_3P$), which slowly reacts with water to produce reduced forms P^{3+} (PO_3^{3-} phosphite). On the contrary, on Earth, phosphorus occurs in minerals with its stable form P^{5+} (PO_4^{3-} phosphate). Besides the mineral supergroup apatite (the primary source of P), this element is a common constituent of organic molecules (DNA, RNA, and so on...) (Pasek 2017). On the Earth's surface, a possible mechanism for reduction of phosphate species may be lightning, as shown by simulated discharge experiments from Glindemann et al. (1999), and from the studies on many fulgurite specimens by Pasek and coauthors. For example, Pasek and Block (2009) report that the reduction of PO_4^{3-} species to reduced forms occurs commonly in many target soils, and they provided two potential causes of P reduction during fulgurite formation:

1. the reduction of phosphate to phosphide, and a later reaction with water; this process is analogous to that of meteoritic schreibersite that reacts with water to produce reduced forms e.g., H-phosphonic acid H_3PO_3 (e.g., Bryant et al. 2009);

2. the direct reduction from PO_4^{3-} to PO_3^{3-} species; thermodynamic calculations indicate that phosphates ($CaHPO_4$) could be reduced to phosphite species ($CaHPO_3$) at temperatures ~2000 K, under oxygen fugacity conditions of ~−5 (as $\log(fO_2)$) (Pasek and Block 2009).

The phosphide species occurring during meteoritic impacts, and especially during the Hadean–Archean heavy meteorite bombardment, have been proposed as a way of altering global phosphorus biogeochemical cycle, and the source of reactive phosphorus required for the processes of prebiotic phosphorylation on the early Earth (Pasek 2008, 2017; Pasek and Block 2009).

Case study: fulgurite. A fulgurite with a dark thick glassy inner core has been used for a series of analyses. The specimen is highly vesiculated, and the inner glass core has a variable thickness up to 1 mm. The outer portion presents unmelted white/grey grains of quartz, and probably salt phases. The specimen has been sliced, and the cross-section polished (Fig. 11). Optical microscope images and Raman spectroscopy analysis have been coupled to SEM/EDS (scanning electron microscopy–energy dispersive spectroscopy) data in order to provide an overview of all the phases occurring in this fulgurite.

The polished slice has been observed with an optical microscope by reflected light, and SEM. The glass matrix of the inner part seems to be very homogeneous, with no relict mineral grains detectable and a few bubbles, if any. Moving from the inner portion toward the external wall

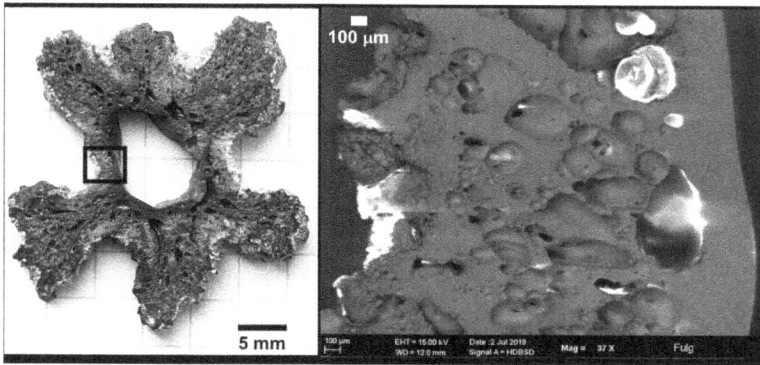

Figure 11. Photograph of the sliced fulgurite (**left**) and SEM image (**right**) of the thinner portion where both inner and outer walls can be observed. The inner wall has a smoother surface, in contrast with the rough outer part. Brighter areas in the right image are residues of carbon coating.

(see Fig. 11), vesicles, mostly elongated, are clearly increasing in amount, and the matrix becomes more heterogeneous. The fulgurite specimen has several outer branches, and many large voids.

Backscattered electron (BSE) images of the fulgurite (Fig. 12) show the increasing number of elongated bubbles, radially distributed in the branches. The matrix is predominantly amorphous and fairly homogeneous close to the inner wall, while all outer branches are more heterogeneous, with completely foamed glass portions, unmelted quartz grains, regions with small zoned crystals, and tiny bright spots, too small to be characterized (Fig. 12).

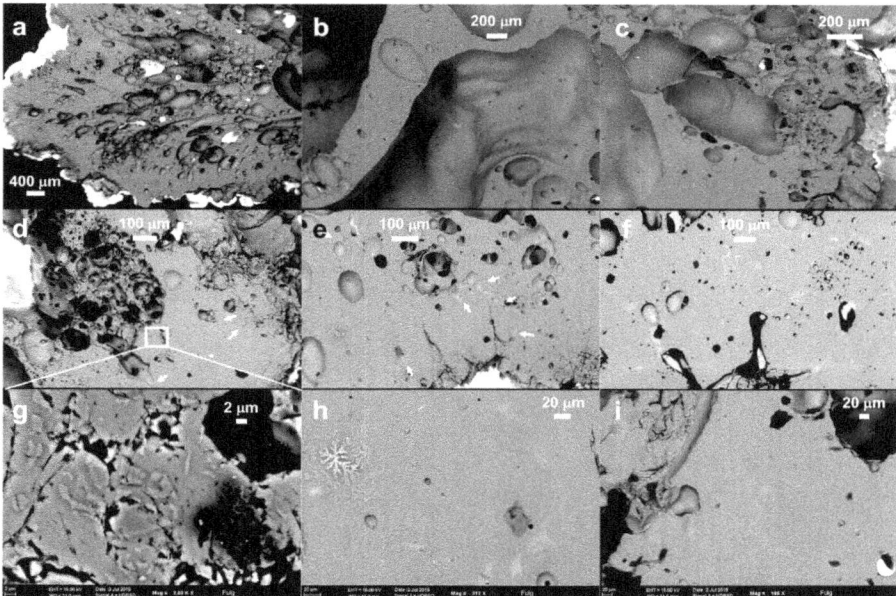

Figure 12. Backscatter electron images of the fulgurite at different magnifications to emphasize the heterogeneity of the sample. Overview of the *a)* outer branch, *b)* inner glass wall, *c)* outer portion in the thinner area; *d–e–f)* outer portions with heterogeneous features, such as foamed area, aligned bubbles, relict quartz grains and pocket and veins of glass with different chemistry (**light grey**; arrows); *g–h–i)* higher magnifications.

EDS analyses of the inner glass portion reveal the homogeneous presence of lechatelierite, with small amounts of Al_2O_3 ($SiO_2 \geq 99$ wt.%, $Al_2O_3 \leq 1$ wt.%). When moving from the smooth inner to the rough outer walls the chemical compositions marginally differ, but without a defined trend. The SiO_2 content ranges between 96.6 and 99.5 wt.% with some random points having higher Al_2O_3 amounts (~2.4 wt.%) and traces of FeO, MgO, CaO and K_2O (all <0.4 wt.%).

The areas where the branches depart and all the branches themselves have areas (pockets) and vein-like features that especially surround bubbles (see arrows in Fig. 12). These portions are highlighted in the BSE images by the different contrast (light grey), and the EDS analysis indicates lower Si contents (down to ~75 wt.%) and higher amounts of Al_2O_3 (up to ~14 wt.%), in addition to the presence of many other elements, including FeO (3–6.5 wt.%), MgO (1–1.5 wt.%), K_2O (0.5–1 wt.%), CaO (~0.4 wt.%) and TiO_2 (~0.3 wt.%). The elemental mapping of one of the regions rich in pockets and veins is reported in Figure 13. The primary matrix is composed almost exclusively of SiO_2 and only 0.5–1.5 wt.% of Al_2O_3. The bubbles are surrounded by large light grey pockets, and veins that contain the secondary melt, which is relatively depleted in silica and enriched in Al, Fe, and alkali/alkaline-earth elements. In a few light grey spots there are also traces of P_2O_5 and Na_2O (~0.3 wt.%).

Raman spectra were collected in several locations of the sliced fulgurite, with a T64000 Jobin-Yvon spectrometer and a 488 nm excitation laser, in order to characterize the amorphous and the mineral phases. The outer whiter grains, supposedly the residual of the target rocks, are mainly composed of quartz and some salts, such as NaCl (from EDS point analyses). The spectra of the outer grains perfectly match the spectra of α-quartz (see Fig. 14A), with the presence of a sharp intense peak at ~464 cm^{-1}, and a broad band at ~205 cm^{-1} (A_1 modes), and also the E modes at ~127 cm^{-1} and 263 cm^{-1}. The differences in the relative intensity of the E modes between the reference quartz spectrum and the spectrum of the outer grains (point 1 in Fig. 14A) is due to sample orientation. Similar spectra were obtained in some of the branches, although with some of the main peaks shifted (point 2 in Fig. 14A). Indeed, all peaks match vibrations of α-quartz, but with shifts of the vibrational modes that indicate the occurrence of shock pressures (Table 4). For instance, while the main peak moves ~3.1 cm^{-1} toward lower frequencies, the broad peak at 205 cm^{-1} shifts more than 10 cm^{-1} downward, and the small peak at 354 cm^{-1} remains unchanged.

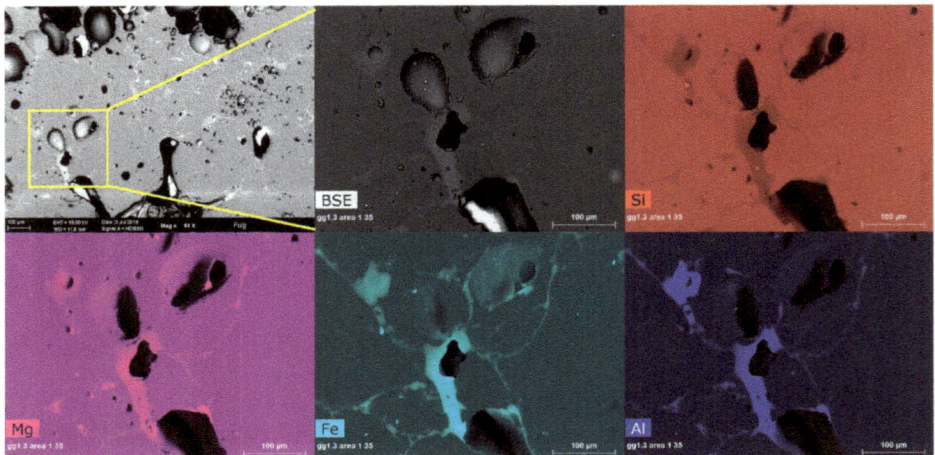

Figure 13. Elemental mapping of one of the regions rich in pockets and veins. The primary glass matrix is composed of ~99 wt.% SiO_2, and voids and bubbles are surrounded by a secondary melt, which is relatively depleted in Si, and enriched in Al, Mg and Fe. The scale bar in all images is 100 μm.

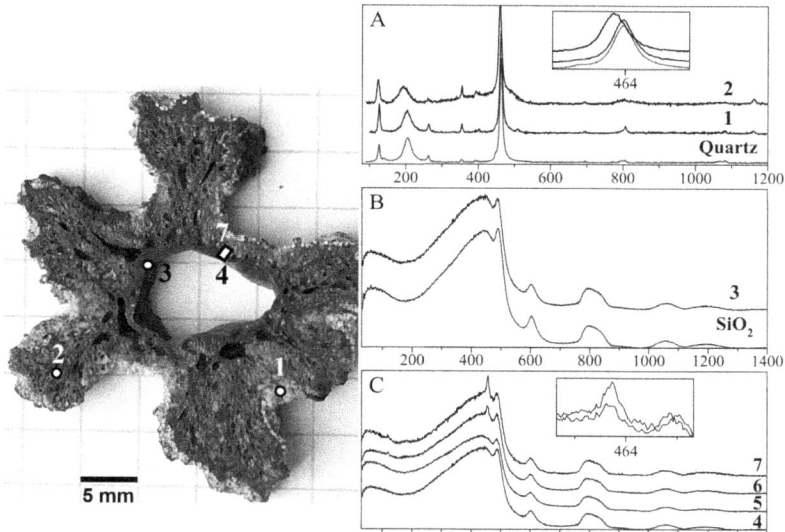

Figure 14. Raman spectra of different portions of the fulgurite (numbers from 1 to 7) and crystalline references. The outer unmelted grain and a small grain in the branch present vibrations associated to alpha-quartz (2). The glass on the inner wall (3) is very polymerized and the spectra is very similar to that of a pure SiO_2 glass. Moving from the inner wall toward the outer part (from 4 to 7) the homogeneous glass matrix starts to be slightly more heterogeneous with the presence of a small amount of quartz (sharp peaks in 6 and 7 and inset).

These uneven shifts of the quartz peaks indicate the presence of shocked quartz, and based on the experimental work done by McMillan et al. (1992), the shifts observed provide a shock peak pressure between 25 and 28 GPa, similarly to what reported by Carter et al. (2010).

The Raman spectrum of the glass on the inner wall (point 3 in Fig. 14B) has the clear broad features of amorphous silica and nearly perfectly match the spectrum of SiO_2 Suprasil® CG, used here as reference (SiO_2 in Fig. 14B). These spectra agree with the SEM/EDS analysis and the occurrence of a homogeneous lechatelierite glass in the inner wall. When moving from the inner wall toward the outer one (points 4 to 7 in Fig. 14C), the shape of the Raman spectra are very similar, and small variations can be attributed to the different concentrations of Al_2O_3 in the amorphous matrix. Moreover, the two last spectra show the presence of small sharp peaks that indicate the occurrence of microcrystalline quartz. Based on the position of the quartz main peaks here the shifts suggest higher shock pressure of ~30–31 GPa (Table 4).

Trinitite or nuclear glass

Within glasses formed by highly energetic events, there is a glass formed by a nuclear explosion. During the first atomic bomb test in Alamogordo (USA Trinity test, 1945), the detonation of the nuclear weapon on the desert sand created a "Lake of Jade [...] a lake of glistening incrustation of blue-green glass 2400 feet in diameter" as reported by Time Magazine (Science: Atomic footprint, 17 Sept 1945). The green amorphous material has been called trinitite, or nuclear glass. Another nuclear test that produced amorphous material is the first underground nuclear explosion, the so-called Rainier test (1957).

The study of these glasses has mainly focused on the radioactivity of the materials, and on the incorporation and distribution of radionuclides, although, more recently, more detailed chemical, mineralogical and geochemical investigations have been reported (Bellucci et al. 2014; Eby et al. 2010, 2015; Fahey et al. 2010). Moreover, because of the mechanisms of formation, trinitite could be regarded as an analogue for impactites.

Table 4. Frequencies of the main vibration modes of α-quartz (reference phase), and of the crystalline phases occurring in the fulgurite specimen (all values cm⁻¹). Numbers refer to the sampling spots reported in Figure 14.

Reference α-quartz	Spectrum 1	Spectrum 2	Spectrum 6	Spectrum 7
126.8	126.8	124.1	120	120
205.1	204.5	194.1		
263.1	263.6	262.1		
354.2	354.2	354.2		
463.6	463.6	460.5	457.8	457.8

In the following, we discuss the event associated with the formation of trinitite, and we will provide information on the different geochemical aspects of this glass.

Trinitite glasses are a record of the first atomic bomb blast on July 16, 1945. The detonation of the 21 kiloton bomb and the resulting fireball melted/vaporized the arkose/arkosic sand desert and formed a large crater glazed with green fused silica sand. The explosion released an extremely high amount of energy, with an estimated yield of 9-18 kt, and average temperature of 8430 K (Hermes and Strickfaden 2005). The heating rate was extremely high with a very short duration of ~3 s and this explain the glass formation and why some minerals (zircon and quartz grains) were only partially melted (Hermes & Strickfaden 2005).

The green fused silica sand was first described by Ross (1948) who, based on the optical properties, reported the occurrence of two different melts. In general, he described trinitite as a highly vesiculated pale bottle green glass that exhibits distinct flow lines. One of the glass phases has a refractive index (RI) between 1.51 and 1.54 and derives from feldspars, clay fraction, and accessory minerals (calcite, hornblende, augite). The second amorphous phase has a lower index of refraction (close to 1.46), indicating that this material is nearly pure SiO_2 glass derived from the fusion of the quartz component of the arkosic sand. Ross (1948) observed also the occurrence of small glass areas with glasses red or grey in color, both containing copper, and suggested that it was a contamination from the explosion device.

Since the nuclear explosion, most of the studies have been done on the radioactive nuclides present in the materials (e.g., Atkatz & Bragg 1995; Eby et al. 2010; Parekh et al. 2006; Wallace et al. 2013). Indeed, the migration of actinides at historical test sites is considered very important, since it is closely related to the waste management and the storage of high-level nuclear waste (Hu et al. 2008; Pacold et al. 2016; Tompson et al. 2002, 2006). However, for a long time, the study from Ross was the only mineralogical investigation available on trinitite (except a magnetic susceptibility study by Glass et al. 1987; see below). Only in the last decade has interest in these materials been raised again, mainly because of the nuclear forensic applications, and many studies on the mineralogy, petrology, geochemistry and on the average Fe or Ti redox state have been published.

Eby et al. (2010) report the occurrence of four different types of glass:

1. glassy and vesicular green top surface coated with sand grains (the top part of the trinitite layer) (Fig. 15);

2. a Cu-rich glass (red trinitite) containing metallic chondrules;

3. green trinitite (scoriaceous fragments);

4. trinitite ejecta, which include aerodynamically shaped droplets, beads, and dumbbell glasses.

The latter glass types were discovered a few years after the detonation as mm-sized specimens. Supposedly, part of the molten material was transported by the hot gas cloud and spread over a wide area into beads and dumbbell shapes (Eby et al. 2010). The red trinitite contains several metallic chondrules having Cu, Fe and Pb deriving from the explosion (blast tower and trinity device), confirming the hypothesis of Ross (1948). The most common material is the green trinitite that consists of vesiculated and amorphous scoria-like fragments, and it is extremely heterogeneous.

Figure 15. Photograph of a green trinitite specimen.

Light microscopy, SEM and BSE analysis show the presence of euhedral heated sand grains, large flow bands and strong chemical variability (Eby et al. 2010, 2015; Fahey et al. 2010; Bellucci et al. 2014) (see Fig. 16). The more recent analyses confirmed the early observation done by Ross (1948) on the bulk chemistry. Indeed, Trinitite glasses have two major glass compositions, one essentially consisting of lechatelierite, and a second one, depleted in silica, with a much higher variability (SiO_2 range = 55–80 wt.%). Moreover, Eby et al. (2015) report the occurrence of planar deformation features in quartz grains, a typical evidence of shock metamorphism in impactites and fulgurites.

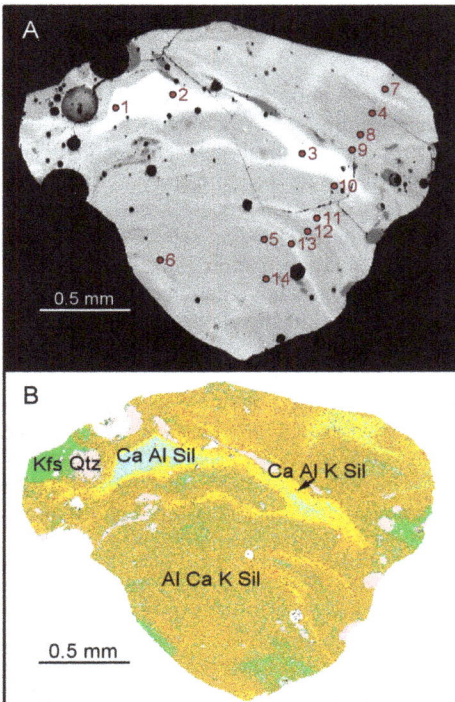

Figure 16 (a) Backscattered electrons (BSE) and **(b)** QEMSCAN (Quantitative Evaluation of Minerals by SCANning electron microscopy) images for a green trinitite fragment. Besides the partially melted K-feldspar (Kfs) and quartz grains (Qtz) there are evident flow structures, and the occurrence of different glass compositions. The brightest areas are enriched in calcium. Numbers represent the point analysis. Figure from Eby et al. (2015) MSA copyright.

Because of the resemblance between trinitite ejecta and distal impact ejecta, trinitite glasses have been compared to tektite and microtektites (Glass et al. 1987). The similarities between the two products were confirmed, especially regarding the water contents (between 0.01 and 0.05 wt.%) and the predominant reducing conditions ($Fe^{3+}/Fe_{tot} < 0.1 \pm 0.1$; Glass et al. 1987; Giuli et al. 2010b; Eby et al. 2015). Studies of the mineralogy and chemical variability of the glass samples have constrained the conditions occurring during the atomic detonation (Pressure–Temperature (P–T) regime), with temperatures on the order of ~1870 K and pressure ≥ 8 GPa (Eby et al. 2015). A recent study highlights also the occurrence of nano-structures in zircon grains, and in particular the presence of baddeleyite fibers surrounding the core of unaltered zircons (Lussier et al. 2017). These fibers that seem to irradiate from the zircon grains are not in physical contact with the core. Based on Raman, TEM and electron diffraction data, the P–T regime was constrained to $T > 1770$ K and $P < 10$ GPa (Lussier et al. 2017). Because of the disagreement between temperatures estimated by trinitite glasses and fireball (~1800 K vs. ~8000 K, respectively), it was suggested that the very short duration of the nuclear event has a major impact on the physical properties registered by glasses and minerals (Eby et al. 2015).

The study of the glasses formed during nuclear events provides information on the type of device that was detonated and its origin, thus it has nuclear forensic applications. Recently, Molgaard et al. (2015) produced synthetic nuclear glasses, comparable with trinitite glasses, as surrogates that could be used to simulate a variety of scenarios and could be used as a tool for developing and validating (nuclear) forensic analysis methods.

GLASS PROPERTIES

The study of the glasses described in the previous sections can provide important information on the formation mechanisms of these amorphous materials and on the related events. In particular, the information provided by the study of shock metamorphism materials have implications well beyond earth sciences, since they also involve planetary sciences, biogeochemistry, and materials science.

By looking at different scales, it is possible to obtain information on the glass average redox conditions during fast cooling, on the short and medium-range silicate network, on liquid immiscibility phenomena, and on the macroscopic properties (optical, physical, magnetic and so on...). All these data can be used to constrain the regimes of temperature and pressure during formation, evaporation, and condensation stages, the timescale of the processes, and the thermal history of glasses. Altogether with the data from experimental studies (e.g., shock experiments) they are the base for simulations and numerical models of crater formation. In the following, we report some of the characteristic of the amorphous materials and a compilation of Fe redox state and Raman spectroscopy data, collected on a variety of non-magmatic glasses.

Liquid immiscibility

A common feature of many natural glasses is the occurrence of two coexisting different melt phases, with features similar to liquid–liquid immiscibility. For example, in the previous sections, the presence of compositional heterogeneities on a variety of scales has been described for several impact-related glasses (e.g. Libyan desert glass, Darwin glass), fulgurites and trinitite. Moreover, the occurrence of ferromagnesian droplets coalescing into a felsic amorphous matrix has also been reported for pyrometamorphic materials (e.g., Capitanio et al. 2004).

Multicomponent silicate systems are stable over wide ranges of P–T–X–fO_2 (pressure, temperature, bulk composition, oxygen fugacity); nevertheless, they might exhibit immiscibility at particular conditions (e.g., Roedder 1992; Hudon and Baker 2002; Hamann et al. 2018). Silicate–silicate liquid immiscibility is a common phenomenon observed in many natural igneous melts on Earth, and in lunar mare basalts (as a late-stage immiscibility), including meteorites (e.g., Hudon et al. 2004; Roedder 1978), and is a fundamental research field in the glass and ceramic industry (phase separation; see Vogel 1994; Shelby 2005 for a review). The silicate–silicate (or carbonate) fluid immiscibility arises from differences in physical properties between two separate phases, and usually, the immiscibility is characterized by coexisting felsic and mafic components; however, the chemical distribution is highly dependent on the duration of melting, the cooling rate, and ambient atmosphere (reducing or oxidizing conditions) (e.g., Roedder 1992; Hudon and Baker 2002; Hudon et al. 2004; Hamann et al. 2018). In highly energetic events, there are dynamic pressure and temperature conditions that allow the physical and chemical interaction (respectively, mingling and mixing) of very different target materials; however, because of the timescale of these processes, equilibrium conditions are hardly achieved, as opposed to the case of magmatic processes where a complete homogenization can be attained (for more information on Magma mixing see Morgavi (2022, this volume).

Liquid–liquid immiscibility arises from the different structural roles of the cations in the glass. For instance, the main cation building the network in natural compositions is Si^{4+} (altogether with AlO_4 tetrahedra), which is bonded by four oxygen ions to form a tetrahedral coordination; each oxygen is shared by neighboring tetrahedra (bridging oxygens) to form a three-dimensional network held together by strong covalent bonds. The other cations present in a natural composition have different structural roles and may either aid or disrupt the silicate network. Thus they are called, respectively, network-formers or network modifiers (see Henderson and Stebbins (2022), Drewitt et al (2022) and Neuvile and Le Losq (2022), all this volume). According to Veksler (2004), the cations that most influence the liquid–liquid immiscibility fields are aluminum and alkali/alkaline earths, and in turn, the peraluminous or peralkaline nature of the melts; additionally, cations that present high field strength tend to concentrate in the Si-poor phase.

Liquid immiscibility seems to be a very common feature of glasses deriving from highly energetic events, and recently, Hamann et al. (2018) discuss the importance of liquid immiscibility in impact processes and the mechanisms that lead to the formation of melt heterogeneities and unmixing of silicate phases. In particular, besides the classical liquid–liquid immiscibility (rising from phase separation of a melt), the authors define two other main mechanisms that could produce the heterogeneities observed in silica impact glasses: i) mingling or incomplete mixing due to the short time-scale associated with highly energetic events; ii) emulsification of incompatible melts (deriving from the target rocks and impactor), without mixing. Moreover, Hamann et al. (2018) report that despite the enormous differences from phase-separated volcanic rocks (both in terms of composition and T–P–time conditions), impact glasses show a very similar major element partitioning between phases.

Heterogeneity in glasses from high energetic events have been described in detail for Libyan Desert glass, blue zhamanshinites, and also trinitite and fulgurite glasses. In all specimens the occurrence of amorphous nanospherules, as a result of silicate–silicate liquid immiscibility, have been observed (Zolensky and Koeberl 1991; Pratesi et al. 2002; Feng et al. 2019). Other immiscibility like features can be found also in suevite glasses from Ries crater where mixture of carbonate impact melt and silicate melt have been reported (Graup 1999).

Reduced iron species

As discussed in the previous sections, the highly reducing conditions of glasses/rocks formed from highly energetic events have many repercussions, not only on the formation mechanisms of the samples themselves, but also on the understanding of the processes that occurred during the heavy meteorite bombardment, the evolution of the early Earth, and planetary impact cratering processes and differentiation. Among the characteristics common to all distal ejecta is the occurrence of very high amount of reduced iron species Fe^{2+} in the melts, and several mechanisms have been proposed to explain the presence of reduced species, and its preservation during the "travel", and the quench (e.g., see Engelhardt et al. 1987; Lukanin and Kadik 2007; Ganino et al. 2018).

Cicconi et al.

Iron is by far the most common multivalent element in all natural amorphous materials, with two valence states (Fe^{3+}, Fe^{2+}) that are stable over a large range of P–T conditions. In the study of glasses derived by impact or other highly energetic events, the occurrence of reduced iron has become one of the common markers to differentiate those samples from volcanic rocks. Moreover, in order to constrain some of the conditions of formation of natural glasses, the study of iron valence has been one of the most common tools. An overview of the average Fe valence in these samples is provided in Figure 17, where $Fe^{3+}/(Fe^{2+} + Fe^{3+})$ (hereafter, Fe^{3+}/Fe_{tot}) redox fractions are reported for many natural glasses. The data for tektite and microtektites were obtained by several techniques (wet chemistry, Mössbauer, X-ray absorption spectroscopy; see below for details).

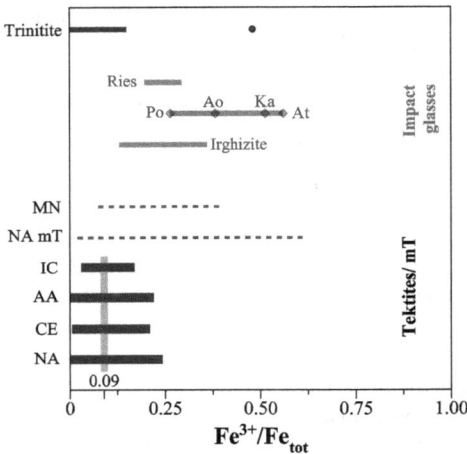

Figure 17. Ranges of Fe redox state (as Fe^{3+}/Fe_{tot}) of tektites and microtektites from the different strewn fields are compared to the Fe^{3+}/Fe_{tot} fractions of impact glasses (proximal from Zhamanshin, Popigai. Kara, Aouelloul, Atacama and Ries craters) and trinitite glasses. For tektites, both the extreme values and the average one (~0.09) are reported. North American microtektites and Muong-Nong tektites are the only distal ejecta specimens that show a broader range of Fe redox states (see the *Glass Properties* section), with values that overlap the proximal impact glasses.

Tektites, from the different strewn-fields, aside from some extreme values reported in the literature, all show an average Fe redox fraction Fe^{3+}/Fe_{tot} ~0.09 (± 0.10; vertical grey line), thus indicating a similar thermal history and similar P–T–fO_2 conditions, despite their different geographical distribution. In Muong-Nong-type tektites (MN) and North American microtektites (NA mT), the large range of Fe^{3+}/Fe_{tot} redox fractions is, in the first case, typical of the layered tektites, and in the second one, a unique case among microtektites (see section *Impactites*). Distal ejecta have considerably lower Fe^{3+}/Fe_{tot} redox fractions than impact glasses, or Fe valences associated with average crustal target rocks (the possible precursor material).

Several techniques have been used to infer the Fe oxidation state. The earlier works reported quite broad ranges regarding the Fe^{3+}/Fe_{tot} fractions with values between 0 and ~0.5. The analyses were based or on magnetic susceptibility or, more commonly, on wet-chemistry methods (e.g., Thorpe et al. 1963; Bouška and Povondra 1964; Schnetzler and Pinson 1964; Cuttitta et al. 1972; Schreiber et al. 1984; Bouška 1994). Early spectroscopic data (Mössbauer) were obtained for a large dataset (tektites from various locations by Evans and Leung 1979; Fudali et al. 1987) finding no detectable ferric iron in one case, and up to 0.12% in the second case. Lately, very similar results on the average Fe redox state have been obtained (e.g., Dunlap 1997; Rossano et al. 1999). Mössbauer spectroscopy, compared to wet chemistry, provided additional information on the Fe local environment. The early studies reported a coordination environment for ferrous iron around 6, with a minor presence of 4-fold coordinated species, but more recently presence of ferrous sites with coordination between 4 and 5 (trigonal bipyramids) were reported (e.g., Dunlap et al. 1998; Rossano et al. 1999; Dunlap and Sibley 2004; Volovetsky et al. 2008).

Another spectroscopy technique that is element selective is X-ray Absorption Spectroscopy (XAS). The study of the X-ray Absorption Near Edge Structure (XANES) region, including the small feature before the whiteline called the pre-edge peak, and the extended energy region (EXAFS) provide information on elements average valence state and coordination number. This technique has been used to study Fe, Al, and Ti structural environments in many natural amorphous materials, and provided insights on pressure and oxygen fugacity conditions during cooling. Because iron is the most common multivalent transition elements in natural melts, it has been extensively studied both in natural and synthetic systems. Thus, accurate information can be derived by the study of the Fe K-edge XAS spectra of tektites, impact glasses, and so on. XAS spectra related to reduced Fe^{2+} and oxidized Fe^{3+} glasses show differences in their energy position, as well as in their shape, with the spectrum related to oxidized species shifted toward higher energies (Fig. 18A). The small feature before the absorption edge (inset in Fig. 18A) is the pre-edge peak that presents different energy positions and intensities depending, respectively, on Fe oxidation state and average coordination number (and local symmetry). The deconvolution of this small feature for many crystalline compounds has allowed the creation of the variogram reported in Figure 18B, and hence the derivation of quantitative information on the structural role of Fe. The intensity of the pre-edge is related to

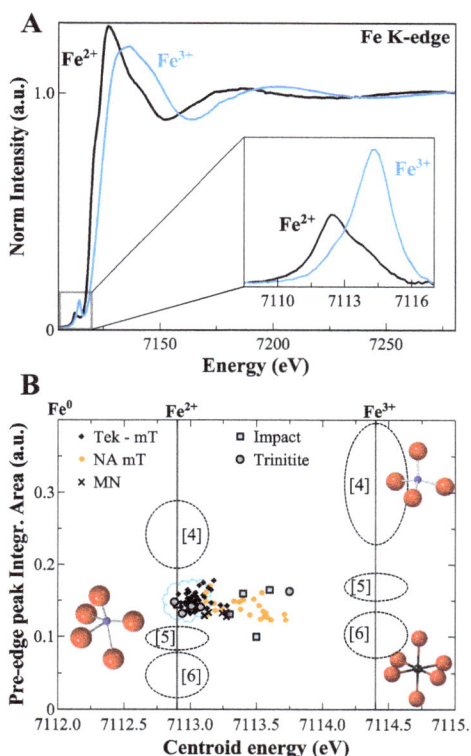

Figure 18. A) XAS at the Fe K-edge for two silicate glasses having reduced and oxidized iron species. The inset shows in detail the background-subtracted pre-edge peaks (data from Cicconi et al. 2015). **B)** Centroid energy position against the Fe pre-edge peak integrated intensity. The **dashed ovals** represent different coordination environments of the iron species in minerals. This variogram is obtained from the study of the Fe K-edge pre-edge peak of several tektites (tek), microtektites (mT) from the different strewn-fields, impact glasses, and trinitite glass. Nearly all the distal ejecta cluster in a small area, here highlighted by the cloud shape. Data from Giuli et al. (2002, 2010a,b, 2013b, 2014a,b). Figure modified after Cicconi et al. (2020a)

the number of first neighbors and site geometry, whereas the centroid position is related to the average valence. For a review on the technique and on studies related to iron in silicate glasses, melts and minerals, the readers are referred to Calas and Petiau (1983), Wilke et al. (2001, 2004, 2007), Giuli et al. (2002, 2012), Mottana (2004), Henderson et al. (2014).

The results obtained from the study of the Fe *K*-edge pre-edge peak for an extensive dataset of tektites and microtektites provide very consistent results, with all distal ejecta having similar values in both average oxidation state and coordination environment. The pre-edge peak variogram in Figure 18B shows that tektites and microtektites cluster in a position that is compatible with the presence of highly reduced species ($\geq 90\%$ Fe^{2+}), and an average coordination between 4 and 5. Trinitite glass as well has a similar behavior and the iron structural environment resembles that of distal ejecta. Impact glasses from different localities have much higher centroid energy values, corresponding to higher amounts of trivalent iron (data in Figure 18B from Giuli et al. 2002, 2010a,b, 2013b, 2014a,b). Besides the analysis of the Fe pre-edge peak, the analysis of the extended region of the Fe *K*-edge XAS spectra allowed derivation of average <Fe–O>distances of 2.00 ± 0.02 Å, values compatible with the presence of ferrous species with coordination between 4 and 5 (Giuli et al. 2002).

Glass structure

Most of the natural materials discussed in this chapter are SiO_2-rich aluminosilicate glasses. The study of the silicate network arrangement can provide information on glass polymerization, temperature and pressure regimes, as observed in Figure 3. One of the most studied phases is SiO_2, either the amorphous phase or the crystalline polymorphs, and for sake of clarity, we compare the properties of amorphous SiO_2 with those of the other crystalline phases in Table 5. There are 11 known SiO_2 polymorphs formed depending on the different pressure and temperature systems, and here we report the most common ones in natural glasses. Quartz, tridymite, and cristobalite are low pressure polymorphs, whereas coesite and stishovite are the high pressure ones (see Fig. 3). The arrangement of the silica is always the same in all polymorphs, namely having a silicon surrounded by four oxygens arranged in a tetrahedron (4-fold coordinated), except stishovite where silicon is surrounded by six oxygens.

Raman spectroscopy is a very powerful technique that allows derivation of information on the silicate melt short- and medium-range orders, thermal history, pressure regimes, as well as providing an overview of sample heterogeneity. Moreover, Raman has been widely used on obsidian specimens in order to get information on the provenience (determination of the

Table 5. Some properties of lechatelierite glass and crystalline SiO_2 polymorphs.

Name	Density (g/cm³)	Refractive Index	<Si–O>(Å)
Lechatelierite	2.20	1.458	1.62
α-quartz	2.65	1.549	1.61
β-quartz	2.53	1.537	1.59
α-tridymite	2.25	1.471	1.59
α-cristobalite	2.32	1.484	1.60
Coesite	2.92	1.596	1.61
Stishovite*	4.29	1.81	1.78

Footnote: * silicon coordination is 6.

geographic origin) of archeological obsidians (e.g., Bellot-Gurlet et al. 2004). Besides magmatic glasses, some of the first studies that exploit the Raman technique to understand the origin of other natural glasses were performed as early as in the 80s, and by using polarized Raman spectra the authors tried to derive information on the thermal history of Libyan Desert Glasses (Galeener et al. 1984), or proximal and distal ejecta (Jakes et al. 1992). Despite these early investigations, it is surprising that in the last decades, not many studies have used this technique to understand possible modifications in the glass connectivity, but rather to determine the crystalline phases present, and the peak shock pressure (for instance, see case study in the section *Fulgurite*).

In the following, we report a series of Raman spectra collected for several glasses described within this chapter. A short description of the main vibration modes of silicate glasses is reported for the sake of clarity; however, for a comprehensive description of Raman spectroscopy and on the silicate network short- and medium-range order, the readers should refer to dedicated texts (Neuville et al. 2014; Henderson and Stebbins 2020; Drewitt et al. 2020, both this volume). Raman spectra for other natural glasses, including magmatic ones, are reported in Cicconi and Neuville (2019).

Raman spectra of silicate glasses have the main vibration modes in the 10–1300 cm^{-1} frequency range, which can be divided in different regions: a very low wavenumber region (<250 cm^{-1}), which provides information on tetrahedral arrangement (see Neuville et al. 2014 and references therein for more details). The frequency range between ~400 and ~650 cm^{-1} is related to vibrations of the tetrahedra-rings (Fig. 19), with the main asymmetric band (*R* band) usually assigned to symmetric vibrations of T–O–T in ≥ 5-membered rings (T is the Si^{4+}- or Al^{3+}-centered tetrahedron). Depending on glass polymerization, two additional sharper bands can be observed (Fig. 19), the so-called defect bands, D_1 (~490 cm^{-1}) and D_2 (~600 cm^{-1}), assigned to breathing modes of 4- and 3-membered rings, respectively. The frequency position and the shape of the *R* band strongly depend on the silica network. For instance, it shifts toward higher frequencies by densification of SiO$_2$ glasses (Deschamps et al. 2013; Cicconi et al. 2020b), or by changing glass connectivity (e.g. by adding network modifier cations) (Neuville et al. 2014 and references therein), thus by varying T–O–T inter-tetrahedral angles, and T–O average distances. The high wavenumber region, extending from ~900 to 1300 cm^{-1}, contains T–O asymmetric stretching vibrations; usually, these stretching modes are labeled according to T polymerization: Q^{0-4}, where Q represents the Si(Al) centered tetrahedron, and 0–4 represents the number of bridging oxygens (BO). This region is also called Q-range or Q-envelope (Fig. 19). By decreasing glass polymerization (e.g., by adding network modifier cations), the Q-range strongly increases in intensity and shift toward lower frequencies because other bands, related to different Q species (Q^3, Q^2) appear in the 900–1200 cm^{-1} region (e.g., Neuville et al. 2014).

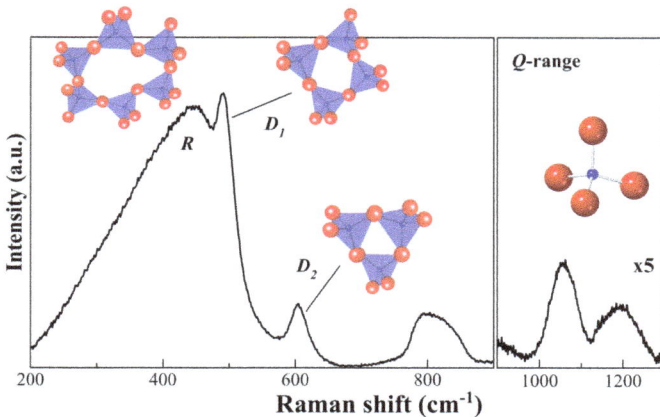

Figure 19. Background-subtracted Raman spectrum of a commercial pure SiO$_2$ glass. The different bands can be attributed to specific Si–O–Si and Si–O vibration modes. See text for details on band labels.

There are several studies dedicated to the modification occurring to the structure of pure silica glass upon compression by hydrostatic pressure (i.e., DAC–diamond anvil cell) or shock experiments or indentation. These investigations report that there is a permanent increase in silica glass density (densification), and for example, the density of a SiO_2 glass increases steadily up to values 20–21% higher than ambient density, approaching the density of crystalline SiO_2 (Bridgman and Šimon 1953; Cohen and Roy 1961, 1965; Susman et al. 1991). For applied pressure below 28 GPa, this densification is not due to changes of silicon coordination, but rather to structural changes, mainly related to variations in intertetrahedral angles distribution and the ring statistics, without variations in Si–O coordination (Sugiura et al. 1997; Deschamps et al. 2013). Okuno et al. (1999) carried out shock wave experiments on a SiO_2 glass by impacting glass with a steel flyer (speed up to ~2 km/s) for pressures in the range of 17.8–43.4 GPa, and they observed the maximum density increment (maximum changes in silica network) at 26.3 GPa. The study of the Raman signals of compressed or shocked glasses has allowed access to much information on the variations that may occur in the silicate network after high-grade shock metamorphisms, and assists identification of the peak pressure values.

A common feature of many natural glasses is the occurrence of lechatelierite that is amorphous SiO_2, or glass phases enriched in silica content. For example, different Libyan Desert Glass (LDG) clear fragments have been analyzed by using a 488 nm laser excitation (T64000 Jobin-Yvon spectrometer), and all Raman signals present vibrations with frequency positions and intensities very similar to that of a lechatelierite glass, including the two sharp peaks related to the breathing modes of 4- and 3-membered rings (defect lines). Some authors used the relative intensity of the defect bands in the Raman spectra to get information on the thermal history of natural glasses (Galeener et al. 1984; Champagnon et al. 1997). For example, Galeener et al. (1984) reported a fictive temperature (T_f) of 1000±50°C for LDG, and suggested that the molten material cooled down to temperature below T_f in a period ranging between a few minutes and days.

The whole region in the frequency range 700–1300 cm^{-1} of the LDG Raman spectrum has a higher background, and all fragments analyzed present a contribution around 940 cm^{-1} that is not present in the commercial SiO_2 glass, or the lechatelierite spectrum (Fig. 20). Because this band is not related to luminescence (electronic transitions), and unlikely is related to Q^1 species, it must be related to other cations present in the Libyan Desert Glasses. Stretching modes related to 4-fold coordinated Al^{3+} will mainly influence the relative intensity and frequency position of the band at ~1200 cm^{-1}, and the intensity of the defect lines because of changes in the ring statistics (Cicconi et al. 2020c); hence, the higher background and the band at ~940 cm^{-1} might be related to other (relatively) abundant cations, such as Ti^{4+} and Fe^{3+} species (~0.1 wt.%).

Darwin glasses are chemically varied, and the Raman signals reflect the heterogeneity of these samples. Indeed, depending on the portion analyzed it is possible to get vibrations almost completely related to amorphous SiO_2 (i.e., lechatelierite) or vibration modes that indicate a more depolymerized silicate network (Fig. 21). For instance, the layered structure of the Darwin glass shown in Figures 7 and 8 have vibrations related to a highly polymerized glass in the light layering, whereas the darker layers have lower intensities of the defect lines D_1 and D_2, and stronger bands in the high-frequency region, most probably related to higher amounts of non-bridging oxygens (NBO). The frequency region between 860 and 1400 cm^{-1} in the dark portion has an integrated area almost three times higher than in the light layers, and these modifications can be attributed to change in the bulk chemistry (see Table 3). In the dark glass portion, there is also the appearance of an additional contribution around 940 cm^{-1}.

Distal ejecta and impact glasses have high SiO_2 contents (typically >65 wt.%; see Tables 2 and 3), thus the typical vibrations of silica-rich glasses are clearly observed, even if sometimes the defect lines are barely visible (Fig. 22). The high-frequency regions show further contributions, and the spectra are very different from the lechatelierite glass, or the SiO_2-richer LDG sample.

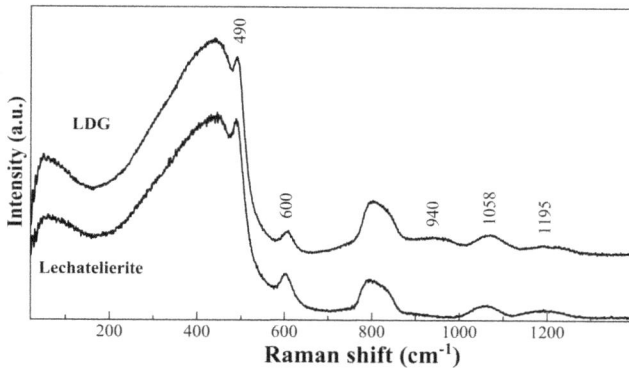

Figure 20. Raman spectra in the frequency range 20–1400 cm⁻¹ of lechatelierite and Libyan Desert Glass (LDG). The frequency position of the main vibrations is reported.

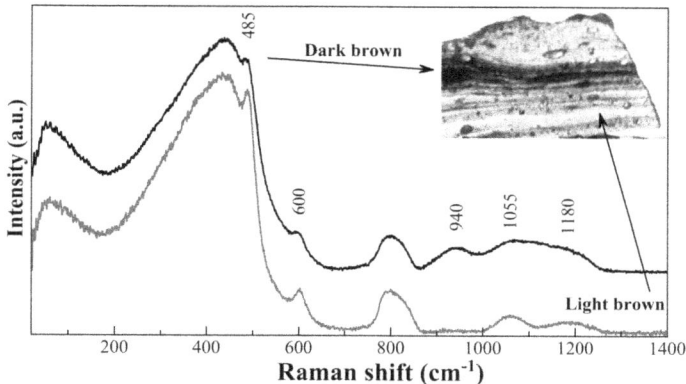

Figure 21. Raman spectra in the frequency range 20–1400 cm⁻¹ of the light and of the dark brown layers in a Darwin glass specimen. The frequency position of the main vibrations is reported.

The atacamaite sample is the most different one, with strong variations in the frequency position of the vibrational bands, and an intense Q-envelope. These modifications on the Raman spectrum of atacamaite glass reflect the average bulk chemistry of the glass, and especially the relatively low SiO_2 content, and the high Al_2O_3, FeO and CaO contents and increased NBOs.

Figure 23 reports the Raman spectrum of a green trinitite specimen, and only the most homogeneous inner glass portion was considered. At the microscale, trinitite specimens are chemically highly heterogeneous (see section *Trinite or nuclear glass*) with SiO_2 contents that range between ~90 and 60 wt.% (Eby et al. 2010). In the specimen available, the Raman spectra do not show Si-rich portions, and the majority of the data collected points to an aluminosilicate glass composition with SiO_2 contents ≪75 wt.%. The high-frequency range has very intense and broad bands, with the highest contribution peaking at ~1030 cm⁻¹. Depending on the glass portion considered, the intensity of the shoulder at ~940 cm⁻¹ slightly varies, and the R band decreases its intensity. These small variations can be related to the different proportions of network modifier cations, and in turn, to the SiO_2 content variability.

The Raman spectra shown here are just an overview of the variability of the glass structures and demonstrate the potentiality of using Raman spectroscopy, especially for spatially-dependent investigations at the microscale. Despite the chemical complexity of a

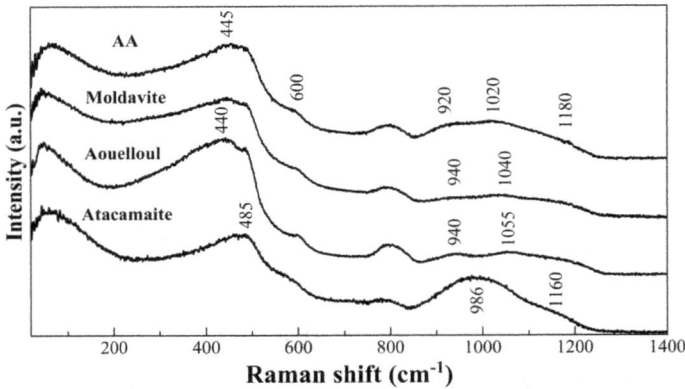

Figure 22. Raman spectra in the frequency range 20–1400 cm⁻¹ of several impact related glasses: two tektites, respectively from the Australasian (AA) and Central European (moldavite) strewn fields, and two proximal impact glasses (Atacamaite and Aouelloul). The frequency position of the main vibrations is reported.

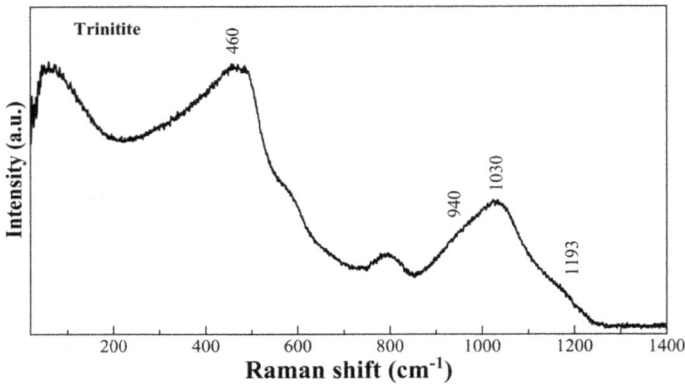

Figure 23. Raman spectra in the frequency range 20–1400 cm⁻¹ of a trinitite sample. The frequency position of the main vibrations is reported.

multicomponent natural glass, the vibrational bands can be interpreted based on the study of simplified synthetic compositions, and it is possible to correlate the degree of polymerization to the Raman features. All changes occurring in the low-frequency region can be mainly associated with changes in the bulk composition, and more specifically, on the Si content that drives the glass network polymerization. For an almost constant bulk composition, it would be possible to use the glass-memory effect and study the variations of the intensity of the defect lines, and the variations in the position of the main R band, in order to relate them to the pressure or the thermal history of the glasses. The high-frequency region contains the T–O stretching modes related to Si^{4+}, but also of other network former cations, such as Al^{3+}. The variations observed in all glasses, compared to the bands of lechatelierite, could be related to the different proportions of network modifiers that creates higher amounts of NBO, and charge compensator cations, that will influence the covalency character of the bonds. Moreover, other cations, such as the 4-fold coordinated Fe^{3+}, can also provide stretching vibrations in the high frequency region. For instance, in aluminosilicate and silicate glasses, both natural and synthetic, it has been reported that the bands between 800 and 1200 cm⁻¹ are very sensitive to the average iron valence state (e.g. Magnien et al. 2008; Di Muro et al. 2009; Di Genova et al. 2016; Le Losq et al. 2019). Thus, in the cases of Darwin brown glass and atacamaite, the contribution observed could be related to the presence of higher amounts of 4-fold coordinated Fe^{3+}.

CONCLUDING REMARKS

We have described several natural glasses that are not derived from volcanic processes but from highly energetic events. The glasses formed as a consequence of these energetic events share many characteristics, such as:

- Compositional similarities (e.g. the presence of lechatelierite)
- Liquid–liquid immiscibility
- Highly reduced species
- Shocked minerals

The study of fulgurites and trinitite glasses may serve as a proxy for impact-induced mineralogic and petrologic changes, and to better understand the formation of glasses from highly disequilibrium conditions.

ACKNOWLEDGMENTS

MRC would like to thank Dr. G. Giuli (University of Camerino, I), who introduced her to the fascinating world of tektites. The authors thank Dr. F. Moynier for the atacamaite samples, and the IPGP-PARI platform for SEM and Raman facilities. JMC was supported by IPGP Visiting Professor Fellowship (Programme 2018). The authors thank Prof. Henderson for all suggestions.

REFERENCES

Aboud T (2009) Libyan Desert Glass: has the enigma of its origin been resolved? Phys Procedia 2:1425–1432
Alvarez LW, Alvarez W, Asaro F, Michel HV (1980) Extraterrestrial Cause for the Cretaceous–Tertiary Extinction. Science 208:1095–1108
Anderson AE (1925) Sand fulgurites from Nebraska their structure and formative factors. The Nebraska State Museum Bull 7:49–86
Artemieva N (2002) Numerical modeling of tektite origin in oblique impacts: Implication to Ries–Moldavites strewn field. Bull Czech Geol Surv 77:303–311
Artemieva NA, Wünnemann K, Krien F, Reimold WU, Stöffler D (2013) Ries crater and suevite revisited—Observations and modeling Part II: Modeling. Meteorit Planet Sci 48:590–627
Atkatz D, Bragg C (1995) Determining the yield of the Trinity nuclear device via gamma-ray spectroscopy. Am J Phys 63:411–413
Barnes VE (1958) Origin of tektites. Nature 181:1457–1457
Barrat JA, Jahn BM, Amosse J, Rocchia R, Keller F, Poupeau GR, Diemer E (1997) Geochemistry and origin of Libyan Desert glasses. Geochim Cosmochim Acta 61:1953–1959
Bates RL, Jackson JA (eds) (1987) Glossary of Geology, 3rd edn. Am Geol Institute
Bellot-Gurlet L, Bourdonnec F-XLe, Poupeau G, Dubernet S (2004) Raman micro-spectroscopy of western Mediterranean obsidian glass: one step towards provenance studies? J Raman Spectrosc 35:671–677
Bellucci JJ, Simonetti A, Koeman EC, Wallace C, Burns PC (2014) A detailed geochemical investigation of post-nuclear detonation trinitite glass at high spatial resolution: Delineating anthropogenic vs. natural components. Chem Geol 365:69–86
Bentor YK (1984) Combustion-metamorphic glasses. J Non-Cryst Solids 67:433–448
Bentor YK, Kastner M, Perlman I, Yellin Y (1981) Combustion metamorphism of bituminous sediments and the formation of melts of granitic and sedimentary composition. Geochim Cosmochim Acta 45:2229–2255
Beran A, Koeberl C (1997) Water in tektites and impact glasses by Fourier-transformed infrared spectrometry. Meteorit Planet Sci 32:211–216

Beyssac O, Rouzaud J-N, Goffe B, Brunet F, Chopin C (2002) Graphitization in a high-pressure, low-temperature metamorphic gradient: a Raman microspectroscopy and HRTEM study. Contrib Mineral Petrol 143:19–31

Bigazzi G, Michele V (1996) New fission-track age determinations on impact glasses. Meteorit Planet Sci 31:234–236

Bohor BF (1990) Shock-induced microdeformations in quartz and other mineralogical indications of an impact event at the Cretaceous–Tertiary boundary. Tectonophysics 171:359–372

Borucki WJ, Chameides WL (1984) Lightning: Estimates of the rates of energy dissipation and nitrogen fixation. Rev Geophys 22:363–372

Bouška V (1994) Terrestrial and lunar, volcanic and impact glasses, tektites, and fulgurites. *In:* Advanced Mineralogy, AS Marfunin (ed) Springer Berlin Heidelberg, p 258–265

Bouška V, Povondra P (1964) Correlation of some physical and chemical properties of moldavites. Geochim Cosmochim Acta 28:783–791

Brachaniec T, Szopa K, Karwowski, Ł. (2014) Discovery of the most distal Ries tektites found in Lower Silesia, southwestern Poland. Meteorit Planet Sci 49:1315–1322

Brauns R (1912) Die chemische Zusammensetzung granatfuhrender kristalliner Schiefer, Cordieritgesteine und Sanidinite aus dem Laacher Seegebiet. Neues Jahrb Min Beilage-Bd 34:85

Bridgman PW, Šimon I (1953) Effects of Very High Pressures on Glass. J Appl Phys 24:405–413

Bryant DE, Greenfield D, Walshaw RD, Evans SM, Nimmo AE, Smith CL, Wang L, Pasek MA, Kee TP (2009) Electrochemical studies of iron meteorites: phosphorus redox chemistry on the early Earth. Int J Astrobiol 8:27–36

Bunch TE, Hermes RE, Moore AM, Kennett DJ, Weaver JC, Wittke JH, DeCarli PS, Bischoff JL, Hillman GC, Howard GA, Kimbel DR (2012) Very high-temperature impact melt products as evidence for cosmic airbursts and impacts 12,900 years ago. PNAS 109:E1903–E1912

Calas G, Petiau J (1983) Coordination of iron in oxide glasses through high-resolution K-edge spectra: Information from the pre-edge. Solid State Commun 48:625–629

Callegari E, Pertsev N (2007) Contact metamorphic and associated rocks. *In:* DJ Fettes J Desmons (eds) Metamorphic Rocks: A Classification and Glossary of Terms, Cambridge University Press, p 69–81

Canil D, Mihalynuk M, Lacourse T (2018) Discovery of modern (post-1850 CE) lavas in south-central British Columbia, Canada: Origin from coal fires or intraplate volcanism? Lithos 296–299:471–481

Capitanio F (2005) Comment on: The Ricetto and Colle Fabbri wollastonite and melilite-bearing rocks of the central Apennines, Italy. Am Mineral 90:1934–1939

Capitanio F, Larocca F, Improta S (2004) High-temperature rapid pyrometamorphism induced by a charcoal pit burning: The case of Ricetto, central Italy. Int J Earth Sci 93:107–118

Carmichael IS E (1979) Glass and the glassy rocks. *In:* The Evolution of the Igneous Rocks, HS Yoder (ed), Princeton Univ Pres, p 233–244

Carter EA, Pasek MA, Smith T, Kee TP, Hines P, Edwards HG M (2010) Rapid Raman mapping of a fulgurite. Anal Bioanal Chem 397:2647–2658

Castro J, Manga M, Cashman K (2002) Dynamics of obsidian flows inferred from microstructures: Insights from microlite preferred orientations. Earth Planet Sci Lett 199:211–226

Cavosie AJ, Koeberl C (2019) Overestimation of threat from 100 Mt–class airbursts? High-pressure evidence from zircon in Libyan Desert Glass. Geology 47:609–612

Champagnon B, Panczer G, Chemarin C (1997) Differentiation of natural silica glasses using Raman microspectrometry. Chem Erde 57:290–296

Chen J, Elmi C, Goldsby D, Gieré R (2017) Generation of shock lamellae and melting in rocks by lightning-induced shock waves and electrical heating. Geophys Res Lett 44:8757–8768

Cicconi MR, Giuli G, Ertel-Ingrisch W, Paris E, Dingwell DB (2015) The effect of the [Na/(Na+K)] ratio on Fe speciation in phonolitic glasses. Am Mineral 100:1610–1619

Cicconi MR, Neuville DR (2019) Natural Glasses. *In:* Musgraves JD, Hu J, Calvez L (eds) Springer Handbook of Glass, Springer Nature Switzerland, p 771–812

Cicconi MR, Le Losq C, Moretti R, Neuville D (2020a). Magmas, the largest repositories and carriers of earth's redox processes. Elements 16:173–178

Cicconi MR, Khansur NH, Eckstein UR, Werr F, Webber KG, de Ligny D (2020b) Determining the local pressure during aerosol deposition using glass memory. J Am Ceram Soc 103:2443–2452

Cicconi MR, Blanc, W, de Ligny D, Neuville DR (2020c) The influence of codoping on optical properties and glass connectivity of silica fiber preforms. Ceramics Int 46:26251–26259

Clark BH, Peacor DR (1992) Pyrometamorphism and partial melting of shales during combustion metamorphism: mineralogical, textural, and chemical effects. Contrib Mineral Petrol 112:558–568

Clayton PA (1934) Silica-glass from the Libyan Desert. Mineral Mag 23:501–508

Cohen HM, Roy R (1961) Effects of ultra high pressures on glass. J Am Ceram Soc 44:523–524

Cohen HM, Roy R (1965) Densification of glass at very high pressure. Phys Chem Glasses 6:149–155

Coombs DS, Beck RJ, Adams CJ, Bannister JM, Paterson LA, Roser BP (2008) Paralava produced by combustion of dead gorse near Colac Bay, Southland, New Zealand. J Geol 116:94–101

Cosca MA, Essene EJ, Geissman JW, Simmons WB, Coates DA (1989) Pyrometamorphic rocks associated with naturally burned coal beds, Powder River Basin, Wyoming. Am Mineral 74:85–100

Cuttitta F, Carron M, Annell C (1972) New data on selected Ivory Coast tektites. Geochim Cosmochim Acta 36:1297–1309

De Blasio FV, Medici L (2017) Microscopic model of rock melting beneath landslides calibrated on the mineralogical analysis of the Köfels frictionite. Landslides 14:337–350

de Boer CB, Dekkers MJ, van Hoof TA (2001) Rock-magnetic properties of TRM carrying baked and molten rocks straddling burnt coal seams. Phys Earth Planet Inter 126:93–108

Debenedetti PG, Stanley HE (2003) Supercooled and glassy water. Phys Today 56:40–46

Deschamps T, Kassir-Bodon A, Sonneville C, Margueritat J, Martinet C, De Ligny D, Mermet A, Champagnon B (2013) Permanent densification of compressed silica glass: a Raman-density calibration curve. J Phys: Condens Matter 25:025402

Deutsch A, Ostermann M, Masaitis VL (1997) Geochemistry and neodymium–strontium isotope signature of tektite-like objects from Siberia (urengoites, South-Ural glass) Meteorit Planet Sci 32:679–686

Devouard B, Rochette P, Gattacceca J, Barrat JA, Moustard F, Valenzuela EM, Alard O, Balestrieri ML, Bigazzi G, Dos Santos E, Gounelle M (2014) A new tektite strewnfield in Atacama. *In:* 77th Annual Meteoritical Society Meeting, p 5394

Di Genova D, Hess K-U, Chevrel MO, Dingwell DB (2016) Models for the estimation of Fe^{3+}/Fe_{tot} ratio in terrestrial and extraterrestrial alkali- and iron-rich silicate glasses using Raman spectroscopy. Am Mineral 101:943–952

Di Muro A, Métrich N, Mercier M, Giordano D, Massare D, Montagnac G (2009) Micro-Raman determination of iron redox state in dry natural glasses: Application to peralkaline rhyolites and basalts. Chem Geol 259:78–88

Di Toro G, Pennacchioni G, Teza G (2005) Can pseudotachylytes be used to infer earthquake source parameters? An example of limitations in the study of exhumed faults. Tectonophysics 402:3–20

Dos Santos E, Scorzelli RB, Rochette P, Devouard B, Gattacceca J, Moustard F, Cournède C (2015) A new strewnfield of splash-form impact glasses in Atacama, Chile: A Mössbauer Study. *In:* 78th Annual Meeting of the Meteoritical Society 1856:5074

Drewitt JWE, Hennet L, Neuville DR (2022) From short to medium range order in glasses and melts by diffraction and Raman spectroscopy. Rev Mineral Geochem 87:55-103

Dunlap RA (1997) An investigation of Fe oxidation states and site distributions in a Tibetan tektite. Hyperfine Interact 110:217–225

Dunlap RA, Sibley ADE (2004) A Mössbauer effect study of Fe-site occupancy in Australasian tektites. J Non-Cryst Solids 337:36–41

Dunlap RA, Eelman DA, MacKay GR (1998) Mössbauer effect investigation of correlated hyperfine parameters in natural glasses (tektites) J Non-Cryst Solids 223:141–146

Earth Impact Database (2019) Earth Impact Database. Retrieved from http://www.passc.net/EarthImpactDatabase/New website_05–2018/Index.html

Eby N, Hermes R, Charnley N, Smoliga JA (2010) Trinitite—the atomic rock. Geol Today 26:180–185

Eby NG, Charnley N, Pirrie D, Hermes R, Smoliga J, Rollinson G (2015) Trinitite redux: Mineralogy and petrology. Am Mineral 100:427–441

Ende M, Schorr S, Kloess G, Franz A, Tovar M (2012) Shocked quartz in Sahara fulgurite. Euro J Mineral 24:499–507

Engelhardt W (1967) Chemical composition of Ries glass bombs. Geochim Cosmochim Acta 31:1677–1689

Engelhardt WV (1972) Shock produced rock glasses from the Ries crater. Contrib Mineral Petrol 36:265–292

Engelhardt W, Luft E, Arndt J, Schock H, Weiskirchner W (1987) Origin of moldavites. Geochim Cosmochim Acta 51:1425–1443

Erismann T, Heuberger H, Preuss E (1977) Der Bimsstein von Koefels (Tirol), ein Bergsturz-`Friktionit'. Tschermaks Mineralogische und Petrographische Mitteilungen 24:67–119

Essene EJ, Fisher DC (1986) Lightning strike fusion: extreme reduction and metal–silicate liquid immiscibility. Science 234:189–93

Etiope G (2015) Seeps in the ancient world: myths, religions, and social development. *In:* Natural Gas Seepage: The Earth's Hydrocarbon Degassing. Springer International Publishing, p 183–193

Evans BJ, Leung LK (1979) Mössbauer spectroscopy of tektites and other natural glasses. J Phys (Paris) Colloq 40(C2):489–490

Fahey AJ, Zeissler CJ, Newbury DE, Davis J, Lindstrom RM (2010) Postdetonation nuclear debris for attribution. PNAS 107:20207–12

Feng T, Lang C, Pasek MA (2019) The origin of blue coloration in a fulgurite from Marquette, Michigan. Lithos 342–343:288–294

Ferré EC, Zechmeister MS, Geissman JW, Sekaran NM, Kocak K (2005) The origin of high magnetic remanence in fault pseudotachylites: Theoretical considerations and implication for coseismic electrical currents. Tectonophysics 402(1–4 SPECISS):125–139

Ferrière L, Osinski GR (2012) Shock Metamorphism. *In:* Osinski GR, Pierazzo E (eds) Impact Cratering. John Wiley Sons, Ltd, Chichester, UK, p 106–124

Ferrière L, Barrat JA, Giuli G, Koeberl C, Schulz T, Topa D, Wegner W (2017) A new tektite strewn field discovered in Uruguay. In: 80th Annual Meeting of the Meteoritical Society, p 6195

Firestone RB, West A, Kennett JP, Becker L, Bunch TE, Revay ZS, Schultz PH, Belgya T, Kennett DJ, Erlandson JM, Dickenson OJ (2007) Evidence for an extraterrestrial impact 12,900 years ago that contributed to the megafaunal extinctions and the Younger Dryas cooling. PNAS 104:16016–21

Folco L, Rochette P, Perchiazzi N, D'Orazio M, Laurenzi MAA, Tiepolo M (2008) Microtektites from Victoria Land Transantarctic Mountains. Geology 36:291–294

Folco LU, D'Orazio MA, Tiepolo M, Tonarini S, Ottolini L, Perchiazzi NA, Rochette P, Glass BP (2009) Transantarctic Mountain microtektites: Geochemical affinity with Australasian microtektites. Geochim Cosmochim Acta 73:3694–3722

Folco L, Glass BP, D'Orazio M, Rochette P (2010) A common volatilization trend in Transantarctic Mountain and Australasian microtektites: Implications for their formation model and parent crater location. Earth Planet Sci Lett 293:135–139

Folco L, Bigazzi G, D'Orazio M, Balestrieri ML (2011) Fission track age of Transantarctic Mountain microtektites. Geochim Cosmochim Acta 75:2356–2360

Folco L, D'Orazio M, Gemelli M, Rochette P (2016) Stretching out the Australasian microtektite strewn field in Victoria Land Transantarctic Mountains. Polar Sci 10:147–159

Ford RJ (1972) A possible impact crater associated with Darwin glass. Earth Planet Sci Lett 16:228–230

Fredriksson K, Degasparis A, Ehmann W (1977) The Zhamanshin structure: chemical and physical properties of selected samples. Meteoritics 12:229

French BM (1998) Traces of catastrophe: A handbook of shock-metamorphic effects in terrestrial meteorite impact structures, Lunar and Planetary Institute, Houston, TX

Frenzel G, Stahle V (1984) Uber Alumosilikatglas mit Lechatelierit-Einschlussen von einer Fulguritrohre des Hahnenstockes (Glarner Freiberg, Schweiz). Chemie Erde 43:17–26

Frey FA (1977) Microtektites: a chemical comparison of bottle-green microtektites, normal microtektites and tektites. Earth Planet Sci Lett 35:43–48

Fudali RF, Ford RJ (1979) Darwin Glass and Darwin Crater: a progress report. Meteoritics 14:283–296

Fudali RF F, Dyar MD D, Griscom DL, Schreiber HD (1987) The oxidation state of iron in tektite glass. Geochim Cosmochim Acta 51:2749–2756

Galeener FL, Geissberger AE, Weeks RA (1984) On the thermal history of Libyan Desert glass. J Non-Cryst Solids 67:629–636

Ganino C, Libourel G, Nakamura AM, Jacomet S, Tottereau O, Michel P (2018) Impact-induced chemical fractionation as inferred from hypervelocity impact experiments with silicate projectiles and metallic targets. Meteorit Planet Sci 53:2306–2326

Gentner, von W, Zähringer J (1959) Kalium-Argon-Alter einiger Tektite. Z Naturforsch A 14:686–687

Gieré R, Wimmenauer W, Müller-Sigmund H, Wirth R, Lumpkin GR, Smith KL (2015) Lightning-induced shock lamellae in quartz. Am Mineral 100:1645–1648

Giuli G (2017) Tektites and microtektites iron oxidation state and water content. Rendiconti Lincei 28:615–621

Giuli G, Pratesi G, Cipriani C, Paris E (2002) Iron local structure in tektites and impact glasses by extended X-ray absorption fine structure and high-resolution X-ray absorption near-edge structure spectroscopy. Geochim Cosmochim Acta 66:4347–4353

Giuli G, Paris E, Pratesi G, Koeberl C, Cipriani C (2003) Iron oxidation state in the Fe-rich layer and silica matrix of Libyan Desert Glass: A high-resolution XANES study. Meteorit Planet Sci 38:1181–1186

Giuli G, Eeckhout SG, Cicconi MR, Koeberl C, Pratesi G, Paris E (2010a) Iron oxidation state and local structure in North American tektites. Large Meteorite Impacts and Planetary Evolution IV, Geol Soc Am Spec Papers 465:645–651

Giuli G, Pratesi G, Eeckhout SG, Koeberl C, Paris E (2010b) Iron reduction in silicate glass produced during the 1945 nuclear test at the Trinity site (Alamogordo, New Mexico, USA) In: Geol Soc Am Spec Papers 465:653–660

Giuli G, Cicconi MR, Paris E (2012) The [4]Fe3+–O distance in synthetic kimzeyite garnet, Ca3Zr2[Fe2SiO12]. Euro J Mineral 24:783–790

Giuli G, Cicconi MR, Eeckhout SG, Koeberl C, Glass BP, Pratesi G, Cestelli-Guidi M, Paris E (2013a) North American microtektites are more oxidized than tektites. Am Mineral 98:1930–1937

Giuli G, Cicconi MR, Trapananti A, Eeckhout SG, Pratesi G, Paris E, Koeberl C (2013b) Iron redox variations in Australasian Muong Nong-type tektites. Meteoritics Planet Sci

Giuli G, Cicconi MR, Eeckhout SG, Pratesi G, Paris E, Folco L (2014a) Australasian microtektites from Antarctica: XAS determination of the Fe oxidation state. Meteorit Planet Sci 49:696–705

Giuli G, Cicconi MR, Stabile P, Trapananti A, Pratesi G, Cestelli-Guidi M, Koeberl C (2014b) New data on the Fe oxidation state and water content of Belize Tektites. In: 45th Lunar Planet Sci Confer, p 2322

Glass B (1967) Microtektites in deep-sea sediments. Nature 214:372–374

Glass BP (1972) Bottle Green Microtektites. J Geophys Res 77:7057–7064

Glass BP (1978) Australasian microtektites and the stratigraphic age of the australites. Bull Geol Soc Am 89:1455–1458

Glass BP (1990) Tektites and microtektites: key facts and inferences. Tectonophysics 171:393–404

Glass BP (2016) Glass: The geologic connection. Int J Appl Glass Sci 7:435–445

Glass BP, Pizzuto JE (1994) Geographic variation in Australasian microtektite concentrations: Implications concerning the location and size of the source crater. J Geophys Res 99(E9):19075

Glass BP, Zwart MJ (1979) North American microtektites in Deep Sea Drilling Project cores from the Caribbean Sea and Gulf of Mexico. Geol Soc Am Bull 90:595

Glass BP, Senftle FE, Muenow DW, Aggrey KE, Thorpe AN (1987) Atomic bomb glass beads: tektite and microtektite analogs. *In:* Second International Conference on Natural Glasses, p 361–369

Glass BP, Huber H, Koeberl C (2004) Geochemistry of Cenozoic microtektites and clinopyroxene-bearing spherules. Geochim Cosmochim Acta 68:3971–4006 Glass BP, Simonson BM (2013) Distal Impact Ejecta Layers. Springer Berlin, Heidelberg

Glindemann D, de Graaf RM, Schwartz AW (1999) Chemical reduction of phosphate on the primitive Earth. Origins Life Evol Biosphere 29:555–561

Goderis S, Paquay F, Claeys P (2012) Projectile identification in terrestrial impact structures and ejecta material. *In:* Impact Cratering, John Wiley Sons, Ltd, Chichester, UK, p 223–239

Golubev YA, Isaenko SI, Prikhodko AS, Borgardt NI, Suvorova EI (2016) Raman spectroscopic study of natural nanostructured carbon materials: shungite vs. anthraxolite. Euro J Mineral 28:545–554

Gomez-Nubla L, Aramendia J, Alonso-Olazabal A, Fdez-Ortiz de Vallejuelo S, Castro K, Ortega LA, Zuluaga MC, Murelaga X, Madariaga JM (2015) Darwin impact glass study by Raman spectroscopy in combination with other spectroscopic techniques. J Raman Spectrosc 46:913–919

Gomez-Nubla L, Aramendia J, de Vallejuelo SF, Alonso-Olazabal A, Castro K, Zuluaga MC, Ortega LÁ, Murelaga X, Madariaga JM (2017) Multispectroscopic methodology to study Libyan desert glass and its formation conditions. Anal Bioanal Chem 409:3597–3610

Gómez-Nubla L, Aramendia J, Fdez-Ortiz de Vallejuelo S, Castro K, Madariaga JM (2018) Detection of organic compounds in impact glasses formed by the collision of an extraterrestrial material with the Libyan Desert (Africa) and Tasmania (Australia). Anal Bioanal Chem 410:6609–6617

Grapes R (2010) Pyrometamorphism. Springer, Berlin Heidelberg

Grapes R, Zhang K, Peng Z (2009) Paralava and clinker products of coal combustion, Yellow River, Shanxi Province, China. Lithos 113:831–843

Grapes RH, Müller-Sigmund H (2010) Lightning-strike fusion of gabbro and formation of magnetite-bearing fulgurite, Cornone di Blumone, Adamello, Western Alps, Italy. Mineral Petrol 99:67–74

Grapes R, Sokol E, Kokh S, Kozmenko O, Fishman I (2013) Petrogenesis of Na-rich paralava formed by methane flares associated with mud volcanism, Altyn-Emel National Park, Kazakhstan. Contrib Mineral Petrol 165:781–803

Graup G (1999) Carbonate–silicate liquid immiscibility upon impact melting: Ries Crater, Germany. Meteorit Planet Sci 34:425–438

Greshake A, Koeberl C, Fritz J, Reimold WU (2010) Brownish inclusions and dark streaks in Libyan Desert Glass: Evidence for high-temperature melting of the target rock. Meteorit Planet Sci 45:973–989

Greshake A, Wirth R, Fritz J, Jakubowski T, Böttger U (2018) Mullite in Libyan Desert Glass: Evidence for high-temperature/low-pressure formation. Meteorit Planet Sci 53:467–481

Hamann C, Hecht L, Ebert M, Wirth R (2013) Chemical projectile–target interaction and liquid immiscibility in impact glass from the Wabar craters, Saudi Arabia. Geochim Cosmochim Acta 121:291–310

Hamann C, Fazio A, Ebert M, Hecht L, Wirth R, Folco L, Deutsch A, Reimold WU (2018) Silicate liquid immiscibility in impact melts. Meteorit Planet Sci 53:1594–1632

Heffern EL, Reiners PW, Naeser CW, Coates DA (2007) Geochronology of clinker and implications for evolution of the Powder River Basin landscape, Wyoming and Montana. GSA Rev Eng Geol 18:155–175

Heide K, Heide G (2011) Vitreous state in nature-Origin and properties. Chem Erde 71:305–335

Henderson GS, Stebbins JF (2022) The short-range order (SRO) and structure. Rev Mineral Geochem 87:1-53

Henderson GS, de Groot FM F, Moulton BJ A (2014) X-ray absorption near-edge structure (XANES) spectroscopy. Rev Mineral Geochem 78:75–138

Hermes RE, Strickfaden WB (2005) A new look at trinitite. Nucl Weapons J:2, 2–7

Herring JR, Modreski PJ (1986) Unusual, high-temperature, iron-rich, mineral phases produced by natural burning of coal seams; analytical data

Hildebrand AR, Penfield GT, Kring DA, Pilkington M, Camargo Z A, Jacobsen SB, Boynton WV (1991) Chicxulub Crater: A possible Cretaceous/Tertiary boundary impact crater on the Yucatán Peninsula, Mexico. Geology 19:867

Hörz F, See TH, Murali AV, Blanchard DP (1989) Heterogeneous dissemination of projectile materials in the impact melts from Wabar crater, Saudi Arabia. *In:* Lunar Planet Sci Confer Proc 19:697–709

Hosgormez H, Etiope G, Yalçin MN (2008) New evidence for a mixed inorganic and organic origin of the Olympic Chimaera fire (Turkey): a large onshore seepage of abiogenic gas. Geofluids 8:263–273

Howard KT (2008) Geochemistry of Darwin glass and target rocks from Darwin Crater, Tasmania, Australia. Meteorit Planet Sci 43:1–21

Howard KT (2009) Physical distribution trends in Darwin glass. Meteorit Planet Sci 44:115–129

Howard KT (2011) Volatile enhanced dispersal of high velocity impact melts and the origin of tektites. Proc Geol Assoc 122:363–382

Howard KT, Haines PW (2007) The geology of Darwin Crater, western Tasmania, Australia. Earth Planet Sci Lett 260:328–339

Howard KT, Bailey MJ, Berhanu D, Bland PA, Cressey G, Howard LE, Jeynes C, Matthewman R, Martins Z, Sephton MA, Stolojan V (2013) Biomass preservation in impact melt ejecta. Nat Geosci 6:1018–1022

Hu QH, Rose TP, Zavarin M, Smith DK, Moran JE, Zhao PH (2008) Assessing field-scale migration of radionuclides at the Nevada Test Site: "mobile" species. J Environ Radioact 99:1617–1630

Hudon P, Baker DR (2002) The nature of phase separation in binary oxide melts and glasses. I Silicate systems. J Non-Cryst Solids 303:299–345

Hudon P, Jung I, Baker DR (2004) Effect of pressure on liquid–liquid miscibility gaps: A case study of the systems $CaO–SiO_2$, $MgO–SiO_2$, and $CaMgSi_2O_6–SiO_2$. J Geophys Res 109(B3):B03207

Izett GA (1987) Authigenic 'spherules' in K–T boundary sediments at Caravaca, Spain, and Raton Basin, Colorado and New Mexico, may not be impact derived. Geol Soc Am Bull 99:78–86

Jakes P, Sen S, Matsuishi K, Reid AM, King EA, Casanova I (1992) Silicate melts at super liquidus temperatures: reduction and volatilization. Lunar Planet Sci Confer 23:599

Jessberger E, Gentner W (1972) Mass spectrometric analysis of gas inclusions in Muong Nong glass and Libyan Desert glass. Earth Planet Sci Lett 14:221–225

Johnson BC, Melosh HJ (2012) Formation of spherules in impact produced vapor plumes. Icarus 217:416–430

Johnson BC, Melosh HJ (2014) Formation of melt droplets, melt fragments, and accretionary impact lapilli during a hypervelocity impact. Icarus 228:347–363

Jones BE, Jones KS, Rambo KJ, Rakov VA, Jerald J, Uman MA (2005) Oxide reduction during triggered-lightning fulgurite formation. J Atmos Sol Terr Phys 67:423–428

Kenkmann T, Hornemann U, Stöffler D (2000) Experimental generation of shock-induced pseudotachylites along lithological interfaces. Meteorit Planet Sci 35:1275–1290

Khoury HN, Sokol EV, Clark ID (2015) Calcium uranium oxide minerals from Central Jordan: assemblages, chemistry, and alteration products. Can Mineral 53:61–82

Killick AM (2003) Fault Rock Classification: An aid to structural interpretation in mine and exploration geology. South Afr J Geol 106:395–402

Koeberl C (1988) Blue glass: A new impactite variety from Zhamanshin Crater, USSR. Geochim Cosmochim Acta 52:779–784

Koeberl C (1992a) Geochemistry and origin of Muong Nong-type tektites. Geochim Cosmochim Acta 56:1033–1064

Koeberl C (1992b) Water content of glasses from the K/T boundary, Haiti: An indication of impact origin. Geochim Cosmochim Acta 56:4329–4332

Koeberl C (1994) Tektite origin by hypervelocity asteroidal or cometary impact. Large Meteorite Impacts and Planetary Evolution, Geol Soc Am Spec Pap 293:133–151

Koeberl C (1997) Libyan Desert Glass: geochemical composition and origin. *In:* Proc Silica '96 Meeting, p 121–131

Koeberl C (2013) The geochemistry and cosmochemistry of impacts. *In:* Treatise on Geochemistry: Second Edition, Vol. 2, p 73–118

Koeber C, Fredriksson K (1986) Impact glasses from Zhamanshin crater (USSR): chemical composition and discussion of origin. Earth Planet Sci Lett 78:80–88

Koeberl C, Ferrière L (2019) Libyan Desert Glass area in western Egypt: Shocked quartz in bedrock points to a possible deeply eroded impact structure in the region. Meteorit Planet Sci 54:2398–2408

Koeberl C, Sigurdsson H (1992) Geochemistry of impact glasses from the K/T boundary in Haiti: Relation to smectites and a new type of glass. Geochim Cosmochim Acta 56:2113–2129

Koeberl C, Brandstätter F, Glass BP, Hecht L, Mader D, Reimold WU (2007) Uppermost impact fallback layer in the Bosumtwi crater (Ghana): Mineralogy, geochemistry, and comparison with Ivory Coast tektites. Meteorit Planet Sci 42:709–729

Koeberl C, Crósta AP, Schulz T (2019) Geochemical Investigation of the Atacamaites, a New Impact Glass Occurrence in South America. *In:* 50th Lunar Planet Sci Confer 2019, p 2132

Kokh S, Dekterev A, Sokol E, Potapov S (2016) Numerical simulation of an oil–gas fire: A case study of a technological accident at Tengiz oilfield, Kazakhstan (June 1985–July 1986) Energy Exploration Exploitation 34:77–98

Krider EP, Dawson GA, Uman MA (1968) Peak power and energy dissipation in a single-stroke lightning flash. J Geophys Res 73:3335–3339

Kuenzer C, Stracher GB (2012) Geomorphology of coal seam fires. Geomorphology 138:209–222

Kwiecinska B, Pusz S, Krzesinska M, Pilawa B (2007) Physical properties of shungite. Int J Coal Geol 71:455–461

Le Losq C, Berry A, Kendrick M, Neuville D, O'Neill HSt (2019) Determination of the oxidation state of iron in Mid-Ocean Ridge basalt glasses by Raman spectroscopy. Am Mineral, 104:1032–1042

Lee M-Y, Wei K-Y (2000) Australasian microtektites in the South China Sea and the West Philippine Sea: Implications for age, size, and location of the impact crater. Meteorit Planet Sci 35:1151–1155

Legros F, Cantagrel J, Devouard B (2000) Pseudotachylyte (frictionite) at the base of the Arequipa volcanic landslide deposit (Peru): Implications for emplacement mechanisms. J Geol 108:601–611

Lin A, Shimamoto T (1998) Selective melting processes as inferred from experimentally generated pseudotachylytes. J Asian Earth Sci 16:533–545

Lo C-H, Howard KT, Chung S-L, Meffre S (2002) Laser fusion argon-40/argon-39 ages of Darwin impact glass. Meteorit Planet Sci 37:1555–1562

Lukanin OA, Kadik AA (2007) Decompression mechanism of ferric iron reduction in tektite melts during their formation in the impact process. Geochem Int 45:857–881

Ma P, Aggrey K, Tonzola C, Schnabel C, De Nicola P, Herzog GF, Wasson JT, Glass BP, Brown L, Tera F, Middleton R (2004) Beryllium-10 in Australasian tektites: Constraints on the location of the source crater. Geochim Cosmochim Acta 68:3883–3896

Maddock RH (1992) Effects of lithology, cataclasis and melting on the composition of fault-generated pseudotachylytes in Lewisian gneiss, Scotland. Tectonophysics 204:261–278

Magnien V, Neuville DR, Cormier L, Roux J, Hazemann JL, De Ligny D, Pascarelli S, Vickridge I, Pinet O, Richet P (2008) Kinetics and mechanisms of iron redox reactions in silicate melts: The effects of temperature and alkali cations. Geochim Cosmochim Acta 72:2157–2168

Martin Crespo T, Lozano Fernandez RP, Gonzalez Laguna R (2009) The fulgurite of Torre de Moncorvo (Portugal): description and analysis of the glass. Euro J Mineral 21:783–794

Masch L, Preuss E (1977) Das Vorkommen des Hyalomylonits von Langtang, Himalaya (Nepal) Neues Jahrb Min Abh 129:292–311

Mathews WH, Bustin RM (1984) Why do the Smoking Hills smoke? Can J Earth Sci 21:737–742

Matsubara K, Matsuda J (1991) Anomalous Ne enrichments in tektites. Meteoritics 26:217–220

Matsuda J, Maruoka T, Pinti DL, Koeberl C (1996) Noble gas study of a philippinite with an unusually large bubble. Meteorit Planet Sci 31:273–277

McCloy JS (2019) Frontiers in natural and un-natural glasses: An interdisciplinary dialogue and review. J Non-Cryst Solids X 4:100035

McMillan P, Wolf G, Lambert P (1992) A Raman spectroscopic study of shocked single crystalline quartz. Phys Chem Mineral 19:71–79

McPherson D, Pye LD, Fréchette VD, Tong S (1984) Microstructure of natural glasses. J Non-Cryst Solids 67:61–79

Meisel T, Koeberl C, Ford RJ (1990) Geochemistry of Darwin impact glass and target rocks. Geochim Cosmochim Acta 54:1463–1474

Melezhik VA, Filippov MM, Romashkin AE (2004) A giant Palaeoproterozoic deposit of shungite in NW Russia: genesis and practical applications. Ore Geol Rev 24:135–154

Melluso L, Conticelli S, D'Antonio M, Mirco NP, Saccani E (2004) Petrology and mineralogy of wollastonite- and melilite-bearing paralavas from the Central Apennines, Italy. Am Mineral 88:1287–1299

Melosh HJ (1989) Impact cratering: A geologic process. Oxford Monographs on Geology and Geophysics, Vol. 11, Oxford University Press, New York, NY

Melosh HJ (2011) Impact cratering. *In:* Planetary Surface Processes. Cambridge University Press, p 222–275

Melosh HJ (2013) The contact and compression stage of impact cratering. *In:* Osinski GR, Pierazzo E (eds) Impact Cratering: Processes and Products, John Wiley Sons, Ltd, Chichester, UK

Melosh HJ (2017) Impact geologists, beware! Geophys Res Lett 44:8873–8874

Melosh HJ, Artemieva N (2004) How does tektite glass lose its water? Lunar Planet Sci Confer XXXV:1723

Melson WG, Potts R (2002) Origin of reddened and melted zones in Pleistocene sediments of the Olorgesailie Basin, Southern Kenya Rift. J Archaeol Sci 29:307–316

Mizera J, Řanda Z, Kameník J (2016) On a possible parent crater for Australasian tektites: Geochemical, isotopic, geographical and other constraints. Earth Sci Rev 154:123–137

Molgaard JJ, Auxier JD, Giminaro AV, Oldham CJ, Cook MT, Young SA, Hall HL (2015) Development of synthetic nuclear melt glass for forensic analysis. J Radioanal Nucl Chem 304:1293–1301

Montanari A, Koeber C (2000) Distal ejecta and tektites. *In:* Impact Stratigraphy. Springer-Verlag, Berlin/Heidelberg, p 57–99

Morgan J, Lana C, Kearsley A, Coles B, Belcher C, Montanari S, Díaz-Martínez E, Barbosa A, Neumann V (2006) Analyses of shocked quartz at the global K-P boundary indicate an origin from a single, high-angle, oblique impact at Chicxulub. Earth Planet Sci Lett 251:264–279

Morgavi D, Laumonier M, Petrelli M, Dingwell DB (2022) Decrypting magma mixing in igneous systems. Rev Mineral Geochem 87:607-638

Mottana A (2004) X-ray absorption spectroscopy in mineralogy: Theory and experiment in the XANES region. *In:* Beran A, Libowitzky E (eds) Spectroscopic Methods in Mineralogy. Mineral Soc Great Britain Ireland, p 465–552

Müller O, Gentner W (1968) Gas content in bubbles of tektites and other natural glasses. Earth Planet Sci Lett 4:406–410

Nestola F, Mittempergher S, Di Toro G, Zorzi F, Pedron D (2010) Evidence of dmisteinbergite (hexagonal form of $CaAl_2Si_2O_8$) in pseudotachylyte: A tool to constrain the thermal history of a seismic event. Am Mineral 95:405–409

Neuville DR, de Ligny D, Henderson GS (2014) Advances in Raman spectroscopy applied to earth and material sciences. Rev Mineral Geochem 78:509–541

Neuville DR, Le Losq C (2022) Link between medium and long-range order and macroscopic properties of silicate glasses and melts. Rev Mineral Geochem 87:105-162

O'Keefe JA (1969) The microtektite data: Implications for the hypothesis of the lunar origin of tektites. J Geophys Res 74:6795–6804

O'Keefe JA (1994) Origin of tektites. Meteoritics 29:73–78

O'Keefe JA, Shute BE (1963) Origin of tektites. Science 139:1288–90

O'Keefe JA, Lowman PD, Dunning KL (1962) Gases in tektite bubbles. Science 137:228

Oberdorfer R (1905) Die vulkanischen Tuffe des Ries bei Nördlingen. Jahreshefte Verein Für Vaterändische Naturkunde Württemberg 61:1–40

Okuno M, Reynard B, Shimada Y, Syono Y, Willaime C (1999) A Raman spectroscopic study of shock-wave densification of vitreous silica. Phys Chem Mineral 26:304–311

Osinski GR, Pierazzo E (2012) Impact Cratering : Processes and Products. Blackwell Publishing Ltd

Osinski GR, Haldemann AF, Schwarcz HP, Smith JR, Kleindienst MR, Kieniewicz J, Churcher CS (2007) Impact glass at the Dakhleh Oasis, Egypt: evidence for a cratering event or large aerial burst? Lunar Planet Sci Confer 38:1346

Osinski GR, Grieve RAF, Tornabene LL (2012) Excavation and impact ejecta emplacement. *In:* Impact Cratering, John Wiley Sons, Ltd, Chichester, UK, p 43–59

Ottemann J (1966) Zusammensetzung und Herkunft der Tektite und Impaktite. *In:* Kosmochemie, Springer-Verlag, , Berlin/Heidelberg, p 409–444

Pacold JI, Lukens WW, Booth CH, Shuh DK, Knight KB, Eppich GR, Holliday KS (2016) Chemical speciation of U Fe, and Pu in melt glass from nuclear weapons testing. J Appl Phys 19:195102

Paquay FS, Goderis S, Ravizza G, Vanhaeck F, Boyd M, Surovell TA, Holliday VT, Haynes CV, Claeys P (2009) Absence of geochemical evidence for an impact event at the Bølling-Allerød/Younger Dryas transition. PNAS 106:21505–10

Parekh P, Semkow T, Torres M, Haines D (2006) Radioactivity in trinitite six decades later. J Environ Radioact 85:103–120

Pasek MA (2008) Rethinking early Earth phosphorus geochemistry. PNAS 105:853–858

Pasek MA (2017) Schreibersite on the early Earth: Scenarios for prebiotic phosphorylation. Geosci Frontiers 8:329–335

Pasek M, Block K (2009) Lightning-induced reduction of phosphorus oxidation state. Nat Geosci 2:553–556

Pasek MA, Pasek VD (2018) The forensics of fulgurite formation. Mineral Petrol 112:185–198

Pasek MA, Block K, Pasek V (2012) Fulgurite morphology: A classification scheme and clues to formation. Contrib Mineral Petrol 164:477–492

Peretyazhko IS, Savina EA, Khromova EA, Karmanov NS, Ivanov AV (2018) Unique clinkers and paralavas from a new Nyalga combustion metamorphic complex in Central Mongolia: Mineralogy, geochemistry, and genesis. Petrology 26:181–211

Piepjohn K, Estrada S, Reinhardt L, von Gosen W, Andruleit H (2007) Origin of iron-oxide and silicate melt rocks in Paleogene sediments of southern Ellesmere Island, Canadian Arctic Archipelago, Nunavut. Can J Earth Sci 44:1005–1013

Pierazzo E, Melosh HJ (2012) Environmental Effects of Impact Events. *In:* Impact Cratering, John Wiley Sons, Ltd, Chichester, UK, p 146–156

Pierazzo E, Kring DA, Melosh HJ (1998) Hydrocode simulation of the Chicxulub impact event and the production of climatically active gases. J Geophys Res: Planets 103(E12):28607–28625

Pinter N, Scott AC, Daulton TL, Podoll A, Koeberl C, Anderson RS, Ishman SE (2011) The Younger Dryas impact hypothesis: A requiem. Earth Sci Rev 106:247–264

Potuzak M, Nichols AR L, Dingwell DB, Clague DA (2008) Hyperquenched volcanic glass from Loihi Seamount, Hawaii. Earth Planet Sci Lett 270:54–62

Povenmire H, Cornec J (2015) The 2014 Report on the Belize tektite strewn field. *In:* 46th Lunar Planet Sci Confer, p 1132

Povenmire H, Harris R, Cornec J (2011) The new Central American tektite strewn field. *In:* 42nd Lunar Planet Sci Confer, p 1224

Pratesi G (2009) Impact diamonds: Formation, mineralogical features and cathodoluminescence properties. *In:* Cathodoluminescence and its Application in the Planetary Sciences. Gucsik A (ed) Springer, Berlin, p 61–86

Pratesi G, Viti C, Cipriani C, Mellini M (2002) Silicate–silicate liquid immiscibility and graphite ribbons in Libyan desert glass. Geochim Cosmochim Acta 66:903–911

Rakov VA, Uman MA (2003) Lightning : physics and effects. Cambridge University Press

Rattigan JH (1967) Phenomena about Burning Mountain, Wingen, NSW. Aust J Sci 30:183–184

Reimold WU, Gibson RL (2005) "Pseudotachylites" in Large Impact Structures. *In:* Impact Tectonics, Springer-Verlag, Berlin/Heidelberg, p 1–53

Reimold WU, Jourdan F (2012) Impact! – Bolidea, craters and catastrophes. Elements 8:19–24

Reimold WU, Koeberl C (2014) Impact structures in Africa: A review. J Afr Earth Sci 93:57–175

Rietmeijer FJ M, Karner JM, Nuth JA, Wasilewski PJ (1999) Nanoscale phase equilibrium in a triggered lightning-strike experiment. Euro J Mineral 11:181–186

Rocchia R, Robin E, Fröhlich F, Meon H, Froget L, Diemer E (1996) L'origine des verres du désert libyque: Un impact météoritique. C R Acad Sci Ser II: Sci Terre Planets 322:839–845

Roedder E (1978) Silicate liquid immiscibility in magmas and in the system K_2O–FeO–Al_2O_3–SiO_2: an example of serendipity. Geochim Cosmochim Acta 42:1597–1617

Roedder E (1992) Fluid inclusion evidence for immiscibility in magmatic differentiation. Geochim Cosmochim Acta 56:5–20

Rogers GS (1918) Baked shale and slag formed by the burning of coal beds. US Government Printing Office

Ron H, Kolodny Y (1992) Paleomagnetic and rock magnetic study of combustion metamorphic rocks in Israel. J Geophys Res 97(B5):6927

Roperch P, Gattacceca J, Valenzuela M, Devouard B, Lorand JP, Arriagada C, Rochette P, Latorre C, Beck P (2017) Surface vitrification caused by natural fires in Late Pleistocene wetlands of the Atacama Desert. Earth Planet Sci Lett 469:15–26

Ross CS (1948) Optical properties of glass from Alamogordo, New Mexico. Am Mineral 33:360–362

Rossano S, Balan E, Morin G, Bauer J-P, Calas G, Brouder C (1999) ^{57}Fe Mössbauer spectroscopy of tektites. Phys Chem Mineral 26:530–538

Rost R (1964) Surfaces of and inclusions in moldavites. Geochim Cosmochim Acta 28:931–936

Rowan LR, Ahrens TJ (1994) Observations of impact-induced molten metal-silicate partitioning. Earth Planet Sci Lett 122:71–88

Schmieder M, Kennedy T, Jourdan F, Buchner E, Reimold WU (2018) A high-precision $^{40}Ar/^{39}Ar$ age for the Nördlinger Ries impact crater, Germany, and implications for the accurate dating of terrestrial impact events. Geochim Cosmochim Acta 220:146–157

Schnetzler CC (1970) The lunar origin of tektites: RIP. Meteoritics 5:221–222

Schnetzler C, Pinson W (1964) A report on some recent major element analyses of tektites. Geochim Cosmochim Acta 28:793–806

Schreiber HD, Minnix LM, Balazs GB B (1984) The redox state of iron in tektites. J Non-Cryst Solids 67: 349–359

See TH, Wagstaff J, Yang V, Hörz F, McKay GA (1998) Compositional variation and mixing of impact melts on microscopic scales. Meteorit Planet Sci 33:937–948

Seebaugh WR, Strauss AM (1984) A cometary impact model for the source of Libyan Desert glass. J Non-Cryst Solids 67:511–519

Senftle FE, Thorpe AN, Grant JR, Hildebrand A, Moholy–Nagy H, Evans BJ, May L (2000) Magnetic measurements of glass from Tikal, Guatemala: Possible tektites. J Geophys Res: Solid Earth 105(B8):18921–18925

Sengupta P (2017) Natural glasses under extreme conditions. *In:* Materials Under Extreme Conditions. Tyagi AK, Banerjee S (eds), Elsevier, Amsterdam, p 235–258

Shelby JE (2005) Introduction to Glass Science and Technology. Royal Society of Chemistry

Shoemaker EM, Chao EC T (1961) New evidence for the impact origin of the Ries Basin, Bavaria, Germany. J Geophys Res 66:3371–3378

Sigurdsson H, D'Hondt S, Arthur MA, Bralower TJ, Zachos JC, Van Fossen M, Channel JE (1991) Glass from the Cretaceous/Tertiary boundary in Haiti. Nature 349:482–487

Skinner HC W, Jahren AH (2003) Biomineralization. *In:* Biogeochemistry. Treatise on Geochemistry. Volume 8. Schleshinger WH (ed) Elsevier, Amsterdam, p 117–184

Smit J, Klaver G (1981) Sanidine spherules at the Cretaceous–Tertiary boundary indicate a large impact event. Nature 292:47–49

Sokol E, Volkova N, Lepezin G (1998) Mineralogy of pyrometamorphic rocks associated with naturally burned coal-bearing spoil-heaps of the Chelyabinsk coal basin, Russia. Euro J Mineral 10:1003–1014

Sokol EV, Volkova NI, Stracher G (2007) Combustion metamorphic events resulting from natural coal fires. Rev Eng Geol 162:373–378Sokol E, Novikov I, Zateeva S, Vapnik Y, Shagam R, Kozmenko O (2010) Combustion metamorphism in the Nabi Musa dome: new implications for a mud volcanic origin of the Mottled Zone, Dead Sea area. Basin Res 22:414–438

Sokol EV, Kokh SN, Vapnik Y, Thiery V, Korzhova SA (2014) Natural analogs of belite sulfoaluminate cement clinkers from Negev Desert, Israel. Am Mineral 99:1471–1487

Spray JG (1992) A physical basis for the frictional melting of some rock-forming minerals. Tectonophysics 204:205–221

Spray JG (2010) Frictional melting processes in planetary materials: from hypervelocity impact to earthquakes. Ann Rev Earth Planet Sci 38:221–254

Stöffler D (1971) Progressive metamorphism and classification of shocked and brecciated crystalline rocks at impact craters. J Geophys Res 76:5541–5551

Stöffler D, Grieve RAF (2007) Impactites. *In:* Fettes DJ Desmons J (eds) Metamorphic rocks: a classification and glossary of terms, Cambridge University Press, p 82–92

Stöffler D, Langenhorst F (1994) Shock metamorphism of quartz in nature and experiment: I Basic observation and theory. Meteoritics 29:155–181

Stöffler D, Artemieva N a., Pierazzo E (2002) Modeling the Ries–Steinheim impact event and the formation of the moldavite strewn field. Meteorit Planet Sci 37:1893–1907

Stöffler D, Artemieva NA, Wünnemann K, Reimold WU, Jacob J, Hansen BK, Summerson IA (2013) Ries crater and suevite revisited-Observations and modeling Part I: Observations. Meteorit Planet Sci 48:515–589

Stöffler D, Hamann C, Metzler K (2018) Shock metamorphism of planetary silicate rocks and sediments: Proposal for an updated classification system. Meteorit Planet Sci 53:5–49

Stoppa F, Rosatelli G, Cundari A, Castorina F, Woolley AR (2005) Comment on Melluso et al. (2003): Reported data and interpretation of some wollastonite- and melilite-bearing rocks from the Central Apennines of Italy. Am Mineral 90:1919–1925

Storzer D, Koeberl C (1991) Uranium and zirconium enrichments in Libyan desert glass: zircon baddeleyite, and high temperature history of the glass. Lunar Planet Sci 22:1345

Storzer D, Wagner GA (1977) Fission track dating of meteorite impacts. Meteoritics 12:368–369

Stracher GB (ed) (2007) Geology of coal fires: case studies from around the world. Rev Eng Geol 18, Geol Soc Am

Sugiura H, Ikeda R, Kondo K, Yamadaya T (1997) Densified silica glass after shock compression. J Appl Phys 81:1651–1655

Susman S, Volin KJ, Price DL, Grimsditch M, Rino JP, Kalia RK, Vashishta P, Gwanmesia G, Wang Y, Liebermann RC (1991) Intermediate-range order in permanently densified vitreous SiO_2: A neutron-diffraction and molecular-dynamics study. Phys Rev B 43:1194–1197

Svensen H, Dysthe DK, Bandlien EH, Sacko S, Coulibaly H, Planke S (2003) Subsurface combustion in Mali: Refutation of the active volcanism hypothesis in West Africa. Geology 31:581

Swaenen M, Stefaniak EA, Frost R, Worobiec A, Van Grieken R (2010) Investigation of inclusions trapped inside Libyan desert glass by Raman microscopy. Anal Bioanal Chem 397:2659–2665

Switzer G, Melson WG (1972) Origin and composition of rock fulgurite glass. Smithsonian Contrib Earth Sci 9:47

Tancredi G, Ishitsuka J, Schultz PH, Harris RS, Brown P, Revelle DO, Antier K, Pichon AL, Rosales D, Vidal E, Varela ME (2009) A meteorite crater on Earth formed on September 15 2007: The Carancas hypervelocity impact. Meteorit Planet Sci 44:1967–1984

Thorpe AN, Senftle FE, Cuttitta F (1963) Magnetic and chemical investigations of iron in tektites. Nature 197:836–840

Thy P, Segobye AK, Ming DW (1995) Implications of prehistoric glassy biomass slag from east-central Botswana. J Archaeolog Sci 22:629–637

Tompson A, Bruton C, Pawloski G, Smith D, Bourcier W, Shumaker D, Kersting A, Carle S, Maxwell R (2002) On the evaluation of groundwater contamination from underground nuclear tests. Environ Geol 42:235–247

Tompson AF B, Hudson GB, Smith DK, Hunt JR (2006) Analysis of radionuclide migration through a 200-m Vadose zone following a 16-year infiltration event. Adv Water Resour 29:281–292

Tulloch AJ, Campbell JK (1993) Clinoenstatite-bearing buchites possibly from combustion of hydrocarbon gases in a major thrust zone: Glenroy Valley, New Zealand. J Geol 101:404–412

Uman MA (1964) The peak temperature of lightning. J Atmos Terr Phys 26:123–128

Uman MA, Krider EP (1989) Natural and artificially initiated lightning. Science 246:457–64

Urey HC (1955) On the origin of tektites. PNAS 41:27–31

Urey HC (1963) Cometary Collisions and Tektites. Nature 197:228–230

Van Hoesel A, Hoek WZ, Pennock GM, Drury MR (2014) The younger dryas impact hypothesis: A critical review. Quat Sci Rev 83:95–114

Vapnik Y, Sharygin VV, Sokol EV, Shagam R (2007) Paralavas in a combustion metamorphic complex: Hatrurim Basin, Israel. Rev Eng Geo 18:1–21

Veksler IV (2004) Liquid immiscibility and its role at the magmatic–hydrothermal transition: a summary of experimental studies. Chem Geol 210:7–31

Vogel W (1994) Glass Chemistry, Springer Berlin Heidelberg

Volovetsky MV, Rusakov VS, Chistyakova NI, Lukanin OA (2008) Mössbauer study of tektites. Hyperfine Interact 186: 83–88

Wallace C, Bellucci JJ, Simonetti A, Hainley T, Koeman EC, Burns PC (2013) A multi-method approach for determination of radionuclide distribution in trinitite. J Radioanal Nucl Chem 298:993–1003

Wasserman A, Melosh H (2001) Chemical reduction of impact processed materials. 32nd Ann Lunar Planet Sci Confer #2037

Weeks RAA, Underwood JR, Giegengack R (1984) Libyan Desert glass: A review. J Non-Cryst Solids 67:593–619

Weidinger JT, Korup O (2009) Frictionite as evidence for a large Late Quaternary rockslide near Kanchenjunga, Sikkim Himalayas, India — Implications for extreme events in mountain relief destruction. Geomorphology 103:57–65

Weidinger JT, Korup O, Munack H, Altenberger U, Dunning SA, Tippelt G, Lottermoser W (2014) Giant rockslides from the inside. Earth Planet Sci Lett 389:62–73

Wilding M, Webb S, Dingwell D, Ablay G, Marti J (1996a) Cooling rate variation in natural volcanic glasses from Tenerife, Canary Islands. Contrib Mineral Petrol 125:151–160

Wilding M, Webb S, Dingwell DB (1996b) Tektite cooling rates: Calorimetric relaxation geospeedometry applied to a natural glass. Geochim Cosmochim Acta 60:1099–1103

Wilke M, Farges F, Petit P-E, Brown GE, Martin F (2001) Oxidation state and coordination of Fe in minerals: An Fe K- XANES spectroscopic study. Am Mineral 86:714–730

Wilke M, Partzsch GM, Bernhardt R, Lattard D (2004) Determination of the iron oxidation state in basaltic glasses using XANES at the K-edge. Chem Geol 213:71–87

Wilke M, Farges F, Partzsch GM, Schmidt C, Behrens H (2007) Speciation of Fe in silicate glasses and melts by in-situ XANES spectroscopy. Am Mineral 92:44–56

Wittke JH, Weaver JC, Bunch TE, Kennett JP, Kennett DJ, Moore AM, Hillman GC, Tankersley KB, Goodyear AC, Moore CR, Daniel IR (2013) Evidence for deposition of 10 million tonnes of impact spherules across four continents 12,800 y ago. PNAS 110:E2088–E2097

Zähringer J, Gentner W (1963) Radiogenic and atmospheric argon content of tektites. Nature 199:58

Reviews in Mineralogy & Geochemistry
Vol. 87 pp. 1015-1038, 2022
Copyright © Mineralogical Society of America

22

Silicate Glasses and Their Impact on Humanity

Randall E. Youngman

Science and Technology Division
Corning Incorporated
Corning, NY 14831, U.S.A.

youngmanre@corning.com

INTRODUCTION

Glasses, especially those in the silicate family, are ubiquitous throughout modern culture and technology. Given the influence of glasses on modern civilizations, the idea of a Glass Age has been proposed, drawing inspiration from the classical three-age system of the Stone, Bronze and Iron Ages. As will be highlighted in this brief review, silicate glasses have truly been transformative throughout human history; from natural glasses used over 75,000 years ago, followed by the discovery of soda-lime silicates some 70,000 years later, to recent centuries where glasses were designed and made for specific technological applications. Silicate glasses touch on every aspect of human existence, and through their use, have impacted advancements in culture, technology, science, and the overall existence of our species. In each case, these silicate glasses have profoundly benefitted humankind, and as will be shown, it is indeed likely that we are currently enjoying the benefits of living in the Glass Age.

Historians, archeologists and anthropologists spend tremendous effort understanding human civilization and the major societal developments throughout the history of our species. This has led to the fairly widespread adoption of labeling distinct eras in human history, categorized using the three-age system solidified in the 19th century by Christian Jurgensen Thomsen, who developed a scientific basis for subdividing prehistory into the Stone, Bronze and Iron Ages (Trigger 2006). These chronological eras represent the main periods of human development, defined largely by the materials featured prominently in the tools, weapons and culture of those eras. While mainly a European-centric view of cultural anthropology, examples of these materials on cultures around the globe have been identified, providing a strong connection between these specific materials and most cultures, and expanding the three-age classification to include the archeology of the Ancient Near East.

The Stone Age, ending sometime between 8700 BCE and 2000 BCE, was a period of human history where stone was the most widely used material in most aspects of civilization. It is likely that the Stone Age existed for more than three million years (Harmand et al. 2015). Most prominent was the use of stone in tools and weapons, allowing humans to forever alter their interaction with the natural world and other peoples. Stone was featured in architecture, as evidenced by some examples of free-standing stone structures, and was the primary artistic medium during this timeframe. During the Stone Age, art was in the form of petroglyphs and rock paintings. The former involved carving images into stone, for example exposed rock in caves. Painting also meant using stone as the canvas, as in cave paintings discovered in France, Spain and India. The Bronze Age followed, roughly defined as the period of time from 2800 BCE to 1200 BCE. This was an era where metallurgy supplanted the widespread use of stone in human civilization. Bronze, which was stronger and more durable than copper, which was probably the first metal mined and smelted by humans, became the material of choice for tools, weapons and other aspects of culture. This led to significant advances in transportation,

1529-6466/22/0087-0022$05.00 (print)
1943-2666/22/0087-0022$05.00 (online)

http://dx.doi.org/10.2138/rmg.2022.87.22

aided by carts, chariots and sailing vessels, new methods for agriculture and a new medium for art. Eventually, the Iron Age and our use of iron and steel displaced the use of Bronze, and humanity entered the last of the classical ages. The Iron Age, from roughly 1200 BCE to 500 BCE, as with Stone and Bronze Ages before, marked a historical era where a single material had a widespread and highly impactful influence on human civilizations.

While the three-age system, which is roughly illustrated in Figure 1, does indeed serve as a mechanism to divide human history into three distinct time periods, with a recognizable chronology, it is most certainly deficient when describing the past several thousand years, which by all accounts are after the end of the Iron Age. One might ask then about the other "Ages" since the end of the Iron Age, and humankind has certainly offered up a variety of options to fill this gap in our history. We have the Age of Enlightenment (18th century Europe), the Middle Ages, the Dark Ages, the Renaissance, the Age of Discovery or Exploration (16th to 18th centuries), the Gilded Age (late 19th century US), the Machine Age (19th and 20th centuries), the Space Age (post 1957), the Information Age (post 1971) and others. All of these resonate with key developments in technology and culture, but unlike the original three-age system, they fail to adequately encompass a broader view of entire cultures. Some of these, including for example the Atomic Age, mark a cultural period defined by the prominence of a specific item or technology, but again lack a more comprehensive view of human civilization.

So while none of the more recent "Ages" have truly captured the essence of humanity since the end of the Iron Age, future historians will certainly identify other key periods in time which are defined by the widespread adoption of a particular material, like those in the aforementioned ages of Stone, Bronze and Iron. However, as will be shown in the following historical perspective on silicate glasses, we can be certain that we are presently living in the Glass Age—a period of human existence where glass is the transformative material, widely used in cultures of the past several hundred years and perhaps even longer. Glass is now the material of choice for much of modern human endeavor and certainly represents a prominent technological advancement in human civilization. Amazingly, the omission of an age of glass from Thomsen's advocation for the three-age system in the early 19th century was already criticized by some of his contemporaries (Trigger 2006), and thus our bold assertion of a Glass Age is not just a 21st century construct.

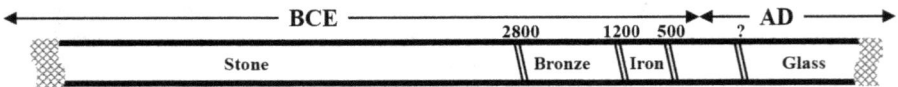

Figure 1. Schematic showing estimated timelines for the three-age system and a proposed addition based on glass. The Glass Age is a recent construct and may not yet have a definitive starting point. **Hatched regions** to the left and right of the timeline denote these currently undefined starting and ending points.

HISTORY AND IMPACT OF SILICATE GLASSES

Glass before and during the Age of Antiquity

Before humans learned how to make glass, ancient peoples during the Stone Age (predating ~3300 BCE) discovered natural sources of glass. Obsidian, pumice (e.g., ignimbrite), fulgurites and tektites are some of the naturally occurring glasses found on Earth (Fig. 2), and in particular, obsidian was used perhaps as early as 75,000 BCE to fashion weapons and tools (Kolb and Kolb 1988). These ancient civilizations found that obsidian could easily be chipped and flaked to create sharp edges, resulting in knives (Fig. 2a), arrowheads and other implements of incredible utility to those civilizations. Obsidian and pumice are silicate glasses with volcanic origin, while fulgurites and tektites are silicates generated by lightning

Figure 2. Examples of natural silicate glasses which were used long before the Glass Age, and in some cases, were prominent in the cultures of Stone Age civilizations: **(a)** assortment of obsidian blades; CMOG 62.7.1 A-E. Gift of the Santa Barbara Museum of Natural History. Image licensed by The Corning Museum of Glass, Corning, NY (www.cmog.org) under CC BY-NC-SA 4.0; **(b)** pumice, "Pumice" by Mauro Cateb is used without alteration and is licensed under CC BY-SA 3.0, **(c)** large sand fulgurites, CMOG 2009.7.1. Image licensed by The Corning Museum of Glass, Corning, NY (www.cmog.org) under CC BY-NC-SA 4.0; and **(d)** moldavite (tektite), CMOG 2013.7.2. Gift of Robert W. Kincheloe. Image licensed by The Corning Museum of Glass, Corning, NY (www.cmog.org) under CC BY-NC-SA 4.0.

(high heat) and meteoric impact (heat and pressure), converting silica-rich materials into glass. Pumice (Fig. 2b) appears to have found ancient uses as a grinding material, as well as something which could be carved into simple shapes—an artistic expression manifested in glass! These silicate glasses have existed as long as any other material on Earth, and provide important examples of how glass already has impacted human existence.

Making glass, rather than harvesting this material from natural sources, was a development in human history that overlapped with the Bronze Age (3300—1200 BCE), a period of time in which civilizations were exploring new materials for tools and other uses. There are varied accounts on how glass making was discovered, but one of the more common stories, told by the ancient Roman historian Pliny and probably too fanciful to be entirely accurate (Ellis 1993), suggests that Phoenician merchants built fires on sandy beaches, using sodium-rich rock to support their cooking implements (Pliny). The resulting combination of heat, sand and fluxes like sodium, generated a new liquid that was likely the first example of soda-lime silicate glass (Harden 1933). Another suggestion as to the origin of glass-making is based on the close development of glass with copper mining and processing in and around ancient Egypt (Macbeth 1908). The oldest glass artifacts, found exclusively in Egypt, are typically blue in color, an indication that they contained substantial amounts of copper—see for example the oldest glass in the British Museum collections (Fig. 3a). The belief is that copper being mined in the Sinai area was processed over fires resulting in copper slags, which were further processed with hotter fires and/or fluxing with alkali.

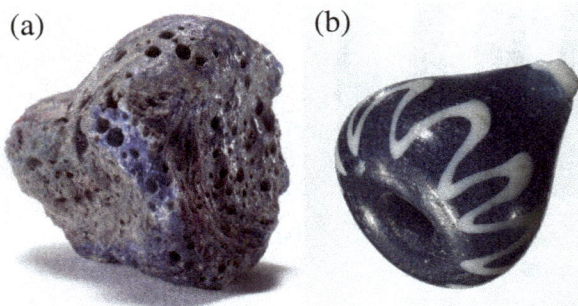

(a) (b)

Figure 3. Some of the oldest examples of synthetic glass: (**a**) a blue glass from Egypt, dated to 2050 BCE (©The Trustees of the British Museum), and (**b**) an ancient Egyptian glass bead from 1400–1300 BCE. CMOG 54.1.146-1. Image licensed by The Corning Museum of Glass, Corning, NY (www.cmog.org) under CC BY-NC-SA 4.0.

Although we cannot be certain of the details of these stories, sometime in the historical period between 5000 and 3000 BCE, people in the eastern Mediteranean somehow discovered a way to make glass. Egyptian artifacts, dating back to before 3000 BCE show glass as a glaze, with entirely glass artifacts being dated back to ~2500 BCE (Maloney 1968). Egypt and nearby cultures, including the Phoenicians who were prominent in the stories of Pliny, were the center of glass making for much of the early history of silicate glasses, but the newly generated materials were difficult to make, and were exclusive to the ruling and wealthy classes. Glass was originally used in decorative form, for example as beads (Fig. 3b) and figurines, but with the use of clay and sand molds, vessels like cups and vases could be fashioned from glass. Already, silicate glasses were being used in a number of growing aspects of human civilization.

The next major milestone in the history of silicate glasses was the invention of glass-blowing, atttributed to glass artisans in or around Syria in the 1st or 2nd centuries BCE (Phillips 1948). The Roman Empire developed the technique and because glass blowing was much faster and less costly than glass casting, brought glass to the masses. As a consequence of Roman glass advances, cups, dishes and other implements became widely available (Fig. 4a), and glass was no longer restricted to the wealthy (Zerwick 1990). The more elaborate glass articles, for example vases (Fig. 4b) and early windows (Fig. 4c), were still considered a luxury, but demonstrated the growing utilization of silicate glasses in human civilization. These advances in glass production and widespread availability throughout the Roman Empire solidified Rome, and regions conquered by Rome, as the center of glass for the first four centuries AD. In addition to further refinements in glass production, namely more decorative vases for the wealthy and production of some of the first, relatively clear silicate glasses, the first written records on window glass suggest flat, somewhat clear glass for windows (e.g., Fig. 4c) were made in the late 3rd and early 4th centuries AD (Phillips 1948). It was also during this time that significant advances in decolorization of glass were made, incorporating specific elements like antimony and manganese to ultimately alter the redox chemistry of glass and remove much of the color originating from other multivalent ions found in common raw materials (Bidegaray 2019).

After the fall of the Roman Empire, there is very little known about silicate glass technology in Europe, at least until the 11th century. However, other regions of glass production continued to flourish, including especially the Byzantine Empire. Here, skilled artisans persisted to advance the artform of colored glass, and sometime in the 7th century AD, stained glass as

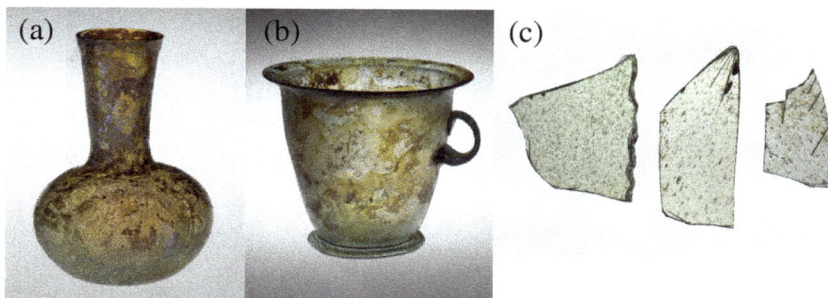

Figure 4. Examples of blown glass from the Roman Empire: **(a)** blown Roman bottle from 200 to 500 AD, CMOG 54.1.75. Gift of I. C. Elston, Jr. Image licensed by The Corning Museum of Glass, Corning, NY (www.cmog.org) under CC BY-NC-SA 4.0; **(b)** blown Roman mug or measure from 1st or 2nd century AD, CMOG 52.1.56. Image licensed by The Corning Museum of Glass, Corning, NY (www.cmog.org) under CC BY-NC-SA 4.0; and **(c)** fragments of blown glass window panes from 300 to 400 AD, CMOG 57.1.10 A–C. Gift of Donald B. Harden. Image licensed by The Corning Museum of Glass, Corning, NY (www.cmog.org) under CC BY-NC-SA 4.0.

we know it was developed in Turkey (Maloney 1968). Stained glass windows were used in churches in Constantinople, eventually spreading to France and England, where such windows were considered required decoration and a necessity in any church (Phillips 1948). Often, these brilliantly colored glass windows were used to teach Christianity (Zerwick 1990), enabling the education of the masses. This particular use of silicate glasses, perhaps overlooked in terms of key milestones in glass history, certainly cemented this material as a key architectural, artistic and educational material, especially when considering the enormous impact and importance of organized religion during the past two thousand years.

Silicate glasses during the Middle Ages

The early Middle Ages witnessed some of the most important and lasting developments in the relationship between humanity and silicate glasses. The next major development in silicate glasses, and again a turning point in how these materials advanced human civilization, was the mastering of glass making and processing by the Venetians (Fig. 5). Clear silicate glass could be made using higher purity (i.e., cleaner sand) or cleansed raw materials like potash, and with continued addition of certain decoloring agents like manganese (Verità 2013), and it was the Venetian glass industry which solidified our fascination with clear glass. By the 13th century, Venice was the new center of glass-making in Europe, and Venetian glass was so highly valued that in addition to safety reasons, glass artisans were moved to the island of Murano to protect the invaluable trade secrets of this industry (Chopinet 2019). Venetian glass, famous for its clarity and ability to be formed into a variety of shapes and figures, also eventually led to an explosion in glasses used for optical applications. Window glass, which had been only somewhat transparent before this time, was now made with higher quality and transparency, letting natural light inside buildings but keeping weather out. The first corrective lenses for sight (spectacles) were said to have come from lenses made using fine Venetian glass (Murube 2014), and this allowed those with poor eyesight to once again read and write. Imagine a learned person in the early Middle Ages no longer being able to read as they aged due to degradation of their eyesight. As we now take for granted, corrective eyewear was indeed a miracle, made possible with glass.

The lens industry spread throughout Europe and eventually led to other key scientific inventions. Magnifying lenses were common by the 1400s, and in the late 1500s, the compound microscope was invented and further refined by Dutch scientists. A different Dutch lens maker, Hans Lippershey, put glass lenses in a tube and invented the telescope in 1608. As a sign of the growing interest in optics, when attempting to secure a patent for his telescope,

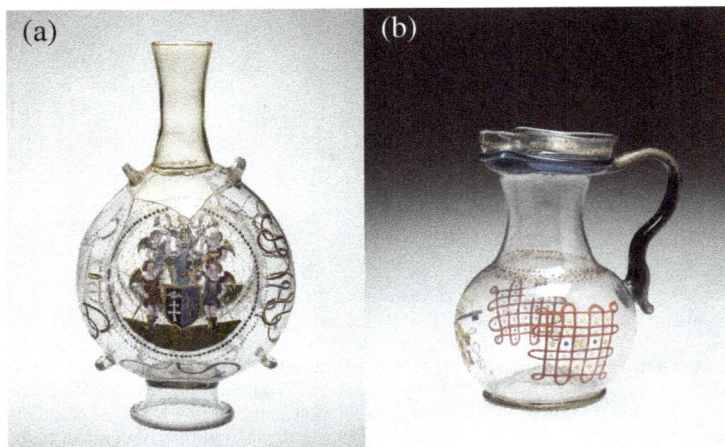

Figure 5. Examples of elaborate blown and decorated glass from Venice: (**a**) an early pilgrim flask (likely late 15th century) with an Italian Bishop's coat of arms, CMOG 59.3.19. Image licensed by The Corning Museum of Glass, Corning, NY (www.cmog.org) under CC BY-NC-SA 4.0; and (**b**) jug with a Medici Pope coat of arms (early 16th century), CMOG 2005.3.28. Purchased with funds from the Houghton Endowment Fund. Image licensed by The Corning Museum of Glass, Corning, NY (www.cmog.org) under CC BY-NC-SA 4.0.

Lippershey's claim was rejected on the grounds that two other Dutch lens makers had also claimed similar inventions (Nascimento and Zanotto 2016). Galileo famously made his own telescope using Italian glass lenses, followed soon by Kepler in Germany (1611). The importance of glass lenses cannot be overstated, having forever altered science and technology since these early examples. Bolt provides an excellent historical review of glass in the context of telescopes and microscopes (Bolt 2017). Figure 6 contains examples of various implements which benefitted from silicate glass optical elements, even before optical properties of glasses were fully understood (see discussion in the next section). The simple botanical microscope in Fig. 6a, highly dependent on glass lenses, allowed one to study the microscopic details of our natural world. French scientists relied on accurate models of the human eye in order to study and correct vision issues, as reflected in the design of an optical model of the eye in Figure 6b. This device incorporated different silicate glass lenses to reproduce the effect of near sightedness and other vision problems. One of the major scientific disciplines to benefit from glass development in the Middle Ages was chemistry, particularly in designing laboratory devices and vessels like beakers, flasks and other glassware. An example of an early graduated cylinder (Fig. 6c) highlights some of the benefits of silicate glasses as the material of choice for these uses. Clear glass allows one to see the contents, something humanity has cherished for millenia. The ability to make glass containers with high precision enabled measurements to be made, as shown by the volumetric markings of this vessel. The precision of glassware will be an important theme in some of the later 19th and 20th century developments in glass. Furthermore, glass in the 17th century featured prominently in the majority of major scientific instrument innovations of the time: the telescope and microscope (both discussed above), as well as the thermometer, barometer and airpump (Bolt 2017). All leveraged our fascination with the optical properties of glasses and prompted the desire for a deeper understanding of glass composition and properties.

Another key milestone in the development of silicate glasses, and perhaps one possible indicator of the imminent Glass Age, was the invention and eventual manufacturing of lead silicate glass. Lead containing glasses were known already, and evidence for their existence can be found throughout medieval artifacts in Europe (Mecking 2013). Some of these silicates were even considered "high-lead glass," having 70 to 75% lead, which substantially lowered

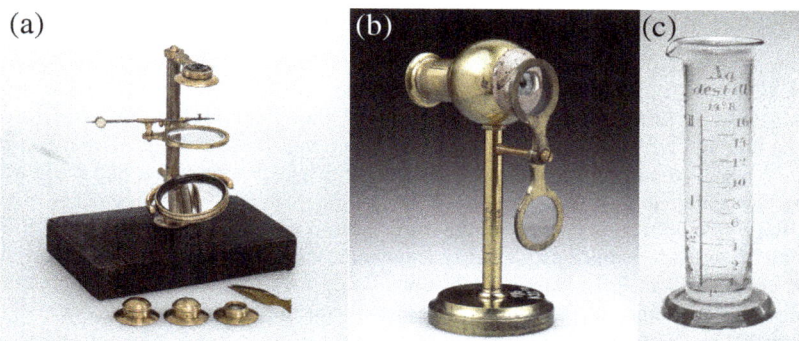

Figure 6. Some examples of glass in use for scientific applications. While the telescope may be the most well known example from the Middle Ages, other uses included: **(a)** lenses in microscopes, like that in this Ellis-type Aquatic/Botanical simple microscope (circa 1770), CMOG 2016.8.1. Image licensed by The Corning Museum of Glass, Corning, NY (www.cmog.org) under CC BY-NC-SA 4.0; **(b)** medical devices like this Optical Model of the Eye, allowing one to study vision ailments (1800–1899), CMOG 2004.3.40. Image licensed by The Corning Museum of Glass, Corning, NY (www.cmog.org) under CC BY-NC-SA 4.0; and **(c)** an example of a graduated cylinder from the early 19th century, CMOG 72.3.185. Gift of Benedict Kolthoff. Image licensed by The Corning Museum of Glass, Corning, NY (www.cmog.org) under CC BY-NC-SA 4.0.

their melting temperatures. In patenting a new lead glass in 1674, George Ravenscroft was credited with fabricating a new type of glass that was much more brilliant than even the finest glasses of Murano and other centers of glassmaking, and something that was relatively easy to work with. Although the actual invention of such a glass may have occurred well before Ravenscroft (Brain and Brain 2016), he undoubtedly improved production and quality of a lead silicate, which resulted in an explosion of fine glass for artististic and daily use, as shown by the example of fine crystal goblets in Figure 7. Ravenscroft's invention also demonstrated our ability to alter the composition of silicate glasses, and to leverage this incredible flexibility to derive a material having specific and favorable properties.

Figure 7. English lead silicate glassware from the early 18th century. From left to right: Wineglass (about 1700–1720). CMOG 54.2.9; Wineglass (about 1730–1740). CMOG 79.2.118. Bequest of Jerome Strauss; Ale glass (about 1700–1710). CMOG 63.2.2; Wineglass (about 1720). CMOG 79.2.122. Bequest of Jerome Strauss; Wineglass (1700–1715). CMOG 55.2.3; Goblet with royal arms and monogram of Queen Anne (probably 1707). CMOG 2005.2.8; Wineglass (about 1720). CMOG 79.2.77. Gift of The Ruth Bryan Strauss Memorial Foundation; Wineglass (about 1720). CMOG 79.2.129. Gift of The Ruth Bryan Strauss Memorial Foundation. Image licensed by The Corning Museum of Glass, Corning, NY (www.cmog.org) under CC BY-NC-SA 4.0.

Silicate glasses in modern and contemporary history

If one were to trace the impact of silicate glasses on humanity up to this point, this amazing material was used for tools, art, architecture (windows), science and medicine, and as containers for foodstuffs, perfumes and beverages. Already, glass had profoundly and irreversibly benefitted human civilization, and perhaps pushed us into the Glass Age. The modern Glass Age really began in earnest in the 19th century, during which significant and rapid advances extended the utility and impact of silicate glasses on many different segments of society. This shifting paradigm can be traced to the introduction of science to glassmaking, effectively converting glass from a novelty to a highly engineered material (Graham 2001). Starting with mid-19th century advances in glass compositions for optics to contemporary uses in consumer electronics and medicine, the last two centuries have witnessed a truly amazing era in glass, and as outlined below, led to a wide variety of lasting cultural and technological advances.

Optical materials and the start of the Glass Age. The development and production of glasses for optical applications may have been the first indication of human civilization entering the Glass Age. Up until the early 19th century, glass in optics relied on simple compositions, though often with elegant applications as highlighted above. Beginning in the early 1800s, scientists actively started to seek solution for the optical glasses and systems in use at that time. One of the key issues was optical aberration, which had been addressed through use of multiple glass lenses with different optical properties. In fact, the widespread production of lead silicates by Ravenscroft (above) provided the flint glass to accompany the crown glass lens in achromatic optical systems. Unfortunately, this solution still suffered from poor quality (inhomogenous) glass with striae, and was also expensive and cumbersome. These issues were a hindrance to the growing demand for optical glass in the 18th and 19th centuries.

The beginnings of the Glass Age may be traced back to 1805, when Pierre Louis Guinand introduced stirring of glass melts to increase homogeniety (Hartmann et al. 2010). Shortly thereafter, Joseph Fraunhofer became the first to address the deficiencies in optical glasses on a manufacturing scale. His work in Bavaria included studies of glass composition and its impact on optical properties, and as a result, the implementation of measurements to study these optical properties (Hartmann et al. 2010). Substantial progress was made during this time, which led others to also investigate ways to make better optical glass. Michael Faraday, in England, and later Ernst Abbe, Otto Schott and Carl Zeiss, in Germany, began to consider other compositional solutions for some of the shortcomings of crown and flint glasses used at the time (Youngman 2021). Their solution was to add boron oxide to the glass, yielding more homogenous glasses with better optical performance. Another Englishman, William Harcourt, was experimenting with glass compositions and discovered that a wide variety of glasses could be made by incorporating boric oxide or phosphorus oxide, but lack of homogeneity and poor durability limited their practical usage (Houghton 1915). The efforts in Germany, capitalizing on improved glass homogeneity and unique compositions, launched an entirely new industry of silicate-based technical glasses for optical applications, solving the problem of chromatic abberation and marking a revolution in optical glasses (Hovestadt 1902). Within a few years after founding the Schott and Associates Glass Technology Laboratory in Jena, their catalogue featured dozens of new, commercial silicate glasses. This followed the successful establishment of two glass companies in England (now Pilkington Specialty Glass) and France (now Corning France), both of which can trace their roots to the early 19th century (Hartmann et al. 2010).

The enormous contributions out of Germany in the late 1800s also gave us glasses which found applications enabled by their unique thermal properties and resistance to chemical reagents (Houghton 1915), and eventually led to the invention of PYREX® by Corning Glass Works. As discussed below, this single glass composition, as well as similar glasses from other companies, forever changed many of the ways glass was used in science and by the consumer. However, PYREX® Glass was also favorable for some of the modern day advances in large, terrestrial telescopes. The photograph in Figure 8 shows rail transport of a large piece of PYREX® made for the Hale Telescope on Mount Palomar. As a low coefficient of thermal expansion glass, large telescope blanks could be made with assurance that size and shape were not affected by changes in temperature, as might occur at the high elevation locales for these types of telescopes. Silicate glasses were now being designed specifically with the properties necessary for even the most stringent technical applications, another indication that the Glass Age had begun.

Figure 8. The 200 inch PYREX® mirror blank from Corning Glass Works being readied for train transport to California. Photo courtesy of the Corning Incorporated Department of Archives & Records Management, Corning, NY.

Other silicate-based glasses have been used for similarly challenging optical applications. More recent terrestrial and space-based telescopes, including the Hubble Space Telescope, use low thermal expansion glasses as the mirror substrate. One common version of this glass is Corning's ULE™ (ultra-low expansion) Glass, which is mostly silica, with a small amount of added TiO_2. For over 75 years, ULE™ Glass has been the standard for astronomical applications, given that it has very little dimensional change over large temperature variations (Sabia et al. 2006). The lightweight mirror in the Hubble Telescope, as well as other large mirrors (up to 8 meters in diameter!) in the Gemini and Subaru telescopes, have allowed scientists to explore some of the far reaches of our universe. The examples in Figure 9 show ULE™ Glass being assembled into the 4 meter diameter Discovery Channel Telescope mirror.

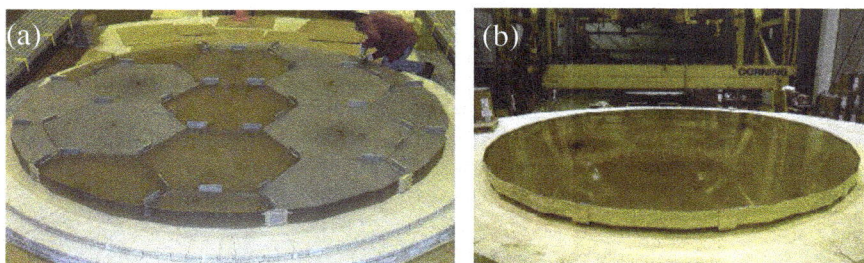

Figure 9. Corning ULE™ Glass segments for the Discovery Channel Telescope (**a**) before and (**b**) after fusing into the 4 meter monolith. Figure reproduced with permission of the Society of Photographic Instrumentation Engineers (SPIE) from (Sabia et al. 2006).

The thermal stability of high silica glasses, including the ULE™ Glass described above, also enabled the computer age, where these glasses are ideal for nanolithographic manufacturing of computer chips. As computer technology advanced, in accordance with Moore's Law, chip manufacturers have relied on increasingly shorter wavelength optical systems to produce chips with smaller features and higher density. Pure silica glass, as well as various doped silicas (i.e., containing fluorine), are highly transparent in the ultra-violet, making them perfect for some of the early nanolithography lens materials, an example of which is shown in Figure 10. Even today, with semiconductor chips being made with the extreme ultraviolet lithographic (EUV) technology, where glass transparency starts to be problematic, the complex optical systems require mirror materials with exceptional dimensional stability, and thus another important use for silicate glasses in modern optics.

Figure 10. Silicate glasses used in modern semiconductor industries include Corning® HPFS® Fused Silica. Image courtesy of Corning Incorporated.

New optical uses of glass include augmented reality, where silicate and other glasses with high refractive index are being used to design approaches for connecting humans with their surroundings. Some of the key design requirements include high transparency, relatively large refractive indices and exceptional durability, which as shown throughout this review of silicate glasses, are material characteristics unique to silicate glasses and thus the future in optical materials depends heavily on glass and our ability to tailor its properties to any given application.

Silicate glasses for transportation. Human transportation is a part of our culture and history which has been greatly impacted by the materials used. As mentioned earlier, some of the great advances during the Bronze Age were enabled by invention of chariots and sailing vessels, allowing civilizations to spread both in size and influence. The same is true for glass and its impact on transportation. Starting in the mid-18th century to this very day, the Glass Age is constantly improving how we navigate land, water and the skies, including current and future space travel. All of these examples can trace their success at least in part to the use of silicate glasses.

The first and lasting impact of glass on transportation was not actually a specific composition of material, but instead was probably the invention of the Fresnel lens. Augustin Fresnel, a French physicist, worked on ways to focus divergent light into an intense, horizontal beam, inventing in 1822 a lighthouse lens comprised of a stack of lenses surrounded by prisms. Some of these prisms were reflecting, while others were refracting, and the order of their stacking provided the focusing effect (CMOG 1999). This Fresnel lens immediately benefited all ocean-going travel, increasing the safety and effectiveness of lighthouses for sailing near shore. The lens design is still in use some 200 years later, found in automotive headlights, traffic lights and airport beacons.

Another lasting milestone in the application of glasses in transportation, besides some of the sailing implements needed by Middle Age explorers, was probably the lantern and signal lights developed for train and boat travel. Both forms of transportation can involve sudden and significant changes in weather, and thus these early lights needed a durable and thermal shock resistant material to encase the light source. Borosilicate glasses became one solution to this problem, as their thermal properties were ideal for the design of lantern chimney glasses—one of the important commercial products of Schott in the late 1800s (Houghton 1915). In a similar vein, one of the earliest commercial successes of Corning Glass Works was the signal lantern for railroads. Early glasses for this application were a lead alkali silicate composition, but with improved optics and colors, resulted in substantially better visibility and safety in railroad travel (Graham and Shuldiner 2001). Similar advances were made for boating and eventually exterior lighting for automobiles. As one example, consider the lanterns in Figure 11, which were constructed using silicate glasses and incorporated advanced lens designs (e.g., Fresnel-type lens) for enhanced visibility and standardized colors. By 1908, color standards, both red and green, produced by Corning Glass Works were adopted across the railroad industry in the US.

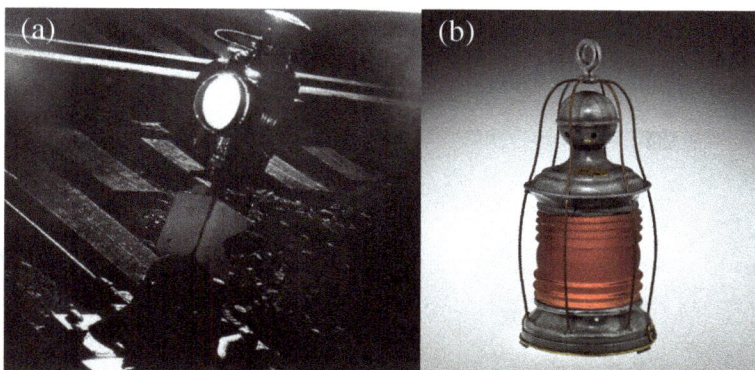

Figure 11. Examples of some early glass applications in transportation. Durable, shock resistant borosilicate glasses, in a variety of standardized colors and incorporating different lens designs, was critical in **(a)** railroad and **(b)** marine applications. Image in (a) courtesy of the Corning Incorporated Department of Archives & Records Management, Corning, NY. Image in (b) is CMOG 2012.4.135. Image licensed by The Corning Museum of Glass, Corning, NY (www.cmog.org) under CC BY-NC-SA 4.0.

The invention of the automobile eventually incorporated a wide variety of silicate glasses in the design and operation of this mode of transportation. In addition to lighting needs on the exterior, including headlamps and various taillights, automobiles use advanced glasses for windows, windshields and sun/moon roofs. All of these have depended on glasses with specific properties, and without which, modern vehicular travel would not be possible, or at the very least, not very enjoyable. Safety in travel is a common theme throughout the incorporation of glass in automobiles and other modes of human transport. Our desire to see the surroundings, and be aware of hazards, can only be achieved through a large quantity of tranparent window glass, made possible by the Pilkington float glass process (see the section on manufacturing) and glass strengthening via thermal tempering and lamination. Glass has been leveraged for these many exterior applications, and now is being incorporated into the interior of vehicles to advance the human experience in transportation and entertainment. Glasses with built-in displays are being added to new automobile design, increasing both safety and information content for the driver. And designers are adopting glass as a medium for upgrading the interior appearance and feel of vehicle interiors, furthering the use of glass in modern transportation.

While these examples are mostly available to the general population, there are other key examples of glass in transportation which are much more specialized and also much more advanced in both their use and impact on humanity. When humans first started to explore space, manned spaceflight required a safe, secure environment to protect humans from the hazards of being in space. This was readily done with other materials and engineering features in the design of space capsules (Fig. 12a) and eventually returnable space vehicles (e.g., the US space shuttle) and now in semi-permanent space stations. However, what would the experience in space, and to the moon and other celestial bodies be without our ability to see the surroundings? So windows in these various spacecraft were a beneficial and arguably necessary component of their design, and silicate glasses were and remain the top material solution for such applications (David 2000). The first US space missions, based on the Mercury capsules, incorporated heat-resistant glass windows from Corning. All subsequent US space flight used similar silicate windows, including the famed Apollo 11 moon landing, where silicate glasses allowed astronauts a safe and spectacular view of the moon (e.g., Fig. 12b). Both of these NASA images were captured by astronauts viewing their surroundings through a robust and sophisticated layer of high homogeneity, low thermal expansion silicate glasses, both to protect the occupants from radiation and debris, but to also provide a direct means by which to see their surroundings. Further advances in large, high-purity silica glass led to even larger and better windows in the International Space Station (David 2000), resulting in lower distortion and wider viewing angles and ultimately the ability to view and study most of Earth's surface and population.

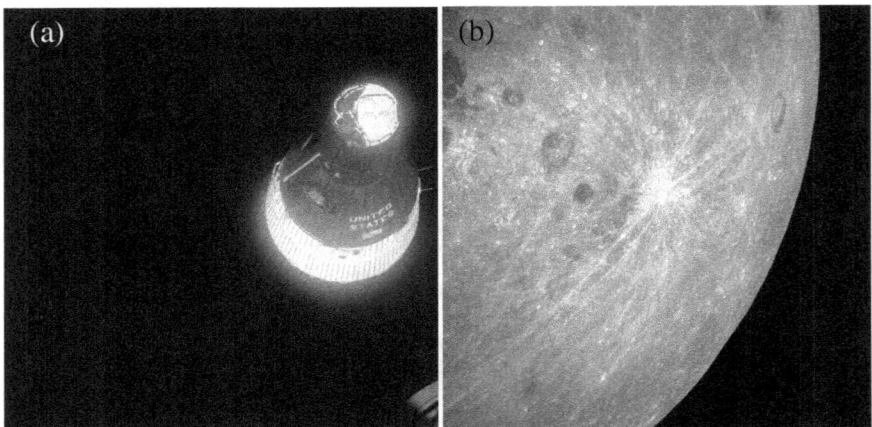

Figure 12. Enabling space exploration with glass: **(a)** view of Gemini 6 spacecraft from the Gemini 7 spacecraft (1965) (NASA ID: s65-64040) and **(b)** photograph of lunar farside crater taken from the Apollo 11 spacecraft in 1970 (NASA ID: as13-60-8675). Source: NASA.

Silicate glasses—The ultimate building material. Glasses in architecture, especially as windows, have been around for millenia. Even with their relatively poor quality and transparency, the ancient windows of Rome and the time before the Middle Ages, described above, demonstrated the strong desire for humans to allow natural light into buildings. This spirit continues to this very day, leading to substantial use of architectural glass over the past several hundred years, as well as advances in how glass can be used in this manner (Eskilson 2018).

The use of flat glass in buildings was only possible from the invention and production of flat glass in large sizes. Clear panes of glass were known from the Middle Ages, where Venice and other glass-making centers were known to have used such material, even though it was still relatively rare and quite expensive. The situation changed with advances in production of flat glass, and especially as society shifted towards the use of glass in more aesthetic architectural applications. The turning point in our use of architectural glass, and potentially another indication of the Glass Age, was the fabulous Crystal Palace (Fig. 13)—an enormous building constructed mainly of glass for the 1851 Great Exhibition in London. Comprised of 900,000 square feet of glass, the design of Englishman Joseph Paxton was widely considered one of the top three architectural achievements of that time (Diamond 1953). In addition to featuring an impressive amount of glass in the walls and ceiling, the contents of the Crystal Palace also showcased a variety of artistic and technical glasses, providing a view of what was possible with glass. Corning's predecessor, the Brooklyn Flint Glass Company, contributed glass in the form of some cut glass tableware for the exhibition. As written over 30 years ago, the Crystal Palace and the 1851 exhibition in London, which featured thousands of glass objects, may "have ushered in the modern Glass Age" (Kolb and Kolb 1988), one of the earliest documented statements about a Glass Age and certainly a foreshadowing of how we would come to view the relatively recent history of glass in society.

Figure 13. The front entrance of the Crystal Palace, Hyde Park, London that housed the Great Exhibition of 1851, the first World's Fair. Contemporary engraving from Tallis' *History and Criticism of the Crystal Palace*, 1852.

Architectural glass in modern society is most visible in the form of skyscrapers and our fascination with enormous, glistening buildings encased in glass. Not only do these forms contain huge amounts of window glass, but now we see glasses with additional features: insulating properties, self-cleaning glass, mirrored surfaces, glass which turn opaque or transparent with an electric field, etc... The use of flat silicate glasses in building applications appears endless.

Other examples supporting silicate glasses as the ultimate building material are abundant. Glass fibers as insulating materials were made possible by new methods to manufacture thin filaments of glasses with specific compositions and properties. Architectural blocks fashioned from silicate glasses can be assembled much like ceramic bricks to form transparent or semi-transparent walls, often used in decorating interior spaces of buildings (Fig. 14). Glass in

Figure 14. Glass blocks used for design and function in architectural applications. Courtesy of the Corning Incorporated Department of Archives & Records Management, Corning, NY.

the form of strong, transparent flooring, continues to be deployed, from observation towers high above cityscapes to viewing platforms suspended over natural panoramas like the Grand Canyon in the southwestern US. Glasses enable the application and enhance the experience of humans with their surroundings.

Lighting—Only possible with glass. Silicate glasses and their impact on lighting was briefly described above in the section on transportation, with the need to encase light sources in a durable and transparent material. However, the greatest invention of the late 19th century based on silicate glasses may have been the "glass envelope" for Edison's electric light. As the story goes, Thomas Edison required a transparent, hermetic material to keep his electrically heated filaments in vacuum. A Corning lead silicate glass, and the skill of their glass blowers, enabled Edison to see his incandescent light bulb to fruition. Glass, arguably the key ingredient to this life-changing invention, "provided the vacuum, protected the filament, and enabled the light to radiate outward with minimal loss" (Dyer and Gross 2001). The electrification of lighting, and especially the mechanization of light bulb manufacturing (described below), was an enormous advance in human civilization and represented a key milestone in the intimate connection between humanity and glass. The iconic view of Edison's electric light, as shown in Figure 15, showcases the simple, yet critical silicate glass which enabled such a change in our lives.

Other examples of glass in lighting applications include some which feature prominently in the culture of the 20th century, including neon lighting. For this application, glass tubing is used to contain the noble gases, but even more importantly because of the incredibly processability of glass, can be formed into a wide variety of shapes and lettering, leading to use in advertising and other signage applications, as exemplified by some of the examples in Figure 16. Glass remains the material of choice for lighting, even as technologies transition from incandescent to fluorescent to light emitting diodes (LED). As was true almost 150 years ago, glass provides the best way to package these light sources, both to protect the contents but more importantly to extract and manipulate the light for human benefit.

Glasses in science and medicine. The Glass Age would certainly not be possible without the advances of silicate glasses in the fields of science and medicine. This was true long before modern times, with the clear glasses of Venice and other glass-making centers used in a variety of scientific glassware and implements, including the thermometer and barometer (discussed in the *Silicate glasses during the Middle Ages* section). While some of the most lasting scientific advances were made during the early eras of glass, the last two centuries have provided a

Figure 15. Example of the electric light bulb, similar to the 1879 invention from Edison, using silicate glass to provide functionality and long lifetime. General Electric, *Electric Light Bulb* (1929). CMOG 2000.4.1. Gift of Helen Hilbert Peterson. Courtesy of The Corning Museum of Glass, Corning, NY.

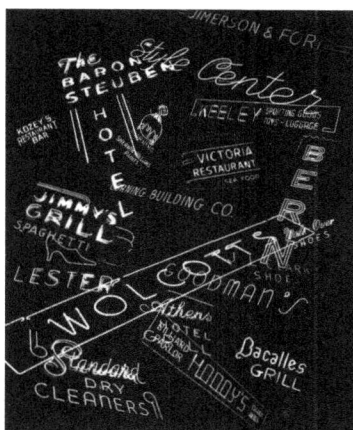

Figure 16. Neon signage, in all shapes and colors, facilitated by silicate glasses. Courtesy of the Corning Incorporated Department of Archives & Records Management, Corning, NY.

wealth of opportunity and impact in science and medicine, facilitated by the many advances in glass science and technology. The first of these was related to the explosion of work on different silicate glass compositions, including the heat-resistant borosilicate glasses from Schott (Houghton 1915) and eventually Corning Glass Works. PYREX® was commercialized in 1915, partially as a result of a shortage of good-quality laboratory glassware during the 1st World War, and it quickly became the glass of choice for scientific labs. Heat and chemical resistance of this special borosilicate glass enabled more reliable and safe explorations of chemistry and biology, continuing to be important to these fields even today. Examples of some PYREX® articles from the early part of the 20th century are shown in Figure 17, demonstrating a variety of different sizes and shapes of laboratory glassware, all with specific functionality.

Some of the other seminal developments in science powered by silicate glasses were described above in the discussion of optics, where for example modern astronomy would not be possible without highly technical silicate glass. Additional examples are found throughout some of the other topics in the *Silicate glasses in modern and contemporary history* section, including communication.

Figure 17. Examples of PYREX® laboratory glassware, which forever changed chemistry and related physical sciences. Image courtesy of the Corning Incorporated Department of Archives & Records Management, Corning, NY.

Medicine is another beneficiary of silicate glasses, beginning long before the Glass Age when early microscopes allowed for observation of microbes and other sources of disease, benefitting all of humankind through better understanding of microbiology. Modern medicine has seen similar benefits from glass, including glass imaging fibers which allow physicians to see into the interior of the human body. Silicate glasses have also been the main packaging material for drugs and medicines, relying on sterility of the glass and preservation of the contents by providing a durable and hermitic seal from the outside world. Today, advances in glass packing, leading to even more damage-resistant glasses, has significantly benefitted medicine. New pharmaceutical vials from Corning (Fig. 18), are specially designed to be chemically inert, but also resistant to damage during filling, transport and use of the vial contents. Other advances of note, including an ion-exchanged silicate glass for the EpiPen®, substantially increase the reliability and safety of these critical devices for human health.

Figure 18. Pharmaceutical packaging in the form of vials, for which chemically strengthened silicate glasses have enabled new and robust components to ensure patient safety. Image courtesy of Corning Incorporated, Pharmaceutical Technologies.

Also of note, glass is being used in the human body to facilitate wound healing, disease treatment and as a scaffold material for implants. These applications, launched with the invention of Bioglass® in 1969 (Hench 2006), have brought the human experience with glass into the body, leveraging composition and material properties in amazing ways. Glass spheres containing radioactive drugs are able to deliver their toxic contents to a specific organ, for example in the treatment of liver cancer (Li et al. 2010). Recent studies of phosphorus- or boron-containing silicate glasses has shown promise in treatment of open wounds which do not usually heal well. Bioactive glasses are incorporated into toothpaste to interact with the human body and strengthen the dentin on our teeth. Bioactive glasses also can dissolve and facilitate rebuilding of hard tissues like bone. The field of bioglasses and their impact on human health is in its infancy, and the possibilities going forward appear endless (Jones 2013).

Modern communication at the speed of light. In terms of modern technology, the use of silicate glasses in communication is probably the most advanced. Invented in the early 1970s, optical fiber, made with incredibly high purity silica and other additives, was designed to carry optical signals over very long distances. As technology in telecommunications grew, cables containing multiple optical fibers (Fig. 19) were replacing standard copper wiring, greatly expanding the bandwidth of these connections. With the advent of the internet and now the "internet of things," bandwidth and overall data transmission across telecommunication networks has exploded. Facilitating this hypergrowth in communication are the thin strands of silica glass which cover large expanses of terrestrial networks, and now connect continents using thousands of kilometers of undersea cables.

Figure 19. Strands of highly pure silicate glass carry optical signals over large distances. Image courtesy of Corning Incorporated.

At the heart of this invention was the idea that the purest glass might be able to carry light over large distances. The initial goal was 1000 meters without too much degradation of signal, a comparable metric for copper wire, but even this was a challenge during the early days of optical fiber development (Dyer and Gross 2001). In the early 1970s, Corning scientists, using chemical vapor deposition as a mean by which to generate the purest form of silicate glasses, were able to make silica based optical fibers that could transmit light over tens of kilometers without the need for repeaters—a feat that meant silicate glasses could easily outperform copper (Dyer and Gross 2001). The technology is much more complex than just a pure glass strand, as these long distance telecommunication fibers require guiding of the light within the core of the fiber, necessitating multiple types of glasses to generate refractive index differences, and resistance to fiber breakage required careful optimization of the mechanical properties and coatings used to turn strands of glass into optical fiber cabling. Even today, the search continues in making optical fiber with transmission even closer to the theoretical limit, a challenge for glass scientists, manufacturing processes and optical physics alike.

Predating the dawn of the internet, communication amongst humans took a variety of forms, including through televisions and motion pictures. Thus cameras and their finely crafted glass lenses, as well as devices to display broadcast video signals, were an important part of the mid to late 20th century. Much of this technology can be traced to the work of Karl Ferdinand Braun, who in 1897 invented the cathode-ray tube oscilloscope, the forerunner to television and radar tubes (Keller 1991). The first commercially practical CRT for television was offered in 1931. In the 1940s, CRT displays were round and formed by centrifugal casting. As technology advanced, the shape of these displays evolved towards being rectangular (e.g., Fig. 20), which presented enormous challenges in manufacturing. Mass production of television tubes by Corning and others lowered the price for such devices and accelerated adoption by large portions of modern society. Figure 20 contains a photograph showing some of the different early CRT

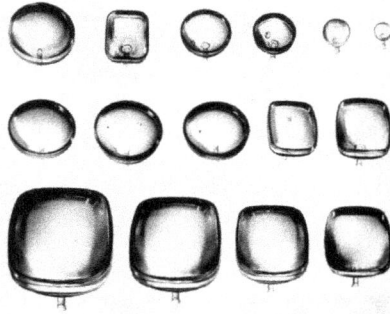

Figure 20. Examples of early television glass funnels, courtesy of the Corning Incorporated Department of Archives & Records Management, Corning, NY.

products made by Corning Glass Works. Different sizes enabled different television sizes, and advances in this technology led to successful development of color television. The glass, which by weight was an enormous part of each television set, was also the key enabler of this technology.

Disruption of CRT-based television technology was achieved late in the 20th century with the invention of liquid crystal displays (LCD), which are much lighter and thinner than standard CRT-based televisions. To meet the challenge of this new form of communication, glass makers rapidly developed specialized flat glass as a substrate for LCD (and plasma displays). Corning's early entry in this commercial area was based on their code 1737 and related glasses, which were multicomponent silicate glasses made using the Corning fusion or overflow downdraw technology, achieving thin, large panels of glass without any contact with either surface of the sheet. Pristine, and incredible flat pieces of glass like 1737 and newer versions, continue to enable new technological advances in this form of communication, and continue to alter the human experience in home entertainment and in other social settings involving visual displays.

The most recent advances in human communication involve wireless technology—basically the widespread adoption of wireless networks and handheld cell phones. Glass has played a key role in this evolution as well, providing clear, aesthetic and durable coverings for the displays on modern smart phones. Corning® Gorilla® Glass is one example of an engineered glass, also a silicate glass made stronger with chemical tempering or ion exchange, and is found on devices used across the globe and by most of the global population. These, and related glasses, enable touch technology on mobile electronics, and are now being used not just in smart phones, but also in other applications like automotive interiors and home appliances, where interaction between a person and information is desired. Changes in the way we use glass are accelerating at an incredible pace.

New frontiers in energy. Glass in energy applications takes many forms, from generating energy and disposing of wastes, to helping conserve energy. The latter is one of the most widespread uses of glass in modern history. Glazing of window glass is done to reduce energy use when heating or cooling building interiors. Window and other architectural glasses are now designed to be transmissive or reflective by electrochomic activity, all with the idea of controlling the amount of sunlight and infrared radiation that can pass through the material. Glass as an insulation material, discussed below in the section on mechanization of glass production, helps reduce energy costs in our homes, as in fiberglass insulation, but is also found as insulating materials in household appliances.

Glasses have also served an enormous role in new sources of energy (i.e., non-hydrocarbon sources). Glass panels are the substrates used in photovoltaics, specifically as solar panels for harvesting the energy of our Sun. Strong, durable and transparent glass is needed for this application, and obviously the glass needs to be inexpensive to minimize the cost of this energy source. Glass for this promising technology can be the normal and inexpensive soda-lime silicate glass that humans have known and used for thousands of years, or it could take the form of more specialized glass compositions. As with any application of glass during the Glass Age, compositional flexibility enables optimization of properties favorable for a particular application. In the case of photovoltaic glass, preserving the function of the photovoltaic material encased by the glass means designing compatible compositions, while maintaining manufacturability, low or modest cost, and providing structural integrity for the assembled panel. Such stringent design and technological criteria can be met with the creative implementation of silicate glasses.

The other strongly beneficial use of glass in energy is the encapsulation of nuclear wastes. Countries with nuclear energy production have been faced with long-term disposal of their radioactive wastes. This waste is often in the form of solutions and sludges, which corrode and lead to leakage if improperly stored, even in metal or concrete containers. One elegant solution, based on our long history of using glass, is to vitrify this waste in the form of chemically stable glass that can then be stored in geological repositories. Tons of nuclear waste from energy (and weapons) production has been converted to glass by fusing this material together with traditional glass ingredients like sand, to make a silicate glass that will stand the test of time, resistant to both weathering and radiation from the encapsulated radionuclides. Solutions like this are one potential way to secure the future of our environment, important for humanity now and for the generations to come.

The modern home—An abundance of silicate glass. Numerous examples of glass in the modern home can be found throughout this brief summary. Some glasses are incorporated into the structure of the home, namely windows and insulating materials. Others are found throughout the many appliances and devices which bring function to our homes. Stoves have glass-ceramic cooktops and heat-resistant windows on the oven doors. Televisions, larger than a person and yet lighter than ever, can be mounted on walls and barely occupy any space. And the image resolution in these large displays continues to advance, requiring new technologies and new glass compositions that can be made in large size with incredible flatness, uniformity and dimensional stability. Computers and handheld electronics are ubiquitous throughout the modern home, and of course also depend on glass for their displays, and in the case of touch screens, advanced glass solutions. We also see this in the design of new technologies for the home, including interfaces between humans and technology in the form of digital information on displays, appliances and other surfaces. And throughout the history of manufactured glass, humans have incorporated glass vessels and implements in their homes. While perhaps mundane and too common to garner much attention when compared to some of the other highly technological uses, the sheer volume of glass articles in a typical modern kitchen is enormous. Inert, transparent, possibly brilliantly colored, and widely accessible, the utility and conveniences of household glass materials continues to enrich the human experience.

Advances in glass manufacturing. The Glass Age, if one agrees that this really started in the mid-18th century and continues today, was only possible because our history and experience with glass was expanded beyond the ancient and medieval uses as art and containers and even early window glass. Adoption by and impact on a broader range of human life, a necessary requirement for glasses to be accepted by cultural anthropologists as meeting the definition of an "Age", was only possible due to the many advances in manufacturing of glass. Some of the great inventions in glass making contributed to the Machine Age of the 18th and 19th centuries, including several which are still in use today. The first advance in glass manufacturing during the modern Glass Age was in plate glass production for architectural use, perhaps signified by the great Crystal Palace (Fig. 13).

Bringing glasses to the masses was achieved principally by mechanization of the glass industry. This affected several industries, including window glass production, when in 1903 the "Lubbers Machine" was built and signified the first machine to partially replace the labor-intensive cylinder method for making window glass (Kolb and Kolb 1988). The basic idea was to mechanically generate glass cylinders which were larger and made faster than the rate of hand-blown cylinders that had been used since the inception of this technique by French artisans in the 12th and 13th centuries (Diamond 1953). Larger cylinders also meant that larger pieces of window glass could be make by cutting and unrolling these cylinders of glass into flat glass sheet, the latter of which was still done by hand. The windows generated from this approach, though still not completely flat and thus causing some visual distortions, were immensely cheaper than before, and therefore more accessible to the general public.

Fully mechanized production of window glass occurred in Europe and the US around the start of World War I (Diamond 1953). A Belgian glass maker named Fourcault invented a process by which molten silicate glass was forced upward through a slot and then cooled while being rolled into a continuous ribbon of glass. The American engineers Irving Colburn and Michael Owens—the latter now recognized as a prominent figure in the history of the American glass industry, designed a similar process, producing 30,000 square feet of glass every day. These continuous sheet-forming processes were used for several decades to produce all flat glass, revolutionizing the use of window glass, until in the mid 20th century, when the float glass process was invented by Alastair Pilkington in the UK (Pilkington 1969). By floating molten silicates on a bed of molten tin, very flat and highly uniform glass could be generated in very large sizes. This continuous melting/forming process remains the single largest method of making window and other architectural glasses, and most float glass is made using the age-old soda-lime silicate composition.

Another milestone in the production of glass for everyday use in modern society was the automatic bottle machine, an invention of Michael Owens which mechanized the process of glass blowing for bottle making. Up until the turn of the 20th century, glass bottles were made by hand, using a process that can be traced back two thousand years. This meant that bottles were expensive, variable in size, shape and quality, and thus limited in their use. Owens invented a machine with fifteen arms which each picked up gobs of molten silicate glass, and were then blown into a mold using compressed air. Soon after, bottles could be produced at the rate of over one million per week. The Owens bottle machine was indeed revolutionary, and "bottles began to provide safe, sanitary, inexpensive and reliably standard containers for the world's medicines, for its milk and many other beverages, for the sauces, fruit and multitudinous items that appears on a grocer's shelf" (Diamond 1953). Thus a seemingly simple glass-making machine expanded access to glass articles—compelling evidence of the Glass Age.

A related invention, which would forever change the lighting industry, was Corning's work in lightbulb production (Graham 2001). As with glass bottles, the glass envelopes for incandescent lighting were made by hand, as even the glass pressing technology used to make solid glass articles was unable to accommodate the narrow opening and the need for a thin, blown glass. Similar to Owens' machine, the ribbon machine (Fig. 21) increased the rate at which blown glass articles could be made (Ellis 1993). Ribbons of molten glass were allowed to sag into holes of specific size, then using a puff of air from above, a suspended glass bubble was formed and molded to the desired shape. This machine originally produced 500 bulbs per minute, then improvements led to a doubling of this rate, as well as other general improvements including inner surface frosting and the ability to fashion many other shapes and sizes.

Continued improvements in melting and forming of glass during the 20th century led to other inventions and uses of glass in modern life. Fabrication of glass fiber in the form of wool, an invention of Russell Slayter of Owens Illinois Glass Co. in 1933 (Nascimento and Zanotto 2016), resulted in fiberglass, used widely as an insulating material in construction.

Figure 21. The Corning ribbon machine for making lightbulbs. Introduced in 1926, this single machine was capable of producing 1000–2000 bulbs per minute! Courtesy of the Corning Incorporated Department of Archives & Records Management, Corning, NY.

Fiberglass has also been used as a reinforcing agent in plastics, similar to how steel can provide reinforcement of concrete. This type of fiberglass can be found in all types of applications, from automotive panels to the hulls of boats.

A paradigm shift in glass manufacturing occurred in the 1930s, when Corning's Frank Hyde developed the vapor deposition process. Instead of melting naturally occurring raw materials like sand and sodium carbonate in a furnace, highly pure starting materials from chemical sources could be burned and deposited as a soot, for example as a way to generate very pure silica glass. This soot deposition process is now widely used in the manufacturing of optical fiber, but has been used by Corning for decades in production of HPFS® and ULE™ Glasses (Graham 2001). Although more expensive than melting sand, the incredibly high purity afforded by pure chemical sources has empowered all of the highly technological uses of silica and doped silicas, including transmission of optical signals over incredible lengths (Fig. 19), or the amazing transparency of these glasses in the deep ultraviolet, necessary for modern lithographic production of semiconductors (Fig. 10).

Summary and perspectives on the Glass Age

This discussion of silicate glasses was not meant to capture the full history of these materials, nor the complete impact of glass on humankind. Instead, it is hoped that a brief review of the history and examples from modern and contemporary history has been useful in demonstrating the incredible usefulness and high value of silicate glasses to ancient and modern human civilizations. There are many more examples of glass development in other cultures, and of course numerous non-silicate glasses which are also important in technology and human existence. For more on non-silicate glasses, the reader is referred to Möncke (2022, this volume).

The history of glass is fascinating, having captured the imagination of scientists and artists alike. Silicate glasses have spanned the entire known history of glass, and the impact of this material on humanity is almost too large to comprehend. Much of what we know about the use and development of silicate glasses mirrors some of the other well-known anthropological descriptions of human civilizations, namely those in the Three-Age system: Stone, Bronze and Iron Ages. Because of the close relation between human usage of silicate glasses and some of these more prevalent materials, there are recent pronouncements that we are now residing in the Glass Age. Although this concept is perhaps only recently becoming part of the lexicon, enhanced by the declaration of a Glass Age by Corning Incorporated (Morse and Evenson 2016; Corning 2020) and the *International Journal of Applied Glass Science* (Pye 2016), there have been additional proponents of a Glass Age, reflecting the enormous impact of this material on human civilization. An article titled "The Glass Age" appeared in 1986, possibly the first documented reference to this era (Sennett 1986). As mentioned earlier in this overview, even the formal adoption of the three-age system in the early 1800s was met with criticism in omitting other ages, including that of glass (Trigger 2006). This chapter was focused on describing this impact, highlighting a few of the many examples in which silicate glasses have transformed human existence, with some discussion on how we might define and justify the addition of the Glass Age as a key moment in the cultural anthropology of human civilization. Perhaps we can now amend the timeline of human civilization, as illustrated in Figure 1, to include the Glass Age. While perhaps a bold assertion on our part, history will certainly judge the necessary details defining the beginning and extent of the Glass Age— hopefully without waiting millennia to acknowledge what we now know to be true:

> *"GLASS! It is older than recorded history, but it is as new as tomorrow. It is hard to imagine what the world would be like without it. And there will undoubtedly be uses for glass in the future that we have not even dreamed about. Glass is a unique material. What we can do with it is limited only by our imagination."* (Kolb and Kolb 1988)

> *"Glass has shaped the world more than any other substance, and in many sneaky ways, it's the defining material of the human era."* (Main 2018)

Furthermore, even other "ages" which have appeared in the lexicon of the past several centuries, can be considered simply as subsets of the Glass Age, similar to how the Copper Age is widely viewed as an early period of the Bronze Age. These may be defined simply as a period of specific human progress, or a cultural period involving a specific material or technology. The Age of Discovery or Exploration, which captures the era of oceanic exploration of our planet between the 16th and 18th centuries, was enabled by simple optical devices like the sextant, of which silicate glass lenses guided the way. The Age of Enlightenment in 18th century Europe was a period of time during which advances in science and medicine would not have been achievable without glass to magnify and reveal the surrounding natural world. The Machine Age of the 19th and 20th centuries widely accelerated manufacturing and technology, and of course glass for all types of applications benefitted from and even contributed to some of these advances. Even something as specific as the Gilded Age in the US (late 19th century) was defined by mechanization of industry and advances in transportation based on railroads as the major growth industry. Some of the key inventions during this era, described earlier, directly contributed to this short period of US culture. And of course, the more recent Space (post 1957) and Information (post 1971) Ages were only possible with glass. Strong, transparent windows for manned space flight, and glass used in some of the most advanced space telescopes today affirm the role of silicate glasses in our exploration of the solar system and beyond. The Information Age, generally describing the advent of computers and rapid transfer of information across the globe, owes much of its impact to silicate glasses. From optical elements for UV lithography and the manifestation of Moore's Law, to optical fiber that connects the world, and even to portable electronics like tablets and cell phones, glass is the material which made all of this possible. The Glass Age indeed transcends these shorter, more specific ages in human history, having contributed to their impact on our lives.

Although silicate glasses are firmly entrenched in modern culture and technology, there are numerous opportunities and challenges which will largely define the extension of the Glass Age into the future. New opportunities in glass are being aided by advances in measurement technology and computer simulation techniques, increasing our understanding of glasses at the atomic scale (Höche 2010; Pedone 2016; Gin et al. 2017). Much of these efforts will lead to additional foundational knowledge building in glass science, connecting composition to structure and ultimately the many beneficial properties of these amazing materials, and of course driving new opportunities to enjoy and benefit from glass. Machine learning and artificial intelligence efforts are underway to maximize our discovery of new and beneficial glasses, and though still in their infancy, such approaches to materials design are already impacting commercial and academic pursuits in glass science (Liu et al. 2021).

In order to keep the Glass Age relevant for future generations, the following challenges must be addressed. Energy usage and CO_2 emissions in glass production are high (Schmitz et al. 2011). New furnace designs and improvements in energy management are continually being made in order to reduce energy consumption. Raw material availability, while seemingly plentiful in many cases (e.g., sand), will present future challenges in the production of glass. Rare earth elements which bring functionality to certain optical and laser glasses, are in some cases becoming scarce and more difficult to safely extract from their sources. Lithium minerals, which have become invaluable in energy applications (including glasses for batteries!) and chemically-strengthened glasses for consumer electronics, have increased in cost due to high demand from these industries. Fortunately, innovations in glass will continue to help solve these issues, and as we all know, glass is one of only a few materials which can be readily recycled, potentially reducing the demand on raw materials and aiding in the energy management necessary to make and process new glasses. Waste glass is also being used in other industries to mitigate their high energy costs and gas emissions (Jani and Hogland 2014)— another example of glass as a potential solution in the future sustainability of humanity.

The end of an "age", as discussed in the Introduction, is brought about when a new material supplants the prevalence of an older one in civilization. The end of the Glass Age, whenever that may occur, will similarly be caused by the adoption of some new, advantaged material which relegates glass to something of lower importance. There have already been materials which compete directly with glass, for example plastics. While these organic polymers have certainly enjoyed a century of wide-spread use, and can be found throughout modern life, they have not prevented the development and use of glass as described in this chapter. In fact, there are even several examples where glass and plastic compete directly in the same technology area and glass perserveres. Modern television technology has evolved from the cathode-ray tubes of the 20th century (Fig. 20) to liquid crystal and plasma displays to organic light emitting diodes and now quantum dot technology. Glass has played a central role in the development of these display technologies, and while plastics may be attractive for certain attributes like weight and cost, glass is the material solution for the past, present and future of this particular technology. Furthermore, with society focusing more on a green economy and stewardship of our planet, plastics continue to fall out of favor in preference for other materials like glass. To this point, there has been a systematic shift away from plastics for food storage, looking again at glass vessels as a sterile, chemically inert, and recyclable material for use by all of humanity.

ACKNOWLEDGEMENTS

Many thanks to Anne Young (Corning Incorporated Department of Archives & Records Management), Suzanne Abrams Rebillard (Corning Museum of Glass) and Katie Greene (Corning Incorporated Science & Technology) for their invaluable assistance. The author especially appreciates the substantial input and support from Amy Lang and Kathryn Allen (Corning Incorporated), as well as the excellent suggestions and improvements resulting from reviews by Doris Möncke and Edgar Zanotto.

REFERENCES

Bolt M (2017) Glass: The eye of science. Int J Appl Glass Sci 8:4–22

Brain C, Brain S (2016) The development of lead-crystal glass in London and Dublin 1672–1682: a reappraisal. Glass Technol: Eur J Glass Sci Techol A 57:37–52

Chopinet MH (2019) The history of glass. *In:* Springer Handbook of Glass. Musgraves JD, Hu J, Calvez L (eds) Springer, p 1–47

CMOG (1999) Innovations in Glass. The Corning Museum of Glass

Corning (2020) The Glass Age. https://www.corning.com/worldwide/en/innovation/the-glass-age.html

David L (2000) Window on the World. Air & Space Magazine. May Issue

Diamond F (1953) The Story of Glass. Harcourt, Brace and Co.

Dyer D, Gross D (2001) The Generations of Corning. The Life and Times of a Global Corporation. Oxford University Press

Ellis WS (1993) Glass—Capturing the Dance of Light. National Geographic 184:37–60

Eskilson S (2018) The Age of Glass—A Cultural History of Glass in Modern and Contemporary Architecture. Bloomsbury Academic Press

Gin S, Jollivet P, Rossa G, Tribet M, Mougnaud S, Collin M, Fournier M, Cadel E, Cabie M, Dupuy L (2017) Atom-Probe Tomography, TEM and ToF-SIMS study of borosilicate glass alternation rim: A multiscale approach to investigating rate-limiting mechanisms. Geochim Cosmochim Acta 202:57–76

Graham MBW, Shuldiner AT (2001) Corning and the Craft of Innovation. Oxford University Press

Harden DB (1933) Ancient glass. Antiquity 7:419–428

Harmand S, Lewis JE, Feibel CS, Lepre CJ, Prat S, Lenoble A, Boes X, Quinn RL, Brenet M, Arroyo A, Taylor N, Clement S, Daver G, Brugal J-P, Leakey L, Mortlock RA, Wright JD, Lokorodi S, Kirwa C, Kent DV, Roche H (2015) 3.3-million-year-old stone tools from Lomekwi 3, West Turkana, Kenya. Nature 521:310–315

Hartmann P, Jedamzik R, Reichel S, Schreder B (2010) Optical glass and glass ceramic historical aspects and recent developments: a Schott view. Appl Optics 49: D157–D176

Hench LL (2006) The story of Bioglass®. J Mater Sci: Mater Med 17:967–978

Höche T (2010) Crystallization in glass: elucidating a realm of diversity by transmission electron microscopy. J Mater Sci 45:3683–3696

Houghton AA (1915) Contributions of the chemist to the glass industry. J Ind Eng Chem 7:290–292

Hovestadt H (1902) Janear Glass und seine Verwendung in Wissenschaft und Technik [Jena Glass and its Scientific and Industrial Applications] Everett JD, Everett A (trans) MacMillan and Co.

Jani Y, Hogland W (2014) Waste glass in the production of cement and concrete—A review. J Environ Chem Eng 2:1767–1775

Jones JR (2013) Review of bioactive glass: From Hench to hybrids. Acta Biomater 9:4457–4496

Keller PA (1991) The Cathode-ray Tube: Technology, History and Applications. Palisades Press

Kolb KE, Kolb DK (1988) Glass Its Many Facets. Enslow Publishers

Li S, Nguyen L, Xiong H, Wang M, Hu T, She J-X, Serkiz SM, Wicks GG, Dynan WS (2010) Porous-wall hollow glass microspheres as novel potential nanocarriers for biomedical applications. Nanomed Nanotechnol Biol Med 6:127–136

Liu H, Fu Z, Yang K, Xu X, Bauchy M (2021) Machine learning for glass science and engineering: a review. J Non-Cryst Solids 557:119419

Macbeth GA (1908) The history of glass making. the ancient origin of glass. Scientific American April 25, 1908, p271

Main D (2018) Humankind's most important material. The Atlantic April 7, 2018

Maloney FJT (1968) Glass in the Modern World. Doubleday

Mecking O (2013) Medieval lead glass in Central Europe. Archaeometry 55:640–662

Möncke D, Topper B, Clare AG (2022) Glass as a state of matter—the "newer" glass families from organic, metallic, ionic to non-silicate oxide and non-oxide glasses. Rev Mineral Geochem 87:1039-1088

Morse DL, Evenson JW (2016) Welcome to the Glass Age. Int J Appl Glass Sci 7:409–412

Murube J (2014) The early use of glass for optical correction and discovery of the microscope. Ocul Surf 12:162–166

Nascimento MLF, Zanotto ED (2016) On the first patents, key inventions and research manuscripts about glass science & technology. World Patent Info 47:54–66

Pedone A (2016) Recent advances in solid-state NMR computational spectroscopy: The case of alumino-silicate glasses. Int J Quantum Chem 116:1520–1531

Phillips CJ (1948) Glass: The Miracle Maker. 2nd Ed. Pitman Publishing Corp.

Pilkington LAB (1969) The float glass process. Proc Roy Soc Lond A 314:1–25

Pliny (the Elder) The Natural History of Pliny vol 6, Bostock J, Riley HT (trans) H.G. Bohn, 1857

Pye LD (2016) Editorial: Arrival of the Glass Age Affirmed. Int J Appl Glass Sci 7:407–408

Sabia R, Edwards M, VanBrocklin R, Wells B (2006) Corning 7972 ULE material for segmented and large monolithic mirror blanks. Proc SPIE 6273 Optomech Tech Astronomy 627302

Schmitz A, Kaminski J, Scalet B, Soria A (2011) Energy consumption and CO_2 emissions of the European glass industry. Energy Policy 39:142–155

Trigger BG (2006) A History of Archaelogical Thought. 2nd Ed. Cambridge University Press

Verità M (2013) Venetian soda glass. *In:* Modern Methods for Analysing Archaeological and Historical Glass, Vol. 1. Janssens KHA (ed) John Wiley & Sons, p 515–536

Youngman RE (2021) Borosilicate glasses *In:* Encyclopedia of Glass Science, Technology, History and Culture. Richet P (ed) John Wiley & Sons, p 867–877

Zerwick C (1990) A Short History of Glass. Harry N. Abrams, Inc.

Reviews in Mineralogy & Geochemistry
Vol. 87 pp. 1039-1088, 2022
Copyright © Mineralogical Society of America

Glass as a State of Matter—The "newer" Glass Families from Organic, Metallic, Ionic to Non-silicate Oxide and Non-oxide Glasses

Doris Möncke, Brian Topper, Alexis G. Clare

Inamori School of Engineering at the New York State College of Ceramics
Alfred University
Alfred, NY
USA

moncke@alfred.edu; BT13@alfred.edu; clare@alfred.edu

OVERVIEW

In theory, any molten material can form a glass when quenched fast enough. Most natural glasses are based on silicates and for thousands of years only alkali/alkaline earth silicate and lead-silicate glasses were prepared by humankind. After exploratory glass experiments by Lomonosov (18th ct) and Harcourt (19th ct), who introduced 20 more elements into glasses, it was Otto Schott who, in the years 1879–1881, melted his way through the periodic table of the elements so that Ernst Abbe could study all types of borate and phosphate glasses for their optical properties. This research also led to the development of the laboratory ware, low alkali borosilicate glasses. Today, not only can the glass former silicate be replaced, partially or fully, by other glass formers such as oxides of boron, phosphorous, tellurium or antimony, but also the oxygen anions can be substituted by fluorine or nitrogen. Chalcogens, the heavier ions in the group of oxygen in the periodic table (S, Se, Te), on their own or when paired with arsenic or germanium, can function as glass formers. Sulfate, nitrate, tungstate and acetate glasses lack the conventional anion and cation classification, as do metallic or organic glasses. The latter can occur naturally—amber predates anthropogenic glass manufacture by more than 200 million years.

In this chapter, we are going to provide an overview of the different glass families, how the structure and properties of these different glass types differ from silicate glasses but also what similarities are dictated by the glassy state. Applications and technological aspects are discussed briefly for each glass family.

1. INTRODUCTION

Glass is not confined to a specific material, such as silicates, but rather is a state of matter (Zanotto and Mauro 2017). In theory, any molten material can form a glass when quenched fast enough and with new high temperature techniques, such as laser levitation melting, the glass forming range is extended all the time (Weber et al. 2005; Masuno et al. 2011; Nasikas et al. 2014). The latest definition of glass designates it as any material that lacks a long-range order, can be prepared in bulk rather than only in an amorphous film, possesses a glass transition region, and exist in a non-equilibrium state according to classical thermodynamics" (Zanotto and Mauro 2017).

Today, most people equate glass with window glass or beautiful vases and objects, reflecting a long history of humankind's use of the first natural glasses, such as obsidian

1529-6466/22/0087-0023$05.00 (print)
1943-2666/22/0087-0023$05.00 (online)

for blades and drills, which in the second millennium before the common era (BCE) were replaced by artificial glasses prepared from sand, plant ash and minerals (Schweizer 2003; Henderson 2013). For thousands of years, many developments in glass production were limited to advances in technology, e.g. the invention of the clay-core technique for the preparation of hollow vessels (Tite et al. 2002), followed around the beginning of the common era (CE) by the invention of the blowpipe that allowed the production of larger vessels in shorter time (Schweizer 2003; Henderson 2013). Advances in raw material selection (Brain and Brain 2016), homogenization as by the new stirring process used by Fraunhofer and Guinand (Vogel 1992), furnace technology (Charleston 1978; Chopinet 2012, 2019), and the emergence of large scale flat glass production spread the use of glassware, as did later automatization of production (Uusitalo 2010; Axinte 2011; Chopinet 2019), see also Youngman (2022, this volume) for a review of the history of the silicate based glass industry. However, until the 19th century, all glass manufacture was based on seven oxides only: SiO_2, Na_2O, CaO, Al_2O_3, K_2O, FeO and PbO (Zschimmer and Cable 2013) with traces of additional oxides from impurities or intentionally added as colorants (CuO, FeO, CoO, MnO) or opacifiers (SnO, $CaSb_2O_6$, $PbSb_2O_7$) for a variety in appearance (Möncke et al. 2014, de Ligny and Möncke 2019).

The first systematic studies on new compositions involving other modifiers or even glass forming oxides, can be traced to the Russian renaissance man Lomonosov who lived in the 18th century (Leicester 1969). During the 19th, the pastor Harcourt (19th ct) (Kurkjian and Prindle 1998), introduced in England 20 additional elements into glassmaking. His glasses were subsequently studied for their optical properties by Stokes. However, the optical quality of the glasses was poor, and many were hygroscopic. At the same time, the German chemist Döbereiner attempted to introduce large amounts of barium and later strontium into glass, hoping to affect the refractive index, but neither of these experiments resulted in new applications (Vogel 1992). A short while later, in the years 1879–1881, Otto Schott melted all types of borate and phosphate glasses, experimentally working his way through most cation oxides from the periodic table of the elements. He mailed his glass samples to the physics professor, Ernst Abbe in Jena, Germany, who then measured their refractive index (Kühnert 2012). Otto Schott overcame the problem of poor optical quality by employing a stirrer during melting and was soon invited to Jena where in 1884, Abbe and Zeiss installed him in an experimental glass laboratory, which later would become the Schott company (Kühnert 2012). Despite their quest for new optical glasses, the first commercial success was the development of heat resistant glasses for gas lamps and up to this day, laboratory ware is made of this low alkali borosilicate glass composition. With this feat, the modern glass era began. Based on glass science, new glasses of varying compositions and varying properties are developed and produced industrially for optics and photonics (Weber 2006; Richardson et al. 2010), architecture (Pariafsai 2016), or container glasses (Schaut and Weeks 2017). Newer applications include thin display glasses for smart phones (Wang and Zimmer 2015), glasses for use in biomedicine (Hench 2006; Jones 2013), nuclear waste immobilization (Ojovan and Lee 2011; Day and Ray 2013), solid state batteries (Ravaine 1980), and hybrid materials (Jones 2013; Ediger et al. 2019).

While many of these glasses are still based on silicates (see Youngman 2022, this volume), silica free glasses, often based on borates, phosphates (Musgraves et al. 2019) or tellurites (Barbosa et al. 2017), are commercially produced today. Even oxygen free glasses are not uncommon, see for example chalcogenide glasses for infrared optics or fluoride glasses as laser glasses (Lucas et al. 2018). Oxynitride glasses and glass films are studied for their hardness and high ion conduction (Hampshire 2003; Garcia et al. 2016). Halogen containing glasses are also studied for ion conduction (Machida et al. 1992). Nonlinear optical as well as magneto optical properties are known for glasses with high loads of highly polarizable post-transition metal ions and rare earth elements, respectively (Weber 2006).

Early glass scientists, such as Zachariasen (1932), postulated rules for oxidic glass forming systems, that helped to define pure glass forming oxides such as SiO_2, B_2O_3, V_2O_5, Nb_2O_5, Ta_2O_5, As_2O_5, As_2O_3, Sb_2O_3, P_2O_5, and BeF_2 as the only fluoride. However, as more glasses were prepared, mixed systems that do not contain any traditional glass former were also shown to exist in the vitreous state, such as for example Al_2O_3–CaO (Akola et al. 2013).

The network theory by Zachariasen and Warren requires the formation of low-order three-dimensional networks for a glass. Such networks are based on network formers which are usually three- or four-fold coordinated, such as oxides of Si, B, P, Ge, As, Be (Be with F), etc. Network modifiers, such as Na, K, Ca, or Ba ions exhibit usually higher coordination numbers. Intermediates may mimic either role, in low coordination of four, they might reinforce the network while in higher coordination they might loosen the network, though on their own, intermediates will not form a glass (Vogel 1992; Shelby 2005).

Goldschmidt et al. (1926) proposd that inorganic non-metallic glass formation occurs for simple compounds for which the ratio of the radius of the cation versus the radius of the anion, r_C/r_A, lies between 0.2 and 0.4. This rule holds for many glasses, but assumes a higher ionicity of bonds than actually present and for example fails for BeO which is not a glass former, though the r_C/r_A value of 0.221 would classify it as such (Paul 1990).

Based on Goldschmidt's conditions, a series of compounds were predicted to theoretically form glasses—some even with different anions than oxygen. While some of these glasses were practically available at the time, others were only projected to give a glass. Classified with increasing anion valence, glass formation was for example predicted for monovalent fluoride glasses (see Section 4.2) and for divalent oxide glasses, including silicates (see other chapters in this volume), borate (see Section 5.2), phosphate (see Section 5.1), germanate (see Section 5.4), arsenate (see Section 5.5.1), chalcogen and chalcogenide glasses (see Section 4.1), etc. Predictions extended even to trivalent nitride glasses, which were only prepared much later (see Section 4.4), and to tetravalent carbide glasses, as realized for example in sol–gel oxycarbide glasses, discussed by Rouxel et al. (2001).

Invert glasses or ionic glasses do not possess a covalent three-dimensional network. Many salts can be quenched resulting for example in sulfate (Sen et al. 2006), nitrate (Duffy and Ingram 1968; Ingram and Lewis 1974), tungstate (Duffy 1977; Montanari et al. 2008) or acetate (Duffy and Ingram 1969; Ingram and Duffy 1970; Wilk Jr and Schreiber 1997; Stokes and Schreiber 2006) glasses. All these materials show a glass transition and a lack of long-range order, though they do have a well-defined short-range order. These ionic glasses, as listed above, as well as invert glasses such as orthosilicate (Nasikas et al. 2011), orthophosphate (Machida et al. 1992), or orthoborate systems (Winterstein-Beckmann et al. 2013, 2015), are made of negatively charged anions of oxo-complexes that are cross-linked by ionic bonds to positive charged cations. The coordination numbers are higher than the ion charges—considering NO_3^- or SO_4^{2-} and PO_4^{3-} as pseudospherical anions with delocalized negative charge. Empirically it can be said, that the higher the variability of cations in size and charge, the better is the glass formation for ionic and invert glasses.

In other glass forming systems, even the division of anions and cations is overcome, and predominantly covalent bonds prevail. These materials include all organic glasses such as amber (Zhao et al. 2013) or plexiglass (Keshavarz et al. 2016), but also chalcogen glasses (Shelby 2005). Like amorphous sulfur (Yuan et al. 2018), organic glasses can also occur naturally—amber predates anthropogenic glass manufacture by several hundreds of millions of years (Bray and Anderson 2009) .

Metallic glasses on the other hand have metallic and covalent bonds, where the metal cations are embedded in a shared electron cloud with some covalency (Gaskell 1982; Wright et al. 1985; Chen 2011; Kruzic 2016). Just as in any other glass, metallic glasses lack a long-range order, though superstructural units in the form of large clusters are known. And as any other glass, metallic glasses do exhibit a glass transition temperature rather than a fixed melting point.

This chapter intends to provide an overview of the many different glass families. The discussion will focus on how the structures and properties of these various glass types differ from silicate glasses as well as what similarities are dictated by the glassy state. Applications and technological aspects are discussed briefly for each type of glass system.

1.1 The glassy state of matter

Too many exceptions to the old definition of glasses as undercooled melts demanded an update on the definition of glasses, the most recent given by Zanotto and Mauro (2017):

> *"Glass is a nonequilibrium, non-crystalline condensed state of matter that exhibits a glass transition. The structure of glasses is similar to that of their parent supercooled liquids (SCL), and they spontaneously relax toward the SCL state. It is proposed that their ultimate fate, in the limit of infinite time, is to crystallize though no proof of this exists."*

While there is some discussion on the ultimate fate of glasses, authors agree that glasses lack a long-range order or translational or rotational order, are not in a thermodynamic equilibrium, and exhibit a glass transition range (Narasimhan et al. 1990). Especially at low temperatures, kinetics overrule thermodynamics and the glassy system cannot find its global potential minimum, that is, it would take an infinitely long time for the system to crystallize (Narasimhan et al. 1990; Zanotto and Mauro 2017).

The consensus in glass science is, generally, that a solid material that can be prepared in bulk, exhibits a glass transition and is defined by the absence of a long-range order, is a glass, regardless of the material or preparation route. Glasses may be viewed as amorphous solids but not all amorphous solids are glasses. Both may be prepared in bulk and possess no periodicity; however, an amorphous solid does not necessarily exhibit a glass transition. Zanotto and Cuthino (2004) estimated the number of possible glasses that can be made from the 80 "useful" elements of the periodic table and derived at a number of $>10^{300}$ glasses, by far exceeding the ca. 2,000,000 reported glasses in 2004 when they wrote their letter. See Figure 1 for an overview on glass forming elements.

Most people would define a glass as the ubiquitous, transparent silicate-based material used in spectacles, windows, and container or stemware. However, considering the much more robust definition above, the scientist is well off viewing glass as a state of matter rather than a certain composition.

1.2 The glass family tree

Considering the above definition of a glass, we can expand our view of the modern glass family significantly. Figure 2 gives an overview of today's classification of different glasses.

In the following we want to discuss briefly the structural characteristics and properties of each of the glass systems listed in Figure 2, starting with organic and metallic glasses, both systems being the least similar to the silicate-based glasses discussed so far in this volume. At first glance, glasses based on organic polymers, oxides, or metallic alloys appear to represent very distinct structural families. The former types range from covalent "framework" materials with very open structures to mixed oxides with considerable ionic bonding, while metallic alloys, appear to be close-packed though most demonstrate some covalency, with bonding dominated by delocalized "metallic" orbitals. There are, nonetheless, definite similarities in the structural questions relating to each class of amorphous solid. For example, the notion of local structural units is a concept generally accepted in covalent glasses but has been disputed in amorphous alloys (Gaskell 1982)—though might be present in glasses, as will be show in Section 3.

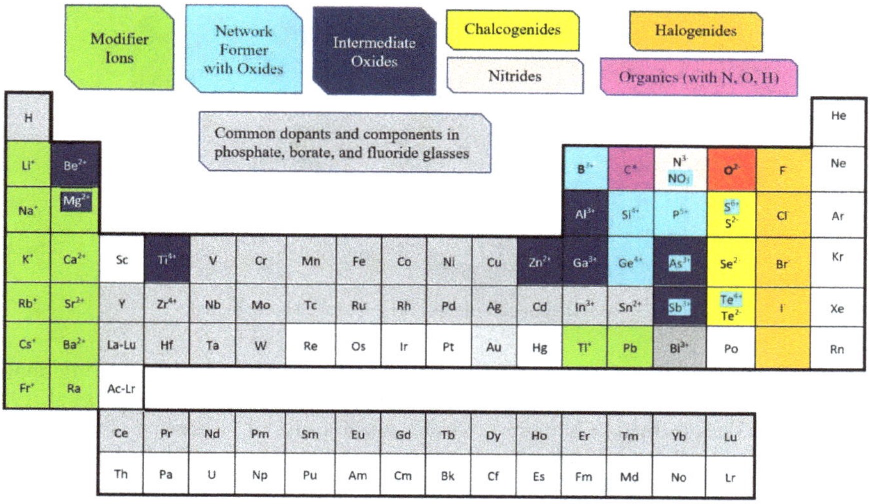

Figure 1. Elements of the periodic table that are commonly found in glasses. Some elements can play different roles in different coordination or oxidation state.

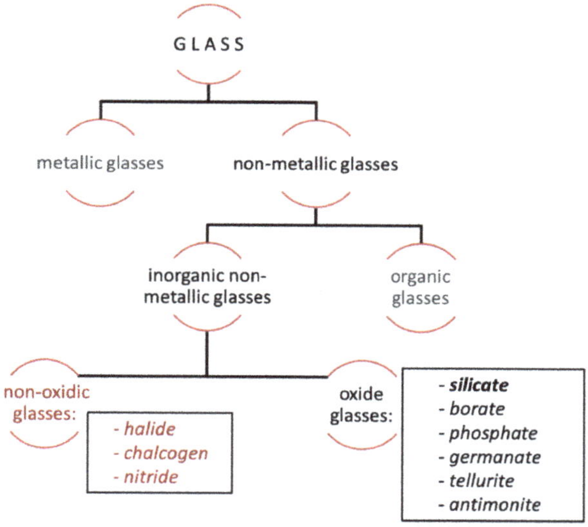

Figure 2. Schematic overview of modern glass families according to material groups, Special glasses often consist of any combinations of the shown glasses.

1.3 Glass models

Since so many different glass types exists, very different models are used to describe favorable conditions for glass formation. We mentioned earlier the glass forming rules from Zachariasen for oxide glasses (Zachariasen 1932). However, as a general rule, any process that slows down crystallization can be seen as advantageous for glass formation. For materials that contain long chains, the random coil model applies, as depicted in Figure 3. This is seen for metaphosphate glasses (Inaba et al. 2015) as well as for organic polymer or elemental chalcogen glasses (Zallen 1998).

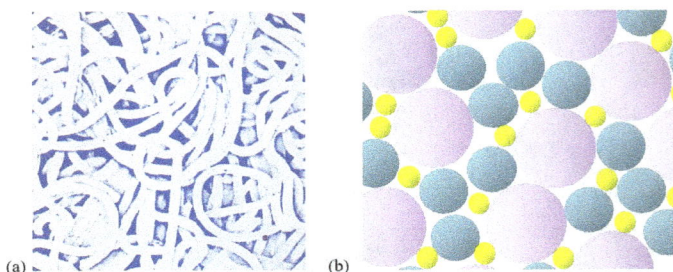

Figure 3. (a) Random coil model; **(b)** random packing model.

In metallic glasses, the random packing model with some near range ordering applies though metallic glasses exhibiting covalent structures have been reported (Wright et al. 1985; Miracle 2004). Glass formation is more favorable when a mix of larger and smaller atoms, ions, or molecules create such a disorder, that easy crystallization is prohibited (Greer 1993; Chattopadhyay and Murty 2016). Differentiation via length scales is important for structural discussions in all glasses. For the local or short-range structure, nearest-neighbor correlation is considered, most prominently the distance to and number of neighbors, and the site symmetry, with an implicit distance scale from 3 to 5 Å (Gaskell 1982). The intermediate range extends roughly from 5 to 20 Å (Gaskell 1982). Depending on the glass type, the intermediate range gives well defined superstructural units such as the boroxol ring. The transition to a disordered long-range environment, length-scales larger than 20 Å, might be fluent, as large polyanions have been described for some systems (i.e., by Möncke et al. 2016).

Glass formation usually is favored by strong, directional, predominantly covalent bonding in well-defined polyatomic clusters, which link together to form three-dimensional networks, as is typically seen for oxide and chalcogenide glasses. On the other hand, bulk amorphous alloys can resemble closely packed arrangements of more or less spherical atoms with some covalency, for example, in Dy_7Ni_3 (Wright et al. 1985).

Examples of vitreous materials that have no three-dimensional network are glasses based on the random coil model, such as metaphosphate, organic polymers or amorphous sulfur or selenium. An example for a two-dimensional amorphous material is glassy carbon. However, despite the name and many glass-like properties such as conchoidal fracturing and a lack of long-range order, vitreous carbon does not exhibit a glass transition temperature and is prepared by pyrolysis and no liquid of similar structure exists. Therefore, while glassy carbon is strictly speaking an amorphous solid, it is not a glass.

The bonding type can vary from strictly covalent as in selenium or sulfur, to mixed ionic-covalent bonding in silicates and most oxide glasses or in the hydrogen-bonded organic glasses, to almost purely ionic forces in mixed alkaline earth nitrates, sulphates, or heavy-metal fluorides to finally, delocalized metallic bonding in amorphous alloys (Gaskell 1982).

2. ORGANIC GLASSES

Many polymers can form glasses or have large amorphous regions. Chain entanglement prevents long polymer chains from forming a well-ordered lattice; however, they can crystallize by chain folding. This glass model, as depicted in Figure 3, is called the random coil model for glass formation and can also be observed in some inorganic glasses, e.g. chalcogen glasses (Section 4.1) or metaphosphate glasses (Section 5.1).

Polymer glasses have the advantage that they are often very inexpensive, and that preparation is fairly uncomplicated. Optical effects are often very strong, e.g. non-linear optical properties like SHG (Weber 2006). However, these advantages are often offset by a low long-term stability, e.g. degradation under irradiation or higher temperatures. The properties of polymeric glasses can easily be fine-tuned by variations in the composition or mode of preparation; however, reproducibility is often a problem when working with such large macro molecules and the accuracy of data on plastics, including mechanical or chemical resistance properties, are generally lower than those of optical glasses, depending strongly on the supplier (Weber 2006). See Table 3 for the properties of plexiglass in comparison to other inorganic glasses. Weber summarized the ranges of selected optical, mechanical and thermal properties of polymers compared to inorganic glasses based on data from Cook and Stokowski (1986). It was found that for example the refractive index is on average lower for polymers than for inorganic optical glasses, though a wide overlap exists. The dispersion range on the other hand is similar for inorganic as for organic glasses, while the refractive index change with temperature is much higher and the transmission range is on average slightly smaller for polymers. The thermal expansion coefficient and heat capacity are higher, the softening temperature lower for polymers than for inorganic glasses. The density of polymers is significantly lower, as is the Young modulus when compared with inorganic glasses.

The nomenclature of organic glasses and polymers differs in some respects from inorganic glasses, as for example various aspects of length scale are important qualifiers. The monomer is the smallest repetitive unit formed by linked elements. Most common elements in polymers are carbon (C) and hydrogen (H), but also oxygen (O), nitrogen (N) or sulfur (S), while halides and other elements may be incorporated as well. Connected monomers form a macromolecule, in which repetitive units of monomers form statistical segments. These segments are often around 1 nm in length and can be ordered in different ways. A random coil region might have a chain radius of gyration around 10 nm (see Fig. 3).

Organic glasses have a much lower transition temperature (T_g) than most inorganic glasses, covering a range that extends even below room temperature. As expected, the properties of sub T_g and supra T_g glasses differ substantially (Gaskell 1982). Hard plastics such as acrylic glass (including the trademark plexiglas) are based on polymethylmethacrylate (PMMA) or polystyrene and have a T_g that is more than one hundred centigrade above room temperature (Jadhav et al. 2009). Elastomers such as, polyisobutylene are normally handled above their T_g, giving them their rubber-like, soft and flexible nature (Jadhav et al. 2009).

Linear structures or chains of hydrocarbons are often linked by weak van der Waals forces. Branched structures reduce the packing density compared to linear polymers and polymers based on branch units often have a lower density relative to comparable linear polymers (Fig. 4). Chains can also be connected via covalent bonds as chemical crosslinking occurs

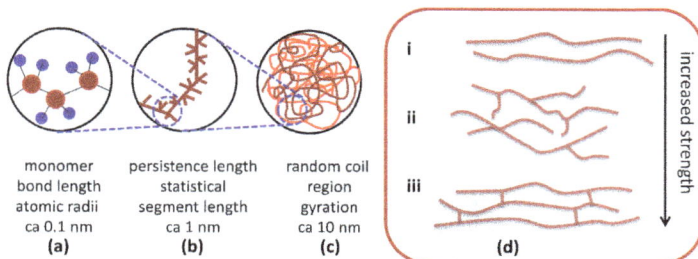

Figure 4. (a) short-, **(b)** medium-, and **(c)** long-range order in polymeric glasses, and **(d)**: schematic how varying interactions between polymer chains impact the strength and transition temperature, in increasing order **(i)** linear polymers, **(ii)** branching polymers and **(iii)** cross-linked polymers.

between chains (see for example Möncke et al. 2011). The mechanical strength and T_g of such polymers depend on the flexibility, that is, on chemical crosslinking as well as on van der Waal forces and dipolar interaction such as hydrogen bridges.

In the following we want to differentiate between mostly artificial polymeric glasses and their properties and applications, natural amber and metal–organic framework glasses.

2.1 Man-made polymeric glasses

Polymeric glasses, like inorganic glasses, have attracted interest for their optical properties, since the absence of grain boundaries removes scattering centers, improves macroscopic homogeneity and provides transparent materials (Ediger et al. 2019). Additionally, polymeric glasses are often mechanically more stable than their crystal counterparts, have a better formability, and the composition can be tuned by a higher flexibility of mixing different components over a wide range. Many polymers are glass-ceramic like, containing crystalline spherulites. Typical applications with OLEDs (organic light emitting diodes) are conventional display screens in which vapor-deposited glasses of organic semiconductors are the active elements (Ràfols-Ribé et al. 2018). These layers allow for the transport of holes and electrons, and whenever a pair of such opposite charges recombine on an organic emitter, light is produced (Ediger 2017).

Like any other glasses, polymeric glasses are usually isotropic, though anisotropy can be induced by preparation, e.g. cold vapor deposition of rod or platelet formed molecules (Ediger et al. 2019), or in extrusion above T_g, when the partially order like alignment of chains in pulling direction is frozen in (Batterman and Bassani 1990). Anisotropy can be measured by the glasses' birefringence, since the latter is closely related to the molecular orientation and in solids, the index of refraction depends upon the direction of polarization of light, see Figure 5 (Batterman and Bassani 1990; Ediger et al. 2019).

Polymers that contain many platelets, like benzene rings or large conjugated networks, can be stacked randomly, or in various directions. The type of stacking can affect many properties such as birefringence or thermal conductivity (Meille Stefano et al. 2011). Likewise, for rods or chains, the partial alignment via chain folding can induce anisotropy. See also Section 5.1, on phosphate glasses, here, as for chainlike polymers, anisotropy can be enforced, for example by directional pulling above T_g, and freezing in of such anisotropies below T_g, even after removing the applied stress. Such polymers might however show frozen-in residual stresses— as displayed in Figure 5 (Meille Stefano et al. 2011).

Applications include photorefractive materials, exploiting properties that include photo-charge generation, photoconductivity, charge trapping, and nonlinear or a birefringent optical response. Often, the active component is dissolved in a polymer host or chemically attached to the polymer (Lundquist et al. 1996).

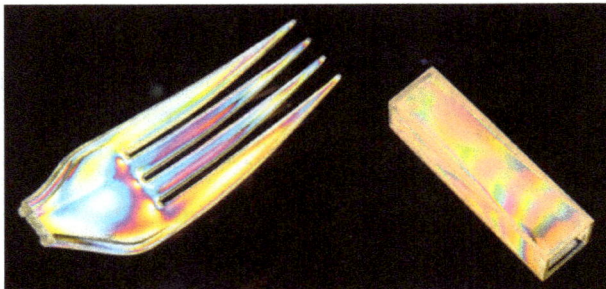

Figure 5. Birefringence is apparent in many plastics when viewed under polarized light, usually as molecules and segments of chains are frozen in while in a stretched formation during molding and processing, on the left a plastic fork and on the right a plastic cuvette (Photo Topper).

2.2 Amber

Ambers constitute a class of fossilized sediments of organic matter, terpenoid resins from higher plants such as conifers (Lambert and Poinar 2002; Bray and Anderson 2009). As a gemstone, amber is unique due to its organic origin; amber is found around the world, with some deposits as old as 320 million years (Bray and Anderson 2009; Zhao et al. 2013). While pieces of amber can be found at the beach, large underground deposits are exploited commercially. When terpenoid compounds in the resins polymerize after exposure to light or air the well-known amber ensues. The properties and color vary with age and thermal history (see also Fig. 6; Anderson et al. 1992).

Figure 6. Photograph of a 2.3 kg piece of amber found 1992 in La Harve in the river Seine (France) and at display at the Inselgoldschmiede & Schmuggelkiste, Langeoog, Germany (Photo Möncke). The chemical formula of the basic repeating units of a regular labdanoid as is typical for Baltic Amber; R can vary e.g. $-CH_3$, $-CH_2-OCH_3$, $-CH_2OH$, $-COOCH_3$ (Bray and Anderson 2009).

Different classification schemes of ambers can be found in the literature (see Table 1) (Bray and Anderson 2009). The most abundant and perhaps best-known ambers are those from the Baltic Sea and adjacent regions. They consist of regular poly-labdanoid ambers, containing succinic acid and are classified as Class Ia (Bray and Anderson 2009). The schematic in Figure 6 shows the structure of regular poly-labdanoid, which with various derivatives is the basis of Baltic amber (Lambert and Poinar 2002). Another classification is based on structural differences as analyzed by NMR or GC/MS (Anderson et al. 1992), finding similar major groupings that can be correlated to certain trees and geological regions and times.

Like other glasses, ambers lack a long-range order, though the short-range order of the basic molecular units can be well defined. Ambers do exhibit a glass transition temperature, and a very low fictive temperature—which however cannot be directly linked to the age and subsequent relaxation of the ambers (Zhao et al. 2013). The Boson peak is evident in the Raman spectra, as it is for all organic and inorganic glasses.

2.3 Organic metal framework glasses

Organic metallic framework (OMF) glasses are a relatively young glass family that so far only attracts academic interest (Bennett et al. 2016, Zhou et al. 2018). Organic complexes of metal cations can form glasses after melting. Relatively low temperatures of thermal decomposition are the main problem, and most organic frameworks are actually crystalline solids. Since the metal–ligand connectivity is retained in the glass, MOFs are distinct from the other glass families, that is from organic, metallic, and the large group of inorganic non-metallic glasses. However, it should be noted, that some hybrid materials that contain organic molecules in an inorganic oxyfluoride glass, or glass particles in an organic polymer, exist as well (see Section 5.1.7.3 or the field of sol gel glasses, Shirosaki et al. 2012).

Table 1. Classification of different amber types according to Anderson et al. (1992), group classification after Lambert and Poinar (2002), with suggested region of findings, or when available time of origin and finally main polymeric components.

Class	Group	Area	Age	Polymers of
Ia	C	Europe	Tertiary	Labdatriene carboxylic acid, communic acid, communal, succinic acid
Ib	A7	Worldwide, New Zealand and Australia,	Cretaceous	Labdatriene carboxylic acid, communic acid, can contain communal, no succinic acid
Ic	D	Americas and in Africa		Labdatriene carboxylic acid, ozoic acids and/or zanzibaric acid
II	B	India across the Pacific to North America	Tertiary	Bicyclic sesquiterpenoid and others
III		New Jersey, Germany		Polystyrene
IV		Moravian resinite		Sesquiterpenoid, cedrane skeleton

Up to now, preparation occurs mostly in DSC pans, the little crucibles that hold the sample during differential scanning calorimetric measurements. The corresponding crystalline MOFs might be used for gas sorption and separation, catalysis, drug delivery, conductive, and multiferroic applications. The wide variety of combinations between different metal nodes and all kind of organic molecules of varying functionality opens almost infinite possibilities of properties and applications. For vitrification, especially zeolitic imidazolate frameworks (ZIFs) are studied, mostly because they prove to have a high chemical stability and display many structural similarities to classical zeolite networks. $Zn(Im)_2$ has for example a melting temperature of 590 °C and a transition temperature of 292 °C (Bennett et al. 2016).

3. INORGANIC METALLIC GLASSES

Metallic glasses exhibit a transition temperature and can be prepared as bulk material, despite their very high crystallization tendency due to their metallic bonds (Miller and Liaw 2008). Like any other glass, metallic glasses lack a long-range order and subsequently possess no metal grain boundaries. However, a topological and chemical short-to-medium range order has been shown to occur in alloys, due to their high atomic packing density and the varying chemical affinity between the constituent elements (Cheng and Ma 2011). Compared to their crystalline counterparts, metallic glasses possess a slightly lower density. For example, metallic glass ribbons, which are rapidly solidified, are 2–3% less dense than bulk metallic glasses, which in turn are about 0.5 % less dense than metallic crystals (Miracle 2012; Suryanarayana and Inoue 2017). Some elemental amorphous metals can be prepared by evaporation or in some cases by other very fast quenching methods such as ribbon quenching on a cooled fast spinning wheel; however, these materials are often relatively unstable and crystallize well below room temperature. Studying binary and ternary alloys was often more successful, many of which can be prepared by quenching from the liquid at high cooling rates. As stated by Gaskell, the process of melt quenching often limits the thickness though there seems to exist no natural upper thickness limit (Gaskell 1982). All bulk metallic glasses (BMG) are alloys, since one component metals tend to crystallize, even though glass scientists love to say that all materials can be made into a glass when quenched fast enough (Zanotto and Coutinho 2004).

Even though the structure of metallic glasses has often been associated with random packing of spheres (see Fig. 3) that is, large and small cations are randomly packed in their shared electron cloud (Yue et al. 2017), it should be noted that in analogy to crystalline metals, which are not always close packed and have a degree of covalency to their bonding, the same is true of metallic glasses. Some similarity to the structure of fluoride glasses, also described by random packing models, has been suggested (see Section 4.2).

The main difficulties in the study of metallic glass structures are related to the small sample size and the fact that even larger samples are complicated, multicomponent alloys. The latter allows for larger bulk samples, prepared by more conservative cooling in which the disordered structure that ensues is believed to be due to the "confusion principle" inhibiting crystal formation. Thus, any structural technique adopted must either be able to measure small volume samples (if compositional simplicity is required to be able to analyze data) or be able to analyze multicomponent samples if larger samples are required. Here, we want to briefly expand on some details taking the example of a Ni–Dy glass from an early study by Hannon et al. (1991). For a careful structural study, it is best to have a limited number of elements, and the Ni and Dy alloy could be prepared and studied in bulk. Combining the isotopic and magnetic dependence of neutron scattering structural studies showed that the structure of the glass was more ordered than a random packing of spheres. Unlike-atom near neighbor distances were not found midway between the two like-atom neighbor distances, which suggests some degree of covalent bonding, which is not unreasonable given that both Ni and Dy are metals with relatively high polarizability.

Advances in structural characterization and analysis of prototypical BMGs are discussed in a review paper by Cheng and Ma (2011), focusing on structural models and fundamental principles, as well as the correlations of thermodynamic, kinetic, and mechanical properties with the MG structures.

Mechanical studies of glasses focus often on metallic glasses, one of the few properties where BMG are directly compared to non-metallic inorganic glasses (Rouxel 2007). The high ductility and elasticity of BMG set them apart from the largest glass forming group, the inorganic non-metallic glasses, because metallic glasses do exhibit extreme plastic deformation, evidenced by extensive shear band formation and are not as brittle as non-metallic glasses, for which shear band formation is observed to a far lesser extent, as recently reviewed by Januchta et al. (Miller and Liaw 2008; Greer et al. 2013; Januchta and Smedskjaer 2019).

Inorganic metallic glasses have found their way in mainstream production, e.g. as high end softball bats or golf clubs, but they have also found their way into the core of electrical transformers for their potential to better minimize losses in comparison to polycrystalline metals (Chen 2011). Furthermore, metallic glasses are more corrosion resistant compared to their crystalline counterparts (Scully et al. 2007).

4. INORGANIC NON-METALLIC NON-OXIDE GLASSES

4.1 Chalcogen(ide) glasses

Chalcogenide glasses have been studied for over 60 years (Glaze et al. 1957; Savage and Nielsen 1965; Phillips 1979; Gaskell 1982; Feltz 1993). Despite the name, this group of glasses focuses on materials that do not contain oxygen but are comprised of any one or any combination of materials that contain the other elements from group VI. Glasses following the random coil model (see Fig. 3) are easily obtained from the elemental chalcogens sulfur or selenium; however, the metallic nature of tellurium necessitates rapid quenching to achieve a glass and requires additional elements to be stable (Bureau et al. 2008, 2009; Varshneya 2013). Interestingly, Se_xTe_{1-x} will form a glass of short Te-chains interacting via van der Waals forces with Se-chains, with minimal heteropolar bonding. The broader family of chalcogenide glasses are comprised of the divalent chalcogen combined with one or more elements from group IV or V. Common examples include As_2S_3 (Abdellaoui et al. 2018), As_xSe_y (King et al. 1995), and $GeSe_2$ (Musgraves et al. 2019).

Most Se- or Te-based glasses are semiconducting and appear black or have a metallic shine (see Fig. 7). Chalcogenide glasses display excellent transmittance in the IR wavelength range which makes them prime materials for use in fiber optics, night vision, thermal imaging, and satellite telescopes. Compared to SiO_2 for which the IR window closes at 3 μm, chalcogenides are transparent down to 15 μm (Feltz 1993; Lucas et al. 2018). However, the extended transparency in the IR range goes in hand with a loss of transmittance in the visible wavelength range, starting as low as 1 μm, see Figure 8 for a comparison of transmittance windows (Feltz 1993; Lucas et al. 2018). Sulfide glasses might be partially transparent in the visible range and are also used as a matrix for rare earth elements (Adam and Zhang 2014). The band gap in the visible range, which decreases with increasing polarizability of the ions from S > Se > Te, is due to the non-bonding electrons, which form an intermediate level between the bonding and antibonding levels.

Figure 7. Arsenic selenide glasses, inside the melting ampule and cut samples after melting at 600 °C for 18 hours (Photo Topper).

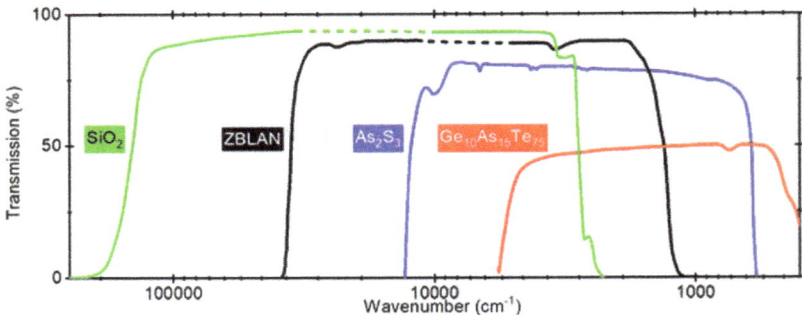

Figure 8. UV–Vis–IR transmission range of a selection of different glass systems, modified after data from suprasil, SiO_2, $d = 10$ mm (Heraeus Quarzglas GmbH & Co. KG 2016); ZBLAN, $d = 2$ mm (Ledemi et al. 2013), As_2S_3, $d =$ ca. 2mm (Verger et al. 2012) and $Ge_{10}As_{15}Te_{75}$, $d = 1.4$ mm (Yang et al. 2010).

The evolution of the structure in chalcogenide glasses is best understood by first considering the glass formed by Se alone. As a group VI element, selenium, like oxygen, has six outer electrons, of these, two are bonding, forming strong covalent sigma-bonds, while the other four electrons form two lone pairs (Lucas et al. 2018).The two bonds and two lone electron pairs can be viewed as a pseudo-tetrahedra, linking via corners to form a glass. Amorphous selenium forms thus a one-dimensional glass consisting of puckered chains with some 8 member rings, interacting via van der Waals forces with one another, resulting in a bulk structure resembling a bowl of spaghetti (see Fig. 3; Lucas et al. 2018). Due to the lack of crosslinking and the rotational freedom of the sigma bonds, selenium and sulfur glasses are very soft and the T_g might lie below room temperature, leading to the claim of having formed the first inorganic flexible bulk glass (Yuan et al. 2018).

The difficulty in forming a glass from tellurium alone is most probably due to the more metallic character of tellurium. This character leads to a delocalization of the electrons that form lone electrons pairs in sulfur or selenium and the formation of a metallic π-bond along the chains. The chains are more stringent, losing the rotational freedom, and instead of a structure that can be described by the random coil model, the material crystallizes. In ternary Ga–Ge–Te, the lone pair deficient Ga can trap the mobile electrons of the Te. Similarly, as shown by Lucas and Zhang (1990), addition of halogens (such as Cl, Br or I) can eliminate the π-bonding as the halogen ions "capture" the free electrons of the Te-chains. Thus, were born the TeX glasses, which show a transparency as low as 20 μm. These glasses are part of the appropriately named chalcohalide family (Sanghera et al. 1988). Two-dimensional chalcogenide networks consist of the higher coordination number group V elements, typically arsenic or antimony, and one or more of the chalcogen ions S, Se, or Te. These 2-dimensional chalcogenides possess higher glass transition temperatures making them more appropriate for technical applications. The divalent chalcogenides are usually linked to only two direct neighbors. Take As_2Se_3 as an example, the network is formed of pyramidal $AsSe_3$ units with a lone pair of electrons on the As-atoms. The same structure would arise if S was substituted for Se. If the composition is chalcogen-rich, the pyramidal $AsSe_3$ units are linked by Se chains. As_xSe_y can take on a more rigid structure when the arsenic content exceeds the stoichiometric quantity of As_2Se_3, leading to the formation of As_4Se_4 cage structures and the presence of As–As homopolar bonds (King et al. 1995). The addition of group IV elements, most frequently germanium, increases the connectivity of the network to 3-dimensions. $GeSe_2$ forms a vitreous network of corner sharing $GeSe_4$ tetrahedra analogous to the SiO_4 tetrahedra in vitreous silica. The presence of the tetravalent element increases crosslinking further resulting in an increased glass transition temperature. These glasses display more suitable mechanical and thermal properties for various applications compared to their lower dimensional counterparts (Lucas et al. 2018).

Of all combinations, binary sulfides and selenides of As and Ge have attracted the most attention on account of their large glass forming regions. The ternary Ge–As–Se system provides an illustrative example of the multi-dimensional bonding accessible to chalcogenides that allows great compositional flexibility. If the composition is selected to have excess chalcogen, the bonding consists of divalent Se atoms forming chains which are then crosslinked three-fold at the As atom sites and four-fold at the Ge atom sites (Zallen 1998). Given this flexibility in network formation, chalcogenide glass compositions can be readily tailored to have specific properties. Such as for example the glass $Te_{20}As_{30}Se_{50}$ (TAS), was developed for the best combination of large IR transparency, durability and good mechanical properties for use in IR spectroscopy. TAS glass has been optimized to develop optical fibers for medical and biological applications (Lucas et al. 2018). Interesting are problems related to the structure of nonstoichiometric phases which, while well known for crystals, do not appear to be a significant problem in other glasses (Gaskell 1982)

Manufacturing high purity chalcogenide glasses, free of impurities including oxygen and water, has received much attention for undisrupted infrared transmission (King et al. 1995; Seddon 1995; Danto et al. 2013). The melting of high-quality chalcogenide glasses is typically carried out by purifying the components, sealing them off in an ampoule under vacuum, melting in a rocking furnace to homogenize the melt, quenching the melt, and annealing the glass in the ampoule. Melting under a controlled atmosphere furnace is another preparation route. Once formed, the chemical durability of chalcogenide glasses is quite good (Lucas et al. 2018).

Chalcogenide glasses can be formed by molding, extrusion and drawing into fibers from a preform (Lucas et al. 2018). The window of stability, $\Delta T = T_c - T_g$, must also be carefully considered when processing these glasses for applications, which typically demands heating the system above the glass transition temperature in order to achieve the viscosity required for the particular forming process (Musgraves et al. 2014; Svoboda and Málek 2015).

Chalcogenide glasses possess a relatively low glass transition temperature which presents additional challenges for implementation in devices and instruments (see also Table 3). Additionally, they have displayed structural relaxation below their glass transition leading to aging effects where the properties of the glass change with the passage of time (Jensen et al. 2012). A variety of photoinduced phenomena take place as a result of the lone electrons on the chalcogen atoms (Zakery and Elliott 2003). Indeed, this nature was first employed for use in Xerox machines and later DVDs (Zallen 1998; Wuttig and Yamada 2007).

4.2 Fluoride glasses

Fluoride glasses challenge some traditional concepts of glass formation. Fluoride glasses—like metallic glasses—possess high crystallization tendencies, this time due to their strong ionic bonds and the very low viscosity of their melts (Ehrt 2015). Coming from oxide glasses, it was natural to describe glass formation in fluoride glasses based on structural criteria, that is, by defined polyhedra and their connectivities resulting in a random three-dimensional network (Poulain et al. 1992). In this scheme, glass forming ability is correlated to the possibility of shaping a disordered network of relatively small polyhedra (trigonal or tetrahedral). Applying the same approach to fluoride glasses is challenging, since the network forming polyhedra may be much larger, such as AlF_6 octahedra (Ehrt 2015), ZrF_7 or HfF_8 polyhedra. Thus, a contrary approach was to describe glass formation in fluoride systems as random ionic packing (Poulain et al. 1992). In this description, lager cations (Pb^{2+}, Ba^{2+}, Na^+ etc.) take the place of an anion in a random packing (Poulain and Maze 1988). The authors of this study emphasize that the atoms are not fully ionized, and binding electrons are located in the molecular orbitals that link neighboring atoms. In fact, the minimization of electrostatic energy leads, in most cases, to the minimization of the interatomic distances and is consistent with a very dense packing. In these conditions, the two models (i) random packing and (ii) network, reflect the different and complementary aspects of the same material. It appears that glass formation in fluoride systems requires at least one cation of mediate field strength (e.g. Al^{3+}, Be^{2+}, Zn^{2+}, Zr^{4+}, Hf^{4+} or La^{3+}) and that multicomponent glasses are often more stable than binary glasses, following the well-known confusion principle as it has been formulated for metallic glasses (see Section 3; Poulain and Maze 1988).

The transparency window of fluoride glasses depends naturally on the cations within the glass, high field strengths ions such as Be^{2+} or Al^{3+} provide in combination with fluorides a very high transparency in the UV, at even highe energies than pure SiO_2 (Ehrt 2018), while heavier, more polarizable cations such as Hf^{4+}, Zr^{4+} or Pb^{2+} will lower the intrinsic absorption edge. The wide transmission window is even more pronounced on the IR side of the spectrum, as the vibrational stretching modes of ZrF_4 are much lower in energy than of Si–O (Möncke and Eckert 2019; see Fig. 8). Depending on the sample thickness and composition, heavy metal fluoride glasses can be transparent beyond 7 µm in the IR. The presence of lighter cations can reduce the infrared window. The IR edge shifts to higher energies in the series: $AlF_3 < ZrF_4 < HfF_4 < ScF_3 < GaF_3 < InF_3$ (Poulain et al. 1992).

The transition temperature is relatively low, starting around 200 °C for alkali free glasses but reaching 500 °C in some alkali and barium containing glasses (see also Table 3), while the thermal expansion is relatively high, ranging for most fluoride glasses between 140 to 200×10^{-7} K^{-1} (Poulain et al. 1992). In fluoride glasses, the melting temperature T_m and transition temperature T_g can often be correlated by the 2/3 rule (Poulain et al. 1992).

For most fluoride glasses, atmospheric moisture is far less detrimental than exposure to liquid water. Some fluoride glasses such as certain fluoroaluminate and fluoro–indate glasses are even relatively stable when immersed in water (Poulain et al. 1992). BeF_2 glasses have very low refractive indices and high partial dispersion (see Fig. 9).

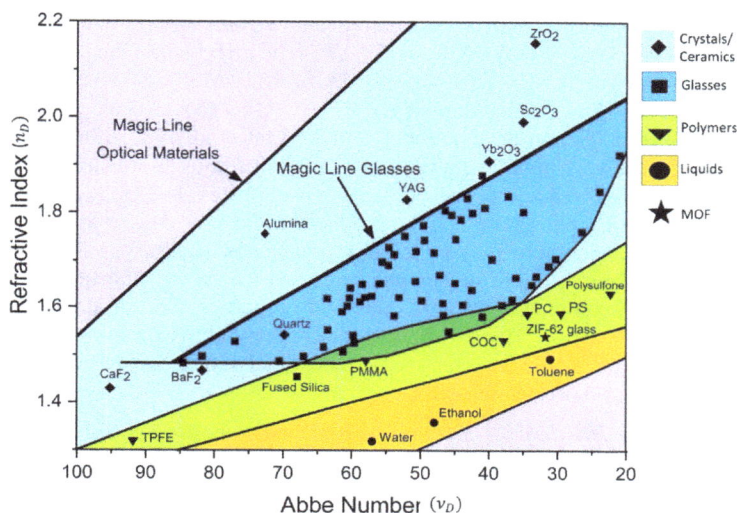

Figure 9. Refractive index n_D and dispersion coefficient or Abbe number v_D ($v_D = \dfrac{n_D - 1}{n_F - n_C}$) of various glass systems modified after Vogel and Hartmann (Vogel 1992, Hartmann, Jedamzik et al. 2010).

4.2.1 BeF$_2$ glasses as a weak model of SiO$_2$. Interestingly, Goldstein's condition of glass formation is also fulfilled for BeF_2 (Vogel 1992), which was therefore studied for a while as a model compound for isomorphous crystalline and glassy SiO_2. Exploiting the weaker binding forces in the fluoride compared to the oxide compound, and the divalent Be compared to tetravalent Si, many properties were more easily accessible, including the viscosity and melting temperature, as well as other transport related processes such as crystallization or diffusion, which are subsequently accelerated in Be–fluoride glasses. For example, BeF_2-glasses have 21% homopolar bonding which is less than half the value than the 50% expected in the bonds of silicate glasses (Mackenzie 1960).

Just as BeF_2 can be seen as a weak bonding model of SiO_2, alkali–beryllium fluoride glasses can be seen as doubly weakened models of the respective alkaline earth silicate melts (Vogel 1992). BeF_4 tetrahedra form the main glass building units, just as SiO_4 tetrahedra do in silicate glasses (Wright et al. 1989). See for comparison the schematic in Figure 12. Be-fluoride glasses show similar regions of immiscibility, with asymmetrical immiscibility gaps as binary alkali or alkaline earth silicate glasses, but the extent of phase separation is much more pronounced, by up to an order in magnitude for the fluoride systems, thus starting the research in phase separation of glasses (Vogel 1992). Fluoro–beryllate glasses are known for their low refractive index and low dispersion. The big drawback of this glass system however is the toxicity of beryllium ions. Therefore, much research focused on the development of other, Be-free, fluoride glasses.

4.2.2 Fluoroaluminate glasses. The earliest reports we found on fluoroaluminate glasses based on AlF_3 and MF_2 (M = Mg, Ca, Ba) were written by Videau et al. (1979) and by Ehrt et al. (1982). A chain-like structure based on AlF_6 octahedra was suggested, based on results from NMR and vibrational techniques. According to Poulain and Maze (1988), only very few AlF_3-based glass systems can be prepared in bulk with a thickness exceeding several mm or and only few systems even allow the drawing of fibers. The few fluoride compositions that allow processing include: AlF_3–BaF_2–YF_3–CaF_2 (Y–ABC), (Kanamori et al. 1981); AlF_3–BaF_2–YF_3–ThF_4 (BATY) (Poulain et al. 1981); CdF_2–LiF–AlF_3–PbF_2 (CLAP) (Tick 1985); AlF_3–NaF–MgF_2–YF_3–ZrF_4 (Izumitani et al. 1987) and CaF_2–AlF_3–MgF_2–BaF_2–YF_3–SrF_2 (CAMBYS) (Kucuk and Clare 1999). As already realized by Ehrt et al. in the early 1980s (Ehrt et al. 1982, 1983; Ehrt and Vogel 1983; Ehrt 2015), the very high crystallization tendency of pure fluoroaluminate glasses can be drastically decreased by adding small amounts of phosphates (1–2 mol%). Fluoroaluminate glasses with 3–15 mol% phosphates can be produced in large scales for applications in high performance optics, and as active laser and amplifier glasses. Contrary to fluoride–phosphate or fluoro–phosphate glasses with higher phosphate content, these fluoride rich glasses are dominated by a fluoroaluminate glass structure and properties (Ehrt et al. 1982, 1983; Ehrt and Vogel 1983; Ehrt 2015). The low phosphate content helps to stabilize the glasses and prevents crystallization, while asserting only a marginal impact on the optical properties. See Section 5.1.7.3 for more details on FP glasses.

4.2.3 Zr–Ba–La–Na–fluoride glasses (ZBLAN). The more accidental discovery of the first fluoro–zirconate glasses around 1974 lead to increased research into heavy metal fluoride glasses (HMFG) and the development of the first industrially produced and traded vitreous fluoride glasses and fibers.

Most fluoride glasses require fast cooling or quenching during preparation and therefore offer limited potential for applications. One system that can be processed and that allows fiber drawing are the ZBLAN type glasses (Poulain and Maze 1988). Like HfF_4 or ThF_4 based glasses, systems based on ZrF_4 were studied for their use in infrared optical materials. ZBLAN stands for a complex composition such as $60ZrF_4$–$20BaF_2$–$4LaF_3$–$6AlF_3$–$10NaF$ (Poulain 1981). Zirconium fluoride does not exist in the vitreous form, however, together with LaF_3, ThF_4, BaF_2 or SrF_2, binary glass can be prepared by fast quenching of the melt (Poulain et al. 1992). In ZBLAN, ZrF_{7-8}, AlF_6 and LaF_8 are the glass-forming polyhedra, whereas Ba^{2+} and Na^+ are classical modifier cations. LaF_3 can be substituted by other, optically active, RE elements allowing the preparation of optical amplifiers or fiber lasers which show more emission lines than silica glass (Miyajima et al. 1994; Lucas et al. 2018). In order to minimize devitrification on reheating, the composition has to be adjusted, and an increasing number of glass components are added, according to the already mentioned "confusion principle" (Poulain et al. 1992).

Hafnium can substitute for zirconium, but despite many similarities, there are some differences in the optimized glass composition seen between fluoro–zirconate and fluoro–hafnate glasses (Poulain et al. 1992).

It can be shown, that some glass forming principles such as the similarity between the structure in the melt and the glass, as postulated in the definition by Zanotto and Mauro (2017), is also valid for fluoride glasses. The coordination number and bond length of the crystalline phases are close to those observed in the glassy phase of many fluoride systems that contain the same cations, thus retaining the same short-range order between the glass and the crystal. For example in fluoro–zirconate glasses, zirconium cations are usually coordinated to eight F atoms while the coordination number of Ba^{2+} varies between 10 and 12 depending on composition (Poulain et al. 1992).

While one viewpoint describes the glass structure as a random network of ZrF_8, AlF_6 and LnF_n polyhedra, that share corners and edges, trapping large cations, e.g. Na^+ or Ba^{2+} between them, another viewpoint focuses on a random packing of large F^- ions and Ba^{2+} ions, between which the smaller cations Zr^{4+}, Al^{3+} or Zn^{2+} are randomly inserted. This approach was especially successful when discussing mixed fluoride–phosphate glasses as laid out in the review paper by Möncke and Eckert (2019). For more on mixed fluoro–phosphate and fluoride–phosphate glasses see Section 5.1.7.3.

4.3 Other ionic and molecular glasses

Fluoride glasses are the most important of the halogen glasses, though other systems with high fractions of chloride (Chen et al. 2018) or iodide (Lefterova et al. 1997) have been prepared successfully. While some simple systems form glasses (Mackenzie 1987), mixed components are studied more often for ion conduction, e.g. in phosphate, borate or tellurite halogen glass systems (Lefterova et al. 1997).

Other salt-like glass systems include mixed alkaline earth nitrates (Duffy and Ingram 1968; Ingram and Lewis 1974) or sulphates (Thieme et al. 2015; Nienhuis et al. 2019) which can be formed by relatively slow quenching from the melt (Gaskell 1982). The $ZnSO_4$–K_2SO_4–NaCl system for example has been studied in the context of immobilization of salt rich nuclear waste (Nienhuis et al. 2019). Contrary to fluoride glasses, no strong anion-cation complexes are formed. The large ions Cl^- or I^-, like multiatomic anions such as SO_4^{2-}, NO_3^- or acetate (H_3C–COO^-) are charge balanced and crosslinked to cations without forming a three-dimensional glass network, thus coining the term "molecular" glass (Gaskell 1982; Thieme et al. 2015; Nienhuis et al. 2019).

4.4 Nitride glasses

In nitride glasses, oxygens ions are replaced by nitrogen ions, which form similarly strong covalent bonds. However, nitrogen ions are in group V of the periodic table and can form three bonds, resulting in a higher degree of crosslinking and thus a stronger glass network when compared to the two bonds oxygen can provide. This extra bond results in a higher T_g, and higher mechanical strength in nitridified glasses compared to the respective oxide glasses (Larson and Day 1986; Loehman 1987; Becher et al. 2011). For each nitrogen bond, the maximal linkages of silicate tetrahedra can increase by +1, and nitridification of SiO_2 can give rise to previously unknown Q_{Si}^{4+m} units—even though the silicate tetrahedron retains its four-fold coordination (see Figs. 10 and 11). The factor, $+m$, describes the additional nitrogen-bridged silicate units that form at the fully linked nitrogen atoms. Koroglu et al. (2011) shows evidence for some $SiON_3$ and possibly SiN_4 species in nitrogen-rich Y–Si–Al–O–N glasses.

Due to the increased crosslinking of nitrogen ions compared to lower valent anions, high melting temperatures are often needed in the preparation. For example, the melting temperature of Si_3N_4 is $T_m = 1900\,°C$, which poses a significant strain on the crucible material, which in turn might dissolve into the glass or react with the nitridification agent (Wójcik et al. 2018a,b). Refractive index and polarizability are also higher for nitride than for oxide glasses (Möncke et al. 2019). In the review by Becher et al. (2011), several studies are cited that indicate that DC conductivity and electrical constants increase in nitride containing systems, a fact, that was also discussed by Mascaraque et al. (2013) and Wójcik et al. (2018a,b).

4.4.1 Alumo-silicate based oxynitride systems SiAlON. The most studies of oxynitride glasses are based on silicate and alumo-silicate systems (Loehman 1987; Becher et al. 2011; Ali et al. 2015; Garcia et al. 2016). The early research was driven by Si_3N_4 and other nitride-based ceramics, and glassy phases within the grain boundaries of Si_3N_4 (Jack 1976). Typical systems include MSiON and MSiAlON where M stands for any alkaline earth or rare earth element, though yttrium can be found in these glasses as well.

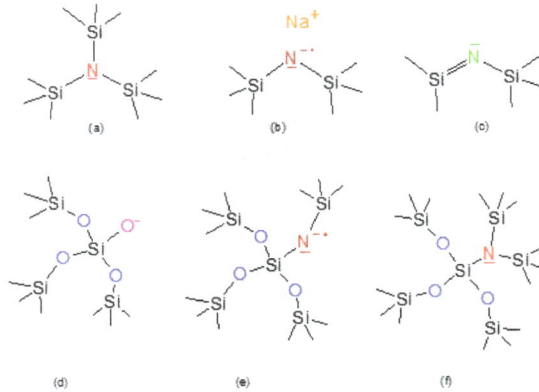

Figure 10. Depiction of basic glass forming units in silica–oxynitride glasses, after Becher et al. (2011). **(a)** bridging nitrogen linking three silicate tetrahedra; **(b)** charge balanced bridging nitrogen linking two silicate tetrahedra—being essentially a mix of the bridging and a non-bridging oxygen atom; **(c)** a double bonded two fold coordinated nitrogen, having a higher bond equivalent unit than a single bond, but connecting only two silicate tetrahedra and having no need for charge balance; **(d)** a classical Q^3 unit with three bridging and one non-bridging oxygen ion; **(e)** the corresponding $Q^{3O(N1)}$ unit with three bridging oxygen and one bridging nitrogen ion that also has one non-bridging site at the nitrogen ion that need to be charge balanced by one modifier cation; **(f)** $Q^{3O(N2)}$ unit with three bridging oxygen and one doubly bridging nitrogen ion which effectively has the connectivity of a Q^5 unit, linking 5 silicate tetrahedra to the central silicate unit.

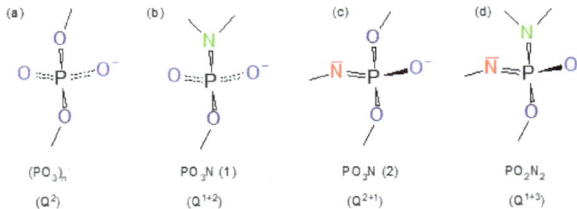

Figure 11. Depiction of oxynitride phosphate glasses (Larson and Day 1986; Bunker et al. 1987; Muñoz et al. 2013).

Good results have been obtained by dissolving nitride components Si_3N_4, and / or AlN in a silicate melt. The first prepared oxynitride glass was made by Mulfinger et al, who studied the dissolution of nitrogen through bubbling N_2, H_2/N_2 or NH_3 in silicate and borate melts (Mulfinger 1966). However, using this method, he was only able to retain less than 1 wt. % of nitrogen in the glasses. As has been shown since, other methods give much better results, for example, when an oxide glass is re-melted with the addition of Si_3N_4 or other metallic nitrides. Another method would be the addition of powdered metallic alkaline earth elements while maintaining an N_2 or NH_3 atmosphere. N^0 or N^{3+} are reduced in a redox reaction, oxidizing the metals to the corresponding nitrides, which in turn will result in an oxynitride glass that sustains a significant nitride fraction (Loehman 1987; Becher et al. 2011, Ali et al. 2015; Garcia et al. 2016). Addition of Si can also help with nitrogen retention and prevents foaming from the melt, as it helps to retain reducing conditions in the melt (Baik and Raj 1985; Loehman 1987).

While alkaline earth, rare earth, and lithium oxide are stable in silicate based oxynitride glass, heavier alkali oxides may decompose in nitride melts according to the following example of the reaction of potassium oxide in an oxy-nitride silicate melt (Loehman 1987):

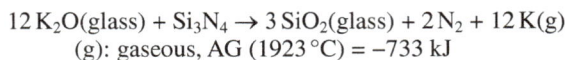

$$12\,K_2O(glass) + Si_3N_4 \rightarrow 3\,SiO_2(glass) + 2\,N_2 + 12\,K(g)$$
$$\text{(g): gaseous, AG (1923\,°C) = −733 kJ}$$

In contrast to most other glass systems, the concept of corresponding salt pairs is used for the description of MSiON glasses. A four-component system such as Si–Al–O–N, is represented as a square plane with the corners represented by the two metal ions (Al and Si) and two anions (N and O). For multicomponent systems with three cations, the five-component system can be represented by a 3D Jänecke prism, where the oxide ternary is plotted at one end, and the corresponding nitride ternary on the other end of the prism (Hampshire 2003). Instead of mol % the components are given in at% or as equivalent fractions, e O/N.

Other techniques for the preparation of oxynitride silicate glasses with high nitrogen levels include hot isostatic pressure for bubble-free YSiAlON glasses, or sol–gel preparation (Loehman 1987).

4.4.2 Phosphate based oxynitride systems. The phosphate oxynitride system behaves quite differently from the silicate system. This is partially because of the high chemical solubility of nitrogen in phosphate melts. This effect causes the often unwanted re-boil effect, when in seemingly homogenous phosphate melts, suddenly myriads of tiny bubbles form during cooling, since the chemical solubility of nitrogen is so much higher at melting temperatures than at lower temperatures (Von Jebsen-Marwedel 2011).

Preparation of P-based oxynitride glasses differs from alumina-silicate oxynitride glasses in several aspects. For one, thermal ammonolysis at relative low temperatures is the preferred preparation method (Bunker et al. 1987; Muñoz 2011). NH_3 is bubbled through the melt at temperatures below 800 °C, resulting in oxynitride phosphate glasses with up to 20 at% nitrogen. In order for this technique to work, it is important to control the viscosity of the melt, e.g. by selecting compositions chosen to be close to the eutectic of the respective metal oxide-phosphorus pentoxide, and having low field strength cations, such as alkali metals (Loehman 1987; Muñoz 2011). The following liquid–gas chemical reaction was postulated by Munoz:

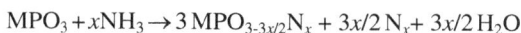

$$MPO_3 + x NH_3 \rightarrow 3\, MPO_{3-3x/2}N_x + 3x/2\, N_x + 3x/2\, H_2O$$

Early reviews on nitrogen rich oxynitride systems appear in the 1980s. Marchand et al. (1983) describes alkali P–O–N systems prepared by the reaction between alkaline polyphosphates with ammonia at 700 °C. Up to 25 % nitrogen was retained in the glasses. The use of ammonia however leads to a water rich glassy phase, that can be ascribed to the H–P–O–N system. Isotopic substitution of ^{17}O-NMR studies on NaPON glasses allowed to distinguish non-bridging oxygen atoms on PO_4 and PO_3N or PO_2N_2 sides (Muñoz et al. 2013). Elemental analysis is important since nitrogen incorporation can vary significantly, depending on composition and preparation technique.

4.4.3 Mixed anionic systems. Recently, Mascaraque et al. (2015) studied fluoride containing LiOPN glasses, 30 years after earlier attempts to introduce nitride into fluoride (oxide) systems by Vaughn and Risbud (1984) and Fletcher et al. (1990).

Mascaraque et al. (2014) also prepared a thio-phosphorus oxynitride glass electrolyte, showing the versatility of glass formation on the anion side.

4.4.4 Applications of oxynitride glasses. In contrast to chalcogenide and fluoride glasses, which found their niche for high specialty applications in optics and photonics, such applications for bulk oxynitride glasses are still not fully developed. Preparation of larger batch sizes and processing of the melts is difficult and so far, improving mechanical properties in silicate glasses is often easier. However, applications that show promise are the fusing of high strength high temperature ceramics, exploiting the natural tendency of Si_3N_4 to occur in grain boundaries (Loehman 1987). Another application might be protective or active coatings on thin films (Ali et al. 2015).

5. SIMPLE INORGANIC OXIDE GLASSES

Non-silicate oxide glasses are technologically very important, as they provide a combination of properties that are hard to derive when using only silicate glasses. This is not only supported by the early systematic studies by Otto Schott (Kühnert 2012; Chopinet 2019), but also by the prominent role non silicate glasses have in the general understanding of glass formation (Zachariasen 1932). Many of today's products are based on all type of oxide glasses, covering applications that range from lasers, to ion conducting, or biomedical materials, as well as nuclear waste vitrification (Musgraves et al. 2019).

Classically, a glass former has been defined as an oxide that can form a glass without additional components, for example B_2O_3, P_2O_5, GeO_2, TeO_2, As_2O_3/As_2O_5, or Sb_2O_3. The first two, B_2O_3 and P_2O_5, are probably the most important of this series, with GeO_2 and TeO_2 fostering more interest for their optical properties and various special applications. Contrary to SiO_2, B_2O_3 and P_2O_5 are very hygroscopic and the simple vitreous oxides are only of academic interest. However, both borates and phosphates react readily with many oxides of the periodic table and have large glass forming regions (Vogel 1992). Both result in a wide range of binary glasses (not counting the oxide as third component), andd many of these glasses are quite stable—depending of course on the second component and the exact composition.

Both borates and phosphates can form glasses with each other, resulting in boro–phosphate glasses, or with other glass formers, such as SiO_2 or Sb_2O_3, as well as with intermediate oxides such as Al_2O_3, rare earth oxides or transition metal and post transition metal oxides (though regions of immiscibility or high tendencies of crystallization occur for various compositions). This extensive flexibility in the glass composition allows the design of glasses with distinct combination of desired properties, from optical to mechanical properties, to dopant solubility, and chemical stability to name just a few.

In the following we will give an overview on phosphates, then borate glasses and afterwards briefly on the less common germanate, tellurite and antimonate glasses.

Both phosphate and borate glasses, as typical network former oxide glasses, show many similarities, but also many distinct differences from silicate glass systems. One is the lower network connectivity of the pure oxides B_2O_3 and P_2O_5 compared to SiO_2 (see Fig. 12), as both elements in fully polymerized pure systems are only connected via three bridging oxygen atoms.

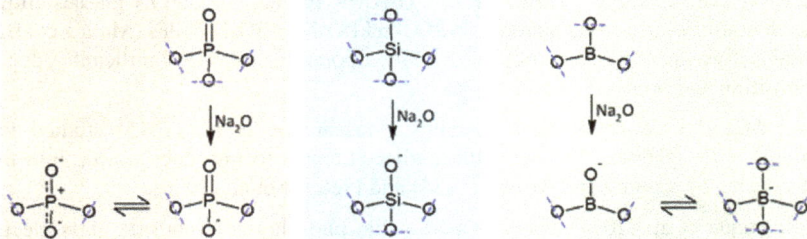

Figure 12. Basic structural units in (**a**) phosphates, (**b**) silicates and (**c**) borate glasses. The top species is found in the fully polymerized simple oxides P_2O_5, SiO_2, and B_2O_3.

5.1 Phosphate glasses

Pure SiO_2 is made of silicate tetrahedra that are connected via all four corners (Q^4). Phosphorus is in the 5th group of the periodic table and therefore has a higher charge than Si^{4+}. In pure vitreous P_2O_5, P^{5+} is at the center of a phosphate tetrahedra with three oxygen atoms that are bridged via a shared corner to other tetrahedra, while the fourth oxygen atom consists of a double bonded terminal oxygen (P=O) (see Fig. 12), resulting in Q_P^3 as the uncharged fully bonded unit (the Q_{P+}^4 unit is a special case, see Fig. 13)

As a consequence, transition temperature T_g and mechanical properties of P_2O_5 reflect the less connected network compared to SiO_2 (see also Table 3 for comparison). Addition of network modifiers such as alkali oxides will depolymerize the network further, much as in the case of silicate glasses.

5.1.1 Delocalized double bond. A peculiarity of phosphate compared to silicate glasses is the delocalization of the double bond to all non-bridging oxygen atoms of the same phosphate tetrahedron. However, this is typical for phosphate compounds in general and not specific to glasses. Details can be found in any book of inorganic chemistry (Wiberg et al. 2007). Since the delocalization of the double bond strongly affects the P–O bond strength, we briefly review the basics in this paragraph. The P=O double bond can easily be identified in pure P_2O_5 by XPS or Raman spectroscopy as well as by other methods. However, in modified phosphate glasses, the P=O double bond cannot be distinguished from a P–O bond containing a non–bridging oxygen atom, as the electron of the second bond will be delocalized over all bonds that contain a non–bridging oxygen atom P–O$^-$ (Brow 2000). The bond strength, given as valence units, vu, increases therefore by the effect of this additional contribution of the delocalized electron (see Fig. 13). Thus, for phosphate glasses, it is more precise to distinguish bridging oxygen and terminal atoms, instead of trying to further discriminate the double bonded from a non–bridging oxygen atom. Having said this, there are rare cases in which a double bonded oxygen atom gives a separate signal from a nbO atom (e.g. by Raman), for example for the high field strength Zn^{2+} ion, which so much prefers bonding to a nbO, creating a Zn–O–P bonds, that the P=O bond can form on non-ligand sites (Brow 2000).

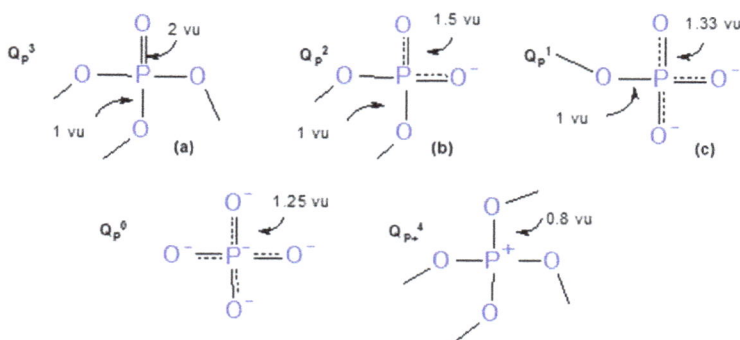

Figure 13. Delocalized electron on the basic phosphate units with increasing depolymerization (a) ultraphosphate, (b) metaphosphate, (c) pyrophosphate, (d) orthophosphate, and (e) the special case of the fourfold coordinated phosphate tetrahedra, bond strength values are given in valence units (vu) after Brow et al. (1995).

5.1.2 Chemical stability of phosphate glasses. Pure P_2O_5 is so hygroscopic, that it has long been used as a drying agent in chemistry (Wiberg et al. 2007). Not surprisingly, pure vitreous P_2O_5, as well as ultraphosphate glasses with a O:P ratio <2 (metaphosphate composition) are often very hygroscopic and often contain H_2O as a third oxidic component (see Fig. 14; Gray and Klein 1983; Hudgens 1994). These P-rich glasses need to be prepared in sealed ampules under dry conditions to exclude the presence of atmospheric water (Brow 2000). For this reason, literature data on ultraphosphate glasses scatters widely, since the water content of these reported glasses varies (Brow 2000).

As mentioned earlier, the different water content from melting in an open crucible or in closed ampules will result in different water uptake and subsequently glasses with different properties.

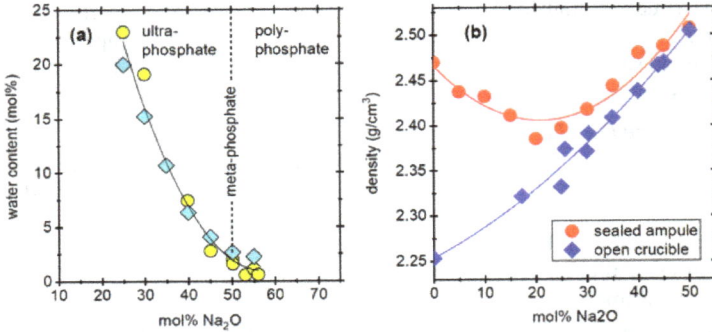

Figure 14. Water content in sodium–phosphate glasses xNa$_2$O–$(100-x)$P$_2$O$_5$ **(a)** analyzed water content for glasses melted in open crucibles (after Brow1990 and Gray and Klein 1983). Reprinted from Journal of Non-Crystalline Solids, Vol. 116, Brow RK et al., The short range structure of sodium phosphate glasses I. MAS NMR studies Vol. 116, pages 39-45, Copyright (1990), with permission from Elsevier and **(b)** variations in density for glasses melted in open Pt-crucibles and in sealed glass ampoules. Reprinted from Journal of Non-Crystalline Solids, Vol. 263, Brow RK, Review: the structure of simple phosphate glasses, pages 1-28, Copyright (2000), with permission from Elsevier.

Figure 14 shows the dependence of water uptake on the Na$_2$O content for the ultraphosphate glass series (Hudgens 1994; Brow 2000). Ehrt showed a similar high water content of 20 mol % in the 60SnO–40P$_2$O$_5$ system (Ehrt 2008). This tin phosphate glass system is overmodified in regard to the meta-composition and was therefore expected to be more stable. Two glasses were melted from the same raw materials, one, melted at 1200 °C, resulted in a water free variant, while the glass melted at only 450 °C was water rich. The glasses varied in their properties as shown in Table 2 in so far as the glass melted at 1200 °C was chemically stable over an observation time of more than 7 years while the glass melted at low temperatures devitrified completely in the same time. Sn^{2+} was the predominant species in both glasses, but the second glass contained 2 wt% or 20 mol% of water while the first glass was water free.

Table 2. Properties of water free (I) and water containing 60SnO-40P$_2$O$_5$ glasses (II) (Ehrt 2008).

Sample	T_m (°C)	T_g (°C)	CTE (ppm/K)	density (g/cm^3)	refractive index
I. water free	1200	300	12	3.554	1.733
II. 20 mol% H$_2$O	450	200	27	3.455	1.748

5.1.3 Binary phosphate glasses. Addition of modifier oxides will not only improve the chemical stability, but also strengthen the glass against mechanical impact and may even result in an increased T_g. The exact properties vary strongly, not only with the amount, but also the type of added modifier oxide (Ehrt 2015; Ehrt and Flügel 2018). For example, a sodium-metaphosphate glass NaPO$_3$ will show significant corrosion over days, while small additions of Al$_2$O$_3$ will make the glasses stable over years (Möncke et al. 2018a). Strontium metaphosphate glass Sr(PO$_3$)$_2$ is relatively stable over years, and iron pyrophosphate glasses are even candidates for nuclear waste immobilization (Day and Ray 2013), reflecting on their stability, and exploiting the ease of incorporation of other large ions within such a glass matrix. Laser glasses with a high long-term stability are prepared industrially (Zhang et al. 2018). Phosphate glasses with controlled dissolution rates are used in fertilizers (Kosareva et al. 2006) and bio-glasses (Sharmin and Rudd 2017).

Metaphosphate to pyrophosphate are very typical compositions and a multitude of publications can be found in this compositional range (Brow 2000; Hoppe et al. 2000; Ehrt 2015). Phosphate reacts readily with many other glass formers and intermediate oxides. Many multicomponent systems are known and can only briefly be mentioned in Section 5.1.7.2.

First, we want to discuss the basic chemistry and structure of simple binary phosphate glasses and the characteristic peculiarities of phosphate glasses. Compared to silicates, binary phosphate glasses are easy to prepare, especially of the meta- to pyro- compositions. Metaphosphate glasses consist of infinite chains and rings made of $(PØ_2O_2)^-_n$ entities (Ø denotes bridging oxygen atoms). Here, as discussed for organic polymers, the random coil model supports glass formation, as it is hard for the chains to crystallize in a well-ordered long-range structure (see Fig. 3). Rings of varying size can occur together with those chains, though smaller rings are not well-defined n-membered superstructural units as those known from silicates or borates (Velli et al. 2005).

Interestingly these chains are a very stable unit and do not easily disproportionate—branching Q-units seem to be the least stable phosphate entity and thus a "no-branching" rule was postulated for metaphosphate glasses (Van Wazer and Holst 1950). Cross linking of the phosphate network by modifier cations may overcome the effect of a high network connectivity, with high T_g and low CTE, as well as higher chemical stability and strength for ions with a high field strength, high force constant for metal–oxygen bonds (Brow et al. 2000). See Figure 15 for the dependence of T_g on the cross-linking strength of the M^{2+}–oxygen bonds as determined from the cation motion bands that appear in the far infrared spectra (for details on far infrared and cation sites see for example Yiannopoulos et al. 2001; Möncke et al. 2016). This effect is also seen when low T_g AgPO$_3$ glasses are melted in Al$_2$O$_3$ crucibles (see Section 5.1.4).

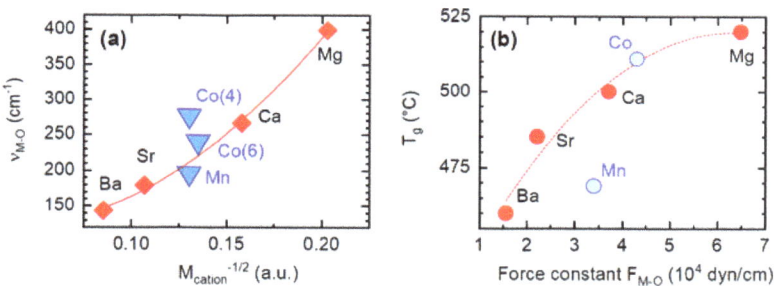

Figure 15. (a) Correlation between the far infrared frequency of the cation-motion band with the inverse square root of the mass of the cation and (b) dependence of the transition temperature T_g on the force constant of divalent cations in binary metaphosphate glasses. Data from: Velli et al. (2005), Konidakis et al. (2010), Griebenow et al. (2017) and Möncke (2017).

Often, the pyro- composition (Q_P^1) is even more stable than are metaphosphate (Q_P^2) glasses. Here, we have to look at the intermediate range order and the stereochemistry and its effect on hydrolysis (Musgraves et al. 2019).

5.1.4 Crucible material. The right choice of crucible is crucial for phosphate glasses, which typically react readily with many crucible materials. Platinum crucibles can experience reduced lifetimes due to catastrophic corrosion or the "phosphorus-burn" (see Fig. 16), when a P–Pt alloy forms under reducing conditions. As long as the melt is conducted under purely oxidizing conditions, the Pt-crucible might not be harmed, but in the presence of polyvalent and reducing elements, Pt-crucibles should definitely be avoided. A superb overview on destructive melts for Pt-crucibles was compiled by Lupton et. al. (1997). On the other hand, small traces of platinum can be dissolved in the phosphate melt, which, depending on the glass composition and application, might result in an insignificant additional absorption in the UV–Vis range due to absorption by Pt^{2+} or Pt^{4+} ions (Click et al. 2003); or be detrimental to laser glasses, if Pt-particles form which will scatter the light, and when heated by the laser energy, the mismatch of the coefficient of thermal expansion of the metallic particles and the glass might lead to the scattering of the laser glass itself (Hayden et al. 1988).

Other crucible materials could be Al_2O_3 or SiO_2. While both are cheaper options compared to Pt, both crucible materials will dissolve even more strongly in a phosphate melt and become part of the glass. The higher the melting temperatures and longer the melting times, the more crucible material will be taken up by the melt (Konidakis et al. 2018; Sawangboon et al. 2019). This process is not necessarily negative, SiO_2 and Al_2O_3 both help to stabilize the new glass and improve its chemical durability. However, not all papers give quantitative details on incorporated crucible materials, a fact that can add to significant scattering of published data of many properties. For example, as shown by Konidakis et al. (2018), the T_g of $AgO–P_2O_5$ glasses might increase by 60 °C when melted in Al_2O_3 crucibles at 900 °C. These papers showed that Al_2O_3 dissolution from the crucibles depolymerizes the phosphate network but adds additional cross-linking strength by forming strong P–O–Al bonds (Palles et al. 2016; Konidakis et al. 2018) or similar studies (Sawangboon et al. 2019).

Amorphous carbon crucibles are another option, though they require an oxygen free atmosphere. As can be seen from Figure 16, phosphorous can be reduced to the elemental state when melting under strongly reducing conditions in a glassy carbon crucible. Even small traces of reduced P^{3+} or colloidal P^0 significantly shift the absorption edge from the UV to the visible wavelength range (Ehrt et al. 2000). The polyvalent nature of phosphorous might also help to stabilize low oxidation states of polyvalent elements in phosphate glasses.

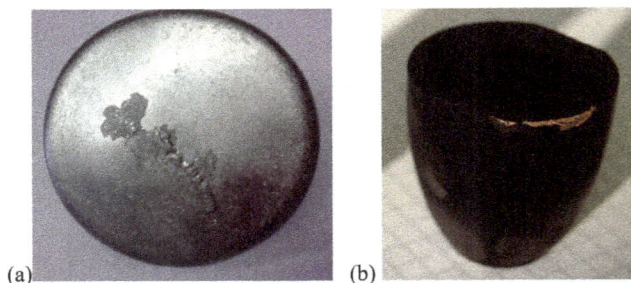

Figure 16. (a) A light case of phosphorous brand on a Pt crucible (Photo Möncke) and **(b)** reduced elemental phosphorous on the rim of an amorphous carbon crucible (Photo Ehrt, Fibikar).

Niobium crucibles are likewise dissolved and Nb-O bands can be identified in the Raman spectra of phosphate glasses melted in Nb-crucibles (Wójcik et al. 2018a,b).

5.1.5 Structural analysis. The structure of phosphate glasses has been studied by many spectroscopic techniques, including ^{31}P-NMR (Kirkpatrick and Brow 1995; Brow et al. 2000; Hoppe et al. 2018), Raman and IR (Hudgens et al. 1998; Brow 2000; Velli et al. 2005; Palles et al. 2016), photoelectron spectroscopy (Brow et al. 1994), neutron and X-rays diffraction (Hoppe et al. 2000, 2018), and even chromatography for the chain length determination (Saitoh et al. 2017; Hoppe et al. 2018).

^{31}P-NMR provides information on the number of non-bridging oxygen atoms; however, if the phosphate tetrahedron is linked to alumina or silicate units, the resulting shift might be so strong, as to overlap with the next more polymerized Q_P-unit (Sawangboon et al. 2019). Nonetheless, the newest NMR correlation studies provide in-depth information on next nearest neighbors and connectivity (van Wüllen et al. 2000; Zhang and Eckert 2006; Sawangboon et al. 2019). Raman and IR spectroscopies also provide information on the degree of polymerization and the intermediate order and connectivity (Brow 2000). Like NMR signals, the vibrational bands shift with the field strength of the connecting modifier ion (Velli et al. 2005; Palles et al. 2016).

Hoppe conducted extensive diffraction studies on phosphate glasses focusing on phosphate units and the coordination and bonding of the modifier cations (Hoppe et al. 2000). For the phosphate groups, no change in the average P–O distances were found initially, but upon deeper analysis, it was found, that the shorter terminal oxygen to phosphor distances grew faster than the longer bridging O–P distances, as the number of terminal oxygen atoms increased. As is apparent from Figure 13, the delocalization of electrons over more bonds decreases the valence units per terminal bond, explaining the elongation of the P–O bonds with depolymerization.

For the cation environment, Hoppe et al. postulated the Hoppe model of coordination, distinguishing three different regions or stages as the coordination of oxygen atoms changes with the type of modifier cation. That is, a change in the modifier cation coordination, especially the number of bridging (bO) and non-bridging oxygen (nbO) atoms, and in the number of oxygen atoms that needed to be coordinated to more than one modifier cation. Please recall, that a Na^+ ion, with a coordination number of 6, is not coordinated to 1 non-bridging oxygen atom only, but to a number of non-bridging oxygen atoms. Depending on the composition, the sodium ion will also share its coordination of one terminal oxygen atom with other Na^+ ions. This partial charge compensation is especially good to understand with phosphate glasses, where the delocalization of the double bond over all nbO creates more sites with a nominal negative charge at all terminal oxygen atoms.

5.1.6 Mixed alkali glasses. Phosphate glasses are relatively stable for high alkali loads and have been studied intensively for ion conduction (see also Section 5.1.7.4). Thus, much research on the mixed alkali effect focuses on phosphate glasses, such as for example (Tsuchida et al. 2017), or for the mixed alkaline earth effect (Griebenow et al. 2018).

Very interesting is also the work by Inaba et al. (2015), who showed that they could induce anisotropy into metaphosphate glasses by aligning the phosphate chains in a mixed alkali metaphosphate glass by fast fiber pulling. Polarized Raman spectra show distinct anisotropy as the orientation of the metaphosphate fiber is changed (see Fig. 17 for the schematic of oriented metaphosphate chains).

(a) Isotropic   (b) anisotropic   (c) ionic (invert) glass

Figure 17. (a) A rare example of anisotropy in glasses showing the isotropic random coil model on the left and **(b)** the oriented one-dimensional structure after fiber drawing, from (Inaba et al. 2015); **(c)** shows the ionic sulfate–phosphate glasses where pseudospherical anions SO_4^{2-}, PO_4^{3-}, $P_2O_7^{4-}$ are crosslinked by Na^+ and Zn^{2+} cations (compare with Thieme et al. 2015; Nienhuis et al. 2019).

5.1.7 Multicomponent phosphate glasses

5.1.7.1 Phospho–silicate glasses. Binary phospho–silicate or silico–phosphate glasses, as well as ternaries are often prone to significant phase separation into (modifier) phosphate rich and silica rich phase (Chakraborty and Condrate Snr 1985; Vogel 1992). Small amounts of phosphate can be dissolved in a silicate matrix, usually as isolated ortho-phosphate units (Jones 2013), and small amounts of SiO_2 can be dissolved in phosphate glasses, as six-fold coordinated silicate that is connected via six bridging oxygen atoms to phosphate tetrahedra while any network modifier forms non-bridging oxygen on the phosphate network (Miyabe et al. 2005; Venkatachalam et al. 2014; Sawangboon et al. 2019). This observation has recently received more attention as multicomponent phosphate glasses are studied for biomedical applications (Sharmin and Rudd 2017). Here, IR and Raman spectra might see octahedral Si^{4+} as modifier ions, while variations in the glass properties reflect the additional cross-linking of silicate octahedra with 6 bridging oxygen atoms (Wójcik et al. 2018a,b).

Chakraborty et al. (1987) studied the 1:1 SiO_2–P_2O_5 glass by vibrational spectroscopy and determined that the structure varied significantly from the corresponding SiP_2O_7 crystals (Chakraborty and Condrate Snr 1985). Contrary to the crystal, no six-fold coordinated silicate groups could be distinguished for the binary glass. However, this changes with the addition of modifier oxides as the coordination of Si^{4+} increases to six, when added at low concentrations to phosphate glasses.

An Al_2O_3 containing phospho–silicate glass xAl_2O_3–$(30-x)P_2O_5$–$70SiO_2$ was successfully prepared by Aitken et al. (2009) and characterized by Raman and enhanced NMR spectroscopies. Melt quenching did not allow formation of a glass for an Al:P ratio of one ($x=15$). When phosphate units outnumbered aluminate units, aluminum was either 4-, 5- or 6-fold coordinated and preferentially found near phosphate groups. In P-rich glasses, higher coordinated Al-species would charge compensate terminal oxygen ions at Q_P^2 units. However, $[AlØ_4]^-$ tetrahedra would link with $[PØ_4]^+$ tetrahedra or Q_P^4 groups, that are similarly found in $AlPO_4$ crystals and isoelectronic to SiO_2. This Q_P^4 species (see also Fig. 13) is also dominant in Al-rich glasses (Al/P > 1).

For other mixed network glasses see borates, Section 5.2.9, tellurites Section 5.3.2, or germanates, Section 5.4.2.

5.1.7.2 Phosphate glasses with high amounts of intermediate oxides. Phosphate glasses form readily with many elements, and in many systems with intermediate network forming oxides like WO_3 or Nb_2O_5 as exemplified by Šubčík et al. (2010) in their study on the glass forming range and glass characteristics. In the frame of this chapter we can only mention briefly a nonrepresentative arbitrary collection of interesting systems. Binary phospho–titanate glasses form only in the TiO_2-rich region with 55–70 mol% TiO_2, which corresponds to the orthophosphate stoichiometry (Hashimoto et al. 2006). Multicomponent titano–phosphates form on the other hand over wide compositional ranges, as reported by Hashimoto et al. (2006) who also studied multicomponent titano–phosphate glasses as ecologically sustainable optical glasses with a high refractive index around 1.9. AgO containing TiO_2–P_2O_2 glasses were found as promising nonlinear optical glass and photocatalytic glasses with self-cleaning properties, the K_2O–TiO_2–P_2O_2 system excels in its athermal properties.

Shen et al. looked into Na_2O–WO_3–P_2O_5 glass for photochromism under gamma-ray irradiation (Shen et al. 2015) while color bleaching and oxygen diffusion was studied by Ghussn et al. (2014) in a niobium phosphate glass of the composition $23K_2O\cdot40Nb_2O_5\cdot37P_2O_2$. Interestingly, the colorless d^0 ion Nb^{5+} could be reduced to blue d^1 ion Nb^{4+} when heated close to T_g under a reducing atmosphere, and bleached under an oxygen atmosphere, the kinetics of the oxidation and reduction processes following diffusion laws.

Recently, a multi-technical study of the structure and properties of $xZnO$–$(67-x)SnO$–$33P_2O_2$ glasses showed a change in the phosphate speciation as Zn^{2+} changes its role from modifier to a $[ZnO_4]^{2-}$ glass former (Hoppe et al. 2018; Saitoh et al. 2018a,b). This in-depth study includes Raman, infrared and NMR spectroscopies, neutron and X-ray diffraction studies, as well as chromatography and the correlation with thermal, optical, mechanical and photo-elastic properties.

See also Sections 5.2.7 to 5.2.9 for similar mixed boro–phosphate glasses.

Fluorophosphate and fluoride phosphate (FP) glasses. Fluoride phosphate (no P–F bond) and fluorophosphate (with F–P bond) glasses combine many favorable aspects of fluoride and phosphate glasses and are currently studied as matrices for applications in optics, photonics and for energy storage (Musgraves et al. 2019). For review papers on the structure and properties of FP glasses we refer to (Ehrt 2015; Möncke and Eckert 2019). Any of the fluoride systems discussed in Section 4.2 can be combined with phosphates. Even when batching fluoride components, fluoride is easily lost from the melt as either HF, POF_3 or even SnF_2. Maintaining a fluoride rich atmosphere, e.g. by the addition of $NH_4 \cdot HF_2$ to the batch, can reduce the otherwise inevitable losses of fluoride during melting. The ready reaction of fluoride with water introduced by the batch materials, ensures that FP glasses are not likely to contain significant water bands, which is important when foped for photoluminescence.

For applications in laser glasses and photonics, the low phonon energies of the fluoride network increase radiative transitions and up-conversion efficiencies, while the phosphate content increases the solubility of active ions such as transition or rare earth elements (Seeber et al. 1995; Ehrt 2015). The ligand field can be tuned by variation of the fluoride versus phosphate components.

The structure of FP glasses consists usually of phosphate monomers and dimers, that are connected via cations to the fluoride components such as fluoroaluminate $Al(O,F)_6$ or fluoro–zirconate $Zr(O,F)_8$. For example, in fluoroaluminate–phosphate glasses Al–O–P bridges link the fluoroaluminate units directly to the phosphate tetrahedra, while bridging fluoride atoms are only found in Al–F–Al bridges (Möncke and Eckert 2019). P–F bonds decompose at high melting temperatures and are therefore only found in lower melted FP glasses, such as the fluoro–stanno–phosphate system (Anma et al. 1991), or the model system $NaPO_3$–AlF_3 (Möncke et al. 2018a).

Since the major application of FP glasses based on fluoro–aluminates with phosphates is in the area of optics and photonics and includes interaction of the glass with low- and high-level irradiation, irradiation induced defect formation has garnered significant attention (Ehrt 2015; Möncke et al. 2018b).

5.1.7.4 Ionic glasses containing halides or sulfates. Phosphate glasses can also contain high halogen levels, such as in the ion conducting $AgPO_3$–AgI system (Rodrigues et al. 2011; Palles et al. 2016). Here, silver iodide addition will not shorten the metaphosphate chains, but the large iodine ions act as spacers between the chains and also contribute additional mobile ions to the glass (see Fig. 17c for a schematic). In contrast to fluorine, heavier hadlides form no direct bonds with the phosphate network, but like orthophosphate or sulfate ions are ionically crosslinked to diverse cations. Following the confusion principle as discussed for metallic glasses (see Fig. 3), glasses might form even without the presence of a higher order network. Perhaps one of the better studied systems of such ionic invert glasses are the sulfate phosphate glasses (Mamoshin 1996; Thieme et al. 2015). Structurally, phosphate free sulfate chloride glasses fall in the same category of ionic glass systems (see Section 4.3).

5.1.7.5 Natural polyphosphates (Holm 2014). Zhang et al. (2007) describe residual phases of glassy phosphates from the Keluo area, Heilongjiang Province, China. Small samples of natural phosphate glass have been identified by Raman spectroscopy. The authors suggest a P- and F-rich melt/fluid played an important role in an upper-mantle metasomatism, where minerals can change their composition by slow dissolution and reprecipitation processes. The assumption was that glasses formed by rapid quenching of this mantle material, and that such samples are widely distributed in the lithospheric mantle. Similar phosphate rich glasses have been identified in France (Rosenbaum et al. 1997).

5.2 Borate glasses

Boron has an unusually small coordination number for a glass former. The basic glass forming entity of pure vitreous B_2O_3 is made of trigonal $BØ_3$ units (Ø denotes bridging oxygen atoms), so the same network connectivity of three ensues for pure boron oxide as for pure phosphorus oxide glasses (Krogh-Moe 1969). Like P_2O_5, pure vitreous B_2O_3 is highly hygroscopic and readily forms boric acid, $B(OH)_3$ (Ehrt 2000, 2013).

The $BØ_3$ trigonal unit is a strong Lewis base, and readily accepts a fourth ligand, forming borate tetrahedra (Wiberg et al. 2007). Thus, upon addition of modifier oxides, instead of the formation of non-bridging oxygen atoms at these borate trigonals, the coordination initially increases to four, as negatively charged $[BØ_4]^-$ tetrahedra are generated. Fully bridged borate tetrahedra drastically improve the chemical stability of borate glasses. The cross-linking capability of the modifier cation contributes as well. Tetrahedral metaborate units $[BØ_4]^-$ are in an equilibrium with the trigonal $BØ_2O^-$ groups, the relative population depends strongly on the type of the charge compensating modifier cation (Möncke et al. 2016). Higher modifier oxide concentrations form significant numbers of nbO (Wright et al. 2014).

5.2.1 Vitreous B_2O_3. B_2O_3 readily forms a glass which might be explained by the multitude of crystalline polymorphs, and the competing restructuring of the melt upon cooling, not settling on one crystal form, thus resulting in an amorphous material without any long-range structure (Ferlat et al. 2012). Interestingly, the structure of pure vitreous B_2O_3 consists of 75% boroxol rings B_3O_6 (Hannon et al. 1994), and the remaining 25 % represent "isolated", that is non-ring $BØ_3$ groups that are connected to these rings and are therefore not actually "isolated" at all–even though this term is found frequently in the literature. The ring breathing mode of this boroxol ring can be easily identified by Raman spectroscopy through a sharp signal at ca. 805 cm^{-1} (Galeener and Geissberger 1982).

B_2O_3 glasses possess some aspects of 'sheet-like' properties that are akin to graphite, such as partial π-bonding (Duffy 2008), with strong B–O–B bonds within boroxol rings, one of the many well-defined superstructural units, but weaker van der Waal's bonding between the rings. The importance of the boroxol rings on the glass properties was also highlighted by an in-situ Raman study of borate melts by Yano et al. (2003a,b,c), who found that the low T_g of borate glasses correlates to the break-up of boroxol rings.

5.2.2 Boron oxide anomaly. As indicated in the beginning of this section, the initial increase in network connectivity with modifier oxide addition has an impact on many borate glass properties, from density to T_g to chemical stability, which often display an extrema around the maximum in network connectivity (Vogel 1992; Brauer and Möncke 2016). N_4, the ratio of $B_4/(B_3 + B_4)$, with B_3 and B_4 the population of three and four fold coordinated borate respectively, is therefore an important parameter for borate containing glasses. Pyroborate ($BØO_2^{2-}$) and orthoborate units (BO_3^{3-}) are usually found as trigonal borate entities, though tetrahedral borate pyroborate ($BØ_3O^-$) and orthoborate units ($BØ_2O_2^{2-}$) are known to occur in crystals (Wright et al. 2014), and possibly in glasses (Winterstein-Beckmann et al. 2013, 2015). See the schematic in Figure 18 for a depiction of the various borate species.

The N_4 fraction depends not only on the molar ratio in binary $xM_2O–(1-x)B_2O_3$ glasses, but also on the type of modifier cation. High field strength cations such as Mg^{2+} and Zn^{2+} prefer the highly charged sites of non-bridging oxygen as found in $BØ_2O^-$, and even support disproportionation into $BØO_2^{2-}$ and $BØ_3$ (Möncke et al. 2016). The small Li^+ ion on the other side prefers $[BØ_4]^-$ tetrahedra over $BØ_2O^-$ trigonals for charge compensation (Kamitsos and Chryssikos 1991).

A second anomaly was found for ortho-borate compositions in the $SrO–MnO–B_2O_3$ and the $SrO–Eu_2O_3–B_2O_3$ systems (Winterstein-Beckmann et al. 2013, 2015). Both glass systems

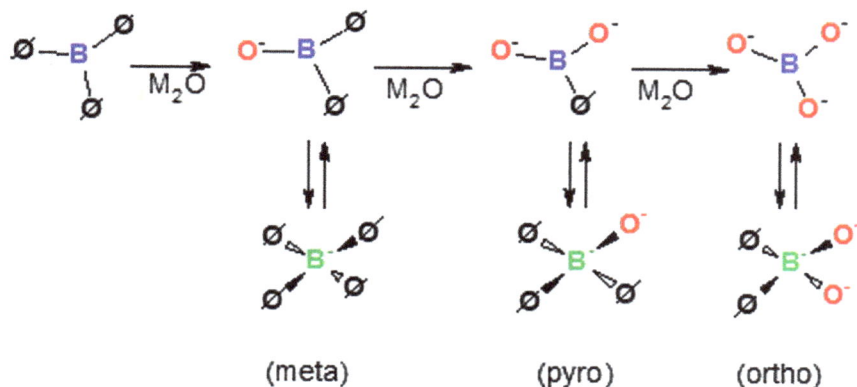

(meta)   (pyro)   (ortho)

Figure 18. Depiction of the various borate species that are known to exist in glasses and crystals. While all trigonal species, as well as the tetrahedral metaborate in the bottom row, are typically found in alkali or alkaline earth borate glasses, the tetrahedral pyro- and ortho-species are less typical though they might occur in lead or bismuth metaborate glass for the pyroborate and in Mn- or Eu- containing orthoborate glasses for the orthoborate unit (see the text for more details).

had initially been selected for their high loads of optically active manganese and europium ions for possible Faraday rotator materials, which renders the glasses paramagnetic and prevented further NMR studies regarding the structure. However, Raman and IR spectra are consistent with the formation of tetrahedral orthoborate units at high modifier concentration and both systems show an increase in T_g for the lowest borate levels, both signs for the occurrence of a second boron anomaly (see Fig. 19). The tetrahedral orthoborate species (depicted on the lower right side in Fig. 18) can form rings of the type $[B_3O_9]^{9-}$ with three bridging oxygen in the ring, and 6 non-bridging oxygen atoms sticking out (see Fig. 19).

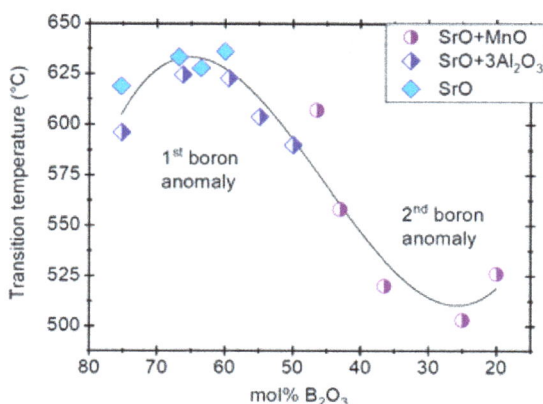

Figure 19. First and second boron anomaly for SrO–B$_2$O$_3$ glasses. Diamonds denote data from Ohta et al. (1982) for the binary system xB$_2$O$_3$–(100−x) sytem with **light blue colored symbols** for the Al-free system and **blue and white split diamonds** for the glasses conating 3 mol% Al$_2$O$_3$, the maximum reflects on the equilibrium of trigonal and tetrahedral metaborate. The **round symbols** are from Winterstein et al. (2013) who found a second boron anomaly as trigoanl orthoborate is in equilibrium with tetrahedral orthoborate units in the ternary xB$_2$O$_3$–xSrO–(100−2x)MnO.

5.2.3 Superstructural units. Borate glasses are fascinating for their variety in medium range order or superstructural units, including various chains and ring type modifications. Many of these superstructural units are known from borate crystals (Wright 2010). Structural analysis depends heavily on ^{11}B-NMR, which can not only distinguish the coordination, but also the presence of ring and non-ring entities (Deters et al. 2011). Raman spectroscopy is further sensitive to even small numbers of boroxol rings with their sharp ring breathing mode near 805 cm^{-1}. Even the substitution of one or two B$\varnothing_3$ units by tetrahedral [B$\varnothing_4$]$^-$, is reflected in separate bands in the Raman spectra. The metaborate or orthoborate ring can further be identified by distinctive bands, as can larger super structural units with strained rings such as the penta-diborate (Kamitsos and Chryssikos 1991; Yiannopoulos et al. 2001; Möncke et al. 2016). Some crystals, such as lead borate or bismuth borate, are characterized by large superstructural units with well-defined structures, such as 20 atoms containing polyanions (Kamitsos and Chryssikos 1991; Wright et al. 2010; Wright 2014; Möncke 2017). Of the many relevant reviews on borate glasses, we suggest the in-depth review on the structural differences between borate and silicate glasses written by Wright et al. (2010).

5.2.4 Non-applicability of Loewenstein rule. The original Loewenstein rule from 1954 had often been wrongly applied to borate glasses where it was dubbed [B$\varnothing_4$]$^-$-avoidance rule. Loewenstein (1954) explained the absence of directly linked [Al$\varnothing_4$]$^-$ tetrahedra in zeolites by the third Pauling rule: *"anion bridges between polyhedra with lowest possible number of coordination, though formally possible according to the electrostatic valence rule, must be expected to be unstable whenever alternative structures with higher numbers of coordination in a part of the polyhedra are possible"*. This means that alumina tetrahedra will prefer to bond to other glass forming units or take on a higher coordination number in order to avoid linking two [Al$\varnothing_4$]$^-$ tetrahedra directly. However, for borate glasses, the tetrahedron has the higher coordination number and therefore the analogy between [Al$\varnothing_4$]$^-$ and [B$\varnothing_4$]$^-$ does not apply (Möncke et al. 2017), as borne out by the many superstructural units in glasses and crystals that do contain linked [B$\varnothing_4$]$^-$ tetrahedra, the most charged one being the tetrahedral orthoborate ring (see Fig. 20). Loewenstein never applied his rule to glasses, nor to B-containing compounds and he did not rule out the tetrahedral linkage on electrostatic grounds or charge repulsion, as often wrongly referred to in the later literature.

5.2.5 Glass formation and phase separation in binary glasses. Like phosphates, borates have a tendency to react with many network former, intermediates and modifier oxides and readily form glasses (Ehrt 2006; Bengisu 2016; Möncke et al. 2016; Milanova et al. 2019). However, some metaborate glasses crystallize (Yiannopoulos et al. 2001) or clustering and phase separation (Ehrt 2013; Herrmann et al. 2019) might occur. ZnO or MnO-metaborate glasses actually display such a pronounced liquid–liquid phase separation, that the heavy MO–B$_2$O$_3$ (56:44) sinks to the bottom of the crucible, while the light B$_2$O$_3$ / H$_3$BO$_3$ phase swims above, preventing even the slightest oxidation of Mn^{2+} (Ehrt 2013).

Lead and bismuth borate glasses form PbO and Bi$_2$O$_3$ pseudo phases embedded in a connective tissue, as described by Ingram for silicates (Ingram et al. 1991). Raman and XPS spectroscopy can distinguish these embedded micro phases from the surrounding phase of undermodified borate compositions (Möncke et al. 2016). The oxygen within the PbO and Bi$_2$O$_3$ regions would correspond to the free oxygen introduced by Henderson and Stebbins (2022, this volume) for silicate glasses.

5.2.6 Modifier cations in borate glasses. The strength of glass network depends not only on the degree of polymerization—and therefore the N$_4$ value with N$_4$=B$_4$/(B$_4$+B$_3$) with B$_4$ and B$_3$ being the fraction of tetrahedral and trigonal coordinated borate species—but also on the cross-linking capability of the modifier cations. In analogy to the discussion in Section 5.1.6, for phosphate glasses, the metal-oxide bonds in the far infrared can be used to calculate the force constant between the modifier cation and the oxygen atoms of the network. Since borate related bands are found at higher energies, more studies have been conducted on the far IR region of borate than most other glasses (Yiannopoulos et al. 2001). Figure 21 shows the correlation of

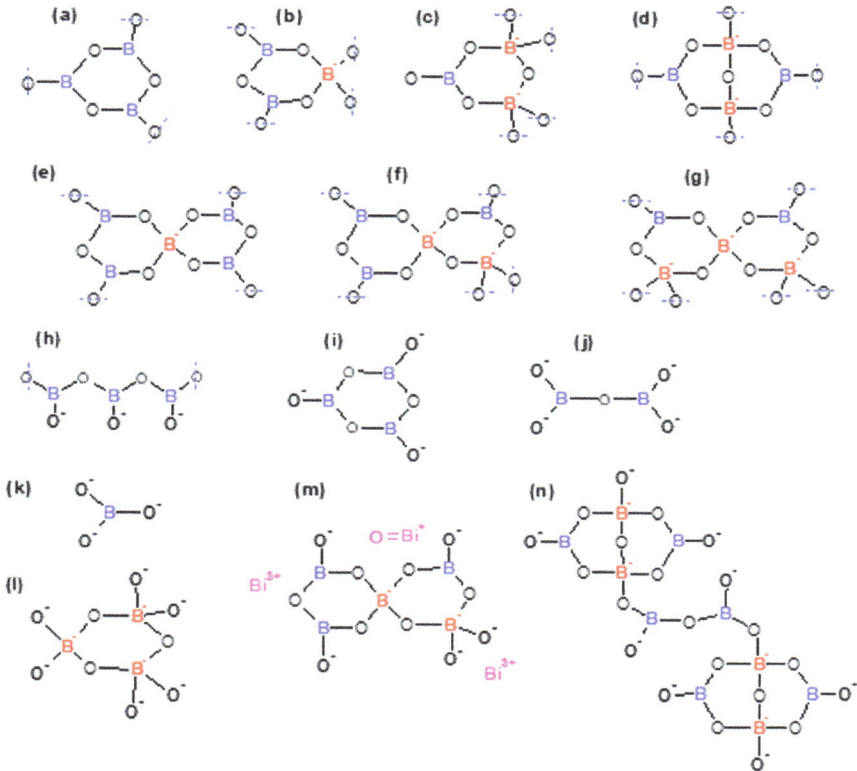

Figure 20. Depiction of typical superstructural units found in borate glasses and crystals (**a**) boroxol ring, (**b**) triborate ring, (**c**) di-triborate ring, (**d**) diborate unit, (**e**) pentaborate unit, (**f**) di-pentaborate unit, (**g**) tri-pentaborate, (**h**) metaborate chain, (**i**) metaborate ring, (**j**) pyroborate, (**k**) orthoborate, (**l**) orthoborate ring with three linked $[B\text{\O}_2O_2]^{3-}$ T_d-orthoborate units; and finally two large polyanions, known from crystals and suggested to exist "(**m**) as di-pentaborate anion $[B_5O_{11}]^{7-}$ in bismuth-metaborate glass or (**n**) as bi-diborate polyanion $[B_{10}O_{21}]^{12-}$ in lead-metaborate glass (Kamitsos and Chryssikos 1991; Wright et al. 2010; Wright 2014; Möncke et al. 2015; Möncke 2017).

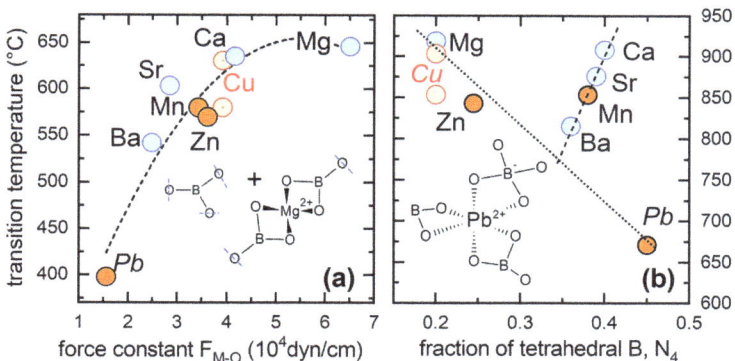

Figure 21. Dependence of transition temperature on the modifier cations in binary MO–B_2O_3 glasses. (**a**) shows the dependence on the force constant of the M–O bond as derived from far infrared measurements, and (**b**) the fraction of tetrahedral borate groups (N_4) as different cations favor different sides of the metaborate equilibrium or disproportionation. Modified after Möncke et al. (2016).

the transition temperature for a series of transition metal and post transition metal cations with the degree of polymerization (N_4) and with the force constant. It is apparent that both factors are equally important. Not shown is the similar correlation with the hardness (Möncke, et al. 2016), results that were recently verified by a mechanical study of alkali and alkaline earth alumoborate glasses (Frederiksen et al. 2018). The dependence of the modifier cation on the optical properties or the viscosity of binary borate glasses is also discussed in (Ehrt 2006).

5.2.7 Multicomponent borate glasses. Borates readily form glasses with other glass formers, intermediate oxides, transition metals or rare earth elements. While some of the binary systems are prone to phase separation or crystallization, many compositions do form glasses, especially when combined with a third or more components (Vogel 1992; Musgraves et al. 2019). For example, the upper limit of glass formation in the $SrO–B_2O_3$ binary lies at $47SrO–53B_2O_3$ while the nominal metaborate 50:50 composition of $SrO–B_2O_3$ crystallizes. Adding a third component to $SrO–B_2O_3$ allows glass formation beyond the metaborate stoichiometry. As seen in Figure 19, the addition of 8 mol%–60 mol% MnO permits glass formation in the Sr–borate system well beyond the metaborate, into and beyond the orthoborate composition. The combination of B_2O_3 with modifier ions has been discussed above and the combination with the glass formers SiO_2 and P_2O_5 will be briefly discussed in the following two sections. Doris Ehrt studied the effect of ZnO, La_2O_3,PbO and Bi_2O_3 on the properties of binary borate glasses and melts (Ehrt 2006), while Milanova et al. combined more intermediate glass former in borate glasses, such as in the $B_2O_3–Bi_2O_3–La_2O_3–WO_3$ glass system with La_2O_3 varying from 0 to 10 mol% and WO_3 levels from 0 to 40 mol% (Milanova et al. 2019). Heavy metal containing glasses are studied as materials for radiation shielding, while other applications focus on the low T_g, such as in sealing glasses.

5.2.8 Borosilicate glasses. Technologically the borosilicate glass system is the most important. These glasses are treated in detail in Youngman (2022, this volume). Applications include, but are not limited, to optics and photonics (Ehrt 2018), laboratory ware, display screens or nuclear waste immobilization. The presence of superstructural units, preferential bonding as well as sub-liquidus phase separation make these glasses fascinating to study (Bunker et al. 1990; Möncke et al. 2015, 2017; Herrmann et al. 2019).

5.2.9 Borophosphate glasses. Eckert et al. studied the structure and properties of different series of sodium borophosphate glasses with enhanced NMR in combination with other spectroscopic techniques (Raman, XPS), once with constant B:P ratio (Rinke and Eckert 2011), once with constant alkali levels but varying B:P ratios for the mixed network former effect (Larink et al. 2012).

For the first case of the $(Na_2O)_x(BPO_4)_{1-x}$ ($0.25 < x < 0.55$) system (Rinke and Eckert 2011), ^{11}B MAS-NMR data reflects a dominance of $[BØ_4]^-$ borate tetrahedra (Ø denotes bridging oxygen atoms) while ^{31}P MAS NMR reveals the successive transformation of neutral Q_P^3 into Q_P^2 and further into Q_P^1 units as the Na_2O content increases. Initially, Na_2O additions form $[BØ_4]^-$ borate tetrahedra, while nbO on the phosphate units form only at higher Na_2O additions (for $x \geq 0.35$). All spectroscopic techniques used revealed strong interactions between the two network formers B and P which are reflected in the preferred formation of B–O–P bonds. Variations in T_g were correlated to the overall network connectivity as expressed by the total number of bridging oxygen atoms per network former species (Rinke and Eckert 2011).

Despite the known phase separation of binary borosilicate or phospho–silicate systems is glass formation reported for ternary and quaternary systems containing only network former oxides. The modifier free boro–phospho–silicate glass system, containing only the network forming oxides B_2O_3, P_2O_5 and SiO_2. has been studied with multinuclear NMR Spectroscopy by Uesbeck et al. (2017). The quaternary, which also contains Al_2O_3, has been reported by Liu et al. (2018).

Borophosphate glasses containing transition metal and post-transition metal ions are studied for their large glass forming region, and their ability to stabilize high fractions of highly polarizable ions. Such glasses are of interest for their non-linear optical properties and electric field induced second harmonic generation (Dussauze et al. 2006).

5.2.10 Borate glasses applications. Applications of borate glasses range from shielding materials to sealing glasses to bio glasses (Bengisu 2016). Borate glass fibers are used in wound healing (Zhao et al. 2015), showing great potential for vascular regrowth.

For the technologically so important field of borosilicate glasses, we refer once again to Youngman (2022, this volume).

5.3 Tellurite glasses

Already in 1834, Berzelius (1834) reported on tellurite glasses consisting of TeO_2 and BaO and similar oxides. Other authors followed up with studies on glass formation and since Stanworth (1952), with characterization of the properties (Vogel 1992).

5.3.1 Pure TeO₂ glass. Pure TeO_2 is not a good glass former and shows a great tendency towards crystallization when prepared in Pt-crucibles. However, dip quenching (Tagiara et al. 2017) allows the preparation of bulk samples of vitreous TeO_2. Pure TeO_2 is unusual in so far as it has asymmetric Te–O–Te bonds, a long and a short bond distance without charge distribution over these bridges (Barney et al. 2013). This is closest to γ-TeO_2, one of the three crystalline phases (McLaughlin et al. 2001). The basic glass forming unit is a pseudo-trigonal bipyramid, with the lone electron pair on tellurium at the 3rd axial corner, the 2 other axial corners are formed by the short Te–O bonds while the Te–O bonds on the z–axis are the long distant bonds. For tellurites, as for borates, the coordination has to be considered when using the Q-nomenclature, this is achieved by adding a 3 or 4 as subscript. The fully polymerized TeO_2 glass consists of TeO_4 units.

The interest in tellurite glasses arises mostly for optical applications (El-Mallawany 2018). The low phonon side bands are of interest for fluorescing materials. The high polarizability induces a high third order susceptibility χ^3 and therefore makes tellurites interesting for their non-linear optical applications. In order to keep a high χ^3, combining TeO_2 with another highly polarizable component, such as Tl_2O or Nb_2O_5 (Bertrand et al. 2015; Carreaud et al. 2015; Kato et al. 2016) is most advantageous.

5.3.2 Structure of tellurite glasses. Contrary to silicate and the previously discussed oxide glass formers, tellurite units exhibit a lone electron pair that takes no part in bonding. Like tin or lead ions, the lone electron pair occupies one side of the basic polyhedra, and in terms of network bonding the lone electron pair can be seen as a non-bridging side at one corner of the pseudo-polyhedra and is one reason for the low T_g (Tagiara et al. 2017) (See Fig. 22).

The basics of this structure has been confirmed by Raman (Tagiara et al. 2017), neutron diffraction (McLaughlin et al. 2001; Barney et al. 2013), and NMR spectroscopy (McLaughlin et al. 2001; Garaga et al. 2017). The lone electron pair at the tellurium ion distorts the fundamental TeO_n polyhedral structure and some ambiguity exists on the nature of the shortest Te–O bonds (see Fig. 22 for a depiction of various tellurite species). Depending on the analytical method used, the short bond might be seen as Te=O double bond or as being weakly bonded to a neighboring tellurium ion as in a Te–O– –Te type of bond, as depicted in Figure 22b as TeO_{3+1} unit. Diffraction studies seem to indicate a significant fraction of such terminal units (Barney et al. 2013); however, Raman spectroscopy does not see a Te=O double bond with a signal at ca. 850 cm^{-1} for pure TeO_2 (Tagiara et al. 2017). On the other hand, if the glass is melted in SiO_2 crucibles, the tellurite network is modified and a Te=O bond is indeed observable in the Raman spectra (Tagiara et al. 2017).

Crucible dissolution has been discussed for the case of phosphate glasses before, and as shown by Tagiara et al. (2017), melting pure TeO_2 in Al_2O_3 crucibles will enhance glass formation through the dissolution and uptake of Al_2O_3 from the crucible. The uptake modifies the network significantly creating non-bridging oxygen atoms (see Fig. 22 for the tellurite species seen in depolymerized glasses) but increasing T_g through the strong cross-linking of Al^{3+} ions.

Like in many other glasses, depolymerization occurs with the addition of modifier oxides to pure TeO_2. From a CN=4 fully polymerized tellurite network with four-fold coordinated entities, Q_{Te4}^4, modifier addition creates first TeO_{3+1} (trigonal bi-pyramid) units, where the 4th oxygen is loosely bound to another tellurite entity, while one oxygen atom is bridging and 2 oxygen atoms are terminal, sharing an excess electron and a double bond. Further modifier oxide addition creates three-fold TeO_3 groups with 1 or two nbO (trigonal pyramid, with the lone electron pair at the apex), that are however delocalized with the double bond of the 3rd terminal oxygen atom. Figure 22 depicts the five tellurite polyhedra that are known from crystals and the TeO_{3+1} group that is an important intermediate species in glasses (McLaughlin et al. 2000).

It is easy to follow these structural changes by IR and Raman spectroscopy. Interestingly, the change in the Raman spectra is very different for addition of the intermediate oxide Nb_2O_5 or when adding modifier oxides such as Tl_2O, ZnO or Al_2O_3 e.g. (Tagiara et al. 2017). Thomas suggested that niobium ions have a similar bond strength to oxygen as tellurium ions and that therefore a solid solution forms between Te- and Nb-oxides. This is reflected for the $(1-x)TeO_2-xNb_2O_5$ series in almost unchanged Raman spectra as Nb_2O_5 is added to TeO_2. This behavior is contrary to the $(1-x)TeO_2-xTl_2O$ series, where the weaker thallium ions act as network modifier, changing the connectivity and coordination of the tellurite polyhedra, which in turn is apparent in significant changes in the Raman spectra (Mirgorodsky et al. 2012) When TeO_2 is combined with another compound that exhibits even stronger oxygen bonds than Te–O, TeO_2 now acts as modifier oxide e.g. in $(1-x)TeO_{2-x}WO_3$ (Mirgorodsky et al. 2012) or when combined with borate as in the $Li_2O-x(2TeO_2)-(1-x)B_2O_3$ glasses (Chatzipanagis et al. 2019).

TeO_2 reacts also readily with other typical glass former oxides, though the mixed B-, Ge- and Si-systems often show phase separation which is not described for phosphorus–tellurite glasses (Vogel 1992), but was for example reported for $(1-x)TeO_{2-x}WO_3$ (Mirgorodsky et al. 2012).

Phase separation or a low degree of mixing is often the reason for a strong negative mixed network former effect, as for example observed for $Li_2O-x(2TeO_2)-(1-x)B_2O_3$ glasses (de Oliveira et al. 2018).

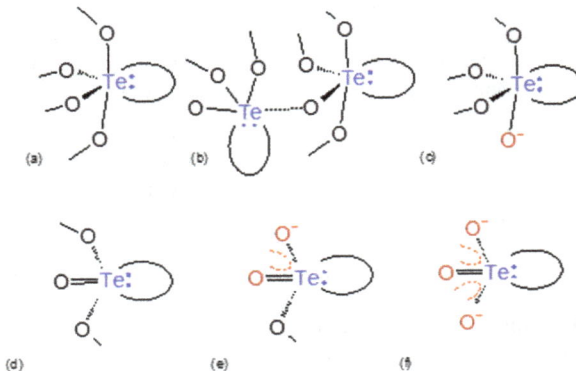

Figure 22. Depiction of the tellurite polyhedra known from crystals and glasses, (a) TeØ$_4$ or Q_{Te4}^4 trigonal bipyramid (tbp), (b) TeØ$_{3+1}$ an intermediate form between the four and three fold coordinated tellurite form with long and short bonds, (c) TeØ$_3$O$^-$ or Q_{Te4}^3(tbp), the non-bridging oxygen can be the long or the short bonded one, (d) TeØ$_2$O or Q_{Te3}^2 trigonal pyramid (tp), (e) TeØO$_2^-$ or Q_{Te3}^1(tp), (f) TeO$_3^{2-}$ or Q_{Te3}^0(tp), after McLaughlin et al. (2000).

When studying the Raman spectra, it should be noted, that Raman spectroscopy has the highest sensitivity for highly polarizable ions such as Te^{4+}, and if combined with low polarizable classical network former such as B, Si or P, those bands are therefore often hidden under the Te-related Raman bands (Chatzipanagis et al. 2019).

For mixed tellurite–germanate glasses see Section 5.4.

5.3.3 Properties and applications. Applications of tellurite glasses are especially focused on optics and photonics for their high polarizability resulting in high refractive index and high third order susceptibility, which is of interest for non-linear optics (NLO) (Weber 2006; Barbosa et al. 2017). Tellurites have a high transparency in the IR, as can be seen from Figure 8, the highest for oxide glasses as low as 5 to 7 µm (Vogel 1992). On the other hand, the high polarizability of tellurium decreases the transparency in the UV-visible wavelength region, shifting the band gap to lower energies (same figure). The low energy phonon side bands are advantageous for fluorescence, as is the high solubility of rare earth ions in tellurite glasses (Wang et al. 1994). The same authors also include a detailed study on the formability of tellurite glasses including fiber drawing and extrusion.

Other applications include laser materials and energy conversion, but also radiation shielding because of the high density and even biomedical applications (El-Mallawany 2018).

Many publications can be found on the structures and properties of tellurite glasses, including The "Tellurite Glasses Handbook: Physical Properties and Data" by R.A.H. El-Mallawany (Vogel 1992; El-Mallawany 2012). The melting temperatures are low, between 700–900°C, and as a consequence they also display a low T_g. Density and coefficient of thermal expansion are high while mechanical strength and hardness are relatively low compared with other oxide glasses (Stanworth 1952; Tagiara et al. 2017). The latter can be improved by crystallization, and transparent glass ceramics for optical applications have been successfully prepared e.g. by complete crystallization of the 75 TeO_2–12.5 Nb_2O_5–12.5 Bi_2O_3 glasses (Bertrand et al. 2016).

Researchers that have worked excessively with tellurium containing glasses might have experienced the typical garlic-odor breath, as the body converts any tellurium taken up, to dimethyl telluride $(CH_3)_2Te$ (Chasteen and Bentley 2003).

5.3.4 Antiglass. The term "anti-glass" was defined by Burckhardt and Trömel (1983) when describing a very small group of oxide materials from the tellurite system. The definition of anti-glass refers to a solid, with cationic long-range order but lacking any anionic short-range order. This is exactly contrary to glasses with short-range order and a lack of long-range order, hence the name. Many reported anti-glass structures derive from tellurites and are based on a CaF_2 fluorite structure with Te^{4+} and other metal ions such as for example Bi^{3+} or Sr^{2+}, statistically distributed at the available cation positions while not all crystallographic anion positions are occupied by oxygen (Bertrand et al. 2015). Reported anti-glass compositions include $SrTe_5O_{11}$ (Burckhardt and Trömel 1983) or glass ceramic from the TeO_2–Nb_2O_5–Bi_2O_3 system (Bertrand et al. 2015).

5.4 Germanate glasses

Germanium is an important element for chalcogenide glasses (see Section 4.1) but plays only a minor role in technological oxide glasses. GeO_2 doped SiO_2 fibers are used in telecommunication, mostly to increase the refractive index in the core of the fiber. $GeCl_4$ is used as precursor in the preparation of preforms by vapor deposition in the same way as $SiCl_4$ (Vogel 1992).

However, since the glass formation of GeO_2 was first suggested by Zachariasen (1932), oxide-germanate glasses garnered so far more academic interest—despite the wide glass forming range. Notwithstanding many analogies to silicate systems, the lower field strength of Ge^{4+} to the smaller Si^{4+} ion results in distinct differences, including a smaller tendency toward phase separation (Vogel 1992).

5.4.1 Simple GeO₂ glass. The structure of pure vitreous GeO_2 glass is in good agreement with vitreous SiO_2. The basic glass forming units are GeO_4-tetrahedra, though the O–Ge–O intra tetrahedral angles seem to be more distorted in vitreous GeO_2 compared to vitreous SiO_2. This fact was attributed to the larger size of the germanium compared to the silicon ion and a higher number of smaller 3-membered rings in vitreous germania glass. A nice review on the structure of amorphous, crystalline and liquid GeO_2 was conducted by Micoulaut et al. (2006). Of interest are also structural variations under high pressure, since GeO_2 has a higher tendency than SiO_2 to form higher coordinated polyhedra (Kono et al. 2016). Not only under pressure, but also in mixed component glasses are germanium ions found in five- and six-fold coordination in addition to GeO_4 tetrahedra. Important differences between GeO_2 and SiO_2 glasses concern the mid-range order and the bond strength. Thus, the T_g of GeO_2 glasses is with only 514 °C significantly lower than the T_g of SiO_2 with 1203 °C, see also Table 3 (Shelby 1974).

5.4.2 Modifier free mixed network former germanate glasses. GeO_2 easily forms glasses with many other glass formers, intermediates and modifier oxides (Haiyan et al. 1986). Shelby studied glass formation and properties of many GeO_2-containing glasses, including the binary B_2O_3–GeO_2 system (Shelby 1974; Vogel 1992).

Germano–silicate glasses are widely used as low-attenuation optical fibers, yielding numerous studies on their physical (optical) properties (Fleming 1984). Studies that focus on the underlying structure are rare and the fundamental question on the homogeneity of germano–silicate glasses still has not been answered, e.g. if clustering or significant phase separation occurs (Micoulaut et al. 2006; Majérus et al. 2008).

Germano–phosphate glasses were studied by X-ray and neutron diffraction and show better glass formation and less regions of phase separation than the corresponding phospho–silica glass (Hoppe et al. 2006).

Germano–borate glasses exhibit a distinct mixed network former effect, displaying negative deviations from additivity for all intermediate glasses for properties related to ion mobility (e.g. viscosity, thermal expansion, but also for density) while a positive deviation from additivity was seen for the refractive index (Shelby 1974). There is no evidence of macroscopic phase separation in these glasses. Raman studies on the binary GeO_2–B_2O_3 and the ternary GeO_2–B_2O_3–SiO_2 system have been performed, and Chakraborty and Condrate (1986a,b) looked into the connectivity and coordination of these glass former oxides in the mixed glass systems.

Germano–tellurite glasses have been studied for their optical properties, especially fluorescence in glasses doped with Er^{3+} and Tm^{3+} ions (Mattarelli et al. 2005).

5.4.3 Modified germanate glasses. Various studies looked into the structure of modified germanate glasses, such as in regard to alkali sites and optical basicity (Kamitsos et al. 2002), including IR and Raman investigations exemplified in the xRb₂O·$(1 - x)$GeO₂ series (Kamitsos et al. 1996). Binary germanate systems such as TiO_2–GeO_2 (Khan and Mohamed–Osman 1986), GeO_2–Bi_2O_3 (Kassab et al. 2019) and GeO_2–PbO (Bahari et al. 2013) were studied for their structure and optical properties after addition of optically active elements such as rare earth ions and silver nano particles.

5.4.4 Multicomponent germanate glasses. Newer, in depth multi-technique structural studies on mixed network former glasses have been conducted by Eckert et al. including the $(M_2O)_{0.33}[(Ge_2O_4)_x(P_2O_5)_{1-x}]_{0.67}$ system with M = Na and K (Behrends and Eckert 2014). The formula uses the term Ge_2O_4 to account for the fact that the same number of phosphate tetrahedra will be replaced by germanate polyhedra. Here, the authors used ^{31}P and ^{23}Na NMR, Raman, and O-1s XPS (X-ray photoelectron) spectroscopy. It was found that heteroatomic P–O–Ge linkages were overall preferred over homoatomic P–O–P and Ge–O–Ge linkages. Alkali ions tend to be always linked to more terminal oxygen atoms from phosphate rather than germanate polyhedra. The preferred association of sodium ions with Q_P^1 groups results partially in cation clusters. The authors deduced indirectly the formation of higher germanium coordination states in this glass system (Behrends and Eckert 2014).

Multicomponent germanate glasses containing Bi_2O_3, Gd_2O_3, Ga_2O_3, WO_3, TeO_2 or even CoO as additional components have been successfully prepared and studied, e.g. by Jewell et al. (1994), Simon et al. (2000), Ardelean et al. (2001) ,Upender et al. (2011) and more recently by Tagiara et al. (2020) and Bradtmüller et al. (2021).

Fluoride containing germanate systems were recently studied in detail, see for example Pereira et al. (2017, 2018).

5.5 Other glasses

In the frame of the current publication we cannot cover the full range of non-metallic oxide glasses that exist in addition to the already mentioned glass forming system. However, we want to mention a couple more examples to show the variety and that even glasses lacking an accepted network former might form glasses in combination with other oxides.

5.5.1 Antimonite glasses. Miller and Cody (1982) successfully prepared vitreous Sb_2O_3 glasses by quenching in liquid nitrogen. Multicomponent antimonate glasses are studied as possible non-linear optical and laser materials. They are in many aspects similar to tellurite glasses (Baazouzi et al. 2013).

Honma et al. studied the binary Sb_2O_3–B_2O_3 system by XPS and optical spectroscopy in order to understand the impact of different polarizability of the trivalent ions in this M_2O_3 systems (Honma et al. 2000). Montesso et al. looked into glass formation, structure and properties of a $SbPO_4$–GeO_2 glass system for possible applications exploiting its high polarizability (Montesso et al. 2018). Antimonates can be introduced into many glass systems, such as tungstates or molybdates see for example Kubliha et al. (2015), as well as germano–silicates (Zmojda et al. 2016). While most glasses contain trivalent Sb^{3+} in the form of antimonite, Mößbauer and photo electron spectroscopy can distinguish pentavalent Sb^{5+} (antimonate) in some samples (Schütz et al. 2004).

This is contrary to arsenate glasses, which usually are found to contain arsenic in the form of As_2O_3–As_2O_5 mixtures (Vogel 1992). Binary As-containing compositions with alkali oxides or intermediate oxides can be prepared as well, though due to volatility and the intrinsic toxicity of arsenic and ensuing safety concerns caused de facto the termination of further studies into arsenite glasses. Moreover, arsenic oxide glasses do not offer any properties that are fundamentally different from other highly polarizable oxides or for that matter, arsenic chalcogenides.

5.5.2 Intermediate glass former (oxides of Pb, W, Mo, Nb, Ta, Al, Zn…). Interestingly, a typical glass former is not always required in order to get a glass. For example, binary glasses form in the CaO–Al_2O_3 system (Drewitt et al. 2012; Akola et al. 2013). For the $64CaO$–$36Al_2O_3$ eutectic, the quenched glasses consist of a topologically disordered cage network with large-sized rings. A coordination number of 4 has been found earlier for Al^{3+} in $50CaO$–$50Al_2O_3$ glass, and generally if $CaO \leq Al_2O_3$. However, a small fraction of higher coordinated aluminate polyhedra form when CaO levels exceed Al_2O_3 levels. With CaO_5 polyhedra, a surprisingly low coordination number is given for Ca^{2+} ions, and depending on the Ca:Al ratio, varies the level of three-fold coordinated oxygen ions. The melt of the eutectic is characterized by a very high viscosity, despite the composition rich in modifier oxides (Akola et al. 2013).

Lead gallate glasses have been studied by diffraction (Hannon et al. 1996) and for ion conduction by Qiu et al. (1996), and lanthano–gallates by Raman spectroscopy (Skopak et al. 2018). Many multicomponent non-traditional network former glasses based on tungstate have also been prepared, a paper on tungstate based glasses by Ataalla et al. (2018) contains also a review on other network former free glasses. For example, Milanova et al. studied diverse network former free glass systems with high Bi_2O_3, MoO_3 or WO_3 content such as the binary Bi_2O_3–MoO_3, or the more complex systems WO_3–ZnO–Nd_2O_3–Al_2O_3, ZnO–Bi_2O_3–WO_3–MoO_3 (Milanova et al. 2011, 2014; Iordanova et al. 2015). Likewise the structure of alkaline earth vanadate glasses was investigated for xMO–$(100-x)$V$_2$O$_5$ (M = Ca, Sr, Ba, $x = 30$ to 55) using ^{51}V-NMR (Hayakawa et al. 1994).

5.5.3 Multi-component inorganic oxide glasses.

5.5.3 Multi-component inorganic oxide glasses. Non-metallic inorganic glasses can be combined in uncountable variations, as was shown by Zanotto et al. (2004). While some combinations show phase separation or are prone to crystallization, a wide range of systems will indeed form glasses.

Mixing different types of modifier cations will often support glass formation due to the confusion principle. Many glass properties will reflect the combined components, such as the refractive index or optical basicity, which is an additive property. Properties that depend on ionic motion such as T_g, viscosity, or ion conduction, but also E modulus, will show deviations from additivity, which is known as the mixed ion effect or mixed alkali effect.

Many intermediate oxides form glasses with any of the classical network formers (NWF) and have been mentioned in each section dedicated to each NWF.

Mixed network glass systems have been treated in this book chapter in chronological order of the first NWF discussed in this chapter. The following table lists selected mixed systems in alphabetical order and gives the respective section number of this chapter where more information regarding each system can be found:

• B_2O_3–GeO_2	germanate glasses,	section 5.4.3
• B_2O_3–P_2O_5	borate glasses,	section 5.2.8
• B_2O_3–SiO_2	borate glasses,	section 5.2.8
• B_2O_3–TeO_2	tellurite glasses,	section 5.3.2
• GeO_2–P_2O_5	germanate glasses,	section 5.4.2
• GeO_2–SiO_2	germanate, glasses,	section 5.4.2
• GeO_2–TeO_2	germanate, glasses,	section 5.4.2
• P_2O_5–SiO_2	phosphate, glasses,	section 5.1.7.1
• P_2O_5–TeO_2	tellurite,	section 5.3.2

Mixed anion glasses were mentioned when the main anion species were discussed, e.g. in Section 4.4.3 with oxynitride, or in Section 5.1.7.4 when discussing mixed FP glasses.

6. GLASS FAMILIES IN COMPARISON

Comparison of such different glass systems can be facilitated by using the concept of optical basicity, a system that is based on the quantification of the electron donor power of the glass matrix rather than on a certain composition. The concept of optical basicity works well for inorganic non-metallic oxide glasses and has even be extended to F and N or S containing systems (Duffy 1992, 2011; Ehrt 2018; Möncke et al. 2019). This concept is most useful for optical properties such as refractive index and band gap; however, the extent of ligand field splitting and redox equilibria of polyvalent ions can also be related to the optical basicity.

Table 3. Properties of monomeric and binary sodium glasses of the type Na_2O–F_2O_n (with F = B, Si, P, Te, Ge, Sb), as well as of selected multicomponent glasses. Density ρ , transition temperature T_g, experimental melting temperature T_m (or when available in brackets, liquidus melting temperature), coefficient of thermal expansion CTE, refractive index n and Abbe number ν (experimental wavelength given as subscript, $e = 546.1$ nm or $d = 587.1$ nm, $D = 589.3$ nm, theoretical optical basicity Λ^1, data in italics are predicted values from SciGlass6.7, calculated with factors from Priven (Mazurin and Priven, Priven 2004).

Glass	ρ g/cm³	T_g °C	T_{m*} °C	CTE $10^{-7}K^{-1}$	n	ν	E_{gap} eV	Λ_{th}	Ref.
SiO_2	2.20	*1203*	CVD (1728)	5	1.48_e	68_e	8.25	0.48	a, b, c
GeO_2	3.64	514	1400 (1150)	77	1.61_d	42_d	5.63	0.70	d, e, f
TeO_2	5.62	307	900	–	2.20_e	20_e	3.37	0.99	g, h
P_2O_5	2.39	393	1000	77	1.50_d	61_d	–	0.25	i, j
B_2O_3	1.84	230–240	850–1000	170	1.45_e	58_e	7.2	0.42	f, k, l
Sb_2O_3	5.11	250	900	*163*	2.10_d	$22_?$	3.33	1.14	k, l, m, n
Na_2O–$2SiO_2$	2.49	460	1600	160	1.51_e	55_e	–	0.60	o
Na_2O–$2GeO_2$	*3.60*	*445*	–	*130*	*1.63_D*	*41_D*	–	0.71	*SciGlass*
Na_2O–$2TeO_2$	4.20	211	800	–	$\sim1.8_d$	$\sim22_d$	–	0.97	p, q
Na_2O–P_2O_5	2.48	275	620	250	1.48_e	66_e	–	0.58	r
Na_2O–B_2O_3	*2.35*	*409*	800–900	*168*	*1.52_D*	*52_D*	–	0.58	*SciGlass*
Na_2O–Sb_2O_3	*4.72*	*255*	–	*180*	*1.92_D*	*25_D*	–	1.16	*SciGlass*
ZBLAN [2]	4.4	260	470	200	1.50_e	78_e	5	0.47	s, t
F00 [3]	3.42	400	1000	–	1.41_e	105_e	8.25	0.35	s, u
FP10 [4]	3.54	440	1000	160	1.46_e	90_e	7.75	0.37	b, s
$73Bi_2O_3$–$27B_2O_3$	7.79	340	1000	110	2.30_e	15_e	2.0	0.98	b
P–laser glass [5]	2.59	460	nda	116	1.51_d	68.4_d	–	–	v
As_2S_3	3.2	454	850	237	$2.4_{1\mu m}$	$160_{3-5\mu m}$	2.2	1.3	v, w, x
Plexiglas 8N [6]	1.19	117	220–260	800_{50}	1.49	59	4	–	y, z

Notes:
[1] Based on recent recommended values as listed by Rodriguez et al. (2011); $Na_2O = 1.11$; values for vitreous P_2O_5 and B_2O_3 from Duffy and Ingram (1976) for sulfides see Duffy (1992) and for fluorides see Duffy (2011)
[2] $53ZrF_4$–$20BaF_2$–$4LaF_3$–$3AlF_3$–$20NaF$
[3] $39AlF_3$–$10MgF_2$–$28CaF_2$–$23SrF_2$
[4] $10Sr(PO_3)_2$–$35AlF_3$–$30CaF_2$–$15SrF_2$–$10MgF_2$
[5] Laser glass, LG-770 from Schott, a Nd-containing alumophosphate based glass (Musgraves et al. 2019)
[6] Plexiglass–polymethylmethacrylate (PMMA), commercial product

References:
[a] Lithosil data sheet (SCHOTT Lithotec AG 2006); [b] Heraeus Quarzglas GmbH & Co. KG (2016); [c] Ehrt et al. (2000); [d] Dennis and Laubengayer (1926); [e] Walker et al. (2015); [f] Shelby (1974); [g] Kim et al. (1993); [h] Tagiara et al. (2017); [i] Hudgens (1994); [j] Kordes et al. (1953); [k] Terashima et al. (1996); [l] Kordes (1939); [m] Doweidar (2015); [n] Bednarik and Neely (1982); [o] Ehrt and Keding (2009); [p] Kavaklıoğlu et al. (2015); [q] n_d and v_d for $40Na_2O$–$60TeO_2$ Vogel (1992); [r] Möncke et al. (2018a); [s] Möncke et al. (2004, 2005); [t] Poulain et al. (1992); [u] Ehrt (2015); [v] Musgraves et al. (2019); [w] Glaze et al. (1957); [x] Fekeshgazi et al. (2005); [y] Evonik Industries AG (2019); [z] Amine et al. (2018).

Actually, the concept was originally developed for the assessment of slags and the redox state of polyvalent ions therein. Importantly, it should be noted that many apparent correlations, e.g. when comparing the optical basicty with thermal properties, is in fact a corrleation with the increased number of non-bridging oxygen ions created by adding modifier oxides of higher basicity to low basicity network former (B, Si or P) and not a direct causation to a higher electron donor power of the glass matrix. This becomes apparent when comparing e.g. the absorption spectra of transition metal ions in a high basicity, highly polymerized tellurite or antimonite glass with a highly modified silicate glass of lower optical basicity. Probe ions like Co^{2+} or Ni^{2+} are indicators of non-bridging oxygen atoms and change their cordination from octahedral sites in low basicty borate, phosphate or silicate rich glasses, to tetrahedral coordination as the optical basicty increases with the addition of modifier oxides. In high basicty tellurite or antimonite rich glasses the octahedral coordinations dominate while addition of Na_2O decreases actually the optical basicity but still supports the tetrahedral coordination of these probe ions as they link to the non-bridging oxygen atoms that are generated by modifier oxide additon, that is preferntial bonding sites of high microbasicity (Soltani et al. 2016).

CONCLUDING REMARKS

Many materials can form a glass when quenched fast enough, organic polymers as well as metals or, oxides, which are the largest group in inorganic non-metallic glasses. All glasses have two characteristic properties: one is the lack of a long-range order and the other is that they all exhibit a glass transition temperature. One big advantage of glasses over many other materials is their formability, due to a viscosity behavior that exhibits a broad softening interval, which allows molding (form pressing, form blowing), fiber drawing or the preparation of large bulk pieces or large planes of float glass. The lack of long-range order allows for an almost unlimited number of compositional variations that do not need to consider stoichiometric compositions as is often found for mixed crystals. Thus, the structure and properties of glasses can be adjusted and fine-tuned on different scales.

The variability of glass compositions cannot be fully honored in such a short book chapter. However, we hope the interested reader got a general idea and will find the given—but by no means exhaustive—references helpful. With applications ranging from everyday materials such as architecture and container glass which are mostly based on silicates, to specialty glasses used in optics and photonics, in biomedicine, or metallic glasses for golf clubs. Even the piece of amber you found at the beach might constitute a very different type of glass.

ACKNOWLEDGEMENTS

The authors want to thank Doris Ehrt and Dominique de Ligny for proof reading and valuable hints and discussions. Doris answered many of our questions to properties by tireless searched in the SciGlass data base. We also want to acknowledge the presenters at the 13 Ecole thématique "*Structural Role of Elements in Glasses from Classical Concept to a Reflexion over broad Composition Range*" - Cargèse, France, 27–31 Mars 2017. Especially for the non-conventional glass former, such as: Invert Glasses and Aluminate Glasses, L. Hennet, Borate Glasses, G. Lelong, Phosphates and Vanadates glasses, F. Munoz & L. Montagne, Metallic glasses, L. Greer, Chalcogenide glasses: structure vs. compositions, A. Pradel, Chalcogenide glasses: properties vs. structure, B. Bureau, Chalcogenide glasses: properties vs. structure, E. Byschkov, Organic glasses, C. Alba Simionesco, Hybrid glasses, N. Greaves, Borates, Silicates and Tellurites glasses, A. Hannon, Tellurite glasses, P. Thomas. Their presentation and slides provided us with valuable references, examples and summaries of the respective fields.

REFERENCES

Abdellaoui N, Starecki F, Boussard C-Pledel, Shpotyuk Y, Doualan JL, Braud A, Baudet E, Nemec P, Cheviré F, Dussauze M, Bureau B, Camy P, Nazabal V (2018) Tb^{3+} doped $Ga_5Ge_{20}Sb_{10}Se_{65-x}Te_x$ ($x = 0–375$) chalcogenide glasses, fibers for MWIR and LWIR emissions. Opt Mater Express 8:2887–2900

Adam J-L, Zhang X (2014) Chalcogenide Glasses, Woodhead Publishing

Aitken BG, Youngman RE, Deshpande RR, Eckert H (2009) Structure–property relations in mixed-network glasses: Multinuclear solid state NMR investigations of the system $xAl_2O_3 \cdot_{(30-x)}P_2O_5 \cdot_{70}SiO_2$. J Phys Chem C 113:3322–3331

Akola J, Kohara S, Ohara K, Fujiwara A, Watanabe Y, Masuno A, Usuki T, Kubo T, Nakahira A, Nitta K, Uruga T, Weber JKR, Benmore CJ (2013) Network topology for the formation of solvated electrons in binary $CaO–Al_2O_3$ composition glasses. PNAS 110:10129

Ali S, Jonson B, Pomeroy JM, Hampshire S (2015) Issues associated with the development of transparent oxynitride glasses. Ceram Inter 41(3, Part A): 3345–3354

Amine T, Libessart L, El Amrani A, Azrour M, Lassue S (2018) Optical and thermal properties of polymethymethacrylate based plexiglass. In: International Symposium: Architectural Patrimony and Local Building Materials, Materiocedia, Cambridge International Academics, Errachidia, Morocco, 2017.

Anderson KB, Winans RE, Botto RE (1992) The nature, fate of natural resins in the geosphere—II Identification, classification and nomenclature of resinites. Org Geochem 18:829–841

Ardelean I, Ilonca G, Simon V, Filip S, Simon S (2001) Magnetic susceptibility investigation of $CoO–Bi_2O_3–GeO_2$ glasses. J Alloys Compd 326(1–2):121–123

Ataalla M, Afify AS, Hassan M, Abdallah M, Milanova M, Aboul-Enein YH, Mohamed A (2018) Tungsten-based glasses for photochromic, electrochromic, gas sensors, and related applications: A review. J Non-Cryst Solids 491:43–54

Axinte E (2011) Glasses as engineering materials: A review. Mater Des 32:1717–1732

Baazouzi M, Soltani MT, Hamzaoui M, Poulain M, Troles J (2013) Optical properties of alkali-antimonite glasses, purified processes for fiber drawing. Opt Mater 36:500–504

Bahari H-R, Zamiri R, Sidek H, Zakaria A, Adikan F (2013) Characterization, synthesis of silver nanostructures in rare earth activated $GeO_2–PbO$ glass matrix using matrix adjustment thermal reduction method. Entropy 15:1528–1539

Baik S, Raj R (1985) Suppression of frothing by silicon addition during oxynitride glass synthesis. J Am Ceram Soc 68:C168–C170

Barbosa CL, Filho CO, Chillcce EF (2017) Photonic applications of tellurite glasses. In:Technological Advances in Tellurite Glasses: Properties, Processing, and Applications. Rivera V, Manzani D (eds) Cham Springer International Publishing, p 93–100

Barney ER, Hannon AC, Holland D, Umesaki N, Tatsumisago M, Orman RG, Feller S (2013) Terminal oxygens in amorphous TeO_2. J Phys Chem Letters 4:2312–2316

Batterman SD, Bassani JL (1990) Yielding, anisotropy, deformation processing of polymers. Polymer Eng Sci 30:1281–1287

Becher PF, Hampshire S, Pomeroy M, Hoffmann MJ, Lance MJ, Satet RL (2011) An overview of the structure, properties of silicon-based oxynitride glasses. Inter J Appl Glass Sci 2:63–83

Bednarik JF, Neely JA (1982) A single component antimony oxide glass, some of its properties. Glastech Ber 55:126–129

Behrends F, Eckert H (2014) Mixed network former effects in oxide glasses: Spectroscopic studies in the system $(M_2O)_{1/3}[(Ge_2O_4)_x(P_2O_5)_{1-x}]_{2/3}$. J Phys Chem C 118:10271–10283

Bengisu M (2016) Borate glasses for scientific, industrial applications: a review. J Mater Sci 51:2199–2242

Bennett TD, Yue Y, Li P, Qiao A, Tao H, Greaves ND, Richards T, Lampronti GI, Redfern SAT, Blanc F, Farha OK, Hupp JT, Cheetham AK, Keen DA (2016) Melt-quenched glasses of metal–organic frameworks. J Am Chem Soc 138:3484–3492

Bertrand A, Carreaud J, Delaizir G, Shimoda M, Duclère J-R, Colas M, Belleil M, Cornette J, Hayakawa T, Genevois C, Veron E, Allix M, Chenu S, Brisset F, Thomas P (2015) New transparent glass-ceramics based on the crystallization of anti-glass. spherulites in the $Bi_2O_3–Nb_2O_5–TeO_2$ System. Crystal Growth Des 15:5086–5096

Bertrand A, Carreaud J, Chenu S, Allix M, Véron E, J-Duclère R, Launay Y, Hayakawa T, Genevois C, Brisset F, Célarié F, Thomas P, Delaizir G (2016) Scalable, formable tellurite-based transparent ceramics for near infrared applications. Adv Opt Mater 4:1482–1486

Berzelius JJ (1834)Untersuchung über die Eigenschaften des Tellurs. Annal Phys 108:577–627

Bradtmüller H, Rodrigues ACM, Eckert H (2021) Network former mixing (NFM) effects in alkali germanotellurite glasses. J Alloy Compd 873:159835

Brain C, Brain S (2016) The development of lead-crystal glass in London, Dublin 1672–1682: A reappraisal. Glass Technol—Euro J Glass Sci Technol Part A 57:37–52

Brauer DS, Möncke D (2016) Structure of bioactive silicate, phosphate and borate glasses. In: Bioactive Glasses: Fundamentals, Technology and Applications. Boccaccini AR, Brauer DS, Hupa L (eds) Royal Society of Chemistry

Bray PS, Anderson KB (2009) Identification of Carboniferous (320 million years old) class Ic amber. Science 326:132

Brow RK (2000) Review: The structure of simple phosphate glasses. J Non-Cryst Solids 263–264:1–28

Brow RK, Tallant DR, Hudgens JJ, Martin SW, Irwin AD (1994) The short-range structure of sodium ultraphosphate glasses. J Non-Cryst Solids 177:221–228

Brow RK, Tallant DR, Myers C, Phifer CC (1995) The short-range structure of zinc polyphosphate glass. J Non-Cryst Solids 191:45–55

Brow RK, Click CA, Alam TM (2000) Modifier coordination, phosphate glass networks. J Non-Cryst Solids 274:9–16

Bunker BC, Tallant DR, Balfe CA, Kirkpatrick JR, Turner GL, Reidmeyer MR (1987) Structure of phosphorus oxynitride glasses. J Am Ceram Soc 70:675–681

Bunker BC, Tallant DR, Kirkpatrick JR, Turner GL (1990) Multinuclear nuclear magnetic resonance, Raman investigation of sodium borosilicate glass structures. Phys Chem Glass 31:30–41

Burckhardt H-G, Trömel M (1983) Strontium-undecaoxotellurat, SrT_5O_{11}, eine CaF_2-Defektstruktur und ihre Beziehung zur Struktur einfacher Gläser. Acta Crystallogr Sect C 39:1322–1323

Bureau B, Boussard-Pledel C, Lucas P, Zhang X, Lucas J (2009) Forming glasses from Se, Te. Molecules 14:4337–4350

Bureau B, Danto S, Ma HL, Boussard-Plédel C, Zhang XH, Lucas J (2008) Tellurium based glasses: A ruthless glass to crystal competition. Solid State Sci 10:427–433

Carreaud J, Labruyère A, Dardar H, Moisy F, Duclère JR, Couderc V, Bertrand A, Dutreilh M-Colas, Delaizir G, Hayakawa T, Crunteanu A, Thomas P (2015) Lasing effects in new Nd^{3+}-doped TeO_2–Nb_2O_5–WO_3 bulk glasses. Opt Mater 47:99–107

Chakraborty IN, Condrate AR Snr (1985) The vibrational spectra of glasses in the Na_2O–SiO_2–P_2O_5 system with a 1:1 SiO_2:P_2O_5 molar ratio. Phys Chem Glasses 26:68–73

Chakraborty IN, Condrate AR (1986a) The vibrational spectra of B_2O_3–GeO_2 glasses. J Non-Cryst Solids 81:271–284

Chakraborty IN, Condrate AR (1986b) Vibrational spectra of B_2O_3–GeO_2–SiO_2 glasses. J Mater Sci Letters 5:361–364

Chakraborty IN, Condrate Snr RA, Ferraro JR and Chenuit CF (1987) The vibrational spectra and normal coordinate alanyses of cubic and monoclinic SiP_2O_7. J Solid State Chem 69:94–105

Charleston RJ (1978) Glass furnaces through the ages. J Glass Stud 20:9–33

Chasteen TG, Bentley R (2003) Biomethylation of selenium, tellurium: Microorganisms and plants. Chem Rev 103:1–26

Chattopadhyay C, Murty BS (2016) Kinetic modification of the 'confusion principle' for metallic glass formation. Scripta Materialia 116:7–10

Chatzipanagis KI, Tagiara NS, Möncke D, Kundu S, Rodrigues ACM, Kamitsos EI (2019) Vibrational study of lithium borotellurite glasses. J Non-Cryst Solids 540:120011

Chen M (2011) A brief overview of bulk metallic glasses. NPG Asia Mater 3:82–90

Chen X, Chen X , Pedone A, Apperley D, Hill RG, Karpukhina N (2018) New insight into mixing fluoride, chloride in bioactive silicate glasses. Sci Rep 8:1316

Cheng YQ, Ma E (2011) Atomic-level structure, structure–property relationship in metallic glasses. Prog Mater Sci 56:379–473

Chopinet M-H (2012) Developments of Siemens regenerative, tank furnaces in Saint-Gobain in the XIXth century. Glass Technol—Euro J Glass Sci Technol Part A 53:177–188

Chopinet M-H (2019) The history of glass (chapter 1). *In:* Springer Handbook of Glass. Musgraves JD, Hu J, Calvez L (eds) Springer, p 1–47

Click CA, Brow RK, Ehrmann PR, Campbell JH (2003) Characterization of Pt^{4+} in alumino-metaphosphate laser glasses. J Non-Cryst Solids 319:95–108

Cook LM, Stokowski SE (1986) Filter materials Handbook of Laser Science and Technology. Boca Raton, FL, CRC Press. Vol IV, Opt Mater Part 2:93

Danto S, Thompson D, Wachtel P, Musgraves JD, Richardson K, Giroire B (2013) A comparative study of purification routes for As_2Se_3 chalcogenide glass. Inter J Appl Glass Sci 4:31–41

Day DE, Ray CS (2013) A review of iron phosphate glasses and recommendations for vitrifying Hanford waste. United States

de Ligny D, Möncke D (2019) Colors in Glasses (chapter 9) Springer Handbook of Glass. Musgraves JD, Hu J, Calvez L, Springer: 297–342

de Oliveira M, Oliveira JS, Kundu S, Machado NMP, Rodrigues ACM, Eckert H (2018) Network former mixing effects in ion-conducting lithium borotellurite glasses: Structure/property correlations in the system $(Li_2O)_y[2(TeO_2)_x(B_2O_3)_{1-x}]_{1-y}$. J Non-Cryst Solids 482:14–22

Dennis LM, Laubengayer AW (1926) Germanium XVII. J Phys Chem 30:1510–1526

Deters H, de Lima JF, Magon CJ, de Camargo ASS, Eckert H (2011) Structural models for yttrium aluminium borate laser glasses: NMR, EPR studies of the system $(Y_2O_3)_{02}$–$(Al_2O_3)_x$–$(B_2O_3)_{08}$–x. Phys Chem Chem Phys 13:16071–16083

Doweidar H (2015) Structural study of density and refractive index of Sb_2O_3–B_2O_3 glasses. J Non-Cryst Solids 429:112–117

Drewitt JWE, Hennet L, Zeidler A, Jahn S, Salmon PS, Neuville DR, Fischer HE (2012) Structural transformations on vitrification in the fragile glass-forming system $CaAl_2O_4$. Phys Rev Lett 109:235501

Duffy AJ (1977) Optical absorption of Na_2O–WO_3 glass containing transition-metal ions. J Am Ceram Soc 60: 440–443

Duffy AJ (1992) Optical basicity of sulfide systems. J Chem Soc Faraday Trans 88:2397–2400

Duffy AJ (2008) The importance of π-bonding in glass chemistry: borate glasses. Phys Chem Glass—Euro J Glass Sci Technol Part B 49:317–325

Duffy AJ (2011) Optical basicity of fluorides, mixed oxide–fluoride glasses and melts. Phys Chem Glass—Euro J Glass Sci Technol Part B 52:107–114

Duffy JA, Ingram MD (1968) Environment of cobalt(II) in nitrate glass. J Am Ceram Soc 51:544–544

Duffy JA, Ingram MD (1969) A spectroscopic study of some transition-metal ions in acetic acid, in acetate glass, melt, and in nitrate glass. J Chem Soc A: Inorgan Phys Theor 2398–2402

Duffy JA, Ingram MD (1976) An interpretation of glass chemistry in terms of the optical basicity concept. J Non-Cryst Solids 21:373–410

Dussauze M, Fargin E, Malakho A, Rodriguez V, Buffeteau T, Adamietz F (2006) Correlation of large SHG responses with structural characterization in borophosphate niobium glasses. Opt Mater 28:1417–1422

Ediger MD (2017) Perspective: Highly stable vapor-deposited glasses. J Chem Phys 147:210901

Ediger MD, de Pablo J, Yu L (2019) Anisotropic vapor-deposited glasses: Hybrid organic solids. Acc Chem Res 52:407–414

Ehrt D (2000) Structure, properties, applications of borate glasses. Glass Technol 41:182–185

Ehrt D (2006) The effect of ZnO, La_2O_3, PbO, Bi_2O_3 on the properties of binary borate glasses and melts. Phys Chem Glass–Euro J Glass Sci Technol Part B 47:669–674

Ehrt D (2008) Effect of OH-content on thermal, chemical properties of $SnO–P_2O_5$ glasses. J Non-Cryst Solids 354: 546–552

Ehrt D (2013) Zinc, manganese borate glasses—phase separation, crystallisation, photoluminescence and structure. Phys Chem Glass—Euro J Glass Sci Technol Part B 54:65–75

Ehrt D (2015) REVIEW Phosphate, fluoride–phosphate optical glasses—properties, structure and applications. Phys Chem Glass—Euro J Glass Sci Technol Part B 56:217–234

Ehrt D (2018) Deep-UV materials. Adv Opt Technol 7:225–242

Ehrt D, Ebeling P, Natura U (2000) UV Transmission, radiation-induced defects in phosphate and fluoride–phosphate glasses. J Non-Cryst Solids 263:240–250

Ehrt D, Erdmann C, Vogel W (1983) Fluoroaluminatgläser: System 2 $CaF_2–SrF_2–AlF_3$ Z Chem 23:37–38

Ehrt D, Flügel S (2018) Electrical conductivity and viscosity of phosphate glasses and melts. J Non-Cryst Solids 498:461–469

Ehrt D, Keding R (2009) Electrical conductivity, viscosity of borosilicate glasses and melts. Phys Chem Glass—Euro J Glass Sci Technol Part B 50:165–171

Ehrt D, Vogel W (1983) Fluoroaluminatgläser: Einfluß von Phosphaten auf die Glasbildung im System $MgF_2–CaF_2–SrF_2–AlF_3$. Z Chem 23:111–112

Ehrt D, Krauß M, Erdmann C, Vogel W (1982) Fluoroaluminatgläser: Systeme; $CaF_2–AlF_3$ und $MgF_2–CaF_2–AlF_3$. Z Chem 22:315–316

El-Mallawany RAH (2012) Tellurite Glasses Handbook, Boca Raton, CRC Press

El-Mallawany R (2018) Some physical properties of tellurite glasses. In: Tellurite Glass Smart Materials: Applications in Optics and Beyond. El R-Mallawany (ed) Cham, Springer International Publishing, p 1–16

Evonik Industries AG (2019) CAMPUS® Datenblatt PLEXIGLAS® 8N—PMMA

Fekeshgazi IV, Mai KV, Matelesko NI, Mitsa VM, Borkach EI (2005) Structural transformations, optical properties of As_2S_3 chalcogenide glasses. Semiconductors 39:951–954

Feltz A (1993) Amorphous Inorganic Materials and Glasses. Weinheim, New York, VCH Publishers

Ferlat G, Seitsonen AP, Lazzeri M, Mauri F (2012) Hidden polymorphs drive vitrification in B_2O_3. Nat Mater 11:925–929

Fleming JW (1984) Dispersion in $GeO_2–SiO_2$ glasses. Applied Optics 23:4486–4493

Fletcher JP, Risbud SH, Kirkpatrick JR (1990) MASS-NMR structural analysis of barium aluminofluorophosphate glasses with, without nitridation. J Mater Res 5:835–840

Frederiksen KF, Januchta K, Mascaraque N, Youngman RE, Bauchy M, Rzoska SJ, Bockowski M, Smedskjaer MM (2018) Structural compromise between high hardness, crack resistance in aluminoborate glasses. J Phys Chem B 122:6287–6295

Galeener FL, Geissberger A (1982) Raman studies of B_2O_3 glass structure: $^{10}B\rightarrow^{11}B$ isotopic substitution. Journal de Physique Colloques 43(C9):343–346

Garaga MN, Werner U-Zwanziger, Zwanziger JW, DeCeanne A, Hauke B, Bozer K, Feller S (2017) Short-range structure of TeO_2 glass. J Phys Chem C 121:28117–28124

Garcia ÀR, Clausell C, Barba A (2016) Oxynitride glasses: A Review. Boletín de la Sociedad Española de Cerámica y Vidrio 55:209–218

Gaskell PH (1982) The local structure of oxide, metallic glasses. Nucl Instrum Methods Phys Res 199:45–60

Ghussn L, Reis RMCV, Brow RK, Baker DB (2014) Color bleaching and oxygen diffusion in a niobium phosphate glass. J Non-Cryst Solids 401:96–100

Glaze FW, Blackburn DH, Osmalov JS, Hubbard D, Black MH (1957) Properties of arsenic sulfide glass. J Res Nat Bur Stand 59:83–92

Goldschmidt VM, Barth TFW, Lunde G, Zachariasen WH (1926) Geochemische Verteilungsgesetze der Elemente. 7: Die Gesetze der Krystallochemie. Oslo, I Kommission Hos Jacob Dybwad

Gray PE, Klein LC (1983) The chemical durability of sodium ultraphosphate glasses. Glass Technol 24:202–206

Greer AL (1993) Confusion by design. Nature 366:303–304

Greer AL, Cheng YQ, Ma E (2013) Shear bands in metallic glasses. Mater Sci Eng Rep 74:71–132

Griebenow K, Hoppe U, Möncke D, Kamitsos EI, Wondraczek L (2017) Transition-metal incorporation and Co-Sr/ Mn-Sr mixed-modifier effect in metaphosphate glasses. J Non-Cryst Solids 460:136–145

Griebenow K, Bragatto CB, Kamitsos EI, Wondraczek L (2018) Mixed-modifier effect in alkaline earth metaphosphate glasses. J Non-Cryst Solids 481:447–456

Haiyan C, Guosong H, Hanfen M, Fuxi G (1986) Structure, Raman spectra of glasses containing several glass-forming oxides and no glass-modifying oxide. J Non-Cryst Solids 80:152–159

Hampshire S (2003) Oxynitride glasses, their properties, crystallisation—a review. J Non-Cryst Solids 316:64–73

Hannon AC, Wright AC, Sinclair RN (1991) The atomic and magnetic structure of melt-spun amorphous Dy_7Ni_3. Mater Sci Eng: A 134:883–887

Hannon AC, Grimley DI, Hulme RA, Wright AC, Sinclair RN (1994) Boroxol groups in vitreous boron oxide: New evidence from neutron diffraction, inelastic neutron scattering studies. J Non-Cryst Solids 177:299–316

Hannon AC, Parker JM, Vessal B (1996) The effect of composition in lead gallate glasses: A structural study. J Non-Cryst Solids 196:187–192

Hartmann P, Jedamzik R, Reichel S, Schreder B (2010) Optical glass, glass ceramic historical aspects and recent developments: a Schott view. Appl Opt 49: D157–D176

Hashimoto T, Nasu H, Kamiya K (2006) Ti^{3+}-Free Multicomponent Titanophosphate Glasses as Ecologically Sustainable Optical Glasses. J Am Ceram Soc 89:2521–2527

Hayakawa S, Yoko T, Sakka S (1994) Structural studies on alkaline earth vanadate glasses (Part 2). ^{51}V NMR spectroscopic study. J Ceram Soc Jpn 102:530–536

Hayden JS, Sapak DL, Marker AJ (1988) Elimination of metallic platinum in phosphate laser glasses. SPIE 895:176–191

Hench LL (2006) The story of Bioglass®. J Mater Sci: Mater Med 17:967–978

Henderson J (2013) Ancient glass: An interdisciplinary exploration. Cambridge University Press

Henderson GS, Stebbins JF (2022) The short-range order (SRO) and structure. Rev Mineral Geochem 87:1-53

Heraeus Quarzglas GmbH & Co KG (2016) Quartz Glass for Optics, Data and Properties

Herrmann A, Völksch G, Ehrt D (2019) Tb^{3+} as probe ion—clustering, phase separation in borate and borosilicate glasses. Inter J Appl Glass Sci 10:532–545

Holm NG (2014) Glasses as sources of condensed phosphates on the early earth. Geochem Trans 15:8

Honma T, Sato R, Benino Y, Komatsu T, Dimitrov V (2000) Electronic polarizability, optical basicity, XPS spectra of Sb_2O_3-B_2O_3 glasses. J Non-Cryst Solids 272:1–13

Hoppe U, Walter G, Kranold R, Stachel D (2000) Structural specifics of phosphate glasses probed by diffraction methods: A Review. J Non-Cryst Solids 263–264:29–47

Hoppe U, Brow RK, Tischendorf BC, Jóvári P, Hannon AC (2006) Structure of GeO_2-P_2O_5 glasses studied by X-ray, neutron diffraction. J Physics: Condens Matter 18:1847–1860

Hoppe U, Saitoh A, Tricot G, Freudenberger P, Hannon AC, Takebe H, Brow RK (2018) The structure, properties of xZnO–$(67-x)$SnO–$33P_2O_5$ glasses: (II) Diffraction, NMR, and chromatographic studies. J Non-Cryst Solids 492:68–76

Hudgens JJ (1994) The structure and properties of anhydrous, alkali ultra-phosphate glasses. PhD Iowa State University

Hudgens JJ, Brow RK, Tallant DR, Martin SW (1998) Raman spectroscopy study of the structure of lithium, sodium ultraphosphate glasses. J Non-Cryst Solids 223: 21–31

Inaba S, Hosono H, Ito S (2015) Entropic shrinkage of an oxide glass. Nat Mater 14:312–317

Ingram MD, Duffy JA (1970) Transition metal ions as spectroscopic probes for detection of network structure in novel inorganic glasses. J Am Ceram Soc 53:317–321

Ingram MD, Lewis GG (1974) Diffusion of trace ions in glass forming molten nitrates. J Chem Soc Faraday Trans 1 70:490–497

Ingram MD, Chryssikos GD, Kamitsos EI (1991) Evidence from vibrational spectroscopy for cluster and tissue pseudophases in glass. J Non-Cryst Solids 131:1089–1091

Iordanova R, Ataalla M, Milanova M, Aleksandrov L, Staneva A, Dimitriev Y (2015) Glass formation, structure of glasses in the WO_3-ZnO-Nd_2O_3-Al_2O_3 system. J Non-Cryst Solids 414:42–50

Izumitani T, Yamashita T, Tokida M, Miura K, Tajima H (1987) New fluoroaluminate glasses and their crystallization tendencies and physical-chemical properties. Mater Sci Forum 19–20:19–26

Jack KH (1976) Sialons, related nitrogen ceramics. J Mater Sci 11:1135–1158

Jadhav NR, Gaikwad VL, Nair KJ, Kadam HM (2009) Glass transition temperature: Basics, application in pharmaceutical sector. Asian J Pharm 3:82–89

Januchta K, Smedskjaer MM (2019) Indentation deformation in oxide glasses: Quantification, structural changes, and relation to cracking. J Non-Cryst Solids: X 1:100007

Jensen M, Smedskjaer MM, Wang W, Chen G, Yue Y (2012) Aging in chalcohalide glasses: Origin, consequences. J Non-Cryst Solids 358:129–132

Jewell JM, Higby PL, Aggarwal ID (1994) Properties of BaO-R_2O_3-Ga_2O_3-GeO_2 (R=Y, Al, La,, Gd) glasses. J Am Ceram Soc 77:697–700

Jones JR (2013) Review of bioactive glass: From Hench to hybrids. Acta Biomaterialia 9:4457–4486

Kamitsos EI, Chryssikos GD (1991) Borate glass structure by Raman and infrared spectroscopies. J Mol Struct 247:1–16

Kamitsos EI, Yiannopoulos YD, Karakassides MA, Chryssikos GD, Jain H (1996) Raman, infrared structural investigation of xRb$_2$O·($1 - x$)GeO$_2$ glasses. J Phys Chem 100:11755–11765

Kamitsos EI, Yiannopoulos YD, Duffy JA (2002) Optical basicity, refractivity of germanate glasses. J Phys Chem B 106:8988–8993

Kanamori T, Oikawa K, Shibata S, Manabe T (1981) BaF$_2$–CaF$_2$–YF$_3$–AlF$_3$ glass systems for infrared transmission. Jpn J Appl Phys 20: L326–L328

Kassab LRP, Miranda MM, Kumada DK, Bontempo L, da Silva DM, de Araújo CB (2019) Germanium oxide glass based metal-dielectric nanocomposites: Fabrication, optical characterization:A Review of new developments. J Mater Sci: Mater Electron 30:16781–16788

Kato K, Hayakawa T, Kasuya Y, Thomas P (2016) Influence of Al$_2$O$_3$ incorporation on the third-order nonlinear optical properties of Ag$_2$O–TeO$_2$ glasses. J Non-Cryst Solids 431:97–102

Kavaklıoğlu KB, Aydin S, Çelikbilek M, Ersundu AE (2015) The TeO$_2$–Na$_2$O system: Thermal behavior, structural properties,, phase equilibria. Inter J Appl Glass Sci 6:406–418

Keshavarz MH, Esmaeilpour K, Taghizadeh H (2016) A new approach for assessment of glass transition temperature of acrylic, methacrylic polymers from structure of their monomers without using any computer codes. J Thermal Anal Calorim 126:1787–1796

Khan MN, Mohamed-Osman AE(1986) Infrared, X-ray diffraction studies of TiO$_2$–GeO$_2$ glasses. J Mater Sci Letters 5:965–968

Kim S-H, Yoko T, Sakka S (1993) Linear, nonlinear optical properties of TeO$_2$ Glass. J Am Ceram Soc 76:2486–2490

King WA, Clare AG, LaCourse WC (1995) Laboratory preparation of highly pure As$_2$Se$_3$ glass. J Non-Cryst Solids 181:231–237

Kirkpatrick RJ, Brow RK (1995) Nuclear magnetic resonance investigation of the structures of phosphate, phosphate-containing glasses: A Review. Solid State Nucl Magn Reson 5:9–21

Konidakis I, Varsamis C-PE, Kamitsos EI, Möncke D, Ehrt D (2010) Structure, properties of mixed strontium-manganese metaphosphate glasses. J Phys Chem C 114:9125–9138

Konidakis I, Psilodimitrakopoulos S, Kosma K, Lemonis A, Stratakis E (2018) Effect of composition and temperature on the second harmonic generation in silver phosphate glasses. Opt Mater 75:796–801

Kono Y, Benson CK, Ikuta D, Shibazaki Y, Wang Y, Shen G (2016) Ultrahigh-pressure polyamorphism in GeO$_2$ glass with coordination number >6. PNAS 113:3436

Kordes E (1939) Physikalisch-chemische Untersuchungen über den Feinbau von Gläsern. III Mitteilung. Binäre und pseudobinäre Gläser ohne nennenswerte Packungsdefekte Z Physikalische Chem B 43:173–190

Kordes E, Vogel W, Feterowsky R (1953) Physikalisch-chemische Untersuchungen über die Eigenschaften und den Feinbau von Phosphatgläsern. Z Elektrochem Ber Bunsen Phys Chem 57:282–289

Koroglu A, Thompson DP, Apperley DC, Harris RK (2011) ^{15}N, ^{17}O MAS NMR studies of yttrium oxynitride glasses. Phys Chem Glass—Euro J Glass Sci Technol Part B 52:175–180

Kosareva IA, Tkachenko KG, Karapetjan GO, Limbach IY, Karapetjan KG, Rozhdestvensky I (2006) Comparative analysis of impact of the complex vitreous fertilizers with microelements on the microflora of sod-podzolic, peat soils. J Plant Nutr 29:933–942

Krogh-Moe J (1969) The structure of vitreous and liquid boron oxide. J Non-Cryst Solids 1:269–284

Kruzic JJ (2016) Bulk metallic glasses as structural materials: A review. Adv Eng Mater 18:1308–1331

Kubliha M, Soltani MT, Trnovcová V, Legouera M, Labaš V, Kostka P, Le D Coq, Hamzaoui M (2015) Electrical, dielectric, and optical properties of Sb$_2$O$_3$–Li$_2$O–MoO$_3$ glasses. J Non-Cryst Solids 428:42–48

Kucuk A, Clare AG (1999) Optical properties of cerium, europium doped fluoroaluminate glasses. Opt Mater 13:279–287

Kühnert H (2012) Forschungen zur Geschichte des Jenaer Glaswerks Schott & Genossen.Wien Köln Weimar, Böhlau Verlag

Kurkjian CR, Prindle WR (1998) Perspectives on the history of glass composition. J Am Ceram Soc 81:795–813

Lambert JB, Poinar GO (2002) Amber: the organic gemstone. Acc Chem Res 35:628–636

Larink D, Eckert H, Reichert M, Martin SW (2012) Mixed network former effect in ion-conducting alkali borophosphate glasses: structure/property correlations in the system [M$_2$O]$_{1/3}$[(B$_2$O$_3$)x(P$_2$O$_5$)$_{1}$–x]$_{2/3}$ (M = Li, K, Cs). J Phys Chem C 116:26162–26176

Larson RW, Day DE (1986) Preparation, characterization of lithium phosphorus oxynitride glass. J Non-Cryst Solids 88:97–113

Ledemi Y, El-Amraoui M, Calvez L, Zhang X-H, Bureau B, Messaddeq Y (2013) Colorless chalco-halide Ga$_2$S$_3$-GeS$_2$–CsCl glasses as new optical material, SPIE

Lefterova ED, Angelov P, Dimitriev Y, Stoynov Z (1997) Silver ion conducting glasses. Analytical Laboratory 6(3)

Leicester HM (1969) Mikhail Lomonosov, the manufacture of glass and mosaics. J Chem Educ 46:295

Liu H, Youngman RE, Kapoor S, Jensen LR, Smedskjaer MM, Yue Y (2018) Nano-phase separation, structural ordering in silica-rich mixed network former glasses. Phys Chem Chem Phys 20:15707–15717

Loehman RE (1987) Oxynitride glasses. MRS Bull 12:26–31

Loewenstein W (1954) The distribution of aluminum in the tetrahedra of silicates, aluminates. Am Mineral 92:92–96

Lucas J, Zhang XH (1990) The tellurium halide glasses. J Non-Cryst Solids 125:1–16

Lucas J, Troles J, Zhang XH, Boussard-Pledel C, Poulain M, Bureau B (2018) Glasses to see beyond visible. C R Chimie 21:916–922

Lundquist PM, Wortmann R, Geletneky C, Twieg RJ, Jurich M, Lee VY, Moylan CR, Burland DM (1996) Organic glasses: A new class of photorefractive materials. Science 274:1182

Lupton DF, Merker J, Schölz F (1997) The correct use of platinum in the XRF laboratory. X-Ray Spectrom 26:132–140

Machida N, Shigematsu T, Nakanishi N, Tsuchida S, Minami T (1992) Glass formation, ion conduction in the CuCl–Cu$_2$MoO$_4$–Cu$_3$PO$_4$ system. J Chem Soc Faraday Trans 88:3059–3062

Mackenzie JD (1960) Modern Aspects of the Vitreous State. Butterworths, Co Ltd, London

Mackenzie JD (1987) Chloride, bromide and iodide glasses. In: Halide Glasses for Infrared Fiberoptics. Almeida R.M. (ed) Dordrecht, Springer Netherlands: 357–366

Majérus O, Cormier L, Neuville DR, Galoisy L, Calas G (2008) The structure of SiO$_2$–GeO$_2$ glasses: A spectroscopic study. J Non-Cryst Solids 354:2004–2009

Mamoshin VL (1996) Theoretical estimation of the possibility of glass formation in sulfate, phosphate, sulfate–phosphate systems. Glass Ceram 53:104–106

Marchand R (1983) Nitrogen-containing phosphate glasses. J Non-Cryst Solids 56:173–178

Mascaraque N, Fierro JLG, Durán A, Muñoz F (2013) An interpretation for the increase of ionic conductivity by nitrogen incorporation in LiPON oxynitride glasses. Solid State Ionics 233:73–79

Mascaraque N, Takebe H, Tricot G, Fierro JLG, Durán A, Muñoz F (2014) Structure and electrical properties of a new thio-phosphorus oxynitride glass electrolyte. J Non-Cryst Solids 405:159–162

Mascaraque N, Durán A, Muñoz F (2015) Effect of fluorine, nitrogen on the chemical durability of lithium phosphate glasses. J Non-Cryst Solids 417(Suppl C):60–65

Masuno A, Inoue H, Arai Y, Yu J, Watanabe Y (2011) Structural-relaxation-induced high refractive indices of Ba$_{1-x}$Ca$_x$Ti$_2$O$_5$ glasses. J Mater Chem 21:17441–17447

Mattarelli M, Chiappini A, Montagna M, Martucci A, Ribaudo A, Guglielmi M, Ferrari M, Chiasera A (2005) Optical spectroscopy of TeO$_2$–GeO$_2$ glasses activated with Er^{3+}, Tm^{3+} ions. J Non-Cryst Solids 351:1759–1763

Mazurin OV, Priven. A SciGlass Information System, version 6EPAM 7 Systems

McLaughlin JC, Tagg SL, Zwanziger JW, Haeffner DR, Shastri SD (2000) The structure of tellurite glass: A combined NMR, neutron diffraction,, X-ray diffraction study. J Non-Cryst Solids 274:1–8

McLaughlin JC, Tagg SL, Zwanziger JW (2001) The structure of alkali tellurite glasses. J Phys Chem B 105:67–75

Meille SV, Allegra G, Geil PH, He J, Hess M, Jin JI, Kratochvíl P, Mormann W, Stepto R (2011) Definitions of terms relating to crystalline polymers (IUPAC Recommendations 2011). Pure Appl Chem 83:1831

Micoulaut M, Cormier L, Henderson GS (2006) The structure of amorphous, crystalline, liquid GeO$_2$. J Phys Condens Matter 18:R753–R784

Milanova M, Iordanova R, Aleksandrov L, Hassan M, Dimitriev Y (2011) Glass formation, structure of glasses in the ZnO–Bi$_2$O$_3$–WO$_3$–MoO$_3$ system. J Non-Cryst Solids 357:2713–2718

Milanova M, Iordanova R, Kostov KL, Dimitriev Y (2014) X-ray photoelectron spectroscopic studies of glasses in the MoO$_3$–Bi$_2$O$_3$, MoO$_3$–Bi$_2$O$_3$–CuO systems. J Non-Cryst Solids 401:175–180

Milanova M, Kostov KL, Iordanova R, Aleksandrov L, Yordanova A, Mineva T (2019) Local structure, connectivity and physical properties of glasses in the B$_2$O$_3$–Bi$_2$O$_3$–La$_2$O$_3$–WO$_3$ system. J Non-Cryst Solids 516:35–44

Miller PJ, Cody CA (1982) Infrared, Raman investigation of vitreous antimony trioxide. Spectrochim Acta Part A 38:555–559

Miller MK, Liaw P (2008) Bulk Metallic Glasses. Boston, MA, Springer US

Miracle DB (2004) A structural model for metallic glasses. Nat Mater 3:697–702

Miracle DB (2012) A physical model for metallic glass structures: An introduction and update. JOM 64:846–855

Mirgorodsky A, Colas M, Smirnov M, Merle-Méjean T, El-Mallawany R, Thomas P (2012) Structural peculiarities and Raman spectra of TeO$_2$/WO$_3$-based glasses: A fresh look at the problem. J Solid State Chemi 190:45–51

Miyabe D, Takahashi M, Tokuda Y, Yoko T, Uchino T (2005) Structure, formation mechanism of six-fold coordinated silicon in phosphosilicate glasses. Phys Rev B 71:172202

Miyajima Y, Komukai T, Sugawa T, Yamamoto T (1994) Rare earth-doped fluoride fiber amplifiers, fiber lasers. Opt Fiber Technol 1:35–47

Möncke D (2017) Metal Ions and their Interactions in Covalent to Ionic Glass Systems—a Spectroscopic Study. Habilitation, Frierdich-Schiller-University Jena, Germany

Möncke D, Eckert H (2019) Review on the structural analysis of fluoride–phosphate and fluoro–phosphate glasses. J Non-Cryst Solids: X 3:100026

Möncke D, Ehrt D, Velli L, Varsamis CPE, Kamitsos EI (2004) Structural investigation of fluoride phosphate glasses. XX Int Congr Glass, Kyoto, Japan

Möncke D, Ehrt D, Velli LL, Varsamis CPE, Kamitsos EI (2005) Structure, properties of mixed phosphate and fluoride glasses. Phys Chem Glasses 46:67–71

Möncke D, Mountrichas G, Pispas S, Kamitsos EI (2011) Orientation phenomena in chromophore DR1-containing polymer films, their non-linear optical response. Mater Sci Eng B 176:515–520

Möncke D, Papageorgiou M, Winterstein-Beckmann A, Zacharias N (2014) Roman glasses coloured by dissolved transition metal ions: redox-reactions, optical spectroscopy, ligand field theory. J Archaeol Sci 46:23–36

Möncke D, Tricot G, Ehrt D, Kamitsos EI (2015) Connectivity of borate, silicate groups in a low-alkali borosilicate glass by vibrational and 2D NMR spectroscopy. J Chem Technol Metall 50:381–386

Möncke D, Kamitsos EI, Palles D, Limbach R, Winterstein A-Beckmann, Honma T, Yao Z, Rouxel T, Wondraczek L (2016) Transition, post-transition metal ions in borate glasses: Borate ligand speciation, cluster formation, and their effect on glass transition and mechanical properties. J Chem Phys 145:124501

Möncke D, Tricot G, Winterstein A, Ehrt D, Kamitsos EI (2017) Preferential bonding in low alkali borosilicate glasses. Phys Chem Glass—Euro J Glass Sci Technol Part B 58:171–179

Möncke D, da Cruz Barbosa Neto M, Bradtmüller H, de Souza GB, Rodrigues AM, Elkholy HS, Othman HA, Moulton BJ, Kamitsos EI, Rodrigues AC, Ehrt D (2018) NaPO$_3$–AlF$_3$ glasses: Fluorine evaporation during melting, the resulting variations in structure and properties. J Chem Technol Metall 53:1047–1060

Möncke D, Jiusti J, Silva LD, Rodrigues ACM (2018) Long-term stability of laser-induced defects in (fluoride-) phosphate glasses doped with W, Mo, Ta, Nb and Zr ions. J Non-Cryst Solids 498:401–414

Möncke D, Ali S, Jonson B, Kamitsos EI (2020) Anion polarizabilities in oxynitride glasses. Establishing a common optical basicity scale, Phys Chem Chem Phys 22:9543–9560

Montanari B, Barbosa AJ, Ribeiro SJL, Messaddeq Y, Poirier G, Li MS (2008) Structural study of thin films prepared from tungstate glass matrix by Raman, X-ray absorption spectroscopy. Appl Surf Sci 254:5552–5556

Montesso M, Manzani D, Donoso JP, Magon CJ, Silva IDA, Chiesa M, Morra E, Nalin M (2018) Synthesis and structural characterization of a new SbPO$_4$–GeO$_2$ glass system. J Non-Cryst Solids

Mulfinger, H-O (1966) Physical, chemical solubility of nitrogen in glass melts. J Am Ceram Soc 49:462–467

Muñoz F (2011) Kinetic analysis of the substitution of nitrogen for oxygen in phosphate glasses. Phys Chem Glass—Euro J Glass Sci Technol Part B 52:181–186

Muñoz F, Delevoye L, Montagne L, Charpentier T (2013) New insights into the structure of oxynitride NaPON phosphate glasses by 17-oxygen NMR. J Non-Cryst Solids 363:134–139

Musgraves JD, Danto S, Richardson K (2014) Thermal properties of chalcogenide glasses. *In:* Chalcogenide Glasses Adam JL and Zhang X (eds) Woodhead Publishing, p 82–112

Musgraves JD, Hu J, Calvez L (2019) Springer Handbook of Glass, Springer

Narasimhan LR, Littau KA, Pack DW, Bai YS, Elschner A, Fayer MD (1990) Probing organic glasses at low temperature with variable time scale optical dephasing measurements. Chem Rev 90:439–457

Nasikas NK, Chrissanthopoulos A, Bouropoulos N, Sen S, Papatheodorou GN (2011) Silicate glasses at the ionic limit: Alkaline-earth sub-orthosilicates. Chem Mater 23:3692–3697

Nasikas NK, Retsinas A, Papatheodorou GN (2014) Y$_3$Al$_5$O$_{12}$–SiO$_2$ glasses: Structure, polyamorphism. J Am Ceram Soc 97:2054–2060

Nienhuis ET, Saleh M, Marcial J, Kriegsman K, Lonergan J, Lipton AS, Guo X, McCloy JS (2019) Structural characterization of ZnSO$_4$–K$_2$SO$_4$–NaCl glasses. J Non-Cryst Solids 524:119639

Ohta Y, Shimada M, Koizumi M (1982) Properties, structure of lithium borate and strontium borate glasses. J Am Ceram Soc 65:572–574

Ojovan MI, Lee WE (2011) Glassy wasteforms for nuclear waste immobilization. Metall Mater Trans A 42:837–851

Palles D, Konidakis I, Varsamis CPE, Kamitsos EI (2016) Vibrational spectroscopic, bond valence study of structure and bonding in Al$_2$O$_3$-containing AgI–AgPO$_3$ glasses. RSC Advances 6:16697–16710

Pariafsai F (2016) A review of design considerations in glass buildings. Front Archit Res 5:171–193

Paul A (1990) Chemistry of Glasses. Netherlands, Springer

Pereira C, Cassanjes FC, Barbosa JS, Gonçalves RR, Ribeiro SJL, Poirier G (2017) Structural and optical study of glasses in the TeO$_2$–GeO$_2$–PbF$_2$ ternary system. J Non-Cryst Solids 463:158–162

Phillips JC (1979) Topology of covalent non-crystalline solids I: Short-range order in chalcogenide alloys. J Non-Cryst Solids 34:153–181

Pisarska J, Sołtys M, Janek J, Górny A, Pietrasik E, Goryczka T, Pisarski WA (2018) Crystallization of lead-based and lead-free oxyfluoride germanate glasses doped with erbium during heat treatment process. J Non-Cryst Solids 501:121–125

Poulain M (1981) Glass formation in ionic systems. Nature 293:279–280

Poulain M, Maze G (1988) Chemistry of fluoride glasses. Chemtronics 3:77–85

Poulain M, Poulain M, Matecki M (1981) Verres fluores a large bande de transmission optique et a haute resistance chimique. Mater Res Bull 16:555–564

Poulain M, Soufiane A, Messaddeq Y, Aegerter MA (1992) Fluoride glasses: Synthesis, properties. Braz J Phys 22:205–217

Priven AI (2004) General method for calculating the properties of oxide glasses, glass forming melts from their composition and temperature. Glass Technol 45:244–254

Qiu H-H, Sakata H, Hirayama T (1996) Electrical conductivity of Fe$_2$O$_3$–PbO–Bi$_2$O$_3$ glasses. J Ceram Soc Jpn 104:1004–1007

Ràfols-Ribé J, Will P-A, Hänisch C, Gonzalez-Silveira M, Lenk S, Rodríguez-Viejo J, Reineke S (2018) High-performance organic light-emitting diodes comprising ultrastable glass layers. Sci Adv 4:eaar8332

Ravaine D (1980) Glasses as solid electrolytes. J Non-Cryst Solids 38–39:353–358

Richardson K, Krol D, Hirao K (2010) Glasses for photonic applications. Inter J Appl Glass Sci 1:74–86

Rinke MT, Eckert H (2011) The mixed network former effect in glasses: Solid state NMR, XPS structural studies of the glass system $(Na_2O)_x(BPO_4)_{1-x}$. Phys Chem Chem Phys 13:6552–6565

Rodrigues ACM, Nascimento MLF, Bragatto CB, Souquet J-L (2011) Charge carrier mobility, concentration as a function of composition in $AgPO_3$–AgI glasses. J Chem Phys 135:234504

Rosenbaum JM, Wilson M, Condliffe E (1997) Partial melts of subducted phosphatic sediments in the mantle. Geology 25:77–80

Rouxel T (2007) Elastic properties, short-to medium-range order in glasses. J Am Ceram Soc 90:3019–3039

Rouxel T, Soraru G-D, Vicens J (2001) Creep viscosity, stress relaxation of gel-derived silicon oxycarbide glasses. J Am Ceram Soc 84:1052–1058

Saitoh A, Kitamura N, Ma L, Freudenberger P, Choudhury A, Takebe H, Brow RK (2017) Structural study of chemically durable BaO–$FeOx$–P_2O_5 glasses by Mössbauer spectroscopy and high performance liquid chromatography. J Non-Cryst Solids 460:106–112

Saitoh A, Brow RK, Hoppe U, Tricot G, Anan S, Takebe H (2018) The structure, properties of $xZnO$–$(67-x)SnO$–P_2O_5 glasses: (I) optical and thermal properties, Raman and infrared spectroscopies. J Non-Cryst Solids 484:132–138

Saitoh A, Hoppe U, Brow RK, Tricot G, Hashida Y, Takebe H (2018) The structure, properties of $xZnO$–$(67-x)$ SnO–$33P_2O_5$ glasses: (III) Photoelastic behavior. J Non-Cryst Solids 498:173–176

Sanghera JS, Heo J, Mackenzie JD (1988) Chalcohalide glasses. J Non-Cryst Solids 103:155–178

Savage JA, Nielsen S (1965) Chalcogenide glasses transmitting in the infrared between 1, 20 µ — a state of the art review. Infrared Physics 5:195–204

Sawangboon N, Nizamutdinova A, Uesbeck T, Limbach R, Meechoowas E, Tapasa K, Möncke D, Wondraczek L, Kamitsos EI, van Wüllen L, Brauer DS (2019) Modification of silicophosphate glass composition, structure, properties via crucible material and melting conditions. Inter J Appl Glass Sci 11:46–57

Schaut RA, Weeks WP (2017) Historical review of glasses used for parenteral packaging. PDA J Pharm Sci Technol 71:279

SCHOTT Lithotec AG (2006) Synthetic Fused Silica Data Sheet

Schütz A, Ehrt D, Dubiel M, Yang XC, Mosel B, Eckert H (2004) A multi-method characterization of borosilicate glasses doped with 1 up to 10 mol% of Fe, Ti, Sb. Glass Sci Technol 77:295–305

Schweizer F (2003) Glas des 2. Jahrtausends v. Chr. im Ostmittelmeerraum. BAG-Verlag

Scully JR, Gebert A, Payer JH (2007) Corrosion, related mechanical properties of bulk metallic glasses. J Mater Res 22:302–313

Seddon AB (1995) Chalcogenide glasses: a review of their preparation, properties and applications. J Non-Cryst Solids 184:44–50

Seeber W, Downing EA, Hesselink L, Fejer MM, Ehrt D (1995) Pr^{3+}-doped fluoride glasses. J Non-Cryst Solids 189:218–226

Sen P, DasMohapatra GK, Ghosh K, Sood N Biswas (2006) Aqueous durability of K_2O–CaO–P_2O_5–SO_3 glasses: Kinetics, mechanism. Phys Chem Glass Euro J Glass Sci Technol Part B 47:294–300

Sharmin N, Rudd CD (2017) Structure, thermal properties, dissolution behaviour, biomedical applications of phosphate glasses and fibres: A Review. J Mater Sci 52:8733–8760

Shelby JE (1974) Properties, structure of B_2O_3–GeO_2 glasses. J Appl Phys 45:5272–5277

Shelby JE (2005) Introduction to Glass Science and Technology, Royal Society of Chemistry

Shen W, Baccaro S, Cemmi A, Xu X, Chen G (2015) Gamma-ray irradiation induced bulk photochromism in WO_3–P_2O_5 glass. Nucl Instrum Methods Phys Res Sect B 362:34–37

Shirosaki Y, Osaka A, Tsuru K, Hayakawa S (2012) Inorganic–Organic Sol–Gel Hybrids. Bio-Glasses, John Wiley & Sons, Ltd: 139–158

Simon S, Ardelean I, Filip S, Bratu I, Cosma I (2000) Structure, magnetic properties of Bi_2O_3–GeO_2–Gd_2O_3 glasses. Solid State Commun 116:83–86

Skopak T, Serment B, Ledemi Y, Dussauze M, Cardinal T, Fargin E, Messaddeq Y (2018) Structure-properties relationship study in niobium oxide containing $GaO_{3/2}$–$LaO_{3/2}$–$KO_{1/2}$ gallate glasses. Mater Res Bull

Soltani MT, Haddad S, Möncke D, Kamitsos EI (2016) Structural investigations of binary glasses Sb_2O_3–Na_2O by Raman, FTIR and optical spectroscopy using Co^{2+} as probe ion. Society of Glass Technology Centenary Conference and ESG, September 4–8, 2016 Sheffield, UK

Stanworth JE (1952) Tellurite glasses. Nature 169:581–582

Stokes ME, Schreiber HD (2006) Copper as a structural probe for acetate glass. Phys Chem Glass Euro J Glass Sci Technol Part B 47:233–235

Šubčík J, Koudelka L, Mošner P, Montagne L, Tricot G, Delevoye L, Gregora I (2010) Glass-forming ability, structure of ZnO–MoO_3–P_2O_5 glasses. J Non-Cryst Solids 356:2509–2516

Suryanarayana C, Inoue A (2017) Physical Properties of Bulk Metallic Glasses. Boca Raton, FL, CRC Press

Svoboda R, Málek J (2015) Evaluation of glass-stability criteria for chalcogenide glasses: Effect of experimental conditions. J Non-Cryst Solids 413:39–45

Tagiara NS, Palles D, Simandiras ED, Psycharis V, Kyritsis A, Kamitsos EI (2017) Synthesis, thermal, structural properties of pure TeO_2 glass and zinc–tellurite glasses. J Non-Cryst Solids 457(Suppl C):116–125

Tagiara NS, Chatzipanagis KI, Bradtmüller H, Rodrigues ACM, Möncke D, Kamitsos EI (2021) Network former mixing effects in alkali germanotellurite glasses: A vibrational spectroscopic study. J Alloy Compd 882:160782

Terashima K, Hashimoto T, Uchino T, S-Kim H, Yoko T (1996) Structure, nonlinear optical properties of Sb_2O_3-B_2O_3 binary glasses. J Ceram Soc Jpn 104:1008–1014

Thieme A, Möncke D, Limbach R, Fuhrmann S, Kamitsos EI, Wondraczek L (2015) Structure, properties of alkali and silver sulfophosphate glasses. J Non-Cryst Solids 410:142–150

Tick PA (1985) Glass forming in the cadmium fluoride–lithium fluoride–aluminium fluoride–lead fluoride system. Mater Sci Forum 5–6:165–165

Tite M, Shortland A, Paynter S (2002) The beginnings of vitreous materials in the Near East, Egypt. Acc Chem Res 35:585–593

Tsuchida JE, Ferri FA, Pizani PS, Martins AC Rodrigues, Kundu S, Schneider JF, Zanotto ED (2017) Ionic conductivity, mixed-ion effect in mixed alkali metaphosphate glasses. Phys Chem Chem Phys 19:6594–6600

Uesbeck T, Eckert H, Youngman R, Aitken B (2017) The structure of borophosphosilicate pure network former glasses studied by multinuclear NMR spectroscopy. J Phys Chem C 121:1838–1850

Upender G, Prasad M, Mouli VC (2011) Vibrational, EPR, optical spectroscopy of the Cu^{2+} doped glasses with $(90-x)TeO_2-10GeO_2-xWO_3$ $(75 \leq x \leq 30)$ composition. J Non-Cryst Solids 357:903–909

Uusitalo O (2010) Revisiting the case of float glass. Euro J Innovation Manage 13:24–45

Van Wazer JR, Holst KA (1950) Structure, properties of the condensed phosphate. I. Some general considerations about phosphoric acids. J Am Chem Soc 72:639–644

van Wüllen L, Eckert H, Schwering G (2000) Structure–property correlations in lithium phosphate glasses: New insights from $^{31}P \leftrightarrow {}^{7}Li$ double-resonance NMR. Chem Mater 12:1840–1846

Varshneya AK (2013) Fundamentals of inorganic glasses. Sheffield, UK, Society of Glass Technology

Vaughn WL, Risbud SH (1984) New fluoronitride glasses in zirconium–metal–F–N systems J Mater Sci Letters 3:162–164

Velli LL, Varsamis CP E, Kamitsos EI, Möncke D, Ehrt D (2005) Structural investigation of metaphosphate glasses. Phys Chem Glasses 46:178–181

Venkatachalam S, Schröder C, Wegner S, van Wüllen L(2014) The structure of a borosilicate, phosphosilicate glasses and its evolution at temperatures above the glass transition temperature: Lessons from in situ MAS NMR. Phys Chem Glasses—Euro J Glass Sci Technol Part B 55:280–287

Verger F, Pain T, Nazabal V, Boussard C-Plédel, Bureau B, Colas F, Rinnert E, Boukerma K, Compère C, Guilloux M-Viry, Deputier S, Perrin A, Guin JP (2012) Surface enhanced infrared absorption (SEIRA) spectroscopy using gold nanoparticles on As_2S_3 glass. Sensors Actuators B 175:142–148

Videau J-J, Portier J, Blanzat B, Barthou C (1979) Etude spectroscopique par la technique d'affinement de raie de fluorescence de verres fluorophosphates dopes a Eu^{3+}. Mater Res Bull 14:1225–1229

Vogel W (1992) Glass Chemistry, Springer-Verlag

Von Jebsen-Marwedel H (2011) Glastechnische Fabrikationsfehler, Einführung. Berlin, Heidelberg, Springer Berlin Heidelberg

Walker B, Dharmawardhana CC, Dari N, Rulis P, Ching W-Y (2015) Electronic structure and optical properties of amorphous GeO_2 in comparison to amorphous SiO_2. J Non-Cryst Solids 428:176–183

Wang C, Zimmer J (2015) Aluminosilicate glass for touch screen. USA, SCHOTT GLASS TECHNOLOGIES (SUZHOU) CLTD O (Jiangsu, CN)

Wang JS, Vogel EM, Snitzer E (1994) Tellurite glass: A new candidate for fiber devices. Optical Materials 3:187–203

Weber MJ (2006) Optical Properties of Glasses. Mater Sci Technol, Wiley-VCH Verlag GmbH & Co. KGaA

Weber JKR, Rix JE, Hiera KJ, Tangeman JA, Benmore CJ, Hart RT, Siewenie JE, Santodonato LJ (2005) Neutron diffraction from levitated liquids—a technique for measurements under extreme conditions. Phys Chem Glass 46:487–491

Wiberg N, Wiberg E, Holleman A (2007) Lehrbuch der Anorganischen Chem, de Gruyter & Co, Berlin

Wilk NR Jr, Schreiber HD (1997) Co^{2+} as a structural probe in acetate liquids, glasses. J Non-Cryst Solids 217: 189–198

Winterstein-Beckmann A, Möncke D, Palles D, Kamitsos EI, Wondraczek L (2013) Structure–property correlations in highly modified Sr, Mn-borate glasses. J Non-Cryst Solids 376:165–174

Winterstein-Beckmann A, Möncke D, Palles D, Kamitsos EI, Wondraczek L (2015) Structure, Properties of Orthoborate Glasses in the $Eu_2O_3-(Sr,Eu)O-B_2O_3$ Quaternary. J Phys Chem B 119:3259–3272

Wójcik NA, Jonson B, Barczyński RJ, Kupracz P, Möncke D, Ali S (2018a) Electrical properties of $Na_2O-CaO-P_2O_5$ glasses doped with SiO_2 and Si_3N_4. Solid State Ionics 325:157–162

Wójcik NA, Jonson B, Möncke D, Palles D, Kamitsos EI, Ghassemali E, Seifeddine S, Eriksson M, Ali S (2018b) Influence of synthesis conditions on glass formation, structure and thermal properties in the $Na_2O-CaO-P_2O_5$ system doped with Si_3N_4 and Mg. J Non-Cryst Solids 494:66–77

Wright AC (2010) Borate structures: Crystalline and vitreous. Phys Chem Glass—Euro J Glass Sci Technol Part B 51:1–39

Wright AC (2014) The great crystallite versus random network controversy: A personal perspective. Inter J Appl Glass Sci 5:31–56

Wright AC, Hannon AC, Clare AG, Sinclair RN, Johnson WL, Atzmon M, Mangin P (1985) A neutron diffraction investigation of the atomic, magneticstructure of amorphous Dy_7Ni_3. J Phys Colloq 46:C8-299–C298-303

Wright AC, Dalba G, Rocca F, Vedishcheva NM (2010) Borate versus silicate glasses: Why are they so different? Phys Chem Glass Euro J Glass Sci Technol Part B 51:233–265

Wright AC, Clare AG, Etherington G, Sinclair RN, Brawer SA, Weber MJ (1989) A neutron diffraction, molecular dynamics investigation of the structure of vitreous beryllium fluoride. J Non-Cryst Solids 111:139–152

Wright AC, Sinclair RN, Stone CE, Shaw JL, Feller SA, Williams RB, Fischer HE, Vedishcheva NM (2014) A neutron diffraction study of sodium, rubidium, caesium borate glasses. Phys Chem Glass Euro J Glass Sci Technol Part B 55:74–84

Wuttig M, Yamada N (2007) Phase-change materials for rewriteable data storage. Nat Mater 6:824–832

Yang Z, Wilhelm AA, Lucas P (2010) High-conductivity tellurium-based infrared transmitting glasses, their suitability for bio-optical detection. J Am Ceram Soc 93:1941–1944

Yano T, Kunimine N, Shibata S, Yamane M (2003a) Structural investigation of sodium borate glasses and melts by Raman spectroscopy.: I. Quantitative evaluation of structural units. J Non-Cryst Solids 321:137–146

Yano T, Kunimine N, Shibata S, Yamane M (2003b) Structural investigation of sodium borate glasses and melts by Raman spectroscopy. II. Conversion between BO_4 and BO_2O-units at high temperature. J Non-Cryst Solids.. J Non-Cryst Solids 321:157–168

Yano T, Kunimine N, Shibata S, Yamane M (2003c) Structural investigation of sodium borate glasses and melts by Raman spectroscopy. III. Relation between the rearrangement of super-structures and the properties of glass. J Non-Cryst Solids 321:147–156

Yiannopoulos YD, Chryssikos GD, Kamitsos EI (2001) Structure, properties of alkaline earth borate glasses. Phys Chem Glasses 42:164–172

Youngmann RE (2022) Silicate glasses and their impact on humanity. Rev Mineral Geochem 87:1015-1038

Yuan B, Zhu W, Hung I, Gan Z, Aitken B, Sen S (2018) Structure, chemical order in S–Se binary glasses. J Phys Chem B 122:12219–12226

Yue X, Inoue A, Liu C-T, Fan C (2017) The development of sructure model in metallic glasses. Materials Research 20:326–338

Zachariasen WH (1932) The atomic arrangement in glass. J Am Chem Soc 54:3841–3851

Zakery A, Elliott SR (2003) Optical properties, applications of chalcogenide glasses: a review. J Non-Cryst Solids 330:1–12

Zallen R (1998) The Physics of Amorphous Solids, Wiley-VCH Verlag GmbH & Co. KGaA

Zanotto ED, Coutinho FAB (2004) How many non-crystalline solids can be made from all the elements of the periodic table? J Non-Cryst Solids 347:285–288

Zanotto ED, Mauro JC (2017) The glassy state of matter: Its definition and ultimate fate. J Non-Cryst Solids 471:490–495

Zhang L, Eckert H (2006) Short-, medium-range order in sodium aluminophosphate glasses: New insights from high-resolution dipolar solid-state NMR spectroscopy. J Phys Chem B 110:8946–8958

Zhang W, Shao J, Xu X, Wang R, Chen L (2007) Mantle metasomatism by P-, F-rich melt/fluids: Evidence from phosphate glass in spinel lherzolite xenolith in Keluo, Heilongjiang Province. Chin Sci Bull 52:1827–1835

Zhang L, Hu L, Jiang S (2018) Progress in Nd^{3+}, Er^{3+}, Yb^{3+} doped laser glasses at Shanghai Institute of Optics and Fine Mechanics. Inter J Appl Glass Sci 9:90–98

Zhao J, Ragazzi E, McKenna GB (2013) Something about amber: Fictive temperature, glass transition temperature of extremely old glasses from copal to Triassic amber. Polymer 54:7041–7047

Zhao S, Li L, Wang H, Zhang Y, Cheng X, Zhou N, Rahaman MN, Liu Z, Huang W, Zhang C (2015) Wound dressings composed of copper-doped borate bioactive glass microfibers stimulate angiogenesis and heal full-thickness skin defects in a rodent model. Biomaterials 53:379–391

Zhou C, Longley L, Krajnc A, Smales GJ, Qiao A, Erucar I, Doherty CM, Thornton AW, Hill AJ, Ashling CW, Qazvini OT, Lee SJ, Chater PA, Terrill NJ, Smith AJ, Yue Y, Mali G, Keen DA, Telfer SG, Bennett TD (2018) Metal-organic framework glasses with permanent accessible porosity. Nat Commun 9:5042

Zmojda J, Kochanowicz M, Miluski P, Leśniak M, Sitarz M, Pisarski W, Pisarska J, Dorosz D (2016) Effect of GeO_2 content on structural, spectroscopic properties of antimony glasses doped with Sm^{3+} ions. J Mol Struct 1126(Suppl C): 207–212

Zschimmer E, Cable MT (2013) Chemical Technology of Glass, Society of Glass Technology